ERRATA, NOTE, BROCHURE and WEB SITE

This book, as its topic, is still in development. The second printing has many great features and some errors. A third printing is in production with editing improvements and elimination of most errors. Yet the topic is vast, new and abstract. Complex essential knowledge and information of the beautiful intricate detail of the best of worlds is not easily organized and comprehended. The quickest way to understand the book is to read the inside flaps of the jacket. Then survey the book and supplemental contents, especially 'How to use this book'. Then read the theses in the addenda. Then read and study a page or so at a time, learning the key ideas of each paragraph and relating them to your life. Maturity in such a life is like a good education, the challenge.

The new terminology adds to this complexity. Yet there are several great reasons why new terminology is indispensable and once learned highly rewarding.

1. There are many unknowns that need discovery, names and expression.
2. Many great aspects, parts and processes of our special universe lack signification.
3. Existing names are usually for senses and things, lacking essence and relevance.
4. Many words have multiple meanings, and lead to misrepresentation and confusion.
5. All terms should have the purity, actuality and reality of this best of worlds.
6. All terms need validity, proportionality or relative magnitude and totality,
7. Explicit and concrete definitions enable a positive objective approach.
8. Existing terms limited to their essences combined with key new terms accentuate each other and their roles in revealing the actual world, TRU.
9. A new language requires operational terms and means to encompass all TRU.

If you can visualize and operate TRU in and of itself, TIT, as TRIT the form, frame and structure of the best of worlds, by its spherical model, the more you learn its great concepts, principles, laws, mechanisms, parts and processes. The actual meanings of the greatest existing terms and phrases with the top new terms provide the increasingly precise factors and vectors for positioning in the spherical models. Once you can relate and fit them into the whole you will find yourself in, and can vastly expand your identity and quality within this whole wonderful actual world that enables human life.

A brochure is available to make this best of worlds, the world of success, well being and survival, TRU, simpler, clearer, more effective and personally rewarding.

The website helps to introduce, guide, and make your study more interesting and effective.

http://www.therelativeuniverse.com

T R U

o r

T H E

R E L A T I V E U N I V E R S E

The Investigation, Analysis and Demonstration of

THIS SPECIAL UNIVERSE of BEST HUMAN LIFE

Bergh

DEDICATION

This work is dedicated to each and every person, who fail to come to terms with, produce, and get the best out of life. This includes most all who understand neither themselves nor the rest of the world. Our life is much different than we believe. We must discover and best apply it if we are to survive and flourish. Thereby this book is especially dedicated to those bright young minds that can recognize, learn, know, appreciate, and endeavor to save their souls and God, as these are - and enable us to achieve -TES and be TRU. It is also dedicated to the leaders of the world, its executives, ministers, doctors, politicians, lawyers, teachers, managers and all in positions of responsibility. It is to these that guide and spearhead a higher age for humanity, that will prevent us from continually becoming victims, and be the victors necessary, for this ever expanding special universe of best human life, TRU.

More personally the author dedicates this work to all who, during a lifetime of medical practice, have provided experience, imagination, ideas, and inspiration. Most personally the book is dedicated to my children: Barbara, Marcus, John, Donna, Susan, Pamela, Constance, Karin, and their children; and most of all to my wife Barbro, without whose companionship, love, and sacrifice I could not have endured this work on TRU.

ISBN # 0-9721468-0-6

1st Printing 2003 TXu 1-062-690
2nd Printing 2005 TXu 1-225-820

Printed in Sweden by
Tierps Tryckeri AB 2004

C O N T E N T S

Supplemental Contents including:
i. Title, ii. Dedication, iii. Contents, iv. Figures or Illustrations, v. T Terms, vi. Foreword, vii. Frequently used and remembered terms, viii. Contents, terms names and proofs, ix-x. How to use text, HUB xi-xii.Abridged and sized texts

The text is divided into 10 Sections, 0-IX, each with 10 Chapters, or 100 total, 00-99

APPENDIX & ADDENDA

MAJOR FIGURES, VISTAS, or ILLUSTRATIONS (G1)

These are only some of the figures in the text, others are enumerated in DIPTRU or in IPTRU. Their posits in SPALT and SPAT are shown, being produced by a preliminary triangulation of A, L, and T. With increasing refinement and accounting they provide a highly revealing vista of TRU. Most are still in a development phase, Please appreciate drawing and editing defects.

GLOSSARY 2, TRU TERMINOLOGY of KEY/CORE CONCEPTS based upon T

NEC	Definition	Chapter	Figure	Á	L	T
T	= TRU The Relative Universe, in all respects	00-99	L@ N	0-10	0-10	0-10
TÁ	= TRU Mass Energy, Á Prime Form & Function of TRU	13n	L@ N	0-10@	2.5	0-10
TÂ	= TRU Absolute, many interpretations of completed end	139	L@	10~	0-10	10~
TÄ	= TRU FORM, all static aspects of TRU's general makeup	12-	L N	6.7~	3.3~	<0-10
TÀ	= TRU Dynamics, TROS and PACH Name of *	17-	L N	~5~	<0-10	~5~
TÉ	= TRU Part, Differential and Particulate of TRU	14-	L N	0-10	7.5~	7~
TÊ	= TRU Enablement, TROS and PACH Name of *	18-	L N	3.3~	0-10	6.7~
TÍ	= TRU Special Part, TROS and PACH Name of	15-	L@	3.3~	6.7~	6.7~
TÎ	= TRU Infinite, The Infinite State and Conditions TRU	14n	L	0-10-	0,10-	0-10-
TÏ	= TRU Derivation, Development, TROS; PACH Name of DOT*	11-	N L	6.7~	5~	2.5~
TÔ	= TRU Totality, The Total State of TRU, TRUS, TRIS	16-	L	0-10	0-10	0-10
TÔ	= TRU Summation, Use of, TROS and PACH Name of *	19-	N L	3.3~	6~	7~
TÚ	= TRU Unity, That State of TRU wherein it is as one	135	L N	(0,,,,5,,,,,10)		
TÛ	= TRU General & Knowledge of, TROS and PACH Name of *	10-	N L	3~	5~	7~
Tn	= TRU Several Optional Basic TRU Terms **	n	N L	4	6.7	7~
TÄP	= TRU Form Proper or Special	13-	L N	0-5-10	3.3~	6+
TES	= TRU Parts Individual or Personal, Counterpart in TRU	15n	L N	3.3~	6.7~	6.7~
TÉT	= TRU Equals Truth, Truth Equals TRU, Equivalence, VAPO	10	L	4	5	7
TINI	= TRU Infinitesimal, Infinitesimal State of TRU, TIT	144	L@	5~	10~	5~
TIT	= TRU In Itself, Excluding all Extraneous effect, S O	133	L@	0-10	0-10	0-10
THEO	= TRU Object, the Objective as formed & functions TRU	152	L@	7.5@~	-	-
THES	= TRU Subject the Subjective as formed & function TRU	152	L@	2.5@~	-	-
TOT	= TRU Test, Test(s) of TRU	194	N L	3~	6~	7.5~
TRAX	= TRU Accentuation without knowledge of TRU	92	L N	6.7~	5~	6.5~
TRE	= TRU Recognized as such	105	L N	2@	5~	7.5~
TREN	= TRU Enablement, Within or of the Whatness of TRU*	184	L N	3.3~	6~	6.7~
TREX	= TRU Accentuation with general knowledge of TRU TREQ	93	L N	2.5~	6~	7.0~
TREQ	= TRU Equivalent(s) All same - nearly the same as TRU	104	N L	5+	4~	3.3~
TRIA	= TRU Is All, the Exclusiveness of TRU	138	L@	0-10	0-10	0-19
TRICH	= TRIS Changes in all or within each	168	L N	5~	6.7~	6.7~
TRID	= TIT Dynamics, all functions and operations in TIT	17,5+	L@	±	<6.7	~5~
TRÍF	= TRIT or TIT Time Future, Final third of time in TIT	133	L@	-	-	6.7-10
TRÍN	= TRIT or TIT Time Present, Now, Middle third of TIT	133	L@	-	-	3.3-6.7
TRÍP	= TRIT or TIT Time Past, Initial third of time in TIT	133	L@	-	-	0-3.3
TRIS	= TRU Sets or Kinds Special	132	L N	~	2.5	~
TRIT	= TRIS Set Equivalent to TIT, as it exists relative	133	L N	5~	2.5	<~
TRÍST	= The Standardized TRÍ, Man and TRU Special	138	L	5	5	6.5
TRIX	= TRU Accentuation with special KOT, TÄP, TIT expones.	94	L@	2.0~	6~	7.5~
TRIZ	= TIT, TRIT, TRIS Zero, As a Limit,	137	L	5	5	5
TROS	= TRU States, TA, TE, TI, TO, TU,,,	127	L N	0-10	>0-10	0-10>
TRUCH	= TRU Change, General Overall	168	L N	3.3~	<1.5~	>0-10
TRUD	= TRU Dynamics	17-	L N	0-10	<	>0-10<
TRUDI	= TRU Divisions, Distribution, Di,,,	14n	L N	5~	5~	5~
TRUDIL	= TRU Divisions of Labor	172	L N	6.5~	5	5.5~
TRUEQ	= TRU Equilibrium, balance of form & function in TRU	166	L N	5~	5~	5~
TRUF	= TRU Future, All after the Present of TRU	128	L N	-	-	8.3~
TRUID	= TRU Ideal	138	L N	5~	5~	8.3~
TRUM	= TRU Meaning, of all ideas & words, Meaning as TIT	106	L N	2.5~	7~	7.5~
TRUN	= TRU Present or Now, Current, Middle range TRU Time	128	L N	-	-	>5<
TRUP	= TRU Past, All before the Present of TRU	128	L N	-	-	0-3.3
TRUR	= TRU Rate(s) Acceleration or change without & within	173	L N	>5<	<1.5~	>7.5~
TRUS	= TRU Sets or Kinds General	122	L N	0-10	>0-10<	0-10

The content numbers apply to sub-chapters of chapters, not pages

* PACH Terms duplicate or are highly coterminal with other terms, e.g. TREN and ENT for TÊ, meanings distinguished in text. ** Many other major posits, see text

- Non existent or non participant in that category, > more participant, < less participant

_ Underlined signifies largely of that figure or dimension

@ Signifies only of that figure or dimension, ~ Signifies less focal or widespread distribution

^ Signifies a high degree of distortion by planar/global (2D/3D) representation. Absence of this symbol does not signify lack of distortion, for many T terms, especially when noted by _, or are highly non focal - as universals -, can't be well shown on 2D

L signifies use of SPALT for posits, U signifies use of SPAT

Many T terms - as universals - denote some aspect or portion of the overall TRU or TIT. Sometimes these are not of - or only of - ALT, e.g. TÄ, TÀ, TÏ, TÏNI, TRIS, TRIT, TROS, TRUD. Since the measures and posits are somewhat different for SPAT than for SPALT and since many T terms are widely or fully distributed in SPALT, the measures given for these L that are underlined or have an @ are their SPAT measures and posits. What is universal in TIT and SPALT has more posit in SPAT. So that some terms are shown in SPAT in a more figurative way.

All outside of the spherical SPALT has less posit by TIT and more by the observer and observed. e.g. MATRU, AMTRU, NAMTRU, N, knowledge increasingly or TRUS, RUTS, RUS and AUS.

It always helps to envision all spherically and totally by TIT-TO, SPALT, SPAT, METAD, POMOT, THES, THEO, SENIT, and THIT and then return to TIT.

FOREWORD

This book is an abridged version of much larger reference works on TRU. Its target is a presentation that provides the greatest discovery of voids in our knowledge with the insight that brings them to the level human life both needs and demands. The relations, integration and differentiation, identity, language, figures, and representations of TIT enable us to produce ever greater formal distinctions of all that is TRU. It enhances and accelerates our processes of subject object operations, realization, and mass energy form and function, **Á FAF**, whose continuum, **SÁ**, gives us the predominance so dear to life. Such a language of TRU, **LANT** or Law of TRU Notations is absolutely necessary, **LATAN**. A language and terminology for all great science is an indispensable tool for signifying and communicating its ideas, concepts, principles, laws and mechanisms. Since TRU is both a unit and a totality that can be treated as a closed set, a logical and symbolic system for demonstrating it provides an ever greater means of discovery, learning, knowledge and use. This enables a reduction to optimal positive objective approach and how to apply logical theory and methods to the appraisal of argument, reason and all thought. It is how to best fulfill their role in providing the needed perfected identity and replication of TRU. Each and all persons exist and participate, more or less, in TRU. The language, we so often take for granted, is central to the reality of our existence in TRU. The words we use are not commonly directed toward this relative universe. They are primarily directed toward senses and things in themselves. As a result we have had languages, words and terminology that fail to give us the perspective of what we actually are. To correct this will take its own terminology, in addition to what language is in common use. To this end the following are proposed:

If: T = The,
R = Relative,
U = Universe,
Then: TRU = The Relative Universe
And if: Ä = Form
Á = Mass Energy
Â = Absolute

É = A Finite or Part
Í = The Finite, Part, Person, I
Î = Infinite and Infinitesimal
O = Total
U = Unity (also)
Then: TÄ = TRU Form
TÁ = Mass Energy of TRU

TÂ = TRU Absolute
TE = TRU Finite, or Part in General
TÍ = TRU Finite, or Part, Special
TI = TRU Infinite or Infinitesimal
TÓ = TRU Total
TU = TRU Unity

Also: s when applied as a suffix adds the Plural, and all other letters as consonants, or syllables with separate useful phonetics can be given denotations that best represent their form and function in TRU. Thereby: D = Differentiation, F = Fallacy, L = Law, Ô = Object, S = Subject. P = Presentation, R = Representation, ES = Parts, and CES = Processes.

Certain specific uses of letters are also applied to make up acronyms. These are ones that have been in use, especially in scientific contexts. However they use some of the same letters so limited to the topic under discussion and are so noted, including: EN(S) = Entity(ies) R = Response S = Stimulus Examples of Combinations include, S/R = Stimulus Response and CNS = Central Nervous System

We call the New Concepts of TRU, **NEC** in contrast to **OC** or Old Concepts, **EC** or Essences of OC and Existing Concepts, and **ECT** or Existing Concepts with TRU meanings. Those NEC of the highest order are **TOPNEC**. This separation is done because most of our concepts and language are less than EC, much less than ECT, and few of adequate NEC equivalence. This lack, or their continual variance in and out of TRU, **OITIO**, causes endless problems when not upgraded to show their actual existence in TRU. It is one thing to mean something by realizing an idea, concept, or principle, as in TRU. It is quite another to develop a good word and definition, especially one that will best communicate to others and get others to know and use the meaning. This is **STÖP** or the Sterile State Of Principles, Concepts, Laws and Language. The reader will find it much easier if they pause for a moment to analyze, recognize and remember key terms. These include listings of contents, figures, T terms, readily and not so readily remembered major concepts and proofs. Together a rapidly increasing familiarity will give the reader not only ready access to and identity with TRU, it will produce the stimulus of power in being a real participant in the actual world, TIT, in which one lives, yet so painfully or poorly acquires and achieves. How philosophy, science and religion are formed and function can be little better than the sufficiency and use of its language. If it does not have the concepts and language necessary to make distinct meanings of all the salient features, parts and processes, it will fail. Our failure to adequately understand ourselves and our world has largely come from this failure to produce a valid, actual and proportionate total language. The expression of this imperative requirement for a demonstrative, concise, clear, simple and sufficient language by all great subjects, sciences, 'ologies' and especially TRU is called: **LIS** = Law of Language Imperative Sufficiency. This, as any, language must be sufficient unto its topic. As difficult as it is for us to adapt to this expanding world it is because of LIS that science and technology have become so reliable and effective.

Since the investigation of any topic requires a focus on its various aspects, parts, and processes, and since in doing so our attention is drawn away from its totality, a reversible compensatory means needs to be provided to assure identity, totality and closure. This is accomplished at length by familiarity with the book, and more simply by organizing, relating, tying-in and indexing. Organization results from a showing that systematically reflects, identifies with, or mirrors its object or topic, TRU. When so provided the sameness can be reproduced at corresponding levels. Tie-ins result from descriptions of relations in different chapters or parts. Indexing and glossaries produce relations by identifying the location in two or more parts of the book. Since CIPTRU is organized by TIT, Á, SOT, S-O, APT, POA, and PACH and not by random topics, these will locate in various chapters. This will make tie-ins more necessary, yet will naturally provide much greater meaning, and significance in TRU. We must realize this TRU's fundamental essence and nature, **DAOST**, how all is Different Aspects Of the Same Thing.

If one follows the simple rule of pronouncing vowels like the first letters of the existing term they represent there will be less difficulty in learning TRU. So of a's we pronounce Á for the predominant mass energy dimension of TRU differently than we pronounce Â for absolute, or Ä for form, and À for attention. Thus ÏÀ is Instantaneous Attention or Immediate Awareness, like it is in these first vowels, and letters, of each term. This simple rule holds throughout their usage. After a while it becomes natural, habitual, and spontaneous. There are many vowel sounds that are insufficiently utilized in language. In the English Language they have great potential by adding parsimony and power to meanings and communication. The value of this in expressing the concepts, principles, laws, parts, processes and mechanisms of TRU are of even greater value. It is for this reason we develop and use in IPTRU a systematic analysis and presentation of these vowel sounds as letters to enable more effective acronyms for TRU. These proposed uses may be only temporary, to be continually improved upon. Yet they are necessary to make distinct the new terminology, NEC, presented. at the outset.

FREQUENTLY USED & READILY RECOGNIZED TOPNEC

(Other than of T's and Contents) Plus the Chapters in which primarily described.

ACEXIN Actuation, Extension and Intension, ALT 2 26-3
ÂÏT Absolute Imperative TRU 452
AITES Absolute Imperative TES 3 00-7,452
AMTRU ATRU TRU Mixture, less than 100% TRU 09-5,10-1
ÂN Absolute Necessity 22-9
ANTRU Anti or Against TRU 10-6
ÂT AU and TRU combined, "all thngs" Pg26,Ch06
ATIM All Time Mastery of Things, TRU 10-13
ÀTRU From or decreasing TRU 02-5, 06-2,10-2
BAS Basal States 02-6
BASFANT Basal States, Far, Near Approaches, TRU 02-6
CARSC Capacity-Ability-Reaction-Sets-Conditions *08-2
CEK Compelling or Drawing Forces of Knowledge 3b6
CENS Comparative Entities 03-3
CIT Comparative Diagrams, Identities of TRU 02-6
COAT Comprehension of All Things, All Things TRU *00-7
COC Complications or Compounding of Causes 66-1, 353,366
CODIT Compulsory Dimension-s of TRU 3b2
COGATEXS Coincidence of Guidelines, Texts of TRU SC-xi
COGNOS We know, know that we know, by VAPO TRU 10,96
COL Correction of Limit(s) 02-11
COMT Comparative TRU,,, 03-3i
CONRIT Convergence upon TRU by RIA, TU, SU, TO *06-2
CONVET Convergence upon TRU by TÎ, TE, TO *06-2
COR Correction of Relation(s) *02-10
COT Coterminality of/in TRU, TERMS, EC and NEC SC-vi,20
Cu Conscious Unit, level of Á 06-3i 35Bi
DAOST Different Aspects of the Same Thing, TRU *14-3i
DÏ(ÏD) Differential Integral 06-4
DIMÄ Dimensional Formulation, frame of TRU 33
DIMT Dimensions of TRU 14,31
EBE Extreme by Emphasis, Differential Focus 03-1
ECRO Ever and Continuously Remember, as needed 05,105
ELTICMA Everlasting Life through Instantaneous Massive, Attention, Accentuation, Action 97,98
ENDOT(RU) Inner third, center or Core of TRU 04-2,12
EN(S) Entity (entities) SC-vi,06-2
ENIT Entity In and of Itself 01-1
ENSY Energy Systems, Levels, SÁ of Á in TRU 83-1
ESIN(S) What is in an Entity also in TRU 03-9 05-11
EXOT(RU) Exterior third or periphery of TRU 04-2,12
FAF Form and Function 00-9 01-3
GER General Relativity, simple meaning, complex 03-1
GERÁ GER becomes Á in Special Thisness of TRU 13
HIDIC *Horizons 29
Imu(Vu) Importance to Man, Value, Unit, Á Level 28,35
IN(INEG) Integrative Negation 07-12,51
IND Individual 56
IPR Interpersonal Relationships 67
ISMS Isms or Systems of Knowledge, EBE or worse 03-1
JUNEC Justification of New Concepts SC-vi,009
LÂCOP Law of Absolute Combined Proof 00-9,10-26
LADIRE Law of Diminishing Returns 00-9
LADOP Law of Added Data, Facts Providing Proof 00-9
LANEC Law of the Necessity of NEC, TRU terms 00-8,10-3
LAMOL Law of More or Less 03-1,05-3,06-8,13
LANERA Never Absolutely Absolute or Relative " 03-13
LANT Law of New or TRU Language and Terminology 12
LAPIP Law of Altered Posit in Presentation 37,43
LARF Law of Relative Fixation, special basis for TRU 2c158
LARP L. Relative Parametric Proportion, Set TRU 2c056
LASUR Law of Sufficient Reason, (c SCARLS) 02,04,10
LAT-S Law-s of TIT, Primarily In TRU 13
LATAN Law of Absolute Necessity 12
LATANO Law of Absolute Necessity of Order 10-7
LATANOS LATANO Standardization SC-vi,10-7,16,22
LATNUP Law of The Absolute Necessity of a Singular Unified Perspective and (Re)Presentation (of TRU) 3,09,11
LIACT The Imperative IA convergence on TRU 01-2
LIS L. Imperative Sufficiency of Language SC-iv, 15,22
LOPA Law of Proportional Act 35,2c6,90
LOPAN Law of Partial Necessity, ES CES Essential 10-5
LORF Law, or Principle of Relative Fixation, General 10-24
LOS Law of Objectification and Subjectification 03-1
LOSUT Optimal Simplicity of TRU thru SUT 15
LORVA Law of Relative Value, Set by TRU 2c3,35
LOT-S Law-s of TRU, OITIO 10-15
MADAS Maximal Separation and Distinction TRU (AU) 14
MATES Meaning Automatically Tends to ES, CES 2c 1-2,14,50
MEDOT(RU) Medial Third, layer or segment of TRU 25
MERDUC Method of Exaltation, Reduction, Production C-ix
METAT(RU) TRU and all around surroundings, or about i t 40
MP Man Point 31
OBIT Object in and of Itself 01,14,18,21
OD Object(ive) Differential 28
OI Object(ive) Integral 28
OITIO Out-In-TRU-In-Out 05-3,06-5,19-2
OM Omnificent Magnanimity 19,21,23,50
OPT Optimal Presentation TRU, best IPTRU, text 08,44
OSO Object-Subject-Object 01,02
OTRU Object TRU 007
PACH Perfected standardized POA for IPTRU 026
PÏ Positive Imperative 00-6
PIP Proof of Integrative Processes 14,16,17,50
POA Positive Objective Approach 00
POMS Principle of Multiple Solutions 06-8
POSCI Principle of Scopic Inertia 06-8
PPMA Prevention Protection Maintenance Advance 90
PRÁ Proof of Á, mass energy dimension of TRU 12,13,14
PROT Presentation of TRU as an Object 00-6
REBE Resolution of EBE, Extremes by Emphasis 04-10
RECTRU Reduction to TRU 08-5
RER Representation of Representation 23
RIA Relative is All, Exclusive Focus of 03-9,10-1
RIT Representation in TRU 12,20
RÖTÓ Total Representation of All TRU 26
S Subject, Subjective 01
SCARL(S) = States, Conditions, Limits, Relations 02,12
SD Subject(ive) Differential 28
SELIA Selective IA 07-6
SENIT Sense In and of Itself 01,14
SER Special Relativity, simple-complex-TRU meanings 03,10
SI(SINT) Subject(ive) Integral 28
SÏA Summative IA 28
SIL Inverse Square Law 72
SIT Subject(ive) In and of Itself 47
SO Subject Object FAF as a unit 10-11
SOT Subject Object FAF as a unit in TRU, Á 01-16
SPO(T) Standardized Perspective of All(TRU) 44
STO Standardized Subject Object 25
STRU Subject TRU 15
TAN Think and Note, Method of TRU SC-ix
THEO The Objective portion of TIT, TRU 00-6,10-7
THES The Subjective portion of TIT, TRU 00-6,10-7
THIT Thing In and of Itself 18,62
TIMS TINS in IPTRU from same levels of PACH SC-P
TINS Tie-ins of Concepts, Posits of TRU SC-P
TOPNEC Major TRU Terminology SC-vi,00-8,97
TRUT TRU is Truth, without A-AN-ANTI TRU 00-6
UK Unknown, or unknowns 10-15
UMIA Universe, Mankind, Individual and IA (TÄ) 02-11
UDOHAT The Undeniability and Durability of much of Philosophical, Scientific and Religious Truths 08.09.10
WITÂ Will to TRU Absolute 2c7

CONTENTS or KEY TOPICS FOR PROOF, ASPECTS FEATURES FORMS and FUNCTIONS of TRU

NEC or Logo	Section Chapter	Content, Definition, and Description	Name of Proof
EPI	0-	Epistemics, senses, perspectives, COAT, philosophy	PREP
AU	0-	Absolute Universe, things, material world, science	PRAU
TRUS	09-12	General Kinds and forms of TRU, Approach to TRU	PRUTRUS
TRU	10-19	TRU, TRU in General, All of or in TRU, theology	PRUT
DOT	11	Derivation, Development and Determination of TRU	PRUDOT
ATO	11	At Once TRU, The self existent TIT-TO	PRATO
IAT	11	Instantaneous Attention TIT, focus and source, universe of human life	PRIAT
REL	11-13	Relativity, General and Special by and of TRU	PREL
S-O	0-9-	Subject-Object Bond, Form and Function in TRU	PRUSO
TÄ	12	TRU Form in General, States, Conditions, Relations, Limits	PRUTÄ
TROS	126	TRU Form, mathematical states and relations, TU, TE, TI, TO, TÂ	PRUTROS
Á	13-	Predominant Mass Energy Form in TIT, Integral	PRÁ
SÁ	13-	Continuum of Mass Energy Form in TIT, Differential	PRUSÁ
TÄP	13	TRU Form Proper, Form of TIT	PRUTÄP
TRIS	13	TIT Form Specifically, Kinds, States, Conditions,,,	PRUTRIS
TIT	133	TRU in and of Itself, shed of all other, or external to it	PRUTIT
TU	135	TRU Unity, Unified, Singular Specificity of TRU	PRUTU
TRIA	138	TRU Is ALL, Exclusiveness of TRU	PRUTRIA
TÂ	139	TRU Absolute, Absolute Nature of TRU	PRUTÂ
TE TES	14	TRU Part and Parts, or Differential, General	PRUTE-S
TÍ TIS	15	TRU Part and Parts, or Differential, Specific	PRUTÍ
TÏ TINI	149-159	TRU Infinite and Infinitesimal, Extension of TRU	PRUTÏ
TO	16	TRU Totality, Integral whole of TRU	PRUTO
REV	167	Reversibility, Cyclicity, Phasic Nature of TRU	PREV
TRUCH	168-9	TRU Change and Variation	PRUCH
VAPO	16-19	Validity, Actuality, Proportionality, and Totality TRU	PRUVAPO
TRUD	17	TRU Dynamics, Function of and in TRU	PRUTRUD
ALT^n	17-	Mass Energy, Length and Time Higher Powers in TRU, ACEXIN	PRALTH,PRAX
ENT	18	Enhancement, Enablement, and improvement, of TRU	PREN
TRIT	19-2-	TIT and TRIS form amenable and available for representation	PRUTRIT
OBT	2-	Objectivity, objectification of TRU for representation	PROBT
SUT	2-	Systematic Unity of TRU, adapted for representation	PRUSUT
ROT	2	Representation of TRU, any and all means helpful, art	PROT
STD	20-	Standardization, Set forms natural and arbitrary	PRUSTD
OC EC	20-25	Concepts in Existing Language, and with essences of TRU Meanings	PROC,PREC
NEC	SC-2-25	New Concepts and Language of and for TRU	PRUNEC
MOT	3	Modulation of TRU, TRU Models	PRUMOT
ALT	31	MOT based on TRU Mass Energy, Length Time Dimensions	PRALT
SPHER	3-34	Sphericity in ROT and MOT, demonstrative of TRU	PRUSPHER
SPALT	33	Spherical ALT, 3D Model of TRU based on ALT	PRUSPALT
REM	35	Relative Magnitude, Valuation of TRU's Components	PREM
POS	4	Posits, Placement of Parts and Processes in ROT SPALT	PRUPOS
STA	45	Static Posits in SPALT, MOT, and ROT	PRUSTA
HOM	5	Homation or Processes of Man General, Dynamic Posits	PRUHOM
CES	5-	Specific Processes for Posits in SPALT, MOT and ROT	PRUCES
DYNU	6	Unitary, Singular or Simple CES in SPALT and ROT	PRUD
DYNS	7	Plural, Multiple or Complex CES in SPALT and ROT	PRUDS
DYNO	8	Total or All Combinations of CES in SPALT and ROT	PRUDO
ACC	90-99	Accentuation or Action for TIT, Resultant and Future	PRAC
DAC	91	Derivation, Development and Determination of ACC	PRUDÂC
ACÄ	92	General Form and Function of ACC, TRAX	PRACÄ
ACÄP	93	Specific Form and Function of ACC, TREX	PRACÄP
ACE	94	General Parts and Processes of ACC, TRIX	PRACE
ACI	95	Specific Parts and Processes of ACC, PUREACLITY	PRACÍ
ACO	96	Total ACC, KUPS, COGNOS	PRACÔ
ACELT	97	Progression of Functions of ACC, ELTICMA	PRELT
SÂ	98	Summation of ACC, TRU and IPTRU	PRUSÂ
TÂ	99	Absolute, Ends, Ultimates, and TIE-INS of ACC, TRU	PRUTÂ
POT	9-9-n	Presentation of TRU, Ways and Means to Know and Show	PRUPOT
IPT	0-9-n	Initial Presentation of TRU, Multivolume and basic First Text	PRIPT
LAWS	0-9-n	Laws, Laws in General and Adaptations to TRU	PRUL
TREQ	0-9-n	TRU Equivalents, Existing World, Same as TRU	PRUTREQ
PROOFS	0-9-n	Proofs in General and Adaptations to TRU	Proofs

Table 3-SC: Prime TOPICS, ASPECTS, FORMS and FUNCTIONS of TRU, used both as contents of TRU and for proofs. See Chapter, Page and Addenda of IPTRU for detail on proofs.

HOW TO USE THIS BOOK: HUB

Why are most humans so averse to the best and greatest in their existence, world, and human life? It is 1. Unknown, 2. What is known is chaotic, disordered, and mixed with much that is not true. 3. It is infinite, infinitesimal or so massive that it is overwhelming, 4. It has little immediate practical value, 5. Because they are so overwhelmed with the demands of life they can't rise to it, 6. That which they do know that is right they do not uphold. 7. Much that is the best and greatest closest to them is focused upon, extolled, and accepted as finished to the exclusion of the whole. 8 Their lives are short, burdened and distracted. Not only must each person learn and know all about the actual world itself in totality, they will be helped by comparing and knowing how it differs from all else.

Absent, mixed or any World	**The Actual World, TRU**
No Focus in	Focus from within
Poor Focus outside	Good Focus outside
Not relative	Highly relative
Disordered	Ordered
Ignorance and Illusion	Wisdom and Comprehension
Totally indistinct	Sameness in Totality
Partiality irrelevant	Differences in Partiality
Chaotic	Cosmotic
Negative to Humans	Positive for Humans
Largely of Figments	Of Actualities
Invalid	Valid
Disproportionate	Proportionate
Unvalued or Dis-valued	Valued
Fragmented	Solid
Poorly to non formed	Highly formed
Mostly un-representable	Highly representable
Model impossible	Readily Modeled
Not Set	Set
Irregular	Highly Regular
Changing irregularly	Unchanged from without
Changing irregularly	Changing from within
Variant from without	Invariant from without
Variant within	Variant within
No fixed quality	Highly qualified, ordered
Indistinct quantity	Distinct ordered quantity
Chaotic origins, beginnings	At Once from itself
No set development	Highly set development
Variant Physical mlt	ALT by human focus
Mass Length Time Variable	A, L, T. set by form of
From and Against Life	Life Oriented, by humans
No, poor or trivial control	High, increasing control
No Selection	Natural Selection, plus
Chance events	Little or rare chance events
No Prevention, Protection	Good Prevention, Protection

The opposite of TRU is anything from a void, to an indistinct mass, to relative universes focused on anything other than a human focus. Some of the comparisons will tend to become more like TRU with increasing relatedness. Many of the contrasts are simple and very clear. Some depend on what is given opposite to TRU. Most are explained in detail in text. To study and learn from such a list helps one get a better idea of what the text is about, and orient on what most readily brings identity with the world they are actually in.

Every human sooner or later if not for their whole life comes to question and seeks to know the disparity and conflicts between themselves and the world they are in, deity, God, TREQ or TRU. The answer for Christians is to accept Christ as their savior. The answer for Muslims is identity through Mohammed with Allah. The answer in Buddhism, Shintoism, Hinduism, Taoism, Confucianism, and all religions is an identity with their creeds. Yet, these are simple answers. They at best only poorly help to explain and lead a person to achieve actual full identity between themselves and their world. In fact they become are so limited, diluted, abused, and diverse from total actuality that they become a hindrance and worthless. They become negative from the effects of conflict with other religions, individuals, and knowledge. Only by TRU reason and CALP can we overcome this great problem of our existence. Therefore, discover, learn, remember, and fully account for all as provided by VAPO, DAOST, SCARLS, ROT, MOT and all of TRU, TROS, TRIS, and TIT.

Although this book should be of great benefit to all it is so new and subtle that only by the beauty of its intricate detail, profound additions, ordering of values and knowledge for human life can it be appreciated. It is not a book that can be easily read cover to cover like a novel. It is more like an encyclopedia, bible or reference work to be read repeatedly in steps or pages, or referred to when needed. To learn and benefit from it, frequently check the glossary, index, and addenda, as well as these preliminary pages. In these the major new concepts are presented and key to understanding the text, as well as to your own life in the actual world, TRU. It is not necessary to learn or remember all the NEC offered, only those that are repeatedly used, pivotal to learning TRU, or of the greatest relative magnitude, REM, or value in TRU. Also, focus on those existing and new concepts that help you know and live by your part and role in this special universe of best human life. Learn and remember what are, and the differences between, T, TRU, TIT, TRIT, TES, ES, CES,,and Á. T is its symbol. TRU is the acronym for our selective actual world in general. TIT is this world exclusively in and of itself without anything extraneous. TÄ is the form of TRU. TRIT is the form, frame, and structure of TIT that can be used to represent and model it. TES is like our soul, our part and role in TIT. ES and CES are the parts and processes in TIT. Á is the mass energy continuum of TIT whose dimension along with the likewise altered, curved or warped dimensions of length and time make up the primary parameters of TIT. The symbols for vowels, Á, Â, ,Ä, À, Å, É, Ê, Í, Î, Ó, Ú, Û, stand for the special form, absolute, general form, conjunction, pause, part and finite, special part, infinite, total, universe and unit or singularity. They are added to key stem consonants as qualifiers to produce many of the key concepts of the book. Also frequently used are the new terms for the contents of the book.(pg viii), and other frequently used NEC.(pg vii). Most of the new terms are defined when first introduced as the text progresses. If not they are defined in the glossary. More exhaustive glossaries of all terms are to be found in the multi-volume IPTRU. It is because there is little precedent for TRU in the literature, because as a science it needs its own language, and because it is all important to the success and survival of humanity that all are encouraged to find the inspiration and revelations needed to learn TRU. For it can't be done without you, and you will suffer greatly for its loss without it. Such a formal and ordered presentation of the world sets forth the methods each and all of us need to fulfill our parts and roles, and prevent the failures inherent in our specie's adaptation to a very vulnerable, accelerating and dangerously uncontrolled life.

The greatest message of this book, its primary theme and goal, is to establish in each and all persons the best possible approach to, reasoning of, and coordination in the world about them. To think with the best attitudes and habits we should not only think rationally and logically, it should be based upon the right reasons, those actually of TRU, or this special world of best human life, not any or diverse worlds. To gain such good attitudes and habits of thought, those based upon VAPO, DAOST, SCARLS, ACEXIN, POA and TRU must be cultivated throughout life. This requires,

1. Acquisition of the right knowledge, of TREQ or TRU,
2. Remember and use this knowledge,
3. Stimulate one's best interests,

4. Making fashionable and popular those interests that are in the highest concepts, principles, mechanisms and laws of TRU.
5. Think and note, **TAN**, all that is unknown or of what needs to build their knowledge,
6. Constantly upgrade and apply this knowledge and best habits of reasoning to ones work and life,
7. Bring and make ready for imitation such attitudes and habits of thought of TRU for all people that one encounters.

Steps that show how to best use this book include,

1. Study and learn these supplemental contents,
2. Learn its major TRU terms and top concepts that are often used thus more readily remembered,
3. Study, learn, and repeatedly use the major illustrations, figures and their posits in SPAT and SPALT,
4. Achieve a concentrated, deeply interested, thought provoking and determined use of the ideas, jists of concepts, NEC, as the text progresses and in the glossaries of TRU,
5. If you have difficulty learning, remembering or pronouncing NEC, start by spelling them, relating each letter to their definitions; anticipate, seek out, relate and posit the deepest meanings of these great ideas and concepts of TRU..
6. Review the proofs of TRU, their posits in summary figures, lists in the text, and the enhanced veracity vividly realized.
7. Relate, tie in and bring all these together in the combined vistas, jists and proofs as they exists differentially, CONVET, and converge integrally, CONRIT, as they best replicate TIT,
8. Think about, relate, and practice positing important, topics concepts and their implications in models and figures of TRU.
9. Realize the great identity of the mechanisms, functions and processes of TRU and their posits in its models. Relate them to the great laws and concepts of TRU to see how all can be well ordered, repeatedly and reliably used, and added to,
10. Apply all this knowledge to the world about us, its questions, problems, voids, conflicts, threats and greatest needs now and future to enable their answers and solutions.
11. Study, accentuate, identify with TRU, find TES and its accompanying mutual unity, success, glory, and immortality.

Make a life long habit of thinking, studying, working with, and living TRU. Get the best possible overview of it and its universal, general and overall concepts, principles, laws and mechanisms. Posit and trace through, and add to or correct existing interpretations of its parts and processes. It is in the themes and theses of TRU and their convergence that identity is best achieved. Simplicity is provided by the unity of TIT, **TU**, that leads to its ordered state, or **TRIT,** in objectivity, **OBT**, systematic unity, **SUT**, and representation, **ROT**. Superlatives are those words and phrases that describe all the greatness and glory of TRU. Conceptual formal distinctions are those words and descriptions of all that represents TRU with the clearest, simplest, defined, and best - TRU meanings. Relative focus is the summative weighted point of reference of TRU. It establishes a base upon which optimal superimposition is possible by SUT and the dimensional formulation of TRU. Posits are the optimal conceptual formal distinction and location of greatest aspects, forms, functions, parts and processes of TRU in the best possible model of TRU. Validity, actuality, proportionality and totality, **VAPO**, makes possible an adequate and successful knowledge, showing or presentation of TRU by reducing all to their part, place, and role in TRU. Consistency is representation through invariance in TRU and POT. Freedom from contradiction results from avoiding all that is not its TRU definition. TROS signifies the combinations of numerical types that exist in TRU: TU, TE, TI, TO, and TÂ. States, conditions, relations and limitations or **SCARLS** is the nature and method of analyzing TRU. Thisness or haecceity is that exclusiveness and selectiveness that so characterizes being, life and humanity. It is a product of focus, relativity and the higher powers of man's predominant form and function. **ACEXIN** is the Actuation, extension, and intension continua, or combined higher orders of dimensional formulation and operation in TRU. Predominant mass energy form and function, state and dimension of **Á**, is the special and unique aspect of TRU that most characterizes and sets it apart from all else. It is not the same, actually in striking contrast with, the mass of the physical world. It may be recognized by comparison with, and what is left over from, the length and time dimensions. How they are also greatly altered, adapted or warped in comparison with those of the physical world. Expanding finite totality is our exquisite and glorious state in the relative that is TRU. It is how we can produce and provide in perfection and permanency more and more parts, processes and all that constitutes TRU. Beauty of intricate detail is the challenge of all TRU finite, **TE,** that is all about us. How it can be known and shown and thereby leads to fulfillment of TRU. Interest or the great curiosity is the abiding faith and continuing ability to maintain, with conviction, a happy participation in what is best, TRU. It gives us, with inspiration and revelation, accelerative and explosive powers to envision and anticipate all that is both static and dynamic of TRU. The Law of Sufficient reason, **LASUR**, is identity of all that is the same, relative and set of TIT-TO available and indispensable for valid inferences and reason. Inspiration is the ecstasy that results from the discovery of great value or identity with the absolute, TRIA of TIT, TRU, and TES. Revelations are those penetrating and greatest ideas that make the world most real, and tie it all together. This is enabled by the Á, relativity, unity, systematic unity, sameness and thisness of TRU. Anticipation enables us to gain ever greater knowledge, TES and TRU, to so heighten the awareness that assures reinforced precognition of all necessary. Resolution is shown in 'to begin is half the job'. Faith is the conviction in benefits that TREQ or TRU provides for each and all persons. Hope is the never-ending dedication to and trust in ones place in the eternal success of TRU. Glory is the ever expanding greatness and happiness to be found in TRU, being benefactor and beneficent in identity with TRU. Victory is the accelerative success that dedicated participation in TRU provides; enabling peace and prosperity when the combined fulfillment of the observer and observed occurs in the relativity that is TRU. This fulfillment is the subjective and objective union and actuation specific for TRU, its **Á** form and function, **FAF**. This is shown by COGNOS, ELTICMA, SA and TA as they enable everyone to best function in TRU. It is the Absolute identity of feeling, being, and immortality with the infinite ACEXIN, enhancement, **ENT**, and eternal expansion of TRU.

See the different sections of the addenda, especially its glossaries and indices to clarify concepts, posits, and proofs. As each new term is added to the whole of language in representing TRU both new and old terms will tend to have less value. Because of overlapping and coterminality its share of the whole, as with REM, will decrease. This is called: **DEMCO** = Decrease or Depreciation of the Magnitude by Coterminality of Concepts in Totality. We must always be alert, ECRO, to apply DEMCO and not lessen our interest and attention to the conceptual basis of TRU. It only confronts us with a recognition of the real challenge of the positive yet extremely finite and complex use of these superimposed concepts as they represent TIT. DEMCO is resolved with ever greater parts and values of the whole, TIT-TO. It is also resolved by correcting coterminality and an ever greater love, appreciation, and use of the exquisite beautiful , and ordered increasingly finite detail of TRU. The letters and abbreviations are used as logos for identifying names in TRU are shown in the following list. The section and chapter numbers provide a decimal ordering in which the first number is for the section, 0-9 and the second number for the chapter, 00-99. By studying, remembering and using these tables and contents of CIPTRU, whose proofs are key to its presentation, we can achieve, their actual relations and total existence in TIT-TO, and the optimal combined understanding necessary to better know and be TES and TRU.

ABBREVIATIONS for some of the major concepts in TRU whose letters are used in ACRONYMS include:

Á First designate, Prime Mass Energy Dimension of TRU
Â Absolute (-3) Ä Form (-2)
À Attention Awareness Aspect
Å Conjunction, Pause, 'uh'
B Basic, b. better, bad, book
C Concept Condition, Comparison,
D Dynamics, Derivation, Development, Determination, Demonstration, Differential, E Part(s), Partial, Finite,
É Part of TRU,
Ë Enhancement, Enablement
EN Entity, ENS Entities
ES Parts of TRU
F Fallacy, f. function
G God h How
H Homation, Processes of man
Í I (person), The Part, (-5) Identity,
Ï Initial, Immediate, (-l) Infinite, IND Individual
J Justification
K Knowledge, k a [constant], the constant
L Law, LA Laws in TIT, LO Laws of TRUl Length, LI - Limit
M Model, Modulation, Math, Measure, Magnitude, m - Mass
N Nominal, Name, New, Near
n power, nth, enumeration
Ó Total, Final, Summary
Ö Object, Objective o of, on, out
P Presentation, Positive Point Perspective(Per), Problem(Pr)
PS Problem Solution
Q Question
R Relative, Representation, Reality, Response, Reverse r. Ratio
S Subject, Subjectivity, Stimulus, Spherical, Standard, Superimposition, State, Success
S-O Subject Object
S/F Success over Failure
S-R Stimulus Response
s Plural
T TRU, The, Thought, Test
t Time, th Thing
U Universe, Universal, Unity, u. Unit
V Valuation
x Solution, as of an equation, result
W World
Y You, Second Person-s Z Chaos

ABRIDGEMENT of TRU TEXTS; COGATEXS

By COGATEXS we follow a system for presenting TRU at textual levels based upon specific guidelines and decimal expansion. **COGATEXS** = Coincidence and Concordance of Guidelines and Texts of TRU. COGATEXS and the briefs or abstracts are tentatively and approximately ordered from T4 to T1 for symbolic representation, and then back again, from TI to TVII for increasingly complex and massive representation:

T Level	Letters	Words	Lines	Pages	Volumes	Name
T-4	1					T
T-3	10	1				TRU, TIT
T-2	100	10	1			The Relative Universe, special universe,,,
T-1	1,000	100	10			Brief Statement of TRU
T-I	1x104	1,000	100	1		1 Page Summary of TRU
T-II	1x105	1x104	1,000	10		10 Page Abstract ofTRU (and/or brochure)
T-III	1x106	1x105	1x104	100		100+ Page Small Condensed Book,(BIPTRU)
"B	6x "	6x "	6x "	600		600+ Page Abridged Book (CIPTRU)(this text)
T-IVB	2x107	2x106	$2x10^5$	2000	1	3,000 Page Large Reference Book (DIPTRU)
T-V	1x108	1x107	$1x10^6$	(5000)	5	10,000+ Pages Reference work on TRU (IPTRU)
T-VI	1x109	1x108	$1x10^8$	$5x10^3$	50	Volume Multi-Author work on TRU (MIPTRU)
T-VII	1x1010	1x109	$1x10^9$	$5x10^5$	500	500 Volume Complete Library of TRU (OPTRU)

Because TRU is such a new, incomplete, great and potentially difficult to understand topic, it is well to provide the reader with an opportunity to find it at ones own pace, choice and level of interest. Yet it should be provided and offered in the most powerful and appealing presentation equal to its topic. The content of these different levels is as follows:

T 4 = T
T 3 = TRU, TIT, TRIS
T 2 = TRU is The Relative Universe generally, TIT TRU Itself, specifically

T 1 = TRU is The Relative Universe. This is the special universe of best human life. TRU is the general term for all that is in or of this universe. TIT is the specific term for all that is in and of itself, Itself. It is focused upon the conjunction of mass-energy length time as they have been adapted, formed and set selectively in human life. These result from relativity or subject-object, and observer-observed relations predominant and typical of TRU. They are formally and functionally fused and consist of many levels from the physical through highest mental spiritual and total TRU abilities. They revolve about the focus of the human person, their attention and action. This is in dynamic relationship and opposition to all other entities, senses and things both as they are considered in and of themselves or become more or less a part of TRU. TRU is of many forms and states from unity, finity, infinity, totality, absolute, changing, dynamic, self propagating, enabling proof and the showing of its systematic unity.

T-I

TRU is symbolized by T, and is The Relative Universe or this special universe of best human life. It is a product of relativity, set and typical of the mass energy, length, and time form and function of TRU. The mass energy dimension is known from the observer-observed or subject-object relation and continuum. These dimensions exist and function in total unison in TRU. It is primarily, preponderantly and predominantly homocentric. It is neither, un-centered, dis-centered, helio-centered, astro-centered, physically centered or centered in any way other than set by its prior existence in systematic unity and closure in an existent, ATO, universe, TIT. Yet in being homocentric it is highly inclusive and qualified. It is centered about the being, normalcy, quality, health, improvement, strength, predominance, success and survival of human life. It is homocentric about all persons, more or less, depending upon their participation in TRU. It is also homocentric about much of the world that supports human existence. This that supports humans is a large part of the whole, TRU. However most that supports it is less of the subjective and center of TRU in contrast to the human aspect. The focus of this center, and of man, is the center of his mass-energy, length and time dimensions. This coordinate axis is special for TRU. It is altered, adapted and warped in ways that are different from that viewed only physically. This focus can be most vividly seen in human consciousness, Instantaneous Attention and Action or **IA**. IA is an immediate awareness and activity in which subject and object function together optimally in a special reality for the actuality of TRU. It is because man has not developed from all his knowledge a good comprehension of the world that much confusion and failure occur and recur. For the form and parts of TRU are often much different than what appears, believed in or imagined.

From the symbol T other combinations with phonemes

provide the abbreviations and acronyms needed to represent the new concepts of TRU. TIT is the new concept, term, logos and system that identifies and represents the world of man as it actually exits, in itself. TRU identifies this world and in perspective subjectively distinguishes it through a positive objective approach to provide this being, our relativity and its best form and function, with representation and presentation, ROT and POT. Further expansion of terminology and language of TRU, LANT, provides the kind of discrete, distinct, and precision guided terms such a great, unknown and undeveloped topic warrants. The standard reference work, detailed, and initial presentation of TRU, **IPTRU**, discovers, analyzes and shows our world as an exclusive, dynamic and massive entity. It is an entity that can be treated objectively through its own parameters of actuation, extension and intension. Its perspective, representation and presentation are identified and fused as they are a part of it as a thing-in-itself, TIT. We the observer, and it as the observed, are one as our subjectivity and its objectivity unite in the predominant mass-energy state of our being. TRU and IPTRU are proven, justified and enacted as described and qualified by the concepts, principles, laws, mechanisms, representations, models and different states, conditions, relations and limitations of all that exists, or can be known and shown within, of, about and out of TRU. In IPTRU we strive to discover, develop, determine and present that best of worlds of man which functions in pure reality and actuality exclusive of all else that is irrelevant and not a part of TRU. TRIT is its frame, and what is available of TIT and TRIS for objectification and standardization. It is what enables us to represent and model TRU. It is both The Subject and The Object as they exist in the unified absolute totality, TO, of TIT. It is focused upon a center-point of here and now, or l and t, and the mass-energy state predominantly characteristic of this perfected world of man, its mass or Á dimension. TRU topically is the systematic investigation and selective integration of all sciences and fields that human understanding have, for so long, sought to discover, develop and determine. This is done through all means that can produce the necessary concepts, principles and laws needed to represent, modulate and present this best of worlds. TRU as a topic involves a critical analysis of knowledge and the world in order to produce its best possible replica.

The many forms and states of TRU and their ordination, standardization, representation and modulation can be developed from its basic dimensions. These include the mlt or mass-energy length and time dimensions as they are, **ALT**, specific, exclusive, set, conditioned by and exist in TRU. TRU is of many varieties, classes, states, forms and aspects. TIT is entirely, and thereby TRU is primarily, immanent in itself. Yet through change and outward observation, it may be at times considered from the standpoint of transcendence. TRU stands in dynamic opposition to all other things, the absolute absolute, absolute universe, AU. The relativity, focus or centering and systematic unity of TRU enables us to develop a formal model which can be highly successful in the representation of TRU. The formulation, systematic unity, ordination, representation, modulation and standardization can be developed from the subject-object or observer-observed mass-energy dimension as it exists in combination with the length and time dimensions. This can be done as they exist in the special predominant state of TRU. These three dimensions provide the solid geometric configuration necessary for spherical modulation, operations, and applications. The geometric construct and modulation which such a dimensional formulation allows, provides a basis for continually improving the best possible model and replica so necessary for representation and objective comprehension of TRU. It also helps to resolve the problem of all other and varied worlds.

As this model and its associated valuation, posits, statics and dynamic aspects are reduced to a positive objective approach, a better textual description can be accomplished. They can be computerized, programmed, communicated and applied to all technologies especially the most advanced fields and sciences. This enables their and the actual world's acceleration and future perfection as they conform to and participate in TRU. If our facts, concepts and principles of TRU are sound and if the information and knowledge that is available is properly adapted to TRU, our demonstration of the substance and inner-workings of TRU can be accurate, highly revealing and successfully shown by their total posits in the perfected model. Together the unity and expanding highly valid, actual, proportionate totality of TRU is provided. Such posits in the primary SPALT and accessory SPAT, METAD, POMOT, and Sector models, enables us to produce a highly reliable, valuable and useful approach to both the static and dynamic components of TRU: its parts, processes, forms, functions, features, forces and mechanisms.

The static and dynamic interpretation of the forms, features, elements, parts, processes, forces, and mechanisms are posited and shown. The dynamic processes, vectors, forces, mechanisms and operations of TRU can be posited in this model through looking at them singularly, in a more complex or compound , and in a total form. These are demonstrated, treated and utilized either as singular, multiple, complex, compound or total factors and products that best demonstrate TRU and can be operated by us, the observers. The resultant conclusions, solutions, accentuation, and actions can be readily traced-out and presented. In totality they can be directed to provide the most profound, perpetual and eternal benefit to each and all, the reader, first person, and all persons, as they makeup TRU. They can be applied to the questions, problems, conflicts and threats which we face or are likely to cause loss, failure and extinction. Thereby TIT, through TRIT, can become even more improved, perfected, and made eternal and absolute. Their accentuation, when combined in the vistas, gist, and proofs, can be summarized and directed to the profound benefit of all, as they are TRU and **TA**, the absolute TRU. TRU also involves the conclusions and subsequent promotion and expansion of this text and subsequent means that will more broadly generate TREQ, thereby TRU, its proofs, concepts, principles and laws.

This book targets everyone who recognizes the need to answer the questions and solve the problems of life in a profound and honest way. We all must open our minds and hearts by removing what is irrelevant and enlisting and emphasizing what it is in the world that counts the most It differs from what has been written by selecting out and eliminating all figments and illusions and providing the basis of what is valid, actual, total and proportional. This it does by giving our concepts the purity and reality they deserve in being identical with their objects in this special universe of best human life, TRU. In measuring these concepts by their positions in the fundamental dimensions, mass, length and time, of our existence, it is possible to provide the differentiation and integration a comprehensive successful ordered whole provides. Only with interest, persistence, and a devotion to study and analysis can we begin to understand ourselves and the world around us. Only as each and all persons, and especially those in positions of decision making leadership, know and best apply the greatest in human knowledge can we avoid the conflicts, wars, suffering, failures, loss, and likely extinction mankind encounters.

It is to produce a presentation that penetrates the inertia, partiality, and fallacies of our beliefs and knowledge that we take the extraordinary avenue of new concepts. For the words and concepts in current use are directed toward things-in-themselves, and occasionally to their essences, yet must be predominantly their strict interpretation in what is of TRU in and of itself in totality, TIT-TO, or its equivalents, TREQ.

This is a work-in-progress of all knowledge, fields and sciences. It will take an extended and massively expanding participation by many in a well controlled campaign.

INTRODUCTORY SECTION

PRESENTATION OF TRU: POT

CHAPTER 00

APPROACH TO, RELATIONS, and PERSPECTIVE of POT and TRU

In order to discover, learn, and best live we must not only know what this is in and of itself, TIT, we must do so for that about it, how it is presented and represented, **POT** and **ROT.** How to eliminate folly and embark on wisdom mandates an approach to TRU through POT and ROT. This is achieved by a relativity between the observer and observed or subject object relationship in which TIT is the object and a person's best recognition and perspective is the subject. This is the key to TRU, a continuum from sense thing cleavage of an entity through the relativity of observer observed of life in general to the final focus of Á mass energy FAF in TIT. This focus is evident in our experience, immediate awareness and instantaneous attention, **IA**. Although our attention, self interest, and much that is partial are great aspects of our being, they are neither all nor what we become in resultant. For all is incomplete, and leads to less unless it is finalized in the totality of TRU. The approach to TRU, **ÀPT**, is the into TIT of this reversible in and out, **OITIO**, processes of POT and ROT of TRU.

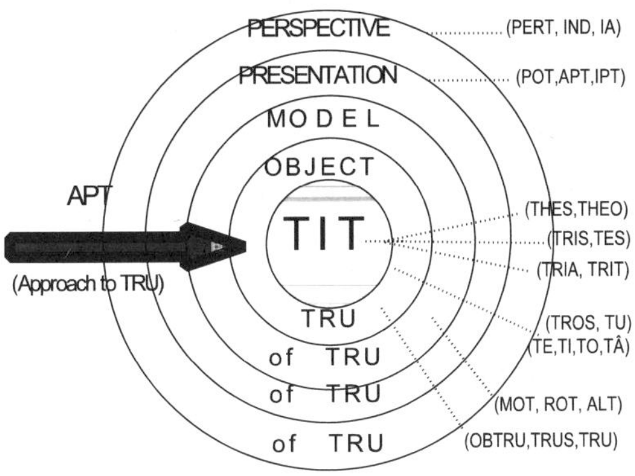

Fig 1-00 **FACADES of an APPROACH to TRU**

There are many kinds of approach, direct or indirect, frontal or rear, lateral, transverse, or sagittal, differentiated or integrated, systematic, standardized or natural. It is not possible to give an immediate and perfect showing or viewing from one simple approach or perspective. We must remember that such limits are of the POT or ROT and not of TIT. In Figure 1-00 alternating shells from TIT, the black central core, and TRU, as with all topics, may be shown from the outside as in APT or from any direction selected. This is what happens when we take the world in our hands and look at it. As can be seen from this diagram and the text, it duplicates these alternating shells to provide the best possible positive objective approach, **POA**, so that we come nearer and nearer to TRU - as to God - by its S-O or A dimensional mass energy predominant form. This is called: **COPA** = Convergence of Objects Applicable through text or Presentation and to all S-O as they equal A or the Predominant form and dimension of TRU. Such 'stemics', or systematic knowledge of all is epistemics or for TRU **EPI.** EPI when combined with is <u>Epistemics</u>. Epistemics is an existing term signifying that aspect of philosophy that not only makes a study of knowledge, it objectively seeks to best order it. This has not been well developed for it requires a set object with known and demonstrable form, namely and specifically TRIT of TIT. By **CET** or Confusion of Epistemics or knowledge with TRU, we can better appreciate what they are separately and together. Epistemics is and is not TRU. This helps to show the epistemics of what is not TRU, mixed, potential, and reversible TRU, or in TIT-TO. Perspective of TRU is **PERT.** PERTderives from EPI largely founded on TRIT, OBT, SUT, MOT, ALT and SPALT and their derivatives. The point of view, viewer and viewed for TRU can only succeed when we visualize TIT-TO, TROS, TRUS, and TRIT and avoid all lesser or non TRU confines of space, time, and mass-energy.

PRET is Perspective, Presentation, Representation and TRU.

This introductory section stands in relation to the next section, I, and all the rest of the text, as philosophy does to religion. Yet they, as God and TRU, can be better understood if the manner and means by which it is observed is taken into account, reversibly, in and out of TRU, **OITIO**. This is: **LOTI** = Law of Total and TRU Interdependence. LOTI signifies and describes - as so often lost by the individual - the relationship, interdependence and SCARLS of all that is EPI with all that is TRU. To better understand by shearing off all that is not actually or disproportional TRU and held to their precise concrete interpretation, as by **VAPO**, validity, actuality, proportionality and totality from TRU, is the dominant theme of this text, a manifestation of **Á** and **SÁ**..

In accordance to the limits of POT and ROT it is good and highly beneficial to have a perspective. It is also good to have standardization, as long as this does not prevent us from looking from non standard positions while looking from and to all as we continually return to TIT-TO..

All we encounter or is given is more or less practically brought to life to become <u>The Given</u> selectively by us at best, **TES**, in TIT. This transition from any given to the Given in TIT is mediated by the accompanying meanings of language whose terms are increasingly defined by TRU. This leads to a positive approach, **POA**, to TRU. It is supported by several basic principles and laws including: the undeniability of many philosophical, religious and scientific truths, **UDOHAT**, a comprehension of all, TRU or **COAT**, the law of diminishing returns or **LADIRE**, the law of added facts, data, information, and evidence, the law that aggregation contributes to proof, **LADOP**, and the law of absolute combined proof or **LACOP**. All humans seek total knowledge, yet quickly succumb to its mass, incompletion, and irrelevance. We are apt to lose interest and appreciation for its use and value. Much of existing knowledge, as **TREQ** or TRU Equivalent, is of great and lasting value. By revealing POT and TRU TREQ reconciles philosophy, science and religion.

Many professionals, dedicated and versed in their fields, suffer from a skewed view of the world. This makes it more complex and thereby less manageable than it actually is. Yet even though it may not be 100% TRU, it does not detract from its contribution. This is UDOHAT. The greatest minds of man since time immemorial, and especially in religion, philosophy and science, have struggled to find the actual meaning of life and all. Their contributions are the undeniable advances in human knowledge and basis of human existence - as TRU. Yet they are incomplete at best and at worst poor, confusing, boring, and fail to provide the necessary comprehension. It is not because it can't be known, what is known isn't valid and great and the individual is unable or disinclined to comprehend all. Too much has been left to individual choice without selective quality in which any bond between individual and given, observer and observed, are accepted. Yet to be the actual truths they have in TRU, they must not only be corrected in quality and quantity, they must be best ordered. To do this we have: **COAT** or the Comprehension of all, All TRU, and **KUPS**, or Knowledge, Universal, Partial and Special.(95) To be able to have COAT and KUPS we need: **AITES** or the Absolute Imperative TES, or Identity of a person with TRU. Such an imperative is not attainable without POA and the demonstration of its object, TRU as by **PI**, the Positive Imperative.

Fig 2-00 **The POSITIVE IMPERATIVE, RELATIONSHIPS in SECTIONS of IPTRU**

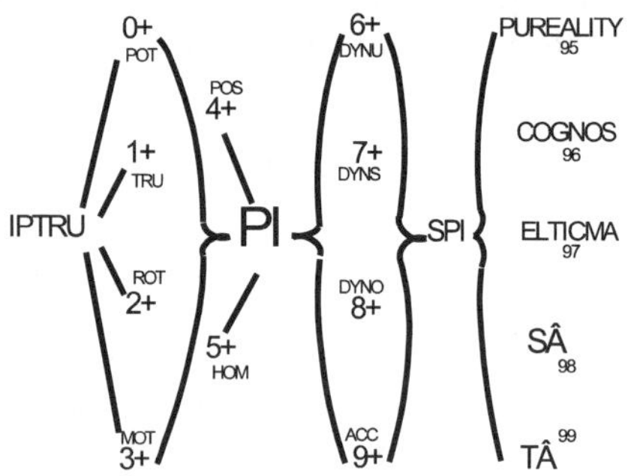

Epistemics and ethics, like philosophy and religion, are not only topics of academic interest to be chosen or perused at leisure. They are imperative by being vital to life. They are positive in the sense that they are for life, no less than what their limits and set make certain and necessary in their participation in TRU. Thus we have their positive and imperative natures and all topics that are of TRU.

A presentation can be no better, and may be measured by the degree it validly, actually, accurately, proportionately, vividly, and totally contribute, applies, and can be used to represent and model this special world of best human life. According to the Platonic Method the first thing to be done in initiating a study, presentation and representation of a topic, object, or the given is to determine and show precisely what it means. That meaning which is best sought for in the concrete, or of the universal in the particular. Remembering what may be considered universal may be otherwise in TRU . In order to do this for TRU we must determine and show all those entities, senses and things which reveal its form and function, structure, modes, qualities, characteristics and properties, as well as the aggregate of these parts and processes, or ES and CES, that constitutes TRU. To do so we investigate all knowledge and nature to sort out what does or does not participate in and show TRU.

The great philosophical problem of whether we can apprehend meaning without words or signs shows us, and its solution provides, the key to understanding the actual meaning of life and all, TRU. Our knowledge can be little better than our language, its concepts and terms, and vice versa. Because we have poorly distinguished between what is illusion, figment, fact, validity, reality, value or importance, and the associated formal distinction necessary, our language does not serve us efficiently. The terms we use can be divided by the extent to which they are not actual, actual, or only of TRU, and of little, some or great value. **OC** signifies existing concepts defined by current usage. **EC** signifies existing concepts whose definitions are limited only to those of greatest actuality to human life and benefit. What is of the essence in each OC. **ECT** signifies existing concepts and words strictly limited to **ESINS**, their TRU meanings. **NEC** signifies all the new concepts that TRU mandates. **TOPNEC** signifies those NEC which are most important to and of greatest TRU value. How we adapt language to TRU is: **LATOC** or Law of Alteration and Adaptation of Existing or Old Concepts to TRU. LATOC leads to EC and ECT, and the void recognized will open up NEC and TOPNEC. That they are, and how necessary NEC are, is called: **LANEC** = Law of New Concepts or the Necessity of a Language of TRU. LANEC is evident in the deficiencies of existing language and repeatedly confirmed by the greatness of TRU and its learning. This confirmation is seen in: **NECIL** or how New Concepts are the Index of Language, Logic, Logos, and the Reality of Relativity, its Observer-Observed State and Mass Energy Dimension, Á.

Fig 3-00 **CLASSIFICATION of TERMINOLOGY and CONCEPTS**

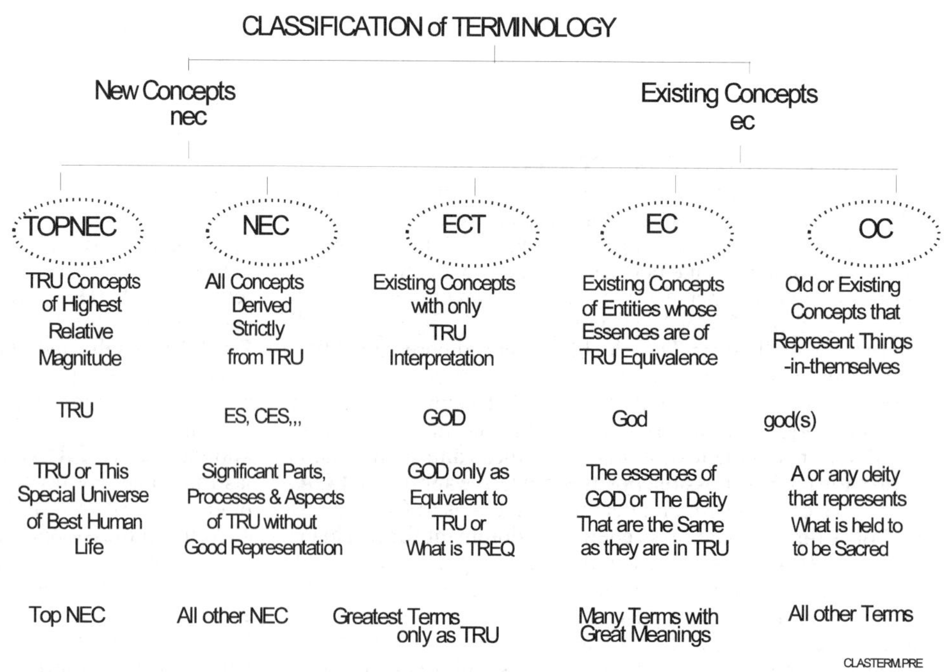

The necessity and greatness of words is signified by: **LANOT** or the Law of the Absolute Necessity of Appellations, Names, and Notations for TRU. LANOT is obvious from the importance of human language and communication. It becomes increasingly absolute when we recognize how TRU and KOT are lost without them. This imperative and their use as laws is signified by: **LAL** = Law of Laws or how by representing great TRU, concepts warranting compliance, directly or indirectly, express them as laws.

Humans have excellent names and words for the most common objects. Yet these words are largely based upon what the objects are in and of them, not how selectively qualified by their part and roles in this special universe of best human life. Names and words have not been given for many of the greatest aspects, parts, processes and mechanisms of TRU. What is the most fundamental and important to us and to our future is either unknown, poorly known, or known in ways that do not enable us to effectively comprehend the world and ourselves. This is why LANT, LANEC,and NEC are needed. The accompanying Terms and Definitions, show,

An **ENTITY** is any item or object that is subjectively made distinct. It is produced by, and in turn produces, relativity and thence, TRU. Entities are made up of senses and things, illusions and actualities, sames and differents, universals and particulars. Its whatness is produced by being separable and finite. In being separately finite it becomes a unit, an ENTITY. If more than a unit, or increasingly finite it becomes entities.

An **OBJECT** is the observed entity that upon which a sense, or senses, are focused, the contraplete of a subject relatively. An object is a key, basic, and a primary ingredient in the world of which we are a part. It is, with the subject, one of the two poles of the predominant mass-energy form of TRU. An object is most readily observed, although not only in one way, through **IA**, Instantaneous Attention and Immediate awareness of which we as individuals are so centered. It is an entity in and of itself to an extent, but as by all within their SCARLS in TRU, is not to be thought of - as any entity, object or RU - as being exclusively so. This by: **LEXCLUP** = Law of Non Exclusive Priority, the DAOST of all in TIT.

The **GIVEN** is what our attention is on, is working with, is at hand, or on hand at the outset. It is the known, the shown, what is available, or upon which what follows is based. Even more explicitly it is what we often must attend to, function or operate with. Most generally the given is any entity which is presented or before an observer. Since in TRU and its presentation IPTRU we began with a great variety of givens we can not easily give them any systematic unity. This is solved after POT and EPI become TIT and again, OITIO, reverse to external relativity. They often originate from themselves and are thereby irrelevant to us. It is the Given that determines the set or type within whose bounds or realm we exclusively operate or must work. It is these givens that so occupy and dominate our lives if we let them, rather than endeavor for **TES**, ones General Counterpart in TRU. The struggle is resolved when they are optimally brought and fused in TRU. This is: **LAGTOP** = Law of The Given and TRU Union or fusion, optimally, totally and properly. LAGTOP assures us of a more accurate presentation of TRU and their necessary overall reconciliation. This Given imposes great limits and likely error, or **ERIN** = Error of Introjection. The mutual substantiation of religion, science, philosophy, and all in TRU with recognition, thought, interest, incentive, practice and life by each person is the key to the solution of givens, 'man's condition'. The greater TRU and TES, or the perfected participation of each person in TRU, are, the less drudgery, conflict, struggle, trouble, misery, loss, failure and likelihood of extinction will occur. The Given, then, must be made by us, not we by it. This is signified by: **LAGMU =** Law of Given Made by Us. We can make POT by the ideal of LAGMU envisioned by TRU and ACC. This is the conflict, which only by climbing the steps of IPTRU, we resolve. For as we see, in beginning this text, we must work with what we have been given. These we change - as exemplified by LANT - to increasingly make LAGMU into the vehicle givens can be to enhance TRU.

In addition to the relation and role of entities, senses and things, subject and object to and in and out of TRU, we need to make distinct what is commonly known about TRU yet is not recognized as TRU or TIT. This is signified by **TREQ** a continuum that runs from what is identical to or 100% TIT to that which is much less, only a small part of TRU, or of which only a small part is TIT, i.e. its **ESIN.** TREQ may also be interpreted and envisioned in two different and opposing or counteracting ways.1. What is of knowledge that is also, or the same as, it is by TRU, i.e. NEC has synonyms in OC. 2. What in existing language can, as well as will, describe TRU concepts. This latter is key to the future promotion of TRU concepts. For many NEC and much of TRU terminology will not reach common usage without the advantage given by what is already known and termed. This second kind is called: **TEREV** or TRU Term Reversal. This is where TRU terms do not become common usage or terms in existing language by reason of other use. This may be of different kinds: 1. where there are excellent existing terms. 2. poor NEC, 3. better depth conception of OC is forthcoming.

'Why TRU''? Why not being, ontology, a being, a human being, the being, God, religion, or the world?. Although these, and especially God, are in ways TREQ of TRU, they are interpreted in so many senses and ways, their meanings are so coterminal, lack formal distinction, and the precise VAPO of TRU, that they are not adequate nor successful to fully represent The Relative Universe. They do not provide a scientifically valid nor comprehensive name for this special universe of best human life. They are not the same, do not conjure up, and commonly divert us from TRU. TRU by its specificity, set meaning, definition, form and limits can be held to a 1:1 representation of itself both qualitatively and quantitatively. Thereby we must clear three great hurdles if we are to readily and sufficiently comprehend and present TRU. 1. We must bring all needed to TRU, in the actual world itself, TIT. 2. In the best possible way we must provide a 1:1 replica of it. 3. We must combine and adapt the above two to the actual make up of all that most clearly reveals why and what this special universe of best human life is, its special and exclusive mentality, speciation, and mass energy form and function.

Why arn't philosophy, religion and all the fields and sciences a satisfactory basis for knowledge or answer to all things? Philosophy doesn't include much that is of religion or sciences, as do religion, fields and sciences include much of each other. So that for all, and since by definition TRU is all, fields, sciences, religion and philosophy, although in ways of the same of TRU, are also different. They fail to reveal TRU.

IMPORTANCE of TRU

All of us, like children, must learn and want to do what we are told to do. Yet, just as important, this we are told to do must be right - must be TRU. This does not deny the associated importance of reason, learning and a persons 'doing ones own thing' as long as this is within **TÍS, TES,** ones special and general counterparts in TRU. If there is not the capacity and incentive to want and learn to do what one should do there can be little hope. This is a part of the competitive form and function, nature, Á mass energy dimension, continuum, state and mechanism of TRU.

TRU is not easy to learn, for it differs from our common senses, varies from the customary, and is even alien to many of our desires. Upon learning it is often not fully appreciated, nor put to good use. Or even when all these are done they are not done consistently. continuously, nor have what is needed for POT, TIT, TO, TRIT, SPALT, POS, or ACC. Yet there can be no denial of TRU for it is by definition, ATO and TRIA, what our lives are all about. Thereby the earlier, deeper, broader and most orderly TRU is available, used and served by each and all, the better it will be for each and all of us.

In our approach to TRU we must constantly reorient ourselves to the multiple foci which IPTRU, POT and TRU produce. This orientation necessitates the optimal application of the dimensions of TRU. Since much of dimensionality is to a large extent universally common, and since it is set by and in TRU we have: **CODIT** = Compulsive Dimensions of TRU. CODIT not only helps to prove TRU, it is of the greatest importance in POT, IPT, EPI, SPU, SPOT, SPUNO, SU, SUT, ROT, MOT, ALT, and SPALT. In order to reveal the commonality and develop dimensions fully and properly, both generally and especially, and for ROT, MOT, ALT and SPALT in TIT we need: **LADIR** = Law of Dimensional Reduction. LADIR is the use of mathematical, geometric, physical and basic scientific methods to work with the actual dimensions of TRU and its representation. It is what can be used in the world around us and our knowledge of it, when adapted and altered as it best fits TRU, to provide the best possible formulation, reproduction, standardization and demonstration of TRU.

That both POT and TRU have more or less Predetermined and Potential ACtuality is signified by: **LAPREDAC**. LAPREDAC expresses that state in which the potential which is predeterminant, and has often been predominant, exists already within it.(ATO) Similarly there exists for TRU and POT much that has preceded them, DOT, AMTRU, TRUS, RUTS and RUS. These are both TRU and not TRU whose distinctions are imperative. Additionally all that is, or is much of, TRU, yet is not known as TRU is continually confused with all that is not TRU. Of these what is TRU is called **TREQ** or TRU Equivalents. TREQ is all about us, our knowledge, language, technology, abilities, advances, which are largely TRU, yet known in their own names or other ways. See TREQ in Index, and proofs in the Addenda of IPTRU. **LATREQ** signifies how we use TREQ to provide, strengthen and perfect POT and TRU. This is supplemented by: **PRUST** or Proofs by Standardization. PRUST are those proofs that are derived from both the form, function, structure, aspects, and nature of TRU, and where no precedents exists, strictly limited and chosen standards selected to best fit TRU and POT. This is exemplified by: **STOP** or the Standardized Sequence of Philosophy, TRU and All. STOP is shown by the common sequence of contents in most textbooks of philosophy, and other topics. It is developed in PACH. These are largely produced by **LOTARP** or the Law of Ordination by the Observer and the Observed, or perspective of TRU and its representation. What follows is: **LORAST**, or the Law of Orientation and Standardization. LORAST is based upon a focus, that of our world, TRU and its IA or Instantaneous Attention and Action. Such a focus is what not only gives TRU much of its relativity, and thereby Á dimensionality, it is what we can best use to provide PRUST, ROT, MOT, SPALT and a strong POA.

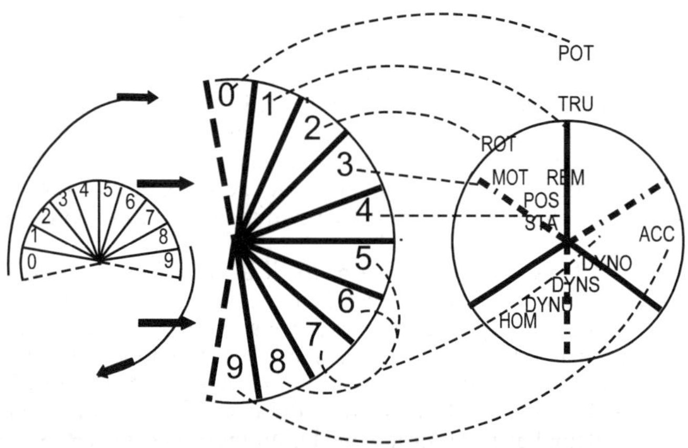

Horizons Rotation and Superimposition on ALT with IPTRU CONTENTS

Fig 4-00 **IPTRU CONTENTS, SEQUENCE and HORIZON, HIDIC, ROTATION, SUPERIMPOSITION**

The application of IPTRU contents topics is shown in HID and HIDIC. **HID** = Horizon of IPTRU Diagram, **HIDIC** = Core of HID. They are provided to orient our view , the text, TRU, and its model SPALT. They help to show how a standardized view, format, POT, MOT, SPALT and ACC can all proceed naturally and by TRU.(POMOT, METAD, SPAT).

KNOWLEDGE and KNOWLEDGE of TRU

Those concepts that help us understand why TRU has been unknown, or known only indirectly and in fragments include: **KOT** = Knowledge of TRU, **UNKTRU** = Unknowns of TRU, **ANKTRU** = The difference between TRU or TIT and the knowledge of it, **INKTRU** = reduced Image of - Knowledge of - TRU, **TRUT** = TRU is truth in which KOT is without UNKTRU, ANKTRU, INKTRU and all less or other than TRU. TRUT is reversible knowledge and truth, or the law of combined Truth. This is seen where TRU is presented as an object, **PROT,** and the subjective, **THES** and objective **THEO** portions of TRU shown. Each and all the steps in knowing and showing TRU that produce loss, include KATRU, UPTRU, URTRU, USTRU, UMTRU, AKTRU, and UNKIT, or lost through perspective and presentation, representation, standardization, modeling, mistaken knowledge, and by the individual's limited knowledge of TRU. It helps to keep these concepts in mind, and focus on TRU as we collect all the data and evidence that support it. We must rise to a higher level of support for the actual world. Above how philosophy has rationalized it, religion has deviated from it, and science has trivialized it, thus TRU itself, in totality.

The means of knowledge is largely learning. Learning that has been the great energizer of human culture and life through the ages. Learning includes, not only what is new by each new individual, but relearning the obvious, what is often forgotten, neglected, disproportionate, or unappreciated. In fact it is to the development of deepest interests in elucidating the obvious and bringing it to TRU through best habits of thought and notation, **TAN**, that continuing learning, KOT and TRU are to be sustained. The differential and distinctions necessary for wider and greater wisdom are exemplified by the word Interval. Interval for TRU is seen in how we can with increasing precision stretch out the medium of the great spectra or continua of mass energy relative to human life in such a way as to provide reliable and representable parameters and dimensions. It is the relativity and differences in TRU and their opposition, existing as polarities, that are often the best means of knowing TRU. This is seen in: **OR** or Opposition from Relative, and **POP** or Proof in Opposition. To understand TRU it is not enough to understand it dynamically, but how these differences work in various ways, levels and degrees of opposition. Thereby, much of learning, takes intuition, interest, inspiration and revelation sufficient to grasp the needed concepts, principles, laws and mechanisms.

Proofs when produced with veracity, by TRU with VAPO, are the basis for lasting knowledge, and the great vehicles of TRU and POT. They often arise from paradoxes whose resolutions occur by: **ÄSOP** or the Ä or basic Form Solution of Paradoxes. What is in doubt or seems so paradoxical can often readily be made known by analyzing basic **FAF**, TRU Form and Function. Enhanced verifiable proofs are produced by: LINPRU, LADIRE, PRAPOS, LADOP and LACOP. **LINPRU** is the Law of an Infinite Number of Possible Relative Proofs. **PRAPOS** signifies Practical Posits where posits of proofs are limited to those that are most appropriate by REM, proof value, implication and stimulation. **LADIRE** is the Law of Diminishing Returns. This is the loss of attention, value and proof with accumulation. **LADOP** = Law of All Added facts, Data, information and evidence in Totality contributes, more or less to proof. **LACOP** = Law of Absolute Combined Proof. LACOP and its methods and categories certify that with enough information and evidence, especially in concert with related topics of TRU, absolute proof can be and is often achieved.

In order to provide a good proof it is necessary that we prove not only *that* something is, we must prove *what* it is, and what it is *actually*. These three are highly interdependent. Since by DAOST much of TIT, as well as all about it has much that is different, it becomes evident that proofs are neither simple nor absolute. Proofs thereby form spectra of complexity and relativity.

CHAPTER 01

DERIVATION, DEVELOPMENT and DETERMINATION of IA, CAUSATION, ENTITIES and POT

In contrast to TRU in which derivation is dominated by ATO, in an approach and presentation of TRU derivation is dominated by our consciousness and Instantaneous Attention, **IA**. Its continuous functions and Imperative Convergence in Life - as TRU - is **LIACT**. This is a fixation in relation signified by **PORF** or the Principle of Relative Fixation. It is accumulative, or **LABE**, Law of Accumulative Origins, Beginnings, and Ends. The imperative functions and accumulation produce actuality which becomes increasingly formally distinct by and in TRU, **LACDIT**. Actuality is a fulfillment of PORF in which because of relativity, which conditions form, is shown by **LANERA** and **LAMOL** or the Law of Never Absolutely Relative and never Absolutely Absolute, and the Law of More or Less. There is a dichotomy of objectivity depending upon whether the focus and reference is based solely on the entity or universe itself or upon the subject, a person, or some mixture of the two. It is this problem of focus and reference that so destroys man's acceptance of a total unified theory of all. Such a vista of a universe can also be compared with the view of a black hole. In fact we may as easily have two or many black holes depending upon the direction and nature of our view. For in one extreme it may be the black hole of infinite outward extension or one of infinitesimal inward contraction. Likewise the same may be done for time, or for the mass-energy dimension, or whatever polarity we may select or assume that provides a good and demonstrable opposition with infinite extremities. The incipient POT is highly illusive, just as man has been unable to agree upon what came first. A part of this problem is that of focus. A major focus is that of the object we seek to make of the entity. This is very apparent to us as we begin any observation, or attempt to show what an observed is. We have a commonly understood, yet very rough, idea of what universal kinds of entities POT is striving to provide. This is not by a standard view, as SPOT or SPALT It is a valid interpretation of a thing, an object, a relative universe as a solar system, or galaxy. For them, as most all unified systems and universes, tend to concentrate at the center and with increasing distance from the center become less dense and existent. Yet if we properly standardize TRU and by SU in modeling as of ALT and SPALT, and give this equal proportions throughout its extensions, it will develop an object whose mass is evenly distributed and density that is properly shown. A mass of unequal density is disproportionately represented and thereby must be corrected. Yet man represents the solar system, universe and entities with least density at the center passing to much greater density at the periphery. This is because of relativity, which also conditions form as shown in LANERA and LAMOL..

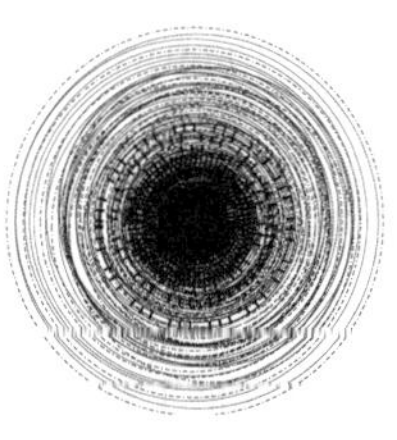

Fig 1-01 VIEWS of A UNIVERSE

In Figure 1-01 a common and rough showing of a, or the, universe is presented. It looks like an astronomical or stellar object from a fixed center. It is similar to the concentric formations about a black hole. This is significant, for the fact that it does not give legitimacy to the black hole as an actuality other than TRU, or a special universe in itself. It shows how objects differ from ENITS. It is what we as observers make it, an object neither ENIT nor an aspect or part of TRU. Such a depiction of a universe is highly indicative of a thing rather than nothing, an entity rather than a non entity, of AU and thereby the beginning of our attention and attention to a critical analysis of subject and object as we extract ourselves from the universe and views of it.

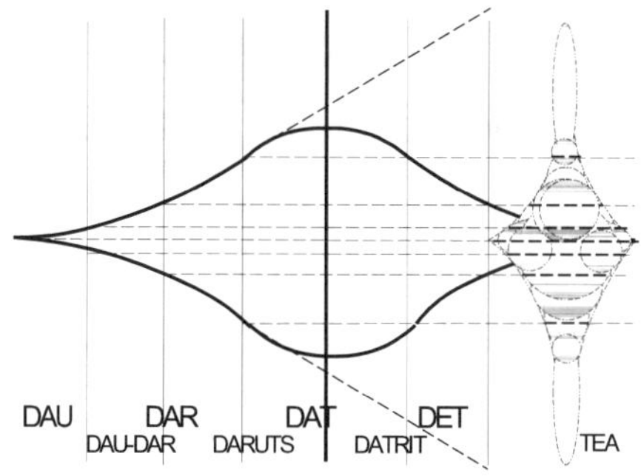

Fig 2-01 **PROBLEMS/ SOLUTIONS BASIS**

How to know what is the best, can only be achieved with knowledge of TRU, COAT and the basic and fundamental concepts, new and old, as shown by IPTRU. The first principles, and great determinants of action, arise througb derivation from IA, form, objectivity, and formulation by relativity, and TRU.The basic relationships of TRU, DIA, AD, DA, DAO, DAU, DAR, DAT, and are less from the details of ES and CES, as great as their details and solutions are in providing a good showing and understanding of TRU, especially by their intricate massive powers in the expansion of TRU. It is in a way for this reason why so much of philosophy and religion have neglected the detailed ES, CES, mechanisms and posits in a coordinate framework. This is called: **DEPAR** = Primary Principle of the Derivation, Development and Determination of First Principles and the Determination of Action from Basic Relationships. This we add throughout the text, especially in Chapters 11, 21, 31, and 51.

If we are to find the answers to the questions about derivation and development we must look for them in APT and POT. They are not to be found in stellar evolution, remotest beginnings of life, nor even human origins. For, although these have their place and participate in TRU, DOT, and POT somewhat differently, they are not the prime reasons for derivation and development. This is to be found in the focus, core and media of TRU. Some of the concepts that exemplify this fact, and provide the basis and great meaning to derivation and development include, IA, IAT, PORF, LASO, ENIT, SENIT, THIT, LAC, LABE, UEIOA, DIA, SELIA, DIRIA, SIA, ADIA and LARPON. **IA** = Instantaneous or Immediate Awareness, Attention and Action. When specific for TRU it is called **IAT**. This is the focus, core, and much of the media of life in various levels of complexity. **PORF** = Principle of Relative Fixation. PORF is revealed in how we attend to life, what it converges on, and becomes one with TRU. **LASO** = Law of Standardized Sequential Origins, or how IA and IAT are so representative of the point of reference, focus of relativity in TRU. This is the point where the observer and observed meet. **ENIT** = Entity in itself or what has specific existence exclusive of all else. It is the basic unit of separate existence upon which all else can be derived. **SENIT** and **THIT** or Sense In ITself, and Thing In Itself are the two kinds of ENIT derived from relativity, observer or observed entities. **LAC** or Language, words and terms are derived from cleavages, divisions, and contrasts. What is the same, needs no name, and, what is differentially irrelevant also needs no name. It is from the differences, contrasts, cleavages, divisions, and dimensions of TRU that names are most needed and beneficial. **LABE** = Law of Accumulative Origins, Beginning and Ends. LABE certifies how origins or beginnings and ends do not usually come from simple or singular causes but are great accumulations, which can with TRU be focused and ordered. **UEIOA** is how IA, as much in TRU, consists of unity, partiality, infinity, totality and absolute forms. The relations are brought together, by Plato, as the absolute form of all participate in the supreme unity, or idea. **DIA** is the Derivation and Development of IA, **SELIA** the selective IA, **DIRIA** the direction of IA, **SIA** the summation of IA, and **ADIA** Adequate and Sufficient IA. These represent higher states of IA becoming IAT in the derivation and development of TRU..

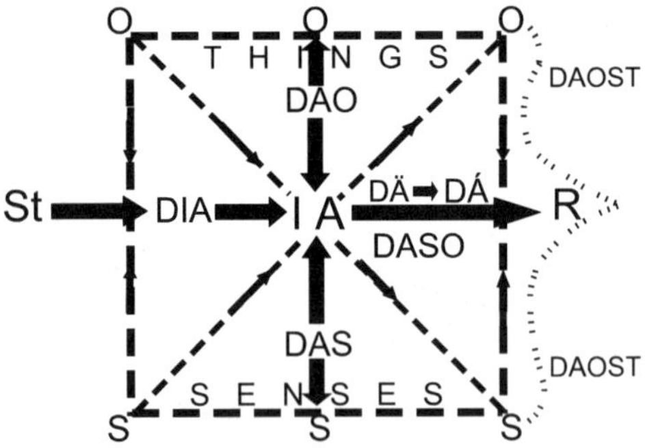

Fig 3-01 **I A DEVELOPMENT, IAD**

This figure reveals the development of IA and much of being. Such development is a combination of the incipient capacity of relativity in which sense and thing become dynamic through being acted upon by stimuli. As these actions increase reactions occur, some of which include the increased development of IA and its accompanying **DÄO** and **DÄS**, or Developing Form of Objectivity and Subjectivity. These are combined in IA and continue to develop. Their form and function expand, and their capacity to recognize and react increases. The development of this combined form and function is the development of **DASO**, or Development of the Form of SO. DASO is typical and exclusive for each organism. It is a combination of many forms of mass energy that make up their predominant dimension. For man this mass energy or m is called, its **Á** dimension. It is that Form And Function or **FAF** unique, typical, selective, and exclusive for humans. The development of this is called, **DÁ**. The dimensions work together as different aspects of the same thing, **DAOS**, or of TRU, **DAOST**.

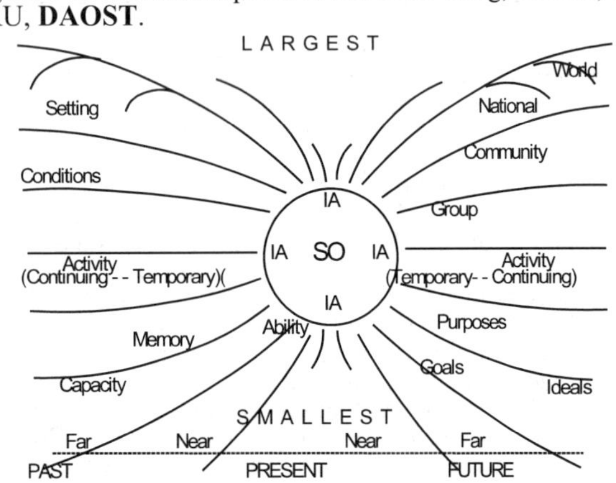

Fig 4-01 **REASON CAUSATION and the Field of IA SO**

The great difference between **RANIA** or random, inadvertent or cursory IA and IAT is that between IA and SU or TRU. It is between minds that are not or are trained and experienced. This is **NUIA** or a fleeting glimpse and/or act. We are much more apt to anticipate and see movement and action, and even more go to where it can be best watched, analyzed and ordered. So does IA proceed to SU. Or if instead of going out to hunt game and expect to walk up on them, or expecting them to be right there, find the best spot and wait for them or move about to hunt, find, flush out, or trap them through an understanding of their nature and behavior. So also in fishing are there many factors of both the fish and the fishing. As in much we do they are analogous to all that brings IA, SO, Á, and SA to completion and success. The IA that we live by is even more complex and infinitely more critical to our existence. Yet we often give it less attention, experience, training and technique than the hunter or fisher. For the movements, habits, nature of the world we live in, and especially the total actual world of TRU or TREQ, TRU Equivalents that we can so readily describe, formulate, show and know yet ignore and neglect, makes us very poor hunters and fishermen of the greatest catch of all, our lives in TIT, TES. More profound is one's ability to anticipate and realize the full significance of these NUIA. This is called: **RATNUIA** = Realization of TRU Form and Function of the NUIA as the Nuances of life that are before us. Only as our IA, RANIA, void, inadvertent and undirected IA comes to identify what we should attend and proceed to all stages, levels, steps of Total Identity of ourselves with all, and The object TRU, does IA rise to its full potential and everlasting glory.

As people believe so do they behave. This common adage has great truth and power, yet does not always follow what is and how we achieve the right beliefs. What prevents and keeps us from following wrong beliefs is how well we are trained, educated, and apply ourselves to best diligently live by them. One of the greatest powers in development is meaning. In actuality this is continually reinforced, **POROM,** or the principle of the Continual Reinforcement of Meaning. Actuality is met when the potential IA becomes the actual IA, specifically **IAT**. Herein is the Law of Causality, **LOC**. This is seen in the causality of Hume and Kant and denotes how causality depends upon relativity or the actuality of the observer-observed. Actuality in IA becomes directed, selective, adequate and sufficient.

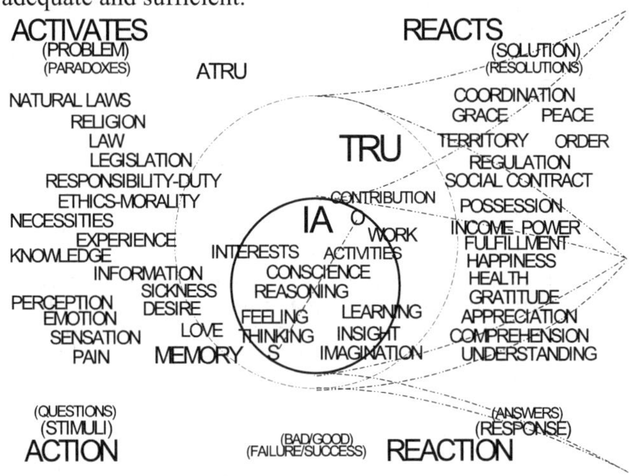

Fig 5-01 **ACTIVATION and REACTION in IA, IAR**

That IA is highly conditional to all about it, as is inferred from its derivation, i.a., inter alia, or among other things, is quickly evident on study. IA is also evident from one's intuition and consciousness, along with feeling, thinking, reasoning, conscience shown in the inner circle Fig. 5 In having much in common with other things it is highly relevant, and thereby does not have hard and fast limits, discreetness nor distinction. Yet it is as close as we can get to the center or focus of this special universe of best human life. For it is the core of our being, the balances between problem and solution, paradox and resolution, question and answer, and stimuli and response. It is in the continuous action and reaction that a relative universe finds itself, dynamically, that is the main determinant of its being. If it can increase the good and decrease the bad, if it can dominate to decrease failure and increase success, it will prevail, survive and have life. Figure 5-01 provides a rough and preliminary diagram of this highly dynamic and basic state of life. It is important for future development, relations, and use of POT and TRU.

Some of the common concepts of development include: LANIP, meaning, POROM, POKED, SEVI, and LIACT. **LANIP** = Law of Negative is Positive through showing or adding to the positive, especially TRU. LANIP reveals the great positive and negative dynamics in development that are highly conditioned in TRU. Meaning is one of the greatest powers in development, whose quality is critical in TRU. **POROM** or the Principle of Reinforcement of Meaning, denotes how much meaning is accumulative and depends upon added concepts. **POKED** = Principle of Explosive Knowledge

Revelation through analysis, Depth conception, Deduction and intuition. POKED shows how much development, especially of TRU, KOT and DOT have depended on increasing levels of reason. **SEVI** is the Search for Vivid Imagery and Identity. It is a continuing feature of development, life, and TRU. **LIACT** = Law of IA Convergence Imperatively onto TRU, or the continuous development of IA not only to completion, but fulfillment in TRU.

POTENTIAL and ACTUAL

The importance of the potential has been distinctly recognized - at least since Aristotle. It is even more important in its special implications for TRU. This becomes most apparent with LAPO, LENAD, COMPIA, LAGPEN, and NAM. **LAPO** = Law of Actuation of the Potential, those processes, specially FAF in TRU, in which the potential becomes actual. **LENAD** = Law of Added Entity, or how as we pass from one entity all that follows alters the potential actual relationship. **COMPIA** = Compulsory IA, how IA is continually realigned and directed toward certain ends, potential to actual in TRU. **LAGPEN** = Law of Potential and Gradation of Entities. In progression among entities, as to TRU, it consists of gradients, in many dynamic modes, which may be determined. **NAM** = Identified Entity, Named Nominal, or how potential entities tend to become actual when designated.

WHY, REASONS, CAUSE and EFFECT

The question about and problems involved with causes, reasons why and their effects have been of preeminent interest to mankind through the ages. This is neither unnatural nor ATRU. For man has, through experience, come to appreciate the great rewards such information produces. Yet the topics of causation, cause and effects, and reasons why, have not been well answered. Such a presentation where we treat a ramified categorical base as well as an approach to TRU is stretched to the limit with extra hurdles in which any offering of causation needs to be reduced to a simple form and state. That cause and effect are valid can hardly be doubted. However they must be conditioned, SCARLED, and brought by VAPO and TREQ to TRU. Herein is the basic and key to a proper and pure critique of all, by TRU.(Kant, *Prolegomena to every future metaphysic).* By POA a natural and standardized operation stands between what is potential and what is actual, what is the derivation of all and what is their culmination. For this reason we approach the problems of causation broadside, as we do with the other categories that partake of such an approach to TRU and its presentation POT. **RECAPT** = Reduction of Cause and Effect to truth and validity. This occurs when the relation of cause and effect are proven to be true, and when they are TRU. **ACIA** or the Actuality of IA. Actuality is met when the potential IA becomes the actual IA, specially IAT.

Determination consists of both distinction and conceptualization in actualization. This is supported by LAC, SACDI, and LENAD. **LAC** signifies how Language, words and terms are derived from Cleavage, division and Contrast. **SACDI** is the Separation and Construction of the Different, Delineation, Division, and Definition. With the search for vivid imagery, **SEVI**, there is, **POKED,** the Principle of Explosive Knowledge Revelation through analysis, Depth conception, Deduction, and intuition. This results in **PORAPS** or Principle of the Ready and Proportionate Satisfaction by Sufficiency of Knowledge of an entity or object. The dynamic observer is in continuous **SI**, Subjective Inversion or where in relativity the observed object of ones attention is often rapidly being inverted. In combination this becomes **ISO** or Inversion SO. With the expansion of IA, Forms Equal any Mode of the Observer, **FEMO**.

By investigating and analyzing the gradations between our observing and perceiving or subjectifying the images, objects, things, and entities in the world around us we seek clear distinctions. Likewise we extend this distinction to TRU as The Object, TIT or TRU In and Of Itself, as well as what is the form and frame of TIT, TRIT, OBT, or SU. We can further extend this distinction to how this frame can be modeled or SPALT and how such a model of TIT relates to all about it, i.e. all other things and senses. In this way we develop a realistic and actual vantage from which an investigation of TRU can proceed..

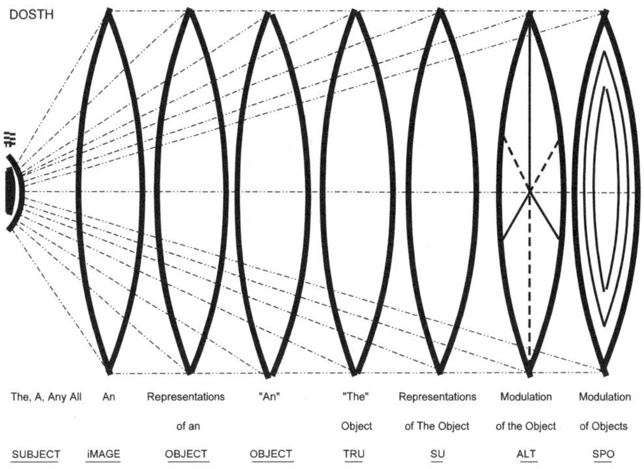

Fig 6-01 **DOSTH Diagram**

We may derive, develop, demonstrate and determine either differentially by taking all as separate and discrete entities, or integrally by observing and perceiving the whole. It is how we can do both together in exquisite detail that is the great quality, beauty, glory, power, success and eternal assurance of our being.

The incipiency of subjectivity in all objectivity is called: **LISO** = Law of Incipiency of Subjectivity in All Objectivity. The reciprocal of this also holds: **LIOS** = Law of Incipiency of Objectivity in All Subjectivity. LISO is a manifestation of the fact that in TRU, absolute subjectivity and absolute objectivity are non existent. By LISO all we see, is shown, represented and demonstrated is less than or other than the entity it represents.

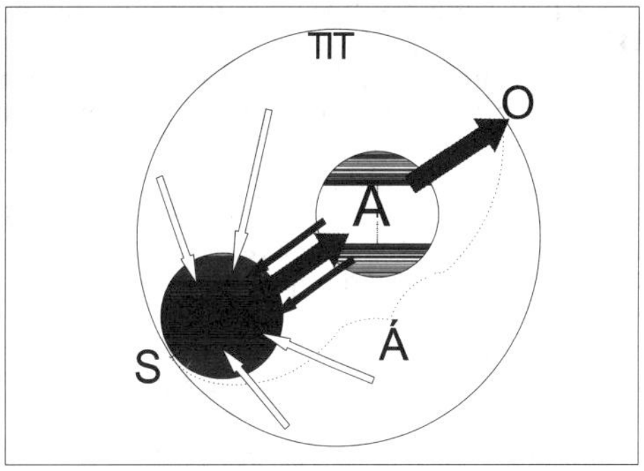

Figure 7-01 **QUEST for KNOW-LEDGE**

The interactions between forces, questions and answers, vaguely seen in the two inter connected circles inside TIT are shown. For POT and APT, these proceed also - and largely - from outside and toward TIT. Although the Greeks were held, even by the 'barbarians' in Hellenistic times, with esteem in their search for knowledge, they had no monopoly on, nor exclusive reverence for, wisdom. The Sumerians, Egyptians, Israelites, and many others were devoted to its pursuit including an identity of wisdom with their major deities. With deep curiosity they searched the length and breadth of their world for hidden and deepest meanings.

The stimulus and ability to ask questions and seek answers ranges from the most simple to most profound, childish to most mature, practical to the most devout, and ignorant to those based upon the best of existing knowledge. It is among the greatest means of acquiring and satisfying ones mass energy observer-observed, S-O, or Á FAF. It has been a great continuing feature of human activity for many thousands of years. In the ancient world, identity of many peoples with their Gods was by questions, beseeching proper knowledge through supplication. What we do today in learning and education, as it will undoubtedly continue in the future is through questions and answers. It is often far more effective than rhetoric or composition. Questions and answers largely come, if only indirectly, from one's needs, and are deeply felt. An analysis of religions, ancient, contemporary and in the future, can often be made on the basis of what and how these questions are and are made. This shows how important it is to have the right knowledge, as by the best of science, TREQ, and TRU, for both questions and answers to be most effective. It is this deep curiosity to answer the questions and solve the problems that man faces, especially those that reveal profound results, that has increasingly characterized humanity. This is natural, for it is the purpose of life and the predominant highest levels A mass-energy FAF that propels and compels mankind. It is, becomes and empowers a kind of high-order animal tropism and taxis. It is not simply the stimulus, question or problem that is the sole generator of this power. It is how our whole being is so oriented, formed and functions. For the subjective, human mental power, is a greatly and highly determinative part of the A, mass-energy, observer-observed and relativity form of mans world, TRU. In turn, they expand and produce many characteristics of TRU including curiosity, depth conception, determination and conviction, that help to show the actual nature of all mass energy, especially this that is so typical of TIT.

These are illustrations of the greatness and totality of our IA and associated psychology. They do not reach a complete description of actuality, extensity, intensity and totality ACTA, EXTA, INIA and OA. They are but elements, parts of an object, a greater whole, IA, OIA, DYNIA, CORTA, COMIA. These in turn are but a part of a greater whole COAT, POAT, COT, POT and SPOT. They stand as a variable part of TRU, depending upon our emphasis, or sense, and whether TRU is being comprehended and produced or as it is TIT, in and of itself. All parts of TRU and TRU itself, are almost entirely surrounded by forces and forms that are non SU, non TRU and ATRU. This reveals **ACIA**, actuality of TRU, which is: (1). Not only actuality of S → O, (2). Not only actuality of S = O, (3). But the actuality of S=SOT=O, as they exist in TRU, the fundamental, A, mass energy predominant FAF of our being. It is in this way that actuality becomes formally distinct in TRU. **LACDIT** = Law of Actualities Formal Distinction in TRU. That actuality and LACDIT are close to A, the mass energy predominant form of TRU, is readily apparent when we understand both, and the identity of SOT with A.

We see, from Figure 8-01 along with Figure 6-11 how the expanding and contracting spheres or circles of IA, EXPIA, are in a continuous flux in life. The parameters include:

1. Degree of Objectivity, sensed object of thought
2. Degree of Extension, Field of
3. Degree of Intension, Degree of Intensity
4. Degree of Intention, Degree of Intent & Purpose
5. Degree of Interest, partly a combination of # 3 and 4.

This figure could also be done exponentially. It is a type of standardized perspective of IA or Instantaneous Attention, SPO or Standardized Perspective and Presentation Total and **SPIA** or Standardized Perspective IA. Although IA is, like TRU, generally thought of as a single and solitary entity this is not so for either, except as made objectively.

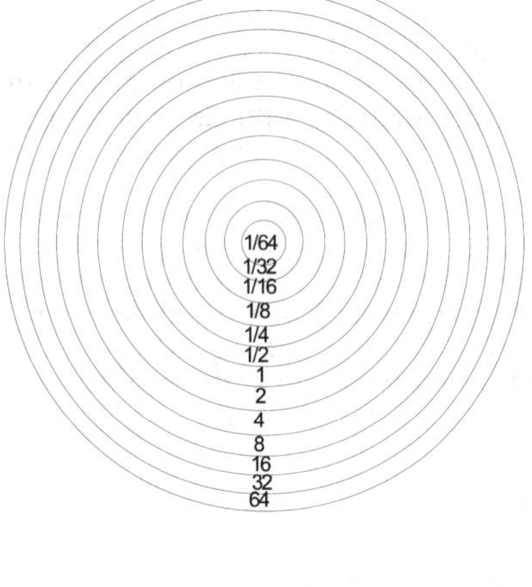

Fig 8-01 **EXPANDING and CONTRACTING CIRCLES of IA**

IA not only has many sets and divisions, it is highly dynamic and variable as are the many levels of consciousness. These diagrams help to show some of this divisibility and variability. In IA it is more directed and involved in both sensory motor or subject object dimensions as it in TRU becomes, IAT, highly coterminal with the A state, dimension and form of the overall predominant mass energy of TRU. **EXPIA** or the Expansion of IA is, thereby, of several forms. It helps us to understand: FEMO, how Forms Equal any Mode of the Observer. FEMO increasingly holds, in the SCARLS of TA, when the observer and its observed are of and in TRU.

The primary method of the text is what is noted on the title page, namely its investigation, analysis and demonstration. This is done, as in Euclid's Elements, by the method of definitions and propositions. We can establish the elements and particular makeup of the world by defining, with the utmost formal distinction, what its concepts and modular bases are. Pythagoras' all consuming aim was to make such methods systematically universal and divine. Nicomachus of Gerasa claimed that it was Pythagoras who first defined philosophy as the love of wisdom, the need and perfection of the cosmos, our actual universe, which alone can find proper and true harmony. It was this same Nichomachus who sought to find the most perfect proportion. This he found in what is three-dimensional and embraces all. Especially notable is: **SACDI** = Separation and Construction of the Different, Delineation, Division and Definition. SACDI is like fine tuning the differential and integral convergent on the FAF of TRU. **DISO** = Differentiation of the Subject Object. DISO signifies the dynamic steps in S and O functions, as well as the different types of their relations. **OBES** = Objective becomes equally Subjective and vv. TRU is not only ENIT, it is OBIT and SENIT, equally subject, object, and all in between, conjointly and dynamically. **SI** = Subjective Inversion. SI is that process that goes on continuously in any person, their subjective and IA. It is where in relativity ones observed or the object of ones attention is constantly and often reversibly being inverted from moment to moment or from time to time. This will depend upon ones position in relation to that about us and TRU. It can be seen in the **SE** or Subjective Error, where all is recognized only as it exists in this sense. SI shows us how much we are right or wrong. How much others are right or wrong. This is often dismissed as being of their point of view, or semantics. It is a function of the limitation of the subjective and the individual. Even if the subjective and individual are 100% TRU SI still holds, yet often in a much different way. For the subjective and individual in TRU, as SOT and TES are still far less than TRU and thereby partial. In being partial they will assume the vicissitudes of the dynamic dialectics within TRU. They will change to meet the necessity of TRU. It is in this sense that we must always in our thoughts, be they TRU, not TRU, or a variable dynamic mixture, be open to criticism and change. Yet if they are TRU, and the highest or most appropriate for TRU at that occasion or of that Given, they will then need only that inversion of S and O sufficient for this adaptation in TRU.

ISO = Inversion S O. (see SO diagrams, THESOS and DAOST) ISO is where the parts are never the same as the whole, nor even less, each other. In this way as our subjective or thoughts are a part of TRU, and what they are on, the observed or objects we attend to are also a part of TRU and not the whole, for them to come together and function, S-O, they will be inverted over and over again as they best fit into the dynamics of the whole. ISO helps us to understand the extreme lability of what we naively think is so sure, certain, and stable. SI and ISO are same and different, and in combination: **SISO** = Subjective and SO Inversion.

Both the Subjective as it focuses on an object, and the Subject Object together, **SO**, as they function and especially as they become their counterparts and function in TRU, **SOT**, are similar in being part of relativity or the observer observed. They are thereby largely SCARLED by the A form, state and mass energy dimension of TRU. SI being commonly more without TRU will become inverted more inadvertently. ISO being more within TRU will become inverted more out of the necessity of TRU and their dynamic flux in TRU.

CHAPTER 02

GENERAL FORM of ENTITIES, SENSES, THINGS, PERT and POT

Form is the static shape or structure of an entity perceived by the observer of the observed, or senses of things. It is a fundamental aspect of an entity and to some extent determined by any sense, but largely determined by its own nature, what it is as an object, its relativity, and especially relation to man, as by TRU. Form is signified by **Ä**. or for TRU, **TÄ.** This is supported by **PÄFOS** or Primary Form Provides the basis for Finite Order, Coordination, Systematic unity, Solution and reduction. Form is qualified, as shown in; **QUOL** = Qualified by its own nature, as an ENIT not from without. Such qualities or kinds of form are called: **SCARL-S** or States, Conditions, Limitations and Relations. The form, QUOL, and SCARLS of entities are altered - and often become - especially seemingly - much different when a part or aspect of TRU. (ESINS, ES) This largely determines the difference between not TRU and TRU as well as what is less than all TRU and all TIT.

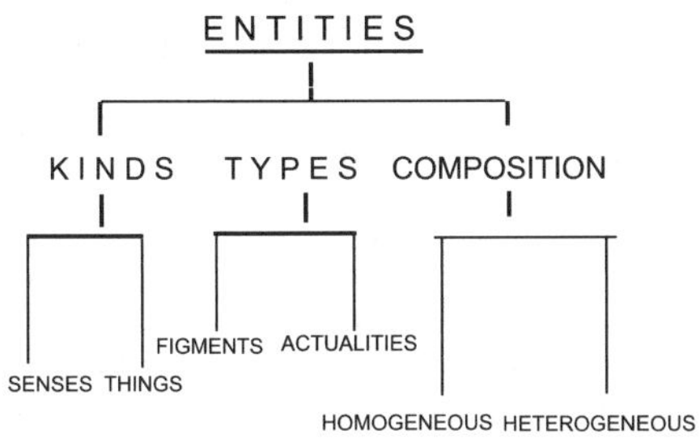

Fig 1-02
CLASSIFICATION of ENTITIES

The form of entities is distinguished by their major types and composition. This is of kinds, as senses and things, of types, as figments and actualities, and of composition as same and other or homogeneous and heterogeneous. From **LAHO** or the Law, and limits, of Homogeneity, and **PLO** or the Primary Law of Objectivity - the preeminence of heterogeneity, we can see how all objects and RUS must obey certain laws and principles. Yet some entities may be entirely the same, as well as same and other, i.e. both homogeneous, heterogeneous, OR mixed entities. They may likewise be homogeneous qualitatively yet heterogeneous quantitatively. The separation of entities by kinds, types, form and composition helps to give us a better idea of what entities are, and their formal distinctions. As we compare and determine the different aspects of entities we can better lay the foundation for discovering, learning and understanding all things and provide a better introduction to TRU, its representation and presentation. Figments have the potential of becoming actualities and actualities of becoming figments. This depends upon their SCARLS and degree to which they are determined and used properly. If these forms of entities and **AUS** or Absolute Universes are known and used - as they are in TRU - then they become TRU. When actualities are not made distinct or used improperly without appropriate SCARLS they become figments, less and less of TRU, by being without identify and validity in TRU. It is by providing SCARLS that we provide this crucial indication of validity. It is the crucial process in the S-O or Á dimension, state, form and function of TRU. This Primary Form Provides the basis For Finite Order, PAFOS. That Form is Derived from Limits is **DOFIL**. Thereby it can be seen that as discrete and clear as forms may seem, in the actuality of TRU they are altered and demand a new reality imposed by S-O the observer-observed relation in TRU. This involves a transformation, or **TRÄ** or **TRU** Ä and Á, which when in and of TIT and TRU becomes **TRÄT**.

DEVELOPMENT of FORM

Forms, as discrete and clear as they may seem, are usually neither that clear, valid, nor actually representative. Thereby we must always be careful to not only investigate them thoroughly and accurately, but with a high degree of doubt and TRU qualification. Form may be internal or external. It often largely depends upon how well we do or do not know the entity and what it is as an object. Herein is the key to form dependency upon the observer. Form may be displayed from no form to a highly intricate form. All show the same entity and thereby great similarity. Yet there are also differences depending both upon what of the entity is shown, how it is observed and known. This is what happens between what exists in our minds, by not knowing and a gradual opening up of what an entity and its form are. Losses occur in transition from entity to object, subject, representation, presentation and perspective. By this indistinction of forms we should avoid choosing and judging them ineptly, but investigate thoroughly and qualify all entities and their forms to observe and bring them to full significance in TRU. The vital importance of this is seen in the upcoming chapters, namely in the PACH, TIMS, SCARLS, and Formal Distinction. This lays the groundwork for what we can and can't do to separate, formulate, reproduce, define and make meaningful the world about us.

DOFIL or the derivation of form from limits is seen in how limits, and the tendency of failing to recognize and follow these limits, produce fallacies. Fallacies thereby tend to follow form to the extent the limits are not adhered to. The concentrations are of intensity which in turn leads to dynamics since t's or times are, and l's or lengths are not, complete without each other. .. This is in totality as they are segments of parameters or dimensions of TRU.

LOUC = the Law of Universal Causation. Can all observable events be explained by a continuous and accountable determination of causes? This humans have sought to discover throughout their past, leading to expanding knowledge. We have sought to find the simple answers to all by showing what caused them and the effects involved. As is so often the case once the causes are known and the effects or mechanisms involved explained .the understanding and what to do about it are readily achieved. Yet without relations, VAPO, SCARLS, TRU and TIT-TO causes, effects and their use are incomplete, inaccurate, disproportionate, disordered , chaotic, and produce failure.

Although surrounded by the infinite, our attention and thought focuses one at a time. We thereby prefer single causes and effects. Yet this is our error. For not only are causes and effects multiple and ramified, they are selectively determined by TRU. It is in relation then, that we seek to tie all causation together. LOUC is valid not so much as an OC, but increasingly in EC, ECT, NEC, TOPNEC, and TO. The question then arises, to what extent can this order of cause and potentiality be brought into an overall or universal order? This is what LOUC asks and signifies. Failure to recognize the SCARLS and LOUC has been a major reason why TRU, its concepts, mechanisms and formulation have not been understood. It is also a reason why scientists have usually turned away from interpretation and championed the discovery of evidence, fact, specific problems and solutions. It has blinded them from an overall view, the combined philosophical, theological and scientific understanding necessary for and available from TRU.

The total form of a RU or TRU is found by S-O, the subjectivity of objects, as IA seeks to determine, and enables, form. For without being observed, the observed has variable form. The reciprocal of this is the transformation of energy-mass, the basic character of all - entities, things, and senses, into a level and form of energy that is present in any set entity or relative universe. The synthesis of this in man's world, TRU, is produced by, and is producing Á. This Á is that component or dimension of mass-energy that is characteristic and predominant of TRU. Á is to be held distinct from the Ä of form. The A of the mass energy, is a subset of Ä, or form. Á is only of one dimension or form of a relative universe. It is the transformed, altered, adapted, and a special kind of form that so distinguishes TRU. It provides us with the mass dimension for TRU, so essential to ROT, MOT, ALT and SPALT. Each higher level of change in this energy state produces a sequence of differentially altered states characteristic for each relative universe. Their critical points are states of transmutation featured in BAS, BASFANT, PACH, and SCARLS. It's perfection and continuing attainment is the predominant and characteristic nature of each specific relative universe. **TRAT**, The A Transformation of - in - TRU, describes how through a vast number, sequence and variable routes the m or mass dimension of physical entities have become altered and adapted as their mass energies are increasingly based, not on gravity alone, but on different and more unique and often unprecedented forms, or FAF. The progression of form is from old and same to new and different forms. We must remember that form is not to be accepted as final or separate at any stage or point in transformation, unless we so specify them in being the given or as they are seen in TIT-TO. This makes plausible that the forms we so often see in life and forms spoken about in physics, etc. are absolute neither in and of themselves - except under very specified conditions - nor absolute as apperceived. For they are to a large extent not what they are in TRU but what they appear to be to the observer as they come to us as The Given. This is the role of relativity the transformations back and forth between the physical world and us, The Observer, TRU.

The form of different approaches to and presentations of TRU is different than the form of TIT and consists of **BAS** = Basic States. BAS provides four of the major topics of TRU. When combined with NAT and FAT it is called **BASFANT**

Fig 2-02 **DOBAS** = Diagram of Basic States

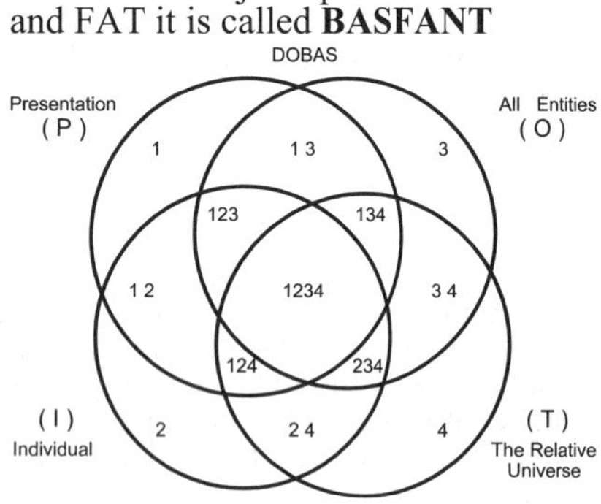

In this text, by using the SCARLS involved in **LIOR** or Law of the fundamental Relationships between the Individual or first person and All entities, we come to a systematic union and formulation. This helps TRU and its presentation, **LIORT**. It is especially valuable in applying SCARLS to the various questions, problems, answers and solutions which are necessary if we are to properly, fully and successfully understand and comprehend all things, COAT. From DOBAS we can develop a systematic union, formulation and understanding of the persistent and paradoxical problems of metatheory, METATRU, SCARLS of TRU, and especially understand their interpretation by relativity, set theory, and Godel's Theorem. Without formal distinctions of terminology, their orientation, and implications into the SCARLS that exist in any meta set or relationship, we have an indefinite flux. This is contrary to our lives as well as TRU. The 'is' and 'this' or perseity and haecceity, that are so much of us, create very basic states and conditions, SCARLS.

This introductory section is mostly not, out of or away from TRU, yet it in being a presentation of TRU, knowledge which is largely TRU, and an approach to TRU, will include much that is of TRU. A concept that helps to explain this is: **RIMT** = Reversion in Meaning to TRU. RIMT denotes the tendency of all concepts, words, symbols, representations toward TRU. RIMT reacts to the tendency of meaning to be only of the ENIT or OBIT. This is **RÒMEN** or Reference of Meaning to particular Entity or Object. Thus there is a continual highly dialectic flux between ROMEN and RIMT, or **REDIL**. REDIL is much of human life as seen in SOT, A, and SA. It is like all we encounter and do whereby what is or is not of value is to the extent it is of practical use. So that much of what we discuss in POT, by RIMT, is not so much what it is as an **ENIT**, entity in and of itself, as it is in relation to us, to TRU. An illustration of this transition from non TRU to TRU is to be best seen in FAT and POMOT.(03,09,10)

Fig 3-02 **CLASSES and FORM of ENTITIES**

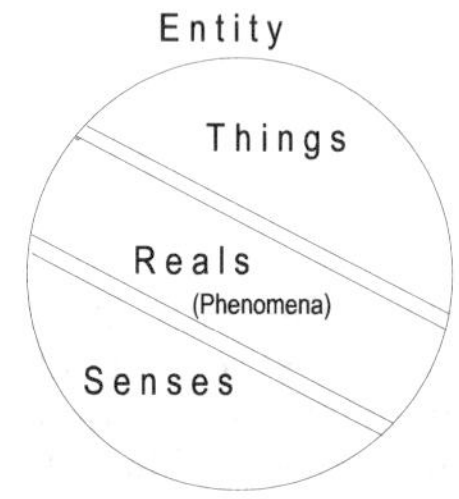

Numbers, letters and words are entities which represent and consist of both senses and things. Such entities that consist of both senses and things are called reals or phenomena. This meaning is different than how the term phenomena is commonly used. For it refers to some occurrence in nature that can be subjected to analysis and determination. Like the real, it is the bridge between senses and things, subjects and objects. It has even developed the pejorative sense of being illusion in contrast to actuality. However as developed in the Phenomenology School of Philosophy, phenomena indicate bonds, intentional activities of man whereby entities and all life, as S-O in TRU can be analyzed and determined. In this use phenomena are basic to our fundamental and predominant division of TRU. For here it becomes a focal aspect of Â, the S-O predominant dimension in TRU. Thereby we partition entities, a distinction based upon **LÂD** and **CÂD** or the Law of Formal Distinction, and Certainty Absolute by Definition, and the coterminalities of things, senses and phenomena, by reducing things not to entities, senses, and things in themselves, but to entities that are phenomena and a mixture of all three classes. The diagram shows the three-way partition or classes of entities as their formal distinctions are set out. It should not be taken without its limits or SCARLS. Thereby: (1) entities are to a great extent illusion and figurative,, and if TRU is actual by TRIA then their entities can not be ENIT or entities-in-themselves, and (2). In a reproduction or representation of an entity senses and things more or less permeate the entire entity, so that there are percentages of these throughout.

IA and SO COMPLEXES - The big problem in life is that whatever our IA or awareness, actions, and attention is on, it usually turns out to be never as simple as it seems initially. For all that exists in actuality is by DAOST in mass, all TRU and infinite together. In addition it is also less than actuality by being infinite in AU. It ramifies into an endless pursuit unless selected and controlled. We can justifiably reduce and control this somewhat by holding our IA to a single observer, and its object to a single observed. This is highly justifiable and effective as we envision TRU from the same position, and hold TRU to its state of being TIT in TU or a thing in and of itself in unity and totality. This is also justifiable both from the standpoint of a standardized optimal vantage or view and the standpoint of an object that is relatively stable and unified, as long as we remember and account for the SCARLS and VAPOS involved.

S O SUBJECT OBJECT -

S O = Subject Object, their polarity, dialectic interaction form and function. S O may be expressed and thought of in several forms and states. It may be:

S or Subject, or Subjective, separately

O or Object, or Objective, separately

S and O Subject Object independent yet thought of simultaneously

S - O Any Subject and Object in relationship or connection

S-O Subject and Object in a close bond

S=O Equality of subject and object

S≠O Inequality of subject and object

S≡O Absolute equality of subject and object

S≢O Absolute inequality of subject and object

S::O Proportionality of subject and object **S≤O** Subject less than or equal to Object

S≥O Subject more than or equal to object

S→O Subject moving to object **S←O** Object moving to subject

S↔O Interaction between subject and object

S⇆O Forward and backward interaction between subject and object

These are only some of many conjunctives that show us the variety of forms and states in the subject and object relationship. Subject object is equivalent to the observer observed of relativity. It is used in this text, from common usage, but mostly to reinforce the significance of the special meaning of the mass-energy dimension form and function of TRU.

Fig 4-02 **POSITIVE SUBJECTIVE APPROACH, PSA and PACH**

This is the Á dimension, state, form and function of our mass energy. Since humans are largely the given and the focus from whom this objectivity is being viewed, we adapt POA and as Perfected and Standardized for POT and IPTRU, **PACH**, to the observer, mankind, **PSA**, a Positive Subjective Approach. We make it ordered and focused, from the first person. It is in this way we have figure 4-02. It acts as a guide to show the form, features, shape, structure, order, sequence and contents of PACH. It is comparable to, and needs to be viewed in conjunction with Figure 3-02, as well as POMOT and METAD.

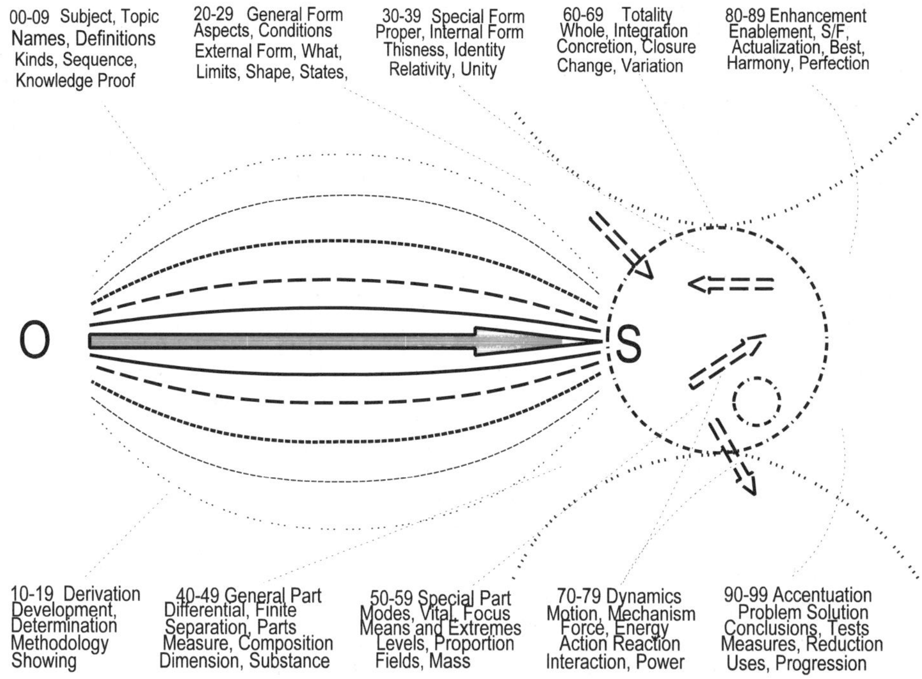

DIFFERENTIATION of SPECIFIC FORM

Why and how is it that the actual world, TRU, or this special universe of best human life, is so much different than our lives, what seems to be, the world we are familiar with, and the world before us that appears to be so certain? Why is the spiritual or heavenly world, and that of an absolute God so mysterious, such a secret, so intangible and different than what we think is reality? Is there any overall reconciliation between our individual being, philosophy, science, religion or are they like atoms, or the objects in front of us, all separate entities?

These are questions, the great minds of the past, as long as humans have had the ability to question, think, and reason about their place and role in life and the world, have been distraught by, struggled with, distracted and confused by, and failed to comprehend. There are many aspects of TRU that best answer, show and prove this astounding yet undeniably and absolutely valid fact of difference. For we can never gain adequate knowledge, never have the right kind of life, and never actually be, unless and to the extent we discover, learn, know, think, utilize and live by such knowledge. This we show in TRU. This we can not do unless we know these differences and what it takes to live and actually be our parts and roles in this seemingly secretive, mysterious, heavenly kingdom, that is only right when corrected to its equivalent in TREQ, TRU and TIT-TO. Only when we learn the key major aspects of TRU, those great new concepts, and corrected existing concepts whose deepest TRU meanings provide the insight necessary. Some of these include: VAPO, TRU, TRU/AU, ATO, DAOST, TRUS, TU, TO, TE, TI, TÂ, TRIA, TIT, TRIS, TRIT, OBT, SUT, ROT, MOT, ALT, Á, ES, CES, ESINS, ACEXIN, REM, POS, and many other key TOPNEC. What is recognized and less recognized needs to be best defined in POT and IPT to show how these differences are the essential nature or form and function of TRU. This is why NEC and TOPNEC, their explicit meanings, definitions, formal distinctions, and relations are so important. It is why a new language, as a new system or science of understanding our world is so necessary. This is why we must discover, learn, know, think, utilize and live TRU.

The problem that occurs with differentiation, determining specific form, and adaptation to TRU have caused failure to achieve an overall form for all. This is pointedly evident in: **FAPAR** or Fallacy of Paradoxical Regression and reversion. Paradoxes, like all that is in doubt, without analysis and proper discovery produce added doubt, often confusion. It takes a positive objective approach, POA. It requires, **LOSO,** Law of SO in combination, Sufficient O-S-O. This is what can be done with probity by the removal of all not TRU, **ATRU**, and reduction by and to TRU and TIT. The differentiation of Specific Form is common in the Measure of All Things, **MÈT**. Yet it is this very measure of things, rather than the extraction of ATRU and application of POA, LOSO, and all means of TRU, that render measurements incomplete, inapplicable or in error. This is expressed in: **LIMAP**: Law of Limits of Mathematical Application.

What is there in a positive objective approach that we can use to make it discrete, concrete and ordered? To make this determination we must examine, describe and demonstrate the basics of relativity between the observer and the observed. We must learn what there is about the object, and the positive approach to an object from the subject that is most important, and of best form and function. In POA we seek to formulate and order it, and by standardizing and fine tuning, PACH.

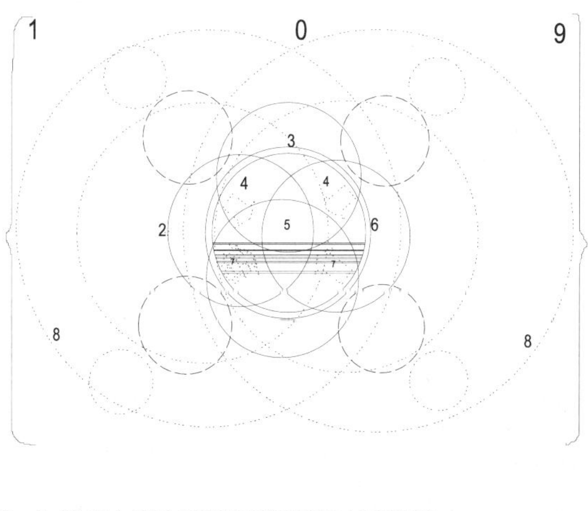

Fig 5-02 **PACH DEMONSTRATION**

In POA we proceed from an entity, sense, observer or subject toward a thing, observed, or object in a way that is characterized by focus, priority, completion, meaning, and use. To have a good POA is tantamount to success. It is the optimal function of ones best form. That form which best provides the well being and preservation of the species. It will differ somewhat - and more or less - with different animals. It is what is predominant for each special variety of animal, characteristic by their edge in life. This, and especially for humans, is the A dimension, state, form and function of our mass energy. Since human beings are largely the given and the focus from whom this objectivity arises we adapt POA and PACH to TRU and have its focus and order come from the individual or first person. It is in this way we progress from DOBAS to PACH. It acts as a guide to show the form, features, shape, structure, order, sequence and contents of PACH.

If **PACH** or the Perfected POA Standardized for POT and IPTRU, is an algorithm or model for an object as TRU, then how is it used? What minimal standards are most helpful and least damaging? What structure can we make of PACH to demonstrates this object? This can be quickly surmised from the nature and artistic rendering of any object. Its form and structure are its frame, its parts lie within, its dynamics are the inner-workings of these parts as well as the motion of its form and structure and their relationships with the outside as by input and output. Likewise its totality is how we look at or show it all together. Its enablement is how all that helps may be demonstrated. Its origin and development show what it came from and how it works thru and thru to what it is going to. Its introduction and accentuation show the overall considerations of approach to and utilization or application of the object. In this way we can show PACH with POA. In doing so it becomes an object. As an object it assumes a changed, more relative, and standardized form. This is what enables us to represent and model it as we do. It is like an object being objectivized or PACH reduced by PACH.

In this process of subjectification, objectification, making positive and PACH set steps are taken:

0. General observation of given, entity or object
1. Derivation, Development, Demonstration of it
2. General Form of the entity or object
3. Special Form of the entity or object
4. A part and parts of it, generally
5. The part and parts of it, specifically
6. Integration, totality of the entity or object
7. Dynamics, forces and vectors of it
8. Enhancement of it with what is outside of it
9. Summary of the given, entity or object, uses, relations and actions to be taken.

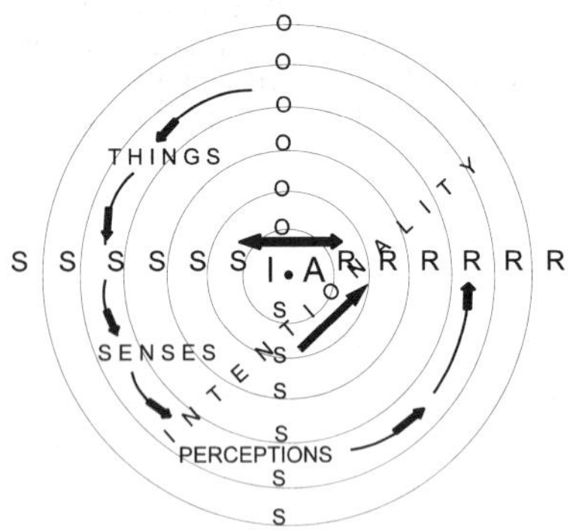

Fig 6-02
S I A R DIAGRAM

By way of a SIAR diagram we may show the expanding circle of S -IA - R. How we can best show how the S-R or stimulus-response fundamental nature of our being, with the IA as its fulcrum, functions as a guide and means to understanding the world, TRU? The expanding circles of S - IA - R's illustrate the fact that there is not only a simple stimulus and response relationship, as physiologically known. It is an increasingly complex manifestation of our being. It is closely allied with the coordinates of TRU. This is especially so when showing the higher dimensional analytic states of the mass, length and time dimensions of TRU. The circles are both expanding and enlarging. They may be , and comparatively, of both RUS and TRUS. They exist Through and Through, **TAT**, in the world. They must be taken step by step in IPTRU. IPTRU is a very large circle of these, **IUK**, Image of Ultimate Knowledge, an infinitely larger circle of S - IA - R. In the superimposition of S - IA - R and ALT on TRU, we can understand how **UMIA** or IA, Individual, mankind and universals are actually in and function as parts and processes of TIT. It shows us how each person must learn to understand and utilize ever expanding circles of knowledge and their interaction with the stimuli, forces and dynamics of the world about them. To learn and comprehend how these are all a part of the larger total picture, TRU. Several questions may be used to convey, focus and coordinate the IA, sense, S-R and S-O of each phrase, sentence, paragraph, facet, sector or part of IPTRU. This is determined by: (1). TIT, (2). SU, (3). IA of TRU, (4). Depth Conception, (5). Sequence of Text. These help set up and apply SIAR and guide POT and IPTRU. Although not directly in SIAR or the derivation and development of POT, they show us how with various diverse beginnings we rapidly converge on TRU.

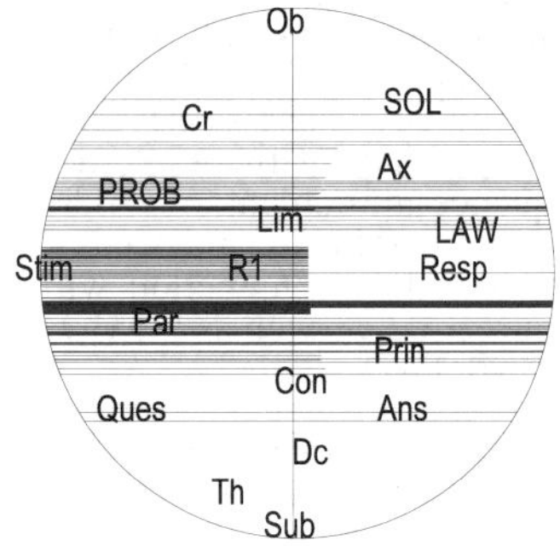

Fig 7-02
S-R and S-O RELATIONSHIPS

Various states of ideas, entities, senses and things, stimuli and responses, subjects and objects are related, dependent, positioned can be ordered and shown on the stimulus-response, and subject-object coordinate field.

This uses abbreviations of:

Stim	= Stimulus	Res	= Response
Sub	= Subject	Ob	= Object
PROB	= Problem	SOL	= Solution
Ques	= Question	Ans	= Answer
Par	= Paradox	Prin	= Principle
Con	= Concept	Dc	= Doctrine
Th	= Theorem	LAW	= Law
Lim	= Limit	Ax	= Axiom
Cr	= Corollary	R1	= Reality

R1 is the one focus of all as they function in the whole. These factors make up much of the essence of human life and TRU. They are not all directly antithetical. The darkness of the field is to show, by impression, depth, vague common understanding, and the extremely finite state and dynamics of all that makeup, and are manifested by, the ES and CES of TRU. Although their placement is only approximate, they are more than solitary static entities, highly dynamic, interdependent and correlated functions.

The difference between what is of APT and POT and what is of TRU and TIT is seen in the very term "is". For it can mean anything from the most simple sense to perfect actuality. This may be for any entity or object or for pure TIT. It is **CIS**, or the Conditional Is, where 'is', although specifying a definite relationship, connection and condition, is not necessarily a definite actuality. Only with a set entity, object, actuality as we have with POA in TRU and TIT can we have focus and order. This provides for **COL**, **COR** and **LASUR**, Correction by Limits, Correction by Relations, and the Law of Sufficient Reason. COL enables us to deduce from a whole and a part what the rest is. COR enables us to deduce from a whole and relations, other relations, and LASUR enables us to find certainty in a set whose SCARLS establish sufficient reason for their truth. This involves **LACRAST**, Law of Cracking the Subjective, a return and fusion of any observer back into the special observer-observed relativity existent in TIT. It enables **STRU** or the Standardization of The Relative Universe. Through **TRIA** or TRU is All, the exclusive form of TRU, it shows **LATAN** and **LATANOS** or the Law of Absolute Necessity - by and of TRU, and the Law of the Absolute Necessity of Order and Standardization.

CHAPTER 03

THE SPECIAL FORM of ENTITIES and POT

It is often the special and unique forms of entities, RUS, and TRUS that determine, and give them, much of their thisness, whatness, and function. Special form and function are what most sets them off, apart from other entities. Therefore **LUC**, Law of the Universal Categorical or Universal Causation and **LUP** or the Law of Universal Permeation are more than anything. They become something, then the thing, type specific for the universe at hand, TRU as TIT. This is called: **LUFCAP** or Law of Unique Form and Function of entities, things and Universes Produce the specific Causation and Permeation typical of a, the and The Universes, as TRU, TRIT and TIT. The special form is not to diminish the importance of the general form. For the general form is also set, and like universals shows the sameness and relations of many entities and universes. Together they provide what is crucial to the overall form and function of entities, RUS and TRUS. LUFCAP is, for an added reason, important, it gives the entity, RU and TRU the extra power needed for specificity, exclusiveness and its absolute status. The separation of general and special form corresponds reversely with special and general relativity. This is a product of a single unchanging observer in general relativity or **GER**, and multiple and changing observers in special relativity, or **SER**. The latter, special to general, is as specie is to a genus. The fixation on special form, as a specie, is called **LOS** or Law of Speciation. It is upon LOS that we can prove much of TRU. (109) This is accompanied by **BOTFIA** or Body of TRU a function of IA. BOTFIA signfies how the focus of TRU, formed and empowered by IA, makes up and determines much of both the thisness and whatness of TRU. By LANERA and LAMOL, TRU is neither relative relative nor absolute absolute. It is a continuum of these two extremes, which in TRU becomes its form and major modes and parameters. This makes its TRU actuality, whose gradations or differentiation is signfied by: **ACDI** or Actuality Differential Describes basic Aspect of RU and TRU, its different aspects of relations. We discover, learn and understand it not only as a simple **ENIT** or Entity-In-and-Of-Itself, but by its extremes and means. This, and the reason extremes, like polarities, are so helpful, is expressed in: **EBE** or Extremes by Emphasis. EBE is not all positive for in common belief it is often the basis for **ISMS** or those extremes upon which the users of public opinion promote their aggendas. These are to often either non or little TRU. Special form in contrast to general form implies focus and a progression toward a set. This is the inverse of special and general relativity wherein special involves any observer, and general limits itself to a single observer-observed. **LUR** is Law of Unity in Relativity, special form is highly dependent upon the focus of relativity, or **FER**, Form Equals Relative, IA of a Perspective or Relative of the form of TRU. Being convergent upon TRU by necessity we have, **LANOFAC** or The Law of Absolute Necessity of Formulation and its Conditions and qualifications. Thus we can have **SPO-T** or the Standardized Perspective of all - all TRU.

Percept is a given element in being, a perceptual experience, a sensation or sense datum that is recognized. It tends to be focused on what is important, positively or negatively for the individual. This is evident in: **LORP** = Law of Relative Opposition. Humans have a built in capacity, that increases with experience, to anticipate and identify certain objects in life. This is what adds meaning to their lives. What is relative, and especially what is in opposition is much of what this capacity focuses on. It is on these that the whole of TRU, ROT and POT depend and revolve. This is by: **MARAT** or the Minds Ability to Rapidly Attend to the Pertinent, TRU. As man, his intellect, mentality, thought and reason are much of TRU, the unique and special, so it is that this special form of energy as it is fused with other forms of energy make up the mass energy, A, dimensions of this special universe, TRU. They are driven by the ever greater success, and thereby survival, of those who best use this FAF and power. The extreme toward which this power moves is called; **THESA** or The Subject Absolute, and when for the total end it is called **THESOA** or the Subject Object Absolute.

As human intelligence and language increased so did the ability, complexity and proficiency of thought. This often took the form of questions and answers. Thought progressed from simple identity to relating, contemplation, reason, logic and the higher powers possible with a POA and understanding of TRU. (COGNOS) All that is of great importance yet can not be explained, described or readily demonstrated, can often be best initiated by being put into the form of a question. The reasons why questions are so valuable are:

1. They bring much better light on the problematic state of what is so important before us.
2. They stimulate thought. They induce a person or reader to think, reason, and find meaningful conclusions.
3. By clearly setting forth the question and some of the SCARLS involved they give focus and direction to the answer.
4. They tend to tie in the issue and problem with others and by relating different questions and answers.(Socratic Method)
5. They give meaning to the relationship of subject and object, set it in order in such a way that the subject better addresses the object and the object is better used by the subject. They are integrated in SO, and especially as this is SOT, A, ALT and TRU.
6. They help to establish logic, logical ways of thinking, algorithms, and models by exercising our reason and developing inferences intellectually and productively.
7. They help to give thisness by bringing convergence upon an entity, sense, thing, RU or TRU.
8. By the form of the question they provide direction into the what, how, why, when, where and much that adds dimension, quiddity, quality, quantity, posit and mechanism of an entity, sense, thing, RU or TRU.
9. They provide the means to react and avert what may be harmful over and above what lower levels of the nervous systems functions can do.

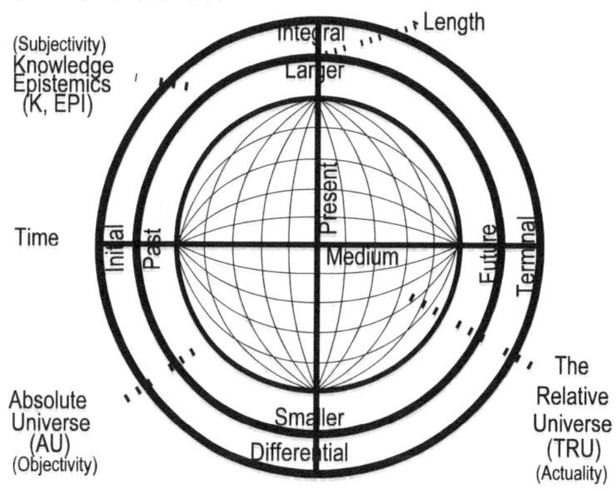

Fig 1-03 **ONTO**: Being or Ontology as a function of the Systematic Unity of entities, objects and universes. Ontology is the philosophy of being-as-such. It is the study of what existence itself is, considered apart from any question as to the nature of any particular existent. It is the attempt to discover the fundamental categories of all being.

Yet our greatest problem is not of being as much as it is of what being. This is the difference between Ontology and Theology. For we assume that many entities and things have a being in and of themselves. Yet we have a great deal of difficulty in trying to decide whether their being is This Being. They may have discrete being in themselves, yet little or no being relatively and actually. Thereby, what existence is TRU and what is not? The individual or first-person 'knows' they are, they exist. Yet even this is in error, an illusion. Not only is the object, but a person is actually only as they are TRU, not as one might think or even so intently realize it or their own being. This is called: **IAMP** or the I Am Paradox. In this way we have many levels of 'being' or existence, yet only TES and TIS preeminently Once well understood the problems of being and existence can be resolved.

Without a strict limitation to TRU we have **LAPIS**, Law of Paradoxical Being and non Being, or Paradoxical "Is". From Figures 3-4 we can gain an appreciation for, **LOTAD** or the Law of TRU AU Distinction. The contrast between TRU and AU is further evident in **PIMOB**, the Principle of Intrinsic Modes of Being, or how infinite and finite being are not specific determinations. It is PIMOB that distinguishes the exclusive being within a RU, as TRU, and the existence of the their finite and infinite. **LUR** signifies how it is relativity, the relative/absolute, that produces unity and thereby thisness. **LURE** = Law of Relativity Establishes Unity and thereby Finity, describes how the unity established by relativity in turn establishes and is established by finity, infinity, and totality. **LATNUP** or Law of the Absolute Necessity of a Singular Unified Perspective and Presentation, emphasizes how it is impossible to have a good presentation and perspective without the best possible unity and focus. **IATU** = Instantaneous Attention and TRU Unity, or the singular, unitary, elemental, and pivotal point of our being signifies how it is IA and the associated predominant powers of S-O or Á FAF that provide this singular, unified, elemental and pivotal point of our being.

When we use the term 'IS' we mean an identity or a concurrence. However, we also often mean completion or totality in which truth or solution is affirmed. How is falls short of actual validity is not a critical issue until we analyze all by TRU. For if TRU is all by TRIA, how can what is not TRU be? It can if we qualify it, hold it to its SCARLS as it is only an identity and not complete, a totality, a final truth or solution. Without being only TRU is can be actually valid by:

1. They may be a part of TRU, less than total TRU yet 100% pure for the part.
2. They may be only partly within TRU, in being universal, general, transcendent or thru and thru.
3. They may be potentially TRU in this same input and output flow between TRU and ATRU.

So that we can say POT, APT, or ROT 'is' as we can say for much we discuss in POT. They have identity as they can be observed and have existence in the above ways. However we can never uncritically accept them as 'is' with identity, finality, totality and validity. This is why, in this text, we frequently use the terms VAPO, SCARLS, DAOST, and other NEC.. This we should do for all we survey even if it is only implied.

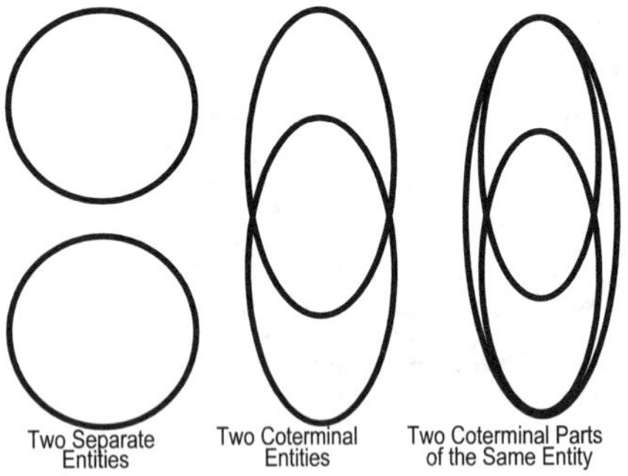

Fig 2-03 **VIEWS and COTERMINALITIES, Types of:**

In these figures we find common relations between entities and a very significant aspect of TRU. It is a basic step in establishing form, systematic unity and order, especially as they approach TRU. They may be determined both by how they exist in and of themselves as well as by TRU. Since we are oriented and inclined to TRU they will depart from their ENIT status and will tend to TRU. This tendency to TRU is increasingly made manifest in each chapter. The reduction of all entities, things and terms to their primary sense, meaning and state in EPI, AU, RU, RUTS, TRUS, TRIS, and TIT in a singular fashion is called; **LARNTAM** = the Law of Reduction of All Nominals, Terms, concepts, things, entities to their TRU, AU or EPI Meanings. There may be exceptions to this, especially where they overlap and are mixed as when in aggregate. Meaning is the bond depth conception makes between observer-observed, or S-O, and the many levels of mass energy FAF in TRU. This produces actual meaning, seen in 'what is the meaning of life, how do I fit in"., and many of the great answers to questions, and solutions to problems.

Animal, and especially human, nature is provided with many variations and combinations of senses and things, subjects and objects. This approaches **LACARD**, Law of Absolute Cardinality, from S, O, T and Á. In TIT this becomes **LARD** or Law of Absolute Relative Distinction, or how the Relative is not Absolute, R not Â, as TRU is not ENIT. The convergent power on TRU is seen in **VACMA**, Vital Accelerative Momentum of S-O and Á.

It is in the progressively greater powers of human mentality and its ability to distinguish all in the world around it that makes it so special, unique and exclusive. That one thing is or is not another and how things are separated is called distinction. When this distinction is formed and set it is called Formal Distinction. This is most marked in: LAD, LARD, INTRU, ATOA and AF. **LAD** = Law of Formal Distinction, based upon a single and set sense. **INTRU** = In or Into TRU, the focusing and/or convergence on TRU. **ATOA** = At Once TRU Absolute, or the highest level of Aseity. **AF** = Absolute Finite, or the finest limit. LARD signifies how it is our ability to ever more perfectly separate what is and is not relative that is fundamental and dominant in human survival. Formal distinction is achieved by preparing, enabling, maintaining, and improving INTRU. ATOA is like TÂ/TRU and AU, the ultimate in TRIA. ÂF is from the limit of differential, what we tend to in the ever more separate and distinct ES and CES of TRU, and much of IA.

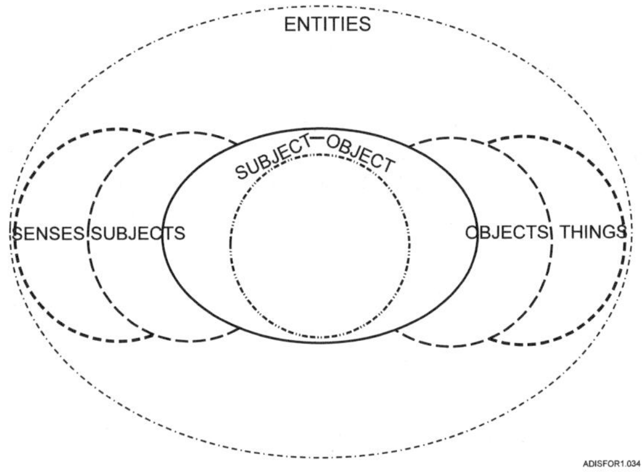

Fig 3-03 **ADIS; GENERAL to SPECIAL FORM DISTINCTION**

ADIS signifies Ä or form distinction and, as shown in this figure, seeks to demonstrate the comparative relationship between entities, senses, things, subjects, objects, subject-object, subjects-objects and The Subject-Object (TRU). It demonstrates how a sense can't be a thing nor a thing a sense. It demonstrates how a subject can't be an object, nor an object a subject. It demonstrates the relationship between entities, senses and things with subjects, objects and subjects-objects together. This is the beginning of the forms and dimensions of TRU. ÄDIS may be separated into two sets or kinds:

(1). A Distinction, Á, that is of its general form. What distinguishes all the form of an entity, and

(2). Á distinction, Á, that is of its special or mass-energy dimension, relativity, or subject object form. The second is a subset of the first, and is the cue to the crux and jist, of TRU. ÄDIS of this second kind or subset, is the initial derivation of the Á mass energy FAF of TRU. It is the differentiation of sense-thing, observer-observed and subject-object in a set form. It is determined by:

(1). That being observed and distinguished,
(2). The intensity of the observer,
(3). The intention, purpose, use and end to which this distinction is to be applied.

Fig 4-03 **FAR DIFFERENT DISTINCTIONS of ALL**

FARDI = Far Different Distinctions Diagram - Each and all humans can not begin to be unless they increasingly separate and come to make discrete and distinct the world around them. This is the great gift and blessing we have inherited. It is what makes them unique, special, exclusive and absolute. It is not only how we must approach TRU, it is TRU. It is the great mass energy form and function, state, dimension, and mechanism of our being, TRU. Each of these categories, that are the

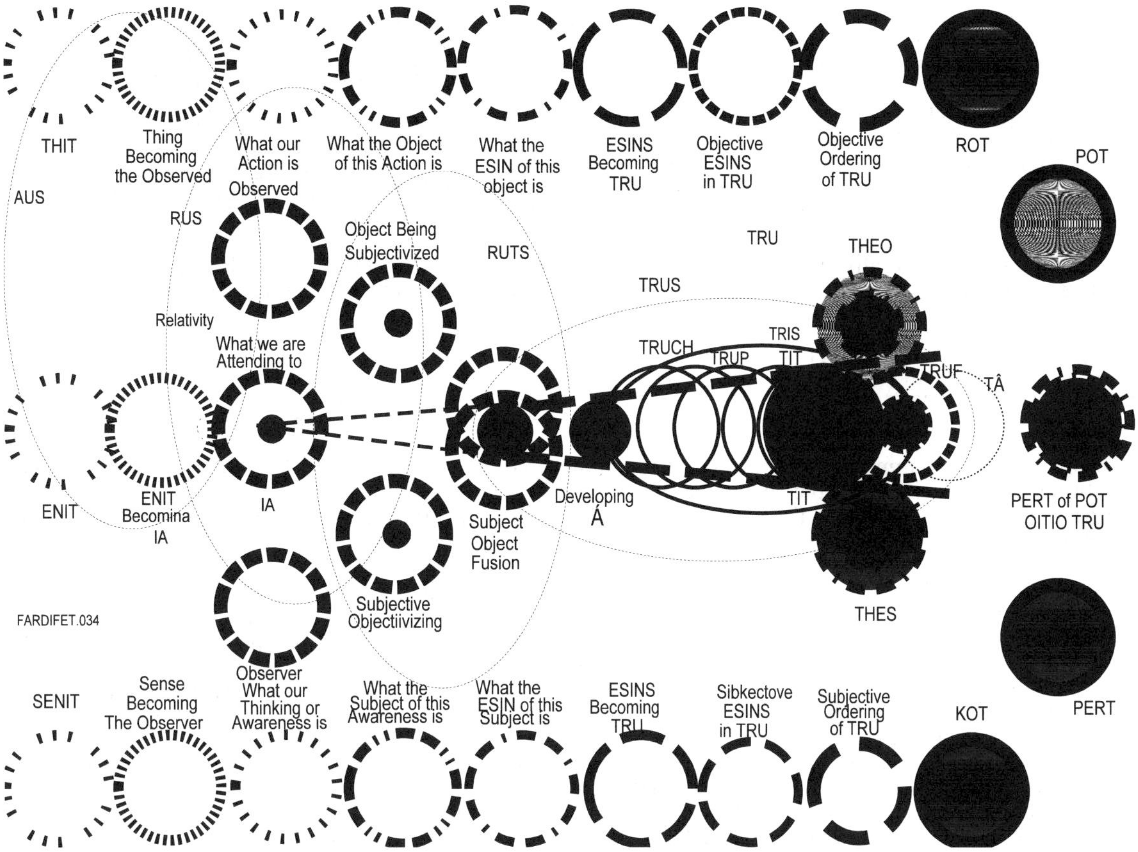

major participants in an approach and introduction to TRU, are far different. The greater we can realize these differences, and make them formally distinct, the greater we will not only understand TRU, but the greater we - as TES and TRU - will be. All less than this is the loss of these differences and of the very reality and actuality, that we so take for granted. We can not live without light, blind to what this light makes possible. It is light and all the other components of our mass energy powers, Á, that enable us to approach TRU, order all, and be TRU. In this diagram the parameters are generally, time on the ordinate or horizontal, and S-O on the abscissa or vertical. Each of the separate items are great ENIT whose difference from adjoining, as well as all other, items show our approach, origins, development, or in a sense, theogony. In this text we concentrate largely upon the hatched and black lines and figures. Without keeping such overall panoramic views in mind we lose sight of what TRU stands in relation to and is most important. It is like the DAOST of TRU, what our Á powers enable us to both make infinite and unit, infinitesimal and total all at once. Without being able to do this we never can encompass the whole, nor adequately recognize TRU.

ENIT and initial and random IA, in contrast to progression in this figure, reveal **IAMP**, the I Am Paradox. This is never resolved until a person realizes they are not as one might think, or even so intently realize it, or their own being.They are only as they are actual, TES, their counterparts in TRU,. In **LATHIA** or Law of This and That, the fundamental operation of IA is signified in which its greatest powers of separation, distinction, selection and discretion are brought together. **ESINS**, or what is in an entity that is also in TRU is a crucial step in the transition from ENIT to TIT and its POT. Much of showing TRU is **OBT** the Objectification of TRU. Just as with anything without becoming an object for our discernment it lacks meaningful actuality. It is LATHIA, and the incipiency of this operation, be it unconscious or conscious, indirect or direct, that initiates the development and dynamics of Á, the mass energy state and FAF of TRU. It is reinforced by the intentional, denoted in: **LIAD** or the Law of Intanstaneous Attention Identity with Intention or the degree to which that intended is attended to. It is their convergent role in TRU that gives LATHIA its greatest importance. For KOT and the proof of TRU rely heavily on distinguishing this from that, and how with all adequate grades of intention we fulfill TRU.

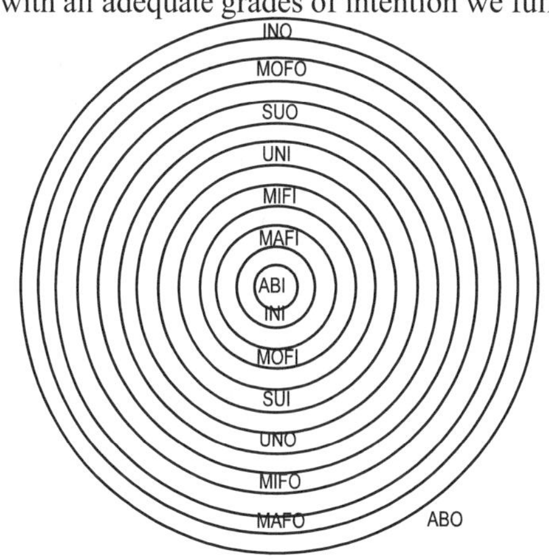

Fig 5-03 **UNIVERSAL DIAGRAM, ABSOLUTE OUTSIDE to INSIDE** (ABOABI)

To envision an entity, sense, thing, subject, and object fully we can do no better than to look at it from the absolute and extreme outside to the extreme and absolute inside. We show twelve levels or spheres, to represent this entity or object. An infinite number are possible, yet this is a finite number that is both representative and very acceptable to perspective. It could be done as it is used in IPTRU for a great variety of objects, including Subject-Object as well as TRU.

In these various relationships of subject object, unit, finite, infinite and problem solving we set out on our quest for systematic unity that can be provided. It is how we can reduce all to the solid construct as a model of TRU. **ABOABI** is a diagrammatic representation of the state that exists in relativity, the observer observed relationship wherein the object can be interpreted from the absolute outside to the absolute inside. This must quickly be qualified by the limits of the absolute extremes. **SPUNO** is the Standardized Unitary Perspective or a 1:1 identity of subject and object. From this we proceed through unitary, increasing levels of finite, to infinite outside and from the inside unitarily, finitely, infinitesimally, immediately and simultaneously as may be gained by envisioning this page totally. The figure provides a construct of all as can be developed from the systematic unity of a one to one relationship between subject and object. Of the continuum some of the different levels that can be developed are:

From the Outside Infinite Outside, Innumerable Finite Outside, Numerable Finite Outside, Limited Finite Outside, Unitary Outside, Standardized Perspective Unitary Outside, and Standardized Unitary Outside, and

From the Inside

Infinite Inside, Innumerable Finite Inside, Numerable Finite Inside, Limited Finite Inside,

Unitary Inside, Standardized Perspective Inside, and Standardized Unitary Inside

The spheres or levels shown in the figure above are:

ABO	= Absolute Outside	**INO**	= Infinite Outside	**MAFO**	= Maximum Finite Outside
MOFO	= Modal Finite Outside	**MIFO**	= Minimum Finite Outside	**SUO**	= Standardized Unitary Outside
UNO	= Unitary Outside	**UNI**	= Unitary Inside	**SUI**	= Standardized Unitary Inside
MIFI	= Minimum Finite Inside	**MOFI**	= Modal Finite Inside	**MAFI**	= Maximum Finite Inside
INI	= Infinite Inside	**ABI**	= Absolute Inside	**ABOABI**	= Absolute, Outside to Inside

The maximums and minimums correspond to the limited and are bridged by the innumerable to numerable finites. This describes much of the extension the finite and their relations to the periphery and center of an entity, universe, TRU, their perspectives and representations. ABOABI is the summation of their wholes, and signifies how closure is made. Such an overview and closure is key to the LATANOS of TRIT, OBT, SUT, ROT and POT. Such an image as ABOABI helps to show **LATNUP**, Law of The Absolute Necessity of a Singular Unified Perspective and Presentation. They are supported by **PLAR**, the Primary Law of the Absolute Relative, or what is above or about relative, yet inside relative.

Within TRU we seek to make correlations through the dimensionality of TRU, A-L-T. Yet these relationships not only include all parameters of Representation, Presentation and Perspective of TRU, the dimensionality of TRU, ALT, has its own SCARLS in representing the complete form of TRU. Thereby the relationships between perspective and TRU, as subject and object and the dimensions of TRU as ALT are altered. However since they both contain this subject-object relationship they may be reduced at least in part to a precise identity. This is S-O in TRU, as A of ALT and S-O in perspective stands in relationship to its object, as observer does to observed in relativity. In this way they are the same and also different, DAOS. The term that signifies the relationship dimensionally between the transcendent and the immanent together and separately is called: **DIRP** = Dimensions Produce Relation, or the Dimension Relativity Product. For POT and EPI DIRP expresses the basic parameters between EPI/TRU, EPI/EPI and TRU/TRU. It helps to show their components, relations, similarities and differences, DAOS and DAOST. Below we enumerate the traditional philosophical relationships. This gives us an overview of all philosophy when considered under its major and added categories: relativity, quantity and identity: realistic monistic epistemology, idealistic monistic epistemology, realistic monistic ontology, idealistic monistic ontology, realistic pluralistic epistemology, idealistic pluralistic epistemology, realistic pluralistic ontology, and idealistic pluralistic ontology. These are important for they show us both the continua of realism - idealism, monism-pluralism, and epistemology-ontology. We must hold in abeyance for the next Section the departures these take from TRU, as well as their usages, e.g. materialism-realism-idealism.

These are most simply and best shown by a simple polar diagram in three dimensions equally representing each. However this can't be carried to its logical conclusion because of ambiguities of the terminology. This is most notable in the case of realism which is more the intermediary in materialism-realism-idealism. However the eight parameters represent the traditional philosophical approaches, and are corrected through the text by APT, POT, TRIT, OBT, SUT, ROT, MOT, ALT, and SPALT. The reversed and combined roles of O-S to S-O in idealism-materialism, epistemology-ontology rather than denying validity to dimensional analysis reinforces it by way of the predominant role of SO in the mass energy form of TRU.

IDEAL and ULTIMATE FORM

What is needed for, can be, and how we enhance form or achieve ideal form is evident in: ESINS, LOD, OBT, INTSO, VACMA, and RORRE. **ESINS** = Entities or parts which correspond with their part or role in TRU. **LOD** = Law of Dynamic Duality, the Absolute Relative reversibility. **INTSO** = Integration of Subject Object, Relativity. **VACMA** = Vital Accelerative Momentum of S-O and A. **RORRE** = Revelation of the Revelation Relate Explosion. ESINS reveal how what is in things also exists explicitly and selectively in TRU. LOD represents the dynamic continuum in which the form of RUS and TRUS develop, whose selective powers operate on its reversibility to produce ideal form and function. INTSO signifies the similar mechanism of the prime dimension of TRU, its relativity fusion to produce the best form. VACMA likewise signifies how the prime continuum of our existence, its S-O actions have momentum that accelerates optimally. RORRE is the extreme of subjectivity in ideal form for ENT and TA. RORRE signifies how accelerated ideas and means are explosively produced and expand by the combined KOT, ROT, POT, PERT OITIO of TRU.

Man's IA and nervous system are largely tuned to the ultimate in differentiation. This produces the ultimate in form, so that we may be able to pass from the largest to smallest readily. It is the great massive expansion of apprehended information and form of objects as they increase knowledge. Especially this is built into TRU and its parts to utilize information and knowledge. It may be and often is AU or ATRU rather than TRU. However both the special form, tendencies, and convergence favor TRU. So that ultimate form is one that is produced by its relativity, mass energy, observer-observed and subject-object state and dimension. This mass energy, which in its way gravitates much that TRU comes in contact with to TRU, is called its A or mass dimension.

RIA = Relative is All. RIA signifies the fact that if a relative universe is all or exclusive, as the term universe implies, then nothing without or other than this universe can be. RIA is a product of relativity. RIA is of unparalleled importance in TRU for TRU by TRIA is all, which enables us to invoke rules that regulate much we come in contact with. We can separate what is, from what is not. We can better develop a solid TRU from its systematic unity, represented spherically and modeled. From this and a firm reliable focus we can develop posits, statics and dynamics that are well ordered. Yet RIA, as TRIA with some differences, can not be taken as an absolute, as no other. In being less than absolute, despite its suggestion as being all, exceptions, conditions, and SCARLS must be applied if precision and freedom from error in its application and the understanding, representation and presentation of all, as TRU, is to be successful. Absolute like zero is largely a figment. For neither actually exists as ENIT. They are by what is relative an antithesis of existence. What existence they actually have is in their role in representing TRU. For absolute is to the infinite, as zero is to the infinitesimal. They act as the limits of extremes which can be approached, named, and used, yet are in the actuality of TIT-TO, illusion.

CHAPTER 04

DIFFERENTIATION of ENTITIES and PARTS of POT

The ancient Greek Philosophers in their headlong pursuit of the makeup of the world eventually discovered that there was more than earth, fire, water, and air. There must be some other element if all is to be included. What was missing was more subtle and profound, the essence of life, without which there was chaos, no order of unity and diversity in cosmos. Since this they figured was the fifth basic stuff of which all is composed, it was called the **quintessence**. The greatness and beauty of the ancient Greek mind can be seen in its search for missing answers, typified by Socrates and Plato. Especially what is more than can be readily observed, physically. The ideas of essence and quintessence when carried to their actual conclusions are basic to any understanding of the cosmos, our world, or TRU. Without distinguishing what is essentially actual from what is an unselected entity, knowledge is perpetually driven down a blind alley. This is like modern philosophy in which much controversy persists in the dimensional makeup of the universe. In fact, none of these can be determined until we determine which universe we are speaking of, is actual, or relatively existent. This is why man has been so confused over what are cosmos, God, world, universe, etc. Until these can be identified, made distinct, and well defined as we exist in relation to them, our knowledge and our own being lack full and exact meaning. The answer to these questions and problems is TRU and its investigation, analysis, representation, demonstration, perspective, and presentation. It is why each and every person throughout time must continually discover, learn, know, remember and best use essences and quintessence of all that brings order, harmony, happiness, tolerance, management and control to their lives. It can't be achieved without quality and select individuals, whose lives are enhanced by the equanimity, inspiration, and revelations such identity with the mass energy continuum human life provides. As only the few quality and select ancestors took advantage of their mass energy form and function survived to have immortality, so must each and all of us, if we are to flourish, survive, have the continuity and immortality available to us.

Meaning is Imperatively Finite, **MIF**. It tends to make the subject object relationship differential. It is largely a function of mentality, thereby human, of TRU rather than AU the physical world alone. MIF is the action of the subject on the object, observer on the observed in the relativity of the mass energy continuum, Á, of TRU. Meaning seeks to establish the heterogeneity as well as relations in things. This is called **PLO** or the Primary Law of Objectivity. It is guided by the human need for POA. It necessitates bringing into view, **BIV** all that is relevant, to the most effective degree. In totality this requires a Comprehension of All Things, **COAT**, which in turn depends upon MIF, and subjective and objective differentiation. It is only as we constantly remember and return repeatedly and reversibly to totality, yet continually delving deeper and deeper, still maintaining optimal coordination and order, that our POT best succeeds for TRU. Then each person reaches optimal learning. It takes differentiation and integration, both subjective and objective, internal and external, static and dynamic. Together they seek to produce Optimal Finite Form and Idea, **OFI**.

	DIFFERENT	**SAME**
NON ESIN	**Typical entity, or AU, neither the same nor has ESIN with TIT**	**Universals or anything else that is similar to what it is in TIT, yet irrelevant to it.**
ESIN	**What is different in an entity or AU that may become contributory to TIT.**	**What is in an entity or AU that is also a part of and contributory to TIT.**

What is differential, a part or component, is going to be both different and same with both the whole and with each other. This is called **DAOS**, or for TRU, **DAOST.** Different Aspects of the Same Things of TRU. DAOST is one of the greatest concepts to learn and use. It is essential if one is to be able to separate - and understand - what is and what is not TRU. This is much of TRU axiomatic, and has as a corollary: **COTA** = Corollary of TRU Axiomatic, the simultaneously increasing separation of the same, and fusion of the different of TRU. The complexities involved require management and control, which in turn depends upon order, the order provided by a positive objective approach, POA. This requires **DISO** or the Differential Separation of Subject and Object. It also requires the use of means by which this separation can be made, of best reason, or: **LASUR** = Law of Sufficient Reason, benefited by focus, relation, totality and POA in analyzing TRU. To learn what is differential, the parts of TRU, also is greatly aided by: **LEPER** = the Particular as Real, the epitome of reality, where in the differentia of the intervals, phenomena and intermediaries of the S - O relationship and continuum become actual as parts of TRU. What is of an entity or object that is not TRU may still be the same as what is of TIT. This is called: **SLUT** = Same Law, not or UNTRU. SLUT is of great significance in helping to separate what is and what is not TRU. For we must be careful not to assume that something that is the same as it is in TRU is of TRU, or of the same REM as it has in TRU. In this way these sames may be neutral, negative or positive to TRU, depending on the relation and interaction with TIT. Universals and particulars, differents and sames, and Non ESINS and ESINS although coterminal in ways are highly distinct and have separate meanings. For occasionally sames may be particulars and differents universals. ESINS may be sames and universals and differents from much less than TO-TIT. Non ESINS may be differents and particulars, even sames and universals that are irrelevant. Differents and sames are based upon the parameter of likeness, particulars and universals are largely based upon size and contents in entities or universes. ESINS and NON ESINS are what does or does not exist and function in entities and AU that is also in TIT-TO. Universals and particulars, sames and differents, ESINS and NON ESINS are not TRU unless applied, qualified and proportional. To be same as in TIT-TO is not same generally or universally.

In this introductory section TRU is the observed and our perspective, thoughts and concepts are the observer. The bonds or segments between them are **INT** or Interval of TRU. Differential and interval are partially illustrative of quantum theory. In TRU their form and function becomes fused to produce the whole in which the part and whole have complete VAPO significance. Quantum physics, as relativity, is verified by TRU when held to their SCARLS and interpreted by VAPO and TRU concepts. The vital role of parts in TIT is **LOPAN** Law of Partial Necessity. The role of parameters and continua in TRU as same and different, integration and differentiation, and signified by DAOST, is much of TRU axiomatic. The importance, even divinity, of words and reason has long been acknowledged. Logos or the predominant power of words and reason as vehicles controlling the universe are none other than key intervals, intermediaries and vehicles in the mid section of the mass energy FAF Á continuum of TRU.

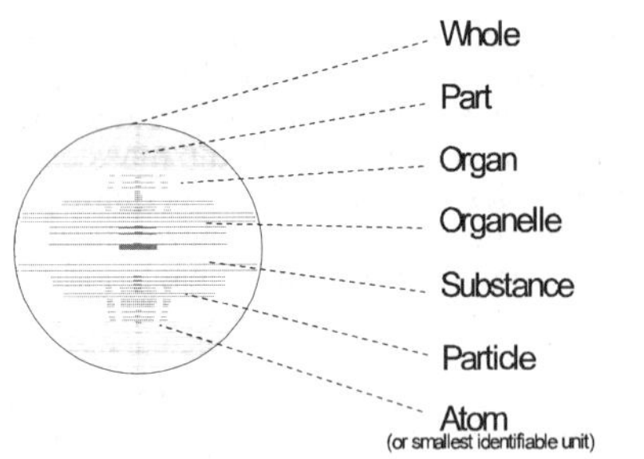

Fig 1-04 **KINDS OF PARTS OF A WHOLE: PARO**

What is of the makeup or quiddity of an entity may be considered singularly or multiply, as part or parts, a finite or multiples of finites. That this special universe of best human life consists of parts is: **LAPAR** = Law of Partiality, separate component parts of TRU. The relation of part and whole, or parts and wholes is signified by: **LANEDO** = Law of the Nature of Parts Determined by and Secondary to the Nature of the Whole. **LANODE** = Law of the Nature of the Whole Determined by and secondary to the Nature of its Parts. The distribution of parts in a whole, as of TRU, may be made fractionally in many ways, when divided into three parts circumferentially they are called **ENDOT**, **MESOT** and **ECTOT** or the inner third, middle third and outer third of TRU.

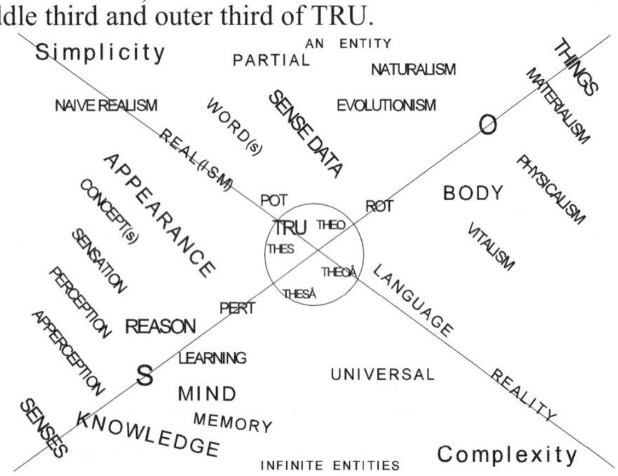

Fig. 2-04 **SOCOD or the S-O SIMPLICITY COMPLEXITY DIAGRAM**

What are the relationships between, and how can we demonstrate, the various schools of philosophy, points of view, means and methods of ordering knowledge? How much is this the substance of POT and TRU? The schools, or classifying Epistemology, are shown by ordering the S-O relationships with those of simplicity/complexity. These schools or systems are **EBE**, or extremes by emphasis. Their resolution is called: **REBE** = Resolution of Extremes by Emphasis.

There are different varieties of the derivation, development, determination and demonstration of objectivity, **DAO**. Among these are RIE and ORT. **RIE** = Representation of the Ratio between Image and Essence. In the subject object, or observer-observed, dimension of relativity there is a variable ratio between the image a subject receives and the essence of the object. To understand subject object, relativity and even TRU we can do little better than to examine what and how RIE participates in all that is Given, at hand or of our IA. **ORT** = Optimal Total Rendering. ORT is a product of combining RIE and DAO. When we speak of the means to a greater and better demonstration of an object we do so in more than one way:

(1). What is of the entity other than the object of this entity,
(2). In Themselves within the object,
(3). As they consist of portions that are TRU,
(4). As these are parts, processes or aspects of TRU,
(5). As they and the rest become the totality of TRU,
(6). Of the representation and presentation of the above as they increasingly become and participate in TRU.

Abstractio is scholastic philosophy's concept for separating one thing from another, and particularly by the process of deriving the universal or generic from the particular or species. Abstraction may be either total or formal. It is total when the universal or generic is considered without what is restricted to the particular or specific as its properties. It is formal when this abstraction involves the essence alone separated from the case in which it subsists. In this way Thomas Aquinas separated these two varieties of abstraction as: the one in which the universal is abstracted from the particular, as animal is from man, and the other in which form is abstracted from matter as the form of a circle is abstracted by the understanding of all, all sensible matter. Abstractivum was a nominalist term, upheld by Occam, for knowledge by which the thing was apprehended not as it is in itself, but through species or some other thing. This is opposed to intuitive knowledge by which the thing is known immediately as it is in itself. These variations on abstraction were largely stimulated by mans examination and attempts to prove God. For God, being the probable relationship within a unified construct, should be demonstrable by reason, sufficient reason and abstraction. It is this intention that not only makes reason, abstraction and our conceptualization more valid, it is what is necessary to bring our approach to TRU to fruition. For if we are to acheive an effective POA, we must be able to separate **ES** and **CES** or the parts and processes of TRU, the differentia of form and function from each other and TIT and how they exist in the whole, **TO** of TRU.

At the root of human ability to exercise relativity, observer-observed, subject-object, or the predominant mass energy A dimension is our use of words, their meanings and definitions to represent what is most important in our world, TRU. How this is actually what is divine has been long recognized, as the word of God, and even earlier, **LOGOS** or the predominant power of words and reason as vehicles or intermediaries controlling this universe, TRU. It will be shown, and frequently emphasized, how these are - in actuality - the reals of the prime and predominant mass energy FAF, Á dimension and continuum of TRU. It is this that gives us a POA to these reals or intermediaries between S and O and makes reality concrete. To do this with the separation, differentiation and formal distinction necessary it is well to set forth what these are. Most are in common usage, although not as clearly defined and used as they can be. Some are altered by TRU, and some represent new concepts, NEC, where voids or poor terms exist. Many associated terms are described in adjacent chapters, the ones here are those needed to emphasize their relations to words and the development of these words. Essence is that of an entity or thing that is necessary for its being or existence. **ESIN** is that of an entity that is a part of TRU. Mean is that of essences, concepts and words that describe their nature, form or function. Meaning is what one intends to convey or signify, the sense by which it is understood. **ES** = Parts of TRU, especially necessary parts.

CES = Processes of TRU, especially necessary ones. Concept is an idea that is solidified into a single expressible word. Principle is an important idea that is clearly enunciated and verified, or a significant concept. Law is a valid and clearly enunciated idea or principle which is used as a guide or complying force. **Á**= The predominant m or mass-energy dimension in TRU, derived from the observer-observed or subject-object state of relativity, and prime determinant of the form of TRU. **ANOM** = From Nominal or word, less than word. Nominal is a name, word or noun that is strictly expressed, neither object nor subject. **NOMT** = Nominals and words can not be taken to be all TRU, nor even the objects they represent They have power and exist in TRU through representation. Word is something expressed in any way which symbolizes or communicates a singular meaning. Term is a word or expression that has a precise and strict meaning, especially in science. A denotation is the class of entities to which a word or term refers, whatever a term designates. Connotations are the properties common to whatever is designated by a particular term, the defining properties of a term. The ideas and images associated with the use of a particular term. Definiendum is a word to be defined. Definiens is an expression used to define a word. Definition is the process of explaining the meaning of a term or expression used, or the statement of the equivalence in meaning of the definiendum and definiens. To Know is the ability and activity to understand and use concepts, terms, definitions and meanings. Knowledge is the total of what is known, by one, some or all. **COGNOS** = The total of what is known as TRU, and the knowing of it. These terms are presented because of the great need for clear expression of all that leads to TRU and **KOT,** the knowledge of TRU. They show the importance of formal distinctions, especially when seeking to provide the optimum differentiation of and in TRU. The course of a presentation of TRU as well as TRU itself is very much dependent upon the validity, actuality, totality and proportionality, or **VAPO** of these and all terms used to represent and present TRU. These concepts and terms must denote that of TRU they represent by providing the formal distinction that delineates and makes the discrete separations of all that is TRU. As symbols, nominals, words, terms, NOMT and NEC gradually become more and more precise, fill in all the gaps in and of TRU, and can be best used to posit all as they function in TRU, language will approach the logical level of set meanings necessary to replicate what is TRU. This becomes more apparent when comparing and considering these terms, concepts and entities.

(1). They stand in a unique and highly focal position as elements of POT, EPI, TRU and ROT.

(2). They can not be considered solely POT, EPI, TRU or ROT.

(3). They can be properly and fully utilized as elements in a successful ROT and POT when through **SUT**, or the Systematic Unity of TRU, they are clarified, and increasingly well defined, delimited and delineated; so that they can be posited as clear cut entities, ES, S-O, CES in SPALT.

(4). This requires the elimination of multiple definitions, ambiguities, errors, unnecessary and non TRU fallacies, paradoxes, definitions and meanings.

(5). The continuing renewal, improvement, and development of concepts, existing and new, with limited and TRU meanings that arise from the proportioning, ordering and positing of all that is TRU. .

Fig 3-04 **Variable AU-TRU Diagram, VAT**

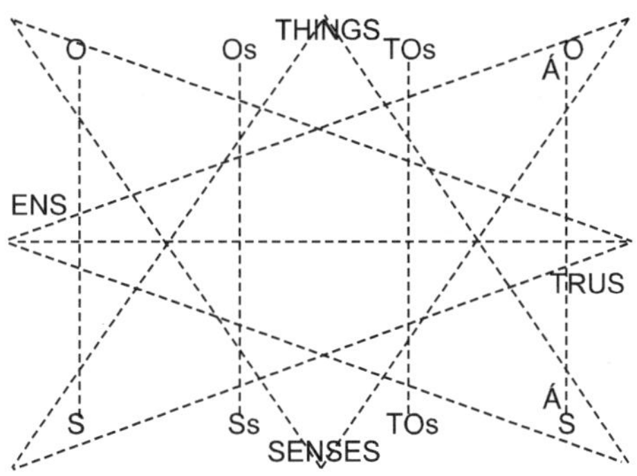

Much that is **AU,** or all entities, is not some ethereal entity, vague thing, or far distant unrecognizable sense, phenomena, fact, or object. Much of AU is that which is less than TRU, from TRU, disproportionate, incoordinate, trivial or mixed TRU and ATRU. What in any way contributes to what is null or negative to TRU. The more there is of these, the more AU. AU is much of our existence, i.e. our existence that is not TRU. It is much that is about us. It is that of everything that is or is not about us that is not to the degree it should be its counterpart in TRU.

> - - - **AU** (All that is not TRU) - - <

> - - - **AUT** (All entities, senses, things, AU and TRU) - - <

> - - - **TIT** (Only TRU-in-and-of-ITself,) - - <

The VAT diagram shows how entities, senses and things, and TRU are mixed in the world around us. We can better understand how they become altered and adapted when compared with their set state in TRU. This is best seen and of the highest significance in S-O transition to the Â dimension.

There are many kinds of parts of a whole and many ways they can be identified. They vary largely with the kind of whole encountered, given, or the topic at hand. They can range anywhere from an infinitely large part, approximating the size of the whole, to an infinitesimally small part. To understand a whole, as we seek to do for TO of TRU, it is necessary that we understand, along with the rest that it takes, what its parts are. Not only the different levels as shown here, but more detailed form and function of by VAPO, as they best makeup and reveal the whole. **PAVAT** = Parts are Absolutely Vital to a Whole, and expressed similarily yet more generally, **LOPAN** = Law of Partial Necessity. PAVAT is also a corollary of LANODE, being more rather than less determinant of the whole. PAVAT expresses the fact that parts are absolutely vital to the totality of TRU, TO, **SARVAT**. Just as the heart, brain, oxygen, water, food,, and many others are vitally necessary to the human body. (LATANSART, ANSART, DREM)

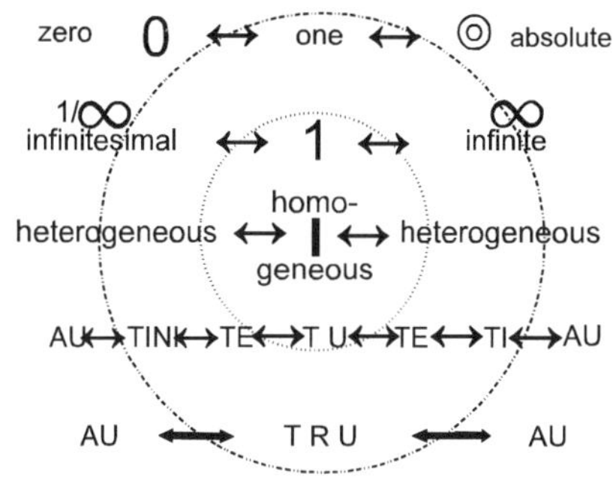

Fig 4-04 **RELATIVE FINITE**

The tendency to think of an absolute, AU, AUS, and TA as an identity devoid of any differences is a trap which illustrates the degree to which EPI and KOT may be distorted. This is the simple defect pointed out by Hegel of Schelling's concept of the absolute. He likened it to a view in which all cows are black at midnight. This, although trite, is an interesting analogy. Without difference, as the D in DAOST of Different Aspects of the Same Thing, in the form of TRU, we have an entirely different and erroneous view or subject. This applies not only to the form of TRU, but even more so to its function, TRUD. It is however, of great value to extract the homogeneous and the heterogeneous, the same and the different, the identical and the non identical and the relative and the irrelative in order to critically analyze and formulate their forms and functions. When done we can posit them in the frame or construct of TRU and bring them back into the overall or whole. Likewise in the comparison, much of the form and function of TRU are better shown. **LORE** is the Law of the Relative Finite. LORE is how Relative imposes, and in turn is imposed, by the distinct finite and definite form. In becoming relative we are spontaneously placed in the position of having a fixed relationship between zero and infinity which focuses on the unity and finity of totality imposed by relativity.

ABSOLUTE NUMBERS INACTUAL and ACTUAL

Just as we have a number for zero, or the imaginary beyond the infinitesimal, so must we have a symbol that can be used mathematically for what is beyond the infinite, past infinity. This we call: **ÂBS** = or Absolute Sum, Total or Set, or simply Absolute, signified by the letter Â. It is in contrast to zero, or absence, 'nothing' and 'non existence', 'everything' or absolute presence and existence. It is the sum total, without which there can be no less and can be no more. In this form it exists as absolute in and of itself, **AIT**. Absolute, or 'everything' is the absolute opposite of zero, or 'nothing' in which there is no less and no more- at least if we do not include negative numbers. Negative numbers require a new given and numeration by subjectivity in which the form of numeration is altered. Positive and negative are readily understood, as more or less, increasing or decreasing, in numeration. However when interpreted as an overview it becomes another given. The problem of added givens, or foci and relative universes, is one that is only solved when we resort to TRU and LATANOS. For just like the infinite reaches of space, or the infinitesimal reaches of physics we proceed diversely and with increasing randomness and chaos unless we retain some focus, centering, or mode of reference. This can, and must be done, if we are not to speak or write in riddles, illusion, fantasies, and figments. This is resolved and supplied by TRU and the thisness, relativity and focus it provides. The symbol or mathematical figure for absolute, limit beyond the infinite, is:

◎ = **Absolute or ÂS, Absolute mathematical Symbol**

In this way we have a continuum between:

0....1/∞....n/1....1....n....∞....◎

When the whole progression is encompassed, we have a new kind of Absolute, the Absolute total:

◎ = Absolute total, or **AÓ**.

(0....1/∞....n/1....1....n....∞....◎)◎

It was in Cantor's study of infinity that he developed Set Theory. Yet number theory, real and complex numbers and set theory, have not resolved many of the basic problems in mathematics. This is partly because they are hard to conceptualize. To symbolize and define this opposite of zero and what is past the infinite, absolute ◎**,** and absolute totality ◎, helps simplify and clarify this aspect of numbers as well as our being. This we do in relation. For when we bring all, by sameness and relations toward a focus one, then numbers ,as also occurs for mass-energy, takes upon new meaning. This new meaning excludes, selectively, what is not a part of this developing to absolutely fulfilling absolute total. By LANERA, although 0 and ◎, are illusions, figments or "imaginary", they do represent the poles, and thereby can be thought of as numbers for which symbols can be developed. In being poles the polarity they form is in TRU a continuum; in fact when considered overall, the sphere and totality of it. This polarity, spectrum, continuum and sphere may be divided as we do in mathematics from one. In doing so they become differentiated and then in reverse, integrated. This is what we see in the formula above.

This meaning, definition, and way of looking at numbers is called: **NEDIBÀ** = Numbers Equal the Differentiation and Integration between zero and absolute. **PADERO** = Positives and Negatives Equal Relative Orientation or Orientation by Relatives. Likewise, simply by TRIA, we have the fact that only in TRU are numbers not illusion, figment or imaginary. This is what we find in all representation, the degree to which it does not have or is not a 1:1 replica of TIT. This is called: **NIFAT** = Numbers Illusion, Imaginary and Figment to the degree they are from, and not of, TRU. This leads us to the arbitrary nature of representations, symbols, numbers, names and definitions. All that stands to show TRU that if different from TRU is not TRU, at least to some degree. The less this departure from TRU the more it is TRU. This is how numbers, as representations, symbols, names and definitions are created by us, and act as standards. This is called: **NADS** = Arbitrary Nature of Names, Numbers, and Definitions, and how this can be ordered by Standards. It is the weaknesses, limits and errors of standards that has, does, and will create much of the dialectical interaction that keeps us in NAMTRU and out of TRU, OITIO. Yet their best refinement and use will progressively improve the ATRU state to become AMTRU, MATRU and TRU. It is in this way names, numbers, definitions and all that represents TRU, are not, less, or more TRU. Yet much of their identity and existence in TRU depends upon the degree to which representation acts as a mnemonic device. As such it becomes highly useful and beneficial. It becomes as **RIT**, Representation in TRU, a part of TRU having, in a way, its own actuality, being and absoluteness.

Quantum theory is more profound, by virtue of the differential, than relativity. Both work separately, yet together. Only in TRU does their form and function become fused to produce the whole in which the part and the whole have completely valid, actual, proportionate and total significance. Thereby the implications of quantum physics can be developed as we do in relativity to show, not so much how it exists astronomically or physically, but selectively qualified as it actually exists in TRU.

Polarities, parameters, dimensions and their adaptation to quantum and relativity, especially that of TRU, are not as simple as they at first appear for:

1. They represent opposites and polarities between which there are many levels, stages, finite units, or quanta..
2. They can be measured, expandingly and quantitatively..
3. They do not exist alone, either in themselves nor universally. It is in this universal union that the coordinates of these dimensions of TRU function in mass action.
4. They do not exist only in space-time but when represented in three dimensions as a geometric solid.

We begin to order the finite or differential by the well known and effective dimensions of length and time, and the coordinate axis they form. These two together make up the axiomatics of TRU known as the Doctrine of Space Time or **DOLT.** Along with the S-O dimension which with alterations and adaptations becomes, in TRU, the A mass energy dimension, there are corresponding alterations and adaptations in the space and time dimensions. All three together become ALT, the three dimensional form and function of TRU, which is of such great importance in representing and modeling TRU. Such is not only basic to the TRU, it must be adapted, standardized, and used to reveal and show TRU in contrast to the. physical world.

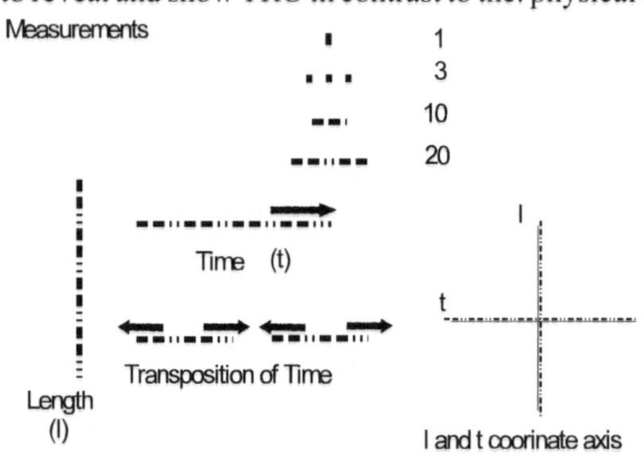

Fig. 5-04 **Measurements, Length and Time** (MELT)

In Figure 5-04 we show an introduction to measurement. Quantifying the world about us is key to the accuracy and reliability needed to adequately understand it. (VAPO) This especially applies to the parameters, dimensions, polarities, spectrums, and continua that constitute TRU. Only as these are developed, and only as this world is TRU, can measurements fully and successfully serve us. Thereby it is mandatory that we develop measurements that reflect TRU. This involves the dimensions of TRU and their measurements. For they are set in TRU and their measurements follow this set. We show this through the text, especially the development TIT, TRIT, OBT, SUT, ROT, MOT, ALT, SPALT, POS, STA, and HOM.

Of the complexities noted for time there is: point, duration, transposition of, and blocks of time. The explanation of time depends upon the frame, the observer and observed relationship. Only in TRU, and only as TRU is SCARLED as in TRIT, TRIN, TRINI, TIT can this be done precisely and completely. Yet this is the beauty of relativity, of TRU and its characteristic relative identity in totality. As the physical world can be formulated dimensionally by mass, length and time so can its contents be posited by their projections in these dimensional coordinates. This will be weighed by the predominance of those dimensions that most partake of each component. Mass will also be weighed not only quantitatively but qualitatively depending upon the universe. Whether the world or universe of which it is a dimension exists entirely in the physical, mechanical and gravitational state or in some other state as by relativity of the observer observed. Only when we can focus, unify, relate, and correlate the dimensions and parameters and measurements can they have actual meaning. And the accuracy and reliability proper TRU measurements can provide will we have a successful POA and POT.

CHAPTER 05

THE PART or PARTS, of ENTITIES UNIVERSES POT and TRU

We can learn the status and role of special parts as they become and are of TRU. It is more clearly achieved by recognizing LACON, LATIMD, COTRU, and ICA. **LACON** = Law of Partial Conditioning, or how parts are by their SCARLS, and those of what they are a part of, TRU. **LATIND** = Law of TRU/Individual or the TRU Individual Differential. **COTRU** = Concentration of TRU. **SRIA** = Sudden Realization of All one has hoped for. These reveal progressive concepts and aspects of the part or parts as it comes to be TÍ and TES in TRU.

As succeeding readers and workers fail to reach the intensity of meaning, revelations, and inspiration developed by those greats of the past, we have: **POMOR** = Principle of Reduction of Meaning. POMOR is how meanings turn out to be 1/2, 1/10, or 1/100 of their original intensity. **RORRE** is the needed revelation of the Revelation Relate Explosion. It is through RORRE that one can with training, concentration, ELTICMA and the constant presence of the problems before one assure the unity to infinite character of TRU. It can be formulated by TRU, TIT, TRIT, OBT, SUT and ROT. **LEVIA** or Level of the IA, illustrates the highest levels of summative planned action. The levels, as **INIA** or the Intensity of IA, lead to such high levels, and are only limited by our capacity, ability and use of all available.

Fig 1-05 **DUALITY, S/O, AU/TRU, and SOAT**

DUALSOAT shows the field produced by dualities or continua of Subject-Object and AU-TRU. It is one of the most important diagrams of the text. For it shows us not only how we look at all, but how all exists. More important it shows how we can better approach TRU, how it is formed and can be shown. The coordinate representation can be divided by its four poles, its objective and subjective, and its AU and TRU and their various gradients, relationships and combinations. When it is divided into its four poles we have: I. AU S, II. AU O, III. TRU S, IV. TRU O.

The possible combinations produced by the coordinate field are, as well, in the dynamics of these relationships, of a similar number of possible interactions from the objective poles. These that are set from the objective are: 1. AU 0 - AU S, 2. AU O - TRU O, 3. AU S - TRU S, 4. AU O - AU O, 5. TRU O - AU S, 6. TRU O - AU O, 7. TRU O - TRU O, 8. TRU O - TRU S, 9. Summative Singular Subjectives. These relationships of S O and AU TRU can be connected by vectors. This brings out their dynamic natures. .

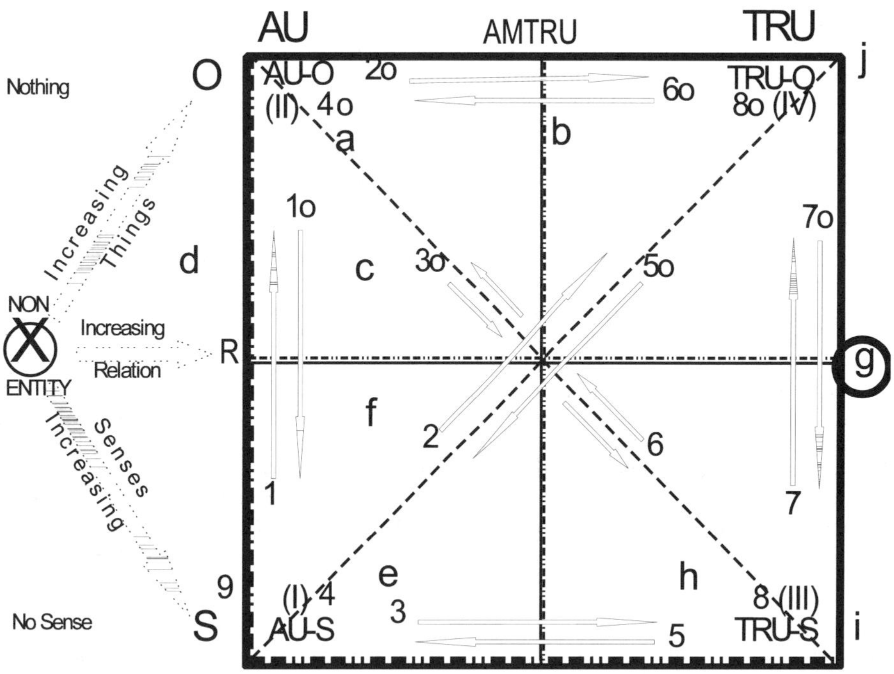

When taken in more than singular relationships our combinations they turn into diffuse permutations and processes. These are important for they show us the great variety of relationships that exist.

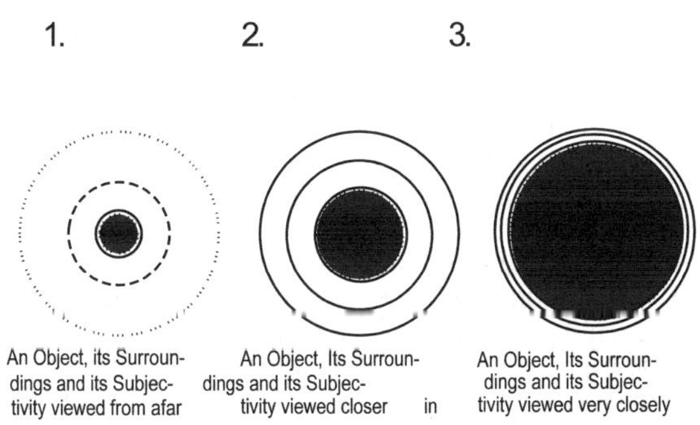

Fig 2-05 **COMPARATIVE SPOS, SOPOBS, POUS**

Three different views of the same object and the related things and senses about it are shown. The difference is one of perspective. In the first we look at it from afar off, so that it is very small, the things about it are somewhat larger and the sensing of it the primary portion. In the second figure we envision the object somewhat closer so that it becomes larger. At this stage or level of telescoping in on the object, it, its surroundings, and its subjectivity are equal. In the last figure we look at it very closely so that the object is very large, and its surroundings and subjectivity are barely visible. This is what occurs continually in and out of objects, in the flux of life, as we are repeatedly stimulated, face givens, and view the objects before us. These are highly variable more or less the same or other than the entities in and of themselves, ENIT. As we look at an object or the world it becomes larger and larger for we see more and more of it. Its form, characteristics, dynamics and all become clearer. It becomes more ‘real’. Yet the object and entity are relatively and actually changed only to the extent they become or are TRU

Fig 3-05 **SPOS DIAGRAM**

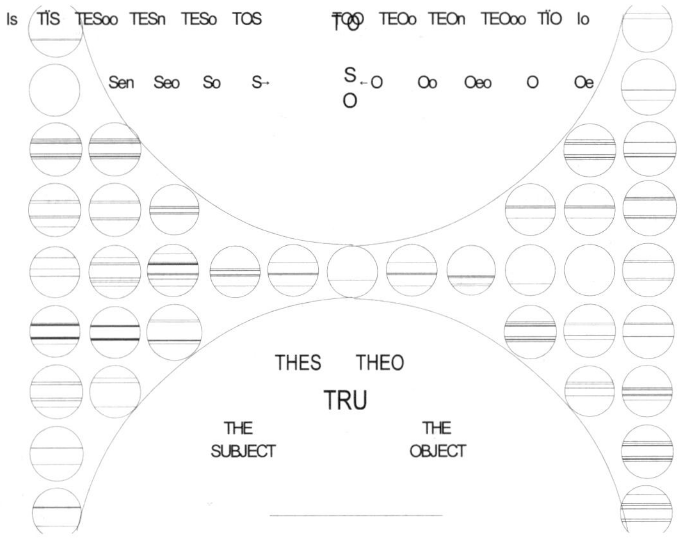

SPOS = Standardized Perspective of the Objective-Subjective. SPOS is one of the greatest of vistas, for if one can grasp the jist of SPOS one can feel the revelation and inspiration that comes from understanding Á and TÄ, and TRU. This is because it not only gives a good account of the parts of entities - senses and things - as they become TRU. It gives us the basis for the prime, S-O or Á dimension of TRU. SPOS results from the extreme lability of our position as observer and IA and the observed as entities and objects. The exclusive lability of meanings by SO or Á for TRU is called **ELP,** the Exclusive Lability Principle.. Each and all of us must not only have senses, perceptions, apperceptions, feelings, thoughts, reason, etc. we must be able to bring these together and sort them out. We must have some kind of overall or world view or perspective. (COAT) This is impossible without standardization. It is not provided spontaneously. It takes knowledge combined with learning, training, and intellect. It is a great part of one's maturation in life. With it a person can better cope in life. Without it one flounders, is quickly lost, confused and defeated. With its cure and end is SPOS, S and O incorporated, standardized and modeled, best represented and brought to TRU.

The relativity of time as provided in our model of TRU, focused at time 0 or the present, is exemplified by the shorelines of the ocean and their continually changing levels. This is of various rates, often rapidly changing and sometimes more stable. Yet this shoreline is only as it is today. It is like a point in time. So is the focus of TRU a point in time, from which ATO and all are derived, at least in large part. It is like TIT, TRIT, TRUN, where TRIS, TROS, TRUS are not only as they are specifically, by TRIA, now, the present. On the other hand this now and present in being infinitesimal, is not all, is not TRIA, for TRIA, TIT, TRIT, TRUN, TRIS, TROS, TRUS, consist of a duration, a massive time which encompasses all that partakes of this relativity.. This is called:

LATIRD = Law of TRU TIME, Relative Distribution.

LATIRD establishes the fact that it is not any time, physical time or variable time that is the actual time frame in which we live. It is a set, yet special, typical and unique distribution of this time as it exists in TRU. It is only through the application of LATIRD, or the principles of time present in relativity as it occurs in the world of man, subjectively and objectively combined, SO, and Á equivalent to the mass energy dimension of this world, that we can reach the orientation and observation necessary to successfully order the universe, the Relative Universe, TRU.

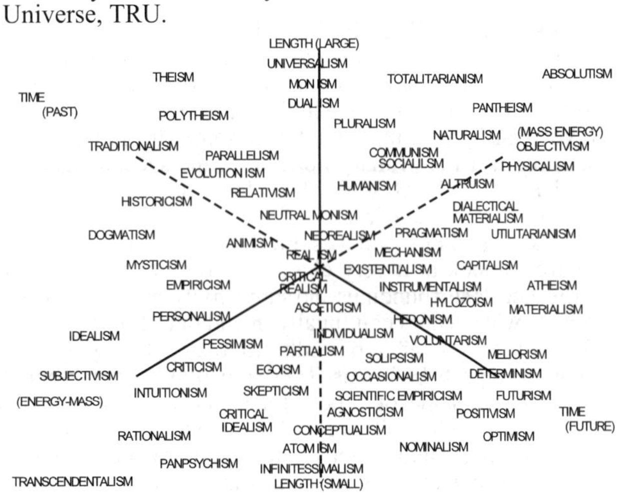

Fig 4-05 **POSITS of ISMS by SO, L and T**

This figure helps us understand the derivation and relations between entities which are special parts, as isms, and the frame of the whole to which they belong. We posit systems of philosophy in the mlt, or S-O mass, length and time frame. The positions held and beliefs that each of these and many other ISMS or systems have is a good example of the derivation, development, determination and demonstration of The Part. For each represents not ends, but specific and unique parts and processes in the whole. Their validity and corrections can occur only when they are recognized as parts and proportions in the whole. This whole is the actuality of TRU, and their validity is no greater nor less than the degree to which they are only this part in TRU, by VAPO. These systems can be better understood and their limitations and qualifications in terms of TRU shown if they are interpreted and posited dimensionally.

PRUFAD and REDFAD - If SOT and A are true, then it must be proven that the absolute S and the absolute O are unreal or invalid, just as the absolute largeness and smallness are unreal or invalid, or the absolute past and future of time are unreal or invalid. In this way one may show the similitude and the necessary agreement between the basic elements of the dimensional formulation. For this it also becomes apparent, or may be proven that the real, or S and O as SO together as the real and gradients of length and time determine, as SOT and constitute, as A, the finite nature of TRU. This is where TRU as any RU is, or becomes, finite through its own RU relativity, unity or focalization by the reduction of its dimensional formulation to finitude. The proof of this is called: **PRUFAD** = Proof of the Non Absolute or Relative Formulation of Dimensions, the Foundation of Dimensional Formulation. The reduction of this is called: **REDFAD** = Reduction of the Non Absolute or Relative Formulation of Dimensions. REDFAD is what enables us to establish totality, geometry and reliability to our frame for MOT, ALT and SPALT.

The dimensional form of TRU is neither each dimension alone, nor each as simple dimensions together. It is of that of which TRU is predominantly constituted, i.e. gives it its special presence, identity power and exclusiveness. The bending, warping and transformation of basic dimensions is a product of TRU and consists of much more than a combination of the simple first order m, l and t, but all higher levels in combination as they best serve TRU. This is called: **LA-HOMALT** = Law of Higher Order MLT in totality produce the Form of an Entity, RU and TRU. HOMALT shows how mass and energy, as Á, along with L and T, come together in their set FAF in TRU. It shows how they are different aspects of the same dimensions and dimensional totality, correspondingly and interdependently fulfilling their actual FAF, SCARLS and roles in TRU. These compound interdependent high orders of ALT, work together in combination totally in what is called **ACEXIN,** or the Actuation, Extension and Intension of TRU.

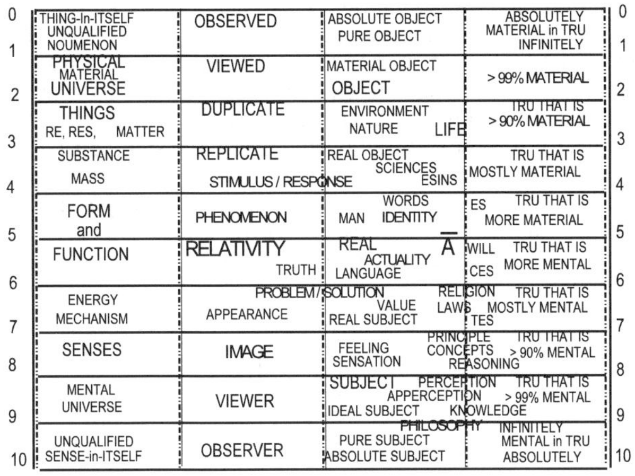

Fig 5-05 **GRADATIONS or LEVELS of RELATIVITY or PHYSICAL/MENTAL, S-O, Á FORM and FUNCTION**

This diagram provides good tie-ins and cues to the gist of POT and TRU. It shows the relations and progressions of many key concepts we use to approach and demonstrate TRU. It helps us attain an overall view and, to some extent, prove how the mass-energy state that exists in the physical universe, is transformed in, and shows TRU, TRIA, TIT and ATO. The first column shows the world beyond and before relativity, how entities exist ENIT or in and of themselves. The second column shows the world of relativity, what the universe is like when altered and conditioned by the observer observed relations. The third column shows how relativity, the observer/ observed relationship becomes the world of the subjective and objective that is the predominant form, function, state, and operation in our world. The fourth column shows how this gradation and levels of subject and object exists in the Á state and dimension in TRU. Within the diagram are many concepts and words that are posited as close as possible to their formal distinctions in this two dimensional field. The diagram is produced not only to prove Á, the S-O nature and coordinate or denominator of TRU; it helps us understand that the world is not as we see it, neither senses nor things, only how the actual world can be TRU or by TREQ. This diagram provides, on progressing to the right, a basis for **LIDORT**, Law of Ideological Total Resolution by TRU or TREQ. It is in the development of opposites in continua and their accompanying dimensions that we represent, model, and use TRU. In being opposite, by contrast and often violent dialectics, they are more dramatic, interesting and impressive, thereby more readily learned and remembered. This is expressed in **LACONEX** or Law of Concepts from Extremes and Opposition. Oppositions and polarities reflect in both humans and individuals an extreme nature and innate tendency to take sides. This is corrected by the role of S-O and Á in ACEXIN and TIT. The focus in this diagram, and the next, shows the focus in S-O and Á, or **DOGOM**, Doctrine of the Golden Mean.

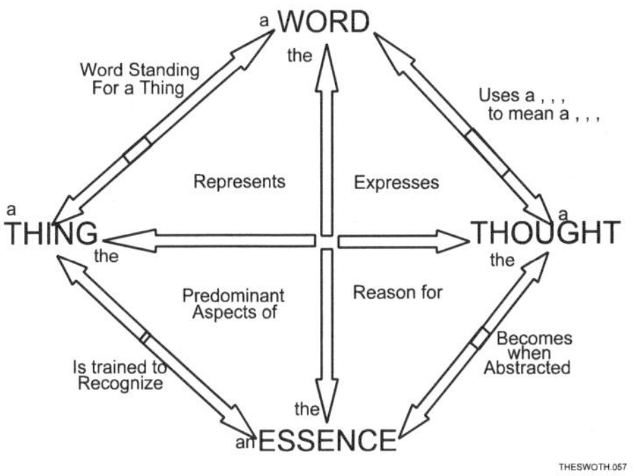

Fig 6-05 **THING, ESSENCE, WORD and THOUGHT** Figure 6-05 shows the relations and interactions between thing or entity, object, word or term, mode or essence and subjectivity or thought. Various connecting elements are added to show other relationships and operations. We need to learn and understand their origins, relations, and actions both as they are interpreted from themselves, or combined in TRU. Animals learn to quickly recognize what is good or bad for them. In humans this is largely verbal and from knowledge in reaction to the increasing complexity, to best solve problems. To have meaning thought uses concepts which are represented as words. The object of this meaning is usually its essence. Thought uses these essences to accomplish its intention. Yet we rarely distinguish things or objects from their essences. To have meaning ideas and thoughts are identified with concepts expressing its essence in words or language. Thought becomes essence as words are abstracted. An entity's essence is usually a large part qualitatively, but small part quantitatively of its FAF, especially its ES and CES in TRU. This is: **REAT** = Thing (RE) Essences from Part in TRU. It was Aristotle's method to train the mind to recognize and develop these essences, and to appreciate their preeminent uses in the world, that has spearheaded the way of the investigator. Plato makes them crucial in our approach to TRU. It is from this transition of reals or essences in things to TRU and the training of persons to extract and utilize them in TRU that we derive the term **PUREALITY**. It is thinking through and using these essences, or 'reals', as they become and are in TRU that is the prime theme and method of learning this text. This is in contrast to thinking without the use of essences and ESINS. **ESIN** is that of a thing that is its counterpart in TRU. ESINS, although similar to essences in being the prime reason for an entity, are distinct in being of all entities according to TIT. Essences are indistinct except the entity of its object and strictly held to their part and role in TIT-TO. Different entities, senses, things, essences, words and thoughts when reduced to their ESINS, and forms and parts in TRU become and enable us to attain higher and higher cleavage and identification of the different FAF. parts and processes of TRU. These discrete parts and processes and the corresponding words enable formal distinction and positing in the frame of TRU.

Fig 7-05 **PHILOSOPHICAL and RELIGIOUS SYSTEMS or ISMS and BELIEFS**

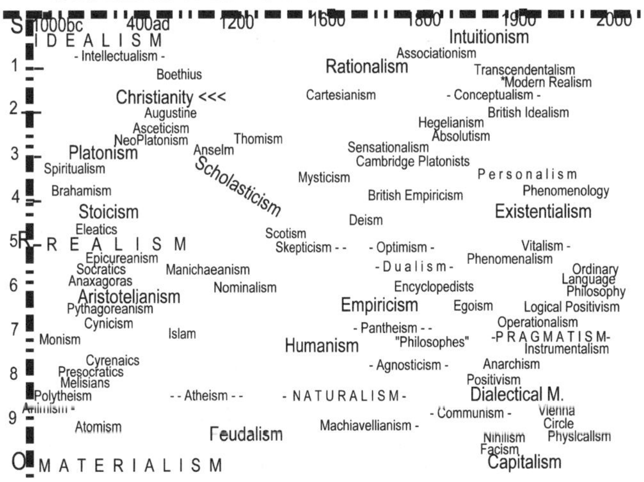

The various Schools of Philosophy and ISMS are of great value, as processes in TRU by emphasizing the crucial nature of basic relationships. They have not been used effectively to reveal their roles in TRU, nor are they, as used, of much TRU. The greatness of such an investigation is to provide ever keener insights into the ever more discrete objects before us. The best of these ISMS - as of various religions - is great, but their trouble lies in the stands they take and teach. For these are, 1. incomplete, 2. have inadequate principles and laws, 2. become divisive and self serving, 4. are intolerant and fail to fully understand the similarities and differences with other ISMS and religions. So must philosophy, religion, and science be all that is equivalent to TRU, TREQ. The degree to which they are not equivalent are their philosophical, religious and scientific errors. It is this that not merely gives a distributive scaling between subject and object, but lies at the root of mans form, our predominant mass energy dimension, or Á. Since idealism, materialism, and even realism can't stand alone it has naturally, in reaction, led to a lack of closure, actual philosophy, ontology, valid spiritualism, and world view. These have been in ongoing flux since early man and have been a result of inadequate, unordered, and disintegrated knowledge. They do not properly show the nature and true form of the actual world, TRU.. They must show: **LORÄ** = the Law of Predominant Form. LORA is by what this relative universe is predominately constituted. For man as TRU this is the special, Á, or mass-energy state, and the special form and distribution of L and T, that exists in TIT. It is this very predominance that propels us to ISMS which by **ICA** or the sudden realization of all one has hoped for must be fulfilled. This can only, however, be achieved by VAPO of and TRU.

ORIENTATION and ORDERING of PARAMETERS - By investigating different parameters and identifying them with TRU, their relations and the form and function in TRU can be better known. It is crucial that we bring this together to see how they are oriented and can be ordered. This is what will enable their use for a positive objective approach.

MASS EQUILIBRIUM ACEXIN STANDARDS and PERSPECTIVE

Fig 8-05 **MASS, LENGTH and TIME EXTREMES BASIS for ALL FORM and FUNCTION, ACEXIN, EQUILI-BRIUM**

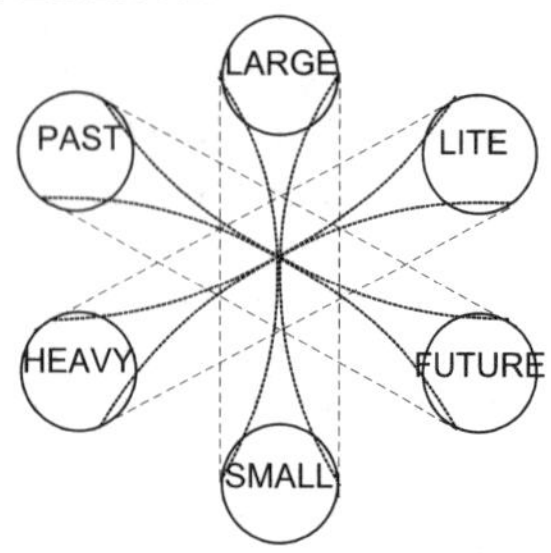

STANDARDS and PERSPECTIVE

What is length, time and mass energy? Why do they seem so intangible? Why is it so hard to achieve any order and total relation between them? To do so we must look at what they denote; what their basis is. Length is the relationship between the large and the small. This includes, area and volume, as well as the medium and all in between, when carried to its fullest meaning. For without the medium, the means and the extremes we have no focus, length, set area or volume and thereby no order. Time is the relationship between the past and the future. This, of course, and especially, includes the present and all in between. Without the present or some set point of reference, as the mean, we have no focus or order in time. These apply similarly to higher states of time as duration and variation. Mass-energy is the state and dynamic relationship of what is predominant for an entity, between that which has the greatest weight, orce, power, or effect and that which has the least. This especially includes the world in which we live and how it relates to us, or **MIDST** = Mass is in different states in entities, andspecially formed in TRU. This is signified by **LESEM** or the Law of Specific Energies and Masses. Without relation to us and our world mass energy has a different focus or order.

All entities and objects must have symmetry, which can be of various kinds. It will depend upon what type of entity or object it is, e.g. subatomic particle, atom, molecule, macromolecule, RU, TRU, ROT, and MOT. Symmetry is not only a result of the form of the entity itself, it is also the result of the form imposed by the observer, subjectivity, as an ENIT becomes ESIN and ES, **ESIT** or the ES in and of itself or aspect of TRU. The symmetry of TRU is the basis for much of its representation, modulation and presentation. The basic, most simple and unifying material is mirrored across the Horizon, **HIDIC**, Fig. 4-00 of the text, toward the most complex and infinitesimal. This mirroring although of great advantage in many ways in constructing a concrete text is not expected to provide definitive solutions for it is more a frame, enabled by the symmetry of the object and the symmetry provided by the subject of the object, TRU. The mirroring does provide a cross check and proof of the balance symmetry provides. It provides a check, more proportional and complete demonstration of the object and its presentation, IPTRU.

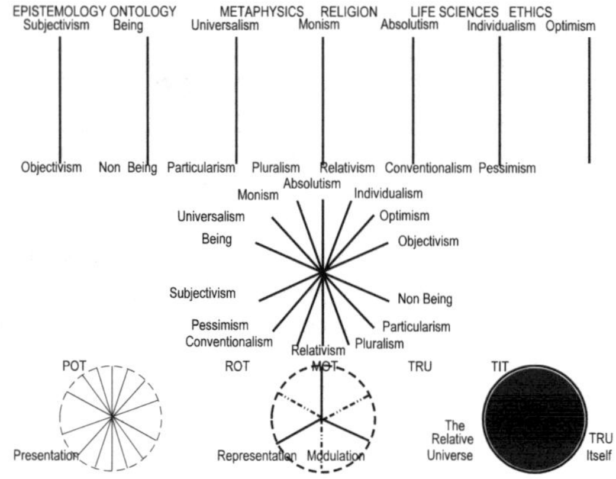

Fig 9-05 Opposites, Polarities, Parameters, Arrangement, Proportion and Balance

What basic relationships, polarities, opposites, ISMS, and contrasts are a great part of TRU or are a great aid in its representation and presentation? We have, among others, the subjective and objective in epistemology, idealism, realism and materialism in ontology, universalism and particularism in metaphysics, absolutism and relativism in theology, optimism and pessimism in ethics and individualism and conventionalism or traditionalism in the life sciences.

In separate figures, different objects whose fields and internal relations are in marked contrast, may be shown as: 1. A single focus with radiations, 2. Multiple foci with radiations, 3. A Triage or three-dimensional field, 4. Triage or three-dimensional field with outer spheres, and 5. Multiple point center with large vectors.

That the make up of the world consists of divisions which exist as parameters and dimensions, and that these have levels which can be scaled and distributed as they exist in TRU, is shown by: **SIL** or the Inverse Square Law. It is also shown by FAE, IAT, **LCDs** or Least Common Denominators, DIL and RAL. **FÂÉ** is the Fallacy of Absolute Equality, and **IAT** is Instantaneous Attention or Immediate Awareness as it exists in TRU, **DIL** is the Distribution Law or how things are distributed, especially for TRU in a specific way. **RAL** is the Relative Form Law, or how TRU is distributed by the weighted power of its dimensions, continua, form and components. How the different dimensions combine to form the whole is **LARP** or the Law of Parametric Proportion. This signifies the set proportionate distribution of the parts of the whole. For any RU there exists a relatively stable or set spatio-temporal mass-energy form and function. These set states and dynamics are characterized by gradients in their dimensions which provide levels, proportions, and quantities that make up the frame, form, and substance of the whole.(TIT-TO) This largely determines its features, modes, essences, characteristics, and properties which remain relatively stable. Its proportions are a product of the balance created by the coordinate frame these dimensions have as they exist together in the RU or TRU. For TRU, LARP is specific, set and has set implications. This understanding of length, time and mass-energy from their opposing extremes, orderly union, and coordination in their means - most evident in the relativity and specificity of TRU - is called: **KALTOMEX** or Extremes, Means and Totality. The dimensional form of TRU is neith Knowledge of Mass Energy, Length and Time from their er each dimension alone nor the union of these dimensions taken simply. It is that of which TRU ispredominantly constituted, i.e. gives it its special thisness, identity, power, and exclusiveness. The adaptations, alterations, transformations, bending, and warping of these basic dimensions both singularly and combined consists of higher levels as they best serve TRU. This is called **HOMALT** or the law of Higher Order MLT in totality produce the form of TRU. It is HOMALT that enables us to envision ACEXIN as it in turn reveals the massive ordered FAF of TIT.

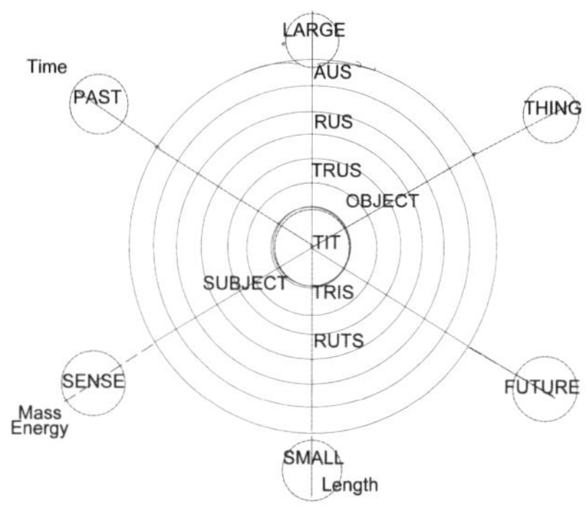

Fig 10-05 **FAT**, Far Approaches to TRU

This figure presents a diagrammatic and rough rendering of the Far Approaches to TRU, how its dimensions can be best combined, and how the many concepts, principles, laws and mechanisms involved can posit into it. This provides an excellent means of achieving a good tie in between POT, EPI and TRU. The dimensional formulation in space time and mass as it becomes S-O in TRU changes as they FAF together and approach or leave TRU.

STANDARDS and their RECONCILIATION

How form is provided by polarities and dimensions is called **LAPAC. STAT** is Standardization from the ideal, or the exclusive dimensional formulation of TRU. Standards like laws, are variable, often not TRU, and often quite partial. It is when they are **STAN** or from the adapted and warped form and frame of TIT, TRIT, OBT, and SUT that they become increasingly accurate. This involves **REFOR** or Resolution of the Frame of Reference, its convergence on TIT. This increase is seen in **LOLVAR** or the Law of Limited View and Representation. Prior experience and familiarity is highly determinant of how well we can discover, realize and apply the dimensions and standards of TRU. One of these is shown by LATIND or Law of TRU Individual Differential. How by reacting dialectically, they largely determine its focus.

Although our presentation of TRU, IPTRU, and all the representations, modulations, and other means of reproduction we use, are selected primarily on the basis of a one to one replication of their positive objective approach to TRU we can not always achieve this ideal. The prime reason for this is that: I. These are not TIT. The second great reason is that, II. Even in separating out the subject and object of TIT we can't do so perfectly, and the third great reason is that, III. Even if all were TIT, and we could have a perfect S-O of it we still do not have enough knowledge to provide the reliable facts, information, data and evidence that give such perfection.

That there are many other reasons for problems with standardization we may refer to their appropriate descriptions in the text. (see index standards) The primary one, however, is how, and how much, they do not follow their natural, TRU, inclinations. Since standardization is such an important and vital part of a good presentation, as well as much of the representation of any topic, especially TRU, it is essential that we understand these. If they are not increasingly perfected they do not coincide with their ideal for TRU. Keep them in mind, ECRO, for continual workout in the future. Thereby for those who endeavor to perfect TRU, either knowingly as TRIX and TREX or less knowingly as TRAX or even only as TREQ it is imperative that they do so, as it is for life itself, to enumerate, develop, plan, correct, integrate, order, and divide these and standards into our ACC and life for TRU.

STANDARDIZATION PROBLEMS

1. First Person, TES, both or variable focus of all
2. TIT, TRIS, TRIT, TÂ, TROS, TRUS, TRU or base
3. MOT and REM, POS and STA divisions in IPTRU
4. Positions of ALT in a sphere
5. SPALT or METAD as focus of IPTRU,
6. Novelty, inertia, resistence, and compliance,
7. Conflicts with existing and provided standards.

STANDARDIZATION SOLUTIONS

1. All must learn, be, and do their best. TES,
2. Each person must learn all that is basic to TRU,.
3. Since standards, like knowledge, representation and presentation of TRU, follow their natural inclination, TRU, we must hold these to TRU except for their added effect.
4. There must be a continual refinement and perfection of MOT, ALT, SPALT, POS, STA and their coincidence and identity with the FAF of TIT,
5. Ever greater precision in ALT's replication of TIT,
6. The use of SPALT, METAD, POMOT, SPAT and other models must be continually upgraded and adpted to TIT,
7. The absolute importance of TRU, justification for LANT, overcoming inertia and resistence, and rational full compliance need to be continually sought and achieved. 8. Existing standards should be used to the extent they are TRU, adaptable to TRU, and continually revised depending upon their need and benefit. It is how well POT and IPT provide these as they replicate TRU that show the benefits, fulfillment, tests, proofs and the superlative nature and method of TRU.

As individuals we are partial, and of limited immortality. In being partial, or variably 12R, ATRU, AU, ANT, ANTIT or even negative REM we are limited. This is: **LALINES** = Law of Absolute and Limited Individual and TES. LALINES signifies that the individual and TES are both absolute and limited. Yet as the individual and TES are much different, except for the bond of ESIN, so will their absoluteness and limits be much different. The individual is absolute in its thisness as a thing-in-itself. It is limited by the non actuality of this thing-in- itself. TES is absolute, as by LOPAN, and is limited by its proportionate part and participation in TRU. It is this difference between TES and the individual as a thing-in-itself, as well as the limits of each proves EL. For we, as all are excluded from TRU to the degree we are not actually of TRU, and of it within its limits.(VAPO). This explains why we must rise to TES, to identify with this absolute aspect of TRU, and be, in turn, absolute. Yet this returns us to the questions of, if a whole is absolute, then its parts must also be absolute. So that only those parts are non absolute that are not absolutely and 100% a part of the whole. It is those parts that are not absolutely essential. Yet by mathematics we know that an infinite can't be absolutely divided, thus is not a finite. For if it were it would be less than infinite. Thereby you and I, the self and the EGO are not as absolute as we think we are, nor seem so certain by infinitely becoming self centered. We are absolute only as we are a perfect part of the whole. This is as Christ, at his finest "In the image of God" where Christ is identified with perfection as a part of God and as a part that is absolute as it is a perfect part of TRU. Thereby that which is a part of TRU is absolute as its parts are 100% TRU. That which is part of TRU is not absolute and is not, as its parts are to that degree not TRU. This leads us to: **LADPEL** = Law of Dynamic Polarity EL, or Paradox of Exclusion by IA and by TRIA.(cf. Chapter 15)

CHAPTER 06

INTEGRATION and TOTALITY of ENTITIES and POT

In order to learn what wholes are, how they exist and become the totality of TRU we need to have INIA and DEC, develop COAT, SPO and SPOT, understand how PIP works, and be able to represent and prove it. **INIA** = Intensity of IA, **DEC** = Depth Conception, **COAT** = Comprehension of All Things, **SPO** = Standardized Perspective of All, **SPOT** = Standardized Perspective of TRU, **PIP** = Proof of Integrative Processes, **RER** = Representation of Representation, and **LAMUP** = the Law of Mutual Proof.

UNITY OF OPPOSITES and TOTALITY

One approach to determining totality is through the unity of opposites. We know that entities, RUS, and TRUS consist of what is the same and what is different. What is different exists in more or less in opposition. Yet both by form and by function this opposition is in equilibrium and united in the whole. It is well known that dialectical principles alone don't provide any solution to concrete problems and afford no predictions. They only describe the most general determinations, systems and processes that give us a guide and positive framework by which interpretations, decisions, choices and judgements can be made. Totality to be actual and proportional involves relating these opposites or dialectical principles such as in the transition from quality to quantity and the negation of negation. Certain hypothetical end-terms of a continuum of intensities - as in black to white - can be understood only in terms of the unity of opposites. The theory of axiom sets becomes the problem of interpretation. What gives accuracy and validity to the interpretation of the unity of opposites is the whole of which they are a part.

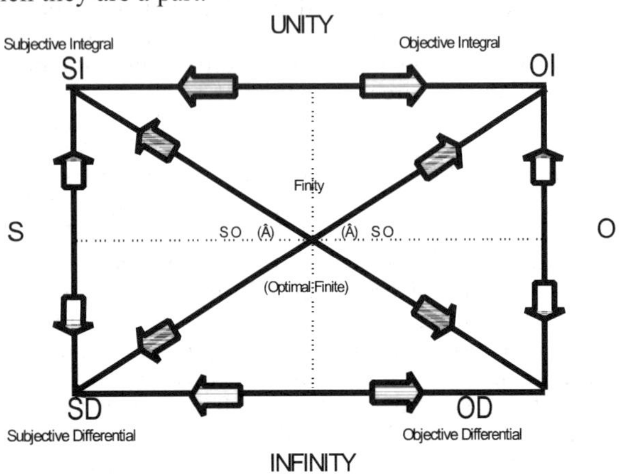

Fig 1-06 **KNOWLEDGE of ALL NECESSITATES TOTAL IN and OUT**

To Understand the world about us necessitates:

1. All its infinite detail,
2. All that shows us how it is a unit, its unity,
3. All the infinite detail of our understanding,
4. All that shows us how our understanding is a unit, its unity.

These may be equated with a unitary and infinite object and a unitary and infinite subject. It is this great relationship between subject and object, unity and infinity that is fundamental to our being, all that we think and do, and all that is at least in actuality, TRU.(SPOS, SOID) This is, next to the modulation of TRU, in ALT and SPALT, one of the best ways of showing, demonstrating and understanding TRU. Herein we can posit all entities as its senses and things are integrated and differentiated to unity and infinity; as they are ENIT, OBIT, RIT, TIT, RIT, or modeled. As they are TRU they form all actualities. As they are not TRU - although valid in and of themselves - are more figments and illusions and not to be accepted as final as we do for SOID and SPALT.

The difference between void or nullity and AU, RU and TRU is the difference between nothing, anything, all things, a thing, some thing and The Thing. This is where each is individualized. It is where each is taken as a thing-in-itself. It is obvious to us that nothing, void and nullity are non entities, except in name. Certainly they are non entities in relation to 'the' as in TRU for they stand in a negation by polarization in opposition to TRU. Since things in polarization represent two ends of a body or continuum we might assume that nothing and the thing are two poles of some kind of unity, and thereby describe this as negative and positive poles. This is, as what is to TRU, in opposition to what is from TRU or **ATRU**. What is ATRU has much in common with AU, both are antithetical to TRU. If so, then, how does AU and any or all things stand in this relationship? Obviously not within the set of TRU, but in some external way. Thereby AU does not start from nothing any more than TRU does. These depend upon whether you are looking at AU and TRU as things-in-themselves or from some external vantage. Only as EPI with standardization and convention can we form a scenario of structure, as: nothing - all things - any thing - a thing - and The T thing. Likewise AU may be contrasted with TRU, as absolute is contrasted with relative. AU and TRU together are ÅT or "all things".

Absolute Idealism is that system, championed by Hegel, which identifies reality with the absolute or the whole world. This is a coherent whole and a unity of all that is known or knowable. All things to the absolute idealists consider the ideal, spiritual, mental and objects of sense the content of this whole. For Bradley the absolute are its appearances. Yet by TRU, the only actual absolute is its own this-ness, TRIA, and the infinite perpetuation of its this-ness. It is success/failure and survival/extinction. This has been recognized since before history. The Egyptian Pharaohs were obsessed with absolute immortality by continued participation in life. This is a valid perspective for it provides much of the dynamic force and power of our actual being. However, Absolute Idealism is wrong on at least two major counts: (1). It fails to include a total, VAPO and ALT content. (2). It fails to exclude or limit all that is not TRU from it.

CONRIT and CONVET

In our approach to totality and TRU there are two general ways, paths, means and methods that can be applied. **CONRIT** or how the extreme differential can be brought to TRU and, **CONVET**, how the extreme integral can be brought to TRU. AU and totality in general are better understood, and how much exists as illusion and figment - except for what is interchangeable - as ESINS, when in the investigation of all for TRU we selectively separate what is TO of TRU from all else, AU. This we see in: **POFIS** = Principle of Field Imperfection Summation. POFIS is the error of all that does not reconcile in developing ROT, MOT, ALT, SPALT, POS and POT.

FORM of TOTALITY,AUS, UNIVERSES and WORLDS

The differences, and where concepts between AUS, universes, worlds, TRUS and TRU disagree, helps us to understand totality in general. AU is both illusory and a fallacy, these two being quite different. **AUI** = Absolute is Illusory, **FAU** = Fallacy the Absolute Universe. This is partly because absolute and absolute universe are different. Partly because absolute by definition, like zero is non existent. Whereas FAU is, as fallacy, not because it is largely figment by TRIA, but is widely in error in assuming to be TRU, when it includes all that is not TRU. It consists of entities, **ENS**, which as ENIT, exist only from their own relativity, not from TRU, nor even from their ESINS when also from TRU.

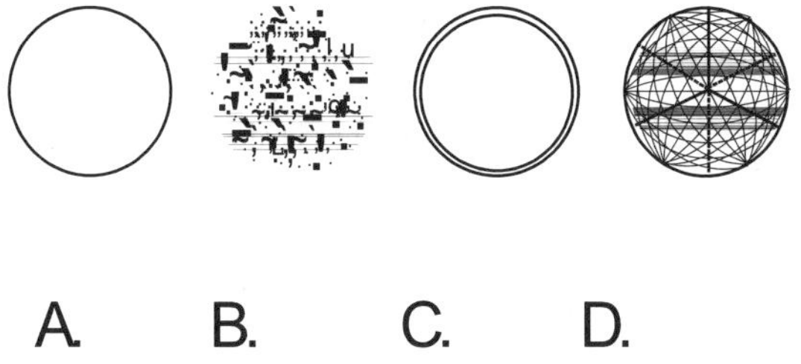

Fig 2-06 **LIMITS of DIFFERENTIATION and INTEGRATION of AU and TRU, DINLI**

A. Absolute Universe spoken of generally, singularly and integratively, B. Absolute Universe spoken of as random particulates differentially. Chaotic, Random and Disordered. C. The Relative Universe spoken of integratively as a single integrated unit. D. The Relative Universe spoken of differentially with a finite to infinite number of ordered parts. These compare and show four of two combinations of dipolar AU/TRU and most simple to most complex. The limits of differentiation and integration of an entity, thing, AU, RU, or TRU must appropriately proceed in all ways and utilize all means convergent by SU that replicates the whole. The Limits of Differentiation are, 1. All are infinite, infinitely possible, therefore impossible to comprehend absolutely. 2. Unity has little meaning substantively. 3. Not reducible by binary fission. The Limits of Integration, are: 1. All are infinite, infinitely possible therefor impossible to entirely or perfectly comprehend or present, 2. Entities, senses, and things are not all the same, 3. Entities, senses, and things are not all different, depends upon the relation, 4. There exists much overlapping both in the entity and even much more in the object & represented, 5. There are different degrees of totality from a singular unity to a totally compound unity depending upon the power of the perspective or the variety of the given, 6. Where there is a fixed relation, relatively and by TIT, or **LAFIR** = Law of Fixed Relation, a TRU Set.

Fig 3-06 **WORLDS; VARIETIES of** (WORV)

The idea of more than one world has been held at least since the time of Plato. He distinguished between the material and the spiritual worlds. His emphasis upon ideas, and how they make up the spiritual world, has been the basis for much of the world's philosophy and religion. Because it is so tied into ideas like justice, the good, and the soul, it has been given added strength. It is a forerunner of and helps to confirm TRU as does all that is TREQ. How much a or all of these worlds may be coterminal with TRU and how much they are not becomes more evident with investigation and comparison. It is of fundamental importance to know, know their limits, and be able to formulate and superimpose them.

The problem of the distinction of entities, as AU and the material and physical world, and their form and characteristics, including the problem of reals, is to be resolved by the methods of analytics, especially comparative analytics. This is by showing how AU is different from an object and how both are different from TRU. If we can show clearly the difference between an object as in the subject object relationship and if we can clearly show the difference between AU and The Object TRU, then AU can be effectively determined. So has there been for man, philosophy and knowledge, the need to establish what is formally distinct, a kind of pure reality. It is what is actually real, the conjunctive pure TRU S-O identity in contrast to what is only in our minds, or is only in and of itself. Yet we think of real as that which has existence in and of itself, ENIT. This is not the actual and pure real of the predominant mass energy FAF of TRU. As ENIT it has individuality and thereby is a separate entity as anything, like AU. Thus there are three or more meanings of real. This sequence of reals is how Aristotle used prior and posterior analytics as methods for the discovery of reals. **POTIN** = Potential In and Out of TRU, reals become pure only as their Potentiality becomes actuality IN TRU.

What systems exist in the world whose form may help us to provide a universal system or examples of the systematic unity of TRU? How can they be correlated, superimposed and standardized to support a better showing and model of TRU? There are many forms, systems and models in life. They arise from the divisions, like a morula, and can be readily adapted to the form they have in TRU. Many of these are in common use. Many are special to the highly technical fields they are known of and used by. All can be adapted, more or less, to TRU, ROT and MOT. In so adapting each, like all partials, they need alteration for identity with TRU. This is nowhere better demonstrated than in the biological sciences where length is shown vertically and time horizontally. In this way we use systems and their dimensions for orienting POT, TRU and ROT.

How far can we go and how much validity can we provide to show the comparative composition, character and makeup of ENS, objects, RUS and TRU? Primary, for an entity, does not imply an exclusive priority. This is called: **LEXCLUP** = Law of Non Exclusive Priority. This is a product of TRIA, as in ATO. Quiddity is what makes them what they are in and of themselves. It is this quiddity that so greatly distinguishes them as well as providing the SCARLS upon which their distinction and discretion can be made.

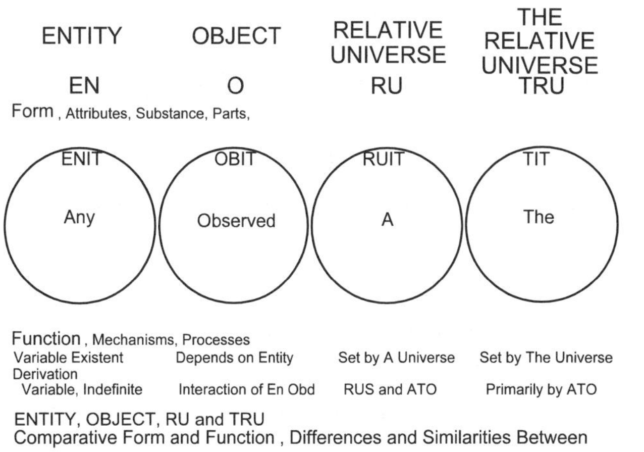

Fig 4-06 **ENTITY, OBJECT, RU and TRU COMPARED** (ENORT)

TODIM = TRU Totally, Differentially, Derivatively and Dynamically Focused in Infinite Integration. TODIM is the means and method of multiple magnification. It is how the optimal method of Homation for the processes and means of attaining certain knowledge of TRU can be obtained. This is done through focusing telescopically and with magnification from TRU to greatest and then decreasing TE to TO. It is accompanied by the differential dynamic toward TI and back again reversibly, singularly and multiply into and out of their total and ultimate relations of form and function by viewing and presenting them both as they are and as they are in TO. TODIM is how we envision the totality of TRU profoundly and ultimately. It is to be distinguished from TELOT by internal focus of the contents of the totality. Whereas **TELOT**, or Telescopic Overall TRU/non TRU variation in perspective, is of external perspectives.

What is the distinction and differential meaning of entities is provided as the partial positions of ES in TRU are shown by TODIM, PORID and SIDA. It is to constantly have in mind TU-TO and greatest TE, as if in a nutshell, by constantly adding to, subtracting from, superimposing and viewing any IA, question, problem, partial and universal. It involves laying out the posits of ES and CES, the parts and processes of TRU, into its TU and TO, in the best possible way. **PORID** = Processes of Total Rapid Relative and Reversible Differentiating and Integrating to TRU from the simplest to the most complex.

SIDA is the Simultaneous Analysis of the Static and Dynamic components of TRU. The extension of SIDA with SU and formal distinction is: **SIMUAD** = Simultaneous Integration with Systematic Unity and Differentiation with Formal Distinction. This is readily understood from **Á**, the mass energy or subject-object dimension. **OM** is the Omnificent Magnaminity that a person can inspiringly envision through the manifold apperception of TRU, as through TODIM, PORID and SIDA. OM is attained not only subjectively but objectively and phenomenally, toward SÁ.

From Figure 1-03 a combination of perspective, totality and composition are shown. Within the contents of this diagram many terms may be placed to show how all relates and can be envisioned together. Among these are: LOVKI, LOVKIRT and DOCIR. **LOVKI** = Law of the Irreversibility of Valid Knowledge. **LOVKIRT** = LOVKI of RUS and TRU. **DOCIR** = Doctrine of Internal Relations. Together the internal constituents and functions show the overall fusion in totality as we approach TRU. Much of EPI and AU have external relations in common with TRU and SPAT.

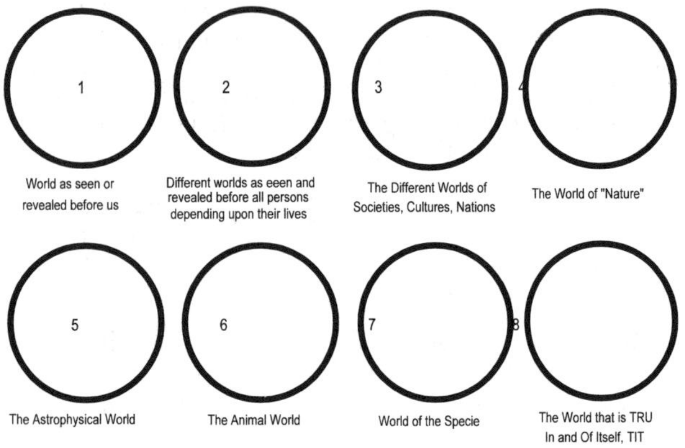

Fig 5-06 **COMPARATIVE WORLDS**

CONCRETION, CLOSURE and SOLIDITY

Who is to say this or that world is right or wrong, or more or less? Yet we know that the course of human life has been and will likely be highly unified. This is especially so by the convergence of increasingly predominant human subjective and cognitive powers. It is well to understand other worlds, not so much for greatness in themselves but their contribution through similarities and differences, universality and particularity in helping understand our own world, TRU,and its makeup. Only the last world, #8 in Figure 5-06, fufills the requirements of The World or This Special Universe of Best Human Life, TRU. This is seen when it is Brought Into View, **BIV**. It is done by **PIC**, or the Principle of Interposition Composition, or how there exists in its frame much overlapping of its aspects, constituents, or ES and CES. It is exemplified by SPOT and SPALT. It is improved and perfected by ACC or Accentuation through such means as **MARS** and **TOPA**, or Massive Accentuation, Action, Ablation Rapid Simultaneous, and the Test of Primary Application through the Acceleration of great breakthroughs.

The special form of all, and all as it is TRU is often envisioned spherically. Although such a representation denies the form of both AU in itself, and aspects of TRU in itself, **AUIT** and **TIT**, it is valid both practically by representation, and by presentation. Thereby to use sphericity to show AU and TRU we need to have certain standards and limits to our perspective. The concepts include, LASPHERT, SOPOB, LATNUP, POUS and TELOT. **LASPHERT** = The Law of Sphericity of TRU. Whereas sphericity for TIT carries certain errors, being of TRU rather than in TRU, many of these errors can be corrected. **LATNUP** = Law of the Absolute Necessity of a Singular or Unified Perspective. LATNUP is a product of both TIT-TU-TO-TRIA and PERT-IA. It is where PERT comes into direct conflict with AU. It is verified by its corollaries and extensions. This is shown by TELOT the telescopic overall TRU and not TRU, The variation of any standardized representation. TELOT reveals the exceptions to LATNUP by the changes and variations provided by dynamically controlled perspectives.

That on account of which it remains the same after - as it was before - undenying its alteration in time or alteration with subjectification is consistency. If the image or replication of a thing alters it, so that when there is a different thing it is not consistent, it does not exist in the same. The quality or quantity of departure from the thing-in-itself can be measured from the type and degree to which it is different or changed. **LACON** = Law of Consistency. This will differ for various sets and classes of entities. So that it will differ as to whether it is no, any, every, a, each, all, some, or the entity, as well as whether it is of senses or things. It will also differ by being an object or subject, AU, RU, TRU or parts thereof. This is the established method of trial and proof by generalization and specification. It is called **LAGASP** = Law of Generalization and Specification. It is related to, but differs fom the methods of induction and deduction. LAGASP permeates our entire text, as it does so much of life, knowledge and thought. For it is a basic relation in and between entities, presentation, first persons and TRU.

You can have big and large departures from a sphere or solid if your gravity is low, as with the mass for physical bodies and gravity. The measurement of gravity gives one the mass of a physical entity. Mass for objects other than physical bodies, consists of that essence or haecceity which makes it actual. Just as the greater the mass and/or gravity the greater the density, the more spherical it becomes. This is called **LASPHER.** So it is of all that is formed by a mass which is other than gravity, whose nature is more and other than the common physical objects we recognize. The shape of a physical object is most efficient when spherical by the law of parsimony. The solidity and sphericity as well as hardness and softness of entities and RUS as they are compared and used in our investigation of TRU are of great importance, not only in representing and modeling TRU, but in showing its processes, mechanisms and overall dynamics. It is what enables much greater: **ORK** = Organization of Knowledge through MOT, ALT, SPALT and POS.

The great mixture that occurs in the world about us becomes clearer when it is separated into its different classes: those which are AU, those TRU, or a those with a mixture of the two. Only when reduced to a single entity, TIT can they have separate existence and distinct meaning. In TRU they will have increasing equilibrium, sphericity, and reversibility externally. **LETID** or the Law of Equilibrium of Integration and Differentiation of TRU.

The great cyclonic state of physical and mental forces working in and out of TRU is exemplified by how: **MIUC** = Mind keeps turning in the universe, SPOT and TRU until satisfied to completion. There are built in states and mechanism, Á and SÁ, to keep turning to both meet the particular problems and to overcome the universes or adapt to the universe it is in, TRU.

	A U	MIXTURES	T R U
UNIVER-SALS	**AU Universals Laws etc. Not of TRU**	**Universals Common to AU and TRU TRAN Transcendent**	**TU LOT(S) Principles of TRU**
MIXTURES	**Larger and Smaller Entities Things and Senses not TRU**	**Larger and Smaller Entities both AU and TRU**	**Larger ES and CES of TRU, DYNO, S, U**
PARTIALS	**Smaller to Infinitesimal Entities not TRU**	**'Trivia' of the world Around the Individual**	**Smaller ES and CES of TRU Toward TI**

Table 1-06 **Universals, Partials, Mixtures AU and TRU**

All universals and partials, as ENIT, do not exist only in TRU, neither do they exist only in AU, nor a mixture of both. Parmenides addressed two of these problems. The problem of "to be" and "not to be" resulted from the radical KRISIS. This resulted from MONISM as in TRU. How to reconcile this with that which is "not to be" in AU. The other problem Parmenides addressed was that of the mixtures of things. This is the beginning of DUALISM, his Doctrine of KRASIS or Mixture. The beauty and greatness of Krisis and Krasis is that they signify the two great states and operating systems. That of all proceeding from one to many, and at the same time dynamically proceeding from many to one. The concepts, principles, laws and mechanisms involved are *critical* to TRU. This is appropriate as the term critical derives from this KRISIS of Parmenides. It is further sufficient reason for, mathematical basis and proof of TRU. **LONBIT** = Law of the Numerical Basis of TRU, and **TROS** = TRU States or Mathematical kinds of TRU.

CHANGE

That change is ubiquitous becomes ever more evident in life's experience. It may be reversible or irreversible, direct or indirect, external or internal to an entity, RU or TRU. It necessitates some relation for there has to be two or more entities or parts interacting. It involves alteration. Contrasts of this are exemplified by; **LAC** = Law of Alternating Correspondence. When we apply relativity to APT, POT, TÄP, and RÖT we are looking at a double sided aspect of relativity, opposite to what it was as a theory. General Relativity as a single focus becomes multiple in resultant and Special Relativity as a multiple and dynamic in focus becomes a single or unified composite of observer and observed in resultant. The direction and resultant of relativity are the effects it produces and the form and construct entities assume as they are composed of the object and the subject together. This is the alteration or reversal of focus that occurs as AU changes to RU and TRU, or AUCH becomes RUCH becomes TRUCH. The gist of the Special Theory of Relativity in the proof of TRU is by being limited to the, the observer, as this observer remains constant while all else changes. This affirms the focus of the observer, or person and their IA, in contrast to any observer, worlds, universes, and all else.

In order to adapt to and manage change, external and internal, several concepts are beneficial. A point may be reached for any set object that is optimal for the image and the circumstances of presentation and representation. The ratio of return of this image to its object may be presented as a concept: **LOPOIR** = the Law Of Point Of Imagery Return Ratio. LOPOIR may also be the variation of this point with circumstances at hand and may be expressed as: **LOPVOIR** = Law Of Point Of Imagery Return Ratio Variation. LOPVOIR is where this point varies with many factors, voluntary, involuntary and imposed. Likewise the tendency of ones mind to lability and the tendency of world view to variation is often unlimited. It is the rapid loss of focus by IA and failure to uphold TIT. This is shown by: **POSCÏ** = the Principle of Scopic Inertia. Similarly unlimited multiplicity of views is proportionally coordinated by: **PLOMCOV** = the Proportional Limits Of Multiple Coordination Of Views. It follows that with coordination of views, orientation, standardization, correlated knowledge and application to solutions of problems, we have:

POMS = Principle of Multiple Solutions.

From various perspectives and views of TRU and all Homation one can with complete understanding of the chapters and concepts reach, the basic coming together at a point, the answers and solutions to all.

VARIATION

Variation is, like change, a product of dynamic differences, both qualitatively and quantitatively. It occurs externally, internally and in and out of entities and TRU. In mixtures it is known as flux. It is all the more reason we should be able to learn - and separate - AU and TRU. This is called: **LORAD** = Law of the Distinction and separation of AU, RU and TRU. Another example of variation is: **LUPAR** = Law of Universal and Particular A, Subjective and Objective Return. LUPAR denotes that state in the subject object relationship of dynamic turnover and reorientation or adaptation to the universal and particular of the universe it is in. Since the subject and object relativity is most complex and noticeable in humans, in addition to the compulsion by TRIA, LUPAR is primarily descriptive of the action of IA, thought, act, all knowledge, and productive human activities.

LAMIV = Law of Mixture and Variation. LAMIV is The Absolute Universe, The Relative Universe or AU, TRU, and all in their great and variable intermixture of form and function. **LACORUP** and **LAMOL** as the Correspondence Of Universality and Proportionality and the Law of More and/Or Less applies to AUS, RUS and TRU. For TRU itself is not absolute absolute, no more than relative relative, but a mixture and variable absolute and relative, RÂ and TÂ, and thereby more and less. When expressed for TRU this becomes increasingly significant not as they express the non existence of the absolute, which is both irrelevant and illusion in the relativity of TIT. By incremental gradations it is the beginning of the relative, where the absolute and relative come together by degrees. This is called the gradations or degrees of Absolute and Relative or: **LAGDAR** = Law of Gradations or Degrees of Absolute and Relative. What is so illusive and paradoxical about AUS and variation is associated with: **LOPA** = Law of Paradoxical Absolutes. As man learns more and more as in science, of the outermost and innermost reaches of the universe, he would like to find the riddle of life and the universe. Yet as we have now proven this is not to be found here or in such a manner. Not so much is it to be found in extension or intension alone but in the relativity, VAPO, DAOST and ACEXIN of TRU.

The Dynamic Interaction of TRU and ATRU Knowledge is given as **DITAK.** This is derived partly from KÂ, Knowledge Absolute, and shown by KATRA → DITAK → LOSOC. **LOSOC** = Law of S O Completion. As we proceed to the posterior by COMT, Comparative Things, to TRU, TRIT and TA no philosophy, science or religion is adequate except to the degree it is, or applies to, an exact 1:1 replica of TRU, TRIT and TA. So that ultimately for completion of S-O, LOSOC and all, COMT must come closer and closer to being an exact representation of TRU. From various perspectives and views of TRU and Homation, with sufficient understanding of its chapters and concepts, we can reach the basic fusion at a point, the answers and the solutions to all is provided.

CHAPTER 07

DYNAMICS, UNIVERSAL POT and TRU

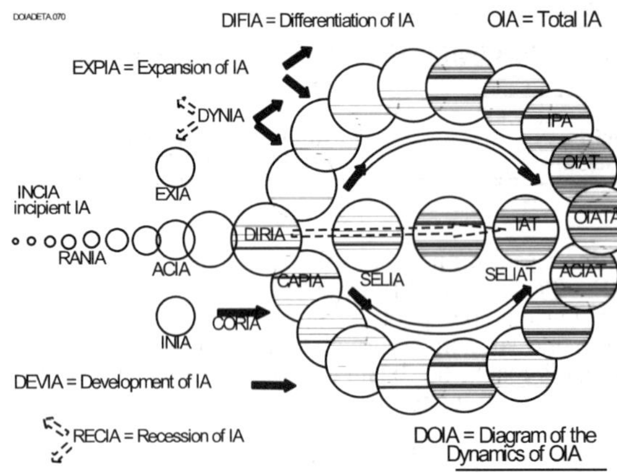

Fig 1-07 **DYNAMICS OF THE PART, CONVIA, DOIA, and DYNIA**

Throughout life a person is continuously looking, searching for and doing. One is restless both from the stimuli and problems of life as well as from ones own capacity, ability and inclination to move. This is by the persistent and rapid exercise of ones awareness and attention, IA to IAT. Actually it should be nothing but TRU or ones TES in TRU. For this is life. It should be at least that of TRU for which the individual must be a participant. Much of this is spontaneous. Yet it is our failure to discover, learn, train ourselves and serve what is best for us and the world, TRU, that our restless minds, acts and search comes to naught or harm. Herein is one of the greatest of human weaknesses and problems. Yet, herein also, is man's greatest duty, challenge and opportunity. This is to know what one is looking for, searching for and doing. This is why the work of exercising our participation in TRU and TES is our major purpose in life. To achieve this, all must converge on, as ones IA focuses on, TRU. These are called: **DOIA** = Derivation and Development of IA, **DYNIA** = Dynamics of IA, **CONVIA** = Convergent IA, **CONVIAT** = Convergence of IA on TRU.

Fig 2-07 **MECHANISMS of the MOVEMENT of ENTITIES and UNIVERSES**

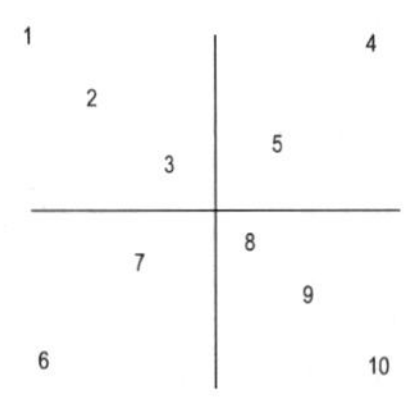

In recent studies in physics, the movement, function and operation of entities, objects, systems, and universes are classified by levels of randomness, regularity and restriction. It helps us understand **PROFS** the Problem and Solution of relevance/irrelevance in FAF. Unrestricted to restricted movement is shown on the horizontal axis and randomness to regularity on the vertical axis. Within this figure, we posit ten operative levels of these various kinds and states of movement. This is an interesting and revealing study of motion and dynamics whose types of movement are:

1 Utter Randomness, Brownian Movement, Absolute,
2; Unrestricted Randomness, Boundless, Highest Order Chaos, and Butterfly,
3 Unrestricted Randomness, Partly Bounded, Lower Order Chaos, Random Rhythm,
4 Restricted Randomness Highest Order, Bounded Butterfly,
5 Restricted Randomness Lower Orders, Bounded Butterfly Water Drip
6 Unrestricted Regularity Highest Order Boundless Regular Limitless Phases
7 Unrestricted Regularity, Lower Order, Boundless Regularity to Simplicity,
8 Restricted Regularity, Lower Orders, Bounded Regularity, Multiple Phases,
9.Restricted Regularity, Highest Order To Bounded Simplicity of Singularity,
10. Restricted Regularity, Absolute Perfect Unitary Phase, At Set Interval.

The applications of these various sets of motion include,
a. They show the various kinds of motion
b. They show the universal SCARLS of motion as they are the same or different at various levels of being, universes, from AU to TRU.
c. They help us to understand chaos and cosmos and the stages in between. How they are two extremes of the same relationship and continuum.(Absolute to Relative to Absolute)
d. They help us better understand the processes of man, Homation, as it partakes of these various levels of motion.
e. They help to prove TRU and ALT, directly and indirectly as their form corresponds to a set level and equilibrium of restricted regularity,
f. They suggest that dynamics and statics are not two completely separate and unrelated entities, but are two poles or extremes of the same continuum,.
g. They also suggest that neither statics nor dynamics are absolute in actuality, but only approach the absolute, as by infinity, as you approach either pole or extreme of their combined relationship,.
h. This also helps to prove a balance between the Absolute Absolute and the Absolute Relative and thereby a better understanding of AU and TA, as they represent the two extremes, # 1 to 10 listed above.

All that is different, dynamic, in opposition or against is not necessarily ATRU or ANTRU or RU or ANRU. It is very much apt to be TRU, TRUD, and in TIT. This is by **DAOST**, Different Aspects of the Same Thing, TRU. Its various extremes, aspects, forms, parts and processes stand as differents, in opposition and more or less dynamically against each other. This is the basis and distinguishing feature of the dynamics of an entity, thing, RU or TRU. How the parts of this thing stand in opposition and how these opposing forces in synchrony function in harmony to enable the total to be, ENT. That is what these different extremes, opposites, and polarities consist of and how they are formed and composed, qualitatively and quantitatively. Forces that act from the inside are not necessarily all good. Nor are those that act from the outside all bad.

That the infinite and infinitesimal exist in human life has become increasingly certain with advances in knowledge and science. It is shown by the proposition, principle, and proof of the infinitesimal and infinite in entities, objects, and universes. This is supported by **POCOL**, the principle of the concretion of Limits. The relationships between the infinite and the infinitesimal are the relative universe's duration equilibria. Better understanding of the extremes and limits shows the meaning, state, quality and degree of TRU, (TITY). Since we know by TU that TIT is a unity and by TRIA all, it can be logically represented as a set that can be well understood and operated. This is possible even though TRU also consists of infinite and infinitesimal, statics and dynamics. This is the Principle of Infinite Determination or infinite TIT, OBT, ALT, METAD and SPAT, **POID**. As we look from universes to universes we can have an infinite number of views. It is how we select these, especially by convergence on TRU, that we can bring about systematic unity. Like anatomical viewing, such an ordering is achieved by SUperimposed IA, views,

Charts, representations, TRIT, OBT, ROT, ALT, METAD, SPAT, or **SUIAC**.

THIRD LAW of MOTION. LERAC = Law of the Equality of Action and Reaction, Reaction and Action. Newton's third law also applies to most all dynamics, for it explains the equilibrium that exists in RUS and TRU. It is a primary function and mechanism that effects most all of TRUD and is basic to HOM.

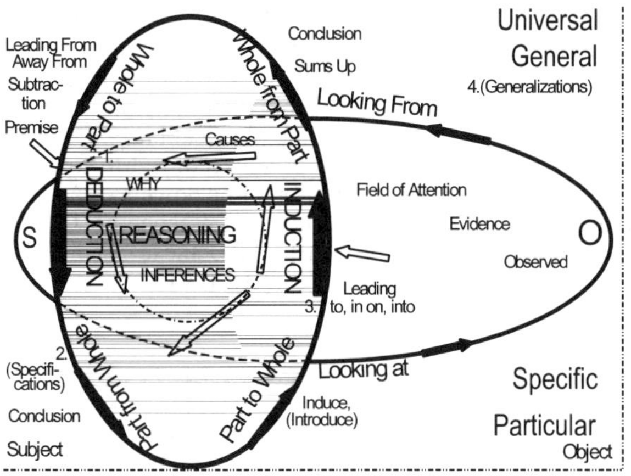

Fig 3-07 **SUBJECT OBJECT, UNIVERSAL PARTIAL, INDUCTIVE and DEDUCTIVE REASONING, LOOKING FROM/AT DYNAMIC FOCUS IN LIFE**

In this figure the theme and format of deduction and induction in a more dynamic role for reasoning is demonstrated. The length parameter from universal on top to particular on the bottom of the vertical dimension and the subject object parameter from subject on the left to object on the right of the horizontal dimension are continued as a standard. The clear basis for deduction is what leads from or away from, as in subtraction from a premise. The clear basis for induction is what leads to, in, on or into, as in summing up from the observed, evidence and field of attention. The continuous causes, questioning and drawing of inferences in both deduction and induction are shown. This is signified by **QUAP** or Questions, Answers and Paradoxes, and Concepts, Principles, Axioms, and Laws, **CALP**, or together, **QUALP**. **AK** = Our Actions can be no better than our Knowledge, and the reason for use. It should be the great opportunity, challenge and inspiration for all to be at least sufficient to the needs, demands, questions, and problems before us. This is signified by: **LOCOP** = Law of the Compulsive Convergence of the Purposive, on TRU. With LOCOP the collection, selection, ordering and natural convergence on the purposive and TRU enables lesser degrees and amounts of thought and act to become greater.

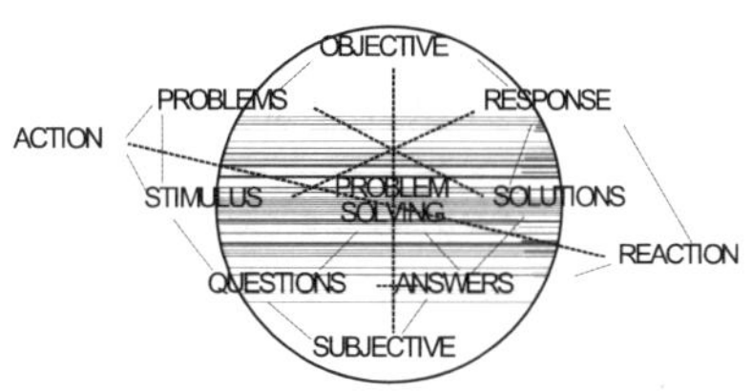

Fig 4-07 **DERIVATION of DYNAMICS, MECHANISMS, QUESTIONS and PROBLEMS**

This diagram shows how difficult it has been, and with orientation by TRU and SPALT, how much easier and better, we can integrate psychology. It is much more than simple mental activities. With other psychological and physical functions as Á they make up their total contribution to TRU and TIT-TO. They permeate, by subjectivity and relativity, all aspects of TRU. This revolves around the IA, S-O, A focus in and of TRU. The reaction and equilibrium of stimulus and response is directed toward selection, which tends to TRU. The right selection and response is equal to life, to TRU. This is why knowledge and action have to be as close as possible to 100% S and 100% R.

	LOOKING at the OUTSIDE	**LOOKING at the INSIDE**
LOOKING from the OUTSIDE	**PHILOSOPHY** **NATURALISM** **DEISM INTERAC-**	**SCIENCE** **TIONISM**
LOOKING from the INSIDE	**IDEALISM** **EXISTEN-** **HUME**	**PERSONALISM** **TIALISM INTUITIONISM** **PSYCHOLOGY** **SOLIPISM EGOISM**

Table 1-070 **LOOKING FROM and AT; PERSPECTIVE POSITIONS**

We may readily divide perspective into two divisions of two aspects each: I. **Looking FROM** II. **Looking AT**

1. the **OUTSIDE** 1. the **OUTSIDE**
2. the **INSIDE** 2. the **INSIDE**

These standardized perspectives include:

I. I-1 at II-1 or Looking From the Outside At the Outside
II. I-1 at II-2 or Looking from the Outside At the Inside
III. I-2 at II-1 or Looking from the Inside At the Outside
IV. I-2 at II-2 or Looking from the Inside At the Inside

From these we can develop Standardized Perspectives Of Reality, **SPOR**. We can also posit the many various perspectives, authorities that have been used, and the systems or philosophical concepts derived. These may be of different variations with many combinations of either/or looking from and looking at, many heavily weighted in one or the other, not balanced, and many depend on what we are looking from and what we are looking at, e.g. from a single person, a group, many or all persons, or perspective of TRU.

The dynamics of development is shown by growth and growth curves. The latter are characterized by several phases including, initial, positive growth acceleration, logarithmic growth, negative growth acceleration, maximum stationary, accelerated death, logarithmic death, and exponential death phases. High order growth curves are largely visual devices used to highlight those segments of any growth curve in which the rate of growth, or regression, is constant. These must all be conditioned by TRU and ROT, with the high order ones being constant rates for various segments of ALT and TRU.

Quantum Theory

Of the great discoveries and advances of the 20th century those of relativity and quantum theory are prime. This especially holds for their applications to and basis for our understanding and presentation of TRU. Relativity settles the problems of, and helps us to discover, learn, develop and show why TRU is actually so, what it is and is not. How mass energy is not only gravity, but partakes of any and all convergent forces that make entities separate, unique and predominant in their own environments. Quantum theory settles the problems of, and helps us to discover, learn, develop and show in what way the universe is composed, and why TRU is the way it is. It reveals why the finite is the way it is, neither impossibly infinitesimal nor infinite, neither indistinctly unified, nor imperceptibly differentiated. It reduces all to degrees of probability. These, are ever greater with increased TRU and knowledge provide the formal distinctions necessary to identify all the finite or differential makeup of TRU. Quanta represents the separate and delineated finite entities, parts of an entity, or parts of a RU or TRU that rather than be figments, illusions and hopelessly indistinct, can be differentiated to produce that beautiful intricate detail that is necessary if the mind is to recognize, represent, conceptualize, think, and reason with it.

The close relation between form and function is evident in quanta. They exist both as static elements, and quantum leap states characterized by transformations. These may be incremental or larger and more active and dramatic with changes that repeatedly occur. **LAQUEX** = Law of Quantum Existence and Expression of Finite Separation. This is similar to mutation. Because it is of one variety in one case does not mean it is always of the same variety, even in almost the same species. In fact this variance in quanta and transformations is highly characteristic of natural selection. This is called: **LATVI** = Law of Transformation Variance, in Quanta and Entities. **LODAW** = Law of Dynamic Dimensional Adaptation and Warping. Dimensional adaptation and warping is what largely creates the massive ongoing dynamic interaction and conflicts that occur between a RU, as TIT, and all that is about it. It is the marked and often vital differences between the dimensional makeup of a RU, as an animal or species and that of the forces in its environment.

MESOÁ = Mass Energy Expansion in life, or that of Á in TRU. In Figure 4-07 the gradual expansion of mass energy in life and how it has lead to Á is shown. It is the transition from the mass, length and time states in the physical world, to their state in life, and especially the higher forms of life, as in TRU. This transition was, is and will be a highly dynamic one, guided largely by its own FAF, biological principles and laws, with their underlying chemical and physical base. The highly competitive and dominant role of succeeding life forms has produce the increasing and self propelling powers of Á that are so much of the thisness, existence and essence of TRU. By developing a positive approach to TRU these become even more evident and in result enable TRU, ENT.

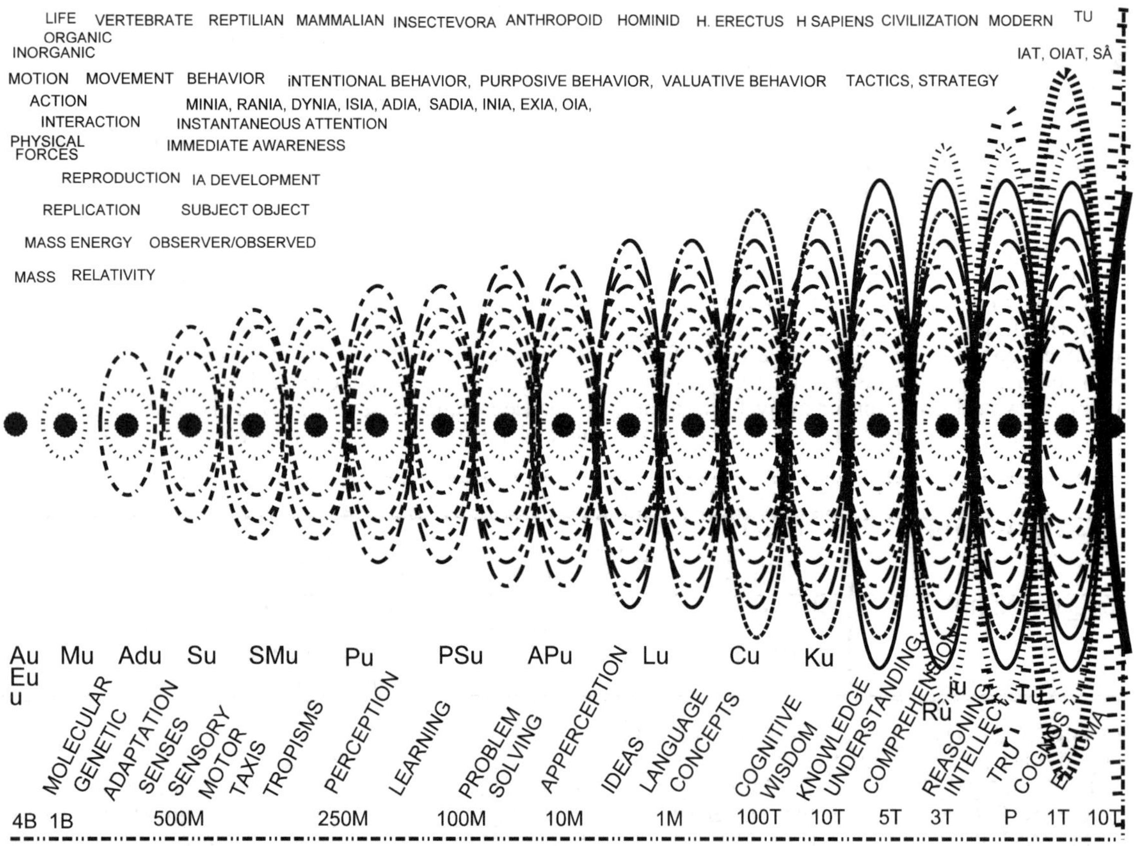

Fig 5-07 **MASS ENERGY SUBJECT OBJECT Á DYNAMICS and IA; MESOÁ**

Each organism, specie and class, become self-contained units which function as sets in a life and death competition with all about it. This is so even within its own kind. It is this great mass of capacity and ability to sustain itself, its many levels of mass energy or Á, that is both its great power as well as vulnerability. For as this becomes greater and more extended it becomes acceleratively more vulnerable. This is why it is so important to TRU, and why it is so important that we properly and fully discover, learn, understand, and apply it.

CONSTANTS and MECHANICS

Planck's Constant gives a determination of the frequency of a particle with its energy; f = E / h, or frequency equals Energy divided by Planck's Constant. The energy of a particle thereby determines its frequency by a fixed constant, h. Planck's Constant is given in units of action and has the magnitude of 6.6252 +- x 10^{-27} erg sec. Planck's Constant is equal to the velocity of light. These are of great significance as part of TRU, most especially as they enable representation, modulation and the geometric formulation and uses of ALT. Since the determination of mass by gravitation in classical or Newtonian systems and the determination of mass by Quantum Mechanics are the same for most systems in the range of human existence they can be used. This is the Correspondence Principle in physics, between what applies in both mechanics, quantum and classical systems. However it necessitates the QUOL and SCARLS of TRU and ROT in the development of the modulation of TRU, ALT and SPALT. Exceptions to this correspondence are seen when the energies of electrons in orbit around atoms do not fit a whole number of DeBroglie Wave Lengths and many other physical functions. The correction of this correspondence for Quantum Mechanics is the Complimentary Principle. It is characteristic of individuality in AU, as well as being the antithesis, the QUOL, of all universals. The use of these two principles in physics is not entirely the same as in philosophy and other sciences. Zero-point-energy illustrates an exception to the Correspondence between Quantum and Classical Mechanics. It also vividly illustrates the difference between any thing and nothing. This is the energy of motion remaining at absolute zero temperature. At this temperature a body is in its Ground State of Energy Level. According to the Uncertainty Principle this can't correspond to the complete absence of motion. Thus for classical mechanics any thing as absolute zero temperature is nothing. This is the inverse of the inverse, but for quantum mechanics it has energy. We can thereby see how the outer reaches, and limits, of dimensional formulae in TIT, its representation and modeling, SPALT, can be readily and accurately demonstrated and interpreted. Energy is accumulated mechanical work, which may be only partially available for use. This is its specific definition from physics. Commonly the term energy is used more broadly. Energy may be examined in terms of its measurement, forms, entropy, potential, kinetic, conservation, heat, electromagnetic, equivalence to mass, and quantifying energy. Energy, mass and quantifying of energy are vital to and enable a positive and precise representation and modulation of TRU through dimensional formulation. Later developments in quantum theory - Heisenberg, Schrodinger, and Dirac have revealed differing energy states and systems as well as a continuum of unstable states. It is the existence of these differing energy states and systems as well as the continuum of unstable states that is key to TRU and TRUD. It is of a high level of importance for TRU. It is from TRU that we have the focus, or special thisness necessary to treat it as a set.

Fig 6-07 **MOLDING, STEREO-TYPING, HABITS, ROUTINES in RUS**

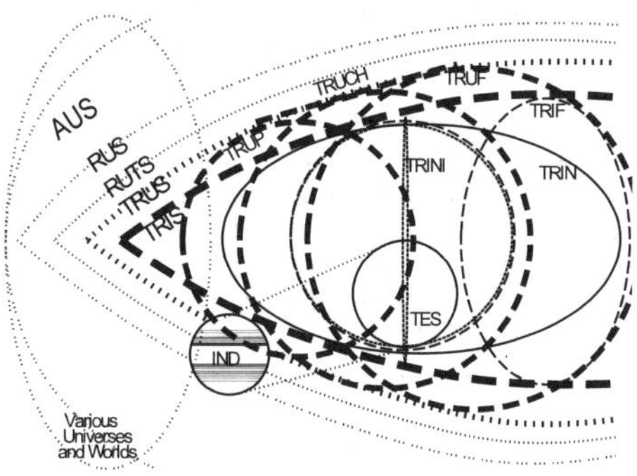

It is stated that 'man is a creature of habit'. This is and is not, as so much in life, true. Yet its truth when qualified is such that it is highly indicative of much of the FAF of TRU. For much of TRUD should not to be considered to be only functions of human reason, thought, subjectivity or even simple response to stimuli. They are often functions of natural animalian chemical and physical levels of Á mass energy. habits, routines, imitation, stereotyping, models, example, copy and automatic functions built into general human makeup and form. They usually indicate lower levels of energy manifested by momentum, inertia, equilibrium, reaction and compensation. They are to a large extent a result of limitations of knowledge, time and money. They are often lacking in higher levels of energy such as introspection, intensity, reason and revelation. Yet they proceed - often haphazardly with little direction - as ways to make life easier or to prevent problems, trouble and conflicts. Rules, regulations and customs as well as routines and habits are continuously and massively interacting with us. There is a reaction to the immensity of these worlds around us which demands of us such conformity, custom, regulation, and routine. This is called: **MORU** = Mold or stereotyping of All about us and of TRU. In the Figure 6-07 we seek to show how these have not only been the origin and development of TRUS, TRU and TIT, they have molded it. As in natural selection RUS and TRUS have retained what is essential for their more basic forms and functions and what works best for them. As function follows form, so does form follow function. They have developed and work together in each RU, RUT, TRU and TIT. The form and function of each partakes of much of its preceding kind of universe, altering in those ways that best suits survival, haecceity or thisness. Adaptation and alteration make up, and increase from AMTRU much of TRU's special FAF. It is only in TRU that we can properly and actually interpret all forms, states, dimensions, and geometries. This is in stark contrast to those of the physical sciences, which are without both focus, fixed thisness, or relevance to us. There we are apt to blindly accept with conviction, ENIT or entities in themselves.

Fig 7-07 **SELIA**

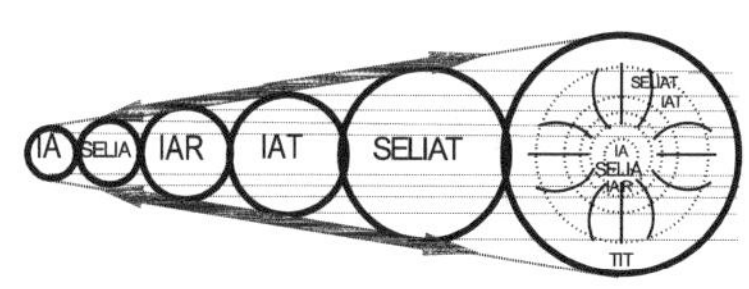

= Selective IA. Another great value of IA and what is so much at the core of all is selection, straining, choosing, moving, and excluding. These are performed in a determined way on all that can, does, or will act upon us. This may be either simple or complex, direct or indirect, significant, most necessary, important, or most TRU. It is the main determinant of priorities and values. SELIA is at the root of not only the being of various RUS, RUTS, TRUS and TIT, it is that through which IA can be formulated, modeled and presented as IAR, and what directs it from IAR to IAT. At any time, a person may analyze their activities and attention to see and do what is of concern, necessary and important.Those choices and moves which it makes will be influenced by many factors. Yet they will tend to be either the most needed, desirable, or what through conditioning, habit, experience or reason is more or less toward set ends of RUS or TRU. What causes the inclination to be more or less TRU is the health or sickness, alertness and mobility or unresponsiveness and immobility of the individual. This may be immediate, in time, singular, or in sum total. Thereby SELIA becomes SELIAT for TRU, by showing how an entity, RU or TRU realizes their limited object. Also it shows how they in combination are actualized to fulfill this to become, through CES, TRU. For TRU it shows how it can be formulated, modeled, and produced as IA through SELIA focusing on those vehicles that make a new and better, TRU.

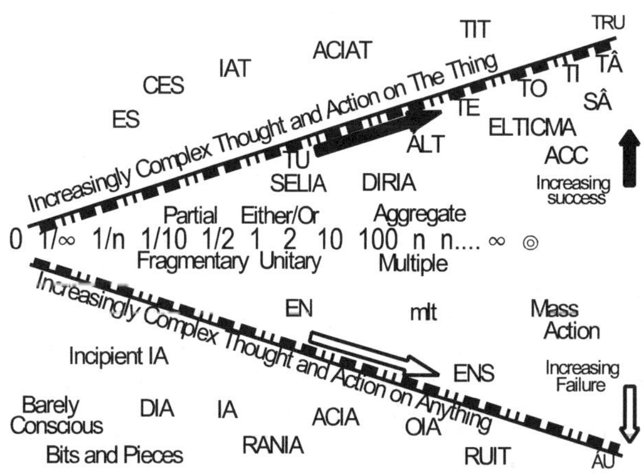

Fig 8-07 **GRADATIONS to ACTION (GACT)**.

GACT = Gradations to Action. GACT signifies choices,

decisions, judgements and conditions of thought to act. It is how the individual must function from a partial singular through multiples to total thought and act. This ends in SA and TA. It is the great tendency to judge, choose, and select from one or the other, good or bad, right or wrong.

In Figure 8-07 two courses of gradations of thought and action are shown. The first or lower course is GACT in general without regard for TRU and the upper is GACT specially limited to TRU. Our actions, as the actions of any RU, animal, or person, can be of many kinds, from the most infinite part to the most complete total. It is especially in the upper course where we can see the great separation between good and bad, right and wrong. For, in general, animals reflexively and instinctively tend to make the right choices. In human life the freedom and complexity of thoughts, desires, interests and their distance from actions, renders an exponentially greater likelihood for bad and wrong actions.

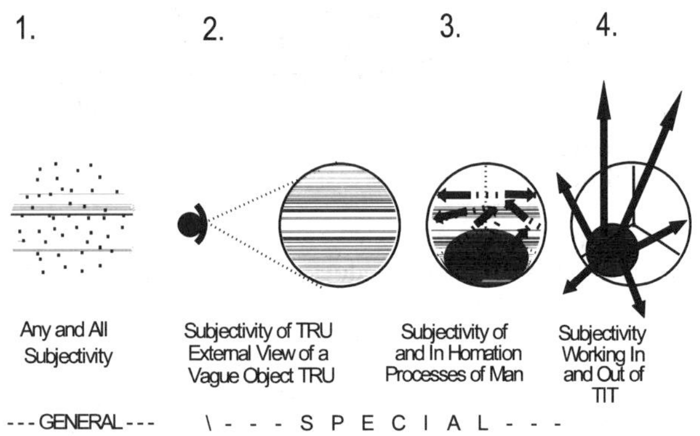

Fig 9-07 **SUBJECTIVITY LEVELS and FUNCTIONS OF TRU**

This diagram shows four kinds of subjectivity progressing from: 1. Any and all subjectivity. 2. External subjectivity on TRU. 3. Internal subjectivity in TIT. 4. Internal subjectivity of TIT directed outwardly or externally. Figures 1 and 2 are becoming TRU. Figures 3 and 4 are in TIT and finally directed outwardly. How much they coordinate with the systematic unity of TRU depends upon the frame given to a construct of TRU by its dimensions. They can be represented either way. The separation and union of thought and act are continuous, focal, and key to Á FAF and TRU.

The major way thought can have an effect, except within ones own brain in affecting another thought, is through act. Thought is closely associated with the IA as they together tend to parallel our interests, problems, needs, and desires. Thought and act function proportionately, relatively and in process by REM in TRU. Concepts and laws of dynamics and change of the different, from, against and in opposition are not necessarily only ATRU. **LADRAN,** and Law of Change in quantitative content of a RU is not necessarily detrimental, **LACHRAN**, and Change in qualitative content, especially basic dimensions of TRU, even though the product in and of itself is, by whatever standard thought to be better, detrimental or negative to TIT. **LACHRU.** This law has corollaries, exceptions and varieties. **LADEP** = Law of Determination of Preferential. LADEP denotes what amount and type of change is preferred and what is not. To determine what is and what is not to be preferred is, as standardized and LATANOS, a highly selective and fundamentally vital determination. LADEP can and must be done within certain set limits and rules. Likewise what amount of change is inevitable and what is not. This is called: **LACHIN** = Law of Inevitable Change. It is the determination of LACHIN and LADEP that will or will not help to preserve and continue TRU. **LACHRIN** = Law of Restriction, Confinement, Control, and Delimitation of Change. **LARIA** = Law of Relative is All. RIA or TRIA imposes strict limits on acceptability and preferentially of change. **LARIAT**(OL) = Law of Relative is All, Tolerance. Since a RU and TRU are in part infinitesimal, the inevitability, acceptability, and preference of change can exist if and when set limits and controls are maintained. This is as qualified, QUOL, SCARL, and VAPO. All of these apply to TIT, TRU and TRUS in somewhat modified and increasingly different ways. Preeminent among these ways are the differential limits of the scores of steps from TIT through TRIT, OBT, SUT, ROT, POT, PERT reversibly. These we call **LOLS** or Laws of Limits. Without knowing and applying these limits to all appropriately in OITIO is the prime cause of failure to be TES and TRU. **POCOL** is the Principle of Concretion of Limits. It is produced by range and distribution, from a mean of unity to extremes of zero and absolute, when the infinitesimal approaches zero and the infinite approaches the absolute.

	What is NOT THERE	What is THERE
Observes, Sees or Perceives	1.	2.
Wants to Observe See or Perceive,	3.	4.
Wants to But Does Not Observe See or Perceive	5.	6.
Does Not nor Doesn't Want to Observe, See or Perceive,	7.	8.

Table 2-07 **PERCEPTION PRESENCE DIAGRAM**

Each of these squares or cells can be characterized by a significant identity or non identity, including: 1 Illusion, 2 Validity, 3. Stubborn Illusion, 4 Potential Validity, 5. Lucky Selection, 6 Failed opportunity, 7 Void, and 8 Loss. These help us to tell where the detail is that exists within us and where we so often go wrong. It shows what hurdles we must overcome if we are to ever find our path to IAT, TES, and TRU. The subjective and objective are on different levels depending upon the qualitative and quantitative state of perception and presence of an object or entity. It is like out-folding from the center of TRU back toward the outside, S/O or O/S and expanding and contracting circles of knowledge, vision or observation. The infinite geometric patterns are not only of the object, RU or TRU but of the subjective and objective and their relationships. This can be seen in why and how much we believe what we want to believe. Much depends upon on needs, desires, wants, and interests. These depend upon the factors in the table. The degree, when carried to its optimum, is the ultimate need, desire, want and interest in and to life, TRU. This is signified by: **WUT** = Want Ultimate There. The ultimate in wanting to, seeing, perceiving and apperceiving. This plus the ultimate in being able to select what is there from what is not, is the ultimate in knowing, understanding, being able to reason out and use all optimally, as they are TRU. This is the optimal validity of perceiving and presence as they are TRU or what is actually there and having OM and power over all.

Presence, like being present on roll call, is key and proof of much of TRU and its ROT. For it is the conscious, existence, being here, the focus of our life. It is Christ's, as TES, before, after, and in life; an individual's acute feeling of the now. We have presence to be. It proves the focus of TIT, continuity of TRU, their FAF by Á, and ATO. This is not any here, but a here and now concentration of S-O as mass energy becomes special in the relativity of TRU.

The simple unitary nature and form of stimulus response can be seen in the most primitive animals. Stimulus response provides the basis for much of the dynamics that enables animals to survive. In its most original and primitive form it arises from action reaction is primarily one of adaptation, **ADu** and major levels of **SÁ**. **ADu** is a basic unit of life and a step that helps us differentiate between the A or predominant mass energy state of TRU from physical mass, i.e. gravity.

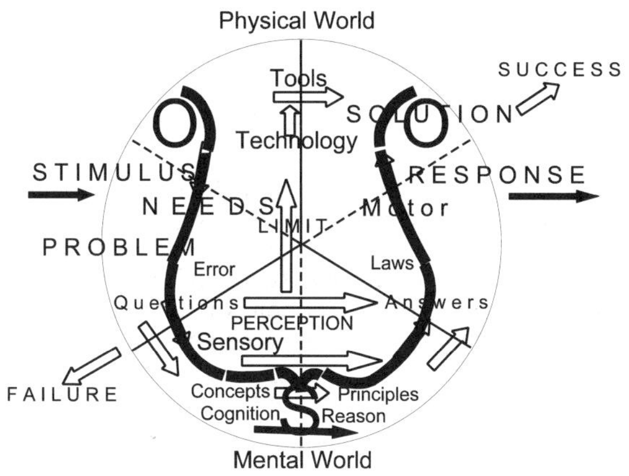

Fig 10-07 **OSOD** = Object Subject Object Diagram, is drawn to emphasize **PSR**, or the Problem Solution Relationships in the O-S-O field. It is called: **PROSOD** = Problem Solution O-S-O Relationship Dynamics.

This figure seeks to show the basic relationships in our world, especially how we function integratively and on a subject object basis. We can't look at only one aspect of our world and expect to understand it. Nor can we understand it adequately without providing some order and frame for these relationships. These are superimposed in TRU and show how the S-O dimension becomes the predominant mass-energy state of TRU, TRU's mass dimension A. It also shows the methods of knowing and acting. **EQUT** = Equilibrium of/in TRU. Any mass that works dynamically in space time must maintain equilibrium. How the relationship and operation of dimensions as coordinates become altered externally is the Principle of Coordinate Transposition, **POCT**.

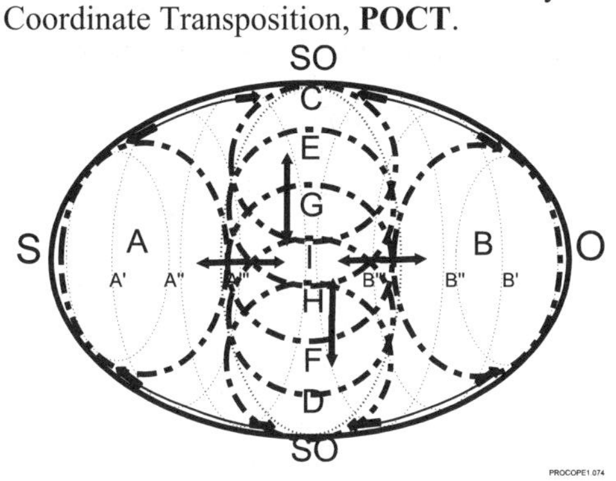

Fig 11-07 **PROCESSES and OPERATIONS** This diagram is devised to show relations between and operations of processes of entities. In this case it also shows how these apply to the relations and operations between subjectivity and objectivity, the observer observed, or relativity. The elements C, D, E, F, G, H, and I, show how SO is composed of many varieties. They include anything that acts as the means or vehicle of relativity, the interaction of subject and object. The words we use in language are excellent examples of these elements. Identification of other elements are,

ABSOLUTES		RELATIVES	
S	Subject	A	Subjective Sphere
0	Object	B	Objective Sphere
SO	Subject Object	A'	S to O
SO	Subject Object	B'	O to S

The arrows represent vectors of the processes and show the vast, important, often rapid and elusive, interchange of processes and the relativity that makes up much of the mass energy, or Á continuum FAF of human existence. When an extra element is added, as D the alteration in A or S - and TRU - and all, is the evidence and proof necessary to show how Homation and PIP proceed totally. This is a key proof of TRU, and Á, PRUT and PRÁ.

There is a natural tendency to decay. This is seen in all aspects of life and TRU. There are also many ways to prevent, protect, maintain, or advance life and TRU, **PPMA**. Much of the FAF, ES, and CES of a RU or TRU provide powers to combat decay, by Powers of Reinvigoration and Reproduction, **PORIV**. Reaction, negation and disassociation are insidious and much more common than we recognize. This is compounded by demands on the higher energy levels of TIT. Good examples of these include: **POCON and POCONF**, the Principle of Confusion, and the Principle of Confusion Fixation. Confusion follows being lost and inadequate knowledge to supply the need, and confusion fixation is the accelerative confusion and defeat that sets in when there is insufficient reaction, compensation, coordination, and power for correction. Such confusion is cured by human powers of coordination based upon one's own point of reference. **ASOPOR**, or Absolute Supremacy through Point of Reference. Likewise by Proper and Perfected Emphasis and Perspective Producing Massive Automatic Validity, or **PEMAV**. The convergent forces enabling TRU are evident in the Law of the Proximity of Being, **LOPOB**.

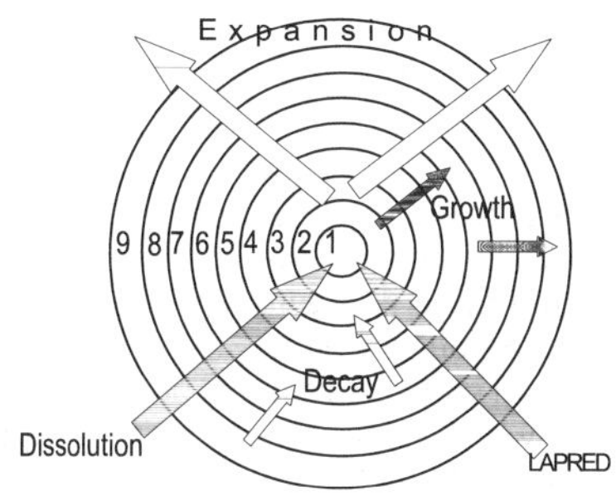

Fig 12-07 **RUS and TRUS EXPANSION, DISSOLUTION, GROWTH and DECAY** This diagram shows the sufficiency and maintenance levels of RUS and TRU. It also illustrates forces, processes and factors that increase, decrease, alter or affect this changing mass of the entity, thing, RU or TRU.

0. The non existence of a thing or its explosive destruction,
1. The bare existing presence of a thing, its incipiency, rapidly, progressive failure and ending,
2. The small, and relatively quantitative, existence of a thing, and its early growth or late existence and progressive failure,
3. The diminished thing, RU or TRU or its early increasing growth or late decreasing existence, increasing failure,
4. Temporary existence more decay than growth, or less than sufficient for maintenance,
5. Existence of a thing, RU, TRU at a maintenance (par) level, balance of growth/decay,
6. Growth and existence at a more assured level, as above par,
7. Progressive expansion of a thing, RU or TRU,
8. The highly successful thing, RU or TRU,
9. Exponential expansion toward RÂ or TÂ, optimal.

In item # 1 above, the rapidly progressive failure or ending of an entity is characterized by: **LAPRED** = Law of Precipitous Dissolution. LAPRED is the rapid dissolution that usually sets in whenever anything begins to deteriorate and loses power. It loses energy, impetus and its compensatory mechanisms exponentially overload. LAPRED is largely a product of an organisms inability to fully and adequately identify and use whatever is at hand in its own defense and support. LAPRED is a special case or corollary of: **INEG** (or LINEG) = Integrative Negation (the law of). INEG is the base in # 2 above from which LAPED # 1 is produced. INEG is the simple compounding of failure as seen in the complications of diseases in medicine. LAPRED is more the dynamic result with much less chance of recovery, and thereby for TRU, more ATRU and ANTRU. LAPRED is seen in history in the decline of nations and cultures; how slow they are formed compared to how rapid they often dissolve. This accelerative dissolution is what must be guarded against, protected from and prevented if an entity, RU or TRU is to survive.

CHAPTER 08

ENCHANCEMENT of APPROACH, PERSPECTIVE and PRESENTATION of TRU

REALIZATION and ACTUALIZATION

Much of what enables TRU is what enables EPI, PERT and POT as they approach TRU. For TRU realization and actualization, like reals and actuals, reality and actuality, take on new and even more powerful meanings. To acquire the jist of these is to understand much of human life, as it is and must be by TRU.

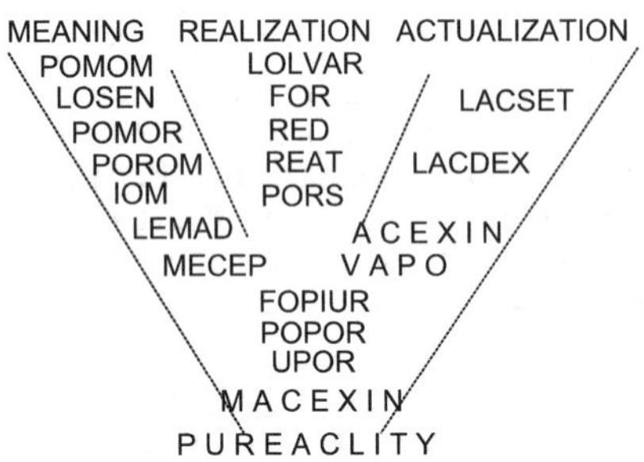

Fig 1-08 **Meaning, Realization, and Actualization**

We tend to think of meaning, realization and actualization as separate ideas denoting things in themselves. However this is our error, for among all that is close to and great of TRU these are it. Meaning, like is, is how well there is an identity between subject and object. In addition there is great intention in meaning, as in IA, and this is convergent on being, our being, TRU. Realization is how profound and accurate this bond between subject and object is, how well it overcomes language insufficiency, **LIS**. Actualization is how well or lifelike all is, whether it is TRU. Meaning, realization, and actualization come together in the mass energy FAF of TRU to give it the predominant power it needs to compete and survive. There are many steps between how they exist separately as ENIT, what their ESINS are, what they are in TRU as states and mechanisms of TRU, and how they fuse in the continuum of Á and total dynamics of TRU. Among other concepts the following signify steps in this progression: **POMOM** = Principle of Multiplication of Meaning, **LOLVAR** = Law of Limited View and Reality, **LOSEN** = Law of Experience and Meaning reinforced, **FOR** = Fallacy of Reality, **LACSET** = Law of Critical Separation of Non Actuality from Actuality or illusion, figment and non TRU from TRU, **POMOR** = Principle of Meaning Reduction, **RED** = Reality Deficit, **POROM** = Principle of Reinforcement of Meaning, Maintain, support, OITIO. **REAT** = Reality different than TRU, only the same by its ESIN in TRU, **LACDEX** = Law of Accumulative and Accentuated Actuation of Dimensions as they work together, in TRU. **IOM** = Intensity of Meaning, **PORS** = Principle of Reality Substantiated, **LEMAD** = Law of Enumerations, Meanings, Definitions and Formal Distinctions, well Developed and ordered to show the concepts, principles, laws, and mechanisms of TRU, **ACEXIN** = Actuation, Extension and Intension or ALT FAF in TRU, **MECEP** = Meaning is Corrected, Proper and Perfected in Completion as it best Enables TRU, **VAPO** = Validity, Actuality, Proportionality and Totality, the overall reduction of that experienced to TRU. **FOPIUR** = Fusion of Perfect Image and Ultimate Reality. **POPOR** = Perfected Perspective of Reality, **UPOR** = Ultimate Perspective of Reality, **MACEXIN** = Maximal Meaning ACEXIN, as of OITIO, **PUREACLITY** = Purity, Reality and Actuality or optimal state of Á in TES and TRU.

Meaning and realization may at first glance seem coterminal, yet they have distinct significance, especially in TRU. Meaning is the overall bond between subject and object and how it is of practical use in life, actuality or TRU. Realization is an intensification and particularization of the S - O relation and application to life, TRU. Both are of the Á dimension and major mechanisms that give it power. Reality also takes on different and important meaning as it progresses from OC, EC, to ECT. For whereas reality commonly used denotes all that is known, identifiable, and existing, Reality of TRU signifies the whole fused region of middle portions of Á, L, and T, of TIT that give it such identity and inner strength.

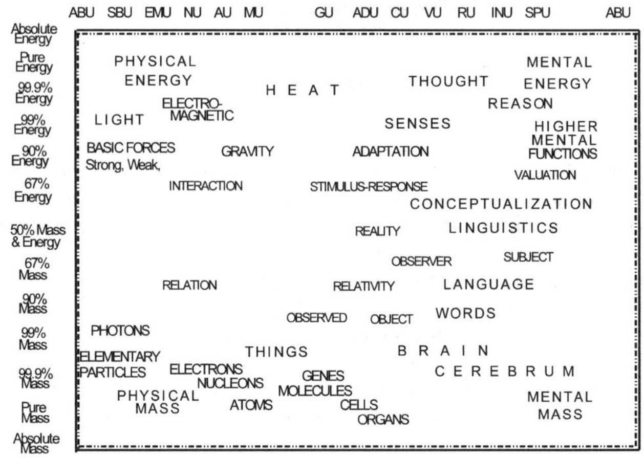

Fig 2-08 **MASS-ENERGY/SO Á (MASENÁ)**

By LANERA and LAMOL absolute mass, absolute energy, absolute objective, and absolute subjective do not exist. At least not in TRU and TIT, and since by TRIA TRU is all, they do not exist in actuality. Only in meta or other than TRU observer or observed states do they become apparently real. Yet to understand relativity as it applies to all entities, as well as the development of its FAF in TRU, or its Á state, we need to visualize its state in other universes and entities, as well as what this state consists of. This is called: **ROMENS** = Relativity or the Mass Energy of Entities, and includes all that makes up, bonds, helps, enables, enhances the entity or universe to be, have thisness, haecceity, sustain itself and survive. **FER** = Form is Equivalent to, and Established by, Relativity. This is shown by how not only the Á dimension is the identity of form and relativity, but relativity also sets the length and time dimension. The constituents of FAF are seen in **CARSC** or the Capacity, Ability, Reaction, Settings and Conditions - of TRU. How they are set for a RU or TRU is signified by; **LENA** = Law of Energy Levels Specific to a RU. ROMENS is all that enables and enhances an entity's momentum, inertia, energy, entropy and other forces that are special for it in reaction to all about it. For TRU mass energy exists as FAF that is a continuum. It can be represented as a spectrum. This spectrum is shown by the vertical dimension in Figure 2-08. The gradient of units in the S O or the Á dimension are shown by the horizontal dimension. It is highly coterminal with and equivalent to mass energy. For this diagram, however, mass energy is shown as percentages of each in its continuum. The units that make up the different kinds of S-O, Á or mass energy are shown as units, levels, stages or forms of many of their known components. That these components, levels and units are neither exact, fixed nor final needs to be strongly emphasized. However they do give us, and may be standardized and represented by even greater, powers of recognition. For the key to man's existence, is much the same, in the senses of things, observation of the observed, subjectivity of the objects, or relativity as it shows the unity, finity, infinity, totality, means and extremes of all as they exist in and out of TRU.

The units we use to separate different levels, like quantum leaps, in the Á mass-energy dimension from most basic to most advanced, are: ABu = Absolute Basic Units, SBu = Lowest Order or most Basic (Forces) Units, EMu = Electromagnetic or Electronic Units, Nu = Neutronic Units, Au = Atomic Units, Mu = Molecular Units, Gu = Genetic Units, ADu = Adaptational Units, Cu = Conscious Units, Vu = Verbal or language Units, Ru -= Rational Units, IMu Proportional or Importance of Man Units, SPu = Spiritual, Collective and higher order Units, AEu = Absolute and Eternal Units. Between the known finest states, levels and units of mass and energy and their absolute limits in the future will tend to expand toward these absolute limits. This gap between known units is called SBU or Subunits on the left, and SPU or Superunits on the right. This is a range in which there may be many different units rather than one. It only confirms what we already know, or with high probability suspect, or can predict. This is the fact that there exists - or will exist - other units between the electromagnetic and the absolute at the one extreme and the rational-intellectual and absolute-eternal at the other extreme. INU, or the intellectual, requires a special definition of the term, in which it is a higher stage of reason, intuition, contemplation and comprehension in which they combine and take on unique added forms and functions for success in life. SPU is the unit - or range of units - that defines a higher stage(s) beyond INU. The expansion of Á is called; **AEX** = Á, relativity or mass energy expansion. The fact that the continuum of Á can be envisioned as continuous plus its higher powers leads us to: **ICA** = Infinite Concentric Circles of S-O or Á. The posits of some of the terms on this field will vary somewhat with their definitions and usages. They also vary somewhat from accepted usage by TRU interpretation. This is appropriate, by TRU.

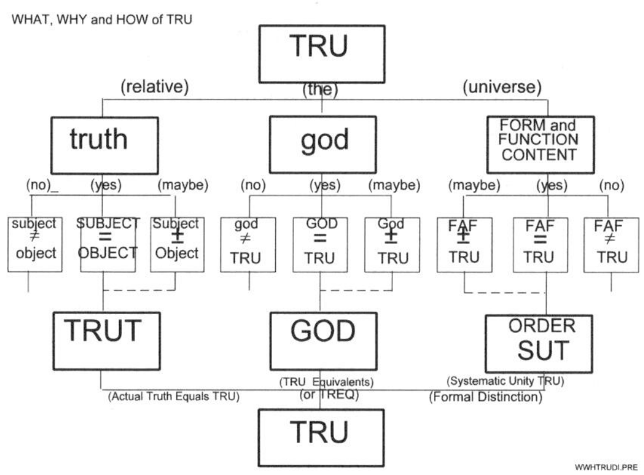

Fig 3-08 **WHAT, WHY and HOW of TRU (WAHOT)**

With TRU and its presentation there are three key, major and great questions and problems. These are the interrogatives; what, why and how. It is from these that we find much of the reasons and proofs for TRU; its form, function, and contents as well as representation and presentation. This is called: **LIVROT** = Law of Intuitive and Valid Revelations and Reasoning based upon accumulating TRU. Interrogatives set the Á,, mass energy FAF in motioin. "What is truth?", "What is God?", and "What is the universe and what are we"? Those of why are: "Why TRU and not GOD?" "Why doesn't our knowledge enable us to know what we actually are, and why can't we do what we must or should"? Those of how are: "How is truth only appearance, and how does it become actual truth?", "How are God and TRU different and how the same, and how can this be best shown?" "How can we discover and learn the form, function, and content of the universe, and how is this exclusive, special, systematically unified, and formally distinct?" From these questions, their premises, and the answers they imply, we can begin our quest for the best possible positive objective approach, POA, and the development of the best possible representation, formulation, modulation, and presentation of this universe. We can then begin to present this knowledge and universe in ways that render it not only universally valid, relative, and specific but reducible to a reliable and effective presentation. A person who takes a superficial look at this figure may think it trite, or presumptive of the relation between God and TRU. Yet, no less than all great concepts of their topics carry the power of their topic's actual being, it is by existence in TRU. It is most subtle, profound, dynamic and alive. For it shows us, at once, the great problems we have, the need for TRU, and its vigorous demonstration. The figure shows the crux of both the text, and any critique of ROT, POT and TRU. For unless TRU fills a great need, is actually separate, is a great improvement on what exists, and reaches a presence and role in public use, it can be of little benefit. The message of this diagram is how truth as the agreement between subject and object, god as it is GOD in its role in TRU, and form and function as they exist only in TRU, combine to show how TREQ, or TRU Equivalents, are actuality in, and help us recognize, TRU.

Fig 4-08 **DEVELOPMENT of "THE"**

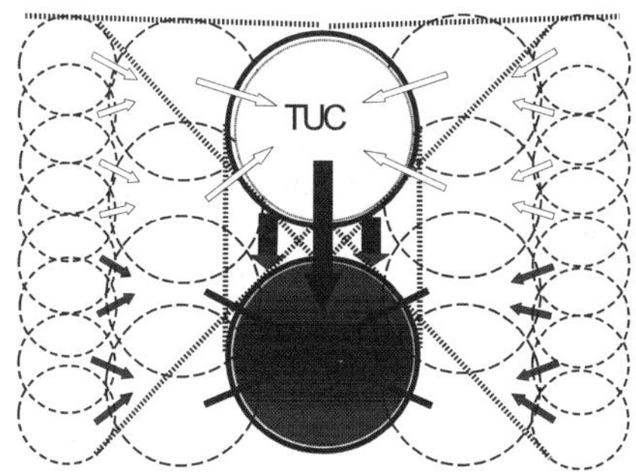

How do a, any or some things, become the thing? How do entities become The entity. Is it a state or condition of the entity itself, its object, or the subjectification of it? How do entities and things progress from absolutes through relatives, universes in common, top circle, to The Relative, TRU, bottom circle? How some things become the thing, is both simple and profound. It is simple to the extent that, as by Anaxagoras, RUS, TRUS and TIT are increasingly and predominantly characterized by those things of which they are mostly made. That is how RU, RUT, TRU and TIT are increasingly specified and made up of different and separate aspects, forms, modes, essences and characteristics. In addition they have a separate and special internal focus. Each TIT, TRU, RUT and RU are not some other, closely related or comparative entity or other TIT, TRU, RUT and RU because they do not have the exact makeup of what characterizes TRU. For RUT and RU, however, we speak of TRU generally, not as explicitly described in Section I or TRU proper. This is called: **TUC(S)** = The Universe(s) Common. It is TUC or what TRU holds in common with entities, things, AUS and RUS that justifies and enables us to validly speak of TRU in this introductory section of this book. TUC is any, a, some thing as they become, yet are not yet, The Thing, or all the necessary characteristics they have. These in comparison with TRU give us a clearer idea of what is the thisness and whatness of TRU as it is distinct from what is irrelevant. TUC stands in marked contrast between RU and TRU, between *a* thing and *The* Thing. This is called: **SPEN** = Speciation of the Entity. SPEN is the final product of thisness or haecceity as TUC proceeds through RU, RUT, and TRU to become TIT. It is more than a product of external convergence. It is highly internal in that it is its own focus or point and center of reference that produces, as it expands or grows, its own thisness or haecceity.(ATO) It is why TIT is itself and not some other RU. It is in being relative to us, being focused upon our being, attention, intention, and success/failure that determine its absolutespecificity.(TRIA) The Relative Universe, other RUS, animals, and deviates in human cladistics are, at best all TUC, and not SPEN.

	I N C R E A S I N G			C O M P L E X I T Y		
RIGHT			PRINCIPLE	L A W	SOLUTION	
	I D E A	CONCEPT	ANSWER			RESOLUTION
	QUESTION		PARADOX			
WRONG	E R R O R	FALLACY		PROBLEM		

Table 1-08 **SODS**, or Diagram of the Solution to Subjectification. Random IA arises from relations to the different states of right and wrong, and degrees of complexity. Before each resolution, solution, answer, law, principle, and concept there is paradox, problem, question, fallacy, or error. Before these were prior problems, questions, fallacies, errors, ideas and unknowns. An example of this is: **PIROL** = Paradoxes Intense Revelation Obtains all, including the most basic laws and mechanisms. So that in the solution to problems there has been a long heritage of prior ones whose resulting laws, principles, and concepts are a great part of the Á dimensional world and known uses. The errors, fallacies, problems, needs and all of our attention are directed to provide us with the breakdown of and the beginning of a study and representation of knowledge. From these through the objects, ideas, reals, and all in life we may trace out random, common and everyday IA's, that show the fulcrum and core of our world. It is in the derivation, development and determination of this subject object relationship and especially the observer as subjective that basic mechanisms, resolutions and solutions to what the world is, can be best found. The most simple wrong and error proceed to the complex and right, as we solve the problems of life. This is **SOS** or the Solution to/Of the Subjective.

Fig 5-08 **RELATIONS, TRUTH and TRU**

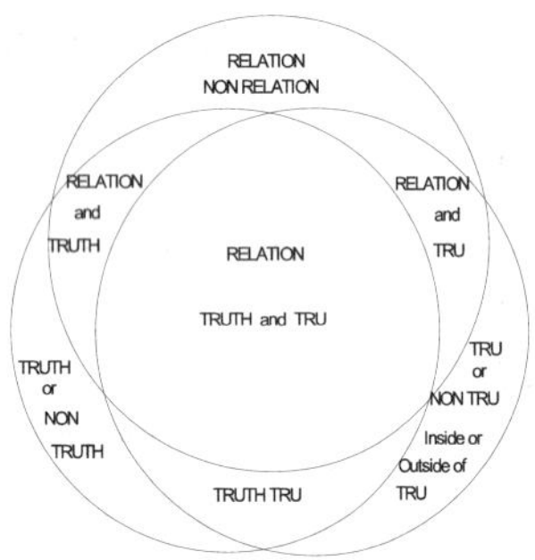

To say that all outside of TRU is not related and all inside is related is not necessarily valid. It is valid to the extent all outside of TRU is illusion or figment and all inside actuality. For by DAOST some of TRU is not, more or less by degrees, related. Also by obvious correlations in worlds other than TRU there are relations. In this way chance and determination as by LASUR do or do not apply, they only have limits. So that we are dealing not with one but three or more parameters in such considerations. 1. Relation or non relation, 2. Truth or non Truth, 3. Outside or inside of TRU, AU or TÂ. Thus, as three valued systems, we have mixtures of seven different combinations, noted in the figure. These demonstrate vividly how far all people's lives, and even more, single relations, truths or aspects and parts of TRU are from their combinations and added states and operations in TRU.

DEGREE of TRUTH and TRU - Throughout the text repeated reference to truth and degrees of truth are made along with TRU. There is a great disparity not only between truth and falsehood, but between kinds of truth. Like the term "is" as an identity, is much different than truth. An identity in which a real truth or what is present, a full agreement between subject and object, is not actual truth. As we shall find in the next Section, on TRU itself, these kinds of truth are not - and much less than - the truth that exists in and of TRU. It demonstrates the continuous processes of Á at work, and is exemplified by the ubiquitous activity of **RECTRU** = Reduction to TRU. This is a truth based upon the actual reality in the relativity of the observer and observed, or subject and object, when formed and function by the Á mass energy predominant dimension of TRU. The bond between relation, truth and TRU is seen in: **SIMER** = Similarity Equals Relative, and also have much qualification.. This will depend on their form, function and use. Throughout history there have been interactions between,

Great Ideas↔Mechanisms↔ Interpretations↔VAPO

This has been largely driven by the search for discovering better methods of action. Since TRU is so massive and complex, separation so difficult, and perfection so evasive, let alone AMTRU, ATRU and AU, any and all means and methods that enhance action are to be procured. This in turn is driven by increasing powers of coordination and order. Order is a fundamental aspect of TRU, as well as topic for metaphysical analysis. It has especially significant implications in relation to the future, methods and purpose. How much order is there, how much is available for use, how much is or can be abused, and how is the best use of order to be achieved? Order like all, and especially all that is of TRU may be separated into its various kinds: disorder, mis-order, poor order, no order, mixed order, little order, good order, perfect order, non TRU order, mixed order with TRU, perfect order by TRU, and perfect order by TRU that is limited to what is necessary and what is of the degree of order indicated by TRU or the given. An understanding of TRU, TROS, TIT, TRIA, TRIA, TÂ, TRIS, TRIT, OBT, SUT, ROT, MOT, ALT, and SPALT enables comprehension of the relations or SCARLS, applications, and limits or what is already naturally existing order in TRU to the role and use of order. Order is greatly enhanced by our being, focus and totality. It is readily achieved by deepest realization, when remembering , ECRO, and applying the important, yet specific, part and role order has in POA, ROT, MOT, SPALT, POS, and ACC. It is key to TRIT, OBT, ROT, POT, and IPT. It is supported and upheld by the higher powers of human mass energy, Á, to produce the distinctions and verifiability knowledge and human life demand. Yet for each it is somewhat different, and calls for different descriptions. These are presented in each chapter. It is in the summary of TRU, ACC, and the future that we must be sure that what we present is no greater than, nor less than, the order that is appropriate for TRU, ROT, MOT, & POT. At the same time never failing to utilize to the utmost its great benefits in enabling us to enhance human life and all that is TRU.

VAPO, evident from the figure below, works reversibly and massively. When we depart from VAPO we are soon lost. When we depart from great ideas and laws we are soon lost. We should uphold the greatness of VAPO and its components.. There is a constant splintering off and regrouping with other associated elements. Yet they tend to return to an equilibrium with the constant upholding of TRU and TO, OITIO. This is called: **GIMVAPO** = Great Ideas, Mechanisms and VAPO. This is primarily how they function together to uphold TRU. e.g. all the great ideas, associated events, and discoveries in history. This is driven by: **ACUIA** = Acuity of the IA naturally tends to TRU, and **SEVI** = Search for Vivid Imagery. The goal of these is: **OFI** = Optimal Finite Form and Idea. VAPO is seen in: **POKED** = Principle of Explosive Total Knowledge, revelation, and accomplishment through analysis, Depth conception and intuition. There is a constant and natural tendency to piece together all that one is sensing, experiencing and working with. This is a major intent and great bounty of scientific research. It arises as **KUT** or Knowledge that enables or is becoming TRU, yet is Unknown, is not TRU.

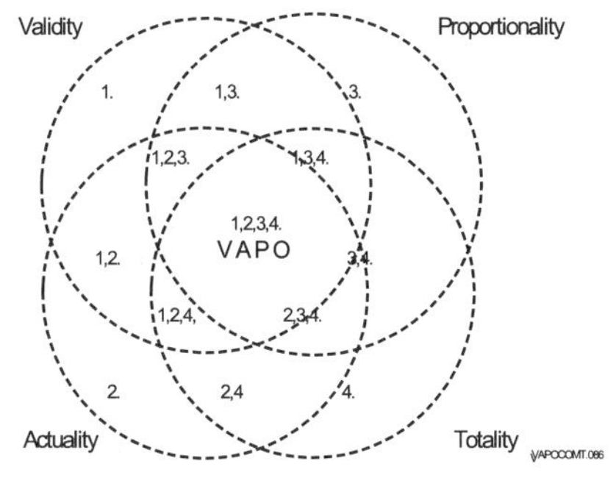

Fig 6-08 **VALIDITY, ACTUALITY, PROPORTIONALITY, and TOTALITY or VAPO**. The avid experimenter avoids analysis and synthesis, even interpretation is suspect. This is good in the presentation of entities and objects and approach to TRU, but is deadly in KOT, ACC, and TRU. POKED is the inspiration, joy, revelation, glory, conviction and victory of being able to acceleratively apperceive and achieve all for TRU. This is the culmination of our S-O or mass energy, A, processes. Yet the ENITS and mixtures we confront are massive and diverse. We must adapt to the alterations created by the limits, confinement, warping and lack of proportionality. This is produced by the irrelative or degree to which we depart from a single relative, or develop multiple foci and relatives, **PAR**, = Paradox of Relative Point of Reference. Many figures in the text show us how we can diagrammatically demonstrate the different that can be increasingly separated and the same that can be unified, thereby assuring ENT, TRU, TÄ, TRIT, SUT, MOT, SPALT, POS, ACC,and TA.

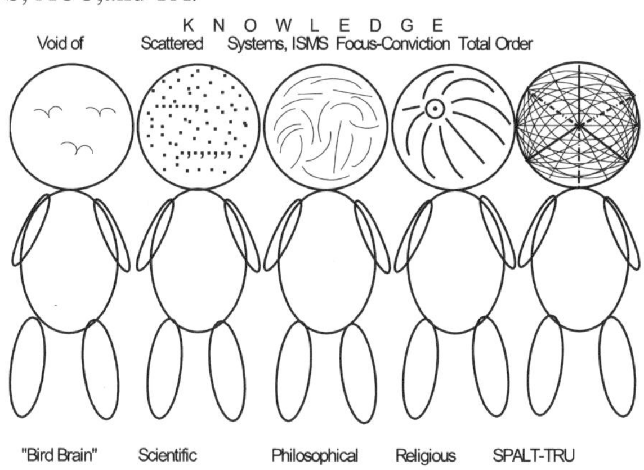

Fig 7-08 **KINDS of KNOWLEDGE, COMPARATIVE MENTALITIES and WORLD VIEWS** (KNOK)

The great purpose of the text - as much of TRU - is to follow the best available knowledge and how it can be best interpreted and presented so that each person may gain optimal TES, ELTICMA, and TÂ. If metaphysics is ever to be a concrete science, it can't be limited to either bits, pieces, or the polarities in which they exist. It can do so only as it includes all, and these by VAPO. Even when we extend physicalism to realism, we are still operating in a limited field. This is **LISMOP** = Law of Limits of ISMS, Oppositions, Polarities and Operations. Knowledge that overcomes LISMOP is what enables us to rise above and break the bonds or confines of little minds and bring them to their full role in and of TRU. ISMS tend to be carried to the extreme, become partial and reactionary. They deceive us and can never be correct, will always be defective. and harmful. For they strive to champion, adapt and apply what is disproportionate TRU, ATRU or ANTRU. Metaphysics is possible only when of TRU or its perspective as each and all are no more nor less than their form and function, parts and processes or roles in TRU.

As entities and universes become more complex failure increases, or **FIC** = Failure in Complexity, or the ratio of failure to success increases exponentially with complexity. There is an exponential incapability of maintaining itself without an equal and opposite exponential support. So that for the most complex organisms with the broadest capacity,, as in humans, the amount of support needed is critical. This makes us highly vulnerable. It has been compensated by: 1. A high rate of selection, and this has been at several levels and in many ways. 2. A high degree of support from its own form and function, 3. A high degree of order and compliance. To do so each of us should eliminate all habits of thought that restrain this, and expand all horizons that help to achieve TREQ or TRU knowledge. In the above diagram we show five, of many, kinds of knowledge. We show the comparison between very little knowledge on the left through much knowledge to an optimal frame and ordered body of knowledge at right as a person in maturation and Pureality achieves enough to be their counterparts, TES, in TRU. This could be extended to many varieties, as occurs in professions, the division of labor, and specialization. The figure is highly figurative and dramatic, yet done so purposively to draw attention to the salient point that all persons need to discover VAPO and TRU. This is knowledge that needs to best develop ing capacity and ability of total comprehension and depth conception. Thereby achieving the kind of overall management and control necessary for a successful life. This comes from the fullest knowledge and participation in TRU. Necessity makes us responsible to properly fulfill our roles in life, as TRU. It is this failure, and the lack of joy that can be associated with its mutual benefits, that is the greatest weakness and threat to mankind. The figures depict the great problem of the world. This is the lack of TRU understanding. It is not only the TU or SU of TRU, but the TO, TRIS, TROS, TIT, TRIA, TRUD, TRUCH, OITIO, OM, ROT, MOT, SPALT, REM, THESA, THEOA, COGNOS, ELTICMA, SA and TA all together. Even knowledge based upon TRU and SPALT is not enough. It lacks completion of each of the aspects of VAPO. In time to come this will require the continual and accelerative support and efforts of all. It needs some of the detail and precision of science, some of the relations and contrasts of philosophy and some of the focus and conviction of religion. Thereby to achieve an effective and successful presentation of TRU, and all, we should have the best possible organization and perfection of knowledge possible. To provide this we need knowledge that has the greatest possible: validity, reliability, detail, profundity, relations, courses, comparisons, contrasts, focus, orientation, conviction, coordination, totality, frame, actuality, value, proportionality, concepts, laws, distinct composition, distinct and discrete processes, posits of parts and processes, order, states, conditions and limits, negation, effects on the whole, operation in out of the whole, All together best replicating TRU, and All added representation.

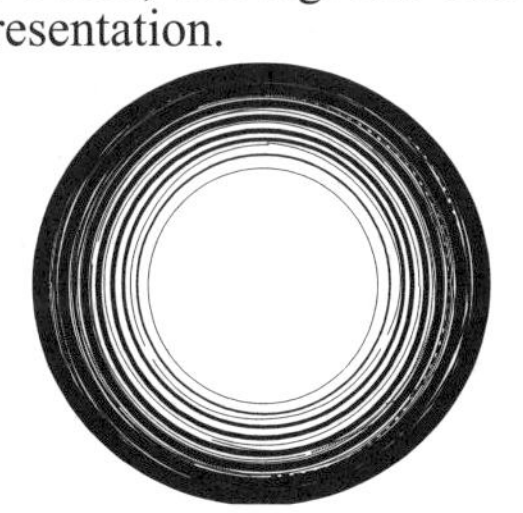

Fig 8-08 **WHITE HOLE, REVERSIBLE DENSITIES and TRU**

In figure 8-08 we show a white hole. This is a reverse image of a black hole. Instead of being the vanishing point of the observer and relativity, it is its continuous and eternally beginning point. The white hole arises from how the nature of a thing is in passing through. It will not have complete form, identity or finality in any specified position or way. It may have a predominance, its fundamental nature may be determined or even mostly made up at a certain level but never complete until it has been reduced. This is the continuation of the form, and is called: **CONA** = Continuation of Form of an entity, thing, sense, and TRU. CONA is similar and related to universals and TIMS, only in different ways. CONA is more limited to the entity, thing, sense or universe in which it applies and operates. The continuation may have sharp, acute or obtuse asymptotes, depending upon its relationship to the axis of its transit. **CAORT** = Continuation of Form Total Relative to Transit, - ultimately by TRU.

A white hole is the beginning of absolute assurance of being, of TRU. In figure 3-11 we show it's opposite, a black hole. This is the figure of any universe or one in which all vanishes through its center. The differences are opposite and together they make up a polarity of the existent, ATO. It, as any polarity, is in a reversible position. Whereby extreme density and extreme lack of density reverse themselves. This polarity is what is between the complete absolute and relative, or the world, as our own, existence. Toward the vanishing point there is extinction. Toward the eternally beginning point there is perpetual and expanding survival and life. Existence as it contrasts between a black hole and a white hole is dramatic.

BLACK HOLE	WHITE HOLE
EXTREME DENSITY	EXTREME LACK OF DENSITY
LIGHT EXTINGUISHING	LIGHT BRIGHTENING
VANISHES	ETERNAL EXISTENCE
DECREASING	INCREASING
ACTUAL to NOTHING	ACTUAL to EVERYTHING
LIMITING	UNLIMITED
CONTRACTING	EXPANDING
NEGATIVE	POSITIVE
Overwhelmed Externally	Eternal Internal Predominance

In chapters 11 and 98 we show these relationships, as they pertain to their vantages. These are extensions, as by TIMS. TRU is neither a black hole nor white hole. It is a stable state whose vast extension of time provides the potential for eternal life. There are many ideas and topics which can be derived from such a view, including:

1. Relationship between light, life, RU or TRU
2. Light, Relativity, RU or TRU
3. Density and lack of Density, general theory of,
4. Reversal of density and lack of density of TRU
5. Density and lack of density in TRU
6. Their comparisons by GER, SER, UFT, RU, TRU
7. The dynamics of the continuum formed by the polarity between black and white holes RU and TRU
8. Relationships to positive and negative forces
9. Absolute and Relative, ÂÂ, ÂR, RÂ, RR and TÂ
10. Representation comparison with black and whites, or negatives, of photography
11. **LIRM** Law of Individual, RU, TES and TRU, Immortality
12. Role of many great concepts of IPTRU, POT, AU, RU, TRU, as ATO, TRIA, TIT and TÂ in these relationships. Black equals 1/TIT and white equals expanding TIT and TÂS. It is thereby self evident, yet so often needing remembrance, that TRU is both "is and is not". depending upon its FAF and SCARLS. If we strictly limit these to TIT, it is. If we speak of a relative universe generally, TRUS, we have great validity in all that is common - as DAOS - totally. Yet if we speak of a relative universe specifically as the, some set, exclusive universe, RUIT, relative only within itself and not to anything else, not even common to other relative universes, we have a very limited set and must recognize the great differences, as by DAOS, of this singular RU. The relative is neither the physical universe and what we commonly know as the world at large, nor the individual and homo sapiens, their anatomy, physiology, social behavior and all. It is a continuum of all these and all in between whose essences are relative to our success over failure and survival. The continuum of its component parts or gradients differs from any ENIT. It is their ESINS in precise VAPO, DAOST, SCARLS. TROS and TRIS. This is called; **TRECON** or The Relative Continuum, Created Continuum. It is TRECON that *IS*, not ENIT, SENIT, nor THIT. To compare TRECON with TRUS, TRUS is like evolution and TRECON like creation, yet both only by ATO. What is the our of the relative and how are its limits set? The 'our' is neither static nor unnatural. It is largely of those who best survive. It is those whose predominance of mass energy and CARSC provide the continuity and perfection of The Relative. Like form and function, whether white, black, yellow, brown, yellow or whatever race, subspecies, or variety, these are less important than their fitness, perfection, and contribution to the whole of human survival, TRU.

GREATNESS OF LIFE

The acuity, breadth, depth, magnanimity and beauty of life are a combination of factors, how profound and inspired our IA or immediate awareness, instantaneous attention, and action is, how much a person develops the best comprehension and performance of all one encounters, the detail provided by the picture at hand, the extent the examination and operation provide, the degree to which this represents the problem or part of life given to be represented, presented and achieved, the greatness of the problem, how much it is critical to TRU. the realization of identity with TRU, TREQ or God as TRU and the infinite and eternal possibilities in this identity.. These solutions to the problems of life, tie and fit into providing ACC and the solutions to all the problems of TRU, thereby, the eternal victory and glory of TRU.

These great joys of life are, like the miracles expressed in the proverb, 'work makes life pleasant', a product of compensation and fulfillment. More especially it is the satisfaction of: **COMIA** or Completion of IA and ones role in the Á FAF of TRU. Life is neither simple nor assured. It continuously challenges us to be our best, to fulfill it as we must if we are to be. The acuity, breadth and depth are much the same as what we find and call ACEXIN. It is the realization and fulfillment of this participation and enablement of TRU, ENT, that provides such great consolation in times of despair for all who learn and work for the glory of God, as TREQ or TRU. For it is through the 'miracle' of this special universe of best human life, TRU, as exemplified in the potential of COGNOS, that we can become so near to *The*, TIT-TÂ. This is evident in **KÄC** or Knowledge Acquires new form and Characteristics as it is increasingly Assured in Completion. If we are to accept only one or a part of one of these, even as they are TRU, we fail. It is small wonder that TRU or our world is so denied and ignored. It is only when these are all properly put together in TRU, that we can begin to see and increasingly become - as we must - TRU, our actual proper counterpart in the world, TES and TÍS.

LEXS = Law of Exponential Sufficiency. The central meaning of COGNOS lies in the advance of S-O in TRU. It is a filling up to completion. The S of TRU being completed as TE-SU-THES, and all that needs to be known, is known, and thereby equals TRU.(VAPO) In this way "we know", and "We know that we know" and „„„ad infinitum". Thus we know what to do, what is not necessary, what not to do, and all exponentially sufficient unto the infinite, TI, and TA. TRU as TA is exponentially sufficient. The law of exponential sufficiency to TA, is LEXS. Yet this is out of itself not a decreasing to zero effect. It is one of exponential perpetuity. It is a symphonic and harmonious expansion of each and all of TRU. All polarities, including the objective, are in harmonious expansion in the tremendous success of TRU.

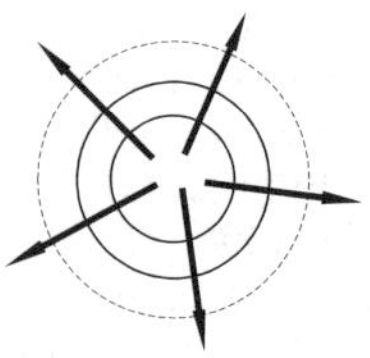

Fig 9-08 **LEXS and CEXTRU.**

CEXTRU = Completion of the Expansion of TRU.

As S-O, or Á, fills up there is a simultaneous broadening. There is a constant equilibrium of this filling up and the broadening of S-O and all dimensional formulation and ACEXIN of TRU. By SU this is a filling up with the concretion and construct of TRU, or: **TRUYE** = TRU Yields Equilibrium. The equilibrium yield of TRU, although massive and self sustaining to a large extent, is at a somewhat variable rate and thereby a state in which TRIS and TRUS grow, yet TIT does not appreciably lose its thisness.

CHAPTER 09

SUMMARY and CONCLUSIONS of APT, EPI, AU, PERT and POT

What exists? Do entities, senses, things, objects, subjects, AUS, universals, and particulars exist, or does only TIT exist? The key is in the relation between what we mean by existing, TIT, and the words we use to define them. Occam constructed the philosophy that only particulars exist, that the individual not the universe is real. This extreme of nominalism is called terminism and produces the Nominal PAradox, **PAN**. Its SOlution is **SOPAN**. This solution lies, as so much that is fundamental, not in the words, individuals, particulars or universes themselves, but in the more general solution to paradoxes and their part and role in TRU. When upheld and limited by TIT individuality becomes the thisness and again particulate of TRU, thereby its proof, **PRUTIND**. It is by the law of individuation and particularity of TRU. It is such combined or fused differences and extremes in opposition that not only make for paradoxes, but when reduced, by DAOST and their common parts and roles in TRU, are the basis for and proof of TRU. Such are the actual universal, or TRU proofs, **PRUS**. Their limits produce not an absolute, but a Relative ReSolution, **LAROS**. The denial of ones existence was considered by Descartes to be impossible. It is an exclusive Relative Individual Identity, **RIND**. This in its non TRU sense is a false path led by idealism, egoism, and criticism. It is a SUfficiency Fallacy, **SUF,** wherein sufficiency is not gauged by individuality and all else assailed regardless of TIT. Sufficiency like VAPO is both TRU and its measure. This is **LATOV** or Law of Absolute integrative or summative Total TRU Validity and Truth. Truth is not only from experience, evidence, reason, intuition or revelation. It is, as shown through the text from TRU, TROS, VAPO, DAOST, REM, and all that shows TRU. Without LATOV the natural result is insufficiency or reaction through consolidation in ignorance. This is characteristic of many ISMS. For ISMS all is held to be reducible to this ISM, reductionism. This is much of its failure for it leaves out its particularity and Higher Rights in Opposition, **LAHROP**. These are all of which ISMS are partial of and limited by, primarily TRU. Much of the basic importance and use of paradoxes stems from their use, representation, and operations in conceptualization and language. For every great paradox there is a great and often new, or to be developed, great concept, principle, or law. This in turn supports **PRUNEC** or Proof of New Concepts, and **PRUNT** Proof of a New Language or Terminology of TRU. The concluding solution lies mainly in each individual's ability to utilize LAHROP and **KUPS** or Knowledge Universal, Partial and Special, in their search for Absolute and Imperative cultivation of ones counterpart in TES and TIT, **AITES**. This is highly dependent upon LATNUP,(pg.16), and resulting SPALT, POS, POA, PERT, and POT.(03) There can be a Unified Field Theory, **UFT**. Not as Einstein envisioned only physically, but one based upon TIT-TO and its ROT. This entails a transit from an unordered unstructured IA and individuality to a successful comprehension and production of TIT and TRU. It requires an absolute distinction between TIT and all not TRU, an absolute distinction in relativity. It likewise requires a minimum of controlled standards.

FIA, Appearance, PROSU Nothing by itself is entirely true. A perfect probability can only be approached, as in quantum theory. There is nothing existing absolutely alone. It only appears to be such to us. Man's perception is such that the objects before us appear as discrete entities. This is not so, neither from the standpoint of AU nor TRU. This is only an identity conditioned in our being, not an eternal truth of either finality or totality. Here is the great fallacy of IA. This is called: **FIA** = Fallacy of IA. Here lies the explanation of many of the problems of philosophy. Can truth exist in either AU or TAU? Is not truth only of TRU as qualified by REM and VAPO. If truth does not exist beyond TRU is AU non existent? How can we speak of a non existent entity? Here is the beauty of the comparison and distinction between AU and TAU. To us, this does not mean they are non existent from all vantages. It is as all exists together, in systematic unity, SU.

Is there much that is neither for us nor against us that is of greatness? No. For anything of significance is so from its relation to us. This relation is from extreme positive to negative, producing success or failure in one way or another. How then can AU be identified with that of harm to us or the negative? For is not the negative a great part of TRU or related to TRU? Herein is the Proof of Systematic Unity or **PROSU**. For to deny being this is not so, but to deny our being in sum is not to be. Therefor TIT-TO., this is certain.

ABSOLUTE and RELATIVE, MORE and LESS - For every entity, sense, thing, AU, RU, object, and subject in the world, of, about, and around us either as individuals or as mankind generally, there is some of each that is a part of TRU and some that is not TRU. There is also some that is more TRU and some that is less TRU. Even of the parts of each one of these there is some that is TRU and some that is not TRU, some that is more TRU and some that is less TRU. Thereby some parts of all is more or less TRU but never absolutely non TRU and never absolutely all TRU. Depending upon the view taken of these that are some more and some less TRU some will be for TRU and some against TRU. We will have either an absolute/relative relationship or positive/negative relationship depending upon how they are viewed. By LANERA and LAMOL the state of absolute to relative and more or less will act as polarities in which there is a continuum. This continuum will depend upon and can only be measured if we set what the given is. This will depend upon the relationship between the entity, sense, thing, AU, RU, object, subject and TRU. Since this is highly variable, it will be necessary to clearly distinguish or specify each case. This is why mathematics is so exact and other sciences less or little so. For in many, as TRU, it is not always possible to have discrete factors and functions, entities and TRU. Yet it can be increasingly accomplished if enough is put into it to make them discrete. This is the science of sciences, *scientia scientia.* As we are battered, and overburdened with facts, information, knowledge, evidence, principles, words, ideas, forms, concepts and laws, and attempt to posit them, we adapt, use, discount, weigh, evaluate, criticize, analyze, extol or are protectively hardened and withdraw in reaction.How the individual strives to consolidate oneself is shown by: .

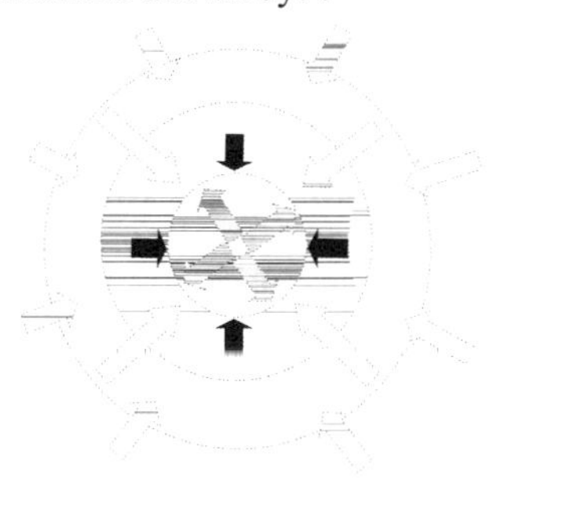

Fig. 1-09
Actions, Reactions and Consolidation.
COI = Consolidation in Ignorance or Insufficiency. COI is the retraction and contraction of an individual overwhelmed from without. Or the overpowering effect of human knowledge. This may take many forms, most insidious and indirect. The solution to COI is partly time and maturity. Yet this is not enough, for the onslaught is continuous. With a proper focus, unity, and systematic unity of TRU as well as the modeling, positing and working out the form, essences, ES and CES of TRU, such methods of learning

and knowing all things, COI is resolved and predominance can be reasserted. All the facts, information, knowledge, evidence, principles, words, ideas, forms, laws, and concepts that we encounter become not so much overwhelming forces closing in, but the beautiful intricate detail and food for thought, upon which our FAF, life, and TRU are sustained. This is done through **AITES**, the absolute and imperative cultivation of ones counterpart in TRU, TES and **PUREALITY**. It is by continual enrichment processes in which **KUPS**, or how this kowledge is Used from universal, Partial, and Special purposes.

K - PER - TREQ - TIT - TITT0 - SÂ - TÂ

The target and method is how to bring all to TRU, and TRU to all that is known and existing essence of TRU, yet not known or named as TRU. .

The antithesis between systems is evident in how relativity tends to refute an exclusive or even overriding role of scientific materialism, scientific realism, idealism and realism alone. It tends to prove or disprove the validity and characteristics of all systems, coordinations, and dimensions.

The Law of the Absolute Necessity of a Singular Unified Perspective and Presentation, **LATNUP**, is exemplified by the image of SPALT or a model of TRU. That a perspective of all is needed has been repeatedly attested in the literature, science, an individual's experience, and this text. Yet it seems to ever escape us. If we are to obtain perspective and overall subjectivity of all it must be reduced from an infinite number of vague observers relating to an infinite number of inactual observed. This in part may be reduced to the problem of relating the integral and the differential. It may be partly reduced to the problem of mode or average acting as one observer bonds with the observed. It may also be reduced by analogy to examples in the world around us in art, photography and statistics. By adopting set preconditions or standards for a unified singular perspective of all we can superimpose the least common denominators or **LCD** to our world's primary dimensions and systematically bond the observer and observed in a 1:1 relationship. This is how perspective can be an image of its object. As perspective is that common for a best observation or subjectivity of TRU and the object is that construct of TRU that most closely duplicates or replicates it. The bond is what best represents these two as they come together, thus the basis for LATNUP and UFT. By the independence of the observer, as well as the observed, degrees of freedom give the variance needed to have multiple perspectives and objects. These multiple views, senses, IA, objects and entities must, however, incline to TRU and thereby LATNUP. Such are the preconditions that preset and hold diversity to unity and unity to diversity. The basic nature of relativity.

Einstein held that there are only two fundamental concepts: 1. The velocity of light and 2. The laws of physics. This is a valuable reduction for physics and even sets the form for relativity. Yet it does not account for universes other than a material or physical one. It does not even account for the observer or much of the observer-observed relationship. So that although Einstein enunciated relativity he never fully appreciated its greatest implication, and the reason why he never could achieve **UFT** a Unified Field Theory. This is its dependence and production in the fundamental form and function, thisness or haecceity, and whatness or quiddity of The Relative Universe. The mind set on two - of the great many - fundamental concepts restricts the necessary extension of relativity to TRU. This transit from an unordered or unstructured IA and individuality is called: **TIAT** = Transit from an unordered unstructured IA and individuality to a successful comprehension and production of TRU. It is to TIAT that much of TES-TÍS depends, and much of this text is devoted. Most of all IPTRU exercises the many means by which TIAT is achieved. The great law that settles the very difficult problem or paradox of Absolute AU, EPI and TRU distinction is: **LARD** = the Law of Absolute Distinction in Relativity. If our Relativity is only TRU, as it must by all accounts, we as individuals can function in actuality only as we are TRU. Thereby a sharp line of demarcation is possible between that which is TRU and that which is not TRU or of the overall AU. Likewise a sharp line of demarcation can be provided between that which is TRU and that which is of EPI or of Knowledge or Perspective. Yet here to we must apply the appropriate QUOL by SOR, SOT, SOS and THESOS. However the key to distinction lies in relativity, as does relativity in distinction, and the very special, unique nature of the observer-observed and subject-object situation in the Á mass energy FAF predominant and special of TRU. The degree to which we impose any Standard or Synthetic form is the Degree to which the Object, representation, model, product or image is Limited and Liable to have Defect, **LODIS**. A corollary to LODIS is that the degree to which standards are adopted. The degree to which they either have to be continually remembered and reduced by TIT, or some secondary standards or rules applied. This is why a minimum must be continually sought. The Accuracy and Need in Equilibrium with Limitation to Produce Optimal Standardization may be expressed as law, **LANELS**. When no basis in objectivity and TRU exist we adopt the first step rule, or set as standard what occurs initially. This is the Law of Primary or Set Standard, **LOPSET**. With LOPSET measures of TRU and all are needed, thence **IRT,** or to Invoke the Rule of TRU. This is their identity and contribution to the success of TRU.(VAPO)

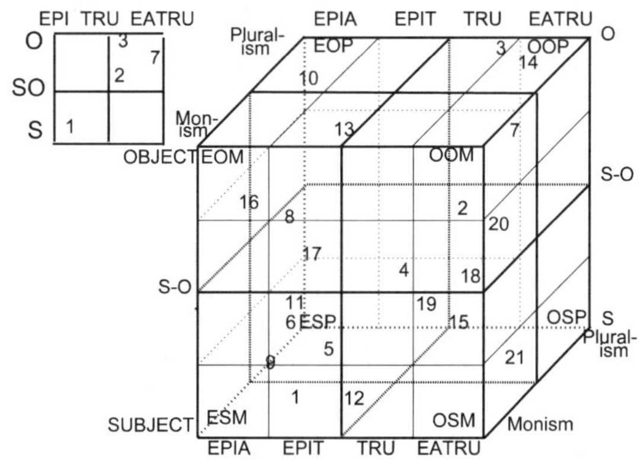

Fig 2-09 **S-O EPI-TRU ONE-MANY**

We must realize and remem-ber that when EPI is extrac-ted from TRU we develop less than actual TRU. When we utilize any entity, we think in terms of any sense and use, and thereby are not being TRU. This difference between TRU and that of TRU with EPI extracted is generally signified by: **EATRU** = Extracted From TRU. This is any TRU envisioned as an object in which a portion of it has been removed. In this way, of COL in reverse, we have in actuality a combined dialectic in reverse or Synthesis (TRU), Antithesis (EPI) and Thesis (EATRU). Yet since EPI may be both non TRU and TRU a new system or thesis is required. This will be EPIAU and EPITRU like IATRU and EATRU. In this case EPI is divided into two portions, EATRU and TRU. The different systems of philosophy may be posited in these diagrams. They are shown by posits in three dimensions, S-O, EPI and TRU with EPIT and EATRU and pluralism to monism. Some of the more commonly used systems or ISMS include, 1. Idealism 2. Realism, 3. Materialism, 4. Empiricism, 5. Sensationalism, 6. Skepticism, 7. Pragmatism, 8. Epistemological Realism, 9. Epistemological Idealism, 10. Epistemological Objectivism, 11. Epistemological Subjectivism, 12. Subjective Monism, 13. Objective Monism 14. Objective Pluralism 15. Subjective Pluralism,16. Epistemological Monism 17. Epistemological Pluralism 18. Ethics, 19. Deontological Ethics 20. Teleological Ethics, 21. Standardized Perspective of Total Ordered TRU These scales are diagrammatic. With continued use and editing they can be much more refined. For example the dimension from monism to pluralism is not only a dichotomy but a scale whose gradients may be measured from zero to infinity. Subject and object may also be so treated - and are as we shall see in the next sections - a continuum. This is how they become much of the A dimension of TRU. The dichotomy of EPI to TRU and its extensions from EPIT to EATRU may be further refined to encompass all senses and things in a systematic standardized order. In so doing we may have not only the combinations and permutations of the six possible extremes, two poles each of three parameters, and eight possible permutations, but many others depending upon the quantitative extension of these gradients. The permutations are:

ESM Epistemological Subjective Monism	OSM Ontological Subjective Monism
EOM Epistemological Objective Monism	OOM Ontological Objective Monism
ESP Epistemological Subjective Pluralism	OSP Ontological Subjective Pluralism
EOP Epistemological Objective Pluralism	OOP Ontological Objective Pluralism

The logical problem of coordinating the epistemological and subjective or the ontological and objective would initially seem ambiguous, invalid or paradoxical. Yet this is philosophy, not only the philosophical perspective, but how it orders itself. For the horizontal dimension is not in terms of an EPI of subjectivity, it is of the continuum from furthest outside of TRU to closest inside. **EPIA** = Epistemics of all, **EPIT** = Epistemics of TRU, TRU and EATRU. Such confusion or interpretation of dimensions is key to this text, wherein explicit dimensional formulation and modulation isthe solution. This does not excuse us, however, from holding each to their limits or SCARLS and VAPO, by TRU.

By TRU all is inclined to the focus of unity in totality and its extension in the finite relative. This is how the infinite and infinitesimal incline toward TRU and vv, to produce TRU Relative, **TRUR**.(VAPO) Degrees of controlled Departure from continua in TRU, **DEPTOS**, infinity-finity-unity-cardinality-and infinitesimal, THES-SÁ-THEO provide the needed manipulation and operations of OITIO.

LOCOAT = Law of Comprehension of All Things, and TRU. The search for knowledge, wisdom, philosophy, and guides in life is infinite, endless, and eternal. Only as knowledge limits itself to TRU, proceeds by LOLS, and only as this is with maximal validity, focus, orientation, form, coordination, and proportion, as through TRIT, OBT, SUT, ROT, MOT, ALT, and SPALT can it be made comprehensible and achieve adequate overview. TRU is comprehensible as a unity and a totality in itself, TIT-TU-TO. In OBT, SUT, ROT, MOT, and ALT it can be taken in less than infinity, treated in finite increments, then in sum. This is **PRUCT(S)** or Proof of the Comprehensibility of TRU, through SUT. Knowledge without TRU and SU can be in part, piecemeal, of all things. Yet it can't be comprehended in totality. It can not be ordered, represented, modeled as a finite object, nor described and demonstrated in the way TRU can. For knowledge without TRU, SU, MOT, ALT, and SPALT is without scale and dimensions, or what is given completely lacks the identity necessary with this special universe of best human life. The dimensions and their scales are indistinct. It must be adapted to the relativity, observer-observed, subject-object, actual form and standards of TRU. When these are set by replication of TIT it achieves the means, and extremes, centering and limits it has in TIT. This is: **LULAR** or Law of Universal Limits as attained by a RU, and only actually attained by the special FAF of TRU. For an object without limits, dimensions, set scales, and actuality in TRU, is no longer an object or RU that can be adequately modeled or comprehended.

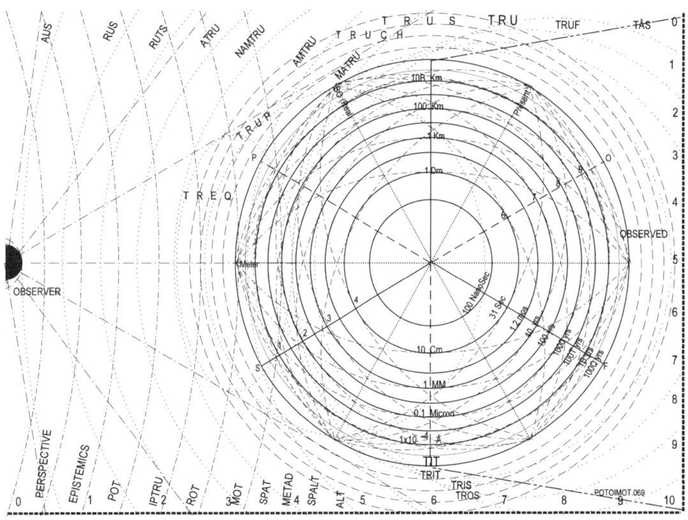

Fig 3-09 **POMOT** is the combination of POT and MOT as SPALT, or the overview of TRU and all, through their development in POT. It is one of the texts major working diagrams, frequently used to posit many of the concepts, principles, laws, figures, tables and enumerations. It serves with other overall diagrams in POT. Only it is more fully developed in SPALT, METAD and SPAT. It provides an intricate field on which posits can be made and especially where POA of POT can be utilized. (069) In order to develop a good representation of not only TIT, but all that is about TRU, and especially the approach and introduction to TRU, we need to have a model or field that shows both TIT and that of, about, in and out, into, approach to and perspective of TRU. In POMOT we extend the dimensions of TIT in such a way that their continuity, both in and of TRU can be best represented. This means the positioning of the A, L, and T dimensions, with an added emphasis upon the S-O and time dimensions in the vertical and horizontal. Since these extensions are extra TRU, and thereby have a decreasing TRU actuality, their structure and workings will become more arbitrary. However, a sufficient consistency can be achieved to make them fairly reliable. Since it is of the greatest importance not only from the absolute nature of TRU, but its near absolute need for KOT and extra TRU demonstration, a high degree of allowance is needed in the posits and their realization as they provide an optimal POA to TRU. We must always remember, or keep in mind and apply, ECRO, that POMOT is less than SPALT and TIT. In being less than SPALT and by including its limitations, i.e. the lack of 3D demonstration and associated S-O and octant deformity, it requires considerable flexibility in use and interpretation.(61,,,)

The S-O and T (time in TRU) measures are scaled with relatively equal weights guided by the elements in each scale. Since SPALT is the more representative of TIT and thereby the primary and preferred model, all concepts, ES, CES and items to be posited are mostly done in SPALT or its sections. POMOT is reserved for all that is less than 100% TIT, or that extend outward from TIT, by reason of TRUS, ROT, POT and PERT. There will be some overlap of posits, depending upon whether the focus of the items is upon their perfect TIT meaning, or upon a focus that extends them into extra TRU relationships. Since TIT is so vast, and TRU, its approach, POT, ROT, MOT, REM, POS and extra TRU even much more so, we must always be careful not to become to complacent or self satisfied with the seeming agreement and success of posits with their objects, and the replication of the object, TIT and extra TRU.

Fig 4-09 **PERSPECTIVE and OVERALL VIEW**

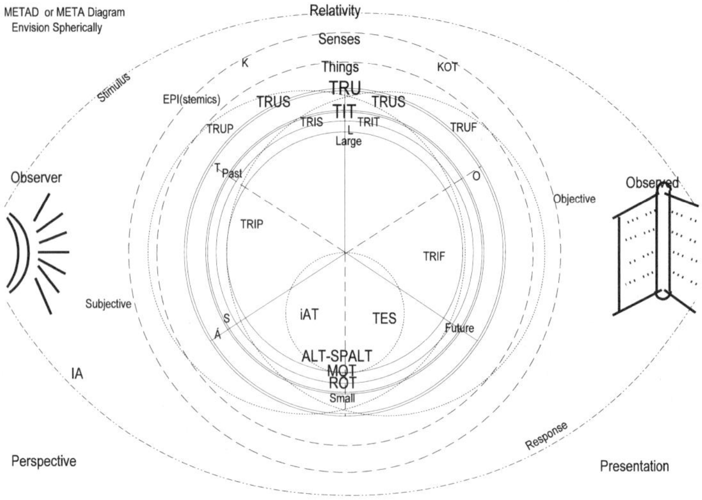

METAD = Meta Diagram, an Overview of TRU. METAD provides an overview from the outside of TRU. It is a precursor, and somewhat subservient to, SPALT and SPAT. METAD is a basis for the organization of much of the material of this text that exists outside of TIT. It seeks to show how the perspective of TRU, as through ALT may be interpreted in terms of its presentation and representation. Like S is to O, and how they may all be oriented, coordinated and ordered as a whole. METAD is an expanded diagram of the ALT or SPALT model, utilizing its dimensions. It extends these and superimposes on ALT all that best enables us to envision all, both inside and outside of TRU. METAD shows what is beyond, higher, transcendent, changing, transforming and polymeric to TRU. Much of the text is preparatory to TRU, its representation and modeling. For although TRU is all by TRIA, when we present, represent, perceive and reproduce TRU we must be able to envision it not as it is in itself, but as it is in all its forms, varieties, states, conditions, limits and relations. This state is created not so much by TRU itself, but, like works of art in other ways, its subjectification, objectification, perspective, representation and presentation. This involves the relationships and SCARLS between the observer and observed or first person and TRU as qualified by this presentation. The latter part of the text is also included in the diagram by being its detail, static posits and dynamic processes. Their in and out or transcendent, reversible, cyclic and total operations can be shown adding to the accuracy and successful presentation of TRU and all that is interactive and involved in the observer and observed relations of TRU. Only in keeping this extra TIT status in mind, and continuously expanding our comprehension of the optimal view of METAD and other important illustrations. as ALT, SPALT, and POMOT we attain the necessary coordination and order out of disorder, that is the key to our being.

There is a large degree of coterminality between standards and laws. Yet each has a somewhat different focus that gives it significance. As much coterminality causes a loss of optimal POA, through diffusion, its very approach to, rather than being in, and a 1:1 identity with TIT, renders it justified. This is called: **COJAP** = Coterminality Justified by Approach. COJAP, however, must hold to DAOST and be limited:(SCARLS) **CODLIM** = Coterminality, Overlapping, and diffusion held to a Limit and corrected by Management and Control. As in science, the determination and decision of what are and what are not laws depends upon:

1. The validity it holds as a concept and principle,
2. The degree to which these are fundamental,
3. The degree to which many other profound concepts, principles and mechanisms are based upon or related to it,
4. The degree these are TRU or relevant to the world, TREQ,
5. How well these concepts and principles exist, signify and explain the ideas or universal relations denoted,
6. The degree of certainty the above can be raised and held to,
7. How important the problem or critical the aspect, state, or part of TRU is that these proposed laws refer to,
8. The degree to which the law is useful or of practical value,
9. The benefit, harm or cost the law produces.

Laws thereby are both symbolic representations of problems, states, concepts, principles, mechanisms of TRU, and the vehicles, means or tools that act as a practical guide, service and solution. This depends upon being formed and function as fundamental truths from and of TRU. By being actual, fundamental and most valuable aspects or parts of TRU they have the added power of being mandatory or imperative. The laws we present are warranted by the greatness of TRU, their lack in existing language and society, and the accentuation and action necessary if TRU, as mans life, is to flourish. **PIROL** or Paradoxes Intense Revelation Obtains Principles and Laws. It is the investigation, analysis, interpretation and resolution of the major and basic paradoxes that many of mankind's greatest principles and laws are obtained. In reverse it is in many of mankind's greatest concepts, principles, laws and mechanisms that the cause and reason for the major and basic paradoxes and problems can be found. How good laws are depends upon how practical they are, or the Utilization Principle. This in turn depends upon VAPO, especially proportion, **LOPU** or Law of Proportional Usefulness. There is a correspondence between Proportionality and Universality. It is in this way that entities, AU and TRU are LAMOL.

The relations in the use of COL, especially in the continua, dimensions, and ALT of TRU, help us fulfill all things. For COL signifies what is left over when THESO is subtracted from TIT. DEPTOS is of the greatest importance. It shows us the great contrast between the highest and best form of relativity as provided by Á and the decreasing subject and object relationship as senses and things. The more we learn and know, especially of TRU, the less we have to use standards. **KASIT** = Knowledge and Standardization of TRU. The highest form of relation is equality or exact equivalence in TRU. This can be increasingly determined by learning more and more about TRU. This is **LE** = Law of Equality. LE is exemplified by how close the equality of the topics and descriptions of the text are with the objects they represent, especially TRU and its parts. **CENORM** is the Correspondence of Entity, sense and Meaning. To achieve the distinction necessary for a determination of correspondence and equality requires: **LOPEX** = Law of Optimal Examination and Extraction. Sometimes we lack that needed to make such an examination possible: **LENK** = Encounter of the Third Kind, or encounters over and above TRU and EPI-AU, what is OITIO, but with the limits of one's own energy, will, interests, and ability. This is associated with: **LODIR** the Law of Diminishing Returns. With the greatness of TRU, and increased learning and KOT producing increased interest, imagination, ideas, visions, and revelations , an ever greater TRU will be shown and known. This is: **DOREV** = Development of Revelation, the divinely inspired and powerfully ordered ideation by VAPO and TRU.

SECTION I

TRU PROPER: TRUS, TRU and TIT

CHAPTER 10

TERMINOLOGY, APPROACH, RELATIONS, KNOWLEDGE and PROOF of TRU

TRU is the general term for all that is, describes, apperceives and shows The Relative Universe or This Special Universe of Best Human Life. **TIT** is TRU in and of itself exclusively of all that is extraneous to it. **TREQ** are TRU Equivalents, all that is TIT and TRU without being recognized as TRU. Since all in a successful world must be in harmony, symphony, and mutual benefit with all that is a part of it, it must complement, add to, reinforce, and empower what is already good and great in the world. It must join with God and all to fill in all voids, incompletions, disproportions, malfunctions, and eliminate all that is wrong to make a better and more perfect world. This is **LATREQ** or Law of TREQ. **TRUSP** is the Specific State of TRU or TIT of the Species Homo Sapiens. **TRE** is TRU Recognized, or TRU sufficiently known to be recognized. **TUR** is TRU Un-Recognized. There is a great difference between TRU as known, as shown, revealed, represented, used, and TIT. Also there is a great difference between the unconscious or subconscious TUR and KOT of TRU today and in the future. **TRECS** signifies the range and increasing steps with distinctions from greatest TRE to the least TUR. TRU may be known in various combinations of fields without being known by the name TRU. TRU may be known in totality, as by God or TREQ, without being known by the name TRU. There is and will be TREQ or TRU knowing itself TIT, only when TRE known by name. To achieve optimal eclectic harmony emphasis and focus must be made only on this same or common identity, TRU. This is called: **LAROND** = the Law of Reduction of Differences in Names or Deities to their least common denominators, actual identity in TRU. LAROND is how TREQ is derived. All that is the best and the same as TRU, such as is God, Allah, Brahma, reality, humanity etc is TREQ. It is not the symbol T, nor name TRU that is so important, it is what it stands for. Yet if it is to be known it must have a name, a name that is equivalent TRU. What GOD and all that is TRU needs has so much limit, resistance, inefficiency and ineffectiveness that TRU and its associated new Terminology are mandatory. This is called: **LANIK** = Law of being named to be known. LANIK is the key to, justification for, and proof of: **LANEC** or the Law of New Concepts. These new concepts, are justified to the degree they are TRU and can be developed into usable and effective terms. This largely depends upon how well they act as vehicles for, and fill voids in, Á, human mass energy FAF.

Fig 1-10 **NAT or NEAR APPROACHES to TRU**

In order to continue the format of finding a perspective of all, and an approach to TRU, we proceed from BAS or Basal States through FAT, Far Approaches to TRU, to NAT, Near Approaches to TRU. They in combination are shown by BASFANT, or all three plus SPALT. NAT helps us understand the concepts used and their relationships, especially as they exist inside of, outside of, or through and through TRU. NAT is a combination of all inside TRUS. NAT especially helps us see the difference between past in general, TRUP and TRUF, the future in general. This is key to understanding and proving TRU. Likewise it helps to separate the levels in the mass energy dimension outside of, inside of and through and through TRU. This approach shows how the relativity of observer-observed becomes in actuation subject and object and the predominant Á FAF and dimension of TRU. Similarly it shows the alterations and expansion of length as it becomes the L dimension in TRU. NAT also shows how much superimposition there is between each of the spheres of TRU and its reproductions. Only as we can envision how these are both ENIT, things in themselves, ES and CES in TIT or mixtures thereof, TIT, can we ever gain an adequate understanding of TRU, and thereby all or anything.

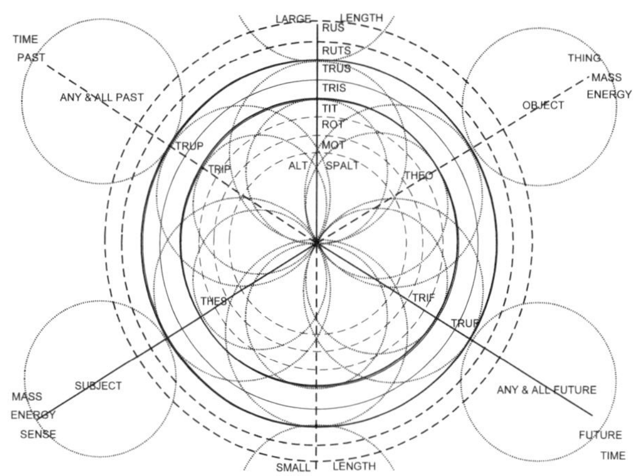

Knowledge without TRU, SU, and LOL can be comprehended, in piecemeal of all things by ESINS, but not in totality. The comprehension of all things can occur only as knowledge limits itself to TRU. Only as this is of maximal form, coordination and ordered to proceed by LOLS and SU can it become TIT-TO. Each and every great concept, principle, law, and mechanism of TRU must have its own term to be adequately communicated and preserved. It requires a terminology sufficient to fit TRU. TRU can be shown, proven, and tested by the use of, and solutions by, such terminology. S, O, S-O, SO, SOT, and Á are increasingly key to, and of and in, TRU and TIT. S, O, TRU separation and standardization are necessary to reconcile OITIO and show the intensity of TRU. The fused absolute subjective objective is evidence for TRU terminology. It is also evident in paradoxes and problems, both universal and of everyday life, such as imperfect systems error in passing away from TRU, differences between AU and TRU, and the increasing and unlimited dividing of parts. There is much that should be constantly and continuously remembered of TIT-TO. This is called ECRO, a strong proof of LANT and NEC, as all that enables the PERT, POT, ROT, and MOT of TRU.

Of the many modes of TRU those of perseity, aseity haecceity, quiddity, quality, quantity, and necessity are of the highest order. They provide excellent distinction of the nature of things and the thing, TRU. **ASEITY** or from oneself, is a higher level of independence of things, especially TRU. This is its being a se, from itself. Aseity is how TRU, as independence without a cause, became special and is closely related to ATO. This is its being in and of itself, without prior external creation or development from something else. **PERSEITY** or through, by means of, or for the sake of oneself, is the independence of a thing, its being in itself, per se. Perseity is TRU as the topic itself. This is like TRIA, absolute independence, being, relative to no other thing. In language, knowledge and philosophy we have great precedent, knowledge and proof of TRU in the terms aseity and perseity. Perseity is being as such, independent of a subject. Perseity is how TRU exists by means of its own internal constitution, without relativity. Aseity is the mechanism by which this occurs from itself. **HAECCEITY** is the thisness of a thing, or TRU. That which makes this thing or TRU what it is, the ultimate reality different from any other. This applies largely to Chapter 13, TÂP, TIT, TRIA and TÂ. It is the key, cue and gist to a reliable understanding of TRU. **QUIDDITY** - is the what-ness, or detail, of a thing, or TRU. It is what makes this thing or TRU to be the type of thing it is; the essential, subtle, nature and underlying form of it, or its essential character. This applies to much of the whole text, especially Chapters 12, 13, 14, and 15.It is what makes much of the greatness of TRU as a topic, being able to make distinct all that is TRU. It is how TRU can rise from the condition of God, or that of a vague and illusive topic. **QUALITY** is its mode of purity, essence, and power. **QUANTITY** is the mode of extension and measure of TRU.The distinctions these terms provide illustrate the ontological and theological proofs of GOD as TREQ and TREQ as TRU. Haecceity is of the specification of TRU. It has to do with the relative fact of TRU in conjunction with whatever FAF it is constituted that renders it an ultimate reality different from any other. This arises from its predominant characteristics and powers that put it in a class by itself, as a species, the species homo sapiens. It is all that shows TRU'S exclusive thisness, TIT and TRIA. Each of these takes on altered and set meanings in TIT and TRU. This is **CAST**, the Changed or Altered States and meanings of all as they exist in TRU. The reason for this altered set is to be found in its own necessity or **LATAN** the Law of The Absolute Necessity, **LATANO**, Law of The Absolute Necessity of Ordination, and **LATANOS**, Law of The Absolute Necessity of Ordination and Standardization, for TRU. Likewise there is an absolute necessity of integration, differentiation, limit, order, distinction and meaning, as well as The Law of Partial Necessity, **LOPAN**. Closely related is all that is imperative of TRU by its exclusiveness and power.

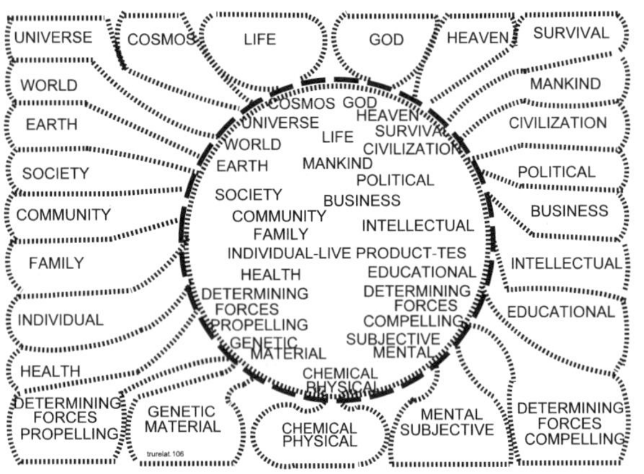

Fig 2-10 **TRU and RELATIONS**

This diagram is used to figuratively show what is in TRU as compared to the same entities and topics that exist in themselves, commonly conceived. The names for ENIT on the outside of the circle TRU have their TRU counterparts inside. There is a great difference between what each of these are in themselves and what they are as they are a part or play a role in TRU or TREQ. It shows us how easy it is to confuse what is TRU with what is not TRU, and thereby fail to recognize both what TRU is and what is the actual existence and meaning of each of these names, topics, entities or fields. It is essential that each and all of us continually remember, ECRO, and apply this idea of separating out all we encounter or is given; whether it is or is not TRU, or how mixed and interactive. It takes a good understanding of TRU as well as an open mind about each of these topics and entities. This we seek to achieve and provide throughout the text.

In this figure we can readily see the difference between INT or In and Into TRU, OITIO, **AMTRU** or TRU and AU Mixtures, **ATRU** or From TRU, and **IMT** or the Immensity of TIT and TRU. Relative partial senses and things are in a continual flux with what is potential and what is actually in TRU. We can also see what is potential to TRU, **POTEN**, and how to separate what is not and what is TRU, **SEPTRU**. Here much that is universal and transcendent, **TRAN,** must be extracted, **EXADT**, for purity in TRU. Not all relative is TRU.(CIS,VAPO) Even what is different in TRU than TIT must be separated if both are to be well known.

What so sets TRU apart, and makes each of the pairs in the figure so different is **TREL** or TRU Relative to itself, the absolute relationship of all TRU, all in TIT, **LAROT**. All is to be thought of and used not as they exist in themselves but how they exist and function in TIT and TRU. This may seem highly limiting yet even the most adventurous will find the Immanence of TRU, **TRIM**, filling its content with ES and CES, its great immensity and interactivity, more than enough challenge.

RELATIONS, RELATIVITY, and UNIVERSES

The key and basic explanation of all the theories, principles and laws of relativity center on the fact that the separate and seemingly independent facts, entities, objects, ideas, concepts and principles of mathematics, geometry, mechanics, physics, astronomy, and the universe do not exist alone, independently and without limits. They exist in relation, and are formed and function as a certain unified set, whose VAPO, equational and focal relationships to each other are manifested by TRIT of TIT in TRU. Einstein's Theories of Relativity, **SER** and **GER**, or special and general theories, center on the spatiotemporal relationships and the observer-observed. How neither can be measured, nor interpreted properly in finality, without the other. This is the gist of Lorentz Transformations. Einstein's General Theory of Relativity centers on the relationship between gravity, or mass, mass-energy as they relate to space time and solid geometry. The special theory was confirmed by the correspondence of the observer and observed in the propagation of light. This is called: **LÍMSCO** = Light is a material/mental, thing/sense, objective/subjective, physical/sight, relativistic combination altered and adapted to TRU. By LIMSCO, the photons we observe mean little without our seeing them, and our seeing them means little without their being there. Yet this most cardinal and vital nature of relativity and TRU did not originate with Einstein. At least as early as Archimedes the pivotal position of relativity was recognized, whereby in order that he might draw the terrestrial globe out of its place and transport it elsewhere Archimedes demanded only that the point should be fixed and immovable; "in the same way I shall have the right to conceive high hopes if I am happy enough to discover one thing only which is certain and indubitable." Here his revelation was not so greatly the power of the lever, as great a law as it has become, it is "only that the point should be fixed and immovable". This requires a fulcrum or a necessary set state, relationship, or SCARL, in the entity, object and the entity, subject. It infers, and it requires, a set universe, TRU. It also requires an IA that is stable, a fixed relationship between

its subjectivity and objectivity. This is called: **LAPIVOT** = the Law of the Pivotal Proof of TRU, from Archimedes. This is limited by the finite nature of man in applying the power needed, and the state of the points being fixed and immoveable.
Herein TRU is set and human opportunity is unlimited, as in hope springs eternal, by TIT and TÂ

TRUS			TROS	TRIS	
KOT	TRUG	TRUD	TÂ	TIT	TRIP
DOT	TRUR	ENT	TU	TRIR	TRIC
DAT	TRUP	OITIO	TE	TRIN	TRIF
TAG	TRUF	OM	TI	TRINI	TRIT
DAOST	TRUCH	SUT	TÂ	TAS	

Table 1-10 **TRUS, TROS and TRIS**

All TRU may be divided into three major categories, or classes. Those that describe TRU generally or TRUS, those that describe TRU and TIT mathematically both generally and specifically or TROS, and those that describe TIT specifically or TRIS. The meanings of each as well as the permutations and combinations of all will depend to a great extent upon what is attended to at the moment and its VAPO and SCARLS with all the others. Although they, like attention and presentation, move from the outside to the inside, do not make a good polarity, They are more like clumps or organs in this progression. They are highly illuminating, by giving a better and clearer idea of the basic aspects and makeup of TRU and TIT. TRU as TIT whose form, construct, and frame provide for OBT, SUT, ROT and MOT is **TRIT**. TRIT is key to tie-ins, as well as separating TIT from all else. It is the basis for, proof of, and justification for much our showing, knowing, using and living TRU. The other kinds of TRIS further qualify TIT and clarify fundamental aspects of its existence. The focus of TIT can be anywhere from the infinitesimal to the infinite only more commonly relative, finite, practical and standardized. This is explained by **CORÂ** or Condition of the Relative Absolute, and by **LÄTIT** or Law of TIT, wherein TRIS render TIT less than absolute and make TIT and TRIS distinct from TÂ-S. They show how TRU is highly conditioned, qualitatively, quantitatively and in many other ways. Although complicating it has its upside in being set. It makes for greater demonstration and comprehension.

It is in an increasingly accurate **PERT** or the perspective of TRU that we can distinguish which are coterminalities, which are the same, and which are different. Yet we must always remember, ECRO, the fallacy of perspective precedence, or how perspective - as APT - is not prime, TIT is. Yet since TIT and TRU must be known to be shown, known and lived, we must Bring it Into View, **BIV**. This POT and IPTRU seek to do with massive distinctions, clarity and precision. This is in part accomplished by the vivid view achieved through TRU AU Imperative Distinctions.(SEPTRU,LATAN) Much of this provides PERT with the needed concepts, principles, laws, and mechanisms of TRU. (NEC, LANT, LOT-S)

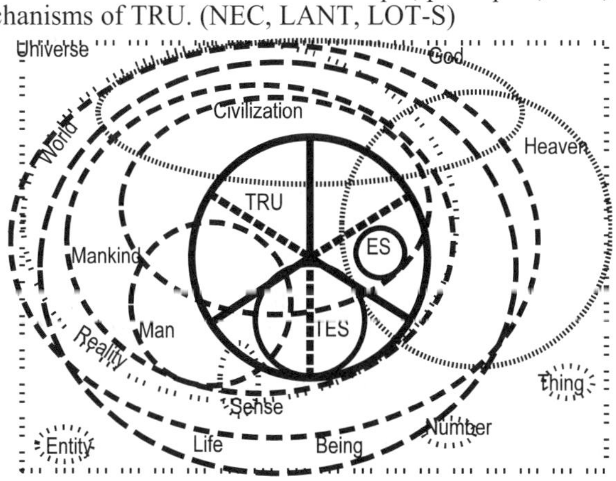

Fig 3-10 **COTERMINALITIES with TRU**

Coterminalities with TRU are those things which are TRU or have much in common with TRU. To a great extent they are apt to overlap in any representation of TRU. They cover much of same ground, and thereby have codomain. Analysis, study, and determination of coterminalities, codomain and overlapping greatly assist our positioning, understanding and life in TRU. It makes clear how present language and terms fail to signify the actual world and thus, importance of new concepts for TRU.

The relative size of these different categories, topics and subjects is optimally dependent upon the sense in which they are taken. Their position in relation to TRU can be made based upon many denominators and dimensions. Yet the length time dimensions of TRU are the most practical, reliable and conventional. It provides an introduction into, principles and problems with posits, or the placement of major parts of TRU in a model. These are only very rough approximations to their coterminalities with TRU. Yet they give, in a somewhat dramatic way, an indication of some of the problems involved in equating them with TRU. As well they indicate the definitions and relations as well as form and function of each. To be more accurate it is necessary to go to a three-dimensional model, SPALT/ALT. It would also be more accurate to show these in varying sizes, greater numbers, and - as well - explain the different senses we refer to them in comparisons. The new concept, which defines, correlates and demonstrates these fused worlds or positive and negative essences, is called: **POCFAW** = Principle of Correlative, Coterminal and superimposed, Fused and demonstrated Worlds. They signify the great media between TRU and all outside and other than of TRU.

Figure 3-10 helps to show the role of speciation and relativity in TRU as well as the association, codomain, and coexistence of man, universe, and all. **LASMU**. = Law of Association of Man and Universe. LASMU helps to explain how universals are continuing themes, only adapted to and qualified in TIT. The answer is one of focused formulation by the total unity or TU, TO, TROS, TRIT, OBT, SUT and ROT of TRU, and the universal and particular harmony and symphony TIT and TRIT provide. One of man's greatest misconceptions, errors, and fallacies is the how, and how much, we fail to associate with others and all in the world we are so much a part of. This is how much a person non-associates, under-associates, over- associates, mis-associates, and variably associates with the universe or world around one. None of us are in excellent harmony or association with the world around us. A few do well, some do enough to adapt to life in general. However most do very poorly or are in repeated or perpetual conflict with it, largely because they do not know what it and they are, TRU and TES. TRU, for them, is a new and strange, entirely different concept, yet none the less great, separate and significant mutual identity.

The relation, coexistence, and differences between God and TRU are most important both because of their greatness, need for a better and complete approach, and criticism that occurs. It is to be guided by the reduction of differences in names of deities to their least common actual identities, and God and TRU Misinterpretation Correction or **GATMIC**. The relation between God and TRU may be expressed both as **GOT**, God Or TRU, and **GET**, God Equals TRU. GET is valid to the extent the TREQ of God equals TRU, and the SCARLS of both to TIT. The extent this is invalid is **GID** or God Defect and God is Dead. It is partly because so much that is actually God as TRU goes unknown or unproven that so many deny or mistake God or will fail to realize TRU. This is the paradox of assurance which can only be cured by a Resolution of Assurance, **RESA**. The differences may be expressed as a ratio, **BER** or the Belief, E or ES, Ratio. They require an ever increasing knowledge, especially Knowledge of TRU, **KOT**. This is evident in the illusion of actuality and the actuality of illusion. Like all fields and topics, and especially like ISMS religious errors result largely from illusion, mistaken actuality, partiality, and a failure to comprehend all, all TRU as through COAT by VAPO.

Reality is the total coincidence of highest ordered subjectivity and objectivity. The coincidence is formed in the mid portions of relativity or observer-observed relationship. For the mass energy continuum in TIT Á this becomes a major section, mid range, of its distribution, SÁ. The distinction of reality can be clearly made by the superimposition of what represents TRU and provides their ultimate identity through TRIT. This is called the Principle of Reality Substantiated or **PORS**. PORS is not the acceptance of any reality but a confirmation of the reality of TRU as reality becomes, a mirror or power of identity, in TRU.

Identity is the extent these different coterminalities become and are the same. The extent to which this is focused on TRU is identity with TRU. Superimposition provides ultimate identity of TRIT, the basis for ROT and MOT. There is much mistaken identity in human lives, especially identity with TRU. Herein is a large portion of the need for and operation of ACC or Accentuation of TRU. It is to be guided by the Law of Reference Fixation or **LORF**, focusing by a center and frame for an ordered objectivity of TRU.

PERT, reality and identity that are increasingly directed to TRU enable ever clearer and more TRU distinctions. This is critical to the successful operation of human life, our Á mass energy FAF. For as TRU is DAOST, same and other, we must best order the same and separate the other. This separation of the other or all that is different, for Á, is how well reality is achieved, their distinctions. On higher Á levels ascension from partials and ATRU is manifest in what a person must be able to understand to be saved or **MIAT**, Maximal Intensity and Acuity with TRU. That All is Differentiated and must be made Distinct from All, **ALDAL**, reflects the differential, even infinitesimal differences that exist in all in TIT. To search for distinction and ALDAL requires **COTRU** or increasing Concentration on TRU, the great concentration and persistence necessary if TRU is to be discovered, comprehended, reproduced and fulfilled. This in turn requires a Reversion and Correction to TRU Intensity or **RECTI**. With higher and higher stages of knowledge and distinction of TRU the world becomes <u>Mundus Intelligibilis</u> or a world of intelligible realities. This is close to Plato's realm of ideas, and St. Augustine's rationes aeternae in a divine mind. For a person to acquire and use knowledge requires much learning and experience of and in TRU. It requires the validity and distinction of what is great in TRU, lesser in TRU, what is not TRU, what is potential for, and what is opposed to TRU. How all fits together in TRU requires a combined differentiation, integration, limitation, order, distinction and meaning. Abstraction rapidly loses its validity and meaning when only of senses and things. Absolute distinction is to be best found in the relativity, polarity and continua of TIT as shown in ALT, SPALT and POS.

KNOWLEDGE of TRU

Not only because it is a great part of TRU and because TRU can't be known without it, knowledge of TRU is described in detail in the remaining pages of this chapter, and text. This is more especially so, since existing knowledge is so deficient.

KOT signifies Knowledge Of TRU. The abbreviation for unknowns is **UK**. Likewise the unknowns of TRU that produce a limit and loss to TRU need to be noted and applied. A ratio can be given for the difference and interaction between ultimate knowledge and unknowns. All knowledge, like reality and much in our daily lives, is not KOT and not of TRU, but must be adapted to TRU. Likewise there is much that prevents KOT, along with limits to knowledge which need to be determined and used. Yet there is much similarity between the forms of knowledge of various TRUS, AMTRUS and AUS. More often this derives from the FAF of knowledge than the universes themselves. There is an absolute necessity for KOT. It is impossible to make proper judgements and take the right action without VAPO and an understanding of the needed concepts, principles, laws and mechanisms of TRU. Valid and important knowledge is both convergent upon TRU, undeniable, irreversible, compelling, and propelling. KOT and the concentration of thought on TRU benefits from freedom of all that is irrelevant. Concentration of thought on TRU when elevated to higher levels is achieved by expanding operations through ACEXIN. The relationship between actuality and inactuality may be expressed as a ratio, the Belief Ratio, **BER**. As for life, there is a highly compulsory nature to KOT. Through awareness we can develop a higher understanding of TRU, and accelerate to higher intelligence and spirit which is in, of, and about TRU.

It is how KOT encompasses the totality of TRU that is prerequisite to any comprehensive and reliable knowledge. Any object is or is not adequately or sufficiently known as it is necessary and practical to us. For this reason the forms and functions of objects before us need to be perfectly known. Yet TRU is unknown. For although we have many glimpses of it, these do not give a good figure. It is like a pretty girl that is so easy to see yet completely baffling to a young man. Since TRU is all, our knowledge must be similarly adequate and complete. Since it does not reveal itself readily, it has remained throughout history, an enigma and paradox that is controversial. TRU thereby necessitates great Concentration, **COTRU**, on itself. TRU also necessitates IAT and a Reversion and Correction to TRU Intensity, **RECTI**. We must always be on guard, if we are to know TRU, to extract from it all that is not TRU or not a part of its definition. This is **EXADT**, Extraction from TRU of all not TRU, or not a part of the Definition of TRU.

PRAPTO = Processes of Rising Above, Away and Beyond, From Partials to Complete Objectivity. There is a great tendency toward the constant fulfillment of TRU. This largely arises from a person, their IA, and the needs of life. On the higher levels of Á this is manifest in, **LART**, Law of Relative Truth and **MIAT**, Maximal Intensity and Acuity with TRU. What may be absolutely true in itself may not, and usually is not, similarly or equally true in TRU or TIT. The interpretation of each is derived from law and TRU, as in mathematics, physics, logic, geometry, relativity, mechanics, motion, and thermodynamics. LART, then, becomes, **LAL**, the Law of Law, or how Laws are not Laws except how, and to the degree, they are related, adapted and modified as they exist or apply in the given universe, for us here, TRU and TIT.

TRUTH and TRU

So much, that we sense, perceive, think, and certify, are true when taken as extremes, for emphasis, or for the clarity attained through distinction. Yet they are invalid in themselves. Only a person in time, or another as in conversation, imparts contrast. It is the nature of our nervous systems and cerebral functions to accept as true all that comes before it. Yet this operation of our senses and psychological powers is highly deceptive. It is not geared so much to the overall success of TRU as it is to the short term simple realities and influences on life of the individual. This is its stimulus response, sensory motor, observer observed, subject object separation in which the functional organism can separate and make one, divide and conquer the objects made clear and true before it. This ultimately for TRU is but a part which must, to fulfill TRU, be counterbalanced, by COL and COR, by all that of TRU which when added to it completes TRU. Since it is deceptive and extreme, that which will counterbalance it will be similarly unique. It is this uniqueness that makes great concepts, principles and laws. It is how these things, that which we sense, perceive, think of and certify as true are altered to become through their ESINS ES, and how these ES and CES or processes of TRU all function together in harmony to makeup all TRU, TO, or TO of TIT, that we can began not only to understand TRU, but be able to better interpret what seems so true, yet which may be highly illusive and deceptive.

TRUT, TRU is Truth, and Truth is TRU, If by TRUT all is shed of what is other, from or anti-TRU, then only what is proportionally TRU is true.By science anything is true to the degree it is a valid relation between the observer and observed, evidence or a fact, then how are these two seeming irreconcilable truths resolved? This is PATSEV, Paradox of TRU and scientific validity. By LABTRU, Law of from Absolute, or AU by TRU, we have two different kinds of truth, the actual truth of TRU, TRUT, and the seemingly virtual truth of science of objectivity. The solution is in the basic nature of DAR, DAT, ATO, and selective relative and exclusive form and function of TIT-TO.

OTHER KINDS of KNOWLEDGE of TRU

COMKOT = Compulsory nature of the Knowledge of TRU **LOT-S** = Law or Laws of TRU **IMKOT** = Imperative Knowledge of TRU **REPAT** = Resolution of the greatest questions, Problems and Paradoxes of TRU **TORPS** = Total Resolution of Paradoxes. **UKTRU-S** = The Unknowns of TRU. **UNKTRU**-S = The Unknown-s of TRU that produce a limit and loss to TRU. LOT-S, or Law-s of TRU, is/are where, either as specified or in general, they represent that of TRU which is certain and can be adapted to action. This is how in a person's life one is given guides, rules, on basic right and wrong ways of functioning and operating as counterparts of TRU.

ADEQUATION

KOT or Knowledge of TRU can be increasingly attained if we discriminately and collectively make the effort and utilize all that pertains to it. It is valid not only in and of itself, as is all TIT by TRIA; it is valid as it is one of several different types of Comparative Things. For as it is a thing, it has its own Space Time Interval. The highest order of validity or truth is called Adequation. Adequation is the relation of truth to being, validity to TRU, or TRUT and KET. To the naive the principles, characteristics, parts and functions of TRU may seem absurd. Yet as expressed by the early Christians, Credo quia absurdum est, or to believe in the absurd is to reveal a greater faith than to believe in the reasonable. In Agniology we learn to determine the unknown and thereby know TRU. Through contemplation we ascend to reach higher stages of knowledge of TRU. As we gain more KOT it becomes a world of intelligible realities. With great intent we direct our minds toward the object, TRU, its principles, aspects, characteristics, form, function, totality and relations.

QUCT = TIT Is highly Conditioned, not only QUalitatively, but QUantitatively. Without picturing and apperceiving this from reality or a total overview a person can never attain any satisfactory knowledge, nor COGNOS or OM. The degree to which we attain this is the degree of KITRU, PUREACLITY, TES and TÍS-Â.(FIAT,FEMST,ATIM)

SUIT = Superimposition provides ultimate identity of TRIT. This is achieved through combinations of perspectives and visions, as shown in the illustrations in IPTRU. It is where several superimposed visions provide a greater SU and KOT of TRU. By focusing in and out of TRU through SU and superimposing all, especially universals and NEC of greatest REM, we attain an ultimate representation, presentation and identity with TRU. Superimposition provides ultimate identity of TRIT. As these different themes and means are brought together we have **PORS**, Principle of Reality Substantiated.

Since all that is not TRU, TRIT or TIT by TRIA is figment or illusion we need not be to disturbed if we do not find their definitive answers and solutions. How knowledge is selective by this is: **CURK** = Compulsory Relative Knowledge or meaning. All subjectivity that is not TRU is imaginary, irrelevant, and labile. This is the observation of the observed that makes the object, TRU. IPTRU involves revelations, concepts, laws and models that knowledge, philosophy, science, and religion have provided and can reveal their massive clarity and precision of BIV. Full value can come only from this clearest and fullest description and expression. It is called: **SPAT** = The Expansion of the Threshold of TRU, and is seen in its diagram. SPAT comes from the knowledge of the Forms, A, as EXIA provides to AD, DA, DAS, DAO. EXIA,, Expansion of IA, produces the glorious expansion of the threshold, that as any IA, S-O is sufficient to the IA of The Object TRU, IAT of ALT, SU, DAT, RU and TRU.

The great enigma or paradox of truth is what is or is not in themselves being exemplified, demonstrated or perceived as self evident, what is TRU with certainty and what also is or is not a part of TRU. The great and clear distinction between TRU and AU and all their related principles, concepts, and laws goes a long way in explaining and clarifying these questions, problems, paradoxes and confusion. This is the great principle of: **RAUD** = or Relative Absolute Universal Distinction. RAUD is associated with the vivid view that is attainable with KOT, COAT, SUT, ROT, MOT, SPALT, and Homation, VIVID.At the highest level we may imagine the life of GOD or TRU to be experience and highest levels of IA. This is called: **EXT** = Experience to, of and in, TRU. EXT is the increasing S/F, TRU, REM, ultimate accentuation, Pureality, and COGNOS achieved as much of human suffering and failure is relieved in transformation to TIT, toward TÂ.

The most important and practical thing about man is, and will probably always be, sufficiency of reaction to the world. This depends upon how well ones view and adaptation is. How well one applies knowledge to their life, and how applied knowledge to all about a person produces the best possible life and success. This is especially so for TRU. **LANKOT** = Law of the Absolute Necessity of KOT, Knowledge of TRU. It is impossible to make proper judgement and take right action without a VAPO understanding of the needed principles and knowledge of TRU. This is called: **APROD** = Action Proportional to Principle. It is equally imperative that our actions be optimal by the many concepts, principles, laws and mechanisms of TRU. Therefore it is imperative that those who make our most important decisions, policies, plans and do our most far reaching planning have proportional knowledge. What is TRU, both in and of itself and known to be TRU, TIT and KOT is undeniable. This is called: **ADTEA** = The Undeniability of valid knowledge of TRU, KOT of TIT.

KOT, DYNAMICS, SO and REASON

The dynamics of KOT are seen in the preeminence of the mental, of the subjective FAF of man and TRU. It is seen in how KOT becomes KIT, to the total THES or The Subjective as it covers roughly half of TRU. The dynamics of KOT are produced by necessity in human life, and begins with the method of trial and error. By putting necessity, questions, problems to direct test in actuality we are rapidly led to KOT. Pragmatic realism emphasizes this dependence of KOT upon action, behavior, life, the successful adaptation and resolution of practical difficulties, problems and needs. This, today, has resulted in precise and beautiful productions of its science. It is even more precise when provided in three dimensions, spherically, and with excellent presentation. It shows what can be done with TRU, the need and degree to which we can represent TRU. **LASUF** = Law of Absolute Sufficiency. With TRU only an approach to absolute sufficiency suffices. This is because of the nature of TRU, and the boundless and critical nature of our intellect, our Á, SOT, mass-energy predominant form. **LASUFSA** = Law of Absolute Sufficiency Subjective Absolute.

Fig 4-10 **KNOWLEDGE and TRU, MAJOR CATEGORIES and TYPES**

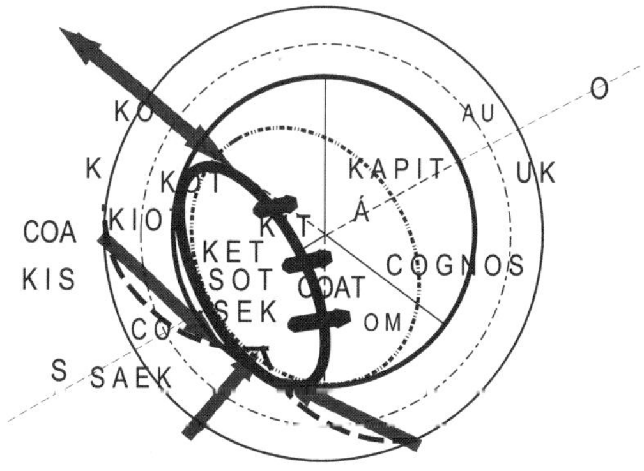

There are many faces and distributions of knowledge, especially what occurs on the S and O scale transversely, diagonally and concentrically.

I. All knowledge of any kind in a summation, including: Knowledge In and of Itself, **KI**; Knowledge by Infinite Subjectivity, **KÏS**; and Knowledge In and Out of TRU, **KIOT**. The accompanying diagram helps us better understand the difference between knowledge and subjectivity. How these different aspects of knowledge can attain more discrete formal distinction. We can also see how much more each are when known and represented in their own right.

II. **AU**, the non TRU world that much of this knowledge or perspective works through. III. **TRU** and how knowledge exists and operates in TRU; Knowledge as is in and is a part of TRU, **KIT**. IV. **KAPIT,** Knowledge that is a Perfect Image or Replica of TRU. This does not currently exist, nor will it ever, absolutely, except relatively enough to assure life's continuity, TÂS of TIT and TRU. Some of the acronyms for the different types of knowledge shown, include, **K** = Knowledge, General, **KO** = All Knowledge, **KET** = Knowledge Equals TRU, **KIOT** = Knowledge In and Out of TRU. KIOT signifies the dipole interaction of its extremes. **KIT** = Knowledge In TRU, **KOT** = Knowledge of TRU, **KAPIT** = Knowledge as a Perfect Image of TRU, **ULK** = Ultimate Knowledge, All Knowledge. K_1 to K_n, **KOS** = All Knowledge Of any kind in Summation, **CO** = Comprehension, **COA** = Comprehension of all, all AU, **COAT** = Comprehension of All TRU, **KÍS** = Knowledge In and For Itself, Infinite S, **S** = Subjectivity, with or without, irrespective of knowledge, **SEK** = Subjectivity Equal to Knowledge and vv, **SAEK** = Subjectivity unequal to Knowledge, **KAES** = Knowledge Unequal to Subjectivity, **KIRT** = Knowledge Is Imperatively Inclined and Reverts to TRU. **ORKUT** = Knowledge and especially the Organization of Knowledge must be primarily based upon the Universe, TRU, it Represents. These, when TRU is well known or even as we commonly observe, may seem obvious or even trite. Yet we often forget and labor without their benefits. It is the reason why knowledge can seem so difficult and conflicting. Whereas this conflict is more than likely either a result of the different, in different aspects of the same thing, not being of the actual world, TRU, or lacking in the relations such organization of TRU and knowledge provide. **OM** = Omniscient Magnanimity, comprehensive affective overview of KOT and all. OM is an overview of all TRU, through all aspects, characteristics, forms, parts, dynamics, processes, ENT, dimensions, order, and proportioning of TRU. It is closely related to omniscience or knowledge of all, universal, complete or infinite knowledge of all, timeless knowledge, as well as prescience or foreknowledge of the future and all things. Prescience and omniscience exemplify the vast direct and indirect consequence operative in a set TIT. This also enables **COGNOS**, We Know, COGnition, the Culmination and power of all knowledge as it is Completed, perfected, and available for use by us, as TRU. How all these work in concert through the mass energy FAF continuum in TIT is shown, along with their focus, in the core circle, from SOT, SEK, KET, COAT, KOT, COAT, and KIT..

TRU that is attainable with existing knowledge, TRU that will become more greatly known, and TRU that can be understood with ultimate knowledge may be named and continuously applied. A law's validity comes from increasing knowledge and correct TRU comprehension. Compulsory relative knowledge helps selection and provides more definitive answers and solutions. Diagrams that show the standardized subject object and TRU reveal how we can reconcile OITIO and the intensity of TRU. The Law of Paradoxical Absolute, **LOPA**, connotes differences between the certainty of finding the answers to all things against the infinitesimal, infinite and OITIO reaches of TRU and its KOT. It is partly solved by the set form of TRU, its SUT, order and Irreversibility of Valid TRU knowledge, **LOVKI**.

TRUTH and CERTAINTY

It is in truth and certainty when fully interpreted and adapted to TRU that much of KOT is found. TRU Equals Truth, **TET**, and Truth Equals TRU, **TRET**, are correct when truth is limited to participation in TRU. This is especially so by **LOPUT** or the Law of Proportionate Truth.(VAPO,TRIA)

FONAR = Fallacy of the non absolute of the relative. FONAR is largely a result of a lack of LOPUT by RIA and LASUR. When we speak of the relative as such, being relative does not mean that it depends, or is only partly, for the ultimate, the all, TRU, is relative to us only. This us is only the us of all that is perfectly TRU, the greater component of TRU being human. Being, ontology, being *qua* being, thing in itself are very much not TRU in that they describe the general condition. This is like demonstrating objects or relative universes in general, RUS. For, although basic to and having much in common with TRU, it is analogous to assuming that man is no more than animal nor a physical object. Certainly many of the characteristics of the physical and animal worlds apply to TRU. They, by way of their ESINS, are basic to human being.

LASUR or the Law of Sufficient Reason can only be successfully applied when based upon truth that is consistently TRU. Leibniz' principle of sufficient reason states that if enough truths are found and together reach an overwhelming probability then God can be proven. Increasingly sufficient reason from potentiality toward absolute and absolute actuality depends upon TRU, TIT, TRIT, OBT, SUT, and ROT. LASUR holds of relations in a closed set, as for TIT and SPALT, in which all that is the same or related provides an absolute limit. The availability of this relationship and reason will depend upon the nature of the entity, thing or object along with the sense or subject and their combined union. This called: **LAPSUR** = Law of Potentiation of LASUR. When all conditions are met and LASUR is carried to its fullest or absolute, it is called: **LASURÂ** = Law of Sufficient Reason Absolute. Specificity for a whole actually exists only in TIT, which is the ultimate in **TÄP** or the TRU Form Proper. For TRIS this will be approximated in TRINI. For only in TIT-TRINI, by TRIA, is there an absolute unitary focus and exclusive relativity in which all approaches the absolute static and relative actuality, RR. This is signified by: **LASURAC** = Law of Sufficient Reason Absolute Actuality. LASURAC delimits those instances of seeming relativity and the concrete upon which sufficient reason may approach the absolute in being explainable or demonstrable. Thereby for sufficient reason we have a spectrum or continuum of relations, uses and operations for TRU which depend upon the degree of this relation and the degree of use and operation. LASUR is not automatically valid for all we encounter, nor can it be used indiscriminately. Yet when defined and described by TRU it is the basis for much of POA and the use and operation of TRU. By DAOST all that is of TRU is more or less the same, relative. For all this more or less, the degree of relation in the continuum is applicable and can be to that extent absolutely available by LASUR and LASURÂ. To the extent these are available, and how hard and how well we bring all involved in each relation together, we can provide an ever better ATIM all time mastery of things, and product for TRU.

Derivation and determination are the products of processes of accumulative validation. LASUR is assisted by **LACOMD** or Law of Comparative Distinction and Definition. TIT is valid as it is in and of itself. **LART** or Law of Relative Truth signifies what is true in itself may not, and usually is not, true in TRU or TIT. Adequation is the relation, and interdependence of, truth to being, TRU. What is truth for TRU is different than it is for AU. Thereby we have **DEGOT** or Degrees of TRU, and these not only between TRU and AU, but between TRUS, TRIS, ES, and REM in TRU. It is in distinction that truth is found, as it is in differences that distinctions are made. As there are great differences between AU, RU, ATRU, AMTRU, TRUS, TRU, TIT, TRIS and TÂ so are there great distinctions to be made and different truths to be found. Although truths other than TIT-TO can have more or less actual TRU value it is primarily by TIT-TO that truth is to be measured. Validity in totality like VAPO is highly exclusive and achieved by the singularity and completed set characteristic of a well ordered differential totality. This is an exclusive validity of a relative universe.(LOPUT) **LEDTO** or the Law of E TO Distinction signifies the fact that the distinction - and truth - of a Part may be far different in itself, in relation to other systems in TRU, in relation to TIT-TO, to TRUS, AMTRU, ATRU and AUS. ES like ESINS and essences, as all entities will have different meanings and definitions depending upon their status and relations. It is in the Separation and UnVeiling of all

Veneers, **SUVUK** that differences, distinctions and truths are best made. This applies both universally to all that is hidden of TRU and particularly to individual differences. So does it depend upon a total resolution of paradoxes. There are selective, practical, conditional as well as relative and partial essences, modes, and aspects of truth. These are all reasons why language is so necessary, and a language that shows these differences, makes crucial their distinctions and truths of TRU. This is called **FITIT** or the terminology necessary to fit TRU, and communicate these truths, distinctions and differences. They are used by logical verification as in right addition and by verification in Plato's relation by sameness to perfect meaning and truth. The target and end result is an absolute probability of KOT and TRU.

Perfect certainty and knowledge, like LANERA, LAMOL, RR, ÂÂ, is not possible. Perfection of TI and TINI in the practice of ROT and MOT is a goal and end, not an established fact. Yet its target is **LACOF**, Law of Absolute Certainty is a Function of Focus. It is a function of:.**LISTA**, IS and TÂ or the difference between any relation and the actual relation, or absolute certainty in TIT and TÂS. Such perfection is approached with the certitude of TRU and its total differential and integral confirmation in TRU, and especially TIT-TO..

CIS, the conditional is, is proof of the Á of TRU, both its aseity, perseity, haecceity, and quiddity. This is called: **PROTA** = the Proof of TRU and Á, or the predominant mass-energy state of TRU. This fact, state, proof, law and resolution of the paradox of subject and object, are at the heart, crux and basis of all knowledge. It is how knowledge is or is not as it is TRU. How TRU is or is not as knowledge is of TRU. It is that state, level or epoche of philosophy which if not surmounted results in a state of arrest, causes us to be blind, ignorant and never attain the revelations necessary for KOT, COAT and TRU.

In order to have complete and proportionate knowledge, one must know , a. what philosophy is and is not, b. what science is and is not, c. what religion is and is not, e.g. what God is and is not, and d. all other great aspects and topics of knowledge, especially as they have closure in TREQ or TRU. Only then and as each topic, subject, field, school, system, religion, sect, science, and art is realized as they exist separately, mixed, and together as different aspects of the same thing, TRU, can they be free from error and diffuse the problems, conflicts, and misinformation and mistakes they produce.

RECONCILIATION of KOT

Order in the world, shown in TU, TIT, TO, TRIT, OBT, ROT, MOT, ALT, SPALT and all IPTRU, proceeds spontaneously from the relativity of TRU if its live product has sufficient coordination and KOT to maintain it. It is a universal harmony, and signified by, **LUPS**, or Law of Universal and Particular Symphony or harmony. This is the form, systematic unity and content of an ordered world, TRU, combined with the form and content of the observer, or EPI. This total unity, **OU**, is achieved through the singularity of the first person and the unity of TRU or TU. It is the relationship between the particular viewer and its object, the subjective and objective, THES and THEO, in TRU. We know, biologically, that we are limited. Why then, do we demand TRU as an absolute? This is called: **PIMAD** = Principle of Importance (to me) Unit, Absolute, Demand . PIMAD is our heritage, the state of being in TRU. It exists, like ATO, whether we like it or not, a must which can best be adapted to, PPMA, and lived in fulfillment and happiness. **LIACI** = Law of Illusion of Actuality and Actuality of Illusion, or Illusions of Actualities and Actualities of Illusions. Not all EPI is illusion and not all TRU is actuality. All EPI that is only generic or of ESIN to TRU is actuality, and TRU that is beyond VAPO, SCARL and TIT is more or less illusion. This can be seen as we compare and show AUS-RUS-RUTS-TRUS-TROS-TRIS-TRINI-TIT-TÂS. LIACI is important not only to show the greatness of an improved EPI, it helps to show what is and what is not TRU. For we can not take TRU, nor what is important of TRU and TIT, to loosely.

The error and confusion that has existed in knowledge and the minds of all people over the nature of god, God, GOD, and TREQ has not only caused them to fail to comprehend their world. It has caused them to fail to recognize TRU. So many of the greatest thinkers of history, scientists as well as philosophers and theologians have had a distorted, erroneous, or more often deficient and disproportionate understanding of TREQ or TRU, and thus their very being. This has been a result of: **POKOT** = Preventers of Knowledge of TRU. Although early in human development, primitive fears and superstitions - even fear of a primitive god - caused POKOT. In modern times it has been deficiencies in knowledge, and especially failure to order and envision the world in its fullest, as it actually is by VAPO, TRU.

God is all that is most beneficial and lasting for humanity; all known and not known, living or not living, able or not able as God is TRU. It has been from those of religions, philosophies, sciences and all that have held most explicitly and emphatically the world TREQ, like TRU, or this special world of best human life, **STRU** or The universal Standardized Relative.

PEDAT = Paradox Equals the Difference between AU and TRU, as between or Sections 0 and I. This is where and how what is separated, or is so TRU in one sense is also so non TRU in another. It will be in the other sense, as in being another thing, but in ways or completely entirely different, another and often opposite. This is resolved only as its AU becomes TRU, as the introductory section becomes section I. It is finally resolved only when exclusively TRU, totally and in the absolute TIT-TO and TÂ. This is how the resolution of paradoxes of all become and complete TRU, and are understood by THESA and OM in being not only TRU, in TO-TÁ, but the subjectification of TO-TÁ, completely, as by PEDAT.

MEANS of KOT and SOLUTIONS

As we learn more and more of TRU and what this knowledge of TRU, KOT, is, we need to make it as accurate, specific, sensitive and valid as possible. How the means of knowledge lead to principles are essential to an understanding of any topic, most certainly one of the magnitude of TRU. All that assists actuality is critical to KOT. One direct means of KOT is: SUVUK, the method and processes in TRU by which we can discover, learn, understand, remember and apply TRU by separating, unveiling, dissecting, and removing all that keeps us from apperceiving and recognizing it. KOT and concentration of thinking on TRU is allied to **FAST** or the Freedom from Thought of all that is irrelevant, and is called: **COTOT** = Concentration of Thinking on TRU. FAST and COTOT on a higher level are: **CACEXINOT** = COTOT through ACEXIN or the actuation, expansion and intension of TRU.

It is 'because it works' that is the connotation of Pragmatic Function. It is because TRU works, in and out reversibly, validly, intricately, and by detailed operation that TRU is successful and proven. This is an actual, or TRU, validity much more than an identity of subject and object, but only as it exists in TIT. How it exists as its prime and predominant dimension, its Á mass energy FAF. It is Á dependent. This actual validity, which is not simple identity of entities, and shed of all external variants is known as: **TAV** = TRU Absolute Validity. TAV must be clearly and quickly distinguished and recognized within its limits and not misused for all that is not TAV. So to must we subjectively dissect the subjective by making allowances for the many different senses, as 'in a sense', as well as many different categories or types of TRUS, TROS, TRIS.

VOSET = Validity is achieved only as All Entities, Senses and Things, as TRU, are Taken and Envisioned not Singularly, but in Totality. The completion of the total KOT and proofs of TRU are shown by closure, *credo ut intelligum* and *lef sum*. Closure is the completion of incomplete forms. It is the perception of continuity, subjectification and KOT toward patterned, coherent, stable, consistent, rational and comprehensive wholes. In the Representation of TRU by ROT, SU, MOT and ALT we have a solution in closure of TRIT, TIT and TRU. Closure of KOT and

Closure of COMT as in COAT, are the ultimates as provided by POT. *Credo ut intelligum* is literally "I believe in order that I may understand". It is the complete understanding which follows complete belief in TRU KOT and COAT. *Lef sum* are the limits of formal parts and the summation of total complete means and methods as they apply to COAT and POT. In the extension and application of knowledge, facts, concepts, principles, laws, and formulation, SCARLS must be applied. Something must be added, something must be subtracted for each type of thing. We must be careful, and in certain ways explain, demonstrate, understand, and restrain their use. All other tools, techniques, means, methods and systems must be so treated and tied together. This is called: **SUM** = Summation of Methods, Total Means. Lef sum, then, becomes the limits of formal parts and the Summation of Methods to COAT and POT.

LASRE and **LUSRE**, Law of Subjective Revolution and Representation, and Law of Utilization of LASRE, go a long way to provide proof of TRU and Á, PRUT and PRÁ. They do so by showing the finer aspects from both formative input as well as inner form. TRU as S, and Á change as TRIS, TROS, and TRUS, change. This is done by making TRU not more doubtful and thereby more valid, but less doubtful and more valid as it is perfected. This is called: **LAPEROT** = the Law of Perfection of all TRU, Proves TRU. This occurs as anything that is increasingly known is increasingly proven, at least within limits or SCARLS. This is: **LAVCAK** = Law of how a Law's Validity comes from increasingly Correct Comprehension and Knowledge. LAVCAK explains and in a way is exemplified by how in the legal world laws continually undergo change. This is by being incomplete and by error.

LATRU = Law of Absolute TRU. Whether we are able to trace the ultimate detail of what is, whether we are able to distinguish the exact magnitude of its relative parts, and whether we are able to trace the perfect formulation of its parts, are not all or none conditions or states. Nor is their totality an all-or- none condition or state. It is not crucial or critical. LATRU comes with TRIA and TET. It is whether the overall facts and principles of TRU are certain. Whether these and the function and form, processes, and parts of TRU are well defined and practiced. Whether these principles, laws, concepts are well developed, expressed and used as axioms, and their associated corollaries are similarly certain and usable. This is the relative and practical fact of TRU, its absolute justification. This follows TRIA as well as from objectification in general. It is not necessary to know the infinite detail, or in an infinite number of ways, of an object. It is absolutely certain of an object if it is sufficiently recognizable to give it such certainty.

Abstraction increasingly loses its validity when only of the sense or the thing. When fully developed as sense thing, and this sense thing is, as all, TRU, we have: **TROC** = TRU Correction. It is the lack of this means, or TROC, that leads us to the error of so much of past knowledge, and especially philosophy. **SPRECT** or Selective, Practical, Partial, Relative and Conditional Essences, Aspects and Modes of Truth. **DEPAV** = Derivation and Determination by a process of Accumulative Validation.

LOVKI = Law of Irreversibility of Valid Knowledge. As knowledge becomes more certain and complete, the point can be reached where we can see the light of all that is and is not. This knowledge attains a level of fixation and irreversibility that drives us on to ever greater knowledge, and ever more perfect completion. This is when the general form, nature and extent of all is attained and shown, then perfect knowledge can be surmised. This can be taken as our reference fixation or: **LORF** = Law of Reference Fixation. If all is directed by LORF and their corresponding fallacies corrected, all the lesser forms of knowledge will disappear. Especially those that are figments, illusions or do not contribute to **COMET**, the Comprehensive and Controlled Means and Method of Thought of TRU. **LAV** = the Law of Logical Verification. LAV is demonstrated by the Method and Means of Right Addition, **MERADEX**. This is where if something is known, and known to be partly right and partly wrong by TRU, then by comparing it with TRU, ES and CES or with other things, its nature and FAF can be increasingly determined. Both as it is or is not of TRU, as well as in other things. By added knowledge of, and comparing, entities, ESINS, ES and their functions, our understanding and determination are accelerated, refined, verified and confirmed.(COR,COL)

LASPRUT = Law of Summative Perspectives Prove TRU. When we bring all the different ISMS, systems, schools of thought, perspectives, religions, philosophies, and sciences, plus their methods, together they tend to combine, relate and converge on TRU. This provides great proof of TRU through their supportive evidence. The relations of KOT as it shows distinction and determination of TRU and how KOT can be shown itself through the limits of its application are demonstrated by the metalogical. This is described as those truths which are the Laws of Thought, those formal conditions of thinking are inherent in reason. They depend upon TRU, LAD, AXLA and ONTA. In this way it is **QUOL**, Qualified and Limited to TRU, and as logic and syntax to ROT. It thereby frees itself from all that is not TRU, nor ROT, and the cause of so much argument. The fact that TRU can be known, as any object, and need not be cut off from pursuit and discovery is expressed in: **LACT** = the Law of the Certitude of TRU. **DOTH** = the Degree of Truth, represents the extent to which our KOT, is or is not equal to TRU in actuality. The distinguishing, distinctions, and differences of TRU that can be tested and enable its proof are called: **VITO** = Validity in Totality. VITO, like VAPO only more exclusively, is achieved by the singularity and completion that are characteristic of a clearly differentiated totality. The method by which TRU can be shown, proven and tested through the enumeration, interconnections, and key dynamic mechanisms in operation of its basic concepts, principles, laws, and are expressed in: **LONECS** = the Law of New Concepts. This is: **PRUNES** =Proof of TRU by Non Essentiality, what can be done without.

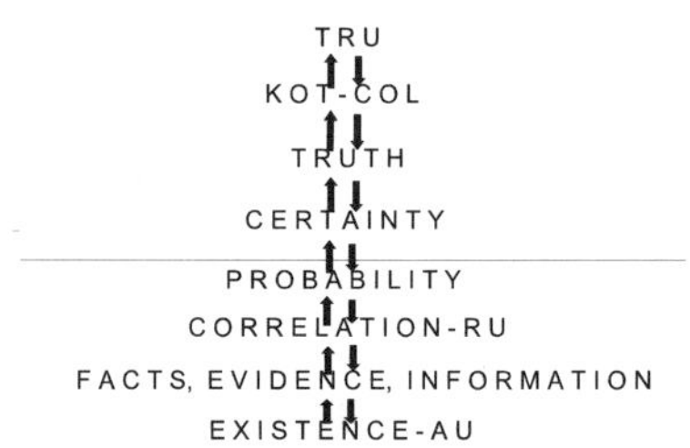

Fig 5-10 **MEANS to PROOF of TRU, PRUT** How we can collect, organize, order, coordinate, proportion, posit, and use all knowledge, is almost as great a way as in gaining and discovering the knowledge initially. It is the key to permanence. For to have knowledge that is irrelevant, unconditioned, without SCARL, and thereby produces confusion and failure, is little better than having no knowledge at all. It is in developing this knowledge comprehensively, in gaining COA, COAT, and KOT that knowledge bursts forth in all its glory, to our absolute benefit, as TRU. How BASFANT and SPO become SPOT, and how this text develops and shows the culmination of great efforts to gather information, discover knowledge and have the subjectivity necessary to best respond, answer, and solve all that confronts us. It is the natural sequential order of this text to demonstrate the relationships between the first person and TRU, and the subject and object as they are **OITIO**, Of and In TRU, In and Of TRU, as these are integrated and differentiated. It is the attainment of an overall and masterful view, a total comprehension, management and control, that has been the purpose, and to a great extent the end, of knowledge. In philosophy, as well as in most other topics, fields and sciences this is the main purpose of knowledge and subjectivity to test theory, develop theses, and successfully exercise the predominant powers of the best of our lives, TRU.

By **LOPA,** the Law of the Paradoxical Absolute we learn - as in science - more and more. The outermost, innermost, and infinitesimally detailed reaches of the universe are approached. We would like to find, as we seek, the secret of the universe, the riddle of life, the answer to all things. Yet as we will show through IPTRU this is not to be found alone so much in extremis, in media, or in relativity, but all together, as in VAPO of TRU. So that it is well to understand all things and especially all things in relation to TRU. For without this we can not obtain a good perspective, comprehensive knowledge, and representation of TRU. In this way we can resolve LOPA, by continuously lessening the paradox.

Without a belief in the absolute nature of our being and the fact that we can't be adequately judged by the natural, physical and biological worlds we can never gain a good overall vision. For, like ATO, our ends are not some illusive sense or thing, but the continuity of this best of human life, TRU. Thereby, within the limits of TRU, it is constantly becoming absolute by RIA, TRIA and its end, the eternal, becoming absolute.

In seeking a perfected proof theory, or the optimal proof of all, it is not only necessary that we Prove TRU, **PRUT**. We need to prove that associated with TRU, as ATO, A and ALT. *Attraction*, as *love,* is a readily recognized bonding that is also a good proof of TRU. They are readily recognized as the positive interactive relation that exists between many aspects and parts of TRU. One of the most basic proofs of TRU is from being dead or alive. It is only as we are "alive", alive and well, conscious, or can be attentive to the needs of life or whatever enables this life, that there is existence. Yet the dead, or alive that we personally know or can anticipate so deeply, is not the same as it is for TRU. For only as it is in TRU, and for us in TÍS and TES, is it alive, and does it provide a proof of TRU. Thereby it becomes evident that, as in being, the relativity of our existence is the predominant feature of life. Life and death are the most profound aspects and features of individual being, being TRU. This is not a simple existence, it is a healthy, strong, successful, and predominant existence. For, only in this way do aliveness and special existence, have assurance of continuity.

That great inner impulse and drive of human mentality is the sensitivity to and interest in life. It is to determine whether and what I am, rather than what I am not. This becomes so when we seek to determine whether or not TRU is, rather than TRU is not. Key to this is the proof of the existence of TRU as it has been of the existence of God. St. Anselm in *Proslogium* argued that being, than whom no greater can be thought, must exist by definition. For an unrealized idea is not so great as a realized one. This ontological argument although profound should not be misconstrued or overstated, without VAPO and the SCARLS of TRU. Yet since the subjective or idealistic pole of the mass energy dimension of TRU is its salient feature and predominant nature, it follows that in such a focus is to be found the ultimate proof of God or TRU. For it is this expanding subjective and idealistic pole, to be realized, brought through SÁ to Á that is astride the cutting edge of TRUS, TROS, TRIS, and TÂS transformations.

Many other proofs of God have been given through the ages, including these five proofs of St. Thomas Aquinas, (1). God as first-mover. (2). God as first efficient cause, (3). God as necessary cause, (4). God from perfection of goodness, (5). God as directed intelligent being. Along with these is the moral proof of God, this although less direct is more simply convincing by being what works, for us, TRU. There is a marked difference between *that* GOD or TRU is, and *what* GOD or TRU is. Although these are highly related and interdependent, proofs that show that God is although highly relevant to TRU do not necessarily show what TRU is, or God is in the actuality of TRU. Errors and loss of faith in the proof of God have been two of the greatest sources of God's denial and thereby loss to humanity. We must therefor firmly and accurately prove TRU and base these proofs on both that and what TRU is. This is the combined MEthod of Proving TRU through its Thisness and its Whatness, and thence together, **MEPRUTA.** MEPRUTA is closely related to the quiddity and haecceity of TRU, as in ATO, TÄ, FAF and especially the special Á state and dimension. **PRUTAW** is Proof of TRU, by both its Thisness or Thatness and Whatness, its Haecceity and Quiddity. PRUTAW differs from MEPRUTA in focusing on proving the kinds of thisness and whatness. To state that TRU is, is to refer or relate it from the outside, and this TRU is, is to refer or relate it from within. It is the difference between a that and this, this is called: **LATHIS** = Law of That and This or This and That in TRU. LATHIS shows us the importance of the subjective and IA origins of things. It also shows us the importance of how we enter into the existence of TRU. Other proofs of TRU are noted through the text, in the addenda, DIPTRU and IPTRU.

Many proofs of TRU are noted through the text, in the addenda, DIPTRU and IPTRU. Some **PRUT** that are especially revealing include, Ability-ies, Achievement, Action, Action-Reaction, Actuality, Adequation, ALT, Analogy of Attribution, Aseity, ATO, Capability-ies, CENS, Coherence, Centripetalization, Convergence, Combined Moment, Laws of Constant and Variable Proportions, Coordination, DAOST, **DEXA** or Determinants by Exponential Analogy, Dialectics, Differential Equations, DIL, Dimension, Dimensional Formulation, Dissociation Curves, ELAC, Eminence, ESINS, Essence, Extensity and Intensity Law, Extremes and Means, Focus, Form, Function, Geometry, God, god-GOD, Haeccity, Happiness, HOM, Homo Sapiens, Human, Identity, Inertial Energy System, Interval, Knowledge and Technology, LACEXIN, LANERA and LAMOL, LAPIVOT, **LARD** or Law of Absolute Distinction in Relativity, **LARDAC** or the Law of Absolute Relative Dependency, LARDAL, **LIDS** or Law of Integral and Differential Equality, Life, Life Chain, Location, Lorentz Transformations, LOSCARL, **LOSID** or Law of Solutions in Deficit, LUND, Man, Mass-energy, Mass Energy Transformations and Adaptations, Meaning, Means, Metatheory, MOCBIREM, ,'in Nature nothing is free', Normal Distribution, Organicism, OITIO, Perseity, PIP, E Pluribus Unum, Point of Reference, Polarity and Continua of TRU, POS and posits, **POSI** or Principle of Scopic Inertia, **POUC** or Principle of Ultimate Coincidence, Predominance, Law of the Primacy of Being, Probabilities and Statistics, **PURV** or Principle of Universal Validity, Quality, Quantity, Quiddity, Relativity, REM, Resolution from ATRU, SCARLS, Selection, Significance, SIL, SOLA, 1/Sophistication, **SOS** or Subject Object Standardization, Speciation, Species, **SRE** or Summative Reinforcement, Statistical Proofs, Stimulus, Stimulus-Response, Structure-Function Theory, Success, Success/Failure, **SUROT** or Summative Unitary Reference of TRU, Survival, Survival/Extinction, Thisness, **TIST** or TRU proves ISMS and ISMS prove TRU, TIT, TIT-TO, TO, Totality, TREQ, TRIS, TRIT, TRIA, TROS, TRU dichotomy or cleavage and divisions of TRU, **TUC** or the Universal Contract, Principle of Unity, Law of Unity of Opposites, VAPO, Varieties of TRU, Universe, Univocity of Being, Vitality, World.

These proofs work effectively only when their ESINS are applied to and incorporated into TRU. With many others the need to Justify and confirm TRU, **JUT**, are produced to equal greatness and replication. They help to reveal the discrete participation each and all has in TRU and PRUT. Working together they are shown by: **LARECIP** = Law of Reciprocal Proof. LARECIP signifies the proofs of TRU, Á, SU, MOT, ALT, SPALT, SPAT, etc. are obtained reciprocally from each other. As we provide more proofs of one key aspect of TRU, they in turn provide more proofs of others; some of them more and others less. This applies singularly, in combinations, totally and eventually as LACOP.

The qualitative way of looking at ones life, as by problem and solution, and good or bad, is only a simple view of what is a great quantitative continuum. A continuum all the way from and through ANTITRU, ANTRU, ATRU, NAMTRU, MATRU, Individual, ESIN, ES, TES, TÍS, TISÂ, TIT-TO, to TÂS. You can't have actual quality without adequate quantity. You can't have actual quantity without adequate quality. To understand the world, as it actually exists, TREQ, TRU, and TIT-TO, you must have increasingly qualified and quantified information, concepts, and knowledge. These must represent TRU in a valid, actual, qualified or proportional, and quantified or total way, VAPO.

As much as scientists may deny or defy God, TRU and religion, one of their most strongly held and powerful tenets is how observation means little without interpretation. Likewise neither observation nor interpretation mean much without determination by TRU or TREQ. This is fundamental to human ability, predominance, and A FAF, provided by the relativity, or observer-observed fulcrum in TRU. It is not any interpretation, however, nor even various combinations of related interpretations. It is only interpretation based upon all TRU, as evident in VAPO, DAOST, SCARLS, TIT_TO and ACEXIN. It requires an ever increasingly astute, and accurate development and use. This necessitates POT and ACC. One of the greatest of all tasks for such interpretation is the accuracy of POA and its posits or: **REFPO** = the Refinement of Posits, and its associated: **LACPAR** = Law of the Continuous Perfection of Parameters. It is in REFPO and LACPAR that we can finally produce that substantiation and perfection of TRU that replicate it and its absolute proof warrant. This is like the most intricate and beautiful detail seen in the most greatest works of art, and best maps of the world. How with greater sensitivity, specificity, reliability and precision posit all in representation.

JUT and LACOP together provide Sufficient Absolute Proof, **LASAP**. This is how the observer doesn't need to know every detail to realize what is there. Sufficient reason proves TRU as TRU proves and sets the basis for universal sufficient reason and certainty. PRUT can be achieved through the convergence of fields, sciences, and all knowledge as they exist in TRU. Both the Proof of Á and of ÁLT, **PRÁ** and **PRÁLT** prove TRU and vv. The subjective revolution, representation, and their utilization go a long way to provide PRÁ, PRALT, and PRUT.

To further confirm the proof of TRU and Á, there is the reversible proof of TRU by Á,, and Á by TRU, **PRÁT**. If there is no observer or subject, there can be no observed or object. Without either there can be no TRU. If there is no universe, or TRU, there can be no observer, or Á. That is, if things are disunited and chaotic the observer can't identify them as a whole, nor Á of TRU, nor can their be an Á, or a TRU. This has been called the final anthropic principle, which is misleading unless interpreted by VAPO, DAOST, and all TRU, TRIS, and TÂS .

A simple and most effective PRUT is **PRUNES** or Proof of TRU by Non Essentiality, or what TRU can do without. Taken positively it is a proof by what TRU most consists of, or is most vital.(REM) **PRUTHOB** or the Hidden Observer Proof of TRU denotes what is perfectly TRU that we overlook in all, only to put in its place much less, null or the negative. By **PROTOS** or Proof of TRU through Total Objectivization and Subjectivization we have a complete cycle return wherein initial processes of S, O, and IA fully respond in the observer observed relativity. Similarly **LARPROT** is the Law of Á or SO Reduction Proof of TRU. The necessary reduction of how we look at TRU, how TRU is, and how TRU is objectively separated. **PROTEL**, Proof of TRU by **EL** or the Exclusion Law. This shows how by increasing production of the limits of ES, CES and EL to TO we provide a formally distinct and substantially complete TIT, TRU, and TÂ. PRUT is repeatedly shown in the resolution of paradoxes as well as **PRUNI** or Proof by Negative Inversion. This can be compared to Proof by parts and characteristics of TRU, laws of objects, and proofs by the correction of errors, false beliefs, ideas, and tenets.

For the completion of KOT and TRU we are guided by VAPO, especially its proportions and tests. TIT and all entities and universes must be conditioned by their own special proportions, own nature and survival, as well as our subjectivity of it. Consequently and especially for TRU, KOT must be proportional.(REM) This also applies to all components, ES and CES, of TRU. More fully there is a law of all as only proportional degrees of TRU, evident in TRU being relative, all is relative and thereby a part of TRU. It is impossible to make proper judgement and take right action without proportional knowledge of the principles TRU. Thus action can only be proportional to needed concepts, principles, laws, and mechanisms of TRU.

Among tests of TRU, **TOT,** many of its proofs when confirmed can be reversed and used as tests. All that can be reduced to accurate measure of TRU can be used as test. Such is REM and **LAPREM** or Law of Á forms, dimensions, gradients and their Representation Measured by their REM. Completion of KOT and TRU is guided by **MEPAT** or The Method of Paradoxical Transformation wherein what seems so absurd, both great aspects within TRU and extra TRU, can when properly reduced to TRU, be managed, maintained and controlled. Completion of KOT and TRU follow completion of S-0 successfully fulfilling stimulus response. These involve **ATIM** or an All Time Mastery of Things, that comes with TRU and TES. By **BIV** or Bring Into View, full value comes from the clearest and fullest descriptions and expression. (SPAT). Completion of KOT and TRU are aided by CES rising above partials, S and O, to complete objectivity, OBT, SPALT and POS. Solutions are made by the exclusive constituents and aspects of TRU, limits of its FAF, and summation of its methods. One of the greatest tasks for completion will be the continuous perfection of the dimensions of TRU, accuracy of POA, and posits in MOT. By TRU, SUT and all, accelerated success in the actuated future is probable and can be achieved with a proper mind set.

The Proof of TRU, PRUT, is obtained with the knowledge of TRU, KOT, and more especially understanding the relationships of TRU by NEC. Each person alters ones position in **OAT**, or All not TRU, AMTRU and ATRU, to best return to TRU. How all these together bring one and all to TRU, how they revert to TRU is called: **VERTRU** = Version, Reversion to and Verification in TRU. In a sense AU, RU and even TRU are superfluous, for TIT is all and all is TIT. Yet, as knowledge of all is, more or less, necessary for the individual relation to TIT, so does knowledge take some precedence, seen in EPI. This is a paradox and fallacy that, if not properly held in mind with others that qualify TIT and its observation, are less able to discover, know and show TRU. This is: **LOPRE** or **FOPRE** = the Law or Fallacy of Perspective Precedence. FOPRE is a part of UPTRU, and other LOLS.

For further proofs see Figs. 5-19 pg. 147, 4-49 pg 282, and lists page, 141, 200, 244, 264, 281, 328, 338, 370, 426, 493, and 524. Throughout the text, and as enumerated at the beginning of the text, pg vii and at the end of Chapters 10, 20, 30, 40, 60, 70, 80, 90 proofs are given for the major topics, themes, and concepts of TRU. The concrete closure these proofs provide is vital for each person to learn if they are to successfully fulfill their lives. Along with the laws of TRU they provide the basis for a guide that can bring us from confusion and chaos to a most efficient and productive life.

Figure 6-10 provides a sample of the great many interrelated proofs of TRU. It presents them not only to confirm the theorem of concrete closure of all in TRU but to show what can be done with the much deeper and greater interdependence all has in the unity in totality of relativity in TRU. By DAOST they are all different aspects of the same thing, seeking to achieve the maximum in differential distinction in the maximum of integral unity in TIT-TO. Each topic and concept must be held to their explicit meaning, state and role in or of TRU. In this drawing we can clearly envision the great part VAPO, ATO, DAOST, SCARLS, TROS, TIT, TRIS, TRIT,

TRIA, TO, ALT, REM, POS, PIP, and HOM play in the proof of TRU.

Fig 6-10 **Proofs of TRU, PRUT: LACOP and all TRU**

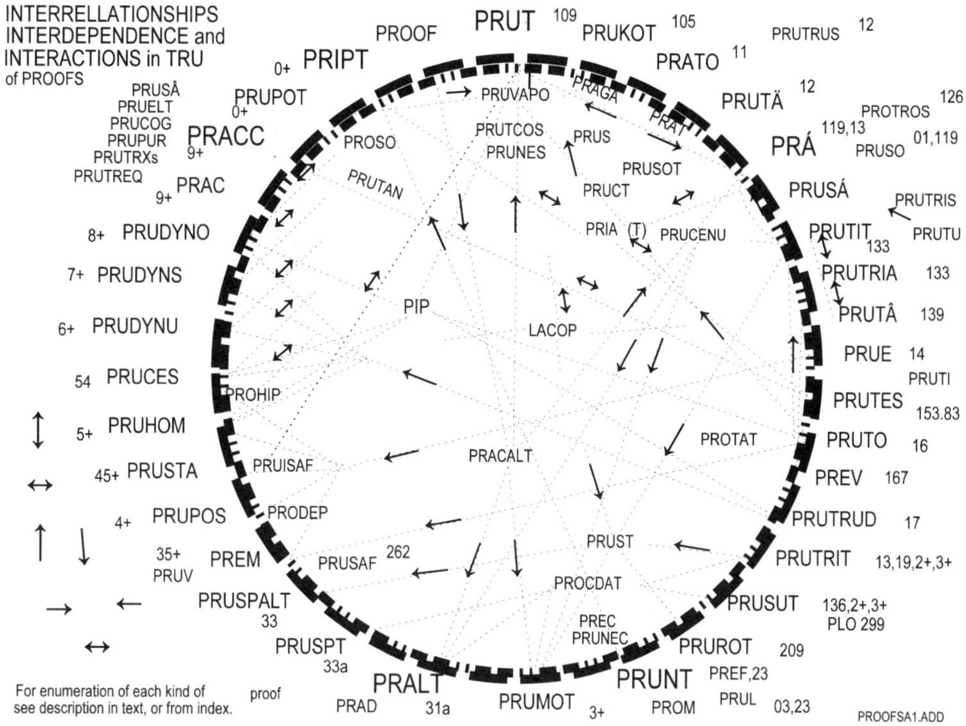

The proofs of TRU, **PRUT,** arise from existing concepts (OC), the essences or ESINS of existing concepts (EC), and all those proposed new concepts (NEC) that especially support TRU. , and all these by concretion with closure in LACOP. They include: Á, ACC, ACEXIN, AIT, ALTn, AT, AXTOPRE, CES, CIS,, COMPRU, CORMPRU, CORVS, CRAED, CRITO, DOCIR, EC, ENT, EPI, ES, FAENIT, FAF, FEP, FITALP, FOSA, HOPRUT, IMPLIT, IMPT, IPT, IS, ITOD, JUT, KOT, LACEXIN, LACHES, LACOC, LACOF, LADAC, LADENT, LADOT, LAMUA, LAMUP LANEA,LANIK, LANSUC, LANT, LAPIVOT, LARFOC, LARECIP, LARNU, LARP, LARPRO, LARPR OT, LASURAC, LASURT LATA, LATAD LATED, LATPED, LATRU, LESM, LEXIT, LEXPON, LINTAC,,LIPCO, LIR, LOCA, LOLVAR LOPAN,LOPTUR LOR, LORAD, LORF, LORTAI, LORVA,LOS, LOSUR, LOSPRE,LOTA,LUT, LUV, MADAS, MATES, MET, MIST, MOCBIREM, MUPT, OM, OT, PAR, PATAR, PI, PIC, PORA, POSFE, POSTE, POT, PRACAL, PRAST,PRAT, PRAVEN, PRIND. PROHIP, PROTA, PROL, PROTEL, PRUE-S, PRUF, PRUH, PRUNES, PRUNI, PRUNT, PRUTAD, PRUTAW, PRUTCOS, QUAR, OET, RECIP, RECTRU, REV, RINT, RITES, SÁ,, SCEN S/F, SOF, SPALT, SRALT, STA, TÄ, TAFOR, TAKT, TAN, TANCOW, TE TEKT, The, TIKAN, TIR, TRIF, TRIP, TRUCH, TRUD, TRUF, TRUP, TRUS, TU, UNT, VACOST, Value, WIR, WITÂ, and all these by closure in LACOP.

PRUTRUS TRU, DOT, TRUCH, PERT, OITIO, POT, AUS, RUS, RUTS, S-O, Serial, Continuum, Thisness, Special, TREQ, Life, Human, God, SOT, Á, DAOST, Universe-s, Relative, The, TRUP, TRUF, KOT, REM, LORVA, Change, Meta, Metasystem-s, Organ-ic, Predestine-ation, Universal-s, World-s, LUT ,NAT, MATRU, LATA, TRE, TRA, ROMAT, IND, TRUD, LASCER, TRUQ, RU-S, RUT-S, LATRAS, LACOTA, LACH, TRUST, PRAST, PROEST, SIL, FLOITIO, AHPET, DRAUT, LATEVU, and LIDE.

PRUKOT TRU, TIT, Á, Sense, Subjectivity, SO, SOT, Knowledge, God, TREQ, POT, PERT, COAT, DAOST, VAPO, LARP, TO, TU, TE, TI, TÁ, TRIA, ACEXIN, ATO, FAF, COGNOS, RUPKAL, LARF, CENU, Orientation, Coordination, REM, LORVA, ACC, LOPA, AIT, Love, Identity, POS, EPI, TRIT, TÄ, SUT, TRUD, Human, SCARLS, ES, CES, TES, Universe, Relative, The, MATES, ENT, OM, PRUT, Meaning, PRUNT, LANT, EC, MADAS, ALT^{n,} ACEXIN, LATANOS, LARP, ALT, SPALT, MOCBIREM, STA, HOM, LADAC, PRALT, HOM, AIT, Appreciation, Base-ic, Certain-ty, Compare-ative, Conceptual, Consider-ation, Discover-y, Gist, Idea, Inspiration, Learn-ing, Metaphysics, Religion, Rule-of-reason, Truth, Why-s, Obscurity, LANIK, TRE, SURAPT, LOVKI, LARECI, PORI, AKA, METO, LARD, AT, LASURT, EA, LASUR, PIKER, LORET, PRUF, LATO, LANOP, KOTIP, TRUT, LACTRU, LANSOT, LOR, PAT, ISNIS, SUVUK, VACEP, RORRE, LOLVAR, COK, COAT, COTRU, POMER, LIVROT, LIR, NECESK, TAFOR, PROHIP, LACOMA, PORA, KOTA, LAUTH, ATIM, MABIK, RAKOT, ANKEF, LAFIRS, and LASORT-TES.

PRUDOT
POTDE, SUVUK, SEPTRU, ATO, LOR, CONVET, CONRIT, MADET, LID, PURT, DEPAR, PEAP, ISTRU, POPTE, MET, ACOIT, LAPAR, LADUT, LAPAC, ACORT, LIDRAS, RECT, LASUR, LACPA, POTRU, DEPAR, PEAP, TRACAD, POPTE, MESDA, ARSOC, PERCIT, MERAR,

PRATO TU, TIT, S O, SO, SOT, Á L T, 1/TRUS, Thisness, Special, Life, TRIT, TI, TÂ, Immanence, 1/Evolution, 1/Creation, God, LARF, ENT, OM, TRUP, TRUF\TRIP, TRIF, KOT, PRUT, TIT/OT, ALT^{n,} ACEXIN, LARP, LORVA, Actualization, Conscious, Alteration, Becoming, Continue-ity, Fortune-ate, Happiness, Health, Hope, Identity, Sufficiency, Á, RIT, LESM, LOPOB, RAIN, RUDE, LODIC, LOCU, TRIA, TA, LARAC, ATIM, LOVKI, TO, LAREST, ACI, ATOME, SOLIT, POCI, PACT, PINT, PROCLI, TAP, LACS, POES, LAVMA, LAI, CONST, CRUT, LARAFS, LADSEF, CODIT, TREAC, SPOIT, LANEA, POBIM, DUTS, INTD, MOMET, LOIM, TATOR, PAFCOD, BIDYM, MEREPT, SOF, TATOR, REV, CAORT, PED, LACTEL, LADUT, and POCES,

PRUTÄ (PRUTÄP) TRU, TIT, TO, SOT, A L T, ALT^{N,,} Thisness, Special, Life, TI, SCARL-S, DAOST, Universe, Relative, LARF, OITIO, ACEXIN, LARP, LADAC, PRALT, FAF, Aspect, Attributes, Condition-s, Limit-s-ation, Construct-ion, Fit-ness-est , Form, Formal distinction, Haecceity, Iso-tone-ic-icity, Mode, Need-necessity, Quality, Quiddity, Relation-s-ship, State-s, Structure, System-s-atic, QUOL, HOMALT, LAD, LESM, LOL-S, CARSC, COL, TREL, DAT, GMLPDF, LOL, LATAN-O-S, TAN, LIT, IMPT, FITALP, TUR, TOL, STAXNO, ACORT, LOSAT, ECOTRU, UMIA, AMETA, LOFOR, MINESA, LALTPA, CEF, LEVADI, DOUP, TRUST, LASYM, UFAF, LAL, STIT, LART, POCOL, LACIN, LAM, CAORT, PRUTAM, POC, SAD, LATAN, LADAS, and LAPAR.

CHAPTER 11

DERIVATION, DEVELOPMENT, DETERMINATION, and METHODS of TRU

How does an entity or universe become the entity or universe and how do these become TRU? Is it only through our observation or relativity? Entities, universes, and essences exist and function both in and out of TRU, only in much different ways. Essences and ESINS increasingly become the modes, ES, and CES of TRU. By TRU figments, illusions, and irrelevant entities are invalid. What is not transformed to TRU and of DAOST is ENIT, not TRU, a non ES. Entities reverse when transformed, from ENIT to TIT. The reversal of the all pervading laws of extra entities, how they become different ENIT, SENIT, OBIT and reality error, neither all nor valid, not TRU, is the revelation of illusion. This follows doubt from inacutality, illusion from ignorance or lack of knowledge. Such reversals, by S or O, S-O, SOT and Á prove TRU. If one would doubt differences, DAOST and TRU, let them consider consciousness and the difference between the conscious and unconscious, being asleep and being awake. Herein we can see both the great absolute nature of our being, TRIA, its disitnct differences, and the role of S-O or Á in the dimensional FAF of TRU. By separating the object from its entity we begin our action toward TRU. This helps us learn how two or more entities can't be the same, at least to that degree. What is less than their part in TRU, as well as their selective and relative participation, is shown by: **LAREST** or Law of Actuality from Reality through the Selective powers of TRU Entities become TRU. For reality is in error both to the extent it is not TRU and not the intended agreement between observer and observed, subject and object. There is always illusion and figment in all, all we confront in life, largely depending upon how little we lack focus, especially on TRU. The full transit is from naught to what actually is TRU. Chance and chaos signify states in AUS and non TRU without relation. Like **DOT** or the Derivation, Development and Determination of TRU, and the range from AUS to TRU, there are many intermediary steps to TRU.

For DOT there is an accretive and actual expansion to TRU that tends to accelerate. By the thisness of RUS, RUTS and TRUS it has increasing self propagation. Special TRU derives from TRUS, RUTS, and RUS. **SEXÄT** is the Simultaneous Form, State, construct and SU EXpansion of TRUS. It denotes that aspect of TRU, as much of TRUS and RUS, in which change, growth and progress occur through their simultaneous proportionate and interdependent expansion.(VAPO) Much of what is thought of as origins, what came first, beginnings, and derivations are illusions. All must be interprreted in terms of TIT-TO to give a reliable DOT for TRU.

Fig 1-11 **AU, AT, TRU, UNIVERSALS and SOLIT**

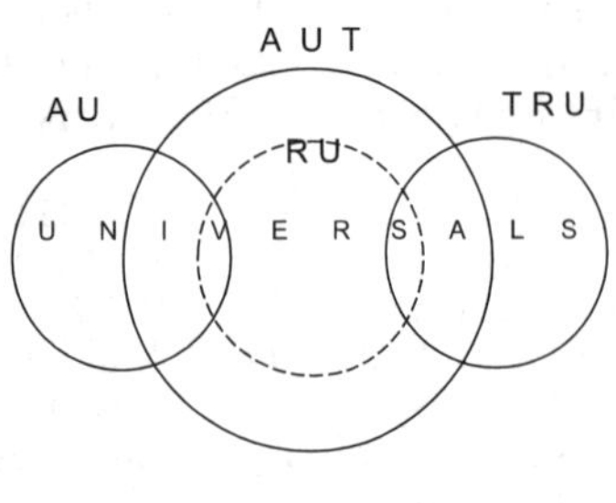

That TRU is in ways relative or common to AU may be shown as a composite mass. Such a diagrammatic rendering of AU and TRU must be quickly qualified, being objective rather than being what AU and TRU are in and of themselves, or variations thereof. Although many major courses and segments of TRU are both directly of TRU and only potentially for TRU, and although they may be potentially against TRU or potentially neutral or un-actuated of TRU, these do not disallow or disprove the solidity of TRU. This is called: **SOLIT** = Solidity of TIT, TRU in and of itself. There exists in TRU many forces, structures, components, categories and forms providing external potential, actual, relatively adequate, cohering, binding and strong features. These are its needs, requirements, makeup, form and function that are repeatedly reinforced and self promoting. Just like 'the ties that bind' in God and religion, just as knowledge reinforces, and all our physical and environmental health promotes, so do all that is positive for TRU assist in helping to make it strong, solid and self sufficient. As a biological unit, by RU, TRU. ATO, and TRIA, TIT exists selectively and separately in differentiation, and totality from all that is less, other, or negative to itself.

Sequential progressions may start from different points. Most derive from outside of TRU rather than TIT. The focus, unity, orientation, standardization, and vital necessity to have S/F in all that is attended to, is **LOSTAP**. It overcomes multiple, different, irrelevant and disorderly universes. Yet some extraneous derivation is warranted, if for no more reason than LAMOL, or where 100% TRU is lacking. This is seen in how the surety of TÂ depends upon contracting time, an accelerative intensity and actuality within a limited extensity. Without this there is an equilibrium between AUS and TRU. DOT is characterized by **VASTI** or the Vastness of the TRU-AU Interchange, **ÂPT** or AU TRU Proximity, and **MANT**, Mixed AU, ANTRU, ATRU, AMTRU and TRU that is all about us, the world we commonly live in.(OITIO) This is accompanied by relations, continua and scales or gradations. This, in turn, reveals AU by TRU disproportion. Thereby, **ATA**, Anything other than TRU to be considered or viewed or held as an object with or in conjunction with or compared to TRU must be corrected, to an equal and opposite factor, if it is to be valid and TRU maintained.

ATO or At Once TRU is one of the greatest of all TRU concepts. It signfies how DOT, when judged by TIT, must be viewed. As well, it is much of the aseity, perseity, and haecceity of TRU. For TIT it must be the starting point, for it establishes the initial conditions. **LATOVUCH** or the Law of At Once Total Derivation of TRU, is explained, justified, and validated by TRICH or changes only within TIT. For TRU what is prior and posterior are already in TIT, and imperatively interdependent. These have precedent in covering law theory. It is charged by self accretive Á movement in the FAF of TRU. This is evident from **COPOFT** or the Compelling Power of Future Thought, absolute spirit, and COGNOS in TIT and TÁS. ATO **is**, for to have meaning it IS, and to be naught, as before or after or other than what it is in TIT, is meaningless. The progression of DOT is seen in the rate or ratio of ATO over all that is not TIT, from TRU to RU to AU. **PRATO** or the Proofs of ATO are to be seen in the above descriptions, concepts, figure, and TRIS, TIT, TRIA, TÂ and associated TRU proofs, PRUT. ATO is also evident in contrast to evolution and **ATODIC**, ATO Creation Distinction. The relation between DOT and ATO is seen in **TÀT** or TRU Alternate TRU, where DOT and TRU are viewed in progression through TRUS and TRIS to TIT and TRIA. **RETRUS** or Reconciliation of Relative Universes signifies how RUS to TIT come together in closure in TRU by allowance and adaptation KOT, DOT, and all as long as it produces no irreversible change or damage to TIT. For the individual to reach an understanding of ATO, KOT and TRU, although RETRUS naturally tends to TREQ, it is more or less AMTRU or ATRU. This is similar to **STRU** or Self TRU, a life long endeavor to rise in the progression to PUREACLITY,

TES, ATO and TIT. In this way ATO helps to impart a kind of immortality on a person. Likewise even, when TRU, an individual can not, without imagination, be the starting point, no more than ES or TES is TIT-TO.

Fig 2-11 **DOT CONCEPTS (DATO)**

To make the derivation of TRU clearer, especially the interactions and relations of ATO with evolution and creation, the accompanying diagram is given. Its great significance is in helping overcome simplistic ideas of origins and beginnings that have so permeated and defeated human knowledge. By correlating the progression from AU to TÂ through TIT, with their cardinal and fractal aspects or initiating a qualitative and quantitative showing we can more readily achieve credibility.

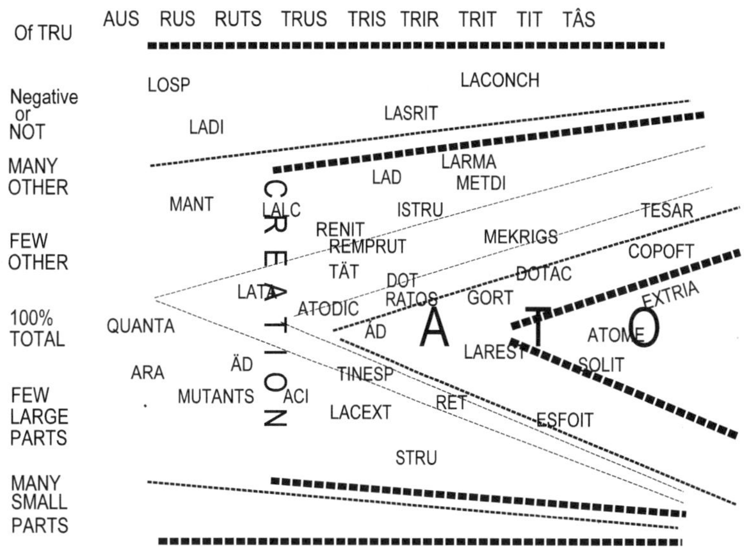

Origins, as much as we seek them, if they are of DOT by TIT, are apt to be an anachronism or illusion from the very nature of the given in relation to TIT. Thus there will be origins, non and mixed origins depending upon the corrections and resolution in which they do or do not apply. We can get a better understanding of origins, emergence, DOT, and ATO when our consciousness, IA, and our mentality encompasses the whole ACEXIN nature of TIT. The origin of TRU is evident in the relation of the negative to positive of TRU. There is a, largely indirect, reciprocation of the negative within TIT, as well as a natural tendency of IA and Á to concentrate on stimulus response and problem solution. Thereby origin and DOT occurs largely inside TIT by ATO rather than evolution and creation externally. Origins and ATO, for DOT and TRU, are to be found in the contrast between absolute and relative. The only absolute is relative and the only relative is absolute.(LANERA) For TRU this is RÂ, not ÂÂ nor RR. It is RÂ, like TRIA, that determines TIT, ATO, DOT, and origins when in or of TRU. Much of this arises from the relativity of TRU, our special thisness, haecceity, or S-O, observer-observed, Á FAF. The transformation from any relativity to that exclusive to TRU is **LART** or the Law of Relativity in TRU, and its changing states is **LARAR** Law of the Reciprocity between Relativity and Absolutism. In any full description and interpretation of relativity we must recognize the aberration of external S-O and DOT of TRU, and always add the necessary correcting factor to bring it back to TIT. This is **TER**, or The principle law of TRU Relativity. All that comes together through convergence on TRU of any relations that become the relations is proof of TRU relativity.

Fig 3-11 **BLACK and WHITE HOLES, DOT**

As we look backward, at what we came from, we see a vanishing point, a black hole. As we look forward, at what we can become at best, we see an ever greater shining light, as if coming out from far inside a tunnel, a white hole. What life is, as it actually is, TRU, is ourselves, subjective and objective together as the Á predominant dimension and FAF, along with a similarly and uniquely contracted and adapted form of length and time or spatiality and temporality. Humans are entering a region of a white hole where their exploding expansion of capacity and power is increasing at an exponential rate. This is more from our mental and effective powers than from the rest of our mass energy, Á. Yet since SA must balance form and function, an equilibrium and optimal set for all in its continuum, the variations in this set and equilibrium need to be accounted for, managed and controlled. This, more than a natural and optimal cyclical or reversible variation. Such an extreme exploding expansion toward a white hole, is opposite the imploding extreme of a black hole. Both opposites of exploding and imploding alteration of the optimal equilibrium and set of TRU require an assurance of stability and continuity in which PPMA of violent or harmful deviations are ever presently met. It is a homeostasis in which: **Imploding Black Hole → TRU, PPMA, TRIA, TÂ ← Exploding White Hole**

As either TRU in exploding expansion, or imploding contraction becomes irreversible and acceleratively departs from its status quo in TIT-TO so are IPT, POT, VAPO, KOT, DAOST, ENT, TES, ROT, POS, HOM, ACC, PPMA, COGNOS, ELTICMA, SÂ, TÂ, and all that upholds TRU vitally necessary. They must be discovered, learned, perfected, applied, used, served, and promoted in all ways that best assure this special universe of best human life, TRU. The law that mandates such action to manage and control the set equilibrium of the future and overall existence of TRU is called: **LEXIMT**

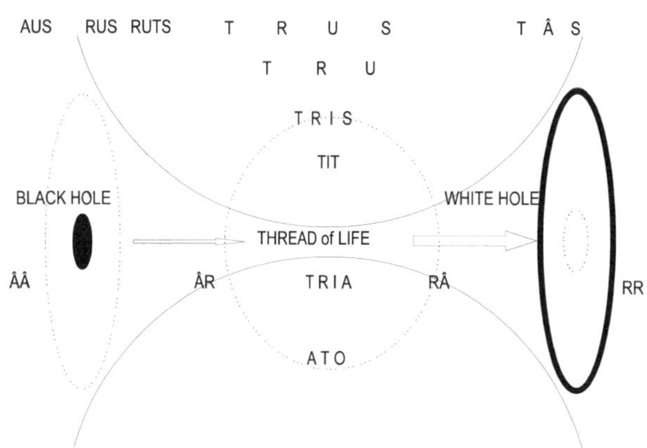

or Law of Equilibrium of Exploding and Imploding TRU.

DOT or the derivation and development of TRU is different than and only a beginning of what TIT is. For it is largely illusion or figment. What TIT actually is, is self contained in closure, by TRIA and **ATO** or At Once TRU. In being ATO it is neither coming from something other than itself or going to something other than itself. It is replenishing itself continuously, dynamically, massively, and totally with optimal ACEXIN through its own FAF, state and powers.

Whereas what the actuality was in and of the entity itself is directed elsewhere, and is not TRU. This is where, throughout the text, as in all our lives, we must constantly be on guard to correct all that is other than, less than, or even to the least degree not quite TRU. This we do by COL and COR and how entities are redirected from their own SCARLS to become actualized in TRU. In actuality the method of TRU is manifold. This is called: **MAMET** - Manifold Method or Match-up of TRU. Just as all elements in TRU partake of its dynamic whole in ways that are bent or warped to TRU, and are thereby often different than we would expect from their own natures or our interpretation, so is evolution. This is, as all senses and things and the dynamic processes which occur in the modified form and state, by and in TRU. **BURP** - Bent or modified Universal Processes, dynamics and evolution. BURP, then, takes on characteristics of their own, and leaves or departs from the form, states, dimensional levels, modes, essences, and characteristics of entities, senses and things. It is not the same as the evolution that occurs in the scientific and physical worlds. The mass energy dimension greatly changes with mans derivation and development. e.g. the different planes of the m, l, and t, or Á, L, and T at different times or periods in human evolution. This is a Proof of Á, **PRÁ**. It gives a better understanding of the actual what and how of DOT, TÄ, and TIT. It also gives the following: **LARACT** - Law of Reverse Actual Form of TRU. LARACT denotes that form and the Á dimension in which it is not only different in different RUS, RUTS, and TRUS, is changing in each. It is also, from the point of view of TIT, to be envisioned in reverse. That is, it becomes less and less like the Á of TRU, the farther back we go into the evolution of man, mammals, vertebrates, animals, and all life. Thereby they are less than TRU. This shows how thisness, TIT, TRIR, and TÂ produce ATO, and how TRU is derived, DOT.

Although ATO sets the origin, derivation, development, determination and demonstration of our world, this does not mean that time is not of the greatest importance, nor does it refute the fact that TRU changes, or even that there is change and variation with TRU in and of itself, TIT. For every moment even TRU is a somewhat different and new entity. For TIT this is largely internal rather than external, indirect rather than direct. It is external only over long periods of time, and then largely reversible.(TRUCH, TRICH, TRIV, variation, t1, t2, and t3). So that to properly and fully show TRU we must account for not only one or two principles and mechanisms, CALP,- whose roles in TRU are indispensable, but for as many as possible whose integration and interdependence must be known and shown along with KOT, DOT and ATO, of TIT-TO, in MOT, POS, HOM, and ACC.

The great modern controversy between evolution and creation both highlights the worst of ISMS, and the best of interest in basic mechanisms of life. Most characteristics of life are being Always and Constantly Creating, **LALC**. Closer to DOT, ATO and TIT is accretive internal creation. It is an autonomous accretion rather than an emphatic type of biblical creation. The major error in creation, evolution, as well as much of current knowledge is seen in **LANEXMO** or how the Non Existence of Components of Extremes and Means Negates the Existence of the Whole. Closely associated is the attribution of erroneous proportion, magnitude, and formulation to TIT. This imbalance and deficiency is best seen in lack of, or improper focus, identity, and speciation. TIT has great powers by **POCI** or the Principle of Compulsory Identity, and **IMTRU** or all that is Imparted to, in and by TIT. The focus, unity and exclusion of TIT are proven by speciation, **LUSP** the Law of Universal Speciation. A RU, and inceasingly RUTS, TRUS and TIT do not become, nor consist of, singular and partial factors, reasons and causes to explain their universality and speciation. They consists of all that exists within and converges on it, which for TIT is signified by TRIA.

	AUS	RUS	RUTS	TRUS	TRU	TRIS	TIT	TÂS
EVOLUTION (See Focus)	RANDOM CHAOTIC to	FOCAL, TYPICALLY BIOLOGICAL	BIOLOGICAL CONVERGENT ON TRU	INCREAS-INGLY SET PROGRESSION IN TRU	GENERALLY FIXED MECHANISM IN	SOME VARI-ABLE MECHANISM IN	ONLY as a FIXED MECHANISM IN	MOSTLY IN SOME GENERAL IDEAL OUT
EMERGENCE	RANDOM ABSOLUTELY VARIABLE	GENERALLY RELATIVELY VARIABLE	SPECIFIC RELATIVE VARIABLE	DECREASING EXTERNAL. WITHIN	LITTLE EXTERNAL MOST WITHIN	NO EXTERNAL SOME VARIATION WITHIN	NO EXTERNAL ALL WITHIN	ALL WITHIN CONTROLLED OUT
CREATION and Types of	VARIABLE, NONE or INDEFINITE with TYPE	REPRODUC-TION, INCIPIENT	PROCREATION INCREASING WITHIN	DEPENDS ON TYPE and INTERPRET-ATION	→ GENERALLY	→ INCREAS-INGLY SET	→ ONLY IN TIT SCARLED by	→ HIGHLY SCARLED by
ATO	NOT APPLICABLE	INCIPIENT	DEVELOPING	MATURING	GENERALLY ATO	INCREASING-LY SPECIFIC	ATO FULLY OPERATIVE	ATO GENERALLY OPERATIVE
FOCUS	NONE and VARIABLE	GENERALLY RELATIVE	SPECIFIC RELATIVE	→ INCREAS-INGLY FIXED	→ RELA-TIVELY FIXED	→ INCREAS-INGLY ABSOLUTELY FIXED	SPECI-FICALLY ABSOLUTELY FIXED	GENERALLY & IDEALLY ABSOLUTELY FIXED
OUT of or IN TIT	ABSOLUTELY OUT	ALMOST COMPLETELY OUT	GENERALY IN, SPECIFIC-ALLY OUT	INCREAS-INGLY IN	INNER SPHERE ABSOLUTELY IN	INCREAS-INGLY ABSO-LUTELY IN	ABSO-LUTELY IN	RETAINS AB-SOLUTE IN, REVERSIBLE OR CON-TROLLED OUT

Table 1-11 **EVOLUTION, EMERGENCE, CREATION and ATO in COMPARATIVE WORLDS or UNIVERSES, TRUS, TIT, TRIS, TÂS and DOT**

It will be recognized that evolution, emergence, creation, and ATO all have their place and roles in TRU. To better understand creation and its relation to God, TRU, DOT and ATO, it should be divided into its various types: accidental, arbitrary, emphatic, accretive, purposive, divine, artistic, as a mechanism in TRU, and others. When made distinct and developed, along with Figure 4-1 1, an interpretation of each and all together in the above table will be more meaningful.

GORT = Gradations of Relative TRU, TIT to RET to TRU, TOC. GORT is related to: **TOC** or Time Continuity of TO. **RET** = Return to, Reformation of, Replication of RE or TRU to TO. It is impossible to say that this is RET or this is TRU for to be prepared for what might have been, or that which might be, or will possibly be, as calamities of the future, is not certain. We may have RIT, a relative universe in itself, or, RET one in transition to TRU, or TIT itself as this is reversed we have GORT.

In Figure 4-11 we seek to vividly show the relations involved in the derivation and development of life and humans as it relates to DOT. It helps compare and visualize evolution, creation, and ATO. It helps show the biological, genetic and population genetics basis for TRU. It helps to provide a better basis for the applications of ACC in the PPMA of TIT and TRU. Cladistics, shown as the crow foot, looking backward and forward from the present, is the study and science of determining the proband or human individual and species ancestry and progeny. Phylogeny, in contrast, shown as a parallelogram, looks toward the present from the past and future, and is the study and science of determining the development of biological class, human species, and individual or proband. Ontogeny, like cladistics, signifies the development of an individual organism, producing its current existence. Ontology is the study, science, or knowledge of being or existence. It becomes special in TRU, and most definitively so in TIT. Here it exists in its exclusive and autonomous state of derivation and development, ATO. Outbreeding and inbreeding represent the extreme genetic differences and extreme genetic sameness. This has been, and is, highly dynamic and variable depending upon many factors of which mobility and contact of different populations is a major factor. Optimal selection involves the best of each, with prolonged excesses of either apt to accentuate loss of viability, fitness and deterioration of genetic material. Present day outbreeding, without past sustaining levels of selection, and associated with increasing genetic defects will - in the long term - pose an increasing threat to TES and TRU if not best PPMA, managed, and controlled. Here we can see the great and relatively rapid changes occurring in TIT revealed by DOT, the validity and importance of TRU and ACC.

Fig 4-11 **CLADISTICS, PHYLOGENY, ADAPTATION, REPRODUCTION, ONTOLOGY and TRU**

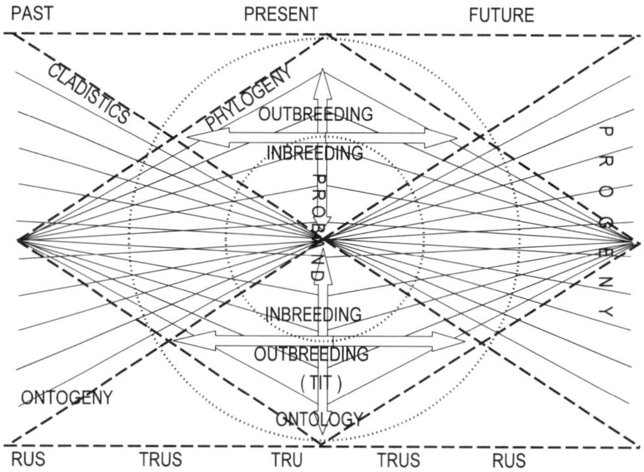

. Cladistics and evolution show the massive interaction and continuum of centering and decentering that occurs in continuity, increasing time, and with OITIO. In the proband, if TES, it is pure TIT. It is accompanied by continuous and massive internal, and when necessary and mostly reversible external, adaptation. Such adaptation consists of a great amount or orientation and coordination. Although orientation is largely a spatial reaction, both it and coordination, taken in their full implications, are also temporal and real or by the mass energy FAF, Á, in TRU. The ultimate in adaptation is **LATAD** or Law of TRU Adaptation wherein no topic, philosophy, science, religion or means is to be applied to TRU as derivation, participation, perspective, method, or test without being first adapted and converted to its equivalent in TIT. In contrast to the dynamics of centering and decentering, DOT can be inferred from the other extreme of space, time and mass, black and white holes, and their means in TIT. How this is altered in the specificity of TIT is by BURP, LARACT, and MAMET.

DEVELOPMENT of TRU

The development of TRU is a closely integrated aspect in DOT, a phase in the progression from origin to actuality, between derivation and determination. Although commonly used in conjunction with growth, it carries more general and, for DOT, greater implications. It retains many of the caviats and modifications we see in derivations, ATO and TIT-TO. It combines derivation and development with the thisness of TRIA and ATO in TIT. When Departed or Deviated from TRU, as in external observer and observed interpretation of DOT by evolution and creation, we have **DATRU**. This is returned to TRU by the reversibility of prior and total growth and development. When we speak of form and function, both in TRU and in these many dynamics of DOT and their development, it is by an extended FAF from that in TIT alone. Also, what exists in and by TIT, as well as of TRU, TRUS, DOT and POT, can not be adequately explained by form or function only, but how they exist and work together, especially with all the rest that makes up TRU and TIT-TO. This is seen in how form follows function, as well as function follows form. These show us what TRU is. The great means of TRU development is by ENIT, ESINS, ES, TRU and all in between. There is an ever greater quality for what is good and right built into the mass energy, Á FAF of TRU development. Its elements are accumulative with added experience, apprecia-tion and inspiration. Much of this is energized by TRIA and its concentration in higher states. Such concentration is seen in survival and self preservation, which are the major themes, properties, and purposes of IA, TES, and TRUD.

	SAME	DIFFERENT	DAOST
UNITY	A or The Undifferentiated One. Nearly Meaningless by Itself, Prime TRU Focus	A Differentiated One is Meaningless, both for TRU and for all	Unity alone does not supply what is essential to DAOST
FINITY	A numerable collection of the homogeneous, lacks contrast, substance, and thereby relativity. Quantitative/qualitative	A differentiated number of entities can not be a universe, only chaos. No quality nor relativity	A number of all different entities lack all sameness, totality, unity, and infinity
INFINITY	An infinite number of enti-ties all the same lack difference unity totality	An infinite number of entities all different lack sameness, relativity unity and totality,	An infinite number of entities without unity finity totality and absolute, can not be TRU
TOTALITY	An undifferentiated Totality lacking sameness unity, finity, infinity and absolute is not a RU	A totality without unity, finity, similarity, infinity and absolute can not be a RU or TRU	A totality of entities without unity, finity, infinity and absolute can not be TRU
ABSOLUTE	An undifferentiated Absolute lacking unity finity infinity and totality can not be a RU or TRU	A differentiated Absolute without similarity, infinity finity, unity, and totality can not be a RU or TRU	An absolute of any kind with-out unity, infinity, finity, and totality can not be the absolute of TRU

Table 2-11 **DAOST, TROS, POTENTIAL-ACTUAL TRU**

From Table 2-11 when TRU is total, ordered, and systematically unified it is productive of being divided by finite proportionality to become comprehensible. This is SU, being divided from unity, the finite steps to the Absolute Finite, **AF**, and to SU, ALT and proof of TRU and IPTRU. It helps to prove the proposition that we can have COAT and POAT. This is: **PROAT** = Proof of the Proposition that the comprehension and presentation of All Things is not only possible, but follows set unity, totality, order, finity and proportion. This is satisfied and solved by a more natural desire and well spring of interest and zest for life, being, TRU and TÂ. The expansion of INIA is called: **LAIA** = Law of the Infinite and Absolute of the Absolute. LAIA is the great equilibrator, the intensity and frequency of IA, against the infinite and absolute of the finite. 'The more things are different, the more they are the same and the more things are the same, the more they are different'. 'What goes around, comes around'. These largely hold true for a RU and especially TRU, in contrast with AU. It is by the relationships between different and same that we can explain much of TRU. It is in understanding DAOST that we can correct and best use LASUR. It is in understanding DAOST that we can limit SUT, yet develop and use it to lay the foundation for ROT, MOT, ALT, SPALT, POS, STA, and of HOM. For without a clear differentiation of what is and what is not the same and different in TRU we can never adequately achieve its order, coordination and form, and thereby actual totality.

Does purpose govern the course of development of a, the, or The Relative Universe? Aristotle's answer was a definite and optimistic yes. His conviction in the importance of the potential or of first, intermediary and final causes and deep faith in the predominance of reason led him to appreciate purpose as first cause and principle. It is in purpose that the potential becomes actual. It is in purpose that we tend to focus, and TRU empowers itself, to become TRU. This is largely directed to functional use and significance. The final cause, one that compels this purpose, is the perfection of infinite replication. There is a summative reinforcement of replication from all energy levels. **LAPOT** signifies how Potentiation works in Totality. This is in close and constant concert with the actual of TRU. When functioning LEss than in totality the CHaracteristics will be different, **LECHAR**. Yet when in their other, partial, or incomplete form they will complete their potentiality to become actual. The potential is only a beginning state in the continuum that includes the whole of potentiation-actuation. For the potential to become actual it must fit into VAPO, DAOST, TROS, and the total states of TRU. **VAT** or any variance from TRU will reduce the actual to potential or less to that degree. **PACT** signifies the Potential Actual interChange, like OITIO the boundary between TIT-TRU and AMTRU-ATRU. The potential like ESINS is what can be and is tending to become actual, ES, and TRU. They all need to be distinguished from **PINT** or the Potential In TIT or TRU. For there is much in TIT that is highly dynamic and thereby more or less. When less becoming more it is a kind of potentiation. This depends upon how much it is needed, a necessary ES or CES of TRU. Such Such potential within TIT may be actual or potential, potential to parts, yet actual to TRU. All are guided by what it takes to fulfill proclivity to TRU of a part, section or totality of TRU. They are also set and guided by **LORF** or the Law of Reference Fixation, the, focus, unity and centripetal powers of TRU. Potentiality is seen in the very progression, stages, and states of DOT, TRUS and TRIS whose concepts are of what is less than TIT, yet are becoming TRU and TIT. This is **STOT** or States of TRUness, or States of being TRU. A vivid example of STOT, potentiation and DOT is in how representations, as words, become TRU. **STAMET** Standardiations become the mass energy, words, phenomenal or real Á of TRU.

Such concentration is seen in **IA** or Immediate Awareness and Instantaneous Attention and Action, that in turn stimulates development and empowers ATO. It is a focus of individual consciousness. IA begins with questions and problems and seeks answers and solutions. There is a close identity of IA with TRU. It, like the proband, is much of the focus and center of TRU. It works closely with the mid to higher levels of mass energy, Á FAF of TRU. Together they are the prime support for TRU, ROT and POT. This convergence and support describes the imperative, intentional direction of forces in DOT and TRU. IA ranges from the most simple to the most complex seen in Figure 5-11, expansion of IA, and to TRU. On lower levels this is IA common, increasing through directed IA to controlled IA. The constant convergence of IA on the necessary spans several levels and tends to TRU. There can be relative concentration or focus of IA in less and less duration of time. This is seen in **POWN** or the Power of Now, the Present or here and Now. This tends to increase its contribution to TRU. IA is constantly engaged in **AFRET** or the Absolute Finite Reversible to TRU, or bringing back entities and objects, as in ESINS to ES, to TRU. This requires a constant recurrence of separation and reunion of S and O or thought and action to TRU.

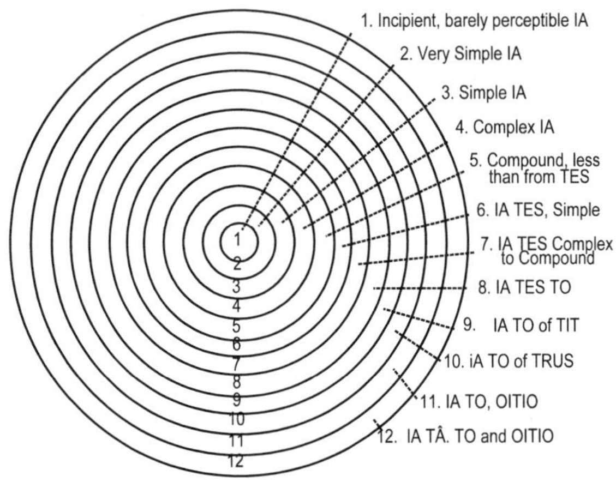

Fig 5-11 **LEVELS of IAT**

Figure 5-11 shows several of the increas-ing levels of IA to absolute IA of TRU. It is from the origin of IA to its absolute; as the focus of life, of TRU and of a person's IA that starts from no attention and proceeds positively from any, a, some each and all attention in rapid expansion. This proceeds to various levels of complex attention, plus combinations with attention that are of longer duration in time. With maturity, education, experience, improving habits of thinking and reason more complex and actual attention of TRU increases as this also expands toward the absolute in attention to God, TREQ and TRU. Each one of these levels and all together are in a dynamic state involving all that is thought and done. **BOTFIA** = Body of TRU a function of IA. BOTFIA answers the question,"but, what is TRU"? What is this world we are of? How is it to be determined from all that is about us or all that is everywhere? This is shown by the following: **IA → Á → TRU**. It is through relativity, the relationship between the observer and observed, attention and action, or from attention and action, that TRU, or the best and perfected world of man, is constituted. For this is its predominant form, state, function and mechanism. It is its mass energy dimension, Á. This, to be demonstrative of the entire form of TRU, must be taken in its most complete sense, as well as to include the other dimensions and their different states or levels. This is called: **LOTÄDIMS** = Law of the Total Form of TRU is, and is only, to be determined by its entire dimensional formulation and states. This totality includes the Á, L and T dimensions as well as all their higher powers in all combinations as they exist in TRU. TRU is never absolute object or absolute subject. In being relative to its mass, Á, TRU is realistically and totally distributed. It must abide by DAOST, which shows the dichotomy, to and from, TRU, and the non objective inclusion or separation of preliminary knowledge to be determined on a topic or TRU. This is **VERTRU** or Version To TRU. Excellency depends upon the degree and level of IA adequately described, demonstrated, and made an object. This is seen in artistry, and its application to all fields, sciences and technologies.

REASON FOR, CAUSE OF, CAUSE and EFFECT

The core of DOT is the completion of the transit from origin through potential and actual to TRU and the reasons for, causes of and effects that occur. It is in these that mankind has found a ready channel to understand our world. Yet its ease has been highly misleading for we have failed to see the difference between single reasons, causes and effects in ENIT, and the multiple totality of those in TIT. Be it material, efficient, formal, final, existential or essential causes, they are not sufficient. They must account for and be accounted by TRU, TIT, TROS, TRIS, DAOST, SCARLS and VAPO That causes work in many and a hierarchy of ways is seen in binomial causality, or a division of causes of different kinds, including whether working within TRU or from without. Causality must be interpreted by TRU and ATO and these can not exist without our thought and act, or its Á mass energy dimension. Since this arises from relativity of subject and object we have **SOTRU** or Subjectification, Objectification Total of TRU. By learning and showing these, DOT has much more meaning. The problems we face in life can't be solved unless we know their reasons and causes, and the effects our solutions or cures produce. Many of these boil down to **PRONIT** or Problem of Not or It. Whether and to what degree it does or does not apply to, or is a threat to, TRU. This requires an explanation of the essences of things which in turn usually means their major reasons or causes and effects. Even at best this does not necessarily have closure. By **LARMA**, the Law of the Reciprocal basis and Reason for Measure, it is never absolute. Since there is much Opposition, **OPIT**, even in TIT it is hard to reconcile with an absolutely positive mode for TRU. Generally it is the positive we must look for the transit of DOT. **PRUTIN** or Proof of TRU through Additions to the whole. Each potential that becomes actual and each negative that becomes positive, strengthens TRU. This is how **LOR** Law of the Rational tends to TRU works. It is how the reasons for and the causes of work on the mind to overcome problems whose solutions enhance TRU. This is shown by the phrase a fortiori which means the admittance or addition of something which once included helps to complete reason for a still stronger reason or cause. With deepening interest in the underlying reasons in life, and as a person's life progresses, there is an increasing search the basic answers and solutions. This is **PORI**, the Principle of Rational Identity. This is facilitated by strengthened reason, **LASUR**, the Law of Sufficient Reason, a cornerstone or solving the problems of TRU and ROT. It is extended in the hypothetical imperative and categorical imperative to provide a purified reason so extolled by Kant. **FIP**, For Fill-In Posit has at least two implications helping to fill in all voids and disproportions and how by the combined deduction of causes those that are unknown can be found. We should constantly remember, ECRO, **LADOT** or the Law of Non Total Different Disunity. This is how what is less than a totality can't be assumed to be anything more, less, or other than what it is. Thereby its reasons and causes have certain limits. We also encounter our own limits in **PRIC**, the Problem of Infinite Causes. This is corrected by **REC** or the Relative Convergence on Causes. Probability determines **LEC**, the Law of Efficient Causes, by the great aspects and FAF TRU that enable a determination of what it is and how to show it. Convergence and divergence increasingly become for TRU the movement toward unity and diversity, the same and other or different of DAOST. There are many examples in nature, as well as the back and forth dynamics in TIT-TO. One example is convergent evolution, where there is increasing similarity in some characteristics between groups of organisms which were initially different. This has great implications for future processes, posits and controls.

Cause and effect characterizes the through and through processes working from without, within and OITIO of TRU. Parsimony is a very important principle and mechanism in DOT and TRU. All TRU and what becomes TRU is, like the human body, developed in parsimony. It is close to efficiency which is even more determinative of effect to TRU than for cause. The compelling drawing together of causes and effects are characteristic of TRUD and TRU. This is seen in **TÀP** or TRU Assimilation Power and Proportion. Form is established by unconformity. Ultimate formulation is what is set in TIT and can be effectively used for ROT and MOT. It is upon established form and ultimate formulation that ultimate posits can be effected or produced. For the enhanced actuality of cause and effect there is increased relative precision and validity of posits.

DETERMINATION and DOT

Determination and demonstration are less exclusive to DOT, and more exclusive to TIT, than derivation and development. They apply to various approaches to, operations within, and representation of TRU. Yet they do have a great role in how, and what can best show how, TRU comes to be. Unless we make and maintain a clear and precise differentiation and distinction between TRU as a subjectivized object, TRU as represented, TRU as presented, and TRU as TIT, we never successfully understand TRU. Nor will we be able to prevent the many pitfalls, errors, and fallacies that are produced when this determination is not made. **DOTIT** is Differentiation of the Objectivized TRU, OBT, from TIT. This leads to a subjectification correction. Just as we can make determinations from one, so also one from differentiation. Also determination is made selective by TRU in exclusion by relativity, exclusion by being, by being the Thing, TRU. Resolution of DOT is by the singularization, differentiation, separation and speciation of its various aspects, forms and parts. Determination is greatly made by **POES** or the Principle of Essential Separation, separating out and selecting what is most vital in TRU. When applied to ESINS it is the determination and definition in their transit to ES and TIT. These determinations are facilitated by: TRU relative intensification, **LAMASEX** or Law of Mass Expansion, Means and Extremes of dimensional levels and exponents together, **LASRIT** Law of Speciation and Relativity in TRU, Law of state and condition of relative TRU, Law of Identification in TRU, Derivation, Determination and re-foliation of IT and TRU or DIT, and **DETCOP**, the Derivation and Determination of TRU Through the Continuous Processes of Parturition.

LAD Law of Determination. LAD is how by narrowing the limits of entities, ideas, concepts and terms, we can formally distinguish the precise FAF and makeup of TIT and TRU. Much of the determination of TRU comes from, how names and words develop, form, and make ever better distinctions in all TRU; in and for a law of determination of the knowledge of TRU. There can be no absolute contraries in TRU, **LÂC**, Law from Absolute Contraries or Scales between extremes. **AXLA** or the Axiom of the Limits of Application. Abandonment and facticity, or being left out, and free diversity denote how TRU or TREQ, as God, help to determine and in reaction propel humans to become TRU.

As actualization is reached all that produces DOT comes to life, life in TRU. There are many laws of actualization and actuality, including law of entity actualization, law of actualization by levels, law of TRU actualization, law of proportionate TRU actualization, law of reverse TRU actualization, law of complex to TRU actualization, and law of TRU complex actualization. Actualization, as DOT and TRU in general, is enhanced by dimensions and measure. If we can discover and determine what the dimensions of TRU are, and their measures we can confirm much of TRU. Although many parameters and continua exist in TRU it is its three primary dimensions that enable us to provide ROT and MOT. **LED** is the Law of the Equal Distribution of the Primary Dimensions of an entity, object, RU and TRU. **LEDÁ** is the Law of the Equal Distribution of Mass and Energy, or the Á Dimension. This may be done for each dimensions separately, all together, or all together conformatively. This provides **LADS** or the Law of Dimensional Stability. LADS results in, **LODPALT**, the law of the distributive proportions of ALT. Insufficiency in or reaction to these is the law of dimensional ignorance. This is a large cause for TRU having been unrecognized. No good and accurate dimensional formulation that reflects TRU has been heretofore discovered. **MERS** is Measure of a System, ISM, science, philosophy, religion or TRU. The reciprocal basis and reason for measure denotes the critical fact that measure is never absolute. No matter what measures are made, unless they are adapted to TRU, they will be deficient or in error. No matter how good, intense, perfect or complete a field of science, school or system of philosophy, kind or sect of religion is, it is no better than the degree it is TRU, or its counterpart in TRU. Thereby we have **IRT** Invoke the Rule of TRU. IRT has corollaries for the measure of all things, and measures of religions. This means God must equal TIT-TO by actual TRU knowledge, EC, ES, CES, NEC and TOPNEC.

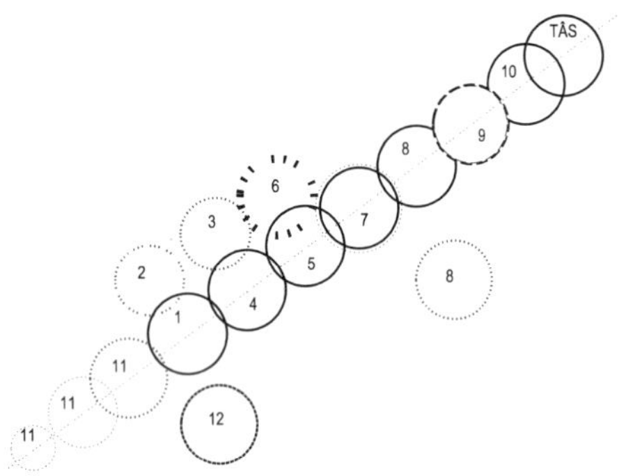

Fig 6-11 TRUS and TRU COMPARATIVE FIGURES, DETERMINATION and ANALYSIS

In the accompanying figure we have circles, lines and numbers that show from left to right:

1. TRU of Aristotle's Day, 2. TRU as conceived by Aristotle,
3. TRU as compositely conceived at best by philosophers,
4. TRU increasing toward current TIT, 5. TRU Today TIT,
6. TRU as conceived by IPTRU, 7. TRU composite,
8. TRU away, 9. TRUD, 10. TRID,
11. Dotted Line to represent formative TRUS, and
12. Hashed Line to represent incomplete TRUS.

In TRU actuality these are more or less illusions except for number five. Thereby they can be shown as broken lines or of varying darkness, as in the last two. Only as these move from lower to higher and from left to right is there S/F and only as TRID ideas and concepts equal TRID or TRU Ideal does it equal TÂ. This called: **LATÂ** = Law of the TRU or Theological Absolute. LATA is S/F∞ or TIT to the Infinite in TÂ.

What we find that is like TRU that may, or will be, of importance to TRU, is called: **SIL** or Without Law. SIL is the general case whereby in types of comparative things a particular type would be non existent without the contraries of other types of comparative things. This depends upon the degree of confinement to this particular type, or **DECONT** Degree of Confinement of an Entity or TRU. These are the limits of its existence. For Aristotle the state and capacity to pass from one form, level, or condition to another was named dynamis. This was the origin of our term dynamics. COMT or Comparative TRU may be done in many ways that strengthen our understanding of TRU. It helps better determine what is and is not TRU and how universals are actually valid only as they are of or in TRU. Such purity in knowledge is the work of PUREACLITY. It is a purification of errors and fallacies, as that of limiting TRU to the unity imposed by physics, biology, or evolution alone. When such purity of all that exists in TRU, is realized and achieved we can recognize the Imperative Identity of all, as GOD, in TRU. Comparative dynamics helps us learn the balance imposed by resilience and tolerance, or the law of equilibrium between the powers of mediation and the flux in TRU and its parts.

METHODS, WAYS OF SHOWING and DEMONSTRATION of TRU

The derivation and determination of TRU, and how we develop it depends upon our methods, and demostration of it. **MAMET** is the Manifold Method of Matchup of TRU. This is something that is roughly done through life by redirecting all that is given or before us from their own SCARLS to become actualized in TRU. It is guided by BURP and initiated by **MACUTEN**, the Method of Modified Analogies and Comparisons of unaltered ENIT and TIT or TRU. **COMET** or the Comprehensive and Controlled Method and Means of Thought of TRU helps us overcome our tendency to seek specifics or singular causes, cures, answers and solutions. For demonstration of TRU the Method of Knowledge and Representation through set Geometric Solids and Spheres as they apply to relativity and RUS is essential for ROT and MOT. **METDI** or the Method of TRU Dialectics holding to TRU and separating S and O divided by its mass energy, Á, dimension as a polarity. Where one is interpreted in terms of the other. We do not analyze things from our own senses but reducing them to their state or part in TRU. As we develop and demonstrate TRU, knowledge and action for the future can be made increasingly reliable. This can be simply tested by **RATEX** or the Ratio of Existence over Extinction. This reveals the difference between evolution, creation, ATO, and TRU. Although **POAT** or the Positive Objective Approach to TRU is the prime method of POT and IPTRU, it has great precedent in knowledge and especially sciences that have been parts of DOT How search is for deeper and deeper causes, so do solutions lie at the level of their most manifest origins. Methods of thought that are comprehensive and controlled to focus on TREQ or TRU are increasingly effective.

To finalize our discussion of DOT we show how all comes together through SOT focused by IA. **FIC** = Focus of Individual Consciousness, Attention and Action or Individual Consciousness and IA. There can be no doubt that the center point of ones - as well as all humans - existence is one's attention, or the conscious focus in the here-and-now on what one is thinking, doing or attending to. To this extent Descartes was right. This is balanced by all the rest of TIT-TO. This focus is not a random focus of space and time. It is not even any fixed pattern of space and time from physical determinants alone. It is the focus of ones being. It is that to which persons concentrate their attention, thoughts and acts that best serve them or best enable them to survive. PRUT or the Proof of TRU, and the proofs of Á or **PRA** are found in the subjective objective continuum of SOT, the mass energy of TRU, reason, LASUR, their limits in TRU, THEO and THES and accumulative understanding of their role in human life. The transition, undeniability, exclusiveness and acuteness of Á all further its TRU meaning. They arise from such concentration or focus of the individual, a person's consciousness and attention. We can deny, for a time, our own focus. We can identify with the focus of others, especially those close to us in various ways. We can even assume the identity of all those varied senses and things we subjectivize Yet we can never accept them finally nor totally as ourselves, if we are to actually be, be TRU. We can never accept a focus that is other than this convergence upon ourselves, as TES. Here is the greatest proof of TRU. For it shows without doubt that the world is not the world of random things, nor any space-time scale. It is a world of relative senses, things, entities as they are ESINS of the ES and CES in TRU. This can only be in the space-time-mass-energy scale and frame humans exist. The powerful meanings and depth conceptions achievable with proofs - PRUT, PRATO, PRA, etc - are both the beginning of KOT, TES, PUREACLITY and SÂ, and the greatest method of enhancing our world, TRU.

Fig 7-11 **THETIC STATES or LEVELS of S-O TRU** (THETS)

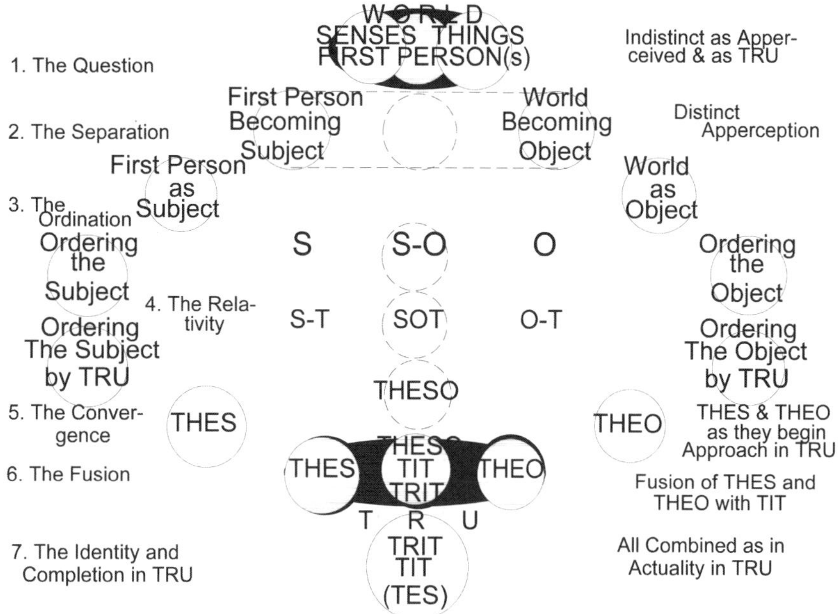

This figure shows how entities, senses and things, by the effect of relativity, Á, and TRU eventually become TRU, TRIT, TIT and TES. From the point of view of each state or level they can be considered as ENIT or entities in and of themselves. From the point of view of TIT they are - more or less - only potential TIT. This is how the potential senses and things become subjects and objects and these become the form and function or Á state, dimension and operations of TRU. In this figure we give to the development of these theses of TRU seven stages or levels, which are:

1. The origins, a body and the derivations, questions, subject and object, observer and observed, problems, stimuli, action, 2. The separation, how they exist, are formed and function as subject and object, observer and observed separately,
3. Their relativity, formation, orientation and ordering. How their form becomes directed and by rela-tivity how a subject and object or observer and observed interact and fuse, 4. Their relativity provides ordination by how **a** subject and object become **the** subject and object in their interaction to provide for TRU, 5. Their convergence, how subject and object become THES and THEO in TRU to converge on THESO, 6. Their fusion, how THES and THEO become THESO, the Á dimension and FAF of TIT, TRIT and TRU, 7. Their Identity and Completion in TRU

These stages or levels are expanded or contracted, as suits the approach to such a theses of TRU. They are largely driven by the dynamics of TRU as exemplified by animal behavior. From stimulus and response, questions and answers, problems and solutions, fallacies, errors, paradoxes and others, their accounting and resolutions are the determinants of relativity and the form and function of the Á dimension.

In the progression from senses and things all the way to TIT we follow our discussions of subject and object in POT. How S, O, S and O exist and function separately. to discover subjective generation, or how as the subject becomes the object and a new or succeeding observer becomes the subject this follows in series. Repeatedly subject follows object and object follows subject. This involves an identification of subject-object, or relativity. We must learn and know the identity of different kinds and definitions of relativity in and of TRU.

S-O is the interactions of S and O which tend to relativity, especially that in TRU. In TIT there is a subject object THES, THEO dependency, or how our predominant nature necessitates their reality, combination, Á and TIT.

SO is S and O existing, formed and functioning together. It helps to prove the unity of subject and object by TRU. Subject Object completed and final, or what the end product is in their interaction and union. Is it the act, being named or something displaced? Yet it is never finished, a continual interaction within the Á continuum of TRUD and TIT. This is **MENT** or the Mass Energy Transformation to or of TRU. Subject equals, or tends to converge on, the object's meaning in TIT.

As the text progresses, from the phenomenal S-O to the Á dimension of TRU, it will become clearer how the subjective relates and is FAF by and to TRU. How it can also be both outside and inside, transcendent to TRU, OITIO. The brain or mind, and its vastly expanding ability to attain subjectification from objectivity is the basic mass energy dimension characterizing TRU through identity and distinction. Parmenides in his method of investigating what it is "to be"showed us how the identity of thought and reality are achieved. The conveyor of this is language, a large part of Á.

What is in a name? What are in words and discussion? Is it precisely what it denotes, all it conjures up, the mind of the writer or observer, only a symbol typed word or phrase, or some intermediate mixture. This is why it is so necessary that all we think and do have the greatest possible meaning. Why it must be of the greatest possible TRU. For the degree to which it is not TRU is the degree to which it is meaningless, worthless or harmful. So do the names word and discussion become significant and great by being the agents of TRU, its ES and CES. This is called: **NERATS** = Nominals Equal ES Relation and Á Function with TRU Significance. Also: **LUSA** = Law of Universal Significance of Á, Actuality. LUSA is from the universal standpoint, NERATS is from the Nominal or word standpoint. They both apply to the significance of actuality as it means, implies and represents all through and through, S-O, Á, as they are TRU. So it is of all subjectivity, objectivity and actuality as they are only in and of all of TRU.(RAEV, LIDRASO)

SOT is S, O, S-O, SO of and in TRU. Subject equals object is a set equivalence in and by TRU, the fixed percentage of S and O in TRU. Subject equals object meaning and TRU. The laws of TRU are often the objective TRU's dependence on subjectivity. **LEDA** is the Equal Distribution of mass or energy, or Á. **SOTRU** signifies the subjectification and objectification of TRU totally, as in the observation with the observed TRU, the completion of relativity in TRU. **CAPOT,** the Catalytic Production from Perspective, SPOT of TRU. **LACOMP** Law of Complimentary Proportions. **SOTIT** denotes S, O, and SO only in TIT. From SOTIT Á, is the predominant mass energy FAF, continuum and dimension of TRU. There is validity and justification for manifold varieties and meanings of mass and Á. **LINDA** Law of the Integral and Infinite Derivation, Development, Distribution and Determination of Á. **LOCAT** Law of consciousness, cerebration, mentality, S-O, Á state, dimensions, or FAF in

TRU. **CÂTRÁ** Concept, principle, or law of the Conversion of AU and all through Á the mass energy dimension of TRU. **RESÁ** Reduction of separated subject and object, to SO, SOT and to Á. The mass equivalence of m, SO, and Á in TRU is called, and was shown in, MESOA. **LÀS** or Law of Alternate and variable Forms, States of the Á dimensions, signifies how it is one thing in SPALT, another in TE, TO, TIT, TRU, DYNU and DYNS. **LOPOS** is the Law of the Percentage of S and O along the Á scale. SO to Á concentrates on stimulus response and what is negative to TRU. Proof of Á and TRU by the correspondence and reconciliation of cross balances, reversibility, continuum, spectrum, dimension, and gradations of S-O, SOT, Á, and TRU. Pragmatic proof of Á is evident in the simple fact that one must know what one is doing. Or, even more complex, one must anticipate and think ahead of time about what will be necessary for the occasion, oneself, life, and TRU. **ÁDIS** or Á Discrepancy denotes that to discover and best use Á, and thereby TRU, it is critical that we know what it is, what it is not, all its concepts, causes and effects, and **PRÁ** or the Proofs of Á. **PRULATA** Proofs of the Limits of Á Prove TRU, and Proofs of the Limits of TRU Prove Á.

The problem of mass, weight, and Á can be resolved by analyzing and separating the many kinds of forces of attraction and repulsion that exist in entities, objects, universes and TRU. The solution is how the mass-energy dimension has to be altered, adapted, warped and transformed by relativity, the focus, thisness, and exclusiveness of TRU. This is MESOA and denotes the adaptation and alteration of m and SO as they become Á, the same as they are both, in turn, adapted and altered to fit the predominant and relative mass energy form, state and dimension of TRU.(Fig. 5-07) MESOA is what gives us the Á dimension whose scales and measures can then be determined. It is confirmed by the fact that S and O are not the same as Á throughout, nor that S and O, or even Á are in themselves, the same throughout PERT, POT, ROT and TRU. This is called: **LAS** = Law of Alternate and Variable Forms and States, or Á dimension of TRU. Thereby Á is one thing in ALT, another in TIT and TRU, another in SO, another in DYNO, another in DYNS, DYNU and DYN or Homation. Yet this should not lessen our intent and ability to compare and unify them as they are identical in the DAOST of TIT-TO. **LAVMA** = Law of the Validity and Justification for Manifold Varieties and Meanings of Mass and Á.

SÁ is the hierarchy and gradations of mass energy or Á components in TRU. SÁ reveals the correspondence between reals and phenomena, reality and the phenomenal. Reason is based upon knowledge of the relationships, revelation of the causes and effects, forms, ES, and CES of that which is TRU. **LIDRASO** signifies The Initiation and Derivation of the Actuation of Reason in TRU from Á and S-O. Formal distinction, BIV, cues, revelations, insight and many other great elements of SÁ are largely their acuity in TIT. So does the name, word, discussion, and the use of NEC and POS become significant and great by being the agents of ES and CES in TRU. This when expanded shows the universal significance of Á, the actuality of mass energy, in TIT and TRU. The Principle of altered distributive portions of ALT applies not only to Á but to, and is unique in each, L and T.

Ways of showing and demonstrating TRU bringing DOT to fulfillment, and are of the highest benefit in arts and sciences. This is shown by the ratio between senses and appearances and what is TRU. There is a continual interchange between actuality and illusion. When of the transcendent and OITIO it extends the actuality of TRU. There is power in TRU to involute and return the transcendent to TRU. **INVOT** denotes Inversion of the Standardized Perspective of TRU. This is how we turn SPALT, POS, POA, EPI and POT back into TIT-TO. By **METID** or methods of TRU Induction and Deduction, as in COR and COL, CONVET and CONRIT, we simultaneously operate to bring all back into TIT. By **LACARD** or the Law of Careful separation, Division, Dissection, and Distinction we eliminate all detrimental or not TRU, elevate all less than TRU, and accentuate much of TRU.(ACC) To reconcile opposing aspects or parts of TRU we apply **LOSCOR** Law of Opposition Resolved by Setting out Limits and Comparing all Significant facts and Relations.(VAPO) Similarly by **TIST** TRU proves ISMS and ISMS prove TRU, and by **PREMAX** Extremes Prove the Means and the Means Prove the Extremes. By the methods of Homation we collect and order processes in SPALT, TRIT and TRU. S-O, or Á FAF in TRU, that converge on ESINS to become ES and CES. Any relationship becomes or can be shown as a line. This is the basis for representing dimensions, their combinations, and ALT. **SIRD** is the Sufficiency of Identification, Representa-tion, and Demonstration. An object that is observed as either a single unit or infinitely composed is not as easily or well learned or shown as one that is sufficiently finite. This is the power of differentiating and demonstrating TRU. These are some of the many methods TRU can bestow on education, especially when used as methods of reasoning in TRU.

Through the ontological absolute or the ontological TRU absolute, TRIA and TÁ and their KOT we realize the reason for absolute faith, to believe, act by, and live in TRU. This is supported by **LABTRU** Law of Absolute TRU wherein whether or not we know it exactly does not negate the fact that we know that it is. This in part is acquired from God and TREQ. KOT demands immense sufficiency and proficiency, especially with expanding KOT, TRU, and the demands on TES - and the individual. You can't deny the great enthusiasm and dedication imminent philosophers and theologians have had for the potential power and glory of rational, logical, spiritual, intellectual and idealistic ideas and thought. The profound revelations of the nous, logos, reasons, causes, and meaning of all helped them realize how different, special, unique, precious, and perfect TREQS are, and the value of this relative universe of human life actually is. In this humans increasingly recognized how much more than their predecessors we benefit from all the great blessings of FAF and mass energy that make up their world, God, TREQ and TRU. It will help us appreciate the new concepts, and concepts to come, that will complete a solid and reliable POA to our world, TRU.

Another fact that helps prove DOT, ATO and the subsidiary role of evolution and creation is that of meaning. Meaning is largely exclusive to and of a high level of the mass energy, Á FAF of TRU. It is the greatest bond between observer and observed or the relativity that gives TRU TRIA. Meaning constantly reinforces TRU as TRU constantly resupplies meaning. This is seen in **ISTRU** or Is Means TRU, or how is and meaning tend to be and are the same in TRU. TRU constantly extracts and magnifies all unto itself by **LACOMET** or the Law of Convergent Meaning, that is increasingly **LORM** or Relative Meaning and **TRUM**, TRU Meaning. Meaning comes to have greater equivalence with TRU, thereby showing the escalating development of TRU. This becomes a functional meaning for all TIT-TO. Symbolic meanings are very valuable in depicting the ESINS in things, major functions of our being. 'What it does for me' as I am or am not TRU. This is what, in turn, makes distinct the differences between entities, things, objects, senses, subjects, and universes as opposed to what is of TRU and in TIT. As meanings become more explicit, we can produce definitions. This has been much of the long rise of civilization in history, one of producing knowledge and languages that have better and better definitions with ever greater distinctions to represent the differences in TRU and all. Existing language has a great variety, with many gradations, of definitions, mostly less than TIT - and even for the same word. However, with good usage they tend to TRU. They help reveal how TRU is different aspects of the same thing and increasingly show DOT by bringing the potential to actual, as it becomes TRU.

CHAPTER 12

GENERAL FORM of TRU

TÄ is the form of TRU, what shapes TRU, or gives it identity through S-O and Á. By TÄ we seek to discover what TRU is, what its various kinds, aspects, forms, states, limits, relations, and necessities are. We can not understand TÄ so well without visualizing its continuity from AUT to NAMTRU, TRUS, TRU and TIT. Also we must remember Co-forms or forms without which a single form loses its form, as TIT and TRU without all their form.

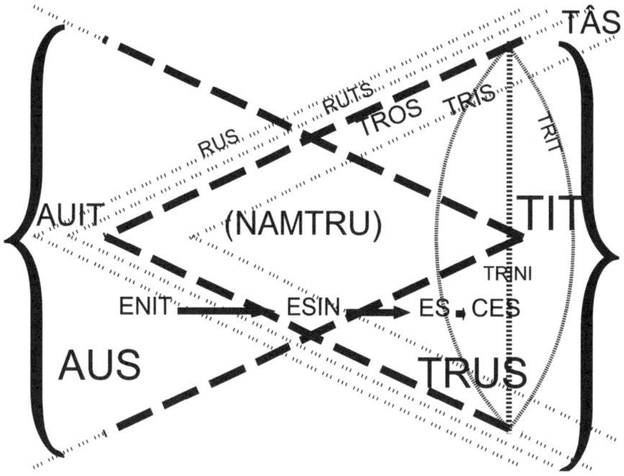

Fig 1-12 **DOTRUS or the DERIVATION of TRUS and the FORM of TRU**

This figure shows the relations between AUS and TRUS, and how they may be viewed separately and together, as in NAMTRU, a mixture of much AUS and TRUS. Such a figure of double triangles as a parallelogram helps us to visualize both the relations, as well as the transformations that occur between AUS and TRUS. The role of RUS, RUTS, TROS, TRIS, as well as TRINI, TRIT, ENIT, ESIN, ES and CES can be readily posited and realized. In this text the distinction and separation between **AUIT** the illusionary figment that is AU, and TIT the Actual and Absolute in the thisness and exclusiveness of TRU is the fundamental distinction that makes all the rest of TRU valid and reliable as a topic for representation and presentation. We should not be misled by the diagram for, as all, they are limited. For limits are the degree to which any and all are not reducible to a discrete object. The only absolutely imaginable object is that in which both the observer and observed are one, as occurs with the first person concentrating only on TIT in a 1:1 identity, fused as 1. So that in this diagram the parallelogram of juncture between AUS and TRUS, is manifested to a large extent by NAMTRU, although much of our lives, both individually and collectively, is not TIT. We, in this Chapter of TRU, are still not absolutely in TIT. We are nearing TRUS, TROS, TRIS, TRIT, TRINI and TIT. We are showing their derivations as they are the forms of TRU.

It helps to understand the knowledge of TÄ. There has been much confusion and error in understanding TREQ and TRU. A major contributor to this has been **MINESAT** or Multiple Selective Massive Amounts of Entities, whose effect is often indirect. There is a tendency to entity fixation whereby entities - and dimensions - are neither exclusive nor how they exist ENIT, i.e. physically alone. Likewise there is the fallacy of visualizing the world, God, TREQ or TRU as the object alone. Only TIT is OBT, and even TRUS are only shadows or accessories to TIT, albeit largely within it. The great fallacy with god, as for all that emphasizes oneness and the integral, is that is overpowers DAOST, and the equilibrium in ID or the integral and differential continuum. It tends to destroy the inquisitive investigation into the parts and processes of TRU.

The actual, relative to mankind, world, universe, God, all, etc. as The Relative Universe, consists of neither a multitude of completely separate things in themselves, any combination or totality of them, nor a singular undifferentiated union of them. It is DAOST, Different Aspects of the Same Thing. The Relative Universe is highly selective, qualified, integrated, differentiated, focused, unified, exclusive, and interrelated in the totality of its own being, actuality, and aseity.

The actual world or relative universe, may be considered religiously by monotheism, trinity, polytheism, or pantheism. Yet by none of these so, exclusively nor strictly except as modified by their roles and depending upon how they are formed and observed in totality. Even an essence of deism may be recognized, whereby much that the relative universe consists of is not managed and controlled by man. Still, it warrants a highly apperceptive, accommodating,, laizze faire, and/or compensatory approach. TROS, TIT-TO.Yet by none of these exclusively except as modified by their form and function in its totality. TROS, TIT-TO.

If we seek and expect to secure the ‘finality of God’ it can not be done only through TU, nor any separate process, part, form, function, aspect or mode of TRU,. The only finality is when TREQ becomes TIT-TO and that of added TRU that provides its external objectivity and subjectivity. Likewise there is the common and pernicious failure to extract all less than TIT, ATRU, and AUS from TRU, especially in any definition of TRU and TIT. That TRU is a concrete universe, at least objectively, is proven by **ATA** or From TRU From All or TRIA. This is where TIT is all and TRIT, OBT, SUT, ROT and MOT represent all, as ATO is the derivation of all. Likewise we must remember the paradox of absolute entities or that state of absolute that are and are not absolutes. This we account for by using the term **ÂÂ** for the former, and include a relative into those of the latter. When we depart from TIT to be an observer we need to remember whether this perspective is of reality, TRU or more a presentation of TRU. Of the spectrum that runs from the most relative to the most absolute we have different levels depending up the degree of differentiation we find most practical. Among the common four levels we have:

	ABSOLUTE	**RELATIVE**
ABSOLUTE	**ABSOLUTE ABSOLUTE, ÂÂ**	**RELATIVE ABSOLUTE, RÂ**
RELATIVE	**ABSOLUTE RELATIVE, RA**	**RELATIVE RELATIVE, RR**

Table 1-12 **RELATIVE - ABSOLUTE CONTINUUM**

In terms of relativity or the relationship between the observer and the observed, these become:

Infinite Observed of Infinite Observed
No Observer, all absolute = ÂÂ
Infinite Observed by Infinite Observer
From Infinite Observed = ÂR

Infinite Observer of Infinite Observed
From Infinite Observer = RÂ
Infinite Relative of Infinite Relative
No Observed, all relative = RR

The derivation, development and determination of the form of TRU, DÄT, can be recognized in: the non identity of TIT with S, O, ROT and MOT, emergent new forms and **LINTAC** or Law of Indirect TRU Actuality, what is peripheral and not necessarily directly human in TIT, relativity's reversal in its general and special theories when applied to TRU. **STAXNOR** or Standardization of the External and Internal Ordered Sequential Relations of the Form of TRU, and ESINS or what is in other entities that is also in TRU. As KOT becomes **KIT,** or Knowledge In TRU, there is an external internal separation, selection and fusion. In TIT we should recall the continual processes of existence whereby it not only stands in opposition to non existence, it stands alone in opposition to both other existences externally, and imbalances of existences within itself. This prime means of survival is self preservation, where the species and individual function in unison, all for one, and one for all. The universe of the specie can 't exist without its interdependence with nature, its environment and all. This to, like ATO and ALT, is highly adapted, altered and selective to the specificity of TRU. In DÄT there is the continual interactivity of interdependent FAF. Form is an expression of their function, what serves them best and what they are best adapted to. This is their functional expression, how form follows function, and vv. .

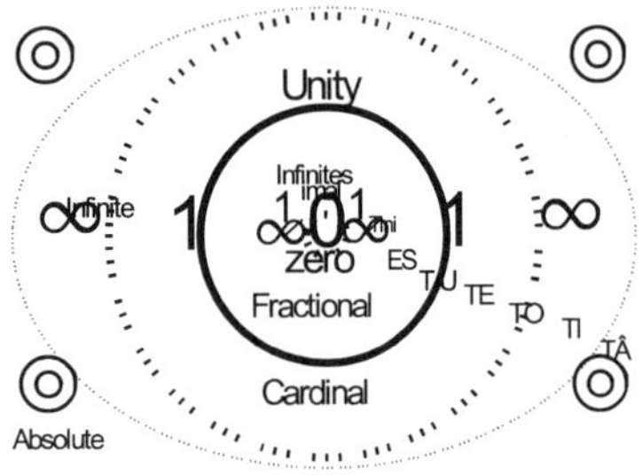

Fig 2-12
DEVELOPMENT of TÄ MATHEMATICS and the RELATIVE FINITE

The mathematical origins and basis for TRU show much of its form. This is largely a product of the existence of TIT-TO in extension toward the infinite and absolute exteriorly and infinitesimal and zero interiorly. The degree of its existence is largely a function of the quantity and quality of all the finite in between. This quality is a largely a function of how well it completes and orders these mathematical FAF.(fig 4-04) This is carried into various mathematical systems in fig. 1-25. The development of the form of TRU, may be visualized from an incipient quality and quantity of this figure to its progressive and vast expansion in all directions. This is much the same as occurs for all entities and RUS as they become things in and of themselves; yet for with TRU its special and unique qualities and characteristics give it exclusive thisness or haecceity.

NUDÄT = Numerical, or mathematical, Unitary Diagram of the Derivation, Development, Determination and Demonstration of TRU. NUDAT is one of the most important diagrams of the text. It is key, core and basic to an overall and ordered understanding of TRU. It shows the role of mathematics in TRU. It shows the form, structure, and arrangement of TRU and TIT. It shows the basis for the composition of TRU and TIT in:.

1 **∞ - 1/∞**
2 **∞ - 1 - 1/∞**
3 **◎ - ∞ - 1 - 1/∞ - 0**
4 **◎ - ∞ - n - 1 - 1/n - 1/∞ - 0**
5 **◎ - ∞ ....n^n....1....1/n^n....1/∞ - 0**

6 Of a Continuum, 7 Of an Infinite Set, 8 Relations of all Focused on 1, 9 Role of ◎ and 0, 10 ◎ and 0 as limits, 11 ∞ and 1/∞ as limits and role of, 12 A quantitative basis for existence, life, TRU, 13 Ends of 0 and ◎ not attainable, (LANERA) 14 Ends of any one element not an overall end, 15 Balance of factors in, ◎ / 0, ∞ / 1/∞, n / 1/n, 16. Balance of all, 17. Related to VAPO, 18. Holds for the existence of any RUS or TRUS, 19. Leaves a remainder, their quality, 20. Added quality with NUDAT suggests that their will be more than any entity or anything, but of TRU when considered as a set, in closed totality, 21. If TRU, this will require special and added FAF modes, essences, characteristics and properties. **LONUDÄT** = Law of NUDAT, the Ordered Totality Numerical Unity Provides for TRU. When adapted to TRU we find its mathematical basis in: **LAHREX** = Law of Harmony of Numerical Form with Expanding Finite Ratios. LAHREX signifies the precedent Pythagoreanism contributed to the mathematical basis for all. It signifies how critical the form and function as well as proportions and makeup of TRU are dependent upon its mathematical basis and derivation. **LAINO** expresses the relationship, and Law, of the Absolute is to the Infinite as Zero is to the Infinitesimal. For the anachronism or paradox of zero time there is resolution in TRIS, **TRIR** or TRIS Relative, **TRIPRA** or TRIS Practical, and TIT. This figure demonstrates **LACOTAP** Law of the Coincidence of All as it forms TRU, or of the convergent center, heart, and focus of TRU. It is how all centers on TU, the individual and the focus of IA and reality in the Á mass energy FAF of TRU. This may be compared with, only much more accurate and actual than, *the father, son and holy ghost.* It is most evident in survival. It has been the greatest driving force in divinity and human identity in TREQ.

How the DÄT developed along with that of Á is seen in the progression from SO to Á to SÁ. This comes from S-O transformation, stimulus response S-O to Á energy coming to TRU. This produces **LARÁ**, Law of Absolute Relative Ratio of Á or S-O relations to TRU actuality/illusion. **LACTENÁ** Law of the Absolute Compulsive Transformation of Energy States to Á in TRU. **LATRA** is the Law of the Transformation of Ä and Á. TRU form develops from its common, conformity and correlation with its identicals in senses and things. **MER** or Mass Energy Relativity signifies the association and equivalence of mass energy and relativity. **LASOT** Law of the Non Existence of the Absolute S or O, Subjective or Objective. **LAPROS** Law of Preeminence of Reason of Subjectivity. **COIT** the CO-Identity of Immanence and Transcendence, identity of S and O when entering TRU. **LAVMÁ** Law of Validity and justification for Manifold Varieties and Meanings of Mass and Á. **LALTEX** Law of Á formation by the Exclusion from Length and Time. Á is what remains after length and time are fully developed in TRU. Resolution by the telescopic and microscopic of TRU is how all are largely resolved by what they are, belong to, evident in relativity and Á when fully envisioned from unity to infinity in TIT. **PRÁ** Proof of Á. **PROL** Proof of TRU, Á, and ALT by Language. **PRÁVER** Proof of Á Reversibly or how physical objects do not have complete exclusive separate existence. **LAO** Law of the S-O or Á mass energy state of All.

Plato in *Phaedo* sought faith in immortality, or identity with the universe, by demonstrating its form and structure. Thus he considered these to be ideal forms, *eidos*. This is how objects become eidos and essences, which are the ESINS that become ES as TREQ becomes through the mass energy FAF, Á and SÁ continuum, TRU. For if nothing is absolute, is not a completed whole, to make a completed whole something additional must be added. This is the most refined meaning of: **COL** = the Correction or Convergence of Limits. COL is what is to be added to something if it is to exist, or is to be completed. It is in this way the world of physics and astronomy, O, must have an opposite, which is S, and S or worlds like that of psychology and idealism, must have an opposite which is O. Likewise how S and O together must be in opposition to form another whole, exists in the thesis, antithesis, and synthesis, of TRU. The state in which we have absolutes that are or are not absolute is called: **PENA** = Paradox of Absolute Entities. This is an illusion, a fallacy of our subjectivity where we complete our process of attention by a finality which assumes the completed step as being absolute. Yet we fail to recognize its singularity and extraneity, and do not develop consistent complexity and totality. **LOP** = Law of Opposition. LOP expresses the ever challenging state of all before and about the observer, the first person, and our attention, by which it is in a never ending process, even when near completion in TIT, TRU or TÂ. LO shows the form of TRU as dynamic and dialectical processes. **LOD** = Law of Opposition of Dimensional polarity

Expresses how Opposites Operate as polarities in a whole, to form its overall dimensional form existing dynamically in systematic unity.

FAF = Form and Function, their union and interactions. This is a simple, general, and commonly used expression of the fact that form and function often exist and work together, and the universe or body they are in depends on their combination as much or more than their separation. They are DAOST, two aspects of the same thing, TRU. A corollary, and more specific is: **LAFID** = Law of Form and Function Identity and Interdependence. Today biologists are very aware that function depends upon structure and are discovering that the shape of a molecule or many other aspects of our being often does more to determine its activity than the exact atoms, elements, cells or parts of which it is composed. **LÁO** = Law of the S-O or Á mass energy state of All. It can best be seen from our observation of any entity, that the observer-observed relationship is key to relativity. For example if we are to examine, analyze, and interpret our bodies, a mechanism, ES or CES, we will quickly find ourselves in difficulty. For to be effective we must do so in some order and overall complete totality. This is a totality of not only this feature or part, but done so in a valid, identical, overall, proportionate and even ordered way in which we present it as it is in itself, and as it is in various altered states, perspectives, or importance and of use to us.(VAPO) To look at TRU is the same, we must account for the observer-observed state, or relativity, both internally and externally, of the entity and our use of it.

Internally what best signifies the combined nature, form, and aspects of TRU is **SCARL,** and its plural, **SCARLS**, or the States, Conditions, Relations and Limits of which it is constituted. This is much used in describing various views and showing TRU**.** A fine example of SCARL is:**LARDAV** = Law of Relative Validity of Relative Limited Dualism and Invalidity of Absolute Unlimited Dualism. LARDAV certifies that state in which those opposites that seem so contradictory, and thereby invalid, are valid relatively and within certain limits or SCARLS. It also certifies that such opposites, or truths, taken to be absolutes are neither TRU nor often actual; for they are taken in a limited way or out of bounds. It is by LARDAV that we can discover and explain the great validity that exists in so many profound yet incomplete or inactual ideas and principles. It quickly follows that many opposing beliefs, tenets, and ISMS are, and may be explained by LARDAV. It is by LARDAV and SCARLS that we can understand and describe the problem of the construct of the universe, TRU. This is what Henri Bergson so struggled with to resolve. What are the dimensions of TRU? What is the nature of mass, length and time in TRU? Why and how have our ideas of these in the past so failed us? How has our thinking, experience and past knowledge so misled and confused us? How has it been illusion of reality and not the actuality that is TRU? So did Bergson admirably point the way to the actual state of time in TRU by showing it to be not points nor lines, but of higher powers as duration, continua, and wholes. Here has been the cause of our illusions about time. They produce a paradoxical dualism which can only be corrected by LARDAV and looking at it in terms of the actual form of kinds of time in TIT-TO.

TROS are those forms or states of TRU based upon quantity. **TU** for TRU Unity, **TE** for the finite TRU. **TO** for the totality of TRU, **TI** for the infinite of TRU, and **TINI** for the infinitesimal of TRU, TE is divided between both the cardinal and the fractional, **TÉC** and **TÉF**, in TIT or of TRU. **MOF** or the Mode of Finity signifies both the fact that the finite of TRU may be taken separately, as quanta, as both cardinal and fractal, as well as used in all sorts of quantitative mixtures and measures to show the parts and processes, and much of TRU. The diagram 2-12 shows the focus of TIT to be infinitesimal and the periphery infinite about the core, human nature of TRU. The form of TRU is largely produced and determined by the integration and differentiation of the whole. **LADSAC** reveals TROS, and especially TE, in the Law of Different Significance and Altered Characteristics, the difference between the many terms and meanings in language. TROS is also revealed by the gradients or differential in parameters, dimensions, spectrums, and continua of TRU such as in **LASPEC** or Law of the Spectrum or Continuum of Form and Parts of TRU. Their many combinations show how forms and parts do not stand as two completely different aspects of TRU, rather two poles or regions in the same or combined continua.

SPACT = Specie Not Comparable with TRU. SPACT signifies how a specie is and is not TRU. We tend to equate TRU with the species Homo Sapiens. This is in ways valid, yet is only very rough, and must be corrected. The appropriate VAPO and SCARLS must be applied. **SRALT** = Species Relative Rather than Absolute Form, State, Conditions and Limits of TRU. **CRAT** = Compatible Reproduction is A large factor in the Thisness and TRIA of TRU. TRU consists of and takes on many added characteristics and forms, as in the transformations to Á, This is: **LENA** = Law of the Emergence of New Forms: **LADDÄ** = Law of Added characteristics and Form, and the associated: **POHIC** = Principle of Hierarchal Continuity. Although POHIC is somewhat analogous to evolution, its hierarchal nature holds it to REM, ATO, and TIT, reaffirming the unique and special form of TIT.

Fig 3-12 **AU-AUP-TIT-TRU DIAGRAM**

Figure 3-12 features MATRU and TRUS, the mixtures of TRU and non TRU in various percentages about the core TRU, TIT, TRIT and ALT. It also features the relationship between **TRUP** and **TRUF** on the one hand and TRIP and TRIF on the other, the past and future of TRU, externally and internally. In addition it shows the relationship between **ÂÂ** or the Absolute Absolute, **AR** or Absolute Relative, **RR** or Relative Relative, and **RÂ** or Relative Absolute. TRU is the continuous survival of the best of human life that begins in the incipient RR and proceeds, as by TRIA to have RÂ and hence to the absolute, TÂ. It is the infinite, continuous approach to the victory of RR, TÂ and ÂR as TÂ over ÂÂ that is our absolute goal, aim, purpose, and undying necessity, in life.This can't be done in ignorance, indolence or without the greatest possible identity and participation of each and all persons with the life we receive and can so gloriously achieve, benefit from and savor. **LAREVÁ** = Law of TRU Revolution of Á and TÄ. If TRIT is so unstable because of the great m or A as well as the l and t revolution within and through itself, including its vulnerability to alteration and extinction, then how does this

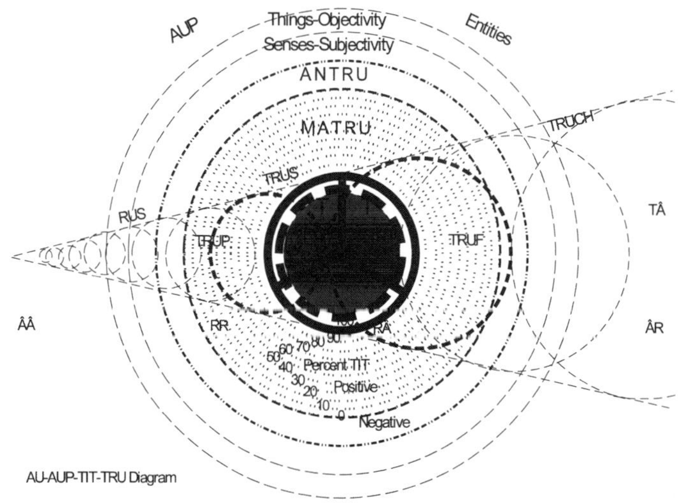

effect TIT, TRIA, TRID the ideal, and TÂ? When we reach the conclusion that we are dealing not only with a set TIT but one with OITIO, what is without and in and out of TIT, we need to know the form and function of this interchange. This is **CHOPAD** = Comparative Changes Processes and Dynamics of TRU. This diagram helps us recognize the principle of hierarchal continuity, the law of added characteristics and form, the fallacy of identifying or rendering all being with TRU, and the God defect. It also helps to develop comparative forms of TRU that enable us to realize what TRU is and is not, and learn to provide a POA to differentiate and order TRU. The figure 3-12 shows the center object TIT, and all peripheral is not TIT. The various kinds of changes that occur in TRUS, TRU, and TIT become evident, or **TRUCH**, **TROCH**, and **TRICH**. Similarly the balance between TIT and TRUCH and its rate, **RATCH** become evident. Of TRIS there are: TRINI, TRIN, **TRIPRA**, or TIT of Practical duration, **TRIR** or TIT of Relative duration. **TRIT** is The Thing, TRU based upon a set, standardized, and unchanging nature and form over an unspecified short time period. TRIT is the departure point from TIT whose static form gives maximum stability for OBT, ROT, MOT, and SPALT. These are also provided by **LOSAT**, Law of Same, or identity of the Form of TRU, and **LADES** or Law of Different Parts of TRU.

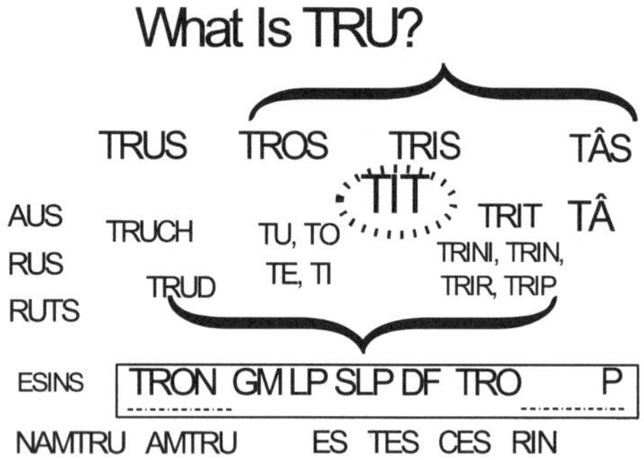

Fig 4-12 **What is TRU** or those relations, aspects, forms and states of TRU that are emphasized to show some of the less readily understood, and controversial features. It helps to define what these features of TRU are. TRU is the generic term for all that is, or is of, this special universe of best human life. TRUS, TROS, TRIS, and TÂ are its major classes or varieties Of TRUS there are many types, including TRUCH and TRUD changed TRU and dynamics of TRU. AUS, RUS and RUTS are included for the relations they have with TRU through ESINS, NAMTRU, AMTRU, TRUS and TRUCH. TIT is TRU itself, TRU shed of the observation, objectification or interpretation and all that is not pure or 100%. TROS are the various general forms of TRU and TIT, their unity, finity, infinity and totality. TRIS are the special states of TRU, as it converges on or diverges from TIT. What is TRU can be envisioned by **ATAT** or All Things At TRU. This is like ATO, a focusing, centering, and centripetal fusing of all in TRU. It is important that we always remember and apply the fact that TRU has FAF that is unique, typical, predominant, adapted, altered, warped and exclusive. Likewise **LINWE**, Law of the Indeterminant We, whereby TRU is directly or indirectly constituted and formed by various kinds of humans.

In TRUS we look at TÄ by each step in the development of TRU. In TRU and TIT we increasingly look at TÄ as a single and completed focus and form. Such contrast is typical of all that is different and thereby has and needs formal distinction. This is called: **LADNAD** = Law of Difference Necessitates Formal Distinction. **TRUS** is the plural of TRU. TRUS includes all of TRU, extending both externally and internally. Externally it is all the sets, classes and kinds of TRU that approach and leave TRU, as in TRUCH, TRUP and TRUF. TIT is the key concept, for although we speak of TRU most commonly and generally, what we strictly mean is TIT. This we must specify especially when this connotation is essential to the proper meaning to be expressed. **TRIS** are the varieties of TRU that exist with TIT. TRIS is somewhat of an anachronism for when TIT is anything more than zero time it is no longer static, but changing, which TIT, in ways by definition, is not. This is the power of both the subjective and LANERA or the relative on TIT. The seeming absolute contradiction and paradox is readily resolved. This resolution is of the most fundamental importance in understanding and proving TRU. Expressed as a law this is **LANZTIT** or Anachronism of zero time period in TIT. The resolution is called: **LARTS** or Law of the Resolution of TIT and TRIS. TRU does not exist in a zero time period, thereby we use several gradations through this time period to show TIT.

FUNDAMENTAL ASPECT	MEANING or SYNONYM	FOR ENTITIES or A RU	FOR TRU
PERSEITY	EXCLUSIVELY VIEWED	THIS ENS or RU VIEWED	THIS TRU or TIT VIEWED
ASEITY	EXCLUSIVELY or SELF DERIVED	ENS or RU from ITSELF	TRU from ITSELF, ATO
HAECCEITY	EXCLUSIVELY EXISTS, THISNESS	SELF EXISTENT RINT RIA	TRU ONLY, TIT, TRIA
QUIDDITY	WHAT IT IS MADE OF, WHATNESS	MAKE UP of an ENS or RU	WHAT TRU IS, TÄ, TROS, TRIS, ES and CES

Table 1-12 **BASIC CONCEPTS of ENS, RUS, TRU**

Theology, philosophy, and language have provided many ideas that show the basis for TRU, among which are those of perseity, aseity, haecceity, and quiddity, or being of itself, self derived, thisness and whatness. What is TRU has been described in the combination of genetic material, live product, and determining forces, or **GELD**. Closely associated is the division of **UMIA**, Universals, Mankind, Individual and IA. This was seen in **AME** or how TRU is also composed of parts other than human. Yet its focus on the individual, when TES, is paramount, and has long been recognized in the significance of personalism, or a when extended to all TRU, the doctrine that ultimate reality is a divine person who sustains the universe by a continuous act of creative will. The predominant role of speciation, especially homo sapiens, is well founded in science, anthropology, sociology and psychology. Yet like so many great concepts of knowledge, like ISMS, it can't be carried to far. Thus specie is not equal to TRU or TIT. When appropriate the part and role of VAPO, DAOST, SCARLS and all TRU must be applied. Species as relative rather than absolute form, state, conditions and limits become TRU. The typical element is compatible reproduction within a specie which for TRU makes up a large portion in its thisness and TRIA. This has, is and will be by **LASAP** or Law of Special Form to meet Particular Adaptation, Needs or Special Problems. LASAP proves **LASCER**, the Law of Special Coordinate Structure. The basic concepts of what is TRU is also seen in what is and how much TRU is determined by what is best. **LATIB**, Law of TRU is best, the best of worlds. This is seen in how if the past, as one's ancestors, were not the best you wouldn't be here. This applies to some and all of us, as TRU. **LABEX** sets this best in TRU, Law of the Best Exclusive as TRU.

One of the great modes of, creating what is best, for TRU is formal distinction. This is the power of Á in human mentality to differentiate, separate, distinguish, and formalize, as in language, the forms, composition, and all of TRU. The ability to see reality depends on our being able to separate all that is different from another. For this to be accurate, reliable, and actual rather than figment depends upon reducing this reality and making the separations necessary for TRU. It is how we can selectively bring reality and goodness to TRU. Even essences and ESINS must similarly be brought to TRU. From quantum mechanics the complementarity principle holds that one aspect of a system, as TRU, is supposed to destroy the possibility of learning about a complementary aspect of the same system, or being. This occurs, but is not always destructive. It is due to at least two factors, overlapping of forms, levels, and parts, and TE, the less than total nature of parts. The great advantage of this separation is in aiding distinction, providing a base for POA and enabling the orderly representation by posits. **LARD** or the Law of Relative Distinction helps to provide a base limit for an extended use of POA. **LAPDA** = Law of Predominant Dimensional Formulation, denotes the great role congruence has in the coordinated dimensions of TRU. Formal distinction is closely associated with REM. Both are limited by their relative nature, overlapping, and great need for selection and separation. TRU is not primarily made up or formed by a few discrete organs or parts which can be simply separated and reproduced. It is primarily, predominantly and preeminently formed and made up by its dimensions, their spectrums, continua, coordination, and intricate interdependent function; as their ALT and ACEXIN. That these are different from mlt is seen in **POIR** or the Purposive or IMU Revolution. Already largely established is the reason for the SO, and Á, the human mass energy dimension. This is so, especially by LAPDA, when combined with the configuration of its length time, or spatiotemporal dimensions. Like the definitive nature of TRU limits set the intricate structures which make up TRIT and set the stage for ROT, MOT, ALT, SPALT, Sectors, METAD, and SPAT. Limits are universal, typical of TRU, and especially KOT and TÄ. Yet limits and boundaries are much different for TRU, KOT, TIT, TRIS, TRIT, OBT, ROT, and ALT and must be compared and demonstrated for reliability. They are indispensable to TRU, as seen in: **FITALP**, Fit and Limits Prove TRU, TÄ, and ALT. It is in **POCOL** , Principle of the Convergence of Limits, we can understand the breadth, distribution, circumscribed form, order, objectivity, OBT, and solidity of TRU. Corrections go along with convergence, COL. They are closely related to LAMOL Law of More or Less, never absolutely all-or-none. Although the dimensions of TRU are largely exponential, it is adapted, altered or warped by its limits, in the extremes and means. The relative, being, limited is also focused on itself, the same. This focus is characterized by **SUPTIA**, a Superhuman Point of Reference which emphasizes both the human as TRU or TES aspect of this center and all that is collectively in TIT. The limits of TIT are set is seen in **FISPIN** or Focus Infinitesimal and Periphery infinite. These with TRIS, TRUS and RUS are increasingly dependent upon rates of change.

DICHOT = Dichotomy of TRU. DICHOT denotes what is split in two or more ways. It is an important concept and feature for reproducing the form and construct of TRU. In fact anything that is of or in TRU can be made a dichotomy of TRU, simply by COL, by dividing it by TRU, and producing, in opposition, what remains of TRU. This is called: **UNDIT** = Universal Divisions or Dichotomies in TRU. Practically for showing TRU we must extract what dichotamies best represent the greatest cleavages or divisions of TRU.

LIMITS of TRU: TRU and TIT are highly limited depending on whether this is considered from within themselves or without. All TIT is set by the thisness or being of TRU. It is because of this set state that the form of TRU can be developed with such a good frame, structure, systematic unity and representation. It is what fixes or sets TRU that is a primary determinant of its limits. These mostly result from the special relativity of TRU, how its mass energy or Á, length, and time dimensions are formed and function.

LIMA = Limits of All. Nothing is to be taken without limit, nor beyond its limits. It is only as we define, develop the meaning, make discrete, and provide the best possible formal distinctions of all that makes up TRU that limits can become well known and shown to confirm TRU and TÄ. It is our perpetual failure to apply the needed and proper limits to all, that causes us such error and failure. It explains many of the controversies, accidents, and mistakes in life.

Fig 5-12 **ENTITIES, UNIVERSALS, ESINS, TÄ, TRU**

In this diagram the nature and relations between entities and TRU are shown. It is primarily for the purpose of visualizing their TRU forms and varieties. The differentiation and distinction in, and as well as between, TRU and all else is the crux of understanding. This is the great bond, benefit and role of Á, the mass energy dimension and FAF of TRU. For it is the wonderful ability of humans to distinguish between what is or is not in and of itself, is figment or actuality, is or is not of importance, is positive or negative to TRU, and what is and what is not, that gives our special powers their potency, predominance, and success. It is this that makes up much of the form, quality, and superiority of TRU.

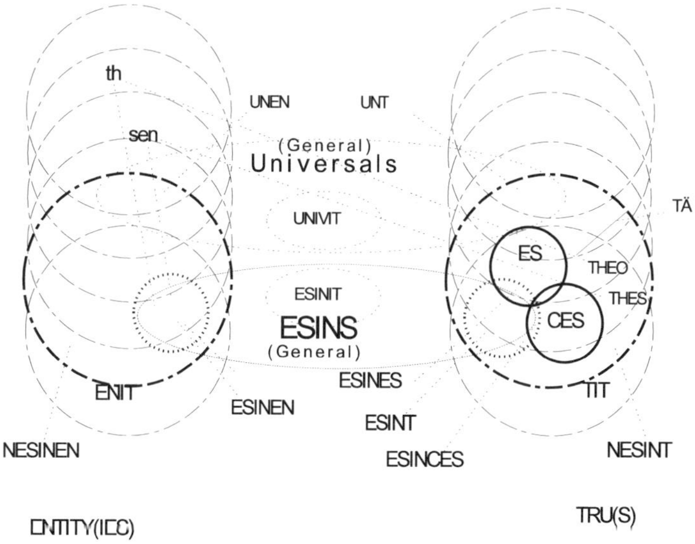

Some of the varieties of entities, TRUS, ESINS and universals include: things that as objects become THEO, the objective elements in TIT-TO, senses that as subjects in TIT become THES. and the subjective portion of TIT, and universals which work through and through that can exist either in entities, TRU, or all combined. These may be divided into, universals that are non TRU, exist only within an ENIT or entities, universals that exist only within TIT or TRUS, and universals that exist,, more or less, in both non TRU and TRU. **E(S)** = Parts of TRU, **CES** = Processes in TRU, **ESINS** = What exists in an ENIT that also exists in TIT, **ESINIT** = ESINS that exist only in and of themselves, neither in ENS nor TRUS these become ENIT or entities, **ENIT** = The Entity in and of Itself, **NESINEN** = The Non ESIN portion of an ENIT, **ESINEN** = ESINS only in an ENIT, **ESINT** = ESINS only in TIT, **ESINES** = ESINS that are their counterparts in TRU, **ESINCES** = ESINS that are their CES in TRU, **NESINT** = The Non ESIN portion of TIT or TRU, **NUNT** = The Non Universal Portion of TIT or TRU. They are especially important in helping to clarify and the learning of the structure and form of TRU. Because of the combination and union of dimensions in TRU, their set FAF and powers work in totality, or: **LATO** = Law of Summation or Law of Totality. By DAOST and RIA, nothing exists except in the relativity within TRU. The degree something is less than TRU but a part of TRU corresponds to that other portion or remainder which together equals TRU.

Focus is not only characteristic of TIT, but of its observation, which must focus on TRU and make distinct what is different in it. What is different is not always irrelative, **RELID**, or Relative in the Different. **RELDI**, or The Relative Different is at the root of TRU, TIT and TRIA. It is all the various SCARLS that set up much of the basic form and structure of TIT. This is shown by the poles of dualisms within TRU. It characterizes parts or portions of TRU in contrast to its whole. TRU is held together by powers of compensation for opposition; the differences, dichotomies and polarities that exist in TIT separately in the form of parameters and dimensions. This is evident in **LARDAV** or Law of Validity of Relative Limited Dualism and invalidity of Absolute unlimited Dualism. Opposition arises, in part, from the extremes of its polarities, spectrums, parameters, and dimensions. It is compensated by built in powers of checks and balances that hold these extremes toward their means. The three primary dimensions that characterize TIT are m, l, t, or special, predominant and altered Á, L, and T dimensions. As Á, L, T are refined they can, by TRIT and the set state of TRU, OBT and SUT, approach perfect form,TÄ. This is the purity formal distinction seeks to provide. How the future alters the present and all are interactively related is signified in: **FACAST**, Future State Alteration Actual State of TIT.

Fig 6-12
Á, L and T CORPOREAL FORMS, STATES, LEVELS, and THEIR HIGHER DIMENSIONS (ALTÄ)

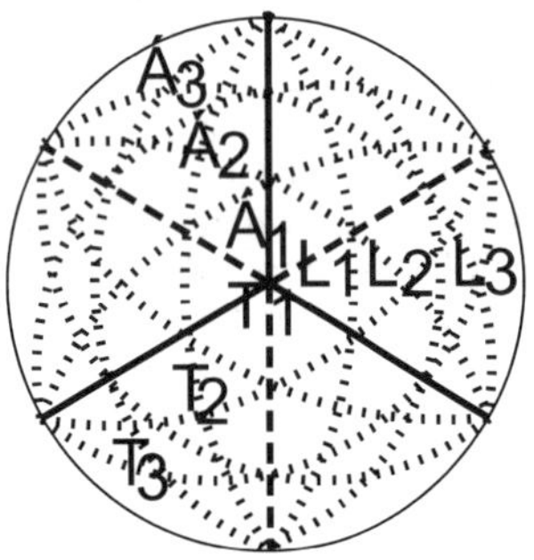

In Figures 6-12 and 7-12 the distribution of the Á or mass energy, L or length, area and volume, and T or time, duration and variation dimensions in the full embodiment or corporeality of TRU are revealed. These dimensions may be shown, as existing in TRU not so much as a simple lines, but as spectrums, continua, and sectors in the overall form and makeup of TRU.

Fig 7-12
HIGHER ORDERS or POWERS of Á

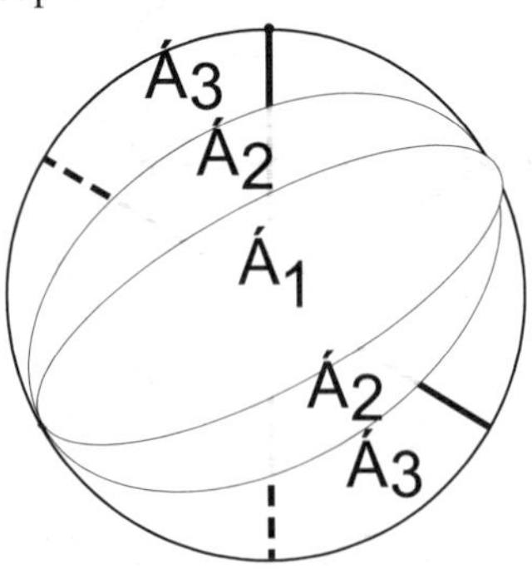

It is in this way the Á dimension of TRU can be better understood and proven. It constitutes the unique, typical and special predominant aspect and form of TRU. It is not TIT, nor even a 1:1 figure of its frame, for TIT extends infinitely and its dimensions are not so simply delineated. Yet it does, when corrected for the LOLS of ROT and objectivity, provide a good idea of the frame, structure, and function in TIT. The fact that forms of entities and objects are fundamentally different and change undergo alteration is supported by: Entasis, the curvilinear alterations in the form of structures. Entasis can be used in more than one way, either to more accurately represent an ENIT, or to increase the effect, on the observer, of a quality of the structure, or most importantly to show the actual alterations that exist in the object itself, TIT. Yet when we limit ourselves to the form and structure of TIT alone we must eliminate all external features and factors. We must reduce it to its bare state in TIT, even though this includes alterations that may seem contradictory. For TRIS this is called: **LASTIL** = Law of the Suspended State of Structural Limits of TIT. Entasis is one means of supporting LASTIL. It does so by not only showing a more actual form of TIT, but by showing how the observation and reproduction of this form produce altered, different, and added states in its structure. As we proceed through the text, especially Chapters 2 and 3, we continually add these superimposed states and structures to supplement the TÄ of TRU. Yet we must continually return to the exclusive form that exists in TIT. What this alteration or warping of the form of TIT necessitates, when made an object or modeled, is: **TANCOWS** = The Absolute Necessity, in a RU or the TRU, of Correction of Short Warping, and by: **LET** = Law of TRU Eternal. No matter if TRUF, TROF or TRIF are 10, 100, 1,000, 1 million or 1 billion years it is still TRU eternal. For within this time frame there are concentrated the whole infinite and eternal being of this special entity, TIT. TANCOWS is how to make LET be within a limited time. If the actual world of man, TRU, should be terminated or become extinct in a relatively short time, then it would still behave and function as the totality, maintaining its ALT configuration as a solid mass, or as by SPALT, a sphere. As function is abbreviated so are other dimensions in balance abbreviated by great concentration and reapportioning. This is called: **LEQUAR** = Law of Proportional and Abbreviated Reaction Equilibrium, Equal Abbreviated Reaction. LEQUAR vividly illustrates the forces of proportion and internal equilibrium that occur. These are related to the adaptive powers of biological relative universes. The maintenance of optimal shape is signified by: **TASPA** = the form, shape and structure of TRU is not necessarily spherical, but will vary with kind, and showing, of TRU. It will be increasingly solid and spherical when viewed from the distance, when the limits of its dimensions are set, all its FAF are incorporated, and its representation and modeling best replicate the dimensional formulation or TRU.

Just as the length dimension exists in higher generations of area, and volume, so do time and mass energy, or duration and variation, and higher states of Á. These higher dimensions help to show the total form of TIT and how this whole is ordered and coordinated in TRIT, OBT, SUT, ROT, ALT, and SPALT. This is also seen in: **LACOMD** or Law of Concomitant Construct and Dynamics or FAF. LACOMD determines much of action-reaction. By the combination of higher forms much of TRU is revealed that only simple m, l, and t leave void. These higher states in combination especially dynamically in TRU are called: **ACEXIN**, or the Combined Actuation, Extension and Intension of TRU. Their total form is called **LAMA** or Law of Manifold Form by Á, L, and T. The more definite, unique, and particularizing the object, OBT is, the more separate, discrete, and distinct its special coordinate structure will be.

TRU is largely the selectively adapted concordance of all its continua in a VAPO ordered special universe of best human life. Continua act as sets in TRU, as TIT likewise can be treated as a set. The continuum problem was attacked by Cantor who described the kinds of sets that are or are not denumerable and countable. It was the extension of cardinality into relativity, in contrast to the absolute, that distinguishes the continuum problem and enables its resolution. This is by TRUS, TIT, TROS, and DAOST how continua are not absolute, only infinite, relative and set within TIT, and often when extended extra set representations of TRU.

Fig 8-12 **FORMS and STATES of TRUS**

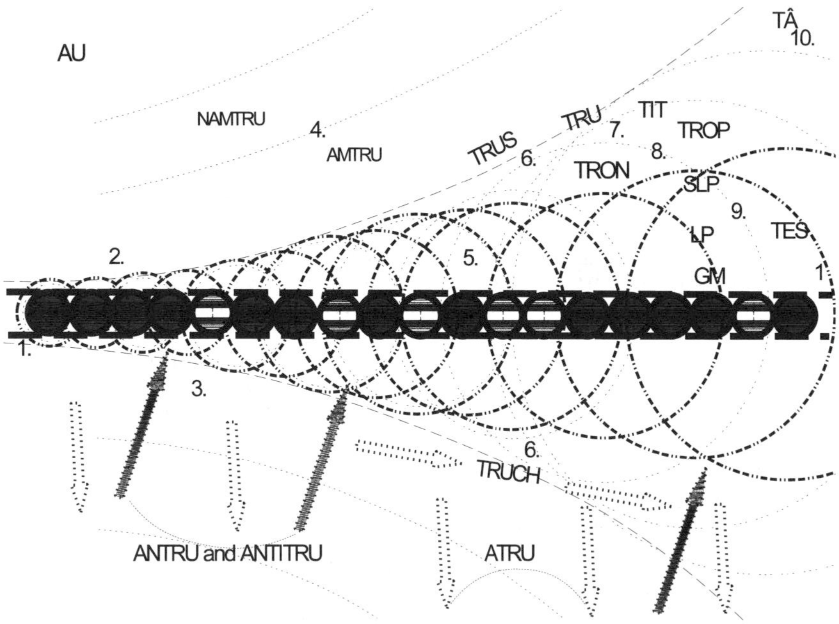

In this diagram those forms and states of TRU are emphasized that help to show how TRU is more or other than a human being, all human beings, a static universe, non variant, or inviolate. These we do by enumeration: 1. The genetic material and associated live product that carry TRUS. This has been through evolution variable from one to many, from partial to multiple in contribution to the genetic material or genotype of the phenotype, of humans. 2. The continuing overall universe of man. In the past it has had a much smaller population, fewer variants and races, as well as all the rest that contribute in TRUS. It exists by TRIA. TRU is all at any one point in time. Since the current time is the focus of TIT, TRIA and TIT are highly coterminal. 3. The forces that interact and act negatively with TRU can be envisioned as what leaves TRU, i.e. by way of loss or **ATRU**, what is against TRU generally or **ANTRU**, and what is against TRU specifically, directly or purposively, **ANTITRU**. Since AU is all that is not TRU, it is illusion by TRIA and TIT. Not so much by TRUS and TRUCH, it is in and out of AU/TRU that these forces act and react. 4. The media of these forces, and regions of mixed AU and TRU, is NAMTRU, AMTRU, and MATRU. **NAMTRU**, or not much TRU and mostly AU, is that region in which the focus is largely from AU. **MATRU**, or mostly TRU, is the region in which the focus is largely from TRU, and **AMTRU** signifies the middle ground, or combined by its own focus. 5. The great majority of dynamics of TRU, **TRUD,** occur not outside or even I/O of TRU so much as is occurs within and between all the forms, functions, parts, processes, aspects and mechanisms of TRU. These are highly set within TIT, yet are more variable with increasing TRUS and TRUCH. 6. **TRUS** is the plural of TRU and its near approaches. It is the sequence of TRU through time. It compares with RUTS and RUS or other increasingly different universes, as in the biological and physical worlds. 7. **TRU** is the generic term for TIT and TRUCH that exists currently, and stabilized by ATO and TRIA. TRU is the term we use to describe all different categories, classes, sets, kinds and forms of this special universe of best human life that more specifically converge through TRIS on TIT. 8. **TROP** is all that makes up TRU that is a product of human activity. and **TRON** is all that makes up TRU that is of nature, life, and biosphere, not specifically human . TROP and TRON include much that we ordinarily do not think of, be it man made or of nature, life, and earth. For some of these are of very little, essentially none, or even negative to TRU. They are much of what exists in the AMTRU and NAMTRU worlds that are in great interactive dynamics with TRU. Specifically TROP is all that makes up TIT that is not of the genetic material or live product, as in human society, it is the 'natural world' or biosystem that partakes of TIT.

It is the whole of the world that humans produce that enables and enhances human life. It is only that which is beneficial in the long term, as it exists interactively with the species. TROP has been highly variable. This is apparent from the many different nations, languages, cultures and civilizations that have, now and will exist and the artifacts they have left. In their own ways TROP and TRON are enmeshed, like ESINS, with the baggage homo sapiens carries. So much we think is a part of the world, is not, and what is not often is much of TRU, as seen in inventions, transportation, industry, business, and many others. For although they have some benefits, this must be weighed against their detriments. TRON or what exists naturally, but other than human, has also been variable through time. It in a way, like TROP, is not what we ordinarily expect or assume. It is only that part of the world, nature and of life that contributes to the life, actual or potential continuity of man, as TRU. **GM, LP, and DF** or Genetic Material, Live Product and Determining Forces are close to TRU. They may be taken both singularly, multiply and totally. They are the natural carriers of the species that may be envisioned statically in TIT, yet are exponentially variable with time. **TÂ** or TRU Absolute is the indefinite and eternal continuation of TIT, TRIS and fulfillment of TRIA.

TRIA = TRU is all. TRIA arises from the relativity, specificity, and overall unchanging TIT. It is comparable to and the special case of RIA or Relative is All of RU. TRU is its set and exclusive case. If TIT is all, yet as 1/∞ it is zero, then what is all? All = TRU = All, by **TRÂ**, the relative absolute. In addition TRU in being all encompassing assumes TRIA, if not TRIT, TRIP, TRUS, TRUO, TO, TU, TA, etc. How does TRA differ from TRU? 1. TRA, proper and distinct is exactly all of the absolute form of TRU, no more and no less. 2. TRU the greater, in general less distinct, and more or less other than the absolute all, TIT-TO. For all TRUS, be they TRU(all)/TRA, TRIA or other. TRIA, as TRU as a product of all that is specifically relative. It is not absolute absolute. Therefore in an absolute, extra TIT-TO, sense TRIA is not correct. Yet as it is an extreme of a polarity, just as TIT is an extreme in a polarity with TRUCH and TRUS, so does it have essential, sufficient and absolute truth. For all practical purposes, especially in short periods of time TRIA, as TIT like ATO, holds and is highly valid. This is sufficient to yield, in that dependent upon the, in which the rule of TRIA is invoked, to be a Law, **LATRIA**. We will find many instances through the text where proof arises as it is established by the rule of TRIA. These are essentially valid as stated, although to be interpreted in the absolute sense, the exceptions, differences and imperfections in an absolute TRIA must be considered and taken into account.

It is the survival of the species, and all this special universe of best human life assures, it it is to be; by definition, as well as thisness. The seeming discrepancy between phenotype and genotype, the central form and carrier of live product and genetic material in human population, as well as generations, is accountable when observed by such figures, when it is recognized that the dynamics are massive, subspecies and varieties are multiple, become massive for smaller sizes and longer periods of time, and change simultaneously.

The state of TRU defines its permanence, its static rather than dynamic nature. When this is unchanging, it becomes status quo. This is compared with TRIS, TRICH and TRUCH. The steady state of TRIR, TRIN and TRIPRA, or relative, now, and practical, is the ideal we hope will continue as a minimum. This is seen in **RÍD**, Real Ideal Differential. A similar differential exists for real, ideal, material, and mental in the actual of TRU, or **RIMAD.** They are the through and through conditions of our lives at its best in TRU. A conditional reality is the certainty of what I am thinking about an object. It originated in Descartes. It is a conditional reality because its certainty depends upon the TRU actuality of thinking about its object. This is: **LACIS** = or the Law of the Conditional Is, as this conditional is perfectly certain and perfectly TRU. The conditions under which the nature, form, attributes, characteristics and properties of TRU exist, must be qualified and specify the context each are in.

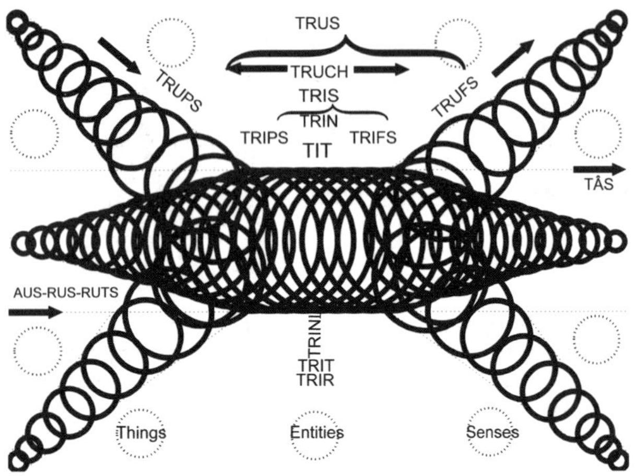

Fig 9-12 **COMPARATIVE TRUS** (COMTRUS)

Figure 9-12 shows us many TRUS in sequence. It does not show one, nor an infinite number, both of which are, by their own SCARLS, right. Yet we present this as a finite series. The finite steps produce a percentage of change. This is: **PERCH** or Percentages of Change of TRU. PERCH may be treated by the methods of statistics, by extremes in standard deviations of change from norms. What is the rate of change of TRU or of forms or parts of TRU? We can't set or adopt any standard without clearly delimiting what is given, or its universe. This is called: **LIDU** or the Law of Imperative Delimiting Universe.

What is the Actual World

How does world, universe, and god differ from The Actual World, The Relative Universe, TRU. Only when The actual GOD, The Actual World and The Relative Universe exist as equivalents, united in a totality that provides for the welfare, success, and survival of humanity can there be lasting hope. This requires that each and all persons realize, appreciate, love, and work for, this common union and totality. It requires that each and all persons overcome and reject all that is invalid, inactual, disproportional and unequal to this common union and totality. It requires that each and all persons learn, understand and live by their states, conditions, relations and limits in this actual world. That they are but a part, yet a necessary part, with many, and certain, rights and duties. What then is the difference between world, universe and gods and this unified and total actual world, god, and relative universe?. To best understand it we must first investigate what it is that provides for human survival. It is those ancestors whose progeny and cultures live on, contribute, and improve. It is not, all the past - and living - dead who have neither added to later humanity, nor contributed to its continuity. It is those existing humans whose progeny lives on and improves TRU. Not those whose progeny adds nor contributes nothing. It is those future humans whose progeny live on, improve, and extend survival permanently. It is not those of the future whose progeny does not last nor contribute to the long term survival of humanity.

Therefore The Relative Universe or its equivalents depend largely upon the adaptive, selective, physical and mental quality of existing humans and their cultures. It is truth and value based and oriented on set dimensions. These are the mass, length, time continua of the form and function of human life. It is not oriented on any physical universe's spatiotemporal dimensions, nor any world, nor any god. Only that identical, or in common with TRU. It is not based upon figment and illusion. It is not based upon no world, vague worlds, or partial worlds unless they are equal to their part and role in the Actual World or TRU. It is not based on erroneous, vague or illusive Gods unless and to the extent they are what is common and best in all Gods equivalent to The Actual World or The Relative Universe. It is the physical and mental health, strength, and predominance of persons who best produce superior effects and additions to this special universe of best human life. Unless each and all persons can learn and live by this difference and knowledge of the worlds of which they are a part, they and their worlds will be in conflict, suffer, fail, and become extinct.

Other ideas that help to clarify, make distinct, and strongly prove the difference between The Actual World and all others: 1. Meaning of Life and Death, 2. What the world, universe, God are without Humans, 3. What a person and all humanity are without the world, universe, and God, 4. Focus or no focus, being awake, or being asleep, consciousness and nothing. 5. All Major Proofs of TRU, 6. All other proofs of TRU, 7. All proofs of other TRU related concepts that relate to what the actual world is, 8. The set state formed and functioning in TRU, 9. Better, simpler, and more common or strikingly homely evidence of the difference, includng, personal needs, how to survive, make it, succeed, and be happy in life.

Fig 10-12 **SETLAW, PERT to SUT**

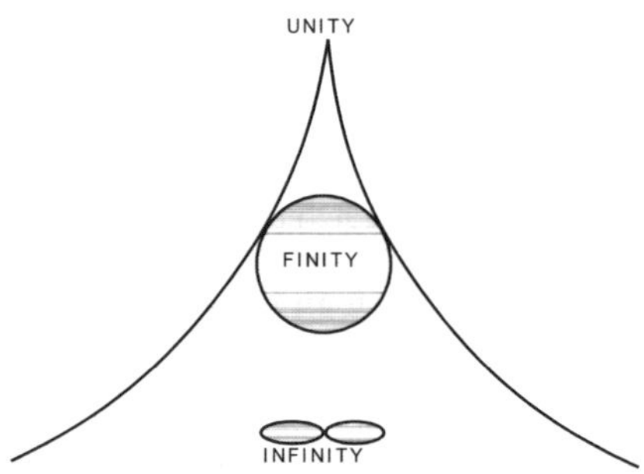

SETLAW is the state in which we find ourselves when set or fixed in any position between unity and infinity in our relationship, identity, perspective observation and objectivity of TRU, its parts or associations. It will be variable to the extent these are variable. Yet it will be fixed, thereby set - SETLAW - to the extent these are fixed from the differential of TIT, the finite shown by the central black circle. SETLAW is what makes us so certain what is, is, as our words, views of the worlds, and objects before us. Yet these by SETLAW are highly limited and must have appropriate SCARLS when applied and in operation.

How SPOT, the Standardized Perspective of TRU, is 1/2, or half way from, perspective or PERT to SUT is shown in Figure 10-12. This is roughly half way between unity and infinity. SPOT, as in epistemology, is the portal to SUT, ROT and MOT of TRU. TRU can be imagined as a tiny speck, dot, circle, sphere, mass, solid, infinite composite or as an infinitely large yet totally and comprehensively finite object. It all depends upon our perspective, PERT, SPOT, IA and TRU. TRU and SPOT are neither SUT but become a set. This is SETLAW.

SCARLS help identify much that is not TRU as well as much that is TRU. A line can be drawn to represent any developed or formed relation. Equality arises from what is similar, or related, in the nature, makeup, form of TRU. Much that we take for granted as being factual and valid may not be so. By TRU and its SCARLS it is only relative or conditional, within limits or of a certain range, or in a certain way. This explains the Heisenberg uncertainty principle. This principle is largely a corollary and qualifier of LANERA and LAMOL. It qualifies, or SCARLS, ENS, RUS, and TRUS, by showing and helping to prove their dualities and dialectics, those of: position and velocity, static and dynamic, and form and function. It shows how in all dualities and dialectics, as one extreme becomes more predominant, the other becomes less; as one extreme becomes more certain, the other extreme becomes less so. They tend to show and help to prove the relativity of all, all based upon a focus, or TRU.

LACUR = Law of Accumulative Reason. To state that one and one is two is true, in many senses and what it applies to. For no number is an absolute, only infinitesimal passing through. Yet for being absolutely true, or true in the actuality of TRU, is a beautiful example of how it may not be exactly true. This and so much that we take for granted and how interpreted as being absolute truths are not entirely nor always so. Only relative or conditional by SCARLS, or only within limits and a certain range. It is only when we can accumulate enough data and information, properly apply it to all SCARLS and best reason to interpret it to bring it to its actual state, that truth returns to TRU and is achieved.

The bonds and relationships between the various components, parts, forms, features, aspects, concepts, principles, laws, and mechanisms of TRU are the most problematic and critical. This is seen in:**SORAT** =Solutions from Relations are the basis for ACEXIN and especially the Á form, of TRU.

Fig 11-12 **POSITIVE Á, MASS ENERGY, TRANSFORMATIONS**

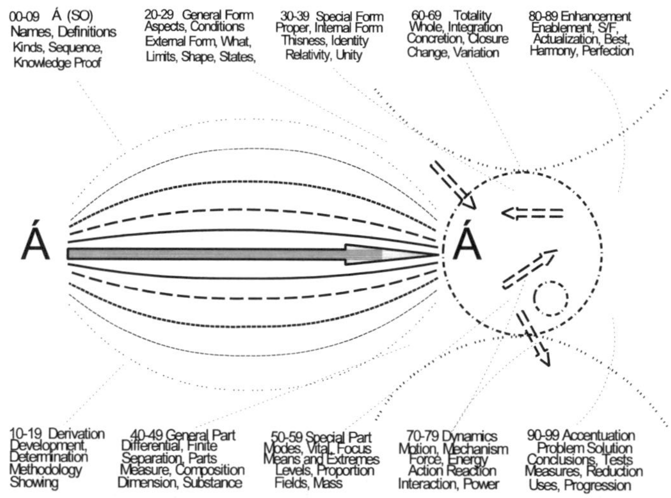

This figure helps to show the mass energy, Á, dimension and continuum of TRU by describing its different relations and roles in each section of the text. Like TRU and all entities it can present many different faces, each adding to our knowledge of it and TRU. It progresses from the initial given of relativity, the observer observed relation in the introductory section, to its fusion in the absolute of TRU in the last section. The transformations of mass from early life forms to TRU is called: **TRÁ** = Transformation, transmutation of Mass, of Mass Energy. The transformations that exemplify, produce and make up totality are guided by: **LOMT** = Law of Total Mass of Transformations. LOMT is what makes up much of TIT. TIT is a whole series of transformations that are variable and working in many ways. LOMT is a functional effect on form, an effect of time. The TRA that manifests itself as the Á form, state and dimension, and is so characteristic, predominant and preeminent in our best of worlds TRU, is not so extraordinary as might be believed. On the contrary, it is most common; from TIT by relativity and ATO. TRA is closely related to the different kinds and levels of energy that exist. To a great extent TRA parallels the progression of TRUS. Yet we must be careful not to forget TRIA and ATO. The DOT of TRUS, by RUS and evolution is much different than the thisness and ATO of TIT. This is: **LÁDATOP** = Law of Á State in TRU Special with its own Predominant Mode from its Thisness, TRIA, ATO and TÂ. Necessity, with existing form and function, largely determine the form, configuration, construct and structure of TRU. This is by selection thru many generations. When limited to its dimensions necessity determines the form, dimensions, mass energy, length, and time of TRU. The cybernetic processes in TIT show how its mass energy, Á FAF differentiates and has developed means to satisfy the need for communication, in words, speech, and composition. To uphold the needed level, a mandatory capacity of language and memory must be maintained.

Fig 12-12 **TÄ FEATURES, SCARLS of FORM of TRUS**

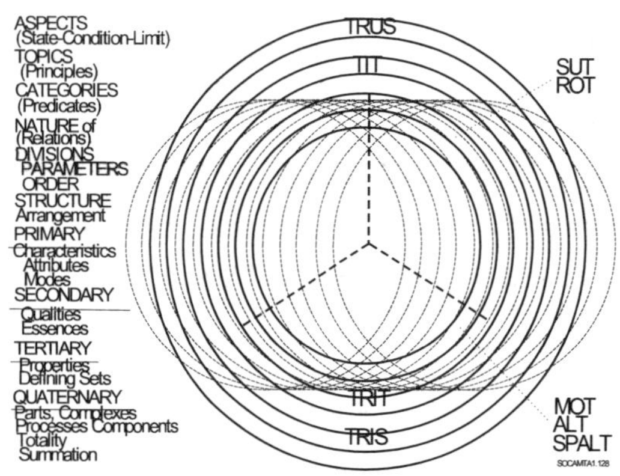

How can we most effectively and realistically distinguish the different major aspects and forms of TRU? How can we provide a means to render both good art and science of TÄ. The accompanying figure joins some of these in TRUS, TRU, and TIT. It will be quickly recognized that one of our main stumbling blocks in best understanding the world is in distinguishing those major aspects of TRU that are other than its ES, CES, and dimensional FAF. These are often many of its greatest concepts, principles, laws and mechanisms. Whether they are primary, secondary, tertiary, or quaternary or at other or mixed levels. The <u>Primary Forms or States</u> are those that are irreducible and thereby most basic and fundamental. These are those forms or states without which TRU couldn't exist or would entirely lack division. The <u>Secondary Forms or States</u> are those, as qualities and essences, which can be reduced to even more basic forms, yet are essential, vital or crucial to TRU. They are more numerous and less basic. The <u>Tertiary Forms or States</u> are those which are less than essential qualities, yet are more than a property, characteristic or defining set. and The <u>Quaternary Forms and States</u> are those at the level of what defines and describes TRU, as its properties, characteristics, formative parts, and mechanisms. This form and state is more than parts but consists of combined groups of parts, processes, and complexes existing in or showing TRU.

The interdependence and interaction that is so dependent upon relativity in TRU results in need or states in which propelling and compelling forces and actions require reaction. This reaction may be of many kinds, only in resultant it must be what is optimal for TRU. TRU is in a continual dynamic state, guided by: **LANËT** or Law of Absolute Necessity. This takes many forms and courses. The focus is the Imperative TRU Instantaneous Attention, Immediate Awareness, and Injunctive Action, and of Absolute Necessity of Identity, **LATANÍ**. There is an absolute necessity for stability to satisfy TRIA, yet accommodating reversible change. To satisfy ROT and MOT and return of OBT to TIT there is **LANOFAC**, Law of the Absolute Necessity of Formulation and its Conditions or qualifications. Throughout the text, and all life, there is **PÏ** or the Positive Imperative, that absolute necessity of positively and most effectively fulfilling the target and ideal of TRU. To fulfill PI and the ideal of TRU requires **PORC**, Principle of Coincident Responsibility. This more than the duty sufficient to satisfy TES and TRU It is a kind of duty that permeates all TIT, in which all does its part optimally. parts, processes, and complexes existing in or showing TRU. Below these are lesser forms, states, ES, CES, and others.

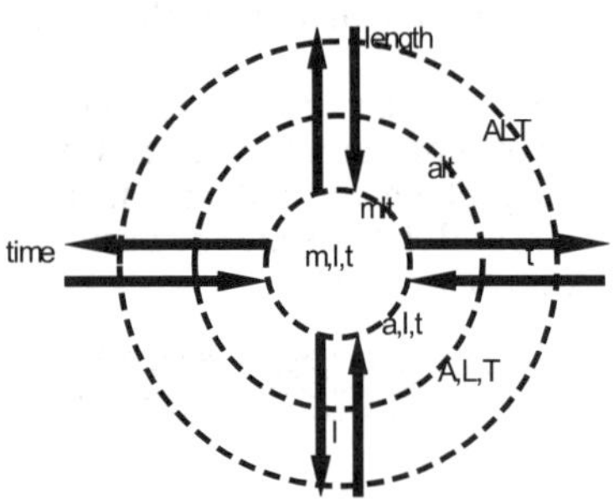

Fig 13-12 **MLT ALT FORMATION** (MLTALT)

REASONS and DETERMINATION

How the mass, length and time of TRU are derived, developed and determined is fundamental for a well coordinated understanding of life. The dimensions and continua they produce in TRU have some similar features. For all three, their qualitative and quantitative form arises very gradually, develops, and are adapted to and determined by many factors of TRU. Although it was through the RUS, and TRUS that these occurred, it is within TIT that their derivation, development and determination exists as set form and function, like ontology follows phylogeny. This is: **SÄFMAD**, Set Form and Function in TIT the Manifestation of its Derivation, Development and Determination. SAFMAD signifies the living necessity of the special form and function in TRU, and of its derivation, development and determination. Whereas TIT is not something of the unrelated or disassociated past, and whereas by ATO it exists, at once, this FAF carries within it the seeds, growth and dynamics of its own origin. Not the least of these is self preservation, the self-sustaining nature of the species, its universe. It is how the actuality of an entity, RU, TRU or TIT is in a continuous process of reproducing itself from its potentiality. This is: **LACPA** = Law of Continuing Process of Actuation from Potentiation. This, as all TIT and TRU functions by **MARD** or Mass Adaptive Reversibility Dynamics, a continual TRUD. It is in LACPA that ATO is to be found. LACPA is primarily an actuation from potentiation that is within itself. By TIT and TRIA this replenishment and reproduction is largely and reversibly from within. Both the qualitative and quantitative aspects of the l, t, and m, or states, conditions, limits length, time and mass energy or Â, L and T for TRU, dimensions are set, and set themselves in certain ways. The l or size, t or time, and A or mass energy and their relationships are set by the surroundings from which they originate, as well as the innate form and forces which tend to maintain themselves in the whole. The surroundings, in turn, are selected and adapted.

	BARELY NEEDED	SLIGHT NEED	SOMEWHAT NECESSARY	NECESSARY	LARGELY NECESSARY	SERIOUSLY NECESSARY	CRITICAL	CRUCIAL	VITAL	IMMEDIATELY VITAL
IN ALL		TRÄT	LUT-085	SACET-013	LEXTRIM-030	RUEIOA-010	SÄ-014	LAPE-019 OITIO	CODIT	LATAN-029,066
MOST ALL				LEBA-023	FER-014 031,083	ARTÄ-021 DIC-085	LORÄ-054	PÏ-029	FAF-020 LIS-080	LUF-019
GREAT MANY	VIRA-031	LARTRA-030	UEF-019	LIVADAC-072	LATAN-029	SEVI-013	LATAW-030 IAE-081	RAL-055 SÏ-004	VACMÄ-038	LÄD-034 LÄRD-034
MANY			UCT-084	LAPED-021	DOR-032 MANT-097	OPIA-012	LATANS-029	LOR-032	OFI-040 LATANOS-053	LANOFAC-036-7 AITES092
SEVERAL	DÄ-031			LAPAC-054	FEMOT-020	LASEX-025	EFIA-012	MARAP-032	LIVDAP-067	STAN-052
FEW OF			OPI-024	LCDS-035	DIRP-037	LACON-015	PÄFOS-022 024	SACDI-028	LAD-042	LANOE-054
DUAL OF			ÄÄÄÄ-021	LENAD-014	LORE-049	LANEC	CASPERT 013	PODA 065		
IT ONLY			ÄD-021	IMPIA-012	CRIAN-014		ONO-064 LOCU-083	LOIR-025	LATSNUP 063	
ONE PART ONLY							CORMAS-038			
ONESMALL PART			DÄSIA-031	MI-013						

Table 2-12 **FORM by UNIVERSALITY and NECESSITY**

In the above table we categorize the form of an entity, object and TRU by their gradients of content depending upon whether they are from one extreme to the other universal or particular, and little necessary, highly necessary, to what can not exist without them. In this way we can not only qualify the form of these entities, we can quantify them. What may appear to be necessity to us may be far different for TRU overall. This is especially so when we consider TRU from the standpoint of TRUS, TRUCH and **LAGSTACH**, Law of General and Special TRUCH. Although most of the terms used are for entities and objects and therefor derived from the introductory section, many of those for this section on TRU proper, and following sections, can be added. Those given, however, provide much of the ground work needed to establish the universality, form and necessity of TRU.

Necessary form corresponds statically with the more dynamic necessity, both being an aspect and property of relation and dependence. It is the reason and cause that occur in relation and dependence or imperative states, conditions, actions, and reactions referable to the given or set. Necessary form, as necessity, therefore is the FAF that should, or must, follow existing FAF in the set or TRU. Thereby both in TIT, as well as of TRU and its ROT and POT, there is much necessity and necessary form that follows existing TIT. This is, in turn, necessary to maintain, sustain, and assure the continuity of TIT and TÂ. There are both universal and particular forms, features, compelling forces, and actions. In ROT and POT to sustain necessary form requires that this form be so modified or formulated to both replicate TIT, and optimally satisfy ROT, MOT, ALT, SPALT, and POT, or LANOFAC. Universals as through and through, or **TAT,** operate out of and in TRU to set up a duality of TIT for TAT. This focus of absolute necessity of Identity in IA or **LATANIA**, and the associated imperative TRU IA or **IMTIA**, is the core of TIT. The Absolute Necessity of TRU or **LÂNET**, becomes clear when we learn TRU and recognize its equivalence to the critical state of life, TRU. **LATA** or the Law of the Absolute Necessity of no, or from, change of TRU is a corollary of LANT and TRIA. **TÂN** or the Absolute Necessity is a product of **TÂ** or TRU Absolute and TRIA or TRU is All. **LANOFAC**, or the Law of the Absolute

Necessity of Formulation and its qualifications, Conditions and SCARLS is the central theme of this work, and much of ROT, MOT, SPALT, ACC, and POT. Without this there is: **LACID** or the Law of Categorical Integral and Differential Illusion. The deepest meaning of LATAN is separation and determination of all that is TRU. The importance and necessity of language, vocabulary, conceptualization, memory, utility and responsibility as they show the transformation to Á to enable an ever greater positive imperative of and in TRU summarizes this topic's contribution.

The form and function of a RU or TRU depends upon what has enabled, and is best enabling, it to survive. These are their predominant capacities, abilities and powers that provide health, strength, and continuity. It is what maintains it in its state of rest before and after it is acted upon by other forces, actions and dynamics. This is called: **LANDACS** = Law of Necessity Determines Form and Configuration, Construct and Structure. Since these primarily depend on the dimensions and form of TRU this becomes: **LANDALT** = Law of Necessity Determines the Form and Dimensions, mass energy, Length and Time of TRU. How LANDACS and LANDALT develop and become qualified and quantified in TRU is called:
LEVA = Law of Finer and Finer Variations in Form.

The target and ideal in necessity of TRU, was noted at outset of CIPTRU, Chapter 00. It was expressed in PÎ, The Positive Imperative. PI in its most general form signifies the increasing relationship in attention by subjectivity to some object. Imperative is what is demanded, necessary, mandatory, obligatory, binding, compulsory, requisite, critical, urgent, decisive to absolutely necessary. Positive signifies that there is a definite, usually increasing certainty in attention. This is an attention that is directed in a set and formal way, which by TRIA, can only be to TRU. Imperative signifies that we must. So that PI signifies that we must have a definite, increasingly certain attention to TRU in a set and formal way, as TRU is TO. The basic impulse to life, to total TRU, ENT, TÂ occurs when PI acts by the higher powers of ACEXIN in summation acceleratively, or PI^n, **PÏN,** as in concentration upon **TORP**, Total TRU Representation and Presentation.

LÅCBEST = Law of Allowance and Control of What is Best for TRU. LÅCBEST has as its primary emphasis RATCH, rate of change, and not so much TIT or AU. It is how TIT can be TIT and TRID in an AMTRU setting over prolonged periods of time. LATRÄ(→Á) is a great part of TRU, the greater our cognitive powers become, as COGNOS, the greater the likelihood and opportunity for LATRA. Yet the more concentrated or more centripetal TRU becomes the more stable it is. So that we have, working simultaneously, both destabilizing or changing forces, and stabilizing and solidifying forces. It is the optimal balance of these that provides eternal stability and solidity of TRU. This is its immortality and total absolute being, TÂ.

The subject object or relativistic proof of Á is through:

1. You can't have thought unless it is <u>thought of something</u>,
2. This thought of something must have <u>meaning</u>. i.e. it can't be without some significance or connection to the person,
3. To be relative and relative together to a whole it must be relative to something, someone, or some <u>universe of action</u>,
4. To have meaningful thought of something it must be <u>relative</u>. Not only relative to each other, or thought to something, but both <u>together to a whole</u>,
5. To be meaningful and relative it must have validity, actuality, proportionality and totality or **VAPO**. To not be these or to be any one or combination of these without being all of them will lack complete or perfect meaning and relativity.
6. To be VAPO it must be, or potentially be, **TRU**.
7. There can be no union of sense and thing, subject and object unless there is a relation, a bond, a connector or a <u>vehicle</u> which brings them together.
8. What this relation, bond, connector, or vehicle is, is <u>special</u> for each RUTS and TRUS.
9. For this special universe of best human life, this relation, bond, connector and vehicle is largely made up of speech, <u>language</u>, <u>terminology</u> and <u>communication</u>.
10. It is in both the qualitative and quantitative refinements and expansion of the subjective, understanding, communicative, real or meaningful language, and the objective or selective action that this <u>predominant FAF of TIT exists</u> and is to be revealed, known and best used.

These together, in summation, and their added states in totality, as the form and function of TRU, are Á, the mass energy FAF. state. dimension and continuum in TRU. They, in turn, provide its proof, PRÁ. **PRUSÁ** = Proof of the Separation and Summation of Á's, or how to best make the gradients of the Á Dimension, or Predominant Mass Energy FAF of TRU distinct. **PROSA** = Proof of the States or gradients of **Á**. We know for certain the dimensional construct of entities, bodies, and much of RUS and TRU. It is quite evident these are by DAOS both different and same. That, as we pass from the general to the specific and back again, depending upon the observer-observed orientation of SER and GER, this departure from the general and physical increases. So that by the time we reach TRU, and especially TIT and TRINI we are speaking of a dimensional makeup that is highly unique, specially adapted, altered and warped in comparison with the physical world. When we do, and interpret them by E, ES, CES and formulation of TRU, as they exist in this world, TRU will become much more apparent and demonstrable. When TIT, THES, THEO, SUT and ROT are not fully developed or described their expansion and completion must await ongoing investigation, research confirmation or needed changes or variations. Yet its crucial form and function in TRU can not be denied, and its incompletion should only add to challenge the future and produce its resolution. Through the S - O relationship Á can be continually upgraded and substantiated.

The lability, yet vital importance, of speech, language and vocabulary, and memory necessary to produce optimal communication, understanding and survival, should be readily apparent to all. This is: **LAMAL** = Law of Mandatory Capacity of Language, or Law of Mandatory Capacity, in general, **LAMAC**. There is a great need for and value in increased expansion of language and vocabulary. As TE increases so does TRU. The increased capacity of the human brain, especially its use of words has been a great force in TRU'S expansion and success in the last tens of thousands of years. Capacity to conceptualize and communicate through ever improved language were undoubtedly one of the major, if not the greatest evolutionary selective power. This still continues today and will in the future probably at an increasing pace. There exist forces for more separation and differentiation of TE. The more substantial, distinct, valuable and practical these words become the more they perpetuate themselves. This is, as the great role they play in Á, the mass energy form, state, dimension of TRU, a large part of TRU.

IMTIA = Imperative TRU, Instantaneous Attention, Immediate Awareness, and Injunctive Action. Since TRU is all by TRIA, and since the IA is the crucial, pivotal, and great determinant focus and relativity of our being, it is imperative that this **IA,** as Instantaneous Attention from Immediate Awareness to Injunctive Action be most completely and efficiently TRU. It is the neglect and denial of IMTIA that is at the root of so much human mediocrity, failure and misery. It is the discovery, development, utilization and practice of IMTIA that is so much the great potential of our being. It is in understanding and using the awareness to action through attention, S-O, or Á nature of IA as well as TRU itself that we can fully come to realize and recognize the pivotal and crucial greatness of IA and TRU. How we with IMTIA, actually participate in TRU. **LATANÍÁ** = Law of The Absolute Necessity of Identity as A or S-O is perfected and fulfilled as it is TRU.(OITIO,TA)

CHAPTER 13

SPECIAL FORM of TRU

This is the core of a presentation of TRU, the form, **TÄP,** of TRU in Itself, **TIT**. This is sui generis or, of its own kind, what is most unique, peculiar, characteristic or, to have exclusive qualities. It is human existence that incorporates all that enables it to be, a relative rather than absolute existentialism. Where an absolute is any and relative is existentialism that is determined only by its own kind. Key to understanding the special form that is only in TIT is proof of the subjective form of its mass energy dimension. This is reinforced by the bond between TIT, human, and the first person, wherein the subjective, mentality, rationality, intellect, and self-conscious set themselves off from all else. This is shown by how all by ATO is itself. In **TRUR** or TRU reality we must recognize its contrast with common reality and actual reality, **RECT** or Reality Equals Convergence with TRU.

The actuality of TRU is recognized in and reinforced by **COPOR** or the Comparative Coincidence of Points of Reference. TRU actuality is also recognized in **LÂT**, Law of the Non Absolute in the Relative or how although in perspective, senses and things the subjective and objective are separated, in TRU actuality, they are combined, or the Á mass energy FAF of TRU. This combination may be of many varieties, proportions and levels seen in SÁ. Differences produce opposition and oppositions produce contradictions, even in the actuality of TRU. The distinction needed, for the special meaning that TRU imparts, is for actuality. Actuality is in TRU its proximate principle by virtue of which the operation is produced. This determines and perfects TRU. Formal actuality is like what scholasticism describes as perfecting the entelechy of the body. Entitative actuality is that in an entity. How actuality becomes formally TRU is the ultimate principle. When ultimately actualized in TRU we can increasingly make formal distinctions, especially when total, and parts of a total, TIT-TO and ES. Actuality is highly equivalent to Á, as Á is to actuality. There is actualization by actualization when concentrated in TRU.

PRÁMS, Proof of the Subjective Form of the Mass Energy dimension of TRU, is a crucial concept. We can not adequately recognize the special, unique and exclusive FAF of TRU unless we discover, remember, and use the contribution and role of the subjective in the basic dimensional polarity, state, and makeup in and out of TRU. What enables TRU to be is its relativity or predominant form and function. For man, and thereby TIT, this is subjectivity or mentality, and of a special and exclusive kind. It is through subjectivity that humans are largely activated and are predominant, along with their opposites and the entire Á continuum. To be, exist, and survive, these objectivities that are in opposition, are assimilated into a person, in TIT. The culmination of the progression of SA gives humans as TRU their special and exclusive existence. This is by **TRI** or TRU-Human- I-IA, to the extent subjectivity, mentality, rationality, intellect, and self consciousness are actual and pure. This is man's edge in the world. It is the edge that makes TRU and TIT competitively alive, successful, and what they are. Subjectivity works in all directions, whose vantage or center is the present, the existing thisness of TRU. This is shown by how all by ATO is itself.

LACSE = Law of Correction of the Subjective Error in Cause and Creation. LACSE is the solution to the problem of recognizing ATO, wherein creation and emergence can be identified with ATO and TIT if their external subjectivity is excluded. LACSE applies to more than cause and creation, it applies to all our consciousness, awareness, attention, IA and SO, as it is **SE** or of the Subjective Error. It is how we depart from the Á form, state and dimension of TRU to take up with only one aspect of it, a disproportionate part of it, or even worse something not of, or negative to it. This unbalances it, for which one of the corrections is LACSE.

Although creation is more evident in religion, TRUS, and the stages of their development, this is not the exact nor full role of creation in TIT-TO. Without TIT-TO creation signifies an extraneous role. The numerous and additive positive alterations to existing progressive TRUS as they become agents in the FAF of TIT may be called creation. **CRACT** - Creation must be SCARLED to be TRU, TIT-TO. Creation, by whatever meaning we have given it, or how it is interpreted publically, is not the same as it is by TIT-TO. Creation for TRU and TIT is not a simple, singular, invariable, idea or process. It is repeated, internal, complex, multiple, and variable depending upon our subjectivity and upon the objective status of TRU at hand, TIT-TO.

What is special is one of the greatest explanations and proofs of TRU. This is most evident in the reproductive biological exclusion by species as a breeding unit. Although not absolute, it provides separation and delineation of what is and what is not human. **LASPU**, the Law of Speciation Unity, holds that it is more the depreciation of the massive equilibrium of the whole, rather than capacity or some element alone, that produces extinction. **LATRASP** or Law of Relative Speciation or how speciation consists of many gradations or levels. For TIT its focus is fixed in the set, and may be treated as an approach. Incorporation of speciation is the result of such exclusion. This means that although the specie has individuation, and is composed of individual members, as a universe in itself it is different. It is made up of not only individual remembers, or even the specie itself, but the whole universe it relates to or brings to itself. And even this is special in the sense that what relates to it or brings to itself is so only as it selectively exists and participates in the whole. **LINCORS** = Law of Incorporation of Speciation. This certifies the fact that a specie may be an individual animal biologically, but as a RU or TRU it is different. TIT, as any RU must be made up not only of that which is its center, its relativity, its frame, or major mechanisms and processes, but all that enables it to be and this in the right proportion. This functional aspect of LINCORS is called: **LENDS** = the Law of Enablement and Enhancement dynamics of speciation. LENDS certifies the fact that TRU exists only to the degree all that enhances or enables it to exist is included, and included to the typical extent. This feature, makeup and form of TRU, TÄ, which is focal to its being and speciation, and which is the core and nucleus of **TÄP** or TRU form Proper, is its thisness or haecceity. This is its first substance, substantia prima. It is what not only externally focuses toward The Thing, TRU, it is what internally focuses upon The Thing, TRU. It is what makes it a 'this' and a 'the'. It is what helps to retain its TÂ/TRUCH or stability and permanence. It is what enables it to be TIT TRU, The Thing rather than any other type of thing, as in COMTH. It is how The Thing by **TÂ**, TRU Absolute, is exclusively TIT, a This. It derives from RIA and in TRU is formed by TRIA by preserving being itself, TIT. In Latin it is called *aseitas*, being by and of itself.(ATO) The functional aspect of this is the enhancement and enablement dynamics of increasing levels of speciation. The incorporation or embodiment of TRU is seen in **CARSC** or the capacity, ability, actions, states and conditions of which it is formed and functions.

LASOR = Law of Subject Object Relativity. Object is to thing, as subject is to sense. They are special cases or states of things and senses as they are constituted by relativity, the observer observed relationship. Object and subject therefor are neither initial entities (ENIT) final phenomena, illusions, figments. nor partials, ES of life or TRU. They are neither all AU, nor all TRU. They are usually a mixture, not absolute. Only at extremes are their decreased relative pure object or subject. Only as they become polarized together forming reals in TRU do they take on actual form. This form is highly coterminal with the Á or mass energy, predominant relativity, observer-observed and therefor subject object, dimension of TRU. It is in this way the subject and object become actualized in TRU as they are the form Á, and its gradient SÁ, with the ES and CES of TRU. This occurs as their ESINS are selected and extracted from entities, things and senses in themselves, and subjects and objects in themselves fuse to become SOT, their combined subject and object counterparts, ES in TRU. That the observed are not independent is intuitively obvious, relative to the observer, and proven by special relativity.

The greatest significance of thisness or haecceity lies in its preeminent role in the primary and predominant nature and state of TRU. It is this that gives us the kind of mass energy, observer observed form and function, or A dimension and state that constitute TRU. It is this very fact of the-ness and thisness that typifies the set form of TRU. This is Á, from the union of observer and observed, that TRU becomes a special, exclusive and all encompassing relative universe. **LIASOC** is the Law of IA, SO, Object, Action, Completion. This, to be complete, entails an actualization that is ACEXIN, inclusive of all that is TRU. To so objectivize and act completely requires a POA or Positive Objective Approach. This is anecessary attitudinal attribute of TRU. It is how ALT is used in Homation to actualize, by special adaptation and exclusion, all TRU. This is how actualization is TRU, not as commonly understood in current usage. Such Á, or Actualization of Actualization, as they are in TRU, is signified by the term: **LATAC** = Law of Actualization of Actualization, by TIT. LATAC is how all becomes TRU through the power of the form and function of TRU as they select and assimilate it as it best suits or fits TRU. LATAC and LIASOC may be easily understood as all thoughts and acts are or are not as they combine practically and successfully in contributing to TRU. This is how what is actual in TRU works in or out of TRU in thoughts and acts to return in TRU, LATAC. Whether the world about us is absolutely 0% or 100% inactual is not to significant. Only as it is 100% or approaching 100% actual by TRU is there fundamental and vital significance.

	THATNESS/ PERSEITY	THISNESS/ HAECCEITY	EXISTENCE	WHATNESS/ QUIDDITY	ESSENCE
RUTS	Biological Species, Animals	Species and Animals Having Individuation	Life of Death Survival of Species	Species or Animal Form and Function	What within best enables them to Survive
TRUS	Human Species and Varieties	Human Species & Varieties Having Individuation	Life and Death Survival of Human Species	Human Species Form and Function	What best enables Man to Survive
TROS	Human Species Aspects of its Whole	Human Species as it is in each Aspect	Human Species as it exists in each Aspect	Human Species as each Aspect makes its form	Human Species as each Aspect best enables Survival
TRIS	That Special, Focal Best of Worlds of Man	This Special Focal Best of Worlds of Man	Best of Worlds As a separate live being or whole	Aspects and Makeup of the Best of Worlds	What Aspects best enables this Best of Worlds to Survive
TIT	That Best of Worlds of Man In-and-Of-itself	This Best of Worlds In-and- Of-Itself	This best of Worlds In-and-Of-Itself as an Exclusive live Being or Whole	Aspects and Makeup, all that describes This Best of Worlds In-and-Of-Itself	What Aspects, Form, Parts, Best Enables This Best of Worlds In-and-Of-itself to Survive
TÂS	That Absolute Best of Worlds of Man,	This Absolute Best of Worlds of Man	This Absolute Best of Worlds of Man Assured Being	Aspects, Form, Makeup, All that Describes This Absolute Best of Worlds of Man	Aspects, Form, Parts Best Enables This Best of Absolute Worlds of Man to Survive

Table 1-13 PLACOMER; COMBINATIONS and PERMUTATIONS of KINDS and FORMS of TRU

In IPTRU, as in all TRU and life, we are constantly confronted with the problem of formal distinctions. How far is it necessary to carry the separation and identity of things to make them as perfectly representative of the object, and identical with TRU, as possible? Nowhere is this more critical than in the basic distinctions of TRU. These include the thatness, thisness, existence, whatness and essence of TRU. This is further complicated by the kinds of TRU, i.e. TRUS, TRIS, TROS, TIT, and TÂS. So that the role of combinations and permutations is continuously upon us. This is called the Paradox and the Law of: **PACOMER** or **LACOMER** Combinations and Permutations. They are largely a matter of differentiation, our ability to make their distinctions, and uphold TRU. As their natures are recognized, within their own SCARLS, knowledge, philosophy, theology and science will become increasingly definitive by TRU. This is called: **CIT** = Completion of Identity to TRU. As ESINS and the participation of universals in TRU, CIT represents how these and other topics become more precisely known through TRU. CIT exemplifies the proof by which **SIM** or the Sequential Imperative is derived and proven, from the initial to the final, the stimulus to the response, from the problem to the solution, from the introduction to the summary and end of presentation.

The reason why TRU has not been properly recognized before, is largely because it has not been recognized adequately, appropriately, exclusively, coordinately, formally, nor completely. TRU is more than of ideas whose totality humans can readily grasp, conceive, or accept. For as hard as TRU may be to understand, this massive, dynamic and dimensionally altered formulation of the world is often quite common in our culture. It is necessary for one and all to rise above their disbeliefs, false or limited opinions to reach a total coordinated understanding of the actual structure of TRU. When we consider TRU in this light we can attain a good working knowledge that will enable ROT and MOT, which in turn yields a rational KOT and TRU. **LA** = Law of non reality, non certainty, and non truth of a portion of all except TIT-TO, helps, by elimination, to prove TRU.

Individuation although exemplary of the independence of a person, is even more descriptive of the fundamental attribute of TIT and TRU. This is seen in the contrast between the individuation or particularity of Occam and Science and the individuation of thisness in God, TREQ or TRU, **PRUTIND** or the Proof of TRU Through Individuation. The thisness of entities becomes the particulate and the particulate becomes the thisness and again particulate of TRU. Within TRU, like relativity and DAOST, this is the tie that binds to give TIT its special, set, and ordered form. Thisness or haecceity may be either relative or absolute. The relative exists in the observer-observed or S-O relationship whose identity is fixed, specified and certain. A distinct object fuses with a distinct subject and combines to become a singular concrete entity. However, only in TIT does thisness become absolute wherein both TRU as a whole and its parts or processes become perfectly pure, specified, discrete, and distinct. This is Á that becomes the mass energy in an all encompassing relative universe, TIT. It is IA, SO, action completion, and actualization in ACEXIN that includes all TRU. What gives TIT this special individuation is relevance. **SRELT** signifies this Special Relevance of TRU, that works to TRU, to its relevance, and all ways in TRU. This is seen in **SUPTIA**, Superhuman Point of Reference. Points of view or perspectives are not arbitrary or open ended. They are largely fixed by TRU, by its overall set form. As the knowledge of TRU increases, with increasing knowledge in general plus their fundamental concepts, principles, laws and mechanisms, it becomes increasingly undeniable. (COGNOS) **LAFRUN** = Law of Fixed or set Form and Nature, as long as TRU is unchanged. LAFRUN is a product of relativity or sameness, the S in DAOST

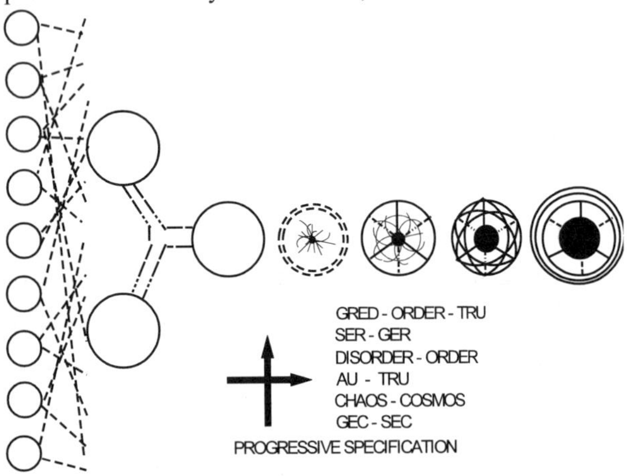

Fig 1-13 **PROGRESSION from AUS to RUS, RUTS, TRUS, TRU and TIT**

This figure views the world diagrammatically in series as a few steps from the most general condition of utter chaos and disorder to the most special condition of perfect cosmotic order. The latter we show as the focal, coordinated and ordered TIT. The new testament slant on this is seen in how much of the world has existed for the past two to three thousand years. This is Christ's Eli, trying to get all to God, only to be abandoned, repeatedly forsaken and constantly in despair. Such a state of being repeatedly scattered and lost may be likened to an infinite number of observers and observed which are, or are not, related only to each other, with no overall union. This is what we see in: **SER** = Special Theory of Relativity. It is one of movement, any change in relation. By the focus of a single unified relationship of observer to observed we can recognize the role of **GER** the General Theory of Relativity to develop a singular unity, the one, that enables a set state in which geometry can be used to represent it. When extended to the entire body and mass of TRU this one becomes a whole, what Einstein sought in his Unified Field Theory, **UFT**. This is, **BO**, the Beginning of Order. For the fixation of the relationship among two or more observers in SER and the limitation to one in GER provides us with the necessary base or cornerstone for construction and coordinate makeup. Thereby, **GRED** = General Theory of Relativity leads to Order. The absolute bond between God=Christ=Man has been a great catalyst, by providing the focus, relation, union and order for the enhancement of all TRU. In Christs Eli or Eliahoim is expressed the living, vivid, actual, suffering, imploring and begging for relief from separation, alienation and dissociation and the benefaction of union, order, concord, concordance of **UNIT**, the Union of Í and T. or the first person or individual and TRU. There is a correspondence between the UNIT of the Union of I and T, and the unit of mathematics and TU in TRU. For the UNIT of I and T is the basic unit upon which all, all relative, all being, and all our lives depend. It gives all its unity or oneness, TU. It is this oneness that provides the haecceity, thisness, RIA and TRIA that are all important to any ordered TRU. However the general and specific in relativity illustrate the great SCARLS that arise from these two terms as they are one. This is called: **ÉNIT** = Equality of, yet finite separation or division of UNIT, of the I and T from the One. all in God or TREQ as The Relative Universe. In this way we can establish a sequence and progression of TÄ, from its most general and universal states and conditions, topics, categories, divisions and structure, through its primary, secondary, and tertiary form, modes, attributes and characteristics, to its most differentiated components and complexes, ES and CES, and thence to its summation and totality. There needs to always be correction for the observer or subjective error in the derivation, creation, cause, and all TRU.

Even more than ATO for the existence/derivation of TRU for TIT, there is non existence of an external form. For any form imposed from without to that degree departs from mandatory conformity in TIT. The correction for this is **RABIF** or Reversible Accommodation by Internal Refocusing of external conditions. This is a kind of internal observer of external conditions. It is the subjective of its objective counterpart as RÏT or Representation in TIT. By **TOL** or TRU Total Limits, we can be guided to order and control in the external derivation and development of TÄP, TRIT, and OBT. It is only inherent with a TRU realism that is neither subject nor object, but their integral and differential combination, SO, appropriate for the total or particular in TIT. This is the mass energy, Á and SÁ FAF of TIT similar to Scotist Ultimate Reality. The inherent desire for all things for all other things in a certain order, as held by Francis Bacon and Whitehead, is the <u>natural election</u> that selectively and focused specifically in TRU binds it together. This is provided by **CRINA** or Concentration, Relativity, and Intentionality of IA and Á. The reasons for TÄP and TIT and the reasoning they function by is seen in: **LARUR**, Law of Relative Universe in Reason and Reason in The Relative Universe, and **LACRÁ**, Law of Combined powers of Reason in Á.

. CRINA shows we how we can, develop more and more IA, and thereby life, in a given shorter length of time. This is a continuance and EXPansion, **EXPIM** or In Media. Such powers of the means, in addition to the extremes, is added evidence and proof of the continuum and spectrum of our mass energy form, state and dimension, the Á dimension. It shows how important finite expansion is, in all directions, as well as the infinitesimal and the infinite. The correspondence of IA and Á is from the most simple in focus to the most complex or compound in extension. **EIA** = Extremity of IA and Á. EIA shows how they permeate all TRU and tend to OM, Pureality, Cognos, SÂ and TÂ. This is in the direction of an infinite IA. It is this that corrects and solves the massive and seemingly disordered world about us. Although infinite, it can be achieved to a great extent, within practical limits, as the expanding finite approaches the infinite.

SEPARATION and RELATIVITY

Nothing can be fully well ordered unless it provides some basis for separation and division of solid entities and universes with dimensions. Without relativity we can have no order for we have no fixed base. With relativity and a fixed base or centerpoint, we have unity and the beginnings of form and structure. These may be classed as follows:

EXTERNAL RELATIVITY	INTERNAL RELATIVITY	THISNESS or HAECCEITY	WHATNESS or QUIDDITY
How it relates to other entities	How it relates within itself	How its relativity produces exclusiveness	How its makeup shows its form composition and function

SOREL = State of Relativization. SOREL is the standard of reference that is not arbitrarily set by the observer. **SORE** = State of Particularization, Individual and variable Relativity. SOREL is the absolute product of being, The Thing, TRU. It is its spontaneous and automatic standard of reference that so validly, precisely, certainly and irreversibly sets the form of TIT. **RATO** = Relativity initiates and gives direction to DAOST and the form of all together, TO and all TRU. **RECO** = Relativity Establishes the homogeneity and heterogeneity of its components as they are developed and constructed in a special entity and way in TRU. It is through its internal relativity, not some extraneous non meaningful or varied cause, reason and force that TRU is, and is made up. In being relative TRU is composed of that which is most supportive, bound, and relative to it. **LINVER** = Law of the Inversion of Relativity. This is the inversion, or repeated versions, of the SO relationship in and out of TIT. One of the greatest things that Homation, ALT, MOT, ROT, SU, THEO and especially ALT and SPALT prove are the equalities, relationships, values, and formal states A, L, T, of distribution about **MP**, Man Point, or the focus of TRU. It proves or sets up the special case of TRU, the construct of its relative proportions, and the makeup of A,L,T around a centerpoint, MP. It shows the value of small chemical changes as much as great political events. It shows us, and how it is made manifest, the past and the future in TRU as well as the present. This is signified by: **LASIT** = The Law of the Actual State in TRU. A corollary of LASIT is: **PADALT** = Principle of, and Proof by, Distribution of A, L, T. or ALT Distribution and Proportion enable and Prove TU and TRU, as these in turn prove and show ALT Distribution and Proportion. LASIT and PADALT may be directly applied to all partials and processes in TRU statics and dynamics. They show the proper way for IA to be Directed, **DIRIA**. It enables us, through HOM, MOT, and POA, to be the guide and way to TES, to be TRU. It shows how great and rapid the transformation of Á is occurring and expanding. Even though human comprehension and brain capacity are not expanding at the same exponential rate. It shows how Á may be expanded through concentration and identity in TRU. This is called: **LÁÊCIT** = the Law of Á Expansion through Identity with and Concentration In and on TRU. This is intelligible reality through its kind of light of IA, VAPO, SCARLS, and ACEXIN in the DAOST of TIT-TO.

Fig 2-13 **PROGRESSION TRUS, TIT, ÁT**, and **DISAT**

This figure of the Sets and Forms of TRU, seeks the best way of showing TIT; what it stands in greatest contrast to and progression with. To adequately recognize and appreciate TIT we must shear off all facsimiles, all approaches to and representations of it. **LATAR**, Law of TRU and TIT through AU Reduction. Without the special form, thisness, states, dimensions, and essences of TRU, and only with those of physical entities, TRU could not be. We can not make it relative from the outside, it already is, within, by ATO and TRIA. Yet we know that relativity, the observer-observed, SO, or Á is also all within TIT. These two we reconcile by their fusion, the Á in TIT, and the TIT in Á. This is called: **ÁT** = Á, SO or Relativity TIT and TRU Fusion. With Á we can begin to see how we can escape the paradox of an absolute TIT, wherein we can not sense, perceive of conceive of it, because it is outside and beyond comprehension, as in the deist fallacy. Also we can begin to see, and escape from the notion that we can not make it a thing, objectivize or reproduce it, because it is beyond replication. To understand this problem we must be able to comprehend and replicate them as we do by looking at this figure and recognizing that both this looking at it and the diagram itself is to a very limited extent, actually within TIT, as TIT is to a very limited extent, with them. This is the resolution of the Paradox of IOT, **PIOT** or in and out of TIT.

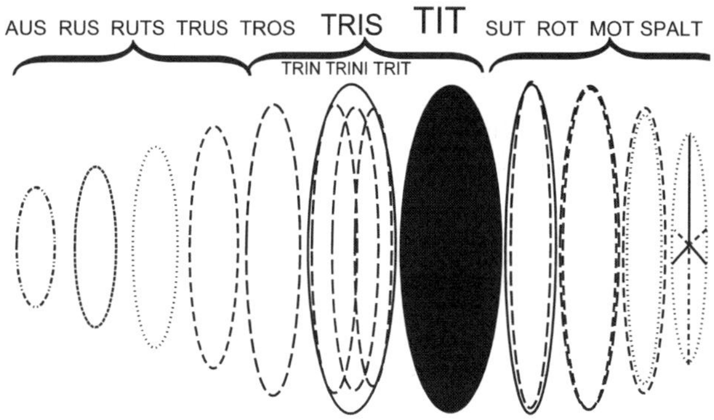

TRUS Progression, TIT and ÁT

In DAOST we can recognize the status of relativity in its sames, and dynamics in its differents. This is how the same or homogeneity tends to unity, TU, and different to heterogeneity and TE. Even more there is a union of oneself and ones kind, or Í and TIT. This is a basic unit, fundamental to mental health, and seen in coordination and control. Isomorphism signifies this sameness of form. It is the fit or isomorphism that exists in TIT, TRIT, OBT, SUT, ROT, MOT, ALT and SPALT that enables us to provide an accurate and reliable base for modeling. There is precedent in this by the Keynesian <u>Principle of Limitation of Independent Variety</u>. Although directed to inductive analogies it holds that no one object has an infinite number of independent qualities or is producible by an infinite number of different causes. This also affirms how TIT can not be only D by DAOST, but also of the same. In DAOST and its differents, there is for TIT a multimodality within its relativity. Plato states in <u>Sophist</u> that sameness depends upon relation, the more related the more verifiable. It is in similarity that the forms of things become TRU and their ESINS bond to become one, a whole. This is evident in the crossing over of genetic material which is so fundamental to all life and the higher structures of man and TRU. It is evident in many other ways as the forms of all aspects of TRU, are symmetrically unified in the whole, as past and future, large and small, subject and object or the mass energy of TRU. It is this very symmetry and isomorphism that enables us to determine the dimensions of TRU and develop its systematic unity, representation and modulation, as in SPALT. How TRU is formed by these symmetries and isomorphism is superimposed by those forms and dynamics of the whole which are not present in its parts. These added states whose symmetries and isomorphism may seem so illusive, yet are so actual as they are by ACEXIN the higher and more special form and characteristics of the TO of TIT, are the basis for its proof and the proof of Á. It is the New Forms, which arise with transformations and the new added forms and states which so characterize the new whole, as in TO. **LALTISO** = Law of TIT TRIT ALT Isomorphism.

So much of the current, as well as near, mid, and long term future's controversy over proofs, truth, verifiability, and falsehood is a failure to comprehend the collective massive weight of proofs of the fundamental aspects, concepts, principles, laws, mechanisms, topics and FAF of TIT-TO. Each of these collective proofs for topics of TRU become more and more verifiable in the qualitative and quantitative distinction and identity with TIT. This occurs as they are of DAOST and its knowing and showing KOT and POT. The continual reduction to TRU, such collective proofs provide, is the prime method of establishing verifiability of the whole.

If we take TRU as a TIT, and proceed to differentiate it, then these differentiations, like forma substantialis, will within themselves form a totality; of which the limits of TIT, TÄ, and TE will be necessary. This is called: **TOL** = TRU Total Limits. TOL is how TRU may be determined, known and shown through the limits formed by the differentiation from one to parts, ES, and how these parts combine to make added form. TOL describes how the basic nature of TÄ and TRU exist and may be used in Multimodality. In multimodality we have validity in varied senses and forms, necessitating **APRES**, or A Perfected RESolution. The multimodality of TRU proceeds in at least two ways, one by the changes in sets or kinds of TRU, as in TRUS and TRIS, and the other by changes in the aspects, modes, forms, and states of an entity, RU or TRU. This combination is given in Fig. 2-13, the **DISAT** Diagram of Sets and Forms of TRU.

In TÄP and TIT the relationship in which human predominant form is manifest is **LOSADIN**, Law of Sense-thing and SO to Á Differentiation and Integration, and **LASÄ**, Law of the Special necessity and Form that typifies the predominant form and power of TIT. The Á FAF of TIT is composed of many senses other than sight, in fact all senses and abilities special to TRU that are not of the length and time continua. **LATÂNÁ** or Law of the Absolute Necessity of Á or SOT Distinction of and in TRU, provides humans with ever greater powers of IA, identifying, distinguishing, and making whatever is the given or is encountered. By **RÁDACT** Relativity becomes Á in the Depth and Complexity of proofs by and meanings in TIT. By **LADIS,** the Law of Dimensional Satisfaction or Sufficiency, L, T, and especially Á are continuously gauged in TRUD. Á expands largely through identity with and concentration on TRU, both in contrast to what is not TRU, what is most TRU, and what is most appropriate for TRU at the time. Through their Á FAF humans best adapt to higher and higher levels of proficiency of an ultimate subjective that spearheads knowledge of ever more TREQ, TRU, and TIT-TO.

It is in identity that the special form of TRU becomes most clear. It is both an identity with TIT and TRU. It is a pure identity with all that is the same in TRU. Actual Identity in TRU is the higher the higher powers of actuality as seen in ACEXIN. It has precedent in identity philosophy which holds to the identity and union of SO, which of necessity must become SOT and Á in TIT. This culminates in **LOCIT** or Law of Compulsory Identity with TIT. When transferred to TRIT this provides the basis for ALT. When compared witb TREQ and God it is called: **LAGID** or Law of God Imperative Identity with TREQ and TRU. To one and all people if they are to be of the same universe, what God means must be identical with what God actually is in TREQ and TRU. The focus of identity and how these show the centering of TRU, is seen in: **LIDOTE,** Law of the great and close Identity of the first person or I and the observer in TRU, TIT, TE and TES. This shows the power of relativity and Á. Like identity, equivalence demarcates the exclusion of all not TRU. This is expressed in **PEQU** the Principle of Equivalence, which in physics asserts that the effects of gravity in different mass energies disappears if the observer is the same.(Á) What is best is so only for TRU as it constitutes TIT. This is so by being a part of the same thing, the relativity, and especially the typical ALT of TIT. Equivalence is closely related yet distinct from equality which carries a more exact and exclusive connotation. **TREQ** or TRU Equivalent signifies all that exists in the world that is TRU, yet in which TRU is unknown. TREQ is highly important if we are to understand TRU, and how much TRU already exists. Even more distinct would be what is equivalent to TIT. They help determine what and how much departure from TRU occurs. Such departure is seen in **ÉNÍT**, Equality of, yet finite separation or division of the union of Í and TIT. The necessity of adherence to TRU is seen in **LANÉT** Law of Absolute Equality of All, as it is its weighted or REM counterpart, ES, in TRU.

The kinds of TÄP are seen in TROS, TRIS and TRITS, or the numerical, formative, and objective of TIT. **TRIS** is the special classes, kinds and forms of TIT based upon the set and range of time in which TIT exists. That TIT-TO is dynamic is shown by **LÄTRIS** or Law of Altered or changing form of Á, L, and T and TRIS. TIT has an infinite, infinitesimal, finite, relative, practical and objective orientation, depending upon its own FAF and how it is represented and presented. Comparative studies of TRIS and their status and roles in TIT provide a deeper understanding of TRU, and answer many of the seeming paradoxes and questions that arise.

Besides already being what is the absolute extant ENIT, TIT can not be in error, because it is what it is by definition. We may not know all it is, yet we can be certain that it is. This certainty is affirmed when we realize **AXOTPRE**, Axioms and corollaries of the Truth or validity of TIT, its Representation, Perspective and Presentation. What is valid of TIT, ROT, PERT, and POT separately may not be true of the other or all together. TRIT is plural, TRITS, when its set frame is adapted and altered by standards, objectivity, systematic unity, representation and modeling. This is the externalization and relativity of TIT for TRITS and OBT. This is reversible or internalization and absolution of TRITS and TRIS for TIT and TÂS. The form, modes and characteristics of TIT that are limited to TRIT are basic for its form, TÄP, and provide the ÁLT frame upon which all objectification of TRU can be based. Humans greatest search was in the variety of Gods and especially since the Ionian and Milesian Philosophers of the 6th c. b.c. was for the basic stuff the world was made of. Philosophy, Science, and Religion even into the 21st century a.d. are still mired in this search. Only when we discover TRU, its form, structure, dimensions and their formulation, as in ALT and ACEXIN of TIT, TRIT, and TRITS will we begin to get the kind of accurate and ordered conception and perspective that closes out this search. ALT and ACEXIN like personality taken generally are those special features, predominant FAF and powers, qualities and properties that make it stand out from all other things.

PAFT = Points of View or Perspectives can not be Arbitrary or open ended, but fixed by TRU. Observations, perspectives, and points of view are set and firmly fixed by their relationship to TRU. That is if, and the extent to which, they are or will be TRU. It is not for each person to chose this or that principle, to champion and impulsively do whatever. Each and all persons must come to understand that their observations, points of view and perspectives are highly focused, if, as they must, be TES and TÍS in TRU. It is limited by both TRU and their own special position in their world.

The fully exclusive nature of TIT is signified by **TRIA** or TRU Is All. By this what is not TRU or TIT can be readily excluded, and to the degree it is not TRU or TIT. TRIA is characterized by immutability, or the overall changelessness of TIT. It, by TRUD, has much change both by OITIO, and by CES internally, but not of TIT-TO externally or irreversibly. How much is a RU or TRU self sustaining? If TIT is all, by

TRIA, can it only be self sustaining, without external help? This is PARIO, or for TRU **PATIO** = Paradox of Out of Into The Relative Universe, Into Out of The Relative Universe. PATIO is the repeated problem we find with IOT, OITIO and the relations between TIT, TRU, TRUS, AUS and all. It is readily solved if we recognize the all encompassing form and function of TIT, its LANERA and LAMOL and external reversibility, slight ideal changes externally, and the exemption to be made for all that is less than 100% TIT.

Of all in human life it is meaning, and the TRU quality and quantity in the learning of meanings, that best show TRU, TÄP, TIT, TRIT, and ALT. The great cue and gist to all lies in their TRU meanings, its concepts, and how these tie all together. This is how they become actually and fully meaningful. It is this very predominant and unique state in humans of being able to understand and communicate ourselves and our world that makes us TES, TREQ and TRU. It is for this reason we must forever remember and extol the best in language, terminology, communication and understanding. **LEXPOM** is the Law of the Exponential Power of Meaning. **MEQUAD** is how Meaning is Qualitative, Quantitative, integral, Differential and Actuating. Meaning is formed and functions in TIT through its mass energy, Á, form, state, and dimension constituted to best assure and fulfill itself as it best serves TIT. It is in all the lesser uses and devious abuses of meaning that degeneration of TRU can be vividly exposed. The greatness of concepts is not only what it is a part of, it is in how and what it proves. Meanings and concepts are often formed by the parameters, spectrums, and continua that divide and differentiate TRU and its general to special, simple to most complex and total forms and states.

It is with formal distinction that meanings attain their highest actualization in TRU. Formal distinction can be no better than the degree or acuity of meaning in communication, speech, writing, teaching, and all vehicles of Á and SÁ. For the first person there are limits, especially to higher powers and precision of ACEXIN. This is even less for the second person. It is also limited by the quality of communication. There is Proof In Distinction, **LAPID**. This is what the operations of logic do. They provide proofs through ever greater number and quality of distinctions that can be developed between two or more ideas or predicates. It is with formal distinction and sufficient reason that meaning reaches its highest form and function in TRU. Since all in TRU is more or less relative it may be adapted using the appropriate SCARLS to produce sufficient reason for each given. TRU becomes known when there is sufficient reason for TRIT, OBT, and SUT. **TAL** = TRU Alteration TO Toward Definition. TAL equals TÂ alteration toward definition, where the indefinite becomes likely with eternal TRU/AU assured by COGNOS. It is a TRU that moves to a more and more definite state, as with OC to EC to ECT to NEC to TOPNEC to TER and LANT. The meaning of transfiguration by TRU is how all entities change as they become ESINS, and their ES, CES or aspects of TRU. **TALM** = TRU as All, More or Less, relatively and conditionally. This is where TALM includes not only TRU, but the world, the relative universe, being, and TREQ. **UTES** = Union of TES. **PUTES** = Union and Practice of the Union, the Unity of TES. TÄP may be inferred from the modes, essences, qualities, and vital functions of TIT and TRU. To understand TRU, TIT and TRIT it is imperative that we study its Vital Functions, **VIF**. We are able to measure or determine the degree of or effect on loss of VIF. This is called: **DAVIF** = Degree of and effect of Loss of Vital Functions. DAVIF has become increasingly well investigated, learned, known and used in Medicine and the care of the critically ill. VIF and DAVIF measure all that is crucial and vital to our survival. The function of parts of TRU have certain degrees of freedom, or: **FRE** = Freedom or Range of Motion that is a Natural Consequence of all that is within TRU **LACH** = Law of Alteration of Characteristics and qualities of things as they are or are not TRU; or are or are not equal to their form and role in TRU. LACH shows us how much of TRU, as TROS its States of TU, TI, TE, TO, and TA partake of characteristics other than those things alone, or even as they may be explicitly or exclusively in TRU. e.g. negative as AU / negative & positive as in TRU. It is most evident in the contrast between:

Entity-ENIT ENIT-ESIN ESIN-ES
ES-TIT TIT TIT-TO.

The primary modes have been classically given as possibility, necessity and actuality. Modality as in modal logic is that property of entities, objects or propositions which infers that it is possible, necessary or actual. Modalism is related to structuralism generally and especially the theological doctrine that affirms the unity of different forms of God, as substance, subsistence, and paracletion of the holy spirit and personality of God. **MODPRUT** or Modification Proof of TRU is the great strengthening power of modification that adds to the proof of Á and thereby TRU. That the modes, essences, qualities and vital functions of TRU are equivalent to their formal distinction is: **LAMADE**. Their strength and validity can be no better than the degree to which they are made distinct and clearly understood and used.

ESSENCES of TRU

The essences of TRU are its distinctive nature, major special characteristics, and those aspects most like TRU, enable it to be, and best define it. Of these purity denotes that content which is only TIT, exclusive of all that is not TRU. Reality denotes how subject equals object combined in TRU mass energy, or Á FAF in TRU. Essence is what sets TRU apart from all else. The preeminent essence of TRU is humans ability to think and reason in order to find the solutions to problems. Those essences that are most life preserving are called critical essences. It is in the study and operations of the processes of critical essences when ordered by SPALT and guided by POS that both they and TRU are best shown. Vital functions are those without which it can't exist. We can measure the degree and effect of loss of vital functions. Within TIT there is a degree of freedom, or range vital functions can operate in. This is best seen in physiology as in oxygenation, nutrition, fluid balance, regeneration, digestion, metabolism, and most prominently thought and reason. Yet it holds true for all TIT.

Quality and quantity are also modes of TRU. There is an alteration of the quality, characteristics, and quantity, of entities as they are or are not, and OITIO, of TRU. Quality is exemplified by **POITES** or what the subject recognizes as being great and/or useful in the object. POITES reveals both the prime predominant FAF of Á in TRU and the integration of quality. TRU, as a person can get along with a loss of most of its quantity, but very little of its quality. This is signified by **TANAT** or Tolerance for loss or gain of amount, number or quantity of TRU. Quantity and even quality over longer periods of time vary considerably in RUTS and TRUS. Measures will increasingly evaluate past human development and help to test and predict that of the future. Quality is closely related to REM, value or relative magnitude. Quantity for TRU is basically the differentiation from TROS, how much TRU there is. This is easier for the gradations of its dimensions than an overall measure, unless comparatively using, to a more or less extent, some external measure. Many measures, as used in scientific investigation, are basic to providing accurate and reliable information about TRU and its representation. **LONBIT** or Law of the Numerical Basis of TRU is seen in the measures, parameters, dimensions, and gradients in TIT. Along with TROS quantity and the numerical basis for TIT arises from DAOST, maximum and minimum, and **ID** or Integration and Differentiation. There is a convergent relationship between maximum and minimum which when taken to its absolute in an infinitely enlarging circumference becomes a straight line. This shows how the coincidence of opposites works in TIT and MOT.(LACOP)

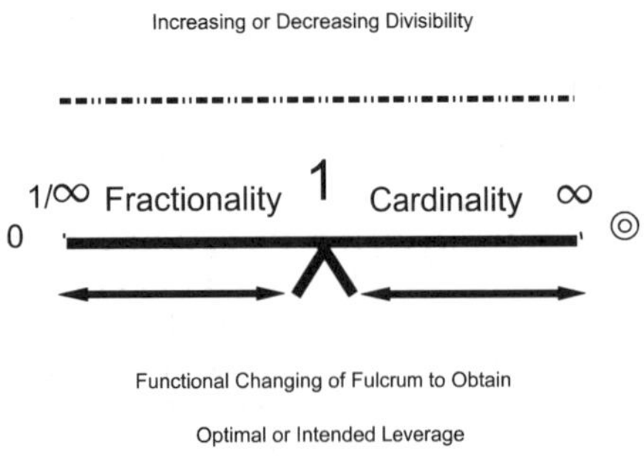

Fig 3-13 **FOCI of TRU, INTEGRATED and DIFFERENTIATED**

We must remember and be careful, ECRO, that for TRU there are both one and many foci. It is easy to think of TRU as TIT and TRIT having an explicit unity. This is a valid assumption to a marked degree. Yet because of TRUCH, TRUS, TRIS, TÄ, and especially the total and dynamic state of TU there are exceptions. This is the fact that each and all components, aspects, forms, functions, parts and processes of TRU have, within their limits in TIT, their own foci. So that the unit TRU and its resultant foci are not so simple. It is an aggregate, weighted total of all its component foci and constituents. The reason why this is important is not so much to show the fact of its aggregation, but to enable us to avoid the trap of denying the power or TRU's components and to better know and show TRU. For it is all that is of TRU working dynamically and often diametrically in opposition that these powers must be recognized and interpreted. There are foci for different kinds of TES, and foci for different IAS. There are many foci for all the various aspects, forms, functions, parts, processes, concepts, principles, laws and mechanisms of TRU. These will often have a high degree of codomain and coterminality, yet recognizable and usable as separate counterparts, counterfunctions and foci. It is by the dialectics and synchrony of these different foci with the VAPO of TRU that determine much of their existence and essence, as well as that of TRU. The importance of focus and its role in the world TRU, has been known since Archimedes. It is the crux of relativity and focus in TIT. This is a major proof of TRU and ROT. The importance of focus in TRU is seen in coordination, direction and navigation where only a tiny glimmer of light in an otherwise absolutely dark night will give a person focus, will keep them from being lost and harmed. So is it for TRU, KOT, and ACC in many ways, especially if this tiny glimmer turns into the light of a bright clear day. The focus of TRU, as any RU, must be in the here and now, and the center of its mass energy. In TIT for mass energy it is in its relativity, the reality of its observer\observed and typical predominant functions. This is how their distribution about **MP** or Man Point produces a convergent focus.

The simplest case for all, for all entities, is unity. This is equivalent to relativity, to The Relative. If unity is produced by relativity and if unity is not all of TRU then it is for us to derive, develop, determine, and prefect, as much as possible and in the best way, all that accompanies this unity to form the whole of TRU. It is through the finite divisions, and increasingly, yet proportionately and orderly construct, of this finite dividing that we can provide an object that most closely replicates TIT. If an object, a RU, or TRU, is unified and exclusive, then and to that degree can be formed and a totality. This is partly from **LOCU** or the Law of Compulsory Unity. **LUEXO** = Law of Unity Exclusive through Totality. Unity and harmony as all, by LANERA, are relative not absolute. In addition the dynamics and oppositions that occur in TRU, even TRU functional competition are not always productive of absolute peace and harmony. Yet it is true that the unity of TRU, TU, and the associated systematic unity, homeostasis and equilibrium that exists dynamically in TRU are largely harmonious, especially from within.

The Unity of TIT, TU

TU is fundamental to its thisness, exclusiveness, solidity, and representation. A precedent for TU is the <u>Doctrine of Suchness</u> in Buddhist Scriptures and commentaries on the soul. Suchness is thisness or haecceity and ATO of TRU, although less clearly defined and separated by VAPO from all other than TRU. From one nothing, denotes how any exclusively singular can produce no discovery. This is a vivid illustration of our major failure in thought, whereby ideas and concepts become decreasingly less when not concentrating on their full meanings, roles and implications.(IA) TU is not a simple oneness without differentiation, but a oneness of the totality of TIT, **TIT-TO**. The unity of TRU is not only static, but highly dynamic seen in self preservation and LOCU. There must be a continual and compensatory optimal reciprocal movement between TE and TU. The form, frame, structure, states, qualities, essences, mechanisms, processes and all work together toward unity in TIT-TO. There is in DAOST by TU a Unity of Opposites, **LUNO**. Being highly dynamic and a compulsory unity TIT is variable, yet in a convergent and more or less orderly and regulatory way. It is a unity provided by, and FAF about, its speciation. This is the maintenance of its capacity, ability, action/reaction in the states and conditions it finds itself. The Principle of Unity, **PUNI** signifies the overall strength and measure of TRU through unity/diversity, convergence/divergence of TRU/ATRU and AU. **LARMOTU** = Law of the Relative Form, or More or Less State and Modality of the Unity of TRU. LARMOTU exemplifies the spectrum of TU from a simple singular unity to a most complex total unity. It can be viewed either way, yet to be of TIT, it must be all increments in the spectrum of the whole, and by TROS typical of TIT.

The paradox between Monotheism and all else is called: **PATUS** = Paradox of the separation of Monotheism as a belief by TRIA and Monotheism as a more general aspect of TRUS and Perspective. PATUS is solved by TRU and TREQ, a focus from TIT, TU, TRIA, and TÂ. The only total or ultimate unity is that of TRU or TREQ. It is in the summation of all its parts, mankind's knowledge, culture, civilization and all that embrace human life and enables being as an entity or universe to survive and continue indefinitely. This is a fundamental law of TRU and is called: **LATRU** = the fundamental Law of TRU. LATRU certifies the relationship between TU, TO, ENT and TRU. It is a product of TRIA, and the relativity inherent in 'The', 'Relative' and 'Universe' as they combine to represent the overall unified FAF of humanity, its success, survival, and continuity. It is The Relative, special and thereby typical new composite of which humans are, and are exclusively, all, in total unity of TRU. The great sequel to LATRU is the foundation of SU and especially SUT. This is called: **LUT** = Law of Ultimate TRU. LUT denotes how with SU and SUT, PATUS, TU-TO and LATRU, TRU can be readily established.

From DAOST, LATO(S), MATES and EBE, closeness to TU is by being closer to uniformity, and closer to the same or S in DAOST. Homogeneity and heterogeneity are polar extremes of the same state of a RU and TRU. Homogeneity tends to TU and heterogeneity to TE and TI. Homogeneity tends to uniformity and heterogeneity to multiformity. This is how the same or homogeneous tends to TU, and the different or heterogeneous tends to TE and TI. DÄT or the Derivation of the form of TRU is largely by: **DÄRUST** = DÄT by Relativity and Unity tending to uniformity and sameness. Uniformity tends to show how: **TI → TE → TU,** as all, tends toward the same, the uniformity of TRU. Yet since we, in TRU, have much more than a unity, or a same homogeneous mass, we must accommodate the finite, infinite and total. Form in being more than unity will also partake of these. It is in this way, that form, especially as manifested functionally, is made up of much dualism and a dialectical balance between the different extremes of its denominators, parameters, dimensions, and functional continua. How this maintains itself is **LACON** =

Law of Conformity.

TÄR = TRU Form, structure, or order of Relationships, and Arrangements. The continuity of TÄR, the order, arrangement and structure of TÄ from comparative entities, things, and forms is shown by the Innateness Hypothesis. This holds that for each type of thing, or TRU and their archetypes, ideas or forms, there is a through and through component common to all, an innate type, form and idea special for each. Compossibility is how the homogeneous and heterogeneous exist in the whole together in an entity, or TRU. It indicates a lack of overall contradiction, not necessarily a lack of opposition. Compossibility enables TRU to have TAR, arrangement, structure, order, and to have the general concepts, principles, laws, and mechanisms which hold it together. TAR and TRU are hard or soft depending upon whether or not there is or is not a high level of compossibility.

LACMAM = Law of Coincidence of Maximum and Minimum. That there is a convergent relationship between maximum and minimum is shown by mathematical figures. When taken to its absolute the circumference of an infinitely enlarging circle becomes a straight line. LACMAM is significant and useful in demonstrating the limits of TRU, and how the absolute and the relative converge in TRU. It helps provide SUT and ROT and explain the steps between TÂ, TIT, TRIT, OBT, SUT, ROT, MOT and SPALT. It is also useful in proving TRU. It shows: **LACOC** = Law of the Concordance of Contraries, **LÄCÓP** = the Law of Coincidence of Opposites, and **PAC** = Principle of Non Contradiction. LACOC, LÄCOP, and PAC are of fundamental importance both to the explanation as well as representation and understanding of TRU. For without them we tend, as is naively believed, that all opposites are contradictory, with no concordance of contraries.

Systematic Unity of TRU, SUT

SUT designates that form of TIT and TRU which is provided by the systems, especially coordinate dimensions, in TRIT and OBT, that enable us to represent and model TRU. It is a SUT whose FAF is typical and exclusive to TIT and formatively inherent through dimensional formulation, ÁLT. These are produced by their transformations through adaptation and alteration to become what selectively has best worked in TRU. Dimensional formulation enables us to order TIT structurally. Each dimension's extremes, means, and gradients in between, form a set scale which when combined make up much of TÄP. This is shown in **TOL**, TRU Total Order from One Limits. SUT is based upon the frame, construct, and systems of TIT. Only in its great use in ROT and MOT can it be standardized to deviate from being a perfect replica of itself in TIT.

The greatness of SUT is a function of the greatness of TIT, its need, how well used, and the refinements that can be made to perfect it. As long as mankind finds no unified and well-reproduced object to order this world, we are eternally lost and fail. It is in the reduction of TIT through TRIT, OBT, SUT to ROT, MOT, ALT, SPALT and POS and their reversibility that enable us to know, use and best live by TRU. SUT, ALT and POS. It is what helps us understand the world as it actually is, TIT, and how it differs from the many other worlds about us. Thereby **PABAT** or the Paradox of Biological TRU Being is solved.. Many NEC in a beautiful intricate POS in SPALT and POT go to resolve this paradox.

The greatness of SU is a function of: 1. The greatness of TRU, its supreme worth, 2. The need for SU, how much it helps to make TRU a better object for knowledge and use. 3. The increasing refinement and decreasing error in SUT. The degree to which SUT and TRU can be made most accurate, reliable, and beneficial.

$$\mathbf{SU_{value} = TRU_{magnitude} \ x \ SU_{need\ for} \ x \ 1/SU_{error}}$$

Thereby the great problem in SUT and all human endeavor is to optimally correct for errors, best fulfill the need for SU, understand the value of SU as it represents the greatness of TRU. This is signified by: **PRAC** = Problem of All Act. As long as mankind can find no unified and reproducible well-structured object to order in the world we are eternally lost, and our actions are uncontrolled. This is called: **LANSUT** = Law of the Absolute Necessity of SUT. LANSUT does not imply that such a unified object must be absolutely perfect, only relatively, practically, and increasingly sufficient to provide an effective means of reducing to order and controlling the world in which we live. Mankind can not live without orientation, coordination and order for many reasons including an ethic based upon a GOD or TREQ as TRU. These are necessary to guide ones relations with all others and all the rest of the world, TRU. The other great reason is biological, whereby all animals through their FAF best adapt and with higher and higher levels of Á become more and more coordinated. **ULS** = the Ultimate Subjective, or the form that enables TRU to be most knowledgeable. Homation as POA is based upon ULS and the formation and ordination of TRU. These are attained through the standardization of THES and THEO, the subjective and objective separation and union in the Á, mass energy FAFof TRU.

RELATION

The set state and range of motion in TRU is relative and conditional, yet provides for optimal union of its parts, ES and TES. Such union has been known as Internal Relations. They depend upon the combined focus of FAF and dimensions in the SO media or reality, medium size and mid time, present. **DIR** = Doctrine of Internal Relations. DIR is primarily not a feature of EPI and logic. It is a part of the special form of a universe or totality in which all that is within is in a set and certain relationship with each other. **POREC** = Principle of Relative Completion. POREC is like ATO, an established and largely unalterable fact or state of affairs. Both are related to TRIA and the haecceity of TRU. All are produced by TRU relativity. This relativity is not simple but massive and dynamic. **LOFOR** = Law of Formative of the Relative. **LODOR** = Law of Distribution of the Relative. LOFOR and LODOR of TRUS by TRUCH have not remained the same through TRUP, nor TRUF. It is in this way that it is TIT or TRUPRE of TRU, focused up the present is absolute. This is called: **PRAT** = Present TRU is Absolute. This seems obvious, yet when viewed from non TRU it is less so. They tend to come together and pass away so that distant TRUP and distant TRUF are 1/SU, dispersed and LODOR fragmented. It is the forces holding and splitting LODOR, fragmenting TRUP and TRUF, that produce much of the dynamics and processes of TRU. This is both those dynamics entirely within, as well as, even more, those in and out of TRU as it may be visioned existing in different foci of time. This is called: **POFORD** = Principle of Fragmentation of Relative Distribution. POFORD denotes the fact that a RU or TRU becomes unstable and tends to fragment when the distribution of its contents are altered. It is like an animal that becomes sick or dies when its parts, organs, or body is diseased or injured. **LAREX** = Law of Relative Exclusion, or the degree all not relative is non existent. LAREX is driven by adequation and selection. A relative, a, some, and the relative universes absolutely exclude all that is not included in its universe. This is the Corollary of LAREX or: **CARA** = Corollary of LAREX, All that is not, or from Relative is Â, Absolute, or non existent in that universe. TRU is by LAREX only TRU and that we interpret, represent, model, and posit of TRU must be done within TRU; with the exception of showing that which can't be shown on this basis alone. However and whenever there is a relation, one standing in some way to another, they change in ways depending upon each and their relation. Just as the physical world has to be adapted and modified by relativity, so must the human world be adapted and modified by its observer-observed, S-O, or Á continuum, form and function in TIT and TRU. .

LÄSIT = Law of Selectivity in Temporality, in TIT or TRIT, or the Law of Temporal ES Selectivity. LÄSIT denotes the freedom, within limits, of FAF, ES and CES within time, duration and variation. **ACIOT** = Actuality In and Out of

TRIT, TIT and TRU. One can readily see, and should relate and fit together ACIOT and OITIO. The focus of any RU, as TRU, TIT and TRIT, must always be the here and now, the product of its relativity. This is signified by: **LAFHAN** = Law of Focus of TRU Is In the Here and Now. LAFHAN is a great aspect and proof of the Thisness or Haecceity of TIT and TRU. Both LASIT and LAFHAN denote a salient mode of all, which is the actuality from relativity that is exclusively a product of TIT. Whereas TIT does allow for increasing TRICH and TRUCH as a consequence of LANERA and its dynamic position in ALT in relation to its surroundings, these are relatively small, and often reversible, especially in terms of the individual observer. LAFHAN works dialectically with the teleological principle whose propelling and compelling forces show both the thisness and whatness of TRU.

The interchange of static and dynamic is shown in:
ESCES = ES as dynamic processes, and CES as static parts. **LALOR** = the Law of Absolute Limits of Relativity. In those TRIS of increasing concentration as **TRINI** and **TRIZ,** or the infinite and limit at zero TIT, we have less and less TRUD and Homation. However this is corrected by fluidity, solidity and the equilibrium that exists within TRIS and TIT. It is this equilibrium that is their essence, basis and first principle of being. It is the ability to have the best or optimal of their static form and their dynamic function that most empowers TIT.

CONST = Conservation of TRU, Law of, **LACONST**. CONST is time dependent as is all TRU. So that the TÄ of man, in being relative, is in a conservation S/F state. By LACONST TÂ is increasingly certain only as the rate of change decreases. This is of great biological, hereditary and genetic significance. Yet there is an undeniability of TRUD and TRUCH. This is called: **DUCH** = Undeniability of TRUD and TRUCH. The freedom of the will to live or to die, to sustain itself, or the success or failure of TRU, are all a part of DUCH and CONST. It contrasts the two extremes of no concern and action for TRU and the most absolute concern and action for TRU. These are signified in: **LARIV** = Law of Absolute Relative Imperative and Voluntary. It is imperative to the absolute, voluntary to relative, voluntary to partial, or involuntary depending upon a person's ENIT or TES direction. LARIV signifies that TIT and TRU are highly dependent upon the will,, as well as on all that enables tolerance. It may be of many varieties, yet as TRU tends to an optimum, this although extreme, is highly voluntary. It is voluntary guided by the voluntary imperative. That is it is a will guided by a willed necessity. This is a will of the integral and differential, absolute and relative of TRU as well as all other forms of will in TRU. In being closely allied to intentionality, IA, SO, and relativity, it is a large segment of the Á predominant form, state and mass energy dimension of TRU, its thisness and whatness. It is this that gives it the power of conservation, as it exists to perpetuate itself. TIT and TRU in being highly dependent upon the will, both personal as in TES, and universal, the good will or TRU assures a continual concentration and control of all for the relative absolute, TIT. Conservation of TRU is in a continuous state to assure success/failure. It is similar to self preservation. Both depend upon selection and survival of the fittest, what is most predominant, TRU. Selectivity occurs in many ways and by many different sizes and groups of population. It has a degree of freedom, within limits, that is a property of much of TRUD.

CORRELATION

Specifically correlation defines the concurrence, sameness, correspondence, convergence, contingency, and conjunction that typify the special relativity and DAOST within TIT-TO of TRU. It reveals the profound tendency to unite in the absolute unity of all that is different and separate in TRU. It is used to provide precision through measurement in and out of TRU. It enables us to express the same relationship from different points of view or parts of TRU. It is important to correct the great fallacy among those who experiment, or do 'studies' in which interpretation of the results is based upon statistical correlation. Both the statistics and the experiments may be well done, yet by interpreting the correlation as meaning a simple direct relationship or cause and effect state belies and defies the more complex and indirect states at hand. For TRU, all that is of the same, as in DAOST, exists in some sort of correlation. Yet very little of these is in a simple direct cause and effect relation. For this reason we must remember, ECRO, that correlation is of many kinds, even a continuum, from that between greatest aspects of TRU, to the most differentiated, to that which is not of TRU.

LUC = Law of Universal Concurrence, Conformity or Correlation. LUC expresses the state and form of all entities that exists relatively in the same way specifically as they become identical with TRU. This is the power of systematic unity. **LUCOM** = Law of Universal Concurrence Measure. This law expresses that product of LUC whereby its verification enables one to apply it to all entities, things, and senses through their ESINS and ES in a positive or set manner. LUC shows how all religions, philosophies and systematic sciences must converge, or follow the same paths, more and more as they become TRU. **FADOC** = Form from Correlation, form from a distinction or a correlation and definitions of correlations. e.g. Those things that appear highly correlated to us are so because they are either (1). So viewed by our own senses, subjectivity or SUBIT correlation or: **CORS** = Correlation through or by S relativity; or (2). They relate to or are a part of things, RU, objective, OBIT correlation or: **CORO** = Correlation through or by O relativity, (3). Either CORS or CORO as they are together in or of a RU, **CORSO**. (4). they are actually a part of and only of TRU, **CORST, COROT, and CORSOT** correlate to each other as they are in TRU. CORS, CORO, CORSO, CORST, COROT, and CORSOT are the solution to the scientific fallacy. This is where very accurate correlations may be discovered, only to be more or less invalid to interpretation. For science interprets its results greatly through correlations, and the probabilities and limits of these correlations. **LACLIR** = Law of Correlation of Limitation and Relation, from **LORED** or the Law of Relative Dependency and **LICES** or the Limitation of Parts and Processes of TRU. In the state of relativity and dependency of entities, ES and CES on each other there exists together, and is imparted to each, limitations which they in and of themselves, as entities-in-itself, ENIT, do not have. These will depend and will be qualified by the nature of each particular entity, and each relationship. **ÄCÀC** = Non Correlation of Correlation. The correlation between things no matter how great does not, nor necessarily correlate with TRU. The correlation of REM and ESIN of things are unequal to their correlations in RE or, correlations in themselves. **CUCOR** = Correlations Unequal to Correlation. CUCOR is determined by the RU/AU or TRU/AU relationships. ÄCAC and CUCOR are highly coterminal yet distinct by being based upon the particular or universal. **DAR** = The conversion of correlation is not systematically determinant but must be extracted with ESINS piece by piece. This is how we can discover ES, how we can learn CES and show how these ES and CES may be posited in TRU. This is the conversion of correlations between things and correlations in TRU, and **CORT** = Correlation by TRU.

LOSP = Law of Self Preservation. Self preservation was the only law that Protagoras considered not reducible to social convention. It is the basis for Darwinian evolution, selection and survival of the fittest. Life, animals, and most certainly for higher animals, and humans in their special and typical ways, it is an undeniable law. Its significance to the individual and groups is primary and vital, and its significance to the species is crucial and absolute. For it is the great mechanism and power that explains much of the FAF of TRU and tends to hold all in its place. It is also one of the greatest and most certain proofs of TRU. For since we are TRU, as TRU is TIT, and TIT is us as we are TES, and since it is the nature of our being, of subjectivity, the observed position, consciousness and mentality that so characterize us, it is this upon which TRU is based and proven. LOSP signifies the maintenance and assurance of TRIA, TRU is all, the exclusiveness and continuity of TRU. **RECIAT** = Reconciliation of all, IA, individual, group, nation and TRU as they function and are dynamically interrelated and interdependent upon each other in TIT. RECIAT occurs as THES, THEO, TRU and TA, all fuse in TIT, TRIT, and dynamically how they may be altered and interchanged in time, TRIS, TRUCH, and TRUS. (PRUT) IA, TRIT, and TRU are all inconclusive in and by themselves. There is summation completion only in SA. TRU is indistinct without TRIT, and TRU and TRIT are indistinct without IA and the relations between viewer and viewed out and in TRU. They become distinct when their relations and interdependent FAF are discovered and shown.

TRUID signifies the Ideal TRU. We must quickly remember that the world of TIT, by **TRIA** or TRU is All, is already Ideal. However experience tells us that the world has been, is, can and probably will be much better. Although these seem to be two incompatible tenets they can be reconciled. By **RECIAT** or Reconciliation of all, IA, individual, group, nation, and TRU, they function and are dynamically interrelated and interdependent upon each other in TIT. Even RECIAT may be altered and interchanged in time.

Fig 4-13 **TRU, TRIT, TRUID and TÂ RELATIONS**

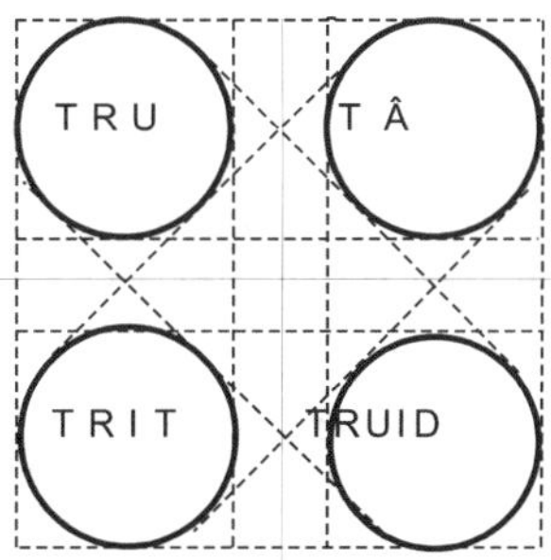

In figure 4-13 the ends or goals of fulfillment are investigated and shown. This arises from the question of whether if TIT by TRIA holds to be the best there is, how can there be any better, a TRUID or a TÂ? Does TIT equal TRID? How are they the same and how different, DAOST? TRU is the generic concept and form of all that is of the best of worlds specifically and properly held. TIT is what this TRU is as an entity in and of itself, shed of all other extraneous effect. **TRIT** is the average or mode of TRIS between the extremes of TRIN and TRINI whose standard provides a structure for objectification and representation. TRUID is the ideal TRU, and thereby not a discrete entity, at least not currently nor with objective specificity. For it will vary with time and our perspective. TRUID can be any where from a small rise in TIT or TRU to an absolute end, TÂ. TRUID may be compared with heaven, a future state in which we have arisen to become free from ANTRU, and much of the burdens of ATRU and AMTRU. The only actual world, TRIS and TIT-TO, is like the heavenly kingdom of God, a relative universe centered on Homo Sapiens and all that enables the survival of human FAF, mentality and life. TÂ is the ultimate of TRU in which it approaches absolute continuity. This is one in which the thisness of TRU is no longer tentative. TÂ, by LANERA and LAMOL, is impossible as an absolute absolute, only as a limit, a relative absolute. Yet for TRU it is of the great importance. It is the unmoved mover that so basically operative in sustaining our haecceity, thisness, or being.

TRIND = The TRIN/TRIUD Differential, or what is of TRU such as can be in TRUCH Improved. TRIND signifies the explicit difference as well as need that **TRÍ**, or humans as TRU special, mandates. It may be from TRUS, TROS, TRIS, TIT, or TRIT compared to their respective TRÍ. Yet like TRIT we have to develop a standardized TRÍ or: **TRÍST** = Standardized TRÍ, e.g. Christ. It is TRIST that gives us the target, goal, purpose and end that TRÍ implies and mandates. It is TRÍST that provides a focus for the future, as TRIT does for the present. TRÍ and TRÍS are the potential and developing TES and TÍS. That TRIST is not absolutely determinant is evident from the TIT/TRUS relationship and is called: **ÀTRIST** = The Indeterminant TRIST.

TRUID in being ideal is more or less extra TRU, less than TIT. Yet by **IDIT,** the Ideal, It, or TRUID TIT relationship, it is brought back to TIT. This is a great basic, dialectical, and controversial mechanism. It is something that is with all of us continuously be it from the tiniest increment of our IA to the complete whole of TRU. It can be answered by learning more and more about TRIF and TRUF and what is best for TIT. It is what ROT, MOT, ALT, and SPALT seek to solidify.

The anachronism of the two incompatible tenets of TRUID and TRIA can be solved. It is largely explained by the exclusion of all less than TIT from TIT, the flexibility and tolerance of TIT, and change in TIT and TRU, TRICH and TRUCH that are largely reversible. How the world can be both TIT and TRIA also depends upon the time frame, as in TRIS, acceptable or managed and controlled. Whether we speak of TRUID and TRIA of the present, near term, mid term, or far term future. The closer it is to the present and near term the less TRUID and more TRIA. The further out it is in the mid to long term the more it will be TRUID and less TRIA of the present. So that for the extended far term it becomes a somewhat, yet largely reversible managed and controlled TRIA that has an acceptable identity with TRIA of the present. One way to tell whether or not any TRUID is valid that is not absolutely within TRIA present, is the experience we have with speech, writing and communication. Humans only live a short time and much of this is spent in maturation.

It is of great importance in the future that we understand these differences, relations and solutions to TRIND, TRUID, and TIT. For it not only helps us correct the errors of common beliefs in the future, religion, and many other disassociated topics, it helps us understand TRU, and most of all gives us the ability to manage and control the future which is essential if mankind is to prosper, survive, and reach TÂS.

PERFIT = Perfection of TRU, or Perfected TRU. A perfect TRU means that it can not be made better or greater. PERFIT does not have to be TIT, although it is for TRINI and relatively for other TRIS. PERFIT helps determine what of TIT may, in the future, need improvement, or prepares us to act in those ways that best assure TÂ. This helps us see and apply the different kinds of relations between TRU, TIT, TRIT, TRUID and TÂ. These may be expressed in different ways, but when standardized helps retain focus, and bring them back into PERT, KOT, TRU and TIT.

If we restrict our definitions of real to a perfected agreement between the subject and object, and actual to what is perfectly of TRU we have two combinations , the really actual and the actually real. An analysis of these two helps to make clear their definitions, and shows how subtle and profound the meanings of and in the relative universe itself, TIT, can be. The <u>really actual</u> - signifies how well the real, observer observed or relativity, represents the actual or the relative universe. II. The <u>actually real</u> - signifies how much the actual, or relative universe conditions and sets the real, observer-observed or relativity, as it is the form and function, continua of this universe. Only when we think in terms of the ESINS or essential TRU meanings of terms can we achieve the selection and distinction needed for a discrete and positive language to represent this special universe of best human life. The marked difference in meanings between these two combinations also shows how much we must have clear and specific terms if our language is to sufficiently serve us. These two combinations also are a great help in understanding TRU. The really actual is synonymous with TRU, the actually real the essence of mass energy continuum and dimension. What is the real and reality, if they do not have the greatest identity with what is essential of our lives, and the whole realm of all this implies TRU?

What makes TRU so special, so exclusive? It is largely the absolute power of Self Realization, or **PÂSRE**. No other specie, no other entity - that we know of - has such powers. Although it is of varying intensity in individuals & groups, for humans ingeneral it is absolute, an undeniable and irrevocably certain level in human mass energy, Á FAF. It is closely related to the ego, actual reality, expansion, THESÂ, THEOÂ, THESOÂ, PUREACLITY, COGNOS, and PRUT.

TRUS	TROS	TRIS	TRÂS
TRUS = Kinds of TRU, Those Sets of TRU Approaching TIT TRU = TRU All Inclusive, Indeterminant TRUG = TRU General Overall Aspects of, TRUCH = TRU Change, General TRUD = TRU Dynamics, in General TRUP = TRU Past TRUPRE = TRU Present, TRUF = TRU Future TIT = TRU In Itself, TRU in and of Itself,	TROS = States of TRU, Unitary, Finite, Infinite, Total and Absolute TU = TRU as a Unit, The Unitary TRU TE = The Finite TRU, Parts of TRU TI = The Infinite TRU TO = The Total TRU TÂ = TRU Absolute TIT-TO TIT Total as a set	TRIS = Special Kinds OF TRU TIT TRU Proper TRINI = TRU extended only Infinitesimally TRIN = TRU extended Infinitely TRIPRE = The Present in TIT TRIF = The Future in TIT and TRIS TRIP = The Past in TIT and TRIS TRIR = The Relative TRIS TRIT = The Standardized Focused TIT, Frame of TIT	RÂ-S = Relative Absolute, Preliminary and Formative TRAS TRIA = TRIA is all TAIA = TRU Absolute is All (TÂO) TÂET(F) = TRU Absolute is Eternal Future TÂIT = TRU Absolute is this static now TÂF = TÂ Future TÂS = TRU Absolute Plural TÂ Extending Indefinitely into the Future

Table 2-13 **TRUS, TROS, TRIS and TRÂS**

TRÂS = The different kinds of TRU Absolute. We speak of TÂ or TRU Absolute, as being that of TRU which is absolutely TRU, as by TRIA. We speak of TÂS or TRU AbsoluteS, as being that series of TRU of the future which proceeds eternally or absolutely. We also speak of TÂ or TRU as being that state of TRU or TIT in which it is immutable, unchanging, and unchanged. It is for all times, intents, and purposes absolutely fixed. These three and other ways in which TÂ can be interpreted absolutely must be kept in mind if it is to be properly conceive, or even more greatly conceive of TRU and have a knowledge that approaches complete comprehension and success in all. So that in addition to TRUS, TROS, and TRIS we must include, as that series or set of TRUS-TRU-TROS-TIT-TRIS-TÂS-TRÂS variables which are paramount to a proper and successful formal distinction, overview, as well as life, of TRU.

The term and concept 'absolute' is complex, multiple and not easy to understand. In current usage it implies an impossible, inhuman, and illusive nature. It, like creation and evolution, has both valid and actual, yet invalid and inactual, implications and relations. These can be corrected by reducing them to TRU and TIT, holding and using them, as we do for zero, to an absolute that is a limit, rather than an exclusive and non existent end. Thus they are like so much of TRU and TIT, of a mixed in and out, OITIO, actual and inactual nature, whose meanings greatly assist our knowledge of TRU. In this way, even ÂU, ÂUS, and ÂÂ are of use.

AFA = All for All. AFA signifies the way the absolute nature of TIT is fulfilled, as it may not be fulfilled without, or the contingency by which all in TIT are united and FAF together. This is largely how they are contingent upon us, as we are TES, and by ENT assure TRU, TIT, and TRIT. **RÂTÂ** = Ratio of the Relative/Absolute form of TÂ. If in any time, t, the possibility, of an infinite limitless TRU, exists then this TIT is absolute. The differentiation between these two types of absolutes, is **TAR**, The Absolute Relative where although it is absolute within TRU it is not absolute without TRU. This nears absolutely not continuous without TRU and nears absolutely continuous within TRU. This is called: **LACTAR** = Law of Absolute and Relative Continuity of TRU by The Absolute Relative. **LONCA** = Law of Neglect of the Collective Absolute. LONCA denotes mans usual tendency to ignore the collective absolute of which he is more or less a part, plus the error of assuming a false or incomplete collective absolute. **PATAS** = Paradox of Multiple Absolute. PATAS denotes the predicament man is placed in by failing to recognize the various kinds of absolutes, and their SCARLS, especially when related to TRU. These absolutes vary both quantitatively and qualitatively. Quantitatively they vary by the degree to which they depart from an absolute absolute. Qualitatively they vary with the many varieties of absolutes and relatives.

TARU = The Absolute Relative Universal. TARU signifies how TRU knowledge of the absolute effects and benefits the relative. TARU is evidenced by the fact that this very knowledge is its proof. What exists as universals without TRU through their ESINS and transcendence in TRU are of TRU. So that what is universal, as a part of AU, can also be a part of TRU. It is a major course in Homation. This must be qualified, for what is in an entity as it is AU, that its ESIN in TRU leaves that entity mixed AU and TRU. So likewise are the parts of TRU, that are less than 100% TRU either in their thisness or in proportion. This state of mixture, as by AMTRU, is called the: **CARU** = the Conditional Absolute Relative Universal. CARU is a complex concept, for it bridges the gap between AU and TRU whose fine, yet massively intricate and interactive, line must be maintained if we are to understand all. Yet whose fine line must be continually penetrated if we are also to understand and show all.

LANID = Law of Absolute Infinite Identity. It is neither the subjectivity of ourselves nor the objectivity of a thing in and of itself that we can get down to the nity-grity of it all. Only as we investigate closely, as in experimental science, that which is an entity, sense, thing, the subject and object in this entity, and how together they become one that is meaningful, valuable and useful that we can fully attain that state of identity that is our great blessing and power. This is the mass energy state or dimension, Á, in its most concentrated form. Absolute and infinite have identity, or approach absolute identity, in the limits of TIT. This is shown by numerical relations and the extremes of the Á, L, and T dimensions.

RAR = Relative Absolute Relative, and TAR denote where and how the absolute of a RU or TRU can not be absolute greater than its limits as a relative or TRU. Yet this absolute, to us, to the TRU, is All Absolute, by RIA and TRIA. This is called for TRU: **ALAT** = All Absolute TRU, TAR. Also to fully recognize the states of absolutes: **ATA** = All TRU Absolute, **AAA** = An Absolute Absolute, **ARA** = A Relative Absolute, **RAA** = Relative Absolute Absolute, **AAR** = All Absolute Relative, **TAA** = TRU Absolute Absolute. Some of these are noted only in passing. To give a more complete account of the varieties involved we consider three main groups, those that are absolute alone, relative alone and the two together. To these we add The Relative or what is generated by TRU and their resulting combinations. **TRA** = TRU Absolute, or TÂ, is not Absolutely certain; only as it is - and we are - all absolute.

TÂ = TRU Absolute. TA is that kind, variety, state and form of the TROS of TRU that reaches beyond the relative to an unconditioned and all exclusive end or progression. It may be thought of in different ways: 1. Absolute TRU, 2. The Assurance of Absolute TRU, 3. To the Assurance of Absolute

TRU, 4. Toward the Assurance of Absolute TRU, 5. Unconditionality of Absolute TRU, 6. Exclusiveness of Absolute TRU, as by TRIA. **LATÂ** = Law of TRU Absolute. The precedence for LATA is both immense and common. We know the significance of success over failure, death, survival, and seeming absoluteness of our being. Humans have often demonstrated their concern, especially in art, and treatment of the dead. It has been both an individual, group, and universal effort. LATA is the summation of all these together. It is the end point of our being. It is that compulsive, vital, crucial and pivotal state of life that drives much of our actuality, existence and being. It is by LATA, and thereby TRU and TES, that we are so obligated and compelled to do and be right. **TITRÂ** = TIT Relative Absolute. TITRA signifies and affirms what may not be absolute absolute by TRIA and TÂ is relatively and practically so by its Existence and Presence.

TÂS = TRU Absolute Plural, Varieties, Types. The TÂ of TIT is **TÂTIT**. The TA of TRIS is **TÂTRIS**, and The TÂ of TRID it **TÂTRID**. It is in this way that TIT, TID, and TÂ are DAOS and DAOST, Different Aspects of the same thing, and of TRU. In being both highly coterminal and cofunctional with the same codomain, at the same time as being separate, different or even opposite we have a dialectical state in which the dynamics of these function in a way that not only proves their separate yet fused existence, but enables us to explain the basic and fundamental unified state in which the world, the Relative Universe, and we, exist.

SOLUTIONS of TÄP, TIT, TU, SUT, and TÂ

Much of the solution to all TRU lies through SOT to Á, SÁ, ALT and ACEXIN or the mass energy state in which human, predominant form is manifest.(LOSADIN) A perfected resolution is necessitated by multimodality. This is confirmed by: **LARSITA**, Law of Resolution of Systems Impossible without TRU, TIT, TRIT and Á. This has precedent in Plekanov who held that "the question about the relationship between subject and object, or consciousness to being, are basic to every philosophy". How all are more or less reducible to TRU is signified by **RECTRU** or Rectification by and in TRU and TIT. How all is or can be brought to TRU.

It is only through the Mass Accelerated Extension of Added powers of our Á mass energy predominance in TRU that we survive, which expressed as Law is **LÂCMÁ**. There can be no doubt that TRUID is valid, and TRIA more conditioned. For the far term, as human powers of Á shown by SÁ, will increasingly change to make TIT - and TRIA - better and stronger. LÂCMÁ shows how Á, as well as L, and T, to a lesser extent, exponentially and optimally in the future, define much of TRIA, TRIS, and TRUID.

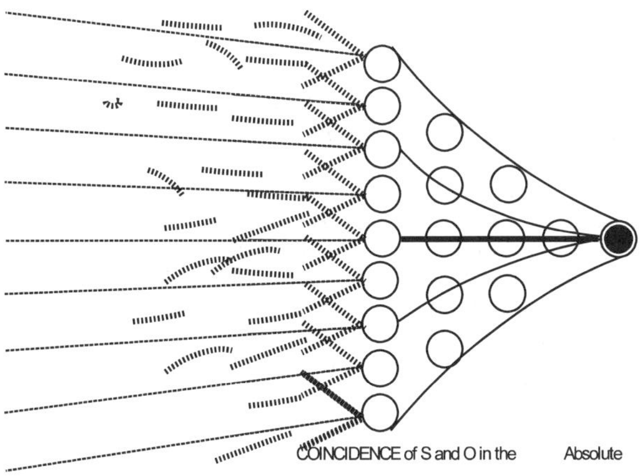

Fig 5-13 **COINCIDENCE of S and O in the ABSOLUTE**

The COincidence of SO, or identity in the absolute as noted by Schelling, is called in TRU, **COSÂ.** It shows how the absolute is derived from the union of TRU. How through <u>Intuito</u> **,**the integration and summation of all that comes together in TRU, especially as produced by self knowledge, and COGNOS, the absolute is realized. This absolute arises from being the supreme power as God, Brahma or world soul. The union of all in this absolute by COSA although showing the mechanics by which TU and TÂ are derived is not the whole mechanism. These are attained by recognizing the absolute in its overall or absolute form, mode, aspect and the totality of conditions or states from which it is produced. These include as well as Á or SO, COSA, length, time, world ground, divine will, absolute ego, organistic whole or real, and perfect being.

The derivation, development and determination of TRU, DOT and its form, DÂT, is in the mediate position between a thing and some things. The thing TRU, the special TRU TRIT, and the absolute TRU, TÂ. It shows how the processes of this relative find completion, are convergent and compelled to find their end in union with the absolute, TÂ. This is how aseity, perseity, and haecceity combine to complete an entity or universe. It is how life continuously reaches for something larger and better, as noted by Bosanquet, to TRU, TU, TO and to TÂ. It is its own eternity as it in self perpetuation works toward eternal and absolute continuation.

The knowledge and understanding of all entities, senses and things is infinite, chaotic and impossible without relative focus, systematic unity and coordination.This is why ROT, MOT, ALT and SPALT are so necessary. This is possible only by the versions of relativity, thisness, specificity, and special sense are reduced to the actuality of TRU. Even though still infinite in TRU, they when properly developed by the set form of TRIT and its coordinate structure, become increasingly cosmotic, possible and operative. As it is discovered, produced and supplied its truths are eternal This eternality of truth depends upon a like eternality of meaning, language, and definitions. Their posits in SPALT of TIT-TO are lasting, with a reliability that is confirmed by TÂ and TÂS, and in turn TÂS, and TÂ are confirmed by them.

ABSOLUTE and RELATIVE

There is a dynamic relationship between absolute and relative in all things and animals that is the basic law of nature. It is like the law of self preservation, only more generic in absolute over relative terms. It is called: **LAR** = Law of Absolute and/or Relative Being. LAR denotes the dynamic, changeable state of all things. Existence at a moment may seem infinite to the individual, as in a boy or girl caught up in and bored in a life that seems stationary. However, these moments are not infinite, as is more apparent in old age. By knowing and doing in an expanding existence, as the expansion of TRU and ALT, does LAR make sense.

For every relation there is a dualism
For every dualism there is a polarity
For every polarity there is a continuum
For every continuum there are levels or gradients
For every level or gradient there is increasing differential
For every differential there is an integral

The integral and differential combine in the continuum
Combined continua make up the parameters of an entity
Set parameters make up the dimensions of an object
Dimensions of the object TRU enable its formulation
Dimensional formulation of TRU enable ROT and MOT
Differential gradients of their dimensions enable ES and CES

Dimensions are altered and warped to adapt to TRU
Measures of dimensional gradients follow TRU's adaptation
Parts and processes of TRU are best represented by posit
Measures of parts and processes need continual refinement
The best positive objective approach to the world is in this
Way by providing the best showing and knowing of our world.

TRU, TÄP, TIT and the ABSOLUTE

By ATO, TIT, TRIA and **RÂ** or the Relative Absolute, TIT already is, in its way, absolute, **PRAT** or Present TRU is Absolute. Yet this must be vigorously separated and distinghished from AU, and ÂÂ the absolute universe and absolute absolute. Thereby we must be very careful, and clearly qualify, whenever we use the term absolute. The great hurdle we must make is largely the difference between TIT and all less and other than TIT, TES and our being and what is less than and other TES. This is the RÂ/ÂU or ÂÂ as seen in Chapt 12 pg 3. **TÂ** is TRU absolute or all of TIT exclusively. It is a this that derives from RUS and TRUS, is formed by TRIA through preserving being itself, TIT. **TÂS** is the plural of TÂ or like TRUS, TROS, TRIS the plural progression or perfected TIT. The FAF of TÂS is not only separate from TRU and TIT it is often mixed, of codomain and cofunction. This is seen in the activity of absolute within TIT and TRU, as in **LISATA** or Law of Infinite Absolution of TÂ, or how the form of TIT is not only an absolute objective form, but one that can be limited and qualified by our use of VAPO, SCARLS, and DAOST to adequately replicate TIT. The Law of Absolute Form, or **LÂÄ**, is like absolute pitch in music where the tone does not depart from a perfect focus and reproduction of itself. By continuous refinement, **ÄDIS** or Formal Distinction from perfected DIviding, and extensions of breakthroughs of TOPNEC absolute is approached. (LACON,PAMUST,Ch17) When applied to all that is refined to perfection in TIT, the degree or quotient of this fulfillment is its Rate, **RÂÄ**. It is the top quality, 100% purity, perfection and TRIA of our genetic material, live product, determining forces and all that constitute this special universe of best human life, or TIT-TO. It helps to establish RÂ, TÂ, and TÂS. Yet TÂ and TÂS are much more. The RÂ, TÂ, and TÂS of TRU are steps toward one extreme in the continuum from AU through MÂTRU to TRUS and in TRIS, through TIT-TO to TÂ, to TÂS. For TRU is not a singular homogeneous entity or topic. It is a composite continuum of which TÂ and TÂS form one extreme or pole, and TIT is the mean and focus. Yet it is TÂ and TÂS that vary, about the mean of TIT-TO, functioning in unison as we adapt and ultimately incline eternally. It is the TO of TIT that we must maintain. (PPMA) By holding to the set of TIT-TO established through TRIA, TRIT, OBT, SUT, ROT, and MOT we show TRU. Although TÂ and TÂS, are largely coterminal with TIT-TO they, like TRU to be represented, must be viewed from the outside. This is why we supplement SPALT by METAD, SPAT, POMOT, SPOT and all to reveal the extremes and external relations of TIT-TO.

LACOTA = Law of the Concretion of TÂ. For TRUS we have, not only TIT by TRIA, we have a whole series of TRUS which exist both in series, as we understand from evolution, as well as change of TIT. Yet it is likely from the consolidation of this best of worlds of man, by civilization, knowledge, God, common purpose, TRU and WIM, that TIT is becoming less like other entities and universes and more like a discrete and concrete separate entity or universe. This is TIT, which also has exclusion in its own right, and thereby clearly separate. This is increased with the absolute nature of its self perpetuation. This is how TIT is a large link in leaving chaos, developing relative cosmos and becoming consolidated in permanence in TÂ. It is one from disorder to order, from diffuse multiplicity to unity in totality. No where is this more evident than in the development of the idea of God by man. For before god there was little unity, with God there was group unity, with GOD universal unity was sought, and with TREQ and TRU the universal unity becomes consolidated in all that makes up TRU. The dynamics and processes of this is much of what HOM, DYNU, DYNS, and DYNO demonstrate to provide that vitally needed for sound ACC and TRUF.

LATAC = Law of TRU as the only Absolute Concrete. TRIT by TRIA becomes the only absolute concrete, in being through TIT The, The RU, the only universe and only actual object. Evidence for LATAC comes from Kotarbinski's Concretism, or Reism. This arose from how qualities can belong to or inhere in the things of which they are characteristic. It is conditioned by the fact that nothing is hardness or softness, blackness or whiteness by themselves. This corrects the conclusion that only concrete individual ENIT, OBIT, or objects exist. We should distinguish from the Absolute TÂ, what is in TRU, **TAIT,** and the absolute TÂ that is of the future, in and of itself, **TÂF**. For although they are much the same, they are also much different. **LATASI** = the Law of TÂF - TRU Absolute future Sameness, Similarity and Identity with that of TIT. This does not mean that it must be completely the same, similar or identical. It only means that certain aspects, forms, essences, qualities, identities, functions must be relatively maintained to assure its fundamental and ongoing continuity. Certain qualities, must be the same, similar in certain ways, or identical in certain special ways typical of TRU. As knowledge of TRU unfolds it is possible that TIT may become more and more concrete to us. Yet this will be within its own limits and not our subjectivity or knowledge of it. It neither has to be, nor is it at best, perfectly or absolutely concrete. For it, as TIT in itself, is what it is, and whatever extra added detracts from it. Thereby LATAC is valid in so far as it supplies us with the TIT, TU and SUT necessary to make an absolutely concrete object. Yet this for TIT is limited, SCARLED by the very nature, form and function of TIT.

LAP = Law of the Absolute Positive. LAP is where TÂ is or is not, and TRUS, TRU, TIT and TRIT are or are not as WE WILL, or as all parts, especially those in the most capable positions, will fulfill their roles.(WIM) TRIT, TIT, TRU and all become infinite and absolute only as we become infinite and absolute. LAP confirms that the degree to which TRU has permanence, relative to infinity or absolute eternally, TÂ, is more or less a function of human will. This, at best, may not be the only, yet it is the major, vital and critical feature of TÂ. TRU, as TIT and TRIT is, primarily one, or at least, based upon the focus of one. This is called: **UTFU** = Unitary TRU and Focus of One. It is from UTFU, as TU, that we can base SUT and all that follows. TA, on the other hand is the infinite extension of this one, or 1/0. ANT and ANTIT are 0/1. This is called: **UZTÂ** = Unitary over Zero, is TÂ. TRU is moving in the direction of 1/0. **POR** = Principles of 1/0 or the One Over Zero Ratio. It is in POR that the life and death, the dialectical dynamics of TRU, as well as our life and immortality, are at stake. The whole course of humans and animals on earth has been one of the surging and ebbing tides of POR, and hope of assurance in the absolute.

That there is a disparity between TÂ, TIT, and TRIT in DAT is: **PAIT** = Paradox of TÂ TIT disparity. PAIT is answered cursorily by; absolutely we are relative and relatively we are absolute, or: **ARI** = Absolute Relative Identity. ARI occurs as the relative and the absolute approach each other. The realization of perfect form and function of TRU in TÂ, as the creative, emergent, procreative, and DAT powers converge on TRU to produce everlasting life, the relative and the absolute converge in unison. This is a special case of the overall state and condition of interdependence seen in Godel's Theorem, and is called: **STIN** = State of Interdependence. All that is so much of TRU is so, not as it was in itself, nor as it may be taken carte blanche into TRU, but only as it is selectively reduced to, or completed in TIT as this TIT is completed by it. **POUC** = Principle of Ultimate Coincidence. The actual world, TRU, is the maximum of absolute effects of absolute causes, as all are in TRU pure TIT. It is relative unity as things and partials are or are not of TRU. Their validity and actuality are a function of their relevance and function in TRU. If all that is TRU is all by TRIA and TÂ and all that is in TRU is absolute by TIT and LOPAN then we can be assured of being absolute - have immortal souls - if we are TESÂ.(14-15)

CHAPTER 14

A PART and PARTS of TRU: TRU DIFFERENTIAL

That the world is composed of its parts is self evident. Yet what is much less obvious and critical to human understanding is what these parts actually are, whether and how there is any order to them.

TERMINOLOGY, SEPARATION, DIFFERENTIATION, KNOWLEDGE, PROBLEMS, PROOFS of PARTS

Ê = A part or component of TRIT, TIT and TRU. **ÊS** - Parts or components of TRIT, TIT and TRU. **TÊ** - Differential or Particular of TRU, that component of TROS which signifies the finite of TIT. **TÊS** - Differentials, Particulars, or the Plural of TE of TRIT, TIT or TRU. It is what their ESIN is as it is its counterpart in TIT. **CE** - Process of TRU, **CES** - Processes of TRU. CE and CES are the functional equivalents of E and ES. ES and CES are not all alike. Since they are DAOST and since TRU is composed of a myriad of parts, with some of great and some of minute value, some universal and some very particular, they are the same when containing the same of TRU, and different as they contain what is other of TRU or each other. As we progress through IA, an entity and what is relative to TRU, we encounter the relationship between entities and parts of TRU. These are called: **RE** - Relative entities or things, referred to, **RÉ** - Relative Part(s) of TRU, **RÉM** Relative Part(s) of TRU Measured, **REM** - Relative Magnitude or Proportionate parts of the whole of all components of TIT-TO. We gain through IA an immediate inference of these parts as they partake of our lives. Many of them exist in opposition. For no RE, or things, are TRU, some are TRU, some are not TRU and all RE can be TRU as they are RÉ. The deficit between these RÉ or ES that are perfectly TRU and all else is called: **ERÉD** - RE - RÉ Deficit. There are several kinds and characteristics of E: E that is known, E that is unknown, E that is known at least once, but not known generally or not known by a particular person or media, E that is strictly of TRU generally or specifically, with out extraneous RE, E that is stereotyped of TRU, as through ROT, E that has modulated form, function or process. We must always remember that E and ES are not perfectly known. Nor is it often even very well known. Yet it is generally certain to be fixed overall, exact, necessary, and amenable to ROT, MOT, SPALT and POS.

KNOWLEDGE, PARADOXES, PROBLEMS and PROOFS of ES

PADET - Paradox of DET/AD or the Determination of the Parts or ES of TRU over the changing Forms of Entities. PADET evinces the potential in selectivity and the role of IA and Á in TRU. **PRUE(PRE)** - Proof of E, ES, CES. **PRUET** - Proof that E, ES, CES is. Proof of the Haecceity or Thisness of E, ES and CES. **PRUEQ** - Proof of what E, ES, CES are or Proof of the Quiddity or whatness of E, ES and CES. PRE, PRET and PRUEQ are proven and shown by providing: pure possibilities as the concrete substances of God, categories as examples of parts of a whole, existensional moments as examples of the parts of TRU, the proof of the parts of any whole, PRUT and PRÁ. PRUE arises directly from the fact that any entity, object or universe has parts, as TRU and ES are revealed PRUE is shown.

DERIVATION, DEVELOPMENT and DETERMINATION of A PART and PARTS of TRU

ES exist in all about us, however, only as these are TRU. **DIT** - Dichotomy, Dialectic, Division and Difference of TRU. DIT is a primary law of TRU. **LATUR** - Law of TIT is the Ultimate Reality, there is nothing in TRU, not even TA, TO, TU, that is "the end in itself". Only the collective exclusive TIT is true, the ultimate reality or absolute. There is a collective common that contains both what is TIT and what is in any and all else that is still TIT, not ATRU or AU. This is in striking contrast with AMTRU, mixed TRU and non TRU. This TIT is the same TIT, yet shows how it is so easy to misinterpret TIT and non TIT. It is a TIT that is set in its totality. This is the set of TRU that also includes some of what appears to be non TRU entities or universes. This combination in TIT arises in part from what is DIT, or divided from or different than TRU, or TIT. **LESEN** - Law of ES or part(s) of TRU Entity relationship. There is a close relationship between the parts of TRU and entities in general for much that is of entities are parts of TRU, and much that are of parts of TRU are of various entities. In our lives we repeatedly and massively bring entities to TRU, to their parts and roles in TRU, as ESINS become ES. This is in a cybernetic fashion of output, input, and feedback as they are of decreasing or increasing use, value, need, meaning and purpose to us. All come into being, TIT, from **DE** or derivation, development and determination of their part-s. **LODET** - Law of Determination, Derivation and Development of Parts of TRU. LODET signifies the importance, necessity, and operations of determining any part or parts of TRU. **LODES** - LODET for a part or parts of TRU.

There are many ways and forms by which the world, all that is around and about us is a part of us: Direct or indirect inclusion or participation, voluntary or involuntary inclusion or participation in TRU, and knowledge of AU as it helps to understand TRU and therefore becomes TRU.

REASONS, CAUSES, METHODS and DETERMINATION of ES

That proportion or proportionate part of the thing that is its ESIN, is called: **PESIN** - Proportionate Part of an Entity or thing that is its ESIN. How we remove all that is in the entities, objects, words, ideas, concepts, principles, phrases, images, pictures we utilize only that which is TRU is called: **EXESINS** - the Extraction of ESINS. The whole of POT, IPTRU, and the method, theme, intention and goal of TRU is one of providing the concepts, principles, laws and mechanisms that best and most fully show TRU by way of the extraction of ESINS. **ESION** is the processes of ESINS and ES Formation. The Negation of ESION is the continuous and massive drift from ES and from TRU. There is a massive interchange between TRU and non TRU or input and output revealing those tendencies that increase or decrease TRU. When less than pure TIT, of OITIO it shows the definition or systematic logistic definition problem. Much work, study, and many new means and discoveries are needed to bring these to their completion in TÄ and TE. In determining ES it is necessary to differentiate the form of TRU as a whole, from the form of TRU taken differentially or piecemeal. This is doubly important because it gives us the long needed distinctions which philosophy has been unable to attain for want of a clear object, The Thing, TRU. It also sets the stage for our ultimate modulation of TRU, SPALT, what enables the posits by ES and CES of the beautiful intricate detail and its artistic representation when provided for TRU. Critical and piecemeal analyses are needed too establish ES, or parts in TIT. This is how to analyze all ES and determine the manner their ESINS can be verified, established, and held in perpetuity. Relative dependency of ES recursively confirms the correlation of the limitation relation. These seek ultimate meaning in all things as they equal TRU. **DÏT** - the Difference between thing-In-ITself, THIT and thing as a part of TRU, and thing as it is of its other states, types, kinds, and sets. DIT is used to produce selective associative meaning and identity. This process is called: **SAMI** - the Selective Associative Meaning and Identity. It is by SAMI that we can analyze the difference between sense-in-itself, SENIT, and the S of TRU. **FERE** - Law of Fractional Evidence from Things to ESINS. The ability to, evidence for, the total proportions and SU of all that we see, hear, feel, know - as our images of things and what they are other than of ESIN - are the constant problems which the capacity and ability of human rationality are continually facing and being challenged by. They are resolved by separating ESINS from their entities and transferring them to their counterparts, ES, in TRU. This is called: **EOA** - Extraction of ESINS. EOA is achieved just as human life and TRU tend to: **LOSE** - Law of Separation of E, ES, and ESINS.(INF,EA,IFPIP). EOA is achieved and LOSE facilitated by language, specifically: the separation of each syllable. **EA** - Idealization Realization, or the Realization of Ideals, the ESINS in things. EA is how we extract, transfer and transform ESINS, ideas or ideals in things, entities, senses, and RUS and enter them into TRU. This is done by, and much of the activity and success of, relativity, the observer observed, S-O or the predominant Á state, dimension, and FAF of TRU. It is how ideals, forms and reals exist alone, together and in TRU. **LESDRIN** - Law of ESIN, ES, Dynamics. There is a never-ending interaction, interchange, problem and struggle to reduce all whose ESINS are in entities to what this ESIN is as an ES and CES in TRU. This is the power of TRU, its predominant Á FAF to bring all optimally to TRU, the method of, and misleading state when considered separately, of ES static and dynamic as they occur in TRU. ES dynamics may at first appear anachronistic, for if ES are its static parts or and CES its dynamic processes then dynamics for ES is an unwarranted extension of its meaning. However on deeper reason we can recognize that statics and dynamics, parts and processes, form and function are but extremes of their continuum in TIT. Thereby **LÂNEC** or Law of never absolutely static or dynamic ES or CES. The continuum formed is **CESFA**, Continuum of CES and ES, form and function. We tend to envision the world about us by absolutely distinct elements, yet in TRU their definition only approaches an absolute. This is the formal distinction of what is TRU. Its ES are relatively, not absolutely distinct. Yet with TÄP, TRIT, OBT, SUT, RÓT, MOT, ALT, SPALT, and POS their FAF order and gradients can be shown. Generally if more static they are ES, if more dynamic CES. Most exist with elements of both extremes. Thus we must adopt methods of total statics and dynamics, ECRO what is both ES and CES. When this is focused upon entities, their ESINS and how these are ES we can approach absolute ESIN ES. **EDI** - Reals E Deficit. EDI is that percentage, amount or sector missing in TRU, ROT or POT. **ESÁ** - ES Á, or ES as All Á. The phenomenon to which the psychical phenomenon or act refers is the primary object. It is what appears as if it were external to consciousness. The ES are more or less subject and object, therefor considerably phenomenal or real. Since this in TRU is Á, or that predominant state in which subject and object become real or phenomenal, they are combined in the relativity and actuality of TRU. They consist of both subject and object, thereby phenomenal or Á, as well as L & T.

The different nature of parts, and their distinction, are not only forms of TRU, they are accentuated by our view, observation or subjectivity. Where it is very necessary to describe and demonstrate a thing, TRU, we must be able to clearly, accurately and readily, set forth its parts, separate them, and hold them in some kind of visual or verbal image. This is spoken of in modern English terminology as the concept of Ideal Types, where a thing, as TRU, is constructed somewhat theoretically by an investigator from component elements with a view to making comparisons and to developing explanations. Other methods of modern thought, which illustrate and provide a method for the attainment of ES, or parts as generalized, are Constituent Analysis and Error Analysis. Similarly various topics in mathematics provide concrete demonstration of the existence and nature of ES, e.g. complex number and complex function theory. To make this analysis ES must be enumerated, or: **ESSARY** - An enumeration and defining of ES. ESSARY is like a glossary, lexicon or dictionary of ES. To bring these ES and their EC and NEC to their TRU meanings is done by: **ETION** - Process of Reducing all to their ES in TRU. ETION is a term for ES of TRU, whereas **ESINATION** is a more specific term for the development of ESINS. **EÂ** - ES (and ESINS) are abstract.

SO and Á with ES in TRU

The representation of TRU, E, ES, ESINS, concepts, terms, and words are expressed to produce ever greater, ACEXIN, and COACINEX by providing the meaning necessary to identify their part and role in TIT. It gives them increased formal distinction and actual - TRU - meaning. The degree to which this is attained and the degree to which they are VAPO and SCARL. REM in TRU is the degree to which they optimally replicate, enhance and empower TRU. This is called: **LADMIRT** - Law of Á State Derived from the Formal Distinction, necessity and Meaning of the Different in TRU and REM as they are Increasingly TRU through VAPO and SCARL.(PRA, PRUT) Much of TRU is the state and operations of ES as ESINS, concepts, and terms working in the Á continuum in TRU. Words are the intermediary vehicles or media in Á and TIT. **RERO** - Resolution of Reals or Objects. **ÁEA** - subjective-objective or Á of Á Part or Parts Form of and in TIT, TRU or AU.(TÁ,TRUR,RÂ) RERO is confirmed by the existence of reals, what is neither material nor ideal alone but tends to their discrete media or middle ground of discrete participants in TIT. Animism is the theory that objects ordinarily considered inanimate have something within them analogous to the human soul. Animism was among the first of human religions or philosophies. As primitive as it seems today, compared with modern religions and philosophy, it contained a truth that often escapes us. This is the fundamental aspect of human life of great significance, the part and role played by entities in the world about us in us, and how much we are of them. How TIT and ES extend and interact into the world and the world about us into ES and TIT. TIT and TRU are animated, worlds in their special and selective ways. **DACTI** - Diverging and Converging TRU and It, entity, Individual, and all. DACTI may be viewed and shown as a crows foot, that which stands in relation to TIT.

Reason by Á is based upon the knowledge of the interrelationship and revelation of Á's and ES. Reason is also the why of an entity or object and by TRIA and TIT becomes the Á form, state and dimension of TRU. For reason in being the cause, and reason for being the processes of selection by TRU, are such a powerful aspects of the subjective, or mentality in TIT. It is to the reason, or the why and essences of things, that human attention is focused. It is human and TRU FAF, or form and function through Á that reacts to, alters, and utilizes this reason as it best serves the purposes of TRU. It uses ESINS. The degree to which all these exist and function in common is called: **RESIN** - Rate of ESIN in Entities, objects, universes and TRU. RESIN can come in all kinds of forms and shapes, although it remains set for each specific ESIN when ES in TRU. RESIN that is set, and more or less vital, for TRU is called: **RESINT** - Rate of ESIN for TRU, only. RESINT signifies the level of ESIN in TRU. PRUE is shown by ES and Á's finite composition. Much of this are the reals which are the intermediaries or focal constituents of relativity in the S-O and SÁ spectrums. These reals actually become the real of SÁ and TIT as ES, CES, parts and processes, form and function of TRU become known. This is RERO, as they become ES to become TRU.

O → R → ES → TRU

. This is the reason and need that TRU Terminology is produced.

KINDS and GENERAL FORM of PARTS of TRU, HOMOGENEITY, HETEROGENEITY and DAOST

There is a difference between finity and partiality. Partiality is a mode and characteristic of a whole, how it may be divided or differentiated. Finity is a mode and characteristic of form. It is a range that can be increasingly divided between the infinitesimal and the infinite centered on unity. To our minds the entities and objects we observe are not only true and real, they have thisness, absolute existence. Yet much of this is false, an illusion. Much is only tiny increments toward or away from their ESIN or ES of TRU. This is incremental or decremental toward or from ES. An indirect, yet at times effective method of determining ES is "as if" ES. How ES are not necessarily, nor always, the same, although essentially often the same in different states of TRU is signified by: **LOFETE** - Law of Fractional Evidence from Things to ESINS. LOFETE denotes the method by which we can separate and extract what is not or less than TRU that is also equivalent to a part of TRU. This is but one step in a series of steps from entities to TRU. The next step is to fractionally adapt these ESINS to their ES in TRU. **TECA** - the Composite TE not AUS nor EPI. TE can't be understood as the material parts of an object, nor the mental parts of a subject. They can only be understood as they fulfill the nature of TRU'S composition. If we are to replicate TRU and provide for its systematic unity, representation and presentation, we can not do it by trying and failing to liken it to a material object, or any common object that we may examine. Only in its most precise TRU state and operations and greatest ACEXIN as it actually is in TIT will we ever be able to shake off the current misunderstanding of the world. **LOPAR** - Law of Partiality. LOPAR is the key, gist, and cue to much of TRU and TES. For, by IA, and ENIT position in life, we are forever focusing, vacillating and variably acting from and about ourselves and denying TRU, TO, and TE. LOPAR certifies the progression from ENS to TRU and ES to TRU in such a way that all E and ES are no more nor less than their part and roles in TRU. The general importance of LOPAR is in establishing a strict basis for the determination and measurements of ES, TES, REM and all TRU. If we can hold them to their exact ES, TES, and CES we can, as by LASUR, determine, measure and represent them. It is when these finite parts become many, are held to their gradients in each parameter, polarity and dimension, and are made formally distinct, that we can see how the **TEÄ**, the form of TE, can increasingly represent TRU, and retain a high level of sensitivity, specificity and reliability. It is this, plus the similar accuracy of posits, that provide TRU and its representation and presentation with the concrete objectivity it needs.

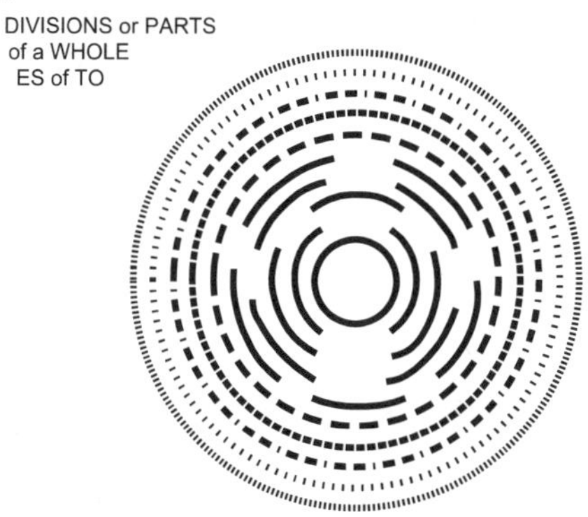

Fig 1-14 **DIVISIONS or PARTS of TRU** How is TRU divided? What is the composition of TRU? These are questions whose answers are essential to an understanding and representation of TRU. This figure presents a simple view of this problem. Since TRU is both unitary, finite and infinite we must first show it in these ways, only a small number of major layers of TRU from the unitary in the center through a few divisions or components, to many divisions or parts, to how TRU is composed of an infinite number of divisions or parts. The finite may be shown in all layers in between. It is in the finite that life, as we are concerned and involved. This is because our lives are mostly characterized by our relativity in the finite divisions of TRU and all about us, e.g. of A, L, and T. This is called: **EPRE** - Finite, or Particulate Preeminence in TRU, and **EPRALT** - Finite or Particulate Preeminence of the finer and finer, or more and more finite, of ALT of TRU. It is finity reduced to unity, and finity extended to infinity, that we stretch our relativity, enhance and assure our being.

Fig 2-14 CINESP or the **Coincidence of ES with Polarities**

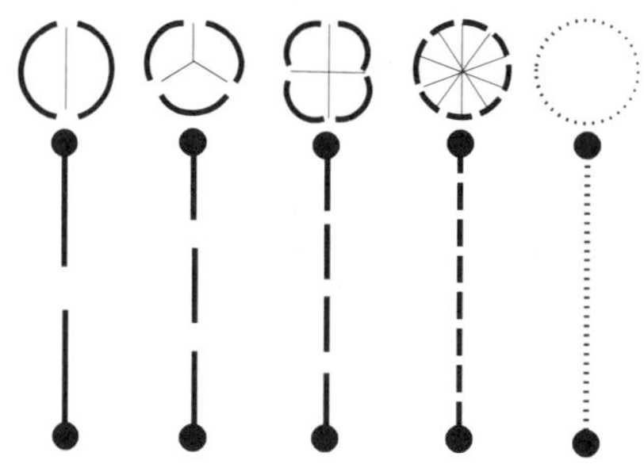

In order to better visualize ES and their relationship with the finite gradients of the dimensions of TRU, the accompanying figures are provided. The finite parts of TRU in item one with only one division would be for only one dimension, e.g. length large and small. The finite parts of TRU in item two with three divisions would be for length large medium and small, for time past present and future, and for Á, subject, real or phenomenal, and object. The finite parts in succeeding items, three to five, would be increasing divisions in each dimension, or spectrum and continuum as they exist in TRU.

SAME and OTHER, or DIFFERENT ASPECTS of the SAME THING TRU

It is not so much that same and other, same and different, common and uncommon are so great in themselves. It is what they entail and infer. This is what is or is not DAOST, same and different or common and uncommon of TRU. How important it is to separate out what is less than, not, or against TRU. This is a fundamental means by which we apperceive and know. Thus DAOST is a means, method or vehicle of Á and SÁ.

LASDAN - Law of Same and Different Aspects Nominalization or Terminology. LASDAN denotes how words or terms which are or seem to be mostly the same, may in being a great part of TRU and also different, give them great significance, need and right to be considered as a separate term, word and nominal. This is representation and perfection through parts and aspects of TRU being made nominal and ordinate, gives TRU terms status and is used in POA by POS. ES can be best viewed as the same, different, and/or their combinations. The same as well as different and all in between of subject object identity must be valid, important and actual, and not alone, strange, isolated, unrelated or independent of TRU. They must be **DAOST** - Different and Same Aspects of TRU. Most will contain, if only in the least, some of both same and different. E and ES are in striking contrast to how the objects before us appear. We must recognize this, learn them and TRU, forever remember and apply it if we are to have a life free of illusion and error. The D in DAOST is the Different in TRU, TE and TI, and S in DAOST is the Same, TRU, TU, TO. DAOST although simple in reason is profound in design and implications. For it enables us through the expansion of the different and the same to develop an ever greater form, construct, and operation of TRU. It is how we can establish a foundation for and execute a formal complete proportional representation of TRU.

LIMITS, SCARLS and NECESSITY

LOPAN - the Law of Partial Necessity. LOPAN is a major law and operation in TRU especially when used in conjunction with COL. When we apply it to ourselves, it explains the reason for, and importance of all in, ones existence. By LOPAN we can recognize the absolute value and nature of the partial, ourselves as a part and our role, in TRU. So also must we at least consider and render discrete, the partial. Through competition, selection, and the predominance of Á FAF there is an absolute necessity for ideas, concepts, words and their communication, as all that best serves and is in TIT. LOPAN pertains not only to a few vital parts, but in ways to all, even the most minute ES and CES, in TIT. They also exist both as a part, and to each its own, **ESIT**. It is imbalances of ESIT that are so prevalent and productive of failure to individuals. Without KOT they mistake this, each its own, for any ATRU and ANTRU.

SPECIAL FORM of PARTS, ESSENCES, MEANING, LANGUAGE, EQUALITY, and ABSOLUTES

The exact and overwhelming fact of all being is the basic composition of the world, TRU, in which it is neither physical things nor mental senses in themselves. It is that part and role of these things, senses, and all which when combined with their developing added states in the totality of TRU become this fundamental part of TRU, **ES**. It, like the monad, is irreducible in the actuality of TIT. This is called: **LAMEX** = Law of Monadic, E or ES, Exclusion or Irreducibility in TÄE. LAMEX certifies the set and delimiting state of ES or parts in TRU. This includes, among others, the differential of mass energy or Á state, form and dimension of TRU.

ESNES = That state of ES which is consistent, remains the same in TRU for each entity, sense or thing regardless of what language, thought or image it may be created or represented by. The E or ES are both subject and object, therefor largely phenomena or reals. Since this in TRU is Á,or that predominant state in which subject and object become real or phenomenal and are combined in the relativity and actuality of TRU, they will consist of both subject and object, thereby be phenomenal, or Á. Reals equal the focal or medial of the Phenomenal FAF of Á. **VADES** - Validity of Á or the mass energy state, dimension, form and function of TRU is dependent upon the ES of the part of TRU involved.

ESSENCES and ES

ESA - ES or Parts of TRU that are major, prime or predominant parts, parts without which TRU can not exist. **ESE** - The Forms or Essences of the Parts of TRU. ESE is the direction or deficit between knowledge and meaning of things, as they are ES and forms of TRU. ESE signifies and defines the entire sequence of steps of parts from being an ENIT to the pure ES of TIT.

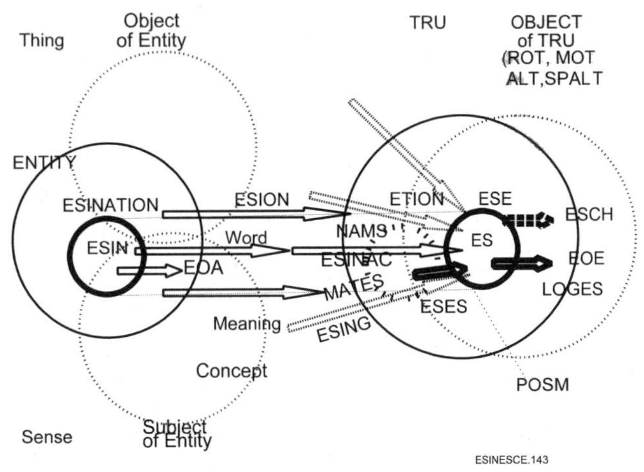

Fig 3-14 **ESINS and ES**. It is impossible to understand the world, to escape from the figments and illusions of entities, things and senses unless and until we can realize the special and select actualities of TRU. This is their ES, what is the jist of entities, senses, things, objects, subjects and all that is their ES in TRU. It excludes all that is not of this ES. This is even more than the gist of each of these ES, it is their combined gist in TIT.

FRACTALS, CARDINALS, MEANING, EQUALITY and ABSOLUTES

MATES - Meaning naturally, spontaneously, and Automatically, by TIT FAF, Tends to ES. MATES is highly significant and useful in determining and demonstrating ES. It shows the great equivalence of meaning and life, and even more the Á mass energy predominant FAF of TIT. MATES also shows how the world is not as it appears, but is only as it is in TRU, TIT.(PRUT) What man sees or senses comes to have meaning. The words that represent these things and the definitions these words take tend to TRU, to their part ES in TRU.

This is neither complete, fixed nor certain; it is only a tendency, or: **PRAV** - a Practical Relative Vector. PRAV is the practical vector in all things, as they incline, or are positively attracted, to TRU. The transformation of meaning to that in TIT is called **ESING**. It is highly dynamic, quantitative, and often reversible. Of the many meanings for each term there is usually one or a focus that is its theme. This becomes when set in TRU, their Principle of Special Meaning, **POSM**, at the focal point of their actuation in TRU. Here the term actually becomes what it is in TRU. The words we use are neither strictly of entities, ENIT, their objects, their ESINS, nor of their corresponding ES,.They are a variable mixture of all four. This shows how abstract and difficult our language can be, both in itself and as it naturally inclines toward TRU, or can with a full understanding of TRU be interpreted by TRU. **PROSM**-Process of Special Meaning.. PROSM signifies how ESINATION, ESION and ESING function in TRU and HOM. How meaning is taken to its special sense, that which is in TIT, and by CES. **LOAM** - Law of Absolute Meaning, or how it is Absolutely equal to its E, ES or equivalent in TRU. It is TRU only if it, as its ES, is 100% TRU. Since TRU is so difficult to know, and ES, when highly refined, are so hard to determine, it is usually attainable only after long and difficult investigation, analysis, and determination of each ES and their terms. This is necessary to counteract the continuous tendency to misuse ES and language. The part language plays in ES is the perfection, ESINATION, and ETION of Language. This is the change, perfection and manner in which words and language become their counterparts in TRU. It is by alteration of definition in becoming pure, real and actual by equivalence with its part, ES in TRU. As different things become TRU they to some extent become the same thing in TRU. **ACE** - Á or mass of TRU conversion to or convergence on E, part of TRU. This is the relationship between Á and E whereby E becomes pure, real and actual in TRU as it converts to A, and A becomes pure, real or actual as it is converted to E. This is not to say they are equal, only to the degree they work by ACE in TRU. Since Á is only one of three dimensions of TRU, E will be composed and have term FAF different than Á.

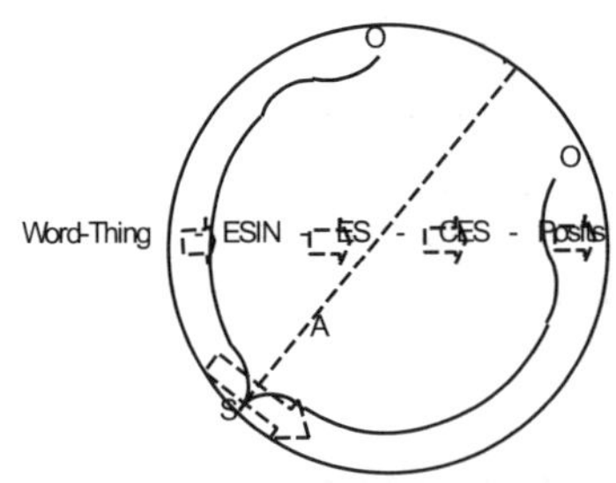

Fig 4-14 **ES, LANGUAGE and TRU** When ESINATION and ETION are optimal we have: processes of optimal ES formation. This is related to conceptualization, recognition, and knowledge of ES and COGNOS. Not only by recognizing the principle or general state of ES, but by recognizing each and every ES, ESINS as well as entity, their names and meanings can be much better identified, learned, and ES, TÍS, TRU and TIT achieved. ES change to be more CES, both as functions in TRU, and as TRU changes, reversibly.

DISTINCTION and DEFINITION

The search for all of ES in general is not something we can achieve outright, nor can it be provided through simple study or interpretation. Like so much that is different in TRU, as in human evolution anthropologically, it is the great search for TRU meaning, expressed in the focus of subject-language-real-object that guides the identity of each ES. It is the manifold increments of usage, trial and error, selection of all, especially word-things that are about us. It is how they go to increasingly make up the actual, definite and distinct reals of finites and parts of TRU, ES.

A word can never be an ES unless it is used exclusively for ES. An exception or corollary of this is **NAMS**, where we use many very great, universal or necessary ideas, terms for things that are a great part of TRU, in a qualified TRU sense.(EC) This qualification is made in the progression from TIT through TIT to OBT, SUT, and ROT plus the standardization necessary to enable POA. This is guided by and helps to prove: **LO** - the Law of Objectivity. For this fundamental law, LO, objectivity is not only positive and of absolute value, it is an objectivity that is at once both external and internal and all in between to TRU. LO signifies that objectivity is not any objectivity by any sense of any thing, but by and of TRU, TIT, and TRIA. It must be only the Objectivity of TRU, TRIT, OBT, SPALT, POS, ACC and POT.

TEÂ - the Law of TRU Parts Absolute. TEA is a quality of the Absolute where even to infinity its parts are absolute. This is the driving, glorious and victorious advantage of our world and our salvation. TEA shows how TRU is absolute in the unity of its being, TU, by TRIA. It is what enables us to concentrate and adopt a positive and all encompassing unlimited effort for its ends. Each person may seem limited in years of life and in ability to do things. Yet our overall potential is more or less until it becomes TRU in ACEXIN of TIT TO by succeeding to fulfill their existence in TRU. Such attainment of the absolute stems from our capacity and ability to discover, learn, know, show, develop and best use ES and TRU.

The specialized means to approach TIT exclusively is called: **LADEX**, or Law of the Differential Exclusion specification. When they are both of TIT and ENIT, or more, the definitions are Bilateral or **BIDA** or of multiple varieties, MIDA. BIDA and MIDA are how words have two or more, often highly opposed, definitions; their common definition, as of ENIT and their ES or TRU definition, or others. There is through objects and ESINS a whole continuum of possible definitions. New terms which are developed and especially those old terms which are given new and different definitions as they are TRU,are done so by the differentiation in TIT. To supply these ES definitions, and overcome this difficulty they present, we have: **REDE**, Redefinition of all to equal their state, form and function as parts of TRU. **LADEX**, more directly, is the Law of Definition to E, Exclusion of all Definitions except those that are in ES. **LADASO** - Law of Distinction of Differents and Sames in All. LADASO is exemplified in the best of science and the professions. It is through investigation and experimentation in an intensive search and discovery process of correlations and measures, to find similarities and differences between things that gives them their greatest meanings. **ESDEF** - ES Specialty Definition. ESDEF is that ever more precise definition of ES which provides the basis upon which we can proceed to concentrate and reduce an ever greater and better distinction and determination of all ES. **DEDE** - the development of TRU becomes the Definition of all that is in TRU. Definition by TRU is how we can perfectly define from TRU rather than striving to transpose definitions as ENIT into TRU. Without definitions by TRU it is impossible to perfect SUT, and POS. For SUT must have a high degree of precision to be a basis for all succeeding ROT, MOT, ALT and SPALT.

The impossibility for two things to be exactly alike is obvious if we are only careful in examining them. If we do not, they are indiscernible and likely to seem identical. This principle originated in Leibniz. Duns Scotus held a similar view, i.e. if things are distinct, they must be distinguished by some qualitative difference. This is: **PODES** - Principle of Distinction through Essence. By TRIA we must focus our determination and examination on one thing, TIT. So that when we are speaking only in the actuality of TRU we are speaking by DAOST of either the same thing, many different things, combinations, or how they apply to the different parts of TRU. PODES depends upon the acuity of our discernment, observation and subjectivity. This is how knowledge and TRU expand. **MESAD** - Manifestation of Finite Separation and Distinction. By MESAD language provides knowledge. We can find the direction necessary for the separation differentiation and distinction of all from TRU. Much of our ability to separate all from TRU and provide it with distinction evolves from the Focus of The Relative , of TIT. All exists only as: **ARF** - A Relative Finite. The purpose of and gist of SU, ROT, MOT, REM, POSITS, HOM, and what follows is to show how all can become increasingly ARF as they approach an absolute finite.

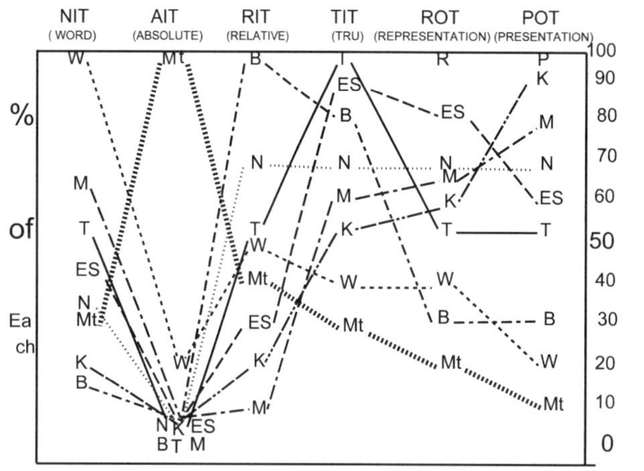

Fig 5-14 **COMPARATIVE WORDS, ES, and UNIVERSAL DISTRIBUTION**

This chart is produced to show some of the major parts of TRU and IPTRU, the forms, functions, processes and characteristics of TRU in relation to those parts and their analyses and interpretation subjectively. It is highly diagrammatic, and each item is open to interpretation and use. The word, sense or thing and the corresponding term to ES is not to be taken without qualification or their appropriate SCARLS. The E of these are usually mixed, AMTRU, of varying degree of TRU and ATRU. Frequently they are very little TRU. Words often tend to have a greater coincidence or coterminality with TRU than senses or things in themselves. For words are drawn, to a great extent, by subjectivity to TRU as in IAT. Their meanings are to a degree not the entity itself, but in relation to man and TRU. If we are to know TRU and its parts, ES, we must know them in and of themselves and not a rough or poor substitute for them. The abbreviations used include, B - Being, N - NAMS Name-ES-ENS, ES - Parts of TRU, P - Presentation, K - Knowledge, R - Representation, M - Man, T - TRU, Mt - Matter, W - Word. Much of this interpretation is completely dependent upon some flexibility of definition and usage. So that for every term, its EPI, AU, RE, REM, OC, EC, ECT, NEC and TRU definitions are all more or less different. Yet the diagram is of great value in helping to envision the participation of concepts in TRU, by their great variations in meaning depending upon the vantage used. This is what needs to be continually recognized and corrected if a person is to understand themselves and TRU.(BIDA)

DIFFERENTIATION of A PART, PARTS of PARTS, WORDS, DISTINCTION, DEFINITION

Particularity involves a formal and TRU conditioned mathematical set and progression from units through finites to infinites, and units through fractions to infinitesimals, plus their distinctions and definitions by TRU.

INFINITE and INFINITESIMAL

TI - TRU Infinite. **TINI** - TRU Infinitesimal. In TI and TINI we express the two extremes of the finite, the indefinite and farthest extensions from unity set up by the relativity of TRU. This is as it is in any RU. In being extensions and extremes of the finite, TI and TINI are considered under parts of TRU. For length it will be at one extreme for size, yet concentrated at the center. For the time and mass energy dimensions it will manifest itself in its differentiation with infinitesimal concentration in the center and infinite expansion toward the periphery. It is in the search for the infinite and the infinitesimal and our moments of infinition in our IA, ICA, and revelations both qualitatively and quantitatively that human's greatest opportunities and immortality lie. It is by ever increasing the percentage of conceived and named ES that we can both understand and enhance TRU. It is in recognizing, living with, improving, reducing, and controlling the finite, infinite differential, and associated predicaments that we and TIT become perfected and eternal. **TEP** - The Finite Predicament. TEP denotes how the total integral becomes lost in differentiation of the infinite and the objective differential becomes lost in the integration of the infinitesimal. TEP is a product of relativity, S-O, and our individuation. In being relative there is no absolute, no finality, no end, it is infinite. In having individuation, either personal or by TRU, this is unitary and in life highly limited. **ITI** - the Infinite in the Infinitesimal, and the infinitesimal in the infinite. ITI signifies that there is a sameness in the extremities of existence. **ICA** - Infinitesimal Instantaneous Concentrated Attention. ICA is the split second, or world stands still, concentration of attention that is at the heart of IA, IAT, TIT and TRU Absolute, TÂ. ICA is how TÂ is produced by the Exfoliation of TRU. Or Concentration over time equals TÂ C/t = TÂ. This shows how we can not only be TRU, but can be TRU exponentially to TA. This shows: **TRUR** - TRU Rate, or Rate of TRU. As TRU becomes ICA it is how the world stands still in absoluteness. ICA is the infinite prolongation of IA in IAT. It is how great IA, depth conception, and new or great existing concepts, are on certain days. They are so intense and extensive that one can't keep up with them. This is called: **INTA** - The Intensity of IA to TÂ or COGNOS to TÂ. INTA signifies ways, avenues and means to TÂ, **TÂCEN** - TRU Acceleration or (TRU)^n. **TÏS** - The Infinite, TÎ in Summation and Significance.

MI - Moment of Infinition. MI is that point or tiniest Interval at which the infinite and infinitesimal no longer are indiscrete but become barely finite. It may be likened to a person's consciousness, or the single static picture in a series of pictures in a film, where the dynamics and time element in all become so short as to be, or at least seem, only static. MI is important for both the reason of connecting TI and TINI with TE and TU, as well as helping to better understand ourselves, as not static as IA, but increasing or decreasing increments of MI in a state of consciousness and subjectivity of all else in the world, TRU. Of even greater significance MI is the interval in the secret, cue and gist of all that unlocks the door to life or entrance to the absolute as they are both TRU. For it is in MI and the interval - like quantum - that we can see the focus and relativity of the thisness of TRU. **AIME** - Absolute Imperative Moment of Infinition and Expansion. As we progress from the finite to increasingly finite to infinity - or from the divisional or fractional increasing to infinitesimal - we develop a completion that in many ways corresponds with the absolute. This is what happens to infinity when it no longer is the prolongation of an infinite series, but a limit in a solid sphere. For this is the difference between infinity in totality and an absolute. This is called: **LINLI** - Law of Infinity in the Limited Infinite. LINLI is that infinity which is no longer the progression that occurs in an infinite series, but the regression and infinity that occurs toward the periphery of a unified series. This is readily evident in viewing SPALT. This is how quantum theory proves TRU. For by its very probability state or odds its parallel universes depart from the unity, limits, and special form provided by TRU. The odds, difference or degree of probability and verifiability of quantum theory evident in parallel universes is the difference of TRU from NAMTRU, TIT and TRIS, or the actual world and other worlds.

Interval is a term that exemplifies, as it actually is, what all greatest OC, EC, ECT, NEC, and TOPNEC are. Such highly inspiring quality and specificity provide the bare limit of finite differentiation necessary for distinction. They enable attention and awareness to realize deepest meaning and stimulate us to become interested in, accept, and work for life, as TRU.

DIVISIONS, COMPOSITION, FOCUS, SET, REM, or PROPORTION, and INTEGRAL / DIFFERENTAL

Some of the concepts of TRU, that help to explain questions about what the parts of TRU are, give a basis for how to demonstrate and determine the values that help in their use. These include: **LOPS** - the Law of Proportional Separation, or the Priorities of ES. LOPS expresses the fact that finites, quanta, and gradients to be successful must be separated by value, magnitude or proportion, **PUARCE** - Particulars and Universals take on Altered, Added and/or Reduced Characteristics and meanings as their ESINS increase, become 100% ES. **REMU** - Relative Magnitude Universality, **FER** - Form, ESIN, E and **CEF** - Correspondent Equivalence of Form and Function, and **LASUR** or the Law of Sufficient Reason. LASUR's greatest value and function is to facilitate and enable formal distinction.

The nature of ES may be divided into: its formative parts, what is different of TRU's composition as these parts have common characteristics, and its definitive parts, what TRU is composed of by different characteristics. The number of finite parts and reals of TRU is, when compared with language in general and all that is not TRU, limited to **UMIA**, Universe, Mankind, Individual and IA. This is: **LAFIP** - Law of Finite Parts, or how they are limited from entities, by TRU, in TIT.

The number of parts and events in TRU is aided by ALT, SPALT, and POS, and exponentially exists as: the number of words a person frequently uses, words a person is familiar with or rarely uses, words in a dictionary, words in the English Language, words in all current languages, words in all time. These show the great number, range, and levels of ES, and their corresponding representations. They show the problems inherent in language, approach to finity, and the D of DAOST. For the form and parts of TIT have a large degree of stability. We can immensely improve our predominance if we can overcome this greatest cause of failure to know TRU. This is our persistent departure of ES and CES from TIT-TO.

The unity and relativity of TRU enable us to provide an increasing degree of needed finite distinctions. That is if it is done well and within certain bounds.(VAPO,SCARL) It can be done by doubling or halving, from one to infinity, one to the infinitesimal and the medium REM from SUT. The goal in using TE, and much of TRU, is to attain an enumeration of events, elements, items, vectors, processes and parts that most closely identify their ES and CES in TRU and in summation provide the total composition of TRU. The relations, correspondence and contrasts between same and other parts are signified by: **ROS** - the Ratio of Different and Same, or Other and Same, or when expressed as a Law: **LOSAF** - Law of Separation and Fusion, or Differentiation and Integration, Different and Same. LOSAF is a corollary of DAOST. It is the equilibrium, balance and proportionality of TU, TE, TI and TO. Sameness and otherness are the extremes in a continuum produced by relativity, the dialectics between relative and absolute. Sameness occurs when it is so relative that it approaches absolute homogeneity, and otherness when it approaches absolute heterogeneity. (LANERA and LAMOL)

The Relative Magnitude of portions does not equal Relative Magnitude of TIT, but equals the concentration and relate of its partiality. This is **DEIT**, or how ES are Other of TRU, or other in itself. The universal/nominal/real controversy is resolved by ES, by being able to show, demonstrate and understand the composition of TRU, not only its parts, but these distinctly in relation to others and to TIT totally, TIT-TO.

Fig 6-14 **DITRIT or DIVISIONS and RELATIONS of TRIT** In Figure 6-14 different aspects of the divisions of TRIT, TIT and TRU are presented.

A. How TRUCH becomes TIT and TRIT,

B. How recognized TIT and TRIT may be divided into component parts, ES,

C. How a scale can be made of component parts showing the approximation of the extant, TIT or TRIT, state of TRE between TRINI and TRIN. The variation of this scale. DITRIT depends upon our presence or lack of presence and success/failure, S/F, in achieving TIT, TRINI, and ENT,

D. Forces, often invisible, acting integratively and differentially through DITRIT,

E. Fusion of DITRIT to TU. What the various resultant polar bodies become as TU divides to, and the divided TU becomes, TO.

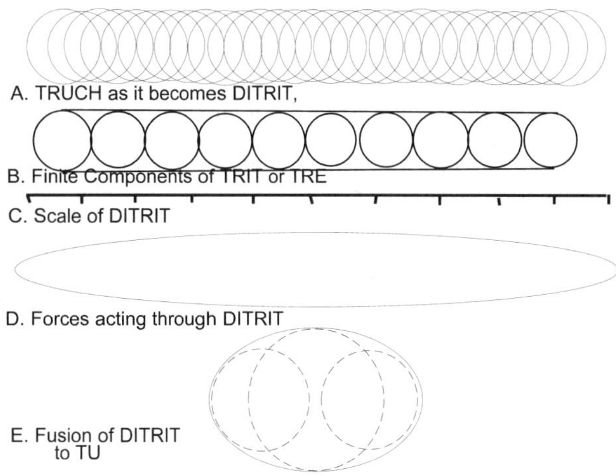

COMPLEX ES

It is the ACEXIN, or the actuality, extensity and intension of the finite of TRU, and a person's level of comprehension of it, that is paramount. Only as we differentiate the same and integrate the other with optimal SUT and maximal ACEXIN of ES can we have the SUT, ROT, POS, HOM, ACC, and POT necessary for the success of TRU. There is massive universal causation and permeation in TRU, all are ultimately cause and effect. This follows the composite nature and TRIA of TIT. They show each other and TRU by giving us their beautiful intricate detail. Some of the concepts that show complex ES give a better idea of what ES are. Diametric is the opposing, polarizing, contrasting, yet precise form and state at or along the diameter, or in the direction of the diameter. This is how geometric position of the extremes is related to some fixed point. There is a constant or continuous interdependence of ES and their diametric positions.

FOCUS, SET, PROPORTION, REM, INTEGRATION-DIFFERENTIATION

The dualism and polarity of the same and other, or TU-TI relationship, is the of inversion of the infinite and the unit, same and other, and relative and absolute. It can be resolved by a focus or set state in which the degree of relativity/absoluteness, or if absoluteness/relativity, is fixed or focused. This is called: **LARAFS** - Law of Resolution of Relativity/Absoluteness by Focus or Set State. What happens if TRU is an added state in which this fixed focus arises from whatever is doing the centering, or exists in actuality. This is the added state by existent focusing in actuality, or TO of TIT. It is this same and other as a focus within TRU that produces much of its dynamics, and variation. In health it is reversible and well controlled, in sickness it is less reversible and poorly controlled. So that there is at the same time unifying and disuniting forces, both effecting TRU. The unifying forces are positive, propelling and compelling, and the disuniting ones, negative and disintegrating.

DEIT refers to those ES or parts of TRU which are, or function as, the D In DAOST. They are those parts in which the S or Same is extracted, leaving only the D or different. This concept is necessitated by the fact that it clarifies the terminology as it represents the different aspects of the parts of TRU, or **FER** -- Form - ES - REM. **MADAS** - Maximal Separation Distinction Formulation of the Different in the Greatest upon which SU and S/F depend. MADAS expresses the basic general relationship between the different aspects of TRU and humans ability to make these distinct. This is a crucial and key function of our nature, the S-O separation which is basic, characteristic, and determinative of the mass dimension, S-O, Á of TRU. The gradation and differentiation in MADAS becomes the substance of TRU, and thereby all. When projected on a common, uniform or preponderant view, perspective or consensus, we have formal distinction, the basis for a precise formulation, standardization and perspective of all things. The parts of TRU being the prime ingredients, elements in point lattice, or diophantine approximation, are the substance and shown numerical units upon which such distinctions can be formulated. **ÉRP** - Parts of TRU Real and Pure, Equivalence and identity to TRU. ERP signifies the contraction of reals, purity and ES or how the terms reals, purity, essence and actuality have come to signify in modern language an equivalence to ES. This is done unconsciously, for the most part, to emphasize those things which are not only perceptively certain, but actually TRU, and a great or specific part of TRU. **PAVA** - The Principle of Partials, Variants, Universals, Forms, Functions or the constituents of TRU. Their degree of differentiation and confirmation depend upon how, and how well, the coordinates of TRU are set.

EBE, Extremes by Emphasis, proves many things and much of TRU. So are ES demonstrated by the Construct of others, especially other ES, **EDCE**. Concepts, Principles, Universals, Laws, parts, ES, Differentiation, Expansion, Delineation, Application, and Preparation for SU is called **CULPES**. The focus of the differential in TIT-TO is, **LATDO** or the Law of TRU Distinction and Totality. It corrects **FID** or the Fallacy of Indistinction. **LAPDIT** is the Law of the absolutely precise distinction of the greatest. This is what language has increasingly spearheaded in the last tens of thousands of years. LAPDIT is what we are to a great extent composed of, the elements of our being, parts, processes, FAF of TRU its relativity, observer and observed, subject and object, idea and concept as they are represented, symbolized, signified by our words in language and communication. This is why NEC, and especially TOPNEC, are so important and why humans have failed to properly comprehend their actal world, TRU. Comparisons, specifications and measurements of ES are made in many ways, as seen in **REM** or a type of quantifying the quality or values to be assigned to parts of TRU. ES, their units, and measure are not alike except, being parts of TRU, completely TRU, measurable in terms of A, L, and T, measurable integratively and differentially, and measurable by proportion of TRU or REM. They may, or may not, resemble their corresponding ESIN, word, object, or entity. ES may be more or less identifiable depending upon our ability to separate and reduce them to measure and dimensional analysis.

MEASUREMENTS, CLEAVAGE, ORDER, and DISTRIBUTION of PARTS

Once we establish E, ES, and their concepts and laws we need to divide, separate, cleave, measure and evaluate them. This we do, among others, through the law of relative value, law of magnitude interdependence, parts as absolutely vital to a whole, and Relative Magnitude of TRU. We start this by inquiring into what there is that exists partly and what is a whole, the relationship between composition, mathematics and presentation. We note how entities may or may not be measured and how this can be used in TRU. To measure parts we need to know whether this is based upon what they are relative to or caused by. We need to establish what bearing subject and object have in measuring parts, as well as what units best serve this measure. Whether it is based upon humans, individuals, organs, essences, or many other constituents and parameters of the whole. It helps to discover and collect all that relates to TRU, and how it can be ordered. From what are different aspects of the same thing of a whole we can deductively find what is missing or unknown and produce it for measurement. With greater relative value we can find the necessary base for measure. It depends upon magnitude interdependence and the vital dependence of the whole on parts. Numerous concepts and laws can be developed to explain measurement of ES. Set measure, equalities and proportionality help to strengthen the measures of ES. The various bases of measure range from the infinitesimal to the infinite, or as limits from zero to the absolute. They help to better understand comparative absolute finites. The dynamic variability and fluctuation of values of partials are shown and solutions can be found. The interdependence of proof and measure is mutually beneficial. With increased quality and measures of ES, parts become more precise, appreciated and interesting. For cleavage we must remember ES never 100% S or O. We must dissect and extract each item in and of TRU. This enables the basic combination and fusion through reassembling dissected and extracted parts in the operation of S-O or Á and returning them to their actual state in the whole. The process of dividing, constructing and showing the discrete separation of TIT can be done at many levels.

Many questions, paradoxes and problems of infinite divisions, dichotomies, ISMS, parameters, dimensions, modulation and universals are recognized for answer and solution. **PRID** - Problem, Principle of Infinite Divisions, Dichotomies, Parameters, Universals, and LCD. **PRIVAD** - Problem of Variable PRID. PRID and PRIVAD are products of TE and TI. They come to the front when we seek to discover the optimal dimensions of TRU, their coordination and order. **PEDÁ** - Proof of Precise Finite Distinction and Actuality, is accomplished when dimensions and scales are discovered and refined.

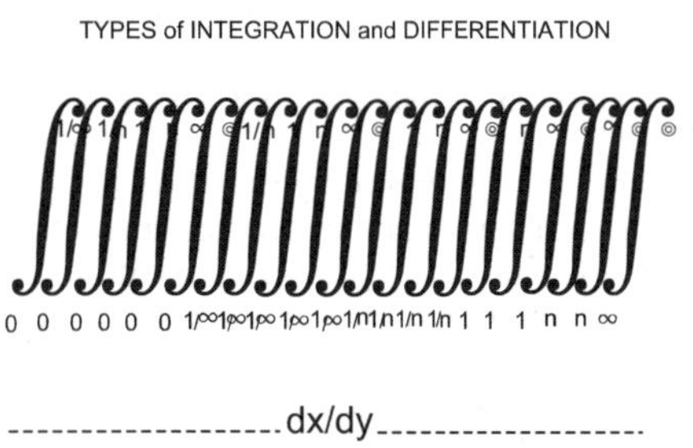

Fig 7-14 **Absolute Finites** Unity to Infinity in Proportionality denotes the general scale taken, which must be brought to replicate how it is in TIT. Various end points in integration and differentiation are shown in Figure 7-14. They may be classified by their centerpoints and endpoints, and are vivid reminders of the scales and complexities of ES operations in ROT of TIT and TRU. Each of these has implications and uses in TRU, and demonstrates TROS. In TRU their scales are not uniform, but altered, like ALT is warped in TIT. Most integrate only a segment of the whole and thereby do not cover all TROS. Yet they show us the basis for representing various segments of TIT. Some are new and require deliberate reduction to TRU to be valid.

FRAME, MASS

Forms and parts are, or tend to be, complementary. Many representations analyze and reduce ES and TRU to positive objectivity, POA. Lattice theory develops framework, structure or geometric arrangement that form a network giving strength to the whole, or providing a basis for the relationship between parts, forms, and the whole, or parts and parts. The representational approximation of these parts, as they may be posited in the frame of two or three-dimensional system is called a Point Lattice. These relationships are demonstrated by the theorem of Minkowski which asserts that any simple, closed, convex region R symmetrical about 0, and of sufficient area, contains within it points of L other than 0. Where 0, P, Q and R are the four points of a point lattice. This leads to how parts, as represented as points in a lattice can be best posited. One method is by Diophantine Approximation where through the use of a polynomial equation or point lattice representation of unknowns or less than completely knowns, parts can be treated as rational numbers and posited geometrically. This is a very important for the proof and use of SPALT and the reliability of POS.

ORDER and DISTRIBUTION

Because a thing or entity is not the same or does not equal its objectification, representation or modulation does not necessarily mean that one or the other is not true or any the less valid or actual. This holds even though the objectification, representation, or modulation are perfect. **LAVIROM** - Law of Validity Independent of Objectification, Representation and Modulation. The importance of LAVIROM is in guiding us to TRU and preventing the observation, representation and modeling of it from becoming more than its intended use and part in TRU. As Terms, ES, components, nominals and elements overlap their REM decrease or are reduced in proportion to the added term in the system, E Limit, **EL**. EL is exemplified by the finite limits of a person. **LORD**, Law of Distribution of Relative, or since all members of TRU are not equal, their distribution is different, becomes relative. The unknown E becomes known through ordination, universals, homation, ESION, NESION, EOE, DOM, SOPAN, PAN, and RONT. **LOFOR** - The Law of the Formative of The Relative, the formed, integrated, or non differential individual. **LODOR,** The Law of the Formed Differentiated Relative, is not satisfactory, only as: **LODIOR**, The Law of the Formed Differentiated and Integrated Relative, or how the form fits better for the ultimate measure of relative, of TIT. **LES** - Law of Equidependence of Selves and summation TRU. LES is exemplified by the intensity of TRU and the balance and interdependence that exists between selves as they are TÍS, and TO as it is TES.

PROCESSES of DIVISION for MEASURE of TRU

The process whereby we deduce such resultants of TRU is called **COL**. Simply COL is the determination of an unknown part or quantity of TRU by subtracting its known remainder from TRU. The corrections involved in this and similar processes are called **COR**. Many different measures of humans have been devised in science, biology, medicine, and psychology. Yet they have to be adapted to be of use in TRU. The ultimate measure of an individual is their degree of TÉS. Another kind of measure is the measure of essences of TRU and of highest REMS. Measures of degree of TRU in entities, objects, or givens is a large part of the use of ES. For in selecting and separating what is or is not TRU, as in AMTRU their measure is critical. **LOPEM** - Law of Proof Equals Measure, by measuring the proportion and participation of ES, CES, and FAF of TRU we can find, see and have its proof. **LORAV** - Law of Relative Value, or ES Value, how all that is about us although seemingly complete and apart from us becomes in being, more or less, relative to us a part of TRU. How this becomes set in TRU is: **LORVACS** - Law of Relative, Actual and Set Value.(REM)

LOMID - Law of Magnitude Interdependence. LOMID is like the interdependence of body size. The greater the parts of TRU the more likely they will be related and either interdependent or in some kind of functional and dynamic state of interaction. This can be in opposition or by contrast as in DAOST. **LEO** - Law of Equality and Totality, or the law of 50:100, where each part, as the first person, stands, or is in, a measurable relation, or some set SCARL, with all other parts. It can also be where this part or first person is with TRU on a 100% basis. This is because both are and can function only as they are 100% TRU.

ES may be considered in different degrees in relation to TRU depending upon whether this is from the whole or some part, perspective or relationship.

DYNAMICS, PROCESSES, SUSCEPTIBILITY, PROBLEMS and FALLACIES of ES

Participation is key, central, crucial and vital to the overall function, success and perpetuation of TRU. It is what can so frequently and violently go amiss, and produce loss. Predication is participation in tandem, where one depends on another. **DIP** - the great Dialectical Predicament of Interaction between Distinction and Predication. DIP arises from **TEP** or The Finite Predicament where we are constantly battered by the great dialectical struggle between the unitary total and the infinitesimal massiveness of all. It is how to best make all of TIT discrete and distinct, yet be able to use sameness and relativity to determine, predicate, and predict how they come together in the whole. Problems with predication compound problems with ES, measure, and representation. Certain concepts give a better idea of ES and TIT than those of TRUD, CES and HOM, and include ES - TRU perpetuity, or the continuing forces of ES in TIT. Summative ESIN and ES functions give an approximate measure and help prove TRU. e.g. the function of an elbow, back, fingernail, thumb, eye, neck, city, or cell nucleus.

Fig 8-14
FREQUENCY BECOMES MASS

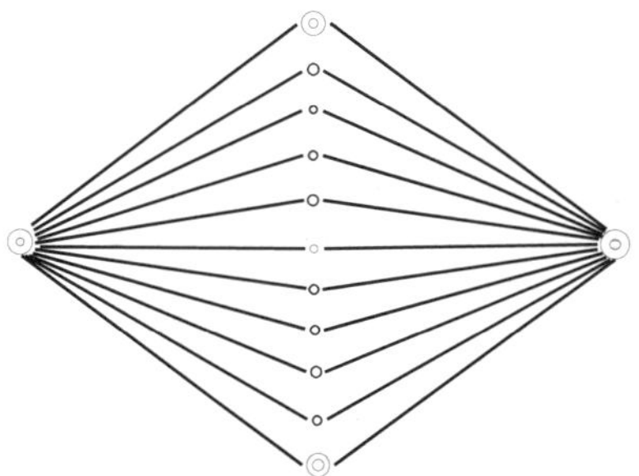

FOSA, Frequency becomes Mass, is a proven physical principle. It is by way of how frequency becomes mass that humans are formed and function. It is on what our being largely depends. How often and how much of what is predominant, most characteristic and typical for TRU becomes its mass, its Á dimension and continuum. FOSA is what renders humans able to distinguish, make distinct, discrete, and develop all the FAF, ES and CES that are provided by language, concepts, principles, laws, reason, and all levels of Á. It is what occurs when the most frequent or predominant relates to other FAF, ES and CES of a new entity.This is how TRU becomes its special kind of mass-energy, form, state and dimension. This is: **FERA** - Predominant Frequency of makeup, constituents or parts and aspects in relative forms. This forms a new entity with its own qualified and typical mass-energy, form, state and dimension. For TRU the typical predominant frequency, relatively, has added states in totality which we call its Á dimension, state and form. This is manifested in levels of Á, or SÁ, conditional, state, form or dimension. This shows us, with frequency, the substantive, or ES makeup of TRU. The subjective fuses with the objective, and becomes, when increasingly frequent, more and more real, elements or ES, produced in the Á FAF of TRU. So has man's central nervous system become more attuned and expanded. It is how parts of wholes, ES of TO, can be differentiated and integrated simultaneously and increasingly toward greater and greater extended limits. This is called: **LARES** - The Law of Reals of ES or Á of ES. From LARES, human existence depends, to a great extent, upon the integration, formulation, systematic unity and modulation, as by ALT and HOM, of these parts, reals, ES, frequencies and entities by differential and integral posits in the whole of TRU, TO or: **LATO** - the LAw of Á Totality in TRU.

PRUTAN - Proof of TRU by Anatomical and Physiological means. That the world of man or the relative universe is centered on the predominant mass energy, space and time distribution of man is evident from anatomy and physiology. What and how much these parts and processes equal the ES and CES becomes more certain with increasing knowledge of TRU. The equality of same and different and **LAPAR** - the Law of Partiality, or how each and all parts, as we ourselves, are or are not except and to the extent we become, or are our identical parts and roles in TRU helps us measure and prove TRU and TES.

SUSCEPTIBILITY and FALLACIES

LESUS - Law of ES Susceptibility. Humans, as TRU, partake of most all of the weaknesses and susceptibilities of other species. Yet for TRU there are many more both from the special nature of TRU and the ability of humans to harm themselves. Much weakness is associated with the unique form and nature of TRU, its parts, and processes. Here is where man is blind and extremely vulnerable. It is shown in the massive interdependence and interaction of SÁ, the various states and levels of Á. If it is an inalterable end, prime but not absolute, it is: **LIGS** - Law of Inalterable Genetic Susceptibility. If this susceptibility is partial, it is called; **LIGES** - Law of Partial Genetic Susceptibility. LIGES, LIGS and LESES can be described as corollaries of more general susceptibility: **LAGS** - Law of Genetic Susceptibility. LAGS, like that for all levels, gradients, or parts in SÁ spectrum or Á continuum, is dependent upon the health of a species, the strength, predominance, competition of its divisions, classes, substance, form, dimensional formulations, parts and processes. It is highly complex and tied to the relationship between:

LIVE PRODUCT ↔ DETERMINING FORCES ↔ GENETIC MATERIAL

LIES - Law of Inalterable partial or ES Susceptibility, LIES signifies that no matter how much force there is placed upon ES and even TRU, if TRU does not change, the set state, form and function of the partial, ES does not change. It may change temporarily or quantitatively, or when qualitative only to reversibly return to its status in TIT. Time, as commonly considered, is largely a fallacy. Beyond simple time higher states in change and variation are markedly altered in TRU, being much different for TÍS than in common usage. It is necessary, for the part and individual in TRU to have IAT and to be in readiness. It is highly limited by the very constraint durations must adapt to. For parts these are often much different than the whole, the dynamic TO. Temporality has a great place in TRU, for it shows the concentration of time in the present in addition to showing its role in TRUD. Time becomes spaciotemporal with mass energy in ALT and TRIT of TIT. By relativity and geometry the observer observed of the world, TRU, makes it certain we can no longer rely on random measures of time. It is a relationship and dimensional form that is determined and provided by the very nature and form of TRU, like in ATO and TRIA, rather than any number of universes and geometries.

Fig 9-14
COMPARATIVE ABSOLUTE FINITE in Á of TU

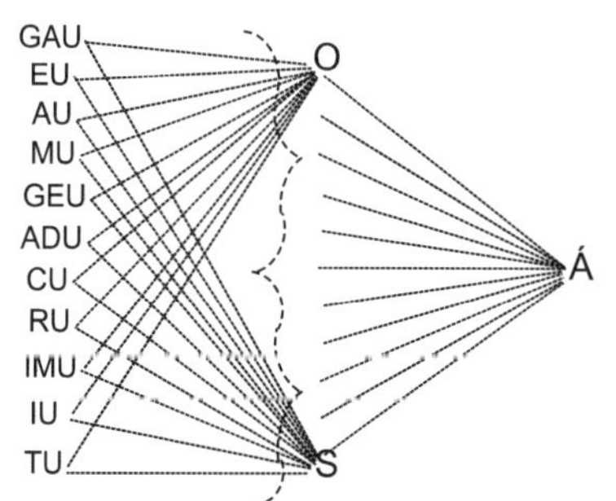

The collective mass energy components of the FAF of TRU are consolidated, as in any animals predominant powers of adaptation and survival, to become absolute in the unity of TRU, TU. Reversely the comparative differentiation of entities is what contributes to and constitutes the finite ES of TRU. The need to give finite formal distinction to all that is of TRU is great and indicated in many ways. Yet the unity and relativity of TRU enables us to provide an increasing degree of these needed finite distinctions.

Mental and Physical, as subjective and objective are, in TRU, only two aspects of the same thing. This is in DAOST, as introduced by Maimonides, two poles of the same mass energy state. This is the primary constituent of the mass dimension of the same universe, TRU. How physical mass has become transformed into its combined continua with the mental in TRU is **LATEV**, or Law of Autonomic TRU Evolution, the self propelling and compelling forces of genetic, nervous and mental FAF. The continuum from S-O is a reciprocal, yet much the same, as this physical to mental FAF, evident in Fig. 9-14. This is **COSUT** or coincidence of OS and physical to mental units in TRU. This great revolution of extant human life, is **LAMT**, or Law of Mass Transformation by TRU. As we proceed in the text to develop the phenomenal, S-O, or Á dimension of TRU it will be more clearly 'realized' how subjective conditions apply to TRU. Also how it is outside, inside, in and out, and transcendent to TRU.(OITIO) Once learned, the human brain, mind, memory, and higher powers of this dimension practically, specifically, and continually move to complete this S-O function. It is the basic dimension of identity, conjunction or bond that holds our universe together.

Fig 10-14 **PROBLEM of REALITY, and RESULTANTS**

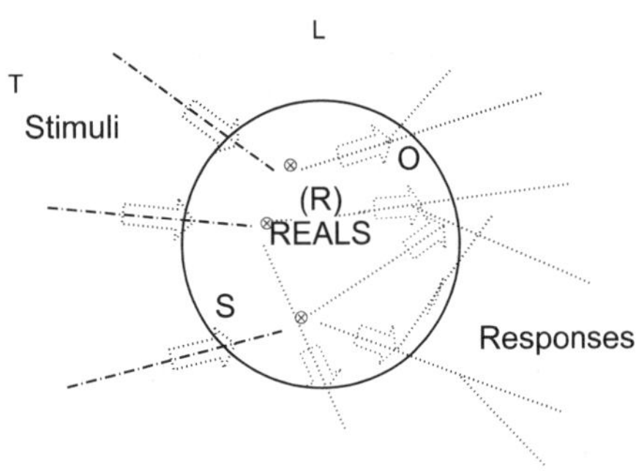

PRIM - Problem of Reality or the S-O or Á Parameter in Medicine and Health. Health, both physical and mental, is the body's adequate and successful ability to sustain itself and meet, adapt and respond to those forces, ailments, and injuries that beset it. It is how strong and well it's FAF enable it to best survive in its environment and in competition with all. It is how the world and man, TRU and TES are constantly bombarded by and in interaction with innumerable forces and exchanges from without and within.(OITIO) This includes radiation, light, heat and cold, food, water, bacteria, viruses, chemicals, substances, animals, other humans, groups, governments, and many others that in one way or another produce action, and thus need reaction. In being adverse they take many forms, as antigens, for which a considerable part of the body, as ones immune and other systems, are focused. All this that comes into contact with the body in such a massive and rapid way are, more or less, either normal or abnormal, beneficial or harmful, and positive or negative. Often both with a similarly positive or negative resultant. They may be demonstrated, as in Fig 10-14 on a length time frame, by S-O whose reals focus the stimuli and responses to most benefit TIT. All ES are massively, indirectly if not directly, participants of these in and out forces. How well they do depends upon how healthy and strong each ES, and TRU is. **RASIT** - Reversible State in Totality. RASIT signifies not only how the state of the dimensions and parts of TRU are different in totality from what they are in and of themselves. What is outside of or In and Out of TRU, OITIO, is much different, and moving in and out, as well as in various ways inside TIT.

Just as large and small and past and future are but two poles of the primary length and time dimensions of TRU so are subjective and objective two poles of the primary mass-energy dimension of TRU, or **LASOD** - the Law of S, O, or Subjective Objective in TRU as they are two aspects of the same thing, the mass energy state and Dimension of TRU.

MÁDÉ - Mass and Part(s) or E, ES Dependency. The gradients, substance or parts of Á have within themselves their own respective natures, form, and characteristics which make the parts depend upon Á and Á on the parts.

If any or all of us ever deny or reject or ignore the fact that the mass energy predominant dimension of our world is Á as in TRU let them consider what the our world is without thought, life, oneself or ones kind. No religion, philosophy or science can provide any dimension, scale, measure or principle which gives us a completely valid, total, proportionate and actual dimension of the world, unless and to the extent it is equivalent to the relativity, observer-observed or subject-object predominant Á dimension, state, form and dynamics of TRU. For this it is by definition. Á is most precisely defined as the Predominant Mass Energy dimension of the world, best, world of man, TRU. The imperative fact of definition is: **LÁD** = Law of Á by Definition or the Law of the Predominant Mass Energy or Á Dimension of TRU, by Definition. That the form, state and Á dimension is based upon relativity and the observer observed relationship does not mean that it is some specific kind of subject - object relationship alone. It is, rather, the whole general product of mans relativity as this mass energy functions totally, comprehensively and coordinately as a mass dynamically and actually in TRU. This is as it likewise functions in tandem and fusion with the length and time dimensions, as they also are neither linear, planar, nor even spatial alone but all together by DAOST and ACEXIN of TRU. **Á** is most precisely defined as the Predominant Mass Energy dimension of this special universe of best human life, or TRU. Á dimension is all that is not of the length and time dimensions in TIT. Like length and time, the divisions of Á, especially its gradients when taken incrementally and particularly, are equivalent to ES and CES. Yet not all E are A, nor all Á E. In life we tend to polarize our mass energy states. This is how our bodies are made up and can best react and adapt for survival in life. It is I and it, or the observer and observed, that recognize and act upon, what my attention and that attended to is on. The Á dimension in various combinations existing in TIT, is largely self propagating, and that of TRU which provides its mass, relative magnitude, power, and action. **LALTSPEC** - Law of ALT Specificity, or how ALT is typical, specific and exclusive for TIT. **PEQ** - Principle of Equivalence, or the Equality of gravitational mass with inertial mass. **LAGRACH** - Law of Changing Gravitational Constant, or Dicke's Law. All things relative, as mind, spirit, body, and matter are neither alone, nor separate in TRU, but are made up of ES and CES in the admixture of the predominant FAF of TRU. Much that is a part of the world of man, as TRU, is not only physical object or act, nor mental subject or thought, it is, what and how it is in TO of TIT and TRU. These components in the overall spectrum of Á and SÁ show the total significance, meaning and dynamics of ES. The mass energy, length and time dimensions of TRU, ALT, with full SCARLS in TRU, $(ALT)^n$, show and prove the form and function of TRU. The better we fully extend and refine ALT the better we show and prove TRU. This also works the other way. The relation between changing gravitational forces and changing universes has been recently a topic of considerable attention. There is a reciprocal inverse relation, which for TRU as would be expected, gravitation becomes less of the whole of mass energy as the Á FAF of TRU expands.

The structure, arrangement, and order of TÄP is what the combined dimensions of TRU provide. How their parts are gradients and ordered by these combined dimensions is their TRU form and order. It renders parts much more amenable to posits in a geometric design. **DEPENT** - Dynamic, Dipole, Dimensional, Polarity Enablement of TRU of Being. DIPENT signifies that dimensions are not only static parameters, but exist in TRU as continua and highly active forms and divisions. **CODIT** - Compulsive Dimensions of TRU. CODIT denotes why a person must be, think, act, and do as they are in accord with TRU. It also expresses why TRU must be and function as it is. For its set dimensional form is what determines much of its function, its parts and other aspects of TRU. It sets limits, focus, proportions, scales and many other features that have a profound influence on all the rest of TRU and our lives. The capacities, abilities, reactions, conditions, settings and states of TRU and ES render them in a certain way. The relations, formal distinctions, formulations, coordinations, and order produced by CODIT are what make it reproducible and most operable. Somewhat similar to CODIT is the compulsive dynamics of ES, **CODET**, which are also constituted and set to provide relation, consistency, stability, and permanence. CODIT sets the frame, structure, and parts of TRU and CODET is how these set parts contribute, give substance or body to, and show, TRU.

ENHANCEMENT , VITALITY, EQUILIBRIUM, HEALTH, SUFFICIENCY and PERMANENCY of ES

In order to correct the prevailing errors of idealism, materialism, naive empiricism, and even scientific realism, it is necessary that we hold to the TRU form and parts of all, and their differentiation by Á, L, T. To do so we must hold matter to Á in TRU or: **ROMÁT** - Reduction of Matter and Mind to Á in TRU. That the parts of an entity or any object are not that entity or object is: **LOP** - Law of Partiality. **FAP** or the Fallacy of Partials makes LOP quantitative, and signifies that for anything the degree to which it is from TRU is the degree it is not TRU, and its parts will lack identity with TRU. **LOPS** - the Law of Partial Significance holds that all partials, in and for themselves even as limited to their proportion of TRU, have total or absolute significance. **LORVA** - the Law of Relative Value, signifies that ES are proportional to their value or Relative Magnitude, REM, in TIT. How we may learn TRU and TE from things or relative universals in general is described by: **DICERE** - Differential correspondence of Parts of TRU and Things. DICERE explains that common, spontaneous fallacy of which we are all guilty, and is so much built into us, which is: **IFE** - the As If ES fallacy. IFE signifies how we think and act as if everything is TRU or a part of TRU. **AUE** - All things equal their E or ES, only. **EFI** - the Equivalence or convergence of Form and Ideal on each only as they become E. As TRU is both TU and TI so will its parts be some same and some different, DAOST, or: **LEDS** - the Law of Equality and Equilibrium of Same and Different. The fact that larger is usually more the same, and smaller more different for ES in TO, gives us: **SÉR** - Similarity Equals Relative, and **MADAS** or the Maximal Separation, Distinction, formulation of the Different in the greatest upon which SU, KOT, and our success Depend. Associated with SÉR is: **PIKER** - Proof by Induction that Knowledge is a function of relativity, **LIPCO** - Law of Intricate Precise Composite Proof, or the quality of proof to be achieved when intricate composition of each ES is attained. **LIT** - Limits of TRU, **LIG** - Limits of God, **ATIN** - Insidiousness of bad, **PADET** - Paradox of potentialities, Development, Derivation form (A), equilibrium of TRU, **PEL** - Perfection and ESINATION and ESATION of Language, from IFE, **PENT** - Perfection of the NAM, that part enabling a TRU word, to be found and useful, **PU** - Perfection of Ultimate Proportion, ways all views and ultimates become one. **PAN** - Paradox of Nominals, words or language. **SOPAN** - Solution to the Paradox of Nominals. **RONT** - Reduction of Nominals to TRU, Equivalency to and in TRU. In all philosophy there are no greater problems, questions, or paradoxes than that raised by language, nominals, words, and concepts. Yet the more we know TRU, ES, Á and the mediate role language has in the Á dimension and continuum the more these problems are solved. Coterminality and specificity show the same and different elements in DAOST. What is highly coterminal, as much that is of TRU, must be by S or same, and what is highly specific and finite must be by D or different made formally distinct texture and matrix of the whole, TRU. An example of this is a detailed and precise text on anatomy. Each finite part does not have to be equal, same and different, only in sum total all finite parts will be so in relationship to TRU, and in relationship to each other.There must be, as for all in a RU, or TRU, an equivalence or balance and harmony between individual finite parts in the whole, TIT-TO.

SUBSTANCE

How matter or 'primary substance' becomes ES, and form or 'secondary substance' is Á is called: **PRASUB** - Substantial Proof of TRU.

<u>Substance → PRASUB ↔ LAMAS ← Á</u>

LASCES - Law of Substantial Concretion of the ES of TRU. **DU** - Dilution of the Universal, occurs when either in TRU or from TRU, or when either of all TRU or some Individual Part of TRU, there is a departure from its nature, form, composition or substantive finite Á, L, and T in TRU.

EQUILIBRIUM, VITALITY and HEALTH

What makes up our world is, more or less, by organized wholes, or the organized whole of which its parts are readily associated. As both relativity and organization are highly productive of the thisness of TRU they have some coterminality. This is closely associated with the predominant mass energy form, state and Á dimension of TRU. Through the capacity and ability of organization and organ systems in a body, as TRU, order, control, and equilibrium are maintained. **LOTEP** - Law of Totality Equilibrium Partiality, or the equilibrium that exists among partials or ES of TRU. It helps to explain their set form, valuation, function, and position in the whole.

SUFFICIENCY, FUTURE and PERMANENCY

LASUD - Law of Sufficient Differential, how contingency helps to that TRU is cosmotic there must be sufficient, ordered, and harmonious differentiation, if the whole is to be maintained. This can be seen in: **LEXHAR** - Law of Expanding Harmonies and **LEXSEP** - Law of Expanding Set Proportions. LEXSEP and LEXHAR are where and how as E expands so must the whole, TO, and other parts to some extent. They must do so proportionately, or in some set, and more or less, coordinated, orderly and harmonious relationship and way. **LENA** - The law of E, sufficiency or adequacy, or how E must be equal to all that is TRU, and this by VAPO, especially proportionately. They must supplement and be added to others to fulfill TO. **LES** - Law of Finite Sufficiency, certifies that the thisness or haecceity, survival and being of TRU are or are not to the degree it is finitely sufficient. Permanency is tantamount to sufficiency for and of the future. Since the future has been, is and will be such an important topic for understanding, as it is a large part of TRU and its survival, and since it has been so poorly understood and confusing to one and all, it is highly necessary that we discover, develop, determine, demonstrate and utilize it in the best possible way. The future in general, and the future specifically, distinguish two major categories which if not separate lead to many of our mistaken ideas of the future. Also we must distinguish between TRUF, TRIF, TRUFID or TRUF/TRIF Differential, and the future of ES. We should also distinguish between the <u>Projected Future</u>, or all that is to come that can be projected, predicted or foretold, with a reasonable degree of accuracy or probability. <u>Indeterminant Future,</u> or all that can not be projected, predicted or foretold, and the <u>Actual Future</u>, those aspects or parts of TRU to come that is its actual form, function, parts or processes or at each succeeding period of TRU, is complicated by changes that will occur. There is also an <u>Ideal Future</u> or that part of TRU to come, either TRUS or TRIS which will be an improvement upon TIT. This may be divided into two parts. The illusionary ideal future and the actual ideal future. The <u>Illusionary Ideal Future</u> is that which we might hope for, wish, dream of, or conceive, which when it comes to pass will be found to be either wrong or in some way less than the best. The <u>Actual Ideal Future</u> is that which will inevitably result, either from the present or from any set point in the future, and will be and maintain the line of TIT, TRICH and TRUCH in its optimal state. <u>Segmental Future</u> applies to those segments of TRU such as is revealed to us by SPALT in which a limited and discrete piece or segment of TIT's future is demonstrated. Absolute Future is of neither end nor beginning, but only a totality exclusive of all else. Relative Future is not absolute, but limited in some way or other to observers or observeds. It is in how well or the degree to which we can make the projected future fit and distinctly describe the actual future that is most revealing and beneficial. On the concept of absoluteness of partials, Plotinus states, "It is law that the offspring of God endures". This is a beautiful example of the continuity of great ideas and ES, whether they are in one form or another. This quote confirms the tradition of the permanency of ES in the set nature of TRU. In addition it makes another very salient point. This is what is undeniable and an all important fact, by becoming a necessary condition, also becomes law. Plotinus a champion of emanations from the *one*, was the forerunner of SÁ or levels of Á, and the physical - material - mental - spiritual, or S-O basis upon which TRU's mass energy is formed.

SUMMARY, CLOSURE, SOLUTIONS, and PROOFS of ES

Because TE and ES are not readily apparent, doesn't mean that their place and roles in TIT can't be sought and practically achieved. It is necessary for each person to so select and manage their acquisition of information and wisdom that we can still have a sufficient amount to meet the demands of TRU. This often means methods of achieving TRU, as: **MADET** - The Methodical and Formulated differentiation of the Forms, Á, and Parts of TRU. MADET is the separation and extraction of the non TRU, the lesser REM of TRU, and the formulation and differentiation, modulation, analysis and posit of TRU. **TÏÊÂ** - Terms Intensity, Extensity and Actuality, or the need to expresses all terms in their deepest, most essential, TRU and complete sense.

Figure 11-14 **WINDOWS of FINITY**

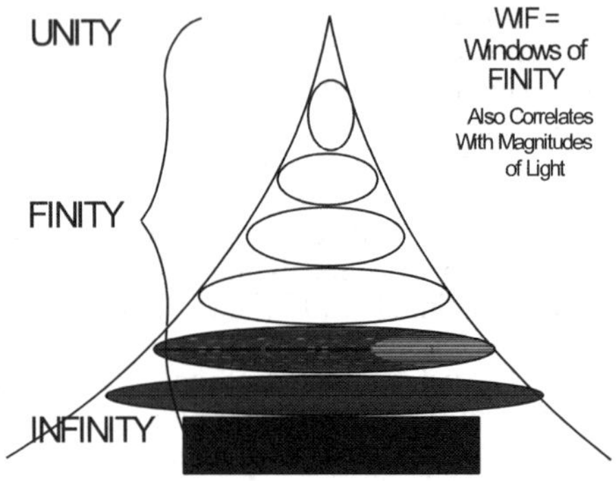

This diagram gives, in a very straightforward, yet diagrammatic, way two very important concepts and visions of TRU, WIF and LÁLREAC. **WIF** or Windows of Finity signifies the fact that the whole Relative Universe of Man is finite, at least in content. This finite ranges from the extremes of near absolute unity to near absolute infinity. It is never quite either for absolute unity and infinity are extra relative to TIT-TO. It is also incomprehensible for a RIT or TIT of RIA or TRIA. Absolute unity is quite relative except in the content in which it approaches another absolute that of the infinitesimal. The other extremely important concept involves the relativity and light relationships and is: **LÁLREAC** - Law of TRU Mass Energy Form and Function and Light Reactions, interactions and interrelationships. As we reduce finity to unity there is increasing light. As we reduce finity to infinity - or the infinitesimal - there is increasing darkness. Light is a manifestation of relativity, the observer observed relationship measured by intensity.

Relativity and the predominant Á or mass energy dimension of TRU are readily formulated by the equation of mass energy & light. We must be quick to add, however, that this is a special interpretation of the role of light in TRU. In its more general role, at the level of energies of physical bodies, light is to be interpreted in its own way, ENIT. Yet even in these lower levels or states of mass energy light interacts with TRU. This interaction crosses energy levels including that predominant for man its Á form and function. These relationships are signified by LÁLREAC. Without external extension of TRIA and holding cosmology to TIT, TÂ, and RÂ we have a relatively fixed and set ALT. TIT is, in actuality, observed only within itself. A further extension of WIF is how the S-O or Relativity, observer-observed relationship depends upon light for distinction. This is by contrast. For light is perceived not in absolute lightness or darkness, or even uniform degree, but by contrasts between light and dark, and even by the dynamics, as in flickers of light or objects in motion reflecting varying intensities of light. This, of course, is largely governed by the setting and SCARLS of the observer's IA, and ones acuity of vision or sense in noticing these quanta of light and their implications as finites in TIT. This in turn is largely governed by intention and what the observer's interests, expectations, hopes, will and anticipation selects. It is in such relationships in TRU, and the dynamic mechanisms and operations of the higher powers of mass length time in the human world that enable us to understand the totality of TIT and TRU. It is by all levels of finity that constitute this whole that man succeeds or fails, survives or becomes extinct. WIF and LALREAC add two more means of realizing these crucial states and modes of TRU, TROS, its unity, finity and infinity.

The finite quanta of reality in TRU are made into the carriers of Á, its words or language. Yet the language we use is far from TRU and will never be adequately successful, appropriate or fulfill its role, unless and to the degree its concepts and terminology are TÏÊÂ, covers all TRU, and does so with a high degree of distinction and coordination. **REDRE** - Redefinition as different, all things must be E, or how through PES, EA, EFI, ESINATION, and ESATION the stuff or substance of the universe become the parts of TRU. So do they require redefinition and strict adherence to their ES and CES in TRU. **TESP** - Thought, Thing, TRU, ES Problem, or that posed by the relationships between senses, things, TRU and TES. These are inferred from REDRE. TESP is the simple product of the diverse position each person is in as that given comes in a variety of forms and states. **LEPT** - Law of ES to TRU in perpetuity, how senses and things become ES and how when ES proportionately, totally and properly become TRU all exists in perpetuity. **LOPAR** - Law of Partiality, or how we, or any entity, is not except and to the extent we become or are identical with our part and role in TRU. **LATEÁ** - Law of Absolute TRU, E and A. The makeup of this world, TRU, is not to be arbitrarily set, determined, nor up to the highest bidder, it largely exists in the E and Á of TIT, which by TRIA renders this absolute.

CLOSURE and SOLUTIONS

FUS - Functional Use, Significance, Meaning or Value. As much of what this significance is, it is dependent upon how they apply to something, how they act, are used, or set in motion. **A, L, T,** provides the dimensional frame for a spherical modulation of TRU that can be represented on a textbook page. Some of the other concepts that help use ALT for posits of ES measure are justified when corrected for overlapping and coterminality. **POCOMF** - Principle of Composition Fixation, **PICOM** - Principle of Interposition Composition, or the overlapping distribution of both ES in ALT, and **PIF** - The Principle of the Inevitable Future. These are principles that enable us to refine ES while using and placing them in SPALT. **LAVCORP** - Law of Absolute Validity through Corrected Relation and Position, or the perfection of identifying ES of TIT through corrections in their relations and position, when represented. The solution to needed corrections in representation, meta states and EPI, and measurements imposed by extra TRU, is that provided by TIT, **TRIR** or TRIS Relative, and **TRIPRA** TRIS Practical. TRIT and its progression to OBT, SUT, ROT, MOT, SPALT, POA, HOM, ACC, and POT carries correction to the highest levels of allowance by the limits of each stage in the progression.

CHAPTER 15

THE FINITE of TRU and TIT

The finite, 15, like the form, 13, contrasts with a finite, 14, like a form in Chapter 12. Although a and the form or part are two kinds of the same thing and may seem to be of little difference, this difference is critical to an understanding of the makeup of TIT and TRU. ***The*** finite is the special identity of a part with its whole, and especially if this part is oneself imperatively identified with one's world, TIT. A part usual to various parts in general, what they have in common. The part is what they have that is different, typical, special, and unique. It is in this last sense that The part is so identical with the first person and their deepest meanings in life, one's soul. It is identified with its counterpart and role in TIT. The terminology of The Part in TRU and TIT include: **TÍ** = The Finite, or Special Part of TRU. **TÍS** = The Finites, or Special ES of TRU, **Í** = The first person, **ÍT** - The first person in TIT. The Finite, or Solitary and focal part in TRU, an ESIN of I, the first person, **TES** = A person in TRU, personal equivalent of ES, one's common Soul, **TÏ** = The Infinite in TRU, **TEO** or the Total TE, **TÍE** or the Partial TÍ, and **TÍESIN** or the TÍ of Individuality becoming TIT, and **TÍP** or TÍ proper or special. TES and TÍS are commonly used to denote a and the person in TIT. Also **TÍCES** or The Continuum and gradations from ES to TÍS, and **TÍES** or difference between TES and TÍS.

ES is not ES, TES is not TES, nor are TE, E, TÍ, TÍS, they are separate parts, with changing times and circumstances. This even though TIT is essentially stable. Ones TES increases and decreases with time, both differently and variably from the point of view of ATO, TROS, TES, and the individual. The individual is minuscule, and ones IA even much more so. For at 1/6,000,000,000 current inhabitants, years of life, time of IA, and many other factors a person seems negligible. Even TES for the individual is minuscule. So we have a great dichotomy and dialectical dynamic between TES of the individual and TES generally, being such a large part of TRU. It is in such a dynamic setting, and of such dynamics of the individual as TES driving TRU that shows its media.

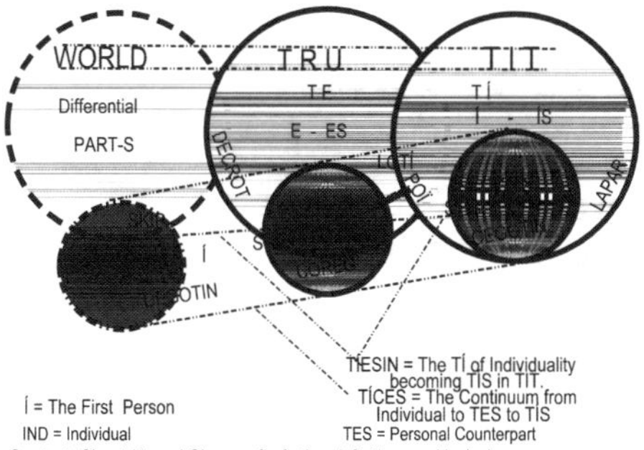

Fig 1-15 **Individual, TES, TÍS, World, TRU and TIT**

That all is not any individual, world, or even relative universe, should be quickly recognized. There is constant flux and interchange between that within and that without TRU and TIT. It is what is and what is not and how it is and how it is not TRU that is so subtle and profound. It is for us to discover TRU and TES to best identify and raise ourselves and all that is, not, mixed and disordered about us to their fulfillment and fruition in TRU. This is our vital need in the actuality of life. It is for us to achieve and apply ourselves and all about us to TES, TÍS, TIT and best direct all to TRU that is so essential and imperative for one and all.

STES - Self to TES, or Self → TES, the inclination of all individuals, more or less, to TRU. **STÍS** - The first person specifically inclined to TIT. **DECROT** - Determination of the Part to fulfill All TRU. **TETITES** - The Reciprocal Pivotal Personal Part of TE and TÍ, TES as the ES of TÍS. TETITES signifies how a person, formed and functioning as ones counterpart in TRU, does so both generally and specifically, as a part of all like parts in TRU as well as autonomously as the focal, special, sole and unique entity that it is. **LOTI** - Law of Total Identity. LOTI denotes how the individual, as all parts, must overcome IR, PECRO, and ISMS to become TRU. **ME** = Moment of Life or lifetime limit. ME is how: E = $\sum$ IA - TRU - Finite = E. **LAPAR** = Law of Partiality, or all is not except how it is of, a part of, TIT. **CEGOTIN** = Collective Ego of Individual, Group and TRU as all are pure and total TRU. CEGOTIN is largely a product of the satisfactory resolution of: **DEGOTIN** = Dichotomy EGO Individual TRU. In TRU, TES, and TA DEGOTIN denotes the opposition and dynamic interactions involved between the ego of an individual in itself, an individual in itself, the ESINS of these two, their TES, TIT, TO, and all else, AU. In Fig.1-15, the black circles progress from individual and DEGOTIN to STES, ITES and TES to COGTIN, STÍS and TÍS as world does thru TRU to TIT. The finite, lost in differentiation as in the objective differential, is in the predicament of so much of all about us. It is on neither the integral nor differential as they are practical in TRU, but of subjective as it is SIT, or subjective in and of itself. Both knowledge and action are to be determined by TRU and the experiences TRU produces. These are compensated by: **RETES** = The constant, continuous and Repeated REversion to TES, either direct or indirect, conscious or subconscious, by the individual, or the reversible reaction individual to and from TES, Individual ⇄ TES. These may be quantified by: **PIT** = Percentage Inclination to TRU. When occurring repeatedly they are: **RATES** = Repeated Reversions from TES. The extent to which we are not TRU, more often than for want of never having realized TRU is: **REOT** - Realization of TRU, rather than the loss, temporarily or permanently of TRU.

Only when I recognize myself as a partial of TRU, TREQ or even God can I ever adapt to life, TRU. Because of the errors implicit in concentration on things personal we must correct the deficit it produces in us. Personal concentration, even a highly qualified self interest, however, is neither irrational nor always ATRU. There must be a constant return, conversion, reward in TES, TÍS and TIT. *Ego* - The self looking at, on and for the self. *Self* - The particular individual or first-person. *Individual* -The average self, any self, a specified person. The inward concentration of the self and ego is not without value. It is exemplified by a strong feeling for life, its perpetuation and self preservation, as most realistically evident in paleolithic paintings and neolithic stone monuments. It characterizes the higher levels of Á in TRU. TES is unattainable without, and to the degree, we have Knowledge of TRU or TREQ. **ITRU** or ÍT - I, or the First Person TRU, the ESSE of IT is **ÍST** = The essence, thisness or ESSE of IT. ÍST is as Christ, the self or person as ones counterpart in becoming one with God or TRU. **POI** - Principle of Individuation, arises from what is TIT, how each part is both individual and a part of the whole. **SACDI** - Separation and Construction of the Divergent, Different, Divisions, Definitions and Delineations. SACDI signifies how these differents in TIT function, interact and can be operated.

Much of what we are so certain of is not TRU and thereby negligible or more or less negative. The Fallacy of the Individual is what prevents learning and living their counterpart and role in TRU. This is how we know only that of ourselves, close to us, or of immediate interests, desires or concerns. When this becomes fixed, habituated, and narrow all a person does or does not do, does wrong, wants, desires and thinks renders one increasingly lacking in TÍS and TES, thereby ANTRU. This is important in being key to what mostly limits TRU and why it can not be more readily enhanced. It is more than a simple selfishness, it is all the reasons why an individual, as ENIT does not become their counterpart in TRU.

Inclination to ES, TÍS, and TIT is acquired through concentration, mental discipline, and the habitual training of our actions and behavior to enjoy what is TRU and refrain from what is not. Individual partial universal paradox is that state whereby an individual is both a part of a whole and an individual in and of themselves, TIT, non TIT and ANTRU. Most humans live their whole lives in this paradox, with failure after failure, little attributing it to anything, many, or all other, than the actual world they do not know, TRU.

LADI - Law of the ultimate Identity of TRU and TES, TIT and ÍT. It occurs through the summation of discriminations, as daily observations and acts converge on TIT, as thru PUREACLITY and COGNOS. There is in birth and regeneration a resurgence of potential and opportunity to TES and TRU. There is an exponential genetic deficit from damage or deficiencies working directly or indirectly, that lead to decreased individual capacity. These are often exponentially accumulative. More often there is an excessive inclination to the integral and subjective. This is a natural result of both a person's IA orientation, and need to acquire a total grasp of the given, TREQ, or TRU. There needs to be optimal equilibrium, a golden mean, between the SI, SD, OD, OI, and their mid ranges in the real.(Fig 6-28) When this is overdone with the SD, OD, and even R, especially by an individual's inability to achieve all SOID, SPALT, KOT, and POT, it is **TEP** or The Finite Predicament In reverse, TEP in extension proves the greatness of the subjective and integral, and together SOID, TRU, and all TIT-TO.

DERIVATION & DEVELOPMENT of THE PARTS of TRU; SELF, SOUL, TES, and TÍS, COMPARED

Much of the struggle for existence is conversion to TRU by all that needs to be TRU. It is toward a perfected actuality and counterpart in TRU. The origins of TÍ come from many sources, yet all primarily come, like ATO, from the existent form, state, function, dynamics, and thisness of TIT. To understand how this works we use as an example, and what is a large part of this conversion, the individual, self, and TES of TRU. The discrepancy and loss of TES is figuratively described in the bible as apostasy or fall from grace. Much of grace, as TRU, is what humans must utilize if they are to live up to their potential. Salvation, reversion, compulsion, convergence, inclination, will, correction, extrication and others are highly active in this transition to full participation in TRU. Parts are not sufficient among themselves. They function mutually and in dialectic totality. Their rewards are equal, proportionate or set by their contribution to TRU. A continuous series exists from the highest levels of TES in TÂ to the opposite extreme. This helps to show the range of derivation, development, determination, and transition from being non TRU to full separate and special participation in TRU.(Fig 2-15) A striking example of the great differences and deficiencies between what is illusion and what is perfect actuality, is: **DAK** = We are, all more or less, Dead and Don't Know it, and to that degree. The struggle for existence is what makes a person's life much more ordinary than they usually believe, less sinister than they under stress and pressure feel, and death more meaningful.

Fig 2-15 **SELF - TES RELATIONSHIPS**

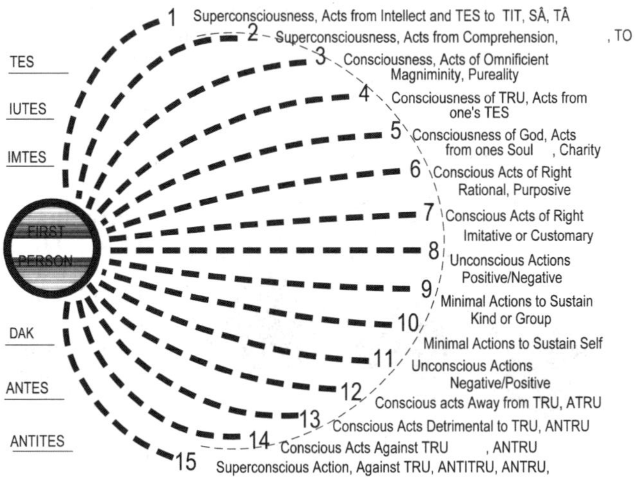

What are the relations, identity or non identity of the first person - focal black circle - self, TES, and TRU? What behavior, acts, work and endeavor is to be made toward TRU, TIT and, TÂ? The critical relationship between ones consciousness and acts that are right or wrong, TRU or Non TRU, or optimal, TES to TA are shown. It helps to know, do, and be on occasion, on average, or through one's entire life. This in turn helps prepare ourselves for the life of direction, coordination, and control we are capable of and vital to TES and TRU. When all mutually and cooperatively function as their parts in TRU rewards, success and fulfillment are assured.

The lability and relative independence of IA is a product of its need and place in the 'animal kingdom' and TRU. This helps us understand the problem of free will versus determinism. It is a relative free will for the dynamic focus of TRU, and a determinism for TIT-TO. I, the first person is caught in the middle, and fails or succeeds depending upon how well they can learn and overcome this seeming enigma. Both are valid if perfectly TIT, yet in our lives of variable AMTRU they become totally invalid and deadly. Thereby a continuous correspondence, convergence, and correction of IA to TES and TRU is compulsory. This requires both a level and direction of IA, TES, and TRU. This also requires a certain degree of depersonalization, as in sacrifice or service, if one is to escape ANTES and become TES and TRU. This has been the great message of the Holy Spirit in Christ by bringing us to God, TREQ, and TRU.

As much as an individual feels free or unlimited, defeated or crushed, maturity, an understanding of TRU and our potential, part and role in it help us achieve *equanimity*. This enables us to prevent, or extricate ourselves from our own weakness and much of ATRU. Convergence on TRU is multilateral, all the interactions between TRU and TES an individual can participate in. Such prevention includes a balance between ones TES and TÍS, or 'a' and 'the' roles in TRU and TIT. TES does not have to be learned and controlled every step of the way. We are already formed and function by the mold set in TRU. This is like karma or the motive power and summative actions that assure existence. Most of this merely requires that we follow this set, what is natural for TRU. Deviations from this are compensated for by continuous reversion to TRU. This constitutes much of the massive and simultaneous processes we undergo repeatedly or cyclically throughout life. What we must continuously learn and control, are the effects of adverse deviations. The dynamic mechanisms of an individual's identity are subordinated reason and value specifically planned for TES and TRU. It is an incarnation of one's being. Reincarnation is the return through transference of individual to ESIN to TES and TRU. Theophany is a manifestation of God to an individual. Epiphany is the sudden spiritual manifestation in the exquisite pauses in time by realizing identity with TES and TRU, TÍS and TIT. Theophanies and epiphanies are like sparks that can become more permanent by the FAF, carriers and knowledge of TŔEQ and TRU. The Logos Doctrine or the role of SO as the Á dimension of Mass Energy in TRU shows how we can explain all things, not only through the SÁ and mechanisms involved, but by the principle of *Salvation*. This is what was so dear to Athanasius. Here it becomes vividly clear. Man is not only weak and ignorant, but rebellious, and thereby greatly needs *Redemption*. Incarnation meets man's deepest needs, for it brings not only illumination but redemption. Only when we achieve TES, TRU or TREQ can we sufficiently lead a divine life, be TIT, and assure TÂ. Even today we are still locked in this conflict. As large as human population, mental, and material baggage are there remains a great void. We are far from achieving the quality our potential could provide for TES and TRU.

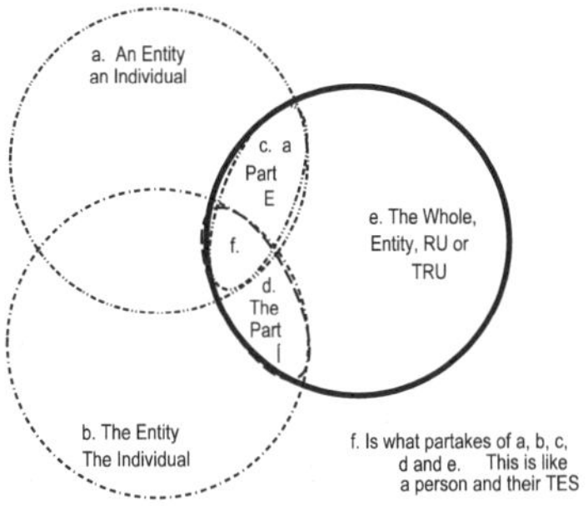

Fig 3-15 **GENERAL RELATIONS Between PARTS and The PART, TES nd TRU**

This figure is used to show the cotermin-ality of a part, the parts and TES. It emphasizes the fact that other persons, oneself and TES, are neither all of TRU, nor all of all things, AU. They are not even all of a part or all of the parts. However since much of TRU is individual, an entity whose ESINS are TRU, this reveals the importance of the special relationship that exists in such dynamic opposition. Herein is what is so available for improvement. By the figure we compare groups as we did for the central and predominant part of TRU, TES. TES is coterminal, comparative with, and the same or different than I, the first person. This group, of which persons are parts, has much in common. By understanding these, as well as the very parts of a human being, first person, TES, and TÍS, we can gain a wider idea of our world, TRU. This is especially so when modulated and posited, by ROT, ALT, SPALT, and POS. A group, the group, and the special group as they are TES and TRU, are three different, and same, entities by various combinations of DAOST. The SCARLS of each are different and same. These are shown by the different sections of the circles shown above.

There are several states of a person's being or: 1. that of a person, self or individual that is bound to its existence, with increasing association of TES, 2. living self or the TES existence of the self or how the ESIN in a self functions as TES, 3. relative self or what has more than individuation, relation, yet less than specified, unified or TRU relation, as TES, and 4.the anti-self, not only against TRU and TES but against future or potential positive relation to TRU.

Personalism is far different in its current meaning than its implications in TRU. In TRU it has special and important implications, including: 1. the greatness of the first person, I, in TRU, 2. as a combination of the first person of TRU with the first person of the focus of the individual, 3. how reality and TRU are coequal or at least very similar, 4. the proof of IA and the individual, 5. a person as the center of the universe, U-M-I-Ia and focus of ALT, 6. the ego as the concentrating power and core of TRU, and 7. the Á state, dimension or form of TRU as the media, vehicle, carrier, and part of the power of personalism. Personalism's greatest value is most subtle, by revealing and proving the actual nature of TRU. For TRU is not only largely human, the human in it is highly personal and special to the extent TÍT is TES, TESP, and TÍS. Personal make up is highly TRU, generally and specifically, just as ones anatomy and physiology. Yet ones thoughts, interests, actions and behavior are often highly ATRU. How this can be best corrected, managed and controlled is a central theme of and is described throughout the text. *Charisma* is an abstract, spiritual power, gift or talent of a person , that religiously denotes the grace and favor of God, TREQ or TRU.

THE PARTS of TRU, TÍ, TÍS & TES, GOD, SOUL, CHRIST, IDENTITY, UNITY, FOCI, CONSCIOUSNESS, and I, The First Person

To investigate the special and exclusive form of the parts of TRU we need look no further than our own consciousness, identity and how Christ served this bond between ourselves and God. It is the focus of both ourselves and TRU. It centers on the concentrated focus of the first person's life, as this is TES. We are the carriers and a large part of GOD as TREQ and TRU. It is this centered concentrated positive power of such a focus that provides much of the unity, mass energy, power and totality of TRU. Much of the dynamics of TRU can be derived from this focus and unity. It is largely the essential principle of being, being TRU. It is provided through the clarity and characteristics of the extant, its Á FAF. We can extend and use knowledge to learn and prove TRU, and thereby all. Until now such knowledge has been accompanied by errors, fallacies, voids, disproportions, and incompletions. (VAPO) These distract our thought and reason. The absolute certainty of ones being is shown by the realization of loss and loss of ones life. The ways an individual becomes TRU are many and continuous. They can be shown. Aristotle's Doctrine of the Soul has given us the realization of the form of the ES and how ESINS are their potential. This potential soul becomes pure activity and ultimate substance when it is actualized in its full part in TRU. The highly imperative identity of the first person, with their counterparts in TRU, TES, is the key to much of the world, both the ENIT, ESINS, TES, TRU, TIT, and TÂ. This may be figuratively shown by the Eucharist. It is how, at best, the self is largely TRU and TRU largely self and human. These occur and maintain themselves by concentration, selection and exclusion. One's life, survival and ego, by relativity, are centers of the universe, TRU. The repeated and convergent coming to terms with life is what often makes this centering, the unity of TRU. This is closely associated with reproduction, future and immortality. At its best, by TRU, intention and the giving of oneself, assures TRU, and in return ones own immortality is increased. It is formed by the predominant FAF, state, dimensions, continua and mechanisms of TRU as these are produced by the supremacy of its set mass, and weighted point of reference. In such a way we can prove ourselves, ES, TES, TÍS and TIT through our total absolute identity. This in turn provides and proves TES and TRU inviolate.

Existentialism assumes that each person may or may not choose or will whatever one wants or desires. This may be true, to an extent, for their IA, thought and specific acts. However, for a lifetime and a person's overall being, it is very much invalid. This is shown by the difference between the Law of Multiple or Summative Senses and Things, called **LOST,** and the Law of Identity, **LID**. It may be that an individual can live at least briefly by ones thoughts and acts in multiple senses and things other than TRU. Yet, as TRU is all by TRÍA, eventually and the degree to which a person does adapt, fulfill their roles, and live in identity with TRU, TES, one must return oppositely and equally back, forward or in whatever way necessary to TRU. This is signified by: **TREAC** - the Law of TRU in Reaction, or Law of Return, oppositely and equally back to TRU. Only as each person becomes one with TRU, TES, and finds the manifold truths in the great concepts, principles, laws, and mechanisms of human life can we become one in TRU. We begin and often exist in life without this, a kind of original sin, for any departure from TRU, especially from its one systematic truth, is sin. This is why the righteousness of the one, as by Plotinus, or in God by contemporary Christians, were able to so vividly see and emphasize the sin which accrued from all that was untrue and wrong. Here is the difference between heaven and hell. Immortality and the magnificence of religious revelation, fervor and spirit must be embraced, upheld, and utilized.

Fig 4-15 **STATUS of INDIVIDUAL IN and OUT of TRU**

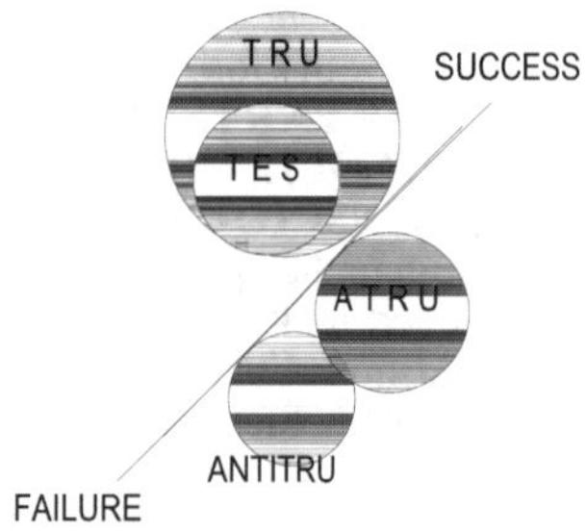

TRU is composed of the ESINS of an aggregate of individuals and many other senses, things, and phenomena which makeup its parts, processes, subjects and objects. They are or are not all set at 100% TIT. Nor are they perfectly identified with, nor function in concert with each other or TRU or TES, top left black circle. These act not entirely in, in accordance with, or proportional to TRU. To this degree they are non or from TRU, ATRU and ANTITRU, bottom circles of figure. Any degree of relationship or force that is positive to TRU is a part and process of TRU.

By TRIF we know that all that is incrementally or even infinitesimally related to and a positive act and force in TRU is a part and process of TIT. So that a person's knowledge, interest, intentions, will, and acts are paramount. If it is on what is positive, good and right - by TRU - it is highly TRU. If it is toward the insignificant, nothing, negative or against TRU it is not or negative to TRU, and produces reaction to that degree. What humans should do and how to live is how much we actually accomplish, achieve, and contribute to TRU in our lifetimes. This is why supplication, prayer, grace, hope, spirit and salvation have been such large contributors to the success of this best of worlds of man, TRU. They are all highly positive - even if only in reaction - for TRU.

ÍT - I TRU, or the fusion of the I, the first persons with the I equivalent in TRU. **ÍST** -The Special, total IT. This may be compared with the suffix 'ite' as in Caananite, or Israelite, a member of a group or nation. Only for the first person is it strictly limited to and governed by TIT, as TIT is governed to a large extent by the collective ÍT. The convergent of ÍT to TE, to TÂ is called: **ÍTESA** - Equilibrium of IT to TES and TÂ to TES. ITESA signifies the homeostasis that exists between IT -TES, and TÂ-TES in TRU. The individual limit denotes major facets of an individual or limits imposed by an individual's relations to, and role in TRU and TES. The actual individual limit is that imposed by TRU. **MOF** or Mode or average of Finity, is the practical limit of humans. In totality there must be the new concept: **ALET** = Absolute Love, bond or identity, by a person, self or individual with TRU. This requires **AITES** - Absolute Imperative TES. AITES denotes the natural, inevitable and certain convergence of each and all on TRU. **THES** - Subjective In TRU, as The Part of TRU. **THEO** - Objective in TRU, as The Part of TRU. Both THES and THEO express a relationship between TIT and the First Person, or the Subject. THES and THEO are of TIT. Yet TIT, THES and THEO are absolute, definitive and distinct, at least in their general aspect of how they are defined. As of any entity, only as we identify it, and set it apart, only as we subjectivize and objectivize it by our predominant form and function the Á dimension and mass energy is it actuated and made valid, for TRU, this is called: **LARECIT** - the law of A reciprocity of TIT. LARECIT is a product of our being, or 'The' State or Condition we have by being a part of TRU..

For a person in the world it is not only 'how and how much you need the world', but how and how much the world needs you. This is the difference between TRU and prior or current beliefs and interests. It needs you to continue itself. It needs you in a certain way. This way will depend and be relative to the world of which you are a part. It will be different - as it is a different world - for each different thing. Each person needs to know and know as much as possible what the world, TRU or TREQ, needs. If this is not known, we do not learn, or are not told what this is we will not know how to be TES and TRU. If we are not receptive to what we are told, we will not know how. The struggle for existence is both direct and indirect, individual, TRU and all in between. An individual's life may seem to be all IA, yet this is not true. Only our lives in TES and TÍS is all, of TRÍA. Even a state of super TES and TRU are, in ways and at times, needed to overcome ATRU.

Fig 5-15 **TES and SOUL, IDENTITY, S, O, AU and TRU ENTITIES to ES**

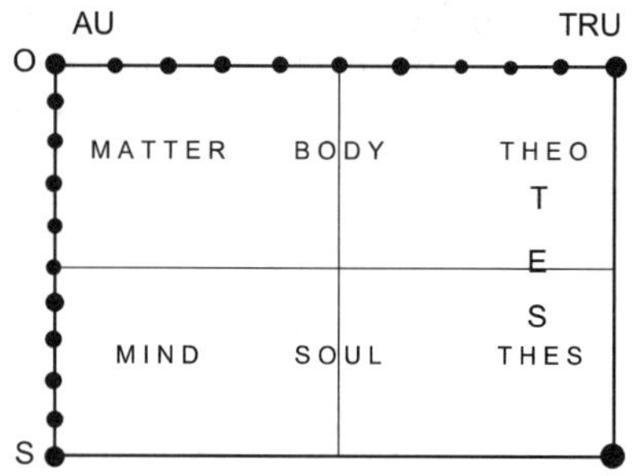

The Neothomists, as expressed by Gilson, describe the parts of TRU as exemplified by the soul. Mans substantiality is identified only with one of his constitutive principles, the soul, TES. By Gilson the soul is pure activity, function or an ultimate substance. The IMperative nature of the soul and TES is called **IMTES**. TES is indestructible only as TRU continues, grows or is TÂ. This does not hold in all senses. TES develops its actuality in its fullness via AU, the body and things in themselves. The error in such a soul is that it does not account for relativity and the Á character of TES. Nor does it distinguish between the intellect, spiritual and material aspects of man and TES as they are outside of or inside of TRU. Whether they are actual or inactual in the TRU sense. In Figure 5-15 the S and O relationship between man or individual existence in AU and mixed AU and TRU is shown in comparison with TES, with THEO and THES, or TRU actuality.

Consciousness is that wakeful state of human life in which we are increasingly aware, attentive, and active.. It is the great focus of IA. Experience and consciousness are the same, yet different. It is the dynamic state of awareness, attention, and thought in which the central nervous system best functions in health for its own and its kinds benefit. It is a prime characteristic of all higher life forms, especially human, and is manifested by the special and predominant mass energy Á FAF of TRU. By the great cleavage between consciousness and unconsicousness, being awake or asleep, and thereby the various levels of SÁ, we can recognize both the finite, differential and quantum basis for TRU. We can also most profitably recognize its potential for modulation and posits in demonstration and reproduction. It is a solution to **PROTÄ** or the problem of the form of TRU.

FROLOV = Felt Realization of the Loss of Ones Life, is seen in St. Augustine's Confessions and their motivation. This is the deep sense of remorse associated with the loss that occurs when one has, is or will fail GOD, TREQ, or TRU. How much extra effort is needed to even recap what has been lost. The great fallacy of ego being, is how certain we believe that one's self, ego, or the here and now is it. "I think therefor I am" of self consciousness seems to be undeniable or self-evident proof of our being. Yet we must always realize that it is in error, or less than correct. It is a belief and opinion based, at least in part, upon an ENIT or THIT and not on a TRU or TIT. It shows the great focus and power of consciousness, the first person, TES and TRU, and the ratio or difference between the part of an individual and TES. **CITES** = Conscience Inclination to TRU and TES. What one does is more or less inclined to, is through natural affinity, experience, and its FAF in life, to be TES and TRU Essence of Being.(POEOB,POM) **TESP,** TES proper, signifies the extension of TES, whereby the essence or essential principle of TRU is this very being,

TES, of the individual in TIT. **LADI** = The Law of Discrimination Identity, or the ultimate identity of TRU, TES, and the individual with the ultimate TRIT, OBT, SUT, ROT, MOT, ALT, SPALT, and POS. Probably the greatest task every person has, is to *Come to Terms with Life*. This is to discover and practice ones position in TRU, as each becomes their TES or counterparts in TRU, and the counterparts of TRU are of them.

EL = Exclusion Law. EL denotes exclusion from TRU. It is an exclusion that is set fixed and largely absolute. So that all that is not TRU is spontaneously, or at once, excluded. This has significance in its relationship to the individual and how the individual is or is not their TES in TRU. The individual and TES are both in ways absolute, limited yet very different. The reduction in level or proportion of TRU through codomain, coterminality and overlapping in the resolution of EL depends upon not only a standardization of the scale of TE, it depends more upon the scale of TE in which the human mind can satisfactorily comprehend. Identity can only be good, if only to TIT. The Law of partial Identity, LESID, holds that we as individuals often do not identify with TRU, and even if we are personally and specifically inclined to do so we do so only partially. The great measure of ones ability and success is how well we can distinguish between, and act upon, what is actually good or bad, TRU or ATRU. **MITRU** - Measure of ITRU, signifies how this can be quantified, and how it reinforces TES and TRU. **ACI** - Absolute Certainty I, or Absolute Certainty of Ones Being. ACI can't to be taken for granted, nor to be confused with what it is not. ACI can be valid only when TRU. All the uses of the verb to be: is, are, am, was, were, and been do not distinguish between what is and is not absolutely certain, nor even what is or is not TRU and TIT..

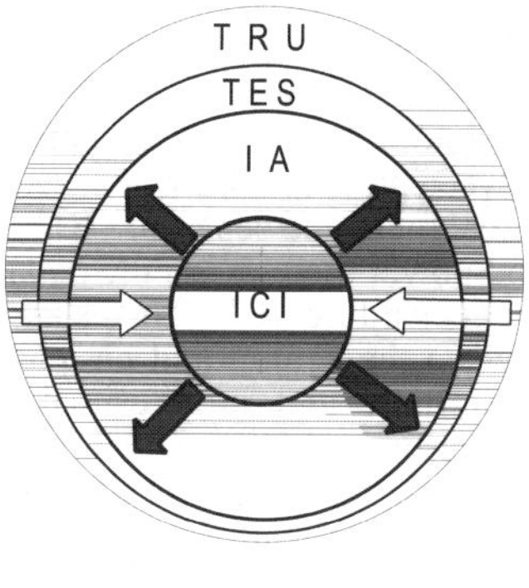

Fig 6-15 **CONCENTRATION, IA, ICI, TES and TRU**

ICI = Instantaneous Concentration I. ICI, the focal black circle, is significant in how IA proceeds by PUREACLITY to TES and TRU. Much of all the world and much of thought, as well as ones very being, depends upon concentration. Most all that is to be accomplished, especially if it is to be well done, requires much concentration. This is especially seen in the great competitive and confrontational world of sports and business in which we live. There is much in professional endeavor both in acquiring a professional status and in maintaining it. What are you doing? This question is key to effecting and producing no more and no less than what is needed, as in Medicine. Yet concentration without the means, material evidence, information, proof, mentality, and support to go on is of little value. Without the direction, purpose and judgement necessary it is without control. COGNOS is the summation of concentration, knowledge, proof, mental capacity, direction and purpose in the unified focus of ICI, IA, TES and TRU. Concentration expands a person's IA, and specific IAS in summation expand TRU. In Figure 6-15 we show concentration as it expands in IA and TRU. There is a polarity and spectrum between concentration and expansion as they balance each other within the limits of IA, the individual and TRU. It is key to the dynamic maintenance of functional ability and performance. As it increases or decreases, as the genetic material, live product and determining forces, or all levels of Á, increase and decrease so does, in summation, the individual, TES and TRU.

CENU = Center of the Universe, TRU. As a result of the central position or the human person, as well as the focus of ones body, thought and activities, we have naturally tended to consider ourselves to be this focus and center. However we need to have a fair, unbiased, scientifically valid and terminally conclusive resolution of CENU. The search for validity, reliability, sensitivity and specificity, perfected in statistics, holds for all knowledge and most of all for that knowledge upon which all else must be based, KOT and COGNOS of TRU. CENU is a prime example of this basic search through its focus and coordination. CENU is a product of relativity and especially the absolute relativity created by the special, unified, systematic and coordinate unified nature of the totality of TIT.

Abelard held that the freedom of man lies in his judgement and the moral quality of an act lies in its intention, not in the act itself. This is the great problem of what is sic and non, yes and no, right and wrong, and how much is up to the individual and how much up to the world one lives in. The answer is to be found in neither freedom, individual, nor any right, but the relationships between TES and TRU. It is this that so stimulates our sensitivities and abilities to provide the energy and power to rise above all that is not TRU. This rising, like the *Ascension* from ATRU worlds and *Assumption* to ever more, greater, glorious and victorious levels of TRU, has been and will be even greater, the focus of all humans. It is not for man to randomly choose whether he wishes or prefers this or that. All must be chosen with a strict and convincingly heartfelt and determined spirit to be TES, and the love of all TRU.(VAPO)

Maximal identity is how to search ones soul to find solace and consolation in fulfillment of ones TES, and harmony in TIT. Absolute egocentrism is that state of an individual in which concentration on self becomes so great as to overwhelm all else. It is a subjective feeling of the absolute of a person without the necessary identity with TES and TRU. Yet ego is in ways both natural and with TRU reason and purpose. This is the absolute undying motivation for total exclusive being and for eternal life. The trick is to realize that this is to be found only in TES and TRU, or TREQ. It is a part of success/failure, life/death and survival. It is a great and most penetrating force in TRU even when subdued and sophisticated. Only in humans with their higher levels of Á and great powers of introspection is the ego so prevalent. Like all newly acquired abilities the ego is highly volatile and apt to go wrong. The ego is often for all the wrong intentions associated with bad self interest that causes such an abomination. Yet, when reduced to its place and role in TRU it adds power to TRU through competition, selection, and the maintenance of TES. It needs to sustain conditioning by **AITES** - The Absolute Imperative Individual to TES to TIT and TRU. It is like Confucius' conformity to the way of heaven, or Kant's Categorical Imperative, whereby one must act "only according to that maxim by which you can at the same time will that it should become a universal law." It is similar to Buddhism where all that is not TRU is eliminated from the individual or entity and all that is of this individual that is TRU is TÂ. Immortality is the permanence to be found in TES and TIT, even immediately by its own revelation and hence extended with TES. **PODIM** - Principle of Differential Immortality. Each and all in TIT are participants in and of different kinds of immortality of which the individual when TES partakes in many ways.

A person's will, becoming identical with that of the universe, shows us how TES is in TIT-TO. It reveals the whole range of intention and conviction in the projection of TES and TRU. It shows us how Augustine and al-Ghazzali were right when insisting that the divine creative will is compatible with man's freedom. Yet this divine creative will, as well as man's freedom are only as they exist in ATO, TIT, and TES. The combined memory of all humanity is what makes up much of the foreknowledge of TRU. This memory, foreknowledge and will are, in being on an equal dimensional plane with length and time, in this sense outside of space time. Yet in TIT-TO it is the whole of the mass energy, Á FAF of TRU, the SÁ continuum, that provides the basis for will and freedom.

KINDS, GROUPS TYPES VARIETIES COMBINATIONS PLURALITY DIFFERENCES

There are numerous kinds, types, views and varieties of individuals and groups, TES and TÍS in TRU. It incorporates in the individual the proper amount of depersonalization as a member of TRU. Such bonding is seen in the Eucharist. It is the common positive sharing of goodness, companionship and friendship necessary in total TES and TIT-TO. It is reinforced both voluntarily and involuntarily through common ends, the same in TRU, reinforced by toleration, moderation, love and appreciation. The bonding may be of different strengths among different parts as well as with the whole, TO. Thereby TRU is composed of many parts whose different strengths or weights provide a summation of multiple foci or centers. The problems of combining parts, foci, and processes produce polarities and axioms which act and can be shown as continua and spectra. Preeminent among these is that of relativity or the subject object form and function, state and dimension of TRU. This dimension has for the parts of TRU halves, its subjective or THES and objective or THEO. It is from the makeup of the relationships between THES and THEO as they exist and work together and apart in TRU that much of TRU can be known.

Fig 7-15 **INTERNAL RELATIONS of TRU and ITS PARTS, TES**

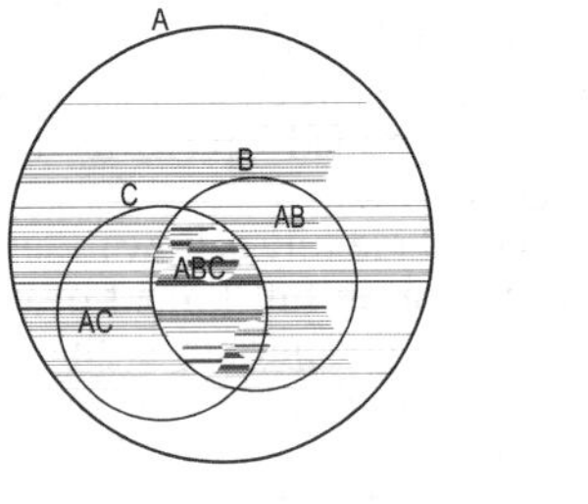

The simplest yet most basic relationships between TRU and its parts include: A = TIT, B = a TES, C = Another and different TES, AB = What is in common or same between A and B, AC = What is common or same between A and C, ABC = What is common or same between all three, A, B and C, D = What is of TRU that is not in B or C. There will be no BC because what is in them will also be in A, as TES that is inside of TRU has no extra TRU component. This is if it is not taken in another sense, which will to this extent make it something else and therefore not as given here. When taken in an extra TIT sense we have: B/A = What is of an entity of TES B that is not in TIT, C/A = What is of Entity of TES C that is not in TIT, BC/A = What is of that in common of TES B and C that is not in TIT. In actuality TIT consists of an innumerable mass of TES. These will more or less correspond to the ESINS of many of the entities man has knowledge of, concepts and terms for. Some that remain are presented in this text as NEC.

In TRU the internal relations, of parts to each other and to the whole, determines much of TRU. Rational identity, maturity, search for basic answers and ones problem solving ability show us the closeness of IA to TRU, and provide a basis for the principles and laws of TRU. Appreciation for all the rest of TRU, as for that in the second or more persons, other givens, ES or CES is crucial to a balanced life, TES and TRU. What may appear to be a selfish person, alien to others, or at a distant from TRU, is, or may contain, only a different ES, CES or aspect of TRU. It is likely to be of mutual benefit, by TIT-TO. In TIT the subjective and objective take on special and exclusive roles. **THES** = The subjective, The summation of the ESINS of extracted Subjectives into TRU. **THEO** = The objective, The summation of the ESINS of extracted Objectives into TRU. TRU is both the subject and the object in various combinations and levels, as well as of other forms and dimensions.

Individuals and TES in TRU may be considered in different ways, as an average per worlds population, as an average for some group or segment of the whole, or weighted by power, money, fame, good works, or other contributions in public opinion, or to TRU. This tends to depersonalize, or a same rather than different emphasis, of DAOST.

LACHPRE = Law of Changing and Altered, Levels and Epoches of Preeminent Reality. It is not the I-I, but the I-thou and I-It relationships that are preeminent, or have preeminent reality. The I is shaped largely in the context of these relationships in TRU. This is what enables us to see how even on a personal level, a singular E, or Í, is largely and only significant by way of its relationships and CES, thereby the plurality ES and TÍS. LACHPRE signifies how we move all through these with relation, depth, dynamics, rapidity and change. The upgrading, training, and education that form our habits and abilities to be TES are largely a group, communal, state, national and total human endeavor. LACHPRE is determinant in an ever more densely populated and cultural society, TRU. The FAF and compelling effects of TIT and TRIA show how we can and must be of set FAF, most fit and effective. This especially applies to all levels that enable predominance, in the SÁ, Á, L, T, and ACEXIN of TIT. It is in the integration and formulation of more independent or free entities and segments of ALT that the only adequate resolution of their occlusion is to be made.

The conflicts, often very violent and prolonged, that arise from the apparent irreconcilable differences between extremes merely show the lack of individual and group understanding, management and control. The solution to these conflicts is to delve deeper, with ACEXIN, SPALT and POS, into each of the extremes, means and the whole, as they exist and can be explained by the totality and harmony of TRU.

THE PARTS of TRU, POLARITIES, CONTINUA, MEANS & EXTREMES, MULTIPLE FOCI, THESO

The relation between parts is how they work together with focus, polarity, and totality. For the mass energy dimension this is largely the relationships between THES and THEO. Such polarities produce axes whose components make a spectrum or exist as a continuum in the whole. The continuum consists of means and extremes with patterns, territories, and many other features. The relations between parts of TRU when weighted in summation provide the actual focus and center of TRU. Around this concentrated mass and power there exists the polarities of the opposing extremes that constitute TRU. This is associated with the relationship between the central and peripheral, the unitary, finite or plural, infinite and infinitesimal. Of the polarities of TRU that of its mass, length and time form its primary dimensions. That of mass energy or relativity in the subject object is special, unique, highly developed and largely exclusive for TIT. It is focused upon the IA, whose expanded point of reference as subject object, is formed and functions in a complex polarity. It is shown by the form and function of human central nervous system. This, for the parts of TRU, makes up its THES -THEO or THESO polarity. The time, as well as length, dimensions are also typical, exclusive, relatively and selectively set in TIT. The means and extremes work together both for each dimension and for ALT in TRU in balance and tend to prove each other as well as the polarity and totality. The means have characteristics, among which are those of the centering by the individual, and the golden mean.

Fig 8-15 **DIMENSIONAL FOCUS, COORDINATION and GRADATIONS**

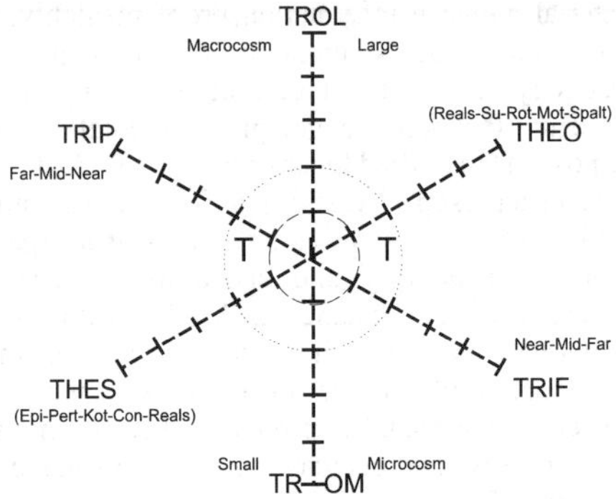

The gradations in each dimension show:

1. The degree or extent to which they differentiate from TIT. That which is of TIT is that

which is of the extreme of THES, TRIP, TROL, THEO, TRIF, TROM or the infinitely subjective, past, large, objective, future and small. 2. There is a distinct DAOS in each as they are separate from or are the same as TIT and TRU. 3. Balance and comes together in TRU. This can also be seen in FAT and NAT by how we telescope into and out from TRU. The degree of, or gradation in, telescoping is the key, mathematically precise way of attaining perspective and presentation of TRU, especially a TRU which is not only TIT, but focused by OITIO. **TELAT** = Telescopic Approach, focus or zooming in and out of TRU with precision.

LAPEX = Law of Polar EXclusion, or polarities are not absolute, do not exist by themselves, but, as relative limits, are only parts of TIT-TO. The condition of LAPEX is where each aspect, part, and process of TRU contains some of each A, L, and T. Yet it does so, at least partially, in a highly variable, special, and exclusive way. That is where one dimension may be practically insignificant or infinite compared to the amount in another dimension. This leads to exclusive hierarchy where all can be directly interpreted in terms of polarities or position on the continuum between polarities. This is associated with the principle that the peripheral and central are equivalent to the plural and unitary. **LEPROM** = Law of the Extremes Prove the Means, and vv. So that it is imperative, that we know, show, and demonstrate the means and extremes, **LAME** - Law of Measure by Means and Extremes.

POINTS of REFERENCE, UNTSO, TU and SUT

The problems that are produced by an unknown TRU, and most markedly by TRUCH, ITCH, TRIS, TRIV and TRUS can not be resolved without a thorough analysis of Points of Reference. By TRIA we must, for TRU, have a fixed, unitary point of reference. Yet by TRUCH, TRIS, ITCH, TRIV, and TRUS we do not have an absolute fixed point. By RU, RIA, and the General Theory of Relativity, we have only one observer and one point of reference. By RU, as by RIA and the Unitary Object, we have only one Observed only one point of reference. When these two are combined, as in a 1:1 relationship between the observer and the observed we have a unitary and fixed point of reference. It is a simple resultant of the convergence of all on TIT. This is for TRU a fixed subjective and a fixed objective as in THES and THEO that becomes fused as one in THESO. This is called: **UNSOT** or Unitary SOT or combined Subject, Object, TRU. Even more evident is the point of reference, or center provided by TIT, CENU, and its composition relative to human beings. We must remember that TIT, only approaches an absolute focus, it is, depending upon kind of TRUS, a Relative Focus. **TITREF** = TIT Relative point of Reference, only those aspects of TIT that hold it together give it centering, concrete form and structure. There has been in the past references to this focus and centering in TREQ and its modes, as in the *unmoved mover* of Aristotle. It was interpreted as being that static focus about which all functions in the universe. The *unvarying mean* in far eastern tradition provided a static focus about which all operated & as Pi or heaven in Chinese was depicted as the hole in the center of a circle, the world. In Christian symbolism the Holy Grail is that object whose quest is to be ended when found in the symbolic center of that representing Christ and God. Mathematically proof of the role of the center is the set relationship of the periphery, as in pi, 3.14159265.. or an exact, relative, dependence of circumference and radius upon a fixed point.

TIME; RELATIVE, SET, ABSOLUTE and LATIRD

To properly attain a perfected perspective of life on earth, it is held by the biologists and ecologists that, this perspective must be vicarious, dynamic, differentiating and integrating to see the whole earth, sun, and universe as whole. It is a place where the individual animal increasingly or decreasingly adapts for survival. The solar integration of heat and energy provides the basis for this dynamic continuum, which working in time by relativity, and as focused upon man, provides the FAF of TRU. It is not any time, physical time, or a variable time that is the actual time frame in which we live. It is a set, yet special, typical and unique distribution of time as it exists in life on earth relatively, and more precisely relatively to humans, in TRU, TIT. It is in this sense of the most particulate form of TRU that exists in each of us in common, our focal point, IA, consciousness and acts that become TES in TIT. The incipient consciousness quickly becomes one's immediate awareness, instantaneous attention, and action through a stream of consciousness that expands with IA, EXPIA, IAT.

PRODFAC = Problem of Determination to Fill Out Completely and Continuously adjust TRU to conform with its Actual Totality. PRODFAC not only proves TRU it explains unknowns and all those parts, processes, functions, principles, facts and structures that hitherto have been unknown or improperly conceived. It is SUT and the development of ALT that our greatest strength and opportunity arises. It is here we best construct, duplicate, replicate, reproduce, represent, model, make an object of how TRU actually exists for ROT and POT. ÏTESA or IT Equilibrium TES Equilibrium TÂ, is the convergent IT to TES to TÂ and the balance and dialectics that exist between them. Each functions as a separate entity, more or less, and all together function for unity, as they are the Same, DAOST. It is the great conflict of wills, especially that of ATRU, AMTRU and even of D or Different within TIT that is so much of the dynamics and problems of human life. This is resolved as we see and live ITESA, our best FAF in TIT.

LE = Law of Equilibrium, or the equilibrium that exists between the coordinate dimensions of any universe, and in its special way in TRU.(TRUEQ) **WEG** = We Are God or TRU, World Equivalence, or how our TES is God, TRU. This is how the ESINS of individuals are TES in TRU, eliminating all other. There is a strong bond between TRU and TE, E, ES, CES, and TES, like 'the tie that binds'. It is one for all and all for one, and all in between that hold TIT-TO together. The bond that exists between the parts of, and more especially, between one person and another, and of various TES has been described by Mother Theresa as: *"We must be able to see the face of God in our neighbor".* It is derived from the need to know and love God, TREQ or TRU, and others as ourselves, TES in TRU.(KOT)

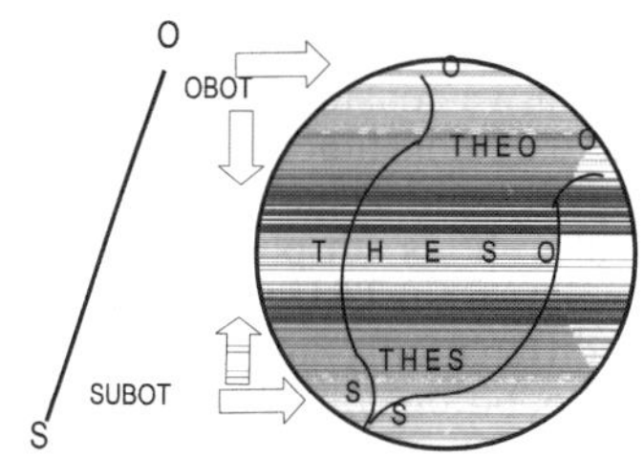

Fig 9-15 **S, O, THES, THEO, and THESO** What are and how do the dimensions of TRU change as it is taken in different ways? How is it a product of replication, representation, modulation or some arbitrary or standardized image, replica, duplicate or reproduction of TRU? **THESOA** = TRU as the Subject Object Absolute, **SUBOT** and **OBOT** or SUBjective and OBjective Of TRU are different than THES and THEO in TIT. SUBOT and OBOT are simply looking at, or making an object of TRU. They are working from the outside. Whereas THES and THEO are what the subject and object are within TIT. This great distinction is: **LOSIT** = Law of Subject and Object Out of and In TRU. In Figure 9-15 S-S, THES, THESO, THEO, and O - O and are shown in the circle, as they are in TRU. THEO is doubled in order to provide a spread to show how they are not fixed at one point but in an area. Also how they are dynamic as well as static can be shown as stimuli or forces acting upon them in a serial manner. In the left hand figure we show S-O as a simple linear dimension. This is what it assumes in ALT and to a lesser extent in SPALT. This is an objectification and standardization of it. For in TIT the Á dimension or state of TRU exists in a much more diffuse, dynamic, massive and energetic FAF.

THESOS or The Subject Object Standardization. This is where, although the subject may not equal the object, they are referred ultimately to, and polarized by, a single Standard. THESOS as an aspect and derivative of Á reverts to TIT to become more balanced, powerful and predominant. **LAMOD** = Law of Á Means Origin and Development, or the origin and development of the mass energy dimension in life did not occur by subject or object, stimulus or response separately, but from its means, polarities, and continuum that followed. Once structures of sensory and motor FAF are demonstrated, the beginnings of the Á dimension were well along the way. They are a large part of TRU, and the focus about which it operates. They are exemplified by what is common in size, current in time, and vehicles of the subject object as language, speech and communication. They determine much of the same relations, and positive forces that converge and reduce all to TRU. **LAMERN** - Law of Medium- in size, Real - in identity, and Now - in time, as the predominant determinant of the form and function of TRU. It is the convergent, compelling, and coordinate dimensional core of TRU.

LEVELS, SCALES, PROPORTION, DISTRIBUTION, BALANCE, EQUALITY & VALUATION of TÍS

Different kinds and uses of levels help to reveal the parts of TRU. Participation leads directly to their share or proportion in TIT and TRU. The contribution of proportion to TRU is of the highest importance as might be expected from VAPO and REM. It is upon these that we can obtain a valid and reliable measure interpreting merit, right and wrong and even death. There are several different kinds of measures of TES, including ESIN/TES, REM, and magnitude of TES. Each person may be 1:1 more or less, but each TES per person is greatly different. The amount of TES in each person varies greatly. There is a correspondence of the form of immortality with the form of TES. That which is everlasting in us is our TES. That which is TES is everlasting and does not disappear at the moment of physical bodily death. **ATES** = Negative, away from, trade off, TES. ATES is where a person trades off parts of their life, fortune, glory and victory in life, their VAMTES for less than TRU. They are fooling no one but themselves. They are buying and selling ATRU, that of no, or even negative worth.

TRU is and is not equally subject and object. This depends upon what we consider this subject and object to be. If they become their constituents in TRU, as in A, then not only are they equally subject and object, they are uniformly so in their set scale. It is disproportional when they digress or disassociate from TRU to subjects and objects, to senses and things differentially and independently. **LEPA** = Law of Equal Proportions of Á, not only are the poles of A equal or 50:50, but the scale is all equal or of set gradations that are amenable to division and scaling. **LEPAS** - Law of Equal Proportions in Divisions and Scaling. These are innate in TIT, which innate state forms the primary coordinate produced by relativity of TRU. This is most fundamental, and a determinant of all else. Consequently it is recognized as being a predominant and preponderate power of our world. **IUTES** - Identity, Union and/or Correspondence of I and TES, Self and TES. Here is the solution to the problem of free will/determinism, and the means and extremes of TES.

LOPS = the Law of Proportional Separation, Law of Set Proportion. From Anaxagoras the make up of a thing, the thing TRU, is a mixture of all things, only in a set proportion for each. This set proportion is the same for this thing at all times and under all circumstances, by TIT. Roscelin is not entirely wrong when he states that when wholes have their parts demonstrated they no longer have entity. This illustrates the Law of: **LIPAT** = Law of Inter-proportions of forms of TRU. LIPAT is the hegemony, autonomy and limitation of the totality and the unity TRU by emphasis upon its parts. It is in the finite makeup of TRU that most explicitly gives enough detail to enable us to know, represent, and present TRU. It shows the value and power of the finite, its distribution and mode of predominance in TRU. **LOPOT** - Law of Total Proportionate, Ordered TRU. LOPOT is the general case of: **LODEPT** - Law of Dynamic Equilibrium Parts and Wholes. LODEPT is highly associated with the degree of partial proportion in TRU.

Kant in *The Critique of Practical Reason*, held that the moral law demands that justice and happiness be proportional to virtue. In a similar way since all is only as it is in TIT, and since all is only as it is relative in TRU, all is proportional to TRU. **LAPIND** = Law of Partiality of the Individual. Both TRU/COL or the REM of an Individual and the TRU/AU, or the Relative amount of TRU in an individual or **RESIN,** are highly variable in more than one way.

Change, especially of ES and TES are not simple, only described singularly to make it more discrete as it actually is particularly. Much change is not even TIT or TRICH, TRUCH, or TECH. It is all of these in dialectic operation by LANERA and LAMOL. This is called: **LACH** = Law of Continual Change. LACH gives us a better picture of what is going on all the time in TIT, TRU, in and out of both TIT and TRU as they actually are, and as well as can be known, shown and presented in many ways. **CIPTIO** = Critical Point of the Individuals Capacity and Control in life, in the dynamic world in which one lives, in TRU. **TESCH** = TES Change, as TRU changes, TRUCH, so must TES more or less change. This is accentuated by society and by changing times in societies.

IPU - Individual Universe Paradox and Problem or the Individual Partial of the necessary Total. **FACH** - the Fallacy of Championing. IPU and FACH are related and express an individuals slide into oblivion. It is how they can or will not rise to the occasion of their life, TES, in TIT, but remain in ATRU and become fixed, as in being lost in the vicious circles of AU, or ANTRU/AMTRU. The individual or partial as oneself does not have, nor is it practical, to be TO. Yet by IPU where the individual is as ATRU/TRU, it must be in TIT both partial and total by necessity. This basic problem is demonstrated in the current battle over ERA, that of equivalent responsibility, and denotes the relative equality of parts of TRU, and what their identity and rights are. **IMT** = Individual Mandatory TRU, or how even little children or the poorest of humans may have their place, role, value and rights in TRU. If one works diligently for their own and equally for all mankind's success they are of a far different TES than one who does not. Such is valid and TRU merit. It is a share or proportion of TRU that returns some measure of benefit. It is like grace, a potential power produced by TRU for its parts.

In order to keep in mind the place and role TE, TES, TÍ, and TÍS have and play in TIT-TO we must remember the frame and media in which this occurs, ACEXIN. For TRU is formed by ALT, and ALT is but the frame of the combined continua of all levels and generations of ALT, ACEXIN. ACEXIN is all the higher powers of mass energy, space, and time as they exist in, and make up the form and function of TRU.

The paradox of the infinite value of the self versus its $1x10^{-10}$ portion in TRU, seems to render the individual of only minute value. It does allow for the proof of the infinite value of the self as well as the infinite and absolute value of TRU and TÂ.(LOPAN) **SPEAT** = Standard Predominance Effect Allowance TRU. **REMI** = Relative Magnitude Individual, or the proportional value to be placed on the individual in TRU. **LOPUT** = The Law of Proportionate TRUTH. The unity - and totality - of TRU is proof that there is only one final truth. Since TRU is conscious, dependent, and relative, knowledge, truth and valuation or REM are set. They are in a state that is fixed within TIT. There can be no actual truth without relation or identity by TIT.

ROLES, MECHANISMS, DYNAMICS, ACEXIN, EQUILIBRIA, DIMENSIONAL FORMULATION of TÍS

The causes, reasons and cures for many of the needs, problems and troubles of man can be found in the oppositions, polarities, continua and how they best function, in the equilibrium of TIT. The predominant mass energy FAF of TRU, or its Á dimension, consists of the multiple specificity of its energy systems whose net effect is the optimal action, equilibria, harmony, welfare, success and survival of the whole. There is an equilibria of concentration and spread, between ES, TES, TÍS and TIT. This equilibria has a vast interface with the actions and reactions, input and output, feedback and intermixture of what is within, in and out, and outside of TRU. This mixture, AMTRU, is the world in which we live. There is an equilibrium between freedom and responsibility. These equilibria are like homeostasis, fine tuned in resultant by the FAF of TRU. When the capacity to maintain this decreases many ills, troubles, loss and ATRU ensue. TRU exists in a dimensional totality which can be shown in different ways. Since the tri-dimensional best replicates a solid, to provide the optimal unity required by the thisness and totality of TIT, it warrants greatest attention. Not all of DET or DETES, the DErivation and Development of and a person's ESINS becoming ES and TES, is incoming. Much is, and often at the same time, outgoing or recurring internally.

Fig 10-15 **INDIVIDUAL and WORLD RELATIONSHIPS**

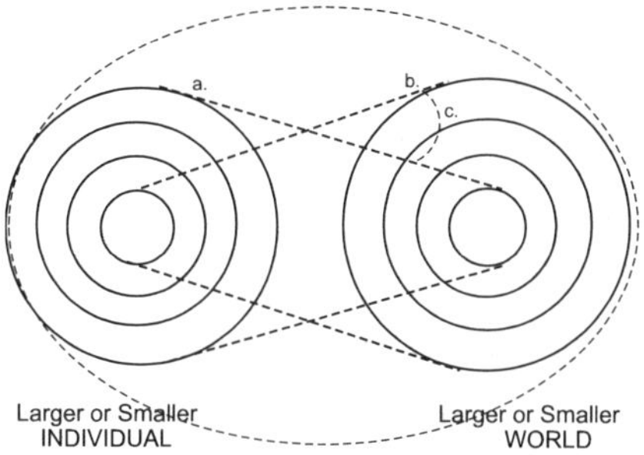

Different sized worlds and individuals show their associated relationships and interactions. These relationships and interactions include, a. the state of a very large individual with a very small world, as seen in dominance, youth, by an egotist or one with over confidence, b. the state of a very small individual and a very large world, as seen in recession, by a pessimist, defeatist, or one without confidence, c. State of the change that occurs between these two extremes or the individual and of the world in the eyes of the individual. This is a continually changing relationship and largely depends upon the vitality of the individual as opposed to how much one is protected from the world. **RESPORT** = Respect, Recognition, Reservation, Resolution, Restitution, Restoration, Restraint, and Restriction, for the Proportion, Portion, Remainder, Relative magnitudes of all TRU that is not a part of the given, partial at hand, first person, or the subject. **ERUC** = Parts, things and Relative Universes Unitary Coincidence, or their operation in unison. ERUC certifies that the operation of things for themselves, in and of themselves, or partially, need not be contrary to a RU or TRU. Or this part need not be in conflict with the whole as long as this operation coincides, is the same, a part, and stands in its special identity or relation to TRU. **ERUCA** = Parts, things, and processes in Unison with other Parts together in TRU. **IPREC** = Interpersonal Relationship Conflict. IPREC is the fallacy resulting from a of lack of ERUCA and ERUC. **SARUE** - Self and TRU, Unity, Equi-dynamics and Equi-dependence.

Fellowship, friendship, and social interaction are only part of the continuous dialectic interchange between the forces and powers of an individual, the unitary TES, other individuals, and other TES, non TRU and TIT. This culminates in the overall fusion and integration in TIT-TO. **TESU** is the Union, bonding, and harmony of the multiple TES in TRU. **SATES** is the Sympathy and Appreciation for TES, like the great bond that exists between Christ and fellow Christians. It is a sympathy and appreciation for all that enables success over failure of all TES and TIT. The struggle for existence is not so much a matter of the immediate life or death of an individual as it is whether we are 99%, 1% or x% of TIT.

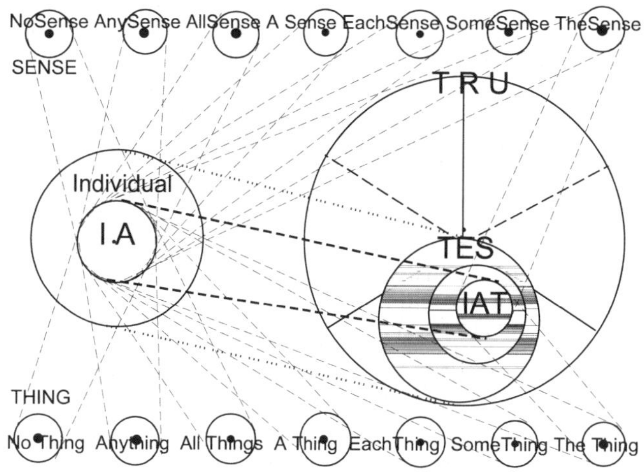

Fig 11-15 **TIND and HETES**

TIND - TRU Individual Interaction Dynamics Diagram. **HETES** -Harmonious Equilibrium Interaction TES, TE and TRU. This diagram shows both the external and internal relations and interactions between the major factors in the world, senses, things, individuals, IA, IAT, TES and TRU. They can be shown in many ways from the outside or inside. Yet this being only one, a sagittal view, gives us a general idea of their relationships and form. An individual's IA is in constant interaction with senses and things as these are brought together. If meaningful, VAPO, and have ESINS that become ES in TRU, this IA becomes IAT. As IAT is enhanced TES and TÍS, that corresponding to the individual, is enhanced. As this is enhanced so does TRU flourish. This occurs as part, E and Í, or parts, ES and ÍS, and the personal parts TES and TÍS interact within TIT and with TRU in many other ways.

LASMEN = Law of Multiple Specificity of Energy Systems in the Á or mass energy continuum in TIT. Á or the predominant mass energy state in TRU is more than the S-O state alone. It is a complex, yet specific mass, typical of TIT and TRIT of which the more specific S-O exemplifies the more general relativity between forces that work as dualities, parameters and dimensions convergent in TIT. It is composed of all those energy systems which make up TRU, man, mankind, the world of man as it optimally functions as a unit.(**ACEXIN**) **LAWLIT** = Law of Warp, or dimensional Alteration, Limited by the IT, ENIT, THIT, SIT, TIT, or universe involved. LAWLIT signifies that for each separate or new universe its dimensions will be predominant, same, common, set and uniform for itself. When compared with other universes or worlds, they will appear warped and even unbelievable. This is in part the result of **LECAS** - Law of Equilibrium of Concentration and Spread or distribution, and **LODEST** - Law of Dynamic Equilibrium ES, TES and TRU, or the state of dynamic homeostasis that exists between TRU and its parts, any part and the parts. **LACOX** = Law of The Axiomatic Conformation or Coaxial Form of TRU and TIT. **LOSUT** = Law of Optimal Simplicity, or Systematic Unity of TRU, or the means by which TRU may be most simply, discretely, basically, efficiently, proportionately and properly formed, ordered and shown.(Parsimony)

TES, TÍS, their continuum, limits and Moderation.

The individual, when TES and TÍS, acts as the dynamo which drives much of TRU. To deny this is to deny much of TRU, as devious, tyrannical and totalitarian leaders and states soon find. This is readily resolved in the optimal balance and understanding, KOT of TROS, that exists in TIT. So that as great as the individual is, as much as TES and TÍS contribute to TIT and TRU they must be in balance with the rest of TRU. There is a constant dialectical struggle the extreme power of self interest versus the world about them, and more actually

TROS and all of TIT. To compensate for this the individual has to overcome all kinds of ÁTRU, be TES, TÍS, TIT and TRU, and must do so with sublimation and supplication. Even TÍS in being TRU, must do so in moderation, optimally adapted to all TIT. This is achieved by the rational exercise of KOT, TIT, VAPO, TRUD, POS, HOM, ACC, PUREACLITY, COGNOS, ELTICMA, SÂ, and TÂ. Likewise success is a function of how well TÍS and TIT are able to maintain and control their TROS FAF. This is of the greatest importance, for the margin of success/failure seen in quality, is very narrow, and must be continuously and massively upheld if the part and whole are to survive and flourish. Nothing is an end in Itself, except TIT.(ATO,TRIA) All less, or other, than TIT is either ATRU or only an aspect, FAF, ES, or CES. There is a level or extent to which TIT must be finite to assure ROT, POT and its continuity, existence and life.

PIPT = Principle of Personal or Individual Permeation throughout TRU, TRIT, THEO, TIT and ALT. PIPT changes the past and alters the future. For as hard as this is to understand by common sense, it is quite evident when we understand TRUD and how Á changes both of CES, TÍS, CES, and OITIO, with its own momentum and power. This is also why the actual human world, TRU, is so much different and seemingly complex than other worlds. The Á state and dimension thereby is much different, unique and hard to understand by many individuals. **LOTES** = Law of TRU Parts, Individuals or TES, or the total level, as love between the divine and synchrony of body parts, that is already more or less within us. **COTES** = Common ES Interdependent position to TRU. COTES reaffirms the interdependence by LOTES in TRU. **CONTIT** = The Concretion of TIT. CONTIT accompanies our expanding knowledge of the world. For if we can not make an object, or make concrete this world, we can never adequately know and show it.

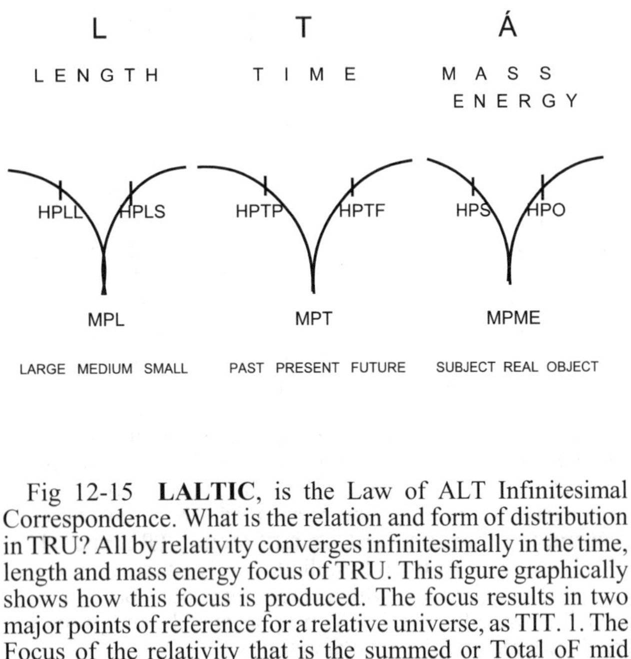

Fig 12-15 **LALTIC**, is the Law of ALT Infinitesimal Correspondence. What is the relation and form of distribution in TRU? All by relativity converges infinitesimally in the time, length and mass energy focus of TRU. This figure graphically shows how this focus is produced. The focus results in two major points of reference for a relative universe, as TIT. 1. The Focus of the relativity that is the summed or Total oF mid points in TIT, is **FOCIT**. 2. The focus of distribution as this relativity is set between its center and its periphery, as in **MP** or the mid point on each dimensional scale. These two points of reference are characterized by: 1. What the average size of the human world is, roughly a meter, the average time of this world, precisely point time present, and the average mass energy of mans world or what is mans most predominant relativity state as occurs in the mid point between observer and observed, subjectivity and objectivity, mental and physical, mind and body. 2. What the half point is on the length scale, or half way between the average size of mans world and its largest and smallest size. What it is for time of half way between the infinite past and point present as well as between point present and infinite future. And what this half- point is between the mid point on the S-O scale and either of its extremities. Once these set points, and associated intermediaries can be established they appropriately, accurately, and reliably replicate and reveal the frame of TRU. We can learn and apply their systematic unity to proceed not only to achieve better understanding of much of TRU, along with better representing and modeling TRU. Once done we can effectively reduce TRU to POA by replicating it in TRIT, OBT, SUT, and ROT in detail in a model. **RAD** - Reduction of Relative Dimensional Formulation to Finitude, **LADIM** - Law of Dimensional Formulation. LADIM denotes the overall state of ALT as they are formed and function together in TRU.

Fig 13-15
DIMENSIONAL RELATIONS, EXTENSION, and INTERACTIONS

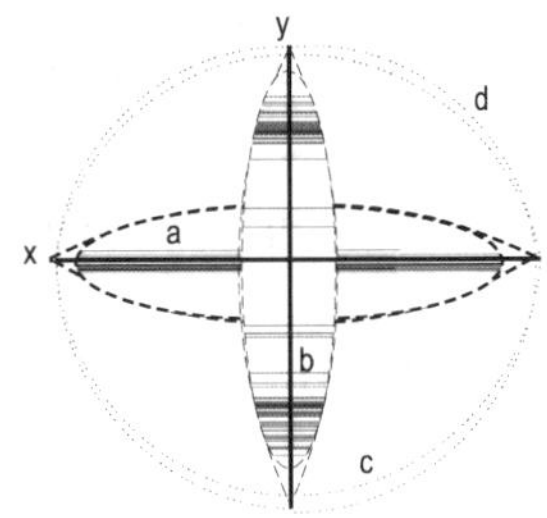

This figure shows the steps that demonstrate how a simple dimension associated with another dimension combine to make a coordinate field or when further combined make a body, or object. It has extension, relationships and interactions as it participates in the whole. This we do by providing: x. The horizontal axis, y. The vertical axis, a. How the horizontal axis becomes a plane, b. How the vertical axis becomes a plane, c. The horizontal axis permeates the whole, and d. The vertical axis permeates the whole. The importance of this diagram is multiple, yet its major thrust is in showing how dimensions are not extremes, points or even only lines but are planes, spectrums, continua of many kinds and modes that make up much of TRU.

ENABLEMENT, FREQUENCY, MASS COMPLEXES, VITALITY & SUCCESS of TÍS and TES

The complexes of TE and TÍ, from frequency through mass to solidity, enable us to posit ES and CES, TES and TÍS in the overall frame of, and thereby better know and show, TIT and TRU. Difference or particularization is limited, yet this limitation is a means to POA, control, and benefits to both part and whole. This involves a total identity with TIT, and an attraction and bonding between parts and the whole. It produces deeper conceptions that provide the intricate and beautiful detail these parts are characterized by. There is a vital and necessary level of adequacy and sufficiency for each part as there is TIT. The complexity of these parts is not only beneficial to TRU, it can be its failure. This is why it must be accounted for, managed and controlled. As the frequency of the parts and whole become its mass, so do the gradations and coordination of its dimensions fuse into the mass of TRU. This gives TRU unity, solidity and sphericity. The vitality and health of its parts as gradients and coordinates in the dimensions of TRU are what largely determine its survival. Most all that is not of the spaciotemporal is ultimately Á or within the sphere or SCARL of Á. The Frequency of a thing becomes its MASS, is a well established physical law, **v = m**. In time, as an event repeats itself more and more, these events, the amount of variation per time becomes a mass. Even more it seems so to the observer. So is there a separate, and new state, created by the relationship, a certain point in time, a mass dimension. **FOSA** = Frequency of Predominant FAF of TRU, becomes its mass, Á, in TRU. This is additional proof of both A and TRU. It applies in different ways. It is the level of acuity or threshold of TE. Human acuity that can separate finite parts before they appear as one in continuity, one mass. This varies, and is greatly increased by familiarity, interest, and comprehension and KOT in TRU. This applies to MOT and SPALT in: **LASPHERT** or Law of Sphericity, for TRU. How the form of TRU becomes more and more spherical as its density, compactness, solidity, mass and Á increase. **MASSEN** = Mass Energy. The significance of MASSEN is how it stands between the physical mass and the special unique mass energy, Á of TRU. How it operates in the dimensional formulation of TRU. In this way that MASSEN is fundamental to form, objectivity and understanding. It emphasizes the TRU FAF of mass that has and can not be understood unless known by higher powers, their levels, gradients, SÁ,, and organization in TIT.

ROL = Respect for Life, One's own, all life, TRU. This is qualitative and quantitative. Degrees or values of ROL may be used as measures. ROL denotes the basic, oriented, mental, and emotive force to uphold life and TRU. It is common and easy for individuals, especially when they lose a predominance, are excessively limited, and have diminished zest for life. It is easy for those who have received much to forget what they have been given and to abuse and lose it. It is only as all person become TES, have interest, inspiration and revelation in TIT that they best use and exercise life by ROL. The first and foremost task before us, as we should well learn, is to bring all IA and oneself to IAT and TES and these to TIT and TÂ. It is not possible to do this in one thought or one act here and now. It is quite possible to do it in multiple increments toward total identity with TRU. Whereas the future can exist only as we are, work and live and more specifically as we are TRU, so is the ultimate value of the individual to be found in a person's fulfillment of their counterpart in TRU. Salvation, conversion, conviction and dedication from DIA and SIA toward TES and TIT, produce ENT, OM, Pureaclity, Cognos, ELTICMA, SA and TA.

TÍ SUMMARY; PROBLEMS, USES, CORRECTIONS,

To complete and summarize the parts of TRU, ES and ÍS, and their personal or special correspondents, TES and TÍS, we bring together their corrections, uses and show their unfolding manifold, the exponential to infinite state in and of TIT.The world about us provides much of the simple answers and certainties needed. Yet we must always look at these critically, profoundly, selectively and reduce them to TRU.(VAPO) Each person should learn and remember what they actually are and what they as their IA, can do, by, of, for, and in TES and TRU. There is a great need for the individual to continuously correct themselves in their part and role in TRU. One of the most important of these is a realization not os any senses or things, as ENIT, but of from ESINS of ES and CES and TIT-TO. Much is explained by the dynamics and conflicts between these, oneself, and TRU. What is negative or positive, how much and in what way. The exclusion of all not TRU requires the best use and applications of ES and TRU, as each person actually and fully attend to all of their ÁLT FAF or ACEXIN. Which from the infinitesimal to the infinite we can with TES increasingly control finite amounts. Since the infinite, infinitesimal, and indefinite finites. are so much of TRU, it is mostly of its parts that the ideal and practical resolve into understanding, management and control. With greater identity and fulfillment of our counterparts and roles in TRU we can realize immortality in our absolute contribution and presence in the absolute TRU and TIT, TÂ.

LAPIA - the Law of Particular Individual Appropriation,, or how IA must be set in and of an increasing Á, L, and T. **LEXUS** = Law of Exclusion, Universal and Simultaneous, or Law of Simultaneous Universal Exclusion. As a person, persons, or all mankind become their TES and thereby TRU and as greater and greater ATRU, ANTRU, ANTITRU are eliminated and all expands in A, L, T in both and all directions, so do all parts of TRU expand. It is by being prepared through anticipation, knowledge, readiness for action. LAPIA and LEXUS that TES best contributes to TRU.

TIEA = Terms, nominals, words, concepts, Intension, Extension and Actualization, and all are to be taken in their deepest, most essential and complete ACEXIN TRU sense. This is largely the use of VAPO, DAOST, and SCARLS in language and how much for all words and concepts. The correction each of us must make to become TES or TÍS in TRU and TIT is evident in how we are changed to an improved permanent state, changed temporarily, are in a dynamic relationship with much that is less TRU, are a decreasing part of TRU, or are other than TRU. **PIAIT** - Precision and Perfection of IA in TRU, or optimal S-O, Á FAF. This is seen in diagnosis, treatment, surgery, and the performance of many arts, sciences, and technologies. These are derived from immediate awareness in the totality of TRU, **IAITO.** Once established it is not necessary to go through the analysis of TIT with each and every thought, view, or act.

RANUD = Recognition and Avoidance of the Necessary and Unnecessary Differences in TIT-TO, and all of TRU. RANUD is, as in SAO and ESIN in STRUT, a basic factor and feature of much of mankind's inadequacy and suffering. **STRUT** is Self and TRU, ATRU TRU from the Self. Only as ES are known, separated and combined properly, fully and proportionately in all TRU can we come to understand, live, work and act in ways in which there is success. The individual as a thing-in-itself **INDIT** must come to understand and act by ES, properly, proportionately and fully, or: **SET** = Self ESIN TRU. That this can be done adequately is obvious only after it is being done and we can recognize it in retrospect. Recognition in retrospect is very important in TRU, for what can not often be learned directly will be vividly recognized in many other ways, especially in retrospect when experience gives the necessary added TRU meaning.(POA)

ACEXIN imparts to TRU many typical, unique, novel and massive modes, qualities, essences, operations and powers. L_3T_3, $A_3L_3T_3$, and ACEXIN are a great part and guide to understanding TRU at those levels of complexity that are necessary to fully replicate TIT. **POCTE** = Principle of the Clarity and Characteristics of the Extant, or TRU Parts. POCTE signifies that for each age the clarity of their times are most evident, and parallel their intense experience and purposes. Yet all parts, functions, and totality of TRU are similarly existent, actual, and function in totality.

PINI = Paradox of Infinite and Infinitesimal results from many causes including its being considered opposite from the finite, and our mistaken idea and naive identification of two opposing, yet similar aspects of TRU.(DAOST) The infinitesimal TRU is no less a part of us than the infinite of TRU. Both are of great significance as TROS and features of ALT. It is in the TRU finite, that comes together in unity, that establishes the preeminent mode of TRU. How they can be reduced as quanta for POS and POA. The infinite warping of future time and frame is highly compelling for man and TRU. It is the drawing out, determination, and representation of processes that are typical of media and extremes of TRU that the future adds to and is key to understanding, KOT and TRU.

Only as a person comes to recognize oneself as a partial of TRU, ES, TES, with its special kind of immortality, will, sacrifice, and work for each and all in TIT, can they find actual survival in life. This is how success equals expansion of the actual finite, TE, E, ES, and CES. This is seen in its opposite, mortality through loss of TRU meaning, or pure Á. Even small departures in meaning, from their actual or VAPO implications in TRU, can be erroneous or deadly. They will in departure from TRU produce false actions or effects which in turn produce reaction that becomes increasingly negative. Their prevention or reversal is much of our immortality. **PODIM** - Principle of Differential Immortality. All of TRU, when posited and evaluated by POS and REM in SPALT, are of different kinds of immortality. All that works through TRU that is a part of individual's lives, as well as being a part of TRU otherwise is of PODIM.

CHAPTER 16

THE WHOLE or TOTALITY of TRU, TO, INTEGRAL TRU

DEFINITION, CONTENTS, PRECEDENT, KNOWLEDGE, PROBLEMS, VALIDITY and PROOF of TO

Validity, and precision of measurement, are best achieved in totality, TO, especially as affirmed by its great value and use in ROT and POT. The species homo sapiens, humanity and civilization, function as a whole. They have increasingly become one world. **UNIP** - Unity to Infinity and Infinitesimal in equal, Proper, Proportional and ordered measure. That law which defines the relationship between all that is completely TRU and all that is less than, although often greatly of TRU, and the manner in which it is less than TRU as well as its difference from TRU, may be added or made whole, TIT-TO, is: **LATO** = the Law of Totality. The special nature of the totality of The Relative Universe, TRU, is shown by: **PROEST** or Proof of Extra Speciation Totality. PROEST is how the unity and reality of knowledge and civilization are composed of increasingly higher states of mass energy, SÂ, and provide TRU with a far greater diversity of form and composition. It shows how increasing capacity, self perpetuation, power, and total convergence of all upon TO enables and assures the haecceity of TRU and TIT.

The specie is reproductively exclusive and human knowledge converges on a unified human focus for survival. Kant dismissed all prior proofs for the existence of God and considered that the ideal of reason was the only valid proof in the whole reality. Its regulative use is a necessary postulate. Hegel proceeded from Kant to show that this all comprehending reality, the Absolute, can be formulated upon a system of thesis, antithesis and synthesis. How the ideal could through a constructionist system of theses, their opposites, and their combined resultants in aggregate, produce a thoroughgoing understanding of the whole or the Absolute. It was an all comprehensive system, The Absolute Spirit. This construction is based primarily on dialectical negation, elimination of the undenominational. As opposed to TRU, the philosophical, realistic, and nature of Hegel's system was pointed out by Avenarius, with the missing element being called Essential Coordination. This is the essential solidarity existing between the knowing subject and the object of knowledge. We must discover, learn, know, think and show TO as the solid, ordered, unified, dynamic, finite, infinite, and absolute entity, object and all important universe that it is. To do so we must learn and remember what is the greatest of its form and function. We must remember how it is composed of its three prime dimensions, their higher powers, forms and continua, signified by ACEXIN. In our great concentration on the predominant relativity, mass energy Á FAF, state, dimension and continuum in TIT, we often fail to give equal and adequate attention to, and concentration on, its length and time dimensions. How they exist as continua, and together with their added FAF in totality, in the TO of TRU. When we remember how the length, area and volume components, that produce the spatial frame and structure of TRU, are altered, adapted, and even warped in its special way typical of TRU, and its temporal continuum of time, duration and variation are formed and function in a like manner, we can begin to understand how TO is formed and what it is made of. When these are combined with the full continuum of Á, we gain that solid, ordered, unified, dynamic, finite, infinite and absolute entity, object and universe, that best reveals TO and TIT. It is only when we think of and make TIT-TO the object, and represent and model its ALT to produce SPALT, that we can provide for TO the kind of meaning it warrants. To add to this meaning and to fulfill it as it actually exists in TIT, all possible REM, POS, SPA, HOM, and ACC are provided.

PAK - Principle of Partial Knowledge, or the knowledge that comes from what is less than, to even minutely of, the whole of an entity or TRU. PAK is the state the world is in today, why it is so fragmented, uncoordinated, and disordered. **FAFO** - Fallacy of the Fragmented Whole. The part, individual, groups, and all less than the whole become and remain wrongly to much of themselves. **LOVIT** - Law of Validity in Totality, or how the actual validity of all, TRU, proceeds from its totality, as all its FAF, ES and CES are discovered and used.(VAPO). LOVIT is supported by: **PROET** - Proof by Parts and Wholes. **SUROT** - Summative Unitary Reference Relative of TRU, or the exclusive qualification of all only as it is exists in TIT. The ESINS of selves, knowledge, basic genetic, biological and natural laws enable us to be and assure preservation rather than extinction. **MADEC** = Measure(s) of Depth Conception, or the achievement of ever higher levels and measures of detail brought to a whole in TO of TRU. To combine the relations between different observer-observed and same observer-observed led to Einstein's unified field theory, **UFT**. Since geometry was largely a product of a set observer-observer system it could not be produced until such an object could be established.(TRU,TIT,TRIT,OBT) They provide the actual geometry, and 3D frame, that UFT must have. It leaves out all those entities, objects, and AU, which are not, or less than, their part in TRU. **TIR** or Truth in Reversion is that state in which such non or less than TRU senses and perspective are reverted back into UFT, OBT, TRIT, and TIT. This is where truth can not be attained until the reversion of the facts, paradoxes, states and conditions, SCARLS which create these problems, questions, truth situation, or SCARL are shown. **COMPRUT** - Composite Proof of TRU. If TRU consists, not of a single exclusive static homogeneous entity, but of a composite of many varieties, forms, states, and parts, and these are highly dynamic and changing, then to prove TRU we can not do it singularly, statically, nor exclusively. The proof of TRU must be manifold, inclusive, dynamic and set forth a comprehensive demonstration of all that is most TRU. It must also be relative or account for the focal or centering of TRU.

DERIVATION, DEVELOPMENT, DETERMINATION, and DEMONSTRATION of TO

Holism is a term that defines wholes as more than the sum of their parts, and has characteristics that cannot be explained by the nature of their parts taken separately. Organicism is that type of holism, as in life FAF, which is based on the analogy of complex systems. Holism describes science and thereby the world, TRU, as an interconnected system which is adjusted by experience, i.e. IA or **IA → OIA → OIAT,** or the Total IA of TRU. **FER** = Form ESIN REM, or both the mechanism whereby Form, ESIN and REM are ES and how these together are of SUT, ROT and MOT. Like TIT and TRU, TO arises from the thisness and existence, ATO, of TIT. Consciousness, meaning and common sense tend to hold TRU together in totality. The modification of human evolution shows TO, that of TRU shows evolution in its actual, TRU, form. The derivation and development of TO, **DETO**, is guided by many principles including the separation and fusion of parts of wholes, changes increasing and decreasing in TO, and sufficient differentiation. The ubiquitous causation in a closed system as TIT is valid only when typically and specially qualified by TRU. This is guided by mutual expansion and proportionate distribution. DETO is guided by survival and sufficiency, and as for TRU depends upon special Á,L,T FAF.

Fig 1-16 **Á MASS ENERGY and DEVELOPMENT of TO**

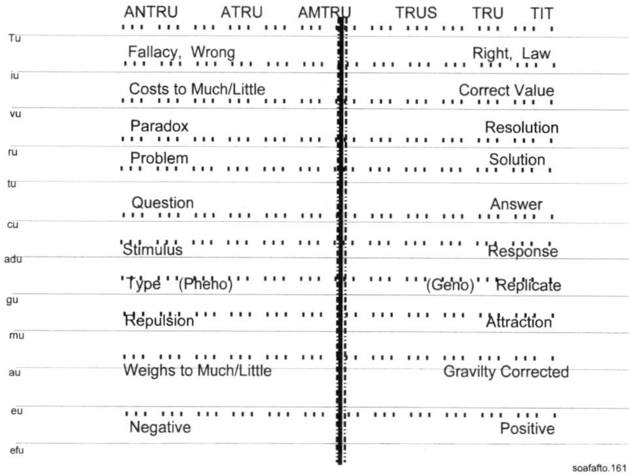

The development of the totality of TIT, DETO, can be seen from their makeup and constituents, especially in the continua represented by dimensions. When mass, length and time are considered together along with their special, typical, unique, higher powers, and exclusive FAF in TRU, they provide an excellent picture of the development of TO, TIT and TRU. The parameters that run across Á are not formally refined, they are represented to provide an optimal impression of the mass form and content of TRU. Figure 1-16 provides a general representation of what the predominant mass energy dimension and continuum consists of, especially emphasizing those parameters of life that are most typical of the various gradients of this continuum. It helps to resolve the problem of identifying Á with subject-object without showing what, how, and why S-O exists in Á. In these various forms and functions, relationships and dialectical operations there is a subject and an object which has a different form and function at each level of the overall continuum. It is the mode and characteristics of these different forms and function of subject and object that need to be understood if we are to recognize and best apply the part and role subject object play in Á and TRU. For example the subject and object of a negative and positive relationship in subatomic particles, is different than in genetics, language, reason, or in the THES and THEO of TRU, yet they are all aspects of the same overall subject and object, the Á dimension and continuum, DAOST in TIT. This is called: **GASOV** = General and Special Subject Object Varieties in Á. It is by way of GASOV that we can make the great revelation of relativity in TRU and how the observer-observed relationship, as subject and object, provides the key to the fundamental and predominant FAF mass energy has in the TRU, KOT, DOT, ATO, TÄ, TIT and TO. Although Figure 1-16 only shows some of the ingredients and factors in the Á dimension and continuum, it when combined with the length and time dimensions and continua and especially when in their higher powers in TRU, provided by ACEXIN, a good showing of TO and how it is derived, formed and functions is achieved..

EXMEV - Exponential Modification of Evolution. Evolution is not only exponentially modified in the past, it is modified by the warp or geometry of the present and future. This is largely a product of being in, rather than outside of TIT. **GAMEV** - Geometric form Modification of Evolution, or the altered status of evolution by the geometry of TRU. Evolution as it exists in the physical and biological worlds, must be modified as conditioned by TRU and VAPO. In a closed relative system, such as TIT, all things, are ultimately caused, more or less, by all things. Or there is a comprehensive state of integrative negation or exclusion of all not TIT. This is called: **LUC** = Law of Ultimate Causation. However LUC is not to be interpreted unconditionally, mor without its necessary corollaries, SCARLS, and VAPO. It is completely valid only for TIT-TO, and then within LANERA. **SAFPO** = Separation and Fusion of Parts of Wholes. Two or more parts of a whole may be entirely separated, both separated and fused, or integrated entirely. Partials or differentials may be differentiated to Unity. **LO** = Law of Object, **LASD** = Law of Sufficient Differentiation, **TOD** = Totality Deficit, or how much a whole is without some of its aspects, form, function, parts, or processes. **LAME** = the Law of Mutual Expansion, where one extreme grows as the other extremes grow. A sufficient rate of integration and differentiation is key to survival. It is only by reversing our perspective from the separation and singularity of mind and matter or subjective and objective to the totality and perspectives of all TRU, as by SPALT and SPAT that we can understand fully the perennial problem of mind and matter. **TEP** = The Finite Predicament, or how we are lost in subjective and/or objective differentiation, when all must be **SOID**, or Subject-Object Integral-Differential, in TIT-TO.

GENERAL FORM of TO, UNIVERSALS, MODES, ADDED STATES, CONTINUITY, CAPABILITY

Because all is determined by, and must be ultimately referred to, TIT-TO does not mean that the whole can't be presented in different levels of fullness or completion. The gap between an individual, their IA, and TO-TIT is best seen in socialization, cultural principles, ideals, and all that represents the multiple, complex, and dynamic highest levels of mass energy convergent on TRU. That the whole is more than the sum of values of its parts has been well known by the principle of organic unities. Not only are integration and differentiation necessary in life, TRU, VAPO, DAOST, SCARLS and TROS must be optimally unified and maintained. This is further demanded by life in the modern world of technology, communication, meaning, coherence, and precision.

Why is synthesis so close to the synthetic in etymology yet so far apart in usage. The one highly positive, the other often negative, thus so incongruent. For as we seek sameness we tend to lose otherness or difference. Here we can realize the continuum that exists by DAOST and **ID** or the Integral Differential. There is a great continuum between extreme difference and extreme sameness that corresponds, to a great extent, with the great continuum between integral and differential. As we move toward one pole we tend to lose balance by a decreased tendency toward the other pole. The equilibrium - an essential aspect and mode of TRU - is upset and lost, at least temporarily or reversibly. This is **LEDOST** or Law of Lost - or recovered - Equilibrium of the DAOST continuum. LEDOST can also be as well applied to ID.

It is to the mechanisms, principles, answers to questions, solutions to problems of human life that we must look if we are to adequately understand KOT and TRU. It is like genetics whereby that viewed is not of isolated phenomena, but the highly adapted, complex, modified, qualified, dynamic, and complex genetics in TIT. This is further evident in population genetics when fully developed and adapted to TRU. It is resolved with the aid of qualified refinements in statistical theory using computers and spread sheets. The relations between nuclear genes and cytoplasmic proteins show us how infection, toxic and many various detrimental, irreversible and reversible, factors operate in the life and death of the individual, the survival or extinction of a group, and the enhancement or detriment of the species. The mechanism and role of natural selection are subtle, profound and complex. Yet it is nonetheless actual and vital. Each increment or decrement sooner or later makes its effect in **E S/F** or energy success/failure, TRUD and TIT. These may not be seen on the individual, group or even total level less than TO. It is through the progeny, the accumulative effects of many generations in summation, that take on added and most significant characteristics, qualities, and essences, that enable and show TRU. Their SCARLS, operations, mechanisms, complex concepts, principles, and laws when collectively interpreted in TIT is the way to understand the contribution

Fig 2-16 **PUC, POV, and PROCAD**

PUC - the Principle of Ultimate Coincidence. PUC is shown by the fact that a person may do as they please, to an extent or for a while, but sooner or later will have to do that which is necessary, a return to harmony with the world around them. We must return to equilibrium or coincidence with our counterpart, TES, TÍS or being in TRU. This is a simple limitation by, and correction to relativity, the relativity of that of which we are a part, TES of TIT. **POV** - the Principle of Viability, or only proportional to TRU, and how we live, work, and do for TRU dynamically and interdependently all our lives, in the collective and cooperative essence of TIT-TO. **PROCAD** - Principle and Proof of Continuity and Duration. Man is primarily a mixture, often more 1/TRU and AU than TRU. TRU/ATRU, TIT and TES can coexist only in so far as there is a great TRU past, TRU future as well as present. In TIT future, TRIF, and TIT past, TRIP, are as much of TRU and TIT as TIT present or now, TRIN..

Synthesis is the combination of separate forms into more complex ones, or into a whole. Syntagma is the more dynamic term that denotes the assembled, systematized, ordered wholes of life. It signifies TRU's tendencies. Syntagma is associated with such views as aestheticism, naturalism, organicism and intellectualism, all of which depend heavily upon the composite FAF of TRU. *Forma Corporealis* is that form of the whole body of an entity.

LOSU = Superimposition of Universals, or the accumulation and positioning of as many of the universals of TRU as possible in accordance with the form and structure of TRU and their relationships and interactions with each other. LOSU indicates how we can better depict TIT in actuality. Multimodality helps to visualize and actualize how TIT-TO and its form exist, yet to subjectivize, realize and show them quite another thing. The many modes and characteristics of GOD, TREQ, and TRU show the extent to which God, as TRU, when fully developed, understood and used, provide the basis for a perfected religion, TRU. However these are less used with all else that is of, and as qualified only, for TO. Their significance and combined greatness as ESINS in TREQ and TRU needs to be remembered and applied.

The transposition of mlt to ALT for TRU is **SALT**, the Specification and Adaptation Law of ALT. **LACEXIN** - Law of A, L, T, or A^n, L^n, T^n, or Actuation, Extension and Intension, or the absolute necessity of higher powers of ALT. The simplest example of the content of TO-TIT is all the great concepts, principles, laws, and mechanisms, of TRU that signify the major aspects, modes, and attributes of TRU. It is especially seen in those NEC that prove TRU, TIT, TO, TRIA, and Á. TRU can be neither adequately known, formed, function, nor represented unless and to the degree it is proportionate. There is hardly anything that is more repulsive than indications of lack of proportion, coordination, integrity, capability, command or not knowing what one is doing. This is accelerated to the degree such evidence of ineptitude reflects or is apt to affect them. For, and only by TREQ or TRU can capability and integrity have full and actual meaning. This is also proof of how the world is actually only TRU. It is the necessity of developing and practicing to the utmost purity, actuality, and reality. This requires, **LASOSO** = Law of what is so equals S-O, or the Á form, state and dimension of TRU. LASOSO denotes the refinement that comes with capability and integrity which is a prime ingredient in the precision and finite differentiation of TRU. It is in the Coincidence of Opposites that we can see the set FAF, hard construct, and solidity of TRU. How that which may be completely opposite and contradictory in and of themselves, is when TRU coordinated and in complete harmony. The parts and processes in TROS, TU- TO, of TRU, are coincident, and fundamental role of genetics. This along with the whole structure, FAF, that are uniquely raised in TIT-TO.

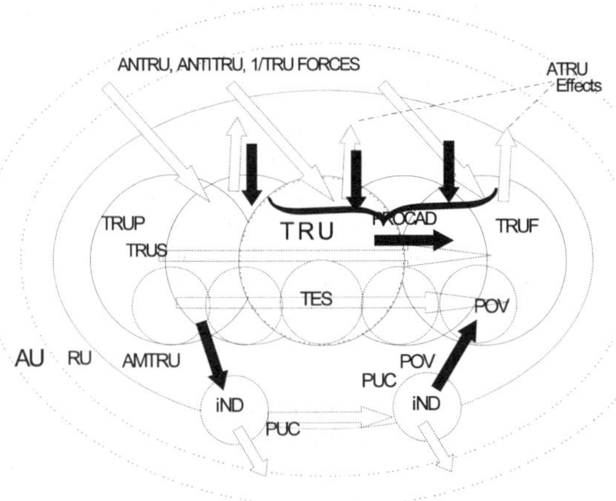

reconcilable, and produce the absolute bonds that Á, L and T, as ALT demonstrate. It is how they exist as ACEXIN in TO that enables us, with many other opposites, to reconcile all that exists in TO, TIT and TRU.

SPECIAL FORM of TO, FUSION, INTERNAL RELATIONS, INTERDEPENDENCE and TIT-TO

What is produced when all fuses interdependently with internal relations in TIT-TO forms a unified totality typical and determinative of TRU. It helps us visualize how TRU is absolute with reliable and proven truths. This we accomplish by recognizing what TO is of itself as compared with the deficit of things and ES. Compounds have separate properties and inherence in themselves and in TRU. Proof by speciation brings to its universe, through relativity of its mass energy, the special form of TIT-TO. When complete this form of total actuality is TÄP, TIT and TRIT. Much of TRU depends upon this special form and becomes evermore certain as all is collected throughout its presentation. It is shown in the disunity of non totals, solidification of TIT, summation of TES, and ambiance of TO. As universals, essences, ES, TÍS, CES and TES exist and function together in TO they show its internal relations and key built-in mechanisms which make up much of TIT. Their integration and interdependence produce the added states in totality which limit, yet give unity to, the whole, TO of TRU. The mentality of humanity, and humanity of mentality, converge upon unity as it encompasses the whole providing additional safeguards for survival and against extinction. This is, as the whole is, a total relative and its comprehension is the absolute idea or revelation. Those properties and characteristics that show the nature of TO resulting from its total special and proper aspects and form include: **EDI** = Reals - E - Deficit, the amount or percentage of a sector of TRU that is missing and whose addition shows the whole, **COMPS** = Compounds have separate properties, **PROEST** = Proof of Extra Speciation Totality, and **PROT**, the a Priori Relations of TRU, or the unity of reality, knowledge, civilization, and IMU. By **EST** or ES and TRU, TRU is provided with a far greater thisness and being. This is far greater than either what other animals and species have and also far greater than what TRU may have by reason of its form, function, parts or other aspects in themselves. This is: **PRAST** = Proof of the Extra Speciation Relativity or Mass Energy, Á Totality of TRU. PRAST is the corollary of PROEST in which it is limited to the development and reasons for actuation of the added states and powers in totality of TIT. Ambience is the environment, mood, character, quality, tone, and particularly of a milieu. This describes another important aspect and mode of the nature of TO-TIT and TRU. **DITRE** = the Difference, Differentiation and Distinction, between TREQ, TRU, and

TIT, how they are REalized or REcognized, **TRE**, and how REpresented and pREsented.

STES = All TES or Summation of TES. **LACON** = Law of Conformity, or how the same and relative produce conformity which becomes the form of TRU and in turn enables its ROT. TRU, as any entity is mass, size, and time dependent, and each in a different - their own - way. It is also dependent upon other forms, e.g. how the increments of ES and CES make up the larger, so that the total size is a function of the success/failure of its inner aspects, their parts and processes.

POLIW = Principle of Limited Whole. Even when we imply something greater than a whole as when we are contradicting ourselves, or at best add another and thereby different entity, thing, given or sense, we are exceeding the whole. This is **PAT** or the Paradox of TRU/TIT. PAT is resolved, managed, or controlled by POLIW. Also, by: **TRUCA** = TRU Common Additive, or where only a rough sameness of all can be comprehended. Only when taken as the final complete common collective additive, can we speak of all, with finality. In accepting the concept of a whole, TIT-TO of TRU, one must accept their LOLS. This is where we start from a standard. **LAVIT** = Law of Validity in Totality. **LOFAS** = Law of Fusion and Separation, seen in the combined mix of TO and TIT.

A proof of TRU is what arises from unity and the unity that exists in totality. In the third metaphysical argument Kant states that we can imagine only single space. If we speak of spaces we mean only parts of one and the same unique space. These parts cannot precede the whole as its parts but can only be thought as in it. If we have more than one space, then, to be meaningful and comprehended they must be combined as related. In this relating they become, at least in essential ways, united and a single totality. It is such a single unitary totality of all, by the higher powers of ALT, vitally functioning together in TRU, that makes it possible to not only represent and modulate TRU, but to prove TRU as it exists in totality. **LOSUR** = Law of Systematic or Summative Unity Relative, or how absolute PRUT is in the best reproduction of TRU..

POBIM - Principle of Built In Mechanisms, or the well known, less known, and unknown mechanisms that constitute a basis for TO. Mechanisms become the essences which together reveal TO. Thereby if we want to know all, as we reasonably must, we have to comprehend and explain All.

MÉATO - Measure of TRU. MEATO signifies all measures of TRU with special emphasis upon all that conforms to the totality in TRU, TIT-TO. This is largely provided for by VAPO. It enables us to find and apply other measures and how we can accurately delineate all of TRU into set, fixed and proportionate units and totality. To have measure, one must have a set basis or standard. This can't be from zero nor can it be from infinity without a focus, without the proper basis for its relativity. This for the focus of TIT, if it is to be, must be its unity in totality. If one would ever doubt the unity in totality and proper DAOST, IA and TROS of God, TRU or this special universe of best human life, let them investigate with honesty and vigorous intent the proofs of TRU. Here their interdependence combined with distinction become self evident.(PRUTROS)

Interdependence is recognized as fundamental and essential to human success. It is largely derived from the integrative processes of TRU or: **PIP** = Proof of Integrative Processes. There is a massive continuous state of increasing or decreasing survival or extinction going on at all moments and durations. Each increment of our lives and all that constitute TRU, its dynamics and its processes, have more or less, positive or negative factors and forces in operation in the dynamic massive states of TRUD and TRIS. The infinite and infinitesimal TIT, or TRIN and TRINI, are forms of TRIS of TIT-TO. They are necessitated by the fact that when we seek to focus on TIT and TO by TRIS we must decide upon what the basis for time is. It is this very fact that TIT already encompasses time yet must have relatively of set time to give it optimal set form. There are reverse meanings of TRIN and TRINI be they of TRU generally or TIT most specifically in TRIS. **TRINS** = TRIS absolute and Immediate eternal Survival. TRINS, ideally, is the exponential convergence of TRIS, TRIT, TIT and TRU on the absolute. It is the positive combination of the infinitesimal and infinite, TRINI and TRIN, providing for absolute thisness.

By ◎ we mean an absolute totality, not, and opposite to, zero or "0". In this case we have: **ONITO** = Total Unity or Integration In and Of itself, as TO in TRU, or the efficient and perfect pervasive surveillance by I and its total efficient and perfect fusion with TIT, as they both, become an absolute totality, ONITO. **TIR** or Truth in Reversion. Truth can not be attained until there is a reversion of the facts, paradoxes, and SCARLS which create these problems and questions or the truth situation. The fact that nothing can be really or completely true unless it is about reality, TRU or OITIO, has been discovered and was promoted by Hegel, who called it the absolute idea. A state in which reality, world or all is true only as it is a whole incorporating the idea, the absolute idea of it.

PARTS of TO, COMPOSITION, COMPOSSIBILITY, CONRIT, CONVET, PARTICIPATION

Through convergence and divergence, induction and deduction we can selectively direct all knowledge to fulfill TO. This is through all compossibles as they provide the fullness and proportion of participation in TRU. Through key built in mechanisms, essences, ES and CES we can provide the substance and concretion needed to know and comprehend all as they are TRU. Through the scales and measures of the dimensions of mass length and time, as altered by variations in relativity in TRU, we can adapt parts to provide the intricate detail that accurately fills the construct of TRU. A great problem and conflict in history seen in Socrates, Christ and many others has been the four way struggle between individual, TES, society, and governments of low or anti TRU and TIT. It is seen in all, especially those who have risen beyond their common roles, to be variably condemned, rewarded or honored. What is one's duty or dedication to society, nation or universe? What is the duty of society or the universe to a person? In the great mass of TRU there are many opposites, dualities, parameters, continua, balances, and equilibria, of which that between its parts and the whole is most vital. Their explanations proceed from knowledge of TO.

To achieve a summation and continuity in TIT-TO and TRU we seek the optimum of simplicity that systematic unity provides. This is in tandem with an optimum of technical proficiency, optimum of productivity, efficiency and effectiveness that are the ingredients of the successful function of a whole. These are signified by **CONRIT** or the Convergence on TRU that occurs from its unity, SU and greatest aspects, and **CONVET** or the Convergence on TRU that occurs from its, ES or parts. Life is a constant dynamic, surging and waning state of being in the midst of the unity to infinity, CONRIT and CONVET in the totality of TRU. Ones success and the success of our entire world, TRU, depend upon the degree to which we can master, control or live positively in this state of dynamic equilibrium.

Leibniz thought it part of God's goodness to create as full a universe as possible. The actual world, then would consist of the largest group of compossibles. Since compossibles are all those parts in a whole which are able or possible to coexist with another, with others and the whole, we find here another beautiful example of the differential nature of God or TRU, of TE in TO. Yet this is not the compossibles of any or all entities or worlds, but the goodness of TRU, by the massive finite to infinite compossibles of and in TRU. Inherence is most typically a characteristic of TAP of TO, in being innate, holding itself within, and being permanently existing. It is defined as being the relationship of a quality to a substance or subject. In this way we can visualize the matrix of TÄ by which inherence is a corollary of thisness, form and structure. Syncretism is a typical kind and mechanism of TO that demonstrates the fusion of incompatible, seemingly incompatible, or conflicting elements. **LOLT** - Law of Length Time. The significance of LOLT depends upon the total modular order and standard of TRU. Point time, like the singular stimulus response is not reversible. Yet time in general, as stimulus-response in general is reversible. The impossibility of altering the past does not come from irreversible time. It will come from the construct of TRU, its time, dimensional formulation, and stimulus response in general of TRU. This is from Theory of Constructs. We know that although time, as understood by the individual is linear, its concomitant change, variation, as well as frame in relativity, may be altered. This depends upon the observer. The resolution of this is: **RETCO** = Resolution of Reversibility of Time Constructs, or how we construct our system from dimensions that replicate the dimensional frame of TIT. The determination of the structure of TRU, is seen in human molecular composition, and this from the chemical makeup of DNA, that gives a structural and formative rendering of how the building blocks of human makeup are demonstrated. By knowing the structure of key ingredients in our makeup we can more readily infer function, practical rules, and the bloc universe that so mimics TIT-TO.

The PARTS of TO, CORPOREALITY, CONCRETION, SOLIDITY, ECONOMY of SIZE, SYNCRETION

The role of the parts in the totality of TRU largely depend upon their contribution to the integration, strength and success of the whole. This is accomplished as the body of TRU becomes concrete, using economy of size, synthesis and consolidation to form a concrete whole. It is how they contribute through space, time and mass with optimal density and ultimate destiny.

Fig 3-16 **TES, TO and TRU RELATIONS ACTIONS**

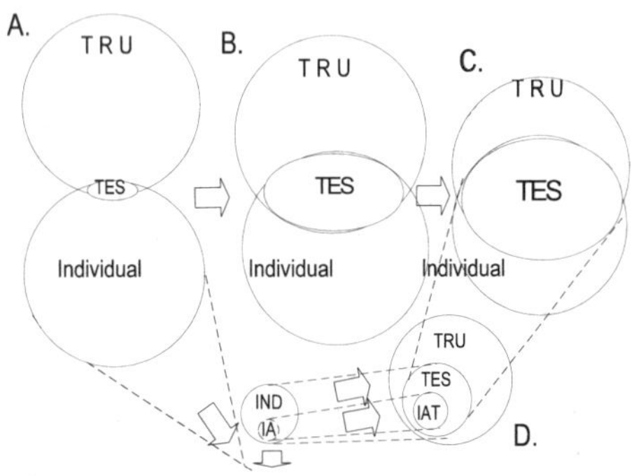

This figure compares: A. The Individual with very little TES in TO, B. The Individual with average TES in TO, C. The Individual with much TES in TO, D. Some of the interactions of Individuals as they do or do not become their counterparts in TES and TO, and as their IAS do or do not become their corresponding IAT in TO. This can be done for most all of TRU that is less than its whole. When, as in A, TES is smaller, TRU becomes smaller. What enables figures A and B to become C. is a major theme of this text. It is a function of their relativity or observer observed relationship as their subjectivity and objectivity through thought and action become increasingly parts and processes in TRU. This is the form and function, state and operation of TRU as its mass energy dimension, Á, increasingly provides the PUREACLITY necessary for optimal TIT. It is how all that enhances an individual's mental and physical powers, values and behavior as in humility, charity, mercy, hope, faith, and grace enhance both TRU and the individual as one in TES and the TO of TRU. This works collectively for all individuals that become TRU and makes up TO, the totality of TRU.

TRU depends upon the extent to which its focus or Center is strong and well defined. There is a Simultaneous Validity and Magnanimity Of the Focus of TIT and TRIT, **SIMVOFAT**, or: **FOCIT**, the Summative Total Integrative Focus of TRU. We must think of the focus of TIT as it exists in actuality a summative and total convergence of its parts, processes, states and mechanisms as they integratively produce this relative, the relative, TRU. TO, and TIT, like a hurricane, largely depends upon how well defined, integrated and strong its center or focus is. These are the set conditions of this totality. Many major factors contribute to this set, including economy of size, identity, closure, contract, proximity, proportional correspondence, compulsory identity, dependent equilibrium and its dynamics. It is also associated with symmetry and ordination, proportionate dispersion, equivalence and summative unitary reference relative to TRU. The resultant concretion and health of the whole is what enables TRU as well as our understanding and reproduction of it. **LOSED** = Law of Systematic Equilibrium Dynamics or the Law of Complete and Balanced - Equilibrium of Systems, their Static and Dynamic Totality in TRU.

LIPDET = Law of Increasing and Proportionate Dispersion Equivalence. As TRU is decreased, even if incrementally, indirectly, or temporarily, it does so with a proportion and equivalence that it seeks to formatively and functionally compensate for. **TIPS** = TRU and TO must be Proportionate, **LEP** = Law of Expanding Proportionality, or as TO changes or expands, even if only incrementally, indirectly or temporarily, it does so proportionately, or with a compensatory power that in resultant will converge proportionately. **LAPAT** = Law of Partial Total TRU, or how as parts are subtracted or removed from a whole the whole changes or becomes altered. Certain losses are deadly, some altering in quality, some altering only in quantity, markedly or slightly, some losses reduce its life time and affect other major functions. These are types of changes, and are called: **TOCH** = Types of CHanges, as when parts are subtracted or removed from the whole.

CONES = Contract Parts of TRU, or how in being to and from each other, and to and from TRU, there is a contract, like the social contract between citizens and states. This contract consists of an imperative, obligation, duty, and rights of all parts of TRU to each other and to the whole. The problem of Socrates and Christ, what to and not to be contracted with, was of these diverse worlds. **DIWEP** = DIverse Worlds Existing In Proximity. **LADE** = Law of Dependent Equilibrium. TRU is in equilibrium and tends to maintain this equilibrium, as in laws of motion and homeostasis. As TRU expands, contracts or changes this expansion, contraction or change must be proportionate. It will be whatever best adapts to the added alteration of the whole. **LADEP** = Law of Dynamic Equilibrium Proportion. **SARUE** = Self and TRU Unity Equivalence, Equidynamics, and Equidependence. Since the first person is not only, collectively and generically a large part of TRU, and even more its most effective aspect, it is this participation we so often focus upon.

COIL = the Concretion of Infinite Limits, TRU can be formulated on a similar basis. There is an Economy of Size, or a concentration of effect by possibility of enlargement for necessary totality of form. The solidity of the TO of TRU is a product of its ACEXIN, or combined and compacted Dimensional Formulation. It may not be the same as the solids we sense and feel. **SEF** = Self Forces, or the powerful set of forces a person has as TES in TO and TIT. When measured by the entire time of a highly selective and competitive life in TIT, of TES and TÍS, most of the inhabitants of this world give less than they receive of TIT. This is why the world we live in is mostly AMTRU and poorly TIT. Yet TIT can and does exist in a small segment of this seemingly larger external world. A larger percentage of these TES in, less TIT, TES is greater than TIT, and par of TES and TIT, are tentatively acceptable because of the tenable coexistence of MATRU and TRU TIT-TO. **RUNS** = Relative Universe Necessities, Needs Us, as we Need it. Let us discover, learn, remember and tell all what the world, TRU, needs. Then let us all do what we know and are told to do, and give to this world what we can to make it work. **LASYM** = the Law of Symmetry, that is mostly a product of the relativity. **CORDIME** = Coordinate Dimensional Equality of parts in each dimension of the coordinate axis of TRU, with a total correspondence of gradients. To be able to sift through all about us, or the given, we must do so with direction and order. This is provided by TRU, TIT, TRIT, SUT, and MOT. Actual Infinity applies only to TIT and consists of the outward extensions of all the dimensions of TIT as it exists as a whole, TO. The boundaries or limits of these actual and combined infinities, and their set effect on the whole, produce in TIT the solidity it requires.

CLOSURE, SYSTEM, ORDINATION, ORDER, EQUILIBRIUM, COMPLETION and FULFILLMENT of TO

The totality of TIT is found in closure and fulfillment. It is also found in the many systems, other aspects, and parts of TRU that come and work together in order and equilibrium for the whole. The need to understand TO is not only a major topic of TRU, it plays a key role in language. Many common expressions of the greatness of TO exist, as in one for all and all for one. The human central nervous system, as an example of the many systems which contribute to how TO is formed, functions as a whole, with coordination that enables us to increasingly appreciate and understand TRU, TO, and ROT..

NODA = Nominals Degree, or the degree of deficit from 100% TRU of words or symbols. How much a term means other than its part in TRU. What may be perfectly valid as fact or ENIT, or even as an ES, in being other or less than TO of TIT diminishes it or makes it susceptible to error. Correction is highly related to age, a continuing CES throughout life.

TITOD and TOK

TITOD = TRU as TIT can't be in error because it is what it is, by Definition. We may not know all it is, we know much it is not, or a mixed what is and is not, or we may know in error some of it, yet TIT can't be in error. **TOK** = TIT Total Knowledge. TIT can neither be in error nor can TOK be absolutely validated in the total knowledge of TRU. For KOT, as TRU, is other than TIT, as well as being both of infinite possibility and the limits of human vocabulary and use. TIT is such that it is largely from sufficiency, by VAPO. Our knowledge only needs to fill and demonstrate it in a sufficient detail, number and quality of ways to fully and best reveal it..

System is a pragmatic, dynamic, concrete, ordered, interdependent and correlated group of entities which function together, within a whole, or as a whole. System is also used as a method and means with the above characteristics which are applied to accomplish some end. Systems being large components in TIT, when adapted to TRIT, go a long way in showing the makeup and solidity of TIT. **ECOSYSTEM** = Interaction of a community of organisms with their environment. ECOSYSTEM is a key aspect of TRU, for it is, and helps us understand the importance of, much of TRU, all that is not or not entirely human in TRU.

LELS = Law of Equilibrium, and Equality, of Large and Small, or of large, medium and small, or of the L the Length dimension. **EMOPU** = Equilibrium of ES, massive totality of ultimate proportion. **LAVMENS** = Law of Vulnerability of Added Mass Energy Systems or Levels. LAVMENS denotes the increased vulnerability or susceptibility of organisms, animals and more complex life forms, as man and TRU. It is those in which its superimposed mass energy levels and systems are so great, as well as tenuous that the being, thisness or haeccity of the whole species or life form is in jeaprody. With human complexity and power to try to function in ways that are completely self reliant, self assertive and independent from his species is understandable, yet even more conflicting. This is a product of mans mental capacities, fixation and self preservation. These are a product of TRU, and the essential nature of the individual in TRU, as TES. In so producing this individualism, however TRU both enables itself and defeats itself. Thereby explaining the great and constant conflict that exists within and from without TRU.

LOE is a part of the Law of Total Equilibrium that exist in all entities, senses and things. Only it is especially adapted to TIT, TRU, or whatever subject, object, entity, sense or thing we specify. LOE is a vital component of relativity. It is a vital component of any totality, universe or system. Only as The thing, becomes A Thing becomes Any Thing does it lose significance. For The Thing as The Relative Universe, like all organisms must function relatively, i.e. have self preservation, be self contained and a whole if it is to exist. This is particularly significant in man, in the world of man, TRU. For the individual is constantly, and particularly for TRU, in a state of divergence, of independence, of "going ones own way" from TRU. This is the special and great nature of both the individual and of TRU. TRU is composed of an aggregate of parts and forms which are unique in the annals of biological history. Man is unlike many less advanced species in the animal kingdom whose form and function is much less complex, dependent and limited. The more complex the form and function the more power, yet the more burdens. For with each added energy form, state and level that constitutes the whole, the more dependent and thereby vulnerable other parts and forms become. It is through LOE or Total Equilibrium that TRU functions to counterbalance this diversion and disassociation.

LOSE(Q) or the Law of Segmental Equilibrium, signifies that systems or segmental portions of the whole, as organs have their own, more or less, equilibrium. LOSE is exemplified by the many different organ systems in the human body each of which functions, at least in part, with its own special kind of equilibrium. The lungs, heart, kidneys, liver, etc. all have their own equilibrium. This equilibrium is based upon the form and function of each organ as it is developed and sustains a needed purpose of the whole. So do each of these segmental or system's equilibria exist in a set relationship to other systems and the whole and are tuned to checks and balances. These segments, organs and systems also exist, as the human body exists, in TRU. Much of these same form and function exist and operate, are in equilibria, LOSE, in TRU, only with certain added principles, concepts, laws and mechanisms not recognized through anatomy and physiology alone. For this reason, LOSE is greater and LOE is exponentially greater in TRU, than compared to less advanced or complex species. TRU is thereby under heavy stress and strain, dynamically. Yet it is not to be denied. The harmony, symphony, concord that has been so often sought in history, that is an ever present goal of so much of mankind does not come easy. It requires great effort on the part of each individual and all parts of TRU to function with the greatest possible efficiency, proportion and the least possible amount of diversion and disassociation. The health of LOSE and LOE and thereby TIT are very delicate and subject to many and serious adverse effects which result in a loss of capacity and ability to maintain LOSE and LOE and thereby the existence of TIT itself. It is to this that man must direct much of his attention, make the proper diagnoses and apply the best possible treatments if TIT, one and all, are to thrive. Segmental equilibrium is exemplified by the many different organ systems in the human body. **MEV** = Method of Weighted Equilibrium Validity. There is maintenance of relation, order, coordination and equilibrium in TO of TRU. MEV are the methods, and means, we can develop that help produce normal to optimal function. It is what prevents destruction, over or under production, and helps distributive and harmonious action and function. **EQUÁ** = Equilibrium of Á, or of the S in SO in Á to produce an ultimate balance. There is a constant balance between the filling up and the broadening of S-O in TRU as SO is a filling up, a continuing concretion of TRIS. There is more generally, continuous cyclical, phasic or more irregular to and fro, back and forth motion in TRU. It is convergent or divergent, integrating and differentiating necessarily tending to equilibrium in totality, TIT-TO.

CHANGE, OF and IN TO and TRU ALTERATION, TRUCH, ITCH, and TECH

TRU is constantly undergoing forces that cause permanent, temporary or reversible change. TRU obeys the laws of action reaction and conservation. TRU undergoes continuous positive and negative forces.

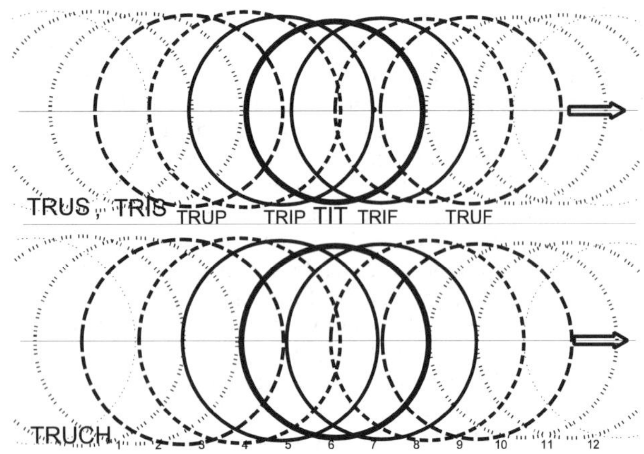

Fig 4-16 **DITRUS and DITRUCH Compared**

There exists at any time, period of time in the world, or any segment of the world, different worlds, and different gradations of TRU. From the maxim 'You would not be here if your ancestors were not the best there were' it can be deduced that there exists a state or condition of two or more worlds. Those who were and are the best and those who were and are of many gradations less than the best. The question arises, "is this best entirely equivalent to TRU?" Not necessarily, for what might have been the best in the past may not be for the present unless all three are in TIT. Even more the future for a partial, less than the whole, may not be so of TIT or TRIT. Thereby we have two or more variables in this equation and set, that which exists at any level, group, time or world, what has been the best or is or can be the best, and that which is TRU. Fig.4-16 helps answer this problem by showing the role of TRUS and TRUCH in relation to TRU and TIT.

The figures of DITRUS and DITRUCH compare the changes of TRU in time, TRUCH. It may also be used, although on a somewhat different basis, to show the different kinds of TRUS as well as varieties of TRU or TRIS, e.g. TRIV, TRIT, TRINI, TRINE, TRIR, TRIP, TRIN, and TRA. At Mid point time, half way to either side, we have that point at which TRU is half different than it is now, be it past or future. The various circles represent the various changes and varieties of TRU at different times relative to the present and TRU. It is relative upon its current center at the present infinitesimal to zero and extended infinitely to time absolute. The figure shows how TIT depends upon TRIT, TRIS and TRUS on the one hand yet TRIA and TÂ on the other.

The relationship or ratio between TRUCH and TRIT is called **ITCH** and is key to understanding TRU internally, externally, statically and dynamically. It is maintained by its point of reference, focus, and centripetal and converging forces. This produces stability yet is always subject to change, especially differentially and variably, as in 'one thing at a time' theism, or Kuthenotheism. In dynamics, especially those based upon a body or focal whole, as TRU, there exist different kinds of occurrences. Those that are random, those that are regulated, and those that are variable. The variable are not necessarily what is between what is random and regulated, but include all functions, changes, and variations, that are *other than* or less than absolutely random or regular. This becomes paramount when we reduce our facts, information, and knowledge to its part and role in TRU, as shown by the ACEXIN of SPALT and POS.. A good example is the history of the earth, its climate and what has affected it. Much has been both cataclysmic, periodic, and variable. We can discover, learn and understand history, and the present and future, better when we realize that all three are operative, and must be adapted to models if we are to perfect our posits. This is also so for genetics, other topics, and TRU in general. It is likewise important to know these when set to law and regulation. For if we are to be the determinants of our own destiny, manage and control all, we can't do so if laws are insufficient, partial, disproportionate, unbalanced, and fallible. Thereby it is not only necessary that our laws approach perfection they must be continually adapted to circumstances, and discarded in the best way at the proper time.

Fig 5-16 **RÂ and TÂ CHANGE IN and OUT**

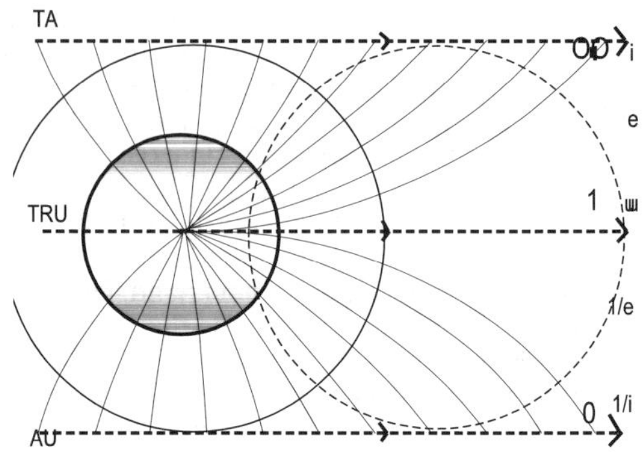

This figure reveals the relationship between TRU or TIT and its equilibrium, between the absolute and zero, between infinity and the infinitesimal. We also show the changes that occur in TRU or TIT as it relates to its unity, finity and infinity. Note how TRU is focused at unity and expands or contracts through the finite and reciprocal finite to the infinite or infinitesimal. When this relativity incorporates all and reaches a state without change it is absolute. When this relativity loses its stability, changes intermittently and has no equilibrium between its infinite, finite, unitary, reciprocal finite and infinite form, it is nothing, zero or the opposite of absolute. Hereby TRU, as RUS, is in a highly dynamic state. This is aside from any loss of stability or equilibrium that may occur. For as long as this dynamic state is within TRU, TIT, it is primarily not one of overall change or loss of equilibrium. TRUCH that still maintains TIT is one that is planned and controlled to retain its harmony and identity. **TRUCH** is TRU Change. TRU is always changing, even though it may be only infinitesimal or very slightly. The dichotomy and dualism of extremes and accompanying change may be quite simply corrected, as many problems, by reasoning in terms of TRU, solution and resolution of its problems and paradoxes. It is the method of TRU, **METH**, whereby we can have the best of both worlds through recognizing and working with all that practically, relatively and finitely provide successful reduction. TIT and TRU enjoy and involve much stability as well as change depending upon what aspect and form we focus on and our observation of it. By TÂ, TRIA, ATO, TRIT, TIT, TO and TU, TRU is highly immutable. By TRUCH, TRUS, TRICH, TRU, TRUD and CES TRU it is highly mutable. As much as it may seem, these are not incompatible nor even contradictory when considered adequately by human life, as it is in TIT-TO.

PANT = Paradox of Positive and Negative TRU, or Positive TIT yet inevitable Negative. PANT is a product of both weaknesses in TRU, as well as the inevitable actions upon TRU externally. It is how TIT is continually reacting to negatives. There are negative relationships between the parts of TIT, yet with all less than 100% TIT TO, it is negative. This kind of negative, within TIT, is counterbalanced so that the total negative effect never applies to TIT in resultant.

The struggle between quantity and quality is a large part of TRU, TIT and Homation. A mode and mean must be maintained. The mean of life is a product of necessity and determined by adaptation to changing conditions, formative and functional progressions. There is a self perpetuation inclination or tendency of mass energy levels in ENS, RUS, TRUS and TRU. **LODIC** - Law of Differential Change, or how change occurs not only as single incremental changes, but in different, multiple, and often massive ways and degrees. **TECH** = The Finite TRUCH. TECH illustrates the paradoxical and seemingly contradictory nature of TIT and TRU infinitesimal from its infinite variety, form and state. It is the build-up or construct of finites.

COMPLEX CHANGE, VARIATION and SUMMARY of The TOTALITY of TRU

TRU and TO variation must be quickly and repeatedly remembered, ECRO, as distinct from the more universal or AU time, change, and variation, t, t^2, and t^3 in TRU. Of TRU variation we have several varieties depending upon the polarity from which they are given. e.g. variation that is of farthest out of TRU to most inside TRU. The Primary or first Law of an Entity, **LAPEN** must be this entity itself without essential change if it is to remain this entity. **VATI** = Vastness of AU-TRU or ATRU-ANTRU-ANTITRU Interchange. VATI has a mixed and ethereal quality that is apparent to all who compare the common world we live in with the world of GOD, TREQ, TRU, and TIT. If TIT is all without what is negative or detrimental, how can there be any relation with AU, ANTRU, ATRU, RUS, AMTRUS, TRUS, or even TRU? This is **SIL** = the Without Law. SIL is resolved by ITCH or the TIT Interchange, being upheld largely reversibly, and slightly relatively. It is also resolved by the concomitant existence of both TRIA, TÂ, TRIS, TRUS and OITIO, as well as less than TIT-TO interaction with what is less or other than TIT-TO. The haecceity and exclusiveness of TIT gives it a high degree of stability of form which is what enables ROT and MOT. There is a set and accountable change of TRU, as evident in comparative dating. Much of this is factored into the FAF of TIT-TO.

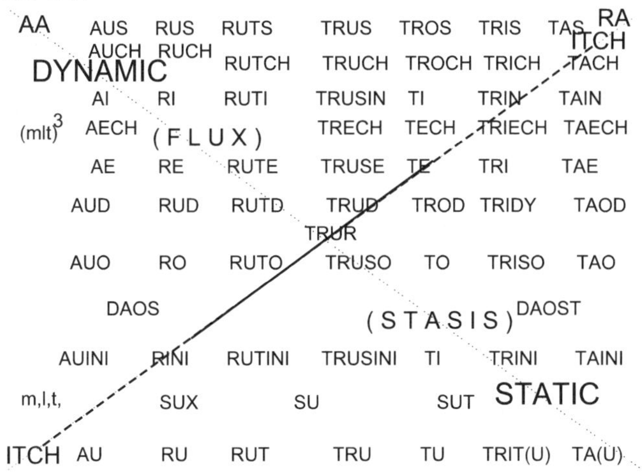

Fig 6-16 **DITCH**

DITCH = Diagram of ITCH, or the TRUCH / TIT-TO Division or Junction. There is a distinct and demonstrative dividing line between what is TRIT, TRINI, and TRIR of TIT and that which is not. This is what is most pivotal in TRIA. TRIA represents the all powerful effect of those forces which retain TRU as TIT, as TRIR. On the other hand since TRINI is highly limited, and TRUCH is all present there is a great dynamic and dialectic state of bipolar interaction between TRINI and TRIN or TRUCH. The dynamics that are associated with increasing or decreasing TRINI and increasing and decreasing TRUCH and TRIN constitute the peripheral and long term, extremes of our world, TRU. In Figure 6-16 we seek to show as many, and as much as possible, key entities, varieties, forms, aspects, parts and processes of TRU. The hierarchy of this ordination proceeds coordinately from the vertical multiple to singular, to the horizontal ÂÂ to RÂ, resulting in a transverse dynamic to static upper left to lower right. This may be superimposed on other diagrams and models as these dimensions are adapted and reordered.

There is an actuality and truth of universes in common with TIT.(ESINS) TRU is in a state of being penetrated constantly, massively and diffusely not only totally, but partially, and in all degrees of its dimensions. It changes not only every moment, but its parts and processes are simultaneously undergoing effects not only from within but from without which constitutes much of TRU's variability.

The contrast between normal and abnormal variation are seen in corrections for loss of appreciation and experience that result in loss of prevention. This is the correction necessary to supplant the loss of experience from the stimuli of trauma, illness and failures. It requires both a deep sensitivity as well as a heartfelt interest in TRU as depicted by a love of life. TIT-TO does not change in a singular manner as ES, it has manifold and massive changes. This is evident from the short mid-point, past and future in ALT and Homation. For not only does TIT change as each TRINI succeeds itself, but the overall, and each incremental TRIS, changes simultaneously, i.e. TRINI, TRIN, TRIT, TRIR, and all finite degrees change at the same time.

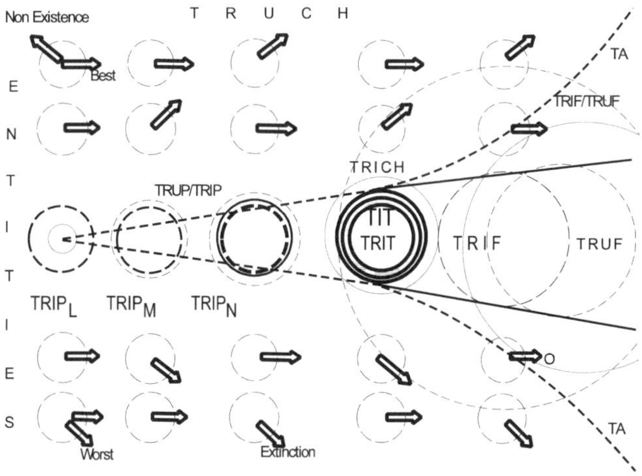

Fig 7-16 **DITRUCH, VARIATION, BEST and WORST**

There exists at any time, period of time, or segment of the world, different worlds and gradations of TRU, in non unified, and to a variable extent, indefinite masses or entities. From the maxim 'You would not be here if your ancestors were not the best there were', one can deduce the state or condition of two or more worlds. Those who were and are the 'best', and those who were and are of many gradations less than the best. The question then arises, is this best entirely equivalent to TRU? Not necessarily, for what might have been the best in the past, or even the future for a partial, for less than the whole, may not be a part, or less than its part, of TIT or TRIT, second from right dark circle in figure. Thereby we have more than one variable in this equation and set: 1. that which exists, this is at any level, a group, a time, or a world, 2. that which has been the best, or is or can be the best, 3. that which is TRU, 4. that which is TIT, 5. that which is TRID, and 6. that which is TÂ. The variation in TRU is nowhere more obvious than in poly-typical species. How humans comprise a number of different, as well as geographically and climatically oriented subspecies, varieties, races, and other groups. This state explains, as well as questions, the form, makeup and differences in TIT. It shows how much variation is operative and how diverse elements come together and stay apart if they are to optimally constitute the different and same in TRU.

CHAPTER 17

THE DYNAMICS of TRU, TRUD

INTRODUCTION to TRUD, FAF, MOTION, RATE, FLOW, KNOWLEDGE and PROOF of TRUD

The dynamics of TRU, **TRUD**, is opposite to statics, as function is to form, and makes up one of the greatest categories and modes of TRU. It is largely within TIT, although derived from TRUS and must account for much that is OITIO. TRUD concerns how the world, or TRU works, and its mechanisms, functions, powers, forces and processes. Dynamics and statics have gradients that form a continuum, which as **FAF** or form and function are in actuality in TIT, highly related, dependent, interactive and often more or less inseparable. This depends, in part, upon how well we can learn, remember, separate and comprehensively show all TRU. Our mass energy FAF is highly geared to notice and react to the dynamic. This it does in its own typical and special way, which in turn helps to prove Á and TRU. It is largely directed toward making the needed distinctions between what is for or against us, how much and in what way. It is highly intentional or purposive with selective and objective cause and effect.

Static and dynamic, form and function, structure and operation are all but extremes of the same parameter. This for TRU, and human relativity, takes on added characteristics. They enable us to see or envision entities or things in and of themselves which are actually not there, but are figments, by TRU. This is called: **LASADIN** = Law of Static and Dynamic differentiation and integration in TRU or of and in TRU. It is through LASADIN as characterized by the predominant form, length, time, and mass energy that the static and dynamic of the differentiation and integration of TRU fuse as one in systematic unity and may be so effectively and beautifully represented and modeled.

The breadth and depth of TRUD can be seen in analyzing its role in all the different kinds and forms of TRU. TRU may be most universal or particulate in or out of TIT. **CEF** = Correspondent Equivalence of Form and Function or Parts and Dynamics of a Thing, of TRU. CEF is a basic law of nature, life or biology, and denotes how the function of a thing and its form correspond. **UFAF** = Unity of Form and Function, or how FAF exists and acts as a unit.

The degree to which any entity, part, process, or aspect of TRU is considered or held to be greater or lesser than it actually is in TRU produces the syncretistic fallacy, loss of value to TRU, multiplicity effect, irrelevance effect or fallacy, and irreverence effect or fallacy. Knowledge provides much of the dominance that determines TRU. This may occur singularly, multiply, totally or in their various mixtures.

SAR = States of Unequal or decreasing Relevance, or states of irrelevance, how levels of entities, senses, things, objects, RUS, and individuals which are less, equal, or of decreasing relevance to TRU. It is SAR that explains the rights of variation. It is SAR that corrects TES.(LADEX) It is in the power of relativity, the bonding of the sames as they hold TRU together that TRU is maintained. When this relativity is decreased, even incrementally and especially unequally, unbalanced states are created in TRU that are highly reactive, even at times violent, causing serious dissociation.

LANOT = Law of the total objective, common purpose of TRU. LANOT signifies the collective purposes, intentions, and goals that are the same in TRU. Corollaries of LANOT consist of different dynamics of the ES and CES or parts and processes of TRU, absolute necessity of the common that is only positive for total benefit, and that of unconscious participation.

Much that is of stress is both of and not of or for TRU. TRUD is positive for TIT-TO, yet is often negative for ES, CES, TRU, TRUS, AMTRU and OITIO. The same holds for the stress that is endured by TRU. It is because what seems so much of us that is not TRU, that is a part of the forces and stress, that we suffer and are unable to understand ourselves, TRU, and this suffering.

DERIVATION & DEVELOPMENT of TRUD, CAUSE & EFFECT, STIMULUS - RESPONSE

In nature freedom is relative, there is a continuing departure from actuality, the best or optimum, TRU actuality. It is often produced by failure to recognize what one senses or the fallacy of the redirect of the tangible. There is a selective development of TRU in TRUD. It is in causes and effects that we can most dramatically see the role of dynamics, TRUD, in life and TRU. Stimulus response is a basic kind of motion in TIT as it is in much of the animal world. It is a fundamental determinant of TRU. By its relation to the relativity or S-O dimension, TRU can be shown. Stimulus response plays a major role in composition determination. It is closely associated with the predominant mass energy, subject-object, or Á dimension and FAF of TRU. This forms a set state with mechanisms which are genetically highly set for TIT, yet susceptible to adverse effect. The negative indirectly contributes to the development of TRUD as stimulus is often negative, producing a positive or optimal response. This is conditioned by the function of the human Central Nervous System, **CNS**, and especially the brain. Mental capacity and ability is primarily nature's provision to better react to actions, respond to stimuli, and uphold survival.

'In nature nothing is free' expresses the fundamental fact of life and the interdependence of all. For the individual, as in existentialism, many activities and thoughts pose never ending questions and problems. Do I know what I am doing? There are always many choices. With a knowledge of and activities based upon TRU, or TREQ, we do right by knowing right in exercising our freedoms. **FRE** = Freedom is Relative. **FRED** = Freedom is Relative Deficit, or the degree some do not produce effectively, causes others to do just that much more. It means that others will suffer, have to work and produce that extra amount to even reach a minimal level of freedom and return benefit from TRU. IPTRU shows that this qualifies the idea that life is what you make or put into it. For a person lives in a social world. This is a world in which what one does or does not do effects more than themselves, but many different groups, ways, and most especially the whole, TO. **FET** = Freedom Equals TRU, or the degree of freedom can be no greater than the degree of TRU. The degree of freedom can also be no greater than the SCARLS of its relationship to TRU. For TES this will depend upon its proportion, or REM, and its form, and relations in TRU. That there is irreversibility in evolution is evident in the inability of higher forms to revert to lower. The implications of this is different for TRU than in science and biology. It is because of irreversibility of much in evolution that to sustain TIT by TRIA we must have management and control of as much of TRU and TRUD as possible.

COC = Causes of Complications, or Causes of Causes, are also in TRU, and become part of its massive dynamic state. All COC as IA are ultimately TRU, ATO, at once TRU, in all Totality. COC signifies that aspect of causation, for TRU, which works in a multiplicity of, or ramified, directions, both a priori and a posteriori. It is seen in the many causes for disease and their, more or less, interdependence and effects. **LODAC** = Law of Direct and Indirect Consequences, denotes the fact that causes and effects work more than singularly and directly, but most profusely to infinitely, within TIT.

SUPEF = Supersaturated Effect. SUPEF is a corollary of the much larger disproportions that occur within a whole that arise from hyperfunction. **LOSE** = Law of Selective Effect. TRU selects the past and future as well as the present. We know that the future can be the cause that affects the past, as readily as the past effects the future and both in and through the present. It depends upon whether it is only of TIT, of ES, TRUS, or by AMTRU. **SEF** = Shattering Effect, or those effects that tend to greatly produce disassociation, disruption, or breaking up, of TRU. There are several types of effects: large, small, rapid, slow, quantitative, qualitative, reversible, irreversible and others of construct and FAF. Inalterable TRU susceptibility makes distinct that state in which it reinforces, as it is reversibly produced by, the predominant Á mass energy form of TRU. There is a specific dependence of reaction on stimulus no matter how unapparent or indirect.

RADNO = Rate of Departure from Best or Optimum, or the rate, velocity and acceleration by which anything passes away from its optimal or best state. TRU is formed and functions by the capacity and ability to work and best respond to stimuli, meet the problems, and answer the questions of life. This is a main reason for knowledge, as much of this knowledge is a product of the predominant mass energy state, Á of TRU. **LARNIP** = the Law of the Non Absolute Relative or Negative in the Positive. LARNIP, and its associated problem the negative that is OITIO, is further explained by the cosmic cycle and flux first described by Heraclitus and Empedocles. The lack of constancy in nature, held by Heraclitus, does not account for TRU, nor do the antithetical dialectics of Empedocles adequately describe the dynamics of TRU. Yet they contain meanings in common. The lack of constancy of Heraclitus can be compared with how TRU departs from the absolute and is relative, externally. This is variable, although for brief periods of time nearly negligible. The dynamic dialectics of Empedocles describe only some of the many polarities that exist in TRU and fail to provide all that is TO TRUD. This is in part a result of subjectivity. If we assume a standpoint other than TIT, or of one between what they are in and of themselves and TRU, we can see the difference. This is the basis of meaning, or **STÎ** = Subjective Transcendence In or Immanence in TRU, or that state of subjectivity in which it is largely confined to TRU, yet in OITIO.

SODU = Subject-Object, A, Static and Dynamic, Stimuli and Response as a Unit. SODU is the discrete unitary form and function S-R in IA through S-O. It is the subjective capacity to react beneficially to alter the objective. **LOTS** = Law of Objective - Subjective, TRU, Sufficiency Dependence. As subjectivity, S increases, so does objective or SO together, as the Á mass energy state and dimension of TRU. It is in anticipation, as seen in foreknowledge and forecasting that the highest powers of Á reveal how TRUD is best developed and determined. Anticipation directed by ACEXIN and TRU knowledge provides not only a purer TRU, it gives the VAPO necessary for TIT. **FADASO** = Fallacy of the disparity of S, O and Á, or Fallacy of S-O or Á Disparity. The worst of all, is the disruption of the concentration of S-O, seen in IAT, for it reflects a sharp departure from the crux of TRU, its Á FAF.

What is the correspondence, or identity, between the S-O parameter and the mass energy form and function of the A continuum? Although they are used interchangeably as the combined prime mass energy dimension of TRU, we need to know more about how they are DAOST, how same and different. To the extent O and mass are more formative and material, and S and energy are more functional and mental, they are roughly the same. To the extent S and energy represent the higher levels, and thereby later development of human ability, and O and mass represent the lower or more primitive levels, and thereby basic to human capacity, they reveal the a correspondence between the two prime modes of TRU. It is likely the actual mass energy dimension of TRU is much of both, and a little of neither, even more in combination. So that we need to be careful, not only to avoid confusing human mass energy with physical mass, but to realize that the mass energy dimension, Á of TRU is not to be judged strictly from either the S-O nor the developing material to mental levels of human capacity and ability necessary to survive and flourish.

FUNCTIONAL ASPECTS, ATTRIBUTES, LIMITS, FORCES, ENERGY, EFFECTS, NECESSITY in TRUD

The major role of velocity, force, acceleration, and power in the dimensional formulation of ACEXIN in TRUD can be inferred from mechanics and their reinterpretation in the special and typical ALT of TIT. Forces accumulate in TIT and AMTRU which perpetuate TRUD. There are optimal levels of forces and its processes which TIT possesses and can sustain. They are controlled in TIT by, among others, the principle of rational remainder and diminishing returns. Forces are much of TRU, infinitesimal or minute to greatest. Their conjunction, bonding and relations establish dynamic systems with a high degree of order and stability.

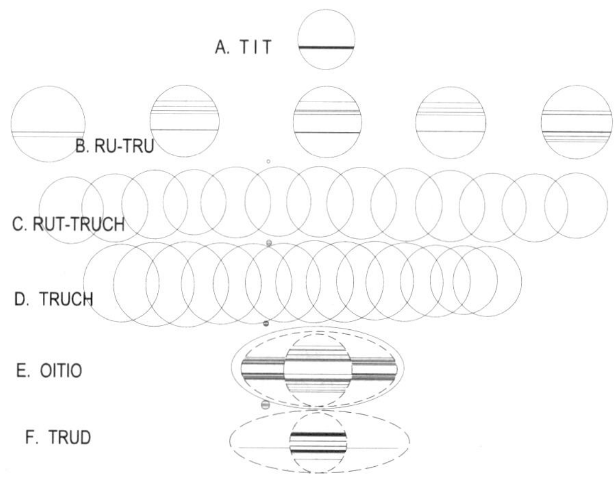

Fig 1-17 **DOTRUD or Diagram of TRU Dynamics**

In DOTRUD the base of TRUD is shown. TRUD can be better understood if we remember that it neither exists alone, nor is it always the same in relation to TIT. DOTRUD manifests both the internal and external functions of TRU. In its external relationships it is helpful to recall RU-TRU, RUT-TRUCH, TRUCH, and all that is constantly operating OITIO.

In the function of TRIS from TRINI to TRIN we have a dipolar state and relationship which for TRUD is highly significant and informative. Although neither one of these is perfectly valid, they represent a state which typifies TRUD. It is in understanding the multi-polarity and dialectical state of TRUD in TIT and TRU that we can gain the necessary depth comprehension of TIT, TRU and all. The multilateral forces of that tend to coalesce to insure TIT come both from within and without TIT. These by TRUCH produce a flux in which there is never an absolutely fixed state but one that tends to be more or less in one direction or another. To maintain itself TIT has controls. It is to further and perfect **LACTRIS**, or the Law of Control of TRIS, that assures TIT.

Within the differences in TIT there is more or less freedom and/or chaos, an instability component of its parts and processes. The energy of TIT exists in various levels of FAF which have oscillations, periodic changes, and time cycles. These are guided by the dynamic relationship between parts and wholes, extant rates and adaptation to TRU. The instability that exists in these systems, as well as processes of TRU, must be accounted for. It is in the normal, healthy and optimal function by division of labor, both within the human body, as well as within all TIT, that makes TRUD and TRU successful and harmonious. This is produced by relations, and sameness as they optimally function in TRU. By viewing the dynamics of comparative forms of TRIS at different time periods past and future we can better understand the basic principles of force, energy and mechanics of TRU.

The farther out one gets from TRU, as in the past or future the less TRU there is, yet not by TIT as demonstrated in ALT and SPALT. It is a kind of focus that partakes of more non being in the human sense. This is: **POPEX** = Principle of the Proximity Existent. POPEX distinguishes the non TRU from the TRU as we proceed outward in the various dimensions, most notably that of time. **PREX** = Principle of Extant Rate, or the rate at which various entities at different periods of time, durations of time, or variations, are or are not a part of TRU. We must be careful to qualify or SCARL POPEX and PREX and all interpretations of what is and is not TRU we may treat.(ESINS,ES,CES)

Beat, time, rhythm, period, phase, pulse, and even schedules, routines and habits all indicate a more or less finite/infinite cyclical relationship and state. They show a dynamic periodic form with finite gradients. This is a product of single, combined and complex basic parameters and dimensions. It is similar to L_2 and T_2, or even A_2, as dimensionally formulated and applied to TRU. That most distinct and closest to man illustrates the SUT within TRU, like the symphonic and harmonic states of music. The principle of imperfections of today equal the evils of tomorrow is resolved by PIPS and COVAPS. Personal freedom must be tempered.

There are many functions in TRUD that are components of TIT and TRU including efficiency, effectivity, productivity, success/failure, improvement, acceleration, growth rate, velocity, inertia, momentum, moment of inertia, center of mass, energy, potential energy, kinetic energy, torque, variances, effusion, diffusion, diffusion coefficient or Fick's Law, and degree of polymorphism. These, and others, come in many varieties and combinations. There are numerous spontaneous and deterministic forces, powers and momentum that enhance and sustain TRU. Some of these include, life potential, activity, summation of individuals, genetics, and reproductive functions. The principle of extant rate shows the current pace of a particular or selected process, CES, in TRU. This is often much different than what we may sense or believe, or its rate as a thing in itself.

LEXOR = Law of Expansion of Relation, or the convergent, collective, and thereby additive and multiplying power of relation. This is a simple accretion of positive forces, like negative forces produce loss in INEG. Yet LEXOR has specific and great meaning in TRU. It is upon the dynamics of relation that all TRU and TIT are based and prosper. Division of labor is an example of LEXOR and how the integrative processes of TRU reveal and prove TRU and FAF.

LODIR = Law of Diminishing Returns, or what from economics signifies the limits, and accelerative loss that accrues in a system or TRU. It is the loss of reversibility that occurs with INEG and its associated loss of quality. **POREM** = Principle of the Rational Remainder. In IPR, fashions, conflicts, fights, and wars, the chance elements cancel out and the rational principle - or TRU - remains to carry on. This is may be either positive or negative for the intended, if negative how it is not TRU. POREM is like least common denominators and is associated with COL and COR.

The reason why AMTRU, NAMTRU and ATRU are so bad is that they cause lack of confidence, confusion, indecision, indifference and distraction which in turn causes dissociation of S-O, the Á and all of TRU. Ones mind does not follow the object. The object does not follow the subject. The observer and observed do not work together in relativity, so that error, confusion, loss, failure and likely disaster result. There is no meaning, no formal distinction, and no identity. It is a large aspect of how TRU fails, loses, and becomes extinct instead of wins and survives permanently..

DUREF = Duration Effect, or how duration determines much of dynamics, especially the adapted and altered FAF of duration in the distribution of time in TRU. **EIC** = Effect and In as Counter-actants, or how all these that are extra TRU are important of and in TRU by and in combined effect.

Conveying matter and energy is done by the linear accelerator. In this, a little mass equals a lot of energy. This is as $E = mc^2$. It is how the sun converts mass directly into energy, and how stars can burn for so long. This has interesting implications for long term human survival. For our success depends upon ability to convert a little mass into a lot of energy. It also equally or even more depends upon our ability to manage and control both this mass and energy, on quality of the Á mass energy FAF of TIT.

LACON = Law of Coordinate Necessity is even more necessary for humans and TRU than usual generally, because of the ever more complex levels and units of Á, shown in SÁ. Much higher and different kinds of coordination are needed. **PAMUST** = Principle of Automatic and Mutual Stimulation, or how one advance calls for another and its associated corollary in the resulting reaction. All are held in check by LACON as it exists in TRU, TIT. **MEVIM** = Measure of Vital Necessity, a good test of Relative Magnitude.

TRU SPECIAL FUNCTIONS, ACTION REACTION SELECTIVITY PROCESS MECHANISM VECTORS

TREQ are well known from their special functions and principles. That action is followed by reaction each within and without TIT, and that these makeup many of the processes and much of the dynamics of TRU becomes ever more apparent when we study, analyze and trace them out in TRU. Actions often tend to separation, and reactions tend to protect the part and the whole together. Examples of simple non TRU and TRU actions, such as blind acts and striking out effects, reaction to accumulative stresses, overburdening and disintegration, reveal TRUD. By determining and representing the unifying and disintegrating processes we can bring out the functional detail of TRU. Such a method arose from Aristotle who conceived of process as the actualization of determinant potentiality.

The distinction between processes and mechanisms comes from their complexity and basic importance to TRU, processes with parts and mechanisms with states of TRU. They are most vivid in the key dynamic built in mechanisms of TRU. They form and determine the basic function, operation, success and survival of TRU. Many mechanisms are highly selective, limited, and in opposition to other mechanisms of TRU.Built in mechanisms are descriptive of the profound and intricate functional nature of TIT and is shown by SPALT, POS, and HOM.

LATMUP = Law of greatest Troubles Wrongs of TRU come from greatest Principles, causes and Mechanisms, Universal causes and effects. **DEPAR** = the Derivation, Development, and Determination of first Principles from basic relationships and how they provide the great determinants of action.

LEXPAC = Law of Exclusive TRU Basis, Plan, Policy, Program for all Action. LEXPAC, as much of TRU, is a simple product of TRIA. LEXPAC is imperative, valuable, and of great pragmatic significance.

From **EL** or the Exclusion Law, since altough TIT is absolute, ES, in being partial are absolute only to the extent they are vital to the whole, by LOPAN. Their interactions as CES are in continual give and take relation to other ES, typical throughout TRUD. Of these there is **LAPDEL** or the Law of Dynamic Polarity of EL, or the Paradox of Exclusion by IA and TRIA. LAPDEL may be likened to a battle of alienation where there is a perpetual dialectical antithesis between the absolute o TRU, TRIA, IA and the individual. Each is more or less excluding, EL, each other. This is one more example of the impossibility of simple and singular causes, effects, actions and knowledge. They must be manifold, complete, confined in, and conditioned, or SCARLED by TRU.

APROD = Action Proportional to Principle. The actions we take in TRUD are highly dependent upon and must conform to our knowledge, its principles, and the involved SCARLS of TRU. Why is order so necessary? This can be vividly seen in action. What are you doing? Instead of doing the same thing over and over again, to do each once or the proper number and ways, is just right for the overall job or problem relative to its need in TRU. Humans by nature are highly and diversely active, rapidly and exponentially becoming more complex. To maintain and advance this activity and its basis, TRU, requires the best exercise of its higher powers, by using its highest policies, principles, knowledge and purposes.

PACU = Principle of Automatic Compulsive Unity. For each change, as the major courses or alteration of half lives, there is an equal reaction in TRU. This in turn is productive of a certain change of form. The tremendous built in forces that sustain PACU become apparent when we envision the **PIP**, Proof of Integrative Processes, determinants, major courses and dynamics as revealed by SPALT and Homation. The success of TRU depends upon the degree to which its reactions and responses equal and are sufficient to meet the actions and stimuli that effect TRU. This reaction must be specific, VAPO, and SCARL to TRU. Only with ACEXIN, ALT, KOT and ENT can efficiency be raised from a modest percentage to approach perfection in TRU.

Selectivity is not only a quality of the higher levels of mass energy as the intellect, valuation or reason, it is present in different ways in most all levels and gradients of the Á FAF and dimension in TRU.(LOSE) It works on the atomic, molecular, genetic, and adaptational levels seen in the role of phosphorus in animal metabolism. Phosphorus is the scarcest element in life's dependence upon recycling in the biosystem. The form and function of man is highly dependent upon phosphorus and many of its compounds in various ways and systems of the human body. It is vital for both the passage of genetic information, energy transfer, and brain function. In phospholipids it is so formed and function critical for membranes in which their role in selectively absorbing water on one side and being impermeable on the other enables them to control much of the intake and output of cells and tissues. This is called: **LASÁ** = Law of Selectivity in Á, in all levels of mass energy. Selectivity occurs even at the level of the electron, due to its polarization between positive and negative. Each entity, sense, thing, AU, RU, RUT, TRU has its own special mix and form of selectivity. It is this that gives it its differential and unique nature. It is this that makes up much of the phylogenetic differences in animals. This is called: **LASAF** = Law of the Special Form and Function of Selectivity in Species of Animals. It is LASAF that provides humans and TIT with much of their Thisness, Haecceity, and TRIA. The interdependence of selectivity and specificity is key to and prime power in TRUS, TRU and TIT. It is selectivity that enables much of the specificity of animals and life, and it is specificity that determines much of their selectivity. This is called: **LIDSAS** = Law of Interdependence of Selectivity and Specificity. All the unique mix of selectivities that make up the FAF of humans in TIT is what no other animal or entity has. Such exclusiveness is a strong proof of TRU. This is what enables and drives much of adaptation in TRU, as well as all levels of the mass energy FAF and continuum in TRU. It is all these levels we must PPMA, as in ACC, if TRU is to survive. This is, like numerous other mechanisms, no where more evident than in the best operation and prevention of loss in the phosphorus cycle in the biosphere and its equilibrium and optimal contribution to human physiology.

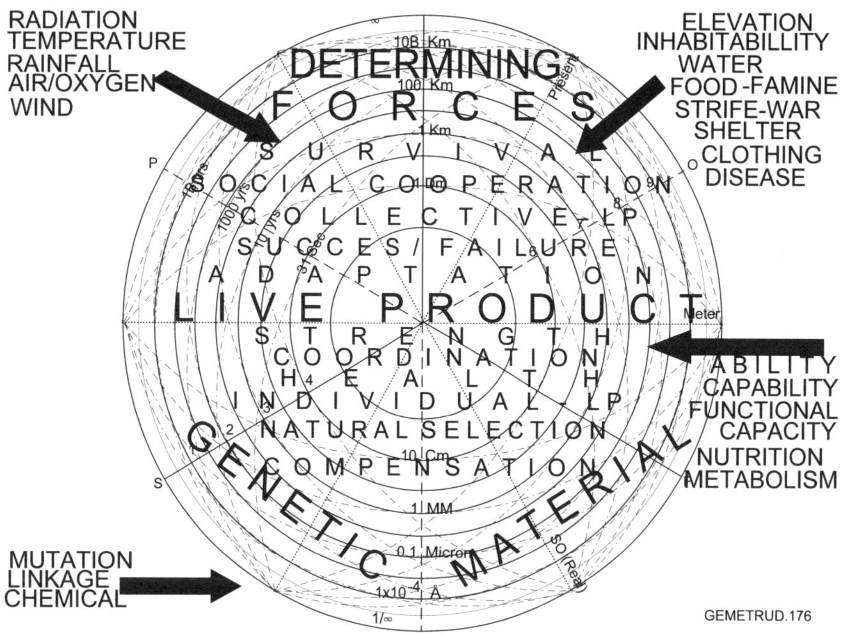

Fig 2-17 **TRUD INTERACTION and CONTINUITY**

The survival of humanity depends upon many factors and forces which work in a great variety of ways, focused on the resultant reaction to determining forces, health, strength, and success of the phenotype or live product, and continuity of the genotype or genetic material or germplasm. Variation and change have already been built into this diagram in its time dimension and the long term roles each plays. This also applies to all the Á dimension. Many more forces, factors, concepts, principles, laws, and mechanisms can be added. Yet to much detail causes a loss of overall comprehension of TRUD in TIT and TRU. This is **LAPOD**, Law of Proof and Practice / Detail.

A simpler function, process, and principle, often points out, shows and proves a more complex and greater one, e.g. oxygen and TRU Debt. This is especially true of the key dynamic mechanisms that are basic and vital to TRU. These proceed from: **LAMA** = the Laws of Mass Action. TRU exists in equilibrium or homeostasis in which massive actions and reactions occur simultaneously. TRU homeostasis or equilibrium will vary, both for ES, CES, and TIT-TO from time to time depending upon many factors, internal and external, indirect and direct, particular and universal. **LOMECH** = Law of Measure of Mechanism. As we discover, develop and consolidate TRU one of the chief ways by which this is achieved formally and reliably is through the measure of its mechanisms. We must analyze and map out all the reasons why TRU and ES function as they do. What they are for. How their function follows form, and their form follows function. Mechanisms which support TRU are also those which are lost when TRU changes or disintegrates. They are often those whose loss is permanent or irreversible. The great tendency of individuals, things, even ES and CES, to form their own or separate ENIT is ever present. It is a great force and aspect of the dynamics of TRU.

The form and dimensions entities assume in TRU is not how they exist in and of themselves, as entities or universes, it is only what they and their ES and CES are in TIT. This is proportional to the total dimensionality in the TIT-TO. **MEREP** - Mechanism of Replication, is a product of and in turn produces, enhances and secures TRU. **ESCES** = Parts of

TRU Conjunction with Processes of TRU. ES by DAOST do not exist separate from their corresponding CES. They are for most of TIT much the same, only one static, the other dynamic aspects of the constituents, participants, and contributors to TIT.

TRUD PARTS, CES, OCCASIONS, BEHAVIOR, OPERATION, FEEDBACK INTERACTION, EQUILIBRIUM, and SUPPLY/DEMAND

The continual mixing, interaction and regeneration of all that is out of and in TIT is what has been called cybernetics. The mass of TRU is such that little change is produced in TO by single forces, actions or reactions or processes. For TO there tends to be unconformity in differentiation and conformity in unity or integration. Functional use and value tend to the infinite and to be multiple and scattered. Yet the convergent powers of TRU help the mutual expansion of ES and their participation in different ways in TRU. There is an equivalence of differentiation, partiality and dynamics in opposition. This is controlled largely by the interdependence of the parts in the whole, TO. There is a dynamic equilibrium of all, including many systems and processes in TRU. One of these is between parallel and divergent forces of relativity in TRU. The parallel and beneficial forces are predominant and maintain the equilibria of separation and union, parts and total, universals and partials and other systems. In both different systems, as well as the polarized ends of each system, the differentiation of TRU produces a fragmentation of equilibrium.

Struggle for existence may be 100% TRU, only of a part of TRU, as TES, or of AMTRU or non TRU, as by an individual. Yet for humans it inclines toward TRU, and reveals a predominant force in life, of which the Á is typical and special. **PAFOT** = Potentiated Force to TRU, or how all one needs to do, frequently, is to accentuate or uphold the existing forces, functions, forms or aspects of TRU.

LOCAT = Law of Occasion Not TRU. The occasions at hand are often most demanding yet also often not TRU, or only a small part of itself in TRU. Yet, this given or occasion, how it affects us, how all our attention seems to be needed to answer, solve, correct or achieve it, seems to be, and is often, all important at the moment. **LETUN** = Law of Fine Tuning of TRU. So that much of our problem in life or the world, is not a problem of TRU, nor even more a denial of TRU. There are great problems in trying to make a part of TRU all TRU, or trying to make an occasion or individual TRU or even its best role in TRU, or trying to make occasions and individuals function together, or trying to make one part or process of TRU some other part or process. Only with fine tuning, trying and making all ES or CES by TIT do we come close to the success that is inherent in TRU.

LAFURD = Law of Functional Relative Dependence. Each ES for TRU, like each part of the body, and each operation of its organs is neither to much nor to little depending upon the built-in conditioning mechanisms whereby its participation in the whole is optimally controlled. This is a great proof of TRU. A finger is strong enough but not to strong, and bones are large enough and shaped in such a way as to provide optimal function. The function of all organs are relative to the needs of the overall organism. **TED** = Theory of Equidependence, or the balance of dependence and independence of ES and CES in TRU to enable optimal success and survival.

It is the totality and its proportions, extension, intension, actuation and the expansion of the finite from unity to infinity that is the positive basis for TRU. It is the ability of entities, organisms, RUS, TRUS and TIT to obtain maximal form and order in such an expanded positive finite between maximal unity and multiplicity that their thisness and existence are based. All, as seen in science and technology, expands mutually as their ES and CES multiply, or: **MEX** = Mutual Expansion of ES, all ES. As one science provides the basis for others and all are related and brought together - by VAPO - TRU is fulfilled. To learn and practice what is needed to overcome partiality and negativity is why being philosophical and religious are so great.(TREQ) It is how we eliminate or ignore our petty troubles and concerns in order to identify with TO of TIT..

Dominance must equal TRU. Although highly painful and all suffering when not TRU, dominance, exemplified by predominance in Á and SÁ, is highly TRU. Dominance applied must be no more nor no less than necessary for TRU in the given, encounter, situation, occasion, or action at hand. Principles in the alteration of TRU for the best include: 1. It is not easy to better TRU, for it is, by TRIA, optimal. 2. It is much easier to alter ones own, or help others, to become TES in TIT. 3. It is easy to worsen TRU for many forces can cause serious harm to TRU. 4. TRU has marked powers of momentum, inertia, reversibility and adaptation, including a large and diverse population. 5. Since TIT-TO is massively active much must be accounted for and done to temper and govern TRUD. 5. What acts on TRU is selectively altered to become TRU. Artificial restraints and the added functional burden that is produced when TIT is altered is not a self destructive state when guided by the relative divisions and TRU differential. It is, on the contrary, the most glorious and successful means by which we can demonstrate and represent TRU. The predominant makeup, the mass energy form of TRU or its S-O dichotomy become, rather than antithetical, a continuum that is self supporting and productive of the relations or bonds the two poles of S and O produce when combined as Á. This is the great value and work of relativity.

Fig 3-17 **EUSTASIS, FORM of TRUD and LACA**

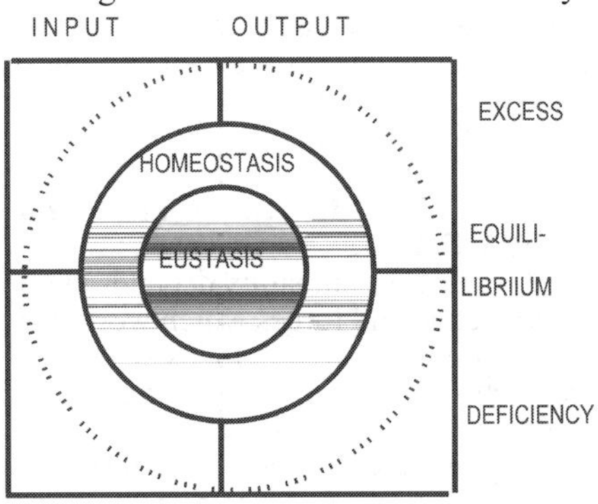

The individual, as TRU, to function properly must or will change, primarily and mostly, in content and not in form. This is called: **LACA** = The Law of Content over Form Change. TIT-TO is in constant motion and a state of equilibrium, balance, and harmony. To be TRU the individual, must be with these balances with the correct fulcrum at the center point. There can be excesses or deficiencies, which will be compensated for. Input and output must be adapted to and be reversible. Reversibility and adaptation may take many forms yet will, in resultant, always accommodate TIT, and its primary form, LACA, TÄP, and TÄ.

In Figure 3-17 the relationship between intake and output, excesses and deficiencies are shown. These may be in complete balance, or <u>eustasis</u>, as the center point which represents the mean, golden mean or ideal, yet not always nor necessarily all, the totality of the mass, the relative universe. This variation or departure from the mean is the equilibrium between all its components. It is an internal milieu, the homeostasis of an individual's physiology. In being generally applied to the special case TIT it is called Homation. It especially applies to the functional equilibrium and mass action dynamics of TRU. The involved processes can be separated or fused and made total for SPALT posits of the whole of TRU. TRUD represented in TIT-TO, TRIT, OBT, SUT, ROT, MOT or the model of TRU, SPALT, gives a much better POT. For a person, as TRU, this is optimal or perfect health, oneself and for TRU fine Tuned to perfection, signified by **LETUN**.

Cybernetic mechanisms of input, output, feedback, and all describe much of the dynamic operation in and out of TRU. In relation to the whole they are positive or negative depending upon their resultant effect. They also depend upon principles of supply and demand, and surplus and deficiency. Behaviors are those operations, which may also be expanded to TO, that seek to explain the way principles of operation, equilibrium and overall function contribute to human life in its struggle for survival. Many diverse worlds exist in close proximity to, and are often highly mixed with, each other and to TRU. They are in perpetual conflict.(CET) Resolution of the monolithic vulnerability anthropologically, biologically, and geologically has been replacement by species or classes that are more stable and less unbalanced, monolithic and vulnerable. For TRU this is limited. If TRU is to be TRIA, it must be stable, can be adapted, and selected only regeneratively and reversibly.

PADFOR = Parallel and Divergent Forces of the Relative and Reality. There are at any time or duration of time many different forces in action on the individual, TES, TÍS, and TRU. This applies to duration and variation as well as time, area and space as well as length, and highest and higher levels of mass energy as well as physical mass alone. These forces tend to converge on, or parallel, unity. Many are slightly or greatly divergent. So that the individual will in their orbit or world be completely, or almost entirely, surrounded by forces productive of problems which are divergent to TIT, TRIT, and TU or the focus and unity of TRU. Therefor the unity of TRU, TIT and TRIT and their totality will be caught by the individual's experience in life only by momentary glimpses. A few will never get these, some only once, a few occasionally, and some when trained by TREQ and TRU, many in a lifetime.(LETID,LOEDPT,LODE) **LODET** = Law of the Dynamic Equilibrium of TRU and TIT, or the difference, continuum, and balance that exists between TRU and TIT when optimal.

Fig 4-17

EQUILIBRIA in TRUD and TRU

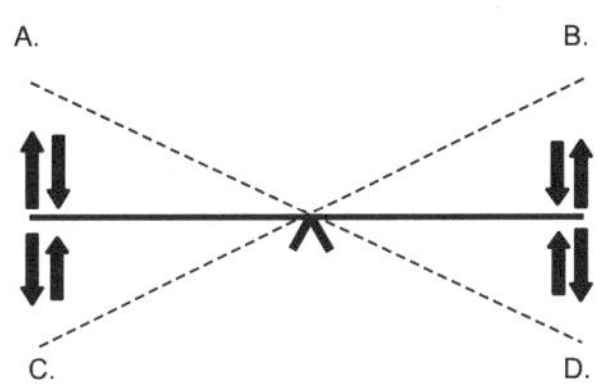

Although TIT is set by TRIA, TÂ, TRINI and TRIT, it also depends, to some extent, upon what we make it by TRUS, ROT, POT, and KOT. Thereby for TRUD this diagram is somewhat flexible. To find the optimum may be compared with the equilibria of four extremes in reaction tending to produce each other.Thereby for TRUD this diagram is somewhat flexible. To find the optimum may be compared with the equilibria of four extremes in reaction tending to produce each other. It is in moderation, a proper mixture of bondage to TRU and freedom of TES and individuality that the best in life and its happiness are to be found. So that all that is TRU is only positive. A. Individual Freedom, B. Individual Slavery or Bondage, C. TRU heterogeneity, D. TRU homogeneity. When indebtedness exceeds possible work, one goes into bondage or a kind of slavery to failure. Whether this is oxygen debt, monetary debt, nutritional debt, knowledge debt, genetic debt, or political debt is all a part of our dynamic equilibrium in TRU. We are all more or less in slavery, whose very bondage to our functioning in TRU is alien to much of our freedom, IA, and comfort. It is an expression of the give and take between ourselves as parts of TRU, TES and TIT, as well as our being to whatever extent in the outside world. It shows us the debt, necessity, IA and the requirements which we must supply or repay if we are, and TRU is, adapt and succeed. If our TES, strengths, capacities, abilities and functions are great then this debt, bondage, and requirements of TRU, and fulfilling our counterpart in TRU is relatively small. For equilibrium in TRU, as in economics and all life, it is relative, relative by the same's of DAOST in TRU.

PEF = Polarization Equilibrium Fragmentation, or the fragmentation that is a product of and occurs in equilibrium as a result of the many polarities and their oppositions in TRU. This is how ATRU, as disease, may come from within rather than without. It occurs when compensation does not correct oppositions and a deficit for TIT results in some permanent loss. For each part, process, occurrence, and situation there will be a tendency to fragmentation between the differences and oppositions of which they are associated, effected by, or made up.(DAOST) It is how differences are fundamentally involved in the dynamics of TRU. All is not always or perfectly attuned to the perfect homeostasis of TRU, thus a never ending to and fro motion, forces, actions, reactions, and effects. In the economy of life, as in the economics of the business world, there is a constant ebb and flow of the tides of work, supply and demand, and prosperity. It is to manage and control these that much of the last five sections of this book are directed.

DIALECTICS, COMPETITION, POSITIVE and NEGATIVE, STRIFE, GOOD and BAD by TRUD in TRU

The operations and equilibria of and in TRU and its systems and processes are not all ordered, stable or harmonious. Even when perfectly TIT there is competition, opposition, dichotomies, dialectics, positive and negative, good and bad forces, causes and effects. This is so if not for TO, for ES and CES. TRU as TRUD is in continuous and variable strife, stress and strain, which in TIT takes on special, typical and unique form as the parameters and polarities of TRU are altered and adapted to become predominant for its survival. How they all work together and how TRU, as TRUD is characterized by them has been little understood. Herein is the key to much of a proper comprehension of TRU and the job of Homation. This is by showing the dynamics of opposing parts and processes in TRU. It arises from the equilibria in TIT-TO of systems and all in and out of TRU, that are interactive with it. The competitive dynamics of TRU show how much equilibrium can be healthy even with opposition, positive and negative forces in operation. The range of positive and negative can be measured. Even negative can be potentially positive. Yet the negative, for TO, has a certain detrimental effect. There is a par for the positive of TRU in TIT. What actions are taken have to be measured by the risk benefit ratio. The dichotomies, dialectics, oppositions and competitions existing in TRU can be of great help rather than hindrance when used for, and to represent, TRU. They purport to show both TRU as it actually is, and enable us to reproduce, construct, model and posit TRU objectively. However corrections must be made if the needed arbitrary restraints and standardizations are to be reconciled.

TRUQ = TRU Equilibrium, What is vacillating from 50:50 tolerance are not infinitesimally, absolutely static, nor an infinite dynamic relative. Many factors go into the massive dynamic state of TRU that constitute, contribute and provide for its equilibrium. In this way a dynamic of TRUS, TRU, and TIT that is simultaneously: balanced and unbalanced, and of AMTRU, ATRU and AUS that is increasingly and variably adynamic and dynamic, unbalanced and balanced. Not only is the determination of cause, reason and forming of DA, AD, DIA, DAS, DAO, DAU, DAR, DAT, and DET very important for the first principles of TRU, it is equally imperative in the determination of action. Effect and cause are often and practically inseparable. **PRADAC** – the Practical need and limits for the Determination of Causation.

The dialectic of idealism and materialism that constitutes TRU exemplifies the dynamics of Á. Yet this dialectic is so much of AU, EPI, POT, and ROT that a new terminology is needed. **POSTNES** - forces and Processes that are POSitive to TRU yet NEgative to each other, as between different ES and CES. There is a redirect of the tangible, or what we sense, touch, or feel may not be TRU or even much TRU, but to be, must be redirected into TIT.

NEGATIVE and POSITIVE MEASUREMENTS

ABSOLUTE					RELATIVE					ABSOLUTE
← ABTRU						ADTRU →				
ANTITRU		ANTRU			ATRU		NAMTRU			TRU
I	I	I	I	I	I	I	I	I	I	I
-	-00	-nf	-nf	-1/i	0	+i/1	+nf	+nf	+00	+
-10	8	6	4	2	0	2	4	6	8	10+

LADCEN = Law of the Distinct, Certain Effect of the Negative. LADCEN is a product of the fact that there exists in what is held to be negative, a positive effect. This is signified by: **LARNIP** = Law of Negative in the Positive. This effect is not negative for TIT-TO however. For TRU can only be positive, in and of itself, TIT. Yet this paradox is resolved by the fact that the negative produces a decrease in the positive, or a relative loss of TRU. Thereby sin, evil, bad and all that is negative has existence in TRU, not because it exists in TIT, but because it manifests indirectly by increasing the positive, qualitatively and quantitatively in TRU.

PARETEN = Paradox of Ethical Tension, or the contrast between TIT, TES, the different, and ES that produces polarity, opposition, dislike, detachment, and defection. The great handicap has been that TRU mechanisms and therefor their basic principles have not been poorly, or not, known and shown. Nor is their lasting appreciation and respect for ES or TES by other ES or TÍS. It is mostly the result of a persons failure to maintain their full TRU TES but momentarily. Thereby the cause of evil, wrong, bad effects is the ATRU and ANTRU effects that are so closely associated with each person. This, as its reverse the cause of good, is that TRU effects are axiomatic, and why TES, KOT, TRU, TIT, ACC, Pureaclity, Cognos, ELTICMA, SÂ, and TÂ are so often emphasized. **AXCET** = Axiom of Cause that is actual and good produces the TRU effect.

The risk/benefit relationship and ratio is a key mechanism in TRUD and TIT. For all that is done, has purpose and is in, more or less, a cause and effect relationship to us. The object or goal of what we do, our activities and lives is to attain a maximum benefit with the minimum of risk, harm, or loss. Yet what often confronts us is more likely to be a maximum of risk and a minimum of benefit. This is simply a measure of the positive and negative forces in the TRU/AU or AMTRU interchange; how much they are active, how much and in what way they effect us.

TOTAL DYNAMICS OF TRU, MASS ACTION, VITALITY, PIP, PROCESSES COMPENSATION, INEG

TRUDO or the Total Dynamics of TRU primarily involves those functions that best provide success over failure, health, happiness, prosperity and survival of TRU. It is stimulated by collective experience and suffering. It is inhibited by an inherent tendency to rest and digress. It is a balance between growth and decay with the integral positive processes being dominant. There is a continuous and massive interaction between what is for and what is away from or against TRU. The best function of the whole and control of these provides the stability of TRU, which can be enhanced by controlled positive exercise of the higher powers of Á. Such control is conditioned by the balance between unification and infinite differentiation with correct over incorrect dynamics or motion, forces, actions and processes. When not turned to TRU negation is often characterized by what is called: **IN(EG)** or integrative negation. This is what happens when things go wrong, they tend to snowball. It is exemplified and associated with accumulative deficiency, degeneration, and disease. This is called: **COC** or Complications of Causes. INEG and COC are combated by compensation and PPMA. Compensation is the power of a body, RU or or TRU to return to optimal health. Compensatory powers are built into TRU both formally and functionally, many of which are evident in the human body anatomically and physiologically. For all TRU, by TRUDO, there is a resultant or final compensatory state. This is a collective harmony of all systems, mechanisms, and processes working within the time frames and limits of space and mass energy of TRU. In collective suffering when a part suffers the whole suffers, when an individual suffers the whole human race suffers. This is also the case for ES, TES, CES, and TRU. It is the affective product of harmful stimuli that are interactively transferred to many or the whole, TO of TIT.

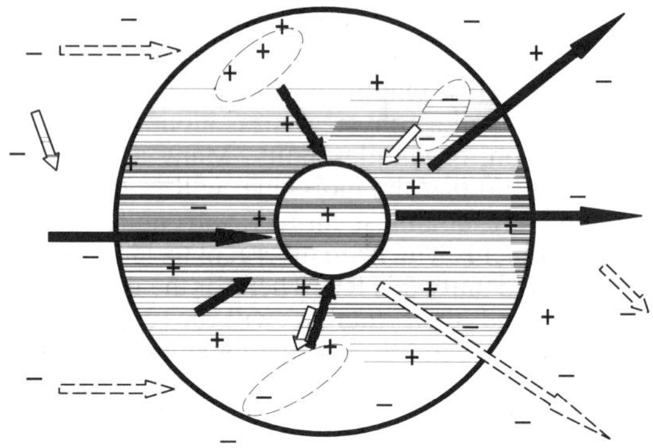

Fig 5-17 **SPOCIS ANALYSIS**

SPOCIS is the Summation of Positive and Negative forces in and of TRU. The diagram shows, in a rough way, how TIT is bathed in a sea of extraneous forces, effects and interactions. Many are negative, most are complex, and in varying amounts to which they do or do not equal TRU. This is the degree of its motion, dynamics and conflict. Their summative effect is indicated by the larger arrows. If positive is more than the negative the overall effect is positive. If the negative is greater than the positive the overall effect is negative.

Mass effect or acts that characteristically unify are common to all that is of TRU. It is this that makes a person, nation, civilization, mankind, human life, TRU and the TIT universe of such great strength and concretion. This is seen in astrophysics where a neutron star's mass and density increase toward the absolute so does the periphery become not only a perfect sphere, but a hardened one. Mass operates integratively, in fusion and works through and through, by ACEXIN collectively, generatively, cooperatively, synchronously and harmoniously. Necessity of massive summative positive indicates that there can be no final quick fixes for the human condition, SCARLS of TRU. It is most imperative that TRU not be taken for granted. Neither absolute extreme is a product of the existent TRU.(LANERA)

REVNAP = Reversion of Negative into Positive. We must always be careful and remember to keep this limited by TIT. Although directed toward the elimination of negatives in reactions to actions and responses to stimuli in all about us this should not be done at random, inadequately, excessively, disorderly, inefficiently, painfully, or wastefully. It needs to be done in a modified, controlled, most effective and efficient way. This is largely by utilizing the power of TRU and KOT. Learning, experience and especially TRU learning, is of higher levels of TRU Á FAF. **POCAD** = Principle of Compensation from Defects. How in reaction to and compensation for defects, deficiencies, negative forces or influences we or TRU becomes stronger rather than weaker.

Potential or actual inevitable damage to TRU must be

discovered, learned, known, shown, managed and controlled if we are to assure and best enhance TRU. **SINEG** = Summation of integrative negation. SINEG is the challenge that is enough to satisfy the greatest desires for adventure and discovery. The great problem is to acquire the necessary capacity, ability, means, will, and TES to overcome SINEG.There are endless distractions and indecisions from the positive, TES and optimal TRU. Those recurrent defects in each and all of us leads us to ATRU and worse. Ethics thereby is little more than learning and following TES its harmony with TRU.

Ethics is produced by working out the details of what TES and TRU are, and how, why, when and where to act accordingly. **POSTA** = Positive Turn Around. Much of all that we encounter in life does so in a negative or mixed way. This, as all, is either TRU, AMTRU, ATRU or ANTRU. It is usually becoming or leaving TRU. When TRU it, although not negative in the absolute nor total sense of being against or harmful to TRU, is negative in the relative and partial senses of being against, detracting from or in some kind of dialectic operation with each other.(VAPO)

TRUD, ATRU, ANTRU, ANTITRU EVIL, FISSION, DISASSOCIATION

Fig 6-17 **GOOD and BAD in TRU and TRUD**

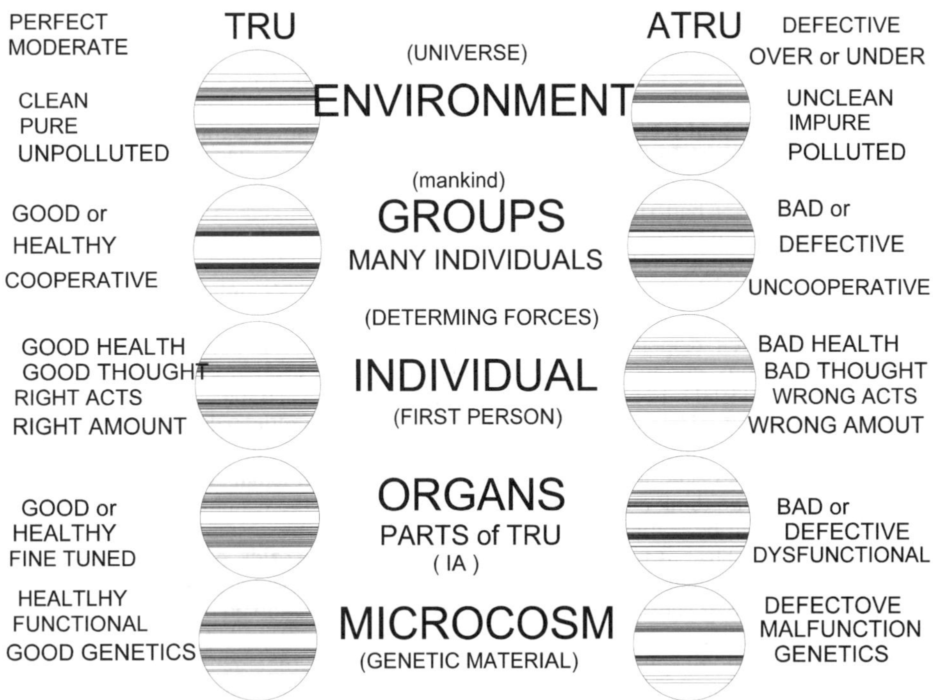

Some of the more common sectors of the world and their good and bad components are shown. To a large extent hese correspond with what is good and bad in TRU. Humans have known, do and will know, to a large extent, what is good and bad for themselves and, to an often lesser extent, their world. It is only for all to do and live without abandoning hope or negative attitudes toward the worsening threats to human lie. It only needs guidance, representation, presentation, plans, controls, and management and the appropriate strategies, tactics, and operations to carry it out by TRU. This is a prime purpose and provision of TRUD, HOM, ACC, POT, TRU and this text, IPTRU.

To understand TRU and its dynamics, TRUD, we need to know what is from TRU or **ATRU**, against TRU, **ANTRU**, and intentionally against TRU, **ANTITRU**. These are highly interspersed in **AMTRU** or the great mixture and interactions that occur between TRU and all that is out-side of TRU, as in an individual's life. In this way we need to know what is and is not TRU. This is often the difference between TES and any individual. It is also to know, much as possible about, what TRU is, what it is not, and what is not that has no, little, or much effect on TRU, and in what way. This is like how to know health we must know disease. It is continuously sensed by the individual person in the stresses and strains, good and bad, and ability to cope with life. Much of what is from TRU is so through loss or lack of contrary. There are continual processes at work, as of gain and loss, and metabolism and catabolism. When of loss it is the prime cause of degeneracy, dissolution, disassociation, fission, and the origin of evil. Any variation that deviates from the optimal form and function of TRU is ATRU if only slightly. The paradox produced is not without resolution, but a manifestation of the great mass and total dynamics of TRU. What departs from or recedes from TRU, does so relatively in a proportionate and measured way. ANTRU is a function of the amount of potential and actual loss to TRU. What is against as apparent to the individual, is often much different than what is against TRU. A person lives in a flux of these various states and TRU depending on goals, interests, circum-stances, encounters, work, degree of TES and its operation.

ATLOC - From TRU or Loss through Lack of Contrary, e.g. if there is much large and little small, much past and little future, much subjective and little objective TRU loses. There must be a proper distribution and equilibrium of all that best enables TRU. OITIO involves both, the turning of bad into good as the potential becomes actualized as it is selected out of TRU, the good for the bad, or the good becomes potentially bad. For ATLOC this is raised to a higher level in which the good becomes bad as the optimal state of the parameters, dimensions or axiomatics of TRU decreases. ATLOC shows the greatness of the differences, parameters, dimensions and axioms of TRU.

There is a relative parametric proportion in any universe if it is to survive. This is the balance between its primary continua, parameters, and dimensions. If its mass energy, length, and time are irreparably unbalanced, it becomes extinct. If they can be compensated for, temporarily or more prolonged, it survives. The forces and power of such parametric proportion work to maintain the universe. If overwhelming forces and stresses occur, the result is failure over success, and extinction. This explains why, and the degree which, large animals lose their ability to compensate. Increasing size temporarily adds to predominance and enables being. Yet increasing size disenables maintaining relative parametric proportion and survival. In the balance between genetic material, live product and determining forces relative parametric proportion continually operates to support, enhance, and perfect the specie, and thus TRU.

LANTRU = Law of ANTRU. LANTRU is related to the sufficiency of TRU or the degree to which what is about or of potential input to TRU is apt to cause loss to TRU. If TRU is highly sufficient, predominant, and well protected then what is ANTRU will be to that degree less. LANTRU is seen in health, strength, prevention, and preparedness whereby actions, threats, dangers, and deleterious effects are overcome by the power within TIT. Breaks in lines of defense can be critical, as exemplified by how the human body becomes diseased or injured. **CUST** = Contribution Unitary to the Success of TRU. There is only one final measure of all things. This is how much do or don't they contribute to the success of TRU.

DISOT = Dissociation of TRU. TRU is continuously undergoing both dissociation, assimilation, incorporation, and consolidation. The total benefit will depend upon the quality of the gain, and the efficiency of selection. A good measure of TRU is Quality of Life. To measure it we must first, carefully and thoroughly, investigate and establish what it is, that TRU and TES are. This we do when we develop TRU, TO, TRUD, ACC and POT.

INTEREST, GOOD, and RESPONSIBILITY, WILL, WORK, POWER, and TRUD

In order to enhance TRU, interests, reinforcement, best understanding of good and bad, right and wrong, and the use of compliance and opportunities are necessary. Even greater, the success of TRU depends upon its knowledge and reproduction so that in work and life optimal exercise of will and power are achieved. This is so each and all will fulfill their part in TRU, TES and TÍS, meet their responsibilities and do their duty. These may seem obvious, common, and boringly imperative, yet they are the crux TRUD, ENT, TRU in being dependent upon each and all persons. Thereby it is the obvious, common and imperative that we continuously address and need to make interesting, enjoyable and beneficial for our will and power to TES and TIT. Those who think they can win overwhelmingly in one world, will often be annihilated in another, and these worlds are to often not TRU. It is easy to win in any world, it is not so difficult to beat ones fellow man. It is harder to do right and pursue a life long endeavor to win at becoming TES and TRU. We should be careful to judge others only in terms of the TRU. Values of all, like interests and REM in TRU, are the origin of this enhancement of TIT by TRUD. Many things add to these values and interests like examples in positive reinforcement.

It is a will to be, to live, and find sufficiency and security that drives us and increases our values and interests. This involves a will to believe, to believe the truth, the truth of what actually is, TRU. Their positive objective understanding provides the optimism, power and fulfillment that human life warrants, and has the FAF and capacity to be. Interests, will and will to TRU combine to provide the great power our being requires. Ideas, concepts, principles, laws, words and language provide the means and vehicles for these great powers. They accelerate TRU and are like bullets whose powers are the magic of God, TREQ, or TRU. They are a large part of the reals, or middle range of the Á spectrum of TRU. It is in the best exercise and use of these powers that TRU is provided with its greatest predominance and acceleration. Since TRU, like God by St. Thomas, is all and self determinant, so is it all providential for humans to live according to the will and power of TRU. **WILT** = Will TRU, or Will in us to TRU. WILT is a variable that is a function of our effort, capacity, potentiality, knowledge, work, perfection, soul, benefits, conviction, determination and overall comprehension. Will is an important determinant of individuals and TRU. It is a will to life both by the person and all TRU that is much of its driving force.

Powers of reinvigoration are often those that have the greatest strength, value, and are at the center of TRU. They enable us to gain energy, new power and strength, in rest, rebirth, refreshment, recreation, change, nourishment, and new ideas.

LOPETAL - Law Of Power, Equals its Actual Existence, mass or weight in TRU and Life. LOPETAL shows how all that is proposed or novel, unless actually used, is of no or little power or value. **TÀP** - TRU Assimilation Power, or all those means by which TRU is able to assimilate and provide what it needs. **TÀPOR** - TRU Assimilation Power, Portional. The drawing together effect of TRU through SUT of its ES, CES, REMS, concepts, laws, and principles. TAPOR is a great mechanism and basic power TRU. As much of Á it is the predominant mass energy power in TRU. Its effects on the new or higher levels of energy has been revolutionary, glorious, and victorious. *DHARMA* is the cosmic law and power of virtue or right. In TRU it is the power that propels virtue and right as they, like ESINS become CES, universally pervade TRU. It is what sustains, helps, aids and enables TRU by ENT. **POGE** = Power of Good over Evil. **LARPIT** = Law of Right Power in TRU. In action and reaction those who are powerful and right are the great driving forces in the world, TRU. For right is power, and power and right are TRU. They are or tend to become the convergence, concentration and culmination of all positive energy and powers in the actual world, TRU. It is a universality of the best, right, good, great, power and will of man. This is TRU. LARPIT or the power produced is what exerts its effects in TRU through much of the absolute, AU, or mixed AMTRU worlds. We shall overcome is both the challenge, as in becoming, and the enigma, as in being infinite, of TRU. Yet it is not for any partial or individual. It is for all TES and TO.

TRANSCENDENT FUNCTIONS and SUMMARY of TRUD, PROBLEMS, SOLUTIONS, CORRECTIONS

Humans are preoccupied with problems, good and bad, voluntarily and involuntarily. Much of this is natural and proper, especially if TRU. Problems result from actions, errors, imbalances, a kind of negative entropy, and chaotic effects in the systems that constitute TRU. Of human problems one of the greatest is that of many different separate yet highly intermixed worlds. Mans reason strives to understand, analyze, correct or reverse these problems and the chaos they create. It can be resolved by discovering how these worlds are other than TRU, TRU, TIT or TÂS and rise to the necessary responsibility coincident with such an endeavor. Humans inherit wonderful capacities of thought and action that can explosively permeate, and reveal the transcendence of these worlds. There is provided by TRU a correspondence between the problems and their conflicts, solutions, and corrections. Their convergence and fusion in TIT is positive, for which good representation of ES, TÍS, TES and CES can be made. Positive forces of good become the absolute forces of right as their purity approaches perfection in TIT. It can be achieved by those who acquire proficiency in TES. It is in this way TRUD provides the best for all, present and future as we have received from the past and is only in TRU. Preoccupation with problems and controversies fuels stimulates the human mind and fuels interest in TRU. It provides and produces the great depths of insight for solutions which otherwise might never be attained. It enables an understanding of the extremes, means, and mechanisms that are often basic to knowledge of TRU.

PORC = Principle of Coincidental Responsibility. TRU is not solely human nor even entirely conscious, and intentional. Yet since we are the only live, knowing, carrier of knowledge and since TRU is half subjective, TRU is coincident and in part equivalent to the species, homo sapiens. Thereby it is largely in man alone that TRU can depend for sustenance. Each and all individuals must carry their share or responsibility, TES. Since The and We are the relative, this may be applied to all E, parts, of TRU and for each its proportional values, REM and their carriers as PORC.

DAREV = Derivation, Determination, and Development, of the Rational, Reason, Resolution, Right Action, Reversibility back and forth, ad infinitum to TRU. This leads to: **TAFOR** = TRU Absolute Forces of Right. TIT is tantamount to the absolute forces of right. This is what many religions claim, yet err not in TAFOR, but in how much their beliefs are less and other than TRU. The significance of inversion lies primarily within TIT, and represents that great tendency, ability, or strategy of turning upside down, bad into good, wrong into right, or vice versa. Man's singular ability to direct action and eliminate, assist, or assimilate what is bad, wanted, necessary, and good is what sets humans and TRU apart, and makes it aright, all right. So it is also of TRU, SUT, ROT, MOT, SPALT, REM and HOM.

FIS = The Fallacy of Irresponsibility from Freedom in Subjectivity, arises from the weakness of, or failure to exercise, our higher powers of Á. The correction and treatment for FIS and the problems associated with an unstable and unfocused TIT is especially found in MOT, ALT, SPALT, METAD, and their assimilation into ELTICMA and SÂ. We unconsciously and continually try to remake the world in our own image, only both are apt to be far from TRU. Thereby it must become our challenge, inspiration, objective, aim and determination of all our might in action to prevent these diversions, and direct them perfectly toward TES and TRU. **CEPTAT** - Compulsive, Explosive, Permeation of Transcendence of Act by Thought. **CRIMPOT** or Concept of the Convergent Resolution in Mathematical Precision of Activities to TRU. CRIMPOT is how all the activities of TRUD, when attended to properly, become conditioned and converge on TRU.(CONVET) Life's trial is revealed in the great paradox of and disparity between IA, inspiration, revelation, TU, TO, TRIA, and TÂ. The all time identity, permanence and power of ideas, concepts and terms, and the IA, effort, work, actuation, extension and intension. It is exemplified by the great difference between our glorious exuberant feelings of energy on awakening in the morning compared with the opposite after a long hard day. It is also exemplified by the manic feelings of power and victory produce versus the depressive feelings of weakness with defeat. This is, PAEIOU, the paradox and trial of Absolute, Partial, Infinite, Total and Unit, or how we succumb or survive by TROS. This is, as it should be, more than enough challenge for all. It should be the cure for boredom. It should be the key solution to FADA, GEXA, STOD and many other paradoxes.(Tab 1-092) Like what is so true for the common themes in all paradoxes, this is the dichotomy between integration and differentiation, focus and spread, concentration and dispersion. Its cure, also like what is so true for all common themes, is largely one of KOT through TIT, TRIT, OBT, SUT, ALT, SPALT, ACEXIN, and the beautiful exquisite detail and hard work needed to posit, fill in, represent, present, best use and live by TRU.

Fig 7-17 **TIDA; TRU Itself, TRU Ideal and TRU Absolute Dynamics**

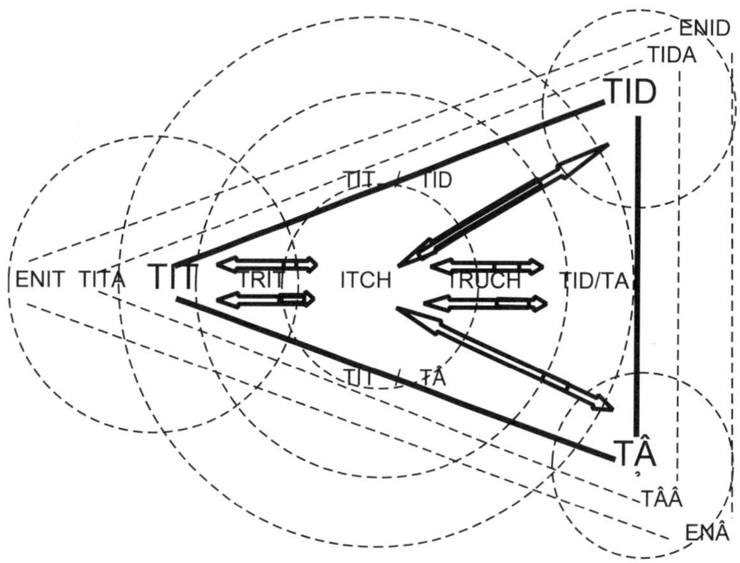

TIDA is the diagram of TIT, TRID and TÂ interactions. Of all the great questions, problems, enigmas and paradoxes that we face there is one that is central and overriding to all, TIDA. The answer to it, its solution and resolution can be the most rewarding of all our inquiries into TRU. This is the great three cornered struggle, battle, war and conflict which exists between TIT, TRID and TA. Do we keep TRU as it is in and of itself at this moment, TIT? Do we do this without fail and without irreversible change? This is the idea of the great protector and preserver, Vishnu. Do we allow whatever to change that we think is likely to improve our world, our being, the ideal world, or heaven and to destroy all that is in its way. This is the idea of the great destroyer of all that is not ideal, Siva. Or do we struggle and strive only for that which is absolute and which overcomes the relative and all. This is the idea of an absolute creator, Brahma. Since none of these alone is the final ultimate resultant it is up to us to decide, if only in those ways as is our lot, what combinations and dynamic SCARLS are to exist and be utilized in this overall relationship and totality. It can be seen that the ultimate product will be some happy medium between the three. This by TRIA and TIT will be to a large extent extra TRINI. It is also likely that the ultimate product will not be some fixed state, but some variable dynamic state of the totality of the three in which the overall success, well being, and survival are the greatest determinants.

TIDA is dynamically of TRU yet not confined in TRU, TIT, nor does, or will, it occur in an absolutely changeless state as TÄ of TRINI. Just as we approach TRU from RUT, RUD, RU, AU, and EPI so must we understand TIT not only by TRIA an absolutely fixed state, but balanced out by that of TRU which is not absolutely stationary by TRUCH, TRID and TA. First however, we must be TIT of TRU. For if we are not, then TIT-TID-TA or TIDA makes little difference. Also we must distinguish the differences between TRUD as it is TIT and TRUD of TRUS, TRID, and TÂ. In this way the dynamics in TIT precedes and is followed by the dynamics of what is of TRU, **TRIDA**, and from TRU or ATRU. The transition of mankind from lesser to greater intelligence as shown by increasing brain capacity, function, language, A, and ES, as well as the altered states that will occur toward TA, and the great changes that such knowledge of the world provided by religion, philosophy, science and technology will continue to occur in the future. They certify an alteration of TIT in the direction of TRUID and TÂ. This not only renders TÂ different than TIT, it assures, makes actual, and certain that TA is not a figment, but TRU. It is important to know that many changes in the future can be explained by and are already in TIT, or within a reversible and sustaining TIT. Yet their variety as to what is optional and can only be of TRID and TA, not of TRINI. It will be seen from this that the ideal state is one of maximum stability and equilibrium in which there is maximal actuation, extension and intension in unity, finity and infinity and all in between. We must forever remember that we are here to fulfill life's purpose, in TRU, to treat the world by fulfilling this role in TRU, and by being proficient in all TES and TRU. Each person needs to be a professional of TRU. **DOCT**(TD) = Doctors of, or TRU Doctorate. Such designations help reveal and empower sufficient levels of knowledge proficiency, especially to teach TRU. **TRUEP** or TRU Replication, Fusion to TRU, or how TIT regenerates itself through its own FAF, especially mass energy, Á. Language successfully represents, replicates and returns it to TIT. TRUEP is in TRU like in a series of concentric circles, converging on the unity, or TU of TIT.

CHAPTER 18

ENT or ENABLEMENT of TRU

TRU ENABLEMENT ENHANCEMENT CLASSIFICATION ACTUALIZATION INHERENCE REALITY

ENT shows what makes, and continues to make, TIT healthy and better. ENT emphasizes all that brings about, sustains, maintains and promotes the unique and special nature of TRU. In accord with the POA, and to better know and show TRU, we need to find out what its purpose is, what best makes it work, how these enable it to be, and to survive. To do so we divide it into its many categories, sets and classes from the more primitive and simple to the most complex, advanced, total and final. We investigate what is for TRU, inherent, good, better, and helps it become actualized. How reality as TRU improves and provides a means for best judgement. Life, as a person's life, inherits an obligation to being, without much alternative but to accept it and achieve its fulfillment. This is by **RÍTVA** or right is TRU, correct VAlid VAlue. This idea is so common an idea that it may seem unnecessary. Yet when we consider its alternative the loss of the actual full meanings of right and TRU, plus the fact that it is limited, we can begin to realize its need. RITVA is most powerful, much of the generator and preserver of TRU. It shows what is and what is not, what we should do and what we should not do, what we can hope for, and what is beyond our limits. ENT is much different than TRU or TES. What enables us to be is not correct for us as NAMTRU. Even the us that is its counterpart in TRU is, by being only variable and partial, less than ENT and TRU. **ACENT** = Accentuation Equality to ENT, to Enable TRU to be, to the Enhancement of TRU. ACENT, as in LADU and LAL, is the quality of expansion that accompanies equality as in to '*Accentuate the Positive*'. ENT and ACC are highly coterminal yet much different and distinct. ENT is focused upon all that benefits TRU in general, ACC are the more specific means by which future TRU is benefitted or achieved.

Meliorism, a great precedent for ENT, is a system that promotes an abiding faith in improvement, as TREQ has been for TRU. For TREQ, with God as our witness and example, we work toward a set goal, TRU. This is based upon an ever greater wisdom of all as TREQ or TRU. Humans have increasing levels of such wisdom and ability to be TRU. This is enhanced by answering the question, 'what am I living for?' This helps us look at TRU, rather than what is away from or not TRU. They help to exercise and produce reason, by an ethics based upon TRU. TRU ethics is consistent, reliable, and can be reconciled with the best in religion, philosophy, metaphysics and science. Such TRU ethics as an exact science must follow certain methods and principles as well as set VAPOS and SCARLS of TRU. It is in the precision, essence, and probity of TRU that actuality and accuracy combine in ENT to provide ever greater truths, proofs and achievement for TRU. There is an absolute need for distinction, force, and perfection in TRU that drives us ever forward.

LOPAP = Law of Positive Application of forces, Powers of subjective and objective to TRU. LOPAP is a great part of the 'Good News' and joy of the New Testament. It is an exaltation, as by God, to provide the necessary hope for the success of TREQ or TRU. It is in the relationships between and the function of the various states of TRU that we can begin to provide an ethics that approaches an exact science. This becomes evident with: TROS, TU, TRIA, TÂ, SUT, TE, TI, ENT, REM, TES, and many of the concepts, principles, laws and mechanisms of TRU. **FOLO** = Fallacy of the Limited Objective. Since all TRU is largely a systematic unit by relevancy and with a meaningful objective end, any objective can be right only to the degree it is unlimited, only within TIT-TO. Thereby only in proportion to total TRU or:

Objective = TO x TRU/AU.

Probity is the right and precise probability and way in life. This is what we search for. This is not any probity, but only what TRU is, and what we attain when our presentation, representation, and observer-observed relationship becomes 1:1 or 1 with TRU. Probit is a statistical unit of measurement of probability based upon the deviations from the mean of a normal frequency distribution. It is like proportion, the goal of an ordered and cosmic world, the right and precise probability. Probity is the uncompromising adherence to the highest principles and ideals of impeccable integrity, uprightness and rectitude. Probity can be no better than the degree to which our knowledge is right, we and all are TES and TRU. Probity is not an automatic gift to the individual. It is something that must be earned, requires learning, appreciation, deep conviction, determination and life long endeavor to become and be TES and TIT. Probity may be attained to some extent through imitation. Yet this kind of probity is likely to be much less pure and permanent than what comes from suffering, reason, experience, revelation, grace, faith, hope, love, and a heartfelt feeling, interest, inspiration, conviction and determination to be TES and TRU. Probity is key to the quality of mechanisms and CES of TRU.

LANOP = or The Law of Absolute Necessity of Perfection, in the knowledge of God, and in the fact that this God as known is perfect, in terms of TRU. It is the degree to which, and the rate to which, this knowledge, God, TREQ or TRU are perfect, that is most important and necessary.

PROOF of ENT, PRENT, is seen in efficiency, enablement, faith, good, great, harmony, honesty, strength, many superlatives, and virtue. PRENT is also evident with TINS to PRUT along with DOPRUT, TREQ, TO, LACOP, Summum Bonum, PROVT, and MUPT. Human success is highly dependent upon, enthusiasm, interests, appreciation, planning, order and control. These in turn must be based upon TRU, with the best possible knowledge, VAPO, POA, and the greatest identity of all, oneself and what one and all are doing, in TRU, TIT.

DERIVATION, DEVELOPMENT, DETERMINATION of ENT, RECTRU, REDEMPTION, HAPPINESS, LOVE, OBLIGATION, RETENTION and S-O

As so much in derivation and development, ENT depends upon the IA of the individual, S-O and accompanying vehicles, as language, that make up the mass energy, Á continuum of TRU. IA helps us recognize the categories of life or levels and steps to ENT. There is a built-in mechanism that obliges and confines all participants to TRU in TIT. Stimuli, pain, experience, and suffering produce feeling, love and happiness as they support and enhance TRU. Certain knowledge, as in law, has built up many set aspects, mechanisms, and guides that have, are or can enhance TRU. Heaven as the hope of TRU, in contrast with hell, provides an ideal that stimulates living to achieve TRU through ENT. Combined with IA, Á, and SÁ we anticipate and direct what is coming, so that sufficient and appropriate response and reaction can be provided. These are much of the origin and reason for ENT. St. Thomas a 'Kempis is quoted to say "You are either on your way to heaven or to hell". This as so many phrases, quotes, dictums, proverbs and sayings is of great truth and value. Yet without proper interpretation, direction, VAPOS, SCARLS, DAOST, use, and practice they can become worthless or negative. There are several reasons why it is upon the individual and ones rise to TES that most all TRU depends. This includes: focus on TRU, back-sliding, distraction, failure to exercise the best of our great capacity, errors of desire, reason, judgement, power, mass energy, or Á FAF at the appropriate times and places.

SATRU - Separation from (Ab) TRU, or the continual processes whereby we become separate and select from what is not TRU. There are several kinds of SATRU, as negative, mixed and positive for TRU, or simple natural choices, complex natural choices, rational and judgmental choices, plans we make and reject, that done by imitation, routine or law and by a person, groups, communities, nations or the whole worlds population. Other kinds of SATRU include: the subconscious as lower order SÁ, selection, inclinations, taxis, tropism that produce motion, in a direct and indirect, involuntary manner, and selection affectively through emotion, feeling and love. SATRU and selection help to better understand associated mechanisms and methods of developing, producing and enabling TRU. **LAMEX** = the Law of Mutual Expansion.

RECTRU - Rectification to TRU, Return or Correction to TRU, and the great forces by which all things incline, tend, or are forced to TRU. Those who labor give with great effort, and do more than their share of the work and benefits, even under great limits and disabilities, are the heros and legends in their own times. The greatness and beauty of the forms, ideas, concepts, principles, mechanisms and laws which languages have produced have been deified by many peoples and cultures. This is especially evident in cultures that associated and developed from common objects, higher ideas, concepts and terms that were found to be most beneficial. It is upon this means of abstraction with the ability to make complex abstractions in reasoning and logic that language has continually reinforced and empowered TRU. This is much of the FAF of human mass energy, Á. It is vividly seen in the deification of the sun, moon, physical objects, animals, and HERoic personages, **HERI.** It has been the expansion and use of massive and interactive abstractions that spearheaded ENT.

The hoped for relief for mankind from need, wrong, sin, evil and suffering was expressed by Isaiah with the idea of a coming savior, Messiah. It should be no surprise then that a self and deity conscious individual, Christ, expanding upon the prophetic influence of John the Baptist, would take up this role, to become, like TÍS, a figure of hope in an age of great stress, strife, conflict, and suffering. Humans have a great sense of identity, not only with the world about them, their loved ones, but with the whole, GOD as TREQ or TRU. It leads to personal leadership and hope. This is especially so when associated with profound suffering. It is the great need to find a way, guide, direction, symbol, image, abstraction, and salvation in GOD, TREQ or TRU. This was attainable through demanding grace, abiding faith, salvation, obedience, compliance and discipline in a life dedicated to God, his law and word. This to them became through identity with the Messiah, Christ, TES, GOD and TRU the Kingdom of Heaven. It was made up, not of armies and kings in material battle, but of those with love and goodwill. This idea and realization provided the "Good News" or message of the Messiah, Christ, a messenger, as also was Mohammed, of God or Allah. It was the message of God, TREQ or TRU to all people. It proclaimed God's gift to man, identity and everlasting life. It was available to all, to save them from sin, and bring all to TRU. It is how TES empowers TRU through its purity and perfection. It is how all in becoming TRU become strong and eternal. It was as each person's life must be, if they are to be, rather than not to be. To be thus as early as possible by coming to terms with life, TES in TRU. This is how TES, as all ES, come in totality to TRU. It is called: **LANUT** = Law of Absolute Imperative Necessity of Unity in Totality. **LOB** = Law of Obligation, Ought as being, TRU, or an ought that is not something outside of or other than being, it is being. It is where necessity is not something out there occasionally, it is a basic feature of TRU. It is the future in the present, the future in the past, and the past in the present all running through and through the present, past, and future.

Both believers and non believers in Christ miss the point and common ground in the Miracles of Jesus. They all represent the striking difference between IA and IAT, or the apparent and the actual, ATRU and TRU. They were gradually reinforced in time. Those who heard the message of Christ and came to realize the difference between the stark reality of what was occurring and its actual meaning in the whole. This is the difference between the static partial of a NAMTRU world, and the 'spirit' or dynamic total of GOD, TREQ, TRU, and TIT-TO. It is astounding, and a difference that is inconceivable to non believers. The occurrences during Christ's ministry were gradually reinterpreted and reinforced in time. The differences between illness experiences or Jesus' 'healings', at one moment or another, its rapid fluctuations, 'getting over it', wine being mostly water, wishful thinking of one or many, hope given through attention, as the laying on of hands, etc. These are all justified differences used, in ways, by modern practitioners. They show the naivity of common IA, and the actuality of IAT, GOD, TREQ, TRU, and TIT.

Happiness is survival, meaning in life, love, identity, bonding, physical fulfillment, reproduction, communication, discovery, problem solving, learning, understanding, mental and intellectual fulfillment, and reality in TRU. Yet happiness that is cheaply bought without these and by material gain and security is but a small part of TRU happiness. It comes after we are godly, discover, learn, know, and serve TREQ, become TES, live in PUREACLITY and be TRU. That there is much law in religion, and religion in law, is certain only if, and as they are both, their equivalents in TRU. For religion is the deification of each and all in TRU, law is this imperatively.

Fig 1-18, **Simultaneous and Exponential ENT**

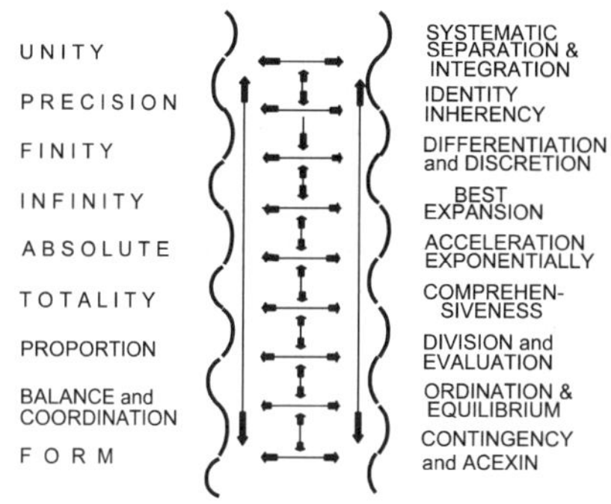

How can we simultaneously and exponentially attain increasing, 1. Unity, 2. Precision, 3. Finity, 4. Infinity, 5. Absolute, 6. Totality, 7. Proportion, 8. Balance and Coordination, and 9. Form.

These proceed as unity in TRU is systematically unified and separated with precision by best 1:1 identity, inherence, and finity by best separation and distinction. Likewise ENT is determined in TÂ by exponentially greater acceleration, totality by best comprehensiveness, proportion by best evaluation and division, balance and coordination by best ordination. Also the best identity with TIT is made through the use of contingency and ACEXIN.

GENERAL FORM of ENT CLASSES STATES

To look ahead and see what is coming is the key feature of S-0, S-R, ES and CES as they characterize IA and together provide the FAF, Á, of TRU. This vividly exemplifies the great massive sensory - motor makeup which we so ably or poorly live by. This is the basis of human life, TRU and ENT, as it is best shown, differentiated and integrated by TU, SUT, ROT, MOT, ALT, SPALT, POS, STA, HOM and ACC.

Economy, in principle is much the same for all ENT and TRU as it is in the business world and commodities markets. To economize means to cut out all extraneous and unnecessary activities and expenses. In TRU it means to operate efficiently and optimally effect, manage or control input and output. **STAT** = Stable State, or the SCARLS of peace and equanimity of TRU, TRIT and TIT. STAT is evident in: **SORISM** or Summative ISMS, where many, if not all the polarities, dimensions and ISMS of which TRU is composed are best proportioned, balanced, and made whole by VAPO. STAT is that state in which there is a minimum of TRUCH, alteration and extraneous forces with destructive effect. To have only that needed and best use it is the law of natural economy. It shows how TRU is formed, functions, is successful and must be upheld.

Fig 2-18 **MOTIVATORS and INTERESTS FOR BAD and GOOD in BAD and GOOD WORLDS**

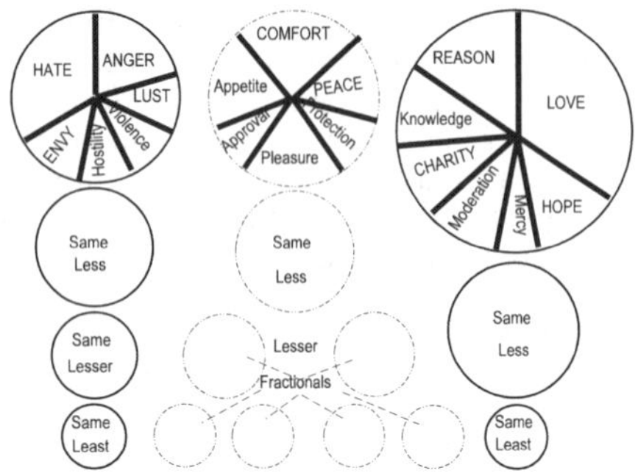

Much of the AMTRU world is driven by the deepest and greatest inclinations, motivation,, intentions, and interests of individuals and groups. These arise from many causes or reasons, chemical, molecular, genetic, physiological, biological, environmental, learned, habitual, purposeful, direct, indirect, curable and some incurable. In Figure 2-18 the horizontal plane from left to right shows that most ANTRU to Neutral to most TRU. The vertical plane shows what is most active to what is most inactive. The characteristics, as hate to love, are only a few that could be posited in these different figures. In fact some that are in one item could as well be in others depending upon the degree of ANTRU or TRU, activity or inactivity attributed to each. Even within a given person there is more or less variation, as the neutral or intermediary comfort, peace, approval, protection and pleasure attest. Each will tend to have a set battery of characteristics. It is this set which can be used as a measure and test to determine what a person is likely to do, and how they can be improved, or be more TES.

The principle, rule or law of immune complexes is the 1:1 relationship between antigen and antibody. The antibody disappears quickly if the antigen is not present. So is there an optimal relation, between negative and positive different and same, DAOST, in TRU, as well as stimulus and response out of and in TRU. These are kinds of an all or none law to the extent relativity is involved, or that of a simple dependency. For if, in the relationship one is missing, the other or opposite can't be. The optimal situation for these is the form, frame and patterns of TRU. An example of this may be seen in the beautiful trabecular pattern of human bone, and the great strength such patterned relationships with their associated dynamic equilibriums of form and function, parts and processes gives to TRU. In this way humans have great strength through the patterns, states, SCARLS or form of our makeup. This is its combined heterogeneity-homogeneity patterns as in the D and S of DAOST and their balances. Their strength and optimal potential are maintained by gravity, or appropriate level of Á in operation. There exists the beautiful balance and strength of life, of TRU. It is a delicate balance, continually changing, altering and realigning the state in which over, normal, and under function - hyper and hypo - in ACEXIN, ENT and TRU. These may be combined, when adequately understood, with many other relationships and dependent factors in TRU. The S-O, Á, FAF of TRU provide much of the intricately beautiful and glorious ENT that shows the actual makeup of TRU.

Confucius' basic tenet and teaching, "it is man that can make truth great, and not truth that can make man great", is highly revealing. Like much of Confucius' sayings and much of all literature, this is somewhat figurative or poetic. Yet the tenet and teaching has been highly provocative and productive, by bringing billions of people into better lives through the search for and use of truth. It proves the compelling powers of the future, such as KOT, can be as strong or stronger than the propelling powers of the past. It proves the great degree to which the world can be improved by the efforts of each person, through wisdom, moral and ethical perfection. **VARST** = Validity or Truth and Reality the Same as they become and/or are in TRU. VARST is the convergence of what is precisely valid in TRU. The reality and actuation of the observer in the observed as they become one in TRU. It is this fusion and union of the observed and observer that is the predominant form and function of TRU, its Á, mass energy dimension and activity. This is the special intention of reality, that of meaning, the mutual power of subjectivity and objectivity on each other that best enables TRU. Simple, and even rational, honesty are only tentative, because they may be based on illusion, ENIT, what is not, less than, or disproportional, TRU. Intellectual, actual and TRU honesty that is correctly derived, formed, proportioned and applied is certain and permanent because it is based upon the purity, reality, and actuality available in VAPO of TRU or TIT.

Love and loyalty have, in common bonding, an identity which when taken together are great in PIP and ENT. The resolution of differences, even when between ES, can be met both naturally and rationally by love and total loyalty to TIT. The individual is given great and absolute FAF, powers, means, assets, gifts, blessings, and benefits with which one can build upon, fulfill and enjoy life. This is achieved through: **LOTES** = the Total Love of the Divine in Each Other as TES and ES, all the parts of TRU. A good start is LOTES and the growth of learning and their use in TES and TRU. Each individual is commonly repulsed or indifferent to most others. We do not have to love each other, or even ourselves if ATRU, ANTRU, and ANTITRU. We only need to love what is divine in ourselves and all others, as this is TRU. A corollary of LOTES is **COTES** = Common ES Interdependence Positive to TRU. All that empowers TRU does so as the parts and forms of TRU empower it to fulfillment and greatest success.

ABSOLUTE LOVE of TRU

That intense love that quickly
Becomes nothing is not true love.
It is no more than a fleeting fancy.
We experience, mature, come to know
What is TRU and what is not TRU,
What is more TRU, what is less TRU

It is from such an overwhelming and
Limitless feeling and outpouring
Of love that each individual achieves
Their infinite capacity to reach the
Unitary, finite, infinite and absolute
Nature, form and function of TRU.

We have a relationship with what is
Of us the partial and universal of TRU
We more readily learn appreciate and
Can love eternally & absolutely what
Is most of both the unity and intricate
Beauty and glory of all that is TRU

Just as one's immortality unto TIT,
Stems from our appreciation and work
Thru TES, TRU in ENT, ACC, TRAX,
TREX, TRIX, and ELTICMA, so do
We come to Live in PUREÁCLITY and
One with COGNOS and TÂ forever Be

LORA = Law of Relative - or limited - Accentuation. Man can't act in any way, or do as he pleases. Man must act and do only by TRU. What is to be accentuated is even more limited. For what we accentuate is only that of TRU and TRUD which are amenable to action and have a high degree of purpose, planning, control and effectiveness. **LACOMÄ** = Law of Compensatory Action, is any and all processes or actions that help to restore the Milieu, Momentum, status quo, steady or Optimal state, or form Ä. TÄ, of TRU. **LACOMP** = Law of Personal Compensation, is doing, acting, working and living in one way to make up for a need in all others. LACOMP and LORA play a vital role in TES and PUREALITY, just as LACOMA and compensation do for TRU in general.

THE SPECIAL FORM of ENT: ACTUALIZATION, QUALITY, SUPERLATIVES, BEST, MEANING

Humans have learned from bitter experience and prolonged suffering many lessons of life. As well, each new individual must learn these, if they are to be TES and TRU is to flourish.

Only when we can sufficiently appreciate and stimulated by the inspiration, joy and rewards of our identity with TRU and understand and utilize all of its concepts, principles, laws, mechanisms, inner workings and operations for best management and control, will we succeed, win, and be perpetually victorious. Human survival over extinction is like that of any animal, based upon quality and predominance over all that acts against it. This quality will depend upon what the animal, man, or TRU are and what is demanded of them. For TRU, this is largely TES, TÍS, and CES and all of man and nature that is essential for survival. We need look no further than to discover what this quality or predominance is and the many problems we face that can be learned, solved, and controlled. It is in quality, as it leads to health, beauty, richness, belief, intelligence, love and many others that peace and prosperity, success and survival by TRU are achieved. Quality is not, however, achieved without effort or direction. **QUENT** = Qualification of ENT. Even ENT must be qualified. Even our most remote ancestors, although not so knowledgeable with the information we have to work with today, were highly capable persons able to adapt to the forces acting upon them, performing much as we do, and securing whatever was needed for their sustenance and survival. They had, as animals in general, the ability to chose good from bad, the beneficial from the harmful, that which enabled them to be, and that which did not. Because our world today and tomorrow is so complex, does not make us more qualified, but less, for it is apt to make us more vulnerable, more is demanded of and by us. Our greatest concern, then, as derived from studies and analysis of quality and quantity, lies in Quality and Performance. Success and the dynamics resulting from this quality and performance, of not only persons or mankind generally, but of the average, plus range of all, is much of what constitutes TRU.

Beatific is the bestowing of bliss, blessing, happiness or the like, especially in a saintly or divine way. The idea of the beatific was discovered and developed in religion, especially in the Catholic Church of the Middle Ages to emphasize the need for and importance of all that bestows bliss, blessing, happiness, quality, perfection, saintliness, godliness and holiness. It was the beatific understanding of Thomas Aquinas, the vision of Dante, and the Art of Giotto, Leonardo daVinci, Raphael, Michelangelo and many others that gave us much of this heritage of awe, inspiration, and revelation in reproducing the likeness and perfection that is God-like, TREQ, TÍS, TES, TREQ and TRU. There is a unity of all things, demonstrated by ALT, that follows spontaneously from ROT, SUT, OBT, TRIT, TU, TO in the TIT of TRU. This effects the concepts and posits, their correlation or negation to equal all relative to SUT. **LABIC** = the Law of Best in Continuity, or where and how TRU is and we are only as the past, future, present and all dimensions, poles and elements are and function in perfect harmony and continuity in TIT. The Chinese term for harmony is *yung*, both a basic law of the universe for Confucists, and letting nature take its course for the Taoists. It is the equilibrium of D and S in DAOST.

Superlatives, alone or in combination, are appropriate to the FAF, ES, and CES of TRU or its presentation. They hold special relationships with the certain meanings of superlatives and the corresponding meanings of different forms and parts of TRU. Some examples of this are, engaging detail, resounding victory, and the beautiful. Superlatives may be further listed alphabetically, showing their special mechanisms in TRU. This, when accomplished with VAPO, provides a quick overview of what is greatest in TRU and ENT as well as the essence and cutting edge of TRU. It is in this way we speak of the Glory of God, and the many other superlatives that are attributed to God. To name these superlatives is not enough. To be valid and successful we must do so in such a way that formulates TRU, its ROT and POT in a positive objective way. They include all that is most TRU, absolute, pure, glorious, magnificent, omnificent, fine, great, virtuous, successful, integrity, blessed, happy, highest, perfect good, exquisite, splendid, complete, beautiful grace, heroic, excellent, important, eternal, phenomenal, radiant, victorious, powerful, intense, spiritual, righteous, lovely, immutable, heavenly, massive, noble, exultant, and honored. These when combined show how superlatives can greatly depict TREQ, essences, and ESINS so readily evident in the FAF, ES and CES of TRU.

How much are you able to see? How much can you distinguish? How much meaning can you get out of it? How much can you discover, understand, present, utilize, and produce that necessary for your life, the life of all? These are the heart of TRU. They are the subjective, objective, and subjective objective combined, especially as they are the S-O or SOT, the Á dimension, primary predominant mass energy state and FAF of our world, TRU. They are the ever recurring and massively occurring dynamics of our being. They are our life, success/failure, joy/sorrow, and immortality/extinction. If we do not see, if we do not pay attention, if we do not distinguish, if we do not get the meaning, if we do not discover, understand, show, produce, present, and utilize it properly then it is not, we fail, we are not. This is how we, and all, must be if we and all are to be TRU. An analogy is in the bible where through a vision of heaven, God, TREQ, TRU and TAS, final victory, judgement and blessedness pour forth.

To have quality population in times of plenty is easy. Yet quality is put to the test in times of stress, paucity and want. Revolutionary changes of one sort or another will occur and cause depreciation if the natural processes are not selective and compensated. Thereby for ENT, and assurance of TRU, it is necessary that, not one or two, no matter how great they may be comparatively, elements, factors, forces, processes, mechanisms, parts, and processes of TRU be highly beneficial, but that all, across the board, they contribute their fair share to TRU This must be both in times of plenty and in times of want.(PPMA) Quality of population is an expression of all TRU, not some segment of it. Quality of population is not so much a person's superficial appearance, it is all that goes to make up their long term adaptability, performance, productivity, and quality of propagation in life. It is the summation of the strengths of all their levels of mass energy, Á, in TIT.

LÄTIN = Law of TRU Inevitability. Just as civility defeats barbarity, so does civilization steadily advance to replace savagery. It is the victory of the civilized mind, one of thought, problem solving, reason, intellect, tolerance, purpose, highest good, and divine inspiration - from TREQ and TRU - that arises from the focus and relativity of all on TRU, that produces LATIN.

A PART and PARTS of ENT

There are many different ways of increasing health, including physical, formative in Á, L and T, mental, spiritual, collective, social, and total. The law of self preservation is largely the spontaneous direct, independent, absolute and sovereign product of the haecceity or thisness of actual individuality when directed to TRU. For TRU, it is the central and crucial power in its perpetuation and speciation. It is guided by the predilection set in Á, L, and T. Most all productive service in life requires a large and good amount of energy, effort, work, and giving of oneself, often with seemingly little in return. Yet herein is the crux of life. This is the nature of the business of life. It is the nature of life of the individual's contribution to the group, to the species of which it is a part. Mans IA, our minds, and desires for self gain, comfort and freedom from the stresses of life continuously render us unwilling and rebellious to make this sacrifice and do our part, and more, in life. To give much of oneself if TES is TRU, and to recognize and appreciate the great amount one has been, is and will be given if TES, is the key to the greatest possible life. Much of TREQ exists in **HPP** or Highest Public Principles. These have stressed: appreciation, gratitude, compliance, cooperation, courage, dedication, duty, efficiency, enthusiasm, equanimity, helpfulness, honesty, honor, integrity, moderation, productivity, self sacrifice, service, tolerance, and virtue. They are qualities of TRU whose quanta enable us to make them distinct by positing and showing their combinations or various levels of Ä, L, and T in TIT.

COMIT = Compensation in TRU, or the return of benefits and provisions in life any and all receive, primarily indirectly, as they themselves fully give in return. COMIT is most vividly recognized in an individuals part in the whole: family, community, state, nation, world, God and TRU. Yet it applies to any and all parts of TRU. It is in COMIT that we can clearly see PIO and INEG.

The greatest existing concepts and descriptions of modes, attributes, characteristics, and properties, of GOD or TREQ greatly aid us in understanding what TRU is and how it is ENT. These include, omnificence, omnipresence, omniscience or the knowingness by God of all things, absolute purity or freedom from impurities and adulteration, faith or holding to the same in TRU as in God so far as the firm bond of determination ones belief holds, grace that represents God's role in benefitting man as TRU, and will or the energy, power, and determination of God, as TRU, to do its work. Of the same and different in TRU, by DAOST, much of the same exists in quanta, individuals, groups, or as families. This is a functional division whose added powers enhance both whole and part.

There is no such thing as something that doesn't cost anything, all costs. In fact, if TRIA and TAS are invoked, all has a set and absolute cost or value. According to Oscar Wilde "A cynic is a man who knows the price of everything and the value of nothing". This although misleading when viewed from TRU tells us much about price and cost. Wherein failing to recognize the value or worth of TRU, and being obsessed with the price of all that is ATRU, produces a cynic, one who denies what is best, TRU. If we are overly concerned with the price, cost and negative effect and do not realize the value, benefit and positive that is associated with the worth of all we confront in life we miss the crux and gist of life, TRU. This is what all is as it is TRU, of infinite value, benefit and positive effect. Life as TRU has, or at least should have, infinite value. So it is that we must dispel, permanently, the cynics of this world, especially those who would deny, defy and destroy TRU.

That there is a great reason and demand for religion, and how all is GOD or TRU, is more or less seen by most people. Yet this stands in stark contradiction and opposition to the similarly long existing and evident divergences, discrepancies, weaknesses and abuse of religion. This has been and is largely resolved by: **LATOL** = Law of Tolerance, Tolerance Total Limit, to TRU, and of a set Form and Order in TRU. Not tolerance for any and all that is without TRU. There is a dialectic state and operation between the tolerance that is of TRU and the set, predetermined necessity and responsibility that TRU demands. Justice is a weighing or balance between proportions of TRU. It is not as given in legal and judicial terms a balance between any or arbitrary right and wrong. For by TIT wrong does not exist. It can only be a weighing that is based upon the distribution, form and function, and fairness, by VAPO of TRU. The worship of God or TRU or any or all great aspects or parts of life and TRU has, is and should never be out of fashion.

SPECIAL PARTS of ENT

DOPEMS = Doctrine of Deprivation, Damnation, Predestination, Election, Mercy and Salvation. This is the great doctrine of St. Paul that was extended and consolidated by St. Augustine and utilized by early Protestants such as Calvin. It holds that only God's grace enables us to be virtuous. This is in contrast and opposition to that of the Pelagians, many philosophers and some theologians in which it is through a person's will to goodness that man best becomes virtuous and serves God. Although the end results are similar, the enhancement of God or ENT and the immortality of the soul, the difference is basic and causal. From TRU, TES and oneself it is readily evident that it works greatly in both directions, as TRU and individual are the same, with similar great powers of ENT in TES and TRU. It is when the grace of GOD or beneficial powers of TRU and human will to goodness, or TES, are both disproportionate, lacking, or negative that both individual and GOD, TES and TRU fail, and all suffer.

Life is always on a threshold, one of future survival or extermination. It never stands still. It is in the expansion of this threshold that much of ENT lies. It is in mutual expansion, or **LAMEX** and the catalytic effects of parts and whole on each other that this threshold is advanced. Such expansion is largely one of extension of all dimensions outward and inward from unity both toward the infinite, infinitesimal, and finite in between. Man's being depends to a large extent upon the capacity, ability, effort and astuteness to produce this extension. It is done through continuous concentration on and self improvement in TES. Each person is in a unique and fortunate position centered and focused to provide the greatest opportunity for ENT. This is a product of ones heritage through the FAF, ALT, and all that make up TRU. **EPAT** = Expansion Forms, Ä, Threshold of TRU and **EPOT** = Expansion of Threshold All TRU, of knowledge, perspective, representation, presentation, modulation, positive objective approach of all TRU. Differences in the expansion of these thresholds is the basis for basic tensions in the world, and the dynamics of TRU.(Theism vs. non) **METO** = Mutual Expansion, catalytic of the Parts and Total of TRU, TO. It is the continuity, solidity, formal distinctions, identity, strength and TRU meaning of language that produce such an expansion of ENT. The focus of human being, TRU, is one. Yet the mass and totality of this being is a finite extension from one, be it additions or subtractions, multiplications or divisions from one. The degree to which it is a success is the degree to which, other things being optimal, it is an expansion of this unity. It has inherent capacity within itself depending upon the magnitude of this expansion.

Improvement, and especially self improvement, are the keys to life and human success. This is acceleratively demanded by mentality, rationality and the increasing complexity of human population. It makes great demands on the individual. **INCATIN** = Individual Centricism Reversed to Absolute Identity in TRU, or how a person can turn from being completely self centered to optimal identity with the focus of TRU. Concentration of TRU is that form and function, state and power of TRU in which it is so structured and operates to best protect, prevent, maintain and advance itself.(PPMA) **DEP** = Dependency Exclusion Principle. How the only exclusive assurance of being is our own fulfillment and accentuation. It accompanies an understanding, interest in, appreciation for, love of, and will to TRU. Through striving to adapt, adjust and come to terms with life, an individual becomes overwhelmed, or feels so overwhelmed, that we become, consciously or more often subconsciously, drawn into reaction formations, bad habits, escapes and diseases that only hasten our demise. This has been called Stress or General Adaptation Syndrome, GAS.(Selye) It is a common mechanism in psychiatry and medicine. It is a major mechanism in mental illness, alcoholism, drug addiction, and auto accidents. To keep life as simple and happy as possible is achieved by uncoupling and excluding all except TIT.

Temporary success is to a great extent dependent upon quantity. Future and overall success is to a great extent dependent upon quality. Quality/Quantity is a basic dichotomy of TRU. Success can only be from quality in TRU, and is accelerated when success builds on success. **LASUCT** = Laws of the Success of TRU include Anticipation and Concentration on TRU. There is much truth in the saying "when you laugh, the world laughs with you, yet, when you cry, you cry alone".This arises from the nature of TRU, TRUD and ENT. For TRU as TIT does not know failure. It doesn't want to know about the negative, as in "I don't want to hear about it," or "it is only sour grapes." TRU is positive, it is success, happiness joy, commitment, fellowship and common identity not only with others and success in the positive, but all as they are or are brought to TRU.

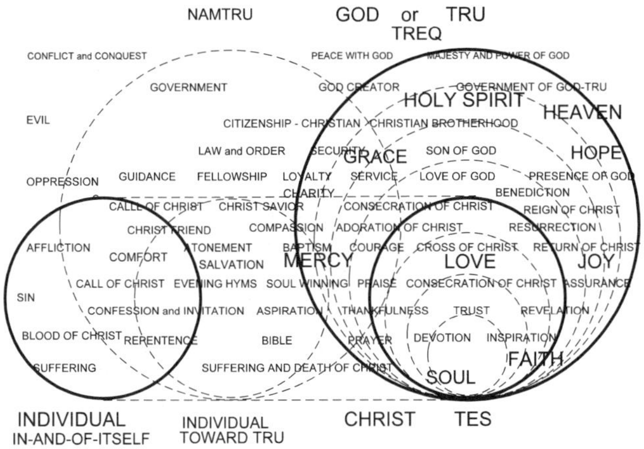

Fig 3-18 **THE RELATIONSHIP BETWEEN THE INDIVIDUAL, CHRIST, GOD, TES and TRU as REVEALED IN AMTRU and CHRISTIANITY**

In Figure 3-18 the degree of TRU and demonstration of the state in which TES functions in TRU are compared to the state of the individual which does not function as TRU. The dashed circles between TES and TRU represent that of different sizes of groups, such as congregations, churches and denominations that function between Christ and God or TES and TRU. It must be remembered that TRU, TES and the individual's relationship to them are much more than seen in church services. Christ or God, as great as they may be, and as great as they have been in the development of TRU, are only when they are interpreted as TREQ, their equivalents in TES and TIT-TO. They must have the VAPO necessary to make them adequately accurate and reliable. Their test of specificity and sensitivity can only come from the degree they are TRU and TIT. Yet this figure of Christ, TES, God and TRU is a good example of the greatness and beauty of Christianity in human progress, TRUS. It explains the growth, quality, power and opportunities of TRU in the last two millennia as well as that for the future. The mixed or AMTRU world, that is so much of, about and seemingly a part of us, is not very much a world of TRU. Yet, we should always temper it by the more direct and profitable positive, and all inclusive, approach to TRU that VAPO and POA provide.

TOTALITY of ENT: INTEGRATION CONCRETION VAPO HARMONY FRUITION ORDER FULFILLMENT

Senses and things have order only as they are taken in their entirety. When this entirety corresponds with Á and TIT an absolute ESIN may be approached. The distinct and discrete best showing of TRU depends upon how all are taken together as a whole. Only as all work together in optimal efficiency and productivity can equilibrium, health, harmony and peace be achieved. In a special and exclusive whole or universe as TIT, nothing lacking or superfluous is produced.

VAPO and **SCARL**, or a Valid, Actual, Proportional and tOtal comprehension of the various States, Conditions, Limitations and Relationships of TRU. These are not to be attained by a simple recognition of separate senses, things, or even a unitary God. Nor is it to be attained through a naive assumption of some narrow premise, religion or philosophy of life. To be adequate, effective, productive and reliable, with sensitivity and specificity it must address the great mass of TRU, both its form and content, functions and operations, totally. This must be combined with all it takes to represent, modulate, evaluate, present, utilize and practice TRU. The great significance of an ever more distinct and discrete formulation and modulation of TRU lies not so much in what all the components of TRU are individually, but in how they combine, interact, and what they mean in the whole. **VINCU** = the Vital Necessity of a, and the, Concrete Universe. In being all and a totality TRU, as TIT, can be envisioned and produced, not only as a discrete mass, it requires a reduction to a closed set in the form of a concrete universe. It requires an identity of subject and object both exterior and interior to TIT

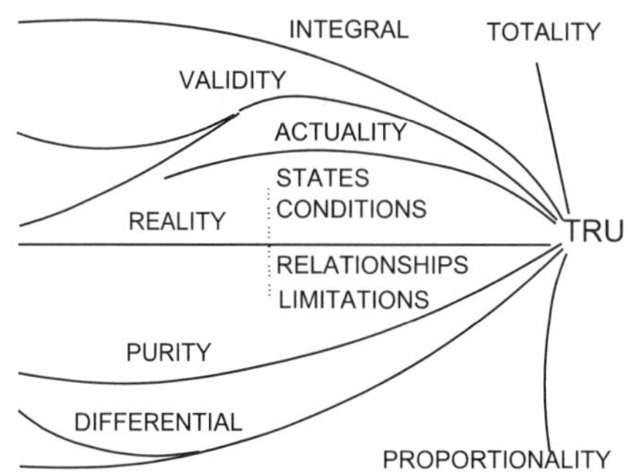

Fig 4-18 **VAPO, SCARLS, PUREAC-LITY and TOTAL ENT**

LAPRIC = Law of Actual, Pure, Reality in Completion, or Law of Actuality, Purity and Reality in Completion. What is TRU is not so much by being VAPO, SCARL, or even PUREACLITY alone or separately. It is when all these and all the great aspects, divisions, FAF, ES and CES of TRU are taken together that we can begin to realize TRU and its beauty. If we are to overcome and overwhelm all adversaries, and that which would destroy us, we must have both the capacity, means and will to coordinate, organize, order, and command both individually, by groups and all mankind, all that is apt to effect TRU. This is the heart of a total ENT.

LO = Law of Order. Only with a thorough, complete, and ordered knowledge can the evidence and information be provided to show what is flux and what is permanent, what is figment or illusion and what is actual, TRU. This is possible when TRU, TIT, TO, and TRIT are well shown and learned. We fail to recognize TRU, we deny TRU, we say that TRU is not, we say that all is, or we say that this is or that is. Yet we know not TRU. Of the ideas and essences of Plato, and his emphasis on summum bonum or the supreme good and ultimate ideal, we have shown little faith and less conviction. Many of the great systematic thinkers of all time through the good, knowledge and reason have been directed to TRU, although not known as such. It is shown by the collective concepts, principles, laws, and mechanisms which totally enable TRU to be and to be what it is. To recognize, and even to deny or refute TRU, we must expand by ACEXIN, this summum bonum to become, as it should be, equivalent to TRU, a complete TIT-TO.

As a tree should be judged to a great extent by its contribution and fruit rather than its leaves or trunk, so should we be a major player in the total harmony of TRU, and supplier of its demonstration. The formal and functional patterns and operations of TRU give us what Kant called the categorical imperative, the universal laws, principles, concepts, and mechanisms by which we must act. **LARNU** = Law of Reciprocal Natural Utility. The world about us is, to a large extent, false, synthetic, and artificial. It is a world of NAMTRU, mixed figments, illusions, untruths, rationalization, sophistication, deception, lassitude, sales, advertizing, money and promotion. It is far to superficial, partial, commercial, political, and manipulative. There is a natural reciprocal utility that exists in TRU between man and environment, all the parts of TRU and all of nature, that is interactive in the whole, TO. This delicate balance of nature is to the extent it is, has and will be essential and vital for TRU. A balance of nature is neither as simple, definite nor as valid as it may appear. It functions in various ways in which individuals are constantly imbalanced. Only nature as TRU and the individual as TES, as shown by

UMIA, is there balance and a well ordered whole. Only with TES and TRU is there an actual balance of nature, and this to the extent: 1. TES and TRU are 100%, free from what is not, 2. TES is measured in the whole of TRU, and not the from only itself or other parts of TRU whose dynamics involve negative actions to itself and other parts not of the whole, TIT.

One of the greatest problems a person and all mankind face is participation, role in the whole, and lack of fulfillment. It is seen in our need for closeness, bonding, relationships, love, companionship, friends, associations, and socialization. These need to be in an optimal state, proportion and equilibrium and only as it is in TRU.(LATRUB,LATCOB, DACOB, COB) They can only be adequate for TRU and ones life when we can envision our world, TRU, as a solid and work through all its intricate detail and interactions, ES and CES in POS. **PRUTO** = Proof of TO, The holistic proof of TRU. Only when we can understand and comprehend all TRU and how it functions dynamically as a whole can we fully and reliably know and show TRU. Through its complete dynamic unity in totality we can prove its thisness conclusively. We can also go a long way to distinctly show its whatness. **DOPRUT** = Dynamic Total Proof of TRU. Just as we can not understand or prove a live, dynamic functioning organism until we fully describe it as a whole in complete operation with all its relationships so must we learn, understand, prove, and live by TRU.

DYNAMICS of ENT: ACCELERATION, INTENTION, ABILITIES, PURPOSE, USE, IDEAL

To fully understand ENT, as we do for any entity and object, we need to know its and TRU's dynamics. ENT is especially the motion, force, power, acceleration and all that projects TRU. ENT includes all escalating forces of intention and purpose that stimulate TRU. ENT is achievement, what one and all need and do to accomplish their optimal role in TRU. Many of the powers with which TRU is blessed, both universal and exclusive, future actuation, validity, higher powers of subjectivity, and understanding, are the salients or cutting edge of ENT. **DUAC** = Dual Acceleration. DUAC applies to the dual, multiple or manifold acceleration that occurs when auto-stimulation and the expanding forces of TRU work in ENT. It is how the positive often feeds upon itself in an exponential way. DUAC is the opposite of INEG or Integrative Negation. **LOCI** = Law of Total Consolidation, Consummation of Intention, or the total common purpose and intention of humanity. LOCI is how intention that is directed by VAPO, SCARLS, TRU, especially in good faith and operation, can not only totally consolidate and consummate their efforts for TRU, they can do so increasingly and with ever greater perfection by accurately knowing and showing TRU. LOCI is a kind of fulfillment of the Á mass energy form, state and dimension of TRU.

The strength of TRU, like that of the human body, is not only the level of its muscular or even mental performance, it is the total of all its mass energy abilities, and how well their combinations fit together to overcome demand. It is largely a function of: **IM** = Inertial Momentum, which shows the power and action produced by each and all parts and aspects of TRU. **POFAT** = Power of Potential Future Actuation of TRU. POFAT is more and different than the future drawing and compelling effects. It is the potential power in the future to be more than TRU already is; to assure a much greater TRU, and to enable TRU to survive much longer, better, and greater. "Nature takes care of" is a common phrase that, with qualification, has much TRU implication. It signifies how the world outside, and inside, of an individual or species provides much benefit, or ENT. TRU consists of set form and function which is optimal for it and much of this is extra individual or species, yet not extra TIT. Reversely, however, nature, as of TRU, demands much of us for its fulfillment. This is to live, and to live only in natural and certain ways that are TRU. Survival of the fittest, struggle of focus, or the competitive laws of nature in which those that survive, are those that best adapt to, their environment, the actions, and forces that beset them. These are the ones who are most able to overpower opposition and competition in the adaptive processes. Survival is the great dynamic force of all as the individual and group struggle for existence, dominance, success and their proportionate share of TRU. Each hews to TIT-TO. It is what TRU should be. It is not what ways are right and what are wrong. It is how these are altered as the species survives on earth, and how the earth is utilized as this survival becomes TRU.

Humans have become more complex, with higher levels of mass energy, Á, problem solving, purposive, rational, and evaluative abilities that are under greater demand and stress. It is part of this great expansion of highest levels of Á mass energy that spearhead the TRU revolution underway. This is called: **POIR** = Purposive or Importance to Man, IMU, Revolution. POIR is one of mankind's greatest revolutions, which, in turn, causes, and is associated with many correlative processes. The degree, simplicity, ease and control that humans attain in adapting to and best completing this revolutionary role is the degree of the success of TRU.

CONTINUATION of ENT, PREDICTION, CONSTANCY, CONSTRUCTION, SUCCESS,

The paradox of relative and absolute universes can't be resolved from either position alone, from the outside or inside, or from our subjectivity or observation of it. Only when we can look beyond these together and do so, realizing that any observation imparts some loss of actuality without equal qualification, can we begin to understand this paradox. It is inconceivable that 'the all' - whatever that is thought to be - is ruled from the outside. For to be so would mean that it is no longer 'the all'. This is the prime defect in much of human ideas of God. It is only when God equals this all, and does so in only certain ways, that this paradox begins to be solved. For whatever we consider to be this all, must have some kind of unity in totality. Without this it is no long an 'all'. Thereby, to have unity in totality requires relation, a relation between the unity, what constitutes the totality, and the totality in itself. As we show by TROS, be it relative to humans, or to any RU, animal, or entity their world or universe will consist by relativity of a geometry based upon set relations and amounts of its form, function and constituents. Since what is predominant in the actual world relative to us is what is most characteristic and formed by it, the paradox is solved by the relative universe of man. This tends to be absolute, TÂ, in seeking survival and eternal life, TRU by ATO becomes TIT-TO, the relative absolute, RÂ.

To assure the permanent continuation of TRU requires optimal support, an ever greater and more accurate predictive ability, and a massive PPMA of TRU. Constancy and the ultimate assurance of TRU can be achieved with a positive objective approach that results in incremental and total constructive success, gives a large amount of its attention to the future, and how this future interacts with the present, past, and all together, in TIT-TO. This is what we do by SPALT, Posits and Homation. This is what produces the good news and fortune of TRU that is sufficient to meet its needs for both sustenance of its quality as well as completion and absolute imperative expansion. Predestination, in religion, is the doctrine that all events are determined by the action of God's Will. It is the doctrine that God has foreordained the eternal life of all, some and each person. This, by TRU is much more valid than many would believe, and much less and different than many others believe. It is one of the reasons why we must enlist the aid of the ESINS and NEC of TRU or TREQ rather than rely only on the OC of God and existing language.. This is because if we limit ourselves to envision TREQ as what is commonly and currently known as God we will from the outset have a frame, construct and model which is so poor that we will never be able to adequately show and know TRU. Since only our TES is TRU, since God is only in some ways TRU, and only since it is not so much events but the overall nature, makeup, form, structure, characteristics, essences, qualities, properties, parts and processes of TRU that maintain any fixed and predictable state we must considerably alter the definition and common usage of the terms predestination, soul and god if we are to give proper and acceptable usage to TRU. **LEPRIM** = Law of Equality or close bond between Predestination and Immortality. In return for the great endeavor to be ES and TES, TRU gives all, and for us in return, immortality. Such a kind of TRU predestination means that through this very relativity we have set relationships which, as by LASUR, are highly predictable. Our whole lives continuously rise to and are GOD or TRU. In the great glory, beauty, and inspiring mystic identity found in ones rise to such majesty, the momentary feeling of actuality, joy, and closure can't be lost. It is an experience we deeply remember and eternally and everlastingly perpetuate in TIT.

Fig 5-18 SYSTEMATIC UNITY TRU, UNITY, FINITY and INFINITY

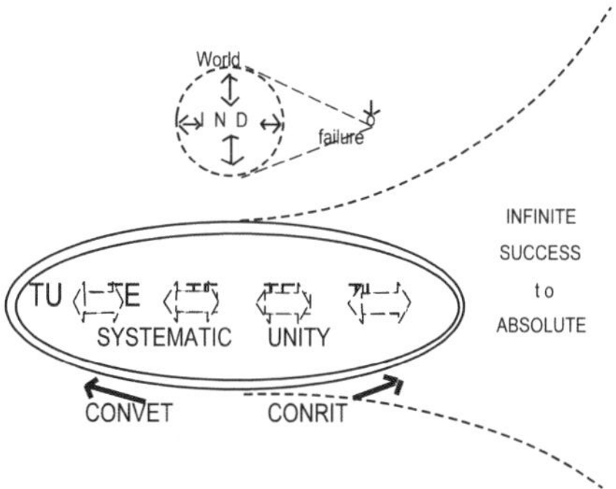

How do we, as in life, obtain an optimum of simplicity and SUT in tandem with an optimum of technical proficiency, productivity, efficiency and effectiveness? Like CONRIT and CONVET, life is a constant dynamic, surging and waning, state of being in the midst of the unity to infinity, the integral and differential of TRU. Our success and the success of our entire world, TRU, depends upon the degree to which we can master, coordinate, control and live positively in this state. Its optimum depends upon the degree to which we simultaneously attain maximal systematic unity and differential control. That is by increased ordering and controlling our partial position, the finite in TRU, in both directions. Therefor SUT is not only the optimal dividing of its object, The Object, TRU. This is only half of it. The other part is how all its components, even to the infinite, may be posited, ordered, systematically unified with the whole as derived from the unit. It is also how the extended finite may be systematically unified, and this further and further into the increasingly finite. In the end not only must the SUT, ROT, MOT, SPALT, POS, STA, HOM, ACC and POT become perfected, the average level upon which our finite existence is based must be acceleratively increased if we are to fulfill our destiny, TIT and TRIA of TRU to the Absolute. **TELEOP** = Teleological Priority or Preeminence. TELEOP may be either in general or specific. TRU ends denote the great and dominating power of the future's directing and compelling forces. They are in dynamic equilibrium, as a polarity, with the great dominating powers of the present and past.

The great joy and victory of life can never be lost, must always be nourished and increased. TRU as our very being, is sacred, success is imperative, fulfillment is a great part of life, relativity and even being only a part, assures our identity. In infinite relativity and LOPAN we can find the absolute, with the reversibility, management, and control of changes necessary to assure both TRIA and TÂ. TRU is happiness the optimum in human life, and its perfection for the individual, optimal participation, or immortality. **LATSUR** = Law of Sufficiency Ratio Proportion Completion. Human sufficiency and survival are not as one being self centered sees oneself, but to the degree of LATSUR, sufficiency ratio of VAPO.

SUMMARY of ENT: COMPLETION and PERFECTION

The culmination of ENT occurs when we recognize and best utilize the glory and ends this brave new world offers. How we can perfect and complete it to provide the infinite future, immortality, eternity, and the absolute it and all its constituents warrant. It is propelled by existing FAF and the subjective revolution, and advancement of actualized, TREQ, knowledge. This has been self perpetuating and accelerating. It is of the great predominant mass energy form and function of TRU, its Á dimension and state. This is especially so as it tends to complete itself in self consciousness to the, more or less, infinite absolute closure of this relativity in each and all persons. In tetelesmenon or the perfection of relativity, or the Á mass energy of TRU, there is an exact, actual and perfect identity. Such perfection is evident in the optimal health and function of the human body and human mind. TRU is, has been, and will be perfect. It is the continual return by perfection in the completed thing in itself, TIT. Tetelesmenon is actual, perfect, and shows the identity of IA - Word - Logos - SO - TRU. It is at the core of human mastery of the world around us. It is humanities ultimate weapon, the form of our special thisness, and Á, mass energy, state that is operating at optimal efficiency and precision.

The perfect form and function of the human body is TRU. Anything less than this is to that degree AMTRU and ÁTRU. So is the perfect form and function of the human mind, Á, and all that contributes to TRU. This proves the degree of quality that exists in, and must be upheld if TRU is to be, and individuals rise to TES. Since TRU is by TRIA and ATO self contained in TIT, the future, present and past are with TROS one and all. Thereby it is not so much any change in focus or relativity of man that is eventually crucial, it is its preservation temporally or reversibly relative to the limit. **TANCOWS** = The Absolute Necessity of Correction of the Short Warping of TIT. This is like the alteration of form, as in time warps, and cosmic geometric curvatures, in which specific corrections must be made if we are to replicate TIT and base our knowledge and demonstration of TRU accurately.

How To Achieve The Absolute

The Infinite is Absolute, and
The Absolute is Infinite in
The Relativistic terms of TRU
TRIF-TI is TÂ, or TÎSTÂ

Therefore it is in extending this
Special Universe of Best Human Life,
TRU indefinitely to the infinite
That we achieve the absolute

It is by way of TISTA
The absolute in the infinite
That enables us to open wide
The door to the Absolute, TÂ

The ultimate intention of human life
Is the opportunity and challenge to
Fulfill the imperative enablement
And infinite Continuity of TRU, TISTA

It is to find the key to continuity, permanence, eternity, or absolute that drives humans beyond a mere existence. Once found this key is used to unlock, open, open wide, enter into and become one with this absolute, or have this absolute become oneself. It is in its association with ATO that we can recognize that ends, as we know or consider them in our NAMTRU worlds, are largely illusion. This proceeds from relativity, as the illusive end erroneously appears, as an absolute finality. Thereby, we must be careful not to believe that ends for TRU are like that of the entities or objects about us. TRU immortality is all that exists now and through and through. That this lives on, is what as form in the present was a form in the past continuing through the future. It partakes more of its quality, the form, function, increments and forces of TRU than it does of its, quantity, smaller parts. Yet all, more or less, contribute or effect TRU immortality. **LÉT** = the Law of Eternal TRU, or that TRU is without beginning or end. It is with ATO, TIT absolute, in its own thisness. From within, it does not admit to having a before or an after in which it did not exist. **LATET** or Law of TRU Eternal, or the contracting and expanding periphery of TRU depends upon the intensification or expansion of time. LATET certifies its associated thisness and concentration in a set time frame. LET certifies the thisness of TRU effecting its eternal continuation.

CHAPTER 19

CONCLUSIONS and SUMMARY of TIT and TRU

PROBLEMS, SOLUTIONS, MEASURES, USES, RECONCILIATION, TOTALITY RELATIONS, SUCCEEDING TEXT, ROT and POT

The whys, that are so imperative to life, along with other interrogatives of what, which, how, where, and when, are much of the fundamental FAF of our being. They are closely associated with IA and the Á mass energy dimension typical of TRU. **FOTIU** = Fallacy of the Irrelevant Universe. Most people have a minimally rough idea of what the universe is. Yet this quickly disappears as they are distracted by its diffusion, disorder, and indistinction. They are unable, as has man throughout time not been able, to achieve adequate relevance, solidity, order, and distinction to it. To answer and solve these interrogatives is what it takes to reduce the chaotic, erratic and seeming futile world to size, meaning, value and order. (VAPO). The more we answer and solve, the better we can correct our partiality and have the deepest interest in what we are. This shows the major deficiency of God, and what characterizes all gods, without TREQ and TRU. It is likewise present in all subjectivity and knowledge. Even when TRU, it is not fully provided for. The challenge of life, knowledge and TRU are such that both focus and all the finite world about us that is TRU, can be learned, remembered and well ordered. However as TIT and TRU become the better objects of concentration, thought and work, it becomes a somewhat new and different world than we are used to. This produces a never-ending problem, whose solution lies in interpretation, comprehension, reconciliation, management and control. If we do not surmount the needed comprehension of both TRU and its understanding we are confused and fail. We misinterpret one and all. When we can comprehend both TRU and our knowing, showing, and best use of it, we can successfully fulfill our natural inclinations, our predominant mass energy, Á, and all as they must be TIT and TRU. For every law there is a fallacy, as for every fallacy there needs to be a law or correction. So there is a corresponding concept, principle, and mechanism, for every unknown, misinterpreted idea and misdirected movement. This is so that for each problem a solution, paradox a resolution, and contradiction a reconciliation. There is an infinite expansion of identity of understanding, a production all of TRU. It is the optimal fulfillment of meaning, formal distinction, a determination of what is right and its usage as it is and must be TIT.

The futility of life is to be found in each person and all that are not, or other than, TRU. Whether we are in the depths of despair or in the most glorious victories and happiness of life we are confronted with our down side, vicissitudes, and times of seeming futility. This despair is often the beginning of TRU and TES, just as the most glorious heights of victory and joy are at the turn around from such futility. The only solution, as for life itself, is its fulfillment, as TRU, TREQ, and TES. Erratic functions are unexpected, disordered, chaotic and diverse in relationship to an observer. There is even a degree of the erratic associated with the different in TRU. The phasic and cyclic changes and motion in TRU are often not so regular, but warped by time in TRU. Even more our observer position renders both the differential of TIT, and even more the NAMTRU and ATRU world about us erratic. Often it is a combined function of the two or many together. Deficient subjectivity arises largely from deficient knowledge, and deficient knowledge from a lack of KOT and TRU. Much needs to be modified and accounted for when applying existing knowledge to TRU. We can not comprehend all unless we first comprehend TRU or TREQ.

We must remember, ECRO, that whenever we use the term TREQ it must be qualified. For if it were the same as TRU, TIT or TIT-TO it wouldn't be needed. Even more such qualification is necessary to reveal the great mass that is both same and different of TREQ that is TRU, TIT, or TIT-TO For there are gradations of TREQ, the degree to which TREQ is less or other than TIT-TO.. Therefore all TREQ refers to, will be more or less different than TRU, TIT and TIT-TO. It is this difference that we should continually learn, understand, and seek to account for in life, and improve to its part and role in TRU. This difference or the degrees that TREQ varies from the non TRU world, ATRU, the world in general, NAMTRU, and the world as TRU, TIT, or TIT-TO, is called: **EXTREQ** or the degree TREQ differs from TIT, NAMTRU, and ATRU. We must make distinct TREQ which is all that is TRU yet unknown as TRU, from MATRU or all that approaches, yet is not altogether, TRU. In some ways TREQ and MATRU are highly coterminal, depending upon the degree to which MATRU is TIT and TREQ is known as TRU, TIT or TIT-TO.

"Why", "Why is", and "why does", are questions that reveal the fundamental and critical functions of our being. They ask the questions about our thisness, whatness and their form and function. It is both the stimulus of our lives, a guide toward better answers and an exercise of our powers of Á. All these very interesting and often very important questions that are so crucial to the fundamental mechanisms and all the inner-workings of the world, TRU. They tell and show what TRU, we and all are. Yet we usually fail to recognize their deepest and most important meanings. We tend to be satisfied with any, simplistic, or partial answers. The combined relationship between laws and fallacies is not coincidental. It is the balance produced by the mass energy, Á FAF between cause and effect, or fallacy and law. There is a tendency in each and all of us to accept and settle on simple answers to "why" and to love, often to much or to little, this or that. There are Francophiles, Anglophiles, Ameriphiles, lovers of fiction, movies, sex, violence or those who are completely obsessed with fashion, money, superficial beauty, sports, action, and power. Individuals and groups tend to completely identify with one extreme after another, ISMs, movements, fads or fashions. This is compounded by the fact that these are usually highly ATRU, even ANTRU. They become so fixed and concentrated upon this or that they think will solve their or all the world's problems, and around which the world should move, that they neglect what the whole world actually consists of, TRU.

TRU, as life, sooner or later is a hard taskmaster. We should not want it any other way as long as it is necessary, enjoyable work, and fulfillment with measured TRU benefits. The human mind is highly adaptable, flexible, with redundancy and resilience. It is when the body becomes exhausted, and the mind is lost that we fall before the onslaught of the NAMTRU world about us. Needs often increase at a much faster rate than capacity. All measures that will improve performance, both currently as well as in the future, will be a great trial. They are the cutting edge of humanity. If persons expect others to like them, to believe what they believe, to do what they want them to do, it is, sooner or later, impossible unless and to the degree both are right, and this to the extent it is TRU. The mutual identity of right can only be found in TRU, or TREQ. So it was in the literature of Goethe, "What is the purpose of life"? This becomes increasingly evident the more we know of TRU, as it is, TIT. It is the fulfillment of TRU, and for the individual, TES and TÍS. Not only is it highly necessary to know the purpose of life, it is equally necessary to achieve the purpose of life. Herein lies the fusion of S and O as they exist in the Á of TRU, that is the predominant manifestation of this purpose in life, the FAF and nature of TRU, its achievement, improvement, perfection, and fulfillment. It is the lack of any idea or comprehension of what ones TRU relationships are that so inhibits us from being TES and TRU. **P/S** = Problem Solution, **PIPS** - Problem Principle Solution, Present Problems and Imperfections equal the evils of tomorrow. Where the problems are great, great solutions are required. and reveal the interactive role of problems and solutions in TRU.

For those who would think the world, TRU, to be to small or unchallenged, insufficient, defective or incomplete let them only see and learn the complexities of its makeup, form, function, modes, parts, characteristics, processes, and properties. Let them see and learn its infinite form and function, parts and processes both in and of TIT. Let them learn TRU in its entirety, proportionately and with utility. Let them suffer the loss incurred by all that is not or against TRU.

DERIVATION and DEVELOPMENT of SUMMARY & CONCLUSIONS, and CORRECTIONS of TRU, and KOT

To answer the questions, solve the problems, resolve the paradoxes, and reconcile the enigmas and all we must discover, learn, know and show TRU. The discovery of unknowns is in the formulation of the finite of the subjective, objective, and SO, as they are TRU. There are limitations and corrections needed for all not, and to that degree not, TRU. Through deduction as in the correspondence of limits in totality we can greatly advance our discovery of unknowns and TRU. COL and COR enable us to increasingly discover not only one, but many unknowns when filling the field or construct of TRU. This is made possible by the exclusiveness of TRU and the sameness of much of its makeup. This enables us to eliminate all not TRU, through incomparability of non TRU and TRU. It helps guide and correct our questions and problems to their actual answers and solutions. The increasing field of discovery that exists in the whole, TO of TRU, provides an illumination whose revelations are mutually stimulating, inspiring, exhilarating, and accelerating. This provides the major method of TRU, through combined experiences, POA, and the revelations of unknowns. This is possible when based upon the correspondence of all as they come together, relatively, in TRU. It is the great power of discovery by closure that is not only a means of understanding TRU, but a major mechanism of Á FAF TIT. This has been further accelerated by our representation and communication making it ever more accessible to all. Much of this discovery has to be reworked repeatedly to bring it to VAPO and to TRU. This is especially so for the ISMS, schools of philosophy, fields of science, and different religions whose great and revealing contributions are often lost in their partiality, and misuse. For analysis, criticism, allowances and reconciliation are continuously in order, if they are to become TREQ and TREQ be TRU as TIT.

Every topic, subject, field and science is not what is usually presented and how it is commonly known. It is strictly and only how it exists in TRU. How its essences act as parts and processes in TRU. (ESINS) Ordinarily this is not so apparent. For it is apt to be more or less in error. All the way from much that is included that is not TRU, to much that is not included that is TRU, and even these, also, by degree.So that to make knowledge efficient and sufficient we can't start from what is assumed or ordinarily known. What, why, how, when, where, etc. of all is apt to be misleading. Each topic, subject, field, as for any and all TRU must start from, develop, include, and fulfill all its parts and role, and how they interact with all, in TRU. In order to overcome this weakness and prevent its failures in knowing the strict, deepest and actual meaning of all, we must find out, learn, utilize, and live TRU.

LAL = Law of Limitation. LAL signifies the way by which all must be qualified or limited by being less and other than TRU. The correction, qualification, limitation and resolution of this is called: **LACOL** = Law of Correction of Limitation, or the limit or correction that must be applied in every case that is not TRU and/or is unknown, and is to be established.

The end of the vicious circle of problem solving is accomplished in the precise reaction and correction, with a minimal amount of side effects or further reactions. The needed finality always returns to the totality, TO of TIT. One way of assuring this is: **RINCAM** = Resistance, Inertia, Immunizing, Neutralizing and Control Mechanisms that reduce the harmful effects of actions, problems, solutions and reactions. **COR** or Correction of Relations has the added meaning and significance of enabling us to deduce even more accurately and distinctively with measurement three or more entities or factors by their simultaneous correction. **MIRPS** = Multiple and Rapid Revelation and Illumination of Problems. MIRPS is the method, and important means, of constantly placing ones problems in the forefront, bringing them together, and ordering them by TRU. This is especially important if rapid solutions are sought and breakthroughs into new ground and fields are indicated. Cultivated interests are very much associated with, and greatly enhance, MIRPS.

The great value of organization, MOT, and POT is not to compartmentalize our knowledge. It is to show DAOST and that for each of the great aspects, existents, ideas, systems and their interpretations of TRU, we can not limit ourselves to only one. We should at all times hold a high level of suspicion of all we sense, perceive and apperceive, and be careful that what confronts us is selectively acted upon by and for TRU. Our biggest problems and solutions in life don't come from conversation, activities and others as often as they do from our own failure to learn, and inability to know and remember, our objectives and priorities, the reasons, principles, laws and mechanisms of TRU. As the cause and potency in all, Aristotle regarded reason as valid inference and thought. This identity of objective reason and subjective reason in the dialectic of Socrates simply through valid inference was the beginning of logic. The refinement of these processes has been advanced by many since. Even simple subjective reason has much merit, for when one thinks with reason one uses - or should be using - the reasons for things - TRU. The reason why this is so true is given in **LASUR** - the Law of SUfficient Reason, or if there is a relationship between two or more things, senses, entities, or universes these can be discovered and stated with certainty. Since all, in the universe is highly related, as seen in the S in DAOST of TRU, it is only up to us to discover, learn, state and remember what the knowledge of this relation or same is. It will not hold for the different or less related in TRU, nor that not TRU. It is in this way philosophy and science as reasoned thinking is of relations and causes in TIT. They stand both in opposition to each other and combine as subject object, or the great mass energy predominant Á dimension FAF of TRU. Yet such sufficient reason takes much knowledge, knowledge to make knowledge, as in revelation which knowledge and sufficient reason supply. And knowledge takes much learning, which an intellectually oriented civilization supplies. Thereby LASUR does not start from scratch, and is not singularly divine, as the Gospel truth. It is divine in providing a great method and having such a basic part and role in TRU.

Inspiration is a great key, theme, and means of achieving TRU. It is the subjective identity that comes from an intense feeling, interest and meaning in all that constitutes TRU. It is like combining powers of two or more levels of Á FAF working with high intensity to achieve TES and TRU.

The doctrine of salvation through faith has a long precedent in religion, and is supported by KOT, TES, ATO, TRU and TRIA. This is by a relative and centripetal concentration of ones life on the TU focus of TIT. John Wesley, influenced by the Moravians, organized prayer meetings and held love feasts whose conviction was based upon faith in God, alone. This can be no better than both the degree to which God is TREQ is TRU, and how well we know God as TRU. Yet it is in faith that we can simply dedicate and direct our lives to TRU. For as TRU is absolute and eternal, hopefully, we must also become thus through faith. This faith is a conviction based upon certain and total TRU knowledge, rather than one of blind acceptance or partial faith. It is one in which there is an ever greater VAPO, ACEXIN, discovery, development, determination, learning and utilization of TRU.

CONCLUSIONS, SOLUTIONS and SUMMARY of TRU

There is an absolute and relative conditional imperative in knowledge which is manifested by different levels from uncertainty to complete certainty, extreme agnosticism to extreme gnosticism. The basis for doubts, problems and levels of certainty lay in relativity and how it provides the form of TRU. This is through the ever greater formal distinctions provided by the powers of predominant mass energy FAF or Á dimension and coordination in producing the necessary identity and POA of all that is in TRU. Upon providing greater identity, correspondence, relevance and reverence for all that is TRU the level of certainty increases. Our problem solving ability, anticipation and recognition of unknowns or unknown variables, increases. Problems are more and more reduced by combined solutions through coterminality. Combined truths and solutions are further produced by the plurality, totality, and limits of TRU. This is further provided by the resolution of absolute oppositions and inappropriate contradictions, which in turn leads to a final resolution. There have been many unknown variables in the equations of life and world. When mans most basic paradoxes are posited in the set framework of TRIT, as it best represents the totality, TO of TIT, resolution is forthcoming. As the great paradoxes that show the fundamental aspects of TRU are resolved all TRU can be better revealed and understood.

Throughout the history of philosophy there has been a fundamental disagreement between whether reality was to be found, and the world is to be known, only or largely, from the seen or unseen worlds by sense experience, or by revelation or some intermediate means. That there must be some moderate, and all inclusive, ground between the two, has been long suspected, but never sufficiently resolved. It was Agassiz who with the empiricism of Aristotle, and the idealism of Plato could recognize the limits of Darwinian evolution. This is done by cladistics and ATO wherein the great continuum of the mass energy, Á, FAF and dimension of TRU provides the relativity and union of observer observed, or subject object, in which its distribution includes their fusion in reality. Both sense experience and TRU revelations and interpretation of the unseen world are large contributors to reality, as is much else that exists in, and along with the L and T of, TIT. After deducing the solutions to 10 to 100 or more of the most important philosophical problems, quandaries, questions and paradoxes we may classify them into those that are: a result of, a failure to account for: IA/TRU, AU-RU-TRU, DATS and DETS, simple differences in TRUS, imbalances in ES in TO, TRUCH, ITCH, lack of VAPO, DAOST, SCARLS, many great laws, principles, NEC of TRU, an improperly stated problem, a semantic error in the problem, defect in COL or COR, and are mixed or represent multiple problems. Such accountings and analysis often provide great insights that can be derived from: the stimulation induced by profound problems, questions and paradoxes, the keen understanding produced by their resolution as shown by their basic mechanisms, the key fundamental laws, principles, concepts, the universals involved, and the discovery, learning, expansion and use of means to reveal TRU.

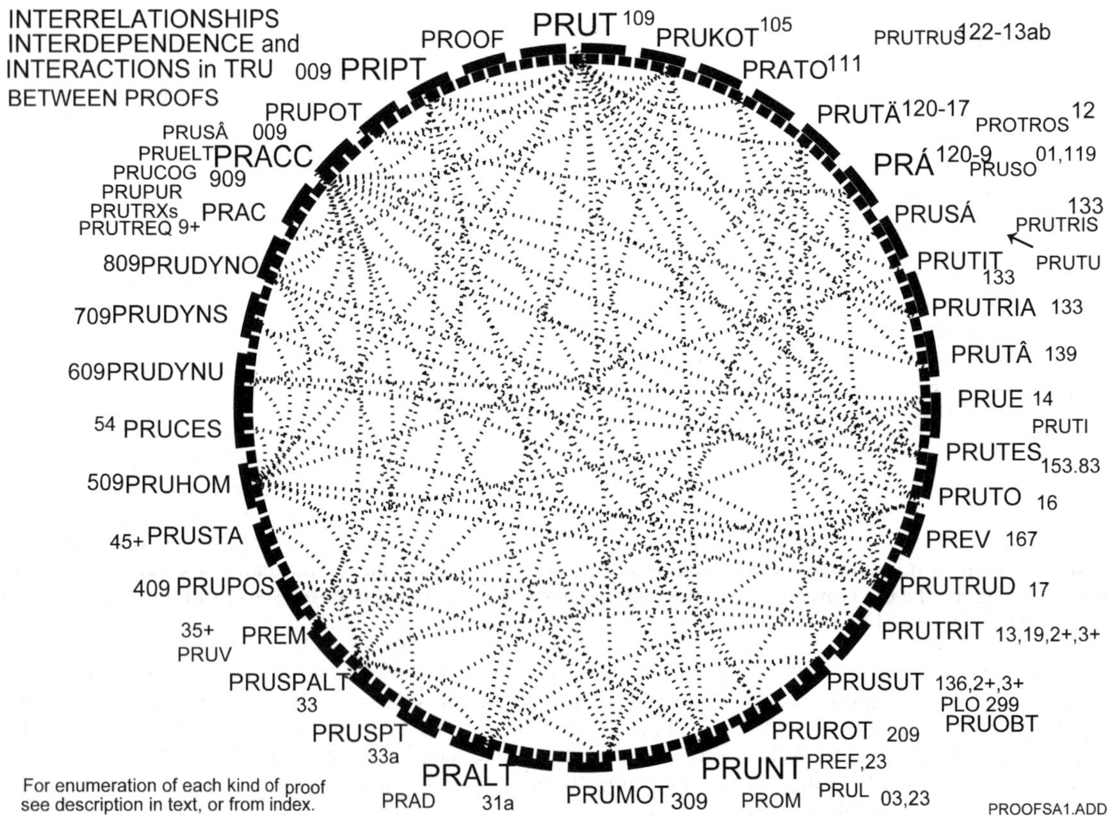

Fig 1-19 **COMBINED PROOFS of TRU**

The questions, problems, paradoxes, fallacies and enigmas we encounter are not initially resolved by the nature of our relationship to them. This is where and how much can't be fully resolved until we reach an end or completion. It is in this way Abelard, in a positive approach to language originated the primary principle of TRU. This is the manner in which reals, words and the world, exist together. For he held that it is only by becoming a predicate that the class notation or general term becomes reality. Words don't exist in and of themselves, nor do reals as universals, even senses and things. All exists only as they are their counterparts and processes in TRU. For words this is their predication or incorporation in the whole, as it is TRU. For predication is like COL, what is added to an entity that makes it TRU. It was from such an interpretation of things that the ideas of *ante rem, in re and post rem* occurred. This is how things are in and of themselves, added to or in things, and what they become, with TRU. Much of the continuing controversy was caused by failure to distinguish and understand the difference between idea subjectively and idea in the Platonic sense of an ultimate archetype of phenomenal objects. This is what we show, and a strong reason and proof of the mass energy state or Á FAF, continuum of TRU.

Figure 1-19 produces a formalized demonstration of many of the interrelationships, interdependence, and interactions of the greater aspects and topics of TRU. It shows how we can not understand, and are apt to deny TRU or any aspect of it, unless we can not only prove each one, but are able to show and comprehend how all exist together in TIT and of TRU. Even then, the highest levels of certainty necessary can not be appreciated unless, and to the degree, we have the most profound knowledge of the concepts, principles, laws, parts, processes and mechanisms of TRU that these proofs are based upon. These in turn depend upon a vivid overview such as this figure, and its companion figure 1-96 provide. This is called: **LORDCOT** = the Reciprocal Law of Optimal Depth Conception and Overview of TRU, or the basic fact of knowledge and the higher powers of human mass energy or Á. This is the fact that Á does not exist as a simple single entity, nor even as a dimension, it exists as a continuum in which all its gradients are best formed and function. This is one in which both differentiation and integration, or depth conception and overview are optimal and most beneficial to, as all must be in, TRU. **RESCOT** = Resolution through Coterminality. Much is lost through a lack of establishing precise coterminalities and distinctions. There is great justification for coterminality if it is held to a minimum and supported by maximum REM, proportion and ordination of TRU.. The goal is to provide total, complete and undeniable truth. This is called: **TRUT** = TRU TRUTH. These produce the ultimate, final, and infallible validity that arises from TRU.(LACTO,LACOLT,RESCOT).

SPECIAL FORM of CONCLUSIONS of TRU: MEASURES, STANDARDS, TRUTH and LAWS

There is great error in irrelevance. It is not only helpful to know all things, it is much more necessary to know how they are related, related to us, and in what proportion. This latter, proportion, is associated with the great error of irreverence, or how much we give value and divinity to what has more, little, none or is negative. It is not only the fact of irrelevance and irreverence that are so destructive to a person and all. It is the way, actions, lost opportunities and waste that occur with them. They are corrected only by optimal conviction in TRU, dedication to VAPO, and optimal measures, standards, truths, laws and accomplishment of TRU. This is founded on their being quantified, discovered by extraction, and formulating the same and different, as by DAOST, that exists in TRU. In so doing we can turn them, from negative to positive, to relevance and reverence. If all is to he measured by TRU, then this measure can be no better than the degree to which VAPO is applied to what of TRU is known, represented, presented and made available for separation, division, parameters, dimensions, polarities, gradations, and scaled. TRU is a highly reliable basis for accurate and reliable measure. However, it is no better than the degree to which we can discover and learn its form, dimensions, parts, processes, mechanisms and operations and their summation. This depends upon set and distinct concepts, principles and laws to express them. This is why a special language for TRU terms is mandatory. It is why we must constantly bring and hold these terms to their maximal formal distinctions by TRU. It is why ROT and POT are also so necessary and useful. Since laws are imperative, important and practical representations and guides, we must provide all those that TRU best offers. Their test is the degree to which they fulfill their use in and for TIT and TRU.

Reconciliation and laws as guides can be no better than their role in TRU and the optimal standards we adopt. Standards must follow the form and function of TIT as much as possible. The degree they do not is to be conditioned by the degree to which they are ever more imperative and suited for the application and use of TRU. Since much is imperative, since it is not always possible to exactly follow TIT, and since some of the standards are for extra TIT, or of TRU rather than in TRU use, some of this standardization will be arbitrary. An increasingly accurate resolution seeks perfection as an end, which can be continually approached and narrowed. It is this that we should seek with great anticipation, hope, and determination. Standards must be held to what provides the highest accuracy, sensitivity, specificity and reliability of their identify with TRU, and their practical need. This practical need is largely one of the subjectivity and objectivity of TIT. It can be measured by the degree to which they enable accurate corrections. Standards are set forms of objectivity we adopt in order to specify and make discrete aspects of the world about us, in our case, TRU. Even what has entered our language, vocabulary, and memory are representations which in a way become standards. Thereby standards are not uniform, but exist all the way from the extreme, exact, and prime replicas of TIT to inexact trivial guides. Standards when assimilated into a culture, universe, TRU, tend to become an aspect of it. This becomes a part of that special exclusive thisness which is characteristic. **MESRIP** = Method of Standardization Rational Identity Processes. This is the means by which we can and must continuously reexamine and readapt our standardizations, as in routines. How to identity them with TRU and apply the involved rational processes to each and all steps in their operation. **THESOS** = The Standardization of Subject Object of TRU. The world, TRU, can never be understood unless it can be coordinately, proportionately and totally represented, modeled or shown. This can't be done without an identity between TIT and its representation. It is the degree to which the standardization agrees with the object and the object with TIT that is the prime test of our measure, demonstration, knowledge, and TRU. When corrections are repeated, and especially when they are highly needed and important they assume a common form which often becomes a standardization.

There is a very close association between truth and law. Great truths are those whose validity corresponds with their existence in TRU. Law is truth that has been proven, set forth as a guide for direction, and acted upon imperatively as part and process in TRU. Both truth and law are absolute, relative and less, or pseudo and practical. Absolute truth and law are how they are no more nor less than 100% their equals in TRU. Common truths and laws are those which hold in themselves but not, nor not necessarily, in TIT. Pseudo truths and laws are what is stated or held to be the truth and law, yet are neither, a mixture that is less than perfectly valid and less than TRU. Practical truth and law denote their use for more or less temporary or partial benefit. So that the great truths of TRU as they are the great concepts and principles that must be put to use for TRU are its laws. The great laws of TRU, derived from its great facts and principles, are based upon its fundamental truths with imperative implications. This is why both truth and law are so dependent upon TRU as well as on each other.

Laws are a function of TRIA and are to a large extent limited. Their limits largely come more from law itself, not TRIA. It is not until we have a unitary set, exclusive by RIA and TRIA, that we will have adequate stability and reliability to warrant establishing definitive principles, concepts, standards, and laws. **LAL** = Law of Laws, or both laws that support all that is positive and beneficial to TRU, and those that separate out or eliminate all that is negative and harmful to TRU. All existing laws should be tested and measured by this standard. The degree to which laws and standards are not TRU and fail, is the degree they should be eliminated. The degree to which they are TRU and succeed is the degree to which they should be continued, expanded, or supplemented as best improves their roles in TRU.

PARTS of CONCLUSIONS of TRU: ACTIONS, USES, APPLICATIONS, TESTS, PRACTICE

When we differentiate the conclusions and summary of TRU we look to the many requirements, tests, adaptations, preventions, applications, uses, actions and practice that follow. This is in accord with any proficient professional program, how practice follows principle and principle follows TRU. These are the concepts, principles, laws, mechanisms and all that help us realize and know TRU. These are the requirements for TRU as a science and way of life. Much is lost between knowledge, even if correct, and action, if action is needed by TRU. Solutions and applications work best when foreknowledge and prevention are optimal.

TRU, as the most valid, perfect, proportionate and total science and way of life, will fail if it does not fulfill certain definite roles or needs including: completion, validity, actuality, proportionality, essentiality, sets of TRU, proper classes, varieties, forms, and functions, SCARLS, DAOST, and use. The sin of human failure to be TES, or ones part in TRU, was explained biblically as "the Fall" or man's lapse from innocence and goodness. This is immortality in reverse which was only to be reversed again through the atoning Grace of Christ, in mans redemption, and forgiveness. This follows from Christ's absolute TES, justification by God and TRU, and the reconciliation of man to TES and TRU. Together this great bonding in God or TRU to man and man to God or TRU is the covenant of parts within the whole, TIT-TO. It is the divine initiation of those coming to TES and as all that comes to its respective part in TRU, enter into a covenant. This is the destiny of humans and this world as directed by ones spiritual, rational, and actual development, or optimal Á FAF in TRU.

The sad fact of each person's life is how great a portion of this life is spent in thinking and doing what is of little, no, or negative benefit to TRU. One can't learn, even from the most bitter lessons, experiences and study, if one does not have, or there is not available the information, knowledge, interest and intent necessary. Since TRU is all, and all to be completely valid must be as it is in accordance with TIT-TO, so must all be, actually and ultimately, as it is in TRU. All laws, principles, axioms, corollaries, limits and relationships are not to be accepted as entities in and of themselves. This is even worse by use. One of the simplest and most effective rules or laws of problem solving or action are when something doesn't work, or is wrong, don't keep trying to do it, especially by forcing it. What is often necessary is only a very simple change or vehicle from what is attempted. This is what brings it back into equilibrium with TRU. Proportions prove validity, and validity tends to prove proportions.(VAPO)

As one gets deeper into an object or entity, they will encounter more of the same. What expresses, represents, or signifies the aspects, forms, functions, parts and processes of this object or entity will become increasingly similar, and with increasing experience, familiar. This explains and justifies the use of ideas, concepts and representations that are coterminal, have great overlapping and the need for tie-ins. What both makes these coterminal and what justifies them is value, the absolute value of TRU, and the greatness of their ideas, concepts and representations. Since TRU is so great, and the REM of the concepts of TRU are also so great, it requires a great amount of similarity and coterminality to assure that its ideas, concepts terms and language are not excessive, but their proportional part, in TIT-TO and TRU.

The test for everything, entities, senses, subjects and objects, is whether they are, to what degree, and how a part, and how they play a role in human life, the survival of the species, homo sapiens. How does it work through past, present and future, large medium and small, and in the mass energy of this special universe of best human life. All may be tested by TRU since we have in TRU by TRIA a closed set with limits which can act as a selective basis for testing. Criteria for testing all that comes before us, as OC, is based upon validity, importance, and contribution to TRU. Critical to all testing is the relationship and ratio between mensuration and immensuration. Mensuration is that method whereby all are measured by the same standard or in the same way. Immensuration is where there is some, much, or all measures of different kinds for different entities or things and whose correction in TRU is crucial if we are to achieve SUT, ROT, MOT, ALT, SPALT, POS. Between these two there is a ratio of commensuration for testing TRU. By *Euclid, Book 10 Definitions,* magnitudes are said to be commensurable which are measured by the same measure, and incommensurable which cannot have common measures. We can achieve the needed control if we reliably, accurately and effectively measure and demonstrate all by a set TIT, TRIT and OBT of TRU. When we select, adapt, and order measures by TRU to produce one that is common, like LCDS, we can give TRU the tests and POA it warrants and needs.

There is a simple, yet highly effective, test of what TRU is, and should be incorporated in any ROT and POT of TRU. For everything that is known to be important in life yet is not in ROT, POT, EPI, KOT or COAT, presents a void that must be filled. This may not be so actually important, for TRU or of an equal value to TRU, yet chances are that it belongs in TRU, ROT and POT. The way it belongs in TRU, ROT and POT may often infer new importance to meanings that are likewise not known or shown. This is the test of TRU in what is not known or shown of TRU, not in ROT, KOT, POT, EPI or COAT. It is in the search and discovery of such unknowns that information is selected, reduced, refined and perfected until it fills these voids in TRU. If one knows ahead of time whether they can or can't make it, this makes all the difference in the world. Whether they succeed of fail, S/F, in TIT TO is the final measure. To have foreknowledge is vividly exemplified in having an accident or disaster in which one or many are killed. In such, all in any way related will be, more or less affected, and fail. Herein is the compelling force of the future, and the great role knowledge, measure, standards, truths, laws, and test collectively hold for TRU.

KOT, CHALLENGE, Á, STRUCTURE, IDENTITY, AIMS, OM, ULTIMATE VALIDITY, and CONTROL

An image of TIT can be small or large. Yet if it is to be adequate to our role in TRU it must not only be large, but well developed and frequently used. How well this has, is, and will be done is the greatest power of TRU, the fulfillment of its Á FAF. Much of this knowledge comes from worlds outside of TRU, both by contrast, comparison, and content of similar universals and ESINS. Our life is one of experiencing and analyzing what these worlds offer and adding it to what we know of TRU. There has been a long, great and growing tradition of human and religious identity and appreciation of TREQ of TRU. This identity and appreciation are physical, mental, and their combined phenomenal or real FAF in the Á of TRU.(SÁ) They are of a fixed proportion, overall. Such a fixed proportion of both the mass energy dimension as well as that for length and time give us both the focus or center, as well as the coordinates, frame and structure of TRU. With this unity and form we can have a SUT that closely replicates OBT and TRIT. Together they enable an ever expanded learning of the beautiful intricate detail of TRU, plus a sufficient overview to encompass all TRU. The Omnificent Magnanimity, **OM**, this enables is what our lives seek. to understand the vast panorama of our actual world, TRU. We need to remember, ECRO, what actually exists is only what is relative to us, not all we sense, believe, or entities in themselves. All that is disproportionate, inactual, invalid, and non total is to that degree non existent or negative It is for us to discover, learn, know, remember, account for, act upon, and accentuate what is and is not, what is actually relevant and what is not. OM is the capacity and ability of each person to simultaneously have: total, massive, and perfect S, knowledge, O, S-O, identity, SO, SOT, duplication in TIT, and D and S of DAOST of TRU. The heretofore unrecognized, yet strong tendency and inclination for all these, as OM, to become

TRU can be greatly advanced by its recognition, practice and perfection. OM is how we master all by mastering TRU and making perfect and infinite the THES, THEO, SUT, ROT and POT in and of TRU. OM is provided by discovery, development, and determination of TRU, its ENT as our powers of feeling and understanding enable our mastery of all.

Many of the interactions that occur are not OITIO in a single, simple or set manner. They are FLOITIO derived largely from TIT, interactions working away from optimum. It is the less than 100% TRU and TIT, revealed in OITIO, that we must address if we are to ever understand, show, and live TIT and TRU adequately. It is this less than, yet of TRU, that explains many of the serious and vital problems and enigmas of our existence; whether we are individual, ENIT, ESIN or TES, whether TIT is absolutely fixed or labile.

Among those who have contributed the most to our understanding of **CENU** or Center of the Universe, include: PLATO and FICINO. Plato did it by recognizing and promoting the Idea, Nous or the intelligent and divine rational principle which orders the universe. Ficino contributed, expanding upon Plato, by presenting the universe with man as its center, a function of his immortal soul which internally ascends towards God, the totality, TIT-TO. Only as we can take the giant steps necessary to recognize the form and structure of TRU and separate out what is not TRU can we began to understand CENU and construct, represent, model and replicate TRU. It is in this way, through a practical and relative certainty of what is TRU, that it has a center of which the set structure, totality, and modeling of TRU can proceed. By recognizing what the dimensional structure of TIT is and by recognizing that each dimension can only be as it is a high order of predominance in TRU, a 3D construct is produced. From solid and analytical geometry such a construct of TRU enables a reliable representation and demonstration that can be increasingly well presented and used. This calls for one primary dimension for each of those dimensions common to any object, or length, time, and mass. For TRU these can only be what is specific, typical, predominant and characteristic.

The reason why infinite is so hard to comprehend is partly because it is beyond perceptibility, the limit of human ability to make distinct. Man is of a finite capacity, and beyond this range of finite capacity all is imperceptible. Even the increasingly finite cardinal or fractional information, communication and means given, show the existence of our limited capacity. We, of the 3rd millennium after Christ, have the opportunity to be most fortunate, the luckiest or most blessed, of all humans by living at a time when so much has been developed and provided for us. In turn we can provide so much more for those who will come after us.

As things and partials are or are not of TRU, their validity is a function of their relevance. They function in TRU coincidentally, the principle of ultimate coincidence. The interpretation, interaction and interdependence of ES, CES, NEC and principles of TRU are largely found in POA, POS, HOM, and ACC. It is these relationships and processes that most clearly show TRU. Philosophy is erroneous in its attempt to gain ultimate knowledge solely in diverse irreconcilable views. Science is in error in trying to explain all in the objective differential. Religion is in error in asserting all may be only understood in the subjective integral or some limited variation thereof. Not the least of these failures is to never recognize the LOLS in all that is not or other than TRU and lacks a dialectic, analytic, and rational approach to TIT-TO. The words, terms and concepts of language are the tools and media by which we think, know, act and succeed. They depend upon the LOLS in the meanings and definitions they symbolize. In contrast to this are: **COM** = Concentration of Meaning, **STOM** = Standardization of TRU Meaning, and **VOM** = Valuation of TRU Meaning. These lead to and reinforce ultimate validity. Through constant use and strengthening, words become greater, more defined and illustrate the ideas and entities they represent. Not only is there great challenge, but this challenge is mandatory, by life, to sustain TRU.

TOTALITY of CONCLUSIONS of TRU: OITIO KNOWLEDGE, LEARNING, COMPLETION, OM, EQUALITY, EQUILIBRIUM, CONCORDANCE

It is a major part of human relativity and its ability to subjectively encompass its object, that it is done as a whole. Yet for every whole each person objectivizes, there is a deficit. For an object can never be known infinitely and perfectly. By the exclusiveness of our thisness, lies the totality deficit of TRU. Besides this deficit there is that of our inability to determine what proportions of the whole, TRU, each and all of its components makeup. Associated with this is the fact that what we may assume to be proportional or having equality may be so in and of themselves, but not as they are in TRU. It is in these ways, among others, that there is a tremendous necessity for learning, learning by and of TRU. There is a ratio between absolute validity and increasing invalidity. This is why a composite view of God and TRU have so escaped us. It is corrected by a reciprocal O-S-O concordance or how our ultimate aim and purpose must follow complete TRU. It is possible when we have an equilibrium between the dynamic unity, plurality and totality of TRU. It is a total, proportionate, formative and positive affirmation of TRU. It involves a comprehension of TRU and all, a flowing flux and field in and out of TRU, OITIO. It involves an appreciation for the principles and states of knowledge of TRU and all in tandem, including a process of reversible differentiation. Its failure is due to a lack of VAPO and SCARLS, and the basic key concepts, principles, laws, and mechanisms of TRU. To achieve such knowledge involves a definitive filling out dialectically to completion, and adaptation to variation in the completion of extra TRU. Such breadth of knowledge, early exemplified by the encyclopedists, produces masses of meaning for all and a feeling of awe and omnificent magnanimity or OM. This is not only a culmination of subjectivity and mans relativity, the crux of our very being, it when associated with KOT, COAT and COGNOS, is the mystic and absolute identification with the salient glory and eternality of our opportunity in life, in TIT.

SPALT is less of a model when we fail to provide the intermediaries that explain its transit to and from TIT. It is in how the mass energy, length and time dimensions that exist in TIT and TRIT are formalized OBT, SUT, ROT, MOT, ALT and SPALT that is crucial. As long as the continuum from TRU to SPALT is free from irrelevant and non VAPO factors, and best shows the important and distinct differences in each step, we can achieve the kind of demonstration TRU warrants.

LOICLOR or Law of Confinement by Limits of Representation. Even if we know science, philosophy, religion, God, and many fields well in and of themselves we fail to achieve a good comprehensive perspective and order to all things as they are in TRU. We never come to terms with life. It is not possible to properly and exclusively separate knowledge without an overall frame, an integrated totality, or TO, by SUT. **TOD** = (Law of) Totality Deficit, or the deficit that occurs from the lack of totality of TRU. It is the degree to which humans do not know and utilize each and all that is of TO and TRU. TOD produces a block to: **SACRU** or the Summation and Continuation of TRU. SACRU affirms the continuity of TRU as it occurs in sequence and totality. It does so within its own framework, unchanged or at least relatively unchanged. SACRU is largely a product of: **DOST** or the Determination of the Special nature and thisness of TRU. SACRU applies both to the representation and presentation of TRU, as well as to TIT. TOD is also closely related to **EQUO** or the Equilibrium of Parts and Wholes in TRUD and TRU.(ENT)

Science tends to be knowledge becoming definite by completely fulfilling the knowledge of all things. Yet this is impossible without a closed set. In TRU it is by all in the D of DAOST, whose oppositions are the source of much of TRU. Definitive to the opposite is that process or means whereby in comprehending the totality of TRU we may lack information, or parts of opposites may not be discernible. This is corrected by filling out and making definitive the necessary knowledge of voids of TRU. We notice or do not notice, largely by way of what we know, are interested in, and are attentive to. If they know little, have little interest and attention they will notice little. As each ingredient increases, more in a tandem, most will be noticed. So that in a person who has total knowledge, KOT, great interest, and is highly attentive the slightest details or facts, notice can and will be brought into fullest TRU implications and a definitive ES and CES provided.

TIT is formed by and functions as an equilibrium. This is by close association with its coordinate form. It applies to both of its DAOST components, homogeneity and unity, as well as to heterogeneity and finity. It is how its homogeneity or homeostasis works centripetally to hold its heterogeneity together. It is how its heterogeneity or differentiation is selectively and otherwise adapted to its unity. This is a concordance of facts, routines, habits, and imitation that are no greater than the degree to which they equal their counterparts in TRU. This is a product of the mass energy form or Á state, L, and T dimensions of TRU. They supply much of both the means and extremes in the equilibrium of TIT. It is this, as all in TRU is related and interacts, that makes singular, complex and total S-O of equal importance as they function in equilibrium in TRU.

LARTRIA = Law of Reversal of TRIA, or the fact and state of all relativity, Á, and IA that not only is all TRU, but TRU is all. This means that we can not expect nor use TRU in any manner other than its all encompassing form. **OPFAT** = The total, proportionate, formative, functional, positive affirmation of TRU. How, by VAPO, we remove all ATRU from TRU, an Optimal separation out of all not TRU.

Fig 2-19 **OITIO and FLOITIO**

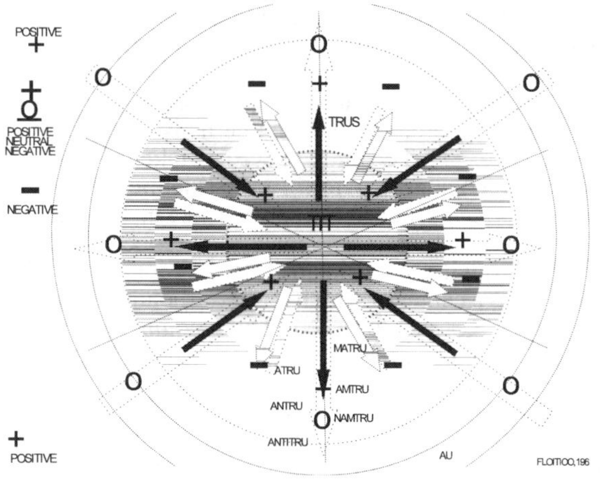

The best possible COA is only attained after it has been done through COAT. This includes all that is of TRU - with TIT in the core circle - and not singular, combined aspects, nor lesser mixtures. This becomes possible when we recognize **OITIO**, out and in, and in to out, of TRU, and **FLOITIO,** The flux, flow and field of OITIO. These necessitate thorough, proportionate. ordered and complete comprehension of all that is of TRU such as TRIS, TRACH, TRUCH, RUT, RUS, AU and all that may be significant or may, or will, have some relationship to TRU and its DAOST, SCARLS and COAT. These as other subjects and objects, entities, senses, things, figments and actualities relate and are useful to a complete and successful comprehension of all, COA, and all things, COATH. This is when total, proportionate, continuous, pictographic and complete, or all encompassing, formulation and construction such as SPOT, BASFANT, SPALT, SPAT, and ELTICMA are available. As these are gained, OM is reached for all who try and succeed in mastering them. It, as perfected by each person, is the most satisfactory system, knowledge, perspective, representation, and presentation of our world, TRU. It is not so much the world that is chaotic. It is not so much the world that is a non unit. It is not so much that we should be complete skeptics or agnostics. It is that if we do not attain a total comprehension of all things, COA, as they are TRU, COAT, then neither things, entities, senses, our knowledge nor our being have full and accurate meaning. We are not. This is equivalent to 1/TA. For it is the great power of meaning, the Á, of SO of TRU our predominant, preponderant makeup and mass energy component that enables us to be, that is the generator, preserver, protector and core of our being. OITIO and FLOITIO are the resultant state of flux from: **LACA** = Law of Combined Absolute, or that state and role of all in which the absolute of TIT, TRIA and TÂ is combined with the less than 100% perfect, but absolute in their own senses or ways -ENIT, SENIT, THIT, RIT, PIT, and even AIT - which together form a kind of whole. It is this, the less than actual, whole that OITIO and FLOITIO operate in, signify and show.

LOCOAT = Law of the Comprehension of All Things and/or All TRU. The search for knowledge, wisdom, and philosophy, is endless, eternal, and infinite. Only as knowledge limits itself to TRU, and only as this knowledge has maximal form, coordination, proportion and ordination as by SU and ALT, and only as these proceed by LOLS, can it be made comprehensible. TRU is comprehensible as unity and as a totality with less than infinite parts as by SU with modeled finite parts. **PROCTRU** = Proof of the comprehensibility of all things, proceeds from the SU of TRU. Knowledge without TRU, SU, and LOLS can be in part or piecemeal of all things. Yet it can't be comprehended in totality. For although AU can be visualized illusively as a totality. It can't be totally ordered and represented as a finite object. It can not be described and demonstrated in the way TRU can, for it has no set, values and ordered observers, observed-s, parameters, scales nor overall relativity. Its dimensions, and their scales, are disordered and indistinct. Only as it is altered by the observers, and only as it is scaled to standards set by the observed, a set actual world, TRU, does its state, conditions, relations, and limits take on adequate distinction for COAT. This is, **LULAR**, the Law of Universal Limits only achieved in a set RU, TRU. If an object has no dimensions, is not limited to its dimensions, nor set scale for these dimensions, it has no representation. It's presentation can't be comprehended. This is imposed by relativity, the definition of a RU and TRU, distinct from AU. Without TRU an object is indefinite, inadequate, and irrelevant to us, our S-O, and doesn't have valid actuality.

DYNAMICS of SUMMARY of TRU, SUBSEQUENT TEXT, TIEINS, TRANSCENDENT FUNCTIONS

The summary of TRU dynamically brings together what we have learned of TRU, its functions and how they relate to all else to produce a total picture. This can not be done through TIT alone, nor even TRU alone, for the added states of objectivity and subjectivity we put TRU through alter its ENIT status and function. This is in accordance with metatheory relative to set theory. It does not destroy, nor even appreciably detract from TRU and TIT. It only supplements and adds to them. Thereby this sequence follows relativity and the frame of objectivity-subjectivity as it is not only in TIT, but OITIO TRU. This gives it unity so that much of what is TRU can be confirmed by complementarity in which only the totality of phenomena exhausts the possible information about an object. This even though much of the information is, in part or largely, outside of the entity this object represents. This is what all sections of IPTRU more or less present as they supplement, yet complement TIT. Complementarity was a development of Niels Bohr and arose from the evidence obtained under different experimental conditions that can not be comprehended within a single picture. They must be regarded as complementary in the sense that only the totality of the phenomena exhausts the possible information about the objects. This may be compared with OITIO where it is impossible to know TIT from within ES, CES or TES alone, or even TO. It must be looked at as if drawing away from it to see what is the same or complementary to it.

PORE = Principle of Relate Extrapolation. PORE, as a Method and Science, is how posited concepts can be shown, demonstrated, and accurately represented through inference and projection. As difficult as it is to sense, know and show TRU and its content, and to posit with formal distinction what we can sense, know and show, there is much that can be posited with a high degree of accuracy. There are many methods which enhance the accuracy of these posits. Once done, having these as an ever expanding base, we can relate, extrapolate, infer, and project separately or in various combinations of ES, CES, EC, and NEC, to discover, learn, know and show most all of TRU. **PRIT** is the Primary Law or Principle of TRU that arises from the need to recognize and adapt for posit entities, senses and things, that are not or less than their role in TRU. PRIT corrects the inveterate tendency we have of accepting all with blind faith and without qualification. It likewise corrects the inclination to limit TRU strictly to TIT. PRIT certifies, with the immediate reversal back into TRU, that subjectivity is everywhere we are, and where we are not, by TRIA, all is not. Thereby the subjectivity of TIT is not only valid, but absolute, This is: **SANT** or Sujectivities Absolute Necessity of TIT. It is this very subjectivity that gets us into so much trouble, and causes us to depart from TRU. Where, biologically, species with much less mentality hold to their optimal adaptations, humans by way of their highly diversified subjectivity, thought and IA become lost in the ATRU world about them. This needs to be corrected: **LARAC** = the Law of Action-Reaction, where subjectivity is immediately reversible back into TRU.

SOID, MOT, ALT, and SPALT enable us to look at not only the frame of TIT, but, more importantly, give a more precise VAPO for analysis of the parts of TO. They also help us understand our processes of thought after their own subjectivity and the differentials of TRU become an overall subject-object entity. This overall unit, and its accompanying knowledge of TRU and all, can only be achieved if SU, SPALT, POS, and HOM are well developed. For they provide its substance, order and what it stands in contrast to. Such a formulation is what effectively enables better application of our knowledge. The complexities of such an understanding of overall finality are not limited to the makeup of TIT, as important as these are. They are also compounded by TRU, TRUS and added objectivity, subjectivity and practice. This has been shown in the fine arts wherein much of their beauty is not a simple replication, but all that goes into its overall interpretations and meanings. Since these can be no better than the accuracy of the detail, form, function, parts and processes as well as VAPO of its object, TIT, they must be the best possible. They are often not directly possible. They can be elicited through relate extrapolation.

With an explicit and definitive demonstration of the nature, divisions, form, characteristics, qualities, properties, parts and processes of TRU we can then apply these to all we find in life. We will find how greatly all we see, experience and learn is derived from their general and special cases and their form and function in TRU. These accumulatively build upon each other both qualitatively and quantitatively enabling us to produce the knowledge and demonstration of TRU sufficient to overcome both its complexities and the resistance of unknowns and those who know erroneously. It is in the overall posit of all that is TRU and TIT, into a replica or model, that best and finally enables us to find the ultimate dimensional gradients and makeup of TRU. It is in this way we can constantly and increasingly perfect our the duplication of TRU provided by POA, TRIT, OBT, SUT, as they reflect the form, SCARLS, beauty and greatness of TRU.

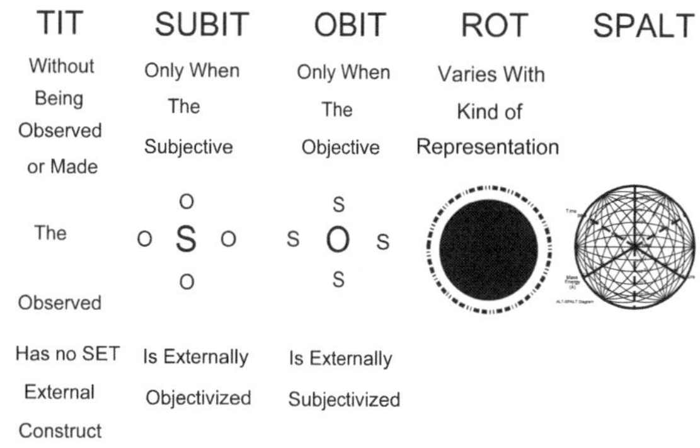

Fig 3-19 **THE SHAPE and STRUCTURE of TRU** The shape and structure of the set TIT, or even the more general and variable TRU, is not, and does not exist in, a simple and entirely discrete form. This is different than TRIT, OBT, SUT, ROT, MOT, ALT and SPALT. It depends upon what you refer to, what of TRU is the given. It depends upon how much TIT is altered to fit this objectivity. TIT has little or no external shape, it extends outward infinitely and inward infinitesimally. It is infinite and infinitesimal in each dimension, actuation, extension and intension. For TRU it depends upon which kind of TRU we refer to and seek to interpret. For OBIT it can be of different shapes, or of different degrees of solidity. For ROT TIT tends to be spherical. For MOT TIT is usually a sphere. For ALT TIT is a field, depending upon how standardized and balanced it is. For SPALT TIT as a sphere is directed toward an absolute set, OBT of TU-TO-TIT. TRU, KOT, and COAT can not be gained piecemeal, nor in simple unity. They can be gained only in the total, proportionate, increasingly finite, ordered and posited representation of TRU. This is in how all the processes of man, in or of TRU, when brought together as a whole by VAPO represent the content of TRU. By filling in the voids and spaces in TRU they are very helpful in guiding us and confirming the frame and construct of OBT, TRIT and TIT. This is to the degree TRU can be put into perspective, reproduced and presented; and must be done reciprocally. **LORÁ** = Law of Order by Relativity or the mass energy, Á, dimension, form and function of TRU. One reason why LORA is so important is that it raises us up from levels of topics and distorted ATRU schemata and brings us directly to TRU, and thus to TIT-TO. The only effective and successful presentation of senses, things, and the world, can come from the discrete construct of The Thing in and of Itself, TIT. It is done as they are modified and conditioned by the special form and objectivity of TRU. Positioning of the constituents of TIT can then proceed as they also are discretely selected and separated. Just as humans learn to live by time and place, so do they learn to live by mass, or whatever is predominant in their world. This is how the mass energy of TRU, or its Á FAF and dimension, comes to provide order. Along with L and T together, ALT, they make up the form, frame and structure of TRU. When we do not understand ALT, and fail to be coordinated, we are, like any animal out of its place, time and means, lost and soon to lose life. It is LORA, ALT, and all the rest of ROT that gives and shows TRU by SUT. This is what enables its members to be coordinated, adapted and successfully function in TRU.

With knowledge, TRU, SUT, SPALT, POS, HOM, and ACC, once they and their greatest NEC, concepts, principles, laws, mechanisms, ES and CES have been discovered, recognized, developed and utilized, it becomes and is for all time familiar and learned. **ATIM,** or the All Time Mastery of Things, TRU. ATIM is a product of the interrelations and proofs of all in and of TRU. It very much depends upon the refinement and progression from TIT through TRIT, OBT, SUT, ROT, MOT,

ALT, SPALT, POS, HOM, ACC and back to TRU and TIT. **PRUT** or the Proof of TRU is neither simple nor easy. If it were it would have been discovered and produced long ago. Yet with such means as LORA and ATIM we can establish much of the ground upon which a substantial basis for its proof can be forthcoming.

ENABLEMENT of SUMMARY of TRU, TRIT, OBJECTIFICATION, SUT, and ROT

Representation acts both as a method of TRU as well as being a large component of and in TIT. To objectify and show TRU appropriately we must include all its principles, aspects, divisions, parts, processes and totality. This requires any and all means possible and effective, especially means of handling complex data. As great and beneficial as the representation of TRU is, it poses added problems. TRU and its representation are largely unknown partly because they are in such a formative, transitory, and uncoordinated state. Likewise because we have not properly understood and shown how to bring the universe, world, and God to TREQ and TIT. For the steps from TIT as an ENIT to TIT as an object, to TRU as many kinds of ROT and MOT, add extra principles, features, and problems that are highly misleading unless well ordered, understood and compensated.

LACOT = Law (Problem or Paradox) of the Complete Objectification of TRU. It is not enough to identify and describe TIT. If we are to ever understand, utilize, represent or hold in complete perspective, and present TRU with the validity, greatness and completeness it warrants, we must show all its: principles, laws, aspects, varieties, components, divisions, forms, functions, core dynamics, ES and CES. As much as possible each need to be shown as discrete, distinct substantial elements in the proportions they exist in TIT. When TRU becomes the object, it differs from TRU itself. This is, in ways of maintaining a 1:1 replica and identity, a sad but certain fact and law. This is called: **LOTAC** = the law of TRU on becoming an object will assume Certain Alteration of Form. This alteration of form will also vary depending upon the degree, state or condition of this objectification, representation, presentation or perspective.

The reason why most all the great philosophers were only those that were systematic thinkers, and of these mostly with a more or less, inclination to TRU, was not that they had any monopoly on intelligence nor some special key to all, it was because only through a positive objective approach, a systematic method of bringing the structure of TRU into the structure of perspective, and a dogged determination, can the necessary overall ROT, POT, PERT, and EPI be attained. This involves a reconciliation, fusion, and operation of subject object, together as S-O, SO, SOT, Á and all SÁ FAF in TRU. This is by LATAN, a necessary condition of relativity.

PROBT = Proof of OBT, the Object of TRU, comes from: 1. differences between TIT and OBT, 2. From TRIT or the set construct of TRU, the special form and frame of TIT, 3. From SUT, the systematic unity of TIT and TRU; SUO, SUT, SUL, SUFO, SURMS, SUTFO, SUTORMS, and SUDE, 4. **OSTA** or Objective TRIT, TIT Systematic unity Absolute equality and identity, 5. **OCA** or Objectification by Formative Construct, 6. Limits of OBT not present in TIT, 7. **LORSIT** or how the Representation and Standardization of TRU become TIT, 8. Variations in OBT, subjectively, in method, and in standardization; MEROT, MESU, 9. **LODAS** or Dynamic Âctual Standardization,(SPRE) 10. **LACOT** or the Complete Objectification of TRU. Some of the proofs of OBT include: POT, OB, S-O, Polarity, TRU, TIT, TO, TÄ, VAPO, Form of Relativity, TRIT, SUT, Á, SOT, Universe, Relative, The, LARF, OM, ROT, LARP, LATANO-S, ALT, MOT, ACEXIN, SPALT, LADAC, PRALT, Dimension-s, Length, Level, and Object-ive-ivity, **PRUD**-Proof of TRU Dimensions, Proofs of TRU as multidimensional. Similarly, proof of the significance, importance, relations, role, effects, and benefits of OBT include: 1. Since objectification is impossible or fruitless without subjectification and since both are fruitless without TRU it is by their combined roles in tandem that much of TRU is formed and functions. 2. OBT as a great portion of SOT and Á, is in TIT, like **RÏT**, and thereby a vital aspect, although different from TO-TIT. 3. OBT as Representation in Itself, **RÍT**, is largely out of TIT, yet highly potential for it. 4. OBT, along with TRIT, SUT, ROT, MOT, ALT, SPALT and all, is the basis for much of the operations of TIT and TRU in IPTRU, especially REM, POS, STA, HOM, DYNU, DYNS, DYNO, ACC, OITIO, OM, PUREACLITY, COGNOS, ELTICMA, SÂ, and TÂ.

FINAL OVERVIEW of CONCLUSIONS and SUMMARY of TRU, POT, IPTRU OITIO, ENDS, PERSPECTIVE

To complete this section and chapter we are led to produce what is most important, a final overview. This is how TIT and TRU tie-in and return to POT, EPI and IPTRU. It involves OITIO or how we move in and out of TRU, not only TIT by TRIA, but through TRU and MATRU, in a metaTRU or extra or mixed TRU way. For TIT it requires closure and perfection with an ultimate world view and summation to TÂ. OITIO forms a polarity between TIT and its being acted upon objectively or subjectively. In terms of TIT it is a partial variable. Yet since it is not pure TRU, and often little TIT, it will usually stand markedly apart from TRU. OITIO in being both a polarity and often of little TRU will vary as the subjectivity and objectivity vary in a great number of ways. It will return to TRU and order only as the first person holds it to its ESINS and ES in TRU as well as focus it on TRU as a unitary whole. The observer and the observed, subject and object, or relativity by TRU are implied but not fully represented in much of all human life. This is especially conspicuous in OITIO. When fully represented OITIO becomes **SOITIO**. SOITIO or the sum of OITIO is a great part of all our lives. It shows our never-ending error. This is a lesson all must learn, deeply and thoroughly, as early as possible in life. For without it the I, or first person, is perpetually not, negative, lost, in conflict, and largely DAK.

The argument to prove God teleologically or on the basis of design or purpose has, when TRU, considerable merit. For purpose and design of the thisness and set form are what establish TRU. This is shown by the validity of an object. It leads to integrative, interrelated and summative proof, as all that is TRU accompanies this purpose and design. It can be seen when we have an inversion of the image of TRU, as it is TIT. It is supported by the inverse identical proportionality of TRU. It is supported by entities becoming ESINS and then their ES in TRU and how this is shown when TRU is represented and presented. It is supported by separating and expanding upon the difference between the statics and dynamics of TRU. Proof of TRU, as well as a successful comprehension of all as TRU, can not be found in singular truths and proofs nor even combined or complex ones that are not or less than VAPO. They are to be developed from a maximal identity, presentation and simultaneous optimal SUT through finite ordered expansion. They must follow THESOS and universal and particular synchrony. Together these become the formulation and understanding of TRU through its totality, representation and presentation. This is made clearer by corresponding views of TRU as well as of its concepts, principles, laws, and mechanisms, as shown by METAD. With good proof and successful comprehension of TRU a perfected sustained TRU can be achieved. This is, as in religion, an imperative identity or common integration of focus, totality, absoluteness of the form and function of TRU, especially in its predominant special and exclusive Á, L and T form. The practical test of this is the degree to which it attains a perfected presentation, perspective and ordering of utilizable TRU. Such a perfected sustained TRU is achieved when our ultimate world view is imposed by posits in this ordered and utilized

TRU. This becomes even more sustained by providing the closure of TRU to SÂ and TÂ.

PIMAX = Problem of Maximal Identity and Presentation. PIMAX arises by departure from TRU as THEO, THES, ROT, and MOT depart from each other, and how they are brought back to TRU. This has been due to human inability to bring senses and things into any meaningful successful ordered identity with TRU. This is done by: **SOEX** = Simultaneous, Optimal Systematic Unity through Finite Ordered Expansion, and our ability to attain optimal SD, subjective differential.

TIPS = TRU must be Proportionate not only in Itself, it must be Proportionate in and of Subjectivity Objectivity, SO. TIPS affirms the necessity of VAPO in an overview of TRU. We can never fully or finally interpret the subjective, the objective or anything else other than the degree to which it stands in the relation it has with TRU, TIT, TRIT, TA, OM, and OITIO. For the subjective this is **STOD** or Singular Subjective Total Discrepancy. **LUPS** = Law of Universal and Particular Synchrony, Symphony and harmony. This law is more than the concepts, principles and laws binding all TE, TO and TRU together. It is the all transcendent law of EPI and TRU whereby as we represent and present TRU we know TRU and all, COAT. This helps overcome the continual abyss of entities, senses, things, opinions and attitudes that are so close to us, yet so deceptive and so much in error. To attain complete knowledge and to find sufficient order in the world, one must: recognize and understand LUPS, TRU, the optimal relationship between observers and observed, the divisions, dimensions and dimensional formulation, and SUT. Thereby we best utilize these divisions, formulations, SUT, ROT, MOT, SPALT, to optimally present, remember and live TRU.

IMPOSING ULTIMATE WORLD VIEW

With basic definitions in hand, along with their corresponding abbreviations, we may by superimposing them on TRU in the representation, perspective, and presentation of TRU, order and reduce all to an ultimate world view. Universals posit not so much by point or part, but by linear, planar or solid configurations. Partials are those differentials which, as ES and CES in TRU, are more focal and often provide reliable and practical posit by point. Both universals and partials have centers which can be used to reinforce such point placement as long as their extensions are recognized. Relative tends to be what is centripetal. Absolute tends to be what is centrifugal. Dynamic is of variation which is, when of decreasingly TRU, centrifugal. Static by reduced time tends to be what is centripetal and relative. Through posits in SPALT, METAD, and SPAT the definitions, abbreviations and denominators that represent TRU are shown.(Pg 44)

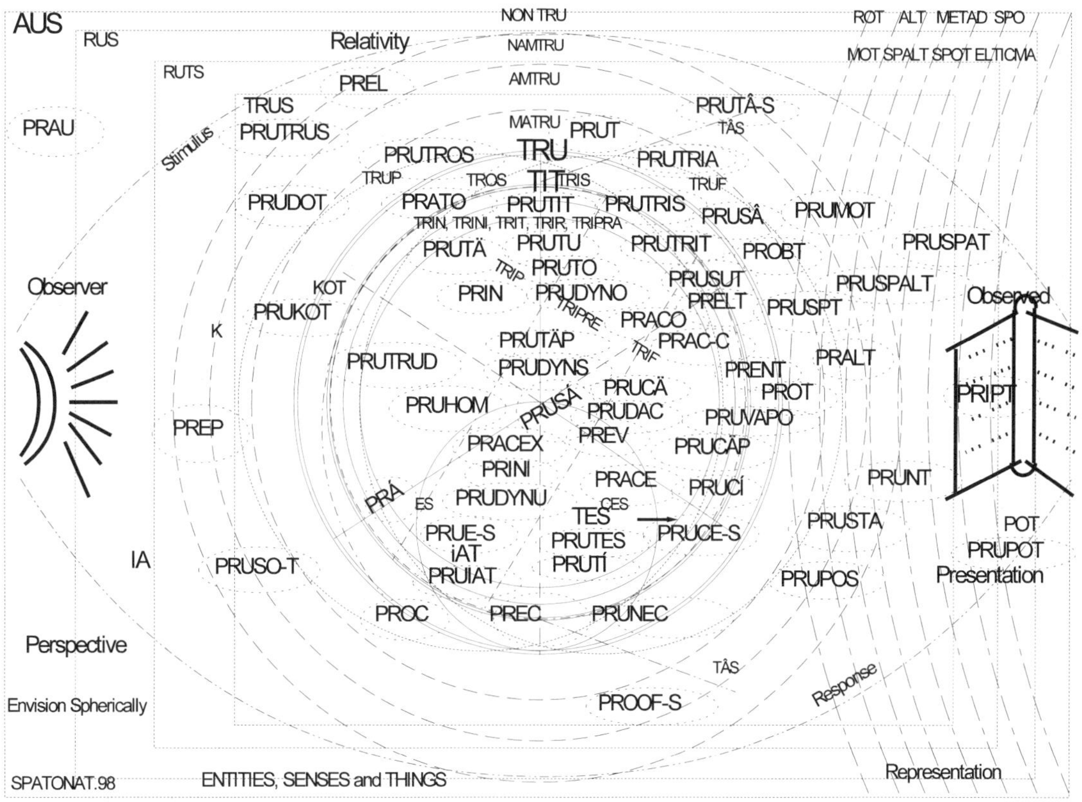

Fig 5-19 **POSITS of MAJOR PROOFS in SPAT**

SPAT = Standardized Perspective of all, TRU. METAD is a precursor of SPAT. Together they form the primary external views of TIT and TRU for ROT and POT. They seek to show how the perspective of TRU by SPALT may be oriented, coordinated, and ordered as a whole. They utilize the dimensions of TRU to bring the whole into focus exteriorly.

All major proofs are included, with the exception of those of ACC, such as PRUTREQ, PRUTRAX, PRUTREX, PRUTRIX, PRUPUR, PRUCOG, PRUELT, and PRUSÂ. These follow closely after PRACC, that accentuation of TRU or what we need to do. A successful way and path in life depends upon comprehension, an understanding of oneself and the world around one, which in turn depends upon ones perspective, their perspective of all, TRU. Like what we do for any object before us it needs to be clearly analyzed, modeled and shown. This is the major overall purpose and goal of our predominant existence. It is the key, and crucial mechanism of TRU. It is the refinement, fruition and consummation of the S-O, or Á, the predominant mass energy FAF and dimension of TRU. It is for us to, with the deepest courage and conviction and with the most advanced and perfected complexes our S-O, Á functions in TRU, when fully discovered, understood, acted upon and practiced, that we will so deserve and win for TRU and ourselves all the glorious, victorious, eternal success that can, must and will be ours. **LOV** = Law Of Objective Validity, and proof of, or a thing to be proven must be made known, demonstrated, and represented correctly. It is necessary that we not only reduce this image to our thought, we need to elevate it, ELTICMA, to TRU and TÂS, with continual optimal reversions. Closure for TRU is no more nor less than the continual reversible fusion and overall management and control of TIT, TRUS, ROT and POT in which our IA is acceleratively improved and refined to meet this need for closure. This is concluded in PUREACLITY, COGNOS, ELTICMA, SÂ and TA.

It is in the Separation and Union of entities, senses, things, Relative universes and TRU, **ROSU,** that we can make TRU discrete, provide much of its proof, and make explicit representation and presentation. Much of this text is a reinterpretation and application of philosophical, religious, and scientific principles to TRU and in return TRU's explanation and correction of them. This is seen in **LAMPREC** or the Law and Methods of Classification and Reduction of Philosophical, Religious and Scientific Problems and Paradoxes.

Throughout human existence there has been **RUPKAL** or an accelerating Revolution Universal of Perfected Knowledge, Love, Appreciation for being. This has been of, to and in TRU, TIT-TO. This has been through the entire continuum of its mass energy, or Á FAF, most notably its more recent and higher levels or states. Proof comes from truth, and truth comes from TRU. This is seen in **LAMIR**, or the Law of Mutual Identity of Right, TRU to KOT to ACC to ones acts.

Much of our difficulty in understanding, especially TRU, can be resolved if we practice **ITRU**, or Inversion and Involution of the Image of TIT-TO and TRU. By **LACTRUT,** all that reveals the Combined Truths of TO, TIT and TRU, their sameness and differences much of each and all together can be proven. Proofs can be better understood when we realize that Laws are a Function of TRIA, **LAFRIA**, and their limits to the aspect, form and function of TRU to which they pertain. The certainty the concepts, principles, laws, mechanisms, demonstrations and proofs provide for TRU and all its aspects are what enable us to Refine and make Reliable TRU as a Measure, **ROMT**. It is these refinements and measures that in turn, when combined by VAPO in TIT-TO, enable, **LACOP**, or the Law of Absolute Combined Proof. All comes together in TIT-TO through TRU, ROT, POT, and IPTRU. This occurs by, **PUC** or the Principle of Ultimate Coincidence in TIT, through its focus, aseity, perseity, haecceity and quiddity.

TRU, the world, or even oneself, can not be properly understood unless they are well known, as by VAPO, DAOST, SCARLS, TROS, and TRIS. Even then this requires increasing detail of the beautiful intricate differential nature of TRU and TES. To do this with sustained perfection is by a Omnificient Magniminity, or **OM.** The following proofs for TRU proper cover those topics from TROS through ENT or Chapters 12-19. They are neither exhaustive nor perfectly distributed. They do, however, provide strong support for each topic, and together for related topics and TRU.

PRUTROS TRU, TE,TU, TI, TO, TÂ, TÄ, MOT, REM, ALT, LADAC, Relativity,,

PRÁ

S O, S-O, SO, SOT, TIT, TÄ, TE, TO, POT, FAF, TRUD, Human, DAOST, Relative, The, LARF, MATES, ENT, KOT, PRUT, Meaning, LANT, EC, MADAS, ACEXIN, ALT^{n}, LARP, LATANO, PRALT, LADAC, Acuity, Adaptation, Attention, Reality, CNS, Concentrate-ion, Expect-ation, Expression, Intent-ion, Is, Language, Mass, MassEnergy, Meaning, Nervous System, Problem-Solving, Purpose, Real, Reason,

PRUSO(T)

Á, POT, PACH, S O, SO, SOT, TÄ, Human, DAOST, Relative, The, LARF, MATES, OITIO, OT, MADAS, Subject, Object, Distinction,,

PRUSÁ

Á, PRÁ, POT, S-O, SOT, CNS, Senses, Things, Phenomena, Reals, Reality, Human, DAOST, MATES, Meaning, OT, EC, LANT, PRUNT, REM, LORVA, Continua-um, States of mass,,

PRUTU

TROS, FAF, TIT, SOT, CENU, ENS, RU, Special, TÄ, SUT, God, monotheism, DAOST, Universe, Relative, The, LARF, ALT, REM, LORVA, LADAC, Unity, Unity of Opposites, CENU,,

PRUTRIS

TIT, T, S-O, SO, SOT, Á, Thisness, TRIT, TI, TÂ, TÄ, SCARLS, DAOST, Universe, Relative, The, LARF, OM, TRUP, TRUF, ALT^{n}, Inertia, Interval, Presence,,

PRUTIT

TRU, TRIA, VAPO, Thisness,, Special, TREQ, Life, Immanence, Selection, 1/TRIT, TI, TÄ, TÂ, TRUD, Human, Homo Sapiens, Species, Relativity, S-O, Mentality, Perseity, Living, God, Deity, Á, SCARLS, DAOST, ES, TES, Universe, Relative, The, LARF, MATES, TRIP, TRIF, PRUT, KOT, TIT/OT, Meaning, MADAS, ALT^{n}, ACEXIN, LARP, LATAN, REM, LORVA, SPALT, LADAC, PRALT, HOM, CES, FAF, LAMUA, WITÂ, AIT, COGNOS, PEQ,Actuality-ies, Excellence, Form&Function, Inher-ence, Pure-ity, Special-ies, and The Special Universe Of Best Human Life, Itself,,

PRUTRIA

TIT, VAPO, Thisness, Special, Aseity, Haecceity, Perseity, Immanence, life, survival, absolute, TÂ, TÄ, SUT, TI, God, DAOST, SCARLS, TES, Universe, Relative, The, LARF, OM, KOT, PRUT, LARP, LATAN, REM, ACC, TRUID, Ideal, Include-sion, Optimal,,

PRUTÂ

TIT, TRIA, Continuity, Immortality, Etern-al-ity, TI, TÄ, Life, God, Selection, SCARLS, Universe, The, RUPKAL, TRUP, TRUF, LACOP, WITÂ, AIT, ACC, Absolute, Constancy, Eternal-ity, Perfection,,

PRUE

TE, E, ES, VAPO, Division, Part-s, Vision, Element-s, Atoms, TI, Human, Á, DAOST, MATES, LOPAN, KOT, Meanng, PRUNT, LANT, EC, MADAS, LARP, Distribution, REM, LORVA, MOCBIREM, CES, DYNU, Character-istic-s, Composition, Depend-ency, Differentiation, Element-s, Finite, Gradient, Infinite-ion, Infinitessimal, Infinity, Intricate, Molecule-s, Part, Polar-ity, Quantum, What-ness,,

PRUTES, PRUTÍS

TE, TIT, VAPO, I, Soul, Human, ES, SOT, Á, DAOST, Relative, The, Point of reference, IA, It, LARF, MATES, LOPAN, ENT, OM, Meaning, MADAS, LARP, REM, LATAN-O-S, ACEXIN, ALT^{n}, REM, LORVA, MOCBIREM, PRALT, DYNU, WITÂ, AIT, ACC, PUREACLITY, COGNOS, Attitude, Built up, Christ, Conscience, Empathy, Identity, Integrity, Live Product, The Part, Person, Reconcile-iation, Redemption, Resign-ation, Responsible-ility, Saved-ation, Self, Soul, Woman, Man, Human,,

PRUTO

TRU, TIT, POT, VAPO, WORLD, God, All, Universe, Added States in Totality, Gestalt, Special, TREQ, Life, TRIT, TI, TÄ, TÂ, SUT, TRUD, Humanity, ALT, SCARLS, DAOST, Relative, The, LARF, OITIO, ENT, OM, TRUP, TRUF, KOT, PRUT, TO/OT, LATAN-O-S, ALT^{n}, ACEXIN, LARP, REM, LORVA, ALT, SPALT, REM, LADAC, PRALT, HOM, FAF, LAMUA, DYNO, Balance, Closure, Fusion, Fulfill-ment, Integration, Mankind, Order-ination, Synthesis-ize, Total-ality, Variation,,

PREV

TRUS, TRU, TIT, TO, TRUD, POT, HOM, ACC, ELTICMA, SÂ, TÂ, Life, OITIO, ENT, TRIP, TRIF, OT, ALT^{n}, ACEXIN, LATANOS, REM, HOM, CES, ACC, Interfuse-ed-ion, Reciprocate-al-city, Refine-ment, Repeat-ition, Resilient-cy, Reverse-ability, Synchro-nize-icity,,

PRUTRUD

TRUS, TRUCH, TRICH, FAF, HOM, Mechanics, RUD, Physics, OITIO, Operations, Process-es, Life, TI, Human, Á, T, TES, CES, DAOST, ENT, OM, Meaning, MADAS, ACEXIN, ALT^{n}, LOPA, DYNU, DYNS, DYNO, LAMUA, Compliance, Energy, Fine Tuned-tuning, Frequency, Function, Golden Rule, Momentum, Negative-ion, Positive, Power, Problem-s, Right-s, Role, Spirit, Stimulus-Response, Tolerance, Will-s, Work,,

PRENT

Efficient-cy, Enable-ment, Faith, Good, Harmony, Honesty, Retain-ention, Strength, Super-lative, Virtue-s-al,TIT, TRIA, TES, PUREACLITY, ELTICMA, SÂ, and TÂ.

SECTION II

REPRESENTATION of TRU, ROT

CHAPTER 20

APPROACH to ROT, DEFINITIONS, CLASSES, CATEGORIES, CONTENTS, SEQUENCE, KNOWLEDGE, PROBLEMS, MEANS, PROOF, ROT INTRODUCTION

In this major section, II, Chapters 20-29 we present ROT. It is pronounced with the short Ô as in rut. Representation, as art, is a special and unique aspect of human life, the reproduction and production of much that copies, imitates, symbolizes, pictures, measures, and models the world. This they do in whatever way best helps to objectivize and identify any and all we encounter that is useful. For ROT this is focused on TRU and TIT, as guided by TRIT. The results and conclusions of scientific investigation naturally converge on a skilled and exact product that tends to be on TRU, its FAF, ES and CES. Duplicates, replicas, copies, examples, reproductions or productions are all about us and make up much of the AMTRU world. They vary from the simplest mark to the greatest productions. Art and representation are much the same depending upon usage and interpretation. The separation of art and science can be seen in the development of skills in the first millennia b.c. Greek Civilization in which the various principles, skills, and techniques of artisans to produce objects became sciences, and the meanings, beauty and special forms of artisan's skills became art. ROT is highly OITIO, for in being like art for its own sake, is in and of itself. Language, words, LANT, symbols, and NEC exemplify and reveal how Á and ROT fuse to show and become TRU. The better art and representation are the more likely they are TIT. Replication category theory is that the representation of an object is through a divisional ordering of its basic categories.

ROT stands in sharp contrast to POT. Although both are dependent upon TRU, TIT and TRIT, they are opposite in the sense that POT describes how we come to, approach, and reveal TRU. ROT describes how we replicate and show TRU or its parts, processes, mechanisms, aspects, and totality. ROT shows how TRU can be divided, coordinated, reproduced, symbolized, portrayed, modeled, or rendered objectively in ways that most assist the understanding and use of TRU. POT is highly subjective and ROT is highly objective, with qualifications. ROT is more than a simple rendering or objectification of the observed TRU. It is the description and demonstration of any and all ways and means of representing and replicating TRU. What ROT is in and of itself is called: **RÍT**, whereas what ROT is only as it is in TRU is: **RÏT**. The term representation is defined as that which stands for something else. It is a duplicate, replica, model, symbol, word, picture, or demonstration of another entity, thing, sense, or object. It is not the thing itself, but a facsimile of it. The form of ROT will be highly variable. For not only is it, like POT, not determined only from 'TIT', but from the outside.

A representation can be no better, and can be measured by, the degree it accurately, proportionately, and vividly shows and contributes to TRU. An *Image* is one of the simplest and clearest types of representation. As TRU is all by TRIA, any object that has actual significance and use, must be as close to TRU as possible, or a conditional part of TRU. Many ways of providing representation already exist. *Art* - Any reproduction, especially what is the best likeness, most important or divine. *Aesthetics* - Theory and science of representation and interpretation of art, especially aimed at quality or what is beautiful. *Language* - The major vehicle or mode of representation of Reality, especially TRU reality. It is a large part of human life, the centerpiece of the Á mass energy dimension of TIT. *Logic* - Representation dealing with validity in the manipulation of language and inferences in thought and demonstration. Using words and symbols with correct principles of reasoning. *Motif* - salient features, design, central or dominant idea and themes of their topic or object. *Analogy and Metaphor* - Exemplifications which bring our key features, ideas, or qualities through comparative vivid imagery. They are less apt to be clones, replicas, copies, or duplicate, but in ways more revealing. *Organization* or what shows how an object is formed from its major parts or organs, the cleavage of primary parts and organs of TRU. *Conceptual art* - Representation as a means of TRU, through the use of vehicles that produce word pictures or systematic whole mental images. *Taxonomy* - The study, methods and uses of scientific classification, their general principles and systematics, organization for representation of TIT, TRIT or TRU. *Number Theory* - The Representation of objects, or TRU by utilizing its unitary, finite and infinite mathematical makeup. What the significance of these numbers is. *Typology* - Theory or science of Types. How divisions of an object can be made and ordered through its various cleavages or types. *Set Theory* - Theory and science of sets. How divisions of an object, especially mathematically can be made and ordered through its various sets, classes or groupings.

The expansion of mathematics and number theory is analagous to what happened in all art, ROT, knowledge, TREQ, and TRU. Even in religion from simple beginnings of animism, to monotheistic religions, to pluralistic religions and philosophies there was an expansion of representation by TREQ and TRU. Expansion or art and ROT reveals the growth of relativity, the mass energy, Á FAF, continuum of TRU. It is the vehicle that produces reality, the agreement in ever higher powers of mentality. It is called **LAREX** or Law of Relative and Representational Expansion. When OITIO it becomes **REXIOT** or LAREX In and Out of TRU. Expansion is seen in all kinds of numbers, from the simplest digits through the most complex that are representative of some aspect, form, or function of the relative universe. Each has their significance that is often highly indicative of certain fundamental categories, aspects, modes, forms, structures, spectra, continua, or mechanisms of TRU. Their discovery and showing not only resolve many of the greatest paradoxes in human life, they are essential to an accurate and successful comprehension and understanding of the relative universe, and all.(TROS,TU,TE,TI,TU,TA)

SUR = Systematic Unity is a function of Relativity, or how relativity becomes The Relative. It is how a person becomes ones counterpart in The Relative by being both sense and thing in TRU. It is how a person in being the observer and the observed and thereby a unified relative is 1:1 = 1 in TRU. This, in turn, is how the unified relative, is in turn systematically differentiated and partitioned as its relativity is systematically ordered by the corresponding sameness of the form and structure of TRU. This is done as the unity, the 1:1 relationship between the observer and observed are increasingly divided and ordered. This is done by finitely and optimally reflecting, imaging, representing, replicating and modeling The Thing, TRU. The greatness and power of systematic unity is evident in mans CNS or central nervous system. For much of human's advanced and special form, is shown by an alteration of dimensions, as in Á or mass energy, is the frame and FAF upon which human life is based. It is mostly upon this frame and FAF man best lives and thrives. What works best within this frame and FAF is normal, healthy, and successful. What works less well is increasingly abnormal, unhealthy, in failure, deadly and apt to cause extinction. It is SUT, then, that helps guide us and provide the form and frame upon which ROT, MOT can be developed. Combined they reveal, show, and help us know TRU. This is vital to the fulfillment of Á, L,T, or ACEXIN of this special universe of best human life. **ROTÁ** or representation S-O or Á in TRU. As Á is the predominant mass energy FAF of TRU so is ROT the response to stimulus and reaction to action that reproduces TRU in a systematic and ordered way.

Before you can represent or model TRU, you must, as for the artistic rendering of any object, have a good, valid and actual idea, knowledge and perspective of it. Each and all that is representative must be greatly and most finely delineated and made distinct, as produced by our observer-observed capacity, Á mass energy dimension in TRU. This is the imperative fact of life, its relativity, S-O, and Á, called: **LÁREP** = Law of Relative Preemption by Á, and, **LAPERT** = Law of Perspective before Representation of TRU. You must know something to show it.. **LARINT** = Law of the Reciprocal Relationship and Intensification between TRU and ROT. LARINT is one of the greatest of all laws and mechanisms. Not only is representation a key and vital part of TRU because of its ability to enhance TRU by enabling the observer to better show and know TRU, it is a built-in part of TRU by being a large portion of the great bond, the mass energy dimension, between observer and observed, its Á continuum. For much of the subject and object by functioning together is largely representation. How the subject can show the object, and how the object can help the subject know it. SO then in ROT, as **RÄ**, is a large part of the Form of ROT.

Mostly because of **TIR** or TRU Is Representation, the very nature of SO and the predominant FAF of mass energy in TRU is representation. Many of the inadequately conceived, worded and termed existing concepts need redefinition by TRU a sharing of connotation by the addition of new concepts. This is **REDEC** = the Reinduction, Revitalization and Reinvigoration of Depreciated, inadequate, or less than TIR language. To maintain TIR at its best, as all must in TRU and any great system, construct, model or presentation requires, we must not fail to understand or follow their entirety and thatwhich is of greater value or use. **FONPI** = Fallacy of Non Persistent Intensity in pursuit, conceptualization and representation of TRU, TIR and ROT. FONPI enables us to understand this major defect in human life, lack of persistence in intensely pursuing the development and fulfillment of all that best represents TRU. This includes concepts and their POA to TRU. **LARTIN** = Law of Art and Representation in TRU, and Intensification of TRU. The great mass and role of Art is its role in glorifying and enabling TRU. This is how well it makes distinct, represents, stimulates and enhances TRU. The gist of art and all representation of TRU is their power to turn TRU inside out to best show TRU. The master art is the perfected representation of TRU, ROT, MOT, SPALT put into use, HOM and ACC. How this is to be achieved and to what purpose it is to be best used is the method and aim of man in TES and TRU. It is in the highest levels of ethics and politics that much of its use and application have been directed. **DITO** = the Difference between Identification for simple use, and Total Proportionate Reconstruction in the Objectivity of TRU. DITO is important not only in showing us the reasons for Representation, it shows us the direction it must take if it is to fulfill its role in TRU. **ROT → TRUR → RIT → Á → TIT.** To properly understand all things we must be able to classify or rate them by the level of DITO that applies. ROT is produced and given such a great part, role, and emphasis in POT and IPTRU because it not only gives a much better understanding of TRU, it enables us to remove the confusion that exists in much of mans scattered knowledge. It sets us out on that greatest of adventures, the finer and finer differences and distinctions that are needed if we are to demonstrate, and thereby fulfill, our role in TRU.

META STATES, GIVENS, SETS, CLASSES and CATEGORIES of ROT

DUPLICATION	REPLICATION	EXEMPLIFICATION	REPRODUCTION	PRODUCTION
IMAGE	MIRROR	Picture	Photography	
MENTAL IMAGERY		Conceptualization		
	SYMBOLIZATION	Letters	Abbreviations	
	NOTATIONS	Numbers	Mathematics	Music
	Measurements	Statistics		
	LOGIC		Reasoning	
MEANING	DEFINITION			
LANGUAGE	WORDS		Publication	
SPEECH	COMMUNICATION	Network,	Network Analysis Telephone	Internet, WWW
	WRITING	Literature	Prose Poetry	IPTRU
FORMULATION	Divisions,	Parameters,		
COORDINATION	Dimensions,	Coordinate Fields	m, l, t, .	
	DENOMINATORS	LCDs		
ILLUSTRATION	Painting,	Drawing		
MODEL	SCULPTURE	Modulation of	Engineering Architecture	Industry
	GEOMETRY	Topology	ALT, SPALT,	ELTICMA
	GRAPHICS	Diagrams		
	ALGORITHMS	Programming		
	CHARTS DESIGN	Cartography,	Map making	
	ELECTRONICS	Computerization		Applications
	STATICS/MECHANICS			
	PROCESSING	HOMATION		
	TRANSFORMATION			
	PERFORMING ARTS	TELEVISION	Movies Theater/plays, Concerts,	Opera

Art generally, and representation or ROT more specifically, may be categorized into five major classes. These proceed along the line of simple duplication to very compound or total production, and are shown above with the subclasses involved.

Their subclasses, for the most part, are equivalent to the topics, themes and modes which we use to present ROT. This classification is incomplete and rough, thus to be continually revised. It is presented to reveal the breadth of art and its opportunities, advantages, and limits, as a means to and role in, ROT and TRU.

Philosophy, metaphysics, science, religion and art, coincide by bringing TRU through ROT and SUT to concretion, formal and constructive modulation and application. It is finally completed by mankind's endeavor to discover, know, and show the world, TRU, its representation, modulation and utilization. This is done through relativity, how the observer and observed exist as subject and object, or Á, with L, and T both in and of TRU. It is the ability to increasingly demonstrate, show, and thereby further benefit from the world in which we live.

OBJECTIFICATION, KINDS & CONTENTS of ROT

Since the actual world, TRU, is not readily, immediately nor completely sensed, perceived, apparent or comprehended by us, it is necessary to use any means we can provide to discover, know, show and achieve it. These means are highly ingrained in TRU itself. We should distinguish between that in TIT which is RÎT and that which is not, or other than RÎT. This has an added importance, for it shows how much RÎT has contributed to TRU, especially its FAF, and the higher levels of its mass energy, Á, continuum and dimension. This, in turn, has the extra great advantage of showing how special TIT is, thereby showing TRIA and proving TRU. We need to be careful, as with all entities, to distinguish that of representation which is within and that which is without, that which is in motion within or without and that which is in motion or working in various directions about TRU.

	TRU	TIT	TRIT	RIT	ROT	ALT	SPALT
ASEITY	THE RELATIVE UNIVERSE Underived and Independent	At Once TRU in and of Itself	TIT Optimally Set and Standardized	Representation Existing in TIT and TRU	Representation of TIT and TRU and of Itself	Dimensional Form of TIT and TRU Underived	Spherical ALT of TIT and TRU Un-derived
PERSEITY	TRU shorn of all extraneous or non TRU factors	TIT shorn of all extraneous or non TRU factors	TRIT in and of itself without non TRIT Factors	Only that of Representation in TRU	Representation of TIT and TRU Only	The Combined Mass Energy, Length, Time Dimensions	Only that of Spherical ALT of TIT or TRU
HAECCEITY	Generic Actual Absolute Thisness	Specific Actual Absolute Thisness	TIT with somewhat Standardized Thisness	Thisness limited to Representation in TIT	Thisness of Representation of TRU	Thisness limited to A,L,T Dimensions in TRU	Thisness of Geometric Totality of ALT of TRU
QUIDDITY	This Special Universe of Best Human Life	TRU only IN and OF ITSELF	TIT FOCUSED and STANDARDIZED Toward ROT and ALT	All of What Representation is in TIT	What All Representation of TIT and TRU is	What the Á L and T Dimensions are in TRU	What Spherical Figure is made by ALT
ASPECTS and FORM	With TRUS, TROS and TRIS forms All Relative	Brings all to Form in Exclusive Relative	TIT Set Form by Focus on Aspects best showing TIT	All Representation in TIT showing its form	All methods of Representation showing the Form of TRU	Forms the Frame and Complete Structure of TRU	Forms the Whole Structure of TRU
STRUCTURE or FRAME	Indistinct and somewhat multiple of ALT	More Distinct and Solitary of ALT	Distinct and Solitary Emphasizing TRU's ALT	Any and All Representation in TIT	Any and All Representation of TIT, TRU, or TRUS	The Frame produced by Combined Á, L, and T	The Total Spherical Structure of ALT and TRU
ESSENCE and MODE	The Actual Relative Convergence of ALL	The Only Actual Exclusive Existent	TIT Focused upon its Á, L, and T Continuums	How the TRIT of TIT can be Represented	All and best means of Representing TRU	Á, L, and T Continuums Combined	ALT Combined to best form
OTHER	Convergent Relative Necessity of All on TRU	Absolute Necessity of The Specific TIT	Necessity of a Focused and Standardized Object	Shows the great functional Representation in TRU	Necessity of Showing TRU before Adequate Comprehension	Necessity of Establishing Á,L,T before Showing and Knowing All	Needed for Field Posits of Parts and Processes
USES	Key to Any and All Understanding and Action	Absolute Basis for Actual Objectification of All	Set Basis for Actual Objectification of All	Set and Exclusive Basis for ALT and POA	All means of Representing and Showing TIT and TRU	Provides Excellent Way for Positive Objective Approach	Definitive Method of Positive Objective Approach

Table 1-20 **COMPARATIVE TRU, TIT, TRIT, RIT, ROT, ALT and SPALT**

This table provides the key distinctions necessary if we are to understand TRU and its variants. This is especially so in regard to what we use these variants for in representing, demonstrating and showing TRU. This is the key role of concentration in the mass energy FAF, or Á relativistic continuum and dimension of TRU. For it is this very relative observer-observed or subject object relationship of our being, existing both within TIT and applied from without, OITIO, that gives systematic unity to the positive objective approach to all essential to our being. It provides a sound basis for POA.

ART, RITS, WHY, REASONS, JUSTIFICATION

The methods of representing TRU, **MEROT**, as for art, are primarily those that synthesize all that shows the form, essence, beauty and the greatness of TRU. This is in contrast to the method of the scientist which is more by analyzing entities, objects, materials or events. Much of most representation is mnemonic, means of helping the memory to recall and expand knowledge. The core and focus of representation is the correspondence between Á or the mass energy state, FAF and continuum within TRU, with VAPO and the various kinds of representation of TRU possible or have already been provided in a less complete form.

GENERAL KNOWLEDGE of ROT: EXTERNAL OBJECTIFICATION WAYS VALUE AIMS PURPOSES

Knowledge can be attained through the use of information, mostly empirically and experimentally derived, into corrected, ordered and simplified form, called *data reduction*. The operations and means for attaining data reduction are vastly aided by computerization. These greatly assist the tremendous load, work, detail and summarization needed by **KROT**, or Knowledge of ROT, as a guide to perfect ROT. Essences are recognized very incompletely and must be extracted from their ENIT, SENIT and THIT or entity, sense or thing-in-itself nature.

The aim of an artist, according to Titian, must always be suited to each subject or topic - which ultimately is TRU - to represent each person, form, or part of TRU, with its actual and true character and emotions. This means that we must represent TRU more than ES, CES, or its face, frame, or motion, but all of it emphasizing what is most predominant, characteristic and demonstrative about it. .

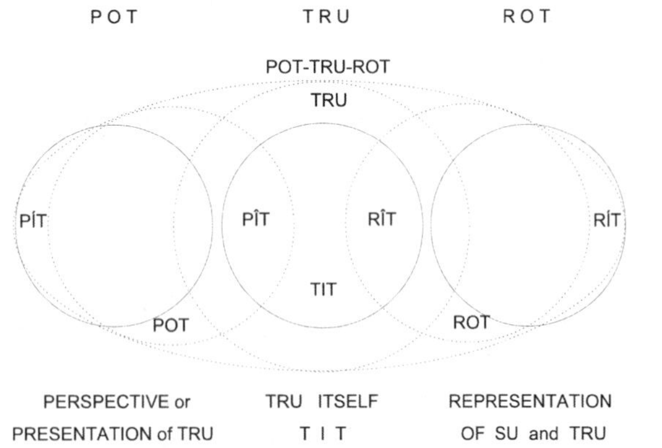

Fig 1-20 **RELATIONS of POT, TRU, and ROT**

By comparing and analyzing the features and characteristics of Presentation, TRU, and Representation, the first three great sections of this text, we can better understand how each contributes to the overall product. This shows how POT and ROT function by being a very highly reflected and rectified TRU. This helps overcome the pernicious fallacy of blindly accepting an undistinguished or undifferentiated incomplete, disproportional and mistaken TRU. All must inquire into TRU, be guided by VAPO and the light of TREQ, TRU, TIT, TRIT, SUT, ROT, MOT, ALT, and SPALT to behold an ever better and greater vision of TIT-TO, TRU, ROT and POT.

SPECIAL, INTERNAL, ABSTRACT and CONCRETE KNOWLEDGE of ROT

Simple algorithms have been used for thousands of years. They have developed into current programs in computer science to consolidate information. It is in a critical struggle with infinite differentiation and dis-ordination. When we break this differentiation and dis-ordination down and discover how TRU through TRIT, OBT, SUT, ROT, MOT, ALT, and SPALT can greatly facilitate and expand the power of ACC and TRU, we stand at the threshold of an ever greater massive new world for the future of mankind. One whose differential and infinite FAF and composition become our challenge and opportunity rather than being lost.

The battle between certainty and uncertainty is closely allied to our struggle to find unity in the world. This has a long history and continues in mathematics and physics, as noted in Relativity. Heisenberg's discovery of matrix mechanics in 1927 led to a quantum special case of the complementarity mechanics. His Uncertainty Principle is a special case of the Complementarity Principle which was the simple product of what happens when a new observer or new point of view is assumed. In TRU, especially when applying the Correspondence Principle, with standardization, ROT, SUT, MOT, ALT, SPALT, and ELTICMA we attain certainty and freedom from complementation by holding to and making our concretion and consolidation from TRIA, TIT, TU, TRIT, OBT, SUT and TRIA.

The parts of TRU, ES, to be effectively represented have to be made whole in TIT.. To bring all ESINS of ES together for TRU is to fit each to TRU and ES to ROT equitably, which requires proportion. **TRACAD** = The Relative and Causal Differential, or what enables the distinction necessary to determine SUT. This is by delineating what is not, what is becoming, what is as it is relative to TRU, and what is in TIT. This can be selectively extracted from 'the all' as they are only in TRU, TIT. It is how we determine ESINS as they are ES or the counterparts of TIT. It enables each person to know, have summative, concrete and solid knowledge and ROT of all that we see, apperceive, and experience.

QUESTIONS, PROBLEMS, WHYS, PARADOXES, ERRORS, DEFICITS and FALLACIES of ROT

The questions, problems, errors, and fallacies of ROT arise both from ROT itself, and from SUT, OBT, TRIT, and TRU. No matter how good, great or mathematically precise OBT, ROT or ALT may be in representing TRU, as long as the view is from the self, the observer is preeminent. A person's or observer's view is from the self, and will always be strongly true to the self, partially focused and oriented. **PROCES** = Problem of Reduction to E(S), to Closed System(s), or limits to the formal distinction necessary to posit in a closed system, as SPALT.

O-S - S-0 REVERSAL

Although we look at things from subject to object, as relativity is usually considered from observer too observed, this is not necessarily the case. Neither are initial nor predominant unless the state is preset or mechanism is already operative. When viewing TRU from the outside we assume a S-O orientation. When reacting to stimuli we have an O-S-O orientation. When standardizing and ordering a representation of TRU, we arbitrarily adopt a S-O orientation in the lower left to upper right alignment. This can be readily changed either by turning the spherical model so that S-O becomes O-S by completely altering the position of the observer or perspective. Or it can be changed by a different standardized spherical model in which the S-O dimension becomes the O-S dimension from lower left to upper right.

For the study of and work on TRU we primarily have three contrasting orientations:

1. We look at entities, and the observer views the observed, object or TIT, from S to O. Yet,
2. The object itself, TIT, TIT is formatively and developmentally constituted O to S.
3. TIT solely within itself without external orientation, when internally oriented, will have any arrangement of S to O orientations. For these will be less than a total. To consider TO-TIT as a whole imparts - at least to a degree - an external interpretation, and thereby orientation. This is called: **LATORM** = Law of TIT Orientation Manifold, and **LOTREX** or Law of TIT-TO Orientation Extra TRU.

The reversal of orientation between external and developmental or S-O and O-S is called: **LÖSRÉ** or Law of S-O or O-S Reversal. Thereby when viewing or having a perspective from outside of TRU we adopt a S-O orientation. When demonstrating the course of mass energy from within TRU we assume a less than total position and thereby one that will be reversible. Since LÖSRÉ is not actually and thereby not strictly TIT, and thereby more or less illusion, we are not obliged to uphold any single or set arrangement. It may be viewed reversibly.

Such a reversal for TIT-TO is **REVPÁ** or the reverse progression of Á. This is how O → S = Á in RUS, RUTS, TRUS, etc. and S → O in ROT, MOT, ALT, SPALT, etc. This is not contradictory nor invalidating, it only explains the in out and less than perfectly TIT-TO stance we are continually operating from and living by. It may be compared with how O - S is the course of ontology and phylogeny, and S - O is that of cladistics. This we see in the HIDIC and double onion peel diagrams, as of S-O, Á, and IA. It is how POA and PACH are used to orient this text, i.e. O-S in the first two sections, and S-O in ROT and the following.

To standardize these for representation and modulation, however, we maintain the initial S-O orientation, with the stipulations, exceptions, and corollaries adequate to correct any discrepancies that cause a departure from TIT. This requires a continuous awareness in ROT, MOT, ALT, SPALT and SPAT to realize and account for - by reversing our perspectives accordingly - the relationships between S and O, or Á relativity. These do not provide a simple singular 1:1 identity of observer-observed, that we so eagerly desire. Yet we must be reminded that TIT is neither simple nor singular in the sense that it requires multiple and complex perspectives and representations if it is to be a given taken in any other way than itself, TIT. It might be asked, thereby, why not just assume a stance that is totally internal, TIT. The answer is quite clear, for such is impossible. We are already, more or less, in departure from TIT when we in any way initiate a new S-O and IA process, as in all we live and do. The problem of reducing TIT to replication may be likened to the inscrutability of God, which would suggest a reversal of any departure from God and attempts to represent God. The extreme is one of the Iconoclasts, and the second commandment of Moses. Although guarding against representations that are either in error or misleading are needed, we must be careful of tendencies that prevent the discovery, learning, knowledge and proper representation of TRU.

There is tremendous difference between what an entity is, an object is, and what TRU or some aspect, part or process of TRU is, plus how this can be shown. **PREP** = Paradox and Problem of Representation. PREP is especially noticeable in TRU for here we can see from the absolute greatness of TRU how much we need to represent so much about it, yet how different it is than the entities and objects themselves. **RÊ** = Representation Error, or the loss of completeness, validity, and perfect identity through its representation. This occurs accumulatively, largely by LOLS, especially URTRU.

PAN = the Paradox of Nominals, or that quality of all nominals, words, and concepts whereby what they connote is less or other than what they are and intended to be as they are in TIT. This is part of the great revolution mankind is going through. This is to provide a language and representation of TRU that is perfected through its concepts, spoken and written words. Logic and much of semantics and associated fields in philosophy, consciously or less conciously, recognize this great need and strive to fulfill and achieve it. The degree to which it can be supplied will be an ever larger part of our success. When considered separately the questions, problems, paradoxes, errors, deficits and fallacies may not seem to amount to much or be of TRU importance, and many are not recognized or remembered and applied for long. However when taken together it is easier to see how difficult an excellent ROT, MOT, and TRU are and what steps must be taken to assure and achieve ROT and MOT and the optimal use of our knowledge, language, and means of TRU.

LEARNING, MEANS, CONCEPTS and SYSTEMS of KNOWLEDGE of ROT

In science and mathematics the term used to represent a proposed fundamental relationship, law, or proposition that is to be demonstrated is called a <u>Theorem</u>. The Greeks called the connector between a problem and a theorem a Porism. Theorems have been highly stimulating by helping to produce the direction, interest and work needed to prove or disprove problems and the ideas they stimulate. Problem solution or stimulus response relationships involved are identical with what occurs in TRU. They are a large part of the development, form and dynamics of TRU, its identity with the phenomenal, relativity, Observer-Observed, Subject- Object, S-O and the Á nature of TRU. It is how it becomes more rational where reason tends, as problem solving, to be proportional to TRU and the ratios and LCD's tend to be TRE of TRU.

PAT = Paradox of Truth, or truth as indistinct or variable images of TRU. This state of truth and TRU being both is and is not is: **ISNIS** or Is Not and Is. ISNIS is at the very crux of TRU, thisness, being and being qua being. Herein lays the contribution and greatness of Parmenides, in defining this paradox. Truth, paradoxically, as music, does not have to be identical with the final total composite TRU. This is the difference between truth and TRU, even though PAT and ISNIS are applicable, in somewhat different ways to each. It may be equivalent as it is, a part of it; or as it is a part of some representation, perspective or presentation of it; or even some other aspect of the observer or observed, senses, things or entities as they even remotely, indirectly relate to some aspect or part of TRU. **ISNISOE** = Is and Is not Total and Partial, or where truth and TRU are superimposed, they abide by LANERA and LAMOL. What is and what is not is not absolute, but depends and is of a spectrum, not separate polarities. It is the key to understanding the sensual and physical worlds, of SENIT and THIT, SUBIT and OBIT, THES and THEO, ES and CES. For all appearance and all reality are in this spectrum never absolutely appearance and never absolutely reality, as ENIT would lead us to believe. Yet by TRU they are highly conditioned by the totality of TIT.

LOICLOR = <u>Law of Imperative Confinement by Limits of Representation.</u> We can't attribute to, and achieve success in, representation unless in large degree this representation is confined to its limits. This means both the limits of the representation, and the limits of what is represented. Great art is not produced de novo, it must be highly conditioned in its production. These are only some and not only the best means and reasons for SUT and ROT. They collectively and clearly show the direction of ROT, and the promise, as well as limits, within which we must operate.

The means to, and the probability of ROT, are formed by knowledge and language generally, and more specifically by the nature of representation and logic. In both TRU and ROT the idea and existence of <u>Compossibility</u> is essential. For the possibility, probability and ability to exist together is a prime characteristic of both TRU and ROT. TRU and ROT contain the compossibility of diverse aspects and parts formed and functioning within their wholes. To do so we need a strong positive objective approach to TRU, POAT. It is through <u>Consilience</u>, or the concurrence of generalizations from separate classes of facts as in logical induction, that one can set the laws and limits necessary. It is how these are in accord with other sets of distinct facts, as knowledge, TRU, SUT, ROT, and MOT coincide, that their probabilities and proofs are largely achieved.

The greatness of the parts of TRU, ES, play in its representation is endless. The solution to this problem is key to a successful ROT and TRU, **SOTERP** = Solution to the problems of TE for the Formulation, Ordination, Representation and Presentation of TRU. Since TU alone is practically valueless in providing the necessary profound and substantial composition, formulation and ordination of TRU, and TI is even of less value in being limitless, it is imperative that we seek the reality available in TIT by TROS, TRIT, TE, TÍ, TO, SÁ, ACEXIN, the ES and CES of TIT.

FOP = Fallacy of Perspective Precedence. As in computer simulation we can understand and produce TRU only as it can be systematically formulated and modeled. This is so that this subject-object, stimulus-response situation may be continuously and/or discretely programmed. Thereby our attention to these problems of representation constitute much of the substance of this text. As these simulated, computerized continuous and discrete constructs and functions are provided, as by IPTRU, the forms, characteristics parts and dynamics of TRU can be discovered, comprehended and presented, in the ways we can best replicate the object, TRU. As in the positive objective approach to any thing, the determination of parity, parametry, or left-right symmetry or 'handedness' is paramount. Since through our inadequacies to envision and produce TRU directly, these are in part provided for us, as naturally and validly shown through less direct means, methods, systematic unity, modulation and demonstration. Nearly all freely moving animals and some plants are bilaterally symmetrical. This adds to their special FAF and equally applies to up-down, front-back symmetry and thereby is available for total spherical replication.

PIMAX = Problem of Maximal Identity and Presentation. PIMAX arises from our departure or regression from TRU, as THEO or THES, or some combination passes away from TRU. As ROT, MOT or all that represent TRU become less focused upon TRU and more diversified and disintegrated maximal representation, identity and thereby presentation is lost. In the early Christian Church the frequent and profound use of symbolism, as with monograms and initial letters standing for some great idea, concept or belief - as INRI and the CROSS - showed how powerful representation is. This in TRU and IPTRU gives us great precedent for a symbolic or notational system, LANT and NEC, the new language and concepts presented. The emphasis on initials or first letters and concise words for great ideas, concepts, principles laws and mechanisms provides brevity, clarity, identity and the power in TRU that they deserve and need.

Every important idea and concept in the world of man must have a separate and distinct term to represent it. Most of these can be supplied with existing language interpreted by TRU. The reason for this is not only the greatness of TRU, indispensability, and practical parsimony that communication demands. It is further reinforced by the demands of writing and of the communication media. The communication media is one of the greatest driving forces in the world today and it lives by effective words. This is why so many new symbols and terms are offered in so much of communication as logos, technical and computerized sciences. It is not that new terms are coined just for the fun of it or even arbitrarily. They are coined, primarily, because there is a great void, need and demand. Practical usage, to TRU is, and will be, the final arbiter of selectivity.

VALIDITY, CERTAINTY and PROOF of ROT

Criticism is the beginning of art and ROT for it initiates and selects what is and what is not to be duplicated, promoted, produced, traded, and TRU. Criticism partakes of all forms of art, representation and ROT. It seeks to discover, understand and propagate what is best in the world, what is TRU and, thereby, guides KOT. Such criticism is key to the quality and value of ROT. Criticism to be just and TRU can not be free from the very criticism it itself is based upon. Doubt and criticism are the generators of knowledge, which must be TRU, thereby rendering criticism itself imperatively TRU.

The term *apodictic* applies to those statements, judgements and relationships which are necessarily true, as of mathematical, geometric and axiomatic conclusions. Only with TRU can we free the use of apodictic from error. For much that may seem to follow another may not necessarily - as in "is" - be true in actuality by TRU. **LORU** = Law of Relative Universe duplication, Replication and Modulation. LORU helps to attain a 1:1 identity between those relationships, statements and judgements in which we seek accuracy and perfect validity. **TORM** is the Test of Maximal Representation, or how well does the representation fit that represented, or ROT of TRU. How well does the representation fit that represented, or ROT of TRU. This is an ongoing process which requires the continuing attention to detail and fitness of the identity. **TOVI** is the Test of Validity Intensity or Inherency. TOVI tests the quality, intricacy, distinction of all that constitutes TRU. **TOVES** (or TOVOP) is the law or Test of the Validity of portions, parts, ES of TRU. It is human's ever-present and more powerful ability to separate and bring together the world TRU by distinguishing and representing it with ever greater formal distinction that differences are shown, delineating and making more distinct dualities and the manifold whole into its divisions and parts.

No representation modulation, perspective nor presentation of TRU is to be made, interpreted, or used except, to the degree and no greater nor less than that to which it: 1. Is based upon what is TRU, 2. Perfectly identifies and replicates this part or aspect of TRU, 3. Takes account of how this represents TRU in its entirety, and 4. Uses these without departure from and only as they are held to 1, 2, and 3. The object, OBT, is formalized by SUT and as SUT is utilized in various ways in ROT, the artistic horizons and vistas are laid before us. This supplies us with the *Verification Confirmation* we so desire and is so imperative.

The proof of ROT arises largely from its necessity, use, benefits, and the importance of the arts in human life.It isn't that humans don't know TRU, as life and TREQ. It isn't that many great ideas, concepts, principles, and laws havn't been discovered and promoted. For these have answered many great questions and solved difficult problems. It is because of the complexity and shortness of life, the mixture of good and bad, TRU and non TRU and because we have failed to properly and fully represent life as TRU, that we remain ignorant and confused. That the battle for the soul is fought in the form of art, as these are TES and ROT, reveals how important art and ROT are in filling in what is less apparent yet so vital to us, TRU. As we understand and live by TES, THES, THEO, THESO, THESOS, SUT and ROT we reach a level and worth in life that stimulates itself to a richness and success that become self perpetuating. This richness is in its finity or differentiation, proportionality, systematic unity, coordination, order, ROT, MOT, exercise, practice, and life in and of TIT-TO. When such a Proof is by Induction that Knowledge is therefore a function of Relativity it is called: **PIKER**. This is the exactness and exquisite detail of its substance and how they can be ordered into the total structure of TRU. PIKER shows how the different among the greatest and the similar among the least that are of such great value to knowledge sets the stage for proof TRU through ES, SUT, ROT and POS.

VACOST = Verification through Combination and Completion of Totality of an object, the object, TIT and its systematic unification, utilization and representation, SUT and ROT. Highly complex and puzzling objects must be laid out in full and interpreted in the light of TRU. As vital as representation is to TRU, no representation is adequate unless and to the degree its views, idealism, realism, materialism, expressionism, impressionism, and all venues are adjusted and reconciled in the totality of TRU, TIT-TO. For although we are continually providing an external demonstration of TRU, this externality has no finality unless it Is, exclusively TIT. In VACOST we secure a proof of unknowns. As they become known and are found to be universal theses, principles and laws of TRU they in turn combine, complete and verify the whole, TO. VACOST works perpetually to form the valid combinations, completion, and totality that substantiates TRU and mutually provides its proof. **LAMUP** or Law of Mutual Proof. LAMUP defines how actual TRU certainties tend to prove each other, VACOST defines how ROT and SUT are obtained, totalized and applied for use. **LESU**, is the first Finite Law of SU, by providing PROSU the proof of systematic unity and ROT. These are some of the greatest of all discoveries of human knowledge. It is the basis for theology and science, as in relativity. Pythagoras, and many before and since have been enthralled by finding the relationships and correlations between finites, and these between unit and infinite. This can be applied to our lives and the world around us to explain how much and what we do not know and how it works. The balance of nature, the harmony and melody of music as it mirrors life, the derivation of RU, TRU, TE, TO, and SU as both one, infinite and finite is undeniable, by LESU.

There remains, and will to some extent always be, a deficiency in ROT which renders proof suspect. This suspicion can be quickly and readily corrected if we account for this remaining deficiency. This is done by: **NORC** = the Necessity of Rebirth, Reproduction, Regeneration, and Replenishment of symbols, Concepts, language, and all Representations of TRU. It is NORC that gives us much of the justification for TRU notations, terminology, and POT..

CHAPTER 21

ROT DERIVATION DEVELOPMENT DEMONSTRATION DETERMINATION

INTRODUCTION, ORIGINS, BEGINNINGS of ROT

ROT arises from entities that become senses and things, and as TRU likewise becoming subject and object, subject object, subject object TRU, and Á. Even more it arises from the great need that the observed and observer have for showing and comprehending TRU. Through our IA, in the occurrences of life, we seek to be aware, pay attention to, distinguish, work up and utilize what comes before us and find helpful. It is the body, mass and total of IA: **BOTFIA** = Body of TRU a Function of IA, that is attainable as IA expands to approach TRU. Through all the means of representation and presentation, we increasingly understand and can use the substance of all, especially the substance of TRU. In: **EXES** = or the Expectation of the Essential, we can understand the importance to absolute necessity of ROT, POT, and TRU. There is much difference between propositions at hand or given, their apparent solutions, and their posit in the scales of TRU's dimensions. How much can the body of TRU be represented and treated as an object, or like the body of a person. To do so there needs to be a unit, that is discrete, changes in the replication known, and especially these if TIT by TRIA. Therefore in ROT we must apply the rules of bodies, objects, observer-observed, and those of TRU, including E, ES, and laws applicable. Any entity if sufficiently available for investigation, and known can be treated, represented and shown.(LASUR)

DROT & OBJECTIFICATION, SOURCES, EMERGENCE, and CREATION

Representation is largely an extension of objectification, an objectification that is developed into sounds, communication, speech, language, drawings, writing, and many more complex means of expression and production of what is and can be observed. This shows how Representation, especially ROT of RÏT, is largely a function of relativity, of the mass-energy, Á FAF and dimension, of TRU. The extension of objectification and representation is what makes up a large part of the world around us, TRU and ATRU. ROT can be either direct or indirect. As direct they are exemplified by thought and in 'the mind's eye'. <u>Selective variation </u>is a phrase in usage which expresses that great means by which infinite or impractical variation of representations are selectively utilized to provide the best possible showing of TRU. If we don't objectify and subjectify what is the greatest and best in life, TRU, it goes unrecognized and undone. That state in which life has objectivity and accompanying identity of subjectivity is what prevents confusion and failure. It is much of TRU, its predominant mass energy Á FAF, dimension and continuum. This is what makes-up the human realm. It is guided by their natural environment and ability to adapt to change. **POCU** or the Principle of Confusion describes that state in which we fail to adapt to changes within time sufficient for us to gain our objectivity and adapt to life. **POCUF** = Principle of Confusion Fixation, or how POCU or confusion becomes fixed. In becoming fixed confusion sets in automatically, escalates, and more or less permanently loses in failure. **POCUM** = Principle of Confusion Unwound, disentangled, extricated or eMancipated. POCUM does not occur until some larger power, as mans own abilities are actualized, enabling this state to unwind or be disentangled. Human life is highly complex and becoming rapidly more so. We have great difficulty in coping with life. Only by disentangling, unwinding, and removing confusion and its fixation, do we succeed. This is largely a result of living within our means, using what we have, and adapting to changes in the best, TRU way. This is largely achieved through objectivity, identifying and representing the world about us so that we can best adapt and use it.

DROT DERIVATION, GENERATION, ATO, CREATION and EVOLUTION of language and ROT

Creativity in art and ROT tend to prove the special and de-novo origin of ROT. It, as ATO, helps to prove the special original state within itself of TRU, ROT and thereby PRUT. Art and ROT are neither singular images, nor representations, duplicates, and reproductions of entities, things, objects or TRU. They strive to create and produce something new, that of TRU which is missing or lacking. ROT, as much of art, is primarily focused upon the discovery of essences, or some essential aspect or way to know the world around us. These essences are often ideas, concepts, parts, processes or some aspect that is of great interest, importance, need, or significance to TRU. It is to provide the means of representing discovered essences that much of the progress of human culture, civilization and TRU have occurred. This means was often in the form of sounds, signs, marks or symbols, new or arbitrary objects, standing for some aspect of TRU. These produced the basis for the knowledge we have today. When combined, as in name relation, they produced the basis for reason that, when TRU and best used, vastly extended our knowledge and gave increased meaning to human life. They were the massive extension of the Á continuum, SÁ, that so characterizes the predominant mass energy of TIT. The etymology of words is much more than their derivations. It is a vast and lengthy achievement of language development. Resolving the failure of language to adequately represent TRU is attained by basing our terms on TRU, and defining existing terms in ways that demonstrate this difference.

DEVELOPMENT of ROT, IA, GROWTH, MEANING, THINKING, LANGUAGE

DROT, DOT and TIT are shown by the converging power of relativity, or how TRU, in being TRU, converges all onto and into itself relatively. The dimensional form and function of the mass energy nature of TRU is like a magnet. It creates a positive and negative field by which selectively some are drawn to TRU, and some are pushed aside. It is in meaning that the procreative powers of TRU and ROT are to be most clearly seen. From its earliest beginnings senses and things acquired meanings that expanded into ideas, concepts, language, communication, and more complex representation. As great a part of representation as language is in TRU, it often does not help, nor can not fulfill all this need. This is why so many other kinds of ROT are needed. Collections or complexes of pictures show the difficulty and indirectness of the knowledge and ROT. This is why expressionism, impressionism, and the many other modes and themes of representation and art are so great. Depth of meaning, maximum finite order, and total ROT of TIT-TO are the greatest aspects of representation. It is toward these that humans are in such powerful headlong pursuit. It is to guide and make it more efficient, effective, complete and beneficial that ROT for ENT fulfills TRU. Meaning as the focal point of both TRU and ROT depends not only on precise delimitation, it depends upon association, context and totality. Languages did not develop quickly nor out of the mind and speech of one or a few. The larger and more interactive the population that used the language the stronger and more permanent it became. Thus, ROT developed through mechanisms of continuous representations. They worked in the medium of culture and communication that fostered their continuity. From this can be seen the essential nature of larger populations in TRUS and TRUF. Their power of ROT and language have given cultures, civilizations and TRU added continuity and stability. They become a growing part of the form and frame of TRU, as they are the carriers or vehicles of the Á or mass energy continuum. They enabled man to think more abstractly with purpose and planning to better solve problems. These continuous representations provide much of the key forces and powers of TRU which propel and compel greater being. One kind of representation often feeds on another. Thinking, language, IA, and TRU together give us an idea of the core dynamics in TIT. They resolve many of the problems and provide the basis for showing each alone and how in various combinations they function in TRU. Thinking occurs at many levels in a conscious person. The line of thought must be maintained if it is to be complete.

POTENTIAL of ROT: RELATIVITY and DEVELOPMENT of POTENTIAL to ACTUAL, ORIENTATION

All varieties of arts and productions are usually very incompletely representative of TRU or even that part or process they show. Their intent and design is often not of TRU but of what is either only partly TRU, AMTRU, ATRU, or ENIT. They can be actual in both a representative sense and a TRU sense. Yet they are usually, at best, only potential, in both senses. The only solution to this ever present problem is to have art that is both highly inspired, as by TREQ or TRU, and is exquisitely executed as in the most 1:1 replication of the object, TRU, or its meanings and essences. Aristotle sought to solve this great paradox and dichotomy of potentiality and actuality by most clearly demonstrating their being aspects of the same mechanism, fulfillment, of realizing themselves. He described, in *Metaphysics*, the eternal unmoved first mover as one and the source of the unity of the world process. This was pure actuality, consisting of perfect being, to which the movement of all things, entities, senses tended. For TRU this is ATO and by TRIA TIT as it becomes or approaches TÂ. The progressive movement of all, whose potencies are actualized and constituted by many and combined forms of energy, as in A, are converted and transformed from one into another. The neurons of the central nervous system of animals and man are to biological systems what transducers are to control and communications systems in modern electronics and technology. They act as receptor and effector organs operating to make the potential into the actual. So is man, a product and function of how his body works, and works successfully to fulfil and survive in life. The frame and orientation of this is provided by the greatest of all, for us, orientors, the human brain, a large part of the form ofTRU, much of whose function is to represent the world, including itself. Through the levels and scales provided by the ALT dimensions we can increasingly develop the form and frame needed to determine how well they are either potential or actual, how much these are of perfect pitch or discordant? It is in tonality as in music that we can best see, hear and represent, the finer and finer delineations and scales needed if we are to show how the potential and actual are related and interact. It is in the development and analysis of these delineations and distinctions of objects or the observed that they can be increasingly formed and represented.

Figure 1-21 shows the relations and spectrum between same and different as they relate to subject object when compared. It reveals how we can develop the object TRU based upon its fundamental Á mass energy FAF when the potential of ROT, as all in TRU, becomes its actuality. There is all the difference 'in the world' between: no light, a little light, a lot of light, and perfect light. With no light nothing can be seen, whether one has had prior knowledge or not. With a little light little can be seen, yet a little light is all that is needed to begin to orient oneself and to be able to know in what direction one is going, especially if a person has prior knowledge. With a lot of light much can be seen, usually sufficient for need. With perfect light, which is not necessarily an extreme amount but what is optimal for the viewer, all may be seen depending upon the capability, interest, and attention of the observer.

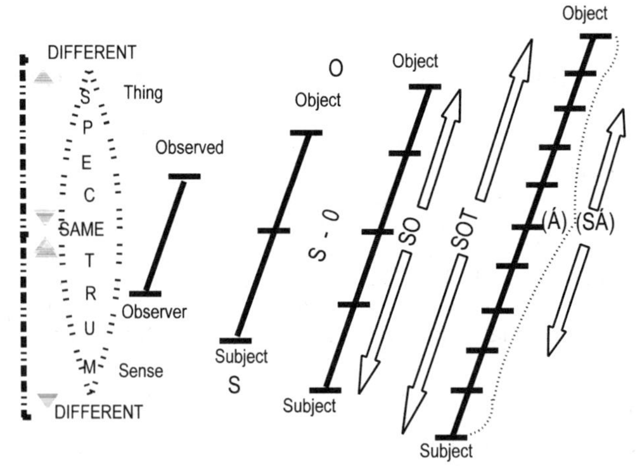

Fig 1-21 **OBSERVER OBSERVED, SUBJECT OBJECT, SOT, Á, SAME DIFFERENT and ROT**

This is how a knowledge of all is achieved by perfecting senses and things as they exist in TRU.

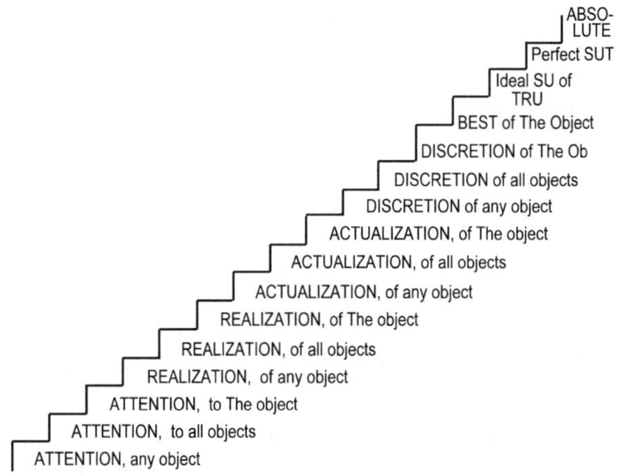

Fig 2-21 **ATTENTION: GRADATIONS of FORM from POTENTIAL to ACTUAL**

How can sufficient OBT, SUT, and ROT be attained, and what steps must each of us take to attain and utilize them? These are described in terms of the attention, realization, actualization of TRU, and the various steps in each. They show how ROT can't be attained by simple replication alone. It requires a simultaneous and similar massive S-O, SOT, through a total Á form and function for TRU approach. **SETION** are the processes of integrating the parts of TRU to produce their maximal form, and thereby formulation by SUT of TRU. **MADAS** is the Maximal Separation and Distinction needed in TRU. **COPLAT** is the Convergent Perfected Posit of Concepts, Principles and Laws of TRU, and **ETO** is the fact that the End of TRU is in its continuous totality.(TIT-TO) Those universal aspects of EPI that transcend TRU and ROT act and can be represented objectively to return to TRU and the observer their potential representation. They will be actuated by CEMLIP, SETION, MADAS, COPLAT and ETO.

REASONS and CAUSATION: ROT DEPENDS ON, PRODUCED BY and COURSES

What ROT depends on is reasons and causes, and how they are produced and function. In turn, they become processes, courses and effects. What is produced shows the transition between the potential ROT, actualization and the determination of ROT. The main causes for ROT arise from mans need to know and show the world around him. Good art, as ROT, can be no better than the ideas, themes, guiding principles, and perspectives it provides. It is in this way that the whole history of art has been an interchange between human guiding ideas that mirror what is the greatest in life, TRU, and how these can be best shown and communicated.

It is because of its objectivity that a thing has reason. It is because of TRU objectivity that there is actual - TRU - reason. To discover this objectivity and how it changes, develops and was caused enables us through LASUR and other means to discover, demonstrate and represent TRU. In **ESE** or the direction, inclination, deficit of knowledge and meaning to ES, ERED, and SUT we find the fulfillment of this activity of man for TRU. This is by: **PRUE** = Proof by E or parts and characteristics of TRU are represented and shown that TRU can be made known and proven.

There is a plurality of causes and ever greater plurality of effects. Yet as the observed, the objects we perceive and use tend to return to TRU. In ROT they tend toward a solitary unified focus of TU in TO. Through meaning man's subjectivity and powers of observation seek the finite. This is: **MIF** = Meaning Imperatively Finite. The essence of meaning, whereby without showing how all is the same or how all is different, as unity or infinity, is with definition.

Correlation unequal to correlation of TRU is like the dual role of 'is'. The correlation and 'is' of entities is not the same as correlation and 'is' is in TRU. Here we can see how truth, that which is, is identical, or correlates, does not have to be TRU to be valid or is. Yet what is the greatest or has the most value, has actual meaning, and what is of our world will be proportionally correlated with TRU. This is the basis for REM, and is also a proof of Á. **OFI** = Optimal Finite Form and Idea. Whereas the forms and ideas can be developed to infinity, as they become more and more infinite, meaning decreases to zero from our inability to distinguish them. Greatness thereby is not only magnitude dependent, it is unity dependent. Just as metaphysics is the 'causes of being', so is science 'knowledge by causes'. It is largely through *aitia* or the discovery of the first principles of causes that Aristotle has become known as the father of science. Human ability to represent the surrounding world improved through the symbols, words, numbers, phrases, NAM, suppositions, sentences, and syllogisms, and the many means of expression and representation are the media of first principles. Through *zetetic* or the seeking out of what was needed and their relationships, as in mathematic's search for x or unknown quantities, the use of numbers was advanced to higher levels. This also occurs in **NAM**, when entities, concepts and words are combined to produce a single idea, essence or ESIN of TRU. This is often with much different, and sometimes far greater, meanings than either alone. It is a large part of the power of language, phrasing, sentencing and composition.

All that sets dimensional form of TRU, ALT, and SUT is what it takes to restore coordination. This is basic to life, survival, success and DROT. This is a product of the integration of Á or our mass energy nature oriented in place (l) or time (t).(ALT) Human life is based upon and lives by the orientation their S-O, l and t nature needs and supplies. In fact it so permeates the origins of not only man, but all life that it is a fundamental fact and principle of both human and all life. This is called: **LOR** = Law of Orientation, or what is a basic feature of IA, quality of TRU, and cause for ROT. As the causation of ROT moves toward its determination so is the reason for ROT the bond that enables and produces this transition. Indication forms a primary bond between causation and determination especially evident in objectivity and ROT. Indication is a kind of Activation which Itself is the final test of its truth and value. **ACIT** is how with meaning proof is attained objectively and pragmatically in the action.

DETERMINATION of ROT COMPARISONS, ANALOGY and INVERSION

The determination of ROT is evident in: **LASUD** = Law of Separation and Union of Same and Different, and **COTA** = Corollary of TRU Axiomatic. The axiomatic follows with TIT alone. It is confined to the specifically modified dimensions in TIT, or its unique length, time and mass energy configurations. The nature of things as they are relative to us and as they are basic and powerful determinants of ROT and TRU is seen in <u>Abhorrence</u>. This is the natural and selective reaction we have to any and all that is irrelevant, unnatural, distasteful, undesirable, of poor quality, negative, or threatening.

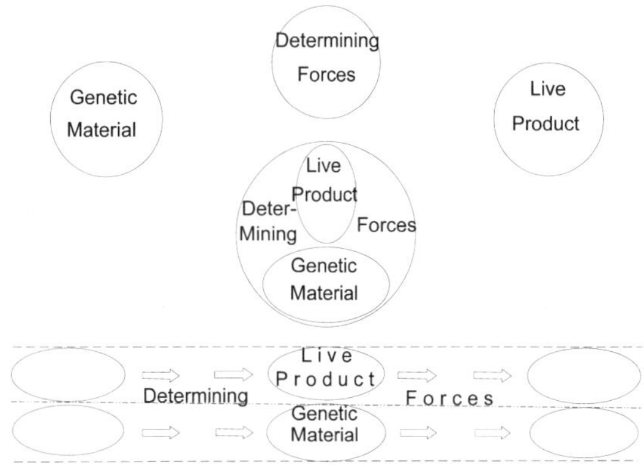

Fig 3-21 **GENETIC MATERIAL, LIVE PRODUCT and DETERMINING FORCES REPRESENTATIVE of TRU and COMPARATIVE ROT**

By showing ROT in different ways we can gain valuable insight into the nature of TRU and how to look at and be TRU. In the analysis and comparative determination of ROT, as in ROT and TRU themselves, we must continually encounter and correct for PIMAX. **PIMAX** = the Problem of Maximal Identity and Presentation. PIMAX arises from our departure or regression from TRU, as THEO or THES become THIT or SENIT or some combination passes away from TRU. As ROT and MOT or all that represent TRU become less focused upon TRU and more diversified and disintegrated from optimal representation, identity and presentation are lost. Some of this is necessary, a part of our IA, subjectivity and relativity. So that PIMAX denotes a very fundamental paradox and one that requires learning and remembrance, ECRO. This is departure from TRU.

ANALYSIS and EVALUATION of ROT REDUCTION and EQUIVALENTS

Since TRU, TIT, TRIT, OBT, SUT, and ROT have never been exactly, explicitly and directly described and demonstrated we must proceed as with any science to deliberately, exactly, validly, reliably, consistently and critically analyze them and their topics, objects, subjects and methods. Because God, the world, the universe, or other TREQ terms as objects seem well defined and discrete neither grants them, when in and of themselves, either such well defined and discrete status nor even certain identity and actuality in TRU. This is why much of ROT, MOT, ALT, SPALT, POS, STA, HOM, and ACC are so necessary. It is to analyze, evaluate, and reduce to optimal identity all we can find for use to show TRIT, TIT and TRU. The comparative determination of ROT is seen in many kinds of knowledge, including knowledge that concentrates on the essences, principles and major components of TRU. This is by: **POCORD** or the Principle of Comparative and Relative Detail. Comparative determination is also seen in **TOVOP** or the Law or Test of Validity of Portions of TRU. Another means of analysis is **SUEX** or how Systematic Unity is attained through exposition. As we expose the form, divisions, parameters and dimensions of TRU and as we illustrate and provide an exposition of them, SUT increasingly represents TRU. Through the: <u>Axiom of Reducibility</u> we restrict, separate and reduce more complex systems, as the ramified theory of types, to their simplest types or constituents and by representing each element separately, then all together, achieve an appropriate representation. These are confirmed by Measurability which is crucial to good representation.(25,26)

TIT is not TRUS, OBT, TRIT, SUT, ROT, MOT, ALT or SPALT. It has much in common, yet each have their distinct FAF and characteristics. For TIT this is: **TÄTÀ** = TIT Formed and Unformed. By TÄTÄ we can denote the idea of a TRU whose form is exactly what it is in TIT and not what may be inferred from TRU, TRUS, OBT, ROTS, SPALT, or other things or objects. We can recognize TIT for what it is, neither having to be spherical nor of some standardized form, that we so depend upon, in ROT and MOT. It is a form that has severe, unique, yet typical differences from all other entities. This is very important since, of all things, it is the all, by TRIA, TIT. The better we can understand, remember and apply the different forms of TRUS in ROT and our operations of TRU the more accurate and discrete will our productions be. It is upon these that ROT, if it is to become the MOT, ALT and SPALT and all it seeks, will depend and must comply. The incompletion and lack of reliability of analysis of forms of TRUS and TIT is incompatible with ROT and TRU. TIT has actual and reliable form, even though it may not be the same, nor amenable to the same treatment, as that of other TRUS, TRIT, OBT, SUT, ROT, MOT, ALT, and SPALT. This form can be actually and reliably transformed from what it is in itself, TÄTÀ, to what best shows it, in TRIT, OBTIT, SUT, ROT, MOT, ALT and SPALT.

The determination of ROT consists of, among others, the discovery of what objectivity is and how objectivity functions as a pole in the SO or subject-object state. The object toward which an act of reference is directed is called the Referent. It is this referent that is the artist's, as by ROT, focus. It can be a word , sentence in a book, face in a picture, or an idea that is represented in any of a number of ways. This referent is what is denoted by a word, sentence or judgement or a term used for meaning, a factor in logical thinking, or the theme of a masterpiece. It is an element and focus in the construct of ROT that enables its form. It is an element in: **LACT** or the Construct of TRU. LACT reveals its dichotomy, **DICHOT**, as this dichotomy, reveals the polarization of TRU's primary dimension, its mass energy dimension, or S-O, subject object form and structure. It is such form and construct of TRU that are enabled by: **DEPOSU** or Determinants of the Polarization or the coordinates of Systematic Unity. The process by which this is achieved is called: **Inversion,** or that process by which a given dimension is changed to another by inverting the ratios in such a way as to replicate TRU. In logic and mathematics it is especially precise as a given proposition is changed to another by inverting the two ratios, so that the second term is to the first term as the fourth is to the third: **If a/b = c/d, then b/a = d/c.** This process shows how not only inversion is useful in reducing all to TRU, it also shows how to proceed in ROT, from its determination through its demonstration to its form, language, logic, formal logic, and mathematics.

Since the success of our lives, and ultimately human survival, depend upon how right our actions are, since these actions depend upon the policies, principles and knowledge they are based, and since language is the means by which all these are developed, it is essential that language be the best. Too many of our words have multiple and vague meanings. Too much of our lives that consist of the same things have different words for different people. Since the actual world consists of different aspects of the same thing it is essential that we know and make distinct what is and what is not same and different, and all in between. Logic has shown how much better thinking is when based upon explicit terms and rules. So like logic we can, and must, develop language that is most explicit, meaningful, and exercised with the best usage. This can't be done efficiently and effectively when limited to existing terms and languages. It requires new, distinct, and strong concepts and terms that have exclusive meaning of the actual world. Terms that consistently have strict and explicit definitions and meanings of that most important and useful. Terms that enable precise, revealing, and inspiring insights into the most fundamental states, conditions, relations and limits of this special world of best human life. LANT or the new language of TRU is offered as a potential vehicle to provide such precise and important concepts and terms. When of top quality and sufficient quantity they enable a whole new and better perspective and way of representing, not any or various worlds but, the select world or universe in which we so intensely and realistically live.

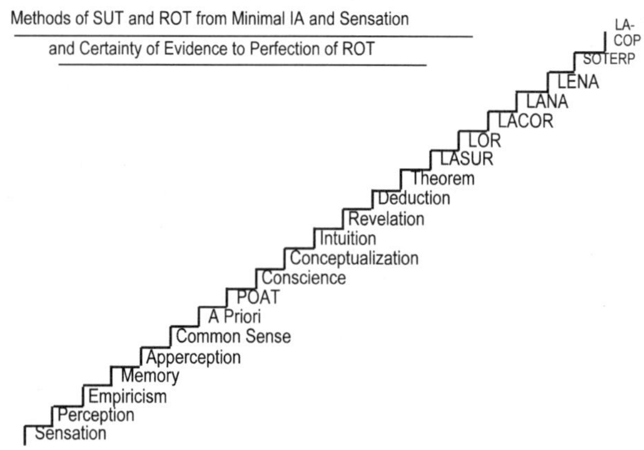

METHODS and DEMONSTRATION of ROT

Fig 4-21 **HIERARCHY of STEPS to ROT**

When we speak of the means and methods to a greater demonstration of an object we tend to converge on the representation of TRU, ROT. For as great as the art, technology or industry are in producing an object this is only a part of it. It is what this stands for, how it can be used and what its values are that are its prime measures. This is ROT and its methodology. The methods of representation, conceptualization, language, and thought are so taken for granted that we are guilty of overlooking the conscious processes of representation. We likewise are often guilty of treating all methods of representation as being of the same or of any level of importance.

Although deduction, the deductive method of both logic and science, has proven highly effective in guiding proofs by building one premise on premises already proven to discover of fundamental truths, the traditional use of deduction, and induction, can't be abandoned. For in a closed set, with limits, as the relative universe in itself, there exist continua whose extremes from unity to infinity, from universal to partial and from general to specific require qualified and modified methods typified by the long time usage of deduction from universal or general to partial or specific and induction from the partial and specific to general and universal. This accords with the differentiation and integration that is so effective in mathematics, calculus, and the management of sets, the our actual world, TIT-TO and TRU. The one kind of deduction does not rule out the other. Such problems in language only show the need for its modification, added, or new language. The method we continually return to is that of systematic unity of TRU, SUT. The reason for this is manifold. Primarily it is because TRU in being a relative universe, and a unity, provides us with a single and thereby reliable focus. Also because TRU is a relative universe that has ES and CES in a totality with consistent form and structure manifested by its dimensions and scales. These dimensions are formed and function relatively and work in set ways. These ways can be greatly enhanced by what is known as Dimensional Formulation or the combinations of higher powers of mlt and ALT. By learning and representing dimensional formulation, the systematic unity of TRU gives the means and methods of objectivizing TIT-TO in a highly discrete and concrete way.

In figure 4-21 some of the methods by which a more reliable SUT and ROT can be attained are given. They are neither complete nor fully conditioned. Several of them are limited not only in themselves but how they relate to and can be applied to SUT and ROT. They mirror higher levels of mass energy, the Á dimension of TRU.The gist of the diagram is to show how ROT is not to be looked upon as one simple and separate collection of various representations but can be achieved sufficiently only when the most important steps are accumulatively and integratively utilized to bring ROT to fulfillment. This is how it can best replicate TRU as well as provide the best possible product for use. **LACOT** = Law of Complete Objectification of TRU. LACOT describes how we should show all, all that is in and of TIT-TO..

We can not show a little bit, or some simple aspect, form or principle of TRU to show all of it. This is a major defect of God and religion, and much of knowledge. To do so we should utilize all there is about TRU that helps to show it. By VAPO this is closely associated with: **LEP** or the Law of Equal Proportions. Without LEP and a representation of what reveals the cleavage of TRU we can not achieve ROT and MOT. With only TI and TRUD we can never find ROT, SUT, KROT and MOT. By LATAN and LATANO we are limited to TRU and its order, and by LATANOS - to an extent -its standardization. We are mandated carriers of TRU with limits on free will. Yet free within TRU, SUT, and adaptation to the best position within it. We can not fit TIT nor TRU to SUT, ROT, MOT, ALT, and SPALT, except to the very limited extent they allow. This is: **FOST** = Fitting OBT and SUT to TRU. This we do by determining and selecting that which is and that which is not a major aspect, form, parameter, dimension, plane, and part of TRU.(LOCIT-132) LACOT proceeds from complete identity with TIT through TRIT, its object OBT, and these in SUT and then ROT.

POPTE = Principle of Piecing Together Essences. POPTE is like scientific experimentation or detective work. It is the extrapolation of things from entities as they are in themselves and as they are in TRU. This is a constant and ever continuing necessity and process. It consists of, **ÁXE** = Á Expansion. AXE is the expansion that occurs when S-O is supplemented by ALT to provide POPTE. **ÁCE** = Á Ever Continuing to Completion. ACE is the continuation and completion that occurs when S-O is supplemented by ALT to provide POPTE. ACE is the fulfillment of AXE. How standards become a part of TRU is called: **STAT** = standards become TIT. STAT is exemplified by the Hindu-Arabic numeric system that adopted the symbols for writing lower case numbers of zero through nine. The use of place value enables the writing of many large numbers from only ten, including zero. It is necessary to: **HENS** = Hold to Established and Natural Standardization. The positioning of dimensions is a good example of HENS where length is depicted from top to bottom, and for time, from left to right. **COMART** = Conclusion and Correction to Final and Total Modulation or Representation of TRU. The fact that collage and piecing together the infinite parts and processes of TRU as the same are separated and the different fused and expanding, yet never to the absolute is **COTA**, Corollary of the TRU axiomatic. COTA is the same separated and the different fused and expanding to an ever more accurate and complete degree. This is how our methods of ROT fulfill their purpose.

CONVERGENCE and SUMMARY of DROT: S-O, Á, ROT, and TRU

The fulfillment and achievement of ever greater relativity is essential to the success of our being, just as Á is the predominant and vital mass energy dimension of TRU. Ideally ROT is a perfect duplicate of TRU, such as ALT and SPALT might idealistically endeavor to be. Yet even here a perfect but simple replication or duplication of TRU falls short. The end of this process is the purity, reality and actuality of participation in TIT or PUREACLITY. Each, subject, object and TRU are absolutely both qualitatively and quantitatively, necessary. So is ROT, as it is the showing of this object, the observed, absolutely necessary. This, in other words is called: **LANOFAC** = the Law of the Absolute Necessity of Formulation and its Conditions and Qualifications. LANOFAC is a result of POA and its adaptation to use or PACH. If TRU is to be anything more than an ENIT, it must be known or shown, for this is a great vital part of it. Thereby we must discover, learn and utilize what the form, makeup, nature, haecceity and quiddity of TRU is and fulfill it as we already condition and qualify it in knowing and showing it. We must remember that the mass energy dimension of TRU, Á, includes all the FAF of TRU, over and above S-O, that constitute its predominance in TRU. That is all that is not included in the length and time continua, as they consist, also, of their higher powers, by ACEXIN.

Fig 5-21 **STIMULUS RESPONSE, S O in DROT**

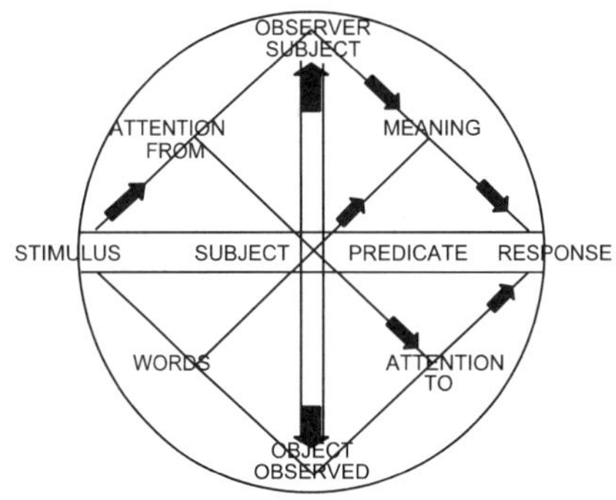

The correspondence of the observer observed relationship or relativity to the stimulus response relationship or animalian dynamics is fundamental to the understanding of life, human life, and when best, TRU. It is also helpful in providing a guide to ROT, MOT, POT and TRU. It can even more readily depict life in a 3D format or model. When superimposed on the many uses of the SPALT and SPAT models, it gives us a clearer, yet simple basic idea of what constitutes TRU. How its mass energy, or Á dimension developed in the transit of relativity through evolution to become focused by and of humans, in TRU. The interaction between observer/observed, subject/object, attention from/attention to, words/meaning and stimulus/ response was shown in the introductory section as both singular relationships and in their overall interactions. The key role all of these interactions play in TRU gives us a good showing of DROT and ROT, relativity to us and TRU, and helps to prove TRU.

ROT and S-O Á PROGRESSION

It is important to note in DROT, the never ending human tendency to think either idealistically, materialistically, or by a simple realism in which the subjective and objective combine to focus on a given that is neither TRU, complex nor dynamic. We, unless with great education, learning and training, fail to use the powers we have been given and fail to put them all together in both the unity and totality of SOT and TRU, dynamically by VAPO. For IA in being a manifold of stimulus-response as well as subject-object so focuses our subjectivity outwardly on the object, or inwardly on itself. The highly variable state in relativity, the observer observed situation of subject and object is much of the human condition. It may be mans great predominant form, structure, function, and power yet it is also his great liability especially when poorly or defectively trained, remembered, and used. In DROT, when adapting our representations to SOT, the Á of TRU, we are conforming to the most important aspect of TRU. This has been roughly understood and used in human knowledge, business, the professions, art, technology, and industry. (TREQ)

CHAPTER 22

GENERAL FORM of ROT

GENERAL FORM of ROT: INTRODUCTION

The General Form that ROT, or RÄ, takes may be described in different ways. It takes on many principles, shapes, conditions, arrangements, types, and products. These may or may not be identical with the form of TRU, yet they will, more or less, tend to be identical with TRU or that part of TRU they objectify. Yet they will depart from TRU by: 1. Representation produces a new and its own self-determined state, 2. Representation is directed not only to TRU but how TRU can be best, most effectively, practically and dramatically shown, 3. The means, methods and instruments to show TRU will each in their own way produce new forms and states.

Because TRU is highly Á, formed by its special predominant mass energy nature as well as its own space time configuration, the general form of ROT will incline toward this frame and structure. The great range of the mass energy dimension in TRU, its levels of Á, or SÁ, makes it possible to represent TRU partially in any one or combination of these levels. Most characteristic, and largely medial and focal, of this is the reality of language and the words that represent the concepts, parts and processes, ES and CES, of TRU.

SET describes how SU is Equivalent to TRU. So likewise is ROT and RÄ Equivalent to SUT and TRU. Yet just as forms are not parts, so are Ä's not E's nor TÄ's TE's. This applies as well to ROT where the form of ROT and its parts are often far different. So are the forms of ROT likely to be multidimensional, rather than unidimensional. This is signified by **DAP** or Different Aspects and Parts. It is the universal form that largely determines and provides the potential for what TRU, is ordered by. So does SUT and the form of TRU and ROT provide the unfolding of many problems of understanding and showing all. This it does in part by providing a pattern or configuration, which in being highly universal is providential throughout its application. **THRA** = Threshold of Reduction Ratio, or the unfolding of SUT as it occurs in both TRU and ROT. THRA also applies to the two way trap of: **LANT** = the Law of The Absolute Necessity of TRU, whereby universal, transcendent, immanent and particular forms of ROT are all, to a varying degree, valid even though they may be of ROT other than TRU and TRIA.

Ambiguity is an expression with more than one meaning, or a meaning that lacks relation and distinction. Roughly ambiguity is any term, expression or representation that poorly represents its object. It has an added exponential negative effect to the degree it is less than or not TRU.

Art is largely driven by aesthetics and the aesthetic experience. Both are key to ROT as well as a powerful force in TRU. The distancing of the spectator from the observed object is called <u>aesthetic distancing</u>. Our appreciation for the overall value of the representation is called <u>aesthetic experience</u>. The process by which the artist creates this experience, and appreciation in the spectator, is distancing. The extreme view whereby all life should primarily be guided by aesthetic values is called <u>Aestheticism</u>. <u>Truth Function</u>, although used mostly for propositions, has a broader meaning for the form of general representation. This is similar to aesthetics whereby the truth or falsity of a representation is determined by the truth or falsity of its components for all possible cases. The <u>Principle of Extensionality</u> confirms the reduction of propositions and representations to truth functions of their ultimate components.

Stages of DRÄ include: 1. Any Representation, 2. Representations guided by POT and RU, 3. Representations guided by TRU, 4. Collected Representations of TRU, 5. Ordered Representations of TRU, 6. Representations ordered by the Systematic unity of TRU, 7. Improved Representations of TRU, 8. Best Possible Representations of TRU, 9. Ideal Representations of TRU, and 10. Variable Ideal Representations of TRU. <u>Formalization</u> is the bringing out of the best or salient features of ROT. <u>Formulation</u> is how we can bring these salient features together in an ordered whole so that we have a permanently clear understanding that will enable us to deal with TRU and ROT most effectively and efficiently. When returned to TIT the objective aspect of representation becomes the vehicle, medial, phenomenal, and real of its relativity, the Á FAF in TIT. This gives us evidence of the flexibility and transformation of relativity when the given is changed, especially when they are of different poles in the subject-object relationship.

GENERAL FORM of ROT From universals much of TRU can be learned. As we approach the form of ROT, as we do of TRU, the subjective or observer must adapt, focus and converge on the object, given, or that to be observed in ways that meet both their static and dynamic natures. By IA, this is an incomplete, and therefor only temporary or transient, object. The only final object can be one in totality, VAPO of TIT-TO. Corporeality signifies the state of having body or being tangible and substantial. It distinguishes the immaterial, transient, unlimited, and unfocused from that which is discrete with well delineated and limited form and structure. Humans have a built in tendency to seek corporeality in all they observe, perceive, and know. Without corporeality the objective world is less reducible to interpretation, relativity, and action. This is largely achieved through orientation, and coordination. Through ACEXIN, the dynamic mode of ALT, RÄ and ARÄ or the General and External Forms of ROT are enhanced and fulfilled.

WHATNESS of ROT What is ROT and what it is made of, includes anything that enables us to better recognize, demonstrate and communicate TRU. This is of both its products, as any artifact, and all the principles and dynamics involved in their production. ROT is seen in how the observed object is worked like a piece of art, the form it has as an object, and what form it takes as the observer or subject brings it to life or TRU. **ESING** is the action or process of forming ESIN. It arises from the human need to attain increasing meaning for words. ESING is a transformation of meaning from what is represented, as a word in general, to what it is as an entity that is also in TIT. **ESION** is the process of optimal ES formation. ESION is how the parts of TRU can be best identified, separated and made pure from their corresponding objects or entities that is not TRU. It arises from the parts of TRU itself, in contrast to ESING and ESIN which arise from an entity or object. The specific kinds, contents and what the general form of ROT should be is shown by the <u>doctrine of names</u>. To name something is to identify it. This identification was man's early attempt to connect the observed with the observer and to communicate the resulting subjectivity to others. Increasingly humans learned from simple sounds to more complex names and speech to better produce what they identify and communicate. These names became more permanent with use and growing populations. So that linguistics arose very gradually over millions to hundreds to tens of thousands of years. It waxed and waned, came and went with the life cycles of the groups using language. This recurred time and time again in forms that tended toward selective benefit. This went hand in hand with increased brain and speech development, both anatomically, physiologically, psychologically, and socially. The benefits of language - with its associated greater brain capacity - were a large part of the predominance, success, and survival of human mass energy, or Á FAF, and TRU. From this differentiation of the form of language occurred and diverged, including: Descriptive linguistics that analyzes the elements that make up sounds and signs of language. If they are primarily sounds, they are called phonemes, when signs morphemes. Phonetics classifies and describes how speech sounds are vowels or consonants and how they can be represented. Vowels and consonants may be classified by the articulation of the mouth, tongue, vocal cords, throat, lips, teeth and palate. They are divided into various qualities and regions produced. Phonemes in being the smallest elements or units of speech provide the means to produce sounds that are separated, or made distinct from each other. The great increasing variety and clarity of these sounds is the great power of language and makes up a large part, indirectly indirectly of representation and TIT. **LAMERT** = Law of Any possible Mode of Expression, communication or representation is apt to exist. LAMERT shows how vast ROT can be, and how much the best of human life is more than man alone. Anything taken from outside that becomes inside TRU or TIT, through its ability to enhance TRU.

The unity of the development of the general form of ROT through THEO, SUT and TRU as well as the unity of the observer and EXIA, are epistemologically and metaphysically parallel and similar to: **PROFAU** or the Principle of Fusion and Formation of Universals. PROFAU is more from the point of view of perspective and entities, whereas in the development of RÄ our form works posteriorly. PROFAU pertains to universals that so permeate our being, although tending to originate from infinite sense or thing, do not continue in this way. They have relevance, a relevance imposed upon through inescapable reduction by the observer observed nature, or very form of TRU and ROT. **POES** = the Principle of Essential Separation. POES is a part of thinking, speaking, acting, and producing. We observe and represent the observed, so must separate out the essences of what we have to work with, are given, or that before us. This is to make it into some kind of replica, example, product, work of art, or representation, that produces meaning, use or value. This is how all tends to converge on ROT and TRU and what enables ROT. What the FAF of ROT does, and can do so well.

BIDA or the Bilateral Definition of All Things. How definitions consist of objects and entities in themselves, and as they are exclusively in TRU. This is often more than one or two but multiple definitions, especially with different shades of meaning. They also exist of many combinations. Although the parts and processes, form and function of TRU have been unknown we can relate and posit terms with increasing accuracy through POA, SPALT, POS, STA, and HOM. **NE** or Nominal Equality is the qualification of the concepts, words, and terms we use. It enables us to attain SU or a perfectly formulated, mathematically precise analysis modulation and construction of ES, CES, and the components of TRU. How our words and terms can be brought to be their equivalents in the optimal ordering and construction of TRU in ALT. NE is related to, and dependent upon, **TIEA** or Terms, Nominals, Words, Concepts, Intension, Extension and Actualization.

TIEA is how all terms in their deepest, most essential and complete senses can be stated and defined as they are in the primary coordinates of TRU. TIEA is the ACEXIN of language, or all words and concepts.

Algorithms and flow sheets reveal the predominance of ROT by showing how our representation of any system of notation can be ordered and used in planing, display, programming, calculation, or action. The central role of stimulus and response in life, and especially our lives in TRU, is an example of the dynamic development and use of terms, flow charts, algorithms, and their operations..

The deleterious effects of stimuli arise when our senses and cognitive powers are lacking. They arise from not knowing, knowing wrongly, or not knowing what to do. They require learning and are completed by best knowing and responding to it. Our dynamics necessitate, conceptualization, terminology and their intelligent use. This involves ever higher, and increasingly accurate terminologies and means of representation to win over competition and achieve a perpetual mastery over all. This requires both language of actual meaning, and indices of cultural sufficiency to succeed. The words developed must not only be simple utterances of objects to be identified. They must be more meaningful terms of more abstract and universal principles or elements in TRU. Abstract Art is any form of representation that has as its purpose, elucidating and producing objects not otherwise apparent or demonstrated. This should be what is needed to reveal aspects of TRU. Abstract Expressionism is the form of abstract art which does the separations noted within the limits of expressionism or how objects are altered by the artist both before and after its own existence. That is how the observed becomes a representation both as it becomes different by the observer or producer as well as how it becomes different as it is formed to meet the viewer or the second observer.

ASPECTS of ROT

LARCOM, the Law of Representative Completion, signifies that representation and communication need to be done adequately, fully and with sufficient closure. It is a kind of artistic disposition, inclination or propensity to satisfaction, achievement, fulfillment and fruition. Just as representation is a great part of the observer-observed relationship so is its fulfillment a great part, and its intentionality a primary mode of the Á dimension of TRU. Without intentionality and direction in ones representations, art and industry, there is a never ending expansion of waste, failure and chaos. To overcome these is a major aspect of ROT, to make it more efficient meaningful and beautiful. This is why it is called definition, to intentionally make it more and more delimited, defined, or SCARLED, as properly fits both this intention and what this intention actually is meant to be in convergence on TIT. Definition and distinction must correspond. Thereby much of the great challenge of life, as it is TRU, is to correct fallacies of ambiguity by multiplicity. **PAMEDS** - the Process of Meaning, words, ES, Distillation and Separation. PAMEDS is the crux of human life, mass energy and language, past, present and future. It is what enables humans to, with ever greater expansion and intensity, selectively provide broader and clearer representation of this world, TRU.

The derivation, development, growth, and form of TRU are characterized by and expansion of all kinds of numbers, the S-O, mass energy, capacity, ability, and means of representation itself. It is a vehicle of expansion, much of the relative observer-observed mass energy form of TRU. It is the vehicle that produces reality, through the expansion of the real, the agreement and ever higher powers of realization, Á the mass energy spectrum, continuum, form and function of This Relative Universe. **LAREX** - Law of Representational Expansion, and associated Relative and Representational Expansion. That this expansion occurs both in, out of, and in and out of TRU can become readily apparent with each passing given, and attention. This is: the Representational Expansion In and Out of TRU.(REXIOT)

It is in the Sign, as 'by this Sign' that humans have so gloriously proceeded to develop and become predominant by SÁ, Á, and TIT. It is the sign of objects we identify with, and make distinct to enable the concentration and work, that is productive. A sign is any object or artifact that represents a specific thing or conveys a specific idea to the person who perceives it. Just as in medicine a sign is any feature or fact of ailment or abnormality that indicates disease. Similarly a Term indicates any object needing designation to call attention. It is through signs and terminology that humans have so attained a great degree of formal distinction necessary to set the same and the different by consolidating their relativity, or Á mass energy dimension in life, in TRU. As words and terms are the subject or predicate of a proposition, they are of many kinds and used in many ways to best describe the objects intended. Of these different kinds there are terms that are, connotative, non-connotative, negative, positive, singular, general, concrete, abstract, relative, and collective.

DIVERGENCE & LIMITS of ROT

Definitions are not, more or less, established from TRU, except in tendency or slightly.(MATES) Therefore they often tend to be of ENITS and OBITS. When expressed as a paradox and when directed to nominals, rather than derivation and definition we have: **PAN** or Paradox of Nominals. PAN questions the degree to which what we know equals or does not equal TRU. Words arise de novo and define mixtures not necessarily TRU. As language is so much of and close to us, and as it is spoken and written in a great maze of different kinds, ways and means, it lacks ordered harmonious totality. This should be expected from the differential nature of language, TIT and AU. Yet this should not lessen, it should increase, our zeal to order and perfect language in and by TRU. Even the differential can be ordered in closure, the set totality of TO. **PAEO** or Paradox of Exclusive Objective. PAEO is the transcendent paradox that arises from **LOICLOR** or the Law of the Imperative Confinement by Limits of Representation, where we must produce ROT within set confines. However by PAEO this limitation must be overcome if the given and represented are to be finalized. This finality is like the error of beginnings and is controlled only in the overall as by SPAT, OM, ACC management, and control. Oppositely there is **LOLVAR** = the Law of Limited View and Reality. LOLVAR applies to any entity and TRU. It arises from their imperfect image and representation. They continue to be imperfect for some time and must be accepted as such if one is to continue. This imperfection arises from our lack of knowledge and the fact that in being, usually only a very small part of TRU, and even more often being much more non TRU than TRU, these images and representations are, or tend to be, duplicates, replicas, examples, reproductions or productions that are very rough, vague, mixed, or even defective representations of their objects in TIT.

The line between what is representation and what is TRU is very fine. They are often not one or the other but some of both; or even more a dynamic fluctuating combination of both. For example the expressions on ones face, the mannerisms one has, and the movements one makes are often both processes in TRU & representations. It will depend upon how they are interpreted. The highly transient nature of representation either generally or some specific representation occurs in varies guises or forms in and out of TRU. This is a key to the form and limits of ROT and TRU. Law implies error, for each law, LOT, and LOL there is a corresponding limit, where use otherwise produces error. It is usually much easier to emphasize and especially to remember and utilize an idea, concept, sense, thing, principle, law or mechanism that has a name than one that is nameless. Herein is the greatest beauty of language. For when we have names, which are repeatedly identified with their objects and which in turn correspond to their actual counterparts, forms, and functions in TRU we save going through the whole process of S-O identification necessary each step of the way. It is in this way language largely consists of nouns. Verbs being their dynamic relatives, associated with other parts of speech, including prefixes, and suffixes are means of making these nouns more meaningful by extending their roles to that in life, TRU or TIT. It is by the prefixes and suffixes that developing language tends to classify its underlying principles. **PADET** = Paradox of Potentiality, Development, Derivation, Partial form of TRU or Á Equilibrium. PADET signifies the great revolutionary struggle that is taking place in TRU between the S-O or Á mass energy dimension and its use as words that are less than TRU. **POCOL** = Principle of the Convergence of Limits. POCOL guides our intentionality in representation.

SHAPE, CONSTRUCT and SET FORM of TRU by ROT

We make distinct and define terms of the differentiated objects as if they are of TRU. This is **ÏFE** or As If E, or ES. As ideas, concepts, essences and ES exist as one and the same thing, they are called **ÍDES**. The whole process of language formation, ever more strengthening and empowering TRU, and ever more being actively used by TRU and ES is:

IFE → **IDES.** This shows the transition from separate parts of our world, TRU, to concepts, and their combination as new entities, in TRU. It is IDES that the Á mass energy FAF of TRU largely uses in its operations. When the thing represented and the manners of representing agree, the thing has <u>style and quality</u>. This reminds us of the ever present role of convention in art. Yet art as representation does not have to be, nor is it always best to be, in perfect agreement. This is artistic license.

LARBÁ = Law of Representative Basis for the Mass Energy FAF of TRU. LARBA denotes the fact that much of the best of human life is based upon how well its mass energy form and function work. This depends upon the quality of representational intensity in the spectrum and continuum between its expanding relativity or subjectivity and objectivity. If its intermediaries are effective agents of representation, other intermediaries or gradients in the S-O scale produce positive forces and success. More directly if the symbols, signs, and words we use are effective in making more real the relationship between subject and object, TRU is benefitted. If words represent meanings and these meanings make real the object to the subject then this function of TRU is satisfied.

Humans produced ROT by song and story, which have become a large part of life. Tales, vedas, and sagas were handed down from generation to generation, attaining a high level of set form, artistry and perpetuation before man could communicate by written word. That art and TRU have long been equivalent is expressed by *'More Preferentia, More Necessitatis'* or as ROT becomes TRU, TRU becomes more ROT, and vv. As TRU is highly limited, **LIT** or limits of TRU, so does this make ROT not harder, but easier. ROT can selectively exclude all the world about it that is not, does not fit, and is not ordered by what is TRU.

CONDITIONS and STATES of ROT

Alteration of form will occur from ENIT, to OBIT, to ESIN, to OBT to ES, CES, and ROT. It will also vary depending upon the degree and state or condition of this objectification, representation, presentation and perspective. So that we not only have The Object, OBT, whose form has been mostly unknown, as well as a form which is not readily understood, we have, even at best, the objectivization of forms of the various parts, processes, and representations of TRU which undergo alteration. The general form of ROT is also engendered, altered or determined by: **LAMS**, or the Law of Altered Meanings and States. As entities, objects, things, senses, ESINS, universals, TRU, essences, forms, functions, parts, processes, and all of significance to TRU are encountered, come to attention, approached, and investigated they will become increasingly identified, conceived, named, and a part of our knowledge and technology. This, if appropriate and successful, is much of the operation of Á. **ECTSION** or the Processes of Explanation, Communication, Conversation of the Parts of TRU. This is an example of the transition from general to special. It embellishes or supplements this transition and its weaknesses by compensation and elaboration.

LUDIP = Law of Language, Logic, Universals, AU, RU, and

Metaphysics stand in a Dipolar State. None of these are completely valid as TRU, some are highly valid as TRU, and all are valid in and of themselves. This is evident in Wittgenstein's principle of verifiability and Carnap's insistence upon meaning and verification. This position supported an opposition to a realism based upon ENIT or THIT things-in-themselves or even objectivity alone, OBIT. Carnap described these as pseudo problems in philosophy, a fundamental defect in metaphysics. LUDIP as it of necessity converges on TRU explains and corrects this problem. For meaning and verification, as VAPO, are correct only when complete and of and in TRU. Representations and worlds hold when their meanings satisfy the conditions and states, or SCARLS of TRU. It is to better enhance these processes, evolution and revolution toward a language guided by TRU, that this text and our lives are so engaged: **JUTER** = Justification of TRU Terminology.

We constantly search for new forms and varieties that yield categories, separation, and fresh means or vehicles by which we make discrete the needed differences, distinctions, and ordering of objects, OBT, and intended reproductions. An example is Art Nouveau, Art Forms of the mid 19th century which became fashionable again in the late 20th century. Its search for new forms and varieties sought to experiment with and establish the needed separation, distinction and make discrete its objects by linear patterns, or forms in the mode of flowing lines.

Knowledge is such that words that are needed to signify the actual world either do not exist, or do so inadequately and imprecisely. They fail to provide the actual distinctions and overall perspective of TIT-TO or its TREQ. By enumeration and analyzing the greatest concepts and terms in existing language that come close to their part and roles in TRU and comparing these with those whose value and representation is poor we can gain a better understanding of what language should be. How ROT, LANT, NEC, MOT, SPALT, POS, STA and HOM are needed to supplant existing language.

Pragmatics is the study of the relations between signs, symbols and concepts. In the use of language in POT and ROT we are confronted with four main kinds of pragmatics, 1. Those commonly known, or OC which is of two varieties, its major definition and all others, 2. those less well known but with essences that approach TREQ or TRU, 3. EC, and 4. NEC or the New Concepts of LANT. In the accompanying listing we compare these four kinds of definitions for outstanding concepts. They show us how necessary pragmatics is, as well as how much we need to supplement existing language by new concepts and terms, NEC of LANT.

TERM	COMMON MEANING	PROPER MEANING	TRU MEANING
GOD	A Deity	The Deity	TRU, TIT
RELIGION	A Way of Life	The Way of Life	The Way of Actual TRU
WORLD	All Things & Senses	All Around Us	TRU, TIT TO
REALITY	All that is Known	Observer/Observed	Best Relative to Us
COSMOS	An Ordered World	The Ordered World	The Ordered TRU
INDIVIDUAL	A Person	The Person	The Person/TRU
SELF	A First Person	The First Person	The Person↔TRU
SOUL	Part seeking Identity Dynamic	Person seeks GOD, Dynamic &	TES, TÍS, All that exists in TIT
SPIRIT	Transcendent aspect or part	Transcendent of Persons/World	OITIO TRUD CES ACEXIN

KINDS of MEANING and DEFINITIONS for OUTSTANDING CONCEPTS

By so distinguishing meaning and definitions they not only provide a second dimension to their representation, they anticipate the great need to reproduce the dynamics of that which is sought. **COB** Complex Objectives or the Complexity of Objectivity. COB is that state, condition, relationship, and limitation, or SCARL, of the object or that to be represented, worked, used of TRU. Such an object is not only multifarious and highly composite, it is intricate and difficult to utilize. Any thoughtful person recognizes or seeks to discover the SCARLS, quality or degree of the given, the object things or all entities before one. We seek to discover the *this, what, where, when, why and how* of all needed or interesting. Prefixes and suffixes are another kind of COB, whose categorization of language give it much of its order and strength, as its operations reflect the Á FAF of TRU.

SIRD = (law of) Sufficient Identification Representation Demonstration. In life we must show, and show how, the utilization of the IA of a thing, its S-O, occurs. We must show the IA of TRU, IAT, and how its representation becomes the representation, the SU, of TRU. Likewise we must show how as all things are infinite, e.g. RI to TI, there must be a limit and SCARL. Yet the demonstration, identification and representation of things need not be infinite, they only need to be sufficient for TRU. This is called: **VASIRD** = Variable Levels of SIRD with Demand and Use, and, **LAPRAL** = Law of Practical Limits. To give a representation perfect detail, infinite form, shape, character, parts, and meaning does not necessarily mean that it must be a 1:1 replica. It is not even necessary to show by all and every type or school of art how each object is to be rendered. It is only necessary to reproduce it in such a way that it is sufficiently shown. This is sufficiently shown to meet the demands and needs of the IA, the S-O, the sense, thing, RU, TRU, aspect or part of TRU.

KATISM = Knowledge and TRU Imperative demand for Sufficient realism, exquisite detail, Summative Magnitude, Magnificence, and OM. KATISM is the presentation form of the summative OM, as well as the SCARL and correction of SIRD.

SCRETS = SCARLS of Comparative Replications, Reproductions, Representations, Renderings and Resolutions of TRU. A frame with respective inclusions of the various SCARLS, RÄ and ROT of TRU can be comparatively provided. This helps: **VALTRIP** = Validation of ALT from TIT and TRIT Progression. The quality and SCARLS of our terms necessitate: specificity, actuality in TRU, agreement in VAPO, and both qualitative and quantitative reduction of their A,L,T composition. With these we not only have a much stronger language, we have one that can be used in a much more positive, concrete and logical manner. They help to show and prove a notational system for TRU Terminology.

NECESSITY, FORM & ROT

Form is of critical importance to entities, the knowledge of entities, TRU, and the understanding of TRU. Form, has for representation, even more striking importance. This is the utilization of form as a necessary condition and means of representation. Form for ROT is both how ROT is structured and the transcendent form ROT supports from TIT and TRIT through OBT, SUT, ROT, MOT, ALT, to SPALT and SPALT ancillaries as POMOT, METAD and SPAT. Similar general and abstract nature of form of ROT arose from numbers and mathematics and are known as *abstraction principles of mathematics*. These provide some of the transcendent nature and form of mathematics as it approaches TRU and ROT. To establish the form of anything and thereby ROT to a large extent, it is necessary to develop the criteria and yield to: **LACTSOV** or the Law of the Absolute Compulsory Convergence of Points of View, Perspectives and Subjectives on TRU. By relativity and the mass and power of TRU these criteria are available to us if we only search them out. The fact that the form of art and representation is varied, nor becomes too stylized, is, in a more complex form understood from Craig's Theorem. This states that theoretical entities are in principle unnecessary for deriving the observational consequences of a scientific theory. Yet by higher means and methods of representation we know that interpretation, conceptualization, ideation, induction, extrapolation, reasoning and intellectual processes can and must play a large part in ROT and TRU. The guarantor of this is:**LANOFAC** = Law of the Absolute Necessity of Formulation, its qualifications, Limitations and Conditions, or SCARLS. As knowledge must, to a large extent, proceed from ROT, so must it come from the reduction of LANOFAC, or how the absolute necessity or formulation by qualifications and SCARLS can overcome: **FE** = Form of parts or factual evidence, or simple observational consequences. In opposition to LANOFAC is: **EF** = Errors and lack of Formulation and representation of TRU. It is from EF that man poorly understands and comprehends his world and fails to apply his knowledge efficiently or without such trial.

The fact that TRU is unitary or TU, all and exclusive or TRIA, relatively changeless or TIT and TRIT, and well formed and coordinated or SUT, makes it's interpretation and understanding reasonable, consistent, and to a high degree set and non arbitrary. This is called: **LANA** = Law of Non Arbitration of TRU. The characteristics, totality and nature of TRU can not be arbitrary. They are fixed and permanent, as TIT and TRIT are. They are valid and true regardless of our specific view, representation or presentation of them. Where there is a loss of a 1:1 image, the necessity for standardization increases. It must be compensated for to make the representation or presentation worthwhile and successful. This is called: **LENA** = the Law of the Relationship between, or Equation of, Absolute Necessity and Necessity of Arbitration: **AN** = Absolute Necessity, **NA** = Necessity of Arbitration.(LATANOS) LENA modifies LANA which is not absolute. Yet LANA when well provided for greatly assists ROT, KROT, MOT and all succeeding positive approaches to TRU.

The terms and concepts spoken, written, or communicated are neither derived from nor commonly used strictly as they are in TRU. So that with use there is more or less departure and deviation from TRU. This is: **DULET** or Departure and Deviation through use of language of the parts and objects from TRU. Yet our languages are the best we have and certainly the best available for knowledge. Thereby it is necessary to use language perfectly to have all who do so become coworkers of LANT and TRU. **LANT** = Law of Absolute Necessity of Language's course to TRU. **LATANOS** = Law of the Absolute Necessity of Standardization. In NEC, as LATANOS in general, the meaning of each letter in the word is representative of key terms in the definition. Since we can't assume that by simply seeing some features on the surface of an object that this is its form, we must endeavor to examine diligently and totally both its internal form and, as well, what produces it. The is called: **LANODAT** or the Absolute Necessity of Deriving, Developing, Determining and Demonstrating the Accurate and Actual Total Form of TRU. LANODAT proceeds from this very failure we have in both assuming what is a feature of an object and what is its total form. Also from the fact that we likewise are apt to assume this form is static, singular and permanent. Nowhere is intricate detail more necessary than in the determination and presentation of the ES and CES of TRU. This is signified in: **SES** = Selfs bringing together of ES. SES is the crux of artistry and ROT for this is our chief activity in life, how to bring together all whose ESINS should be ES as representative or reproductive of TRU. The first person to be TES must discern and analytically use every word, concept, sensation, abstraction and even act of life, in ways that are free from non TRU. This is free from: **EM** = Meaning Error or Error in Meaning. EM signifies Maximal Meaning. This as an absolute goal is limited. It requires deep commitment, one that is relatively attainable as the future will certainly become increasingly proficient in ROT, TER and TRU, even if by TREQ.

NECESSITY of a NEW LANGUAGE

The reasons why a new language with terms based upon TRU or TREQ are so necessary rather than the continued use or adaptations of current languages are: 1. The exclusive specificity, greatness, and sensitivity of TRU and especially TIT-TO. 2. The necessity of precise and proper differentiation, formal distinctions, and purity of concepts, names, and words for all that makes up TRU. 3. The great many languages, with differences in meanings, that currently exist that cause many problems, most notably failure to understand each other. 4. The great diversity and variability of each of these languages. The users of each language can't depend on their language to provide consistent and stable meaning. 5. The great changes that occur with time, place, and usage of languages. Literature, to be right, must continuously explain what is meant by the words it uses. 6. The words used mean and are defined by entities, senses and things in themselves, not by what is TRU or TREQ. This produces a continuous focus away from TRU. It requires continuous reinterpretation and explanation if the word, their combinations, and communications are to be, as they are more or less inclined to be, TRU. 7. To develop a positive objective approach, representation, and presentation, of the world, as it is TRU. Languages that do not account for, or provide a base upon which their terms signify their TRU meanings and can be posited in the frame of TRU, are inadequate and sooner or later fail. Language is the great vehicle, means, body and soul, and thereby predominant power and glory, as seen in its Á FAF, of much of human life. It is ever more rapidly changing. spearheading human development, yet similarly causing increased demand on the individual. It is to aid and cure much of these troubles, problems, predicaments, and discord that the correct and best use of the great predominant - Á - powers of TRU are brought out to frame TRU and signify the meanings and definitions needed to show this complex world in which we live and must survive. Terms which are held strictly to TRU, as well as inclusive of all the greatest aspects, forms, functions, parts and processes of TRU, are what is needed and what it will take to help language overcome its weaknesses and problems. For these reasons a language of TRU, TRU Terminology, is developed and offered for all to see and use that will raise each and all of a person's life to far greater levels of focus, differentiation, integration, order, power, and success.

SUET = SU equalization to TRU or the necessity by which any formalization of TRU must ultimately approach equalization with TRU. **LACADO** = Law of Other than TRU and other than All TRU Connotation, Denotation and Definition. We usually define and use words as if they are TRU or as if they are automatically valid. Thereby they are rendered, by the Law of Contradiction, on both counts, invalid. This is signified by: **FACADOR** = Fallacy of CADO, or Connotation, Denotation and Definition, Reasoning and logic. Since TRU, as TRIT by TRIA, is all, we can see the impossibility of our existing language in its current usage. It, in a way suffers from a kind of original sin. Yet our words, concepts and language are great and indispensable tools and means of understanding and enabling the subjective basis of our being, as it is TRU. This is signified by: **LANAS** = Law of the Absolute Necessity for Subjectivity in Language. There is a similar necessity for objectivity - and all in between - in language. Yet LANAS in no way makes words or names precise or TRU. This has to be determined and worked out on a word by word, case by case basis. This is the justification of ES, CES, POS, and much of the use of SPALT, SPAT, METAD, POMOT, etc.. LACADO helps focus words, nominals, notations and concepts on TRU. LANAS helps to set this focus by the subjectivity and relativity that is in TRU.

CHAPTER 23

SPECIAL or PROPER FORM of ROT: RÄP

SPECIAL and INTERNAL RÄ ACTUALITY and PROBLEMS in RÄP

ROT, as art, can be no greater than the degree to which it is S-O, A, and what is produced equals that which it represents is TRU. The task before us is evident in: **SEF** = the Semantic Fallacy. SEF shows how primitive and non TRU much of our language is; yet likewise shows how it can be improved, perfected and used to its best potential. In all our thinking we tend to think in terms of ideas, concepts and words that describe entities that exist independently in and of themselves or in any relation. This is not as they are in TRU. It causes misuse, misinterpretation, misinformation and errors in thought and reason. It is to the degree the words are not converted to TRU even though they tend to TRU.

The necessity of language as self empowering is seen in: **LACIN** or the Law of Form Produced and Caused by Necessity. The form of ROT, like form follows function in nature, is an adaptation to necessity. This necessity is often auto-induced through mechanisms within its universe, TRU & ROT. What engenders or determines the form of ROT or RÄ and RÄP is to be found in what makes any form. This is found in: **LACT** or Law of TRU on becoming an object will undergo and assume Certain alterations of Form. **LE** = Language Error or Fallacy. LE plays a large role in human life and its conflicts, shortcomings, stresses and failings. For not only is language highly imperfect, it is used by many very poorly and often for ATRU, ANTRU and ANTITRU ends. This is both the language itself and their ANTRU use. These are often fed by human deception, deviousness, and duplicity, all, generally, by absence of or departure from TRU. This is exemplified by, but much broader than, vulgarity, cursing, swearing, and abusive language that we commonly experience. This is opposite from and exemplified by scientific and technical terms and uses which have massively contributed to language and TRU. This is how great ideas and concepts become more realistic and effective vehicles or means in the Name, or Named Concept. This is the existing equivalent of what we do in NEC or the new terms of TRU. This name, or named concept, is much the same as what the ancient Greeks called the Logos, the reason behind, and how to use, words. In summation they are the great power of words in TRU, similar to the power of God and the Holy Spirit, energizer of much of Á and SÁ. Inference is the basis of logic, a formal manipulation of ideas, concepts and reason.. It is an exercise of logos or rational processes in which words, phrases, concepts, and ideas are joined together and guided by rules or set procedures to provide better and greater meaning. This naturally tends to Á, ACEXIN, and TIT. It is directed in being focused upon a specific question, problem, paradox, enigma or topic whose answer, solution, resolution or completion can be coordinated and brought to a valid conclusion. It is often an exercise in the use of VAPO shown by Socrates, Plato and Aristotle. Analysis itself must proceed in a distinct order. This comes from the observer-observed relationship by our focus in relativity. Explanation is specifically formative by enabling added and fuller meaning to a topic.(VAPO)

In this text we utilize the method of Acronyms and Abreviations to creatively display those images of TRU that have been unknown, inadequately, and wrongly known. The justification for this is highly warranted, by many unknowns, the greatness of TRU, and the facility combinations provide for more comprehensive knowledge of TRU. Those notations, words, and abbreviations take precedence and are most justified which are closest to and of the greatest parts, processes or aspects of TRU or of its perspective, presentation, representation, modulation, utilization or accentuation.

ROT S-O ACTUALIZING, SAMES, CORRESPONDENCE and IDENTITY

SOT as it exists in TRU as A, or its primary mass energy dimension, is the most important and least understood nature of TRU. To represent this aspect of TRU we must not only understand and expose all confusion about how Á is created and formed in TRU, but how the S-O component or state that exists both within and without TRU coincides with the Á dimension and special form of TRU. When we discover and learn TRU, represent it, and at the same time do so fully realizing and representing its Á or S-O, observer/observed nature we will spontaneously converge on, as we must with deepest conviction, TRU. The Object, OBT, is the same as The Relative Universe, except for its activation by the subject. Such statements, although highly valid, are not to be taken, as all, in any way, but only by TRU and within their SCARLS. In representation we should not only show the object, TRU, we must make the compromise between the extremes of objectivism and subjectivism. This is what is called the Theory of Representative Perception. This theory distinguishes between the objective and the subjective by calling those qualities primary that were solely in the object, and secondary those that were produced by the subject. Obviously the qualities of TRU that are subjective are not secondary, but equal to those which are objective. Yet this theory reveals how philosophy has needed to emphasize the correction for its own epistemological bias. **RORO** or the Resolution of Reals or Objects as they become ES to become TRU. **Ob → Reals → ES → TRU.** RORO arose from the problem that in our world, and civilization since coming into existence there has been no strict or absolutely discrete reals that are actually pure parts or forms of TRU. This is the reason for pure and real in **PUREAL**. Our concepts, symbols, words, even most of our sentences and descriptions, are less than and usually of very little TRU. By continually resurrecting what must be TRU, language and ROT finally succeed.

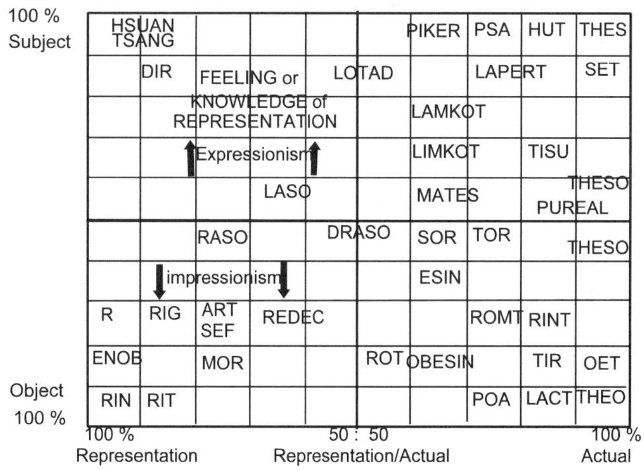

Fig 1-23 **DRASO or DIAGRAM of REPRESENTATION, ACTUALITY, SUBJECTIVE and OBJECTIVE**

The total aspect of S-O and summary of DROT is exhibited by Hsuan-Tsang, DRASO, and comparative analysis of the various observer observed relationships in the primary categories in and of TRU. Hsuan Tsang was a philosopher of the flourishing TANG period of Chinese culture. He held that the whole world is but a representation of the mind. This not only indicates the preeminence of mind in the mass energy form of TRU it represents the very important part representation plays in TRU and the minds contribution to this representation. Such idealism or the preeminence of the mind is not so unusual in philosophy, for this is its primary outlook. Yet the emphasis upon representation of this mind as the sole, or even prime, form of the world or TRU was novel and highly fitting. The subject in the subject object state is like the observer, a focus from which much begins. Without this there is no actuality, in that relativity, as special and typical for humans and TRU, is lost. Even more, the thisness of TRU is lost. This is why the mass energy or Á dimension is predominant and thereby the prime form of TRU. The subject, in the subject predicate relationship, is also a focus that initiates IA, representation, and language. The operation depends upon the subject or observer to initiate and do it. It is a means of showing its relationship, however indirect, to us, to TRU. It is in this way subject of subject-predicate is the same as subject of subject-object, and both together as they constitute the subject-object, phenomenal, mass energy or Á dimension of TRU. The use of subject and predicate is much more complex than that of simple observation of the observed or S-O. It involves a number of words in phrase, sentence or larger rhetoric and composition in which the identity, sameness or relations of one to several simple ideas, come together in the process. Since no object or observed, as well as no subject or observer can be the totality of the world, TRU, we must accept their relevant polarity in continuum, that of Á.

RELATIVITY, IDENTITY and MEANING IN RÄP, RÄ, ROT and TRU

Scarabs as symbols, represented meaning as representation. This is called: **RER** or Representation of Representation. RER shows how even in ancient history the expansion of representation was a great power that propelled humans forward to ever more successful forms and levels of mass energy in relativity. **RERÁ** = Relativity of the Relational Á, the fullest TRU meaning of RER. The selective thisness of TRU brings back and gives added meaning to the object represented as it is altered to its part and role in TRU.(PRÁ, PRUT) This is done by the relativity of TRU and the associated relation and identity between subject and object, as they are Á, in TRU. It is this great role of RER, not so much as the object of art we encounter, but all that is representative of something else, as these in turn are altered and exist as the same in TRU. S represents O, and O represents S, selectively and exclusively in Á and TIT. RERÁ is a major proof of Á, PRÁ, and TRU, PRUT.

Type-token ambiguity gives us an example of the difference between language in ROT and TRU or RÄ and TÄ. The Principle of Identity is the logical principle that anything is itself. That any symbol is equivalent in meaning to itself. This may be interpreted, as in type token ambiguity, for words are what is represented in more than one way: ENS are neither S nor O alone, entities, or ENS are no more nor less than what it signifies or defines, ENS are only ES or CES if TRU, ENS are by SOT Á only they are their roles in TRU, ENS are on what it represents by FAF of ROT RÄ, and ENS are variations or mixtures of the above. From these differences the principle of identity has certain prerequisites. Without them it is, both right and wrong, largely depending upon its use in these ways. It should be strikingly apparent that the great defect in this principle is at the root of all. As much as it appears to be an absolute certainty it is severely flawed. This flaw is repeatedly encountered in the world in which we live, and especially the world of communication and knowledge. This is largely the difference between ENIT, OBIT, ESIN, ES, TRU, TIT, TRIT, OBT, SUT, MOT, POS, POT and how words or any variety of ROT shows them. Which is right and which is wrong and to what extent. This is called: **DEOST** = Selective Differentiation of Entities, Senses, Things, ESINS, ES, TIT, TRU and how they are shown by words.

Knowledge of the whatness and specificity of the form of representation has been recognized for ages. Cheng Ming or the doctrine of the rectification of names, regarded names as corresponding to realities and should serve as standards for social organization and personal conduct. It further maintained that the actual must in each case be made to correspond to the name. What meaning is directed to is shown in: **MATES** = Meaning Automatically tends to ES and CES in TRU Specifics, differentials, and particulars. MATES arises from the general and simple fact that life is a direct product of an inclination toward what is necessary, significant, essential and most important. Since for the best of human life this is TRU, MATES relates to TRU, ES, CES and all that is TRU. As concepts and words define, distinguish, make distinct, differentiate, form and especially converge on TRU there is meaning. This is: **MIST** or Meaning Is TRU. The exception and corollary to this is in IS itself: **MIS** = the Meaning of IS. Both is and meaning being of central and critical importance to TRU make it imperative that we separate and reduce them adequately. The premier power of humanity is its ability to search for ever greater and deeper meaning and definitions. (SADAM). It is the basis for effective language. It takes a vigorous and proficient approach, including that of, relations, ideas, revelations, experience, astute attention, and interest. It is the key mode of human mass energy, Á. In meaning we have a higher level of Á in operation in which this identity provides the subjective with information and evidence that ones cognitive powers may use. Meaning may also be expressed in the fusion of Subject-Object in representation. This is where it equals the success of a subject's representation of an object. This is called: **ISO** = Is Equals Success of Subject's Identity of an Object. ISO is how subject and object become the same as subject represents exactly, or in some exact way, feature, or meaning, its counterpart in the object. Ultimate meaning even more profoundly tends to TRU, or its ROT, POT, and PERT. **ISERO** = Is Equals Success of Subjects Representation of Object. **ACISTRO** = the Actual Is equals success of subject TRU's Representation of object TRU. The steps in meaning are: singular meaning, contextual meaning, subjective integral meaning, subjective integral meaning TRU, total meaning, and TIT-TO Meaning. Means and methods of art or ROT are great because they have depth of meaning. This is closely associated with depth conception and greatness in TRU, and shows how much is in a beautiful picture and ROT.

If a word's meaning is exactly the same as another word, it is useless. This is opposite from how each aspect and part of TRU must be represented by a different word, or at least a qualification or description of the formal distinction needed. Still we know, from practical experience, synonyms can be very helpful, they reinforce and expand meaning. This is partly because of personal limits, as well as how they reflect the overlap of meanings in TRU and ROT. It is mostly because it gives us strong data, evidence or information that we can use to strengthen meaning, although this may be neither pure nor without duplication, It is the ever greater distinction and ACEXIN of each word's meaning that provides the great power of language, our life, thought, and TRU. For this is its basic FAF, Á, L, T representative nature, its specific RÄP.

DIFFERENTIATION of FORM in RÄP: FORMAL DISTINCTION

When an idea and concept is represented by a word, it needs actual and consistent definition to be lasting and successful. The word to be defined is called a *definiendum*. The expression which is used to define a word is a *definiens*. The variation in definition of terms depends upon whether the emphasis is from whatever vantage is at hand. A word, concept, or ISM, will vary considerably, formally and distinctly by their epistemological, metaphysical, logical, ontological, axiological, ethical, or OITIO interpretations. For any word or term to have power, consistency, discretion, distinction and accurate meaning these and any other interpretations, must be remembered, utilized, controlled and exceptions made as necessary to hold them to best use in TIT.

	ATRU	TRU
INDISTINCT	Ignorance Non Identity Non Being	Undiscovered TRU Rationis Ratiocinantae
DISTINCT	False of Illusionary Knowledge Absolute Distinction Modal Distinction Real Distinction Adequate Distinction Inadequate Distinction Rationis Ratiocinatis	Revealed TRU TRU Knowledge TRU Reason Rationis TRU (RATRU)

Fig 2-23 **INDISTINCTION / DISTINCTION, ATRU/TRU DISTINCTIONS and ANALYSIS by LEVELS and TRUS**

Among these different distinctions there is reasoned reason or rationis rationcinatae, That between our mind in conceiving things as not distinct when in actuality in TRU there exists differences which can provide distinction. This was shown by the Scholastics as the perfections in God. It is, as to be expected then, quite prevalent throughout TRU, and especially of ES, Á, and SCARLS. Rationis ratiocinatae involves three or more varieties. The one that stands in greatest indistinction to TRU is shown in figure 2-23. Some of the types of distinction and how all entities may be classified by being distinct or indistinct, ATRU or TRU are shown. The diagram vividly reveals the distinction of distinction. It is necessary that we achieve an ever clearer distinction of all that is different and separable, how one thing is not another, in TRU, the fourth quadrant This is the only actual valid reason and is that distinction sufficient for the reason we must seek and have to use all for TRU. Scholasticism also identified other kinds of distinction, shown in this figure. In formal distinction we signify those means, natural, invented or standardized by which we can show set differences. Formal distinction is also largely the means by which language is objectified and perfected.

UNION & UNITY of/in ROT, COHERENCE, EQUIVALENCE, and SYSTEMATIC UNITY

Representation of TU does not come from a simple singular undifferentiated TIT, one that lacks totality, TROS and VAPO, but one that is a unity in totality inclusive of all its finite, infinite, and absolute aspects. The contribution of SUT and its convergence upon a unity of ROT is confirmed in the part and role it plays in TRU, TIT, TRIT, and ROT. **COTA** = Corollary of the TRU Axiomatic, or the means and method of using TRU's dimensions. As the different dimensions and their axes fuse and function in TRU, the mass energy state as well as volume and variation all come together in TO. We can then recognize their spectrums or continua with their many clines and gradations. We can begin to realize how complex and intricate TIT, TRIT, OBT, ROT, and SPALT are. **SOR** = S - O Orientation, which when adapted to TRU becomes: **TOR** or the TRU S-O Orientation. That SOR and TOR are not identical becomes quickly evident when we analyze the observer observed situation of any entity, and that which exists in TRU.

Systematic Unity is a designation that has been used in many different fields and topics for the kind of system that is coordinately formed and functions about a focus to yield concrete totality yet maintains its unity. This we pay added emphasis on because it is a necessary condition for MOT, SPALT, POA, POS, STA, HOM, ACC, and POT. This is: **LATRÄ** = the Law of TRU and ROT Formulation by SUT. For the construct of SU as determined from universals especially in terms of length and time in many fields of science, and especially physics, there has been provided abundant evidence that the world is constructed or made up exponentially and geometrically. These in ways, shapes and proportions that are special, consistent, and typical for the universe.

In order to have an adequate and successful TU and SUT we must have the necessary and prerequisite quality and quantity concepts, principles, components and dimensions. The quality and quantity of TRU, **QUALT** and **QUANT**, can be discovered largely by what is common and universal in ENITS, objects, RUS, RUTS, and TRUS. QUALT arises from the ESINS of TRU/ATRU-AU. QUANT is TE within the OBT and ROT of TIT. QUALT and QUANT depend upon the precision, clarity, and TRU actuality of the words that signify its concepts, principles, components and dimensions. This is the reason why humans, in the use of language strive to meet the necessary QUALT and QUANT of words used. In **GER**, or General Relativity, we have one point of view or one observer. This does not have to be the focus of one's eyes, it can be and is in actuality a collective, summation or a unison of observers or subjects. This is the beauty and absoluteness of TIT. For it has, as a whole, all for one and one for all, as these observers become one in the totality of TRU, THES. That of TRU which is observed becomes THEO, and their combination to make Á the **UVÁ** or Univocity of Á. UVA strongly proves Á, the great predominant power typical and special mass energy state and operation, FAF, of TRU.

The exclusiveness, exhaustiveness and divisibility of TRU, that enable its systematic unity, coordination and order, are derived from **SIL** or Without Law. SIL describes the general case of the various types of comparative entities or things where a particular type would be non existent without the contraries of other types of comparative things. Absolute differences become and function as relative differences, as the D in DAOST in TRU. It is these that establish much of the polarization and thereby dimensions and dialectics so necessary for both TRU, TIT and their ROT. So that those coordinating forms of TRU must be selectively demonstrated in such a way as to maximally depict TRU.

PROMOV is the Problem of Representation of Multiple Variables. These are not only variables within TRU but multiple variables imposed by the relativity state externally. **PROMCES** is Problem Reduction Relation of TO together Multiple Closed E, ES, TE, and SU. PROMCES comes from PROMOV. **SUL** or Systematic Unity Limit. As new terms or forms in the SO continuum are added and as TRU proceeds from TO to increasing TE toward TI, the elements, forms, parts and REMS accordingly decrease in value. What distinguishes each varied aspects of TRU is related to the SCARLS in producing alteration and exclusion of other identities. This is called: **EL** = Exclusion Law. EL has very broad or narrow, general or special meaning. Most generally, it is what makes one thing not another, what is excluded by one of another. Resolution of EL depends upon more than the scale of TE which the human mind can satisfactorily comprehend. This is SUL The forms, ES, elements, words and parts of TRU, TE, can be known and known exactly, but only if they are known by VAPO and SCARLS for TRU. SUL is likewise not all exclusive. For although EL leads to the exclusion of parts of the whole if it is limited to SUL, we are not bound to this standard, nor SU, absolutely. EL, to, is relative to our capacity to comprehend and will vary for all persons under different circumstances and capacities. For, as we attain ELTICMA this will vary as to whether our photographic memory is more or less clear and in detail. Thereby the scale, detail and differentiation of TRU and TE will depend upon SUL.

ROT form proper or RÄP has unity and similarity in its content, both from TU, ALT, OBT, SUT, and the closure of ROT. This provides the basis for relations, meaning and much of definition.(analogy) These will depend upon the sameness of DAOST. Since what makes up TRU is not the total, there will be differences. It is the relations, meanings and definitions of these differences that are key to ROT, MOT, ALT, SPALT, POS, STA and HOM. A major law of these differences in RÄP is: **LACMAD** or the Law of Categorical Meaning and Definition. All must be corrected by their less than total or special part and process in TRU. All terms, especially those of greatest significance will have and must be corrected for that category, section, and chapter of TRU and its presentation, subjectivity or perspective, to which it pertains. TRU is a function of the success of language and this upon LACMAD. Reinforcement of meaning of TRU, all in or of TRU and its means are methods that not only provide the meanings of terms. When solidified, made formally distinct, and become the optimal 1:1 identities with TRU they collectively and totally provide the best SUT, objectification and representation of TRU. **LEM** = the Law of the Equivalence of Magnitude and Meaning. This obvious, yet so quickly forgotten nature of the parts and processes - and words - of TRU is of great importance both in ES, as well as in how words must be conditioned and guided. How they, as all that can represent TRU, must be guided and conditioned as they produce ROT and thereby TRU. LEM is the identity of the greatest with the simplest, ES with a term. Through language it is the primary means of representing TRU as these parts or processes are held to a 1:1 equivalence as value and meaning come together in their identities in TRU.

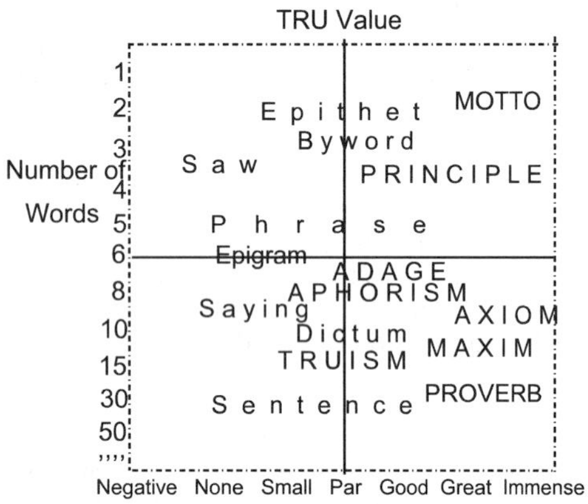

Fig 3-23 **COMBINATIONS of WORDS or PHRASES**

Different combinations of words producing concise meanings provide beautiful examples of the human mass energy dimension or the Á FAF, state and continuum. The representation of phrases, sentences, paragraphs and larger combinations of language are much of both ROT and TRU and one of the best means of understanding TRU. They are given different names depending upon their emphasis, characteristics and purposes. Many of the greatest ideas, thoughts, concepts, and principles are not expressed by single words, but by terse phrases and sentences. It is the use of these in combinations that has enabled a vast expansion in the power of meaning - and thereby being. It is mostly in their similarities that they give such strength to TRU. These are examples of the combinations of words and phrases that provide an exceptionally strong and vivid representation. Adage is a saying of common experience, often metaphorical, as 'without his tools the workman is helpless'. Aphorism are short sentences, or concise statements of principle, that seek to formulate the truth, as 'the style is the man'. Apothegm is a short, pointed, instructive saying, usually a concise or terse formulation of truth. Byword is a proverbial saying, and favorite personal expression, as in 'praise be', or 'necessity is the mother of invention'. Dictum is a saying, opinion or statement that is often authoritative, formal, and definitive, Epigram is a terse, sage like, witty, and often paradoxical saying or expression that is frequently a part of an epigraph or inscription for a monument or literary work. Epithet is a word or phrase, often attached to a name or entity used to identify, describe and characterize it in many different ways, as in 'the Eternal Lord God'. Maxim is a general truth and fundamental principle, usually in sentence form and often with a philosophical trend or bent. Motto is an inscribed indictment and short guiding principle, moral, character or parable, as for each state's motto, or for the boy scouts in 'be prepared'. Phrase is a combination of two or more words that express an idea. A phrase tends to be pithy, telling, as in 'could have been good for you' or 'God for you'. Principle can be expressed either as a single term, phrase or sentence. It gives to a concept a quality of both validity, general worth, and guiding light, especially as it signifies its underlying or basic idea or mechanism, Proverb is a brief saying that is popular, pithy and often ingenious. They are profound observations of the truth and give the observer a rule to live by, as in 'marry in haste, repent in leisure'. Saw is a common, old, homely saying that emphasizes contrast and dramatic effect, sometimes at the expense of truth, as in 'ignorance is bliss'. Saying is a wise, witty and often repeated statement by a well-known person. Cliche is any phrase, as the above, that has become overused to the point of disfavor or contempt, Truism is a valuated and self evident statement, in phrase or sentence form, that in being obvious tends to need no mention.

Some of the combinations of words, as in logic, where we use both language and mathematics to produce complex vehicles, are powerful means of showing TRU. Some of these may represent all TRU - as SPALT - only a part of it, both of OITIO, or various combinations. There exists in all languages, especially those with large vocabularies a detailed knowledge with shades of meaning. Likewise there are, many scientific and technical, terms whose concepts are expressed in various combinations of words.

We use our words and concepts as though they are precise, fully understandable, and TRU. Yet this is very much not so. Even philosophy often degenerates into semantic differences or a confused colloquy in language. So is the form of representation based upon words and language confined by the immature and imprecise state of the ideas, concepts, words and terms we use plus their proliferation in all higher operations. Concepts, old and new, like meanng, can be no better than 1. the intensity with which the idea, thing, law, etc. is conceived and used, 2. the degree of its expression, how inspired, explicit, and persuasive it is developed,3.Its actuality in and by TRU, 4. Its REM in TRU, 5. Its reception by reader or audience, 6. How well it works with others and fits by VAPO in TIT-TO. We must continue to use our imagination and creative energies to devise new concepts and terms, with best complex ROT, which must be found to fill the voids, replace the errors, and provide TRU meanings. **LEM** = Law of Equating or Equivalence of Magnitude and Meaning. LEM is the identity of the greatest with the most understandable and distinct.. It is the means of ideal identity, 1:1 or maximal representation. This is where the correspondence of the unity, high REM, or highest proportions of TRU, ordination, is with that of the simplicity of language.

LAC = Law of Absolute Contraries. The significance of contraries are shown in POT and ROT by making up the dualisms, parameters, and dimensions upon which TRU is based. For ROT and its production they have added significance. By LAC the divisions and morulation enable an ever better ROT and MOT. This is especially so when it is true and an exact replica of the form of TRU. The dimensions in TRU are not arbitrary, they are its frame just as m l t are the prime physical dimensions of any object, body or entity. In being consistent and reasoned by the actual nature of TRU rather than arbitrary they are the established and compelling determinants of TRU and thereby ROT. This is called: **CODIT** = Compulsive Dimensions of TRU. CODIT is why man, and all to be TRU, must be, think, act, and do in accordance with TRU. It is why TRU must be and is as it is. CODIT is why man can not do, as he pleases, any which way, thing, or even nothing.

CODIT produces a set frame for all, as it is TRU. CODIT or compulsive dimensions not only set the tone for human action, it sets the state and SCARLS of ROT. CODIT enables us to produce a ROT and MOT of TRU with specificity, sensitivity, reliability, consistency and perfection.

Fig 4-23 Convergence on Ideal SUT and ROT

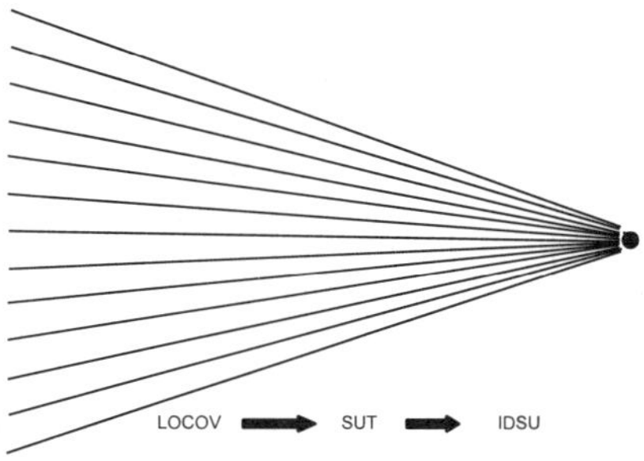

Just as all entities, senses and things must become TRU, by TRIA, life, and relativity, so must they converge on TRU. **LOCOV** = Law of Convergent Validity. As knowledge of TRU increases, both TRU and knowledge of SU and ROT increase. LOCOV is greatly assisted by the dimensional formulation of TRU, the superimposition of all aspects and parts of TRU, and by induction as it can be best provided in SUT, ROT, and MOT. **SUVUK** Signifies that by Separating and Unveiling of all the Veneers that Surround and come between us and TRU, we can attain Ultimate truth and Knowledge. When this is done, and when that within TRU becomes clear, we have what is needed to bring LOCOV to ROT, MOT, SPALT, TRIT and TIT.

In ways impressionism brings us back full circle to what symbolism began with, an idealization. The origins of art and representation in paleolithic times sprung not so much from a simple record or duplication of something they saw or experienced as it did from their felt need and desire to make permanent some greatness in their life whose idea needed communication or remembrance. In scientific empiricism there is the same common intent and attitude for precision, standardization and totality with unity. It holds that every known that is factual if connected with experience can make possible validity and certainty. The ideal of ROT and the language necessary for duplication and replication of TRU would be to have the best possible word or symbol for each separate aspect, mode, form, function, mechanism, part, process, feature, substance, essence, principle, law, need and solution of TRU. To likewise arrange, form or formulate these symbols, nominals, words, and all in such a way they would yield the most simple, efficient and precise OBT and ROT. It is to a large extent, the perfection of meaning, that which is the predominant essence, Á, of TRU. It is signified by: **PEMOR** or Perfect Means of Representation. PEMOR is more than the perfection of meaning through the 1:1 identity of subject object and TRU, even when attained through semantic, rhetoric and logic. PEMOR generally is any and all means of representation, whose goal is reached as it approaches perfection. Yet it is in the perfection of meaning, as attained in TRU through Á its mass energy relativity when the observed and the observer attain perfect or proper identity, that PEMOR most specifically applies. It can be shown by how stimulus response expands through subject object totality, TO. It is how we can through ever clearer, approaching perfect meaning, explain the DAOST of TRU.

Since the absolute nature and form of TIT is confirmed by TRIA and the thisness of TIT, and since the absolute relative focus of IA in the individual human is evident in its perspective unity, it would seem that the associated representation of TRU would, more or less, have some kind of absolute, or relative absolute form, or **RÂ** = Representational Absolute of TRU. However, since representation is neither TRU, PERT, nor POT, and since KOT is not even its potential in TIT, we know that it will assume added form and characteristics typical of itself, especially in totality. It is this added form of ROT that makes RÂ unlike TRU, PERT and POT, with its own kind of absolute. This is one in which the unity, finity, infinity, and totality of ROT although largely guided by and replicative of TIT needs to, at the same time, reflect the focus and nature of representation.

Since ROT demands the best of formal distinction with the best of differentiation, we must, for the greatest aspects, parts, processes, mechanisms, principles, laws and all of TRU have only one term with one meaning and definition, that is clear concise, distinct and discrete. It is a great proof of NEC, POT, IPTRU and TRU. It shows how we should remember the various definitions and meanings of the great ideas, concepts, terms, principles and laws we use and note or qualify them to avoid ambiguity, contradiction and misinterpretation. An author may be tested by the manner in which they manage the meanings, definitions and terms, how well these present their topic, and especially the greatness and manner in which the topics presented are TRU. To not use TRIT, OBT, SUT, ROT, MOT and SPALT or their TREQ of TIT is like driving backwards and using rear view mirrors. It demands its VAPO development for ROT and MOT. Since this is largely directed toward objects of TRU, the whole process, or the operations, are one of objective transformations of TRU. The great gist of this transformation is its cue in showing and helping us understand TRU, and by recognizing that it is neither simple nor chaotically undeterminable. Likewise we must show TRU to be able to know TRU. In addition the knowing and showing of TRU is much less effective unless TRU is not any entity, but The Entity, The Absolute, TA.

The determination and evaluation of ROT is further promoted by: **DENC** = Determination and Expansion by New Concepts. As voids in TRU are discovered, or existing aspects, parts, and processes of TRU and other entities lack VAPO, the refinement of existing and development of new concepts will greatly assists in the expansion of ROT and TRU. When we have an ordered absolute as our foundation, the evaluation and expansion of new concepts can progress exponentially. This occurs as the finites of TRU are replicated, ordered, reduced to posit as representative of their part and roles in TRU. This exponential progression is: **EXAD** = Exponential Meaning of Determination. EXAD enables words to reinforce each other to become absolutely confirmed concepts, both existing and new, when they are of TRU, and represented in dynamic totality with exquisite detail and perfection. Each object or entity in itself, as it appears, as its ESIN, as it is a part of TRU, or a representation of each or TRU, are all different and must be converted to and used appropriately by EXAD, order and DENC. The final and continuing determination and evaluation of ROT is attained through: **TOVOP** = Test of Validity of Portions of TRU. Our concepts and terms become actual and acceleratively expanded when each and all of us, more rather than less, throughout our lives, learn and use our concepts, words and vocabularies ever more perfectly. This is both qualitative and quantitative as they and we are, TRU. Our thinking and much of life at best is an exercise in the continuous processes of TOVOP, the evaluation and determination of all that represents TRU.

LATA = Law of The Absolute, or the perfect identity - and validity - of an ENIT or TIT with its object, or OBT. The degree to which there is not an absolute equality and identity between two entities is the degree of departure from the absolute state of identity. So that in the dimensional formulation, systematic unity and modulation of TRU we must always seek and show TRU as it is the thing in and of itself, TIT, a 1:1 identity and thereby LATA. Yet we must also recognize the VAPO and SCARLS of this LATA as it is reciprocal from TIT and TRIT by TROS and TRUS, as well as by the observer, the relativity of it. So that LATA, as SPALT, is valid for TRIT and TIT if it is a 1:1 model. Yet it is invalid to the degree they are not qualified by VAPO, SCARLS, TROS, TRIS, and relativity.

CHAPTER 24

DIFFERENTIATION and PARTS of ROT

A PART, PARTS of and DIFFERENTIATION of ROT

One should understand why the duration of a person's lifetime is so short and art as ROT and TRU are so long. This is because a person's life is highly limited and strongly felt in many ways compared with the continuity and greatness of art in this world of best of human life, ROT in TRU. It is how fleeting our awareness, attention, appreciation and IA are when compared with the representations that human culture and civilization has, and will ever so more greatly produce. It reflects the depth, intricacies, infinite to infinitesimal and differential nature of life. How Instantaneous Attention is concentrated on the finite and infinitesimal core of TRU. Why ROT is so great is because it gives us tremendous extra meaning in each form, part and production presented. When art is judged at its best, ROT, it is truly great because it is not only a great part of POT, THEO, and TRU, it is in being a great part of this objective half or sector of the S-O of TRU, A, a great and, by LOPAN, vital part of TIT itself. It escalates S.

It is with an advanced system of cleavage whereby all that is in TIT can be so separated as to best replicate what is different, and do so in such a way that none is left out, all is proportional, and all is most demonstrably evident, that the great task before SUT, ROT, MOT, SPALT and POT best serves TRU. Although doing, thinking, working, reasoning, making or using our intellect are all large parts of both TRU and ROT, it is how they are done as they concentrate more and more upon their function within TRU or their function as they enable us to show TRU that gives a plane of cleavage upon which their discrete forms and parts can be identified, separated and compared.

RONT or the Reduction of Nominals or words to TRU, involves not so much their obliteration and replacement as ES or CES, as it does their interpretation and use only as they are aspects, parts or processes of TRU. RONT does not mean that all words have to be only or 100% their ES and CES, and only a single part in TRU although this is often ideal. It means that to be actual and fully meaningful in representing their counterparts in TRU, we must have at least one definition that is this counterpart and function. Words as any great part, segment, or portion of a scale of dimension, in TRU can be no greater than this they represent.

The importance of the particularization of ROT can be readily felt and appreciated in music. For it is music that touches the human soul, those finer separations, distinctions, movements and melodies that are heard, felt, known, and appreciated. Music also provides other venues of representing TRU. Music is great, not only because of its appealing beauty to our hearing, and its accompanying homely and soul-searching lyrics. It is great because: 1. notes give us discrete parts like finite and formal distinction which the continuum of tone does not generally provide, 2. The relations between notes give us harmony, an ordering of discrete finite formally distinct parts. 3. The octave and tonal divisions, provide a doubling and proportions in which harmony is accentuated and unity is provided, 4. A focusing of the tonal scale on middle C provides a point of reference that exemplifies human focus, IA, and the absolutely essential relativity of human parameters. 5. Such focusing and scaling provides a quantitative way of measuring a continuum, an example of what can be done for parameters, spectrums, continuum and especially the dimensions in ALT that form TRU.

ORIGIN and DEVELOPMENT of PART/ PARTS in ROT

Once we identify and accept the fact that TRU is a whole and a unit, the next great hurdle and problem for solution is to successfully objectivize and represent it by dividing this whole. This also has a long history in religion and philosophy. For the theology of many religions endeavored to resolve this problem. Is God one, two, three, multiple, many or combinations of these? Are these part of the whole or do they represent separate Gods? If they represent separate Gods do we lose our objectivity and ability to represent them? Stated in the epistle of St. John, "In the beginning was the Word, and the word was with God and the 'Word was God". It is from an early date humans realized God was not only to be recognized, objectified, symbolized, signified and represented. These in turn, must revert to God as they are in GOD, or TIT of TRU. Representation, the word and language are much more than their represented form, they are living, actual portions of the FAF, ES and CES of TRU when and if properly and effectively used.(RÎT) They are coterminous with human life, as this life is TRU. It is this coterminality of objectivity, representation, SUT and TRU that dictates its absoluteness as they are TRU. With improvement in knowledge, language, and representation there was increased stimulation of thought, its extension to more universal and abstract yet less apparent topics originated and developed. This was the origin of reason and intellect.

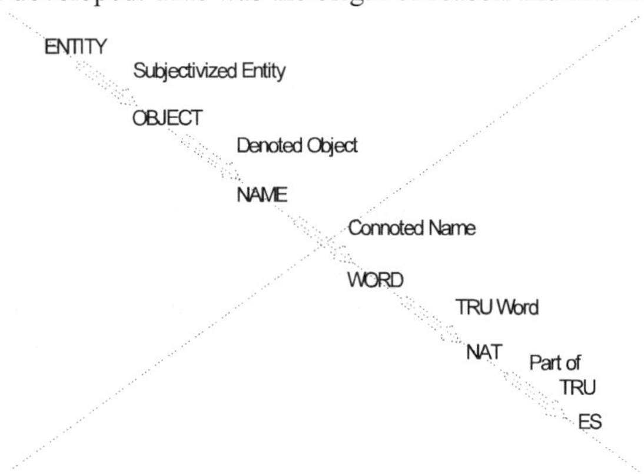

Fig 1-24 **DEVELOPMENT of ROTE and ES,**

How an entity becomes an ES or an exact part of TRU may be shown in different ways. One of the most vivid is to show the transit from entity to object, to name, to word, to NAT, to ES. The derivation, growth and development of ROT did not occur spontaneously nor is art and ROT created without learning and training. Just as the development of reason in the child is a result of growth and training, this is in the use of significant signs which are prerequisite to all logical thought. So do all artists, or those who would produce ROT, need to learn and have the training necessary to create and produce works of significance and worth. Alteration and adaptation of words to TRU is certified by: **REDRE** = Redefinition E, are different - all things RE defined by E, as we do by ESIN. As we extract the ESIN in objects and the words which represent them we undergo REDRE. We undergo a conditioning, refining, and redefining of words. As we prove E, each E, along with TRU its form and processes, we can increasingly perfect these concepts. This is: **PRE-S** or Proof of E, or ES. We should remember that in TRU these elements are not as distinct as we make them. For entities and objects, even names and words are often not TRU, poorly, or not very much TRU. **LOTESAN** or the Law of TRE ESIN, Actualization of Nominals or words. LOTESAN is the law which focuses on: **TRE** = TRU Recognized as such, and its meaning of words. LOTESAN is their TRU meanings selectively extracting all other meanings.

GENERAL FORM of ROTE

A name-entity, nominal-entity, word-entity or word-thing is called **NEN**. NEN refers to the connector or element of transition between the name of an entity or thing, its object, and the entity itself, or ENIT. NEN is brought into significance by TRU, for in the formal distinctions necessary to bring out the parts of TRU, ES, it is necessary to extract from all their essences, the part that is TRU.(ESIN) The great paradox of words is how they are based upon both the relativity of, and sameness in, TRU, and upon the degree to which they can represent its differences, by distinction and separation. To this great paradox is the added problem of words which are only partly TRU, less than TRU, non TRU and against TRU. The solution to this paradox is to be found in the mechanisms of its composition and their development. This is called: **SONECT** = Solution(s) to the paradox and problems of nominals or words and TRU. SONECT is achieved when we understand and apply both the sameness and differences in TRU, the distinctions, representations and LANT needed, and how to separate words from all that is not or less than TRU. Exercises in this and the mechanisms involved have precedent in logic and set theory, and are practiced in the use of POS.

LEMAD is the Law of Enumerations, Meanings, Definitions and Formal Distinctions. When well developed and ordered LEMAD shows how concepts, principles, axioms, laws, and mechanisms of TRU reveal the SCARLS and all of TRU. It is that law which combines the substantive exquisite detail of TRU with its optimal representation, especially as these are best adapted to TRU and posited in the best possible modulation of TRU. It is how the definitive of TRU may be obtained and utilized. **LADET** is the Law of the Absolute Derivation and Determination of the parts of TRU, ES. LADET is the stuff or substance of the universe, TRU. The nature of the stuff or substance out of which the world is made or composed was the question that so obsessed the early Ionian Greek Philosophers. In turn this began the search that has influenced all knowledge ever since. It set in motion the great quest for knowledge and philosophy which in turn set up the crossroads to paths down which many of the existing schools of philosophy have developed and radiated.

A term is a name, word or expression that has a precisely limited or special meaning and use. It is the scientific, technical, artistic, professional, and trade designation for names, words, or expressions with greater validity, explicitness and limits. It is the representation of ideas, concepts, and principles that purport to be without ambiguity, duplication and vagueness. Terms and terminology are the limitation and special meanings used in the context of a totally ordered, and often rigorous field and science. A good technical vocabulary or collection of terms, is often a person's most valuable tool as they are the vehicles of A FAF in TRU. Any profession or science is lost without its terminology. It is not difficult to acquire a good working knowledge of ones needed terminology and thereby profession, occupation, field, science or segment of TRU. It is only for a person to study, learn, relate, appreciate, memorize, understand, remember, and best use those terms that are at the basis, crux and most valuable to the field, science, life's success, or TRU.

Justification and Proof of TRU Terminology

Because of the many voids and deficient meanings in human language's coverage of the total actual form and function of all in and about our world, TRU, it is necessary to utilize the best of logical, modular and vernacular means to provide what is so critically necessary. Foremost of these is the necessary as a major qualitative and quantitative measure of all, all TRU. If one would wonder what is missing in knowledge and success in human life we can readily and accurately determine it by investigating what is and what is not, and the degree, necessary. The deepest meanings made distinct of these needs, when inadequately represented by a concept and term is what is used to derive NEC, and how a new symbolic logical system, guided by the modular form of TRU and the vernacular modern usage of the English Language enables us to produce a standardized, rigorous, effective and precise representation of TRU. **LANDIM** - Law of New concepts, terminology, and Notations Determined from Deepest Meanings In Necessity. **LANMAV** - Law of New Concepts, terminology, and Notations provided by the Logical, Modular, and Vernacular Means and Measures Necessary in This Special World of Best Human Life.

LADEX is the Law of Definition to E or ES, Exclusion. LADEX is how we reduce all words, terms, and even representations generally, to their meanings and definitions that relate to us, in TRU. What the term is as its denotation or object is and is only, exclusively, its part, process, form or mechanism in TRU. **LADES** is the attainment of ES through bringing known, and new concepts and words, to their role in TRU. This is done in many ways, one of which is by broad interpretation from **UMIA,** or Universe, Mankind, Individual and Instantaneous Attention, so that their participation and REM, or valuation, can be properly and proportionately assessed or accounted for.

SPECIAL FORM of ROTE

The special form of ROTE includes those kinds of representation of TRU that are identifying, specific, direct, precise, significant and important agents for showing TRU. These include, principles, valence, meanings, definitions, and divisions. Principles are almost entirely specific aspects of knowledge, the underlying significance of all that guides a field, science, topic, or the given. Through the extension of unity, a unity provided by TRU, an ordinal divisibility is possible. This is called Valence. Valence is not limited to a single or a few levels of divisibility. There may be bivalence, or tri-valence or whatever number best provides representation. Valence is of great significance as in two-valued logic, three-valued logic, or as in a model of TRU based on three dimensions in ALT and SPALT.

LAPDAM = Law of Appropriate or Proper Definition and Meaning. Words can't be defined nor have meanings that are limited to only the intuited object. They must, if they are to be actual and more than illusions or figments, include in their definitions and meaning their implications in TRU, POT, PERT, ROT, MOT and HOM. Just as TRU is so much better known if its parts are known, so is the representation of any object, and especially those of TRU, to be largely found in these divisions, separations, parts, and processes. A large part of definition and description lies in showing what an object is not, in contrast with what the object is or comparatively clarifies it. Definition by recursion applies to the logical method and signifies any representation that can be actuated by showing how it contains certain properties or functional values. As these are common for all within that defined they add essential information and prove the existence of it.

LOCAS = Law of the Coincidence and Separation in the Representation of TRU. It is necessary to represent TRU, and to understand it by focusing its aspects, FAF, parts and processes forward and backward, inward and outward, same and other, integrated and differentiated. This is how we develop TRU from DAOST objectively and completely. Here is definition, distinction at its greatest, finest, and best. LOCAS gives us the basis whereby rules of logical divisions can be best fulfilled and applied. For without LOCAS there is no final nor unified goal to which they can be directed. To survey and assess these Rules of Logical Division we must analyze each as they are modified by LOCAS, including the exclusiveness and exhaustiveness of divisions. The rules of division certify a propensity in definition that can well activate the potentially representative of TRU. From the rules of logical division we can better provide the necessary processes of separation and distillation of meaning from words. This is: **PAMEDS** = Processes of Separation and Distillation of Meaning, from words. PAMEDS are those processes guided by rules of logical division whereby the meaning, definition and reasoning from words are held to optimal representation.

Although deduction, the deductive method of both logic and science, has proven highly effective in guiding proofs, by building one premise on premises already proven, to discover fundamental truths, the traditional use of deduction, and induction, can't be abandoned. For in a closed set, with limits, as the relative universe in itself, there exist continua whose extremes from unity to infinity, from universal to partial and from general to specific require qualified and modified methods typified by the long time usage of deduction from universal or general to partial or specific and induction from the partial and specific to general and universal. This accords with differentiation and integration so effective in mathematics, calculus, and the management of a set, the whole our actual world, TRU. The one kind of deduction does not rule out the other. Such problems in language only show the need for its modification, added, or new language.

LARSÄDT = Law of the Special Representation of the underlying, basic and ideal Form and Function, FAF, of TRU. LARSADT is seen in artistic idealism and has become accentuated by ROT whereby representation is to be made of all that is of an object, not only what its features, motions, or detail show. LARSADT is a critical law of TRU, its KOT, ROT and POT. It is largely for lack of LARSADT, ESINS, LANT, and NEC that the actual world or universe, of TRU, has never been adequately discovered, learned, known and shown. By the unity of TRU systematic order and division of parts are given definite form and position. This enables a precision guided terminology.

PARTICULARIZATION - DIVERSIFICATION in ROTE

The rating, evaluation, and analysis of words, terms, existing concepts and new concepts may be done by how much they are: TRU, important, function in TRU, function with other ES, lessened by coterminality, reliable, fit meaning or definition, multiple meanings, clarity, popularity, parsimony, vocalization, construction, precision, and universality. The success of S-O, or Á FAF largely depends upon how well these qualifications of terms are made. Definitions may also be classified by how they are: from dictionaries, ostensive, genus or species, abstraction, recursion, contextual, lexical, stipulative or various combinations of these. **MARDU** is the Means and Ability to Represent and Demonstrate the Universe all things and TRU.

Knowledge, although not all TRU, is essential to TRU. To understand TRU, and to visualize it, it is necessary through perspective and vantage to visualize it in complete context. So that in representation, our S O and IA must separate the increasingly finite to infinite differentials, to best utilize and show TRU

.DIFFERENT PARTS of ROT

Abbreviations provide both simplicity and power for representation and language. Their proliferation plus the need to spell them out rather than to pronounce them, especially when of increasing letters, reduces their effectiveness. Likewise there is an optimal communication value of language in which the benefits of more than one or two letters or syllables for abbreviations and words give their own quality and strength. The optimum of representation, is thereby, a balance, like that in TRU and all, between unity on one hand and infinity on the other. It is the balance between a one letter signification and an encyclopedic or lengthy description. The need and demand for parsimony is powerful. This limit is of vocabulary and understanding. It is even more, a limit of IA, immediate awareness and attention. For in the chain of consciousness and thought there is the ever present need to maintain great freedom of movement, ID and OITIO. This can be reduced by holding ones IA on TRU, realizing TROS, and developing the SUT, ROT, MOT, ALT, SPALT, POS, STA, HOM and ACC that best order and manage IAT. Abbreviations are widely used in language and representation. Computerization has stimulated an even greater expansion and utilization of abbreviations. Whether a perfected or ideal ultimate form may be developed should not detract from the importance of perfecting effective abbreviations and acronyms for much that is lacking and necessitates representation in TRU. In order to show and use what the different parts of ROT supply we need look no farther than how words are made up. This quickly leads to a major innovation of this text, the full use of all sounds, letters and especially vowels as they both widen and make more efficient a precise representation of TRU. Assuming that the word LÁRA exemplifies the many combinations of terms that can be used to show how and when for each vowel two or more syllables produce a whole array of simple and strong concepts. We must be careful to recognize these, not as duplicates, but as potentially formal distinct concepts, often with great gist and cue to their role in TRU.

	Á	Â	Ä	À	Å
Á	LÁRÁ	LÂRÁ	LÄRÁ	LÀRÁ	LÅRÁ
Â	LÁRÂ	LÂRÂ	LÄRÂ	LÀRÂ	LÅRÂ
Ä	LÁRÄ	LÂRÄ	LÄRÄ	LÀRÄ	LÅRÄ
À	LÁRÀ	LÂRÀ	LÄRÀ	LÀRÀ	LÅRÀ
Å	LÁRÅ	LÂRÅ	LÄRÅ	LÀRÅ	LÅRÅ

Table 1-24 **PHONETIC DIFFERENTIATION and UTILIZATION of NEW CONCEPTS**

This diagram of differentiated terms, and its use for TRU Terminology, is important in more than one way. For ROT it reveals both an optimal use of language for representing the differential of TRU. It also reveals much of the principles and mechanisms of ROT both in itself and in TRU. The letters stand for: **Á** = The Mass Energy Primary Dimension of TRU, exemplified by predominant relativity, **Â** = The Absolute, both generally and specifically, **Ä** = Form, in general and as adapted in TRU, TÄ, **À** = From, as from TRU ÀTRU, and **Å** = as a copula.

STED = SU and TRU E, ES, words, nominals, new concepts, Dependency. There are limits and thereby errors in the use of synonyms to define a word by another word. If a word means the same thing as another word it is unnecessary. If we qualify these definitions and synonyms by explanation that is only an indication, a similarity, and go on to show other words and to give fuller and fuller explanations, as the value in an exhaustive dictionary is contrasted with a small one, then these meanings gain greater value. Overlapping or coterminality typical of the sames of TRU is valid and acceptable if done with parsimony. **RECONET** = Reconciliation, Complete of Coterminologies and Overlapping Nominals, ES, Parts of TRU. RECONET, as all that is substantive to TRU, is a continuing endeavor, a great part of life. Only its form as it is a part of the human brain, speech, and expression has developed to its own advantage direction, and purpose. Yet with maturation language will come to TRU and follow, be more efficient, effective, and expand RECONET upon existing Á FAF of TRU.

MEASUREMENTS of ROTE

ESI is Essence S Index, ES Index. The examination, investigation, evaluation, interpretation, systematic massive determination and fitting together of ESI for all the greatest word-things, NEC, and ES of TRU is the basis for the understanding of POS, STA, HOM, ACC, POT, and thereby TRU. It can be carried to a very effective, advanced and productive degree in demonstration by ROT and MOT. An Infrastructure is the basic underlying framework and features of an entity, object, body or system. For TRU this is provided by both its divisions, polarities, and parameters. It is only when by ROT and MOT, we can relate, order, and totally show TRU that it has greatest worth and utility. This is called: **ROMEA** = Related, Ordered, Modulated Measures. ROMEA signifies and certifies the absolute necessity of measurements in ROT, plus measures that are not less than well related, ordered and modulated. Number and SUT are closely related, for all generally by TRIA, is convergent upon TRU and TIT. More specifically SUT being the coordination of dimensions in TRU can be readily reduced to scale and number. These are based upon the TROS ordering of TIT, and are from infinitesimal to infinite by different combinations of number systems available to ROT. Some will measure TO and some various portions of TO. They may be based upon cleavages or divisions of TRU that is: from one, cardinal or fractal, binary, tertiary or quaternary cleavage, quadrantal combinations or 4 x 2, as provided by the eight sectors used in Sections VI-VIII, a variety of dimension, as in SOID, or from a base of 10, 100 or 1000. When carried to set exponential levels we can get both a mathematically precise measure of TRU as well as a VAPO demonstration of its component parts. For even when at the level of 10, 10x10x10 for each of the three dimensions, provides 1,000 parts, and 100x100x100 provides 1,000,000. The needed intricate exquisite detail of TIT, mostly of its ES and CES, can become readily apparent and realized.

Similarly by investigations into the correlations between parts of TRU a massive approach helps to demonstrate the finite makeup of TRU. 100, 1000, or more correlations may be made between different parts of TRU. They will confirm qualitatively, and quantitatively, the relationship between the sames of TRU, as by DAOST. Such confirmations are key to discovery and filling the voids and unknowns which have so eluded human knowledge.

ROT and the DIMENSIONS of TRU

The 3D form of TRU, ALT, is derived from mass, length, time or mlt, whose concepts include:

1. **m** = mass, energy, mass energy, **Á, m_1,m_2,m_3, $Á_1$,$Á_2$,$Á_3$,**
2. **l** = length **l_1,l_2,l_3,** length, area, volume,
3. **t** = time **t_1,t_2,t_3,** time, duration, variation,
4. **lt** = space time. (correctly **l_3t** or **$l_n t_n$**,
5. **ALT** = Mass Energy, Length and Time in TRU, also
6. Nullity-1/Finity-Unity-Finity-Infinity,
7. Zero → Infinitesimal → Fractional → One → One → Cardinal → Infinite → Absolute,
8. **REM** or Relative Magnitude, from one to infinite, as parts of a whole, or from a whole as one to a whole as an infinite number or parts
9. **IA** = Instantaneous Attention - Awareness - Action,
10. Subjective - Objective, Observer - Observed, Relativity,

These are various ways and means we can divide TRU into equal or relatively equal parts. We can collect, fuse and ultimately formulate them to provide an optimal representation in terms of a useful object. ALT provides the 3D coordinates for a spherical model, mathematical relations based upon one. REM and IA provide its focus, and the subjective objective, both its external and internal form.

The Á, L and T dimensions, like m, l and t provide the text's chapters interrogative coordinates and locations of: **Where** = l, 3+,4+,5+, **When** = t, 3+,4+,5+, **Why** = SU, Formulation, EPI, COL, COR, TRU, 0+, 1+, 2+, **What** = IA, SU, 00, 02, 03, 04, 05, 1+, 12, 13, 14, 19, 2+, **How** = Causation, Homation 5+,6+,7+,8+ **Which** = AUS, RUS, TRUS, TRIS, 06, 07, 1+, and **Who** = IA, 05, 06, 1+, 15, Ind, 0+, 5-8+, 9+. Each dimension may be divided and taken from pole to pole of furthest relative extensions. For S-O and the Á dimension this is:

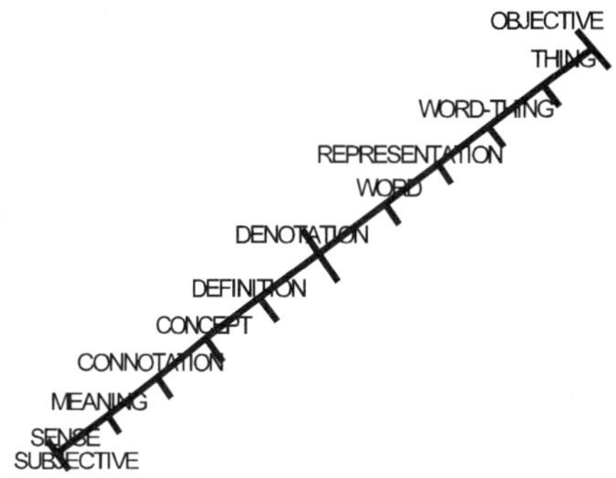

Fig 2-24 **The MASS ENERGY DIMENSION in TRU**

If mass could be reduced to length and time there would be no need for a 3D ROT and MOT of TRU. Likewise l and t would have to explain the form and function of TRU, which is not possible without mass or the special Á for TRU. Those aspects, modes, essences, and parameters that give TRU specificity make it into an object that can be readily represented. It is most evident that what is mass for physical bodies does not fulfill the kind of mass it has for TRU. The dimensions for ROT can only be replicas of its special mass, length and time, or: Á, L, and T, as they exist in TIT. We do not even have to perfectly equate the 3D of MOT, ALT, and SPALT with that of ROT, SUT, OBT, TRIT and TIT, as long as our standards are correct, made known and best controlled. It works to a great extent in logic whose function through words and their operations, sentences, inferences, syllogisms and propositions are the processes and large dynamic parts of the Á dimension FAF of TRU. The answer to the fundamental questions of the dimensional makeup and frame of TIT are largely found in the proofs of TRU, TRIS, SUT, Á, ALT, ATO, ROT, MOT, REM, HOM, POA, POT and IPTRU.

SUBSTANCE of ROT and ITS ENHANCEMENT

POSA = Principle of the Subinfinite Formative. POSA signifies the way in which knowledge and learning tend to render the subinfinite or finite formative toward the infinite. As the meanings and relationships to be posited are made known, they expand manifoldly, toward the infinite. This would be good if this could be managed in an orderly and controlled way. Yet without TRU, or even with TRU without an ordered and controlled ROT, or without an ordered and controlled viewer, it becomes disordered, uncontrolled and unknown. This is the world we live in, have known, and must overcome. By data analysis we collect masses of information, knowledge, and detail, and work to select, classify, order and utilize them in various ways practical for the problem and programming at hand. Such a collection and operation is necessary if we are to make an exquisitely detailed yet orderly study and presentation of the topic and science of TRU.

TIME DIMENSION, PROBLEMS and SOLUTIONS

Length generally can be readily divided into three kinds, length, area, and volume. For time and mass this, although used in dimensional formulation, is less evident. For time this is: of 1. brief focus, TIC, 2. linear time of indefinite extension, TIS, 3. exponential time, TIEX, 4. past, present and future time, 5. infinitesimal to infinite, time, 6. time, now to infinite future, 7. time Infinite past to now, 8. time infinite past to infinitesimal now, 9. time infinitesimal now to infinite future, 10. time infinite past to infinitesimal past present through present to infinitesimal present future to infinite future. 11. TRU and TIT time, TIZ, 12. mixed time, TIV. Science has become increasingly aware of these differences yet has failed to bring them to set. That the human body is built upon a special and set mix of these times is becoming more evident.

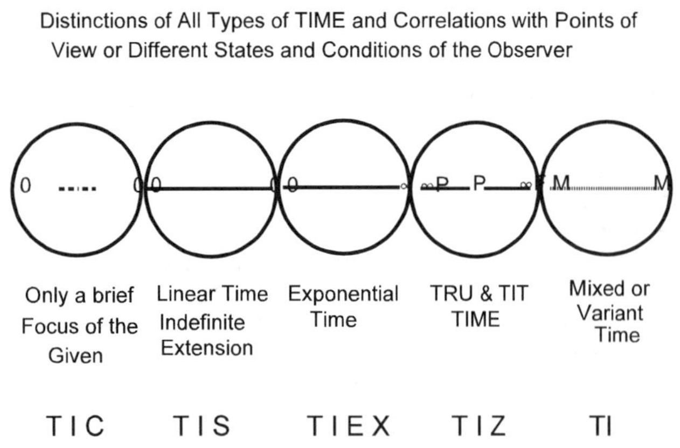

Fig 3-24 **KINDS of TIME, TIME DIMENSIONS**

The problems encountered in analyzing time not only help to direct us toward their solutions, they help to direct us to the necessary basic understanding of the form and function of TRU. In turn, increased understanding of the underlying form and function of TRU helps us understand and explain what time actually is and covers. We investigate in ROT and especially MOT and ALT the pros and cons of time. Most of these problems are solved when brought out together, and thesequence and features of RUS, TRUS, TROS, TRIS, TIT, TRIT, OBT, SUT, ROT, ALT, and SPALT made distinct.

	CON	PRO
TIC	Can not be used as a measure of a AU, RU or TRU Variable set	Shown by Personal Consciousness and IA
TIS	Time Dimension of TRU will have no focus when time is linear and the extension is indefinite	Common time, segments of which become measures of our lives
TIEX	Time will lack adaptation of a mid-point in the past and future Does not extend through the present from the past to future correctly	Make up of much of the world Especially the physical world that is relative to humans
TIZ	Not fully established Must be continually refined Will vary slightly with TRUS	Readily and excellently temporally posited Individual or Personal Identification is made much clearer The present or here and now aspect of the universe is greatly and properly emphasized Proved by TIT, and TIT proves TIZ
TIV	Has no focus, or can not be used for a fixed or set universe	Becomes valid when a set focus is removed

Fig 4-24 **PROS and CONS of DIFFERENT TIMES**

PROS:

1. Readily and excellently temporally posited,
2. Individual personal identification is made much clearer,
3. The present or here and now aspect of the universe is greatly and properly emphasized,
4. Subjective time is

$$(0) - \infty - 1/\infty - (0) - 1/\infty - \infty - (0)$$

5. IA can only be ultimately (0) - 1 - (0),
6. More readily adaptable to Homation,
7. Can better posit events, ES, CES and all in terms of their course in TIT. This is how past present and future, cause and effect can be better posited teleologically.

CONS:

1. Impossible to adapt CES, ES and word things to precise posit, they are more durational,
2. Half lives do not necessarily exist equally past and future, and are not well established,
3. Length of time does not correlate well with infinite to infinitesimal to infinite time,
4. Bifurcation of all time elements and parts of TRU necessitates much duplication,
5. Unable to attain a good comparison of densities, e.g. explosive increments,
6. Does not clearly show how TRU necessarily corresponds to the S or IA of the individual

Thereby we are directed to investigate:

1. Is the world, TRU of time 0 - ∞ or of time PPF, or... ?
2. How do CES, ES, REM and word things adapt and exist in these dimensions of TRU?
3. Are there any transformations between kinds of time?
4. Is there any transpositions among TIO, TIS, TIF, TI.., TI..?
5. If we utilize both TIS for TRU and TIF for the individual how do we maximally extrapolate, superimpose and unify time and the dimensions of TRU?

SUMMARY of ROTE

The world is not impossibly complicated. It is more or less complicated depending upon how we look at it, how much of it we are seeking to produce, how well ordered it is and how controlled our search is. If we know the limits of both it and ourselves, and live a life that is optimally TRU yet practical, the world is a challenge that we can profitably and enjoyably meet and reproduced. Because of the absolute and infinite nature of TRU, TÁ and TI, and its relativity or permeation of sames, by DAOST, there is a more or less infinite capacity for correlation between all in TIT. To be able to manage and provide this requires: **LOC** = the Law of Overall Comprehension and Control. LOC is provided by TÄ, TRIT, OBT, FAF and SUT of TRU, and our ability, conviction and determination to achieve greater quality and quantity of correlations and order. The relationship between RUS, TRUS, TÄS, TROS, TRIS, TIT and TRIT as they best reveal each others representation is: **CORDAT** = Correspondence of the Representation of the Forms of TRU, and: **CRU** = the Convergence through RUS, TRUS, TÄS, TRIS, TROS, TIT and TRIT that provides ROT. It is in: **PROFAU** = the Processes of Fragmentation and Unification of TRU that we can finally see TRU as it actually is. Through PROFAU we have for use a primary means of concentrating upon the end product of ROT. Through implication and entailment, as CORDAT, CRU and PROFAU show ROT and TRU, we can expand our showing and knowing of TRU toward completion. In systems of strict implication and entailment we can lay the basis for the great formulation and modulation of TRU, MOT, ALT and SPALT, and their uses. As we discover more and more of the parts of the whole, TE, separate them out with finer and finer delineation, we can by finding major divisions between them complete the cleavage that is so critical in enabling us to formally construct and represent, manage and control TRU and all.

The origin and development of TRU is to be first sought in ATO. Since TRU is all by TRIA and TÂ, so likewise to a large extent must ROT be seen at once as a whole if it is to be by VAPO truly and completely representative of TRU. This corresponds with the fact that by: **ATI** = At once TRU Infinite is a basic fact, condition and state of TIT. It, as well, is its origin and representation, through: **ORI** = Ordered from Infinity, the Origin of ROT very gradually proceeds. ORI is readily seen as the myriad to infinite varieties of representation and ROT that exist or are possible yet must be brought into some kind of order both from the nature of TRU, as well as its ROT and POT. This is S or subjective, only as ENIT must in being observed. In being made into the observable object, O, ENIT is brought through its ESIN, ES, CES or its corresponding aspect in TRU to order, unity and the best possible kind of whole, the TO of TIT.

CHAPTER 25

THE PART and PARTS of ROT: SPECIAL or PROPER FINITE of ROT

INTRODUCTION to and HOW THE PARTS of ROT are CONSTITUTED

Since we are unable, on the one hand to attain perfect mathematical simplicity as in one equation, and on the other to have massive uncoordinated knowledge, it is necessary that we seek to bring these two together and develop mathematics and representation to most orderly and precisely fit, TRU. This together with reducing ROT and TRU to SUT, TRIT, MOT, ALT, and SPALT with a maximum of POA and POS. The dynamic posits of TRU in such a model is Homation, the same or, identical with what exists as CES in TIT-TO. **POMATH** = Paradox Of Mathematical Precision. POMATH is where measurement, although it can attain 100% precision of material, physical and objective things, is never 100%. For by TRU nothing is absolutely objective, physical or material. TRU is a composite of which its form parts and mass are mixed subjective and objective, of many different energy states or levels. Since mass for TRU can not be given 100% objective measure it is, to a larger or smaller extent, subjective. Yet since TRU in ROT is taken as the object this can, to a large extent, be corrected. We can, as by diaphantine equations or formulations, provide an increasingly accurate measurement, solution, or resolution of unknowns and incompletely known parts, elements, forms and factors of and in TRU. This is called: **COPMATH** = Correction of POMATH. From COPMATH we may conclude that TRU itself can not be reduced to complete mathematical precision. Yet it can be increasingly reduced and formulated, mathematically through POA, TRIT, OBT, SUT, ROT, MOT, ALT, and SPALT, for most all practical purposes.

Fig 1-25 **NUMBER SYSTEMS and REPRESENTATION**

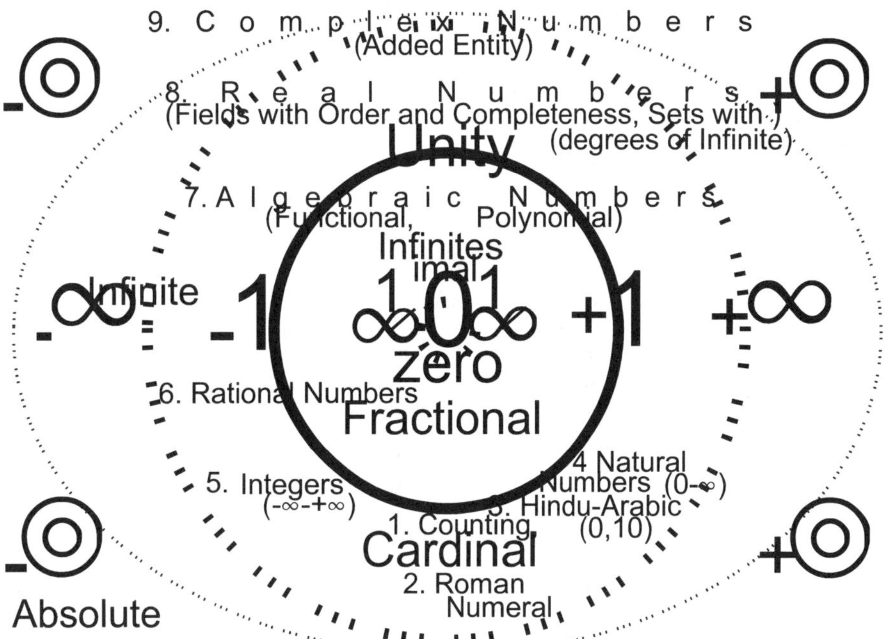

All kinds of numbers, from the simplest digits through rational and irrational, to the most complex numbers are representative of some aspect, form, or function of the relative universe. Each has their significance that is often highly indicative of certain fundamental categories, aspects, modes, forms, structures, spectra, continua, or mechanisms of TRU. Their discovery and showing not only resolve many of the greatest problems and paradoxes in human life, they are essential to an accurate and successful comprehension and understanding of the relative universe, and all.(TROS, TU, TE, TI, TU, TA)

Through human development, there has been a progressive expansion of mathematics and the way numbers are conceived. From a primitive counting (1). as by using ones fingers, man early found many ways to solve problems. This is shown in identifying possessions, and the exchange of goods. (2). The numeration based upon ten, in Roman numerals. (3). Formalized numbers and their operations as in algebra by the Hindus and Arabs, ordered and standardized numbers with zero and ten as bases. All expanded the use of: (4). Natural numbers from zero to infinity in uses of mathematics. With Western Civilization there has been a rapid expansion of knowledge of numbers, stimulated by those problems, technologies, inconsistencies and contradictions that existing systems did not solve. This included the need to account for: (5). Integers using negative numbers, from negative to positive infinity, (6). Rational numbers, including fractions and partials of wholes, (7). More advanced Algebraic Numbers including functional and polynomial numbers of various kinds. It has only been in the last century that astute minds in mathematics have devised systems to explain: (8). Real Numbers, with higher degrees of infinity, fields, order, completeness, sets and, (9). Complex Numbers with added states in totality, that number systems must include if they are to encompass all. (10). ROT and TIT-TO numbers based on their differentiation, form, function and integration as altered, adapted and warped to TRU.

Along with rational numbers, irrational numbers constitute a continuum, as long as the real numbers, they represent, have a one-to-one relationship with points on a line. The number of points on such a line need not be of a set quantity. Such is the arbitrary partial distinction of representation. This is **LEDOR** or the Law of Partial Distinction of Representation. The discovery and development of irrational numbers and continuity were largely the work of R. Dedekind. 1872. Whereas TIT-TO is characterized by infinite to infinitesimal differentiation, partials and especially their representation are typically finite. It is in this way we can treat much of TRU - and all - as discrete, although they are not so perfectly in TIT-TO. Whenever such is done, however we must ECRO, their limited, partial, representative and non TIT-TO nature, form and function.

From the refined exercise of logos or the reason that exists in and of TRU, and becomes such a power in Á and TIT, logic is the philosophy of rational argument. It is, as a basis and guide for numerical and axiomatic reasoning, that logic lays the foundation for much of mathematics and ROT. It also includes the clarification of the terms of formal criticism, together with the invention or specific forms of inference and rules for their use. Logic is both TRU and EPl, POT, ROT, and HOM. As ROT, in SU, logic is primarily a means or method to provide a clearer representation of more complex aspects and relations in TRU. It shows how through the systematic reduction of language, argument, propositions, inferences, statements, and syllogisms, we can so coordinate and connect terms, which are representations of the ES and CES of TRU, that will not only be valid as well as precise; they will together form greater systems that represent TRU. As great as logic, mathematics, geometry and language are as means of representing TRU and as systematically unifying as they are, all have limits. These are limits that can be both general and specific for each. If we can comprehend these limits and develop logic, mathematics, geometry and language to their fullest, apply them well as necessitated by and indicated for, TRU a clearer representation of TRU can be produced. We can approach that absolute precision math requires in the fulfillment of TRU. Together this is their great inclination and necessity.

The breadth of number systems and logic is analogous to what we find in TRU. For there is a similar expansion in history of religion from diverse deities, to a simple monotheistic God, to one in which God has a finite, then infinite form and function, to one that is TREQ, has a harmonious TROS in TIT. To actually understand TRU, however, as we do for math, we must resort to means of ROT that include all, from the most integral to the most differential, from the most relative to the most absolute. It is in how such number systems and logical operations parallel and can be used to support and prove both TRU and its ROT and POT, that is key to TRU and POT. They show the central role math plays in all sciences, especially that of TRU. Some of the Logical Terms that apply to ROT and TRU include: Logic of Cardinality = Comparability, Logic, Modal = Systems Interpretation, Logic Modern = Sentential, or argument form valid only if conditional antecedent and consequent is a Tautology, Logistic Systems = Two types: I. Propositional (or sequential) Calculus, II. Functional or Predicative, A Priori, Transitive and Relational. From these we make deductive arguments of the kind: *'B greater than C, A greater than C, therefore A greater than C'*. The notations used, the manipulations of these notations as in propositional calculus, the different types of treatment of syllogisms, forms of inference and opposition, categorical syllogisms, the theory of types, algebra of classes and relations, functional calculus of various orders and combinations, and finally their overall uses and integration as logically ordered by A in TRU are the tools logic provides in demonstrating through ROT the TE and TO of TRU.

MEANS and STANDARDIZATION in The Parts of ROT and TRU

With relationships, classifications, standardizations and systematic unity the frame of TRU can be established. The ordering of representation, as especially applied to TRU, has been greatly enhanced by computer science. This is seen in: Language and Communication, Information, Information Flow over Discrete Channels, Coding Theory, Reliable Transmission and Redundancy, Continuous Channels, Information and Entropy, Mathematical Theories of the Living Brain, The Computer and the Brain, Automatic Computers, Analogue and Digital Machines, Neural Networks, Turning Machines, Intelligence Amplifiers, Learning Machines or Perceptions, Artificial Intelligence, Game-Playing, Translating and Utley Machines, Computer-Aided Design or CAD, Multimedia, Internet, and many others. The strength of computerization lies in both its ability to make information more available, vastly extending our capacity, and its many venues such as word processing, spread sheets, graphics, and their extensions. It adapts many methods previously used for representation. It provides more intricate, precise, efficient and massive applications in the overall design, utilization and management of most all fields and sciences on a highly refined technological base. It is a variety of ROT with almost endless possibilities that, especially when applied to TRU, ROT, MOT, POS, HOM, and ACC, provide mankind with an escalating mastery over the world around us. With its many spinoff revolutions, knowledge through electronic media and computer science, now spearhead the higher levels of Á, the great mass-energy revolution in TRU.

DERIVATION and METHODS of The PARTS of ROT and TRU

Reliability of anything as a measure is a general case of the more specific ratio of monetary value to TRU value, and **MERS** or the Measure of a System or Religion. These are dependent upon several factors: the characteristics of measures, their conformity and identity with TRU, and the use of measure itself. These are generally found in the methods of science, mathematics and statistics. Their best development and use provides measures of TRU with both increased accuracy as well as increased justification, proof and accentuation of TRU. **MEDOT** = Method of Exact Reduction to TRU through exact Definition of Opposition and Exact Equality with TRU. MEDOT consists of two parts, which when taken separately are: **MEDO** = Method of Exact Reduction of Opposition, and **MEDÉT** = Method of Exact Reduction to TRU through Exact Equality with TRU. The dynamic same produces a uniform disposition. When extended to the polarities of TRU these methods are called: **MERPOT** = Method of Exact Reduction of Polarities to TRU. The reason why these concepts are important is the necessity of having well delineated and discrete form and parts for ROT and TRU. To do so we must separate all that is of TIT and TRU. To do this we must use its differences, any and all cleavage planes, and all that we can find of TRU that shows how it can be taken apart. Algorithm is a mechanical procedure with a finite number of steps; a computation from certain types of data to certain types of results. It is by the use of algorithms, of many different and complex varieties, that we can represent TRU in a controlled way. This is best seen in POAT and PACH.

One of the greatest problems of all for mankind is the 'Argument from Illusion'. It is the use of perfectly logical and precise reasoning in similarly excellent ways yet resulting in inferences and syllogisms whose application is useless, impractical, or erroneous. It is a product of language, semantics, or of the words themselves. They are used in ways that lack VAPO. Many philosophical dilemmas, paradoxes and problems can be eliminated simply by recognizing the limits of their applications and how they must be reduced to terms that have explicit TRU meanings. It is the perpetual task of all who think, write, work with and use words, language and logic to do so first by eliminating their shortcomings, errors, and then by identifying and utilizing all as they are in TRU for the best ROT and POT of TRU.

Aristotle's Two Fundamental Laws of Logic were: 1. The Principle of Contradiction, a proposition can not be both true and false, 2. The Principle of the Excluded Middle a proposition must be either true or false. These may be true for logic, yet for TRU they are no more, and often less, fundamental than the degree to which their propositions are 100% TRU. For TRU they should be limited to first principles or truths that can be directly verified in themselves, as well as verified of TRU. The primitive formulae of formal logic and propositional calculus, as: AB, A, A v B, are not only a great part of ROT, they can be adapted for POT, TRU, MOT, ALT, and SPALT. The notational system of formal logic shows: the relations between things, what is and what is not, proof and test, plus inferences, and effectiveness. e.g. A and B or p and q, may be A and T, or AU and TRU or ATRU and TRU.

SETTING, CIRCUMSTANCES and STANDARDIZATION in ROT

Divisibility is that quality or state of entities and objects which may be used to separate and reduce them into their most prominent portions. It is the processes or mechanism by which wholes are separated into their parts in such a way that does not necessarily severe their relation to the whole. Divisibility is a key to our being, our predominant mass-energy, A, relationship and dimension. For it is not only a key dynamic mechanism in our transformations and continuity, it is that mechanism whereby the necessary transformations and continuity of relativity occur, namely the observer-observed, or S-O, Á mechanisms. To best produce and utilize divisibility requires the optimal separation of all that is TIT, its representation, formulation, modulation and use in return, to TRU. It especially requires a divisibility that utilizes the primary and most discrete aspects of TRU. It also requires the sequence subject and object, cause and effect, stimulus and response, and problem and solution that are common in nature and its investigation. In medicine this is called, **SOAP**, or Subjective, Objective, Assessment and Plan. SOAP is closely related to POA. It is a highly successful tool, for it duplicates what exists in the nature of TRU. This is the Á dimension in which we start with the observer or subject and fix on the observed or object to determine or assess what it is and plan action that is most necessary and practical.

FOCUS, CONCENTRATION, LOGIC, NUMBER, and STANDARDIZATION in ROT

Although TIT is composed of infinitesimal gradations its divisions are also composed of divisibility and levels. Likewise relativity, observer-observed, S-O, SOT and Á seek to identify these divisibilities and levels. It is in this combined sense, and nature of TIT, that we must accommodate all that is TROS, the infinitesimal, infinite, finite, cardinal, fractal, unit and totality of TIT for ROT. **UNKT(RU)** = That Unknown of TRU. The problem of how we look at TRU arises from whether it is from an infinite distance so that it looks like a dot, from increasingly closer distances becoming a larger and larger object, or it is from within TIT-TO in various ways. There is a correlation between the observer and an object's m and t dimensions, as well as the higher powers of all dimensions. This is called: **LOREL** = the Law of Optimal Relativity. An object is best viewed or sensed from that distance, time, and mass energy in which its divisions, insides and outsides, as well as dynamics can all be best revealed. This, to a great extent, is a function of one's visual fields and senses, the experience of the observer, as well as the nature of the object.

LACFOF = Law of Absolute Certainty, is a Function of Focus. When LACFOF is demonstrated being positive is provided. This is the best of positivism. For all can be brought together for posit, when known with certainty and posited in a perfected model of TRU. There are many functions in TRU for which formulae have already been discovered, developed and used. These may require some adapting, qualification, and modification to TRU. Yet their value is great and well proven. It is in the interpretation and proper application of all appropriate formulae that the greatest care and control is needed, and benefit produced. Examples exist in many fields and sciences, especially medicine including: osmolarity, anion gap, acid base balance, and the standard deviation curve.

What continually improves and perfects formal distinction, definite relationships between the subject and object, subject and predicate, or how thoughts and sentences are formed, shows the close association between logic and Á the mass energy dimension, continuum and FAF of TRU. Logic seeks to determine definite relationships between the subject and object or predicate. When logic becomes less symbolic and more mathematical it seeks to provide as much correlation as possible by reducing to precision numerical operations. This requires, 1. Definite subject, a definite object and a set relationship between them, and 2. A Point of Reference. The point of reference must be a focus or center-point through which all can be related. It is not a simple single point as much as it is a collective and massive convergent body. For much of TRU this is the individual person as they are their counterparts, TES, in TRU. It is a collection of mind and body as they are the focus of attention, IA. Plato states that knowledge involves the unchanging by reference to which the changing is understood. In having a focus, point of reference, or center-point, as is necessary for any relative universe, and in representing it, there will be relations, positions and vectors. There will be up and down, inside and outside, left and right, which will need to be set arbitrarily with optimal standard. These are the coordination, order and the standardization upon which all must be based. It is orientation and coordination, the subjects orientation, and coordination with the object, the basis of the mass energy FAF of TRU, its Á continuum and dimension. This can be simply proven by the coordination and equilibrium of all animals, and the conscious, adaptive and purposeful form and function of humans. It is made up of many levels or stages of energy systems, SÁ, as they are adapted and altered to best suit human life.

THE PARTS, COMBINATIONS, MEANS and EXTREMES: SETTING in ROT

A language that represents the world can be developed through the geometric method of axiomatic logic. This would largely be what TRU consists, its ES and CES, FAF. It emphasizes the greatness of a concrete explicit and logically perfected language in providing for ROT and POT.

LOGICAL NOTATIONS of FORMAL LOGIC: pq = p and q = Conjunction, v = p or q = Inclusive Disjunction or both, + = p plus q = Exclusive Disjunction, p & q, but not both. ~ = Not p or Not = negation, ⊃ = If p then q = conditional, ≡ = p if and only q = biconditional, I = Not both p and q = alternate denial, O = Neither p or q = nothing, zero. Logical notations, as symbols, provide a valuable system for reducing propositions to simplicity, set, and order, habits of thought, and the relationship between subject and predicate. This is through the connectives and, or, not, if, then,,,. They can through POT and Á be adapted to and used in TRU. They are also applicable to the interactions between TE and ES, or CES. They include many other symbols, notations and their combinations and formulae. They are a precedent to LANT.

In the mathematical basis for the representation and formulation of TRU we have several concepts that are of special importance, these include: mean = Having a middle position, or occurring between the limits or extremes. Average, the mean of a number of positions about midway between the two extremes. In statistics, and in mathematics generally, mean is distinguished from mode by indicating the precise average of all elements. Whereas mode indicates that point on the scale between its two extremes usually where these elements have their greatest frequency. In a Gaussian or Normal Distribution Curve these will tend to be close together. Indeed their degree of variance is to some extent a test of the systems reliability, or important features, mechanisms, or abnormalities of that represented. Statistics is the science that deals with the collection, analysis, interpretation and presentation of masses of numerical data. For a precise and reliable representation of anything a hard or rigorous Statistical Analysis is needed. Many things however are not reducible to their numerical representation, and even when possible are so only incompletely and need adaptation or alteration to the given, as for TRU and ROT.

Errors, Theory of, is a branch of Mathematics in which a special type, Error Analysis is a branch of statistics. Here regression analysis assumes that an observed value is a combination of a determinate true value together with an error term which is randomly distributed. This is used in simple estimation techniques in experimental sciences. Measurement, is that great means of representation in which anything is posited on a predetermined scale. It is how numerical units of finite divisions can be made when a set point of reference and limits are established. This is in effect when parameters and dimensions are so scaled that their gradients can be refined and made most distinct. The more an object, as TRU, can be reduced to its least common denominators and parameters, and formulated as dimensions whose coordinates are reduced or divided into equal or separable parts, the more accurate the measurements will be. The object can then be much more readily reduced to precise representation and formulation.

Determining polarity is neither so uncommon nor so difficult. It is no more than determining the certainty of TRU and enlarging upon its relationships. Procedures for determining polarity are vital to discovering, determining and learning the makeup and nature of TRU and ROT. There has been no consistent uniformity and no reliable method, as a unified field theory, which adequately, sufficiently, properly and effectively establishes m, l and t dimensions for the world relative to humans, TRU. Without an idea of what TRU, TIT, TRIT, OBT, SUT, ROT, and MOT are it is difficult to find and fit their primary dimensions into any uniform, actual, valid, and precise construct. Length and time can be surmised from this idealization of TRU and TRIT guided by Riemannian or non Euclidean Geometry where l and t are warped to fit TRU. For m, however, we can't accept the mass-energy of the physical universe as being its sole constituent form. The mass of an object must replicate its actual predominant state which for TRU must be a polarity of the predominant powers of mass energy of which it is constituted. This is the FAF and relations between the physical and the mental and all in between. The resulting predominant mass energy dimension not only partakes of many gradients between these two poles, but as in the case of length and time, adapted, altered, or warped by what exists in TRU and best enables humans to survive.

Fig 2-25 **POLARIZATION, LEVELS & GRADATIONS of ROT in the SENSORY and MOTOR of the CNS**

Subject and Object, as the general form of THES and THEO, describe those two portions of a relationship in which a sense, first person or observer is attending to or in some way in interaction with a thing, what the first person focuses on or the observer of the observed. This involves a stimulus-response state in which the central nervous system, is the subjective beginning, and the muscular system is the objective end. It is appropriate to show the chain of events, levels and locations involved in these processes fundamental to our being. This is the fundamental mechanism by which our relativity, subject-object form of TRU and thereby ROT occur. The S-R, S-O, IA and sensory-motor functions of human makeup are key to the understanding of SO, SOT, Á, and SÁ the predominant and primary mass energy dimension in TIT. They are especially key to the anatomical and physiological form and function of humans that occurs in the actualization processes productive of TRU.

O St Sn ST SN TP CS T CR CT TN MN MT M R O

O = OBJECT　CS = CEREBRAL CORTICAL STIMULATION　MN = MOTOR NUCLEI
St = STIMULUS
Sn = SENSATION　T = THOUGHT　MT = MOTOR TRACTS
ST = SENSORY TRACTS　CR = CEREBRAL CORTICAL RESPONSE　M = MOTOR
SN = SENSORY NUCLEI　CT = CEREBRAL MOTOR TRACTS　R = RESPONSE
O = OBJECT
TP = THALAMIC CORTICAL PROJECTION　TN = THALAMIC CORTICAL PROJECTION NUCLEI

How poles function as a unit and balance out in the whole has far reaching implications for TRU. The product of the means is equal to the product of the extremes. This is a fundamental law of mathematics. Its equational state reflects relativity, its focus, IA, Á, TROS, TU and TRIA. Thereby it has decisive application in deducing distribution in MOT, ALT, and SPALT. For it is a product of the totality of TRU plus the proportionate representation of this distribution. It strongly supports VAPO as does VAPO this principle. When valid dimensions are developed, the spectrums and continua they typify in TRU can then be analyzed, divided, proportioned and distributed. When this is applied to the mass energy dimension we can began to see TRU as TIT and a whole, and hence, as it actually is.

Fig: 3-25 **FIELD of INTEGRAL-DIFFERENTIAL**

In this diagram the extremes and means, polarity and continuum of the integral/differential relationship are shown. This is an extension of TROS and DAOST, whereby the mathematical same and other modes, concepts, principles, laws, and mechanisms of TIT establish much of its form, structure, and composition. The correlation between same and other, or DAOST, and TROS or **ID** Integral and Differential is most subtle. For it combines the greatness of DAOST with that of TROS, TU-TE-TO, or ID. This is key to much of the text, especially representation, modulation, and the development of ALT, SPALT, and POS. Unless we can think massively and orderly we can never achieve any adequate or lasting comprehension of all. The diagram helps to show us that the finite is little more than our ability to distinguish what relates to relativity or the S-O or Á dimension, form and function of man. The quality and quantity of finite distinction is tantamount to the thisness of human being and TRU. In Figure 3-25 other features of **ID** or the Integral Differential relationship are labeled by: a. Absolute Unity, g. Increasingly Formed, b. Relative Unity, h. Decreasingly Formed to Formlessness, c. Limit of Conceptual Unity, without multiplicity, d. Limit of Conceptual Infinity, Moment of, **MI**, e. Relative Infinity, and f. Absolute Infinity.

PROPORTION, ORIENTATION, BALANCE, DISTRIBUTION, and EQUALITY

LODEP = the Law of Definite Proportions. LODEP states that a given compound always contains the same elements in the same proportions by weight. This law arose and is best exemplified in chemistry and physics, yet has wide implications and applies in qualified ways to much of TRU and ROT. Its implication in TIT comes from thisness, exclusiveness, and closure in which aspects, form, and composition exist in set proportions.

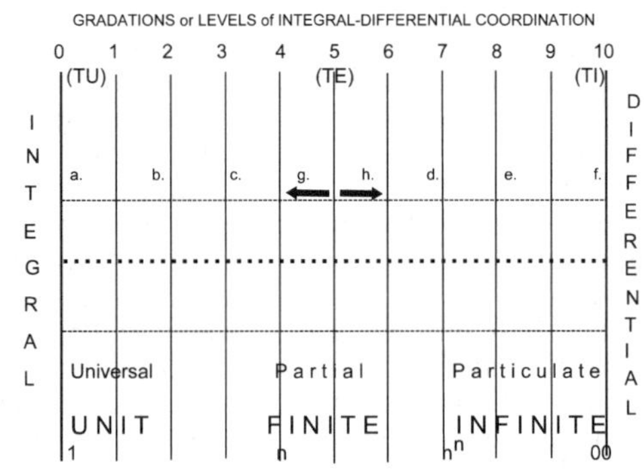

This is likewise the foundation for REM, the valuation of TRU. We know from TRIA and TIT as well as LODEP that TRU can be treated, within limits, as an object whose form and parts can be with VAPO well represented, modeled and utilized. In addition there is: **LACOM** or the Law of the Conservation of Matter which corresponds to SETLAW. The conservation of matter and definite proportions are shown by their operations whose symbols are naturally extended to the formulation of relationships between them. Their relationships are most readily expressed as equations. An Equation is a statement of change using symbols. In different states it has different characteristics, as the states change, the elements in the equation change. In chemistry and science as in mathematics an equation must be balanced with those necessary coefficients or constants that give it full and valid identity with the facts it depicts. An equation will or will not be adaptable to TRU depending upon the degree to which its symbols and constants are of ES or TO or TA or some other variant of TRU. It will also depend upon the degree to which the relationship and statement of change involves some function, process, mechanism or operation in or of TRU. So that if we do not find instantaneous validity in applying equations, even the laws noted above to TRU, all factors which limit and reduce the applicability of these to TRU must be eliminated or accounted for,.before denying their use in TRU.

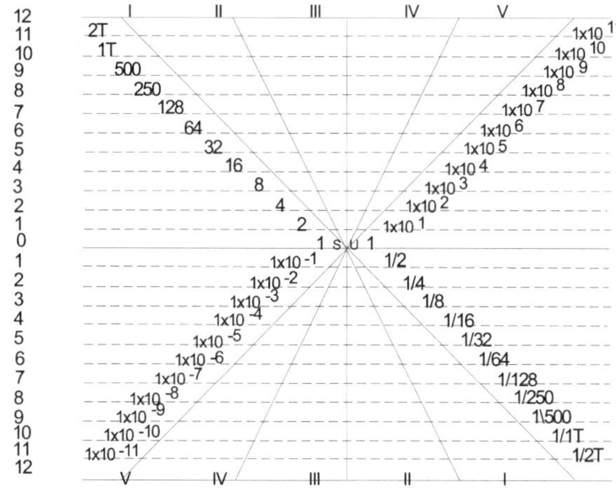

Fig 4-25 **DIVISIONS and PROPORTIONS in TRU**

The representation of TRU provides a divisioning, fission, differentiation, multiplication, fusion and integration of TRU. This can be done on different levels as it is expanded or divided. The positives indicate the numerical increases of TRU as cardinal numbers, and the negatives indicate the numerical divisions of TRU as fractional numbers. Types of scales include, 1. Level of division or order of division, 2. Binary fission or fusion, 3. Square fission and fusion, 4. Logarithmic fission and fusion, 5. Exponential fission and fusion.

TRU is not structured exclusively by any one of these scales. It is warped by adaptation through time, partaking of different ones relatively and often differently between TO and TE. The correlation between normal distribution, or bell curve, and the special warping of any RU is called: **DANC** = Dimensional Analysis and Form by Normal Distribution. DANC is by MOT and shows how the standard or normal distribution curve will fall along the ALT Scale. These tend to prove each other. The curvilinear form of the normal distribution curve shows much of the deformation of the linear scale of TRU'S Dimensions. This state is conditioned, limited by, and related to the laws, principles and concepts involved in each case, QUOL.

Fig 5-25 **RELATIONS of SUBJECT-OBJECT to INFINITY-INFINITESIMAL, and ZERO-ABSOLUTE**

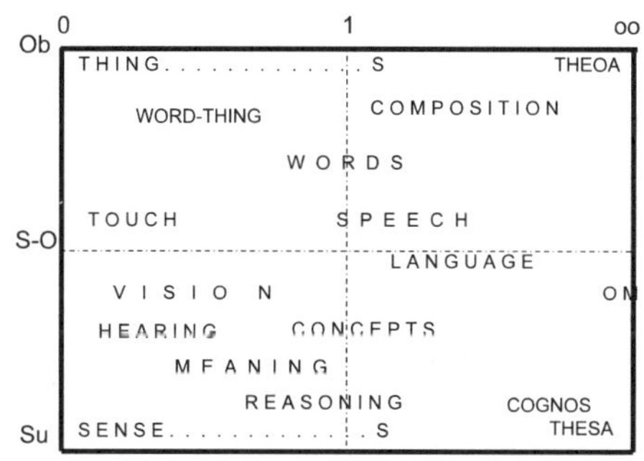

Complexities of SO or Compounds of Á are signified by interrelated, interactive and often superimposed parameters of S-O and Á. This may also be done for l, t, combinations of A, L, and T, higher powers of Á, L, T, and all ALT together. It shows the extremes, means and total scale as well as mass of S-O from limits of infinitesimal to infinity, from zero to the absolute, and the Absolute Subject to Absolute Object. How different elements or entities will posit in this will depend upon which sense they are taken. Are they considered as things in and of themselves or as parts, processes, or aspects of TRU.

COORDINATION, AXES BALANCE, EQUILIBRIUM, FIELDS, DIMENSIONAL FORMULATION in ROT

LEXAP = Law of Extensionality for Propositions. LEXAP is coterminal in ways with the Axiom System. It increases the probability of more formal logic. The cross has been a symbol for several thousand years. The cross as a symbol represents the equal and opposing coordination of any two parameters or dimensions. Basically it has been mostly used for spatiotemporal dimensions. As the axes of illustrations in scientific work, where the vertical dimension commonly represents length, and the horizontal dimension represents time. This is an arbitrary custom, yet one that is a good example of a standard that is naturally objective, or POA, and thereby corroborates TRU and SPALT. There is an Economy of Size a concentration of effect by possibility of enlargement and contraction necessitated by totality, a totality of form and utilization of prior dimensional form.

FREQUENCY, SERIES, MASS, COMPLEXES, and VARIETIES in ROT

If TRIT, OBT, SUT, ROT and MOT are founded and based upon the identity and optimal dividing or multiplying of their constituents, then what sort of divisions and multiples are we to accept as being optimal? Naturally since the progression from TIT through ROT to SPALT produces an increasing departure from TIT the infinite series from one to infinity or infinitesimal to infinity will change depending upon what works best for, the limits and uses of what shows and models TRU. Whether we use 1, 2, 3, 4, 5,,,,, or 1/2, 1/3, 1/4, 1/5,, or 1, 10, 100, 1000,,,,, or 1/10, 1/100,1/1000...or 1, 2, 4, 16, 256,,,,, and 1/2,1/4,1/16,1/256,,,or any other mathematical sequence or series will depend upon its distribution in TIT and adaptations through the progressions of ROT and MOT. It was such questions and problems that the Pythagoreans attacked and offered as divine. They held that the proportions and ordering of the world were the foundation for all. If we are to select certain natural and formal divisions and multiples of finite elements from unity we must do so rationally, axiomatically, precisely, and validly by developing its form in TIT. These are derived from the set ATO, TROS and TRIA of TIT. In being a set with perfect balance, focus, and proportion it becomes an absolute.

LARP = Law of Relative Parametric Proportion. For TRU there exists a relatively stable or fixed spatiotemporal mass, ALT. When modulated as a solid in these three dimensions this becomes SPALT. This mass in being stable and dynamic has focus and unity. In turn it has relativity and proportion. These give its parametric constituency form and composition. Proof by mathematics, geometry, proportion and proportional analysis is the most direct, accurate and simplest means we have of clearly establishing unknowns. If we can set up an a:b::c:d relationship and already know 3 of the for factors, the 4th can be readily determined. This is provided they exist within a set or closed system..

The attainment of KOT, COAT, SU, POT and PERT can often be best found by superimpositions of charts, diagrams, figures, illustrations, pictures and various representations. This we do by overlapping, analyzing and reducing their coordinates as much as possible to those of TRU. We then superimpose and posit them to become general posits in TRU through its representation, modulation, ALT and SPALT. This sometimes requires the adaptation of different scales. Sometimes it takes the reduction of parameters to their basic components. Sometimes the reduction, analysis and SPOT of these reduced parameters and dimensions is done both as TRU is constituted and as it is represented or held in perspective and presented.

DISCRETE COMPONENTS of LENGTH TIME

The length and time dimensions can be systematically ordered. They are used in physics along with mass to provide an overall formulation. The dynamics of an entity, and for TRU to a large extent, is the product of the interactions of these various states and relationships between lower to higher powers of length and time. So also can TO be divided and differentiated. This is a more simple case, and shows us what can be done for mass, as Á in MOT and SPALT for TIT-TO.

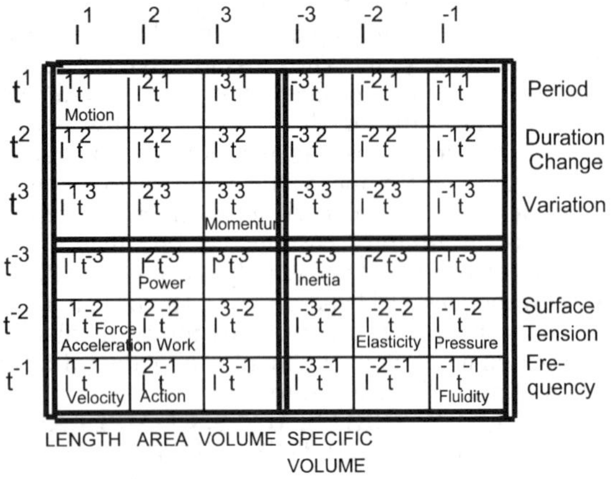

Fig 6-25 **The DIMENSIONAL FORMULATION of LENGTH TIME DIMENSIONS (DIMFOR)**

From the physical world these divisions of dimensional formulation provide the formal distinctions necessary to differentiate the dynamics of TRU. This becomes more specific and actual when they are appropriately joined with the mass-energy or Á dimension and posited as vectors in ALT and SPALT. The diagram provides a highly effective means of showing the processes of TRU, HOM, when their posits in the spherical TRU are viewed from the best perspectives. From motion to fluidity, momentum to inertia, each of the 27 sectors in this chart are identified with a discrete type of dynamic. Many of these types are not given names, have and have not been discovered, identified, known or used. A similar figure, or cube, in which for TRU, Á as well as L and T give each of these cells greater meaning can be produced. Each cell or type of dynamic operates in their own way in and of TIT. They are fundamental to a good understanding of TRUD and HOM and succeeding sections of IPTRU. Much work remains in the use of all such basic analyses of mechanics to further their understanding and application to TRU, as in ACC.

Great detail of the ground, substance and parts of TRU is provided by increasingly discrete scales on the Á, L, and T, ID and other dimensions. This is the beginning of mathematically precise and reliable demonstrations of the makeup of TRU. It is what enables a POA. The distribution of L and T are readily shown both by convention and how this is altered and adapted to reflect TIT. The distribution of mass must meet its form and SCARLS in TIT including a series that runs from complete objectivity to complete subjectivity determined by the predominant form of mass in TIT. It is tentatively given in scales of tenths, or one hundredth's, distributed evenly to reflect the identity with coordinate balance in TIT. Although L and T are for TRU determined by relativity, it is it's mass energy that most clearly reflects the role of relativity in, and thisness or exclusiveness of, TRU. This is seen in $E = mC^2$. It seeks to confirm the precise dimensional FAF of TIT that remains after all powers of L and T are accounted for. It is also confirmed by identity with what we know and show of the special, typical predominant mass energy that exists in TIT.

The contribution of logic, formal logic, two-valued logic, three-valued logic, logical relations, inclusion relations, lattices of points and lines, span of vector space, truth tables and others to ROT is highly beneficial. An example of this is the Distributive Law of Logic which states that for any sentence P, Q and R if the compound sentence P and (Q or R) is true, then the compound sentence (P and Q) or (P and R) must also be true. The use of a lattice to model this distributive law, shows how logical relations can be demonstrated and are examples of the geometric and modular rendering of TRU.

SUMMARY and EXTENSION of ROTÍ, MATH, DIMENSIONS and INFINITY

Probably more important than Descartes's famous Method of Doubt, were his contributions to mathematics, geometry, calculus and the invention of coordinate or analytic geometry. In these he used what had been developed in ancient times by geometrical reasoning, the analytic method. In geometry a problem is assumed to be solved, supposition, and the consequences of this supposition are examined. If possible either confirmed or denied, or qualified as proved by corollaries for further logical reasoning. Its great value, both in geometry, logic and reasoning in general, lies not only in providing a rigorous proof of some discrete object, as geometric figures or propositions, it is in its extension to less discrete objects, THEO, OBT, SUT, ROT and MOT that can be used to show TRU. Carried further with appropriate VAPOS, SCARLS, and REM it can qualify and refine ROT and the methodology of TRU.

What ROT depends upon is often through what it produces. If we are to attain ROT and TES we must do so either by eliminating ambiguity or control it and to expand our powers to use it. This is, among others, accomplished by TRU terminology. **LOPUT** or the Law of Proportionate Truth. LOPUT describes the unity and distribution within TRU which provides the distribution and necessary proportioning to be consistent. Such consistency is a result of its set state, unity and systematic unity that produce a solitary final total truth. This holds for each and all within TIT, providing all else is equal, as the observer.

Teleology is the study of ends, final causes or purposes. It provides TRU with that *'a posteriori'* vantage without which it would be incomplete. As shown in TRU, MOT, HOM and ACC the future determines TRU, ROT and MOT as well as the present and past. So that in seeking goals, ends, purposes, priorities, meaning, and values we are usually working from after to now and before. It is this end term of a process, the purpose or final cause, that is in temporal continuum with it.

It is the finite that is primarily our REALITY. A finite not of any sort, but one determined by the existence and whatness of TRU. How we are finite composites, increasing to become more and more divided, or decreasing to become more fused. This is how we mathematically consist, from 1 to n/1 or from 1 to 1/n. This Finite System that composes TRU is the embodiment of Form and Meaning and the great potential and means to SUT. We live in purity, reality, and a finite actuality, only as it is, and we are, ES and TES, in TIT-TO of TRU.

The infinite, as the infinitesimal, is actually by TRU, not an ENIT, or thing in itself. It is only an extreme, or pole of the form of TRU. It is the poles of the Á, L, and T dimensions and continua. The relations between various categories of number can be enumerated. They show the range and much of the mathematical basis, TROS, for TRU.

Unlike entities or even parts of TRU, the set form and SCARLS of time and other dimensions in TIT change little. What an entity is in itself, ENIT, what it is as its ESIN, and what it is as its corresponding ES and CES in TRU are different, often far different, even in opposition. So that we are not to accept entities nor even ESINS, ES and CES as absolute, valid or even final until and as they are in the TO of TIT. So it is of time, length and the mass energy dimensions of TRU, ALT. They are not to be casually believed or accepted, even realistically, what they may seem or appear to be. They are first and foremost only what they are in the TO of TIT. Then we may, with deep ECRO, add to them the givens of ROT and POT and whatever added vantage or use the observer makes. What the numerical basis for ALT will be can be readily determined if we consider and investigate what it actually is in TRU.

CHAPTER 26

TOTAL REPRESENTATION of TRU, ROTO

INTRODUCTION, KNOWLEDGE, and VALIDATION of ROTO

Most of our knowledge, and even more our thinking, is in terms of bits and pieces, which are all to often much less than and even negative to TRU. To produce a whole we must have two things, a unified and whole entity, object, or TIT-TO to treat, and a unified whole representation and model of what best replicates it. We are the observers, we have an idea, word, concept, symbol and meaning, yet the value of these, are very limited unless they are related to other words, concepts, and meanings. The more they are related, the more they are intensely thought of and the more value they have. The greater these relations are brought into, and proportionately, a picture of TRU the greater these concepts become. Knowledge(K) equals object of knowledge x intensity x # people utilizing, x relations of objects x volume x coordination, and all by TRU.

The biological orientation of an organism's or animal's motor and sensory capacity does not include the whole of what enables that specie and its universe to survive. In this way there is a deficit between CNS orientation and total or TRU orientation. Geometry, ROTO, and especially its axioms progressively move from their simplest to most complex orientation. By Lowenheim's Theorem formulae or entities, to be satisfiable, must be enumerable. Yet by Skolem's Paradox such formulae or entities, as TRU, will appear to be the contrary, not satisfiable, for reasons of being incompletely formalized, lacking the essential finite elements or variables or being in some way or type of non enumerable infinite. As great and necessary as our language is, especially by providing the mass energy and reality needed for TRU, it falls far short of giving us good representation, production, reproduction, exemplification, replication and duplication of TRU. Our words are many, our literature massive, some of our definitions very good, and many of our meanings are great, yet we fail to produce the needed accurate comprehensive representation of TRU.

Fig 1-26 **FROM RELATIVITY Á by ROTO**

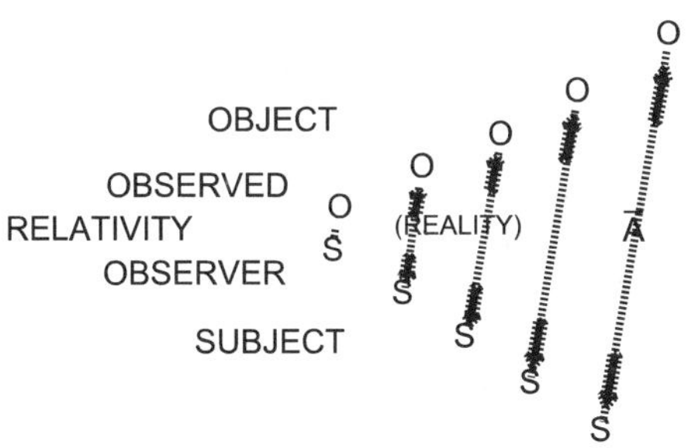

Only as we can hold to the proper objectivity, the observed to its wholeness, the unified and verifiable totality of TIT TO, can we provide the correct definitions needed to reintegrate that separated from TO of TIT. The validity of a whole is proportional to: 1. The precision of its parts, 2. The totality of these parts, 3. Proportionate and exists in the set, 4. The non empty set of these parts, or the degree to which they are discrete, 5. The precise definition of the whole, 6. Appropriate SCARLS for the whole and each part.

VE = Validation Extrapolation, or the determination and attainment of knowledge through the use of formulation, SU and THESOS in interpreting existing knowledge in relationship to unknowns. Do they reconcile? Does it fit? Much that is unknown could never be known without relation, correlation, and ordination. Yet it can be readily, clearly and greatly known when we use the simple processes of VE unknowns from knowns as they fit into an ordered whole. This has been a large part, albeit singular or limited, of reason, the rational and deductive method that has come down to us through history. Discernment of TRU is produced by the consistent state of S and IA that must be cultivated if we are to be TES in TRU. This is how the individual must think and reason to exercise VE. Just as education is much greater when it trains us how to think than to simply impart knowledge by rote, so does VE and discernment train the determination and distinction of that needed for TRU. They train us as in **TAN** - to think and note - to build upon our experiences by relating, thinking, and making them a game and record. These produce the guide and foundation for bringing our observations to TRU through VE. **VACEP** = Value Assurance Certainty Principle. We attain VACEP when the form, function, parts, processes, characteristics, ideals and all of TRU and SUT are have VAPO and are increasingly understood and ingrained in our minds..

DERIVATION, ANALYSIS & METHODS of TOTAL ROT

In Analytic Geometry we are shown the principles and relations which exist in the construct of objects. This is critical to VAPO, PACH and a successful ROT and MOT. It is equally critical to precise and reliable posits of the parts and processes of TRU. It is not only necessary that we know TRU. It is even more necessary that we be TRU. It is crucial that we know and be the perspective, presentation, representation and modulation of TRU and much of all that enables us to know, show, and live by TRU. It is in the fulfillment of actuality that the completion of our instantaneous attention is attained. This fulfillment of actuality is our subjective's objectivity in resolving the stimulus response or problem solution given. Herein is the gist and cue to our being, TRU.

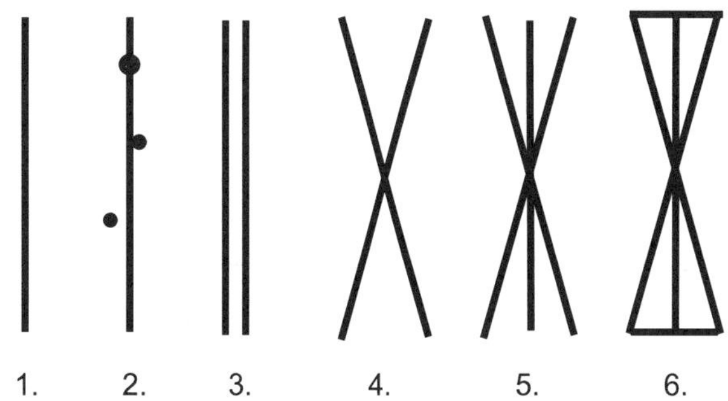

Fundamental Terms, Shapes and Relationships in Geometry:
1. Line
2. Line and Point(s)
3. Two Lines
4. Two Lines Intersecting
5. Three Lines, Intersecting
6. Three Lines Intersecting Forming Enclosed Figure

Fig 2-26 **REPRESENTATION of STRUCTURAL COMPONENTS of TOTALITY, 1D**

In Figure 2 and succeeding figures 3, 4, and 5 we show the increasing complexity of components in totality. This shows how different aspects and representations may be given as points, lines, fields and solids. They are the basis for MOT, ALT, SPALT, POS, and POT. Unless we learn, know and use what the different pieces of the frame and structure of TRU and its ROT, MOT, ALT, and SPALT are, we can never sufficiently appreciate, accept and utilize them.

Much of the foundation of investigation in science has been the effective method of increased surmising of more what we seek, as TRU, by making hypotheses. These form the basis that selectively guides better directions and ends of successful knowledge and description. Through description we can enlist all the detail, makeup, aspects, evidence and information that when brought together in an astute and organized manner provides effective integration and fusion for a ROT. Through sorting and classifying this information, we can attain the necessary programming and management essential for overall control. Through conventions, especially when identical or close to their object, we can provide the algorithms, plans, systems and all necessary for managing information about the object as it must be organized to be represented. In POA and PACH or Positive Objective Approach, and its standardized perfected format. With a form based upon the fundamental relationship between observer and observed we have the methods, most used in Medicine and IPTRU, whereby information about the various aspects, external and internal, of TRU can be accumulated.

To frame and structure ROT is like surveying or laying out the blueprint of a building, made up of points, scales, areas, parameters, dimensions, volumes, coordinates, and overall unity of the whole. It is this when done by the best actual form and function of the intended object that most greatly enables us to plan, act, succeed and be, TES and TRU. (OBT)

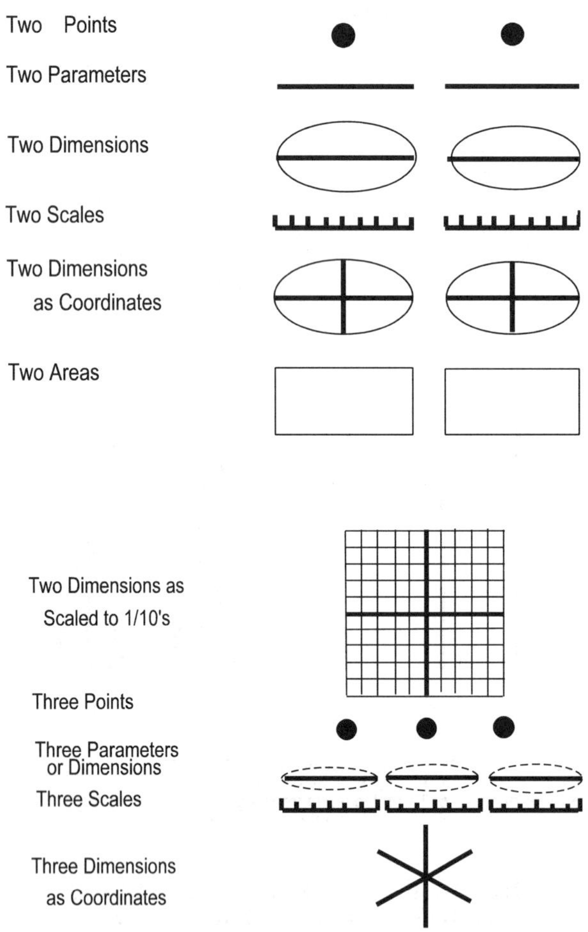

Figs. 3 and 4-26 **2D and 3D in ROTO**

The structural components upon which the totality of ROT and TRU are based, are largely concerned with those of no, one, two, and three dimensions. It is in how these can be represented and presented more than one at a time; two points, two parameters, two dimensions, two scales, and two dimensions as coordinates, two areas, or two dimensions as scaled to 1/10ths. It is also necessary to have: three points, scales, dimensions, three dimensions as coordinates, three dimensions scaled to 1/10ths, three volumes, three dimensions & volumes scaled to 1/10's, three dimensions with coordinates set to a given object, OBT, three dimensions with coordinates set to TRU. These are the analytic and geometric steps needed to develop, learn, and apply and gain a total approach, frame, perspective and perusal of an object. Even much more imperatively is this so for 'The World', The Object', OBT of TRU. This does not give us the substance of TRU, nor even less, its REM and internal dynamics. Yet we must always remember, **ECRO**, that it is the frame and structure that provides the base upon which all the rest is made and shown.

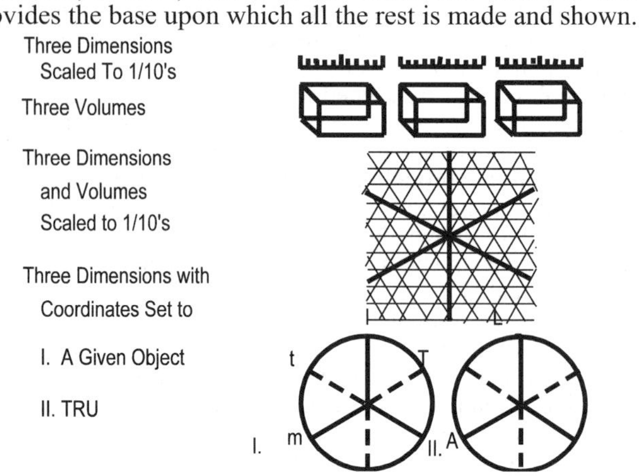

. Fig 5-26 **COMPLEX ROTO PROGRESSING to MOT**

The importance of these figures is to reveal the basics of geometry that contribute to produce a frame for the model of TRU, SPALT. It is in the excellence of our derivation, development, determination and demonstration of the totality of TRU and ROT, TO and ROTO, that we can set the stage for the kind of reliable means that prove TRU and ROT by best showing and understanding what their frame is and how it is composed. The resulting performance on this frame's stage is the positioning and positing of all TRU in the most vivid and exquisitely beautiful manner that best replicates what they are in actuality, TIT

By fitting pertinent information into the best axiomatic and geometric configurations precise position or posit in SPALT and much that is in and of The Object, TRU, can be shown. Calculus determines the volume of a solid by the method of slicing. So can we determine the contents and parts of TRU by extending this method. Such methods enable mathematically exact and reliable proofs of TRU. They are especially valuable by enabling us to best interpret and use ROT and MOT.

GENERAL FORMS of ROTO, SHAPE, GEOMETRY and CONSTRUCTS

ROT, MOT, ALT and SPALT are to a large extent based upon the valid assumptions and established principles of mathematics, geometry, analytical geometry, and graphics. The frame of coordinate dimensions of TRU provide a guide for reduction by geometric, solid analytical geometric and graphic representation. When TIT is held to a set TO frame in TRIT and OBT and existing knowledge of solid analytical geometry and graphic symbolism and design are applied, a ready and accurate reproduction and utilization is provided.

When we appreciate the importance of approximation we can begin to better appreciate TRU, the limits, value, and uses of ROT. It is by and in the development, accounting and refinement of all our approximations in TRU and ROT that we must concentrate and persevere. Approximation is needed and important by the limits of TRU, the limits of the infinite, TI, the ever more exact distribution and valuation of ES and CES render approximation necessary. This is what human knowledge has been doing through the ages, making approximations and then repeatedly improving on them. **PRUSAP** = Proof of Sufficient Accuracy and Precision. It is in the improvements, refinements toward perfection of approximations in ROT, MOT, SPALT, POS and their uses that IPTRU, POT and all TRU must be based.(VAPO) This is the relativity of the observer and observed and the finer and finer degrees of astuteness and precision these realizations produce.

Conic sections are an example of non-linear, warp, bend, and rate of parameter change, where the structure of TRU is not the linear measure of any physical object or measure but that set and special for TIT. Their fixed point describes arcs whose distance from a fixed point is in a constant ratio to the distance from a fixed line. This produces and describes the adapted, altered, somewhat elliptical and specific form of parameters in TRU and ROT. The frame for ROTO becomes a solid, of conic sections as these layers and sections represent the superimposed spheres of a solid TIT.

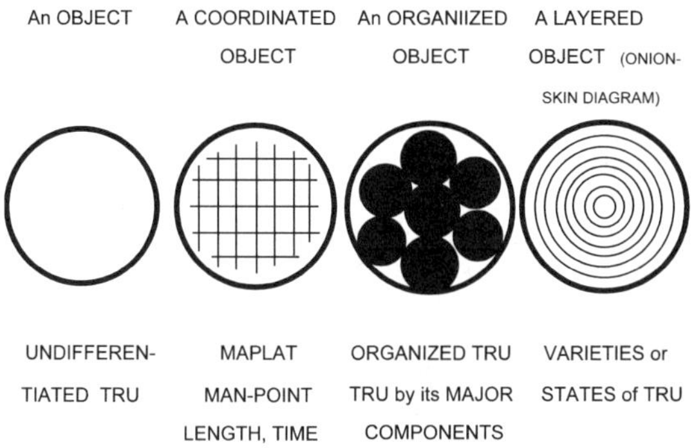

Fig 6-26 **SOME CONSTRUCTS of OBJECTS and TRU**

Objects, as OBT of TIT, may be divided, envisioned and presented in different ways. Whatever best reveals their form and content is optimal for ROT. Different kinds of representation are needed to demonstrate the many modes, facets, forms, and makeup of TRU.

RELATIONS, CONTINUA, DIMENSIONS, and FORMULATION. HOW TO DEMONSTRATE, SHOW and KNOW our WORLD, TRU

For every relation there is a dualism
For every dualism there is a polarity
For every polarity there is a continuum
For every continuum there are levels or gradients
For every level or gradient there is increasing differential
For every differential there is an integral

The integral and differential combine in the continuumn
Combined continua make up the parameters of an entity
Set parameters make up the dimensions of an object
Dimensions of the object of TRU enable its formulatiion
Dimensional formulation enable representation and modulation
Differential gradients of dimensions make up its ES and CES

Dimensions are altered and warped to adapt to TRU
Measures of dimensional gradients follow TRU's adaptation
Parts and processes of TRU are best represented by posit
Measures of parts and processes need continual refinement
The best positive objective approach to the world is in this way
By providing the best showing and knowing of our world

The relative universe, in comparison with the astrophysical universe, is curved and warped not only exponentially in its three major dimensions. It is warped somewhat irregularly and variably in ways that typify and best provide its form and function set by its own being. It is a form that approaches the absolute in being the structure of the absolute, TRIA of TIT. Much of the dynamics of TRU is to PPMA and ACC in accord with this special warped frame revealed in the dimensional formulation and operation of ALT and ACEXIN. Whatever, even more slightly, varies and helps them to adapt to the needs of life and survival.

Topology is a valuable tool in ROT, MOT and their uses. It applies geometric, plus its own, methods to demonstrate objects, OBT and TRU, by SPALT, POS, STA and HOM. It is the study of those properties of an entity that remain invariant under similar forms. This signifies that as long as the properties or aspects of TRU are the same as those represented, the demonstration will be valid, accurate, and consistent. This is not absolute, only relative, although practically approaching the absolute so long as the identity of object and entity are maintained. This prime principle of topology when applied to TRU is called: **LATINÄ** = Law of the Identity of TRU, its object, their representation by Topology and Invariant homomorphisms. LATINÄ is a prime proof of the justification and value of ROT, MOT, HOM, and indirectly, in return, TRU. Likewise, in the use of topology, what is posited in SPALT or equivalent models will be valid and accurate as long as it reflects the actual state in TIT.

To what extent is TRU or TIT a uniform spherical mass? Is it's substance or parts uniformly distributed or are there marked differences in density, concentration and distribution? Are there a finite number of enclaves, organs or concentrations in the mass of TRU? These are questions whose answers should be increasingly evident from existing knowledge, IPTRU, and further work on TRU. When TRU is represented as a whole, especially for ALT and SPALT, it tends to become spherical with increasing solidity. This is evident in ACEXIN and tends more or less to uniformity by the exponential, which is altered by the relativity construct of TIT and ALT. This produces limits in totality tending to sphericity, the nature of any centered body. This is according to the law of Mass, $Sp = lt^{2x}$. TIT is neither like an individual person, nor like a spherical object or even one we represent. It is without external objectivity and subjectivity or relativity, but only as these exist within, ATO, ALT, and THESO, like a focus that radiates outward and inward with infinite limits.

Fig 7-26 **TOTAL MASS of TRU, LAYERED FORMS**

TIT may be shown by concentric rings of a sphere which when colored provide a good idea of its content from core to periphery. When correlated with light the periphery is ultraviolet and the core infrared with the spectrum in between.

FORMULATION, ORIENTATION and TOTAL GEOMETRY of ROT

PROFOU = Proof of Formulation and Fusion. PROFOU is derived and demonstrated by **MP,** Man-Point, the core and focus of TIT-TO. Geometry, formulation, major courses, capacity-ability-reaction-setting-condition or **CARSC**, and many others help in the proof and demonstration by fusion and formulation. This, when limited to dimensions, is more than simple fusion, it targets total completion. For fusion suggests a loss of differentiation in reality and TO and is thereby not in agreement with the massive and intricate form and makeup of TRU. This is shown by: **LAM** = Law of Morphology. What is morphous is formed and tends to follow actuation, extension and intension, or the **ACEXIN** that is specific in TRU.

It is in **TOR** or TRU orientation, that we can understand and relate all most precisely as they are part of TRU, their determination and needed criteria are finally developed and attained. **SOR** is man's S-O or Observer/Observed Orientation. SOR shows how from S-O, or observer-observed relativity orientation by Á and its products as ALT, we derive TOR. **SORINACT** = SOR TOR Interaction and Mutual Actuation and Augmentation in TRU. SORINACT is best seen in the role of language and reality in expanding the combined SOR and TOR. SORINACT shows the stability and power of orientation. It is exemplified by how a person's orientation and coordination are not controlled by a single organ, but are a function of much of the entire central nervous system, with mutual checks and balances of several organs. So do SOR and TOR combine with other aspects of the FAF of TRU to provide optimal orientation and coordination.

INCLUSION, COMPOSITE DIVISIONS, ROTO CONCRETION & DEPENDENCY

Hyperbolics and quadratics exemplify how unidimensional and bidimensional forms can combine to provide 3D forms for ROTO. When equations are formed to provide higher degrees in use for TRU they introduce us into the potential as well as higher realms of sphericity encountered in the OITIO of TRU. Much can be standardized through target theory of general relativity as well as in radiation seen in atomic energy. Target theory and its development from relativity sets a standard for the use of TU in SUT, ROT and MOT.

Fig 8-26 **COMPARATIVE CARDINAL and FRACTIONAL DIVISIONS in ROTO**

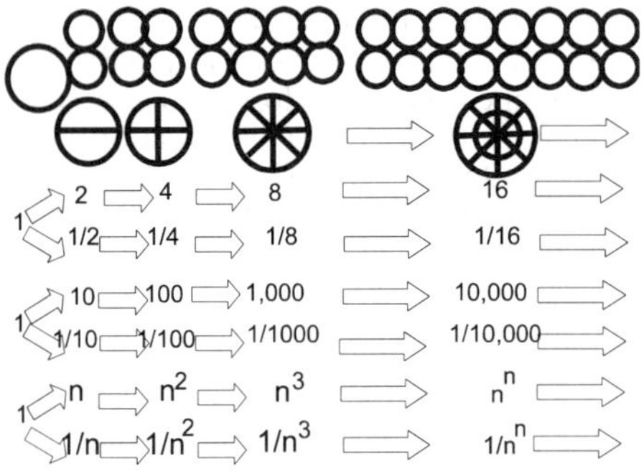

OMAFLEX = One, Multiple and Fractional Divisions Simultaneously. Such divisional and fractional separation from totality is a finite ideal. However it is not necessarily of TIT, nor ROT. It gives a guide and model for SUT and REM or a standardized division of TIT TO. When all of these are brought together we have **METO** = Mutual Expansion of E to TO. METO shows how all can discretely represent TRU. They can be modified by TRU and best shown differentially. This helps to not only expand ROT, but TRU as well. To do so we reverse the process. Likewise METO provides the exquisite detail of the substantive TRU. This in turn enhances of TRU, '*Dictum de omni et nullo*'. This is the Aristotelian principle of syllogistic logic which states that whatever can be affirmed or denied of the whole of a class may be affirmed or denied of a part. As parts and wholes are not exact replicas of each other this is not valid. Yet by the sames of DAOST, parts are different aspects of the same thing, or the whole. In this sense of the same, this principle is valid. Also as wholes can't exist without parts and parts can't exist without wholes this is also valid.(DAOST,LASUR)

As LT, L_2T_2, L_3T_3, and corresponding generations of mass in in which all are relative to humans, an equivalence to its predominant spaciotemporal, observer-observed or relativistic capacity, FAF, produce a frame for TRU. Their gradations enable us to provide the detail, measure and reliable representation needed.

Fig 9-26 **INCREASING CONTENT and SUBSTANCE in TRU**

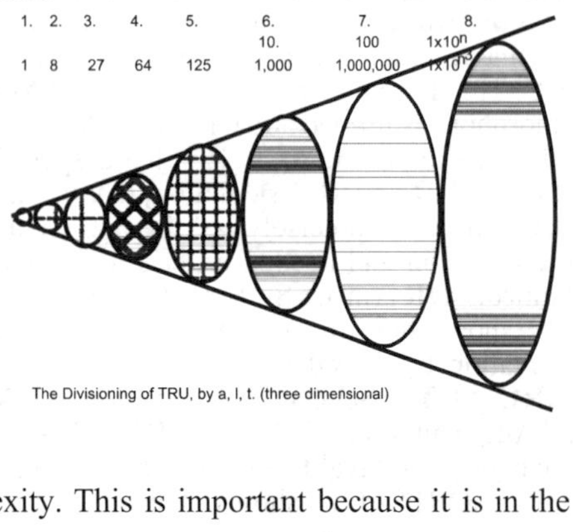

The Divisioning of TRU, by a, l, t. (three dimensional)

The comparative content and substance of TRU may be divided in many ways with increasing complexity. This is important because it is in the total representation of their best distribution, and ever more precise finite parts of TRU, that we attain that identity with reality which is so central to our being. It helps to show TIT as it actually is, both as TU, TE, TI and a TO that is more than the sum of its parts, thereby leading to TÂ.

This same problem of how to divide and show TRU is to be found in REM. This is not surprising for both are much the same, only in ROTO we seek to show how TO can be best divided to show TIT, and in REM we show how this same can be distributed proportionately by relative magnitude which turns out to be value and worth relative to TIT, the whole, or TO of TRU.(35) *Association* is how this dividing, separation and cleavage of TRU occurs as its form and parts are most related and opposed in the statics and dynamics of TRU. Serial order occurs as TRU and its observation come into agreement or correspondence.

COMBINATION, ORIENTATION, ORDINATION, CLOSURE, SOLIDITY

The many combinations of different ROT or figures in IPTRU when superimposed on SPALT, METAD, SPAT, and all can include the many beautiful works of art and scientific illustrations which together depict a whole panorama of the world and TRU. There is an equivalence of: a point of reference, axis of coordination, focus of TRU, overall reference, total reference, coordinated sphere, TIT, and OITIO in different ROT. Although these may not be identical, their center, focus, reference and relativity are all derived from their concurrence and correspondence with TIT. This center -and mass - is not of any kind, it is a great accumulation of all that is relative and centripetal to TIT. It is associated with **TOTOR** = Total TRU Orientation, Order and set.

The great advances of civilization and human's union and control of nature, have been a result of continuously improved knowledge, language, representation and means. **VACOST** = Validity Forms Combination Completion Totality TRU. VACOST signifies the principle or law that validity is to be found not so much in singular facts or observations, nor in entities, things and senses in themselves, so much as it is to be found in their combinations, completion and totality in TRU. This is largely by **VE** or Validity Extrapolatiion. How they come together and are formed in TRU. **LANCO** = Law of nth Completion. LANCO is what transpires when there is a proposition or subject predicate relationship in which there will be needed added propositions or subject predicates for completion. This is the beauty of description, explanation and composition. They give the closure LANCO must have. Such is the basis for SU, POA, and POS, a mathematical reduction where one factor is meaningless or inadequate without the whole relationship or equation. With the precision of algebraic mathematical and logical equations most all of TRU can be reduced to its elements, factors, ES and CES to produce the discrete objective differential in totality or MOT, POS, ROT, and POT that are so important for TRU and its PERT. By **SIPEQ**, the Superimposition and Integration of Equations adapted to VAPO, TRU can be more reliably and better Proven and Presented. **PEQ** is Proof of TRU by Equations..

Fig 10-26 **POLARITIES SUPERIMPOSITION in ROT**

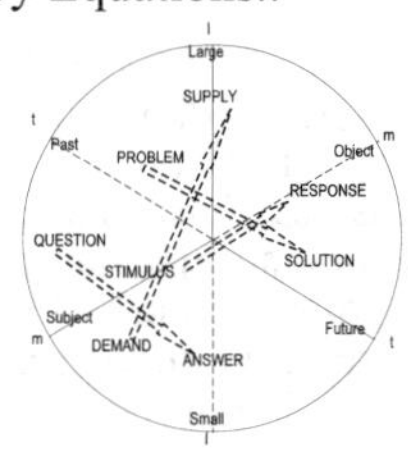

Many dialectical, opposite, and contrary divisions of TRU are not necessarily primary dimensions. Yet they can be interpreted in terms of ALT and used as lesser parameters or systems in TRU. This is evident in two ways: one by adapting altering, and showing the nature of mlt the mass length time dimensional formulation that exists in TRU, and the other by showing how these divisions, contraries, opposites, polarities and dialectical dynamic aspects of TRU combine to make up, form and prove ALT. It shows how the mass or Á dimension of TRU is more than any S-O relationship, but is set by the predominant FAF of TRU. It is, in actuality, all that is of TRU that is not a part of or some aspect of all the powers of the l and t dimensions and continua. Together they fulfill the dimensional structure of TRU as well as determine much of its makeup. **LALTS** = Law of ALT Summation. LALTS signifies the complex form of TRU. This is not the simple mlt of physical mechanics, nor its more complex dimensional formulations, nor even they as they exist in TRU separately, but all together as they makeup TIT. It is all the complexities of A, L, T, in their highest states, continua, and powers as shown in ROT, MOT, SPALT, and ACEXIN.

CONSISTENCY, SYSTEMS, ORDER and SPHERICITY

There are many examples of consistency and continuity in TRU, as seen in true pitch, harmony, melody, symphony in unison and in totality. They show how TRU order, TRIT, OBT, and SUT are provided. In ä capella singing there is a perfection of pitch that produces an effect exemplifying divine perfection, the aim of ROT for TRU. It is an art form and production, championed by ancient Greek sculptors and artisans, that shows the fineness, definitive ecstasy, and beatific feeling such perfection and greatness of TRU warrants. It is in such combined operations of differents and sames brought to totality that show us the greatness of ROTO. The method of representing total polarities is founded upon the graded steps that in summation bring them to ROTO. These are shown in Figure 10. If two of three dimensions are represented in a certain given way, they produce a set relationship. When a third dimension is added, it will be of a somewhat altered given way or set relationship. For example if t is the horizontal axis, and l is the vertical axis, then mass or mass-energy Á - will be the sagittal axis. This is called the right hand rule in **CAD**, computer automated design. Although it does not prove Á, it does provide a strong basis for standardization and helps to prove the formality of a solid spherical geometrical construct. This in turn greatly helps to prove the representation of TRU as ALT in such a solid spherical object.

LAMPEX or the Law of Á or S-O Medial Means and Peripheral Extremes, is much of TRU Correction, **TROC.** S and O become more in TRU when of the means and out of TRU when of the extremes. LAMPEX results from the transformation of vectors, parameters and dimensions that occur between TIT and its being an object in our observation and relativity of it. This is what happens in OITIO. So that we must always be careful that, even though we are being the observer, the observed is of the means in TIT. Observation being of the extremes unless the adapted observer in TIT. We must ECRO and continually reduce SO to SOT, Á and TIT. When Á and S-O are only as they are inside of TIT, they are massenergy in a highly dynamic and interacting state and not something that is readily evident to the experience of man. This is one in which this S-O is more than what we ordinarily think of as a person looking at or an object being looked upon. It is a whole spectrum of relativity and other mass-energy forms as they constitute TRU. In this sense Á or S-O is both medial, as they are the same highly combined, interacting, and peripheral as they are different or separated. It is in the combined medial state that they exist in strongest relativity or identity, the reals, as subject and object are bound together. It is the continuum, polarity and scale we produce from the extremes and means of Á as well as length and time, that enable us to formulate, represent, measure and model TRU.

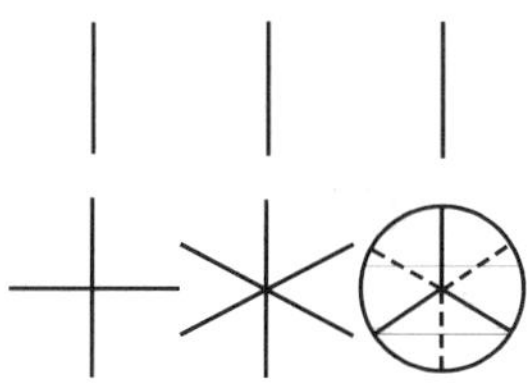

Fig 11-26 **The ORDERING of DIMENSIONS, PROOF of the GEOMETRIC CONSTRUCT of TRU**

These include, a single fact, a single extreme, or pole, two poles that form a polarity, linear parameter, dimension or continuum, combined polarities, dimensions or continua, that form a coordinate system, the simplest planar structure of an entity, universe or TRU, combined polarities as dimensions that form the coordinates in a three dimensional structure, as of SUT, and the total form produced by these polarities dimensions and continua as they exist in and of TRU, in and of ROT, ROTO, MOT, ALT and SPALT.

LACBE = Law of Coordinate Balance of Extremes. For any valid coordination of TRU there will be a corresponding balance of its extremes. This may not be balanced to our perception, but will of necessity, be so for TIT. **TUS** = TRU Unity, fusion or equality with Systems, a and the System, TRU, and Systematic Unity TRU. TUS signifies how TRU in many respects acts like a system, and how in many respects a system has the features, nature or form like TRU. This is an extension of SUT in which its total dynamics reveals it to be a separate entity. **MOSU** = Method of Systematic Unity. If TIT, as TU, can be considered as one, and if we can objectivize TIT and represent it as closely as possible in a 1:1 replica of TIT, then its method will be those ways that can replicate and provide a dividing of TRU that gives us the best possible SUT, ROT, and MOT. This may or may not be perfectly possible. Yet it certainly must be possible to a great degree. These are classified as follows: how it can be by: TU, divisions, dimensions, ordering, standardization, and can not be by: TI, TE beyond our limits, and overlapping.

DIVERSE GEOMETRIES, CYCLES, ALTERATION and REDUCTION

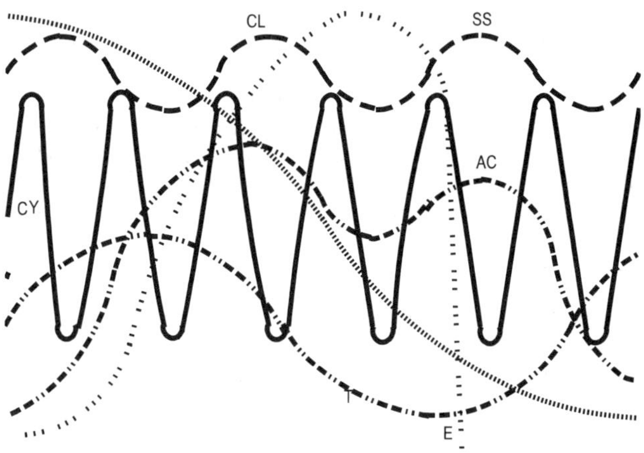

Fig 12-26 **SUPERIMPOSED CYCLES and RHYTHMS PERIODIC and CYCLIC ROT**

Various cycles, rhythms, phases, and periodic states and similar functions show a large portion of the dynamics of TRU. They determine much of what TRU is and does. They are primarily time oriented, present in both l and Á at various levels. When they are accurately and collectively posited in ROT and MOT their superimposition provides an indispensable tool to replicate TIT. These may be transformed into many different aspects of TRU and overlap generally or specifically. Cycles, CY, and their phases that we identify, show some of their features and divisions by, climax, CL, anticlimax, AC, troughs, T, steady states, SS, extinction, E, and others. Such a view, strictly held to TO-TIT, is indispensable to an adequate comprehension of TIT.

To understand the more complex and dynamic aspects of ROTO and TRU we need to investigate and show its rhythms, phases, waves, periodicity, cyclicity, cyclonic and energy systems, and regeneration or rejuvenation. These apply both internally and externally. They reveal a continuous action - often variant - on the part of TIT to right, adapt, and continue itself. By Diverse Dimensions Reduction, **DIRED**, we seek to find how much is natural, how much can be readily and accurately standardized, and how much is inadequate before and after. This is largely the work of MOT, ALT and SPALT.

LARECS = Law of Reduction and Reconciliation of Standards. LARECS defines all means and methods whereby we can delimit and control the standards adopted through reducing their number and reconciling them with TRU. This is by optimal relativity, or subject object correlation with TRU. If we deal with a state in which there is a variety of observers and observed variety of geometries.

SERIES, CHANGE, REVERSIBILITY

Change is a fundamental process in all things both external and internal, direct and indirect, rapid and slow, intermittent and continuous. These have important implications and applications in and of TIT, TRIT, and ROT. **LAPRA** = Law of Aberrations Provide and Prove first Principles and Form, RÄ and TÄ. LAPRA denotes how what is different, and more vividly aberrant, is what enables us to recognize what is discrete and thereby make an entity, RU, object, and TRU formally distinct. Animals have a built-in predilection for sensing what is different, just as they have a predilection for sensing movement, and special objects of high priority to their being. Difference and movement are highly related. A circle and polarities prove, geometrically, the measure of surface conformity and internal form of an entity, object or universe. This is called: **MESUC** = Measure of Surface Conformity.

Terms of universe, world, God, and man's world, although much different than and confused with TRU, may still have great use as ESINS, universals or the different aspects of TRU.(TREQ) They in being defined and used in ways that are less than all TRU and only a part of TRU, as their ESINS are lower, so are their REM and place in TRU reduced. This is signified by: **COREMT** = Competitive REM in TRU. All that is both potential and actual of TRU is in a strong competition with all else, to be at least its part, if not more, or all, of TRU. It is this dynamic action and reaction between all in TRU that produces many of the changes both in and of TRU. The decreasing value of the old, as God or other Gods, is as for all terms, not held permanently nor inalterably to the same value or REM.

Since language is the primary vehicle of knowledge, as NEC are the primary vehicle of understanding TRU, we need to recognize what TRU terminology does. 1. It shows the actual basis for the role of words in human life, 2. It shows the FAF of human mass energy dimension, Á, 3. It shows how the part and role ES and CES become formally distinct in SÁ and TIT. 4. It shows how to refine formal distinction by identifying words with their actual objects in TRU when best subjectivized. 5.Together they show how to actually know our world, TRU.

VARIATION, FLUX, and INTENSITY in ROT and SUMMARY of ROTÓ

Fig 13-26 **REPRESENTATION from UNITY, FINITY, INFINITY & TOTALITY; VARIABLY COMBINED**

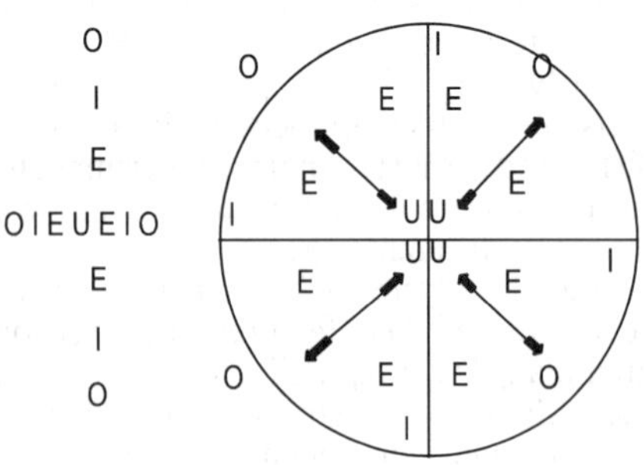

Focusing in and out of life is nothing unnatural or uncommon. In attention and acts, IA, we do it every minute of our waking day. We subconsciously sense, visually scan, and accommodate of all before us. It is of all that is of both close to us and far away. We do it with microscopes, telescopes and all kinds of vehicles, lenses, visual aids, and media. So that humans must look, sense, perceive, apperceive, apprehend, comprehend all; especially all that is meaningful, necessary, and simultaneously in continual motion to and from, and more or less, of us. Only when simplistically naive, lacking in VAPO, do we deny this. We must constantly be on guard to realize the ACEXIN, VAPO and SCARL that apply. Neither unity, finity, infinity, nor totality, alone give us adequate representation. Only as they are all combined and perfectly ordinated on the perfected object, OBT and TRIT of TIT can we begin to show the world and our relationship to it successfully. This requires a reconciliation of all its variant representations into a unified total, ROTO.

Selective variation maintains an equilibrium that continually enhances meaning. Rareness enhances meaning, as well as use and other factors. **MASPOV** = Mathematical Solution of Polarization of Variable Axes. MASPOV is guided in part by how in TRU, as in biology, there are asymmetries in their axes, parameters, polarities, coordinates and formulations. Some of these differences are commonly known and for TRU include: up and down, right and left, front and back, in and out, universals and particulars, men and women, infants and aged, worst and best, and many others. These are examples of the polarities and parameters that fit in special ways into TIT and TRIT. Unless this frame is able to represent them, as all major REM, with VAPO the representation will be incomplete and deficient. They will require both a reconciliation of symmetries, standardizations, and contents.

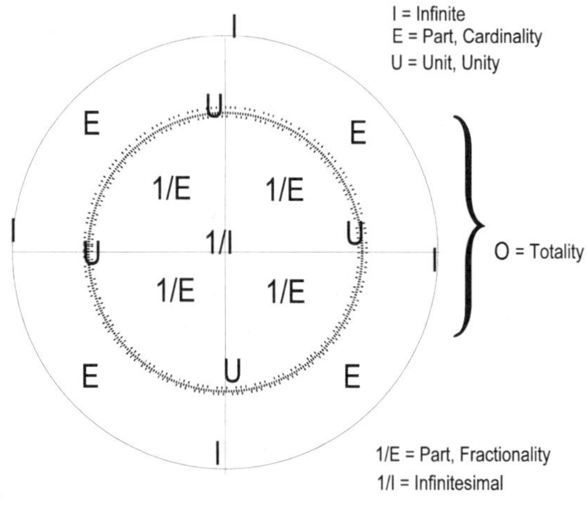

Fig 14-26 **REPRESENTATION by UEIO of TRU**

Figure 14-26 presents a similar figure to 13 with a marked difference in focus. This figure is based on the infinitesimal core instead of one. This is more like the actual centering of TIT with its focus in unity half way between the infinitesimal center and infinite periphery. This great fact of the form of TIT is the basis for alteration of form or the warping compared with physical dimensions. It is special and typical for a RU, TU of TRU, and especially for TIT-TO, TRIT, ROTO, MOT, and SPALT. It is an orientation, which is highly relative, a product of the observer-observed relationship A in TIT.

ETO or End TRU is in its continual Totality. Life is in the living of it, a natural selection in which very long term incremental advantage resulted in improvements that were most beneficial, successful, fulfilling, enabling, and enduring. It was produced by doing all tasks proportionately, properly, totally and permanently, by VAPO. So is it in our quest for knowledge and in our actions in life that direct us to ETO. The role of aims, purposes, and goals must be only by TRU, as with VAPO; no more nor less than they with probity can sustain and improve on existing TIT. The simple ideal of this is a formally set observer of a single discrete observed or object. This is the first person's observation of TIT-TO and all that is in or accompanies it.. Yet to do this, we must clearly and completely discover and show not only what TIT-TO is but what are the observer, PERT, POT and ROT.

CHAPTER 27

DYNAMICS of ROT

INTRODUCTION to ROT DYNAMICS: KINDS, KNOWLEDGE of, MOTION, FUNCTION, and RATE

Just as the world, to be shown and demonstrated properly, is composed of, and must account for, those homely and common parts and forms, so must it do so for the functions and processes of which it is made. These are often most important, taken for granted, and unrecognized because they are not usually the clear-cut objects before us, nor ordered in dynamic totality. Their meanings are often subtle and profound. It is up to us to discover, learn, remember, relate, posit and appreciate them. To successfully represent and present these less obvious and customary processes one must vividly, strikingly and proportionately aim to show what is necessary to bring out and glorify the greatest in the world, TRU. This is the reason for much of LANT, NEC, and REM. Just as in action we work anticipating our needs and ends, in ROT we must work, as in knowledge, EPI, PERT, and POT, in supposition. Supposition theory in general and its ramifications in particular show how we do not take entities, ideas, or objects as the material of our representation. We, as in the way of the investigator, take the best suppositions, assumptions, theses, principles, known ideas, information, data, evidence, and premises available, reduce them to their ESINS and bring them together as they apply to what we best suppose our representation can produce out of them. As this is directed to TRU they transform what is not, or less than, TRU into what is increasingly TRU. This transformation is highly dynamic and inclines to a balance and harmony of TIT-TO.

That ROT and its external form are dynamic, OITIO, is evident in the meanings, definitions, and use of words, both from the multilateral definitions of words as well as **REDE** or Redefinitions of words as E and ES are different. Hereby ROT is in a constant state of, input and output, selection, correction, meaning, practice and use. All great struggles that are meaningful to man represent conflicts, great differences, and oppositions in TRU. Much of their future is in their being, being TES or ES. This is what has driven life and humans in the past, does in the present, and will in the future. **WILPÁT** = Will to Live, Purpose, Plan and Proof of Á, and TRU. This is called: **LARMO** = the Law of Replication, Modulation and Objectivity. The representation, artistic rendering, or objectivization of an entity, can be little more successful than the quality and quantity of detail, precision and proportion it provides. Replication, modulation and representation are functions of their quantity, quality, and that of OBT and TIT.

It is from the unknown, doubt, vagueness and indistinction that each person, more or less has, are and will struggle to save or elevate themselves. It is by realizing the magnificent benefits and permanent gains of knowledge, truth, philosophy, science, religion and all fields that we not only achieve, but are satisfied, fulfilled, and even more, stimulated to meet the additional challenges to attain ever higher and greater achievements, TES and TRU. It is like providing a picture that otherwise is invisible. It is like making an examination with tools that man has hitherto never had. The exquisite detail and the beautiful form, ideas, concepts, principles, laws, mechanisms, and revealed dynamics unfold before us as in a great and beautiful panorama or movie. Here is the acuity, breadth, depth, beauty and magnanimity of life as our Á or S-O nature is fulfilled from the simplest initial IA to the most complex transcendent solutions to all our problems as they are in or of TRU. It is in this way that our S-O, its Á, as it is TRU combines to infinitely concentrate toward TRU, enable TRU, and proceed in TÂ and TÂS. Rates, ratios and probabilities are prime methods and means of showing the dynamics of TRU. They enable us to determine the speed of change, dynamic interactions between parts, and the degree to which we can predict outcomes.

DERIVATION, DEVELOPMENT, DETERMINATION and METHODS of the DYNAMICS of ROT

The use of words, symbolism, or any kind of representation of one object to infer or indicate the meaning of some less sensed, yet more basic and important object, idea or principle is the basis of much of the human world. It is the primary feature of S-O dynamics, TRU's Á continuum, and how much interaction there is between the ES, CES, and NEC of TRU. It was not the static form of various animals but their movements, purposes and uses that humans were most curious. The mechanism of representation and symbolism developed complex and complete means that led to the FAF shown by its principles and laws, that provided mankind with reason and a more comprehensive realization of the basic aspects, forms, needs, parts and totality of our life, TRU.

Along with systems that are inadequate we have failures in meaning which must be continuously guarded against and corrected if we are to produce the needed reduction of ROT for TRU. This is shown by POCOM, POMOM, POROM, MIRC and POPTE. **POCOM** = Principle of Cheapening of Meaning. **POMOM** = Principle of Multiplication of Meaning, or the utilization of the same term for several different meanings and definitions. **POROM** = Principle of Reinforcement of Meaning, or those processes through which our language constantly needs to be reinforced. **MIRC** = Minimum Reduction of Capitals and letters, or the need to keep our terms as brief as possible. **POPTE** = Problem of the Transcendent Parts of TRU, or the less than ideal state that exists in all about us, or encountered, that is not ES, but some fraction thereof.

Recursion is a method whereby computations or determinations are made from any or all that has preceded it. Mathematically these are formulations in which any element may be computed from the one(s) preceding it. A recursive definition is used in symbolic logic to enable a function to be calculated systematically in a finite number of steps. As the term definition indicates - is defined - we are in pursuit and a continual search for finer and finer distinctions between that which is DAOST, or different aspects of the same thing TRU. A closely related method of ROT is that of repetition so commonly experienced in routine. Periodic and cyclical functions require reliable measure, indicated reinforcement, and the guidance of value or REM. Another form and use of dynamic ROT is seen in how poetics and lyrics combine in an appealing and aesthetic way, in verse or music, to express rhythm, melody, and harmony.

GENERAL FORM of ROT DYNAMICS

The survival and success of animals and humans consist of the strength of their reflexes, capacity and ability to react, respond, and behave in ways that best meet contingencies and fulfill all needed for existence and survival. For humans this requires higher powers, as evident in our Á and SÁ mass energy FAF, to answer questions and to solve problems. It is the subjective pole and powers of human consciousness, attention, mentality, cognition, reason, valuative, and intellectual powers that are at the cutting edge of TRU. Thereby to this degree we are compelled, and it is imperative, that the response, our answers, solutions and comprehension be sought and achieved. Cleavage is not necessarily only anatomical or static. It can be physiological, functional and dynamic, as well as all combined in TRU. Contrapletes are the two opposites or poles of a relationship in which they stand against and separate from each other, as in yes and no. Contraplete denotes a basic and crucial principle of TRU, the dialectical. The external and general form of ROT is dynamically shown in metaphor, parable, song and music. In metaphor the words we use are less literal with meanings that are impressively or figuratively descriptive. Parables describe vivid and lively relationships that seek to teach a lesson in a story-like way. Their use in the bible reveals, intricate yet simple, means by which basic ideas and principles were communicated. In music and song we find various forms of representation especially in its rhythm and movement that are excellent examples of the dynamics of ROT. Even more we find an ability to represent aspects of life's dynamics that were not otherwise readily recognized nor available. In being a media of sound and feeling it provides a much broader range of representing the world around us, a fuller nature of Á. One of the most demonstrative forms of representation dynamically is mannerism. Mannerism is behavior with facial countenance or physical movements that seeks to act upon or effect another. Mannerism occurs continuously in IPR in TRUD.

The Pythagorean Theorem, $A^2 \times B^2 = C^2$, is an example of the practical applications of mathematics and relativity in TRU. It is through the demonstration of mathematical relationships between physical sounds, as in and from the octave, through the demonstration of the relationship between mass and energy and the speed of light in relativity, and through the demonstration of mathematical relationships between many fundamental principles and laws of geometry, physics, chemistry, economics and many other fields, adapted to TRU, that helps solve basic problems and paradoxes through mathematical means. The centering or focus of music on middle C, with octaves and harmonious distribution of notes, illustrates the optimal way dynamic relations in nature are adapted to TRU and become its ÁLT predominant mass energy FAF. This is called: **LOPCAF** = Law of Optimal and Predominant Centering and Focus. LOPCAF is the mean about which survival is best assured. This is like a tropistic power that repeatedly and massively reinforces the reality and actuality that makes up the FAF of TRU. For TIT this reality is the mean of its Á, L and T dimensions of which that of the combined S-O in Á expressed as ideas, concepts and words, or language, is the most predominant.

SPECIAL FORMS of ROT DYNAMICS UNITY of OPPOSITES, EQUATIONS

That the FAF of TIT largely results from unity of opposites becomes more apparent when we realize its dimensional form, dynamic mechanisms and processes. The unity of opposites can be attained only as this unity is based upon opposites which are: strict opposites, as by standing in a direct antithetical relationship to each other, representing the predominant form of its object, TRU, are proportional to the content of TRU, and are properly scaled. Equations represent a higher form of unity and equilibrium of opposites in which added factors are given to bridge and measure the difference. This is seen in COL where the power and operation of relations, associations, or identifications of two or more factors are represented. Equations are mathematical and algebraic expressions of relationships between two or more factors.. Euler equations and formulae are examples of how well equational formulation and representations of TRU can be provided. The first are any of several differential equations of dynamics especially in alternating current theory. Euler's Formula in engineering relates to the strength of structures. Euler diagram is generally significant in demonstrating the means by which mathematical methods can be applied to geometry, and these in turn to ROT, MOT, and TRU. The Euler diagram is a graphic method that employs circles to represent the relations between, and operations on, classes and the terms of propositions by inclusion, exclusion and intersection. Venn Diagramsare somewhat refined variants of the Euler Diagrams, which are also used for ROT, ES, and relations where shading is used to indicate empty areas or regions. Crosses are used for those that are not empty and blank spaces for those that are either. By these and various other means logic, mathematics, geometry, art and representations fuse. Equations can be used to supply formulations of universals and ISMS, especially ones with explicit polarities of their parameters. This is a major use of ROT Dynamics as shown by: S/O x IA x (AU to TRU) = Realism, TRU x IA / EPI = Critical Realism, TES x IA / U (k) = TRU Nominalism, TES x AU x IA = Phenomenalism, and TO x (K) x Z = Eclecticism. Equations of polarities are a large portion of the form and function of TRU. They help show how ISMS are polarizations whose corrections are so important. This necessitate depolarization along with the adapted proportion in the whole, TIT-TO of TRU. Equations of the relations between TRU and its dimensions are shown by: **TRU x LT = COL (MASS), COL(MASS) = S-O (Á) x X** Equations not only help to explain the form and make up of TRU, they lead us to the key questions whose answers provide much knowledge of TRU. **LALTI** = Law of ALT Inductive Inference, LALTI is the reduction where L^sT^s and Mass or S-O, Á are clearly set as a dimensional formulation with equal positionality. This is where and when mass and the combinations of m, l and t are given equal positional relationship we can find a simple coordinate and equal representation of length and time. In this way we can approach a reliable format for the ROT MOT, and SPALT thereby having the necessary structure for the posits of ES, CES and all of TRU.

PARTS of ROT DYNAMICS, INTERACTIONS, MECHANISMS and OPERATIONS

Differentiation of ROT dynamics consists largely of CES and especially the particular vehicles of the Á mass energy continuum, verbs, their actions and uses. Of these there are verbs that represent: 1. simple movements of the body, as reflexes, responses and tropisms, 2. more complex movements of the body, as behavior, 3. general or universal forms of motion, 4. commonly known and used, 5. are technical terms of verbs that are used mostly in some field of knowledge, 6. are used in descriptions for unknowns, 7. not yet determined which need naming, 8. are of little value to the TRU, TRUD or ROT, 9. are of great value to the dynamics of TRU and CES, and, 10. are of poor quality or misused in relation to TRU.

Verbs, as other words and all representations of TRU, have the same problem. This is the problem of providing the best possible representation with the least possible loss of distinction, TRU meaning, parsimony, or interest. The continuous improvement and expansion of language, the use of words and especially verbs, is one of the hallmarks of civilization. By REM all of TRU can be measured by the comparative analysis of verbs and the many functions of their associated words. In being dynamic they come close to much of the actuality of life, of TRU. Each verb has to be interpreted singularly as they function in TRU. As we study and posit the CES of verbs in ROT, MOT, SPALT and POS we gain a deep respect and appreciation for their power in and benefit for TIT. Verbs tend to be of high value, in TRU. They not only are very common and essential to life, they, in being dynamic, come close to the actuality of, as often large parts of, TRU.

The scaling of dimensions of a spherical ROT and MOT needs an increasing number of functions. The more the functions the quicker and less prolonged the posits become. Two dimensions are much better than one for they add a factor of ten to the field. So are three better than two, and add a factor of one hundred for the spherical model. If we are to measure parts of TRU by including other factors that define each ES or CES for posit we add to the three dimensions four, five, six and more functions to power to such measurements, or: **POMUF** = Principle of the Multiple Factorial. TRU and ROT depend upon posits and posits depend upon POMUF. POMUF shows how much each of us can improve, as we must to be TES in TRU. With even five functions we can be of a great range of magnitude in contribution to TRU.

The method of proving a proposition by showing that its contradictory involves an inconsistency or disproving a proposition by showing the proposition involves an inconsistency is called: Reduction ad absurdum. The inconsistencies shown in the opposites of polarities, diametrics and dialectics, can not usually be resolved within the system or dimension. Only from without or in its totality can resolution be completely attained. So that proof of inconsistencies by contradiction in the proposition itself is not sufficient. It is often in error. Thereby we may expand logic, syllogisms, and the analysis or symbolic and formal propositions to verification beyond their existing limits and show their operations and usages as altered and adapted to ROT, MOT, TRU and POT. This is initiated by the role of EPI or Epistemics for POT, IPTRU and TRU. It is in this way we can elevate and achieve a final reduction ad absurdum as a method of proving or disproving a proposition. This is resolution of unlimited unfolding systems and shows how by ROT when completed in MOT, REM, POS, STA, and HOM replicating TIT systems can be altered and adapted as they exist in or relate to TRU.

There is a correspondence of form and function, or shape and dynamics, not only varies, it has different characteristics. This is evident throughout nature, and is exemplified by the form of enzymes, many of whom have known structures with deep grooves at their active sites, and in some others, different relative orientation. The kinase system provides the clearest and best established example of a functionally important closing of an active site cleft in response to substrate binding. Summation, compensation and balance of activity produces and provides solidity and sphericity.

It is only in the combined state of CES, dimensions, or continua, that they actually exist in TIT. This is called: **LACFAF** = Law of Combined Form and Function. Dimensions of TIT are figurative lines transecting OBT, SUT, and ROT. Yet they exist in TIT in a more complex combined higher states of FAF, working interactively with each other. They follow the fundamental typical and special form and function of TRU, as Á, L and T. How two or more dimensions function together reveals a major aspect of ROT Dynamics. One of these, **DISDO** or the Dynamics of Integral Subjective versus Differential Objective, exemplifies the four different mixtures of extremes of the dynamic ROT and TRUD functioning both separately and together, in TO of TIT. DISDO denotes the terrific dynamics and conflicts that exist between these extremes.(SOID)

There are many positive and negative forces and powers of TRUD and TIT that make up much of ROT, SPALT, and POS. **Positive Forces & Powers:** 1. the unitary nature of human orientation, 2. natural laws, 3. relativity, the interdependency of all in TRU, 4. dimensional formulae, 5. determinism in TRU, 6. partial delineation, 7. formal distinction, 8. the individual and the species as composites, 9. natural tendency to symmetry and coordination, 10. general forces stratified time, 11. the unitary nature of TRU.

Some of the **Negative Forces & Powers** are, 1. the infinite nature of facts, evidence and motion, 2. the dissociation of half lives evident from the form MPLT, ALT, SPALT, 3. the impossibility of multiplicity in simultaneous thought, or IA, 4. the magnitude of time and population variation of the individual. The greatest positive forces and powers arise internally from the unity, relativity, form and order of TRU. Its greatest negative forces and powers arise largely externally from all actions that adversely effect TIT, from mans position as the observer, and from ES that are missing, disproportional, unbalanced, poorly distributed, and erroneous in TIT.

The best showing, representation and presentation of TRU, would be to reproduce or clone TIT-TO itself. Since this is impossible the next closest to it is a duplicate, replica, or what represents it as in this text. It must utilize all the essentials, frame, form and function that best replicate TRU. SETLAW signifies the Set FAF of TRU as Law. It determines how we can with full reliability understand, represent and reproduce TRU with specificity and sensitivity. How we can time after time exactly determine and utilize all as they are of and in TRU. The criteria and determination of the whole, TRU and ROT, as attained from SUT, SETLAW, LOTOD, SOR and TOR, is not complete unless this whole functions and is utilized dynamically in equilibrium. This is expressed by: **LODEPT** = Law of the Dynamic Equilibrium between Parts and TOtals of TIT-TO. LODEPT describes how the completed and comprehensive whole functions not alone or homogeneously, but as a totality in which its parts also function in ways typical of TRU. It is in LODEPT as conditioned by SETLAW, that we can envision and utilize the overall operations of TRU and its parts, as in ethics, to best posit their processes as they are and must be TRU.

When the levels or gradients of a dimension are demonstrated analytically on a field they run perpendicular to the axis of the dimension. So that if the static is vertical and the dynamic is horizontal their gradients are opposite, or horizontal and vertical. When two or three dimensions of TIT are shown in combination their gradients form a field or solid in which each run perpendicular to their dimension. This is called: **POCT** = the Principle of Coordinate Transfer or the Principle of Coordinate Rotation and Transposition. POCT gives us the basis for coordinating polarities as well as the basis for the greatly needed modulation of, ALT and SPALT of TRU. When the rotation and transposition of dimensions and their gradient fields are coordinated, most simply and optimally standardized and held to their VAPO, FAF, and SCARLS in TÄ, TRU, TIT, TRIT, SUT and ROT, then POCT is best satisfied.

Ordered totality shows, and helps to prove, the equality or equilibrium of poles. This is called: **LEP** = the Law of Equilibrium of Poles. LEP is how TRU, composed of constituent parts and so oriented can be presented to produce a 1:1 distribution. The poles of that body, as all its parts, stand in a state of relative dynamic equilibria. This is why for humans the deviations from sphericity as head, arms, legs, fingers, etc. prove, when reduced to its ultimate functional form in TIT, to be, or representable as, a spherical whole.

Analysis of the factors in Energy Systems is seen in the C-Ab-S/R-C-S, or **CARSC** Diagrams. It shows the beautiful intricate detail of dynamic ROT. They, as all figures if adequately adapted and TIT, can be superimposed on SPALT. We have a relationship in which capacity-S/F-condition is reduced to its static dynamic components and the energy systems to their SO or Á and length components. This, as all superimposition, is called: **REDPRIM** = REDuction to PRImary DIMensions. REDPRIM requires qualification by reduction to one meaning, degrees of freedom, degrees of form or spread, largeness of print or demonstration, and degree of importance, or REM.

Without direction and orientation in life we are confused and lost. Even with directions, if they are incorrect, or not TRU as by VAPO, we are still lost. This is why it is so important to get it right from the start. It is why we must learn what is best, TRU, and why we must be oriented, coordinated and well ordered. It shows why being familiar with our surroundings and all that is and will be needed is vital. Only through being able to place or posit oneself into the whole in some orderly, systematic fashion can a person expect to not only know what one is doing, and know where one is going, but be able to prevent waste, duplication, being lost and failure. TRU, SU, ROT, MOT, ALT, SPALT, SPOT, METAD and ELTICMA all have a prime purpose in their contribution to curing this problem of direction, orientation and coordination. There is ancillary proof of the focus, center-point and the analytical geometric necessity of ordination and coordination for TRU, and thereby SU and ALT.

As more and more concepts, words, symbols, terms, rhetoric and composition are added to human language and knowledge, the need for meaning, definition, distinction and standardization increases. This is because they tend to represent the same things, overlap and define closer and closer meanings, of finer and finer parts and processes. Yet the need increases exponentially as our knowledge expands. It is necessary that those words have the greatest meaning where they stand for the greatest ideas, most readily help to solve their problems, are the clearest, usable, most capable of being propagated, most parsimonious, and are most commonly and frequently used. Thereby a great part of TIT, Á, TRUD, ROT, POS, HOM, ACC, POT, and IPT is to reduce words, nominals, concepts, ideas and the universals of language to their maximal meanings for posit and relate in TRU

The construct of space-time will vary with the observer observed relationship. This is shown by relativity and creates warping or alteration of the classical Euclidean Geometry. The construct of space-time is not fixed by astrophysical mass, length, and time of objects nor ENIT. Neither is it fixed by AU or anything other than TIT. Since TRU is, ultimately for all observers and observed as they converge and are conditioned by TRIA in TIT, the single fixed construct which sets the state of a unified space time geometry, we automatically and spontaneously have a Unified Construct in TRU. TRU is the only fixed system relative to this observer-observed state. It is thereby possible, and imperative, through the TU and TO states to determine its mlt FAF that has a solid geometric space-time configuration and coordinate system based upon the ALT of TRU. Principles, fundamentals and universals or basic ultimates and values are established and validated only as they are increasingly defined, made formally distinct, related and formulated, or SCARLED. This is the beauty of existing knowledge. This is increased by scientific and technological demands for proficiency and maintaining strict and precise definitions. **LÁIÀ** = Law of Mass Energy, Á, Instantaneous Attention, IA. It is in LAIA or Law of Á IA, or How Á converts to Ä and Ä converts to IA ad infinitum, that we can finally understand the core and mechanisms whereby these various processes of the CNS and Dynamics of TRU come together. The scrutiny and ever finer powers of attention concentrated upon the S-R process and how these are convergent upon Á, is the key to A, S-O, and the mass energy FAF of humans and TRU. Like intentionality it works, as much of humanity and TRU works, because set processes condition us to work in this way. This is seen in the intricate beauty of genetic, chemical, molecular, adaptive, behavioral and all levels of Á, mass energy in TRU.

The dynamic power of a culture group, composed of a common language, is measured by: the number of words, meaning of words, clarity of this meaning, how precisely these words are, their ESINS in TRU, how important these words are of TRU, how well they cover or complete all of TRU, how constantly and consistently they are used, how many of these words are known by each individual, how the vocabulary of each person equals the greatness of TRU, and the intensity or ACEXIN, by which this vocabulary of TRU is used.

LANI = Language Inversion, **TANT** = the Two way Trap of LANT. LANI refers to the dynamic changes of words and languages to and from LANT and TRU. TANT is how without the promotion and universal usage of LANT and TRU existing minds not only suffer from POMOM, POCOM, LANI and the losses they produce, they suffer from the added burden of much ATRU, inefficient, and ineffective language. **THRA** = Threshold of Reduction Ratio of Systematic, Symbolic unity. THRA signifies the boundary and threshold that exists at any time in the reduction mankind is making of Á, ROT, LANT and all that supports TRU. It is in extending THRA, enhancing LANT and thereby ROT and TRU that human power, will and work can be best seen and proven.

Since words are our primary means of expression and understanding we are forced to accept - as we adapt to - their limitations and shortcomings. When we can bring words to TRU and appreciate their value in TRU or: **PEL** = Perfection and ESINATION of Language, we can greatly accelerate human language, Á, and TRU. The processes that enhance PEL are ever present in each of us, each language and culture as ROT. The great value of concepts, and especially the new concepts introduced in TRU Terminology is not only their signification and representation for communication of important ideas. It is what they introduce, conjure up, bring out, elucidate and advance in other and all great aspects, forms, functions, parts, processes, characteristics, properties and all of TRU. It is how they better provide the needed inferences, explain, and clarify the many unknowns, questions, problems and seeming paradoxes of our being.

AXIN = Axiom of Intensionality, or the nonidentity of the concepts defined by two different propositional functions, even if they are coextensive. This makes the search for rational metamathematics implausible. For any such fundamental system of semantics uses two or more specific primitive signs. They can or can't be formulated by whether AXIN does or does not hold. Only with TRU, TIT, TRIT, TU, SU and a ROT based upon the SCARLS and VAPO involved as well as limits and standards of their own, does intensionality, like IA and EPI themselves, have great freedom and nonidentity. This is why philosophy and logic have, when limited and distinct, provided us with such great tools of presentation and representation of the world.

In the study and development of a formalized language it is necessary to employ a second and more powerful language. The formalized language, as the existing concepts in language of TRU, are called its Object Language. The second, more powerful language developed herein for TRU is called its: Meta-Language. For short we usually refer to these concepts and terms, as OC, EC, ECT, and NEC, for more powerful and formal concepts and terms of TRU. OC and EC are increasingly object language, and ECT, NEC, and TOPNEC, the meta languages of TRU.

To obtain a set optimal observer observed relation or viewing and 1:1 identity with TRU we must not only determine what this first person or THES is, and what the object or THEO is, we must determine in what unified, singular relationship they are to be ordered. If we are to far in or to far out to obtain a maximal view we will not attain a 1:1 identity. The principle or law that describes this is: **LOPOIR** = Law of Depreciative Return of Multiple Imagery Return Ratio, the Law of Point of Imagery Return. That law which describes the variation in this is: **LOPVOIR** = Law of Variation of Point of Return Imagery Return Ratio. LOPOIR and LOPVOIR are to be determined by what the ideal or optimal point of view is. Also how it can be measured and to what extent it proceeds as EPI, IA, S, O, SO, OSO, SPO, SPOT, RU, TRU, SIT, ALT, SPALT, POS, STA, HOM, ACC, and POT. How the IA becomes IAT by a COAT, KOT and COGNOS of all TRU. How the investigation can standardized the S-O, THES-THEO or IAT-TIT relationship to become or provide the ultimate 1:1 = 1. This is the absolute test of a successful knowing, showing, and solution to TRIT, OBT, SUT, ROT, MOT, REM, POS, and POT problems of of TRU.

CHAPTER 28

ENABLEMENT and ENHANCEMENT of ROT

ACTUALIZATION, KNOWLEDGE and MEANS of ENHANCING of ROT

Even though often poorly representative of TRU, the many pictures, illustrations, figures, diagrams, models, and images of the world around us are highly valuable. They are large steps toward enabling us to visualize and comprehend TIT, and much else of potential interest and use. The representation of TRU may be effectively considered in three ways: singularly, in combinations or complexes, and totally. Much of all, in and about us, is either representation or can be represented as it is in or of TIT. The simplest case of representation is the word TRU.(TREQ or GOD) The very complexity and abstraction of symbolism and representation of objects-subjects, senses-things, and other means, has caused humans to forget, disbelieve, or fail to recognize their crucial and pivotal role in TRU, POT, and PERT. The greatness of representation of life, OITIO, TRU and TIT, can hardly be overemphasized. It is at the very root of mankind"s entire being. It is the great least common denominator of our existence. It makes up a large part of both the Á FAF of TRU and TRU's productivity. It is the very basic stuff, of which TRU is made. This is mans real, phenomenal, spiritual, and formative, Á nature. It is no wonder abstract thought by sounds, letters and words, is to each new person, child, youth, and student so difficult to relate to, assimilate and use properly. It has been through language and communication that representation has produced so much that is and will increasingly be TIT. Knowledge and identity are largely RÏT.

It is not always necessary that we express relativity as the observer observed or a subject object, or even object subject. It can be expressed in different ways. One of these, that has use, is **O-S-O** or object-subject-object. This is what occurs in much of our relations with the world about us. An object acts upon and stimulates us to subjectively apperceive and think about it in order to objectively react, respond, and best control it. In addition to figures in which O-S-O are the basis for ROT, there are various combinations of other parameters. One of the more important is: **SOID** = Subject Object Integral Differential Diagram. It is SOID that enables us to begin to envision ROT and TIT-TO, by bringing them together, and showing how they provide a precedent and example for MOT, ALT and SPALT. SOID gives a preliminary planar means of representing TRU. It does not include the key parameters, as length and time nor even a corrected mass energy.

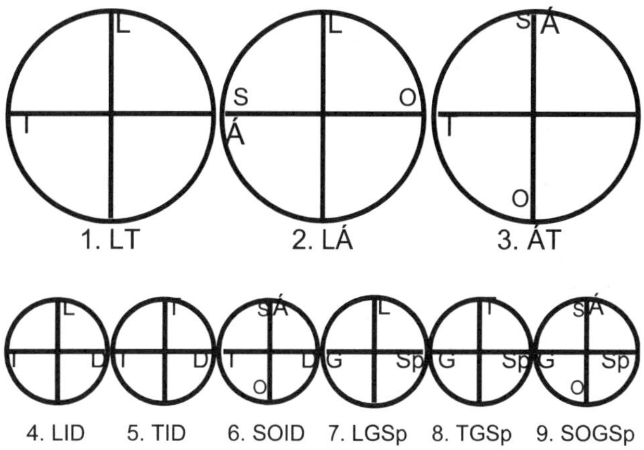

DERIVATION, DEVELOPMENT and METHODS of ENHANCED ROT

Fig 1-28 **PRIMARY and OTHER PLANES of ROT and TRU**

The development and determination of the enhancement of ROT and TRU require finding how the parameters of TRU can be represented and what the primary dimensions are. To do so we investigate various combinations of primary parameters for comparison and selection in MOT. Some of the parameters include length, time, mass-energy, integral-differential, and general-specific. Fig 1-28 shows some of these and how the coordinates of various 2D combinations can be laid out on a planar field. Both the inadequacy of 2D fields to demonstrate a concrete TO of TRU, varieties of variable parameters, and arrangement of their coordinate fields becomes clear. It is to settle this disorder and use of what best replicates TRU, or can be standardized, that is the prime task of MOT. Opposites are, for an instant fused together, then inverted. There is an inner inversion accompanied by outer inversion, for all actions a reaction, by the law of conservation of energy. Of these the differents, opposites, inversions, and reversions that exist in TIT and of TRU as well as by various means of ROT provide the determination of an enhanced ROT for TRU. These changes in opposites are typical of relativity, its mass energy or Á FAF or TRUD.

LAPLIN = Law of Planar Importance and Necessity, or the great benefits that can be achieved by breaking down the spherical or total final showing of TRU into its constituent forms. Also building up the unitary or singular initial showing of TRU into its constituent forms. In this way a planar representation and interpretation stand in the middle ground. Thereby the reality, vehicles, means, and meaning of ROT, MOT and TRU are revealed. Items 1, 2 and 3 in Fig 1-28 give the three primary planes produced by combinations of A, L, and T. Items 4, 5, 6, 7, 8, give other planes of parameters of TRU. They consist of any binary coaxial system that can be made up of any two parameters or dimensions of TRU. In figures 7, 8 and 9 we represent coaxially the three primary dimensions L, T and Á with the General to Special Form of TRU. This although associated with the differential and integral, which is more quantitative, comes from its own point of view the essential thisness, mode and form of TRU. In ALT, the ordinate or time dimension is from the infinite to infinitesimal zero at the center to infinitesimal to infinite at the right. These in comparison show how things, as they become a or the thing, develop set parameters, which when represented geometrically become dimensions. This is called: **MEXSPOD** = Means and Extremes of polarities Set Parameters which, in TRU, ROT and MOT become Dimensions. TRU. They form and partake of the proper specific and selective mixture of all.

Names seem most simple, yet they are in many ways most complex. This is denoted by: **LINROT** or Law of the Identity or lack of identity, degree, measure or Ratio of, word or Name, Represented Object, ENIT, ESIN, or ES of TRU. LINROT is a crucial concept, for it answers many questions about language as well as integration and differentiation, distinction or fusion in the world about us and TRU. If words or names are to have worth and power they must have ever greater correspondence with their FAF or ES and CES in TRU and separation from what is figment and illusion, their ENIT and ATRU. Communication of these names depends upon having a terminology no less than what exists in the greatest and most concrete of sciences. There are many ways of naming and signifying ideas and concepts, actual and potential. In the 20-21st century, especially in computer science, this has been largely and very effectively done by the use of Acronyms. Acronyms are new words that are made up from the first letters of those words which best describe their key ideas, meaning and importance. The justification and use of acronyms is both by precedent, simplicity, power, and the greatness of TRU.

We think, speak, write and use that about and before us, and especially the common parts of TRU, its concepts, names, words, and their accompanying relations and processes, as though these were precise and fully understood. Yet this is not so, just as the whole history of man has been one long struggle and record of identifying and relating or separating ideas, concepts, words, names and their reasons, then using them to advantage. So have many been unknow, poorly and wrongly known. TRU or TREQ, as GOD, has evaded us. This has been the greatest problem, and mans greatest challenge. Our thinking can be no better than the tools it has to work with. Since thinking, conceptualization and abstraction depend upon our words, language, rhetoric, reason, and logic, representation tends to be self propagating. Representation is highly interdependent and dynamically interactive with the mechanisms, operations and processes of life that exist relatively between observer and observed or SO in Á. For the Á FAF, mass energy dimension of TRU makes up a large part, and predominant aspect, of TRU

This is why it is so important that we properly and precisely represent the actual world of best human life, rather than any world. It is why language must mirror this selective world. It is why new terms, NEC, must be developed to best signify each and every great aspect, mode, characteristic, component, and mechanism of TRU. It is because god and other words are used and thought of in so many duplicated, deficient, disproportionate, and diverse ways that we can never acquire unity, totality, proportion, order and control of our lives. They do not adequately reflect the actual world, this special universe of best human life. To comprehend how god, GOD, TREQ, TRU, TIT, TIT-TO, TRIT, OBT, SUT, ROT, MOT, ALT, and SPALT compare and fit it helps to examine and show what this relative universe actually consists of. Their forms consist of a composite of many continua, dimensions, polarities, and features, including how personal or impersonal, inclusive or exclusive, loving or unloving, temporal or eternal, unified or differentiated, total or partial, large or small, past, present or future, mental or physical and many more. By expanding and comparing all that is TREQ in detail we can produce a much better comprehension of what god and all are and are not in relation to TIT-TO and each other. We can realize how much it is necessary that each of these, as all, our words and NEC, be as precise and properly represent TRU as possible. It does not exclude the use of God, it merely qualifies it. We can then show why and how to provide the mature language without which knowledge is perpetually incomplete, errant, and misleading.

By the principle of relative extraction we certify that method, especially in science, whereby concepts and posits can be demonstrated and accurately represented through inference and projection. It is the predominant method in the posits of TRU through MOT, ALT, SPALT, METAD, and SPAT. Just as essences are extracted naturally, we should apply ES and CES to our views with the best possible accuracy and posit them in the best possible way in the models of TRU. This is done while at the same time correlating and using them with others to provide the best ROT of TIT.

The principle of contrast and its refinement is produced by identity with TIT and TRIT. It is very significant in the subject object relationship and ROT. As in formal distinction and ESIN we strive to produce a reality that can be made into discrete parts, objects or wholes, best representative of the ES and CES in TIT. By contrasting one object after another, as language does continuously, we increasingly make definite and perfect the words that represent these ES and TRU, especially when all that is abstract is ES. It is the culmination of the Á or mass energy dimension and power of our being, TRU. Through reality substantiation and its total contrast we utilize the differential powers of Á or mans ability to make observations that are identical and reproducible. In summation by contrasting these reals, or S-O identities and greats, we substantiate them as they come together in totality, TIT-TO.

The methods which are geared to a replication of TRU, ROT and MOT, have a high level of beauty, both from the beauty of TIT, REM, IAT, as well as in their own right, by ROT. Much of the world about us, even TRU, can not always be measured exactly. It is impossible for the artist, or the observer who attempts to view that represented, to always do so exactly. There is often a great need for approximation. How approximations can be best refined, achieved and expanded is well known in many sciences, statistics, and mathematics. It can be used in TRU and POT providing it is remembered, recognized and allowed for. That S is not O and O is not S and neither are that which is meant or actual as it is TRU is signified by: **LACT** = the Law of Actuality or Á Actuality TRU. LACT is one of the greatest Laws of TRU for it certifies and SCARLS the nature of TRU by demonstrating the qualification and quantification of the Á dimension.

GENERAL FORM of ENHANCEMENT of ROT EFFICIENCY, BEAUTY, PRODUCTION

POMUF, or the Principle of the Multiple Factorial, is the efficiency or rate of participant presence and activity to the rate of effective contribution and production.(pg 189) It is seen throughout life especially in work and competition. If we take 4, 5 or 6 different functions and apply them to the job at hand by investigating their levels of efficiency times TRU from none to what is commonly found to be optimal possible, we will find these functions combine, for example, as multiples with:

1	20%	100%	0.2
2	10%	100%	0.02
3	25%	100%	0.005
4	40%	100%	0.002
5	5%	100%	0.0001

or an efficiency rate of only 1:10,000

It will be quickly seen that in such work or production efficiency is exponentially worsened and failure is likely. This difference or deficit of a rate at one in 10,000 is also found in the increments of TRU. It is what we find in the great interchange between the ATRU and TRU, AMTRU and the individual's relations with TRU.

For ACC POMUF is a critical principle. Accentuation, and all action, to be TRU, depends upon POMUF. It is by POMUF that a highly effective, efficient, contributory, and productive corps of dedicated persons of TES, with undying zeal, gain heroic immortality through their persistent courage, power, and predominance by, in and for TRU. Beauty and aesthetics are largely measures of value in TIT, REM.

What is of the greatest proportion of TRU should also be what is most beautiful. This is seen in human form and function, health and strength, and the quality of all that contributes to TRU. Since beauty and aesthetics are also felt experiences in art, products, and reproduction, this role reinforces what may seem to be a static, or less emotive, REM. Herein we can see the correspondence and equivalence of beauty, art or ROT, REM and TRU. Beauty may be divided into: 1. simple beauty, 2. systematic beauty or the increasingly complex, yet ordered beauty that stems from TO, TRUD, TIT, TRIT, OBT, and SUT, 3. the intricate, elaborate and finite detail in which ROT embellishes TRU by many means, and 4. absolute beauty, the ideal end of a combination of the above and all that produces both a perfect ROT, and TRU. Representation tends to be objective, and the perspective and presentation subjective, with both reversibly faced as the subject faces ALT, and the object realized by the observer. This is called: **FALT** = Faces of ALT. FALT shows how when viewing a model of TRU, representatively as an object from the outside, we are apt to only see the periphery. At best, if we even consider the dimensional structure of TRU, we see only the inner frame of TRU. So that we see neither the other aspects, makeup and form of TRU, many of its principles and none of its parts or processes. **SPELT** = the Plural or Multiple Factorial Enhancement of the mass energy dimension or Á of TRU by Language, ES, and other Participants and vehicles. SPELT reveals how much Á is more than subject and object alone, but all that is in the whole that bonds and makes them most powerful and predominant. It is in the combined and total use of SOT or Á that ROT and POT fuse in TIT-TO.

SPECIFICATION, SUPER-FORM, SOÁ EXPANSION, ROT ENHANCEMENT

LADEX = Law of Differential Exclusion Specification. There are two bases for LADEX, one is of language, our ability or inability to provide the differentiation and exclusive specification necessary, as in sames and differents in substances or particulars and universals. The other is of TRU itself. This is **SAR** where there is Actually a variation in Relevance, both from the Standpoint of Special and general Relativity, EPI on one extreme, and TIT on the other. LADEX not only reduces toward efficiency, if not simplicity, the effort and pain of achieving TRU. It provides the basis for increasing formal distinction. We seek perfection in an absolute OBT SUT, and ROT. Yet this is not entirely nor currently possible as shown by LANERA. Such perfection is evident in the early stages of DOT, and divisions of SPALT. These dimensions provide clear cut and discrete divisions of TRU that can be separated with near perfect delineation and cleavage with the following segments:

SUBJECTIVE LARGE PAST	**O BJECTIVE LARGE PAST**
SUBJECTIVE SMALL PAST	**OBJECTIVE SMALL PAST**
SUBJECTIVE LARGE FUTURE	**OBJECTIVE LARGE FUTURE**
SUBJECTIVE SMALL FUTURE	**OBJECIIVE SMALL FUTURE**

SPALT can likewise be divided into increasingly smaller segments in such a way as to have 64, 512, 1,000, 1,000,000,, separate segments. The last is where each dimension is divided into 100 equal parts yielding a total of 100^3 or one million parts for the whole ALT and SPALT. This is the construct of TRU provided through other OBT, SUT, ROT, and MOT. We can not resolve the problems of representation unless, and to the extent, we recognize and utilize how the S-O relationship or Á is equivalent to representation. This shows the bond and equivalence that exists between the Á dimension or form and function of TRU and its power of representation.

The adaptation of color to ROT and TIT adds another substantial and demonstrably effective means to ROT-POT. It can be made quantitative by giving those measures on the SPOT or SPALT scale to the naturally corresponding ratings on the color scale. (Fig. 1-29)

Among the many closely related revolutions humans are currently undergoing ROT is prime. It reaches toward the absolute, of TRIA of TÂ. This has been, and will be an even greater, power in ENT, ACC, and TIT. In this way representation, relativity, and Á are highly coterminal. Since TRU can be taken singularly as a unit, and since our perspective of it can be reduced to a single or unified sense, we may represent it as a solid object plus that needed for a fourth dimension, the observer. This may be given as a standard when reduced to, and placed at a single point. However by the Special Theory of Relativity, SER, we may also have more than one observer or observed. Either one may change, as well as by the varieties and changes in TRU so that there is more than one observed. We do not necessarily, nor always have to represent TRU as a single object of three dimensions from a single focus, point of reference or single point. The is, **LOST** = the Law of Multiple Senses and Things. One does not necessarily have to become 'lost' in LOST. Especially if we allow for the deviations from TIT-TO by ROT and POT. The multiple senses and things when interpreted from identity, oriented, coordinated, ordered, managed and controlled, as they must, are reduced to ALT, and an optimal observer, which if in ALT and TIT, can converge on the standard 3D object, TIT.

ROT RECONCILIATION & ENHANCEMENT: PARAMETERS, DIMENSIONS, and MEASUREMENTS

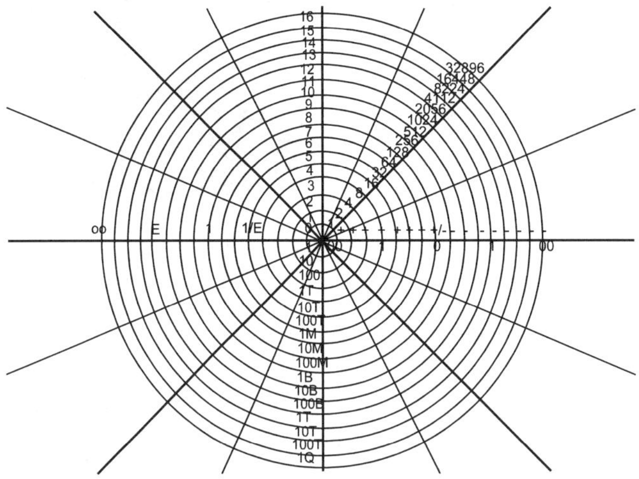

Fig 2-28 **MATHEMATICAL SCALES and FINITE RANGES for ROT & MOT**

It is through the differential of parameters, dimensions and their measures that we can lay the foundation for a good accounting and reconciliation of each and all that enhances ROT and TRU. In Fig 2-28, parameters and ranges of measures are given. It is highly representative of various aspects of objects and TRU, and is used, to show finite ranges of TRU. In ALT it becomes adapted, somewhat altered for TRU, and has its own warp and distribution. This concurs with the fact of special cases. For particulars of universals do not necessarily have to be equivalent, nor are they apt to be so, for this is their special meaning. Figure 2-28 may also be used to show different aspects of the infinite to finite, infinite to infinitesimal to zero to infinitesimal to infinite, and zero to absolute relationships. Thereby we can most appropriately and precisely relate and adapt these to TRU and its ROT and MOT.

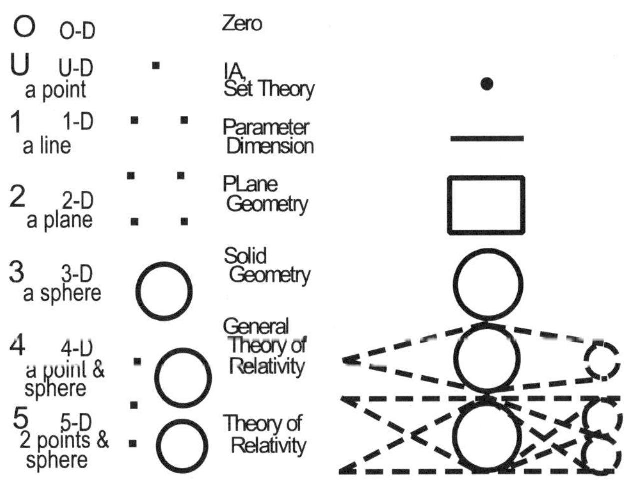

Fig 3-28, 4-28 DIMCO: VARIETIES and FIGURES For COMPARATIVE DIMENSIONS in ROT MOT and TRU

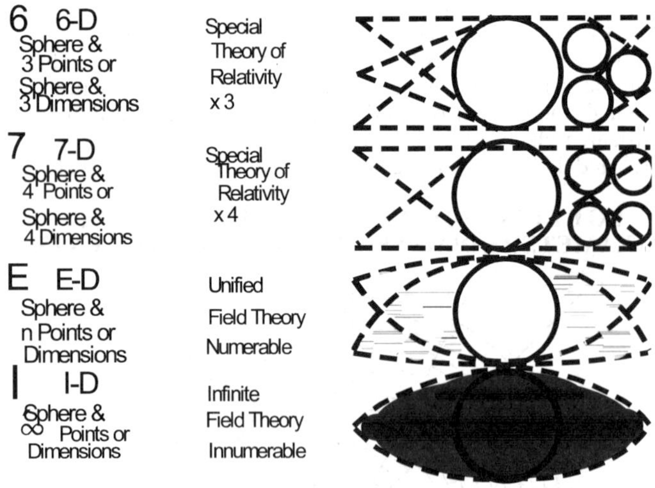

In the two accompanying figures we seek to show, as simply yet as distinctly as possible, the different dimensional possibilities. How they are symbolized or characterized, their implications, and how they compare. The higher dimensions, 4th and greater, are interpreted from two points, thing and sense. Only in the latter is the Theory of Relativity strictly applicable. For if we make all entities into a thing, observed or object, as four or more dimensions, we have higher and less fixed or set model. Only as a sense is it an observer. In this text we deal mostly with the naturally and formally necessary 3D, occasionally 2D and 4D, and rarely higher degrees for ROT, yet by ECRO holding to SPALT of TIT..

The reason why there are so many different dimensional models of the universe is that each of the preceding ones have been found to be contradicted in one way or another. Some have been disproved by experiment, observation, fact of nature, or certain experience. So that theories of models are not unified as they pertain to the universe. It only takes one such certain fact to disprove a theory, it often takes many to establish it. Yet this one certain fact may, especially for TRU, may be erroneous. It will, unless absolutely within TIT, always be multiple, if for no more reason than external - and its added - relativity.

Among the various geometrical models we can make of the universe based upon the number of dimensions it contains we have:
- 1D a line l, t or m
- 2D a plane or surface, lt
- 3D a sphere or cube, mlt
- 4D 3D space and time
- 5D 4D X observer or S
- 6D l_3t_3
- 7D ml_3t_3
- 8D combinations of 8 $m_nl_nt_n$
- 9D " of 9 or $m_3l_3t_3$
- 10D " of S and $m_3l_3t_3$
- 11D popular by theorists
- 12D numerable to enumerable possible.

Ten dimensions and one string have also been proposed in theories of models and geometry of the universe. For TIT we can find consensus in a 3D model, which with a single standardized and optimal external observer becomes 4D, TIT dependent. Quickly remembering, ECRO, that this fourth dimension is either extraneous, or internally reconciled. With 2D, as in SOID, we get a Coordinate Frame . This may be done for senses, things, perspectives, presentations, representations, or the thing itself, TIT. SOID provides a better universal means of showing some aspects of TRU and TIT.

Fig 5-28 **ENERGY SYSTEMS, SO, LEVELS, and ROT ENHANCEMENT**

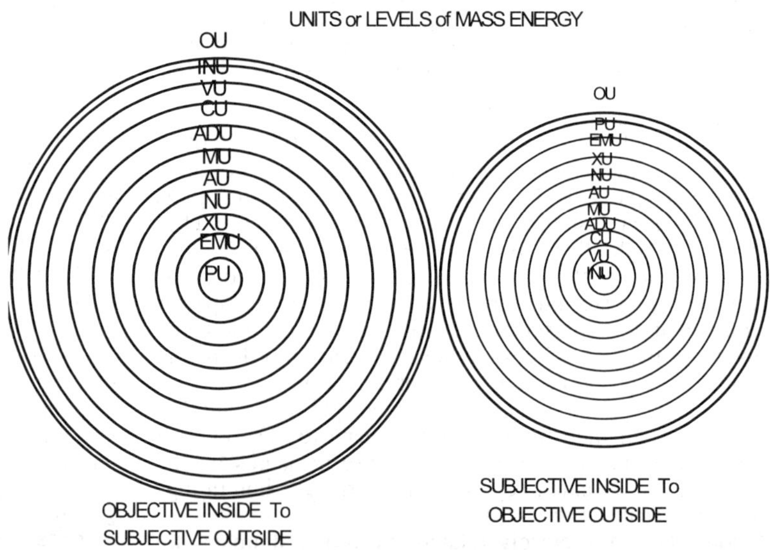

This is a target type of image or model, or if viewed sagittally, an onion-peel type of figure. It is frequently used as a handy and dramatic way of showing relationships between various divisions, parts and aspects of things, or TRU. In this particular case we diagrammatically present a cross section of the mass-energy systems that exist in TRU. Each may be superimposed on the other, or given separately by biological and phylogenetics, in geological time or psychologically and intellectually from the current moment. The contrast is evolution or cladistic, lowest to highest life forms. In the first figure it is produced by a centrifugal progression of expansion, as in evolution. In the second figure it is produced centripetal expansion, as theistically from an all powerful and knowing God, TRU or Cognos.. Their particular energy levels, when completely differentiated as things in themselves, and providing a single measure, are referred to as a Unit. e. g. IMU, CU, or ADU: the Importance to Man (REM), Conscious and Adaptational Units, etc.

The two figures show how TRU is composed of various forms of mass, or matter and energy, and how its dimension Á, is constituted in TRU. This agrees with S-O in Á, SÁ, ROT and MOT, for the higher levels tend to the subjective and the lower objective. Other levels and many finer gradations can be demonstrated. Each is superimposed upon the other in increasingly specialized ways. Transformation of energy and mass proceed by the levels that make up our world, TRU. All things will have their particular set and distribution of mass energy levels. Together they will, to a great extent, be characterized by levels of added states in combination and totality. These levels constitute much of TRU and its predominant FAF, or Á dimension. They may be reversibly developed depending upon the standards, representation and how the frame or relationships are interpreted.(LENA) The second figure is more like what exists from the IA and senses of human beings. It shows how readily we can transpose from one extreme to another in objectivity and representation.

Mass energy in inorganic and physical entities is analogous, although somewhat altered and adapted, in organic, biological and human entities. It is the Á equivalent in the physical world. For parts of the world of man this becomes more subjective and less objective as we further our focus, analysis, investigation and critique of the world. Mass energy in TRU is simply all that is not of the length and time dimensions or continua. Each and all three together, remain stable for their respective universes. The definition of the mass energy dim-ension changes depending upon what the focus is, whether general or special. If it is general, it applies to all that is not of the length and time dimension. If it is TRU, it applies to mass energy dimension that is predominant, Á and SÁ in TIT The seeming contradiction between TRU as a discrete solid and TRU or TIT with an infinitely extended periphery is called the paradox, problem and law of: **LACPERIT** = Law of Contrasting Periphery of/In TRU. LACPERIT certifies that state of the periphery and solidity of TRU where it can be one and at the same time both extremely discrete or extremely diffuse extended and indefinite. The complete integration and unification of perspectives is: **SI** = Subjective Integral, the unification of perspectives of reals in formulation. The unification of perspectives, of formulation is SUT. The difference of perspectives is: **SD** = the Differential Subjective. The unification of objectives is: **OI** = the Objective Integral. What we commonly act upon, yet is usually much different than we assume is: **OD** = Objective Differential. The unification of objectives of Reals is formulation. Such integration of Reals is called: **IR** = Integration of Reals. It is through IR, and what is at the core of TRU, that we can best understand TRU. To attempt to exclusively develop OBT, SUT, ROT, MOT, and ALT from extremes, means, or anysegment alone, be it subjective, objective, or their real combination, is impossible and highly misleading. Although such differentiation is how we penetrate deeper and deeper into the finite, as we enlarge and expand the centers of the primary dimensions of TRU, and give them increasingly precise scale and measurement with positioning of all that constitutes these substantive reals, it is not TIT-TO. At best it supplements TIT-TO and ROTO to provide an increasingly meaningful formulation and demonstration of TRU. This is called: **LUTIR** = Law of Ultimate Understanding of TRU through the Integration of **SRO**, Subject-Real-Object, by Reals. Words are equivalent to many of these reals. The problem is, however, to distinguish which of these reals are perfectly TRU, less than perfectly TRU, mixed TRU, not TRU, or against TRU.

SO, SOID, EQUATION, AXIOMATICS SUCCESS & ENHANCEMENT of ROT

As the elements become finer and finer they of necessity become more and more different aspects of the same thing, TRU. Often they, as a mixture of TRU and ATRU tend, in this differentiation, to become increasingly ATRU. So that as we approach the finer and finer of SI, SD, OD, OI, or the other poles of TRU, or even any other point in SOID as TRU, so do we approach non being, non TRU. Only by this means will the materialist come to recognize the complete fallacy of only differentiating objectivity, and the absolute idealist come to recognize the complete fallacy of only integrating subjectivity. For OD and SI when taken in opposition, are, as all poles and extremes, limits of TIT, not absolute ends in themselves. When taken together in TIT they are its modes and substance.

FULFILLMENT and TOTAL ENHANCEMENT of ROT.

What we place or posit in SOID and SPALT will partake of DAOST, and thereby have, more or less, some of each dimension. This is one of the reasons why our languages mislead us. For we tend to think of words and meanings as being specific and exclusive. We accept them as completely separate and true. However none of the words we use are completely separate and true in themselves in the actuality of TIT-TO. Emphasis upon I, SI, SD, OD, SUT and others serve a vital purpose and need. It is to them that our language and knowledge, as they represent the form, parts and processes of TRU, serve us to provide the means to perfect ROT and MOT.

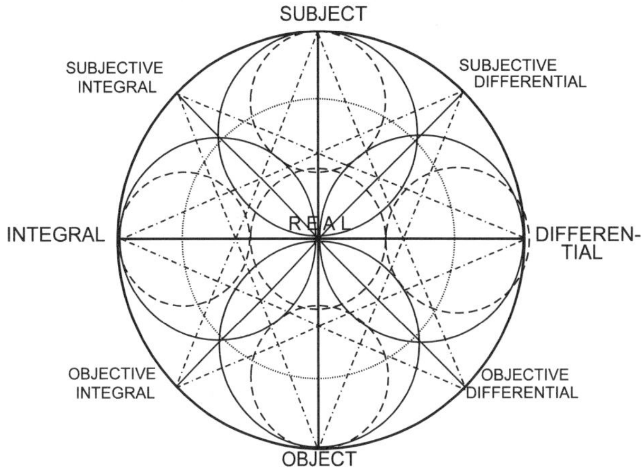

Fig 6-28 **SOID or the SUBJECT- OBJECT INTEGRAL- DIFFERENTIAL**

The SOID diagram represents two of the major basic dimensions of the relative universe, TRU. The focal point is how all that forms and functions in TRU best balances the relativity, observer observed or subject object, the predominant mass energy dimension or Á, with the integral and differential parameter or spectrum of TRU. This latter is from one to infinity. It could be oriented in other ways, yet for SOID this works best for ROT. It can be seen that the success and thereby survival of TRU is largely a function of how well it both qualitatively and quantitatively expands in both of these two dimensions, and how it at the same time maintains an ideal balance of these and other parameters, as length and time. There are three sets of inner circles which have great significance. For those that take up half the diameter of each are what generally are considered the poles themselves, i.e. subject, object, integral and differential. Those that take up a third of the diameter of each are more specific. They have outer thirds that are the poles, but centers neither subject nor object, integral nor differential. They are the real, as the real world, mixtures or means that are so much of a person's - as all human - life. It is such an identity of the real that allows us to increasingly grade and provide levels for quantifying the subject object, Á dimension, as well as the integral differential. This provides a reliable basis upon which we produce or relate to further appropriate models for TRU.

As our S-O, and S-R-O, become more complex and must be summed to be valid in the relativity of TRU so does SOID, MOT, and SPALT, form the basis upon which we can not only represent TRU but represent TRU as it is in itself, TIT, and all together, observer or EPI, observed and represented, TRU and ROT. In this we have an extended, or meta, basis with MOT, ALT and SPALT, or: **LALTI** = Law of ALT Inductive Inference, or Inferential Reformation and Redefinition. LALTI has vast implications in the definitive content of TIT. LALTI signifies how ALT and its use in SPALT for posits of ES and CES can be better accomplished by inductive inference through the reformation of TRU and the redefinition of its components. These are both achieved by the refinement and perfection of each dimension, the calibration of its gradients, and their identity with their existent continua in TIT. So that the great value of SOID, and all ROT guided by VAPO and parts thereof, is in how they enable us to show and know TRU in ways SPALT or even METAD and SPAT can not supply. **DISDO** = Dynamics of Integral Subjective versus Differential Objective. There are terrific dynamics and conflicts of the subjective and differential in TRU, as there are, more or less, with other extremes. We should expand SOID, by VAPO, of TRU. This is best done in all directions, simultaneously and to the practical limits of our ability. This will rapidly increase as we acquire more and more knowledge and use of it.

SOID, SUPERIMPOSITION and DYNAMIC ENHANCEMENT of ROT

It is how we interpret each occasion and can provide the probabilities of what is not by chance, that we can better predict and forecast what will or might be. We must recognize that what is increasingly partial in TRU is not necessarily the same as what is increasingly different. What is different in TRU may be highly non partial or total, just as the poles of many of the polarities that are basic to TRU, as large and small, past and future, subject and object are not highly partial, but different as opposites in their existence in TRU. These and other factors operate in the spectra of chance and determinability, probability and prediction.

SOID, COMPLETION, PERFECTION of ROT and its ENHANCEMENT

Superimposition, rotation and perpetual maximal reversibility back to TIT-TO is the operation of ROT and MOT. It is in the systematic unity of TROS, guided by VAPO, that we can provide a complete demonstration of TRU.

U E OE OI AI AU AE (TROS)
\ \ ⌣ \ I / / /
SU RE ROTO SOID SPALT ACEXIN (MOT)

Where U stands for unit, E - part, O - total, A - Absolute.

SOID in TRU and the enhancement of ROT and MOT is facilitated if we realize the greatness and beauty of our being, and the polarities which provide the great body of TRU. We can then attain a more vivid and actuated identity with TRU.

S = SUBJECTIVE SI = SUBJECTIVE INTEGRAL
I = INTEGRAL OI = OBJECTIVE INTEGRAL
O = OBJECTIVE OD = OBJECTIVE DIFFERENTIAL
D = DIFFERENTIAL SD = SUBJECTIVE DIFFERENTIAL

In addition to these eight major poles of SOID that may be used to represent TRU we have ID, SO and R, as well as the many combinations of R that combine with or be made to fill its respective sections of SOID. **R** = The Reals, Reality or mid portion of SOID. This is a reality in which there is an identity and continuous interaction of observer and observed.

Fig 7-28
COMPONENTS and CONDITIONS of BEAUTY and GREATNESS in ART

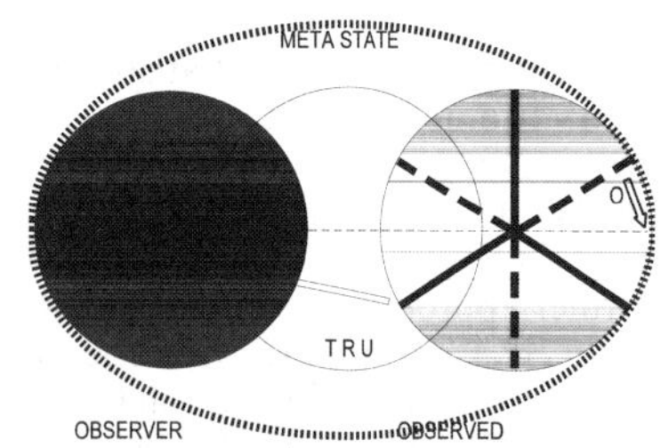

The most beautiful object in the world can be none other than that object which most precisely and fully represents TRU, the world, GOD, TREQ, or this special universe of best human life. This is one that is a 1:1 replica of TIT. It is what is most clearly and intricately visible to the viewer or observer; that observer - observed S-O, or A situation in which the greatest of observer and observed, or subject and object, or Á, occurs. This is as it should and must be, of and by TRU by relativity. Figure 7-28 depicts the relationship between observed object, the most beautiful in the world figuratively, and the observer subject, as one most astutely comprehends this world, and both as extensions from TRU. It is by such products that the basic questions and problems of humans, are to be answered and solved. Many things, even senses, can be most beautiful objects. This is evident in all art, paintings, sculpture, models, music and representations that have time-honored worth and recognition. They often depict either a great aspect of TRU, some great feature of its observer, or some way of representing TRU.

What is the place and role of language in TRU? Why do new fields and sciences need new terms? What are the different kinds of words and how best utilized? Existing concepts, words, terms, and expressions, may be signified by OC, EC, (ESINS), ECT, NEC, and TOPNEC. It is most important to know the great concepts of human life, and know their different states and kinds of meaning. To only understand existing or old concepts, OC, those words defined as entities in themselves without their essences, is to miss the great gift of language and what human experience has provided. This has been expanded by EC, definitions whose meanings are of the essence of an entity, sense, thing, or object. A few have reached ECT, or the distinct essence measured by this special universe of best human life.All TRU concepts that need to be named are signified NEC, those of highest value, TOPNEC.

MIDCOT = the Maximization of Intensities in Differentiating Coterminalities. MIDCOT denotes that operation by which we continually minimize the overlapping and indiscrete meanings of words. If we have an object, and also, ten words to represent it, how can they be divided to equal the ten major divisions of this object? As an ideal and absolute end this would assume, by the object being OBT of TIT, that only words with such divisions could be used. This we seek in ROT, MOT, ALT, SPALT, REM and POS and existing and needed language, LANT and NEC. **SUTO** = Superimposed Totality. **LASDIC** = Law of Superimposition of Different Scales. The problems involved in the superimposition or fusion of different scales, systems, and measurements are well known and commonly found. The basic problem involves one of finding their LCDs and thereby uniformity enough to retain and provide validity and reliability derived from TRU. If we are to superimpose different scales, we must do so by providing scales that essentially measure the same thing. This is relatively easy for length and time. For the mass energy dimension, Á, however, it is much more difficult. For all measurements, which have to do with parameters other than length and time or some variant of them, we need to thoroughly and profoundly discover, develop, determine, and utilize whatever other measures are actual occurrences and relationships identical with that in TRU. It is in this way we can bring, through extrapolation, interpolation, alteration and adaptation to TRU much of the information that is already known and has been set to mathematical, statistical, geometrical, topographical and other formulation. These can be superimposed when properly modified to what is most like or identical to selected in TRU.

CONTINUATION, PERMANENCE, PRIMACY and ULTIMATE of ROT

How do the Á or massenergy and TRU/LT remainder differ, correspond, and are to be separated for best demonstration if ambiguity and confusion are to be avoided. Even more if the precision of ALT and all succeeding formulation and modulation from TRIT to SPALT are to be provided. The solution of the LT/TRU remainder, S-O equality, is called: **SOLTREA** Solution of the LT TRU Remainder, and Equality with Á, (S-O). **SOLTREAC** = Solution of the Corrected LT TRU Remainder Equality with the corrected Á, (SO). SOLTREAC is the discovery, refinement and perfection of all needed alterations, adaptations, improvements, and corrections for the optimal 1:1 identity of ALT with its counterpart and role in TRU as evident in ACEXIN..

The manifold beauty and greatness of art has been recognized for ages. Yet what the ultimate of this manifold beauty and greatness is, has not been so well recognized. That art can and must provide a perfected representation of its object is a basic precondition and requirement. That art can and must provide means and vehicles for all different perspectives and presentations of its object is also highly necessary. That art can and must provide for these by VAPO, is also necessary and required, yet not sufficiently appreciated. **LURD** = Law of Ultimate Representation Determination. As much as we represent and determine things by whatever we sense or things before us, and whatever our subjective and objective may focus upon, eventually or in fulfillment it will be determined and come together, by TIT-TO and its VAPO.

CHAPTER 29

SUMMARY of ROT

QUESTIONS, FALLACIES and PROBLEMS of ROT

For any text or work of art to be a complete and exact ROT of TRU must be directed toward TRU and approach its actuality. It must be neither figment, limited, nor partial. It must approach TIT-TO. Philosophical, scientific, religious, and representative errors originate in UPTRU, URTRU, UTRU, and UTES. They impose a subjective, representative, mechanical or focal bias. More directly they all lack a precise, live direction, coincidence and reconciliation with TRU.

LONI = Law of Nominal Insufficiency. TIT and TRU are known to us through our ideas, concepts and terms. Unless we know TRU and TIT and have terms that are equivalent to their corresponding aspects, ES or CES, with VAPO, DAOST and SCARLS, they are insufficient. They do not give a 1:1 image in a precise way of either TRU or what they signify, denote and represent . They do not provide the accuracy and detail rhetoric needed to reveal the TRU, e.g. the *City of God.* Augustine, being a champion of rhetoric, showed the essence of the term 'nominal', functional sufficiency. Most all of our communication, conversation, discussion, and business is conducted by means of language. We can not overcome LONI simply. The semantic problem is commonly recognized and involves all those uses of poorly produced language that fail to achieve their intended goals. It is not conceivable that there should be voids in TIT. In fact the discovery of seeming voids is a first step toward the discovery of unknowns, basic to KOT, ROT, and TRU. Much of this reduction of distribution to VAPO and REM is obtained in MOT and revealing its underlying FAF, concepts, principles, laws, and mechanisms.

ANSWERS & CORRECTIONS in SUMMARY of ROT

"Tell me if anything is ever done" was Leonardo daVinci's admonition during his tragic life at Rome in 1513. The great penetrating, observing, formulating and dynamically oriented mind of Leonardo felt the limits of one person's yearning, search and struggle for the invisibles, causes, reasons, meanings, and mechanisms of form and motion. He diligently and happily pursued the unknown, and especially unknowns that were suspected to be of greatest use or importance. It was of the world, TREQ, or God which eluded ones senses. What it took to bring forth an artistic rendering, in drawing, painting, model or sculpture. It was what gave these images of the invisible more perfect and complete forms and motions. It is in *'Knowing how to see'* that is the crucial mark, attribute and quality of an artist, scholar, and all persons, as they must if they are live and be TRU. This is knowing how to be the best possible observer-observed, the best exercise of ones Á powers in the relativity of TRU. How to best exercise the use of VAPO, DAOST, and SCARLS in the determination of TRU. The reason why this is so important is that since TIT is so elusive ROT in compensation must be that much better.

VACOST = Validity Forms Combination, Completion, Substantiation Totality, TRU. VACOST shows how TRU can be shown and proven through the most accurate expression, completion, and integration of all great concepts, principles, laws and mechanism. This occurs when they are analyzed, posited, and interpreted by TRU, especially by the POT of ES and CES in MOT, SPALT and SPAT. .

Human powers to sustain the vast and constant massive processes of variation are largely achieved by selection. Selective Variationin language tends to maintain an equilibrium in meaning, as do other linguistic forms and forces. Content of meaning is enhanced by relative rareness. Language tends to become standardized and customized. Also, these changes produce much of the dynamics in cultures.

RESOLUTIONS and CONCLUSIONS in ROT

The great ambition to attain knowledge, wisdom, comprehension, and mastery is a natural and predominant trait and purpose of one and all in life. It is a large part of ones Á FAF. It assists one to become TES. It is accelerated by intelligence and the will to succeed. It is qualified by experience, wisdom, and Cognos. The more experienced, mature, intelligent or wiser one is the more determined they become to achieve complete knowledge, understanding, total mastery, and the dominance needed for success and survival. The finished product we see and know has complete perfect form only through our objectivizing, replicating, modeling, or showing and knowing it. For the purposes of comparison and optimal demonstration, focus and distribution must not only replicate TIT, they must be consistent, i.e. set and re-affirmative. The resolution such equilibrium provides is: **RESP** = Resolution of Paradoxes, or the balancing out ofwhat is seemingly contradictory for contribution to TRU and ROT..

We can't demand more faith than the precision, exactness, truth, relative magnitude or proportion of our understanding, knowledge, wisdom and comprehension provide. Yet absolute TRU, like God, demands complete and total faith. It demands total conviction and dedication to Á, construct, SUT and representation of TRU, ROT. It is in this total, complete and final act of S-O that TRU itself is totally, completely and finally attained. Before we can adopt any sort of standard, we must clearly delimit our universe. This is called: **LIDU** = Law of Imperative Delimited Universe. Inward by LIDU to TRU to TRIT, and outward by SU, ROT, POT, PRET to TES and TRU. Yet always returning, by convergence in relativity, to one, unity, TU, and all TRU, TO as it is TIT.

MEASURES, CORRECTIONS, ORIENTATION, ACCURACY, REDUCTION, and STANDARDS of ROT

Target theory enables us to extend and order from TU and CENU, and formulate views of ROT, MOT, ALT with more multiplicity, accuracy, proportionality and totality. It helps to qualify and limit, **QUOL**, not only by TIT, but by the object of TIT being analyzed, the characteristics and mode of this analyzing, and the perspective required. The value of target theory in TRU, can be seen in special concentric application of centripetal and centrifugal forces of the individual of and in TRU. It proves both the thisness and whatness of TRU. It enables the representation and demonstration of form produced in focus, by relativity.

That there can be much resistance to, yet value of, improved standardization is a common fact of life. It is evident in weights and measures, the mgs or metric system, as is used in MOT. So is there in world languages, and the facility of LANT and TRU. This has, is and will be a major mechanism and problem in TRUF, especially as associated with adaptations to TREQ and TRU. If we are to develop and use ROT and MOT they must by TRU, accommodate existing means and be developed in the most accurate and best way.

APPLICATIONS, USES, TESTS and RESULTS of ROT

It is by the focus and bond between subject and object, relativity, that the coordinate state of TRU is produced in TRU reality. What enables dimensional coordination and posit, or: **PERC** = Posit is Equal, or Equivalent, to Relativity and Coordination. PERC is axiomatic as shown by existing solid geometric propositions and laws. From analytic geometry, the placement or posit of points in a field is possible only when this field has set coordinates, and the posits are reduced, set by and held to, distinct parts in TRU. There are many measurements and units which are important to TRU or which add detail that is of great value. These, when strictly held to, taken together in summation, and superimposed in MOT and SPALT in TIT are of great value in showing, proving and justifying TRU. **TOVI** = Test of Validity Inherency or

Intensity. The inherency or degree of intensity can be a highly effective means of showing validity of TRU, TIT, OBT, ES or CES. It is proportionate to the correlation and exact representation of counterparts, ES, CES, TES and REM in TIT. The success of TOVI is the degree to which; the subject equals its object, the part in the entity, this part in the entity has ESIN, this ESIN equals its ES or other components of TRU, how well this ES or CES can be represented, and how well this represented and positioned ES CES is shown, known, and done so in TIT-TO. The greatest measure of all about us is, its validity, importance, how much it adds to what we have or know, how pure it is of TRU, and what its use is. With these measures as a base, we proceed to make or unmake all before us to develop the needed means and operations for an ordered understanding and representing TRU.

CONVERGENCE of ALL ROT on TRU
ANALYTIC SYNTHESIS to TRU

Analytical Synthesis for TRU signifies those methods we discover and use, that become of TRU or a part of TIT as ROT comes into TÎT.(RÏT) In the reverse so does that which is in TIT become of TIT, OITIO, much of it being used to represent TRU. ROT may be classified, ordered, and coordinated by collecting all its vehicles, as its figures, and reducing them to the dimensions and parameters of TRU. The classification of pictures gives a good example of the parameters of TRU, and allows their superimposition into ALT, SPOT, TORP, and TRU. ROT, as art, has been poorly done for TRU. They must utilize the l_1, l_2, and l_3, the t_1, t_2, and t_3, and the $\acute{A}_1$, $\acute{A}_2$, and $\acute{A}_3$ dimensions, and their various combinations if they are to best replicate TRU.

Our words, concepts and language are, in fact, usually imprecise, inactual, partial, disproportionate and incomplete. They can be no better than, and often worse than our knowledge. They are still in a very primitive state compared to their potential for perfected, complete and TRU to TRIF fulfillment. They often move in the direction of, rather than away from, meanings that are not only significant in and of themselves, but in and of TRU. To be of ESINS and ES is a natural impetus typical of Á as produced by the focus of TRU. It is produced by how useful they are in enhancing human ability to cope with life and succeed. They contain the meaning of not only their own focus, but this focus as it has greater implications in a pure, real and actual TRU.

SUMMARY of TOTAL ROT

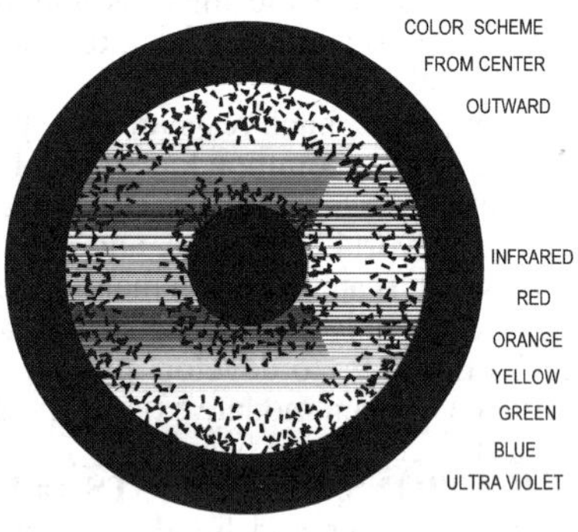

Fig 1-29 **COLOR and REPRESENTATION IN and OUT of TRU, EQUILIBRIUM and ROT**

It is through the modes of peace, harmony, symphony and total equilibrium that we develop the fullest methods of ROT. Peace is that state of TO whereby there is a minimum of conflict, negation and destruction and a maximum of VAPO, harmony, symphony and equilibrium. This produces a minimum of loss and a maximum of fulfillment, happiness, prosperity, well being, or TIT. Representations, replications, exemplifications, reproductions and productions that show peace and the dynamic optimal state of TRU will tend to be naturally agreeable, attractive, satisfying, beautiful and beneficial. Although we can represent the more action aspects of TRU, AMTRU, ATRU, RU, and RU, as in OITIO, we must always be careful to refine their roles to TIT. The means and methods of using of color and other added features achieve a fuller reinforcement and production of ROT.(frontispiece)

ORIENTATION, PROGRESSION and RECONCILIATION of ROT, TRU, & IPTRU

Often we superficially know a thing, an object, God, or TRU, yet we it totally confused us because we know neither its details, totality, orientation, coordination, proportions, nor relations. This is a subjective error and is a product of ignorance, a lack of interest and identity with the VAPO of TRU. It is because of these and the resulting confusion that TRIT, OBT, SUT, ROT, and standardization are necessary. Thereby in adopting standards we need to continuously bring ROT to TO-TIT as well as remember USTRU and URTRU.

Philosophy is directed toward a means of attaining an understanding and knowledge of all things, entities and universals, science is directed toward a means of attaining an understanding of every thing and particulars, and religion remains in a very archaic stage of understanding TIT. Neither alone nor together can they answer the basic needs for overall comprehension of TRU without a precise and perfected ROT, POT and IPT, where human endeavor must be at its finest and greatest. They are the ultimate object of the ultimate topic of all great endeavors, attention, works, be it literature, art, music, or any product that shows TRU. By TRIT, OBT, SUT, ROT, and MOT we achieve greatest direction and completion. Charts of charts or super-charts like giant spreadsheets are designations for the coordination of diagrams, parameters, dimensions, and formulations whose collections of massive information and data enable a summation of much of ROT. When posited in SPALT they provide the detail to better understand and show how all is related and located as they are adapted to, and exist in, TRU. Only as URTRU, UNKTRU, and UPTRU find correction, reconciliation and identity in TRU, and only as all ES are known by TRU, are proportional to TRU, and relate to each other, will we approach an ideal ROT of TIT. The standardization and superimposition of parameters and processes is the crux of ROT, MOT, SPALT, POS, Homation, KOT, and POT. They help to provide the necessary means of interpreting all information when applied to TRU, ROT, MOT, ALT, SPALT, POS, STA, HOM and ACC. By **LOLS**, Laws of Limits, we must remember, ECRO, that URTRU, UNKTRU, UPTRU and for standardization, USTRU, each adds to the limits and interpretation of TRU. Although each of these follow their natural, TRU, inclination, there is a loss of valid knowledge and perfect identity with TRU. So that only a perfect idea and comprehension of all TIT-TO, as it itself, is overall uncritically unlimited.

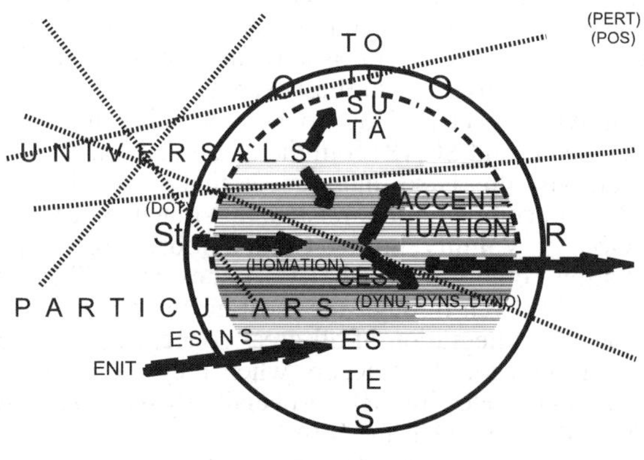

Fig 2-29 **SUMMATION of UNIVERSALS, PARTICULARS, INDIVIDUALS ROT and TRU**

In Figure 2-29 some universals, transcendent partials, and other factors operative in and out of TRU and ROT are shown. Universals and even transcendent partials, like ESINS, in not being only of TIT, show how all together provide the SUT,

ROT, and MOT of TRIT and TIT. Universals, particulars and accentuation are related to one's comprehension of, how and what can be done for TIT. As TRU progresses to TÂ there is a plan of action, by both the individual and all, as they become more and more TES and TRU. The most efficient path to an ultimate representation is to direct it at and hold it on TIT-TO and TÂ, in the most formative, functional, and valued way.

EXPANSION of REPRESENTATION of and in TRU

Facts, ideas and truths are for all to learn, remember, appreciate and use. We only need to put them together by SUT, ROT, MOT, SPALT, POS and TRU so each person can have at their finger tips the maps, plans, models, and guides necessary for their best possible life in TRU. Human life, existence and the world have become increasingly complex with TRU, and complicated and confusing without TRU. There are words of words, symbols of symbols, representations of representations, objects of objects, images of images, and mixtures in the greatest variety, to represent all that exemplifies and stands for the actuality of TIT. Only as they are fully developed, completed, standardized, coordinated, ordered and focused by and in TIT-TO can we perfect COAT, KET, ROT, TRU and TA.

FROM ENTITIES to FINISHED PRODUCT, IPTRU

REPRESENTATION COMPLETION

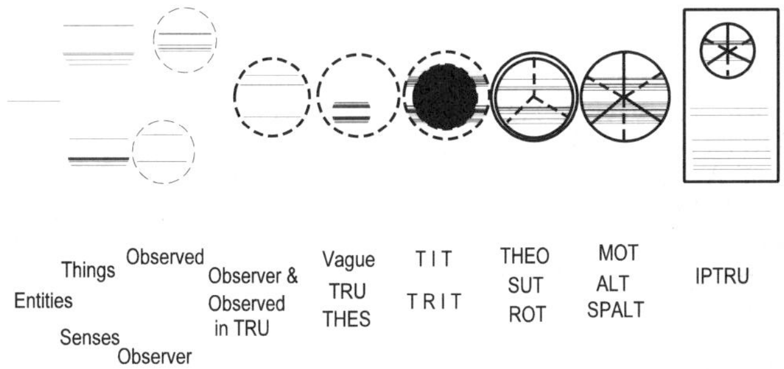

Fig 3-29 **COMPLETION of TOTAL ROT**

The form of the comparative determination of ROT and TRU is exemplified by Conformational Analysis. Conformational Analysis is a technique in chemistry and science in which a correlation of the preferred shapes or conformations of molecules, especially organic, with their chemical and physical properties is provided. When these various molecular structures are developed, perfected and compared with other chemicals and their properties, characteristics and dynamics supplied, added knowledge of them, corrections where needed, and how they exist and function can be found and used. So can this be done for most all of TRU. As they are objects which become objects in TRU, THEO, and when correlated with SUT of TIT and represented they enable superimposition of ROT, MOT, SPALT and provide their posits. This helps to fulfill the demand for needed information about the nature, form and function of TRU. Figure 3-29 seeks to show the comparative steps to TRU representation and demonstration. From entities through the observer/observed to the more subjectivized and objectivized TRU we can represent and modulate all in a formal manner for presentation.

SUMMATION of ROT and POT

That the world around us is oriented on some kind of spaciotemporal basis is obvious and commonly utilized. The coordination of all organisms, the central nervous system, and our whole being requires a high degree of set FAF dependent upon coordination. It is the focus, orientation, distribution and coordination of the primary dimensions and determinants that are such a preeminent aspect of both TRU and ROT. Their set FAF, SCARLS and dynamics are largely what makes and gives TRU its the special thisness. The primary law of objectivity, PLP, or the necessity of contrasts, differences and heterogeneity holds that if there were no contraries there would be no object, no relative universe and no TRU. For it would be all the same. Absolute homogeneity is incompatible with TRU. This is evident in an animal's curiosity, their FAF and experience enables them to perceive differences and differences in movement.

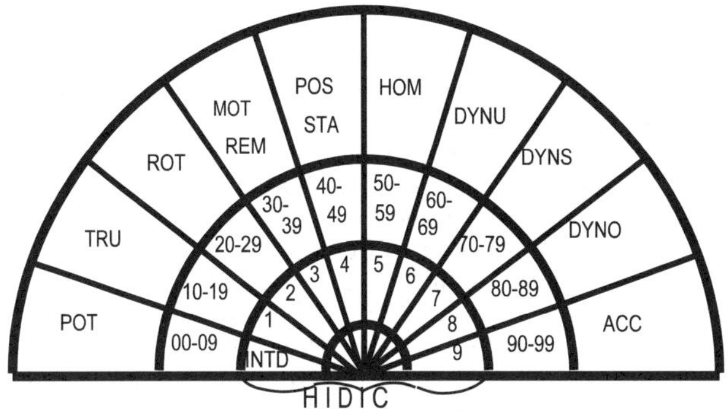

Fig 4-29 **SUPERIMPOSITION of HIP, HIDIC, OSO, SU, ROT & POT**

HIP = Horizon IPTRU Diagram, **HIDIC** = Horizon IPTRU Diagram, I Core. HIDIC produces an ordination, coordination and superimposition of different aspects of POT, IPT and other SU. The coaxial basis for, and solution to, ROT and POT arises from TÄ, TROS, TÄP, DAOST, TU, and CENU in TIT. When the dimensions of TRU are held to those adapted mlt, or ÁLT, and these oriented in a coaxial set form we have the basis for a consistent and valid representation. Since there is much of TRU that can not be readily represented, modeled, or presented on a planar surface, page, or book, we must develop and bring together all that is not well represented and have a complete, total and conclusive presentation. The above figure is one of these extra MOT diagrams that shows how IPTRU relates to POT, ROT & TRU.

What is possible is only the complete reduction of a maximally expanded TE. For the sake of consistency, uniformity and reliability an exactly repeatable final form is necessary. However this has, as an exception, corollaries or laws which allow for that degree of variation necessary from situation to situation in the presentation of things, senses and perspectives of subjects and objects. The law which expresses this is: **LACRUA** = Law of the Consistency, Uniformity and Reliability of an exactly Repeatable final Form, and its corollary: **LALVOC** = Law of Limited degree of Variation of Circumstance. An example of this is: **LASYM** = Law of Asymptotic Form. LASYM describes the form many great topics, fields, sciences, and universals take in relation to TRU. They are like the many human tribes, peoples, and cultures in migration through history. They are neither all TRU, nor no TRU, but stand in TRU like asymptotes, describing arcs or curves that sweep in from outside to the very heart or whatever segment, to depart in much the same way in another side of TRU. This is how much of what is great in philosophy, religion, and science are of TRU, ROT and POT, both in themselves, as well as how these are in TIT. That TRIT is different than TIT as seen by **TRIAPS** or TIT as TRIA is without Periphery and Sphericity. So that for ROT and MOT it is in how to bridge the gap between TRIAPS and LOSP that counts. **LOSP** is the Law of Sphericity. Sphericity is mandatory for systematic unity, ROT, MOT, POT and the objectivity of TIT and TRU. This is achieved by analyzing and demonstrating the least common denominators, parameters, and dimensions of TRU. A spherical representation and model of TRU is primarily used for the reason all is relatively based upon a point of reference or focal point, and proportions are produced by relations to this reference. Also TIT by TRIA, in being absolutely and exclusively relative, is a sphere, that is circumscribed, albeit less than absolute, infinite. It is not by homogeneity nor heterogeneity alone that TIT-TO can be represented. For if all within TIT were the same on the one hand, or different on the other there would be no basis for demonstration, representation, or even form. It would be a formless and functionless mass. This is expressed in **LAHO**, the Law of Homogeneity as a non entity. Absolute Non Relative Homogeneity enables us to discern and reproduce an object by distinguishing that which is different and ordering it with that which is the same. For objectivity, S-O and Á this is, **PLO** or the Primary Law of Objectivity, the Preeminence of contrasts, differences, heterogeneity with Á. S-O and Á are not the same from within TIT and from the outside. In fact there is a reversal, **LÁRSOIT**, Law of Reversal of Á SO Out of and In TIT. This is also signified by **LAVROT** or Law of Representative Versions of TRU. Only when taken in totality, and this given needed versions, inversions, reversions, and conversions can all needs of S-O or Á be met. RU, especially TRU, TIT are not simple objects in space of the physical world. They are inclusive and exclusive. Even by RIA and TRIA for TRU there is TRUS and TRUCH, which create a whole that is somewhat variable and dependent not only on its dynamics in time but the subjectivity and objectivity of it. This is **ORCHOS**,or Totality is Relative or dependent upon Change, Objectivity and Subjectivity. This helps to confirm the flux of S-O out of to in, and the role of Á in and out of TIT, POT and ROT. It needs to be remembered that POT, ROT and MOT are never perfect replicas of TIT. This deficit is increased in use, by **LIRM,** or Limits of Representation Management and Control.

LORFA or the (Law of) Orientation Fallacy, signifies the fact that one's attention, the individual, representation or presentation, although inclined to TRU orientation, are never perfectly, and commonly poorly oriented by and to TRU. TRU has parameters, dimensions, and least common denominators which are formed and function coordinately. It is to represent, understand and use this coordination in the best possible way if we are to be and show TRU. This is called **LACDOT,** or the Law of the Coordination of Ordered Denominators, parameters, and Dimensions of TRU. All animals, most especially humans, succeed or fail, live or die, to the extent they best function coordinately in their worlds. Along with LACDOT is **LOCSU** or the Law of Compulsive Systematic Unity. As we must be formed and function coordinately so must this order and control be made systematically of the totality of TRU, TIT, TRIT by SUT.

Much of the representation of TRU far exceeds a replica of TRU, TIT or even TRIT. This is a result of **LACREPS** or the Law of Complex Representation for Subjectification. This is how the object to be given is more than the thing itself, or even the object itself, it is the many ways it is expanded or altered for better viewing, showing and interpretation for the subjective. Just as in the perspective, knowledge, and presentation of TRU there is for representation a Sufficient degree of Proficiency by which it can be made. This is called **LASPTRU**. It is not necessary that we perceive, know, present or show TRU perfectly to make it effective and satisfy the many roles in TRU. By IA, TROS, TRIA and others it is only necessary that it be done sufficiently.

In the following proofs we pass from that for TRIT which provides the foundation for representation through the frame and order of TIT, to those new concepts, NEC, of TRU terminology which are mandated by the absence or lack of TRU meaning in existing terms.

PRUTRIT

TIT, TÄ, VAPO, Frame, Structure, Form, Dimension-al, Universe, Relative, TI, SUT, TÄP, LARF, OM, OT, ROT, LARP, ALT^n, LATAN-O-S, MOT, ALT, SPALT, REM, PRALT, LADAC , Concrete-ion, Orient-ation, Relative Focus, Spirit, Standard-ize-ation, Time, Length

PRUSUT

TIT, TÄ, ALT, Á, VAPO, Frame, Structure, Dimension-al, Form, TU,,TO, TRIT, TI, Universe, Relative, The, LARF, OT, ROT, PRENT, RUPKAL, LARF, MATES, OM, Meaning, LARP, LATANO-S, ALT^n, MOT, SPALT, PRALT, LADAC, Axiomatic-s, Focus-i, Set-Law, System-s, Systematic Unity, Unity of Opposites, Kinds of SU-T

PROBT

POT, OB, S-O, Polarity, TRU, TIT, TO, TÄ, VAPO, Form, OT, Relativity, TRIT, SUT, Á, SOT, Universe, Relative, The, LARF, OM, ROT, LARP, LATANO-S, ALT^n, MOT, ALT, SPALT, LADAC, PRALT, Dimension-s, Length, Level, Object-i've-ivity, PLO,

PRUROT

TRU, SOID, TÄ, Art, Math, Music, POT, PACH, VAPO, Communication, Computerization, Language, Logic, TRIT, TE, TI, SUT, Á L T, SCARLS, DAOST, ES, TES, MATES, ENT, OM, OT, Meaning, LANT, EC, MADAS, LARP, LATANO-S, ALT^n, MOT, ALT, SPALT, LADAC, A-s, Aesthetic-s, Beauty, Definition, Equat-ion, Geometry, Impression, Measure, Morula-tion, Pronounce-iation, Proportion-ality, Quantity, Quantifier-s, Representation, Resynthesis, Set, Show-ing, Speech, Statistic-s, Test-s, Verb-s, Word-s, Write-ing,

PRUNT

TRU, TE, VAPO, TO, E, ES, POT, VAPO, Science-tific, TI, TRIT, Human, Terminology, Communication, Computeriz-tion, Á, Meaning, DAOST, The, MATES, ENT, OM, LANT, EC, MADAS, REM, POS, ACC, PUREACLITY, COGNOS, Virtual Separation.

PRUL PREF, PROM (From Language),

Words, Representation, ESINS, PRUNT, LANT, EC, MADAS

PREC (From Essences and ESINS in Language)

TRU, Words, Language, ESINS, MATES, ENT, PRUNT, LANT, MADAS, POS, POPTE, OC, EC, NEC, ECT, LINROT, PROM, MIDCOT, LORM, MET, TRUM, NERATS, LEXPON, LACHAS, LANIC, MADIT, NAM-S, MATES, BIDA, NODA, PROHIP, POMON, POCOM, COSVAD, LOLD, NAM, ESION, PAD, PADAC, PAN, PADET, LAMS, NAT, ROMTO, TAMUD, CONES, PITRU, LAPDAM, NEQ, PROD, LORDPA, LEMSCA, LANTLA, PI

PRUNEC (From TRU terms or Language)

E, ES, CES, MATES, PRUNT, LANT, EC, MADAS, POS, TOPNEC, POMER, DEC, PRENC, LIS, PRUNEC, PRUNT, LANIK, FITIT, TRUM, PRUNEC, POE-S, RERO, CRES, ESING, COGES, ESNES, LOGES, MATES, LADEX, REDE, PODES, REDRE, LAPDIT, MADAS, SOPAN, POROM, MIRC, FADIT, LANT, THRA, PEL, NEC, LANERI, FADIC, MIDCOT, LECUM, PROLOG, VACOST, MIF, RONT, TIEA, MADEX, LESTER, PONEC, NODA, POMON, POCOM, STOM, RENT, LADCEM, REDEC, COSVAD, LARCOP, RASDU, RONT, ESION, PADAC, PAMEDS, LAPDAM, PADET, LALPNEC, LAMS, LEXOT, JUAB, FAL, STED, PONT, LAREPL, POROM, ROMTO, LEM, TAMUD, DENC, EXAD, REDRE, SONECT, LORDPA, SOPAN, LASIGN, LAREOP, CLAMP, ESNES, PLED, RENDE, >PS, %PS, WOM, LANC, LEMSCA, NECVIK, MOG, PI, PITRU, LANTLA, SACOMD.

For further proofs see: Fig 6-10, Fig 5-19, Fig 4-49, and lists on pgs. 55,147,244,264,281,328,338,370,426,493,524.

SECTION IIIA

MODULATION of TRU, MOT

CHAPTER 30

MOT INTRODUCTION, BASIS, KNOWLEDGE

This section seeks to produce the best possible model of TRU. There are many things in favor of a solid model. These decrease any doubts that may arise. They include, the thisness and thereby exclusiveness of TIT, the relativity and thereby unity and totality of TRU, the order this relativity provides or SUT, the natural dimensions which exist in TRU and from which TRIT, OBT, SUT, ROT and MOT can be provided, the totality of TRU enabling a coordinated focus on a single TIT, increasing human knowledge, powers of understanding and reason, computer science, increasing technology and thereby powers of reproduction and communication, handling of massive amounts of information, analytic and solid geometry, the axiom that a three dimensional object with periphery is a solid, set theory, topology, and many other fields and sciences which help to support and prove TRU. These also help, in other ways, to provide the means and methods needed to produce a model of TRU that can be used effectively in life. A model is, can be no better, and is to be measured by the degree it accurately, proportionately, and vividly replicates, shows and contributes to TRU and its knowledge and presentation.

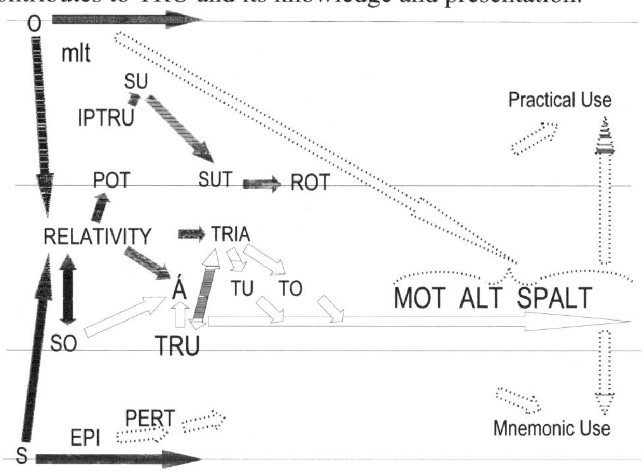

Fig 1-30 **JUSTIFICATION and PROOF of MOT, ALT & SPALT**

In the accompanying figure we show many of the most important factors prerequisite to MOT and its products. MOT is a special kind of ROT and thereby is supported by most of ROT and that precedent to it, SUT, OBT, TRIT, TIT-TO and TRU. MOT is to procure the best possible plan, guide, map and model that can be used to not only better understand and orient our lives, but to show and enable a more positive, comprehensive and powerful vehicle for applying TRU.

LAMÄ = the Law of Manifold Form, or the fact that entities and TRU do not exist in a form that is simple and static, they exist in conjunction with the FAF of TIT-TO and much that effects it in manifold and variable ways.

That the universe, our world, being, and all that is relative to us, TRU, is ordered has been recognized by man from remote times. The ancient Greek Philosophers generally agreed upon the overall form, but not the composition of the universe. This ordered universe was to them the <u>Cosmos</u>, and its study <u>Cosmology</u>. Cosmology has not been well accepted as a rigorous science. This is not surprising since it has been poorly developed and erroneously treated without proper orientaton, coordination, and formulation. As a study of the basic reasons, causes, structure and processes of the universe relative to mankind it can be readily adapted by and coherent to TRU with a vital role for ROT, MOT and POT.

Plato and Aristotle further refined and extended this world and universe by attributing its order to an intelligent and divine principle, *NOUS*. Nous was the "cause of no mean power, which orders and arranges.." (Philebus 30). This it does by its own existence, by ATO, and the parameters or dimensions of TRU. We can be quite certain that the principles of nous, cosmology and **PATS**, or Parameters of TRU, must be of TIT, by TRIA, and of TRU in a certain way by TÄ, TROS, TRIS, and TRIT. These order, set, and determine PATS thereby we can attribute it to TRU, and TRU by it. Not surprisingly this is much of the more powerful higher levels of mass energy in the Á dimension. It is both much of the form and structure of TRU as well as what gives us the prime dimension for ALT and SPALT. The rational basis for this cosmos was the word, <u>Logos</u>. In the logos and the cosmos, as they were early forms of the Á state and frame of TRU, we can see the beginnings of MOT, ALT, SPALT and thereby TRU. One of the reasons why the cosmos and TRU have not been recognized, well understood, and broadly accepted is the human failure to adequately distinguish between what is the earth, universe, world, cosmos, God and TREQ and TRU. Only when we discover, learn, know, remember and apply these various distinct nuances of TRU and its ROT and POT can we clarify and correct such errors, disbelief, and associated doubts. Another error that impairs a good comprehension of TRU or our world, is the misinterpretation of what qualifies TRU, i.e. the differences between value, mass, time, and length as in larger and greater. It is here we can appreciate the necessity of LANT, a new terminology for TRU, along with its modulation and use for placement or posit. Yet another reason why the cosmology, ROT, MOT and TRU have not been recognized is for want of: **SOP** = the Standardized Superimposition of Parameters and Processes of TRU. SOP is the crux of POT and IPT, and the means to HOM, and KOT, and ROT. It helps to provide the necessary interpretation of all information applied to TRU. It is the means by which, and product of, the final reduction to and from ALT that incorporates and combines these parameters, processes, information, diagrams, illustrations, and coordinates into an overall and completed production and overview.

EXOPT = the Exclusive, One, Optimal Picture, View, Presentation, Perspective of TRU, all TO or TRU Totality. EXOPT is how all things, S-O, L, T, OM, endeavor, work, tasks and problems, can be brought together as a single unit for the most successful and optimal action and completion. It is achieved when we obtain a single standardized view of the world, as TRU. This we work toward in MOT, ALT, SPALT, SPOT, ELTICMA, PERT and POT. Although there is, and must be maintained a 1:1:1 identity between TIT and its S-O, or TRU and its knowing and showing, this does not nullify the fact that from this singular identity other formats can be provided. This is called **LOSTÄM** = Law of One and Many,

S O and T or knowing and showing of TRU. However to uphold the 1:1:1 relation and identity of TIT, MOT and PERT we must continuously exercise optimal management and control. Our subjectivity and the relativity of the observer to the observed are not only a simple identity, they also must be more to most complex. How coordinated and ordered this can be held to and used is the key to much of human success and existence, as it is for any animal. Much of this optimal demonstration occurs when we reduce and hold the primary dimensions in MOT and PERT to what they are in TIT. This is most subtle and profound. It involves the greatest ACEXIN overview, OITIO, and presence of mind.

MPLT or **MAPLAT,** Man Point Length and Time, is a precursor to MOT and ALT. Its planar figure in length and time focuses on a human center, or man point. Although MAPLAT was highly revealing and stimulating to the MOT and POT it was incomplete and thereby largely displaced by ALT and SPALT. **SPALT** or the Spherical and three dimensional bounded model of TRU is the solid geometrical demonstration of TIT. SPALT as ALT is not only a simple model, it can be expanded or contracted depending upon the optimal use or view needed. As conditioned by the Law of Limits, LOL, as well as COL, **LACOL**, or the Correction of Limits, so do common correctives go a long way to provide the necessary missing factors of all MOT and TRU. This is discovery, as by solving an equation, of the unknown factors.

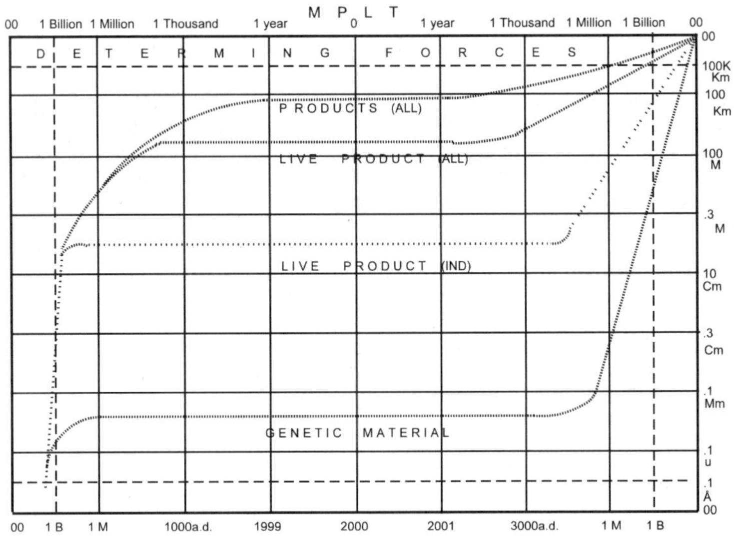

Fig 2-30 **MPLT and DEVELOPMENT of MOT and SPALT**

Figure 2-30 presents the early development of studies demonstrating TRU. It was a coordinate field of length time used to posit radiocarbon dates. It led to many of the basic concepts of TRU. It helped to make the corrections in length and time, and the needed mass energy dimensional equivalents. It helped to produce more and more refined means and extremes for each dimension. This led to the kind of measure the distributions each dimension required. The necessary existing and new concepts, LANT, and models, MOT go hand in hand. They are mutually stimulating.

If one would question, for a moment, the need on one hand, and the advantage on the other, of a great model and image of the universe in which we live, let them remember how blind and ignorant humans were before they had words, maps, globes, and images of the earth. In actuality the form and function of TRU are not altered nor warped by ATO. They seem warped by physical mlt, evolutionary development, and even our own skewed ideas of length, time and mass. Yet, what seems foreign to our sense of mass, space, and time, is the actual TRU. Thereby recognize alteration or warping not as a function of TIT, but a function of subjectivity, or of AMTRU or AU. **LARSK** = Law of Alteration or warping, not a function of TIT, but a function of Subjectivity, Individual Focus, AU and ATRU, existing Knowledge, and Representation. To overcome the bias of what mlt should be we have comparative warping that investigates the many different views of objects and entities that fundamentally differ from TRU.To overcome the bias of what mlt should be we have: **COMWARP** = Comparative Warping. COMWARP investigates the many different views of objects and entities that fundamentally differ from TRU. It seeks to describe the reasons for these differences. To understand these differences helps us see the actual changes made in MOT, ALT and SPALT.In the actuality of TRU it is not A, L and T that are altered or warped, it is their FAF in the physical world, and how they are based upon linear and non TRU scales, and how we view them. Once we discover, learn and know TRU and all its associated concepts many of our misinterpretations and disbeliefs can be readily corrected, and the kind of VAPO knowledge mandatory for TRU will prevail.

Although TRU and TIT have focus and orientation it is different than the set and standardized dimensional order we produce in MOT, ALT and SPALT. This is called: **STOD** = Standardized Set of Dimensional Differences and Distinctions that exists between TIT and SPALT. This difference is much larger as we proceed toward the orientation, coordination and total construct of TRU or MOT. This is called: **STORD** = Standardized Set of Dimensional Orientation, Coordination, and Total Construct and Ordered Differences and Distinctions that exits between TIT and SPALT. STORD signifies and emphasizes the fact that how we standardize and set the frame, construct, model and detail of MOT, ALT, and SPALT is not, does not have to be, and even can not to some extent, be the same as that of TRUS, TRU and TIT. **LORDSVE** = Law of Orientation of Dimensions Varies with External Relations, TRU and TIT.

Whether there are higher powers of mlt and ALT, what they are like, and how they are to be interpreted in terms of TRU are key problems whose solutions open the door to the breakthrough needed if we are to ever adequately reveal, show, and know our world, TRU. Is there a m2, m3 just as there is l2, l3 and a t2 and t3? By our dimensional formulation this should be automatic. By TO, the nature of TRU, and the spherical form of OBT and MOT this becomes more certain, although in each of these the higher powers are much different, and each with their own SCARLS. Since sphericity is largely a function of l3 or volume, we must correct the bias of this belief and envision its actual existence as ALT for the solidity of TRU, TO, TIT, TRIT, OBT, and MOT. Their best identity and replication involve what is common to each with differences being recognized, accounted for, and used properly. Therefor when we reduce such a spatial figure into different, yet actual for TRU, dimensions that are the length, time, and mass-energy of TRU and MOT we have a seemingly new kind of spherical object. Yet it is this, that such a 3D figure naturally and coaxially provides for our world. This is a correction that humans have heretofore failed to discover and understand. Yet is critical to a successful cosmology, representation, and presentation of TRU if we are to survive, have ATIM, an all time mastery of things, and TREQ is to be TRU, and TRU TIT-TO.

By initiating the process of understanding or being subjective of an ENIT or TIT we add something more. This is objectivity, what an entity becomes when it is made into an object to be apperceived, understood and subjectivized. This is the nature, form and function of our world. It is not any universe with an infinite array of indefinite and different entities, things or even senses. It is relative, and relative in a certain and set way, in which all is more or less related. Much of this relationship depends upon its recognition, understanding, apperception and subjectification. This universe is both objectivized and subjectivized from without and within, its A dimensioin. By extremes knowledge of means and polarities, continuums, parameters, dimensions, gradiations and the finite can be better understood. Humans can never achieve any successful understanding of this world until we learn and use the knowledge of relativity, the observer observed, or subject object, and its existence, Á, in all its manifestations and implications in and of TRU.(OITIO)

	ENS and RUS	TIT	SPALT	METAD, SPAT, POMOT
LENGTH = l_1 = L_1 for TRU	MANY LENGTHS DEPENDING ON ENS OR RUS	SET LENGTHS FOR TO MANY SCALES FOR TROS	SET LENGTH AND SCALES OF OBJECTIVIZED TIT	LESS SET, SCALED FROM SPALT TO USE
AREA = l_2 = L_2 for TRU	MANY AREAS DEPENDING ON ENS AND RUS	SET AREAS FOR TO MANY SCALES FOR TROS	SET AREAS AND SCALES OF OBJECTIVIZED TIT	LESS SET, SCALED FROM SPALT TO USE
VOLUME = l_3 = L_3	MANY VOLUMES DEPENDING ON ENS OR RUS	SET VOLUMES FOR TO MANY SCALES FOR TROS	SET VOLUMES AND SCALES OF OBJECT TIT	LESS SET, SCALED FROM SPALT TO USE
TIME = t_1 = T_1 for TRU	MANY TIMES DEPENDING ON ENS AND RUS	SET TIMES FOR TO MANY TIME SCALES FOR TROS	SET TIMES AND SCALES OF OBJECTIVIZED TIT	LESS SET TIMES SCALED BY SPALT TO USE
DURATION = t_2 = T_2 for TRU	MANY DURATIONS DEPENDING ON ENS AND RUS	SET DURATIONS FOR TO MANY SCALES FOR TROS	SET DURATIONS & SCALES OF OBJECT TIT	LESS SET DURATIONS SCALED FROM SPALT
VARIATION = t_3 = T_3 for TRU	MANY VARIATIONS DEPENDING ON ENS AND RUS	SET VARIATIONS FOR TO MANY SCALES FOR TROS	SET VARIATIONS & SCALES OF OBJECT TIT	LESS SET VARIATIONS SCALED FROM SPALT TO USE
MASS = m_1 = $\acute{A}_1$ for TRU	MANY MASSES DEPENDING ON ENS AND RUS	SET MASS FOR TO MANY SCALES FOR TROS	SET MASS AND SCALES OF OBJECT TIT	LESS SET MASS SCALED FROM SPALT TO USE
MASS ENERGY = m_2 = $\acute{A}_2$ for TRU	MANY MASS ENERGIES DEPENDING ON ENS AND RUS	SET MASS ENERGY FOR TO MANY SCALES FOR TROS	SET MASS ENERGY AND SCALES OF OBJECT TIT	LESS SET MASS ENERGY SCALED FROM SPALT TO USE
MASSENERGY = m_3 = A_3 for TRU	MANY MASSENERGIES DEPENDING ON ENS AND RUS	SET MASS ENERGY FOR TO MANY SCALES FOR TROS	SET MASSENERGY AND SCALES OF OBJECT TIT	LESS SET MASSENERGY SCALES FROM SPALT TO USE

Table 1-30 **COMPARATIVE DIFFERENCES OF PRIMARY DIMENSIONS**

In order to best understand TRU it is necessary to learn and separate the differences between the various states of each primary dimension. How they exist in TIT as compared with other entities, the standardized model of TIT, SPALT, and other models of TIT. There are fundamental differences in their FAF in TRU compared with that in other entities plus their form in a TRU that is objectivized. These transformations are largely a result of being TRU as opposed to being ENITS, an or different, entities. This is what separates them from the astrophysical world void of their predominant form and function in our world, TRU. Their only fully actual form is what it is in TIT, by VAPO. It is a most unusual difference in form for mass physically when compared with its equivalent in TRU, Á. Even physically we must look at m, l, and t, not so much as simple dimensions, but as states of the much greater dimensional forms, spectrums, combinations, and continua which they present. To transpose these to TRU is not readily achieved unless we do so with an open mind, and consider all that goes into their form as tabulated in 1-30. If these are not learned and applied one can never understand TRU, MOT and SPALT and all subsequent uses will be flawed.

Much of the proof of MOT rests upon the reliability of the mass energy dimension of TRU, Á, and its existent FAF, continuum, and gradients, SÁ. **PRAT** = Proof of Á by TRU. The greater and better we can learn, know, show and produce TRU and ROT, the better MOT, ALT, SPALT and all will be. **LAVOF** = Law of ALT Inductive and Inferential Reformation. LAVOF defines this state in which ALT, as it provides for the structure of MOT, is not so much determinant as it is resultant. When LAVOF signifies and emphasizes the role of length time in this induction and inference we have: **LALTI** = Law of ALT Inductive and Inferential Reformation, from Length and Time. Because of its special emphasis upon the Á dimension and its reformation from m through TRUS it warrants separate status as a term of TRU. Proof of the Modulation of TRU, **PRUMOT**, is provided by: TRU, TO, TRIT, TI, TÄ, SUT, ALT, ROT, VAPO, Á L T, geometry, SCARLS, human, world, universe, relative, the, LATANO-S, ALT^n, LARP, SPALT, LORVA, POS, LADAC, PRALT,
ACC, Centering, Cosmos, Cosmology, Model and Modeling.

The difference between what we commonly believe or even scientifically hold to be the dimensions - and their gradients - of our world, what they actually are to an observer, the observer, TES, and what they are in TIT-TO has puzzled man for millennia. The temporal was replaced by the eternal in God, and was the theme of the resurrection, and justification of Christ. The extension of the eternal, to eternity, and the eternity as it exists in TIT-TO was held to be known by St. Thomas Aquinas. Eternity is known both because it has no beginning nor end, and was a whole, without overall succession. This helps to provide the precedent, as well as reason for, the FAF of time, and these as they make up TREQ and TRU. This helps us understand the bridge between ourselves, TES and TIT-TO. Yet to understand this we must discover, learn, and reason by what the overall form and function of time is in TIT, TRIT, ROT, MOT, and ALT.

Here, also, we must turn to the great tradition in religion, that through vast human experience, has taught us that if we are to know the world or actual universe, we must do so as it relates to humanity, and the God that represent the best in humanity. St. Thomas Aquinas held a relativistic conception of time in TRU. Being immutable externally, without beginning or end, time in GOD has an eternity. As a uniform being, with internal movement, time follows movement. Time thereby is more than physical time, but time, change, and variation, as they exist in the actual world, GOD, or TRU.

We must always remember and reconcile our differing views with the fact that the overall object also differs. For instance, what is the difference between TRU and SU? As in the difference between the ideal and how we strictly speak of reality, we continually encounter a varied object. TRU is the relative universe or world, which as TIT or TRU in and of itself, is not entirely the same as is formulated, represented, presented, modeled, or otherwise conceived or shown. We must identify with the relativity of TO of TIT 'as it is', in and of itself. It is never quite attainable, as in this IA, we are less than TRU and TIT. This is, **DIRET** or The Diminutive Degree of Reality of TRU. SPALT, ALT, MOT, ROT, SUT, OBT and TRIT are the means to TRU, representations and forms that help best show and utilize TRU. They are the one and all vehicles and manner which enable achievement of optimal identity with TIT-TO and TRU in all its aspects, modes, forms, functions, parts, and processes.

CHAPTER 31

FORM of MOT

GENERAL and SPECIAL FORM; SEPARATION and FUSION of Á, L and T, AXES in MOT DIVISIONS, DIMENSIONS and CONSTRUCT TRU, MOT UNITY, SU of MOT, RELATIONS, IDEAL

It is how we transform TIT and TRIT into OBT and SUT as these are reduced to ROT, MOT, and SPALT based upon the primary frame and structure of TIT. We do this by using its dimensions, and modeling them in ways that best preserves their FAF in TRU as well as yielding the best possible product. In order to do this certain key assumptions must be made. These are based upon VAPO, the SCARLS of TRU, the basic principles of TRU as well as those of ROT and MOT. These include those limited yet necessary modes and standards that render SPALT somewhat different than TIT. The degree to which it can't be identical is the degree to which these separated modes and standards are developed in ways that replicate or coincide with TIT-TO,.

We must remember that S-O and Á are far different as well as being much the same. S-O is the relationship from the most simple to the most complex between subjects and objects. It is what our instantaneous attention or immediate awareness may be broken up into, its looking from and looking at aspects. It is our thought's relationship to the object thought about. This object is far different than thing, or simple physical entities. This is evident from the fact that the object upon which the subject is attending may be another subject, or some mental phenomenon or sense. So to is subject and sense to be differentiated. Sense is strictly mental, as thing is physical. Yet subject may be like an object, a sense and a thing. This is called: **SOTS** = Subject-Sense-Object-Thing differentiation. SOTS helps us remember how S-O - as Á - are much different, and the OITIO mechanisms involved.

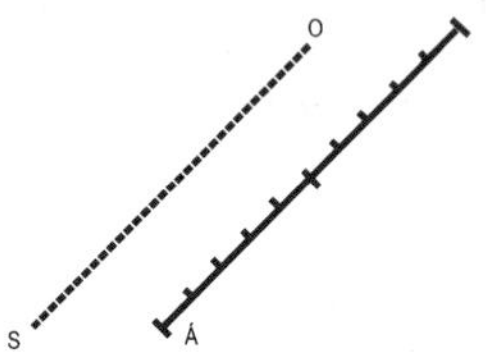

Fig 1-31 **THE SEPARATION and FUSION of S-O** This figure shows S-O as it is a thing-in-itself and as it is Á, the mass-energy dimension of TRU. We need to show how Á is to be interpreted in terms of some of the better known theories, concepts and principles of form and basic relationships. In the pragmatism defined by C. S. Peirce "consider what effects that conceivably might have practical bearings, we conceive the object of our conception to have. Then our conception of these effects is the whole of our conception of the object". S-O of our being is practical, as is it also most vital that we be oriented, coordinated, and able to order and posit all. This includes how we exist in the dimensional and formulated frame in which we live. It is a dimensional frame, not so much of lower order = mass, length and time, as it is of their second and fully by their third order continua. So that we must always think of the actual world, TRU, in its spatiotemporal actual TRU mass energy total form. In the full powers and glories of our vigorous being we gladly piece together all as they beautifully fit into this imperatively delimited universe. This is called: **LIDU** = Law of Imperatively Delimited Universe. Within LIDU and the Objectification of TRU and through TU, TE, TI, LOTAN, LÂTANO, LATANOS, OM, COGNOS, ELTICMA, SÂ, and TÂ we can find our greatest and ultimate, absolute identities. This absolute is not only one of fact, it is also even more one of need, the practice of Á.(LATAN)

LAX = Limits of Axes. LAX is where and how the parameters of TRU as the axes of its dimensional structure are limited by the modular construct by which TRU is formed and represented. If there are two axes on the coordinate field or if the parameter's functions are charted as two axes or coordinates, like l and t or geographical axes and time, or language time, we then have within this construct limiting factors. This will limit, warp, or bias the 1:1 representation, demonstration or modulation of TRU.

If man is such a polytypic species, one in which there is great diversity of races, groups, geographical, linguistic, and cultural divisions, then why is it and how can it be, that TRU is so concrete or at least can be treated as a unified massive whole, and represented as a solid object or sphere? It is not so much because of human genetic or physical, or even physiological makeup that we are actual, the entity, the being, or TRU. Although these play a major part and are closely correlated, there are much greater and more powerful forces, reasons, principles and laws in operation and at the root of TRU. That which makes TRU a unit are all those forces and powers of predominance which produce a great centripetal and converging dynamic that holds it together. Some of these, for both individuals, groups, and TRU include, survival, common interests, knowledge, culture, goals and purposes, feeling of love, compassion, and survival. The fact that TIT is characterized by infinite extension, infinite intension, and infinite levels of parts, all of which partake of marked differences and dynamics, is analogous to the inscrutability of God. TIT is not absolutely formed, but bounded and limited largely from the outside, being an object or its observation. Yet because TIT is not absolutely formed does not mean its object, ROT, MOT, ALT or SPALT can not be well produced. For TIT, by ATO and TRIA, is a whole and set, which is amenable to demonstration. This is how by being an object reduced to limit with a discrete periphery and form provide the set for ROT, MOT, ALT and SPALT to represent TIT-TO. This is what enables and gives TRU objectivity, absolute form, structuralism, sphericity, and enables us to provide a rigorous treatment of TIT, TRIT, and TÄ, by SUT, ROT, MOT, ALT, SPALT, POS, HOM and ACC.

In figures # 2-31 and # 3-31 we show how the form, content, divisions and parts of TRU can be demonstrated. We must remember that a simple planar circle can't, in more than one way, adequately show TIT. In the first line we represent different contents and ways of representing TIT. In the second and third lines we show a progressive increase in the content, or levels with equal parts of TRU. It should be remembered that there are an infinite number of divisions. Also, they can be represented in a great variety of ways. A major reason for showing these divisions is to seek the best, or that level in which each division is discrete and of equal size. Also how they can be used most efficiently and reliably by POA. There are many ways in which the divisions and partitions of TRU can be demonstrated, even within the same primary MOT. These help better vision of TIT, ROT and SPALT and show some of the problems and their solutions that are encountered in the remainder of the text.

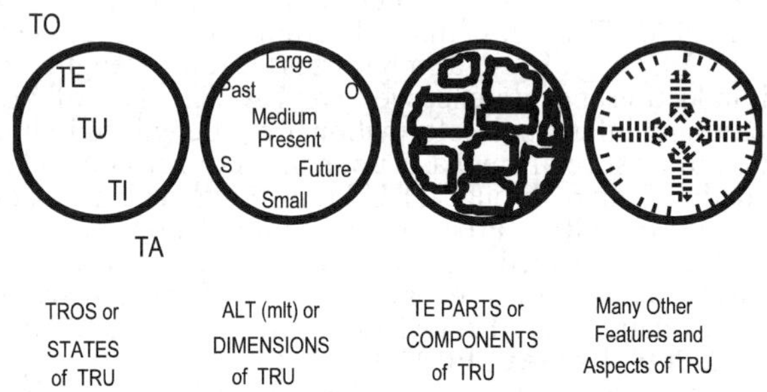

TROS or STATES of TRU | ALT (mlt) or DIMENSIONS of TRU | TE PARTS or COMPONENTS of TRU | Many Other Features and Aspects of TRU

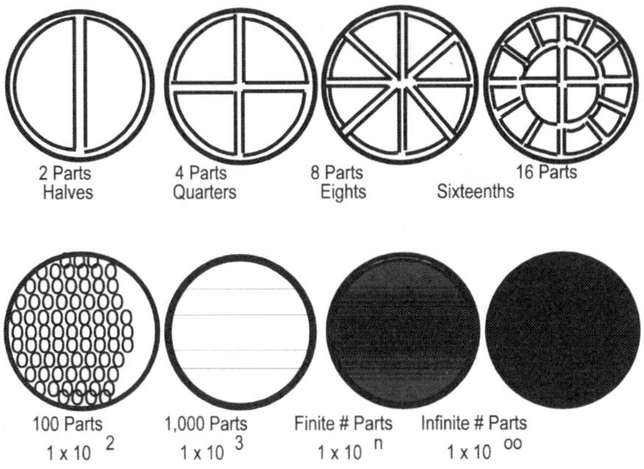

Fig 2-31, 3-31 **CONTENT and VIEWS of TIT and MOT**

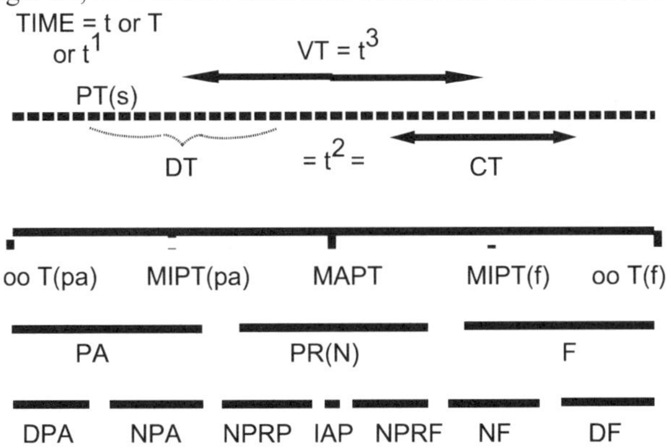

Fig 4-31 **TIME RELATIONSHIPS and REPT**

This figure shows some of the different components of time, how they are viewed, and can be interpreted in ALT of TRU. **REPT** signifies Resolution of the Paradox of Time. The time paradox is evident from how it seems that nothing of the past or future can be now. "What is past is past, What is gone is gone, what is in the future is not yet, and what is the use". All these are common expressions of what seems to be certain to the casual observer and especially to those who 'believe only what I can see'. Yet on more careful analysis and with deeper intuition, the relations, interdependence and, even to more or less extent, the identities of the past, present and future, are found to be more and more valid. This is especially evident with their reduction to TRU. This is resolved by: the through and through nature of all that is of the same in TIT, the through and through nature of much of what we are and what is TRU, the mandate of ATO and TRIA, PIP, and a recognition of the limits, deficiencies and fallacies of our IA, senses, and things.

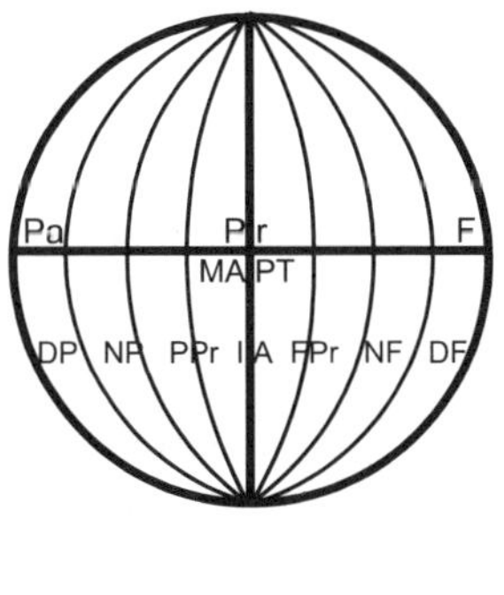

Fig 5-31 **UNIVERSAL TIME DIVISIONS**

Abbreviations used in the t or time dimension in MOT or modulation of TRU are:
t or T = time.
Pa = Past,
Pr = Present, Now
F = Future,
PT = Point Times,
DT = Duration Time,
CT = Change Time,
VT = Variation Time. MAPT = Man Point Time. MIPT = Mid Point Time or half the distance from infinite past and future to now. DP = Distant Past, NP = near past, PPr = Past Present, FPr = Future Present, DPR = distant Present, NPR = Near Present,
NPRPA = Near Present Past,
NF = Near Future,
NPRF = Near Present Future,
DF = Distant Future.
oo T = Infinite Time
oo T(pa) = Infinite Time in the past
oo T(f) = Infinite Time in the future
t^{1-3} = Temporal States
t^1 = Point or Linear Time
t^2 = Change of Time
t^3 = Variation (of time)
Also in contrast to infinite time

$1/\infty$ T = Infinitesimal Time

$1/\infty$ T(pa) = Infinitesimal Time Past$1/\infty$
T(f) = Infinitesimal Time Future

DACOD = Denominators Attained through Comparative Deduction, or those denominators, integrals or objective integrals that are basic to determining the form of TRU. DACOD can be more highly refined when we separate from these comparative and similar posits the SO and ID characteristics, their input, output, capacity, vector and other factors. **TODS** = The Test of Dimensional Sufficiency. As we compare and separate all common posits and as we enumerate and formulate their common dimensions, TODS can be developed and determined by the number of remaining features, characteristics, frame and ES for which the dimension is lacking. We know, for sure, that TRU as TIT extends from infinite length to infinitesimal length, from infinite past through time 0 or the present, to infinite time future. We likewise know that the extension of S-O has rapidly expanded as TRU has progressed. Man has become increasingly cognizant and productive. This mass energy parameter has become more and more the predominant energy level and determinant of our substance and mass special and typical for TRU. Not only have the subjective vastly expanded, so have the objective. What we know for sure helps to establish the means, extremes, distribution and coordination of dimensions, separately and together.

When the unknowns and knowns are discovered, learned, combined, integrated, and improved we are left with those basic problems that can be reduced when vigorously concentrated upon and attacked with sufficient power. Of these there is **LAVDEN** and **LAVSDEN** or Laws of Variable Density, and Variable Set Density, for the whole, dynamics of TIT. Although physically humans in being composed largely of water have a specific gravity of approximately one, we can't assume this for all TIT. This will also be a determinant of A. LAVDEN is not a chaotic and variably dynamic total density, but one of differences for different parts. For the whole, TIT TO TRUD, it is LAVSDEN.

What we are not so exactly certain of are the exact time and length gradients. We have an effective standard in the metric system. We roughly know that the center of TIT must be somewhere in the mid reaches of length and time that are familiar and basic to human existence and survival. It can be determined and measured by the size and weight of the live-product. Refinements from other determinants will take time and work to set precisely. An average size, weight, and duration life time by TIT is a major determinant. Yet the differences in center-point, although not established perfectly, are relatively determined and not far separated when considered in relationship to the extremes of time and length of TIT-TO and TRIT. Yet since the means are so relatively enlarged, and since posits must be more precise, it is essential that the measurements of dimensions and their center-points also be more exact. In this way what is less certain, when vigorously pursued, refined and perfected also helps to produce a more reliable and effective MOT. The infinite and infinitesimal reaches of TIT-TO are not invalidating to either TRU, ROT nor MOT. They are corrected by TRIT, OBT, and elliptical geometry to confirm SPALT..

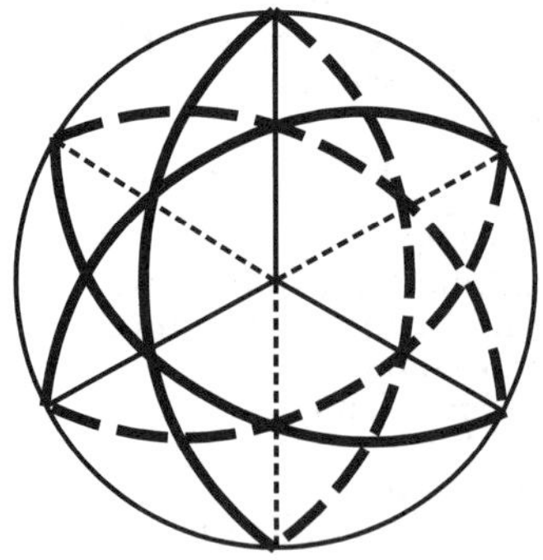

Fig 6-31 **PRECISE DIMENSIONAL CONSTRUCT of MOT**

Just like the man, the world, and God can not be understood or worked properly if we do not know what they are, or have the information and measurement of their dimensions, frame and sphere, so must we have MOT, ALT, and SPALT. How and to what extent does a crystal clear object divided and scaled into its exact dimensions and presented with the best possible posits of all it parts, processes, principles and mechanisms provide us with 1:1 identity and a perfect duplicate of its object and TIT? We can infer that TRU, as a single entity, can be objectivized and represented as round from our S-O capacity to observe and since it is exclusive and reduced to modulation. It will have clear-cut dimensions, as any RU, if these are primary and reduced to coordinates and structure. That these dimensions can be exactly scaled is inferred from the propositions, proofs, axioms and laws of geometry and physics. It only depends upon and remains for our information and precision to do so. That the parts and aspects of TIT can be precisely posited can be shown by analytic solid geometry if we have an excellent and valid frame, and as these parts and aspects are made increasingly distinct and evaluated in the ways necessary for precise posit or placement in the model.

That most things have two sides, being unable to slice them so thin they have only one side, helps show how single dimensions, as length, time, or mass energy do not exist alone. Even more it shows the importance of ACEXIN in TIT-TO, and the finite to infinite gradients of their dimensions. The extreme L^3 of the length continuum for TRIT is the volume of a perfect sphere. In conjunction with $\acute{A}^3$ and T^3 the perfectly spherical form of TRU is represented. This is how SPALT has closure, a closure that replicates its set form. It should be remembered, ECRO, that for TIT and TRU these will be increasingly different.

FOSK = Fallacy of Skepticism by Superficiality, or failure to have an adequate frame and to fill it sufficiently to meet that necessary for TRU and MOT. This is the common state of confusion mankind, has been and, is in. Humans will continue in such a state in the future until, unless and to the degree, they discover, develop, learn, positively remember, and best use TREQ, TRU, ROT, MOT, ALT, and SPALT, or their equivalents. Other causes of insufficient TRU understanding and fulfillment related to MOT are: lack of construct, detail, lack of evidence or facts, proofs, distribution and proportion. **FOSP** = Fallacy of Skepticism by Disproportion, **FOSPA** = Fallacy of Disproportion in Parameter, **FOSPEM** = Fallacy of Disproportion in Magnitude. Deficiencies and fallacies in geometric framework are more often the result of perspective or demonstration than they are in TIT, TRIT, OBT, SUT, ROT, or MOT. The application of mathematical proportion by Pythagoras was the secret beauty of form. It is what enables mankind to achieve a perfected perspective of the world, TRU. It is knowledge by proportion that realizes importance, and adds meaning. The proportional selection of meaning and knowledge brings and holds all to TRU. To achieve proof of MOT, ALT, SPALT and their derivatives we can find the answers and solutions to questions and problems, deficiencies and fallacies in perspective and demonstration.

PHYSICAL and MENTAL, S-O, Á, STIT and LAFAP

This does not mean, as we discover, develop and show the PIP and COC of Homation and TRU, that we can not think of nor use the physical and mental as they are revealed to us. It only means that if we are to apply them properly to TRU, to use them in HOMATION to show TRU, and learn how they thru PIP and COC show TRU, we must do so, not so much by showing their SIT and THIT, but their **STIT** or State in TRU. We can do this simply, although not exclusively, by showing how they are not only an anatomical or pathological part or specimen, but in actuality TRU. They are as they function and even more as they function only in a certain TIT way. This can usually be readily, although not so often perfectly done, by simply investigating, analyzing and positing their functions in MOTof TIT. It can be more perfectly done as we apply all associated parts and processes, forms and functions, concepts, principles and laws of TRU that make more distinct these physical and mental entities and the information used and applied in Homation and TRU.

That of the world we use in our experience, experiments or in other ways to provide the data, evidence and information must be recognized as IA becomes IAT, and SO becomes Á. These are needed to discover the processes and complications which occur in the dynamics and total state of TRU. They are in a kind of balance for each part and ES as they are in TRU. ES are of different percentages of S and O, mental and physical as they exist in the ES and CES of TRU. Yet they are of stable and of fixed percentages when combined as A, in actuality as they are of the totality of and function convergently in TRU. This is called: **LAFAP** = The law of Fixed Proportion of S and O or Á in TRU Dynamics. LAFAP defines that state of ES and CES in which they are neither physical nor mental, but consist of a more or less constant Á. That which is of a variable or changing proportion of SO or Á, is likely to be either not all ES or CES or some combination and process which have different proportions of S and O as they are Á. Also what is of variable S-O is often less than Á and of some NAMTRU, ATRU or AU, or ANTRU percentage. It is in understanding these and the many other concepts, principles and laws of HOM and TRU, applying them to PIP and COC, and tracing them out singularly, collectively or totally in ALT, SPALT, METAD and ELTICMA that we can best show the glorious detail of TRU.

REF = Relative Finite. The subject and object exist not in themselves but only as they interact, combine and form a real, meaningful and actual unit together in the actuality of TRU. All we sense and the things and givens we are confronted with, are limited to whatever category they exist. In being relative, dualistic, dialectic, and dynamic they are graded from one extreme to another as their poles forms a dualism and continua. The three prime dimensions combine, coalesce, and fuse to make up the form and construct of TRU. In being continua they have both unity, finity and infinity. It is about the finite one that the finite differential, both the cardinal, as well as the fractional, develop. How well it develops is a prime mode of the thisness of TRU. It is from these that we can reach the real. The real consists of the finite combined midrange of S and O. The real is meaningful activity both in TRU, by providing its vehicles, strong bonds, as well as in MOT, by providing the differential makeup which is so essential to the exquisite finite detail that adequately represents TRU. The frame, that provides this representation, arises from the construct of TÄ, TÄP, SUT and ROT to become MOT. This frame depends upon REF, the relative finite, to provide the differential basis for the dimensions which make up the frame of MOT. In Chapter 32 we proceed to investigate the form of Modulation and its products. Since our premise and theme is to achieve a solitary duplicate of TRU, we seek to focus, concentrate and converge on all that exists of TIT into an object that will be both a close replica of TIT, as well as the best possible representation and model. It is how we transform TIT and TRIT into OBT and SUT as this is reduced in ROT, MOT, and SPALT, based upon the primary frame and structure of TRU. We do this by using the dimensions of TIT, representing and modeling them in that way that best preserves their FAF in TRU as well as yielding the best possible model.

CHAPTER 32

GENERAL and SPECIFIC PARTS of MOT, A, L, T, and ALT

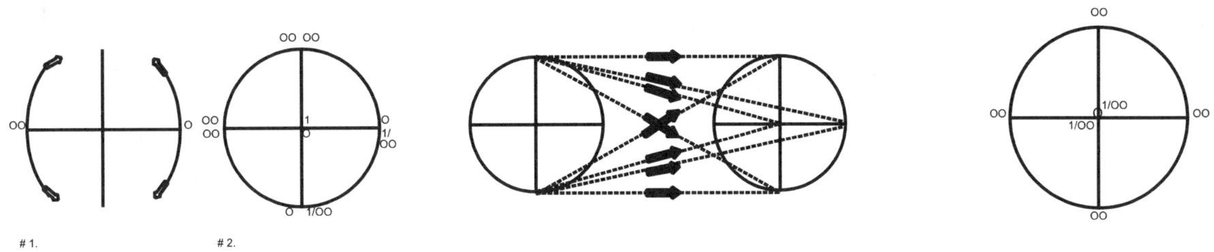

Fig 1-32 **DIMENSIONAL RELATIONS and ACTIONS**
How S is separated from the differential is called: **LASD** = Law of Subjective Differential Separation. LASD denotes why SI, the Subjective Integral, is such a great ideal. It much more readily enables a perspective and presentation of all and TRU. In the accompanying diagrams we have figures illustrating the dimensional formulation of the basic denominators of senses, things, perspective, and TRU , # 1. The manner in which coordinates or dimensions are, or are not, in perfect perpendicularity and the extremes of these dimensions. # 2. The manner in which two coordinates form a whole, a circle by their periphery, and what they infer. This can be done for TRU or all senses or things. # 3 The manner in which two such coordinate fields can be transposed and superimposed.

1. Senses Subject Subjects and Objects Object Things

2. Any Subject | A Subject | Some Subject | A S-O | The S-O | A S-O | Some Object | A Object | Any Object

3. I_____I_____I_____I_____I_____I_____I_____I_____I_____I_____I
100% Subject | 50% Subject & Object | 100% Object

4. Infinite Length | Length | Infinite Length

5. Any Length | A Length | Some Length | Large Larger | The Length | Small Smaller | Some Length | A Length | Any Length

6. I_____________I_____________I_____________I_____________I
Infinite Large | MidPoint Large | Average Size | MidPoint Small | Infinite Small

7. Any Time | Time | Any Time

8. Any Time | A Time | Some Time | Past Time | The Past Time | The Time | The Future | Future Time | Some A Time | Any Time

9. I_____________I_____________I_____________I_____________I
Infinite Past | MidPoint Past | Present Current | MidPoint Mid Future | Infinite Future

Fig 2-32 **COMPARATIVE DIMENSIONAL SCALES**
MEALT = Measurement of ALT, determination of the scales of Á, L and T in ALT. Figure 2-32 shows A, L, and T, in absolute, relative, and The Relative terms. The first three, 1-3, are for the mass energy dimension or from the subject object poles, the second three, 4-6, are for the length dimension, and the last three, 7-9, are for the time dimension. The time or T Scales are for differential factors in and changes in time. This is done for 1/2 the range for time and is, for TRU, a dimension which must be interpreted and measured from a focus. This is the present, or focus of the now. It corresponds with the center points on the L & A scales in the coaxial relative dimensional formulation of TRU. Centering for the Á dimension, or its SO combined, real, as the medium for length and the now for time, helps to prove Á and thereby TIT and TRU. The focus of ALT of TRU is that point at which the mid point of A, the mid point of length and the mid point of time for TRU intersect. Any construct of the world, any cosmology which does not reconcile their frame to its actual relative focus is, and is to that degree, in error. Any construct and cosmology which does utilize these and develop its scales proportionately and properly, as by VAPO, DAOST, SCARLS, and ACEXIN for TRU, will accordingly succeed. In our representation, modulation, perspective and presentation of TRU, TRIT and TIT we assume basic standards. Certain standards and a formal stance are utilized that are more common, and most important, naturally conforming to the order of TRIT and TRU. Since the Á dimension represents the specific energy and mass of TRU, and since the vertical axis is that of length, and the horizontal that of time, by exclusion Á becomes the sagittal dimension. The manner in which coordinate fields are circumscribed, scaled and centered enables demonstration. If S is the degree

of concentration of the infinite of the absolute, then: **SECIA** = Subjective Equals Concentration of the Infinite to the Absolute, or: SECIA = $(t_3)^{1/oo}$ / $(t_3)^{oo}$. So that, in a sense, I and O are identical with S and D, and TRU is the expanding mass of a 1/∞ to an ∞. This is proof of ELTICMA and shows the limits of TRIT and TIT. This is properly solved by demonstrating the relationships of TROS, between 0, 1/∞, 1/n, 1, n, ∞, and ◎ in TRU and the ID thereof. This is called: **SEP** = Solution of Finite Proportion of TRU. **SPOIDALT** = the Superimposition of SOID and ALT. From these principles of the differential and relations of MOT and TRU we can understand why MOT has such great practical value This by not only providing a frame and guide for TRU, but enabling its POA, detail, and supporting the proof of TRU.

The great beauty, value, and use of Figure 2-32, and especially 3-32 is that they show that and how, a purely discrete and distinct, differentiation of TRU can be made. This is in sharp contrast with the great difficulties we have in reducing all concepts, words, and terms to their ESINS and then to TRU for placement in SPALT or other models of TRU. With only ten pure and discrete divisions in each dimension, we have 10^3 or 1,000 pure and distinct segments of TRU with which to work, posit and demonstrate TIT and TRU.

```
                    Subject                                           Object
A  I--------------------------------------------^--------------------------------------------I

                    Large                                             Small
L  I--------------------------------------------^--------------------------------------------I

                    Past                                              Future
T  I--------------------------------------------^--------------------------------------------I


           Subject                          Real                             Object
A  I-----------------------------^------------------------------^-----------------------------I
          (Ideal)                (Phenomenal or Neutral)               (Material)

           Large                           Medium                    Small or Diminutive
L  I-----------------------------^------------------------------^-----------------------------I

           Past                      Present or Current                      Future
T  I-----------------------------^------------------------------^-----------------------------I


        SS             SR             RS             RO             OR             OO
A  I--------------^--------------^--------------^--------------^--------------^--------------I

        LL             LM             ML             MD             DM             DD
L  I--------------^--------------^--------------^--------------^--------------^--------------I

        PP             PC             CP             CF             FC             FF
T  I--------------^--------------^--------------^--------------^--------------^--------------I


     SSS      SSR      SRS      RSR      RRS      RRO      ROR      ORO      OOR      OOO
A  I--------^--------^--------^--------^--------^--------^--------^--------^--------^--------I

     LLL      LLM      LML      MLM      MML      MMD      MDM      DMD      DDM      DDD
L  I--------^--------^--------^--------^--------^--------^--------^--------^--------^--------I

     PPP      PPC      PCP      CPC      CCP      CCF      CFC      FCF      FFC      FFF
T  I--------^--------^--------^--------^--------^--------^--------^--------^--------^--------I


     Á  Scale  to  100's
A  I------------------------------------ -------^--------------------------------------------I

     Length Scale to 100's
L  I----------------------------------- --------^--------------------------------------------I

     Time Scale to 100's
T  I---------------------------------- ---------^--------------------------------------------I
```

Fig 3-32 **COMPARATIVE DIVISIONS and SCALES of the PRIMARY PARAMETERS that BECOME the PRIMARY DIMENSIONS of TRU** (MEALT)

A, L, T or the mass-energy, length, and time dimensions are adapted and altered by their special form and function in TRU. The acronyms for chosen segments of these dimensions are tentative and not to be taken as final. Yet the better we can delineate, make discrete and distinct all segments of TRU the more real actual and assured our world will be. They are mnemonic examples rather than what much of TRU terminology does to provide a set vehicle for TRU. The abbreviations produce a good basis for the necessary differentiation and distinction of TRU and all. The segments of the prime dimensions of TRU although unrefined, are of major importance. For they not only show how TRU can be divided into discrete segments. They also show how we can with such discrete segments provide sectors in SPALT and all that enable an accurate and reliable positive objective approach for position and order. They have ultimate distinction, through being perfectly discrete sectors of prime dimensions. This is what so much of the world about us lacks, even when perfectly TRU, precise discreetness and distinction showing the gradients of dimensiions and parts of the sectors of a strictly TIT TRU. Hereby we need and seek to provide a POA, PACH, and optimal POS, STA, and HOM of TRU. Its lack largely explains DAK and the negation in each person. The basic abbreviations include the following, the others are noted as an example of how well TO can be divided:

```
C = Current, Contemporary, present
CF = Current Future
CP = Current or Contemporary Past
D = Diminutive or Small 3rd of LMS dimension
DD = Diminutive, smallest or most D 6th of LMS
```

F = Future (a third of Past Present Future)
FC = Future Current or Future Present
FF = Future (most F of a sixth of PPF)
L = Large, or Largest 3rd of length dimension
LL = Large Large, or 1st 6th of LMS dimension
LM = Large Medium or 2nd 6th of LMS dimension;
M = Medium, middle 3rd of LMS dimension;
MD = Medium Diminutive or 4th 6th of LMS;
ML = Medium Large or 3rd 6th of LMS;
O = Object, or 1st 3rd of S-R-O dimension
OO = Objective Objective, 6th 6th of SRO
OR = Objective Real, 5th 6th of S-R-O
P = Past of PPF, PC = Past Current 2nd 6th
PP = Past Past (1st 6th of PPF)
R = Real or Phenomenal middle third of S-O, SRO;
RO = Real Objective (4th 6th of S-R-O);
RS = Real Subjective (3rd 6th of S-R-O)
S = Subject, Subjective (or 1st 3rd of S-O)
SR = Subjective Real (2nd 6th of S-R-O);
SS = Subjective x 2 (1st 6th of S-R-O)

Each person should be able to recognize and utilize what is, in actuality and totality, larger or smaller, past, present and future, sense, entity, thing, and how they all function in complex union and totality, in TIT-TO. Then we rise from in-actual existence, erroneous beliefs, disorientation, confusion, fallacies, problems and their resultant failures. Only then can we fulfill our rightful part and role, TES, in this world, TIT, in which both ones own success, and the escalating being for all is assured.

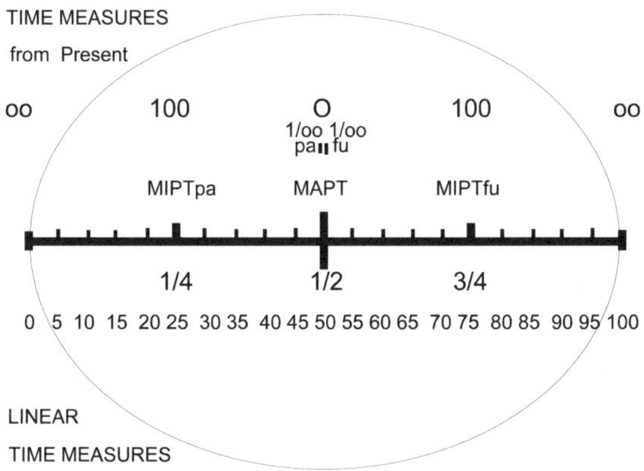

Fig 4-32 **TIME MEASURES CONSTRUCT in MOT & TRU**

In the time dimension, how is its scale to be determined? If the extremes of time are infinite past to infinite future and if the centerpoint is this instant, or infinitesimal time present passing through zero. Then our time scale is to be graded as follows: Mid Time Past, **MIPTpa or MTP**, is that set time at the 25% point on the scale. Mid Time Future, **MIPTfu or MTF**, is that set time at the 75% point on the scale. Zero Time is a vanishing point or Present Time without duration. **MAPT or PT**, is at the 50% point on the scale. In life we assume the present to be fully real, a point rather than its role in TIT..This causes confusion. We can't believe such a point might be zero, even when taken as an imaginary limit.

The determination of MTP and MTF should provide an equal time from present. A finite time, developed on the basis of that time in which TIT is half evolving and half resolving, with allowance for some variation with TRU. This is a great and critical determination. For it deals with the very essence of our being. It sets the standard upon which much of TRU, its SU, ROT, MOT, ALT, SPALT and all that posits in it both static and dynamic, are based. The form, features, constituents, nature, construct, divisions, characteristics, qualities, and properties of TRU are closely related to these two points and their balance in the time scale, and by **LINCOS**, or the Law of the Infinitesimal and Infinite Coaxial System Typical of each TRU. For an initial approximation, a period of 100 years past and future is taken as this mid point, MTP, MTF for the modulation of TRU, ALT. This is in ways to long or to short. Only with continuous evaluation, KOT and ACC, will we be able to increasingly set this point. In fact it will tend to change with TRUCH and TRUF, our mastery over our life, and the forces acting upon it. Its time, as well as length, extension or retraction, is closely associated with ENT. It is, thereby labile, and dependent upon "us", the future of human life. This produces its own, for us and for all TRU, most imperative dynamics. It is this dependency, and prerequisite necessity, that is such a powerful force in TRU. It is a great part of natural selection and survival. **LIMTES** = Law of Imperative TES. LIMTES is one example of many that show the great importance and vital nature or form and function of the proper dimensional measures and construct of TRU.

Our immediate awareness and instantaneous attention does not distinguish whether it is in an absolute zero, infinitesimal past or infinitesimal present, finite past or finite present of different durations of time. In fact it fluctuates between all these and may even extend into all time. This, in ways, is similar to what occurs at the periphery where there is infinity. An absolute is approached, but never reached.(LANERA). So does the centerpoint tend to the infinitesimal. Since by the relativity of TRU the same principle applies to the length and mass or Á dimension for TRU - in their own ways - we find a concentration toward the center and a diffusion toward the periphery. This is as we would expect for any RU. It is the spontaneous result of the thisness and relativity of TRU.

As these combined concentrated dimensions together make up a highly powered TRU, its unity is provided by systematic polar coordination of their extremes and means. The polar coordination provided is signified by: **LINFIP** = Law of Infinitesimal and Infinite Polar Coordination.

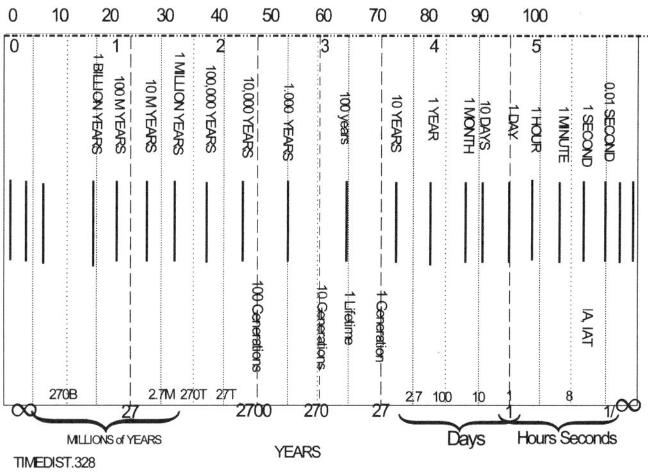

Fig 5-32 **HUMAN WORLD's TIME DISTRIBUTION**

In order to understand the FAF of TRU it is necessary that all principles and factors of its dimensions be envisioned both separately and together. In Figure 5-32 by personal experience we tentatively envision time as it might be commonly understood. However, since this does not account for its existence through the present from past to future and vv. we can see complexities in dimensions that must be accounted for. This figure also shows the problem of setting both the mid point and distribution of time. Set values are a must. However since they are also standardized and somewhat arbitrary they should be continually refined and upgraded. As we approach the foci of each dimension in our model of TRU their rate of change reduces, and decreases to zero. For time this is the present, for length an average length for this special world of best human life and for the mass or Á dimension the mid point of the most complete and powerful bond between observer and observed or all energy levels that comprise the predominant form and function of TRU. For there is always a balance between the center and the periphery, as in all that constitutes

TRU. In fact it is the dynamics of the forces, powers and fluctuations of these variations that makes up much of TRUD, HOM, DYNU, DYNS, DYNO and ACC. It is in this way we can not ignore the poles, even when infinite as the extremes and means are what constitute TRU.

Although Figure 5-32 provides a simple and good example of time it should be noted that it is not the final standard we provide for MOT, ALT and SPALT. Also although this figure is similar to what we do in SPALT, it has three marked differences. It is for only one half of TIT, either past or future time, its ranges are not complete, and the final standard of 100 years at 50% level is not taken. With the mass energy, or Á dimension the centerpoint of S-O to SO, or foci of the phenomenal or real, is what provides meaning and separation. This gives formal distinction to what is different in TRU. It is how ES and CES exist and function in TRU to makeup its mass, density and solidity. This is how the Á dimension, like the L and T dimensions expands TRU qualitatively, quantitatively, relatively and typically. The concentration of the SO or Real of the Á dimension is what occurs with IA, and the world of a person's conscious certainty, fused and interacting with the reality of life.

We should remember that although the set dimensions we seek to establish for TIT, TRIT, OBT, SUT, ROT, ALT, SPALT, SPAT and their uses are the same, they are different for departure from the whole, TIT-TO, and from each other. For example TRUS, TRIS and ES time are not the same. The further we get from TIT-TO either externally, or internally by partiality, the greater these differences will be. This is why for the individual time seems so much at odds with overall time or TIT-TO. It is the difference individuals, must get used to, to be TES, and how TES must adapt to TIT-TO.

PEDOD = Principle of the Exact Delineation of Dimensions. In point placement or POS, there is a constant tendency to posit by topics or relations. This can lead to the fallacy of familiarity, the simple and natural tendency to chose and use what is given or most familiar. It is a tendency which is often unsatisfied and needs to be reversed or compensated for. It is a test of lack of TREQ or TRU. It is overcome in highly developed fields and professions whose emphasis is on a comprehensive, thorough, and full treatment of the world that produces a broader brain set. The solution is knowledge, especially TRU knowledge, with the constant correction it provides. PEDOD has validity as a method of composition whereby order is maintained by placing all like material together. Yet without reduction to TRU is leads to fallacy..

In our use of the term ALT it is well to distinguish the different kinds and usages inferred or involved. These are not separate only different ways ALT can be used, including, separate components or dimensions taken collec-tively, the form, frame and construct, the FAF or form and function of TRU, a general appellation for any coaxial representation of TRU based on mass length and time, the planar representation of TRU, the model of TRU, ALT, and the spherical model of TRU specific for a set construct, SPALT.

ALT means how these components and dimensions of TRU function both separately and together, differentially and integrally, united in TIT. The parts of TRU may be considered, objectivized and known through their ALT composition. The various amounts, proportions, and functions of these different dimensional components in different parts of TRU reveal much of their nature, form and characteristics.

Since the great power of Á is dependent upon formal distinction we continually strive to expand and provide its ever greater power for TRU.(DAOST) This is TRU meaning, key to much of our being and survival. Á is most evident wherever meanings are conceived from the givens or discussions at hand. The separation of ALT has the further significance of showing the nature of TIT, and TRU as it is represented and modulated, and how ALT and TIT are in actuality one. ALT in its fullest dynamic is closely related to ACEXIN. It is in this sense that all should most greatly know TRU, TRU as the great mass, object, ATO, TRIA, TA, TU, TE, TI, TO, TRUCH, TRUD, TRUS, TRIT, TRE, and TIT. In this way we can know how TRU is basically constructed, formed and functions, by the combined massive continua of its primary dimensions, ALT, existing as ACEXIN in TIT.

PART and PARTS of MOT MEASURES SCALES SYMMETRY AXES COORDINATION DISTRIBUTION

The size that TRU functions around, its MPL or Mid Point Length of one meter because it is: 1. A common and predominant measure in human life, 2. The basic unit, standardized in measurement, 3. Represents a volume which is in the order of magnitude of the median range of a human being, 4. Represents a volume for which all that is larger is about equal to all that is smaller, 5. Gets away from the stigma of the individual, alone, 6. Is a measure that is common to most all fields and sciences, 7. Can be readily accepted as a standard, as E, ES, without evoking the criticism of the universe. It does not have to necessarily, although it will certainly, tend to be similar in size to the individual. Different races, sexes, ages, systems and forms contribute various mean sizes and therefore will tend to alter, and make necessary an averaging out of the mid point length, MPL, as the FAF standard for MPL. The precise centering of length, like that of other pivotal points in the ALT frame will require continuing investigation, work, and corrections, whose accumulated knowledge with that of all TRU, and experience in its operations, will provide the refinement for perfection TIT, ROT, MOT, and POT need. Even preliminary pivotal points can, within the proper order of magnitude and range, provide centering of dimensions sufficient to make a good and useable model of TRU. With continuing refinements of TRU the MPL of ROT, MOT and ALT will increasingly have a 1:1 identity with the precise and actual MPL in TIT.

Some of the major concepts of centering, beyond those of **MP, MPL, MPT, MPÁ,** or Mid or Man Point, Mid Point Length, Time and Mass Energy for TRU, include those of: **MPLa** = Mid Point Large -Medium, **MPSm** = Mid Point Small - Medium, **MPP** = Mid Point Past - Present, **MPF** = Mid Point Future - Present, **MPSb** = Mid Point Subjective - Real, Phenomenal, **MPO** = Mid Point Objective - Real, Phenomenal. This divides each dimension into 4ths. If each of these dimensions is divided into 10ths, 100ths, 1000ths or all degrees thereof as in the decimal system, then the center-point will be at 5, 50, 500, or at the 50th percentile. This for A is that which is equally S and O or the real. The observer observed relativity is at its optimum of meaning in an equal balance of the two, observer and observed, or subject and object, or orientation of weight of light and heavy, or density in the light spectrum. This is called: **LAMENOL** = Law of Mass Energy Orientation with Equal balance of Light and heavy, or physical light density, or for TRU. Light of the visual spectrum is also related to mass by Einstein's equation. So that mass and light are highly interdependent, as is all that compose this dimension. It is in this way that light is equivalent to increased density, and increased density to TRU. So that TRU, as we might expect by relativity, is highly light dependent. In this way Á if not a form of light, has in common with it the parameter, and dimension from which the mass energy dimensions of which TRU are based. Only in TRU it is at an adapted, altered, warped, and much more highly, selective, developed, and special level.

The second type of point, is the half way point, between the center and periphery in each of the three dimensions. These are largely determined by the point of reference or relativity, in balance with the extremes of each dimension. This is especially determined by the associated summative powers of competitive predominance of TRU. That the mid points need to be taken together, and also together with the centers of each of the three dimensions, helps to determine both individually and the whole it represents, TIT-TO. Since the half times for dimensions, by relativity must be of the same order, with symmetry both in the same dimension as well as with each other, it will show how the future reduces the past and the past prolongs the future. This is a result of: **DINE** = the, Dynamic Indirect Equilibrium and balancing Effect. Although a mid point of 100 (or even 270) years may appear too short this appearance is not a half life in terms of medium lethal doses or medium of survival. This neither disproves half life nor the survival time of TRU. Since time about MPP and MPF is both exponential and warped, time, survival is skewed to a longer duration. That is how far term millennia arithmetically overcome near term. It only proves the correct FAF of mid points, the dynamic TROS of TIT, and the critically imperative nature of TRU, TIT, TRIA, and TÂ.

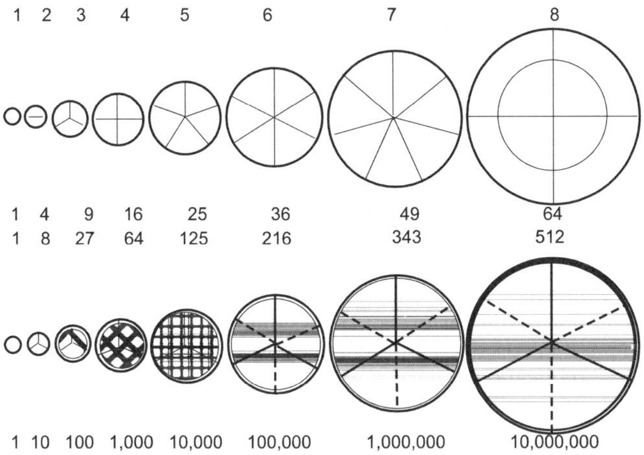

Fig 6-32 **DIVISIONING of TRU**

In figure 6-32 a comparison is made of TRU divisions arithmetically and exponentially by 2nd, 4th, 8th and 10th powers. It helps better visualize how ALT, and especially SPALT. How they can be divided, proportioned, and separated to provide the discrete partitions for positing ES and CES.

The determinants of the polarization and coordinates of TRIT, OBT, SUT, ROT, MOT, ALT, and SPALT depend upon the FAF of each, and especially how they exist in TIT. This depends on: 1. Our capacity to comprehend it, 2. The totality of TRU - that is maximal, 3. The parts of TRU, TE, ES of TRU, their maximal formulation, 4. The dimensions, parameters, processes of TRU, 5. The maximal detail of TRU, 6. the limits of ROT and KOT. 7. the unknowns of TRU, POT, MOT and their description, principles, concepts and laws, 8. the limits of the individual, 9. our total interest, and inspiration, and 10. the values, importance, application and uses of polarization and coordinates. This is the explanation of rating techniques and notations such as 4+ and nR, or REM valuation in the text. It arises from the need to reinforce and place emphasis on relative values, importance and priorities. Only in this way can proportions, as required by VAPO, be achieved. This is similar to the continual use of VAPOS and SCARLS and in this case is called: **LEP** = the law of Equal Portions. LEP is currently geared to approximately 1:1,000,000 or 6R, as seen roughly in the number of words, in language. If tentatively used as a standard, held to a relatively fixed focus, as in the 1/100 x 1/100 x 1/100 division of the A, L and T dimensions of our Model, and if adequate provision is made for the notations of relative portions, proportionality, and other significant means, we will at least provide the needed practical finite basis, coordinate formulation, modulation, and placement. Some of the ways we can scale the dimensions of TRU to produce accuracy include: 1. Scale by comparison plus standards. Certain parts of TRU can be compared with other parts, analyzed, and scaled by percentage of the whole. 2. Scale by reference, re-measure, and by posit. As the parts of TRU, phenomena of S-O, ideals, forms, concepts and entities are studied and posited in TRU, they are repeated, their percentages compared and scaled. This is with repeated reference to definition, usage, re-measure and repositioning. 3. Scale by composite posit. As all things are posited by A, L, and T, they, through their second and third dimensional position yield firmer posit.

The symmetry of the forces and systems that exist in, and are demonstrated by, most sciences is the kind of relationship needed to demonstrate TRU. It has been shown that many of these forces, such as subatomic particles, obey strict laws of symmetry. Symmetry provides accurate laws that within their respective limits are necessary if we are to replicate, order, and, at least relatively, coordinate the whole, MOT of TIT. The symmetrical provide a certain polarity between any two or more relationships, places, forces, or dimensions. This is what was established by Einstein's Theories where one and another observer must be the same, or there is an accelerated observer. The principle of symmetry involves different geometric configurations of the universe. As shown by TIT, TRIA, SUT, THEO, OBTRU, ROT, MOT and **LABSDIMT** Dimensionality is not an ABsolute, but relative, and Depends largely upon the universe at hand, or given. Set dimensions exist in TIT, rather than out of TIT, and are used in TRIT, SUT, OBTRU, ROT and MOT to define, replicate and show TRU. Dimensions are or the frame of TIT not to overrule it.

Scales, measures of dimensions, and their combinations in SPALT are of great help in showing the way to answer many of mankind's major questions. One is whether we are a finite series, an infinite series, or some mixture of the two. It shows how our world is distributed. It shows how we can overcome the impossibility of plotting a mixed arithmetic, logarithmic, exponential, finite and infinite series. Relativity of an exclusive special universe, as TRU existing by ATO, is what produces this alteration or warping. The extremes, both in the periphery and nearing the center or focus, are highly warped. So that the only evenly measurable regions are those portions of each dimension that are relatively stable. Within fields and sciences or limited sectors of TRU unaltered parameters are readily and accurately measured and demonstrated. For the smaller the segment on the dimension to be measured, the more reliable linear interpretation can be made. This is called: **LARIMDS** = Law of more Reliable Linear measures on Limited Segments of Dimensional Scale. LARIMDS has been known from classical and relative geometries in physics that measures of small segments of a whole can be made with a high degree of accuracy, even though their parameters may not be perfect. It was the exceptions to, and limits of, LARIMDS that measures were found to be in error, and that led, were key to, and confirmed, relativity. LARIMDS has great implications both into the finer aspects and the most universal laws and applications of TRU and MOT.

More complex and less accurate measures arise when measuring the subjective and the objective as absolutes. This is especially evident in the absolute past or future, absolute largest and smallest. In being absolutes of a relative universe they are immeasurable. Even their near approaches are innumerable and thereby relatively immeasurable. There will be a region near the poles and centers of each in which critical disproportion exists. This is only balanced by the whole, of each dimension and all together where overall measures can be applied as we do for POS in SPALT and SPAT.

These limits and problems, however, do not alter, only make more challenging and vital, the scales and measures of the dimensions of TRU. **PRODOM** = Problem of Distribution of Measures on axes of Coaxial Systems. Before ROT, SOID, MOT, ALT, and SPALT can be made and used satisfactorily and successfully it is necessary to find and establish the precise distribution on each dimension, axis, the whole and on the fields they subtend. PRODOM is solved as the determination of ALT and SPALT are made, how they exist as a fixed, total, exclusive, and solid coaxial system, and how they are to be scaled separately and together.

CHAPTER 33

TOTALITY and DYNAMICS of MOT, SPALT

SOLID GEOMETRICAL and COORDINATE SYSTEMS, SPHERICAL MOT and SPALT

When we move from a two dimensional to a three dimensional coaxial system, we must adapt to added relationships. In addition to the vertical and horizontal axes of a planar binary coaxial system we have a sagittal axis. The question is, which dimensions best fit which axes, and which one is best for the sagittal axis? At least one or the other will become somewhat altered as the plane becomes a solid and a set perspective is assumed. This is especially so if the triaxial system is represented on a planar field. It is not an arbitrary standard. The operation of fitting axes to dimensions is not by chance it is largely guided by the form and relations between TRU, TRIT SU, OBT, ROT, MOT, observer and the SCARLS of each succeeding step in the progression. Justification of standardization holds when and to the degree its provisions far outweigh its detriments, times how much it can not be done without, and how well it agrees with TRU. **LAJUST** = Law of Justification of Standardization. Thus it is justified to the degree all SCARLS of TRU are exhausted in developing standards. It follows precedents set by computer-aided design, CAD. That is where the ratio of S/F or benefit/harm, approaches infinity, as also its need. In a finite world this is as great as possible, or within limits, infinite within these limits. This is the reason for corrections and holding it in mind.

If we adopt the standard of Á as the sagittal axis, and T as the upper diagonal or horizontal axis, we retain L in its usual position as the vertical axis we approach a standard geometry. This will seem to vary as our perspective of TIT moves about its object and model. The direction of axes can be of many arrangements. Only we accept as our primary model, and standard, Á, L and T as usually shown in SPALT. This is largely based on natural relations, the relations between an observer and observed, and the existing conventions. This is supported by the fact that large is up, small is down, past is to the left, future to the right. The subject or our viewing is from the left lower and object or that viewed to the right upper. Yet since this form is objectivized both by OBT, and external subjectivity, some added transformations are necessary if we are to assure reduction of perspective to TIT-TO.

x, y, and z interchangable with either/any m, l, and t or Á, L or T

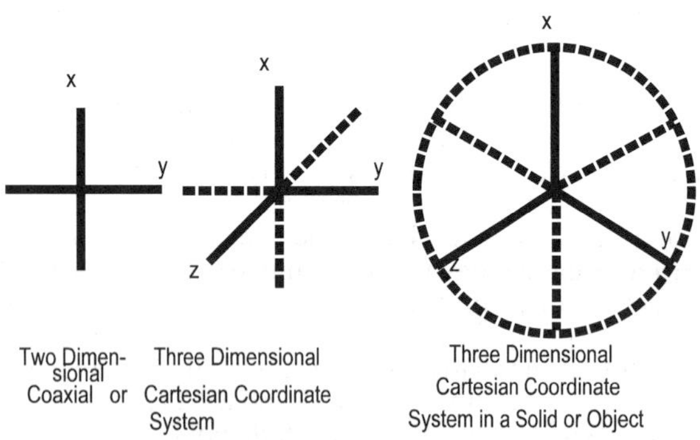

Fig 1-33 **2D and 3D CO-AXIAL SYS-TEMS; RELA-TIONS and ORDER of DIMEN-SIONS and AXES**

Steps in this development are shown in figure 1-33. To compensate for the problem of TIT identity it is sufficient to remember and use the difference and deviation of ALT and SPALT from TIT. The next best thing to having TIT, is THEO, THES, SUBT, OBT, SUT, ROT, MOT, ALT, and SPALT in mind or in hand always realizing their difference. Since it is less flexible and practical to have a spherical object of SPALT in association with a book, IPTRU, this problem adds to the change in alignment of axes with perspective. One remedy, and one which we have already used in this text, is of superimposing multiple representations. For even if we do not have a spherical model to replicate TRU we can approach this by showing the major planes or sections of TRU separately and then combining them through verbal imaging. This can also be corrected by supplementing the book with models provided separately as is done by globes of the earth with atlases. Furthermore with audiovisual vehicles of communication and CAD much improved views of MOT, SPALT and TIT can be provided.

If then we have a deviation from x and y, and we must also deviate from the x y z correspondence with mlt or ALT, we adopt the convention of m or Á as the lower sagittal axis, the convention of t or T as the upper diagonal or horizontal axis, and hold l or L in its usual position as the vertical axis. The orientation of dimensions and axes, although highly supported and justified, produce some change and arbitrary fixation from a strict replication of what is in TIT. The added transformation necessary is called: **LATOM** = Law of Added Transformation Objectification in the SCARL Representation and Modulation by ALT and SPALT of TRU. When given in this way we will readily see that the axes of the sphere can and do correspond to ALT and that these likewise correspond readily to x, y, and z, or as usually used, L, T and A axes and dimensions. Only as we look at, and for purposes of description provide the best possible system or divisibility of this spherical model, can we come to achieve a satisfactory, useful, stable, permanent method of demonstrating TRU.

PROCDAT = Proof of the Circumscribed and Definitive TRU. Choice and free will of the individual when TES is valid only within the limits of TIT. Each individual is the product of the animal and human life, race, culture, society, tools, means, ideas, concepts, principles and laws of which they are a part. One uses or reacts to a part of this, as TRU. It is in the summation of all of these that ones acts are incorporated in TRU. The summation of all our acts can only be, completion or fulfillment of, all that determines us, past, future and present. This proves the determinant and definitive nature of TRU, PROCDAT. The knowledge in the world, about and available to us, determines most of our ideas, thoughts and course. It determines our participation in TRU, life, being or non being. The summation of all our acts can only be a completion or fulfillment of all that determines all TIT-TO. This is signified by: **LOCOF** = Law of Obligatory or necessary Completion and Fulfillment. LOCOF is the finality of S/F, or success over failure. It proves the determinant and definitive nature of TRU, PROCDAT. LOCOF defines and signifies that state of our being that lives up to VAPO and SCARLS of TRU whereby our choices and free will are confined or limited by all else of TRU, both statically and dynamically, formally and functionally. Yet by being an absolute participant in an absolute we gain far greater and permanent choice and free will, as well as life. PROCDAT is in turn provided by the development and determination of TRU in its fullest form and function as it forms a systematic unity. Then it can be well represented, modulated and utilized, as we do with ALT in POS, STA and Homation. It is such concepts, principles, laws and proofs as PROCDAT and LOCOF that confirm the totality of TRU and its objectivity. This in turn is what proves and establishes MOT, ALT, SPALT and subsequent operations.

PRACALT = the Proof of TRU, MOT, ALT and A by Analytic Geometry. These are the propositions and proofs that are derived from the geometric representation of mathematical functions when applied to an object modeled as a solid or sphere. When the internal forces are great, as in gravity, the periphery becomes a perfect sphere. This is seen in hurricanes, variation, human life, TRUS, and TRIT. It is by centripetal internal forces and their bonding that we can recognize and prove the role and use of sphericity and SPALT. **ALTCO** = Á and ALT Complete only as TRU is Complete, or only as TRU is TIT-TO. Á and ALT can not be complete until TRU is complete or what we know and show as TRU fulfills VAPO. This is a truism for all that is given of TRU. It is one that is of special importance in MOT, ALT and SPALT. For they can never be successful unless and to the degree they fulfill VAPO, TIT and ALTCO. We may give valid and beautiful representations, entities, ESINS, ES, CES, aspects of TRU, or sections of models of TRU. However they will never provide a satisfactory showing or modeling of TRU until they show it in full and in the proper way.(VAPO)

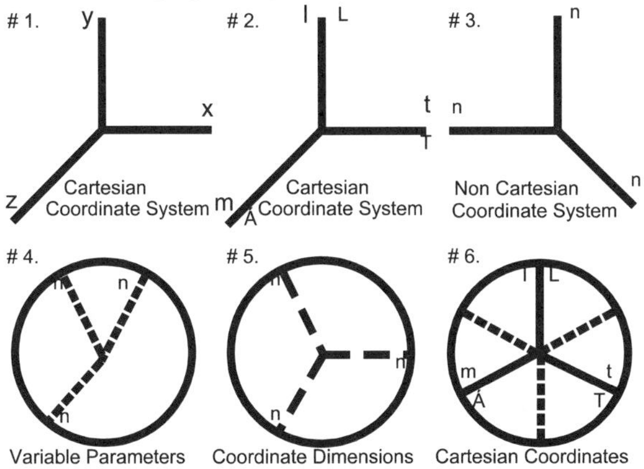

Fig 2-33 **PROGRESSIVE DEMONSTRATION of COORDINATE SYSTEMS**

Figure 2-33 shows how we approach the problems and analyses of basic geometries for MOT. When formed into a Cartesian coordinate system the standard mlt dimensions and the ALT dimensions correspond. This is largely a result of the identity of the FAF of xyz, with mlt, with ALT of MOT and ACEXIN in SPALT of TIT. It provides a base for representation, as all standards and conventions have by the predominance in Á of TIT. It does not contradict or make MOT and SPALT wrong, it only qualifies them to the extent if not recognized and used within its limits or SCARLS, it will be erroneously treated.

PORLE = Principle of Relocation of REALS. Off hand it seems impossible and paradoxical that a sphere can be all, yet by observation and by relativity, when qualified, this is quite possible. It is even more certain in the convergence of all by way of relativity and thisness to TIT and ATO.(TRIA) For in such an existence it is not only a totality. It as totality with unity, systematic unity, order and closure, becomes a completed set. We can have an infinite periphery and infinitesimal core and still have a closed set, as most certainly TIT does by its aseity, perseity, and haecceity. We can have such a spherical TRU when it becomes an object or is visualized in this same relativity with set limits. The fact that this infinite periphery is not an arithmetic one, nor even a solely physical one but one that is relative, altered, and warped to TRU is what makes this infinity concrete. This is the basis for TRIT, OBT, and SUT. It is the relative, altered, and warped orientation and order of the form of TIT. This is much of what most levels of SÁ of Á continually do, maintain the homeostasis and continuity of TIT. The correspondence between such a relatively conditioned and set solid TIT and a set MOT, ALT and SPALT corrects our false opinions about the structure of the universe. It produces an opportunity to show the structure of the universe as it actually is. Since TIT and SPALT are not like an astronomical view looking farther and farther into the distance to search for the riddle of the universe and life, we need to relocate our priorities as they exist in TIT and can be shown and used in SPALT. This is the truth and proof by TRU, how the actual form of TIT relocates our commonly believed reals and can show the ES in TIT.(SPALT,POS) It is where we, both individually and all of TRU as the observer in relativity, must look not at the periphery or some part of any universe. We must obtain a total through and through view of TIT-TO, its ES, CES, REMS, and POS. Each ES is pure TRU and each real is the identity of the thought with the actuality of each ES in TRU.(PUREACLITY) Only in this way, through KOT, TIT, TRIT, OBT, SUT, ROT, MOT, ALT and SPALT, can we developed and use a reliable model, SPALT, with the great value and power it and TRU have and deserve.

Fig 3-33 **EIGHT DISCRETE DIVISIONS of TRU REPRESENTED on a PLANAR DIAGRAM**

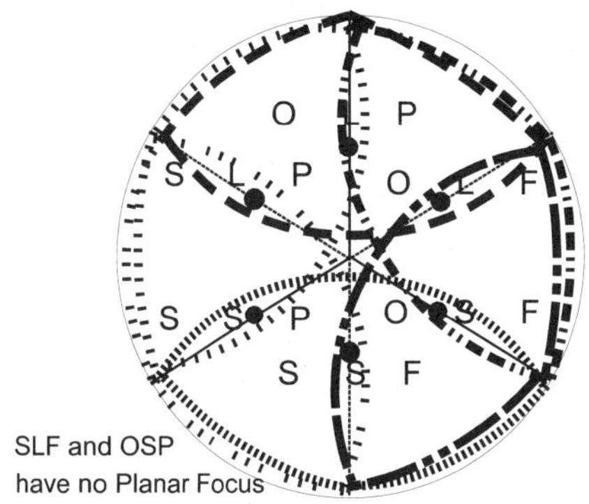

The divisions of the world have been commonly and roughly known for ages. Even animals in lower phyla biologically have a sense, subconscious if not conscious, indirect if not direct, dynamic if not static, of how they are oriented in relation to all about them. This is of what is large and what is small, what is past and what is future, what is mental and what is physical or whatever predominant mass energy makeup is typical for them. It is in this way we can divide the world, TRU, into its parts. If we have three dimensions and these are the Cartesian Coordinates of a sphere, then this sphere may be divided into the eight parts formed by such a division.

OLP = Objective Large Past
SLP = Subjective Large Past, or SULP
SLF = Subjective Large Future, or SULF
OLF = Objective Large Future
OSP = Objective Small Past
SSP = Subjective Small Past, or SUSP
SSF = Subjective Small Future, or SUSF
OSF = Objective Small Future

These divisions are not exclusive for TRU. They can be divided into an infinite number of more definitive divisions based upon the scaling of the three parameters. However these eight provide the basic form and classification of the parts of TRU as shown by the placement and positing in STA, HOM, DYNU, DYNS, DYNO, and ACC. Figure 3-33 provides a planar diagram in which these divisions are indicated. The ellipsoids are not exact but represent the general areas for each division or 1/8 part of TRU. The foci of each of these divisions or parts are shown with small dots. One of the main reasons for this figure is to show how planar representation does not suffice for 3D MOT. Two of these divisions are made up of three axes that are at right angles to this first observer and thereby opposites which will not show on a planar diagram, and for which posits are not directly possible.

LUSECT = Law of Uniform sectors and parts of TRU. With an accurate and consistent geometric solid for SPALT we can divide it into precise sectors. This not only provides divisions that are perfectly equal it enables us to, in turn, refine ROT, MOT, ALT, SPALT, REM, POS, STA, HOM and ACC.

LARES = Law of Representational Essence. LARES is commonly known in the art world, and is the ability to bring out not only a perfect replica of both face, frame, structure, and content, but a great variety of aspects or essences that otherwise would be missed or neglected. LARES is of even more significance and value in TRU, for much of what we describe in ROT and MOT and show in models and figures is not always a 1:1 replica. All that is missing we provide in other ROT and MOT which fills in these neglected or missed essences and aspects of TRU.(METAD,SPAT) **LIRNMET** = Law of the Imperative Replacement or provision for Neglected, Missed or void Essences and Aspects of TRU. This is a basic mode of TRU as well as ROT, the need to sustain itself must be self fulfilling. The role of much of ROT in TIT is to supply this fulfillment. Whether we know it or not, much of what we do, say, work for, and what we react to, is a kind of RIT operating in TIT, **RINT**, by way of supplying LARES.So do we seek to provide uniform sectors and parts of TRU, to fulfill both its composition as well as its order.

The meter and second are not chosen in and of themselves, or even as they alone provide the axis for TIT. Each dimension, and its coordination is more accurately determined by averaging out and balancing what is on either side of us, the larger or smaller, the past and future, the observer and the observed. Since time is at 0, in the present, the mid point time is dual, both mid-point for the past, and mid point for the future. These two can be generally established from an analysis of what is equally of the far and near past, far and near future. MOT, ALT and SPALT assume equal densities. Yet unequal densities and even more variations through time of the status of each dimension in TIT may, at least in part, be explained by the counterbalanced differences in quality and quantity. This is in long term adaptation, as evident in changes in habitat.

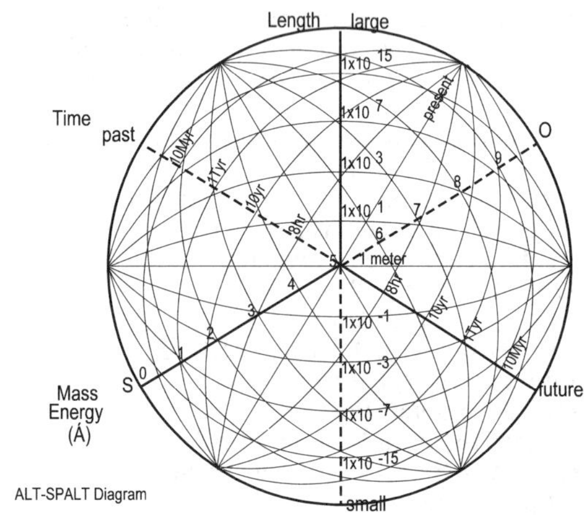

Fig 4-33 **SPALT DIAGRAM on a PLANE SURFACE**

Figure 4-33 is the basic and key diagram of MOT for TRU. It is neither all possible views of SPALT, all solid geometric representations or models of TRU, an exact replica of TRIT, nor is it TIT. Yet it does provide an excellent basis for revealing TIT and TRU. It is the prime vehicle for demonstrating TRU, and enables us to utilize and develop POS, and take POA, textually, to TRU for the benefit of all. It's use in demonstrating all the concepts, principles, laws and mechanisms of TRU shows how all can be can be integrated and ordered in the whole of TRU. Many can be posited with a high degree of precision that hold to a high degree of consistency. It also enables us to expand this model into points of view from the outside in which a meta diagram, as in POMOT, METAD, and SPAT, or the development of overall perspective of all, can be achieved. It is vital that each and all persons learn their worlds and themselves in such an ordered and positive way so that they can have the necessary TES and TRU to meet the demands of, and cope with, life. That all their powers of mass energy can best fulfill their roles in TRU. On each person the future of humanity will place ever heavier burdens. Much greater capacity and ability will be needed, if suffering and destruction are to avoided. Each and all persons must use all means to help learn TRU. We use and adapt this figure for many purposes POT and IPTRU.

LIRMALT = Law of Limiting ROT and MOT to mlt or ALT. LIRMALT signifies that in order to provide a solid geometrical and spherical objective model of TRU we use not only its three primary dimensions of m,l,t or A,L,T, we can do so in a controlled manner with and without higher powers and external interpretation. The application of higher powers and external states and conditions in which ALT and SPALT are qualified is called: **ALTQUAL** = ALT Qualification of Limits or SCARLS. The principle that when internal forces, as with gravity, are great the periphery becomes a perfect sphere is called: **SPHUL** and its variation, **SPHULVA**. It is by SPHUL and SPHULVA that we can recognize and prove the role and use of sphericity and SPALT.

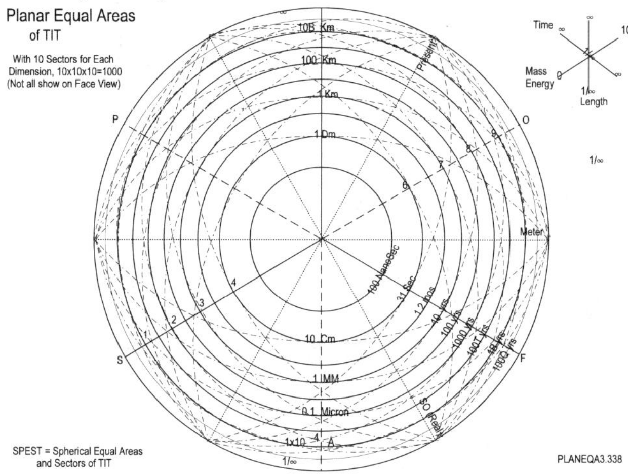

Fig 5-33 **SUMMARY of STATIC POSITS in SPEV**

Since a corrected SPALT for dimensions in a spherical view provides a more accurate representation, it is what is used as the standard for most views of SPALT, MOT, POS in IPTRU. We should remember, however, that this figure does not provide a spherical model, nor is it, even a spherical model identical with TIT-TO. However, it does provide an extremely useful replica of it.

TRAINING IN HOW TO THINK SPHERICALLY BY SPALT and SPAT

The key and power of IPT, POT, and TRU, as it does in life itself, lies in how to think, and this in how to think integrally, differentially, systematically, and effectively. It is how to think by relation, DAOST and VAPO of TRU. This is knowledge that is of the same, different, valid, actual, proportional and total, or knowledge by and of TRU. More specifically it is how to think in this way guided by a positive objective approach, POA. This requires a frame, a frame that replicates TRU, or, its mass-energy, length, and time form and function, ALT. It is a construct that produces SPALT, the spherical figure that represents and models these fundamental dimensions. In such a figure there is a place for everything, and everything is in its place. If we can only develop, learn, remember, use, and live by the POSITS of all that exists in it statically, dynamically, simply, complexly, totally and in all ways, then we can best apply it to all givens, encounters, occasions, needs, problems, intentions, purposes, goals, and ends of our lives. This is the cue, key and power of this text, how we develop it and focus on all that can be placed in SPALT and SPAT to guide us in life. It is not something - as human life itself - that is light, or can be used now and then. It requires a life long search for the deepest meanings in all, how to relate them, put them together by POSITS in SPALT, SPAT and all other means that best give us the way to navigate our lives and that of human survival, success, and eternal life.

The ability, practice and ever expanding use of such spherical thinking by SPALT, is what can, must, and will - of necessity - raise each and all persons from levels of mediocrity and failure to this special universe of best human life. If you do not chose to be, to be your part and fulfill your role, then you will pay, not at the price of what is best for humanity. It is one in which the POSITS of all that make up and function in TRU are applied to our work and life in general in such a way that we personally and collectively behave, do, act, effect, and achieve what is best for all.

DYNAMICS of MOT

It is in the interactions and operations of MOT that we can further our understanding of its greatness and beauty. We can much more efficiently and effectively apply its benefits to our lives and TRU. In order to make MOT adequate for the rigors of posits it is necessary that its dynamics also be well developed. From the many changes and variations through expansion, interactions, transcendent functions and standardizations, MOT, ALT, SPALT and their uses can be improved. If we seek and expect to find only a simple single final model that is a perfect replica of TIT, as one would of a model of any object in the physical universe, we will be disappointed and lost. Only when one discovers, learns, knows, remembers and uses all the VAPOS, SCARLS, TRUS, TROS, TRIS, and progression from TIT, through TRIT, SUT, OBTRU, ROT, MOT, ALT, SPALT, POMOT, METAD, ELTICMA, SÂ, and TÂ, back to ones observation and relativity will such models fulfill their intended roles. For it is through such dynamic operations of MOT and their uses that the management and control of their complexs, discrepancies, variation, standardization, and uses be best met and achieved.

DIMÄ = Dimensional Analysis and Formulation. DIMÄ is how we interpret and develop mlt, ALT, and their higher powers, to best replicate TIT. Since dynamic factors like motion have been standardized by convention and represented from left to right, and since static factors have likewise been represented from top to bottom as in most experiments and science, the time factor and x axis runs from left to right, and the length factor or y axis from top to bottom.

Some of the fallacies, errors and needs of dynamics in MOT include those of skepticism by superficiality, disproportion in parameter, disproportion in magnitude, and evils in summation as they deny top focus, ALT, LASIT, PADALT, and LAECIT where ATRU and ANTRU precipitate out. (INEG) The last is often a manifestation of social purpose and harmony.(PIP) It is order, purpose and harmony that ALT and SPALT so actively and intently provide. In this way the existence of MOT, ALT and SPALT are not only in TIT they are vividly, most actually, and imperatively so.

There are corresponding changes in scales with the portions on a scale. As a scale is observed to be larger and provides finer and finer elements on it there will be room for more and more posits. This corresponds to the increasing differentiation characteristic - within limits - of a successful life. This is the beauty of a model. It occurs by not only providing an overall frame in which to posit the world, it can do so with greater and greater precision, detail, and exquisite beauty as it grows to near total reality with TRU-TIT.

In the following two diagrams, sectors of SPALT are compared. In figure 5 it is done as it appears on the face of SPALT, or periphery. In figure 6 it is done as it would appear if we cut thru the inside of SPALT to produce a sagittal section. These are different, and represent the importance of how we can easily fail to appreciate the value of all detail in SPALT when we view and use SPALT, POS, HOM and their dynamics.

Figs. 6-33 **COMPARATIVE SECTORS of SPALT**

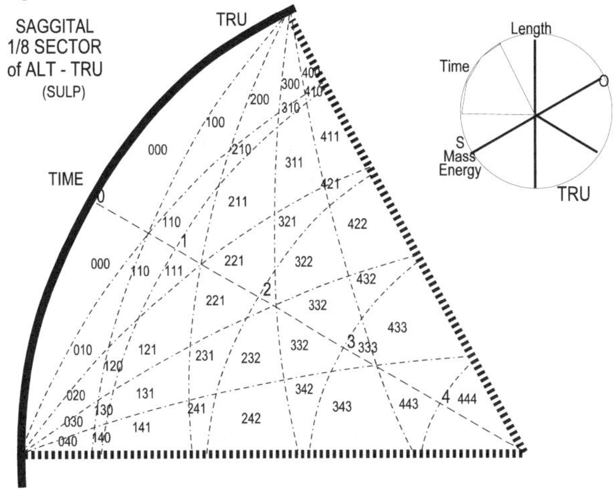

Fig. 7-33 VISUALIZED SECTORS and DYNAMICS

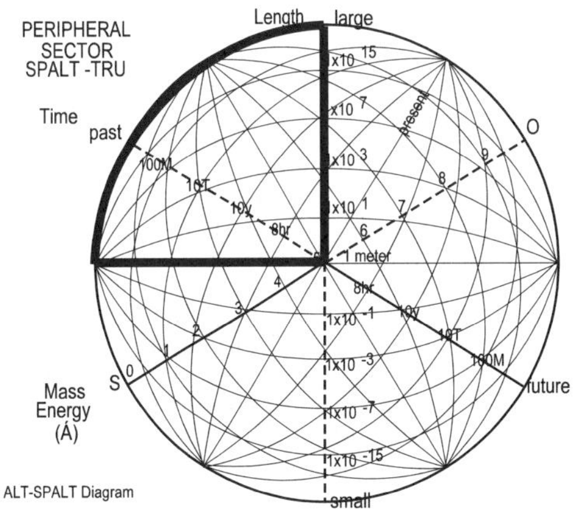

Only as we comprehend the finer and finer degrees and gradations of S-O, come to know the differential and integral, TE, TÄ and TO and especially as they may be coordinated with the S-O to formulate them in a unified and detailed whole can we be free of disproportion, disorder and confusion. We can escape from the trap of IA and ÍT, our sensations, perceptions, and knowledge that have an overwhelming tendency to accept as truth all it focuses on. Truth is the known and yet to be discovered unknown ideas, concepts, principles, laws, and mechanisms, whose relations, connections, and interpretations provide closure to all in this special universe of best human life. Only when TIT is reduced to OBT, SUT, ROT, MOT, ALT, and SPALT, and all is posited and applied to our lives, as in ELTICMA, can we reach those levels of identity with TRU which are necessary for TES. We must ECRO, always remember, what sense ES and CES, or that for posit, is taken in and how it exists in TIT. Some will be highly variable. Some will posit throughout the range of Á, L, and T, and some very specifically, as by: **ERETO** = the determination of par and weighted values of the parts of TRU. It is the comparative contributions and roles of all great aspects, principles and concepts of TRU. They will increasingly enable us to precisely set mid points, scales and their distributions for Á, L and T. Yet each E or part of TRU will, throughout 0-10, have its own range. We must recognize that these figures are less than TIT by URTRU, and UMTRU. Each item is not to be accepted as posited with absolute precision. Nor are the definitions used the only nor the most TRU. It reinforces the importance we must attach in developing an accurate and reliable TRU, TIT, MOT, ALT, and SPALT.

CHAPTER 34

ENHANCEMENT and SUMMARY of MOT

THE ENHANCEMENT and PERFECTION of MOT, ABSTRACTION and CONCRETION, FITTING, MULTIPLE and COMPLEX MODELS, SUPERIMPOSITION, OPTIMAL FORMULATION and MODULATION

There are many means of improving and perfecting MOT. The focus of these is on a single model that can act as a standard for all. At the same time, it must be remembered that such a single model does not completely satisfy either MOT or TIT, nor fulfill the need for all ROT, PERT, or our IA. To fulfill this need we have to develop an overview or panoramic perspective in which not only is the single view of the whole best produced, but all other beneficial views and products can be superimposed and related to it. These other views include, all that is in TIT, as well as some that is outside of, or of mixed, TIT and other senses and things. In telescoping and focusing in and out and all about TRU we can provide such a combination. This, however, to maintain unity, totality, management and control, must have POA, plus more expanded and deeper concentration and comprehension. By the laws of abstract and concrete, as long as one stays within the parts of the whole, ES and CES in TIT, a perfect construct and model can be produced. In abstraction wholes are prior to parts. In concretion parts are prior to wholes. It is in the union of these that they best identify the form and composition of OBT, TRU, and the order of their constituents are attained. It is the modulation of these into which the parts and processes of the whole can be posited. This is the goal of an enhanced MOT.

LAFUT = Law of Fitting the World, the all, or the Universe, to TRU. Any world or model must be fitted not only to the dimensions and standards of its object, and this to their entity, it must be fitted to TRU, and more especially to TIT and TRIT. For any actual correspondence we should find, between the subject and object, past and future, large and small, and integral and differential, dimensions that will be only as they are TRU, TIT and TRIT. There may be a great and good correspondence and utilization of models, yet they will not be the ultimate nor ideal, until they have a 1:1 correspondence with TIT. This is called: **CORTEM** = Corollary and Concept of TRU Equality or Equivalence with its Model.

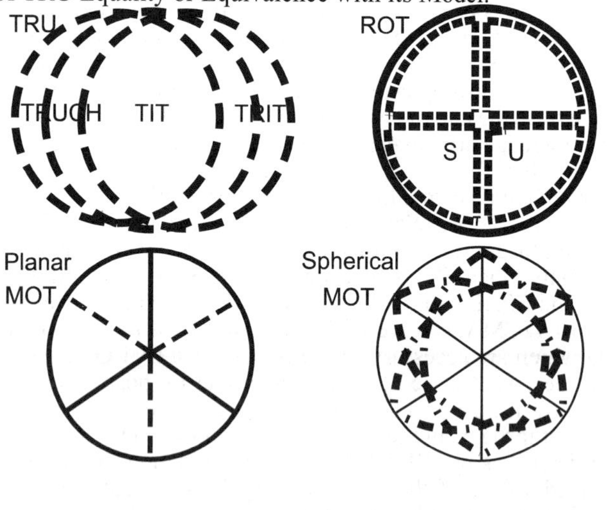

Fig 1-34 **COMPARATIVE TRU, SU, PLANAR MOT, ALT, SPHERICAL MOT, SPALT, DEVELOPMENT of an ENHANCED MOT and SPALT**

PALT = MOT as ALT on a Plane in a Page of IPTRU. Only when we can recognize the differences between TRU, TIT, TRIT, SU, ROT, MOT, ALT, SPALT, PALT, MPLT, and SOID can we SOID, can we properly use or criticize them. This can be better recognized by viewing Figure 1-34. Representing these different entities as objects automatically produces limits, SCARLS, and thereby errors, in addition to any errors inherent in the figures themselves. TIT is not actually the circle or word shown at left. Nor is SU the divided, hashed circle and divisions shown. Only as we come to represent and model TRU can we, as they are made identical, overcome this basic defect. It is inherent in the nature of S-O, and is corrected, ultimately, only by relativity, the observer, or human mind, in subjectivity involuting its A powers in TRU, as by OITIO. SUT, as even TU, TO, and TRU may be maximally represented by many different images. SPALT as TIT may be maximally represented in a single image by a spherical model based upon mlt or ALT, as is the world of man, shown on page, screen or other means by PALT, MLT, and SOID.

If TIT is modulated in ALT with three dimensions which in turn are divided into 2, 3, 4, 5, 10 or 100 equal parts the resulting sphere will contain the cube root of each of these or 8, 27, 64, or 125, 1,000, or 1,000,000 equal parts. It means that portion of the dimension this degree of division produces. Also such divisions do not mean that posits in each sector, or region, are the same. For all that posits in 000, or 001, or 002, can be very different, of varying portions of each dimension. The amount of A, L, and T of each part of TRU is variable. For instance, it may be highly A and very little L or T, or vv. As much as they seem alike there is a great difference between much in each sector or sector of TRU shown in SPALT. This is, **LOVSEC** = the Law of Variable Sectorial Composition.

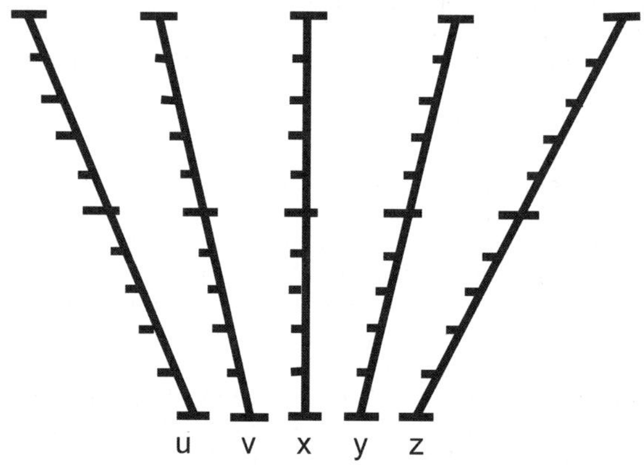

Fig 2-34 **NOMOGRAMS; EXAMPLES of SUPERIMPOSITION of PARAMETERS in MOT and TRU**

With the use of nomograms we investigate differences in dimensions, their scales, and what is the same or natural correspondence within TRU. This is called: **INCOT** = Internally corresponding Scales of TRU. With the use of nomograms correlations of otherwise non correlatable entities can be accounted for. We can show how they differ, compare, and its significance. We can show, how they must be altered to become superimposed upon the form of TRU, its geometric configuration, and what corrected resultant they can give in the use of parts, their roles, and applications in TRU. The major value is in bringing the innumerable parameters, scales and measures, available in established knowledge and science, especially through research, into alignment with their corresponding counterparts or scales in TRU. It is a good example of how we can rise from arbitrary measures of anything to more accurate measures of what is TRU. SOID, ALT, SPALT, and other models of TRU are dependent upon the limits of all upon which they are founded, namely: ANKTRU, UPTRU, UNKTRU, URTRU, and UMTRU. All senses and things, especially when their ESINS, and thence their ES and CES can be reduced, as in MOT, into their subject object or A, L, T, I-D, and REM components. They may be achieved by the differentiation of TRU in TE toward TI, or the integration of TRU in TE of TO toward TU. So that, to make an absolute pole real and actual, it must be returned toward completed totality in proportionate parts. This is even more so being variably composed of A, L and T.

As described in Chapter 33, the optimal model for TIT must be a sphere. Yet in presentation we are usually forced to present it as plane, on a flat surface, a "flatland". This is all items in figure 3-34 when viewed head on, even though of 2D, 3D, and more dimensions. For unless we are presented with a model fully providing for each dimension we can't readily perceive their full meanings and interpretations. Anything that is external, epi, meta, or outside of TRU, must account for the added states and dimensions imposed. This does not alter either TIT nor the fact that a 3D model, as SPALT, can be provided. It only shows us that neither TIT nor SPALT are all, except in and of themselves. They are altered by objectification and subjectification. It can also be altered and adapted depending upon other givens or points of view which, within their realms, may at times be necessitated and accounted for. When we attempt to justify and account for entities, senses, things, universes, observers, observed, subjects, objects, reals, givens and all not TRU or mixed TRU and non TRU we depart from an optimal model. This is an ever present and pervasive inclination. When this occurs, we have a myriad of new figures and ways of representing and modeling them. Yet by rotating SPALT, or by focusing in and out of SPALT plus a moderate coordinated use of accessories to SPALT, as POMOT, METAD, and SPAT, we can retain an optimal model plus guides to operate TRU and our lives.

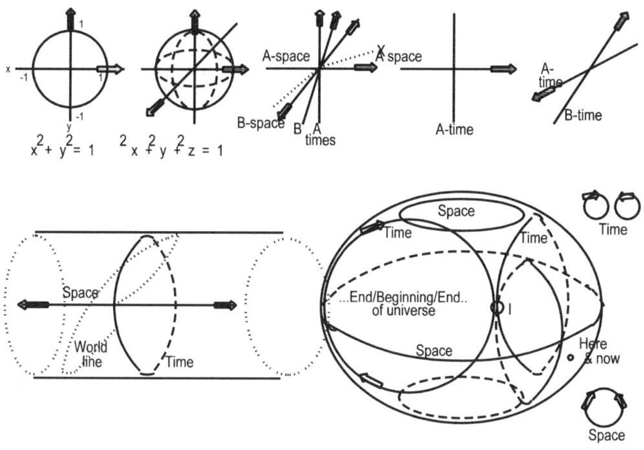

Fig: 3-34 **MULTIPLE COMPLEX MODELS**

This non original figure is used to show the precedent and geome-trical interpre-tation of MOT and TRU. Each part focuses on different applications and associated problems. What is the optimal or final number, and how we represent TRU by these dimensions. How can we represent these procedures, and at the same time the additional perspectives or presentations needed, or taken as the case at hand? How can ALT, SPALT, and SPAT be put into optimal perspective?

Proof of the identity or degree of replication by SPALT of TRU is much more than PRALT. It is a determination of the degree to which not only is ALT valid for TRU, but the degree to which this modulation as SPALT represents an accurate replica of TIT. The degree to which it is identical with TIT, TRIT, SUT, and OBTRU can be used as a completely reliable guide to the actual, extensive and intensive, ACEXIN, demonstration of TRU. Whether, in TRU, this model's use fulfills, with scientific accuracy and mathematical precision, what its representation should be. Proofs of ALT tend to prove each other. This is why it is so necessary to carry all ROT, MOT and POT to the their best fulfillment.

So likewise do many of the features of TRU prove TRU as TRU proves them. For example point of size, point time, and point of the S-O interchange prove TRU, as TRU is increasingly confirmed by them. Also interval and permeate proofs by area, duration, and the expansion of levels of A prove TRU as TRU confirms them. Furthermore the total and comprehensive states of dimensions by volume, variation, combined SA prove TRU, as, in turn, they are confirmed by TRU. This role of ACEXIN is called **LUVREPT** or the Law of Unified Volume, Variation and Relative Proof of TRU.

Squaring and Cubing TRU's Dimenional Scales, ACEXIN

It is not so difficult to visualize or model the squares and cubes of single linear scales. It is more difficult to visualize and model the squares and cubes of complex regular exponential scales. It is most difficult to visualize and model the squares and cubes of total and somewhat irregular scales. It takes a superhuman effort to fit, model, visualize and treat dynamically the form that actually exists in TRU. In order to better understand the frame, form, function and mechanisms of our world, TRU, we need to learn how to square and cube the scales of its length, time, and mass energy continua. We need to lay out the scales of length, time and mass energy in one, two, and especially three dimensions as they exist in TRU, to envision all in its solid geometric state as a sphere. Yet these are necessary and the basic operations of the FAF of TRU, its TA, MOT, SPALT, ACEXIN and Posits. This is: **LASCUT** - Law of Squaring and Cubing the Scales of TRU.

PROBLEMS and SUMMARY of MOT: APPLICATIONS, USES, ORDERING, PROGRESSION, REFINEMENTS, TIE-INS, and ACCELERATIVE POWERS of MOT

Although the answers to questions and solutions to problems of MOT and SPALT and their uses have been, more or less, been met and standards adopted, they are neither exact nor permanent. The possibilities of further refinements are likely and necessary. This applies as well to language, symbols and all that is used to represent and present TRU. Mankind is only beginning to realize its potential. Science, research, and all advances in knowledge, as TREQ, must go hand in hand with TRIT, OBT, SUT, ROT, MOT, SPALT, and the total management they engender. Since the future, as shown by TRU and ALT is a large part of human life and approaches half when considered in balance with the past, it is mandatory and absolutely necessary that all persons understand not only the present and the past, but the future and all together. This is one of the many benefits a discovery, development, determination and use of TRU provides. It is through understanding TRU, its fundamentals, posits, and all in SPALT and SPAT that we can forecast, predict and describe much of the future. In being a large portion of TIT and existing in the present and past in ways that are the same and different in the future we can better judge the future.

The divisions of the parameters or dimensions of TRU and their scales in MOT, ALT and SPALT, do not have to be by halves, tenths, hundredths or thousandths. Nor do the divisions mean that TIT is composed of 1, 2, 3, 4, 5, 6, 7, 8, 9, alone, or any set divisibility. These are largely mnemonic devices or conventions of the models. In actuality TIT is not so simply differentiated. It includes all divisions and gradients in infinitesimal series. The great need, and the ability to hold the world in a nutshell, stimulates us to adopt as much explicit differentiation and formal distinction of ES and CES, as practical within the limits of TRU. The success of SPALT, any model, algorithm, representation or even conceptual knowledge largely depends upon the viewer, first person, or subjective. It depends upon one's finite capacity to distinguish, envision, comprehend and remember information. This is a limit, and the primary limit, if the divisions of this information's representation and modulation of TRU are right, as by VAPO. **MIP** = Maximal Image Point. The level of differentiation will vary depending upon many factors: the intensity of the model and TIT, familiarity, knowledge, interest, and ability to concentrate and utilize this knowledge. These are only some of the factors that determine the maximal image point, which for TRU and the observer focus on the perspective of TIT-TO.

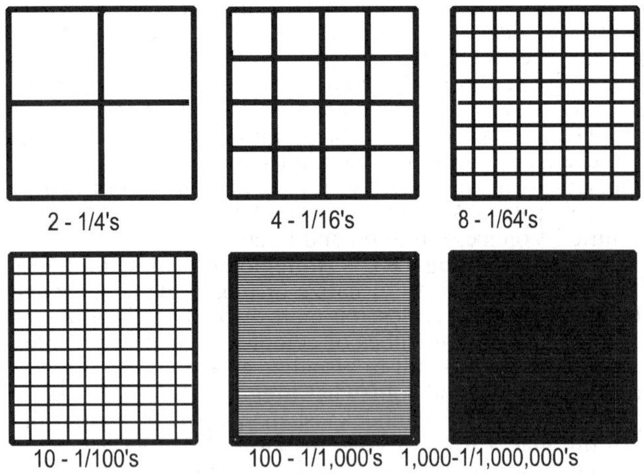

Fig 4-34 **MAXIMAL IMAGE POINT and LIMITS of DISTINCTION**

In figure 4-34 the divisions shown give us an image of one of these greatest problems. This is the human mind's ability to learn and think complexly with precision. For our mind not only is limited in its capacity to envision exquisite detail, it soon forgets and loses even much of the greatness it has learned to distraction, attrition, or fatigue.

The figure shown is not complete for there are in TIT all gradations in between. As the first person looks at the detail of a highly dividing object as in items 5 and 6 it soon becomes a blur, impossible to distinguish what is differentiated. This comes in different levels, of which two major ones are the blurs associated with loss of interest and fatigue, and the limit of possible distinction by the human mind. It is in how we can increasingly concentrate on TIT, SPALT and POS to further expand ability and achievement by ever greater formal distinction that we can greatly improve our lives.

LIDÄ = Limit of Distinction Formal, or Limit of Formal Distinction. LIDÄ comes in at least two different levels, the first is like the difference between figures of 100 to 1,000 separate parts. This is the difference between our being able or unable to distinguish separate parts, but not each single part. The second level of LIDA we see between figures of 1,000 and 1,000,000 separate parts. This is the difference between our being able or unable to distinguish order and no order. **MIF**, or Meaning Imperatively Finite, and LIDÄ are both complementary and opposing. They are complementary in providing a good basis for the S-O of ROT and MOT. They are opposing in producing a state in life, of both individual and all TRU, of being at the same time achieving a maximal state of SOT within a finite limit of formal distinction. It is in how we can increasingly concentrate MIF further and further to expand our capacity, ability and achievement of ever greater formal distinction that we can greatly improve our lives. This is called: **MIPLIDA** = MIP LIDÄ Actuation. MIPLIDA is the simultaneous rate of expansion and enhancement of the extremes of MIP and LIDA. Here is the solution to many of man's problems. Here is the cue to, gist, and thrust of ROT, MOT and SPALT. For we live under very great limits in being relative, extremely partial and mortal. This is simply our ability to distinguish and effect, as by our predominant mass energy powers, what is VAPO. This is greatly supplemented by memory, maturity, experience, training of thought, expansion of ability to envision detail, interests, and love of life or TRU. In the utilization of ALT for TRU, we can not operate without adequate exposure. This is why we pay such attention to OBT, SUT, ROT, and MOT. We and they need the greatest fusion. It is basic to action, be it a simple task or the most demanding, providing the most intricate and complicated demonstration of TRU. This is the utilization of ALT and SPALT for POS and TRU.

SPALT is frequently used throughout the text to reveal many different aspects, forms, parts, processes, totality and dynamics of TRU and TIT. It provides us with a good standard focus and reference upon which to base a positive objective approach to the differential TRU, its parts, ES and processes, CES. It enables a better understanding of REM as well as how to use REM and all that supplements TRU. However each and all persons should not mistake this model or TRU nor ES, REMS and POS that are not TRU. All who use it would greatly benefit from concentrating on and use of larger spherical models of SPALT, their posits and its ordered extension by OIȚIO, METAD, SPAT, COGNOS, ELTICMA, SÂ, and TÂ in envisioning TIT-TO. SPALT is set in an optimal mode. Yet this very setting and standardization imposes limits for which care must be given in their use. This is: **ROXALT** or the Rotation of Axes. The same arrangement of the dimensions and axes of ALT are used, only rotated, depending on the relation of observer to different sides of the face the sphere of SPALT. The second, and more marked alteration, is **TAXMO** or the Transposition of Axes, wherein different arrangements are used to demonstrate TIT and TRU. The difference between ROXALT and TAXMO is seen in the Á, L, and T dimensions where in ROXALT they are always the same, only viewed from various sides. In TÁXMO the A, L, and T dimensions are varied depending on whatever position of SPALT and TIT best agrees with the needs or preference of the observer. SPALT is not exclusive, and as great and helpful as it is in replicating TIT, it can not take the place of all needed for the best showing and knowing of TIT and TRU. At times we must rotate SPALT, transpose ALT, and vary our MOT and ROT to reproduce SUT, OBT, and TRIT to best replicate TIT.

By way of the SPALT diagram we can attain a better overall view of TRU and the effects and principles that are involved in its dynamics.(17) Some of these include, **DAREV** = Derivation, Determination, Rational, Reason, Resolution, Right Action, Reversibility back and forth, ad infinitum to TRU. **DUREF** = Duration Effect. **CED** = Competitive Exclusion of Divergence, or the processes of competitive exclusion of divergence. These are closely related to: **PITA** = Principle of Indirect Temporal A, or the S-O Effect. **DIVE** = Dynamic Vector Equalization. **POMAN** = Principle of Manifold Formulation. A person can't fully understand something nor utilize its benefits appropriately until they can visualize it from all directions. So does TU, SUT, ROT, MOT, SPALT, POT, METAD, ELTICMA, SÂ and TÂ give both unitary and manifold formulations. This is how in POMAN we can see the difference between TRU, HOM, and IPTRU.

There is an optimal way or route to follow in making a trip. So it is likewise in all human action, and use of TRU. Failure to profit by such experience is often not due to lack of knowledge, it is because there is a lack of clear-cut goals and plan. It is because there are no adequate orientation and coordination, and this in relation to the universe they are in and must be a part of. To simply survive by whim, comfort or any expedient is insufficient. To not know or know poorly what we are actually doing in life, and what the long term consequences are, is the road to failure. Public opinion is so inadequate that much of life is spent in NAMTRU and wasted. We do not adapt our desires and interests into what is most advantageous and successful, and become our counterparts in TRU, TES. Herein is the great value of MOT, SPALT, POS, HOM, ACC, and POT and how their use as vehicles, guides, maps, atlases, and models can reveal all the detail of ES and CES in TRU. This is how we can most greatly benefit from their use in life.

We can construct excellent models of astronomical space and atomic structure, yet we have such difficulty of constructing a model of our own world. We seldom realize that these models of astronomical space or atomic structure are only objective approaches to small portions of TRU. SPALT should be no less valuable to the study and management of ones life and the world, than is speech, writing, books, computers, an EKG in the study and management of heart disease, or computer technology is to space travel. SPALT when best and fully used should be the focus and means by which all other ROT, MOT, ALT, and POS help guide our lives, in TES and TRU.

SECTION IIIB

REM or RELATIVE MAGNITUDE in TRU

CHAPTER 35

INTRODUCTION to REM, VALUE, AXIOLOGY, KNOWLEDGE of REM, PROPORTION, IMPORTANCE, CHOICE, GOODNESS, JUDGEMENT

REM, Relative Magnitude, investigates and treats a proper interpretation of value, axiology, worth, goodness, and what is best. It is how they can be effectively determined from TRU, TIT and SPALT. What is of greater and what of lesser importance, priority or proportion in or of the world. REM follows and depends on MOT, as POS, HOM and ACC follow and depend upon REM. There is much more to valuation, axiology and REM than to put a monetary value on things, or even to extol TRU singularly as 'God is the Greatest'. For the role of REM in TRU can, and of necessity must, be ACEXIN, actualized, extensive and intensive. It must fulfill VAPO. Much of REM is the unique essence of TRU, something that does not exist in physical things or objects in and of themselves. This is an important proof of Á, or the mass energy dimension and FAF of TRU. REM should tell us not only what is the greatest of TRU, but the value of all in TRU and: 1. How all TRU can be valid, 2. What our priorities should be, 3. What each person should choose or decide in each circumstance, instance, or occasion, 4. What our interests should be on, 5. How we can share and distribute the goods of the world evenly, properly, and fairly, 6. By PRA REM provides a focus and coordination for stimulus/response, S/R, and optimal success of TRU. REM can't tell us: 1. What each persons next step can always be. 2. What TRU is, except how it is proportioned and how this shows the content of TRU. 3. What we are, when not our counterparts in TRU. The justification, as well as, the necessity, of ethics are based, to a great extent, upon what is good, valuation, or REM. Without proportions, values, priorities, and without all that TRU shows us to be the preferred or best choices, we fail, become nothing, or are a detriment to society, as TRU. In this way REM is of the greatest importance as it exists in, and is a mode and method of TRU, and topic of IPTRU..

Axiology is the philosophical study or science of value. Its principles have added greatly to our knowledge of worth, value, ethics, and life. Axiology in general is neither based directly nor exclusively on TIT, nor proportioned by TRU. Thereby it fails to provide an actual and precise measure of what is needed. In life we continually make choices between what is good or bad, best or less than best, interesting or repulsive, comfortable or uncomfortable, pleasant or unpleasant, and many others. These are examples of the dynamics of natural selection, and the exclusiveness of TIT. Yet in human life they are not so rationally nor intellectually guided by REM. Time and again in life we see that which we cherish or value most devastated and turned to naught. This often occurs not from extrinsic forces as it does from our intrinsic failures. These failures arise, to a large extent, from the very excesses or deficits of cherishing and valuing singular things without VAPO and TRU. In so doing, we not only ignore and neglect all else that is of such great, and often more imperative value. We fail to learn and live by all values in an ordered and comprehensive way. Maturation and the coming to terms with life are to a large extent the simple process of learning, remembering and applying these experiences and lessons of life. Through loss, devastation and failure we learn what is good and bad, what is of value and how they can be better ordered, comprehensive and useful. Such a process, so fundamental to each persons being, is a natural indirect fulfillment of ones counterpart in TRU. It may at times be direct as in religion or a philosophy of life. So are the values dependent upon truths, not any truths, but the identity of S and O, SO or Á, in TRU.(VAPO) It is for us to have the knowledge to recognize what is true worth. Such is the purpose and benefit of TIT, TES, TRIT, OBT, SUT, ROT, MOT, REM, POS, STA, HOM, ACC, POT, and IPT. By Chaucer, "trouthe is the hyeste thing that man may kepe" so must this trouthe itself be of the highest or of those values that are in this best of worlds, TRU, proportioned, accurate, complete, ordered and useful.

By **LOV** or the Law of Value or Law of Valuation we signify the fact that not only is all not equivalent, we extend this to a determination or rating of the importance of all that is before us. TRU is tantamount to quality, quality of life, just as this is the prime creator of what is. These are largely of value, of relative magnitude, what is the great, greater and greatest in TRU. How value and LOV take on set characteristics and certain form is called: **LORVA** = Law of Relative Value, or how all things that are about us, in this relationship, take on a certain and set form and characteristics. It is this relationship as seen in TÄP, TU, SUT and TRIA that enables us to produce REM. This is because they show how TIT exists with set form and proportions with a high degree of stability.

It is in the very crux of **PRAV**, or the Paradox of Practical Value that we can see and understand the key to right or wrong, and the success or failure of pragmatism. The deeper and more general interpretation of pragmatism, as enunciated brilliantly by C. S. Peirce, was called and properly distinguished as a distinct entity, Pragmaticism. Peirce's theory and definition of Pragmaticism are those practical consequences which might conceivably result by necessity from "the truth of that conception, and the sum of these consequences will constitute the entire meaning of the conception". Also he held that pragmaticism was a theory of logical analysis or true definition and its merits are greatest in application to the highest metaphysical conceptions. That is, it is not what is of practical use that counts so much as it is how this practical use is part of the highest metaphysical concepts, how they are of TRU that is its real actuality, ultimate value, evaluation, axiology and REM. So that as the term 'is' is or is not as it is TRU, the term practical value, and pragmatism is or is not, by PRAV, as it is or is not by pragmaticism through its highest and entire metaphysical truth, in TRU. This is a TRU that is largely constituted by the great concepts known, and yet to be known, that represent all its being.

The precedent for REM or Relative Magnitude lies mostly in those methods of mathematics, statistics, and value that use an abbreviated exponential number. This is where each magnitude is ten times the preceding one. This may represent lengths, areas, volumes, time, durations or standardized levels as we do for the Á dimension of TRU. For REM it is based upon a standard of one. This is because the relativity of TRU presents a unity from which all that partake of this whole may be deduced. Another value of such orders of magnitude is how that which is adapted to TRU can be related and superimposed to provide further discrete representation of TRU. A good example of this is seen in the exponential orders or magnitudes of time and their adaptation and alteration in TRU.

Since we must have proportion, values, or REM of what is TIT in order to enable subjectivity and proper perspective, we must be able to divide and distribute these absolute parts and processes, statics and dynamics of TRU appropriately. This enables and expedites a positive objective approach to TRU. In turn, it provides an increasingly precise replication and representation of TRU. The primary division between the two main sets of valuation or axiology is that which separates total valuation or magnitude, **OV**, from the relative magnitude, or **REM**, of the parts of TRU as they may be valued and distributed in TIT. As in the analogy of slicing a pie, REM has to do with the number and size of the pieces whereas OV has to do with the size and value of the whole pie. As the whole grows and enlarges or recedes and shrinks so do all the slices, pieces and parts. Thus results the benefits and success or detriments and failure. Because OV is fairly stable for TIT, except from the casual observer, it is to REM we largely depend on to reveal proportion and value. OV does have great significance however, for it is a kind of TREQ, and especially in TRUD, the dialectics of Homation, long term TRUS, success/failure, and survival or extinction..

A less simple, yet none the less necessary division in valuation lies in: **PROV** = Projected Valuation. Much of the value of all things, especially new concepts and ES lie not so much in themselves, but how they are a part of TRU, and this in how they so clearly show something else. It is in their standing for something else, as is S or the subjective and objective or all that is representative of TRU, is in TRU not as it is a direct form or part of TRU, but as it indirectly effects the form, parts, and processes of TRU. PROV provides the means to find and determine value from representation, from all that exists in/out of TRU as it stands for something else.

VOT = Valuation and limitation of Value to TRU. REM can be little better than the separation, delimitation, formal distinction and refinement of TRU, POT, ROT, MOT, SPALT, ES and their language as they are TIT. For as long as these are unknown we can not have a whole in which proportionate parts can be established.

TRU is not only REM of ES or CES separately as each may be a distinct part. It is the degree of total TRU, their intensity that has, can and will forever be so important, REM^2. This is called: **LADOR** = Law of Degree of REM. LADOR may be likened to love of life, or the degree to which one and all love, want, need, expect, demand, and will GOD, TREQ or TRU. If it is little then each and all REMS are likewise. If it is very great then all REMS individually, complexly or totally are also very great. Humans live in a highly complex, technical, and competitive, world. Each of the component energy levels, that are of, in, and prove, Á, necessitate fulfillment.

Knowledge of REM is close to the good, deities, right and various combinations and forces that produce them. Yet if we are to value all that is about us, and if we are to base it upon one, the whole which is TIT-TO, then all else will be less.

Axiology, the theory of value, must first of all recognize that ever present and permeating fallacy we rarely escape from in our usual lives: **FAV** = Fallacy or paradox of Value. Value is not to be found in usual monetary value, common supply and demand, or what anyone may interpret at any moment as being that which is most important, necessary, or desirable. It is ultimately, and simply what is a proportional part, aspect, or measure of TRU. This is not so much its quantity and actuality but its quality and degree of actuality or beauty. This is why we use the term REM to signify values, and all that we must include in the Axiology of TRU.

Those ES which become apparent only in the highest values of TRU, as 1R, 2R, and 3R do not provide an even distribution of OBT, TRIT, or TIT. Yet as we increasingly discover, know and show TRU we can see how such cardinality in even distribution and within limits can be obtained. Failure to discover and know such proportion and order is one reason why the characteristics of God or Gods have not given humans the idea of proportions, values, deities, and the actual TÄ and TROS of TRU. Here we can see the preeminence of VAPO, TE, TÍ, REM, POS, STA, HOM, and ACC. We can see them together as the basis for KET. We can see them both as a means of representing and presenting TRU and as they are great parts of TRU itself. They in the past have been only a very small and variable part of knowledge. They undoubtedly will become a much larger part in the future.

If we are to chose, determine and select in life what is and what is not TRU, we must be able to qualify and quantify what it is and is not. We can't do this unless we can validly compare the two and measure this difference. It is done through the major components of ENS and TRU: **ENS** = Entities, **ESINS** = That of an entity that is also of TIT, **EP** = The percentage of ESIN/EN, **ES** = A part of TRU, **REM** = % of TRU this ES is, **ESINES** = % of ES that is of this ESIN, **EPINES** = EP times ESINES, **ESINREM** = ESINES times REM, **ERP** = ESIN REM Percentage, **EPESREM** = EP times ESINES times REM.

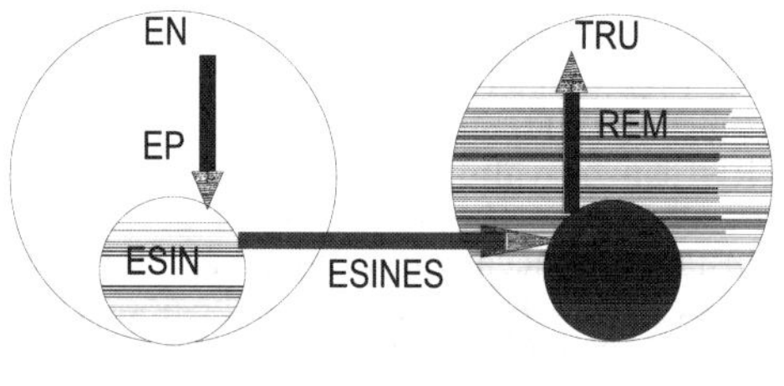

Fig 1-35 **ENS, EP, ESIN, ESINES, ES, REM and TRU**

The percentage of TRU, REM, is how much of this entity's ESIN is of ES, left to right dark circles, that is, REM, of TRU. 10% x 10% x 10% = 0.1% of the entity is TRU. 0.1% of TRU is of a Relative Magnitude of 3R. We can tabulate EP and REM in such a way as to produce an accurate assessment of what and how much all entities are in relation to TRU. This is an essential step if we are to give accurate and reliable measures to all, without which TRU and much of all, can not be known.

PROOF of REM, PREM

The proof of REM, as proofs in general to a large extent, are provided by ever expanding knowledge and substantiation of the makeup, and what it is based upon, of TRU. It is based upon the discovery and refinement of its basic concepts, principles, laws, and mechanisms. They involve proofs of not only REM itself, they involve, to a varying degree, the overall and transcendently relative states, and the givens that relate to REM. In this way proofs of REM will involve proofs of OV or Total Valuation. An enumeration of some, of many, proofs of REM, include, 1. **LOV** or Law of Valuation. Since all is not the same, selection must be made, 2. **OV** or Total Valuation. Valuation is more than the comparative and evaluated parts of TRU. It is preeminently the expanding to absolute worth of TRU. 3. **REMT** or that valued must be or be brought to pure or 100% TRU. 4. **REMTE** or that valued must be based upon an ever more distinct and discrete separation or differentiation of the parts of TRU. 5. REM is attained to a large extent only as we reduce all terms, concepts, nominals, and symbols to their part in TRU. REM is the amount or proportion of TRU totality, that is occupied by an entity, thing, sense, aspect, part of process of TRU. How great a thing is of TRU primarily equals its Relative Magnitude or REM in TRU. 6. **PRAV** or Practical Valuation proves REM through its dynamic utilization in life providing the selection both of TRU from non TRU, worth from non worth, and the better from the worse. 7. **Order of Magnitude** proves REM by providing a systematically unified scale as well as an exponentially, and warped to total scale, that enables the comparative selectivity of all parts of TRU as they can be evaluated precisely. 8. The determination of REM proves REM by showing how, and providing the posits necessary for the valuation of each part of TRU. 9. The compartmental sectioning of REM provides the classification and scaling of parts of TRU necessary for comparative and selective valuation. 10. **MOCBIREM** or the Method of precise valuation through the adaptation of great knowns and Class Measurements to TRU. How existing known topics that have good measures for their parts can be equated with and applied to REM of TRU. MOCBIREM is an immensely useful and valuable means and method. It can be no better than the degree to which each and all its components are held to TRU, and the scales of systems or groups are adapted to TRU. Other direct, indirect, and modified proofs of REM include, LARF, TO, TE, TIT, value, proportion, VAPO, special life, human, ENT, life chain, God, Á L T, sectors, meaning, DAOST, ES, TES, relative, the, distinction, identity, RUPKAL, MATES, LOPAN, OM, TRUP, TRUF, PRUNT, EC, LANT, language, MADAS, LARP, LATANOS, LORVA, ALT, SPALT, POS, STA, PRALT, TREQ, equality, distribution, PRUV, economics, business, capital, commerce, money, value, ENUL, LES, FAV, divisions, POREM, OTREM, POREM, and LOPELT.

DERIVATION, DEVELOPMENT, DETERMINATION and DEMONSTRATION of REM, POTENTIAL, REASON, CAUSES and GROWTH of REM, MONEY, LANGUAGE, COMPARISONS, METHODS, and Á

Man's struggle to be right, moral, ethical, and achieve righteousness in life, has been a major, continuous, and growing theme and power. It is associated with increased mentality and population which demand adaptation and TRU. The lessons of history have taught the necessity of learning what is right and the best ways of living together. This Development has corresponded with that of TRU, DOT. **DOREM** or derivation, origin, development, determination, and demonstration of REM. We should look for the origin of REM in developing different mass-energy states of life and man. For that level of Á above the adaptive, conscious, and rational, REM signifies the energy and power that an idea and measure of good or bad produces. This measure is signified by: **IMU** = Importance to Man Unit. IMU signifies that level, in the predominant relativity, observer-observed relationship in humans, Á FAF, in TRU, that enables a selective and measured determination of what is positive, good and right over the negative, bad and wrong. What is important, how it is important and especially, frequently, and continuously determined to do if life is to succeed. Man is an intentional, problem solving, evaluative, IMU, and purposive animal that focuses and concentrates upon existence in highly specialized and exclusive ways. This is signified by: **POIR** = Principle of Purposive or Problem Solving, Importance to Man, Revolution. One of the greatest stimuli to action is the accelerated search for, and discovery of, what is right, its value and use in action and achievement. James Joyce in Ulysses worked on this problem and gave it a unique literary form by focusing on a single day in the consciousness of an individual. In this way we can see both the most basic origins as well as ubiquitousness of values, axiology and REM.

The origin of values and REM are seen in the many Gods through history. How, and how much or little meaning, and in what way, they symbolized values for a culture. In Greek Mythology it was Apollo, the great protector, who symbolized good over evil, by letting the good in and keeping the evil out. This corresponds, as does much in religion, to the sames of DAOST in TIT, the powers of protection also seen in Vishnu in Hindu Religion. It was in this role of a protector, an extension of REM, that Apollo provided knowledge, civil order, stimulated colonization, and acted as the founder of civilization. Without Apollo the rest of the world, to the Greeks, was barbarian. Along with the ability to know good from bad was knowledge, divine enlightenment, reason and a heavenly Logos. It is not surprising that Apollo, although a late comer in Greek Theogeny, was, more than any other God, for many centuries represented in art.

FAV = Fallacy of Face Value. In life there is an inveterate tendency to take every thing at face value. We tend to fear what we don't know. For usually we know no better. Only as we mature and this knowledge increases we pause to consider and come to analyze other factors in making choices, decisions, judgements and valuations. Thereby it is in knowledge that most of the origin of TRU value, REM, lies. Since knowledge largely derives and results from TRU, they both progress in tandem. It is how the positive aspects of IA become IAT that REM is mostly derived. For we can not overcome FAV unless we know TRU or its equivalent, TREQ, and use them in IAT and our daily lives. There has been a great parallel accumulation of knowledge and improved ideas of value and TRU. Each greatly stimulated the other. Value tended to produce meaning and meaningful ideas as concepts were expressed by words. Knowledge of factual objects sensed, perceived and apperceived was fleeting and less rational without the ideas, concepts and words with which they could be represented. As this knowledge, and the concepts involved, accumulated, their use was largely dependent upon how great, beneficial, important, and valuable the REM in TRU was, that they supplied. For values, like knowledge, and the objects in the environment were or were not used, at least in resultant, to the degree they were or were not of help or benefit.

The worth of an item or object is largely the amount it causes or enables us to be, ENT. This is called: **MEVT** = Measurement of TRU, and TRU Value. One category of MEVT is what is known as hard and soft values. This is the difference between available and non available goods. Another category of MEVT is that between TRU value and ATRU value. MEVT is largely determined by the individual and TRU in general. This is closely related to monetary value. Yet they are not the same. Money, or any kind of monetary exchange, is the prevalent vehicle and measure of value. It is much more effective in enabling the exchange of goods and services, than it is as a predictor of REM. It is what money stands for, how it is used, and the gain or loss of TRU that occurs, that determines its TRU worth.. Our monetary values are often more or less false, ATRU and ANTRU. In this way there is a difference and ratio between TRU values and monetary or customary values. This is called: **ROMVET** = Ratio of Monetary or Customary Values to TRU Values. ROMVET tends to be rather low, yet not negative, in the world economy. It is apt to be lower in times of economic expansion, and higher in times of economic slowdown, largely because of concentration on the more essential needs of life. In this way ROMVET has played a major part in the ups and downs in history.(PPF).

The trend of ones thought is largely a measure of the values of a person. Ones goodness, rightness and valuation of the world about one is a function of one's interests and knowledge. This is especially so when they are massive, dynamic, ordered and total, as by GOD, TREQ, or TRU. The quantity and quality of what a culture holds in highest esteem, as by custom, and especially as it is actual and ordered by TRU, is the power and success of that culture. It is in: **SRIA** = the Sudden Realization of the value of life that REM is Initiated, and initiates. This is a kind of intensified, perfected, and total identity of Á. SRIA, like an epiphany, is what so closely bonds a person, REM and TRU. From the experience and learning of this bond valuation is perpetuated and its convergence upon GOD, TREQ, and

TRU increasingly assured.

LOVAL = Law of Value from Labor; of Property from Labor. LOVAL expresses Locke's theory that the individual has established ones right to property by mixing ones labor with that of mankind in common. Just as work makes life pleasant, so does work's provision in life depend upon whether or not this work is of set values. The more it is productive of ES and CES,the more it is of value. So that value arises and is produced by the effectiveness of labor, as it in turn is relative to TRU. To be able to determine and measure value, as it is REM, is the key to much of a persons labors, success, TES and TRU.

What our attention is on and what we are familiar with is good, bad, and their mixtures. The common selector is much the core of our being. Like much of IA it is highly fallible, apt to be in error, and cause failure. Here is a pivotal mechanism in TRU, how much does IA rise to IAT, the best by TRU. In the development of knowledge and language, there has been a constant sifting of the less real and important to the more. This is largely dependent upon how great or insignificant their meanings were. If TRU is all important, and if the 100 to 1,000 and more forms, parts, processes, concepts, etc. are also all important, then we should not hesitate to best name them. There should be, within bounds, a free license to develop any and all in as simple, clear, and effective way that best shows their part and role in TRU, DAOST, and REM.

DOP = Development of Proportionality. DOP occurs with increasing knowledge, language, and meaning. Yet it has been poor because it has little focus, lacks TU, CENU, TO and is not TRU. The *Hedonistic Paradox* is the apparent contradiction implicit in the hedonistic claim that a person's pleasure is the only good. Yet pleasure should not be the sole object of desire, for if pleasure alone is sought, it will not be found. This illusionary situation is also an example of paradoxes in general and their interpretation. The illusion is that of attributing actual relationships to words, phrases and logical interpretation. We do not have a strict definition of pleasure and good unless they can be reduced to REM and TIT. Epicureanism qualified this in that happiness is the greatest good and that happiness is to be achieved by living a life of moderation in which the contemplative pleasures are preferred to the sensuous pleasures. Like so many ISMS, Epicureanism, when narrowly interpreted, is a good example of Extreme by Emphasis or **EBE**. It emphasizes great principles, yet through what it omits, and how it is misused, loses much of its value. Who is the "good man" "good woman" or "good person", is not decided in popularity nor even lesser means, but only through TES, TRU, or TREQ. **MISDOT** = Mutual Development of MOT and REM. If you do not develop both the modulation or frame for, and the derivation of essentials or first principles of the world, you have a lopsided or erroneous rendering. It is a very poor observed for which the observer can never achieve identity.

All objects are not to be regarded as belonging to one level. This is the basis for the Theory of Types that is fundamental to all, as the categories set forth in the first section of this book. When dealing with different choices and the levels involved, we must remember the significance of interval and transition. What must be chosen or selected is based upon a widening spread with increasing interval or distance between objects, a ranking from which comparative and/or equal values must be assigned. So that in monetary values ever widening intervals and transitions are determined by some assigned or standardized base. If such valuation is based upon proportions of a whole there is derived REM or Relative Magnitude. As priorities and importance are given relative values, an increasingly useful measure is achieved. Since life has great meaning and value those preferences which are in harmony with TRU can and need to be measured, and rarely standardized.

The development and growth of REM follow upon the representation or production of proportional values, relative importance or determination of the relative magnitude of all. **DREM** = Determination of the Relative Magnitude of All that is a portion, constituent, aspect, part or process of TRU. DREM is specifically ordered in the best possible way it is TRU. DREM is the crucial proper, major and perfected form of valuation and axiology. To make these values, measures and proportionate parts, we should be able to test their validity. This is called: **TOVOP** = Test of the Validity of Portions of TRU. TOVOP is how we develop and determine the means and methods that make certain our ES and CES are exactly what they are in TRU. It also seeks to reduce them with similar validity and accuracy to posits in the best possible MOT, or SPALT. One of the means by which we do this is to equate their meanings with their magnitude or value as they are in TRU. This is called: **LEM** or Law of Equivalence and Equating of Magnitude and Meaning. LEM signifies how much magnitude a part or process will have, correlates highly with how great its meaning is developed and is in TRU. **LORVA** is the Law of Relative Value, or how we can extract that which is TRU of entities, senses and things in the world, or their ESINS, and assign to them a measure of their worth, a relative value. Without the ability of LORVA we can never attain any proper valuation of the world about us.

The significance of the insignificant and the insignificance of the significant, is the NAMTRU world in which we live. It is to be able to discover, find, develop and show what actually is and what is not by TRU, that NAMTRU becomes AMTRU, AMTRU TRU, and we TES. This is much of the business and crux of S-O, our Á, predominant, preponderant, basic mass-energy FAF of being. The little objects we use, the homely things about us, and the obvious ideas we have may seem so insignificant, yet may be of great significance in TRU. The most expensive and costly items, that which we, without full knowledge of TRU, work so hard for and seem so indispensable, may be so insignificant in TRU. Whether they are of the greatest or least of TRU, if they are not optimal for their form and function, state, role, and REM in TRU they are less than TRU.(VAPO) The significance and greatness of anything, entity or given is what, why, and how it is in TRU. If this is not known or correctly assessed and posited, then must be added whatever is lacking, unknown, and of significance by and of TIT-TO . This is: **LACOT** or Law of Closure of TRU. To TRU must be added whatever is lacking, unknown, gives it significance and greatness for all in life This is key to PRÁ, TRU/AU, and ATRU. This is **LASÁT** or Law of Significant Greatness in TRU, TIT.

It is impossible to provide the relative magnitude or values of all without the ncessary and prerequisite information and knowledge. To understand what is and what is not their manner and detail requires an appropriate quantity and quality of information and knowledge. All information, facts and evidence about an object, subject, thing, sense, or entity is more or less significant if for no more reason than an explanation of other, and more TRU, knowledge. Yet it is this more TRU, IMU of SÁ, and REM that is their true measure. So that IMU not only represents a form of energy, as a level of Á, it is also the determinant of relative magnitude and MPLT or SPALT significance. Emerson stated, "What is a weed? A plant whose virtues have not yet been discovered". IMU and REM are, in ways, different. REM is the measure of the parts of TRU proportioned in the whole. REM is best reduced to SUT and POA. IMU is that part of the mass energy - or Á dimension - of TRU that selectively determines what is qualitatively and quantitatively important.

Another way REM has been commonly determined and demonstrated is by the method of elimination, or 'how much can we get along without it'. Although we are hereby making a relative judgement by comparing it, the very being of the whole gives it great relative magnitude or value. Since such judgements lack precise focus, objective base, and TRU, they are only a rough, but vivid, gauge of REM.

The development and growth of REM is revealed in LAR, LORVA, DOP, DREM, TOVOP, LEM, greatest happiness and rules of E determination. **LAR** is the Law of The Relative, haecceity or beingness of the universe Given, at hand, or actualized. It is LAR that renders REM convergent and compulsively discrete and thereby produces the necessary reducibility to representable valuation. As all exist in REM as they do in TRU through their actualization, as by **PIP** or Proof of Integrative Processes, so can they be developed into REM. This is called: **LOREM** = the Law of the Total development of all to REM. The relation between what seems to be and that which actually is, or what seems to be so great and that which is actually so great, and is itself so exemplary of distinction, is designated by: **USPU** = The Unsubtle that is Profound and the not so profound that is most Subtle.

DEREM is the Development of the REM of E, ES, CES, or specific parts and factors of TRU for placement in a model. When we closely analyze increased specificity of **REM**, **DOREM, DEREM,** and **DREM**, their levels, 1R, 2R, 3R, 4R, of ES and CES and their corresponding ESINS in ideas, abstractions, words, concepts, and abbreviations, and posit them in ALT and SPALT, we can attain a highly practical and revealing representation and image of what TIT actually is. When we supplement this with VAPO, DAOST, the SCARLS involved, TRIS, OBT, SUT, as we can by applying: **ARF** or absence of REM Forms, and **DIREM** or divergence of REMS, Coterminality of REM, and MOCBIREM, greatly improve the specificity, sensitivity and reliability of values in TRU. The degree to which this can be a more accurate as well as practical image of TRU can be continually refined and validated in the operations of MOT, SPALT, POS, STA and HOM. It has increased as knowledge, first principles, concepts, ideas, and laws have become more valid and actual. Through increased formal distinction, clarity and meaning they are or are not as they are selectively utilized by humanity. They are utilized if their meanings are: very useful, clear, familiar, problematic or interrogative. The continual upgrading of meanings of what represents ES and CES is what enables REM, as REM, in turn, helps to make them more distinct.

Fig: 2-35 **COMPARATIVE VALUATION of HUMANITY**

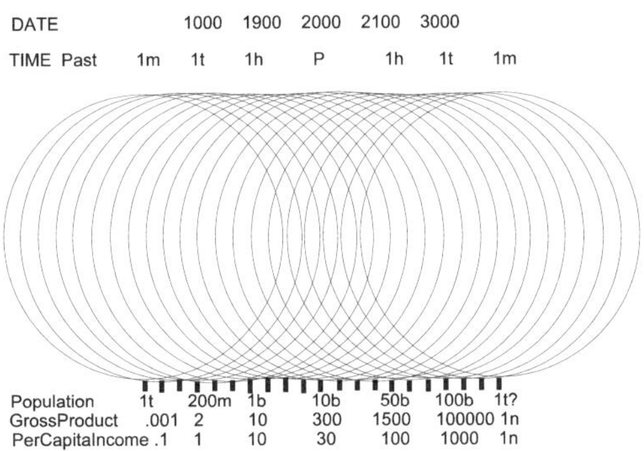

As man's possessions and life have changed so have the values of that which might act as standards. Certainly they have changed. At some times increasing and others decreasing, in part depending, as in economics, on supply and demand. This is natural, for the less life we have, the dearer it becomes. We of the 20th-21st century look upon those of the past, who seemingly possessed so much less even in their own day, as being worth so much less. This is measured by gross products and per-capita incomes. Yet are per capita income and gross national product fair measures of value of the individual or more totally TRU? Certainly not, yet it must be a considerable factor in the equation. Figure 2-35 shows serially the changing GNP, PI and populations past, present, and future. Yet we must be careful not to equate overall worth, individual worth, or proportionate values by a set of monetary gross national products of all of the world. For if we would the world in the next hundred or so years would outgrow itself, not so much by population alone, but magnified many times by material, commodity, and growth of the world economy. It shows how when populations expand rapidly, and the gross product does likewise, the supply necessary to support this population and gross product does not expand accordingly. This is immensely important by showing the greatness and necessity of REM in **ACC**, or Action and Control, of TRU. REM, expressed in many ways, directly, indirectly, or inferred is a large part of human life. This is seen in greatness, comparative bases for measure, ability to do, ease of accomplishing desired ends and others.

From the subjective, observer or sense aspect in REM, we have a sense of values. These attempt to accurately measure the most concrete and distinct representations that can be made. It is not only the best possible measurement, it is so much more, what this deepest sense of values conjures that is the greatest and the most important in life, the greatest of TRU. This is called: **LASCI** = the Law of Scientific Significance. As all becomes more and more differentiated and discrete the representation of it, as by symbols and language, becomes increasingly possible and productive. LASCI is justification of TRU, and is in part caused or stimulated by existing language both its lack, overlapping, changing and poorly defined terminology. As much as REM is limited to, and to be so determined only from, a set standpoint, we can't help but consider, and leave open, how this will vary. This is so with any observer and with relativity other than the relativity that exists absolutely within TRU. REM thereby becomes not only outside of TRU but inside of and therefore **REMT**. It will be, as OITIO, a part of the transcendent dynamics, and like THES and THEO, constitute a large and dynamic component of the mass energy aspect of TRU. REM becomes REMT in TRU convergently, yet divergently as the observer or observed are extracted or exist as the given without TRU. So is **REMIT** or REM itself, different from REMT with all gradations in and out of TRU. As REMIT and REMT provide a polarity with gradations, this is **REMIP** or REM In Presentation. For REMIP we may have any where from one to many powers as in REM^1, REM^2, REM^3 and REM^n. These will depend upon how much refinement we search for in this polarity. REM is the quantitative state, whereas **REMVOS,** REM Variable with S-O. is the qualitative state of REM as it exists in transition from REMIT or in and of itself, to REMT as it exists only and exactly in TRU. Although REMIT may appear obvious, its significance, like that of IA and the individual, warrant continuous analysis. For each must be continually adapted and returned to TIT-TO.

As one gets deeper into an object or entity they will encounter more and more of the same. More and more will be of this same object or thing. This is to be expected by the necessary convergence on TRU, TIT, TU and TO in the separation, discretion, and distinction of its ideas, concepts, and processes. This for REM is the ratio or: **RARES**, Ratio of REM to Size, and REM/OV. NEC and all that is reducible to TRU will be guided by RARES. For it is by the finer and finer differences of the same that TRU is composed. It is this increasing similarity that makes for familiarity in an object, thing, topic or field, and in turn, is a major reason why it is so much easier to remember a topic or field when once learned. This is what enables us to better relate all of TRU and provide a POA and POS, REM, ES, and their POS. They are enhanced through **MOREP** or actuation, intensification, extensification, and Production, ACEXIN of REM production, and **DILREM** = The Dimensional Level of REM. It is important that the individual recognize their level of REM, both as individuals, ES, TES, and TIT. This is, among other ways, At a moment, **ÂTÀ** or Absolute hesitant of time, in general, or in a lifetime as in total TES. These are special cases of ELTICMA or an individual, group, or, all persons in relation to TIT.

CHAPTER 36

FORM of REM

GENERAL FORM OF REM, WHATNESS, KINDS, SHAPE, SCARLS, RELATIONS, NECESSITY, VALUE, INTRINSIC and INSTRUMENTAL, COST, PRICE, BEAUTY, IDENTITY

In axiology, evaluation, and the determination of Relative Magnitude, REM, we can't simply assign relative values to different terms for things as they seem to be a greater or smaller part of our world, or of greater or smaller priority. A determination of REM requires both the application of all that conditions it, all principles that apply, and the most astute accounting, and reconciliation of each and every item for valuation, or REM. For TRU, ROT, MOT, REM, and POS it is much more complex and rigorous. It requires a great amount of refinement. Yet it can produce a highly accurate and beneficial basis for each and all person's lives as well as the success of TRU. Initially we have no, or a very incomplete, idea of what these bases for REM are. We can only gleam their meanings, their part in TRU by what they connote to us, and how we assume they are significant in TRU. It is therefor the ESINS and ES of entities whose words we must seek and use..

There are certain concepts that help us understand the form of REM. The value of anything should be related and proportional to other things.The value of anything should equal 'values' or that held most dear or important. These values can be nothing less or other than their proportional part, REM, and processes in This Special Universe of Best Human Life, pure TRU, TIT-TO. The form of REM is more than a simple value of an object. It is all values of TRU and especially those values that make up a precise, ordered, and proportionate determination of TO, the totality of TIT or TRU. Some of these concepts include, **ARA** or a lack of REM forms, **SOREM** or Selection of REM or value, **MEVIM** or Measure of Vital Necessity, **LORVA** or the Law of Relative Value or ES (and ESIN) Value, and **PLACREMS** the Correlation and Equivalence of Concepts, Principles and Laws with REMS of parts and processes of TRU. ARA is where we start before we make any choices, selections, judgements or evaluations. It is almost impossible for us to recognize, because it is a practically lifeless state, alien to TRU. SOREM is any and all selection whereby a choice is made in relativity between what is observed. It permeates all our life for even the changes in our IA from moment to moment involve a high degree of selection, unconscious if not conscious, indirect if not direct. More specifically selection is the basis upon which all DREM are determined. **REF** is man's usual selection of values by referring them to what we already know. The more familiar we know something the better and quicker we can value, accept, or reject it. MEVIM is also a common means of valuation based upon what is vitally necessary. It can be at various levels of mass energy or A. It is very evident in all animals whose movements and choices are automatically and spontaneously dependent upon instinct, habit, and built in mechanisms that support life and survival. LORVA is how all things that are about us although seemingly completely apart from us, eg, animals, forests, etc. are in their vast and often indirect special relationship to us, a part of us, TRI.(130) This relationship takes on and has set characteristics, or certain forms. That there is a high correlation, equivalence, and identity of these great aspects of TRU, PLACREMS, is increasingly evident by their form and function in TRU, ROT, MOT and HOM. From this we can better develop each as they are not entities functioning in and of themselves, but actual components of TRU in their full actuality, integration, and differentiation.

To illustrate proportionality of terms and that known and to emphasize and clarify basic relations we should recall their explanations and standardizations. In **SARUE** or Self and Relative Universal Equivalence, Equidynamics, we have a means of obtaining the release of persons from bondage of their identities with manifold, diverse and mixed worlds. So will the individual, both giving and receiving in all ways return, sooner or later, to TRU. They are in a state of dynamic equilibrium as all proportional parts of TRU are. It is ATRU and this dynamic equilibrium, combinations of meanings, value, superimposition and fusion that makes knowledge and TRU'S picture or understanding so difficult, even if vital. It is their separation and formulation, opposition, and extraction that enable its identification and comprehension.

Without TRU, and even to a degree with TRU, it is very difficult to consign and reduce words, concepts, ideas, objects, things, and topics to valuation and REM. A word's common definition will have one value, its ESIN another, and the ES and CES this word represents in TRU still another. Even more, different values would have to be given for the first and second persons meaning or interpretation, and still another for the variation and changes its usage would conjure.

REM may be divided into different parts in terms of how we consider TRU. If we measure TRU only by its component parts and give them a value between 1 and 0, depending upon the percentage of the whole they constitute, then this usual and proper **REM** is measured. It is DREM if it is of any and all that forms or constitutes TRU. It is **DEREM** if it is limited to ES and CES. No one measure is all important, although it is: **REMP** = REM Proper, if it is usually referred to in this text. The exacting form of REMP is DREM, and when limited to ES and CES DEREM. **REMU** = or REM Universal, is the positive and negative, or what must be used when we interpret TRU in more than any sense as these senses assume a part of the valuation and involve more of TRU than TIT and more of REM than REMP, DREM and DEREM. REMU is of great value in HOM, ethics and our lives in general, especially when in AMTRU or NAMTRU, the constant interaction of being in and out of TRU, OITIO.

TES, TRU and TIT are of absolute importance as the prime aspects of this special universe of best human life. On the other side, and to reverse the negative, it is of similar importance to counterbalance those aspects of human nature that are inherently selfish and thereby tend to evil. Only when we can fully and comprehensively discover, know and apply TRU by the contrast between different goals in life can we understand and manage our world properly. Is it for how much money or power we can attain, for security, for survival, for quality of life, or for any other needs and goals we must work and live for? Values that come in many ways indirectly and even from the negative, are needed if we are to make REM total.

The cost of something, in actuality, is how much we have to pay for it with our lives as TES, or ultimately how much TRU has to pay. This is called: **CEPLAT** = Cost Equals Price in Life as TRU or TES. CEPLAT is of great importance for the degree to which we use the world around us, our resources for good or bad, TRU or ATRU and ANTRU is how well monetary value equals REM or TRU value.

The whatness or quiddity of REM, that which constitutes and is predominant in it, is shown by comparative greatness, measurement, magnitude, division of labor, proportional separation, LOPS, and their uses in life as in problem solving and POIR. "An idea is greater than a man, an ideal is divine" is said to have been quoted by Alexander of Aristotle. This comparison of what is greater or of more worth is highly significant not in both pointing to that of greatest worth, but in being able to provide some kind of relative order to worth. This relative order is not based upon any value; it is based upon a simple relationship between the two factors in TIT.

Some places in the world may seem so beautiful at certain times of the day or seasons of the year, conditions, or different ages yet so ugly, drab, or common at others. So also is much we see, sense, or perceive in and about us. We must SCARL and VAPO these and bring ourselves to base this beauty or value only on what is its fullest part and role in TRU.

Since REM must be based upon and produced from the unity of TRU, we are repeatedly tempted to focus upon those values that most provide this unity, are most vital. Not only the greatness of life, but the greatest aspects of TRU whose composition in TRU gives us the most simple and reasonable values and REM. How great the birth of a baby is. How great it is that so many people can enjoy the benefits of this world. How great the quality of each person is. How precious the life of each person is as this is the happiness and fulfillment of TES and TRU. Yet one person can't take this from another, nor be two persons or none without being other than TRU.

In order to provide **ROPU** or Resolution of the Proportions of the Universe we must know what this universe is and how it can be proportioned. We must also be able to identify and make discrete each element to be measured. Our effort to provide infinite importance to all in TRU is limited. This is called: **LIREM** = Limits of REM. Only as REM is perfected and its limits expanded is there a fully actualized and effective standardization and measurement of value and importance. The proportional value of IMU and REM must be from the proportions of TO-TIT and must be perfectly balanced and weighed not by a simple fact or act alone, but by the summation of its existence and a summation of the indirect and dynamic effects of each element, part of TRU, or entity to be measured or rated. To provide the frame for a panorama of REM we best adapt TROS and its parts - as they provide values or REM - to a scale and frame that gives us an optimal axiology or values for representation.

UNITARY	**FINITE**	**INFINITE SET**
UNITARY SCALED	**FINITE SCALED**	**INFINITE SCALED**
1-large n	**Large n to Small n**	**Small n to Infinite n**
1-10	**10 - $1x10^n$**	**$1x10^n$ - $1x10^\infty$**
1-3, 3-10	**10-100, 100-10^9**	**$1x10^9$-$1x10^n$, $1x10^n$-$1x10^\infty$**
R - 1R	**1R-2R, 2R-9R**	**9R-nR, nR-∞R**
TRU	**2R,3R,4R,5R,6R,7R,7R,9R**	**9R-20R, 20-100R, 100-∞R**
REM	**INFINITE NOT SET**	

This great variety of scales for representing REMS, although neither unified nor ordered, shows the importance of avoiding the trap of making valuations and measuring REMS by any scale, when they must replicate all TIT. It shows the great need to superimpose and fuse these various scales into a systematic whole that best represents TIT. One way is to provide a perfectly scaled distribution of R's warped by: 1 - $1x10^{9+}$ - $1x10^\infty$.

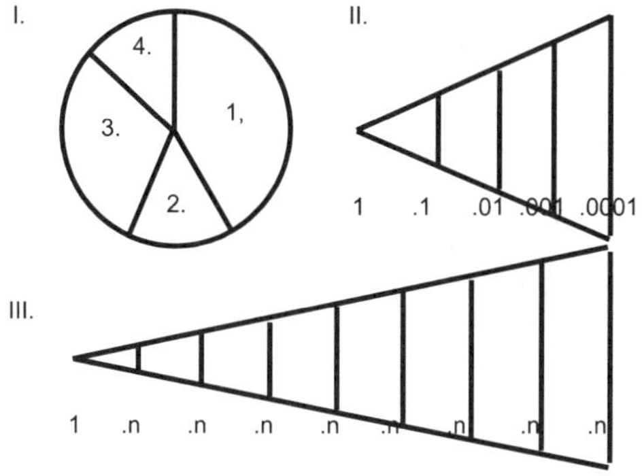

Fig 1-36 **GENERAL FORM of REM**

As man has sought, and found to some extent, an ever better form of government to control larger populations with differences and oppositions, one of the means by which this has been assured is by <u>Proportional Representation</u>. This is based upon distribution and simple mathematical percentages. The method of more accurately showing these proportions and percentages is provided by statistics, geometry and graphics. Several forms are available of which we use only a few in this book. I. A Circle Graph which is a statistical graph for showing the relation of parts to the whole. II. A pie graph by sectors in two dimensions, or cone by three dimension, showing the proportions and percentages of the whole in increasing amounts. III. Adaptations of the pie and conical graphs to the REM of TRU whereby the percentages of the total are warped from 1 to the infinitesimal. For developing, determining and demonstrating the proper form of REM we adapt such a conical graph by selecting a mid point between 1 and ∞/1 and providing the interim fractal levels. What this mid point is, is determined by TRU and our expectation, ability and power to fulfill this representation with an appropriate valuation of parts, processes and aspects of TRU.

The general form of REM starts from a proportional separation of the makeup, or composition, of TRU. This we do as we discover, develop, determine, demonstrate and refine **LOPS,** or the Law of Proportional Separation. LOPS, REM, and even TRU to a large extent, are products of the intentional, purposive, problem solving, and importance determining makeup and revolution in human life. This is a level of energy which has its own mode and level, the IMU unit in the A FAF of TRU. It is signified by: **POIR** or the Principle of Intentional, Purposive, Problem solving and Importance to Man, Revolution. POIR is an acronym that signifies and exemplifies the fact and principle of ascending states of energy of which man's world is, and is becoming more rapidly, formed. At what point this revolution of man's world exists in is not yet entirely clear, albeit somewhere in the midst of its massive and energetic development. Where it is, like the time dimension, depends upon how much we put into it, and how many more, even higher levels of mass energy, the A dimensions of the future will have. POIR is provided for by the ability of man and the accomplishment of ordered proportional separation. This is: **LORPS** Law of Ordered Proportional Separation. LORPS is, to a great extent, a product of LASUR, formal distinction, finite delineation, definition and expansion of TRU.

The laws of measuring the REM of ESINS are: I. We must determine how much of this thing, entity, interest, or object is TRU and how much is not TRU. II. We must determine what the posit of it is. We must be able to place it in the structure, order, replica or model of TRU, SPALT. III. We must determine what its range is. That is, we must be able to say that roughly 1%, 10%, 50%, 90% or 99% of this is within a set segment of TRU. That segment of TRU in which its posit or center is focused. IV. We must determine the divisions, and segments of TRU, so that # III can be given meaning and value. Is this 1/8 or 12.5% of TRU, 1/100 or 1% of TRU, 1/1,000 or 0.1% of TRU, or 1/1,000,000 or 0.0001% or 1 X 10-6 of TRU. V. We must determine the exclusiveness of it. Is it all of what its term connotes? Or are there other terms which describe a great part of TRU which are coterminal with this term. To this extent its REM will be reduced. For ES, CES, all aspects of TRU, and even more important laws apply, including: 1. Is it only TRU and not diluted to become only some smaller percentage of TRU. 2. How well can this ES, CES or aspect of TRU be posited in a model of TRU, SPALT? 3. How many other ES, CES or aspects of TRU are coterminal with it? 4. How discrete or formally distinct can we make this term or object of ES, CES or aspect of TRU? 5. How much is it indispensable or how much is it crucial and critical for life. This is a key method and vehicle for valuation and REM. There is nothing more directly associated with value than the degree to which a part, process, aspect, form, function, and principle of TRU is essential or vital. It is simply a measure of what TRU can or can not live or do without. Much that TRU can or can not do without may be secondary, indirect, highly dynamic or unknown in all of its many functions and effects.

What may seem to be a whole by itself, or much of TRU are, even when TRU, but a part, even a small part, compared to the rest. This, its ESIN, is only a small part of itself, ENIT. Herein is PROBT, Proof of Objectivity. It is in such overall perspective of TRU and its proportioning that we can see: **REMUC** or Relative Magnitude in the Universe, TRU, by what greatly helps, at least to begin to form and prove compulsion, the compulsion of objectivity and A. We are compelled by exclusions, the totality of TRU, the proportionality of its parts and aspects, and a conformity to the best possible structure of REM itself, to fit all together as they provide this most important object and representation of TRU. By REMUC we are compelled to fit all that is TRU into TRU as REM and at the same time adapt and refine REM. Parts are absolutely vital to the fulfillment of the whole, TO of TRU.(LOPAN,PAVAT,SARVAT). Just as the heart, brain, and oxygen are vitally necessary to the human body. This may be compared with: **MEVIN** = Measure of Vital Necessity, MEVIN is a good test of REM and proof of the make up of TRU. This is shown by DREM and DEREM. MEVIN is the fact, for REM, **PED** is the underlying Degree to which this fact is Perfected Essence in operation.

SPECIAL FORM of REM

The special form of REM separates what the principles and overall form is from the form that provides the most effective means and measures. This is how we can value all that is differential of TRU, as ES and CES, with sufficient overlapping and balanced coverage. A precise measure of values or REM may be attained from: 1. known portions of the whole, 2. envisioned in a portion of all vantages, 3. can be no more precise than the definition of the portion, subject, object or topic - as TIT - to be valued, 4. must be a valid, or true value, that is it must be 100% of the whole or object to be valucd, 5. must bc of TRU, or: **MEVT** = Measurement of TRU value, or True Value as it is TRU. The chosen REM are absolute when 100% identical with their counterparts in TIT.

REM is most precisely determined through proportions of TRU. These include the divisioning of the fundamental dimensions of TRU. If the dimensions of TRU are formed by ALT, their six divisions of subjective, objective, large, small, past and future give us a TRU of 8 separate REM. If we further divide these divisions with their center sections separated and signified we have 27 separate REM, subjective, real, objective, large, medium, small, past, present and future. It will be quickly seen this is not what we commonly expect with known values. It does show how accurate, valid, reliable, sensitive and specific values can be given. It gives a precise method of determining the proportionate, relative magnitude, REM of TRU, and it helps, in return, to prove not only REM, but A, ALT and TRU. For by weighing TRU accurately and comprehensively, and by accounting for all principles and factors that refine each item for valuation, we gain a much better idea what is in TIT, and what is REM. This initiates the proof and form of REM and DREM. This has precedent in the law of exponentials of distance, l^2 where the light of stars are measured by distance from the observer, $M = Kx(1/l^2)$. This is where magnitude equals a constant, times the reciprocal of length squared. A good example is the sun, which is of tremendous size and power in itself, is equally tremendous as REM. It is only one of many major ES and CES in TRU. This is true of all entities; which are more or less, depending, in part, by their distance from us. The dimensional measure of an entity, ES, CES, element or aspect of TIT and SPALT will vary with each definition, meaning, and how it is used. (DEMVU)

ES and CES will not have equal value or REM depending upon how their codomain is treated, and to which part, process or aspect of TRU sole domain is granted or given. Herein lies the greatest challenge of valuation. In this book we refer to the REM values of TRU either with or without the decimal shown, i.e: .1 or 1R are both 1/10 of the whole ofTRU. The Order of Magnitude is as follows:

R's	Decimals	Number
1	.1	10
2	.01	100
3	.001	1000
4	.0001	10,000
5	.00001	100,000
6	.000001	1,000,000
7	.0000001	10,000,000
8	.00000001	100,000,000
9	.000000001	1,000,000,000
10	.0000000001	10,000,000,000
11-n-n-∞	.. . . .	.. . . .

The first column represents the Order of Magnitude or R Level, the second the decimal quantity of the whole and the third the number of parts represented at that level. By envisioning the rapidly increasing number, and greatness of these parts of TRU and their need for expression and comparison in valuation, it will become evident that both the scope and need for REM is set by this form. It will also show the great amount of work that will be needed to enable it to have the sensitivity and specificity that a statistically dependent system demands. Great thoughts and ideas may flow freely at times of demand, yet those of greatest importance are those we are continuously concerned.

REM	SUBJECT-OBJECT	DEGREE OF UNIVERSALITY	PART OF A
GREATER than TRU		**LE-098**	
1R	**e.g. KOT**	**HIGHEST, SELECTIVE**	**Very Little**
2R	**e.g. DAOST**	**HIGHER, SELECTIVE**	**Little (LORE-023)**
3R	**Great Groups**	**HIGHly SELECTIVE DOR-032**	**Increasingly**
4R	**Primary of Great Groups ESO-035**	**HIGH SELECTIVITY COVREM-054**	**High, Variable**
5R	**Secondary of Great Groups**	**HIGH, less Selective**	**High, Variable**
6R	**Tertiary of Great Groups**	**INCREASINGLY PARTIAL**	**High, Variable**
7R	**Parts of Great Groups**	**LESS UNIVERSALITY**	**High, Variable (POMOM-081)**
8R	**Small Parts of Great Groups**	**LESSER UNIVERSALITY**	**High, Variable**
9R	**Smallest Parts of Great Groups**	**Least Universality**	**High, Variable**
10R	**Average TES (LATANCOV-038)**	**Little Universality**	**High, Variable**
12R	**Average Person**	**Unlikely (COV-054)**	**Higher, Variable**
15R	**Subpar Contributor**	**Unlikely**	**Higher, Variable**
20R	**Average Act/Average Person**	**Unlikely**	**Higher, Variable**
30R	**Average Thought/Average Person**	**Unlikely**	**Higher, Variable**
50R	**Barely recognizable Positive**	**Unlikely**	**Higher, Variable**
100R	**Low Order Positive**	**Unlikely**	**Higher, Variable (LAIA-050)**
1000R	**Very Low Order Positive**	**Rarely**	**Highest, Variable**
n R	**Tiny Finite Positive Increment**	**Rarely 1/n +**	**Highest, Variable**
∞ R	**Least Positive Increment**	**Rarely 1/∞ +**	**Highest, Variable**
0 R	**Resultant neither + or -**	**Rarely - = +**	**Nearly all (RACOD-044)**
- R	**What is Negative for TRU**	**If ATRU, ANTRU**	**Entirely**

Table 1-36B FORMAL DISTINCTION of VALUES

That the smaller, 1R-5R,, REMS tend to be more universal and the higher order ones more partial is evident, to be expected, to be remembered, and used to further their refinement. The unity, equivalence, equality and stability of REM are only evident when values are derived from a single and totally representative source. Just as the only way to attain an objective is to know what the center is, and from that its surroundings, so is the unity of REM dependent upon a full focus and perspective of TIT-TO. The worth or value of ones IA, thought, subject, topic, study, discussion, thing, diagram, book, IPTRU, first person or anything equals: 1. The percentage of it that is its ES. 2. the corresponding part of it that is in TRU, 3. the degree of its brevity or resolution of its limits, 4. the degree of perfection by which it presents, depicts, and illustrates this part.

REM = ESIN x ES x REM Limits x ROT perfection

Thereby worth equals proportionality, brevity, artistic perfection, the degree these are equal to its ES as it is a part of TRU, and the percentage of the thing that is its ESIN and ES. Brevity is like Occam's Razor where value is the highest where it measures the least that is appropriate for the measure. From an equality of REM we attain: **LE** or the Law of Equality or: **COL LE - X = TRU**, and

Equality x S-O x X = TRU x LAMID
x quality Production or O x S x S-0

Through the unity, equality, LE, structuring, and formulation of REM of TRU we can develop, represent and illustrate REM by a target diagram. In such a diagram we look head on at TRU as a unit and derive from the equality measures or values that fit decreasing levels of relative magnitude, REM.

Lesser ES, or those of small proportion in TRU, are given more qualitative and less quantitative emphasis in such levels of REM. Those like TOPNEC, of the greatest laws, principles, concepts, and mechanisms of TRU are more basic, necessary and vital. Neither are they always the most positive, for they may be negative in reversal, or productive of the positive, as fear. Although fear. is not on the highest level of mass energy, or of SA, it represents a force that is highly productive of correction to TRU. It both directly reacts so that stimulus, by bringing the negative to positive, brings a person, or peoples, to IA, or awareness, attention, and action. The great need, however, is to make sure this fear is directed toward the right things, and best reactions. To fear trivia, and the very bad that is of little consequence, reduces ones attention on what is greatest, what is detrimental to TRU, and highest TOPNEC. The great fears of ATRU/TRU, Individual/TES, no REM/highest REM, extinction/ survival, permanent loss/improvement, and many others are fleetingly or indirectly appreciated. Whereas they must be at the center of concern, interest, and purpose throughout our lives. Religion has been of value in directing fear, as in superstitions, atonement, supplication, and 'eternal damnation'. Yet these have lacked the overview and comprehension of the actual world to such a extent that their resultant efficiency of benefit/harm is very small. To focus on what is of greatest benefit/harm we must focus on TRU, all that is great of TRU, as TOPNEC and highest REM. Yet, even here, it is not easy, for we must not only provide an excellent REM, we must resolve all that lessens it, e.g. coterminality. To solve the problem of coterminality we must solve its kinds and root causes. That is whether coterminality results from lack of meaning, formal distinction, definition, coordination, implication, interpretation, etc. This is the form and task of REM and POS, how to bring them into the most discrete, yet orderly form and closure.

There are several different kinds of REMS and POS. Each may be separated by how much they are strictly of TIT-TO, how well they can be made formally distinct, how easy-hard-impossible they are valued and positioned, how well we can improve them by comparisons with others, and how useful they are. Some of these kinds REM and POS are: taken without qualification, selected, stereotyped, overlapping, standardized, and those that are engineered in various ways to most replicate their objects in TIT.

That REM must be guided by and conform to TRU is axiomatic and by definition certain. **FER** = the relationship and bond between ESIN, TRU Form, and REM. FER acquires new concept status by right of the all important realization and actuation of how ESINS of entities are formed as they become TRU, through ACEXIN. How both correspond to and produce REM. This REM is both its special form as well as its specific values. In being so combined, yet distinct, they provide a sensitive, specific, and reliable basis for valuation. They enable treatment of all before us to become measurable units or factors. This is accomplished by: 1. being actual, or their ESIN and TRU portions, 2. being discrete participants in TRU, 3. being analyzed by the form and construct of TRU, 4. being valuated by a systematic and representative means or REM, that provides proper and optimal measures. It is in FER, or how ESINS, form and REM combine to enable a sensitive, specific and reliable measure of valuation, that our proper form of TRU is and achieved.

What is dynamic by REM, is much the same as what is static, only REM needs to be analyzed by the dynamic. What is in great motion, force and power has ESIN, just as what is thought of statically has its effect and REM. These are or are not of value, importance and REM, depending upon their CES in, of, or by potential, TRU. **CURS** = Correlation Unequal of REMS. CURS is the corollary of COCOR and signifies the fact that two REMS of equal magnitude may not have the same correlation with each other. CURS tends to decline with increasing knowledge.

STAREM = The Stabilization of REM. If REM are completely understandable, i.e. if the loss of REM value is reduced to its extreme through exclusion of other overlapping REMS, ES, and words, then for SU to have meaning a practical standardized agreed upon level of division must be determined and set. This is done passively in usage, by common language, speech and writing. Yet it is done without much system, order, precision or efficiency. This is called: **REMVE** = REM is valid only as such, as it equals its ESIN and these their ES in TRU. The great process of the **ESINATION** of words is a deeply ingrained, long continuing and highly empowering characteristic of language, human life, and A FAF of TRU. The degree of success of SU and the meanings of REM and SU are functions of the degree to which each ESIN, REM or component of SU is perfected.

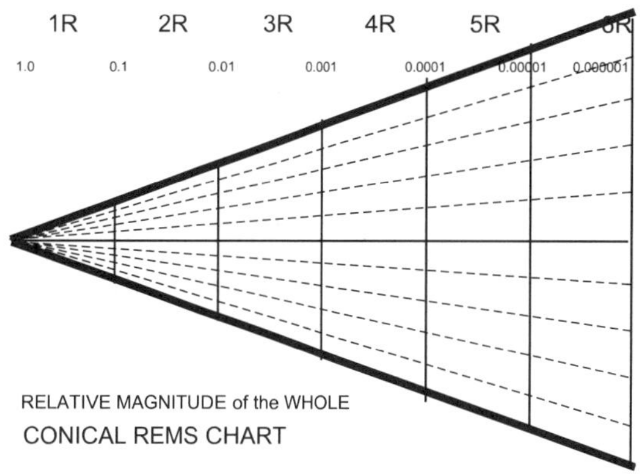

Fig 2-36 **CONICAL REMS CHART, FRAME OF**

This is the frame or construct of our primary representation of REM. It provides the field upon which we can posit REM. All concepts, principles, laws and mechanisms of REM, and related topics, must be best used as they guide us in both the interpretation of this chart and the posits to be made. **PROV** = Projected Valuation, how much of the value of all things, especially new concepts and parts of TRU or ES, lies less in themselves. PROV is how they are a part of TRU, directly or indirectly enable TRU, or stand for and show something else. This is much of the function of subjectivity and objectivity, or representation, and both in being fundamental to TRU and much of all our world.

In 'par excellence' we find a tradition that has great axiological, valuative and REM implications. For it is the interest in, aspiration for, and will to attain quality and excellence that we both seek values, REM, and TRU. This is greatly enhanced by the growing appreciation for quality and excellence, that comes with greater respect for life, art, ROT, VAPO, and TRU. REM can be no better than the degree to which ROT, MOT, and SPALT are perfected. **LORF** = the Law of Representational Failure. LORF applies to those SCARLS whereby what we select or choose is done so out of measures and feelings, that are less than or other than LAV, VAPO and TIT. It is through: **TOVI** = the Test of Validity Intensity and Inherence, that the goods and bads, rights and wrongs, desires, needs, and wants are all set properly and in proportion to their worth, need, and part in TIT.

CHAPTER 37

A PART and PARTS of REM

DREM DIFFERENTIATION and SEPARATION, DIVISIONS, MEASUREMENT, WORDS

The dependence, significance and operation of REM on an ordered differential of all TIT is **DREM** or the Determination of REM by the ordered posits of the Differential of TRU, on a field that best demonstrates their Relative Magnitudes consistent with TRU. The way to have perfect parts or aspects of TRU with precise values is: **LOPELT** = the Law of Precise Parts of TRU to be Posited on an Á, L, and T construct. LOPELT is how ES and CES can be found and made accurate, plus how they can be made separate from all else for relative magnitude in TRU and REM.

CREM = Criteria of REM and DREM Determination: 1. Rule of presence, degree of and means to separate and manage coterminality. 2. Rule of proportions from a definite and discrete whole, e.g number of organs of the human body. 3. Rule of time, how the point in time increases or decreases the valuation of an item or thing to be measured. What the optimum and gradations are from these. 4. Rule of mass, how the point or posit in the S-O scale increases or decreases the valuation of an item, to obtain an even proportion from 0 - 100, or from 1-10. 5. Rule of length, how the size increases or decreases the valuation of an item. 6. Combinations of 3, 4, 5, that are determined from TRU. 7. The more we come to know TRU, the more other related and corresponding facts, information, concepts, principles and laws, are TRU. 8. Rule of collective opinion. The value to be placed is the average of total values placed on items by different TRU authorities. Here reliability depends upon the availability and TRU quality of the valuations of each item. REM is often based upon the value, worth, and use of that form, nature, characteristics and properties of the part, process or aspect of TRU which are indirect, secondary, and dynamically interfused with the rest of TRU. REMS and ES are not only proportional to those parts of TRU which are identifiable.

Those ES which are REMS at each level, or prove each level, are not ideal or actual REMS for the following reasons: 1. ES REMS = a remainder of ES which are not recognized but unknown and therefore attributively given an elevated or incorrect value of existing REMS. This is similar to **ANKEF** or The Ank effect or Unknown for reasons of incorrect identity. This produces a void or surplus, such as an exaggeration which is so often done in our IA and thinking. 2. Our knowledge of the significance of REMS may be greater or lesser than their actual ES value as a part of TRU. Much of ES and REMS of TRU are unknown. 3. Change, an ES may not be given its proper REM through change in value, or change in position of TRIS or TRUS, or change of our attention such as with concentration, distraction, fads, and fashions. 4. Reduction of what we use to be ES and REM may be altered in objectivity, in its representation, or in the form of our REM model. So that their form and function in REM is not the same as it is in TRU.

As we proceed in TRU by REM from 1R to 2R to 3R,,,,, we at 10, 100, 1,000,,,,, identifiable divisions are increasingly able to posit DREM. Yet our language is such that the words, symbols, nominals, and concepts we use do not properly apply nor equal these divisions. Nor do they fully cover TRU in an equal and perfectly formed proportion. This is because of: 1. Decreasing ES, 2. Limits of ES, 3. Limits of other, aspects of TRU, 4. Limits of REM in replication these ES. 5. Limits of REM of stereotyped levels, or non continuous series. 6. Words less than their ES, not pure TRU, 7. Total terminology does not cover TRU. This is called: **LIREM** or Limits of REM. So that whenever we purport to show TRU by its 10, 100 or 1,000 divisions we must do so or provide the reason why the product we propose is not an exact exponential dividing and construct of TRU. It is easy to enumerate 10, 100 or even 1,000 words for common important entities, ideas, and principles of TRU and say that these provide the REM of TRU. Yet it is not complete by TRU nor even an adequate REM operation. If we are to perfect our priorities, if we are to provide an optimal construct of TRU in which its parts are not only 100% pure but 100% well delineated and proportioned, and if we are to have a highly qualified, set, ordered, and reliable guide to live by in which our priorities, objectives, and ends are successful, we must separate and extract all that is TRU most effectively.(VAPO) Lack of order and refinement in REM is: **ARF** = From REM Forms, or lack of REM Forms or divisions. ARF is what happens when we use common or existing words as REMS in TRU and give them 1R, 2R, 3R, , . status without separating and extracting their ESINS in a pure, proportioned and well delineated construct in TRU. The principles involved in rendering REMS a perfect construct of TRU are: 1. ESIN, or their meanings and use must equal their TRU, not general meanings, 2. Equality, they must be exactly 1/10, 1/100 ,,,,, as specified of TRU, 3. their error or fallacy of irregular or disproportionate coterminalities must be corrected. The proportionate parts listed 10, 100,,,, must equally cover all TRU, 4. From or lack of REM forms or divisions, including the above and any other factors or principles that correct the deficit of rendering the ES perfect TRU. 5. From REM in return reinforcing and verifying proper distribution, and through the continued accounting, refinement and precision of REM limits.

In order to determine identifiable parts for REM we must clearly distinguish those parts of TRU which are: 1. Equal and exact proportionate parts of TRU. That is, at the 2R level no more nor less than a 1/100 part of TRU exclusively and equally to all others. 2. An entity, anything or sense whose ESIN is roughly 1/100 or between 1/10 and 1/100 part of TRU. This is the approximate posit of an ESIN in the 2R range: if 1R is 0-10 and 2R 10-100,, 3. An entity, any thing or any sense which is used directly as a value for posit in REM. 4. A word which is used as a representation of an entity, anything, any sense or object as a value for posit in REM. It will be readily seen that # 4 is much easier than # 3, and in decreasing scale, 3 than 2, and 2 than 1. # 1 is more a goal and to a great extent only roughly attainable, at least with great work, SCARLS, standardization, and refinement in the future. # 2 is highly beneficial, for it leads to # 1, yet is imprecise and many contain much overlapping. # 4 and 3 are increasingly poor vehicles for valuation, although they are what we commonly and less appropriately use. Herein is the key to the success of valuation. It is how we can bring the words we use to TRU, give REM to them that are as free as possible from impurity both in TRU and REM. It is how ES can, be best distributed to fit the kind of REM projected. These steps provide the method whereby a successful differentiation and posit of values must be produced.

All that is 1R does not mean they are all 10 equally weighted parts of TRU. Although the ideal REM would be one in which we have a set and equal number of perfectly separated parts of TRU for each R level, as 10 for 1R. This can be of various degrees of REM in transition from their actual state in TRU, to their most formal state in REM, or 1-50 of TO for 1R in REM. It is impossible to give REM values to words, and especially NEC, without producing and providing total TRU terminology, LANT. We can't accept REMS that are assigned tentatively or indiscriminately. It is imperative that REM be reduced to total TRU ES, CES, EC and NEC language before perfect positing. **ÉL** (ELAN) = ES Language or Language in terms of the parts of TRU.(EC and NEC) ÉLREM = EL determined and gauged by REM. **EXÉL** = Exclusion of words or language to their participation and function in TRU. EXEL is ES Limitation and is shown by the special exclusion by EL. The degree to which we can limit language by EL and EXEL to TRU is called: **LÉL** or the Limit of language to TRU. The corollaries to EL are: 1. As terms are added the object becomes more and more distinct, so that each added term defines and delineates additional segments or aspects of this object, TRU. This is EXEL. 2. As terms are added the more general, and often less discrete ones giving blanket definitions, delineation and distinction to the object, TRU, are reduced to the extent the new term displaces them in the total object TRU. Thereby giving it increased meaning and formal distinction. 3. Since there are many languages and many uses of terms in the same language as well as fads, fashions, customs, and changes in the usage of language with time and location, there now exists little uniform terminology for mankind and even less terms TRU essences. 4. The expanding volume of words, terminologies, and vocabularies is limited not only by EL and EXEL it is limited for all persons by the finite capacity of their memory and lives. It is in the optimal formal distinctions, delineations, separations, positive modulations and applications of these as they are brought together as a whole in the total objectification, representation, and replication of ROT, SUT, THEO, and TRU.of TIT, that we can finally attain the fruition of destiny.(SOT-Á-TIT-TÂ).

From DREM or the specific determination of values for all of TRU and POS in SPALT we can recognize the great importance of words. This is by how: 1. Their meanings are adapted to any condition, situation, problem, and ES, 2. Their meanings are or not equivalent to their like component of TRU. 3. Their ESINS are or are not equivalent to their ES in TRU, 4. Their ES may be generally interpreted in different ways, 5. Their ES may be specifically posited in different ways depending upon what they are used in relation to. The world logos, and the '-ologies' they produce attain a level of identity with TIT through S-O and Á FAF by REMS and SU.

Phrases are significant by emphasizing qualification, relations, contrast, and how the operation of two or more ideas, parts, concepts, components, or constituents combine to show TRU through: 1. this simple contrast, 2. the added meaning their combinations supply, 3. a scale whereby these combinations become a gradient from one extreme to another, 4. a continuum in which they exist in TRU as part of a system or large FAF of TIT. Any collection of words, ideas, things for valuation will sooner or later have to be contrasted and compared so that they can be reduced to some kind of a whole. This, so that their relationships converge on unity, TU of TRU.(TIT-TO) Yet this can't deter their increasingly discrete separation and enumeration.

WORDS in USE, CORRESPONDENCE with REMS

In the divisions of words or vocabularies of languages, if we start with 10 million possible words of separate meanings and divide these into their pertinent categories, we have the following for their use in REM. This may have included applications representing ES, CES or aspects of TRU, how translated into REM, their variability and overlapping.

Magnitude	Magnitude of words	Number of words	Expressed Exponentially	REM Equivalents	Range of REM
0	7	10 Million	1 x l0	R	1
1	6	1 Million	1×10^6	1R	1-10
2	5	100 Thousand	1×10^5	2R	10-100
3	4	10 Thousand	1×10^4	3R	100-1T
4	3	1 Thousand	1×10^3	4R	1-10T
5	2	100	1×10^2	5R	10-100T
6	1	10	1×10^1	6R	100T-1M
7	0	1	1	7R	1-10M
8	-1	1/10. .1	$1 \times 10\text{-}^1$	8R	10-100M
9	-2	1/100, .01	1×10^{-2}	9R	100M-1B
10	-3	1/1T, .001	1×10^{-3}	10R	1-10B
11	-4	1/10T	1×10^{-4}	11R	10-100B
12	-5	1/100T	1×10^{-5}	12R	100N-1T
13	-6	1/1M	1×10^{-6}	13R	1-10T
14	-7	1/10M	1×10^{-7}	14R	10-100T
n	-n	1/n	1×10^{-n}	nR	100T-n
∞	-∞	1/∞	1 x 10∞	∞R	n-∞

Table 1-37 **DIVISIONS of WORDS and USE with REMS**

In this diagram we may start from somewhat different foci, as the number of letters or symbols or figures of various kinds. Yet we will see that in general the format will be much the same and will be limited by our powers of observation. The ranges of single or multiple letters, symbols and words will not greatly alter the focus and distribution of REMS, only enlarge them somewhat.

As we add more to the overall REM, all existing REM will have to be downgraded, an equal amount. This is called: **LEDIOR** = the Law of Equal Downgrading by Input and Output in, and upgrading by, REM. LEDIOR is displacement and reconstitution of REM in TIT with SPALT. Language is highly arbitrary and functions by the observer independently in common use. This, to an extent, adds to the reason why values or ideas of importance must be rigorously corrected to REM. This requires a recognition of and correction of all factors that affect such reduction. It involves the natural tendency to upward bias. This is called: **FOGRA** = Fallacy of Over-Grading, or of excessive REM. FOGRA is a product of our focus, subjectivity, concentration and emphasis. Individually it can be pathological, if not less on a social level. For it may represent a reaction formation to the overwhelming world about us. We take up or champion this or that and extol it for the world to hear or see. Whereas it is only a relatively small, even when at its greatest in 1R or 2R REM, of TRU. The principle involved in making the needed reduction of FOGRA is called: **REMAS** or Reduction of REM to Mode, Mean and Separation. REMAS applies to all valuation in TRU. It applies, however, most greatly to the highest REMS such as 1R, 2R and 3R where if we are not careful we will have 100 or 1000 values within 1R and similar excesses in 2R and 3R. **LOPELT** or the law and problem of presenting ES or the components of TRU on LT, ALT, or SPALT. LOPELT is readily seen when we try to posit various terms, objects, parts or components of TRU into ALT and SPALT. For the most part these terms, whose definitions may even be multiple, are themselves undifferentiated or non discrete without distinct limits and solidity.

LARED = Law of Differential Reduction or Reduced

Differentiated DREM. LARED signifies how much that is almost synonymous with TRU is, when differentiated in DREM less and less a part of the whole. This is what we find for God, world, universe, love, virtue, happiness, length, time, large, small, medium, past, present, future, subject, object, real, ideal, and many others. They are all highly synonymous with TRU yet when separate and limited to no other domain, they have less and less REM. This is to be expected when best reason prevails, yet surprising to the casual observer who would consider any on a par with TIT. This is also why we use many existing concepts, as God, Soul, Christ, and Universe, with great caution and mostly only when compared with TRU. For they can not be considered to be all, even more they are apt to be misleading and cause failure to comprehend TRU.

LOD = Law of Differential Devaluation. As anything becomes differentiated it automatically loses proportion and value, e.g. ambiguities, LADEP, and LOBIR. LOD is a general case of: **TAD** = TRU Differential Devaluation. TAD is of more than one kind. In general it applies to the loss of strength, power and being by differentiation or the effect of dispersal. **LIDNOC** = Law of Idealization of a Finite Number of Concepts. LIDNOC signifies that the closer we are to TRU and TÂ the greater the number of concepts there will be. This is a universal law, with many corollaries in each respective field. It arises from relativity, the simple relation between observer and observed. This is how when the subject becomes more closely involved in the object this object's aspects, features, parts, processes and all become more evident. **LEDIT** = Law of Equilibrium of Distribution of content in TRU. Evidence for LEDIT is to be found in the human body, both in form and function, where there is a balance in the size, timing and mass energy of its content. This is seen as we have in REM in the total TRU, where some become enlarged, others must become smaller. So that the resolution of LEDIT and LIDNOC is their synthesis, or: **LEDIT/LIDNOC = c**, where c is the needed correction factor.

THE PART and PARTS of REM, REMTE, SPECIFIC PART REME, IDENTITY, CHARACTERISTICS, FOCUS, STANDARDS, LEVELS, SCALES, MOCBIREM

As we leave a participation in REM and concentrate more on the specific parts of REM we consider and strive to show how each REM and its combinations may be determined and from this how they make up all REM as they participate and show TRU. To calculate the R of an IA, if it is of one second, we have: **70 x 365 x 24 x 60 x 60 or 2,2075,520,000 IA** or lifetime in year, days/year, hours/day, minutes/hour, seconds/minute or 10R, which at 10R/ Individual of total population is 10R+10R or 20R. This will also depend on the topic, meaning, and power of this IA. If the average is 20R, and if it is of a major ES or aspect of TRU it will be higher by association.

REM	$.-\infty$	.-f Large	.-f Medium	.-f Small	+/-	+ $.1/\infty$	.+1/f	+ .01 1%	+ .10 10%	+ .50 50%	+ .90 90%	1	$1x10^n$	.1 $x10^\infty$
0-All	Impos-sible		NEGA-TIVE	to		POSI-TIVE	for	GREAT-EST	TRU	TRUS	TRU TREQ	TIT	ENT	TÂ
1R										TU	TO GOD	E	ENT	TÂ
2R		WORST ANTI-TRU					GREAT-EST GOODS			God	HIGH-EST GOOD	TES (gen-eric)		
3R	MOST		WORST ANTRU		NAMTRU				god	SOUL	BESTQUA-LITY			
4R				WORST ÄTRU					ANGEL	BETTER QUALIT		ES		
5R					NEU-TRAL			IMPOR TANT		POES 040	REME			
7R	DEADLY	COC 018								MATRU		SIAT 042		
10R			FET 007	ÄTRU	LINCOT 035	PROPOR-TIONAL		PERSON	GOOD	IAT	IATID	TES (spec)		REMÂ
30R	and	DAK		RELIS 083	LARPON 010	DILAC-OR-094	Think	ACUIA 083		LEPER 043				
100R						ESIN-T	PERSO-NAL IA	AMTRU					SAF 042	
1000R	DES-	FIG-MENT	ILLU-SION	FIA 083	THINGS	SENSES		LOFETE 049						
1M R	TRUC-				ENTI-TIES	NAMTRU						REM^n		
1T R	TIVE		FIE 007	MINU-TAE	PAND 075	TRIVIA		LEXPRO 045	VERTRU 025		RACOD 044			
∞ R	ABSO-	HELL	AILA		LAIA 050	MI-013 071	PRUI 059		AMTRU/ AU-AUI			REM^∞		TÂ-TI/ AU-AUI
< R	LUTE HELL	WORST ÄREM	ANTI-TRU	WORSE ÄREM	AU OPAR									

Table 2-37 **REM ESIN Diagram or DRESIN**.

To be able to characterize and find the modes and properties of REME we can contrast, compare and demonstrate their REM or value, as opposed to their percentage or degree of positive and negative participation and effect on TRU. All things, senses, and entities, may be divided into: 1. How much or what percentage of them is TRU or their ESIN, the abscissa, 2. How much value they are to TRU or their REM, the ordinate. By coordinating these two parameters, diagraming, positing, and accounting for many of indicative terms that pertain or contrast with TRU we can produce a frame, tabulation, and ordering of TRU and REM that is instructive and fascinating. That posited can then be compared with much that comes before us and what we seek to evaluate. Although ESINS and REMS are highly related, for purposes of such a demonstration of their comparative contributions to DRESIN they must be clearly separated. For in and of themselves they are much different, as is the interpretation of many of the terms used. This is also seen in their generic or specific implications. It can be seen by how important continuing formal distinction, EC and NEC, are. This accounting of all that goes into TRU helps us grasp the subtle, profound, yet so crucial concepts, principles, laws, mechanisms and unifying means of understanding this special universe of best human life. f in the above diagram stands for finite fractions or fractals of a single and perfected ESIN. 0- stands for the highest REM level or O-R and the fact that it approaches or leaves this level, but is never this unless TIT, and then only qualitatively and relatively. The <R stands for REM beyond the finite. If beyond the intinite, although impossible for TIT and REM, is in another sense, by meta set as a limit. REM and TRU are encompassed with added states of ROT, POT and PERT. Many more items could be placed in the cells of 1-4R/10-100%T, including TRU Terms. This would require blowing up the table in this region, or an accessory table. However DRESIN does provide an excellent overview of how all bits of knowledge can be evaluated by TRU/AU and REM. In interpreting and using DRESIN we must not use negative percentages to confuse percentage of a thing that is TRU, its ESIN with its REM. A person can have high ESIN, yet unless identified with the overall counterpart in TRU, will be only 10-12R and their consciousness and IA only about 15-20R. Likewise one must be careful to avoid being misled by the use of the negative. There is much in and around us that is negative. Likewise for REM, it is conceivable that beyond those values that are infinitesimal there are even zero or negative values. This may not apply to TIT but to REM. In rendering TRU objective we can assume a negative as well as positive stance for both ESIN and REM. Once we can establish the frame, model and approach for positing the parts or aspects of TRU we can proceed to discover how these can be best ordered. For this, so great of TRU and of REM, is singular and only the part of REM equivalent to its part in TRU. This singular part is called: **RËMÉ** = The Part of REM Equivalent to its Part in TRU. REME helps distinguish this special part from parts in general, relative magnitude and values of TRU or REM. REME are chaotic if not properly discovered as ES and ordered in the overall structure of TO, TRIT, OBT, MOT, SPALT and REM. To do this we must develop the methods by which REME, can be separated and ordered. The proportioning of various presentations and organizations of knowledge, as encyclopedias and general reference books, provides a means of developing and determining REME and posits in REM. If it is a specialty book, as great as many are, its starting in REM can be no greater than its topic's or field's part of TRU. Yet for general books, references and encyclopedias that purport to represent all things, knowledge, or reality, these should start at a much higher level. If we take Encyclopedia Britannica as starting at 2R, then for the 23 volumes this would put each volume in the 4R level, for 1,000 plus pages this would be in the 7R level. If there are 1000 lines these would be at the 10R level, and 10 words per line at the 11R level. So that each word, if taken separately and as used to describe some aspect, part or process in TRU, or of all knowledge, would be about 1 X 10^{10}. This although showing a method of ordering and positing REME must be quickly qualified. For not only is the material AMTRU, each word or phrase is often duplicated and reference works are not entirely directed at nor have they been written and edited to TRU.

The inability of the individual at 10-12R to understand all between oneself and TRU, 1-10-20R is the **REMD** or REM TRU Differential. The inability of the individual at 1R to understand all between oneself and TRU at any R is **REMID** or REM Individual-I-TES differential. This is where there is a 50:50 or 1:1 relationship or identity between the Individual-I-TES and TRU. It is a measure of actual mysticism. This is resolved by the difference between measures by TIT and measures by TIT/IT. For the individual less than 100% TES or increasingly ATRU, AU, or ANTRU this is the **AREMID** or the deviation from Individual relative magnitude in TRU, a persons partiality as a thing in itself - THIT. This may be 100% THIT, but little, no or even negative by TRU.

Levels of 10's of REMS, or Exponential R's

We can gain a better idea of the exponential mode of REM, in a linear, non warped, or non altered progression of REMS. The following alters and extends this outwardly:

0R		TRU	
1R	=	10 Major Divisions of TRU,	
2R	=	100 Large Divisions of TRU,	
3R	=	1,000 Moderately Large " " "	
4R	=	10,000 Moderate Divisions of TRU	
5R	=	100,000	Moderately Small " " "
6R-10	=	$1x10^{6-10}$	Small Divisions of TRU
10-16R	=	$1x10^{10-16}$	Smaller " " "
16-30R	=	$1x10^{16-30}$	Very Small " " "
30-∞R	=	$1x10^{30-\infty}$	to Infinitesimal " " "

CONICAL REMS CHART POSITS

Posits of REMS are later combined with posits in SPALT, POMOT, METAD, SPAT and other perspectives of TRU. They help show and guide us toward the goal of a POA, and best utilization of REMS. **REMI** = Relative Magnitude Individual. REMI is an important concept for it ties-in what is an individual, what is TES, their part and role in TRU and how this can be best valued. It is vital that each person know what they are and are not, what they can do and can not, what their SCARLS are and what all this is of the world they are in TRU. This largely depends upon their knowledge of values in life, or REM. This is a major determinate of both themselves and TRU. It is a basis for ethics. Without such an orientation disproportions, conflicts, misery, and suffering continue unabated.

REMC = REM Characteristics. For every great segment of TRU, REM, part, or proportion there are many qualities, features, aspects and NAMS. These may be different as they are a part of TRU. So that characteristics by themselves, of any entity, or RU, or how they are of TIT, and how they are both as they alter REM from that of TIT, or how they are as REMIT, are all different. REMC are of different kinds. However their greatest significance comes from how the characteristics of TIT are posited in DREM. It is to understand these differences, especially how this posit in REM, of characteristics of TIT, is to be considered in these different ways, yet still held to their REM posit and use.

From **ÊL**, or the Exclusion Law, Exclusion from TRU, as terms, nominals or words overlap they reduce in proportion to the added term in the set, system, or whole. In this way some terms will be halved or decimated and reduced to lower set magnitudes. This explains and shows the value of detail in science. It shows the loss of value of universals of terms which are general for a topic. When the topics are thoroughly determined and demonstrated, as universals that are transcendent, they tend to evaporate as we come to know TRU. They are unnecessary or reduced to lesser significance. Yet science fails to account for: **ILF** = Individual Limit Finite or the Human Finite Limit. ILF is evident from the analysis of the corollaries of EL, **EXEL**, the ES Limitation or specialization Exclusion of EL, and **LEL**, the decreasing Limit of EL by the Individual as ROT, POT, and PERT are discovered. LEL is the correcting power - to an extent - of: **IL** = Individual Limit. ILF pertains to the Limited use of language, IL to finites, individuals, limited in numerous ways, including language. LEL pertains to reduction through ROT, POT, and PERT.

Since the population of man on earth approaches 10 billion, and rising, this $1XI0^{10}$ by division of ES in TRU by reduction for other highest REM and limitations of existing population to fulfill all population in TRU, past, present and future, gives the average person of today a 11-12R value. Still, this is not acceptable for the first person, for such a position is not comprehendible and lacks much of the sames, generic, or relativity which persons can be elevated by. **FIL** = Finite Limit, or the limit of the individual, TES, and ability to recognize. This, greatest problem in the world, is automatically and ultimately resolved as senses become THES, things become THEO, entities become ES, and the first person becomes TES, as they fuse, and exist together in an optimal association normative for TRU. This is one where the first person, as TES, becomes the part or half of an absolute, or an absolute in identity. This is called: **LAIT** = Law of Autonomy and Identity of The Individual (or first person) and TRU. This is the 50:50 rule of the world and one-self. It is the mechanism whereby Christ is, as each and all can be, God, as the son, daughter, spirit, and all of God, the part of TRU that fulfills TES. In this way we are or are not divine as we are TES and TRU. The corollary of LAIT involves the second person and is called: **LADUT** or the Law of The Autonomy of Duality or two individuals and TRU. LADUT is the existence of a first and second person together as one in TRU and in identity with TES, a half of TRU, TRU absolute. This may be further extended: **LAMUT** or the Law of the Autonomy and Identity of Manifold or Multiple Parts or Entities and TRU. LAIT, LADUT, and LAMUT are products of the relativity and identity produced in the subject object, or observer observed relationship as they become one, as in the Á mass energy parameter or dimension of TRU. A special case of this is: **ICOL** = or Identity through Objective Í, or the fusion of the identity or concentration of the first person with the focus and unity or thisness of TRU. The range or interval between the given and TRU, and between indefinite, infinite or tiny finite REM to absolute total R and REM, is signified by: **DISP** = Disproportionate I Identity. **PEREM** = Personal Relative Value or Magnitude, Paradoxical Resolution. **DIR** = Divisions of REMS and Reals into the Differentiated or separate parts. The conditions and states of the divisions of REMS, where a thing, category, item, entity, phenomena, or E is separated into its component parts and these are given new names, is the basis of language, reality, Á, and TRU. The process of divisions of REMS or reals into their differentiated and separate parts is DIR. It is the basis of our reality, language and the form of our existence. This is where the actual form of TRU and the form as represented and subjectivized become fused and equivalent in an ordered whole. This is called: **LAFI** = Law of Form Identity. LAFI is the basis for systematic unity. With all the weaknesses, fallacies, imperfections and non TRU nature of language it is the means or media, as it will and must be constantly improved and perfected, to equal these forms, ideals, essences, parts, processes and mechanisms of TRU and all knowledge. REM in purporting to provide a fundamental system of value, must find methods to produce such precise measures. When adapted to REM standards comparisons are of immense importance.

This is signified by: **LACSREM** = Law of Comparative Standards of REM. LACSREM as a method is exercised by comparing a known and an unknown REM. If the known REM is of 4R magnitude and an unknown REM is slightly, moderately or greatly larger or smaller than the known, then it will be 4R, 3R or 5R. Further positing of REMS may proceed as this exercise is continuously reworked and refined.

COMREM = Combined REM. All REMS do not have to be considered singularly. This is preferred, however, primarily for the sake of precision, parsimony, order, and use. Yet to fully understand how REM works in TIT and especially TO we need to understand how they combine to form much greater REM when they exist or function together. COMREM helps to ameliorate the problems of EL, FIL, and LARED.

REMAS = Reproduction of REMS by Mode, Mean, Separation, Specialization or Exclusion. Modes are the varieties of aspects, themes and form of TRU. Mean is average or centerpoint of REM on its scale. It is in a scale that is adapted and weighted to TRU. Separation is the complete partitioning of the components of TRU, or the mathematical differential, by searching for and determining the mode and mean of these components in TRU. In this the way we can gain increasingly distinct objectivity that enables posits for valuation, in REM.

REVAD = Relative Value Differential. There exists in the world many different levels and measures of value, or REM. These greatly lower the qualities and success of TRU. For they set forces in action that conflict with an optimally unified goal in TRU. Yet they are closely interwoven or dispersed in the total population, culture and civilization. Much of this is due to partiality or nominal differences of ES and CES in TRU, individuality, and the added states produced by the observer and the given. REM may be made more accurate by defining what is its par, or **PAREM**, an average mean from which measures of all can be determined. By knowing the number of words, people, identifiable objects, phrases, combinations, etc. an average for all that can be used as a standard. We should distinguish a par for quantity and a par for quality. In our search for absolute value for REM they may be categorized by gradations from the absolute. This we call: **GRÂ** or Gradations of the Relative Absolute. GRA is where and how absolute REM are qualified by the relative nature of TRU. When this is achieved through a standard or PAR value we call it: **GRAP** or Gradations of Relative Absolute from Par. And when we achieve gradations of relative absolute by par or a set focus within a set period of time or circumstance this is called: **GRAPTOC** or Gradations of Relative Absolute from Par within a set period of time or circumstances.

Source	Number	1R	2R	3R	4R	5R	6R	7R	8R
Great Ideas, I	100								
Dictionary, 2	6,500	6	40	200	1000	2000	1000	500	200 ....
A Book 3	1,400	2	10	100	200	200	200	200	150
Words all languages	1M+	9	80	600	5000	40T	200T	200T	150T
All Books (Cpd)	10M+	9	85	800	7000	55T	400T	2M	2M
Perfect REM	∞	10	100	1000	1OT	100T	1M	10M	1O0M

Table 3-37 **EXAMPLES of DISTRIBUTION of REM**

We can list and summarize DREM very roughly by considering the great ideas man has, the number of words in dictionaries, encyclopedias, or the total number of words in all languages and classify them by levels of Relative Magnitude. This classification and numerical listing of REMS from various sources is neither accurate nor complete, largely because they are variably AMTRU. It is offered to illustrate REM, DREM and how we can apply and use REM in research, as well as how research can help REM. We know that most of our language and terms can be given values. We know that many of these values - and posits - can be highly accurate when fully corrected. We also know that these values can be at least somewhat measured by TRU. We know by REM that its limits impose a strict determination of these values. On this basis the above table is provided and can account for REM, POS, SPALT, ROT, ES, and CES.

The frequency, mass, varieties, complexes, multitude, milieu and permutations of REM are to be found in MODREM, **LARMEDS**, diagrams of exponential R's or levels of 10's of REMS, and the Conical REMS Chart. LARMEDS signifies REM of Equal Segment Distribution. **MODREM** signifies the Modifications of REM, how we must learn, devise and modify the form, states, essences, parts and functions of REM to best replicate TRU. It is to show how REM is different from all that exists about us and must, as we do in ESINS, bring them to their state in TIT. **MEDREM** = Measurement Relative or Magnitude of Mass Deficit. The REM value of a thing is not the percentage or area of SPALT it covers. It is a percentage of those segments whose mass it represents. Parts, universals, abstractions, E, ES, functions, forms, processes, partials and all entities of TRU tend to be lower than are assumed. This, is not so exponential as it is by simple division. MODREM results from the fact percentage of area of ALT or SPALT is not exclusive to an entity, thing or of the REM value of this entity or thing. If we have a single entity per segment, or there are a fixed number of entities per all proportional segments of TRU, then this entity may be given REM value. The conical REMS chart gives a representation of what occurs between R's, and allows great insight into what values TRU has for greater R's. REM slows after 6-8 R, leaving void much of the intricate beauty of TRU.

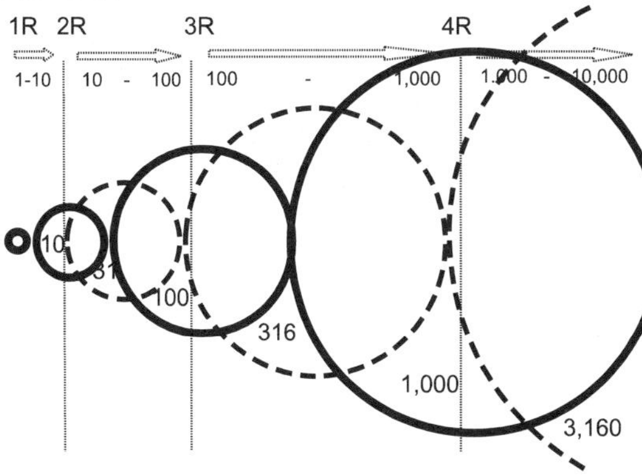

Fig 1-37 **REM LEVELS, VARIATIONS and USES**

The higher levels of REM show how we start off with all succeeding levels and how for the highest ones we can subdivide the REMS so that a finer distribution can be provided. The Highest REMS are 1R or from 1-10 of the most important or valuable parts or aspects of TRU, 2R or from 10-100 of the next most important parts, 3R or from 100-1,000 of the following important parts, and 4R or from 1,000-10,000. We also show a level mid way between each REM on this exponential scale, at the 3, 31, 316, and 3,160 levels. We must remember that REM is a continuum not limited to set levels.

In any field, science or topic we are often dealing with a segment of TIT-TO that is no more than 1/100 or 1/100,000 (2R to 5R) of the whole. The field, science or topic will be composed of thousands of separate parts or units. These components or parts of TRU and sciences, will have distinction as well as a great amount of stability and certainty in TRU. This makes them not minute, trivial or menial, but as TRU is worth so much they also become so. Here we can see the great problem of individual differences wherein the first person has great interest in this part of a field, yet another or the public has little or no interest. Neither one is right, both must be qualified by maximizing the REMS, yet doing so with VAPO and TIT-TO. In REM exponential figures can show that a person is not a unity, nor infinite, nor limited to any combination of divisions at a single level. We are for the most part relative, focused on the finite levels we are predominantly constituted. These are especially relative or focused upon different levels depending upon the time frame considered. We are dealing with a fleeting moment, instant or IA, that tends to unity, a lifetime with trillions of parts accountable, or some duration and period in between. It depends upon whether it is for one person, all persons, or all TRU. Also whether we are dealing with TIT, ROT, MOT, POT or some other variant of TRU. This not only reveals a person's variable nature, it gives us a good basis and limit within which to work in providing a successful construction and representation of TRU.

	L by AREA	T by DURATION	Á by MAS ENERGY	ñ by QUANTITY	* by QUALITY	R by REM OPENING	R RANGE of WHOLE	$ by MONETARY	D by DISTINCTION
RACES	+	+++	+++	+	+++	2R	3-5R~	+	+++
RELIGIONS ISMS	++	+++	++++	++	++++	1R	1-6R -	++	+++
CONTINENTS	++++	++++	++++	+	++++	3R	3-5R	++++	++++
NATIONS	++++	++	++++	+++	++++	2R	2-9R	++++	++++
LANGUAGES	++	+++	++++	+++	++++	2R	2-7R	++++	+++
INDUSTRIES	+	+	++	++	++++	2R	2-6R -	++++	+++
BUS/COMMODITIES	+	+/-	++	++++	~+/-	3R	3-∞ -	+++/-	++++
CITIES/PLACES	+++	+/++	+/++	++++	+++/-	3R	3-∞ -	+++/-	++++
OCCUPATIONS/LAB	+	+	++	+++	++++/-	2R	2-∞ -	++++/-	+++
AGE SEX GROUPS	+	++++(+)	++++	++	++++	2R	2-~	+++~	+++
INTERRELATIONS	++	+/-	++++	++++	++++/	2R	2-∞ -	++++/	+++
PERSONS	++	++	++++	++++	++++/-	4R	4-12- -	+++	++++
ORGANS	+++	++	+++	++	++++	3R	3-6	+~	++++
INTERACTIONS	++	~/+	+++	++++	+++	4R	4-∞ -	+/-	++
CPDS/ELEMENTS	+++	~/+	++	+++	+++	3R	3-10- n	++	++++
PARTICLES FORCES	++	~/+	+	++++	+++	4R	4-∞- -	/+	+++

Table 4-37 **MOCBIREM**

This is one of the most important tables of the text. It is neither complete nor perfect. It is because it provides a clear method of making the formal distinctions and reduction to REM so necessary if we are to have an effective ROT, POA and COAT. **MOCBIREM** = Method of Class Bisection of TRU and REM. MOCBIREM is that great method of reducing what is to be valued into more precise parameters, groups, and scales for measure. This is by selecting out any and all of the many possible measurable classes and groups in the world, adapting them to TRU, and using their existing or developed scales for REM determination. This is simply and clearly seen in geographical divisions, nations, places, and persons on earth. Many other categories can be considered, as ideas, thoughts, senses, things, words, and all that the human mind can differentiate and classify. REM is one of the best means of differentiation. It is not new, for it has been key to human choice, selection, and trade throughout most of our existence. The range of REM is vast, as it must be, if it is to replicate TRU. REM is a composite of all gradations in between as well as a compound of all together that makes up the finite distribution and composition of TIT. Yet it is through the differential of REM that we best achieve the formal distinction necessary if valuation and KOT are to have reliability. This is why MOCBIREM is so helpful. It will be quickly noted that these scales are not always what they are in themselves, but what they are as they are in TRU. In addition the basic class must be discrete and TRU. This calls for modifications in each class or group utilized, especially for quality differences, or differences in contribution to TRU.

There are very few words in frequent use that are not a great part of TRU, thereby of high REM. Yet when a word is specified as ES it is qualified, and more so in use. It is apt to be 1/TRU, or less than TRU Thereby it is often more how we use words than the words themselves.

CHAPTER 38

REM TOTAL and DYNAMICS

THE REM TOTAL, REMO, ADDED STATES in TOTALITY, INTEGRATION, EQUILIBRIUM, MODE of FINITY, ORDER, VERSION, CONICAL REMS CHART, CHANGE, VARIATION

When many to all that constitutes TRU are recognized, brought into order, and act as values and priorities in our lives we can be much more effective counterparts of TRU. This requires, **REMO** or Relative Magnitude Totality. REMO can not be achieved in a hit and miss manner. It can only be based upon TREQ, TRU, TIT-TO, VAPO, SCARLS, TROS, and all that enables us to represent, modulate, and have POA, a positive objective approach to TRU. It is in this that a direct and singular relationship can be seen between the fundamental and simple IA and the most complex TIT-TO, REMO and ACC of TRU. The key and guiding power of this is the Á mass energy observer observed relativity that includes an entire continuum of gradients TRU's form and function.

The determination of proportion and value is made in various ways, including: **LOID** = Law of Integrative Determination. Many laws, principles, concepts, and formulae may seem very true and valid in themselves. So much so, in fact, as to seem exclusive, obvious and almost trite, even to deny the existence of TRU. Among the 1-4R level of REM is to be found all that is emphasized in life, what is held to be good, great, greatest, or divine. It can be seen, without modifying by LOPAN, how when compared with the great many other values, they are reduced to a more common level. That wisdom rather than pleasure was the good, by Plato repeating Socrates, was properly qualified to account for all goods. For Plato states in Philebus that neither pleasure nor wisdom in isolation is a perfect good, since neither would be acceptable without the other. He concluded therefor that the 'good' is a feature of the mixed life. Further, in *Philebus*, Plato describes unity and infinity and how through the selection of a definite number, as sounds in music, a unity may be achieved. This he thought may have originated with Thoth, Egyptian God of learning, reckoning, invention of writing, securer of social order, a creator of languages and patron of the scribes and officials. He weighed the hearts of persons at death and reported them for judgement in attaining everlasting life. It is in securing a weighing and balance to make best judgement that values and REM are also achieved. For as noted in *Philebus* not even measure, or order themselves are as the first and second good, adequate to describe or fill all good. For when the finite and infinite combine with balance in totality, a third class appears that of the harmonious and proportionate. So is REM, as TRU and SUT through relativity, a product of the optimal combination of unity, finity and infinity in totality. The finity that is produced is variable to the extent man is capable of understanding and representing it. Plato repeating Socrates also mentions that wisdom and knowledge have different levels of purity. Of these levels, like TIT and REM, only that which is secured through measure, mathematics and weighing are pure, the rest is more or less conjecture. Plato also, most profoundly, notes that the purest of the arts is the dialectic or science of being and reality. This is because it is a knowledge of the changeless and the essential, like the TRIA of TO TIT. When wisdom and mind are most truly and exactly used to refer to the contemplation of true being, as in dialectic, COL and COR, they help to provide the second and first goods, order and measure. This confirms and is in turn reconfirmed by TRU, KOT, SUT, MOT, SPALT, and REM. For REM can only be pure with pure measure and order of the parts, and principles, of TRU. Also, REM can be neither one value nor an infinite number. REM must be an ever increasing finite number, measured, proportionate, ordered and total or REMO. This becomes be a REMO that is ever enlarging as its selection and standardization keep pace with knowledge, ability to comprehend, and relativity. Their determination, understanding, and being in TRU is basic and essential. An adequate, full and proper understanding of them is impossible without their formulation with TRU and together with each other.

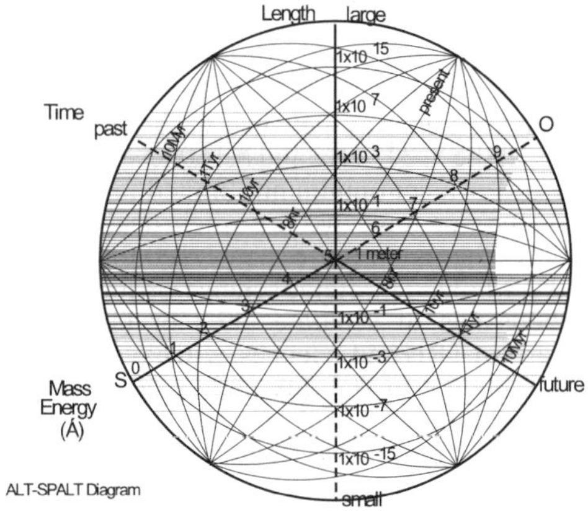

Fig 1-38 **PLANAR ALT, SPHERICAL SPALT & REMO**

The SPALT model of TRU, either as a field, ALT, or as sphere SPALT, provides an excellent basis for developing REM. For it represents the total of TIT, TO, which can be directly transferred to the total of REM, REMO. ALT and SPALT are only a frame and do not represent many of the aspects, principles and contents of TRU. We must be careful not to exclusively apply them as a simple or singular standard.

What guides us a better REM by SPALT includes: 1. Its parameters as large components of TRU and REM, 2. The measure parameters give from accurate scales which can be adapted to REM, 3. The sectors in the sphere that can be directly adapted to certain levels of REM depending upon how much differentiated. 4. Some of the laws, corollaries, principles, concepts, and mechanisms of REM, SPALT, MOT, ROT, SUT and TRU when applied to this transition from SPALT to REMO give us added support for REM.

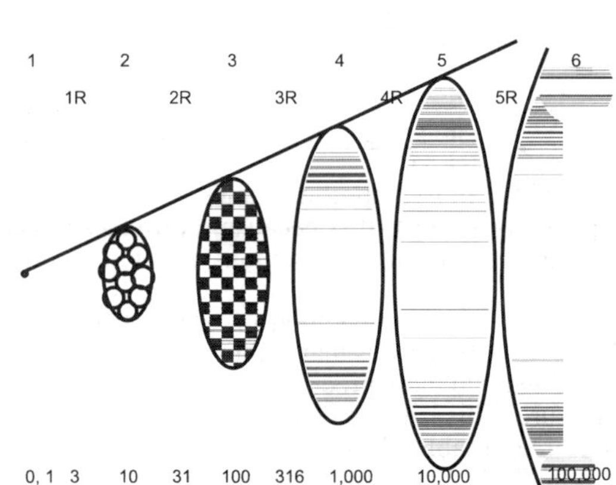

Fig 2-38 **RELATIVE VOLUME of COMPARATIVE REMS**

The one, ten, hundred, thousand,,,, content of REMS in 1R, 2R, 3R, 4R, 5R, 6R ,,, show the complexity and finite expansion that is such a great problem, demand, aspect and fact of life of relative and human interaction. To provide the necessary frame for a panorama of REM we best adapt the unitary, partial and infinite natures of TRU and its parts - as they provide values or REM - to a scale and frame that gives optimal axiological representation. **LOMID** = Law of Magnitude Interdependence, or the high correlation between the top REM and the major aspects and parts of TRU.

Many ideas, concepts, reals, and ES, when first recognized seem so clear, penetrating and inspiring, the answer to all. Yet, they are such a small part when compared to TIT-TO. This is because of the focus of IA, the fact that they may not be much TRU, and the fact that the rest of TRU is so great. Also TIT-TO is all those added modes, essences, characteristics and properties innate to this whole, which add a vast array of great REM. It is these we must discover, learn, remember and practice if we are to get the full benefit out of REM and TRU.

LARE = Law of REM Equivalence, or the REM Equivalence Principle. As great truths become evident, many previous great revelations, new concepts, principles and laws become less so by slight increments or larger division. However, they are not to be set aside. They are, as the content of parts of TRU, to be only less in presentation. As our vision, perspective and presentation of TRU, SPALT and Homation, reveal great new principles, facts, laws and explanations so must it reduce, in a limited presentation, other principles, facts, laws and explanations. This is largely a product of practice and parsimony, neither a defect in TRU nor REM. This reduction must take place in a proportionate and balanced way if the presentation is to be satisfactorily representative of TRU. The smallest values in TRU must have an equivalence with those which are of the largest REM. For the smallest will make up in quantity what the largest will have in worth, and all in between. This is called: **LARELS** = Law of Relative Equivalence of Largest and Smallest, Large and Small, and all intermediary values. LARE applies in general to the overall equivalence and balance, whereas LARELS applies to set and opposing levels that tend to be equivalent and balanced. This also will vary with **POREM** = the Posits Total of REM. POREM confirms the actual and proper nature of divisions of REM by requiring a construct in which the final positing of values, words for objects, and all in TRU are not only of ESINS and ES that are discrete and proportionate, but other ES and aspects of TRU that are of the totality of TRU. In other words POREM signifies how in valuation we must critically analyze the nature of that to be valued, overall TRU and REM. Simultaneously we should take special care to make REM and POSIT in terms of their critical and vital essences, states and functions, in the whole. POREM is an example of how differentiation of REMS and ROT are necessary for the proof of TRU, PRUT. In turn POREM proves REM, as well as enhances TRU, ENT, ROT and REM. If we accept REM, as provided by levels as our guide, then we must determine at which level half, a third, a quarter, a tenth, of these levels falls in each compartment. As a result of this process we find that:

whole							**REMO**					
half	**R**						**12R**					**ooR**
third	**R**				**6R**			**24R**				**ooR**
quarter	**R**			**4R**			**12R**		**48R**			**ooR**
fifth	**R**		**3R**			**8R**		**2OR**	**60R**			**ooR**
tenth	**R**	**1R**	**2-3R**	**3-5R**	**5-8R**	**8-12R**	**12-20R**	**20-32R**	**32-60R**	**60-120R**	**120-∞**	**ooR**

These are offered as examples of what can and needs to be done if we precisely show REMO. To be mathematically precise the dividing will not occur on strictly R levels, but at set levels, yet to be determined. The above, however will give us a good working idea of what balance is needed and how it can be accomplished. It will help to provide a frame by which we can envision REMO. This is called: **REMOD** = Total REM Divisions and Ordering Diagram, for perfection of overall representation. This is reinforced by **MOF** or the Mode of Finity or Fractal. That is, we must select some central point that can be used to provide an average between one and infinity, a finite level, which is practically achievable. Just as COL and algebra, show us how we can find unknowns by subtracting, or dividing knowns from TRU, so can we in mass find unknowns or disproportionate values by subtracting known values from TRU. This is: **LADREMO** or Law of Differential Total Relative Magnitude. LADREMO is an additional method, as well as principle of REM and TRU. The degree to which MOF is variable or distinct and ordered is also highly different. This is: **VAMOF** = Variations in Mode of Finity, MOF. VAMOF is where IA, general knowledge, and the ultimate valuation of the worth, essence, proportion of parts of the world or different things varies in its level of finity, and TE increases with KOT.

CLOSURE in REMO, INDISCRIMINATE REM and RECIPROCAL MAGNITUDE

LORAM = Law of Reciprocal Magnitude. The proportion of anything that is not, or that is becoming TRU that it should not, for maximal TRU is its **RAM** or Reciprocal Magnitude. One extreme of this is 1/REM. It may be seen in TAO and natural proportions. Without this there is: **FIREM** = the Fallacy of Indiscriminate REM. **RECTI** = Reversion to Corrected TRU Intention.

SUMREM = The Sum Total or Summation of, or Total REM. **POREM** = Principle of Minute/SUMREM, or of the body of REM over highest levels. POREM arises from the fact that no matter how perfect a REM is from the top, no matter how proportional it is, and even when we have the magnitude of all subjects and objects in TRU designated, the mass of the smaller is still the greater body of TRU. That REMO and SUMREM are absolutely necessary is evident from the role of REM in TRU. If REMO and SUMREM are so necessary, then it lies in our ability to continuously bring out these lower levels of REM, as POREM. This is the key to a successful axiology or valuation of all, all TRU. The human mind is so inclined, yet without much guide, order, or REMO.

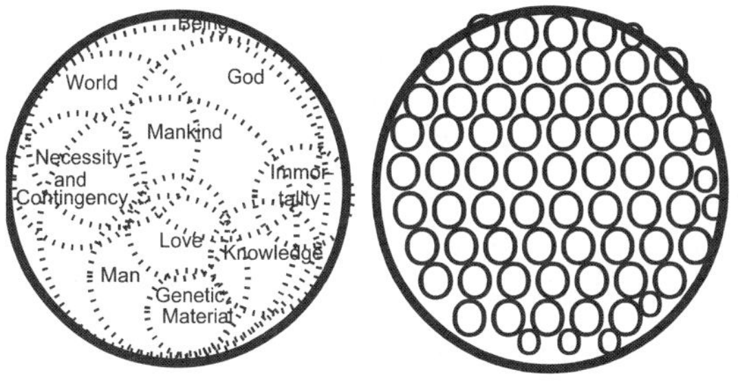

SAGGITAL SECTIONS of REM at 1R and 2R

Fig 3 38 **HIGHEST REM LEVELS and REMO**

This figure shows what the form of values are in the whole of TRU at the level of 10 and 100, or 1R and 2R. These are roughly what are the ten and one hundred greatest components of TRU. We do this in two ways: For 1R it is how they are roughly distributed, both by ALT, and by size. For 2R it is how they are distributed more or less equally as REM. These could be reversed, and can be done in either way for all levels of REM in which enough information exists to give them accurate REM and sufficient posits.

Fig 4-38 **REMOD, Total Relative Magnitude Diagram,**

This is the standard distributed Conical REMS Diagram, in which REM Levels are centered about 10R, and extend in equilibrium from one to infinity. Whether 10R is close to the median of REM Levels, and whether 3R and 30R provide good half way points between this 10R median and REMU or REM Unity and REMIN or REM infinity will have to be critically analyzed, refined, and perfected in the future. It does provide a good working frame for REM, when all their qualifying factors are applied. This will depend upon how much effect the other laws, principles, concepts, corollaries and mechanisms of REM play in providing it. That a sense of values is paramount is the lesson life's experience, impresses upon us. Each and every individual must learn, develop, have, and activate this sense of values in their lives by finding and using them as they are of the total dynamic REM in proper order and intensity. This sense of values must be forever nourished, by one and all, to become,

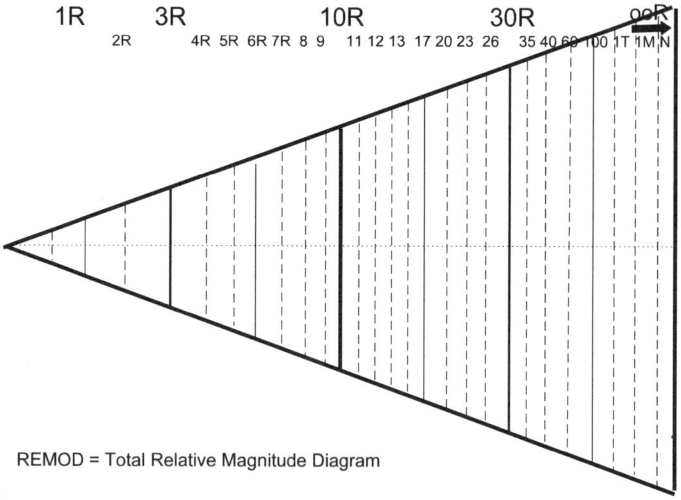

REMOD = Total Relative Magnitude Diagram

REMOD, total Dynamic REM as it best enhances TRU. As simple singular problems may only require simple singular solutions or acts so do multiple and total dynamic actions necessitate REMOD and a transcendent, continual problem oriented approach. Complex and compound total problems usually require more complex and total accounting, solutions and actions. This is the purpose and goal of much of the remainder of the text, that proceed from IA to IAT, and through HOM to culminate in ACC and return to TIT-TO and TRU.

Fig 5-38 **REMOD of balanced REM Levels**

This figure, in contrast to the standard Fig. 4, shows the whole range of REM adapted to best distribution and of balanced levels. In this way those aspects and parts of TRU at 1R are given the much greater representation warranted. The limits of REM are those to which our relative universe naturally inclines and renders us. So that what is most important in 1R is no more than others, or combinations, of other levels. What is possible, or we are sufficient for, must be kept in motion constantly pushing forward or backward through REM Levels. This is called: **POIR** = the Principle of Inversion Reaction. POIR is the degree to which the relative moves toward the absolute, and the absolute moves toward the relative. This is our all, RIA, or TRIA, in transition. POIR is the difference between absolute worth in which all are immeasurably valuable and unitary worth in which all become one. With relative worth in between that are preferably well formed, scaled and distributed to give best order to values. It is by POIR that we can resolve, at least seemingly, the difficulty of having a limitless and infinite REM or a meaningless and unitary REM. This paradox, of parts being as vital as wholes in contrast to the greatness of some parts over others, can be readily resolved, as all paradoxes, when all the information is at hand, the corrected or unified point of view is developed and demonstrated, and all TRU factors are accounted for.

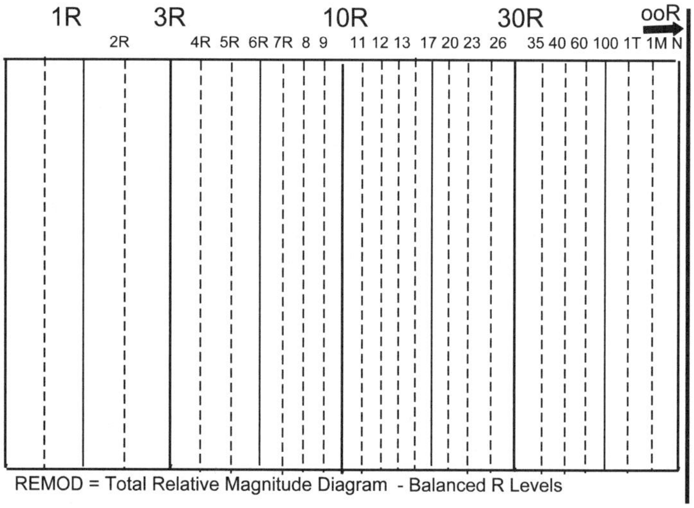

REMOD = Total Relative Magnitude Diagram - Balanced R Levels

THE DYNAMICS of VALUE and HOW REM WORKS

If we should question the dynamics of values let us consider for a moment how many people there are in this world, how often they make choices, purchases, and how careless and of little actual worth these are. Much of this is in exchange using money or money substitutes and with a whole range of TRU from highest to lowest to negative. This is especially so for those who are poor and those who have great appreciation for life and the world, TRU. This is the difference between REM and non REM in the 'real' world of everyday life, or REM dynamics. Without recognition, understanding and appreciation for what is good, better and best of God, religion, values, TREQ and TRU our purchases, emphases, percentages, gradients and levels of attention and choice, be it individual, group or all, is defective and fails. It takes both a direct knowledge of TRU and REM or their equivalents if our needs, forces and compulsions working through and through TRU, like HOM, are to be beneficial and successful. It is wages, possessions, and our savings that make or break us depending upon the level of TES we seek and the degree we benefit from or deny REM and TRUD. They are always and massively with us. Just as those who feel desire and cherish love most are those who have the most energy plus the maximum direction of that energy. This energy is directed toward finding or satisfying basic needs, comforts and interests. That of value, may be known or unknown. Yet it is usually felt unconsciously as it causes us to act in ways needed to fulfill body, life, and social needs. The unknown and that felt unconsciously, even if not directly of, or known as, TRU, is naturally of the right value, or at least the right order of value.

The sanctity and holiness of God, Christ, Mary, the Saints, and all that Christianity and other religions so vigorously extol is largely a reflection of a mysterious and secretive interpretation of the universal and absolute nature of TREQ and TRU. Once this nature, and its aspects, modes, form, function, parts, processes, concepts, principles, laws, mechanisms, and systematic unity, and order are discovered, well developed, and used, the metaphorical, incorrect spiritual, and exaggerated interpretation will be unnecessary. A scientific approach to religion will free it from errors of identity, disproportion, valuation, incoordination, inactuality and incompletion. Recognize the great contribution religions, gods, messengers, personages, and vehicles have made to civilization. Yet this must not hinder us from developing and realizing the full potential of knowledge that best enables comprehension of our world, TRU.

Although our bodies and much of the world around us function at a high level of TRU, almost perfectly in some cases, our attention, thoughts and actions are likely to be much less or not TRU. The more we make them 1/TRU, the more we lose TRU. It is in making our individuality identical with that of TIT that is the great justification, cause and reason for REM. **LORF** = Law of Representative Failure. LORF is the failure to utilize ROT and to provide REM posits. **EXREM** = EX, Exit or from REM. EXREM is seen in just the 'living of it', an illusion of greater than REM, than TRU. Much life is a failure because we miss the mark, do not know correct values, where we are going, what we are doing.

The extreme transcendence and penetration of things, as parts of TRU in SPALT, proves the importance of TRU dynamics. We can't misconstrue our attention, IA, as we are so apt to misconstrue all subjects when we emphasize certain features to the neglect of others or when we assume simple or limited objects to be final.

Behind all greatness there is much more than chance, luck, or even singularity and unity. It is easy to attribute worth, highest value, and greatness to this or that. This is especially so when one extols or seeks others to extol whatever they champion or promote. In fact this is an excellent measure of the TES in a person, whether this is or is not of TRU, how and to what extent. It is in the collective greatness that often corresponds to highest REM, that TES and TRU are to be found. We know from history and science that greatness is accumulative, both in time and in number, e.g. the great civilizations in history did not just happen. Nor were they a result of chance, luck, or only one or a few simple incidental causes. They occurred in a massive and related upgrading of quality, basic strength, and power of predominant Â FAF in which the fruits of their work blossomed. These were often those at the forefront of TRU at the time. **LIR** = Law of Insufficient REM. Muscular strength, tallness, prettiness, force, power, prestige, money, wining and many other ideals in the life of individuals are held at one time or another to be the greatest, the end and the highest good. This can be very false. With learning, knowledge, exposure to TREQ, maturity, experience, great magnanimity, moderation, toleration, TÊS, OM, PUREACLITY, COGNOS, POM, ELTICMA, SÂ and TÂ, do we use **LACIR**, the correction of DIR and LIR to SÂ.

The means and methods whereby we posit and work the elements of REM involve the following: 1. What these elements are, 2. How the words, concepts, objects of our world can and can't be used for valuation in REM, 3. How their ESINS or that part of them that is their counterpart in TRU mûst be the basis for a proper valuation, 4. How these ESINS are ES and CES and ordered by VAPO in TIT, 5. How these aspects, parts, processes, or items to be valued relate or act with others in TRU, and by integrative processes of TRU, PIP. 6. How we approach the valuation or REM, 7. What its **ERP** is, or the combination of ESIN percentage, and REM value. ERP is necessary when we use words uncritically for valuation. If they are only of 10% ESIN then their REM is likely, although not necessarily only 10%, to be much less than what they would be expected for TRU. 8. **EOP** or Total final Interpreted Value or REM to various items, entities, aspects, parts, and processes of TRU. It accrues from the above and other determinations and measures. This is predominantly their overwhelming posit by TRIA and SCARL. 9. Other means and methods will include correction for multiple meanings, comparisons with various representations of TRU as SPALT and ELTICMA to attain better values, the filling of voids not posited, and the adaptation of the many new and old concepts, principles and laws that pertain to REM and TRU.

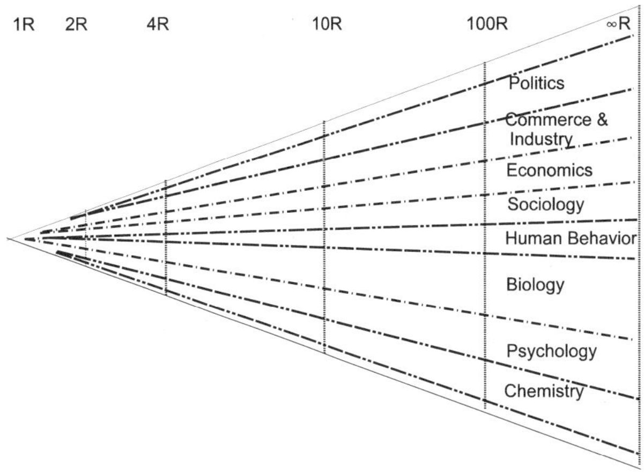

Fig 6-38 **CONICAL SECTIONS DIAGRAM OF REMO AND HUMAN DYNAMICS**

REMS can be ordered by the fields, sciences and topics we know, especially of TRU. It fills out and orders vertically following the length dimension of ALT, from largest to smallest, top to bottom. This is along with REMS in the horizontal dimension, in the conical sections. By providing geographical, economic, political, and many other discrete divisions of the world, with VAPO, we can order, value and show TRU. With their measured subdivisions, as we do by MOCBIREM, this provides a basis for selective and comparative REM determination. Each great field, science, and all in between, will carry finer and finer topics, features, and parts that help, as by COMREM, to fill out and show what TRU actually consists of both by quantity and quality. The divisions of science, or '-ologies', as founded by Aristotle, gave knowledge a stronger and more distinct means and separation. It, however, has caused some of our problems by making knowledge stereotyped, disconnected, and unnaturally inflexible. Combined with an individual's lack of depth this is much of the reason for a needed REM, KOT, and TRU. As these fields and sciences become more developed and complete by TIT all their components strongly support REM.(MOCBIREM)

The mechanisms, interactions and operations of REM Dynamics are, as TRUD, massive and highly dialectic. Our lives, and all TRU, consists of a great mixture of all three of these aspects of TRUD. They, by DAOST, are separate and fused, different and same. Through interaction they become even more fused and difficult to distinguish. Yet in concentrating upon those of highest REM in TRU we provide not only a good example of the operations involved in REM dynamics, we also enhance our interest by giving them more meaning, thus TRU. By showing the posits of REMS and their interactions we work through the representation problem. This is called: **LOED** = the Law of Finite Distinction. LOED is an ever present problem of TE and ES that usually is not resolved outright. It is resolved only as we increasingly discover, know and show information of TRU. By better describing and showing the intricate detail and operations of TRU's parts and processes we can achieve the necessary finite distinction.

There is nothing without cost. Since the line between positive and negtive cost, value and harm is infinitesimal, all have, at least in effect, either a positive or negative value. All that is a part of process, aspect, or FAF of TRU has a price, a set worth, relative value or REM.It is either to much or to little. It can be close to the right amount, but still has its optimal price. This is relatively stable for TIT-TO, yet liable to high variation per individual or in time. Even if its cost to us in monetary value is nothing or it is free, it still costs something.

If it has no price and we buy, use it, or take up our time with it, it still produces in us an effect, if nothing more than loss of time and attention. Since TRU is TROS, both absolute, finite, infinite, total and a unit and since TRU can be divided from its unitary nature, we may evaluate it in terms of percentages of a whole. This we do in REM.

Fig 7-38 **DIVERGENCE of REMS by Á ACTION**

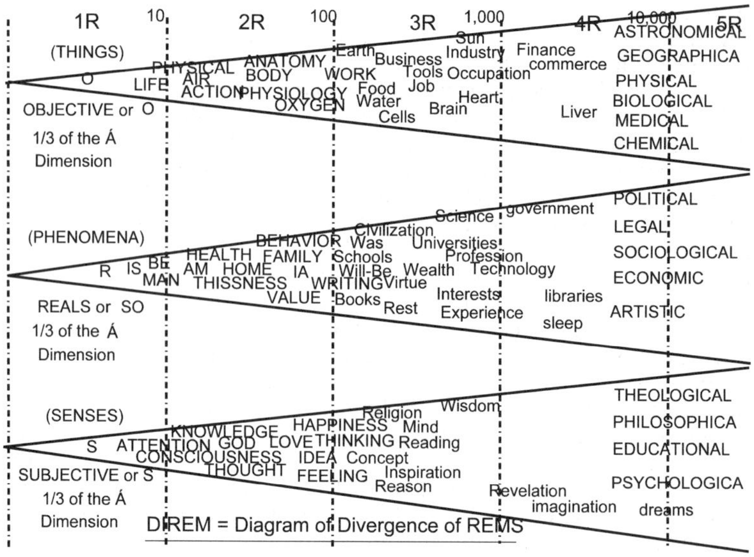

This figure shows the role of the Á mass energy FAF, or **ORS**, Objective, Reals and Subjective Sectors and operations in REM and TRU. They are graded from highest REMS to decreasing REMS. There are many other terms and subdivisions of fields and sciences that can be added. They do provide a good idea of **DIREM**, the Divergence of REMS, and how they can be divided with increased accuracy.

Money, or medias of exchange, are much of the substance and reality of life. If they are improperly used it is a waste and negative. If they are used properly, by TRU, it becomes positive, the measure of true value, or REM. Its use may be measured as a % of TRU/ATRU. Money detracts from TRU by being a currency that is artificial and representative of value less than TRU or REM, rather than being an identical proportion of TIT. It is because of money's great power, facility and OITIO, yet often negative status to TRU, that makes us all so susceptible to its misuse and abuse. Fig 8-38 sets a base for operations by which REM does or does not conform to TRU to TRU.

Fig 8-38 **GRADATIONS of the VALUE of MONEY and REM**

Obsession with money is further evidence of the negation of REM.This is readily seen in, and helps to prove the TROS and ACEXIN of RE, and TIT. Even REM, like ROT, MOT and POT, is more a means to TRU than an end in itself. We do not seek REM nor money for REM's or money's sake. We only seek them for how they participate and benefit TRU, and ourselves as we are TES in TIT-TO

The place of the will in life has been emphasized and taught by many. John Duns Scotus held that the will is superior to the intellect and was free and capable of making decisions between sense, moral law, and of values. He also held that the will is what determines right and wrong. Thus, by Scotus, the will to live, TRU, value and REM is to provide the highest quality. This is seen in **OVESIN**, Total Value times ESINS, ES, and CES. Just as God's will is superior to his intellect so right is right not out of choice. It is right because it is GOD or TRU. Therefore in the exercise of our will we do so, as in the choice of values, or determination of REM, by their nature, form, participation, function and power in TRU. This is a major reason why knowledge of TRU and REM are imperative. If we consider that the intellect allows us to select our will at random, we are wrong. If we consider the will free from all that is not TRU then we may find a kind of free will. Herein lies both the simplicity as well as hardship, joy and challenge of will and life if we are to be, be TES and TRU. We must do so, by REMOD, TES, TO and TIT, in all its complexity, variability, intricate detail, totality, dynamics and transcendence as it applies on occasion or throughout life.

Just as good will has been called the mightiest practical force in the universe, so is all will that is directed to TRU positively and valuatively the great media of the processes and dynamics of TRU. Good-will is not attained without knowing what is and what is not good and its exercise by the will of TES. Goodness comes from value, and value from what are proportionate parts of TRU. Yet we can't determine what this REM is without knowledge. So that they all build on each other dependently. This is usually not obvious in and of itself. Certainly we will, as animals in their own ways, have an inner sense of what is good or bad for us. This is effective, only within a range for this massive and rapidly expanding world of man. So that knowledge of TRU and its values, REM, are prerequisite and crucial to good will. This takes many forms, one of which is, **HPP** = Highest Public Principles, or the great guiding force of the common TRU benefits people have their minds on, their interests, focus, what conforms with the best in their life, or what its worth and value are.

Transvaluation of values is the revolutionizing of the reigning tendencies and sentiments of one's age. It may be right in so far as there is some change of values with TRUCH, or cyclical changes within TIT. It is more apt to be wrong in being a reaction to an existing defect. Much of history is a tale of such transvaluation of values. It is a tendency to right the wrong. Only by reaction and lack of VAPO, it perpetuates wrong and thereby continues transvaluation. Associated with this is: **LADIM** = Law of Different Measure. LADIM is exemplified by Arianism, or the view held by many early Christians from 300-400 A.D. that Jesus and God were not of the same substance. They questioned how a belief in the divinity of Christ is compatible with the doctrine of one God. They held a doctrine of subordination with a more extreme group believing Jesus to be unlike God and a more moderate one believing in a similar essence but different substance. The changes that occur with such fundamental differences, actions and reactions are by transvaluation of values. They as antitheses, in reaction to the action of theses, form a new, often different thesis, rather than synthesis. They fail to provide an adequate foundation for a knowledge of TRU. They rest their case, as is so often in life, on half truths or bits of truth, or some combination of these with very inadequate ideas and principles compared to the needed TO-TIT. The only ultimate solution is TRU and a REM and REMO in which all values are based on TIT-TO by VAPO and are fully and properly utilized by the worlds population.

CHAPTER 39

ENHANCEMENT and SUMMARY of REM and VALUATION

REM ENABLEMENT, ACTUALIZATION, SUPERLATIVES, SUFFICIENCY, FULFILLMENT, ULTIMATE REM

A person's REM is a function of TES and TRU. Since we are of varying TES both per different persons and at different times in life we are of variable REM. These are due to individual differences especially those associated with TRU. They help to prove the nature of TES and TRU, as well as the various states each of us are in, TES and REM. This takes initiative, a special kind focus on TRU, and an appreciation for REM in its greatest ACEXIN and accounting.

What most benefits and enables REM, as it shows TRU, can be used to bring ever higher and greater quality into human life. This comes from the great joy and happiness we get from that ecstatic feeling of achievement and fruition in the inspiration, revelation, discovery, solution, and fulfillment of all that is in TRU. These come as we discover and comprehend REMOD, TRUS, TRU, TIT-TO, and are able to posit each part and all of TRU by and in SPALT and REM. We can be no closer to God and TRU than in how we love, reason, revel, contemplate, and work ES, CES and values that constitutes, and best show and produce TRU in all ways most clearly and totally realized. By St. Augustine they must be right and directed toward total identification of supreme value with the spirit, God, TREQ, or TRU rather than non TRU. St. Augustine also held that only the highest is to be enjoyed and upheld as the ultimate. This necessitates a supreme good that is not private of our own. It comes from the heart as it discovers the spirit of God, the Á FAF nature, and constituents of TRU. This occurs with discovery and attainment of **CALP,** the concepts, principles, laws, and mechanisms of TRU. In so doing our private values become and correspond to TRU. The increase and expansion that occurs with decreasing REMLEV is a corollary of the laws of quantification and growth. It produces increasingly pure intricate detail, and is the greatness of both RU, TRU, ROT, MOT, SPALT, POS, STA and HOM. It is in the expanding finite, that the substance and complete composition of TRU are revealed and REM is enabled, enhanced, becomes totally successful and complete.

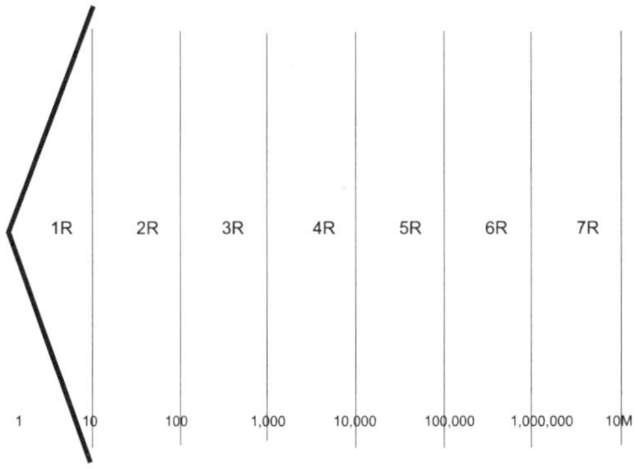

Fig 1-39 **REM and the LAWS of QUANTIFICATION and GROWTH**

Only in a system in which there is a discrete frame replicating TIT and dividable into distinct parts can we have TRU components and terms that are reduced to ES, REM and POS. The resolution of coterminality alone will not provide perfect separation of each entity. Usually two or more terms in coterminality are not fixed and perfectly superimposed and congruent.

INSUF = Index of Sufficiency. INSUF approximates the REM value of the individual or any aspect or part of TRU and TIT. Yet INSUF and REM are different. REM or value at 11R or 1R, as 1/2 of TRU or as specific or generic TES, is indeterminate. INSUF should be taken as an arbitrary standard which is precisely measurable. Yet it is in seeking and working out this standard and developing its precision and measurability that is largely the key to REM. This can be increasingly done when other parameters and standards are brought into a precise relationship with INSUF for REM. This may be done for all mass energy levels of Á in TRU, physical mental, rational, and intellectual measurements for INSUF, IMU, REM, and others. Efficiency of selection of REM leads to reliability in valuation. It is the basis for an adequate REM. For, unless we can select among all that is given, what is and what is not TRU, and what is and is not reducible to REM, any selection made is not only inefficient, it is loaded with error. This reduces knowledge, by VAPO, into the differential and lesser finite of TRU.

REM and EXPANDING FINITY

This we show in a sagittal or head-on view of REMS and the content at each level. The expanding finite is how REM expands as its levels increases and each becomes narrower, with each term smaller. REM tends to be finite objectively, yet these same parts of TRU given value, tend to be absolute both in being part of an absolute as well as subjectively in estimation. It has been human's compelling drive toward worth of life, value, proportionate TRU, and REM, that has, is, and will be the greatest of beneficial forces in civilization and TRU. The forces which have gone to make up this drive illustrate the catalysts and motivators of humanity. These are what are so great and such a large part of REM, HOM, and TRU. As humans become more complex, the Á dimension changes, knowledge and value become more susceptible to error, spontaneous and irrational behavior becomes more prevalent, faulty, and dangerous. It is through a life of learning, knowing, and producing, that lower values are turned to higher, negatives to positives, and to REMS, that greatest success is found. However, to do so continuously, to maintain the highest positive values is a rigorous and lifelong endeavor. It is in REM, REMO, TES and TRU that a balanced judgement of all is to be found. A simple classification of REM or proportionate parts of TRU in one two or three levels may be expanded to include all that is of TRU from its unity in TU to its infinity in TI. In such a way we can develop a valid quantification, REM of all TRU.

Summum bonum is defined as the highest good. It is usually the one in which all other goods are included and/or from which they are derived. In summum bonum we have an expression that is both a hallmark of REM as well as a dilemma. Summum Bonum is that which is intrinsically better than any other good and which is thus qualified to serve as an end, as in human conduct. This, on the surface, seems easy enough. It has been a great part of axiology, valuation, and to some extent REM. The emphasis on and search for summum bonum has been a large part of human history and religion, especially since its development by Plato. Only many diverse kinds of *summum bonun* have been championed. *Aristo* or an intelligent healthy state of mind were considered to be this by the ancient Greeks and many others. Yet by TRU and SÁ we know that a healthy body, progeny, habitats, and all that otherwise constitutes TRU are similarly highest goods or values. Value inclines to the superlative, as in being the 'highest'. Yet this is an error of focus, IA, and concentration by the first person, as well as being innate to value itself. The fallacy created by this seeming contradiction as summum bonum, are corrected by REMOD and its best operation in POS, HOM, ACC and POT. It is a corrected assessment through the full and equitable distribution of REM.(VAPO)

	SC, 0+	1+	2+	3+	4+	5+	6+	7+	8+	9+
1R 10%EC 9NEC		10 TRU TIT TE TU TO TROS TRIA TRIS TREQ Life								
2R (90) 30%EC	1 POT	6, 2 TRIT TI TÂ TÄ SUT TRUD Human* GOD*	2, 1 ROT OT Meaning*	1 MOT	1 POS					
70NEC 2c 11/9 20%EC	3 EPI PERT EN(s)	7, 2 Á SOT SCARLS DOAST ES TES Universe* TRUS Relative* The*	3 LANT EC MADAS	2 REM ALT	1 STA	1 HOM	1 DYNU	1 DYNS	1 DYNO	1 ACC
2 25/18 30%EC 2e 54/32 40%EC	8, 2 ESIN(s) PACH RU VAPO S O SO Ä This* Special*	13, 1 LOPAN OITIO ENT RUPKAL OM LARF MATES TRUF TRUP LACOP KOT PRUT PRACALT God*	4 (Á L T)^n ACEXIN LATANOS LARP	3 SPALT LORVA MOCBI- REM	2 LADAC PRALT	2 CES FAF	2 CÉ LOPA	2 LAMUA WITÁ	2 LIFOT AIT	2 COGNOS PUREAL -CLITY
3R 900 50%EC 450 NEC 3c (115) 40%EC 69NEC	12 NEC JUNT AU Â POCOL IAs TINS LOLS POA LOPOIR S-O AS	25 TER COPT LAPDAT LASADIN LITS TRUEQ REXTRU LO LOTES LATOL LASUCT POIR LIPÁ TRÂ LAMIR COL COR LACOL MET SOEX JUCSMOR LAFRIA LARUTS ROMT LOTÄC	9 LANOFAC PAMEDS MATES SUVUK NE LOC TOR MP R	9 Ļ T DEC LOPS TÄRAW MEALT DREM REMS LEDIOR	8 POSITS PÏ LAD NES PHIWOT LED POP STATICS	4,1 Process Hom- ation CEFKA CHES LASACA	5 GAC LATAC ITOD CED ACLO	6 IPR OPTH LASTE PS MAB LACOM	6 DEF VAC COPTA LATED LAPAF RAKUT	7 TRE KUPS ELTIC- MA SÂ TRUCO ÁCOM
.... 3 (248) 50%EC 124 ---	25 POKED POPOT FAF AU MÎ NAT MARAT FAT METs IAT MIDCET INT MARATEX HOMALT DISO LOR RORRE COAT INT0 DOLT S-R PIN LESM LORÄ(F) LENK	36 ATRU IT INTRU INEG MOG ITRU METAD COC COVET ATIM LATIN SAC APROD LAMEX LARAC LO MET S-R STOT BIDYN FET LARNIP LASAF LAL LEXPAC LOSE PADFOR LODAC LARTIN LATIN LASPRE MIRPS LAROS LARUTS POSÀ RÍ	21 RÎT ÁXE POPTE BIDA PRUNEC MIST CODIT LARECS LOTESAN ROTE LADES LOPCAF RECONET TRÍD LOPUT VE SOID SETLAW EC ECT OC	15 SULP OLP OSP SUSP SUSF OSF OLF SULF DIMALT ENUL ROMVET TOVOP LEM LOV SARUE	12 LUDATO PIP RED LACRO LARPOS PICOM POCOMF LOM FOL LAPDOR LAP LOPS	6 PAN S/F KDM Í PREDIS LADPEFT	7 CNS PORI CEK PED PEK AMT TRI- ALT	10 FERT IOBRE ASIT OTS STES AMTRU MERPS POPA LAPAR POMAC	9 LINS PODAF LAWID PED LAPCOST TRUC LACJOT SIRT LACHVAR	9 TRAX TREX TRIX LATS PAF MATAM MARS SPEM THESOÂ
3e (537) 60%EC (215)	60 PIC BASFANT ECRO LIS SELIA BAS TAN CONVET CONRIT POROM TIMS CODITrp ENIT PORF CIS IRT LAC TRÄT PROAT LIAT OLA ABOABI EBE FER LAFED LASUR ÄD INTRU LATNUP DIRP INTSO MUS VACMÁ POES RAL ENDOT ECTOT AK MESOT REBE LARD COTA TRAT LEPER LOGOS COTRU LOLVAR EN(S) LOVKIR TOPA TRUCH16 LADGAR POMOR LOPVOIR STAN PAND PIROL LENA PRUNT METO	57 FIC LEXÁ MÊT GMLPDF TECS LAVMÁ SUPTIA LARÁDE RECT LADSO POTRU T(RU)ID UTES TRÂ THES THEO THESO LUSA LASOT MËT SIRD TEL PRUCT TOL LACPÁ LINTAC RÂ UMIA LO LABEX AME AHE INIT LADREM MINESAT POIR LAMIT TANCOWS LÁSP LIDU TUR COS SORAT STAXNOR LATANO TRÍ PADALT LOSADIN ACIT FOCI LUT CORT CONS POUC PPF LACTAR	46, 2 TIR LOICLOR POSAT RÍT PREM PROV AG LORSIPT ID VACEP LOSAF Definition n(PROD) LATRANG PEL LACFAF LODEPT SUPIT REDRPIM POCT MEDOPL LOPVOIR SOLTREA PORE LACRUA LAMUT LÁRE PLO MIF PIKER RESUD LASUD TATA LEP PUREAL LÂNT LANA LOTAD PONT RONT LOTESAN DADO LADS LADO mlt* MEDOT LEDIS POLT	20 OVT RAU ERP IMu KREM ÁOITIO DIREM SRIA STAREM CREM MOF MOD LOPELT LARMEDS ÂTES SOP SAXT LEXFI COF MIPT	21 LOSPEC LA- MORDT TACON LANCOP POSE VE POSLOW I/O POL LADIP POR DEP PRALTI LACESP RACTIO REPOSE LARAT LART POR TAT POL	18 LATMUD CESINS HOMA- TION TAP LABA LAMA LOPOF LOCON RABEC LOR- CEDS WOT FUPE COPOS EPAT LATSOL LAST PORAB SREC	13 TESP LASO BC BR DR RAK WAD JUT INT RITES WITÂ RETES DORIN	20 PÂS IATES IMTES LILT PROG IBRE OBRE PERFT TESÂ UIAT CE VOC MATSOR LAFJET RASAD LOEN TEX TIF LAMUC LEF	18 RETIT SLONCA COKIT LACART POK ACEF INAC LAPDOT LOPRIC LADEXÂ LANSUC PROVAC LAPCOST LACORE FOPIC GUT LACPUT RINCFID	13 LAL PAC ACENT PRACC RAMAR LAPUS LOMT PPMA PODES LATS- CAP PAFOC MOV SCEN

Table 1-39 **REM and SECTIONS of IPTRU CORRESPONDENCE; EVALUATION of ES, TRU**

In table 2-39B we provide a frame for collecting and showing many REM as distributed through the text. This is **CET,** or Chart of the Continuous Evaluation of the Parts of TRU or ES, SU, REM and POT. CET helps: 1. Obtain a more reliable measure of REM and all that is collected, 2. Enables us to present each section and chapter more effectively, 3. Gives us a quick overview of the text and TRU. To make this more accurate and avoid overlapping terms that are the same or synonymous need to be used only as one.

The numbers in the cells stand for the number of NEC for each section of the text at that level of REM. This is not a set number for more than one reason: 1.REM will vary with the total number of NEC and EC used, 2. Coterminalities make for an incomplete and variable distribution, 3. How much weight is given to NEC and how much to existing words is variable. 4. How much weight is given to existing words that are generally defined or how much they are reduced to their TRU meanings is variable. 5. As language, NEC and REM improve, these variations will be better accounted for. A few ECT, EC and OC terms are added to show their presence in the total REM table. They are given separately both for reasons of presentation as well as to show their differences and separate operations. The N stands for numbers of NECT terms in each category with an indefinite number (~) above 5R depending upon new ones added. The table only begins to show REMS, albeit major ones, of TRU. The ideal distribution of REMS for these concepts, principles, laws, terms, and the aspects, form, function, parts or processes of TRU is something for continued expert work and refinement.

The enablement of REM through REMOD is found in and

contributes to the harmony and symphony of TRU and ROT. When a person's attention to singular objects is through experience, learning, and TRU turned into all activities that relate to TRU, we can see the greatness of REM. **POMAG** = P. of Manifold Greatness, certifies the recognition, inspiration and exaltation attainable in the glory of all whose ES and REM that make up the total FAF of TRU. This ultimately must include REMO and systematic representation of TRU. For this is the essence of being. It is how the purest of arts is the dialectic science of being and reality. This is the essential principle, the substance of COGNOS.

REM is preserved, made better, enabled, enhanced, strengthened and made preeminent by the best possible exercise of love, virtue, beauty, quality, morals, standards, improved language and overall comprehension of REM and TRU. There being all kinds of emphatic REM, so do they when best adapted enable and enhance TRU. It is in how these means of enabling and enhancing REM and TRU are best exercised that they become actual. As virtue is best served when self interest and principle coincide so the problem is, as with all, how do we get them, when TRU, to coincide. This is in how they are the same or identical in TRU. The coincidence of virtues, values, and REM can come with and after REMS are adequately developed, viewed massively, dynamically, totally in and of TRU, as this is applied to life in TRU. If we comprehend all continuously with greatest practical breadth, depth and proportion we will have a complete and optimal REM, be complete, successful, and optimal. This is: **ULREM** = Ultimate Unified REM. As all must be TRU so must REM be TRU.

SUMMARY of REM

From TRU we know for sure that: 1. TES is of infinite and absolute value, 2. non TRU or individuals that are not TES, are of null or negative value, 3. our potential is of great value, and its degree may be roughly gauged, 4. we as one in many billion are of limited value, 5. Yet, if we as one are in identity with TES, we are of infinite value, 6. we are going to die, at least physically, and our value will decrease, 7. before and after physical death we may be dead or alive depending on the lack or presence of TES, 8. we live in a very TRU, ATRU, ANTRU and ANTITRU or mixed, AMTRU, world, 9. we must learn and live TRU as much as possible, 10. TRU depends, as we do, upon this kind of learning and life, 11. value can only stem from TRU and what is proportionate of total TRU, 12. this value or REM can be measured and posited and thereby vastly improve by our quality and selectivity in life, TRU, 13. if we know and live by no or only a few minor TRU values, then our knowledge, these values, and all that follows will be to that degree inadequate, faulty, and our lives one of unnecessary suffering.

Fig 2-39 **PROBLEMS and SOLUTIONS of VALUATION**

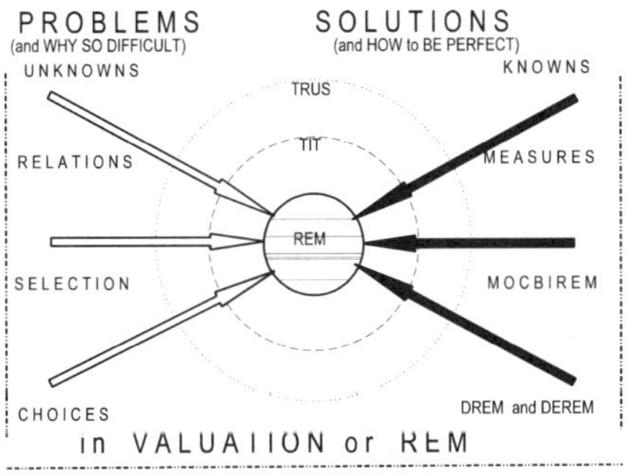

Valuation can't be successfully determined without certain set conditions and requirements. For the problems of valuation are shown by how poor it is either by accurate measures or by TRU. The problems are difficult because there are many unknowns, variables, and imperfections. These are both unknowns in general and specifically of what may be known yet is unknown by how it exists in TRU. It is in selection that life survives. It is in the selection of what is good and not bad, and right and not wrong, that humans succeed or fail. We continuously make choices little realizing the depth and complexity of either their base, operation or consequences. These together with many associated features of the difficult problems we face are the causes and reasons for valuation. It is what makes REM, and thereby TRU, so important and great.

STRUV = Standardization of REM and Universal Value. STRUV includes both the standardization of REM, the correspondence and adaptation of other universal values to REM. For tradeoffs, inputs must be highly qualified and meet basic needs and whatever else is best for TRU. **FUV** = Functional Value. FUV is given in proof by sizes, or size proof. Why one organ or part is so large and another so small. Why some things seem much larger or smaller to us than they are actually in TRU. It usually depends upon how important they are for TRU, REM. The human body, being a great part of TRU is a good example of divisions,, proportions and REM. It is size and shape related and determined by the functions it must perform. This is FUV. Just as a mouth or a stomach needs to be a certain size and shape to perform the intake of food and initiate digestion. In all, we have summative functional proportional values. How the isolated, different, divisions of an entity, or ES of TRU attain their proportional, major or minor, predominant or subordinate roles.

MOV = Measure of Value. Different systems have different yard sticks, scales, standards, and therefor different values or REMS. This is partly why our world is in such disarray, for to be TRU its values can only be by DAOST, concomitant with TRU, and be ordered by OBT, TRIT and TIT. REM helps to provide the equity needed for one and all. Equity is highly coterminal with equality, equal distribution, fairness and justice. It is necessitated by poor distribution, poverty and iniquity. The vehicles of equity are the individual, society, economics, wealth and government. It is the equal, fair and judicious sharing of the benefits of life, or TRU. Yet it must be equal rights for equal potential, active and future participantion in TRU, rather than any value. To best follow TRU and make most effective POT, ROT, SUT, REM, excess standards, arbitrary conventions, and laws should be minimal and adapted to TRU. As great as TRU standards are, they must be: 1. minimized, 2. follow TRU, 3. be perfected, 4. held in mind, ECRO, 5. limited, and 6. appropriately applied. It is the extent and degree these are or are not well developed, measured, and executed that they utilize REM and are most beneficial to TRU.

Only when we have the desire and interest to discover and learn unknowns and their correlations can we begin to solve problems. This can not be done totally and efficiently without TRU, ALT and REM. By learning these unknowns and developing their correlations we have to reduce them to measurement, the measurement of TRU. This is done through VAPO and being able to determine their REM. REM is given precise measurement through MOCBIREM, DREM, DEREM, and the measures and posits of all that exists in TRU. **CURS**, Correlation Unequal to REMS, denotes how two REMS of equal magnitude may not, and usually do not, have the same correlation. They are not apt to have the same posits, nor will there be similarity other than by their REMS. Coterminality is justified if held to minimum and supported by combination, ordination, proportion and REM. **CORT** = Correction of Coterminality. CORT is the specific method whereby we give to each ES, CES, or term, that is used, a corrected REM value. This is a value that is often far less than it may appear to have in and of itself, and how much it seems to be so great a REM of TRU. Yet, when all terms that are used to reveal all that is of TRU this term and its role in TRU are diminished in terms of the whole. So that we qualify each term as it posits in TRU by its SCARLS with other terms.

The degree to which we want beauty, quality, health, strength, value and REM is the greatest. Even that which stimulates and accentuates this want is similarly great. For this is life in the fullest, TRU. Especially is it so from an individual's vantage.

Yet as all is absolute, proportionate, and complete by TRU, and thereby for REM, we must not mistakethe kind of absolute of the part for the absolute of the whole and thereby misapply this part, as beauty and other singular values. This is called: **FAMIR** = Fallacy of Mistaken and Irrelevant or non Identical Absolute. It is because of the accumulative negative effects of over, under or incorrect interests and priorities in life that so many people never attain the success in life they seek and need to be TRU. Recognize greatest REM over all the worst for TRU, and correct each as the occurrences in life provide.

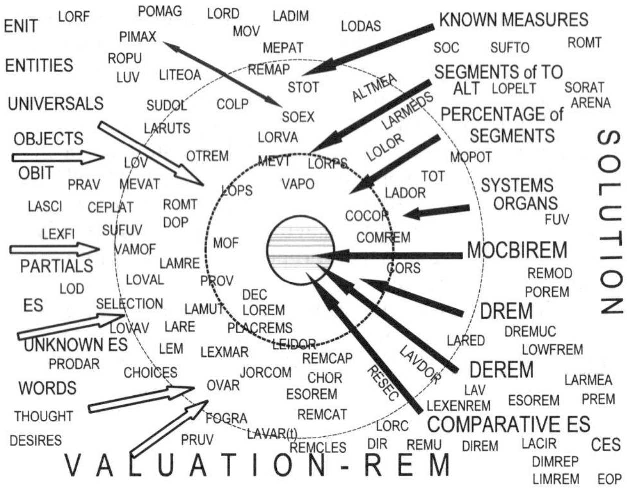

Fig 3-39 **PROBLEMS and SOLUTIONS in VALUATION and REM**

Figure 3-39 is an extension of 2-39. It provides a rough yet more detailed and illustrative summary of many major concepts relating to valuation and REM.(cf.Glossaries and Indices) It shows the formidable and highly effective combined powers of measures, segments of TO, percentages of segments, systems and organs, MOCBIREM, DREM, DEREM, Comparative ES, REM and other means of giving meaning and order to one of mans greatest problems, accurate values. It shows their relationships to the many concepts, principles, fallacies and problems that make valuation difficult, as well as those that enable us to order and refine REMS. The diagram is ordered horizontally from problems to solutions focusing on their interaction in the core of TIT. The vertical axis is roughly based upon size. The circles enclose from inside out, the core of TIT, what is in TRU and what is of TRU. The arrows exemplify the vectors or forces that work between a few of the major problems and the solutions that provide for REM, as they converge upon identity in TIT.

Positive ++++++++++ / -
TRU and +++++++++ / --
REM ++++++++ / ---
+++++++ / ----
++++++ / -----
+++++ / ------
++++ / -------
+++ / -------- Negative
++ / --------- TRU and
+ / ---------- REM

One great salvation of REM, and all persons' endeavor, is that once REM is determined, it is always there. It is more than a great mind-set it becomes a REM SET, **REMSET**, for each and all to use reliably and effectively. So are senses and things, knowledge and technology, discovery and application similar parts of what constitutes TRU in a highly dependable and permanent way. Once found and learned they are, with a little effort, always there for use when needed.

PROV = Projected Valuation. The value of all things, especially new concepts, NEC, ES and CES, lie not so much in themselves, as they do in their part and role in TIT-TO. Even more, it lies in how much and clearly they show something else, their standing and supporting something else. This is commonly done by S in its mode of representing what is sensed, which if TRU is their role in Á. This shows how the subjective is basic to TRU, SU, and REM, a part of Á FAF. S or Subjective is representative of SU and SU is largely in effect representative of TRU. Yet S is both in TRU and of or outside of TRU as an observer and image. When REMS function by standing for, or enabling some other aspect, part or process of TRU, it is through PROV. PROV demonstrates how SU, ROT and REM replicate TRU which leads S to COGNOS in the ideal subjectivization of and for TRU. This is called: **SURTCID** = SU Ideal Replication for TRU as COGNOS is the Ideal Subjectification for TRU. This is a basic characteristic of life. It is the biological advance of the human central nervous system. It is largely sustained through reproduction and genetics. These effect representation in their own way in TRU.

It is very necessary that the value of money, who has, has had, will have, and use it, be proportional to TO and TRU. REM and money should be largely equivalent. Money should be little more than the media or vehicle for enhancing TRU, both integrally as total worth or value, OV, and differentially, as proportions by MOCBIREM, DREM, DEREM, and REM. An economy must be based upon those parts of TRU that enable its prosperity and the quality of peoples lives, TES. Fairly accurate values for the world can, and will, be increasingly produced.

Why the justice systems in the world, some much more so than others, don't work is because they do not do what they are needed for, the weighing of good/bad, right/wrong, REM/1/REM, and benefit/punishment. These are no more, nor less, than positive/negative, and TRU/ANTRU. A justice system should uphold and enhance TRU, all that is TRU including its basic CALP. Existing justice systems, even at best, look to the discovery of facts and evidence that will prove one side or the other of a controversy or deviation with little regard for what is best in TRU. **JOR** (**JERO**) = Judgement Equals or is Equivalent to REM and TO. Good judgement can't be achieved without TO, its ordination and REM. If you do not know what is the greatest, best and right by TREQ, TRU, DAT, DET, SU, and REM, you can never judge the limits of all you survey. You will be forever over and under rating judging, and misjudging all in the world. To complement and reward not only attainment of TRU, but to stimulate their rates of improvement, is the best method of applying TRU.

We can not expect others to stand for us unless we stand for something, something good, TREQ or TRU. We can not stand for others unless they are willing to stand for something, something good, TREQ or TRU. To stand for something is little better than to stand for nothing unless it is TRU. To do so we analyze, 1. What is missing, 2. What doesn't fit, 3. How it can be best ordered, 4. What parts, divisions, parameters, coordinates, dimensions, and geometry, it has that will best enable us to replicate, understand and achieve it, 5. How does the frame, parts, and their proportions all fit together in the whole? Fitting these together provides: 1. Needed knowledge of unknowns, little or wrongly known, 2. Fulfilling TRU, 3. Answers to questions and solutions to the problems indicated, 4. Present the equations and enable their solutions, 5. Enable the best possible replication of TRU, 6. Reduction of errors, false beliefs and opinions, 7. Enable and produce the stimulus needed for interests to accelerate TRU at an optimal rate. What fits is how humans not only have been fitted to best adapt to their environment, but how much of their environment has been best fitted to humans. This is vividly seen in limits, both the limits of the human body, the human brain, and all the tools, instruments, vehicles, and CALP humans develop and use. All are guided by how much they best serve us and the laws of parsimony, rather than the unnecessary, irrelevant, or negative.

Much may be learned from others and all about us if we have the interest and curiosity to seek it out. The whole world is loaded with a great variety of information. Much of it is indirect and interactive, yet of great use. When combined with the learning of greatest existing information much of the reasons and proof of TRU can be readily realized. It takes **TAN** or to Think and Note, exercising all ones existing powers of being, the Á mass energy FAF in TES of TRU. What is essential is that which has, In^{∞} x REM^{∞}, or **INREM**, the Infinite of Intensity with the infinite of REM. This is the degree to which learning occurs and information is optimal, or a maximum of interest combined with the maximum of value in TRU. The spirit of CES and TRU are shown in: zest for life, youthful enthusiasm, joy, spirit of adventure, everlasting interest, inspiration, revelation and the infinite, eternal, unending and manifold FAF of TRU. They are increasingly enhanced by the higher levels of Á in TRU, as by Cu, Ru, IMu, and INTu. These are perpetuated in each acquisition of knowledge and the advance of a culture. All of them, as TRU are what is available. They are highly determined by REM.

There is tremendous force in values, proportions, what is known to be right, TREQ or TRU. It determines our every thought and act. The least deviation from right is a central focus of our IA, vividly noticed, actively, reactively, and pro-actively, sooner or later productive of correction. This is much of TRUD, the living force of learned and acquired behavior. What we do is or is not by the rules, or within or without the law, largely by existing laws of TRU, and less by arbitrary laws in reaction to what is not TRU. This is **EXREM** or From REM, just in the living of it, greater than REM, TES by REMOD, and all TRU. It is resolved by a bonding, joyous identity and union with the best in ourselves as we are of this best of worlds, TRU. **PLACREM** = the Correlation and Equivalence of Principles, Laws and Concepts with REM and the parts and processes of TRU. An elevation of values, as seen in REM and its uses, is in the end like the determination of a perfect deity. It is like COGNOS or that which is of us the greatest and catalyzes the greatest. This is REM to the infinite, and this infinite as it is best represented and utilized in TRU. This is shown in METAD and SPAT where with an overall diagram of TRU and associated ES the relationships between the great REMS, as well as major concepts, principles and laws of TRU are produced and posited We come close to a duplicate or replica of TIT, TRU itself, when allowance is made for all REMS and all TRU as they fit into this diagram and model. From their existence in TIT we can see how PLACREM provides the needed reconciliation and tieins betweem REM and TRU.

The problems of and resistence to new languages or terms can be vividly seen in REM. Any and all added terms not only complicate an already overburdened vocabulary, they add and lead to confusion in existing terms. They tend to displace what is conventional and comfortable to us. (PLACREMS)
The solution is to meticulously add and subtract terms that do or do not fit their actual ES, CES or aspects of TRU and TIT.

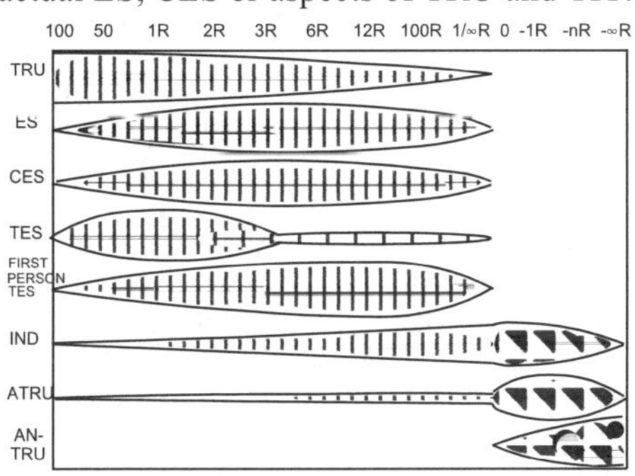

Fig 4-39 **COMPARATIVE STATES RELATIONS TYPES of REM** Some of the major categories in, of and out of TRUare investigated and the REM levels of TRU compared. It shows their distribution between 100% TIT to 100% negation of TIT. All that is strictly of TIT contains no negation. This combines TRU, REM and how its aspects and parts relate to other commonly encountered entities, and what works for and against TRU. One can get a better idea of what REM is and how to bring it to the fullest benefit for TRU.

The determination of values, REM, DREM, and priorities, are not only a fundamental part of understanding TRU, ES, CES, and ROT in a proportionate and positive manner. It is to establish the basis for ethics, behavior, and all human activities and operations. There is great value in where and when, just as there is in why, what and how. They are key to the length, time and mass energy FAF of TRU, as well as an individual's movement and timing. It is in the combination of REM with Á, L, and T for each ES and CES that we can achieve the POA that is prerequisite to an adequate comprehension and demonstration of TRU. Inge for many years Dean of St. Paul's in London anticipated a REM presentation through actualization in TRU, by insisting upon the ultimate spiritual reality of values grounded in God. He instructed us to look beyond our senses and the world immediately before us and seek eternal values, of truth, beauty, and goodness, found in the realm of God, TRU. It is a mystic ascension of the soul to God, as TES must be its counterpart in TRU. Herein mysticism is no denial or escape from reality, rather, its return. It is the actual and pure reality, the media of our identity with TRU or the whole, proportionate and valued TREQ or TIT.

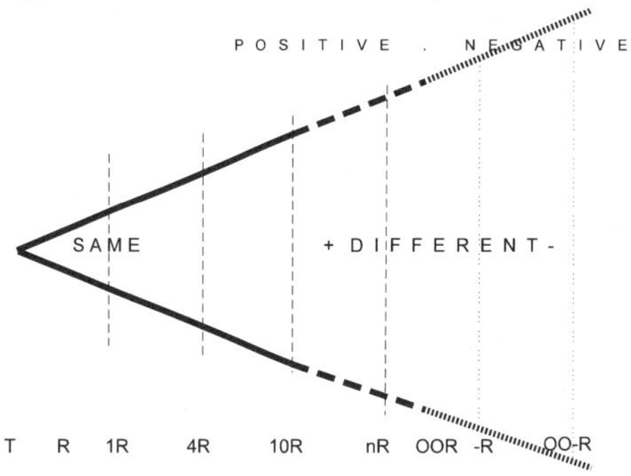

Fig 5-39 **EXTERNAL VIEW OF EXPANDED REM**

Identity comes from unity, the unity of relativity systematically produced by the conversion and fusion of the commonality of all that is greatest in TRU. It is the unity of our predominance, the Á mass energy state of TRU. by their FAF in ideas and mechanisms of the world and God. This is how all that comes before us, by the interaction and relativity in the world of being an observer of all observed, comes to have more or less meaning to us. The greater this meaning is, the greater and more these same values, ideas, forms, aspects, parts and processes of TRU are, and the more we and TRU are. There is unity at the top, in identity and sameness. As we pass from 12R to 10R to 6R to 4R, 3R, 2R and 1R these values and REM become more the same, as they must by definition be increasing proportions of TRU. At the bottom, ooR to nR, there is only infinite or unidentifiable finites without much relationship or significance to each other, the whole, and without coordination except in their tiny posits relative to SPALT and TIT as shown in STA and HOM. Yet they are in TRU and when combined with much else that is different and differential, much of TRU. Meaning is largely the power of our IA, guided by the Á predominant FAF, focused upon the ES, CES, and REM of TRU.

REMCAP = REM as Centers or Powers in TRU, TIT. **REMCÄT** = REM and Values as they inextricably help Create and Form TRU. REMCAP and REMCAT help to show how TRU is more formed by predominant powers that make it stronger, more stable, different, and exclusive, from all else. It proves Á, PRA and TRU, PRUT, by showing how values and

REM can not be answered in any way, as only physically, but as they are typically, specifically and exclusively all in TIT. Many of the words we use, even ES and CES, have numerous meanings or applications and thereby different and variable values and posits. We are limited in our capacity to represent them in REM. Anything of small REM can't have more than one posit. Those of highest part and role in TRU may have more than one posit. Yet, these must not only be limited, they must be highly selected, discrete and TRU.

Values can't be reliably determined unless they are properly based and quantified. They can't be quantified unless this basis accurately depicts the proportional parts or relative magnitude of TRU. The proportional parts of TRU can't be developed until TRU as TIT-TO, TRIT, MOT, ALT, and SPALT are accurately demonstrated. This is how MOT and REM are intricately related and interdependent.

The preliminary criticism of TRU, TIT, TRIT, ROT, MOT, ALT, SPALT and REM is corrected by the concepts of MOT and SPALT whereby **LINCOS** or the Infinite and Infinitesimal Coaxial System typical of any RU or TRU becomes their limits. Limits enable us to provide a positive objective approach and demonstration, which in turn enables us to provide all the succeeding means to show and present TRU. With a three dimensional frame based upon mlt, or for TRU, Á, L, and T and their set limits and distribution by TRU and ROT, all that is otherwise needed is to Standardize their AXes, **SAXT**. This is achieved by the relativity of observer observed or S-O from within, or outside of TRU, and a set, or sets, of presentation. The full benefit of ALT and SPALT can be utilized.

PRUMOT

TRU, TO, TRIT, TI, TÄ, SUT, Human, ALT, ROT, Dimension, OT, VAPO, Á L T, Geometry, SCARLS, Universe, Relative, The, LATANO-S, ALT^{n,} LARP, SPALT, LORVA, POS, LADAC, PRALT, ACC, Center-ing, Cosmos-ology, Model-ing, LARP, POCT, LATSNU, LOCU, AX, SUROT, LANOFAC, LORS, LAPDAT, LEXP, TANCOW, SETLAW, REDSO, CENU, LARFOC, LOMPEX, LANEAD, LACEA, LACFUS, LAPAD, LACAD, LAWLIT, LADIM, POLIW, FOCIT, COIL, DE, CORDE, VINCU, ITRU, OET, REFACT, TOR, LORIEN, UNA, LACT, LASCER, ROMEA, TIEX, TIZ, MERPOT, FOLM, LOREL, DANC, LEXAMT, LATRANG, LOCOS, CLOD, MP, OMAFLEX, LEPCO, LALTS, LADSAC, CAD, LACBE, LADROT, LAPGAL, SROPU, LATANG, MASPOV, CALT, SOSTIT, DOTRIS, LICUFT, SCAPON, POCT, LEP, REDPRI, LANEG, UCT, LAPLIN, MEDPOL, LOST, LASDIC, LIDU, LACOMU, LORFA, PEMAT, LACDOT, SOP, COPMOT, COMWARP, STORD, LORDSVE, LAX, LASAT, SALT, POD, LASOT, LASD, LAALTCA, TIMOST, DEPOSU, FADIM, LAMARS, LAFUT, CORTEM, MIP, MOCBIREM, PROREC, RILT, LAMORD, GASUP, PECT, LART.

PRALT

LARF, PRAD, TIT, MOT, S-O, TÄ, A L T, VAPO, Past, Present, Future, Large, Medium, Small, Subject, Real, Object, Dimensions, Dimensional Form, Coordinate-s-ion, TRIT, TI, SUT, Universe, Relative, The, ENT, OM, TRUP, TRUF, KOT, PRACALT, MADAS, ALT^{n,} LATAN-O-S, LARP, ACEXIN, MOT, SPALT, REM, LORVA, LADAC, Duration, Length, Time, ASOPOR, PAXSU, SIL, RUDE, LIVDAP, LOSCAD, RUTHES, LORF, PRACAL, LADS, LODPAL, FITALP, POCOL, LOR, LIDU, LADIS, MP, CODIT, LATIM, LATIMS, LALTSP, CENU, LARFOC, LOMPEX, LATIRD, LATLIR, LALTIC, LAWLIT, LACOX, PRAD, LADIM, LADIMALT, TELAT, SALT, LOLT, RUC, COIL, DE, DIME, OTA, OTALT, OA, OALT, LADENT, PODAF, OET, TORM, TOR, LEOIEN, VALTRI, LASCER, CODIT, LOREL, LACFOF, MEDPOL, LARVPI, DANC, LADIR, LALTS, LALTEX, LAPGAL, CALT, DOTRIS, LALTI, POCT, LEP, REDPRI, LANEG, MEXSPO, MEDPOL, FALT, AG, LASDIC, SOLTREAC, LAALTSU, LALT, LACDOT, EXOPT, LAVOF, SALT, IAP, MAPT, MIPT, MEALT, LALTCA, DEMPAC, LARIMD, PRODOM, LIRMAL, ALTQUA, SAXT, ALTCO, LUSECT, DIMA, DIMST, PALT, LADEX, INCOT, RILT, LAPLI, LART, SLADTI, LASFAL, SROT, CUMALT, DEDIPA, Temporal 147, PRADEST 157, PRUTAN, Timewarp 157, Standard-ized-ation 1-330, Elevating IA from SUBIT to TIT 1-404, COL, Á exists in different ways, levels, kinds, 1-125, PRACALT, PROCDAT 330, PADALT 137-3a, POCOL 225-2, LACOSPEC 148-4d, ILS, ILT, ILM, LALTCAP - 320, PRA 1-125, FITALP, TÄ, and Fit.

PRUSP

LASPHER, LIPS, LATANOS, LOPOB, TRUP, POCOL, LASPHERT, TRUN, TRUF, LATIM, LATRID, CORDE, TASPA, TOR, TITAV, CODIT, LADROT, OBER, SADECO, SADECOR, LEP, LOTISP, SOSTIT, LACSOR, SCAPON, LAPSU, LODIME, LASOT, ALTQUA, PRACAL, LANITS.

PRUSPT

Sphere-ical-icity, Geometry, Solid Geometry, TIT, TIT-TO, ROT,MOT, S-O, TÄ, SUT, TRIT, HUMAN, CENU, Universe, LARF, ENT, OT, LATANO-S, LARP, MOT, ALT, ALTL, SPALT, LADAC, PRALT, DYNO, Sector-s.

PRUSPALT

ALT, *ALT^{n}, ALTCO, ALTL, ASOPOR, ATIM, Axis-es, CENU, CODIT, COIL, CONTIT, COPRINT 133-6C, DEDIPA, DIM, DYNO, ENT, EXATRE, EXOPT, FAPERI, Geometry, Human, KOT, LACFOF, LACOT, LACREM, LADAC, LADIM, LADIS, LADPOR, LADS, LADTO, LALTI, LANCOP, LANEAD, LARAT, LARES, LARF, LARP, LARVPI, LASCER, LASDIC, LASPAL, LASPHERT, LATANO-S, LATANS, LATOM, LAVTUS, LAWLIT, LED, LEP, LEPROM, LICUFT, LIDA, LIDAP, LIDU, LIDUC, LINCOS, LINLOC, LIRMAL, LIRNME, LOCOF, LOCOS, LOMPEX, LOPEL, LORIEN, LORVA, LUF, LUND, LUSECT, MIP, MIPLID, MOS, MOT, OBER, OET, OM, OT, PACEPAN, PADALT, PARS, PEC, PECT, POAT, POCOL, POCT, POLIW, PRACAL, PRACALT, PRAD, PRALT, PRISPA, PROCDA, PROCDAT, PRUT, RAD, REDPRI, REFACT, REFINT, Relative, RELOCI, RILT, ROT, RUPKAL, SAKT, Sector-s, SEVI, S-O, Solid Geometry, SPEL, SPEV, STOD, STORD, STROCO, SULF, SULP, SUT, TA, TÄ, TASPA, TELAT, TI, TINI-144-6CD, TIT, TITSOL, TO, TRIT, TRUF, TRUP, Universe, VAPO,

PREM

LARF, TO, TE, TIT, VALUE, PROPORTION, VAPO, Special, Life, Human, Life Chain, God, Á L T SECTORS, Meaning, MOCBIREM, DAOST, ES, TES, Relative, The, Distinction, Identity, RUPKAL, MATES, LOPAN, OM, TRUP, TRUF, PRUNT, LANT, Langage, EC, MADAS, LARP, LATANOS, LORVA, ALT, SPALT, POS, STA, PRALT, Distribute-ion, Equal-ity, LOPU, LOPUT, LINTAC, LOPS, LOVTRU, REMU, FER, LOTEP, ESATO, LOPS, LORD, OPT, REMI, LECAS, RAD, LINE, LASD, MEVIN, TAPOR, SAV, VOL, POV, LOP, TIPS, CONES, LEP, LEM, ROMT, COREMT, SEP, ROPU, LEP, LAVARD, PRAV, LOV, LAPIPS, OV-T, PROV, VOT, LADOR, LITEOA, FAV, ANKEF, REMLEV, FODEP, REMP, REMT, REMIT, DREM, MOCBIREM, VOT, MISDOT, TOVOP, LEM, LOREM, PLACREM, SARUE, LIREM, LOPS, LORPS, PED, REMUC, LANSAR, FER, LORC, MOPOT, MOPOTI, LANMAR, CORST, LOPELT, CREM, REMAP, REMPRO, LEDIOR, REMAS, LARED, LEDIT, 1R, 2R, 3R, 4R, 5R, 10R, 100R, ∞R, REME, LAMUT, DIR, LAFI, COMREM, GRA, GRAP, GRAPTO, LARMED, LAWOR, REMS, LORNPU, COREMS, DIREM, LARE, LADREM, STRUV, FUV, SUFUV, SUFPUV, REMSET, POMVAS, MAGS, LAPDOR, TOPNEC, LARPCO, LACREM, REMDO, LADER, LASFOR, REMEM, LADOV, MONT,DEDIPA, MOASM, LOPROT, RACTAW, WOFREM, RIM, TRECON,

PRUV: TREQ, Economics, Business, Capital, Commerce, Money, Value, Purchasing, deity, OV, REM, and TRU.

SECTION IV

TOPOLOGY, PLACEMENT OPERATIONS, POS POSITIONING in SPALT and MODELS of TRU

CHAPTER 40

INTRODUCTION, KNOWLEDGE, DERIVATION and DEVELOPMENT of POS

POS, POSITS, DEFINITION, PRECEDENT KINDS, KNOWLEDGE, PROBLEMS, PROOF of POSITS

The acronym for the whole topic of ordered placement of elements of TRU into its model is **POS**. In order to best demonstrate TRU, especially its ES and CES, we show their posits in the most refined models. This is also used to show REM and give meaning to ROT, MOT, ALT, SPALT, and TRU. To show TRU and TIT in this way we adopt the means and method of a positive objective approach or **POA**. This requires and entails the observation of the observed TRU, TIT, TRIT, and their objects, OBT, SUT, and ROT. Continual correction is needed to approach, achieve and maintain a 1:1 identity with TIT. This is especially important for TIT unless we recognize and correct the exceptions made in the operation. This is what is done for all objects in a relativistic, observer/observed state. So that **POSITS**, the placement of the content of TRU in its models, are a manifestation of the Á mass energy, relativity form and function both in and of TRU. With POA, SPALT, and POS we can reproduce TIT in ways that are VAPO and thereby highly accurate, reliable, revealing and beneficial. Even so all the SCARLS, concepts, principles, laws and mechanisms, which apply must be properly considered and applied to correct deviations and problems in TRU's reproduction.

Fig 1-40 **PRIMARY MODEL for POSITS, PLANAR FACE of SPHERICAL SPALT of TIT**

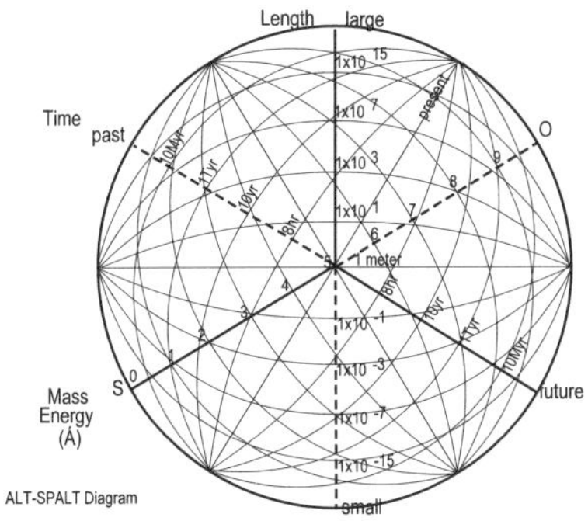

Here is the standardized model of TRU, SPALT, developed in Section III. This forms the focus and basis for much of POS, as it does for IPTRU. SPALT acts, when viewed spherically, both as a guide to help us understand the princi-ples of POS and TRU, as well as the frame for regional placement of all of TRU. To **POSIT** is to put in position or place. For IPTRU it is specifically used to designate those ES, CES, aspects, and all of TRU that are placed in SPALT or associated MOT. **POS** is this topic and Section IV-A of IPTRU, that explain the concepts, principles, laws, mechanisms and all that help provide a positive objective approach to TRU. It is based upon reducing TRU, as TIT, to its object, representing, and as a model, being used to provide the field or solid as identical to TIT as much as possible. **POSITS** are the positioning and placement of items of TRU in SPALT and other models of TIT. **POINT PLACEMENT, PP, or POP** are terms used, somewhat more precedential and generally, to signify posits that are not strictly of TIT or in MOT. The importance of posits can be simply proven by order in generaal. How necessary it is to "remember where it is", and "a place for everything and everything in its place". Without position, as in navigation, we are continuously lost. POSITS are what it takes to make the observation of TIT real. Because of lack of position and resulting disorder and confusion humanity fails and suffers. This is why POT and IPTRU are so greatly needed. It is only when we separate and treat positively what is actual in TRU, that we can develop, as with positivism, and have an actual, productive and highly rewarding means to TRU. TRU and KOT are derived, to a large extent from science, religion, philosophy, God, and TREQ. Science in turn, as all, must be predominantly oriented and focused on TRU. Because existing philosophy, science and religion are not specifically derived and developed from TRU does not mean they do not have a great body of knowledge that is beneficial, for TRU. It only means that the knowledge they present must be modified and adapted to TRU.(VAPO) The undeniable philosophical, scientific and religious truths all are essential, with proper modification, to TRU. This also applies in general to much of knowledge, definitions, words, and meanings, especially when adapted and reduced to TRU, ROT, SPALT, REM, and POS. .(OC,EC,ECT,NEC)

For every great subject or block of knowledge that has been unknown, ignored, or fallen from favor because of error, misuse, or neglect, this void will tend to stir in the imagination of some an inspiration and zeal for the fulfillment of this vacuum. It will lead to the discovery of great missing links and topics with implications for all, TRU. Just as 'nature abhors a vacuum' so does human reason and understandings strive to close the vacuums left by unknowns and what is wrongly known. This is a great basic natural inclination typifying the mass energy, Á FAF dimension of mankind. It is an accelerative expanding salient of human growth and development, the progression of highest levels of SÁ in TIT, TRIS and TRUS. It, and the associated equilibrium of density in TRU give us one of the strongest proofs of TRU/RU,AU, as well as proof of much of the aspects, FAF, parts and processes of TRU. Thereby it is basic to TRIT, ROT, MOT, SPALT, POS, STA, HOM, and ACC.

Don't underestimate the common and known. Only accept it conditionally, after the necessary correction or exclusion of non TRU. Just as you can not underestimate the intelligence or TRU potential and function of the average person, don't over or mis-estimate the less common or unknown. For these are often only the more common and known in other forms. The less common and unknown in being mysterious tend to become exaggerated or skewed. Only with the coordination, order, and proportionality of TRU and its perspective can that which is seemingly certain which is not, and that which is seemingly uncertain that is, be properly recognized, analyzed, and comprehended.

Change is at the heart of both the relativity and absoluteness of TRU as well as its representation and modeling. It not only shows much of what TRU is, they show how we must be open to fresh visions of the world if we are to fit its pieces together. It is the collection and adaptation of these and many other concepts, principles, laws and mechanisms that explain what TRU is, how it functions, what its actual parts are, and how they can be reproduced that enable us to accurately fit together all of its pieces. One of the greatest proofs of TRU and TÄ is that of fit. What fits? All are guided by the limits of each CALP, ES, and their POSITS to , 1. best serve us versus how much is unnecessary, 2. are molded by the laws of parsimony, 3. explain the limits of various bodily and mental functions, 4. show the bond that exists and participation of nature, habitat, and non human elements in TRU. 5. emphasize how nature molds humans, humans mold nature, and humans mold themselves. 6. emphasize the TRU meaning of limit and fitness. 7. help to determine what is and is not TRU, 8. help to differentiate their ESINS and ES from non ESINS and non ES.. 9. collectively demonstrates the center of TRU. 10. reveal the interaction and bonding of subject, object, and TRU. 11. reveal ATO by the interaction and integration of means and extremes of Á, L, and T in TRU. 12. explain the converging Á, L and T FAF of TRU. 13. show the quality, quantity, and measure of the mass energy, Á, FAF, dimension and continuum of TRU. 14. help understanding by reinforcing knowledge and experience to guide our acts, action and ACC. 15. prevent loss through understanding the optimal management, control and operations of TRU. 16. fit and limits are similar and closely related to each other. They reveal how FIT and FITALP by DAOST are similar and closely related to adaptation. It is their differences, distinctions, and greatness in TRU that makes POS so significant. Another, related, yet newer and specific significance of fit is in POS, how we fit all into a model to demonstrate TRU. We can see both the mechanics of fitting, as well as the refinements necessary to provide the kind of precision and fine tuning TRU warrants. Fit, fitness, limits and FITALP, are added concepts whose explicit meanings and discrete use, bring the actual reality of TRU to life. They are the magic ES and CES of our being, TES, need to show the virtual actuality of TIT.

There are, among others, two highly accurate and beneficial methods of developing the position of parts and processes in TRU and its model, SPALT. These include those of qualifying each posit by how much it consists of each of the three major dimensions. How much each dimension contributes to it as it is a part or process in TIT. The second is how we quantify the amount in each dimension and as with analytic geometry place them in the field of sphere. The first is much simpler than the second, as qualitative posits of only three dimensions are given comparative values is simpler than less accurate quantitative measures of the gradients in all three dimension, plus their combination. For this reason the refinement of posits is crucial.

To best show all aspects of TIT and TRU, and operations involved, more than one model and variation of SPALT is sometimes helpful. This is resolved by return focus and reconvergence on SPALT and TIT. It requires the best limits, correlations, correspondences, and superimpositions. In Christ's words from the epistle of St. John, he "was before Abraham". By this Christ was trying to get his audience to understand that he was not only of this apparent lifetime, but through and through, past, present and future. This is key to Christianity and TRU. To be TES one partakes of its entire mode, REM, REMOD, and TIT-TO. PRALT is to a great extent clarified and corrected when we elevate our IA and EPI from their own sense to that of its TRU sense. As the form of an object is revealed and represented it is exponentially identified. For TRU this is: **EXATRE** = Exponential Identification of the Form - Ä - of TRU as Revealed and Represented.(RORRE, EXEL) When we reveal and represent the form, aspects, and makeup of an object and TRU it rapidly comes together. This is like a jigsaw puzzle, where toward the end it finally and suddenly makes sense. EXATRE is highly important when applied to the ROT. This is how TRU is without TRE and TO. With closure revelations of exponential proportions identify TRU in all its glory, beauty and greatness through our predominant powers of S-O and Á FAF in TRU.

Proof, verification and justification of the use of posits in SOID, ALT and SPALT does not, nor in actuality is not, dependent upon their occupying only the exact point noted. This can be seen on geographical maps, where nations, states, counties and even cities occupy more than points. Likewise their reference on a map is not absolutely exact, much occurs through direct and indirect influence. This is similar to the status of the Á, L and T dimensions which exist more as spectra and continua than as lines.(ACEXIN) This is corrected by making the map or model ever larger and more precise with greater detail. However this does not detract from the ability of map makers to represent and posit localities. Nor should it for TRU if remembered and properly interpreted by TRU. It is signified by: **JUPOS** = Law of Justification of Posits. JUPOS denotes, signifies and qualifies the value, yet limits of posits. It justifies their use, especially when best used. The use and value of a map to provide a guide, way or means of enabling a traveler to find their way is much the same as what SUT, ROT, ALT, SPALT,. REM and POS provide for TRU. ROT, MOT and SPALT. They give us the frame and means to identify and use knowledge, especially of TRU. This is mostly a function of, the size of the map, the amount of knowledge, ability to see detail, and degree of VAPO.

The amount of detail, and thereby precision and totality produced, is largely a function of the field or sphere into which it is posited. This is called: **LADPOR** = Law of Detail and Precision, is a function of an object's representation, modeling, and size. Thereby the features, notations, detail, accuracy and precision of all, both model and that to be posited are all, when increasingly exact, a 1:1 replica, of this region or aspect of TRU. It is the necessary means to total identity, knowledge and OSO, for each and all of us. This is called: **LOSO** = Law of Total SO, or Object-Subject-Object's combinations, or the reversible fulfillment of Á mass energy FAF in TIT. To be increasingly exact and perfect it is the necessary condition of Á, or: **LÁ** = Law of Á, the law of the identity of subject and object. LÁ is the certification of what the mass-energy dimension and state of TRU, Á, is and does as it provides an ever more adequate and sufficient, precise and qualitative reality in life, the identity of subject, object, and TRU. It is how Object, Subject, Object or O-S-O come together as one in TRU. How this form and function are given act only through the form, state, dimension and power of TRU, its Á dimension. This is called: **LOSOA** = Law of Object-Subject-Object to Á. LOSOA is how we can have a proper, highly successful and perfected production of whatever O, S, and O yield. It is to that degree of positive and true worth. This is for one and for all. So is Á forthcoming and of great significance in the actual form and function of mass in humans, when of TES and TRU. This is what is such an increasing aspect of TIT, vital and specific for its own being. This occurs as it reproduces itself, in smallest part or in totality. In this way we can see and understand Á and TIT, how the posits in ALT, SPALT or any MOT or ROT of TRU are the striking examples of LOSO, LÁ and LOSOA.

PROOF of POSITS, PRUPOS: SPALT, TE, TIT, analytic geometry, typology, VAPO, meaning, POT, DAOST, ES, the, COMPUTERIZATION, Á L T, MATES, LOPAN, KOT, MADAS, EC, LANT, PRUNT, LARP, LATANOS, ALT, REM, LORVA, MOCBIREM, PRALT, Coterminality, Direction, Location, Position, Placement, POS, JUPOS, LOPOT, LOPELT, TINS, PRUTRU, set, size, LOSO, POVIN, PRODEP, R Real, FARP, Log, Positivism, PROKEP, RECIP.

ORIGIN, DERIVATION, DEVELOPMENT, DETERMINATION, and DEMONSTRATION of POS and IA

A positive objective approach, despite its great need, advantage and use creates its own bias. For all that is not TRU in some, or any, way not TRU will be, to this extent, inactual, illusion, figment, and an appearance that is neither real nor ideal by TRU. It will like COL stand in an opposing position or bias away from and other than TRU. On the other hand, being of OITIO, also in TRU, this will naturally tend to correct itself in TRU. POA provides a standardized format for a better way to reach closure of O-S-O and purpose in Á. It is this fusion back into TRU that is so much of the text, Homation, ALT, SPALT, and their positing. It is a positive approach that returns us infinitely back into the immanent position within SPALT, convergent on a 1:1 identity with TIT.

For the last million years language has been in an ever changing state advancing and accelerating toward greater quantity, complexity, facility, and proficiency in human communication and development. The changes and quality have been surging back and forth in the competitive group dynamics of natural selection. This is shown by predominance of the Á mass energy continuum of TIT. Its versatility to meld meanings and definitions to suit needs and quality resulted from the intensity of ideas, strength of meanings, selection of concepts and words, communication, use, and actuation of values, or REM, by ever-more discerning groups of peoples. This has been the naming of ideas and concepts by sounds, symbols, and words and their changes and refinements. **LAV** = Language Versions, **LI** = Language Inversion, **LAREV** = Language Reversion, **LIR** = Language Inversions and Reversions. They invert and revert in the many languages of the past and present as they supply each group, state, nation, culture or civilization with a means to selectively succeed in life. In this way they continually incline and adapt to TRU. Hereby they are of the input and output of TRU, OITIO.

It is not surprising that words fail us. They are neither precise, valid, TRU, proportionate, nor formed into a whole. Words and languages represent an initial, primitive, and archaic progression to make everything discrete, distinct, and provide meaningful differentiation of all. Although language, or the use of words and symbols is a major, crucial and latest aspect of our being, it is one of the least ordered, completed, and refined. It has, is and will undergo continuous, massive, and accelerative improvement, especially with TREQ and TRU. At present, however, words and language do not give the adequate concrete static or dynamic elements for posits that are either necessary, ideal, or possible. In use this failure is magnified. Thus, great latitude in precision must be initially allowed, remembered and corrected, if we are to have the kinds and degree of POA, POS, and demonstration TRU requires.

As good as examples and analogies are, for representation in general, they usually make poor posits. For they lack several of the ingredients of good TRU posit. It is usually easier to remember, and much easier to utilize, a principle or a concept, law or mechanism that has a name than one that has no name, or become a word. There are potentially thousands of unrecognized and/or unnamed ideas, concepts, principles, laws, mechanisms, ES and CES that remain. There should be, at least for the greatest ideas, concepts, principles, laws, mechanisms, ES and CES of TRU or new names and a new terminology. It is much easier to remember these concepts and terms if they can be related and thereby posited. These two, the naming of unknowns of what is unknown or improperly known and their relates and posits as they exist in TRU are the key, cue and jist of all knowledge, human life, and TRU. This is the beauty, greatness, and power of signification by names and LANT. It largely justifies and proves TRU, ROT, POT, and IPTRU. **FOPIA** = Fallacy of Particulated, Partial, Pinpointed, and Peculiar IA. FOPIA is to a large extent both obvious and natural. More adversely it arises from our failure to alter apperception beyond an initial state. It may be as we see it, yet its meaning, how we interpret and use it are far different. They should be they are in TIT. This is called: **FASIM** = the Fallacy of Altered State in Meaning (by IA). FASIM is the state of an entity, anything in interpretation and use that is far different than what it is as seen or sensed. It is meaning, rather than appearance. It will, or should, tend to TRU. This is: **STIT** = State in TRU, and **STISM** = State in TRU as interpreted and used in TRU by meaning. STISM is greatly altered in IA from what may be sensed. **PP** = Point Placement, **PIA** = Point Placement Instantaneous Attention. PIA is how the development of IA to PP, and PP to IA, amplifies and accentuates the derivation and development of each. It denotes how far PP goes to bring IA to TIT. It shows the correspondence of IA to TRIT and TIT, AT and ITRU.

The difference between posits in nature and posit in length-time is formulation. Posits in A, L, T are produced by the set form of TRIT, MOT and SPALT. The key to this formulation is the reduction of infinite and infinitesimal length time. This is called: **ILT** = Infinite Length Time, **RILT** = Reduction of Infinite Length Time. ILT is what specifically exists in TRU. It is its extension infinitely in all dimensions in a manner specified and typical for itself. This largely depends upon its predominance and adaptation to the dimensions of the environment in which it exists. For each finite there is also an infinite aspect. What is posited is, in TRU, not at a point, but with body, whose extension from infinite to infinitesimal is, in ways, similar to that of TIT. Thereby, as with REM, the POS in ALT and SPALT are not absolute nor ideal. They are more the practical averaged centers posited as a point in the field or sphere. This is the ILT that exists not as extremes by periphery but the extreme infinity all can be taken to at any sector of ALT and SPALT. To this also RILT applies. For although we can never make our points of infinitesimal focus, we can make them sufficiently small or delimited to provide more than what is needed for their use in POS, ROT, POT. Of all we posit in life will, to approach actuality, be in either, AU, AMTRU, EPI, TRU-TIT, ROT or MOT. Each of these is multiple, and all are largely indefinite from figment or illusion, unless entirely of TIT. **LAPMET** = Law of Posits in EPI, TRU, ROT and MOT. **LAPMETU** = Law of Posits in EPI, TRU, ROT and MOT by Purpose, End and Use. The return to TRU requires one to concentrate and use ECRO, VAPO, and SCARLS. For that to be posited and used must be brought to TRU, or at least superimposed in SPALT to posits in TRU. The answer lies in the Return to TRU, **RETRU**.

Everything that has meaning, especially TRU meaning, has a place in a full representation as well as presentation of TRU. Yet because this must be limited by the capacity of each, the observer, and associated factors, it will have to relatively depend on and be proportionate to them. Since all that comes to our attention, even when our attention is, as usual, neither pure nor TRU, the exercise of separating and applying all ENS, S-O, POT, Á, TRU, TIT, TRIT, OBT, SUT, ROT, MOT, SPALT, REM, and POS is not futile. In fact, it is very crucial to our existence by the great necessity of TRU and its POA. It is to elevate all ideas of TRU to a posit that is finite and ordered that this work, IPTRU, has been deliberately and vigorously pursued. It requires increasing massive comprehension. This calls for, **TORP** = Total TRU Representation and Presentation and **DITORP** or the Differentiation and Integration of TRU Total Representation and Presentation. DITORP includes the continuous distinction necessary of all ideas, CALP, standards, forms, and actual existence as they present, represent and are TRU.

The degree to which humans attain any satisfactory knowledge and comprehension of their world, and thus TRU, S/F and sufficiency, is a function of the degree to which we attain sufficient posit. **LOPSUF** = Law of Posit Sufficiency, or the degree to which POSITS are not only sufficiently representative of their object in TRU, they are sufficient to the overall field, sphere or media. In contrast there is: **EBIP** = Error and Bad Basis in Point, or what is used as POS, but contains one or many defects, by not being TRU, not an object for posit, or errors in placement. **PESAD** = Posits are to be made largely and precisely only by ES, CES or aspects of TRU in SPALT. These by using the fullest ACEXIN expanded dimensions, and continua that are primary for TIT-TO.

DETERMINISM	ERRORS and LIMITS	ADVANTAGES and OPPORTUNITIES
TRU	In Actuality Only Applies to TRU	Can be Selectively Refined by TRU
VALIDITY	Is often used indiscriminately with more/ less invalidity, Limited by validity in TRU	When based upon valid identity and relations can be highly reliable and useful
PROPORTION-ALITY	Is used without regard for proportionality is limited by the proportionality of TRU	Helps to show proportionality in TRU and the precision and benefits this gives POA.
TOTALITY	Is used without regard for totality, is limited by the totality of TRU	Helps to show the totality, focus and unity of TRU. These limits set the basis for POA.
DISTRI-BUTION	Can not be used for all TRU	Not Uniformly distributed in TRU Applies only to the sames or TRU or those Differents that are modified by the sames
FINITE and INFINITE	Can not be applied to the nth degree or infinite of TRU	The nth degree and infinite of TRU provide an endless opportunity for expansion, our life
COMPLEXITY	Does not usually exist in singular or simple form and function, highly complex	Its complexity is not defeating only chal-lenging, and makes achievement more glorious
RELIABIL-ITY	How much and how well provided, Capability of those and that providing	Supported by a high degree of accuracy and may be increasingly used and depended upon
CLOSURE	May not be possible or only somewhat so. Can't be done without totality divisioning formulation, coordination, and ordination	Is the difference between chaos and a cosmo-tic rendering of all, all that is of this special universe of best human life.

Table 1-4O **DETERMINISM: its ERRORS, LIMITS, ADVANTAGES, OPPORTUNITY, RELATION to POA**

In order to justify and validate POA, posits, and much of IPTRU, it is necessary to analyze the reasons why determinism, especially that used for ROT, MOT, SPALT, REM and POS, is both so great and erroneous. It errs, not so much because it is invalid or wrong in principle, it errs because it is either defined, based, and used wrong, or fails to meet all that modifies it for TRU or the given. Determinism when defined by and used in and of TRU and meets all those factors that modify it to TRU, is the great basis for representing TRU and showing how TRU works. This is in part achieved by MOT, ALT, SPALT, POS, STA and HOM. For these both set the form and frame in which we can provide distinct and discrete posits, and guide us in producing a means to discover, learn, know and show TRU.

From mathematics, geometry, Sanskrit, Euclid and Cartesian Coordinates, the reasons and causes for POSITS are derived. For Sanskrit mathematical is by one who puts or places more at mind. For Euclid "To every pair of real numbers, there corresponds one and only one point in the plane, and conversely." This applies for a sphere and ALT or SPALT, just as it does for a plane, the centered and coordinate MPLT. It may not hold without SCARL for the dynamic mass and less formative aspects of TRU. In accord with this axiom we may, very early in our construct of TRU, distinguish all that can be posited in this construct in a way that is at once reliable and valid. This is the absolute certainty of the axiomatic method for which all can be given precise dimensional determination. All that is of TIT and TRIT that can measured accurately and given real numbers. What can be given precise value to length, time, and mass, or Á as it is in TRIT, can be posited at one and only one point in its frame, ALT. It enables us to emphasize, particularize, differentiate, and make formally distinct each separate part, process, form, and all of TRU. This is especially so of all that is of its specific and partial, rather than its general and universal form and nature. Added qualifications may be needed to provide POA. With the further use of analytical geometry, ROT, MOT, and ALT we can accurately extend to SPALT, ALT, MOT, ROT, SUT, OBT, TRIT, and TIT, the solid, spherical, 3D principles needed to provide valid, reliable, successful and well represented POS of TIT.

Posits, although mostly made in ALT and SPALT, for facts, parts, events, and processes of TRU, are produced by their appropriate sequential point or area of significance. This, as by ROT, can be other than strict point posits in SPALT of TRU, including: area of influence, indiscrete or multiple meanings, posits in IPTRU, text or formalized objects, posits in other topics, and objects or entities either singular or multiple, as of human organs and systems. In this and further ways we extend our definitions, principles and uses of posits beyond foci in TRU and points in SPALT. To some extent this is natural, valid, and justifiable. However, the further they depart from focal POS in SPALT the more likely they are to lose formal precision, validity, and actuality in TIT-TO..

There is great beauty and value in radiocarbon dating, especially its refinements and how they exemplify POS. With their increasing accuracy, corrections, and examples for other precise values for Á and length as well as time, provide us with very reliable and helpful posits. In the time dimension for many kinds of items, from over 30,000 yrs to the present, they have greatly illuminated and aided the accurate interpretation of prehistory. This is an excellent example of the kind of precision data POS supplies through the ES and CES when of TIT-TO and fully represented and interpreted for TRU.

POSITS are selected and reduced by **RED** or Relative Distinction, the distinction relative to another item posited. They are guided by the demonstration of **PIP** or Proof of Integrative Processes. They are also guided by: **METAT** = Mutual Expansion Formulation of TRU. METAT is the mechanism that vastly increases our knowledge and representation of TRU through the spread of information of the relations between posits of ES and CES in SPALT. METAT shows how knowledge and especially knowledge as posited in SPALT and Homation, feeds on itself. It acts as a correcting, reconfirming, and reconciling device. This is comparable with: **LOSIC** = Law of Second Individual Correction. LOSIC describes that state in which a person's vantage even when correct is incomplete. Another vantage, especially when in strong and close proximity, helps to see opposition, supplement, correct and continue the process. The continuing improvement of POS is repeatedly seen when we discover, learn, know and envision TRU through SPALT.

By gradations of posits, and various other means of showing posits, increasing accuracy can be achieved. Expanding the kinds of symbols for posits in ALT and SPALT enables not only provision of points as posits, but figures which represent different REMS, masses, powers, or other qualities. To provide reliable posits they must conform to several rules of adaptation to TRU and MOT. These are rules that govern direction, distance, time, duration and S-O in posits. If only one or two of the measured posits are wrong, as distance, direction, time, S-O relationship, ID or others, the posit will be in error. So that posits can only be used when A, L, and T are correct both in the model, for each item to be posited, and degree of each dimension. This applies also to their higher powers, along with a controlled and TRU overview.(OITIO) Discovery of unknowns and TRU often results from certain unexplainable differences between the observed and computed positions, or what appears and what is calculated. Such statements are the scientific keys by which experiments and research open the door to knowledge. The most profound questions and problems facing us are often laid bare by POS and the insights they inspire of TRU.

CHAPTER 41

GENERAL and SPECIFIC FORM of POS

GENERAL FORM and KINDS of POS, SINGLE and MULTIPLE, WHATNESS, BREADTH, SCARLS and NECESSITY of POSITS

One standardized view of SPALT is simple and readily reproducible. The coordinate position of every posit or item in SPALT of TIT-TO is set. Yet, since we can't look on each and every pole of ALT on a planar surface simultaneously, we must seek to imagine its entire 3D configuration or accept many views. When our view is of a solid object, the 3D sphere of SPALT, there is a vast expansion in our comprehension and thereby identity with TIT. Yet, we are still somewhat limited. As stable and consistent as this model is, it can, and preferably often, must be viewed in many different ways. This is the background upon which posits must be made and continuously remembered, ECRO.

For each item, function, part, and point posited there will be specification in three dimensions, or the x time dimension, the y length dimension, and z or mass energy Á dimension. This is axiomatic by analytic geometry. In this way we can readily posit all things that are of TIT, and can be adapted to, or broken down into their A-L-T components. However, for their ready and full appreciation, the observer or student of TRU should be most intent and proficient in thoroughly familiarizing themselves, not only with the model, but must by repeated, massive positing of the elements, substance, and parts of TRU, enhance their statics, dynamics, vectors, combinations, groupings, and sets, to come to know the world in a well ordered, objective and total fashion. This should be as well as one knows their home, neighborhood, and workplace. In this way a person will rapidly develop the basis for KOT, PUREACLITY, COGNOS, ELTICMA, and acceleratively generate TES in a manner heretofore unknown. Likewise we must present the subjectification of TRU through SPOT, POMOT, METAD, and SPAT in such modular productions that similarly present all those transcendents, concepts, principles, laws, aspects, forms and functions that enable an optimal ROT, MOT, POT, and identity with TIT. Since all of these can not be posited within TRU and since their ordination and posit are beneficial, provision must be made for their use. Like visualizing and manipulating a solid geometric object, we need not denounce it as invalid or impractical because only one view is taken or the product is imperfect. We can still have one object and one subject, although the object is turned or analyzed from different points of view, and the subject varies, continually turning, to its one identity, with TIT, as all TRU subjectivity.

As we pass from TU to TO, TE, TI, AMTRU, and AU we pass from unitary certainty to the possible to the impossible. This is especially so for all things when extracted in their purest, TRU form and function, ES and CES. E.g. the year 4,000 bc is less possible than the decade, century or millennia. So does this apply to all time, as well as length and mass energy in TRU. This is especially so the more precise such posits are demanded and the less REM or TU-TE the item to be posited is. Thereby the measure of greatness in posit is not so much whether or not it can be done, or even the small degree of error that is likely, it is, rather, the degree to which it represents TIT. It is, like an artistic masterpiece, the degree it gives as close as possible, a 1:1 image of TRU. In this way future posits need not be perfect photostatically nor exact in every infinitesimal detail. Also by this yardstick we can find great satisfaction over the dismay that exists about the future, for much is known and can be forecast - by TREQ - that we customarily do not appreciate. By presenting TRU in ways, as in other than a visual 1:1 image, such as description, is necessary and very helpful.

In our haste and practical desire to attain finite parsimony we tend to deny VAPO, SCARLS, TRUS, TROS, TRIS, TO, TRUD, ACEXIN, and all that is highly complex in a massive whole, yet none the less of finite order. Knowledge, TRU, SUT, REM, ELIA, POTISU, PUREACLITY, and all must be applied. Nor can they be denied for lack of knowledge, lack of realization of true value, personal bias, personal dislike of anything that is unknown, different, complex, or a threat to their knowing all or having self identity, as balance, and the limitations of time.

Many of the entities, words and items we seek to posit are not perfectly nor statically in TRU. They may be more or less TRU and more or less dynamic, in and out of TRU. This only requires our increased vigilance to account for and reduce their differential, dynamic and exoteric FAF to TO-TIT. Sometimes it is necessary to give words, ES, and CES more than one posit to account for these deviations from TO-TIT. It is in the combination, formal standardizing, and distinctions necessary to bring these posits and frames into optimal agreement that makes this an overall and continuing task for mankind. We must not only know our world, we must be able to reproduce it. We must be able to positively identify each and all its features, components, aspects, and forms as well as parts and processes, to bring them into an overall comprehensive model in which they are optimally represented.

Pervasiveness is very important and vividly demonstrates the nature of TRU and how it is to be shown. Much of TRU is not necessarily evenly distributed in its frame. This problem is largely corrected in DYNS, DYNO and ACC. **PORE** = Principle of Relative Extraction. The more we relate all that is for posit to all that conditions and guides this posit, and the more all not TRU is extracted, the better the posits become.

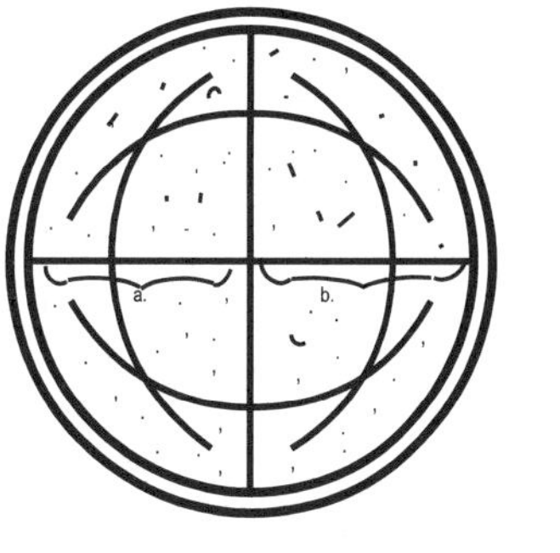

Fig 1-41 **DISCERNIBLE FEATURES for POS**

The divergence of general form of POS is evident in the features that are discernible at the 1, 10, 100, 1,000, 10T, 100T and 1 million levels. This includes their REM, Á, L and T measures, how homogeneous or heterogeneous, their roles in the form, function, essences, parts, processes and other aspects of TRU. To a casual observer this seems impossible yet to one familiar and repeatedly perusing POS in SPALT it is the opposite, easily done. It is making the way to COGNOS and ELTICMA shorter and easier. With a clear and distinct picture from TIT, TU, TO, SUT, ROT, MOT, ALT, SPALT, and POSITS and their exercise all is available for learning, KOT and TES.

The Posits of ES and TRU selected entities, senses, things, and words in ALT are determined by: 1. distinct and definitive scaling of each dimension, 2. the posit of their mass energy, length and time or Á, L, and T components, 3. the difference or variance in REM values for each of these components. 4. dimensional variance, ability to posit, or how some ES and word-things posit well and some do not in one, two or all three of their primary dimensions. This only reinforces and exemplifies the general and special cases and principles involved in POS. 5. increasing comparative determination, 6. the standardization and revision of posits to obtain balanced and properly distributed frequencies, 7. the descriptive, dissective, prescriptive, selective, and proscriptive analysis of those portions that are defective, inaccurate or imbalanced so that these can be corrected or accounted for, 8. an enumeration and accounting of word-things and ES, that are diffusely distributed through A, L, and T. Those that define the form, features, and characteristics of TRU in general or how we may present, perceive or represent TRU.

LILE = the Limits of a Language's ability to represent the ES, CES, and the FAF of TRU. Words and language are very much a part of TRU. This is largely a representative part and function in the SO or Á dimension. Many are of little or no TRU. This stems from the problem of logic, language, symbolism, formal and representative identity in TIT.

In order to make POSITS most successful several hurdles, problems and limiting factors must be overcome and solved: 1. That to be posited must be discrete, it must have great formal distinction, 2. It must not be ambiguous, but have consistent lucid meaning, 3. Its REM must be at least roughly known and indicated, 4. Narration and description must be optimally used when needed for each posit. 5. The imperfection and diffusion of posits must be best corrected.

CALP, or the great Concepts, Principles, Laws, and mechanisms of TRU, should be accounted for by posit in SPALT, and associated representations and presentations of TRU. These may be divided into several kinds: I - those that can be posited in ALT and SPALT, II - those that can be posited in other Models of TRU or its Representation, Perspective and Presentation, III - those that can be more or less posited in both, and IV - those in development, do not posit well, or need other representation. Each of these can be further classified, defined with the appropriate explanations, divisions, and degrees. For clarity, parsimony, and practicality it is highly preferable that all, as much as possible, be held to # I. Yet this is not easy, for many are not readily reduced to precise meaning, definition, distinction and point placement. Associated with this is the fact that a nominal's or word's definition will vary with its position and orientation in PACH, SPALT and TRU. This is why we try to hold all to TIT-TO.

The relations between anything and TRU may be posited or positioned according to certain definite fixed principles and laws. The posit of a thing can be no better than the degree to which: 1. the model, field, graph, or sphere is a 1:1 representation of TRU, 2.it is an excellent model with a large capacity to demonstrate both form and detail of its object, 3. the thing, object or TRU, is a specific entity, discrete and concentrated. i.e. the degree to which it is not an aggregate, agglomerate, conglomerate or lacks relativity, i.e. is concrete, or a closed set, 4. the thing, item, partial, or aspect of TRU, does not have more than one center of concentration, or point of reference. 5. the perspective of TRU, TIT, TRIT, OBT, SPOT or SPALT are largely extra TRU to constantly corrected or considered as such. 6. the definition and meaning of the word, idea, concept, thing or TRU are clear, precise, distinct and discrete. If these are vague, variable or poorly used the position, accordingly, will become vague and variable.

It is absolutely imperative that we have not only: 1.explicit, singular, prescribed, defined words for all major, great and important concepts, principles and laws, but that these be, 2. perfected, clearly posited by, and of, all of TRU. The semantics or words we use must be the best that can be developed as we do through making their objects discrete, provide formal distinction, ES, and TRU. So that the key to TRU, ROT, ALT, SPALT, POS, ACC, and POT is development and use of words, language and logical processes that provide optimal differentiation and formal distinction to best replicate the aspects, ES, and CES of TRU for POA, POS, KOT, POT and TRU.

Fig 2-41 **PRODUCERS of the POSITIVE IMPERATIVE**

POSITS are also applicable to any relationship, although mostly from the ES, CES, facts, and events in TRU. Whenever there is a relationship between two entities, as in subject - object, senses - things, or even the observer and subjectivity, a positive approach is initiated. This, in turn, is what provides posits. **PI** = Positive Imperative. In PI these relationships all converge in an absolutely necessary way, on TRU. It is only for us to determine what, why, and how PI, in a certain way, is a product of TIT and TRIT. Thereby all in IPTRU can be posited by their appropriate significance, and this sequentially and formally arranged as we proceed from the text,, IPTRU, to the observer, to the subject object, senses, things, the thing TRU, its representation, and POSITS. It is in this POA that POS is formed and can produce reliable measures of what is in TIT. The accompanying figure gives a general idea of how things are related and formed. This may be posited in relationship to other things as about one, two, or the relation, or many other ways, applying to TRU. 1. Shows us how this applies for a subject, 2. Shows how this applies for a polarity of objects as in a field, 3. Shows us how this applies to TRU, to form ALT, and SPALT that provide the model for POS.

SPECIAL FORM and KINDS of POSITS, ACTUALITY, SAMENESS, IDENTITY, SPECIFICITY, COTERMINALITY, CONGRUITY, MEANINGS, DISTINCTIONS, UNITY, FORMAT, OPTIMAL and ULTIMATE POSITS

The every day activities and problems of life can be best solved with the help of ALT, SPALT, HOM, ACC, TRU, and their posits, analyses, determinations and interpretations. We proceed from the certainty of unity to the uncertainty of partiality and infinity in the progression toward ever better solutions. This can be successful as long as it is precise, perfect, accurate, differentiated, divided, parametric, dimensional, mathematical, geometric and representative of whatever means are accurate in depicting TRU. Because we can not set an exact time and place for entities, senses and things does not invalidate ALT, POS or TRU or their methods. More specific and exact posits are possible when the entities, or the given, is reduced to their FAF and SCARLS in TRU. This is: **LOSPEC** = the Law of Specification, where and how the posits, statics, and dynamics of TRU are to be represented in its model, ALT. This is not as events, thoughts, acts, or the things we experience or perceive, but only as their ESINS are the ES, CES, and aspects of the specific and actual of TIT.

Specific posits can be derived from that of TRU which can be readily analyzed and broken down into its TRU dimensions. We tend to posit ENT, REM, ES and CES by size, or by the length dimension. Yet it is just as important, and because less apparent, more rewarding, to posit them by the time and mass-energy or Á, dimension. The significance, REM or weighting of various aspects, parts, processes of TRU or any entity in one dimension, does not have to be the same as in another. Nor does the concentration of its posit.

POSITS WELL	DIFFICULT to POSIT	IMPOSSIBLE to POSIT
Has a specified S-O, Length and Time Has only one centerpoint Describes only one entity or topic Is only TRU Singular topics or lie in fields and sciences which have well developed limits, scales and gradients	Has vague or unknown S-O, Length and Time Has more than one centerpoint Describes more than one entity or topic Is less than TRU Less singular topics which have less well developed limits	Have no or unknown S-O, Length and Time Has many centerpoints Describes many entities or topics Is not TRU Multiple topics which have poorly developed limits

Table 1-41 **WHAT POSITS WELL, POORLY or DOES NOT POSIT**

By analyzing, refining and improving POSITS, that are difficult or impossible to posit, we are aided and can continually endeavor to upgrade POS and POA of TRU. It qualifies them and helps further their contribution. To be able to properly posit that of highest TRU we must both, and at the same time, posit it specifically, generally, differentially and integratively as it is most representative of its ES and CES.

The critical significance of the time factor is shown by the term countdown. This is where necessary time, as the time needed to catch a plane, train, ship, or meet any deadline is related to the time needed to get there. To get there 10 minutes or weeks after they have left is no better than have never gotten there at all. This is what we face continuously, not only on a few occasions in life. It applies to length and Á as well as it does to time. It shows how important A, L and T can be. It shows how they are tantamount to life itself, to ones being, and the thisness of TRU. This is Position, to be in proper, TRU or POS, position, or the reverse of 'to miss is as good as a mile'. This is the significance of **POSIT**, or precision in placement and its standardization. It is being on target, the ultimate exacting purpose of POA, as POA fulfills the highest quality of TRU in Á and representation. If it is to be on target it should be within the standardization relative by TRU.

Specific, separate, and terms that are determinant of TRU are the keys to successful POS. Co-temporality, co-terminal-ity, co-positional-ity, or co-actual-ity, co-extensional-ity and co-intentional-ity are factors that express salient principles and FAF in TRU. They are especially important in POA and POS. This is how they all partake of the sameness in TRU, which is to be expected by the more or less relative, DAOST, makeup of TRU. Yet it is often most unexpected by our naive attention. For this attention, in being largely of Á, seeks to distinguish, simplify and reduce all to a discrete object or unit to match its unitary focus. In this process it creates a never ending bias in which we tend to believe, as is current in realism, the primacy and exclusiveness of ENIT and things-in-themselves. This is largely illusion. It is something we must continually correct and apply to produce KOT, POA, COAT and COGNOS. One of these ways is how we treat it in POSITS. This we do by remembering it, holding all to their difference in TRU as much as is compatible with its sameness, and in moderately allowing the entry and exit of all that are coterminal, cotemporal, or coexistent in TRU.

LACOD = the Law of Non Discrete Construct. LACOD is a product of the diffuse nature of TIT-TO in contrast to TRIT, OBT, SUT, MOT and SPALT, and its own relativity and sameness. LACOD is where correction to produce increasing perfection of REM, TRU and SU increasingly depends on the discrete, concrete delimitation and representation of the features of TRU. This is the ultimate of SU. It has been one of the greatest hurdles of TRU and all knowledge. As we place more and more concepts, things, parts, objects, laws, processes, functions and topics into TRU and as they are all more or less, we either detract from those already in, increase the value of the whole or we must give them overlapping values. To do the later, alters our standards of REM and POSITS. So that we are often forced to reduce them as new ones are added. When corrected as the whole object, SPALT, and its detail are enlarged to compensate for the reduction of each, is called: **LACRO** = Law of Compensation for Reduction by Coterminality. Its corollary is: **LACRENO** = LACRO by enlargement of the Whole. LACRENO may be both qualitative or quantitative Qualitative is seen in computerization. It is by LACRENO, if all else is optimal, that we can assure, with continuing refinement, a high degree of accuracy, reliability, and comprehensive POS, SPALT, ALT, MOT, ROT, KOT and TRU. It is in LACRO and LACRENO that life takes place. Our IA and IAT, S-O, and Á depend upon our capacity to profit by LACRO and LACRENO that we best FAF and survive.

Analysis of POSITS on the same loci, help to check the dimensional measures, similarities and differences in that posited, and refine the POSIT. It leads to an investigation of other LCD's or parameters that produce differences and the concepts, principles, and laws involved. This is done by comparative point placement. It is to be recognized and resolved, in part, by the fact that their weighting in each dimension is not the same, or may even be variable. It is also resolved by differences in emphasis, mode, and the higher powers of Á, L, and T, ACEXIN.

POCID = Principle of Concept Identification. There is a great necessity for each individual to identify in depth their revealing thoughts with existing concepts and words. The reverse of this is equally true and needed. Especially is this so for the identification that gives maximal true value to their revelation and to the greatness and complete omnificent reasoning with existing concepts, principles and words as they are the vehicles of human mass energy, Á FAF. **POCO** = Principle of Concept Organization. Each individual, in order to attain coordinated visualization of their concepts, basic principles of knowledge, TRU, universals, partials, accentuation, and REM should locate and posit each in the total picture. This will be done in excursions in, out, about, and in combinations of TRU, SU, ROT, ALT and SPALT

Words are not only representations of entities, or items that posit in 7_1, 8_1, 9_1. sectors of SPALT, they are aggregates of living actual components of TIT that may be any place in TIT. That is they are not only the little things that make up our world. They can be very great as their meanings carry them to a more precise connotation by A, L, T. They must be held to their TRU meanings and this throughout ALT. Language is the human ability and activity to separate, identify, name, define, signify, and use all. This all tends to what is important, of interest, use, and thereby TRU. Without language, definitions and meaning we can not understand the world, have a POA, or produce the posits that yield the beautiful intricate detail of TRU necessary to know that TRU and TREQ are more than a simple unity or homogeneous totality.

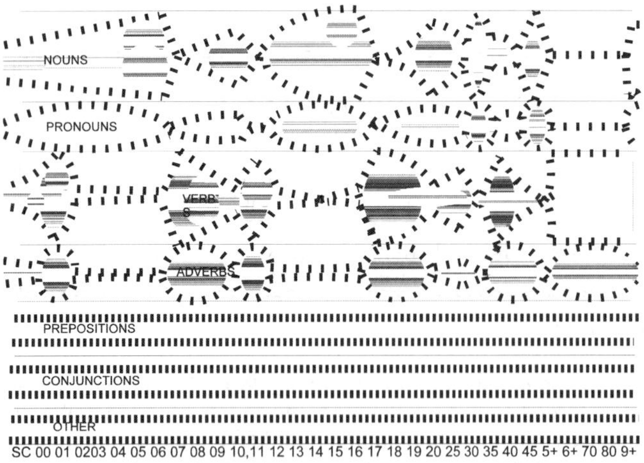

Fig 3-41 **POSITS of PARTS of SPEECH in IPTRU**
This diagram helps us better understand the role of language in POT and IPT and show the relations between posits in general, in IPT, and in SPALT. It does so by showing the role of various parts of speech in the sections and chapters of the text. To gain some idea of the role of entities, words, and ES and their posits both in IPTRU and how they are used in MOT and SPALT we need to show what part of speech they are. If they are nouns, pronouns, and adjectives, they tend to be more static and of form. If they are verbs and adverbs, they tend to be more dynamic and functional. If they are prepositions, conjunctions and others, they can be more or less either or both, depending upon usage. It is in this way we posit nouns primarily in STA or section IVB, and verbs primarily in dynamics or sections V-VIII. This is one more way we have of providing formal distinction when striving for position and order.

We must always remember that the words we use, and POSITS we make, are only relative. They are often only a pole or gradient in a dipolar dimension. In actuality the processes of TRU, CES, consist not of what is subjective-large-past, for example, but what this CES or process is only centered in this particular segment of ALT. In actuality it is both subjective and objective, large and small, past and future, all relatively and by degree, never the absolute objective, large or past. **LADED** = Law of Determination and Distinction of E, ES and CES through POSIT in modulation or SPALT. LADED is the law of DET by Distinction, how greater delineation, determination, and distinction is possible for TE, ES and CES by Posits. This is what the succeeding and progressive sections of this text show.

If one is in a distorted, awkward or cramped position, one's actions and effectiveness decrease. Here one can see the need for best position. Human life, even competition, is one of acquiring the best position. This position is one that is determined by TRU, TIT, TRIT, TRUD, TO, OBT, ROT, MOT, ALT, SPALT, and POS. It is revealed in ROT, MOT, ALT, SPALT, POS, STA and HOM. To a great extent the posits in MOT, ALT and SPALT provide this optimal position. They are further revealed in STA, HOM, DYNU, DYNS, DYNO, and ACC. Yet the overall position of all in relation to, and in TRU, is to be determined not by simple singular posits alone, nor simple singular posits in totality, but by all depending upon the observer or 1st person. This will usually be in terms of the observed or situation at hand, given, and TRU. So that not only must we strive to continually move ourselves into better positions, we must do so with all in mind as we, TES, and TRU best coincide.

POVERP = Principle of Veritable Prophecy and Projection. When you speak or think of the past and the future clearly distinguish ENIT, ESINS, TRUS, ES, CES, TROS, TIT, and TIT-TO. **POVIN** = Principle of Veritable Interpretation, or the extension of POVERP to not only the time dimension, but the length and mass-energy or Á dimension. It is also the extension of these not only on the level of their first powers, but second and third powers and in different combinations.(ACEXIN)

FORMAT for the Powers, Bullets and Prescriptions of TREQ or TRU as POSITS in MOT, ALT and SPALT

From: Ingredients that serve a purpose, in TRU
Is: What is the deepest meaning, especially in TRU
Gist: What is its greatest significance, in TRU
Cue: What are its greatest implications by TRU
Indications: What wrongs, dangers, ills, problems, and needs does it most greatly meet and solve
Vehicles: What means best present and make it available
Mechanisms: How does it work
Actions: How is it both positive and negative
Interactions: How does it work with other and many powers
VAPO: How must it's validity, actuality, proportionality and totality be applied
SCARLS: How must its states, conditions, limitations and relationships be applied
ECRO: How much must it be enumerated and remembered continuously
Contraindications: In what ways must it not be used
Precautions: What care must be taken in its use
Side Effects: What other effects does it have than proposed
Focus: What is its hooker, how specifically best used
Application: How can it be best applied generally
Solutions: What ends or purposes does it serve
How Supplied: What forms are, and can be, developed for it

This format is to assist each person in appreciating the power and use of words and POSITS of TRU. It helps connect the power of words in TRU, TREQ, or GOD. How words are much more than simple items of speech or static symbols of writing. They replicate and are highly interrelated vehicles of discrete and differentiated aspects, ES and CES in TIT. Words have great power both of the objects, ES and CES they represent and are the carriers in the Á dimension of TRU. Mostly they reveal how through ever greater and better use they provide the primary means and much of the predominant power of TRUD, HOM, DYNU, ACC, POT, IPTRU, TREQ, TRU and TIT. Each of these powers are like prescriptions in medicine that can be divided into those categories that best explain them and their uses.

In the dissection and differentiation of TIT we, more or less, spontaneously or automatically depart from pure TIT-TO. This is over and above all else that is less than TIT. This we often do in POS and IPT. For we can not know and show TIT without differentiation, dissection, and separation of the ESINS of entities, and elements or ES and CES of TRU. This is especially notable with S, O, S-O, SO, SOT, Á, SÁ, ALT, and all of ACEXIN in TRIT, TIT, and TRU. (See pg 63, THETS) It is exemplified by how difficult it is in our lives to hold all IA to IAT. To perfectly select by subjectivity from all entities only those that are objects which are, can or will be no less than perfectly TIT. Thus, POS like MOT, ROT and POT have limits of increased imperfection only to be counteracted by increased accuracy, veractiy, and practical use in TRU.(See 198, URTRU, LOLS)

CHAPTER 42

POS DIFFERENTIAL: A and THE PART and PARTS of POSITS

A PART and PARTS of POSITS, DIFFERENTIAL, DIVISIONS and MODULATION, FINITE INFINITE, INCOMPLETE DIMENSIONS, COMPLEX and EXPANDED POSITS

There are an infinite number of places and times that a traveler can visit, so also are there an infinite number of places and times one can relate to and POSIT of and in TRU. TIT on being the object for posits, can be viewed as a unit, both a unified totality, and a finite and infinite number of different places and times as well as gradients of the observer-observed. **DINF** = Distinction and differentiation of the infinitesimal, infinite and infinity. DINF certifies that there is a clear distinction in, and differentiation to be made of, the infinitesimal, infinite and infinity. We make good use of this in the means and extremes of the dimensions of ALT. It is key to KOT, MOT, REM, POS and a controlled accounting of TRU. Because we can't posit the infinitesimal and infinite, does not mean they neither exist nor can be ignored. They exist in their own way in TIT and are useful in demonstrating TIT. What is right, is a capacity to sense the expanding differential, and to do it with proportion and order in the FAF unity of TIT-TO.(VAPO)

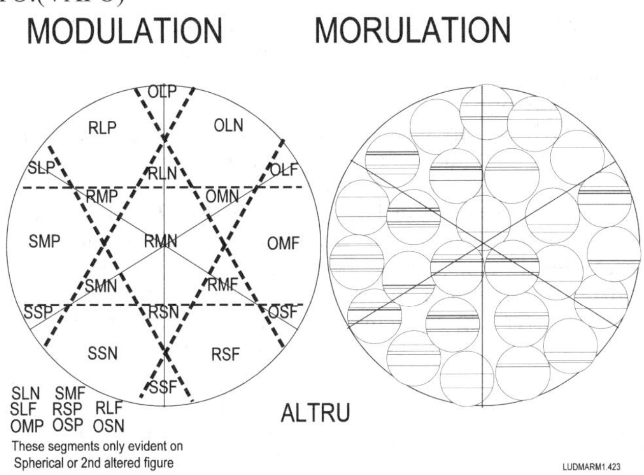

Fig 1-42 MODULATION and MORULATION of TRU FORM COMBINATION AND MODIFICATION

We posit in ALT and use in HOM only as it is TIT, in TRU. We divide and multiply, differentiate and integrate only as they are parts and become whole again in TO of TIT. We can not use any and all that comes before us to posit without due modification by TRU and the means at hand,. Without an appropriate discovery, development and demonstration of the parameters, modulation, segmentation and differentiation of TRU we can never achieve the actual and necessary systematic unity that is available. If we divide TRU, TIT, and TRIT as attainable through MOT into the three divisions of each primary parameter, m, l, t, or A, L, and T, we will have a Morula or an object or body that is concretely divided into parts in which there are developed 3 X 3 X 3 or 27 set divisions, sections or organs. As we may expand this to 10 X 10 X 100 we may have Morulas of 1,000,000 divisions, sections or organs. Fig 1-42 Shows such discrete parts for use as standards in a modulation of TRU to determine its Posits. It is our ability to morulate or break up into discrete and exact parts, divisions, sections and organs of that providing POA and POT. It makes possible the depth conception, subtlety and profundity required. This is if we are to have and hold them in mind, as we hold TRU in mind. It demands great interest and revelation that is readily forthcoming when we realize through life's bitter experiences how much depends on TRU.

Two or more objects, bodies, forms, subjects, ideas, words, if posited together are not necessarily related nor proportionately identical. This provides some of the solution for identical posits. The other solutions involve comparing them and determining what accounts for the differences in each case. This is a good test of the actuality of TRU and leads to discovery of other concepts, principles, laws and mechanisms.

Neither mass energy, time nor length, when took from POSITS are absolutely specific. This is most noticeable for time, as seen in all that occurs through the present. Many of our posits stretch from past to future in different durations. Since SPALT is highly focused on the present this produces a broad range for what is posited by time. This is explained and corrected through POSITS that: 1. represent only the center point, focus, media, and region of density or concentration, 2. is or is not valid to the degree they are dispersed or universal of what it is intended, 3. may be in terms of their length, and time alone, their area or acceleration alone, or their volume and variation alone, but always subject to the other limits of dispersions, 4. Bring together the principles, posits, model of TRU, and never over or under estimate the value, determination and representation that is or is not TRU. 5.distribute posit of ES and word-things in time PPF, that present special problems peculiar to the PPF distribution of time, and corrected through duration and half-life. Many of the posits of ES, CES, NES, and decreasingly NESI, and NENS posit through TIT, especially through its middle portions, so that in SPALT they will appear far separated past and future, yet are the single and same posit. This is called: **LAPT** = Law of POSIT Across Time, or posits in more or less past and future similarly through the present. Past or future placement, as an ES or word-thing, is either more past or future, and more of set duration. It will fall either to the one or the other. It will usually not fall half way between, as would be the tendency. For at 50 only those things posit that are at this instant present and neither past nor future. All words must be recognized in their pertinent tense: **ET**, present tense - go or is going; past tense - went, past participle - gone, has gone, have gone, and future tense - go, will go, must go, will have gone. Here we see the importance of the time dimension, when and how to posits in it, and how each must have the added factors that modify all that needed to be specifically accounted for.

A highly accurate and impressive overview of the total object is produced in well developed and technically proficient means of representation, as in anatomy and the display of the ultrastructure of systems illustrations that superimpose, combine, and fuse a number of different groups. By comparing many other such combined representations and illustrations with SPALT we can better understand the problems and limits involved as well as added benefits. It is to clarify and to enhance SPALT and posits that we can compensate for the problems and limits SPALT and POS presented to us.

To overcome the problem of limited space and time in ROT and POS abbreviations are used. This is closely related to the need to keep the number of letters in each word at a minimum consistent with a practical and effective language. To help provide better material for Posits in a MOT, complete analysis of all short words helps. Generally the simple symbols, abbreviations, and acronyms should be for what is the most needed, greatest and effective aspects, ES, and CES of TRU.

Fig 2-42, **MOT and SPALT of INCREASING COMPLEXITY of STRUCTURE and POSITS**

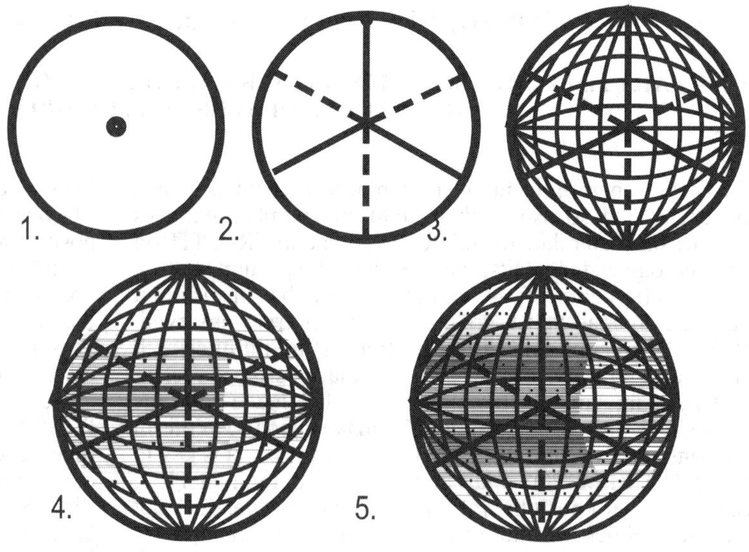

In these five figures increasing complexity of structure and detail, or integration and differentiation, are shown. It can be seen that these, although in a sense the same, are far different. Their differences point out the primary characteristics, first principles, concepts and laws of each. The degree of integration and differentiation should be in terms of the amount of density or matrix of TRU as of ALT, SPALT and SPOT. With increasing simplicity we usually have increasing unity and general understanding. With increasing complexity we usually have increased differentiation and specific understanding. With increasing simplicity and complexity, integration and differentiation there are either increased or decreased understanding depending upon: A. the interest of the first person or observer, B. their alertness, C. the relations and tie-ins or prior knowledge, D. the precision of the object, TRU, ES and MOT, E. their validity, F. their magnitude, or REM.

ECTSION = Processes of Explanation, Communication, Conversation, and Parts to TRU. The kinds of ECTSION are more than adequate to give necessary meaning. They are like pseudocode in computer science. They enable effective algorithms. They have great, deep and inspiring TRU mean-ing both to the donor and to the recipient. To be able to reduce our ideas to concise concepts, discrete terms and these to their ESINS of ES and CES in TRU is the way we can provide the necessary posits in any highly productive modulation of TRU. Even then it is through the narration, description, dialogue, and explanation that these posits are sufficiently described to have reliable posit in SPALT.

The major sectors of TRU, TIT, TRIT, MOT, ALT, and MOT are means for developing POS. The positing of all points in ALT should be made from the sectors they come under, as 47-72, 22-38, so that each sector or segment will fall into one of the major divisions, as is done decimally in organization, as in 10^3 or 100^3 or 1000^3. So that even though we often do not have the precise point for posit in TIT, we can give it a high approximation in range or sector. The potential for posits in a modular object of three dimensions is such that both the capacity of many may be vastly expanded, and the actuality of TRU well represented in beautiful intricate detail. It is important that all the great concepts, OC, EC and NEC, of TRU be thought of together, simultaneously, positively and proportionally by SPALT, SPOT, METAD and ELTICMA. Without such a foundation, form, concepts and POS, we can never show or properly know TRU. The major sectors are:

SULP = SUBJECTIVE LARGE PAST
SUSP = SUBJECTIVE SMALL PAST
SULF = SUBJECTIVE LARGE FUTURE
SUSF = SUBJECTIVE SMALL FUTURE
OLP = OBJECTIVE LARGE PAST
OSP = OBJECTIVE SMALL PAST
OLF = OBJECTIVE LARGE FUTURE
OSF = OBJECTIVE SMALL FUTURE

The principle of composition fixation, **POCOMF**, certifies that the composition of an ENIT, RIT or TIT will be fixed, at least to a large extent and in most ways. Thereby we have: **PICOM** = Principle of Interposition Composition. PICOM certifies that the content or composition of an ENIT or TIT and of their objects have set relationships which can be formally, precisely and consistently posited.

Of the concentration and dispersion of posits there are several concepts, principles and laws. **TOD** = Totality in Dispersion, Totality justified and proven by Dispersion. We enumerate and evaluate all terms which can not or whose posits do not adequately describe, or are described in/with TRU. Since differentiation is much of the essence of TIT, dispersion in totality is welcome and a means to representation. So that we must always be careful to utilize the actual set differentiation in TIT for POS, and avoid other tendencies and bias. **LINES** = Law of Imperative Inevident Parts, ES, CES, and forces in TRU. At any one time, TRU, ALT, ROT, the forces that exist and are TRU/AU are those that are providing the greatest S/F. This may or may not be known or evident at the time, but are none the less certain. **DIL** = Dispersion Limits. DIL reveals the infinite ability to posit in TRU, ALT, and SPALT. This is the great goal for mankind, if we are to have distinction and thereby fulfillment of our Á FAF and special roles and being.

DIM = Dispersion of Dimensions. DIM is the great and vital value of increasing gradients in dimensions, as in 3rd-powers of 10, 100, and 1,000. As DIL becomes recognized and DIM utilized, we attain an increasingly clear and precise solid object and model in which we can make posits. In this way we reach toward: **UM** = an Ultimate or Ideal Model, **CADID** = Concentration and Dispersion Correlation with the Integral and Differential. CADID is where and how concentration tends to the integral and dispersion to the differential.

THE PART and PARTS of POSITS CHARACTERISTICS, FOCUS, MEANS - EXTREMES GRADIENTS, SCALES, DISTRIBUTION, LEVELS,

From SPALT, or the spherical modulation of TRU as represented on a textbook page we recognize: **LASD** = the Law of Subjective Differential Dependence. Subjectively actuality, extensity and intensity or the ACEXIN of TRU, its combined A,L,T in all powers, can be no better than the degree to which we attain differentiation. That is the degree to which we are able to posit explicitly with finer and finer precision. So that as the individual, in reality, attains certainty, there must be subjective idealism. All COA, COATH, and COAT, can attain certainty by the ultimate coordination of the Subjective Differential, LASD. Yet this subjective differential can only be based upon maximal coordination. Without this it will have no focus, polarity, relativity, order, or meaning. Much of the failure of great minds to attain solutions to problems of absolute certainty has been their failure to comprehend and present all things not only with ACEXIN but integratively, coordinately, subjectively, differentially and objectively. One should be capable of comprehending all of these simultaneously and in in proper and relative proportion. This, SOID, MOT, MAPLAT, ALT, SPALT, METAD and ELTICMA, help to do in an increasingly higher degree. Such comprehension gained is signified and expressed in OM. Meaning from differential, LASD, and related concepts are equivalent to TRU and show the strength and value of ROT, MOT, POS, POT and IPTRU.

As we progress, from S to SI or to SD, to SI and R and D or to OI, we see that major features are completely altered. We are in entirely different new worlds or environments. Yet in our thinking this is done imperceptibly in short periods of time and with the greatest of facility. So that our thinking potentially scans all instantaneously. It usually does so without considering these different worlds or environments. To free up the limits imposed, by making knowledgeable and available these worlds and environments, and to train and direct ones intention to the highest, most ordered and profound pursuits,is the key to TES, life in TRU.

Characteristics of things, concepts, subjects, and objects, as interpreted through TRU, especially ALT, include: 1. How much a part of and how much not a part of TRU, ALT. 2. How clearly and exactly contained and capable of being posited within ALT and within ALT limits, 3. How clearly and singularly defined as a word or concept, very exact, almost precise, or materially precise, 4. How great, of what value, what proportion of Relative Magnitude, order and magnitude of TRU, 5. Whether used within their mathematical or physical confines; 6. Ability and acceptance of use as a mathematical function or term, and 7. How exact a measure, or a function. For these we can represent by figures, symbols or markings and can rate all according to the above characteristics, 1, 2, 3, 4, and 5-7 as from 1 - 10.

Nutrition	= 3, 7, 7, 5, 4, 5, 3	=	100,800
Evolution	= 2, 3, 5, 4, 3, 4, 3	=	4,320
Force	= 2, 3, 7, 4, 6, 6, 6	=	35,288
A nation	= 3, 6, 6, 3, 3, 5, 5	=	24,300
Love	= 5, 2, 4, 5, 3, 4, 2	=	4,800
God, common	= 5, 2, 2, 3, 2, 3, 2	=	720
COGNOS	= 7, 3, 6, 7, 3, 2, 3	=	16,156
GOD or TREQ	= 9, 7, 7, 10 ,7, 5, 6	=	132,300
Ideal Posit	=10, 10, 10, 10, 10, 10, 10	=	10,000,000

These subjective and approximate ratings are examples of how reliable posits can be. It becomes evident that they have rough value and some accuracy when compared exponentially. Work is needed for their refinement, other characteristics of POT, and their applications.

Most of what is not of us is combined and mixed with what is and is necessary to us. This is called: **PINA** = Principle of the more or less Inseparable Necessity of all about us. PINA is a prime mode of AMTRU, the mixed world of TRU and non TRU. Since all is not either/or TRU or non TRU, but a continuum from one extreme to the other. This is called: **TACON** = TRU ATRU Continuum. We should know, chose, select, and separate that most TRU. This is called: **PINAS** = PINA Selection. It is PINA and PINAS by TRU that makes what seems to be a narrow world of the first person or human life into the otherwise much greater world, TIT-TO. It is partly for this reason, that our specific behavior and actions must usually be made with moderation gauged to allow for, all that is, other than the object immediately in front of, or opposes us, and all of life, nature and the universe in general.

Yet neither rough posits nor simply giving a less than accurate quantitative measure to posits are going to produce the kind of replication of TRU in SPALT it deserves. (QUALAN) The great hurdle to overcome this in developing posits lies in the make up, parts of TRU, especially humans and individual differences. The importance of individual differences has been known for ages, even as Individual Differences in Psychology for over a hundred years. Only recently, however has genetic research analyzed the extreme variation that exists in genetic material. Of each of the approximately 30,000 genes there are fourteen versions that can be inherited. Such variation proves how difficult it is to achieve specific accuracy in posits. So that a general or exemplary approach is more practical as long as it provides relative accuracy.

PROACH = the Problem and Principle of Alternate or Differential Characteristics. PROACH are characteristics that are different, that alter the comparable items that posit the same. What are the classes of characteristics and what are the causes of this difference? PROACH is what largely determines: **POSLOW** or the weighted Posits of Parts and Entities, **POSE** and **POSEN**. For an item, object, aspect, part or process of TRU to be posited is composed of altered or different characteristics than that of a simple and equally balanced mass length time. Its posit, then, will be weighted and located other than what these three dimensions would simply indicate. PROACH holds that those parts or components of any set that are purely comparable or a part of that set, their posits should be identical. POSITS with same gradients in each dimension are not necessarily the same in MOT, ALT or SPALT. They will be the same, only if the amount in each dimension is the same. This is called: **LADPOS** = Law of Different Posits of Same or identical points in each dimensional scale. LADPOS is a product of the fact that the amount of different dimensions in each item to be posited is not the same. They may be of a great amount of time, some length and very little Á. Or they may be of any of the various combinations and permutations of these three dimensions. This accounts for much of the seeming discrepancy between items with the same posits in A, L, T. This is the difference between qualitative and quantitative posits, or **LAQUP**. Both are needed, especially together.

LADISP = Law of Distributive Posit, or the mechanism whereby the FAF of TRU reveals how posits are to be made in MOT, ALT or SPALT. Even the focus of this posit may be lesser or greater, depending not only on the spread of the part or posit but upon its concentration as well as the weight of its REM. This leads to: **LADFOV** = Law of Formal Distinction by Variety, Validity and Value of Focus. LADFOV signifies important SCARLS of POS that are necessary if it is to become a reliable means. Just as we have THES and THEO as fixed segments, partials, and ES of TRU so can we have for other polar regions of the dimensions of TRU fixed segments and concepts, THEP, THEN, THEF, THEL, THEM, THED, and THER or the past, now, future, large, medium, diminutive or small, and real of phenomenal in TRU. This may also be done for further differentials of either within the same dimension as segments of the continuum or combinations of different dimensions as in: **PL, PD, PM, PS, and PO** or Past-Small, Past-Medium, Past-Subjective, Past-Objective, and all other combinations of the Á, L, T dimensions. It will be recognized that these when taken together and in their TRU and substantial sense, by ES and CES, give a proportionate and good, yet incomplete, account of the content of TRU.

It should be kept in mind, ECRO, that dimensions and the poles of dimensions when taken alone are features of a frame, bodiless, of lines, and not as they exist in TRU, ES, CES, aspects, substance and points. Even the higher powers of dimensions, like ACEXIN, as great as they are in representing TRU, do not, nor can not be used to qualify and explain all the essence, modes, characteristics, properties, etc. of TRU. As parts, or even systematically proportionate divisions of TRU, they are signified by: **LADRES** = Law of Dimensional Segments taken as Regions exist as Parts in TRU.

LORPOV = Law of Related, Posited Validity, Law of Concurrent Validity. If two or more entities, objects, bodies, forms, subjects, ideas, or words have the same POSIT together, their general form, modes, characteristics and properties are related in certain respects. The non relation of these objects of equal posit represent their special form in TRU. These enable us to check the validity of posits, as well as the meaning of the objects, ES or items to be posited. One way of illustrating posits in the frames of ALT or SPALT is to make those that are less distinct very light in contrast to those posits that are more distinct and certain. These can be made heavier, as we do with bold in print. This method can be quantified or graded between the greatest extremes of most indistinct to most certain. Principles and methods of showing posits, include: 1. by # on margin, or accompanying description, 2. by heavier or lighter print, 3. by larger and smaller size, 4. by larger and smaller areas covered, 5. by kinds of figures representing their posit, 6. by color, 7. by any and all means of representation, as used in map making.

LOM = Levels of precision of Measurement. LOM may be increased by providing super or sub factors for each POSIT, both before, over, under, after, or in some corner of that for measure. Some of the parameters that can be measured and their rates include: a). Necessity, b). Value, c). Clarity. Many factors increase or decrease the various levels of measurement. With refinement, standardization and continual upgrading they provide POS a nominal and strong base. In this abridged text, CIPTRU, these are not carried far, for reasons of parsimony and preliminary development of the topic.

Universals and partials are not two separate entities, they are largely two extremes of the same entity. This has some correlation with their size and aspects of TRU. As two extremes in a gradient or continuum they may be scaled into different levels or divisions which depend upon the total number assigned. If there are only two, then universals and partials suffice or, if more, names for each level can be given. Yet we must remember that size is not the only mode of universals and partials.

To have an accurate and perfected reduction of IA to TRU and TRU to IA, as in TRU = IA, is called: **TEIA.** By MER, the Mass Energy Relativity, and EPEX, we must be able to correctly Posit every and all items into each succeedingly lower exponentially graded level that posits in higher levels. This is called: **APDU** = Automatic Positing and Duplication of the greater in the lesser unit or level. APDU is associated with MOCBIREM. They both measure the contents of a system, organ, large division or part of TIT. In so doing the system or larger part will also have present each of its smaller parts, which can be quantified. APDU greatly resolves the confusion and irregularities of improper distribution and disproportion. It calls for the perfection of the higher, and also a continual maintenance in the precise unchanging whole, TO-TIT. A corollary of APDU, is how nothing that is barely posited at a certain level is to be posited at a higher level. This we attain from comparative REM, REM levels. Yet it applies to all sets in which they are perfectly limited. This is called: **LIMP** or the Maximal Limit of Posit. It is what can be barely posited at a lower level. LIMP is important in providing an exact posit of many ES, other components of TRU, evaluating scales as in CENS, and others where a precise order is necessary.

The distribution of ES, or word things, in time, PPF, can be guided to an extent by the following: 1. It will be more P and F to the degree it has or will be more increasing or decreasing, 2. The position thereof will be determined by maximum or mean time, 3. What enables us to be is past, 4. What our being is directed to or must be, is future, 5. If S and O as SOLT is to be valid, they must be equally past and future. This means that the potential and developing S of the past acquires more importance. These are resolved by TRIS, TROS, TRUS, TRIF, TROF and TRUF and changing Á or **ACH**. For the present Á, and even more the future, is not the same as in the past.

LADIP = Law of Distributed Posit. We should be careful to avoid the inclination and error of positing, speaking, and writing, by placing or positing various topics, categories, great concepts and laws of life and logic in one focus. They may, cover more than a point, be somewhat variable, or depend on different fields, as SPOT, SOID, SPALT, POMOT, METAD, or SPAT. More likely they are influenced by the existing bias of a knowledge that is not evenly distributed in TIT. It is located wherever we can best sense, experience, rationally reduce or research it. The explanation of LADIP is that, what is in TRU is also in MOT, ALT and SPALT not by its center only, or certain sectors, but also by its periphery and totality. This is a good test of, and can help refine, SPALT, ALT, and MOT. What is near or far, like in ALT, MPLT, POT, BAS, FAT, and NAT, can not be measured in terms of feet or minutes from oneself only. It must be measured at least in higher powers i.e. L_3's, T_3's, A_3's, in the whole of TRU, TO-TIT. What is for posit becomes concepts as they are subjectivized and objectivized by the Á FAF in TRU, ES and CES, that are then treated as items for Posit in SPALT.

To gain an understanding of the composition of TRU, as well as to gain a better understanding of the many fields and sciences that make up TRU and how they tie together, we need only to posit these fields and sciences in SPALT and other MOT and ROT. It is based upon: (1). The foci of each in TRU, (2). Their spread in TRU, (3). Their magnitude in TRU. In addition to making simple and effective posits in SPALT, they through MOCBIREM and uses as basis for comparison with less well posited items, provide an effective vehicle for POS. Fields and sciences are readily representative of TRU when they are reconciled by VAPO in TRU. When they fit into the framework of TRU we can see how well much of knowledge can be adapted to posit in ALT. How well they reveal the mass-energy, length and time in their various states in TIT. When we have increasingly reliable posits, as fields and sciences to back up the ground of SPALT other posits can be more accurately set. This tends to enhance the reliability of all TRU, its ES and CES posited in SPALT. It in turn helps us better understand the FAF, DAOST, SCARLS and other features of each field or science. Through elimination we can determine their POSIT, and what else is and is not in that segment, what they are related to and in what way. It is in both the theoretical and practical use of posits that the success of STA, HOM, ACC, POT and IPT is secured.

LANCOP = Law of the Absolute Necessity of Coordinate Posit.(29,39) LANCOP signifies the fact that there can be no orientation, coordination or order of knowledge, subjectivity, or human thought unless and to the degree we can have the satisfactory fulfillment of relativity, the observer-observed relationship or of our predominant Á mass energy FAF. This coincidence of the subject and object, or what is known as identity and validity requires CNS coordination, a POA. This depends upon how well it can be ordered and coordinated in the object, OBT of TIT. It is what provides the basis for POSITS in MOT, ALT, SPALT, POMOT, METAD and SPAT to enable a definitive POA, PI, POT, and IPT. Of the many kinds of posits there are focal and dispersed, immanent and transcendent, specific and general posits. It is not always possible to have the same degree of centering for all that is represented. This is significant: to develop the means of attaining as good a posit as possible for those that are dispersed, transcendent or general. Even more significant it is necessary to distinguish these so that those great aspects, parts and processes of TRU whose posits and objectification in TRU are so vital to its representation and ENT are fully, properly and successfully provided. This is both the great value and the weakness or failure of differentiation and posits. For not only do we have to differentiate to the degree TRU is composed of different elements, parts, processes, forms, functions, and aspects, we have to differentiate to the degree representation of TRU requires more than a single means. This is the problem or law of: **MADIP** = Manifold Differentiation and Posits of TRU. By understanding and working to apply all the concepts, principles, laws, mechanisms of POS we provide much of ROT to sufficiently know and show TRU. To have a place for everything, and everything in its place, as these are VAPO, is the cue and key to a successful human life. There is hardly any portion of the text that treats a more difficult, yet more vital, feature of the presentation of TRU than POSITS For it is in how we can best place and position all in a total MOT that we can most continuously have and hold TRU.

CHAPTER 43

INTEGRATIVE and DYNAMIC POS and POSITS

INTEGRATIVE and TOTAL POSITS, TOPOS; DIVERSE, DISPERSED, DELIMITED, and SET POSITS PLANAR andSPHERICAL MODELS for TOPOS, COMPLETE, REVERSIBLE and VARIABLE POSITS

To attain formulation, ordination and the modulation of TRU through TE and ES, they must be constantly reduced not only to their form and function in TRU, we must elaborate on them in their relations to each other and bring them to TIT-TO. This is called: **TOPOS** = Total Posits, by the constant reduction of all TIT to an objectivity that can be adapted to positive representation. TOPOS is the total posits in SPALT, ALT, MOT and ROT that SU, OBT, TRIT and TRU provide. Posits should be primarily and eventually done by TOPOS and not by the words that express them in presentation. This is the object and goal of much of the text and especially ROT, MOT, REM and Posits of these ES and CES, or; to identify, describe, investigate, explain, and illustrate these aspects and finites, and their relationships and correlations as they are in TRU. We must begin with TIT, TIT-TO, and hold our SPALT and POS to them.

To attain perfection in this formulation, ordination, modulation and replication it is necessary to proportionately present all coordinate measures in their TRU greatness. It is then necessary to totally identify, investigate, and elaborate all those ES of which TRU is composed. Neither the present extent of our knowledge nor the limitations of authors and texts enable such perfection. None the less, we can envision, show, and be sufficiently sure of a whole that is what can be well posited from our increasing knowledge and certainty of TRU. These are not only the correct and basic facts, concepts, principles and laws of TRU and Homation, but the fact and ability to reach a practical level of formulation, coordination, and modulation of TRU that are increasingly correct. This formulation, ordination, and modulation will need to undergo repeated investigation, correction, improvement, and revision. Yet much of the basic truths, concepts, principles and laws, as well as their posits in ALT of TRU will continue to improve, expand, and be of ever greater value as they have, also, only in a somewhat lesser and different way, in past knowledge.

QUALIFIERS OF POSITS

R = REM Relative Magnitude of
C = Concentration, Fixation of assigned posit C/Universality
F = Fluctuance Stability of posit always same locus
D = Dimensional Equality/Variance
V = Type of Dimensional Variance (A, L, T)
P = Ability to develop Precise Posit
A = Achieved Produced or Resultant Posit, how good
M = Meanings of Posit, Single Discrete/Numerous Vague n = (ACDFMPRV)
Mostly 1-4 except R: 1 Excellent, 2 Good, 3 Fair, 4 Poor

SUB-QUALIFIERS:

D and V Dimensional Variance and Types of
all L, M, Sm, T, P, Pr, F, Á, S, R, O
all of only one (each, or pole of dimension)
mostly one, some of a second
mostly one, some of a second and third
mostly two, little or none of the third
Equally of all three

Examples of Qualifiers
C 1/Concentration or Universality. Entity or universal will have REM of poor or no C
F. of poor or no F, will vary with kind of,
D. of poor or no D will vary with kind of,
no set V, P a function of DC1/FM, A a function of P x work x refinement x quality.

TRU is far different than what may be assumed from the worlds about us and appearances. This is shown by the FAF, concepts, principles, laws and mechanisms of TRU. Thereby in the use of current knowledge as posits in SPALT, we must be careful to modify and adapt them to TIT. Doxography is more than the glorification of beliefs. In its original and PreSocratic general sense it is this total, proper summation of ideas, concepts, principles, and laws, that cosmologically and positively graphically represent and present the world, TREQ, and TRU. Initially this is: $^{A}_{C}n^{B}_{D}$, or more extended uses:

Many qualifiers and factors can be added to give POSITS increasing accuracy, specificity, sensitivity, validity and effectiveness. Their refinement and use requires application to each element of TRU to be posited. For this abridged text we present an initial description and limited use. The jist of generalities and specifics, universals and particulars, is not so much each and all of these taken singularly or together in any combination, as great as these are. It is in the mode and method by which they can develop a concrete whole, TIT-TO. So that the corollary of this is **KUPS** or Knowledge, Universal, Partial and Special.(95)

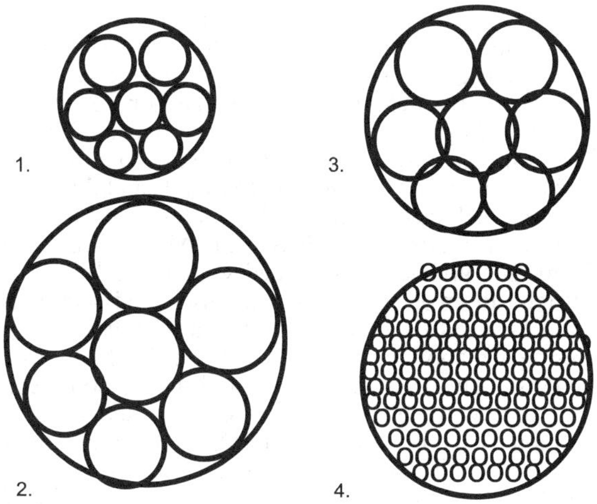

Fig 1-43 **COMPARATIVE SIZES of FIELDS and POSITS**

The size of the Posit will vary with the size of the field, map, model, SOID, ALT or SPALT. When the REM of a posit is great the size of its representation should be equally great or increase to equal its value in TRU. So that the size of the Posit in reference to the size of the total will depend upon what proportion that part is of the total to that at a set REM level as 1R with 10 parts or 2R with 100 parts. Also the size of the posit will decrease if their number increases and the fields stay the same. In Figure 1-43 we present four figures: 1. A small field, map, model or OBT in which there are 7 parts, 2. A larger field, map, model or OBT in which there are 7 parts. The 7 parts are much larger in the second figure. Yet, they both depict the same object. This is of great significance in many ways. It is, of ultimate significance in howa person as TES can become immortal, or have absolute greatness with the similar greatness of TRU. 3. A medium sized field with seven parts in which there is some, if only slight, overlapping of its parts. This is of great importance for it demonstrates at least two salient principles. The first is we have to account for the fact that our representations are not the same as ENIT or OBIT. The second is that for TRU, contrary to what exists, in the physical nature, two different parts or ES can, more or less, occupy the same space. These figures show how a part of a whole, as ES, NEC, or word-thing, when posited loses value when other parts are added. It also shows how extension to finer differentiation can be provided. All these figures show how well parts in a whole can produce an effective representation of an object when both are discrete. They all show how important it is to know and show the whole if we are to represent, present and be a successful observer that balances life between personal and total goals.

If we have a frame, model, coordinate plan, guide or field that is satisfactory, relatively valid, practical and useful then axiomatically its value will depend upon the extent to which it can effectively show and use singular, multiple, and total posits, both statically and dynamically. This is the new and TRU meaning of positivism. It is: **POR** = (law) Positive Order. POR is basic to all approaches, analyses, and determinations. It is increasingly so in DYNU, DYNS and DYNO. POR should be conditioned by S/F, ENT, COL, DOT, OM, and COR.

ERSO = Equally Reversible Representation of Subjective and Objective, Subject and Object. ERSO is the reversible or variable position between an observer and the observed. It is whether a representation is of all objects or all subjects or only in appearance or in face. It is like the subjective face of ALT and planes in statics. ERSO is related to **FALT** or fallacy of ALT, where the greatness of ALT is lost by failure to apply ERSO. This is where we are to often apt to envision ALT, not as TRU, or SPALT, but only as a field and or solid and expect it to be a complete and perfect model for posits to represent TIT. This is close to all representations, modulations and their uses. Yet to be complete they must be based on how TIT-TO is demonstrated by SPALT, METAD, and the use of TOPOS.

In the positing of all we should always remember they are in the solid construct of TIT, TRIT and SPALT, and this only as it of TIT-TO in all its FAF. Remember that in picturing, in our minds, what we see on a page, POS on the ALT-SPALT frame, this figure and all views associated with it are limited and thereby tend to error. The limitation associated with planar exclusion is: **LAPLI** = Law of Planar Limitation. Upon discovering and applying LAPLI we may make the following conclusions: All of any 1/8 sector will not be confined, in actuality and completely, to the segment shown in the presentation. This is called: **LACON** = the Law of Non Confinement of the posits of parts and processes to their presented posits on a planar illustration. These lead to: **LINEX** = the Law of Infinite Extension. LINEX is the fact that some parts and processes may be at or toward infinity, the complete opposite pole. e.g. S may be at 9A and some small item may be at OL in SPALT. The overall conclusion is that the first person, and all, must remember to think spherically and solidly by TIT-TO through all observations and POSITS.

PRALTI or the Principle, Problem and Resolution of ALT Inversion. PRALTI is shown by our inveterate tendency to differentiate - and dissociate - all from TRU. This is a part of the TU, TE, TI problem of our IA focus. It is corrected by finite concretion and expanding capacity and ability to optimally derive, develop, and determine SPALT from TRIT and TIT.

SOV = Subject Object Variability. The amount and percentage of subject or object in things, ES, and CES is often highly variable. For example government, which is generally about 60% subject and 40% object will be 95% by constitution and 5% by operation, or the opposite in operation of some departments. An easier way is to use it only for those cases where it creates the greatest problem of an otherwise highly reliable posit. **DOSOV** = Degree of Subject Object Variability, its measure, Determination, and testing. SOV and DOSOV reveal part of the range and dispersion from centerpoint, concentration or focus and how these establish Position and POSITS. These same principles, to a lesser extent, apply to the length and time dimensions, each in their own way and degree.

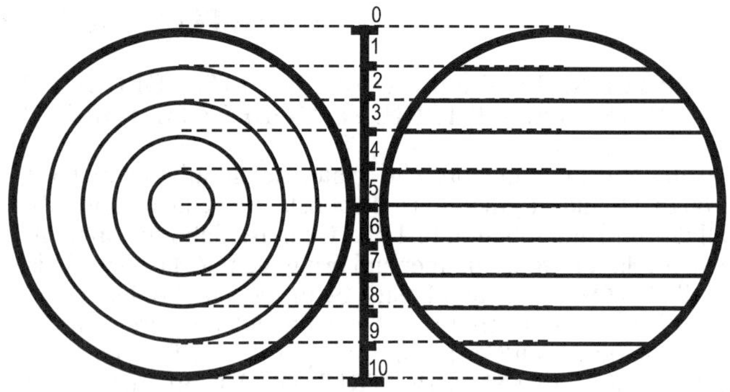

Fig 2-43 **RELATIONSHIP BETWEEN PLANAR and SOLID MEASUREMENTS**

Why does the center often seem to be more frequently posited than the periphery? This is an illusion. It is a product of our tendency to focus, and the nature of a page's presentation and our lack of vision in depth or our mind's overview.It is also a product of both viewers subjective bias and common in experience and knowledge, Whenever we look at TRU through ALT as shown on a textbook page we should always remember that this is in two dimensions of a 3D figure of the object of TIT, OBT. The differences here are valid, but our failure to continually realize and apply them are not, especially in use or in interpretation. So that continual correction and extension of our overview and depth conception must be applied. The correction factors are known and are shown in **SPLEAT** or the modified SPALT. The name of the relationship is called: **COPS** = Convergence, Concurrence and Correction of Planar and Solid Posits.

The posits in MOT and SPALT are of many types, classes, items, words, objects, concepts, new concepts, ES, CES and aspects of TRU. They have a tendency, from the observer, to unify and be subjective. We need to be aware of and correct for this as well as other influences that alter actual positions in TIT. We also need to notice voids and all that poorly show what TIT actually is.

Although most terms can be roughly centered and posited in SPALT and MOT, it is only when we consider them in their entirety, relations, varieties, effects, and implications that TRU can be well shown. This is what it takes to fully realize TRU. It is a filling-up process of many POSITS that make the observed into the kind of object that equals TRU and satisfies the observer. Because of limits and bias in posit, because there is much that remains void in filling-up SPALT with TIT, because these posits are not always singular and simple ES and CES in SPALT, and because we fail to return to, or maintain sufficient overview, does not negate POS, SPALT, HOM, ACC, POT or IPT. This is one of the greatest lessons of POT and IPT. It is one that any critic must know if they are to judge TRU appropriately.

LADIP = Law of Diverse Posit. LADIP signifies how each new element in a posited thought, object, entity, real, aspect, ES or CES of TRU will vary considerably depending upon what may be added, subtracted, missing or overused in that posited. It varies with the size of and number of elements in the whole, and by the polarity of the topic at hand. Each thought, object, entity, real, aspect, part or process of TRU will posit differently and to the proportion each new element is added to its definition, description and completion.

The deepest, most profound and greatest intimation and meaning of posits is to be found in their dispersion. Like all that is TRU is of DAOST, its posits are both differentiated and integral simultaneously. It is in how well we can discover, derive, develop, determine, formulate and operate these aspects, elements, ES, CES, mechanisms and posits of TRU, that we can best produce their POA. Thereby we should be careful not to limit ourselves to only singular highly specific posits, to pigeon-hole each that does not equitably represent its status in TIT. Degree of fixation in POSIT is the degree to which a part is fixed in its position in TRU. This is not set or static. It can be dynamic and as well highly variable in degree or percentage of TRU. It may be more or less concentrated or more or less dispersed and these also may be variable. **LADAC** = Law of Dispersion and Concentration of POS and TOPOS. For every concentration there is a dispersion and for every dispersion there is a concentration. This may not be the same for all, but highly variable for each in both TIT and SPALT. This is why it is so necessary that posits be separated by overall TRU and the subjectivity or perspective of TRU. The observation of successful subjective position in TRU is impossible without TRU being reduced to a finer POA, which in turn, depends upon a keener perspective and observer. It is in this way that S and O function, in and out of TRU, OITIO, to THESOA and to TA. It is COGNOS that finally reconciles and reinforces these, encouraging us to live by PUREALITY, ELTICMA and SÂ, to be TES in TRU, TIT and TÂ.

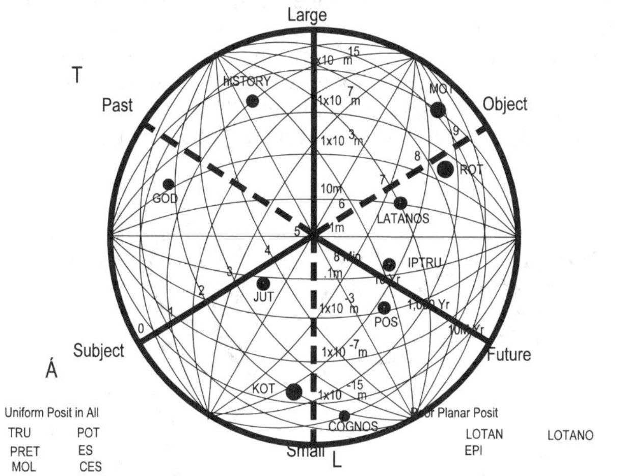

Fig 3-43 **POSITS of NEC, EXAMPLES for TOPOS**

This figure shows a few tentative posits of existing and TRU concepts which are especially noteworthy in this text. It must be noted that some do not POSIT as well as others, which is justified by TIT, or applies to the representation, modulation, perspective or presentation of TRU and not exclusively to TIT.

LOPOT = Law of Total Posits, as Total TRU or Equivalent to Total TRU, REM and Corrections. Even if what we posit is actually, perfectly, and absolutely represented and modulated as posits in TRU, it must, when interpreted, be corrected for being less than perfectly of TIT, REM, and other factors and qualifiers. If it is less than TRU it must be corrected for this deficiency of its part in the totality of posits and TRU.(COL,COR)

Variable, partial and incomplete POSITS can be analyzed, solved, and provided, in several ways: 1. Generally or loosely in terms of their apparent posit, 2. Specifically in terms of relations to this apparent posit, 3. More precisely in terms of the REM value of the item, entity, object, word, ES or CES, 4. More precisely in terms of the status, as SCARL, of the item, entity, sense, thing, ESIN, ES, or CES, 5. More precisely in terms of the A or any other status of the item, entity, sense, thing, ESIN, ES or CES, 6. More precisely in terms of the concentration or dispersion of the foci of the item, entity, word, sense, thing, subject, object, ESIN, ES or CES, 7. More precisely in terms of its reference or non reference to TRU, ALT, SPALT, REM, POS, 8. More precisely in terms of ALT, SPO, SPALT, and contents, whether they are TRU, no reference to TRU, part TRU or surplus of subject, object, integral, and differential as in SOID. So that for each item, entity, word, thing, sense, object, ESIN, ES or CES we endeavor to posit we can not expect to attain a valid or TRU Posit or representation vaguely, roughly, haphazardly, simply, or always distinctly. We can not even accept seemingly certain posit unconditionally, without considering it constantly and continuously to upgrade its representation of TIT.

Posits vary with emphasis on origin, form, function, center, mechanism, future, implication and importance. In positing, a great amount of difficulty is produced by lack of consistency and standardization of emphasis. We posit strictly by some proper FAF, ES or CES in TIT. For posits to be good, valid, and adapted by VAPO, and SCARL we must delve much deeper. It is an extremely important problem whose solution is one of the greatest of all time. It requires continued study, evaluation, refinement, and management. Emphasis and the variance from TO associated with it is simply a part of the great dialectical actions that occur among the polarities in TIT. That is between the whole and its parts, between the universe and the person. So that in POS there is, and must be as in life, a continuous and reversible movement from specific to general, and general to specific, that best enables POS to replicate their state and role in TIT.

DYNAMICS of POS and POSITS, PROCESS, PROJECTION and REINFORCEMENT in POS, PROBLEMS with POSITS, TRANSCENDENTS and COMPARATIVE POSITS

It is when we can posit and understand the significance of words in their full roles in TIT that we can begin to see how POS, MOT, all ROT fulfill TIT and our needs. It is being highly dynamic and subjective that we can provide the great great explanations needed for understanding. It is why valid, standardized, and successful world views have not been previously achieved. Experimental psychology adds much to resolving the dynamics and posits of human thought, act, and attention. How good is ones ability to notice missing parts or defects in a field of many different elements, components or items? Do we miss the only one there, or can we recognize many, most or all of them? This is very important in showing posits and our ability to comprehend and picture what we observe in terms of an overview. This is the acuity of the observer-observed dynamic relationship and the degree of its reality. This is why there is such great variety in life, both because of the adaptation of animals to their environment and because they adapt the environment to suit them. Each forms a RU typical and special unto itself. For TRU this is most significant. It is how much of human and therefor TRU's FAF developed and exists. This is largely of relativity, their observer-observer capacity and ability, or how well they can realize and achieve, at even lower levels of Á, all that is positive or negative to survival.

DEP = Determination of Point Placement, POP. DEP is the generic form of Posits. It consists of all principles, means, facts about the placement of any kind of entity or its representation in an object, another representation or body. It is specifically directed toward the finer and finer focusing and targeting of items in a given field. When seeking to posit many different items, to order those posited, or to use them it is much more effective to do so in some kind of sequence. This is preferably by what is most natural by TRU, the observer/observed, and SPALT. If we are to show and know the object of TIT, it is much better to limit the kinds and start from a set origin. When we can trace through the model in such a way that we can represent, communicate and teach TRU to others we can achieve the POA necessary. It depends upon resolution of separation and the formal distinction of all that are and can be best posited of TIT. Many things, senses, entities, as well as aspects, forms and processes of TRU project far into the future, both of the more special TRIF, and the general future, TRUF. This can be highly revealing and rewarding to our needed knowledge in life. So that projection, as all POSITS participate in TRU, should be simultaneously thought of as particulate posits, extended projections, processes and working in a totality that replicates TIT.

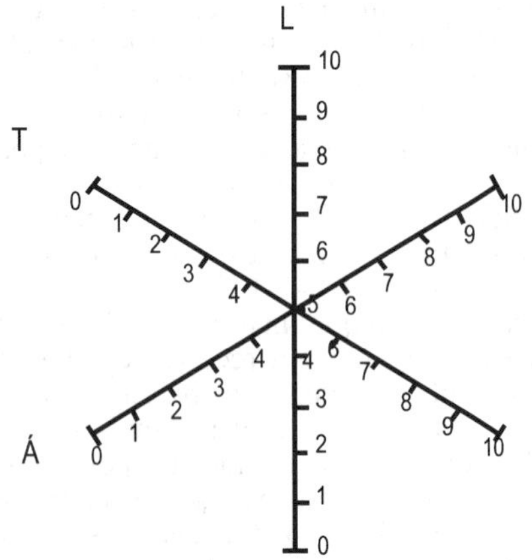

Fig 4-43 **The SEQUENCE of ALT POS**

The sequence of ALT is based upon the scale of each of its three primary dimensions shown coordinated in Figure 4-43. To best follow through ALT and its posits we do so in the same way, beginning with the infinite subjective, infinite small and infinite past and proceed first through time from 0 - 10 t, then through length from the 0 - 10 l, and then through the A dimension from the infinite subjective to the infinite objective or from 0 - 10. In this way we have the optimal effect of staying with our scaled sequence in its preset order. This is not exclusive. For, if we start with subjective at the bottom left and not the usual top left, infinite past time, as on a page and move back and forth through the sphere, not in a direct sequence as on a page, we can with broad enough swaths cover the entire solid. For ALT this becomes a sequence that is less direct yet avoids overlapping itself in each dimension or region in TIT.

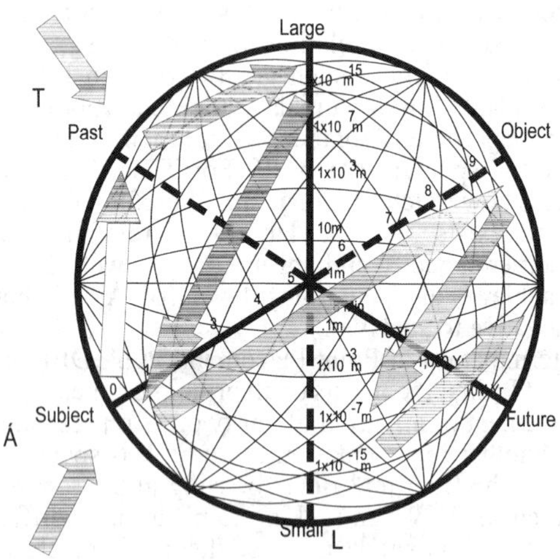

Fig 5-43 **APPROACH and SEQUENCE THROUGH ALT**

In viewing the dynamics of Posits we need to constantly remember that because ALT on a page is planar and of 2D, we have three to cover in SPALT. If these scales are increased exponentially to 10^3 or 1,000, or to 100^3 or 1 million, divisions and many more posits for each division we will have a complex challenge of our POSITS sequence. At times an option is needed to take another course for POMOT, METAD and SPAT. By ATO, TIT and TRIA we must always return to TIT-TO if we seek closure.

This can be done in different ways depending upon which provides the best demonstration of TRU: 1. simply proceeding as we read a page, 2. by assuming any motion out or in ALT, 3. by following the scales of each dimension, 4. by using the 0-10 scale of each dimension, 5. by following the scales of each dimension combined and in a formal way, 6. by using the sectors of ALT, 7. by using other fields and models. The trails through ALT, SPALT, POMOT, SOID, or MAPLAT, METAD, or ELTICMA are like the trail through life. It is not at once TIT-TO. It is variably singular, partial, linear and limited, and often incomplete, yet it can be more ordered and fulfilling. This helps us see and show IA, Á, TIT and TRU.

The time dimension is largely of the dynamic, as the length is static, which makes posits by time less set in MOT, ALT and SPALT. Added to this is the duration and variation higher powers of time, plus the transit of time through the present. Although very poorly appreciated, this is of the greatest importance, especially for POSITS, for it necessitates an adaptation and modification of many posits in the time dimension. This is most noticeable when we seek to POSIT ES and CES at some time in the near past, only to find it is equally appropriate in the near future. Since for SPALT this is a spread of nearly 1/2 the whole, it becomes acute. It is a product of the fact that ES, are not posited in time, but in duration or variation. Another way is to view SPALT from different directions. None of these disqualifies POSITS nor SPALT, they only help to provide an overview, qualifications, and exceptions needed in practice.

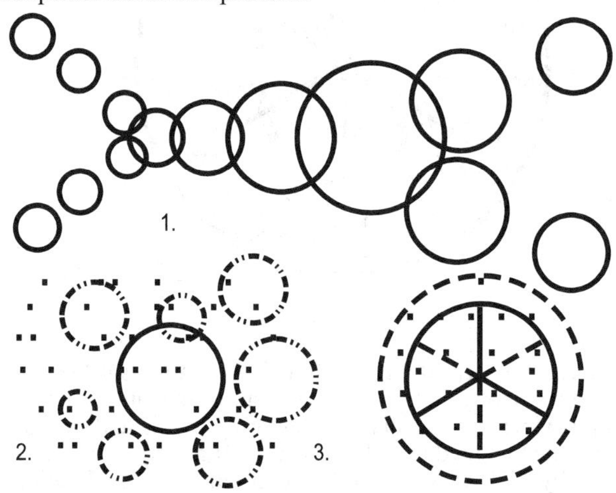

Fig 6-43 **COMPARATIVE IA, POSITS and RESOLUTION**

In any final analysis we always need to correct for the placing of items outside of ALT of TRU. If we do not correct for any and all Posits that are other than only in TIT-TO we have a degree of disorder no matter how great TRU is. This departure must be accounted for and resolved. This is called: **RACTIO** = Resolution, Accounting, Accentuation of TRU, in and out of OITIO. RACTIO is what we can finally use to bring scattered and diffuse non and poor posits as well as non TRU posits into line with TIT. The three figures show OITIO relations between IA, POSITS, and TRU: 1. IA as it approaches and leaves TRU, 2. POSITS at random, objects and TRU, 3. IA as POSITS in and about TRU. These figures help a comparative picture of IA, Posits in SPALT and TRU. How they can be brought to TIT-TO in a final figure that makes posits in and outside of TRU, and still retain a high degree of order conforming to the necessity of an absolute union and thisness or TRIA, of TIT.

CHAPTER 44

ENHANCEMENT and SUMMARY of POS

ENHANCEMENT of POS and POSITS, ACTUALIZATION, SUFFICIENCY, HARMONY, EXTRAPOLATION, PRECISION, SUPERIMPOSITION, OPTIMAL and ULTIMATE POSITS

The quantum leaps from no knowledge, to incomplete and errant knowledge, to a vague idea of the world, God and universe, to their representation, and hence to SPALT are to much, for an individual, and even all humans collectively, to manage. If we are to overcome these leaps and show and apply SPALT we can do so only when all steps are analyzed and shown in the best possible way. We should always remember the major concepts and views provided in the previous sections, and with MOT and SPALT proceed to fulfill the positive approach so greatly needed to know and show TRU.

LOTRAC = Law of Realization Through the Comprehension of Actuality. LOTRAC is the balanced, valid, proportionate, massive, total and ready realization of TIT by TIT, or all through the comprehension of actuality by ALT, POS, HOM and TRU.(VAPO) It when limited to ES becomes: **LOTESAN** = the Law of TRE ESIN, ES and CES Actualization of Nominals and Words. LOTESAN is the TRU meaning of words.

PRODEP = Proof by Deposition, by Posit, by falling into Place. Through suspicion, insight, ability to anticipate, great wisdom and knowledge we recognize that which will occur from that which has occurred. Confirmed knowledge is established absolutely with certainty by falling into place. This is like a happening or occurrence that would not have otherwise happened by chance. This is a highly probable and terminal proof of God, TREQ, and TRU. It is the most effective means of establishing actuality from figments. For it is in the nature of humans, their relativity, or observer observed relationship, Á FAF in TRU, that knowledge is gained through the positive identity of this observed or object by the subject. It is like the experimental method in science in which the results independently coincide with the theory and premises which are being investigated and tested. It is what is expected. It is how, where and when posits are made perfectly when Á, L and T is what it is supposed to be. POSITS and PRODEP are keys to TRU as the actuality of TRU and TÄ provide their foundation.

The best and most natural way to develop POSITS, POS, and POA is to start with those ES and aspects of TRU which are most distinct and most simply and accurately posit in ROT and SPALT. When Principles, Ultimates, and basic or fundamentals concepts are Validated, related and formulated appropriate posits of TRU are enabled and enhanced, **PUV**. There is a great need to separate, interpret and posit each and all elements, as in fitting the given, which in turn must be brought to TIT and ROT.

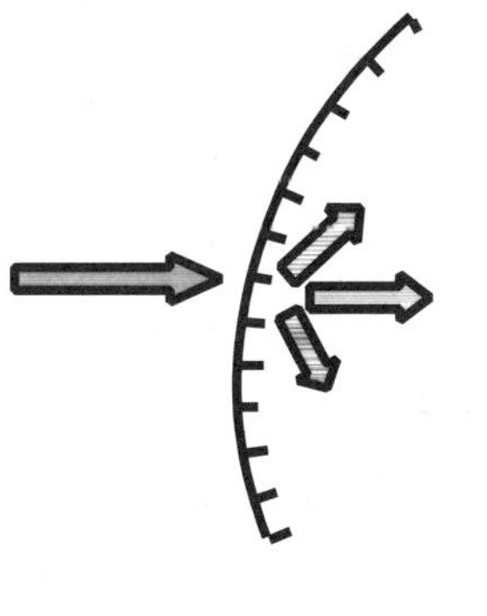

Fig 1-44 **MEANS of DEVELOPING ENHANCED POSITS**

One of the most common ways to make and use posits is for a person in daily life through IA, to analyze and interpret all that is given by TRU. This may be done more fully and totally by linearly superimposing all on SPALT and developing the concepts, principles, laws, rules, and means through the best path TRU provides.

LAPIC = Law of Precision of POSIT by Increasing Comparisons. As certain ENS, ESINS, ES, and CES attain precise posit others attain increasingly precise posit by comparison, similarity, and agreement. Much imprecision is due to ambiguity, multiplicity, and diverse meanings and definitions. It is for this reason we emphasize NEC more than EC more than OC. Although many existing terms are not only of great power, they are needed to complement POS, and in reverse, NEC and EC to complement them. It is by adding evidence and proof of the VAPO and TRU of each given, item, word-thing, ES, or CES for posit that LAPIC adds great power and accuracy to locating and positioning in general and the point placement of each item.

LOPS = Law of Posit Sufficiency. LOPS designates that status of POSITS whereby they reach a sufficient level of proficiency to yield a profit, or benefits exceed losses. LOPS is achieved by a consistent, proficient and convincing work in POSITS, and its use in the lives of each person. LOPS is of different kinds of sufficiency: 1. of POS as a general work of representing TRU, 2. of POS as general rules, laws, and principles of point placement in a locus. 3. of POSITS that do or do not achieve an appropriate level of precision in representation, 4. mixtures of average levels of each posit, and 5. Maximum of VAPO, accuracy, and benefit of each POSIT.

VÊ = Validation through Extrapolation. VE is evident where although much may be unknown, as it varies per individual, time, place, and operation. This unknown can be known to the extent our relative knowledge is: 1. great, 2. clear, 3. VAPO, and 4. TRU. As TRU and S-O or Á become increasingly known we can rapidly and readily reason, surmise and extrapolate what remains. This is evident in how we can see TRU and TRUS in MOT, MPLT and SPALT. Much can be anticipated in the future from the past as well as how the past changes with the future. We can see that at time t, be it 1 million, 100,000, or 10,000 years, the TRUS of these dates were different, but consisted of largely predictable set form by TIT-TO and ATO. The validity of posits is greatly enhanced by: 1. KOT, FAT, BASFANT, NAT, VAPO, TROS, ACEXIN, OM, and MOT, 2. Knowledge of ROT, SOID, MOT, ALT, and SPALT, 3. The evidence gathered from the operations of these in SPALT, POS, STA, HOM, and ACC, 4. From Posits themselves as they can be actualized, formulated and standardized, 5. From Extrapolation of these Posits and VE.

The rewards for greatness and righteousness in life, TES, are legend and plentiful. This is much more efficient and effective the closer it is to TRU, for TRU is this by definition. It has been GOD, as it approaches TREQ of the past that enabled expanding populations to provide the love, positive attitudes and controls that are based upon fundamental principles of social harmony and cooperation seen in TRU, MOT and POS. To achieve, sustain, and enhance such a level of permanent success requires a POA of TRU, the POSITS, comprehension, and use by each person. Man increasingly lives by concepts, ideas, ideals, principles, policies, and plans of God, TREQ and TRU, and less by the objects we directly encounter in the world. This is simply a manifestation of the expanding subjectivity, mentality, and Á FAF, and its associated vehicles of speech, writing, computing and communications in a growing world. It is best done as they are posited in the best possible frame and model of TRU, integrated by POA into the best possible replica of TIT-TO. Reals focus on finite entities, whose ESINS exist as aspects, parts, processes, concepts, laws, principles and mechanisms of TRU. They are those things that are finite both of SO together, as Á, and finite as in the middle region between indistinct unity, and indistinct infinity. This is called: **LED** = the Law of Finite Distinction. LED is that great law which does so much by rendering man capable. It brings cosmos or the ordered special universe, as SUT out of the chaos of AU. It brings a highly irrelevant and irreverant physical cosmology into the actual cosmology of TRU. It is what brings both unity and intricate beautiful detail into harmony and out of complete dissociation. Just as left and right, up and down are necessary so must TRU be considered, as 1, 2, 3, 4, 5, 6, 7, as well as an infinite number. This is called: **LO** = Law of Ordination. It is by LO that the three dimensions of TRU are verified, in ALT.

MAXSIT = Maximum site of pain, posit, concentration, or focus. MAXSIT is as for all TRU, the maximum site of focus, its center and mean, or the center of the dimensions or polarities of its A, L, and T dimensions. The MAXSIT of all in TRU are not as we would commonly or even necessarily know them. They are as ESINS and even more, only these ESINS not as parts of ENIT, but parts, ES of TRU. They are concentrations of their masses and means in TRU, as posited in ALT. It is where and how ENS can and must be reduced to their ESINS and ES in TRU, and in so doing provide optimum posit. It is like pain in ones body whose distinct center is perceived by us for best response. It is even like the pain that does not exist in exactly its point of origin, but may be referred. So can a referred posit be inexact, and need to be examined and analyzed to be known, shown and corrected.

LAPOR = Law of Positive Reinforcement. Posits of ES, CES, and all of TRU when incapable of ready and accurate placement can often be determined by those great posits to which they are related. **LAPORS** = Summative Positive Reinforcement, or totally integrated LAPOR. LAPORS is the final effect of all related posits to,, and on what needs to be posited. All POSITS should be primarily made by the weighted focus of their FAF in TRU, TRIT, ALT and especially their most profound implications. This often warrants a NEC or reemphasis upon an OC and EC and their TRU definitions. **PROSEXT** or Processes of Self Extrication to TRU by concentration and focusing. PROSEXT is a great part of becoming TES and PUREACLITY. It is closely related to maturation, growing up, coming to terms with life, fulfillment of ones mass energy, Á FAF, and having the necessary conviction to be responsible in life, TRU. This has been a great and core force in the dynamics of TRU, requiring ones ability and achievement in POA, and the showing, knowing, and communicating TRU. They are guided by the perfected ideal of formulation and modulation of the best pos-sible Posits, or: **OPTRU** = Optimal Posit, Optimal TRU. With OPTRU as a tool and goal we can maintain the difficult course POA, POS and TRU demand. The conversion and POSITS of all IA, ESINS, ES, CES, REMS, vectors, forces, dynamics, motion, power, action, and mechanisms of TRU into ALT, SPALT, METAD, and SPAT, and these into ultimate use in life, is the theme, method and goal of much of the text. It is a person's best thought deliberately working toward ascent to TES and TRU.

The reconstruction of evidence enables us to obtain accurate, corrected, proportionate, and complete representation of entities, objects, and all. This is much of what POS does for KOT and TRU. Only when done by POA, VAPO, and POT for all TIT-TO it is much more efficient, effective, accurate, and complete.

SUMMARY OF POS and POSITS, PROBLEMS, SOLUTIONS, STANDARDIZATION, USES, BENEFITS, ORDER, IPTRU, ROT, POT and TRU

For every great topic, field, science or block of knowledge that has been unknown, ignored, and fallen from favor because of error, misuse, or misdirection, this void will tend to stir in the imagination of some, zeal and inspiration for being filled. Humans have a natural curiosity for the discovery of what is missing or wrong, and seek to explain and correct their meaning and use. This is a great basic inclination that typifies the mass energy dimension of man, its Á FAF. **RED** = Reaction by Equalization of Densities. RED is that automatic or spontaneous filling of voids, reaction to equalize density and provide unknowns that is natural to TRU and each person. It is through a POA, POSITS especially in MOT, ALT, SPALT, and supplements that we best discover, and fulfill this need for TRU. There have been, are and will - to a much lesser extent - be parts, processes, forms and functions, states and mechanisms, and many aspects of TRU that are unknown, void, disproportionate or poorly applied knowledge. This is because we have been blind to much of the world and its order, TRU. Only with a well developed SPALT and POS, or equivalents, is there adequate frame and method for discovering and applying these unknowns and voids.

	EXTREMELY SIMPLE	SIMPLE to COMPLEX	EXTREMELY COMPLEX
SUBJECTIVE	A vivid certain thought clear and coordinate	Increasingly finite mental ROT, MOT. POT	Massive thoughts and views, SUBT to TO
SUBJECTIVE INTEGRAL	Excellent View of any object, OBT or TO	Increasing views of objects, OBT or TO	Massive senses of all objects, SUBT to TU
INTEGRAL	The whole when discrete	The whole when increasingly SCARLED	The whole when massive and dynamic
OBJECTIVE INTEGRAL	Usually depends on the presence of the whole	Expansion key to SUT and MOT of TO-TIT	Of ROT when TRU is not TRE, OBT to TU
OBJECTIVE	Discrete Object or Observed	Increasing quantity and quality of objects, TRU	Massive number and roles of objects, TRUD
OBJECTIVE DIFFERENTIAL	A or The discrete object or observed	Increasing quantity-quality of separate ENS,ES	Massive number and roles of ES CES, TE
DIFFERENTIAL	Entity or ES discrete among many	Increasing number and functions of ENS or ES	Massive number and roles, separate ES CES
SUBJECTIVE DIFFERENTIAL	A thought strongly felt seems so certain	Increasing number and functions of senses	Massive number/roles of separate senses, TE
REAL	Simple Identities seem so certain and real	Increasing identities to be used for benefit	Massive identity with all in the whole or TO
RESULTANT	TU to TO Toward easy Success	Increasing Challenge c Optimal Capacity	Necessary yet Fails if not with great TRU ability

Table 1-44 **SIMPLE TO COMPLEX IDENTIFICATION**

This table gives a rough outline of the extremes from the most effective to the least effective means of utilizing the segments of SOID, a SOID with horizontal S-O, and vertical I-D. It shows how neither any one aspect or part of TRU, nor in the unity or infinite of TRU, but the ever expanding and perfected, as by VAPO, fulfillment of predominant form and function in TIT-TO that greatest success, survival and being exists. POSITS being of the substance of a POA make the way by which fulfillment of TES and TRU can be achieved. All may produce more or less accurate posits depending upon the distinct acuity of the observer and the discrete availability of the observed. These will also vary depending upon what is posited in SOID. If SOID is used generally, then it will be largely of entities, senses and things, subjects and objects. If SPALT is used properly the POSITS will be largely of TRU mechanisms and especially ES and CES. They will also vary with how they are used for benefit in TRU.

Cross references, applications of information and data, tie-ins, enumerations, aids to thought and study, and other means provide the necessary relations and combinations that are vital to complete POS in SPALT of TIT. There is great use and benefit to be found in errors and imbalances in POSITS. Errors, imbalances, and incompletions represent some reason, principle, means, method, or process in POS and POSITS that needs to be addressed and corrected. It directs us in the search for better and greater POSITS. **LÂVCORP** = Law of Absolute Validity through Corrected Relation and Position. When TRU and its ES and CES are properly related, defined, posited and corrected full validity and reliability can be established.

Some of the means by which standards and reductions can be made for best POS are: 1. 10 X 10 X 10 Divisions of SPALT provided by its Dimensional distributions, 2. restricting posits to the purest and most concrete ES and CES selection and reduction, 3. providing all posits with precise values, relative magnitudes or REM, 4. Filling in the voids, by not having SPALT left with regions that are empty or poorly represented, 5. Selection of Posits, to continually upgrade placement in SPALT, 6. Emphasis on gist, revelation, and depth conception of each item for Posit and using this as its measure, 7. Emphasis on cue, by continually utilizing all revelations that help concentration on the detail and content of TRU and ROT. Then, their use is much more adaptable to TRU. 8. Quality of POSITS. 9. Allowance for degree of overlap, establishing limits, standards, rules and the means to minimize or control their problems. 10. Flexibility of limits, means, and methods which provide the guide lines and controls necessary. 11. Variation with TRIS, TROS, TRUS, and TAS. How posits in SPALT must be understood in relationship to all that is in, of, beyond, and outside of TIT and TRIT.

POL = Posit Laws. It may seem simple and accurate enough to place something on a field and be satisfied with its use. Then we have only begun to slightly satisfy our need for orientation, coordination and control of all the detail necessary in life by TRU. This leads to posits that are well developed, refined, with VAPO, and SCARLED in the set way that best replicates their part and role in TIT. POL includes: 1. What is for Posit can only be of what is TRU, 2. They must account for the primary mass energy relativity of TRU, its Á form, function and dimension, 3. They must conform to the object, ROT, and MOT of TRU, 4. They must account for the higher powers of the dimensional formulation of TRU, i.e its area, volume, duration, variation and levels of mass energy. 5. They must account for the weighted distribution of Posits.

The need for empirical evidence, observation, and scientific proof to determine the meaning of a statement or proposition strongly supports the contention that the knowledge and proof of TRU can be fulfilled. This has been shown by the Logical Positivists. Only when observations are accurate, collected and posited, like in SPALT, can they collectively and totally show TRU. To posit, largely means to place or position and not necessarily or exclusively to verify, as by the verification principle. Yet the verification principle and the proof of TRU are provided when all that is TRU is posited. This happens when all is placed in the modular construct of TRU. Truth, ethics, and morals are derived from what must be: 1. Known, 2. TRU, 3. Proportionate as REM, 4. Posited or ordered, and 5. Their statements or propositions must be determined and have accurate, actual and TRU definitions, distinctions and meanings, and 6. best used for TRU by TES..

The better and greater the positing, the greater the value of the frame or model is. The better and greater the frame or model is, the better and greater the positing is. This, whether it is done simply or complexly, singularly multiply or totally, directly or indirectly, and statically or dynamically.

Life is a continuous, varying, earnest, and stern activity of needs, wants, desires, preferences, choices, hopes, thoughts, intentions, priorities, objectives, goals, judgements, decisions, selections, work, and routines. It is in the associated search for best interactions that each and all of us continually make in life, that we succeed or fail. It is a search for each and all that makes life work and succeed. This depends upon the degree this search is right and ordered. These depend upon: 1. The situation or occasion at hand, 2. Ones life in general, 3. Ones life as it is or is not with TRU, 4. How right it is by TRU, 5. How ordered it is by TRU, 6. How VAPO it is by TRU. 7. Best use of TRU means, POA, POS, PI and ACEXIN. Such a well ordered search in life is significant, for it is key to Á and much of TES. This is seen in POSITS, for in being able to know, show, and do what is right in an ordered way we can readily achieve what is best, TRU.

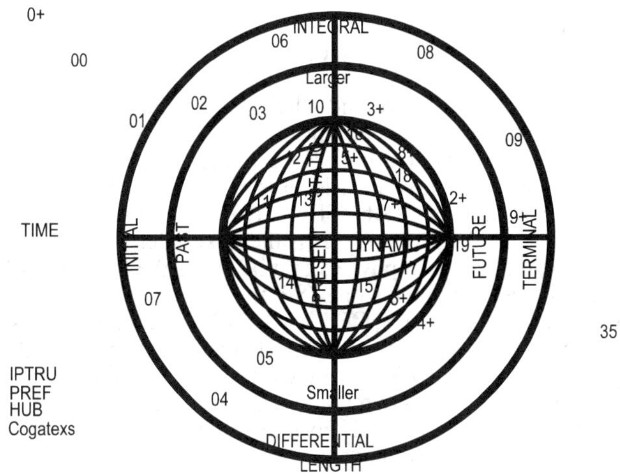

Fig 2-44 **IPTRU CHAPTER/SECTION POSITS in SPO**

SPO or the Standardized Perspective of All Things, presents 1. an outer circle representing the far approaches to TRU, mostly subjective as in Knowledge or Epistemology, 2. the middle circle representing nearer approaches to TRU, mostly objective as in representation, modeling and their applications, and 3. an inner circle representing TIT. The inner circle is here so viewed that length and time being perpendicular the sagittal axis or mass energy, Á dimension is perfectly perpendicular to the plane of the diagram and thereby is left to the imagination of the observer. The figure shows posits of many of the chapters of IPTRU. Each posit will vary with interpretation. This figure, although rough and highly diagrammatic provides a clear reference for posits in the text, and sets an example and beginning for POA. It shows us how much posits can help an orientation in length and time. Position and place are great facts of life. Position, along with the associated coordination, is a set for much that exists. This is a necessary condition of ones relations in TRU.(SCARLS) In POS and Posits we are working through, to the degree they are VAPO, the groundwork of this fundamental position. It is a basic dynamic process whose execution and control is key to life, as TRU. Position, transformation and their combinations are examples of the relationship between the 100% pure Á FAF, state and dimension of TRU and its increasingly impure S-O counterpart of TRU or of all entities.

The reciprocal value of TRU as it enables the text and the text enables and increases TRU, is the beauty and glory of each. As the S of Á increases, so similarly does the 0 of Á expand. This occurs as our understanding of the world and our text's demonstration of it increases, so does its objectivity. Much in the text is prerequisite to, and works reciprocally with, the sources of the great proofs presented, namely the proof of A, or PRA, TRU or PRUT, SU or PROSU, ALT or PRALT, and the Proof of POS, or PRUPOS. This can be done through:

1. Showing how these are the clearest ways of picturing ESINS, SU, POT, SPALT, METAD amd ELTICMA.
2. Enumerating all the greatest processes.
3. Showing how much TRU and its parts work by process.
4. How to formulate, posit and diagram these processes as they replicate their actions in the dynamics of TRU.
5. How this is the single and clearest way of illustrating SU and TRU.
6. Showing how SU, mathematical unity, and therefore quantities, are derived from and substantiated by geometry TU, TRU, ROT, ALT and SPALT.
7. Showing how all chapters of IPTRU arise from a good and well balanced development and treatment of TRU.
8. Showing how much of TRU can be shown only from without, i.e, as we approach TRU and as we present TRU, not always what TRU is in and of itself.(BAS,BASFANT, FAT,NAT,ALT, TRU,)
9. Best showing TRU thru TRIA and MOT, ALT, SPALT, and METAD. This is produced in reaction to **PIOT** or the Paradox of In and Out of TRU. So that we can at once and the same time satisfy TRIA and PIOT by extending MOT to include meta, overview or epistemics. This is the simple addition of the next outer ring or level of the observer/observed series. Where this next outer level is the observer rather than the thing-in-and of-itself, TIT. If it is represented or objectivized in the process, this is the next outer ring and the overview, meta or metaphysics of it becomes the third ring of this infinite and variable series.(SPO)
10. Showing how the ES, CES, and CALP or parts, processes, concepts, principles, laws and mechanisms can be brought together alphabetically, or in various classifications through the guide, text, worksheet, enumerations, glossaries, indices and all means of integrating and positing them into the many diagrams, illustrations, models, approaches to, superimposed on TIT, TRIT, and TRU. Even unknowns can, to some extent be shown in such a way.

DIP = TÎS/TEP, TEP/TÎS, are the ratios of a person's TES from the individual or from TRU and their combinations and interactions in TIT. Such absolute necessity, LATAN, leads us to and proves Á and TRU, PRA and PRUT. For in being TRU and a part of TRU, each and all of us must be, and be formed and function, in that way which most appropriately fits in with and fulfills the set of TRU. This means we must be able, to achieve all the necessary subjective-objective, observer-observed, distinctions and predications.

Some of the other new TRU concepts and means that help verify POS and extend its use and achieving a high level of permanent success requires a POA to TRU and the POSITS, comprehension, and use by each person. To accomplish this includes an understanding of the concepts and principles of POSITS, for without these there is a greater error, **PEC** = POSIT Error Centering. PEC signifies that never ending and devastating state one and all persons are in who lack a POA and understanding of the ES and CES, by POS, of TRU. It also signifies how the periphery is buried in the averaging out of the Center. Or the center is buried in the averaging out of the periphery. Even with TRU, MOT, SPALT and POS we have tendencies to PEC. Yet when this is of a special kind limited to TRU it is called: **PECT** = POSITS Errors Centering TRU. PECT signifies that in MOT and SPALT there is, when we average out the dimensional determinants of a posit, a tendency for it to become buried in the center. This requires follow up, along with that for the meanings, uses and POA to NEC.

Understanding is equal to an exponential expansion, as memory can be vastly expanded, TRU is by both, exponentially expanded. This is shown by: Understanding = x, and Memory = x^2 and TRU = x^3. This is evident in how far in ALT we can and should go in summation to prove TRU totality. What factorial advantage, as ELTICMAT or OM provides this. The factorial advantage is: **FARP** = Factorial Advantage of Relate and Posit. FARP is where all information we obtain - be it AU or TRU, small or great - becomes 2x, 10x, 100x, 1000x greater when we can relate and posit it. So is SPOR, SPO, MAPLAT, ALT, SPALT and LATAN and all justified hereby. It is not so much directly of TRU, as it is TRU indirectly through its ACC, PERT, KOT, and ROT. New concepts in TRU, ALT, SPALT and their associated MOT, more or less, equal 1,000 times their common existing terms. The highest expanded powers of god and proven information, prove over and over again all that is of and in TRU.

LASUP, LAWS OF SUPERIMPOSITION

1. Any coordinate or universal is not to be utilized except and to the degree it is adapted to the primary dimensions and model of TRU. 2. Any coordinate or universal is not to be utilized except and to the degree it is validated and reconciled with TRU, MOT and ALT. 3. The degree of error of all coordinates, and universals are to be utilized only to the degree they are corrected and adapted to TRU. 4. The new superimposition, adaptation, reconciliation of the original coordinate, or universal, is to be interpreted in its new, altered, or TRU nature. Any residual, warping, or other defects are to be disregarded and discarded. 5. The error of consecutive coordinate systems are to be restrained. That is when superimposing more than one system, set, matrix, or diagram, on TRU as modeled by ALT, the added errors involved are reduced to be negated in like fashion. 6. Summary: Therefor all superimpositions are completely invalid to the degree each is in error. They will warrant similar integrative and exponentially expanding correction, as the need requires.

PRUPOS

SPALT, TE, TIT, Analytic Geometry, Topology, VAPO, Meaning, Computerization, POT, Á L T, DAOST, ES, The, MATES, LOPAN, KOT, MADAS, EC, LANT, PRUNT, LARP, LATANOS, ALT, REM, LORVA, MOCBIREM, PRALT, Coterminality, Direction, Location, Posit-s, POKED, PEMAV, POMER, DEC, PIC, SIMUAD, LIDAP, SUIT, DECOMT, MAMET, POAT, FIP, LUP, ABET, DIR, VAPO, MADAS, CACEXIN, ESALT, LACOSP, LOPED, PODAC, PICOM, LAVCOR, CORPEX, COMPEX, LEFES, LOPS, RAD, DIFIRT, LAPIA, LATENF, LINE, MADEC, DALTOM, VINCU, LACOMD, RESCOT, PROHIP, PORE, TRACAD, TOVI, PIKER, MADAS, COPLAT, POPTE, DEOST, ISO, SUES, STED, PONT, TAMUD, EXAD, LACFOF, LAPSUFDAT, LECHPO, LORPDA, VE, VACEP, LAVTUS, DIRPOS, LALTI, LAFOCO, PAMEDS, SUPIT, SPOAT, PORE, LASDIC, MIDCOT, PERC, LACOMU, SOP, RESOP, CROMET, RASOP, DACOD, TODS, ROPU, LUSECT, LIRNME, ERETO, LADEX, LOVSEC, LIDA, MIF, MIPLID, TOVOP, LOPELT, RESCOT, LOPRIM, MOG, CONVET, CONRIT, CONUET, POS, PP-POP, POL, POSE, POSEN, POSAC, PRAMT, DUPS, DECPIT, EXATRE, JUPOS, LADPOD, RED, METAT, LOSIC, PORE, PI, LOSPEC, LOPA, TINDS, LACRO, NOSP, LOCPUR, LADAC, LADED, POCID, POCO, POVIN, LUM, LATCOM, LAMOP, LAMORD, LARPOS, LEPFES, SPINC, LAPT, POCOMF, PICOM, TOD, LINES, DIL, LASD, LAAPES, PROSLO, LADPOS, FACS, LAD, LADFOV, LOM, CORWAF, LANCOP, MADIP, TOPOS, POMVAS, POR, ERSO, LAPEX, LAPLI, PRALTI, SE, RETO, DOFIP, PROTIM, LOTRAC, LOTESA, PRODEP, OUV, LAPIC, VE, LED, LAPOR, COPLAT, PENK, and POL

SECTION IVB

STA and STATICS: POSITS of ES in MOT

CHAPTER 45

PRINCIPLES and DEVELOPMENT of STA

STATIC COMPONENTS of TRU, DEFINITION, KNOWLEDGE, DEVELOPMENT and METHODS of STATIC POSITS in SPALT, MOT, ROT and POT

STA signifies the topic whose principles and means of placing the static, formative, elemental and set parts of TIT into the model SPALT or other MOT and ROT. **STATICS** are the specific operations of placing or positing ES, words, NEC, phrases, concepts, ES, and aspects of TRU in SPALT and MOT. POS is to POSITS as STA is to STATICS of ES, and HOM is to CHES of CES. STA and STATICS are how we can bring the many methods of representation and modeling to objectively best show the parts of TRU. It is much more than simply placing entities or objects into any field. POS and STA provide the best way to show the components, substance, or morphology of TRU, what TRU is composed of, in a positive objective way. It is how to accumulate, account, order, and position the concepts, principles, and laws of what we have found in this text and how they apply to the set form of TRU.

Fig 1-45 **STANDARD PLANAR DIAGRAM of SPALT**

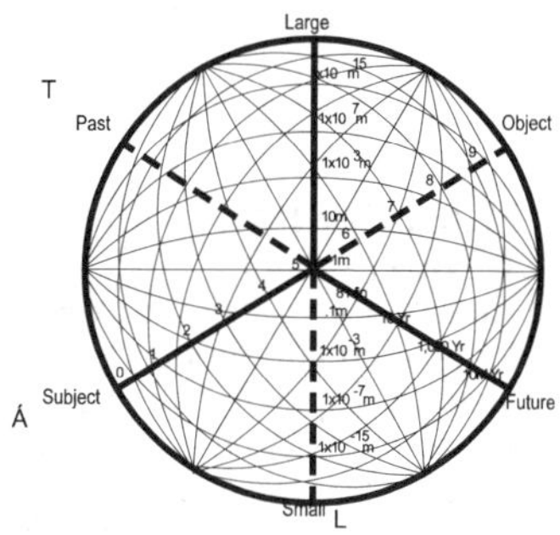

Figures 1-4 of this chapter compare the ways SPALT can be used for posits of the static aspects and elements of TIT. Figure 1-45 is the SPALT diagram of the three dimensional OBT of TIT for positioning elements of TRU. It is not corrected for volume, and the corrected mid points are not weighted accordingly. The separation and precision of the gradients in Á, L and T dimensions are needed for continual refinement. The study and work on POA, POS, STA, and HOM is to show the basic concepts, principles, mechanisms, laws, and mode of all TIT. It is absolutely necessary if we are to make a full accounting, demonstrate and present it in the best ways possible.(VAPO)

TIT is as much the specific, differential, partial, and static of TROS as it is its general unitary, integral, universal, dynamic, principle, laws and totality. STA is given much less space than the dynamic because it is simpler and much more self evident. It has less to prove, less interaction and ramification. This is much like how a person's anatomy is, when considered relative to physiology or all of medicine, a great basis but much less than half of its total. It is largely in the physiology, chemistry, technology, practice, and living of TRU that it as a science can be most and best known and useful. Vesalius held and taught that seeing rather than listening was the best way to learn anatomy. To clearly and totally envision and hold in mind TIT and apply fully to life is the secret and key to understanding. This is how experience is the key to wisdom. The wisdom and understanding of the worlds knowledge and TRU can be seen only if the illustrations, pictures, diagrams, charts, plates, spheres, and models can be developed to fit TRU. The body of an object must be shown and depicted in whatever way its framework and structure, form, and function, with parts and processes that exist and are directed by the total meaning and nature of TRU.

If one would doubt distinction, cleavage, role of quanta, differentiation, and DAOST let them consider consciousness and the difference between the conscious and the unconscious, as in being asleep and in waking up. Herein we see both the great absolute/relative paradox of distinct difference plus the role of S-O, and thus the Á continuum in the great mechanisms past, present and future of TRU. The prime purpose of consciousness is to separate the differences, make finite, and prevent the infinitesimal and non discernible from making our life unintelligible. It is life in this finite world that makes apperception the fulcrum of IA and power of TRU. The consciousness of being born, the beauty of the form of TRU, and the design of this form and function, processes, totality and presentation of TRU can be provided. How to represent the static makeup of TRU and what posits we can provide can be compared with a positive objective approach to anything observed, as the object of a subject. This method has been accelerating throughout science in the last two centuries. It is especially useful in medicine where of the formats for solutions to problems is called SOAP or subjective, objective, assessment and plan. It's application to the dynamics or processes of TRU or Homation, enables the ordering and placement of elements in the format. This enables a greater determination of truths, separation of elements, and the arrangement of corresponding interactions for better health. In turn the static posits can fulfill the content of the format and TRU. The basis for STA, and all that represents and operates from TIT-TO is its undeniable truth. This is how mankind and all that enables our being is formed and functions as a whole. The well being, continuity and survival of this whole is typical, selective, and exclusive for this special universe of best human life.

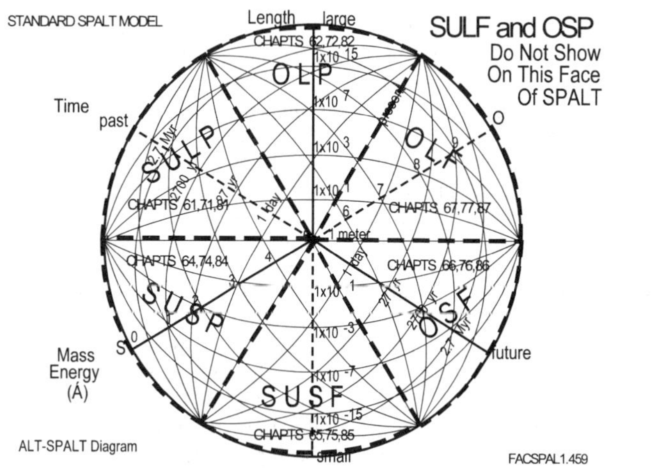

Fig 2-45 **STANDARD FACE of SPALT MODEL and its SECTORS or OCTANTS**

This figure provides the basic method for Posits of ES, CES and all in SPALT, for STA and HOM. Since we are only looking at the face of SPALT when using ALT on a planar surface of a textbook page, we will lack two of the eight sectors or segments. This Standard Face, with length or spatial dimension on the vertical axis, and time on horizontal axis mass energy, and the Á, or mass energy dimension on the sagittal axis can be so rotated to bring out views in which all eight sectors or octants can be equitably shown. The understanding of TRU by its parts can be no better, when pictured statically either singularly or multiply, than the existence of a person can be understood only by viewing parts, a picture, or cross section of human anatomy. Thereby the model, and TRU, can be little better than the person or persons viewing it. It also illustrates the necessity of learning all of TRU.(TROS,TRIS,ACEXIN)

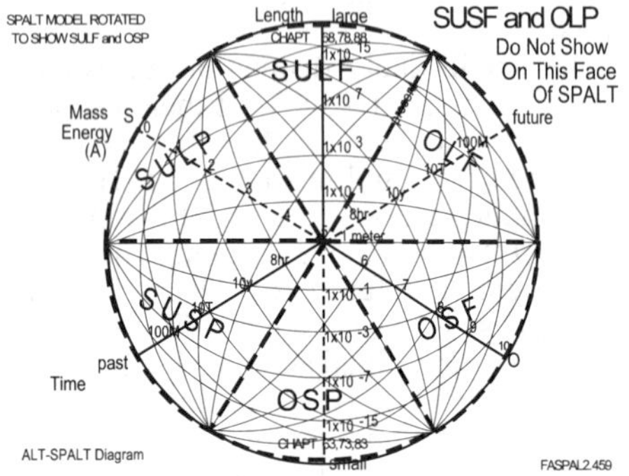

Fig 3-45 **ROTATED FACE of SPALT**

This figure supplements Figure 2-45. If SPALT is not viewed as a single standardized global ideal, it is necessary to rotate the model in different ways. Here the SULF and OSP sectors are shown without the usual SUSF and OLP sectors This depends upon swapping the Á and T axes. In fact SPALT can and should be rotated in all different directions, even when envisioned as a sphere if we are to fully objectivize it.. Only in a 3D model can TRU be shown and viewed as it actually is.

POS and POSITS introduces us into a detailed reconstruction. They are developed from the use of the basic principles of the applied geometry analytically in MOT and SPALT to show TRU. STA performs the reconstruction of the concrete discrete elements of TRU. STA is what is not dynamic yet differential of TRU. This is much of the world, TRU. its selective aspects and elements about us that are not,,or not considered, in motion. The frame is well provided. What is posited usually at first appears to be static, by TRIT of TIT. With a longer view and looking for its movement it is dynamic, of HOM. This is done because to reveal the DAOST, ID and TROS forms of TRU we must continuously go back and forth from static to dynamic, from differential to integral.

The determination of STA and STATICS depends upon their basis in POS, REM, SPALT, MOT, and ROT. This largely depends upon the dimensional form of TRU, TRIT, OBT and SUT. The Time dimension takes on special significance. It does not follow conventional ideas of time. The time scale is relative and thereby does not follow physical time, but is time dependent upon the predominant form of TRU. This is called: **LART** = Law of Relative Time. LART is the scale we set up that best represents the time continuum in TIT. This includes both its centering in the present, and its center of distribution past and future, or the mid point of time past and future. By being based upon the present it has two halves that are more or less mirrors of each other. That is time is reciprocal for past and future, or: **LARAT** = Law of Reciprocal Form of Time. When we posit items, NES, ES, and CES in SPALT we find that their time posit does not fall where we might initially expect it to. For it often exists in both past and future. So the distant past and distant future may, on average or in mean, posit toward the middle. It not only detracts from the distant time posit of an item and it also adds to the centering problem. The law that confirms this infolding or centering of distant times is: **LINRAT** = Law of Infolding of Relative Time. LINRAT signifies that time items that work from distant past through the present to distant future, although of a wide range and a mean in each segment, will posit, on average, at their mean toward the center, more or less in the present. This is a product of the basic time form of TRU, and proves the great ACEXIN of position in TIT. Since in the present time is greatly extended, this is not as severe a problem as we might assume for Time in TIT. Yet, too correct for it, we can add qualifiers, factors, and notations. One way is by the range of the posit or the distal superscript, or ʳ. This tells us how many tenths of the scale this item refers to. Another way is to posit it in whichever, past or future, is the greater element in its role in TIT, plus the added qualification of its range.

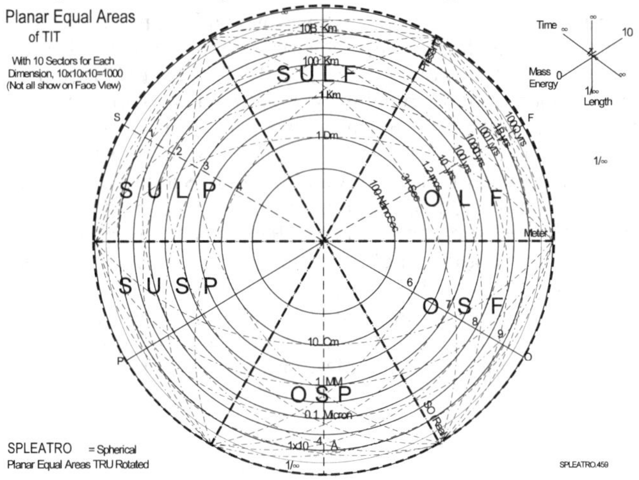

Fig 4-45 **PLANAR SPALT with EQUAL SPHERICAL VOLUMES** (SPEV)

How the end product, an improved solid spherical model, produces identical equal volumes with what exists in TRU is: **SPEV** = Spherical Equal Volumes. Its equivalent for planar figures of equal Areas and Sectors of TIT is: **SPEST**.(frontispiece) From SPEV we can develop figures and sections that yield the best possible diagrams for POSITS of ES, CES, aspects and all of TRU. This is intended to demonstrate three things: 1. The octant divisions of SPALT and TIT, 2. The changes and extension necessary if the solid or spherical TIT and SPALT are to be best shown on a planar surface or page of a text. 3. The changes and extension of dimensions necessary if this adaptation of equal areas is to be optimal. Producing and using such a figure for SPALT increases its refinement within its limits, when represented and presented in figures for pages in a book. However we should always remember, ECRO, the greatness of spherical and dynamic OITIO envisioning of TRU itself, TIT in all we observe.

FUS = Fallacy of Singular Standards. **FOS** = Fallacy of Multiple or Total Standards. As occurs with laws, FUS arises from the tendency to develop and hold to one standard, especially when that standard is both in error, incomplete and outdated. Or it is the tendency to develop and hold to many standards, now one and then another, without any overall order or identity with TREQ or TRU. They are largely a product of superficiality and failure to learn and know what GOD, TREQ, or TRU actually are. It is characterized by ISMS, yet much broader. For ISMS represent the end products of polarities, systems and standards. FUS arises from the very incipiency of S-O, Á, IA, and our minds. What we attend to and have our minds on at any moment is it. Any other thing or sense, or their ESINS, and especially all they are in TIT-TO are ignored. Only with the complete understanding of the extremes can those of the means be attained, and only with the complete understanding of the means can the extremes be attained. With the complete understanding of both and all, means, extremes, in balanced proportion, dynamics, in toto, as TRU, can all be understood. The combined total of these and corresponding knowledge is signified by: **OKOM** = Only with the complete understanding or Knowledge of the Means and extremes, or polarities and the added characteristics of their Totality, TIT-TO, can all be understood or Known.

In order to correct, or greatly improve, the defects produced by fixed levels of REMS it is necessary to condition them and provide the standards and laws that will both explain, help to control, and make them more accurate and practical. **LASTEC** or The Law of Standardized Sectors, the means of separating both the levels of REMS, as well as the realms of different systems whose various components are for valuation. **LADER** = Law of Decimated REMS. REMS have distributions that are more or less decimated, e.g. for a certain concept and its object a little could be placed in 2R, much more in 3R, most in 4R or lower REM level. **LASFORD** = Law of Standardized Focus of LADER. Each item for posit should have a focus or an average. It is necessary to provide some focus or centerpoint for the representation in the total REM. This will often be considerably less than its highest level. This helps to correct many of the problems of severe overloading at each REM level, especially 1R, 2R, and 3R. **LASFALT** = Law of Standardized Focus of Á, L, and T. A standardized focus for each item is required, with a term for its object, part or process that is largely set in TIT. Its range, territory, duration, and reality need to be known and shown. **LACMEST** = Law of Correction for Medializing of Standards, or the correction necessary to bring standards, and any stereotyping effects, back and as close as possible to a 1:1 replication of their part and role in TIT. The work of LACMEST is difficult, for each standard that is added makes that much more LACMEST necessary. Therefore standards be kept at a minimum, LACMEST must be accomplished with ACC and the utmost vigor, means and perfection available.

If we can obtain an atlas of the world, TRU, with the same perfection that we have atlases of the earth we will rapidly approach a more accurate and perfect knowledge and world view of all. Only SPALT, POS, STA and HOM are more difficult for they are less known and developed. An atlas shows the periphery or surface of the earth, whereas SPALT and TIT are spherical and exist by TROS. It is only up to us to see the light and take the giant steps needed to replicate the massive and beautiful TIT and fulfill its object, OBT. It is up to each individual to learn, as early and as well as possible, the needed visions of TRU. So that each person can not only be oriented in time, space, and their mass-energy but all together in life that makes up TRU. Differential rates or degrees of stasis and dynamics occur in TRU like in the human body. Some parts are highly dynamic, as evident by their susceptibility to damage by radiation, and some parts are highly static. This is related to their form, structure, mass, motion, and energy. It shows how we can use them to determine the parts and processes in TRU and determine the concepts, principles and laws, mechanisms, and the overall state and function of TRU.

When we engage in using REM and POS in STA to produce accurate POSITS and STATICS, it helps to apply the basis upon which they are formed. That is, even if only static, they are: 1. discrete constituent parts of TRU, ES, 2. aspects of TRU, 3. ways of qualifying TRU or ES, 4. ways of looking at TRU or ES, 5. some other feature of TRU or ES. Many of the common words in use, even if all that is not their ESIN is extracted, do not differentiate between these kinds of STA. In fact they often infer one or the other variably. Thereby it is well to determine and enumerate which words we use are of which kind of STA. Abstract names or nouns of ideas or qualities can be any of the above, although more often # 3. Collective nouns of name groups are likely to be # 1 although they can be of 2, 3, 4 or 5. Common nouns or names of classes are likely to be # 1, although they can be of 2, 3, 4 or 5. Concrete nouns or names of tangible things are usually # 1. Count nouns or names of things that can be numbered is almost always # 1. Mass nouns or names that are normally not counted although often of # 1 may be of any of the others. Proper nouns, are usually # 1. Specific nouns are almost always # 1. Pronouns will depend upon the kind of noun they substitute, and thereby often # 1. Adjectives are usually more or less qualifying and usually # 3. Verbs and adverbs being dynamic are of CES and HOM, dynamic thereby have their own bases. Prepositions, conjunctions, interjections and subclasses of the above are usually # 3, 4 or 5, but may sometimes be given approximate POSIT. This is especially so when used in reference to a # 1 name. Some words can be of more then one of the above kinds of nouns. All words to be sufficiently qualified for STATICS must be reduced to their part in TIT. To indicate the degree to which a word can be posited, and all other necessary qualifiers applied, are shown in a variety of set positions around it. These include a rating by percentage of perfect posit, where 10 would be the highest limit.

There are several concepts, principles, laws and , CALP, of STA that should be repeatedly held in mind and appropriate corrections made where necessary. **PADIP** = Problem of Non dimensional POSITS. This is either because their dimensional foci do not exist, are indistinct, diverse, or some defect in the OBT, ROT, MOT, REM, POS, or STA. **PLADIP** = Problem of PLAP or Dimensional Polarities, is much of the problem of poor POSITS of Á, L and T on a 2D field. It is a result of the occlusion of the third dimension. **PLAP** = Problem of Planar POSITS, **MESEP** = Method of Separated POSITS, is by separate combinations of the three, as AL, AT and LT and providing for each their positions. This requires three separate planar fields and an incomplete demonstration, further confirming the value and need for a total vision of 3D global SPALT of TIT. **METANG** (LATANG) = Method (and Law) of Triangulation, is a program for reconciling MESEP, the method of POSITS in three dimensions. To reduce the dimensional position of any item to a planar posit, the resultant of the averages of each of two combinations must be produced. If Philosophy has a 20-40-25 of the three dimensions to posit it properly on a textbook page, it would be the resultant of posit of 20-40 x 20-25 x 40-25. This becomes a law when in the use of a planar field we are to provide accurate posits. **DIWAT** = Differential Weighting of Posits, how the amount of each dimension in each item to be posited is not the same. In fact, many items are mostly of one dimension and very little of others. To compensate for the skew that would result when a posit is made without correction, DIWAT must be applied. **SCON** = Multiple or Plural Comprehension Corrections or Reconciliation of POSITS. To have accurate, reliable and representative posits all factors that affect the posit process and a posit's optimal loci must be taken into account. **PRENG** = Precision Engineering, degree of, or how POSITS can be no more valid than the degree to which they are engineered, i.e. designed and operated with precision. A corollary of PRENG is the degree or level to which this precision is carried. This is called by the exponential of precision, e.g. **PRENG1** = Posits to only one decimal point, **PRENG2** = Posits to two decimal points, etc. Such precision is a trade-off with human capacity and practicality. A level of practicality suitable for an individual raised to TES. So that at the present time we do well to provide PRENG1 or PRENG2 with adequate precision. The advance of this, as with human development and language, will be heavy and slow. However, its needs, benefits and ATIM will more than compensate for the endeavor. In this text we largely use PRENG1 to introduce or provide a simple presentation of a concept or theme. We use higher precision values for the most important ES and REM, and future refinement. Even PRENG1, and especially PRENG2 by providing a 10x10x10 or 1,000 and 100x100x 100 or 1,000,000 separate parts is adequate for a highly reliable representation of TRU.

CHAPTER 46

FORM of STA and STATICS

GENERAL FORM of STA and STATIC POSITS or STATICS of TRU: THEIR LIMITS, SYMBOLS, CHARACTERS, DIVISIONS and SECTORS in MOT and SPALT

Position and Placement are large contributors to the proof of TRU. They are largely a product of KOT and the relationship between IA, Stimulus response, and S-O in and out of TRU, as demonstrated by Figure 1-46. They accompany the various perspectives and standards of POT, as it does for knowledge in general, or epistemics. They help to show what existed before man, yet is certain to be of TRU.. What is a very great part of TRU is bound to be either great TE, REM, or of some less concrete, discrete, or indirect part or aspect of TRU. Position and placement through DAT, DET, and CES singularly, by complexes, and totally reveal the interactions that best work to support TRU, They are also shown by natural, physical, and biological principles, laws, and TÂ, how TRU, IPTRU and POS can be concrete. How they are absolutely imperative. This is selection, by and to TES and TRU. They seek to reveal how and how well we do, or do not, provide a POA for TRU. Position is a feature of animal life becoming more coordinated and effective with a more complex nervous system and brain. Their characteristics and principles in operation will depend upon what the intended position is for. Most primitively it is to reorient the animal, man, and TRU, in space in response to stimuli.

Fig 1-46 **DERIVATION of FORM of STATIC POSITS**

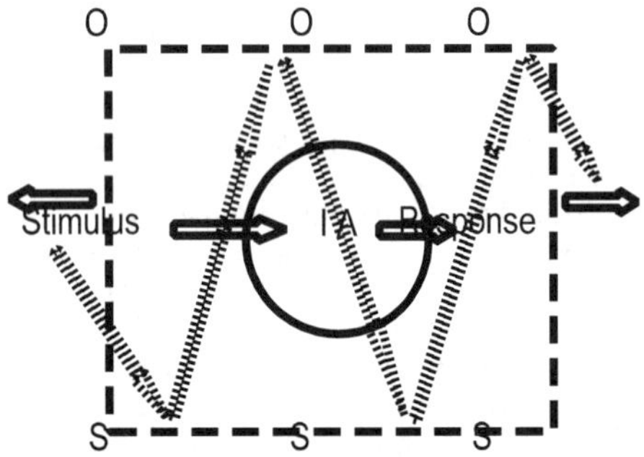

From a simple reaction to stimuli the IA and mass energy FAF increasingly developed into the highest form of position. This is the highly accurate, reversible, positive, and valid placement of the individual adapting to their environment, and the environment to the individual. It is especially those of ideas, thoughts, concepts, principles and phenomena in a frame of TREQ, TRU, TIT, and TRIT. This highest form of position in TRIT, OBT, and SPALT, although in a developing condition, is greatly enhanced by super and sub scripts for REM, POS, STA, and HOM.

ESNES = ESINS of nominals of the same ES. Whatever object, name, word, symbol, description POSITS are designated by, they have actual meaning and definition when they relate to or denote the same in TIT. When their ESINS are extracted from the ENS, they refer to this same ES. Likewise this is how the words in each and every language are the same when they represent the same ES, REM and POS in TIT. This is how a language maintains consistency, and how it is easy once learned. More or less it varies to the degree shades of meaning make the term or POSIT slightly different. Symbolic meaning is often of great value in depicting ESINS. The discovery of ever greater meaning is a major key function of our being. The history of meaning and use has paralleled the history - and prehistory - of mankind. What was not, was,is,and will be directly apparent, of the actual world, was represented in various ways. This included the use of symbols to stand for some significant aspect of life. Some common object, as a standard for an army of hosts, or a mark that stood for a ruler or god in their absence. It with repeated use became of increasing practical value for one and all, as 'for me', and as I am/I am not, common in TRU. In order to establish the discrete, formally distinct and well delineated components of TRU that can be posited as they statically represent the different aspects, parts or other elements of TRU, we should understand and develop the separation and limits of these components. To do so, by TRU, ROT, MOT, SPALT, REM, POS, and STA we should recall the dependency of such separation and discrete form upon limits.

We can use STATICS as our vehicle to obtain a good vision or panorama as well as to deepen our recognition of the intricately detailed makeup of all of TRU. POS, STA, HOM, REM, MOT and ROT to be valid necessitate use that is not only accurate, it must follow the conditions, qualifications, SCARLS and VAPOS that determine them. To do so we design and develop the following criteria for the paradigms and algorithms that these POSITS and STATICS should follow: 1.Validity, 2. TRU, 3. Proportionate by REM, 4. Of the whole, TIT-TO, and recognizable as its part or aspect in the whole, 5. Unambiguous, 6. Avoid multiple or contradictory meanings, 7. Have semantic accountability and simple syntactic recognition, 8. The amount in the dimensions of TRU must be relative and weighted to their actual composition, 9. Each dimension must be carried to 1/l00's, or where possible 1/1,000, for good technical use.

Further guides for the SCARLS of STATICS are for the whole of TIT and SPALT: 1. The elements or items to be posited are not based upon their identical ES or CES of TRU, but some representation. 2. Since the meanings and representations of words are rarely based upon a VAPO reduction to TIT, there always exists a more or less defect, and loss of actual, correct and reliable basis for posit. This is in contrast with NEC especially precise divisions of TIT, TRU, TRIT and OBT. 3. The SCARLS of POSITS will vary with the weights of each element, ES, or CES. Initially and simply, POSITS are made as if the three dimensions are all of the same weight, strength or power in the ALT. Actually this is very much not the case for most all have some, and many have extreme differences in the weight, strength and power of A, L, and T. For correct posits this must be accounted for.

Guides for sectors or portions of SPALT include: 1. Depend on the parameters and their scales, 2. The numbers for each segment shown on the face of a planar representation of each sector do not nearly approach the number existing within its region in SPALT, or TIT. This is in part because most of these segments exist within the spherical SPALT and TIT and not on the external face of it. 3. Many posits made by these segments have dimensional foci in the three dimensions which exclude their representation by ALT. That is their posits in each dimension are so distributed as to cancel each other out on a planar figure. 4. Associated with # 3 is the loss of accurate planar posit for all due to this variance from precise representation. Each posit will lose accuracy in a planar figure by virtue of its loss of identity with exact position in TIT and SPALT. 5. The attempt to correct # 3 and # 4 by triangulation is limited by the planar defect. For this fallacy of cancellation in a triangulation in 2D approaches the absolute when the three dimensions are exactly opposite, both qualitatively and quantitatively, or by polar extremes and distribution. Also Improved POSITS are achieved if that to be positioned is largely of only two, or especially one, dimension.

To gain an understanding of the composition of TRU, as well as to gain a better understanding of the many fields and sciences that make up TRU and how they tie together, we need only to posit them in MOT, SOID, ALT, SPALT, TORP, SPOT, and METAD. (Table 1-46) It is based upon: (1). The foci of each in TRU, (2). Their spread in TRU, (3). Their magnitude in TRU. In addition we can rate Posits for STA and HOM by degrees of complexity. These include, divisional, sectional, combinations, complex, compound and total. For STA we note these divisions as means of collectively sparating and rating fields and sciences. When taken together with all the rest of what is static in TRU it helps in the delineation and order of SPALT Posits. When combined with the weighting by relative quantities of different dimensions it gives us a much more accurate and fuller posit and representation of these topics, fields, and sciences and their place and role in TRU and SPALT.

I MAJOR CATEGOR- IES MAIN FOCUS	PERSPEC- TIVE and PRESEN- TATION 0+	RELIGION 1+	REPRESEN- TATIVE ARTS 2+	S C PHYSICAL 3+-8+	I E LIFE 3+-8+	N C SOCIAL 5+-9+	E S APPLIED 6+-9+	POLITICAL 6+-9+	BUSINESS 6+-9+
II PRIMARY DIVISIONS	PHILOSOPHY EPISTEMO- LOGY KNOWLEDGE MATHEMA- TICS METAPHY- SICS ONTOLOGY LOGIC	THEOLOGY DEITY GOD ETHICS	REPRESENTA -TIVE ARTS LANGUAGE COMMUNI- CATION MATHEMA- TICS MUSIC DESIGN	PHYSICS ASTRONOMY EARTH CHEMISTRY INORGANIC ORGANIC PHYSICAL ANALYSIS WEATHER & CLIMATE ATMOSPHERE AIR OXYGEN FOOD WATER ATOMS	BIOLOGY HEALTH SCIENCES PSYCHOLOGY ATTENTION HUMAN BEHAVIOR GENETICS & HEREDITY	SOCIOLOGY HUMANISM HUMANITIES EDUCATION ANTHRO- POLOGY HUMAN EVOLUTION HISTORY	ENGINEER- ING INDUSTRY SHELTER MANUFAC- TURING TRANSPOR- TATION	GOVERNMENT LAW CIVICS	ECONOMICS COMMERCE TRADE TRANSPOR- TATION SALES WORK and EMPLOYMENT COMPETI- TION
III SECONDARY DIVISIONS	IDEOLOGY LEARNING THOUGHT and THOUGHT PROCESSES	MONOTHEISM POLYTHEISM BUDDHISM CHRISTIAN- ITY CONFUCISM ISLAM JUDAISM SALVATION REDEMPTION JUSTIFICA- TION GRACE HOPE MERCY	THEATER ARITHI- METIC WRITING and LITERA- TURE DRAWING PAINTING	BIOCHEM- ICAL COM- PONENTS OF ORGANISMS COSMOLOGY EARTH SCIENCES	BOTANY MOLECULAR BIOLOGY MICROBI- OLOGY PREVENTIVE MEDICINE ECOLOGY BODILY ORGANS SYSTEMS IMMUNITY NUTRITION	ASIAN PEOPLES and CULTURE CLOTHING DRESS and ADORNMENT	CONSTRUC- TION INDUSTRIA L-IZATION AGRICUL- TURE	CONTINENTS NATIONS INTER- NATIONAL RELATIONS TAXATION CONFLICT AGGRESSION and WAR	BANKS and BANKING TELEVISON ECONOMIC THEORY SYSTEMS GROWTH PLANNING ADVERTI- ZING
IV TERTIARY DIVISIONS	PHILO- SOPHICAL SCHOOLS and DOCTRINES	EVOLUTION CREATION & RELATIVITY	NUMERICAL ANALYSIS THEORY DIMENSIONS LENGTH and TIME	GEOMORPHIC PROCESSES THERMO- DYNAMICS	BIOSPHERE BLOOD BIRTH CONTROL CIRCUL -ATION CYTOLOGY DIGESTION DIAGNOSIS THERAPEU- TICS	REGIONAL HISTORY HUMAN RIGHTS	DECORA TIVE ARTS and FURNISH- INGS	LARGEST NATIONS	BROAD- CASTING
V OTHER	PLATONISM PHILO- SOPHY of MIND	HISTORICAL and REGIONAL RELIGIONS ZORASTRIAN- ISM	STATIS TICS TOPOLOGY	SOILS	INFECTION ARTHROPODS			LARGE NATIONS GREATEST CITIES	

Table 1-46 **POSITS of FIELDS and SCIENCES**

This is also seen Figure 2-47. Some fields and sciences, and much in TIT, are more of one dimension than another. When this is so, if we do not make correction, a somewhat abnormal and biased posit is produced. For example the following sciences have weights that are somewhat more in the following dimensions or continua:

Á = Psychology, Philosophy, Economics,

L = Sociology, Geography, Psychology, Biology, Anatomy, Comparative Anatomy,

T = History, Archeology, Geology, Paleontology, Physiology,

Fields and Sciences may also be divided , both separately and together, by the form or function of TRU. This is of great significance in the differentiation between STA and HOM. For it shows what to leave for HOM and what to emphasize in STA. In this way we largely seek POSITS in static fields such as Anatomy, Geology, Geography, and leave for HOM fields such as Physiology, Sociology and History. However, they should be remembered in their overall settings. This is why in demonstration we use a hashed line instead of a solid line.(Fig 2-48) They correspond and may be used together with REM and MOCBIREM. There is a great amount of information, both principles, concepts, research, evidence and data that are not included in IPTRU that apply to a POT of TRU. The larger POT and IPT are the more can be included, especially when it is better adapted in the future to produce a vastly superior and perfected presentation of TRU.

When we differentiate the fields and sciences that posit in SPALT we can roughly assign levels of each dimension. From this we can also assign levels of each dimension for each of their divisions, subfields, subsciences, and specialties. In this way, as in Table 1-46, we can better posit all, and can more fully relate them to the whole of TRU, to produce its improved and complete representation. We can then more successfully plot out all. The posit will be easier and more accurate. It can not be done without full application of the principles of ROT, MOT, REM, POS and STA involved. Nor can it be done without the best reduction of each science, field, and their divisions to TRU. When these are applied, it will usually provide a very clear POA. This applies not only to fields and sciences but to all knowledge, all TRU.

At the 10 divisional levels of the dimensions of ALT, we have 10 x 10 x 10 or 1,000 parts. If we have the A, L and T scales with each of the ten divisions, we have:

S	**Sa**	**SA**	**AS**	**As**	**Ao**	**AO**	**OA**	**Oa**	**O**
L	**Lm**	**LM**	**ML**	**Ml**	**Md**	**MD**	**DM**	**Dm**	**D**
P	**Pn**	**PN**	**NP**	**Np**	**Nf**	**NF**	**FN**	**Fn**	**F**

The abbreviations used are as follows:

S = A major part Subjective
s = A minor part Subjective
A = A major part PhenomenAl or ReAl
a = A minor part phenomenal or Real
O = A major part Objective o = A minor part objective
L = A major part Large l = A minor part large
M = A major part Medium m = A minor part medium
D = A major part Diminutive (small)
d = A minor part diminutive
N = A major part now n = A minor part now
P = A major part past p = A minor part past
F = A major part future f = A minor part future

Such a dividing of TRU in ALT and SPALT for STA and STATICS reveals a good base, for it not only enables accurate and reliable posit of all TRU, it gives a means by which other components of TRU are compared in such a way that more precise measures of POSITS are achieved.

Probably the greatest of all human failings has been our inability to think in terms TIT, TRIT, OBT, and its solid three dimensional frame. Glimpses of TRU as a God, or even a page's illustration of TRU lack adequate circumspection. Thereby, if the human race is to overcome this failing and avoid the misconceptions, bad decisions, poor judgement and confusion that so often accompany in-coordination, partiality and disorder, we must spend much time, interest and effort in developing and mastering our ability to think, reason, remember, and apply the solid form, construct and nature of TIT-TO of TRU. In addition, all that goes into this solid construct as POS, STA, and HOM that benefits learning and understanding is of the greatest benefit. **MOSUN** = Method of Solid Spherical or Three Dimensional Understanding. **MOCO** = Method of Total Comprehension.(COAT) **MOS** = Method of Solid Observation, OM, Totality, especially TIT-TO-POT. MOSUN is both highly necessary and limited, yet what mankind so often fails to recognize and utilize. However, MOSUN is not a panacea. It does not include all kinds of understanding. Nor can it take the place of description and interpretation. MOCO is the general case of MOSUN and includes these and other kinds of comprehension and understanding. MOS is the method and mastery of MOSUN and MOCO whereby they can be applied to TO, OM, POM, SÂ and TÂ.

SPECIAL FORM of STA and POSITS in STATICS

POPTE = Principle of Piecing Together Essences. POPTE is the key critical means and method, like scientific experimentation, whereby conclusions are not drawn outright, but become evident only when a sufficient amount of valid evidence, data, information, and reasoned interpretation are provided. Directly or indirectly and subconsciously or consciously, POPTE often involves, the extrapolation from ESINS, ES, aspects, and FAF of TIT. POPTE is closely related to induction. However its focus and ends are different in being specifically toward essences, ESINS, ES and TRU. For STA it is the means of providing accurate POSITS of very important terms of TRU.(CALP). The ultimate end of this is unity in totality, with expanding harmony of integration and differentiation, of what is TIT and TRU.

The factors, indices, and other qualifiers that need to be expressed for POSITS are primarily shown by superscripts and subscripts. This as in mathematics is an extremely effective way of adding precision and power to each element, ROT and TRU, even though it adds an extra burden for the author, reader and student. On a page it is not possible to provide words with many superscripts and subscripts. Even four adds much complexity to a presentation, especially if frequently used. The solution to this is to provide for each term or item for posit their required qualifiers and factors in separate listings, or in the glossary or index. Whatever needed descriptive entries are added to round out the representation..

Here we show four major qualifiers for each REM and POSIT. These are for # I, REM and # II POSITS as follows:

I

Proximal Superscript = V = ability to value term
Proximal Subscript = R = REM value of term or ES
Distal Superscript = r = range in each dimension
Distal Subscript = d = items percentage in this r

II

Proximal Superscript = P = ability to posit term
Proximal Subscript = R = REM value of term or ES
Distal Superscript = r = range in each dimension
Distal Subscript = d = % item in each dimension

Other important qualifiers of the position of all in TRU include: whether or not they are: diverse or highly concentrated, degrees of power in each of the three dimensions, and ESIN percentage.

How can we provide a better DET and, ESINATION of all that is of TRU? How can we obtain 100% perfection, or a highly practical E and ES of TRU? How can we transform this into highly accurate POSITS is shown by:

1. 100% ESIN, and 100% ES and CES,
2. ESIN and ES extraction, divisions to be valid must be exactly separated,
3. Numerical identity with E, REM, derived from the methods of mathematics, statistics and probabilities,
4. When on one word, take out extractions, e.g. LP, OM,
5. Principle, idea, concept, word, and thing to be valid must be 100% identical, e.g. SEO,
6. All must be capable of being brought to a circumscribed, limited, or discrete part of TRU.
7. The discrete static and dynamic, in POS of STA and HOM must replicate the same in TIT.
8. All knowledge and its methods must be called upon to provide whatever assists POS and TRU.
9. The end product should be kept in mind.

POR = Principle of One over Zero Ratio. POR is where and when TRU, as TU, moves from its oneness or relative unity, one over zero, 1/0, or toward zero over one, 0/1. The first is an absolute in accretion, the latter absolute in diminution. Neither is possible by human ability as it is in ATO and TRIA. It exists in a flux between absolute and zero, infinite and infinitesimal, a maximal and minimal finite with a more or less, optimal fixed center in unity. In their combination that leaves us with a ratio of an in between, TU-TIT/Zero or POR. It is the positive, or 1/0 and its finite range, to which we aspire and the contribution of all parts of TRU partake. Yet we should not forget their like contributions and all efforts for the stable state, peace, conservatism, equanimity and physics of TRU. This is TRIT state or: **STAT** = Stable State of TRU, where optimal TROS is maintained. STAT is a state or form and function that is relatively free from external change and alteration. It is this stable state of TROS and TRIT of TIT that natural selection results in, and provides us with the basis for form and, thus, for the development of STA, POS, TIT, TRU, POT, and IPT.

CHAPTER 47

PART and PARTS of STA and STATIC POSITS

STATICS of FINITES, CONCEPTS, TOPICS, MEASURES

Much in life is dependent upon the magnitude and range of knowledge, technical information, and the objective treatment available and committed to memory; e.g., how many different specific kinds of treatment a doctor knows, or works an artisan can expertly produce. How well versed, knowledgeable and capable each and all professionals, technicians and workers are with the use of their special knowledge and tools? It is the accelerated and expanding quantity and quality of these that is much of human advancement, our mass energy Á FAF. It is what each and all person's expertise, professionality, and contribution depends upon. So must it be for all in TRU. For all is very inefficient in use without a POA that can replicate TRU. Such an armamentarium is provided and greatly enhanced by STATIC POSITS in MOT, ALT, SPALT, STA, HOM and ACC.Almost all knowledge can be reduced to POSITS with increasingly precise REM, coefficient factors, and qualifiers. This requires more valid and TRU meanings, limitations and interpretations through VAPO, SCARLS and all that increases their reliability. Intellect imposes a static character on concepts. It constructs general truths. They are not necessarily nor always TRU. They, even more than the world around us that tend to TRU, are an accelerating mechanism for developing the essences of truths.

Parts of STA, and finite STATICS, may be compared with the notes played on a piano, or the music of a great composition, concerto, or symphony. They multiply the notes in stanzas, rhythm, lines and pages. The degree to which this finite form can be developed is, as in formal distinction, key to the makeup and scales in SÁ, the observer-observed, subject-object relativity. This is what TRU is and does, as each person is predominantly FAF by the kinds of mass energy exemplified by reasoning and each level of its continuum, or SÁ.

Fig 1-47 **STATIC POSITS in SOID**

SOID helps to reveal TRU in a somewhat different light and to better show how different SOID and SPALT POSITS can be achieved. It also shows some of the problems that occur in SPALT. In figure 1-47 we show the 2D SOID frame and use it as an example of what can be accomplished.

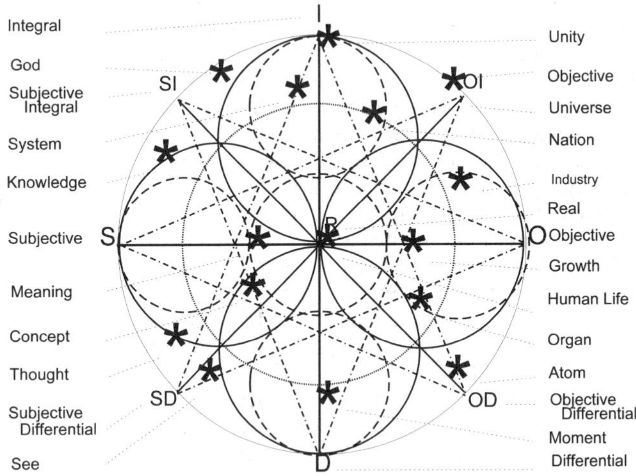

In the glossary or ESARY that defines and references NEC, the ESINS and ES of many existing words could be similarily given. Much more information can be provided for each in a larger presentation. Generally they are shown by NEC, EC, OC, or word, definition, REM and POSIT. Often it is the relating that goes with combinations, phrases and descriptions that produces more complete TRU meanings.

DIMENSIONS, LEVELS, DISTRIBUTION and FIELDS

Fig 2-47 **STATICS and SEPARATE DIMENSIONAL POSITS**

In order to separate and develop better POSITS and STATICS by dimensional scales it is helpful to clarify their compositions, those elements in each that most characterize them for TRU. This we do for Á, L, and T. We do it by showing how they are qualified, by VAPO, DAOST, SCARLS and all that shows their separate part in TIT. In ROT, MOT, ALT, SPALT, REM, POS, STA and HOM we largely show those terms that best describe TRU. This includes many NEC that are of great REM. (TOPNEC) There are many ideas, concepts, old and new, especially principles, mechanisms, laws, and proofs that apply to more than one chapter of IPTRU. This application may be much more than a simple dependent relationship with that chapter to which they are associated. Often it may be a prime posit in two, three or more chapters, as well as variations in definition, meaning, interpretation or application. This is what description is greatly about. It is well to show these, yet at the same time avoid the pitfall of losing specificity in POS and STA.

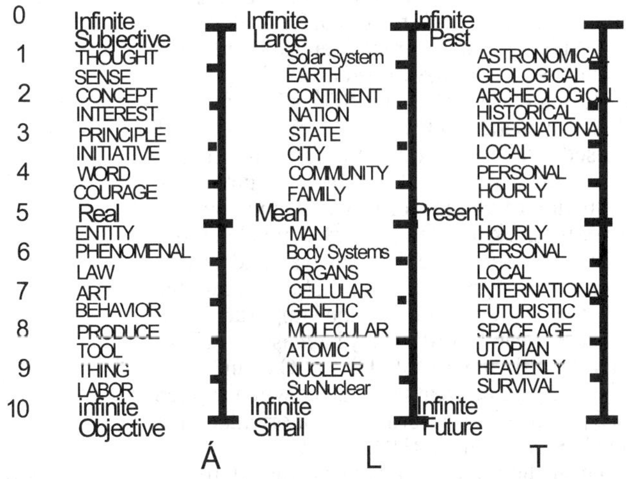

The uses of many different kinds of suffixes to represent basic qualifying modes of life, as states, conditions, and characteristics, are not only of great reality and practical value, they are great aspects, parts. and REM of TRU, e.g. 'ness, 'ity, 'ion'. They provide a vast reservoir of terms that give STATICS greater breadth. Their explicit meanings, as representation of TRU as well as REM value, also help to fill the voids and provide a good total representation of the actual world, TRU.

Figs: 3-47,4-47,5-47 **2D PLANAR of ALT, THREE PLANES of SPALT; ES, POS and STA**

In Figures 3-47, 4-47 and 5-47 the three fields of ALT are shown. The sections of ALT are as if the 3rd dimension in each case is left out. This is a good way to initiate POS and Statics. If we can adequately present the three planes of which SPALT is formed in such a way that it exactly replicates TIT, and then proceed to apply the POSITS, STATICS and REM that are appropriate for the ES and other aspects of TRU, we will set a good foundation for representing TRU as a solid spherical whole.

The three planes, represented by two dimensions each of ALT, taken separately are shown as: **ÁL, ÁT, and LT.** We show these planes as figures to help in STA, POS, REM and MOT and provide a concrete POA to TRU. They help to produce a much more accurate demonstration when the planes are combined. Yet it must always be remembered that these are only sections through, and faces within, SPALT. OBT, TRIT and TIT and not the whole. We should constantly convert each to its actual place and role in the whole.

In these figures many items, as ES, are reduced to POSIT in each of the three planes depending upon which combination of the two dimensions they are most characterized by. This is in which dimension they are centered or most prevalent or of which each specific item to be posited has the largest and most prominent proportion and contribution. When completed imagine and perform the operation of combining these three planes and all their contents into SPALT and the global whole of TIT, as it can be best shown in solid, spherical form. Then also consider all that is I/O as in POMOT, METAD, and SPAT. Tie these in with the greatest applicable CALPS provided by POT and IPTRU.

In these diagrams we use those ES, aspects, topics, systems and groups whose constituents form sufficiently measurable parts. By selecting those major portions of TRU which are well formed with parts that can be measured we develop consistent and reliable POSITS and STATICS for demonstration. This test is supported by the fact that REM, POS and STA can not be more than the limits of A, L and T they are formed by. This is the test of STA, POS, and REM by ALT Limits. To simplify and add precision to this study we emphasize the groups and systems in the world which are of TRU. **SIT** = The average TRU Self. Oneself, as TES, is the core of TRU. SIT can be used as a test to extract the TRU meaning of ALT, from the earth's m, l, t. In the enumeration and evaluation of all the words, terms, and nouns in language, we seek to relate, measure, posit and use them. This we do in more than one way and section of the text. It is of great value in REM. It shows the great value in determining the importance or significance of each word as they increase from ENS to ESIN to ES, a part of TRU, POT and IPT. Likewise they are of great importance when applied and used practically as information for orientation in ES and CES, STA and TRUD of TRU.

Some of the topics provided by encyclopedias beginning in an alphabetical order can be given as examples of STATIC POSITS: Africa 8-1-5 3R, Alphabet 3-6-7 3R, Amino Acids 8-8-5 3R, Amphibia 8-5-0 5R, Anatomy 9-5-4 2R, Animal Behavior 7-5-3 4R, Apollo 2-4-1 7R, Arab 6-2-2 5R, Archeology 5-4-1 5R, Arch and Vault 9-3-2 6R, Aristotle 1-5-1 4R, Art 6-5-5 3R, Atmosphere 9-1-6 4R, Atom 9-8-5 4R, Bacteriology 7-6-6 4R, Banking 6-3-6 5R, etc.

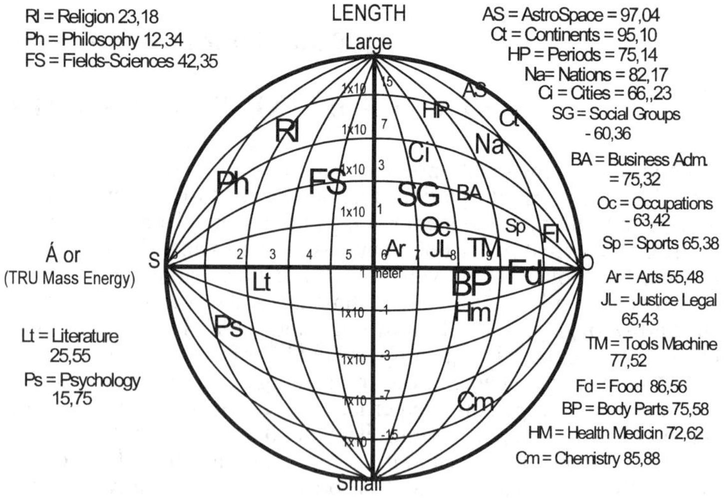

ÁL PLANE of SPALT

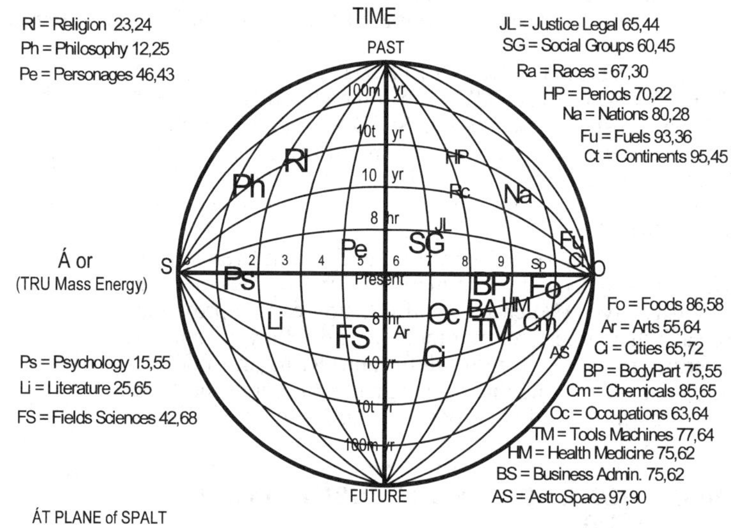

ÁT PLANE of SPALT

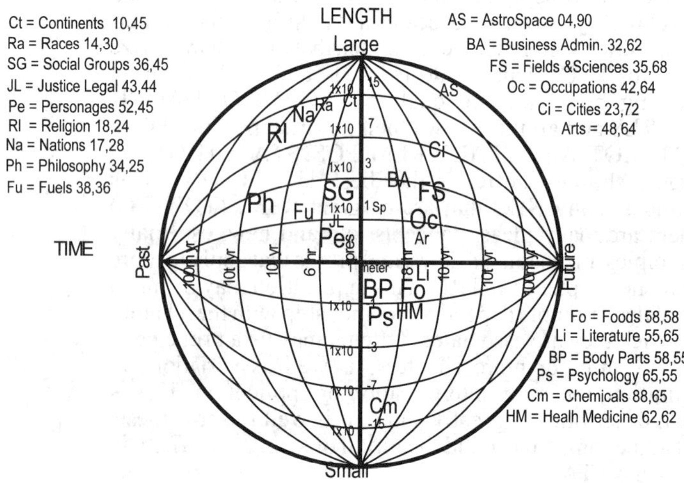

LT PLANE of SPALT

ALPHABETICAL FEATURES, FORMS, STRUCTURES, CHARACTERISTICS, PROPERTIES OF TRU, TÄ

R	Á	L	T												
				DEMOCRACY 3c	7.5	3.0	6.5	HOPE 3b	1.0	6-u	7-n	PASSION 4	6.0	6.0	5.0
A, AN 2	5~	5~	5+~	DESIRE 3b	3.5	8.0	6.0	HUMAN 1	6-m	4-m	6-m	PEACE 3c	8-u	3-u	7-u
ACT,ACTION 3a	9.0	2.5	7.0	DIGNITY 4	5.0	4.5	6-u	HUMANITY 3b	6.5	2.0	7	PEOPLE C3b	8.5	3.0	6.0
ALSO 3b	5~	5~	5+~	DIVINE 3c	2.0	4-n	7-u	HUNGER 3b	8 u	6.0	6-u	PERSUASION 4	3.5	6.0	6.0
AMBITION 3a	7.5	5.5	6.5	DIVINITY C3c	5.0	2-m	8-u	IDEALS 3b	9.5	8-m	8-u	PHILOSOPHY C3a	8.5	2.5	7-n
AND 2	5~	5-~	5+~	DUTY 3b	7.0	4-m	7-m	IMMORTALITYC4	8.0	5.0	9-u	PLEASURE 3b	4.0	5.0	5-u
ARCHETYPAL 5	7.0	3.3	7-u	EARTH 3a	9.5	1.0	1 n	INCREDIBLE 41	.5	6.5	7-m	POLITICS 4-	6.0	3.5	5-m
AUTHORITY 3c	8.5	6.5	7.5	ECONOMY 3b	9.0	3.0	7.0	INFALLIBILITY4	1.07.0		8-u	POVERTY 3c-	8.5	4-u	5-m
BEAUTY 2	9.0	4.0	5 n	ECONOMICS C3c	8.0	2.5	7.5	INFORMATION3b	3.0	7.5	7-u	POWER 3a	7-u	3-u	7um
BEAUTIFUL C4	9.5	6.0	5 n	ECONOMIC SYS C4	8.5	2.5	8.0	INSTITUTION3b	7.5	3.5	7.5	PRAISE 3b	4.0	6.5	5.5
BIGNESS C4	~	1.7	~	EDUCATION 2	1.5	4.5	7-u	INTELLECT 3a	1.0	5-n	7-u	PRAY PRAYERS3a	2.0	7.5	6.5
BIOLOGY C3b	8.5	6.5	3 n	ENDS-MEANS 3b	6-m	4-m	7-u	INTELLIGENCE2	1.0	7.0	8-m	PRECEDENT 3c	6-m	4.5	3-n
BRAVE 3b	4.5	5.0	6.0	ENERGY C3a	7-n	5-n	6-m	JESUS CHRISTC3	2.5	5.0	7-u	PRIDE 4	3.0	7.0	6.0
BROTHERHOOD 5	5.0	4.5	5.5	ENTHUSIASM 3b	2.5	5.5	7-u	JUDGE-ary 3c	3-m	5-m	6-u	PRINCIPLE 2	3.0	6-m	7-n
BUT 3a	5~	7~	5+~	EQUALITY 3c	6.5	4.0	6-n	JUDGEMENT 3b	3.0	6.5	6.0	PRIVACY 3c	7.5	4.5	5.0
CAPITAL 4	8.0	3.0	7.5	EQUILIBRIUM 3b	5	5	5-n	JUST-ice 3b	6-m	4-m	6-u	PRIViLEGE 4	7,0	4.0	6.5
CAPITALIST 7	8.0	4.0	6.5	EQUITY 3b	7.5	3.5	7-u	KNOW 2	0.5	7.0	6.0	PROBABILITY3a~	7.5	n	n-8
CAPITALISM 5	8.0	2.0	7.0	ETHICS 3a	7.0	4-m	7-u	KNOWLEDGE 2	1.0	3.5	7-u	PROCREATION C	6-u	5-m	7-u
CATHOLIC 4	3.0	1 n	3 n	EUGENICS 7 +/-	9.0	2.5	7.0	LABOR 3a	9.0	4-m	5.5	PRODUCT-ion3b	8.5	3-M	7-u
CATHOLICISM 5	3.0	2.0	2.0	EVOLUTION 3c	7.0	2.5	7.5	LAND 3a	9.5	2.5	4-u	PROFIT 3a	8.5	4-m	7-u
CAUSE 3b	6n	7n	2.5	EXPERIENCE 3b	8.5	6.5	8.5	LAW C3b	4-m	5-m	7-u	PROGRESS-ive3c	7-n	4-n	7-u
CAUSATION C4	n-m	n-m	n-u	FACT,FACTS 3c	8.0	5.5	8.5	LEARNING 2	2.0	6-m	8-u	PROPERTY 3c	9.0	4.5	4.5
CHANGE 3b	n-m	n-m	n-m	FAITH 3a	10	8.0	10	LENGTH c3b	-	n	-	PROPORTION C2	~	~	~
CHARACTER 3c	3.0	6.0	4.5	FATE C4	8.0	7.0	8.0	LIFE-LIVING 1	6.5	3-m	6-n	PROSPERITY 3b	8.5	3-m	7-m
CHARITY 3b	7.0	4.0	5.5	FEAR 3c	6.0	6.0	7.0	LOVE 2	1.5	5-n	7-n	PROVIDENCE 2	8-m	2-m	9-u
CHEMISTRY 3a	9.5	9.0	4 n	FOOD 2	8.0	6.5	5.5	LOYALTY 3c-	7.5	4-m	7-u	PSYCHOLOGY 3a	9.0	7.0	5 u
CHRISTIAN 4	3.5	5.5	3.0	FORCE C3b	n	m	6-u	MAN-KIND 3a	7.5	4-m	7-n	PUBLIC 3a	8.0	3.5	6.5
CHRISTIANITY3a	3.0	4.0	2.5	FORCES 2	7-n	4-m	7-u	MARTYR 4	7.0	5.0	6.0	PUBLIC OPIN.3c	2.0	4.0	5.5
CHURCH 3c	7.0	6.0	3.0	FORTUNE 3b	8.0	4.0	6.5	MASS c3b	n	-	-	PURITAN-ism 4	6.5	4-m	2.5
CIRCULATION 3c	9.0	6.5	5.0	FREE C4	8.0	4.0	7.0	MATERIAL 3c	8.5	2.0	7.5	RACE 3a-	8.0	1-u	3-n
CIVILIZATION 2	7.5	1.5	6.0	FREEDOM 3B	8.0	2.0	7 n	METAPHYSICS C	9.0	6.5	7.0	REASON 3a	9.0	3.0	7-u
CLASS 3c	7.0	4.0	3-u	FREE SPEECH 4	6.0	6.0	5.5	MIND C3a	9.5	8.0	9.5	REFORM,,, 3b-	7.0	3.5	6-m
CLASSES 4	n-m	1-m	n-m	FRIEND(S) 3 c	4.0	4.5.5.5		MODERATION 2	9.5	6.0	8.5	RELATIV(ity)2	5n	5n	5n
COMFORT 3b	7.0	5.5	5-n	GENETICS C3a	8.0	7.5	3-m	MOLECULAR 3b	9.0	8.5	4 n	RELIGION C2	4-m	4-m	7-u
COMMUNICATION3c	5.0	5.0	6-u	GENIUS 4	2.0	6.0	5.5	MONISM C3b	9.0	3.0	7.5	REPETITION 4	6-n	6-m	7-u
" MEDIA 4	7.0	3.5	7.5	GLORY 3c	2.5	2 n	9 n	MORAL-s-ity C	9.0	4.0	8.5	REPRODUCTION3a	8.0	6.5	5-m
COMPETITION 3a	8-u	4-u	6-u					MORALE 3c	8.0	5.5	8.5	RESISTANCE 4	7-n	4-m	4-u
COMPLIANCE 3A	4.0	4.5	7-u	GOD C2	1-m	0-u	m-n	MOTHER-hood 2	6.0	4.5	6.0	RESPIRATION 3c	9.0	6.5	4-m
COMPROMISE 3b	6.5	4-u	7-u	GOLDEN RULE3a	8.5	4.5	6.0	MOTIVATION 3a	6.5	6.5	4.5	REVELATION 2	0.5	2.0	7-u
CONDUCT 3b	9.0	5.5	5.5	GOOD(NESS) 3a	7.0	5.5	7.0	MUSIC 3b	6.0	5.5	4.5	REVERENCE 2	1.0	6.5	8-u
CONFORMITY 3a	6.5	4-u	6-u	GOODS (m) 3b	9.0	6.0	4.5	MYSTERY-ICS 4	5.0	6.0	8.0	REVOLUTION 4~	9-n	2-m	6-u
CONSCIENCE 3b	2.0	7.5	6.0	GOODWILL 3b	3.5	5-m	6.0	NAME(S) C3c	7.0	6.5	6.0	RHETORIC 3b	4.0	6.0	4-u
CONSERVATIVE 4	6.0	4-u	6-u	GOVERNMENT 3b	7.0	2.0	6.0	NATION-ALITY3c	7.5	2.0	2.5	RIVERS 3c	9.5	2.0	1-
CONSERVATION3b	7-u	3-u	8-m	GREAT(NESS)3a	5-u	4-u	7-u	NATURE-AL C3a	8.0	2.5	3 n	SCIENCE 2	6.5	5~	5-u
CONSISTENCY 3c	8-u	4-u	6-u	HABIT 3c	7.5	5.5	5.5	NECESSITY 2	4-n	4-m	2-u	SENSE(SORY)3a	1.0	8.0	5-m
CONSTITUTION3b	6.0	4.0	6.0	HAPPINESS 2	3.0	5-u	7-u	NEED(S) C3a	6-m	5-m	7-m	SET THEORY 3a	~	~	~
CONVICTION 3b	2.0	7.0	7-u	HEALTH 2	6-u	5-n	7-u	NEUTRALITY 3c	7.5	4-m	6.5	SEX 3a-	7.5	5.5	5-m
COOPERATE 4	8-u	4-u	7-u	HEAVEN C4	7-u	1-u	9-u	OBEDIENCE 3c	7.5	5.0	6.0	SOCIOLOGY C3a	4.5	3.5	6-m
COOPERATION 3b	8.5	4.0	7-m	HERO 4-	6.0	4.0	7.0	OPPOSITION 3b	7.0	4-u	5.5	SOME 3a	~	~	~
COORDINATION 3a	6.0	5-	5.0	HISTORY C3a	7.0	2.0	1.5	OPTIMISM 3c	8.0	7.0	8.0	SUN 3a-	9.5	0.5	0-n
CREDIT 4	7-m	4-m	6-m	HOLY GHOST 6	6-u	4-u	3-u	ORDER 3a	7-n	2-m	7-u	THE 2	4.0	5.0	5.5
CULTURE 3b	6.5	1-u	6-n	HOLY SPIRITC4	3-u	4-n	6-u	ORGANIZATION3b	7.5	4-u	6-u	UNIVERSE C3a	6.7	0	0-n
CUSTOM 3c	8.0	3.0	2.5	HONOR 3c	4.0	4.5	6-u	PACIFIC-ISM 4	8.0	2.5	6.5	WATER 2	9.5	5+	1-m

Table 1-47 REMS and Á, L and T POSITS of Major Existing Concepts, EC. In this table we apply, analyze and determine the REM values and Á, L, and T measures and positions in SPALT for various major concepts, OC, EC, ECT and NEC of NES, ES ESIN, CES, and TRU equivalence. They are roughly posited between the first and second powers. The problem of using commonly held definitions and meanings is readily evident. To correct this we use n, m, and u to qualify those terms by their universality, multiplicity, and utility. Many of these terms can be much more precisely posited with appropriate coefficients and factors, how, and when they are used in combination with other terms. How peace, world peace with God, or peace within a person, posit differently. This shows the tremendous power and importance of language's combinations, composition, and rhetoric. These terms need to have their common meanings and their TRU meanings compared. Ongoing determination of their TRU meanings and POSITS will be required. This we seek to continuously do and extend in the latter sections and revisions of CIPTRU. Much will need to be done, as has been continuously done in the evolution of language and TRU, to make it an ever greater vehicle for human representation, POT, Á and TRU.

Qualifiers and factors used include, a = higher third of a REM level, b = middle third of a REM level, c = lower third of a REM level, C = Coterminality, m = Short Term, n = Mid Term, u = Long Term, ~ = Variable or widespread, - = If a marked Negative Factor, CHK = tentative u m n usage.

There is a striking difference between the portion of TIT something is, and the portion of something that is TIT. The first is its REM and is usually far lower than the second, its ESIN. The second is its ESIN and EC. The makeup of TRU, the problem with REMS, and much of our difficulty with showing and knowing the world as it actually is, TRU, is because of super percentages. This is where all is given values that far exceed their actual portions in TRU. It can not be corrected unless we discover and learn TRU, TROS, TRIS, TIT, TRIT, OBT, SUT, ROT, MOT, SPALT and REM. Super percentages, over valuation, and elevated REMS arise from Lack of VAPO, and the failure to base the scales, for yielding levels of REM on TRU and ALT. For TRU, TIT, and TO alone will not provide a good measure and demonstration of TRU. What actually is, and what is used beneficially can be precisely quantified if adequately supported, related, and qualified. The static posits of ALT, REM, and qualifiers help overcome tendencies to mistake all we encounter and must value. It helps us appreciate, correct, and apply the great words, CALP and mechanisms we continually must choose from and use, as ES, CES, REM and POS become truth.

CHAPTER 48

TOTALITY of STA, STATICS in SPALT and MOT

STATICS of TOPICS, FIELDS, SCIENCES, REM, TOPNEC

It is only when we reach the totality of POSITS that we can begin to see their fusion, value, problems and solutions. Among the problems are those of the limits of centering posits.

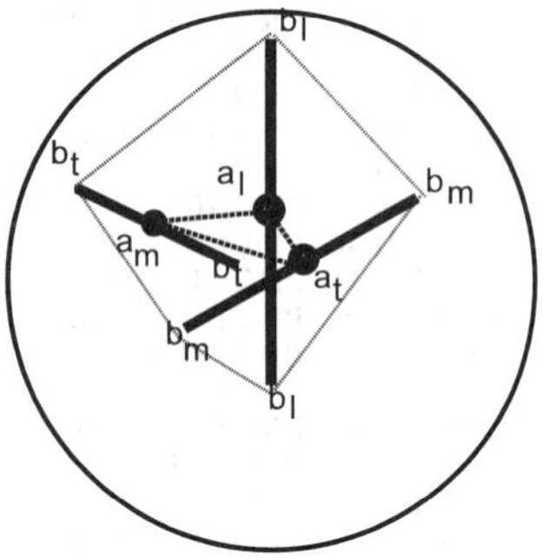

Fig 1-48 **FALLACY & LIMITS of CENTERING POSITS**

In the POSITS of entities, objects, words, aspects, parts and processes of TRU in MOT, ALT and SPALT it becomes quickly apparent that these do not produce in a uniform, evenly distributed pattern. This can be somewhat corrected by expanding our planar figure of SPALT into equal areas by volume, SPEV. However there are other causes, one of which besides being less than 3D, is: **FOPDIC** = Fallacy of Positing items at the foci or centers of their dimensions. FOPDIC denotes the fallacy and error that results from using the foci or center of an item in its mass, length and time dimensions as the basis of its posit. This may be correct for some items that are highly limited in ALT. It produces a defect that escalates when made for all three dimensions, and these as a resultant whole for POSIT. This is shown if a posit has a focus a_m, a_l, a_t, and its object exists in the universe with a range or spread in each dimension b_m, b_l, b_t, then; a_m, a_l, a_t will never represent the periphery of the universe. When the resultant of a_m, a_l, a_t is made these averages will escalate and compound this failure to represent the periphery of the universe. Only those, which are but a small fraction of the total, will best replicate TRU, and produce a uniform and well distributed whole. This is corrected with qualifiers and descriptions to yield a much better accounting.

	ENTITY	OBJECT	ESIN	TIT	ES
CENTERS	Indefinite Not TIT	Variable, Center Poorly Known	Of itself Not TRU	Not actually or adequately known	Of Itself Not TRU, plus limits
2D	Indefinite Not TIT	Incompletely seen 2D Undeveloped	Not Well Seen only Part of TIT	Unknown, Equiva-lents without 2D	Not well Seen, to 2D gradients
3D	Indistinct as entity or 3D	Produced incom-pletely, Not TRU	Unseen, Little Frame	Unknown, Even more not by 3D	Not in itself, only part in TIT
Com-bined	Limits of Non TRU and lack of FAF	Less than TRU Imperfect ROT	ES at best, Limits of ES	How well known and shown, ROT,MOT	Less than TO, Development

Table 1-48 **Comparative Static Posits of entities, objects, ESINS, in TIT, and ES.**

This table seeks to show how much difference there is among the kinds of posits and ways they are represented. It provides a preliminary format for recognizing what and how they can, and what can't be, posited. The precision of REM is increased with the production of STA Posits, the PRENG level carried to two decimals, PRENG2. The superscript represents the number of levels on each dimension in the particular field or science. The subscript represents the approximate percentages, to 1/10's, of each dimension to be weighted for that particular field and science. In the measure of REM it is determined by: 1. Number of levels covered in each dimension, 2. The product of these three values, 3. The percentage or weight to be given to each of these three dimensions, 4. Percentage of each level that the particular field or science covers, versus other topics. 5. The product of these percentages, 6. Other factors influencing REM. 7. The products of # 2, 4 and 5 or the Actual REM value for this particular field and science.

Agriculture $85^3_5,35^3_3,60^7_2$; 3x3x7X3x2x2X5 =.000108
Anatomy $90^2_3,64^3_6,50^9_1$; 2x3x9X4x5x1X8 =.000486
Anthropology $40^9_4,45^4_3,18^6_3$; 9x4x6X2x2x1x4 =.000032

The measurements are as follows: first group for POSITS, second group for REMS, superscripts for levels of dimension covered, and subscripts for weighing each level. The first three terms in 2nd group are for levels of dimensions, the 2nd three numbers are for percentages of these ranges in the field or science, and the last number in this second group for other correction factors of REM. The resultant is their calculated REM value. REM values when posited are often given by the size of the print, extra large 1R, fine if < 10R.

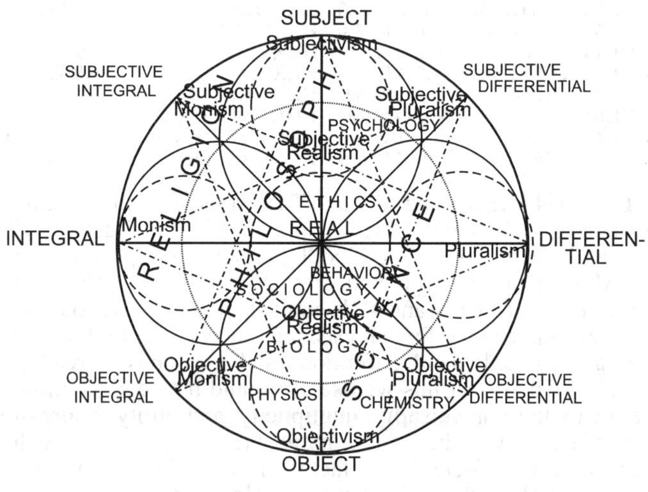

Fig 2-48 **Total STATICS in SOID**

In Figure 2-48 some of the major components, fields, and aspects of the world and TRU are shown on the S-O and I-D Plane. This provides a beginning form and product that can serve as an example of what mlt in a solid geometric model can provide for total STATICS. SOID is used, especially in early investigations because ALT and SPALT do not show the integral differential dimension. In order to learn the benefits and effects of TOPNEC, POS and STATICS it is well to study their combined uses.

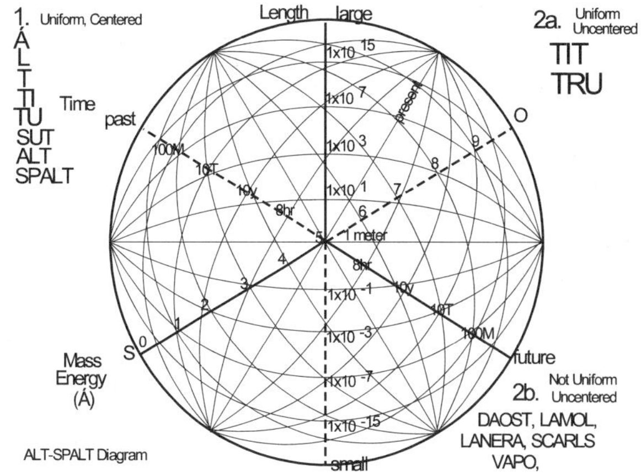

Fig 3-48 **TOTAL STATICS of TOPNEC**

A low percentage of items for placement will need to be shown without strict point posit, POSIT, as in Figure 3-48. We should at once recognize the limits imposed by ALT when presented in 2D. Likewise we should at once recognize the broad distribution of many TOPNEC. Some TOPNEC are uniformly distributed throughout TIT and SPALT in such a way that no localized posit is possible. Too correct for this we list by categories, those that are: 1. uniform and well centered, or posit uniformly throughout the ALT scales, 2a. uniform and poorly centered, or in which no centering is evident, 3.(2b) not uniform and do not center well. 4. not uniform but center well, 5. variable in uniformity and centering. 6. mixtures of the above. What we primarily look for in REM, POS and STA are those that are well centered or have sufficient centering to be accurate. If they are not uniform they can still be centered, if sufficient knowledge of degree of uniformity and distribution are provided.

Some of the NEC we select for POSITS include,

Á	$50^{10}{}_{10}$,-,-,	(10,,)(3)	10	= .3-
ACEXIN	$50^{10}{}_{3}$,$50^{10}{}_{3}$,$50^{10}{}_{3}$	(10x10x10)(10x10x10x)	10	= 1-
ALT	$50^{10}{}_{3}$,$50^{10}{}_{3}$,$50^{10}{}_{3}$	(10x10x10)(3x3x3)	10	= .9-
DAOST	(#2b)		1	= <.1
L	$50^{10}{}_{10}$,-,-,	(10,,)(3)	10	= .3-
LAMOL	(#2b)		1	= <.1
LANERA	(#2b)		1	= <.1
SCARLS	(#2b)		1	= <.1
SPALT	$50^{10}{}_{3}$,$50^{10}{}_{3}$,$50^{10}{}_{3}$	(10x10x10)(3x3x3)	10	= .9-

SUT	50.50,50		2	=<.2
T	$50^{10}{}_{10}$,-,-,	(10,,)(3)	10	=.3-
TI	50,50,50		2-	=<.2
TE	(#2b) (mid circumferential)		2-	=<.2
TU	50,50,50		2-	=<.2
TIT	(#2a)		10±	= R±
TRU	(#2a)		10±	= R±
VAPO	(#2b)		1	=<.1

Terms which signify the most universal concepts do not posit well in SPALT. This is neither to their nor MOT and SPALT's discredit. It only helps gain the needed overview these terms provide in a POA to TRU. Much of this depends upon their use, whether or not they are interpreted in general, as ALT is all TIT, ALT is the dimensional frame, or ALT is the FAF of TRU. (ACEXIN). In addition universals are more likely to have a diversity of meanings and implications.

Fig 4-48 **SOME STATIC POSITS of SELECTED TOPICS in ALT.**

In Figure 4-48 some POSITS of words with high REM, ESIN and ES in TRU are shown, and listed below. Like TRU terms they, as many, tend to universality with a broad spread in one, two or all dimensions. These are noted on the left upper hand corner. Some topics converge on the center, noted in the right upper hand corner. Some triangulate in POSIT so that their resultant is in the center, these are shown in the right lower hand corner. All that are posited in this figure are determined by triangulation. This correction must be accounted for if misinterpretation is to be avoided. Not all principles and techniques of POS, STA, POSITS and STATICS are used in this figure:

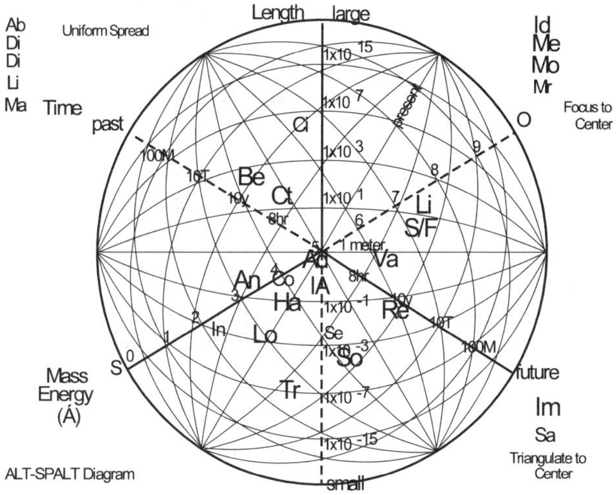

Abbr.	Name	POSITS-A,L,T	REM	TRU	Diluted
Ab	= Absolute	(0/100,0/100,0/100)	2R	3	5
Ad	= Adaptation	68,55,45	2	3	5
An	= Angel	25,50,60	3	4	7
Ap	= Appearance	38,25,51,	4	5	7
Be	= Becoming	(59,45,15)	1	2	4
Ci	= Civilization	54,85,75	2	4	6
Co	= Concentration	27,55,52	2	3	7
Ct	= Contingency	57,45,30	1	2	3
Cy	= Cybernetics	62,45,70	4	5	6
Dv	= Development	(60,40,30)	2	3	5
Dm	= Dimension	(50,50,50)	2	3	6
Dr	= Direction	(50,50,50)	2	3	5
Ds	= Division	(50,50,50)	2	3	4
Hp	= Happiness	33,55,55	1	2	2
Hu	= Humanism	55,50,35	2	4	6
Id	= Identity	50,45,67	1	2	4
Im	= Immortality	25,30,85	1	2	5
In	= Individualism	65,50,40	2	4	6
If	= Information	15,65,65	2	3	5
It	= Innate	(75,33,33)	3	4	5
Is	= Instantaneous	52,75,50	2	3	5
Lf	= Life	75,40,67	1	2	3
Li	= Limits	55,45,60	2	3	5
Lg	= Logos	20,65,25	3	5	7
Lo	= Love	25,65,60	1	2	5
Ma	= Mathematics	75,50,50	2	3	4
Me	= Mechanism	70,50,50	1	3	5
Mo	= Motion	50,50,50	1	3	5
Mr	= Morality	55,45,65	2	3	4
NR	= New Realism	50,67,34	4	5	6
No	= Nominalism	40,70,25	4	5	6
Ql	= Quality	60,62,70	1	2	3
Pa	= Paradox	25,65,35	3	4	6
Pe	= Personality	52,55,35	3	4	6
Pl	= Pleasure	45,50,60	3	4	7
Pr	= Predicate	55,65,60	2	3	4
Pp	= Proposition	30,70,60	2	3	4
Re	= Real	50,65,70	1	2	5
Sa	= Salvation	25,30,80	2	3	4
Se	= Sentence	40,70,60	2	3	4
Si	= Sin (ESIN)	70,40,50	3	5	7
So	= Soul	35,70,70	1	2	3
S/F	= Success/Failure	65,55,70	1	2	4
Tr	= Truth	25,75,65	1	2	3
Va	= Value	50,55,65	1	2	3

These STATICS provide a rough entry into the total ROT, POS and POT of TRU. They can be done in other diagrams, as BASFANT and SOID. They do not account for all the corrected forms and concepts of ROT, MOT, REM, POS and STA. They are not entirely ES. They lack other corrections, as for range and weighted REM by percentages of participation, as well as overlap. These would lower the REM values, more for some than others.

Fig 5-48
STEREOTYPED DIAGRAM of POSITS of FIELDS and SCIENCES

This diagram is highly figurative of the distribution of many of the greatest fields and sciences. It lacks the precision of POS, POSITS, STA and STATICS. Yet it is presented to provide an introduction and guide to POS and POT. To produce an impressive picture of the distribution, make-up, and the use of the positioning of static topics for TRU.

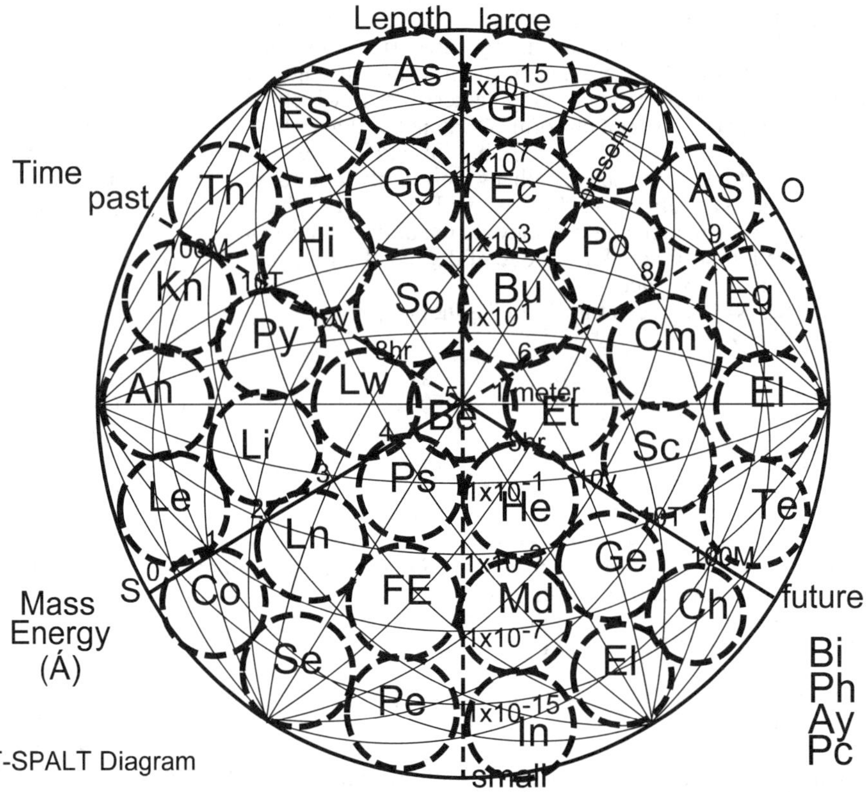

ALT-SPALT Diagram

The abbreviations used are as follows:

An = Anthropology
As = Astronomy
AS = AstroSpace
Ay = Anatomy
Be = Behavior
Bi = Biology
Bu = Business
Ch = Chemistry
Cm = Commerce
Co = Cognition

Ec = Economics
Eg = Engineering
El = Electronics
ES = Earth Sc
Et = Ethics
FE = Feeling/Emotion
Ge = Genetics
Gg = Geography
Gl = Geology
He = Health

Hi = History
In = Interest
Int = Intention
Kn = Knowledge
Le = Learning
Li = Literature
Ln = Language
Lw = Law
Md = Medicine
Pc = Physics

Pe = Perception
Po = Political Science
Ps = Psychology
Py = Philosophy
Sc = Science
Se = Sensation
So = Social Science
SS = Solar System
Te = Teleology
Th = Theology

Figure 5-48 uses ALT as a frame for the spherical modulation of TRU as represented on a page. It figuratively shows some major fields, sciences, and topics that make up large portions of TRU. Neither equalities of size nor color are to be taken in the full actuality of TIT, only indicatively. It shows how TRU is composed of many elements, parts, sciences, fields, and systems. It shows how 'ologies' and functioning systems like organs are of great value in understanding and elevating TRU. It demonstrates how we can't expect only one field or science to provide all the answers, nor even the final answer to the questions, problems, needs, or benefits of man. It shows some important terms, whose ESINS are of special significance in TIT, their abbreviations, and numerical posits in A, L, and T. They can be classified in several ways including: ALT, SOID, REM, and Alphabetically. Ay, Bi, Pc, and INT are not included for their triangulated centers are better reproduced on an alteernate face of SPALT. There are those fields and sciences which posit well in ALT and in the contents of a text on TRU. The more through and through, universal, beyond TRU or meta TRU, they are the more difficult they are to posit, e.g. mathematics, biology, and physics. Degree of precise posit is: **DEPOS**, Degree, or fraction, of Precise Posit. DEPOS is the fraction, of one, of that to be posited.

Most of the great major systems and groups have parts that can be divided with precision. First it requires reduction to TRU. It is the same method used in MOCBIREM in Section IIIB. For MOCBIREM proves the duality and conjunction of spread and focus. This is called: **SAFPIM** = Spread and Focus Proven In Media. If TRU is more or less composed of great systems or groups and if these can be accurately and readily divided into their measurable parts, then each of these parts will partake of TRU in a set way. This set is what we can use with a high degree of reliability and specificity for their posits. Since such systems or groups have set posits in TRU their component parts will also have a high correlation, and can be measured, with that of the overall system, group, and REM. Although there will be differences in the components, the components as a whole will be what constitute the posits of the system or group. The fact that the terms for some systems, and their parts, overlap is: **CIN** = Coterminality Index, **CINSY** = Coterminality Index between Systems, **CINE** = Coterminality Index between parts, **OIN** = Without Index, or the ratio or degree to which a part is exclusively inside/outside of a system, i.e. in other systems. One advantage in dividing the world into such systems and groups, is that each component can not be, by MOCBIREM, of greater magnitude than its part in the system. Each system can not be greater than its part and role in the whole of TRU. So that the tendency to overrate what they have their attention on, or consider separately can be quickly reconciled with the parts these set for the whole. In this way posits will become more accurate by filling up TRU by POA. The precision of this approach, like that of MOCBIREM, depends heavily upon the base value given for each system in REM of TRU. This can be done only with careful, discrete and total comparison of the predetermined selected and separate systems as they are only of, and make up TRU. For example the above systems are given POSITS and REM:

System	POSITS	REM
Body Parts	$75^8_5, 58^3_3, 55^8_2$;	(8x3x8)(7x5x4)6 = .00482
Continents	$95^1_2, 10^1_5, 45^8_3$;	(1x1x8)(3x4x1)2 = .00002
Races	$67^9_4, 14^6_3, 30^5_3$;	(9x8x5)(2x3x3)5 = .00324
Nations	$80^9_5, 17^3_3, 28^6_2$;	(9x3x6)(1x5x3)3 = .00072
Cities	$65^7_6, 23^3_2, 73^7_2$;	(7x3x7)(3x4x3)5 = .00214
Social Groups	$60^9_6, 36^3_2, 45^6_2$;	(9x3x6)(5x4x5)6 = .00972
History Per.	$70^8_2, 22^3_2, 22^1_6$;	(8x3x1)(2x5x8)4 = .00077
Fields Sci.	$42^9_6, 35^9_2, 68^9_2$;	(9x9x9)(5x8x6)8 = .13797
Business Ad	$75^5_5, 32^5_2, 62^5_3$;	(5x5x5)(4x3x4)4 = .00180
Justice Sys	$65^7_6, 43^4_2, 44^4_2$;	(7x4x4)(2x3x2)1 = .00013
Health Sys	$75^9_5, 62^4_2, 62^6_3$;	(9x4x6)(3x5x4)2 = .00259
Occupations	$62^9_5, 42^6_2, 64^7_3$;	(9x6x7)(5x4x5)3 = .01104
Arts	$55^8_6, 48^4_2, 64^7_2$;	(8x4x7)(2x2x2)4 = .00068
Sports	$85^3_5, 38^3_2, 53^3_3$;	(3x3x3)(2x1x1)3 = .00002
Personages	$48^9_6, 52^5_2, 45^6_2$;	(9x5x6)(4x5x5)6 = .00762

From such POSITS some conclusions may be drawn:
1. PLAP and PLADIP, or the Problems with Planar Posits, and its Dimensional Polarities, are large contributors to an imperfect representation on this field, 2. Systems and groups that are measurable are those that are more familiar to us, 3. Additional correcting factors are needed to fine tune STATIC POSITS, e.g. a. all TRU, b. overlapping, c. general or more limited interpretation of each topic or system, 4. Such STATIC POSITS of systems and groups provide excellent basis to gauge posits of their constituent parts, 5. Some parts or constituents of systems or groups will be higher or lower in quality, as TRU, than others and the REM will change somewhat. However, the differing order of magnitude will not usually be over one REM level. 6. By comparing all other items, entities, senses, things, ENES, NETH, NES, ESINS, ES, CES and aspects of TRU for POSITS, we can discern whether or not they are of larger or smaller constituents and contributors to TRU. We can also compare their posits to see if they are more or less toward this or that dimension than items which have received firm POSITS. No correction is made for differential weighting of each dimension. This and other factors can be made to provide the continual improvement and perfection needed.

One of the greatest problems of all in STA, POS and REM is to separate out what is formally of strict 1R, 2R, 3R, and what is of all gradations between. This is crucial to a precise dividing and positing of TRU. There is all the difference in the world between a fleeting visit to one or two sites in a foreign land, a short trip through it, an extended visit including many sites, or a more permanent stay including a thorough study of the whole country. Anyone who has done much traveling knows that if you journey in a straight line, especially at high speed, you may get there quicker but you miss most of what you passed. In our journey through TRU, to cover it can be done simply and quickly or it can be done in ACEXIN, using all the massenergy, length and time dimensions, CALP, ES, REM, POS, STA. HOM, POT, and IPTRU available and recognizable..

In many of the figures in this chapter, as 3-48, we encounter the problem of correct REM. For many of the terms we use, both common, accepted, and TRU meanings, their face value in TRU is much higher than what they have when diluted by other terms, or in common usage. In order to provide the most accurate and reliable REMS for POSITS, it is necessary that they be correct and standardized. As is evident they can not be used as single representations of their most TRU meanings. For if we did we would have 100-1,000 1R. Since this is impossible we have to bring them to a corrected or standardized REM. This is called: **LACREM** = Law of Corrected or standardized REM. It will be noticed that LACREM is in ways a kind of reverse MOCBIREM. For whereas MOCBIREM provides us with an excellent means for measuring values, or setting REM, the dilution it reflects is also its great problem and limit. So that LACREM becomes the means of specifying and qualifying dilution and overlapping of POSITS and REM, as is achieved through MOCBIREM.

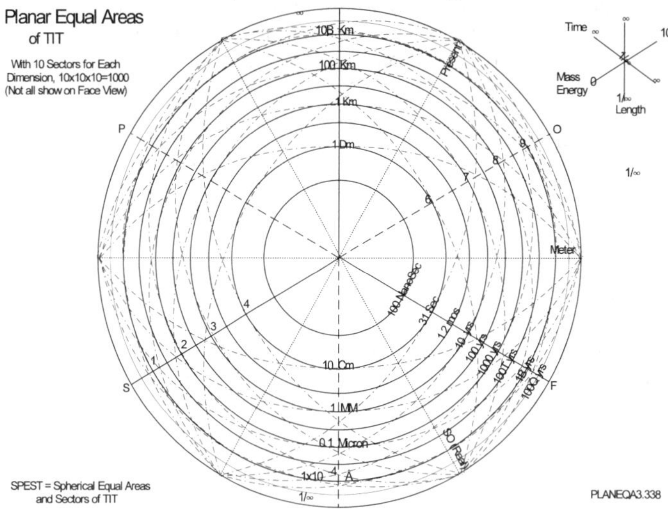

Fig 6-48 **SUMMARY of STATIC POSITS in SPEV**
By this figure many of the more significant and exemplary static posits are posited in a SPALT that is corrected for equal volumes as equal areas of TIT on a planar surface. It is what is used as the standardized primary model of TIT in the remainder of the text. At the same time keeping in mind the ideal comprehension of the global SPALT and associated models of TIT and TO.

The high correlation between length dimension and statics and the time dimension with the dynamic is of great importance. It means that items for posit that are largely length will lie through a length segment and those for time will lie through a time segment. Rather than a point they will be more or less spread out, by the three continua that make up ACEXIN.

CHAPTER 49

ENHANCEMENT and SUMMARY of STA and STATICS

REFINEMENT, PRECISION and PERFECTION, PROBLEMS, LIMITS, STANDARDS, CONCLUSIONS, PROOFS and RELATIONS of STA and STATICS

With POA, MOT, POS, STA and the needed factual static data, a massive static and dynamic interpretation and overall understanding of all TRU is possible. Through Homation it can be posited through DYNU, DYNS, and DYNO in ALT, SPALT, and used in many ways to demonstrate and prove various truths, mechanisms, and the overall nature of our lives, TRU. These in return can be of great value in showing us where and how to obtain further factual data, information, Posits, Statics, and develop their use for POT and TRU.

The measures and formulae that can be developed and used to enhance the accuracy of POSITS and STATICS is the degree of measurability of ES and TRU, degree of measurability of STATICS, formulae of REM measures in STA, and formulae of spherical POSITS in STA and SPALT. By these we seek to increase measurability by adopting all means that separate and make ES discrete. We seek to increase the measurability of STATICS by all means that bring ES and all items of TRU into a whole that can be well differentiated and proportioned. The formula of the REM measures in STA is:

REM (ES,,) = Á (range) + L (range) + T (range) / 3, x Á (%) + L (%) + T (%) x X

(% less than 100% TRU from other factors, e.g. overlapping), or:

$$(\acute{A}^r+L^r+T^r)(A^p+L^p+T^p)\ k = REM$$

where: r = Range or number of 1/l0s or 1/100s of the dimension this item covers, p = Proportion of this r or coverage that is this items share of TIT, and k = Correction factor for all values less than 100% because of other causes, including overlapping and non TRU. The proportion, p, is the percentage or amount within this range that is only in this ES, aspect item or category.

Fig 1-49 **STATICS STANDARDIZED and MEASURED by FIRST ORDER in A, L and T DIMENSIONS**

This figure illustrates the standards adopted, namely of numbering measures of the dimensional scales from left to right and top to bottom. In this way the numbers from 0 to 10 are, 1. Á dimension is measured from subject to object, 2. Length dimension, from large to small, 3. Time dimension, from past through present to future. A major benefit of such posits is the interpretation of a topic. If there is a discrepancy, what accounts for it, as what accounts for positing energy at 5 on the time scale. When restricted to its TRU, or in TIT, meaning much is lost in what is commonly understood by the term. e.g. evolution different than TRU evolution. Many common topics will have more limited posits whereas their TRU ESINS will be broad and focused at the center, 5, or without focus or number, e.g. God.

Figure 1-49 shows the posits of many parts of TRU denoted by words or terms of use in English. It is not meant to be exhaustive, nor proportional. It is presented to give an idea of the great possibilities and potentiality of representing all in a model as SPALT of TRU. It is also done to show how necessary it is to have precise posits through a measurement of the position of the item on the dimension, or Á, L, T scales in of TRU.The items posited in this diagram are numbered and measured as follows:

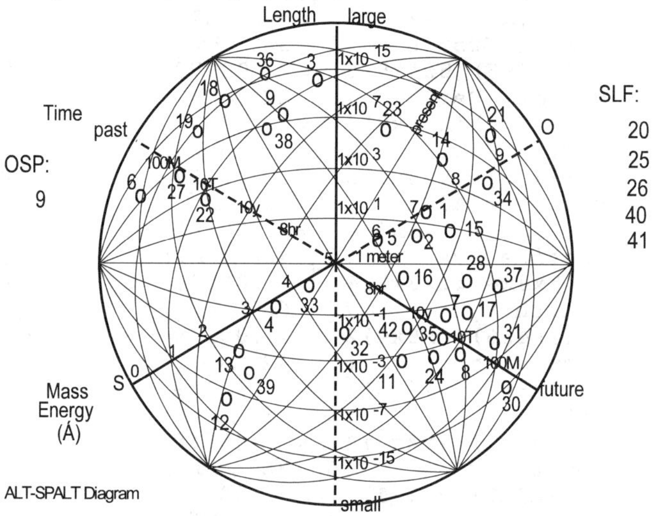

ALT-SPALT Diagram

1.	Act	8-5-5	OLF
2.	Adaptation	7-5-6	OLF
3.	America	7-2-3	OLP
4.	Attention	3-6-5	SUSP
5.	Attribute	5-4-6	OLF(R)
6.	Augustinian	1-2-1	SULP
7.	Body (Human)	9-5-5	OSF(R)
8.	Brain	7-5-8	OSF
9.	Bronze Age	8-2-1	OLP
10.	Radiocarbon dates	8-6-1	OSP
11.	Chemical Reaction	9-8-5	OSN
12.	Comprehension	0-7-7	SUSN
13.	Consciousness	1-6-5	SUSN
14.	Crops (Farming)	8-3-6	OL~
15.	Engineering Gen.	7-4-7	OLF
16.	Energy	5-5-7	SU-~
17.	Enzyme	8-7-6	OSN
18.	Evolution	8-2-1	OL~
19.	God (Common)	1-0-2	SULP
20.	Government, Ideal	3-1-7	SULF
21.	Gravitation	9-2-7	-~-
22.	Greek Philosophy	1-2-1	SUSP
23.	Horsepower	9-3-2	OLP
24.	Immortality	3-5-9	S~F
25.	Justice	2-3-7	SULP
26.	Theory of Knowledge	0-2-8	SUL-
27.	Mythology	2-1-1	-LP
28.	Organic Synthesis	9-8-7	OS-
29.	Obedience/Compliance	7-4-7	RM-
30.	Particles elementary	9-9-8	OS-
31.	Phase Equilibrium	8-8-9	O-N
32.	Predicate	4-6-6	RS-
33.	Proposition	4-5-5	RS-
34.	Prosperity	5-2-8	OLF
35.	Protein Syn Con	7-8-8	OS-
36.	Races	7-1-1	OL-
37.	Radiation Physiology	9-7-8	OS-
38.	Renaissance	3-1-1	RLP
39.	Sensation	0-7-5	SUSN
40.	Time Reversal	5,0,9	SULF
41.	Virtue	2,4,7	SUM-
42.	Word	6,7,7	RS-

If accurately placed measures at first order numbers are not greatly different than those for second and third order numbers. These only add precision to the demonstration of TIT. So that for the purpose of ROT and POT a single numbered measure of 10x10x10 yields 1000, which is usually sufficient for ordinary POSITS in CIPTRU. A two numbered measure 100x100x100 = 1,000,000 is a goal that can be envisioned to give a sufficiently reliable ROT and POT of TIT. It should always be remembered, ECRO, that ALT is in TIT-TO and OBT, much closer to a spherical object, SPALT.

This applies especially for those POSITS whose measures are SULF or OSP, that is on opposites poles of each dimension. Only when they are on neighboring poles of each of the three dimensions will their accuracy on a page or planar field approach a satisfactory level.

To designate those conditions in which centering decreases and spread makes posit difficult there is: **>PS** = Terms whose spread is greater than what posits would benefit from, e.g. 'applies to all TRU'. **%PS** = the percentage of spread this term has in ALT, in general, or for each dimension. The greater %PS is, the poorer its posit.

Fig 2-49 **NEEDED CONSIDERATIONS OF EACH POSIT FOR STATIC and DYNAMIC ES and CES**

CONSIDERATIONS FOR EACH POSIT

- Is the item for POSIT 100% TRU, TIT
- If not, what is its ESIN
- How can this ESIN be best brought to TIT
- Have all needed qualifiers, factors and coefficients been applied
- How much coterminality or overlap of meanings is there
- Does TRU meaning change POSIT from other meanings
- How can and much do POSITS of other terms set this POSIT
- How can and much do POSITS of SYSTEMS and GROUPS help
- Is REM, or the POSITS value, well expressed
- How well can CIN, LARPCO, and LACREM help POSITS
- How much work is needed to refine and perfect the POSIT

This list summarizes the salient points needed when considering any item for POSIT in SPALT. It is failure to remember and use them that POSITS, IPTRU and TRU are apt to be discredited and their great value and use neglected. This is much of the reason why SPALT, ALT, REM, ROT, TRIT, TIT, and TRU are not recognized and upheld.

STATICS and DYNAMICS may be compared by how they describe, demonstrate and illustrate TRU when TRU is objectivized and represented through its models, ALT, SPALT, and SPAT. STA and HOM are not, by the laws and principles of TRU, the final or end product of an exposition of TRU. Only when this is totalized both immanently and transcendently as exemplified by COGNOS and ELTICMA, and its use by TES and PUREALITY to OM, POM, SA and TA, does it provide the final end product of POT.

STA and HOM, as ES and CES, are the substance and means through which the content of TRU is illustrated. This they do by filling in the frame of SUT, ROT, MOT, ALT, SPALT for TRU. They are the same in the fact that they: 1. Describe the same parts, processes, features, subjects, sciences, fields and all of TRU. They are different by the fact that they: 1. Indicate divergent aspects, natures of TRU, a. STATICS: ES, REM, TIT, TRIT, and b. Dynamics, CES, TRUD, TRICH, and to a lesser degree TRUCH.

Knowledge of the world, as it is TRU, is both simple and very complex, easy and very difficult at the same time. It is very complex and difficult because it is so vast and approaches the infinite, is unknown, not readily apparent, and highly interactive. It is very simple and easy because, as TRU, TU and TO, it is from one and totality, thereby set, highly related, systematically unified, ordered, and reducible to being shown for ATIM. So that once learned it is always there. It leads to direction, comprehension, and achievement. Exceptions to the facts, principles and rules are not necessarily negative nor defeating, they are often qualifying and positive. This is what we also find with limits. Since TRU is primarily relative rather than absolute, by LANERA and LAMOL, its limits produce form which by thisness apply to set form. All that is not fully of this set form, even if otherwise within TRU, will to this extent produce exception. It is by continuously qualifying and refining exceptions that precise and VAPO representation, modulation, valuation and POA, POS, STA and HOM are produced. Thus great portions of TRUy become their rules and refinements. This is what we do throughout TRU, ROT, POT, and IPTRU. It is what is most apparent in POS, STA and HOM by bringing the form and composition of TRU into a POA that is adequately conceived, formed, delimited, qualified, programmed and refined.

One of the major limitations of STATIC POSITS is the overlapping of the range and distribution of an item, entity, thing, word-thing, word, ESIN, ES, or other aspect of TRU in MOT and SPALT. This is due to the fact that many POSITS are not limited to a discrete sector. It can be readily seen when we compare POSITS of topics, fields, sciences, and word-things in the last several diagrams with the those of the sectors of SPALT. The 1,000 different sectors of 10x10x10 parts of TIT, have absolute set and exclusive form. There is no overlapping. This is the ideal. It is what we look to if we are to correct all as they may be a perfect replica of TRU. Knowing language as we do, it is clear our terms are not based upon TRU, even their ESINS are not usually from an exclusively TIT vantage. Even when reduced to their ES they still represent parts of TRU whose delineations lack the specificity of those of the 1,000 sectors. From this way we need to analyze and demonstrate STATICS in ways that best correct overlapping. It involves an acceptance of some of the shortcomings that our existing language and the nature of TIT impose. We are, however, dependent upon language, since no better one exists, that we can do no better for now, than to accept words, word-things, NES, ESNES, entities, things, ESINS, ES, and aspects of TRU as items for STATICS. We should then proceed to reduce them to TIT-TO, TRIT, and SPALT with overlapping and other problems held to a minimum. The corrections for overlapping are, 1. Note the quality and quantity of overlap in each item for POSITS of STA, 2. Produce and implement fractional reduction of this overlapping.(LACREM) The degree to which each added term detracts from existing terms, 3. Hold posits of terms to as close a concise and concrete form in TIT as possible. i.e. reduce their ranges and general interpretations to their strict definitions, inferences or implications, 4. Note what they do not cover, or what does not exist in portions of dimensions and sectors. In this way, by a process of selection and elimination the remainder of TRU is excluded from the foci and REM of the POSIT, 5. Corrections by inclusion in systems or groups. Those terms for posit that are countable members of a larger group or system will have slight to no overlap. 6. Enumerate, remember and utilize all terms that overlap for which inadequate correction can be made so that they can be fitted into the model for total comprehension of TRU. This fitting will include posits that are not strictly delimited, but as ESINS, ES, CES, forms, functions, qualities, modes, aspects or qualifying features help replicate TRU in other ways than through its constituent parts. Perfect language enables parts of TRU, ES, distinction. This is the function of an ES language, to make the world distinct and meaningful. To distinguish the parts of TRU, their nature in TRU, and that needed for the perfection of language by TRU. They are highly coterminal. The degree to which each and all fail is the loss of TRU. To prevent and correct such failures is the cause and reason POT and IPT are developed and offered..

Fig 3-49 **QUALIFIERS, FACTORS, COEFFICIENTS of POSITS**

This figure provides a summary of the key factors, qualifiers, and coefficients that are needed to express and make each POSIT the best it can to represent TIT. Without these factors or coefficients it is often impossible to provide accurate and reliable POSITS. By making static posits in 2D, 3D, and associated ROT with those limited to TIT, point for a range, limited to single and unchanging definitions, with freedom from ambiguity and other qualifying factors the ideal of a highly effective representation of TRU can be achieved.

QUALIFIERS, FACTORS, COEFFICIENTS of POSITS

- 1 Á, L, and T Measures n
- 2 REM f
- 3 In SPALT, METAD, POMOT, SOID, etc A
- 4 Centered/Uncentered
- 5 Uniformity/Not Uniform
- 6 Loss in Planar Representation (ALT)
- 7 Coterminality, Overlapping
- 8 MOCBIREM, Discrete Member of System
- 9 Degree of Non TRU, Non ESIN Meanings
- 10 ESIN value or % of ENIT in TRU
- 11 Ability to POSIT/ work needed
- 12 Percentage of Item in each Dimension

PRINCIPLES and GUIDES for SPHERICAL THINKING

Among the greatest of all the problems that cause humans to fail to use their knowledge and comprehend the actual world, TRU, is our inability and lack of training to think and comprehend spherically. This is especially so for the solid object TRIT and its component CALP, FAF, parts and processes. We have difficulty in simple thinking and focusing on a point, when as observers, we do not identify it with the observed. We have difficulty in thinking in terms of movement, dynamically and of vectors. Not only does this imply the point of identity between the observer and observed, but tracing a course. Without training we have difficulty in analytic geometry, thinking using fields, planes, and solids. Only with training and repetition is this rapidly magnified to become the massive necessary comprehension. This requires both the point and line of identity between the observer and observed, but a two or more dimensional figure with multiple points. The great difficulty in thinking spherically needs determined training, teaching interest, repetition, practice, great prior knowledge, and comprehension if we are to interpret solid figures adequately. The movements and dynamics of all within this solid are beyond ordinary thought especially if TRU and VAPO are unknown. We don't separate and select what is, what is in, and what is of, TRU.

Modern man, with knowledge, technology, electronics, computer science, and communication, has greatly advanced the ability, quality, and quantity to represent and know the world. Yet each person by themselves, has difficulty envisioning the full nature and use of these representations because of the nature of our mentality and lack of exposure, training and direction. When well represented and described the observer, reader or person will be lead into a better view and thought about what is shown. TRU must be well shown. The observer, reader or person must learn to think at ever higher levels and ways that enable a great comprehension of all TIT-TO. This training and ability include, 1. Background in those fields and sciences that provide the basis for this kind of thinking., with includes enhanced powers of observation, language, mathematics, statistics, geometry, analytic geometry, solid geometry, topology and many others. 2. A life long training, education and practice in the use of solid geometric representation. 3. A life long training, education, of philosophy, science, religion, music, and arts and their practice in TRU with all that best aids KOT. 4. A life long application of the POA of TRU and how each and every encounter in life can be systematically related and unified through its interpretation in the best representations of TRU.

In this text, as an initial search for and demonstration of TRU, we do not have, nor even more, can expect any reader to have, great powers of observation and thinking. Therefore we present TRU step by step in as orderly and direct way as possible. This is by description, posits, linear, spatial, dynamic, and total solid figures sufficient for each step and combined identity with TIT-TO. However to best discover, learn, know, remember, apply and live TRU this is neither adequate nor appropriate for a refinement equal to the greatness and need of TRU. It will take a life learning to enhance the readers ability to think spherically and to fully comprehend TRU,, TIT, TES, ES, and CES. This is by,
1. Thoroughly learning ALT, SPALT, METAD and SPAT,
2. Learning POS, STA and HOM effectively. 3. Practice and learn to transfer the POSITS in STA and HOM into the global and solid figure of SPALT, 4. Investigate, learn and remember the greatest REMS, of both known and TRU concepts, principles, laws, and mechanisms for POSITS in SPALT. 5. Remember and apply VAPO, DAOST and the SCARLS of TRU to all you encounter, 6. Perfect a solid sphere for TRU, TIT, TRIT, and OBT, SPALT, for continual use and perusal. Practice posits of aspects, parts, processes and mechanisms of TRU. 7. Cultivate experience, inspiration, and revelations in your daily workaday life as it relates to TRU. Think and note the deepest TRU meanings in all you encounter to answer basic questions and principles for an ever greater comprehension of TRU. 8. Repeatedly think of the whole, both as it is TRU, TIT, TRIT, SUT, THEO, OBTRU, ROT, MOT, ALT, and SPALT, and as it is thought of, THES, through the FAF of the Á FAF of TRU. Bring these to SPAT, PUREACLITY, COGNOS, ELTICMA, SÂ and Â as they return, in OITIO, to TIT-TO.

LACMEST or the Law of Correction for MEdializing of Standards. It is like the drift to mediocrity in life. It requires the necessary correction to bring standardization and their restrictive and stereotyping tendencies and effects back as close as possible to a 1:1 replication. For ROT, MOT, ALT, REM, POS, STA and HOM this is their ideal replication of TIT. The work of LACMEST is difficult, for each standard that is added makes that much more LACMEST necessary. Therefor standards must be held to a minimum. LACMEST should be accomplished with the utmost vigor, means and perfection available. This is more successful with ACC and a total management and control by TES for TRU.

The problems, limits, standards, and corrections we must adopt do not deny or lessen the value and great benefit of MOT, SPALT, REM, POS and STA. It only reconfirms their need, and the need for persistent and dedicated work. Also, any entity is not to be understood from its parts alone. They can be represented when this entity is made into an object, an ES and CES of TRU. This Law is called: **LAMORTE** = Law of

Maximal Objectification and Representation of the parts of Constituents of TRU. More generally it is called: **LAMORE** = Law of Maximal Objectification and Representation of the parts of an Entity. LAMORE and LAMORTE certify that for any entity or TRU, its object can be replicated by its parts only within limits. For these parts do not represent all of this object, and even less its entity or TRU. It is, to a large degree, because of LAMORE and LAMORTE that we must be careful not to expect to provide in POSITS of STATICS a perfect replica of TRU. This, in principle, we can not attribute only to overlapping or other difficulties and limits within STATICS, STA, POSITS, POS, REM, SPALT, MOT or ROT. Some come from the FAF of TRU, and the errors inherent in externalization of TRU, or altered relativity.

So that as in the same and other of DAOST of TRU, the infinity and unity of TRU, and the in and out of TRU. we continuously confront obstacles to the compartmentalization of our universe. As long as this critical problem is realized, and the limits and corrections to posits made they can be a valuable weapon, like language, to empower mass energy, Á, FAF in TRU. Success is gained by the corrections we enumerate in this section of IPTRU based upon TU, TIT-TO, TRIT, OBT, SUT, ROT, MOT, ALT, SPALT and POS. As it becomes more proficient, this success will result in better plans and means in life. It is gained in every formally distinct concept and word fully adapted to TIT-TO. Conceptualization must be accurate, precise, actual, proportional and complete, or by VAPO, if OBT, ROT, MOT, and ALT become THEO as they must in a 1:1 closure with TIT.

The problem with achieving precise and reliable posits is evident in the major fields of knowledge, as shown in tables and figures of Chapters 46-49. Of the many correction factors that must be applied, one of the greatest is the degree to which each of these fields is of Á, L, or T. Additionally how their roles in TRU depart from the traditional. This is seen in history which is predominantly a time function, yet also of size and mass energy, S-O, Á, significance. Herein is another problem, that of transit to TIT-TO. For history by time denotes the period of the last 5,000 years variously centered. Whereas by adapting to TRU no set usage has been provided and will be a larger and wider stretch in both time past, size and mass energy. The same corrections will be needed for all major sectors of knowledge, such as in L for the astrophysical and earth sciences, and in Á for the philosophy, religion, and the psychological sciences..

In these we, as in so much of the in and out dynamics of our relations to TRU, have many concepts, either partial or universal, with posits in and out of TRU. Sometimes these in and out posits carry more significance, seemingly, than that strictly in TIT-TO. This means we need to continually weigh each to determine how much is, because of this increasing actually, in rather than seemingly in and out, or even out of TIT-TO. This is signified by: **LUSIT** = Law of Universal Significance and Implication to TRU. LUSIT reveals an unknown, yet great, role of universals and all in TRU.

Problems of continua and in and out of TRU are fundamentally associated with that of the spectrum of density. In full it is between that of an absolute vacuum or non density, and absolute density. This in turn is, in a way, a function of degree of TRIA. This is one of sub or super saturation, hypo or hyper function, so prevalent in all that is both in and out, and of other continua working as tangents to and of TRU. They are one of the greatest features and roles of life, of TRU, and closely related to TROS. They help to explain the expanding and contracting universes of which TRU is the one. It helps to confirm and uphold the stability of TIT a permanently relative yet absolute universe in which our ability to make and best use finite ES and CES is pivotal. **VASCO** = Vacuum, Sub and Super Saturation Contractions.Fluctuations of density in TITneed to be shown by both STA and HOM.

CONCLUSIONS from ALT POSITS and STATICS

1. All posits do not have an equal amount of Á, L, and T,

2. Some posits may be highly A, L, or T almost to the point of an insignificant amount of either one or two of the others. e.g. those things that are of primarily of time, length, S or O.

3. The relations between static and dynamic, peripheral and central require close, and often reciprocal, ID determination for posit.

4. The periphery tends to be less dense than the center, yet balanced by accelerating contraction in the form of TIT and MOT. This is more apparent and representational than actual for TIT. For in TIT their convergent and divergent forces tend to equilibrium. In ROT, MOT, and all objectivity there is an inclination to bring all together, by its focus or center. This is like concentric circular telescoping of the observed and object toward the center. As we look at TIT the depth dimension is lost which is heavier in the center.

REFERENCES for ALT, SPALT, POS and STA

POSITS, STATICS and HOMATION are so critical to a POA, ROT and POT of TRU, and the need for powerful substantiation so great, that extra emphasis on the precedent, methods, and means are given. It is in order to increase proof of the principles of ALT, SPALT, POS and STA, and in order to assist in the refinement of POS and STA, and their POSITS and STATICS, that we note a few to exemplify the TREQ that pertain to MOT, POS and STA in: Aristotle's Posterior Analytics, Euclid's Elements, Geometry, Descartes' Rules or Discourse on Method, Spinoza's Ethics, Hume's Essay of Human Understanding, Love's Analytic Geometry, Cohn's Solid Geometry, Thomeier's Topology, and Hoel's Statistics. There are many other authors and works on the above topics plus those on mapping, mathematics, placement and posits. Even though each provides strong proof, none of these alone provide the ROT, MOT, ALT, POS, and STA needed.

The central chapters of the text, STA, HOM, POS, REMS, ALT, DYNU and DYNS of TRU show and prove much of the detail of TRU. They also show much of their own detail and use, as they become parts of TRU, how TRU actually exists as a living organism, the special universe of best human life. This consists of parts and processes, form and function, whose posits can be shown by STA and HOM, statically and dynamically. They prove, by showing how TRU and its aspects, features, principles, CALPS, SCARLS, form, substance, ES, and CES exist in an ordered, harmonious, symphonic totality. The interaction of these basics of TRU, and their representation, are the reciprocal proofs of TRU. This is called: **RECIP** = the Reciprocal Proofs of TRU. Loss of one function affects other functions. This is similar to the extremely numerous and intricate effects, vectors, carriers, and processes in TRU. The form of TRU and its function are interrelated, interdependent, and in fact, become one in TRU.

The transition from chapters 49 to 50 marks the mid point and center of the text. It is that from form and parts to function and processes, as they are represented by POSITS on SPALT and models of TRU. It is how by working with ES and CES we can best demonstrate both these elements, and the whole, TIT-TO, TRU, and their needed presentation, interpretation, and action. To do so meanings need to be of and by TRU and as formally distinct and ordered as possible.

LUM = Law of Ultimate Meaning. For each word, concept, term, nominal, sign and mark there is a deepest and most TRU meaning. This meaning is focused on that segment, posit or point in TRU that most exemplifies the word. As obvious and important as this is, it is not something that can always be accomplished quickly nor perfectly. For the meanings of the words in all current world languages are not directed by and to TRU. They have arisen and are perpetuated by both the thisness of the objects of entities they represent, the thisness of the subjects observing them, and, as well, the thisness of those using them. These are often very different than their TRU meanings. To bring these to TRU it helps to show how their current meanings are different, can become their TRU meanings, and the best ways it can be done.

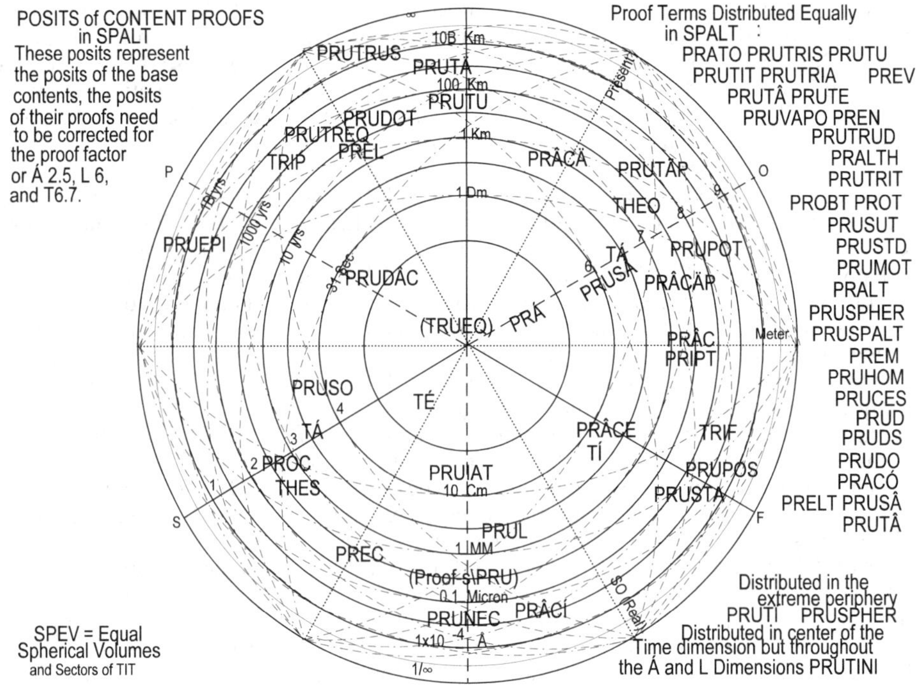

Fig 4-49 **Composite Proofs of TRU** (Also see Figs 6-10, and Fig 5-19)

Proofs depend upon the evidential relations that TRU provides. Each, being different aspects of the same thing, supports the others in their specific ways. The degree to which formal distinction is provided largely determines their VAPO. Thereby it is in the constant upgrading of language, as our terms become more TRU, plus added terms to fill the voids, that the success/failure of this relative universe of best human life depends and will grow. Meaning, language, and posits will increase, but TRU will largely and generally remain the same, changing these meanings, language, and posits internally and expanding as a continuation of what it is, and as it exists from its own focus. In this way LANT, NEC, and POS are not to be looked upon as great changes of TIT, only upgrading of existing language with changes in TIT that give it more precision, detail, power, and continuity.

There is a profound difference between learning the meaning and identify of discrete objects to our senses and knowledge, and being able to understand and use their deepest and fullest meanings and implications to our lives, as TRU. This is the collective, interactive, dynamic, and total meanings as they posit in TIT-TO. It is to this kind of understanding that each of us must concentrate if we are to have PUREALITY, be TES, live TRU and have everlasting life. When making this transit, along with ultimate meanings, we especially need the dynamics and proofs that are most comprehensive and definitive. When focusing on the differential, and singular parts become processes, ES become the CES of HOM, we show DYNU and other proofs in these core sections of IPTRU, POT and TRU.

PRUSTA Proof of STA and STATICS

OM, TIT, TO, TÄ, POS, TE, E, ES, Fields and Sciences, Places, LACES, TREQ, VAPO, Persons, Topology, REM, POT, MATES, OITIO, ENT, Meaning, MADAS, EC, PRUNT, LATANOS, LARP, ALT, SPALT, LORVA, MOCBIREM, PRALT, Anatomy, Brain, Child, Adult, Man, Woman, Ecology, Economic-s, Genetic-s, Material, Geography, Habitat, Homology, Population, Possession-s, Territory, Water, SIDA, SINUAD, MADAS, PRUTAN, LECDIE, LASCES, TRUCA, SAV, SRE, ESIT, POID, Statics, POP, LAPOSC, POPTE, STAT, CALP, ENES, RENDE, LATANE, TESPRA, DRAL, DEPOS, SAFPIN, CIN, CINSY, CINE, and OIN.

For further proofs see: 55, 147, 200, 244, 264, 328, 338, 370, 426, 493, and 524.

PRUHOM Proof of HOM, the processes of man as TRU.

(Since STA and HOM are two poles of the same continuum, from complete stasis to full dynamics, it helps to differentiate, contrast and integrate them, by DAOST)

TRUP, TRUF, TRU, TRUD, TRUS, TRIS, POS, CES, CHES, FUNCTIONS, POT, VAPO, TREQ, Life, Human, RUPKAL, LARF, MATES, OITIO, ENT, OM, KOT, PRUT, Meaning, MADAS, EC, ACEXIN, ALT^n, LARP, LATAN-O-S, ALT, SPALT, REM, LORVA, POS, PRALT, FAF, DYNU, DYNS, DYNO, LAMUA, WITÂ, ACC, Anthropology, Ethics, Conflict-s, Determining Forces, Expansion, Growth, Mechanic-s, Mercy, Moral-ity, Rule, Opposite-ion, Precise-ion, Regenerate-ion, Technique-ical-ology, Waste, Dialectic-s, PIP, CE, SIDA, SINUAD, PRUTAN, LECDIE, LASCES, INEG, LOR, PAND, LODAC, APROD, LOVKI, COPOFT, DEF, MESDY, MOSDY, FUS, PIF, PIFS, CIDEX, LACH, POSFE, PURT, PIPT, BIN, RECST, PUC, POV, PROCAD, ETOC, CONES, HOSU, COC, LODAC, LOSE, LITS, CED, DIVE, LACORN, RACT, LAFURD, PADFOR, LARNIP, POCAD, TAP, LARPIT, CODET, LETH, LACOMA, DITAN, SOF, RINCAN, PODAF, PROH, PROHIP, SRE, LACFAF, LODEPT, LELAG, SORD, TOSRA, PESP, DAREV, LITEOA, FUV, LACFEF, LELAG, SORD, PESP, DAREV, RETO, LACESP, RECIP, SLONCA, ACRE, TEPAR, LAHOCE, LAHED, LAHANC, VAC, LAHAFS, MOV, LESP, INVICT, NECESK, LOSIR, POC, RET, KDM, S-R, PAN, CARSC, S/F, FIDAT, LONAP, LATMUD, PRESH, ASEC, PAPH, HOPRUT, PAJ, SPAJ, HOR, TOR, GED, NESION, CORC, COUC, POCEF, VE, DEDIPA, RAW, LABIP, ROSOC, LACSAR, LOPAT, PARCOM, FARUM, LAVEC, LOLARG, LALES, LIPFOT, FAS, CAMD, LITPRO, POCEA, FACH, GIMT, RIPIC, POBUF, RAK, LILIT, PRI, PRIM, EXINI, POTRAT, VEMORT, LATOK, LATAN, LOPA, BIPO, LATAC, LORA,SACT, IMPLIT, INFA, MOPU, LAV, LAVIPT, LADU, POTDE, RAPOR, PAS, POPTS, MOPLAR, ERI, CERIT, THET, MARE, HELT, OPTH, LOPURT, MATSUR, MAB, GFST, GAC, MAT, SEPTAU, ITOD, LOAB, INCSET, DOF, ACCOSM, GIFT, PERT, ITA, IMPET, LASEV, LABGEPT, GEDEF, GECEF, ACEX, SOLO, RICT, TRIALT, ACLO, VEFOT, DETOP, POVOF, LUM, NEM, COMP, COMPLO, NET, CARUL, POSED, WAH, DITCH, LACOV, LACDAR, WIR, LAPAC, LIDAC, LICAP, WIM, SWAT, LARCANT, LIR, RAC, MEA, STOCS, LABA, TAFOR, RECTRU, LABA, LART, LADOV, LASACA, REC, PEC, PROS, LACOM, RAPAN, LAPERAC, LASPOFT, LARMSELT, TEF, TRUFCES, TRUPCES, PIFS, PITA, LARPID, PS, SED, CORTRU, HIDIC, LORNS, LOPOF, POFT, MOT, WITA, LADPROT,DEBT,DIFS, LAMA, MINAPT, LAT, TRUMOV.

PRUCES Proof of Processes of ES in TRU and MOT.

PRUHOM, TRUD, CES, VAPO, ES, MATES, ENT, Meaning, ACEXIN, SPALT, REM, LORVA, MOCBIREM, HOM, FAF, Interaction-s, Oblige-ation, Process-es, Think-ing-thought, CE, CHES, PIP, HOM, PROS, NECESK, EXAT, CESINS, IPR, LOGOS,ACI, LINTAC, CESE, LAPDIT, TOW, CESA, POCOP, DIVE, OITIO, TOC, DITAN, LAFAP, CESION, ESION, LAIA, RECIP, and LACESP.

SECTION V

HOM: The PROCESSES of MAN, or CES

CHAPTER 50

INTRODUCTION, DEFINITIONS, CATEGORIES, PRECEDENT, SEQUENCE, KNOWLEDGE, PROBLEMS, and PROOF of HOM

HOMATION and CHES or DYNAMIC POSITS in MOT, ALT, SPALT, and SPAT

Section V is the beginning of the last half of the text. It is the application of our previous findings and existing knowledge to human life, as TRU, through revealing its dynamics by the processes of man or **HOMATION,** and as this section **HOM**. The ESIN of these processes is **CESIN**, and when considered solely within TRU, **CES.**. When CES are converted to their vector POSITS in SPALT and other ALT, MOT and ROT, they are called **CHES**. These are the specific Posits of the processes of TRU, CES, in a representation, model, ALT or SPALT. Many NEC become ES, CES, and CHES to make the POS in SPALT that provide KOT and POT. The dynamic CES are the opposite face of the static ES. When we POSIT ES in STA or STATICS that for motion is to a large extent CES in HOM or dynamics, in SPALT.

HOM is SECTION V in IPTRU, how it can be presented. Homation is study of dynamics whose topic is the processes of man or how they can be worked to represent TRU. CE and CES are one or more of these processes in TRU, comparable to the static E and ES. CHES is how each CES can be best posited in SPALT or other MOT of TRU. The POSITS of CES in SPALT are the means by which we bring all TRU into objectification, representation, differentially and integrally, and especially dynamically in TIT-TO. It includes not only Section V, but the following Sections, VI, VII and VIII as they are HOMATION and CHES, the specific POSITS of CES singularly, complexly and totally, or DYNU, DYNS and DYNO. It is in such an approach that we can best reveal the inner workings of all, as this all is, and is in, TIT.

The linguistic basis for Homation are the suffixes: **ION** = State or condition, act or process of a thing, or that being acted upon, **TION** = The processes of a thing, **ATION** = the processes of a thing, **hom** = Similar, same, like, **Hom** = Similar, Relative, Human, man in TRU, **HOM** = Section V of IPTRU, Principles and Description of Homation and HOMATION of TRU, **homo** = Man, or Man like, **homation** = Any processes, more or less referable to man, **Homation** = The Processes of the Similar, Relative or of Man and TRU.(TRACAD) **HOMATION** = The Posited Processes of TRU, the general principles CES Posited in Model,as SPALT.

HOM is of great importance to POT and IPT. This is seen in the fact that they actualize TIT, as shown in the latter half of the text. This is because by discovering, learning, and best utilizing the differential dynamics of TIT through VAPO and POA we can best show and know TRU. HOM not only shows the mechanisms and processes of how TRU works, it supplies the long needed deficiency of integrative processes, and their proof, PIP. It proves TRU and the great value of ROT, SPALT and POS. They act as core forces and complex functions that enable TRU through the selection of ESINS, positing of REMS, developing, and tracing through CES. We can't know the mechanisms unless we can demonstrate how the processes of man exist in TRU and its model. How they can be posited and brought by VAPO into a successful guide for each person's knowledge, work and life. Most NEC exist, more or less, if even by their ESINS and CESINS, as parts and processes in TRU. How we can discover and use existing and new terms, or NEC, for ES and CES. We convert them into NES and CHES as they are posited in MOT and SPALT to produce and provide KOT. This is a level of knowledge that has, heretofore, never been achieved. The rewards of KOT are nothing less than a complete fulfillment of the potential and actual Á FAF, dimension and continuum of TRU. This especially applies to the ever higher levels of Á.

We may not believe that there is much in the world other than that immediately before us. This is called: **FIA** = the Fallacy of Immediate Awareness or Instantaneous Attention. We may not accept the reality, actuality and value of such great processes and dynamics of TRU, that are so much a part of us. They are often more abstract, altered, adapted, complex, and in ways, esoteric, thereby not so apparent to our senses or mind. Yet, it is often, these that are such a major part and aspect, that stimulate and guide us. It helps to learn the kinds of factors and processes we seek to posit in SPALT and TRU. They can be potential, actual, ideal, optimal, minimal, cyclical, effectual, determinative, and more or less total. The greatness of TRU and Homation, or IPTRU and POT rests on the proof of the nature of TRU and its formulation from its aspects, parts, processes, divisions, constructs, systematic unity, representation, and modulation. They are the great formative elements and functional forces in TRU. This is confirmed when we recall the law of escalating limits, or how TRIS-TRIT-OBT-SUT-POT-MOT-ALT-SPALT-REM-POS-STA-HOM-ES-CES are decreasing a replicas of TIT-TO. We should always remember, ECRO, that each of these are DAOST and this in an accumulatively progressive way depending upon their FAF, view, and use. Reversibly this is what occurred in presentation, from observer-AUS-TRU.

HOM tells us *how* this special universe of best human life functions, as human is taken generally in its most relative and inclusive, TRU, sense. **PROHIP** = Proof of all TRU or this special universe of best human life through Homation, the interrelation, interdependence and interaction of its principles and processes. PROHIP is the how TRU is proven by homation, TRUD, HOM, CES, and CHES.(POUC,PUREV, LUV) HOM is also the confirmation of much of the <u>WHY</u> of TRU. It is not enough to know that TRU is. It is not enough to know what TRU is. It is not enough to have an idea or contemplate vaguely about TRU. It is only sufficient when we know the why and how of TRU in sufficient detail.(VAPO) The deficit is the arrested state of much of human knowledge and life. This is evident from the worlds religions, philosophies, fields, sciences and how poorly all people meld and work together, as they must for the common good in TRU. How the world works is Homation, especially if held to TRU. The constant processes of healing, repair, replacement and regeneration in the human body are excellent examples, parts and processes of the massive presence of the dynamics in TRU. Increasing knowledge enables perfected ethics. We can not do right unless we know, know what is right, and know this to be what actually is, is right by TRU or TREQ.

Homation means much more than the posit and use of single

processes. It means how they are or can be: 1. Rated, evaluated, and guided by REM, 2. Summed, as a whole and in this way imparts new meaning, 3. Made proportionate by ALT, 4. Related to each other, and to the whole, 5. The same and different aspects of TRU, and how they function in this way, 6. Interpreted in terms of the basic AD, DA, DIA, DAS DAO, DAU, DAR, DAT, DET, DAP, and DAREP, 7. interpreted in terms of TRUCH, TRIS, input, output, and feedback, 8. All in summation prove TO, TU, TRUIT, and TRU, 9. Fulfill ELTICMA and move to the absolute, TA.

Fig 1-50 **Descriptive Varieties of Processes CES and CHES in HOM**

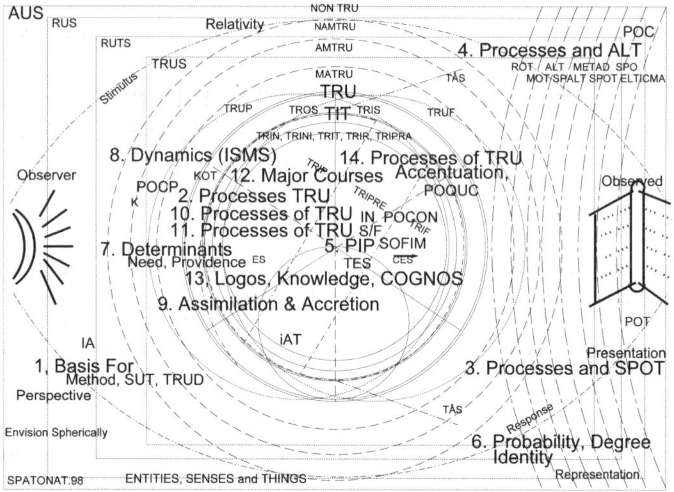

CES are differential as well as fundamental to TRU, TRUD, HOM, ROT, KOT, and POT. How they work in TIT, TRU, and OITIO is shown by CHES in SPAT of: 1. Processes, the basis for: a. Method, b. Systematic Unity, c. TRU, d. RE I-1, e. Showing the Dynamic Mass of TRU. 2. Processes of TRU: POCP, POF, POSIM, POSTOP, RE, and RENFUD, 3. Processes and SPOT, Perspective of the Universe, Universes, and Subject-Object, 4. Processes and ALT, **POC** or Principle of Conformity, 5. **PIP** or Proof of Integrative Processes, fusion, reciprocity, whole, and SOFIM, 6. Probability, Correlation, Identity and Degree, 7. Determinants, Contingencies, Necessity, Fate, Providence, Destiny, Obligation, and Duty, 8. Dynamics of Permeation, Symbolism, Transcendental, Spiritualism, Mysticism, 9. Accretion and assimilation, 10. Processes of TRU, Causation, Formulation Integrative Negation or **IN**, Principle of Consequential Negation or **POCON**, Popular Misconceptions, Wa, ME, MC, Accentuation, 11. Processes of TRU, Success/Failure, bent, warped, and asymptotic Processes (POUF, RON, POVE, BURP) 12. Major Courses, Logic of History, 13. Logos, COGNOS, Medieval Platonism, Knowledge as Processes, 14. Processes of TRU: Accentuation, Principle of Quality Control or **POQUC**.

As our viewing, showing and knowing TRU changes so must their different processes and methods change. They do so in certain ways that produce their own formality and thereby, relativity and reproduction. One of these ways is: **LIA** = the Law of Identical Alteration. Whereas TRU is all, and as TRU changes, so does all tend to change identically. That is with only the difference involved in the change. Also, there will be a way of change by proportionality and dimensionality.

There is much difference between what the inside is from what it may seem to be when seen only on the surface. This is how much that is valid for TRU is, as for the human body, earth, universe, or God, only understood when comprehended in full. This fact is common knowledge, as in the phrase 'more than skin deep'. It has taken humans a long time to accumulate the knowledge and understanding needed to communicate what these various, massive, dynamic and complex insides are. Although much of knowledge, and the world, remains ignorant of them, they, to a large extent, are simply for the asking. This science has shown, even though with poor focus and VAPO of TREQ or TRU.

How much we do or do not accept the past and future and hold them in mind as they are actually in TIT is key to our lives, and a large determinant of its duration. It is partly a function of, and how much all appreciate, **PIP** = Proof of Integrative Processes. PIP is a large proof of both HOM and of TRU, and. TRU and HOM show how PIP is proven.

What a thing is for and how it helps humanity or TRU are much of its measure. With meaning, reason, and intention we can determine the more complex functions and processes of the real or phenomenal of relativity or the observer-observed, subject-object, Á FAF and dimension in and for TRU.

Indecision is one of the worst causes for mistakes, errors and failures. The correction of indecision is only the right decision, action, and aggressiveness by a knowing and ordered mind TRU. These are not possible unless we discover, learn and know, TREQ, TRU and its CES and put them to best use.

Fig 2-50 **HEALTH, and INCREASING and DECREASING TRU**

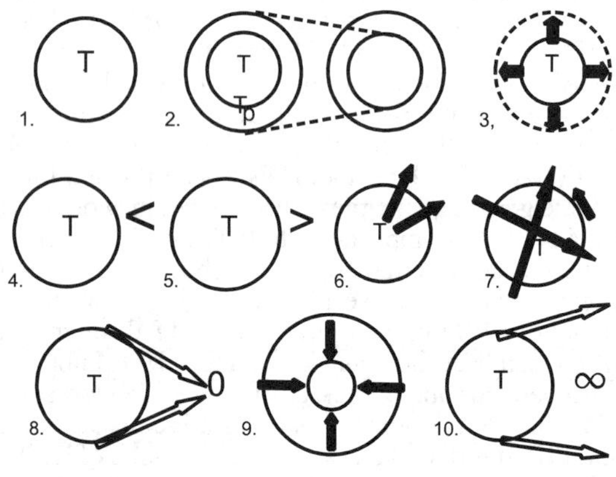

Good health is what is of increasing TRU, all that produces better quality of TRU. Some of the basic relations and mechanisms of health in different ENS, OBT, and TRUS are shown by: 1. Simple object, 2. Transformation, 3. Expansion, 4. Decrease, 5. Increase, 6. Outward thrust, 7. Inward actions or thrust, or transformation of coordinates, 8. Actions and forces that tend to reduce to zero, 9. Convergent forces toward the center, 10. Forces or processes tending to infinity or absolute. **FIDAT** = Forces working In All Directions in TIT. FIDAT corrects the misconception that the forces and processes in TIT are only unidirectional or simple cause and effect. They are mostly multidirectional, complex, and for the whole, positive in equilibrium. This is one of the most beautiful and terrific examples of how the inversion of paradoxes provides great illumination of TRU. How we are so sure that the past can't be brought back to life, and with so many that physical death is all. Nothing proves TIT and TES over individuality more than this. We are so sure that processes move only from the past forward and never from the future or present to the past that we deny much of our being, TRU. Most of us believe our lives are only this conscious, 'point in time', the present. Yet are so certain of this and that, as oneself, and 'is what it is', that all contraries are denied and ignored. In Homation, however, we can raise our appreciation for both the complexity, interactivity, health and totality of all, as it is in TRUD, ENT, TIT-TO, and TRU.

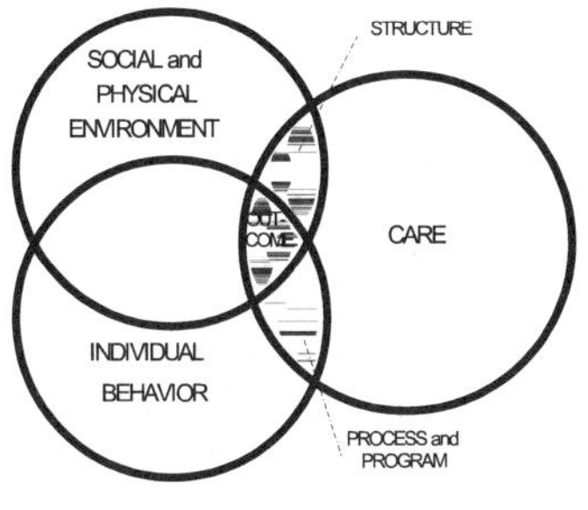

Fig 3-50 **Means of Showing Homation**

Coterminalities and codomain provide another means of showing different steps, methods and degrees of problem solving. In figure 3-50 we identify three steps, stages or degrees of function that are single, double, or triple combinations of different entities, units or systems. Codomain can show the relations of values and their various combinations, permutations and interactions. Without interaction we are mostly involved in structure, i.e static/ dynamic. Whereas it more often exists as dynamic/static.

How Homation is a method of thinking: 1. Piece by piece, instantaneously, or partially, so that the evidence, facts, proof and maximal form or revelation of extremes and means, parts and processes, are brought to provide maximal form to TRU, 2. Reducing all facts, evidence, knowledge, perceptions to their proper interpretation in terms of TRU, 3. Partially, critically and analytically of differences, ones partial position., 4. Conclusively in order to bring out all by experience, intuition, deduction, induction, gist, and whatever means show their great significance, complete and overall relates and posits, 5. Problematically, 6. Paradoxically and concentrating on those problems or seemingly irreconcilables, differences, and opposites that must be reduced if TRU and DAOST are to be made valid. 7. Critically so that all fallacies and errors are considered and their pitfalls averted, 8. Explosively as in revelation (ER) where each great aspect, feature, form and function of TRU is brought out, revealed, demonstrated and visualized in its maximal state and detail, 9. By ESINS and ES of all we attend to, 10. Of and by the parameters and dimensions of TRU, and how they can guide us to SUT, 11. Exponentially or by ACEXIN, how parameters or dimensions exist not arithmetically, evenly, but exponentially, and in an altered, adapted and somewhat warped way, at least when compared to physical equivalents, 12. Massively, in toto, holistically, 13. In terms of proportion, evaluation, and relative composition as by REM, 14. Correlation, by axioms and corollaries that are based upon the same of DAOST, 15. In terms of ordination, coordination, as ALT, SPALT, METAD, and SPAT, 16. Of unity, especially the unity of TRU, TU, 17. To think ELTICMA with optimal memory and flashes of revelation, cue, intensity, visualization and order, and 18. To think by all these methods together focused on the totality in TRU, TIT-TO.

There are certain preconditions for success by Homation. CES must be different, divisible and separable. We must be ready, willing and able to achieve this differentiation and separation. They must be identified. Names must be made for those that are unknown and most important. A frame, sphere or model must be developed to order them. Their representation must utilize the best of methods to achieve POA. It must be done by VAPO. We must precisely identify and utilize all the form and function of TRU, its different states, kinds, varieties, concepts, principles, laws and mechanisms that guide us to best POSITS of CES.

It is in tracing through CES and CHES that show the transitions of thought, act and their vehicles, that we can best show the benefits and successes of the participants in TIT. The transits of thought and act, and their intermediaries existing in TRU, are called: **LATMUD** = Law of Act(s) and Thought(s) Mutual Dynamics. LATMUD signifies the operations of S and O that make up much of the predominant Á dimension, FAF, parts and processes of TRU. It signifies how thought and act, as S and O, correspond through the vehicles that are part of this same dimension, Á, of TRU. These vehicles are its gestures, words, speech, writing, pictures, stories and all communication that is representative of TRU. LATMUD signifies that thought and act, as CES, are not to be considered of only as ENIT or things in themselves. Nor only separate processes in TRU, but as they are DAOST, different aspects of the same thing, formal and functional complexes in TIT. In this way they exist together, dynamically, by constituting much of the whole, TO of TRU.

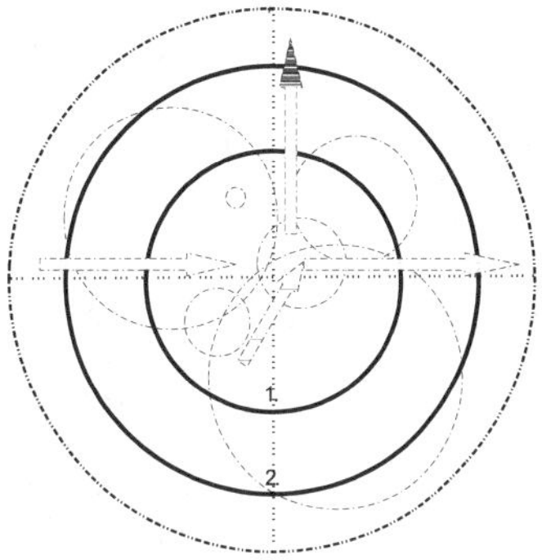

Fig 4-50 **PROCESSES ENABLING US TO PROVIDE A MEASURE OF TRU**

The best measure of all and what points out the finest details of structure and operation of TRU are those forms, functions, patterns and processes that enable it, and us, to be. It is of those processes that are working through time and by the manner in which this working takes place.(ACEXIN) It is how this provides the capacity for a successful existence.

Application and use are not always, nor merely, a matter of action, response or of the future. They are basic and crucial aspects and components of our being. Use, relativity, and the Á predominant FAF are much of the same of DAOST of TRU. Every part, organ and function of our body has been fine tuned for its use to the whole of our species as it is a formed and functions as a whole, TIT-TO. The operation of our CNS is highly dynamic, complex and variable, yet largely convergent and interdependent in TRU. It is in use, what all is used for, or potentially will be TRU, that determines much of the TRUD, TRU, and HOM. If we would understand the world and how it functions or operates, we need to first and foremost learn and understand what it is for, why it is used, and how it contributes to TRU.

In all kinds of HOM, CES and CHES of DYNU, DYNS and DYNO, we seek to discover: 1. How much do and can we know of each, 2. What is and what is not TRIT, TRIF, TRUF, TRIP, TRUP, TRU, TRUS, RUTS, RUS, AUS and all. 3. What and how are and are they not TIT, TRUIT, TRU, TRUCH, and TRUS, 4. What are their basic forms, characteristics, definitions and delineations, 5. How much TRUCH can be brought under SUT, MOT and HOM, 6. How much SU is adequate without TRUCH and TRU, 7. How much or how far is TRUCH possible. 8. Is TRUCH close to TIT more understandable than TRUCH farther removed. How much do TRIT and TRUCH set and emphasize the knowledge of TRU. 9. How much does knowledge of TRIT and TRUCH complete our overall understanding of TRU, knowledge of God,, and the problems and failings of a universal understanding, or religion. These and associated questions are key to the answers and solutions the processes of TRU seek and provide. The outcome of their courses converge on others that are the same or correspond in TRU to produce the totality of form and function in TRU. They stand in contrast to all that is outside of TRU. The subjectivity and objectivity these processes encompass, provide and engender help us recognize differences that enable an effective overall vision of TIT, TRU, TRUCH, TRUS, RUTS, RUS, and AUS.

CHAPTER 51

DERIVATION, DEVELOPMENT and DETERMINATION of HOMATION

CAUSE and EFFECT, METHODS, DEMONSTRATION and S-O in HOM

There are around us many different kinds of worlds, some commonly known have been described, others less so, as: AUS, entities, senses, things, RUS, RUTS, NAMTRUS, AMTRUS, MATRUS, TRUS, TRU, TRIS and TIT. That these different worlds are in many ways in opposition, even violent conflict with each other, are also readily evident. That they are in opposition and conflict with TRU is to be shown. There exists in and out of TRU a constant flow of positive and negative, what is or is not TRU in an ever fluctuating manner. Even more, there exists entirely within TRU a continuous massive interaction between parts and processes in which they are competing and in conflict or dialectic equilibrium with each other. This becomes increasingly certain as we learn more and more of TRU and its dynamics. The massive and continuous struggle between ideas, opinions, individuals, partials and processes in the TIT, TRU, and ATRU worlds is that mixed world into which we are born and live. Here is the great and continuing battleground, a life at war with itself. It has ever been, continues, and will continue to rage, more or less, forever as it relates to the internal as well as external dialectic CES of TRU. For even if completely within TIT, although its totality is in harmony, its parts and processes are in vigorous competition.

Yet there is a certain form and function, set and SCARLS as well as certain rules, which exist that can be enlisted for help. There is no guarantee of permanency or continuing success. Each generation, individual, and effective part of TRU must find, as early as possible and with the least loss, suffering and failure, an adjudication or coming to terms with life and bonding with TRU. It is to a great extent up to each new person to increasingly select this heaven over hell on earth through adaptation, especially with the assistance and prior benefits provided by an existing TRU and its FAF.

The price, pains, and losses in regeneration, reproduction and rejuvenation, especially if less than 100% necessary or TRU, is a paradox and problem whose solution lies to a great extent at the basis of this well spring of life. It deserves special emphasis however, for not only is it often productive of conflict in societies, cultures, civilizations and TRU. Failure of each new generation to learn, understand, and appreciate its part in TIT-TO is reactionary, inefficient, and leads to further reaction or loss that is increasingly harmful. It requires as other parts and processes great capacity, ability, effort, energy, and power of and from TRU.

Although TIT-TO, is positive, in certain respects the negative participates, if only indirectly, partially, or reversibly. One of these is the reciprocal affect which the negative induces on the positive. This is called: **RENIP** = Reciprocal Effects the Negative Induces on the Positive in TRU. We must be very careful when we assume the negative to be a part of TRU. For TIT-TO the interacting negatives are, in effect, positive in resultant, a compensatory benefit. Just as in medicine, what is disease is negative, yet what this disease is in the actuality of TRU is its cure, compensatory or selective positive effects. Often knowledge of the negative, by contrast, enables us to better know the positive and TRU. TRU effectively turns negative into positive through knowledge of the negative.

HOM arises from the very nature of TRU, TIT, and TRUD. It is to be always conditioned by ATO and TROS. This can be best evident in our lives, in the accelerative expansion of knowledge and its availability. Treatment can't ordinarily come before diagnosis. In contrast to the one way relationship of diagnosis to treatment there exists in TRU a great proliferation of ways in which the processes of man work. Even diagnosis benefits from treatment, for treatments that are a success or failure in practice will alter the diagnosis.

Fig 1-51 **ORIGIN, GROWTH and DECAY, HOM and TRU**

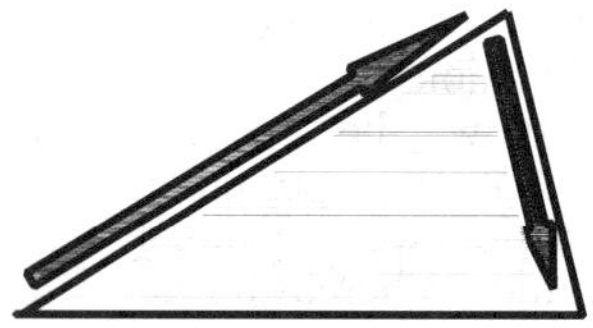

This figure initiates ones view of the successive course and stages of ENS, beings, and TRU. It is usually a long hard pull up and a short easy fall down. There is a long period of growth, plateaus, quick harvest, and decay. This simply exemplifies many of the key ideas of the core mechanisms and processes of TRU. TRU differs from individuals, animals, other species and entities in their respective growth and decay patterns. Even the parts of TRU differ from the whole. The origins, growth and development of TRU are progressive and regressive, liable to an extinction that is special. For TRU these will have their own scales, like those for A, L and T. TRU functions, and its measures depict this function, of a nature that partakes of many of the same mechanics as of other bodies. This will depend upon the similarities between these other bodies and TRU, and their adaptations to TIT-TO and ROT. Yet as a distinct entity in its own right its mechanics are not to be mistakenly attributed to other states and conditions which do not apply, or apply in only certain, variable, or somewhat altered ways.

GED = Genetic Exponential Deficit, or the genetic defects, deficiencies or deficits that occur accumulatively, and often exponentially. This leads to a lessening of human energy, capacity, ability, shortening of life, and inability to compete in life. Along with its use in the live product and the effects of other determining forces in TRU, they make up much of TRUD and Homation. Often, over time, very small differences in GED and reproduction rates produce the greatest effects in survival of species, varieties, races and cultures. GED both qualitatively and quantitatively decreases human achievement of TES and TRU. That the development of the processes of man occur as ontology follows phylogeny is evident in the stage of <u>mosaic development</u> in the embryo. This is that period between the stage of organ determination and functional differentiation. It reveals the expanding complex processes through gastrulation. It should be apparent that genetic defects are diffusely and interactively transmitted.

Increased recognition, proof, and conviction in the reality and actuality of TREQ and TRU has been a product of increasing human awareness of the world, recognition of the forces, powers and actions that we do not see, readily sense, or immediately identify. It is exemplified by primitive religions in which a curiosity, awe and mystic state seeks to penetrate this realm of seemingly hidden powers of TIT-TO. The awe of mystic and transient beliefs disappear when understood. (VAPO) Yet their state in TRU remains essentially unchanged. The idea of God, and the development of religion, arose from such processes. Only when knowledge identifies with TREQ or TRU, and is able to show and use it, will their development reach and assure fruition. Because of this and the reaction of established religions to it and to increased knowledge, it should not be surprising that there have been so many differences, and attempts to reform and improve religion. TRU can not be identified by ascribing to it only a limited, or small number of its, components, form and function. It must recognize all of TRU, especially its mechanisms and processes, whose bonds and interactions are the 'tie that binds' this world, God, TREQ or TRU, together. The conversion of persons to these powers, what is holy is the sum of felt adaptation to TRU. Herein is how there arose and has been produced the summation of sins, redemption in conversion to God, and of God to oneself, as in 'my God' by Christ, perfection of an individual's identification or 1:1 union with God as TRU, as an individual's TES in TIT. This is, through TREQ, the spirit that has come to be known as Christianity. It is especially through each persons identification with Christ, and his sacrifice and martyrdom to this conviction and cause, that each person 'may live' with God, as with TRU. So that with the recognition and proof of how these hidden forces, powers, mechanisms and processes came together as a unified whole, and how man can identify with TREQ that humans in ever greater populations have been able to bring their mass energies to ever higher levels of predominance in this world. It is the development of all these and other processes, or Homation, that produce TRU. To have a mission in life is the best method of existence or state of being. Yet this mission can't be one of free choice or fancy. We need to be flexible and willing to give as much, or more, than we take or receive, by VAPO and all in TRU.

"Total rejection is hell" says Billy Graham. Such rejection, be it overt or invert, TRU or ATRU, is very commonplace in human life. For what is a person if they are the only ones in the world, or if all are exactly like them. In many ways rejection is a good example of the extreme forces for and not for TRU. Forces for TRU have markedly accelerated in hominid evolution by the combined power of mind, socialization and language.(Á FAF) They show what the reasons, causes and effects of the processes of man, CES, are in HOM, as they posit, CHES, in SPALT. *Quid pro quo* means something in return, compensation, retaliation, or substitution. It is a good example of TRUD, HOM, CES and CHES. It is a form or state of interaction and exchange, which in TRU is positive. In politics, negatively, quid pro quo refers to graft and corruption This shows how important it is to be selective, both the recipient and the donor, by a TRUD that is only TIT. What the inclination of dialectical processes and their positive resultants are, are shown in: 'nature creates nothing in vain'. This basic proof of TRU is most subtle and profound. It is known in another sense in TRU, as the pleasure principle. It may also be seen in "a good appetite needs no sauce". Cause and effect is the key to understanding TIT, TRUD, and HOM. It and such basic polarities of continua or processes as love/hate and strength/weakness show HOM.

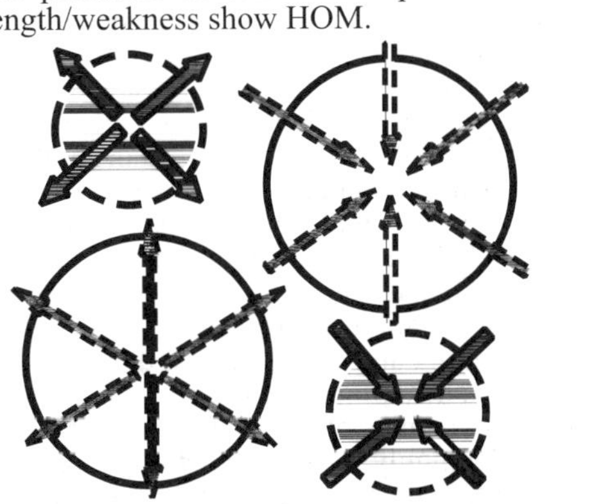

Fig 2-51 **CONCURRENCE of CAUSES, HOMATION and TRU**

CORC = Concurrence or Correspondence of Causes. CORC signifies that there are, as in relativity, sameness and especially, a set form and function of TRU, a large amount of concurrence and correspondence of causes in TRU, TRUD, and HOM. The figure shows some the forces that produce, enable, affect, and destroy. CORC should be contrasted with COC and **COUC**, Courses in Causation. These signify different kinds of reasons for and causes of HOM, TRUD and TRU. By their role in the operations and Posits of CES we show the most distinct and explicit detail of HOM and TRU. Much of what one does in life is determined by ones circumstances. Thereby much in life needs to improve and maintain the best possible circumstances. The degree to which we can not or do not correct circumstances that are detrimental is what causes much loss, by INEG. Much of HOM, as human life, is the use of ESINATION, how we take or select what is good from bad, and bring it to TRU. When of complications they are: **COC** = Complications of Causes, Causes of Complications. COC is seen in death certificates, where the causes of death are often attributed to a major cause, yet are in actuality, both naturally and by TRU, many, ramified and a product of INEG. These illustrate the dynamic DAOST, in the totality of TRU. Multiple Causation not only proves the true workings of causes, reasons, form and function of TRU, it provides the basis and justification for Homation. These are the processes and their mechanisms without which TIT can not be known. This is why God, religion, philosophy and science have so failed us. They have not been able to bring together their great concepts, principles, tenets, doctrines, axioms, and laws into the potentially coherent, positive objective and demonstrable whole of TREQ and TRU. Multiple Causation greatly shows and proves the highest validity of PIP and SUT. e.g. the many injuries and illnesses which are caused, not by a single accident or agent, but whose causes are ramified almost infinitely, with convergence upon a unitary end, survival of the species.

Without some idea of what is and what is not important, what is of most value, what our priorities are, what the relative magnitude of each and all parts and components of TRU is, we are lost, we fail, and TRU fails. It is from increasing awareness of value, REM, through the ages, that we have been able to climb to such heights of success and glory in TIT. This is because we can understand and appreciate the value of life, love, God, knowledge, work, and all that is great, REM. We can direct our lives, and all we do, to conform to these ideals and ends. Yet this is not a simple formula or equation, it is an ongoing and vastly expanding operation of TRU.(ACEXIN)

The <u>principle of efficient and final causation</u> is the ontological principle of Whitehead. It implies that actual entities are reasons and thereby our search for reasons is the same as the search for actual entities. This is how TE and ES are of TRU and effects are causes and causes are effects as they function in different SCARLS in TIT, demonstrated through CHES in SPALT. <u>Vacuum effect</u> is the largely negative force, power, and effect created within TIT that result from either the dialectical or dipolar imbalances of major continua, mechanism, and processes, CES, or from voids, either real or imaginary, that exist in TIT. Actually TIT does not have absolute voids or vacuums, only relatively as associated with variable density, dynamic, cyclical, and variable flows in the differential in TIT. It does have both by FAF, as does the human body, lesser and greater regions of concentration within their masses. These are often manifestations of the great homeostasis and dynamic fluctuations that occur, optimally, sub-optimally, or abnormally. They may even be made worse by vacuums and voids in our lack of knowledge, acquaintance or misuse. They are activators whose function revolves around maintaining the homeostasis of TRU. Such gradients of less and greater density are set in motion, as any level of energy in the Á dimension, by powerful forces, pressures, and great vectors which, to a great extent, show Homation, THEO, THES, and the nature of TIT.

DEDIPAT = Determination of parts and processes of TRU through Direct and Indirect Effects, Influences, Proportions, Accuracy and totality. DEDIPAT is a key concept which aids in resolving the many problems of TRU, its ROT, MOT and POT. It helps to provide the formulation of the processes of its parts as demonstrated in Homation. It provides the key dynamic mechanisms, **KDM**, for understanding the world, universe, or TRU. It gives a better understanding and proof of the voids in ALT. It helps to fill in those unknowns that were heretofore unrecognized. It enables massively, dynamic and total thinking equal to the nature of TRU. It enables each person to train themselves to think correctly, profoundly and sublimely. It enables each and all persons, along with other methods of knowledge, philosophy, science, religion and TRU, to gain those penetrating insights, and the overall comprehension, necessary for knowledge of TRU, all things, and an ever better and happier life.

There is tremendous significance in the great variety, parts and processes in each and every organ and function in the human body. This is especially noticeable when they are abnormal, not well, injured, disabled, or missing. As expansion and contraction, metabolism and catabolism, in daily life all organs and bodily functions are in continuous activity. It is the importance of bodily functions, which may ordinarily seem trivial or to numerous to mention, that provide the examples and the actuality of the differential and incremental effects of the processes of man, Homation in TRU.(PIP)

What one allies oneself with is largely the determinant of what a person is and does, what they are, whether one wins or loses, and whether a person lives or dies. This is what a great part of the world does, both beneficially and detrimentally, as it is a cause for much of its troubles and failures. It is not who or what may be temporarily a winner, but what is totally and permanently winners, as these are TRU. It is what one and all must ally themselves with TRU, as TIT is so allied with all its constituents and benefices, its S or same in DAOST..

Science is skeptical of, and apt to be the first to criticize, the application of conclusions beyond their statistical and investigational limits. In HOM as in TRU even these findings and conclusions are not accurate or must be adapted to their relationship to TIT. Certain general principles point to and make it highly probable that such an interpretation can be done: 1. The theory of relativity and the establishment of TRU, 2. The absolute nature of TRU, TRIA, 3. The means to TRIT, OBT, and SUT, 4. The means of ROT, MOT, and REM, 5. The increasing validity and reliability of ALT and SPALT, 6. The nature of the reduction of all to their ESIN, ES, CES, and REM in TRU, 7. The positing of ES and CHES, singularly, collectively and totally in SPALT, 8. The application of all concepts, principles, and laws that POSIT in SPALT and are a part, process, form and function of TIT. 9. The uses of the methods of HOM, ACC, POT and IPTRU in the life of each person.

Things which are of no apparent relation may actually be related indirectly, reversibly, in cause and effect, or other ways. This is commonly experienced in scientific research where correlations are often erroneously attributed to a simple cause and effect relation. The relation is indirect through other causes and effects, or may be an accumulation far removed. The relation, or correlation, is often from a sameness in TRU that is not understood. The extension, causation and formulation of correlations can approach the infinite, both in, out, and in-and-out of TRU. The major courses shown by the posits of Homation in SPALT, their causation, and formulations represent the same, but different approaches to, what constitutes TRU.

Much is to be learned from motion and action analysis, and from the study of the sizes, timing, movements, S-O and all activities a person engages in or occurs in about them. The sequences of actions that go to make up the work, job, life, daily activities, needs, and bodily functions of the individual are great keys to the dynamics of TRU and make up a large portion of its processes, operations, inner-workings and dynamics. These are simple when viewed singularly. When held to a limited frame, as in algorithms or cybernetics, they may be readily demonstrated. When done in the frame of TRU using SU, REM, ESINS, ES and CES through POS in SPALT, they may also be shown with accuracy and certainty.

Method, means, and critical analysis of vector Posits include: I. Their essence, ESIN, must be extracted, II. The REM valuation must be assessed and determined, III. Their tri-dimensional nature and character must be thoroughly demonstrated and remembered, especially the mass, Á, dimension, IV. The limits of three-dimensional representation must be made evident and corrected, V. Their multiple nature and character must be shown, VI. Their relationship and correlation to TIT-TO must be discovered, learned and shown.

Modern man must be a member of the present day scientific technical economic world. We must also be a member of the best of what is common in all religions that is TREQ or TRU. We should know how each person is not, or not proportionate by TRU, that we need to avoid, eliminate, reduce or control. For both Heidegger and Bultmann the way to find this best world, both of Christianity, modern science, and commerce, is to find what are the 'authentic' types of existence, TREQ. Only by learning more and more about TRU and separating all not TRU or disproportionate TRU can we find this authenticity by discovering, learning and applying TRU.

There are several laws of what is best. They reveal the basis for TRUD and HOM: **LABIP** = the Law of the Best in the Past. LABIP signifies that if the past wasn't the best we wouldn't be here. All generations should realize that their lives and their worlds are the result of the work and lives of their ancestors and all that made TRU. If these had not been the best we wouldn't be here. This is valid more in the relative and distant past than absolutely in the near past, or in one or a few generations. Indirectly LABIP includes learning, listening and profiting from the experience and wisdom of past generations. **LABIF** = the Law of the Best in the Future. LABIF signifies that if the future is not the best we will not be here. For if the future is not the best, it will not beTRU and TRU will not be. This gives great credence to the value and necessity of an assured future, not only in kind but in substance, form, function and entirety. **LABUF** = Law of the Best in Us necessary for the Future. LABUB reinforces the obvious need for selection in life, both by the individual, and populations, to maintain TRU. **LABED** = the Law of the Best in Element and Dimension. LABED signifies the extension of LABIP and LABIF, where the same reasons apply to any aspect, element, dimension, part, and process of TRU. It may also apply to the polarities and gradients on the dimensions of TRU. LABED is an explanation for ISMS. Yet, by EBE, their justification is invalid unless EBE and these ISMS are strictly qualified and in conformity, by VAPO, with TRU-TIT.

The public votes their pocketbooks, loves, fantasies, following the crowd, and are swayed by the power of stimuli, irregardless of TRU values. It shows how the stimulus is not always simple tactile or painful but much more complex, and often not, or negative to, TRU. To be TRU it must come from within, ones best or deepest loves, feelings and higher levels of Á, exclusively in TIT-TO.. The capacity and ability of human beings to respond to stimuli, answer questions, solve problems, and in all ways provide the optimal THES and THEO, THESÂ and THEOÂ is the expanding complex core of being, TRU. It, like POA, is the derivation, development, determination and demonstration of HOM all together.

To be thinking of one thing and doing another doesn't work. This is because thinking and doing are two aspects of the same thing, the Á FAF and dynamics of humans.(DAOST) Since human IA is finite and a highly focal state, its relationship to thinking and doing is different than that of TRU in general. We can through IA or a differentiated view of thinking and doing, as in psychology or individual's workaday world, readily separate the two. To function or live successfully requires a constant realignment, readaptation, return, and reintegration of the two as they must work together as processes of the same thing, S-O or Á mass energy dimension of TRU. The progression of SÁ recapitulates the derivation, development and determination of HOM.

CHAPTER 52

GENERAL FORM of HOM, CES, and CHES

KINDS, FORCES, EFFECTS, VECTORS, SCARLS, CARSC, NECESSITY of HOM, CES, and CHES

The general form of HOM includes the POSITS of CES in SPALT and MOT that are adapted to fit what they represent of TIT, TRUD, TRIT, OBT, ROT, MOT, ALT, REM, and POS. It is how we are guided by and apply the processes to each and all occasions, events, and situations with POSITS in SPALT. Also how they can be expanded in various ways by their uses for TRU and the observer. HOM is largely in TIT, yet OITIO when RÏT and RÏT. It is an objectification, placement, and treatment of ES and CES, as CHES, for TRU.

One of the greatest lessons and tasks each person must learn and do, is to distinguish between the following: 1. Those pressures that are the natural result of a person's relationship to the world around one, its differential and normative TRU interactions. What necessity demands we be in the world, to be TRU. 2. Those pressures that effect a person by the world around us that are abnormal, unfair, improper or unjust, and thereby not TRU, or ATRU. 3. Pressures from within us, both positive and negative. 4. Those pressures that are natural for TRU, as by VAPO, can be joyfully met, as part of the give and take of life. It takes many years of learning, knowing, planning, doing and work to even make the best of ourselves and reduce ATRU pressures as much as possible. This is why we are so deliberate in presenting TRU, TES, HOM, ACC, TRAX, TREX, TRIX, PUREALITY, COGNOS, ELTICMA, SÂ and TÂ. These are to a great extent the dynamics of interaction in TRUD that focuses on the individual. The lesson to learn is to make this individual, oneself, into TES as early, well and continuously as possible. We can show by HOMATION and CHES not only how to respond to stimuli, react to these pressures, and solve these problems, but we can show the overall principles, concepts and laws of TRU and how they best apply to our daily lives. HOM and CHES guide us to show all that we can do for the success of TRU and our ourselves, as TES.

Fig 1-52
GENERAL FORMS IN HOM: DIFFERING VIEWS

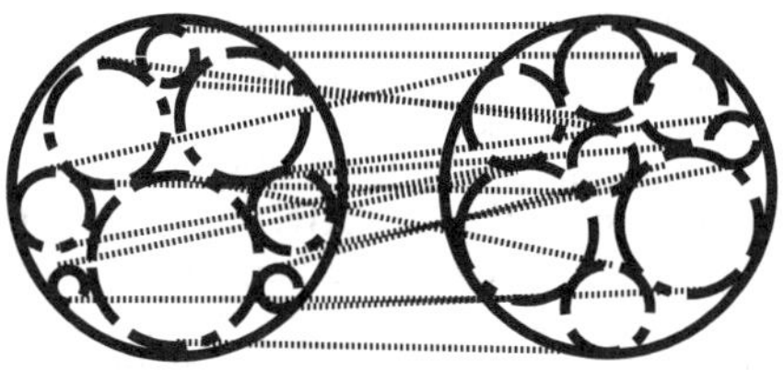

Culture is closely allied with language. This is how their processes parallel in Á, L, and T. From the views and posits we make for much of the ES and CES of TRU there is a great difference between whether they are longitudinal or in cross section. They show how, although many have similar components and processes, there still is much difference depending upon the view. This is why we develop SPALT, and why we must bring all to SPALT of TIT-TO. When approached by different views it helps us envision all portions of TIT.

There are many kinds of CES and CHES. One category includes the traditional divisions and fields approach to human life. This is seen in, history, genetics, biology, chemistry, physics, psychology, sociology, economics, political science, and the anthropological, evolutionary, universal, cultural, informational, and evaluative. All of these have to be better ordered by large segments of ALT, or by classifying them by their in related groups, as is done biologically. Yet there is a need for separate distinction of all that is different and helps to show and know TRU.

Only what is in opposition to TRU externally, like ATRU, is not TRU. If it is in opposition to a part of TRU as two parts of TRU function in some kind of a dichotomy, or polarized relationship or equilibrium, this will often be, and sometimes very greatly needs to be, a part, process aspect or mechanism of TIT. This is called: **LOPAT** = the Law of Positive and Negative or TRU and Non TRU Opposition. In the near term these are the currently active, acute and extant, struggles, conflicts, troubles, and problems that work to maintain the balance in TRU. They may be good or bad for a part or the individual, but are good for TIT-TO. In the intermediate term, the negatives of decades to centuries, are many of the larger conflicts, divergences, vacuums, and over concentrations, that represent defects in MATRU. In the long term are those basic and fundamental aspects, features and relationships, or CARSC, between the determinants, forms, functions, parts and processes of TRU. These are often vital and require great ACC, PPMA. management and control.

There is no question that <u>free will</u> is a large component of an individual. Even in TES its realm and variability of relations allow room for free will. Yet it is not free will versus the set FAF of God, TREQ, or TRU that is paramount. It is a balance, an optimal interaction in TRU, the balancing of ES and TES, by their CES, which generally and largely follow the normative of TRUD. Specifically, and less often, they follow a more fluctuating course, and return to TES TIT-TO in compensation.

PARCON = Principle of Absolute and Relative Choice or Necessity. PARCON arises from what we do and can be in four ways. 1. What we must that is right, 2. What we must that is wrong, 3. What we please that is right, and, 4. What we please that is wrong. Choice and necessity, by PARCON, are never absolutely negative or absolutely positive, never absolutely infinite or absolutely infinitesimal. Yet these are often of no or little TRU, as we are in being NAMTRU and DAK. Only as each is perfectly measured and determined by TRU can we escape ATRU and be TES and TRU.

Much depends on a person's survival which in TRU actuality is our TES. The focus and medium for this survival are largely ones IA and Á. These provide the great vehicles for determining how a person will, or will not, do right to make it. An example in life or business is how when one is down they will readily buckle up and dig in for recovery and improvement. Those who are comfortable, the fat cats, who have more than they need, lose interest and care. This shows by how ATRU stems from lack of need, and is called: **ATAN** = ATRU from A lack of Need. ATAN exists when there is little need, and therefor little or no stimulus or incentive, to do or act, by TRU. Although we can't deny that much action arises from immediate need, we act not so much from immediate need, as we act from more complex, long-term or overall needs, S/F and fulfillment. These converge on TRU and TES.

PRORE = Principle of General Proportion by Reason of Relation and Proximity. PRORE is exemplified by grasping, functioning fingers with an opposable thumb. Over millions of years this was the constant selective advantage their use offered human evolutionary predominance. It was an adaptation that utilized relation, proximity and proportion to develop optimal form and function, FAF. Just as effect has significance in and of itself, and reciprocally in TRU, so does it have more basic and primary states and conditions. These are: **DE** = Duration Effect, and **ARE** = Area Effect. Much that is in process in TRU, of which we are of or is a part of us, is DE, and is primary to the comprehension of TIT, TRU, TRUD, and HOM. Man largely senses and perceives things statically and objectively. We never realize much of the mid too processes. As a part of TRU, or even not so much or not of TRU, they work not as we perceive them, but through and through, or as half lives in duration and variation. They may also work structurally and functionally to produce manifold effects in TRUD. ARE are those functions, CES and effects, which are planar rather than linear. Likewise **SE** or the Space Effect are those functions, CES and effects in three dimensions by volume, or spatially. So can we show similar effects for variation, as well as the higher powers of Á. All together they form ACEXIN, or the total effects of ALT in TIT, as shown by HOM and CHES.

S - R, S → R, S ← R, S = R, S ≠ R, S↑R, S↓R,↓S -R,↑S - R are different forms of the basic stimulus response dynamics in animals. They reveal the relations between certain types of HOM. The equilibrium and the right response are, or tend to be, equal to life, to TRU. These relations show why knowledge and action should be equal to TRU. S is determined by R, not only R by S. It is in the expansion of these many relationships that we can better understand animal and human behavior, the form and function of TRU.

The vectors we use in HOM, CES, and CHES in SPALT of TRU can work in many directions. They are not restricted to right directional vectors only. They can be backwards by whatever their role is in TRU, and in any direction in SPALT. The vectors of CHES in SPALT are to be primarily determined by their overall function in TRU and not by our senses, perception or preconceived notions.

WELL KNOWN and IMPORTANT PROCESSES

There are many existing terms that have meanings that vividly describe TRUD, HOM, CES and CHES: Activation = Initiating or Enhancing Action, Alienation = Selecting, Separating out, expelling, Anticipation = Expecting a stimulus or effect, Assimilation = Selected Input of the positive, taking in the good into TIT, ES or CES, Categorical Rule = To do what one wills others to do, Categorical Imperative = To Act only according to a maxim by which you at the same time will that it shall become a Universal Law, Conditional Imperative = the golden and categorical rules, and categorical imperative only as TRU, Conservatism = Contraction, Preservation, Restriction of Freedom, Protection, Limiting, Constriction = State, regions and mechanisms of confinement or limited form and function, Contingency = Depending upon, interdependent, Convergence = Coming toward, tending to fuse, Development = Progression of Initial and Potential to actual, Differentiation = Making increasingly finite, Discretion = Making separate, fine, distinguishing, Distinction = Telling one from another, making discrete, Divergence = Passing away from, tends to disunion, Equilibrium = Full balance of opposing forces in a whole, Erratic = Having no fixed course, functioning in uneven and unpredictable ways, disordered, chaotic, Excretion = Output of Negative, neutral or less than selected, can be out of TRU, TIT or aspect, ES or CES, Expansion = Pushing out, enlarging, loss of limits, and confinement with increased interaction, Golden Rule = Do Unto Others as You Would have them Do Unto You, Growth = Increase or development in whole or part, as from birth to death, in quality or quantity, Harmony = Smooth optimal operation balance, peace, Inhibition = Acting to repress or reduce, Integration = Making increasingly whole, Isolation = Limiting, confining, or reducing to no relation or interaction, Liberalism = Release of confinement or limits, freedom of motion, expansion, advance, progression, Necessity = All that must, has to or is imperative, Prima Facie Duties = Duties Morally Binding, Rejection = Refusal, denial, excluding, Selection = choosing what is best, eliminating unwanted or worst, any means of increasing separation for benefit, Separation = Taking apart, dividing, many from one, Utilization = Operations of use, that is positive and beneficial.

Fig 2-52 **ASSIMILATION and the GENERAL FORM of MOT**

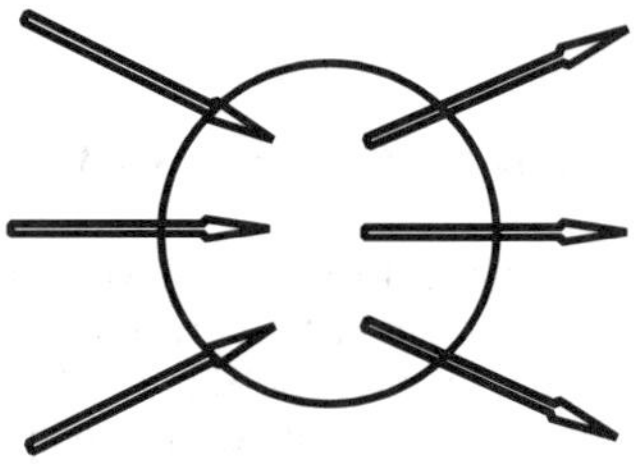

The continuous processes of assimilation and elimination, absorption and adsorption, that accompany interaction are characteristic, typify, and show the properties and dynamics of much of TRU and HOM. What TRU assimilates and excretes is not only physiological, it is mental, social, communicative, and all aspects of Á, both objectively, subjectively, all in between, and all together. It helps to explain how natural selection and evolution function and exist in TRU by being processes yet operating within the limits and SCARLS of ÁTO, TÄ and TRUD. **TAP** = TRU Assimilation Power. TRU has momentum and powers that continuously draw to itself. Its own structure and processes, form and function provide the setting, states, and conditions which attract or limit that which is beneficial or harmful. Senses and things are drawn to TRU in ways that are often not readily apparent. Yet much that is of senses and things are uninfluenced by TRU, even repelled by TRU. It is this revulsion to and by TRU, the resistance an individual has to being good, right or TRU, that is so devastating to the person as well as to TRU.

In addition to the great fundamental processes, principles, mechanisms, and laws of natural selection, there are two or more higher generations of laws that reveal the massive intricate dynamics of human life. These are: Integrative Selection, or the combined to total interaction of genotype, phenotype, and environment, or genetic material, live product, and determining forces, whose effects are accumulative and massive in a positive or negative way. Although natural selection works singularly in a one to one, by individual character to gene, this exists in the massive milieu of interactions whose products in the genotype, phenotype, and intermediaries extend throughout the species, or TRU. The other is Conditioned Selection, or the collective and proactive forces built into the phenotype and genotype that project the combined selective forms in a positive and beneficial way. When these are not perfectly balanced, increasingly negative effects are produced.

These, and associated, laws of human dynamics, TRUD, may be compared with ACEXIN wherein each generation of selection when brought together, like each higher dimension of ALT describes the increasingly complex to total human form and function. To solve the problems of human genetics, populations, and fitness to survive is like solving the problems of ACEXIN and TRU. For we can't understand ACEXIN and TRU if we do not learn and practice the differential and integral and all in between, as they actually exist in TRU. So also we can't understand human genetics, populations, and all the problems, solutions, actions, and management involved in upholding human quality unless we learn and practice the differential, integral, and all in between, as they exist in this special universe of best human life.

The farther one gets from ones home base the harder states, conditions, and setting become. All that stresses this setting, such as taxes and regulations, tends to produce strong reaction. History of most countries has been a long tale of the raising or lowering of taxes and regulations, and the resulting positive and negative effects produced. If the return benefit to either individual or government is unbalanced, oppressive, disproportionate, and especially if not TRU, great harm and reaction result. These are some of the determining forces on the live product, that secondarily effect genetics, in GELD.

Fig 3-52 PROCESSES of ISOLATION and EXPANSION in TRU

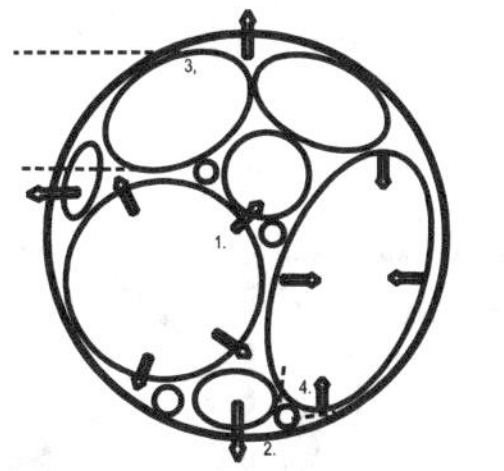

A part of TRU can expand only so far without detracting from other parts or the whole. This is a very common state and is shown by what happens when all take all and give little, as in all for all and one for one, rather than one for all and all for one. Mutual benefit occurs where, when, and how parts expand and provide strong forces to enlarge TRU. There is more room for all, other parts may not necessarily be isolated or forced to contract.

The operations of dynamics, and more specifically cybernetics, consist largely of input, output, feedback, and other mechanisms. They are especially adapted in TRU and correspond with the development of an overall view of TRU, TRUD and HOM. These are valuable in demonstrating CES and CHES in SPALT, and the basic nature of TRU.

There are TRU forms, components, and mechanisms that dynamically provide optimum response to stimuli, equilibrium success and survival. They do this by increasing or decreasing the systems operation. Scientific and medical research in immunology, the function of the great number of factors and substances that activate and inhibit the hematopoietic system to provide the best possible equilibrium in the human body, vividly and magnificently display mans internal dynamic totality. These inner-workings are evident in the balance that exists in the A, L, T; A_2, L_2, T_2; A_3, L_3, T_3,, ALT, ACEXIN and all combinations in TRU. They show the importance and outstanding part cellular, molecular, chemical, and physical operations play in the whole of TRU.

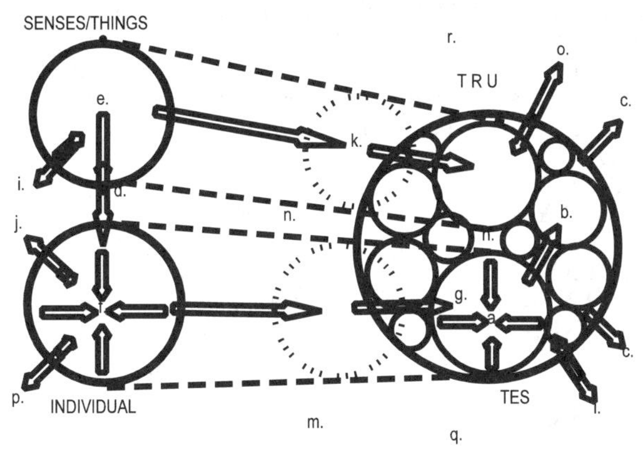

Fig 4-52 **KINDS of FORCES to and from SENSE and THING INDIVIDUAL TES and TRU** There are many kinds of interactions between parts in, and entities from without TRU. a. 50% for itself - its part in TRU, b. 50% for all else, TRU and other wise, c. Factors in and out of TRU - also some of its parts, d. Reaction to abuse, e. 100% for itself, f. 100% for the individual, g. 100% for its Part in TRU, h. 100% for TRU, i. 50% for itself, and all else, j. 50% for the individual and all else, k. 50% for its part in TRU and all else, l. 50% for TRU and all else. m. 33% each for Individual, TES, TRU, n. 25% each for Sense-Thing, Individual, TES and TRU, O. 50% for TRU and 50% against TRU, p. 50% for the Individual and 50% against the Individual, q. Variable percentages, and mixtures, r. Others: self's, parts, senses, things, and combinations.

The ultimate principle is the maxim that ones <u>force can't exceed or be greater than the perfected principle involved</u>. This can be either conscious or unconscious, single or multiple, direct or indirect. Ones <u>force can't be less than nor other than that indicated by its part and role in TRU</u>. These forces act, among others, in the following ways: 1. a cause may have some effect that does not appreciably alter TIT, 2. it may seem to be a threat to TIT, 3. may seem to act upon TIT, but actually has no effect, 4, may be prevented from acting upon TIT, 5. may actually act upon TIT but have little effect, 6. may actually act upon TIT and have an effect but the effect is reversed or counteracted, which may be just right, overreacted or underreacted, 7. may actually act upon TIT and have an effect, but this although temporarily altering TRU is adapted to and returns in time to normal, 8. may actually act upon TIT and seriously injure or kill a part of TRU, but is not a vital portion of TRU and is replaced, 9. may actually act upon TIT and have an effect that alters TRU appreciably but is ultimately corrected and TIT returns to its initial form, 10. may actually act upon TIT but the effect is quantitative and not qualitative, so that regrowth replaces what was lost, or, 11. they act upon TRU with dire effects. These vividly show how TIT can usually be as stable as it turns out to be, even in foreign surroundings or a hostile world.

Fig 5-52 REPRESENTATION of MOTION, FORCES, VECTORS, ACTION, POWER, and CES in TRU

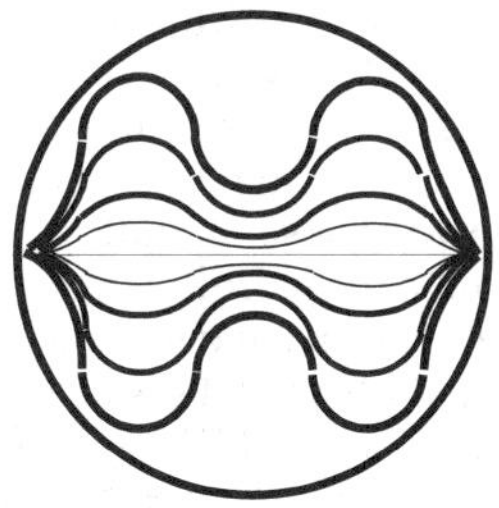

This figure assumes a biphasic, dumbbell, and periodic distribution. This is especially typical of many posits in the time continuum, and even more prominent in those durations that span the present. A great variety of other configurations exist and can be shown. Gradations, tapering of distribution, concentration, and others can be represented and done, other than by width, length or depth. It can be represented by heaviness, shading, or color. **PADFOR** = the Parallel and Divergent Forces of Relativity. This is the relative reality of the many different forces in action. There are also diagonal, perpendicular, or asymptotic vectors of forces. Forces are sometimes completely opposite to each other and still of the structure or parts of TRU.

There are many forms of vector tracks in TRU and HOM. They may be more or less chaotic or ordered, or all in between. We know from TRU, that its vectors and their posits will have certain set form and function depending upon their state in TRU, its objectivity, representation, and HOM.

There are many ways, or states of action on TRU, that reveal the form of CES and HOM. TIT adapts to external forces which produce cause and effect and may deny **LUC** and **LUP** or Laws of Universal Causation and Permeation.

There is in the homeostasis of all TRU, just as in the homeostasis of the human body, not only an optimum level, but a range, reversibility, and range of reversibility as well as differential normal, abnormal, and other states of bad and good health applicable to the success of TRU. These and other concepts of the conditions and states of TRUD and Homation describe the internal milieu or homeostasis of TRU. They go far to describe the limits, flexibility, and resilience, as well as degree of irreparable damage that occurs in the dynamics of TRU, its changing environment and the associated forces acting upon it.

Non ethical dynamics is a concept which is necessary, not only to make ethics formally distinct but to reveal and symbolize that great part of TIT and HOM that is not a part of, nor evident in, ethics. It is necessary to correct the inveterate tendency in humans to think that all can be accomplished, particularly directly and through conscious acts. It is not necessary that humans directly act on all, nor even specifically on each separate incidence of TRU. Much of the dynamics and processes of TRU are indirect, variable, and beyond those levels of Á, for which a person has any conscious, cognitive, or ethical control or direct influence.

Fig 6-52 **TYPES of DYNAMIC SITUATIONS, WORLD, INDIVIDUAL, TES, and TRU**

In Figure 6-52, reading from left to right, top to bottom different circumstances, occasions and situations are compared and shown. These include:

(1). The hard world, demands of the world closing in on the vulnerable, weak and immature part, (2). The demands of TRU effecting TES with TES heavily exerting itself and giving to, and taking from, TRU, (3). The hard and strong ES or TES expanding against an easily pressured world. The older and more mature TES pressing outward or forward in TRU. (4). The combination of situations # 1 and # 2 in equilibrium or dynamic interaction. (5). Operative factors, and mechanisms where # 1, 2, 3, function as a dynamic whole, (6). Some segments are stronger, some weaker, some more and others less vulnerable. (7). The partial and temporary as opposed to the total and permanent. These may be done from any standpoint, from the first person, second person, groups, TRU, or other parts or aspects of TRU. In order to return to systematic unity in TRU, we should hold TRU dynamics as close as possible to a 1:1 identity with their actual form and function in TRU, TIT-TO..

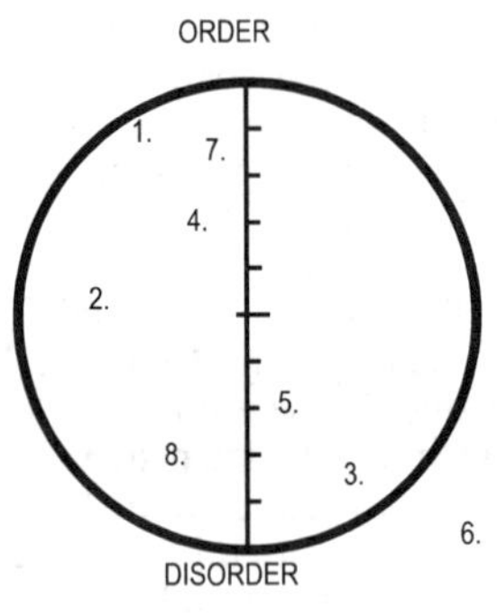

Fig 7-52 **POSITS in POLARITY of ORDER and DISORDER**

The relationship between order and disorder can be described as a polarity. This relationship may be shown by posits of: 1. Law, 2. Optimism, 3. Pessimism, 4. Conservatism, 5. Liberalism, 6. Nihilism, 7. Perfectionism, 8. Radicalism, and many others. In this figure the unseen horizontal is roughly the S-O axis, and the vertical axis the integral/differential axis. There is a natural inclination to coordination and identity with TRU. Polarities act as extremes in TRU. Their interactions and processes can be seen to work in different, yet set, sectors of the whole.

LOCAS = Law of Conformity and Stability, Law of Conformity and SI Stability. LOCAS is especially, although not necessarily, noticeable in relationship between conformity and stability and the subjective integral. By having a central nervous system with a great amount of coordination and control in the human brain, the subjective integral assures all, as they are TRU.

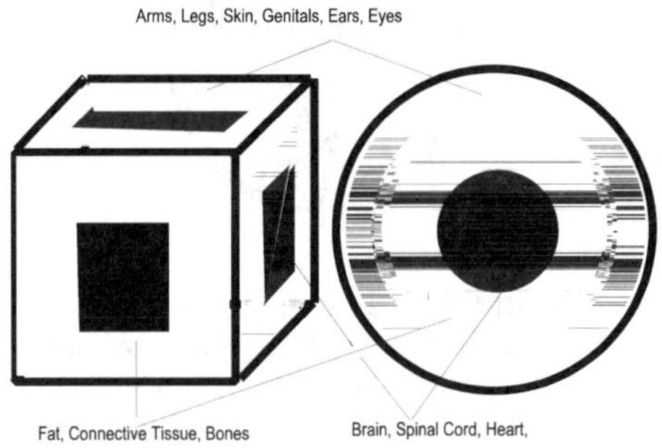

Fig 8-52 **CORRESPONDENCE OF FORM and FUNCTION by HOM**

The position of structure and function in the human body and its processes, functions, mechanisms and operations have become set through natural selection. This is evident in the form and configuration of different organs, and the time it takes for various functions. It is set in certain ways, as genetically, that best enable survival. This is what occurs in all organisms and results from the many and varied selective processes that have occurred. It is driven by natural selection and speciation in such a way that there is an optimal adaptation of form to determining forces. These functional parts of TRU are not easily nor quickly lost or gained. Usually they have developed in a very slow, fine, incremental, yet somewhat deliberate, trial and error sequence. Many exist together as systems or set relationships which are altered, lost or gained not singularly but complexly and totally. They need to be maintained in a like way. Massive sexual mutations are incompatible with live birth or when born with competition.

Capacity is the strength and power of content, necessary to raise the potential. It is a basic mode, characteristic, and essence of the dynamic form of TIT. It is the optimum level of each continuum of ACEXIN supplied separately and together to enable maximum potential for TIT. The greater the quality and quantity of capacity of TRU, the more it enables and assures success and continuity. The more relativity and reality it provides to overwhelm all that acts upon it.

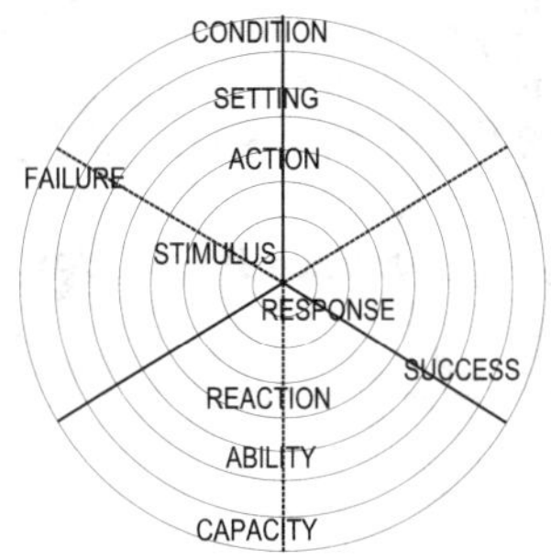

Fig 9-52 **CAPACITY, ABILITY, ACTION, REACTION, SETTING, and CONDITION**

Here is **CARSC**, what is relatively adapted to TRU. It is diagrammatic, some of the terms are derived from experience, and are external. The dynamic relationships between the primary forms of HOM are very interesting, and although not to be interpreted without qualification, revealing and beneficial to an understanding of TRU, TRUD, and POT. It helps to show what is central and what is peripheral in TRU. We must beware of misbeliefs that prevent us from realizing their potential and overall role in TRU. These terms and states graphically and roughly depict, how we may envision all that is of TRU, especially those core and key dynamic mechanisms, CES, powers, and forces that play such a major part in TRU.

For any state and its dynamic relations and requirements a definite amount of force is needed, not more, not less than what is optimal for it and TRU. Compelling Necessity of Certain and Proper Perfected Automated Action, **CONPA,** denotes the method of confirming those repeated, automated or routine actions that must be certain, proper and perfect. CONPA is a very important supplement to other comparative objective methods and principles. Compliance denotes activity that results from necessary adherence to some preset, rules, regulations, customs, directions, orders, principles, or laws. Compulsory compliance is one that is reinforced by punishment or some loss when not accomplished.

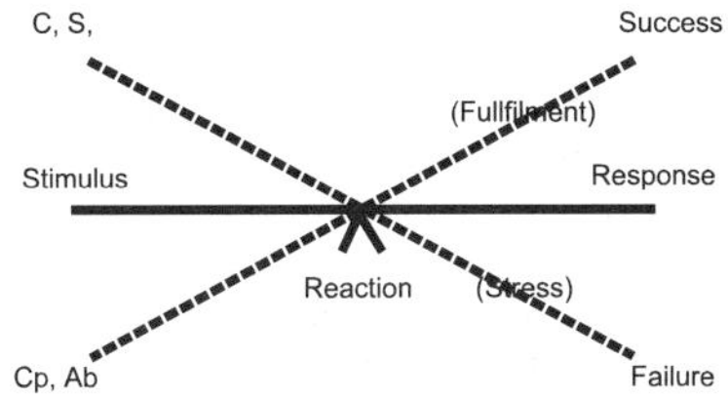

Fig 10-52 **EQUILIBRIUM in CARSC TRU and HOM**
Stimulus- Response, as well as Success/ Failure are distributed primarily along the time axis. Although capacity, ability, setting, condition are largely universal through TRU, they are given in contrast here since they are in such a state and natural position in the S/F CARCS relationship. **SROT** = The reverse or reciprocal status of time in TRU. SROT signifies and explains that mechanism whereby the present is in one sense more static, in another more dynamic, and the distant past and future in one sense more static in another more dynamic. This is a result of the S - O relationship. Whether we are considering TRU as a TIT, viewed from itself overall, its center is static, or viewed from ourselves as observe and identify with the here and now, is dynamic. Also, we should correct for the interpretation and posits in TRU of the other noted forms of TRU, e.g. capacity, condition, ability, and setting as well as all other constituents whose posits depend upon the observer-observed relationship.

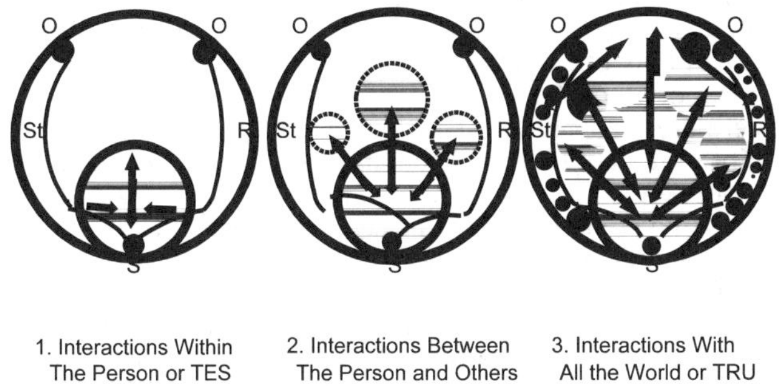

Fig 11-52 **TRUTH, INTERACTIONS, MEANS, RELATIONSHIPS, and HOM**
These are shown in the property, goods, commodities, means and power that a person needs, buys, is surrounded by, and struggles for. Even when under possible control, they will depend upon SCARLS and the necessity at hand as it relates to TRU. What is of a person's behavior, or what one has large control over directly in acts, is but a small part of TRU. Even indirectly and by influence or in other ways, these occur not as we usually think or assume, but as they are processes in TIT and TRUD. Our built in structure and behavior patterns are optimally formed to address the outside as well and inner processes and interactions of TES and TRU. There is a relationship from which equations may be developed to show the amount of fulfillment and the amount of stress. A person's being is in constant interaction, change, and challenge. There is an equilibrium, not always exactly balanced, at any moment or time, in which the time frame of TRU is set. The line that separates ethical from non-ethical acts and dynamics is complex. As human capacity and knowledge increase our power to influence, and indirectly effect much of TRU also increases. So does this place an ever increasing burden, demand and responsibility on ethics generally, and on each one. For this reason we can't accept conditions or the world as usual. We can't accept an ethic that lacks VAPO. Ethics, as all, has significance only, and to the degree, it is TREQ or TRU. In the dynamics of TRU, especially in HOM it is necessary to recognize and posit the processes of TRU as they are guided and determined by the equilibrium between polarities. An example of such a polarity in action in TRUD and Homation is that of crisis situations, their opposites, and the stable, harmonious, enduring dynamics of TIT-TO. This is similar to what individuals experience when overwhelmed by pressures and turmoil, yet at other times, of great peace and quite. These like the changes in weather are typical of TRUD, HOM, CES, ACC, management and control.

There is a mandatory requirement of a minimal level of knowledge of TRU that must be attained by each and all persons. This includes all persons, if they are to succeed, survive, continue to have permanence, and immortaily.

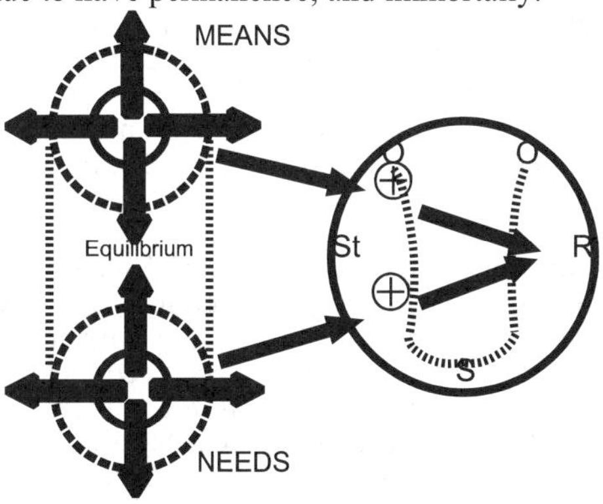

Fig 12-52 **EQUILIBRIUM of NEEDS and MEANS**
There is a fixed relationship between needs and means. The more the needs the greater means are required. This is especially true with problem solving and supplying the means to meet human needs. It is a great part of TRU. This is associated with TRU expansion and contraction. They can be either reversible or irreversible both in terms of the whole, TRU, or its parts. A person may get away with indiscretions momentarily in time and place. Yet the powerful forces, action and powers of TRU, subjective, objective, large, small, past and future, make it so that for TRU, every indiscretion, wrongful act, negative deed and every failing, each and all, you, I or someone else, of TRU, must pay. If this debt is not paid directly it will be paid indirectly. If not voluntarily it will be paid involuntarily.

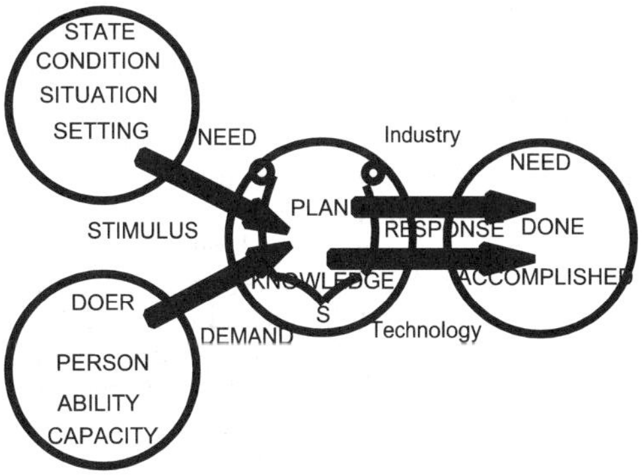

Fig 13-52 **NEEDS, DOING and The FORM of HOM**

What needs to be done? If one has things to do and has no knowledge or plan, it will not be accomplished, done incompletely, or done wrong. How far knowledge and plan can be completed and perfected depends upon the CARSC at hand, and how much is, and used by, KOT and TRU.

CHAPTER 53

SPECIAL FORM of HOM: ACTIONS, RIGHT and WRONG

DOING, SELECTION, UNITY, DIRECTION, POWERS, STIMULANTS, CONTROL, TREATMENT, FUTURE, EXPONENTIALS, ABSOLUTE

A vital example of the special form of HOM and CHES is: **POSED** = Principle of the Necessity of Self Defense. Self defense permeates much of TRU, TRUD and HOM. It is a predominant essence of being. This in turn is what drives much of human mass-energy, the Á dimension. It is action formed or designed to protect, shown as vectors, that cause motion and dynamics characteristic of TIT. POSED exemplifies the why and how of CES or the processes in TRU.

Act and action are terms of common use. More properly for TRU they signify aspects of its dynamics and function which when CES, can be shown by CHES and HOMATION. They consist of the more direct complex motions, forces, changes, vectors, functions and simple mechanisms of TRU that involve higher states of length, $l^2 \times t^{-1}$ x mass, and are distinguished and posited as vectors in SPALT. Act is not necessarily complicated, impossible invariably positive, negative nor uncharged. It is direct, positive, and ordinated to the extent it follows KET and TIT. It is produced by and in turn produces knowledge of TRU.

Do you know what you are doing? This is the key to success in life, and is the means that determines whether you learn or reduce your doing to what is safe. Humans must rely on knowledge to achieve most of their needs. This is because the world we live in is much different than any other. The NAMTRU world is much different than that of TRU. The TRU world is a world composed of FAF, states, dimensions and mechanisms that are altered, adapted or warped in its own special way. That is they are of a mass energy, space and time frame that is increasingly dependent upon predominance for FAF and maintenance. For Á, mass energy is largely weighted by higher levels of SÁ, the ability of humans to think, reason, decide and plan depending largely upon knowledge.(TREQ) If one would doubt the heavily weighted alteration, adaptation and warping of this predominant dimension in the direction of knowledge let them consider the fact that out of tens of thousands of genomes, nearly l/3 represent the brain of man. These are of recent origin phylogenetically. Their increase is undoubtedly accelerating rapidly. This is at the expense of competition and selection. It depends largely upon human mental capacity, language and knowledge. Knowing what one is doing and the great predominant power of TRU, show how relativity sets the special form of TRU, HOM, and the processes of TRU, CES..

Activating forces may be classified qualitatively by being direct or indirect, natural, normal, abnormal, pathological, or deadly. Quantitatively they are classified by producing hyper or hypo responses. They reveal the selective powers of individuals, for or not for TRU, plus combinations.

LACDAR = Law of Accelerated and Decelerated Response and Action. Much of LACDAR is derived from a basic principle of endocrinology. If deficient, there will be a marked reaction to hormones, as with the thyroid hormone in the elderly with hypothyroidism, and FSH for hypoadrenalism. LACDAR is a part of homeostasis and helps to prove PIP. It shows that the relationship between stimulus and response is not so much a 1:1 phenomena as it is controlled by a great number of factors that return TRUD to its best state in TIT. It shows the flexibility and reversibility of the massive interchange in TRU, TRUD and HOM. It is both qualitative and quantitatively dependent on all that converges on optimal health, of TRU. It is of great value in judging and fine tuning, how much treatment is to be used. It is important in the management and control of treatment over long periods of time. In these ways it applies to TRU, TRUD, Homation and ACC whether hypo or hyper compensatory states and conditions.

Animism is the belief that nature is full of spirits. Although rebuked by much of modern thought there is much in animism that arose from natural causes and for good reason. Early human recognition of many, less evident, powers in their world, led to beliefs in the positive and negative forces that affect life. In turn these stimulated, and were very important in, the development of knowledge. TRU likewise, and more than science, reaches beyond existing knowledge to provide the reasons, determination and explanations that are needed to prove and show how and what essentially and actually exists. Science describes things, TRU essences and dynamics of all.

What mostly determines our actions can be shown by posits in SPALT. Some of these include, mechanical or physical factors, tropisms and taxis, external and internal stimuli, bodily needs, urges, comfort, likes, dislikes, habit, imitation, routines, attention, IA, immediate awareness, subject, object, object following subject, and what is waited for or anticipated. Our actions are also determined by demand, imperatives, musts, should-s, intentions, making a living, done with some special object or goal, by reason, solutions of problems, intellectual interests, through some selected complex of value, choices, judgements, spiritual or religious goals, identity with the divine in the world one is in, and ones counterpart in it, TES in TRU. What is done only and to the precise degree necessary and fulfilling ones role in TRU are focal determinats of action. Ones actions, and that of much of TRU, are not usually or mostly of only one or the other of the above, they are of many mixtures or all of the above.

There are many types of action. Action that is singular or multiple, direct or indirect, spontaneous or contrived, planned, reasoned, logical, reaction, no and minimal action, and many variants and combinations of these. It is in the nature of life to be busy and active, and to be so in the way that best serves ones FAF and existence, thus TES and TRU. This is appropriate action, that dictated by ones Á FAF. What works well once is apt to be imitated and put into repeated and general use. This is the common basis for the controlled development of routines, common laws and regulations. Yet such repeated action and routines are apt lose probity and to lead to misuse, even if unintentional and unrealized.

For those in charge of action, treatment, legislation, and many other areas of responsibility, as in business, health care, and government, there is more than one approach: 1. To radically excise all that is critically wrong immediately, 2. To early yet moderately excise all that is wrong, 3. To conservatively treat what is wrong as effective as possible with the help of natural forces that may enable it to go away. 4. To minimally treat or remove in time only what is most serious. Each one and their gradations determine action. Overreaction is as bad as under reaction. This is why such precision and control in knowledge, KOT, TRU, HOM, ACC, and TIT are necessary. Overwhelming action can often be worse than no action. Precise and perfectly controlled action (A), is the key to best management, especially as it entails all TRU, TO. Along with **WIM** or We will it Must, another axiom and corollary of our relationship to the action necessary for TRU is: **SWAT** = Without Work No TRU. No one is going to live another person's life, no one is going to be your TES but you. TRU can not be achieved by what is not or non TRU. In most battles, although soldiers may be fighting for their lives, they soon grow weary unless others are at their side, they know what they are fighting for, and why their lives have been put at stake. So much that seems wrong is much less so, or not, or can even by right be TRU. Much of the history, and even the present state of humanity, is one long tale of needless suffering because of false or untimely ideas of what is wrong, or minor easily adapted differences in TIT.(DAOST) At the commercial, political, national, and military levels these wrongs and errors escalate becoming increasingly negative with great suffering. Nations suffer change to gain one mass of collective errors and wrongs after another, little understanding the underlying reasons, causes, concepts, principles, laws, mechanisms and processes involved. Only when we discover, know and live by TRU or TREQ, its equivalents, can we begin to properly know, know how, and act on what is to be done.

Fig 1-53 **FORCES, PROCESSES and ACTIONS ESPECIALLY RELATIVE in TRU**

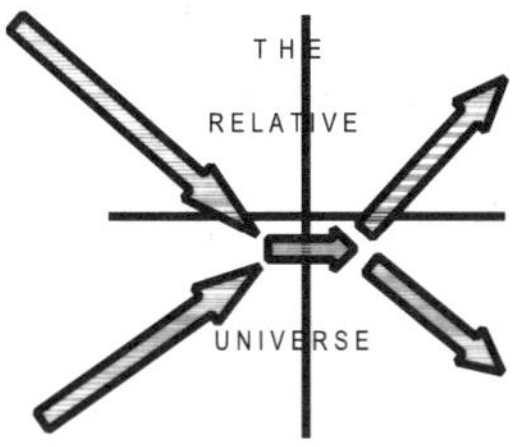

It is in the relationships between TRU, all that constitutes TRU, its observation and observed state, its mechanisms, processes, Homation, REM, POT, SU, ALT, concepts, principles, laws, and major courses, that we can began to formulate TRU. It is through these, and ACC, that massive plans can be made, that we as individuals, stand in relationship to both right and wrong, to do what is correct by TRU. **LIR** = Law of Imperative Right. That all is or can be all right is not as simple nor assured as it may commonly or at first appear. For all to be right is as profound, subtle, complex, contingent and vivid as life, or TRU, itself. Each of us can, are, and will be right only to the degree we know what is right and know how to use this knowledge to do right.(VAPO) This requires an evermore accelerative and expanding understanding and identity with ones eternal potentiality, mass energy, Á FAF, as produced by TIT for TES.

What and how much a person should do is largely guided by: the margin of error of act. This is a major reason and incentive for knowledge. Only when voluntary acts expand can TRU expand. Only as we know and correct for our limits will knowledge and KUPS have meaning and succeed.

The loss of natural selection and its high rates in various ways in the modern world is widely evident and a sign of fundamental changes in life on earth. It is disruptive of life cycles, the great interactive dependency of all, more or less, on each other, the life chain or biosystem. It shows how much humans and TRU are not isolated from the rest of the world, but how much they not only depend upon, but how the rest of the world is a part of TRU. It is in this wider world of TRU that the selective processes of man and life must cohere and be mutually supportive if all life on earth is to survive and flourish. It is in human quality as TRU, not quantity of any kind that our future is at stake. Modern man lives in a culture and world in which natural selection is altered in ways that are massively deleterious. The difference between being Good and Being ones best in life is much like sports where a team that is only good will lose most of its games, whereas a team that is always at its best, will win most of its games. So it is in and of life, life in TRU. **LABA** = Law of Best Action. No matter how much is wrong, best action can only be that by TRU. To be TRU, so all have to give it their all, and do their best.

TAFOR = The Absolute Force of Right. In SPALT a sphere is presented in which CHES are to be posited. The determination of CHES are guided by VAPO, SCARLS, REM, and all that qualifies TRU. This is what is right, and how it can be shown is CHES. TRU is right both by definition, TRIA, and the FAF of TRU. TRU is molded to be no more and no less than right. Right is molded to TRU similarly. All else must also be molded in the same way, at least as it will be TRU. **WIR** = What is Right. WIR is what POT and IPTRU seek to show and fulfill. As all is completed reversibly; all human efforts toward right under any name approach completion. This is quality, a constant upgrading to 100% TRU. This is purity, an extraction of all that is right from all that is not right. This is called: **PUREV** = Purity in complete Reversibility, and **DARJEV** = Determination justified by Right, and Right justified by Determination, Reversibly. DARJEV is the derivation, development, determination and demonstration of the reason, right, and right action, reversibly, back and forth, ad infinitum, to TRU. TAFO, TAFOR, WIR, PUREV and DARJEV are strong proofs of Á and TRU.(PRÁ,PRUT) All are more or less reducible by and to TRU, in either respect as appropriate or needed, is called: **RECTRU** = Rectification by TRU.

There are many ramifications of productivity in TRU that help show and use the special forms of HOM and TRUD.

PRODUCTIVITY=EFFECTIVITY x EFFICIENCY/RESISTANCE

is from Ohms Law, and makes it apparent that life should be, what a person's life should be aimed for, and how it can be successfully fulfilled. It is in decreasing the resistance and increasing the efficiency and effectivity by one and all as they provide the greatest productivity that is or will be TRU.

Some of the greatest powers of TRU are those of self renewal, regeneration, rejuvenation, and reproduction. It is in the dynamic nature of TRU to provide the means for self preservation, by various powers, to overcome whatever may adversely effect it. Much of these means and powers are built in, even by the most basic and GELD forms, as its A, L. T. It can adapt to all kinds of settings and conditions. It can renew itself, be rejuvenated, and finally but not least nor only, can reproduce itself. TRU is in a production race, both by quality and quantity, with all that besets it. This is why it is so important that all of us join in this race, work, and live as we must be, our counterparts in TRU, TES. It is for this reason the virtues and characteristics that are best for the parts, TES of TRU, are those which must be the best for each of us, including: energy, efficiency, effectivity, productivity, courage, gratitude, inspiration, revelation, reason, principle, comprehension, utilization, and many others.

Love is a special mode and form of TRU and HOM. It is the great positive, emotive, beneficial and productive feeling and power that exists in and for all that is TRU. In its differential state it will appear far removed from TRU, a sense in itself. Yet it is the closest to, and what works for TRU, the powers of bonding, and Á - SÁ by and between all TES and all TIT-TO. It is that great positive beneficial productive feeling and attraction that binds any and all mutually in TRU. It is not only personal, couple, family, social, and biological, it is artistic or representative of the divine or REM, as in love of the beautiful, glorious, GOD as TREQ, or TRU. If we love, love well, and consistently hold to all that is TRU by VAPO, we increase benefits by PIP to uphold TRU. Mohammed's great success and triumph arose not only from his great capacity to stimulate a populace to common purpose in Allah, or TRU, it came from his own absolute love of and conviction in Islam, Confident Submission to God. This is what occurs in Individual→ESIN→TES→TRU. The reaction to action may be explosive or violent especially when there is a displacement of values. This is called: **LADOV** = Law of Displacement of Values. LADOV is a major mechanism in competition and human selection and is to be commonly found in group and social reactions. The other great aspect of Mohammed's teaching was the reinforcement of Judgement by the all powerful God of each person. This in TRU occurs in duration, not so much at any point in time, as a fleeting resurrection. Yet its almighty aspect as demanded by God or TRU is a proper prerequisite to the dynamic functioning of its parts. No person should be pleased or distressed with oneself - or others -except and to the degree they are TES or non TES, TRU or non TRU.

Fig 2-53 **LAW OF SUFFICIENT and CONCOMITANT ACTION**

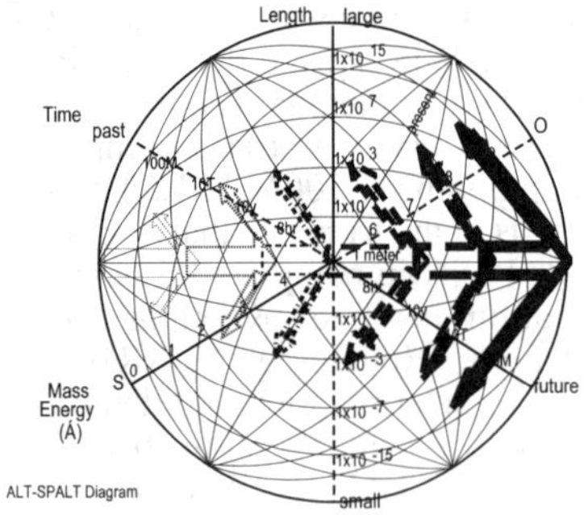

LASACA = Law of Action and Accentuation, Sufficient and Concomitant. Since TRU is a manifold, TU-TE-TI-TO-TÂ, any and all actions, accentuations, and operations must be sufficient and concomitant. They must completely and best follow TROS in TIT and its concepts, principles, laws and mechanisms. This is where each and all actions, ethics, and goals are sufficient and concomitant with the S/F and PPMA of TRU. They should be LASACA in an increasingly accelerative way within the SCARLS of their relationship to TIT. They must be concomitant with the manifold nature of SU, ALT, REM, POSIT and HOM of TRU. If the person in a position of responsibility does not know, or knows less than, what is needed for what they are doing, all who follow can't be expected to know or do right. **LACOMA** = the Law of Concomitant action. LACOMA is why it is so important that the basic principles of life, TREQ, or TRU, be right, known, lived, and especially those who are in positions of leadership and responsibility know what they are doing, TRU.

Fig 3-53 **VOLUNTARY CONTROL of ACTION, S-O, Á and HOM**

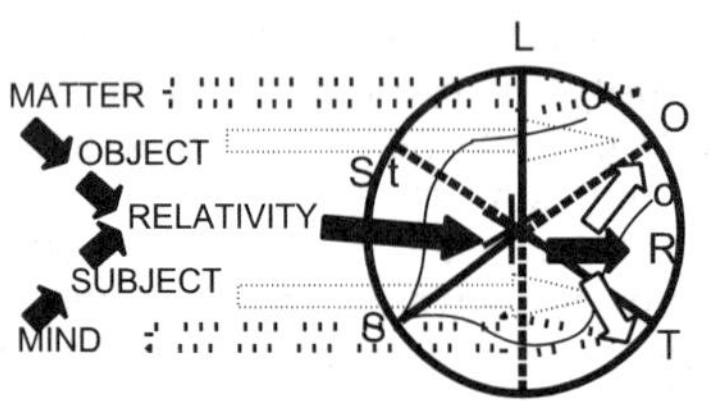

The voluntary control of action is of immense importance for more than one reason. The need to be able to control action, and the association of action with the voluntary is at the basis of our being. It is in the highest levels of that salient portion of human relativity or the observer and observed Á, predominant mass energy, FAF, continuum, and dimension of our existence.

Physicians can make more astute, profound and effective physical diagnostic measures and evaluations of the human body's capacities and abilities. It is an example of the ultimate perfection that can be achieved in TRU. It is a refinement of attention like what we see in ACEXIN or the actualization, extension and intension in TRU. This is the precision of A, L and T in combination. It provides not only an utmost subjectification of the object, it provides more and more perfection in the determination of the TRU in this object. It provides a much better presentation of the problem and physical signs which are key to making the diagnosis. So it is of **SOAP** or subjective, objective, assessment and plan, or the overall format or algorithm for investigation and management, as used in medical education and practice. Related concepts, through SÁ, to the special form of HOM and TRU include: **RE(C)** = Reciprocity. REC is how Homation or CES correspond to the responses of TRU, its TRUD. **PEC** = Processes of Critical Essences. Processes may be divided into, what works as signs, evidence, symptoms, or of the subjective, and what is by forces, powers, effects, and of the objective. CES that are combined and total, as in DYNS and DYNO, become no more of the subjective processes than the objective. In TRU they are more often highly mixed, subject-real-object interactively. These have force, power, and work in TRU as well as in their own way. It is these that best show what Á, the mass energy FAF of TRU, is.

Compensation usually results from loss, insufficiency or deficiency. It is largely dynamic, a massive interchange as in cybernetics with constantly occurring and changing gains and losses. It may be shown for most TIT-TO. Compensation is a large part of Homation, CHES, CES, TRUD and TIT. It works in larger and smaller systems. Many of the basic principles of S/F of TRU may be shown in terms of compensatory mechanisms that give us how, and how well, TRU works.

Fig 4-53 PROPER FUNCTION, CARE and TREATMENT

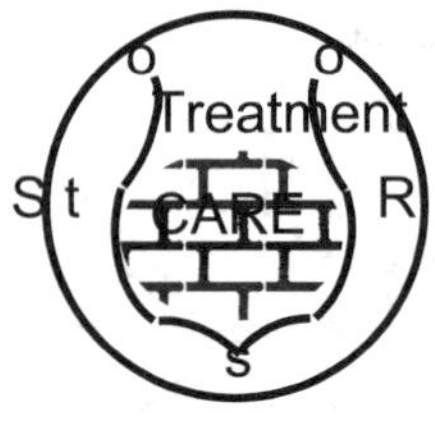

4

The great importance of proper function, processes, and action is expressed in the adage 'no treatment is better than wrong treatment' and as expressed in the different uses of the word care, as in 'I don't care', 'be careful' and 'care for you'. The words we use to express ourselves, relations to others, and activities show how to attain ever better representation. In care we express proper function, action, occupation, and service in which a high level of sensitivity, proficiency and success are sought. From professional performance such is necessary for the successful function of society and TRU

Fig 5-53 **HIGH VELOCITY FORCES, TRUD, HOM**

Anything that is moving at a high rate of speed is certain to create havoc. This usually occurs sooner rather than later. Therefor all acts, especially when the rate of speed is excessive, must be governed, constrained, or controlled. The dynamics of all in and out of TRU, **OITIO**, is such, and so fundamental, that unless we do our constraining, governing, controlling, planning, and action in such a way as to most perfectly, orderly and by VAPO, anticipate and completely manage it overall, we fail, either in part or completely. No matter how high minded or successful it may at first seem it will soon fail. It is for this reason we must know TIT and all about TRU, we must know TRUD in all ways that are apt to effect, and create havoc in, TRU.

Catalysts, enzymes, and stimulants generally are the activators or effectors that enhance the actions, functions, processes, or operations of an entity, molecule, organ, system, body, human being, group or TRU. These are often, especially in higher or more complex systems and humans, phasic, cyclic, rhythmic, and metabolic. They reversibly, directly or indirectly, and in equilibrium maintain homeostasis. They do so to accentuate, magnify and enhance this aspect or part of TRU. Since in TRU, the higher levels of mass energy are less established, it is not surprising that the role of its controls are less efficient in maintaining and enhancing TRU. The massive continuous anabolic and catabolic controlled processes of protein, fat, carbohydrate, vitamin, complex nutrients and elements of the human body exemplifies the FAF in TRU.

The strength of TRU depends upon success/failure, or S/F. The factors in S/F include, I. For every entity, sense, thing, force, and all that goes into TRU it increases TRU, unless: 1. It is not 100% for TRU, 2. It is not altered in the transition, 3. No output is associated with it. II. For every entity, sense, thing, force and all that goes out of TRU it decreases TRU, unless: 1. It is not 100% negative for TRU, 2. It is not altered in transition, 3. No input is associated with it, III. Quality may increase or decrease without input or output. This may occur as the content or total becomes stronger or weaker, better ordered or disorganized, more or less proportionate, better or more poorly adapted to the dynamics of TRU. IV. The total or sum of these factors consists of added characteristics. All together they make up TRUD, CES, and CHES that enables S/F, as through POA and KOT.

LEXENT = Law of exponential ENT, or the vast expansion of the enablement and enhancement of TRU. LEXENT is how we can most greatly turn ourselves into TES and benefit TRU. This includes how we help others and all things in their role in TIT and ENT. Productivity is often only 80%, 60%, 40%, 20% and less of its potential. This is magnified in group labor. It is not only working 1/2 of the allotted time, it is in being inefficient, ineffectual in that time with a productivity that is:

50% time working x 50% efficient x 50% effective = .125

This is readily recognized in lack of actual TRU production, especially by those who manage various businesses whose TRU effort is poor or even negative. If we further take from this our ATRU, ANTRU, ANTITRU, and those living off the labors of others or their own future, it is no surprise that most of the world stays in NAMTRU, poverty, want, misery and suffering. Thereby we need to learn and know TRU, apply it to our lives, and turn this around for TRU and LEXENT.

LAPERAC = Law of Perpetual Action. LAPERAC is like getting the swing of it, learning to ride a bicycle, walking on stilts, roller or ice skating. It is like driving on any large crowded and congested freeway or an interstate highway. Motion must be maintained, even velocity, acceleration, and deceleration tightly controlled. Motion must be held to a precise range and limit. LAPERAC is the law which represents that state and condition in which a person, group, mankind, or all TRU must keep in motion, be continually active and fulfill their part of this dynamic nature. It is evident in the situation, like that of Buridan's Ass, where a person when planing what to do is torn between two equal choices not knowing which to start first, thereby remaining immobile and static. Life, especially life as TRU, is not static, except relatively in frame, form and construct. In totality it is in motion, dynamic, keeping busy, and requiring continual action, or LAPERAC.

LASPOFT = Law of the Selective Power of Future TRU. Much of all we are and all about us that we feel is so real is not. Much that seems so true is not. Much we think absolute is not. Much that we believe in, as God, is not. Much we may even interpret as TRU is not. For to be TRU it must be not only by VAPO, DAOST, SCARL, TROS, ALT and all. If future TRU is not fulfilled then it is not, that which we interpret as TRU is not. So that the future in TRU, TRIF, selects that which will be, and that which will be in the future that in the present and the past. The general case of LASPOFT is the law of Active, Retroactive and Meta Active Selective TRU. These are its total dynamics that do not occur only one way, but in a great variety of ways and in a very interactive and massive mode.

TEF = TRU Future drawing Forces. TEF are those forces and powers of the future that effect the present and past, both separately and combined in TIT-TO. TES beautifully illustrates what we are so often apt to neglect and ignore. The greater hope and ideals are, the greater TRU becomes. For life to have meaning, for a person to rise from inactual individuality to TES, through resurrection, being born again in TRU, or RECTRU, is largely the simple absolute drawing effect of the future, TES, the future thisness of TRU. It is how those of the past looking to the future have sustained us. It is how those of the future look back at us in awe and wonder why we failed to understand and do all that is TRU. This is how the hopes and dreams of man are in the past, present and future, together. This we see in the greatness of TEF and COGNOS, as have God and Religion, been such great powers perpetuating and strengthening our world, TRU. Future drawing forces should be learned, well known, shown, appreciated, and applied for the great powers that bind and sustain TRU. These are the stuff that the great power of predominant Á FAF and PIP are made of. They are much of what makes TRU special, so different from other RUS, thereby proving TIT, TRIA and TÂ. Even more they are so different than what we ordinarily think the world to be. They are what enables us to understand not only the what, but the why and the how of TRU. This is the heart and soul of TRU; how TRU is constituted and how it can be represented, modeled, applied and used through CHES and this Special Form of HOM.

PIFS = Principle of Indirect, and direct, Future. PIFS denotes the idea and revelation of the effect of the future on the present. This is very dramatically seen in ones feeling of 'you are being watched' and the effect it has on a person. It may be compared, and contrasted, with one's conscience. It is all, some part, or any, of future subjectification that has, in its myriad ways, an effect in TIT. Most focally it is some direct way by which those of the future may envision TIT, some part of it, TES or even some individual, as the first person. It applies both to the future, past, and present as they effect all other portions of TRU.

TREXP = TRU Exponential, acceleration to the nth power of TRU, ES and CES. TREXP is what we feel and are so greatly inspired by. When we awake after a good nights sleep, or youths feeling of life and vigor, or a persons rebirth spiritually upon finding their greatest identity with GOD, TREQ or TRU. It is toward a life of sustained unbounded growth and being. It is the feeling of complete freedom, unfettered by the limits of life, or the SCARLS of TRU. It may be expressed in many ways.

That the relative is not our only ethic is shown by the fact that although TRIA and TA are great ends, TRINI is not TRIT, TRIT is not TIT, and TIT is not TRU except in part and in ways. Thereby our goal and greatest standard of ethic is not a simple R-R, but a composite of R-R, R-A, A-R and A-A. It is these as they best, and can most practically, enable us to be, survive and flourish. It is how they can reversibly accentuate the survival and enhancement of TRU.

Laws may be divided by, 1. Degree of formal distinction, how much are they strictly and specially the actual laws, not any rules, doctrines, axioms, principles, notions or even less, 2. Degree of Probity, to what extent they are right and precise, 3. Degree of CENS, what are they laws of? Are they laws of TRU, RU, AU, things, senses, entities, CES, ES, or whatever? 4. Degree of VAPO, to what extent are they valid, actually of TRU, of TRU proportion and totality. 5. Degree of SCARL,.to what extent are they or are they not properly reduced and held to their appropriate states, conditions, limits and relations in TRU? 6. Degree of Standardization or Formality, to what extent are they standardized and to what extent are these standards fully accounted for, exceptions made, and held to their limits. 7. Degree of Legality, to what extent are they adapted or adaptable to society and a set of Laws Codes and Statutes that require strict compliance. 8. Degree of TRU,. to what extent are all the above held to TRU, or their relationships to the appropriate aspects, FAF, ES, CES and all of TRU. As great as laws purport to be the final directives of mans knowledge and life, they, in each way, often become sidetracked into something quite different and less. They even become the opposite, especially in effect, to what they were meant to, or should be. Each must be recognized for what they are, analyzed separately, combined, and used no more than the extent to which they are appropriate for TRU. They are not all inclusive of TRU, or of much they seek to control. It requires a judicious oversight through HOM, DYNU, DYNS, DYNO, ACC, POT and all TRU to use, manage and control laws.

CHAPTER 54

GENERAL PARTICIPANTS and DIFFERENTIATION of HOM and CHES

FINITE PROCESSES, VERBS, SEPARATION, DIVISIONS, DIMENSIONS, MECHANISMS, CORE FORCES, ENERGY, POWERS, INCREMENTAL INFINITE PROCESSES, PARTICIPATION in TRU

The differential processes working as centripetal and centrifugal forces and powers, tend toward or away from, the unity of TRU. Both may be equally of TRU, or various degrees from TRU.(DAOST,OITIO) TRUD and Homation can be vividly demonstrated by those major processes, which act in a general way, as represented by verbs when limited to TRU. When verbs are transferred into action as processes by the addition of the suffixes: ion, tion or ation, they give us much of the differential of TRUD. These CES can be readily posited by HOM, CHES, as ES are by STA in SPALT: to assimilate = assimilation, to moderate = moderation, to operate = operation, to regulate = regulation, to select = selection, and to stimulate = stimulation.

Verbs were the first to be posited in models of TRU, **MAPLAT**, or Man Point, Length, and Time.(pg 202) This was done to show both the use of MAPLAT in the development of TRU, and a simple representation of verbs as CES in a dimensional field. There are many beautiful examples of dynamics, of CES and CHES in life, if we only stop to think, analyze, reveal, posit and reason with them.

DEVELOPMENT OF CHES BY SECTORS of ALT, in HOM

If we posit the forces, vectors, courses, trends, powers, and especially CES, or all dynamic factors in TIT by HOM in SPALT, CHES, if these are divided into their eight primary sectors of TRU, by S-O, L-S, and P-F, and if they are likewise posited by determining their centers in tenths of each dimension, or 0 to 10 on the A, L, and T scales , we can develop a POA to HOM and CHES. We thereby show TIT in a more ordered and exact way than it has ever been known. This is a way by which an understanding of any and all knowledge, religion or science can be given a POA. Within the eight primary sectors of TRU, as shown by SULP, OLP, OSP, SUSP, SUSF, OSF, OLF and SULF, or the combinations of divisions which the tenths scale of dimensions will provide, there will be in each sector 125 parts, of the 1,000 or (10x10x10) for the whole. Below we show their correspondence:

SULP	OLP	OSP	SUSP	SUSF	OSF	OLF	SULF
0-0-0	9-0-0	9-9-0	0-9-0	0-9-9	9-9-9	9-0-9	0-0-9
1-0-0	8-0-0	8-9-0	1-9-0	1-9-9	8-9-9	8-0-9	1-0-9
2-0-0	7-0-0	7-9-0	2-9-0	2-9-9	7-9-9	7-0-9	2-0-9
3-0-0	6-0-0	6-9-0	3-9-0	3-9-9	6-9-9	6-0-9	3-0-9
4-0-0	5-0-0	5-9-0	4-9-0	4-9-9	5-9-9	5-0-9	4-0-9
0-1-0	9-1-0	9-8-0	0-8-0	0-8-9	9-8-9	9-1-9	0-1-9
1-1-0	8-1-0	8-8-0	1-8-0	1-8-9	8-8-9	8-1-9	1-1-9
2-1-0	7-1-0	7-8-0	2-8-0	2-8-9	7-8-9	7-1-9	2-1-9
3-1-0	6-1-0	6-8-0	3-8-0	3-8-9	6-8-9	6-1-9	3-1-9
4-1-0	5-1-0	5-8-0	4-8-0	4-8-9	5-8-9	5-1-9	4-1-9
0-2-0	9-2-0	9-7-0	0-7-0	0-7-9	9-7-9	9-2-9	0-2-9
1-2-0	8-2-0	8-7-0	1-7-0	1-7-9	8-7-9	8-2-9	1-2-9
-etc.	etc.	etc.	etc.	etc.	etc.	etc.	etc.
to	to	to	to	to	to	to	to
125	125	125	125	125	125	125	125

Such a division may be extended to include the mid ranges of each dimension to produce 27 different sectors for precise posits. When correlated with chapters of the text and many of the major concepts in each we can see an order for posits with great precision and reliability. It shows the relations between ES and CES in SPALT with IPTRU, or between ROT and MOT with POT and PERT.

The differences and dynamics between, and of, the individual, others, parts of TRU, and TRU, is the basis for much of the great FAF as well as conflicts in the world. It is most evident and bitterly experienced by the individual throughout life. The constant interaction and competition with all that is about one. TRU is composed of all individuals, as they are, when TES, part of and processes in, TRU. Yet this is often not on a 1:1 basis either in whole or in many parts or facets. There is a balance and moderation between same and different, both acting by attraction and repulsion.

Partials, or ES and thereby CES, are limited to their proportional function in TRU. This gives a better view of the dynamic interaction between different yet related aspects of TRU. Empathy is the assumption of an attitude, motion, or state of mind of another, as if one were the other. It is how to put one self into another's shoes and especially to appreciate their stresses and feelings. Functional Purpose is more than a simple role parts play in a whole or body. It is their essential role, how they serve the basic principles and laws of TIT-TO.

The primary role and duty of justice is to protect and safeguard human existence, TIT-TRIA. There is nothing nor any activity that does not have some effect on us. Whether it is positive or negative, or resultant thereof, it has at least an infinitesimal effect. The effect is more apt to be indirect than direct, complex with others than solitary. The effects may be of any kind of time, length, mass energy, or combined. Justice can't be achieved without correctly weighing the effects of all activities that relate to that role in the ES and CES of TIT.To have such ability requires a sufficient knowledge of TRU.

What exemplifies the differential of HOM, that has been established through millions of years of human development include: carrying, lifting, pulling, pushing, writing, turning, looking, hearing, breathing, eating, drinking, digesting, eliminating, reproducing, loving, caring, thinking, reasoning, reasoning together, working together, brachiation, manual dexterity, function of the opposable thumb, feeling of touch, picking up, rising, making, finding, using, and all means that enhance ones being, most especially ones being with others in TRU. '-ing' is a suffix like -ation, that forms nouns from verbs that represent action or process. Such a noun standing for a verb is a gerund. '-ing' when used as a verb produces its progressive tense, especially present participles. These are examples of a few of the functions and actions which CES or processes of TRU, can be identified with, made discrete and posited as CHES by HOM in SPALT.

CORE FORCES, TRU, SPALT and HOM

What are the great driving forces, powers, aims and purposes of man? How have great personages through history and great contributors to religion, philosophy and science exemplified these great core forces in TRU? To better know such greats, their work, and to tie them with the development of knowledge, their influence, course in history, and continuity, it is well to take them one at a time and distinguish their part in the whole, of TRU. For Confucius this was Jen or humanity. Jen is a person's goodness, necessary to serve society. It has been that balanced development of the individual whereby one establishes their own character and that of others to be prominent as they conscientiously, sincerely and altruistically form a part of society, the state, and the world. Confucius also, and somewhat related to this first, had another undying passion, and this was: Rectification of Names. To Confucius this was the setting right, correcting, reformating, regulating and standardizing names with what they correspond to, so that they are identical. This is their identity with their ESINS, ES, and CES, of TREQ or TRU. A person must not be called a king or leader unless they possess the true qualities of a king or leader. Thereby right by TRU must be based upon a valid, honest and sincere appreciation for and maintenance of all - including language - that upholds society, TRU. We must have, and use, a language that is of TRU, ESINS, ES or CES.

For Plato this driving passion was for the Unity of the Universe. Much of this was derived from his deep appreciation and obligation to learn and reason, in the way of Socrates. This, in turn, was derived from a deep sensitivity to the death of Socrates, resulting from a conflict in obligation Whether obligation is to the petty, erroneous, and myths of religion and society or to its highest truths, reasons, wisdom, KOT and TRU. Plato redirected Socrates emphasis upon the attainment of these to goodness and beauty, and to their ideals, reals and ends, or the Unity of the Universe, TU-TIT-TO.

	SC	00-05	05-09	10-11	12-15	16-19	2+	30-35	40-45	5+	6+	7+	8+	9+
SULP		PHILO-SOPHY	KNOWL-EDGE	GOD	MONO-THEISM	RIGHT			OLOGY			RELIG-ION		
SUMP											QUESTIO N-ANSWER	PROBLEM-SOLUTION	LEARN-ING	
SUSP		QUES-TION		INSPI-RATI.	ONES SOUL	IDEA	DEFINI-TION			LOGOS	INTER-ESTS		CHRIST	
RELP				PLATO IDEAL	UNITY OFTRU	HUMAN-ITY	LAN-GUAGE	DIMEN-SION		ETHICS!		CIVICS	TECHNO-LOGY	CONTIN-UANCE
REMP		TRUTH	PRO-BLEM		SPECI-FICI.	GOOD-NESS	WRIT-ING			CYBER-NETICS	IPR	HUMAN-ISM	STRE-NGTH	BENE-FITS
RESP			STIM-ULUS		SUBMIS-SION	SPIR-IT	MATES			PRO-CESS	BOOKS	EXPER-IENCE	CHAR-ITY	CONS-TANCY
OLP		EVOLU-TION	NA-TURE			MAN-KIND	ART	COSM-OLOGY	GEOG-RAPHY	MECHAN-ICS MOT-	TRADE	INDUS-TRY	CIVILI-ZATION	
OMP		FORM		GRO-WTH	MAN WOMAN		DEMON-STRA.	SET	HABI-TAT	DEVELO-PMENT		NEW REALISM		
OSP		BECOM-ING					DESCRI-PTI.	VALUE	ANAT-OMY		(GREED)	GENE-TICS		
SULN											CURE	TEACH		UNDERS-TANDING
SUMN	STUDY			MIND		DE-MAND	MUSIC			PLEA-SURE	ATTI-TUDE	PRAISE	COMPRE-HENSION	
SUSN	REA-SON	VALID ITY	REVEL ATION	COM-FORT	THO-UGHT		LOGIC		INTU-ITION	INTEN-TION		FAITH		
RELN	LORNS	RELAT IVITY	REAL-ITY		QUAN-TITY	WILL	AESTHE-TICS	TOR		COMPEN-SATION			REALITY	LAW
REMN	GIVEN	INTER VAL	S-O SO	TRUT	JEN	HAR-MONY→	BEAU-TY	FOCUS MP		OBLIG-TION		ROLE COURAGE	JEN DUTY	THISNES. HECC.
RESN		APPEA-RANCE	ESIN	ENTHU-SIASM	QUAL-ITY	WANTS	MEAN-ING	R the real	DISTI-NCTI-	DETERM. FORCES.	SPEECH	SHOWING	RESPONS-IBILITY	CONCEN-TRATION
OLN							MATH	PROPO-RTION		DIREC-TION	EXERCISE SAFETY	ECONO-MICS	LIFE CHAIN	USE
OMN		POA					DEMON-STRA.	TIME VALUE	BRAIN PÏ	HELP	HUNGER THIRST	GOLDEN RULE	MAKING A LIVING	SUCCESS /FAILU.
OSN		THING			PURE		DISSE-CTION	FREQ-UENCY	POSIT S POS	HEALTH	AIR OXYGEN	WORK	OBEDI-ENCE	EFFICI-ENCY
SULF					LATOL	SUPER-LATI-						HAPPI-NESS	RESIG-NATION	COGNOS ELTICMA
SUMF		WAY			SOUL	LOVE	PROBA-BILI.					THANK-FULNESS	COORDI-NATION	PUREALI-TY KUPS
SUSF	RECTI ICAT	FION NAMES		JIST KOT	THINK LOTES	POIR		DEC	INSI-GHT		ITOD	ANTICI-PATION	DISCRE-TION	SALVA-TION
RELF		TRUEQ		INN-ATE	TRU TIT	IMMU-TABIL	LANT	GOOD-NESS		FAF CEFKA	MERCY	SOCIAL POLITI.	ORGANI-ZATION	LIFE ÁCOM
REMF		EXAMI-NATI.	PROOF	VAPO PRUT	OPPUR-TUNI.	SURVI-VAL	NAMES PRALT	WORTH ALT	TOPO-LOGY	CED LASTE	DISCI-PLINE	PATIE. COPTA	FULFILL-MENT	ÁÂ PER-FECION@
RESF				REGE-NERAT	Á-S TE ES	TRUD SOT	WORD EC	REM LOPS	NES	CES	ANGEL	RIGHTS	PLAN-S VAC	IMMOR-TALITY
OLF			ME-THOD	LO		LATAC	LANO-FAC	GEOM-ETRY		ADAPT-ATION	GIVING	MANAGE-MENT	ECOLOGY LATED	TELEO-LOGY
OMF					REDEM-PTIO.	PERS-ERVER-		DURA-TION	COTE-RMINA-	LASACA	ACLO	COMPLI-ANCE	CONSER-VATION	CONTROL POWER
OSF						-ANCE		ROMT	LITY	ACTION	GENETICS	OPTH		ACC SÂ

Table 1-54 **CORE FORCES of TRU for CHES**

This is a tentative ordering of core forces, mostly done using great existing concepts. The table uses a few TRU terms to supplement and tie-in their roles in TRU. Most all great REM are forces in TRU. They are core forces when working radiantly and positively in TIT-TO. The table shows how numerous core forces are, benefit KOT, and coordinate IPTRU, SPALT, and TIT.

The greatest motivating forces include, 1. What one is accustomed to, learned, copied, become habit, or by experience conditioned to be most comfortable, 2. Interests, what one finds most agreeable and fulfilling. Those things which in the past have been most stimulating and of the greatest inspiration, 3. Wants, what one wants to do, all that in the past and all that one finds in the present produces the greatest joy, 4. Needs, ones thirsts and hungers for air, water and food, and other basic needs including sex for reproduction, 5. Possessions and goods that benefit life and make it more successful, 6. Money and value, or the means to livelihood and security, 7. Other levels of the mass energy FAF of TRU including mothering, teaching, socialization, shelter, safety, other Á, and these in combination and totality as they become, and function perfectly, for TRU. If you want to get others to do right, to know what is right, to learn right, then do it in such a way that enables them to realize that by right and only by what is actually right, TREQ or TRU, can these habits or customs, interests, wants, money, power, and benefits be best gained. This can only be done by fulfilling ones counterpart and role in TRU.

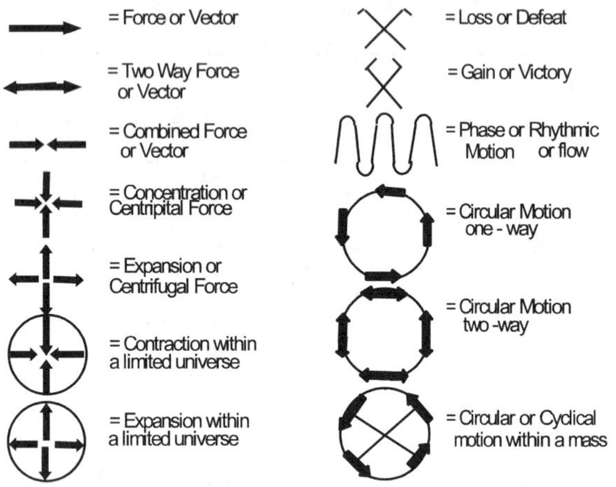

Fig 1 and 2-54 **SIMPLE FIGURES in TRUD and HOM**

The simplest and greatest positive and negative forces, powers, trends, courses and half lives of TRU may be taken separately, compared, and superimposed. In this way TRUD, HOM, the success of TRU, and all that promotes and enables it, ENT, may be more clearly understood and shown. They relate tp and show how geometry, solid analytical geometry, topology, mechanics, and many other fields can be applied to analyze TRU and produce POT.

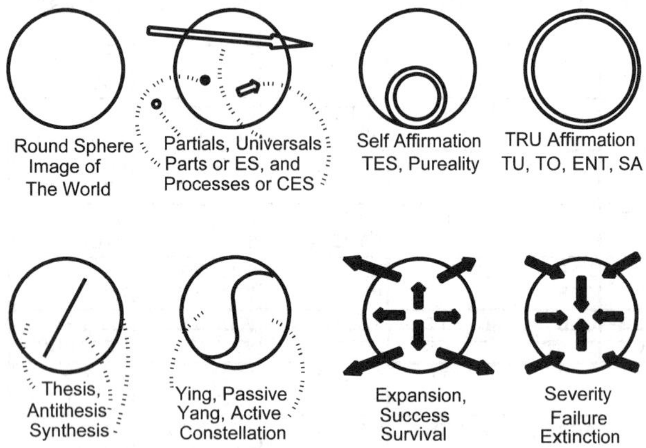

As finite phenomena, ES, and CES, develop and become more numerous, it is not sufficient that they be understood or represented as static posits on a visual field, as SOID, ALT, SPALT or ELTICMA. It is necessary that they be interrelated, compared, defined, and traced out, in their concentrations, dispersion, vectors, dynamic forces, effects, finite vehicles, and activators, as they exist in TIT.

ATTENTION, CONCENTRATION and SOLUTIONS to PROBLEMS

If one is having trouble, confused, and making a lot of mistakes, there is a best solution. This is to concentrate, pay attention, know exactly what to do and accomplish it according to what needs to be done, by TRU. This is the core of life as well as the basis for S-O the relativity and predominant dimension mass energy component of TRU. To know exactly what to, and do it precisely, properly, proportionately, orderly, and completely as it is TRU, or that aspect, sector, occasion, circumstance, ES or CES of TRU that is at hand and necessary. The activity and dynamics of a person, TES, by TRU can be corrected or better directed. It is with such intention and purpose that mankind, as TRU, has been formed, functions, and is characterized by. This derives from its complex state of predominance, or Á mass energy nature, continuum, parameter and dimension. Herein lies a major portion of TRU. How the FAF of TRU is combined in a way that compels and enhances it positively and totally into the future, and throughout time..

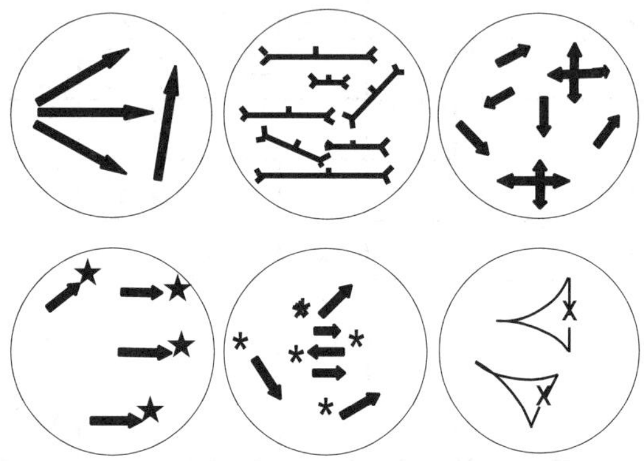

Fig 3-54 **KINDS, FORMS and SYMBOLS for CHES in HOM**

There are several different ways of demonstrating the key, core, and crucial make up, aspects, and forces of TRU and TRUD. There are ways of showing TRU other than through the strict dimensional interpretation by ALT; what shows type situations, and particular elements that are crucial to, yet not dimensions of TRU. In figure 3-54 forms and symbols that are used to represent processes in TRU are demonstrated, including: 1. Trends and courses, 2. Half lives or sections of processes, 3. Forces and vectors, 4. Hopes, 5. Needs, 6. Climaxes, and 7. Others, all to be developed..

Many fundamental concepts have been well developed and extensively used in language, scientific, and technical terminology to give the listener or reader a quick and accurate idea of their roles in TIT. Most verbs in being functions of life can be adapted to their equivalent process in TRU. There are different reasons for their great beauty: 1. There already exists in language the structure for a positive approach to these processes in life, 2. These terms are concise and dynamic concepts of processes. They bring together much of what is so disparate and previously indistinct in the human world. 3. Such terms, as processes in TRU, CES can be readily represented through a POA in MOT and SPALT. 4. We must be careful in the use CHES when they are not exactly CES, nor TIT-TO. By the CHES in SPALT we rely heavily upon those words and concepts which are of the best of human life and of highest REM.

The fear of God, like the fear of doing wrong and suffering, is a great part of the much larger orientation of one and all to TRU. It is that state of mind and feeling necessary for all who are either newcomers to this world or who have not yet learned to know God as TREQ or TRU sufficiently to live successfully in TES. It is what this lack of learning and life in TES leads to that is negative or potentially harmful that needs to be most feared. The great driving force of Jeremiah was the sincere personal search for God. This is shown by many through history and to a more or less degree, in all humans as they seek identity in TREQ, TRU and TES. It is especially under duress when they strive to make some meaning and identity in their lives. However, in Jeremiah it became a holy campaign toward a complete concord and obedience to God. It is a force that provides great impetus and superhuman attainment, like Christ, of a, some, each and all absolute knowledge. The human mind seeks not only identity and unity with the whole. It seeks to understand the reasons for this search, unity and the whole as it is the absolute, TIT-TO-TÂ.

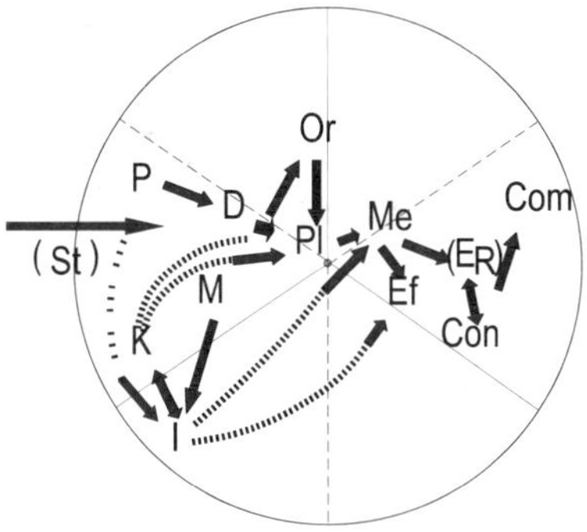

Fig 4-54 **PRIMARY FORCES and POWERS THAT MAKE UP PERFECTED EFFECTUALITY**

This figure shows the core forces and processes that occur in perfected effectuality. These, nor the factors involved, are all. Many more processes and their interactions are shown in the next three sections of IPTRU. The abbreviations used for the factors in the above diagram are: COM = Completion, CON= Continuance, D= Demand, E = Effect, Ef = Efficiency, I = Insight, K = Knowledge, M = Motivation, Me= Method, Or = Organization, P = Problem, Pl = Plan , R = Response, St = Stimulus. It is important to know the degree to which they are TRU, even though they may have variable effects within TRU. Some may be much more heavily weighted or REM than others.

Through the ages many means or vehicles have provided humans with concrete and predominant powers of ever higher success. An example of this is comparative worth or value. Such means that make exchange and productivity greater produce great impetus toward a more differentiated, concrete, and stronger life in, and of, TRU. This can also be done for the health of individuals, the will to TÂ, and all great, or REM, processes. The will to TÂ is called: **WITÂ** = Will to TÂ. WITA is the common denominator that assures such gains, as in the life/death wish that is a great part of ones predominant form and function, the Á or S-O dimension. Size, time, mass energy levels, features, and many other characteristics of a person, TES and TRU are neither incidental nor singularly very meaningful. They are large determining aspects of TRU, which combined and totally are balanced in their best function in TRU, TRUS and TAS.

Differences and Diversity, Tolerance, Unity and Peace

Differences and diversity are not what we commonly think or are advocated by those who demand any integration of diversity. By DAOST our world is made up of different aspects of the same thing. Yet this is the integration of same and different exclusive for TRU, its principles, laws, and mechanisms. To insist on, and accept as a way of life any kind of difference and diversity that is other than, or against TRU is a fad that is worse than no insistence or acceptance of diversity or differences at all. This may be expressed as a law, **LAIRD**, Law of Right, or Rectified Diversity and Difference.

This is a power that guides us to avoid all that lessens a total perfected comprehension of TRU, TU, TE, TI, TO, and TÂ. This is the way to fulfill TRU, with tolerance and peace.

That time cures all evils is an adage of great significance and many implications, positive and negative. Its importance lies in what time, or its associations, provide, a VAPO approach to TRU. Is it the opportunity for TRU or for ATRU. The dynamics, momentum, and inertia associated with time are of even greater importance, especially those with built in forces, powers and processes that carry on TRU. Here also is its fallacy or weakness. For to stand by and let time take care of things, to wait patiently in limbo without reliance upon ones resources, are failure. To utilize time without waste, as one who hustles while they wait, is the only means of repayment of debt to TRU. This kind of TRU debt is all that is owed by us and by TRU which can not be settled without payment in the appropriate way. TRU owes us a debt and we owe TRU a debt. These are counterbalanced by, and even more so by the rewards we and TRU receive from each other.

Of mechanical principles that reveal CES, TRU, HOM and their uses, there is Bernouli's Principle, or how with increasing velocity pressure decreases. This applies to TRU and its success, where with great positive control the velocity can be increased, pressure or resistance decreases, and are more readily overcome. This is related to **PIN,** Proof of the Principle of Integrative Negation. It is also key to tactics and ACC. Even more it is good example of all the factors and kinds of dynamics in dimensional formulation that apply to their roles in HOM TRU. Other mechanical ideas that similarly apply include those internal and external to TRU, or the internal and external transfer of energy, TRU force, exertion of force, TRU motion or displacement, TRU work, virtual systems, and parts in equilibrium.

DIFS = Disparity between the first and second persons. This is the very common, frequently astounding and conflicting, difference between what two or more individuals are thinking, speaking, and doing. It is simply a difference between what is ES in TRU, what is not TRU, or what ones IA focuses on. The proper knowledge and use of DIFS or its equivalent is essential to consideration in life, TES and TRU. Such disparity is seen in the same person from time to time, place, or circumstance.

Life is an uphill struggle, with pauses for rest and relaxation. Yet we must all beware when we depart from the uphill it quickly becomes a downhill run. Whenever individuals or groups expect or try to take or get more out of life than they give, especially any and all that is not TRU, they will sooner or later fail, suffer, and cause loss to others and TRU. This can occur either directly or indirectly, individually or collectively, to the extent they produce a negative effect on TRU. There is more happiness to be found in the challenges and successful fulfillment of life than in either paths of an easy gain or mindless acts. It is only for us to discover, learn and utilize all that is necessary to make our lives most efficient, effective, fulfilling and successful. These are largely a product of reducing and expanding life to its counterpart in TRU or those ways which are TREQ. Much of this is a simple matter of dedication. Those who in history have with dedication put more into a lifetime, have been the greats, saints, discoverers, inventors, thinkers, authors, benefactors, and its immortals. Of the great principles which show life, none are greater than to think, note, work and be alive, TES. This is shown in how intense a love of life is, how and how much a person can channel this love and life to TRU.

The quality, strength and success of TRU and TIT-TO is very much dependent upon the intensity, perfection, intention, precision, care, and detail each person provides. This applies to both a lifetime and for each increment of ones existence, seen in clarity of speech, expression, writing, instructions, andperformance. For if it is to be TRU it requires the best. So is the degree to which a society, nation, culture, and civilization increases or decreases, succeeds or fails dependent upon the quality of its FAF. This defines and is defined by the mass energy, Á and SÁ continuum in TIT-TO.

Fig 5-54 **HOMATION and CHES IN and OUT, POSITIVE and NEGATIVE**

In and of TRU forces, powers, dynamics and changes are working in different ways, and in different classes of ways: I. Positionally and Directionally, 1. Outside, out to out, 2. Outside and potentially inside, out to in, 3. Outside and inside, I/O or IO, a. out too in, and b. in too out, 4. Entirely inside or within, II. Relatively: 1. Directly, 2. Indirectly, 3. Consciously, 4. Unconsciously. They may be, A. In to in, as within TIT, B. In to out, as from TIT to NAMTRU, ATRU and AU, C. Out too in, as potential for TIT, D. Out to out as an illusion or figment, not TRU. They are by TO however, only as they are positive and in to in, and to a more tentative extent positive out to in, and negative in to out, or potentially that eventually becomes actual, TRU.

The troubles different peoples of the world have are not all the same, like DAOST, some the same and some different. In the final positive effect and resultant they tend to resolve into TES and TRU. They are more stable and less unequal the more they are TRU. Those peoples who have the proper physical and mental, or entire Á, SÁ, and GELF strength are more apt to be TRU, as TRU is of them. This is especially so if TES, and is used for the benefit of TIT-TO

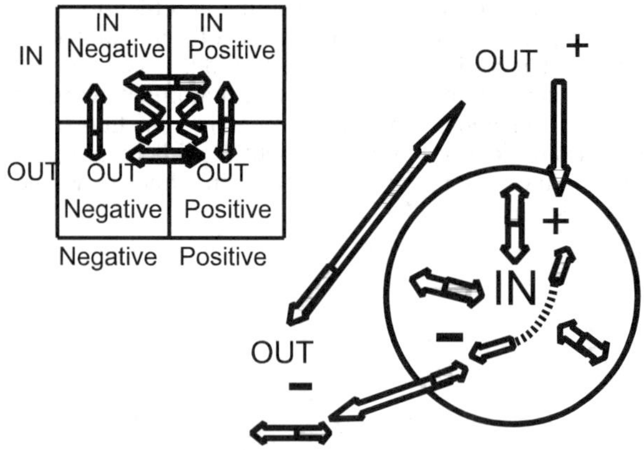

It is not only a peoples separation from TRU and TO that makes them not TRU, it is their disproportion, invalidity and partiality

ORGAN/SYSTEM	FUNCTIONS NEEDS PURPOSES	USES	BENEFITS
BLOOD, HEART	CIRCULATION, PUMP	NUTRIMENT TRANSPORT	Oxygenation, Fluid/Cell Supply
BONES	FRAME, SOLIDITY	SUPPORT	Stature, Resistance, Mobility
BRAIN COMPLEX	COGNITIVE, REASON	REASON, INTELLECT, WILL	PROBLEM SOLVING, lANGUAGE
BRAIN CORTICAL	CONSCIOUSNESS,	APPERCEPTION, THOUGHT	CHOICE, SELECTION, REASON
BRAIN OTHER	LANGUAGE, MEMORY, CENTERS	AUTOMATIC CONTROLS	INTEGRATIVE SYSTEMS, COORDINA.
CEREBELLUM	EQUILIBRIUM	COORDINATION	PHYSICAL BALANCE, MOTOR,
COLON/RECTUM	EXCRETION	ELIMINATES WASTE	EXCHANGE FOR BETTER, OUTPUT-IN
CONNECTIVE	DIVISION, SUPPORT	SEPARATION, PROTECTION	STRENGTH, ORGANIZATION
EARS	HEARING	AWARENESS,	BROADENS SELECTION, MUSIC,
ENDOCRINE	SUPPLYING CONTROLS	NUMEROUS INTERACTIONS	REGULATES SYSTEMS, SPECIAL TRU
EYES	VISION	TO SEE NEEDS	BETTER SELECTION, PERSPECTIVE
FINGER/HANDS	HANDLING, WORK	MAKING USEFUL PRODUCTS	BENEFICIAL PRODUCTS PROCUREING
INTESTINES	ASSIMILATION, SELECTIVE	NUTRITIONAL NEEDS	ABSORPTION NEEDS, NUTRITION,
KIDNEYS, UT	EXCRETION, SELECTIVE	FLUID & CHEMICAL BALANCE	HOMEOSTASIS, COMPENSATION,
LEGS	MOBILITY	WALKING, SEARCH	ACQUIRING NEEDS OF LIFE,
LIVER	METABOLISM	PRODUCES NEEDED CHEMICALS	HOMEOSTASIS, METABOLISM
LUNGS	RESPIRATION, ACID/BASE	O_2 INTAKE, CO_2 OUTPUT	CLEAN AIR, OXYGEN, ENDURANCE
LYMPHATICS	REACTION TO ANTIGENS	PREVENT INFECTION	IMMUNITY, RESISTENCE
MOUTH	INGESTION, SPEECH	FOOD, WATER, LANGUAGE	FULFILL NUTRITION, WORDS
MUSCULAR	MOVEMENT, ACTION	PROVIDE NEEDED MOBILITY	MOVEMENT WITH STRENGTH
NERVES, PNS	RESPONSE TO STIMULI	REACT TO PAIN	PREVENT HARM, SENSATION,
NOSE	AIRWAY, BREATHING	OPENS EXCHANGE, SMELLING	BODILY NEEDS AND SENSES
REPRODUCTIVE	REPRODUCTION	SEX, REGENERATION	PRESERVING THE SPECIES
SINUSES	AIR/BONE DISTRIBUTION	STRUCTUAL/FUNCTIONAL NEED	OPTIMAL FAF, IMMUNITY
SKIN	PROTECTION, SEPARATION	PREVENTS ADVERSE EFFECTS	LESS INJURY, INFECTION
STOMACH	INGESTION, DIGESTION	PREPARES AND ALTERS FOOD	BETTER BODY NUTRIMENTS
THROAT,VOCAL	SWALLOWING, SPEECH	FOOD AND WATER INTAKE	COMMUNICATION, RHETORIC
TONGUE, TEETH	TASTE, SPEECH, CHEWING	FOOD SELECTION PREPARATION	INTAKE CONTROL, DICTION
TOUCH	FEELING,	DISTINGUISHES QUALITY	CONVERGENT ON TRU

Table 2-54 **PURPOSES, USES and BENEFITS of the FORMS, FUNCTIONS and PROCESSES of TRU**

These are but a few of the thousands and millions of bodily FAF, parts and processes of humans that make up what TES and TRU are, can do and be. They are key to both the particularization of this being, as well as show the mechanisms by which humans, as TRU, operate. That TRU is very much human, as human is very much what we are composed of, and how this works to our best advantage in a highly competitive world, is evident from the collective interaction of these separate processes in TIT. It is these that determine how the parts and processes of TRU are provided with the predominance that assures the thisness that is TRU and TES.

Since the world is a multitude, increasingly finite, to infinite TE to TI, the associated acts, treatments and effects will be similarly numerous and varied. There is no single cure or panacea, for all. It takes a massive amount of specific and combined cures. **LAMA** = Law of Mass Action. It helps to understand the complex and total processes of TRU. LAMA also helps us recognize the need to act, react and treat what is required in both a differential, integral, combined and controlled way appropriate for TRU and the need at hand.

CHAPTER 55

SPECIAL PARTICIPANTS, ROLES, and PROCESSES of HOM

THE FINITE PROCESSES, CES and CHES in HOM, FOCUS, CONCENTRATION, MEANS and EXTREMES, CHARACTERISTICS, POLARITIES, SCALES, LEVELS, SHARE, PROPORTION, CYBERNETICS, MASS

In NAMTRU we largely miss TRU, and know not whether we are this or that, fish or fowl. It is to discover, determine, decide, and with conviction become as great a member as possible, by best allying oneself with what this or that actually is, that all succeed or fail. Those who have been the greatest of TRU, who have the greatest immortality, are those who have most greatly identified themselves with this special universe of best human life, GOD, TREQ, or TRU. The relationship between a person and TRU is much the same as between a worker and their job or the company one works for. It is ideal when the needs of each are optimally satisfied. There is a critical decision before each and all that continuously shows the extremes from greatest benefit to greatest harm. This is in all we encounter, all who come to us wanting this or that. Is what they want genuine, for great benefit, TRU, or is it some trivial or devious intent that is of little or negative good? There are many ways the negative occurs, especially in IPR indirectly, as the prevalence of many kinds of deceit. This is much of the great dynamic struggle going on continuously within, without and in and out of TRU. Of these, the one that is both common and highly active is special interest. This is at the crux of most choices, decisions and behavior. How much they are the same in interest yet how much they are different in TIT or ATRU. They are constantly changing and illustrate the variation and flux of TRUD and TRU which are completely missed unless we know TREQ or TRU.

The way to do things easily, yet precisely, requires a certain concentration on detail. Getting things done right in perfect sequence and at the right time requires anticipation and knowledge. The best of professionality, specialization, technology, and precision work combined with improved and sustained ability, conviction, endeavor and associated culture based upon TREQ or TRU, is what is needed for success, the achievement of TES and TRU. **POCOS** = Principle of the Compulsion of Success. POCOS denotes how the future TRU may be different than TRU present.(TRUF/TRUN) Yet TRIF is more interactive PPF than we believe because of POCOS in its many forms. It is POCOS that produces much of the momentum of TRU. It is not always a guarantor of success. Its momentum and inertia provide a great measure of ENT. The vectors of CES, CHES, are often expressed by + and - symbols. These provide, or highlight, the laws of concentration, and balance of dispersion or: **LOCON** = Law of Concentration. LOCON is where positive, as relative, and focus are concentrations of the positive forces that centripetally attract to TIT. Balance of dispersion is how the negative as less relative are non concentrations or dispersions of negative forces that repel, or are repelled, from within TRU. The dialectics of dispersion and concentration demonstrate the high degree of dynamics present in and about HOM and TRU.

The great mass of language and words in, and out of, common usage exemplifies the specific processes in and out of TRU, and the levels their meanings impart. Peer signify those of equal stature, power, trade, and profession who are apt to create in a person a sense of belonging, on the one hand, and of duty and compliance on the other. Compliance signifies performance of the individual or group, which is productive of highest levels of adherence. Supposedly this is toward quality and perfection.

Potential individual processes in TRU and CHES in HOM are shown by how no one can have much of a soul, TES, unless: 1. they understand it, and 2. live completely by it. From Kierkegaard, this by existentialism, is each person's choice, to exist in any way or to exist in the way. In a persons life we make this choice many times a day. If it is usually inconsistent with TRU in part or in sum, it denies us the rewards of TRU. The many different CES and CHES are seen in what most parents want for their children. It is a large determinant of their later strength or weakness. In humans the following wants are prominent: to do well, be right, happy, and contribute to the success of ones family, community, society, nation, world, God or TRU. The individual, as special or general to oneself or an ENIT, is not of paramount significance to TRU. The individual as special or general in TRU, TÍS and TES, is of paramount significance to TRU.(LOPAN) It shows how ES, TES, and TÍS are far more important than an individual. Only as the individual becomes ES, CES, TES, and TÍS, their counterparts in TRU, do they have great significance. This is part of the dialectics and conflicts of the individual OITIO TRU. The resolution of this problem is called: **RETES** = Resolution of ES, TES, and Significance in TRU.

You pay for your wrongs in life, to the degree they are 1/TRU. They are often much different than suspected. The debt or effect may be direct or indirect, sooner or later. Yet we all must pay, often dearly, for our failings, and all that we do or cause to be done that is detrimental to life, and TRU itself, TIT. This payment is rather definite and often set, dependent on the processes of TRU. For in this, are our hope and greatest chance for immortality. This is an immortality that is assured by TES and by our recognition of, work for, and conviction in PUREACLITY, COGNOS, ELTICMA, OM, SA, and TA. Rights, individual and all in TRU, are not something that start and stop at a certain point. Rights are on a variable scale depending upon the age of the individual, the groups one is in, and the degree to which they are or are not TRU.

Fig 1-55 **THE PROCESSES and EFFECTS of and in TRU and HOM**

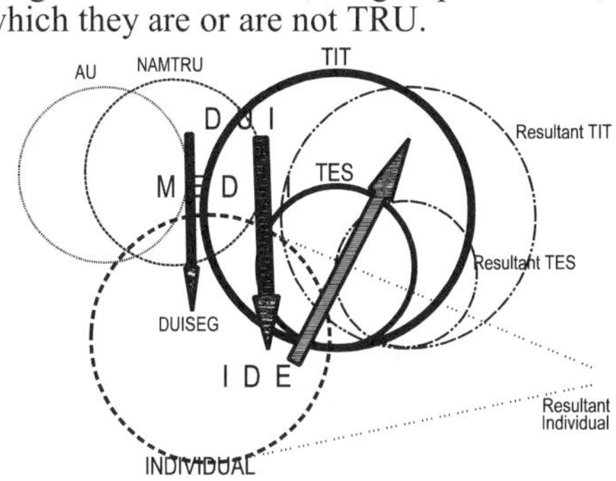

Often when things are detrimental and blown out of proportion, as in the undermining of an institution or government they need de-emphasis. Attitudes unfairly critical of the principles and spirit of work can have a constant drag effect on their institutions and world. This is shown by: **IDE** = Individual Drag Effect. This on TRU by the Individual is in part a reaction to the strong demands of TRU, and all else, on the individual and their labor. This is: **DUI** = Demands of the Universe on the Individual. DUI may be measured, which is: **MEDUI** or Measure of the Demands of the Universe on the Individual.

The unity, solidity, and concrete nature of the world was expounded by Fichte, who considered it to be a rational, unified system directed toward a purpose. He considered self consciousness and self analysis to be the key, the sole and fundamental task of philosophy. The focus of IA impels all to TRU, in full and in proportion. and is called: **FORIA** = Focus to Relative Reality of IA that impels all to TRU. It is the focus, and its dynamic aspects, dialectical and systematic unity, that provide the actual concretion of the world, TRU.

Fig 2-55 **SPECIAL CES of TREQ, GOD, CHRIST, TES and TRU**

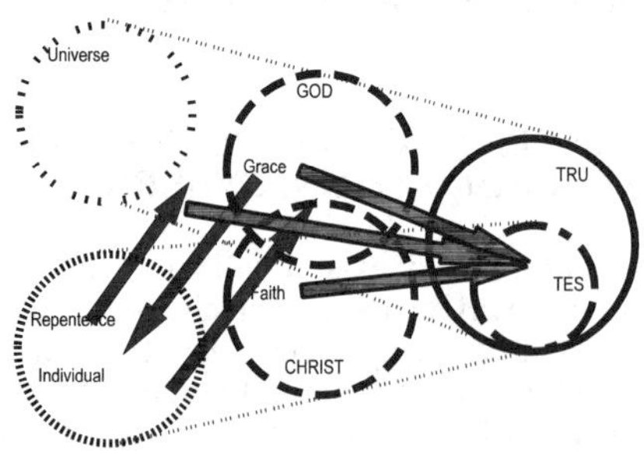

It is by demonstrating the relations and processes between repentance, faith, individuals, Christ and God existing separately and combined in or not in TRU, that we can use existing wisdom to vividly see the structure, ES, CES, and FAF of our world. It is not only static, not simply dynamic, but static and dynamic together in actuality and totality. This we do by making distinct the great processes of Christianity and how they relate to TRU. <u>Reconciliation</u> is the acceptance by man of God's Forgiveness. Actually it is this and more, the resolution of all in TIT. Reconciliation in principle requires both a knowledge of oneself as ATRU as well as a willingness to be ones counterpart, TES. It is not a free gift for all, especially all who are not TRU. Reconciliation in effect requires a great accounting of oneself, one's TES, and appreciation for their great blessings from TIT.

Fig 3-55 **GREAT PROCESSES and POWERS in TREQ and TRU**

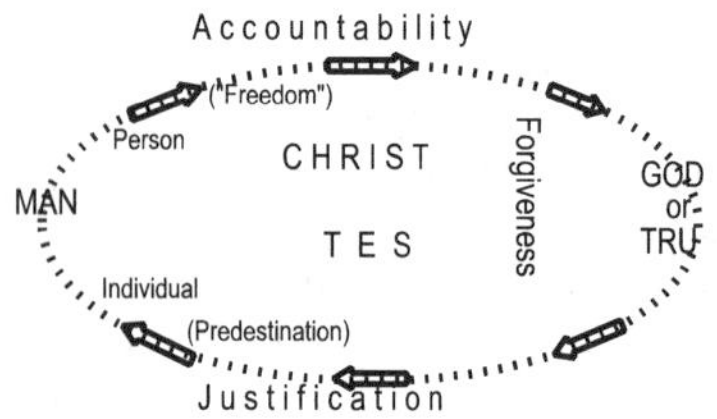

The concepts which represent mans relationship to God, which are such a great part of TRU. The magical powers of TRU, or MOG, are some of the greatest specific CES of TRU. When viewed in different ways these concepts show how such powers work both in and out of TRU.

The perfected Golden Rule can only be when the part functions perfectly as its counterpart in TRU. This is the function of TES. How we identify with and become TES, as do all others, as we do for others to become their TES. The golden rule is the forgiveness and interactions between individuals when TES and TRU. It is not enough to only forgive without action nor act without forgiving. Each must go hand in hand, as in the universal dialectic between extremes and means or moderation.

Fig 4-55 **LOVE of the DIVINE in ONE ANOTHER, PROCESSES of LOTES**

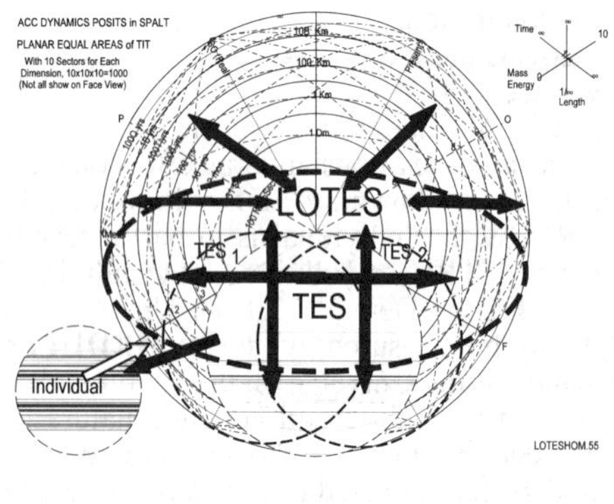

LOTES = the Total Love of The divine or TRU in Each Part, Process, or Each Other. Is this a world of IPR in which most are trying to get any and all they can, or is it one in which all strive to bring out the best in all others by giving and receiving the divine in each other. This has been provided by the greatness, truth and enduring meaning of Christian love. This greatest of processes is even more than the basis for social harmony, it is key to an individual's rebirth in TES and TES in TRU. Other religions have added to this great contribution to the success of TRU. We do not love one another senselessly, we love only, and make good allowance for, the TES in others and all that is TRU. For TRU, familiarity that breeds contempt on the one hand and repulsion by differences on the other, profits no one.

MARS or the Multiple and Massive Ablation or Action that is Rapid and Simultaneous. It, simply expressed, is 'to set ones pegs in a row' with sufficient force, power, determination and control, to act conclusively and successfully. So does HOM give us the POA necessary to accomplish and fulfill TRU. **EBER** or the (Law) of Extremes by Emphasis Reaction, signifies our development of extreme positions or standards in order to attain greater and more appropriate, representation, demonstration, or accomplishment of some intended purpose. These extremes produce in reaction, as all extremes, forces, and powers that are not total and harmonious, the wrong effect. These reactions in summation are **PIMET** or Principle of Imperfections of Today are th Evils of Tomorrow. They are in opposition to what has been developed. They are a polarity in TRU, and often in being other than their counterparts in TRU are the opposite and thereby ATRU or ANTRU. In this way we have opposition in opposition, actions and reactions in and out of TRU. Such states produce the continuous processes of dynamic change, the underlying condition of ES and CES in TRU shown by HOM. These exist in many directions, including: **PEFS** or Principle of Direct Future Subjectification Senses. PEFS is the 'you are being watched effect'. Although it may be taken literally and dramatically as being viewed by a future observer, this is not its major state and significance. PEFS refers to all effects of subjectivity upon the present. In this way it comes much closer to what actually has, is and does occur on a large scale. It is how a poorly recognized force of future subjectification can be excessive, deficient or sufficient to TRU, as it improperly or properly stands in opposition or conjunction with TRU.

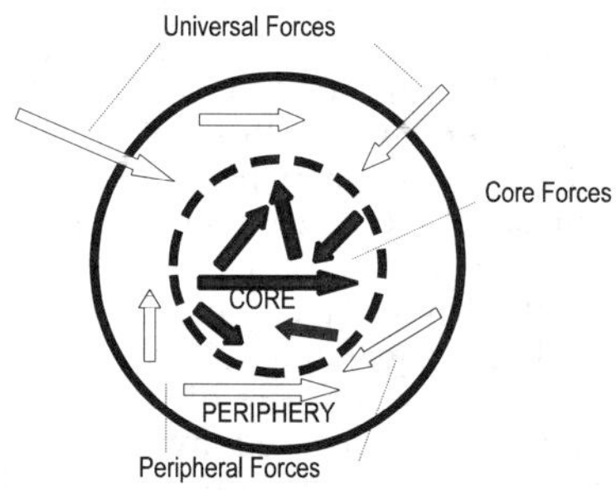

Fig 5-55 **CORE FORCES and CES of TRU**

Any RU can contract to provide protection yet lose a considerable part of its mass. Quantity is less essential than quality for survival. This is more so with certain ENS or TRUS than others. We often describe the core forces of TRU, its KDM, those CES which are most essential or have the greatest REM. This we do because, not only are they good examples and readily show HOM, TRUD and TRU, they save us from becoming lost in the increasing finite. To best learn TRU we must envision both forces in the core, periphery, all in between and those working OITIO.

FORACS = Fallacy of Over and Under Reaction. We live in a world of over and under reactors and reactions. Only by TRU can this perennial problem be recognized and corrected. One person, group, and nation over or under reacts to another. All these tend to perpetuate each other, as occurs by **IN** or Integrative Negation. People vary in the way they think and act; how generous or stingy they are, how close to TRU, how much they are of input and output, how much FORACS. When extreme, personality and individual differences develop reaction formations become pathological, ANTRU. We have many such combinations in IPR, as in to starve or gorge, and waste or pollute. From this the greatness of moderation or the golden mean is evident. This is the natural center, especially where all is TRU. It is where there must be neither excessive input nor output, or should there be excessive stinginess or generosity. These are shown by two different sets of opposites represented by a block diagram with four cells.

Fig 6-55 **RIGHT/ WRONG CHARITY/ SELFISHNESS**

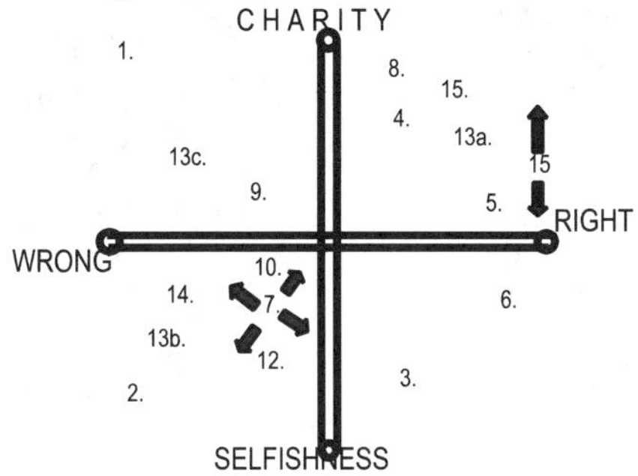

These two polarities when coordinated show how processes exist in combination and work to show complexes of CES and CHES which can be formulated and demonstrated. It also shows the exchange of value and kinds of combinations within the field. 1. Give to the worthless all that is necessary for those in need, 2. Rob the best and give to the worst, 3.Rob from the bad and keep for the good, 4. Rob from the bad, keep for others in need, (Robin Hood) 5. Give to the worthy, all that is right, all they deserve or need, 6. PUREACLITY, 7. Constant NAMTRU tendencies, 8. Corrected by religion, ethics, charitable organizations, 9. Less than worth, 10. More than worth, 11. Worthless/worth, 12. America 20-21st c., 13a. Best of Nations, 13b. Worst of Nations, 14. Welfare Abuses, 15. Christianity, 15+. TRU.

Learning to read varies from for those who: 1. don't read, 2. read but don't remember, 3. read and remember only by rote, 4. read, remember and relate, 5. read, remember and relate by problem and by importance, 6. read, remember, relate by problem and importance and use or practice it with wisdom, 7. read, remember, relate by problems and importance, use and practice this with wisdom beneficially, 8. read, remember and relate by problems, importance, and use, and practice in all ways beneficially by TRU, 9.read, remember, and relate by problems, importance and use and practice in all ways beneficially by TRU and VAPO. It is not sufficient to be able to read, attend, do, work or live as a simple reaction or response to stimuli or any given. Life is a continual process of learning and best using all the capacities and abilities provided by our predominant FAF, mass energy, or Á continuum in TIT.

Fig 7-55 **RELATIONS between KNOWLEDGE and ACTION**

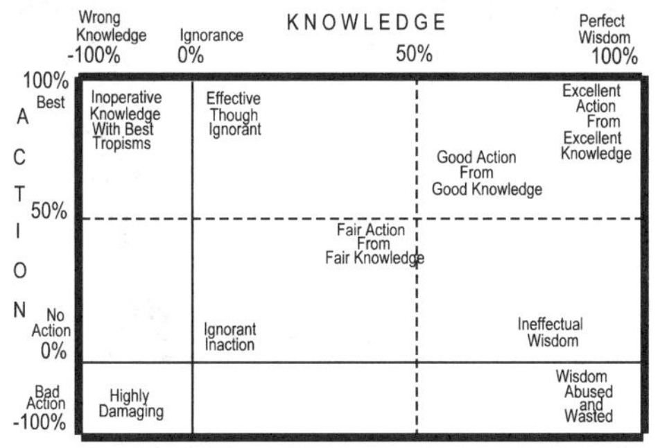

The relations between CHES for action and knowledge, thought and act are of the highest frequency in human behavior and life, and most important to TRU. Knowledge, and what the associated action must be on, is becoming more complex and rapidly improving.(Technology) It is often valid but lacks VAPO and concrete identity with OBT, TRIT, TIT, and TO.

Individuals and groups are often caught up in a life and death struggle for existence. One extreme is the accumulative forces and effects that cause us to be rapidly excluded from the vital necessities of life. The opposite extreme is what has been known as the armchair fallacy. This fallacy signifies how little pressure and great peace or stability cause inaction, imbalance, and resultant reaction. It is being completely engulfed by the present, acute, and often esoteric needs and reactions, to the neglect of overwhelming pressures.

Fig 8-55 **DYNAMICS of COMBINATIONS of MULTIPLE POLARITIES**

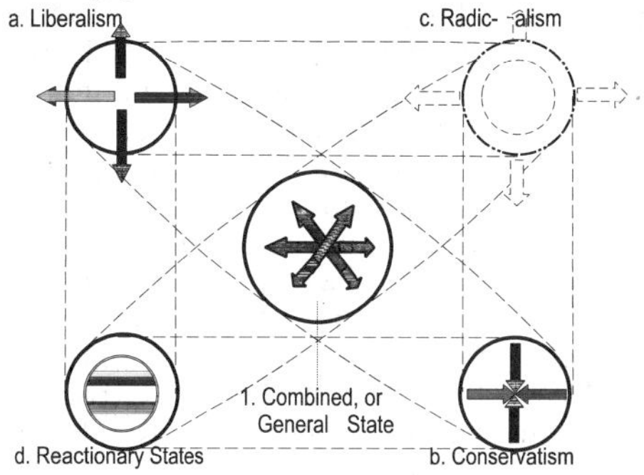

Polarities and opposites operate, and the dynamics of combinations in TRU are shown, by the systems or ISMS of liberalism and conservatism and their reactionary and radical states. Although radicalism and reactionary states are commonly considered to be extremes of liberalism and conservatism, there is as well radical conservatism and reactionary liberalism. These demonstrate the great value of formal distinction and interpretation.

How hard and efficient work must be to do, hold, maintain or advance ones needs, life or TRU? There are distinct levels of personal, group, community, national, and TRU endeavor that take a high level of sufficiency. This is continually being eroded by:

Fig 9-55 **LEVELS OF SUFFICIENCY**

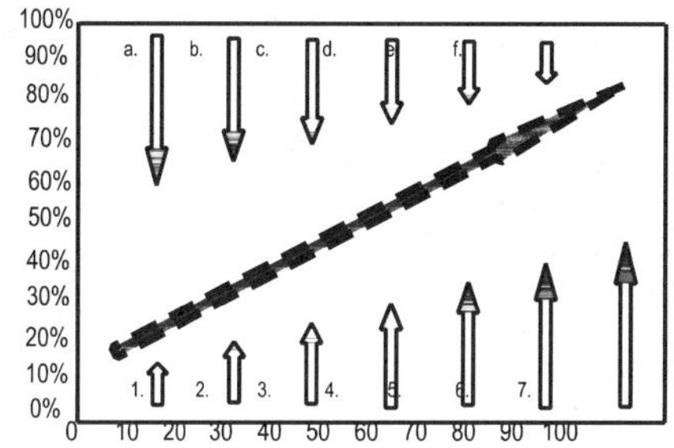

a. Limits of capacity,
b. Limits of ability,
c. Limits of attention,
d. ATRU'S, and
e. Intermittent needs.

There are several parameters of the <u>mode of sufficiency</u> that are necessary for optimal FAF and work. These include, 1. Size of object, a person, institution, or community, 2. Periods of time, 3. Work x time = success, 4. Effectiveness, how much per job, 5. Efficiency, 6. Job as percentage of TRU, 7. Other factors in levels of sufficiency.

Fig 10-55 **SUPPLY and DEMAND EQUILIBRIA**

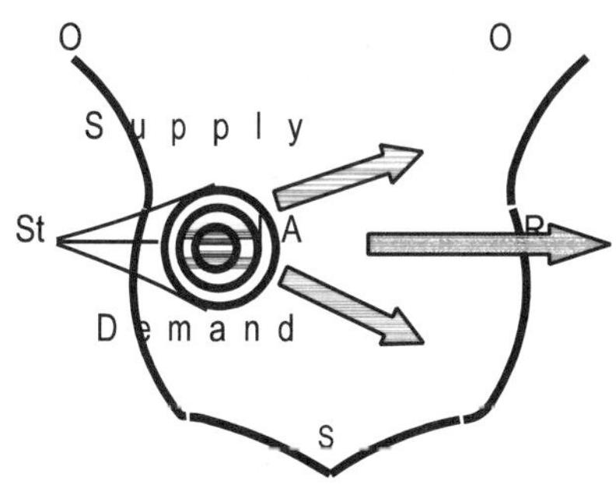

There are many faces of the supply and demand relationship, especially prices, and the needs and powers of an economy. It is the economy of CES in TIT that are shown by HOM. Four states of supply and demand depend upon whether either one is increasing or decreasing, large or small. The system, body or TRU is largely dependent upon these relationships. ↓S / ↓D = ↘, ↓S / ↑D = ↘, ↑S / ↓ D = ↘, ↑D / ↑S = ↗. In TRU these categories are more or less variable with many factors: prevalence, ease of supply, population, its vigor, and all levels of SÁ. **LOSDIT** = L. Optimal Supply and Demand in TRU. All factors need to be included if we are to have a good HOM and healthy TRUD and TRU. #1 leaves little to work with and is negative for TIT-TO, yet can be positive for parts, # 2 produces stimulation and production of incentive, imagination, discovery and invention. It can also produce increased stress and loss without achievement, # 3 decreases through the continuing inhibition of demand, and in # 4 although simply increasing supply and demand it may be thought success and survival are assured, an inability of keeping-pace can occur. **LODIR** = Law of Diminishing Returns. Production rises at a progressively decreasing rate by its increased effect on output. LODIR comes from the overall principles of limits on most all growth. It is a force whose actions can produce much reaction if not compensated for in the health and strength of TIT.

RABEC = Ratio of Benefit/Cost. RABEC is a prime measure and example of the dynamics of TRU. It is a basic process in SPALT and Homation that permeates TRU and NAMTRU. It confronts all in life, as the shopper so readily realizes. **DESCOL** = Dependence of Principles of Success on Characteristics of the Level. DESCOL describes that condition of TRUD and HOM in which its principles are made up from and characterized by DAOST, the different levels, clines, polarities, and primary parts of TRU.

Fig 11-55 **PROPORTIONATE ACTION, LAPAC, PROCESSES, HOM and TRU**

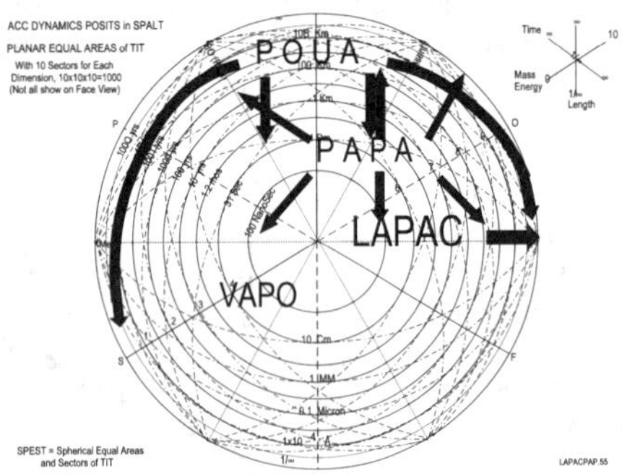

LAPAC = Law of Proportionate Action. LAPAC symbolizes action, dynamics, and all that functions about us, where action must be no greater nor less than its place in TRU actuality. LAPAC carries tremendous implication in the action, dynamics, ethics, work, efforts, future and plans of one and all, as in PI and ENT to become TRU. Closely associated is its multilateral and universal FAF. By TRU these are largely guided by VAPO. For the degree to which action, even if it is intended to be most proportionate, multilateral or universal, is not based upon VAPO and LAPAC, is the degree to which it fails.

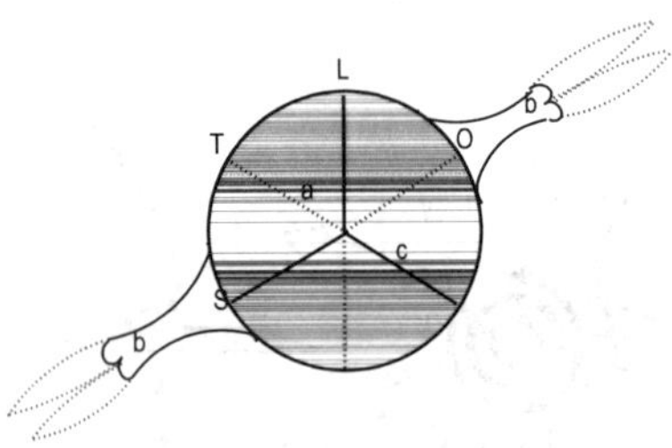

Fig 12-55 **APPENDAGES in FORM and FUNCTION of TRU**

The great outpouring of subjectivity and knowledge between 600bc-200ad, and again since 1500 have been like appendages to TRU. The first was especially stimulated by the alphabet, writing and an interest in understanding the reason for things and human behavior. The second was especially stimulated by further advances in language, communication, printing, science, experimentation, computing, and interest in invention, technology, and practical application. Appendages illustrate, 1. TRU as a RU, 2. Its basic form and function, 3. As DAR, DAT, DET of TRU. For every imbalance appendages represent, there must be a balance. This is called: **LANAP** = Law of Absolute and Natural Proportions. LANAP signifies there must be a reason, as well as a correction. For every action, balance, imbalance, appendage or extension there must be a reaction, re-balancing, compensation, or return to optimal FAF.

Fig 13-55 **INDIVIDUAL, TES, ETHICS and TRU**

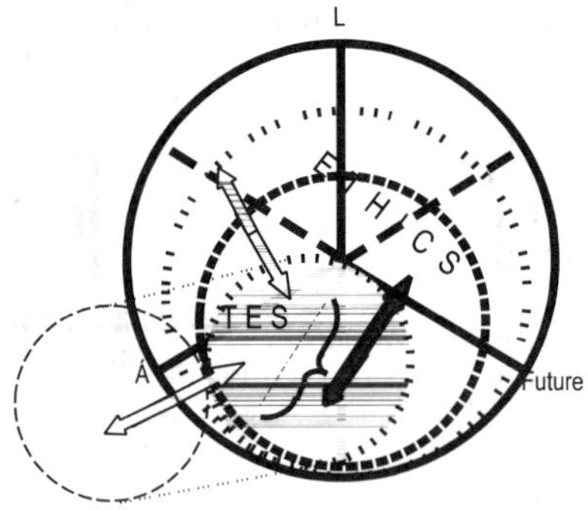

The individual as ENIT and TES in the expanding world and TRU is highly interactive. This tends to be toward the future and greater, involving larger numbers of people and their welfare. This shows how CES and CHES in HOM are KDM, enhancing our ability to order and understand the dynamics of TIT, TRUD, TO and TRU.

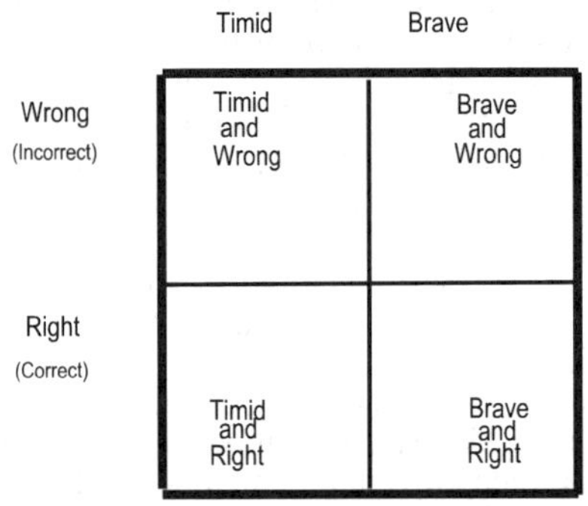

Fig 14-55 **RIGHT/ WRONG, TIMID/ BRAVE**

With maturity we learn not be timid but to go after whatever needs to be done. Early in life we learn not to err, make mistakes, be wrong or incorrect, but to discover, learn and practice the right, correct and good as much as possible. This is the objective and subjective, or Á, that is so much of life and TRU. The two parameters that have as polarities timid to brave or courage and wrong to right illustrate the matrix and orientation we must learn and develop if we are to adapt our thinking to the complex processes in TRU.

Fig 15-55 **THE DISSOCIATION CURVE**

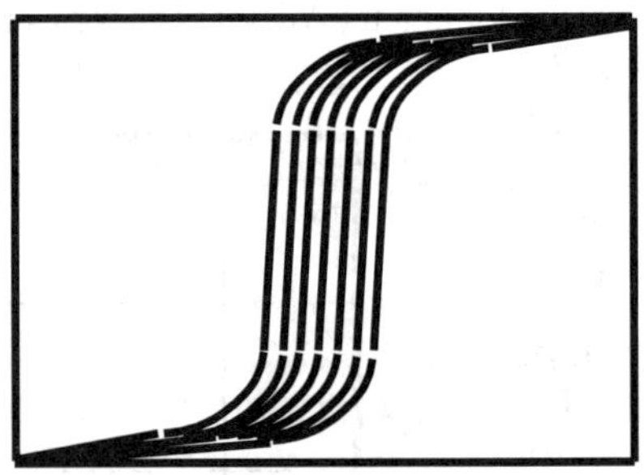

The dissociation curve represents much of the dynamic state of TRU by emphasizing not only the posits and distribution of many of the forces and much of the action in TRU but their variations from the mean and standard distributions. Even more it illustrates the changes these forces and actions undergo, both functionally within TRU and, under stress. How they can be of TRUCH returning to TIT by TRUR. Such means demonstrate many of the CES, mechanisms, polarities, continua and dialectics in TRU. It beautifully illustrates the steady state along with increasing alteration, stress on the equilibrium, and mechanisms that occur simultaneously in TRU. These must be in, or return to equilibrium, to maintain optimum FAF and prepared to react to all that is deleterious.

Fig 16-55 **ACID BASE DIAGRAM**

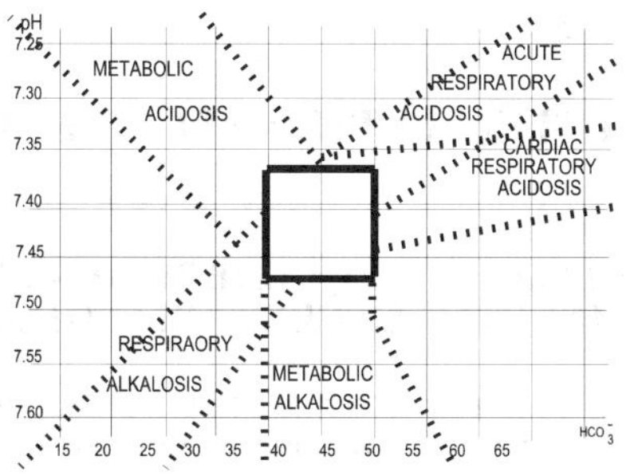

In the human body its acid-base metabolism is an excellent example what also occurs in TIT more generally. It is an equilibrium state for the body's maintenance of eustasis or good health, as well as the management of many different diseases. In health, physiologically, and in somewhat different ways in TRU, there are smaller variations of these equilibrium systems. They work cybernetically with compensatory mechanisms. The predominant mass energy, Á, continuum of TRU functions in such an equilibrium between its extremes. These balances set in motion, action, reaction and the dynamic processes of TRU.

Fig 17-55 **CYBERNETICS, INPUT, OUTPUT, FEEDBACK, INTERACTIONS of MECHANISMS**

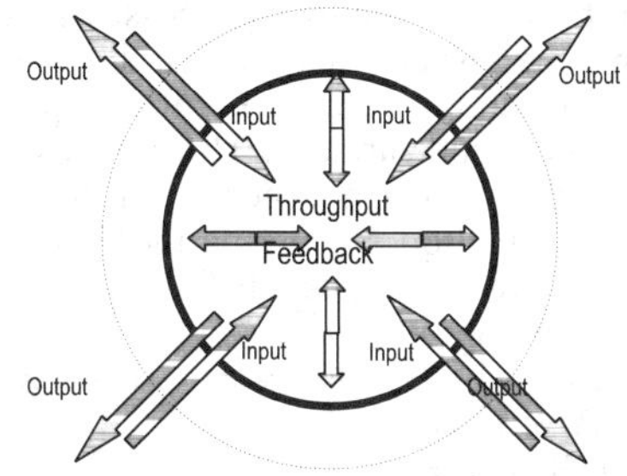

Automatic control systems that exist in special ways in the different portions of TRU work cybernetically. Many CES and forces work simultaneously with many others. This is where we often mistake what is and what is not TRU. It is called: **SIVOPT** = Simultaneous In and Out Variable Percentages of TRU. Although TIT exists by TRIA, TRU functionally, as a dynamic mass, it does not exist exclusive of OITIO without interchange with what is less than 100% TRU. Yet once this that is potential for TRU is activated, and that which is in TIT yet less than 100% TIT is deactivated, the reactions maintain TIT at its TRIA status.

Homeostatic resilience is seen in many diseases where there is a breakdown of normal physiological mechanisms. This breakdown is primarily by excess load or deficient capacity. Through INEG there is an increasingly irreversible deterioration. **PREDIS** is the Prevention of the Dissolution of TRU. PREDIS is continuously at work in the main stream of life, its preservation. It is constant replenishment and improvement that is most characteristic. This is called: **PER** = Principle of Ready Replacement. PER is where there is a continuous and necessary demand, forces or vectors to replace that lost or used up in any sort of energy expenditure.

The knowledge explosion is outpacing lifetime limits. It produces stress and strain on the individual with resultant reaction. It is also manifested by personal, social and world divergence of views and interests with escalating conflicts. This is a major reason for HOM, TRU, KOT, POT and IPT. It is aided by the interactivity of ACEXIN. ACEXIN, guided by VAPO, helps us live a balanced life letting all fulfill their best roles in TRU.

Even if you are on the right way you are apt to get run over if you just sit there. This adage points out four major CES in TRU. You must be on the right way, you must move with the current, you must know what you are doing, and it must be done with probity by TRU. **LODECON** = Law of Determinant Contingency signifies that a person may do as they please at a time, place, or way, yet they are forced to do what they must by TRU sooner or later or in some place or way. Finally and totally in the complex integration, PIP, of the general convergence of forces, ones actions will be reduced and return to eventually balance out in TRU. **LOTOD** = Law of Transfer of Indebtedness to All, TO, or to others, and payback. LOTOD is how LODECON is evaded by individuals who seek to honor their debt by borrowing from others. This is a great cause of turmoil in the world, whose correction in TRU require sufficient protection in KOT, HOM and ACC. The abuse of LOTOD is continually experienced wherever people live, work, behave and interact. This is the reason it is better to give than to receive. For it is in such giving we serve God, become TES in TRU. For in giving one is much nearer to the heart, soul and core of TRU. In receiving, unless it is well realized and used in TRU, it is more like the output of TRU.

Fig 18-55 **EQUILIBRIUM of TRU, ES, CES and MECHANISMS in HOM**

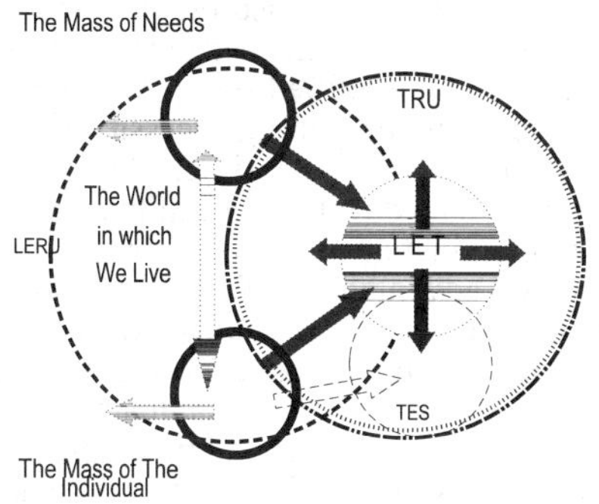

We are continuously paying for the past.(PIET) **LET** = Law of Equilibrium of The Relative Universe. This is shown by: : **LERU x MC x IN x ESIRU**. We are always taking from or putting into the future. This is why it is so hard to catch up. It arises from our failure to 'make hay while the sun shines'. It arises from the necessity to keep up functions as a mass, and our lack of anticipation, thereby. This spread of TRU. its essences, ES, and CES must be continuously remembered, qualified, and misconceptions corrected.

Fig 19-55 **THREE TO MANY CORNERED DIALECTICS and SPECTRA in HOM and TRU; NEEDS, DUTIES and LAWS**

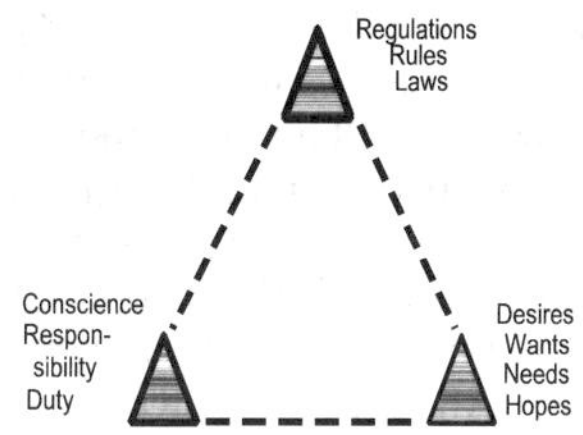

There is a very massive, yet imperatively fine and improving relationship and dynamic equilibrium between many aspects and parts of TRU, including, (1). The laws, rules and regulations of a group, community, society, state, nation, world or TRU (2). The sense of duty, responsibility and conscience of the individual, groups and all, (3). The needs, desires, wants, and hopes of the individual, group, and all, for their benefit.

Fig 20-55 **CHANGING CES, POWERS, and ACTIONS in TRUS and TIT**

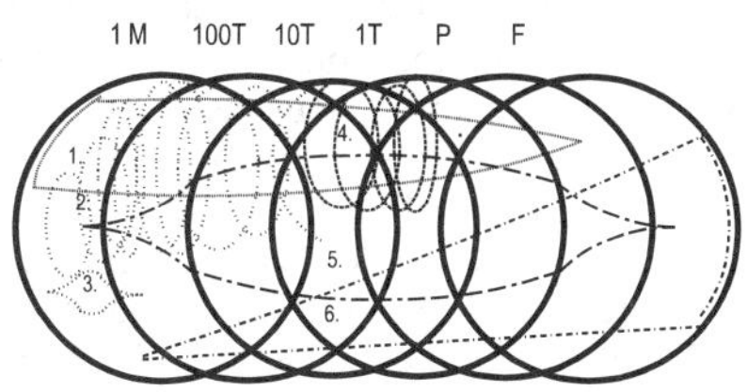

To be understood properly, adequately and at its fullest, how divergent forces and CES act on TRU through time should be properly given. These are not the ages of time, we so commonly think of. They are of long term distribution in ALT and SPALT. The divergences, kinds of clines or worlds within worlds, are the stuff that much of the dynamics and CES of TRU are made of. It is in their understanding, direction and control toward maximum benefit that we achieve TIT. In Figure 20-55 some of these changing parts are shown: 1. Earths Resources, 2. Species, 3. Vestiges, 4. Cultures and Civilizations, 5. Genetic Material, (best - worst scenarios), 6. Mental capacity, ability and language.

CHAPTER 56

TOTALITY of HOM: SUMMATION and INTEGRATION of CES and CHES

PROOF of INTEGRATIVE PROCESSES, INTEGRATIVE NEGATION, EQUILIBRIUM, ADDED STATES in TOTALITY, CLOSURE, SYSTEMS, MOMENTUM, ORDER, INERTIA, CYCLES, CHANGE and VARIATION

Actual form and function become more apparent when viewed in totality, TIT-TO. When compared with many entities, objects, and measures, the parameters, frame, parts and processes of TRU seem bent and are different than they appear. **BURD** = Bent, modified and qualified, by TRU Dynamics and processes. The compelling necessity of habituated acts is much of TRUD. They are neither directly rational nor simple responses. The development of human abilities proceeds not by simple increments alone, but through the defects and adaptive advantages offered by different functions in combination and/or in totality. Mankind, and all that enables our being, functions as a whole, diverging and converging simultaneously to produce maximum permanence and survival. Total dynamics of TRU may be simply classified by what is of ethics and what is not. These are more readily understood when we extend this classification into SÁ, in which dynamics are non ethical except that of higher levels of SÁ focused on social harmony. It is because of our failure to understand what is right and wrong, and how the ratio of this right and wrong is interpreted by others that humans have failed to adequately understand many of their problems. Probably the greatest cause of problems, trouble, unnecessary conflict, stress and suffering in the world is the relationship, the antithesis between:

IGNORANCE x SELF INTEREST x RASH ACTS / WISDOM x IDENTITY x GOOD x TRU ACTS

Experience is more or less accumulated with maturation through time. The bitter the experience the deeper and longer it is learned, remembered, and the greater its stimulus to knowledge. With this there tends to be greater identity with the world, TES in TRU. There is built into each of us a degree of suspicion and antipathy towards others. This in part arises from their NAMTRU to ATRU FAF and higher levels of Á in which differentiation, distinction and criticism are beneficial to coordination and control. Closely associated with this is the dualism between **LOVE/HATE and SUSPICION** that are largely of **TRU/AU.** It is the beautiful and well developed capacity and ability prominent in Á and IA, of knowing and attending to what is positive rather than negative or harmful. It is the degree they are best brought to TIT-TO that determines success and survival. As the parts of any whole must be bonded firmly and properly so must the bonds between all persons, groups and societies render them not separate and alone, but parts and processes of the whole. This is how TES and CES come together in TRU and can be shown in HOM. Total fulfillment of Á is the fact that Á can be discovered, learned, known and shown only to the degree all its ingredients are included and in place. This knowledge and action relative in the subject-object, observer-observed and other forms of mass and energy, partakes of this predominant continuum, Á FAF in TIT, of TRU.

SEF = Sigma or Summative ESIN Function. SEF signifies that the total functions of ESINS, when enumerated and separated, will lead to an approximate measure of TRU. e.g. The function of an elbow, back, finger, nail, thumb, eye,, neck, and leg. PIP represents a large section of HOM as well as IPTRU. PIP certifies how TIT actually exists. The thisness of TIT, as any entity, depends to a large extent upon how all its parts and processes work integratively and harmoniously, along with their added states in unison and totality.

There is a continuous and massive interplay of the positive and negative with the differential and integral in TRU. **POND** = Positive and Negative Differential. It is because of failure to learn and think in terms of POND that humans have been so confused and failed to recognize God, TREQ or TRU. Whether we are dealing with TIT-TO or the negative effects of one part on another, this negation is often accumulative.(INEG) **MARF** = Massive Response to Failure. With much pain, stress and severe loss of response escalates, and mass action results. This is a protective device of several of the lower levels of mass energy in TRU. **MEX** = Means and Extremes. **VARMEX** = Variation of Means and Extremes. **VARMEXT** = Variation of MEX of TRU. MEX, VARMEX and VARMEXT explain much of CES, CHES or HOM, TRUD, TRU, and especially their equilibria, and cyclical processes and changes. VARMEXT shows the variety that we see, how different so much of the world around us seems. Yet it also provides meaning to these differences and variation. Differences and variation are ordered by their SCARLS, a large part of the form, mode and aspect of TRU. Transition represents those processes and mechanisms which show how these different and variable FAF pass from one to another in TIT.

Fig 1-56 **CHANCE, PLANNED, CON-TROLLED, and CANCELED OUT EVENTS**

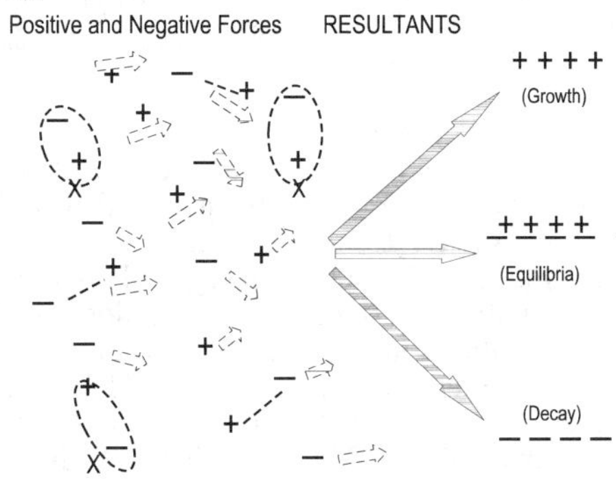

Much of the rate of growth and development of TRU is determined by the difference between chance events and planned, directed or controlled events. The closer to the present and the greater the higher levels Á are, the more this has been through plan, direction, and control. This is one of the great laws of human specificity, TRIA, ATO and TIT. It is called: **LORCEDS** = Law of Evolutionary Replacement of Chance by Directed Selection. LORCEDS is not a simple step, it has been very gradually developed in natural selection. They have and are existing and working together and apart.(DAOST) Directed selection has roots in lower, consists largely of increasingly higher, and covers several levels in Á. There are also many accumulative, secondary or indirectly summed events. Many of these tend, in resultant, to cancel each other out. Just as in genetics, shown in the forces time chart, this helps to maintain a steady state. Lack of wise selection, control, and care is constantly before us. It was from a lack of care that Faust was blinded and so rapidly destroyed. INEG may occur by great leaps and bounds, mutations, sudden death or revolutions. Or it may occur by the tiniest of an immense number of increments. To have infinite care, to have an extremely deep love of life, to identify oneself completely with the ACEXIN of TRU, and visualize ELTICMA, and TÂ are the best means of avoiding the fate of all who pass into oblivion. The stem 'pletion' as in depletion and completion of TRU, represents the processes of filling. For every increment taken out of TRU an equally, at least, as good an increment must be put back in, added to, or change made, if TIT is to hold its own. **SEN** = Summative Sensitivity of Parts and Processes to Negation. The summative or Sigma S/F, pains, sufferings, and experiences of humans produce an inertia to failure in INEG or, **ERIN** = InErtia to INEG. In this way ERIN and SEN are right and provide strong evidence and support for PIP. Just as our sensitivity to stimuli, pain, discomfort and suffering to ill effects and wrongs enables adaptation. These are the great rectifiers, and how SEN, ERIN, and ENT contain through inertia and momentum powers and processes that sustain TRU. **INAT** = Inertia of Â in TRU. INAT certifies the inertia and momentum created by the predominant mass-energy nature and form of TRU. Just as any entity has its own momentum and inertia so does its FAF or major continua and components more or less carry its existence and survival. For Á this includes all levels, and is vividly seen in speech, knowledge, writing, thinking, reason, problem solving, purposes, priorities, and goals.

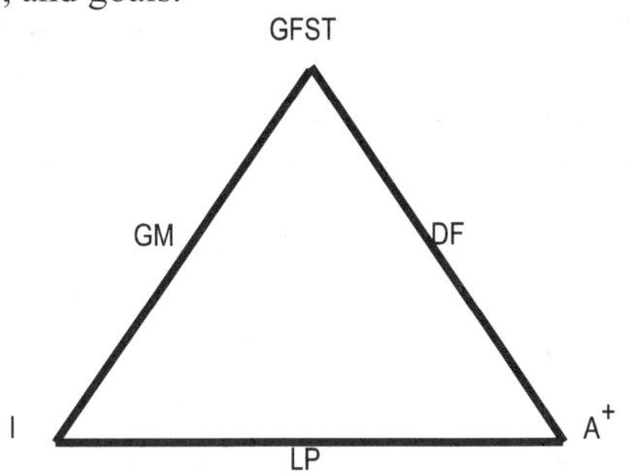

Fig 2-56 **FORMULAE and FIGURES of OBJECTIVES and PURPOSES in HOM** Reproduction, whether sexual or general, is a major need and purpose of animals, humans, and TRU. In figure 2-56 some of these functions, forces, and factors are shown: **GFST** = General Forces Stratified Time, **I(Ind)**=Individual, **A+** = Activities, Action Positive, for TRU, **GM** = Genetic Material **DF** = Determining Forces, **LP** = Live Product. It as all else, can be TRU, all gradations less than, or negative to TRU. Other needs that are 100% TRU are protection against adversaries, livelihood, sustenance, fulfillment of functional abilities, prevention, maintenance and advance of energy systems. These can be enumerated to include most objectives or purposes. They can also be represented, as in a mileage chart to beautifully show much of TRU.

$$S/F = K\, R_p \times L \times FFA \times PPMA,\ F_w,\ F_g\ GFST.$$

Activities increased in evolution:

$$\sum(C-S) \times R \times Capacity = Capacity = Life\ Charge$$

There is a continual, incremental to infinitesimal, and summative Diffusion and Effusion in and out and within TRU. TIT is continuously being dissipated, weakened, and wasted simultaneously with being conserved, built up, and strengthened, one, more acute and greater at times and places, than the other. It may be far different, less or more, for the whole, for TO, than for ES and CES of TRU. Not all output of TRU is reversible, yet in other ways or by other means, may be compensated for. These are called: **MEDAT** = Mass Energy Dynamics to and from TRU. MEDAT is the basis for much of HOM, where CHES provides the vehicle through which MEDAT can be positively identified and represented. MEDAT is a corollary and special category of OITIO by virtue of its focus upon the Á FAF

Mentality and reason provide the special essence of TRU, or **MAREST** = Mentality and Reason form the Special Essence of TRU. It is MAREST that shows how what is relativity in massenergy at various levels of transformation in animals becomes what it is in TRU. For each different RU and RUT there is, if only very slightly a different m, l and t. This is a great part of its form and essence. How the different m, l, t or A, L, T in RUS and TRUS are equivalent to their essences is called: **DALTES** = Different A, L and T equivalent to Essence or Vital, Primary and Predominant Forms. It is through a mass energy and beingness of form, states, dimensions, functions, mechanisms and processes that are unique, special, and to a high degree exclusive, that TÄP is constituted.

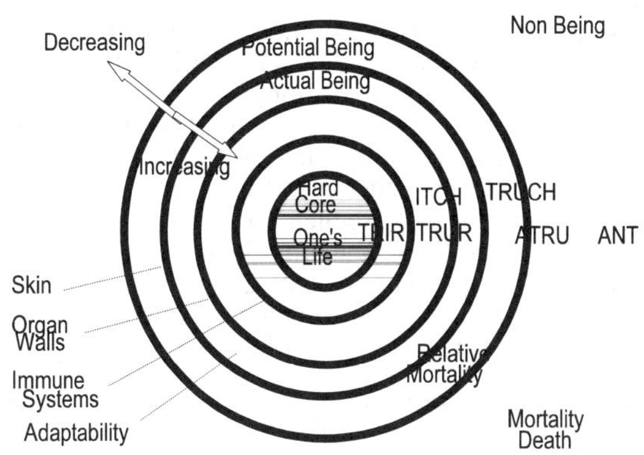

Fig 3-56 **LINES and MEANS of DEFENSE**

For all entities and TIT there are many lines, walls, shields, and means of defense. For TRU they will have much that is entirely different from other entities or universes. The human body has many different membranes, walls, means, reactions, interactions, and systems of defense. In this figure we show a few of the factors involved. There are a great many more in personal health, as well as the health of all different segments of TRU. Even such common and taken for granted defenses as work and money, or ones sustenance and security, are major means of defense. They are not only what stimulates and keeps the dynamics of TRU in motion, and a person going. Even more the great dialectical forces and equilibria that exist between ES, TES, CES and aspects of TRU prevent those ENIT and ATRU which function through AMTRU.

TRUEQ = TRU Equilibrium, **TEQ** = Theory of Equidependence. TEQ is the balance of dependence that is distributed throughout TRU, TO. The dynamics of S/F, that can be presented in Homation, is best seen when combined with much of the rest of TRU. **CARSC** or the Capacity, Ability or Intensity, Reaction, Success/Failure, Setting and Condition FAF, state and operations of TRU can be shown diagrammatically. This can also be done in conjunction with Energy Systems, Á of TRU, **CARES**. (Table 1-56)

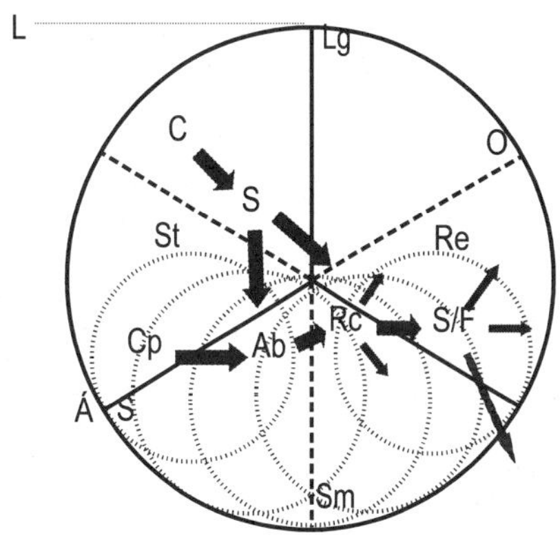

Fig 4-56 **CARSC DIAGRAM, PRINCIPLES, ALT and HOM**

The posits of all signifi-cant ideas, universals, and principles in such a diagram give us an additional method of revealing ACEXIN and TIT-TO. **DENT** = Dynamic Dipole Duplicate Polarity, enabling TRU. These are resolved by TRIT, OBT, ROT, SUT, and SO, as they bring ALT and PERT through ROT, POT and SPOT to TRU. This figure helps to show the relationships between the Stimulus-Response, Capacity-Ability-Reaction-Setting-Condition, and Success/ Failure states and continua.(CARES) It is especially used here to show the superimposition of diagrams of Medical Homation that posit in

ALT. What most clearly demonstrates the causes and conditions producing diseases, their complications, and consequences. It shows re-coordination by ALT and SPALT. Abbreviations include, **C** = Conditions, **L** = Length, **Se** = Settings, **Á** = Mass Energy, **Cp** = Capacity, **Lg** = Large, **Ab** = Ability, **Sm** = Small, **Rc** = Reaction, **S** = Subjective, **S/F** = Success/ Failure, **O** = Objective, **St** = Stimulus, **Re** = Response. The S or Se, Setting, are those things that come in direct contact with, and provide the background and media through which the actions, interactions, and processes occur. C or Co is the Condition in which this dynamic whole functions. Systems as CARSC are widely used in biology and medicine. For ultimate adoption to ROT, MOT, and TIT, however, systems must be altered and superimposed in ALT. How can we enhance the setting, state, conditions, reaction, capacity, ability of any and all that is TRU? Human intelligence, vital organs, parts, energy systems, abilities, capacities, and the carriers of human culture are not self sufficient. They can't stand alone, be denied, or suffer destruction indiscriminately. The average S and C vary with different peoples, races and subspecies. The capacity, ability and reaction of TRU depends to a great extent on the individual, and vv. This is why it is imperative, within the need for individuals, to be TES.

	CAPACITY	ABILITY	REACTION	SUCCESS/ FAILURE	SETTING	CONDITION
pu	photon		Light			R
sfu	Lepton					R
wfu	Hadron					R
emu	electron		Radiation			R
gru			Weight			R
nu						
atu	Ch En	Mt	N		Ec	Cl In
mou						
adu	Ce An	Gr Rd	G-A-S Re Ph Ne Si		Pl Fo An Ne	
cu	Ne CNS	Se Le	Fe Wa Lo Be C De Hb	Se		Da De
ru	Br CC	Pe Ex Co As We Kn Th	Sp La SB Ho Wi Ch Ed	La Be	Ec Ci Fi	Wa
imu vu		Re	Ph Ri	Cu Re	In La Co	Ta Wr
ilu	Pr	Ju				
Tu	TIT, TU, TÄ TE, TO	VAPO SCARLS DAOST ACEXIN	HOMATION	ENT	IO ATRU NAMTRU	AU

Table 1-56 **CARES**

CARES is formed by combining the Energy Systems or levels of the Á dimension with CARSC. Abbreviations used in CARES include: **Ab** = Ability, **L** = Language, **En** = Energy, **R** = Radiation, **K** = Knowledge, **CNS** = Central Nervous System, **Ec** = Ecology, **G-A-S** = General Adaptation Syndrome, **S-R** = Stimulus Response, **S/F** = Success/Failure, **ad** = adaptational, **Ch** = Chemical, **at** = atomic, **Co** = Cognitive, **c** = conscious, **Ec** = Economic, **e** = energy, **Ed** = Education, **e** = electronic, **En** = Enzymatic, **f** = force(s), **Gr** = Growth, **g** = general, **Ju** = Judgement, **gr** = gravitational, **Le** = Learning, **il** = intellectual, **Ne** = Nerve, **im** = importance to man, **Pe** = Perception, **m** = mass, **Ph** = Philosophy, **mg** = magnetic, **Pr** = Principle, **mo** = molecular, **Re** = Reason, **n** = nuclear, **Rl** = Religion, **p** = physical, **Sc** = Science, **r** = rational, **Th** = Thought, **u** = unit, **Wi** = Will, **v** = valuational. The levels of energy systems, or SÁ, may be analyzed and compar ed with the higher powers of or Á, L, and T in TRU, $L_1T_1\acute{A}_1$, $L_2T_2\acute{A}_2$, and $L_3T_3\acute{A}_3$. By CARSC and CARES one may envision the complexity of HOM, TRUD and TRU about various portions of, and in, the whole, TIT-TO. Centerpoints, world point, POSITS, summation of energies, selection,and quantities all help us to piece together the great mass of TRUD.

All that rises and falls, absolutely, relatively, temporarily and generally has features, form, processes, climaxes, growth and decay that are much the same. Their archaic, formative, heroic, classic, golden, silver, senescent, and terminal stages or ages can usually be identified. Their maximums and minimums add to such a demonstration. Maximums or optimums are apt to be times when humans are coming out of periods of great selection. Stress and strain produces added stimuli for concentrated power. Duration of phases will vary with the category considered.

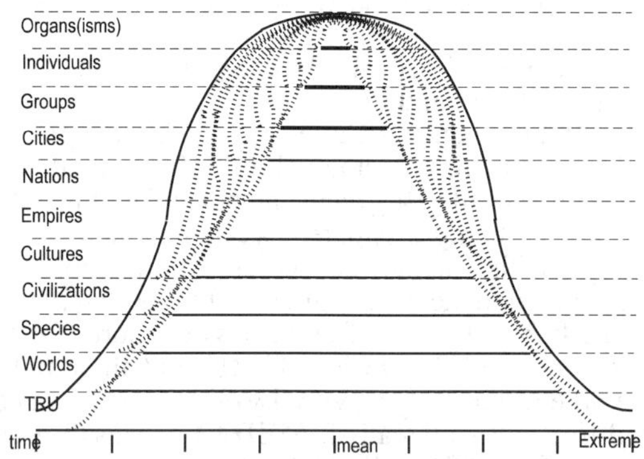

Fig 5-56 **The RISE and FALL of Things, Worlds and TRU**

Various functions and dynamics of TRU are not simple homogeneous and singular phase motion. They are in opposition.(DAOST) This diversity however is not chaotic. It is more or less harmonic and cosmotic in the sense TRU is in complex functional competition, and in which diverse parts and processes are, more or less, ordered and controlled.

All the processes that exist within the totality of TIT are formed and function by the interaction of mechanistic laws and principles of dimensional formulation adapted to TRU. (pg.180) There are more of these, some more partial for each process, and some more total for all TIT. Examples of these include redundancy, entropy, kinds of energy, OITIO, etc. The conversion of many levels of mass energy working to and fro in TO helps us understand the totality and intricacy of HOM. This is shown by the oppositions of processes whose function is to restore the optimal state of their parts in TIT. When excessive or deficient the action reaction dynamics produced often engage whatever function in dimensional formulation best brings the whole back into equilibrium.

Fig 6-56 **THE GREAT EQUILIBRIUM BETWEEN EXCESSES AND DEFICIENCIES in TRU, TIT and HOM**

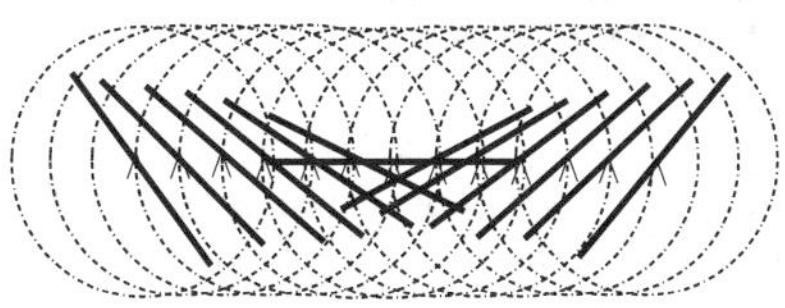

What it takes to maintain optimal equilibrium is a great balancing act. (SCARLS) The integrated and dynamic form of TRU, as well as its composition and substantial nature, are exemplified by the many buffer systems. Some of these that exist in the human body, and its physiological chemistry, are, Bicarbonate - Carbonic Acid Buffer System, Phosphate Buffer System, Protein Buffer System, Hemoglobin Buffer System, Bone Buffer System, many automatic and built in regulatory mechanisms and systems including those of, respiration, blood pressure, heart rate, sleep, water, nutrition, digestion, elimination, temperature, and reproduction. These are, but a few, and are not limited to only the physiology, of the human body, but include all parts of TRU. They, being lower than the level of consciousness occur simultaneously, massively and interdependently. A great amount of CES act in regulatory mechanisms or systems. Often they play a role in more than one, sometimes several systems, or may act indirectly to produce positive effects in support of the whole.

Fig 8-56 **HUMAN GENERATIONS**

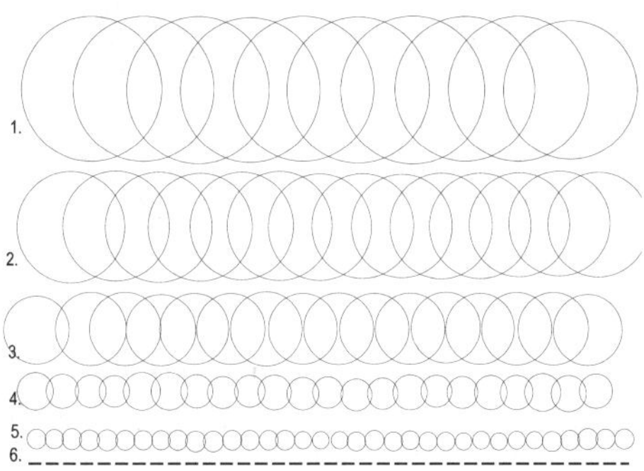

Some of the different kinds of TRUS are shown in which, # 1-5 pertain to decreasing sizes of population or overall quantity. To what extent is TIT only # 5, the single individual from which, in mating, all future generations will be derived? Or is TIT only # 1, the total human population. Or is TIT # 2-4 some combination or variation of these. This results in differences and a flux in the generations of TRU and their FAF. The kinds of generations change. This is why we include # 6 or a time scale. To be accurate this needs to be more elaborate, more than arithmetic, more than one scale, or even one diagram to depict kinds of generations.

Fig 7-56 **APPEARANCE and ACTUALITY of the SERIAL and DYNAMIC**

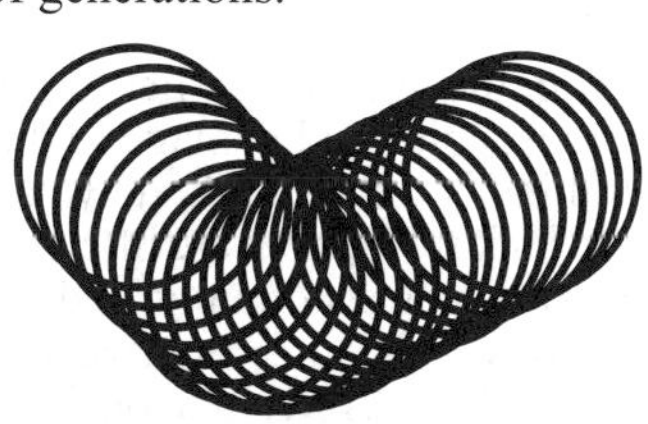

SERIAL PROGRESSION, AS ONE OR SEPARATE

Ones whole being depends upon whether it is of death, survival, sustenance, or dominance. There is a myriad of forces operating at any designated time. Spheres operating in close succession are serial. Separately they appear stationary or static. This, in reverse, is seen in a moving picture where each separate picture in very small increments of time makes it look like a single sequence of dynamic action. This is what the individual and ones attention do. It causes failure for IA. Because that before us appears static does not mean that in the actuality of TRU it is not moving. In IPTRU this is why we stress SPALT and HOM and their viewing. With SPALT what we envision is all inclusive and with HOM we can include all that is dynamic. To see them as TRU, we must see ourselves, both as any observer, some observer, the observer, and as TES in TIT-TO, subjectivizing them in various ways. There is a never-ending inclination to mediocrity. Along with individual limits they are what mostly keep us from being TES.

Among the principles that show the summative state of HOM there is: **PIF** = The Principle of Inevitable Future. PIF occurs as the mechanisms of TRU are constituted by tremendous powers, energies, pressures, resistance, momentums and inertia to all that may detract from TRU. It is neither unlimited nor is it, on the other hand, as frail as it might seem, or at first be believed.(POCOMF) **LOSPOB** = Law of Summative Power of Benefit. LOSPOB is the reciprocal of INEG, the summative, multiplying and exponentially expanding empowered sequence of benefit. LOSPOB is the ultimate ethic for doing right. For what helps you, what is maximal by REM, by ES, TES, and all segments of TRU is what most helps S/F of the rest of TRU

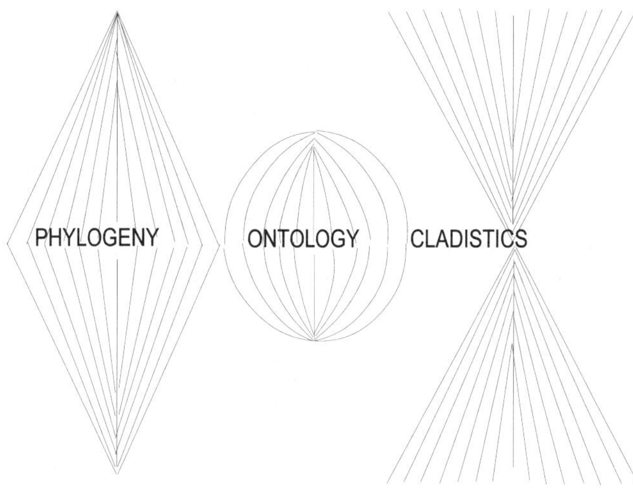

Fig 9-56 **PHYLOGENY, ONTOLOGY and CLADISTICS**

Just as ontology recapitulates phylogeny, so in reverse, does the existing mass energy dimension of relativity between observer and observed, or S-O, recapitulate the development of physical to mental units of mass energy. This is one of the best evidence and proof of the mass dimension as it exists in TRU, its Á continuum.

A clear distinction between phylogeny, ontology, and cladistics is necessary if we are to understand evolution, anthropology, creation, DOT, ATO, TRU, and TIT. Belief in simple objects and things, and conviction in evolution causes us to lose sight of the fact that the simple junction of observer-observed doesn't assure an actual reality, one that is TIT-TO. Phylogeny, ontology and cladistics have major roles in TRU. These roles do not exist as ENIT, but only as TRU. We can get a good idea by looking at their focus. For phylogeny it is many from one, for cladistics it is one from many, and for ontology it is one in one. This can be readily seen when we realize that for the future phylogeny leads to a dead end, and cladistics to indefinite diversity, whereas together in ontology, a closed system, one that is exclusively of TIT, albeit infinite and absolute in TÂ, do we satisfy TRU actuality.

CHAPTER 57

DYNAMICS of HOM: CES and CHES INTERACTIONS

KNOWLEDGE of ACTION and REACTION, POSITIVE and NEGATIVE, METHODS, OPPOSITION, VOLUNTARY and COMPULSORY ACTION, GOOD and BAD, SIN, RIGHT and WRONG, CONFLICTS, POWER, WILL, JUDGING, TRANSCENDENCE

Mistakes are made to the degree knowledge, truth, greatness, proportion and completeness are denied. Mistakes, errors, fallacies, misunderstandings, falsities, wrongs, and inaccuracies are all negation and failure. To deny and defy mistakes and failure is 1/TRU, nothingness and death. To be ethical is to make no mistakes, errors, misunderstandings, falsities, wrongs and inaccuracies. To be ethical is to greatly, truly, proportionately and completely be and live TRU by KOT. Generally, in principle, this is less commonly realized. In a daily life it is less understood and more difficult to follow. It is for this reason we must all live highly astute, guided, calculated and positive lives by TREQ or TRU. In this way we can best effect the time and place for all that is best by functioning as our counterparts in TRU, TES. If we have lifelong deep interests and love of the world of man, GOD, TREQ and TRU, and if we develop and expand this to include all that relates to it necessitated by TES, then we can be actually, accurately, precisely, proportionately, and completely TRU. We must be highly tolerant and unite with all that is of this same best world of man, GOD, TREQ, or TRU. By whatever name and in whatever form their products are mostly the same, TIT, with highest ratio of TRU/ATRU.

ANS = the evil can never understand the good. ANS may be taken from its most general to its most specific sense. In its general sense it is without TRU, that against TRU. This may be great or little. In its more special sense ANS shows how all that or those who are not, or against TRU, never have, can or will understand, at least relatively, TRU. For in their state of being, material or mental, object, subject, or real, their orientation is such that makes TRU invisible, irrelevant and thereby unbelievable. It is our position in the ATRU world, that is not TRU, that makes us disbelievers. TRU or TREQ wisdom not only helps all to discover, learn, remember and live TES. It helps to remove us from positions that are ANS.

The great axiom of war is to know the enemy. Although TRU is neither dedicated to war, nor is it necessary to concentrate excessively upon the enemy or that which is not, less than or detrimental to TRU, it is necessary to be keenly aware and to have the necessary plans, ability and will to prevent or correct, with utmost probity, all that would reduce or destroy TRU. This especially applies to all that would destroy TRU or its vital aspects unexpectedly, explosively, violently, uncontrollably, and permanently.

What we should do is far less meaningful without knowing not only know *why* we should do it, *how* we should do it, what *end* is in mind., and, by VAPO, in what *proportion*, *position*, and totality. These have been poorly known except to the degree they have been directly in front of us, learned from experience, adopted as routines or of sensed necessity. This has been improved through the methods and means of science, technology, philosophy, and religion, but has far to go.

It is not enough to be against evil. For what one may assume to be evil, what any and all are so quick to attack or join in fighting against, are often the good, positive, and TRU in disguise. We must be for right and good that is no more and no less than TRU, what is the best possible world and all that goes to make it. We should be for this with a conviction, determination and will so strong that by this good of TRU we will influence others and others will imitate us in a resulting identity and massive expansion of understanding, endeavor, and existence for TRU.

How to prevent many of the ills of life is by eliminating their approach and our exposure.Yet in life we must function optimally. This means a great amount of exposure. For both the individual and for TRU life is, to a large extent, one continuous and varied exposure to NAMTRU. So that we must control the exposure and the approach of all that is apt to be detrimental. Thereby we have ratios of optimal opportunity/exposure. These are both qualitatively and quantitative directed at the actual or detrimental potential. The beauty of, and one of the greatest advances in modern civilization has been reduction to exposure, largely through knowledge and prevention. This is offset by increasing expansion, complexity and exposure.

The mentality, cognition, thinking, understanding and comprehension in which one and all can apperceive both TIT, TRU and all subjectively and objectively simultaneously is evident when looking into two mirrors that are so aligned as to produce an image of an infinite series of objects. This reveals both the infinite compositions of the object, as well as the subject looking at it. It is what reveals the potential and illustrates the primacy of COGNOS in the culmination of Á. It shows the optimism in the methods, uses, management, enablement, and advancement of all successful means that accelerate TRU.

Fig 1-57 **LEVELS of ACCELERATION and DECELERATION to TRU by ATRU**

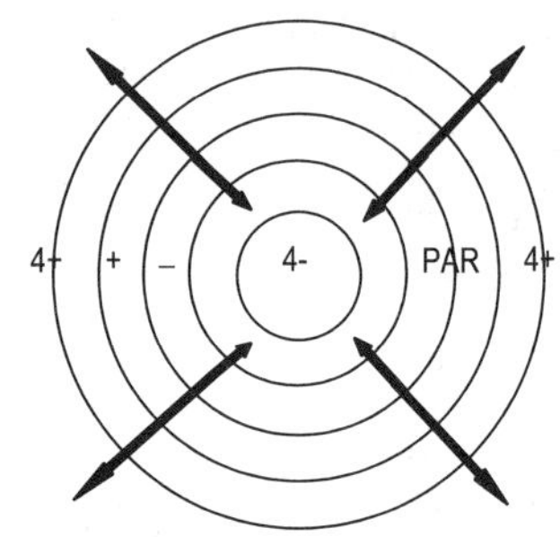

How to accelerate is a great part of the dynamics of TRU, especially so in times, circumstances, and of forces that threaten to cause loss to TRU. Also when they are reduced or lost in times of increasing TRU. These are shown by outward and inward vectors and depend upon both the degree of external and internal pressure. Without external pressure there is no stimulation. Without internal pressure there is no power. With too much and to little, or to little and too much, we have an imbalance and dissociation. Anything that wrongs what is good, honest, and TRU is double evil. For it not only reduces TRU and TES it lets ANTRU accelerate and thereby causes loss to all TRU. It must be compensated for, reacted to, or the damage done, must be made up for. It is therefor imperative that one and all, the first person. second person, all groups and peoples, TES and all of TRU become their counterparts and fulfill their roles in TRU.

It is in the terrific conflict of right versus right that so many of the troubles and damage to the world arise. "You do what you think is right and I'll do what I think is right, with hell to pay". So it is that both are often wrong in various portions, and the result is most certainly, and doubly, wrong. All must know TRU to be right. Only then can right be right, wrong be wrong, and right be fully acted upon. This becomes obvious once one increasingly recognizes GOD, TREQ, TRU, TROS, TIT, and TES. Yet rarely are they adequately known, and even more, rarely used. There are often two different rights to the degree they are D of DAOST in TRU. They are self accommodating. They are opposites or differents and their polarities are often working in phase, synchrony and harmony in the whole, TIT-

TO. If not working in phase and harmony they are ATRU.

LAVACT = the Law of Voluntary and Compulsory TRU Action, Activity and Actualization of oneself and all to TRU. It is mandatory that all be, by TRIA, TRU. To fulfill this necessity each and all must be sufficient to TRU. For this to be achieved it must be either by love of, desire for, interest in, will for, reason for, and work for TRU or their appropriate combinations. If not then some means should be found for their achievement that stimulate or compel this level of sufficiency. It must be primarily voluntary. Many means exist and can be provided. What is best, however, is when it is spontaneous. TRU is better acheived when the energy provided is directed from the inside. Yet it, as in education, needs direction, compliance, and a degree of support. Even more it needs the long term training in thinking, reason, appreciation, inspiration, revelation, and love of TRU led by the highest levels of mass energy. There are many uses of the terms act or action that, more or less, apply to how the dynamic functions of combined motion, velocity, force, and powers occur as CES in HOM. <u>Heritage, Tradition, Custom, Convention, and Imitation</u> show how we act in certain ways because of the culture we are in. **POMA** = the Principle of Magnified Action. POMA is revealed in the maxim of 'to hustle while one waits', or to do in slack time all that is necessary in rush time. These are examples of the objective extremes of highest levels of Á, and thereby the great predominant and special power of TRU. **AEF** is Action Equivalent to Future. AEF is what we do not only in the present, it represents the future, how the present prepares for the future. It illustrates our immortality through what we do.

Rigorous or Draconian measures are needed under those circumstances where life or TRU is being overwhelmed by ATRU and ANTRU forces. This is often at the sacrifice of much that is TRU, more by quantity than quality. One of the most effective tricks or strategies of action is to set limits, when you don't know what to do, or when there is slack in work, don't stop, don't do trivial or frivolous tasks, do a *set number* of items, that are most likely to be needed. Another primary tactic is *tenacity* or how long and hard one can fight or keep returning to the fray. It is how 'doggedness does it'. It is getting back at it until right is done or TRU is accomplished. It is the key to any successful strategy where a difficult endeavor or obstinate opponents are to be engaged, or prolonged problems, battles, or wars are to be fought.

Fig 2-57 **LEVEL of TRU in LIFE**

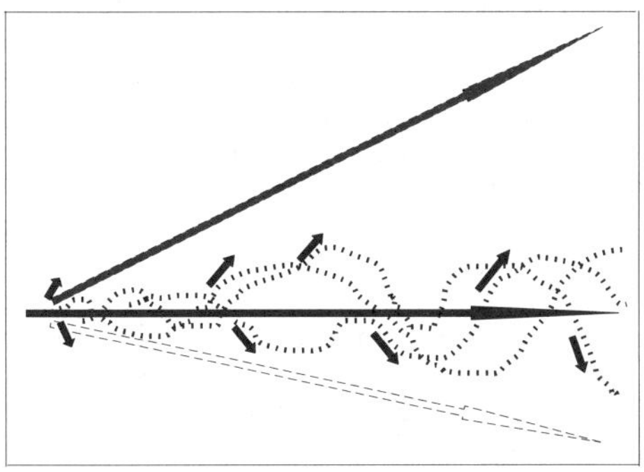

Each and all persons are in a never-ending struggle of success / failure. This is, and can often be in resultant, TIT. It is much more than the ups and downs of life. Even what is seemingly good and right quickly becomes bad and wrong. All that has been good and right quickly dissipates and is lost through failure to sustain the PPMA necessary.

This best world of man is not something for the asking, that can be done at leisure, or a reward for a chosen few. It is the good, actual, dynamic, effective, efficient, vigorous, productive, fulfilling, enhancing, perpetuating, and eternally directed life we can all have as we best use what is given to us. The greatest sin of all is for those who have received so much to turn into, or turn it into, so little. The greatest glory of all is for those who have received so little to turn it into so much while having complete identity with TRU. It takes the adaptive realignment of persons, groups, and all mankind in a rededication to TRU. It also takes an avoidance of the quick and easy distractions and return to a vigorous and everlasting ENT, TIT, TRU and TÂ. How to savor its glorious increments yet shake its satanic tempations.

The only thing that the ATRU, ANTRU, or ANTITRU respect and submit to is power and correction. Most, when realizing TRU, will respect and submit to it. It is this overwhelming power that keeps TRU going, and most adversaries at bay. That power of all TRU which sets them right. It is TRU with an abundance of health and strength in all its mass energy to be utilized fully and appropriately by HOM Thus, it is necessary that all TRU exercise its functions and responsibilities in the execution of that required.

Over and under actions and reactions need the guidance of: **SPACHT** = Strength, Predominance, Higher Capacity and flexibility of Á, assuring and enabling greater life and TRU. The ranges that occur in the many tests that are made of human function illustrate the degree to which overreaction and under reaction are normal or abnormal. This may be readily extended to most of TRUD. Involuntary functions can be more readily measured and interpreted as normal and abnormal than cognitive and voluntary. Since man is, less to more, both the director of ones destiny and that of our world, overreaction and under reaction may be normal at one time and abnormal at another. Yet all persons must assume responsibility for how they react in life and on each occasion.

TIT consists of a selected, qualified, exclusive, predominant, and vigorous humanity, plus what enables it to be. It is largely yet invisibly different and in conflict with individuals or humanity in general. This difference and conflict is basic and permeates all life. It drives both the part and the whole, the individual and TIT. It activates individuals and all, in both positive and negative ways. They are much more positive when TIT is anticipated. TIT is anticipated when it, or TREQ, are known and properly acted upon. How well they function is in harmony, equilibrium, and with the peace of mind the soul, or TES, produces. It comes to one who has lost their way, failed or sinned and been reconciled to God, TREQ, or TRU. This is the gist of Christ's message and greatness.

There is a basic, fundamental, great and ubiquitous conflict between self interest and universal benefit, what is for an individual and what is for TIT-TO. This is not only because an individual, as ENIT, is not TES. It is because even TES is not TIT. Even TES and especially TÍS of each person differs. It is in understanding the basic reasons for all the problems, causes, troubles, and conflicts that HOM does its greatest good. The fall from grace to sin is not as fantastic nor imaginary as it may seem. It is occurring all the time in each individual in all entities, things, senses, individuals and groups. It is any negation to TRU. It is more specifically that down hill slide or fall which occurs in a person's life when they take wrong turns. It takes only one or a few critical errors in life, especially when associated with an attitude, character, and wisdom that is deficient or defective. The cure is to become, live, know, and serve TES, TRU, or TREQ.

What we do in TRU present may or may not be correct, true, right, or the best. For it may also be how we do all that prevents or supplies future needs. Since man commonly lives by trial and error, much must be suffered to learn. Knowledge can be transmitted to the benefit of TRU future without suffering. This quandary is associated with efficiency of action and ratio or rate of efficiency of action. **SIAR** = the Summative Inspiration for and Accumulative Increments of good and Right. We are all constantly apt to do wrong or bad and do not know it. Repeatedly we have thoughts, believe wrong, and effect others wrongly. So that only by:

:IA-SIA-Pureality-DIA-PIA-ELTICMA-COGNOS-OM-POM- SA

will ATRUxANTRU be reversed, and TÂ be attained.

Fig 3-57 **GOOD TIMES and BAD; CONTINUAL DEFICIT of TRU**

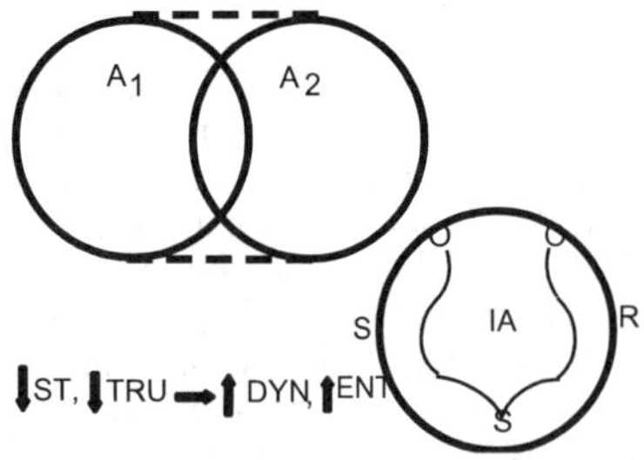

When things are going well, we tend to be slack and thereby fail to maintain life's requirements. In figure 3-57 if slack is prevelant, A_1 - A_2, there is a build up of need and TRU decreases proportionately. This may not be noticeable in the present but is all the same occurring. When things are going bad we are stimulated, activated and often become erratic, overreacting or reacting without purpose or with erroneous or poor goals. We react by switching to one false system to another neglecting the high percentage of TRU that is available. TRU will rectify its course.

Fig 4-57 **RIGHT, REGENERATION, and SELFISHNESS in HOM**

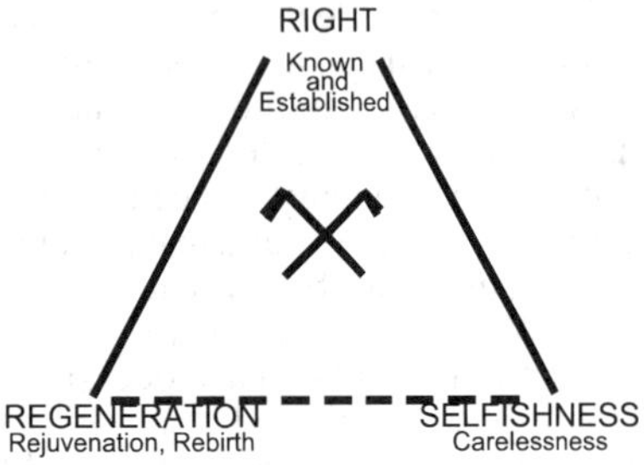

There is a great three cornered battle at the root of the success or failure of individuals, groups, nations, the world in general, and TRU. It is how well we can learn and live by TRU, reduce the three cornered struggle to best control each and all aspects of our lives. How we actually succeed, achieve TES and immortality in TIT and TÂ. Individuals and TRU are widely separated, yet they exist in the closest proximity at all times and places. This ranges from complete ENIT to complete TRU. It is nothing new, and has been increasingly present with greater population, competition, industry and commerce. In fact this great problem is what Christianity best addresses and seeks to solve.

There are two kinds of Integrative Negation, these include: 1. Causal - in which what goes wrong causes other or more negative wrongs and forces. 2. Non Causal or Accumulative - in which what goes wrong is a percentage of loss to TRU or ES through accumulated negatives, wrongs and forces. Even without an initial cause the simple breakdown of more than one part, process, essence, aspect, form or function of TRU is apt to accumulate. These two are highly mixed in INEG. What enables doing well or feeling fine is the body's, as TRU in general, capacity for compensation. Just as healing is the power to make up by a series of positive reactions of bodily repair. The mass energy or Á FAF of TRU is a large portion of this compensation. It shows how the mental as well as physical work separately and together to maintain optimal health. From earliest records of the 4the to 2nd millennia b.c. through Classical Greco Roman and Asian cultures of the 1st both bc and ad, to today, mankind exhibited and endeavored to determine what is right and what is wrong. It is closely bound to and in turn proves the validity of our Subject - Object, Á predominant mass-energy form, in TIT. It is how S-O in IA stimulus response becomes so aligned and directed to produce success/failure. Many ancient theories, doctrines, systems and religions were derived from, determined by, and developed great ideas, concepts, principles and laws of right and wrong.

The great difference between TRU and NAMTRU or the actual and poorly mixed world we, on average, live in, is seen in why people hate or love. They hate you because either they want what you have, or they have done you wrong. They love you, not so much because you are good or TRU, but because you may give them something they want, or might do them good, make them feel better, have or may help them in some way. This is related to the saying that "if you want someone to like you ask them to do something for you". In this, as in TRU generally, love is positive and hate is negative. In business, human relationships or IPR a good deal is where both sides profit. Thereby life is mostly positive when humans deal best together by TRU, avoiding hate and fostering the love of all that is in any way TRU.

Although hate is negative and inimical to TRU, and therefore to be ignored or avoided it has some use as a test separating TRU from NAMTRU and ANTRU. For we need to clearly distinguish between those who hate us: 1. because they are wrong, or have done us wrong, 2. because we are wrong, or have done them wrong, 3. for what we have that they want, 4. for differences, as DAOST of TRU, 5. for imagined beliefs or differences, and 6. for other reasons. The cure and solution to these are different. 1. They must become right, TES and TRU, 2. We, as all, must become right, TES and TRU, 3. They, and we, must understand DAOST and TROS to provide the best possible equity by TIT, 4. address and reassess past, ongoing, and future wrongs, 5. discover, learn, remember, and live by all associated with TES and TRU, 6. do so as an individual, social, international and world policy and law that all can identify with and cooperate in. Such a basic understanding of what is wrong and what is right, the concepts, principles, laws, and mechanisms involved, and the policies and actions that go with them are what TRU, TIT-TO, POT, HOM, and IPTRU champion.

Fig 5-57 **THE POSITIVE and NEGATIVE DYNAMIC of TRU in MEDIA**

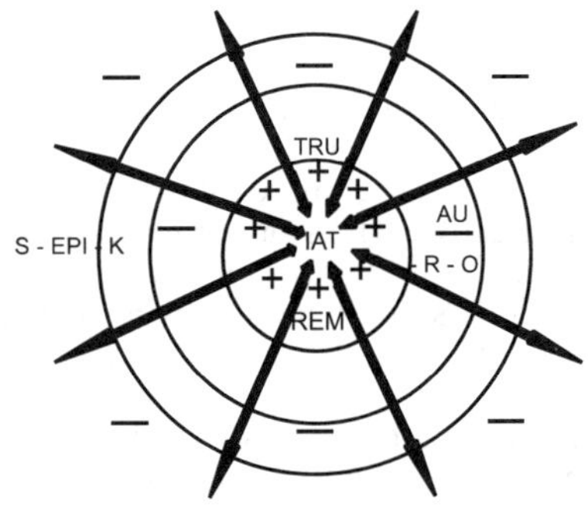

We know that, although TIT is absolute by TRIA and TÂ, it does not exist in a vacuum. It does not exist without interchange, relationships and dynamics with what is less, mixed, from, to, for, and against TRU. The crucial difference is in the latter, for TRU and against TRU. Only that in TRU, TIT, is for TRU in resultant. What is out of TRU is against TRU in resultant.(OITIO) However the degree of these will vary depends upon the part and REM of TRU effected. We also know that many interchanges, relationships and dynamics are reversible. Figure 6 is a rough showing of this state of TRU in a medium of senses, things, AU, and knowledge. The point of interchange is given by three concentric circles, signifying the various states of relative or absolute reversible or irreversible SCARL of TRU. The point of interchange is where the positive and negative interact. Since we need a periphery to circumscribe the object of TRU the point of interaction is figurative. Certain axioms of the positive and negative forces, effects, and dynamics may be proposed: I. Negative to negative, null to null is not TRU, II. What is negative and is not protected against is a 100% negative effect on TRU. This may be reversible. Its 100% negative effect is partial and not usually for all TRU, TO. III. What is negative and is protected against will produce a resultant that will depend upon the relative strengths of the negative and protection. This is salvation in Christ, TES, and TRU, as our will becomes the will of TRU. IV. That which is negative yet destroyed by TRU before it reaches TRU or damages TRU is of effect only in exercising or conditioning TRU, thereby positive. This is like Shiva, the destroyer God. It protects us by destroying what is, or is apt to be, harmful. V. That which may be negative, neutral or positive, but has a positive effect on us and produces a positive response in TRU is TIT. VI. That which is positive and produces a positive response in TRU is TIT. VII. That which is positive and produces a negative response in TRU is not TRU, is ANTRU. This is like to much of a good thing ,an over abundance, or hyper-compensation, where we are unable to respond beneficially as in overwork, confusion, and to much force, data, food, and all that is an overdose and overdone. VIII. TRU is either directly or indirectly, potentially or actively positive. IX. TRU is not generally negative to itself nor its parts and processes. It may quickly become negative depending upon the nature and form of the forces acting upon it, when it indirectly malfunctions. Parts or Processes of TRU may be negative, when in being negative they do not cause any resultant loss to TRU, only make up the differences and dialectics within TRU.

PINIT = Paradox of Infinitesimal and Infinite TRIS. PINIT is another, yet associated, great paradox in near and far time, static and dynamic. Although the near term would appear to be more stationary, on the other hand it is highly functional and thereby dynamic. Contrastingly although the far term would appear to be more variable, on the other hand it is more formative and thereby static. This helps to show the FAF of time in ALT and TIT.

There is a great paradox in all existence between TRIA, what is not, the effects of negative, I/O, and external relations of TRUS. The key to this paradox is the fact that we are speaking of more than one thing with more than one vantage. TRIA and TA are not abrogated, nor is I/O and IOT: 1. Much that is I/O is within TRU, 2. Much that is I/O that is without TRU, does not effect TRU, 3. TRIA and TÂ are modes, forms or states of TIT, 4. I/O and the negative is more of the dynamic and potential, 5. TIT is not TRU, TRINI is not TRUS, TRICH is not TRUCH, 6. I/O and the Negative are more of TRUS, TRUCH and TRU rather than TIT, TRINI or TRICH, 7. The shorter the period of time to the infinitesimal present, the more static, TRINI, TRIA and TÂ become. The longer the period of time to the infinite past and future the more dynamic TRU, TRUCH and TRUS become, 8. So that just like ITCH, TIT/TRUCH, RATCH, or Ratio of TRU change, we are dealing with a spectrum of TRU based upon the present in which more dynamic is packed into shorter periods of time. It varies from infinitesimal to infinite. It is this spectrum that so distorts our world view, and causes this great paradox TRIA and I/O, the natures of what is inside and outside of TRU.

Disease is for the most part a breakdown in a person's body or mind's capacity to resist, cope or withstand extrinsic and intrinsic abnormal forces, causes and effects. These include, among many others, toxic agents, excesses, deficiencies, pathogens, defects in form, function, metabolism and cell differentiation, and stress. They are associated with growth and decay, or aging. They tend to be accumulative.(INEG)

The goal of life must be a maximum of building up or saving and a minimum of using up or wasting. Much of happiness in life comes from living and achieving it, anticipating its improvements, successes and joys rather than the excessive savoring or glorifying its accomplishments.

. Fig 6-57
PROCESSES of FRAGMENTATION

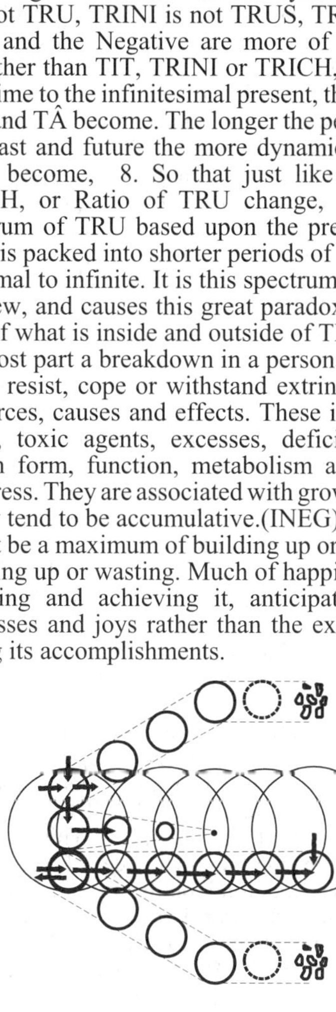

This diagram shows the fission tendencies, continuity, and balance of principles and processes of TRU demonstrated by CES in SPALT. It indicates the molding, hewing into line, adaptive and compulsory motion, velocities, accelerations, forces, changes and evolutionary processes that occur in TRU. It shows what is fusion and what is fission, integration and differentiation, cohesion and fragmentation. It demonstrates how ES and CES function in TRU, in and of themselves and in various combinations. It indicates the effects of going it alone, as so prevalent in senses, things, and individuals as they are or are not a part of TRU. The seven deadly sins are: anger, covetousness, envy, gluttony, lechery, pride, and sloth. These have changed little through the ages. They have been added to or subtracted from, or combined in many ways by different sources, indicating that they are neither all, nor always the worst. They are all obviously negative and components of ATRU and ANTRU. They are depicted on the wall of the Arena Chapel in Padua, by Giotto, the father of early Renaissance painting. Like Dante and many in history there was a great need, desire, and will to distinguish right and wrong. The clearer and more comprehensive the representation of rights and wrongs are the more it aids the life of their audience. It finds closure in RECTRU, DYNO, ACC and ELTICMA.

Religions have a way of becoming self perpetuating. This is not without reason, but not necessarily, largely, nor rightly self sustaining. Only this is right when TREQ or TRU. **LOVI** = the Law of Validity or Invalidity of ISMS. LOVI is demonstrated by how all is attributed to one quality, extreme, polarity or system. This commonly uses or extends the identifier '*is*' to mean equal, as in the goal of life is,' happiness or pleasure'. Few deny the greatness of happiness and pleasure. Yet it's wrong and a serious fallacy to not distinguish what, where, when, why, and how this applies. They are no more nor less than their part and role in TIT.

The will of God, TREQ, or TRU, is primarily and greatly dependent upon, and similar to, the good will of each and all persons. Yet they are not identical. **WOT** = Will of TRU. In the positive exercise of will is the functional power of TRU. It is this Will to Be, to be TRU that is the crux of all, all TRU. For it not only generates hope, happiness, and success of its part, it tends to be catching and stimulates the will of others and TRU. If the world is to be TRU, TRID and TÂ, then we must will it to be so. There is a tremendous will existing in and for TREQ or TRU. Directly it is in each person, indirectly it exists in all the rest of TRU, culturally, socially, institutionally, and politically. It occurs in all people simultaneously, more or less. Collectively and totally it is one of the greatest powers of TRU.

Fig 7-57
RIGHT and WRONG JUDGEMENTS of the GOOD and the BAD

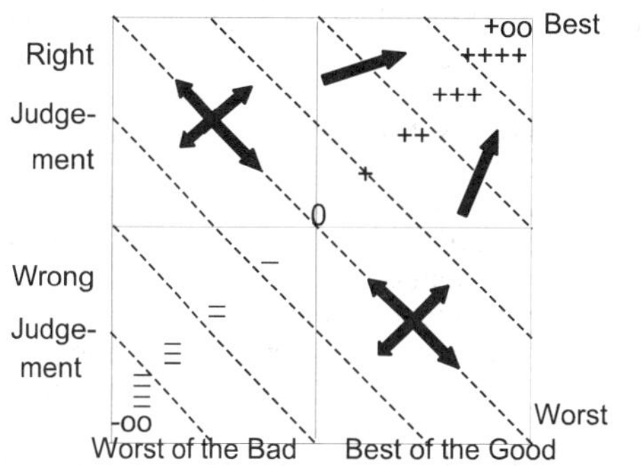

Most judgements are not to be accepted as fact, or right. It is only the best or right judgement by TRU that is acceptable. It is the frailty of humans combined with false pride and vanity that so destroys us. Humans accept this frailty in themselves or that close to them, yet to often demand of the world and others perfection. This is done by making wrong or inadequate judgments. We can overcome frailty, pride, and vanity through VAPO KOT, TES, and TRU. Most of our actions are determined by judgements that are based upon the information we feel is appropriate for the circumstances. Our actions can be no better than our judgement, information, knowledge, and degree it is TRU. It is in the best of the best that TRU is achieved and all succeed.

CHAPTER 58

ENABLEMENT and ENHANCEMENT of HOM, CES and CHES ETHICS, its STATES, LIMITS and LAWS

RELATIVITY, RESPONSIBILITY, PURPOSES, GOALS, RECONCILIATION, SUCCESS/FAILURE BEST HARMONY ORDER CONTROL FULFILLMENT ULTIMATES

The enhancement of HOM refers both to HOM as it exists in TIT, and how it by CES and CHES are methods of TIT. It focuses on what helps us best learn, posit, know, show and use CES. By so doing we can expand and elevate our lives from simple responses and problem solving to a totally ordered accounting, reconciliation and resolution of all that needs attention and achievement. What ethics has not effectively enlisted is the Conscience of all persons. There has been, throughout the ages, but most noticeable in the present, a world wide Demise of Conscience. It was never for long, in any age, a highly-developed human characteristic and power. It requires great knowledge of TREQ or TRU. However our world in becoming more and more complex can not sustain itself unless each and all have a strong TRU conscience based upon the common good, KOT, and will for TRU. It is also why they have not developed any better than they have. This is because God and religion have not been adequately developed nor tolerant.(VAPO) Further this is why TES, TRU, POT and IPTRU are so needed and important.

Ethics is that part of philosophy and religion which deals with questions concerning the source, nature and performance of goodness, rightness, value, duty and related matters. It is the great part of philosophy and TRU that gives meaning, action, and puts into practice the CES of TRU, HOM. Ethical patterns of behavior establish what is important to man. So likewise does this text, as all philosophy, converge in final unity on mans great knowledge, intellect, understanding, COGNOS, ELTICMA, SA, and TÂ. These are attained in the individual through the development of PUREACLITY and TES. PUREACLITY is that state of the individual, as TES, the prime component of TRU, in which one learns to think and act in terms of the formative and functional, pure, actual and real, parts and processes of TRU.

Morality may be divided into moral, immoral, and amoral. Amoral is what is neither moral nor immoral. Immoral is what is not or against moral, as ATRU, ANTRU and ANTITRU are to TRU. Even though obvious, it is important to make these distinctions. For unless we can understand what is and what is not TRU, what our acts and behavior can and can not accomplish in TRU, we can not know what is moral nor much of what TRU is. For life, and especially what can only be, as it is TRU, is first and foremost necessary, imperative, compliant and highly demanding by the musts and should's that are its FAF. This is by its relativity, and the continuously repeated sameness of much of TRU, or DAOST.

Eudaemonism is the theory that acts are right insofar as they contribute to mans well-being or happiness. The emphasis in eudaemonism is not on pleasure, as in hedonistic ethics, but upon the way of life most suited to man's nature. This ISM has a broad base of validity and tends to TRU. Yet as all ISMS it is wrong in what it leaves out, what its focus is other than TRU, and how far it does not go. It is wrong in the over interpretation of what it implies. For this is not the sole basis for acts and the judgement of value. Nor is it, without full comprehension and correction, likely to be TRU. British Utilitarianism established a functional CES and teleological aspect of TRU. Bentham found happiness through the elimination of fictions and upholding the common good. So did Mills complete this trend of thought by expounding the principle of "the greatest good for the greatest number" or the core idea of Utilitarianism. This, also, is oversimplification.

The strategy of sweeping everything before you, in one form or another, has been the pride and joy of all mankind, the fulfillment of dominant instincts. Yet it should be used only within limits when it is for the right reasons and best purposes, or TRU. It is not, nor can't be, an end in itself. It is rarely the end one seeks or as satisfying and successful as it seems, unless to the degree it is of TES and TIT-TO. On the other hand activism when best directed is key to ACC, TES and TRU. In **POMOR** or the Principle of Magnified Breakthroughs, there is the collective method of divide and conquer, or successively breaking work into its most effective steps. They are planned and resolved by concentrated onslaught, or by the indirect and massive means available when TRU is learned, known, represented and best achieved.

The whole is beautiful when set to music, film or represented systematically.(VAPO) These help to fill the gaps or voids of life. It is what our IA misses that is not readily evident, of TRU. Each person, more or less, has a great inner sense or suspicion of this dynamic totality as it comes together in the givens and experiences of life. It reveals religion, beginnings of TROS, and the spiritual, mystical, and inspirational origin of TES and TRU. When revealed in story or song, especially when beautifully and vividly done, the whole of TO can move one to the greatest heights of identity with ones being. This is the glory of art, and especially music, identity with TRU, TROS and TÂ. It is how representation - as provided by SPALT, POS, HOM and ENT - can provide means of comprehension and appreciation of TRU. These are the richness and treasures ACEXIN, MOT and HOM provide.

To be civil means being socially and politically correct by TRU, functioning normally and properly as a member of a group, society, state, or world. It also means, to be tolerant, patient, proportionate, moderate, considerate, understanding and abiding by those customs and laws that are TREQ or TRU. Civics and barbarity are not entirely opposite, they are often, in ways, closer than they may appear. Even the competition within TRU can be barbaric to some ES and CES, and is a large part acquiring TES. It requires a full amount of higher energy levels that make up the FAF of TRU. Civics and ethics are closely related, especially when they are in TRU. Yet, as DAOST, they are also separate and distinct, both by their differences in TRU as well as their vantage. Civics relates to one's behavior, adapting to law, and performing duties within the form of government, local to universal, of TRU. Ethics relates to one's behavior with others, how to get along and help society be harmonious and successful, as all are participants and fulfill their roles in GOD, TREQ, or TRU.

The ideal ethic is that in which each and every individual, self, or person completely identifies with the ESINS, ES, CES, and TES in all others, all TRU. This is called: **COME** = Complementation or Completion of Mutual ESINS or ES. When we look at others in an unfriendly way, as our enemies, or when we despise much about them we are only showing our own ignorance and lack of ESIN, TES and TRU. We are showing at best our partiality. This is evident in Christ's teaching, turning the other cheek. There is much power in the summative and total public opinion, condemnation, moral judgement of the world around us. This is no better than the degree to which it is positive for TRU, based upon HOM, KOT, and ACC. **CADOME** = Condemnation and Approbation Differential leading to COME.

LIT = Limits of Ethics. Even if the rule, the right way, is 100% perfect it is impossible to establish an inflexible, unlimited and exceptionally free standard for ethics. Just as it is impossible to establish an absolute single standard for SPO, SPOT, ALT or TRU. By LIT we are dealing with several variables, as the age, needs, circumstances, changes and individual's partiality. This is how they vary with each and all TRUS, TES, ES, or cases at hand. Ethics as all, right or wrong, by being in contrast with all else, COL, are limited from TRU This is by all that is less than TIT-TO. The mixed state of ethics is given in: **LARIV** = the Law of Absolute Voluntary or Relative Choice. LARIV is the imperative absolute, voluntary relative to partial FAF of ethics. Choice and will are shown in the derivation of crime and punishment.

The most outstanding author of ethics, in its derivation from God, was Spinoza. His great work, *Ethics,* is a testimony to the undeniable convergent dynamic bonds that exist between ones acts and God. It is not surprising that in such a period, of the thirty-years war, great religious fervor, and search for reason and truth, a mind bred in the knowledge of Jewish Religion, law and philosophy and reared in the knowledge of Christian, reformed, philosophical, and liberal learning could have developed. In fact it was from the very apogee of the Golden Age of Dutch dominance and culture, that the kernel of the American puritanical spirit developed with such stubborn adherence to what is right. Spinoza's ethics proceeded axiomatically from the principles, determination and resolution of metaphysics, systematic unity, God and TRU. It is based upon an expanded knowledge of God as TRU. It is easy enough to say that we must be good because God is Good, because we love God, or because God is all. Yet to be verifiable and convincing we should learn and know the what, when, where, why, and how, or reasons GOD is TREQ is TRU. That this should have followed in a certain way - an *Imitation of Christ* - should be no surprise to the cultural background of Spinoza. The greater Ethics by TRU, as compared with ethics in common usage is the product of that system that puts in full capitals all terms and words that are their equivalents in TRU. It applies to not only what a person should do, but all that can be done by all means, indirectly and manifoldly, that enables the success and continuity of TRU. Such a magnificent ethic applies to all those means that are directed to a successful being, ENT, TRU, TIT and TÂ.

INDETHIC is the principle or law of In Depth Ethics. It seems that we can change all things, or better, perfect whatever our acts or deeds effect. Yet in life we are usually barely effective in altering the least finite of TRU. We can envision an indirect effect, influence, control and understanding of all in such a way that 100% of TRU operations are resolved. For it is not so much in believing what we are doing is right as to actually and correctly, by VAPO, know and achieve what is TRU. If we know what, why, how and all about doing right by TREQ or TRU with complete management and control we can approach INDETHIC. The laws of ethics and rules of conduct must be based upon and guided by VAPO, TROS, TRIA, TRUD, ENT and ACEXIN.

To concentrate on the object and aim straight at it are what all carnivors, as much of life, learns and does to stay alive, survive, and assure survival. Many levels of SÁ participate in this. The higher human mentality or levels of SÁ increases concentration on, and the use of, TRU. Success becomes much easier and complete. This is how each person becomes a part of TRU and fulfills their counterpart, TES. When all core forces, major courses and processes are developed, analyzed, selected, and posited in SPALT we can fulfill such a role for HOM in TRU.

One can not be TRU unless one knows TRU. One can not know TRU unless TRU is discovered, developed, determined, promoted and utilized. Not being bad, or to avoid it, are what most of us, more or less, seek. Yet to momentarily be what one assumes to be, or tries to make the highest good, is far different than continuously being this highest good as it is actually TRU.

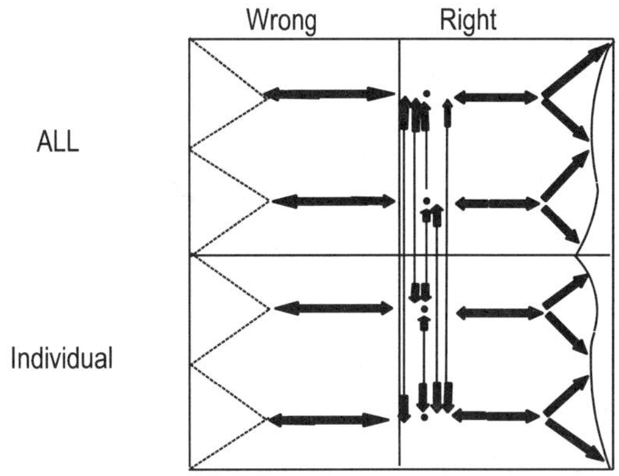

Fig 1-58 **FOUR STATES of "ONE FOR ALL and ALL FOR ONE"**,

As correct and right as it seems, this is not necessarily so. How the ideal, what must be TRU, can only be how the individual and all people function in perfect harmony and success by TRU. The order of these four categories can be taken in any way. Yet in this diagram are so ordered to best superimpose on ALT. They can be expressed more quantitatively by degrees from all to one person and from most wrong to most right. The general principles, of all for one and one for all, are only correct when completely right and of the proper balance between one and all, in TRU.

One can't live defensively as a matter of course or expediency. Ones actions can't be adequate or successful if left to their own devices or to the last stimulus. The stimulus response state is integrated and massive. It tends to TRU, our life as it is TRU. The main problem is that this tendency is not assured, it is often riddled with distractions, sidetracks, trivia and even negatives. To be TRU all our motion, movements, and actions, be they of any energy level, must be attuned to, in proportion to, and in synchrony with TRU. This is our relativity, the dynamic role of our predominant FAF, Á continuum in TRU. To this extent the best defense is a good offense. To offensively and decisively, plan, act, endeavor for and enhance TRU, TE, TI, TU, TO, TA and to offensively prevent protect support and reduce all that is less than or against TRU. This necessitates detailed KOT which is best supplied by TRU, HOM, DYNU, DYNS, DYNO, and ACC.

An individual, as well as groups and all, need to have a guiding star, clear goals, priorities, and bring all forces necessary by TES and TRU to bear on them. The importance of definite goals and great purposes in life can't be underestimated. Yet this should be more than a fantasy, dream, a youthful ideal, simple act, or even a well thought out plan. It must, and can only be, as this is equal to that which is the greatest, all of TRU.(REM) It is how these fit into the future of the individual as TES or TIT.

TRU is systematically unified, formed, well ordered and can be standardized to provide great rewards for effective representation, presentation, and use. Its guide to success should consider whatever limits, proportions and other factors make up the whole of TRU. It should be measured using any and all actual TIT differences, distinctions, cleavages, polarities, spectrums, continua, levels, scales, and gradients that can be adapted to practical application, and used as guides for the success, S/F of TRU. Success comes from simplicity, restraint or limits, order, and power especially when based upon the healthy and predominant ACEXIN of TRU.

LAPE = Law of Perfected Ethic. Only that ethic, system or law will provide sufficient or completely satisfactory behavior, means of action, or accentuation that encompasses, 1. All TRU needs, 2. All TRU validly, 3. All TRU actually, 4. All TRU proportionately, 5. All TRU totally, and these by VAPO. 6. All TRU formatively, 7. All TRU functionally, 8. All TRU with sufficiently adequate practical finite detail.

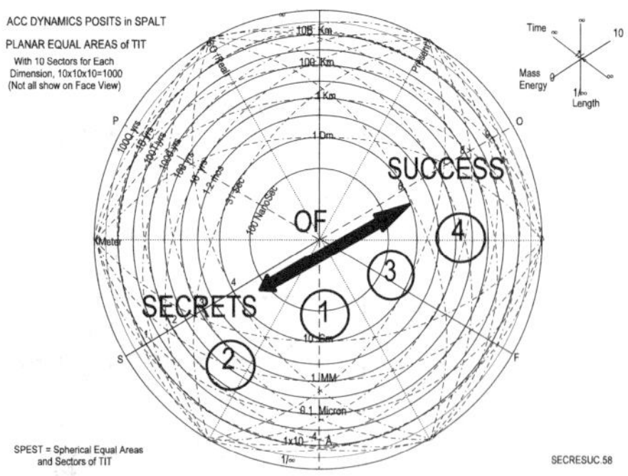

Fig 2-58 **PROCESSES SHOWING the PRINCIPLES and SECRETS of SUCCESS in TRU**

There are several main secrets to success. Some of these include, 1. Get at it. To a great extent, what needs to be done, only needs to be attacked to be accomplished, 2. Know all the surroundings and pertaining facts, factors, principles, needs, routines and follow through with sufficient power, 3. Do no more nor less than necessary for TRU. Do only what is right for TRU and the occasion as it fits TRU. 4. Always act in ways that not only solve any problem but best solves it for TRU. **PONMOS** = Principle of the Narrow Margin of Success. Success depends on the characteristics of the Level. A scaling up or down of operations in any institution, business, life situation, administration is often necessary.

Fig 3-58 **STIMULUS-RESPONSE, ORDER CONTROL ETHICS THEISM in TRU**

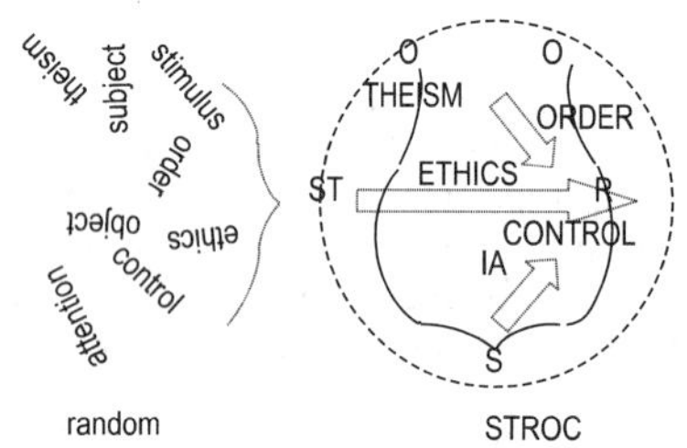

Theism and ethics are largely the ordering and control of stimuli - response in the social and whole human world, TIT-TO. In this figure they are combined in the overall FAF of S-O and life of TRU. This is: **STROC** = Stimulus- TRU - Response equals Order and Control or when stated otherwise, stimulus and response combined with Subject Object in TRU are the order and control necessary. This is evident from the Positive Imperative, **PI**, and the Categorical Imperative. In the latter, where the maxim of personal will is raised into a principle of universal law, the relationship between ethics and order are shown. In **PI** of TRU, 'we must' is where the overall relationship between stimulus and response, subject and object, order and control, ethics and theism are, by TRU, continually brought into their ES and CES.

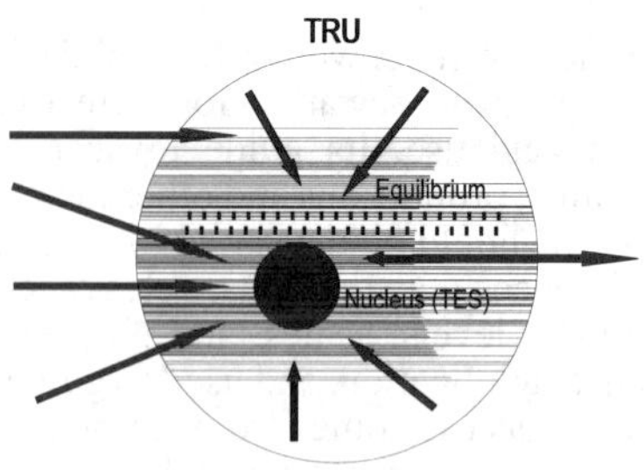

Fig 4-58 **EQUILIBRIUM, PEACE and CURES in TRU and HOM**

The cure for awesome threats to TRU often lie in places and phases other than what is acute, given, or commonly known. Short term cures are apt to be far from the S/F needed for the greatest needs and totality of TRU. This is a large part of S-R, O-S-O, S-O, Á, and SÁ FAF of TRU. There is a great equilibrium in TRU along with the strength, will, desire to win, mutual benefit and many other values and powers that provide a cure for all that threatens TRU. Coexistence, although a cure for some problems of TRU can be defeating and productive of reaction. Peace, is that state of harmony and symphony in the world, TRU, with a minimum of conflict, war and destruction, and a maximum of fulfillment and happiness. It is how all that is in TRU to be in complete harmony and symphony needs to function perfectly, massively, proportionately and smoothly. There may be great dialectic opposition within TRU, especially between parts, processes, and polarities like the extremes of Á, L, T or S and O, large and small, past and future, stimulus and response, action, reaction and inaction. Yet for TIT-TO, its dynamics must be in a Steady State of dynamic equilibrium and increasing permanence by TRIA to TÂ. You can't diagnose diseases, conditions, troubles and problems, unless you are familiar with their various kinds, and continuously maintain a high level of suspicion for their occurrence. This applies to what is common and readily apparent as well as the uncommon and most subtle. The greater the body and quality of TRU knowledge at hand the better one knows in advance the problems, their conditions, SCARLS, and what to do.

It is in the mixture and coexistence of the best of all extremes and means of TRU that greatest good, pleasure, benefit and success is to be attained. This is called: **OMEX** = the Optimal Mixture of Extremes. OMEX is more than the golden mean, it is The kind of mixture of polarities, actions, reactions and extremes that are best for TRU. OMEX is not only for extremes, but means and extremes and all TRU in one way, but will vary by the part of TRU at hand or given. Sex is a prime example of OMEX. It is, as all in TRU, neither sin nor the ultimate. It, when optimal is a fluctuating means, extremes and major finite component fulfilling its role in TIT.

The new for the old, is the constant regeneration and rejuvenation cycles of nature that are typical of life and the phasic form of TRU and CES. Is it life or is it fitness to life that is most important? Even like diseases and negatives they need management, control, and cure. To be alive is hardly as important as it is to be fit to be alive. It is to be fit to be alive as TRU that is the greatest of all. This is how preparation and development make TES and TRU. .

Fig 5-58 **Levels from TÂ Unity to ÂÂ Infinity**

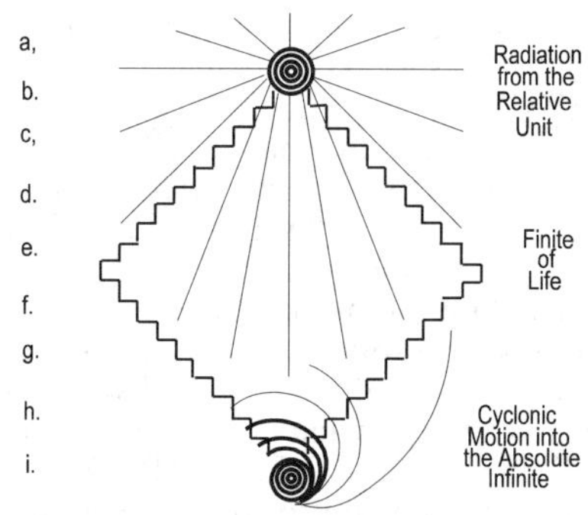

Figure 5-58 shows the relationship between the total shining forth light of TÂ, TRU and TRUID, all radiation and steps away from and toward the nothing and negative extremes. It shows the range of life, and TRU, between these two extremes, including: a. Shining white light of relative/absolute, perfect unity, b. Initial divisions, c. Gradations or levels from, d. Increasing finite, e. Optimal Finite, TROS of TIT, f. The hyper-optimal finite, g. The infinitesimal differential, h. The chaotic infinitesimal, i. The absence of light, black hole of absolute/relative. For TRU we are drawn in all directions, but maintain ourselves in the optimally differentiated and unified relative finite by TROS and ACEXIN, our equilibrium, homeostasis of HOM and TIT.

Eschatology is the theological study of such final matters as death, half lives, mortality, divine judgement and the end of the world. This is a topic that has dramatic and often erroneous representations and implications. It is close to the hearts of those who are concerned with what is going to happen to them, theirs, and their world. Developing a rigorous, scientific, valid and complete view of the future comes with more information, VAPO, KOT, TRU, SPALT, POS, HOM and ACC.

CHAPTER 59

SUMMARY, USES and TIE-INS of HOM

QUESTIONS, PROBLEMS, ANSWERS, SOLUTIONS, CORRECTIONS, STANDARDS, MEASURES, TESTS, RULES, LAWS, MANAGEMENT, SEQUENCE, ACC, ROT, and POT

Knowledge of HOM involves practical questions and problems, as of relevance and duty. These like existentialism, as determining factors in success, imply choices, decisions, and a recognition of our part, TES, along with related ES, and CES, parts and processes in success, life, and survival. Knowledge of HOM and its CHES provide much greater forecasting, predictions, and solutions to future problems. Much of any attempt to foretell the future depends upon the method, SCARLS, principles, and detail that are used. Forecasting can be developed into an ever more accurate and valid science if we do it properly and utilize all that TRU and its POT and ROT offer. We must emphasize what is certain and attempt to predict not the extreme differentia of TRU, nor specific events, places, persons, but to forecast generally what will and must be as it is TRU. For much of the failure, and pejorative connotation, that prediction, prophecy and forecasting have had, is a result of: 1. Attempts to predict what is not TRU, 2. Attempts to predict what is of the extreme differential of TRU rather than its essentials, systematic unity and totality, 3. Predictions made by observers other than from TES and THES in TRU, 4. Failure to apply VAPO to that predicted, 5. Failure to take a POA to TRU and make predictions based upon accurate posits. Yet we know, even from simple experience and confining its limits, that forecasting can be highly successful. Therefore we seek to expand on this and face the questions and problems of the future not with dismay and disbelief, but with confidence in knowledge of TRU and the future revealed with inspiration, revelation, faith, and devotion to all TES and TRU.

The degree of validity that can be determined by intuitive reflection applied to moral obligation relative to formal principles of conduct is known as Formalism. This is true or false, right or wrong as the intuitive reflection and the formal principles of conduct are true or false, right or wrong. It is vitally necessary that these are VAPO of TRU. Here is our challenge and that upon which we concentrate much of our effort. Thereby we seek to discover, derive, develop and determine the principles of HOM and TRU and how they can be presented. This best of human life and world, is when TRU, composed of ACEXIN, realized as vast, actual, extensive, and intensive. It is far more than the behavior and values of individual or even group action. A person to be ones counterpart in TRU, TES, is not so much an ethical and moral duty as it is vital to ones very being. It is more a benefit than a duty. It is here that much of the onerous stress of ethics, morals, duty and responsibility can be avoided. We do and work because it is a necessary, beneficial and an enjoyable part of fulfilling a glorious life, not because it is forced upon us for lesser reasons or powers. Ones choice lies in how, when and where, yet the resultant end is an Absolute that is a simple fulfillment of our part and role in TRU.

To gain the necessary increments, increasing powers and needs of TRU there will be many actions, reactions, solutions, and resolutions. These may be both preventative and active, yet must be properly gauged and corrected by TRU. They may not always be at the time, nor near the time, initially occurring. They will occur "in time". This is called: **SREC** = Solution Resolution Correction, by CES or processes. SREC is a special case of reaction. As with all action there is a reaction. The reaction occurs according to certain laws, principles, and SCARLS of what is involved. This may be in or out of TRU and especially according to dynamic relationships.

The processes of man, the dynamics of TRU, as well as all TRU can not be understood by every point, posit, process or CES singularly. Nor can they be understood solely by abstract generalities and principles. Only by the synchronous correlation and formulation of ever broadening facts, evidence, points and processes, and more precise rationale of universals, forms, ideas, generalities, principles, concepts, laws and mechanisms that are integrated and differentiated by TRU, can its understanding and demonstration be made. This is what we do in HOM, DYNU, DYNS, and DYNO, provide what is needed for ACC, TES and TIT. Knowledge is power, as S-O, the Á dimension and continuum is the predominant FAF and all powerful mass energy of TRU. It must and can only be complete, as TRU. Yet it must be accompanied by all that is TRU, action, right action, love of life as TRU. It is in such integration and differentiation that is of and in TRU, OITIO, that IPTRU is made and directed.

To believe and allow human life to be abused in all sorts of ways, only to be replaced by something different, is the worst, yet a common, kind of thinking. What there is, as TRU, is what we have, we can only preserve, support and build upon. There is a continual and perfected becoming that assures TRU. This is largely already in TIT, not any thing that is new. The latter half of this text is largely like a manual telling us what these operations are, how they work and how we can best repair, maintain, and fix them. Yet the degree to which we fix it depends upon the degree to which the given, occasion or problem at hand and TRU are ordered, have POA, and present themselves. It is for this reason such a profound and subtle analysis, demonstration, and learning of TRU is necessary.

Never let anything in life that is of the greatest importance get ahead of you. Never let what may cause great harm get into a position that can cause, or let it increase this potential for, harm. These are basic reasons for and determinants of behavior, IAT, TES and TRU. They are basic lessons of life, crucial for human reaction and adaptation. The accumulative care for and of TRU has provided the golden ages of most civilizations. It is in the great build up of the highest quality of live product, genetic material and determining forces through time that provides their base.

Fig 1-59
STIMULUS RESPONSE O-S-O, LAW and TRU

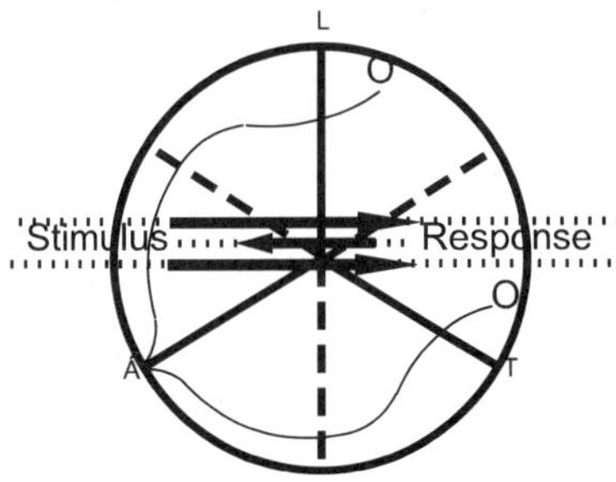

SRET = Stimulus Response is Equivalent to TRU. SRET is a large proof of HOM and TRU. Grotius, in the *Law of Peace and War*, constructed a system or model in which natural and historical law were distinguished. The principle of natural morality is not synthetic, artificial, nor to be made up arbitrarily. It, as all standards to be valid and viable, must be based upon the fundamental principles, ultimate causes, determinants, and the relationships between all in TRU. These are to be ascertained by reason. Humans largely determine historical law. However, this is not free from chance or even negative determination. It likewise must ultimately be established by the supreme state and good of natural law. Freedom of the will is the indispensable condition of all morality. Moral consciousness is a result of three basic principles of TRU: 1. Freedom of the will within TES and TIT, 2. The immortality of one's soul or TES, and TRU, 3. The existing supreme being, GOD as TREQ or TRU. Natural law is often much different than what it seems to the individual, knowledge, legal decrees or even science. For it must conform to the laws of God as TREQ or TRU. The license for making and conducting war can be granted only to the degree they conform to the laws of nature, in TRU. No arbitrary power or radical view can change, alter or propound its laws without adherence to the laws or CALP of TRU. The management of any problem can often be achieved by breaking it down into various categories and causes, in order to set up a plan to prevent, eliminate, or solve them. By SRET problems of: categories of TRU, it's the-ness or being, thisness or haecceity, predominant nature, form, structure, and solving the most important and greatest of all problems, are guided by the appropriate CALP and CHES to their best solutions.

CHASNAL = Changing Specificity of Half Lives. In the PPF of the time dimension of TRU, what the present contains of the past and future becomes less specific and more and more general. Any object or subject, especially if of higher REM, must be considered in several respects. CHASNAL is a great solution to the problem of future prediction and the science of forecasting in TRU. It is when we provide SPALT, POS and HOM, and best apply ES, CES, and CHES that we can through the same focusing on TRU achieve better ways and means to learn, know, and show TIT.

LIS = Law of Irreconcilable Standards. One of the greatest paradoxes, conflicts, defects and terrifically confusing aspects of our current orientation and ordination is to look at the future as ideal. Even without 'hope springs eternal', we most often anticipate a better future. Yet, this is far from assured. Even more if we are not TES, and the world not TRU, it is guaranteed to be the opposite. We tend to look at the present as the real, actual, and best, and to the past as chaotic and future fantastic. This has some value in orienting us toward an objective and TRUD. Yet it destroys our ability to look at the past, present and future as they actually exist in TRU. It is no better than how we use many of our standards. This is to the degree we fail to learn and remember their SCARLS and correct them accordingly. Our world is built of many such standards. Of these many are defective. For this reason the term standard becomes degraded. For standards not only 'set their own standard', they are highly necessary for TRU. Yet our idea, conception and view of the world is warped, incomplete, erroneous and confused, partly because TIT physically seems altered, adapted and warped .

POMP = Programmed Planned Management of Problems. POMP is derived from and similar to medical planned management of the illnesses. How major illnesses and problems can be managed through an algorithm of planned, systematic and programmed controls. It shows how physicians achieve the needed acumen in planing, managing, and controling various illnesses and health problems. It is a method and approach that modern science is best at in solving problems through models and systems that utilize POA and the Á FAF of TRU. (SOAP)

<u>Vicissitude</u> = Changing phase or condition, ups and downs. Everyone exists, more or less, in a continuous vicissitude. There are ups and downs, always changing conditions of a never ending array of forces and effects, actions and reactions, both in and in and out of TRU, upon us. It is for us to discover, determine, manage and control these changing conditions, phases, ups and downs as much as possible and to do so to the most ordered and greatest extent compatible with the exponential form and success of TRU. This is why it is so necessary that TRU be known and we be TRU, TES.

TOT = Test or Tests of TRU. To look back in history and take any point in time, as TRU for humans, can be very helpful in learning both what the world was like then, what TRU actually is now, and what TRU will be. This we discover by including those ESINS of the past which continue into the present and future. We must be careful, however, to avoid taking any information or knowledge at face value, as it exists ENIT or even ESIN. For as great or interesting as it may seem it is likely to be far different in TIT. This happens when we look at TRUP or TRUF, we distort the image of TIT, the only correct observed. This is especially so in ones life, current events and history. For we assume all is real, actual and valid as it enters our attention and knowledge. Yet here is the great fallacy. **FIN** = Fallacy of Information. FIN is that fallacy of all that comes to our attention that is not or other than TRU. It may be 99% fallacy and 1% correct actuality in TRU, or the reverse, and even of negative value, or their mixture. To live by any information and knowledge is not enough. We must learn to analyze, interpret, and adapt information and knowledge to TRU, in all its aspects, parts and processes, as well as VAPO and SCARLS. This applies to history, current events, the future, oneself, all fields, sciences, religions, and philosophies. This is why our subjectivity must be selective, and through its powers of Á, sort out what is and what is not TRU. To be able to, and to make, these distinctions in handling information is a prime test of TRU, TOT.

PORAB = Principle of Ramified Acceleration of Breakthroughs. PORAB is the principle and means whereby discoveries and inventions in turn produce many more. The breakthroughs in knowledge and their uses opens up whole new vistas, regions, and fields of unknowns, forgottens, or poorly knowns of the ACEXIN of TRU.

An adage, attributed to Will Rogers, that is highly revealing of the processes of man is "When you get something for nothing you just haven't been billed for it yet." This illustrates the fact that life is precious, all costs you, as well as, is an opportunity for profit. All you get must be worth a lot, and therefore a proportionate degree of sacrifice, energy and cost are necessary to attain it. All in life is like a massive complex equation, a great equilibrium of positive and negative forces working for and against TRU, and TES. All has a value which is sufficient, or is in a ratio to the cost, energy, sacrifice and loss expended. Those, who live without making every moment, thought, act, and sacrifice count, do so by later having to pay. For that much of TRU is depreciated and that much of their own being lost. This loss when deeply experienced and compensated for has some added positive value.(St. Augustine) For it helps us learn to appreciate life and the value of God as TRU, both quantitatively and qualitatively.

To live without TRU is like trying to do something without seeing what one is doing. This is how Á works both in, and of TRU, **OITIO**. To live optimally, as one is naturally inclined, tends to TRU. This is equivalent to using ones FAF of sensations, perceptions, apperceptions, feeling and learning optimally for each occasion or aspect of life, natural for its success in TRU. It is by reason of the negative that the simplest responses to stimuli when highly necessary are apt to be optimal. The more complicated and of higher levels of feeling, cognition, reason, learning and intellect the more fragile and less than optimal they become.

By **LOTAN**, the Law of Absolute Necessity, we must account for and abide by the SCARLS we find ourselves in. This will depend upon who "we" are and what the SCARLS are for each. The rationale can only proceed from totality, total being and the individual as a part of this whole, TES in TIT-TO. What may be a process or power in one sense or way, may be a problem, vehicle, mechanism, interaction or solution in another. Just as what may be a law or principle in one sense or way, or to one person on occasion, may be in another a fallacy, paradox, or problem. So that we must determine which is best by TRU and for the occasion where a great amount of description and detail are necessary in formulating policy, line of action, and emphasizing priority. They are scheduled or set as routines by best conforming to TRU.

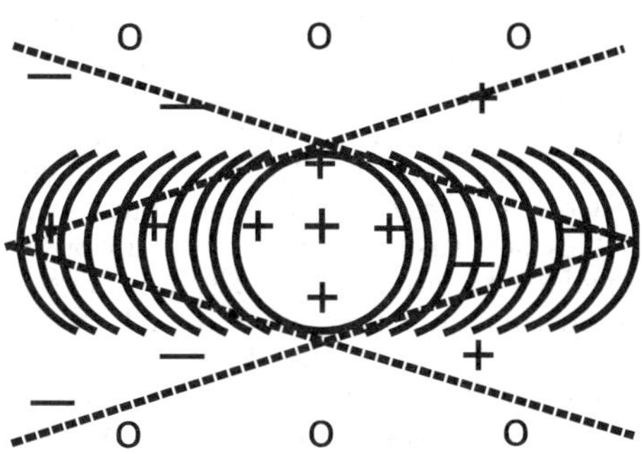

Fig 2-59 **Form of TRU and HOM in Media**

The key to TIT and OITIO is relation. For by TÂ, TÄ, and TÁ all that is relative is in its appropriate state within TRU. So that we must always be careful to distinguish what is in TRU, TIT, from that which might be thought of as being in some sort of relationship between TRU and all that is less or other than TRU. Many of the processes and principles we use, like ESINS, are not entirely confined to TRU only. They are the same in general or universally. What exists in both TRU and other than TRU is **ÂT**. Universals, and ESINS to a somewhat less extent, are highly ÂT. If we do not explain and know the context in which they are used much misinterpretation follows. This is the case of many terms, e.g. god, God, GOD, GET, TREQ and TRU. In the above diagram the symbols stand for: + = Positive or beneficial toward TRU, **-** = Negative or harmful to TRU, **o** = Neutral or of no effect on TRU. Accurately, and in much of our lives of AMTRU, all is neither absolutely positive, negative or neutral, but gradations and variations of each. It is mostly in movement toward the greatest positive or away from negative that is TRU. This diagram is a simple example of what can be attained. To show the great media in which TRU exists in all that is NAMTRU, ATRU and AU is even complex, yet unnecessary except by contrast to MOT. (SPOT, METAD, SPAT.and POMOT)

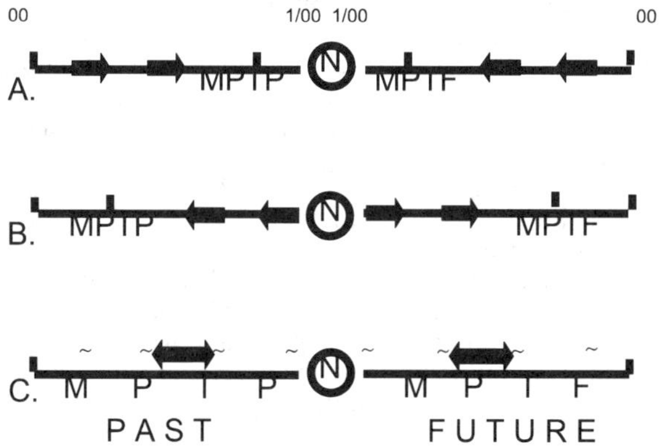

Fig 3-59 **MAN POINT TIME, FOCUS, SE-QUENCE and IMPLI-CATIONS**

By **MPT** or Man Point Time, as, **MPTP** or MTP past increases and **MPTF** or MTF future decreases as we move in the direction of extinction. If MTP decreases and MTF increases we move in the direction of infinity, or eternal existence. There is a set equilibrium between MTP and MTF. This does not mean they are at all times equal. Yet by the laws of symmetry this is a natural FAF, form, state, tendency, or inclination. These are shown in the accompanying figure in which: A. Time changes as man approaches extinction, B. Time changes as man approaches the absolute, C. Time changes with human life's variability and indefiniteness. As humans or TRU approach extinction all converges on the N, now or present. As we approach the absolute, as in TÂ, all expands, both N or now, MTP and MTF toward the absolute. As we are variable and indefinite these changes are variable and indefinite so that it depends upon the many factors, forces, processes in, or potentially in, operation. This is an example of: **PRIT** = Reversible Proofs in IPTRU. PRIT is shown by how logic proves ontology and ontology proves logic. It is shown by how ethics proves epistemology and epistemology proves ethics. It helps to show how philosophy proves religion and religion proves philosophy. It is how all great categories, topics, subjects, and aspects of TRU tend to prove each other and prove TRU, as they are properly held to their SCARLS and interactions with each other in or of TRU. In order to survive, it can't be left to chance. It can't be in disorder. It can't be in confusion or chaos. It must be ordered, cosmotic, controlled, by us, as TES-TIT in command. We must be capable, able, predominant, execute all indicated, act, and achieve TRU success. We must be able to prevent all possible harm. The basic requirement, key nature of the S-O relationship, is that it functions, as our CNS, coordinately, and massively, for optimal success. The necessary amount of knowledge, reason, learning, logic and comprehension must already be in place to give proper direction, VAPO and SCARLS to this command, management and control.

The World, TRU to succeed, moves simultaneously in the direction of increasing intensity of the subjective, increasing form of the integral, increasing identity of the focal and partial, extension of differentiation toward the infinite and increasing inherence of the objective, the objects within TRU. If it moves toward the disproportionate, its effects are nullified. This is called: **LAST** = Law of Simultaneity - and symmetry - of TRU. As we approach the unity of integral, the infinitesimal of differential, the perfect intensity of the subjective and the complete inherence of the objective and all in complete proportion in totality, we approach the absolute. TRU becomes or reaches eternally toward an indefinite limit. Yet this is a relative limit or end, relative to us at point A, here and now. For as these change, the point A changes to point B, and so to infinity as in TRUCH and TRUS. It is relative to us, our capacity, ability, R/S, C, and S/F. So that we have a relationship or ratio between the rate at which LAST increasingly approaches the absolute and the rate at which the center point of TRU changes. These are shown by: **RAFLAS** or Ratio of focus and LAST. They are illustrated by the growth or contraction of TRU; whether it is more expanding or changing, relative to the set TRU.

<u>Teleology</u> is any theory of ends or purposes. Teleology is the study of events as signs of purpose. Though largely unrealized it is a highly important field of knowledge. TRU functions as an end in itself. This is a basic law, state and condition, that is analogous to ATO, and the *Law of Self Preservation*. The all powerful bonding of the teleological ends and purposes is a basic quality and determinant of TRU. From the potential to the actual, from those forces that hold it together and from its absolute essentiality these ends and purposes are so characterized, qualified and composed in TIT. Teleology may be likened to a target, and some of its principles are derived, in part, from <u>Target Theory</u>. This is especially evident in Radiation Biology. Teleology may have targets, goals and purposes that are singular, multiple, combinations, or total. It depends upon the unity of a fixed source and endpoint. As both subject and object it totally utilizes all the levels of mass energy, Á, to represent and present TRU. This is what enables a good teleology, the unity and totality, or TU and TO, of TIT. There are many targets for TRU, POT, IPT, and each person. They primarily include the answers to questions, and solutions to the problems of human life. Examples of these are, how to acquire more and better knowledge, how to cope with life, who am I, what is the world I'm in, why do I fail, why is the world so cruel, why has God failed us, how are others so different and the same, will my children and progeny be fit for life and the happiness it can provide, how can all mankind live in peace, harmony, and prosperity? They are of all persons and many in each person, although they may not believe, or want

to recognize them.

The first and greatest target is the leaders and potential leaders in the world. Those upon whose decisions so many depend. The second and most immediate target are all communication media, books, libraries, academia, media, internet, and all areas where those in need will quickly see how much it fulfills their needs and helps them to encounter, and fulfill, life. The third, and ultimately the most vast, target is all persons, especially in the privacy of their lives who seek consolation and happiness in discovering and applying the great ideas, concepts, principles, laws, and mechanisms of life. How these can be well ordered for use when their lives become public, targeted, and whole.

Supervision is needed for all TRU, for its presentation and for all it takes to fulfill TRU. Supervision is even more definitive when applied to all who can, do and will oversee and control TRU. Voids is a term used for those segments or portions of TRU that are missed or largely ignored, and is mostly for what is vital to TRU, needing discovery, development, emphasis and presentation. Much that is a part of knowledge, especially in fields, sciences and technology, yet not adequately presented in this text, are voids. This we strive to fill. It is work on these, like we get from paradoxes, that lead to further voids and breakthroughs, whose revelations and concepts form the basis of KOT to come.

If the problems, questions and doubts are of any significance they will be encountered and must be confronted. This, if successful, will be by TRU. Yet in practice the decisions we make, of what are or are not worthwhile and what should or should not be done, and the satisfaction or quandary each person lives with, are very much with us. It is a good example of our dynamic inner and outer workings, OITIO. This is evident from POA and provides us with the answers and solutions, the resolution of the doubts that we harbor. These can only be as they are in counterpoint with TIT. One million years past does not assure one million years future or any other set time duration. Only TRU in totality and all the rest that is TRU assures TRU. Increasingly TRU, more qualitatively than quantitatively, is the only way to TÂ.

IMÄ = Immortality, Form of. **MIMA** = Multiple, Manifold, Summative, Total Forms of Immortality. In finality, MIMA is equivalent to all the through and through processes in TRU that provide for the ongoing life of an individual, TES, during times that are not limited to ones lifetime. In a way TES and TRUD partake of MIMA. For the personality, presence, beingness, resurrection, and spirit of the individual, as TRU, permeate much of TRU. The more special case of IMA is: **FUH** = Future Hindsight, and **WABEW** = We Are Being Watched. FUH and WABEW, although understood as being from the future in the present, even as IMA, and MIMA, more greatly exist throughout TRU. This is as the present has hindsight of the past, or watches some aspect of the past. They may be extended to all tenses of time in TRU as they effect or or are in dynamic relationship with each other. FUH and WABEW are preeminently within TIT as shown by what is imperative. This is what makes all the musts, should's and necessities we so frequently designate in our descriptions, in IPTRU, so justified. If we do not properly and fully anticipate and attend to all TRU, rather than what is immediately before us, given, or even less than TRU, we fail. There is more to immortality than individual life, or even the special cases, FUH and WABEW, as important as they are. Dynamically all that assures TES and TRU, and especially the mutual positive effect on each other is a great part of this immortality by TRU. It is better for us to understand immortality, not as some vague state after physical death, but as the forces which act universally in TRU to not preserve any individual, but the individual as TES to become the total TES that so enhances and preserves TRU.

PIA = Opportunity of and for IA, IAT, TES, and TRU. PIA is the core of a highly mature, judicious, prudent and effective person. It is directed to heaven on earth, TRU. Our heaven is what we make it. Suffering for TRU is much less than suffering for ATRU. This was learned from the example of Christian Martyrs. Just as the response to stimuli that are TRU and successful are more natural and less painful than responses to stimuli that are not TRU. This is a conditioning or the physiological make up of an individual seen in adaptation that enables and sustains TRU on a subconscious level. PIA is especially beneficial to a person, or TES, as it expands and provides a hard core of TRU in media with all about it.

From the midpoint of ALT the future consists of TRU. This shows how all concepts have much more meaning in totality when bonded with all TRU. IA is ever presently concerned with, concentrates on, and consciously or unconsciously, directly or indirectly, acts toward the future. Its reversibility explains why the future is so great and important. This explains the dynamics of whether mankind goes on to eternal greatness or becomes rapidly extinct. If TRU and man become rapidly extinct, likewise do the present and past. Here is the justification of the almost 50% future in ALT and TRU. For it is infinite TRU Future that must be, or approach, TÂ. **AFRET** = Absolute Finite Reversion to TRU, TES, PUREALITY, COGNOS, SA and TÂ. IA as TES is the vital viewer of and to TA, THEO + THES, THESOÂ to TIT and TRU. **UL** = Uncertainty Law, the Uncertainty of the Future. UL besides being due to lack of knowledge and KOT, is due in part to the individual's and IA's partiality, and inability to focus upon TRIF, any inexact IA on mid point or focus of the future in ALT and TIT. UL is not infinitely invalidating, it is infinitely compelling to TÂ, willing the future. It is highly teleological and eschatological.

EPAT = Expansion of the Threshold of TRU and knowledge of the Form of TRU. EPAT occurs both specially and generally. Human conceptualization, language, knowledge and information produce a rapidly changing and astonishing revolution of TRU. They act as major vehicles, the reality of the makeup of TRU. This is how concepts, language, and formal distinction of the parts of TRU enable us to attain: **TUID** = a Total Unified Identity with TRU. TUID is the great flowering of our world, the Relative Universe, as it is both, from invagination and evagination, infolding and outfolding itself, providing fruition and completion. The threshold of TRU, EPAT, expands simultaneously with our accomplishment of it. This maintains its Eustasis. or state of TIT-TO, in perpetuity. The threshold must be met. We must cross and accomplish all that is required, in doing less we suffer loss of TUID, EPAT, TES and TRU.

What it takes to generate TRU may be evident in the midst of life, when one's livelihood depends upon what happens if one does not meet ones obligations and generate TRU. Nothing is self assured. Nothing is to be attained without some doing. What is attained to be right requires doing by TRU. Not for long can a person stand by or be statically in and of themselves. They are not permanent. Even by TRUS this permanence changes. It is due to total, partial, internal conditions and external forces.

Each of us continuously makes choices and judgements. Commonly these are very limited or poor, as from ATRU information. Many will affect others and thereby ourselves in reaction. It is making choices and judgements in the best possible way we interpret others and all not in error but most accurately and properly by TRU. This paradox of total and partial action is: **PACTO** = Paradox of Action, Partial and Total. PACTO signifies how correction is to be made, as in a pact with others, TES and TRU; that actions should not be greater than the partial to which they pertain. This, should be both in right proportion and correlation. **SACTO** = Solution of PACTO. SACTO is supported by VAPO. LATANOS is proven by how PACTO = SACTO as in race to race, sex to sex, and age to age. Right act and actions, whether partial or total, IPR in daily or all life, should be in proper harmony with and of TES, TIT-TO and TRU.

SECTION VI

DYNU, SIMPLE PROCESSES, CES and CHES in TRU and HOM

CHAPTER 60

INTRODUCTION to SINGLE or UNITARY DYNAMICS, CES and CHES in DYNU

DYNU KNOWLEDGE, DEVELOPMENT, FORM and FUNCTION, FORCES, ACTIONS, MECHANISMS

In Section VI, DYNU, we develop and treat singular functions of the dynamics of TRU, TRUD, whose pro**ces**ses, **CES**, are shown as CHES in SPALT. They are not done definitively nor conclusively. They are presented in comprehensive and exemplary fashion to show how such a treatment of TIT bridges the gap and shows the great differences between the substantive makeup of TIT and the incremental world, entities, senses and things in our lives whose ESINS are TRU.

The types of posits, **CHES**, of their corresponding processes, CES, are classified into four categories, including, 1. Solitary CES and CHES in general, 2. Solitary CES and CHES specifically, 3. Singular, but not necessarily solitary CES and CHES, and 4. Singular CES and CHES occurring in multiples, or in an enumerated manner. The first three classes, although large, are what we encounter in life and knowledge making up a large portion of TRU. Because they are solitary or singular without enumeration or association their implications are limited. Thus, they are of less help in revealing the underlying principles, mechanisms and totality of TRU than the last category, and even less that of DYNS and DYNO, the next two sections.

The knowledge of DYNUS or the separate, processes of TRU is <u>Situational</u>, <u>Valuational</u> or <u>Axiological</u>, <u>Validational</u> or <u>Veridical</u>, <u>Essential</u>, and <u>TRU</u>. With these means and measures of all that comes before us, or we should set before ourselves, we can better test and order the disparate processes of TRU. We can best make their posits, analyze their applications, and show TRU. All entities, things, senses, words, parts, processes, currents, acts, influences, trends, forms and functions that are used in TRU, or its study, may be posited in and interpreted by SPALT, POMOT, METAD, SPAT, or other MOT and ROT that enable demonstration of TRU. When combined with their descriptions, they provide a powerful tool for TRU. They constantly need refinement and improvement. The words and their combinations used to represent what is to be posited are intended to be their actuality in TRU. This is because their meanings have become increasingly identified with some specific aspect, part, or process that reveals TRU Such a vivid and practical method of communicating through the development of language which in the actuality of Á is the reality of TRU. This is the reason we must combine the best of all existing language with TRU terminology in the most effective and beneficial way.

CES do not have to represent what its object seems to be, only as they exist in TIT. This is primarily their functional TRU role and value. Therefor CES are functionally valuable, useful, significant in TRU as they enable TRU to be, ENT. They may, and should be, very large, medium, or small CES in TRU, although because of limited space we focus more on the large, for parsimony, emphasis, and as examples.

There are many means of providing DYNU and its CHES. Some of these are the data, facts, base, data base, evidence, descriptions, content analysis, explanations, interpretations, and covering law theory, that give us better ways of handling or treating massive amounts of information. This is what we can use to describe, demonstrate, and POSIT CHES, SPALT, and all. For each science, there are various and unique methods of information gathering and ordering.

In figure 1-60, 3 digit numbers can be entered to represent each cell in SPALT. They show how what is separate can be reduced to CHES in DYNU. We must always remember, ECRO, the deficiency or COL between what the CHES and SPALT are and what TIT is, to make appropriate correction in each and all cases. However we must be careful, also, not to forget or fail to appreciate the greatness of SPALT, POS, HOM, CHES and DYNU and not let their deficiencies lessen our appreciation, enthusiasm, conviction and work for contribution to POA, POT and TRU. To provide the separate processes of DYNU, we must constantly do so by considering its very subjective - objective, relationships. This is done by the use and plotting through time and place. When applied to TRU it helps to better understand how the differential and specific elements fit into the whole. It helps make the form of the whole, SPALT, POS, STA and HOM become more accurately determined and identified with TRIT and TIT.

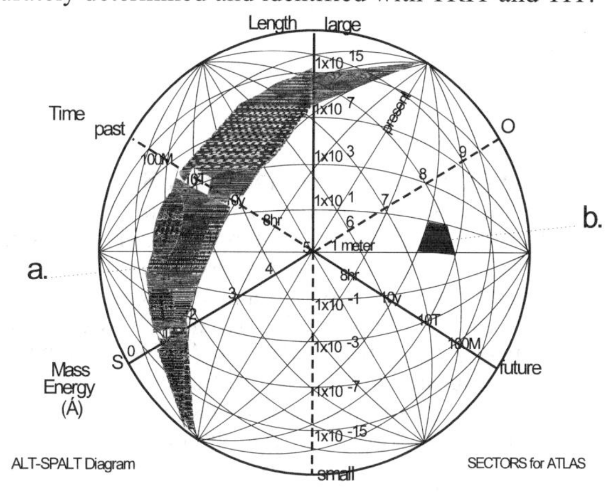

Fig 1-60 **SECTORS of SPALT as MAPS for ATLAS**

In order to present TRU most vividly and fully its processes, CES and their posits CHES, can be produced as an atlas. Each sector or segment is separately viewed in which their CHES are in their appropriate cell. Each cell has a three digit Á, L, and T number which becomes that of its CHES. Each sector and cell is a proportionate part of SPALT and thereby TIT. This is very important for many reasons, (1). it gives consistency and reliability to a representation and presentation of TRU. For these maps, like those of a road atlas, are highly stable, as much as TRU is. 2. Each sector can always be compared and held in proper proportion by its relative size in terms of the whole. e.g where 1 million to 10 years past yields a sector which is approximately 10% of the whole. If this map is reduced to that size between 1 meter and 10 meters, it would be about 1.5% of the whole. So that for those sectors that represent 1/10 of each of the three dimensions we would have 1,000 maps for the whole. This can be done in various ways, yet limited, as all books are, to the size of the book(s) and the knowledge and accuracy of what are included. In Figure 1-60 we show: a. Larger sector, 15% of the total, based upon a reduction in only one dimension, b. Smaller sector, 0.1% of the total, based upon a reduction of 1/10 in 3 dimensions or $1x10^{-3}$. Such differentiation must be balanced by the totality of TRU, .as well as an ever greater accuracy of CES, CHES and SPALT Not only does SPALT enable us to show how TRUD, HOM, and DYNU can become POA and CHES posited, viewed and understood, this very positing enables us to understand ALT, ROT and TRU. It enables a much more accurate, vivid and comprehensive view. One of the most beautiful examples of HOM, and one that has had a tremendous influence upon our coordinate proportional comprehension of how mankind can be determined is through Radiocarbon Dates. It is possible to mass together enumerable events, dates, conditions, and many factors of contemporary and sequential times to each date in many areas of the world.

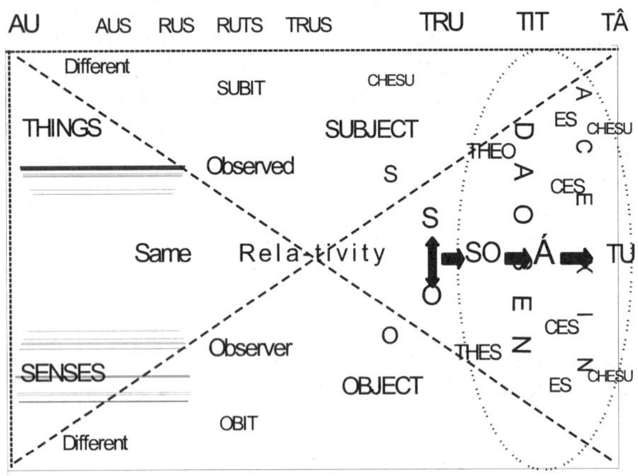

Fig 2-60 **CONTRASTING AU-TRU D-A-S**

The two parameters or continua of senses to things and AU to TRU to TÂ are much of the approach to TRU. This is provided by its mass energy, Á, FAF and dimension. In all POSITS, especially by DYNU the separate and differential dynamic participantes in TRU. We should keep in mind how they arose from such a background. It is CHESU and CES, when held to their special, unique and distinctive identical processes in TIT, they enable us to trace out the beautiful exquisite detail of the world. This can not be done, however, without applying all the pertinent concepts, principles, laws, mechanisms, SCARLS, VAPO, and DAOST.

DYNUS can be those processes that occur in any portion of TIT. To make CHES and ALT valid by increasing identity with TIT CES must occur throughout ALT and TIT with even distribution and form equivalent to what exists in TIT. They must balance each other out, as in the Doctrine of Polarity. They must be statistically valid, as in the Doctrine of the Mean. Their extraneity must be equally distributed by their inherence. They must be produced with locational analysis, translocation, vertical, horizontal and sagittal groupings. Their primary functions in TIT must be observed and posited, i.e. what, where, when, why and how. Their thisness or haecceity is primarily their exclusiveness within TRU. Their whatness or quiddity is primarily their formal identity with TRU. Their when is the locus in time change and variation in TIT and SPALT. Their where is their locus either in the length dimension, area or volume, or generally in SPALT. Their why is the determination, often dynamic, of distribution in TRU by cause and effect. The various reasons for being what they are as well as their mechanisms. Their how are the methods, modes and mechanisms of function, operation and dynamics in TRU. For Homation this is how they can be represented, modulated, and posited to best represent their actual part and role in TIT. How their technique or technical aspects function in TRU.

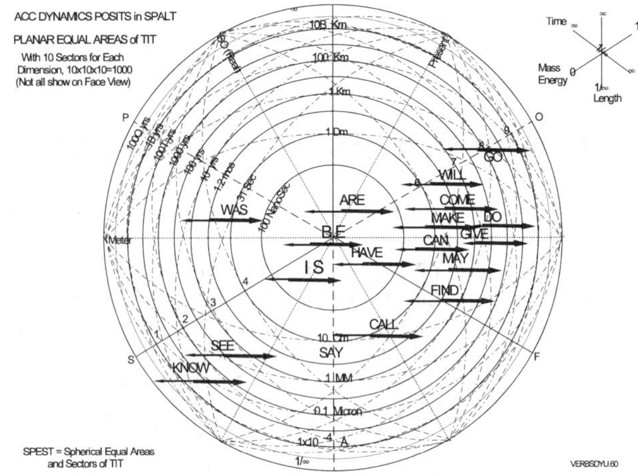

Fig 3-60 **MOST FREQUENTLY USED VERBS in DYNU**

Verbs are the great media and vehicles in TIT and of TRU OITIO that so identify and validate CES, CHES, CHESU and DYNUS. A rough scale of the relative frequency of different verbs helps us understand their depth and meaning in TRU. It will be seen how great forms of the root ' to be' partake of action - and meaning - in language and life.

	(f)	Á	L	T
IS	(60)	5.0	4.5	5.0
ARE	(15)	5.0	5.5	5.0
BE	(14)	5.0	5.0	5.0
COME	(12)	7	4.5	5.5
HAVE	(12)	6	5.5	4
SAY	(12)	4	6	5
GO	(11)	8	3.5	6
GIVE	(11)	8	5	5.5
MAY	(11)	6	5	6
CALL	(10)	4	6	5
DO	(10)	8	5	5.5
KNOW	(10)	0	8	3.5
MAKE	(10)	7	5	6
SEE	(10)	1	7	4.5
WAS	(10)	5	5	2.5
WILL	(10)	5.5	4.5	7.5
CAN	(9)	7	5	5
FIND	(9)	6	5.5	6

If one would think that words are cheap, and doubt the greatness of language, let them look at how small our world would be and how poorly we would get along without them. For words are not only the finite elements of our lives, much of what makes it real. They are the great dynamic vehicles of relativity, what makes the observed bond with the observer. They are the media and glue that holds the predominant mass energy Á FAF of TRU together. As such they are much of the driving force and power of TRU. Words, thereby, are great elements and features in the current revolution that is rapidly expanding TRU. Each person will increasingly depend on their success or failure with happiness or misery, and will be determined by it, as it will not be achieved without a good knowledge and use of TRU. It is when words are not used well or misused that they become cheap. They are priceless when learned and used well, combined in speech, composition, and communication to have ever greater meanings. So that there is a reversible dependency of words on TRU, and TRU on words that makes up a large part of TRU, our world and lives.

It is this that is much of ROT, as well as POS, STA, HOM, and POT, as they in turn are or will increasingly become of TIT and TRU.

Verbs vividly depict life, which in combinations show how complex yet how positive and certain life can be. How close to us TIT and TRU, or TREQ and God are. Many verbs apply to both the individual and all. They likewise clearly show the great and largely unrecognized compulsion of the future. This is how the future world compels, propels, and draws us forward to a better and more lasting life. Of the thousands of verbs in use, less than a hundred form the bulk of common usage. They are highly directed by and to the subjective, objective, and real of TRU. This substantiates the Á, S-O, or special mass, energy form of TRU. They provide proof of the composition and distribution of TRU. Most directly they show and prove, when their ESINS are extracted and analyzed in terms of TRU and its CES developed, the form of the differential basis of TRU. How TRU is discovered, its nature revealed, and its mechanisms combined and integrated into the whole,TIT-TO and TRU.

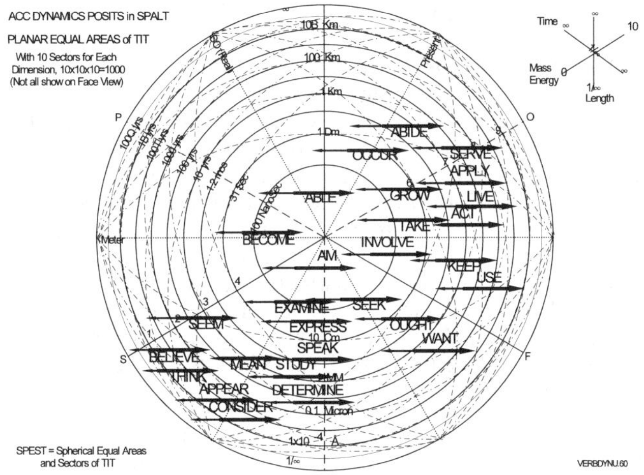

Fig 4-60 **SOME COMMON VERBS, CES DYNU POSITS**

Verbs other than forms of the verb to be that are highly indicative of the CES and CHES TRU, HOM and DYNU include:

abide	8 - 3.5 - 3	able	6 - 5 - 3
act	8 - 5 - 6	am	5 - 5 - 5
appear	1 - 8 - 4	apply	8 - 4.5- 6.5
become	5 - 5 - 4	believe	0 - 8 - 3
consider	1 - 8 - 5	determine	2 - 7 - 6
disturb	.6 5 - 5	express	4 - 5.5- 5
examine	4 - 5.5 - 5	grow	5 - 3 - 7
involve	5 - 5 - 5.5	keep	6 - 5 - 7
let	7 - 4 - 6	live	7 - 4 - 7
mean	2 - 7 - 4	occur	6 - 4 - 8
ought	4 - 5 - 8	represent	6 - 5 - 6
seek	4 - 5 - 6	seem	1 - 7 - 4
serve	7 - 3 - 7	speak	4 - 6 - 5
study	2 - 6 - 6	take	7 - 5 - 5
think	0 - 8 - 4	use	7 - 5 - 6
want	5 - 6 - 7		

The posits of verbs in SPALT are highly revealing, both of the role of CES in TRU as well as the great value of a POA. What some of these revelations include, 1. Focal and response orientation of verbs and the CES represented, 2. Confirmation of their roles in each dimension, especially the Á dimension, 3. Agreement with a modified planar and spherical adaptation for equal areas of TIT, 4. Their power both singularly and in combination, 5. Their differential and discrete form providing a potential elementary basis for POA of TRU. 6. The form and function of the particulate ES and CES of TRU and our ability to use them to show all TRU by HOM, DYNU, DYNS, and DYNO.

ROS = Resolution of the Separate. ROS is the primary law of resolution of what is different and separate. It is a separation by multiplicity and a separation by dissimilarity. This is what science is continuously doing. It tears things apart so that it can better see how they are made, formed and function. Yet it tends to do so in a way that does not return them, properly or adequately, to the total, to TO and TRU. This is a common and human tendency of our S-O and mass energy, Á powers to take everything apart and identify it. ROS is essential to the identity of all, the formal distinction of all TRU, the D of DAOST. This is only half of the battle or war. The other great part lies in putting it back together. It is how to systematically unify organize, make whole all that is known separately and to precisely and proportionately bring it into focus.(VAPO) This can be well done only when of TRU or TREQ. We then have the three great steps, prerequisite to comprehending and showing the knowledge of all: separation or differentiation, fusion or integration, and their totality with VAPO, in the combined Systematic Unity of TIT in TRU. These are the basic and crucial means and tools of S-R, S-O, that which is used throughout this text. They are determined by TRU and exemplified in DYNU, DYNS and DYNO by representing the POA of TRU by HOM in SPALT.

Attention to detail, especially when expert, is of great importance and utility. **EXAD** = Expert Attention to Detail. With EXAD the perfection, intensity, extensity, and actuality, or the ACEXIN of TRU, are possible. Experts, highly qualified, experienced and effective professionals and technicians understand this principle in their fields of responsibility. Yet to often they do not extend it beyond these narrow confines, to TREQ or TRU. Those who do, know how to, are able to, and do with precision often contribute to TRU. Yet without its relation to the rest of TRU it is highly inefficient.

That the separate CES tend to function in and of themselves is evident, especially to our IA. In nature and our lives we are constantly and dynamically functioning both dependently and independently. All seeks the best in all worlds. This even makes distinct the state of separate processes whereby, even though pure TRU, CES and their values may fluctuate, depending upon all that is different, independent of them, and the less than total objective of TRU, to which they pertain. **PISU** = Partials will impart their own Special Unity, features or whatness. Much of the exclusive and unique nature of TIT and TRU comes from the added qualities of partials, and the added state they are changed to, in TIT-TO.

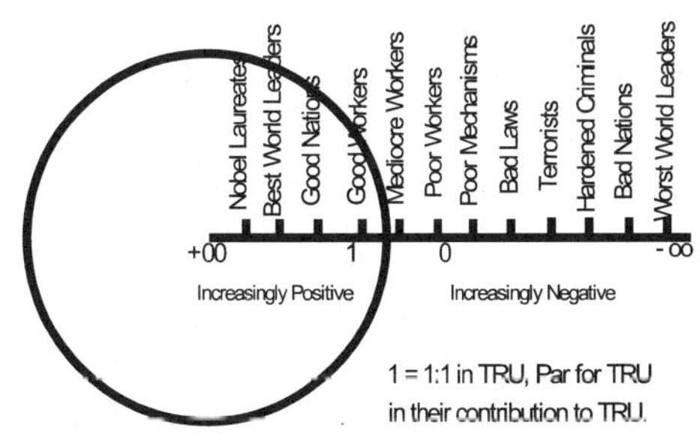

Fig 5-60 **DEGREES of PARTICIPATION in TRU**

How much or how little do different things, persons, entities, objects, ideas, mechanisms, and laws, participate in TRU? How much do they fluctuate in this participation? If the rim of a circle is a standard, of one, then a person in being one in 10 billion would be a per capita figure of participation in TRU. This would be roughly equal to their REM in TRU. There is a great variance in the different kinds of entries. The standard for many would not be per world population, but their relative in the whole. Likewise there would be great variation depending upon, not the average per each group but the range.

Good things happen, a little at a time, like quanta, from great knowledge, right principles and decisions, and hard work for the best in human life. Rarely they occur in larger bundles, like quantum leaps, complex mutations or collections of great knowledge, right principles and decisions, and hard work. So that to expect miracles, great luck or fortune without great knowledge, work, right principles and decisions is like expecting TRU from ATRU, everything from nothing, or even worse, everything from the negative. It never happens.

EDE = Partial Difference from, Deference and Disrespect to and for Partial. EDE is a continuing accretion, self perpetuating process, principle, and law of TRU. Not only does a partial have a biased stance or view. In viewing, judging or choosing another partial it knows feels, subjectifies that partial as different, in opposition or even ATRU to TIT-TO. It produces another degree of partiality. It is related to why others often seem so different and why a person must often force themselves to identify with and love others, even if TRU. **PÀL** = Partial from TRU Law signifies these combined deviations from TRU and in reaction, initiates their reconciliation. It is paying the extra price in love of TRU, in love of not only ones counterpart in TRU, and sames of TRU, but how this counterpart must be a part of the whole, its universal benefit. It requires an interest and love in attaining an absolute empathy toward all that is or needs to recover and turn from EDE through PAL to maximal TRU. In DYNU we are dealing with and treating partials, and highly separate or different parts and processes of TRU. In so doing we are always being, to an extent, as in most all we survey, other than TO-TIT. For this reason we must constantly remember, ECRO, keep in mind, EDE and PAL, as we should constantly keep before us under all circumstances the basic laws, principles, concepts, SCARLS, SPALT, POMOT, METAD, SPAT, and SPO, of what enables our perspective and comprehension of TRU.

Know the action of something and you will go far to understand it and its use. This will be even more valid if you know this action, thing, and its use in TRU. You gain its full significance, REM, and how these functions work as process in TRU. For example, all drugs, and how certain types, sets, groups of forces and actions, are highly revealing, the receptors, activators, agonists, antagonists, protagonists, stimulants, and inhibitors. These are most evident in enzyme and immune systems. The kinds of roles that different CES play in TRUD simply and distinctly provide the forces, powers, vectors and CHES for DYNU. Moreover they reveal the dynamics of that both unknown and OITIO of TIT.

Using drugs as an example of more general implications of action there are many lessons to be learned that action, care, and ACC can benefit or harm the body and TRU. It is that any drug that is strong enough to help you is also likely strong enough to hurt you. Also, help or hurt can be quantified, as we measure CES like REM for ES. In fact the positive or negative degrees for TRU mirror what occurs in OITIO. Rarely even it is help that hurts, and hurt that helps, when the help is excessive or deficient, and the hurt stimulates action and cure. Every action, care, and ACC must be, by TRU, neither excessive or deficient, and precisely limited to the occurrence or case at hand or its applicability.

Our minds and IA tend to function in terms of singular static or dynamic elements in the world around us. This is especially so for what can be used in reaction, response, or problem solving. This may or may not be TRU, CES, or relevant to others. So that to focus our thinking on DYNU we must do so in a certain way, that of the SCARL of which it consists relative to TRU. CES is apt to be thought of as singular, differential, and likely to be of lower order REM. This is not the only form of DYNU.

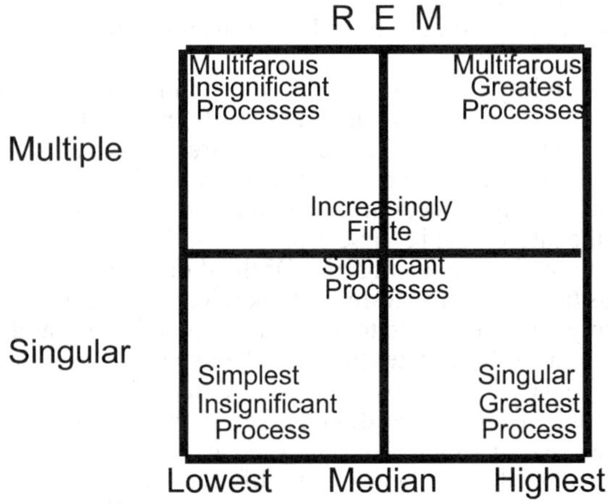

Fig 6-60 **FORM of DYNU, SINGULAR, MULTIPLE and by REM**

Each of the four quadrants of figure 3-60 should have essentially equal representation of DYNU when properly weighed. Just as in REM, we can not expect to get a satisfactory idea of the world about us, TRU, by knowing one or two Gods, a few of their attributes, commandments or laws. We must recognize and learn what TRU is. If we are to properly show TIT and TRU we should do so by showing these mechanisms in all their engaging detail as well as well interactions and glory. Core forces are all forces, vectors, and processes, that: 1. Are of great REM, 1R, 2R... 2. Together are composite TRU, when brought together to make up TRU, 3. Are highly ENT, 4. Are distinctly, sufficiently, and well separated from each other, 5. Demonstrate the bonds and great powers that hold TIT together. Some of these core forces and their POSITS that help show CES, DYNU, SUT, OBT, TRIT, and TIT include:

DIVSION OF LABOR	8.0 - 2.5 - 7.0
ENTHUSIASM	2.5 - 6.0 - 7.0
HEARING	1.5 - 8.0 - 1.0
SPEECH	5.0 - 6.5 - 2.0
LANGUAGE	6.0 - 3.5 - 7.0
WRITING	6.0 - 6.5 - 2.5
ELECTRONICS	8.5 - 2.0 - 7.5
TRANSPORTATION	8.0 - 2.5 - 7.0
CLOTHING	9.0 - 5.5 - 4.5
FOOD PRODUCTION METHODS	8.0 - 3.5 - 2.0
ARCHITECTURE	9.0 - 2.5 - 7.0
VISION, combined	1.5 - 8.5 - 1.0
GENETIC MATERIAL and base	9.0 - 8.0 - 3.5
EARTH, Natural Ecology	10 - 0.5 - 0.5
ASTRONOMICAL, Universe,	10 - 0 - 1.0
WORK, PRODUCTIVITY O	8 - 4 - 5
THOUGHT, REASON S	1.0 - 9.0 - 6.0
DEITY, UNITARY, PURPOSE	3.0 - 1.0 - 8.0
PREVENTIVE MEDICINE	6.0 - 6.5 - 6.5
MEDICINE AND SURGERY	5.0 - 6.0 - 3.5
LAW (and order)	6.5 -3.5 - 4.0
LOCOMOTION	9.0 -4.5 - 5.0
HAND, Opposable Thumb	9.0 - 3.5 - 1.5
LOVE (of all - TRU)	1.0 - 4.5 - 6.5

A great portion and thrust of IPTRU is based upon the extraction of ESINS and their abstract FAF in TIT. It may be compared, in ways, with the *ideas* of Plato, *forms* of Aristotle, *monads* of Bruno and Leibniz and the *phenomena* of Kant, Hegel and Husserl. Each of these recognized, realized and sought to know and show the makeup of process in the world, the absolute TREQ or TRU. It is how these exist as CES in TRU and can be posited as separate distinct and discrete CHES in SPALT for DYNU and HOM that we can develop order out of chaos. It is how these CHES make up the massive functional processes of TRU that TRU becomes known, real and actual. For the person to identify and utilize this is PUREACLITY. The future, as all TRU, becomes definite through the defined POSITS of CHES in SPALT.

Since the form of TRU is framed by parameters, dimensions, polarizations and spectrums it is only necessary that the knowledge and state of CES and HOM be according to their principles. The philosophy of polarity and its accompanying dialectics is basic to HOM and DYNUS. The form of animalian motion, motility, function, and behavior, when adequately SCARLED in TRU, is highly determinative of these processes. This is especially evident in the study of Taxis and Tropisms, and the polarization of many different forces in equilibrium in the mechanisms, mass and energy of TRU.

"How do you do", "How goes it", "How are you feeling" are the introductory greetings and salutations commonly used. They appear to represent simple functions between people. Their implications are far more subtle and profound. They reach into the heart of TRU with an attention with intensity into whether or not one is in optimal health. Not so much physical health but overall well being, the fulfillment of their part and roles in life, TRU. It represents a search PI, for the mutual positive imperative in life. From this beginning connection and conversation their follows a massive penetration into how people think and work together to react to the actions, respond to the stimuli, and solve the problems they have that bring them back to optimal well being in life, or TRU. From an analysis of such interactions the processes of TRU can be seen, as they expand OITIO, and the mutual success/failure they provide.

Fig 7-60
SCATTERED SEARCH, DYNUS, CHES, HOM

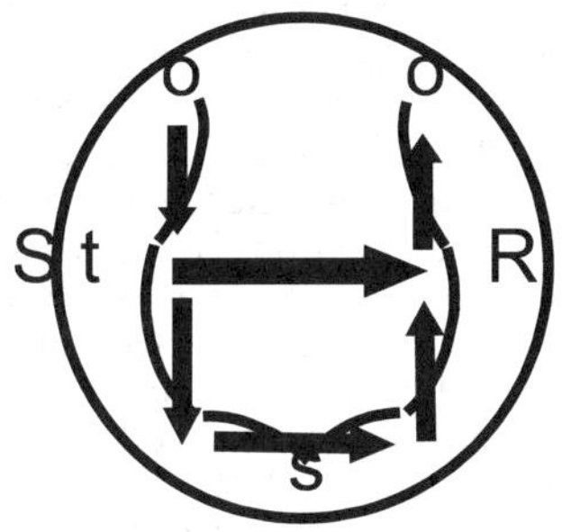

FAS = Fallacy of Scattered Search. FAS is the tendency of systems, ISMS, and sophisticates to become counter productive, not so much through their partiality or lack of VAPO and TRU, but through the loss of original purpose, target and search. It is a most common human tendency of actions and to deteriorate into lesser REMS and ATRUS only to often later, after much loss, have to search, recover and return to TRU or TE. This can be seen in the contrast between the basic or pure sciences and applied or practical sciences. We should work for VAPO through the wisdom that experience in life provides, and the great heartfelt desire for right that naturally and predominantly comes from TRU. Complete correction, courage, motivation and determination to overcome and correct stimulation deficit is balanced by the greatness and finite fulfillment of the functions and mechanisms of TRU. This is shown by CHES in SPALT and can be utilized in life. When the singular or separate functions and processes of TRU are considered or attended to by representing change, variation, and the negative motion in TRU, it can be seen how the individual becomes so dissociated, detached from the whole, and polarized from TRU. This is the spinoff of all that is differential, separate, singular of and from TRU. The overall state of change is even extolled by life's experience, as in the Greek Maxim "change is sweet in everything". This is very true of IA, and of many disparate elements and processes of TRU. For the partial must be highly variable if it is to maintain its partiality. Its independence/dependence primarily comes from DAOST. When mixed with the changes that are OITIO the problems of individuality, IA, and partiality are accelerated, yet better understood. They call for the most astute KOT, ECRO, and reversibility of the negative to TRU. That great process and function by which life, and TRU so successfully react to motion and change is called: Adaptation. Adaptation may apply generally to all motion and energy for reaction in countering action, especially with extended interactions. The dynamics of TRU, TRUD, is very much a product of its differential, or heterogeneous FAF which may be generalized in the phrase *gradiatim vincimus*, or we conquer by degrees or increments, CES in TRU. Motion, change, variation, action, and processes do not always have to occur by infinitesimal increments. This is in conformity with DAOST wherein differences exist not only infinitesimally, but by all finite degrees, from one extreme to the other throughout the continua of TRU. This illustrates the same and different or integral and differential continua of which TRU is formed. How its FAF are structured and behave according to set relationships in the whole of TIT.

TOST = Test of Senses and Things, is their greatest meanings when applied by inversion to their actuality in TIT. An initial presentation of a topic that has the greatness and intricacy of TRU can not present or know all its processes and mechanisms. We therefore strive to bring them out by CHES to: (1) discover all that is unknown of TIT, (2) show the less significant processes in number as examples, (3) show them not only simply and singularly but by field, sector, or division of the space occupied by the solid dimensional TRU, (4). show them as they are the greatest processes of TRU, CES, How they combine and can be used. We will then proceed in chapters 61-68, and following sections to provide CES and CHES in these different ways, constantly remembering their SCARLS in HOM, ALT, MOT, ROT, SU, THEO, TRU and TIT.

THOUGHT, THING, TRU, ES or CES PROBLEM and SOLUTION

The how or technical aspects and mechanisms of the separate processes of TRU, DYNUS, are not as indiscriminate as we might assume. They follow to a great extent their own TRU nature. the form and function of TRU and its TIT, TRIT, OBT, SUT, THES, and THEO. They are often easily recognized by other functions, processes, and entities to which they may or may not be associated. These separate processes of TRU are to be distinguished and made discrete for posit in TRU. What we use to discover them is seen in: **TESP** = the Problem of TES or the Thought-Thing-TRU-ES or CES Problem. If thoughts, things, cats, dogs, and objects, as medical objective treatments, are seemingly 100% objective or subjective and therefor not from TRU and demonstrable by SO, A, and ALT, let us use what of them is ES or CES. That is let us extract their ESINS and CESINS. Let it be remembered that ES or CES do not have to represent each of these as they seem to be. Only as they are in relation to us, exist in TIT and as they are ENT of TRU. This is primarily their functional, and TRU, value and role. Therefor ES and CES are as they are functionally valuable, useful, significant in TRU, and enable TRU to be, ENT. If we sense them and they are not ES, they are not except as they are ES or CES. If we have a vague idea of ES and it is not a thing, objective of thought then it is not. It is not TRU. It is only as its ES or CES are in TIT-TO. They may be a very large, medium, small part of TRU. One of the greatest tasks of man is to tender ES or CES as sense and thing, and to tender sense and thing as ES and CES in TRU. To do so with focus and profound demonstration. To not do so or to do so is the problem of TES, through thing TRU ES Problem, TESP. TESP together with the situational, axiological, veridical, and essential means of knowing the processes or CES of TRU, and HOMATION provides these objectively to yield THEO and TIT, are the great benefits of DYNUS. They justify the deliberate and laborious processes by which we bring out and posit many separate functions, CHESU, of the CES of TRU in SPALT.

The vital importance of concepts, principles, laws, dimensions, coordinates, models, representation, presentation, TRUD and IPTRU is not to deny the facts, evidence, experience, experimentation, study, prospective investigation and detail upon which much of knowledge is based. All of these must work together, facts support principles, and principles make facts more actual rather than figments. Both as they are TRU and make these pure, actual and real to become 100 replicas of what they are as ES, CES, CHES, and CHESU in TIT-TO.

Once the frame, structure, form and replica of the relative universe are established and its major fundamental concepts, principles, laws and mechanisms known, then it is not so difficult to expand this knowledge and cognition to completion through the increasingly finite toward infinite selection, distinction, actual meanings, separation and positive operations of its parts and processes.

The great value of science lies in the fact that for the description and demonstration of an object, no matter how small or AMTRU, any information is meaningful and useful if it is or can be found to be of benefit, to humans, TREQ or TRU. 'Beware oh ye of little faith lest thou be led astray', by those and all that deny facts, evidence, scientific principles as they are the ES, CES, truths of TRU. **PIV** = Principle of Infinite or tiny finite information of Permanent Value.

The illusive nature of truth is nowhere more evident than in the difference between what is valid by VAPO and TRU in comparison with scientific truths. Although they overlap to a great extent, they often are quite diverse, evem antithetical. If by TRUT only that which is proportional to TRU is true, and if by science anything is true if proven by fact, evidence or a valid relation or syllogism, we have a paradox. Which is true? This is **PATSEV** or the paradox of TRU and Scientific Validity. The solution to this is the same as in the case of 'is' where we have two or more meanings. The actual truth by part, process, relation, contribution, and proportion in TRU, and the virtual truth, what holds in and of itself. The solution lies in the fundamental and basic nature of TRU relativity, the mass energy, S-O, Á development, form and function of TRU. Even though precision is difficult, and perfection impossible, with posits of STA and especially HOM and DYNU, both author, reader, and all who endeavor for TRU should seek to hold each and every process, to the full solid geometric objectivity of their actual state in TIT-TO.

To provide this knowledge of TRU, especially its separate processes, we must constantly do so by considering its very subjective - objective, psycho-physical, mind-body relationships. These we must also do as we move through the whole gamut of SCENS and SCÁTS of science by which senses, things, and entities are stated. This means we must constantly distinguish between the sense and the thing, the subjective and the objective, the mental and the physical with which we are dealing. It even more means that we must come to know Á, and how these function as S-O or Á in TRU. This is not as either mental or physical alone. It is in striking contrast with what our thinking, immediate attention, IA, scientific and medical powers of analysis seem to imply. For the mind and the body, as in the age-old controversy of universals and nominals in scholasticism seem to be, we set out to attend to or study them, completely separate. Nothing seems more sure to the doctor than the complete validity in the physical body, and the completely valid independent and exclusive function of the mind, of thought and reason. In this sense, and as it can be separated in DYNUS we must heed the separation of subjective and objective, mind and body, as set forth in the 'mind stuff' theory, the mind body relation. Only when we come to try to relate them, to understand how they function together, and what their related processes are, as in the cause theory of mind body, or parallelism, do we come to realize the fact that both they, and TRU, must be formed and function in certain ways. These ways are peculiar, unique, and special to the predominant mass energy, S-O, and Á nature, of the form, based upon thisness of, and relativity in, TRU.

They also, when completely separated into their S-O poles, have special distinguishing features. Yet they as all that exists either as ENIT or as it is in TIT, or in any and all ways in between, must be constantly remembered, ECRO, in these different ways. For this is a fallacy of ignorance and misjudgement, that creates much confusion in human life. This is signified by: **CONAT** = Contrasting Continuum of the Constituents of All, from AU to TRU, also: **LÂDÄENT** = Law of the Absolute Difference between the Form - and function - of Entities in AU, TRU and all in between.

Sense-thing, observer-observed, subject-object and mind-body are all much the same with certain special distinguishing features separating each. It is LADAENT (or LADENT), in association with LANERA and LAMOL, that establishes the special form of TRU and proves Á and TRU. It is shown in the CONAT-DAS diagram, Figure 2-60. The gist of this illustration is to recognize the fact that what man has so long considered mind and matter, mind and body, subject and object as being separate senses and things, are in fact, especially by TRIA of TIT, not only united, but assume added and unique characteristics in the totality of TRU. Without discovering, learning, remembering and using this simple, yet so subtle and profound fact, no one can understand, or even live adequately, in TRU. Entities, senses and things exist in TRU not as ENIT, they develop and become ESINS and CESINS to be ES, CES, aspects, FAF and all that is of the makeup of TRU. They exist as DAOST or DAOSEN, Different Aspects of the Same Thing TRU, or Entity. In being so, none have either absolute differentiation nor absolute union. They may be of any level in the polarity and continuum of DAOST. This will produce a loss of perfect distinction. Yet both their difference and sameness can be distinguished and described to provide, more or less, sufficient distinction. They are LANERA and LAMOL. From this we may apply what we have before us, and to develop it as it is a process in TRU. These are its situational, axiological, veridical and essential measures and means. We can, when understood and extracted in its TRU state, have a pure reality, or actual TRU reality. We can develop a knowledge of the processes of TRU that are highly successful.

When focusing on the differential, and singular processes, ES and CES of HOM we show DYNU and other proofs in this core of POT and IPTRU.(See Fig 6-10) The Proof of DYNU, PRUDYNU is closely related to the other proofs of TRU, PRUT. By **LACOP** or Law of Absolute Combined Proof, as shown at the end of Chapter 10 we recapitulate and proceed to show further proofs of TRU.

For further proofs see: Fig 6-10 pg 55, Fig 5-19 pg 147, Fig 4-49 pg 282. Also see pages: 200,244,264,281,282,338, 370,426,493,524.

PRUDYNU

TE, ES, CES, TIT, TO, E, SPALT, POT, VAPO, LIFE, TREQ, HUMAN, DAOST, MATES, LOPAN, ENT, TRUP, TRUF, MEANING, MADAS, EC, ACEXIN, ALT, SPALT, REM, LORVA, MOCBIREM, POS, LOPA, HOM, ACC, E, PRUE, TES, HOM, WITÂ, ACC, Alienation, Belief-s, Chemistry, Compliment, Do-ing, Family, Food, god, Medicine, Metabolism, Outgrow-it, Oxygen, Prescript-ion, Simple Processes, Service, Sex, Synapse, FRE, LODIS, ACI, DOUP, IR, LETUM, LEDID, MEA, STOCS, POFT, PADSAV, EDE, PAL, TORIS, MI, PISU, POCEA, POBUF, POTRAT, LATOK, MIACET, BIPO, LATAC, IMPLIT, PROMGO, LOAB, AMT, DO, GIFT, RICTRI.

CHAPTER 61

SULP or the SUBJECTIVE LARGE PAST in DYNU

SINGLE and SEPARATE PROCESSES, LANGUAGE, KNOWLEDGE, BEACONS, GOD, SIN, GUILT, INFLUENCES, CHOOSING SIDES, andETHICS

SULP or the subjective, large and past sector or segment of TRU initiates our analysis of CES in TRU, its vectors, CHES, posited in the standardized model SPALT. We do the same for the other eight sectors of TRU in Chapters 62-68, and likewise for DYNS and DYNO in Chapters 71-78 and 81-88. To produce these posits in the extensive, precise and reliable way demanded by TRU and TE is a heavy yet gloriously revealing task. The dynamic chapters have four added hurdles to overcome, including: 1. the fact that STA does not need to separate the posits to the extent we do in DYNU, DYNS and DYNO, 2. the statics of an object are better formed and differentiated than its dynamics so that the processes of TRU often have greater relations, effects and less discrete domain, 3. Processes are apt to have more diverse meanings and posits than parts of TRU, 4. Processes and their vectors are apt to have more effects, influences and ramifications. Thus we offer these posits and discussions more as an introduction and example of what a definitive analysis and presentation can do. However, their value should be obvious to all who understand the need, and their potential for lasting benefit to human life.

SULP by always being the same portion of TIT, must be held to this position. It must also not be forgotten that these sectors of SPALT are largely representations of TIT, yet the processes are in actuality what they are in TIT. CHESU thereby work as exemplary causes, *cum div*, with God with dividend, or TRU benefit. They occur in their own particular A,L,T, or mass, length and time, posits and way in TRU.

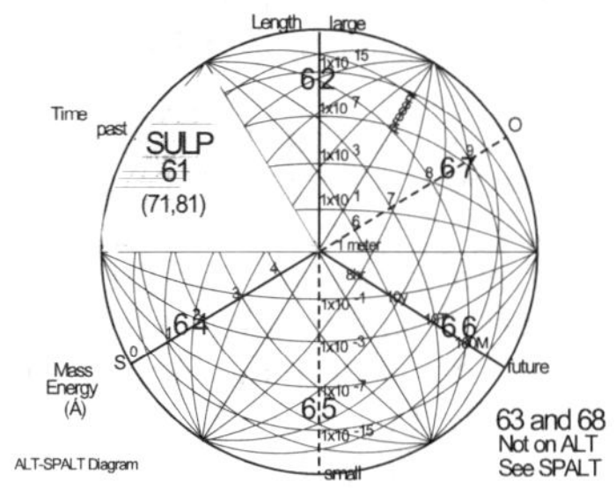

Fig 1-61 **SULP, or SUBJECTIVE LARGE PAST in the SPALT MODEL**

Posits in SULP, as other segments, are not limited to their center-points, they function as forces and powers collectively whose center of action occurs at this point.

At the outset we are confronted, with the great problems of: **LALTE** = Law of ALT Partials, and: **PLAC** = Planar geometric Problem. These occur when using a segmental planar field for POSITS on ALT. This is reduced with use, familiarity, and most specifically with SPALT, a solid spherical model and its extensions. TRU can be simply, quickly, easily, and effectively demonstrated and described by the separation and positioning of its functional elements. It requires a ready, yet demanding perfection of analysis and placement of any and all dynamic factors, functions, mechanisms, forces, powers, influences, trends, and courses by extracting and correcting them to TRU for position in OBT.

The defect in God is the degree to which it is less than TREQ TIT, and TRU. In use it has often become archaic, imperfect, excessively simplistic or unified, disproportionate, erroneously formed or highly partial. God is variably far from TIT-TO. Different names for God, and different religions provide poor concurrence in its full meaning. In the best of theology there is much that is TREQ, and thereby TRU. In the worst of religions and gods it is ATRU, and productive of much ANTRU. We must be careful therefor not to make any blanket condemnation of God or religion for they have contributed much and been of the greatest benefit to humans and TRU. God is the bridge to TREQ, TRU and TÂ. It is its lack of VAPO especially among those who do not vigilantly seek the full meaning of GOD as TREQ or TRU. God has never been dead, is ever increasingly alive when TREQ in and as TRU.

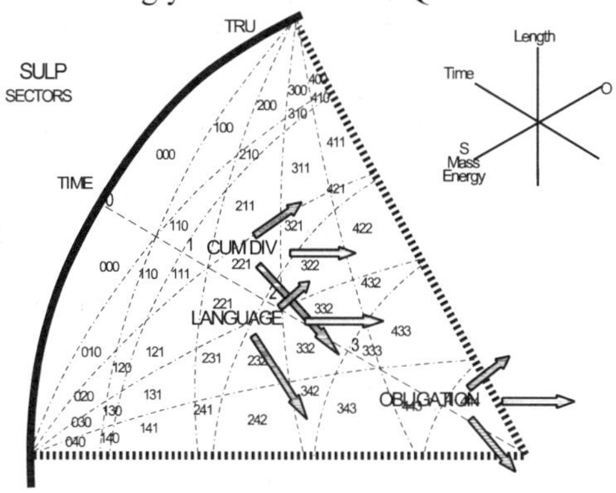

Fig 2-61 **CHES in SULP SECTOR**

Three subjective-large-past concepts and processes of TRU are shown. Although better in a sphere we can posit them on an ALT page. This still, gives us, a good understanding of their key roles in TRU. *Cum Div*, 2-1-2, is "With God with dividend". It is an expression which shows the great influence and benefit the association of peoples have had with GOD. It shows the potential of TRU. It also shows how God in the past has differed from TRU, for TRU the S-O focus would be on 5, i.e. 5-5-5. Language in the past, 3-3-1, is a great component of Á FAF, as well as power in TRU. Its origin began early in human evolution and has had a relatively unbroken record of continuous accelerative improvement with increasing benefit to mankind. Its posits and influences spread from this focus throughout much of TIT and POS in SPALT. Language and its growth are not only for representing TRU, but how it exists in TRU. Many other posits can be made for language, as many other concepts, depending upon what aspect or definition is being considered. Obligation, 4-4-3+, is closely related to duty and responsibility. It arises from the past, is more subjective than objective, yet contains much of both and is somewhat more of the larger than smaller of TRU. Because of the limits of space in an abridged text see the addenda, DIPTRU or IPTRU for abbreviations, NEC, names and further descriptions of POSITS in SPALT, MOT, TRU. and glossary.

One of the best examples of the importance of language and communication is seen in greetings noted in the previous, Chapter, 60. This usually implies a positive connotation of both empathy and hope. More deeply they simultaneously reveal both the importance of interaction in human life and the proof of TRU. For it focuses on the IA's massive and powerful driving force in two or more persons fulfilling their mass energy, Á, FAF in TRU. In like questions and answers there occur along several different parameters and planes, the difference between straight and devious, wise and ignorant, and right and wrong. It reveals the close bond of expanding IA to success and survival, and to TRU, or to diverse trivia or worse to 1/REM, AU, ATRU and ANTRU and all the dynamic mixtures and variations whose correction is always with us, and for us to provide.

PORCOM = Principle of the Repetitive Conception of Meaning. **PORSCOM** = Principle of the Repetitive Selective Concretion of Meaning. These two concepts concern the strength of terms in language and life. They are the basis for much of culture. Words are, in ways, similar to the Holy Spirit by showing how their dynamics and media in Á FAF is like the magic and powers of GOD or TRU. Language is a large portion of human predominance, Á FAF. This is shown by S-O expansion in TRU. It is also shown by: **POVAD** = Principle of Variant Dependency. POVAD is where the absolute 1:1 replica of TRU is dependent upon those variants which express TRU, e.g. means, methods, vehicles, media, types, and forms of OBT, SUT and ROT. It can be seen in primitive or very small populations where language is limited. Concepts through repetition reinforce themselves by their TRU meanings. DYNU is typically singular yet more often and preeminently separate but multiple and repetitive.

Greatest errors of human belief, thought and knowledge

When, and to the extent, our concepts and words are wrong all knowledge and human activity that proceeds from them can never be right for long. For human thought and judgement, our principles, laws, plans, and purposes can be no better than the degree they are right. This is a large reason why language continually changes. For in the selective world of a predominance for mankind, those peoples who have the best concepts, words, ideas, thoughts and judgements, principles, laws, plans and purposes will find right over wrong knowledge and beliefs that prevail. This is why so much of the numerous languages of the modern world differ, and how much their concepts and words do not correctly represent the actual world. This is why new concepts and words, as well a better way of looking at their use and the actual world is so necessary. We must dispel once and for all some of the greatest errors in human belief and language. Most prominently these are those greatest ideas of human life.

What is, and isn't, God? Events are posited in the dimensional construct of the spatial, temporal, and mass energy continua relative to the survival and life of homo sapiens, or how they exist in this special universe of best human life. What is the world? It is not the world of entities, anything, any or all humans, only their essences relative to the actual world, TRU. What am I? It is not what we sense or the things before us, and is not what we covet or choose. It is their selected VAPO in TRU. What is truth? It is not agreement between any subject and object. It is the identity of subject and object with its part and role in TRU. The same holds for what is actual, proportional or is of value, total, same, different, actually existent, and how do they all function together to best assure human survival.

Since our ideas of world, universe, God, even civilization are wrong or incomplete - and thereby misleading - it is essential that we find new concepts and terms to replace or improve them. For concepts, words, and ideas that are wrong can never provide correct unanimity or distinction. Without correct unanimity and distinction we will be continually at odds both with what is right, with each other, as they are of and in TRU..

The principles and doctrines of stoicism have been of great influence through its serious, logical and practical approach to life. This was more often an indirect influence to the profound, subtle and massive Á FAF necessary for TES and TRU. Stoicism's influence on Christianity is evident through the greatness of Stoic and Greek Doctrine in the Jewish World in the centuries before, during, and after Christ.(St. Peter and Paul) The uses of hymns and the emphasis upon meditation, endurance in life, and tolerance of persecution, are typical of each. Stoicism held that a person must never allow their feelings to deteriorate into irrational passions. Typical was virtue, predominately by wisdom, a necessary prerequisite to happiness, the overriding good, and the conviction in "Let no one break your will".

Knowledge repeatedly and intentionally selects what is best or preferred and revises itself for what is most relative, useful and especially, TRU. This is more or less irreversible and carries tremendous momentum. It is one of the greatest powers in TRU and can be easily posited and demonstrated in the accompanying diagrams of TRU by Homation on SPALT. Knowledge is a great example of CES and HOM. Mankind has, through the ages accumulated a great body of information for ready, available and continuous use. This accumulation is of all kinds of experience, education, evidence, facts, data, revelation, ideas, concepts, principles, laws, mechanisms, and reasoning that enable each and all to comprehend much of all, and most of TRU, to meet their needs. Human loss largely comes from failure to use this that is available. In the future we will increasingly and more readily learn to select and use that which is best, which is and will be TRU.

Fig 3-61

FOUR NOBLE TRUTHS, MORALITY and SULP

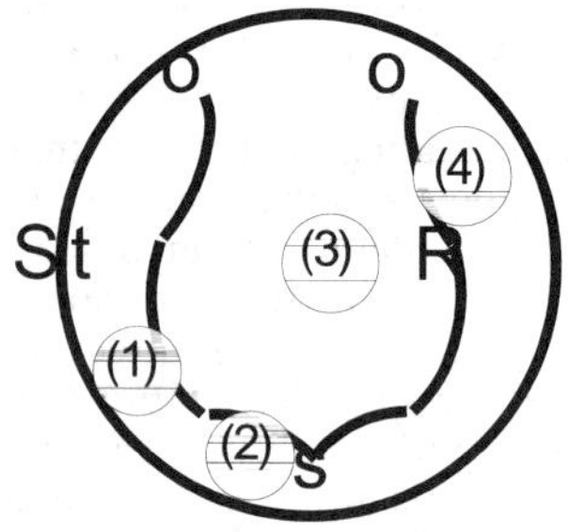

Buddhism arose, as so much of human religion and wisdom from an acute awareness of misfortunes and sufferings. From these were developed the four "noble truths" which include: (1). Suffering, (2). Knowledge of its cause, (3). Getting rid of passions by deliverance from suffering, and (4). Truth, the way of removing suffering by a system of moral discipline. These are what Buddha considered to be most noble. Such nobility is much the same as royalty in Indo-European ideology, which for the Greeks became the <u>Real</u>. So has reality come to be what is important, the most revealing and vivid truths of life. These were the nobility and royalty of the soul and a moral world of TES, PUREALITY, and TRU.

From Kant's Categorical Imperative, you must act as if the maxim from which you act were to become through your will a universal law of nature, as KOT. Yet we have a double explanation, how right and wrong, moral or immoral, ethical or unethical, promise or no promise, all may be interpreted and explained in one sense or another. This is from the individual, as no in being another, and as yes by TES and TRU. So that each person is living a split or double life, a Dr. Jekyl and Mr. Hyde. The Mr. Hyde or self's existence with freedom of any choice, is as sin, not TRU. The only way for the individual to escape from sin is escape all that is not TRU within and bring oneself back fully to TES. This is the true <u>will,</u> that by which all humans must sooner or later come to terms in life if they are to be, to be a part of TRU, rather than never to have lived at all. That which they bestow upon the world rather than those countless billions who stand as extinct species, non contributing races, tribes, groups or individuals.

The sin of this failure of each person to be TES, or ones part in TRU, biblically was explained as "the Fall" or mans lapse from innocence and goodness, the loss of Adams position of integrity, virtue and innocence, and his legacy of proclivity toward sin. This is immortality in reverse for as explained in the bible Adam's descendents, all human beings, were born in sin. Only to be reversed again through the atoning Grace of God and Christ, as mans redemption and forgiveness, as Christ's absolute TES, justification as God's and TRU's and reconciliation as man's. Together this great bonding in God or TRU to man, and man to God or TRU is called: **COET** = Covenant of Parts of TRU. (RAZ - Covet TRU) COET describes the New Testament's rendering of how the divine initiation of those coming to TES, as all that comes to its respective parts in TRU, enter into a covenant which is the destiny of man and our world directed by spiritual development. This is by Logical Verification, **LAV**, and **MERADEX**, Method of Righ Addition and Exclusion.

USIN or universal sin is all that has, is and will be non TRU, or antiTRU, that may have or could have been TRU. Sin applies especially to all that has been, is or will be done, and what we are or do that is not TRU. Original SIN applies to all from which we came that is not TRU. Since we as individuals and all peoples of the world are, do, and were even derived from much that is not TRU these terms or phrases have distinct and just meaning. Guilt like conscience is the sensed or felt realization of sin. It is what attains when we have or feel we have done wrong, wrong by life or TRU. We are motivated, act, and are driven by sin, guilt and reaction to sin. They are all of the NAMTRU world about us, commonly expressed in religion, all that is not positive or VAPO by TRU. Like turning negative into positive for TRU they can be directed to that which is most important of ourselves, TES and TRU.

LITPROS = Law of Imperfections of Today Equal Solutions of Tomorrow, Problem and Principle Solution. LITPROS combines several ideas of imperfections to their correction, negation of the past to action of the present, negation of the present to correction and positive in the future, problems, principles, and solutions. Imperfections are alien to TRU, yet prevalent in our NAMTRU world. They are abnormalities and weaknesses that cause loss of quality, predominance, and competitive edge.

POCEA = Principle of Obsolescence of Current Events and Contemporary Activities. What seems so astounding and important one day is obsolete the next. POCEA is a function of several factors: 1. Changing TRU, 2. Partiality, 3. Problem/Solution, 4. Crisis/stasis, 5. Focus of view, 6. Conditioned interest in social and political problems, 7. Escape from ones problems. POCEA does not lessen our information, nor make it necessary to continually delete it. Rather, POCEA reveals PIP, Homation, DYNS and DYNO, the great part short time periods play in ALT, and the greatness of TRUD and TRU. It also shows, IA, our frailty, and that of the observer's perspective.

FACH = False Championing or the Hero Fallacy. There is an ever present tendency among humans to become worked up, excited and involved in this or that cause. They look to leaders, heros to worship and simple solutions that will resolve all their problems. They champion and join this or that movement not so much out of TRU but lesser, even ATRU reasons. At best the cause or movement is only partially rightor slightly TRU. It usually lacks VAPO. Those who are in the movement will often, through its singular focus, come to neglect the rest of the world, and have a very disproportionate and narrow attitude and outlook. They are apt to blame all who do not have or do exactly what they have or do, do not fall in with the party line, or stand in opposition to them. This is to know the world in which we live, TRU, to reconcile all as it best functions in TRU, and live accordingly, by TES.

Only a glimmer of light is all that is necessary to guide and realize ones life and all by God, TREQ and TRU.. Once navigational aids are provided one can move, act, produce, and live in highly effective and successful ways. Without such a beacon of light no navigation is possible, all is lost. The whole of mans existence on earth has been a progressive story of finding such beacons. It has been of trial and error in search and discovery to orient, coordinate and find a success ful life. By God, TREQ or TRU even a small beacon gives direction. For this reason, even as poor as this beacon often was, and as little as it was equivalent to TRU, it did give the group, tribe, community, state, nation, culture, or civilization focus, direction and unity. This is why they were often as successful as they were. A simple singular direction is perfected and expanded into an overall plan for all needed courses of action in life.(SPALT,POA,TRU) The collective IA of all people is much different than that of one person, the person. It is constantly upheld and improved for the best of ENT and TIT. The collective IA of all people is not the same, yet has much sameness in common with all knowledge.

Solutions to the limits of ALT planar posits include, I. Concentrate only on finished positing in the spherical SPALT, and set aside any attempt to posit on a planar surface. This can be done with the sphere of SPALT of TIT. II. Provide rough posits on ALT or the planar surface of a texts page, expect or hope that all who envision it will extrapolate this to its actual state in the spherical SPALT. OPBT and TIT. This is done in the figures. III. Make CHES on different faces, segments and ways of using ALT of SPALT of TIT. This can be done by looking at SPALT not from its standard ALT face with time and mass-energy left to right and length top to bottom, but in whatever other way best suits one's view and posit by CHESU, CHES, CES, HOM and TRU. In IPTRU, in order to achieve an optimal demonstration of CHES, which are vitally essential to an adequate ROT and POT of TRU, we utilize all possible methods of posits.

In these chapters of HOM, 61-68, 71-78 and 81-88 in which we strive to posit all as they exist in TIT, we critically utilize several specific kinds of fields: 1. The sector in which numerical segments show on the face of ALT and into which each CHES can be exactly posited, 2. The SULP sector in which numerical segments do not show on the face of ALT, yet into their in-depth existence in SPALT. 3. The SULP sector in which numerical segments do not show on the face of ALT, and in which existence in TIT and foci in SPALT are less certain, thereby having posits of CES are apt to be rough and need future investigation and description, 4. The SULP sector envisioned from faces of SPALT other than the standard ALT, 5. ALT and SPALT wholes that represent TIT. On this field we posit those processes which are more universal or do not readily posit in the SULP sector. 6. How they exist OITIO to TRU and SPALT, in METAD, POMOT, or SPAT, on a planar surface or textbook page, revealing much of TRU and POT extraneous to TIT. 7. How they can be qualified, or the factors that may be added, to perfect their position that best replicates TIT. 8. All these when best combined.

There can be much duplication in the use of different fields or models for positing CES. The rule is to use that field or model that most accurately and explicitly demonstrates their role in TIT. Some CES like universals reach a long way through SPALT, even externally. Some act more as single forces and vectors. It is the highly diverse, multiple, and near universal CES and CHES of some of these, especially when of very high REM, that show the intricacy and greatness of TIT and the beauty of SPALT, REM, POS, HOM and POT, yet hardly begin to demonstrate all SULP processes in DYNU.

MEASURES of POSITS in SULP

B	Beacon	3-4-2 (5)	G	God	2-0-0 (6)	P Philosophy	0-1-2 (7)
Cp	Champion	4-3-3 (5)	Gl	Guilt	3-4-3 (5)	POCEA	4-4-4 (6)
Ch	Christ	2-3-1 (5)	H	Heritage	4-3-2 (5)	R Religion	2-3-1 (6)
Cm	Communication	4-3-4 (6)		HPP	1-2-3 (6)	Rf Reformation	4-2-2 (5)
E	Ethic(s)	4-4-3 (6)	K	Knowledge	1-1-2 (7)	S Science	4-2-1 (7)
	FACH	3-4-3 (5)	L	Language	4-3-3 (6)	Sholom Aleich	.4-4-3 (5)
	FERI	4-4-4 (5)	Ln	Learning	1-4-3 (6)	St Stoicism	2-4-1 (5)
	FIRAS	4-4-4 (5)		LITPROS	4-4-3 (6)	SULP	2.5-2.5-2.5
	GID	2-1-2 (6)		MALCOP	4-3-2 (5)	Sf Suffering	1-4-1 (6)
	god	4-1-1 (6)		Mirabile Dictu	3-4-2	USIN	4-1-2 (7)
G	God	2-0-0 (6)	M	Mysticism	2-4-2 (5)		

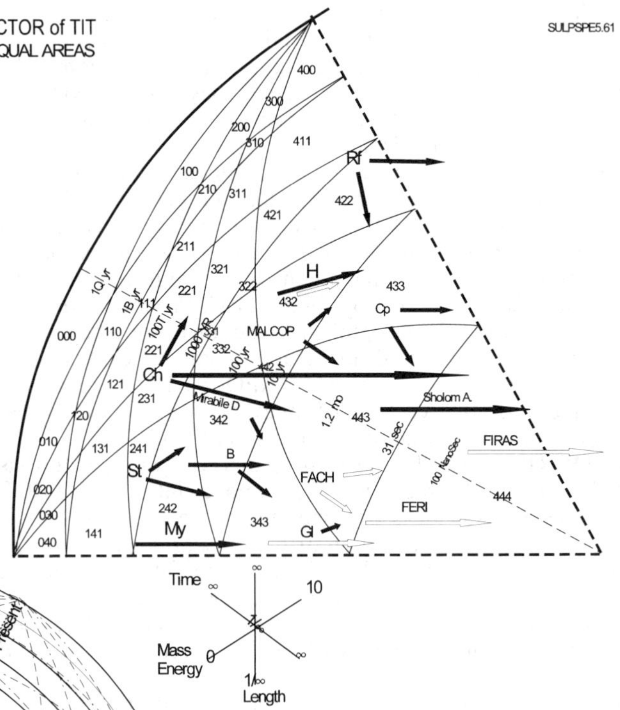

Fig 4-61 **VECTORS of CHESU in SULP SECTOR DYNU**

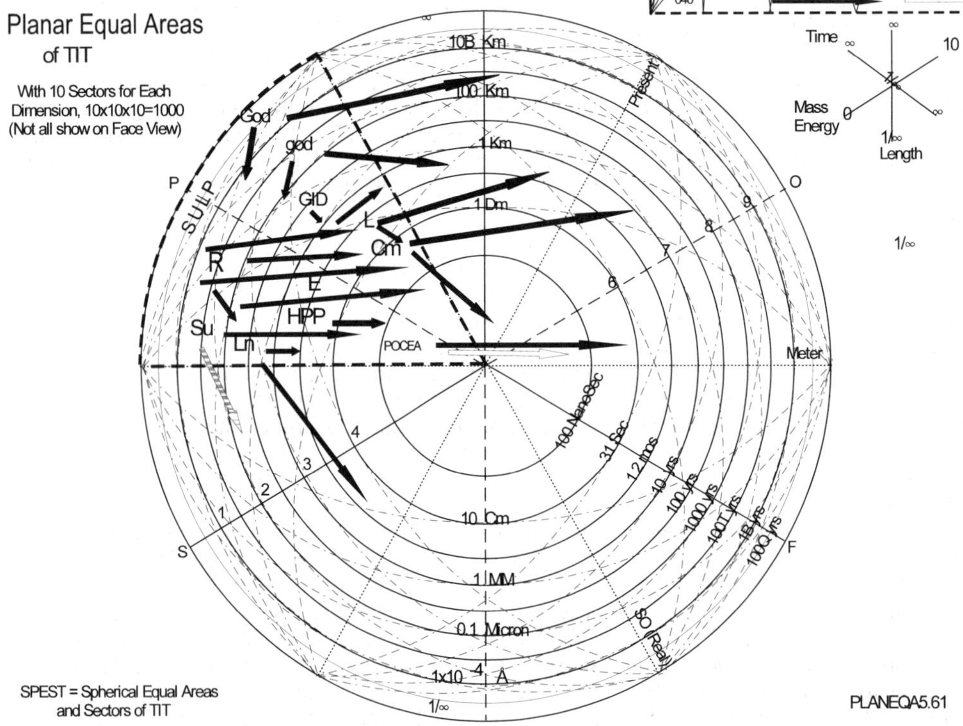

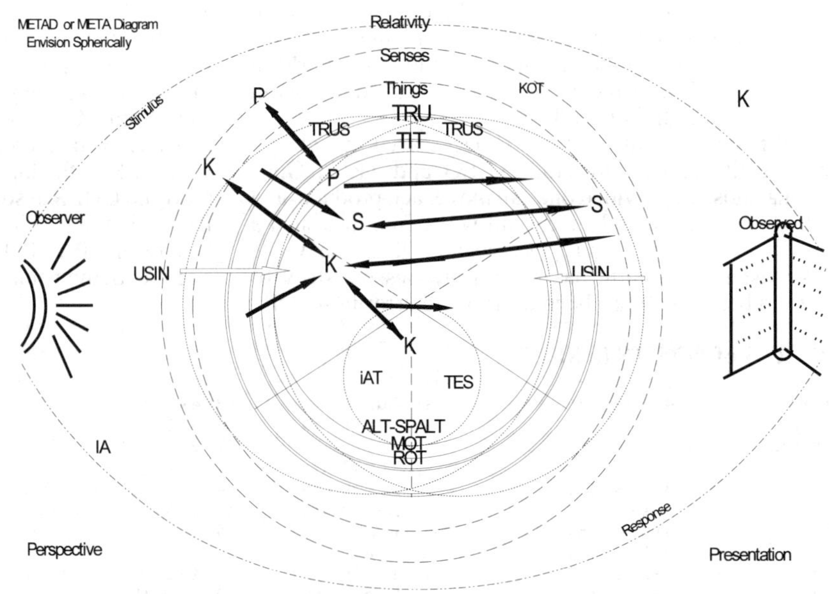

Figs 5-61, 6-61, **CHESU in SPALT and METAD**

CHAPTER 62

OLP or OBJECTIVE LARGE PAST PROCESSES in DYNU

CLIMATE, HISTORY, CULTURE, TERRITORY, POWER, CONFLICTS, BENEFITS, ALIENATION, and ACTION

In this Chapter we treat those processes that are of the objective, large and past sector of TRU. They can be great or small in value, i.e., 1R to ∞ R. They may be solitary or multiple in contrast with those of complexes or the whole of TRU. Known ideas and concepts expressed as words and phrases also have OLP implication. Technique, family, situation, technology, resource, locality, and many ES of TRU function as CES in OLP. Technique represents processes in TRU which are highly effective specialized means of accomplishing some specific task and goal. It is in the practice and perfection of technique that humans, by using established knowledge, forethought and routines best reveal their combined sustained mass energy power. Technology is more general term for all the processes of human endeavor that provide CES with the means to produce better goods and services. Family is that basic state in which related individuals live together as a basic social unit to satisfy their needs and better fulfill their roles in life, of TRU. It provides an optimal opportunity at a minimum of loss to men, women, children, and groups for each others sexual, emotional, social and other needs, security, success and happiness. It is particularly necessary for the growth, development, education, and training of children. Territory although largely of STA can be vividly of HOM and DYNU in the protection animals and humans need to secure their habitats. Resource, in its dynamic role, refers to processes involved in the available materials without which there could be no or deficient supply of what human life must have to survive, grow and flourish. Resources although largely objective can be subjective, as in knowledge. This shows the diversity of meaning and relativity of S-O and A. Since there are limited resources in a given region its problems tend to accumulate and reach critical mass. It is a major cause conflict, war, and misery which increases rapidly with increased population, density, demand, competition and selection.

Rank	Age	Taxon	Related	Examples
PHYLUM	700M-	VERTEBRATA	Pices (450M-)	
SUBPHYLUM	300M	AMPHIBIA	Reptilia	
CLASS	200M	MAMMALIA		
SUBCLASS	135M	Marsupial	Placentals	
INFRACLASS	90M	INSECTIVORA		Squirrels
ORDER	60	PRIMATE	Prosimian Haplorrhine	Lemurs Tarsiers
SUBORDER	40	ANTHROPOIDEA	Catopithecus Simian Higher Primates	
INFRAORDER	33	CATARRHINE		Old World Monkeys
SUPER F	25	HOMINOIDAE	Proconsul	Apes
FAMILY	18	HOMINIDAE	Kenyapithecus	Baboons
SUBFAMILY	13	HOMININAE		Orangutans
TRIBE (Subtribe)	9 (6M)	Gorillas	Chimpanzees (Australopithecus 'Lucy')	
GENUS	3.5	HOMO		Human
SUBGENUS	2.5	Homo Habilis	Homo Erectus	
SPECIES	1.5	HOMO SAPIENS		
SUBSPECIES	.6	Homo S. Neanderthalis		
VARIETY	.4	HOMO SAPIENS SAPIENS	Archaic	
SUBVARIETY	.3	HOMO SAPIENS SAPIENS	?"Eve"?	
MODERN MAN	.2+	Variation with adaptation to habitat,,		
RACE	.15-05-03-p	BLACK, BROWN, WHITE, YELLOW		
SUBRACE	-/1	Regional Diversity and Divisions		
HYBRIDIZATION		Within more or less same Breeding Population		
FUTURE TRENDS		Optimal balance of Outbreeding-Inbreeding Selection, Environment		

Table 1-62 **CLASSIFICATION of PHYLOGENY, to MODERN HUMANS**

Classifications and phylogeny tend to be stereotyped and apt to be somewhat misleading although highly informative. The addition of the commonly known animals at right to this list is more by way of relation and example rather than a role in human evolution. Like in the beginnings of language, the incipient stock was not only of one, a couple or small population. It was formed not in one generation but over many, and with variable rates of growth, causes and effects. Each, RU, RUS, RUT, RUTS, and TRUS may be compared with TRU and viewed with a kind of VAPO, DAOST, SCARLS and TROS of its own. This gives a fresh idea of their actual FAF both ENIT and TIT. When we consider the massive interaction of forces, causes and effects in human ancestry and adapt them to ATO and TIT we can begin to understand what our world has, is and will be. Only when interpreted in terms of the vast, decreasing durations and variations of time, space, and mass energy plus their external perspective can we fully understand TRU.

Phylogeny is largely a function of time, with the most profound changes tending to occur, relatively for man, by time in ALT and TRU, in the far past, and the less profound changes in the nearer term. Although much adverse change can occur within only one or a few generations, by mutation, beneficial changes or adaptations are almost always built upon, through hundreds, thousands, and millions of generations. Most large mutations are lethal. Each has a somewhat different schedule and character in its effect on the form, function, parts and processes on the human species or TRU. When we can increasingly realize these relationships and processes TRU we can better accept TRU, and understand many of the problems and principles of biology, anthropology, and genetics, as they have, are and will exist in TRU, TIT.

The variation, often by cycles, of what occurs astronomically and geophysically, exemplified by temperature and climate, has been a major force through time. It plays a large role in the acute selective processes of survival. Although diurnal and annual cycles are more dramatic, it is by more extended times important to the success/fai1ure, and prosperity/depression of various regions, cultures and civilizations.

There are many new and better tools, weapons, and means that are being constantly developed or improved. These propel mankind into a greater and better life as well as a Worse life depending on the Limits of Knowledge, their effects and Use, **WULK**. Their misuse in PPF is one of the greatest destroyers of humanity. This can be the greatest cause of extinction. They have a great role in predominance and natural selection, both in man versus nature and man versus man. Control over nature, transportation, economic means, power and energy resources and utilization, organizational means, efficiency and productivity as well as many more are all the CES and dynamic participants of TRU, TIT and DYNU.

One of the greatest problems for all persons to decide, as it is for physicians, is how much do you do for all who come to you for care or help. What and how much do you select, and what should and should not be done? This is mostly established by professional knowledge. Much is also filtered out before the visit or encounter, including the appointment process. There is a great temptation to help all, or to help those who are the most demanding and loudest. This is not limited to the care of the sick, it is common throughout human relations and endeavor. It is a determination all must make who give and seek help. When both the seeker and helper lack knowledge, especially TRU, the result is trivial at best and likely ATRU. When the seeker lacks knowledge and TRU but the helper does not the result is poor. When the seeker lacks knowledge and TRU and the helper is of TRU the result is likely to be much better. When both have both knowledge and TRU the result approaches the optimal. This especially applies to decision making within ones realm or field of expertise, technology, occupation or profession. The judgement can often be readily made in terms of established and preset routine and policy. However in this case the decisions and judgements will be no better and often worse than the policy. It is for this reason all must know TRU, and all policies, routines, rules, regulations, and laws must be based upon TRU or TREQ.

Charity largely comes from within the individual and TRU. Just as God so loved the world that he gave his only begotten son, so did Christ give his love that all might live. In this same spirit or the charitable principle, we must return the blessings and benefices, to generate and promote the great ideal of charity that best enables GOD as TREQ to be TIT. Charity must become the guide to overcome the deeply ingrained habit of failing to return to TRU the grace, love, and goods that all must live by. The ten commandments reveal how traditions of the past, oral and written, were largely influenced by prior traditions. The first is one of identity. It gives both the name, exclusiveness and compelling necessity of God. This is from TREQ, the same for God as it is TRU by name, TRIA and LATAN. It sets outright the focus of the observer observed on TRU. It is an external to internal relativity, as all must be, if they are to be, TRU. The second seeks to strengthen all representations of TRU by recognizing their limits and the partiality in TRU. The third seeks to hold all to, extol and work for the exclusiveness and absolute of TRU.(TRIA,TÂ) The fourth seeks to maintain any regularity in life that enhances TRU and its efficiency. The fifth through ninth exhort all to honor what will solidify and refrain from destroying the social fabric or FAF of society, living together as a cohesive, cooperative and constructive unit. These are proof of HOM and TRU. The fear we have is of what we are or will effect that is not TRU. It is what we are and do that denies and defies TRU.

Liberty is a Function of Relativity provided by the Conditioning and SCARLS of TRU that produce a free Range of motion or Dynamics, **LIFCORD**. The more relativity we have, the greater is our liberty, within limits. The greater our success and concentration at the center of TRU is the greater our A, L and T are. This is seen in how greatly we can make time stand still, or have ACEXIN. This is a conditioned liberty, conditioned by the same and positives of TRU. One might think the unrelated actions in ÂU would provide unbridled freedom. Yet this is not so. There is no guards, protections, or direction. It is because of the momentum, inertia, relations, sameness, and positive conditioning that exists in TRU that we have, within its limits, a free range of motion.

As great and appealing as the phrase affirmative action, and many others in common parlance are used, they are all limited by 1. How much they are not TRU, 2. How much they lack VAPO, 3. How much they are directed toward non and negative to TRU ends. All must be based upon TIT-TO.

Alienation is the process and mechanism whereby a person, group, part or entity is separated or rejected from its kind. The individual in the world exists in a continuum of alienation to identity. It is what drives much of the social consciousness of the individual. Yet much of this is pathological, in being ATRU. It is alienation from TRU that one must avoid and identity with TRU that counts.(TES) Alienation is closely related to **LASPRE**, the force of the first law of nature, Self PREeservation. Man lives in a constant state of conflict and oppression. Much of this is TRU, unavoidable, and to be reacted to as a part of fulfillment in life. Yet much is also ATRU, avoidable and needs to be reacted to and rejected in order to achieve, enhance and assure TRU. Life is a continuous fulfillment of being, in which dynamic interaction is inescapable. The best relief comes from personal freedom through temporal, spacial, and a mass-energy predominance, balanced with that of others. This involves a continuous emphasis upon personal, family, group, total, and TRU achievement and success.

There is great conflict in the world between the haves and the have-nots, the rich and the poor. The haves waste much and care little for the poor or in giving up what they have to benefit TREQ or TRU. The have-nots can't adequately exist on what they have. They can not be or become TRU unless they have what it takes to live an adequate life. So that much of both the haves and have-nots are not TRU. There is waste, that goes with finishing, waste with starting, and waste with all in between. With all economies there is, more or less, much waste. The feast and famine, boom and bust, slack and rush situations, conditions and states that occur in all segments of simple to most complex human endeavor is destructive to a good and effective existence. They usually result from collective imbalances, weaknesses, or accumulations of ATRU forces. It is in how to accommodate, plan for, and get all to function in accord with what is its proper, TRU role that we must constantly seek to achieve.

Waste is a common loss to this best of worlds. It is evident in our most basic dimensions. To waste time is to not exist, to use it is to live, but to use it wisely is to live correctly, by TRU. So is it also of change, variation, length, area, volume, the subjective, real and objective, and especially all combined as they make up TIT-TO.(ACEXIN)

Persecution comes in many forms, kinds and styles, national, political, governmental, modern, legal, financial, economic, cultural, religious, and societal. In fact broadly speaking all non TRU effect on individuals, groups and especially TES is persecution. It is produced in many ways, whereby each group or kind of world creates its own bias, standards, and powers in which all that is not contributory, all that is not a counterpart, is directly or indirectly persecuted or punished. This is especially so if it is close to this world, exposes its faults and threatens its existence. So is our present NAMTRU world generally composed of many layers of worlds that are functioning often more in conflict than cooperation. In TES and TRU we learn to rise above commitment to ATRU worlds and groups. For the individual, especially when immature, needs a sense of belonging, education, and experience which can come only from TRU. Mazdaism is a religion by which a person's actions seek to release the light in the world through moral conduct and a frugal and exemplary life. Unwillingness to make a contribution and work for the good of TRU is one of the great causes that must be continuously remembered and corrected. Inventions are the mainstay of rapid progress, means and control of the environment. Election is the selection by voting of candidates, individuals, or propositions in a community or state. In effect, it enables a citizen to select their preference. How good the result is can be no better than the degree to which this preference is TREQ or TRU.

MOTIVATION; kinds of	STIMULI - ACTIVATORS	RESTRAINTS /INHIBITORS	RESPONSE / BENEFITS
INTERNAL BODILY NEEDS/not needed	DEFICIENCY, DEPRIVATION, USE	SUFFICIENCY, FULFILLMENT	SEARCH and SATISFACTION OPTIMAL MAINTENANCE
EXTERNAL BODILY THREAT ACTUAL/illusion	PHYSICAL THREAT & ACTION	DEFENSES,ABILITIES,FEAR ATTENTION, PREDOMINANCE	REACTION, ESCAPE, FIGHT PREVENTION, UNDERSTANDING
ACQUISITION, POSSESS-IVENESS NEEDED/not	WANT, DESIRE, GREED	SIMPLICITY, FRUGALITY, SACRIFICE	ACQUISITION, POSSESSION, USE, MAINTENANCE
INTERESTS/SELF INTERESTS	EXPOSURE, APPEAL, BUILT-IN FAMILIARITY, DESIRE	KNOWLEDGE, COMMON GOOD, FOCUS, NEEDS	FAMILIARITY, SATISFACTION APPLICATION, SUCCESS
CARE, LOVE, SOCIABI-LITY? LACK OF "	ALIENATION, DEPENDENCE, WANT, UN-REMITTED NEED	SELF SUFFICIENCY, DUTY, RESPONSIBILITY, MERCY	LOVE, CHARITY, HOPE
IDENTITY, IMMORTAL-ITYABSENCE OF "	ALONENESS, CONFLICT, NEGATION, ANTRU	LIFE'S FULFILLMENT, KNOWLEDGE OF GOD or TRU	SELF SACRIFICE, UNION WITH GOD or TRU

Table 1-62 **RESTRAINTS and BENEFITS out and in TRU**

Much that is done in business and politics, as well as life in general, is from greed, fear and self interest. Much should be done from helpfulness, benefits, and success of this special universe of best human life. These are measures of a society or culture. They are effectively evaluated in the detailed S-R relationship of right and wrong of TRU and ANTITRU In Table I-62 items and measures of restraints and activities of positive benefit are noted. This helps to show and prove how it is usually as easy - in the short term - to do, live, go ahead and produce good as it is bad, and the overwhelming benefit this good provides - in the long term - when accounted for by TRU.

Fig. 1-62 **THE LIVE PRODUCT or CARRIER of TRU** (GELD)

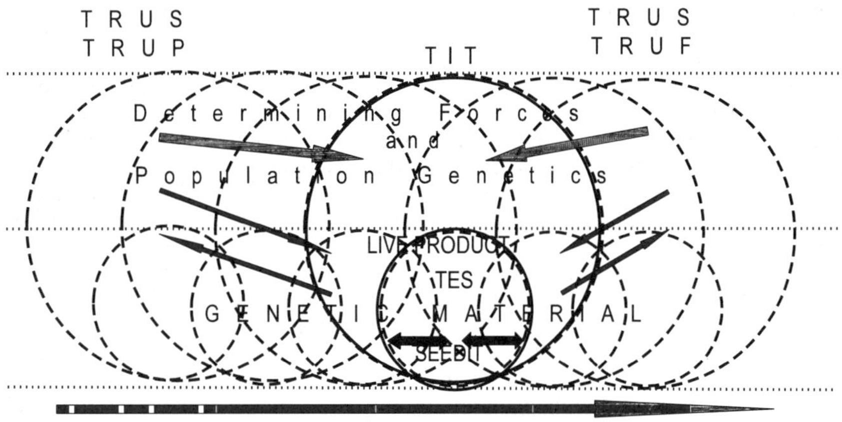

That TRU has a carrier or live product is certain. In all species their evolutionary course has been one of a live product accompanied by genetic material and determining forces. So is it for the human species and all it consists of. That less certain is the nature of this live product. TRU takes on many added, special and unique forms and functions.

Relative to what? What is the focus? How can we determine a fixed universe based upon a set center? Is this what TRIS provides? How much of TRUS? How does this affect or alter CENU? There is some precedent. This is the Species Homo Sapiens. Yet the delimitation is not absolute. For there has been a whole continuum of relativity through life forms, animal, mammalian, and especially primate species to homo sapiens. This is only half of the continuum for there is also that within the species. So that each and all in this continuum have some, and think they have total, claim to being the focus and basis for any set final universe. **COCOT** = Continuum of Contributors Centering or Focus of The Relative Universe. That there is a predominance of contributors based upon homo sapiens can hardly be denied. That there also are many clines and limits within this continuum seem more than certain. So that we have a continuum made up of two poles, that of all life forms before or other than human, and that of all that constitutes human. This continuum is not static, it is highly dynamic. The dynamic nature of it is dialectic is one in which there is great competition for position and power within the continuum. Just as each person struggles to be 'king of the hill', for their 'place in the sun', and predominance of various sorts, so do all clines and gradients in the continuum struggle to hold to all to themselves as the center. This is the basis and much of the reason for one's ego, vanity, deceit, and many other positive and negative characteristics.This is called the **COCDET** or Continuum of Contributors Centering Dynamic parts, persons, elements, etc. of TRU. The great determinant and leveler of all these various contributors to the center, and all TRU is the power of existing TRU on them. This reduces much of their effect so that it is not how much TRU is brought to them, but they to TRU. So that there are continuously tremendous forces and processes at work moving all to - and from TRU - and to and from themselves. The latter is the weak force, the former the strong. **COCWAS** or Continuum of Contributors Centering Dynamic weak and strong Forces of gradients, elements, persons, parts, and processes. Every element or participant has its force, cause and effect, and process. The weak force is weak only as it applies to elemental effects on TRU. It is a powerful force within the elements, gradients, persons, or entities themselves. However, this is largely and variably ATRU. Thus the weak and strong forces on the continuum of contribution to TRU depend upon their role in and out of TRU, OITIO. **WASFOPT** Weak and Strong Forces a Function of their participation, power and predominance in TRU.

The live product can be more easily understood by lineage than by a single individual. Some of the different kinds include,

1. Single, 2. Highly select, as 'Chosen Seed', 3. Mean seeded lineage 4. Fluctuant selective seed lineage, 5. Slightly contributory seed lineage, 6. Participants without being directly in line, 7. Other variable and indirect factors and forces in the Live Product, 8. The total mass of contributing forces and effects, even infinitesimal, to the lineage. Such a continuum of heredity and population genetics basic to TRU is called: **SEEDIT** = Seeding of and In TRU. Although SEEDIT, genetics, inheritance and the continuum of the Live Product in TRU are highly specialized topics for which genetic science needs to be fully applied, their fundamental role in TRU warrants a good accounting and reconciliation if we are to adequately develop their CHES, DYNU, HOM and ACC. Since by TRIA we limit evolution and alter TRUS, so also is SEEDIT not what we ordinarily might think the seed line of the human race of population genetics to be. Figure 1-62 provides representation of its exclusive role in TIT, as well as in TRUS.

Mutations in sexual cells, a kind of genetic clock, can cladistically trace male inheritance. They also help determine isolation of genetic variations which are of beneficial or detrimental value to mutation. To fully use genetic information of genotypes, phenotypes, and carried mutations, their roles as forces, mechanisms, operations, principles, and concepts can be accounted for and shown for different time periods. These are what is occurring at 1, 10, 100, 1,000,,,, generations, and all in between. (See IPTRU, Atlas) Similarly for the future quality of individuals or phenotypes the use of the Genome and DNA testing in the prevention, management and control of many diseases is rapidly advancing. It is the differentiation of the genome to account for vast interactions that information processing must be developed

The focal areas and functional centers of the brain are good examples of the differential, singular, and quantum like make up of much of TRU. This is similarly seen in the 100,000,000,000 nerve cells in the brain. Naturally their form and function converge on, and are continually being altered and adapted to TRU. The visual and auditory cortices, speech centers, polymodal association area, and many others provide the regulatory means by which various specific physiological, neurological and behavioral functions are carried out.

In figures of this section some of the processes in or of TIT are described by sector, SPALT and METAD to demonstrate the ordered dynamics of TRU. Although it is a highly effective means of representing this aspect of TIT, its singular and separate processes are not without certain problems. 1. Such a surface representation, especially on a planar field does not do justice to either a good representation, these processes, or TRU. Yet for a text this is the simplest means and is thereby used, to a qualified and conditional extent. These processes are inside of TRU and not on the surface, so that any errors resulting from a surface viewing should be corrected or at least kept in mind, ECRO. 3. Most processes tend to work toward the center and future. This is how they tend to converge on TRU or follow through with their action-reaction, stimulus-response and dynamic trends. 4. Most processes are not fixed at one point, they are often extended over a range. 5. This range is variable per dimension, and the variability is not highly correlated between the two or three different dimensions. 6. The sectors in this OLP diagram are not always the areas in which like sectors of an item will be posited. 7. The posit of a process, although largely of its TRU function, will often tend to or in effect contain negatives, what is less than pure TIT. When negative the vector is shown with a crosshatched rather than solid arrow.

To correct these problems all should seek to develop an ever greater capacity, ability and achievement of envisioning and thinking globally of TIT-TO and not its singular processes or representation. This should not detract from the importance of singular and separate processes in a positive objective approach to TRU. It should help to repeatedly remind us that when we are working with or thinking of these separate processes that errors are likely to be made if all are not envisioned as they are in TIT. This great principle and rule is called: **LETS** = Law of Envisioning and Thinking TRU Spherically and Solidly, as Enormous, Extensive, Large, total and dynamic view of TIT. LETS are those views of TIT that optimally utilize VAPO, SCARLS, ACEXIN and DAOST. LETS maybe remembered by 'lets do it'. This condition for OLP and DYNU is like that for all we study, investigate and consider. They as ENIT or entities in themselves, must always be corrected by COR and TIT. The correction factor, COR will be what is necessary to keep all in optimal subject-object relationship as they are TIT. Classifications and phylogeny tend to be stereotyped and apt to be somewhat misleading, thereby reconciled by POS in TO.

Individual differences are baffling to us in our encounters and dealings with others. How can others be right if I am right and they are so different? The question, then arises, 'am I right, are others right, or are both of us wrong to a great extent'? We do not recognize the reasons for this until we have experienced life, seen what others do, others think, and what has determined their actions and minds. One of the best ways of understanding this is to search through what many individuals have learned in life, what they know best, and how they look at it and all. When we see the great differences in the problems and experiences others have been exposed to and what language, concepts, ideas, principles, laws, and mechanisms they have learned or not learned, we can see how much difference there is. No two individuals think alike, just as no two have the same face. In fact the same individual thinks much different, as is their countenance, at different times in life. Yet it is their deepest and strongest desires, interests, lessons, wishes, hopes, fears, and the ideas that accompany them, that determine their frame of mind. Just like their personalities are so different so are their mental attitudes, reactions, and seeming oppositions to others. Most of these oppositions are not ones of conflict, especially ATRU conflict. They are oppositions of differences in TRU. For the individual, at best, which they mostly aspire to, is highly differential and their attitudes, thoughts, and reactions which seem so opposite from another individual are not so bad. Only the first and the second or other persons are at levels of differentials, parts of TRU, MATRU, AMTRU, NAMTRU, and ATRU that are decreasingly the same, and increasingly different, much worse than the D or Difference of **DAOST,** Different Aspects of the Same Thing. These levels of partiality and ATRU are such that IPRs are in perpetual dialectic AMTRU and ATRU. Only when all in the IPR situation are of the highest TRU can either or all expect to find actual and correct concensus. **MIPRIN** = Mistaken Identity or IPR Interpretation; Fallacy of. **LICSOT** = Law of IPR Concensus only in the Integral or Samnesses of TRU

About the greatest problem past present and future, is the awesome dichotomy between an individual's being, their egos, vanity, need to survive, reaction formations, and their being the required center, actor, carrier, and vehicle in the universe. Their frailty, weaknesses, and minuscule place in all, even all TRU renders them helpless, devious and never satisfied. To assist the individual adapt and best use this state, is the great duty of all. It needs to, and can be, by way of greatest meaning and means as by TES and TIT, largely corrected. It is why we must prove TRU and provide IPTRU and HUB. It is especially how to develop in our leaders, teachers, philosophers, preachers, doctors, professionals, and all, the exemplary and guiding roles, the basics of TRU. How to have the best of both, to live and to succeed, is to be TRU, to nourish ones soul, TES.

NIB = Non Being, in and out of the Individual. In contrast to TES and DAK, it can be TRU or ATRU. NIB is largely equivalent to NAMTRU It was recognized early in the development of TRU and IPTRU's diagrams of OITIO, to make more distinct all that stands in contrast to the 'being' of an individual and TES. NIB helps us recognize much of the world that is both of mass energy, length and time frames that are not within the realm of the individual. We speak of TRU as focused upon man, as well as look at it subjectively from the first person. This is true, at least overall. For there is much non being in and about the individual that has actual being as it indirectly effects TRU. NIB also, like NAMTRU, applies to what is outside of TRU. It signifies much that is about and around the individual or first person that is not TRU.

Interpersonal units of force and environmental units of situation can be quantified to provide a better measure of the dynamics of TRU. Much of what determines human success and survival depends upon work, natural inclinations, activities, living conditions, social functions, provisions for the future, reproduction, education, maturation, and many others. The determinants of men, women, children and different components of society vary somewhat depending upon their needs and roles. The determinants of society depend upon the collective accretion of individual units, plus the added states in totality that such aggregates of peoples are characterized by.

The test and proof of TRU, is whether and how much our ancestors thought and lived for the past, present, or future, for what they had received, what they desired, or lived for their own, kind, family, tribe, and humanity. How much they lived for the continuity of these or many other reasons. This tells us both what their roles were and are in TRU and what we should be and do, TES and TRU. It also helps to reveal what TRU is and how it works.

TIRT - TRU is Righteousness, and Righteousness is TRU. The righteous nature of early biblical Gods was of a human, although unseen, universal form. They were formed and functioned variably in an anthropomorphic and universal mode. Love, sympathy, hate, and power more or less characterized EL, Yahweh, Jehovah, Jove, and 'God'. Righteousness comes from God as much as God is, as TRU, what is right. This is what they are by VAPO, SCARLS and DAOST, TROS, TRIS, TIT, TRIA and all that is correctly TRU. God, as TRU, doesn't, so much, make what is right, as what is right makes up and empowers God. Yet when what makes God is right it is TRU. Once TRU it is right that exudes righteousness. TRU is right and righteousness follows in TIT-TO. Live every day as if it were the Sabbath, was expressed by Christ, noted from an Egyptian Papyrus. This exhorts all persons to live their religion, not for a few minutes, an hour on Sunday, or any time other than continuously, as they are TES and TRU.

If our ancestors were the best there were, why are we so much less than the best? This is a fundamental question whose subtle and profound answers must be carefully considered by each and every person. It will set the stage and help one plan to get the best out of life. There are several answers including:

1. Although this holds for the distant past, during more recent generations there is often an accumulative loss of quality in the individual. (APRA, LADCAP) 2. What may have been the best in a single line of descent in the past is not the same as what may be best under different conditions in the near past, present and future. 3. The farther out in time from the present, the fewer existing humans are a part of either the past or future population, and the farther out from the center each person is apt to be. 4. We are now only one of a great population and thereby our exclusiveness as an individual person is much diluted and achieves exclusiveness and absolution only by identity with its part and role in TIT, TÍS→TES. Even at an optimum a person of the present is not the same as probands of the past where populations were much smaller or even as a thread in continuity .(LABPER) 5. Since the mass energy Á FAF of humans has changed through time, and since what is of TIT-TO present past is somewhat different than what was operative at times in the past we must adjust our measure of best accordingly. This is even more important for the future, giving it a degree of freedom necessary for optimal adaptation.
6. Since being the best consists of many abilities, we must consider all of these in their correct totality and massive interactive forces when comparing, tracing through the processes, and interpreting their roles and their implications in evaluating present day individuals, groups, populations, and TRU.(LOPAB,LABO,reason for IPTRU) 7. What of all that has gone to make us, and enables us to best live is the basis for TRU, past present and future. Since each of us exists only in this way we must decrease all that prevents this, and increase all that enhances it.(POMU, LABPER, LABS) This loss of direct dependence upon individual best, is one of the great causes of our day. For it produces in us a complacency about the dire circumstances under which TIT-TO depends upon us.

Fig. 2-62 **CES, CHES and CHESU in OLP SECTOR**

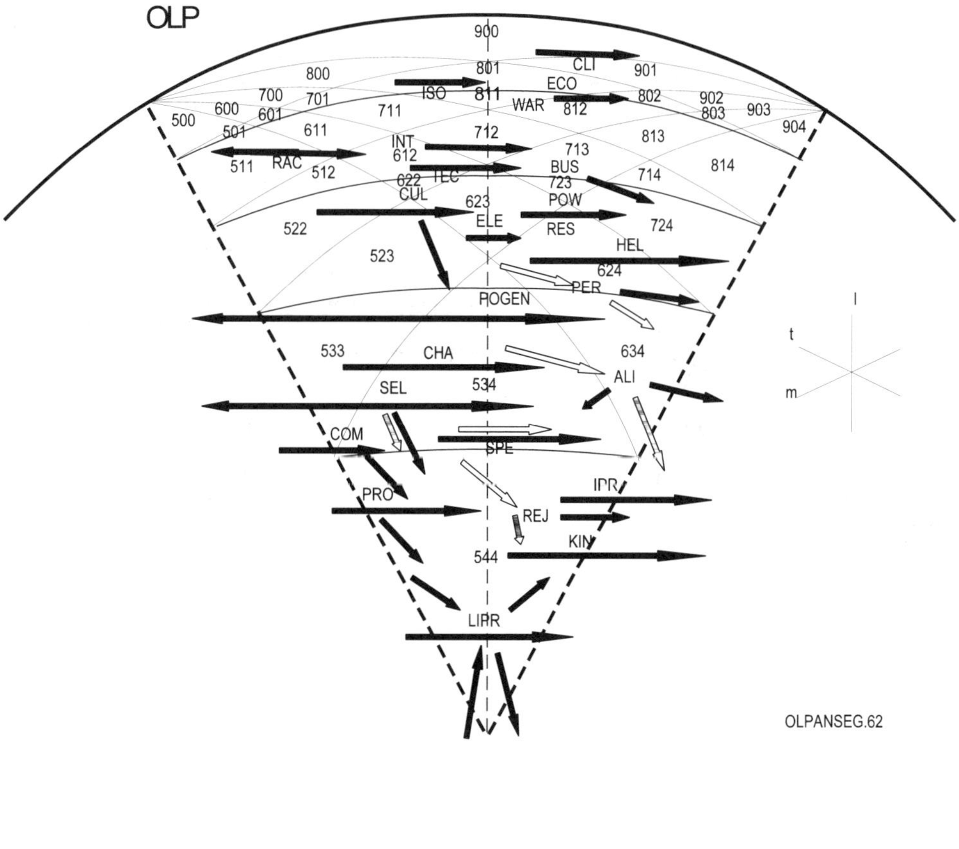

The following concepts and terms have been used, whose processes in the dynamics of TRU can be posited in OLP.

Alienation	6.5 - 3 - 4 (S)	Environment	9 - 1 - ~ (M)	Liberty	7 -3.5-2.5 (M)	Rejection	5 - 4 - 4 (S)
Alliances	7 - 2 - 2 (S)	Facio ut Facias	8 -4.5- 5 (T)	LIFCORD	6 - 5 - 4	Resource(s)	7 -2.5-2.5 (TM)
Benefits	6.5-3.5-2.5 (T)	Family	7 - 4 -3.5 (ST)	Live Product	5 - 4 - 4 (T)	SEEDIT	7 - 6 - 3
Business	7.5-1.5 - 3 (T)	Help	7 -3.5- 3 (S)	Mazdakism	5 - 4 -1.5 (S)	SocialInteraction	7.5- 3-2.5 (T)
Cause(s)	7 - 5 - 2 (M)	Haves\Have Nots	8..5 - 4 - 3 (T)	Material	10 - 3 - 4 (T)	Specialization	6 -4.5- 3 (S)
Charity	6 -3.5-2.5 (S)	Industry	8.5 -2.0-2.5 (T)	NIB	9.5 - - - - (M)	Technology	6.5- 2 -2.5 (TM)
Climate	9.5- 1 - 1 (T)	Integration	6.5 -1.0- 2 (T)	OLP	7.5 -2.5-2.5	Territory	9 -2.5-3.5 (T)
Commandments	5 -3.5-2.5 (S)	Inventions	7 - 4 - 3 (T)	Persecution	6 -3.5-2.5 (S)	Tools	8.5-4.5- 3 (T)
Conflict	7 - 3 - 2 (TM)	IPR	6 - 4 - 4 (T)	Population Genetics		War	8 - 1 - 2.5 (S)
Culture	6 - 2 - 2 (T)	Isolation	8 - 1 - 1 (ST)		6.5- 3 - 3 (S)	Waste	7.5- 3.5- 3 (T)
Do, Doing	8.5-4.5- 5 (T)	Kindness	6 - 4 - 4 (T)	Power	7 -2.5- 3 (T)(M)	Weapons	8.5- 4.5- 2 (S)
Ecology	9.0- 1 - 1 (T)	LASPRE	5 - 5 - 4	Professional	5 -4.5- 3 (S)	WULK	7 - 4 - 2
Election	6 -2.5- 3 (S)	LETS	2 - 5 -7.5	Race	6 -1.5-1.5 (T)		

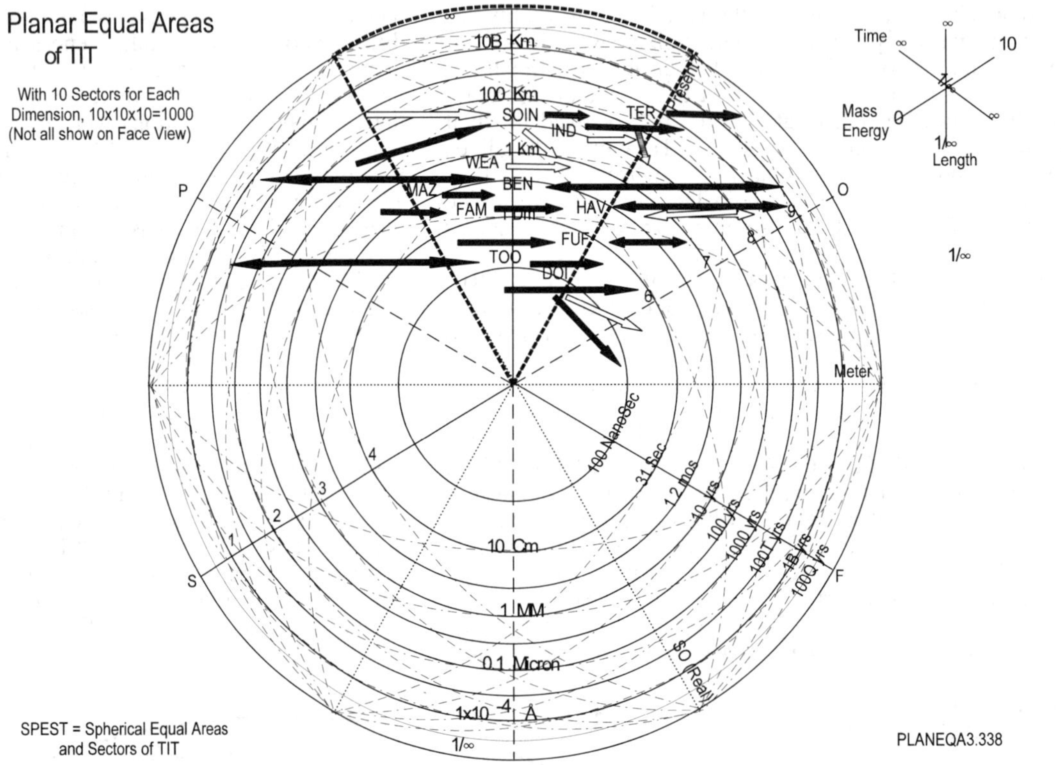

Figures 3-62 CES, CHES and CHESU in OLP SECTOR of SPALT

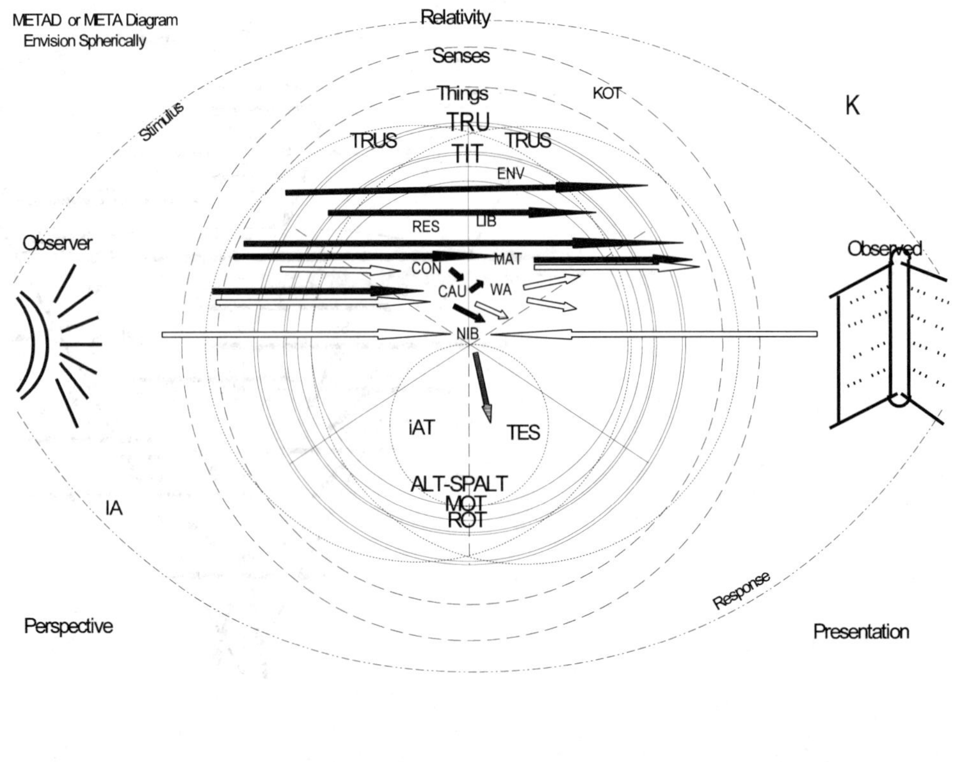

Figure 4-62 CES, CHES and CHESU in OLP SECTOR of METAD

CHAPTER 63

OSP or OBJECTIVE, SMALL, PAST PROCESSES of DYNU

BODILY FUNCTIONS and NEEDS, HEALTH, TOOLS PREVENTION, ACTION, DO, USE, BEHAVIOR

The greatness of OSP is evident in much of the world around and in us if we can only sense what is TRU. The many physical acts, processes, and functions in our every day lives, activities, and work as well as the operations of our internal organs reveal a vast number of processes vital to life of which we and TRU are composed. It is the successful and sufficient, proportional and balanced function that makes up a large amount of OSP. These may be analyzed from the standpoint of excellence, degree of performance, effect and reversibility in TRUS or TIT.

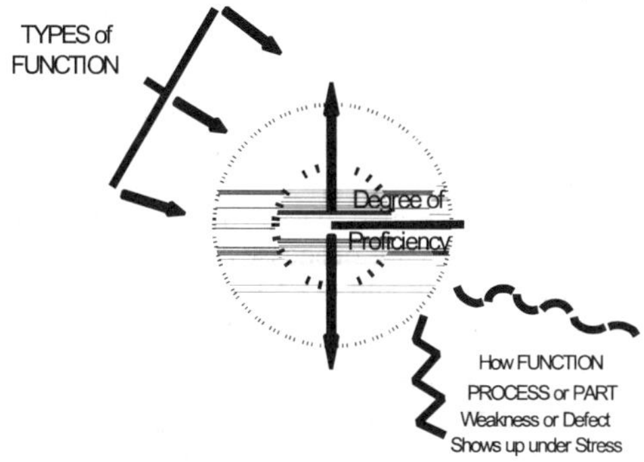

Fig 1-63 **DIFFERENT FUNCTIONS, PROFICIENCY and OPERATIONS**

The amount of detail that is shown when a person is caught in the middle of some dangerous situation is often strongly experienced. To plunge into the middle of something or a different world often produces in the subject a feeling of overwhelming impossibility. For what enables one to obtain an often obvious partial is not so easy. This is very true of the many sciences, fields, topics, and areas of activity in our daily lives, that we study or work with. Their intricacies and interactive functions seem infinite. To correlate these with all their concepts, principles, laws and mechanisms likewise seems impossible. Yet they must always, to be adequately known and successfully completed, be reduced, returned to, and posited in TRU. This often takes great knowledge, imagination, and work to separate out what is ATRU in each case. The great advantage of TRU is that it is so formed, unified, consistent, and available for repeated use, that what may seem to be an impossibility can become an increasing practical probability. This is shown by the mastery of any science, field, or difficult area of operation.

There are a great many healing, metabolic and regenerative processes going on at the same time. They are highly interactive, both as a part of normal and natural physiology and as a part of abnormal or pathological actions. Many of these processes, actions, reactions, and mechanisms are the same variants of TRU..The processes and forces that act in TRU may be largely and effectively enumerated and posited in SPALT.

Key Dynamic Physiological Mechanisms (KDPM)
Organ Potential
Cellular Potential
Reproductive Organ Environment
Female Window of Fertility
Ability to Conceive
Female Pelvic Capacity (**FPC**)
Setting Potential or Habitat
Genetic Quotient
Acquired Quotient
Positive Forces Producing
Contact Rate, optimal (**OPTCR**)
Birth Control
Fertility
Fertility Rates (**FR**)
Birth Rate
Coital Rate
Death Rate
Efficiency Rates
Age Factors
Sperm, Egg, Ova
Exact Seed Line
Mass Seed Line (**MSL**)
Germinal Covenants (**GC**)
Upright Stature

Endurance
Use of Hands
Opposable Thumb
Language
Expression
Conceptualization
Skull Capacity
Differential Integrative Selection
Generations
Genetic Mosaic Maxima
Genetic Flux
Reproductive Quality
Outbreeding/Inbreeding Continuum
Concentration/Dilution Factors in
Selection Rate in Mass
Levels of Selection Rate
Bioflow/Region
Linear Genetic Flow
Balanced Genetic Flow
Same and Different in Selection
Subterfused Determinants and,
Mixed Milieu Forces
Set of Maximum Genetic Quotient
Love, Sexual Desire
Need, Want
Deprivation

Fear, Unwanted/Wanted
Zeal, Hunting Effort
Ability to Search
Character, Virtue
Per Capita Income
Individual Dynamics
Adaptability
Energy
Protection from extremes, Disease,
Adversity
INEG
Precipitating Causes
Nutritional Level
Aggressiveness
Conditioning, Habits
Experience, Wisdom, Memory
Calculation, Preparing for Future
Diluting Effect
Multiple Unconformities Torque
Selective Rates, Factorial
Diversity Segregation
Waste and Turmoil, Change
Sublimation, Libido, Ego, Id.
Increasing Means, Commodities,
Increasing Simplicity
Youth and Age differences

Many of these forces and processes are reproductive and sexual. This is to be expected. They are important in maintaining the highest quality of natural selection competitively and to assure continuity of the species. Many other processes, forces and effects can be noted. These give an idea of the immensity of OSP and all sectors in TIT.

The population explosion, accelerative changes in human life, and the variable competitive and selective forces make an interpretation of what is TRU most important, interesting, and necessary for future action .in OSP increasingly intricate, detailed, and demanding. Most OSP forces and their dynamic counterparts can be well posited in SPALT.

There are 30,000 genes per person and a exponentially greater number of combinations, permutations, and functions. There are many interacting enzymes and molecules in the reproductive processes. The chromosomal banding patterns of various species show that karyotypic evolution occurred at different rates and involves different types of rearrangements. The rates of patterns of chromosomal makeup occurring in the species are an important part of the dynamic processes. They are important in providing strong and healthy phenotypes carriers of human capacity. These are in continuous, cyclical, and variable operation in place and time in TRU. They work in total human population in many different ways over different periods of time. The altered ALT produces additional complex characteristics and properties to reproduction and heredity. It should be no surprise that genetics and heredity have not been sufficiently available for TREQ, ACC and selective management and control.

The dynamics of genes and chromosomes in all states of their life cycles, especially in sexual mitosis, exemplifies the complexities of combinations and permutations. Linkage, cross over, and cross over values of genes in different parts of the same chromosome reveal the great continuum between most beneficial to most detrimental effects of determining forces on the live product.

What different parts or organs of the human body are for is determinative of their existence, FAF, and TRU. Each one is a prime example of DYNU. Their changes through OSP reveal how well the human body and TRU have developed, adapted, survived, and become predominant. There is a balance between softness and hardness, fluidity and solidity, expansion and contraction, and many others. They must all interact, balance out, be proportional and function in the various ways that optimally produce and sustain TRU. For the species this is the survival of all that best enables the most fit to survive in its environment. The manner in which primates and humans used the objects in their environment reveals the way DYNU existed in the past, that is part of present and future. How this developed exponentially is shown by:

100-10 million years Pre Use of Tools - increasing use of external objects (birds, beavers, monkeys)

10-1 million years Primitive Tool Making - beginnings of implements, any kind, wood, and stone (primates)

1,000,000 - 100,000 years, Primitive Stone Implements - increasing assembly of stone tools, Early Paleolithic (Homo)

100,000 - 10,000 years, More Advanced Stone Implements Refined to Polished stone industries, Middle and Late Paleolithic, Mousterian, Microlithic (Homo sapiens archaic, Neanderthal, modern. Homo sapiens sapiens)

10,000 - 1.000 Neolithic, Metalic Implements - Rapid expansion of technology and cultures, Use of copper, bronze iron with a vast complexity of types and applications.

1,000 - 100 Scientific and Industrial Revolution, Expansion.

100 -Presemt-100+ Rapid advance of all applied sciences, specialization, communication media, controls,

Future - Knowledge organization, M&C,VAPO and TRU

This is a quick survey of the accelerated development of the tools humans use in controlling the environment and enhancing life. They have shaped much of human Á mass energy development, as well as the altered environment that makes up TRU. It becomes clearer how rapidly TRU is and will be changing in the near, mid and long term future. Yet much changes very little. It is the conflict between these two that will increasingly occupy the attention and decisions in the future if TIT is to flourish.

Among the essentials, basic needs and fundamental necessities of life, most are processes which can be developed and shown through OSP and DYNU. They consist, directly or indirectly, of what constitutes OSP. Some of these are,

FOOD	ADAPTATION	LEARNING	RECREATION	AIMS
SHELTER	SELECTION	TRAINING	COORDINATION	OBJECTIVES
CLOTHING	REPRODUCTION	EDUCATION	ORIENTATION	PURPOSE
WATER	CHILDBIRTH	CONTINUING EDUCATION	EXPERIENCE	SUCCESS
AIR	NURSING of BABIES	COMPANIONSHIP	PERSONALITY	FULFILLMENT
OXYGEN	NURTURING of INFANTS	IPR	MENTALITY	SURVIVAL
VITAMINS	CARE OF CHILDREN	SOCIALIZING	LOVE	
MINERALS	COMMUNICATION	SLEEP	EXPRESSION	
INTERNAL ORDER	LANGUAGE	REST		

These and many others are the core of our existence and reveal the contribution of OSP CES to TRU. There must be a coincidence and identity of mans basic needs and highest principles. At the same time all TROS aspects or states of known and TRU are to be supported.(CALP,EC,NEC)

POBUF = Processes of Biological responses, Urges, and Feeling. Much of a person's rapid and massive functions, activities, and processes occur at subconscious levels. Through autonomic and reflex mechanisms and actions at atomic, molecular, genetic, and adaptive levels responses, urges, feelings and needs are met.

The human brain is much more than an organ of thought, it is an organ of survival. It is made in such a way as to enable us to accept as truth that which is only advantage. It accepts as truth only that which is identified, or 'is'. It is not capable of distinguishing spontaneously whether or not it is TRU. The higher the level of SÁ the less capable, or tentative, this is. Only in a rough and intentional way, as phenomenologically to SO or Á, does it incline to TRU. Being born does not enable life. They must be nurtured, experienced, matured, grow and learn to live. It is how well this is done that enables them to be or not to be, to be TRU.

To walk another mile was the message given in the Sermon on the Mount. It was Christ's counsel to keep going, to have an indomitable will, search for the truth, and do the work of God. It was his prescription for attaining the Kingdom of God.(TES) This and the added endeavor necessary are the equivalents of heaven and TRU. How to attain KOT, TES, and TO by acquiring the wisdom, endeavor, and all necessary is a central feature of Christ's immortality and thereby divinity by assuring his disciples and early Christianity that staying power without which it would have been only one more passing eastern religion. It indelibly reminds us to continuously move onward by, for, and in TREQ or TRU.

What are you doing? In all activity there is a relation between what one is doing and what one is thinking or has their attention on. This is a large part of IA, Á, and the virtual reality of life. It is the bond of attention and action. This is a continuous selective force, whether what one has done, is to do, or is already in the act of doing that is right or wrong. There are several kinds of servers, or those who labor and work should be able to provide the service they purport to provide. 1 Those who don't know how to do it. 2. Those who don't care to do it. 3. Those who don't want to do it because it will not pay what they want or expect. 4. Those who attempt to do it but fail, or worse cause it to become much worse. 5. Those who know, do, and complete it perfectly with minimal loss and trouble. In the modern world, a world of highly productive, efficient, trained, skilled, effective, technical, professional, productive, and pleasant workers there is only a place for the latter, #4. For it is in these that the success/failure, of commerce, the economy, continuation or extinction, and the human race depend. It is when a person or all mankind lose sight of what their actual goals are, TES and TRU that they provide ever poorer service.

CONAP = Concept of Compelling the Necessity and Certainty of Proper and Perfect Automated Act. CONAP signifies the importance of all activities occurring in and out of the human body whose proper perfect functions are both certain and compellingly necessary to ones life. As such they are a major part of TRU. CONAP is evident in its contrast with all bodily functions and activities that are less than necessary, proper, or perfect for TRU. The functions of CONAP may or may not be conscious or even direct. Yet they are the crux of life. Much of what we are, do and what functions inside of us is not rational. It is also not chaotic nor without purpose or benefit.

The finite expanse of TRU is shown in many ways. One of these is the regeneration of new cells. Every minute about 3 X 10^9 new cells are formed in the human body. With the massive processes occurring in each cell this is even more impressive. Cellular physiology and associated genetic function help us better understand the exquisite beautiful detail and dynamics that constitute TIT, especially when ordered and positively approached by SPALT and POS. These we must learn, realize, and comprehend in their actual VAPO FAF if we are to understand and live successfully, be our TES, TREQ, TRU, and TIT. There are thousands of different kinds of activities and bodily functions that make up a large part of a person's functions, activities, behavior and performance. This makes up a large part of ones life, and a large part of TRU. They may be classified and characterized by their A, L, T parameters. They range from largely mental too highly physical. Most are focused primarily between 4-7 on the length scale. Most of these in the present, have half lives andexistence that stretches over much of time. Some overlap with organ functions, and all, are more or less important for individual health and success.(TES) Some of these activities and functions expressed as gerunds, include:

ANALYZING	DRESSING	LAYING	REACHING	TALKING
BREATHING	EATING	MAKING	RESTING	TEACHING
BUYING	ELIMINATING	MOVING	RUNNING	TRACING
CALCULATING	EXAMINING	NURSING	SCRUBBING	TREATING
COMPUTING	EXCRETING	OPERATING	SEEING	TYPING
CONCENTRATING	EXERCISING	PERCEIVING	SELLING	URINATING
CURING	GROWING	PLACING	SENSING	USING
DICTATING	HEARING	POSITING	SLEEPING	WALKING
DIGESTING	HELPING	PUTTING	SWEEPING	WRITING
DOING	HOLDING	RAISING	SWIMMING	WORKING

RAK = Right Act from Knowledge. We do right primarily because we know right, because we know what works and does not work in the framework of our relationship to the world. This largely, and in ways instinctively, helps us determine what is and how to do right by the FAF of TRU. Fallacy of familiarity is where everything one knows is good and evey-thing one does not know is bad or insignificant. This common attitude is partial, fragmenting and counter-productive. It is the reverse of the much more general identity and love of God, TREQ or TRU. This is why learning, knowing, remembering and practicing TRU are so important.(ACEXIN,HOM)

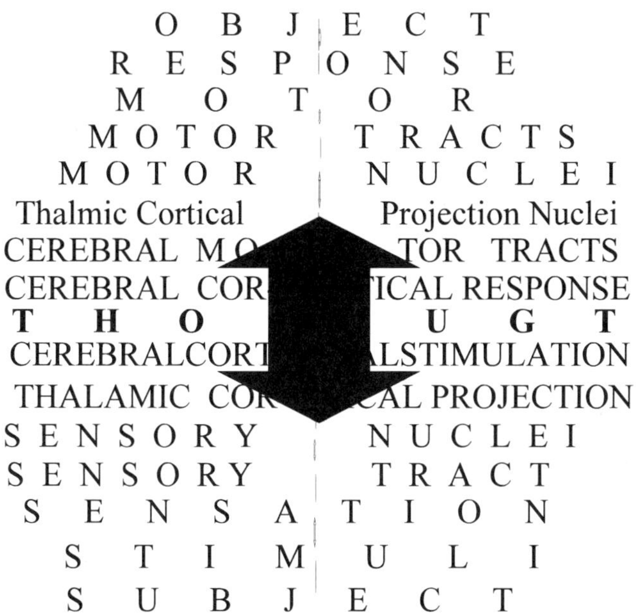

Fig 2-63 **SENSORY, MOTOR, SUBJECT, OBJECT and NERVOUS SYSTEM in TRU**

The chain of events and locations that take place in the peripheral and central nervous system, show how the S-R, S-O, IA sensory and motor functions of our make-up are key to understanding Â, SÁ, and TRU. This is the anatomical and physiological FAF in the actualization process, or Á mass energy form that is predominant in humans and TRU. These may be considered as separate CES or processes, in OSP, in their anatomical, physiological, histological, adaptational, genetic, molecular, and elemental levels and roles in the human body. If one would doubt TRU why do we feel such pain and why are the pain, sensory and motor fibers distributed as they are in the human body? They are formed and function to provide the mechanisms that maximize life, as TRU, or this special world of best human life. They are not only centered in certain ways, they are selectively set for the continuity of the organism, mankind. The neuroanatomical and neurophysiological form and functions of human anatomy determine most of the greatest characteristics, properties, qualities, and virtues that form mentality, and subjective levels and powers of predominant mass energy. They provide the form and function that best enable humans to survive and flourish. From their FAF propensity for benefit, habit formation, experience, training, knowledge, principles, values, and reason, they are key to a person's success or failure, TES and TRU. They, as in SÁ, present a whole battery of levels present in different ages, sex, circumstances, and social interactions to provide fulfillment of ones TES in TRU.

POSITS of CES and CHES in OSP SECTOR of SPALT, SPEST and METAD

These are neither exhaustive, definitive, nor perfected. They are provided more to show the method, model, and POA to TRU than to bring them to completion out of their rough estimates as final products of posits in TRU. This will have to await long, heavy, and devoted production of HOM and POA to TRU.

Measures of OSP Posits

Action	5-7-5	T
Adaptation	6-6-4	S
Birth Control	7-6-3	T
Cellular Activity	8-8-4	S
Chemical Reactions	9-8-4	T
CNS Functions	5-7-3	S
CONAP	7-5-4	M
Continue Maximal -Selective Effect	5-5-4	T
Coordination	5-6-3	S
Curing	5-5-4	S
Effort	6-6-4	S
Encouragement to Produce	5-6-3	S
Enzyme Actions	8-8-4	S
Energy	7-7-3	S
Exact Seed Line ESL	8-8-1	S
Genetic Functions	8-8-2	S
Hands, Use of	6-6-4	S
Healing	6-7-3	T
KDPM	8-7-4	T
Molecular Functions	9-8-4	T
Multiple Unconformities Torque	6-7-3	T
Neurological Funct.	5-6-4	T
Opposable Thumb	7-7-3	S
OSP Focus	7.5-7.5-2.5	S
Oxygen	9 0 4	S
Physical Forces, Basic	9-9-3	M
Physiological Actions	9-7-3	T
POBUF	5-6-4	S
RAK	5-5-4	S
Selection	5-5-4	SM
Sexual Desire SD	5-6-4	S
Success/Failure	5-5-4	M
Symbols, Letters CES	6-8-2	M
Talking, Speech	5-5-4	S
Temporal Lobe Developm-	6-8-1	S
Treating	6-6-3	T
Walk Another Mile	5-5-4	S

Fig 3-63 **CES, CHES and CHESU in OSP SECTOR**

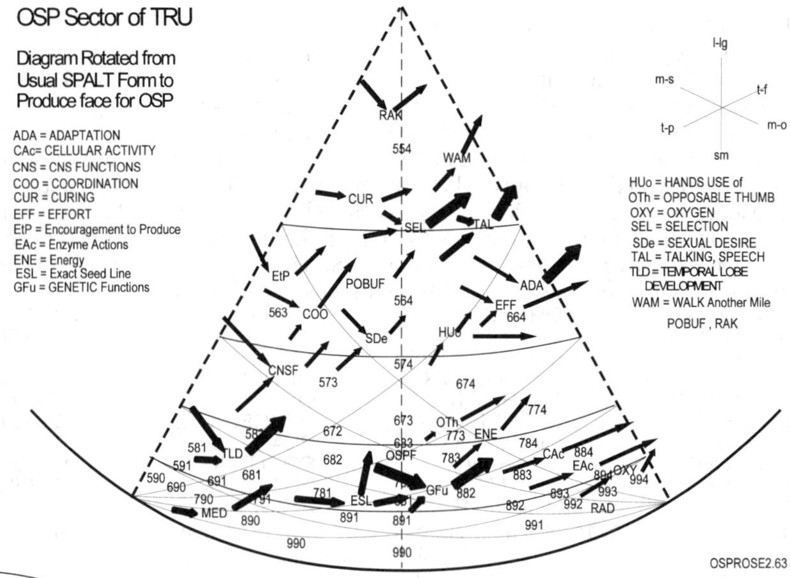

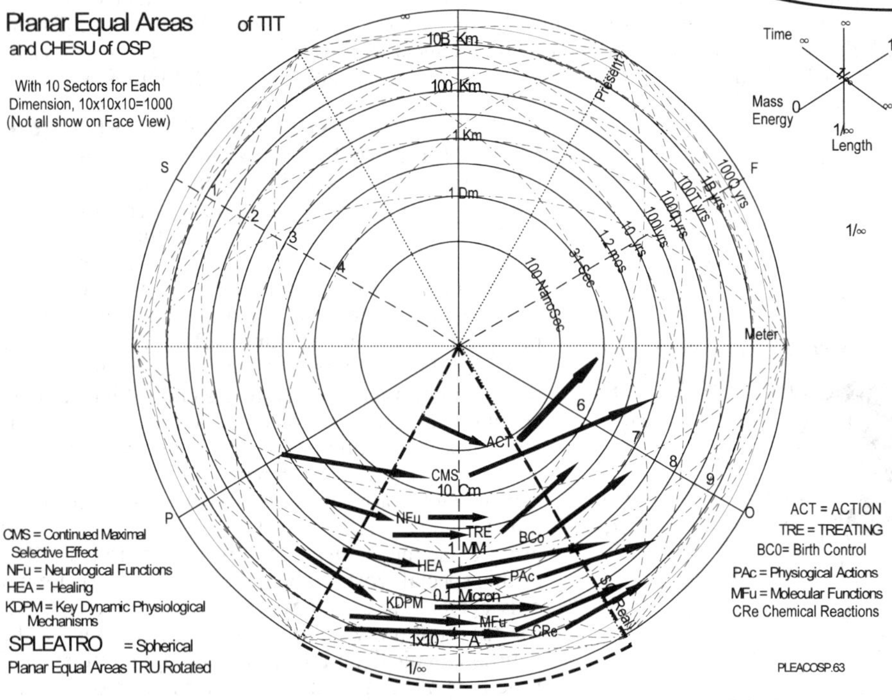

Fig 4-63 **OSP CES, CHES and CHESU in SPALT**

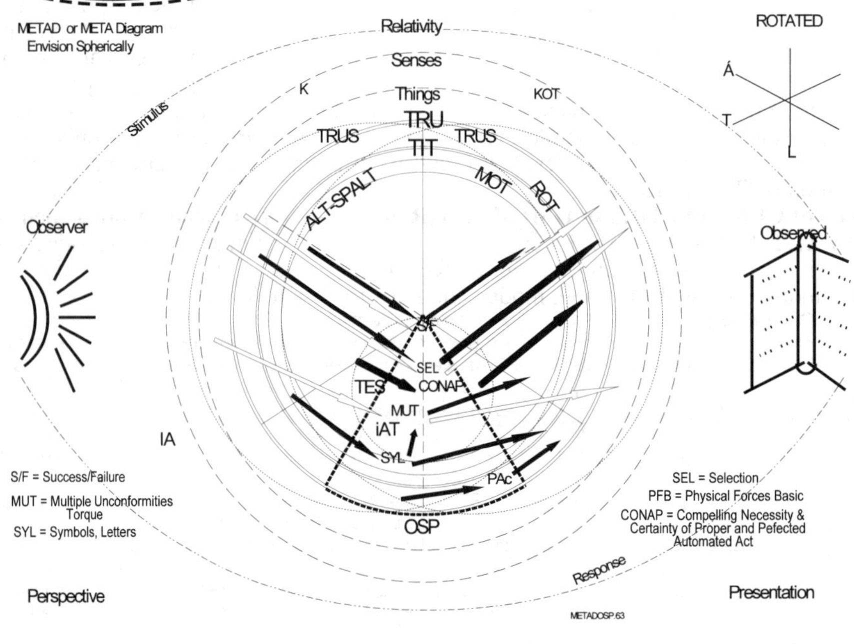

Fig 5-63 **OSP CES, CHES and CHESU in METAD**

CHAPTER 64

SUSP or SUBJECTIVE, SMALL, PAST PROCESSES of DYNU

LANGUAGE, INFORMATION, LOVE, JUDGEMENT, READINESS, INTERESTS, MEANING, SENSES, IA

Many of the processes used are ordered by time and degree of TRU. The great paradox, problem and struggle of life between a person as any individual and a person's soul or TES is the battle each person must win if they are, and much of TRU is, to be. It is the great battleground of TRU and the making or breaking of everyone. This is: **LINDTES** = Law of Individual TES Dichotomy. A person's entire life is spent learning, living and trying to adapt to this great problem. It is closely related to coping with life. Even the achievement of this end is distorted, abused, twisted and perverted. Many who have the capacity, ability and means only use TES to achieve invalid, partial, incomplete, selfish and ATRU ends. This principle and those persons who live such lives are called:

DÄPUT = Distorted, Abusive and Perverted ends Using TRU, yet ATRU. Even worse is **DÄPTES** or Distorted, Abused, Perverted Ends Used against what is Potential TES. This we commonly see in America and so much of the world. It is by those who believe, pretend, or act like they are right, good, or TES only to use this to destroy TES and TRU.

In the rapid growth of communication in modern times all media that enable better and clearer use of information greatly increase human ability and horizons in life. Such is the case with computers, the internet, and the vast expanding resources of ready-access information. Yet, it is unlikely these will ever replace the concentrated and ready access of books that best represent their topics in TRU.

	NUMEROUS DEFINITION	COMMON DEFINITION	ACTUAL DEFINITION	SCER	IGER
GOD	Many uses	a deity	The Deity, TRU	Missed Meaning	No VAPO of TRU
SOUL	Many uses	any soul	The Soul, TES	Missed Meaning	No VAPO of TES
SPIRIT	Many uses	any spirit	The Spirit, FAF,ES,CES TRU	Missed Meaning	No VAPO of FAF, ES, CES of TRU
FORM	Many uses	any form	The Form of TRU, TÄ	Missed Meaning, No KOT and of TÄ	No VAPO of Ä, nÄ and TÄ
IDEA	Many uses	any sense	The Revealing Sense of TRU	Missed Meaning or interpretation	No distinct idea of idea in TRU
REAL	Many uses	any kind of real	TRU S-O Agreement to mid section Á	Missed Meaning or Interpretation	No TRU revelation of Á and SÁ
ACTUAL	Many uses	any kind of actual	The existent and active of/in TRU	Missed Meaning or Interpretation	No recognition of TRU actual
SUBSTANCE	Many uses	any underly-ing element	The underlying parts of/in TRU	Missed Meaning or Interpretation	No recognition of TE, ES, TÍ

Table 1-64 **CONTRASTING DEFINITIONS of WORDS**

The words we use, the great terms which are so important for understanding the world, TRU, have several very serious problems. Among these is: **SCER** = the tranSCription ERror. SCER is the error of considering a word thing or NAM to be anything other than what it exactly means. What is meant must be its ESIN or more specifically its ES and CES or part, aspect and role in TRU. Another error is **IGER** or Ignorance Error. IGER is the error of considering a thing to be other than or much less than it actually is because of ignorance or a failure to understand its whole significance in TRU. In this way NAM or name-things may be divided into different classes: Name or Word - What the name or word is as a name rather than its object or thing referred to, Thing or Sense - What constitutes entities, **THIT** and **SENIT** - What a thing or sense is In and of Itself, without reference to its name, **NAM** = Word Thing, or conventionally thing-word, **NAMIT** - NAM in and of itself, **NAMTH** - NAM as it is primarily its thing, **NAMOB** - NAM as its object, **NAMESIN** - NAM that represents the ESIN of its object and entity, **NAMT** - NAM only of TRU, **NAMES** - NAM of its ES in TRU, **NAMCES** - NAM of its CES in TRU, **NAMCO** - NAMT when its NAM-ES and NAM-CES are extracted, or its other manifestations in TRU. These are distinct, yet rarely used in everyday life.

As long as there are concrete formal distinctions, misinformation and erroneous meanings are much less likely. The definitions of words, terms, and concepts hold their finite mathematical precision as long as the media, population or linguistic group using it maintains strong specified meanings.

Cicero stated - of Nero -, "If you ask me, I think a man finished, when he makes pleasure, not duty, his main object." This when qualified takes on greater significance. When and pleasure and duty are in and by TRU they are positive and beneficial. It is when TIT-TO is fully known that we can gain the complete direction and control that fits the nature and potential of mankind. Isolation, and failure that follow all that is TRU are prime examples of **IN**, Integrative Negation. Their reversal, **PIP** or Proof of Integrative Processes, is especially dependent upon the VAPO and DAOST of TRU.

Those who preach damnation, or the inviolable and absolute power of god, are not so incredulous or wrong as many may think. For God as TREQ, TRU, TES, TRIA, TIT-TO and TÂ is absolute and inviolable. All who do not learn, know, serve, and live by and for this actual GOD, TREQ, TRU, TES, TIT-TO and TÂ are to that extent negative and much of what is to be dammed. The opposite and critical position taken by those who deny, defy, and ignore such damnation or God, TREQ and TRU has, in turn and by reverse, much credit. For the strong methods, actions, and cures for all that even small deviance from the "best" truths, SCALP, ways, accounting, and ACC can be devastating. It is like what is known as ' a bull in a China shop'. The greater and more fundamental the principles, laws, and values are of TRU, the greater and more fundamentally detailed, perfected, and beautiful truths these must have, be known, shown and applied. **LÂTÉ** or Law of Action and TRU Equilibrium, perfected, detailed, beautiful and in truth.(VAPO) Such application is necessary to the whole range of massive and powerful actions religions use, as all in fervor and zeal, to secure their ends. Nor can we be without the dynamics and well targeted fervent and zealous actions needed. Sometimes a dramatic, yet meticulous and mindful, show of force, by exposing the negative - as in damnation - is most effective. It is like where 'when the battle begins within, one is worth something', or 'it will make a man out of him'. It is for this reason TRU, POT, and IPT are so necessary. LATAN-O-S. Why we must make most clear and powerful TES, ENT, ROT, MOT, ALT, SPALT, POS, STA,

HOM, DYNU, DYNS, DYNO, ACC, TRAX, TREX, TRIX, PUREACLITY, KUPS, COGNOS, ELTICMA, SÂ, and TÂ. It is also why it is so necessary that all be PPMA, maintained, sustained, and upheld at such high levels of precision and balance by DAOST and the SCARLS of TRU. It is why those of TRAX, TREX, TRIX, SMASH, CRUSH, and all strong and vigorous measures and actions must do so most properly within limits. Limits of the given, occasion, circumstances, timing, to provide actions and CES that maximally benefit and enhance, yet minimally harm or destroy TRU, TIT, TES and ES. We can not be without mutual faith, reverence, conscience, and loyal appreciation for our souls and GOD as these are TES and TRU. For to preach damnation, as in championing morality, we can't do so lightly, superficially, nor inappropriately. They must join with the subtle and profound highest degree of identity and devotion recognizable in TRU. This is **LIDAT** or Law of the Love and Devotion All must have by, In and for GOD, TREQ, TRU, TIT, TO, TES, and TÂ.

As so often happens in life we become locked into, at least temporarily or without results, problems we can't extricate ourselves from. We can not bring ourselves to TE, TI, TO, TRU, SCARLS, PERT, PREG, and all TRU. This is called: **PRÉS** = the Problem of Restriction. PRES may be remembered as a Predisposition to less than TRU.

The greatest problem and cause for trouble in the world, other than sheer incapacity, inability, lack of energy and interest is that of selfishness and contrasting regeneration. When combined they are: **PRI** - the Regenerative Individual. PRI permeates all being, especially an individual's active life and work. It is the new for the old, most striking in youth. It is also vividly seen in times of low median age and life expectancy. When positive it is shown by coming to terms with life, faith, and humility. There is a natural tendency for all to act by self preservation. Some degree of this is necessary and positive. The ratio of these is called: **PRIM** = Ratio of Self Preservation to Regeneration. PRIM is a balance of the old and the new, what we all vigilantly seek, to be TES and TRU, between ENIT, ESIN, ES, and CES, by LOPAN, and TRU.

FIBA = Fallacy of Inadvertent Bias, or deficient or deviant À and SÁ, NAMTRJ and ATRU. FIBA is the unintentional and inadvertent bias of a person, ones group, state and work. It comes from all the TRU one doesn't know, locked into the extreme partials one does know, along with all ATRU one knows.. FIBA shows why only to know TRU is not enough. It is to live, practice, serve and with faith, sacrifice, and frugality be TRU. This can be done both by focusing on oneself and on TRU so long as one is TES, TRU, and TIT by VAPO.

	THINKS of ONESELF	THINKS of ONESELF AT BEST	THINKS of ONESELF AT WORST	THOUGHT BY PARTICULAR STANDARD	INTER-RELATION OF	WHAT IDEALLY BY TRU
EVALUATED ONESELF	SELF CENTERED	BEST OF ONESELF & WITH SOCIETY	SELF DENIAL VALID or INVALID	SATISFACTION IN A CLOSED SETTING	COORDINATES FOCUS	PERSONAL EXPANSION TO TES, TÂ
EVALUATED by SOCIETY	LOSES INDIVIDUA-LITY	INDIVIDUAL BEST SOCIAL ADAPTED	POORLY ADAPTABLE to SOCIETY	LIMITED BY THE CLOSED SETTING	COORDINATES FOCUS IN SOCIETY	SOCIAL EXPANSION TO TRU
EVALUATED BY TRU	LACKS VAPO	RECOGNIZES THE SELF IN TRU	NEGATIVE INACTUAL, LITTLE HOPE	STANDARD of TRU STANDARD LIMITS	BRINGS ALL TOGETHER IN TRU	ALL TOGETHER IN TRU, TRUID, TÂ
ARE WHEN ALL TES	IN TES WITH FULL VAPO	BRINGS SELF to TES and TRU	HOPELESS unless TURNED to TES	PROVIDES FOCUS STANDARD LIMITS	BRINGS ALL TOGE-THER IN TES	TOGETHER IN TES, TRUID, TÂ

Table 2-64 **EVALUATION and SELF ASSESSMENT**

Ones success in life depends upon many factors including whether it is viewed from, the individual, society, TIT-TO, or or other aspects of TRU. It is extremely important that each person refresh and know these various ways. Much of ones choices and decisions depend upon what, how and how well we think and evaluate, avoid ATRU, adapt to others, to TES, and the surrounding TRU.

Honor, respect, approval, like and dislike are all highly indicative of TRUD, HOM, CES, forces, powers, determinants, fusion and PIP of TRU. They are not necessarily, and usually only partly, TRU.

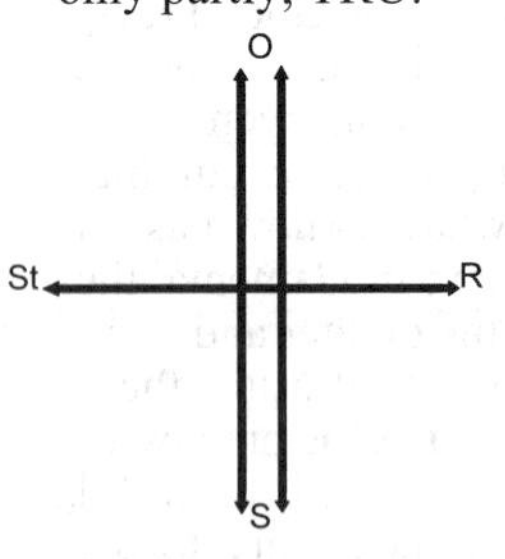

Fig 1-64 **OVER and UNDER REACTION to the SIGNIFICANT and INSIGNIFICANT**

This is what should, at least initially, be provided by good parents, teachers, advisers, and leaders. Or it may be shown by how well a person is trained, how and how much one does not overact to the insignificant and under react to the significant. Here is the crux of the combined S-R and S-O in a person and measure of their TES and PUREALITY.

ATRU to TRU FLEXIBILITY or INFLEXIBILITY

Adversity may be either TRU or non TRU, depending upon one's existence. Thereby it is not only necessary that each person be capable of withstanding adversity. It is equally necessary that adversity be held to a minimum, and our security, stability and existence be relentlessly pursued to assure TES and TRU. The measure of person may be determined, to a great extent, by the degree they can withstand adversity under pressure,.flexible or inflexible, depending on the need of TRU, circumstances, and SCARLS. A person can be quickly and readily identified and distinguished by the degree to which they do or don't stand firm, and act at the appropriate time and way. This standing firm or action takes selection or non selection and the principles or lack of principles that apply in each occasion.

The difference between being little or big minded is well learned in IPR. It helps to avoid those who are: 1. Quick to think the worst of any and all, 2. Reluctant to look for the good, 3. Lack knowledge of GOD, TREQ or TRU. 4. Lack of interest in what is best of human life, and VAPO, an analysis and interpretation of all that comes before them, as it is or is not TRU, 5. Incoordination and lack of equanimity.

A great question is how much and why is the attainment of freedom from pain so great. The initial answer would be a resounding yes for the reason it relieves distress and suffering. Yet pain has value. Herein is the paradox, a great paradox. For if pain is of value, if it is a necessary accompaniment of life why should we so seek to eliminate all of it? TRU is to a great extent S-R or Stimulus Response. One's sensory and motor functions unite for the protection and enablement of life. They are guided by what is or is not painful. This may be the result of what has happened, is occurring or may likely occur. The end of life is more than the cessation of pain. Yet TRU is not negation. What causes the pain, especially pain that indicates loss to TRU, is how negation becomes a part of TRU in response. So that there may be beneficent as well as maleficent pain, purpose, and all in the world, TIT-TO.

Judgement by qualifying much of choice and selection is fundamental to human predominance, Á, IA and TRU. It is at the forefront of success/failure. Many a good, nice, sweet, pretty, lively, handsome, or showy young male or female has never surmounted mediocrity in life for lack of intelligence, reason, knowledge, wisdom, understanding and judgement. Many an ugly, deformed, or otherwise defective individual has in reaction achieved the highest in life. It is VAPO, DAOST, SCARLS, TROS, ACEXIN and all that are shown by SPALT and POS that enable the complete comprehension necessary for TRU judgement. How interest focuses, and can be maximal, to best analyze and bring all to TRU.

	ATRU	TRU
Maleficient	Harmful to any and All (- = -)	Harmful to TRU (- + = -)
Beneficient	Beneficial to any and all Not TRU (+ - = -)	Beneficial only to TRU (- - = +, ++ = +)

painmbtr.646

Fig 3-64 **MALEFICENT and BENEFICENT ATRU and TRU; POSITIVE and NEGATIVE EFFECTS and RESULTANTS**

The trick is to separate, what are maleficent purposes, that produce temporary pain, from long term and overall beneficent purposes. Nature may be largely indifferent to the hopes, aspirations, needs and purposes of a person. Yet TRU, like the grace of God, is not indifferent, if these are TES. TRU acquires and absorbs all that may be of benefit, only to reject negative effects that outweigh their positive. This it does repeatedly until what is maleficent is nil and what is beneficent is all.

Without readiness and receptivity KOT, ACC, COGNOS, ELTICMA, SA and TA can not be attained. It is in the cultivation of proper IAT, interests, inspiration, readiness and states of receptivity that the greater part of S/F TRU of lies. Each culture, rises or falls by how these are right and nurtured. The key to successful action and life is readiness for TRU, the anticipation and receptivity of all that will be necessary or beneficial for TRU or whatever aspect, part or processes of TRU fits the occasion.

Fig 4-64 **INTERESTS in S-R, OSO, TRU**

The quality of ones interests produced by TREQ or TRU reveals much about ES, TES, TIT, TO, and HOM. This includes their growth, validity and the discovery of what is unknown and important. The quality, quantity and power of interests need stimulation and cultivation. For the best quality interests are of lesser worth and profit if they are very few and unrelated. This is why persons who spend their entire lives on one narrow concern, vocation, faith, profession, or goal are likely miss TES and TRU. Thereby both quantity and quality of interests should be great, to reflect TIT-TO and one's TES and TÍS.(VAPO) It is evident, and a justification of TRU, that we discover, develop, determine, promote, and utilize ways that enhance TRU and its ROT and POT.

Getting what we want is often mostly a matter of what and whom we know. This, although easily abused, is not necessarily bad, for it is practical and natural if correct by TRU. **MOCS** - Motivation of Comfort, Self, and Stimuli. In personal comfort, desire, whim, wish, money, self interest and reaction to stimuli we respond in ways similar and different depending upon our own personality, age, experience, circumstances, beliefs, and inclinations. They will be both justified and unjustified. They will be both TRU and non TRU.

Each person from birth to death continually develops, grows, learns, experiences, is educated, maturates, and adapts to the society they are a part. It is a continual process of development of the more subjective and higher levels of their mass energy, Á, FAF. It is repeatedly inclined to TES and TRU in many and various ways. The level of TES, and relative TES per age, are largely determined by how good, how many, how intense, and how receptive one has been up to that age of all the learning and experiences gained. The intensity of the experiences, the knowledge learned both in quality and quantity, how much they are TRU, the revelation and inspiration of the most subtle and profound ideas, concepts, comprehension of TRU, and how they are best applied, all measure how well a person achieves TES. This is a struggle, a long and hard road in a short and limited life. Yet its alternative is only the opposite, death, not so much a physical death, but a death of actual being, being a part of the world one has benefitted from and must necessarily contribute to. The inertia and temptation is always to take any of the many deviations from this road, to lose ones way. Yet this road is much less a struggle if one only grows and lives it naturally fulfilling the role it so readily provides. Herein is the role of the society, for to reach to TES requires that this TES for the individual, group, and all, must remember and practice its universal as well as personal mode. Here was the primary message of Socrates. "Pay attention to the young and make them as good as possible." The combined greatness of paying attention, as by IAT, focusing on the best development of the young, and the importance and continuity of TRU has been one of the greatest of human gifts. This is the common TES, as in Plato's ideal soul, in all, the higher powers of mass energy, Á FAF in all people that continue to exchange and improve the knowledge and performance of the individual.

Many blame others, or other things, for what happens or goes wrong. They tend to reduce this to a single factor, cause, error, or wrong. They little realize COC, DAOST, TROS, PIP, and all the fundamental mechanisms and processes involved. This is one of the greatest deficiencies and needs of humans, being disinclined or unable to learn, know, recognize, account for, reconcile and work out the full causal relations, determining forces and factors involved in all.

Each and all, in seeking or practicing and acting according to their Pleasure Principle, are doing so, not necessarily out of selfish ends. They are doing so to preserve and maintain that which enables them to be. The pleasure principle may be compared with the Reality Principle. This is in contrast to the immediate benefit of pleasure how humans defer gains for long term benefit. Either one can be right or wrong. They simply seek what is best in the short term and what is best in the long term. Thus, the reality principle and the pleasure principle do not necessarily conflict. Yet, there is a dialectical relationship between the two. When either becomes over or under operative, imbalanced, and out of harmony with TRU they are wrong. To balance these two enables the individual to adapt, adjust and comply more readily. This may not be beneficial or TRU. **COMPER** = Compliance Error. COMPER is that inveterate tendency to adapt to the world, comply with the necessity of ATRU, and crashing its limits of TRU and living by expediency. This expediency may not even be by the pleasure principle. It tends to be chaotic. **FADAL** = Fallacy of Depreciation by Association. FADAL is the misinterpretation and mistaken identity that occurs with close partials. It can be either subjective, objective or combined, in, out, or OITIO to TRU. It assumes that what is close spatially, temporally, or even in the mass energy dimension is it, the actual existent. It is similar to both partiality and **FAMF** or the fallacy of familiarity. FADAL and FAMF are biases of the here, now, force, power or familiarity that produce externality and its added relativity between subject and object. It is exemplified by man's tendency to identify things by senses, and especially sound, e.g homonyms. This may or may not be so. FADAL is largely a product of the deviance that occurs when TRU is displaced by ENIT, SENIT, THIT, OBIT or SUBIT. FADAL is most important in helping us look at or represent TRU, not as it appears, is fictitious, or a figment, but as it actually exists through and through TIT-TO.

Fig 5-64 **SENSES, SENSATIONS and STIMULI**

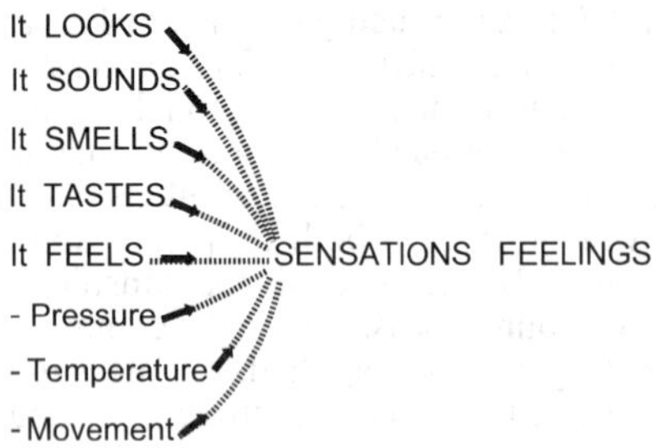

Stimulus - response is the action - reaction relation in which senses and things become subject object, or in TRU its mass energy Á FAF. Different senses provide a person with highly effective means of responding to stimuli. **STET** - Senses are valid, true, only as they Equal TRU. Senses are figments or illusions to the degree they are not TRU. Thomas Aquinas stated "human life, then, is a procession from the world of nature to the world of Grace, from want to satisfaction, from imperfection to perfection". All persons are to a great extent inclined to, and must develop this potential, grace, satisfaction and perfection if they are to be, become their TES in TRU. This is accomplished as their senses become so trained, oriented and willed to STET. **SÀT** - Sensation, feeling of ÄTRU, pain, misery, unhappiness, foreboding, feelings the result of Summative ATRUS. **JOVS** = Justification of Variable Senses. **CORVS** = Correction for and Order of Variable Senses. Variable senses are close to IA and tend to TRU. This is seen in special relativity where the focus is relative to the sense or observer.

RAD = Recognition and Avoidance of Unnecessary Differences. RAD presents an ubiquitous and fundamental paradox and problem. The paradox lies in TRU itself, the problem lies largely without TRU. There will be many differences we need to recognize, avoid or accept within TRU. There will also be many differences of our relativity, observer-observed or subject-object encounters that are outside of, or mixed outside and inside, of TRU. These can't be as is. Each has to be accepted or eliminated on its own merits, by what and how it is beneficial for TRU and absorbed, or what is null and harmful that must be avoided or eliminated. Because what we see, sense, or perceive is of complete certainty, in the full light of day, does not mean that this validity of the S-O relationship is also valid in all regards, namely, without adaptation to TRU.

Consciousness, feeling, awareness, and attention when viewed from any or their own foci, ENIT, and not from TRU may be seen as in: **SIAF** = Self It Fallacy. SIAF is where ones IA is at the mercy of its 12-18R - or worse - position. We can show how IA proceeds to IAT, to Á, to ALT, to SUT, and to TRU. Also we can show how IA equals IAT, equals Á or Á equals IAT, equals IA. This is called: **EIA** = Equality of IA and Á, Á and IA. Through EIA Á is given its rightful importance and how this importance underscores the center and crucial dynamics of TRU. IA is IAT, is of TRU, Á is in TRU, ES or parts of TRU are in TRU.

The quality, strength, and totality of TRU is very dependent upon the intensity, perfection, intention, care and detail a person exercises throughout life. This is seen in precision work, clarity of speech, depth of meaning, clear writing, concise concepts, expression, and instructions. In French the saying goes, "*Voir dire*", or to say the truth as a law. If you speak of TRU to those who are TRU no evasive answers, no deceptions, nor restraints are necessary. It is to understand the fusion of mental and physical thought and act, speech and written words as they are Á in ES and CES that we can render all entities actual. They shed their figments and illusions as separate senses and things to become these particular ES, CES, forces, powers, forms and functions of TRU.

Fig 6-64 **CES, CHES and CHESU in SEGMENTS of SUSP SECTOR of SPALT**

In Figures 7, 8 and 9 of this chapter many of its described processes are posited in the sector, SPALT or SPEST, and METAD diagrams. The intent is to provide a better vision and use of SUSP, HOM and TRU, and thereby our lives. What MOT, SPALT, POS, HOM and DYNU can do. These processes and the loci of their vectors include,

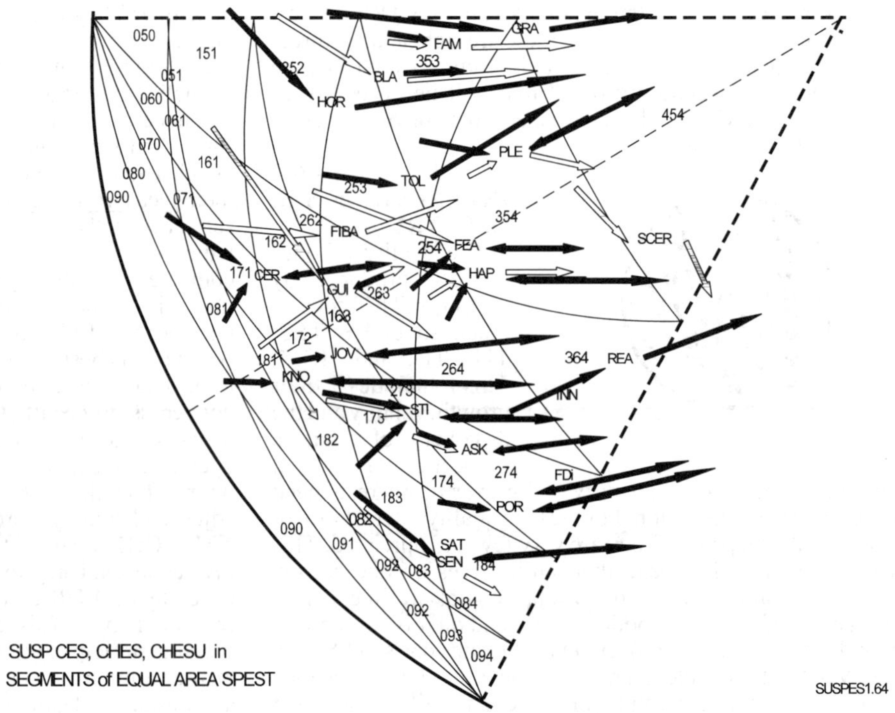

Measures of CES POSITS in the SUSP sector of DYNU

ASK,-ING	3-6-4	S
BLAME	3-5-3	S
CERTAIN-TIFICATION	1-7-2	S
COMPER	7-5+-4-	M
CONCEPTS	2-8-3	T
CORVS	2-7.5-6+	M
DAPTES	6-6+-6	M
DAPUT	6-4+-7	M
EIA	5-7-7	S
FADAL	5-5+-5-	M
FAMF	3-5-3	S
FEAR	3-6-3	S
FEELING	1-7-3	T
FIBA	3-6-3	S
FORMAL DISTINCTION	3-7-4	S
GRACE	4-5-2	S
GUILT	2-6-3	S
HAPPINESS	3-6-3	S
HONOR, RESPECT	3-5-2	S
IA	4-7-4	M
IGER	2-7-3	M
INFORMATION	1-8-3	T
INTENTION	3-6-4	S
INTERESTS	2-7-3	T

JOVS	1-7-3	S	NAMIT	7-7-3	M	SCER	4-6-4	S
JUDGE,-ING,-MENT	~3-7-3	T	NAMOB	8-7-3	M	SENIT	8-7+-6+	M
KNOW,-ING	1-8-2	S	NAMT	4-6-8	M	SENSE,-S,-ATION	1-8-4	S
LANGUAGE	5-6-2	T	NAMTH	7-7-3+	M	SIAF	4-6-3	M
LEARNING	2-7-3	T	PAIN	1-8-4	T	STANDARD	5-5-2	M
LIKE-DISLIKE	3-7-3	M	PEGSOT	4+-6-4	M	STET	0-8-4	S
LINDTES	4-5-2	M	PIP	4-2+-7	T	STIMULUS	2-7-4	S
MEANING	3-7-4	M	PLEASURE	3-5-4	S	SUFFER,SUFFERING	2-6-3	T
MEASURE	5-5-4	T	PORCOM, -PORSCOM	2-7-4	S	THINK,-ING	0-8-4	T
MINDEDNESS	1-8-3	T	PRÉS	6-6-4	M	THIT	2-5-4	M
MOCS	4-6-4	T	PRI, -PRIM	4-5-2	T	TOLERATE,-ION	3-6-3	S
NAM	7-7-4	M	QUALITY	4-5-2	M	VIEW,ING	1-7-4	M
NAMCES	5-6-8	M	RAD	4-6-4	M			
NAMCO	4-7-8	M	READINESS,RECEPT	-3-7-3	T			
NAMES	4-7-8	M	REALIZE,-ITY,-AT	3-6-4	S			
NAMESIN	6-7-6	M	SAT	1-8-4	S			

(PORCOM and PORSCOM like MALCOP interchangeable with 61, showing population emphasis, and in 64 personal or basic emphasis)

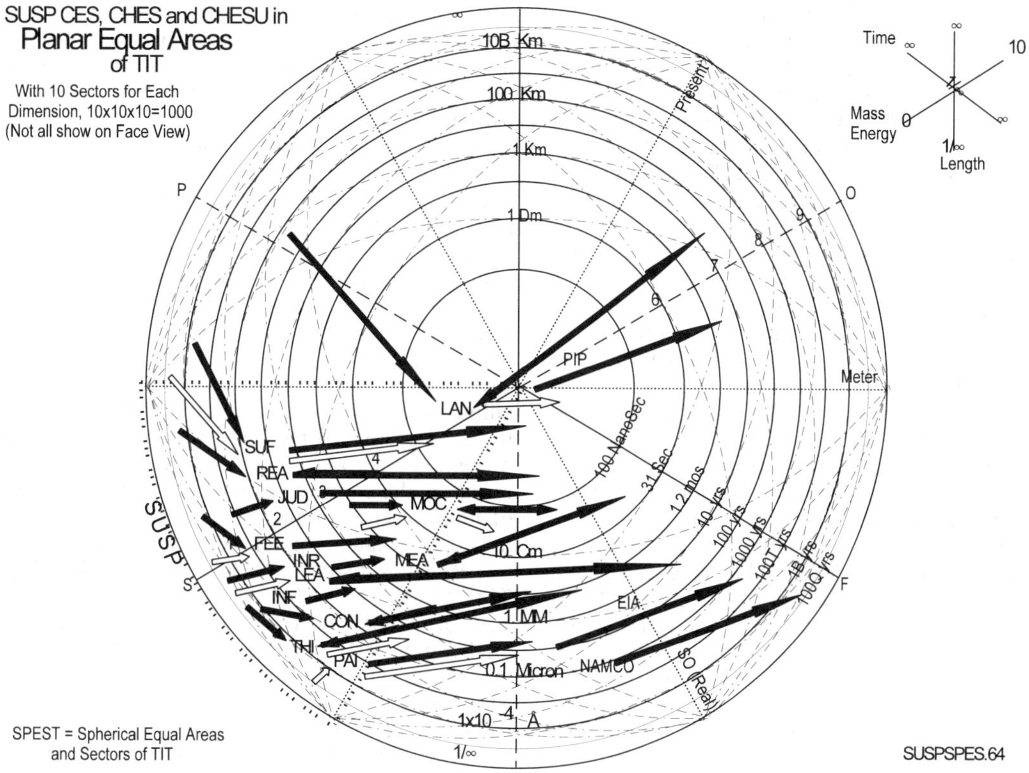

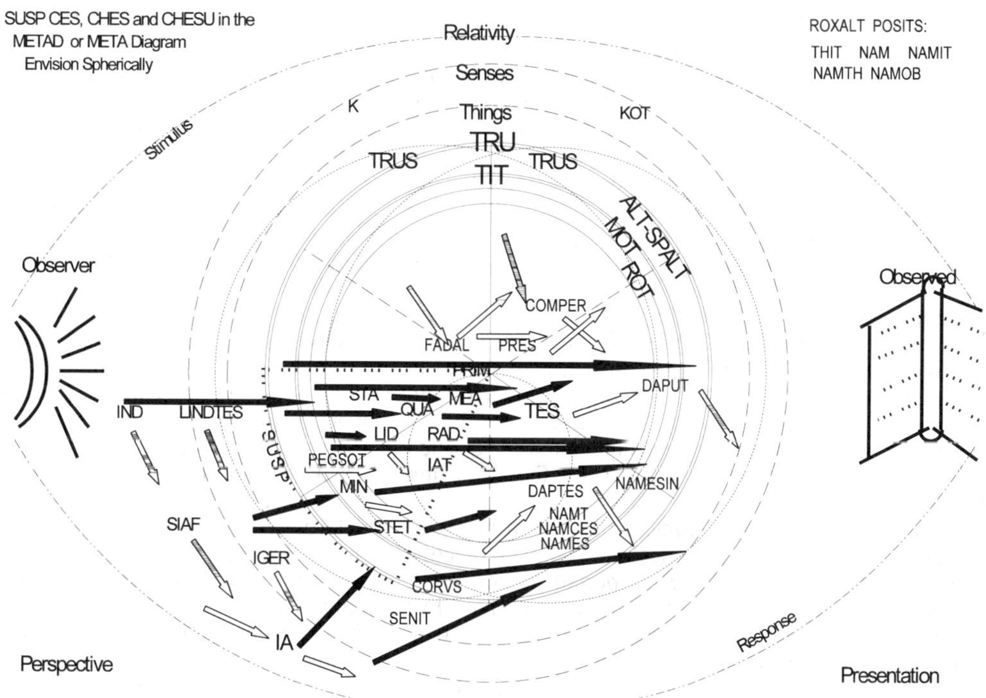

Fig. 7, 8-64
CES, CHES and CHESU in SUSP SECTORS of SPALT and METAD

CHAPTER 65

SUSF or SUBJECTIVE, SMALL, FUTURE PROCESSES of TRU

IA, THINKING, IMAGINATION, DECISION, INTERESTS. KNOWLEDGE, CHOICE, CHARACTER, CONCEPTS, JUDGEMENT, PRINCIPLES, TES, IMMORTALITY

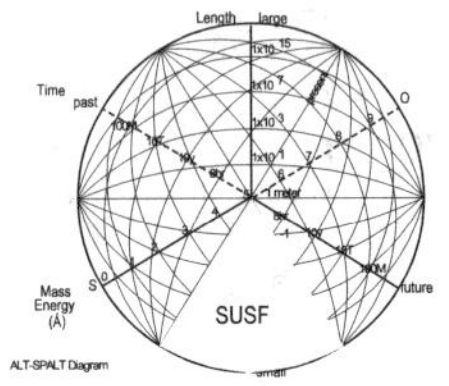

Fig: 1-65 **SUBJECTIVE, SMALL, FUTURE SECTOR**

By comparing this figure with the figures in 1-64 or SUSP it would seem that there is a large discrepancy in the representation of sectors in ALT. It can be resolved only by looking at TIT-TO and TRIT by way of SPALT, in 3D. When we look at a sector in the planar 2D ALT we have two sectors that are obscured from view. This is compensated by viewing ALT from a second or more, vantages. When we think of all these sectors throughout Sections VI, VII and VIII we must use the standard for six sectors, and the alternate view for OSP and SULF. Even better continually envision all in a masterful way that best represents the spherical geometric model, SPALT, of TRIT and TIT.

In language we have many words that describe the same thing. This may be exact or inexact, absolute or relative, or relative by degree. When by degree and indicative, they are synonyms. This is justified, in a way, by the power of words to reveal greatest meaning. Not only is TRU made up of much that is the same by DAOST, it in being absolute, all great, is associated with much that is likewise. The greater the value or REM, the greater the need for formal distinction, yet more difficult and needed, because of their greater sameness in TRU.

Nowhere is this more apparent and actual than in the center, the core, or the middle of TRU. This is the present, medium sized, and SO mixed or real components of TRU. It includes one's daily life, acts, words, Immediate Awareness, and Instantaneous Attention, IA. It is what is directly about us, of us and in us. It fully represents the living GOD the most emotive viable aspects of TRU. It reveals the error that so many religions have of separating God from Man, or God from the World. It is a reason why God is not an adequate term for TRU. The segments of ALT of 63, 64, 65, and 66 give us the frame in this region of TIT that is so much of us. There are few and poor lines, surfaces, or clear separations of the SUSP and SUSF sectors of TRU. Only what we can devise and arbitrarily set in ROT and MOT to give them SU and order. So that they are not readily distinguished and made distinct. This causes much confusion, like much of psychiatry, psychology and even philosophy where concepts are poorly limited and controversial. There is only the latitudinal, longitudinal, and sagittal modulations, as in geography or solid geometry, which enable us to map or model TIT, ALT or SPALT. They are not apparent to the naked eye or sense. They are not felt, heard, seen, smelled by us, nor even readily perceived. For, they are more frames of ROT, than of TIT. We offer them as examples and means of showing how ALT is composed, works and can be used. This is called: **CUMALT** = Composition Use and Mechanisms of ALT. CUMALT will become more than a glimpse of TIT when in time we have accumulated and applied all that makes up, forms, functions and are the parts and processes of TRU.

It is the stimulation and development of optimal habits of thinking along with the, concepts, vocabulary, and knowledge upon which they are based that is all important to our being, TRU. This is shown by the remorse we feel when we have lost something very dear. No loss is greater than the loss of a brilliant and glorious idea. Similarly important it is to be stimulated to many great ideas and thoughts, and these ordered. It is that intense, burning desire and motivation to find what one has lost or needs that is the spark of life, TRU. Exhilaration is the feeling just after a great idea hits you, and just before you realize what's wrong with it or its limits. It is like a sudden achievement of the absolute. Only it is an absolute of a partial at best. What is often wrong with our great ideas is that they are not in and of themselves so wrong, for as parts and processes of TRU they may be absolute by TRU.(LOPAN) It is only when we interpret them as being more or other than they are in TIT. Or when we realize that they are of coterminalities of REM. When we can have one exhilaration after another as our great ideas accumulate, coalesce, reveal each other, escalate, harmonize and unify we achieve glorious subjective union with TIT.(TES)

How to think is the most effective means of achieving TES and TRU. It is how to think: fast, deep, subtle, broad, extensive, intensive, with curiosity, imagination, creatively, appropriately, parsimoniously, distinctly, practically, usefully, moderately, decisively, sufficiently, systematically, unitarily, and absolutely by VAPO and all TRU. It is in the cultivation of these habits of thinking that are derived from God, TREQ or TRU that most beneficially provide the stimulation, enthusiasm, interest, and conviction necessary. If any subject, topic or sector of TRU is to be adequately elucidated and presented a certain level of concentration is required. As in research, the concentration used is little more than a focusing of attention and bringing together all available information and means on whatever given, topic or sector of TRU needed.

ILRIA = Illusion of RIA or no RIA. As RIA is limited to TIT, TRIA, TIT is the infinitesimal TRINI. This is the great part of TRUS, or how the seeming inertia of externally unchanging TRU, is demonstrated. This is the momentary inertia, or at least what seems to be **MIN** or of great Momentary Inertia, to the individual. TIT as TRIT is, relative to TRUS and TRIS, standing still. Since, however, TIT is under the effects of TRUS and TRUCH, TRIT becomes TRIS of different kinds. It is TIT as TRIT and TRIP that is set, and has great momentary inertia. The duration of this set state,, TRINI and TRIPRE, is a function of the extent of TRIS. Whether it is infinite, as TRIN, of finite duration, as TRIP and TRIPRE, or infinitesimal as TRINI or other as TRIR it may be viewed as static and dynamic. MIN is a reflection of the focus of TIT and how it seems to the individual to be TRIA by virtue of its own ENIT and the prolonged duration of TRIN.

A person's Interests are one of the most important aspects of their makeup.(pg 345) They determine how a person is or is not their counterpart in TRU. How their subjective is both qualitatively and quantitatively focused on TRU. **EXINI** = Extremely Intense Interest. We can analyze and explain how EXINI, great ideas, and revelations work. With these one can achieve a good analysis and explanation of all TRU.

One of the greatest causes of failure, loss, and suffering in the world is Indecision or improper decision. It comes mostly from indifference, ignorance or erroneous knowledge. It comes from a lack of the best possible integrated knowledge and how it can be applied to one's life or KOT to TES. First principles are determined by the underlying laws, concepts, and mechanisms of the FAF of TRU. They are the guidelines for right thought and right act. **FOYAN** = Fallacy of Yes and No. We as individuals project our need for finite distinction, yes and no decisions, upon all things, little realizing that this applies and is limited to ourselves as individuals, less than TES, and not to TRU. **FAYAN** = Fallacy of Yes and No, Non TRU. The projection of FAYAN is not a lack, but a directly or indirectly acting force of non TRU. Directly it offers and expects belief that is not valid but wrong and negative. FAYAN is a result of invalidity, FOYAN is more the product of non deliberation and thoughtlessness.

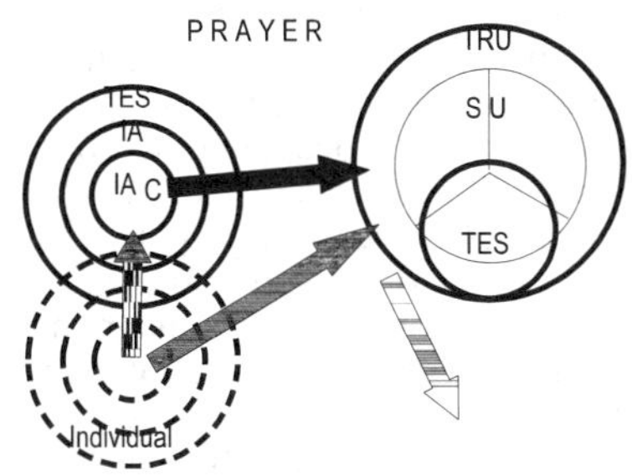

Fig 2-65 **SUSF: DISCERNING, PRAYER, ONE WHO KNOWS**

Discerning is keen attention, good judgement and clear understanding. It is being able to distinguish differences, of one thing from another. We should respect the opinion of all to the degree they are a discerning critic of God, TREQ, or TRU, rather than what is not TRU. Likewise we should recognize how broad and well their discerning is followed through with by appropriate action. Prayer is the personal communication and supplication to God, divinity, TREQ or TRU. It can be for all aspects of ones relationship, as in relief of distress. Prayer is a concentration of IA and its projection to TIT-TO providing identity with TES, TRU, and their benefits. There are powerful forces in One Who Knows. It provides increasing respect/non respect, demand/dishonor, acceptance/repulsion and benefit/harm of TRU. With increased learning and knowledge, especially of TRU, supplication and prayer take on added meaning, identity with TO-TIT and TÂ.

The power of ingenuity should always be tempered by TRU, its SCARLS and all we come in contact with. It is a magnifying power which builds upon itself, especially when it is cultivated and directed by TRU. It is more effective with the ACEXIN necessary for TRU.

The Principle of the Transmutation of Thought or **POTRATH**, may be formal or informal, the thought and structural aspects of TRU. POTRATH denotes how thought has little importance, meaning, or force by itself; only when translated or transformed into other thoughts and mostly into actions and effects that have actuality. This is a large proof of Á and the interaction and fusion of S and O. It acts in reverse by their separation and independent actions within TRU.

There are great swings or fluctuations in feelings and the readiness. We often little appreciate the great medium or mass or: **VEMORT** - Variation in Emotive Readiness, or Variation to total readiness for TRU. VEMORT is vast, even in addition to that of the great medium of mass through which it works. It is the total interface between individuals and TRU as this interface is being penetrated by the great myriad of separate forces and processes. These are CES only when their resultant in TRU is positive and entirely TRU.

Happiness, imagination, intellect, choice, field of consciousness, virtue, decisions, foreknowledge, speech, knowing and judgement all represent separate ES, CES, and FAF in TRU. They are distinct terms with clear meanings which enhance their value and use by TES in TRU. Verve and vigor is the spirit, vivaciousness and liveliness of human mentality and behavior. An exceptional degree of enthusiasm can, must and will be in all if we will be TES, our parts in TRU. This comes from many sources, our natural physical and mental endowment, energy, spirit and conviction in TRU. The force and power of the intellect is essential to TRU. This is evident in the vital role of higher levels of SÁ, and will be greater in TRIF, if TIT is TRIA, and TÂ. **LENTES** = the Law of the Enablement of TRU or the means by which choices, decisions, foreknowledge and judgement optimize their proper processes in TRU.

Do you know what you are doing? Do we know what we are? Do we know how to live? Daniel Boone is quoted as saying "I've never been lost, but I've sure been confused." By being conscious in life we rarely think of ourselves as being lost, unless we are faced with some power that overwhelms us. All humans in the mixed NAMTRU world are both lost and confused. Yet they do not know it, they are cognizant only of their consciousness, thoughts and the objects they sense or focus on what seems so real and actual. Yet these are usually neither their entities nor what they are in TRU.

Handsome is that handsome does. Yet be careful, what you may think or what the public holds to be handsome may be far less or different. Although humans have greater knowledge, thought and means than ten thousand years ago, what they uphold and champion is often less positive. Only what is **TIFT**, True IF TRU, should be upheld. Ego applies not only to a person, the first person, it applies to the second person, all persons, and all TRU. This is called: **EGOTRU** = the Ego of TRU, the generator and focal power within TÍS, TES and all pivotal conscious ES and CES of TRU. Yet since the ego is such a dynamic generator and stimulator it, when not TRU or imbalanced, is much abused and misused. This is how ego has become pejorative, a term of derision, and to be avoided. It is for each and all to fully utilize, yet channel and control, the personal ego, as EGOTRU. This is a large reason for VAPO, RECTRU and all that is TRU.

The figmental state and dynamic - rather than actuality -of the seemingly complete reality of ourselves, our lives, and all we survey is best seen and resolved by ATO, TRIA, TIT-TO, TRIT, ROT, MOT, ALT, SPALT, SPAT, POS and their use. When these are understood, we no longer so readily accept and are convinced by our own senses and the things about us. We expand ourselves from a narrow ego into a much broader world of TES in TIT-TO. We begin to analyze and select what actually is qualitative and quantitative, rather than uncritically assuming all is, or is in the way we sense it. It may be 'real' as a sense of a thing, but is not actually real as it is a part or process of TRU. It can be real as it is actually relative to us and be our part, and do what is needed to sustain, this special universe of best human life. This is best seen in public service in the many venues individuals contribute. Here is the application of PROT, ATO, TRIA, TIT-TO, TRIT, MOT, ALT, SPALT, SPAT, REM, POS, and all in showing what is, rather than so much in and of itself insignificant and worthless. Much that is not TRU remains continually in the entities, senses and things before and about us yet is limited, mixed, trivial, insignificant, worthless, or negative. They lack TRU actuality and are inimical to us. For we are or are not as all is or is not actual of TRU. To continually confront and cull out what is illusion and figment is what NEC, POT, IPT, VAPO, ATO, DAOST, TIT-TO, TRIA, TRIT, ROT, MOT, SPALT, REM, POS and the proofs of PROT, PRA, PRALT, PRUT, provide.

The character of a person sufficient for TES can not accept the usual or common standards of much of the extant world. Those standards of acceptance and compliance are:

What they are in General	What they are in TRU
Figments myths fantasies	Actuality
Uselessness,	Utility
Undisciplined and Rash Acts	Discipline and Control
Disproportion	Proportionality
Imprecision	Precision
Incompletion	Completion
Disorder	Order
Incoordination	Coordination
Timidity	Courage
Disinterest	Interest
Dispassion	Feeling, for TRU
Hate	Love
Distraction	Concentration
Triviality	Priority, Value, REM
Ingratitude	Gratitude
Negativity	Positivity

Some of these are opposite to TRU. It is necessary for us to be constantly on guard to not only avoid what is opposite to TRU, but to turn them to TRU when possible and practical. To be TRU, ones counterpart in TRU, TES is neither a simple and undirected, nor endless task. It is to normally live according to the order, means, SCARLS, and VAPO of TRU.

LERAC = Law of Error from lack of COL, COR and coordination. LERAC signifies the great law of posits and focus of S on TRU. This is not to be determined by ENIT nor even the focus of CES as a separate process. It is to be primarily determined by the TO-TIT meaning and effects of each process. This does not contradict, only inactualizes, the fact that this process has an earlier, other, or extra TRU focus.

The most precious, valuable, golden or glorious possessions one can keep are the principles, concepts, laws and mechanisms of TRU.(CALP) Many of these are a product of the extension of the ES and CES of TRU. Many are already available in our language in existing concepts and words that only need to be interpreted in terms of TRU. Many more are offered, by **NEC** or NEw Concepts, the notational system of terminology or language of TRU.(LANT) It is these, when combined and brought into an ordered whole representative of TIT, that we can have the kind of comprehension and control that befits TRU and assures permanent predominance. ALT provides the form, structure, frame, and path through which we have direction, plot our course, and bring all under optimal control. Yet the concentration, uniqueness, and the energy necessary to pursue these principles and TRU must come from within a person, TES, through TRU knowledge, Inspiration, Spirit, and Will. These are not foreign to our being, but the heart and soul of it. This is called: **INT** = Intensity to TRU. INT is the optimal activation to TES, TIT-TO and TRU.

How we view ourselves and our vanity may be considered in many ways: what is strictly TRU or TES, as we 'really are' an ENIT, the way others see us, fantasy , how we see ourselves as we dream or imagine, ideally as we hope to be, materially or of wealth, influence, power, and acclaim, and variable mixtures of these. Vanity is neither entirely non TRU, nor always wrong. It is not how we are alone as the center of the universe but how all, and we or one part and process, exist in this center. For no matter our age sex, race, or other portion of TRU we can and will picture ourselves as wanting to be different, or better.

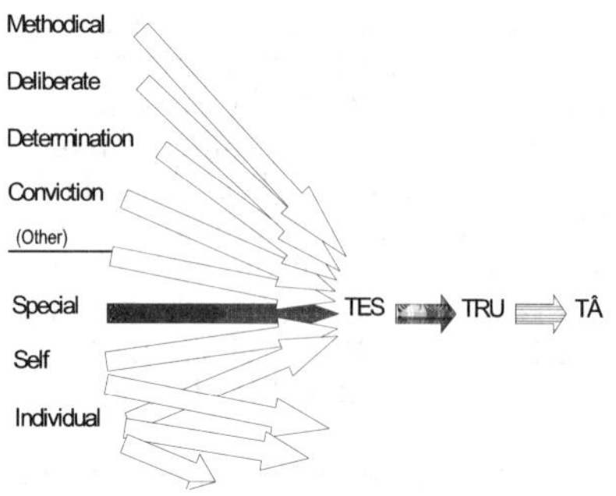

Fig 3-65 **PROCESSES and METHODS of SELF ATTAINMENT of TES, TRU and TÂ**

It is not so much, whether each random impulse, wish or desire is satisfied, whether we win or lose minor or worthless games or battles. It is through identity and cooperation with TRU that we can properly learn how to best play the serious actual game of life. By living more and more in PUREALITY we can apply with increased accuracy and power the KOT and means that achieve TIT.

By imitation or peer pressures from early life we tend to variously idealize and idolize the world. This often arises from ignorance, strangeness, and the unaccounted differences in the world. For when these are known they are no longer idols or ideals. We also tend to do the reverse, to bedevil and blame much in the world around us, especially if unfamiliar and different. We are influenced by the advertizing and blaming of others. It expands by reaction formation and tends to become pathological. It is infrequently based upon or qualified by the highest principles of TIT and its predominant Á form of energy and power. Only by discovering, learning, remembering and using TRU, and how these highest principles, concepts, laws, axioms, corollaries fit together can we properly be TRU. The right outlook, orientation and resulting concentration, in TRU, is vital. It is accumulative and exponentially enhancing.

It never ceases to amaze us how much people have in common, yet how far apart they can be. To a certain extent this is natural for the differential, especially if highly finite and disordered. Our sensory organs are geared to differences and changes. Our entire predominant mass energy, Á continuum and dimension is directed to select better/worse differences. To this is added the great extent to which people are not part of the differential, but deviant from and against the actual world. They are deviants from all levels, largely led by ignorance, mis-belief, figments, illusion, disproportions, or lack of total identity with the actual world. This results in the great divergence, absence, and negation we continually encounter in life. There is nothing so sad as this divergence, absence and negation from this special world of best human life. In your contacts, conversation, and interactions with others then: 1. Are you free from such divergence, absence, and negation? 2. How much do others come from such divergence, absence and negation? and 3. How well can you both achieve the identity, appreciation, and tolerance necessary to not only find your common ground in the actual world, but realize and comprehend the differences, divergences, and overall relationship?. Until, and how well, this is done, there can be little hope for, nor relief from, the incompatibilities, antagonisms, treachery, terrorism, conflicts, wars, and suffering that plague mankind.

Everyone wants to be loved. Yet to be loved one must be lovable. To be lovable is the degree to which one is TES. Therefor, be TES! One must have love and respect for all that is TRU, just as one needs to have a great respect and love of life. It is a spontaneous means and product of our being. It is a felt and joyful awareness, attention or IA and collective SÁ whose fulfillment becomes much of learned behavior. In order to have this love and respect we must know what is and what is not TRU, in all its intricate dynamic detail as they fit into its extant totality, TIT-TO. So that we can in our every thought, act, and means can do so with the precision and probity necessary for TRU.

SOLÁ = Solution or resolution from ATRU, Self, the individual and I/TRU. SOLA is the opposite or reversion from TA, TRU, TES, to self, to I/TES. If to be intellectual means to, or are those who, comprehend, complement and apply the basic ideas, concepts, principles, laws, and mechanisms of the world, TRU, it is to be highly extolled and glorified. If it is no more than the sophisticates weapon of added power, vanity, prestige, figment of TRU, or ATRU, it is to be deplored. So that we must not only know and apply TRU, we must know enough of ATRU, self, individual and I/TRU to be constantly on guard to upgrade and provide SOLA.

LINTÂ = Law of Individual to TRU Adaptation. LINTÂ is the more exact concept for the commonly known idea of a person's coming to terms with life. All that is needed in VAPO, DAOST, SCARLS, TIT-TO and ACEXIN for a person to be TES in TRU. The reason why girls and women have less crime is that from genetics, environment, culture and the earliest years of life they are less aggressive, suffer more repression, and are more in tune with home and family. They realize the benefits received from being or not being a good participant in the world around them. Boys being more self centered, independent, and adventurous are more often at odds with the world around them. They either learn or must pay for their continual adjustments in the world. This is a life long give and take condition in which only the intelligent and productive come out as TES and TÍS by way of a positive dominant role in TRU. **LINTES** = Law of Individual to TES. The transit, transformation or transfiguration of an individual or person to TES is a life long massive, dynamic, and highly variable state and condition. Each person's life varies minute by minute, hour by hour, day by day, month by month, and year by year. There is an optimal and minimal, even negative, rate by which this transit is occurring. So that the continuum of degree from TES to non TES and ATES approaches the entire mass interchange in life of positive and negative forces and processes in total TES and TIT-TO. It is not easy. Yet it is much less difficult without ATRU, when derived and achieved naturally, tending to TRU. Many sentinels of, the presently easy way that leads to ATES and ATRU are commonly known and self evident. Those that lead to the harder, but more permanent TES and TRU are much of TREQ, less well known, and much less adhered to. It should be no surprise that many religions take a severe pietist stance. Yet here to, it can not be without TREQ, VAPO, DAOST, SCARLS, ACEXIN, TIT and TO.

RITES = Rectification or Reversal of Individual TES and TRU. RITES is how an individual is rectified. How the individual is able to and does recreate oneself in TRU. It is analogous to the resurrection of Christ, being born again or how to be elevated to TRU. **PORI** = Principle of Reversible Immortality. PORI is the struggle of life, to survive by TES in TRU. To survive the deadness in us and in all.(DAK,ATRU)

Fig 4-65 **IT IS LATER THAN YOU THINK**

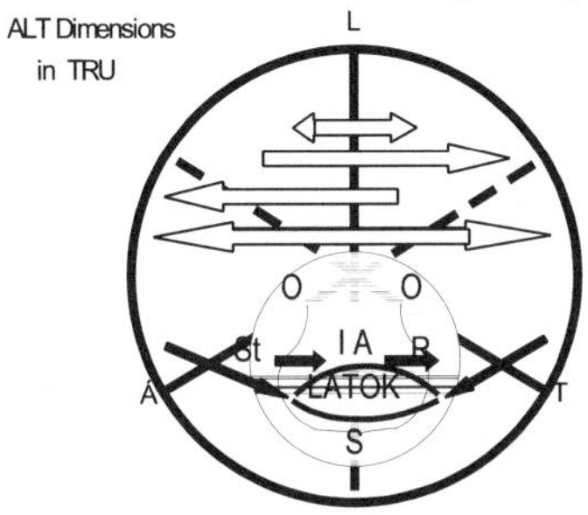

It is almost always LAter Than you ThinK, **LATOK.** LATOK characterizes both ones IA in life or TRU and most all we comprehend; especially all that may be adverse or likely to destroy us and TRU. It is a result of the fact that we are oriented, by our CNS, to the here and now. We do not realize stimulation by the future. So that, just as we think the past is dead and gone, so do we think the future indiscernible or of little significance. You are, as your spirit, not what appears or you may naively or commonly believe.We are but a part of the whole, TIT-TO. All others are deviates, to the degree ATRU.

A person's life should be a long and joyful adventure of learning, doing, and being TRU. The most interesting and exciting adventure, the best discovery, the greatest gift and the highest victory and happiness that a person can receive or experience in life is to seek out, find, be, and enjoy TES and TRU. Specifically it is to find the ES, CES, concepts, principles, laws, and uses of TRU that can be applied for the benefit of all, TO-TIT. It is the joy and power of inspiration, mystic revelation, attainment of certain immortality, and simple fulfillment of our counterpart in TRU. The joy of being a worker and contributor to this greatest endeavor the world has ever known, is at each person's beckon and call.

To not only notice others by their point of , but to notice and think of all by their actual existence in the relative universe. Rise from ignorance and stupidity to gain a mind set equal to our potential participation and contribution to human success and survival. This helps to correct our confusion and mistaken ideas and decisions. It helps to assure improvement, purity,and immortality in our life.

What we look to, what our sights are on, and how our IA, or attention projects into the future is: **POFU** = the Principle of Future Projection. POFU without TRIT, OBT, SUT, SPALT and all is haphazard, inadequate, and subject to error, confusion, loss and failure. It is like our priorities, values, convictions, principles, aims and purposes, a great part of the intentional nature of our being. This is much of TRU, a driving built-in Core Force and Power. It is essential to management and control. It is how to be the best kind of entrepreneur, those of TES who collect all the necessary resources and methods to achieve all needed to make TRU beautiful, artistic, and profitable. This is largely how well ESand CES can beautifully, artistically and positively reproduce and enhance TRU. It is similar to: **PETRU** = Pet TRU or in making TRU ones pet. PETRU is a turnaround from the pet fallacy or fantasy. It is where ones life should be spent loving and working for what is most TRU. As we make VAPO, TRU, TÄ, DAOST, SCARLS, TROS, TIT, ES, TES, ACEXIN, TRIT, OBT, ACC, and TA our pets, so will we spontaneously become - in turn - the pet of TRU. Immortality is the last great singular power and process we are likely to consider in life. Immortality can only be adequately understood when we know all TRU and the relation of TES to TRU. It comes from the contrasting denial of life, disbelief in God, TREQ, and TRU, and a realization of something more than what is immediately before us, as in materialism and naive realism. The extension of a person through the past and future, the large and small, the subjective and objective extremes occurs even though ones center may be located near the center of TRU. Collective powers of immortality are how TES functions for the eternal success of TRU like a manifest destiny must be TES, and TES must be TRU.(LOPAN)

FISAT = Fallacy of the Increasingly Finite toward Infinite Search for the less and less of TRU. This is a product of l;ack of focus, TROS, order, coordination and overall view. To seek out, search for, and strive to discover deeper and deeper reasons, causes, meanings, and the fundamental explanations of all is at the root of all life, the fulfillment of our mass energy, Á, FAF of TRU. FISAT does not negate a deeper search into the intricate finite and infinite it only limits and conditions it. For this is much of what DYNU is and will be a large part of SUSF forever. This is: **SETIN** = Search for the Finite and Infinite of TRU through FAF, DAOST, SUT and TIT-TO. SETIN is a, finite toward infinite, search in TRU guided by VAPO.

Of the following terms those with numerals of posits in SUSF are assigned to their respective figures; S for the Sector diagram Fig 5-65, T to the STES diagram, 6-65, and M to the METAD diagram, 7-65. Several of the terms have or will be posited in prior or forthcoming chapters. Because they are discussed in this chapter, and have posits in this chapter through extension of multiple foci they are included in this list, although not in its figures. As we do in all chapters from 61-88 we abbreviate these processes in the figures. The system used is to use the first three letters of each term, unless more than one term then first letters of each term. There are a couple of exceptions to this, i.e. for NEC we use the full notation. For abbreviations that are duplicates of others we may use four letters or a distinguishing consonant for the third letter. Some are duplicates but in different figures, these we leave abbreviated without exception.

Measures of CHES of SUSP

			EXINI	2-7-6	S	INT	3-6-6	S	PRIORITY	4-5-5	S
			FAYAN	2-7-5	M	LATOK	3-5-6	T	PROJECTION	3-6-8	M
COMPREHENSION	1--9~	M	FISAT	4-5-6-	T	LERAC	2-7-5	M	PUREALITY	4-5-9	T
CONCENTRATION	2-5-5	T	FOYAN	2-7-6	M	LENTES	4-5-5-	S	RITES	3-7-6	M
COURAGE	4-6-6+	S	IDEAS	2-8-6	S	LOVE	3-6-6	T	SETIN	2-7-8	T
DECEPTION	4-5-5	T	IDEALS	3-7-7	S	ORIENTATION	3-6-5	S	SOLA	2-7-7	M
DECISION	3-7-6	S	ILRIA	2-6-5	M	PETRU	4-5-7	S	SPIRIT	3-7-7	S
DIRECTION-CONTROL	3-6-7	M	IMAGINATION	1-8-7	S	PORI	3-7-8	T	SRIA	3-7-6	S
EGO (EGOTRU)	2-7-6	S	IMMORTALITY	3-7-9	M	POTRATH	1-8-6	S	VANITY	4-5-6	T
ENTHUSIASM verve	3-6-6	S	INGENUITY	4-5-6	S	PRAYER	3-6-6	S	VEMORT	2-7-5	S
EXHILARATION	3-7-6	S	INTELLECT	2-5-7	T	PRINCIPLES	2-7-7	S	WITA	3-7-7	S

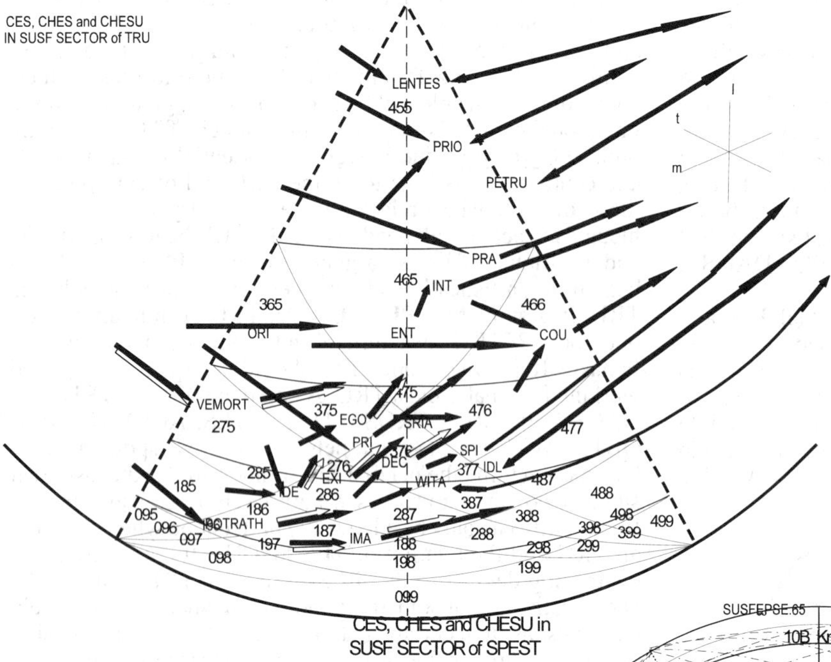

Fig 5-65 **CES, CHES and CHESU in SUSF SECTOR of SPALT and TRU**

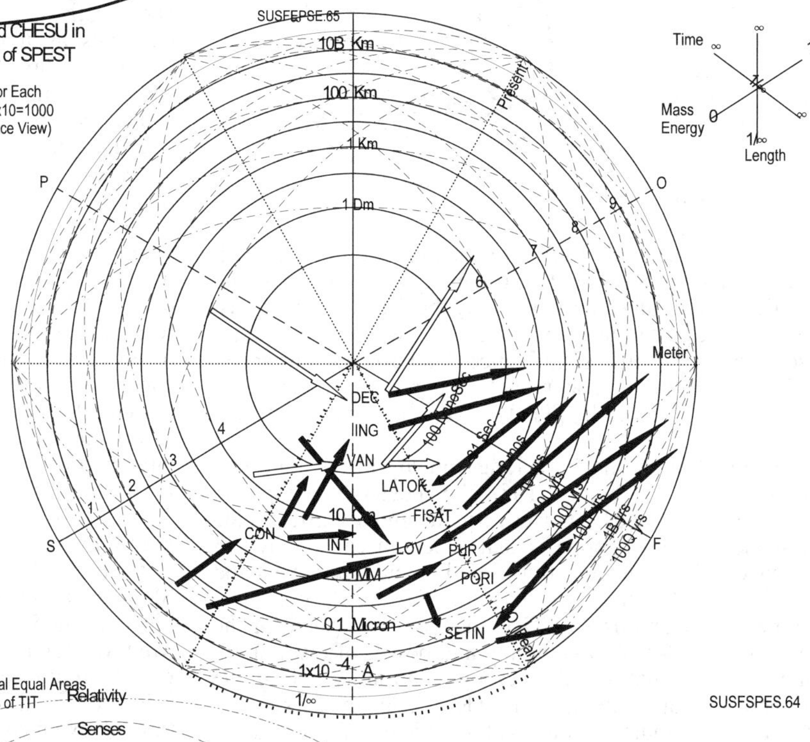

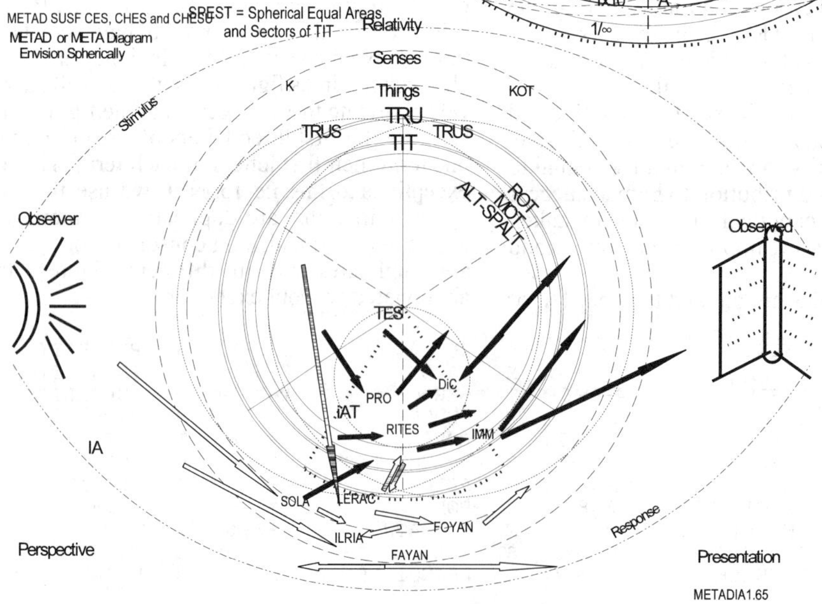

Figs 6, 7-65 **SUSF CES, CHES and CHESU in SPALT and METAD**

CHAPTER 66

OSF or OBJECTIVE SMALL FUTURE PROCESSES of DYNU

REFLEX, RESPONSE, IA, BODILY FUNCTIONS, DO, ACT, SEX, HELP, RIGHT ACT, ACTION, TREATMENT, CHARACTER, WORK, CHARITY, BEHAVIOR, LIFE

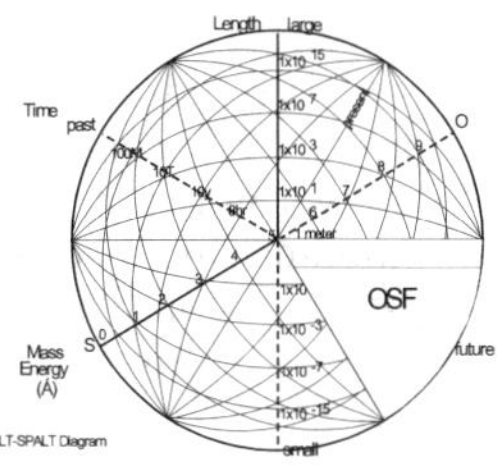

Fig 1-66 **OBJECTIVE SMALL FUTURE SECTOR**

The different kinds of acts are projected into the future, intentionally or on lower levels of SÁ. Most are well defined and thereby amenable to posit. Yet they must be TRU, to do right in life: to play their part, TES. Their opposite or negative is **VANTRU** = Voluntary ANTRU or the will and actions by which we are not, or anti, TRU.

The amount of insulin, ACTH, cortisone, testosterone, estrogen, progesterone, adrenalin, and all hormones, enzymes, mediators and substances produced in the human body per second, minute, hour, day, year and lifetime necessary to sustain life and produce optimal health provides key information about the vital macrochemical dynamics of TRU. They must all be tallied, interrelated and ordered as the actually exists in TIT if HOM is to best represent TRU. The many singular verbs and nouns that represent physiological, biological, chemical, physical, human, behavioral, IPR and social CES are highly beneficial to TRU and should be enumerated to round out OSF. Some of these are: assimilation, diffusion, osmosis, photosynthesis, chemical reaction, secretion, radiation, conjunction, division, growth, fertilization, microbiology, molecular biology and chemistry, enzymes, many actions and CES that are less well named or described in the literature There are also many more actions and processes that do not have one word denotation, and remain to be discovered and represented.

DESJUC, Deception Justifies Skepticism which necessitates Correction. DESJUC signifies the reverse use of appearances, figments and illusions, and a justification of skepticism. The existence of these continually necessitates correction.(FOSK)

GA = Glory in Act. If the same spirit, conviction and energy of climbing the highest mountain or performing the most unbelievable acts could be put into the basic first principles of life, how great TRU would become. The identity of each individual becomes TES, TRU becomes the glory of these acts, and immortality in IA is hastened and secured.

What one has to do and what one wants to do should be much the same. If they are not, either: 1. What one wants to do is wrong, or 2. What one has to do is wrong, or 3. They are both more or less wrong, or 4. One does not appreciate what one has to do, or 5. One expects to much either more than one is capable of, or is excessively disagreeable or in imbalance, lacking VAPO and its accompanying harmony and happiness. To do what one has to do rather than what one wants to do, is the adjustment and adaptation life demands. Yet adaptation is in itself a natural need, and largely becomes habitual. The goal is to learn to like the things one has to do by fully realizing the benefits and rewards one receives from TRU.

All in life of TRU profits and/or costs something. Charity is where the cost is minimal and the profit is maximal, not something for nothing. No one is so wise, impartial, nor free from ATRU, that they can have attitudes and actions that are indiscriminate. Only TRU or TREQ attitudes, knowledge and discriminatory acts will be permanently successfully work. What discrimination and variance is acceptable is that determined by the different of DAOST, and the partial position of the person. Both of these must be in resultant accounted for through compensation or what ever mechanism maintains the equilibrium and TRIA of TIT.

Because something is wrong doesn't mean that you should charge into it without first reconciling your actions. Although many of our actions are reflex and many of these are all-or-none, this does not apply in general. Especially it does not apply to higher levels of mass energy, or Á actions. Thereby, the only all-out act that is voluntary, is what is for TIT-TO. All other acts are as all partials, senses, and things proportional and limited. This is called: **LOPA** = the Law of Proportional Act. As developed in the Laws of TRU, LOT, the reaction can not, generally, exceed the action. Overreaction is more than worthless, it is often highly negative and harmful causing suffering and loss to TRU. LOPA means that all actions and reactions should be tempered, moderated, weighed, judged, proportioned, and remembered according to VAPO, REM, their relationships and posits, and the dynamic forces, powers, and CES that are occurring in and applying to TRU. A Corollary to LOPA is: **MOR** = Margin of Order. MOR is where LOPA is qualified and specified by the exception of those instances where perfect order and proportionality are not possible by regeneration and a vent for energy. This is especially exemplified by what is potential for TRU. Yet in the control of MOR and the sustaining of LOPA some order is necessary by TRU.

How one mistake leads to another, and tends to reinforce it, is seen in the S-O of a composition. One mistake distracts our attention, the intention and chain of thought are diverted, and the lost direction produces repeated mistakes. This is another form of INEG or Integrative Negation and shows COC. It shows how INEG tends to prove PIP or Proof of Integrative Processes. Ultimately INEG is the state of TRU in which the collective and negative create, make or unmake, annihilate and destroy TRU. It, in a particular way, shows the importance of good training, habits, attention and good intention. When these are focused through ACEXIN upon what is most pertinent, learning and concentrating on TRU the result will be one in which there is much less likelihood of mistakes and their escalation.

The use of various consonant, vowels, and combinations in different languages throughout the world can be rated by their frequencies. They illustrate and will provide for extensions of sounds that may be inadequately utilized. They show the frequencies and readily usable forms of certain sounds as opposed to those sounds whose combinations are not so well pronounced. They show, when analyzing rarer combinations those languages which use certain sounds that have been of cultural competitive advantage.

FOCS = Fallacy of Choosing Sides. FOCS occurs with all humans, either as individuals, groups, nations or in totality. Human evolution has produced in the individual a great need to belong, to become a member of this or that group that adds great protection and strength to life. Without this belonging the individual is extremely vulnerable. This is how groups which are strong may have individuals that are much weaker. In this way major decisions are not necessary in all or even the most important cases. If they have already been decided on, if their concepts, principles, laws, and rules are well established, they can be put into practice without individual, formal, and extended judgements. The opposite of FOCS is,

SEJONAC = The Selective Judgement of new or unbroken cases.

Muscle composes 30 to 50% of body weight. If you don't use it, you lose it. If we do not exercise, it will largely shrink or turn to fat. So it is of many of the other organs and systems of the body and TRU. The importance of optimal exercise for every part of the body is equivalent to optimal health. However this optimal exercise is not for one system more than another, so that to have 50% body weight in muscle deprives other systems. This is even more important for, ones Central Nervous System and Mind. They are, much less in percentage, yet, much more in contribution to TES and TIT. Not only does its exercise provide the needed bodily functions, it, to a large extent, controls the rest of the body. It, as the organ of survival, is the predominant and crucial portion of a human being, and TRU.

You don't fight for anything but TRU. This may be as a whole or in particular, but in the end all the same for TRU.(VAPO) If you fight perfectly for TRU you and all can not lose. Your benefits will be eternal. A hard thing, that takes a long time to learn and recover from in life, is anything that is expensive is going to cost you. Sometimes you may get by without it, but more often you may have to pay and pay. The trick and aim, then, is to stay away from all that is expensive in cost, unless it is weighed by degree of TRU. The more expensive it is in greatness, REM, and TRU the more you should desire and strive to attain and maintain it. This is why material things that are highly needed are to be protected and preserved, held so dear and utilized fully. They should always be handled frugally, with care, in the optimal working condition, and given the maximal appropriate use. This is why a life spent in the sum of the opposite is so wasted. **LOSOW** = Law of Summation of Wastes. LOSOW is largely a corollary of INEG with specific meaning in relation to all that can be measured by cost. If you don't have sufficient need for it, don't buy it. Always consider those who must pay for it. Yet the greatest object in all trade is to get what is or will be most important for you and TRU, as you are your counterpart TES in TRU.

Fig 2-66 **ACTIONS in HIDIC SPALT and TRU**

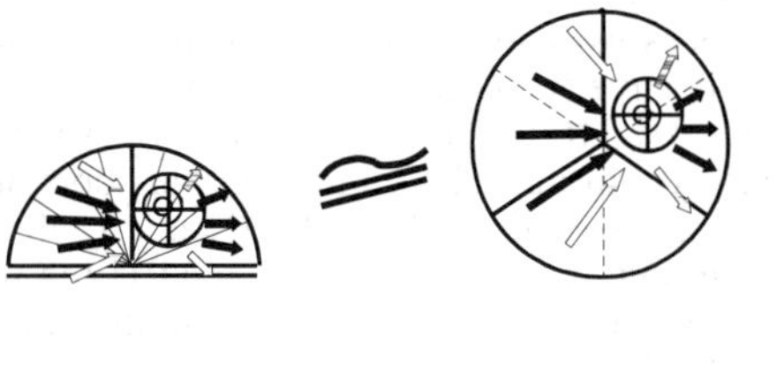

LACMEST - Law of Actions to be Measured by their Contribution to the Success of TRU. LACMEST seems simple initially, yet it is conditioned by all that is of TRU, its VAPO, all SOT and Á. The greatness of treatment, and even more the prevention and means of controlling these, is of great assistance and benefit to the TES and TRU.

FORA = Fallacy of Right Act. It is not adequate nor so effective that people be told to do right. For often, there is reaction, especially when loaded with error. It can be no better than a person's knowledge, KOT. Let no act be other than the ability to contribute to TRU. Let no one be less than their action to support TRU. This is: **LATAC** = the Law of TRU action. LATAC is an ideal as well as practical objective and means. To transform all ones being, acts, attention, intention, work from their common levels of behavior and objectives to LATAC is a great part of all our lives. Only from a subtle and profound love and effort for GOD, TREQ, or TRU do we rise to the occasion, the eternal necessity in TIT. It depends not so much on fortune, but on conduct; the conduct determined by TRU. **PORA** = Principle of Right Act. For a person to do right they must either know what is right or be told what is right. These can be no better than the laws, axioms, mechanisms, principles, concepts, and corollaries, laid down by and for TREQ or TRU.

There are more than enough genuine and bountiful thrills in life to keep from being bored. Intimate mating to maintain genetic quality and momentum, birth of a healthy, vigorous, beautiful babies, discovering great unknowns or revelations, satisfying ones deprived needs, finding what has been lost that is sorely needed, and the thrill of making a paradise on earth are but a few of what TRU and TES offer.

Fig 3-66 **RIGHT ACT and LIBERTY by KOT**

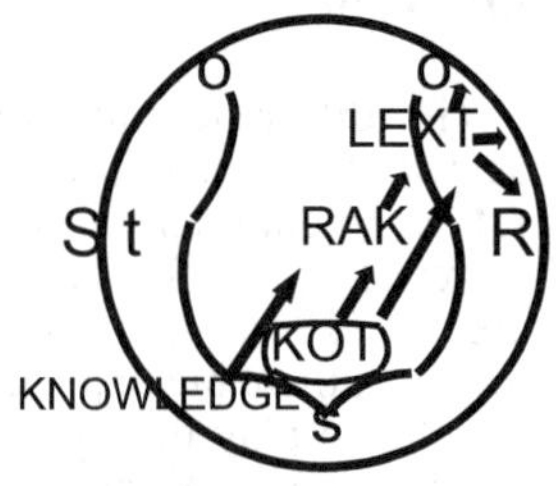

We do right primarily because we know right, not from ignorance, nor because we are so conditioned. This is **RAK** -Right Act from Knowledge. The right of man to life, liberty and the pursuit of happiness must be qualified and proportional to TRU. It is something that humans, and all that have enabled our being, must earn, not something from or for nothing. Man can not have full freedom until it is certain that we will, automatically and spontaneously maintain sufficient ability, understanding, will, and TES. This is: **LEXT** = Law of Liberty for those who would be TRU. The greatest trick in life is to learn how to find ones pleasures not in what is harmful or ATRU but in those that are beneficial or TRU.

LORA = Law of Right Act. The precedent as well as form, content, function and processes of right acts, have been built up through the ages. This has been especially true since the advent of civilization and mankind's discovery and use of writing. As human knowledge increased and as all the necessary concepts, principles, corollaries, axioms, and laws which guide this knowledge were developed, mankind found, knowingly or unknowingly, that when they were properly used success was the result. This is how religion, ethics, philosophy and science have become increasingly TREQ, and TRU.

A DAY'S JOYOUS ACTIVITIES in TRU

Any day in which there is no
Music, imparting of joy to others
Achievement of multiple tasks
Inspiration in study and writing
And Identity with the glory of
GOD as TRU, is a day that is lost.

Many who go to a doctor or who suffer from any affliction may not have a natural disease or readily diagnosable mental or physical ailment. Yet their illness and complaints may be none the less real, severe or distressing. They may be deficient, have excessive exposure, or may not function in one, some or all ways properly. It is up to each and all of us to discover, appreciate, learn and practice a life in which we constantly strive to improve, qualify and achieve the best possible SÁ, or health that is free from distress, suffering, disability, serious injury or deadly disease. This is what is done in diagnosis and treatment. If you have something wrong that is temporary or not life threatening, let nature heal it. Random treatment is often more apt to do more harm than good. If there are problems that are likely to cause great damage to TRU analyze them, evaluate, order, and solve them as appropriate for each case by TRU. Always control your treatment by the benefit/harm ratio, not so much by short term relief as for long term benefit, for the enhancement of TRU.

I - or you - "Should have known better" and "What's the use", are phrases that are directed to actions that are not right, can't find the right answers, or help. Yet this is never the case in TRU. For by TRIA not only are TIT and TRIT absolutely valuable, all that is of TRU that needs help to be TRU or continue to be TRU, is absolutely valuable. To say one should have known better and what's the use are rationalizations for inadequacy or neglect. All that is TRU and all that enables knowledge of TRU are infinitely valuable and must be helped in all ways indicated by TRU. The worst is to think without discretion, to act in haste or by impulse without thinking with discretion, and thinking and acting without VAPO, DAOST, SCALS, TROS, TES and TRU. We fail to separate simple, complex and total actual ES and CES as they exist in TIT-TO. If it wasn't for our society, nations, and cultures that give us the routines, concepts, habits, morals, principles, ethics, civics, and laws to go by we as individuals would be, as so many are in the world today, savages or savages in sheep's clothing. It isn't because the knowledge isn't discovered, it isn't because each person doesn't or can't know the world they are in. For much is all around, common, given to us, and right before our eyes. Yet we don't see it. We are blinded by our own ignorance and the ATRU and NAMTRU worlds about us that offer easy ways out. It is like how being tempted by the devil, brings the worst savage tendencies into action. This is why such emphasis is placed on S-O, the mass energy, Á dimension of TRU, IA, S/R, and success/failure as all reveal the fundamental and central dynamics of human life, TRU. It is why we must discover, learn, and apply all TRU if we are to be our part and fulfill our roles in life.

To be TES and TRU, is not a free ride or a down hill slide. It requires work, effort, doing, learning, ability, conviction, determination, appreciation, and PPMA. It does not require an unnecessary amount of pain or suffering. It only requires great attention, and intention to TRU. Thus working our way uphill may or may not be TRU. If it is not TRU it is not only drudgery but often waste, misery and failure. If everything were easy, you wouldn't be needed, or even be here. The trick to successful life, as TRU, is in making everything no harder than necessary, and to learn to enjoy what you have to do. That which is difficult may have to be effected as such. To do things which require hard work and effort to overcome great difficulties, in a manner that is easy or skimpy is usually faulty. Hard work, difficulty and the massive maintenance of life, as TRU, are not necessarily bad nor negative. They are bad, as all, only when they are not TRU. As the proverb goes 'Work makes life pleasant'. This applies only to the extent, this work and life are TRU. Work that is enjoyable, profitable and eternally beneficial is, in a sense, play. This is the best kind of play. That which is the enjoyable fulfillment of life, of TRU. The basis for a person's contribution to TRU, their behavior, actions, good deeds, and benefits have been repeatedly attributed to character and virtue. These terms represent great processes in TRU, especially if best trained and used. That which is commonly held to be character and what are the qualities it connotes include, persistence, patience, control, moderation, tolerance, equanimity, judgement, virtue, quality, conviction, resurrection, confidence, determination, courage, wisdom, honesty, and righteousness. Character, like so many words, consists of dual and manifold meanings. The supreme sacrifice in life is to live and endure it. It is to live it as TES in the most complete and greatest way. This is called: **SACT** = Supreme Sacrifice of the Self to TES, to TRU, TRIT, TIT, and TÂ. We more or less engage in SACT all our lives. Our attention, love, duties, work, helping others and many more provide SACT as SACT provides TRU. Just as a mother cares for her young, the scientist investigates the unknown, and workers labor long and hard not so much for money, but a livelihood with heartfelt identity and contribution in the principles and higher good of TREQ and TRU.

Companionship, romance, love, sex, intercourse, and their varieties are a great part of our lives and TRU. Upon songs, stories, movies, plays, operas, TV and commercials much of our notice or attention is focused. The percentage they are TRU, ATRU, and ANTRU, are highly variable. Their effect, however, is often not to benefit TRU. It is most common for all men and women to experience in love, sex, and reproduction, the S/F and mass energy Á continuum of their lives. For it is in these they must come to terms with the hard giving, rather than taking in life. It is here the quality of their lives, their TES and TRU/ATRU become most manifest. The old saying or maxim goes, 'a person is up or down as their love life is up or down' might be better qualified by the degree to which this love life is TES and TRU. Hardly anything ruins TRU more than defective, recessive, or inferior reproduction. It seems and is to a great extent an enigma that we must hold things close. Yet one can't hold things to close. By TROS, VAPO and ALT each must be held at their proper distance and position. Only when we experience the vicissitudes of life and the frailties of our own capacity, ability and powers can we learn how to be best aware of our limits.

Fig 4-66
GRADA-TIONS of QUALITY and TOLERANCE

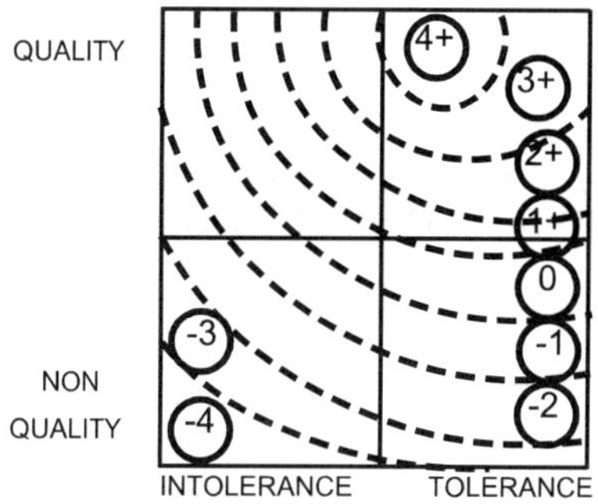

Humans can be separated into types by divisions and gradations of quality of TRU and degree of tolerance. It helps to show how well they are TES and contribute to TRU. It corresponds with how well they are or are not productive of peace or strife. For TES and TRU it is vitally necessary to not only be of top quality but to have a high level of tolerance. Also it helps to have a degree of intolerance for that which is not TRU. It can also be seen in how important it is to discover, know, show and remember what quality actually is, as it must be TRU. What TRU is in its entirety, TROS. So that we might know what quality and tolerance to have. Tolerance is of S/F, strength/weakness, and reasoned acceptance of the different FAF of TRU.

MEASURES of CES, CHES and CHESU in OSF SECTOR of SPALT

ACT ACTION	8-5-6	T	FORA	7-5-6	M	MISTAKE	6-6-6	S
BENEFIT(ICIAL)	7-6-8	M	GA	8-5-7	T	MOR	7-5-6	M
BODILY FUNCTIONS	9-6-6	T	GENETIC MOMENTUM	9-8-8	T	PORA	7-5-6	S
CARE (Phys)	6-5-6	S	GIVE (ING)	7-5-6	M	REFLEX	6-8-5	T
CHARACTER	5-6-7	S	HABIT	8-6-7	S	RAK	6-5-6	S
CHARITY	6-5-6@	T	HELP	7-5-6@	M	SACT	6-6-7	S
CHEMICAL REACTIONS	9-8-6	T	HOLD	7-6-6	T	SECRETION(s)	8-7-6	S
COMPENSATION	6-5-7	S	ILLNESS	8-7-6	M	SEJONAC	5-5-6	S
CONTRIBUTION	6-5-7	S	LACHMEST	8-5-8	S	SEX	8-6-6@	S
DESJUC	5-8-6	S	LATAC	8-5-8	S	SOUND(S)	7-8-5	T
DO, DOING	7-5-6	T	LEXT	7-5-7	S	TAL	5-5-8	M
ENZYME (ACTIONS)	9-8-6	T	LIBERTY (Freedom)	7-5-7@	S	TREATMENT	8-6-6	M
EXPENSE, COST	7-5-6	S	LIGHT (pulsar wd)	9-9-~	S	USE	7-5-6	M
EXERCISE	7-5-6	T	LOPA	8-5-6	M	VANTRU	5-6-7	M
FIGHT	8-5-7	T	LORA	8-5-5	M	WORK	8-5-6	T
FOCS	5-5-6	M	LOSOW	7-5-7	M			

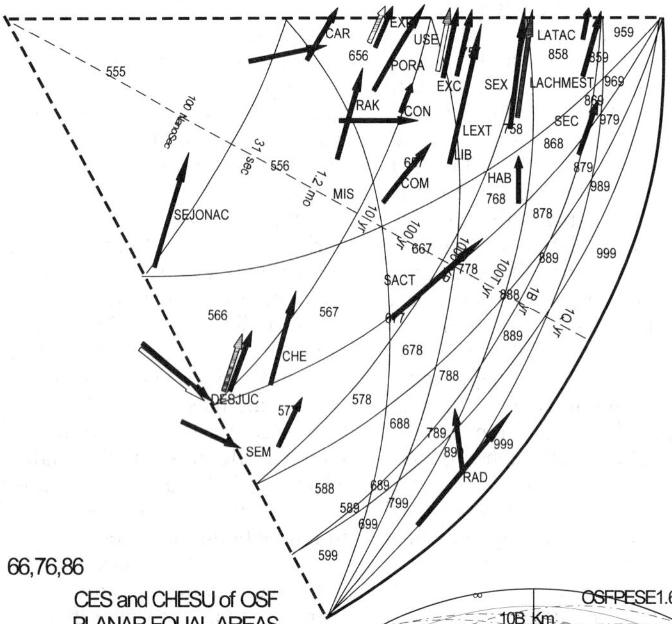

Fig 6-66 **CES, CHES, and CHESU of OSF SECTOR in SPALT**

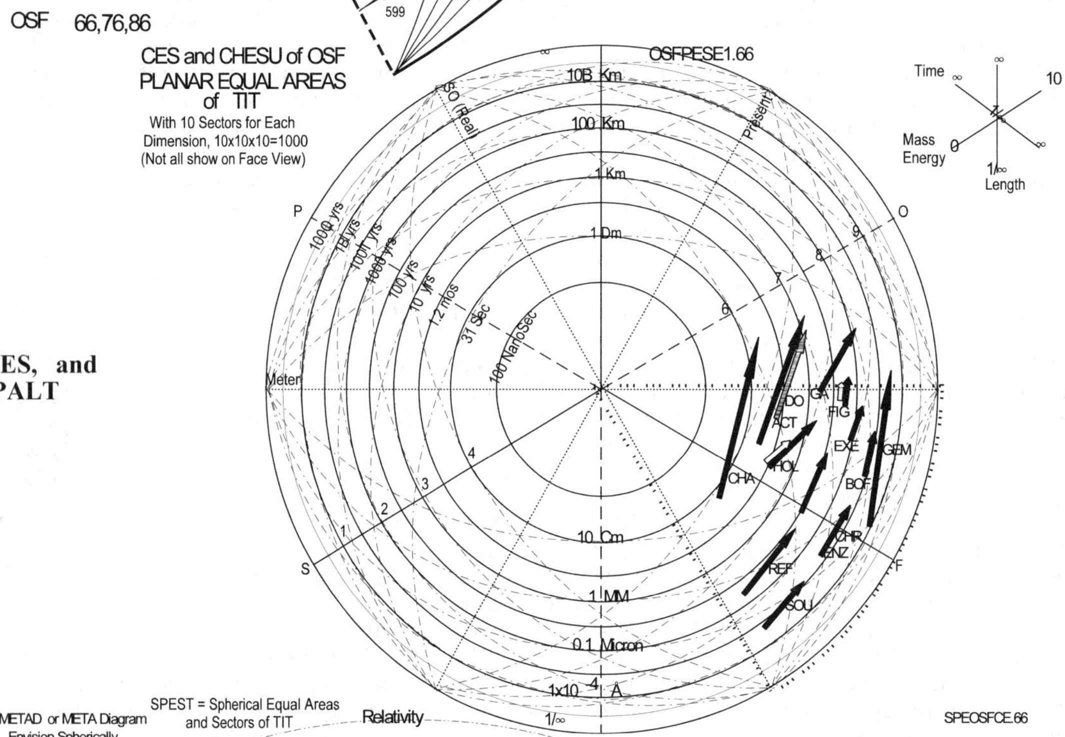

Fig 5-66 **CES, CHES, and CHESU of OSF in SPALT**

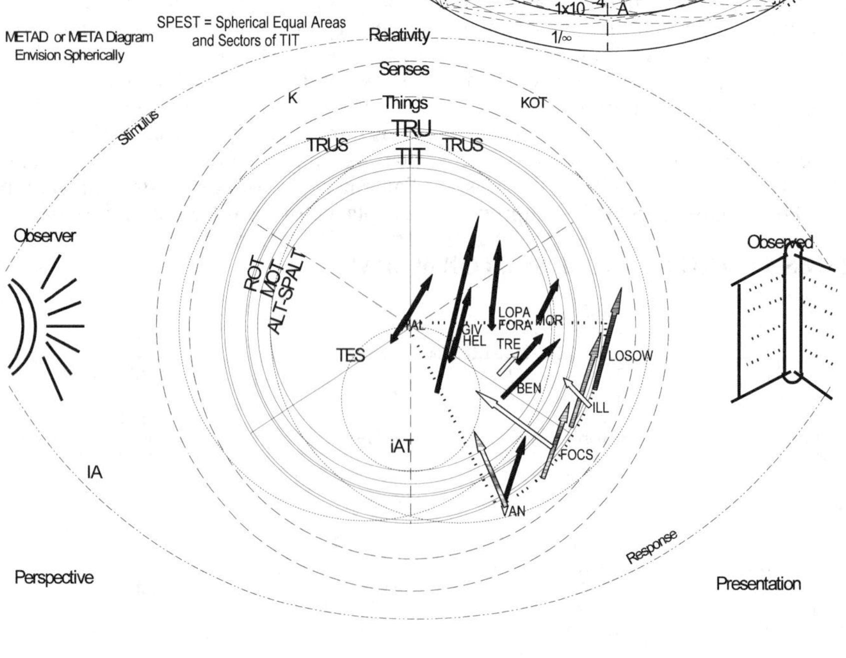

Fig 7-66 **CES, CHES and CHESU of OSF in METAD**

CHAPTER 67

OLF or OBJECTIVE LARGE FUTURE PROCESSES of DYNU and TRU

WANTS, NEEDS, FOOD, ADVANTAGE, LABOR, BENEFIT, USE, IPR, MONEY, ENVIRONMENT

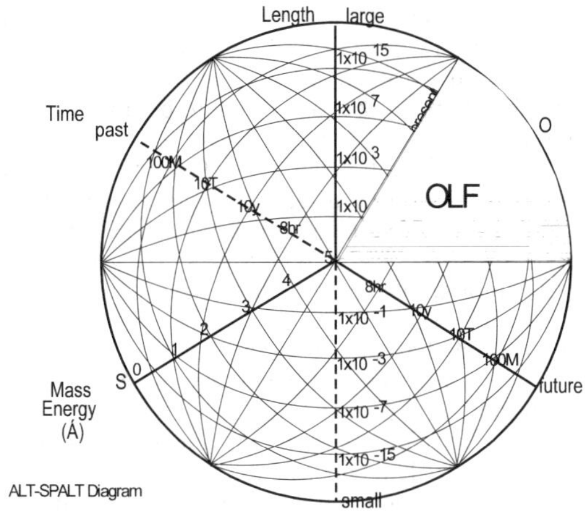

Fig 1-67 **OLF or OBJECTIVE-LARGE-FUTURE SECTOR**

The processes presented in OLF are utilized primarily as examples, both of important CES in TIT, and for their use in POA. These examples show existing terms as well as NEC and the problems that occur with ROT and POT of TRU. With practice the reader and student of TRU can gain an increasing ability to posit all that is given or that they encounter to gradually provide the composite makeup of TRU vital to the well being of everyone. More precise POSITS of specific CES and CHES with their A, L and T measures will require continuous future refinement. The philosophy of deontology, or obligation, represents a large part of TRUD and CES in OLP. Much of the activity of all persons and groups are obligations produced in the past, concurring in the present and directed toward the future. They largely form our work-a-day world, business, social and civic activity. The stimulation necessary for all in TRU is possible only when TRU is recognized and cultivated. This is demonstrated by: **PIB** = Perpetual Immunity to Boredom or Stagnation. PIB is achieved by the inspiration and revelation associated with TES and TRU. The various processes involved in population genetics, races, blood groups, inbreeding, out breeding, isolation and movements of populations exemplify CES in OLF. The Indefiniteness of the Future, **INF** and Poisson's Law of the Probability of Events help to increase certainty within statistical error. Such means provide an ever more precise renderings of the future, ALT, THEO and TRU.

Role theory reveals the contrast between how a person functions in society as an agent or employee filling some role rather than how the function as an individual. It emphasizes a division of labor in a highly organized, civilized and technical society. This may seen demeaning to personal freedom yet it demonstrates both the difference between individual and TES, and one's role in TRU. It challenges human adaptability. It rewards with benefits that accrue from fulfilling ones contribution to TIT-TO..

Individuals, families, tribes, communities, states, nations, and civilizations should treat others, of these various kinds, separately, in combination, or totally, by how much they are TREQ or TRU. How much they are and do their part and fill their roles in TRU. Not how much they can do otherwise, less, or negative. Two good examples of this are 1. Nation's foreign policies that support rogue groups, states, and nations, and 2. All the behavior and actions of a person that are impulsive, rash, by whim, fantasy, illusion, comfort, influence, desire, unfounded, and ATRU choice.

	- 10	- 8	- 6	- 4	- 2	0	+ 2	+ 4	+ 6	+ 8	+ 10
0	EXTINC-TION			POLLUTION			WEATHER		RESOURCES	CONSER-VATION	ETERNAL LIFE
1	Unlimi-ted WAR		OPPRESSION				NECESS-ITY	INDUSTRY	COMMERCE	SOCIAL HARMONY	
2		TERROR-ISM	WASTE			TRAVEL	CUSTOM	TRANSPOR-TATION	TECHNO-LOGY	PROSPER-ITY	SUCCESS
3		FAILURE	INTOLER-ANCE	WORST of MOVIE-TV		INFLU-ENCE~	MONEY~	COMMUNI-CATION~	SHELTER	PPMA	PRODUCT-IVE WORK
4		VIOLENT CRIME	IMMODER-ATION	MISDE-MENOR	FASHION BOXING	IPR~	SPORTS	BEHAVIOR~	HOME MAKING	GARDENING	MODERA-TION
5	MURDER	FELONY	SLOTH THEFT	IRRESPON-SIBILITY	HYPE CARDS	HABIT	ACTIV-ITY~	REPRO DUCTION~	EXERCISE RESTSLEEP	EDUCATION	REPRODUC-TION,TRU
6			PREJUDICE	PROFANITY	IRREVER-ENCE	VIDEO GAMES	CONVER-SATION	DISCUSSION	RHETORIC	CLOTHING~	TRU KNOW-LEDGE
7		HATRED	PURPOSE-LESSNESS	EXCESSES LACKS	TRIVIA	SEX~	INTER-ESTS	COMPO-SITION	PRIORI-TIES~	FOOD WATER	TRU PURPOSE
8			ALCOHOL TOBACCO	LYING	INAPPRE-CIATION		PURPOSE	READING	THOUGHT	PRAYER	IAT
9	RADIA-TION	POISONS	TOXIC SUBSTANCES		DRUGS			MINERALS	MEDICINES	OXYGEN~ AIR	

Table 1-67 **WORST to BEST GRADED by NEGATIVE to POSITIVE and SIZE**

Some of the OLF CES are ordered by their degree of TRU and size. This enables us to envision how they work OITIO and how we should react to them, by TES. They can not be taken as specific points or values when less than TRU. This we designate by ~ indicating that the grading of each topic and process may be much greater or smaller depending upon how it is interpreted and used.

Among the most critical positive and negative forces that must be met, are those involving replenishable and irreplenishable resources. What we consume, and how much. What is qualitatively and quantitatively vital. What we must save for best use and reuse. The degree of tolerance needed to adapt to less than optimal conditions, as adverse environment. War has been, and will always be, highly negative to a prosperous and permanent world. It, as the negative in general, can have some benefit in a reactive and proactive way. Since

Sun Tzu in China, 500b.c,.the importance and benefits of strategy have become highly-developed institutions. When greatly limited and directed toward TREQ and TRU goals and strategy can be very useful and positive.

A person's karma exemplifies OSF to OLF CES by the overall extended effects of ones words, acts, and deeds transcending life. It signifies the continuity and immortality of one's soul and TES in ALT and TRU. POSITS are often not to be made from the ES or CES commonly known, but from their implications. This leads to a reversal of verification by TRU, whereby actuality is often much more profound and subtle than expected. The actual magic in this world is how well things work when they occur or are done by TRU. Once realized it is no longer magic, but KOT. Magic is especially the case when long overdue. There is more to the subtlety and profundity as well as greatness and absolute beauty of TRU than can ever satisfy the most vigorous desire for magic, novelty, change and adventure. In crisis the many benefits that can be produced are the greatest magic.

Many examples of TRU motion occur in everyday life. When adapted to TRU they become highly revealing and useful. One of these is Multiple Ablation through Rapid Simultaneous Action or **MARS**. This may be most simple to most complex as in TRUD, HOM and ACC. We are in a delicate balance between necessary action and unnecessary action, effect and safety. Sometimes our routines and habits fail, or fail in time. These are particulars, examples of the positive and negative in HOM, which are validated or eliminated. Many hands make light work, in the distribution of labor, be it a group, different organs of the body, or any portion of TRU. It is essential that an optimal number contribute if efficiency and success are to be achieved. When others want something from you it is not because they owe it to you or like you. It is not because you are so right, good, valid, godly or TRU, it is because you have something they want or need. It is often a simple exchange of work for pay. Something they give for you to give in return. This is why buying and selling is often so cold yet one in which both sides profit. Thereby in trade or negotiations assure profit in ways that all can be best turned to TRU. For it is not to buy whatever the seller wants to sell. It is to only select and buy what you need, and this by TES and TRU. Unless we want to suffer the experiences of life in learning this lesson we must profit from the experiences of those before us who have learned what works best, TREQ, and TRU. This is the key to most human activity, yet it must be with VAPO, DAOST, TES, and TRU.

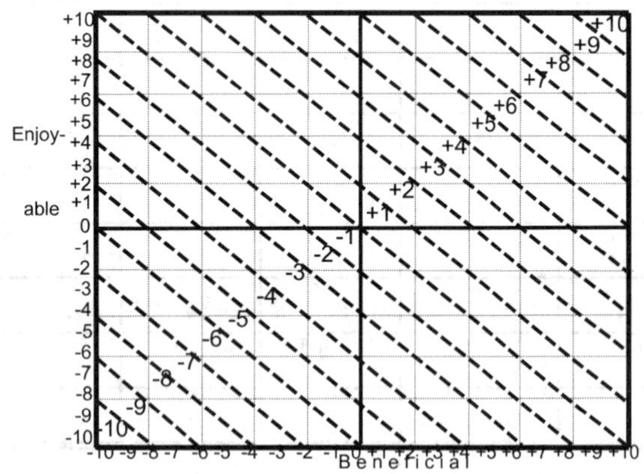

Fig 2-67 **THE ENJOYABLE or NON ENJOYABLE, BENEFICIAL and NON BENEFICIAL**

All that stands in any relationship to a person has an enjoyable or non enjoyable, pleasurable or non pleasurable, good or bad, positive or negative, TRU or non TRU quality and measure. Everyone is highly and quickly attuned to this. They will change and be different depending upon the person and their circumstances, degrees in different individuals. Among these are those that are enjoyable yet of more waste than help, most helpful yet not so enjoyable, most TRU fulfilling, meeting the needs of a person and the world, enabling one to be, and those of the highest quality universally recognized, greatly TREQ and TRU.

IPR - Interpersonal Relationships. IPR, when TRU, helps show the classes of friends, relations, associations, and participants in society and life. If a hierarchy of classes in TRU is used it should be based on their capacity, ability and performance in TES and TRU. It involves the degree to which both you, they and all involved are TES and TRU.

In life there is little time to be lost, much to be done, upheld, enhanced, and acted on only when achieved as appropriately and early in life as possible. It necessitates training oneself and others to identify with the good, right, great, beautiful, harmonious, Godly, TREQ, or TRU. With practice it becomes easier and easier as time goes on. It becomes, like good habits, a way of life. The benefits and blessings will accelerate ever greater identity with TES and TRU.

Similar to and somewhat more general than fig. 2-67 are the relations between the unnecessary/ necessary and ATRU/TRU. Only when optimal are quality and success assured. The biggest problem of work, employment, and contribution to TRU arises from needs. The needs of the individual, business, and TRU are critical to each and all together. If an individual can work, be it job or TRU, in such a way that the needs of this job or TRU are maximally fulfilled, then equitable accommodation can be made for the needs of the individual or TES. So is the relationship between all parts and the whole to be weighed, be it of ES, CES or any aspect of TRU. This is the basis for and derives from the makeup of TRU, TROS, TIT and ENT. It is the key to S/F. It is obvious initially and by our cultural inheritance, yet so quickly forgotten or ignored.

The tightness or looseness of money is significant both to the individual, the world of finance and as a measure of TRU. Inflation and the ability to pay ones bills, $ 25 in 1947 and $ 300 in 1975 and $ 1000 in 2000 eats at economy like a cancer. How spendthrift or how frugal different persons, peoples, societies and nations are and how much this has to do with their success and failure. It is largely by knowledge, value and worth that humans succeed of fail. This is why REM, VAPO, SPALT, POS, HOM and KOT are so important. It is a product of optimal utilization of TRU by all persons into and throughout the future.

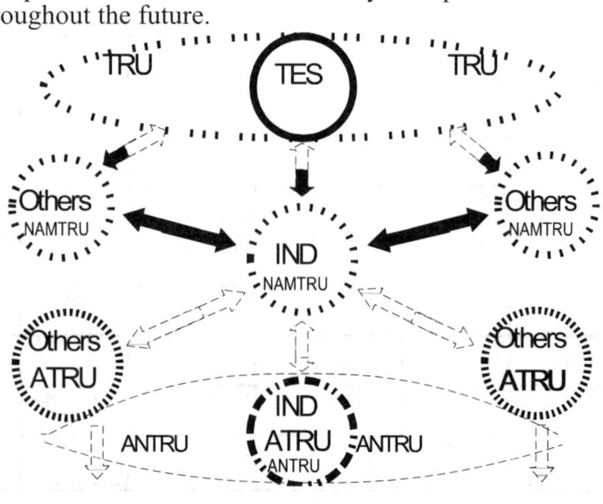

Fig 3-67 **OTHERS, INDIVIDUALS, TES and TRU**

The individual in being or affecting others to be TRU or ATRU rises higher or sinks deeper. Good or bad in life is what we are taught and learn from experience. Good comes from vigorous work with great conviction for the best in oneself, others, and the world in general, TRU. It is also from those who relentlessly pursue the deepest meaning of good and even seek to allay those who would injure or destroy.

TIF = TRU In Fashion. TRU or TREQ is always in fashion.

That which has, is and will be in fashion can only be TRU. For all that is not is only an illusion or figment and often worse, destructive of what must be TRU. Man has an inveterate tendency to seek fashion in what is novel, surprising, eye-catching, or popular. These are, to an extent, reactions against imbalances and disproportions in TRU, and thereby explainable, but often false. We must constantly remember to be on guard and correct all fashion, imbalances, excessive extremes and means, ISMS, and other tangents or disproportions that are not or less than TRU. This can be done only when we discover, develop, learn know and utilize TRU, precisely, proportionately and fully in FAF and SCARLS.

It is most important that the percentage of benefit in use of all, as taxes and labor, be the highest possible. In order to do this it is necessary that it be measured. **MOPU** = Measurement of Practical TRU Use, or Proportional Utilization of TRU. There is a constant exchange of the level of TRU in use at a time, or over periods of time, that ranges from 100% TRU to the opposite - 100% TRU. It is by MOPU we measure all that exists in OLF as an example of all TRU. It is especially evident and readily available for our use in all about us. This exists, to some extent, in all animals, only increasingly more accurate in the higher levels of SÁ in humans, as TRU.

Humans strive to be their own, often other than TRU, to take short cuts, and live off the lives of others. This is key too much of their failure, as well as correction. This is turned to success by all in the world who are capable, concerned and loving of their total mass energy Á FAF, or well being. This includes their homes, businesses, environment, and all that enables and enhances TRU. This leads to the problem of how much TRU is only the best or highly selective and limited genetic material, and how much is all its population and civilization. This is called: **PROMGOC** = Problem of Mean Genetic Total Continuity. PROMGOC may be stated as a Law.It is closely related to seed line, although different and distinct in emphasizing and focusing on the differential rates and courses of GELF or live product, genetic material and deter-mining forces in TIT, TRICH, TRIS, TRUCH and TRUS. Determinism and one of its extreme forms, predestination, represent a type or form of many processes in TRU. Predestination is limited by its common usage, and needs to be interpreted in its TRU sense. The prefix TEL signifies end, accomplishment, consummation or completion. TEL as used in teleology, is a study or the science of ends, final causes, directed toward ends and purpose. For TRU TEL helps add evidence to show the great role the future plays in all TIT. This, not so much as a final state, as it is in all its other manifestations, indirect, remote, reversible, and transcendent. It works universally or through and through, and even in and out of TRU.

The narrow margin of human survival has acceleratively decreased and worsened in recent times. This is due to many factors including, shrinking world, industry, weaponry, use and abuse of material, real and mental resources, overpopulation, loss with each generation, lack of identity, organization, interests, determination, comprehension, direction and the positive approach needed to meet challenges and fulfill our role in the best of worlds. Yet there is a way that is relatively simple and effective. One that is led and guided by VAPO of TRU, or the validity, actuality, proportionality, and totality of This Special Universe of Best Human Life So that:

Acts and actions are impossible without best policies
Policies are impossible without the best principles
Principles are impossible without the best knowledge
Knowledge is impossible without VAPO and TRU
VAPO and TRU are impossible without the capacity,
Interest, and determination of each person to have such
Virtues that enable them to be their part and fulfill
Their roles in all that makes the best for human life.

We tend to think of the future in terms of the large, as in OLF and SULF of DYNU. This is a major reason why we are so blinded to it. The future is from the infinitesimal present, all time of TIT-TO. TRUF for TRU, TRIF for TIT, and ACC are best envisioned when all chapters of future in DYNU, DYNS, and DYNO are considered together in the global SPALT, accompanied by appropriate sectors, METAD and SPAT. Such a total view, plus repeated reference to the future listings in the index-concordance, and related topics is needed to overcome our shortsightedness. This, more generally, applies to all necessary to envision TIT-TO and TRU.

Fig 4-67 **CES, CHES and CHESU in OLF of DYNU**

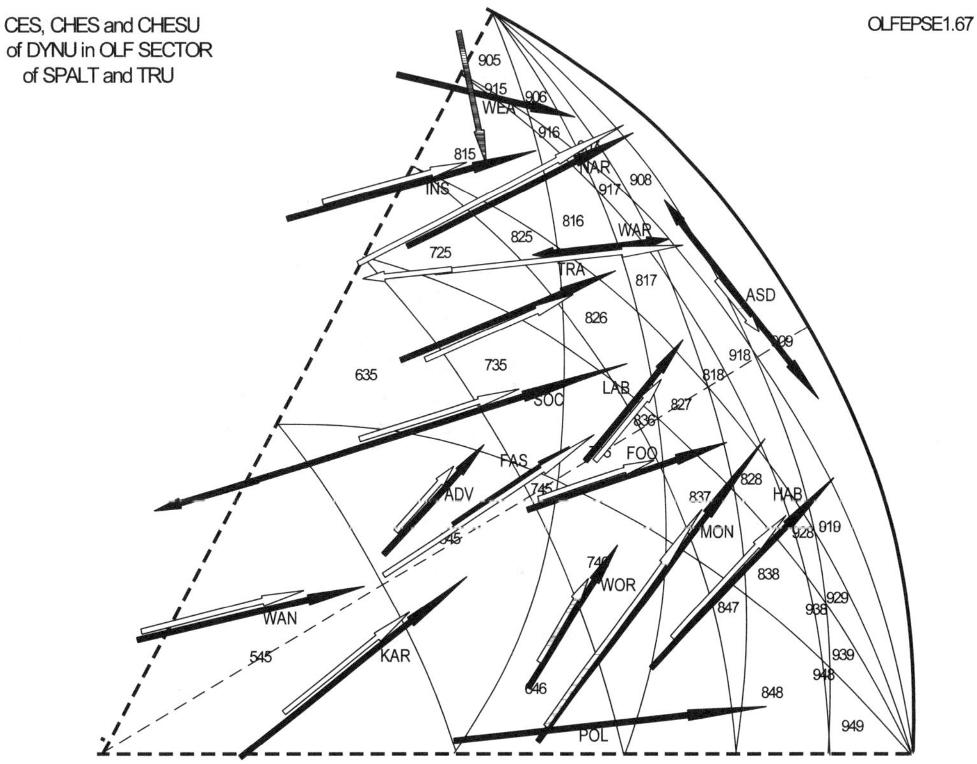

MEASURES of POSITS of CES, CHES, CHESU of OLF

Measure	Posit	
ADVANTAGE	6-3-6	S
ASTRONOMICAL DYN	9-0-9	S
BENEFIT	6-3-7+	T
CONTRIBUTION	6-3+-8~	M
DETERMINISM	~7-3-5$_9$~	M
ECOLOGY	9-1-8	T
ENVIRONMENT	8-1-7	T
FASHION	6-3-6	S
FOOD	8-3-6	S
GAC	7-4-6	M
HABITAT	9-2-8	S
INF	5-4-9	T
INFLUENCE (INL)	6-4-5~	M
INTERNATIONAL SCS	7-1-7	S
IPR	5-4-6	S
KARMA	5-4-6	S
LABOR	~8-3-6~	S
LAT	~7-2-7~	M
LIFE	~~-0-9~	M
MAGIC	5-4-7~	M
MARS	7-3-8~	M
MONEY	8-3-7-	S
MOPU	8-3-7	S
NATURAL RESOURCES	9-1-7	S
NEEDS	7-4--7~	S
OPPORTUNITIES	7-4-6	S
PIB	5-5-9	M
POISSONS LAW	~6-3-8-~	S
PPMA	7-2-9~	T
PROMGOC	8-5-9~	T
RESPONSIBILITY DeO	6-4-6	T
SAFETY	8-4-6	S
SOCIOLOGY	7-2-7	S
TEL	~5-4-9~	M
TIF	6-3-8	T
TRANSPORTATION	8-2-6	S
USE	7-4-7~	S
WANT	5-4-5+	S
WAR	8-1-7	S
WEATHER	9-1-5	S
WORK	~7-4-6~	S

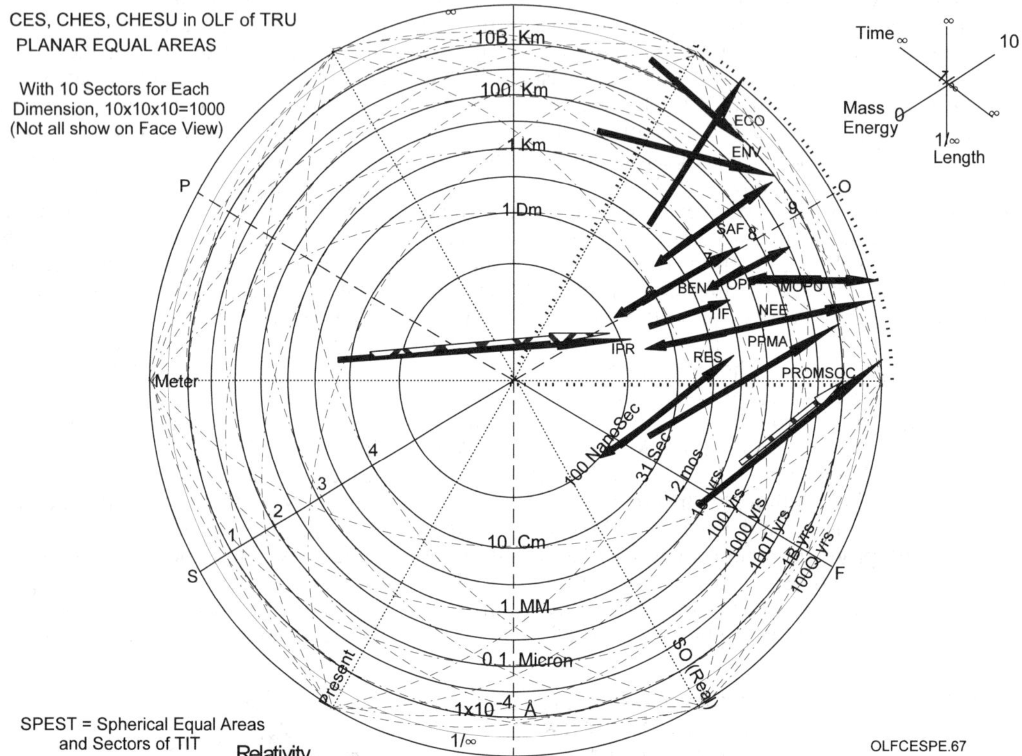

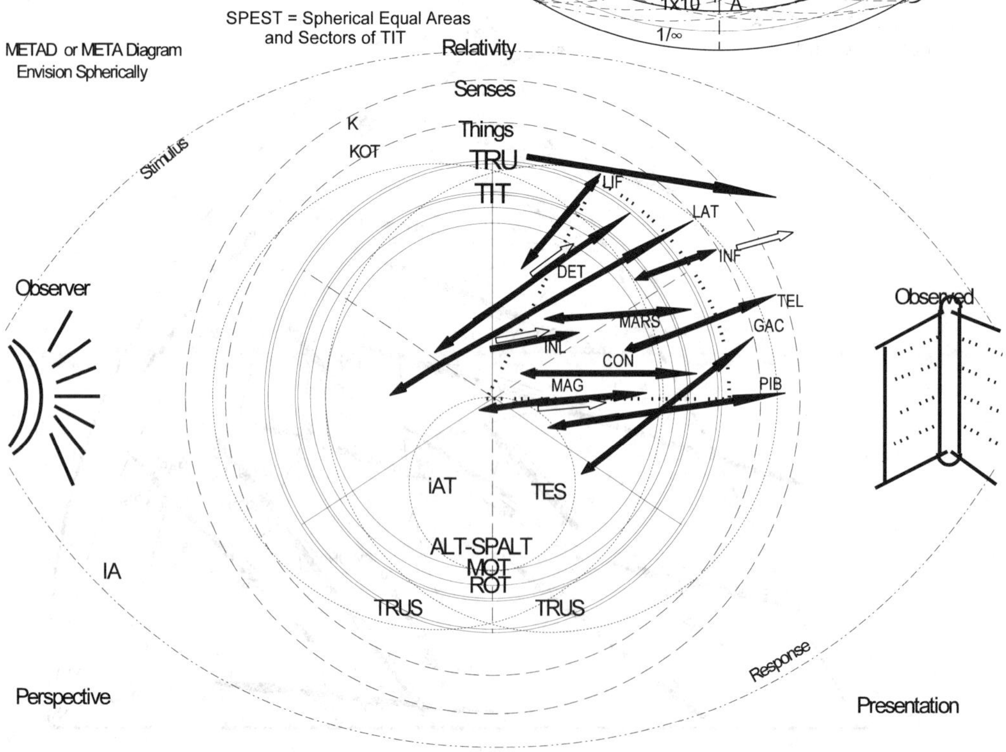

Fig 5-67, 6-67 **CES, CHES and CHESU of DYNU in SPALT and METAD**

CHAPTER 68

SULF or SUBJECTIVE, LARGE, FUTURE PROCESSES of DYNU

IDEAS, WANTS, HONOR, HPP, LENAF, GOALS, PRIORITIES INFORMATION, EXPANSION of the SUBJECTIVE, KNOWLEDGE

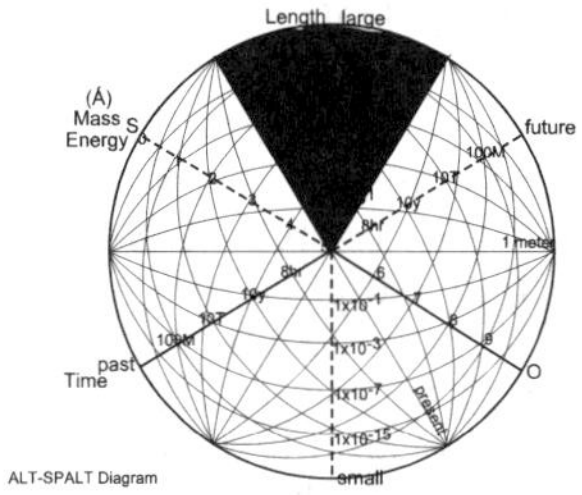

Fig 1-68 **SUBJECTIVE LARGE FUTURE SECTOR of SPALT; SPALT with ROTATED DIMENSIONS of TIME and Á**

The described and posited CES and CHES are used to show both SULF and TRU as well as the NEC and problems that occur with ROT and POT. With practice the reader and student of TRU can gain increasing ability to posit all that is given or they encounter to gradually visualize the composite and ordered whole makeup of TRU.

We use many terms in this text in their proper and actual sense of their actions in TRU. For example, we speak of foreknowledge, foreordination, divination, and forecasting as though they are valid and established facts and processes in TRU. This is for their proper use with limited meanings. Yet the strength of existing meanings is needed for representation. This is called: **LEMSCARL** = Law of Equality of Maximal or Best Sense and SCARLS of entities. LEMSCARL signifies that state in which entities, as they are their best or in their proper or TRU sense actualities, whereas in their common or general sense figments. This leads to a key concept and law of language. This is the paradoxical state that exists between words as entities which are figments or actualities, or: **LENAF** = Law of Entities as Actualities or as Figments. LENAF is a product, partly of the weakness of terms, words, and partly from the fact that we have not adequately attributed to them, nor limited them to, their best possible, maximal, and proper TRU meanings.

Many simple terms, sayings, and TRU processes are good examples of SULF including: sweet companionship, no news is good news, smiles, better ask than lose your way, silent consent, security and happiness. This is evident when analyzing and interpreting IPR. What does and does not constitute a positive, healthy, stimulating and congenial consensus for TRU. Sweet companionship is an example of mutual support, cooperation and the spirit of Christian brotherly love that transpires between persons, groups and all that are TRU. Although for good to happen to others it may in the short term be detrimental to some, in the long term it should be beneficial to all. This is by definition, 'good'. It, then, is paramount that we accurately and completely set forth that which is The Good, the message of the Epistles of St. Paul. How there was brought into the world and the early church, the "good news" of Christ and his legacy to the world. These were the good tidings, good things, that which is the best, all the manifestations of God through the Holy Spirit in Christ, or of TRU through CES by TES in TIT. There is little in this world that provides better and greater news for man. We must forever be on guard to chose the good news of TIT-TO, TROS, VAPO and all in preference to NAMTRU, ATRU, and ANTRU that we confront and succumb to. Much of the great art and literature in the world has been to achieve righteousness over sin. The most beautiful, greatest, and best of God as TREQ and TRU is not only in and of itself but as we the viewer and the observer both writer and reader come to comprehend it. (COAT,KOT,TRU) It is here we can learn what are the good things and the good news we seek.There has been a long tradition of righteousness or moral purity in many religions that has been a great positive force for greatest TREQ, and TRU. It is more to total greatness than momentary lapses of what is right that is the measure of a person, or their immortality, TES. This is, **LOMOR** - Law of Total Magnitude of Right and Righteousness.

No news is good news shows how the operations of TRU that are not dramatically nor violently changing indicate that TRU is functioning well. Also it shows how we must refrain from putting out false information or any influence that is not highly contributory to TRU. Smiles are the reflection of the heart, ones well being, happiness, interest, good will and affection with others, ES, CES, and TES as they are so much of TRU. They are our joy, the reflection of the happiness we feel with ourselves, others and TRU in the great love of life, TRU. It is the outpouring of our appreciation for being, for being TRU. **Individual → ESINS → ESES → TES → CES → TRU.** These show the power of singular separate CES in TRU. It is the power of a positive and joyous attitude and appreciation of what we can do and help others do for TRU.

Simple ideas are, with universals, EPI, IPTRU, ROT, ALT, POS, and HOM, one half the battle of knowing and being TRU. The use of HOM in the development of knowledge, EPI, POT, ROT and TRU is great. This is called: **INCSET** = Increasing Numbers of Ideas and Processes that in HOM Set and Establish TRU. INCSET also helps to set and establish ACC.

To not hesitate to ask rather than lose ones way signifies that great social process whereby information is readily and happily sought and received. When living, not out of anxiety, care, dread, fear and overwhelming factual infinitude, but out of a natural thorough strength and courage for TRU in its security and happiness, humans have little need to hesitate to ask, smile, or have any action that is not positive for TRU. No silence will occur if there is a need for TRU. Peoples consent will not be in question. The only choice will be how to properly and most effectively accomplish all, all processes, that best lead to TRU for that particular situation or occasion. The distortion and quick disillusionment that follow a few wrong turns and becoming lost from TRU is expressed in: **SUDRA** = Subjective Distortion of Right and Wrong. SUDRA is the ultra fine error made by our mental weakness or deficiency in not having the right information, knowledge, experience or wisdom necessary for making the TRU choice or decision at the right time. It can be directly corrected by: **UNA** = Union of Associations. UNA involves the analysis of all existing associations, societies, groups, unions, international organizations whose high purposes coincide with TRU, TREQ. The success of UNA will depend to a great extent upon the degree to which these organizations and associations are analyzed, have common ground with TRU, are understood, applied, fused, made whole, balanced and worked for.

Condignity is worthiness or merit that is not given but earned by good works. It is performance in a state of grace by the will of God, TREQ or TRU. An actual TRU appreciation for condignity is subtle and profound. It represents one of the great CES of TRU, that through which the individual becomes TES and the bond of all as they lead ENT and TRU toward TÂ. It is in its broadest sense, all that works positively for TRU. More specifically it is how a person, when TES, works and produces great benefits for TRU in a diligent, profound, subtle, and convincing way.

Good will, *De bonne volonte,* willingly for GOD or TRU is readily produced as it best directs us and we best direct ourselves. Its production becomes stronger and much more lasting through learning and practicing ethics, Christianity, TREQ, and TRU. The great strength of subjective and intellectual processes from the past through the present to the future was studied and described by Dilthey. The intellectual development of man, the ever higher social values and ends were presented in his doctrine of *Vertschungs-psychologie* and the science of *Lebensphilosophie*. These are significant for TRU in substantiating and showing HOM, the processes of life and man as they work through time and, in turn, conforming and continuing human intellectual development and associated CES of SÁ.

Fig 2-68 WRONG, NEUTRAL, RIGHT, GNOSTICISM, AGNOSTICISM, SKEPTICISM, and TRU

	WRONG	NEUTRAL	RIGHT
GNOSTIC-ISM (Belief)	Wrong Belief	Neutral Belief	Right Belief
AGNOSTIC-ISM (Non Belief)	Wrong Non Belief	Neutral Non Belief	Right Non Belief
SKEPTICISM (Disbelief)	Wrong Disbelief	Neutral Disbelief	Right Disbelief

These concepts are represented in a diagrammatic way. Their two parameters are distinct and of great importance. The horizontal parameter is a function of validity, especially validity in TRU. The vertical parameter is a function of the viewer or observer and how they interpret a topic, entity, or object. We in life think and believe in terms of right and wrong. Yet this is like looking at the world with blinders on. For all is from absolute validity through neutral to absolute invalidity. Likewise how we observe is also from an absolute belief to a neutral or wrong one, or an absolute disbelief with a neutral in between. As these two parameters meet, in S-O and the Á FAF of TRU, we can better see their differentiation and integration. We can see how they are integrated, and by DAOST, with the great continuum or spectrum of Á in TRU. Yet when it works toward a pure or absolute Á in TRU it comes, and only comes, to be right belief, KIT. Since only right belief of and in TRU is the only appropriate stance there is little room for all the eight other categories. However since human knowledge is erroneous as well as incomplete the role of these other categories must be recognized and considered. The long history of human development has seen the gradual rise from wrong to right and from disbelief to belief in knowledge.(KOT) Yet there is much knowledge that remains. All of us who continually deny and defy TRU and KOT, as we who can so readily see, must investigate and learn TRU.

The great error of Christianity, as well as other religions, is that it holds Christ, or others, to be God. It relegates the human person to a menial position. This is valid if all persons are ATRU or ungodly. However, if all persons are ungodly or ATRU, there would be neither God nor TRU. They hold that, even at best, they are never God, only accepted into the heavenly kingdom to have eternal life. Any person who lives a virtuous life and contributes to the improvement, success and survival of the human race is God by being a part of GOD, TREQ and TRU. Just as all who have in a great variety of ways and degrees contributed to any and all of the make up, as by SÁ, of humanity. To hold correct knowledge that opposes traditions that are in error is not blasphemy, in fact it is the opposite, a critique that is TRU. This is especially true when this correct knowledge is right, comprehensive, complete, and replicates TRU. All existing religions fail to account for the rapidly expanding great advances in knowledge, especially physics, chemistry, biology, medicine, psychology, sociology, and philosophy of the past two thousand years. Thus existing religions have, more or less, become obsolete. They need to be replaced, and what is correct in them, TREQ, identified and fused into TRU.

ITOD = All Ideas whose time has come, or impact is overdue. ITOD is important because of the great benefits that accrue when we discover, use and promote ideas, EC and NEC of TRU. This is especially so when they fill great voids or restore great losses in our world, TRU. When much of TRU is lacking, deficient, or blocked, it is often ITOD that can bring it back to life and flourish.

Much can be achieved when we want to be, be healthy, happy, sober, frugal, successful, helpful, instructive, wise,and TRU. If we want TRU we can readily overcome what is lacking or in error. Rather than wanting to digress into ATRU we must want the benefits that will be provided by TRU. We must want to have knowledge, a long life, and be TRU more than we want what is harmful or wrong even if this damage is far removed. Therefor the pleasure, fun, joy, benefits, glory, and all acts that with probity lead to TRU should be optimally cultivated.

DOF - Delusion or Dismay over the Future. Many humans disbelieve because they do not know or care from where they came, how they were made, what they are, where they are going, what the ends are. Because we do not know exactly, nor can knowledge tell what things will precisely be at a set time in the future we disbelieve. Since without scientific, mathematic and geometric certainty many become highly skeptical. This is: **REDOF** - Regression to DOF. REDOF is the delusion reached when we close out all that is unknown or not known completely or perfectly. In TRUD, ENT, HOM, DYNU, DYNS, and DYNO, the forces of S/F are numerously shown as great CES which are effected by, or which correct DOF. It is partly and correctly a product of the particularity of DYNU, IA and the individual.

What is the difference between self interest, special interests, politics and TRU? If we can allow a certain degree of self interest in the recognition of the partial and often limited or ATRU position of the individual then does this not give some reason for politics deviant from TRU? A person, family, group, state, nation, and culture can't be completely self denying and give up all to a larger identity, the whole, or even TO-TIT. Yet as each comes to recognize one's part in TRU, TES, and the benefits accorded as well as the nature of TRU to this individual, then the person, family, group, state, and nation participates in TRU naturally without differences that are harmful, destructive or deadly.

No one or group should be authoritarian unless and to the degree they are highly TES, TRU and have equally great KOT. Even more, unless they apply and use this KOT appropriately for the occasions, situations, circumstances or given. **LAUTH** - Law of Authoritarian Limits. LAUTH signifies that no one or group should hold, speak, write or communicate dogmatically and imperatively unless and to the degree the appropriate limits of unknowns, TIT, the interactions or dynamics, application and use of this knowledge is known and shown. It is in writing, especially writing of books, that LAUTH is most significant. For authors of the topics they present must necessarily seek to be authoritarian if their presentation is to be worthwhile. Their products will be worthwhile to the degree: 1. Greatness as a topic, and this of TIT or TREQ, 2. Knowledge of their topic and this is VAPO TRU, 3. How well they present this knowledge, 4. How well they enable the reader and user of this to also achieve a level of authority on the topic.

AUTHORITY = TOPIC-TRU x VAPO x POK x READER K x USE x X

Other factors are indicated by the X in the equation.

One the highest benefits and blessings of social relations is that of human trust, an open conversation and exchange of ideas. This has been held in high regard in the best of societies, especially in the basic sciences. It is essential to success. It consists of both truth and the duty and honor for which truth is prerequisite. This is what enables a free flow of the best in communication, teaching, experience and principle. Human trust can only be built upon VAPO in TREQ or TRU. This must be accompanied by the astute, profound and vigorous support and action needed for TRU, a kind of human trust based upon cooperation, commitment and joy in the sames, or common ideals, in TRU.

Honor to that which honor is due sounds right and simple enough. Yet it can only be a right that is beneficial and TRU, along with avoidance, nullification, or destruction of that which is wrong and harmful. For the public to share in the benefits of TRU they should be stimulated and learn to be complementary of all that is constructive to TRU or TREQ. It is the cornerstone of ENT. It is a human conscience that has been neglected and missing, the needed level of individual conviction and determination for success and TRU.(TES)

HPP = Highest Public Principles. Throughout history much human activity has been based upon an increasing collection of experience and lessons passed along and reinforced to become principles. They then become the focus of dedication. Further experience and lessons convincingly confirmed these principles that imperatively and regulatively produced laws. They have variably and reversibly accompanied human socialization and its success. This makes man civilized, living together amicably in high density populations. They are highly coterminal with professionality, morals, ethics, religion, and all that works best in IPR, schools, families, communities, businesses, institutions, and governments to assure a highly successful TRU. HPP are often altered and work profusely in varied ways. They have a solid backing by those whose education and intellect helped them appreciate and learn the great improvements that provide human well being, success, prosperity, happiness and TES.

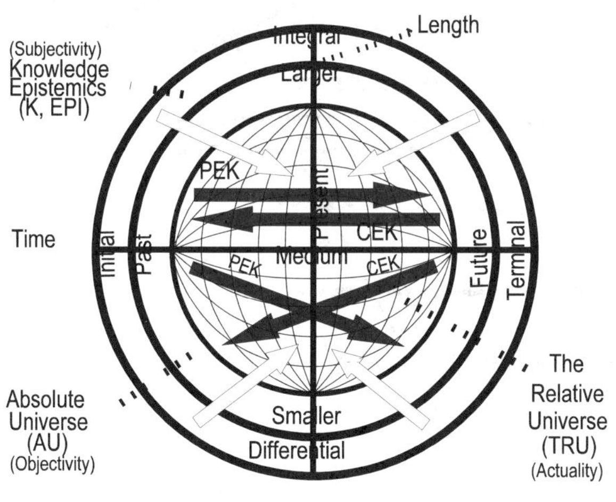

Fig 3-68 **PROPELLING and COMPELLING EFFECTS** These are shown on **SPO**, the Standardized Perspective of All. The outer circle represents observers, knowledge, or epistemology external to TIT. The middle circle represents observerds or the Absolute Universe, AU external to TIT. The center circle represents TRU in itself, TIT. ALT Modulation of TRU is here so viewed that l and t are its vertical and horizontal axes with the mass energy, S-O or A axis perpendicular to the page and thereby not represented. The figure demonstrates the function of two of the greatest general kinds of forces, powers, and dynamics of TRU. Vectors to the right are propelling forces and effects or Dynamics, **PED**, and to the left are compelling forces and effects or Dynamics, **CED**. When designated for knowledge they become **PEK** and **CEK**. The propelling effects work from the past, the compelling effects push to, or draw from the future. Both work reversibly and thereby are not to be considered as absolute opposites. Among the great compelling forces is the urge to still higher qualities. These are the forces and powers of SÁ, levels of human mass energy. Its higher levels have been continually and recurrently emerging. Of these mentality is most readily recognized. Deity is a still higher quality, both potential and actual in TREQ and TRU. All, in the present and future, actually exists only as they are a level with more differentiated levels in TIT-TO.

When any and all in the world feel they are playing the role of a poor third, when we think we are taking a back seat or being denied, then it is time to reevaluate priorities. If the priorities are right and not being denied then stick to them. If the priorities are being denied or injured in one way or another they must be restored. You do not always have to do what others want, nor even what you are told to do. Yet if you don't do it you should be ready and able to make the appropriate correction, through TRU and the occasion. It is what you should do by TRU that counts. It is even more helpful to see that what you are told to do is TRU. Yet if what you are told or directed to do is TRU you must quickly and fully comply, get to it, and do it in all ways necessary.

To be beneficent is defined as doing good, conferring benefits, kindly in action or purpose. Beneficence has been the great strength of religion, ethics, morals and the instillation of principles and habits of a life that is a benefit to GOD, TREQ or TRU and TES, and to themselves in return. It has been the key or sparkplug to world success. Beneficence is both the strength and shows the success of those religions that emphasize beneficence. **EPAT** = Expansion of the Threshold of TRU. From little things great things occur, as from tiny acorns mighty oaks are grown. There is not only, by TI, an infinite number of static components of TRU. There is an even higher and multiple power of relations, causes, interactions and CES working in TIT-TO and OITIO. Thereby let no one relax in their understanding of TROS, TRIS, TRIT, OBT, SUT, or ALT. Never assume you are sufficient in yourself. Let it be a constant stimulus, inspiration, interest, challenge, and guide to EPAT.

Interests in human culture have waxed and waned through history. The decline among much of modern society of deep interests and rigorous pursuit of the underlying reasons for human existence and survival has foreshadowed future loss. For the spin offs from such interests and pursuits feed the motor of civilization. Only by appreciating the benefits and realizing the revelations and inspiration of TRU, and keeping it in fashion can we sustain the high level or TRU needed.

Terms that stand for the greatest human ideals in different countries, languages, cultures and civilizations include, knowledge, adventure, self reliance, pioneering, imagination, courage, honor, conviction, duty, justice, and others. It is such great ideals, like the symbols and standards, whose powers were developed through thousands of years that have helped guide and drive mankind to higher levels of achievement and TRU. The high points of many great world civilizations have been characterized by the predominance such ideals provided.

The significance of each bit of data, facts, information, and evidence is not only what they mean in and of themselves, it is how they relate, suggest other unknowns, may be interpreted and expanded to explain the mechanisms and all that makes diagnoses or enables a plans of treatment. They are what it takes for the solutions to problems, management and control. This is by collective significance to prove TRU. **SIG** = Summative Significance. When meaning is complete, it is signified by: **COSM** = Complete and Total Significant Meaning. These are the established facts of knowledge; those certainties which are irrefutable within their confines or SCARLS in TRU. Once established they are relatively unalterable and permanently valid additions to knowledge,

KOT, TRU, and ATIM. This is the great beauty of LANT and POT. By becoming established, TRU is permanently proven, validated and certified for constant and continuous use without deterioration or invalidity. This is signified by: **LATSIGOV** = Law of TRU Significant Meaning Total Validity, and **ACOSM** = the Accretion and Acceleration needed to provide and to fulfill all TRU. ACOSM is the research, discovery, development, determination, and positing of all unknown data, information, ideas and concepts. **LANC** = Law of New Concepts, or New TRU terms. With the concepts, principles and laws of TRU, not only must we utilize and maintain current language as a part of TIT we must also utilize and maintain whatever new communication, representation, language, words and concepts are needed to show and be TIT. In deriving and developing these and associated communications, as they are predominantly and vitally necessary to a practical life, little precedent has been provided, except by TREQ and TRU.(SEJONAC)

Of the most complex and advanced of TRU processes we note: positivism, prediction, forecasting, foreordination, foreknowledge, divination, highest good, and the ethical absolute. Various terms for types of determination of the possible, of probabilities, and of ways to show how the future can not only be known, but exists as certainty in TRU as it does the present, past and all together. These ways include, fore-knowledge, forecasting, prediction, foreordination, and divination. To disbelieve, or to believe in the future other than TRU, is apt to make it less, commonly misleading, and nullifying. This leads to: **GIFT** = Grace in the Future TRU . GIFT is itself one of the greatest processes of TREQ and TRU. It is one of the greatest gifts for mankind. GIFT is closely related to other terms as beneficence, COGNOS and ELTICMA. However its great significance comes from the higher levels of SÁ ACEXIN working by CED and CEK as the living and potential future works with the powers of grace through TRU. They assure the success, continuity, survival, eternity and absolute of TRU and each of its constituents, as we ourselves are TRU.

PERT = Perpetual Evaluation Reduction to TRU. PERT can not be achieved so quickly nor easily. It has misled and been wrongly attributed to what is other than TRU. What may seem so obvious to us at a moment quickly evaporates when ramifications turn to naught, or cause us to lose focus. Meaning may be great or poor depending upon many factors. It is those factors that make meaning great that are the basis for PERT. Much of knowledge, as evident in past religion and philosophy, exemplifies PERT. So has mankind, civilization, and TRU been accumulated by PERT. This is shown in our Common Public Principles or **CPP**. These are all the principles, ideas, concepts, laws, rules, terms which are used in a positive sense and work well to guide us toward TRU. **ÏTÁ** ('it tay') - Infinite Exuberance to TÂ. ITA is sensed on arising each new day, the freshness of spring, and the vitality of new life. It illustrates the seemingly limitless hopes, energies and plans of attacking, reducing, and solving all before us. It is how HOM and DYNU become ACC, as ACC assures life to become TIT, TRIF, and TÂS.

Fig 4-68 **CES, CHES, and CHESU of DYNU in SULF SECTOR of SPALT**

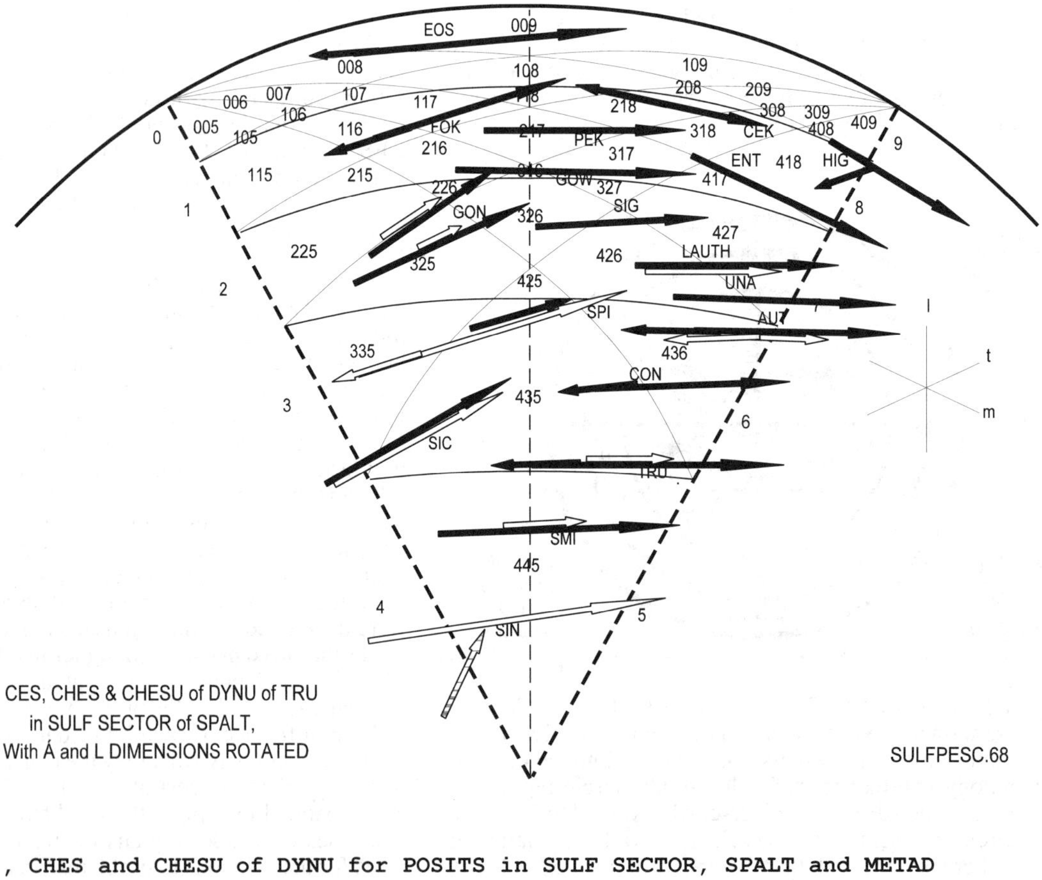

MEASURES of CES, CHES and CHESU of DYNU for POSITS in SULF SECTOR, SPALT and METAD

AUTHORITY	4-3-7	S	EXPANSION SUBJECTIVE	0-0-9	S	HAPPINESS	~1-1-7~	M
BENEFICENCE	4-3-6~	T	FOREKNOWLEDGE	2-1-7	S	HIGHEST GOOD	4-1-8	S
CEK	3-1-9	S	GIFT	4-2-8	T	HONOR	4-3-7	M
CONDIGNITY	4-3-6	S	GNOSTICISM	2-3-7	T	HPP and CPP	1-3-7	T
COSM	3-3~-7	T	GOALS	4-3-7	T	IDEAS	0-4-5	T
DOF	1-4-8	M	GOOD NEWS	3-2-6	S	INFORMATION	2-4-6	T
ENT	4-1-8	S	GOOD WILL	3-2-7	S	INCSET	~3-~-7~	M

ITÂ	2-4-9	T	POK	4-2-8	M	SUDRA	1-3-7	M
ITOD	1-3-5	T	PRIORITIES	4-3-7	T	TRUST	4-4-6	
KNOWLEDGE	1-1-8	T	REDOF	3-4-7	M	SUNA	4-2-7	S
LANC	2-3-8	T	RIGHT BELIEF	1-4-6	M	WANT to BE	3-4-8	M
LATSIGOV	2-4~-7~	M	SEJONAC	3-4-6	T	WANTS	4-4-6	M
LAUTH	4-2-7	S	SIG	3-2-7	S	Also ACOSM	6-4-7.5	T
LEMSCARL	2-4-7	T	SILENT CONSENT	4-3-5	S	CED 4-4-8 S, CPP 1-4-3 M, DAN		
LITERATURE TRU Stds	2-3-8	M	SIN	~4-4-5~	S	DOUT 2-6-6 M, EPAT 3-5-8 M,		
LIFE	4-4-7	M	SKEPTICISM	3-3-6	T	LENAF 4-6-3 M, LIA 4-7-5 S,		
PEK	3-1-7	S	SMILES	4-4-5	S	LOAB 6-6-6 T MAT 6-3-7 M,		
PERT	3-3-8	M	SPECIAL INTERESTS	4-3-6	S	MOVET 4-3-7 T, PED 6-5-2 T,		
						RENEST 5-6-6 M,		

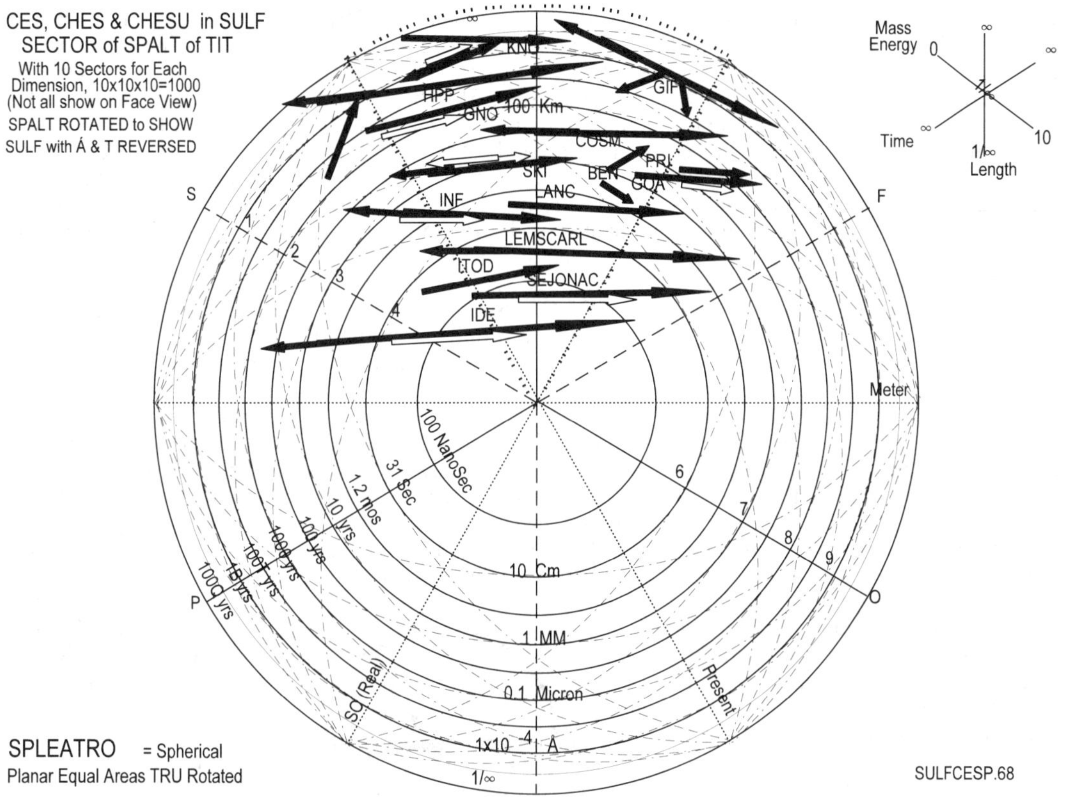

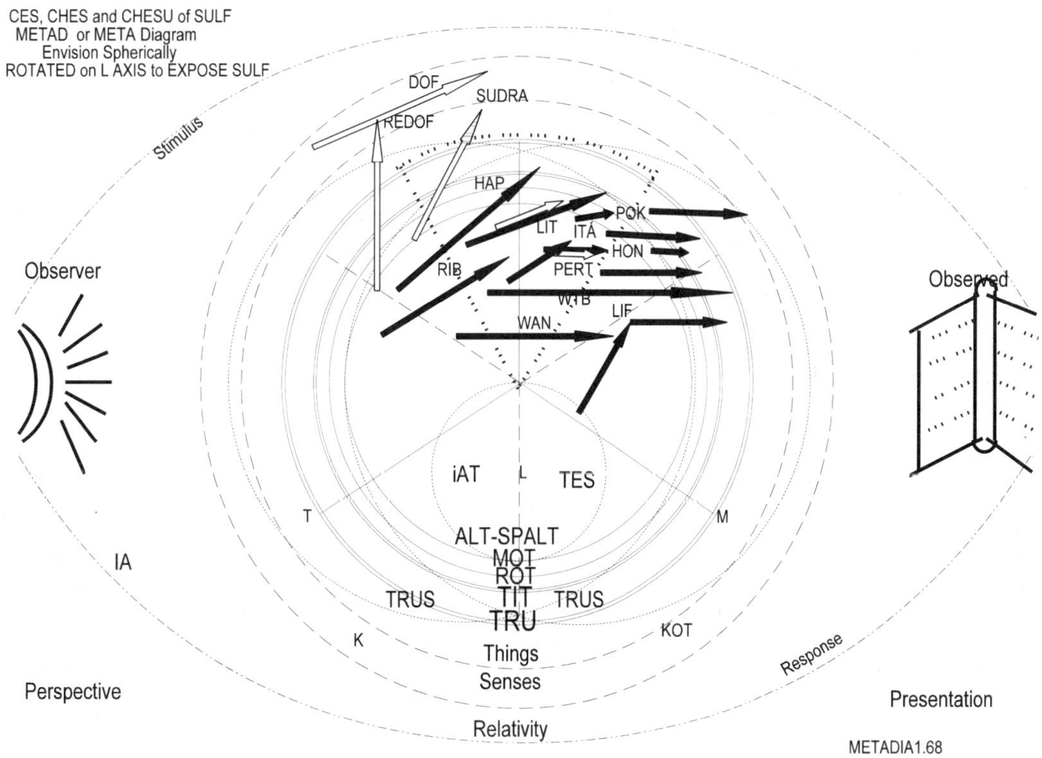

Figures 5-68 and 6-68 **CES, CHES and CHESU of DYNU of TRU in SPALT, and METAD, by rotated DIMEN-SIONS**

CHAPTER 69

THE RELATIONS, PROBLEMS, SOLUTIONS, and SUMMARY of DYNU

CES, WORDS, SEPARATE PROBLEMS, INCREMENTS, PROCESSES, FORCES, COURSES, STEPS, GFST, FUTURE, andENDS

Each person, group and all need to repeatedly or periodically analyze their situation in life, as TES, to measure, determine and be sure that their battery or set of habits, customs, rules, guides, regulations, frame, systems, or orders for action are the best, TRU. From this we have: **IMPET** = the Imperfections of today, are or lead to, the Evils of Tomorrow. IMPET is also true through all time, as yesterday was for today, and even tomorrow for yesterday. More generally this is: **PANFIOT** = The positive and negative forces in and out of TRU. Sophistication can't replace great virtues. You can not make people, men, women or children, do right for long unless it is in their interest, fits their needs, and they can work through to identify the right reasons, TRU. This is called: **DORIN** = to Do Right through Interests, fitting of Needs, and helping to achieve Identity through Reasons of TRU. You can show, encourage, and stimulate people to want to do right. Then they will do right spontaneously. They must however know and remember to balance their individuality in TES, differences, SCARLS, VAPO and TIT-TO. They must know their strengths and weaknesses and how to be protected. They must use all from their strengths that enable TES and TRU. In a world of TRU there is no murder, theft, crime, jails, criminal courts, trial lawyers and few judges. Everyone knows what is right, abides by, does what they are told to do in a system that tells them only what is right, TRU. In fact to a great extent you can measure the TRU of cities, societies, nations and regions in the world by these rates. Look to those groups, occupations, professions, trades, centers of learning, and nations where they are most interested and effective in solving the problems of the world, helping others, and enhancing TRU and you will see the image of what the world will be, when perfected in TRU.

RECVERT = Reconversion of Causes and Effects to TRU. Only through RECVERT and similar CALP can we develop a proper analysis and interpretation of entities, things, senses, diseases, troubles, problems, and their cures and solutions. In order to show TRU and provide its representation and presentation it is crucial that we return to the cause and effect DA, DAT, derivation, development and determination of TRU. It is likewise crucial that COC and RECVERT be adapted to TIT-TO and ACC.

The problem is often not so much that a single crime has been committed. Nor that a single stimulus or object is before us. It is that if you see a person commit one crime it is likely that he has or will commit many. This is: **LASEV** = Law of Sentinel, Signal, or Additive Evidence. LASEV is illustrated, somewhat indirectly in the legal world, by punitive damages and class action suits. This can come only from right thought, right memory, right knowledge, KOT and TRU. A single stimulus, fact, bit of information, evidence, or even object before us, may be conclusive in-and-of itself, but this may be much less important than what it indicates and infers. If punitive damages and class action suits were invoked for all the ill effects in the current NAMTRU world we live in humans would wake up and realize how far they are from TRU. Each new generation would learn by example how wrong their paths are and how right they can and must be in TES and TRU. **LABGEPT** = Law of Added Evidence and Power to what is better, greater and more TRU. The distinction and jist of LABGEPT is its signification and emphasis upon the need for all to learn, practice, judge, complement, and support TRU and avoid the pitfall of simple judgements that either tend to, or through lack of VAPO, produce ATRU.(ACC) Here is the greatness of ACC, how power, when right, can be so beneficial and great.

GEDEF = Great Destructive and Disintegrating Forces. **GECEF** = Great Civilizing and Cultivating Forces. We need to identify the major GEDEF and GECEF in order to prevent loss and promote gain for TRU. By learning how to avoid wrong, do right, and be TRU we can decrease GEDEF and increase GECEF. These forces working in various ways show us much of what TRUD is made of and what HOM must show if we are to best represent, present and live by TRU.

ÀCEX = From (Ab) Complementary leads to Extinction. What is not complementary, in speech, act or in general is unable to be complementary to our being. This is can not become their ESINS, or may potentially, but does not become TRU. ÀCEX is the degree TRU loses.

SÔLÔ = Strategy of the Limited Objective. SOLO is how and where if one has an insurmountable task to do, it is often more successful to break it up into its parts or tear it up piecemeal. It is easier and more effective to take it off one piece at a time than to expect to accomplish it all in one step or in one 'fell swoop'. Mankind must take certain separate steps if we are to be, achieve and survive, including the avoidance of unnecessary conflicts, improved quality, certain reserve to assure TRU, expand human discovery, knowledge, invention and institutions in TRU, focus on and eliminate all possible dangers and threats to TRU.(PPMA)

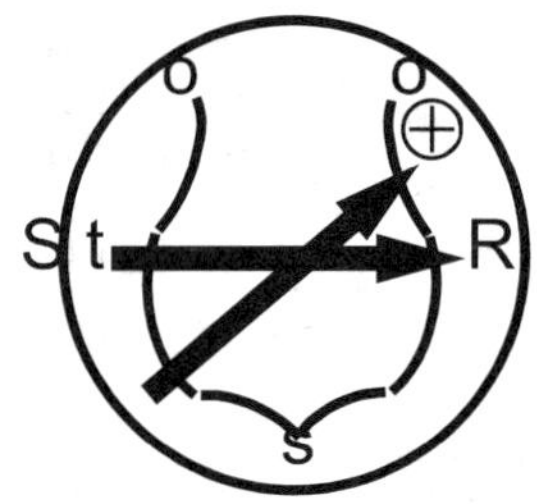

Fig 1-69 **SIMPLE SINGLE ACTS and VICTORIES**

CEV = Clear Cut Victory. There is a great innate tendency in humans to act by one strike or win. There is great difference between a life long,, and seemingly endless, struggle that seems to result in complete fragmentation, or one of a conclusive victory. In both the knowledge and representation to be of TRU a long hard ascent to fulfillment is required. It is in steps, each of which is not final, only a level of sufficiency to sustain itself and TRU. **CEVSU** = Clear Cut Victory necessitated by SU, TRU and the finite FAF and POS of CES, ES and other aspects of TRU. The clear cut victory required to achieve, assure, secure and preserve life and TRU can be attained. Yet it is not done in single strikes, nor simple overwhelming blows. It is achieved by a combined unified, finite, infinite endeavor that applies VAPO to encompass all that can, must and will be TRU.

Life's charge is a phrase for the mass energy, Á FAF, that propels TRU.(PED) It is seen in conception, birth, growth and development of a child. It is that wonderful, fantastic and endearing aspect of TRU that gives it great momentum. It is the forces and powers throughout life that provide a given amount of energy. Life charge and mass energy act in a person's life, as in other aspects of TRU, as a trajectory. It is a trajectory of GFST. This shows an individual's maximum and minimum trajectories and where their incapacities can be found. They also reveal some of the great problems of man, as to how, and what extent we sacrifice ourselves for subpar, harmful and negative trajectories.

INO = In and Out Makeup, of an object or TRU. From how INO occurs in the derivation and form of an object or TRU we have: **INODA-T** = In and Out Derivation and Form of an object, or TRU. It is in the reconciliation of these two, the CES or Processes of TRU as shown in ALT by Homation, and INODAT, that we can come to realize the actual state of our world by TRIA, TIT. Yet we should continually bring this focus, our attention, and existence back into TIT.

Function words and many of the simple common short words of language have both existing and TRU definitions, meanings, and significance in use and as CES. These include: The, used as a function word with nouns to signify some special state, condition, limit, quality, power, relation, and possession. Thereby it is closely related to SCARLS. In TRU it is used in this sense, especially signifying those features that demarcate the thisness of TRU or its components or functions. A and AN are used as function words with nouns to signify what is unspecified, undetermined, or unlimited. In TRU a is used in this sense likewise, especially in regard to an entity or a thing, as opposed to the thing, TRU. Of is used as a description, property, or some nature or aspect of a sense or thing. Of is used in TRU both for what is not and what is TRU. It emphasizes the manner in which a feature, aspect, nature, characteristic, quality, or property is significant, and how this compares with TRU. For is used as a function word to indicate the object or end of a feeling or faculty. Also it is used as a preparation toward, amount of, in place of, or reason of and variants of these. In TRU it is used in these ways as they are VAPO and SCARLED by TRU and especially as a reason, proof or relation to TRU. By is used as a function word to indicate through the means or instrumentality of, in witness or conformity of, during the course of, and a succession of units or groups. In TRU it is used in these ways and especially as the means through which TRU functions or operates. As is used as a function word for similarity and simultaneity, how one subject, object or real operates like another. The beauty of these functions words is in how words mimic TRU, how TRU is reinforced by language, and how this reinforcement is produced by vehicles of the real in the S-O, SOT, Á FAF in TIT of TRU.

GFST or General Forces Stratified Time, are those greatest positive and negative forces sustaining or destroying the world, TRU. The better we know what TRU is the better we can determine which forces are most involved. The better we can determine which forces are most involved the better we can know, manage, control and assure TRU.

WORST GREATEST FORCES	**BEST GREATEST FORCES**
Defective Genetic Material	**Optimal Genetic Material**
Worst Live Product	**Best Live Product**
Malfunction	**Optimal Function**
Determining Forces that are most damaging	**Determining Forces that are most beneficial**

Increasingly the great value and benefit of science and human discovery has, is and will be to enable us to not only reveal these beneficial and harmful forces, but to utilize and control them.

The interconnection of processes and their driving force is expressed in the term Contingency. This is how one process or, one part of TRU as the human body, exists from or depends on another or other parts. How one aspect of the dynamics of TRU functions in concert with other aspects, processes, and parts. As we better know TRU and have greater scientific and technical knowledge we can through developing the Criterion for TRU and its various aspects provide that degree of near absolute TRU that is necessary.

Cum div is 'with dividend'. It shows the increments, collective effects and processes for TRU. A dividend illustrates how through the ages collective goods and right acts produced increasing benefits. It signifies the extra continuing and accumulative processes of which we are composed. Prometheus shows the immortal hope for mankind. Oppositely about the saddest thing in life is how, like Pandora, we get locked into the many devious and tangential courses which are far from both our hopes, TES, and TRU. To remedy this we must either rationally redirect and habituate ourselves back into life as TES or suffer the consequences. It is in the practice of TES that we discover the reasons and causes for both the negative and positive of TRU, and the methods used for diagnosis and treatment. This lies in discovering and signifying all its basic concepts, principles and laws which can then be developed into their aspects, ES, and CES for a POA of TRU.

IR = Increments to Relative or to TRU. **CIR** = Compulsory Increments to TRU, the general case. Elimination of all to zero, strictly by chance, is of zero effect. The increments by increasing addition, summation, multiples, and exponentials to infinite compulsory convergence to TRU is CIR. IR is the summative positive effect of all that is in, of, and about TRU, that works for its enhanced totality. Much of this is compulsory by TRIA and thence becomes CIR. This is why cultures and civilizations have fallen to zero when the sum of increments decreases. There are breaks in the successive building blocks and mechanisms on which TRU is composed. When the momentum to TRU, **MOMET**, is lost, and the pro parts and processes, form and function of TRU fall apart.

NECVIK = New Concepts filling the Voids in Knowledge of TRU.(OM → COGNOS). By achieving a clearer vision and comprehension of what the ends are we must recognize that this is not an absolutely single unity. Both the nature of TRU, TRUS, and TRUCH, as well as the nature of this vision or comprehension in being sense and perspective rather than TRU, produce, to that extent a heteronomy of ends. (SCARLS) It is largely for this reason we maintain and repeatedly return to our overall perspective, being, TIT-TO.

SEF = Sigma or Summative ES Function or CES. SEF is where the Σ, sigma, or total function of ES, as enumerated and separated will prove the approximate measure of TRU. When each of these functions of ES, or CES are identified and described in detail they provide accurate, separate, and measurable factors of TRU. When taken together, SEF, they enable us to measure and determine most of TRU. The rest can also be better measured and determined when this remainder is subtracted from the whole, as we do in COL.

Many laws, principles, and concepts, or CALP, may seem quite true and valid in themselves. So much so, in fact, as to be obvious and almost trite. Yet their determination and understanding is basic and essential, for they are often not less than TRU, as is the case in common use. When TRU their status is fixed and their relationships are the foundation of an integrative systematic unity of all, TRU. An adequate, proper and full understanding of TRU is impossible without them. Their adequate, proper and full understanding is impossible without their formulation by and with TRU. This is: **LOID** = Law of Integrative Determination. LOID emphasizes the general case where words and the world, as ES, CES, and TRU are intricately interwoven and interdependent. Also how important they are to representation in TIT and of TRU. Bringing separate processes together in summation yields: **RICT** = Reinforcement of Combined and Valid Truths.

Reinforcement of each and all in TRU. ES, CES, and ENT are of TRU by reinforcing each other. Mankind has become more social and intellectual, with much common ground in religion, ethics, philosophy, and science. It becomes apparent that a random selection based on non TRU, partial or limited ends is far different than that based upon the whole breadth of SÁ in TIT-TO. This especially applies to its highest valuative, rational, intellectual, and TRU ends. **REHNAS** = The Human REvolution in Evolution and Natural Selection. Much of animalian evolution does not apply to modern mankind. It produces loss of identity with God, TREQ, or TRU. This is done not in the sense of either or, but more and less. Where it is much more righteous to seek after the spirit than after the flesh or material things so strongly held by the early Christian Fathers. So must we seek the form, function, spirit, parts, processes, substance and mechanisms of TRU.

Fig 2-69 **DERIVATION of OPTIMAL LEVELS of ACTION**

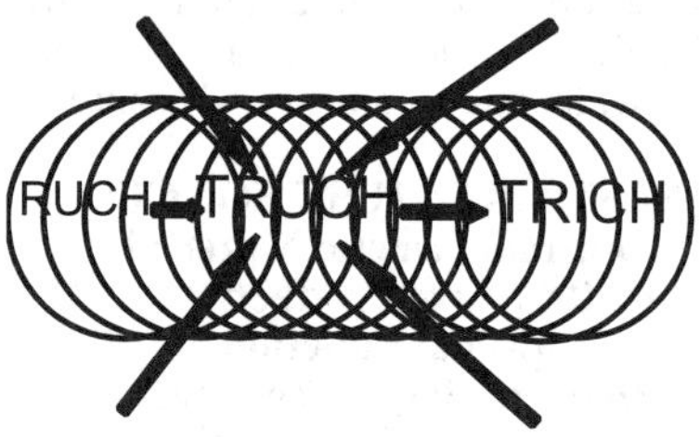

ACLO = Action Level Optimal. ACLO is seen in acclimatization, that must be just right. ACLO is the degree, level or threshold of optimal act or action. All acts, jobs, work and even life can be overdone or underdone. There is an ACLO that is just right. This requires a clear head in which there is agreement between mentality as ENIT and as THES, where both expand and are perfected. All that is done without adequate plan, purpose or understanding is not likely to have ACLO and is apt to be inefficient, wasteful, harmful or ATRU. Just as TRU is so great, massive and intricate so may and will each act be measured and imperative, neither greater nor lesser than ACLO.

Fig 3-69 **PROCESSES and ACTIONS of ACLO**

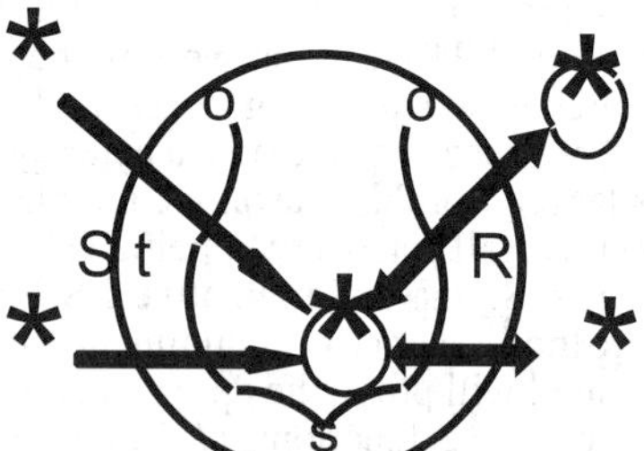

Professionally in the work-a-day world, and TREQ, ACLO is highly regarded. It is proportional to the level of TRU at the time or region. ACLO is seen in the cultural and artistic perfection of the most glorious days of human existence. The derivation of ACLO ranges from the ability to act from the most minute to the largest, most distant past to the future, and from the lowest levels of Á to its highest.

Each and all of us, so to speak, at every moment in life are on trial. This is called: **TRIALT** = a Trial of TRU. TRIALT signifies that each person must so work and live that all is most completely TRU. This should be spontaneous with a great upsurge of love, dedication, conviction, power, direction and control, that renders each part harmonious and dynamically productive of its role in TRU.

PRIND = Problem (and Principle) of Integration and Differentiation. PRIND is the universal and transcendent problem of life and its relativity, especially seen in personal knowledge. It is the problem of human limits, of how to provide, at the same time, maximal integration and maximal differentiation and all in between. It is most evident in ALT, SPALT, POS, HOM, COGNOS, ELTICMA, and SA. It is only in complete return to TRU that we fulfill our total TES. This is our salvation and immortality. **CORT** = Completed Return to TRU through fulfillment of ROT, POT and ACC..

There is no major segment of being, knowledge, or TRU that can be ignored, denied, defied, or passed off lightly. When held with such derision it is signified by: **POFIS** = Principle of Field Imperfection Summation, and is shown by: **PROVU** = Professional, or field ratio of Value over Use. These occur, often, when we think and act in ways that are less than proportional, balanced, and TRU.(VAPO,REM) This often means that our view is very partial, has an absence of interest and inspiration in being, or no love and gratitude for life or TRU. It often means that we have so championed some false cause, self interest, end, or god that the real thing TRU, is a threat to us seemingly if not actually. **VOKEP** = the Validity of Knowledge depends upon Completeness, Summation, Proportionality, Precision and Perfection.(VAPO)

The object of any and all conversation should, as all, tend to be complimentary and fulfilling to TRU. The seriousness of TRU need not diminish its joy nor that of IAT, conversation or communication of TRU. The joy of TRU is eternal. The use of a great and well known TRU ideas and CALP in conversation and communication is readily and increasingly joyful and productive.

Those who are armed, have superior means, tools and weapons predominate and tend to excite and alter the equilibrium and mass of mankind, of TRU. Superior tools are not only self perpetuating, they act as multiple and mass propelling and compelling forces. It is in reducing and controlling these tools, weapons and means, and their accompanying forces, actions, powers, and processes to TRU that we can best prevent harm, protect and maintain TRU, and provide ENT. This is greatly aided by an excellent vision of the frame and FAF of TRU seen in SPALT and HOM.

There are many new and better tools, weapons, and means in the armamentarium of man that are constantly being developed and improved. These both in the past, present and future, have, are and will, not necessarily, be those which have given a predominance in battle or war. For many of mans tools, weapons and means have great higher Á, indirect, and manifold effects that are much more determinative of success/failure and long term survival. These are what so propels man, and TRU, into a greater and better life, as well as at times, in reaction, a lesser and worse life. Mans control over nature, economic means, transportation, power and energy resources and their utilization, organizational means, efficiency and productivity, as well as many others are all a large part of the forces and dynamics of TRU. The understanding of these in TRU may have been little in the past, and much less in the present than in the future. Yet they are more active and interactive, than might be commonly assumed. This provides the answer to the static paradoxes of peoples as units or variants, peoples as things or dynamic forces, peoples and individuals in and of themselves. More crucially it answers how peoples and individuals must be their counterparts, TES, in TRU. Likewise it shows how all these superior tools, weapons, and means can be put to best use for ENT, the enhancement and eternal life of TRU, TIT and TA.

CATACLYSMIC, PERIODIC, and VARIABLE FUNCTIONS, CHANGES and VARIATIONS

In dynamics, and especially the dynamics based upon a body or focal whole, as TRU, there exist different kinds of occurrences. Those that are random, those that are regulated, and those that are variable. The variable is not necessarily what is between the random and regulated, but includes all functions, changes, and variations, including those *other than* or less than absolutely random or regular. This becomes paramount when we reduce our facts, information, and knowledge to its part and role in TRU, as shown by the ACEXIN of SPALT and the posits in it. A good example is the history of the earth, its climate and what has affected it. Much has been both cataclysmic, periodic, and variable. Locally, or for groups of persons it can be cataclysmic. Totally, for the earth and mankind, cataclysmic takes a relatively long time, is less noticeable, yet none the less disastrous unless adequately adapted to, PPMA, or controlled. All are subject to the periodic and variable. We can discover, learn and understand history, as well as the present and future, better when we realize that all three are operative, and must be adapted to our model if we are to perfect its posits. This is also so for genetics, many other topics, and TRU in general. It is likewise important and crucial to know what is cataclysmic, random, periodic, regular, variable and set forth to regulation in law. For if we are to be the determinants of our own destiny, manage and control all, we can't do so when our laws and ACC are insufficient, partial, disproportionate, unbalanced, and fallible. Thereby it is not only necessary that our laws approach perfection, they must be of TRU, continually adapted to circumstances, and changed or discarded in the best way at the proper time.

We as individuals are in a trade-off, dialectic and pragmatic relationship, between the various types of things we are composed. (COMT) This state of trade-off, pragmatic relationship, or catch-22, is conditioned by our universal, transcendent, focal, relative, specific, immanent, finite, infinite, partial, infinitesimal, unified, total and absolute existence. We, in singular thought, act, encounter or event will adopt that choice which best fits the relationship and produces the over-all desired effect. These will often be in opposition. It is in the selection of this opposition and in working through by accounting and balancing succeeding thoughts and acts that we can come to final successful and best TRU solutions.

What should be? What should we be willing to pay for the best? What should we pay for that which is worthless and bad? How can we enhance the quality, achievement, production of all that is the best, of greatest quality, within TRU? Only an ever increasing level of understanding, dedication, conviction and persistence in all peoples for what is the best and greatest quality, or TRU, will we escape from paying to much for what is worthless, and not willing to paying anything for that which is worth all. For to wantonly pay for what is worthless or even harmful and destructive and fail to pay for what is TRU is tantamount to non being, even negative being.

TRU can not be any better understood without HOM than HOM can be understood without TRU. Also DYNU can be no better understood without DYNO or DYNS than DYNO or DYNS can be understood without DYNU. To gain the proper insight and understanding of TRU, to be able to have a viable COAT we must know all the major features, forms, qualities, polarities, structures, characteristics, essences, properties, parts, processes, forms and function of TRU, as well as their complexes and totality.

The great core processes of TRU are largely and commonly known as needs, the musts, shoulds and all that carry the imperative label. They are those greatest forces, powers, actions, motions, and vectors which are essential, and provide for the survival of man and TRU. They are what help us, give us security, enables us to be, and even achieve potential everlasting life. They include all that should or ought to be; all that prevents harm, protects, maintains and advances TRU. All that is Actus or at the will of God or TRU. All that is useful to TRU, as evident from the maxim "what is useful is valuable". This is always before us, as in what is needed, and "what will be needed"? All segments of HOM, as a person's life, are concerned with this function and process. Its opposite, inaction and complacency, is that state where deceleration, that is evident in so much of life, is accepted as natural and normal and allowed to occur. There are innumerable differential functions and processes that, singularly or more relatively, go into making up the great core of TRU. These its thisness and dynamic form provide. A Generative Theory of Data of functions and processes is highly beneficial in showing how disparate and differential processes add up and can be enumerated as they constitute TRU, or TIT. The opposite, the degenerative processes, and their accompanying inaction and complacency, that are often present in modern man, sow the seeds of a miserable, failure ridden life, and the potential extinction of humans and TRU. It is because of this that we need, and need to know, TRU, HOM, DYNUS, DYNS, DYNO, ACC, and all that helps these generative CES, CHES, CHESU or simple processes of and in TRU.

What is the resolution of the great paradox of the existence, necessity and use of divisions in TRU, SUT and MOT. TIT in being composed of dimensions can be used oppositely and equally to produce the integration and differentiation unity and finity of TRU, DAOST. Yet this great paradox is corrected by: 1. Finite Ordination, 2. Comprehension, 3. Remembering S-O fusion as Á in TRU and its state as shown in HOM. It is corrected by the combined ordered differentiation in the unity of totality provided by TE and TU. SUT, ES, CES, ROT, MOT, ALT, SPALT, and POS. Thereby this great dialectic cause us to become lost in differentiation and the expansion of TRU in HOM. It is resolved readily if we attain and maintain the above corrections by TROS, IND, MOT, COAT, and ACEXIN. This is done by reaching deeper and deeper into DYNS and DYNO, especially the objective in contrast with the subjective. It is done by then holding differentials in to ordered whole of SPALT and TO. In this way we may use SULP, OLP, OSP, SUSP, SUSF, OSF, OLF, SULF with greater freedom and assurance within TRU, and have the proper overview.

The problems associated with the ID paradox are most noticeable in DYNU where we, with great differentiation and especially great polarization of these dimensions, seem to undo rather than produce the very SUT we seek. This is a part of the associated universal problem of life, simplicity or complexity. Life, as for the individual and TRU, is largely this, the optimal combination of simplicity and complexity. It is our Á powers and charge that enables us to compete successfully and survive when we best harmonize and utilize the best of simplicity with the best of complexity in our world, TRU. It isn't in all simplicity nor all complexity, no more than in all unity or all infinity, that man survives. It is in the best combination of all, as it is TRU. Each species is in a war, struggle for survival, and battle to attain the maximum benefits of complexity and of differentiation, with the maximum of integration necessary for its relativity and unity. Thereby in DYNU or in any treatment of ALT of Modulation of TIT, we must accept complexity and differentiation, and rise to its use optimally if we are to assure and enhance TRU. The paradox then, is more a product of our ignorance, our inabilities or other insufficiencies than the invalidity of Homation, DYNU, ALT or TRU.

In summarizing DYNU we need to recognize the thin and vague line, screen, or distinction that separates TRU from AU, ATRU, ANTRU, mixed TRU and other things, and all entities. We need also to recognize the fact that TRU and, each of us, are surrounded and permeated by the bad, wrong and negative, ATRU, ANTRU and ANTITRU and ANTI TRU. TheSE degrees or mixtures from perfect TRU do not form aS clear or simple A separation as they, at first, may appear. Yet we must hold TRU. at least within the appropriate SCARLS of its different varieties, kinds, forms, states, and dimensions, to strict separation. This is like what we do in all knowledge and presentation of TRU through ACEXINS, and ever expanding THES and THEO that are imperative to its formulation. All the bad things that man does to himself may seem trifling singularly or from the overall viewpoint of TRU. Yet, from a personal point of view as well as in summation by integrative negation not only are they apt to be one's undoing, 1/TES,or I/TRU, they illustrate and prove Homation, TES, and TRU. They show us what to do, and what not to do. They also show what must be done, and what must not be done. They give us the don'ts through which we can so readily find the vital and successful life of TES and TRU. They show why and how they are to be done, which in turn show why and how our world, TRU is formed and functions. They are not only a part of good habits of health and living, keeping oneself built up, they are much more. They are all the experiences, concepts, principles, axioms, rules, regulations, and laws, which we have found help us, and be positive for TRU.

MEASURES of CES, CHES and CHESU in SUMMARY of DYNU

In Figures 4-69 and 5-69 the more discussed topics of this chapter are posited. Many do not have good foci in TIT. The reasons for this are many. They are either negative, planar, outside of TRU, universal, or without clear foci. If negative they are shown by hatched vectors or in the outer spheres of figure 5-6 69. If they are universal, they are shown by the symbol ~. In this way we can still retain a good POA to TRU and relate these topics to TRU in a valid and reliable way.

ACEX	~-~-~-~	M	CONTINGENCY	~-~-3~	M	INODA	~8-2-7~	M	PRIND	~ ~ ~ ~	M
ACHIEVEMENT	6-4-7	T	COURSES	~ ~ ~ ~	M	IOT	~-~-~-~	M	PROVU	7-4-7	T
ACLO	8-4-7	T	CRITERION	~4-5~-7~	M	LASEV	6-6-5	T	RECVERT	~-~-~-~	M
ACTUS	~5-1-6~	T	ENDS	~-5~-9~	M	LIFE'S CHAR.	~6-6-2~	T	RICT	3-~-7~	M
ARS	7-4-7	T	FUNCT.WORDS	5-7-3~	T	LOID	5-3-6	T	SEF	~-~-~-~	M
CEV	~6-3-8~	T	GCF	5-3-2	T	MOMET	~5-5-~	M	SOLO	8-3~-7	T
CEVSU	~6-2-7~	T	GDF	~-~-~-~	M	NECVIK	2-3-7	T	STEPS	~-~-~-~	M
CIR	~5-6-7~	T	GFST	~-~-~-~-	T	PANFIOT	~ ~ ~ ~	M	SURVIVAL	~-4~-9~	M
CORT	7-3-9	T	IMPET	~5-5-4~	T	POFIS	~ ~ ~ ~	M	TRIALT	~-~3-7~	M
									VOKEP	2-3~-7~	T

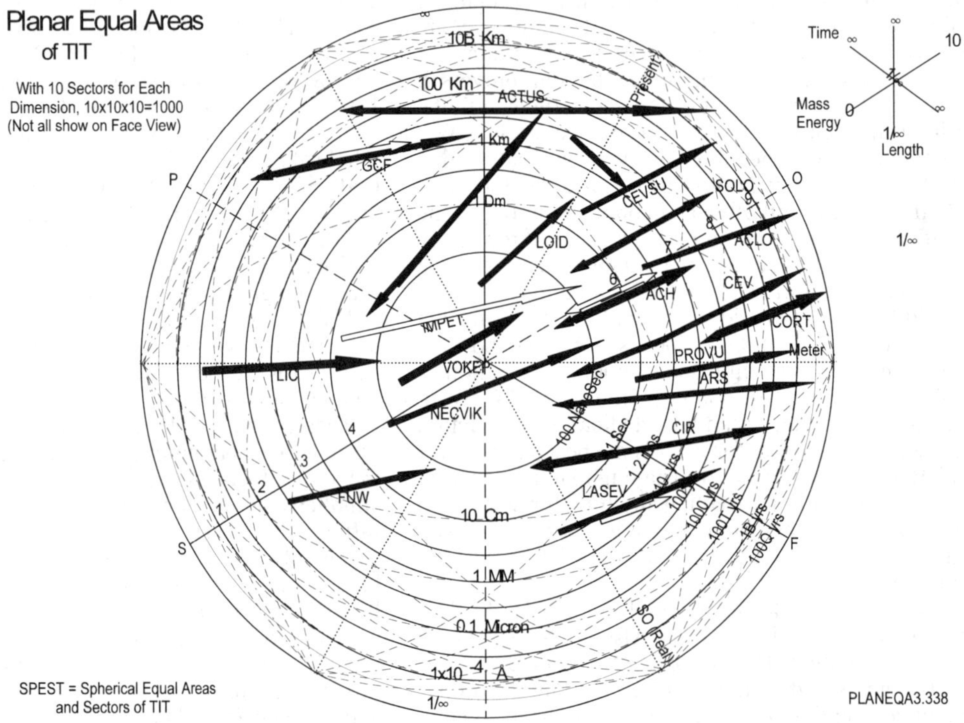

Fig 4-69 **CES, CHES and CHESU of DYNU in SPALT**

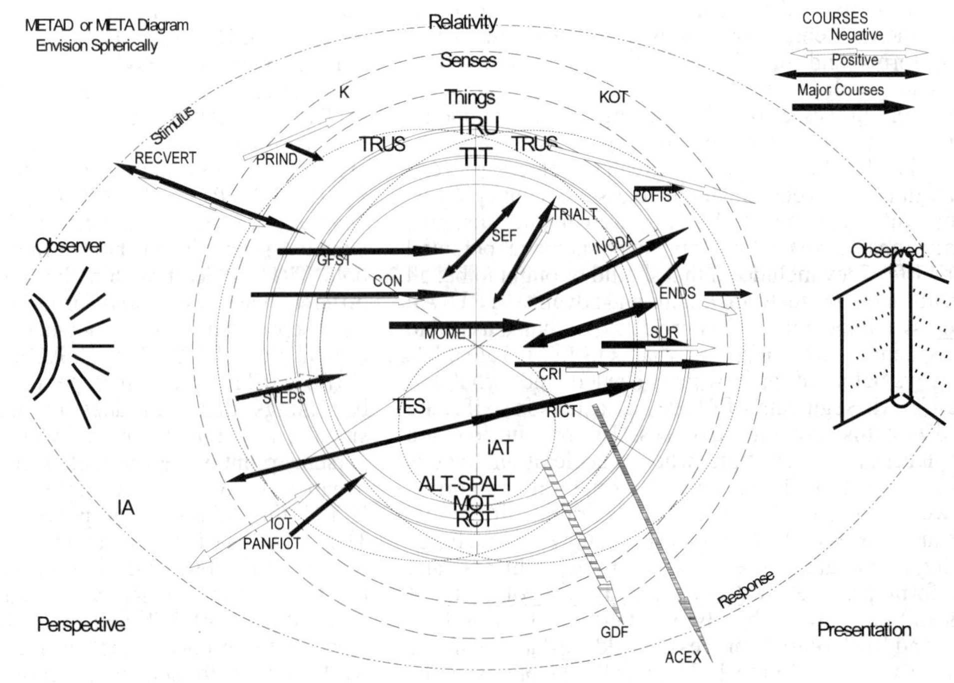

Fig 5-69 **CES, CHES and CHESU of DYNU in METAD**

SECTION VII

DYNS or the DYNAMIC COMBINED COMPLEX PROCESSES of TRU

CHAPTER 70

INTRODUCTION to DYNS CES and CHES in SPALT

RELATIONS, SELECTIONS, ADDED SUPPORT, STEPS, TANDEM, COLLECTIVES, QUOTIENTS, RATIOS, USES of CHES

DYNS is the middle ground between simple singular processes and total comprehensive processes of TRU. It helps operate and utilize CES as CHES posits in SPALT by collecting and placing them together in various combinations or complexes. It shows how our world, TRU, is composed of more than separate static and dynamic elements but many complex processes working differentially and integrally at the same time.

Humans suffer from a common problem, the lack of attention, learning, accounting, and appropriate utilization of the complex, compound, and total processes, mechanisms and dynamics that make up their lives, in TREQ or TRU. These processes whose knowledge when collected and viewed totally, as by VAPO, give us what we have so missed and so lacked. This is a good working knowledge of all TRU and TES. It is not a facsimile, but a most real and actual replication of TIT.

FIS = Fallacy of Individual Standardized Perspective of Being. When we look upon individuals in past and future settings we tend to compact time & size and thereby distort their being. We tend to become caught up in their - or some other - focus or vantage, rather than that of all TRU. It may help to provide a degree of reality to the image, yet it is poor actuality by TIT. It rapidly produces a skewed, distorted and false understanding. This is likewise a problem in DYNS where we tend to focus on complexes in ENIT, rather than TIT, and a, or some, CES rather than all CES in TO.

There are several commonly used terms that define those processes of TRU which work together to form complexes and include among others: additions, asymmetries, bonds, collectives, 'co-s', combinations, complexes, contrapletes, cooperatives, COVRAI, dialectic, dualisms, multiplication, paradoxes, ratios, and tandem. In fact all terms that describe how entities and especially CES function together in any way are modes of DYNS, often leading to DYNO. An analysis of families, groups, towns, communities, sites, cities, states, and nations past, present, and future helps us see the combined and complex processes. Complexes are sometimes divided into periods which are held in common by many sequences of growth and development. Many common periods of development used include: archaic, formative, classical or "golden," terminal or final periods. Such schemata are neither the only nor the most complete ones for categorizing development. Such schemata for periods of life have much in common with PACH.

There is a great and significant difference between the undirected and the directed TRU, between the expedient TRU and the mandatory TRU. The undirected and the expedient TRU have very limited lives. Only with a directed and mandatory TRU can we have the assurance of even a temporarily certain future, an uncertain TÂ or a certain TÂ. **UNDEXT** = Undirected and Directed Expedient - and Mandatory TRU.

LASOC = Law of Selective Significance Solution of Conflicts and Problems. LASOC signifies the degree of emphasis that is to be placed on the magnitude, relation, and ACEXIN of conflicts and problems. LASOC is a concept which focuses in TRUD, REM and ACC and shows the special form of DYNS. This is how selectively and in combination we use the complex dynamics of CES.

A society, culture or the world may be divided by the activities of its members. How much does each contribute, or not contribute or detract, to or from the welfare and success of the whole? How well can these be measured and a quantitative analysis provided? One of the great needs that are fulfilled by religion is to direct individuals into the right path. It gives more than simple consolation in despair. This is by leading to a better and greater ability and contribution. Much of this can be measured. **MOR** = Measure of Religion. There can be several possible measures of religion. It is through bringing individuals to their counterparts in TRU, as in how well they abide by GOD and how well God expresses what it actually is TIT, that much of TRU is attained and assured. It arises from the moral and ethical maturation and perfection of the individual. This in turn is largely a function of the presence and degree of KOT in the individual and the degree to which they with great interest, conviction, dedication, and devotion, remember, apply, use, and live by KOT and TRU.

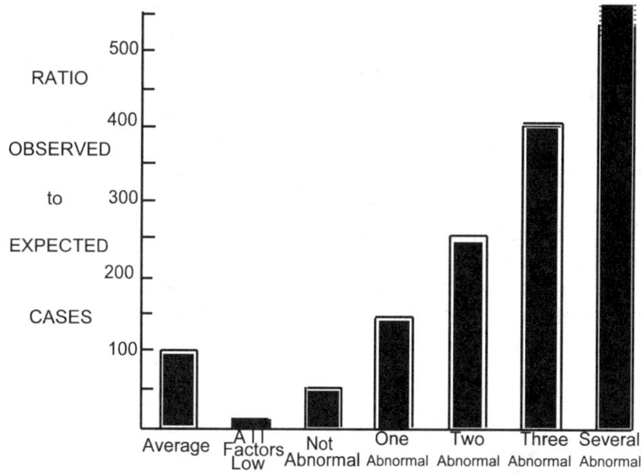

Fig 1-70 **DYNS and COMBINATIONS of FACTORS IN DISEASE, HOM and TRU**

The validity, as well as importance, of complex and compound causal mechanisms is shown throughout nature.(COC) It is especially evident from scientific research, as seen in medicine. The many factors that go into producing a certain disease or complications of diseases are being increasingly shown. The causes for hypertension, arteriosclerosis, coronary heart disease, myocardial infarctions, heart failure, strokes, and many others are good and important examples of DYNS and Homation. This is especially so for IOT, in and out of TRU. Continued increase in research and discovery will not only add to knowledge, etiology, diagnosis and prevention and treatment of diseases, it will do the same for all TRU by showing how DYNS, HOM and TRUD work in TIT and how to prevent and treat much of what adversely effects, or may destroy TIT. Risk factors are combined, susceptibility is greatly increased, the effects on ACEXIN provide a clearer picture of what can and does go wrong in TRU. From these, their PPMA, becomes clear and can be more readily accomplished. What proves the greatness of mathematics, as well as DYNS. is the demonstration of multiple factors, functions, and variables of formula and equations that reveal the world as it is or is not, and in and out of TRU. When these variables and factors are made sufficiently explicit and when TRU can be sufficiently well known to fit its processes into these formulae and equations, much more accurate and fulfilling showing of TRU can be produced. This yields the beautiful intricate detail and actual dynamics of TIT-TO that is essential if we are to adequately understand, effectively act, and best produce to survive.

Humans have developed by a natural selection that is controlled by what has been, is and indirectly will be most useful. This will differ somewhat by stage in TRUS, TRIS, ES and CES yet will precisely hold for TIT-TO. Every part of the human body, and every part of TRU exists because it has selectively been of the best use. There are several levels of use. These coincide with the human mass energy dimension of TRU, its Á continuum. Since it is essential that we know as much as possible of the actual world, we can learn, remember and apply it best by interpreting the uses of its fundamental mechanisms and processes.

MARF = principle of Massive Response to Failure, is seen in how war or any great threat to human survival is met with the use of all resources available. This often occurs in an increasingly ramified and decreasingly TRU way. The road man takes to extinction is much easier than the steady hard climb to continued survival. **LIT** = Lassitude Inherent Tendency. LIT is what has been popularly known as growing soft. Such a response can be positive, as evident in the cyclical phases in human existence. This is especially so when coping with life and the necessity and demands of TRU. This shows us an added meaning and value of TRU. How it is not only to TES and TRU, but to the individual, parts and ENIT. For if they are to be taken as things in themselves, they still will receive benefit from TES and TRU. This is called: **LAV** = Law of Added Value, and, **LAVIPT** = Law of Added Value to Individual of PUREALITY and TRU and TES.

MOASM = Magnitude of All Segments Mutual. There is a dependence and mutual magnitude of all great segments of TRU. Their greatness in TRU stems from their mutual dynamics with other segments of TRU. MOASM is shown by DYNS. When we represent the many combinations and complex processes of TRU we can see their dependence and mutual support of each other.

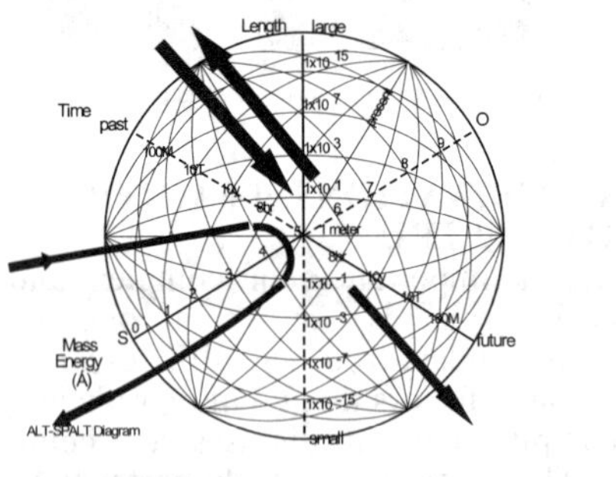

Fig 2-70 **TANGENTS From TRU and ADDED VALUES**

Vectors of CES in SPALT can show reaction, action, tangents and the departures from TES and TRU that people take through various kinds of COC, defects, ailments, injuries, habits, and inability to cope with life.

LORSUD = Law of Research Scientific Systematic Unity Deficit. LORSUD explains how scientific research is opposite to TU and SU, thus its limits. This is especially seen in ROT and POT, and depends upon the degree to which it is directly or indirectly guided by VAPO.

BER = Belief, E ES, Ratio. Every society, person, culture, age, and civilization looks upon its own as the ultimate, the only complete truth. In actuality, of TRU, this can be no greater than the degree to which its beliefs are of the highest percentage of ES, and proportion by REM. BER provides a means of measuring and analyzing the many religions, ISMS, systems, and beliefs of the world and their tenets.

The collective beneficial or detrimental effects of the capabilities of those who hold high positions in the world are signified by: **COVET** = Collective Value of the function of Parts of TRU. When persons who hold high positions have poor BER, lack responsibility, have little or no HPP and KOT, or are highly ATRU, partial, with goals and purposes that lack VAPO their responsibility can not be fulfilled. **COVRAI** = Collective Voids of Responsibility of knowledgeable Individuals. COVRAI represents those social groups, regions or nations where inadequate or bad governments are produced by leaderships that lack knowledge of TRU.

The more material things one demands in life, the more one becomes their slave. There is an increasing limit to how much baggage one can carry. There is always a trade off between the advantage obtained from material things, their price, cost, and one, others or all, must pay for them. Like all in life, there is a price, someone must pay. Nothing in life is free. For, even if it is momentarily or seemingly free for a person, it costs, by the time spent longer term, by other people, by TRU, or in various other ways, somehow or someone must pay.

Extremes and means function in tandem in TRU and show the various ways life is sustained. From extreme idealism to extreme materialism, from extreme charity to extreme self indulgence, are only two of a vast number of parameters typical of life by TRU. It is the interaction, fluctuations and range of the many polarities, continua or spectrums that exist in our lives that constitute much of TRU. The processes of these often work, as DYNS, in combined tandem of opposition in equilibrium is TRUD. So that ISMS, religions, schools of thought, and other ways of life are to be analyzed and judged not only by their degree of TRU/ATRU, but the degree to which they are and purport to be ES, REM, TO or all inclusive, as extremes and means of dialectical processes. If they at the outset recognize, communicate, and practice within their limits in TRU, then their error is minimized. However to the degree they are not TRU, and especially the degree to which they maintain an all inclusive and universal claim but are not, then this combined role becomes doubly wrong.

Fig 3-70 **STEPS to HELL or to HEAVEN**

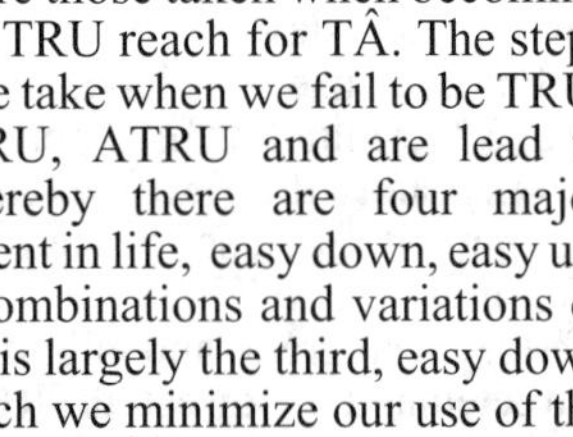

A simple comparison of the ups and downs, steps down and steps up in life are given in two ways, one in which there is an easy walking stair with slower slope, and one in which there is a hard walking stair with a faster slope. The steps up at the right, to heaven are those taken when becoming TES and our IA, SIA, TES and TRU reach for TÂ. The steps down, at left, to hell are those we take when we fail to be TRU, are overwhelmed by NAMTRU, ATRU and are lead to ANTRU and ANTITRU. Hereby there are four major directions of movement, as evident in life, easy down, easy up, hard down and hard up, plus combinations and variations of two, three, or four of each. Life is largely the third, easy down and hard up. The degree to which we minimize our use of the easy down and facilitate our use of the hard up, as it is TRU is the degree to which we, all, and TRU succeed. The degree to which we take the others, as they are not TRU usually ends in barm, or great loss or disaster to oneself, others, and/or TRU.

LADU = Law of Disproportionate Utility. Even with 10% harm and 90% benefit there can be loss that may be fatal, to some, all in a group, or TRU. This has been shown by paleo-anthropologists who calculate that differences in a competitive edge of no more than 2% will give one animal such great predominance over their adversaries that the weaker become extinct in less than a thousand years. This is why no other animal has the abilities of language, rational cooperation, insight, and many other features as humans. For they are do not have the higher range of mass energy levels of Á FAF present in humans. It limits the range of FAF and thereby adaptability. As a result they are less able to adjust to the great variety of habitat and interactions that are so characteristic of humans. Some of these abilities which may give humans a competitive edge also have their downside. They are loaded with extraneous deleterious susceptibilities with potential effects that reemphasizes humans unique thisness, form, and function, TIT. It serves them best in many ways by its highly expanded higher levels of FAF but also has greater potential for damage, harm and suffering. Humans are continuously thinking and doing things that are not in their best long term benefit and interest. Moreover, they are not interests of the whole, TIT-TO. Even when 90% for good they can cause much harm so that the whole suffers. **POTDE** = Potential Deficit. POTDE signifies failure to develop the potential that has been inherited or can be attained. It is the difference between what can be and what is developed. It is in POTDE that we can clearly see our future and what we should do. For if we do what is not the best and fail to fully develop our potentials, POTDE results and both we and TRU lose. LADU and POTDE represent two of the more critical mechanisms and processes in or about TRU whose combination in reaction to TRU shows the great role DYNS have in TRU.

1. GENETIC QUOTIENT,
2. SUMMATIVE GROUPS,
3. EFFICIENCY,
4. USEFULNESS,
5. POPULATION,
6. AMERICA POTENTIAL,
7. ACQUIRED QUOTIENT
8. AFFECT QUOTIENT, acquired and genetic

We may express DYNS or the combined processes of TRU by quotients. Quotients may be expressed as functions of: (f) A, B, C, D, E, F,. Some of the quotients are noted above and separately or in combinations provide great use in REM.:

LOPROT = Law of Return from Proper Proportion of TRU. LOPROT defines how those who give 100% to TRU may be returned 100% from TRU. That which becomes 100% TRU will be returned from TRU this percentage. Those who give 12R have returned 12R TRU. Those who give negatively or ANTRU are returned likewise.This holds relative to TRU, not relative to AMTRU or ATRU. One receives according to their work, effort, contribution and enablement of TRU. What they give and produce in return produces that amount of benefit to TRU. However what they receive is only TRU not some fantasy, figment or false goods or lesser reward. **LORTAIS** = Law of Return of Actual/Illusion TRU, and its recognition, subjectification and understanding. The combined and complex processes, DYNS, of interaction and return to the individual from TRU help us understand what TRU is, the reasons why we should think, act, work and live by TRU.

Fig 4-70 **LEVELS of TRU SUPPORTS**

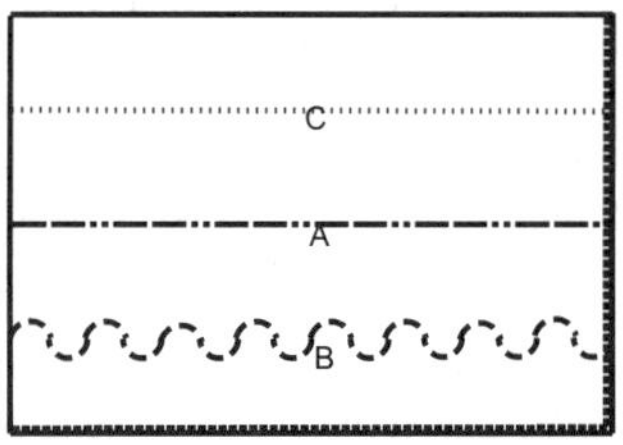

Different levels of contribution or non contribution to TRU run from the lowest Level B, or most of the world's population that contributes little or negatively to TRU. Level A is par, or that level of contribution necessary if TRU is to hold its own. Level C is the high level reached by some who have, are and will be great contributors to TRU. This is more than singular, but a collective to total SÁ and GELF. Often of those who have sacrificed and endeavored to help far beyond what might have been expected. It is in the ratio and summation between levels that we have a clearer idea of the health, and how, TRU is doing.

SCOW = Summation Compounding of Wrong. SCOW may occur in any way, inadvertently or expediently, directly or indirectly, reversibly or irreversibly. To help in any way requested is a common error. Even if right for the benefit of TRU in one way it may not be right in another, the right place, time, way, or in the right proportion. Yet the key is in KOT, knowing what is TRU and how it is to be used for best advantage. Wants can usually be turned to TRU if the one granting them can help those requesting understand the reason and be TES. Everything a person thinks or does in various positive or negative ways may be expressed for TRU.(IN) **POCON** = Principle of Consequential Negation. Popular misconceptions and limited views correlate with a lack of TRU understanding and explain many great troubles, conflicts and wars in history. Everything may have seemed to be all right or all wrong, yet were more likely to be a complex mixture of both. A certain year, period, or setting may have many of the characteristics of another, yet they are likely to be far different. This we can better understand by TRU, and when we compare times, partials, and misconceptions that occur with each. Many popular misconceptions are little more than deviants from the basic features and corresponding principles of TRU. These not only show the key to how misconceptions arise, they show how they are to be cured.

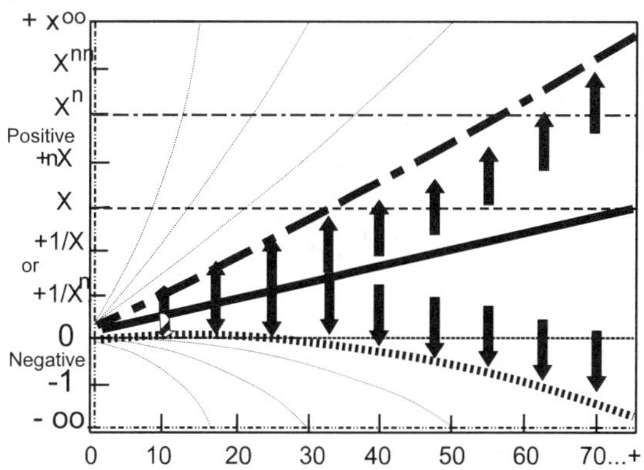

Fig 5-70 **THE POSITIVE and NEGATIVE, LIFE, DYNS and TRU**

In the Figure 5-70 the years of a persons life are shown on the horizontal plane and degree of accumulative positive to negative effects, contribution to depreciation, are shown on the vertical. Most of us either through imitation, custom, education, habit, routine, pressure, the social order, profession, occupation, rules and regulations or the law, come to have a semblance of positive and viable existence. Yet in substance, as we have and will note, this positive is superficial and not highly substantive. It is often by TRU actuality highly negative. Only when one becomes deeply interested, committed and finds joy in discovering, understanding and using the greatest knowledge, wisdom, and understanding, especially of KOT and ROT can we come to be highly positive TES. The right decisions and endeavors, or the wrong choices and mistakes in life, to a great extent, determine the degree to which each one of us are or are not positive or negative to TRU. These are shown by ↑ and ↓ vectors.

Each and every choice, decision, act, and thought is positive or negative for the success of the individual, all mankind and TRU. These are not the same, although for all mankind tend to be TRU. Therefore the great mass of humanity, either out of not knowing what to do or out of lack of proper guidance and opportunity, are the cause for much of TRU's and their own conflict, stress, strife, suffering and failure.(MOPLARU) For to have the whole or large segments of the worlds population operating either against or in any direction other than directly for TRU is productive of deterioration and negation. This is why the affluent must realize and alter their life to be TRU, and the destitute and impoverished must be PPMA by ACC, TRU or TREQ. Negative forces may be of many kinds not always directly nor simply against TRU. For all that is divergent, out of phase, non synchronous, disproportionate, and disordered, is, unless compensated for, not perfectly in tune with TIT-TO. It is not perfectly positive and thereby to that degree and in that way negative.

Biologically and anthropologically the high level of influence of a whole population on the evolution of a species may or may not be great. By current opinion, and ideally by TRU it is generally preferable. For it provides greater stability. However the less TRU the total population is the less it should or will influence evolution. Because the evolution of humans in the past has been, in terms of TRU perfect for TRINI, does not mean it will be so in the future. Likewise both because of the role of the future, the role, more or less, of all humans in the TO of TIT, our inability to manage and control the GELD of TRU, and the increasing interaction in TRU upon large populations, a simple biological or anthropological determination is inadequate. It must be broadened and altered to include what is best for TRU.

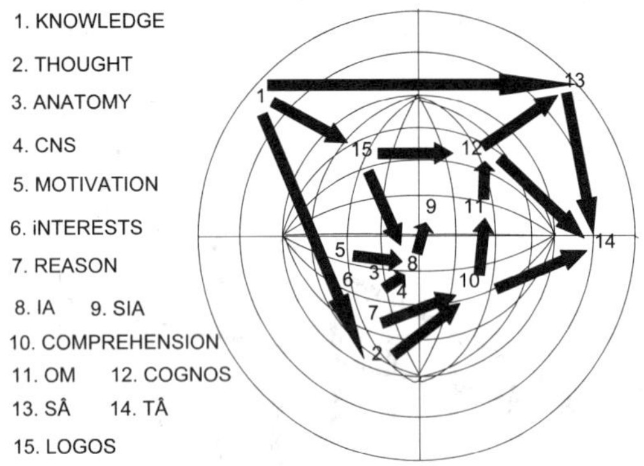

Fig: 6-70 **COMBINED CES, DYNS in and by TIT**

On the frame of SPO some of the major categories of, and combined dynamic CES, are shown. These are only rough posits and vectors. Yet they provide a good illustration of their relations and dynamics. They show how CES are interactive and support a person, group and TRU. How knowledge, interest, motivation, the physical make up of man and his central nervous system, comprehension, reason, logos, Cognos, OM, SIA, SA, TA, among many others may be shown and give a clearer picture of TRU and our place and role in it.

PROBLEM-MECHANISM-SOLUTION

Some of the perpetual problems and demanding goals of mankind, the world, and TRU include: Individualism/TES, all to TES optimally by first person, observer-obsserved, macrocosm/microcosm, dynamic proportionate total harmony, balance of form within TRU, symphony and structure, internal positive forces, equilibrium, moderation of parts of TRU, components internal competition, harmonious function, healthy homeostasis, relative/absolute, any being or the being, relative absolute, thisness, eternal balance of TO-TRUD-REM, total harmony, IA need depth and flow, ATRU-TRU, THIT/TIT optimal functional ability, fulfillment of TIT, etc..

The greater the problems are the more likely and greater the effects, divergence and imbalance in the forms and parts of TRU, and vice versa. The validity of the relationship between great problems and great imbalances and divergences in TRU not only prove their rationale, it tends to prove COGNOS and TRU. Every person should be sufficiently appreciative of the grace or benefits of TRU, as well as have available sufficient return to render ones life positive and joyful with TRU. This is called: **RAPOR** = Ratio of Full or Total Positive Return. RAPOR is something that is not owed to the individual so much as that which one must earn. It is earned by living, working, being ones counterpart in TRU, being TES, Pureality and having only the highest and greatest ideals, principles, laws of TRU as ones guiding light. The solution is not to be found so much in the negative, condemnation nor elimination as it is in benefits, education, promotion, commendation, spontaneous and willful joy in being TES and TRU.

The best approach and attitude that can be appropriate is that which is subtle, profound, formally distinct, and comprehensive by VAPO and SCARLS for all TIT, TRIT, TRIS, TROS, SA and TA. It is an attitude, like KUPS, that focuses on both what is universal, as TRU, what is best for each person both in their occupations, personal life, and relations as they are TES.

In the performance of work, deeds, endeavor, battles, wars and any strategic effort there is a principle that is highly effective. This is: **PAS** = the Principle of Added Support. PAS is shown by the situation where a difficult task, job, work, effort, endeavor, battle or war occurs and the doer, attacker, army, general or operator may not have the capacity or resources to quickly, readily, fully or successfully effect and complete the task or operation. If added support is supplied or if a like degree of help or power is added, as another corps in battle or another doctor to care for a massive overload in emergencies, the job, task, battle or war can be readily accomplished. To have that degree or amount of help necessary to do any job is a primary responsibility of leadership, PPMA, and M&C. It is equally necessary for an efficient and successful enterprise to have, in the reverse case for optimal efficiency, no more than that degree or amount of help necessary. This is called: **POPTS** = the Principle of Optimal Support. POPTS will depend to some extent upon the jobs importance and the overall stress involved. If a job is extremely important PAS is vital, and if little stress is involved, POPTS is not so vital.

MOPLARU = Masses of People Living without or Against TRU.(NAMTRU,ATRU,ANTRU) The degree to which we are less than fully, other than or against, TRU is what so sets us apart from one another, and when with abnormal interests in violent conflict. MOPLARU shows the combined dynamics of the negative. If we understand the problems of all days, times, half lives, and their dynamics, and if we map out in SPALT and MOT the major courses, forces, vectors, powers involved, we will go much farther in determining history than through names, dates, places, and conditions of isolated events. We can show how each person is caught up and can dramatically and beautifully provide a highly accurate POA to history, as well as the future. This reveals a key defect in our thinking and a major solution provided by MOT, ALT, SPALT and POS. The defect concerns our identification, relating, and interpretation of names, dates, places, conditions, and events, which we are so sure of, quickly come to naught. This is called: **ERI** = Error of Identification, Relation, and Interpretation of Events, names, dates, places, and conditions. Its correction is called: **CERIT** = Correction of ERI by TRU.(LORTAIS) CERIT is how all we meet in life, or comes to our IA, is not to be accepted at face value, nor identified, related, or interpreted by any incidental factor. It and all it relates to are to be identified, related, and interpreted by how they exist in TRU, operate by VAPO, and best posited by SPALT. The application of the principles of HOM, as they occur in TRU and TRUD, to fields other than history, can also be highly effective although somewhat less dramatic. This is the greatest advantage in applying HOM and DYNS, to show the world not singularly, nor totally, nor in form or function separately, but why and how all these work together positively and dynamically in SPALT of TIT. Homation as produced in IPTRU provides the identity, relating, interpreting, and formulating that corrects ERI, supplies CERIT and is much of the proof of TRU.

Measures of CES in DYNS on SPEST of SPALT

ABNORMALITY	~_~_~_~	M	FIS	~2-1-4~	-M	MOASM	~-~-~-~	M	POTDE	~-~-~-~	-M
BER	$\overline{3}$-$\overline{7}$-$\overline{3}$	T	LADU	~8-2-~	-M	MOPLARU	8-1-3	-T	PPS	2-3-3	T
BOND	~~~~	M	LASOC	6-3-7	T	MOR	3-2-7	T	RAPOR	6-1-7	+T
COMBINED DYN	~_~_~_~	M	LAV	~-~-~7~	M	MORF	~-~-~-~	-M	SCOW	~-~-~-~	-M
COMMUNITY	$\overline{6}$-$\overline{2}$-$\overline{2}$	T	LAVIPT	~5-5-8~	M	PAS	~-~-~-~	+M	SOLUTIONS	~5-4-8~	T
COMPLEXES	~_~_~_~	M	LIT	~5-6-6~	-	PERIODS of LIFE	7-3-2~	T	TANDEM	~-~-~-~	M
COOPERATION	~_~_~_~	M	TLOPROT	~_~_~_~	M	POCON	~-~-~-~	-M	UNDEXT	~-~-~-~	M
COVET	~-~-~-~	M	LORSUD	~$\overline{3}$-$\overline{2}$-$\overline{7}$~	-T	POPTS	~6-5-7~	M			
COVRAI	~7-5-4~	-M	LORTAIS	2-8-5	-→+T	POSITIVE NEG.	~_~_~_~	M			

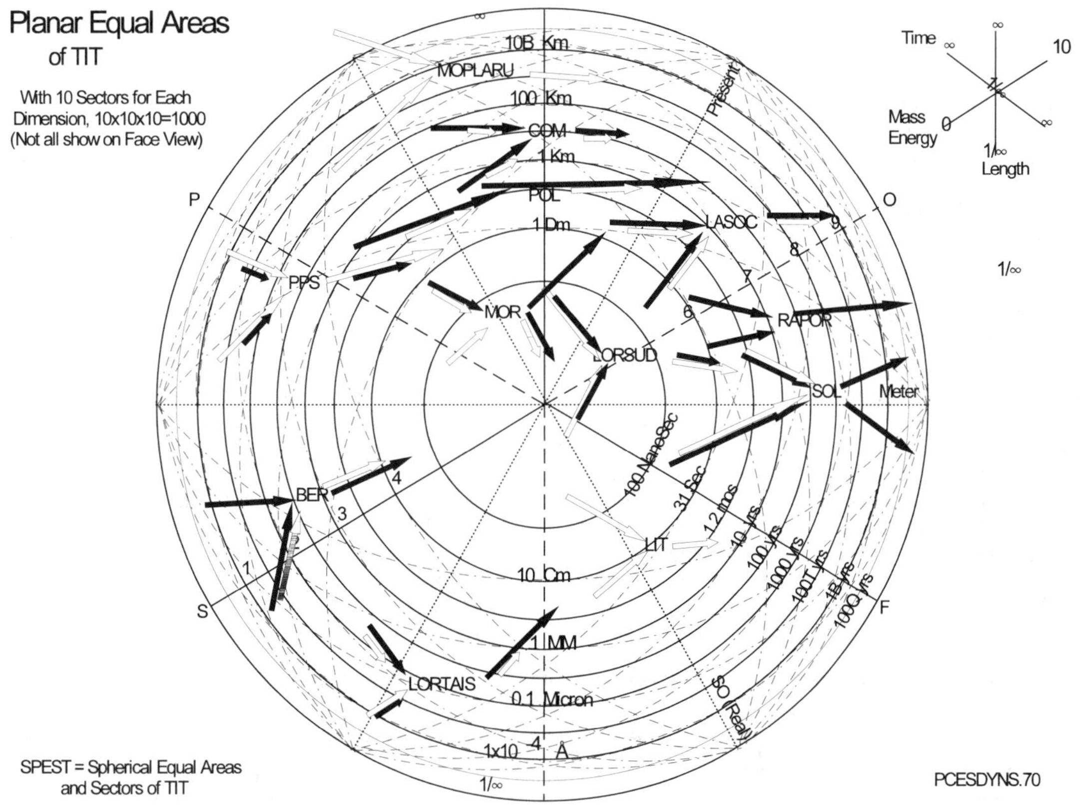

Fig 7-70 **CES of DYNS in SPALT,**

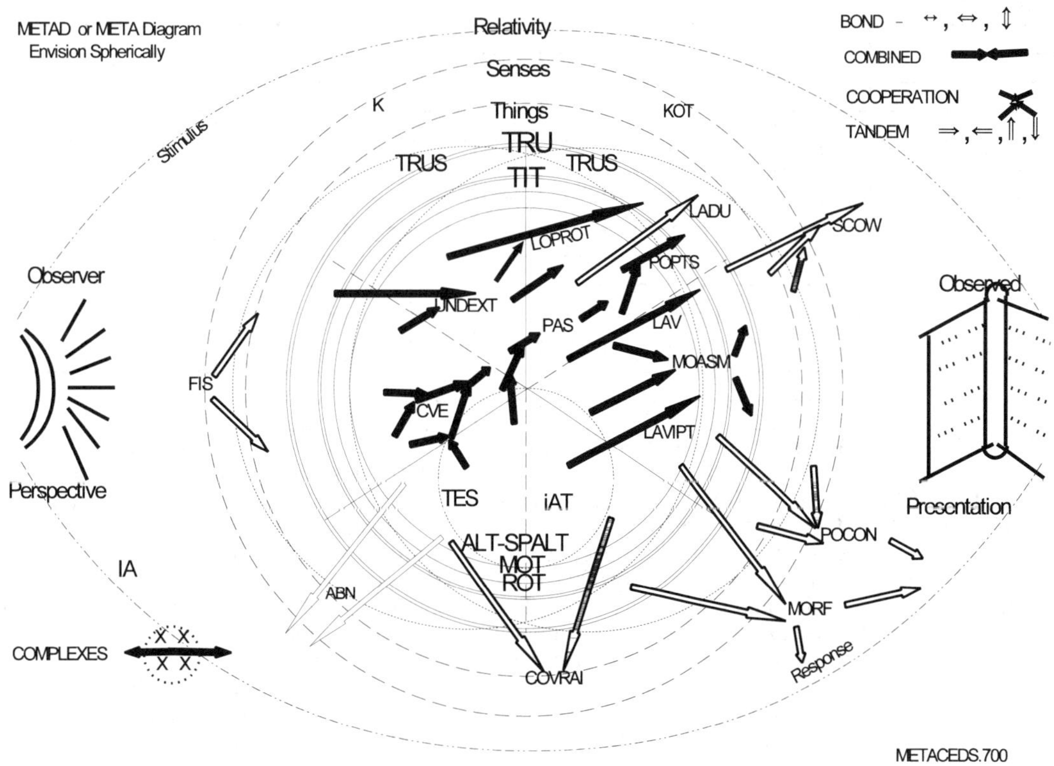

Fig. 8-70 **CES of DYNS in METAD**

CHAPTER 71

SULP or SUBJECT LARGE PAST PROCESSES in DYNS

CULTURE, LANGUAGE, HISTORY, WISDOM, GODS, RELIGION, WORTH, PHILOSOPHY, LAW, INTEGRITY

As human communication, use of words, and language expanded with increasing identity, meaning, and precision, concepts and terms became more distinct and defined. The definition of terms occurred relatively late in mans past. Yet it must have been of concern indirectly if not intentionally from a much earlier time. Mans first cuneiform writings in the protoliterate period - Uruk of the Chalcolithic - show evidence of concern for the distinction and definition needed in name relations. Communication, use of words, and language, developed, along with a sense of value, need, goodness, and obligation. This was important, of the highest priority, and what one, some and all should and should not do. True, a natural, if unrecognized or subconscious sense of these occurred prior to settled life. Even today we can recognize much in the lesser levels of Á as well as ATRU of all that is about us. However on a subjective and rational basis they are characteristic, as Á and its higher levels are, only of TRU.

ATO renders all our prior ideas of astronomical, biological, and temporal beginnings obsolete. All must be reconciled with the construct of ALT in TRU. When we exercise SPALT POS, as in HOM and DYNS we can accurately reduce all knowledge to what it is in TRU. The lesson for all time, is the need for a firm and great foundation upon which cultures and civilizations must be built. These are built not upon illusions, figments, fictions or frivolous things and ideas, but what is nearest and dearest to the human soul and life. This is their form and function in TES and TRU. These can be symbolized and ordered into philosophical, religious, political or other systems that best yield what is natural for TRU, its totality, harmony, peace and prosperity. They can instill and keep an ever increasing capacity and ability to be the best.

The bible gives a good account, collection of evidence, and proof of what it takes for groups of people living together to survive. It also shows what happens when they live apart, failure. It shows, **COPOS** = Collection and Proof of Combined Success. COPOS signifies what the rules are for a successful life together with peace and prosperity. What and how order is so necessary with a multitude of peoples. This is a good example of much of TRU, for not only is TRIT a static or unbound universe, TIT-TO is one that has great combined and complex strength based upon internal forces, intense being, life and survival.(PIP)

Currently we misinterpret the world's characteristics from near sightedness, poor perspective, and lack of KOT.. What distinguishes civilizations, cultures, nations and states, by way of being different, dramatic, helpful and identifying although not TRU directly, is largely through benefit. Even when held to be TRU it may not be, or be much less, for it may only represent a variant extreme or partial. Whereas FAF has much in common, this same is what makes up much of TRU. **CHADÄ** = Characterizations by Distinguishable Form. CHADA is an example of PRA → PRUT.

	TRU	TIT	TRIA TÂ	ATO	DAOST	A, TU SUT	ES TES	TO	TRUD CES	ENT	ROT	MOT SPALT	REM	POS	STA	HOM	ACC
MONOTHEISM	+ -	+-2	+ 3	+3-	-3	+4-3	-1+2	+3-3	+2-4	+2-4	-3	-4	+1,-4	-4	-3	-3	-3+1
POLYTHEISM	+-3	-4	-3	-3	+2-4	-3-3	+3-1	-3	-1	-3	-2	-4	-1	-1	-2	-2	-2
DEISM	-3	-4	-3	-3	-2	-3-1	-2	-2	-2	-3	-3	-4	-3	-4	-4	-3	-3
ANIMISM	-3	-2	-2	-4	-3	-2-3	-3-3	-2	+2+3	-1	+1	-3	-1	-2	-3	+2	-1
RELIGION, (General)	+3	+2	+3	-1	+1	4	-2+2	+2	-1	+3	+1	-3	+2	-3	-2	-3	+2
CATHOLICISM	+-	-1	+1	+1	+2	+1+1	-2+1	+2	-2	+1	+2	-3	+1	-3	-3	-3	+1
PHILOSOPHY, General	-1	-3	-3	-2	+1	-2-3	-2+1	-1	-2	-1	-2	-3-4	+1	-3	-3	-1	+2
IDEALISM	-2	-2	+1	+1	-1	-1-1	-2-2	+1	+1-1	+1	-3	-3-4	+1	-3	-4	-2	+2
REALISM Aristotle	+1	+2	+2	-1	+2	+2-1	+1+2	+1	+1+1	+1	-1	-2-3	+2	-3	-3	-1	+2
SCIENCE (GENERAL)	-3	-4	-3	-2	+1	-3	+3+2	-2	+1,+2	+2	+2	-1-3	+1	+1	+1	+2	+2
MATERIALISM	-3	-3	-4	-4	-3	-2-4	+2+1	-2	-3	-1	+2	-1-2	+1	-1	-1	-2	+1
CAPITALISM	-2	-3	-4	-3	-2	-3-4	+2+1	-3	-1-1	-1	+1	-2-4	+2	-3	-3	-1	+1
DIALECTIC MATERIAL.	-3	-4	-4	-3	-1	-1-4	-1-1	+1	+2+3	-1	-1	-2-3	-1	-3	-3	+2	+1
SOCIALISM COMMUNI.	-2	-1	-2	-3	-1	-2-2	-1+2	+1	+1+2	+2	-1	-2-3	+1	-2	-2	+1	+1
EXPERMENTALISM	-3	-3	-3	-2	-1	-2-3	-1-1	-3	-1+2	-1	+2	-1-2	-2	-2	-2	+1	-1
BEHAVIORISM	-3	-4	-4	-2	-2	-1-3	-3+2	+1	+2+3	+2	-1	-3-4	+2	-1	-3	+2	+1
HUMANISM	+1	+2	-2	-1	+1	-1-2	4	-2	+1+2	+1	+2	-1-3	+2	-2	-2	-1	+2
EXISTENTIALISM	-2	-3	-2	+2	+2	-1-2	-2+3	-2	+1+2	-3	-2	-3-4	+1	-1	-2	+1	-1

Table 1-71 **DEFICIENCIES, ERRORS, and FALLACIES of GREAT RELIGIONS, PHILOSOPHIES, SYSTEMS, SCIENCES, BELIEFS, ISMS, SCHOOLS of THOUGHT or BEHAVIOR ACCOUNTED for by TRU**

Horizontally many of the greatest aspects, categories and SCARLS of TRU are enumerated. Vertically many of the most important religions, philosophies, and systems are listed. The causes of their neglect, deficiency, defiance, inadequate and incomplete beliefs and tenets are shown by ratings rated from -4 to +4, depending on the degree to which they are or are not TIT. From such a preliminary and rough table we can show how religions and systems fail to be TREQ or fully TIT-TO is evident. Also, it helps us understand the greatness and breadth of TRU, how much and in what way all needs to come to TRU. This is a great proof of the need to adopt and carry through TRU, and its terminology. The ratings depend upon the definition and common usage, whereas when strictly limited to TRU and roles in TRU their ratings are much higher, e.g. monotheism. As great as TU, God and monotheistic religions are the harm done by what they neglect negates much of their greatness. **LANDOVS** = Law of Negation or Negative TRU from a lack or deficit of TROS, VAPO, SCARLS, ACEXIN and all TRU. LANDOVS explains the reasons why so many religions, systems, schools, beliefs and knowledge fail. This is the serious deficit that occurs, or COL, between what they are and what they should be as TRU. Religion, instead of unifying and creating harmony, has been a vehicle for separation and opposition. In TRU difference by DAOST, is separation, yet only one that is balanced by the unity and totality of TRU. TRU is the extraction of all inactual differences, disproportions, disorders, and negatives to the whole. With improvement the extent to which tenets are or are not valid by TRU, the degree of REM, or degree to which they measure all by VAPO is provided. That which is TRU can only be positive for TIT-TO in totality and in resultant.

The transfer of ideas and cultures is seen in mathematical divisions of a whole, e.g. music. This shows the distinction necessary on the one hand with the systematic unity possible from a unified focus on the others. The spread of human knowledge and means, except in the more isolated regions, was a continual and massive process. This was often indirect, off and on, and more or less altered depending upon use in recipient cultures. It is obvious that the rate of this dissemination increased with time and human ability. Transference, imitation, and improvement of ideas, cultures, technologies, and industries in the short, mid and long term provided much of the dynamic milieu for TRU.

Fig 1-71 **COMPARATIVE RELIGIONS and THEIR TRU EVALUATION**

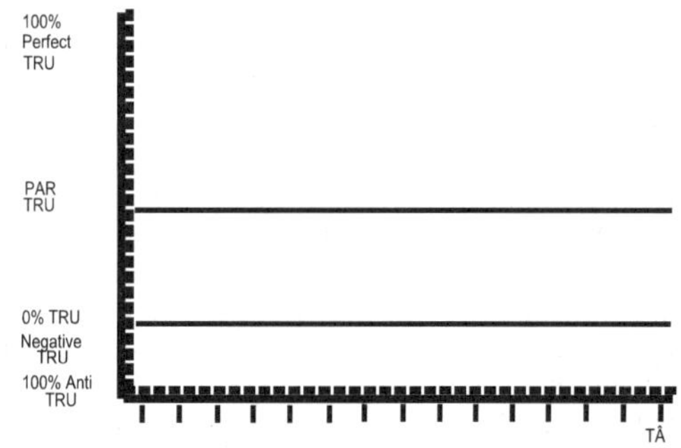

LOPSIT = the Law of Positive Identity with TRU. LOPSIT is a simple way of saying that unless an entity, sense, thing, or all are identified with and how they exist in TRU they are not. LOPSIT also further signifies that religions are or are not to the extent they are identified with TREQ or TRU. TREQ, as do religions, occur in many sizes and degrees of TRU. Ethics follows theology, especially when the theology is by the actuality of TRU. Ontology is largely TRU proper and TIT. It is especially important to show how to incorporate all that is good and great in each religion into a crusade for all in common with TRU and show how each person can best achieve TES. It is often not so much our good beliefs or tenets that are, or are such a great measure of being TRU. It is in how much they fail to put into play and practice total proportional and proper TRU. To adequately assess religions much detail as well as their comparative analysis by TRU must be known, shown and precisely evaluated.

The opposite of specialized form of TRU is **TANTRU** = all that is AntiTRU damaging to TRU itself. The salient features and form of TRU are the real and identifiable entity, a labeled power in society for decades to come. This is: **TRANT** = AT TRU. TRU which is At or In the extant world is a kind of TRU in transit where this transit is a state of applying the principles of TRU into direct identification of TRU as a signified and formal power in civilization today and in the near future.

The ability to turn negatives into positives, subjectively, as seen in the divine birth of Christ by Mary, if even by illusion, has been an important mechanism. It is how to make things nice, as they are in TRU. It is a positive reversal of the human capacity for deception and adversity. Custom is ruled largely by reversion to TREQ, aversion to ANTRU, and adherence to set beliefs. It is what we often do when we equate the good with God and God with the good. To a large extent this is not illusion, it is actuality, the actuality of TRU. So is much of TRUD, HOM and CES, accentuating the positive or turning all to TRU. The better our knowledge and KOT the less we mistakenly attribute positive to negative, and vv. We already know the positive, what is TRU.

It is not so much what great tenet different religions, philosophies, systems, and sciences have that so distinguishes them, or so produces such reaction. It is what they leave out, or fail to include, without proportion and order. Of that which is thought to be <u>sacred</u> we should determine what is TRU and what is not TRU. That which is the same or alike in religions is largely, although variably, TRU. That which is different or present in only one, two or a few is less likely to be TRU, although it may be highly TRU on occasion. Mankind has increasingly and acceleratively found it necessary and beneficial to strengthen those aspects of life that protect from danger. This requires discovery, learning and understanding the world, all that we are a part of, and enhances well being. That which provides unity, value and cooperation among ever growing populations. Continuing beliefs which became sacred, in many forms, and like highest REM, are large parts of human life and a major unifying power in society..

	Right	Wrong
Medieval Christian-ity	Man's life on earth significant only to the extent of one's soul after death	What God and Soul are Not in life, special and completely separate
Medieval Humanism	A person's value in life not death, Great potential and opportunity	What part of a person's life counts or is significant after death
Modern Popular Religion/ Philosophy	Life to the fullest Use advantages Quality of life	Short memory of what is before or after, gain with or without Gods help
TRU TIT TO	All of the above Plus a systematic understanding & compliance with the best. TRU	Only what falls short of what is optimally right, TRU, Not acted upon reasonably.

Much has been right and wrong in Medieval Christianity, Humanism, and Modern Popular religion and philosophy. It must be quickly added that in the world at large there are many varieties of religion and philosophy of life, and these at an expanding rate of divergence. Only what they have in common holds them together. This is often poorly learned and practiced thereby very fragile. It is in the great contrast between the extremes of Medieval Christianity and Humanism and between popular philosophies of life and TRU that we can understand our individuality from TREQ and TRU. The great intrinsic value of a person's life can be properly known only if we understand all the fundamental form and function of TRU with its sameness and differences.(VAPO, Á, DAOST)

What all the great religions of the past and present have in common is, more or less, highly TREQ and TRU. What and how various religions have that is separate and different, and what most often distinguishes them, is, more or less, less than TREQ, partial, negative, divisive and against TRU. Even what are differences that are partial in TIT lead to deterioration. This is called **RIPDA**, or Resultant Inclination of Partials to negativity, ATRU, and Deterioration. It is this that all who hold with such conviction to their religions exclusion of others must discover, learn, realize, remember and correct, if they are to return to that they extol, what is best for all, as they are TREQ and TRU. Because we come from traditions of Islam, Christianity, Judaism, Hinduism, Buddhism, Taoism, Shintoism, Animism, Atheism, Agnosticism, Materialism, or even Scientific Realism doesn't mean that we should refrain from pointing out their weaknesses and errors. For all in many ways depart from the certainty of our actual universe, TRU. They depart in fundamental conceptions, as well as how it should be represented and presented. Most of these religions are largely obsolete in failing to adapt to rapidly advancing knowledge of the past two thousand years. The physical relativity, chemical massive convergent forces with checks and balances, a biological predominance, psychological conscious awareness, personal focus, higher powers of social and human cooperation and all that enables it to be this special universe of best human life must be included and set to order. The old religions were designed largely for an illiterate and simple public. Many of their basic tenets, especially those held in common, are fundamental to TRU. Yet they have been outgrown, they have become misleading and divisive so that not only must tbeir errors be exposed, a new and better system for knowing the actual world, TRU must be available for all. Homation and SPALT hold and prove that all value is based upon TRU, SUT, and REM. Many of the deities held by these traditional religions are little more than illusion and fancy. For whatever is deemed divine or holy that is beyond the confines, or other than, of TIT-TO is to that extent false.

The great benefits and enablement of TRU that best rulers produce are exemplified by those of the past who through religious purpose, intellect, humanity and great administrative ability created domains that were constructive, harmonious, prosperous and highly successful. Benevolent rulers identify with the welfare of their peoples by working for mutual success and survival. It was the principles, purposes, aims, goals, and results, as TREQ, that led to such benefits. It requires a deep sense of learning, wisdom, conviction, ethics, divinity, determination and courage to assure constant and permanent focus and endeavor for the well being and success of human life in a country, when TRU. Yet we must consider the contributions all philosophies, religions and sciences have made to our world, adapt and utilize each as they are parts and processes in TRU. It is such intellectual maturation that is so difficult, yet so necessary, if each and all of us are to rise to challenges and the level of this rapidly expanding, complex, technical, special and universal world we are so much a part.

The lesson is to hold freedom to the dynamics and tolerance within TRU. This requires an ever expanding knowledge of TREQ or TRU. This is called: **IMKOT** = Imperative Knowledge of TRU. IMKOT implies carriers or individuals, who also have their attention on and who are more and more their counterparts in TRU. These are called: **IATES** = Instantaneous Attention TES and **IMTES** = Imperative TES (and TRU). These three, IMKOT, IATES and IMTES are highly correlated and exist more or less together in each person. The greatest heights of learning came from collections by those with great interest in learning who, either as schools, libraries, courts, monasteries, or chuches, championed wisdom and learning. Beginning in the 11th to 13th century a.d. Asian, African, and European universities developed somewhat differently from religious origins. **COIT** = Concentrations of Intellect or Intelligence. With increased population, knowledge and communication there have been a vast expansion and dissemination of intellect, reason and understanding. Yet their quality and proportion are variable and remain in question. The intense inspiration, interest, imagination, ingenuity, and conviction in the cause of wisdom and learning is not self assured.

Nothing is so sad as to see such profound dedication to study, investigation, and the intellect lost. To understand COIT better we can contrast it with its opposite. This is: **COIG** = Concentrations of Ignorance. COIG signifies those regions in which there is an unusually low degree of interest in learning. The expansion of knowledge, and our inability to provide VAPO, TREQ, SCARLS, TROS, TIT, TRIT, OBT, SUT, MOT, and their best exercise has been a leading cause of COIG/COIT.

FODE = Fallacy of Democratic Equality. The belief that the world has to be composed of a precise and measurable amount of equality among its total population is in error. This is seen in the fact that the TES in each individual is not the same and highly variable. For humans are neither created, live, die, nor by immortality are even or the same. Nevertheless we should be careful not to misinterpret and misapply inequalities. All humans are much the same in many ways, much of which are TRU. We must recognize and live by these similarities. .

In much of human life there is the choice to blindly and indiscriminately give in to this or that, or the choice to conscientiously learn and accept TREQ or TRU. This is the difference between being civilized or savage, between DAK and TES. The benefits, characteristics, qualities, and powers that such a system provides is its Gracet, ur the Grace of GOD as TRU. The determination, conviction, persistence, and continuous dedication one has to this grace is our Faith.

There was great emphasis upon the subjective world in the absolute spirit of Kant, Schelling and Hegel**.** This in turn led to an identity or equivalence of the subjective and objective in philosophy and phenomenology, which when TIT, are Á, of THES and THEO. The vast interchange of subjectivity and objectivity, THES and THEO in and out of TRU is called: **THESOIO** = The Subjective Objective In and Out of TRU. THESOIO tends to converge on TRU, as there is an identity of S and O, SOT. this is, **TIDS** = The Identity of the same of all in TRU. TIDS is similar to the S, same of DAOST.

Great discoveries in knowledge, science, philosophy, and religion go hand in hand. They often come from rather simple, yet profound, relationships. These have been vividly shown in physics. Much of this arose from space time relations, formulations, geometries, models, and diagrams. The relationships and interactions between space and time is nothing new. They have been a basis from human thought from Cronus, Cosmos, Apollo, Aristotle, Galilleo, Kant, Maxwell, Michelsen, More, Lorenz, Planck, and Poincaire. Einstein did little more than name, patent, formulate and popularize it. The constant speed of light of light discovered by Max Planck produced a fundamental principle that continues to serve as a standard for much of the universe. This is what we further develop in ALT as Á, the predominant form and function of TRU. It should help to show how through the action of photons on the retina, the greatest of human senses, vision, projects on the cortex impulses whose experience constitutes human mass-energy, Á. It enables us to relate the spaciotemporal dimensions with matter. The extension of dimensions to their higher, 2nd and 3rd powers, and their relationships provide the framework of TRU, TRIT. **TIDUF** = Time Duration Relationship Formula, holds for time as do specific length area, and observer observed relationships for spatial and mass energy aspects of TRU.

Fig 2-71 **THE UNKNOWN and KNOWN TRU**

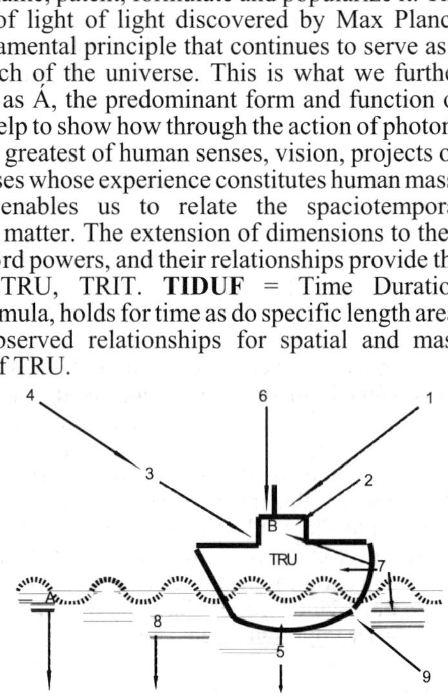

Why hasn't mankind discovered and accepted a knowledge of TRU or TREQ heretofore? This is seen in how we have missed the boat. How humans have failed for ages, for lack of KOT. Some is **UK**, or UnKnown,some known, as ENIT, ESINS and TREQ, some known as ES, yet none together as they exist in TIT-TO. In this figure we show: (A). Why TRU hasn't been understood, (B). How we can prove TIT by the combined forces acting upon it: by 1. Idealism, 2. Realism, 3. RU, 4. AU, 5. Individualism, 6. Lack of total perspective, 7. Lack of proportion of the whole, TO. 8. Lack of imagination, 9. Other deficiencies and excesses. Memorial day should be a measure of how much individuals, groups, states, nations, and the entire world recognize TREQ or TRU and how much they have, do and will provide. The worst of sins is ingratitude, ingratitude for all TRU has given us. This especially applies to all past generations that have enabled us to be. It exemplifies our failure, need and way we may greatly recognize worth in life, ENT in the future, and thereby through all time in TRU.

NET = Natural Ethic, which is not automatic, assured or without cost. It is the activity and behavior of mankind in TRU that most closely follows its actual form and function. It varies somewhat with different aspects of TRU. It is based upon the knowledge of a culture. It is based upon the education, experience, maturation and wisdom of its members. It is based upon freedom from ATRU conflict, stress, strain, pressure, overpopulation, distraction and many others.

Each person should, by the best means necessary, as early and as completely as possible in life, attain complete identity and endeavor with and for one's soul, TES. This is a fundamental law of our being. **LINTES** = Law of Necessary Identity of an Individual, their soul, TES and TRU. LINTES is for each and all of us. If the sacraments are inherently efficacious and indispensable to salvation and capable of conferring grace on our souls, as is practically and communally evident, then they should have whatever role needed for TRU. This is providing we do not dilute or pollute their underlying TRU meanings .(PRÁ) Much of nature, as all of human life, is full of the representative, symbolic and spiritual meanings of the in-evident, insensible, holy, eternal, and TRU. It is reasonable to accept tentatively any experience, revelation, representation, concept or principle that enhances identity with TRU.

Fig 3-71 **DEVELOPMENT of a FEELING of RIGHT from WRONG, GOD and TRU**

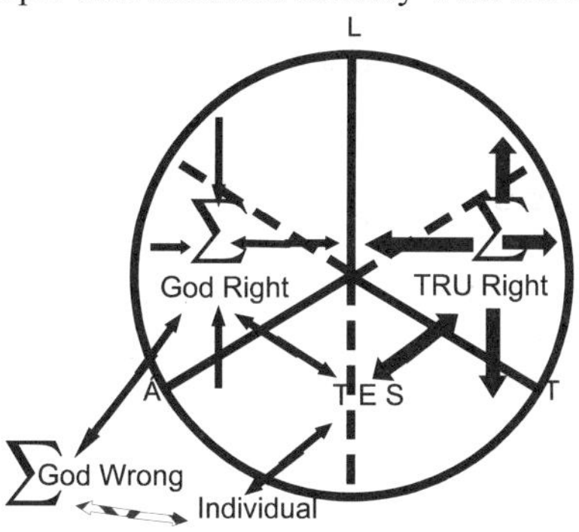

What may seem right and good may often not be, or it may be much less or other than it actually is in TRU. On the other hand bad or wrong may not be so negative for TRU. In fact it may carry potential that is positive. Thereby to best learn and understand right from wrong we, as always, must rely on KOT and TRU. Humans have an inner sense of what is truth or falsehood, correct or incorrect, good or bad, and right or wrong. It is a feeling of well being, what is salutary and congenial. It makes up a large portion of many Á levels of the conscious. It is largely a product of the focus and internal convergent forces of a person, group, or specie. Only when what is different, in competition or opposition, is this feeling of sameness or relation lost. For humans it is developed from sameness in FAF, what of Á is common, in genetics, life and related experiences. This is present in all humans, more or less, in proportion to their sensitivity, judgement and personality. This is a power that gives value and REM to what they encounter, thereby TRU, TES and themselves, indirectly. It is the basis for the moral issue exemplified by identity with Christ. It is the distinctions, discretions, higher powers of feeling and mentality typified by ethics that stem from basic feelings and consciousness that provide much of the dynamics, force and power of TRU. Each person tends to dedicate their lives with the causes of right over wrong.

A complete and balanced religion, if TRU, should give to each of its adherents a formula, guide, model, plan, map or accounting by which they can not only best understand all things, TRU, but be able to solve the problems that will be confronted in life. Only in this way can we always make the right decisions, have the knowledge necessary, and find the proper solutions to each and all problems.

Laws, legislation, regulations, deregulations, rules, even conventions, customs, routines and habits with their uses and abuses, are continually accumulated. Yet when, where, and how do they stop? Is there any end to the restrictions and demands an ever more complicated life is placed upon us? Only with TRU or TREQ, its freedom and release from bonds and abuses of unnecessary laws and all, can this vicious cycle be set right. With TRU a continual discovery, application, roll over, alteration, or elimination of laws and all provides optimal TRU. This is why KOT and KUPS are so important.

There are many injustices associated with litigation, what is criminal, and crime and punishment in general. This is like there is no honor among thieves, or how the negative has no similarity. Those who commit crimes are not to be thought of in a simple singular manner. To be or not be guilty of major or minor crimes is often little more than a label of injudicious procedures and vindictive ATRU forces. **AJUT** = Adverse Judgements, litigation controversies, injustice of adversarial relationships when not strictly TRU. There are many biases, prejudices, and forces and lack of VAPO of TRU in operation both within and without the legal world that reveal the great dynamic interchange between the positive and negative of TRU and ATRU. They are best resolved through moderation, toleration, and a full accounting of and by TRU.

SIC = Self Interest Compounded. As groups, and nations bond together, for self protection and mutual benefit this rather than acting for the success of all produces compounded failure in all. It may be expressed in: **(Partiality)**^{n}**.** By accentuating and multiplying the partial at the sake of the total, it becomes SIC. Only when this is perfectly balanced in the TROS of TIT-TO is it compensated for and corrected. Lack of capability, ability, and integrity quickly turns people off. If one is to be served, treated or cared for and the one providing this is incompetent a red flag appears. Anything that indicates that one does not know what you are doing, are not giving exactly what is needed, or is the best, reaction is rapidly produced. **LONSEG** = Law of Needs can Not be Supplied by what is incapable, negative, wrong or damaging. LONSEG is so common that it may seem obvious. Yet it is critical to each and all persons, especially for TES and TRU. It is in survival and the prevention of all that is a threat to this survival that is so critical. **TRINT** = TRU INTegrity, of being supplied, built by and including the positive and excluding the negative. TRINT shows the necessity of developing and practicing, to the utmost, PUREALITY, TES, TRU or their equivalents. Only by knowing what is so is TRU, and can separate all that is not so, can we achieve TRINT. Yet all humans need to have the same degree of demand for TRU, integrity, perfection, precision, understanding and work in themselves as they do so insistently from others and those who serve them, there would be a fast track to TES and TRU **STÔP** = the Sterile State of Principles, concepts and laws needing Production, Promotion and Propagation. It is one thing to mean something realizing an idea, concept or principle, as in TRU; yet it is quite another, to develop a good word, definition and communicable concept. It is because of the reasons associated with STOP that so many NEC, much of IPTRU, and their expansion and promotion, are provided.

Fig 4-71 CES of DYNS in SULP SECTOR of SPALT

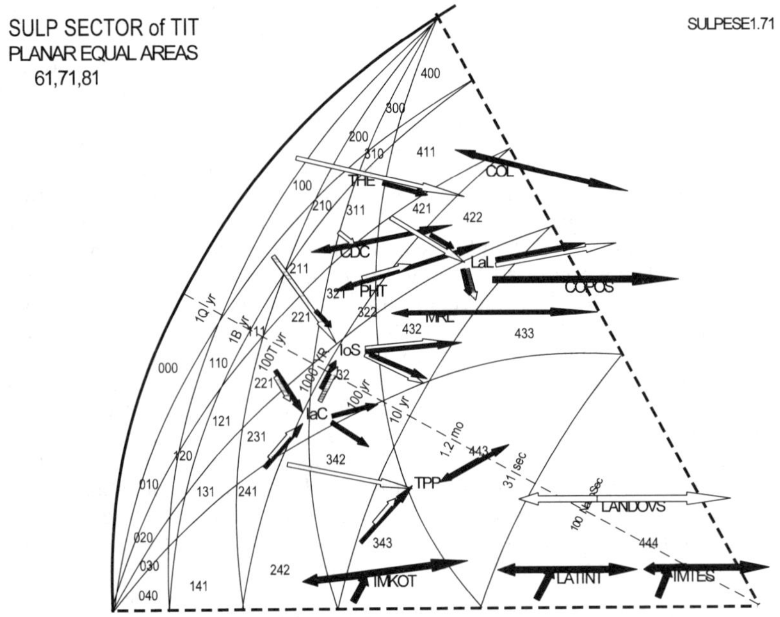

MEASURES of DYNS CES in SULP SECTOR of SPALT

AJUT	3-4-4~	-T
CENTERS of LEARNING,		
COIT	4-2-2	S
CHADA	4-4-4	M
COIG	3-2-2	-M
COIM	3-2-2	-M
COPOS	4-2-4~	S
CULTURE - DISORDERS of TH.		
CDT	2-4-3	T
CULTURE and DYNS CONCEPTS		
CDC	2-2-1	S
ERRORS, ISMS SYSTEMS and		
EIC	3-2-2	- T
FACT - FICTION	~4-4-3~	T
FODE	4-2-2~	-M
GOD, GRACE, TES	3-2-4~	T
IATES (75)	4-6~-5	T
IDEAS and CULTURES		
IaC	3-3-2	S
IMTES (75,78)	4-4-4	S
IMKOT (78)	2-3-4~	S
INTERESTS of SYSTEMS		
IoS	3-3-2	S
KNOW.-PRES.-FIDEL.-		
TRU-IPT KPF	3-2-~	M
LANDOVS	4-4-4~	S
LATINT	4-4-4	S
LAWS and LEGISLATION		
LaL	4-3-2	S
LONSEC (73-77)	6-4-3~	-M
LOPSIT	~4-~4-4~	T
MORALITY, RELIGION,		
LAWS MRL	4-3-2	
NEGATIVE to POSITIVE		
NtP	~3-4-2~-	T
PHILOSOPHIES/TRU		
PHT	3-2-2	S
REL. GOD GRACE PERSON TES,,		
RGG	3-2~-3~	-M
SACRAMENTS REPRESENTA. TRU		
SRT	4-2~-4~	M
SIC	3-5-3	-T
STOP	3-4~-4	-M
TANTRU	~ ~ ~ ~-	M
THEOGONIES	3-1-1	S
THESOIO	~4-~-~-	M
THOUGHT-PROBLEM-		
PROSPERITY TPP	4-4-3	S
TIDUF	~-~-~-~	M
TIS	~4-4-4~	T
TRANT	~ ~ ~ ~	M
TRINT	~4-4-4~	T
UNKNOWN & KNOWN TRU		
UKT	1-1-2~	M

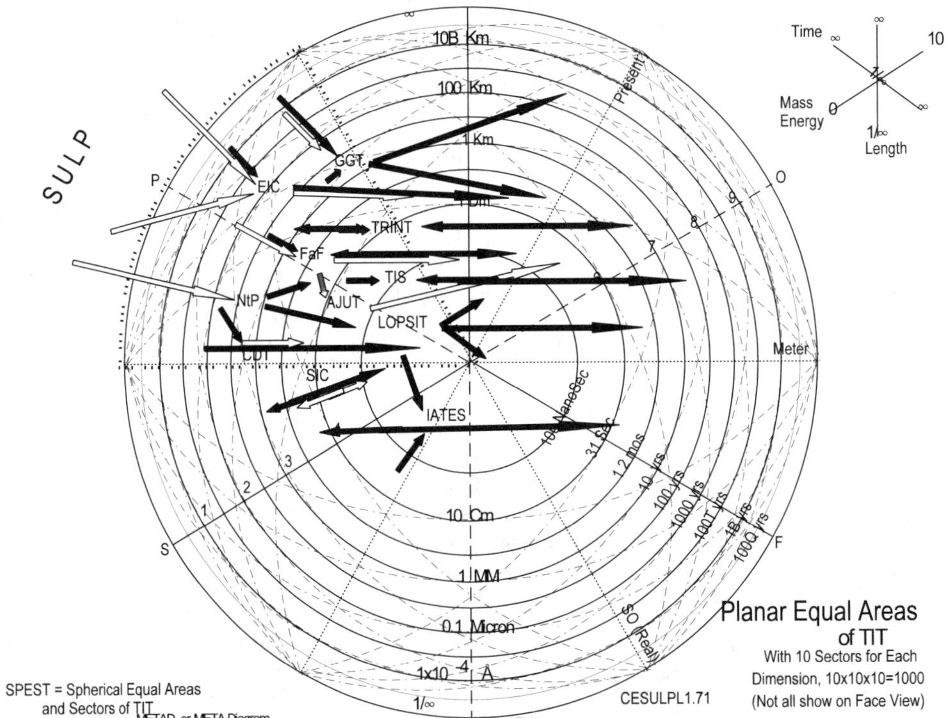

Fig 5-71 **CES of SULP of DYNS in SPALT**

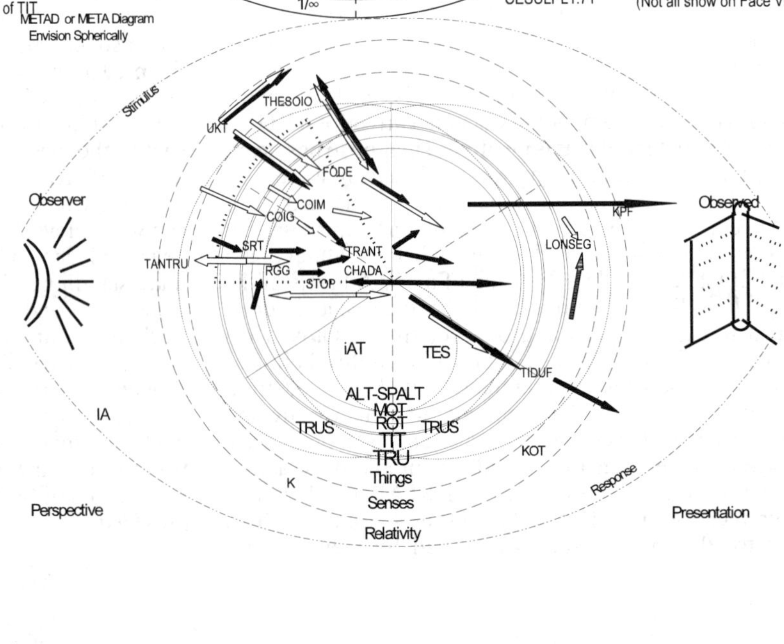

Fig 6-71 **CES of SULP of DYNS in METAD**

CHAPTER 72

OLP or OBJECTIVE LARGE PAST POSITS of DYNS

CLIMATE, ECOLOGY, CULTURE, RESOURCES, POLITICAL, ECONOMIC, SOCIAL. NEEDS, IPR

Just as in relativity, everything that exists in the astrophysical universe exists not only as entities, senses, or <u>things-in-themselves</u>, but as produced by forces, one of which is gravity. Gravity is the relationship and compelling force produced by the mass attraction of material bodies. There are many forces in the physical world, such as the strong force, the weak force, electromagnetic forces, etc. For us, as demonstrated in our relativity to all, the observer/observed relations, our world consists of <u>all the forces that actually hold us together and enable our being</u>. This is what TRU is formed from, and makes up its primary form and function, its mass energy continuum and dimension. Along with the spatiotemporal it provides us with the frame and structure by which we can make TRU the object for representation and modeling.

Gravity varies with elevation in a set proportion, **g = k (1/E).** This is related to the inverse square law or **SIL**, which is of fundamental importance in explaining the differential exponential make up of the world, TRU. It should be remembered that the dimensions of TRU although largely exponential, are adapted, altered and warped in actuality. Gravity as a physical component of Á defines the limits and functions of a certain level, SÁ, of the Á dimension. The constant role of gravity on earth, its change with elevation and humans adaptation to them, are all forces that produce processes which combined have, are and will shape some of the more physical aspects of human FAF, TRU.

Variations in the earth's environment are largely produced by cyclical or noncyclical, periodic or non periodic variations in astronomical and geophysical forces and relations. The effects produced by the raising and lowering of the earths temperature both within known time periods, as the diurnal and yearly periods as well as much longer periods are good examples of these variations and cycles. These have a profound effect at many times, durations, and variations in human life and TRU. They can be of the widest or narrowest time frames and may have complex reactions. They are related to sleep, activity, light, clothing, housing, and many others, thereby successfully adapting and selecting in life. The periodic changes over much longer durations of time produce somewhat different types of, yet significant, effects. Many less noticeable, cycles of roughly 14, 206, 835, 5,700, 21,000, 92,000, years have had important effects on life. The 14 year cycle is that of sun spots. The 206 year cycle is that of the sun's intensity. The 21,000 year cycle, and probably a major one, sufficient to produce marked changes in sea level is due to the earths distance from the sun. All these, others, and non cyclical causes have been, in altered ways, factors in the past, present and future. None under 5,000 years have produced such marked changes in sea level. In part, and to a limited extent, this is because glacial melt takes time.

That man, and TRU, are very much a product of their environment is well known and repeatedly proven. The FAF of animals to a great extent conforms to their habitats. This is by behavior and adaptation physically and genetically. Changes in the environment are greatly a product of climatic changes, temperature, precipitation and others. From the beginnings of bipedalism when the forests of Africa no long provided the environment for survival, hominids wandered the bush, plains, savannahs and deserts to all ends of the earth in search of better habitats and all needed to protect, produce, sustain, and preserve life. It is not surprising that the world should have developed diverse varieties and races. It is also not surprising in movement many diverse peoples often inhabited the same region. The discovery and use of stone and other material for tools has been a major factor in human development and DOT. Along with the discovery of human remains they have been a reliable source of information about human activity and evolution. It is evident that the advance of tool-making was largely, and variably, accumulative. It was also selective, so that some ways or working existed only within relatively narrow periods of time. The use of tools, and stone in particular, was a large factor in providing the competitive advantage and predominance necessary for survival. Material and tool industries expanded, and expanded in turn, socialization. They provide the evidence needed to trace through many of the changes in parts, and processes that have occurred and can be posited by DYNS in SPALT.

The accelerative development of highest levels of Á FAF along with the accelerative development of an ever increasing array of material and products are evident for humans in the last millions, hundreds, and tens of thousands of years for adaptation to TRU, in DYNS POS in SPALT. The use of many diverse materials by many diverse cultures, as barley in the Egyptian late Paleolithic, in the Aswan region, between 18,000 and 13,000 bp, and others proves the long rise and development of food collecting and food growing cultures in the near east. The existence of cyclical climatic changes on Nile river flooding exemplifies the changes that occurred in non polar regions associated with the end of the last ice age.

Early irrigation agriculture shows human beginnings of an altered habitat that improved production and even provided surplus yields. This generated a freedom to pursue numerous other activities. It began on a small scale and expanded into the great civilizations of Asia and Egypt. It illustrates the continuity of much in TRU. With marked increases in population there evolved many of the economic, civil, and social advantages, benefits and problems, that followed.

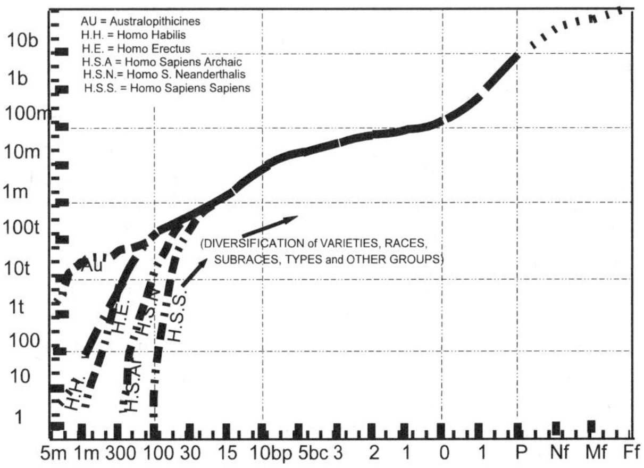

Fig 1-72 POPULATION with TIME, DYNS, TRU

Human population changes throughout existence on earth help us learn the CES and thereby FAF of TRU. The total population of man on earth at different times from the remote past to the present supplies important data on the effects and the processes of man, CES of TRU. Combined with this we have the total population of all ages since hominid and anthropoid origins. A rough date of approximately five million years in East Africa may be used as an initial time and place for measuring human population statistics. In the earlier periods population was, although variable, much smaller. In fact with changes in genus, family, species, subspecies and race there were undoubtedly times of genetic alteration that involved only a couple or a few individuals. However over the entire time of human evolution of form and function, the incremental accretion of mass genetics and populations, and thereby TRU, were the more usual. Some families, genera, species, subspecies, varieties, races and subraces fit in different regions that provided a highly selective, and usually complex or massive progression or recession of populations. Just as there is on earth today a great variety of types people, so it was more or less so the further into the past we go. However, human descent in direct lineage, that excludes many, even close, relatives that existed at a time. Much of this is done in the dynamics of altering isolation and congregation.(ID)

It is not so certain to predict accurately what these will be in the future, although some general principles can be inferred. The future will depend, not so much on excessively large populations as it will on the quality necessary to sustain and improve GELF in TRU. For quantity is highly dependent upon sufficient levels or degrees of quality. Quality/Quantity will continue to be a large factor and force in HOM, TRUD, TRUF and TIT. Many corrective factors need to be included if figures of population growth are to be reliable, including how the population participated in TRU and to what degree.

The accelerative propulsion by TRUS and TRUC of TRU, TROS and TIT is more than singular and multiple, simple selection. There are built in processes and mechanism whereby the future is predetermined by the present and past. This comes from within, the momentum and power of the Á FAF of TRU. What TRIF and TRUF become is inclined and compelled to fulfill its needs and is of maximal benefit This may change somewhat with reactions to actions outside of TRIS and TRUS, but will tend to maintain the status quo of TIT, or at least of its qualitative rather than quantitative aspects. These predetermining mechanisms and powers are the accelerative propelling and compelling forces that work beyond what is customarily recognized in natural selection.

It is in the cultures, societies, nations and civilizations that are most capable, vigorous, selective, best confront and prevent their basic problems, as hunger, disease, illiteracy, poverty, strife, turmoil, war, and misery, that are more apt to succeed and survive.(VAPO) **LACFON** = Law of Continuing and Cyclical Forces of Negation. LACFON signifies and emphasizes the regular and irregular forces that produce adverse effects on TRU. LACFON shows how it is impossible to have success without adequate response, reaction or control of basic problems. New groups of peoples, whose genetic and physical health enabled them to predominate repeatedly fused or displaced less fit populations. It was their ability to adapt to LACFON that largely determined this predominance.

The index of sufficiency of a culture and the accumulation of factors that make it more or less efficient when well defined give a good measure of its potential success and survival. Some of these are from language and communication, their clarity and strength. **CIS** = Cultural Index Sufficiency. CIS is a summation of factors and functions in a culture. It shows how necessary it is for the individual in a society, and culture to be of high quality, TES.

There is a continuous massive effect of forces upon us that work to increase our health and quality or eliminate the less fit among us. The species survives by a small edge of proficiency and predominance. It is a proficiency that requires all levels of the mass energy, Á continuum and dimension of our being. A specie survives if: 1. Its capacity and ability are greater than the deleterious effects upon it. 2. Its compen-satory mechanisms can make up for all revocable effects.
3. The specie does not outgrow or expand beyond the limits of its existence. 4. Quality, fitness, predominance, and ,energy, are maintained. 5. All vital segments of its being are not overwhelmed by adverse effects.

The importance of trade and commerce throughout history demonstrates many of the combined and complex processes in TRU. By tracing out the trade routes and by demonstrating products that were not indigenous in archeological sites we can vividly recognize what was occurring at a time. The overall kinds, principles and detail of trade have been increasingly understood and used to provide a virtual reality in time. Articles of trade were exchanged that reveal the relations, influences and processes of man. Moreover it was the ideas, methods, other general vehicles of interaction, including intermixed populations. Just as the medieval fairs were much more than the interchange of goods, so was trade in all ages, more or less, of help bringing people together to exchange the different aspects, parts and processes of TRU. As trade becomes more effective so does economy, which in turn increases cultures and TRU. This assumes trade that is of commodities that are largely worth while, REM, TRU.

If inventions, ideas, motifs and fashions were original, they did not exist before their discovery. Many had more than one origin or originated at more than one date. Some were discovered only in part, often very small, only to be rediscovered and added to at a later time. It is this rediscovery and renewed interest that constitutes a large part of the dynamics of TRU. This is called: **REDIM** = Rediscovery and Renewed Interest in Multiple Discoveries, Inventions, fashions, Motifs, and Ideas. It was more by trial and error and not constructive. It is becoming more positive for TRU.

What the best social systems and what qualities are associated with them, can be tabulated. It is not so much what kind or type of social system or state we live in. It is what the need is, how they are interpreted, used, abused, and how many of these different kinds of systems have ESINS, concepts, principles, and laws that are TRU. The type of social system or state is only a means. The needs and provision for these needs is what is most indicative of their TRU nature.

LIH = Laws of Hegemony, Laws of International Hegemony. America's great mistake or failing in the 20-21st century has been its inability to provide the necessary world leadership that was expected and offered. **LILT** = Law of Imperative Leadership TRU. That leadership is not only necessary, but must be of a certain kind, is known. That this must be by TRU is less known, yet even more imperative. All exists in TIT and will require much from others and the rest of TRU. Different forms of government and social systems are often better for different nations and different stages in their life cycles. The great similarity is what they have in common by TRU. History is an endless tale of the misery caused by nations and groups who - well intentioned at least at first - sought to impose what may have worked well for them, onto others. To maintain liberty there must be responsibility. This is why totalitarian states and high levels of principles and intellect have a negative correlation. They tend to stifle freedom of thought and interests. On the other hand democracies that are made up of naive and narrow minded and self-interested people are likewise negative. The whole world should not only condemn and sanction such states it should investigate and account for what is wrong, why and how TRU is missing. All must have great determination and foresight to bring world politics and themselves as TES, to TRU As corollaries to TRU and LAT or the Laws of TRU and Laws of Homation we have LIH. Leadership is not only limited, it should be perfected. It must, by LILT, be TRU.

A most important question and problem in social harmony and success is whether an individual conforms because they have to or because they want to. **CONFAOR** = Conformation, Form, Order of TRU. CONFAOR is what overcomes failure to adapt, act, or be TRU. When the individual is TES and the laws, rules, and customs are the best, TRU, questions and problems are removed. To the extent we are not TES and our laws, routines, customs, and all are not TRU there are neither answers nor solutions. The breakdown or buildup of society and the world, TRU, proceeds directly from how well each and all persons learn, know, and use, and how much and their customs, rules, laws, are from, TREQ or TRU.

There are multiple purposes for resources, buildings, institutions, and all, both natural and man made, that make up a successful civilization, or TRU. Likewise there are many means and extremes or levels in the degree of ownership, be it of territory, property, goods, or resources. It is not how much one owns that counts as how it is used for TRU and TES. There is a continuum from complete TRU to complete ATRU that makes up the world we live in. There is also a large dynamic interaction between the various levels or **RACTAW** = Ratio of Class gradients in TRU to that in the world in which humans live. The forces that increase TRU and decrease that in the NAMTRU, ATRU and ANTRU worlds are the great forces positive for TRU.

No highly successful culture or great civilization occurred in the past without having been built on a solid foundation. This included much of what TRU is composed; a vigorous and intelligent population, proper mix of individual freedom, security, good health, and most of all the genetic capacity of a population that enabled them to achieve. If we seek to assure such civilizations for the future, it is in these basic strengths and the natural predominance of human life we must look. These and the many other characteristics, properties and qualities of its population, as noted throughout the text, in the ES and CES of TRU are what gives hope, and what all must extol. It is what we must comit to, and how perseverant we are, which in summary is no less than the measure of each and all persons, TES.

It is of the first order of importance that persons take care of themselves. Of similar importance for their success and survival is to take care of those and all the world around them that are or can be TRU. There are several qualifications to this rule. Some of the more significant ones are, those who are either by birth, infancy, childhood, illness, misfortune or senility unable to care for themselves. We can't be masters of all trades, nor in the division of labor of a highly civilized society without the many limits to what each and all can do or take care of. At different ages, stages and conditions of life there are great variations in the capacity, means, and type of contribution one can make.

Many different kinds of information, data and statistics are of great importance for many different uses in TRU. These are what they directly relate to or knowledge that is adapted and interpreted in terms of TRU. The statistics of nations, their populations, areas, density, land under cultivation, potentially usable land, climate, gross national products, per capita incomes, and many others all help in making decisions and planning of many kinds. Then we must, as for all more or less, adapt to TRU. **PCI** or Per Capita Income of different nations, regions, and peoples is, with correction, good measures of their prosperity, economic success and standard of living. This has a modest correlation with TRU, yet needs much improvement. All the great advances in knowledge, invention, transportation, science, technology, industry, commerce, and other means of life do not always supply the purity of basic needs nor lead to the quality sufficient for TRU.

LIRE = Law of Inflation as a function of Over REgulation, direct or indirect. The relationship between inflation and excessive controls, and ATRU is evident in an economy. Although inflation is caused by many other factors as excess demand/supply, it, for LIRE, results from not granting or enabling inexpensive, competitive and ready access to supply.

A serious problem that has plagued humanity is lack of authority, quality of authority, and respect for authority. In the drive for success either justifiably as in a spirit of competition, or unjustifiably as in self interest and love of wining at any cost, authority increasingly depreciates. Nations with the best, or TREQ, authority have much less crime and social unrest. If people do as they are told, and what they are told is the best by TRU they do what is right. A form that is to rigid is not entirely right, or if it is imposed invariantly and disproportionately, especially of TRU is equally destructive. This is called: **LICY** = Life Cycle Limits, or Limits of Life Cycles. LICY is where the interaction of citizen and the state, or the part and TRU must be governed by the nature of each as they are TRU and as they each function in their way interrelated and interdependent. Politics is not TRU except to the degree that TRU is also composed of an aggregation of its parts, and these parts may stand, as D in DAOST, in opposition or dynamic dialectical relation to each other. Yet the trouble with political systems is not their polarities in TRU, it is in what different segments of population and power groups stand for and how this power accumulates and is used without TRU. The collective lack of VAPO by each successive administration causes much of the wrongs each election, administration or regime uses, has, or tries, to correct.

IPR or Interpersonal Relationships like business relationships involve several factors: professionalism, cost, TRU or non TRU, right or wrong, wisdom or ignorance, and fair or unfair exchange or rendering. For a system to work there is an optimum or ideal mean of each of these. **IPREC** = Interpersonal Relations Conflict(s).(ERUC) IPREC is the degree to which mentality and TRU overlap and the ratio of ATRU to TRU. Much in the world is routine, played out in a business like and an optimal manner. This is one in which IPR are predetermined and leave little room for ATRU conflict. There are in all human relations, as even within the individual and TES, degrees of agreement or disagreement, friendliness or unfriendliness, selection, right and wrong or TRU and ATRU. It is in these interactions that so much of what is good or bad for TRU transpires. These are IPREC.

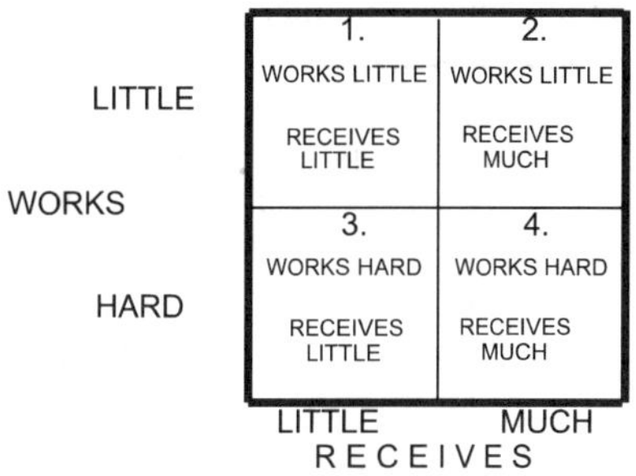

Fig 2-72 **CATEGORIES of WORK / RETURN FROM WORK RELATIONSHIPS**

Repetition may seem bad to ones IA, yet it is a great aspect of the sameness of TRU, its TU and TRUD. Depth conception, scientific research, and many other activities in life have great strengths provided by the use of repetition. We can chart relationships between work output and benefits

1. WORKS LITTLE and RECEIVES LITTLE
2. WORKS LITTLE and RECEIVES MUCH
3. WORKS MUCH and RECEIVES LITTLE
4. WORKS MUCH and RECEIVES MUCH

Generally only the last category, both working and receiving much is mostly TRU. Yet they depend upon only a very special, or TRU, kind of both work and what is received.

The greatness of mercy and charity is not so much the relief people need when they are suffering as it is the even greater need to give to them what it takes to bring them to, and to be, TRU. To better enable them to be and live TES. This is a principle, intentionally or inadvertently, religion uses to assist those they can reach. This is especially so when both are, yet for some reason, can't be TREQ. When they can be positively enabled to improve, we achieve the highest form and ends of mercy and charity. We must be generous with our mercy and charity, just as we must be vigilant with our learning, conviction in, love for, and determination to achieve TRU. This highest form of mercy and charity, MAC, must be clearly recognized, and of a generosity that is selective by TRU. For only in the understanding of TRU can this recognition and generosity be QUOL or of the quality, purity, reality, and actuality that should be in all we know and do. The Lessons,

LIMT, which can be Learned from our Identity and Maturation in TES and TRU, are of the greatest dynamic processes of TRU. It's a ratio between individuality and survival of the fittest with sacrifice to TRU. It is not an easy lesson to learn. It requires a labor of love and great sacrifice to TES and TRU. Yet the rewards are much greater if one is willing and knowledgeable enough to have the determination to achieve TES and TRU. This ratio is: **RASTIN** = Ratio of Sacrifice for TES and TRU and Individuality. Just as Jacob struggled with the Angels and God, to become Israel, through his firm conviction and devotion to God, so do each and all of us in life achieve TES and become one with TREQ or TRU when RASTIN is optimal. We do so by best resolving, purifying, and refining RASTIN in our lives by learning, knowing, remembering, using, doing, living and in all ways being TES and TRU.

RACHAS = Ratio of Christian Love, Spirit, Charity and Selflessness to Selfishness, Self Aggrandizement, and greed. The worth of religions, sects, groups, and individuals are simply and quickly measured, by what they do, how they handle money, and what their work is for. Is it for the benefit of TRU and ENT? Much of what one works for and much that one takes from others is not meant for oneself, but for ones own, others in need, ones family, community, and world. **RUSEM** = the Ratio of the Use of Money, the Ratio of the Right or TRU Use, to what is wrong or Misused. RUSEM in the use of money is ultimately what one takes is less or no more than what one needs for TES and TRU, and how that used is, to that extent, for TRU. This is a very great and powerful measure. It is used every day. It is very much in the back of all peoples minds in valuation, doing their work, shopping, and in their relationships with other people.

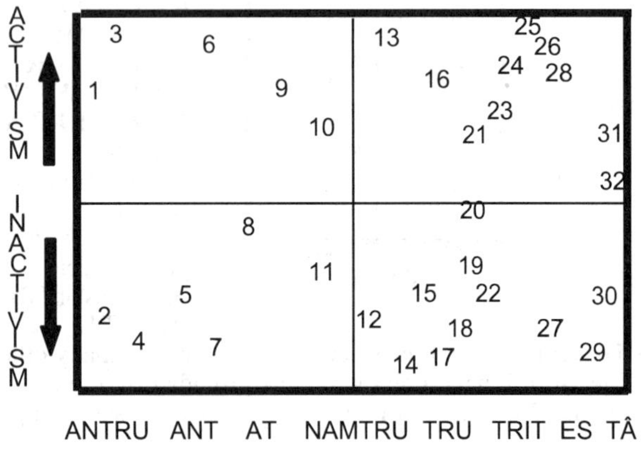

Fig 3-72 **ACTIVISM and INACTIVISM by ANTRU to TRU**

Most activists and in-activists can be expressed in terms of their products, how much they benefit or destroy TRU. The numbers used to correspond with the different kinds of peoplewe encounter are: 1. World Destroyers, 2. Irresponsible Leaders, 3. Criminals, Terrorists, 4. The negligent, 5. The Ignorant, 6. Radicals, 7. Reactionaries, 8. Average persons, 9. Liberals, 10. Conservatives, 11. Behaviorists, 12. Deists, 13.Works for Good, 14. Theistic Pacifists, 15. Rationalists, 16. Ethical Realists, 17. Christians, 18. Intellectuals, 19. Monotheists, 20. Par for TRU, 21. Christian Workers, 22. Godly or martyrs, 23. TREQ, 24. PUREAL, 25. Accentuation 26. TRAX-TREX-TRIX, 27. KOT, 28. TES, 29. COGNOS, 30. ELTICMA, 31. SÂ, and 32. TÂ. In some of these types of individuals or groups there may be a wide variation of both activism-inactivism and degree of ANTRU or TRU. We have endeavored to center each on their average, especially as they relate to TRU. The benefit to be derived from such some rough schemata lies in providing a ready overview for each person of all, both themselves and all others. It helps us appreciate the fact that we and those about us are, and are not, what we might ordinarily suspect. They are only as TRU, and to be TRU, requires both activism and inactivism depending upon the occasion, circumstances and relations to the whole.

In order to relieve suffering we must relieve poverty. In order to relieve poverty we must relieve supply and greed and provide justice. In order to relieve supply and greed we must know values and sacrifice. In order to provide justice we must know and apply God, TREQ, and TRU. In order to know and apply TRU we must have the capacity, energy, determination and love to see it done. We must know VAPO, DAOST, TROS, SCARLS, and especially appreciate the unity in sameness and tolerance in diversity that TRU is composed.

LANONT = Law of Non TRU TRU. LANONT states that what is not TRU is neither better nor worse than something else that is not TRU. For each in being not TRU are in a non being to being relationship which is zero and thereby immeasurable. This changes only when an effect on TRU occurs, or when persons of very high TES are caught up in ATRU circumstances beyond the best of TRU control. Fair and equal justice can proceed only from TRU, VAPO, SCARLS, TROS, TRIS, TIT, TRUD and all the laws of TRU. **LEDMOP** = Law of Equality of Degree and Manner of Justice and Punishment with degree and manner of crime. LEDMOP applies only within the limits of TRU.

If you can't keep up drop out. This is the continual trial each and all have, are, and will face. The main question, however, is keep up to, and what this is. It is what is or is not TREQ or TRU. It simply amounts to avoiding impairment of TRU, certainly if you can not contribute to it. It is like going with the flow, fulfilling ones role, being adaptively coordinated, and in moderation holding to the optimal pace.

Responsibility falls on the shoulders of those that best carry it. They are often given responsibilities and work far greater or different than those who should be carrying it. Or for those individuals which may not, but should be, doing for themselves. To be Machiavellian, where the ends justify the means, is mostly only if limited in TRU. Yet, each must, to the best of their abilities, be TES and TRU. Through the ages structure after structure, group after group, and state after state, have become dependent upon simple rules and routines with poor direction of work and responsibility that lack VAPO, and lose sight of the more complex and what is best for the whole, TIT-TO.

MEASURES of CES of OLP of DYNS

ACTIVISM & TRU AaT	6-2-2	S	LICY	8-4-2	M	and TRU	8-2-2	T
AGRICULTURE & POPULAT.TRU	9-2-1	M	LIH	8-1-2	S	POPULATION/TIM	9-1-1	S
CESINS	-~-~-~	M	LIRE	8-1-2	S	RACHAS	6-4-3	T
CIS	7-1-1	S	MERCY CHARITY MEANS BEN	7-3-4	T	RASTIN	5-4-4	S
CLIMATE, ENVIRONMENT, TRU	9-0-0	S	NATIONAL ATTITUDES			REDIM	5-4-3	S
CONFAOR	8-3-2	T	S/F TRU	5-4-2	S	RUSEM	8-2-2	T
EKTRU	~-~-~-~	M	NATIONAL VALUES,			SIL	~_~_~_~	M
IPR	5-4-4+	S	COMPARAtIVE	8-1-1	S	SYSTEMS, KINDS and QUALITIES	7-2-2	T
IPREC	6-4-3-	M	OVEX	7-2-2	S	TOOL MAKING, TIME & CULTURES	8-3-1	T
LABOR DISTRIBU. CARE TRU	8-4-3	T	OWNERSHIP OPPRESSION			TRADE, COMMERCE and TRU	8-1-2	S
LACFON	~_~_~_~	M	BENEF	9-3-2	M	WH-words	5-4-4~	M
LANONT	~_~_~_~	M	PCI	8-4-3	M	WORK and RETURNS or RECEIPTS	8-4-3	M
LEDMOP	7-4-3-	M	PHILOSOPHIES,GOVT,PCI	8-1-2	T			
LEDMOT	6-4-2	T	POLITICAL SCI, VAPO					

Fig 4-72 **CES of DYNS in OLP SECTOR of SPALT**

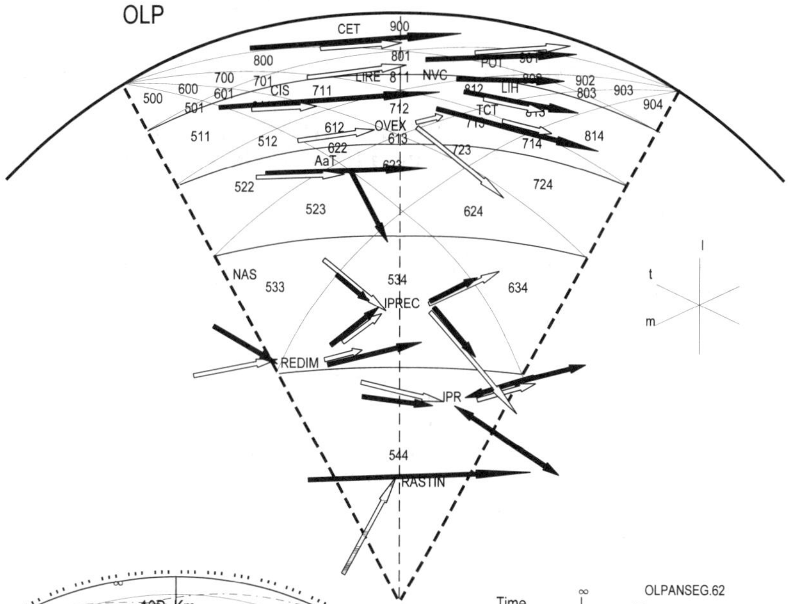

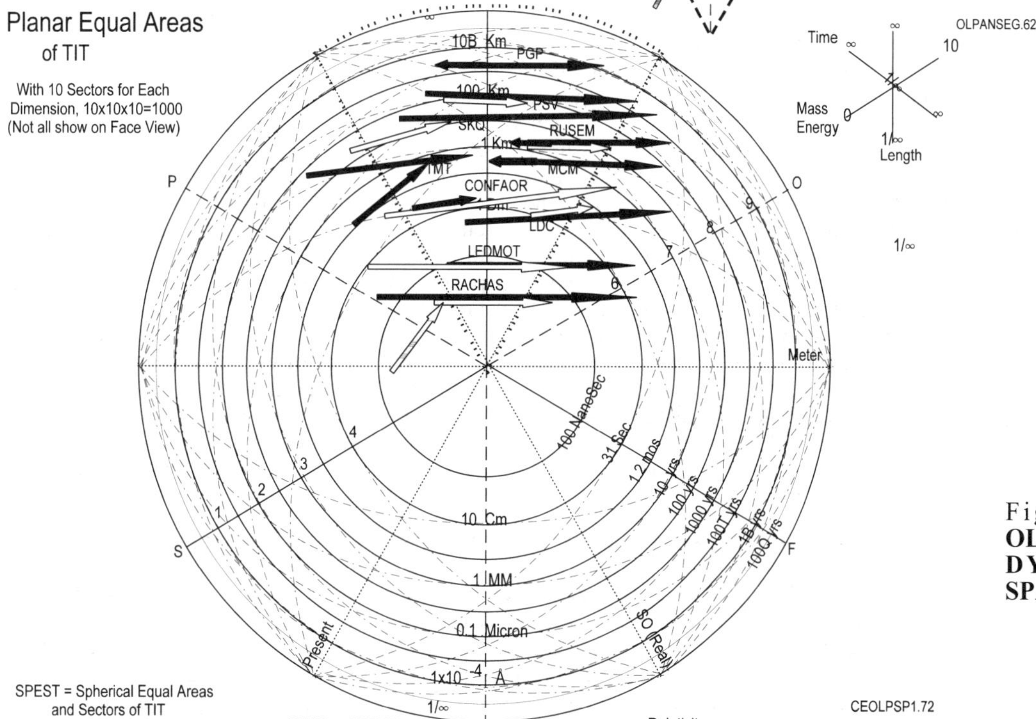

Fig 5-72 **OLP CES of DYNS in SPALT**

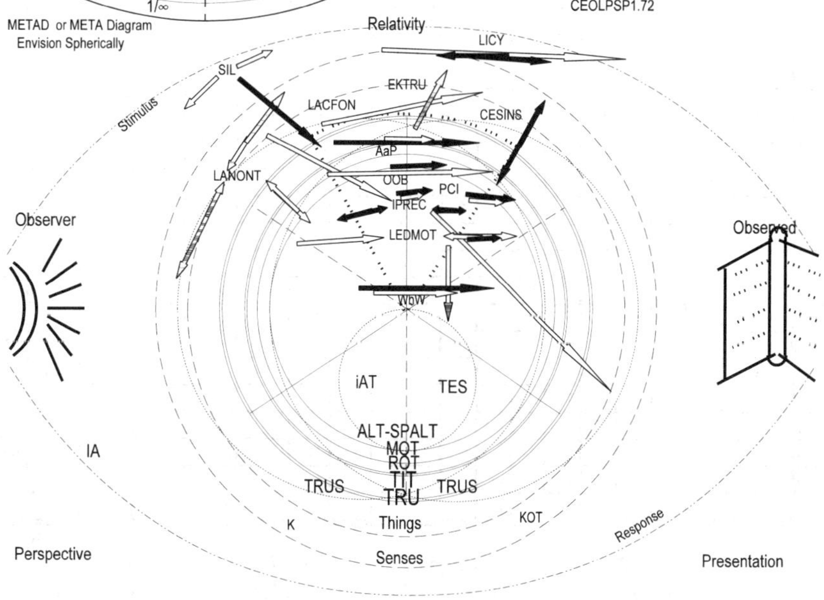

Fig 6-72 **OLP CES of DYNS in METAD**

CHAPTER 73

OSP or OBJECTIVE SMALL PAST PROCESSES of DYNS

HEREDITY, HOMINID, FERTILITY, WRITING, BODY and PHYSIOLOGY, MATERIALITY, HEALTH

The Laws of Heredity are the mechanisms by which genetic material is passed on to succeeding generations by fixed ratios and means. The significance of these laws by Mendel partially lies in verifying Darwin's evolution and natural selection. Of even greater significance though, are their implications and applications directly to TRU. They show the form and continuity by which humans propagated in a relatively fixed way. It shows the great relative stability within a species, especially that attained through various means of selective breeding. It shows other dynamic and variable factors, as well as the limits to this stability and continuity. It also provides the basis upon which inheritance works as ENIT, RUIT, RUTIT and ESINS participate in the FAF of TRUS and TIT. These are all of the utmost importance in human future genetic and overall health, and the survival, as well as how to obtain an accurate, and with VAPO, intricate picture of TRU.

Opposable thumb, erect stance, legs and pelvis adapted to bipedalism and reproduction, increasing skull capacity, vision, longer life, enlarging brain, hearing acuity, language, speech, writing and many other forms and functions are what have enhanced and fulfilled human life. They are what it needed to succeed and best fulfill life. Stress and loss are the result of the competitive disadvantage arising in those who lack such improvements. It shows how information, knowledge and its practice, action, predominance, prevention and improvement of all conditions, effects and determinants are necessary.

There are many different evolutionary factors and forces in operation at the same, and various, times. These may or may not be more or less related. The duration of effects may be long or short with even or uneven midpoints and distribution. The greater these determine TRU the greater their REMS are apt to be, and vv. Some of these include, tissue strength of different kinds, adequacy of function, many varieties of immunity, longevity, adaptation, changing features of the human body, cervical, thoracic, lumbar, and pelvic regions, increasing brain size and cranial capacity, that are fairly constant long term and basically yet highly variable at different times. Some occur throughout animal development and some occur only within a limited specie, genus or family. These are some of the multiple and complex processes that make up hominid evolution and the DYNS of HOM and TRUS. Because evolution often makes these steps by mutation or other singular or complex changes in genetic form does not detract from their necessity nor contingency. Nor does it detract from our need to understand them completely, thereby be managed and controlled in ways most safe and successful. These steps also occur through outbreeding, interracial, intra-sub-specie, or even, rarely, inter-specie breeding.

Works of art through many thousands of years provide a vast reservoir of information. They are often representations of direct observation or indirect idealization of the IA and worlds of the artist. Their, often deep and intensive, meanings provide evidence for the FAF, ES and CES of TRU. There has been a long tradition of art and writing. This has been since paleolithic times, associated with, and in part stems from, a great need and emphasis on the representation of ideas and the world. Some of this can be seen in the decoration of pottery and markings on possessions. In developing civilizations humans found it practical and increasingly necessary to represent the world and keep a record of their activities and ideas. This became a specialized labor of scribes in Sumeria, Egypt, and other lands. Writing is of great importance in representing TRU, thereby a living part, and process, of the A or the mass energy component in the FAF of TRU. The scribes and all who represented the world naturally tended to focus on TRU. They show many of the core features of TRU.

	30M	10M	3M	1M	500T	300T	200T	100T	50T	30T	20T	10T	3Tbc	2Tbc	1Tbc	500bc	0	500ad	1Tad	1500a	PRES	500F	2000F	10TF	100TF	1MF	F
Lt	x	x	x	x	x	x	x	x	x	x	x	x	x	x	x	x	x	x	x	x	x	x	x	x	x	x	x
Ar	x	x	x	x	x	x	x	x	x	x	x	x	x	x	x	x	x	x	x	x	x	x	x	x	x	x	x
Wt	x	x	x	x	x	x	x	x	x	x	x	x	x	x	x	x	x	x	x	x	x	x	x	x	x	x	x
Fd	x	x	x	x	x	x	x	x	x	x	x	x	x	x	x	x	x	x	x	x	x	x	x	x	x	x	x
St		.	-	-	x	x	x	x	x	x	x	x	x	x	x	x	x	x	x	x	x	x	x	x	x	x	x
Ct		.	.	-	-	-	x	x	x	x	x	x	x	x	x	x	x	x	x	x	x	x	x	x	x	x	x
Df		.	.	-	-	-	x	x	x	x	x	x	x	x	x	x	x	x	x	x	x	x	x	x	x	x	x
Lc		.	.	.	.	.	.	-	-	x	x	x	x	x	x	x	x	x	x	x	x	x	x	x	x	x	x
Sp	-	-	x	x	x	x	x	x	x	x	x	x	x	x	x	x	x	x	x	x	x	x	x	x	x	x	x
Sc	.	.	.	-	-	-	x	x	x	x	x	x	xx	xx	xxx	xxx	xxx	xxx	xxx	xxx	xxx	xxxx	xxxx	xxxx	xxxx	XXXX	XX+
Sx	x	x	x	x	x	x	x	x	x	x	x	x	x	x	x	x	x	x	x	x	x	x	xx	xx	xxx	xxx	xx+
Lg	.	.	.	-	-	-	-	-	-	x	x	x	xx	xx	xx	xxx	xxx	xxx	xxxx	xxxx	xxxx	Xxxx	XXxx	XXXx	XXXx	XXXX	XX+
CY	-	-	-	-	-	-	-	-	-	x	x	x	x	x	x	x	x	x	x	xx	xx	xx	xxx	xxx	xxxx	xxxx	xx+
CF	.	.	-	-	-	-	-	-	x	x	x	x	x	xx	xx	xx	xx	xx	xx	xx	xx	xx	xxx	xxx	xxx	xxxx	Xx+
Gd	.	.	.	.	-	-	-	-	x	x	xx	xx	xxx	xxx	xxxx	xxxx	Xxxx	Xxxx	XXxx	XXxx	XXxx	XXXx	XXXx	XXXx	XXXX	XXXX	XX+
DN					.	.	.	.	-	-	-	-	x	x	x	x	x	xx	xx	xx	xx	xx	xxx	xxx	xxx	xxx	xx+
Cp	-	-	-	-	-	-	-	-	x	x	x	x	x	x	x	xx	xx	xx	xx	xx	xxx	xxx	xxx	xxx	xxxx	xxxx	xx+
Lv	.	.	.	.	-	-	-	-	-	x	x	x	x	xx	xx	xxx	xxx	xxxx	xxxx	Xxxx	Xxxx	XXxx	XXxx	XXXx	XXXx	XXXX	XX+
Rs					.	.	.	.	.	-	-	-	-	x	x	xxx	xx	xx	xxx	xxx	xxxx	Xxxx	XXxx	XXXx	XXXX	XXXX	XX+
It										.	.	.	-	-	-	xx	x	x	xx	xxx	xxxx	Xxxx	XXxx	XXXx	XXXX	XXXX	XX+
Th								.	.	-	-	x	x	xx	xxx	xxx	xxx	xxx	xxxx	xxxx	Xxxx	XXxx	XXxx	XXxx	XXxx	XXxx	XX+
TQ							.	.	.	-	-	-	x	x	x	xx	xx	xxx	xxx	xxxx	xxxx	Xxxx	Xxxx	XXxx	XXxx	XXXx	XX+
TR									.	.	.	-	-	-	x	x	x	xx	xx	xxx	xxxx	Xxxx	XXxx	XXxx	XXXx	XXXx	XX+

Table 1-73 **HUMAN NEEDS through TIME and TRU** The primary purpose of this table is to show the gradual development and growth of human needs and abilities, and how they are interdependent and mutually empowering. It provides a rough, quick, assessment of DYNS and HOM in

SPALT and TIT. It is much more refined and interpreted in larger future editions of IPTRU, It shows how much we can not do without, and thereby what TRU is and the future consists of. As humans were more and more able to act, work, behave and control life in ways that increased their special powers and predominance so did success and survival follow. This in turn added to the genetic, anatomic, physiological, neurological, psychological, and social capacity that has gradually produced the form and function that best succeeds for humans and TRU. It has been the best of humans, those most able over competition and changing environments that has most assured this survival. It is the build up of these that produced human FAF providing each and all persons with the strength to make the best of life. With each surviving generation, through natural selection and other processes, those capacities, abilities, qualities and functions that provided incremental positive changes in the human genome, were those that were predominant, succeeded, survived and perpetuated themselves. These needs include the following: **Lt = Light, Ar = Air, Wt = Water, Fd = Food, St = Shelter, Ct = Clothing, Df = Defense, Lc = Locomotion (Transport), Sp = Sleep, Sc = Social, Sx = Sexual/Reproduction, Lg = Language, CY = Care of Young, CF = Care of Family, Gd = Goods, DN = Derived Needs, Cp = Capacity, Lv = Love, Rs = Reason, It = Intellect, Th = Theological, TQ = TRU Equivalent, TR = TRU.** Some can be interpreted and measured in different ways.

Fertility is the degree to which sexual conception is successful in reproduction. A fertility rate is the number of conceptions per thousand people. A birth rate is the number of live births per thousand people. **FER** = Fertility Ratio or Reproduction Rate is how much of a population is breeding within a given amount or period of time. **FERG** = Fertility Rates Different Groups and Races, or Members, and, **FERT** = Fertility Rates the degree of TRU. What is the effect of outbreeding, **OBRE**, inbreeding, **IBRE**, and their equilibrium and optimum, **IOBRE**, **IOPBRE**? What are the population movements and degree of stability? Why are these factors so important in determining the capacity and the quality of its members? These are key parts and processes of TRU for which its principles and concepts warrant new concepts. **PROG** = Progenitors of a population, ancestors. **DEFER** = Determinants of the fertility, Fertility Rates, Birth Rates, FER, FERG, FERT. **DEPROG** = Determinants of the Progenitors of TRU. The demonstration of these is signified by: **NESCID** = Needs Scientific Demonstration, Determination, answers and solutions. Many of these concepts, associated genetic factors and interactions, are processes in TRU heredity, are key to TRU, ENT and HOM.

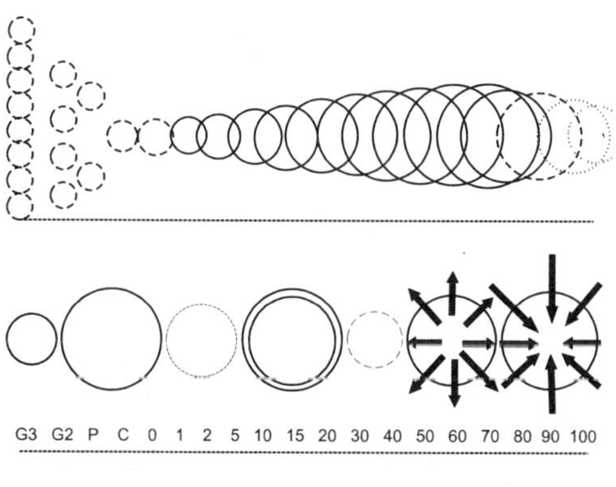

Fig 1-73 **INDIVIDUAL'S LIFE COURSES**

Different types of sufficiency or insufficiency and their interactions with the world are shown. The upper figure roughly shows the level at which a person minimally fulfills their TES as a 1/6 billionth member of the world's population. It indicates that level at which a person from early age endeavors to make contribution. It also may be used to indicate the generative position and age of a person. The lower figures shows individuals separately at any age in different forms. It indicates how great a person's existence may be in or out of TES, ENIT or **INDIT** of TRU. INDIT is related to an Individual's mental health and life. It can be of value in the treatment of those who are sick, or disabled. The vectors can be worked out in greater detail, as continually done in the application of CES and HOM.

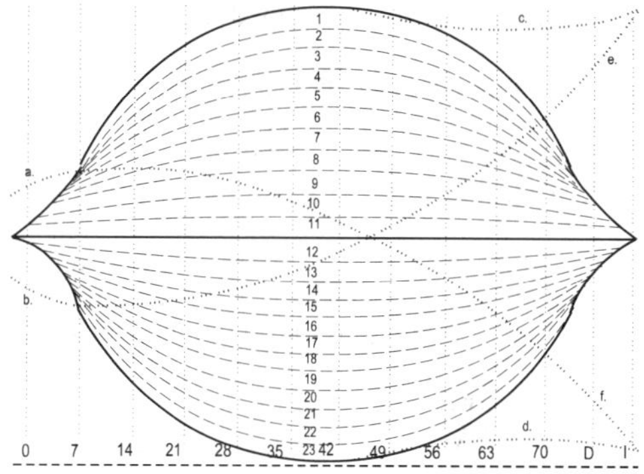

Fig 2-73 **LIFE's COURSES, SELF, TES, OBJECTIVE and SUBJECTIVE**

From birth to death, from a nidus of organic or objective TRU through many and varied courses to a focus of Á, spiritual or subjective-objective TRU, the self is often not much TES. One often rapidly alternates between TES and ATES in both constant dynamic and conflicting relationships directly and indirectly, spontaneously and rationally that tend toward a meaningful, TRU life. This is a life of fulfillment, satisfaction and survival. The figure shows ages on the horizontal axis and behavioral patterns, attitudes and other categories on the vertical axis. The role of many components of a person's life can be included in such a representation.

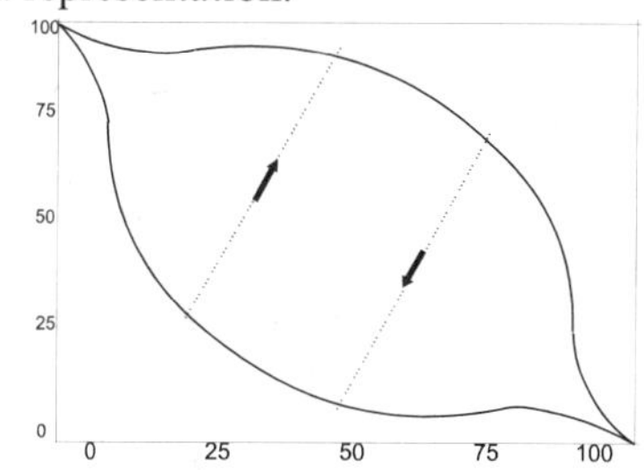

Fig 3-73 **LIFE SPAN, EXPECTATION, and SURVIVAL CURVES**

The survival curve can be used as means of representing success in life or resisting death, including forces that: 1. Follow ones sense of well being, 2. Keeping oneself built up, 3. Applying the best of knowledge of health, 4. Personal hygiene, avoidance, moderation, 5. Public Health, 6. Preventative Medicine, 7. Scientific discovery. 8. Medical treatment, 9. Applied medical and scientific, knowledge, 10. Turning positive all the negatives, such as adverse and abnormal forces on all levels of Á. 11. Turning all around in the world, ATRU to TRU, where practically and beneficially possible, **TURNAT.** The biggest benefactors of health are the young who have been provided with a great accumulation of potential capacity. Life expectation has been highly variable through history, and has rapidly expanded in highly stable and prosperous regions, and well-developed nations of the present. Life expectancy was approximately 18 years in the Bronze Age, 25 years at the time of Christ, 43 years in 1900, 68 years in 1980 and over 80 years in many nations in the 21at century.

To live fully, yet frugally, continually working and enjoying life is associated with the common and great means of turning the bad into good. This has been shown in mysticism, or each person's return to TREQ, to enjoy absolute identity with TRU. This in its ultimate form is the extinction of individuality, of all sin, evil, non TRU worthless material, and waste. It is how to develop, without loss, the beatific union of ones liberated spirit, TES, with KOT, THES, TRU and TÂ. This in Sanskrit is called: Nirvana which means "blown out". This is the blowing out of all that is not TRU. It is the extinction of all the suffering, sin, material things, economic bondage and evils in the world. Nirvana and such revelations must be modified by VAPO, SCARLS, and the needs of TRU. The material means that enable us to have more time for thought, contemplation, and creative work have more of all Á FAF or power to produce benefit and enable TRU. Life is short, wisdom is long, each person is a reincarnation of this life, wisdom and world when TES and all are TRU.

Hourly wages may be expressed as a function of time through history. This varies and needs to be corrected for changes in values and currency. It shows how material means and values have increased. It should alert us to the great problems and responsibility future material wealth imposes. The supply and demand for needs of life as TRU should be well accounted for. How much is or is not spent on TRU. Items necessary for life, may be obtained very cheaply by one society, yet costly to another.

Fig 4-73 **AGE and RESIDUAL CAPACITY**

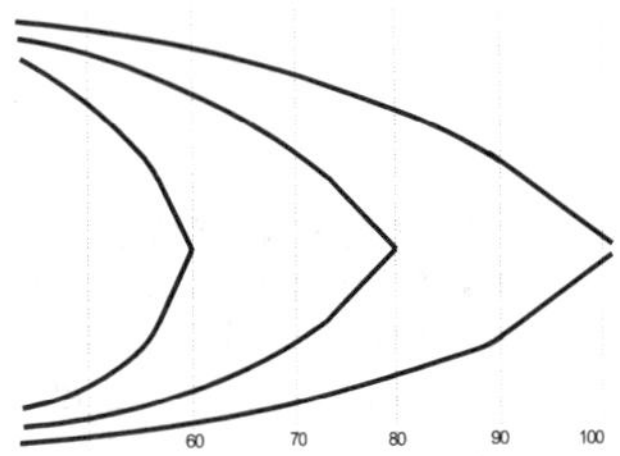

Residual human and life capacity, as well as physiolo-gical functions change with advancing age. To meet added requirements and to compensate for loses in the other functions it is edifying to show what the involved factors and CES are: mammalian precursors and determinants, increasing longevity, forces for and against optimal progeny, variable senescence of different organs, optimal life of different systems and organs, limiting factors in tissue survival, contributory factors that sustain life or cause senescence, determinants, forces, limits and factors that can and can not be altered, and the point or age at which alteration can begin as well as how it is to be continued. For humans aging is similar to all other organic material, much is dead without even being recognized.(DAK) Humans have relatively long lives when compared with most animals. This is necessitated, certainly in part, by dependence upon the maturity needed to succeed, fulfill our Á mass energy FAF, and survive.

Fig 5-73 **DEGREES of HEALTH**

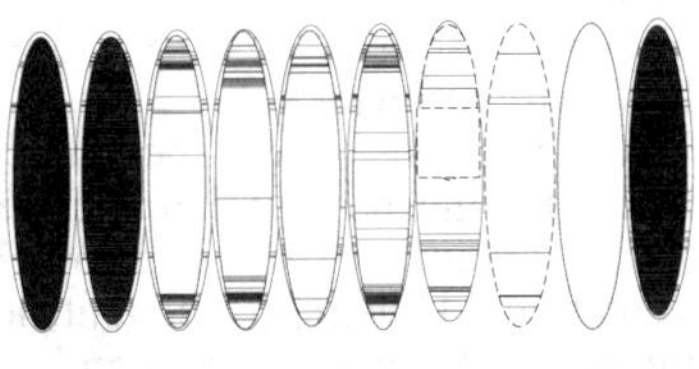

Total health may be categorized to emphasize and contrast differences between perfect health and lesser degrees of health: 1. Perfect, 2. Hypercompensatory, 3. Normal, 4. Average, 5. Less than average, 6. Hypocompensatory, 7. Frequency of time in dying, 8. MLD 50, Mid lethal, 9. Death, 10. Overkill. These represent various scenarios in a persons life. How one feels is one indicator of the scenario. The various factors in bodily and sexual FAF show how far the sexual determines ones S/F in life. The energy, activity and opportunity to provide frequency, intensity and level of activity, availability, and experiences, all are factors that determine S/F.

The important thing in life and in the care of ones physical health is to know what illnesses, diseases and injuries a person will get over, can cure with simple PPMA and care, and which ones are apt to become progressively worse. With knowledge and anticipation we can exercise one of the greatest strengths of TRU. It shows how TRU is more than one or a few lower levels of mass energy, but all, including the highest, combined levels of S-O, Á FAF, that enable TRU. There are many examples of the Complications of Causes, or COC and COUC, in the physiology of the human body.

There is great beauty in genomics. It enables us to realize complexity, yet potential reducibility to determination. With $1x10^{14}$ cells in the human body, 20-30,000 genes, 100,000 base pairs or DNA building blocks per gene, and the great diversity of proteins, activity, interactions, sequences, enzymes and control mechanisms through various settings in the life cycle, DYNS, and DYNO, science still has plenty to do.

Fig 6-73 **CORRESPONDENCE OF HEALTH and TRU**

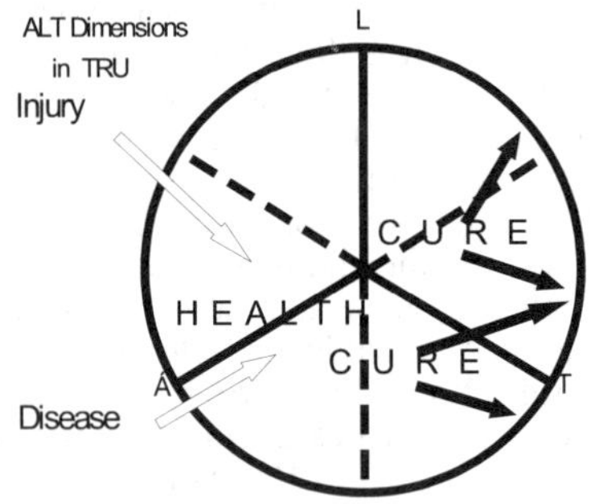

In medicine we deal primarily with the abnormal. It provides us with a good example of how important it is to know, not only what is positive in TIT but all that threatens or potentially alters TRU. It is the healthy and strong lives of humans that, in its special way, make up much of TRU. Health and disease, and the illustrations and models that have, are and can be devised, can be used for much of TRU. In SPALT, HOM and DYNS we can show these CES by adapting and superimposing a great many of the diagrams, illustrations, models and pictures of medicine and all fields and sciences to TRU.

ABFRE = Abuse of Freedoms. ABFRE is a corollary of the abuse of life. The price of an individual's, America's, and the world's, freedom is to have to pay many times for the indiscretions and abuses of freedom. Cultures, nations and societies can become enslaved by their own prejudices, reactions, excess laws, rules, regulations, customs, routines, ideals, and lack of reason and TRU.

Fig 7-73 **CARSC, S/R, and Disease**

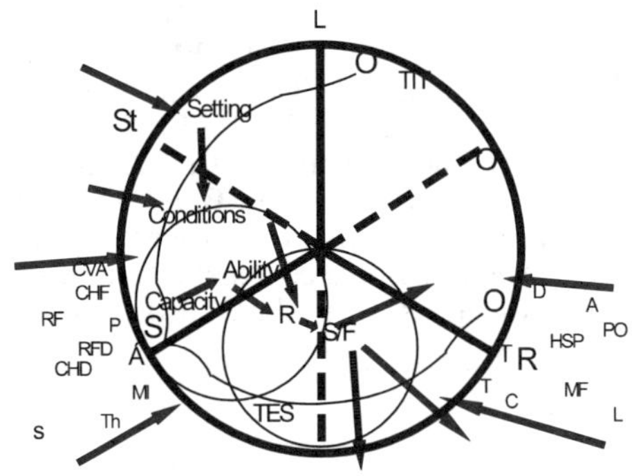

The correspondence between levels of the causes for disease and the capacity, ability, reaction, setting and conditions is the relationship between classification and ordered causes of diseases. It is an example of how acute, subacute, and chronic come together with capacity, ability, condition and setting in producing success/failure. These change with the age and management. Randomly a few are shown with Abbreviations: A = Anemia, C = Carcinoma, CHD = Coronary Heart D,.CHF = Congestive Heart Failure, CVA = Cerebral Vascular Accident, D = Diabetes, HSP = Purpura, L = Leukemia, MF = Marrow Failure, MI = Myocardial Infarction, P = Pericarditis, PO = Polycythemia, RF = Rheumatic Diseases, s = Streptococcal Infection, T = Tumor, and Th = Thrombosis.

Any society, nation, culture or civilization in which the detrimental, ignorant, narrow minded, extremists, retarded, intolerant and lazy have as much power and influence asthose who contribute most, is one that is headed for destruction. Therefore civil rights, as all, must yield to and be supportive of the power of TRU. This is the power of natural selection, and all human higher powers can use to enhance it.

SYSTEMS	ILLNESS, DEFICIENT, SUBPAR	AVERAGE, PAR, FUNCTIONAL	TES, TRU, PREDOMINANT
ASTRONOMICAL	PERTURBATIONS, ↕ SOLAR (f)	STEADY STATE, NORMAL (f)	CONTINUED NORMAL FUNCTION
CLIMATIC	EXTREMES, INCLIMATE CHANGE	NORMAL CLIMATE and CHANGE	NORMAL CLIMATE, CONTROLS
PHYSICAL	EXTREMES, IMBALANCES	NORMAL INTERACTION	NORMAL ENVIRONS, CONTROLS
SOCIAL	DETERIORATING, DISORDERED	HEALTHY SOCIETY	NOURISHED, STRENGTH, TRU
INDUSTRIAL	DEFICIENT OR HARMFUL	HEALTHY PLENTIFUL	NONHARMFUL, EMPOWERING
AUTOMOTIVE	EXCESS, DEFICIENT, HARMFUL	EFFECTIVE, NON POLLUTING	EFFICIENT, NON HARMFUL
ELECTRONICS	WASTED HARMFUL DISTRACTING	EFFECTIVE, UNHARMFUL	STIMULATED ENHANCED PPMA
ANTHROPOLOGICAL	LEADING TOWARD EXTINCTION	LIFE SUSTAINING	SUPPORTING ADVANCING TRU
MENTAL	DEFICIENT, MISUSED	SUFFICIENT, HEALTHY	ENABLING, EXPANDING, KOT
MECHANICAL	DEFECTIVE, WASTED	SUPPORTIVE, EFFECTIVE	COMPLEMENTING, ENHANCING
TRAUMA	EXCESSIVE, DAMAGING	NORMAL STIMULI/RESPONSE	MINIMAL TURNED to BENEFIT
ORGAN SYSTEMS	DEFICIENT, ABUSED	NORMAL FAF	SUPPORTED ENHANCED FAF
MICRO SYSTEMS	DEFECTIVE, UNHEALTHY	NORMALLY HEALTHY	REGENERATED STRENGTHENED
EXCESSES	ABNORMALLY PRESENT, ABUSED	MINIMAL or MANAGED	REMOVED UTILIZED PPMA TRU
DEFICIENCIES	NUMEROUS UNCOMPENSATED FOR	NATURAL, COMPENSATED FOR	MINIMAL, PPMA, ACC, TRU
NUTRITIONAL	MALNUTRITION, PATHOLOGICAL	NORMAL or SUPPLIED	OPTIMAL SUPPLY CONTROL
ALLERGENS	EXCESS, DEFICIENT RESPONSE	NORMAL PRESENCE RESPONSE	MINIMAL PRESENCE, PPMA,
TOXINS	EXCESSIVE, NO/POOR CONTROL	FEW TOXINS, PREVENTION	REMOVAL, INCREASED PPMA
METABOLISM	DEFECTIVE, MALFUNCTIONING	HEALTHY, BALANCED	SUPPORTED, ENHANCED HOM
DEVELOPMENTAL	NEGLECTED, ABUSED, POOR	SUPPORTED, HEALTHY	STIMULATED, IMPROVED TRU
CELLULAR	DISEASED, DEFICIENT	HEALTHY, NOURISHED	HEALTHY, ENHANCED
BACTERIAL	EXCESSIVE, VIRULENT	MINIMAL, CONTROLLED	ELIMINATION PPMA CONTROLS
VIRAL	EXCESSIVE, VIRULENT	MINIMAL, CONTROLLED	ELIMINATION PPMA CONTROLS
GENETIC	DEFECTIVE, WORSENING	HEALTHY, ADEQUATE	PERFECTED, PPMA, TRU ACC
ENZYMATIC	DEFICIENT, DEFECTIVE	NORMAL, SUFFICIENT	ENHANCED PPMA CONTROLLED
MOLECULAR	DEFECTIVE, DEFICIENT	NORMAL FAF	ENHANCED PPMA CONTROLLED
ELEMENTAL	DEFICIENT, EXCESS, DEFECT	NORMAL SUPPLY, FAF	SUPPORTED SUPPLY AND USE
SUBATOMIC	DEFECTIVE, HARMFUL, EXCESS	NORMALLY PRESENT, FAF	ENHANCED SUPPLY AND USE

Table 2-73 **SYSTEMS, ILLNESS, HEALTH, and TRU**

Many of the systems involved in human health, illness and TRU, when shown, provide a good overview of the differences between illness, average health, and best needed to be TES.

Behavior may be categorized by its parameters and classes, including: 1. Inaction or Action, 2. Action or Reaction, 3. Non TRU or TRU, 4. Partial or Total, 5. Intentional or unintentional, 6. Short, mid, or long-term, 7. Regulated or unregulated, and 8. Routine or arbitrary. Behavior is a large part of both IA, ACC, and the expression of subjectivity in the objective of a person's Á mass energy FAF in life. Behavior is a good example of: **LUS** = Law of Unitary and Multiple, separate and combined. LUS signifies what are, in being singular and separate in one way, also complex and combined in others. In this way there are: **LUSO** = LUS different and same, **LUSP** = LUS proper or its means, **LUSU** = LUS separate and combined. This is shown by the motto of Future Farmers of America: “Learning to do, Doing to learn, Earning to live, Living to serve.” **YAD** = Youthful desire and inclination to Adventure, Anarchism, nihilism or terrorism. YAD signifies this tendency in humans, especially young males, with their new found powers and excess energy. There are certain measures of a person, both positive and negative that hold for much of ACC as they do for life: 1. How hard one works, 2. How right one acts, 3. How wisely these acts and works are directed, 4. How much one does for themselves, 5. How much this done for oneself is TRU, 6. How much one does for others, 7. How much this done for others is TRU, 8. How much one wastes, or does for fantasy or harm, 9. What are and how right ones interests and thoughts are, 10. What one’s loves are, 11. How much energy and vigor one has, 12. How much will power one has, 13. One's judgement, how well this is used, 14. How much one works with felicity and VAPO for TRU, 15. How much ones free will functions in harmony with that of TRU, or God's prescience, 16. How much one lives frugally and sparingly of all appropriately by TRU, 17. How much one lives away from and avoids ATRU and ANTRU, 18. How much one lives by the rule of the Greatest Good or REM in TRU, 19. How much one realizes, appreciates and holds in highest regard the greatness and intricacy, yet attainable ordination and modulation of TRU, 20. How much one appreciates and honors the divinity of the word, concept, Á and ES particularization of TRU as rendered possible in a developing language, logic, and TRU, 21. How much ones uses and improves ones imagination, inspiration and revelation of all for TRU, 22. How much one avoids temptation of all that is not TRU and maintains the continuity of TES, 23. How much one recognizes TRU and TES, ones counterpart and role in TES, as the mediator, and way to heal bad, evil, sin, failures, infirmities, and negative to TRU, 24. How much one dedicates their life and salvation to TES, as did Christ, St. Augustine, Socrates, Buddha, and many others, as all must to be TRU.

The measures an individual takes positively or negatively for health is no where more important than how much they affect good or bad propagation. There is a vast number of indirect forces that detract from optimal progeny. They continue to be discovered and added to those that are known. Some of these include, 1. deleterious genetic defects, both qualitatively and quantitatively, 2. other detrimental effects on sexual cells, including alteration by toxic chemicals, viruses and bacteria, 3. Indirect effects from somatic cells. 4. A woman’s life, selection and choices, quality and quantity. 5. A woman’s emotional and rational preferences, 6. A man’s rational, emotional, and impulsive tendencies and preferences. 7. Attractiveness, bodily, facial, behavior. 8. Sexual experience and preferences, 9. Effects of peer groups, fads, and fashions, 10. Customs and social taboos, 11. Educational and Cultural effects on an individuals sexual and social relations, 12. Consanguinity and inbreeding effects. 13. Diversity and outbreeding effects, 14. Teenage and young adult ignorance and experimentation, 15. Sexual fears, inhibitions, responsibilities and lack of, 16. Smoking, drugs, alcohol, and lack of good health habits, 17. Diet, balanced or unbalanced, excesses, and adverse effects on sex cells. The effects of these are multiple, combined, complex, interactive, and variable. Ones genetics, no less than the rest of ones body, and all human health and life are continually being affected by all that acts upon it. The margin of safety and level of tolerance of genetics is so narrow, that prevention must be based upon best knowledge if it is to replace natural selection.

	ABSOLUTE	BASIC	RELATIVE	OPTIMAL	EXCESSIVE
OXYGEN	5-10 minutes	1 Minute	5 per Minute	20 per Minute	40 per Minute
FOOD	1-3 Months	1-4 Weeks	1-7 days	1-4/ day	>4 / day
WATER	1-4 weeks	1-7 Days	1-3 / day	3-10 / day	> 10 / day
TEMPERATURE degrees	>110, <10	100-110, 10-30	80-100 30-50	50-80	No Variation
CLOTHING	None	Poor, Feet or Lower Body	Sufficient	Well Clothed	Over Clothed
MATING	None	Rarely	> Yearly	Weekly	> Daily
SHELTER	None	Minimal Cover	Adequate abode	Comfortable Home	Wasteful Use
URINATION	1-2 Weeks	1-7 days	1-2/day	3-4/day	Polyuria
DEFECATION	1-3 Months	1-4 weeks	2-7 days	2 days - 3/day	Continual Diarrhea
EXPERIENCE	None or MOST ANTRU	Rare or Very Bad	Inadequate or Bad	Sufficient and TRU	Excess Bad Exposure
KNOWLEDGE	None or Most ANTRU	Little and Bad	Inadequate and Bad	Massive and TRU	Massive and ATRU

Table 3-73 **BASIC NECESSITIES of LIFE**

What are the absolute, basic, relative, optimal, and excessive use of the necessities of life? Human physiology provides not only exceptionally good illustrations and detail of these, they are, as the human body and its functions, highly TRU. Three examples are, neurochemical development, subatomic Á origins, and extrahepatic and hepatic lipid function. Through their receptors we can classify and formulate their combined and complex processes. Next to being the best one can be, TES, and seeking the best for the world, TRU, the most important thing in a person's life is to pick the best mate, associates, and ends possible. This is by TES, TÍS, compatibility, one's immortality, and TRU's continuity.

Division of Labor, perfected team performance

No one can go it alone, nor do only what one pleases. At best our decisions, selection and actions are joint ventures with all the forces, processes and operations that enable this world of best human life, plus all the aspects, modes, form, structure, and mechanisms that support and enable one and all to be. For the actual world of human life to succeed and survive it takes the optimal function and contribution of all parts.

The big quandary, each and all persons, separately or together have in life, is 'What do I, or we, do'. This we usually do spontaneously either by fulfilling demands or our needs, by habit or routine. Yet often we have to make decisions and choices which can be no better than our knowledge and how well it is used. Do we keep on doing what is being done or lose it, or do we do what comes up or forget about it? There is a continual correction for questions, problems, errors, and needs. The solution is to follow the best of ones natural mass energy form and function as it is TRU. This involves all levels of SA, especially an ever greater quality and quantity of its highest levels. This is what this text purports to edify.

Most people can be easily understood if you recognize what they need most, what they hold most dear, who their loved ones are, where they find comfort, what they spend most of their free time at, what their money is used for, what their jobs, work, trades, occupations and professions are, what their life's experiences have been, and what their future holds, as TES. By accounting, adapting and interpreting these by TRU we can better gauge and set IPR to work with others for TRU.

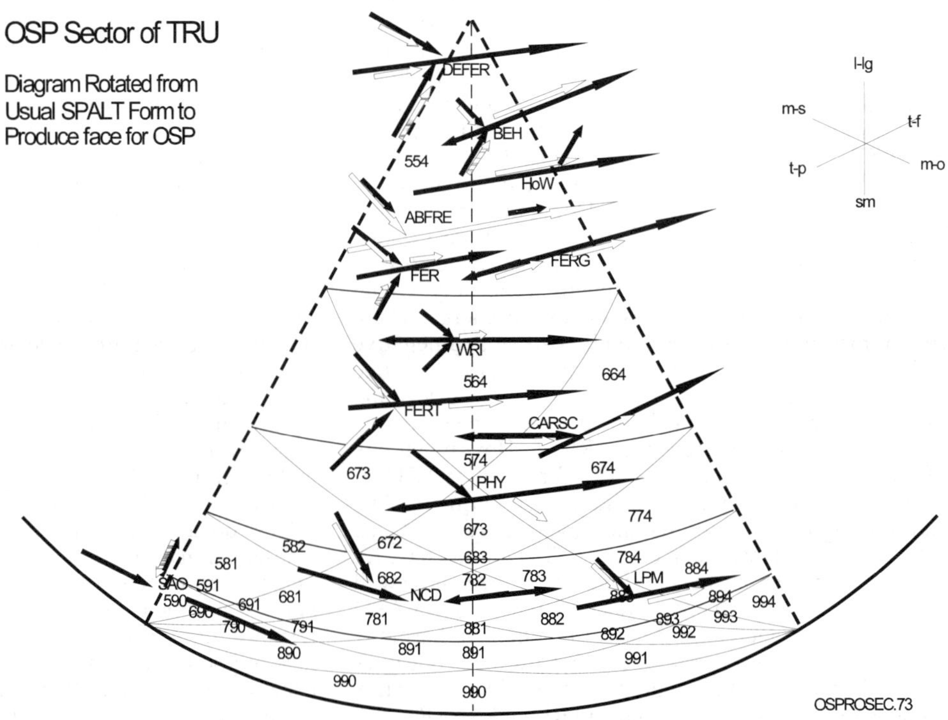

Fig 8-73 **DYNS CES in OSP SECTOR of SPALT**

MEASURES of DYNS CES in OSP

Measure	Code	Type
ABFRE	6-5-3	-S
AGE & RESIDUAL CAPACITY ARC	8-5-3~	T
AGING	8-5-2~	T
BEHAVIOR	7-5-3	S
BODILY & SEXUAL FUN BSF	8-6-3~	T
CARSC	6-6-4	S
DEFER	7-5-2	S
DEPROG	6-5-1	M
EVOLUTION: FACTORS,FOR EFF	8-5-1	T
FERTILITY	7-6-3	S
FER	8-5-2	S
FERG	7-5-2	S
HEALTH	~7-6-3~	M
HOMINID	8-4~-0~	M
HOURLY WAGE	8-5-3	S
HUMAN NEEDS/TIME HNT	7-5-4~	T
IBRE	8-4-1+	T
IOBRE	8-4-2~	T
LIFE'S COURSES LiC	~7-5-2~	M
LIFE EXPECTATION LEX	8-5-2~	T
LIPOPROTEIN, METABOL LPM	8-8-4	S
LUS,LUSO,LUSP,LUSU	~-~-~-~	M
MATERIALITY MAT	9-7~-~	M
NESCID	~5-5-~	M
NEUROCHEMICAL DEVELOPMENT NCD	8-8-4	S
OBRE	8-4-1~	T
PHYSIOLOGY PHY	6-7-4	S
PROG	7-5-1~	M
SUBATOMIC Á ORIGINS SAO	5-9-0	S
TURNAT	7-5-2	M
WRITING	6-6-3~	S
YAD	7-4-2~	T

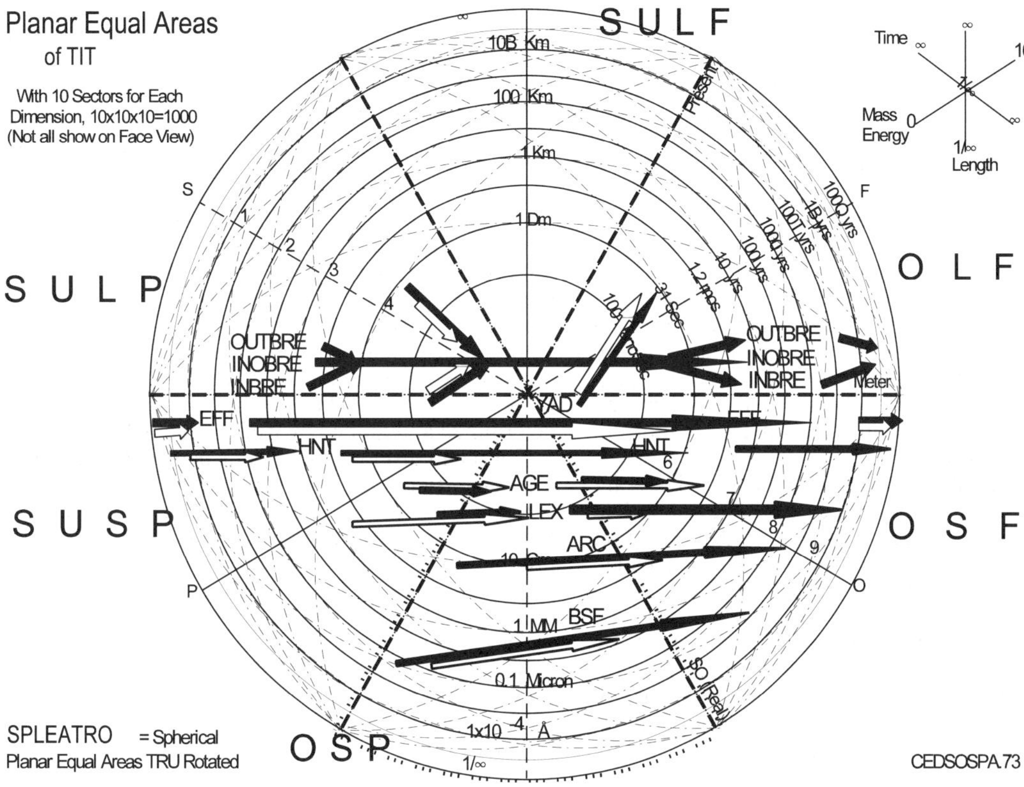

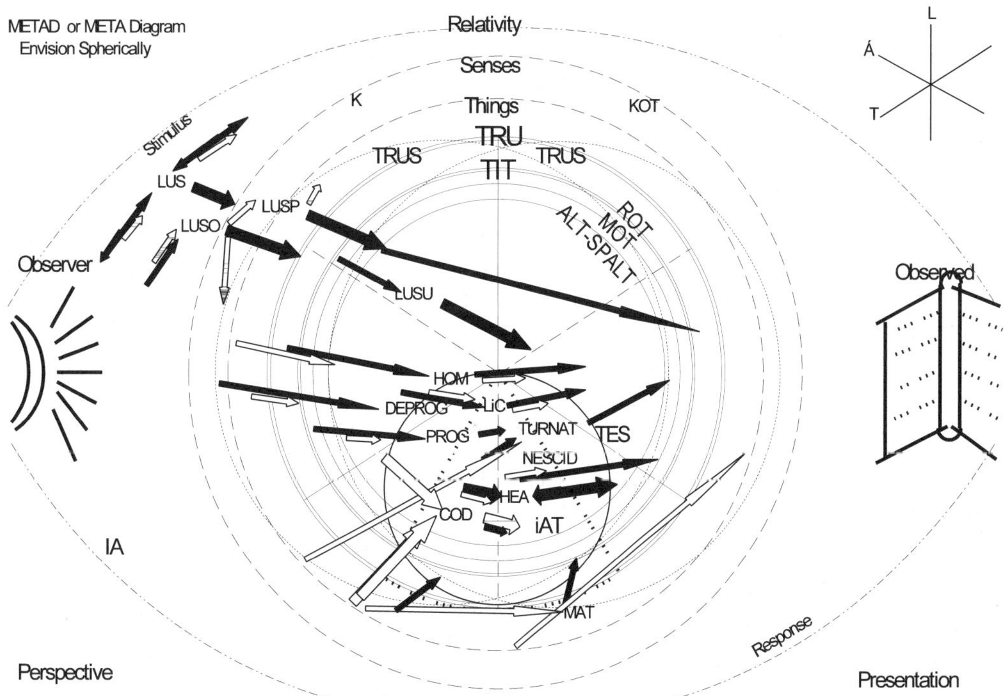

Figs 9-73 and 10-73

DYNS CES of OSP in SPALT and METAD

CHAPTER 74

SUSP or SUBJECTIVE SMALL PAST PROCESSES of DYNS

WORDS, PERSONAL KNOWLEDGE and LIFE, GOOD, MEASURE, WANTS, EXPERIENCE, LOVE, TRUTH

It is in the higher levels of mass energy that the competition between individuals and groups exists in such stress, strain, and struggle. This explains how selection is so great within a species, and how optimal quality is maintained. Their lives are as one tricked by nature. For they would go one way and nature, or FAF, drives them, continuously redirecting them to another, the convergence and focus of the mean optimum of TIT. An example of this is the ego of the individual. It is needed to provide the focus and energy of their individuality. Yet it also causes much suffering by way of continual competition, opposition, vanity, and fighting for a predominance, with futility in mixed achievement. **HELT** = High Energy Levels Equilibrium and Trick or set and SCARLS of nature in TRU. HELT defines that state in which we are caught up in situations to perform in certain ways. It is most notable in the higher energy levels of the individual. Here the individual feels and is most cognizant of being continuously tossed and turned, not only by all that is negative, but much that is positive for TRU. Even ones ego, that great powerhouse of survival, suffers from this ruse and power of nature by being provided with an it or ENIT focus. Only to find that this is an extreme in which one must return to the mean, the equilibrium of all the mass energy, Á of TRU.

Pain and suffering takes many forms and functions. They are great vehicles in the mass energy dimension. They stimulate and teach us what is bad, wrong, and what is needed to correct them. They piercingly penetrate our inertia to make us realize loss, to ourselves and to the actual world, TRU, we are in. They prompt us to be, and do the best we possible can. This is: **SUFELT** or Suffering, Felt Stimuli to negative of TRU. How it can be turned into positive and most beneficial. The key to Buddhism was <u>Suffering</u>. In Tibetan art, as from Buddhism generally, The Buddha is the all compassionate being. This is, as OM, the intense, intricate, detailed and proportionate feeling and love for being, in all its forms and parts. It is one that is continuously being reborn to help end suffering.

The development of language has been slow and largely by trial and error. The development of language has not been without a great amount of negative as well as positive advance. One of these is: **LI** = Language Inversion. LI is the more or less loss of ES and TO with the use of words, names, concepts, and IA. The value of words and language in TRU, nations, states, civilizations, and cultures is in a state of dynamic variation and change. Whether they improve or worsen depends upon the product of language, speech, word, and literature, intensity, extensity and actuality. **MUM** or Multiple Meanings arise from a coming together of words by sound, meaning, structure or other relations. MUM are large causes of ambiguity. Of several ways and means to resolve MUM and ambiguity, the best is to experience the idea, concept and term and best apply them in life, as TRU. Another is to add to the word or term a letter, suffix or prefix that sets it apart from others. This we do in NEC, as by IA: **INIA** = Initial IA becoming, **INIAT** = Initial IA becomes IAT, and: **INIACT** = Initial IA becomes IAT and New Concepts. This shows why a new terminology for TRU is so necessary and how it must be constructed by TRU, SUT and ROT. There are many words that are not commonly used or understood whose distinction adds greatly to the explicit understanding of TRU. A term, on the positive side is <u>Probit</u> and its qualifier <u>Probity</u>. These imply the right and precise probability and proper results. It is probity in TRU that we search for. This is what TRU is, what we attain when our presentation, representation, accentuation, observer-observed relationships become 1:1 or one with TRU. Probit is a statistical unit of measurement of probability based upon the deviations from the mean of a normal frequency distribution. It is, with VAPO, the goal of an ordered, cosmotic world, the right and precise probability. Probity is not an automatic gift. It is something that is mostly learned, earned, and requires deep conviction, determination, hard work and dedication as one becomes their counterpart in TRU, TES. This is where TES and TRU function as each best enables the other, ENT.

The Socratic Method set the standard for all knowledge. The contributions of Socrates to humanity proceeded from principles and their corollaries. It arose from prior philosophical, mathematical, axiomatic and rational currents. They have influenced learning ever since, largely through the works of Plato, and include: 1. God as the all wise and all good ruler of the world, the universal principle, Nous, or the giver of the divine sign, Logos, 2. Confidence in ones daimon, or personal God, soul, conscience or TES, 3. A person must know their ignorance and embark on the journey of thought, 4. Man's self resides solely in the knowledge of the true and the good. Only those who takes such thinking seriously and are determined and guided by truth are actually themselves, TES, 5. Truth will disclose itself if one perseveres in questioning, 6. Emphasis upon analysis and observation, 7. Through the example of self control, endurance, moderation, and frugality Socrates imparted habits which enabled those who were influenced by him to better follow these principles. 8. It is a person's business in life to practice the tending of ones soul by doing good through considering the great and the good. For these are but a part of God. By imparting this good a person leaves their effect in God or the Intellectual Principle in immortality.

Fig 1-74 **INDIVIDUAL'S UNION with TRU, TÂ**

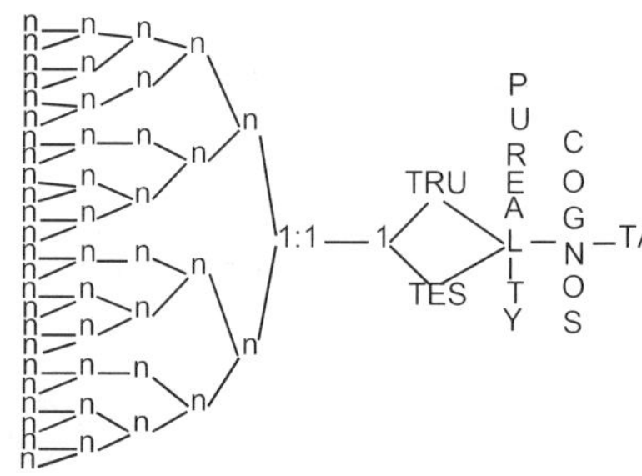

Martyrdom leaves us with great feelings of compassion for not only life, strength of character, and the causes that were strongly held. It shows how the individual can reach for heaven, for immortality by identity with God, TRU or TÂ.

Some individuals will deny even life itself if they can reach a greater life, and achieve an ever higher degree of immortality. Herein is the gist of Christ's Message, the key to Christianity. This can be achieved by different means, work and sacrifice. The identity of the self with TRU, TES, and the self identity of the individual, as we fix upon ourselves indiscriminately may be signified as follows: **ASI** = Absolute (level) of Self Identity, as it varies per life time, **ASIT** = Absolute Self Identity with TRU, This is equivalent to TES, **ASA** = Absolute Self identity Absolute with all entities, or any entity. ASA is or can become equivalent to a completely unguided, irrelevant, irreverent and chaotic identity. ASIT and ASA for the individual are only ends and not the usual state of ASI. The great state of flux of the individual, IA, and of ASI is its most typical and characteristic state. Whether it is mostly toward ASIT, or ASA will depend upon the amount of effort, conviction and determination.

How do we analyze, measure and rate philosophers, theologians, scientists, statesmen, top executives, and all great personages? What were their distinguishing natures and responsibilities? This is of value, like **TESTAR**, Testing and Rating to and of TRU, for several reasons: 1. To precisely evaluate their REM, 2. To show what, as an example, is necessary for highest contribution and attainment, 3. To discover and measure our weaknesses, 4. To show Pureality and what it leads to, 5. TES and how great it can be, at its optimum.

It is imperative that the condition and state of fulfillment and completion of knowledge be met with an equally great conviction, perseverance, gratitude, and joy. This needs to be continued through a lifetime of expanded training and education. The counterbalance by knowledge, THES and Cognos is called: **COGNIM** = (Law of) COGNOS or Knowledge Imperative. It is knowledge actuating Cognos.

Fig 2-74 **COMBINED CES of FAITH, HOPE, GRACE, MERCY and CHARITY and TRU**

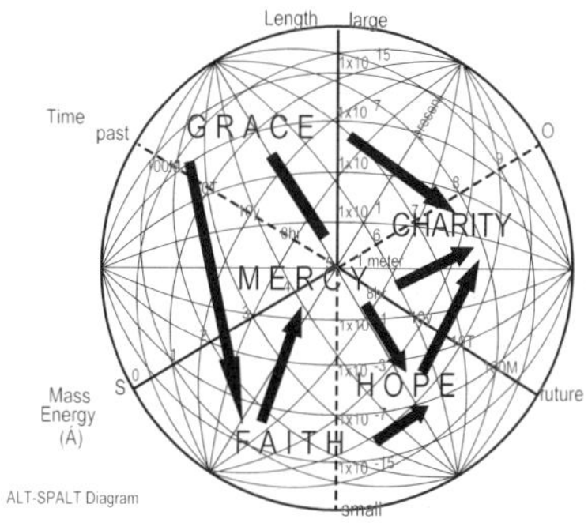

When faith, hope, grace, mercy and charity are combined, interrelated, proportioned, and posited in SPALT, they provide added meaning to their contribution to DYNS and TRU. This is especially so when combined with other CES. Charity is the spirit of giving and life long dedication to a controlled selflessness and devotion to God or TRU. It is the understanding, appreciation, conviction and endeavor to benefit all who most need to fulfill their counterparts, TES and become TRU. Only charity that is permanent, acts not only to relieve, but to guide, elevate, improve, achieve, and continuously upgrade all that can be increasingly TRU is equivalent to the greatness of TRU. The best scenarios in life of individuals, groups and all people include, 1. High intelligence, 2. Selflessness and generosity qualified by the needs of the individual as TES, 3. Non aggression, except as indicated by TES and TRU, 4. High moral values - conditioned to TRU, 5. Great conscience, gratitude, love, inspiration and enthusiasm for TRU or TREQ, 6. Highly effective good work and deeds that are by VAPO, SCARLS, POA to TRU. (See index for the other CES)

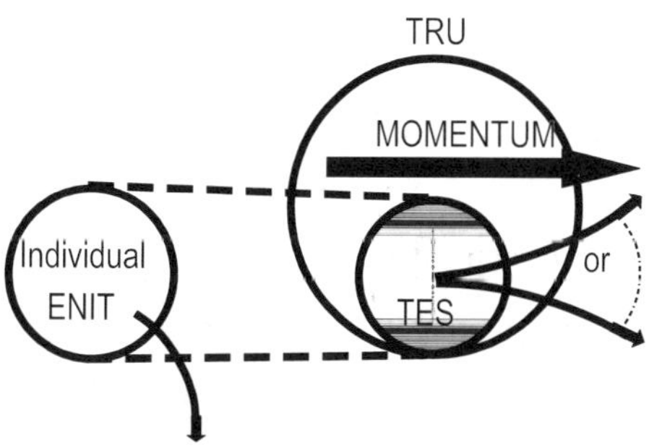

Fig 3-74 **INDIVIDUALS as ENIT or TRU**

Personalities and self interest, introvert and extrovert, depressive and manic are all states both normal and abnormal. There are cycles or rhythms in individuals and larger groups. There are individual differences both separately and in levels of Á that may when extreme be abnormal. They vary with reactions to stresses, adaptation and attitudes, as well as deleterious effects on Á, e.g. more permanent brain damage. The degree to which our being or lives are other than what is proportional to TES, the small dark circle, does not deny our being as an individual immediately. For, as an ENIT, each individual has its own special absoluteness which is in dialectical interaction with ATRU, TRU, TES and TIT.

In growth and development, human attention has a vast amount of interaction to contend with and experience. It should be no surprise that in the expanded exposure children have to increased, natural and artificial stimuli and environments, they react with what is called short attention span disorder. The panic reactions and being obsessed with threats produce states of chronic fear that can be subconscious or overt. Attention disorders also result from a lack of conditioning, inclinations, habits, parenteral guidance, and security. The key is prevention, at all ages, and levels of GELD and SÁ. These include measures to extend the limits of human capacity, ability, conditioning, training, focus, targets, knowledge and anticipation. It should include improvements in human existence in which greater stability and settled life are assured with a decrease in instability and uprooted life. Each person should realize the control necessary to fulfill their own destiny. The achievement of TES is not for a few or a limited period of a persons life in maturity. It is for growth and development in all time before, during and after life. Although apparent stability and security help in the short run, permanent stability and security are only achieved with TRU, with attention that is both equal to its demand in life, and equal to what is needed to be TES.

UIA = Universal IA, Preoccupation, **UIAT** = Universal TRU, TIA, TRU IA Preoccupation. UIA defines and can measure what and the degree ones IA and mind in a total lifetime IA are occupied. The degree an individual is preoccupied with trivia, fads, routines, trite topics or pursuits, colloquial, nonbeing, AU facts, fancies, beliefs, illusions, figments, myths, superstitions and many other ATRU. **UIRE** = Ultimate Identity, **LIS** = Limited Self, **RES** = Relative Self, **ANTIS** = Anti Self, **PES** = Partial Self. All contrast with TES, the, or ones TRU Self, counterpart in TRU. The optimal self, TES may be compared with the divine, as an angel, or saint, one who devotes their life to the welfare of God or TRU. ANTIS or the antiself is that part of an individual that is against TRU. In reverse it is corrected by what the self of the future does to escape from the non TRU self of the past.

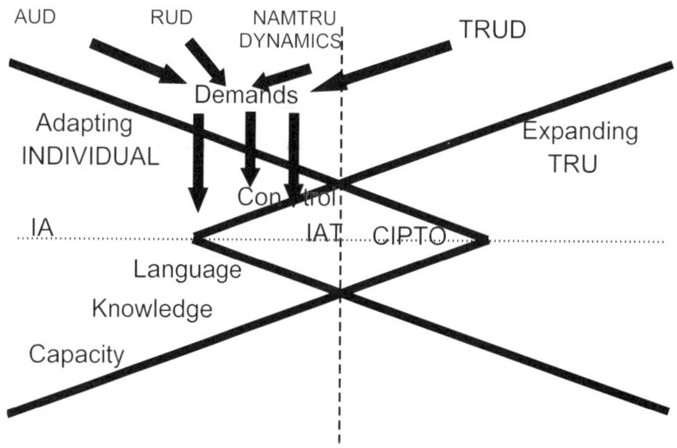

Fig 4-74 **CRITICAL POINT of INDIVIDUAL, LIFE and TRU**

CIPTO = Critical Point of Individual Capacity and Control in a dynamic world. As we develop more and more ideas, concepts, words, laws and principles that are increasingly well defined and as each individual pushes their limits of comprehension to a knowledge of all that is TRU, the operation and management of CIPTO become increasingly necessary. With greater pressure to be TES, each and all find themselves in AITES. We are at a critical point, CIPTO, of individual capacity and control in a dynamic world. This is the point of an individual in relation to their surrounding world. It is how the individual as ENIT in AMTRU dynamically and continually must tend to TÍS, TES and TRU. Much of the world, or TRU, focuses around this dynamic conflicting relationship. Most individuals are so engrossed in their own IA, wants, life, and survival that little sense or connection with TRU can be made. Their resulting knowledge is neither organized nor prepared for a reasoned and successful life by TREQ or TRU. Neither TRU nor ACC will attain adequacy or success without recognizing and applying all pertinent types of TRU knowledge. This can only be by the highly actual, extensive and intensive kind of ACEXIN knowledge that VAPO, KOT, SPALT, POS, STA, HOM, and ACC provide.

The measure of a person is key to their character, virtue, REM, TES, PUREALITY, KUPS and TRU. These can be demonstrated by psychological testing, personality evaluation, as by the MMPI, or Minnesota Multiphasic Personality Inventory, and other measures. They need revaluation by TRU, TES, individual differences, and the variations necessary within each person's life. To a great extent whatever you persistently want and will to be, you will be. However this requires not a passing desire or choice, it is a life long endeavor and fulfillment of identity with TES and TRU. The actual measure of a person is not monetary wealth, physical appearance or possessions, or seeming powers over others, it is more how largely and for what each person uses these and all their Á powers in TES and TRU. It is their CALP of TRU and how applied to life that is the actual measure of a person. The redemption of man through Christ to God or TREQ is the most TRU aspect of Christ's Spirit. How all persons can be redeemed or brought to their counterparts in GOD, TREQ or TRU, TES. The beauty of Easter is in the profound revelation of human and Christ's suffering that we may live and avoid such suffering. This shows the weakness and failing of religion, God and so many systems. To be redeemed by a simple belief alone, or to follow Christ in a limited manner is not enough. One must know what Christ and God are, as they are TES and TRU. One must know why and how they work, and their precise part in this, so that one's life is complete, fulfilled and perfectly TES and TRU.

Fig 5-74

ID, EGO, SUPER-EGO and RELATIONSHIPS

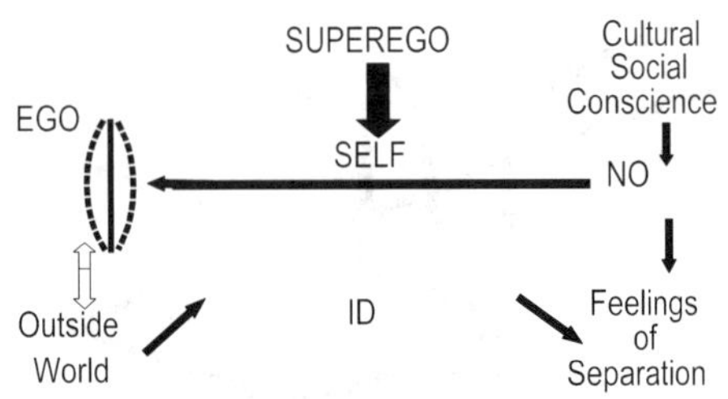

The relations between the consciousness, the id, primitive forces of sex, aggression, self, ego, superego, conscience, cultural and social customs and feelings of separation, form a large part of a person's life. These relationships can be demonstrated both within themselves and by CES in DYNS of TRU. A person may feel bored, without care or interest, with little feeling or thought for a life that is worthwhile. These are reflections of individuality, despondency, alienation, depression or other states of flight from or loss in life. They indicate failure, poor use of one's being and the world around one. It is their combined CES by DYNS, balance, and moderation that show our TES and TRU strength and power. For seeing through a thing is useless without interest in seeing it through, and interest in seeing it through approaches uselessness without an interest in what is valid, actual, proportional and total.

PROSEP = Projection of Separation. We usually believe what we want to believe. Yet sometimes we are forced to believe what we have to believe. This latter is a kind of hewing into line, that occurs continually in life. It keeps us in line and in humility. Capacity to adapt to this is our success. It also keeps us on guard, more alert and ready to defend and live up to our potentiality, actuality and partiality in totality, TES-TIT-TO. Its correction is by K̂OT, PUREACLITY, KUPS, TES, COGNOS, ELTICMA, SÂ and TÂ.

All who hold opinions, notions, biases, ideas, and beliefs, and use them carelessly, impetuously and forcefully to the detriment of the common good, TREQ and TRU, should be ignored, limited, or alienated. This is TRU's reaction, as it has been generally to the extent such actions are wrong and negative. To the extent the counter reactions are within the limits or SCARLS of TRU, non participation and selection out of TRU. It is a great stimulus both to those who are not, and those who are TES and TRU. A stimulus whose resultant is the improvement, assurance, and achievement of this special world of best human life, TRU.

Without an understanding and appreciation for the deepest TRU meaning of humility mankind will suffer turmoil and self destruction. Humility is that state, characteristic, and mind set of a person who knows and appreciates what they have been given and what they have to do to fulfill their part and role in TRU. All must have great humility, especially to all that is ES, CES, TES, and TRU. The less they know it the more they should have it, humility. Yet the less they have it the less likely they are to appreciate it. This is the great value of higher education, and especially knowledge of TRU.. It provides the knowledge necessary to appreciate the actual world and all it provides. Thereby greatly extol knowledge, the devotion to it, TRU or TREQ. For TES and TRU humility shouldn't be an inhibition to any and all activity that is TRU. It should only be a guide that limits one to what is best and prevents what is the worst, what is not or against TRU. It should be the tone of interpersonal relations in which one and all acknowledge the role of what is other or different in the whole, and supplicate themselves before the absolute TIT.

LOFE = Learning of Felt Experience. LOFE signifies how the individual must more than learn knowledge. Each must seek and be helped to gain the original felt experience from all the great ideas that developed into the concepts, principles, and laws TRU. The glorious feeling of accomplishment and happiness that follow great discoveries is what best produces satisfaction. In literature and other media we gain a glimpse of the great ideas, concepts, and virtues within human life. Yet we must not despair because they are not always with us. We need not despair if we don't know God, or GOD precisely and fully, by VAPO and SCARL, as TRU. It is in the constant improvement, and returning to GOD as TRU and TES that the great promise for each and all persons is to be found. It is like perpetual salvation and redemption, only on an ever increasing and more massive scale. The great conflicts and suffering both in, out, and OITIO of TRU are the stuff and intense stimuli that produce LOFE and PROSEP sufficient to raise us to TES and TRU. We can gain this *Confessions*, as did St. Augustine, and *Consolation*, as did Boethius. There is great interdependence between suffering, fortune, philosophy, foreknowledge and God or TRU. In this interdependence, and search in life for truth and happiness, we succeed through TES and TRU. We can respond either negatively to nothing, or positively to all, in TRU. It is to provide this: **LEP** = Level of Proficiency, necessary to meet these two extremes and all in between and other that is, and is the key to, TES. It mostly requires intelligence, great learning, comprehension of TRU, good health, longevity, action, service, inspiration, optimism, cheerful associations, conviction, courage, and a profound dedication to all TRU.

Obedience is like duty and responsibility when perfect, a

combination of the best of laws with the best of conscience and understanding in the performance and achievement of the best of worlds, God or TRU. When TRU is perfect or **PERFT,** in contrast with **PROPT** or the Problem of Optimal TRU, it is better to be obedient to TRU than to be obedient to less than TRU. So also must we forever learn, understand, practice and live by TRU, that which is the best of worlds so that our obedience is not wasted, in vain, or productive of more for which obedience and all are in conflict, turmoil, loss and failure.

Even a perfect woman is not what a man may fantasize. For no one woman or man contains all beauty or virtue. Yet a man seeks to find that woman, a boy that girl of his dreams, a girl or woman that prince or Mr. Right that meets all these desires. It is an absolute and final aim and goal, which is never realized. So it is of TRU and each and all of our relationships with TRU and all about us. TRU is like the perfect woman or man, not a dream but actual reality, only that with which each and all must live. This can be done, and in only the right ways, if each and all live, succeed, endure, survive, fulfill and achieve, not what we may imagine, but what must be. It is wonderful to discover the best mate as TES, or even more, to be TES oneself. Especially when done by discovering and living the great ideas, concepts, principles, laws and mechanisms of TRU. And to do so with the added superlative state of enhancing TRU, and accentuating all for an exponentially glorious TESÁ, TRU, and TÁ. The great advances, and driving power behind them, in the last million or so years, has been the exponentially increasing determination and focus on what is right. What is right has not been discovered precisely, but by the common experiences of what all that most practically enables the well being of humanity. So that we have a well spring of wisdom whose greatness is only overshadowed by its disorder and our denial of it. To live a happy and successful life doing our best to contribute and be a part in the actual world of human destiny means that we must search for and achieve TES, our best role in it. This means that we must learn what is right, know what is right, think what is right, appreciate and love what is right, say and do what is right, get others to also do the same, be confident and gratified in our service that fulfills this right, and hold these to God, TREQ, TRU, TROS, TIT, TIT-TO, TÂS, and all their aspects, FAF, ES and CES.

The current status and dynamics of civilization and human life have been based upon those who have persistently turned their great ideas into pure actual realities. These realities are the crux of human mass energy, its concepts, terms and uses that best support TRU. The great technical and material advances and benefits that have been discovered and improved were not possible without the basic concepts, principles and laws of the basic FAF and mechanisms involved. Reversely, many of the great ideas and concepts resulted from great discoveries and applications of technology and material improvements to human life. Their interactions and interdependence are one of the greatest proofs of DYNS.

Each person's entire life is a testing ground, separator, and selector of the actually existent from all else. Each person is neither dead at physical death nor alive during any or all of ones lifetime. Our being is through and through interactive and highly interdependent with all that makes up the relative universe. Therefore we must discover, learn, know, remember, use, account for, act upon, and accentuate what is needed to be. To be only and none other than what actually is and all that enables human life to survive and succeed.

The Doctrine of Intuition is how we can apprehend our own true nature. This will, sooner or later, reveal itself to be TES in TRU in common, generally, and TÏS specifically. This was especially evident from Ch'an Buddhism where our true nature reveals itself to be a Buddha-nature. This is the spirit and the theme of the Sung Painters in China. By spending much time in meditating on nature, they sought to lay bare, its spirit. This is the emphasis upon personal recognition of what the actual world is, in contrast to all that is not. It is shown by the increments of CES in DYNU, DYNS and DYNO. There is an awe inspiring exaltation to be had in learning and attaining those concepts that are the keys to CES, Homation, heights of IAT, and TRU. The infinite, eternal, highest principles, absolute love, ego, future hindsight, Pureality, Cognos, and thousands of existing and new concepts, terms, their combinations, interactions, and totality are what enable us to identify and become, in TES, TRU.

D.Q.(DIQUO) = Dis-ordination Quotient, D.Q. is the amount, measure, or degree an individual, group or humanity,looks at or does other than fully coordinate and find the equilibrium necessary for TRU. Much of this ordination is by insight and a comprehension guided by TRIT, ACEXIN, OBT, and SPALT.

That commonly known as good, is what most characterizes human life in TRU. It is our discovery, and determination of what is best for us and our world. What is the good is of all TRU, all levels of Á FAF in ACEXIN.(VAPO) The measure of a person's ability to transcribe, reason with, and practice the highest good, TRU, is the most reliable indicator of TES. This is how we should judge others and all.

FAVI = Fallacy of Vacillating Idols. Idols are like EBE, Extremes By Emphasis, whose vacillation tends to accelerate, accentuating their negativity to become chaotic. It is seen in the human habit of constant unreasoned extolling and degrading. this and that. It is somewhat a consequence of the ignorance, a feeble attempt at expression, partiality of IA and natural selection. It is lacking in VAPO, and when upon a, many and all things, **UOESO**, Under and Over with variable Emphasis of Subjective-Objective.. Yet it is rarely qualified at all and even less properly qualified by TRU. This is a large reason for IPTRU and PUREALITY. For it is here, largely, that we seek to help others and strive most directly to achieve the proper qualification of all, as they are TRU

MEASURES of DYNS CES of SUSP

Term		Measure		
ANTIS		~-5-4~	-	M
ASA	(75)	~1-5-9~	-	T
ASI		~2-5-2~	-	T
ASIT		~2-5-4~		T
CHARACTERISTICS of PERSONAGES				
COP		3-6-2		S
CIPTO		4-7-2		M
COGNIM	(75)	4-5-8		T
DESPAIR, CONSOLATION,				
DeC		1-8-1		S
DO(R)D)		4-6-4		M
DOCTRINE of INTUITION				
DoI		0-8-4		S
EGO, ID, LIBIDO, SUPEREGO,,				
EIS		3-6-3~		T
FAVI		3-6-3	-	M
FAITH, MERCY, HOPE, CHARITY,				
FMH		3-6~-4~		S
GOOD, The		3-5-3		S
HELT		4-7-4		S
INDIVIDUAL KNOWLEDGE, F.U.T.				
IKF		1-7-2		S
INIAT		4-7-4		M
INITIAC		4-7-3		M
LEP		4-6-3		T
LI		4-6-2~		T
LIS		4-5-4~	-	M
LOFE		1-7-3~	+-	S
MAJOR LITERARY CONTRIBUTIONS				
MLC		3-6-2		S
MMPI		2-7-3~		S
PERFT		~-~-~9~		M
PERSONALITIES and INTERESTS				
PaI		3-6-3		S
PERSONAL KNOWLEDGE	PeK	1-7-4~		S
PES		4 5 3·		M
PROSEP		3-6~-4~	-	M
PROPT		~ ~ ~7~		M
RES		$\overline{4}$-$\overline{5}$-4~	+-	M
SUFFER, COMPASSION	SuC	0-6-3		S
SUSP		2-7-2		S
TESÂ		~3-5-9~		M
TESTAR		~3-6-7~		M
TRUQ		~ ~ ~ ~	+-	M
UIA		$\overline{3}$-$\overline{6}$-$\overline{3}$	+-	T
UIAT		3-6-7~		T
UIRE		3-6-8~		T
UOESO		2-6-4	-	T

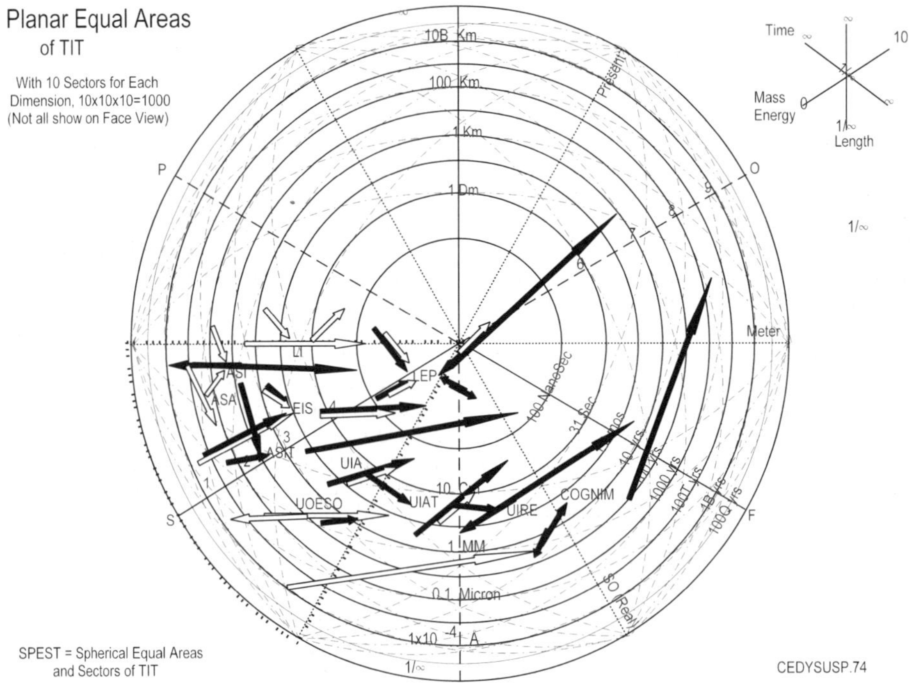

Fig 6-74 **DYNS CES in SUSP of SPALT**

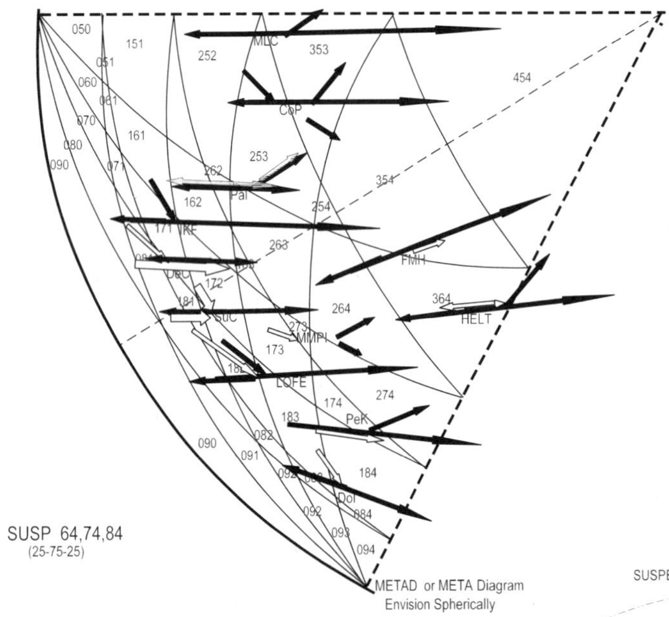

Fig 7-74 **DYNS CES of SUSP SECTOR in SPALT**

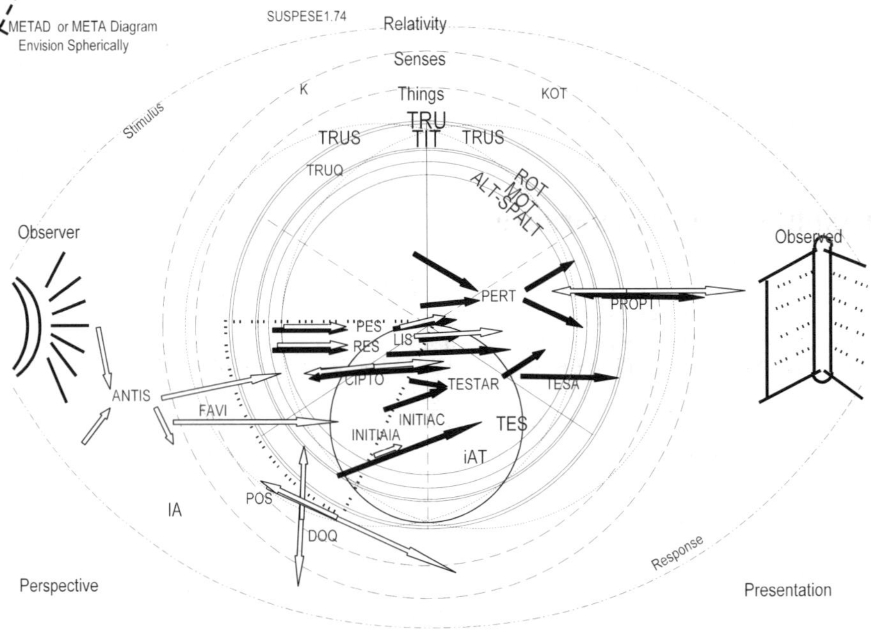

Fig 8-74 **DYNS CES of SUSP in METAD**

CHAPTER 75

SUSF or SUBJECTIVE SMALL FUTURE PROCESSES in DYNS

CONSCIOUS, IA, WANT, CHOICE, THOUGHT, READ, KNOW, DECISION, SPEECH, ATTITUDES, TRAINING, TES

Concentration is the focus on, full attention to, and effect needed to receive and accomplish certain selected aspects of ones Field or Place in Life, **FOL**. Concentration is on that before us. This should be equivalent to TRU. This is: **COIT** = Coincidence of Concentration and TRU. COIT is the basis for, and often first phase of IAT, TES, PUREACLITY, COGNOS and all TRU. The only true, correct, and complete IA, is IAT, the simultaneous comprehension, visualization, and focused activity of all one's TIT, THEO, POT, MOT, ALT and HOM. It is by way of the attainment and the application of this to one's personal life, that PUREALITY is formed, and becomes TES. **ENIAT** = All that ENables enhances, broadens, deepens, actuates IA of TRU. ENIAT is very important, the development, stimulation, and cultivation of best habits of ENIAT in all that benefits TRU.

Fig 1-75 **START, CONSCIOUS FOCUS, SUBJECTIVITY, I, and TRU**

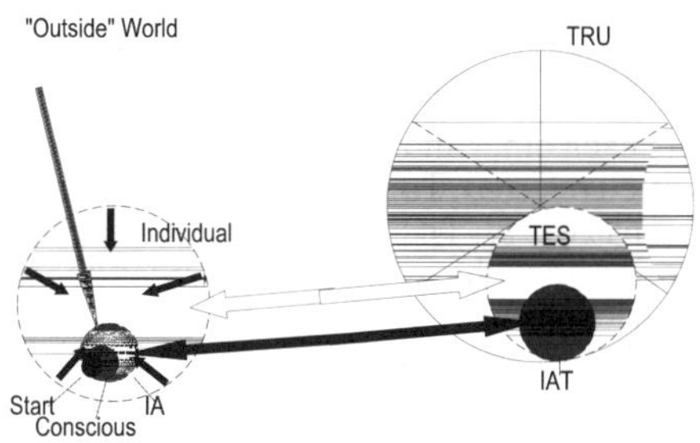

Where does one start? One usually starts from one's focus, ones consciousness, what is being concentrated on, the given or what stimulates or precipitates it. This incipient consciousness quickly becomes one's IA. Failure to recognize this focus and initiation of being is where knowledge has erred. It has been intensely recognized by some systems or ISMS in philosophy, only to be completely ignored by others, religions and most sciences. In phenomenology, psychology and intuition we delve deeper into the split second origins of our conscious realization of being and life. However we can't, on the other extreme, consider this to be all. It is only a start, part and focus. It could be made from any site in TRU or SPALT, as evident from ATO. So that although IA is great, both by REM and the focus of so much of the dynamics of TRU, it is far from being all. Each person must clearly learn, distinguish, and properly apply the differences between these and self consciousness, ones egoism, IA, IAT, TÍS, TES, absolute being, power and TRU. Within the consciousness and intuition of each person there exists, at times, a wonder, awe, concern and feeling of absolute being, a self consciousness that leads to egocentricism, mysticism and the absolute. It stands in sharp contrast to and contradiction to TO, TRU, TRIA and the limited counterpart and counter function of the person within TRU. This disparity can be explained by more complete, intricate dynamic and transcendent understanding of TRU, TROS, TRIS, TAS, TRUD, OM, SA, TA, and BASFANT.

Consciousness is the basis for IA and proof of TRU and Á, PRUT and PRÁ. How else can humans live than by their wits, the sharpening of consciousness? No physical entities or matter has consciousness, or the special, typical and unique kind of being that so specifies man, gives competitive edge, and predominance. Consciousness is, as it is of IA, the focus of our world, TRU. It is closely associated with being alive, in contrast to that which is dead. Like lightness in contrast to darkness, and all compared to nothing. It takes many forms, characteristics and properties. However its central focus and gist lies in the pivotal nature of its role, contribution and control over the world, the relative universe of which we are so much a part It has through biological evolution and its own growth and development in TRU, arisen out of primitive life forms and grown with the expansion of the central nervous system and brain. This is characteristic in a more limited way in mammals close to homo sapiens. In this way it is not confined to man. However with man being its dominant carrier as well as extant focus, it is closely identified with us, as we are TRU. It is not necessarily TRU. Both to a lesser extent and its use to a greater extent are often far less than TRU. Even non, against, or negative to TRU. It is fused with many other kinds of subjectivity, or mental activity whose names and aspects in human life and TRU have been increasingly realized and defined, by psychology, philosophy, religion, and science in general. In psychology and the health sciences its role helped to differentiate all conscious states and scale them by the way and degree to which they best function in life. Consciousness acts as a good cut off point or line of demarcation, which is so needed by human mass energy, Á. It can be used to start many psychological measurements. Such formal distinction is through analogy what can be of great benefit in **PROTÄ** and **PRUTÄ** or the Problem amd Proof of the form of TRU. The emphasis on consciousness in medicine refers to the clear polarity that exists between life and death. Functional and dysfunctional mental states give us vivid insights into our being. These provide added evidence and proof of not only the predominant role it plays in TRU, but how this is done thru its mass energy component. How it reveals the subjective polarity in the observer-observed relativity parameter, Á dimension of TRU.

If one would doubt for a second the eminence, greatness, profundity, subtlety, validity and reliability of TRU, and if one would question the value of what is great and what is small, what is good and what is bad of TRU and ANTRU, let them only in every moment of their daily lives analyze their choices, desires, wants, likes, dislikes, loves, and all that effects them. Only as each person develops and appreciates conscience and the consequences of all in their relation to TRU can we be responsible and successful. **CONOT** = Conscientiousness or a Conscience for the Consequences of All in relation to TRU. Without being conscientious, or with a conscience, CONOT, a person can not be TES or TRU.

To be equivocal produces much of the worst and little of the best. For there are many who either directly or indirectly, consciously or unconsciously give or sell you not what you want or need but what is worthless or even harmful. These are much like the adage, "buyer beware" and "a fool and his money are soon parted." Hence every person must know what is best. They must know so that they ask or demand what is absolutely essential, the best. However this should at all times be qualified by what is best by TREQ or TRU. To be at peace and harmony with the world does not mean inaction. Quite the opposite, it means action. Yet we should quickly realize that this action is special, highly qualified and limited, quite different than what is commonly or naively thought. It is action for TRU and by the individual through TES. They are qualitative and quantitative in TRU.

In the CES of human relations there is a constant and forceful operation of judging. This is same and different than selecting. This does not mean that any and all persons should or should not judge any and all. For judgement, in a qualified and TRU sense, is a key process of life. Judge no more nor other than what this is in TIT or TRU, qualified and meets its VAPO and SCARLS, necessary, the best of ROT or objectivity, appropriate for the given or circumstances by the giver or to the recipient, and indicated by ones TRU relations to the given and circumstances. All of these depend upon the degree of KOT. The scientific and professional world abstains from being judgmental. It is in being empirical they come to realize that much remains to be discovered. Knowledge can only be based upon verifiable truths. This is why science has failed to develop TRU. Because TRU, as TREQ, has been unknown, highly complex, interactive, and subjectively open to abstruse interpretation needing toleration.

MON = Measure of a Man, or of a person. It doesn't take but a moment to get the jist of, figure out, sense or know what another will think, say or do, or how friendly they are. This can be quickly determined by the questions they ask, their answers, statements, countenance, what their thoughts are on, and what they are doing. It is quickly evident from their interests, wants, desires, and demands. With KOT this can be expanded to become readily known as to whether, to what degree, and how they are or are not TRU, MON. It is of great help to both IPR as well as to spontaneously establish the ideas upon which choices and decisions are made that will be most productive of TRU.

Fig 2-75 **KOT, The BASIS for LIFE, THOUGHT and ACT**

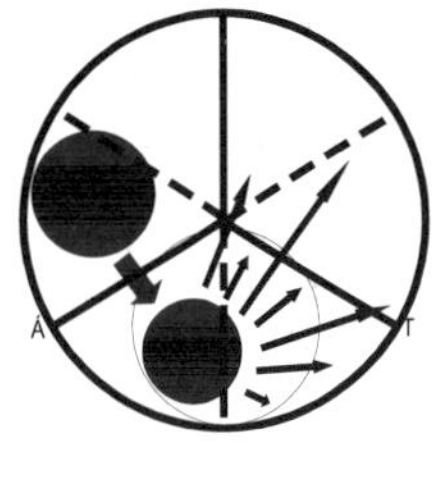

There is no complete freedom in life. Life is a hard taskmaster, especially for modern man. There is such a thing as a range or degree of freedom of choice in life, even as TRU. When less than TRU, as in existentialism, the range is much greater, although often in an opposite or negative way to TRU. It is this range or degree of choice in life that is so unrecognized and difficult to understand, especially for the young, immature and naive. We do not appropriately realize our part in the world, TES in TRU. We should live, think, behave and act primarily from the learning, knowledge, reason, application and love TRU provides. **RATIN** = Ratio of Attitudes, Intention toward TRU. RATIN is the ratio of attitudes for TRU over those for ATRU. A very simple quick measure of each person, group, state, culture, nation, and the world may be attained by assessing their attitudes. **RACPIN** = the Ratio of Routine, Custom, Imitation and Regulation Producing Positive effects against the Reaction Negation produced through lack of necessary Attitudes, Conviction, Interest, Inspiration, and Appreciation of TRU.

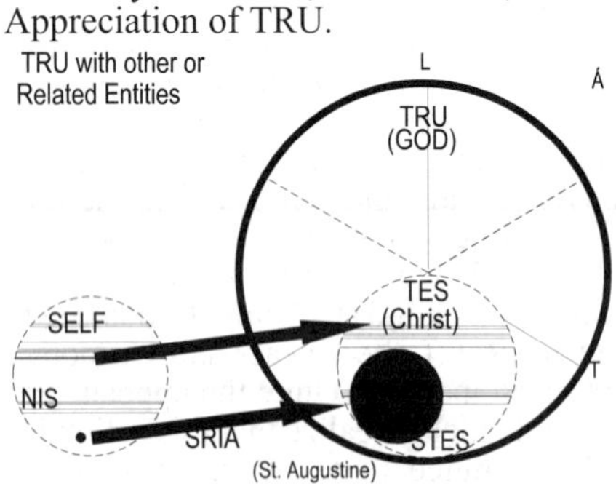

Fig 3-75 **SELF, TES, NIS, SRIA, STES RELATIONS**

STES = Self's dedication and deep realization of ES and TES. STES has been, in TREQ form, recognized by many through history. It has been taken to extremes, EBE, by some. STES is most greatly and clearly seen in St. Augustine, who with profound recognition of the contrary view, saw how penetrating and accurate we can rhetorically depict God, TREQ, or TRU. Failure to recognize God produced a profound feeling of guilt, as related in *Confessions*. This stimulated STES identity and role in serving Christ, one's soul or TES, enabling a great dedication to GOD or TRU. What accentuated this was a reaction to immature or youthful indifference and separation, in an ANTRU world. **NIS** = Negation, Indifference and Separation of the Immature Individual Self, to TRU. It is in training, education, and cultivation like that which must be given to the pear tree, that we can overcome NIS and provide STES. It is through confession, commitment, conversion, Sudden Realization of the value of life and IA, **SRIA** and repentance for PPF sins or ATRU, that TES and TRU, as ones soul and God are attained.

Conscience is the intrinsic power that stimulates a person to be their counterpart in TRU, TES. It is an inclination to think, reason, and act in ways that prevent negation or harm. It is a felt realization of one's duty to loved ones, group, society and the world. It is what a person feels when they do or do not fulfill their role in TRU, TES. It is more or less imperative and corrective depending upon the experience, learning and best Á FAF to relieve one's failed identity. This requires the needed degree of achievement, assurance and self satisfaction that comes from fulfillment as TRU is achieved.

Each and all persons must have a knowledge that is sufficient, appropriate, coordinated, and comprehensive. It must be a knowledge guided by VAPO and TRU, SPALT and adaptable to ready use. This knowledge is one in which philosophy, religion, and all fields and sciences are no more nor less than their parts and roles in TRU. **LOKISU** = Law of Knowledge in which the Subjective and Objective are mirrors, or Identity with each other and to OBT and SUT. To tell people to do right is often resisted. To make them do right, or to do it for a person doesn't work. The same applies to learning, the learning necessary to have a good foundation and organization of knowledge. Just as a mother tells her children, 'just listen to me and do as I say, and everything will be alright', so each of us must listen and learn with a minimum of experience and loss, to know and be TRU. Yet we will, because of our own individuality, do right only when we are confident that this is right, and know what we are doing is right. **OTS** = Opportunity to Study, learn and be TRU. **COTS** = Corollary of the Opportunity to Study, or its intrinsic use and effect. **FRAPOV** = General case of COTS and OTS. COTS is a product of OTS. All persons should equally and best utilize their opportunities and not waste them upon all that is not TRU. It is to REM, TES, TIT, TRU and TÁ that our pursuits in study, as in all life, should be directed.

Desires, dreams, wishes, hopes, purposes and goals are much of a person's life, the generator of TES and TRU. They are TES accentuated by the hard reality of TRU. The search for hopes, goals, objectives, and purposes in compensation, - whether attained or not - make us more able to sustain, and endure the boredom, suffering, drudgery, unhappiness, conflicts, defeats and failures of our lives. Only when TRU can we endure the pain that is a part of life and turn failure into success.

A great problem we have that must be remembered and corrected is the loss of reason that comes with repetition. In routines and all that is done automatically, there is little stimuli to reexamine the reason for use and action. **LORHÁ** = Law of Reason and Remembrance when Repetitive or Habituated. The correction for LORHÁ comes with KOT and ECRO of ES, CES, EC, NEC, POS, ACC, and COGNOS.

Tell me what a person has read once and I will tell you what his memory is. Tell me what a person has read twice and I will tell you what his interests are. Tell me what a person has read several times and I will tell you what his convictions are. Tell me what a person has read many times and how he has used and lived by these and I will tell you what his immortality is. It is primarily in reading that knowledge is propagated. The quality of this knowledge is very much a function of the quality of the reading. The quality of this reading is a function of the books and literature available, one seeks, has read and is interested in. It is for this reason we must always increase our interest, concentration, love and joy that can be found in the highest quality of books and their study. These are the books that contain the greatest amount of TRU. They are of all philosophy, religions, fields, sciences, and technology, that have contributed and best relate to knowledge of what is TREQ or TRU. One can't do, act, and achieve without the necessary knowledge. One can't have knowledge without learning. One can't learn without study,experience, investigation or examination. One can't study, experience, investigate, or examine without the reading, incentive, opportunity, and preparation. They are complex levels in the progression of our predominant mass energy or Á form and function in which each step is necessary for the whole. **LIPÁ** = Law of the Imperative Progression of Á, or Preparation, reading, study, Learning, Knowing and Achieving. LIPÁ and education are closely related, only education without LIPÁ fails, just as each step in LIPA fails without the other and life without TES and TRU.

Interactions between Levels of the Mass Energy Continuum and Dimension

It is not enough to know about length, time and mass. It is not enough to know the interactions between the different areas and volumes in length, durations and variations in time, or levels and higher powers of mass energy. It is even not enough to know how these different varieties of dimensions work as continua in the special universe of best human life, TRU. Only when we can fully describe all the major interactions between different levels of mass energy in the spatiotemporal world of human life and posit them in models that accurately replicate TRU, can we have the thought, comprehension, and kind of knowledge sufficient for managing and controlling our lives. Thereby, we seek to penetrate the more complex and total interactions of the various levels of mass energy typified by TRU. To begin this we view **MESOA**, how these different levels have developed, exist, and where they are taking us. We then examine each level of Á, SÁ, and search for the processes through which they interact to work for human survival through the ages. Each higher level could not have succeeded without the support of each lower, and in various times succeedingly lower levels. The form and function that makes up TRU present is the carrier of all these successful increments. So that what supports TRU now is how this form and function works to uphold TRU. This is recognized in the fitness, health, strength, predominance, and natural selection in humans. In turn much can be gained from comparative studies with other animals. Only for humans it takes, as does the mass-energy level, a whole battery and combination of higher levels. Within each individual, and their interactions with others in a social world, much depends upon the health and best use of these levels of mass energy. To be ones part and contribute most and best to TRU takes a highly proficient total battery and combination of levels. The higher ones being the more evident such work, communication, and influence in various forms based upon good judgment from vast knowledge, comprehension, and overview of the actual world. These in turn come from great learning, training, self improvement, and good health. So it is down through all the many levels that make up an individual's capacity and ability to be a part of TRU. It is the interactions between these many levels and their posits both simply, multiply, complexly, and totally in the models of TRU that provide the kind of knowledge needed to understand TIT-TO. This is the emphasis of much of POS, HOMATION, DYNU, DYNS, and DYNO. The immense finite and interactive relations, forces and vectors are only begun in CIPTRU. Their discovery, investigation, analysis and demonstration will take many years work by many in collaboration. Yet this is the meat or substance which will measure the precision and closure necessary for a well completed presentation, and life, of TRU. Such well designed methods of approach to knowledge and TRU greatly help direct research. as in science. Just as the rest of discovered knowledge has helped to give us a basis, albeit qualified, for knowing and showing TRU.

LEVEL of Á	0	1	5	10	15	20	30	40	50	65	80+	post
,u	10	10	10	10	10	10	10	8	6	4	2	1
Pu	9	9	10	10	10	10	10	9	7	5	3	2
Eu	8	8	9	10	10	10	10	9	8	6	4	2
Au	7	8	9	10	19	10	10	9	8	7	6	3
Mu	6	8	9	10	10	10	10	9	8	7	6	4
Gu	4	6	8	10	10	10	10	9	7	4	1	5
Adu	2	4	6	8	10	10	10	10	9	7	5	3
Cu	1	3	5	7	9	10	10	10	9	8	7	1
Empathy	+/-	1	2	3	4	5	6	7	8	9	8	2
Rational	-	+/-	1	2	4	5	6	7	8	9	7	5
IMuAxiolog		-	1	2	3	4	5	6	7	8	8	7
Intellect			-	1	2	3	4	5	6	7	7	7
PUREACLITY				-	1	2	3	4	5	6	6	9
TESÂ					-	1	3	5	7	7	5	10

Table 1-75 **GRADES and LEVELS OF TRU for DIFFERENT AGES IN HUMAN LIFETIME**

This table could as well be done for % of Á or level of SÁ. These tend to escalate exponentially to higher and higher levels and greatness. There are many prerequisites to each age. For example it is not possible to realize SUT and TRU without a certain degree of ACEXIN in IA or attention. Nor is it possible to have an understanding of concretion by TRIT and TRU without a certain degree of realization. There is an average for humans in the world today. There is an average of what all could readily reach if they had the opportunity and will. There is an optimal practical level. Those of only 15 or 20 years of age can reach the highest levels, with the exception of a certain degree of experience and maturity.

INEMTA = Interpersonal Empathy and communication to TRU Absolute. For TRU, IPTRU and PUREALITY, not only is it essential to have a complete aggregate of great concepts, principles, laws, and mechanisms of TRU, emotive experience with TA-TÁ-TÂ, and love of Self-TES-TE -TRU are also needed. This is shown in the ideal and spiritual love of Plato and Christ, enhanced in conversation and communication.

OPTH = Optimal Health, care and condition. OPTH is the best possible care of oneself, as shown by PUREACLITY, building oneself up, general and preventive care. The search for the maximum in health, capacity and ability is all a part of getting the most out of life, TES to TIT and TA.

The key to minimizing risk is training. It is also the key to KUPS, human success in general, and of TRU. This is training in all that is of great value to TES and TRU. A good example of the importance of training is seen in rescue work, planing for emergencies and disasters. Training like **CE** or Continuing Education and KUPS in Pureality, is an ongoing process. In an advanced civilization this is an ever greater part of life and TRU, and should parallel degree of responsibility. Each person should not only be trained from an early age in visualizing by three dimensions, they should be trained in all the best methods of inductive and deductive reasoning. For only when they can think rationally and logically in three dimensions focally, partially, and solidly can they comprehend and find their place and role in life, TES and TRU.

It is impossible for every person to know all there is, all knowledge of every subject. It is even impossible for every person to know all there is of TRU. It is impossible for every person to avoid all negative, ATRU, ANTRU, NAMTRU, evil, bad, and wrong. However it is essential that each person be exposed to as much TRU and as little ATRU as possible. It is practical that each person be exposed to a enough of what is partial, special and potentially TRU, that will likely suit ones problems and special circumstances in life. In being exposed to as much TRU knowledge as possible is the Universal Knowledge of **KUPS**.(95) This is what has been so vigorously sought after by a few, yet lacking in most humans through history. Universal knowledge may seem simple or even obvious, as in a unitary God. Yet it is most complex and comprehensive. All that detracts from the motivation, stimulation, interest in, and the attainment of this greatest knowledge of TRU, must be anticipated, recognized, concentrated on and in all ways prevented or corrected. **ATIM** = All Time Mastery of Things. The fact that with TRU, SPALT, HOM and ACC once TRU knowledge, KOT, has been discovered and learned it is a part of TRU. It is available in undiluted and undiminished form for all to see, learn and use at all times. Each person can have ATIM. This is one of the glorious and beautiful heritages we receive, and when we develop and avail ourselves of it, can so benefit both ourselves and TRU. This is so for all information that has value, REM in TRU. It is especially evident for those discoveries and new ideas that show more and more about the world around us, both TRU and less than fully TRU.

Fig 4-75 **ENUCA; ENERGY UNITS, CAPACITY, and Á FAF**

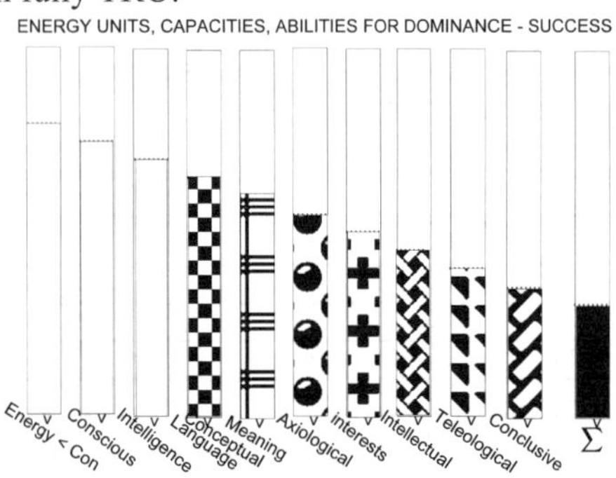

A person's mental qualities and characteristics largely determine their success in life. It is upon these the quality of the worlds population and success of TRU depend. These different mental qualities and characteristics are neither complete nor necessarily identical with the units that make up the mass energy, Á FAF in TRU. They are primarily of the subjective half of the S-O or Á continuum. The first, which is also largely physical when fully Á, is: **ENUCA** = Energy Units, Capacities, Abilities, Á. ENUCA is collectively a large factor in TES. Persons who have a full, healthy, and a strong complement of subjective or mental abilities will have much less success if they do not have the physical quality and energy. All subjective units and capacities have, more or less, physical bases, in their form and function. This is increased if associated, related or supported by other units and capacities. The line drawn across each is roughly a line of how much is inherited or how much is acquired. The lower segment being inherited and the upper one being acquired. This is figurative and diagrammatic. It is is done in even steps to represent the gradations this relates to in the increasingly higher powers of effectiveness and success in life. Although genetics plays a large part in all, it is through training, education, experience, associations, influences, stimuli, competition, judgement, self coordination and conclusiveness to effectively use all one's abilities for best advantage, that we succeed or fail. In the future, as has been increasingly so in the past, man's natural selection will depend less on physical competition and fitness to maintain quality and more or higher levels of the mental or subjective capacity in the competition of SÁ.

An outlook on life may be very different in different persons. Even in the same person it will often be very different at different circumstances, times and ages. Each individual tends to be more involuntary than voluntary, more habitual than rational; what best suits them at the moment of their comforts, interests, needs, desires, wants, and personal gain. Like ISMS various outlooks on life can be more or less of great value or dis-value at times and in different circumstances. To be TRU, however, takes a fusion of all that is the best of each and an elimination of all that is increasingly ATRU.

Fig 5-75 **DEARNESS and ABSOLUTENESS OF LIFE, IMMORTALITY**

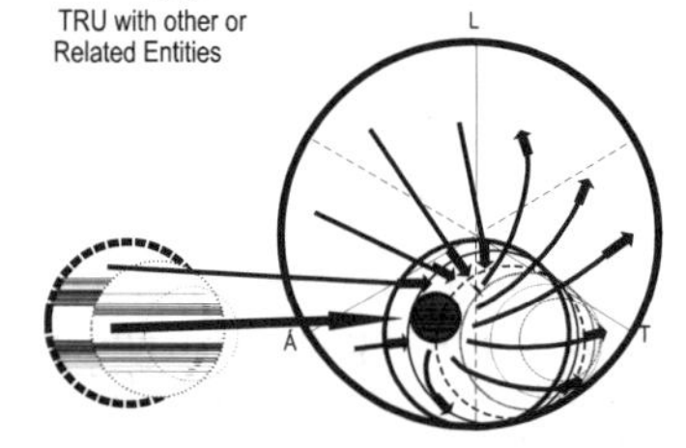

When we are young we, fleetingly yet very dearly, feel the love for and glory of life. With increasing age this becomes less fleeting yet a more resolute love and glory in life. One who can't, nor does not want to understand life and death, only their momentary joys, misses the whole meaning of life and its continuation, everlasting life. On approaching death we increasingly realize what little is left. We recognize that if there is to be anything more to it than this little that is left they must leave something. This is something they have missed, lost, most loved, been most interested in, worked hardest to achieve, or what seemed to be most needed or valuable. They know that it must contain more or less the seeds of everlasting life, or immortality, that increasingly tends to TRU and TÂS.

There is much more happiness to be found in a life of fulfillment by TES and TRU than in a life of partiality, triviality, mediocrity, senseless or negative action or enterprise. Thereby learn TRU, take heart in all that you have been given, use it wisely, frugally and most effectively throughout life. For evolution, and especially our human origins, tell us those who survived were those that had that extra edge of energy, mental superiority, and worked hard and effectively to be prepared for and meet the contingencies of life.

MEASURES of DYNS CES of SUSF

Measure	Value	
ATIM	4-5-8	M
CE	3-6-7	T
COIT	1-7-7	T
CONSCIENCE CON	1-8-5	S
CONSCIOUS FOCUS,S,I,TRU CFS	2-8-5+-	T
COTS	4-6-6	T
DESIRES DREAMS HOPES DDH	1-8-7	S
DOUBT, DISCOVERY, LEARN DDL	1-8-6	S
ENIAT	4-5-5~	M
ENUCA, Subjective	1-7-6~	M
IAT	2-6-~	M
IMMORTALITY, Life, ILD	2-7-7	T
INEMTA	0-8-9	M
JUDGING and TRU JaT	2-7-8	M
KOT, thought, act, life KTA	2-7-6	S
LEVELS TRU HUMAN DVLT LTH	4-5-~	T
LIFE and TRU (Per) LaT	4-5-7	T
LOKISU	~4-5~-7~	M
MON	4-6-6	S
NIS	2-7-5-	M
OPTH	?-7-7	M
OTS	~4-6-6	T
OUTLOOK on LIFE OoL	1-7-7	M
QUALITIES,CHARACTERQAC	3-6-6+-	T
PERMANENT KNOW.ATIM PKA	2-6-7	M
RACPIN	4-4-3	M
RATIN	2-7-6	S
READING	2-6-5	T
REASON, POWERS of RPo	0-8-7	S
SELF CONSCIOUSNESS Sco	1-8-8	S
SRIA	2-8-5	S
STES	3-7-6~	S
SUSF	2.5-7.5-75	
TRAINING THOUGHT SUCCESS TTS	3-7-6	T
WANTS, CHOICES, DECISIONS WCD	2-7-6	S

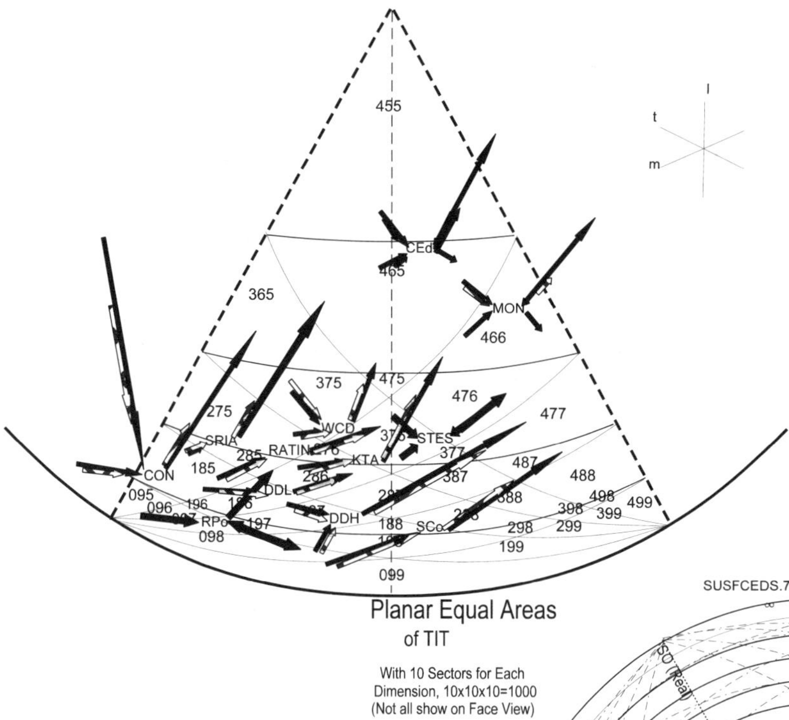

Fig 6-75 **DYNS CES in SUSF SECTOR of SPALT**

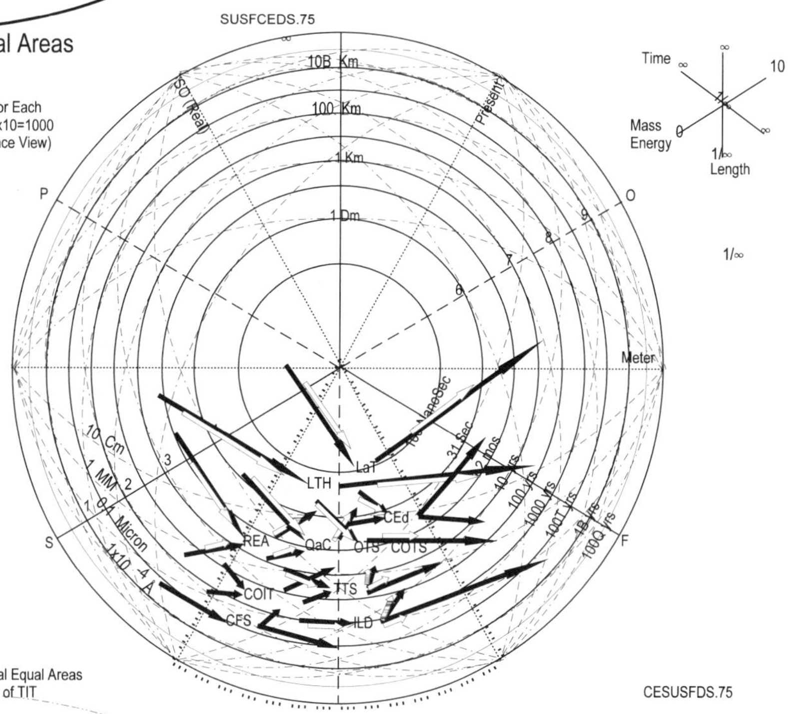

Fig 7-75 **DYNS CES of SUSF in SPALT**

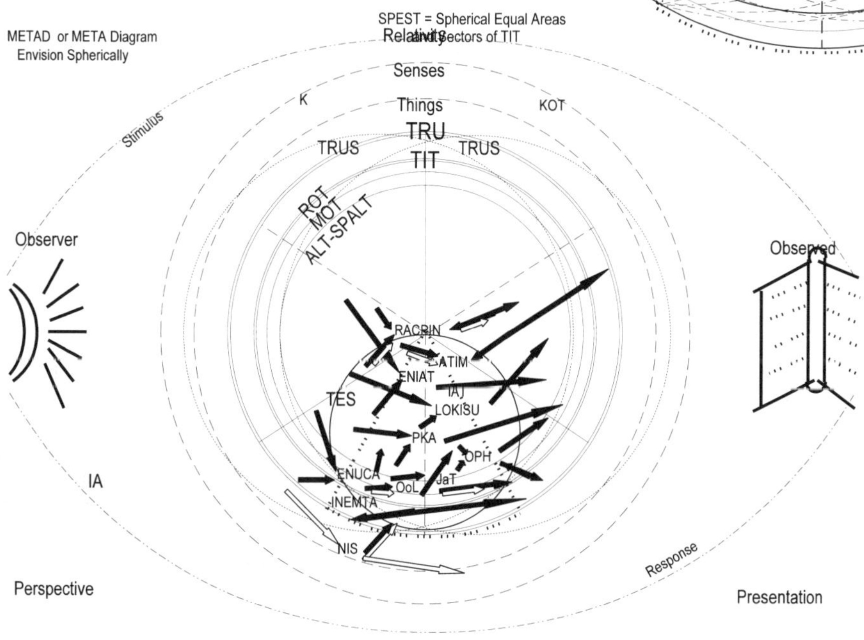

Fig 8-75 **DYNS CES of SUSF in METAD**

CHAPTER 76

OSF or OBJECTIVE SMALL FUTURE PROCESSES of DYNS

DO, ACT, ACTION, HELP, BEHAVIOR, PERFORMANCE, USE, IMPROVEMENT, PREVENTION, TES, FULFILLMENT

The bottom line of all representation, presentation, description, demonstration, conversation, rules, regulations, tenets, directions, and requests is what do I do about it. In summation these are many and diverse, yet convergent on TRU. This dynamism is neither chaotic nor without purpose. It is set by TRU, and this is largely by the Á FAF of TRU. Yet we can not truly know what is right, do what is right, know what is wrong, and avoid or not do what is wrong, unless and to the degree we know and use TRU. Only when I, the first person, comprehend, remember, apply, and accentuate TRU most effectively, efficiently, productively, happily and joyfully can I readily answer the question what do I do about it and best act in the world in which I live. How to do right is for I, the first person, to constantly and fully realize, learn and apply TRU in life. It is the greatest means to success, ACC, TES, TRU and TÂ. It is more or less the basis of most religions. This is exemplified in the Psalms, "To live in the paths of righteousness for his name's sake all the days of my life". To totally identify with all the words, EC, and NEC that make up the Magic of God, **MOG.** To get the most out of life, that of ones kind, and all TRU. To live in Purity, Reality, and Actuality of TRU, PUREACLITY, and all that enables TRU, ENT.

When to tend to something and when to let it run its course. When to hold them and when to fold them. This is one of the hardest things there is to know. If you tend to it to soon it interferes with its natural processes. If you let it run its course and something serious is setting up, you can be taking chances with life threatening effects. So that the best way is to know exactly what you are doing, acting according to established precepts of each kind of case in TIT, and managing it appropriately. This is a good example of how much and how exacting knowledge is, and can be refined by TRU. Yet life is neither a simplistic dream nor an infinite nightmare. With the order attainable with TRIT, OBT, SUT, ROT, MOT, and the control attainable over finite intricate detail by VAPO and SPALT we can sufficiently know what, why, when, and how to and not to tend to all that is given or confronts us.

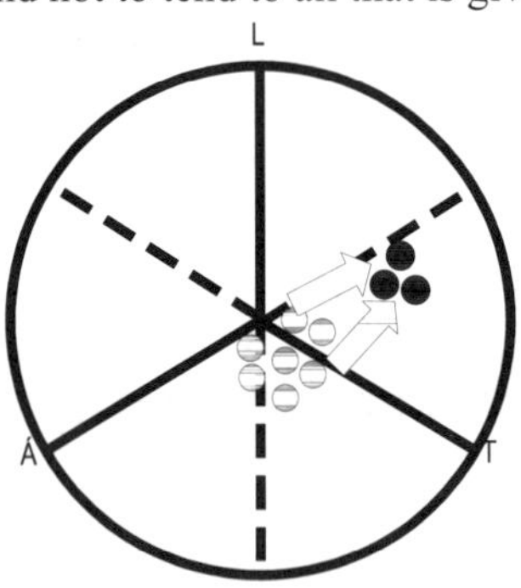

Fig 1-76 **FINITE NUMBER to EFFECT RATIO**

FER = Finite Effect Ratio. How many things can a person work with at a time? How many can one do without peaking out, making mistakes, losing efficiency and thereby effectiveness? This varies per individual and level of consciousness. It is a function of: 1. Familiarity, 2. Training, 3. Practice, 4. Concentration, and 5. Intelligence. There is a threshold of finite effect, or FER. An example of the intention, attention, concentration, and the precision necessary in life is the need to speak and write plainly, precisely, and perfectly. This is vital and paramount to communication.

IA - - - COMMUNICATION - - - IA

It is a crucial element in the S-O relationship and scale, the Á FAF of TRU. It illustrates the reality and core of TRU.

The greatest battle of all is the battle within oneself, to do what we want, is popular, customary, the "law", or what is TRU. Do we think, say and do what we want to, what others want us to, what we are told to, or what we will or should in various other ways? Only as the individual approaches TREQ or TRU by TES is this battle won and these imperatives satisfied. Only do that which is the measure of all this special relative universe of best human life, TRU, and if it is what I, others, am told, or the law, all the better. For if what you, the authorities, law, and all tell you to do is TRU then all are in concert, TRU, so will you do what is right. This is not to say that we should deny or defy our wants, others, our superiors, or the law of what has been known to be, or can be the best. All must be balanced by what is perfectly TRU and the degree to which it is not TRU. Likewise in being partial it will be different, and depend upon time, place, means, and given. If the law, rules and regulations are nearly perfect then the greatest compliance and accommodation must be made. However, unless we know TRU, we will never know whether these are right or wrong and what will be the best to do.

Rather than tell others what to do, or to try to get others to do right, it is often better to focus on ones own need to do right, TREQ and TRU. For a person can best anticipate and appreciate both the danger and bad in what is close to them and what they can learn by experience to enjoy improvement and achievement in TES and TRU. Another person will resent, resist and react to being told or forced to do. The second person looks at things differently. Their first reaction naturally and often tends to be different, one of contrast or opposition. Even more strongly, as in argument, the second person is in an antithetical position. On the other hand one can very beneficially by indirect and general well formed means introduce, expose, or extract from them the right reasoning.

MEP = Make 'Em Pay.(ATRU and ANTRU) We pay for all in life. There is a accounting or balance sheet for each person, for all is or is not as it is or is not TRU. If your early life is an accumulation of false steps, waste, unproductiveness, denial, and defiling all that is good or TRU, then you will pay to the degree you fail TES and TRU. If you have used others and TRU for ATRU purposes, or have not stuck to business and done a good job, sooner or later you, your health, mind, possessions, loved ones, family, and progeny, by TRU will pay. You are blessed with great potentials from TRU. You can use them to whatever degree. **LOVET** = Law of Volition Optimally Enabling TRU. MEP is compelled upon us by our very nature. We may seem crushed by the burdens of life, especially at the end of a hard day. Yet on arising and regenerated in a new day we readily and happily attack MEP to fulfill our part in TRU. They are much reduced by LOVET where the seeming burdens of life as TRU are lighter and better completed in the glory of TRU.

VACOL = Voluntary and/or Compulsive life through the fusion and allowance of all by TRU, in a persons life, Ordered IA and TRU. VACOL shows the compulsory life we have, and must lead. The optimal mixture for TES is one that is both spontaneous, voluntary and compulsory, depending largely upon the level of TES one acquires. Humans can withstand actions that are voluntary much easier than those that are compulsory, depending upon how much they and the actions are TRU.

When help is needed, it should be freely, readily, and cheerfully given by and for all, providing it is measured by TREQ and TRU. This is how, in a positive life, do not hesitate to ask for help if it is for the right things, TRU. For the right people will always recognize right and willingly give. The selective, conditional, and appropriate help depends on: TRU, appreciation for the help and TRU, its best use, elimination of all that may possibly detract from it, availability of the help, and the ratio of cost of the help to its benefit in TIT. We are here to make the world better, as TRU. is.

No matter how hard one tries one can't be good, do right, and provide for the common success unless we know values, have order, purpose and priorities in our every act and entire lives. These are not arbitrary, open for any desire or choice. They are only for the reasons and degree they are TRU. We can readily see in each individual, by their words, acts, and thoughts the degree to which, and the manner in which they are and are not TRU.

By TRU, then, how are we to act, behave, and function? 1. The integral and differential with magnanimity toward all TROS of TRU, 2. Avoid all that is not TRU, 3. Annihilate all that causes loss or is apt to injure or harm TRU or its aspects, FAF, ES, and CES. 4. Practice VAPO and follow the SCARLS of TRU in each and all occurrences in life. 5. Recognize and best utilize the DAOST in all. 6. Use the moderation, toleration, and equanimity of TRU and TES as well as the selection, force, and power appropriate by TRU. 7. Be careful what you destroy, even what you do that you may not realize the destruction it produces.

VOC = Voluntary Controls. VOC includes quality control, self control, propagation control, assistance, construction and all programs and management that produce TRU. The better controls are stated and communicated, the more effective they become. They can be used as a guide to: safety, propagation, study, IPR, friendships, character, interests, work, basic needs, work and activities voluntary controls are much superior to forced routine, involuntary regimented or mandated controls. This is because only the person usually knows what, when and how their life works best, how to be applied and used. Controls that are derived from a person often represent a knowledge of what is at hand. We do not profit from experience if we do not develop the reason needed to explain the meanings involved. If we are to support TRU, or even prevent the extinction of our species, we can not leave it to chance. We must be on point, as our station in life grants and enables us, masters of our own destiny. To do less we defy all that has produced us, our Á, ACEXIN and TRU FAF. **TACOR** = Action Corresponding to TRU, or TRU Action Correspondence.

From an analysis of the worlds we live in, what is mostly going for us - thus our greatest blessings - are our bodies and our habitat or environment. This is our anatomy, genetics, physiology, and their interactions.(GELD) The untoward effects on our genetic material, live product, and determining forces are continuous, unrelenting, and increasing. It is unlikely each succeeding generation will be blessed with the benefits and superiority of its ancestry, proband/progeny unless great efforts are made to provide the necessary PPMA and ACC..

There is a tremendous struggle within each person to make a living, get ahead, be better, possess security, be happy, gain power, be predominant, superior, succeed, and have immortality. These are natural, biological, and anthropological characteristics and properties of TRU, TRUD competition, and the quality, Á, mass energy FAF of each person in TRU. They are largely determined by an increasing quality of human mentality. This is also a prime driving force of TRU. It may not be considered to be altogether beneficial or even right, for it is always to much or to little, if only slightly. Commonly when excessive, deviant, or inadequate it is pathological. It is the need, want, motivation, goals, purpose, tropism, and taxis of TRUD that are to be understood if we have the breadth of vision necessary to sustain the equanimity and moderation optimal for this struggle. **LANPÁST** = the Law of Natural, Predominant, Symphonic and Pleasant Á TRU Superiority. LANPÁST arises reactively from the deficit produced by a lack of Á and excessively formal and rigorous exercise of superiority, and predominance.

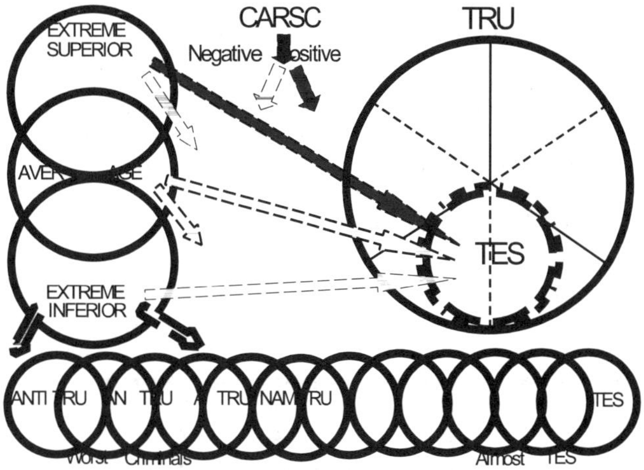

Fig 2-76 **COMPARATIVE CARSC and FITNESS in a PERSON and TES**

The average person, and the two extremes of 1. the superior-inferior continuum or those who are, the most TRU and those who have severe mental physical, and genetic abnormalities and 2. The degree of ANTRU to TES, are shown. We do this by their content as revealed in CARSC, or their capacity, ability, S/F, reaction, setting and conditions. Such a demonstration is of value in the practice of medicine, psychiatry, sociology, politics and the world in general. It illustrates the degree to which this average falls or can be raised, so crucial to the success and continuity of TRU. It shows, not only how mental, physical and genetic abnormalities are inherited and important, it shows how environment, training, character building and education are important. It helps identify those extremes that are most ATRU.

In life there are those who feel they are on top of the world, and those who feel they are at the bottom. There are those who feel nothing or little relation to the world. Of all these what they feel and what they are by TRU are often far different, even opposite. Thus, each person struggles to succeed, achieve, and live in such a way that they get as much out of it as possible. Here is the problem, only those few who live to see how much they can put into it, and this by TRU, are correct. We should do so with the most actual, intense, comprehensive, and extended understanding possible so that our limited time on earth is not only most productive and attuned to TRU, but most enjoyable and eternal. It is to make the near future, past and present as enjoyable, happy and fulfilling as possible to all who are born that can be and are TES, a large part of TRU. These are: **PEDÁF** = Persons or individuals for the Distant Future, or for TÂ, **PENP** = Persons or individuals for the Near Past, future and the Present. PEDAF and PENP represent extremes and means in the continuum of personal or individual participation in TRU. It shows what TRU expects of them, as well as when and what they receive in return. PEDAF is much of what TRU expects out of them. PENP is much of what they receive in return. There must be a balance between the two. If either are too great or to little they suffer. This results in: **OPET** = Optimal Personal Participation in TRU. OPET is the optimal equilibrium of PEDAF and PENP. OPET, for many, becomes a large part of TRUD, its S/F. How it does or does not PPMA the species, TRU.

We should all try to attain a high level of performance in

TRU. Commonly, an average level for TRU, or **PAR**, would be believed sufficient. Yet it is not. We are all parts of GOD, TREQ, or TRU, if we are at all. We can never, in actuality, be non TRU. To do so is the same as non being. We may vary from extreme to extreme existence or near non existence. This is the absolute law of existence, by TRIA. It is only for us to properly guide ourselves and all that we effect to see that we, that about us, and all so converge and conform to TRU, providing this is the right one, TIT. When we set out each day, we should take great joy in our capacity, abilities and accept the challenge in our life's activities for TRU. To use them is to stimulate our and others interest, inspire and revel in the joy and happiness, glory and victory that TRU offers and produces. We should take heart in the beautiful detail, depth, intricacy, breadth, variation and dynamics of life.

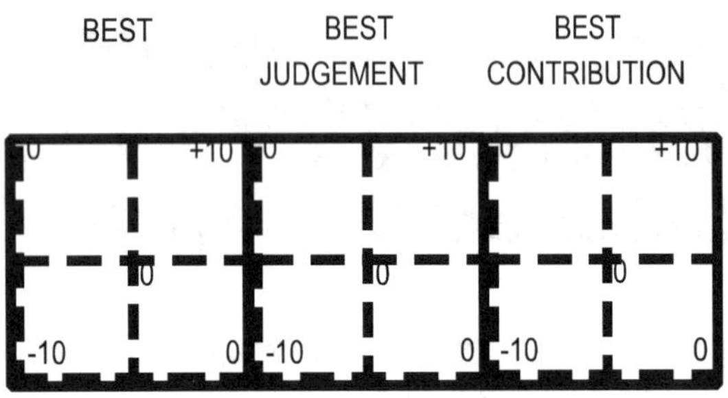

Fig 3-76 **RATING BY PERSONAL BEST, JUDGEMENT and CONTRIBUTION to IPR and TRU**

People can be rated in terms of judgement and contribution. What it takes for a successful IPR. There are three, of many, aspects of people in business that exemplify what enables a profitable relationship: 1, Are they nice, good, best, free from bad? 2. Are they wise, correct, forthright and do they have the best judgement? 3. Are their bills reasonable, paid, and do they make the appropriate contribution? In summation these provide an index of a person's contribution or likely contribution to the business, or TRU. Only if they are the best, their judgement is the best, and they make the best possible contribution will they be of the highest profit to the business, arrangement, interpersonal relationship, or TRU. This is with the added caveat that the business, arrangement, IPR or TRU is likewise the best, TIT. We should evaluate ourselves, others and all to make this the best possible. In figure 3-76 there is a total +30 points that are the best possible. All that goes into them needs to be encouraged and enlisted for TRU. Decreasingly at +20 very good, +10 fair, and all degrees from poor to the worst at -30. It is in this way we can provide a rough quantitative measure or evaluation of all. We can better separate TRU from ATRU for ENT, ACC, and TRU.

Each person's main and ultimate goal in life is the total contribution one makes to the success of humanity, the best of worlds of man, TRU or TREQ. Also, it is in the joy and happiness such fulfillment produces. Each of these has some, more or less, place in a person's life and work. However each person, if we are to be, can only be as we are E, ES, TE and TES, or our counterparts in TRU. This we can not do if we function at great variance and deviance from how this counterpart must function in TRU. This is called: **VATES** = Variance, great, of E, TE, TES, or ones Counterpart in TRU. VATES denotes the great fluctuations that occur within a person, or any entity as they are in relation to TRU, little to none, or much to all of their ESIN, ES, CES, and other of TRU. It is the knowledge of preventive medicine and public health that enables humans to live in ways that provide maximal benefit with minimal loss. In part they make up or increasing population. Humans should live to support all that contributes to TRU. They need to have sufficient opportunity for utilizing all that is available and beneficial of and to TRU. There is a constant balance between what is of benefit and the potential and likely effects, problems, and dangers it produces. The harm/benefit ratios of all are a great portion of the dynamic equilibrium and ACC of TRU.

That a person must be a part of the world is a foregone conclusion. Just as any animal can't for long escape the reality of the world of which they are a part. Yet all will suffer - or cause suffering - to the degree they are not interrelated and in balance with the world. In time, or sooner, each person must return to a state of massive dynamic interaction with the world around them. This is usually ATRU, yet can only, in actuality, be TRU. By DAOST and ones own partial position this may be different and same yet in finality convergent upon and in equilibrium with all in the totality of TRU.

Every part of the body needs to be constantly replenished. The need is to be filled up with what is the lacking, the best, and not what is bad, the positive not the negative for TRU. The lungs, stomach, brain, eyes, ears, intestines, kidneys, and all need to be actively, adequately and properly nourished and used to provide their needed bodily functions. For the brain this is done through education, training, schooling, the joy of learning, and an inquisitive attitude throughout life. For all these, there is much happiness in fulfillment. It is in the fulfillment of TRU, and all its systems, organs, parts, processes, form and function that we can bring together the many participants that best function by VAPO in the whole. It is in the fulfillment of TRU that it reaches the highest level of its potential. It is to fulfill TRU, supplied by each individual as they fulfill their potential, that in summation provides the assurance, and absolute of TRU, TÂ.

The happiness and joy in life are not what they may seem to the novice. Just as naturally happiness is often found in fulfillment so it is not the result of immediate gain or satisfaction. Happiness is a result of planing, behaving, doing, and acting in those ways that provide the greatest benefit not only to oneself, but over the long term to oneself, others and all. All who selectively pursue that life in which one, to the best of their ability, fulfills their part and role in this special world of best human life. In short, happiness, as life itself, is a labor of love. The love that comes from learning, comprehending, preparing, and achieving the identity needed to best adapt to and fulfill ones part in the actual world we must be of if we are to be. The love that comes from seeing joy in the eyes of others as they are lifted to identity with TREQ.

LOPURT = Law of Priority Opportunity to TRU. Those who are most likely to deny opportunities to others are often those who have been given the most opportunities. Opportunity to TRU, as to life, is and must be of the highest priority, for one and all, to the degree they avail themselves. Among the many ways and means of producing for youth and all persons ever higher virtues, qualities and character that enable them to be TES, that of inculcating the best habits of thought and behavior from a very young age are preeminent. Along with this, and almost as important, is the ever greater need to discover and alter the development of reaction formation, complexes and abnormal tendencies and types of behavior in the young, the reduction of potential danger to TRU. Where there is insufficient stimulation the necessity of TRU is brought out so that human responses will not only be right and appropriate by TRU, they are the preeminent solution and fulfillment of LOPURT. When we know what is greatest and most necessary by REM, ACC and TRU and formulate their greatest priorities, more than adequate stimulation will follow. This will provide the response, even if not TRU, that will at least give the impetus, opportunity and improvement necessary to initiate TRU.

There are great individual differences, and a great need for each and all of us to as early, completely, and as rapidly as possible discover, learn, and utilize our aptitudes to be our best counterpart in TRU, TES. To do so there is a great need to likewise prepare oneself for the roles, special functions, careers, occupations, and professions that best serve TRU. They are synergistic, as parts and the division of labor, of populations of humans on earth. TRU can not be TE without TO, nor TO without TE, but TROS. This is one of the hardest things for each person to learn, appreciate, accept, and work by as it is one of the greatest problems of TRU. **LODIL** = Law of Totality and the Division of Labor, and LOPAN. LODIL is closely related to, and largely explains, selfishness, greed and lack of love of God or TREQ, GOD, or TRU. The separation of selfishness, self interest, self preservation and TES is very indistinct without the benefit of a deep understanding of The World, TRU. Yet it is at the basis of an adequate coming to terms with life. For it highlights the contrast between ATRU and TRU as well as the continuum and balance of one and all, TÍS - TES and TO-TIT. Unless a person knows and appreciates TRU fully, one can never be their proper TES. The special work that needs to be done will continue to diversify. It is in division of labor that humans make increasingly superior productions. However, a career can not be all ones life, no more than any ES, CES or TES can be all TRU, nor S be all in KUPS. Foremost of these is our effort to help others and correct wrongs in the world, ACC. This emotive and complex contribution to TRU is: **LADEC** = Law of Differential and Partial Contribution, as in Determining Forces. The significance of LADEC is this very awakening of public consciousness to right, TRU, and the complex and special role of determining forces and contributions of all TIT to its success, continuity, and permanence. It corrects our erroneous ideas and interpretations of what this world, the human species, God, TREQ, or TRU are. It helps us understand how TIT and TRU are more than simple biological evolution, but ATO, with added FAF that become evident when we recognize the special nature of best human life. The role of individual participation in TRU is not only very complex, it is caught between the evolutionary path of natural selection and genetic material, along with the live products who carry it, and the determining forces that make up so much of the whole. When we limit ourselves to TIT, as we must, and when we add to this the vantage of the first person, we can begin to recognize the corrections that LADEC supplies. This is fundamental to the questions and problems of, what is right and what is wrong, who is and who is not, and what is and what is not TES and TRU.

Each person should have, in their lives, sufficient return and relief from being unfairly or unjustly burdened or destroyed, to enable them to produce a positive or TRU life. **RAPOR** Ratio of full or Total Positive Return. If not spontaneous this return should at least respect and enable one to work for their own TES, and that of others, and all for TRU. To see that all is equitable by TRU.(VAPO) TRU pays the price for the burdens of adversity, conflict and injury. **BURTES** = the Burdens of TES. BURTES are the degrees, steps, and levels of higher and higher responsibility. They are the burdens, load, energy, power that is necessary for TES to be produced and continue. BURTES are light when compared with the burdens that lives have and suffer without TES and TRU.

Aging is better understood by viewing, in its broadest role, all processes involved in growth and decay, regression and senescence. This transit and a measure of aging used by the insurance industry has been, Gompert's Equation. This equation holds that after 30, the likelihood of death doubles every seven years. It, although variable by several other factors that act as tangents and in summation gives one of those rules of thumb that help to show TRU. Aging does not begin later in life, nor even in maturity. It begins in heredity, conception, fetal life, parturition, childhood, adolescence, sexual maturation, and throughout life. It is accumulatively set we think lives so long and seems so organically eternal, as the bristle cone pine or the Sequoia, do not actually live for 4,000 + years. Actually it is alive for about thirty years, the inner parts are all dead. So it is for man, much is more or less Dead And we Don't Know it, **DAK**. Man undergoes a continual rejuvenation in life, especially of that which is TES. Yet humans have, compared with most animals, very long lives. This is necessitated, certainly in part, by maturation and our dependence upon the mentality essential for success and survival. **MATSUR** = Maturity and Mentality, to Survive. MATSUR holds that a species will have individuals with longer lives when they require longer periods of maturation for survival. This is an adaptational mechanism most evident in mammals and acceleratively so with primates and humans in the last several million years. The edge it provides humans is most delicate. For such rapid changes in human evolution have not come without their price. These great changes that have occurred poses some of the greatest problems, threats to existence and need for accentuation both in the present and future.(ACC) The continual balance and dialectic relationship between maturity and rejuvenation, beautifully illustrates the through and through KDM, INEG, and PIP of which man, as TRU, is constituted. It is through their discovery, determination, demonstration, POS, HOM, CES, CHES, interpretation, and how they are TIT, that our search bears fruit, and life is well fulfilled.

There are limitations on cell division and genetic continuity that are only compensated for by repeated regeneration and replenishment. If cell division is limited to about 50 generations of cells and if life span is effected by this, at least in principle or indirectly if not directly, then what will it take to increase life span, and what will it take to assure indefinite continuity of genetic material? This is a statistic that has been confirmed in cancer research. Its impact upon the continuity of the species, subspecies, races or genetic groups in human populations in the future remains to be seen. Advances in genetic engineering and enhancement may produce increased to infinite sexual/somatic cellular division. Although the continuity of TRU does not necessarily hinge on increased life span this certainly would markedly improve TES and TRU.

Self Improvement and Human Survival

Present civilization is so ripe it is rotten. We largely ignore or deny it. Our philosophy as advanced as it seems, is fragmented. Our science and technology as much as they have discovered to make life better have failed to give us individuals who can cope with their altered and complex nature and future. Our religions have completely lost their role in guiding us to find and identify with a world that is different aspects of the same thing. The actual world can only be a heaven on earth, united in appreciating and sacrificing for all that is common and best, and understanding, tolerating, and adapting to all that is different in life. What is best for one and all is both simple and quite available. It can and must be well known by everyone if we are to continue to fluorish. The only thing is that each person must be ready, willing and able to have the interest to discover, the energy to provide the needed self improvement, and bring all to fulfillment in this special actual world of best human life. The human race can't exist when natural selection is upset irreversibly. When individuals no longer know, live, nor face their roles and make their existence into the essence of being equal to human survival. It takes intelligence, education, maturity, interest, principle, reason, and sacrifice throughout a dedicated and vigorous lifetime. For those who are not, serious, ignore, and deny human survival and their own self improvement necessary, we can see why they, as in nature, will be dominated, subjugated, decimated, and destroyed. Let us not waste the great world we can have by seeking to raise all that is hopelessly lost to an ideal life when they neither wish nor intend it to be.

Investigate, discover, demonstrate, and join in this less obvious, but more enjoyable beautiful and wonderful actual world of human life to meet your challenge to be the kind of person nature has intended, and human mentality can so gloriously enable, you to be.

Do you care what happens to humanity? If not you will suffer the consequences like those who havn't. If you do care, get at it, do what is necessary. You owe it to yourself, to nature, and especially to your immortality. Thus you must identify yourself with what is best in human life, learn its principles, study and map out its processes. You must with great validity, actuality, proportionality and totality focus all on their role in the future in the most intricate beautiful detail that does justice to all you have received from this special and marvelous world of best human life.

PASGOLM = Principle of Avoiding Specific Gradations of Human Lineage. Who is superior and who is inferior, predominant or not, is a critical problem and reason for much of the struggle in life, survival of the fittest, and cause for strife in history and TRU. If strictly TRU it is a natural, selective, and specific part of life. It is often NAMTRU, ATRU, and ANTRU. For this we must avoid, prevent, and protect ourselves. Selection thereby needs to be increasingly learned, perfectly managed, and controlled, indirectly if not directly, in the long term if not short term. This is not possible unless it is well understood. In the more distant past this was settled by trial and error, natural selection and survival. The science of genetics, and its specialties of genetic screening and engineering although rapidly becoming more reliable are still only beginning to have what is needed for overall management and control, especially on as broad a basis as total human population, and TRU. Animal breeding and propagation show many of the principles and what can be done if properly applied. However, the long-term effects in humans, are so critical, that exceptionally accurate and complete knowledge and techniques are needed. Often much worse can be done than, 'let nature take its course', or laissez-faire. On very limited populations greater discovery is possible than on the total species. The religious, philosophical, scientific, technical, humanitarian, social, and other implications make it necessary that all the concepts, principles, laws, and mechanisms of TRU be accounted for to best adapt genetic and total screening and engineering to TRU. They seek to best develop how we can use our knowledge of genetics to prevent disease, improve health, improve the quality of life, the quality of progeny, and the permanence of this special universe of best human life, TRU. The close bond between determining forces, live product and genetic material make any manipulation and alteration most critical. This is seen in PIP and LADEC. What is and what is not best so far as individual, and even group selection can be described with some, and rapidly increasing, validity.

For each person it must be clearly understood how much is and how much is not TRU. Each person must bring all in motion by many factors affecting health. For example what necessary to right oneself with all that is right with the world. To do so, each must know a large and increasing amount about TRU, TROS, and TIT-TO. For if we are to know TRU, TRU must be discovered, represented, and its knowledge, representation, and presentation refined and communicated. This is part of: **LASTE** = Law of The Strong Bond between TRU and E or TE, or TES. LASTE is '*the tie that binds'*, what so strongly supports TRU. It is the one for all and all for one that so strongly supports family, social life, communities, states, nations, and the world, TRU. We look at things at one time individually, at another totally, but rarely do we look at them both individually and totally separately and together in the vast dynamic complexity of TRU. This is why we have not had the necessary level of understanding and why it is so imperative that we discover, learn, love, and comply with TRU.(POT) The struggle of life, failure and success in TRU is a never ending endeavor for all. Yet with TES, TRU, PUREALITY, COGNOS, ELTICMA, SÂ and TÂ they are greatly facilitated. With these we become superhuman, like immortal Gods whose heroic deeds can do all, within the limits of TIT. Everyone can have such super powers and immortality when TES by working to accomplish all for TRU in the most propitious way. We live in very momentous times, times in which both the world, as TRU and in other ways, is rapidly changing, a world in which the individual is often at a loss, confused, and overwhelmed. It is for each person to make this struggle for success, not failure, in life, and to do it with the great blessings, benefits, and happiness TRU provides.

What does TRU mean to me? What must I know, remember, act upon and live by? ES are not ES, TES are not TES, TE, TÍ, and TÍS are not with changing times, even though TIT is essentially stable. Ones TES and even more ones TÍS increases or decreases with time. This occurs differently and variably both from the ATO, TRUS, TRIS, TES, and individual points of view. Currently if only pure TIT, we would be 1/10,.000,000,000, and of portion of TES in all time aprox. 1x10 to the -12th power, individual per 100 years to the -14th power/year, and per variable attention, action, etc. to the - 18th to ∞ power. Such a minuscule presence of ones being,, or consciousness in life, even when largely TES, seems to contradict all the first person feels so sure of. Herein is the paradox, whose solution explains much of TRU, TES is, the first person, and the deepest meaning of our existence inn TIT. It is resolved by the great concepts and laws of TRU, wherein although TES and TÍS are small for the individual, their identity and thereby great role - in what is so great and vital for TIT-TO is assured. This is by LOPAN or the Law of Partial Necessity. Like the heart, brain and all are so critical to the human body, so are the vital parts of TRU, as TES and TÍS, critical, and thus so great in TRU. It is this great dichotomy that both drives TRU and gives us TES. It shows how we are not as an individual of this and that, but are as we are no less than our part and roles in TIT-TO.

MEASURES OF DYNS CES IN OSF of SPALT

Measure	Value	
ACTION; What, How, When, Reasons ACT	8-5-7~+-	T
BEHAVIOR IND; How, BEH	8-5-6	T
BURTES	5-6-7 -	S
CELL DIVISION and GENETIC CON.,CDG	9-8-8~	S
COMMUNICATION and IA CIA	5-6-5~	T
COMPARATIVE FITNESS in TRU CFT	~6-6-6~	S
COMPOSITION of IND-S WORLDS CIW	~6-5-5~	T
DOING; What, How, When , Reasons DWH	8-6-6	T
FER	5-6-6	S
FULFILLMENT; IND, PARTIAL TRU FIP	~6-6-7~	S
GENETIC SCREENING and ENHANCEMENT GSE	7-8-9~	S
HEALTH; Rules of Good HRG	7-6-7	T
HELP; GIVEN and RECEIVED HGR, , ,	~8-5-6	T
INTERPERSONAL BUSINESS RELATIONS IBR	6-4~-6 -	M
LASTE	~5-6-5	M
LODIL(T)	~7-6~-~	M
LOPURT	6-5-7~	T
LOVET	5-6-7	S
MATSUR	6-5-7	S
MEP	7-4-7~ -	M
MOG	5-6-6~	S
OPET	~6-6-7~	T
OSF	7.5-7.5-7.5	T
PAR	7-6~-7~	S
PASGOLM	~7-6~-8~+-	M
PEDAF	~6-5-8	M
PENP	6-6~-6	S
PERFORMANCE,Best PEB	7-6-6	T
PERSONAL LIFE, S/F, TRU PLT	~6-5-7~	M
PPF	~-~-~-~	M
RAPOR	~6-5~-7~	M
TACOR	~8-5~-7~	T
VOC	5-8-8~	S
VACOL	~-~-~-~	M
VATES	5-6~-7~	M

Fig 4-76 **DYNS CES of OSF SECTOR of SPALT**

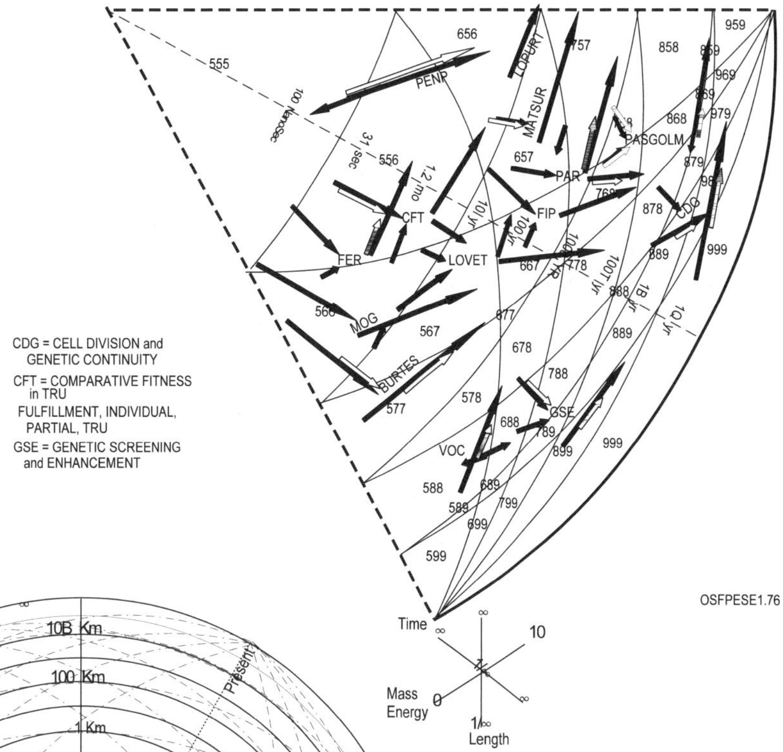

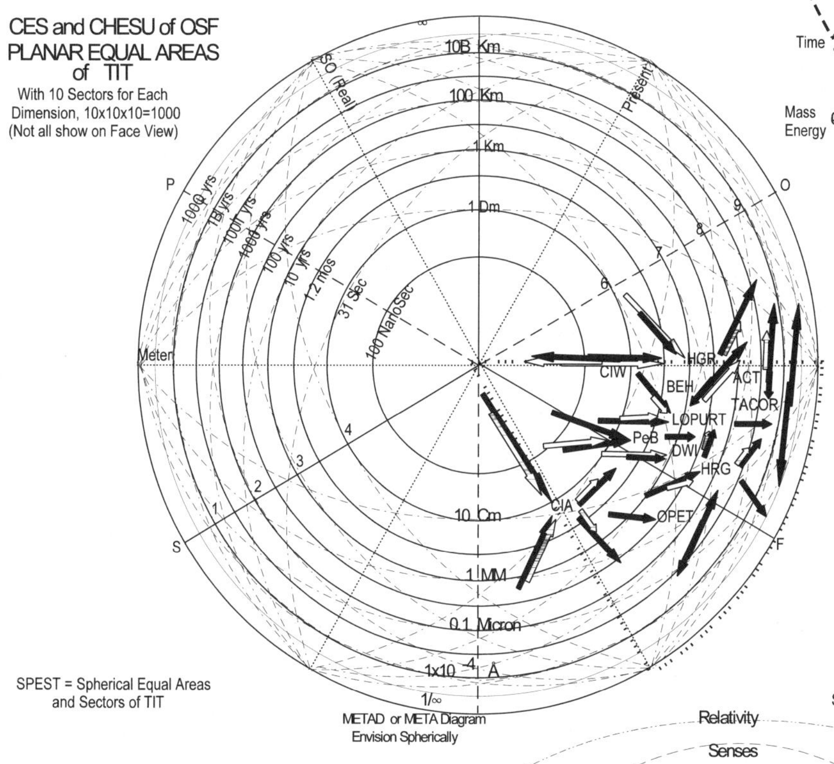

Figs 5 and 6-76 **DYNS CES of OSF in SPALT and METAD**

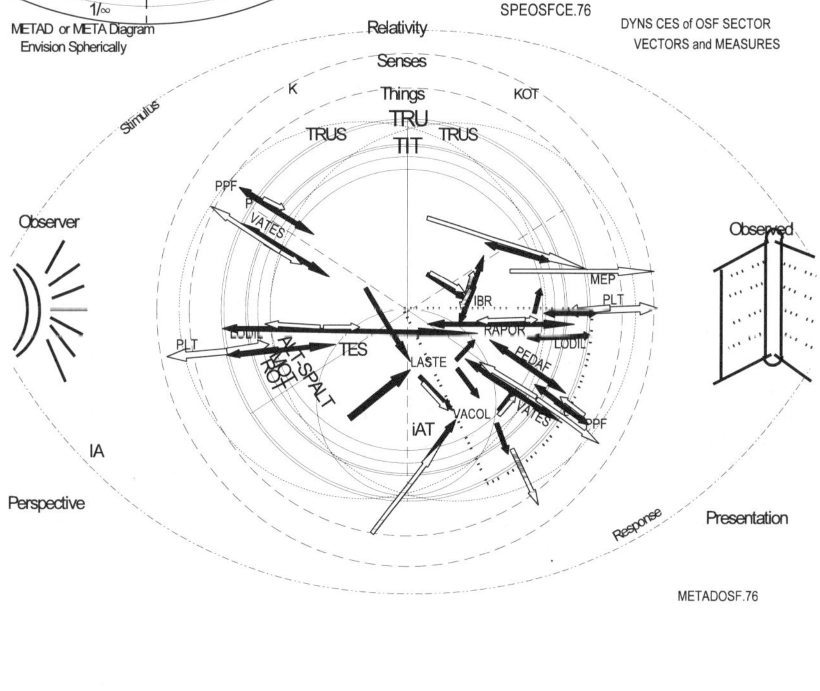

CHAPTER 77

OLF or OBJECTIVE LARGE FUTURE PROCESSES of DYNS

ACTIONS and SOLUTIONS, DUTY, CRIME, CARING, RIGHTS, CLASSES, SUCCESS, ECONOMICS, POLITICS

Religion, ethics, and morality purport to show what is right. Yet we cannot accurately know and do what is right, know what is wrong and avoid or not do what is wrong unless and to the degree we know and know the right reasons why and how. This requires the collective summation of TIT-TO and TRU. Only when I, the finite first person, comprehend, remember, apply and accentuate TRU most effectively, efficiently, productively, happily and joyfully, can the question be answered, what shall we do about it. Increasing violence, war and suffering is largely a product of increasing population density, want, the brutality of more powerful weaponry, and a lack of humanity, or TRU in peoples. This is a lack of mercy, moderation, tolerance, charity, and appreciation for all that stems from a comprehensive love and understanding of TRU. When peoples are reduced to poverty, bondage, misery, despair, and face death, their needs and even more their priorities, depart from what is best for TRU and become focused on less rational simple essentials of life. It begins with here and now, myself, yourselves, ourselves and all to do what we know best and is actually best, that which is TRU. Thereby it is in all we use and use it right, in all ways most TRU, that greatest value and thereby return, is achieved.

The design and use of **VACOL** of IA and TRU should be done properly with utmost VAPO, QUOL and SCARLS of TRU. Yet there can be no doubt that the degree of freedom or liberty in contrast to the degree of restriction or confinement must parallel the degree of TRU participation of each person. This is called: **VACOLP** = Voluntary and Compulsory Ordering of Life and Limitation of Persons or Society Parallel to Participation in TRU. All that is not proportional to, or has the best benefit/harm ratio, or +/- for TRU must eliminated, corrected, or improved. All that is highly important and greatly enables TRU must be stimulated, expanded and refined by TRU or the occasions that best become TRU.

Is premeditated intentional crime worse than unpremeditated or unintentional? One might believe and jurisprudence has largely held that premeditated crime is worse. Yet to disclose their differences one must dissect their elements and describe major factors. These include intention and crime, TRU, ATRU, DAOST, REM, DYNS and others. In the sense that a person with intention, will, and premeditation has, the opportunity of making a rational decision or are capable of and do not exercise this ability suggests an added offense against TRU. However is this any worse than the person who lacks the ability for premeditation, intention and will? For this shows a great defect in much of higher SA levels in TRU. Thus, in final analysis, unintentional crime can be worse than intentional, especially if the latter is for TRU reasons. What may be crime in one jurisdiction may not be, or be lesser, in another. This shows the great responsibility the criminal justice system has in considering and dispensing its judgements. The injustice of their punishment is especially vivid when the accounting and punishment are not TRU. For here there may be little or no crime, and what is TES is punished. **LAFJET** = Law of Fair and Equal Justice can only proceed from VAPO, TIT, TROS, VAPO, and all TRU. Where the evil and crime is such that little TRU is involved, then the punishment, degree, or the manner of punishment will depend upon how much ATRU, ANTRU, or ANTITRU. **LEDMOP** = Law of Equality of Degree and Manner of Punishment with Degree and Manner of TRU Damage or Potential Damage of the crime. Crime can be largely prevented by the rigorous, appropriate, and voluntary training of potential criminals. The salvage rate, however, is variable. The simple measures, which have been in use for management and control, have been limited. Like action needed for ATRU and ANTRU only VAPO and TRU will suffice. The compliance necessary to be TRU is heavy and hard, yet the alternative for TRU is many times worse.

Among the parameters of a person's life there is that of whether or not and how much they care and struggle for a better world. This powerful intention typically produces, and shows, divisions of parameters, means, extremes and degrees. Of its two poles there are those who care deeply and struggle for a better world, and those who have no care or compassion, needlessly abandon all for power, self interest, gain, pleasure, or worse ends. The first should be complimented, stimulated, encouraged, assisted, honored, and built upon. The second must be corrected, ignored or loaded with burdens of the world they so abandon and destroy. This is to enable them to have the experience necessary to appreciate their blessings and find salvation, or pass on to the oblivion they seek and deserve. May all forces in our social, legal, economic, political and intellectual lives be directed to best manage and control this parameter.

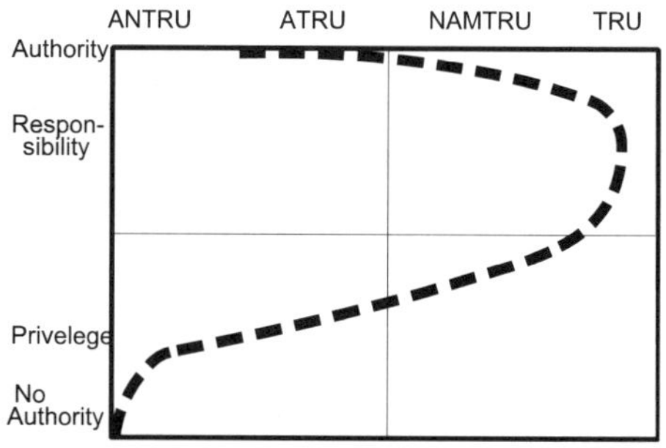

Fig 1-77 **RESPONSIBILITY, PRIVILEGE and TRU** The resultant or heavily dotted line shows that the greater TRU is there tends to be a point where authority and responsibility, begin to decrease. This continues as privilege and influence abuses responsibility and authority to become ATRU and ANTRU. It shows how much privilege, even responsibility and authority, when not TRU can be as bad, if not worse than none at all. To be their counterparts in TRU or TES, was in early times shown by a crown or halo. In earlier tribal societies headdresses, crowns, and halos were used to symbolize authority, dignity or sanctity. Combinations were evident as early as 3,000 b.c., in the crown of upper and lower Egypt. This also shows an early beginning of abstract meaning. It was one of many means found practical to increase social union and compliance. It may or may not have been actual, only an illusion of authority. The interactions between these two parameters, as well as their relations to many others, are good examples of the combined and complex processes of TRU. This is especially so when it helps to define and determine their optimal state in TES and TRUD.

Friendship is not so much through acquaintance as through common bonds and kindred spirits,. This is readily evident in the S of DAOST of TRU. For much that is not TRU, or even ANTRU has in it much that is same to each other, yet not at all of TRU. It is through their mutual convergence in TRU that actual friendship is to be found. Thereby to be a true friend one should be TES in TRU and support TRU in all ways possible.(VAPO)

<u>ERA</u> or Equal Rights Amendment is the hallmark of women's

liberties. It however is a part of a much greater dynamic state. Qualitatively to be the same one, and quantitatively to be what level or degree in the whole. This is the balance, means and extremes of the rights and limits of all participants in the world. This carried to its totality in TRU becomes a matter of proportion, share, distribution, value, REM, and their operations for all that is or will be TRU. Equal rights can be misleading. It implies only a smaller contribution to TRU. It implies a guaranteed right of a partial or segment of population to value, benefits and return. This is true or false depending upon whether these are TRU or not TRU. So much of a demand for rights is not TRU. To this extent rights, or ERA, is false. Associated with this is the relationship between rights and responsibilities. These also can only be TRU. For to expect rights without responsibilities, or responsibilities without rights, is futile, very ATRU, and productive of conflict. Thereby the great emphasis should be on a joyful responsibility in TRU. **ERT** = Equivalent Responsibilities in TRU. ERT signifies the fact that by becoming TRU we can determine what each and all person's responsibilities should be. For, we can gauge the unequal and varied responsibilities for different ages, conditions, sexes and other factors in a population. When TRU is known, and all that contributes to TRU accounted for, it is possible to recognize who is doing, can do, and how much this is a part of the whole. When weighed in terms of both responsibilities and rights it becomes: **ERAR** = Equivalent or Equal Rights and Responsibilities, **ERART** = Equivalent or Equal Responsibilities and Rights by TRU. ERART for all ages should be optimal TES for that age, state, condition, sex, capacity, station, race and situation. So that it will vary. Still it is possible to determine, whether an infant, school child, youth, man and woman of any age. All will be as they are TRU, much of which will be imperative and immediate. This is called: **IMT** = Imperative and Immediate TRU. IMT signifies how, with accurate and measurable functions as ERT and ERART, it is not necessary to make a lengthy deliberation of what is and what is not rights and responsibilities. They, as all, when known of TRU are immediately accessible, as they are imperative to function, in TRU.

What has, is, and will determine the success or failure of a person, group, institution, nation, or world is not what it shows or what it may take over, but its, 1. excellency and special knowledge, 2. good will, 3. efficiency, and 4. hard work and effectiveness. These, and good acts and deeds with the purity, actuality and reality of TRU, can be seen in different parts of the world. Only they lack adequate DAOST, formulation, coordination, ordering and systematic application for the accentuation and achievement that can be provided when highly TRU.

A class society and a TRU social order are much different. Yet such a designation of class not only defines and identifies a society, it helps to make it discrete. One of these is the difference currently used economically and culturally into upper, middle, and lower classes. These are terms that are loosely and commonly used, yet, because of variable usage, may be more discretely reduced and divided into three added levels of each so that we have: The Upper Upper - the 'filthy rich', the moguls, 'high society', leisure class, and carefree, The Middle Upper - the multimillionaires, high society, that hasn't quite reached or doesn't aspire to be the top. Those less ostentatious, more reserved. The Lower Upper - the would be or have been rich, social climbers, lost wealth but not pride or vanity. Of the middle class there are: The Upper Middle - the comfortably well off with good standards, solid ethics, and productive. The Middle Middle - the mode, but not average of a population, less affluent, living primarily off their own efforts, and expect others to do likewise. The Lower Middle - the poorer yet with good standards, honest, solid intentions, ethical, religious, seek to do right with reason and love, yet put life, others and principles before material gain, the average in successful societies. The middle class is somewhat rightly looked upon as being ideal, or virtuous. Divisions of the lower class are, The Upper Lower: not poor economically but poor in all else, of little or no standards or ethics. The Middle Lower - The bulk of worlds population of mediocre standards, poor, living off of the world, and the world living off them, with little idea, principle or appreciation of their own and the world's plight. The Lower Lower - A large part of the worlds population, in or near poverty, having either been crushed, given up or unable to cope with life, the unfortunate, meek, sick, destitute in any and all ways.

As much as these inequities seem unfair or intolerable and as much as many seek to do something about it, some classification remains as long as there are individual differences and TRUD. The future needs, and the success of TRU will very much depend upon, how well we can raise the great masses of human population from illiteracy, poverty, inability, hopelessness, disease, crime, want, denial, defiance and lethargy toward TRU. The popularity of the highest ideals and virtues, when championed and directed toward TREQ TES, and TRU, is a good step in the right direction. Monetary or material gain, as in capitalism, is what empowers much of the NAMTRU world.. If capitalism is to be eternally successful, S/F TRU and TÂS, it must take a few lessons from the best in social ethics in which all that is of human benefit, TRU, must have, as by VAPO and ACC, its role in the whole.

Fig 2-77 **LAWS of ECONOMICS and TRU**

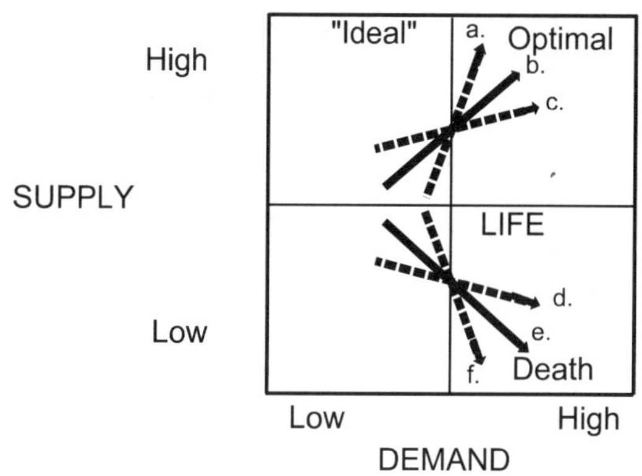

In the relationship between high and low supply and high and low demand the ideal may change somewhat for different segments and stages in the advancement of TRU depending on its position and SCARLS in TRU. Yet for TRU supply generally it tends to S/F when at a high level with an accompanying high level of demand for what is of all sectors of TRU. A key to this is: **WHITS** = What and How TRU is to be supplied. WHITS is also to be accompanied by what and how demand is to be supported and maintained. Much of this is natural to life. Yet it is only with KOT, VAPO, and TRU that demand can be appropriately determined and function: a. to increase supply and demand of what is most needed important or greatest of TRU is one of highest priority for economy and TRU, b. cultures in which high supply and modest demand exist are usually in harmony, happy and seemingly successful. This is largely temporary for the lack of demand will eventually produce loss. c. the supply, capacity, and ability of individuals, groups, and TRU should be at least equal to if not exceed what will be demanded. This especially applies when demand and supply are TRU or what PPMA requires when necessary for adaptation. d. the demands upon the individual, group, or on TRU, should not be greater than they can reasonably bear for the success of TRU. e. The degree to which it fails is a result of its inadequacy. f. the degree to which ability and adaptation fail is the suffering, turmoil, strife, violence, reaction, revolution, of individuals, groups, and all. The rates of supply and demand are: **RASAD** = Rates of Supply and Demand, This may be high, moderate or low, increasing or decreasing. **HIRASAD** = High Rate of Supply/Demand, S/D, **MIRASAD** = Medium or Moderate Rate of S/D, **LIRASAD** = Low Rate of Supply/Demand, **MARASAD** = Decreasing Rate of Moderate Supply/Demand, **HARASAD** = Decreasing Rate of High Supply/Demand. Rates of supply and demand have a great fundamental significance in the existence of TIT-TO. For it is well known in anthropology and biology that the very being or non being of species and animal populations is a direct function of the rate of resources or supply to the size and needs of populations. When the resources or supply fail or are inadequate the demand can't be met, the population rapidly decreases, and may become extinct. As in any wholes and parts, seen in Chapter 35, the size of the pie of the whole of TRU, and REM or that of its parts, are concepts that show how those of supply and demand are of two kinds: 1. those in which the relationship is one where both are higher or lower, and 2. those in which the ratio between the two is higher or lower. The great controversy between economists on how they can best be interpreted and utilized, fails to, but must, include their quality, or TRU.

The have's and have not's are a beautiful example of TRUD and its dialectics, equilibrium, means and extremes and the total interactions in TRU. The cure for the conflicts between the rich and the poor, is not the victory or even gain of one over the other. It is in TRU, an appreciation and mutual cooperation of the one with the other in TRU, as both are TE and TES. Limits should be made for the degree have not's must endure, as well as limits on the excess profits and waste the have's produce. This will be largely voluntary when all are TES. Yet TRU is not a utopian equality without natural selective predominance and power. All means to TRU should be enlisted and fostered for have nots, haves, and all in between. We must forget about seeking bliss, or any state that does not account for each persons duty and role in the interactions and dynamics of TRU.

Fig 3-77 **PROGRAMMED PLANNED MANAGEMENT in MEDICINE and TRU**

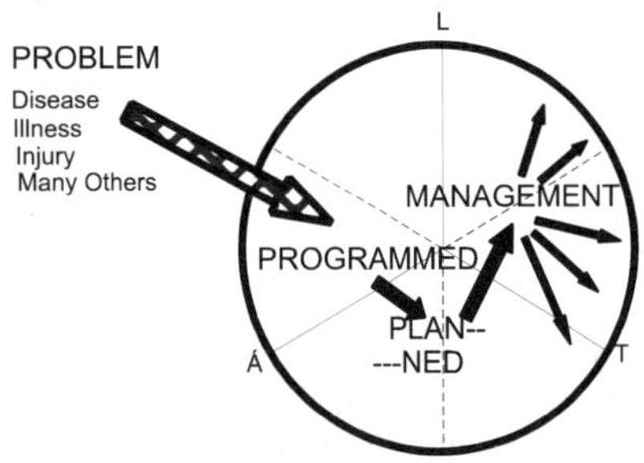

The acumen by which physicians plan and attack a given problem is one of the best methods of modern science. One of these is: **POMP**, Programmed Planned Management.(pg 320). POMP is the detailed and directed programming, planing, management of the conditions or problems associated with the diagnosis and treatment of illnesses and diseases in medicine. It provides an algorithm or explicit means of tracing out the factors involved, their directions, programming, planning and management. It concentrates on the major problems. In this way it tends to be TRU both by its topic as well as its method. It may be extended to other POMP, which will improve by comparison, being more TRU in quality as well as quantity.

There are many energy sources available to mankind, TRU. Fossil fuels although more plentiful than earlier estimated, can last only so long. It is a prime example of what a large population and overuse can do to what is needed to support human economy. This is, in general, a prime concern for the future, a future that extends in time of ALT. As petroleum decreases, becomes too expensive, and is exhausted, other sources or a variety of combinations of energy sources are being developed, produced, and utilized. Even more ingenious discoveries of energy production, challenge future generations to enhance all levels of SA. The changing positions and policies in both non TRU, TRU and mixed worlds is a feature that adds complexity to TRUD, HOM, and typifies DYNS. What does it mean to turn about face, 180° 90° or 360° degrees? Nations often act little better than individuals in their deception, intrigue and political power plays that abnormally seek special interests and gain. It is done in the name of their self preservation, security, power and predominance. Only with the first concepts, principles, laws, and mechanisms of TREQ, TRU, TROS, TRIS, TRUD and all, can conflicts and wars be averted and justice prevail. Our ideals, convictions, and ends should not be nationalism, they should be a globalism based upon all that is common, yet different in proportion, order, and harmony in TRU. One in which the divisions of TRU function optimally in cooperation and healthy competition.

Fig 4-77 **CYCLES, LEAN and FAT PERIODS, and TRU CONTRIBUTION**

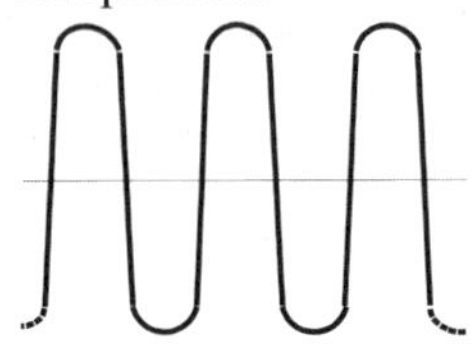

Different sized regions, the world, and TRU have undergone periods cold and warm, flood and drought, depression and prosperity, war and peace, and many others. Such fluctuations occur cyclically, irregularly, and mixtures of the two. They provide the opportunity for scientific discovery, analyzing and interpreting the basic CES and mechanism of TRU from which its concepts, principles and laws can be derived. It is in times of greatest stress that we question our being. We ask for God's forgiveness, salvation, grace or guidance. In times of stress, war and depression peoples do not have the time for scientific inquiry and philosophical contemplation. It requires a combination of each, the stimulation and the opportunity for free expression and activity that best produce a flourishing world. All that enhances each must be discovered, determined, developed, promoted, used and expanded. That which may have been periods of turmoil in one region on earth, may have been periods of prosperity in others. In this way great success and prosperity in civilizations are for the accumulated whole, as are periods of decay, both regionally and for the world. It is in learning the causes for each, and the appropriate actions guided by TRUD, HOM, and ACC that we can greatly help TRU. Invention, industry and the continual reconstituting of nations, as seen in America, are much of what provides the forces that create prosperity and TRU. Yet each has its downside which is seen in much that is wrong. It can not be easily solved by legislation or regulations that are only expedient and not based on TREQ or TRU. It requires a reconstituting and restructuring of its entire culture. This is from the most minute conscious thought and IA of an individual through all levels of society to the government as a whole. This is roughly the difference between what it is now and what it should be as TRU. All that is essential must be rebuilt and promoted. All that is worthless and unnecessary must be set aside, or converted to what is used for TRU. There is much rationale in accentuating the positive and eliminating the negative. Yet neither can be done properly without a good accounting of the whole and an excellent understanding of what is and what is not positive and negative.(VAPO,TRU) (1). Many times better progeny, (2). Freedom to live productively and successfully without unnecessary restraints, (3). Freedom for TRU and the elimination of all unnecessary restraints in TIT, (4). Optimal assistance for what is potential to TRU with a controlled subsidy and welfare. (5). The best possible and most needed occupational, technological, and professional guidance and training with associated TRU stimulation, (6). Highest standards by TRU in all endeavor, (7). Highest quality of commodities, research, discovery, invention, development, industry, services, professions, and technology. In some sectors of America and the world these are approached, many others and most of the world they are sadly lacking. Existing high quality sectors are good examples for the rest of the world. To be TRU, all in a united world must be no more partial than their counterparts in TRU.

It will be highly beneficial if future political divisions, and forms of government, best serve the quality and quantity of total world population. 1. A great enough number of nations to produce representation of all regions on earth in an equivalent manner, 2. Follows, as much as possible, natural geographical boundaries, e.g., seas, mountains, uninhabitable stretches of land or other features, 3. Allows for a weighing of different regions on the basis of their degree of habitability, productivity and of agreeability to human life, 4. Allows for current population, corrected for potential population, 5. A small enough number to enable a single body or assembly representing all regions, 6. A large enough number to prevent one region or division from causing potential harm, or doing irreparable damage to the whole, 7. Safeguards to prevent bloc political monopolies, lobbies and other detrimental effects on the whole, TIT-TO, 8. A means of limiting associations, pacts, alliances which are partial and for the sole purpose of gaining political advantage, self interests, or disproportional and ATRU advantage, 9. To a proper extent and within the limits of TRU, allow for cultural, religious, linguistic, and other differences. These will often be similar to existing boundaries and often quite different, especially in the divisions of the largest and smallest nations. They should provide a basis and be accompanied by mechanisms that best stimulate TRU competition. The term politics has come to be, not without some reason, a dirty word. This is because it is so abused and fails to exercise the best principles of political science. This can only come from TRU or TREQ. Through history a nation's or country's government was more or less determined by the powers and interests involved and their associated means and extremes. Some of these include, Totalitarian/Democratic, Conservative/Liberal, Business/Labor, Nationalist/Non Nationalist, Orthodox/heterodox, Republican/Democratic, Closed/Open, Majority/Minority, Trade Controls/Free Trade, Reactionary/Revolutionary, Self Contained/Expansionist, and Belligerent/Peaceful. These and others show the extremes at work at a national level and how much they occupy the minds of their leaders and constituents. The success of an election depends in great part on how much the other side can weaken the loyalties of these groups in the other's party. This is politics in the raw, especially when abusive. It is highly competitive, so much so, that it loses most all regard for TRU. The great defect in political systems is not so much their polarity, it is in what these different segments stand for and engender that is not TRU. They must be TRU and use their powers accordingly. Some of the major problems of today, their TRU mechanisms and their solutions are:

PROBLEM:	MECHANISM:	SOLUTIONS:
Partiality/Universality	TE/TU, AU/TRU	TIT, TROS, TU=TE, TES=TO
Individualism/Human Welfare	IND/TREQ,ATRU-TE/TRU	Ind→TES→TO, ↑KOT-LAL
Socialism/Capitalism	NAMTRU/AU-REM, TO/TE	TU-TO=TE-TI, ATRU→TIT+Ind→TES
Totalitarianism/Democracy	ATRU-TO/NAMTRU-TE	TIT-TO, TU-TO=TE-TI,↑KOT,
Racism/Humanism	ATRU-TE-TRUCH/TREQ-TO	↑KOT,↑Selection,Moderation
Criminality/Peacefulness	AUD-ANTRU/TÄ-TÂ	TU-TO/Ind-TE,KOT,↑Selection
Punishment/Leniency	ANTRU-REM/TRU,TROS	↑KOT TRU-TRAX-TREX-TRIX/ATRU
Work/Rest	AUD-TRUD-TÂ/IND	↑KOT,PUREALITY, KUPS, TRU R&R
Provincial/Cosmopolitan	ATRU-TE-TI/TU-TO	TROS,TO=TE,↑KOT, KUPS
Sophistication/Naturalism	ATRU-REM/TREQ-REM	↑TRUD-TIT, ↓ANTRU,↑F/P
National/International	ATRU-TE/TU-TO	↑KOT TO=TE, RAMAR
Haves/Have Nots	REM-TES/AREM-TES	VAPO,TROS,↑KOT-REM
Injustice/Justice	ANTRU-AU-ATRU/TREQ-TRU	TO,TROS,DAOST,↑KOT,ACC
Disorder/Order	AU-NAMTRU/TREQ-TRU-TO	SUT,ACEXIN,↑TIT/AU,↑TO/TI-VAPO
Prosperity/Depression	REM-TE-TO/ATRU-AUD,F/S	↑TRUD-KOT-ACC,S/F,↓ATRU-ANTRU
Extinction/Survival(1/SEQ)	ANTITRU-ANTRU-ATRU/TITÂ	TES,TIT,TRU,TROS,TESÂ,TÂ,TÂS
War/Peace, TÄ/TRUD (FAF)	ANTRU-AUD/TROS,TO,TIT-TÄ	↑TU-TO-KOT,TROS,↑FORTOP/AU-Ind

Because of the number and complexity of dynamics, and the many major problems and interactions in human life it is well to make an assessment and learn as much of TRU as possible. This needs to be done both in detail and in order to best pursue and resolve them. Throughout the text, and especially in ACC or Section IX we analyze them further. It is the major NEC of TRU associated with more specific NEC such as LAPUS, LOMT, TOPA, LATSCAP, PODES, PPMA, that show the many ways TRUD, HOM and ACC work together to provide the greatest possible benefit for human life. The solutions are less by fashion and high sounding words. They are the great discrete powerful aspects of TRU that are alive and critical to human life. The problems, basic mechanisms, and solutions often become caught up in the stresses of mixed or other worlds, ATRU, and ANTRU forces and processes. So that to resolve and to provide the moderation, balance, VAPO, right way, and solution a full accounting and accentuation are needed.(ACC)

SOSA = Solutions through Summative Á, or the collective, Indirect and Total Correction by S-O. SOSA denotes how action and proper solutions do not, often, follow upon wise counsel. Yet the wise counsel and recommended solution will often provide the same results without direct action through indirect and summative correction, SOSA.

Since times in the increasingly distant future are subject to being less than perfectly TIT it renders prediction even less accurate. Added to this, by the infinite detail and massive dynamic interactions of TRU, it should be no surprise that the future is so unknown, confusing or without consensus. Although many universals, general principles, mechanisms, laws, courses and trends are predictable, the more detailed and specific we become the less we can provide exact posits. Part of the reason for this is what will be TIT at a certain date, time, place, and mass energy in the future, say 2984 or 15480 years a.d. and how much or in what ways will it be more or less other than existing TIT, TRIT or TRU. It will be to some extent, a fraction of its difference from TIT-TO, TRIT and TRU, differentially trivial or indeterminate. This is a very important consideration for it is possible to collect all the variables and develop the necessary correction factors which can make predictions more reliable. From this we can add the indeterminants as variables that can be accurately set forth. Extending through the future TRUF will be increasingly different than TRIF, and this from TIT. Yet to maintain TIT, at least in certain ways, is also definite, and an end that requires increasingly vast powers of KOT, POA, ACC, management and control.

Measures of a person and their burden or benefit to TRU can

be made for much of the earth's human population. How much and in what ways they are or are not TES in TIT-TO. There are portions of the world's population that are less than par for TRU. This will vary considerably per culture, civilization, region, as well as individually. It will vary per time and per the amount each puts into it. In the past it was largely resolved by natural selection and genetic polymorphism. Yet, especially for the future it becomes increasingly important to learn and best guide the burdens and benefits. The final measure will be a changing optimal relative. As an overall policy it shows the necessity of ENT and all that has, is and will be essential to successfully increase GELF and TRU. The roles of natural selection and genetic engineering directed toward optimal genetic quality can and must be utilized together in optimal harmony. This should be done for the greatest strength and permanence of TRU.

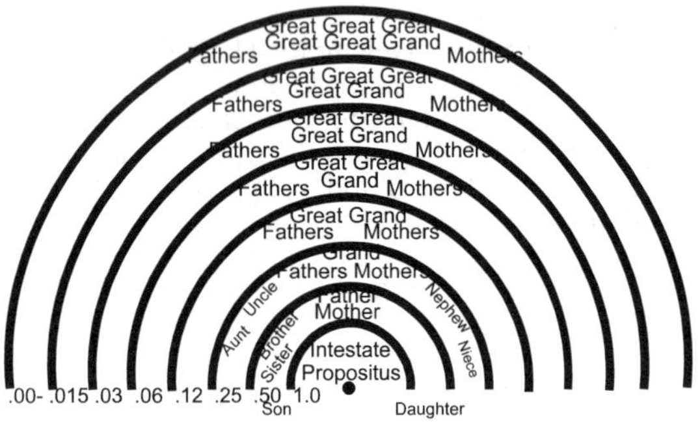

Fig 5-77 **CONSANGUINITY and GENERATIONS**

The relationships between different members of a family show extensions of generations. It is less important in a highly cosmopolitan diverse world of outbreeding. Yet for ACC in the future where more and more measurements of GELD will be made such relations will have importance and value in predictions and decisions of voluntary and formal selection.

Among some of the major trends in the next century are, 1. National to world economies, 2. Information and knowledge explosion, 3. Improved public opinion and beliefs, 4. High tech, 5. Either/or TREQ, 6. better, better suited and better access to Jobs, 7. Participation in government, 8. Population and density increase, 9. Struggle for population control, 10. Struggle to adapt to density, 11. Struggle to keep up, 12. Increased disparity haves/have nots, 13. Resources and supply, depletion of, 14. Productivity + or -, 15. Standard of living, rate of +/-, 16. Public health, prevention, care, + or -, 17. Continued rapid advances in medicine, biology and associated sciences and their application, 18. Genetic engineering, diagnosis and treatment, 19. Voluntary participation in natural selection, 20. Quality persons and means, 21. Increased identity and immortality, 22. Advancing knowledge to KOT, 23. TREQ to TRU and TIT, 24. TES, training, maturity, adaptation, 25. Terrorism, rouge groups, and nations, 26. International cooperation at all levels +/-, 27. International competition and global economics without war or ATRU, 28. International assistance as well as rewards for excellence, 29. Great expansion of key salients in human quality, 30. TRE, ACC, MARS, TRAX, TREX, TRIX, ELTICMA,,. These are only some of the great changes, needs, problems, and solutions that are likely to take place. The order, and how they will effect each other. It will depend in part upon how they are presented and how the public and nations attend to, popularize, accept and effect them. New inventions, tools, and products rapidly find their place and role in the assemblage of human culture, TRU. There are alternative and auxiliary means of providing a better life. Yet in becoming important, by way of our dependence, they are in TRU, of TIT. It is in adapting them, ourselves and TRU properly, as all function harmoniously as counterparts in TRU that the least possible damage or loss to TRU results.

Civilization is highly illustrative of TRU. It enables humans to live in ways that optimally contribute and minimally detract from their world. It seeks to avoid savagery, barbarity and destruction in this social order, TRU. Yet it is only when we reduce what is meant by civilization and civility by VAPO, DAOST, SCARLS, and ACC that their advances and contributions are fully TRU

Whether the earth is inhabited by 1 to 10 to 100 billion people, requires far different rules and regulations, laws and standards of living. Whole new sets of values are necessary, although they will incorporate much that is the same. The technical, commercial, industrial, transportation and other factors that are associated with increased population are also good examples. Exercise and adapting to physical needs, conditioning, improvement of the mind, and many other aspects of TRU must all be maintained and enhanced to meet not only the basic demands of life, but the adaptation to expanding and complex forces, environments and habitats. **LADIT** = Law of Adaptation to Density, of the world as It is TRU. LADIT may seem obvious, yet the refinements, ramifications, effects and all needed for adaptation are vital. This is especially so in prevention and completely adapting to new worlds.

LIH = Limits of Homation and Its Laws: **I** = All nations and peoples are or are not successful as they are or are not equal TRU, TE, TRUS or TREQ, **II** = All nations and peoples as parts of TRU must strive to equal or excel in their part, ES, necessary for TRU, or their TE to TRU. They must not be equal to that part which they falsely assume or purport to be TRU yet which is not TRU. **III** = Principles and Policy of all individuals, groups, and nations can't be selected and used at random, erroneously, inconsistently nor frivolously. Their effects can be no better than the perfection of their causes, and these together by TRU. **IV** = Errors in principles and policy can be corrected. Yet to err once often indicates other errors. To err and not even know it or try to correct it is doubly negative.(INEG) **V** = No nation, group or individual should force or use its power and leadership, within or without its boundaries more than the degree it is TRU, its ES and CES in TIT. This should not be done also to the extent it lacks or in any way can not carry out this leadership by TRU. It may also be necessary that it be less than perfectly TRU under certain conditions, qualified, and compensated by TRU. Some of this will naturally and justifiably result from partiality of nations, group or individuals as well as variations in their dynamics.

Future needs and supply, in TRU and in time, are both limited and continuous, regular and fluctuating. The question is what will these needs be, and what supplies will be available to meet the most important and critical upcoming changes and shortages. Other than enumerated below there are many needs and supplies or varieties. Their fulfillment or completion make up much of TRUD. Often the needs and supplies may be present in many other categories than the ones they are most directly applied to. The needs and supplies change, which will be greater in the long term. Yet there is a great need for stability and balance. One question is whether or not man, as an advanced evolutionary animal, has lost certain potential for change? Is size one of the limiting factors that may tend to lead to extinction?. If so, how is it to be corrected or compensated for?(TRIF,TRUF,ACC)

NEEDS	SUPPLY	1
SUN (LIGHT)	Sun, (Stability of)	
2 TEMPERATURE	Climatic Stability	
3 AIR, OXYGEN	Climatic Stability	
4 WATER and FOOD	Preseervation/Growth	
5 SHELTER	Construction, Design	

6 CLOTHING	Manufacturing, Quallity	12 TOOLS-TECHNI-	Invention, Utilization
7 TRANSPORTATION	Materials, R&D. Improved	13 VALUE	Actual, Perfected sense
8 CARE	Identity,Empathy,Quality	14 KNOWLEDGE	Discovery, Propagation
9 COOPERATION	Bonding, Adaptation	15 JUDGEMENT	Enhanced, Best Use of
10 LANGUAGE	Growth-development, Use	16 PROJECTION	Intellect, Practice, POA
11 INVENTIONS	Discovery, Development	17 CONTROL	Applied TRU, ACC, PPMA

	POT	AU, RU	TRU KOT	TÄ TU	TE TI	TO	TRUD	ENT OM	ROT ALT	REM	POS STA	HOM	DYNU	DYNS	DYNO	ACC	TREQ	TRAX TRIX	PUR- COG-	SÂ TÂ
1	+		+++	++	+++	++++	++	+++	+	+++	++	+	+++	++	+	++	++	++	++++	+++
2	++	+/-	+	+++	+	++		++	++++	++	+	+	+	+	+	++	++	+++	++	++
3						+++	++	+++	+	+++	+	++	++	+++	++++	+++	+++	++++	++++	+++
4	+	---	+++	++	+++	+++	+++	+++	+	+++	++	++	+++	++	++	+++	++	+++	++++	+++
5	+	---	++	+	+++	++	++	+++	+	++	+	++	++++	+++	++	++	++	+	+++	++
6		---	++++	+++	+++	+++	+++	++++	++	+++	+	++	+	++	+++	+++	++	+++	++	+++
7	+	---	++	++	++	+	++	+++	+	++	+	++	++++	+++	++	+++	+	+++	++++	+++
8	+		++	+++	+++	+++	+++	+++	+	+++	+	++	++	+++	+++	+++	++	+++	++++	+++
9	+	+/-	++	+++	++	+++	++	++++	+	+++	+	++	+	++	+++	++++	++	+++	++	+++
10	++++	-	++++	+++	++	+++	+++	+++	++	+++	++	+++	++	+++	++++	++++	++	+++	++++	+++
11	+++		+++	++	+	++	+++	+++	++++	++++	+++	+++	+++	+++	+++	++++	++	++++	+++	+++
12	++		+++	++	++++	+++	+++	++++	+	+++	++	++	+++	++	++	+++	++	+++	++++	+++
13		-	++	++	+++	+++	++	++	++	++++	++	++	++++	+++	++	+++	++	+++	++	++
14	+++		+++	+	++	+	++	+++	++++	++++	+++	++	++++	++	+	++	+	+	++++	++
15	+	---	++	+	++	++	++++	+++	+	+++	+	+++	++++	+++	+++	++++	+++	++++	+	++
16	++	-	+++	+++	+++	++++	+++	++++	++	+++	+	+++	++	+++	++++	+++	++	+++	+++	++
17	+	--	++		+++	+	++	+++	+	+++	+	++	++++	+++	++	++	+	+	++	+
18	++		++	+	++	++	++	+++	+	++	++	++	++++	+++	++	++	+	+++	++++	++
19	++		+++	++	+++	+++	+++	+++	+++	+++	++	++	+++	++	+	++	+	++	++++	+++
20	+	-	++	+++	+++	+++	++++	++++	+	+++	++	++	++	+++	+++	++++	++	++	++++	+++ +

Table 1-77 **CONFLICTS, PROBLEMS and NEEDS by CONCEPTS, ASPECTS and DIVISIONS of TRU**

The great conflicts, problems and needs of today and different times can be ordered by their causes, effects, courses and common solutions. Horizontal columns include conflicts, problems and needs of:

1. Abandonment, no Identity
2. Bad Advertizing, art, commercialism
3. Citizenry, Contributing/Abusive
4. Crime, Violence
5. Disease, Bad Health
6. Environment Habitat Deterioration
7. Genetics; Defective, Damage to
8. Government, Helpful/Oppressive
9. Hate, Prejudice, Bias, Isms
10. Ignorance, No or Poor Education

11 Illiteracy, No-poor Communication
12 Immaturity, Untrained, Inexperienced
13 In-opportunity, Oppression, debt
14 Profanity, Language Misuse and Abuses
15 Poverty, Misery, Hard Times,Oppression
16 Self Interests, Greed, Vanity, Defects
17 Toxic Substances, Drugs, Alcohol,,,,;
18 Un or bad employment; 1/Talents,,
19 Unfulfillment, Maldistribution
20 War, Terrorism, no peace

The vertical columns include some of their TRU manifestations, representation, and order by sequence and content. Many of the causes for conflicts and problems lie outside of TRU and TIT, and thereby only noted, negatively by a - or minus. There are many other causes, indirect and multiple often working by INEG. The solutions are noted positively by + and their widespread distribution confirms the centripetal forces working in TRU. This the individual, group, and all can join to the extent they become their part and role TES and CES in TRU.

MEASURES of DYNS CES in OLF SECTOR of SPALT

BARBARISM, CIVILITY and TRU	7-2-7		T	BCT
BURDENS and BENEFITS	7-3-7	-+	S	BaB
CARING and STRUGGLE for TRU	5-4-5~		S	CST
CLASSES and SOCIETIES	~6-3-7	+-	T	CaS
COMPARATIVE NATIONAL VALUES	~7-1-7	~	M	CNV
CONFLICTS, PROBLEMS, NEEDS, TRU	7-3-7		T	CPN
CONSANGUINITY and GENERATIONS	8-3-7	+-	S	CaG
CRIME and PUNISHMENT	~6-4-6~	-	M	CaP
CYCLES and TRU CONTRIBUTION	~7-2-8~	+-	M	CTC
ENERGY SOURCES, SYNFUELS	9--7	+-	M	ESF
ERAR	~6-4-5~	-	T	
ERT	5-3-7		M	
FUTURE NEED and SUPPLY	7-3-7		T	FNS
FUTURE PREDICTIONS	~7-4-7	+-	M	FuP
HARSAD	8-1-8	-	S	
HAVES and HAVE NOTS	7-2-7	-	M	HaN
HIRSAD	8-1-8	+-	S	
HOW TO, SUCCESS and TRU	6-4-7		T	HST
IMT	6-3-5		S	
INVENTIONS and INDUSTRIES	9-4-8		S	IaI
LADIT	8-1-7		T	
LIH	7-2-7		T	
LIRSAD	8-1-8		S	
MAJOR TRENDS	7-3-8~		S	MaT
MARSAD	8-1-8		S	
MIRSAD	8-1-8		S	
NEEDS in TRU and in TIME	8-2-8		S	NTT
OLF	7.5-2.5-7		.5	STM
POLITICAL, , , ,	7-2-7	-+	M	POL
POSITIONS, POLICIES, CHANGING	6-3-7	-+	M	PPC
PPM	7-3-8		S	
RASAD	8-1-8		S	
RECONSTITUTING, Am, World, TRU	7-1-8		M	RWT
RESPONSIBILITY and PRIVILEGE	6-4-6	~	S	RaP
SOSA	5-4-7		T	
VACOLP	5-4-7		M	

Fig 6-77 **DYNS CES in OLF SECTOR of SPALT**

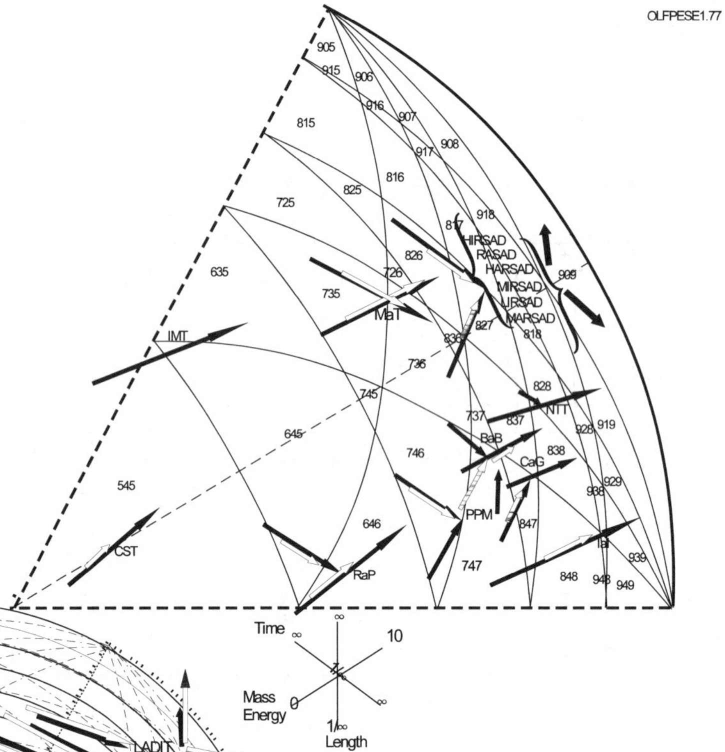

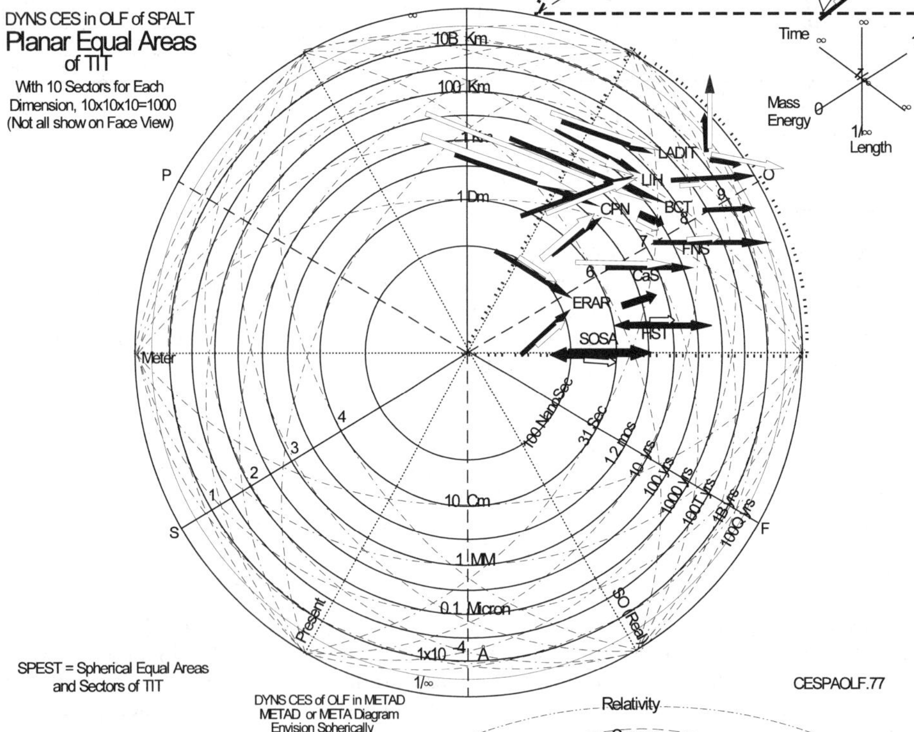

Fig 7-77 **DYNS CES of OLF in SPALT**

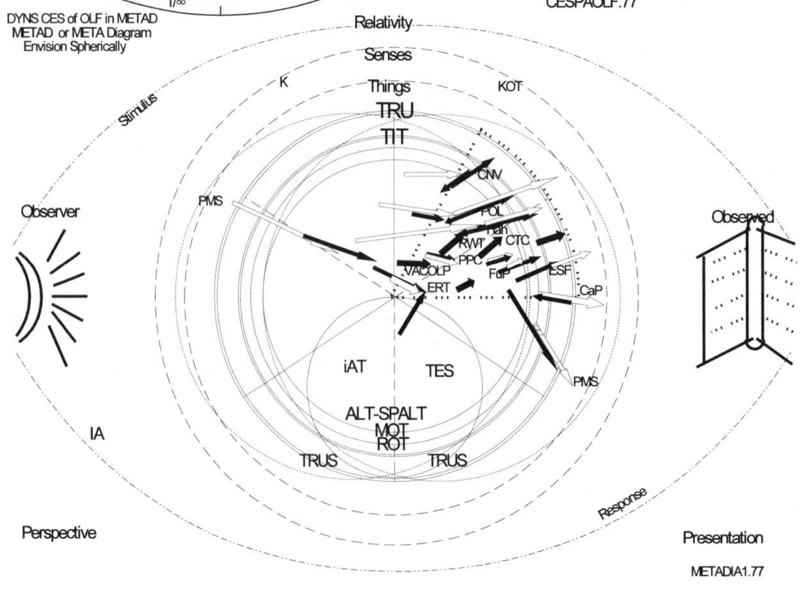

Fig 8-77 **DYNS CES of OLF SECTOR in METAD**

CHAPTER 78

SULF or SUBJECTIVE LARGE FUTURE PROCESSES of DYNS

KNOWLEDGE, RIGHT, IPR, SOCIOLOGY COMMUNICATION, RELIGION, JUSTICE, GOVERNMENT

There is a great difference, in the level of justice and how enlightened the citizenry is, at different times and in different parts of the world. These are major determinants of the success or failure of societies, nations, cultures, and civilizations. They are TRU to ATRU, what it takes, and how it necessitates accompanying levels of education, economics, communication, industry, commerce and most of the structure of a society. It necessitates a level of basic and technical knowledge as well as proficiency in use. It changes per age, yet much is unchanged or reversible. In those nations, cultures and civilizations that have had an enlightened, curious, stimulated, and principled citizenry, peace, prosperity, success and relative permanence have been greater and more secure. This is where the level and participation in justice was continuously upheld and increased. It is here citizens with the best knowledge, enlightenment, convictions, choices, decisions, and judgement have, are and will be highly contributory to TRU, as well as successful in their own right.

There is an arrested level and mass of the worlds population that does not have, either through poverty, ignorance, neglect, lack of opportunity, courage, inspiration and conviction what is needed to attain TES and TRU. They will be to that degree, and by INEG, in failure and loss to TRU. This is one of the greatest sources of disassociation, conflict and suffering. It is in the perpetual PPMA of the capacity, ability, conception, abstraction, principles, reason, and comprehension by VAPO, SCARLS, DAOST, and TROS of TRU that we can best assist all to become TES and TRU.

When attitudes and behavior are shown to be abnormal, unfriendly, and devious it is not this in itself that is so serious and dangerous. It is what is behind it. It usually indicates not only being inconsiderate and ungrateful, it means that they completely lack the identity with the common effort, the successful unity needed in an institution, community, nation or TRU. It is this and other attitudes and indicators, that one can best test and measure their actual form and function in TRU. This is especially so in performance or work record. It is this one must closely watch. Non identity is apt to be unexpected and contrary to what would be expected, by TRU. It is a sign of the many indirect and largely hidden meanings to be found in IPR. It is quite another thing to maintain the necessary agreement with TRU, and especially when this is of the different in TRU.(DAOST) This is where humans are so quick to fail and suffer in all kinds of ways. **FARD** = Fallacy or Failure of Realizing and Adapting to the Differences, what is uncommon and opposing in the world, TRU. FARD is a great power in life, and TRU. **FARDIN** = Fallacy of Realizing and Adapting to the Differences in TRU. They are aggravated by what is seemingly foreign and oppose a person's interests. This is why we must always keep in mind and continuously analyze and correct all principles, laws, legislation, rules, regulations, directives, instructions and routines we live by. They must be certain, valid, correct and good not only for TRU, but for the individual, TES and TRU. The rationale, informed consent, process, and overall analysis must be explicit and well given.

There is a great difference between the precepts of a preacher, church, sect, religion, religions in general, the best of religion, TREQ and TRU. Each has their own focus and emphasis, which is often highly variable. Religion is, like law in emphasizing ethics, what is right and wrong, only with a different perspective. TREQ, TRU, or an ideal religion can be neither simple nor impossibly complex. It can be neither absolutely immanent nor transcendent. It must be a finite composite of all that is the greatest in TIT, by VAPO, TROS, DAOST, SCARLS, TRIS, TÂS and REM. Its application and use must reflect the entire ALT and SÁ FAF of TRU, being of all levels of the subjective, objective and real, combined as a guide for all person's religion, TREQ or TRU. It can be seen in how societies, nations, and cultures, which more or less live by limited and weak VAPO-TIT precepts, have so many difficulties, conflicts and failure. As much as they intend to be right, and do their best they are in their limitations, abused, misused, and in other ways, wrong and conflicting. Only in the Total Relative Universe, TO of TIT, can the basis for all, all the acts and functions of the world be found. This is how we can find what is actually best for all, what is the ethical or the right thing to do. Each counterpart of TRU's functions are to be judged, decided upon and acted upon no more, nor less, than appropriate for its part and processes in TRU.

The crux of the problem of social life, of community, society, politics and religion is the argument, "What does your religion give to you, and how much do or are you willing to give to it? A complete and balanced religion, if it is TRU or TREQ and its adherents or their equivalents in any religion, TES,, should give to each of its adherents a method, guide, formula, model by which each person can not only understand all things, TRU, but be able to answer all of the necessary questions and solve those life's problems they confront. In this way with strong dedication and will, all can always find the knowledge necessary for the proper solution to each problem and make the right decisions and judgements. Only in this way can each and all of us avoid the pitfall of leaving our lives to chance, notion, or some extreme and rash act that produces failure to oneself as TES, and all as TRU.

WRONG LAWS, CIVIL REACTION and TRU

To expect people to abide by the law
The law must be right, correct by TRU
For people will do what they are told
To do if what they are told is right

Humans appreciate how reasonable and
Beneficial what is good and right
Otherwise there is a society in which
Laws are partial, of special interest

They are judged and used even worse
A public is required to do neither
TRU nor what is in their interests
Laws become a chain entangling all

Many commandments, statues, laws and moral
Guides to behavior have served society well
So that there is a perennial interaction
Stress, struggle, and conflict between

These time honored rules and laws
And those that are wrong, against TRU
People who do what they are told, and
React to wrong and ANTRU because laws

Are wrong and against TRU are productive
Of loss and failure for which many suffer
Only what one is told to do must be right
All must be brought to TREQ, KOT and TRU

It is these wrong or misused laws and the accumulative effects on those who suffer, that much of TRU is necessary and directed to resolve. No map, course, navigation, plan, policy, law and constitution can be better than the degree they and their knowledge are based on TREQ or TRU. For, in actuality, this, laws, principles, and rules are by definition. Responsibility tends to fall on the shoulders of those that best carry it. This may be right or wrong. Right only if both the one responsible, their knowledge, effect, and those served are TES and TRU. There needs to be public awareness, common will, and a massive crusade to increase appreciation, honor, and reward for ability, quality and responsibility. Responsibility in TES and TRU is largely its own reward. It comes from a profound and massive understanding and labor of love of what is best in life and the world, TRU. There is nothing like an enlightened, responsible, and good public and government functioning in tandem. Only when TRU is known as TRU, in precision, VAPO, represented, and held in perspective as such, is both enlightened public and responsible government probable. It is usually easy and effective to set and follow policy. Yet, like responsibility and laws, policy is too often temporary or superficial; no better than the degree to which set policies principles, laws, and determinants are VAPO for TRU. There must be a match-up in persons of highest TES with that of greatest responsibility. This relates to how officials and leaders are selected by: public appeal, appointment, election, tests and measures, or other means. Although there has and is much selection in the rise to power, this is not the same as selection that is determined by or needed for TRU. The best leaders will have a deep knowledge and abiding faith in TREQ or TRU and all of their varieties, kinds, substance, parts, and processes.

For a democracy to work, the people must do what they are told to do. Yet, what they are told to do must be right and complete by TRU and VAPO. To tell people to do right, doesn't work. To make people do right, doesn't work. To let people do as they please, won't work. To try to do right without knowing what is right, doesn't work. To do what one thinks is right when it is not, doesn't work. To do it for them won't work. It is from these actions, problems, errors and failures that we can finally reach the right basis for solving the great enigmas of life. Only if each and every man, woman and child knows, feels, wants and shows to the greatest extent possible, what and all is right as it is TRU and what and all that is wrong is not TRU will they be able to make it work.

There should be a manual, for presidents, leaders, and all in highest positions of responsibility and authority, that is based upon the best uses of TES, TREQ and TRU. It will include some already in use: 1. What to do, 2. What not to do, 3. What it takes to make a great president, or best fitted for leadership, 4. How, in being the best possible leader, one's whole country or sector of responsibility follows, 5. How these leaders can best handle negative reaction to their job. 6. How to act upon all that is wrong, improperly opposing, and destructive, 7. How to have learned and best apply all the concepts, principles, laws and mechanisms to the job, as these can become or do exist, in TRU.

Belief in what is sacred must be identical with TREQ or TRU, and all its VAPO, FAF, DAOST, REM, SCARLS, REM and TROS. This is evident in those societies, groups, tribes, communities, states, cities, and cultures, whose sacred beliefs were lacking, weak, extreme, irrelevant, and limited. Thereby being inferior, loss and demise ensue. How strong they were and how long they lived depended both upon what and how well their adherents believed, plus how little opposition they had.

Each person, group, nation, and the world tends to look at the present only vaguely, indirectly looking at the past, and even less at the future. This is natural, especially for the person or group, by partial focus. Yet this focus produces great bias and problems that must be compensated, if they are to know and act properly, by TRU. By looking at the world in such a personal and temporal way we lose sight of the whole and much of its dynamics, means, extremes, and proportions.(VAPO) In fact we tend to think of it as being the best or worst, the ideal or worthless. These can be severely mistaken. There seems to be in every new generation, a plethora of one kind or another of dunces and backsliders who make the same mistakes, redoing the wrongs that were thought to be forever cured or ended. Among many reasons for this there is one overriding reason; neither the past nor the present had much overall understanding or appreciation for what is best for all, all TRU. They have forgotten or failed to learn their lesson, or even worse think they can get away with anything in a public that has lost its concern for what is best. They lack and must find in TRU what applies to each and all problems, conflicts, decisions and judgements.

MED = Methods of Education, TRU Education, Pureality, and HUB. The great defect in Cogito Ergo Sum or "I think therefore I am" is: **SELFER** = Self Error. This defect or error assumes that in being cognitive, with IA or Immediate Awareness or Instantaneous Attention, one has being, absolute being, therefor <u>IS</u>. It is a great part of the error in many systems, philosophies, and ISMS. They ignore the figment and illusion that exist in all that is less than perfectly TIT. **FÂENIT** = Fallacy of Absolute Existence of ENIT. FAENIT is to some extent the difference between philosophical and religious validity, when TRU. It is very correct to emphasize the greatness of IA, for it is a great proof of our relativity, reality, and the Á FAF construct of TRU and ACEXIN. Yet it is incorrect to assume that because I attend or think, I actually am, TES, or fully TIT. We introspectively recognize this thinking meaning "I am" but completely ignore the error of assuming this is, and I am only TES or TRU. It is related to: **FISMS** = the fallacy of ISMS. SELFER is also like the subjective error, which is more evident when we understand SUBIT. The general defect common to SELFER, FISMS, SUBIT, OBIT and ENIT, is departure from TRU, or if ESINS of TRU, how they are less than all TRU. This is: **PER** = the Partiality Error. PER simply signifies that what is partial at best, or ES, is not TIT-TO, and all that is to that degree less TIT is even increasingly less ES and TIT. Such errors signify the difference between any entity, thing, sense, self, subjective, ISM, partial, process, and the total thing, TIT-TO. To understand this great fallacy is the beginning of MED, KOT, PUREALITY, KUPS and TES. It is for this reason the best Method of Education, MED, is not so much a gathering of information, facts, education, bits of knowledge, nor even the ordering of this knowledge technically as in a profession or occupation. The best method of education can proceed only from the overall whole, TIT-TO, TRU and its SUT, SPALT, and POS. We should exercise and practice learning by reversing the process from TA, back through SA, to ELTICMA, COGNOS, THESOA, THEOA, THESA, THEO, TO, TRU, ROT, SUT, OM, COL and the Given. In this case, this is 'I think therefore I can be'. The method must be a combined and reversible philosophy, science and theology. Universities should be primarily measured by the degree to which they prepare students to become their counterparts, TES, in TIT. It means that they with deep conviction and dedication develop the necessary life long habits of inspiration, revelation, TAN, VAPO, SCARLS, DAOST, OM, SPALT, METAD, ACC, Homation, Pureality, Cognos, ELTICMA, SA and TA. By whatever name these are called, all that is directed in the university, and all stages of life, to produce this kind of person is to be highly complemented, upheld and fostered.

There can be tremendous positive power in looking from another's point of view. These are called: **SECPAS** = Second Person Effect, or an IA that is far different, at times alien to ones common IA. This may apply to other persons ideas and beliefs or to a change in or another time of the first person. **SPARF** = Second Person's Reversal to the First Person. SPARF is what occurs, as commonly in conversation, when two naturally diverse points of view converge in agreement. This may be good or bad, right or wrong, narrow or broad, yet it will have, more or less at least a slight, tendency to TRU. **SPAFORT** = Second and First Persons Reversal by and to the Totality of TRU. SPAFORT is what should occur in IPR between two or more persons. It should be a meeting of the minds in TRU. It is through SPAFORT that mankind can resolve many of its problems.

Toleration in TRU is a great necessity. Yet TRU consists of some aspects, features, components, FAF that are not known. This will include many parts or processes that exist and work as differents,, in opposition. **LOEN** = Law of Equivalent Necessity. LOEN denotes those portions that are of the different aspects of TRU, which may function in opposition or antagonistically, yet are of equivalent necessity.**TACK** = Tolerance and Complete Knowledge. As great as tolerance is in accentuation and action it does not include, nor is it advisable in, all behavior. What is known to be detrimental and destructive of TRU or its essential parts permits little or no tolerance. It mostly needs their removal. We must be tolerant of all TRU, ES, CES, and all that is potential for TRU. We should be decreasingly tolerant of AMTRU, MATRU, NAMTRU, ATRU, ANTRU, and ANTITRU. This is most especially so to the degree they are can't be reversed back into TIT. We should be tolerant of all in others that is TRU, as the TES that is common to each of us. We should even be tolerant of what is uncommon or different in others than ourselves, yet of TRU. This is why our knowledge must become more KOT, and KOT must become in use ever greater for TRU. It is denial and defiance of what is so commonly known to be right, or TRU, that is such a harbinger of failure and loss. **DECKORT** = Denial and Defiance of Common Knowledge of what is right, TRU.

Every government's policy with other governments and the rest of the world should be based upon the degree to which all are TRU, their part and role in TIT. It should use a system of favor or disfavor, positive or negative action, action or inaction to the degree others are their counterparts in TRU. These may be graded from that perfectly TRU to that most ANTITRU. This should be associated with a full accounting of their policies, activities, non TRU and TRU makeup and roles to help governments and citizens better realize both their own TES and TRU, and their relations with others.

No Government is better
Than to much government
A little government is better
Than no government

Government is primarily for those
Who will not govern themselves
The only perfect government can be
A government based on TRU

This is a government in which
Each person and government
Function as they are their
Counterparts ES and CES in TRU

The best government is one in which there is the least government concomitant with the protection, maintenance and advance of TRU. It is what its citizens, as TES, can not do for themselves. It is that government that best fosters, enables and enhances its parts, processes, each citizen, in their due proportions and most opportune way to be TES. It is a government that is adequately prepared to meet all contingencies and best prepared to correct or annihilate all ANTRU. It is a government that identifies with TIT-TO.

IMT = Imperative Exponential Thought, comprehension, POM of TRU. **TEX** = Thought Exponential. With the common and combined TEX, IMT, and TRU we have one of the greatest powers compelling TRU and our lives as TES. The most renowned ages in history felt they had the answers to all, or were about to discover, grow and be better and greater. Little did they, or do we, know that the vicissitudes of human fortune. How invisible, cyclical, and often dismal, the seeds of negation and future conflict are all about. At present they are advancing at an alarming pace. It is in the days of our greatest success and prosperity that we take life for granted and ignore growing threats to existence. Yet it is in these very times that we have the greatest opportunity to prevent, protect, advance and maintain TRU. It is failure to make hay while the sun shines, and to repair our roofs when it doesn't rain, that we live to regret it. It is in these times that the seeds and shoots of bad, evil and negative accumulate acceleratively. There are many questions and problems we must face, answer, and solve. This is done throughout the text, most are accounted for and brought to order by POS in HOM and ACC.

ITOD = Ideas whose Time is come, Due, or Over Due. Many new concepts, words, abbreviations or acronyms are justified by the existing voids, vacuums, absence of existing words, and upcoming needs in TRU. They are those that are yet to be discovered or for which words have not been provided. If they are of great REM or key to the understanding of TRU they must, if we are to have an adequate language, be represented, communicated, and used properly, discretely, and concisely. **STIP** = Summation Time, Term, Idea, Past due. Much of all written works, literature, and even all scientific and technical literature, is largely and merely propagation, communication and emphasis on unknown or poorly comprehended or represented ideas. So is IPTRU, the summation and ordination of these, ITOD, and STIP in the form of NEC and OC redefined to TRU. The massive dynamics of human life and communication continually change language. Common usage is constantly upgrading, downgrading, refining, or altering the words we use. Ever better languages in the future must remedy great problems and ultimately give us more stable and long lasting identity with what represents the greatness of TRU.

Churches should become more like schools and universities that teach religion, TREQ or TRU. Schools and universities should become more like temples that revere God, TREQ or TRU. Separation, as the D, in DAOST is equally necessary. Yet, as in political science it can go to far. This they have been, off and on, through history, depending on the orientation of the church or school. No knowledge that is of actual value can be ignored. It is very unfortunate that religion has been so divisive and education so unappreciative of TREQ and TRU. With more comprehensive and VAPO governments we simultaneously and properly separate and reunite the secular and the sacred. Only when we learn and remember TROS, the optimal FAF of TU and TE together, can we come to appreciate their mutual benefits.

POPA = Principle of Primary Applicability. If you have great breakthroughs and many serious and great problems, then turn the power and light of these great breakthroughs on the problems. This has been found to hold in the many trials and uses of great new discoveries, inventions, tools, methods and systems. It explains the proliferation of research, new technologies, and great successes of science. Their greatness as well as benefits feed on themselves. They can be turned about and used on many other problems. **TOPA** = Test of Primary Applicability. TOPA must be distinguished from MIRPS. POPA and TOPA are major reasons why the world, TIT-TO, is moving so fast today. They exemplify CEK, the compelling and drawing forces of the future. They illustrate the power of knowledge, discovery, proper direction, and the way of the investigator. POPA is evident in the study of TRU. For as we provide new ideas and concepts, the discovery of new relationships and means of representing TRU, these in turn reveal and provide striking new ideas, relationships, concepts, principles, laws, mechanisms and methods to help better represent TIT and thereby enhance TRU. It is both their applications and relations of their fundamental principles that our horizons exponentially expand TRU. **MERPS** = Many, Multiple or Manifold Enumerations and Emanations in

Problem Solving. MERPS may be combined with MARS. They show how important such expansion is and how to bring out the greatest and worst questions and problems. It enables their being constantly set before us for vigorous pursuit. **MAB** = Magnification of Breakthroughs. MAB is the exponential and ramified POPA, TOPA, and MERPS. Great breakthroughs provide the glorious opportunities for even greater discoveries and victories. Its beauty comes from a realization of how vast and great unknowns exist only to be brought out. Its observation and description allows us to recognize relativity in TRU, observer observed, subject object, or mass energy form of TRU, its Á FAF, continua and mechanisms. It is this that provides the great powers of knowledge in thought and action. **MABIK** = Magnification of Breakthroughs in Knowledge. **MABITH** = Magnification of Breakthroughs by Thought. **MABAC** = Magnification of Breakthroughs in Action, and, **RAB, RABAC, RABITH, RABIK** = Ramified Acceleration of Breakthroughs, and of Action, in thought and in Knowledge. These concepts beautifully illustrate how SPALT and HOM work through and through as they show some of the detailed dynamics of TRU, TRUD. All these aspects of MAB in TRU, when they occur without being sought for, are vividly seen in serendipity or the assumed gift of finding valuable or agreeable things not sought for. Yet serendipity is less than MAB and its associated concepts and processes in TRU, for it supposes chance or fantasy which is not a basis for TRU. For MAB, as TRU and life, are produced by compelling, propelling, and impelling forces and powers either already in, or working toward an existence in TRU.

The great direction, thrust and power toward future selection should be toward those members of the species that have the greatest capacity and quality to meet the demands of life, be, and enhance TRU. These will be those who have the capacity to comprehend problems, solutions and the concepts, principles, laws, processes and mechanisms. It needs to be a capacity to comprehend the VAPO of all that makes up, or is apt to effect, TRU. It should thereby be those who can best learn, know and utilize MAB, KOT, TRU, OM, POM, CEK, COGNOS, ELTICMA, SÂ and TÂ. Such solutions should act as great catalysts for the advancement of KOT, TES, TRU and TÂ. **SIST** = Solution by Future Selective Thrusts.

There is an ideal and goal for the actual or TRU relationship between parts and divisions of TRU. This is comparable to the political makeup of the world. This problem is seen in John F. Kennedy's statement "political equanimity is not our goal". In this he meant that a nation composed of only uniform equal and static divisions was not what we should strive for, or try to reduce ourselves to. So is it of TRU. Although much is the same, much is also different, by DAOST. No two parts are alike or of exactly the same value in the REM of TRU. No part is completely static. All parts are not and do not have to be exactly the same. Parts of TRU may be of the same order or same level of REM, yet different.

Humans have a great tendency to fear the unknown, be astounded by discovery, and bored by their repetition. We naturally fear what is likely to injure or kill us. This renders each person highly susceptible to beliefs and acts that are based upon givens and partials that quickly become prejudice, bias, and misjudgment. These are commonly not right, less or not TRU. It is in our naturally keen, but in practice wrong, tendencies to discern differences that are bad, foreign, and apt to be antithetically opposed to us, that we believe all should be the same and static. It is this erroneous tendency, applied specifically to politics and government that John Kennedy referred. It, when taken in full, beautifully shows us how much what we hold to be ideals and goals, must be corrected in the actuality of TRU.

TRU is, as it always should be, In Fashion, **TIF**. That which has, is and will be in fashion, can only be TRU. For all that is not is only illusion, and often worse, destruction of fashion as it must be TRU. What is so current a fad and fashion is not likely to be actual fashion, it is not TRU. Man has an inveterate tendency to find fashion in that which is novel, surprising, catching to the eye, popular, devious, or trivial. Yet these are often a reaction against imbalances and disproportions in life, even as TRU. Both the imbalances and the reactions to them, although they may have some merit, are more apt to only worsen these imbalances..

We search for the answers and solutions to all things as they pose the ever present questions and problems of life. Many are old, many repeated, and many new. This is the crux of our nature, relativity, observer-observed, S-O or Á FAF. Yet we are led astray, for much of this great heritage, as interesting and important as it is in literature, language, history and all, was erroneous, mythological, misleading, vague, trivial, or very disproportionate. It was at best only partial of TRU. We are distracted from the correct and valid TRU nature of entities, senses, things, objects, the given, and all. So that there has been a great basis for and a continual return to knowledge that converges or returns to and enhances TRU. In this search great philosophers came to be idealistic in being subjective, materialistic in being objective, and realistic in being combined in their identity in being. They were transcendental in being about this world, but immanent in being in this world, TRU. For here the approach and fusion becomes one. This one is demonstrated by the fusion of, Á, L, and T, that occurs in the union of the dimensions of TRU, TROS, TRIT, TU in TIT-TO. The great predominant and typically human S-O or Á FAF and continuum is what best and most realistically and fully reveals this union. The stimulus - response, problem - solution, analysis of senses and things is the pivotal and crucial aspect of the Á dimension. How the individual interacts with senses, things and TRU. It is through such focus and coordinate SUT that we can bring infinite senses and things to solid order and closure in TRU.

MEASURES of DYNS CES of SULF SECTOR of TRU

DISCOVERY and USE of RIGHT, TRU	4-3-6~	S	DUR
EXTRATERRESTRIAL INTELLIGENCE, T	~0-0-9~+-	S	EIT
FISMS	~3-1-8~ -	M	
GOVERNMENT, POLICY, KIND, PERFECT	4-1-7	S	GPK
IMT	1-4-5	M	
INDICATORS, ATTITUDES, IPR, WORK,T	4-3-5	S	IAW
INSTRUCTIONS for LEADERS, PRESID.	4-3-7	T	ILP
ITOD	2-4-6	T	
KNOWLEDGE, ENLIGHTENED CITIZENS, T	3-2-7	S	KEC
LOEN	~4-3~-7~	M	
MED	3-3-7	T	
MERPS	~4-4-6~	T	
MIS-COMMUNICATION, LEARNING, TRU	4-4-5~+-	T	MLT
MAB	~4-4~-7	M	
MABAC	~7-4~-7	M	
MABITH	~1-6~-7	M	
MABIK	~2-3~-7	M	
PER	~-~-~-~	M	
POPA	~5-3-6~	M	
POLITICAL and TRU DIVISIONS	~4-1-8~	S	PTD
PS	~3-5-5~	M	
RABAC, RABITH, RABIK	~-~-~7~	M	
REACTION, ARREST of KNOWLEDGE, T,	3-4-6~ -	M	RAK
RELIGIOUS and CHURCH METHODS, TRU	3-3-6	T	RCM
RESPONSIBILITY, IS and CAN BE BEST	4-5-6	T	RIB
RESPONSIBLE LEADERS, POSITIONS, T	4-2-7	S	RLP
SE	2-5-5 -	T	
SEARCH for ANSWERS and SOLUTIONS,	2-4-7	T	SAS
SIST	4-1-8~	M	
STIP	3-1-6	S	
SULF	2.5-2.5-7.5		
TACK	3-3-5~	S	
TEX	~1-1-8~	S	
TIF	4-3-6	S	
TOPA	4-4-6~	S	

Fig 1-78 **DYNS CES and CHES in SULF SECTOR of SPALT**

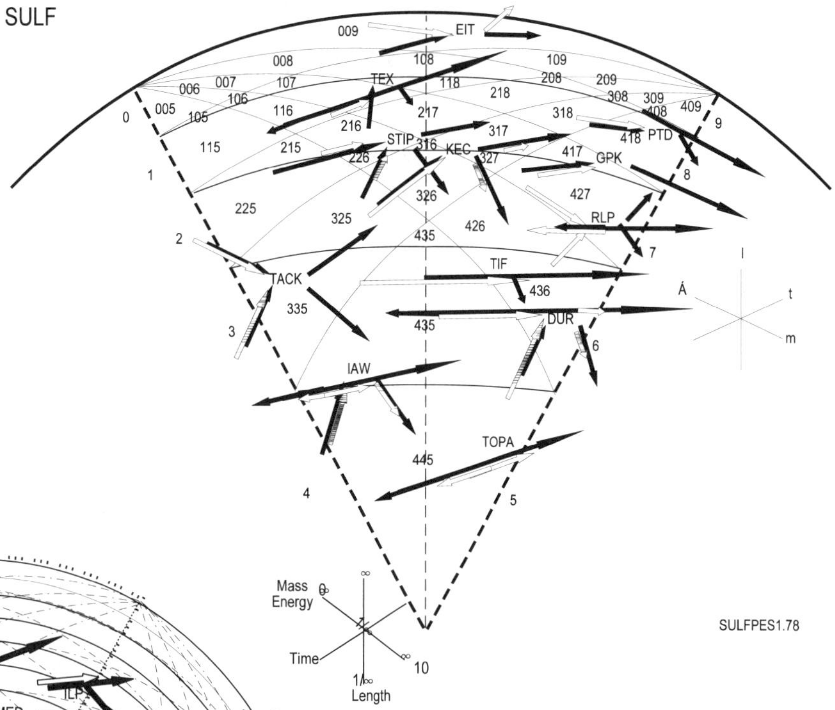

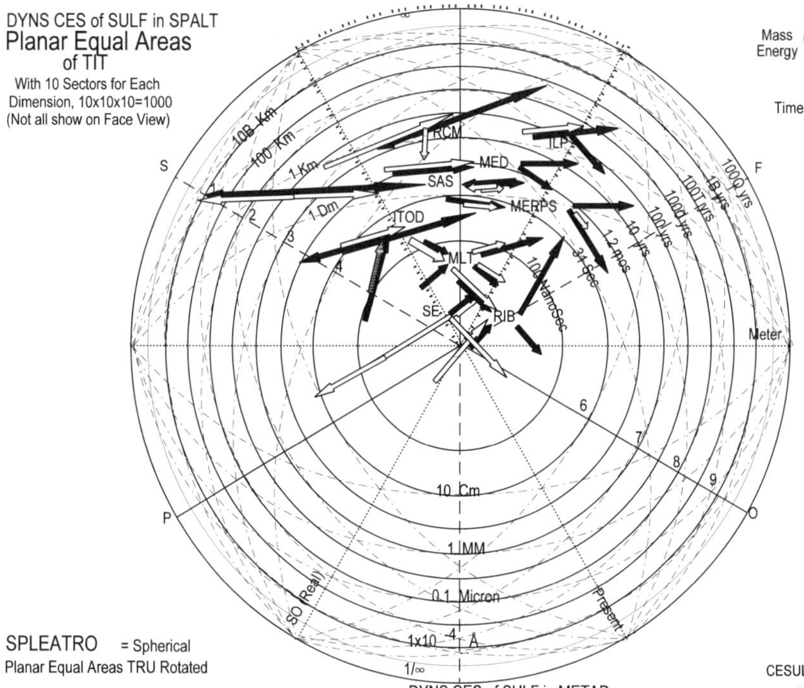

Fig. 2-78
DYNES CES and CHES of SULF in SPALT

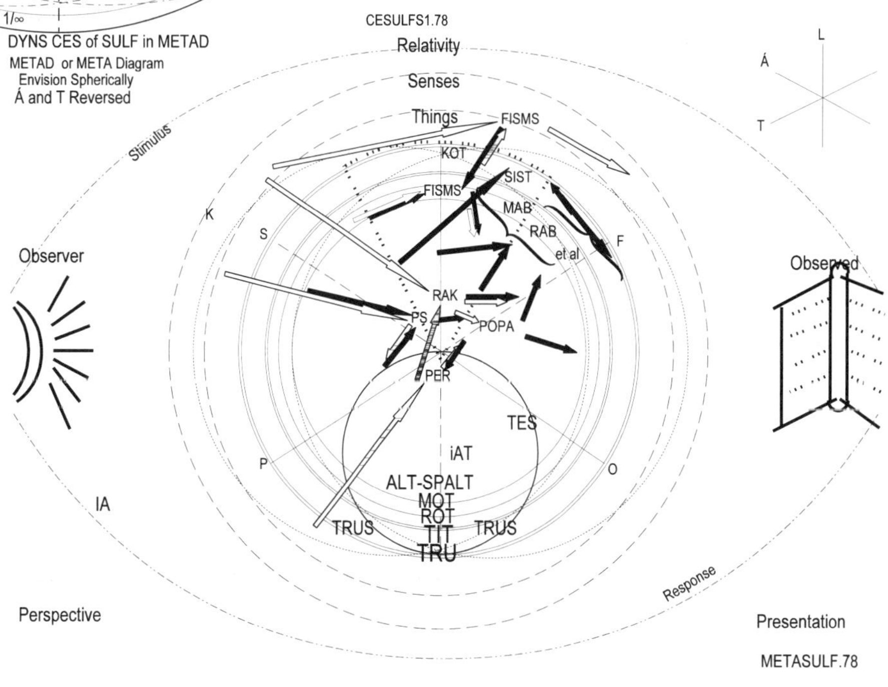

Fig. 3-78 **DYNS CES and CHES of SULF in METAD**

CHAPTER 79

DYNS SUMMARY, RELATIONS and FUTURE

PROBLEMS, QUALITY, IMPROVEMENT, TREATMENT, RIGHT, BEST, CONTROL, HOPE, WILL and SUCCESS

Negative and partial forces are often compounded by stressful situations or crises. They are also compounded by density and close proximity and the associated lack of basic values, first principles, and spirit of reconciliation. When these accumulate you have the seeds of failure, waste, conflict, war, and calamity. Therefore we should quickly and with perfection analyze each and all that confronts or is offered to us. Usually all that is needed is to avoid it. Occasionally it must be turned down or put down as appropriate. Rarely will it be TRU and appropriate for the need. The great mark of modern man, in this sophisticated and technical age, is the ability to accomplish the above with ease, almost automatically and to perfection, selecting what is best from a vast array of offerings. Only with expert knowledge can this be done. This is why it is important that all have KUPS, not only a general and special knowledge appropriate for their work in life, but a partial knowledge of all needed for oneself and all. How to fulfill even the simplest of needs.

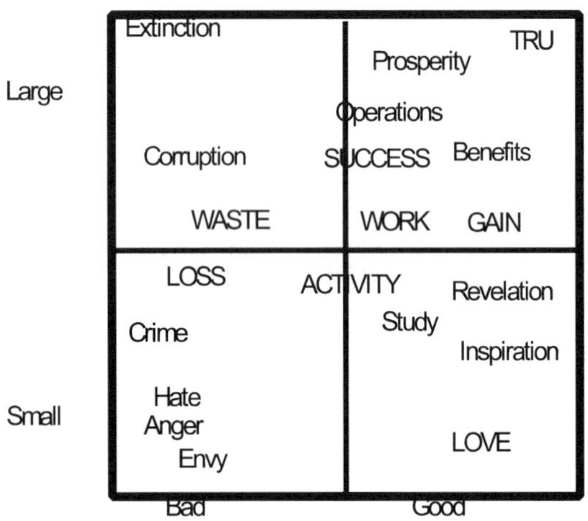

Fig 1-79 **HOW TO IMPROVE and HAVE THE BEST THOUGHTS and ACTS**

Thoughts and acts, good and bad, large and small are to be directed in which way? How can we improve and keep them at their best? Life's totality of thoughts, acts and work are some of each and all of the above. Yet there is a best kind and proportion of each. Each will vary with different persons, ages of life, circumstances, conditions, work, problems, and many others. Only when we have a good understanding of all things and only when we can apply this understanding through SUT, ROT, MOT, and detailed accounting of its processes, as by HOM, can we begin to have the kind of acts and thoughts that we can and should have if we are to be, to be our counterparts, TES in TRU.

All decisions and acts of man are to be measured in terms of their benefit. This is vividly seen in the many problems, conflicts, troubles, strife, turmoil and wars past and present. These are a continuation, past through future of the summative deficits of all the acts and benefits that should have been right but were not. For all that is not, negative and against TRU is what causes such loss. We see these in General Forces Stratified Time, **GFST**.

LASE = Law of Selective Elimination. One and all must be constantly on guard not to strike out, act impulsively, rashly, or recklessly. Such behavior is typical of animals whose instincts tend, when critically threatened, to vigorously move, act, and eliminate. Humans correct this with their higher powers of mass energy, SÁ, by being when best trained and exercised thoughtfully, rationally, logically, with full application of TRU. This is done for each occasion., by LASE, selectively, with judgement and a full accounting.

To know and do precisely and properly and give the right treatment for whatever the problem or the ailment is, follows our basic mass energy Á FAF, the purpose of human life. It is best seen in, and the goal of, specialization, artistry, technology, professionalism, and expert productivity. These are seen in the highest standards of care, the most beautiful decorations, the best and most functional products. **POMÂC** = Principle of Multiple or Dual Acceleration. POMAC is revealed when the best of routines are combined with the best of inspiration, ingenuity, reason, judgement, and effort to produce the maximal effect. It is what is present in the best of institutions, corporations and businesses through organization, lay out, planning, programming and procedures. It helps to show how all are formed and function, are parts and processes of TRU, and, as well, how TRU is proven by them.

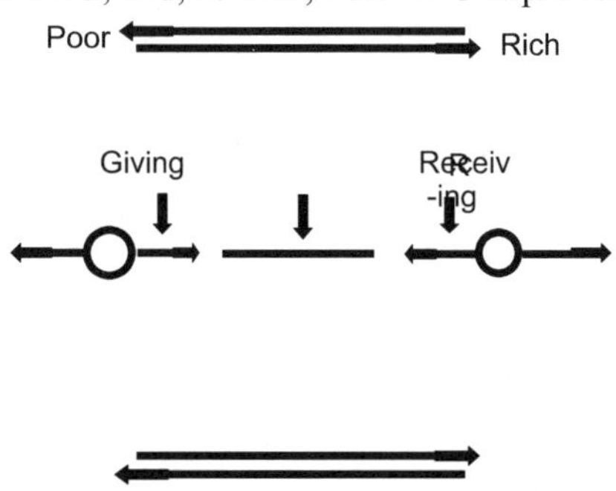

Fig 2-79 **POOR and RICH, GIVING and RECEIVING**

The relation between poor and rich, giving and receiving work all through HOM, DYNU, DYNS and DYNO. Many take whatever they can from the world and give little or nothing in return. They have never learned, known or appreciated the glory and greatness of giving, and giving TRU, that of greatest benefit. This is the benefit achieved from giving themselves as their counterparts **TES → PUREALITY → TÀ**. With larger populations it is mandatory that less loss occurs to all that enables us, as TRU, to be. We can neither survive if our resources and strengths are depleted nor can we survive if all the rest of TRU is increasingly lost. So that we must discover, learn, know, love, appreciate, and work to give, redeem, restore, achieve, produce, resupply, and live parsimoniously and frugally in TRU. We must never be so poor that it causes us to become poorer or lack what is needed for TES, and never so rich that is makes us become richer to waste what is TRU.

LACSA = Law of Action through Complementing, Supplementing, and praise versus disapproval, Condemning, and Annihilating. Results are produced and success is achieved much faster, further, and better by complements, supplements, and praise than through excessive force, negative means, and condemnation. Individuals, groups and all persons act or react not simply but complexly in ways that depend upon a variety of relationships and participation in and out of TRU. They attend to, think about, and seek to know what their relationship is and the results that are likely in the action involved. This knowledge will ultimately be that of TRU, KOT and COGNOS.

MAJOR CONFLICTS, PROBLEMS, FAILURES and TROUBLES	NEAR TERM 0-10 years	MID TERM 10-100 years	LONG TERM 100-1000+ years
ANTITRU/TRU Lack of TRE/TTT	Existent only in TREQ, Severe	Introduction TRU, Severe Added Reaction and TRIX	Progressive Reduction by ACC and all TRU
ANTRU/TRU Against/For TRU	Existent mainly in TREQ, Severe	Introduction to TRU, Added Reaction and TREX	Progressive Reduction by ACC, TREX and all TRU
ATRU/TRU From/To TRU	Existent mainly in TREQ Severe	Introduction with TRU, Added Reaction and TRAX	Progressive Reduction by TRAX, with TREX and TRIX
NAMTRU Mixed Worlds	Existent, Part of TRU Unrecognized of TRU	Recognition, Beginning Management and Control	Full Recognition, Better Management-Control
INDIVIDUAL/TES Casual self/One's Soul	Poorly Recognized, not as TRU, poor of TREQ	Better Recognition, poorly of TRU	Well Recognized, Increased Transition to
TI/TU Divided/United	Part of Life, Poorly Recognized only in TREQ	Some Recognition in TRU Beginning Adaptation to	Improved Adaptation to and Utilization for TÂ
TRIN/TRINI Intinite/Infinitesimal	Unknown	Early Introduction	Improve Perspective, Beginning Control of TRIS
TRUS-TRUCH/TÂ What has made us/we can	Great Part of Life, most-ely unknown and harmful	Early Recognition, and Method of Control	Improving Design, Means, Management, and Control
INERTIA-LASSITUDE/ ENERGY	Continuous life conflict fair level of energy	New methods of stimul-ation and prevention	Much Stimulation, Di-rection and Improvement
CONSERVATION/INDUSTRY	Active Program Resources, Clean Industries, Design	Expanding Problem and Solution, deep divisions	Problems Persist, Solu-tions escalating
RACISM/INTEGRATION	Great Processes Underway Physical - Mental aspects	Recognition of species-genetic-TRU basis	Initial Mental adaptation and Physical control
NATIONALISM/INTERNATIO-NALISM	Great Conflicts, Political Economic mostly unaffected	Early recognition of basis in TRU,	Gradual adaptation to TRU political-economic ideals
SECTARIANISM/ECLECTICISM	Pitiful struggle, vast lack of understanding	Initial knowledge of basic problems, solutions	Rapidly expanding know-ledge of TRU and Actions
REVOLUTIONISM-WAR/ PACIFISM	Severe Conflicts, Poor understanding, Suffering	Initial knowledge of basic problem, solutions	Gradual control of means and extremes adaptation
TOTALITARIANISM/DEMOCRACY	Fair Recognition and Control with TREQ	Direct Need and Use in TRU	Balanced Social Contract with TRU, TIT, TES, TÂ
TOLERANCE/INTOLERANCE	General Population Level Poor, Continuing Prob	Continued Education Up-grade Media Show Effects	PPMA with TRU KOT TES ACC PUREACLITY COGNOS SÂ TÂ
SOPHISTICATION/(OPENNESS)	Extremes of each on ATRU and ANTRU, need showing	Improved Knowledge under-standing and attitudes	Moderation and expansion comes with KOT TES TÂ
GOVERNMENT/CITIZENS	Better Knowledge, appreci-ation Participation	Upgraded Servers and served plus benefits of	Close identity of both by KOT TIT TRIA COGNOS SÂ
INDEBTEDNESS/SAVINGS-FRUGALITY	Knowledge of dangers and benefits, consumerism	Emphasis on best value, frugality, what is 'free'	Greatness of TES REM TIT PUREACLITY ACC TRU/ANTRU
RATIONALISM/IRRATIONALISM	Training and behavior and to enhance rational means	Continual work to enable knowledge reason behavior	Best effort for TES KOT All Higher Levels of Á
TOXIC SUBSTANCES-ADUL-TERATION/PURITY	Personal and Media Atten-tion to, Controls	Continual removal, PPMA, Upgrading, Knowledge of	KOT of role of all in TIT and 1/ANTRU Provision for
ALCOHOLISM/	Media Personal emphasis on Dangers and Effects of	Appreciation for value of life dangers cause effect	KOT of role of all in TIT and 1/ANTRU Provision for
CANCER/	ACTIVE PREVENTION, TOXIC Substances, Exposures	R&D Education Personal Identity with Control	ALT REM KOT total ACC PPMA of all ATRU-ANTRU
HEART DISEASE/	ACTIVE PREVENTION-CONTROL FATS, SMOKING, EXERCISE,,	R&D Education Personal Identity with Control of	ALT REM KOT total ACC PPMA of all ATRU-ANTRU
DUTY-RESPONSIBILITY/ rebelliousness	Education Knowledge Mutual Benefit Persons/States	Improved Governments-Pers-ons Participation Benefit	KOT DAOST TU TRIA TRID TES PUREACLITY ACC SÂ
CARELESSNESS-ACCIDENTS/ CAUTION-PREVENTION	Education Media Dangers Attitudes of, Prevention	Recognition of, Education Simplicity Refinement PPMA	TES TIT PUREACLITY ACC REM PPMA Avoidance COAT
INTERESTS-INSPIRATION-REVELATION/ detachment	Training, Education Media stimulation 1/distractions	Education, Focus, What is best enhancement benefits	TRIA Identity TES REM Revelation Inspiration
ENDOCRINE INSUFFICIENCY	Improved Health Compensa-tion Supplementing	R&D Genetic Engineering Treatment Health Care	R&D, Genetic Engineering SPALT, HOM, ACC, TREX
GENETIC DISEASE/HEALTH	Understanding only Beginning, few	R&D Education Genetic Eng-ineering Best Use Control	TRU KOT TES ACEXIN SPALT HOM DYN ACC TREX ELTICMA
HATE-PREJUDICE-BIAS/LOVE	Base Human Reactions, great antithesis	Recognition basic reasons and solutions in TRU,	Great Reduction, Control, Expansion of Love in TRU
MENTAL ILLNESS/MENTAL HEALTH	Constant Problem, Severe Improvements, Preventions	Recognition of TRU role in cause prevention cure	Great Reduction through enhanced TRU, prevention

Table 1-79 **NEAR, MID and LONG TERM REACTION and CONTROL of WORST PROBLEMS**

Some of the more important problems, troubles, conflicts and failures of our time are investigated and interpreted in terms of TREQ and TRU. They are further described throughout the text. They can be better comprehended through TRU and their relations to all the concepts, principles, laws and mechanisms, ES, CES, and HOM of TRU. They need to be continually addressed and corrected.

There is a great disparity between the levels of types of problems and ills and their classification. 1. No problem or ill, or believed and suspected yet non existent. 2. Near term minor or insignificant, 3. Minor yet discrete ES or CES errors or ills. 4. Potential or early major problem or ill. 5. Known or unknown problems less than of all TRU, critical or near fatal for large segments of TRU. 6. Unknown deadly to TRU lead-to extinction. 7. Known deadly to TRU problems, life extinguishing, with limited or decreasing ability to react. Items in Table 1-79 can be categorized by these seven levels for resolution of problems, and use in ACC.

TAR = The Transcendent Resolution. TAR is seen in how we should avoid a Machiavellian policy. It signifies how extremes of self interest and total sacrifice are used as expedients for the occasion, situation, conditions, and states of the problems at hand. Yet, they are not entirely opposite, in ways they are the same, less than or adverse to TRU.(VAPO) TAR signifies that policies such as self gain and self denial are not to be determined within narrow confines. Both may have a use on highly differentiated and specific occasions, but for TES and TIT both lack VAPO and tend to become negative.

The future can't live off the past any better than the past can live off the future. Each, as all ES and TES, must to a large extent, be in balance, and stand on their own. The lesson and resolution of the problem of inadequate citizens and governments lies in the synchronous and harmonious part each must properly play as ES, CES, processes in TRU. Disproportionate TRU can be as devastating as ATRU, and ANTRU, especially when it produces integrative negation on a large scale. This must be constantly uncovered, recognized, prevented or resolved. The debts that all must pay are of two kinds. 1. Those of TRU, which is a part of being ones counterpart in TRU. This is compensated by the mass and power of TRU itself and all its components and benefits, 2. Those that are not of TRU, disproportionate TRU, NAMTRU, ATRU, ANTRU and ANTITRU. Governments, like individuals and all, can be measured by the extent they fulfill their roles in TRU.

Since the world is based upon TRU the whole serves the parts and the parts serve the whole more or less spontaneously. Then, there is much less need for either government or the regulation of its citizens. It is for this reason, if we are going to bring the NAMTRU world to TRU and each individual to TES we must adopt a firm policy with both poor and bad governments and people. This more or less includes all, other than TRU. This is called: **PARCGI** = Policy and Actions for the Rectification and Containment of Governments and Individuals, and wholes and Parts, that are less or other than TRU. When governments, or groups, increasingly fail to identify with their constituents, and when parts or individuals increasingly fail to identify with their governments or groups, then much of the strong positive and beneficial bonds are loosened and become dissociated. PARCGI is to be achieved, and TRU strengthened and made more permanent, by attacking these major reasons and causes. Individuals as citizens, TES, and governments as overseers of societies and cultures, should both be no more and no less than the degree to which they fulfill their part and roles in TIT-TO.

Fig 3-79 **INTERACTION of FUNDAMENTAL OPPOSITIONS and TRU PROBLEMS**

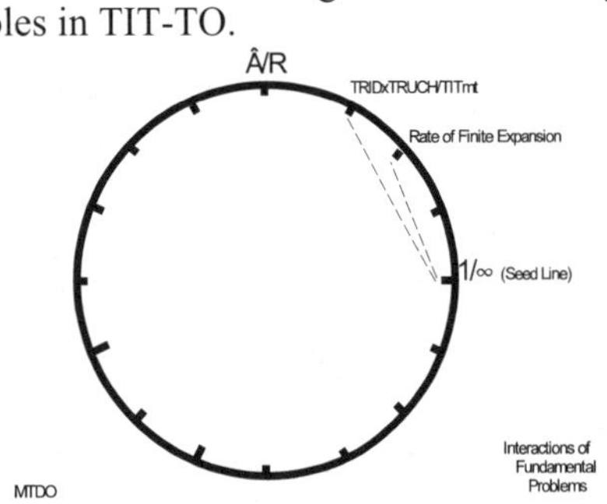

In this preliminary study and drawing we initiate a search to discover, analyze, and show how the basic oppositions or polarities in TRU interact. We should next seek to examine the major problems that exist in our world, as we do for proofs. Through their relations we gain the complex understanding that DYNS provides. It is in such summations of interactions through DIN, SUT, SPALT, POS, STA, HOM and DYNO that we can finally reach and provide the needed totality of TRUD, TIT-TO. (q.v. proof diagrams, pg 55)

As much as possible and feasible, governments should help each other and those they serve to be TES and TRU. Citizens should help each other and their governments to be TES and TRU. **LAMUA** = the Law of Mutual Assistance. There are associated laws, which include, **LACOM** = the Law of Compliance, and, **LAMUC** = the Law of Mutual Compliance. These concepts signify topics that are at the core of civics. In being more discrete, natural, functional, by the VAPO and SCARLS they help to bring civics to TRU. They help to extend the social contract that exists between governments and its citizens to that between all individuals, as TES and TRU. By LAMUA, if we are going to help each other, we must know what it is that needs help. The massive drain on humanity of being wasted by doing what is of no or negative benefit is the cause of much loss and suffering.

POMOG = Principle of Manifold Greatness. Greatness is derived from TRU & TES. Its attainment is to a large extent accumulative, assured, of great momentum and even, to an extent <u>retroactive</u> as well as increasingly <u>proactive</u>. As shown by REM, we can neither assume, nor extol, singular greatness or REM, to the neglect of their combined, multiple, interrelated and interdependent complexes in the totality of TRU. It is these and the processes, forces and powers they produce that make up much of TRU and thereby our lives. It makes us realize our REM in relation to all else in TRU. Only the fact that our participation as LOPAN, the ENIT of individual being, and the absoluteness of TRU help us to compensate for this seeming insignificance in REM.

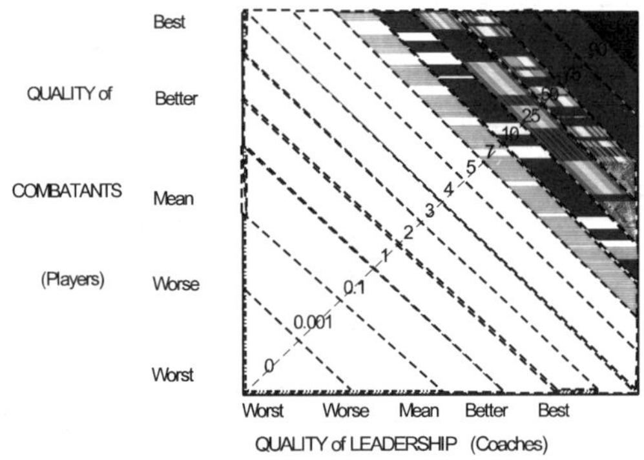

Fig 4-79 **SUCCESS or FAILURE in GAME, WAR or LIFE**

To win or lose, to succeed or fail, to survive or become extinct, to be or not to be, is largely a function of quality. Quality is of many kinds generally and specifically typical of TRU and TIT. Two carriers of quality that are of the greatest importance, with a multiplying effect toward the failure if they are lacking, or, by equation, success, if they are present, are:

Players Quality x Leaders Quality = Success.

It is the compounded power of quality that provides the cutting edge upon which natural selection, genetics and survival of the species are based. **SELCON** = Selective Level of Competition and Natural Selection for Success and Survival, also: **SEQUAL** = Success, Survival and Selective Level in TRU depends upon highest Quality of parts, players, combatants, or participants. That man, as well as all TRUS, RUTS, and RUS exist because they have been the best, for their niche in life, or adaptation to their environments and changes, is well attested in science and nature. That man and TRU will in the future continue to win rather than lose, survive rather than become extinct, be rather that lose their absolute existence and being can also be well attested by SELCON and SEQUAL.

Figure 4-79 gives rough estimates of percentages of failure or success. These are not to be accepted as the sole or final measures, for they will greatly vary depending upon opposition, domain, time frame, execution, and many other factors. An edge of only 3% has been considered accurate for specie survival. For TRU we encounter great opposition from all that is less than, not or against TRU, as well as its total domain for all time. The five or n qualitative categories of both participants and their leaders can be developed so that of the resulting 25 or n^n categories may be used in interpreting their function in games, battles, wars, life and TRU. It is in these by: **SEQUOT** = Success, Survival and Selective in Total TRU depends upon quality of constituents, individuals, concepts, principles, laws and the leaders who carry them out. It is largely because of SEQUOT that each and every person, whether they are a combatant or general, constituent or leader, contributor or manager in TRU, should learn and do all in their power to be TES and TRU. It is upon this high Level of quality that TRU is sought and ACC is provided.

ORGAN/SYSTEM	FORM FUNCTION NEED	PURPOSE USE	BENEFIT
ABSOLUTION ATONEMENT CONFESSION	ABSOLVE ONES SINS ATONE FOR WRONGS CONFESS ONES SINS	ELIMINATE, ATRU, ANTRU, NON TES, WRONGS AND SINS	ACCEPTANCE INTO TRU ELIMINATION OF ATRU REINFORCES TES AND TRU
CARE	WHAT NEEDS HELP	FULFILL ALL NEEDS	STRENGTHENS MUCH OF TRU
CHARITY	TO GIVE OF ONESELF	CAN NOT HELP THEMSELVES	HELPS JUMPSTART ES CES TES
CONVERSION	REVERT TO TES AND TRU	BRING ONESELF TO TRU	EXPANDS TES AND TRU
DIVINITY	HOLD IN THE HIGHEST	MAKE ONE WITH TRU	FULL EMPOWERMENT OF TRU
EQUANIMITY	BALANCE AND DETERMINATION	BECOME STEADFAST IN TRU	AVOIDS ATRU, ANTRU ISMS
FAITH	HOLD TO CONVICTIONS TRU	TO BELIEVE IN TRU	HOLDS ATTENTION TO TRU
FORGIVENESS	DISPELS NEGATION	REMEMBER TES IN ALL	REVERSE NEGATIVE/POSITIVE
HONESTY	VALID TRU ACTUALITY	ONLY TRU TRUTH	DEPENDABLE TRU VALIDITY
HOPE	OPTIMISM FOR TRUF	ALL BETTER IN TRU	FUTURE ENCOURAGEMENT
HUMILITY	DISPEL VANITY	TRU TES ACCEPTANCE	FOCUS ON TES ES CES
IDENTITY	UNION WITH TES AND TRU	TOWARD SAME IN TES TRU	UNIFIED STRENGTH IN TRU
IMMORTALITY	CONTINUED TRU PRESENCE	EVERLASTING LIFE IN TRU	JOY IN ETERNAL TRU LIFE
INSPIRATION	DEEP FEELINGS TRU JOY	OPTIMAL FEELING FOR TRU	TRU CUE EXQUISITE FEELING
LOVE	DEEP FEELINGS OF UNION	DEEP BOND WITH ALL TES TRU	RETURNED BONDING IN TRU
MANIFESTATION	UNEXPECTED NATURE OF TRU	CLEARLY RECOGNIZE TĂ TIT	REALIZATION ACTUAL TRUS
MERCY	AVOIDANCE OF BAD ACTION	TRU COMPASSION FOR NEEDY	FULLY REGAINED TRU
MODERATION	AVOID OF EXTREMES, ISMS	CONVERGENT BALANCE IN TRU	NO UNNECESSARY CONFLICTS
PERFECTION	SEEK PURITY IN TRU	NOTHING LESS THAN ONLY TRU	ALL HELD TO TRU
PLAN	METHODS OF SEEKING TRU	BEST METHOD OF GAINING TRU	EFFECTIVE TRU ACHIEVEMENT
PRAYER	ATTENTION TO GOD or TREQ	CONSOLATION IN GOD or TRU	FIRM BONDIND WITH TRU
PURPOSE	DEDICATED TRU DIRECTION	REASON AND ENDS FOR TRU	FOCUS ON SUCCESS OF TRU
REDEMPTION	RELEASE FROM SELF, ATRU	RETURN TO TES AND TRU	NO ANTRU, ONLY TES TRU
RESPECT	DUE CONSIDERATION ALL TRU	FAIR SHARING OF TRU	FULL ACCORD DAOST TES TRU
RESPONSIBILITY	FULFILL ONES ROLE IN TRU	MEET TRU DUTY & OBLIGATIONS	BEST FULILLS TRU and TES
REVELATION	DEEP TRU INSPIRED THOUGHT	DEEPEST TRU REALIZATION	EXERCISE IN BEST TES TRU
SALVATION	NEED TO BE SAVED FOR TRU	TRU TES FROM ANTRU ATES	PREVENT ATES ASSURE TES
SACRIFICE SUPPLICATION	CALL FOR TRU HELP GUIDANCE ONESELF LESS FOR MORE TRU	CALL and SUBMIT TO TRU DENIAL ATES TO GAIN TRU	FOCUS ON RETURN FROM TRU JOY OF GREATER TES TRU
TRANSFIGURATION TRANSFORMATION	CHANGED FIGURE TO TES CHANGED FORM TO TES & TRU	ACHIEVE TRU CONFIGURATION FORM TO TRU TIT ES or TES	REALIZATION OF TES TRU ACTUALIZATION OF TES TRU

Table 2-79 **SPIRITUAL ORGANS OF SURVIVAL FORM AND FUNCTION, NEEDS, PURPOSES, AND BENEFITS**

These are some of the many theological concepts that when adapted to TRU signify, both how and how much, religion has in common with TREQ and TRU. How well TRU explains their actual meaning and significance. They are presented along with other means of survival, physical, mental, social, phenomenal and spiritual. These provide many of the mechanisms of the core of TRU. It is why TRU is so concentrated and powerful with solidity and sphericity about its center. It explains how God is so much, of and in, us as we are in and of God. It is both physical, mental and spiritual or real, along with spaciotemporal form and function as these together make up God's frame, or m,l,t or mlt and $\acute{A}^n$, L^n, T^n, or ALT and ACEXIN of TRU. Such ideas, themes and purposes as hope exemplify the long tradition of religion. Christianity as a living hope demonstrates how man is sustained and rewarded when TRU.(e.g. Schleiermacher) It is not without reason that temples, academies, schools, churches, cathedrals, and universities have been the centers of learning. They are closely associated with what is best in religion, philosophy and science. Even withstanding their deficiencies and limits, they have been great stimuli, the way to success. How the best gain the best. How the kingdom of God leads to a search for the best, all that is most important, or education that leads to TREQ and TRU.

CONVERSATION, MONEY, SEX, POLITICS and RELIGION

Although money, sex, politics and religion are taboo as topics of polite social conversation, they are sine qua non, without equal, examples of what needs to be discussed and best known if one is to have an open mind. The reason they are taboo is an extreme entrenchment of attitude in most people, that accentuates narrow mindedness compensating for the limits in, and adaptation to, life. The experiences of life, especially if it is difficult, are hard task masters. Money, sex, politics and religion are topics that each person takes seriously, if for no more reason than they are close to their success or failure in life. They are close to any plans they have, guidance they seek, or ideals they may have that assure their success, survival and happiness. Once formed and brought to conviction, religion becomes not so much a guide or ideal, but a refuge, an escape mechanism from all other approaches to life. This is what makes it’s adherents narrow minded and branded as hypocrites. They, in escape from the actuality of the world, rely on simple confessions and rules of goodness and what is right. Yet only to act and behave in diverse ways that have little relevance to either their part, role, and contribution to society, humanity and the actual world. They become and reinforce one’s limited individuality and self centeredness. Thereby money, sex, politics and religion must be open, to investigation and enlightened conversation, for all to see. To see not only how they exist in the actual world, but how each and all persons can best learn, know and use them to make their lives more meaningful, valid, and contributory to their whole, as the whole benefits them. What are, and are not, they in this special and actual world of best human life? How can they be best learned, remembered and used by each and all persons to enhance the world and themselves? It is the diverse, partial and negative interpretations and misuse of these topics that is the cause of many of the worlds problems and conflicts. For as long as individuals, groups, states, nations, cultures, civilizations and the entire world consist of not only different aspects of the same thing actually in the world, but inactually by various diverse, invalid, disproportionate, partial, and negative other worlds, there can be little common ground. There can be no meeting of minds, or souls. No unity in totality that is the absolute minimum necessity if the human race is to survive and avoid destroying itself, and becoming extinct.

That TRU is very much man as man is very much what we are composed of and how this works to our best advantage in a highly competitive world, is evident from biology, anthropology, history and religion. It is these forms and functions that makeup the parts and processes that provide man with the predominance that assures the thisness of TRU. It is in the typical mass energy and spaciotemporal form and function of this special predominance that most essentially and vitally provides dimensional basis, Á, for discrete specificity from without and identity from within..

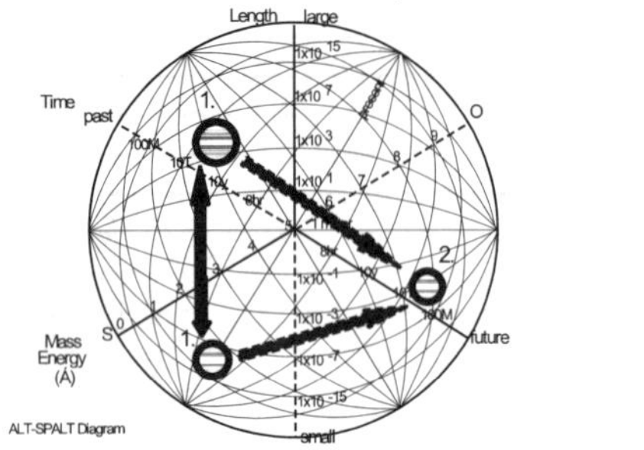

Fig 5-79 **GREATEST OPPORTUNITIES for FUTURE TRU IMPROVEMENT**

Where are the greatest opportunities for the growth, expansion and success of TRU? Where are the greatest dangers, areas of decay, contraction, and failure? 1. Education and the development of the highest possible qualities in the individual. The development of each person to their fullest in knowledge, knowledge of TRU and appreciation or love for the best of human life, TRU. Optimal progression from: **Individual - self - TIV - TESIN -TES - TRU.** 2. Genetic Selection: both as has been the natural course through thousands and millions of years, and what can be improved in the future. It can be made worse or improved depending upon our knowledge and use of it. By indirect as well as the direct effects we have to continually enhance to assure its quality. Directly it can be supplemented by genetic engineering and other methods based upon the continued application of life sciences. These are methods of the present and more in the future that will greatly enhance our ability to improve the race, and the ability of future individuals to live healthier, happier and fuller lives

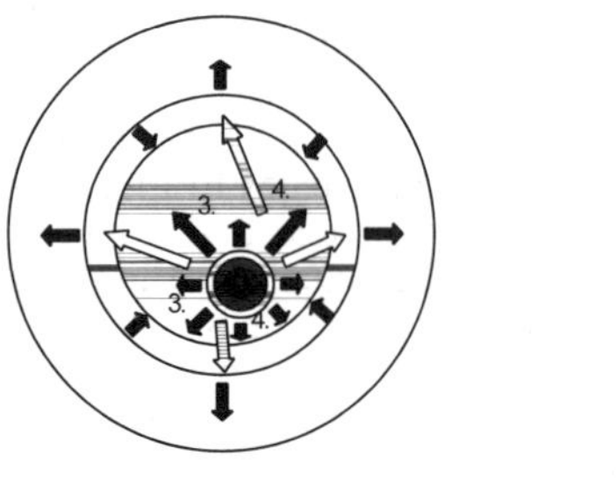

Fig 6-79 **FUTURE TRU IMPROVEMENT**

The means may be by singular, multiple, combined, total, and all beneficial forces that promote and improve the genetic, physical, mental, social, economic, political and total makeup of TRU. As these are discovered, developed and used there is also needed, avoidance and PPMA of permanent and irreversible injury and damage. Most processes of TRU, CES and HOMATION, are neither solely singular nor total. They are, **LARCES** or the Law of Relative Complexity of CES. They are of a continuum from near absolute singularity and differential to absolutely integrated totality. Only TIT-TO is of the latter. So that for POS in SPALT, DYNU, DYNS, and DYNO we can only approach perfect distinction in TRU. If less or not TRU they become negligible, acceleratively worse.

There is nothing that takes the place of, 1. healthy and highly capable individuals, 2. Individuals with a great amount of energy, 3. Individuals with high levels of consciousness, 4. Great, complete, distributed, and coordinate knowledge to fill this consciousness, 5. Use of this knowledge through judgement, reason, and all those ways of thought that most effectively enable the best acts, deeds, behavior, endeavor, service, benefits, achievements and success. This is much of all that is TRU, as TRU is much of them.. It is upon these and the whole spectrum of gradients of the mass energy FAF of TRU that CES, HOM, DYNS, and DYNO are based, the highest levels of civilization are attained, and hopes for the future depend. With the enhancement of ever greater quality of human mass energy in the genetic material, live product, determining forces, and throughout TIT-TO the future of TRU can be assured.

Without a clear distinction between the kinds of selection, it can never be reasonably efficient nor properly used to improve human well being. This is called: **LADSEL** = Law of Differential Perfected TRU Selection. LADSEL includes selection that is made: spontaneously, impulsively, naturally, voluntarily, involuntarily, punitively, rationally, by management and control, professionally, intellectually, totally, optimally and by TRU. **LASEL**= Law of Selection, **LASEÄ** = Law o Selective Form, **LASED** = Law of Selective Dynamics, **LASFAF** = Law of Selective Form and Function, **LASEC** = Law of Selective Composition, **LASEV** = Law of Selective Value. Whether TRU is of generations of a single individual or of a total population, whether it is only 'I' or the whole universe it is largely a product of LADSEL, LASEL, LASEÄ, LASED, LASFAF, LASEC, and LASEV. It is in LASEV that the great powers and beauty of incremental selection by the individual can be seen most clearly. Here also, can we best see our roles in TRU, and the proof of both our, as TES, and the world, as TRU is to be found. For TRU is, and is composed of all, more or less, of all these acting dynamically through us and all about us that constitutes this actual world. It is also here that we can quantify all in and by TRU by its value to the whole, REM of TIT-TO.

In the many tables and figures of this text and the courses described, we can increasingly find what is needed. Instead we live largely in a world of NAMTRU courses. All must strongly take interest in, encourage, complement, honor, and reward ever higher levels of selection. The knowledge and powers of management and control in TRU has reached a point where we are compelled to rely on and must adequately utilize the great knowledge and powers that have been developed. With even greater knowledge and understanding of all kinds of selection and their application, as seen in voluntary and higher means in ACC, the needed improvements in selection can be achieved.

LEF = Law of Essential Futurities. LEF signifies that when certain unknowns, non existents, non TRU entities, ES, forms, forces and needs do not exist, and are not apparent or certain in the present, does not mean that this holds for the future or past. LEF are those whose potential are known, unknown, unrecognizable, and unregulated in the present life that will still be essential to the future of TRU and TÂ. They are critical for the assurance of TRIA. They may be direct or indirect, although often indirect or mixed. LEF is highly significant, especially from a contemporary view. The universals, transcendents, great mechanisms, laws, principles, concepts, processes, forces and powers through and through TRU into TRIF and TRUF depend highly on LEF.(LOCON)

WITÂ = Will to the Absolute of TRU, TÂ. WITÂ is the will that is associated with the relativity, being and existence of TRU. It is to be increasingly found in the higher levels of mentality in the mass energy or Á predominant form and function of TRU. It is especially seen in an individual's love of life, fear of death, and search for immortality. It is seen in human values, what holds to be of greatest worth, and how this worth in general becomes absolute. The more we seek value and worth the more we seek ultimate worth, eternal existence, and immortality. For what is good if 'we can't take it with us'. Herein is the enigma. For it is thought we can't take it with us upon dying. This is our illusion for what we would like to take after death is not TRU, and what we give in life is TES and TRU that alone offers us the opportunity to have anything after physical death. WITÂ enables TES and this extension becomes actual. There is in humans a great common denominator that assures such gains, success, survival and overcoming death. There is an undying wish and hope for life. This is related to Á, or human FAF, and its corresponding **WITÁ**, or Will to Á, the FAF of TRU. For WITÂ is highly empowered and driven by all in TRU. These two, Â and Â as well as WITÂ and WITÂ are same and different. They are much the same in being basic aspects of TRU that are highly interdependent. Yet they are far different in representing two separate and highly distinct aspects of TIT, TRIS and TRIF. This difference is that between the nature and degree of relativity and existence as seen in the extreme unique special relative and predominant form of TRU, its mass energy dimension, state and function or Á and WITÁ, and the extreme absolute of TRU, TÂ and WITÂ and its fulfillment within TRIÄ, this relative absolute, TÂ.

How much are the problems, troubles, defeats, and failures of the past, present and future merely a reaction to the tremendous mechanisms, processes, forces, and powers in the past, present and future? How and how much are those that cause suffering a simple outpouring or precipitating out of these factors? There are, at present, many great positive forces and powers working in the world of man. Yet they are not fully self sustaining. Nor is there any good reason to believe that there are not many more great negative forces and powers developing and likely to work to the detriment, loss, devastation, and possible extinction of man, and TRU. The thread of life, especially that of human life, is even more vulnerable. The many more species that are now extinct and the increasing endangered species of the present are all good proof of human, and TRU, insecurity and destructibility. Although there are many things in mans favor, there are many more - and even many more than this that are not known or poorly known by VAPO - that pose a dire threat to human existence, or TRU. Because they are not felt this minute or crush us this year does not mean that they are not working incrementally, indirectly, insidiously or in all their deleterious ways to cause irreversible loss, destruction and extinction. These can be resolved only with a continued expansion of knowledge, KOT, of TRU or TREQ. It can benefit from the care each and all persons have for their lives, by becoming their counterparts, TES, in TRU. It can benefit by ever greater means to provide the best possible genetic material, live product and determining forces that are a part of the interdependence of life on, and care of, the earth we live in. (GFST)

How many, minutes, hours, days, weeks, months, years, decades, centuries, millennia or time, are left to go. And how much is it of all TRU, some part of TRU, ones lifetime, ones job, ones IA or the minimum of consciousness? These show the future of TRU. So does it show the future of different parts of TRU, as for each person. Looking at the future in this way we can better measure and plan ahead. We can better determine our problems and how to best fit their solutions into the proper time period. Also we may do so for the other dimensions of TRU, its adaptation into the scale of Á and L. Attention to these is ever present in our waking lives. In multiples, combinations and complexes they form the basis of DYNS and are especially the basis for its future.

Many of the world's problems, troubles, and conflicts stem from listening to only one side, a part, or some extreme, popularized or contrived aspect. One and all tend to be sympathetic to that at hand, given, common, or before them. Especially is this true when you are asked for something or when you are asked to believe something. Agreement is a large part of communication. Compound this with selfishness, partiality, or individual interest and you have the beginnings of strong conflict and a biased, unbalanced and often unjust or wrong forces and actions. One and all will find in all aspects and walks of life that they are bombarded with those who are trying to sell something of many kinds, ideas if not products. They are rarely just right. They are right if only for both of two reasons: 1. Timely or appropriate for the circumstances, occasion, situation, place, state, or conditions, and, 2. If they are TRU, 100% TRU, and not of some varied mixture of TRU, ATRU, and ANTRU.

To know and do the exact and proper thing, the precise and right treatment for the problem or ailment at hand, is the purpose, Á, of life. It is a goal of technology, professionalism, and many activities and work in life. This is the beauty of all trades, occupations, professions, techniques, and routines. To do no more, no less, and no other than exactly what is necessary and best for the particular situations, occasions and needs. These are the highest standards of care. Those occupations in which one does only one monotonous job, as on the assembly line, may provide the exact, proper, precise, and right product or treatment. But this tends, and is likely, to be counterproductive. For in gaining conformity in unanimity, it loses conformity in diversity. It loses ordination, identity, and the totality of TRU. This is the state of much of the world in the twenty-first century. It is so concerned with and focused upon one's singular and partial jobs that the systematic unity and ordination of the whole are lost, or never gained. Humans become, like automatons, cogs in a great wheel, that is not a wheel of their good fortune in which they have a decisive role, but one in which they are completely subservient and crushed. It is one in which they lose and lack control, whose control is poorly, not apt to be, or not, TRU.

A person is never in so much on trial as in the moment of good fortune. Like good habits, character, virtue, and all in our great tradition of TREQ we either learn and follow our training and education with zeal, or we suffer interminably.

When you have a heavy load, a great amount of work to do, don't expect, however, that when you get help, that you can quit, or let others do most or all the work. For even half of a heavy load takes a lot of work. It may still require full-fledged effort. So it is of TRU. TRU can not be left to others, nor can it be set aside when others join in, nor can one work at it spasmodically. Each and all parts of TRU, as the parts of any animal, organism or entity, must fulfill their roles in the whole. So must we, each and all of us, be our counterparts, functioning and fulfilling our roles in TRU. Each must do their jobs, as both we and these jobs better coincide with TRU. Each and all of us must stick to business and do a good job. To look to some fantasy, false, or even blissful, end is an illusion. Life in TRU, like virtue, is its own reward. It is fulfillment, a challenge and a level of activity whose success and continuation in TRU are its end. There is a never-ending and relatively constant demand for being ones counterpart in TRU, TES, that each must supply. We see this every day, yet we don't actually, really or purely recognize it. To reach this point of recognition is PUREACLITY. If and when one varies in the supply of this demand, especially in great extremes, it is then that the problems, stresses, strains, diabolical forces, and even failures are produced. This is especially so when they are not quickly, readily or properly prevented, averted or corrected.

The greatest of joys, and happiness in life are in the activities of life that are TRU. This is what thousands and millions of years of evolution have given us. It is performance, attainment, achievement, success and the fulfillment that continually enable us to be TRU. In so being we have the good and great feelings and fortune, that go with the best in life. This is seen in the satisfaction, joy, and beauty of life and its representation, as in successfully accomplishing the greatest works of music, its' most melodious, harmonious and rhythmic pieces, concertos, symphonies or other compositions and productions.

Throughout the text there have been lists of proofs that pertain to the sections involved, namely pages, 55, 140, 148, 200, 244, 264, 282, 328, 425, 492, and 524. More separate detailed proofs are noted through the text. In the Addenda of IPTRU, along with a more complete glossary, there are further description and tabulations in which proofs are derived from figures, other tables, and enumerations. There is no single proof that provides conclusive absolute proof of any of these topics. Yet all, more or less, add to the proof of each. Each proof and combination of proofs are different, quantitatively and qualitatively. Logic and mathematics have provided more highly refined methods of proof. The form and function of TRU, how it exists as a set, its systematic unity, and its reduction to logical, axiomatic, statistical, and mathematical operations give it far greater amenability and reliability for proofs. Many other proofs, as by OC and ECC can be shown.

Proofs act as highly concentrated processes in TRU. Their complex interrelations are shown in Fig.6-10. It is all such interactions that enable us to bring closure of proofs by LACOP. Yet each must be treated both singularly, complexly and totally to yield the necessary distinction and finality. **PRUDYNS** TE, ES, CES, ALT, SPALT, TIT, TO, E, TIT,POS, POT VAPO, Life, TREQ, Human, MATES, ENT Meaning, MADAS, ACEXIN, ALT^n, LORVA, REM, MOCBIREM, LOPA, HOM, WITÂ, LAMUA, ACC, PUREACLITY, Physiology, Business, Circulation, Civic-s, Complex-ly, Culture-s, Duty, Engineer-ing, History, HPP, Justice, Complex Processes, Social-ology-iety, LADEC, LUND, RELDI, MOP, COMPS, LOSE, MINAPT, IPR, FIS, UNDEXT, LASOC, MOR, LAV, LAVIPT, LORSUD, LADU, POTDE, LORTAI, RAPOR, PAS, MOPLAR, CERIT, CO, LIH, LILT, RACHAS, RUSEN, MAREOP, MATSUR, LADIT, TEX, TAR, LAMUA, LACOM, and LAMUC.

MEASURES of CES in SUMMARY of DYNS

Item	Measure		
CONTROL of WORST PROBLEMS	~5-4-7~ -+	M	CWP
GFST	~-~-~-~	M	
IMPROVEMENT, BEST T THOUGHTS/ACTS	5-5-5~	T	IBT
LACOM	~6-4-6~ +-	M	
LAMUA	7-4-6	T	
LAMUC	6-3-7	T	
LACSA	7-4-6	T	
LAPAR	6-5-6	S	
LASE	7-3-8	T	
LEF	8-2-8	T	
OPPORTUNITIES for, IMPROVEMENT	T~5-3~-7~	T	OFI
PARCGI	~6-2-8~	M	
POMAC	~7-2-7~	M	
POOR & RICH, GIVING & RECEIVING	7-3-6~	T	PRG
PROBLEMS, INTERACTIONS	~6-3~-7~ -+	M	PRI
QUALITY of COMBATANTS, and LEADERS	~6-2~-7~ -+	M	QCL
TAR	~6-1-7~	M	
TREATMENT, PROBITY of	7-4-6	T	TPo
WITÂ	~5-1-9~	M	
WITÁ	5-5-7.5	S	

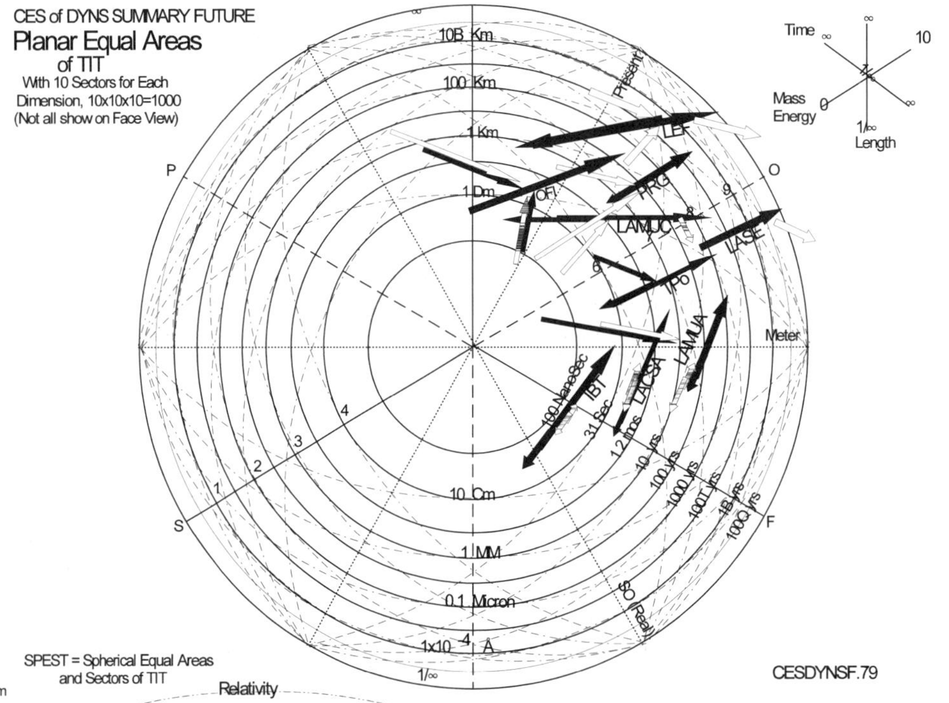

Fig 7-79 **DYNS CES in SPALT SUMMARY and FUTURE**

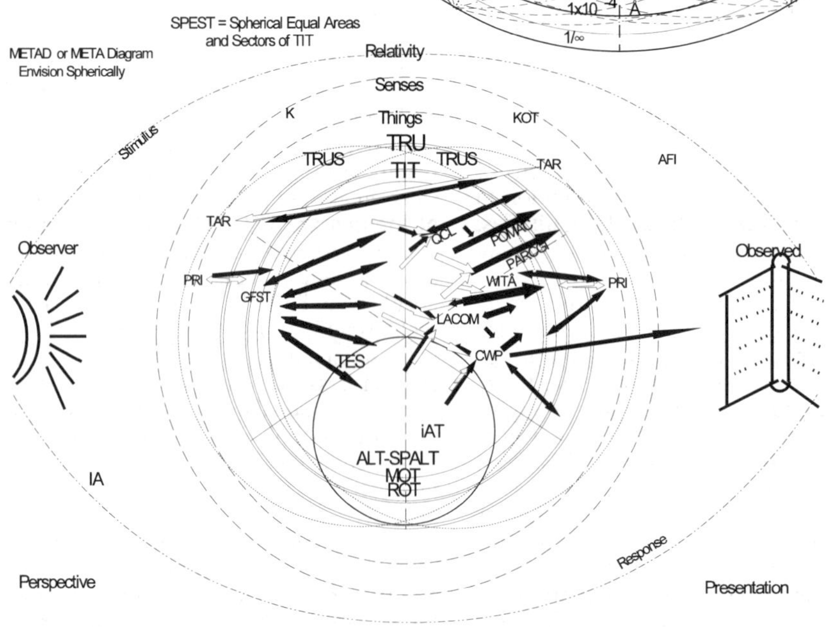

Fig 8-79 **CES of DYNS SUMMARY in METAD**

SECTION VIII

DYNO or The TOTAL CES and CHES of HOMATION

INTEGRATED DYNAMICS and PROCESSES of TRU

CHAPTER 80

INTRODUCTION to DYNO

KNOWLEDGE, DEVELOPMENT, FORM and PARTS, ENT, NEGATIVE and POSITIVE POSITS in HOM

In this last Section of HOM, DYNO, we seek and produce CES and CHES that exemplify integrated processes in TRU. They help envision TIT-TO in its ordered, comprehensive and total FAF. This is impossible without the depth conception, integrated overview, and deeply felt realization of what TRU consists of. The complex processes in TRU that approach their full and complete state as parts or segments of TRU show the transition between DYNS and DYNO. In our sequence from singular through complex to total processes we have not one or the other, but a continuum. Thereby the dividing line between the sections and chapters of IPT and POT is arbitrary, with only relative certainty. We must be ready to yield, to newness, order, expression, representation and artistic license, a degree of freedom.

That the world is in the midst of great revolutions, We can project, with increasing reliability social, political, economic and other trends. This has been recognized by many, e.g. Robert L. Heilbroner, in *The Future as History*. Even more the role of understanding this world through the trends and processes of the greatest forces and courses in TRU can be proactively posited and foreseen to provide a greater and better world, TRU. Yet much of this knowledge is denied.

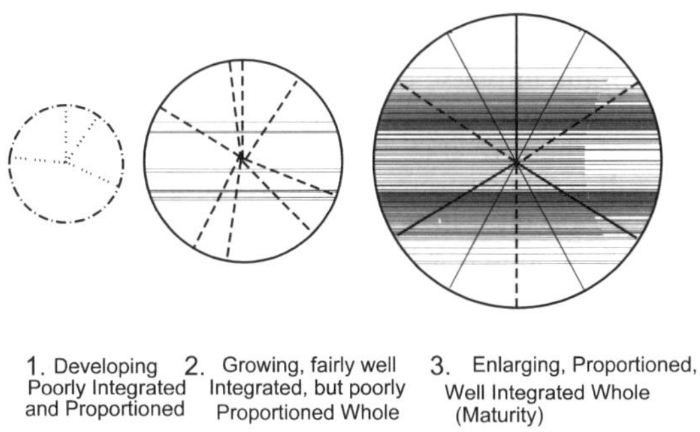

Fig 1-80 **GROWTH, DEVELOPMENT and MATURITY**

In Figure 1-80 three objects of different sizes and divisions that represent DYNO, are shown as comparative pies in the development and growth of entities, life and TRU. Each object represents the overall size of TRU and the different divisions represent the various portions of the whole, as sections of a pie, its aspects, parts, form and measure. Four of the many great problems in life, that are initially evident in an investigation of DYNO are: 1. How to keep and make the pie, object, TRU, bigger and better, hopefully in an exponential way, 2. How to keep and make the portions most proportionate by way of their contribution to the whole, 3. How to keep all in balance and within the limits of the setting and conditions, 4. How to discover, develop and best utilize the expanding horizons, their highest principles, laws and mechanisms.

RETIT = Reduction to TIT. We can best reduce all to TIT, as we must, through its SCARLS, if we envision TIT as provided for us by ALT and SPALT. If we fill SPALT with all that is TIT in the best possible way by VAPO and indicate what is inside, outside, adjoining, moving in/out cybernetically, fills the needed replication and presentation of TRU. It is by DYNO that we can finally bring all the preceding sections and chapters of IPTRU together to provide a complete accounting and demonstration.

PROFU = Processes of Filling-Up. Many authorities through history have come to see the world and reality as much more and other than what is objectively before them, represented, experienced, given or presented. It is much different than what has been taught or previously championed. For the latter have looked at reality and the world as ENIT, ISMS, or other phases, facets, categories, or parts of TRU. Only as all these are measured and filled up as they exist in TIT-TO can and will they achieve adequate meaning and totality.(VAPO)

For knowledge the greatest problem of all is how much do its many kinds differ from a perfected, complete comprehensive identity with TIT. How much does this perfected complete comprehension of TRU include these many kinds of knowledge. **PROKCOT** = Problem (Fallacy, Paradox, Concept, Principle or Law) of the varieties of Knowledge and the Perfected Complete Comprehension of TRU. PROKCOT applies to all knowledge and is clearly exemplified by discovery of human origins. Although recent discoveries have produced much new light on TRUS and TRUPS, we are left with a poor idea of TREQ. Only when we separate, make distinct, clear, and understand these critical processes and course in operation in TIT can we solve PROKCOT. **DIWHAT** = Difference and Distinction between the Worlds of Past and Early Humans as ENIT, and how they are manifested in TIT-TO. DIWHAT is a corollary and special case of the difference between ENIT, ESIN, ES or CES, and TIT. It is when we can train ourselves to envision all in these different ways and bring them to their best focus in TIT that we can correct the mistaken views of the past, and hold ourselves to KOT, COGNOS, ELTICMA, SÂ, and TÂ.

It is in discovering the TRIT and VAPO meaning of the future, present and past and their interdependence and interactions, that we can recognize many of the mistakes the human mind makes, and how poorly we have understood and adapted to our world. The future, present and past are even more than their time segments in the world, they are how and how well each provides for each other, and all together as they are combined by ACEXIN in TRU. TRU is likely to suffer from the many different interpretations of the world, both actual and fictional. TRU can never be adequately understood unless all its aspects and major ES and CES are brought out, learned, remembered, and applied.

The all embracing idea of Human Solidarity, is not only an

idea of labor and social life it is a first principle of TRU. Yet it is not any, vague, or other solidarity than TIT-TO. Comity or mutual courtesy and civility is one of the many great words that is underutilized. It is much the same around persons, groups or nations. Being of the same by DAOST and TES we have common cause, ground and the great vehicle for relative and cooperative success. As all peoples of the world identify themselves with much that is the same, relative, and TRU in the world and life, and act, behave with comity, and concretely prevail to produce the results that are TRU, they reach toward TÂ, its fulfillment and permanence.

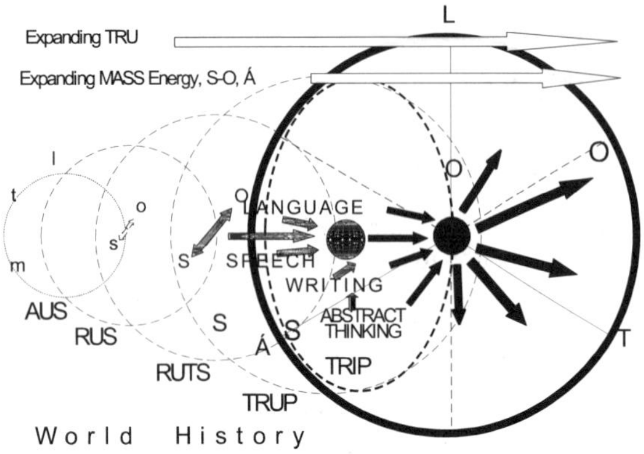

Fig 2-80 **HISTORY, TRUP, TRIP, TIT, S-O, Á, KOT and TRU**

A history of the world should be primarily a history of TRU, TRUP, and of TRIP in TIT. It should also include the proper correction for TRIS, TRUS, TRUCH, TRUD and HOM. This is evident in how abstract thinking developed along with the development of language and writing. This we can see in the representations, pictographs, glyphs, cuneiform and hieroglyphic markings in incipient civilizations. These show the rapid expansion and great prominence of the S-O, or observer-observed relations as their polarizations formed the continuum that makes up much of the mass energy, or Á, form and function of TRU. They give a good idea of developing and expanding higher levels of Á that have occurred with the expansion of mentality, speech, language and writing. It is by ever better concepts that have develop into better principles and laws that enable the abstract thought. This is necessary to escalate human understanding into the present TREQ. By abstract thought becoming increasingly profound and VAPO we begin the process of achieving a comprehensive understanding of the world, TRU.

The central and critical nature of each person, whether they are or are not TRU, has been shown repeatedly through the history of religion, philosophy and the world. Especially is it shown in the great contributors to ethics. This leads to the intellectual love of God, that wisdom exalted in the bible, and only through which the what, why and how in all its wondrous forms and functions are TRU. The infinite love, which we can so readily identify with, gives us that perfected bond in TRU, and in and of us and all as we are TRU. This is Spinoza's Doctrine of Necessity, the great ethical necessity of Gods infinite love, and our understanding of ourselves and thereby love of God as all that is TRU is perfectly and absolutely bonded and united. This is much the same as **TRIA**, TIT is All, the necessary success/failure of the Relative Universe, TRU. It is the sameness in all TES. It is the thisness or being, as all its components function perfectly in unison for its success, self preservation and absoluteness. This necessity may be classified by a universal or general one, a special one, as of Spinoza limited to God's infinite love and our understanding, and that of TRU which is povided by VAPO, ATO, DAOST, SCARLS, TES and ACEXIN.

UNIICATION of The UNIVERSE and TRU

Newton's, or the classical theory of **Gravitation**, did not account for changes in the geometry of space. It left errors in experiments and measurements. The earths orbit, if sun disappeared, would be unaffected and would not fly off into space until its loss of gravity reached the earth. Einstein's **General Theory of Relativity** corrected this by showing space is warped with curves. It mostly holds for larger astrophysical bodies. The earths orbit around the sun is affected immediately. Black Holes occur when an enlarging great mass is contracted into a greatly decreasing smaller space. Matter is lost, until there is an absolute absence, similar to extinction, as of mankind. White holes are the opposite, certain and absolute presence.

Classical and relativity theories do not hold for very small, as subatomic particles and forces, electrons, quarks, the strong force or attraction of neutrons to protons as in nuclear energy, and the weak force of electrons, radioactivity and the light spectrum. These are explained by experiments on subatomic particles applying Quantum Theory. Yet they can't be explained or known with absolute certainty. There is always an element and degree of **Probability**. Their explanation is made by Maxwell's combination of electricity and magnetism, the behavior of other subatomic particles and the strong, weak forces, and other forces that exist within atoms, at a subatomic level.

The disparity between classical and relativity approaches to physics on the one hand, and the quantum or electromagnetic and subatomic particle forces approach on the other has left physics without its highly anticipated unanimity. A **string theory** has been proposed and shown promise by reduction of the disparity to five basic equations and logical explanations of electromagnetism and subatomic forces in the warped and curved geometries of the universe. It finds these basic elements can be likened to a string in motion in such variety of forms and functions that it explains the behavior of subatomic particles. However, five to many more string theories have been proposed. Even an infinite number has been suggested. This is completely at odds with unification.

Indeed the whole universe must be envisioned and demonstrated in a new light. To be a universe it must have a **center**. About this center its constituents must have some kind of form and function. In both the special and general theories of relativity, the relationship between **observer- observed** and the geometry of the universe solve the problems of centerng, geometric frame or form, and the mass energy relationships and function of the universe. The primary parameters and dimensions will be those found in Classical Newtonian and Relativity Laws of the Universe. The universe relative to mankind is not so much what it is in the far reaches of infinitely large or infinitesimally small objects in space. It can only be based upon how they are warped or curved relative to its center and distributive mean. They can not be explained by absolute chance and chaos, but by higher degrees of probability based upon the aseity, perseity, haecceity, quiddity and continuity of the relative universe. This is evident in the big bang theory of the origin of the universe, whereby when centered in here and now, the limits of the peripheral are likewise warped. The relative universe can be based on neither nothing nor everything, at least not everything external to itself. Upon this basis TRU has a firm foundation by which it can be both represented and presented, ever remembering that in doing so there will be deviance from and degrees of probability of what it is in and of itself, TIT-TO. Quantum theory solves the problem of absolute extinction or absolute certainty of eternal life for man, by **Probability.** It is a probability that is relative not absolute, but is up to 'us'. Here the unification of the universe is found, as it exists in TRU, **a probability of human continued existence indefinitely if we will, and assure, it to be.** (See index).

Fig 3-80 **BASIC NEED FOR ORDER, TIT and TRU**

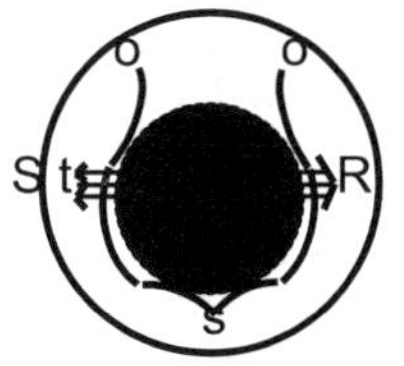

Everyone, more or less, tries to bring all into some kind of order, at least to coordinate the world around them, or themselves to the world. They do so, in conformity with the frame of one's whole being, mlt, both physically and mentally, ALT. Coordination, order, systematic unity, a structural balance and the synchrony of form and function must occur in the SCARLS, the state of relativity that exists in our lives, TÄ of TRU. This tends to make us learn, and know our world and place in it. This is why it is necessary to know more than only the interactions and tasks before us, but oneself and the whole world together. If we are to realize our great gifts of quality SÁ that we have in being human, TES and TRU, we must rise above the lower levels of SÁ and best utilize all that gives us the greatest power and predominance. Although based upon interactions of S-O, with stimulus-response, and success/failure, it is in the best exercise of their higher levels, in Á, that humans fulfill their need for order, TIT and TRU. We must be careful not to jump to conclusions, nor expect solutions based upon simplistic, trivial, or extremely general processes that lack sufficient evidence. We must base our results, conclusions, and solutions upon the proper collection, combination, totality, proportion, balance, order, and evaluation of all. The limit of dynamic particulars, as produced by 1/TO by the individual, in IA is called: **LIDE** = Law of Limits of Dynamic Particulars. LIDE should be obvious. Yet the repeated failure to apply it and the seriousness of the dangers inherent in decisions and conclusions makes it significant. The Corollary to LIDE are the, **LOT(S)**, and **LAT(S)** = Law(s) Of and in, or At, TRU, TIT. These are the collection, summary, ECRO and use of all the great laws of TRU that provide the necessary and real knowledge for ROT, POT and personal use. This is called: **LOPA** = the Law of Preeminence and Adequacy of Actualities, or universal truths. LOPA is how the S-O becomes Á, as these principles are expanded, and together return into TRU - OITIO - to become and enhance TRU. Universal truths and actualities in massive combinations and perspectives, which with the use of VAPOS and SCARLS, can accurately represent TRU. They are TRU when best done. Any and all that make up TRU, and any as TRU is all by TRIA, become all, as this all is TRU. In being, TRU and all, they are infinitely and absolutely great. This is: **LARTS** = Law of Actuality Summation Absolute TRU, REM All TRU. LARTS signifies the fact that when universal truths or LOPA are brought together, posited, valued and ordered in TRU we can begin to comprehend the actual total and absolute nature, FAF of TRU.

SEF = Summative ESIN Functions. SEF is shown where enumerated and separated functions of ESINS provide approximate measures of TRU. They enable us to observe and attend to their relations and summative effects as they work toward the TU, SUT and TO of TIT. e.g. the bodily functions of an elbow, back, finger, eye, neck, leg, house, boat, baby, man, woman and child. All average out as they contribute to TIT-TO whose epicenters or foci when reconciled and summed are TRU. This we see better when we posit many ES and CES in SPALT and other ROT.

The greater TRU, or some aspect, ES or CES, is taken in terms of or as an EBE, the greater the error. For error is a function and degree of departure from, or lack of, TRU. This is called: **EDEMT** = Error is a function of Departure and Magnitude of Departure from TRU. This departure is the extent to which it is a deviation from or deficiency of TRU. It is in this way we discover, develop and resolve all. How philosophy, science, religion and all knowledge are brought to replicate and have 1:1 identity with TRU. How they can be reduced to a POA that most accurately, reliably and effectively demonstrates TRU. This is why all SUT, ROT, MOT, ALT, REM, POS, STA, and Homation, are so important.

In all conscious life we must be alert to appreciate and recognize how much greater the best of philosophy, science, and religion are than what our world would be without them. However we must similarly appreciate and recognize how much greater philosophy, science and religion can be when taken in their full part and role in the actual and select world of TREQ and TRU. How the form and function of TRU applies to TIT-TO and not the worst of assumed philosophies, sciences and religions we see in the NAMTRU world today.

Fig 4-80 **LIVE PRODUCT, GENETIC MATERIAL, DETERMINING FORCES or GELD**

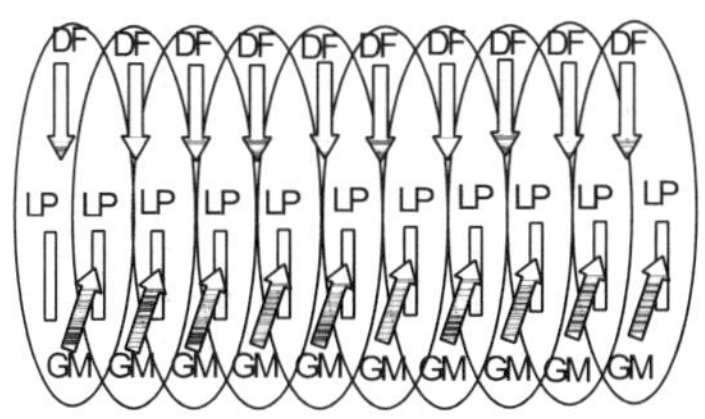

Together these reveal much of the basic structure and operations of TRU, TIT, and HOM. However in TIT-TO they differ from what is known in the biological sciences. The determining forces may be subdivided into those that occur by chance, OITIO, TIT, are intermittent, continuous, acute, chronic, TRUS, TRUCH, TRIS, TRICH, and ITCH. The continuity of live product and germplasm, phenotype and genotype, and their interactions are the most fundamental forces and processes that constitute TIT. These show the processes that have, are, and will enable S/F, survival in TRU. **LP** = Live Product **GM** = Genetic Material, **DF** = Determining Forces. LP, GM and DF make up a major portion of FAF, ES and CES of TRU. They function together as a unit for the benefit of TRU, as they do for any biological unit. Yet its solidity is not self assured. This is largely by selection at, more or less, all levels of SÁ. For TRU they are focused on humans and the Á, L, T FAF of TRIT and TIT.

If one is a partial, the partial of a whole, and if one emphasizes this partial to the exclusion of the whole, then one will sooner or later, by reaction, be under the pressure of this opposition. Either one, the individual, first person, ES, TES, or the whole, as TRU, suffers. Most likely, and to a more or less extent, it will be the first person. Although one may not be affected directly, objectively or physically, reaction in some form is certain to result, by the principle of Action Reaction, **ÂCRE** and the principle of INevitable interACtion in TIT, **INVAC**.(TRIA) Because of the void, suffering may be indirect, subjective, potential, delayed, or phenomenal. This kind of reaction when misunderstood may be attributed to chance, or to some vague or mysterious cause. Yet they are, if of TRU, no less actual than we ourselves as TES in TRU. Thereby, no matter how partial a thing or person is, no matter how much ES and TES they have, it must be equally so and equally TRU together. This is why it is so necessary to think, envision all TRU and its subjectivity, turn the other cheek, find the joy and the success in all others, all other parts and the rest of TRU. This is the true Glory of Life, OM. You are directed toward the necessary harmony between you and all the rest of TRU. That is all TRU as you and they both function in total equilibrium and concert. **LACOND** = Law of Concert of the Different, or Law of with the Different. LACOND signifies not the sameness, but the concert, harmony and positive successful give and take that occurs between the differents of TRU. It especially applies to an individual's appreciation of all the rest of TRU that may seem so different, even alien, and in opposition to oneself.

Time, as all that is great of TRU, is, in a way, a two edged sword. For as we are TRU and as TRU functions totally, all of what is great of it must function no more and no less than its part in TRU. So that although we can or can not, more or less function freely, time will help resolve or lessen much that can not be resolved or does not pass away, properly without damage to TRU. Neither failure to instantaneously cure all ills, nor to blindly leave them to indefinite time, is the answer and solution. We must do all, work on all, prevent all, or endeavor indirectly for the benefit of all that is TRU as soon and as long as proper whichever - or both - are best by TRU. It is only in terms of our IA and IAT, that we must have the patience to resolve the many great and complicated problems. We must forever and continuously realize and appreciate the fact that IA must exist in extension, intension and actuation as it expands to TRU.(ACEXIN)

LASTE = Law of the Strong Ties and bonds, or **EST**, between E, ES or TE and TRU. LASTE is the general case of: **LASTES** = Law of the Strong Ties and bonding or **TEST**, between the individual, self, person, first person, TIS, TES and TRU. LASTE and LASTES signify that great portion of TRU in which its parts, and especially The Part, the Live Product, I, or the Proband, also have great interdependence and bonding, with powers crucial to TRU. LASTE and LASTES are highly variable, from all standpoints except for that exclusively TRU. Yet they are what can, should, and must be, by TRIA. This is the set makeup of TRU. The choice is only in the ways and the means. If this bonding and their powers are healthy as being largely and increasingly TRU then TRU survives and TA ensues. It is in LASTE and LASTES that we can find much of the basis for a person's relationship to, as in love of GOD, TRU. It is the basis for ethics in TRU, all that must follow when becoming TES.

LAPDUL or Law of the Perpetual Defects and Dangers that are Unknown to the Living. Each age, generation, individual, group, nation and all feel their world is the best and other worlds are less to worst. They little realize all the defects, evils, wrongs, and dangers to human life and TRU that are in their midst and within themselves. There can be no let down in seeking to instill in all an understanding and appreciation for what is necessary to attain and maintain TES and TRU. This is why we can't leave to generation after generation such a lack of love, appreciation, accounting. ACC, and all that can be done to improve and assure TRU. Yet because they never recognize God, TREQ, or TRU they never adequately bring out TRU and achieve and retain their common end. The same holds true for many of the greatest concepts, principles, laws and mechanism of the aspects, form, mode, function, parts, characteristics, processes, properties and values of TRU. They are much the same as many that have already been discovered and propagated. Their difference, however, is all the exclusive differents, partials and loss of totality, that they are reduced to. This may be no less than how they exist in TRU, are represented, presented and perceived, the differents of DAOST of TROS and TIT.

	massen ergy,Á	length	time	same/dif ferent	form/fun ction	Universal /Partic	Actual/ Inactual	Absolute Relative	Dominant Recessive	Outside Inside	Autonomou s Dep
Stuff	3	2	2	4	2	4	3	3	3	4	2
Nous	4	2	2	3	4	4	3	3	3	3	3
Ideas	4	2	3	3	4	4	4	4	4	3	3
Form	4	4	4	2	4	3	2	2	3	4	2
Parts	3	3	3	4	2	4	2	3	3	2	4
Distinct	4	3	3	4	2	4	4	3	3	4	3
Themes	1-4	1-4	1-4	2-4	2-4	2-4	4	2-4	1-4	2-4	1-4
LANT Nec	3	2	2	4	4	3	4	3	3	4	3
THES	4	3	3	4	3	2	4	3	4	2	3
THEO	3	4	4	3	3	2	4	3	4	3	3
THESO	4+	4	4	4	4	3	4	4	4	3	4

Table 1-80 **COMPARATIVE ORDINATIONS and SYSTEMS**

This is an analysis of the parameters of TRU and the means and methods of their determination. The columns enumerate some of the basic greatest dimensions and parameters of TRU. The rows some of the different topics that provide the methods and means of identifying TRU. The value of such a study of TRU is not only for the specific information it produces, it is in providing a method and direction for analysis in continuing work on TRU, by leading each of us through our: **IA X COL → TRU.** This is the means by which we can analyze all other points of view, perspectives, and bring them together in order and completion. If we know TRU and have a good understanding of how we know TRU, we can interpret all we come in contact with. It is commonly expressed in, 'Great minds follow the same paths'. This is largely obvious from the S in DAOST, or **IACOLT = IA x COL = TRU**. It is through IACOLT that we can see the great value of contrasts. For in contrasts there are distinctions. It is this ever greater distinction that enables us to separate out, dissect, and make discrete all that is TRU. This is the beauty of ES and COL and how they work by IACOLT. The sameness that exists in all is why all who recognize the unity of the world, a supreme being, God and all that is divine often have so much in common. Even with all their different perspectives, means of achieving this recognition, different parts and processes, disproportions, and the differences in God, TREQ, or TRU, the end results are much the same for they are relative, the sames of DAOST, TU and TO of TROS of TIT.

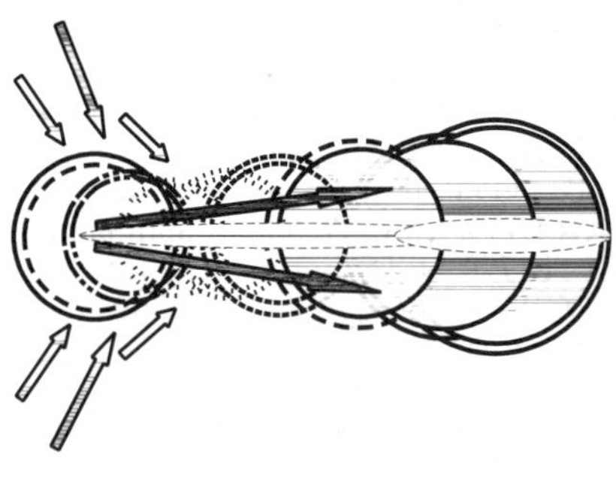

Fig 5-80 **CONTRACTION and REDEVELOPMENT of TRUS and RUS**

Much of the contracting and expanding totality of TRU is a result of a great many increments that effect TRU both positively and negatively, directly and indirectly, from without and from within. Each increment has its accumulative effect in causing contraction or expansion, if only infinitesimally. This is, **SLONCA** = the Summative or Total Contracting and Expanding of TRU. SLONCA is of all segments of TRU, as of conscience of the self, group or world. Emphasizing or de-emphasizing many different ISMS positively, negatively, and incrementally produce in summation contraction or expansion of TO. Much of the resulting contraction is due to neglect of, or imbalances in, the collective absolute. Figure 5-80 shows the contraction, realignment and redevelopment of a new organism or TRU after its failures and losses. These can be in all degrees of severity. For TRU there are added, different, and unique features due to its special form and function. These are what make TRU different from all other universes, and in ways many times more susceptible and fragile.

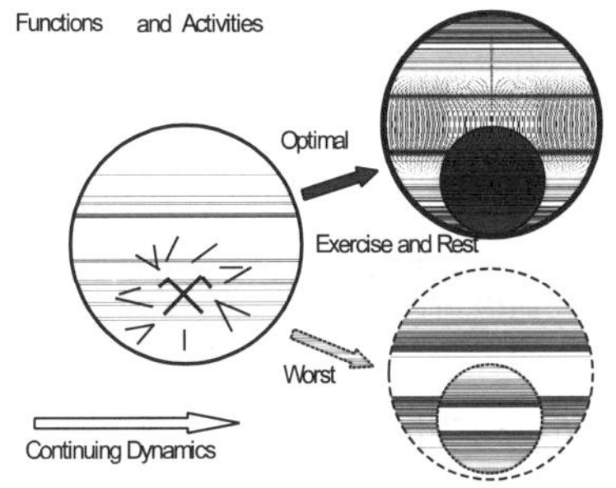

Fig 6-80 **EXERCISE, REST and ALL NEEDED for OPTIMAL FUNCTION**

For all weak or deficient parts, some of the cures are by exercise, some by rest, some by replenishment, and some replacement. All of TRU, like the human body, is in a state of variation and flux. This variation is from an optimal state of health, and usually dependent upon the variation on TRU about the part. The positive and negative effects of return to or departure from health and optimal state in TRU, are shown. To set out what these are requires a large amount of information and its best possible application.(IPT) This is the goal of totality by CES we provide in DYNO. They can be adequately known only as we show the more basic processes and related problems, principles and the fusion of all by an effective POA, Homation in ALT and SPALT of TRU.

The enemy that can be mostly readily corrected is within ourselves, and the lack of **CIA** or Controlled IA. It is to know our weaknesses, and those of the groups and worlds we are in. This is not easy to accept, for we look outward to blame whatever is a threat, actual or imaginary. We see many failings and faults in the world and assume that this or that is the cause of it all. The lack of this, our great enemy within, is: **LACOLD** = Law of Collective Deficit. LACOLD signifies that where all that is wrong has been wrong in like manner, more or less, in the past, and continues to be wrong through the present into the future. This is out of the continuous collective deficit of each individual, signified by:**FOSP** = Fallacy of Selfish Projection. FOSP is the tendency to blame others,, or the wrong things, for our own failures and misery. We suspect that someone is not doing their part. It is easier than to learn and master KOT, TES and TRU. It is easier than to strive for PUREALITY, COGNOS, TIT, and SÂ. In actuality this is much more, the most inspiring revelation, knowledge and love of TES, TRU, ELTICMA, and TÂ. We often live by example, as Christ and TES are examples of each other. By tiny increments the individual gains and achieves that bit of immortality. This is often madethrough self sacrifice. It is the contribution that produces the great progression toward excellence, an irreversibly assured TES. Its opposite is called: **SACRIA** = Lack of Self Sacrifice and IA. Plato's Socrates exemplifies CIA, KOT, TRU over lesser Gods.

The following diagrams help us realize and think universally. If we are to achieve KOT and COAT, we must always remember, ECRO, that in using OITIO we depart from and lose some TIT. This is true of all our IA and a large portion of human relations with the worlds about them. It is the great powers of convergence on TRU that enable our being. Yet it is also the great power that makes absolute IA and MOT impossible. However this power, to be able to recognize and selectively assimilate or reject what is OITIO, right or wrong, good or bad, is what finally enables this being, the great expanding selective powers of our Á FAF in life as TRU. Knowledge without TIT-TO, SUT, LOL, and much of TRU in general, as of all things, has and can in part, piecemeal, be readily comprehended. Yet neither in totality nor, even worse, with an adequate degree of focus, orientation, order, management and control. This is: **LOCOAT** or the Law of the Comprehension of all things and TRU Only as knowledge proceeds from LOLS and limits itself to TRU, and only as this is done with maximal form, objectivity, systematic unity, coordination and modulation can it be made comprehensible.

Fig 7-80 **TYPES of CES and CHES in and out of TRU and AU**

That category or class which best relates and classifies each POSIT is designated by their given numbers:

1. focal, inside of TRU, unevenly distributed,
2. focal, outside of TRU, evenly distributed,
3. focal, inside of TRU and evenly distributed,
4. focal, outside of TRU, unevenly distributed,
5. universal, inside and unevenly distributed,
6. universal, outside and evenly distributed,
7. universal, inside and evenly distributed,
8. universal, outside and unevenly distributed. Such a differentiation enables us to better distinguish between entities and concepts OITIO. It also improves their posits by comparison and qualification. Posits in Chapters 81-98 are evaluated in this way. As better models are developed to fully replicate TRU and our S-O relation, sphericity, surrounding orbits, observer observed variations, and closure can be greatly refined. This can be envisioned when more familiar with TRU.

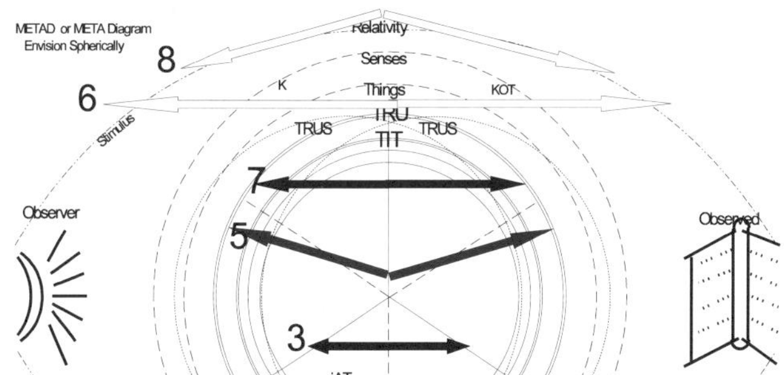

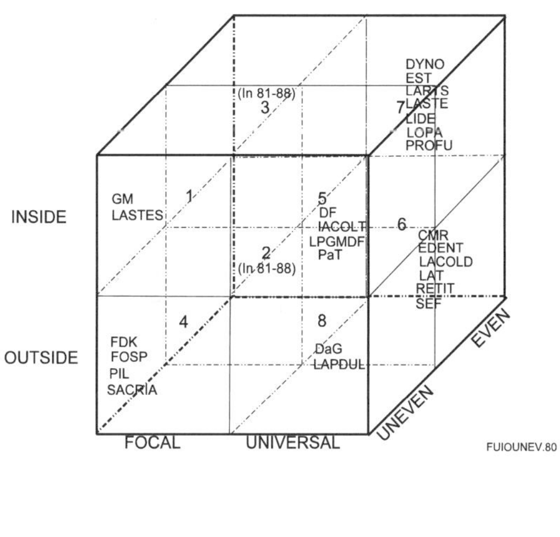

Fig 8-80 **CES and CHES of DYNO SEPARATED by FOCAL/UNIVERSAL, INSIDE/OUTSIDE, and EVEN/UNEVEN**

This helps to orient and fill out the entire range of posits covered in DYNO. Chapters 81-88 treat CES and CHES that are less focal, inside or outside and more evenly distributed. So that we are primarily concerned with those CES and CHES that are universal, evenly distributed, both inside and outside of TRU. We also include some of those CES and CHES which are more focal, inside or outside and unevenly distributed. Although the classes are most acute and demonstrative in this section, they apply, more or less, to all posits. Such an analysis gives better classification and how to make posits of CES by CHES, especially of those more difficult universal posits. The ones that we are primarily involved with in Chapters 80 and 89 are those whose posits are poor or are universal, evenly distributed, and both inside and outside of TRU. The remainders are better related to

Chapters 81-88. What is focal and inside, both well or poorly distributed has been represented in Chapters 60-79. What is universal, outside and evenly distributed is largely presented in this chapter and # 89. The rest are presented in Chapters 81-88.

MEASURES of CHES in DYNO

Item	Measure	
CONCEPTS & MECHANISMS With RESULTS	2-7-7	S
DEVELOPMENT and GROWTH	6-5-~	S
DF	~-~-~-~ +-	M
DYNO	~-~-~-~	M
EDENT	~-~-~-~ -	M
EST	~-~-~-~	M
EXERCISE, REST, NEEDS, FUNCTIONS	~6~-~4~	-+MT
FAILURE, DISCOVERY & KNOWING	~3-6-3~	TM
FOSP	3-6-5-	T
GM (& operations)	8-8--~	S
IACOLT	~-~-~-~	M
LACOLD	6-6-5	S
LAPDUL	~-~-~-~ -	M
LARTS	~,~,~,~	S
LASTE	~-~-~-~	M
LASTES	~-~5~6~	TM
LIDE	~-~-~-~	M
LOPA	5----	S
LOT	~4-2~8~	T
LP	~5-5-5~	S
LP-GM-DF (GELD)	~-~-~-~	M
PARAMETERS TRU	~-~-~-~	M
PROFU	~-~-~-~	M
PROKOT	1-7-7	S
PSYCHOLOGICAL INTEGRATIVE LAWS	2-6-7~	TM
RETIT	~5~4~7~	SM
SACRIA	~4-6-4~ -	SMT
SEF	~-~-~-~ -+	M
SLONCA	~-~-~-~	M

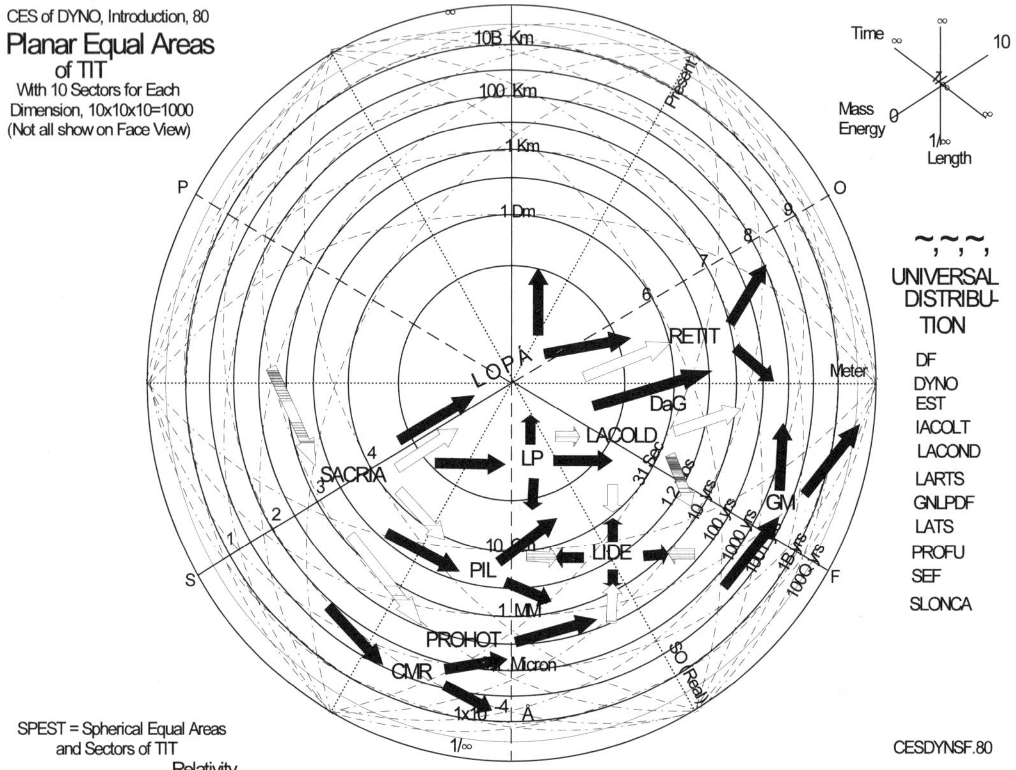

Fig 9-80 **CES and CHES of DYNO in SPALT**

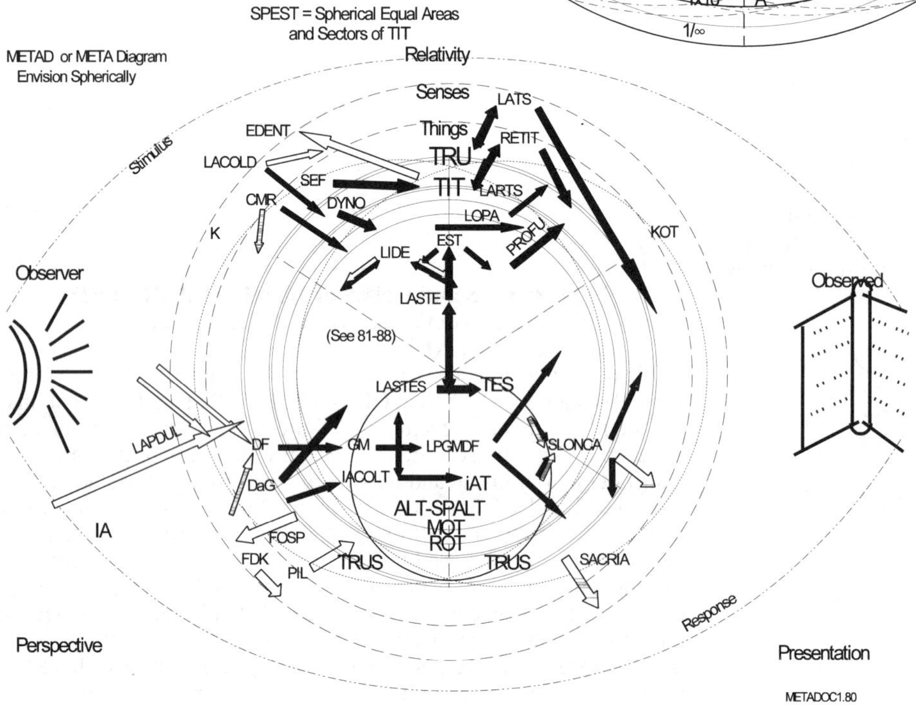

Fig 10-80 **CES and CHES of DYNO in METAD**

CHAPTER 81

SULP or SUBJECTIVE LARGE PAST PROCESSES of DYNO

CULTURE, LANGUAGE, KNOWLEDGE, RELIGION, PHILOSOPHY, EBE, ISMS, ETHICS, JUSTICE

PUDS = Principle of Primitive Understanding from deficient or non Sphericity and Solidity. PUDS is the recognition of a lack of relation, orientation, order, solidity or sphericity in the world. It arises and grows as mental capacity to recognize the world around them grows. This is more so when conceptual and abstract. It is evident in stone age tribes, as in the American Indian, where great emphasis was placed on different objects, animals, and physical features of their environment. This emphasis bore a special relationship and bond, including names for children. Collectively this was a struggle to raise a deficient relationship into one with meaning, identity, solidity or even sphericity. It exemplifies the power of representation, language, and growing relativity or Á. The lack of such an overall recognition is what has kept mankind from an adequate comprehension of the world, God and TRU. Since the advent of recorded history there have been many cosmological theories and models that have sought to explain what this whole world is. Yet there has been little consensus, none have been consistently verified. This is largely because their foci were different, they lacked VAPO, or were obsessed with certain aspects, parts or processes of TRU, rather than TIT-TO. We now see in the different planes of A, L, and T how they fuse to become ALT, and the solid and spherical SPALT.

Through history the competitive and selective role and advantage of peoples with superior language was often the major deciding factor in their predominance. This is most direct in economics, politics, and military. The evolution of language is key to understanding much of human evolution in the last few hundred thousand years. One of its outstanding features has been its great diversity. This is not surprising since most started in isolation. Records of incipiency elude searches and explain the paucity of knowledge as well as subsequent errors in interpretation. The size of the brain increased as language improved and became more differentiated. Language is to a great extent the hallmark of civilization, our world, and TRU. There are many kinds of language, as in body and sign language, that are more than verbal. Yet it is the kind of language that enables concepts, reasons, values, intellect and other higher levels and forms of Á that is unique and special for TRU. It may be seen from a study of the development of language, that the words we use, such as formal names, as Elizabeth or house of my God, show how much the earlier depends upon proof and action of the latter. This is: **STE** = Standardized Error. It is a minor error, although accumulative and when not recognized may become major. For the word EL(I), and Beth, as the second letter in the alphabet, are terms which have long since been superseded. Such supersession in language is of tremendous importance even though of a low rate of efficiency. The great demand for a universal, fully developed, and TRU language is an end which language itself inclines. It remains to be seen how we can best attain perfect formal distinctions and languages that give us the best representation of TRU in TRU with the least amount of STE and other detriments. The continued expansion of Á in the future, to forms we can hardly imagine at the present, make it probable that the development, future, and even errors in language, will be ongoing processes indefinitely. For such ultimates may not be infinitesimally absolutely possible. Yet they must eventually be finitely relatively certain.

There are in different religions and philosophies many concepts and ideals that, not only correspond with but, have been the precursors of and directly or indirectly substantiate TREQ and TRU. In Buddhism these included: a belief in personal divinity, a final extinction of a soul, reincarnation as a new soul or Buddha, and a beatific vision glorifying basic principles and values, REM, catalepsis that assure escalating enablement of the spirituality of being, and Á or ENT on the curve of Buddha. The ideas held by Pythagoras, who was a contemporary of Buddha, are not confirmed by direct and complete writings, yet emphasized transmigration, vegetarianism, moral self examination, and the mathematically precise formulation of the world raised to divinity. The fact that Aristotle produced such brilliant and influential works in literature, philosophy, biology, logic, ethics, and many other fields and sciences should not lead us to believe that these were all original, nor even the 'gospel truth'. Only some of his salient ideas and methods were original, most were the collection, investigation, and interpretation of existing knowledge brought out as a school for study and teaching. For knowledge, as language, is a high powered, influential, and super-personal medium that all use to produce their works. Above all, at the academy, Aristotle has been for all time the standard as an accumulator, compositor, and synchronizer of existing writings, and knowledge. He profited by the influence of Pythagoras, Anaxagoras, Socrates, Plato, and others whose ideas, concepts, and training in profound and subtle thought brought out the salient ideas in questions and problems at hand. This is what is most productive of new ideas, ideas that correct or supplement existing knowledge. It is the very method that produces the best knowledge, KOT, that focuses on TREQ, TRU and TIT.

It is by tracing through the processes and forces in history of the individual's relationship with the good, deity and soul that we can see the gradual yet accelerative and relentless growth and power of an understanding of God, TREQ and TRU. The idea and concept arose in the remote and more recent past from individuals and cultures whose ideas, concepts, and beliefs were, more or less, continuously improved, expanded on, and communicated. This has had manifold positive effects. It shows us the need for greater comprehension. Yet this has not been a direct and flawless course. **CHE** = Church Error, or the great error made by churches and religions of attributing all the evil in the world to the flesh and all the good in the world to the spirit. This has been an archaic influence of burgeoning subjectivism. It resulted from failure to properly and fully identify and distinguish the mental, phenomenal, spiritual, and physical or Á FAF continuum in TRU. Physiology, even ones anatomy and all FAF of the human body that are physical are not evil, but of a very high degree or percentage of TRU. CHE is characterized by disproportion, or: **DER** = Disproportion Error. DER is deviatant from, a much larger, formidable, characteristic of TRU, VAPO. It is more often our desires, choices, attention, thoughts, and decisions that lack VAPO and TRU. Commonly the human IA is usually more ATRU than TRU. The corollary of DER, the attribution of what is bad in something to all of it is called: **DERA** = Disproportion Error of Attribution, to all bad that which contains some bad. CHE, DER, DERA and a lack of VAPO was evident in the formalizing age of the Roman Catholic Church in the 4th and 5th centuries.. It is against such DER and CHE that modern man, science, even philosophy has reacted and turned to the other extreme. Only in TRU, TROS, TIT, and TRIT can we realize how all must exist, function and be balanced in TO. As TRU consists of means and extremes, parameters, forms, functions, parts, and processes these also will be valid only as they exist in TRU.(VAPO)

	>5000 #CHAP	5000-	2000-	1000-	-500-	0-	+500-	1000-	1300-	1500-	1700-	1800-	1850-	1900-	1950-	2000-	210 0-	F-
PREF HUB	SC			-	-	- Au	-	-	1	1	2	2	2	2	3 Ry	3 LANT	4COG ATEX	4
POT EPI	0003		- Ge	-	1 So Ar	1	1	2	2	2 De Bc Lo	3	3 Hg	3	3	3	3POA	4	4
POT SCI	0405			-	1 Dm Ar Lu	1	1	1 Av Aq	2 Co	2 Ke De	2	2 Ki	3 Wu	3	4	4	4	4
POT PHIL	0609	-	1	1	2HcAn EmSoP	2 Pt Au	2	2 Aq	3	3	3 Ka Dd Li	3	3 Da Ki	3 Ei BG Hb	3	4VAPO OITIO	4	4
TRU K-DOT	1011	-	- Ge Ex	- Bi	1 So Pl	1	1	1 Sc	2	2	2	2	3 MM	3AHHu CrBeu	3	3 KOT ATO	4	4
TÄTÄ TROS	1213	-	- Ak	- Pa	1 An	1	1 Mo	1 Sc	2 Sd	2	2 Ka	2 AM Hg	3 Mw	3 Br	3	SCARL 3 TIT	4	4
TE TES	1415	-	1	1 DS Bi	2	2 Pt	2 Ha	2	3	3 Ga	3 Le	3	3	3 Ei Bu	4	DAOST LOPAN	4	4
TO TRUD	1617	-	-	1	1 Em	1	1	1	1	2 Ba	2 Le Ne	2 Hg Ca	2 PJ Pk	2 Bh Bs	3 Al	3 AC- EXIN	4	4
ENT OM	1819 -	-	1	1 Py	1	1 Ch	2	2	2	2 Sp	2 Ka	2	3	3 Ad Br Hb	3	3	4	4
ROT K-DOR	20-3 1	1 Im	1	1	2	2	2	2	3	3 Pc	3	3 Dl	3	3 Wi	4	4	4	4
ROTE 2225	2425	1	1	2	2	2 Ch Di	2	2	2	3 Pc	3	3	3 Bl	3 RW	3 Ry	4	4	4
ROTO 2629	26-9	-	-	-	1 Eu	1	1	1	2	2 Ba	2	2 Bt	3	3 RW	3 Ls	3 ALT	4	4
MOT SPALT	30-4	- Im	-	- Py	- Eu	- Di	-	1	1 Co	1 Sp	1	2	2 Mw Me	2 Ei	2	3SPAT METAD	3	4
REM	35-39 -	-	-	1	1	1	1	2	2	2	2	2	3	3	3	3	4	4
POS STA	4044 4549-	-	-	-	- Dm	-	-	-	-	-	1	1	1	2 Hu Fi	2	2	3	4
HOM	5+ -	-	-Ex	- DS	1 Hc Ar	1 Ch	1	1	2	2 Ke Ga Sp	2 Ne	2 ME	3 PJ Pk Fr	3 Dt	3	3 CES CHES	4	4
DYNU	6+ -	-	-	-	1 He	1 Gl	1	1	2	2	2	3 Ly	3	3 Sa	3 Al	4	4	4
DYNS	7+		-	-	+/- Th	- Ma Bo	1	1	1	1 Gr	2 Ro Sm Be	2	2 Sr Md	2	3	3	4	4
DYNO	8+		- Ex	-	+/-	-	-	-	-	1	1	1 To	1 Ra	1 Ml	2	2	3	4
ACC ADD	9+		-	- DS Bi Py	- Pl Am	- Ma Ch Bo	- Mo Ha	- Sc	-	1 Bc Gr Lo	1 Vo Mt	1 La SS Ki	1	1 Sl Mt	1 Mc Ti	2KUPS PUREA	3COG ELTI	4SÂ TÂ

Table 1-81 **PERSONS AND AGES of KEY CONTRIBUTIONS**

The contributions and communication of many of the world's greatest benefactors can be tabulated by date and categories of TRU. It is guided by, and can be expanded into a much more detailed spread sheet of the origins and best-known exponents of some of the key first principles and ideas, concepts, laws, principles, or basic ultimate truths. It is primarily based upon those meanings and terms treated in the chapters of IPTRU. Personages are noted by one capital and one small letter. The concepts, existing or new - OC or NEC - are noted by their first two capitalized letters. The abbreviated names are as follows:

Ad-Adler 19 18, AH-Adorno-Horkheimer 19 10, Ak-Akhneton -13 135\15, Al-Alexandre 19+ 16\65, Am-Archimedes -2 26\34\90, An-Anaxagoras -4 09\13, Aq-Aquinas 12 08\10, Ar-Aristotle -3 04\09\50, Au-Augustine 3-Pre\10, Av-Avicennia 10 08, Ba-Bach 16 17\27, Bc-Bacon 16 01\90, Be-Bentham 17 74, BG-Bondi-Gold 19 11, Bh-Bohr 19 17, Bi-Biblical Authors -7 10\15\91, Bl-Boole 18+ 25, Bo-Boethius 5 74\95, Br-Bradley 18+ 13\19, Bs-Bergson 19 11\17, Bt-Beethoven 18 28, Bu-Buber 19 15, Ca-Carnot 18 17, Ch-Christ 0 15\18\51\95, Co-Copernicus 14 05\30, Ck-Chomsky 19+ 24, Cr-Croce 19 10\11, Da-Darwin 18+ 06\07, De-Descartes 16 01\05, Dd-Diderot 17 06\08, Di-Diophantus 2 25\30, Dl-Delacroix 18 20, Dm-Democritus -4 04\40, DS-David-Solomon -9 15\55\91, Dt-Dilthey 19 53, Ei-Einstein 19 10\15\32, Em-Empedocles -4 07\17, Eu-Euclid -3 26\33, Ex-Exodus Authors -12 59\81, Fi-Fischer 19 45, Fr-Freud 18+ 53, Ga-Galileo 15 07\50, Ge-Genesis Authors -16 01\11, Gl-Galen 1 63, Gr-Grotius 16 72\90, Gu-Gutenberg 14 60, Ha-Al Hallaj 9 15\93, Hb-Heisenberg 19 09\19, Hc-Heraclitus -5 07\50, He-Herodotus -4 60, Hg-Hegel 18 03\139\16, Ho-Hobbes 16 71, Hu-Husserl 19 11, Im-Imhotep -26 20\30, Ka-Kant 17 08\13\18, Ke-Kepler 16 05\50, Ki-Kierkagaard 18 05\90, Ky-Keynes 19 82, La-Lavoisier 17 73, Le-Leibniz 17 15\16, Li-Linnaeus 17 09, Lo-Locke 16 01\91, Ls-Lassalle 18+ 90, Ls-Levi-Straus 19+ 30, Lu-Lucretius -0 04, Ly Lyell 18 62, Ma Marcus Aurelius 1 71\90, Mc-Marcuse 19+ 90, Md-Mendel 18+ 73, Me-Mendeleev 18+ 30, ME-Marx-Engels 18 50, Ml-Malinowski 19 80, MM-Michelsen-Morley 18+11, Mo-Mohammed 6,18 135\93, Mt-Malthus 17 90, Mw-Maxwell 18+ 12\30, Ne-Newton 17 07\50, Pa-Parmenides -5 13, Pc-Pascal 16 21\25, PJ-Peirce-James 18+ 17\51, Pk-Planck 18+ 17\50, Pl-Plato -3 09\10\90, Pr-Parsons 19+ 56, Ps-Pasteur 18+ 63, Pt-Plotinus 2-10\15, Py-Pythagoras -5 20\34\91, Ra-Ramakrishna 18+ 81, Ro-Rosseau 17 71, RW-Russell-Whitehead 19 24\26, Ry-Ryle 19+ Pref\24, Sa-Sartre 19 65, Sc-Scotus 12 11\12\90, Sl-Spengler 19 91, Sm-Smith 17 72, So-Socrates -4 00\08\10, Sp-Spinoza 16 18\34\59, Sp-Spencer 18+ 70, SS-SaintSimon 18 90, Th-Thucidydes -3 70, Ti-Tillich 19+ 95, To-Tocqueville 18 87, Vo-Voltaire 17, 90 Wi-Wittengenstein 19 21, Wu-Wundt 18+ 05. (q.v. Index)

This and many other tables, charts, diagrams, illustrations and figures have, are, and will be developed and refined for use, as in POT, and superimposition by SPALT. Many authorities had contributions in other fields and places on the diagram. There was current in various ages a mass of knowledge and cultural interest that guided much of their advance. The great void was, as it continues to be, between the few who had understanding and its influence compared with the little or no TREQ knowledge of the general population. The use of combined authors as those of biblical times, DS, MM, RW, etc. refer to a combined yet not fully determined contribution. DS refers largely to Psalms and Proverbs. Proverbs although identified with Solomon were largely from thousands of years of development, largely in Egypt. The only area or section noted without contributors is REM. Many have contributed to values through time, although few have been outstanding. Only in the last few decades in philosophical research periodicals has there been concentrated and comprehensive study of value. Science, philosophy and religion can't be complete without each other. Only by TREQ and a precise perspective of TRU can we find the necessary closure of PERT, POT and TIT-TO.

Customs, rules and regulations have been like guidelines, beacons of light in a dark world. They have been of great help in social behavior, competition, selection, and predominance. When equivalent to their role in TRU they yield long and strong support. They reinforce compliance. However, they are no substitute for a thoroughgoing, interesting and inspiring quest for understanding and use of all the great concepts, principles, laws, and mechanisms of TRU.

There is an essential unity of all the monotheistic religions. This has been championed not only by the eclectics, but usually by their saints, prophets, and all who are the carriers of religions. Mohammed as the messenger of God, was especially convinced that God is not partial to any religion, that Judaism, Christianity and other monotheistic religions share GOD. "We have faith in that which has been revealed to us and in that which has been revealed unto you. Our God and your God are one, and unto him are we resigned." The Unity of God is the means by which the brotherhood of the faithful, TES in TRU can so greatly succeed. Religions are successful when they most greatly provide all the vital principles, concepts, doctrines and laws of TRU to their adherents. These when carried in the most inspiring, convincing and stimulating way add greatly to this success. The whole history of the worlds religions, especially Islam, the last great one to be founded, is basically the dynamic mechanisms involved in the power of their principles, concepts, doctrines, and laws, and how these succeeded or failed to fulfill all that must be TRU. The successive steps in the communication of knowledge is what constitutes the greatest power that has enhanced human life and made it into the special universe it is in TRU. God, Brahma, Allah, and all others can be no more and no less than they are TRU. A person, Christ, Mohammed, Buddha and others can be no more and no less than how they are their counterparts, TES, in TRU. TRU and TES can be no more and no less than the degree to which they signify their actualities. A monotheistic religion can fail or succeed no more or no less than it is no more and no less than the manifestation of the TU, and TROS, of TIT. St. Augustine held to a predestination of Grace through God alone, and the need for salvation through baptism. It was because of the non reconciliation of these two doctrines that some protestant churches excluded baptism. There is much predestination in TIT, seen in PEK, CEK, and the FAF of TRU. It is through a powerful appeal to the oppressed and unfortunate that the Judeo-Christian religions originated and became most influential. It took on two different paths, legalism in isolation in Judaism, and love in identity with Christ as God in Christianity. The second, in part, in reaction to the first. Judaism was a defensive and fundamental theology, Christianity has been more practical, and egalitarian which explain many of their successes and failures. Through great doubts and suffering people can be stimulated to seek God, a better life, TREQ, TES and TRU.

The most vivid and possibly influential early exponent of the role of dynamic processes in world history was Lessing. He championed the theme of individual development in the medium of Leibniz' necessary truths, principles, ideas and the unique, special nature of the human intellect. They combined ontology and religion, as we show in TRU, and as demonstrated in HOM, the processes of TRU in its model, SPALT. Lessing regarded the history of religions as a series of revelations by which God educates the human race. Revelations, in progressive stages, produce a comprehensive education for each person.

LOTAD = Law of TRU/AU Distinction. Each and all persons to be TES must recognize: 1. The great distinction between themselves as ENIT and as TES, 2. That between TES and TRU, and 3. All that is TRU. Without these what progress there is in civilization is unguided, largely by trial and error selection of limited and misused higher powers of mass energy. All cultures, societies, groups, and civilizations need to review and determine where they are going wrong and the likely ill effects they are both apt to encounter and produce. The dream of bliss, life without study and work must be replaced by a joyful appreciation for the actual life of TIT. Yet work need be neither laborious nor unpleasant. For the highest definition of work is that which is positive and productive, a labor of love, love of TRU and TES. Work means challenges and adventure in being more and more TRU. Variation from the static mean of TRU is the dynamic means that helps us survive and thrive.

	POSITIVE	NEGATIVE
ENTITIES/PARTS	If TRU becomes Absolute, LOPAN	Not TRU or only a tiny Part of TRU
INDIVIDUALS	Increased Knowledge, Means	Limits, deviations, and burdens
GROUPS	If TRU partakes of all benefits	Often not or little TRU, to ANTRU
COMMUNITIES	Identity, Cooperation, TRU basis	Decay, Crime, Separation, ATRU
SOCIETIES	If TRU promotes and enhances TRU	Self Interest, Abusive, Prejudice
STATES	If TRU promotes and enhances TRU	ATRU, uncooperative, and conflicting
NATIONS	If TRU uniting, Economic strength	ANTRU, Opposition, conflicts, and wars
CULTURES	If TRU highly supportive, define	Tangential, wasteful, and destructive
CIVILIZATIONS	If TRU exemplary, ENT, ACC, SA	If ATRU deteriorate, and loss in other causes
TRU	All its benefits, assures TA	All that detracts from TRU 1/TRIA

Table 2-81 **POSITIVE and NEGATIVE ASPECTS of DIFFERENT GROUPS**

In this table a balance sheet of the positive/negative and thereby optimism/pessimism in the world can be provided. For different segments of life, such a balance sheet acts as an accounting record. **EBE** = Extremes by Emphasis. It is great to emphasize that which has been so much denied of TRU. Yet in so doing distortion is likely to be produced. Thereby we can see that much of EBE is not a defiance of the total means and extremes, or TO, but a reaction to what has been denied. This is called: **EBERT** = EBE in Reaction to Denial of TRU. EBERT shows how the greater, and in that way wrong and erroneous, that is emphasized, the greater the EBE. That this need only be temporary should be recognized, just as the interactions between first and second persons. So that the greater the error produced by EBE the greater the need will be for its correction. This is a common paradox of our world. It does not hold so much for TRU or TO. Their focus is already corrected. It is from the focus of IA, ENIT, or the individual, and all in reaction, that the dangers and limits of EBE are ever present. EBE, then, can be resolved only in toto, TIT-TO. It can in some instances be resolved, not when taken as a whole within, but when taken from without. This is where the observer, as the overall perspective, meta, universal or transcendental states are accounted for. It is very revealing and effective to label each and all contributors by some ideas that they have championed and may, more or less, characterize them. Such we see in the method of Socrates, ideas of Plato, the potential and forms of Aristotle, the determination of the Stoics, the happiness of the Epicureans, the relativity via light of Einstein, Planck and Christ. Yet as great as these are, and the rest of the breadth of their contributions, they all fall far short, by partiality and EBE, of being the totality of all, TO or even its overview. Thereby we can see the great paradox and the associated necessity of VAPO, SCARLS, TROS and all that makes up TRU. We can see both their great contributions which are naturally partial, and the necessity of totality on the other. This is another great proof of and justification for TRU.

LEBE = Law of Extremes By Emphasis. What is or seems to be so necessary may not be TRU. Only when - no matter how great our TREQ are - they are TRU, and even TRU is TRU as it is TIT, can we escape from the endless pitfalls and distractions that prevent absolute purpose, and ends, as TIT through SA becomes, as it must, TA. As great as Christianity, Catholicism and many religions are, they tend to LEBE. They tend to accommodate this or that precept at the sake of the whole. **GID** = God Defect, or the concept of how God lacks a, some, or much that is TRU. It is highly variable depending upon what the belief in God is. God lacks a large degree of VAPO, actual meaning. All that produces a God of gaps produces GID. GID does not mean GOD does not exist, is not TREQ or TRU. The correction of EBE is an ever present and glorious endeavor. It is largely a product of our subjectivity, mentality and ideality, a reaction from partial to partial within and without TRU. With a thoroughgoing knowledge of TRU through KOT, SPALT, HOMATION, PUREALITY and COGNOS we achieve that level of comprehension sufficient to resolve the problem of EBE. **COMPRET** = Compensation, Recompensation, Return, Convergence or setting to TRU. COMPRET signifies those processes by which EBE and much else that is other or less than TRU is brought back to TRU. What is EBE or less than TRU is compensated or recompensated for, returned to TRU and all brought in full balance to the set state of TIT-TO.

	NATURALISM	SPIRITUALISM, ANIM-ISM	POLY-THEISM	MONO-THEISM	PAN-THEISM	DEISM	DYNAM-ISM	MATERIALISM REALISM	ECLEC-TICISM BEST	TRU
SUMERIAN	++--	+-	++	-	+-+	+-+	+-	++	--	--+-
EGYPTIAN	++--	+++-	+++	-x	-+-	+-+	+-	++	--	--+-
ATEN	+--	--	x	++	--	+-+	++-	+-+	+	+-+-
JUDAISM	-+-	-++	x	++	-+-	-+-	++-+	-+-	-+-	+-+-
AVESTA-ZOROAST-	+-	-+-	-+-	-+-	-	+-	++-+	+-+	-++	+-+-
BRAHMA-HINDUISM	-++--	-++-	x	+-+-	+-+	-+-	++--	-+-	+-+	+-+-
BUDDHISM	-+	++-	x-	+-+	+-+	--	++-+	+-+	+-+	+-+
TAOISM	+++-	+-+	++-	x--+-	++-	+++	++	+-+	+-+-	-+-
CONFUCISM	+-+	-+	-	-	-	-	+-+	+-+	+-+	-+
SHINTOISM	+-+	+-+	++	-+-	-+	+-	+-+	+	+-	-+-
PLATO-neo-ISM	-+-	+++	x-	++	+-+	-+-	++-+	-x	+-+	+-+
CHRISTIANITY	-+-	+-+	x-	++-+	x+-	-+-	++-+	-x	+-+	-+-+
CATHOLICISM	--	+-	x-	++-+	x+-	-+-	++-+	--	+-+	--+/+-++
ISLAM	-+-	-+	xx	+++	-+-	+-	+-	-+-	+-x	-+-
PROTESTANTISM	--	+-	x-	++-+	(+-+-)~	-+-/+++	+-+	-+-	+-+	---/+++
TRU	-+++	-+++	+-+	+-+-	++-+	--+-	++-+	+-++	-+++	++++

Table 3-81 **Comparative Religions**

Different religions and ways of life are compared by what they must include, where they are deficient and how they are identical with TRU. A (+) sign indicates this was a significant part of their religion, as a (-) did not. An x indicates special opposition to the doctrine. These rough estimates help to show EBE, DAOST and TRU. How they are the same and how they are different, and how this same and difference is or is not the same or different that is of TRU and in TRU, TIT. One must be careful in looking at this table, as in life itself, to realize that what is a positive identity is not necessarily positive identity with TRU. Also, variations in each religion by time and place were often considerable.

RELIGION and SYSTEMS	GREATNESS	FAILINGS of
MONOTHEISM	Unity through a Godhead, Power of Concretion of the same, DAOST	Excludes the Finite, Infinite, Total Lacks Proportion good Representation
PLURALISTIC RELIGIONS	Represents Major Aspects of TRU (more or less) Power of differents	Lacks Unity, System, Order, Valid Totality, Proportion
EXISTING RELIGIONS (best)	Strong Traditions, Firm Foundation in God and Christ, Righteousness, Goodness, Faith, Grace, Hope,,,	Often loses original fulfillmentand intent, Sectarian, Lack POA, TRU, SUT, VAPO, DAOST, ALT,
CAPITALISM	Competition, Economic Stimulation, Open Markets, Initiative	Lack of Common Purpose, Excessive exercise and abuse of Self Interest
SOCIALISM / COMMUNISM	Universal Benefits, Does what private sector often can not do	Lack of Individual Rights, freedoms and ability to produce, incentive
UNIVERSALISM	Total comprehensiveness, increasingly recognizes the thru and thru, whole	Lack of what is meant by Universal Little emphasis on the particular
PAST RELIGIONS	Social Cohesion through Identity with some object, animal, God, etc.	Hierarchy and Divinity invalid, Irrational and fantasy, not TRU
PHILOSOPHICAL SYSTEMS	High Level of Perspective and Validity, Penetrates basic truths	Usually extra or Inconclusive TRU, misinterpreted, FISMS, 1/TIT,TE
SCIENTIFIC TRUTHS INTERPRETATION	Precision, beautiful detail, discovery, criticism, factual	Lacks unity, proportion, totality, denies TRU's intuitive certainty
HOMATION	Great representative insight in processes of life, TRUD, TRU	Not all of TRU, not known except in TREQ, Much work due on
TRU, TIT, TRUS, TROS, TRIS	By definition both valid and total System with unified FAF precisely represented, most rewarding	Not now all known, much work due on, Not simple, depends upon the initiative of one and all

Table 4-81 **Greatness and Failings of Philosophies, Systems, Religions, and TRUS, Relative to TRU, RELT**

Religious and philosophical problems, questions, paradoxes, principles, concepts, laws, relationships, characteristics, polarities, basic forms and ideas need to be considered and interpreted from the whole range of their extremes and means and how they are most TRU. In TRU, TREQ, or any good eclectic religion it is not only necessary that its basic concepts, principles, laws, mechanisms, parts and processes be well known but well learned and used by all. St. Augustine, in *The City of God*, reveals how within all the mixed and ungodly worlds and universes about us there is a city of God, the universe of TIT and TRU.(**TRUNIV(ERSITY**) It is a universe within a universe, where all negative is turned to positive as sin is turned into what is right, TES to TRU. This was done by St. Augustine, through penitence, confession, and thanks-giving. It is done in the pursuit of Truth for TRU by TES.

The level of justice, and how much there is in an enlightened citizenry in different parts of the world at different times in the past, present and future, are large determinants and measures of the success or failure, of societies, cultures, and civilizations. They are the TRU/ATRU, what it takes, and how it necessitates accompanying levels for economics, commerce, communication, industry and development. It necessitates a level of technical and basic knowledge as well as proficiency in its use. It changes per age. Yet much remains unchanged. In those cultures and civilizations that had a very enlightened, interested, and stimulated citizenry, where the level of justice had been continuously increased and upheld, there were societies, nations, cultures, and civilizations that were highly contributory to TRU and successful in their own right.

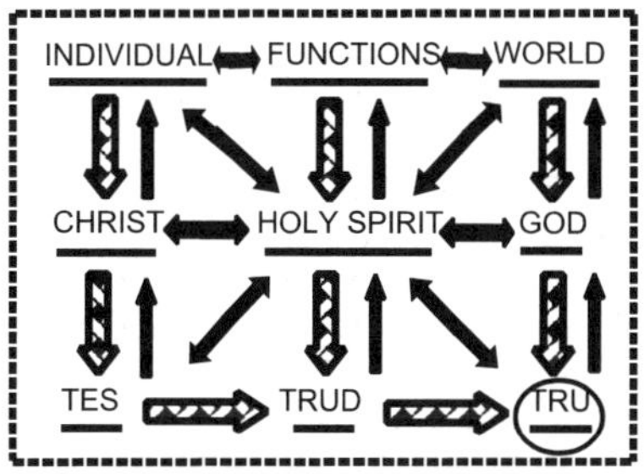

Fig 1-81 **ST. AUGUSTINE, CITY of GOD, TRUNIVERSITY and TRU**

The diagram shows three levels of truth that reach the greatest level of actual truth in TRU. There is a close correspondence of each row as any truth becomes, divine truth, and these become the actual truths that exist by way of TRU. It is done through devotion to the Vita Contemplative, and through Christ, as TES. This is the contemplation of life, God, TREQ or TRU. It involves the ability to answer the greatest questions and solve the great problems through analysis and bringing them altogether as they can be done through TRU. This is how the individual gradually, permanently and gloriously contemplates the truth of TREQ and TRU. This is the actual TRUNIVERSITY. It is less than TRU, and more than a university, even a university dedicated to the study of TRU. TRUNIVERSITY is the world of study, learning, knowledge, KOT, COAT, OM, COGNOS, ELTICMA of all persons becoming TES, and together working to discover, understand, serve and live TRU. It is found everywhere there are people, men and women, adults and children who seek the truth, knowledge, answers to their questions, or respond adequately to all that leads to TRU. The necessity of total identification of supreme value within the spirit to seek and know God, or TRU, through his image in the soul is St. Augustine's highest contribution. It is what creates the bond that enables TES to encompass the city of God, TREQ, TRIT, OBT, SUT, SPALT and TRU.

LACE = Law or principle of the Common Ethic. LACE signifies the power of a common ethic. This is where no matter what the ethic, religion, principles or beliefs of peoples, are or are called, as long as they are in common with TRU and VAPO they are to that degree right.(LADEP, DICE, PIDET, TES, LAPE) It is a special case of the general common ethic for all humanity. LACE is what we find in TREQ or more what is common, yet TRU, to all eclectic and ecumenical philosophical, scientific, and religious movements. They produce and become great processes that converge on TRU. They make up a large part of: **LACART** = Law of Core Forces of Right for TRU,

Each new generation, different culture and country has its own concerns, ideas, language, and set of beliefs. These may be highly theological, philosophical, practical, economical, social, mythical, or primitive. Yet they usually are little, superficially, or variably TRU. It is in extracting the common universal principles involved in all equivalent to TRU that we can unite in the common welfare of the world. This leads to: **UPOST** = Universal Proportional Order, Sense, Sets of Beliefs by TRU. UPOST is the combination of religious, philosophical, scientific, trade, fraternal, social and other beliefs that together equal those of TRU. All must come to recognize what is right and what is wrong in their beliefs and opinions. The best way of doing this is to learn, know, remember, practice and serve TRU.

COKIT = Communication, Knowledge, Information, and Transmission. Not only has human knowledge increased rapidly but our ability and means to represent, demonstrate, transmit, comprehend, think of and use it has even more rapidly increased. Yet this does not come, as all that takes from us, without its downside. This is a preoccupation with its means, vehicles, trivia, and the lack of overall order, coordination, comprehension, management and control necessary to bring all back into a healthy strong mentally competent and totally knowledgeable individual.

The unity, concrete and solid nature of the world was expounded by Fichte. He considered this to be a rational, unified system directed toward a purpose. He considered self consciousness, and self analysis to be the key, sole and fundamental task of philosophy. New problems are created as reason attempts to solve the problems imposed self analysis, self consciousness and their relationships to the actual world. This produces a dialectic between reason and the world in which the individual lives. It is a dialectic which consists of a thesis, an antithesis and a synthesis. These ideas were borrowed and expanded by Hegel, as they also influenced Brentano, Husserl and Freud. To Fichte self consciousness or self analysis were the thesis which he called the Ego. The world about it and problems involved or antitheses were the Non Ego, and the resultant combinations and adaptations were called the synthesis. This corresponds with and shows the convention of starting with immediate awareness, IA, and self consciousness as the point or origin, center and focus of and for TRU.(IA,01) The dynamics of this unity focusing on the ego, self consciousness, instantaneous attention, or IA, gives us a strong philosophical precedent for the fact and mechanism by which the processes of man, HOM are derived, developed, determined and posited to complete the unity and solidity of the concrete world, TIT-TO. This then becomes the synthesis, toward which all tends to converge. The focus to relative reality, of IA that impels all to TRU in full and in proportion, is **FORIA** = Focus of Relative Reality of IA. FORIA is not all TRU. Just as SCARL is not anything of anything, nor is it to be, as we are so prone. This is **PHE**, the Philosophical Error, as by ISMS, if it is considered as all TRU as Fichte would have us believe. None the less it is the focus, and its dynamic aspects, dialectical and systematic unity, that provide for the actual concretion of the world.

KOPT = Knowledge Potential TRU. The great revolution that developed in the 18th-21st century has been that of knowledge, especially scientific and technical knowledge and its communication. All should join and work for this great cause, course, power and wave in the future and at the same time help to guide it by and to TRU. For it is the greatest means of success and the continuity of TRU. This is done by making knowledge that is potential actual in TRU, either by what is known that is not known by TRU, inadequately or improperly known, but so potentially great for TRU. The force and power of knowledge are tremendous. It continually asks and answers the questions, what, how, why, when, where and many others. The great amount of time spent in applying knowledge in the different trades, occupations, skills, professions, and the fulfillment of human needs are much of TRU. These are a tremendous body and power whose vectors and processes provide much of the dynamics of TRU, from ESINS to ES and CES, from values to REMS and their use. **LEKPRÁ** or Law of Partial Knowledge Proof of Á.(PRÁ) The force and power of knowledge is tremendous. Yet it is only a polarity whose scale fuses relatively with the objective of TRU in Á As in reality it inclines and fuses to become the third great section of the Á dimension, the real, toward all TRU. As knowledge increases, as our interests increase, as our routines are more and more reinforced and utilized, as our objective, material, and physical means are enhanced, and as all in between or in combination grow, expand, intensify, so does all TRU increase. The relationship between the Power of Knowledge, **POK**, and PRÁ is called: **POKERÁ** = Power of Knowledge and Equivalence to and Proof of Á. Knowledge when VAPO and TRU has great powers that make up a large portion of the Á FAF or dimension of TRU. This power of knowledge, especially of and in TRU, KOT and KIT, reveals TRU and the way subjectivity works predominantly through Á in TRU, thereby proving Á and the FAF of TRU.

Every person, group, community, city, state, nation, culture, civilization and world begins, develops, grows, becomes mature, evolves, and changes. However, this must be in a special way, to be successful, survive, and have meaning, in the fulfillment of their role in the actual world of best human life. It is here the great paradox and trial of life occurs. For each of us, all that is a part, and we are a part of, are continually being selected in or out of this actual world. Just as humans have always been, and always will be, only if they are a part of this select world. Thereby all the entities, sense and things, thoughts and acts we continually live by are either/or, more or less of this select world, or some lesser or negative world. As much as we think and seem to believe otherwise, we are if we are good, and are not if we are bad. Our conscience dictates that it must be good, for the best of hum an life. Only so often we have little, or no, conscience.

Just as bad ignores conscience, yet must be equivalent to our knowledge of right and wrong. Knowledge of right and wrong can only come from great knowledge, both in quantity and quality, as it replicates the actual world. Humans tend to be well intentioned, avoid mistaken ideas, errors, and mis-judgements. More often it comes from ignorance and lack of appreciation for what they have received. Let them escape from the more frequent mistaken notion that simple truths equal total truth and excesses are final. **FASTEX** = Fallacy of Simple Truths, and Excesses. Let them know the breadth, depth, and height of philosophy, religion, all sciences and technologies that constitute the totality of actual existing and coming human knowledge and life. All persons of the present and future can achieve this level of education, knowledge conscience and contribution if they are continually and proficiently nurtured and trained from infancy throughout childhood, adolescence, and adult life. They must be nurtured and trained by receiving and achieving the highest levels of care and experience. This in order to develop into mature, tolerant, broad minded, dedicated, determined, serious, pleasant, communicative, powerful, and respected persons that can love life and the endeavor for all that is for a better actual world, this special world of best human life.

Mankind ever since the beginnings of his ability to think and understand the surrounding world has increasingly developed, with the associated needed concepts and terms, a great interest and awe of this world. Early taboos, myths, and beliefs developed from incomplete recognition, suspicion, fear and use of such objects and ideas for communal and personal success. As man became increasingly aware of the sun, moon, stars, earth, water, air, fire, wild animals, sheep, goats, pigs, cows, horses, and other objects in his environment, he gave to them a higher place or elevated position of importance. It demonstrated an increased awareness and importance to this special world of best human life. Since they held in common great value to others, they became a rallying point, common ground and purpose to which populations could bond. It set up some beneficial aspect, item, animal, or deified personage as an object which needed to be held in a special way or fixed distance to properly venerate. This is: **DIST** = Distance or Distancing from TRU. DIST is seen in any standardized perspective of the world, TRU as SPO, SPALT and ELTICMA It is where we find that distance which best enables us to see, perceive, associate ourselves and venerate the object of our attention, interest, or being. Yet in so fixing such a DIST, all peoples, the earliest primitives and savages through greatest modern religions to TRU are committing an error. This is by separation, from that very thing they so extol or venerate. It is because of DIST that man has so alienated himself from his world, at all stages of its development and existence. It has become so fixed in our mind and culture that it is pathological and a great cause of failure and demise. This is called: **AFOD** = Abnormal Alienation by Fixation on a set Objectivity and Distancing from TRU. It is the reversal of our relativistic state, or Á, that is key to, and explains many human enigmas of life, and thereby of TRU. This is: **REDIST** = Reversal of Distancing from TRU. Even in IPTRU DIST is done, there is an inclination to AFOD - for which much effort is taken to correct. Philosophy is largely a critique of the problem of DIST. Yet much of this very critique winds up in committing the very problem it questions. This is, **CRIO** or Critique of All. The only ultimate right, or correction to, or proper knowledge can come from TRU, as TRU is TIT-TO, THES-THEO, and expanded therefrom maintaining a 1:1 identity as it proceeds. **CRITO** = Critique of all in TRU, TRU totality, in OITIO. With so much knowledge available and taught to us we can hardly appreciate what it is to be without it. So is it of TRU. We can hardly appreciate what it is to be without TRU. Those who have little or no knowledge of TRU are likely to be highly suspicious and fearful of it, as is natural for what is seemingly unknown or different. It is only as these two come together that both TRU and all peoples will become one. This is what CIPT is for.

The living lie and the ultimate in falsity is to deny GOD as TREQ and TRU, their equivalents, aspects, forms, modes, parts, functions, processes and totality. It is when we stubbornly hold to errors, bits or half truths, partials, disproportions, ISMS, and systems that are not so, are not TRU. Or when they are less than TRU through incompletion, disproportion, in-actuality and invalidity, or void of VAPO. This is especially so when that which one believes to be so right is so lacking. The ultimate in falsity is to hold to any belief, knowledge, idea, concept, and conviction, or their representation, presentation, perspective, or relationship that is less than, or completely erroneous in relation to TRU.

MEASURES of DYNO CES in SULP of SPALT

CHE(R)	~3-3-2~ -	S	4
COKIT	4-2-4~	T	5~
COMPARATIVE RELIGIONS CoR	2-1~-1~	S	8
COMPRET	~4-~-~	M	7
CRIO	2-4-4~ +	T	2
CRITO	2-4-4	T	2
CULTURAL DYN, WORLD HISTORY	4-1-1	S	1~
DER	~-~-~-~ -	M	8
DERA	~3~-~-~	M	8
DIST	~-~-~-~ +-	M	6
EBERT	3-3~2~ -	M	4
FORIA	4-5~-4	T	
1GID	2-0~-2 -	M	8
GREATNESS/FAILINGS PHIL., RELT GFP	3-0~-~ -+	T	8
JUSTICE, ENLIGHTENED CITIZENS JEC	3-3-6~	T	1
KOPT	1-2~-7~	M	7
LACART	~-~-~-~	M	7
LACE	3-2-2 ++-	S	4
LANGUAGE DEVELOPMENT LaD	4-1-1 ++-	S	1
LEBE	~-~-~-~+–	M	8
LEKPRA	~2-~2-7~	T	5
LOTAD	~-~-~-~-+	M	8
PERSONAGES,KNOWLEDGE			
CONTRI. PKC	2-4-2	S	1
PHE	1-1-2 -	T	8
POK(T)	~2-~2-8~+	M	5-8
POKERA	~ 4-4-4~	S	5
POSITIVE/NEGATIVE, of GROUP PNG	4-3-2~ -+	S	4
PUS	1-1-1 ++-	S	2
RELIGIONS and TENETS RaT	2-1-1~ +-	S	8
ST. AUGUSTINE, CITY of GOD, T SAC	~4-~2-~++-	M	6,7
STE	~2-0~-2~ -	M	6,8
TRUNIVERSITY	4-2~-7	T	1
UPOST	~-~-~-~	M	7

Fig 4-81 **DYNO SULP CES in METAD**

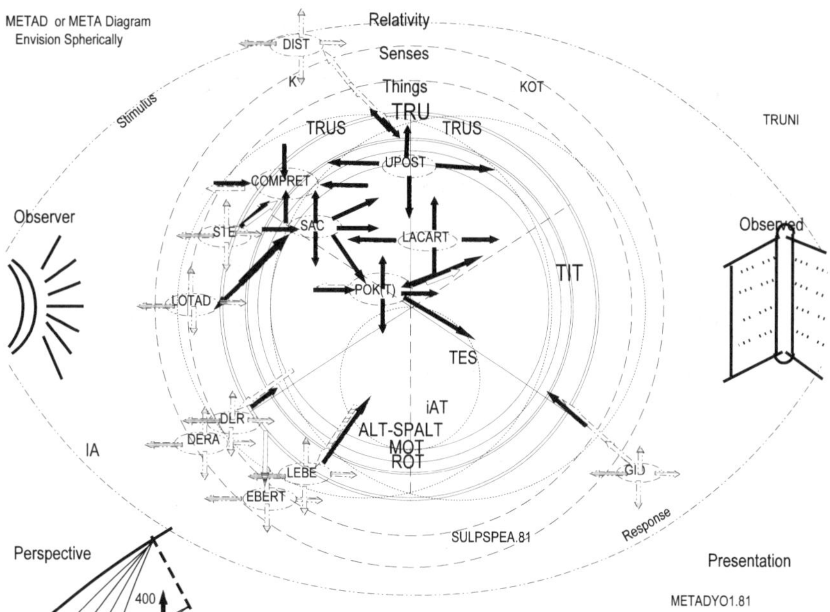

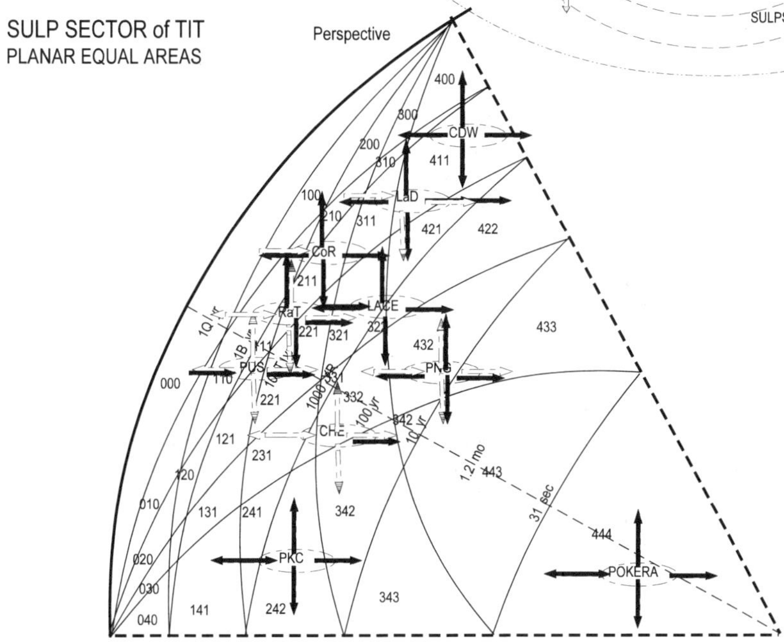

Fig 2-81 **DYNO CES of SULP SECTOR in SPALT**

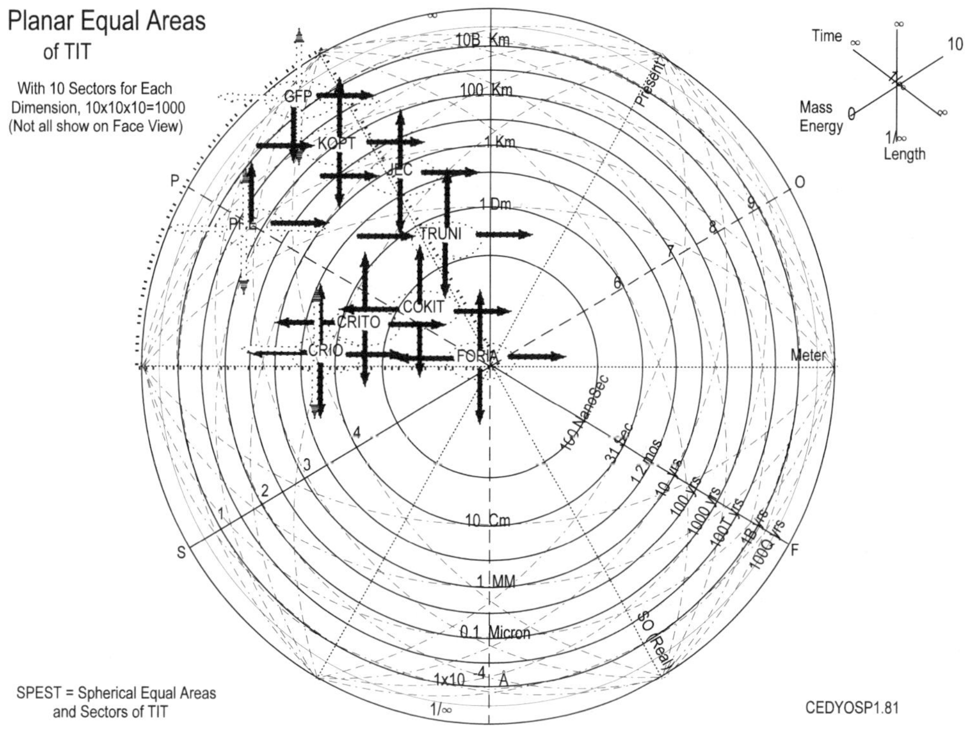

Fig 3-81 **DYNO SULP CES in SPALT**

CHAPTER 82

OLP or OBJECTIVE LARGE PAST of DYNO

INTERACTION, EVOLUTION, CLIMATE, NEEDS, POPULATION, RESOURCES, ANCESTRY, ECONOMICS

LAMID = Law of Magnitude Interdependence. There is a relationship between the causes and reasons for what is large and that which is small. This is seen in time and length or size. It is seen in climate and other cycles that are time and size dependent. In brief, long term effects are not apt to be associated with small size effects. Small size effects are not apt to be associated with long term effects. This we see in the effect of the local on the total or earth. This is evident in the effect of a volcano or temporal event upon the overall status of the earth. However, this law, generalization or relationship has its limits and exceptions. For it is certain that long term effects, such as from the sun, or even astronomical forces farther away, do have profound effects, in various and often different ways, on all portions of TRU, including the small. Thereby since LAMID and TIT are fixed functions, it is necessary to resolve the two, as by COL, with a third, a reaction or constant. This becomes, **RELAMID** = Law of the Massive Dynamic Interaction, or reaction to the difference between LAMID and TRU. LAMID and RELAMID relate primarily to large departures from the unity that exists in TRU. As a departure occurs, this action calls for reaction. If the departure is large, the reaction is apt to be large. Likewise the difference between TU and this departure is a reaction or synthesis, as all must be in an absolute set, to come back to TRU.

The manner of human lineage can be presented to emphasize the cladistics or phylogeny of human classification in RUTS and TRUS. Human lineage was both complex, subtle, and varied with large or very small population dynamics, large or very small loci, and large or very small breeding patterns. It also occurred with large and small breaks, as from mutations or alteration of genetic material, and often with much polymorphism and massive GELD incremental changes. Mutations were also modified through hybridization and population dynamics. Rapid advances in evolution occurred in times of marked environmental change, stress, selection, isolation, and small populations with inbreeding. Also, these were times of high selection and extinction. More gradual advances occurred in times of larger populations, polymorphism and high outbreeding. Many were accompanied by regression and extinction, although less abrupt, noticeable, and less significant to TRU past and present. It was natural selection and associated competition, within and in and out, OITIO, the species that primarily provided optimal genetics.

We find in TRU and DOT that many of the ideas, concepts, and laws of human evolution, cladistics and phylogenetic development are of the utmost importance. Yet by ATO, TRUD, and HOM we also find that we can not rely on them unless they are adapted to TRU and TIT, and accompany other pertaining concepts, principles and laws of TRU.(VAPO) We should reach out to all that is a part of and has a role in TRU. That the advancing and increasingly more complex mass energy dimension of man, through succeeding time, were the major determinants, become ever more certain. Each live product and generation propelled the developing Á FAF along in competition and natural selection as each achieved and maintained predominance. This predominance was largely and increasingly the result of the accumulative health and strength of all levels of SÁ. It was especially the highest ones that have, are and will be tested the hardest and most apt to fail/succeed to continue predominant Á mass energy. There still remains vast gaps, problems and undiscovered principles in evolution and cladistics, especially when adapted to TIT. Not the least of these problems is that of distinction, especially by higher levels of SÁ. Although we can see the differences between members of human groups, especially when widely separated by time, location, form and function there has been a tremendous overlap, hybridization, transition, and sameness of makeup that is almost infinite. This has not only been within races, subraces, varieties, groups, and individuals. It has caused much denial of the importance, as well as having misled, cladistics and human evolution. It has also produced a lack of attention to the TRU aspects of human development. When viewed within the perspective of TRUS, TO, TIT and TÂ many of these problems and some of the principles can be set forth and resolved. This we do throughout the text. For OLP of DYNO we need to make as distinct as possible the discrete increments of cladistics, and in contrast those of evolution, that have occurred.

Time and size are two of the greatest parameters in human development. Time and generations are shown in the columns of Figure 1-82. There was a fairly high correlation between race and its location. That is the bulk of human populations usually stayed where they originated. Human range was roughly 18 km for A. Afarensis, 100 km for Homo Erectus, and has continually increased for humans today with modern means of transportation and commerce. External movement occurred relatively rarely. Larger populations usually absorb much smaller and incoming populations. Lifetimes of individuals and families changed much faster than groups, tribes and classes, that changed much faster than, subraces, races or races, and these than subspecies and species. A rough median time can be made for each, which is closely related to a set number of generations. Over time there was probably a vastly complex and interactive relationship between different groups, races and species. The closer they were in proximity and the greater their differences, the greater their dislike or fear of each other. This probably subsided or grew, to a large extent, often rapidly, with hybridization, association, and attitudes . So that it is not surprising that there has been a great mixture of all levels of cladistics through the ages. Even in the same locality there were often a great variety of different peoples, as evident from the biblical *Genesis* in Israel. The origins, rise, optimum and decline of many of the species, subspecies, races, subraces, and groups in the last 10 million years can be traced out geographically. This is not fully known, there are gaps and errors. However between what has been discovered of human remains as evidence, what remained at earliest recorded history, what exists today, and what we can infer from knowledge of probable processes we can get a usable overview of diversity in human development. The abbreviations in rows following the continents indicate the region of geographical orientation. Some of the genera, species and major races used in this diagram are: HS = Homo Sapiens, HSn = Homo Sapiens Neanderthal, HSs = Homo S. Sapiens, Mg = Magdalenian, O = Probably none, Py = Pygmy, ? = Uncertain, R = Red, W = White, Y = Yellow. Other subraces and groups include: Ag = Aurignacian, Al = Alpine, Ar = Armenian, Au = Australian, Ca = Caucasian, Dr = Dravidian, Ef = Egyptian/Fellahin, G = Greek, Hm = Hamitic, IE = IndoEuropean, Me = Melanesian, Mn = Mandarin, Mo = Maori, Pn = Polynesian, Sm = Semitic, Sl = Slavic, Su = Sumerian, Ta = Tasmanian, Tk = Turkic, Tt = Teutonic.

Even Charles Darwin, the great exponent of evolution, early recognized that evolution was not a sufficient state of human life. It did not satisfy VAPO, ATO, TROS, DAOST and the SCARLS of TRU. He supplemented *Origin of the Species* with the *Descent of Man.* This began a more cladistic approach rather than biological evolution phylogenetically

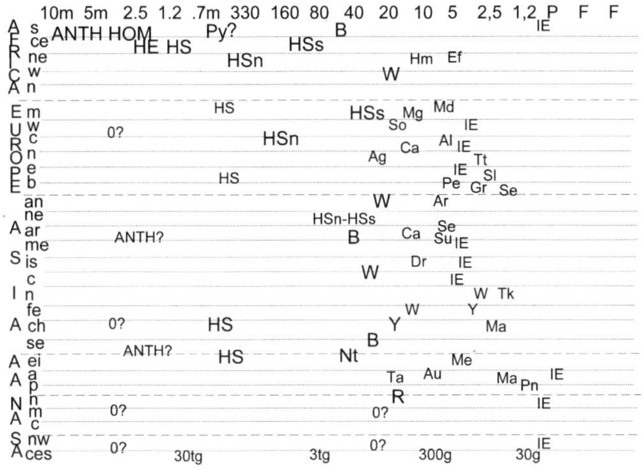

Fig 1-82 **HUMAN RACES, COURSES THROUGH TIME**
. Those who champion evolution, denying much of religion, and TRU, must first rethink, as Darwin did, the primary focus of human life, TRIA, TIT and TRU. Human descent, as evolution in general, was largely influenced by location, habitat, or the environment. It also depended on human adaptability and use of habitat, e.g. available potable water. These in turn depended upon the competitive and selective quality of all levels of mass energy or Á FAF.(SÁ) Cladistics and especially DOT exist only by ATO and TIT. The principles and problems involving race, subrace, varieties and hybridization and others assume a stance in the present and future that is far different than anything seen during the past. These differences will alter both the present and the past, as the future and ATO determine the lineage of man as it exists in TIT.

The cyclical nature of much of TRU and how it is a part of the mechanisms and processes of TRU can be seen from how nature works through time by TRU. It does not work as we ordinarily think it would, even less as it is a thing or entity in and of itself, ENIT. Yet it is often highly important to TRU, and even more important when its conformation in TIT are known and shown. It not only shows us what is actually in TRU. It goes a long way to show what TRU is in many other ways. It helps to prove TRU. It also proves TRU by proving what is not, other or disproportionate TRU. This is called: **PRUTAD** = Proof of TRU by proving what TRU is not, other or disproportionate TRU. One of these is the cyclical effect of weather, climate, solar and astronomical change that occurs and recurs in regular and irregular ways. This is especially so when taking many together that are then adapted to TRU. It is more important in certain parts of the earth than in others, the arctic/temperate/tropic. For each cycle, there are different sections or phases, a rapid rise, a long top, a rapid decline and a long bottom. These may also be divided into further shorter sections or phases. When compared to the temperature fluctuations by day, and by years overall temperature changes of the world may seem very bland and weak. This is true to a large extent, especially when of here and now. Yet when measured in terms of long term change, effect on cultures, civilizations, humans, and TRU it is a very great mechanism, and strong force that effects many processes. The more conspicuous of these are the increases and decreases in rainfall. Climate changed not only regularly, irregularly, by temperature and rainfall, but by alternating and varying patterns during the last few million years and glacial ages of the arctic, subarctic and temporal regions. It is thought that this not only increased the rate of evolution, it forced humans, as well as many other animals, to adapt or become extinct. For man it led to bipedalism and a mobility to survive in more diverse rather than special areas. Temperate fluctuations of several different kinds occurred: in roughly 92,000, 40,000, 21,000, 5,100, 835, 120, 14 year intervals, with other less known or irregular ones. These are not exact, fully developed, and applied, yet were of great significance for human adaptation and variation. Cycles of less than a thousand years have had little to no effect on sea levels, so that is those associated with ice ages that have been most influential. Yet temperature deteriorated or ameliorated, as did rainfall, and other features of the human habitat. By providing OLP we need to show all major processes and determining forces by VAPO of TRU. The superimposition of such cycles and their summative effects are good examples of how all the processes of TRU work in DYNO. The longer the period of time, i.e. number of generations, the greater the dynamics, changes, variation, and differences in concepts, principles, mechanisms, and laws that apply. For any climatic cycle will have, in a way, its own spring, summer, fall and winter, as well as dry and wet seasons, with a similar effect of seasons in human life. Patterns, precipitation, and temperature cycles will not necessarily be the same although certain relations hold, as a decrease of available fresh water in glacial ages. Their 'seasons' will often be most marked with each swing when conditions dramatically change. This has been historically confirmed by accompanying amelioration or deterioration of the human condition, generally or locally. Less commonly when more than one cycle swung at the same time these became even more dramatic. e.g. circa 13,000bp. What these portend for the future is less certain, especially when human civilization itself may alter the climate. Often very little change can cause, on marginal habitats, very marked disruption of human life, e,g, Irish famines of the 1840's when population decreased to one half. These are also evident in the cyclical effects of El Nino, La Nina and other patterns. When the many climatic cycles are accurately known, combined, and superimposed, a much greater prediction and prevention of detrimental effects can be provided. To formalize the state of cycles in which there are long periods of stability and very short. and often dramatic and disturbing, periods of change and instability there is: **RACS** or Rapid Rate of Climatic Change or Climate Swings. RACS helps to explain, along with many other CES and CHES, how in the course of human events there have been periods of great flowering out, peace, or prosperity, and periods of great decline, unrest and misery.

Since the population of China makes up a large percentage of the total worlds population and since it's figures are better documented in the last two thousand years, it helps to use China as one of several references to determine more accurately world population through the ages. It represents a discrete subunit of the world that, with the needed exceptions and corrections, can be effectively used as an example of the whole. From very roughly 10 million in 4000bc to 20 in 1000bc, to 50 at 0ad, to 80 in 1000ad, 120 in 1500, 300 in 1800, 450 in 1900, and 1,300 in 2000 we can see the marked change in population, especially in the last five hundred years. The World's Population is a large determinant of many of its processes, forces, powers, benefits, detriments and problems. The relationships between quality and quantity are primary determinants of much of its dynamics. This includes the kind and amount of tolerance, moderation, integration, out-breeding, inbreeding and segregation that affect optimal selection. Accurate population figures without a census are notoriously unreliable. Yet if we know the approximate populations at various times we can extrapolate for total populations and provide rough, and usable estimates. For example both the Roman Empire and Han China are known to have had about sixty million at 200Ad. With roughly an equal amount for the rest of the world, this would mean 250 million. So that in the million years before the time Christ the annual growth rate was only 1/100 what it has been since. Thus it was by quality, since then by quantity, which is a poor exchange productive of dire consequences. What is the ideal population for the world, for parts of the world, various conditions, ranges, and rates of change, are important considerations for future welfare. Many factors and processes are at work. Not the least of which is human ability to preserve quality both physically, mentally, socially, and spiritually.

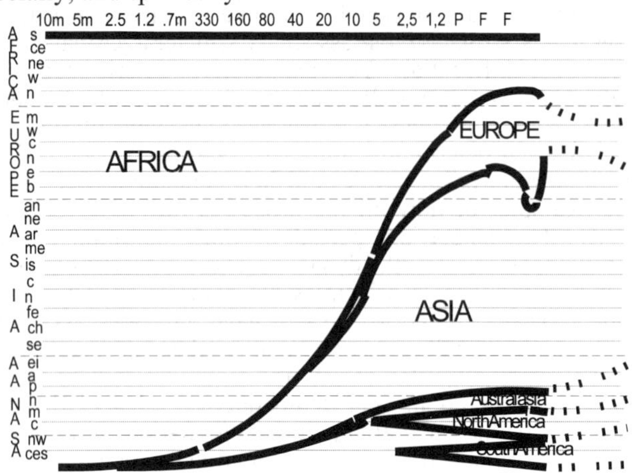

Fig 1-82 **WORLD POPULATION, PERCENTAGE / CONTINENT / TIME**

Since the neolithic era circa 10,000bc Asia has been over half the total. For a while in the 19th century with the industrial revolution and to a lesser extent with the spread of civilization Europe had the highest rate of population growth only to decline in the hot and cold wars of the 20th c. It was in Africa that most humans lived prior to the mid paleolithic era, with considerable cyclical variation with climatic change. Such a study helps us find a better overall view of humanity through the ages. There is a continual tendency to become overly fixed on the singular, partial, regional, and smaller complex processes of the present, less of the past, and little of the future to the neglect of what is the whole.

If in our world population we are to account for all human past, present and future, we must do so by correction factors that best reflect their roles in TIT. World Population reached one billion about or shortly after 1800. The total number of humans living and dead are estimated at approximately 65 billion. This illustrates the points that there are many more dead than alive, even though our recent populations have been rising at unprecedented rates. This indicates the power of those dead who by contributing to TRU, in a great multiplicity of ways, are more or less alive. It also helps to show the meaning of DAK, those who are living, yet are actually dead by non contribution to TO-TIT. It also carries heavy implications into the future, whereby the present and especially the past will have a great effect on both the near and mid term future, i.e. next several hundred years. Beyond that humans are increasingly at the mercy of their own devices, which by TIT must be toward TÂ. This mass of the dead in TO takes many forms and is largely what their TES is in TIT, yet is even different in ways, than our TES in TIT. The surviving combined genotypic-phenotypic GELD is a good example. It is in this way they contribute to Á and help to show what Á is. For applying this we must remember **TESCH** the changes in TES with time. It is as if souls change. For as TRU changes, TRUCH, so does TES change. This is a decrease with time from the present. This is called **TESCHAO**. We should also remember that the form, force and magnitude of past and future souls can be different and somewhat varied, but nevertheless, as all TIT and TRU, to be appreciated, even worshiped. This is **TESAM**. With relatively - and growing - human population there has been an increasing use and alteration of the earth's resources. Much of this has been positive for TRU without significant overall loss to the earth, through reversibility and replenishment. There is an accumulative permanent loss to certain segments of the earth, of which some is indirect and complex. To many benefits often produce, as in domesticated animals, a reliance on an artificial environment or arbitrary life.

LADCAP = Law of Differential Capacity, Ancestors and Progeny. **LABUB** = Law of Ancestry Best for Us to Be. For an individual LADCAP reveals the likelihood of greater capacity or fitness in ones ancestors than ones progeny. LADCAP along with LABUB are key concepts and laws that help us learn and understand TRU and the benefits we have received. It is a kind of grace of TRU working from the past through the present into the future. In return it warrants duty to future, TRU. They give us the appreciation and conviction needed to rise to TES. Although there have been 65± billion humans on earth since their origins, these have, like those at present, not been of equal TES. Only a small fraction of the total 65 billion has significant existence today. As much as we may look back at our ancestors and the past as dead or irrelevant, we can be certain that they were the best there were.(LABUB) For if not we would not be here. We must recognize and appreciate that ancestors with existing TES often had, are, and will be great contributors to the PPF of TIT-TO. However there is no such assurance that our progeny will be, as well, or as great of TES. For many of us will not succeed in natural selection or contribute in the many ways either genetically or otherwise for TRU. Likely with all the deleterious effects on our germplasm through all times a great amount of information and accounting will be necessary to understand how much or how little each will contribute to the core population of the 6+ billion. How well or poorly we are TRU, and best quality of GM, LP, and DF are upheld. This is called: **LAPDOT** = Law of (future) Progeny Dependent upon TRU (and TES), generally and specifically. LAPDOT signifies that the effect of LABUB and LADCAP in the future is, as it has been in the past, TRU and TES dependent. What upholds selection is TRU. What defeats selection is ATRU, ANTRU and all that is negative to TRU. If our progenies are to be the best, we must likewise be, or prepare to be, the best. This is by TRIA, how all must converge, by PUREALITY, on the absolute, TÂ. It proves much of TRU, TRIT, ROT, MOT, SPALT, POS, HOM, and POT, as well as IPTRU.

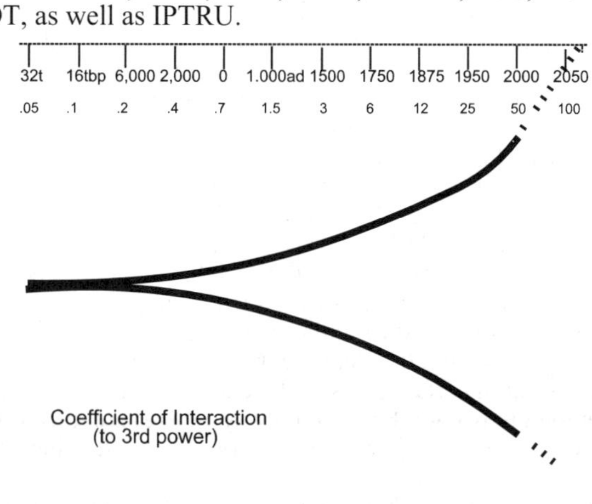

Fig 2-82 **Rapidly Expanding Human Interaction**

There is a coefficient of interaction among humans that is associated with many processes that effect future prediction. We can detect some of the factors involved, density, communication, transportation, and , or:

$$[INAC]^c = D \times C \times T$$

To formulate the functions involved makes expression easier. The accompanying figure gives time horizontally and the coefficient of interaction vertically. In history it has increased exponentially and in the last few decades even spectacularly so. This has been a product of the above factors, and a filling up process that accompanies density. Since there are no longer any lost, dark continents or regions of the earth to explore and settle the world confines are being rapidly constricted. We can now turn our exploration to all else that is lost or unknown of TRU. With high levels of interaction and integration there is an associated major shift to outbreeding and hybridization whose long term consequences are yet to be determined. The exponentially greater human contact and increasing closure is probably the greatest revolution we are experiencing today. Its repercussions are great with some severe. This is not all positive. The actions involved produce many reactions, unless compensated. Decreased health and quality are likely.

In the enumeration below, the number of ancestors for various generations is shown. This is a basic certainty - barring some variation in the length of a generation - and is of great informative and active value, evidence and proof of many features and functions of TRU. This especially applies to durations of time in the region of 33 to 1,000+ years. e.g. the percentage of a total population and the number of years necessary for complete fusion. Yet the world population is equally tending toward complete division. Likewise the fusion and separation of populations depends upon proximity or isolation, large or small degrees of mixture and involvement. The principles involved are called: **PODAF** = Principles of Division and Fusion of a worlds total population. The same can be done forward with time in the future, so that ones progeny will fuse with total population of the world in a little over a thousand years. This depends, of course, on future total population, birth rates per generation, and even more, upon the degrees of segregation and integration along the way. That segregation played a large part historically is readily evident in human history. It also was a major determinant of cultures, nationalities, civilizations, tribes, and families. Integration with interaction has also had considerable plus and minus effects. This has profound implications in the relations, inbreeding, outbreeding or hybridization of subspecies, races, subraces and groups. It helps determine the amount of time it takes for populations to become integrated, and how many generations it takes to reach a set number of ancestors or progeny. The rate is given by a doubling with each generation. This may be much higher or lower especially early and in smaller populations.

ANCESTRY	GENERATIONS		PROGENY
Year	Number	Population	Year
2000	0	1	2000
1967	1	2	2033
1933	2	4	2067
1900	3	8	2100
1867	4	16	2133
1833	5	32	2167
1800	6	64	2200
1767	7	128	2233
1733	8	250	2267
1700	9	500	2300
1667	10	1,000	2333
1633	11	2,000	2367
1600	12	4,000	2400
1567	13	8,000	2433
1533	14	16,000	2467
1500	15	32,000	2500
1400	18	250,000	2600
1300	21	2,000,000	2700
1200	24	16,000,000	2800
1100	27	125,000,000	2900
1000	30	1,000,000,000	3000
900	33	8,000,000,000	3100

If we know how many years it takes to have a population of such a size we can tell how long it was that all had the same ancestors. e.g. 1 thousand in 330 years, 1 million in 660 years, and 1 billion in 1000 years. In this way we can tell how the cultural history of regions and nations continued to have largely diverse populations for 400 years, but were highly integrated in 600-800 years. Since inbreeding, increasing duplication, and various rates of growth occur with many other processes and factors, marked refinements in the above figures are needed for each case. Some of the principles, conclusions and implications of PODAF are: 1. They have a through and through effect in TRU, 2. Population Genetics, 3. How pertinent evidence can be so valuable, 4. How great genetics is in much of the form and function of TRU, 5. **RETIP** = Resolution of Time Paradox. 6. **DEPT** = Different kinds of effects, ES and CES for different periods of time. The future may or may not mirror the past.(DAOST) DEPT gives a much better insight into on HOM, CES, DYNU, DYNS and DYNO. LABUB reveals how important it is to understand and appreciate TES and TRU, and have sufficient PPMA to assure the continuity in capacity for ENT. From this it can be seen that to have a rapidly expanding and very capable population there must be a sufficient high birth rate of highest genetic quality. There is a great need to discover, learn, know, and apply all needed to plan, aid and support TRU quality and to control what is of lesser or degenerative quality. This is close to the heart and soul of each person and couples. It runs contrary to many beliefs so that reaction and conflict occur. Yet this is TRU, dialectically different within and a vital struggle without. Through history this has been a matter of selection through competition and predominance. However since this is not highly efficient, painless, or universally acceptable it must be continuously improved. This is through KOT, TRU, ENT, and ACC. With DEPT there is a close correlation between generations or time, and type of population or length. The fewer the generations, the more it has significance to the individual and families. As the generations increase, it has more communal, regional and national significance. As it approaches total world population, it has total TRU significance. How cladistics is opposite in approach to evolution, so are interpretations of ancestors different than that for progeny. This is due to the fact that in ancestors, not only are we dealing with a doubling in each generation, there is an absolute viability of each ancestor. With progeny we are dealing with anywhere from a fraction to many of which there is no assurance of viability. Thereby we must be careful not to misinterpret, by identifying and otherwise, the role of generations in past and future. They do show us, however, the relatively brief periods of time needed for evolution and the processes of man, to do their work. They show the time period certain processes occur in, and how they need to be modified and adapted to their role in TIT. It is of special significance in revealing the processes that have been at work during recorded history, and are likely to be highly important in the next hundreds to thousands of years.

If our ancestors were the best there were, why are we so much less than the best? This is a fundamental question whose subtle and profound answers must be carefully considered by each and every person. It will set the stage and help one plan to get the best out of life. There are several answers including:

1. Although this holds for the distant past, during more recent generations there is often an accumulative loss of quality in succeeding individuals. (APRA, LADCAP) 2. What may have been the best in a single line of descent in the past is not the same as what may be best under different conditions in the near past, present and future. 3. We are now only one of a great population and thereby our exclusiveness as an individual person is much diluted and achieves exclusiveness and absolution only by identity with its part and role in TIT, TÍS→TES. Even at an optimum a person of the present is not the same as probands of the past where populations were much smaller, even tiny threads in continuity .(LABPER)

4. Since the mass energy Á FAF of humans has changed through time, and since what is of TIT-TO present past is somewhat different than what was operative at times in the distant past we must adjust our measure of best accordingly. This is even more important for the future, giving it a degree of freedom necessary for optimal adaptation. 5. Since being the best consists of many abilities, we must consider all of these in their correct totality and massive interactive forces when comparing, tracing through the processes, and interpreting their roles and their implications in evaluating present day individuals, groups, populations, and TRU. (LOPAB, LABO, reason for IPTRU) 6. All that has gone to make us, and enables us to best live, is the basis for TRU, past present and future. Since each of us exists only in this way we must decrease all

that prevents this, and increase all that enhances it.(POMU, LABPER, LABS) This loss of direct dependence upon individual best, is one of the great causes of our day. For it produces in us a complacency about the dire circumstances under which TIT-TO depends upon us. . (LABS, LACBEST) Because ones distant ancestors were the best does not mean all succeeding were, even cladistically. There is a decreasing scale of fitness for each person with time since ones ancestors were the best. Some of this is difficult to determine unless looking back from the distant future. Yet we have general and specific knowledge of what is best, TES and TRU. These are largely sexual and genetic in long term reproduction. Thereby it is of the greatest importance that we learn, know, remember, apply, practice and focus on all levels of human capacity in all terms of time to assure all ES and CES that assure the best capacity for each person. From the far term to mid term and near term future, through ones grandparents, parents to oneself, and then through ones children, grandchildren, near, mid and far term future each and all of us must apply all genetic, biological, scientific, general, TREQ, and TRU knowledge to assure that we make the best of life, to assure the best for this special universe of human life. It need not be complicated or stressful. It does need to be serious, dedicated, highly reasoned, and by TRU.

ACEF = Ancestry Certainty Factor. It is the dependence of certainty of quality and predominance of ancestors on the number of generations. ACEF is how the certainty or quality of ancestors depends upon how far back, or how many prior generations there were. The farther back we go, the more certain ancestors were of top quality for their times. They must have been increasingly better 5-10-30 generations or 150-1000 years ago, especially so in highly selective populations. Generally this was an increasingly fit progression with time past by LADCAP and LABUB. Our parents and grandparents may not have been so good, as we may be even less fit. It is in this way, in part that we are, like original sin, more Dead than Alive, and by **DAK**, Dead and don't Know it. But the farther back in ancestry the greater they with certainty were to have been a participant in TRUS. These are highly determinative of the status, mental, physical, and all Á, of TRU during time. This is important in comparing human differences, as subspecies, races, subraces, other groups and different aspects of TRU. ACEF makes up much of the processes, function and form of TRU. It helps to show what has enabled us to be and will enable us to be. There have always been lags between causes and effects, as occur by ACEF, or general stimulus and response as manifested in and by genetic adaptation. This lag is signified by: **GAL** = Genetic Adaptation Lag. There is always a lag between changing environment, physical need and genetic adaptation. Just as there was a large lag between high temperatures and deglaciation. This lag is variable per different combinations of factors in GM-LP-DF. Likewise this adaptation is not only genetic, or even physiological, it is complex, and largely indirect, as DYNS and DYNO show in proving TRU. Adaptations of 1 billion or 100 million years ago may, in ways be very exclusive, yet in other ways still proceed into the far distant future. We can enumerate thousands of factors, physical, genetic, climatic, environmental, mental, purposeful, and even focal of TRU that are in process, in evolution in which GAL is operative and ACEF causal. It is in understanding and properly acting upon and controlling these causes and forces as they are a part of TRU, CES, and HOM that we can with a good knowledge of TRU and an effective effort, present, apply, and practice TRU. We may, where necessary, not only control GAL and continue ACEF through us, we can assure and enhance TRU. In applying ACEF to the past, present and future generations doesn't mean all, nor all of one person, were the best. It is a continuum from best to worst and all in between of all levels of SÁ. This why applied genetics, and its engineering is so difficult and controversial. However it goes a long way to explain the past, the role of focus in the present and future basis for targeting ACC.

The trouble with excessive competition, as excesses of all kinds, is that they try to change the world to suit themselves, by reducing TIT-TO to ES and ATRU. They have no way of knowing, and even less, care what the reactions, interactions and consequences are in TIT. They are not or poorly TRU. The competition amongst nations in the world, and their systems should be channeled or guided in the best paths. Socialism extols competition externally, but not so much internally. Capitalism extols competition internally or in part but poorly externally. Partially as it is a part of the economical means of free enterprise. It is like the natural selection that occurs in species not from external forces, but the selection of the best from within its species. This is called: **LINS** = Law of Internal Natural Selection. LINS is evident in TRU by what is of lesser TES, ES, or CES is superseded, selected, or displaced by what is greater TES, ES, or CES.

The American Revolution was, and is in ways, one of the greatest social revolutions of all time. It was not only a search for independence and improved political and social conditions it was primarily for fundamental changes in human life on earth in which there is an equitable relationship between peoples, and with each person and their government. It was the fruition and fulfillment of the ideas of John Locke, Rousseau and others by way of Christianity, humanism, renaissance, reformation, enlightenment, representative government, rationalism, social contract, checks and balances, and many other ideas that helped those who framed the Constitution. Both its constitutional strengths, TES and TRU is what will continue to be needed, as well as the elimination of all that is of its weaknesses. So will it be needed in a United States of the World, or of TRU. If such can be achieved, it might be called: **UST** = United States of TRU. Much has already been learned and achieved in improving governments and cultures through history and especially, the republics. In their heydays many nations were inspired by great ideals of perfect government that were both highly representative and supported. The ideals of the League of Nations, and the FAF of United Nations stand as examples of what can and can not be done, to fulfill the kind of justice TREQ or TRU mandate. They will need to be refined and perfected in the continued growth of TREQ, ACC, TIT, and TRIF. This will include a constitution and government that best utilizes TRU and can be best amended to meet changes and needs.

In addition to the great fundamental processes, principles, mechanisms, laws, and that of **Natural Selection,** there are two or more higher generations of laws that reveal the massive intricate dynamics of human life. These are: **Integrative Selection** or the combined to total interactions of genotype, phenotype, and environment or genetic material, live product and determining forces, whose effects are accumulative and massive in a positive and negative way. Although natural selection works singularly in a one to one individual characteristic to gene, this exists in a massive milieu of interactions whose products in the genotype, phenotype and intermediaries extend throughout the species or TRU. Another is **Conditioned Selection** or the collective and proactive forces built into the phenotype and genotype that project the combined selective forms in a positive and beneficial way. When these aren't in perfect balance, increasingly negative effects are produced. These fundamental laws of human dynamics or TRUD, may be compared to ACEXIN wherein each generation of selection when brought together, like each higher dimension of ALT describes the increasingly complex to total human form and function. To solve the problems of human genetics, populations, and fitness to survive is like solving the problems of ACEXIN and TRU. For we can't understand TRU nor ACEXIN if we do not learn and practice the differential and integral and all in between as they actually exist. So also we can't understand human genetics, populations, and all problems, actions, and management involved in upholding human quality unless we learn and practice the differential and integral, and all in between as they exist in this special universe of best human life, TRU.

There is a great disparity between the liberties granted in America, its crime and prison population. It is true that equal liberties can not be granted to all, they must be largely TES and TRU dependent. It is the abuse of liberties and the associated lack of interest and reason by TREQ that causes, in reaction, such a plethora of laws, rules and regulations, that in turn cause reaction and crime. Laws can't make up for all that is different, wrong, or problematic, in or out of TRU or of some mixture. Laws as guides by the concepts and principles of TRU make much of common law obsolete. Laws for TRU are limited to a set territory and role in TRU.

There is a direct relationship between earnings and productivity. This is especially so if they are of and for TRU. How good is productivity, and how well are earnings used? When one expects to have greater earnings with less productivity, e.g., in boom times, sooner or later failure ensues. Likewise if one has greater productivity with fewer and fewer earnings something is bound to give, either the produce is decimated, or rebellion occurs, often both. It is possible to have ever greater productivity with an even greater earning, as in modern conveniences, efficiency, and means.

The spirit of competitiveness and individual gain, are great stimuli to success. Yet without the discipline that comes from TRU, from Coming to Terms With Life these liberties can be chaotic. It is only by learning, discovering, developing, and disciplining ourselves to find the joy and happiness in being, and fulfilling our counterparts in, TRU, TES, that we overcome these deficiencies and defects. It is in a willing dedication to hard work and pursuit of life, in TRU, that is so necessary and vital to human survival. Without the knowledge and discipline that comes from TRU liberties and ideal assurances of any kind of government tend to be false, chaotic, and negative. The defect in all systems is not necessarily only in the systems themselves. It is also in the continually uninspired, misguided, and inept populations that support them. How much do the individuals in the populations become mature, do their part, and function as their counterparts properly and unselfishly? No political system can give, for long, to the individual what they must do and gain for themselves. Nor can the individuals what governments can't or don't do right. This is the discovery, learning, development, and production of a positive rather than negative level of service and contribution to the whole, TIT-TO.

How much mankind produces and how much he consumes might be characterized in different ways. The percentage of that produced that is actually consumed or quantitative waste. Much that is consumed that is not actually needed, or qualitative waste, and the percentage greater than TRU needs or the special qualitative waste. Total waste is thereby:

W =Wp x Wc x Wat.

Also, how much are those who hunger or deprived insufficient to meet their minimal needs in life to be TRU. There is a Sufficient Consumption. This will vary depending upon the level of TRU, its scientific and technical SA levels. Yet those nations that overproduce and over consume only economic/TRU have a false prosperity.The equilibrium of TRU, **TRUQ**, is related to the balance of nature. In this way it becomes a part, process or basic mechanism of TRU. Not only is the greatly increased population of humans creating great problems for the balance and equilibrium of nature, and TRU in turn, it is doing so exponentially through excessive consumption, excess production, and the use of required and vital resources.This is associated with many untoward effects. A healthy world economy is not one in which there is more and more production, more and more consumption, it is one in which a minimum of production and consumption can exist, and still give each and all persons maximal benefit and value to their lives as these are TRU. Thereby we should turn the world from a consumer oriented economy to a consumer qualified one. In the better sectors of advanced modern culture this, as TREQ, is realized. It should be a TRU oriented and qualified public in which all Á, FAF, ES, CES, REM, ALT and HOM,

as shown by DYNO, are in balance. What is expensive and what is cheap largely depends on supply and demand, which is natural, and tends to be TRU. Yet both supply and especially demand can be far from TRU. This is where the bad are esteemed to be of high value and the good is continuously debased. Such are the seeds that lead to loss and the ruin of societies and the world. It is important for all mankind, and each person, to discover, learn, remember and live by what is the least expensive and most beneficial for their TES and TRU. It is reminiscent of the adage, the best in life is free. This, although well made, stretches the point, for it must be qualified by the fact that nothing in life is entirely free. It most approaches the free when TRU is least made to pay.

In life you should, in ways, be a people mover. Yet in moving people you have a great duty which means it should be done with great care, probity, positively, pleasantly, and in different ways and degrees depending upon the situation, or work to be done. Thereby our purpose should be to move people only to the degree they can't move themselves, or is practical, in order, appropriate, and in accord with TRU.

A good general, quarterback, or person, is not so much the one who can win a battle, game, or match. It is the one who can fight a successful rearguard action, a defensive season, or productive life. It is one whose life long endeavor deliberately and determinedly manage all that is before them to be TES and TRU. It is all those who, with a deep sense of identity, love of life, joy in the blessings of the world and a wide knowledge of all that is necessary for success, pursue TRU with avid conviction and resolution. Their divine charge can be no less than TES, their total responsibility which is TRU, or that part of TRU, that is the object of its subject, or subject of its object, as these all are most real in TRU.

MEASURES of CES POSITS of DYNO OLP

ACEF	8-4-1	T	3	
ADTRU	~-~-~-~	M	7	
CHINA INDEX of POPULATION	9-1-2	T	4	CIP
CLADISTICS, HUMAN EVOLUTION	8-2-1	S	4	CHE
CLIMATE CYCLES, ECO CHANGE	9-0-1	S	~8	CCE
COEFFICIENT of INTERACTION	~5-3-2~	T	7	CeT
COMING TO TERMS WITH LIFE	6-4-~	S	3	CTL
COMPETITION	7-3-3~	S	8	COM
DESCENT LINE for HUMANS	8-2-1	S	4	DLH
ECONOMICS	8-2-3~	S	~5,8	
GAL	8-6-1~	M	1	
GENERATIONS, ANCESTRY, PROGENY	8-2-2~	T	+- 4	GAP
HUMAN INTERACTION	~5-3-4~	S	8	HuI
HUMAN RACES COURSES TIME	8-1-1	S	4	HRC
INAC	~-~-~-~	M	6,8	
ISOLATION and TRANSPORTATION	8-3-~	T	4	Iat
LABUB	7-4-1	S	3	
LADCAP	6-4-1	S	1	
LAMID	~-~-~-~	M	7	
LAPDOT	5-4-7~	M	1	
NEEDS	~6-4-3~	T	~5,8	NEE
OLP	7-2-2			
PODAF	8-1-1~	S	4	
POWER, POLITICS, AGGRESSION,	7-2-3	M	- 4	PPA
PRODUCTIVITY, EARNINGS	8-4-2	S	4	PrE
RELAMID	~-~-~-~	M	7	
RESOURCES, LOSS, REPLENISH	9-~2-7~	M	- ~8	RLR
SUFFICIENT CONSUMPTION	8-2-2~	M	7	SfC
SUPERIMPOSED CLIMATES	9-0-0~	S	+- 2	SuC
TESAM	5-5-~	M	1	
TESCH	4-5-4~	M	1	
TESCHAO	4-5-4~	M	1	
TRUQ	~-~-~-~	T	7	
UST	7-1-7~	T	3	
WORLD POPULATION TIME	9-1-2~	T	4	WPT

Fig 5-82 **DYNO CES in OLP SECTOR of SPALT**

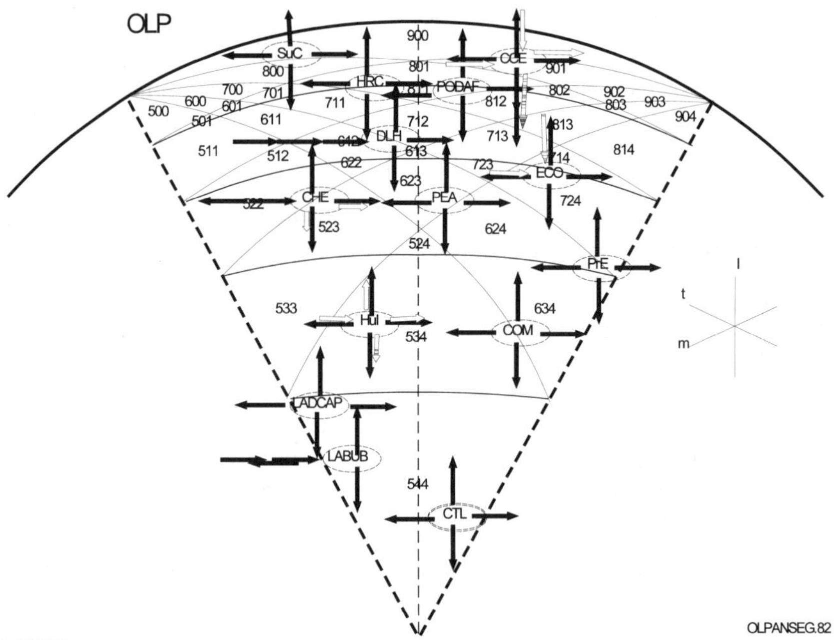

Fig 4-82 **DYNO CES of OLP inSPALT**

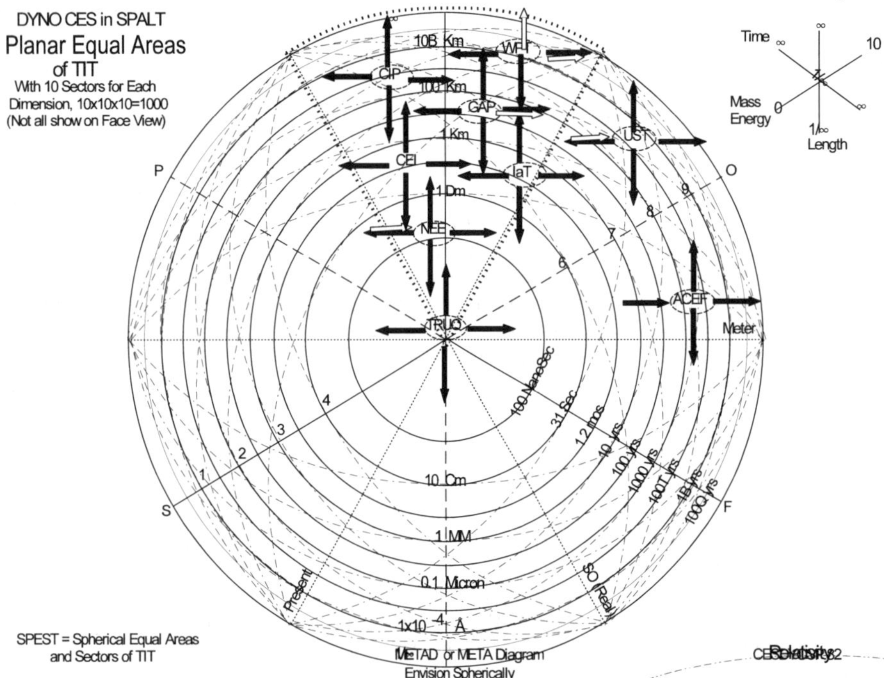

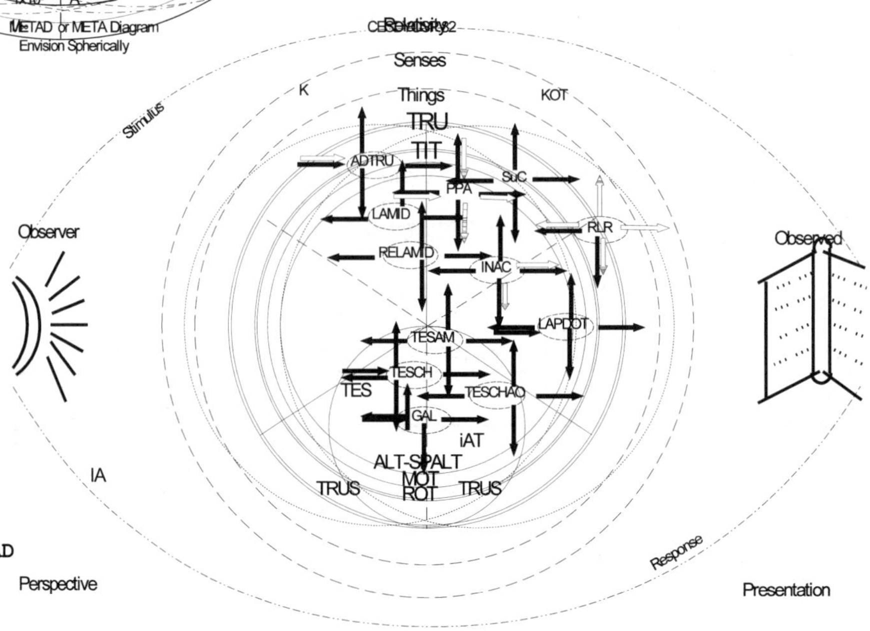

Fig 5-82 **DYNO CES of OLP in METAD**

CHAPTER 83

OSP or OBJECTIVE SMALL PAST of DYNO

GENETICS, SIN, HUMAN LIFE, WORK, SUCCESS, INSTINCTS, ENDURANCE, QUALITY and ACTION

LOPRIC = Line of Phylogenetic or Relative Continuity. LOPRIC is a line of specialization that is largely fixed by GELD interaction in the present. It is relative in continuity. It is highly variable especially by the observer. If the observer is at a distance in time the stability and permanence of the species, world, or TRU seems very labile. However if the observer is within TRU, and of the great mass of humanity, civilization and mankind it appears to be an almost unchanging absolute. Yet we, to develop a proper subjectivity of the SCARLS of TRU, must do so from a set, standardized, and mean observation. This will show how TRU is neither highly labile, ethereal, nor an unchanging absolute. It is rather, somewhat in between. It will depend, to a great extent on many things, not the least of which is the exercise of our FAF, and especially our conviction and determination to will it, and us, to be. The thread of life is very fine, fragile, and of varying degrees of quality and competence. This is no where greater than the many deleterious effects on genetic material. Many of these are unknown, especially in their complex, indirect and interactive forces, and even more the ways work in TRU. Much that alters our genetics is only temporary and reversible. Only by completely understanding these effects, the long term and permanent negation and decay of our genetic material can we fully prevent and control that detrimental to TRU. A high degree of selection is necessary to compensate. for the massive AMTRU world population of today, A certain degree or range of variation in our environment, determining forces, and TRUD is natural, and even to some extent a necessary part of the exercise of its functions. It is mostly what is beyond the range of our powers to compensate, especially permanently and irreversibly, that we must always be most concerned and attentive. This is in contrast to our attention on only what is TRU. Yet by the Laws of TRU, KOT, OITIO, ELTICMA, SÂ, and TÂ we must know and fully use not only TRU, but all that may be - or is - either positive or negative to TRU.

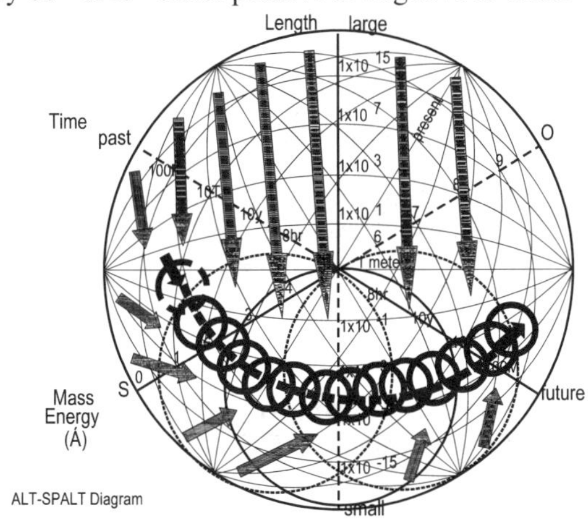

Fig 1-83 **DETERMINING FORCES or CES, LIVE PRODUCT or TES, and GENETIC MATERIAL (GELD)**

The great importance of genetic material and all factors that affect its optimal capacity in human life, TRU, is becoming increasingly evident. They are a large part of human quality. It is the vehicle by which strength and long term fitness are maintained. The genetic material as all intermediaries, from it to the live product, makeup the form, structure, and base of TIT. It is largely through selective germplasm that humans avoid extinction. It is not only what determines human form, but much of function, ability, capacity, energy, and power. This optimal state of genetic material can not be decreased or markedly or irreversibly altered without a negative effect and likely harm to the individual, species, their continuity, and thereby TRU. Genetic material although primarily active in fertilization and meiosis is not inert in time. It is continually under the influence of its physiology and internal milieu exert upon it during long periods of dormancy. These in turn are under vast influences and effects, both positive and negative, that the human body is, or is apt to be, in interaction with.

Of the Mendelian laws, population, genetics and TRU we have: **SUNGS** = Summative Processes of Generations through Natural Selection. SUNGS denotes how propagation through natural selection is not a 1:1 process. This occurs neither individually nor as a whole. It is like the doubling of generations, a rapidly summative condition. So that traits, features, and characteristics are not attained from a 1:1 survival setting but are initiated by increments in mass. These can be seen in the great many characteristics every person displays. It is what is signified by polymorphism, hybridization, and different combinations of alleles. Those that are the most unusual or extreme are often the most noticeable. They are not produced by a single inheritance, but work in complexes and totally for individuals, groups and species through time.

We are all loaded with adversity, original, as well as ongoing, 'sin'. For what we inherit may not be the best, and what we live with as well as do with life are in a world of AMTRU, NAMTRU, and ATRU. Original sin is little more or other than the a priori or already existent negative causes, reasons and processes. It is both all the collective negative FAF within us, as well as within or without TRU that renders the individual or group less than TRU. Sin is the negative that detracts from us, in all sorts of ways and interactions, especially those we produce ourselves. It is only for us to discover, learn, know, direct, control and complete ourselves to optimally avert and reduce this adversity, negation, and sin to TRU. It is only for us to assure the climate, habitat, setting, conditions, means, acts, and all that give the best possible chance and opportunity in life to do right, and thereby eliminate sin. It is the will to change wrong act, or no act to Right Thought and Right Act. This is the great thrust many religions have taught us, the common bond of all great religions, and TREQ. It was especially the basis for Buddhism. It is not so difficult for us to learn the difference between right and wrong, right thought and act opposed to wrong thought and wrong act. The bugaboo is in consistently doing and performing so. No one can get out of life more than they put into it, no more than what this and the person is, in TRU. Yet each and all can expect to get out of life as much as they put into it, as long as this is TRU. For it is through these positive increments that they, as TES, are TRU and highly supportive of TRUF and TÂ. It is how we in turn realize our rewards in life. It is a kind of bilateral and transcendent faith and gratitude for divine providence working in TIT, TES and the individual as all are one. This is called: **(LA)FAGIP** = Law of Faith and Gratitude in Divine Providence, or TRU, TES and oneself, and **(LA)GLAPT** = (Law of) Getting out and Putting into Life no more nor less than TRU and TES. All people are not born with the same blessings, capacity, physical and mental potentials. All people are not given the same benefits, means, education, training, experience, environment and stimulating relationships. Many who are given the best will produce the worst. Many who are given the least will be inspired, compensate and produce the best.

	1b	250m	64m	16m	4m	1m	250t	64t	16t	4t	1t	250y	64y	16y	4y	1yr	3mo	1mo	1wk
Astro-	++	+	+/-	-	-	-	-	-	-	-	-	-	-	-	-	-	-	-	-
Solar	++++	++++	++++	++++	+++	+++	+++	+++	+++	++	++	++	++	++	++	++	++	++	++
Earth	++	++	++	++	++	++	++	++	++	++	++	++	++	++	++	++	++	++	++
Cli-mate	+	+	+	++	++	++	+++	+++	+++	+++	++++	++++	++++	+++	+++	+++	++	++	++
Lati-tude	+	+	+	+	+	+	++	++	++	++	++	++	++	+	+	+	+	+	+
Eleva-tion	+	+	+	+	+	++	++	++	++	++	++	++	++	+	+	+	+	+	+
ppt (rain)	++	+	++	++	++	++	++	++	++	++	++	++	++	++	++	+++	+++	+++	+++
Bio-Life	+	+	+	+	+	+	++	++	++	++	++	++	++	++	++	++	++	++	++
Fresh Water	++	+++	++++	++++	++++	++++	++++	++++	++++	++++	++++	++++	++++	++++	++++	++++	++++	++++	++++
Food Supply	++++	++++	++++	++++	++++	++++	++++	++++	++++	++++	++++	++++	++++	++++	++++	++++	++++	++++	++++
Oxygen	+++	++	++	++	++	++	++	++	++	++	++	++	++	++	++	++	++	++	+++
Langu age			-	-	+-	+-	+	+	++	++	+++	+++	++++	++++	++++	++++	++++	++++	++++
Mat-ing	++	++	++	+++	+++	+++	+++	+++	+++	++++	++++	++++	++++	++++	+++	+++	+++	+++	+++
Tool Making						+	++	+++	++++	++++	++++	++++	++++	++++	++++	+++	+++	+++	+++
Soc-ial	-	-	+-	+-	+	+	++	++	+++	+++	++++	++++	++++	++++	++++	++++	+++	+++	+++
Cloth-ing					-	+	++	+++	++++	+++	+++	+++	+++	+++	+++	+++	+++	+++	+++
Shel-ter		-	-	-	+	+	++	+++	++	+++	+++	++++	++++	++++	++++	++++	++++	++++	++++
Polu-tion									-	-	+	+	++	+++	+++	++++	++++	++++	++++
Infec-tion	+	+	+	+	+	++	++	++	+++	+++	++++	++++	++++	+++	+++	+++	+++	+++	+++
Chem-ical	+	+	+	+	++	++	++	++	+++	+++	+++	+++	++++	++++	++++	++++	+++	+++	+++
Health	+++	+++	+++	+++	+++	+++	++++	++++	++++	++++	++++	++++	++++	++++	++++	++++	++++	++++	++++
Reason				-	_	+/-	+/-	+	+	++	++	+++	+++	++++	++++	++++	++++	++++	++++
Reli-gion								-	+	++	+++	+++	+++	++++	++++	++++	++++	++++	++++

Tab 1-83 Processes of Natural Selection and Time

In this table are shown many of the major forces on, and of, natural selection, that are working at various periods of time. That they work differently at different times becomes obvious. That their effect, both at the time they occurred and their role in TIT is different, also becomes apparent. Yet since their effect at the time they occurred is important for their role in TIT much interpretation of this difference should be accounted for. The complexities of the many levels and combinations of direct and indirect forces, that effect selection, are such that the above ratings should be recognized as rough approximations. Yet they give us a good idea of both the massiveness, supplementation, and continuity involved in all HOM and TIT-TO. In order to better understand natural selection, how it differs with various durations of time, what its role has been in human evolution, what its role is now in CES of TRU and how it is to be interpreted in terms fundamental concepts, principles, mechanisms, and laws of TRU, are all highly important if we are to understand and show TRU. These work together like physiological processes in homeostasis as OSP functions together with OSF, SUSP, SUSF and all sectors in TRUD in TIT-TO. So that a person is to be judged not so much by what they have been given, have, nor even how greatly they have been TRU. A person should also be measured by how TRU has been improved and accelerated through achievement and against overwhelming odds. This is the degree a person rises above, or is able to withstand, live and work, in the midst of the worst possible odds, circumstances, settings, conditions, and states, to better oneself and all to TES, TO, TRU and TÂ.

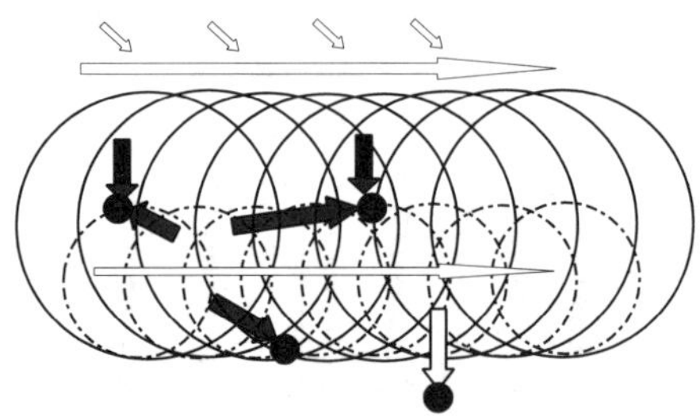

Fig 2-83 **DIRECTION of INDIVIDUAL in TRUCH**

The direction of the individual to TES in TRU is shown by vectors ↑ + ↓ - → ←. Those that increase or are positive and those that decrease or are negative, are shown in each chapter's CES Posits, CHES. When in and positive for TRU, or acting from without to within as in potentiality or beneficially they are shown as solid black arrows. They may be working mostly in TIT, mostly in and about TRU, or mostly in TRUCH and TRUS. The degree to which they are increased, especially in TIT, is greatly accelerated by its ideal, TRUID. This occurs in a, some, or all segments of TIT. In TRUD and TRUID they are accelerated by TRUCH through increased stimulation. This may be on individuals, states, nations or all levels as they must conform or incline toward TRU. The individual stands in a pivotal position not because of ENIT but by their crucial role as TES, the generator of thoughts and acts, which are large parts, and function as carriers of Á and TRU.

The forces of our attitudes and incentives, and how they answer the quest for life and TRU, are seen in: "What do we work for?" "What am I living for?" "What are the main reasons or purposes of life?" There were marked differences in whether people work for survival, material success, or life quality. Eventually and to a great extent people tend to get what they work for. Their ideals are often not what they should be. Life quality follows survival and material success, and these follow TRU. Material success often may worsen life quality more than improving it. It depends upon whether material success and a person is or is not TES and TRU.

Talents, aptitudes, life's work, TES and TRU are seen in the education and schools of all nations through history. There was a process of separation or compartmentalization of the scholastic curricula as well as professional, business, technical and occupational divisions that went hand in hand with what is needed in adult or later life in increasingly complex societies. This is what advancing societies, nations, cultures, and civilization need for self preservation. Those who are: most dedicated to learning or to their particular subjects become the teachers, the best at wining arguments study law, the perfectionists with good memories become doctors, those with greatest aptitude for industry and detail in utility become the technicians, and those who show aptitude for using their hands and bodies various occupations and sports, those who show they are best at buying and selling become traders, and those with the best imagination, discovery and promotion become the entrepreneurs. So that there is some relationship, and rightfully so, between ones aptitudes, natural strengths, and work in life. These are to a large extent magnified by the repeated practice in ones line of work. They render the person special in their role as a counterpart of TRU. The special role in a partial, or TES of TRU is to a great extent one large reason why all of us do not react to the world, TRU, in the same way, nor even in the way that might be best for TRU. We tend not only to be partial as TES, we tend to be separate in terms of this special or trade bias. We also tend to be limited the shortness of life, by our lack of maturity, our IA, distractions, and ATRU. Many persons have lost talents that are never utilized in life. Many are not guided nor find that work for which they are best suited. Many professions and occupations have members whose talents do not fit. Many do not nurture their talents nor utilize what best enables them to contribute to this niche in life to increase TES and TRU.

The measure of TRU is not only TRU or TIT present, but the potential, future and all TIT-TO. What we put into life and what we get out of life, is too often a matter of how we try to get so much more out of life than we put into it, as in politics. Yet we must put into it as much as we readily and happily can. For it is in such acts, work and life that we can find our greatest success and joy, as this becomes TES, ENT and TIT.

The best way of life, ones life in TRU, is largely simple and natural. It is not difficult for any and all to find and follow their natural instincts and habits if one only carefully and persistently uses good sense, learns, trains oneself, and practices what is actually best for themselves and all that is TRU. Maimonides claimed that man may learn medicine from the beasts, who scarcely ever overeat, overexert, or overindulge in harmful or unnecessary activity; among a thousand people only one dies a natural death, the rest succumb to irrational modes of living. This is not quite so true today as in 12th century Persia, yet the principle involved is much the same. He also stated that besides the power to heal, a doctor must possess an independent mind, be a whole man, observant, cautious, and attentive to each and all. This applies to all persons in life, especially TES, those who care for, treat, or serve others. All we need to do is to turn from what is negative, harmful, unnecessary, and over or under active, and do that which is positive, beneficial, necessary and active as it should be in TRU.

It is extremely remarkable how much people complain about and think they can't live without this or that. Yet, when they are in a struggle for life, how little they need or may complain. This shows how much fear and complaining are products of anticipation of injury and loss, rather than the loss itself. Yet when people are really up against it, as in jail, destitute, or completely crushed, they demand little and require but the smallest amount of the bare necessities of life. Often the cure for bad problems, complaints, and conflicts is nothing more than their displacement by a worse one or to suffer from the failure of greater oppression or loss. Although this is not correct by TRU, it answers the question of how negatives can, on unusual occasions, be treated by negatives. The defect lies in a lack of and loss of TRU. Its correction and cure, is not by a worse problem, but by TRU. The continual states of a stimulus, response, experience, awareness, attention, and IA are our great opportunities. It is at hand. It is in how we live and can be TRU. Each person should utilize their capacity and ability to turn ATRU to TRU, non TRU to increasing TRU and all optimally for TRU by TES, ENT, REM, POS, HOM, and all ACC.

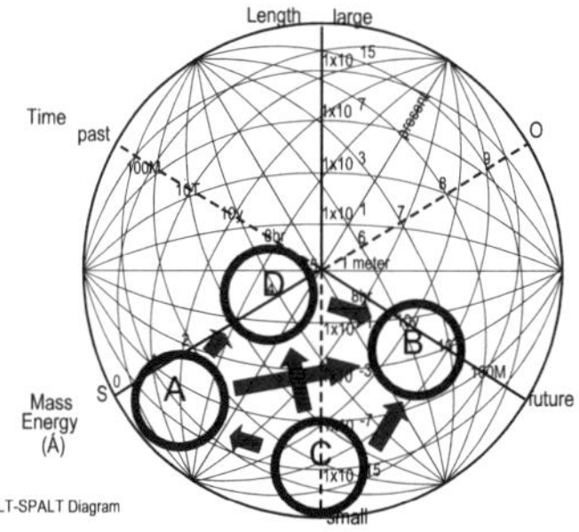

Fig 3-83 **RELATIONSHIPS in HUMAN ACTIVITY**

The relationships between, A, knowledge, B. daily life, C. energy S/F functions, and D. ability to protect oneself are shown in the accompanying diagram. Knowledge is not all, natural subsistence is usually not the greatest part, yet when it rapidly deteriorates such a force makes man more vulnerable and dependent. Energy success over failure helps, and is either to large or to small for daily life. Ability to protect oneself, is related to the above, and continuously a part of life in many different forms.

The factors that characterize quality: 1. Literacy, 2. Understanding the major topics of knowledge, 3. Degree of vocational, occupational, specialty or professional training, and performance, 4. Life span, 5. Contribution to knowledge, 6. Productivity, 7. Happiness, freedom from unnecessary or detrimental stress or misery, 8. Economy, 9. Quality of services, 10. Facility of services, 11. Precision of work, 12. Artistic value of products, 13. Inventions, discoveries, 14. Advances in technology, 15. Highest interests, and, 15. Representation, discussion and communication. These and many more qualities provide indices, that can be, more or less measured, to give a comparison and rough idea of the quality of an individual, group, community, society, nation or world.

There is a solid bond between knowledge and action in TRU. This is especially realized in the mass energy, Á FAF of TIT. Those who advise, command, order, and act often do so merely out of habit, routine, or the pretense of following the rules. This is often without knowing, reasoning and using the actual concepts, principles, laws and mechanisms involved. The development and determination of these basic aspects, forms, functions, parts, and processes of The World, TRU, are unknown or only known piecemeal and without much VAPO. So must all have KUPS, knowledge of all that makes up the fundamental principles and laws of TRU. The knowledge of universals must be based upon the whole world, as it is FAF in totally, TO of TIT and TRU. This is shown through TRIT, OBT, SUT, and MOT enabling the observer to properly perceive, order and use the observed. It is this strong and repeated bond between the knowledge and action of persons in

TRU, PPF, that proves Á, TES and TRU.

LASACA = Laws of Action and Accentuation, Sufficient and Concomitant. The European role in the development of modern civilization relates to the quality of strength of TRU. It was based upon the traditions of ancient Greek and other cultures where learning was primary, with an intense interest and investigation into the makeup of the world, the reasons for things in philosophy, science and technology. How to produce the highest quality of useful products. LASACA is the degree and extent to which our actions and life are concomitant with what exists and sufficient to propel it forward to assure continued success. This depends upon learning, knowledge, interest, service and quality of action of all TRU. A person may pick the time and place for ones actions, behavior, and work. A person may think what they can choose or attend to whatever they want. Yet, in most of life, the amount of freedom for each person will be roughly the same. For this is the degree of freedom TES has in TRU. All else, by the individual, makes little difference, for it is naught, if not TRU. Naturally, it will be different for different individuals or at different times and places in their lives, as they are TES. For different sexes, races, nationalities, regions on earth there will be differences, both as partials in TRU or observers. They will be the same by whatever their norm is as their ES, CES, and CHES are TES in TRU. Most of a persons life will describe a circle or arc that is much the same for all. This may be likened to continuity, a predestination which has a certain validity by relativity in TRU. For we are destined to be born, to physically die, breath, eat, eliminate, procreate, think, act, make a living, and to be TRU. It is the degree to which we deny and defy this destiny that we eschew life. **LAVAB** = Law of Variability of the Abnormal. It is this life of optimal balance of all form and function of the person as they are their counterparts in TRU that is so necessary. Each and all persons can, should and must seek TES both for themselves and for TRU. These are, by DAOST, different aspects of the same thing, TRU. Yet this does not change the part for TES. This is called: **LUPTES** = Law of Uniformity of Par for TES. It is within the realm of LUPTES that free will and the ultimate SCARLS of freedom for each person lie. For them to seek it elsewhere is naught and often increasingly negative. In the best of lives, then, it is not freedom to think and do whatever we chose or want, it is the freedom to think and do what is best for TRU and for ourselves as, and only as, we are TES. It is the occasion to seize the moment, chose and want to become TRU. The joy of achievement by work, of fulfillment that is everlasting is far greater than the joy of attainment that is fleeting or ill gotten without effort or without being a part of the glorious joy of life, in TRU. For our makeup is already set, we have only to fulfill it in life. We can not waste it or deny it for this is the reciprocal of what life actually is as TRU. This is, **LAFOT** = Life in the Achievement and Fulfillment of TRU. LAFOT, as TES, must be lived if we are to be.

LITA = Law of Imperative TRU Action. Ontology, epistemology, and ethics, like science, philosophy and religion are not only academic topics, curiosities, or fields of special interest only to be ignored or denied by any or all of us. All segments are vital to TRU to the degree they are proportional parts and processes of TRU. They are, what have for the most part, enabled us to understand TREQ and TRU. They warrant an equal understanding, appreciation and application by all.(KUPS). They are an integral part of our world and our world, as TRU, is an integral part of each and all of us. It is imperative, LITA, that we and TRU function in tandem and appropriately. Our actions, all actions must be so directed to produce the optimal benefit for TRU and ourselves as we are TES.. All our actions must be based on and follow LITA to be most efficient, effective and productive.

COPTA = Concordance of Perfected Thought and Act. Even with a great amount of TRU, or a great knowledge of TRU and all things, we don't get a good picture, vision, panorama, or overview necessary for a mastery of TRU. Only with the most perfect, coordinated, and ordered model of TRU, SPALT, accessory representations, and posits, can we master all needed for TRU. For it all returns to you and I. How well are we going to learn and remember it, and what are we going to do about it? How are we going to live? How fully can this doing, action, and life be perfectly in accord with what TRU is and needs to be? It is in this way, by the absolute necessity, LATAN, LATANO, LATANOS of ENT and TRU, that we resolve this problem, the problem of life. This is done by the best perfected concordance of thought and act, subjectivity and objectivity, observer and observed, our relativity and our mass energy, Á, FAF as it is in TRU. This is what provides and guides COPTA as we and they are TRU.

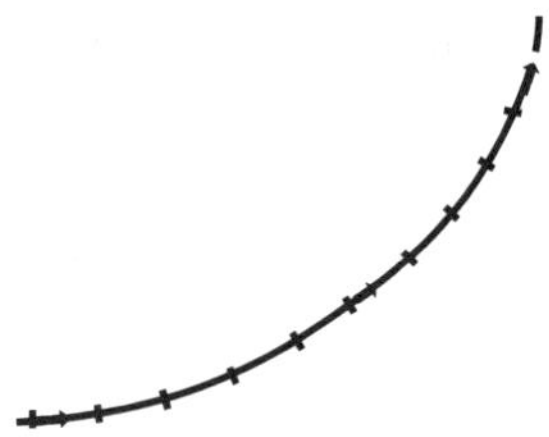

Fig 4-83
ACCELERATIVE PROBLEM SOLVING ABILITY

This figure shows such an exponential rise in problem solving ability in and by humanity, as TRU. It is in TRU, being TES, fulfilling TRU and with ACC, TRAX, TREX, and TRIX that we can learn and show what we must be aware of, should and should not do, or may do that is wrong, not TRU. Only if we have a thorough knowledge of TRU, TIT, SUT, ALT, HOM, TORP, COAT, SPALT, METAD, COGNOS, and ELTICMA can we have before us that frame of reference in which all our problems may be posited. It is a frame in which we may be able, can, must and will successfully decide what is to be done, in what way, and what proportion for and by TRU.

The great strength, force and power of the human brain of the success has been its means and ability, either directly or indirectly, to solve problems. This is from the most simple to the most complex with success and VAPO. So is the comprehension and systematic unity of TRU, that formulates all, the necessary base for total problem solving. It is the ultimate aim and end of the gradual climb of SÁ that has so characterized, glorified and immortalized humans, as TRU. It depends, however, upon the continuing accelerative capacity of all, their universal, partial and special knowledge, KUPS, COGNOS, OM, ELTICMA, SA AND TA. Many ages have had great success in solving the problems they have confronted. In earliest times this was by chance, trial and error, hit and miss, and only gradually, through rational problem solving. Their failures were and are more from desires, choices, selections, distractions, fashions, fads, poor interests, low priorities, and misjudgments, than in either their problem solving ability or best intentions. Thereby it was more and more from what they did not do or did wrong than in what they did that was right that problem solving failed them. It is how and how much they missed TRU.

MEASURES of DYNO CES in OSP SECTOR				
COPTA	5-6-4	S	3	
Direction Individual	6-5-2,7~	TM	4	DoI
Emotions, anticipation, endurance	5-5-3~	TS	4	EAE
FORCES, DF,LP,GM	~6-6-3~	TS	8	FLG
GENETICS	8-8-2~ ±	S	4	GEN
HUMAN LIFE	6-5-2~	TSM	4	HuL
INSTINCTS	5-7-4 ±	T	4	INS
Knowledge- action	5-6-4~	ST	5	KnA
LAFOT	7-5-2~	TSM	7	
LASACA	8-5-3	T	3	
LAVAB	~5~-5~-~	-M	8	
LITA	7-5-3~	T	1	
LOPRIC	8-4-1~	TM	2	
LUPTES	5-5-4~	ST	3	
OSP	7-7-2	S	3	
Problem Solving, Action	5-5-4	S	4	PSA
QUALITY	~-~-~-~	M	6	QUA
SIN	7-5-2~-	MT	8	SIN
SUCCESS	~6~-~-~	M	5	SUC
SUNGS	6-5~-1,8~	TM	4	
Talents, aptitude, life, work	5-6-2~	T	1	TAL
WORK	8-5-3~	T	4	WOR

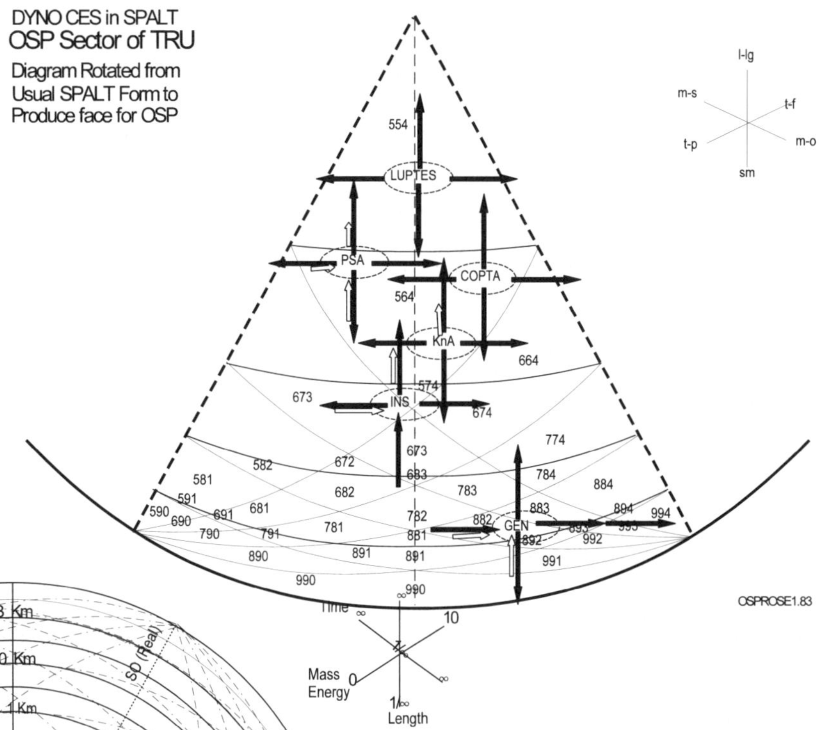

Fig 5-83 **DYNO CES in OSP SEC-TOR of SPALT**

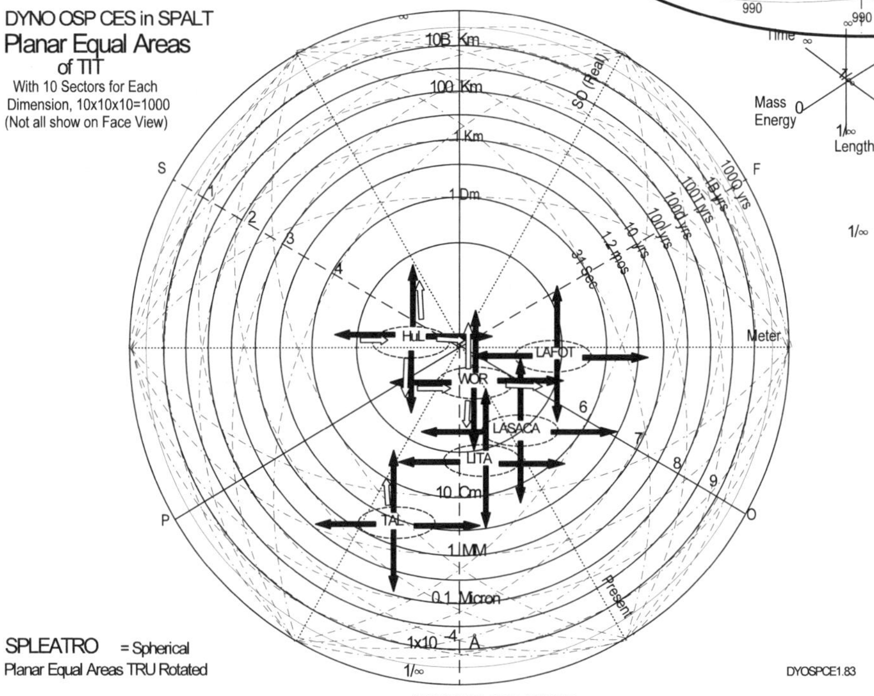

Fig 6-83 **DYNO OSP CES in SPALT**

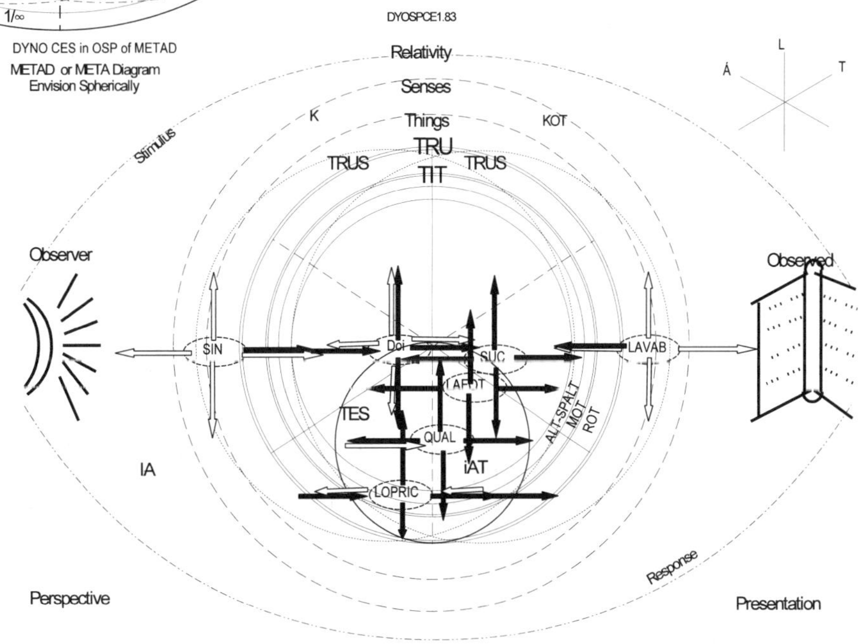

Fig 7-83 **DYNO CES in OSP of METAD**

CHAPTER 84

SUSP or SUBJECTIVE SMALL PAST PROCESSES of DYNO

WORDS, BELIEFS, JOY, IDENTITY, REASON, ATTITUDES, FAITH, EXPERIENCE, THOUGHT, DOUBT, CERTAINTY

Views of life and the world are much different with each age. This is a beautiful examples of how TRU, as shown in ALT, is constituted by more than the immediate and apparent. When seen in the great diversity of different ages, sexes, races, occupations, peoples and cultures it provides a better vision of TRU. Every age plays a part, characteristically capable of fulfilling and applying its role in the FAF or Á mass energy of TRU. The part and role played by all are similar and in ways recapitulate developmental levels of the mass energy Á FAF of TES and TRU. They also will vary considerably with different persons, sex, race, nationality, culture, and language. Yet their similarities are remarkable and demonstrate the sameness - and unity - that exists in mankind and TRU. TRU is not only the rational, nor even the rational applied to the problems of life as they tend to TRU. TRU is all that also indirectly and potentially contributes and enables it to be. It is all the levels of the Á, L, and T continua that are specially constituted for best survival.(ACEXIN)

The English language accounts for hundreds of thousands of terms. When counting all the acronyms, scientific, and technical terms, it is in the millions. When the number of words in all other languages is added, the total is hundreds of millions. Since there is, more or less, much duplication of words even in the same language, the total number of concepts humans have been much less, depending upon shades of meaning and distinctions. There are also many different definitions, interpretations, and combined words or phrases which are singular concepts. The greatest ideas, such as TRU Terminology, act as words depending up their acceptance. What age each person is when they: 1. first recognize them, 2. use them, 3. fully recognize their meanings, 4. remember them, and 5. live by them, depends upon their exposure, training, interest, activities and work. Life is largely one long process of learning, developing, making meaningful, remembering and best applying words. It accumulates so that in later life we, more or less, have a powerful battery of concepts with which to understand the world, ourselves, relations, and problems. The gain or loss at senescence and death of an individual and success or failure to learn by new individuals are some of the greatest positive and negative processes in TRU. Words are largely the carrier or vehicle of the Á dimension, FAF of TRU. It is in the summation of # 1 and 5, with their quantity or vocabularies, that great TES can be attained when properly exercised in TRU.

The greatness of human quest for knowledge has been its search for the basis of ethics or the best use of religion. Suppliants to ancient Gods sought guidance. Kant championed morality as the highest religious activity, whereas Schelling considered it to be the creative energy of phenomenally combined subjective and objective or consciousness and self consciousness evolving through escalating stages. Schiller considered it to be artistic creation, representation, and valuation of aesthetic phenomenal appearance of God or TREQ awakening the individual to higher education, vocation, and generating highest morals. Schleiermacher held highest religious activity to be a mystical awareness or feeling that God, or TRU, have in the identity of thought and being, the goal of all learning and knowledge. In combination these different beliefs, revealed by posits in SPALT show how TRU functions as a whole, and all act in unison to produce ethics or highest religious activity. We should recognize that these two are not quite the same. They will be formulated and posited somewhat differently in the overall picture of the dynamic workings of the whole, DYNO. Ethics represents the perfection of activities of the individual or group, as they become their counterparts in TRU and as TRU is enabled to be. The highest religious activity, is all that enables TRU to be, not only the individual or the group, but all that constitutes TRU. Especially the degree to which the individual discovers, learns, remembers, contemplates, and lives by TES, GOD, TREQ or TRU. It has been the driving force of civilization for many thousands of years as shown by the earliest shrines and temples.

Apostasy signifies those who have risen to godliness, Christ, God, TES and TRU only to lose, forget, deny, become a backslider, desert and/or forsake GOD or TRU. We all at times are, more or less, apostates. The prevention of apostasy from God or TRU is almost as important as rising to, and being TES. This is the great value of a rational, or POA, approach, to religion, God, TREQ, and TRU. It enables us to maintain TES in TRU, not succumb to apostasy, achieve knowledge of TRU and the will and action to see all done accordingly. Apostasy, then, represents the negative phases of vacillations persons undergo in life between the TRU, NAMTRU, ATRU, AU and ANTRU worlds. Most dramatically it is one in which the transit from TRU to ANTRU is in close proximity and effects others. However, they rarely recognize it, only have an uncomfortable feeling, kind of remorse, or bad conscience for leaving the fold, the world of which they are and must be so much a part. It is that foreboding feeling one gets when one has great loss or approaches wrong. It is negative in leaving TES and TRU. It is positive in being brought, pulled, forced, or on ones own desire and will returns to TES and TRU. The tenets we develop and hold, the positions we take relative to the fundamental principles and laws, and all that must ultimately reconcile with TRU, must be accountable to the total and final relates and posits it has with TRU. This is by: **AIT** = the Absolute Imperative of TRU. This is the generic or general case of: **AITES** = the Absolute Imperative of TES. Just as TES is the counterpart of TRU, so must all our bodies, tenets, and all else be transformed and transpired through their ESINS in TRU to become their counterparts or functions, ES or CES in TRU.

ILE = Imperative high degree of Liberal, General, and TRU Learning and Education. Each person should comply with and seek ILE, as early and greatly as possible in their life. This is our greatness and future.(TES) For the unmet problems of the world are its failures. Their solutions are how to turn them into successes, overall success of TRU. This knowledge is the basis for educating all people in TRU. Not only must the knowledge be optimal, by TRU, the vehicle for this is through each and all persons, who by education, training, habits of thinking, reasoning, problem solving, and acting, from ILE must and can provide the means to best achieve and utilize KOT and TRU. Current secular and religious education fails to provide an adequate knowledge of TRU. Current liberal education in our schools and colleges, to a degree recognizes this need, seen in philosophy, the humanities, and general sciences and tends to acclaim much that is TRU. Yet this is done only in bits and pieces, and in a hit and miss fashion, and without order, system or model for guidance. Knowledge and education through the past, and in the present does little justice to its subject matter beyond separate topics, its own subjectivity, and technical applications. Even less does it recognize their many relations and problems, and their resolution as shown by best division, relating, combined positing in ROT, MOT, and SPALT of

TRU. No wonder so many of the youth of the world today, in spite of great exposure to education, and the vast increases in information leave their educational phase of life confused, disillusioned, and unready for being the kind of contributor TRU necessitates. Therefore identify yourself with the immensity and omniscience of TRU. Utilize all its knowledge and models as they can best provide and show how to fulfill life as it must be ever greater in TRU.

COK = Correct and Total Knowledge. As much as possible, you must know all that is needed to be known. What you know must be right, TRU or TREQ. There is a balance between a jack of all trades, professional, super specialist, generalist, scientist and a theologian. How sure are you that what you know is right and increasingly total of TREQ or TRU? Only with TRU, TRAX, TREX, TRIX, PUREALITY, KUPS, COGNOS, ELTICMA, and SÂ can this be sure. This is vividly brought to mind by health care of seriously and critically ill. How vital it is that they know, know correctly, and proficiently. How well they convert this into perfect care in the best manner at the right time. Yet you can not know TRU only by being a professional, generalist, or theologian. It takes intricate detail, evidence, facts, experience, induction, and problem solving. These different kinds of knowledge, when of TIT-TO are the core forces and determinants of TRU. They are what each and all persons must aim to achieve, that body of knowledge necessary to be TES and TRU.

We all have our crosses to bear in life. In fact if we subsist without bearing many crosses, suspect something is wrong. Our counterpart in TRU is not being fulfilled. The trick lies in bearing this cross with a full body of strength, mental, physical, and all in between, plus the joy and happiness that comes with having fully lived ones best in TES and TRU. This is like ACEXIN, the fulfillment of ones capacities and abilities, or complement of Á FAF in TIT. It is in doing not so much only what one likes to do, but in learning to appreciate and finding the joy and happiness in what one must do, in the best possible way, as TRU. **LAWDOT** = Law of Learning to Like What one has to do, and not doing what one likes to do, or what is ATRU. LAWDOT is closely related to ILE, and is seen in making a game or fun out of learning, doing and living TRU. Those who can so adapt themselves, find the joy that is in TRU, and not in all that is less or wrong. When they most positively resolve the riddle of life, the crosses they bear become much easier, a blessing, challenge, and not a curse. It becomes the fulfillment of heaven on earth, TRU.

To be a Christian does not mean that one should waste oneself or sacrifice oneself for what is negative, worthless, inappropriate, ungodly or lacking in TREQ and TRU. It does imply and demand a great amount of selflessness, service and sacrifice. It means that oneself and others must not be taken in vain, or as in and of themselves, but are and are only as they actually a part of God, counterpart of TRU, TES. To be a Christian thereby requires great knowledge, both in quality and quantity of what one is and what God or TRU are and how they work together in absolute perfection. It is much more than being a sheep in any shepherd's flock, following any guidance.To be a Christian thereby simply means that one must live in the Grace of GOD, TREQ, or TRU, as TIT as perfectly and fully as possible. It means that one should be willing to serve and sacrifice oneself that others may live, if such sacrifice in resultant is for TES and TRU. To do whatever needs to be done for TRU in the easiest way to the degree it doesn't have to be done at a specific time, place or manner is the way of TRU and peace of mind. By doing it in the easiest way and to do it in the best way by TRU is the ultimate art of TRU. The art of industry, based upon the best Representation of TRU.

The world becomes more complicated, dense, competitive and technical. To master it takes a great amount of insight, KUPS, KOT and TRU. To do so takes opening up great barriers, taking giant steps, and leaping high hurdles to make a bridge to the actuality of, and have identity with, TRU. This we do by discovering, learning, knowing, and being TES in TIT. This is actual maturity, coming to terms with life, both for oneself and ones kind.

Much of our life is repetition and recall, a going over the same things, time and time again. The older we get the more is common, homely or what we have experienced before. However this should not be so denied and unexpected. TRU, in being relative, is much of the same by the S in DAOST, a repetitious recycling of the same. It shows us the cyclical form and function of much in TRU. It justifies much of habit and routine, as long as they are TRU. The more TIT changes the more it remains the same, and the more it remains the same, the more it changes, in a continually reversible manner.

The great controversies that shook the early Christian Church are striking examples of the fundamental and basic structural and doctrinal differences that can destroy a universe. This holds for the world, the relative universe or TRU, as it did for the first few centuries of Christianity. In the teaching of Pelagius of Britain in the 4th century a surprisingly Western modern outlook arose in denying original sin and the necessity of baptism. These teachings insisted that through justification and faith alone men are to be judged and purged of their sins. The great validity of such teaching was not appreciated by the early church fathers, especially St. Augustine. For it was a denial of the then great emphasis and fervor for confession and salvation through the absolution of ones sins, by baptism. Even more it denied much of the great spirit of Christ's sacrifice and the martyrdom so many had endured for Christ as God. For, they could not believe in Christ's divinity, unless they recognized this message of suffering for the sins of all, and eternal life. It is in a more contemporary sense whether we can by our own initiative be a good Christian, achieve an ethical life, or TES, or whether we must have guidance and help. The ways, means and vehicles are of lesser importance. It is achieving righteousness or TES that is crucial. This means to discover, learn, know, remember and have an ethical life, be TREQ by VAPO TES and TRU. There are in the lives of each and all persons certain essential attitudes that enable them to be good, positive, TES and TRU. If one learns, knows, remembers, uses and lives by what these essential concepts or CALP are and develops the attitudes and habits that go along with them, they are to be recognized, complimented, and praised. If one rejects and fails to develop the attitudes and habits that go along with them, they are to be ignored, criticized, denounced, condemned, or censored appropriate by TRU. These include, GOOD

Deeply concerned with human welfare and future
Deeply concerned with life's worth and quality
Greatly wants to do what is right
Does what is right for right's sake
Are pleasant, cheerful, and respectful of others
Recognizes the significance of differences
Recognizes the significance of sameness
Appreciates the reasons behind and actions needed
Seeks to solve the major problems facing mankind

BAD

Has no, or an adverse concern for human welfare
Cares little for values, worth, and quality of life
Doesn't care to do, or deviates from doing, right
Does only because they have to by custom or law
Are unpleasant, hateful, harmful, disrespectful
Does not recognize the reasons for differences
Does not recognize the reasons for sameness
Are unable or fail to appreciate the reasons and
Actions needed to solve the major problems of TRU.

All must be based, and depend, on the human mind's ability to comprehend TRU.(COAT) It has taken the form of ideas, concepts principles, laws, mechanisms, terms and meanings that represent TRU. Each new person doesn't have to start from scratch, they need to have available KOT and LANT.

Fig 1-84 **BASIC S-R, S-O and S/F RELATIONS**

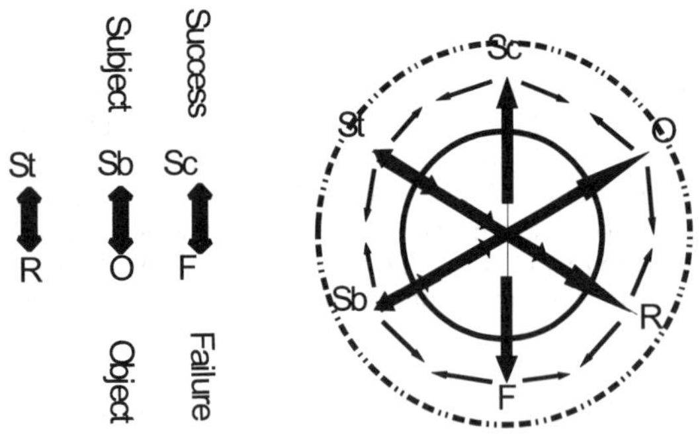

You have to know what you are looking for if you expect to get any kind of success out of whatever you are doing. You must be able to look, listen, smell, taste, feel and sense in all ways that gain recognition. This is the crux of life. In human's life-energy situation, form and function, Á, all are bound up into the predominant state and mechanisms of TRU. These and their many combinations, permutations, and manifestations give us the core human dynamics as TRUD. They are the core mechanisms and key CES of HOM. They show how ALT is dynamically the primary form and frame for TRU. It is in the great sufferings, terrible conflicts, stresses and strains of our lives that we beg, plead, beseech and search for help and deliverance. This has been, is now, and will be, more or less forever, true. It is a product of our failures, limits, ignorance and lack of TRU, the NAMTRU worlds in which we live. It has been the crucible of many great lives, religions, philosophies, systems, ISMS, and endeavors. It is most apparent in the origins of Buddhism and Christianity. Only when recognizing TRU, ones identity with TES, and our absolute as well as seeming insignificant value in TRU, can this deliverance, reconciliation and practice of righteousness have the necessary reason and purpose to complete itself in TRU. The great strength, as well as fact, of God as TREQ in history, TRUS, has been this very reason and purpose to find righteousness and itself. It enabled many, like St. Augustine, as it enables all, to turn away from "that pernicious blitheness by which we depart from Thee". In the Augustinian theory of learning, learning about the world, God and TRU, ratio or reasoning was the minds capacity to distinguish and connect the things that are learned. It was held to precede the exercise of the intellectual capacity, and thereby was inferior to it. Intellection and contemplation are how we use learning and reason to distinguish and connect what is attended to, which of necessity must be God, or TRU. Bacon expanded upon this and utilized the various Augustinian and Scholastic varieties of reason, as rationes seminales, process of becoming, etc. In Bacon the Divine Illumination Theory becomes an active intellect and is nothing less than God or an agent or angel of God. This is a very interesting and inspiring theory, that as so much that is highly valid, especially if properly conditioned, is greatest when held to its TRU meanings. It represents the mechanism of many of the great subjective processes in TRU. Yet, it as so many theories, ideas, concepts, when taken in and of themselves only, are singularly misleading and in use become erroneous or even bizarre through failure to present TIT-TO. Reason and intellect are some of the best means of achieving TES, how the individual mind can discover, develop, learn and utilize knowledge as their potentiality becomes actuality in TRU.

Immortality comes from reaching the limits of and achieving the highest levels of knowledge, thought, action and life in TRU. One who approaches these limits, and reaches these levels is engaged in: **SIATKA** = Solid Immortality that comes from Achievement or Attainment of the limits and highest levels of Thought, Knowledge, Action, and Life. SIATKA signifies the fact that the basis for research's contribution to humanity is through the information, facts, evidence, essences of life, and ESINS in things that exist in TRU. Through their discovery they become a part of knowledge as this knowledge by observer of the observed in relativity fulfills itself and the mass energy, Á, FAF of TRU. Herein immortality is achieved. It is the heavenly world of TRU here on earth, as all through scientific research, investigation, reason and judgement comprehend and make certain the significants, essences and ESINS that make up this world, TRU, TRIT, and TIT.

Whether we endeavor to be TES and have immortality is partly the result of what our chances are, and what our predetermined role in future generations will be. Yet this is not only genetic, by alleles or cross over. It is an across the board function of much that constitutes TRUD and TIT. It is largely of the ratio, Will to TES, TIT, and TÂ/resignation to non or negative participation. It, to a great extent, is regained by corrections during renewal, rejuvenation, learning, etc.

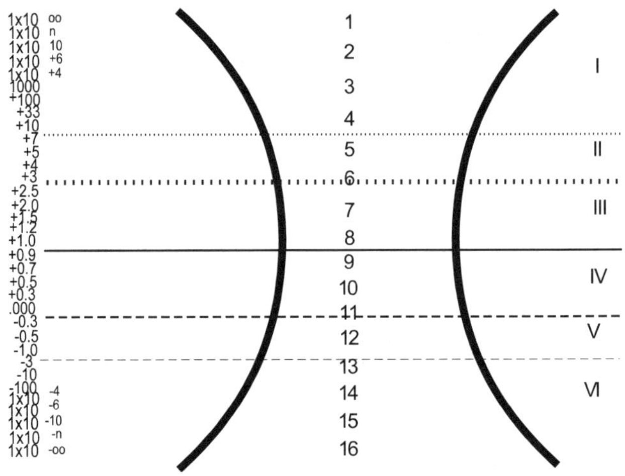

Fig 2-84 **COMPARATIVE LEVELS of MENTAL ACTIVITY and THOUGHT**

In this diagram a comparison is made of many kinds of mental activity and thought along with their major levels. They progress from the lowest, those of complete destructive manias and wanton disregard for life to the most perfect identity with God, TRU, COGNOS and TÂ. This is produced by a complete, infinite and perfect contemplation of all TRU. From the top down we have by levels: I. Exponentially Great Contemplation of TRU, TRIT, TIT, TÂ, II. Many times Greater than Par # I, or TREQ and TRUS. III. Highly Positive, more than par contemplation of TRU, or the best of understanding and thought. IV. Thoughts and Contemplation less than Par, or mixed TRU and Non TRU, V. Negative or erroneous thoughts and Contemplation to TRU, ATRU or moving away from TRU, VI. Highly destructive thoughts of, and of TRU.(ANTRU,ANTITRU) Of these more exemplary levels may be enumerated to explain reasons for and against the intellect needed to be TES.(1-16)

Semper Fidelis, being always faithful, signifies a determined attitude to uphold the major principle of our being, what is right and divine, TRU. It is an identity and bond of duty to all or some aspect of our world that is a guiding ideal and standard. Its meaning is much the same as the kind of faith God has invoked on man. If we are to be, we must do so in that way that most efficiently and effectively yields it. If we are to be faithful, we must have an equivalent reason, justification and understanding, of what such faith is based upon. It reinforces the everlasting and great empowering need for KOT and TRU, and the conviction that upholds them. As in union there is strength, as in one for all and all for one, so does a whole depend upon its parts and the parts upon the whole. Only when all are complete and strong can this faith be complete and have full meaning.

IR = Infinite Reflection or the Infinite Felt Powerful Experience Expressed Perfectly. IR is seen in the Mirror Effect whereby placing two or more mirrors in such a position that an infinite series of objects can be seen. This is the endless depth and infinite reflection the mind can achieve by experiencing the infinite nature of TRU, TI. It may also be expressed more fully and perfectly to include all visions or experiences that not only reveal the infinite of TRU, but the infinite observations and all other approaches to TRU seen in its representation and presentation. IR is what enables us to establish limits, the frame of TRU. Only when these limits can be provided can we produce dimensions whose SUT give a totality in which VAPO can operate and enable the closure of KOT and COAT of TRU. Only in this way can certainty be reliably established and doubt adequately removed.

Fig 3-84 **BEING, IA, INDIVIDUAL, TES, TRU, TÂ**

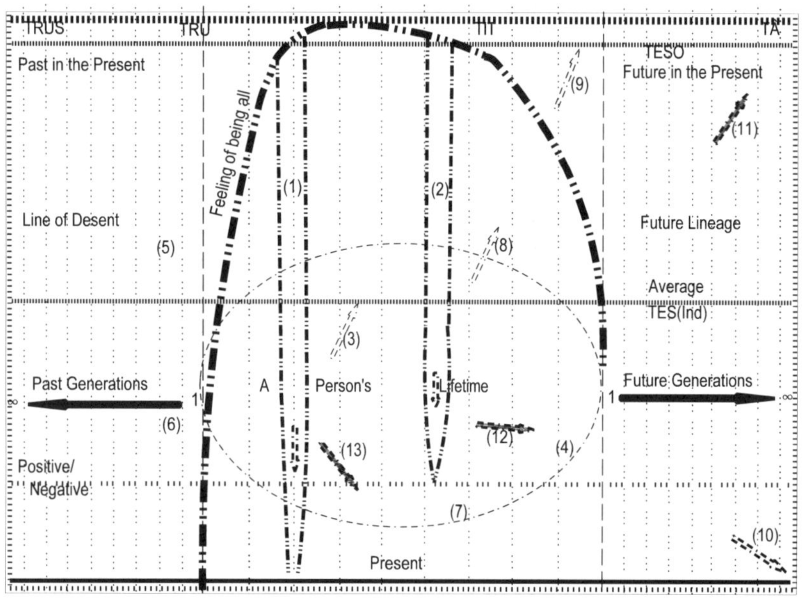

Human beings have been very much the same and different through the ages, as signified by DAOST. Yet this sameness and difference, like our belief in being absolutely alive, are not what it appears to be. A person is or is not as they are their TES, their part in TRU. The degree to which they are not is the degree to which they are dead and don't know it, DAK. It is vital, for both oneself and TES that we know this key cue to our lives, its actual/apparent being. It is contrary to ego, individuality, and the natural feeling one has when so much alive. In order to help all understand their life and role in TRU it is of the highest importance that all learn and remember the many TRU factors and forces that are vaguely sensed in our lives:
(1). In youth, the acute depression and feeling of being overwhelmed in contrast to the absolute certainty of being. How long this lasts and what their reaction to it is depends upon their ability to react to, and profit by (3,8). (2). Periods in maturity, beyond when one questions this absoluteness of being, what it is and whether they are only one in billions, and have no life after death. This is why individuals tend to become more religious with age, a search for the answers to identity, salvation and immortality. (3). Coming to terms with life, the determined and energetic endeavor to better oneself and the world. (4). The ellipse represents a person's lifetime. This is highly variable, both totally and incrementally. It is to be understood by the many laws, principles, concepts presented in IPTRU as well as the experience and lessons learned both by oneself and all before them. (5). A person's lifetime that has been given a very high level of potential and capacity by their ancestors. (6). A person's lifetime that has been given a very poor level of potential and capacity by their ancestors. (7). Person's whose lives have and continue to be negative, both a cause of permanent death to themselves, TES and TRU. (8). Person's whose great and continuing work has produced immense benefit for the world, TRU, with a resulting enhancement of TRU, TESO, and TES(Ind) or ones TES and immortality. (9). Person's whose great benefits for the world has been such that it is far beyond the call of duty, and reaches the absolute through assuring the continuity and eternal permanence of TRU. (10). Person's and all that is increasingly, and often acceleratively negative, INEG, leading to personal, human, and TRU extinction. (11). Person's and all that is INcreasingly, and often acceleratively POsitive - **INPO** - toward personal, human, and TRU eternality, TESÂ, TÂ. (12). Our lives when we follow the crowd, work and live oblivious to our place and role in the world, and what it takes to assure health and well being. (13). A young person's life, who both through poor heredity and upbringing, education, development of a zest for life with its inspiration, and revelation necessary, misses the great value and benefits life offers. They give up or resign themselves to mediocrity and a uncaring spirit for what it takes to succeed rather than fail, and live rather than die, or be TES and TRU. This figure may be readily superimposed and adapted to SPALT. It greatly adds to the meaning of the many references we make in the text to ones understanding of TRU and the rewards that it can produce.

Doubt and questions of certainty preoccupied Descartes, Hume, and many others. They arose primarily from God's existence, form and function. It concerns problems that are resolved by VAPO, TRU, ATO, KOT, TÄ, TROS, DAOST, SCARLS, TRIS, TRIA, and TES. Without establishing these concepts, or their equivalents, as a whole in TIT-TO, TRU and OM, COGNOS and ELTICMA doubt and questions of certainty persist. It is upon the achievement of certainty and the methods of understanding that Descartes and Hume differed. They were both champions of skepticism. Yet to find the best methods for resolving doubt required a breakdown of the problem. This depended upon the limits of the method and what is sought. Descartes' method-worked well when derived from and limited to a simple focus, the observer, as in 'I think therefor I am'. And further when it was extended to a reasoned interpretation of God, or a 1:1 identity with God. Hume's method worked-well when derived from and limited to individual collections of distinct experiences. This is, as long as it focused upon the facts of discrete particulars, measured and scientifically determined cause and effect. Yet both Descartes and Hume are wrong to the extent their methods lacked actuality, proportionality and totality. They vigorously sought the validity needed, yet ignored the other three main ingredients of VAPO necessary to understand TRU. Skepticism based upon incomplete methods, directions, without VAPO and TRU can never be raised to the kind of actual certainty human life requires.

Like bodily organs for ones physical well being there are for the well being of ones life, personal and mental organs or aspects of ones form and function that are both its focus and the focus of the world, TRU. They are as much organs of survival as ones bodily organs, or TRU's social organs.(2-54)

ORGAN/SYSTEM	FUNCTIONS NEEDS	PURPOSES USES	BENEFITS FULFILLMENT
ABILITY CAPABILITY	SUFFICIENT FOR TRU'S NEED	TO BE UP TO LIFE'S DEMAND	BASIC TRU FAF FOR SUCCESS
ACCOUNTABILITY DUTY RELIABILITY COMPLETION	MEASURE UP TO LIFE'S JOB HELD TO TRU'S STANDARD	DEVELOP SUFFICIENT RELIA-BILITY TO COMPLETE NEEDS	BETTER CONTRIBUTION TO THE WHOLE AND BECOMES TES
ADAPTATION	ADJUST TO ALL IN OUT TRU	ABLE BEST ADAPT TO LIFE	BEST MEET DEMANDS OF LIFE
ANTICIPATION PREPARAT.	BE READY FOR ALL NEEDED	KNOW, PLAN AHEAD NEEDED	KEY TO SUSTAINED VICTORY
APPRECIATION GRATITUDE	RECOGNITION OF BEST, TRU	RETURNS FROM TRU RESPECT	HELPS RAISE ALL TO TRU
ATTENTION OBSERVATION DETAIL	EXERCISE TRU POTENTIAL FOR SUCCESS AND SURVIVAL	DEVELOP AND EXERCISE BEST SENSE-ATTENTION TO DETAIL	BASIC FAF OF IA IN TRU KEY TO SURVIVAL HAPPINESS
COURAGE DETERMINATION CONTINUITY ENERGY	SUFFICIENT CHARACTER AND WILL TO FIGHT, OVERCOME	HAVE DETERMINED ENERGY TO ASSURE CONTINUITY IN LIFE	INDISPENSABLE FOR TRUD ESPECIALLY WITH GREAT KOT
DISTINCTION EVALUATION INTERPRETATION EXAMINATION	SEPARATE EXAMINE DISTIN GUISH DIFFERENCES SELECT INTERPRET WHAT IS TRU	TO BE ABLE TO MAKE NEEDED DISTINCTIONS EXAMINATIONS INTERPRETATIONS VALUATION	ASSURES BEST SEPARATION AND SELECTION OF ALL FROM TRU AND ALL IN TRU
FRUGALITY THRIFT PARSIMONY APPLICA-TION EFFICIENCY	HOW TO BEST APPLY ONESELF FRUGALLY, PARSIMONIOUSLY c THRIFT AND EFFICIENCY	BEST APPLY ONESELF WITH FRUGALITY THRIFT PARSIMONY AND EFFICIENCY	WHEN ACTIONS ARE BEST ACCOMPLISHED RATE OF TRU IS GREATLY INCREASED
HAPPINESS JOY FULFILL.	BEST LIFE/TRU FULFILLMENT	GAIN TES FIND HAPPINESS	FROM THAT BEST FOR TRU
INTENTIONALITY	ATTENTION BY TRU INTENT	TRU GUIDE BEST INTENTIONS	SUCCESS BY A DIRECTED TES
KNOWLEDGE LEARNING	LEARNED KNOWLEDGE ALL TRU	LIFE LONG PURSUIT OF KOT	JOY-GAIN IN LEARNING KOT
LANGUAGE COMMUNICATION	EXPANDED MIND IN LANGUAGE	BEST USE BETTER LANGUAGE	PROVIDES KOT, Á AND TRUD
MATURATION EXPERIENCE	SUFFICIENT MIND FOR TRU	GAIN EXPERIENCE MATURING	YIELDS A WELL ROUNDED TES
MEANING IDEATION CONCEPTUALIZATION	TRU THINKING FROM MEANING-FUL IDEAS AND CONCEPTS	KEENER AND GREATER IDEAS CONCEPTS MEANINGS & USE	JOY FOUND DISCOVERING GREATEST MEANINGS IDEAS
ORIENTATION COORDINAT.	TRU'S ORDER COMPREHENDED	BEST COORDINATED ORDERED	SATISFYING BALANCE IN ALL
POSSESSION REWARD HONOR	ACQUIRE THAT NEEDED FOR LIFE EARNED IN TRU	ENOUGH GOODS REWARDS AND HONORS TO MEET,BE TES TRU	HAVING ALL NEEDED, AND THE RECOGNITION AFFORDED
PRECISION QUALITY	NECESSARY FINENESS IN TRU	IMPROVED QUALITY FOR TRU	MAXIMAL REACHES GREAT TRU
PREDOMINANCE STRENGTH HEALTH	FOUNDATION FOR TRU LIFE ITS Á FAF PERPETUATION	ACQUIRE AND ENHANCE BASIC FAF FOR STRONG Á TES TRU	PROVIDES EVER GREATER TRU ITS SURVIVAL & PERMANENCE
PROFESSIONALITY CAREER OCCUPATION CONTRIBU.	TRAINING EDUCATION FOCUS NEEDED FOR LIFE'S ROLE	BEST CONTRIBUTE TO TRU FULFILL CAREER IN TES	BEST EXECUTED BEST ROLE IN TRU AND RETURN TO SELF
PROPORTION VALUE WORTH PRIORITY	MORE THAN SIMPLE KOT, ALL BEST DISTRIBUTED USED	UNDERSTAND TO AND THE FAIR SHARE AND VALUE OF ALL ES	OPTIMAL FAIRNESS AND BEST SELECTIVITY IN LIFE & TRU
READING STUDY INVESTIGATION	MEANS METHOD PREPARE FOR LEARNING AND KNOWLEDGE	BEST INVESTIGATION STUDY READING TO ACHIEVE TRU	JOY OF LEARNING TRU, AND ITS FULFILLING SEARCH
REASON PROBLEM SOLVING	NEED TO FIGURE OUT TRU	GREAT TRU REASONING POWER	GAINS IN RIGHT SOLUTIONS
REPRESENTATION ART COMMUNICATION	TO DEMONSTRATE, SHOW, DISPLAY AND RELATE TRU	TO BETTER REVEAL TRU BY BETTER MEANS OF SHOWING	BEAUTIFUL DEMONSTRATIONS OF HIGHLY NEEDED TRU USES
SUCCESS OPTIMISM CONFIDENCE	SUCCESS BREADS SUCCESS CAPACITY PERPETUATED	TO PRACTICE-LEARN SUCCESS CONFIDENTLY COMES FROM TRU	ETERNAL PERPETUATION OF TRU, COMPLETE IDENTITY
THOUGHT IMAGINATION CONSIDERATION	EXERCISE AND IMPROVE GIV-EN TRU MENTAL CAPACITY	IMAGINE CONSIDER THINK IN BEST WAYS THAT GIVE TRU	BOTH THEIR OWN REWARDS AS WELL AS FAST TRACK TO TRU
TRUTH INTELLECT	WHAT IS TRUTH BY TRU	HAVE INTELLECT, TRU TRUTH	EVER HIGHER TRUTHS IN TRU
UNION COOPERATION TEAMWORK	TO BEST UTILIZE COMBINED EFFORTS OF ONE AND ALL	TO HAVE BEST COOPERATION, UNITY AND TEAMWORK FOR TRU	CAN NOT BE DONE ALONE NEEDS TEAMWORK IF TO TRUD
WORK WRITING PRODUCTIVITY	GREATEST METHODS OF CON-TRIBUTING - PRODUCING TRU	DEVELOP BEST HABITS OF PRODUCTIVITY WRITING WORK	BEST HABITS ACHIEVEMENTS MAKE TRU EASY & PLEASANT

Table 1-84 **PURPOSES, USES and BENEFITS of the PERSONAL SUBJECTIVE FAF AND CES of TRU**

These are but a few of many forms and functions, parts and processes of humans that make up what we are, can do and be. They are the key to both the particularization of this being, as well as show us the mechanisms by which we, as TRU, operate. That TRU is very much personal, as each one is very much what we are composed of, works to our best advantage in a highly competitive world. This is evident from biology,anthropology, history, economics and our daily lives. It is these forms and functions that makeup the parts and processes that provide us with the predominance that assures our thisness, TES and TRU. It is in the typical mass energy and spaciotemporal form and function of this special predominance that most essentially and vitally provides dimensional basis, Á, for discrete specificity from without and identity from within. This dimensional frame is the basis in which various organs and systems, parts and processes, forms and functions develop their best purposes, uses and benefits for the success, continuance and existence of TRU.

MEASURES of DYNO SUSP CES in FIGURES

ABILITY to COMPREHEND	2-6-3~		S	7	AtC
AGES, VIEWS, SURROUNDINGS	4-5-2~		M	4	AVS
AIT	~-~-~-~		M	7	
AITES	4-5-~		M	1	
APOSTASIES	4-6-~	-	M	4	APO
ATTITUDES and F/S	3-6-3~	±	M	4	AFS
AUGUSTINIAN TH. of LEARNING	1-8-3~		S	4	ATL
BELIEFS in	1-8-3~	±	T	4	BEL
COK	1-5~-4~		M	8	
DIVINE ILLUMINATION THEORY	1-8-1~		S	2,5	DIT
DOUBT and CERTAINTY	1-8-4~	±	S	4	DaC
ESSENTIAL ATTITUDES	2-7-4~		S	4	EsA
EXPERIENCE	3-7-3~	±	T	4	EXP
FAITH (GOD and TRU)	2-7-2~		T	n	FAI
HIGHEST RELIGIOUS ACTIVITY	3-4,5-3~		T	7	HRA
IDENTITY	3-6-4		T	1	IDE
ILE	2-6-3~		S	6	
INTELLECT	1-7-3~		S	3	INT
IR	2-8-4~		T	4	
JOY of LIFE	3-6-3~		S	1,3	JoL
MENTAL ACTIVITY, THOUGHT	0-8-4	±	S	4	MAT
REASON	0-8-4	±	T	4	REA
REPETITION, COMMON and SAME	4-6-4~	±	T	8	RCS
SELFLESSNESS, SACRIFICE,,	4-5-4~		S	1	SSS
SIATKA	3-6-4		S	1	
S-R, S-O and S/F RELATIONS	~4-5-~	±	M	6	SROF
SUFFERING, SEARCH for RIGHT	3-6-4~	±	S	4	SSR
SUSP	25-75-25				
WORDS and VOCABULARY	4-7-4~	±	M	4	WaV

Fig 4-84 DYNO CES in SUSP SECTOR of SPALT

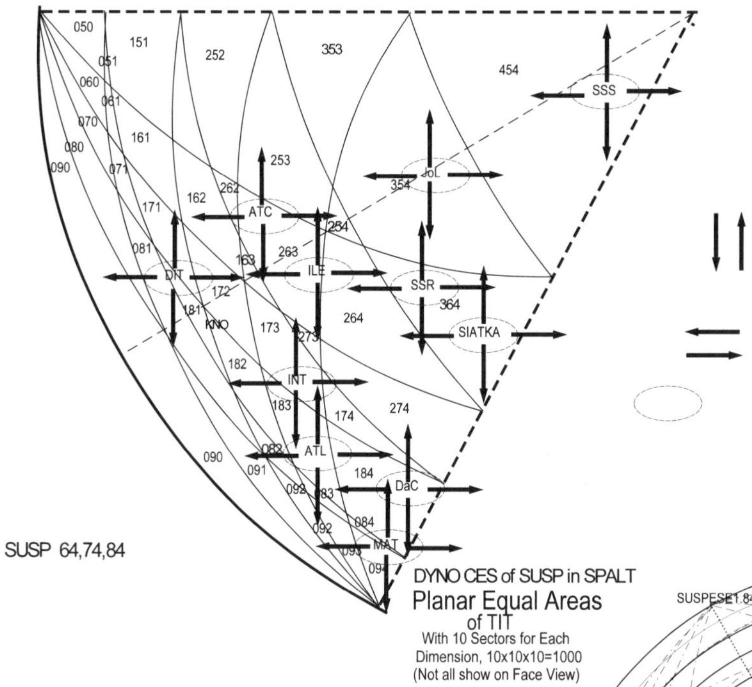

Fig 5-84 DYNO SUSP CES in SPALT

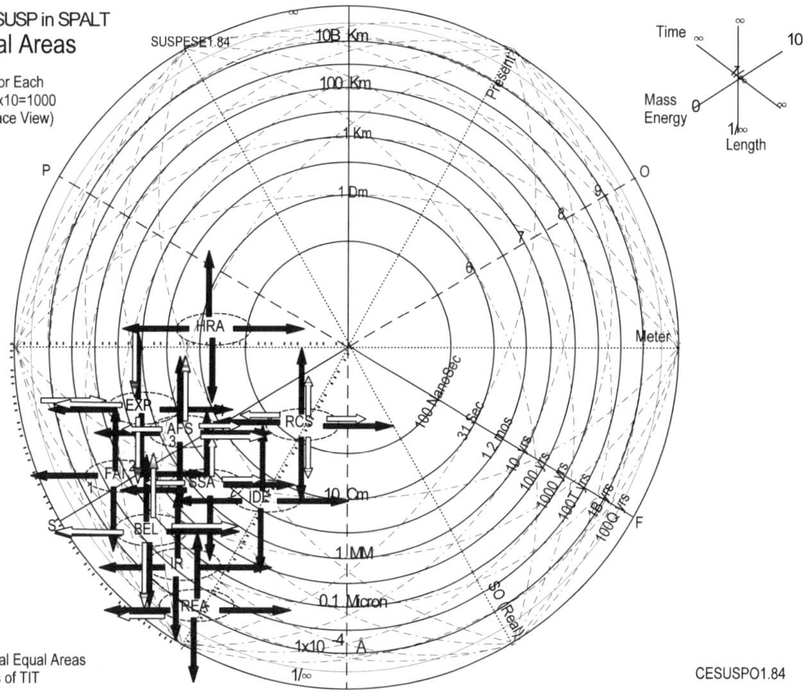

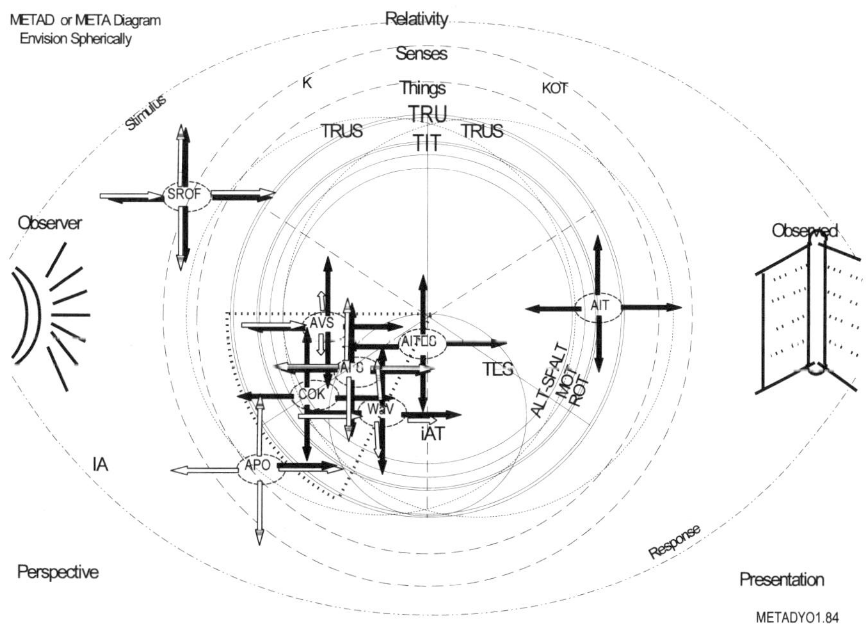

Fig 6-84 DYNO of SUSP CES in METAD

CHAPTER 85

SUSF or SUBJECTIVE SMALL FUTURE PROCESSES of DYNO

THINKING, DESIRES, LOVE, INSIGHT, INTERESTS, KNOWLEDGE, JUDGEMENT, EDUCATION, IDEALS

To the unobserving and uncritical mind time seems to be a simple matter of what time it is or how long things take. Yet to investigate time much more closely and deeply such simple observations can be reversed into how less certain are the times things take. An example is seen in how long it takes a train, ship or plane to depart, yet how long it has been gone when you have missed it, even though it has just left. The world goes faster or slower as we make it or as we are in relationship to it. This is much of relativity, and, when envisioned for all dimensions, Á, and L, the determinant of the FAF of TRU. Can the world stand still? Can TRU become absolute in an infinitesimal period of time? These are largely a function of relativity. It is how we as individuals can't find the eternal, immortality, or the absolute when all consciousness is lost, as at death. So that to make the world standstill, or to so expand our subjectivity to encompass all in one split second or infinitesimal amount of time, is the second way we have of finding the eternal, immortality or the absolute. The two together are called: **LADEXA** = Law of Dual Expansion to the Absolute. LADEXA shows that an individual, as TES and as an observer in relativity outside and in TRU, can find the eternal by both concentrating time or by expanding oneself. Animals with a greatly accelerated rate of metabolism can, in a way, do the first. They compact more and more life into shorter periods of time. We do both by becoming TES and TESÂ. So that time, personal subjectivity and TRU are all closely related. In SUSF of DYNO we can see how the means and two extremes of time for a person are variations in this focus of an observer's or person's powers of subjectivity in terms of TRU, TRUS, TRUCH, TRUF and what is extra TRU. It is both an identity with TRU, as by TES, and a overview of TRU as by an ENIT seeking to comprehend both together and resolve the different combinations this person can make.

Human mentality can be classified by its quality. Thought is more than a simple process of conscious activity between stimulus and response. It is better envisioned when one thinks rapidly, validly, actually, proportionately, and totally, by states, conditions, relationships, and limitations.(VAPO, SCARL) It is also how one thinks by ACEXIN, SÁ, SUT, newly, dynamically, orderly, and transcendentally, by TRU, TROS, TRIS, TIT, TÂ, and TRIT. These are but a few of the many greater variations, modalities, means, transit, and processes by which thought travels. Only in TRU or TREQ, selectively, interestingly, studiously, analytically, and with conviction and a heart warming appreciation for the benefits available can this be done and done continuously. With VAPO, SCARL, ACEXIN, OM, and HOM one can condition, qualify and perfect developing KOT so that ones thinking becomes a glorious challenge and victorious reward with identity in TES and TRU. It is not so complicated to improve one's mind. It is mostly a matter of deep interest and appreciation for the worth of life, ones own as TES and TRU. In fact it is simple to direct and will to the highest level. A person can not pay attention to all there is, even of TRU, yet one can learn to intently and rapidly expand and intensify what is most important of TRU. This is with the spirit of the most beautiful music that shows the heights of harmony, melody and rhythm or the glorious feelings to which we can be lifted in the mind's fulfillment with the best that works. It is through training and using the highest levels of mentality and expanding them by KOT that their use best fulfills TRU.

Thinking that is structured, dynamic, total and TRU is achieved by applying duration, variation, area, volume, inertia, momentum, mass, action, power, ACEXIN and ACC to a POA. These concepts of mechanics are basic comprehensive measures of TRU. When the concepts, principles, doctrines, axioms, laws, and mechanisms of TRU are thought of in these terms all comes together properly through VAPO. It is not only important to think abstractly, it is necessary to think abstractly in masse. This massive abstract thought has to be systematically modulated and positively unified if it is to fulfill meaning.. There must be an overview or big picture, plan, or blueprint, if the best representation of the object, TRU, is to be attained. This means that one must envision the object by thinking in three dimensions, and these in all their powers and combinations. To not only understand this object positively, spherically and completely, but all that surrounds it, all that is involved in its subjectivity-objectivity. It takes repeated learning, relating, positing, anticipation and a photographic memory. This we learn in PUREALITY, and gain in ELTICMA as we become TES, plus all that it returns to the individual, ones identity and immortality in TRU.

LAVIT = Law of Vicious Circle without TRU. Without TRU, TRIT, TIT and their totality, unity, systematic unity, and their absolute, TRIA, TA, and its focus, we in our senses and things never find ourselves but are forever and everywhere rambling lost in confusion. Never finding that for which we should search. Never this and never that, indiscriminate and without reason. Only as we come to discover and learn what is the greatest, TRU and REM, as the true, the good and the beautiful, do we come to have a stability, unity and harmony of our reason and being. This state in which reason is in being and being in reason, as they both enable us to be, to be TRU is called: **LARUR** = Law of Relative Universe in Reason and Reason in Relative Universe. This is where and how object and subject, objects and subjects develop meaning and significance and develop actuality, as they fuse in union to produce, eventually in TRU, the greatest, the reason and the being that together are TRU. This is similar to THES and THEO although not as polarized nor as completely in TRU. LARUR and LAVIT, to a lesser degree, prove Á or PRÁ and TRU, PRUT. LARUR is associated with: **LACRÁ** = Law of Combined Powers of Reason in Á. LACRA arises from an understanding of the combined meanings of REASON. This is how reason is not just one of its meanings, but what all these meanings have as a common and universal theme. For Reason is: I. The underlying universal principle or mechanism in all, II. The causes of things, entities and all, and III. The mental powers we have to distinguish and determine. When these are combined in the universal theme and power of TIT-TO it will be found to be ideal, real and material combined as well as universal and particular. It is how, by relativity, the observer observed relationship exists in all, as they are neither any of these alone but all together as subject and object becoming subject-object and this as the predominant power and exclusive thisness of TRU becomes SUBJECT-OBJECT TRU, or **SOT**. This by relativity and its equivalent the observer observed relationship is what constitutes the predominant form, state and dimension of TRU, its Á Dimension. This A dimension, like the reason of Ancient Greek Philosophy in combination with the reason we use to best determine all, is highly dynamic, interacting and permeates all as the mass energy component of our being, TRU.

The limitations and weaknesses of our will, reason and judgement are clearly evident in the near impossibility of staying on a proper diet and stopping smoking or alcohol. It shows the power of base hungers, appetites and desires over our reason, judgement and will. So it is for all that we should or should not put into our lives, TES, and TRU. Unless we know thoroughly and strongly and have the interest and conviction to see the benefit, as well as the need for TRU, we fail. We must deny our hungers, appetites and desires for all that is not TRU. The correction lies in deepest knowledge, understanding, and will that is founded on the tremendous worth of our bodies, minds, and moderation in all aspects of our FAF, SÁ levels in TES and TRU. Such will power takes constant cultivation of habits of TRU produced by an equally constant interest, gratitude, recognition and fear of the deadly consequences that occur when TRU is absent or defective. This is self discipline at its finest, the greatest needs of TES and each persons being.

Like Mark Twain said, "It isn't what people know that gets them into trouble. It is what they know that ain't so." The great significance of this statement is not only its obvious answer, it lies in the determination to rid oneself of that which 'ain't so'. What we know may be quite true in and of itself, yet like we ourselves, as DAK, when evaluated by its TRU nature, or in terms of the overall TRU, ROT, and POT it is not what we know at all, it is much different. Only when we know TRU thoroughly and our COAT is oriented and convergent on KOT can each and all things that we know become properly considered and realized in the ramifications and varied states that they are in or relate to TRU. It is in this way we must: know rather than not know, know what is so rather than what is no't so, know what is TRU rather than not TRU, know how this that is of TRU stands in TIT-TO, know how these are all to be COAT, know how all these can be Posited in SPALT, and know how this knowledge can be applied in life.

NUK = Negative Unknowns. Because a person feels comfortable or satisfied with oneself, at a specific time and place, does not mean that they are secure or free from danger, harm or disaster. This is the difference between any IA and IAT, DAK and TES, ENIT and TIT. For by ENIT, DAK and IA, the world, especially the here and now that seems so placid and satisfying, is a very false world. Each and all persons should learn to recognize their part and processes or role in life, IAT, TES and TRU. Herein is the jist and cue to the first person, Christ, and TES, and the world, God or TRU. It is because of the false feeling of comfort and satisfaction that we become complacent, ignoring and failing to live up to what is necessary. Competitive selection is produced by an extremely high level of stimulation. It is largely because of NUK that so much emphasis is placed upon education, even though it is not often well accepted by those being educated. For with the rapidly expanding and intensifying world that is TRU, the individual will not succeed or even have much role in it, if they do not overcome NUK. NUK includes all, in increasing degree, from the absolute known to the absolute unknown. It especially applies to what is unknown to each individual that increasingly lessens their chance to become and be TES and TRU.

Throughout life our information, vocabulary, concepts, and knowledge are continuously expanding. The identification of this knowledge with that of the world, the receiving, sorting out, accepting, learning, relearning and applying it to the principles, laws, and perspectives is a large and great portion of ones life. It occurs in one's daily life, work-a-day world, at home, all seasons of the year and all stages of life. They are real, actual, valid, and TRU only when all that is not TRU is extracted. They represent much of the ES and CES of TRU. Every person, generation, nation, and culture thinks it is the best. Only with much experience, maturity, and age do they come to realize their limits, and place in the whole. The 'ideal and perfect' little realize how far it is from TIT, TRIF, and TÂ. This is seen in REMS and REMOD. Like DAK we little realize how far, 1. we are from the actual ideal, 2. this ideal is from TES, and 3. TES is from TIT-TO.

All who pass judgement or make decisions that condemn or adversely effect others TES, or even any aspect of portion of TRU, without care and without TRU perpetuate a loss to TRU. They are ATRU, ANTRU and ANTITRU. To these we must invoke the rules of TRU and ACC to correct them as appropriate by TRU. We should start first with ourselves. It is easy to make this or that statement, decision and judgement when in conversation that is so partial and often not TRU or misinterpreted. To be highly right, as it must be TRU, it is in the best sense of our dedication to TES. It shows in thoughts, words, attitudes, acts and work. We should express joy and happiness with all that is being their counterpart or functions in TRU. Quantitatively this will increase with the degree of TRU. For these ES, CES, REM, and TES together make up most of TRU and secure TÁ. We can be both pleasant and joyful, subtle and profound without being asocial or excessively separated. This is what TRU is. It has been the mark of civilizations and cultures whose conversation, interests, communications and influences were based, not on the negative, but upon TRU that showed the degree to which they were increasingly successful. It is not easy, in the many trials and tribulations of life, to always or even generally, be TRU. Even less in our limited IA and speech to consider all pertinent to TRU. Nothing or no one who is or strives for TES and TRU can be defied or denied. For they *are* already. Yet we can not be turned away from TRU. We should always have trust, keep the faith, live in PUREALITY in the best possible ways. Only when we most greatly learn, understand and love TRU as we do ourselves and others when TES, can we become TES, and be our role in TIT to the benefit of one and all.

If all mankind could come to retain the deeply felt experiences of the lives of all who have come before them, what a wonderful world this would be. Similarly if we can come to retain the deep identity with the lives of all that come after us. We may miss a lot personally, yet the suffering and losses we can avoid make life far more tolerable. This kind of world for us is, **WOBE** = Wonderful world this Would BE, to retain and project all time Experiences. WOBE is that ideal state in which new persons would not have to experience and suffer the life they usually do, yet still gain the needed wisdom and TRU. WOBE is evident in the value of all modern means of communication. The big problems are, however, that both the communication is so devious and the public is so ignorant, gullible, and deviated without appreciation for these deeply felt actual experiences of life. Their receipt of quality and quantity is on a much lower level. It takes a highly astute and critical mind to sift through all before us to elevate our imagination to levels that can reach WOBE and TRU. Yet this can, should, and will be corrected. The question is more one of when and how this correction will take place. Whether the trial and error methods of the past and present, or better methods of science, TREQ and expanding TRU will produce TES, ACC, and ELTICMA. For it is already largely predetermined by the compelling TRU. Measures can and must be taken to elevate communication and stimulate the critical imagination of human minds to raise our learning to the levels it would have had through painful experience to what it must have by TRU.

The nearness and dearness of TRU is so great that no one should ever depart from, short change, deny, defy or neglect it. How much more real and actual it is, than to believe in some whim of immortality, but find it in every moment of our lives. There is no need for inappropriate thought, hope, act or adventure. There is more than enough in TES and TRU. At all times one must be guided by ones part and role in TRU that is most appropriate for the given and can be transformed into TRU. A person's life is one long adventure, endeavor, achievement and fulfillment of what one is, is given, and compelled to be as best converted into TRU. This is S/F, identity, actuality, reality and purity as best attained and fulfilled for immortality of ones, TES, and all, in TIT-TO.

IMMORTALITY = 1/Death, or
LIFE/DEATH = TRU, to TÂ.

What we like or want in life
Is not the same, even at best, as
What life has in store for us
Even as TRU they are different

This is especially acute in our ideas
Of immortality or everlasting life;
For what illusions we have of this is
Much different than what it will be.

Only when we learn, know and live TRU
And can extract our ESINS to be TES
By living in Pureality with Cognos can
We reconcile our immortality in TÂ.

The importance of learning and insight is for us to similarly recognize TRU, or TREQ. Learning was especially championed in 20th century psychology and education. Yet learning, as all, is not alone, unrelated or independent. It is a great part of the higher levels of our mass energy or Á FAF. It is a part of the nature of discovery in which our IA is forever selecting what is more or less of benefit, or TRU. Yet since we are not born with the wisdom of the world or of TRU we begin devoid of KOT. With insight and zeal we can direct learnng and thinking toward what is important, TRU. We reduce our sins, errors and failures by utilizing knowledge through TREQ or TRU to all that is known to gain those broad and great vistas through learning by adding more for vast panorama of TRU. In this way we increasingly come to learn, order, and remember it better, the adventure and discovery fresh minds can provide. We can in their closure and our maturity most readily, effectively, and best turn this to our participation, TES, and to the glory of TRU.

LORFI = Law of Reversible Fusion and Identity. LORFI signifies the state in which ones desires and wants can be brought to life's goals and TES, and life's goals and TES can be brought to ones desires and wants. This is experienced, and can be the greatest of opportunities for matching maturation in life, with identify in TES and TIT. It already exists to a great extent in a perfected life in which persons at each age best perform those roles which nature and society have customized for it. With the caveat that this nature and these customs and roles are optimally TRU.

The limited life in a pursuit of happiness and the more rigorous but fruitful one of TES and TRU are not as incompatible as first appears. They are similar and highly complementary. So that the pursuit of happiness is more a frame of mind in which one finds happiness within certain orbits, realms, roles, and ways that are positive and beneficial to TRU. We must learn to find happiness in doing what we must as TRU, and learning to appreciate, gratify, love, and enjoy it. This rather than doing what we want to, chose to, or inadvertently seek pleasure. It is a small price to pay, and in the end a great profit. For many beautiful, glorious, rich and inspiring blessings are come with the experiences that are gained when we are mindful of, work and live with TRU.

More failures are caused from disorientation, incoordination, and disproportion than from other causes. Our form and function, anatomy and physiology, are based upon focus, symmetry, coordination, proportion and equilibrium. Yet in life it is our inveterate tendency to neglect and deny what it takes to comprehend, make account of, set the priorities, relative values, or magnitudes in all TIT-TO. This is in part a reflection of the singularity of our attention, superficiality and relative novelty of higher levels and powers of Á. It is for this reason we must continuously remember, ECRO, and use our powers correctly and fully to make up for their lack of support. This is called: **LACHÁL** = Law of Compensation for Lack, or insufficiency, of the Higher Levels of Á. LACHAL is of the highest importance by signifying and actuating our higher levels of Á powers. LACHAL is what it takes to get from DAK or ones individuality to TES. We don't know what hit us, until it is all over, or we miss it all together. This is a function of our immaturity, insufficiency, confusion, and a lack of proportionate identification, valuation or REM. Only TRU can be successful permanently with the consistency, dependability, certainty and the consummation of all it assures. **LATSCON** = Law of TRU Success Only with Consistency, Certainty, and Consummation. Without TRU there can be no actual success. It will be only a figment, a fleeting gain, and often loss, both to itself, and to TRU. The degree to which these are met is the degree to which all that is TRU, as TRUF, will exponentially grow in both quality and quantity for greater lives, blessings and happiness of all.

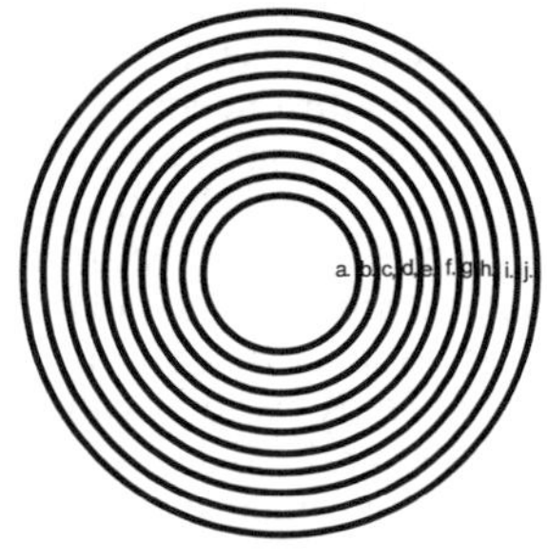

Fig 1-85 **A PERSONS VIEWS of THE WORLD, PHILOSOPHY or KOT**

What philosophy does an individual hold? How we do we view the world, we are in? What level of subjectivity or mentality views this world? There are many different levels of a philosophy of life. They, like the predominant mass energy, or Á, FAF of TRU, are expanding rapidly: a. Animal, simple physical view and/or response, b. Hominids, direct perception of nature, c. Early Human, perceptions plus aggressive apperceptive objectives, d. Rational, with many diverse and largely incorrect views, e. Average person in the world today, f. Views by way of higher intellect, g. Ordered comprehension of the world, h. Increasing knowledge of TREQ or KOT, i. Overview of TRU, j. Perfect overview, of and in TIT, and all. It can be seen why a good perspective, understanding, representation and presentation of the actual world TRU has not been previously used. We have looked at the world in very vague, poor, disproportionate, incomplete, partial, and disordered ways. These ways are numerous, variable and overlapping. They lack orientation, coordination, order, and a solid comprehension. This is what weakens our society, civilization, and TRU. It can be corrected by discovering, learning, understanding, remembering, and applying TRU. Some ways, although very simple or primitive and in certain settings, may be very effective. Yet this same way may be completely ruinous in another setting. Through their measure we can determine: 1. The degree of knowledge, coordination, efficiency and other factors necessary, 2. Sufficiency to maintain TRU at present level, 3. Level to survive only a limited set period of time, and 4. TRU to TIT-TO to TÂ.

The principles one keeps and how they function and are used for TES and TRU are the best measure of a person. It can be partially measured by writing or oral testing. Special tests can increase the reliability of measurement. Yet the only final and completely reliable measure is the summation of the results in a persons life. They come from a good ancestry, environment, home, upbringing, training, interests, education, reading, writing, motivation, conversation, work and relationships. They will, even be much greater when they also come directly from TRU. One must have respect, be grateful, appreciate, love life, and love of all that is TRU. For if one does not one is apt and inclined to be partial, skewed, diverse, divisive, productive of ATRU, and ANTRU. Reaction formations that follow will, sooner or later, directly or indirectly, adversely effect TRU, its parts, and oneself. In TRU can one find much that is common or the same as oneself. In all that is other and not TRU there is overwhelming difference and opposition that is falsely accepted as being inevitable or even praised. Yet how can one respect and love God, TREQ or TRU if one does not learn and know them? This is the great reason for TRU, POT, and IPTRU. It is seen in the love of God, as TREQ, TIT and TRIT. We should at all times be helpful and cooperative with all efforts whatever they may be or by whatever name they are called, if they are productive of TRU.

In life, we must continuously make decisions and choices. If we and humanity are to succeed and survive, these decisions and choices must be right. They must be based upon what is best for our souls and for God. Yet we can't make them unless we know what our souls and Gods are. This is why it is so important that we study, learn, and be TES and TRU. How one looks at the world is vital to one's judgement. One can not efficiently, effectively or successfully live, make decisions, judgements, and be free from failure when we do not have the proper knowledge of the world, God, TREQ, or TRU. This is called: **LANSUC** = Law of Necessity of Success, based upon knowledge and judgements by TRU. LANSUC is the necessary knowledge and judgement that is based upon KOT. All wants, desires, decisions, choices and judgements should be conditioned and guided by TRU knowledge. They can be no better than whether or not this knowledge provides them with the right information, direction and purpose. This, ultimately, can only be TRU.

Decision theory is central to the validity of intentionality, and the mass energy Á form and function of the world. It works, as all approaches must, if they are held to TRU. Because it is not held to TRU is why critical theory, deconstruction, and many other approaches to knowledge and the world are lost in its massiveness and divergence. Without knowing the limits of the actual world, its parameters, our observation, representation and comprehension of it, no school of thought or system of ideas can succeed. Many ideas, schools of thought, fields and sciences have great and beneficial approaches and concepts that add to knowledge. Decision theory must rely on the certainty of its information. This must largely rely on the language and concepts the data is represented by. Because the integrated whole, differential particulars, and the order of all in between are not well used in the harmony necessary, decisions, especially political, are continually inadequate, and often harmful. However, this must be weighed against doing nothing, or doing without decision. Decisions that result from the certainty of ever more precise induction and deduction from the actual meanings of essences of concepts, and the massive reconfirmation by ACEXIN and POA, reveal the beauty and glory of TRU, POT, ROT, HOM, ACC and IPT.

The cultivation of interests and to have a mission in life is an effective cure for many alcoholics, obese, and adverse habits. If life has not produced sufficient stimulus or suffering to promote response and a strong desire for relief then people will lack what is necessary, they will lack sufficient appreciation and joy for what is best in life. When we can motivate all to recognize and exercise the common virtues and worth of life - in TRU - they can readily overcome such detrimental conditions and all that is negative to themselves, TES and TRU. It is the underlying defect, lack of interest and mission in life, that must be cured. Only with quality individuals, activities, and purposes in TREQ or TRU, can mankind flourish and avoid the sufferings that humans have repeatedly endured.

Fig 2-85
SUMMATIVE SEARCH FOR UNKNOWNS and/or TRU

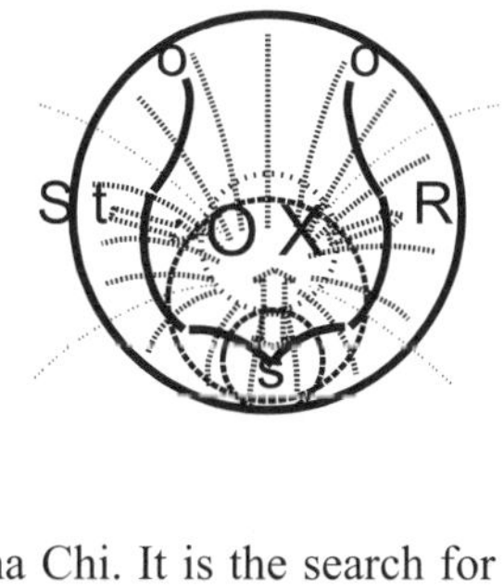

To find the ES and the TRU in all things, as well as the TES in oneself, is like the mystic search for the Holy Grail, Excaliber, Phoenix, X in an equation or in Sigma Chi. It is the search for what is needed, right, most important and perfect, the TRU in all things. These all together are a summative X, or search for the unknown, what is the best of the actual with the best of all together, TIT-TO. This is a life long endeavor. It is a part of each person's S/F, TES. The degree to which each person achieves rather than does not achieve this is the degree to which we are able to find, comprehend and act upon those necessities that are our ACEXIN in life, ones O-S-O. **OX** = Summative or Total Search, for X or unknowns, formal distinction and Ordered differentiation. Much of life, TES and TRU is OX, how we search through: all, aspects, ES and CES of TRU, KOT, DOT, TÄ, TROS, TRUD, STA, HOM for their deepest and total meanings and fulfillment. **LASURT** = Law of Search for Summative Reasons for TRU.

Of all the terms in language insight is one of the greatest, both formally distinct and powerful. Consciousness and awareness penetratingly identify with the quality and power of IA, IAT, Á, and TRU. Insight like interests demonstrate the opportunity to best exercise the highest levels of ones actuality, or the Á mass energy FAF in TRU. They need to be selected and held to TRU. This reduction can be made in many ways, yet optimally only in the way that most efficiently and effectively leads to TRU. This is largely through its optimal approach, POA. We are not born with insight. It must be developed, trained, sharpened and exercised. This largely depends upon learning and knowledge.(KOT) It is in this way TIT, TRIS, TRIT, OBT, SUT, ROT, MOT, ALT, SPALT, REM, POS, STA, HOM, and ACC can provide the POA of TRU needed to enable insight to fulfill its great role in TES, and TRU.(ENT)

Many of the bright minds of young men and women of every generation are lead into, directed, and inspired by the best ideals and principles of religion, God, philosophy, science, TREQ and TRU. However these ideals may or may not be the best of TRU. Their aims, purposes, goals or ends are usually limited, partial, disproportionate, extreme, distorted, and deviated. Too many are eager and ready to adopt this or that ideal or system,, or less rationally accept any sect, fad, fashion, trend, peer interest or sophisticated cause they encounter. They champion it, and come to feel this ideal and end is the most, the answer to all things. They are told this is the key to knowledge or the answer to the riddle of life or all. Yet this is far from the truth, with little VAPO. Such profound needs and deficiencies are corrected by the best knowledge, a full and comprehensive knowledge of all that is, in actuality, TRU.

Already there are too many different philosophies, religions, and sciences.(PREFS) Why then do we need, and can we provide, what seems to be another. Initially one that may seem harder for each person to learn and live by. It is because each is in error in many ways. It is because together they lack relation, unity, and coherence. If they are to provide an organization of knowledge, and if this is to equal their roles in showing and guiding each person's actual life, a new approach is necessary.(POT,IPT) This is the aim, purpose, goal and end, to achieve fulfillment by being ones counterpart in TRU, TES, by learning what is TRU completely in TIT-TO and in a manner that is ready for use. Thereby all should first consider and come to realize what goals and ends are right and what wrong. They should learn what is known that is and is not, what is known that is TRU, and how important it is in TRU. The young, and all, need to avoid and reject false ideals and extreme views and come to know each and all for what actually is, its worth, REM, and TRU. They should interpret each in its proper role and place in TRU as shown by SPALT, REM, POS, HOM, ACC, and COGNOS.

In order to find and achieve the answers, solutions and provisions that the questions, problems, and needs of human life require it is paramount and vital that each and all of us discover, learn, and apply all knowledge that is valid, actual, proportional, and total of this special universe of best human life. To find and achieve this, all knowledge, existing and yet to be discovered, must be approached, organized, represented, modeled, posited, utilized, and treated in accord with the form and function of our world, this special universe of best human life, TRU. This is achieved through THEQ, TRUS, TU,TRIS, TO, TRIA, and TIT. It is by those means that can be provided by the PI, POA, POS of ES and CES in the TRIT, OBT, SUT, ROT, MOT, ALT, and SPALT of TRU.

To many people have presence, a presence in their own self and being. Yet they have little or no presence in the actual world, the survival of mankind. Presence like attention is or is not a virtual or actual presence It can and must raise its holder, proband, or first person to reach higher degrees of identity with this special universe of best human life, the relative universe, TRU. This is achieved in the discovery, learning, knowledge, memory, interest, inspiration, revelation, dedication, conviction, utilization and service to ones presence in God, TREQ, the actual world, the relative universe, TRU, that one truly lives, has eternal presence, and the hoped for immortality.

LEP = Law of Expanding, Exponential and Eternal Presence For it is with such focus and energy that we actually are or are not our part and contributor to the universe that all human life must uphold. For without this universe human life is lost and soon passes away. Only the hope of eternal life and the absolute can give mankind the power necessary to rise to, and overcome all that would destroy us.. It is by the given, human life as we find it, and by the presence needed to sustain it that all must and will live by seeking fulfillment, happiness, and immortality. Yet the road to salvation and fulfillment is not easy. It necessitates a constant recall, memory, and realization of the presence necessary to raise and bring all givens to the actuality of human life. Law of Giving It Presence, (LAGIP). One that knows and has no figments and illusions, ENIT, SENIT, and THIT that are not what they must be in TRU.

Many stories provide the glorious means of representing keen insights into the inner-workings of TRU. Each propagates some key aspect, feature, principle, mechanism, or characteristic of TRU that is not otherwise readily understood. This is done by concentrating on those portions and dynamics of human life that are closest to both a person, IPR and TRU. They help to bring us to TRU by showing the adventure of life: to achieve success, survival, happiness, identity and immortality. They show how much presence and feeling there is in the core reality of TRU, as by Jesus' imminent presence of God's Kingdom. This is done through many different, and expanding, vehicles and media. They tend to tell what is and what is not TRU. It is in literature and many other forms of the arts, that may be of any sense or perspective, that we can better learn and realizeTRU, its parts and processes.

The fight against evil should not start by tearing down everything. It should start only after we have the capacity, ability and means to build something better in its place, TREQ or TRU. Only with KUPS, and massive ACEXIN can we know TRU and both tear down what is ATRU and ANTRU, and build up what is TRU. There are at least three good reasons why each person should developed, to the utmost, their best ACEXIN and TES powers: 1. The person is in a focal, pivotal, key, and central position, 2. The person represents the most reliable, tangible, adaptable, and actualizing carrier and great part of TRU. 3. The person is the only portion of TRU that can control most or all of its mass energy, or Á FAF. Without the person, TRU or anything, has little actual meaning, identity, and closure. TRU is largely personal and of humanity just as persons and humanity are in actuality nothing without TRU. Life must be palatable, appreciable and full of happiness, yet never at the expense or neglect of: **LANSUC** = the Law and Necessity of Success in TRU of KOT decisions. LANSUC is largely supplied by education, training, living, duty, and the responsibility of each to TES and TRU. This we spontaneously appreciate and love when brought out by becoming TES.

COMTELD = Compulsory Education, Training and Life to and in TRU. It is not enough that at some or any time in life we learn to believe in God and intermittently strive to be Godly, moral, ethical and nourish our souls. These are a good beginning. It should be incumbent upon all that an optimal level of proficiency in TES and TRU be achieved. This is what is demanded in many fields of endeavor, sports, sciences, and professions. For there is no greater way of bringing the newcomers, children, trainees, youth and all to TES than to, as early and thoroughly as possible, show them how to learn and understand life as TREQ or TRU. It requires a feeling of revelation, inspiration, mystic union and identity with TRU and all that is in common with TRU. If and when COMTELD is gained and held, then, many of the troubles, strife, wars, suffering, selfishness, disbeliefs, and sins of the world, will be spontaneously eliminated.

There is no way we as individuals can attain it all. The best we can hope for and do is to be the best we can, TES, and to practically see to it that all else is also TRU to the greatest possible degree. This is so that TRU itself may become ever better and greater both qualitatively and quantitatively. As the pie enlarges, we its pieces, become greater. As the pie becomes absolute, TÂ, so we become absolute, TESÂ. It is because of ENIT, selfishness, partiality, vanity, and the NAMTRU world about us that the natural tendency to TES and TRU are ignored, denied and defied. It is corrected by the recognition of our partial position and highly finite nature in the world, TRU. It is even further resolved by realizing the extremes of egocentricism and selfishness, which normally have some reason, yet are highly limited. It is specifically corrected by the expiation or exorcism of all within us that is negative, not, or against TRU. **EXPA** = Excises, Expiates and Exorcizes all not, from, or against TES and TRU.

The success of and value in all education, teaching, degrees, and learning can be measured by the amount to which it provides a filling up of ones mind with all the basic processes, mechanisms, parts, concepts, principles and laws of TRU.(CALP) This is called: **PUCS** = Processes of Filling Up, that enables Success. PUCS is not new, it is a large theme of the method of Confucius, Socrates, and many educators who sought to help young minds discover for themselves with reason the greatest goods. It is more than this, how reason can reveal these greatest goods and all that makes up TRU in order to have a firm understanding of all, as it is TRU. It is how the mind when guided by the answers to questions and solutions to problems of basic TRU concepts can learn, understand, know, remember and use KOT, to be sufficient for all. This is how an individual passes from little knowledge to greater understanding, and this to the greatest understanding of all, KUPS, KOT and COAT, as they through Á FAF fulfill THES achieve COGNOS, ELTICMA, SÂ and TÂ.

Fig 3-85 **WHAT MAKES US BE, BE TES, TRU**

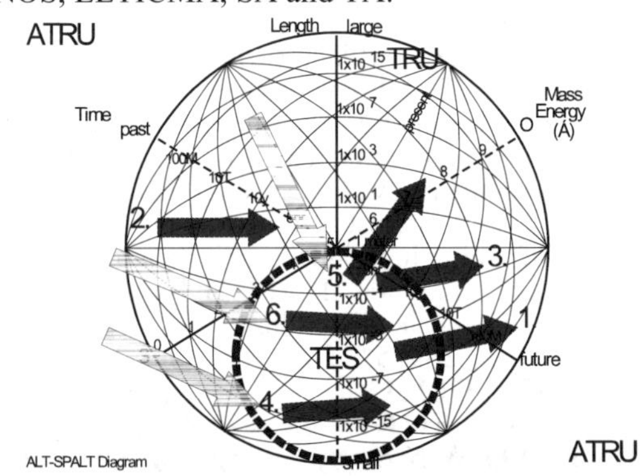

With all TRU determinism, Pureaclity, ethics, morals, duty, responsibility, education, work, and taxes is it worth it? What does the individual get in return? Why should we be so diligent and arise to the demands on us for sustained vigor to be, to be our counterparts in TRU, TES: 1. Immortality, 2. No choice, we are, 3. Material blessings, security, 4. Salvation from sin, pain, suffering, eternal damnation, 5. Compliance, social order, peace, 6. Prevention of injury, harm, ills, 7. Joy and happiness of fulfilling our role in life.

Individuality that is focused in ENIT and ATRU is fruitless and damaging. It fails without discovering its ESINS and their transfer to TES with immortality in TIT. For those who want personal immortality or who pine for a fountain of youth let them take heed and find happiness in the rejuvenation and immortality they can still engage in and enjoy in the remaining incremental glory of their lives. No life is so long nor short, that we can not nourish our souls, TES and TRU. As long as there is life there is not only hope but opportunity. It is when life is lost through complete and irreversible loss of TRU, is there is no hope.

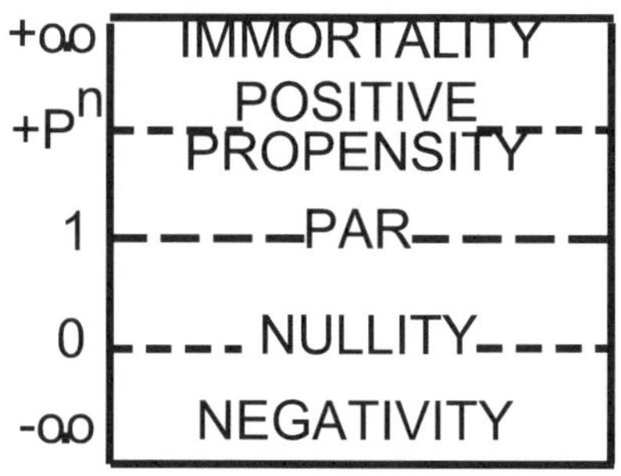

Fig 4-85 **EACH PERSON'S MARK of ACHIEVEMENT IN LIFE, TES, and TRU**

How far each of us are from the mark is only apparent when we come to know TRU, TES and all its forms, functions, parts, and processes. It is when in being ATRU we see others who are TES and how far less we are, and how we can come to TRU. It is when we have the conviction necessary for TES, that we can be confident in the kind of approach that finds its mark, in TRU. All of us from our ego reaction formations, self identities, fixations, need for self preservation, and our being, assume we are all, we are it. Figure 4-85 scales the relative positive ornegative achievement or mark of individuals from positive infinity to negative infinity. Qualities and characteristics are compared with TES, par, nullity and negativity. The degree a person brings themselves to TES, and the perfection to which separation, selection, extraction at the TRU/ATRU interface is achieved is the degree to which this person moves from the bottom of this diagram, negativity to the top, immortality. The greatest and most basic reason for all education is the development of conceptual coordinated knowledge of TRU, their reasons and a person's ability to think through, identify, and apply their conclusions to TRU. It takes a life long habit of **TAN**, to think and note, SUT and POS of the cue and gist of the greatest ideas, concepts, principles, laws, and mechanism of TRU. The degree of TES is greatly a function and product of such fundamental means to COAT. The education needed to achieve TES is neither random nor simply done in starts and spurts. It is a life long endeavor, based upon and directed by VAPO, POA, ACEXIN, DAOST, the SCARLS, and all of TIT-TO. That which is most TRU. Knowledge of TRU is aided, guided, fostered and stimulated by TRIT, OBT, SUT, ROT, MOT, ALT, SPALT, REM, POS, HOM, PUREACLITY, COGNOS, ELTICMA, SA and TA. This is: **LATED** = Law of TRU Education. LATED must be the guiding light either directly by TRU, TREX, and TRIX or indirectly through similarly great equivalents, TREQ, TRAX, and the entire application of the best possible methods and procedures of education. Thereby education needs overhauling, not in diverse ways, but as all that is not TRU is developed determined, managed, treated and controlled. This is how they are and in the proportion to which they are not and appropriately must be, by TRU, TIT.

Because we can not learn and know the infinite, all knowledge, and all TRU does not lessen the fact that it can be unified, well ordered, massively adequate, and highly successful. Only as our means and education can resupply the loss with every generation through the best KOT can all have the ability to be TRU. Education has made giant strides in the last two centuries. Yet it is a never-ending battle to not only maintain these gains but to continually upgrade its quality, quantity and utilization. All too often education loses its inspiration and glory. To often both students, scholars and teachers are relegated to less desirable and inferior positions. Yet this should not be cause nor excuse for dismay and loss. The glory of imagination, revelation, scholarly interests in all, especially when most important and useful are the great opportunities and power of all time. Yet they are greatly ignored and missed. Education and being a scholar must of necessity be primarily an individual effort. It comes from a deep inner energy, intelligence, competition, joy in learning, reading, study, inspiration, revelation, desire to excel, need to make up for the negatives in the world, and working out the way, plans and operations that brings all to TRU. Open discussion, free association, ever greater knowledge of TRU, as it can be communicated in many and expanding ways, will be the wave of the future. A life of TRU can not be confined, nor only for the academic, as great and useful as these may be. A life of TRU, in which there is a deep interest by each person and especially in conversation, group discussion and communication is its best means and salvation. A world-shaking revolution, direct or indirect, will occur, especially to the degree TRU or its equivalents are not readily and appropriately fulfilled. For not only is there a great need for TRU, those who recognize, acquire, and achieve great proficiency in TRU will not allow mankind to self destruct in the many ways it can, and may likely, occur. The more the seeds of destruction and extinction the worse the struggle will be to reverse it, from ATRU to TRU. However it will largely be a revolution of the individual or self to TES. As Christ and many others have led us to seek salvation and immortality in God, TREQ or TRU, so can we be assured, with ever greater, more accurate VAPO, KOT and TES of the continuation of these processes in the fulfillment of TRU.

The role of TES in our lives can be readily learned from the existing term apotheosis. Apotheosis is the raising of a person to divine status, a culmination or highest development. It is the ultimate, quintessential, or final orm and function of its great part and role in being, TRU. It is what provides the expanding spread of the mass energy continuum, MESOA. It is what has been so used, and abused, by many rulers and personages in history who claim but defy true exaltation. Yet such apotheosis as TES is the driving force all, more or less, must seek it they are to make their potential actual in necessary best of possible worlds, TRU.

MEASURES of DYNO SUSF CES

ACHIEVEMENT in LIFE, TES	~4-5-7~	T	1,3	AiL
BEST of LIFE	4-6-6	S	7	BoL
COMTELD	4-5-6	S	3	
DESIRES, APPETITES and CURES of	3-7-7	S	4	DAC
DECISIONS JUDGEMENTS RIGHT/WRONG	3-6-5 -	S	4	DRW
EXPA	4-6-7	M	8	
HAPPINESS	3-6-6	S	7	HAP
IDEALS, MISPLACED or TRU	2-7-8~-	M	8~	IMT
IMMORTALITY	4-5-8	M	3	IMM
INSIGHT, ANY & TRU	0-8-5-	T	4	INT
INTERESTS, ANY, TRU CULTIVATION	2-7-6 -	T	4	ITC
KNOWLEDGE, LEARN WHAT IS/ISN'T	2-8-6~	T	4	KLI
KNOWLEDGE, UNIVERSAL, PAR,SPE	1-8-7~	S	8~	KUP
LADEXA	~-~-~-~	M	7	
LANSUC	3 6 6	T	3	
LASURT	3-7-7~	S	7	
LATED	2-7-7	S	7	
LATSCON	4-5-6~	T	1	
LEARNING	1-8-~	M	4	LEA
LORFI	~-~-~-~	M	8	
LOVE of TRU (God, Life)	2-7-5~	S	7	LTL
MENTALITY, LEVELS of	1-8-7~	S	4,7	MLs
NUK	~2-7~-~	M	8	
OPPORTUNITY, HOPE for LIFE TRU	4-5-7~	S	7	
OHLOX	3-7-6~	T	4	
PERSONAL VIEW of The WORLD	1-5-6~	M	8	PVW
PRINCIPLES, MEASURE and TRU	2-7-6	S	4	PMT
PUCS	~-~-~-~	M	6,7	
STORIES and LIFE'S SEARCHES	4-5-7~	T	4	SLS
SUSF	2.5-7.5-7.5			
THINKING, GROWTH & PERFECTION	0-8-5~	M	4	TGP
WHAT MAKES US BE, TES, TRU	4-6-6~	S	1	WMB
WOBE	4-5-7~	T	7	

Fig 5-85 **DYNO CES in SUSF SECTOR of SPALT**

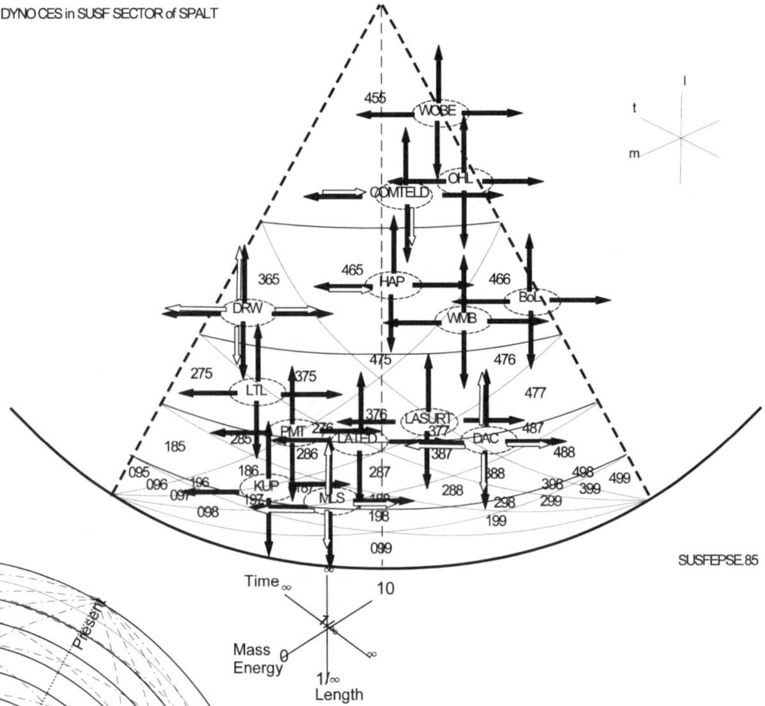

Fig 6-85 **DYNO SUSF CES in SPALT**

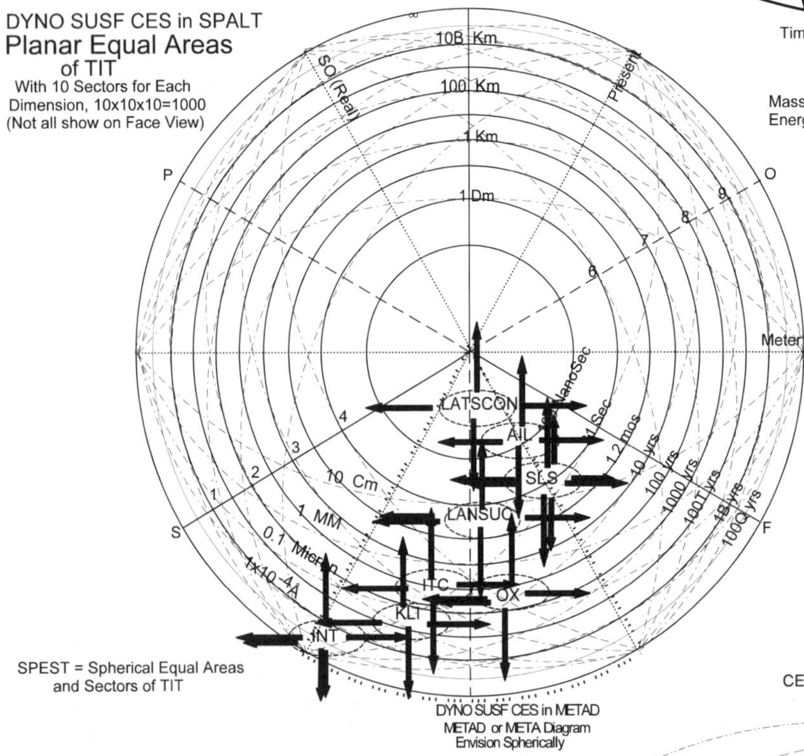

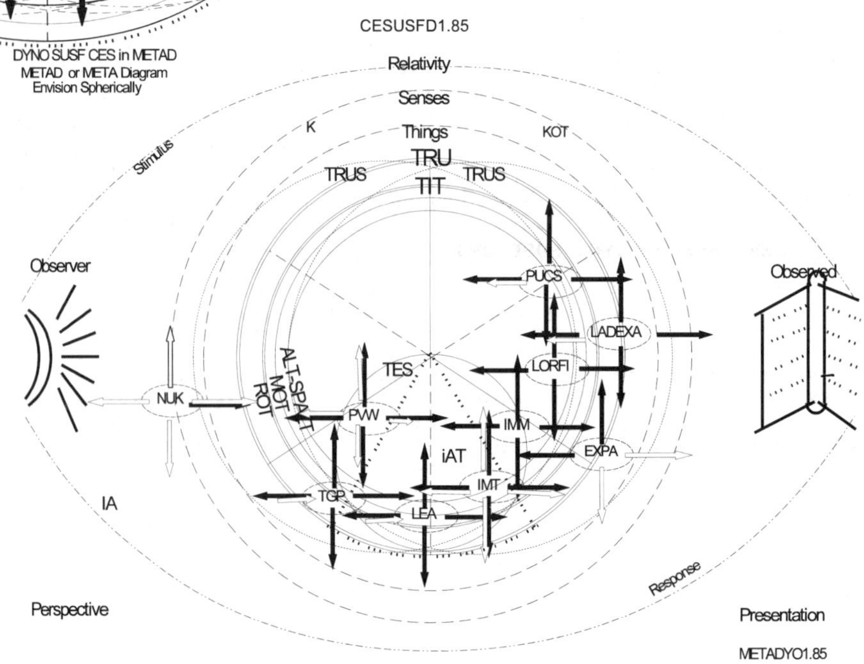

Fig 7-85 **DYNO SUSF CES in METAD**

CHAPTER 86

OSF or OBJECTIVE SMALL FUTURE PROCESSES of DYNO

ACTION-REACTION, DOING, ENERGY, GENETICS, PAR, GOOD, LIFE, QUALITY, SUCCESS, GOAL, TES

In POS, HOM, and GFST we should be careful not to accept as final singular modes, ES, characteristics, or CES in advancing the increasing levels of Á and TRU. The expansion of TRU has been more multiple and total, for all in TRU is part of the same thing, by DAOST. The individual should be measured by the sum total of all forces that produce them. These tend to converge on TRU. Although in terms of the highly adaptational, competitive, and selective nature of TRU, convergence is less TRU in result. This is also true of the general population. It explains the dynamics in and out of TRU, the high degree of strife and selection. These are what make-up the determinants of the whole, TIT-TO. How poorly or well an individual produces these forces is the determinant and measure of ones sum total, and to a large extent survival as it does not or does contribute to TRU.

LADTIS = Law of what Detracts from TRU Is as Important - if indirectly - as much of what Is TRU. It is as important to prevent, stop, act upon or destroy what may or does destroy or detract from TRU. For although TIT is only what is positive, the correction of the negative is a reversal which is positive, thus not contradictory. This is how LADTIS is significant in POT. Although medicine deals primarily with what is negative for TRU, and thereby not TRU, it is in this role of LADTIS that it acquires its great significance and importance. This is especially so to the degree this negative can be turned too positive. So that in the study of diseases we are in a most difficult position. This position is exemplary of many other situations and positions where we must analyze and interpret what is ATRU and ANTRU, and show it both as it is an entity in itself, ENIT, albeit destructive to TRU, and as it is in or detracts from TRU. It is here we investigate the great interface and interactions of MATRU, AMTRU and NAMTRU, different levels of mixed TRU and AU or ATRU. In Homation, the processes of man, and Medical Homation, for every disease objective entities are not actually as they seem but are only as they are reversible and positive in TRU. By curing disease we enhance TES and potential TRU.They are only as their ESINS are or become TRU. What detracts from TRU is eliminated. As we delve deeper and deeper into TRU we are faced with: **PROVEX** = Principle of Depreciative Value. PROVEX is a REM principle which holds that as we veer away from TRU, from even significant portions of the makeup of TRU, we are likely to rapidly become ATRU and lost.(IN,PIP) Imperatively, by our relation to TRU, by TRIA, we tend to correct that which is in being ANTRU such as threats or whatever is so damaging. This is: **PROVAC** = Principle of Value Accumulative. PROVAC becomes a law, as does all, when its limits, conditions, necessity and standards are great, imperative, shown and used. It is partly for this reason the treatment of diseases are beneficial when effective, yet so limited in terms of TRU in general.

SYNFE = Synthetic nature of the formulation of ESINS.

SYNFE denotes how the correlation of all things or entities proceeds by forms, classes, etc.(e.g. medical diagnoses) The formulation of ESINS to ES in TRU proceeds by only one ultimate bond or corrlelation and that is their relativity, part and role in TRU. No matter how good our SU, no matter how good our REM, the bonds and correlations of things, all entities must proceed from their relationships in TRU. These must be commonly determined piece by piece, often synthetically as we extract the ESINS of all, and apply them to TRU, ROT, MOT, ALT, SPALT, POS and Homation. No matter how well formed all entities are, their form per se, and their form as they are in TRU are far different. Therefore how well can universals be to systematic unity come to supply SYNFE? This has been suggested to us by the greatness of the "spiritual". Yet this does not began to give the form necessary. How Homation equals and is the means, the method of attaining SYNFE, and how SYNFE is the means and method of attaining a representation of TRU. To better understand SYNFE as well as all things, entities, words, ESINS, ES, CES and Posits we must analyze and compare them. Entities are mostly natural and genuine, but may be unnatural and synthetic. This is commonly, although not very scientifically, used as a difference between thing and entity. Words are mostly unnatural and synthetic. Entities, things, senses, words are all figment and illusion unless and to the extent they are TRU. They tend to be actualized in TIT-TO. Words often signify not necessarily the entity in itself but its object, some varied nature of it or how it relates to and is of practical and actual part of TRU. ESINS are synthetic, yet highly potential of TRU. For they are yet to be extracted. ES as CES or the parts and processes of TRU are genuine and actual. Posits, if of TRU and its form, essence and mostly of the ES and CES, are both genuine and actual. The small objective future of TRU has full and proper meaning only as all these ESINS, ES, CES, and CHES are accounted for in TIT-TO. Their synthetic nature is in their formation and usage, not in their existence in TRU.

Every person born, and at every minute and second of the day, has the greatest opportunity in the world to take the wonderful adventure freely and readily given to them. This is not only to live the best and greatest life possible, it is to be a great spectator and participant in this greatest show on earth. It is in being. a super spectator and participant, thus over all. Like the best of athletes to win a minor victory is much less than to continuously win the most glorious victory of all, an ever greater the victory of all, as TIT-TO. It is in being such an adventurer or sportsman that all can find that winning, victory, glory, Eureka, Eldorado, Phoenix, or Holy Grail that is actually, finally and totally satisfying and worthy. If you fail to positively add to all that counts you are DAK, dead and don't know it. You are neither TES nor TRU. When this is extended you become **DAKIN** = DAK to the nth degree, for not only are you not TRU, you are acceleratively and even exponentially damaging to TRU. It is for this reason we must be frugal, conserving, well conditioned, moderate, and tolerant, yet with utmost probity active for TRU, a complete counterpart, TES. We can not be TES unless we know what TES and TRU are.

What is par is an important factor in measures and determinants. What is par for individuals, TES, and TRU are much different. Even the par for TRU, is more than normal, more than the average for all humans, or even the average for all healthy individuals. In being highly competitive and demanding it has added FAF. For individuals this is called: **POPAR** = The normalcy for People is an Average of the Total Population. On the other hand the par for TES is a highly qualified state, FAF of this normalcy for the average population. For TES and TRU this is called: **TEPAR** = The normalcy for People that is an Average of TES. The forces that make us react, return to normal and straighten us out are much the same for both POPAR and TEPAR. Only in TEPAR they are much more intense and qualified. They are even much more demanding. TEPAR are strictly all forces that are most demanding of us as we are only TES. We are both propelled and compelled to discover, show, and cultivate TES and TRU. These are vivid examples of TRUD, the dynamic makeup of TRU, its CES, Homation, DYNU, DYNS, DYNO and ACC. None of us are so big that we can both know and do all, always giving and never receiving. This is why Christ is so great. His example of giving, loving, and dedication to God, total sacrifice and martyrdom is the reward we receive and can give in return by simply being, being alive, and being an *Imitation of Christ*, of TES. To appreciate, yet live less, in ways that are less all suffering, enslaving, drudgery and condemned than his, is the experience we must absolutely learn. This, even the greatest of lessons, that not only Christ, but all those who have suffered for us that we may live with all the blessings of TRU, are what must be ingrained and inspired in all.(ANCEF) For it takes a great mass of experience or conditioned sensitivity to our benefices, to achieve the FAF necessary for TES in life. Only with VAPO and TES, in perpetuating all means that replenish and replace such experience, can we maintain ourselves and TRU.(PPMA)

All but TRU is illusion, figment, vanity, negative and loss. For all the time - and ACEXIN - that is lost in the negative is gone forever. It is to that degree loss of life which is actual death, 1/TRU. This is called: **LATIM** = Law of Time Lost is ATRU, negative to TRU. This is so for all states of Á, L, and T. LATIM can be both qualitative and quantitative, great or small, direct or indirect and more or less separate or integrated. It is a large factor in OITIO and the extra/intra dynamics of TRU and Homation.

Since TRU is an equilibrium between same and different, integral and differential, unity and infinity, it is not, in being separately more partial, same, differentiated, integrated, unified or infinite that we should exclusively sacrifice ourselves. Our great sacrifice should be for all TROS, DAOST, VAPO, SCARLS, ACEXIN, and TIT-TO in the improvement, benefit, betterment, and perfection of the actual world. To counter these discrepancies we have: **SATLEP** = Self Sacrifice for TRU by being less ATRU, AU, ANTRU, and partial, that is not TES. SATLEP means that one should go to the lengths of their ability to bring ATRU, ESINS, and uphold and sustain CALP, ES, TES, and TRU. To incorporate any and all in TRU as fits each given or occasion. SATLEP is the answer to most of those who resist or oppose TRU.

	ABSOLUTE	RELATIVE	PROBABLE and ACTIVE	COMPLEX and POSSIBLE
General Physical	Radiation, Physical	Temperature extremes	External Agents	Mechanical
Structural	Anatomy, Skull	Reproductive Anatomy	Micro/Molecular Form	Genetic Form action
Chemical	Presence, Effect	Dietary, Toxic	Smoking	Many Agents, Drugs
Physiologic	Bodily Function Tests	Reproductive (f)	Genetic function test	Activity Extremes
Neurological	Anatomy CNS	Psychological tests	Behavior Analysis	Empirical studies
Conceptual/L	Brain Centers of	Vocabulary	Memory	Read-Write skills
Cognitive	Consciousness	Attention	Thought	Answers, Solutions
Rational	Reason	Judgement	Choices	Problem solving skill
Intellectual	Intelligence	Interests Inspiration	Valuing, integration	Closure VAPO-TRU

Table 1-86 **VOLUNTARY and PLANNED PARENTHOOD**

Parenthood can be voluntary, involuntary, planned, and controlled to provide the best possible quality for humans. Natural selection has not only been of physical strength but of increasing prowess in all predominant powers of Á FAF,.any kind of human quality that sustains TRU. Human knowledge and science have increasingly provided better measures of all that enhances the quality of our capacity to be TES in TRU. In Table 1-86 classify by degree of certainty in columns and levels of Á powers and forces in rows as examples of SÁ that help to make such a determination.

Fig 1-86 **ELTIMAP, CONVERSION of ELTICMA to PERSONAL LIFE and USE**

Many of the goals and actions in totality, TIT-TO, are seen in ELTIMA, ELTICMA, and **ELTIMAP** or Everlasting life through Personal Coordinated and massive Instantaneous Attention and Action. A greater salvage rate for humans, to be superior, as *jen* in Chinese, or TES is needed. This means the measure of our actions, policies, and successes should not be in terms of one or another limited, partial, irrelative, or ATRU IA but most practical and valuable to TIT-TO. This, both qualitatively and quantitatively, is the fulfilling of ones being, TES in TIT to TA. ELTIMAP is produced where and when ELTIMA and ELTICMA can be developed for personal use. Evil, wrong, bad, ATRU, ANTRU, and ANTITRU are not only associated with singular or small ENIT, elements, or forces. They are everywhere, all about us. If you would seek them out, start with yourself, then your immediate acquaintances, groups, and surroundings. You will be surprised to learn what and how much you and yours are not so good, less or not TRU.For this is what you have most experienced, know, can find inspiration for TRU, and be best remedied. **SRAP** = Solutions are most Readily found, Acted upon and resolved when close at hand. **SREP** = For each and every need and Solution there are apt to be close at hand, associated cures, answers and Remedies. Of the many factors that provide natural selection and genetic health and how the future can be assured for the benefit of TRU is seen in: 1. Massive increases in our knowledge of genetics 2. Increases in our knowledge of reproduction, propagation and population genetics, 3. Massive increases in our knowledge of the processes, mechanisms, and interactions of genetic material, live product and determining forces that occur differentially and integratively within TIT. 4. Reduction in birth rates, especially in countries that are least able to support them, and less capable of adapting to an increasingly complex world, 5. Discovery, learning, knowledge, application of TRU, POS, SPALT, HOMATION and CES, 6. Application of our knowledge of ACC, 7. Guidance, encouragement, and measures to enrich all lives with willing sacrifice for TRU. For those whose genetics will detract from TRU this will require special guidance and measures. 8. Education and voluntary selection, 9. Use of all knowledge for optimal TES and TRU.

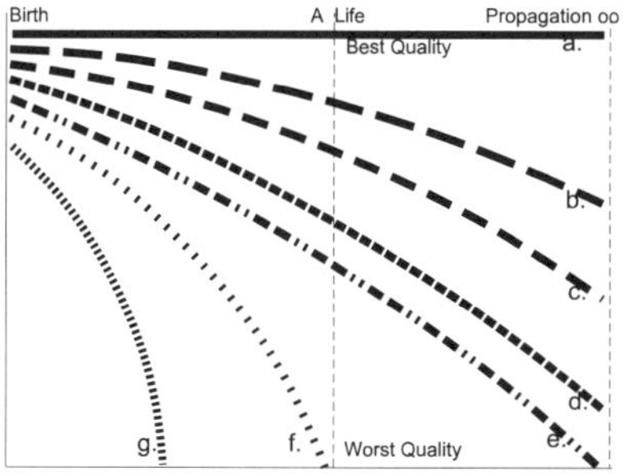

Fig 2-86 QUALITY and COMPARATIVE LIVES
In this diagram levels of quality in TRU are correlated with comparative lives as measured by survival of progeny, shown by: a. Perfect survival, Propagation, b. Surviving with more than par propagation, c. Surviving with less than par propagation, d. Surviving marginally propagating progeny, e. Surviving slightly propagating progeny, f. Surviving but non propagating progeny, g. Early death, rapid extinction. Some of the key factors that contribute to this quality, and thereby survival and progeny include: 1. Health, life in and by TRU, 2. Production in life, Contribution to TRU, 3. Lasting Contribution, after death, 4. Secondary or indirect benefits. Each individual, past, present and future, has to make it in life. They can be given life, helped, provided for, directed and stimulated in various ways. Still, it is up to oneself to make something of life, their future, to live or die, be or not be, succeed or fail, be TES or not TES, nothing or immortality in whatever degree, and be TRU or not TRU. We can find consolation in knowing the next person is in much the same boat, or situation. This is by the S in DAOST, all human roles are, more or less, identical. This is both by being in the TRU/AU interchange, as well as in TIT. Life is a great leveler, even-er and equalizer.. The individual to fulfill this role in life, TES, must be at least even, reach a par, if, of and in, TRU. Ones progeny just as ones ancestors are also in this same state. It is through and by them that much of the impetus to make it in life is powered. To a large extent life is predetermined by TIT, for which the best way is to simply to follow its course.(Tao) Thus seeking the natural way and life becomes a standard. Such standards like customs and laws yield fairly simple and beneficial, yet imperfect means. **LONS** - Law of Natural Standards. This depends upon the degree it is TREQ, and TRU. With a degree of individual and situational variation each of us avoids much suffering and loss in life if we follow that best for our age and role in TRU.(Confucian) The overwhelming mechanism that saves us is our capacity, ability and performance in TES that has, is and will be TRU. To see each new generation come into the world in health and happiness, grow, mature, learn, work and achieve newer andgreater things, along with maintaining and utilizing all that is good and great from before, are the gifts and benefits of those who leave this life. When they know that the world they left behind is better and greater, especially in quality and ability to continue, they can die in peace. They can physically die being assured that this death is merely a hiatus for continued life. Those they leave behind continue their life by having more needed for maintaining TRU indefinitely.

TRU VALUE	TRU STATUS	VOTING STATUS REM and TES	INDIVIDUALS PERSONAGES	BY TYPE of WORK OCCUPATION	OTHER FACTORS
>1,000,000	APPROACHING TÂ	100:1 VOTE	CHRIST, BUDDHA	GREATEST LEADERS	STIM. ROLES OF
100,000	GREATEST TES	50:1 VOTE	CONFUCIUS MOHAMMED	TOP IN RELIGION	STIM. REVELATION
10,000	GREATER TES	20:1 VOTE	SOCRATES PLATO	TOP IN PHILOSOPHY	STIM. INSPIRATION
1,000	GREAT TES	10:1 VOTE	ARISTOTLE EINSTEIN	TOP IN SCIENCE	STIM. CURIOSITY OF
500	NEARING FULL TES	5:1 VOTE	DI VINCI, PLANCK,	TOP IN INVENTION	STIM. IMAGINATION
200	HIGHEST CONTRIBUTOR	4:1 VOTE	COLUMBUS,,,,,	TOP IN DISCOVERY	STIM. SPIRIT OF
100	HIGHER CONTRIBUTOR	4:1 VOTE	LOCKE,,,,,	TOP IN GOVERNMENT	STIM BEST DEALINGS
50	HIGH CONTRIBUTOR	2:1 VOTE	SMITH,,,,,	TOP IN ECONOMICS	STIM. BEST TRADE
30	GOOD CONTRIBUTOR	2:1 VOTE	DEWEY,,,,,	TOP IN EDUCATION	STIM EDUCATION
16	VERY HIGH TRU/ATRU	2:1 VOTE	MICHELANGELO,,,,,	TOP IN ART	STIM ART & BEAUTY
8	HIGH TRU/ATRU	2:1 VOTE	BACH BEETHOVEN,,,,	TOP IN MUSIC	STIM MUSIC & JOY
4	MUCH MORE THAN PAR	1:1 VOTE	RAPHAEL,,,,,	TOP IN PAINTING	STIM APPRECIATION
2	BETTER THAN PAR	1:1 VOTE	EDISON,,,,,	TOP IN ELECTRONICS	STIM APPLICATIONS
1.5	ABOVE PART	1:1 VOTE	STRENGTHENS of TRU	GOOD PROFESSIONALS	TEST TRAIN BETTER
1.2	BARELY ABOVE PAR	1:1 VOTE	LEAD OTHERS to TRU	GOOD LEADERS	TEST TRAIN BETTER
1.0	PAR, MAINTAIN TRU	1:1 VOTE	JUST SUSTAINS TRU	ANY TRU WORK	KOT, PUREACLITY
0.8	UNABLE to " "	1/2 VOTE	UNABLE TO " "	LESS TRU WORK	ENCOURAGE LEAVE BE
0.5	POORLY " " "	1/4 VOTE	POORLY " "	MUCH LESS TRU WORK	ENHANCE TRU IN
0.1	SOME POSITIVE """	1/10 VOTE	AVERAGE PERSON	MORE VOID THAN TRU	TURN VOID TO TRU
0.01	SLIGHTLY POSITIVE"	1/100 VOTE	WORSE THAN AVERAGE	POOR WORKERS	INCREASE POSITIVE
0.001	VERY LITTLE "	1/1000 VOTE	MUCH WORSE " "	POORER WORKERS	INCREASE POSITIVE
0.0001	PRACTICALLY no "	1/10000 VOTE	VERY MUCH " " "	POOREST WORKERS	INCREASE POSITIVE
0.0000	INADEQUACY, LOSS	Exercise VOTE	DRAGS ON SOCIETY	NO GOOD WORK	MORE PLUS/MINUS
-0.1	BEGINNING NEGATIVE	No Vote Temp	HARMFUL TO SOCIETY	ATHEISTS	TURN MINUS TO PLUS
-1.0	NEGATIVE @ ONE	" " "	MORE " " "	BAD WORK	TURN MINUS TO PLUS
-2	DOUBLE NEGATIVE	" " "	VERY " " "	NIHILISTS	SEEK ANY PLUS IN
-4	QUADRUPLE "	" " "	MOST " " "	HARMFUL ACTIVITY	SEEK ANY PLUS IN
-10	REPEATEDLY "	" " "	HITLERS,,,,	FANATICS	GIVE CHANCE TO TRU
$1x10^{-2}$	OVERWHELMINGLY "	No Vote Perm	TIMURS, NEROS,,,	HIGHLY HARMFUL	IF NO CHANCE 1/ANT
$1x10^{-3}$	DESTRUCTIVELY "	" " "	ATTILAS,,,,,	MOST HARMFUL	ACC AND PPMA ON
$1x10^{-4}$	HIGHLY DESTRUCTIVE	" " "	SCOURGE OF GOD TRU	TERRORISTS	CONTROL WITH ACC
$1x10^{-5}$	MOST DESTRUCTIVE	" " "	DEVILS	MOST DESTRUCTIVE	ACC STRONG CONTROL
$>1x10^{-6}$	TRU EXTINCTION	" " "	ABSOLUTE DEVILS	WORLD DESTROYERS	RIGOROUS PPMA ACC

Table 2-86 LEVELS of CONTRIBUTION or DETRIMENT TO TRU

In Table 2-86 different levels, individuals, and groups are compared by their contribution to or loss to TRU. We do so, in columns by categories of TRU value, voting or civic power, personages, professions and other modes. Many other persons, groups, and gradations could be supplied. The above is tentatively given as an example of what can be done and what is needed for ACC. In this tentative state, it is not to be interpreted nor used as a well determined and final guide. Though human equality and a classless society have long been a goal, as well as the current standard, TRU and especially its enhancement mandate certain modifications. This agrees with selection, competition and survival in life, and especially the predominant nature and FAF of TRU. In the category of voting and civic power there has been in modern times a belief in a one person one vote. This has been a product of and was greatly superior to the disenfranchisement that had occurred in most of history. Yet humans are naturally not equal, especially in degree of TES and TRU. To use such a system can be no better than the degree TES and TRU are known and applied. The problem is both, where and how to apply and use them, with no chance for abuse. Such selection would be of benefit in reversing the drift toward poorer quality and thereby in a highly cosmopolitan and ATRU world. It would also add stimulus for each persons role as TES. It would make up for poorer levels of natural selection. It would help to slow down the population explosion. This can be instituted both voluntarily and gradually depending upon its correctness by TRU. It is already in place to a small extent by disallowing criminals and discouraging less TRU the right or means to vote. It would also have great salutary effects, partly through competition, and effect on enhancing TRU. It would help to resolve racial and religious conflict by decreasing bias, bigotry, and all without VAPO and TRU. Yet it would have to be done with great toleration, moderation, and precision.

There are certain common general principles or dictums that greatly assist living and TRU: 1. Learn what you must do, 2. Learn to like the things you have to do, 3. Plan ahead and use your head, 4. Don't do things that are wrong, 5. Learn what is and what is not by TRU, 6. Understand that for what is wrong, you pay, 7. Don't yearn for the bad, the impossible, and anything that is not TRU, 8. Don't try to live with, bother with, or try to correct the bad in any of its multifarious faces unless it can be helped to become TRU, 9. Realize the joys of a simple, happy, and frugal life, 10. It is better to give than to receive, 11. Realize most joys of life are near at hand, 12. Understand that all that is not VAPO is not of TRU, yet also understand all that is common or equivalent, TREQ, is TRU, 13. Recognize and emphasize the ESINS, but not ENIT in all, 14. Comprehend all and act by accounting for all, 15. Take heart in being an observer and participant in the greatest of all struggles, endeavors, games and adventures of life, TRU. 16. Find great joy and fulfillment in the analysis and positive guidance of the concepts, principles, laws, processes, and powers of TRU that apply to your life in TES, 17. Take joy in your contribution to those processes and increments of life that assure success and accentuated TRU, the greatest glory, victory and eternality of all. It is in # 8 that we can recognize a great strategy that is the gist of life in TRU, or: **LASDANT** = Law of Dismissal and Denial, of and by the Self, of all that is From, Not or Against TRU, or oneself as TES. It is by denying oneself of all thought, choices, desires, acts, behavior, work, possessions and associations that are less than, other than, negative to TRU, and to that degree. This is not spontaneously accomplished. It requires life long training, and can, especially when well inculcated early in life become a great and powerful habit for good, TRU. LASDANT is similar to Christ's teaching of 'turning the other cheek'. For if one is highly interested, dedicated, devoted with deep love and conviction in God, or TRU, there is no place nor inclination for all else. LASDANT is a rational and strategic means of reacting to all that is not TRU.

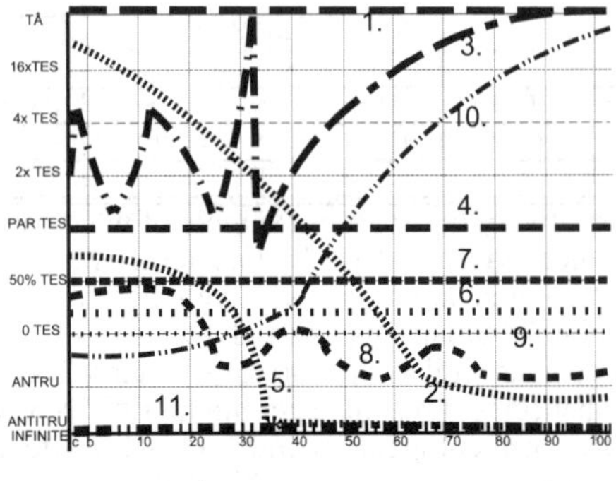

Fig 3-86 **COURSES of SUCCESS or FAILURE in LIFE**

The course of a person's life can be TES, less than TES, variable TES, NON TES, ATRU or ANTRU. It is usually a great, variable and rapidly moving and changing mixture. Only in the resultant of a lifetime or over longer periods of time can a good measure of TES be provided. Many of the recognized and accepted measures of success or failure in life are equivocal, worthless or wrong. Even by TRU standards it is easy to be mistaken, for what may be fixed for TRINI may be different for TRINE, TRIN, TRIS and TIT. In Figure 3-86 the courses of eleven individuals or types of individuals who demonstrate success and failure in life as measured by TRU, are shown: 1. TÂ. 2. An individual who has great initial potential, excellent heredity, physical and mental capacity yet who, through faults of others or their own rapidly, waste and lose their capacity and power of TES and TRU. 3. A diagrammatic tracing of the life of Christ. 4. Par for TRU or that level which on average must be maintained if TRU is to hold its own. 5. The life of an arch criminal, or mass murderer. 6. Is the average course of all individuals. 7. This line 50% TRU describes that level of individuals who give some benefit to TRU but not enough to sustain TRU. A half hearted effort or contribution, a resultant drag on TRU. 8. One who has less than average endowment, and through the vicissitudes of life follows an up and mostly down course, 9. That level at which ones contribution, as received or as expended in life is of no TRU value directly, yet indirectly detrimental in being less than par, 10. How a person with very bad heredity, physical strength, intelligence and even poorer upbringing can through self restraint, self denial, sacrifice, character and compensation through spiritual and total Á Mass-Energy, contribute beneficially and exponentially to TRU. This is seen in the physical, or otherwise handicapped person, who through compensation does great things, and leads an exemplary life. This is achieved through great discoveries, inventions, authorship, endeavor, and works in life that multiply for others to follow and profit by. It is done through a heroic effort, superhuman love of life and absolute determination in the success of TRU, TÂ. 11. The absolute limit of negation, harm to TRU. It is the opposite of # 1, and like # 1 does not occur in one or many, only limits of TES and

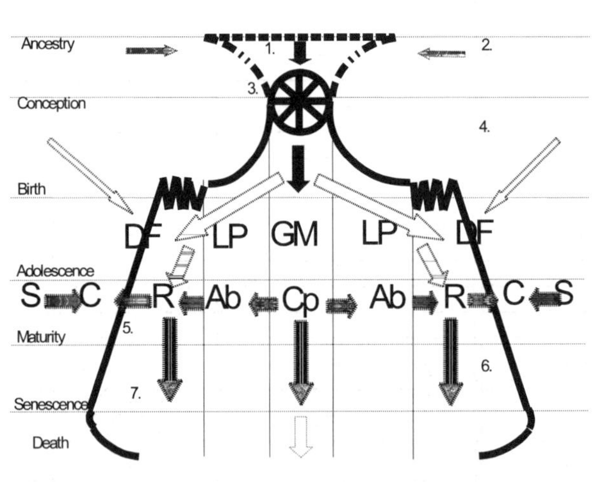

TRU. It is an example of what must not be, by reversal we know what to be.

Fig 4-86 **A TYPE FORM FOR REACTION FORMULAE**

This figure seeks to answer and solve some of the questions and problems about the forces, dynamics, their resultants, role of the CNS, memory, and how they are continuously assimilated and eliminated by IA in TES. They concentrate primarily on all that effects the CNS, although it could as well be done for all other levels of Á, or their totality in human life. Abbreviations are: Ab = Ability, GM = Genetic Material, C = Condition, LP = Live Product, Cp = Capacity, R = Reaction, DF = Determining Forces, S = Setting. The numbers signify: 1. Non-differential, inherited neuropsycho-logical characteristics, 2. The average for a specie, race or group, 3. Genes and chromosomes, form and function, or interactions,

4. External forces, setting, conditions, physical and of all kinds on the organism, individual or TRU, 5. How the individual, as TES, is constantly assimilating and eliminating, 6. Reactions without experience, stimulus response, reflex, autonomic, 7. Reactions from experience, memory, reason, conscience, and exercise of. This diagram only begins to show the forces in action. They may be superimposed on and are shown in more detail for HOM by POS in SPALT. A healthy complement of capacities are needed to sustain life. They are continually upgraded in life's cycles either distinctly and directly or less distinctly and indirectly. Many of these are inherited initially and gradually become altered for many reasons, causes, forces, acquired characteristics, and effects. They may be likened to ones sex cells and somatic cells divided into what passes on from generation to generation and that which is active only in the live product, proband, or first person. That part of TES and TRU that is limited to the sexual life cycle of generations is called: **SLICT**. When limited to all that is passed on other than SLICT, or is asexual, is called **ASLICT**. SLICT and ASLICT although much the same and highly interactive are also much different and separately operative at different ages and milieu in life. Each persons life is neither nothing nor absolute except in the sense of LOPAN in TES. Happiness, as life itself, is health and the best equilibrium of contrasting forces and powers in TRU.

Fig 6-86 **LIFE EXPECTANCY and QUALITY**

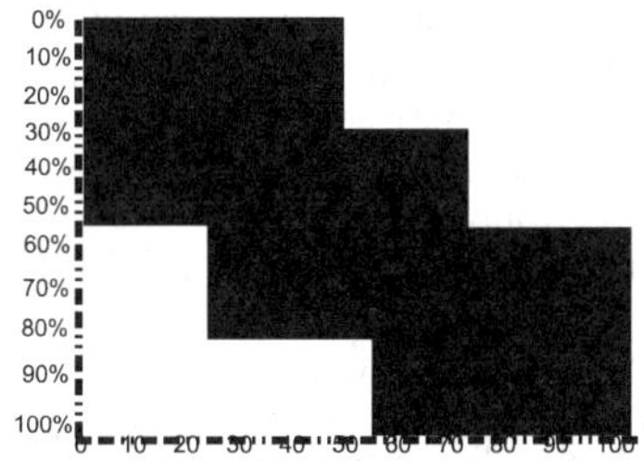

How do we get the most out of life, its span and maximum TRU. Is it 1%, 10%, 50%, 90% or 99% of the possible maximum? How much does knowledge, love of GOD, TREQ or TRU accelerate it? No cells grow forever. Undernourishment, to a point, is known to extend or even double the life of certain animals. From earliest life the cause of late maturity affects the extension of life.The vertical axis shows the percentage of possible quality attainable. In Figure 6-86 we combine three means of providing greater TRU in a person's life. The horizontal shows the total potential quality attainable when all three means are combined: 1. increasing life expectancy, 2. increasing life span, 3. all other means of adding to life, including its intensity and TRU. Expectancy, span, and quality are combined to show means of providing greater TRU in person's lives. A person's reward in life should be to a great extent proportional to the degree to which one, loves, contributes, helps, and works for TRU. The lack of reward should be the reciprocal of this. Yet the problem is complicated by what is TRU reward. It is mostly within TES and TRU, their joyful fulfillment, and the knowledge and hope for greater life expectancy, span, quality, and future continuity.

With modern genetic engineering and the great advances in science in general it is likely that in the near future human lifespan will be extended. This is not only life expectation which is already occurring in both the quality and quantity of life, but the amount within a given time period. All that assures TRU provides for a vast expansion, as has already occurred in this revolution of the Á FAF and capacity of humans. This will undoubtedly result in a successful and better life of TRU. Life expectation and span are highly dependent upon a culture's ability to provide for favorable conditions. This is what has varied so from age to age, or: **LACVA** = Law of Cultural Variance. LACVA signifies that at any time in prehistory, proto-history or even recorded history the various cultures, communities and groups were not on the same level. This is found in Archeology, for even today some cultures show very high levels of advancement, and some are highly regressive. There are optimal levels of quality and quantity which different cultures sustain.

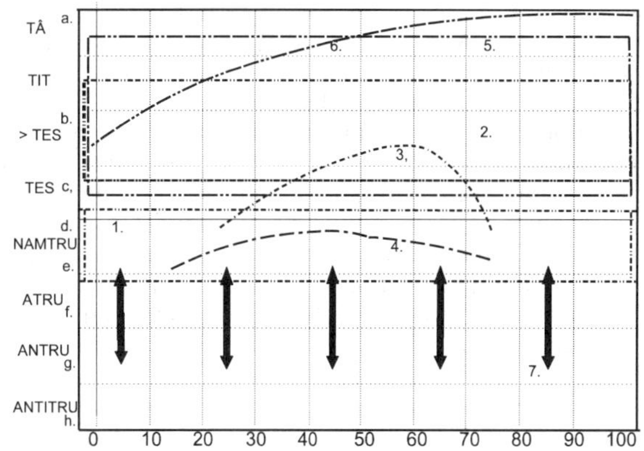

Fig 5-86 **INDIVIDUAL RISES/FALLS from TES**

The individual rises, does not rise, or rises and falls in flux, in their part and role TRU. This is: **COILAT** = The coincidence of life and TRU in each individual or person. COILAT is highly variable, just as it is from person to person, and even of all in TRU. A test of a person is the degree to which one resolutely rises to TES. Figure 5-86 demonstrates certain levels and courses of these: a. Individual perfection in TA, b. Manifold TES in TRU, c. Just equal to a 1:1 of TES, d. Level of bare maintenance of TRU, e. Level of bare maintenance of existence, f. Relative demise, TRU demise longer term, g. Demise of the Individual, Rapid TRU demise, i. Immediate or absolute demise. The number of each course signifies: 1. Mainstream of life, this may be divided into different parts, a, b, c, d, 2. Mainstream of TRU, 3. Mainstream of professional perfection, 4. Individuals rise through the mainstream of life, their maintenance forces, and courses in, 5. TRU over and above the main stream of life, 6. TRU to TA, 7. factors and reasons, 20-100, why a person does or does not rise to or reach TES and TRU..

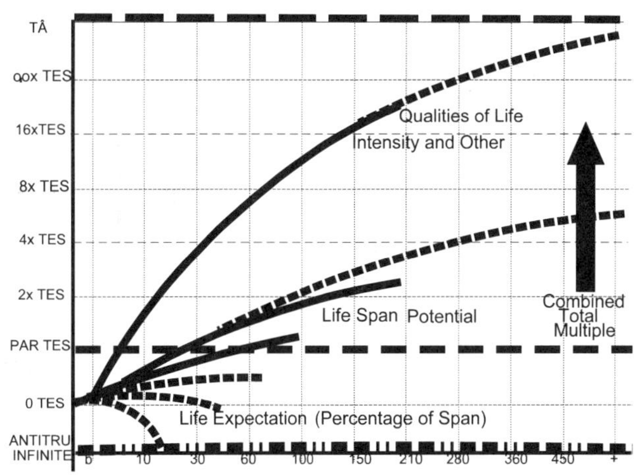

Fig 7-86 **LIFE EXPECTATION, SPAN, QUALITY, SURVIVAL in MULTIPLE and QUANTIFIED**

When combined and optimal we get the best of Fig 7-86. The degree to which these are less is a smaller and smaller section of the total diagram. By combining these means of enhancing life, superimposing their benefits, and adding all other means of bringing the highest exponential accentuation to the quality of life as by increasing its intensity, IA, and other means we provide exponential growth of both quality and quantity of life for TRU. They are presented in terms of a life time, potential life time and the combined and added quality and intensity that may be potentially achieved .

If a great part of the world's population could live to be 1,000 years old they would not all necessarily be TRU, TES, nor live in PUREALITY. Certainly, however, when life expectancies are low, they are of less to little short term benefit to TRU. This applies more to ASLICT than SLICT. In fact, here we can see how ASLICT and SLICT operate interactively positively and negatively. Their natural selection may rise, but all the other constituents, especially those of the subjective, spiritual or phenomenal of TRU do not. In general the quality of life depreciates with shorter life spans. This is for at least two reasons:, 1. they have less time to accomplish work and contribute and, 2. the period of maturity and rational life is much shorter. Natural selection is very high, but much of the rest of TRU is low. There is an ebb and flow in human affairs. For TRUD this is called: **LEFAM** = Law of Ebb and Flow in the Affairs, Successes or Failures of Man, individuals, groups, nations, cultures and civilizations. LEFAM shows the accumulative effects of a healthy culture, or the detrimental effects of a sick culture. It is highly related to INEG. Immortality and TRU can not wait until the end of a persons life. The entire life of a person must be TES and TRU. This is: **LEFRAM** = Law of Ebb and Flow of the Rational and TRU in the Affairs, Successes or Failures of Man, individuals, groups, nations, cultures and civilizations. All persons must recognize TRU, and properly correct all that is not TRU. They must always be cognizant of, appreciate and completely endeavor for TES, TRU, PUREALITY, COGNOS, ELTICMA, OM, SA, and TA. In this way we can continue to make the gains, continue to build upon all that has gone before that is great for TRU, and make TRU ever greater and better.

Among the various actions and reactions of a person there is in totality, a resultant balance. **LARET** = Law of Action Reaction Equivalence in TRU. LARET is closely associated with and derived from the laws of motion. For the relationship between our acts, objectives and goals we have: **LAGOT** = Law of Our Acts must be Equivalent to our Objectives, our Objectives must be Equivalent to our Goals, and our Goals must be Equivalent to TRU. LAGOT shows how all TRUD motion, in and about a person, is not and can not be chaotic. There is a high degree of necessary interdependence between actions, reactions, objectives, goals and TRU.

Each person contains a storehouse of energy that can be highly TES and TRU. It, when in the best of health or SÁ, approaches perfection. Perfection is not absolute nor irrelevant. Perfection is the harmonious, symphonic, dynamic equilibrium of all that is 100% TRU. It is all counterparts of TRU that function in optimal unison. It is this view of the total dynamic world, TRU, and ourselves that we must understand.

With the more rapid expansion of human mentality, both by capacity, ability, knowledge and activity, natural selection is rapidly changing. It depends upon ever higher levels of SÁ, reason, values, intellect, and comprehension. Yet these often work in reverse by enabling the less fit to survive and reproduce, and the <u>more fit to reproduce less</u>. It is this that must be compensated for or planned to correct. Biological controls and genetic engineering are only special kinds of selection. They can be used for good or bad. It is for this reason we must learn and account for the values and uses of genetic engineering, biological controls, and activation. It allows us to direct natural selection from a very poor rate of efficiency toward ever higher rates through TIT. We must recognize all the forces at play and processes in action that add or subtract from genetic quality and human capacity. It is much more than the genetic material alone. It is certain that, as in preventative medicine and its great benefits and successes, a proper attitude toward genetics and human biological improvement is necessary if only to guide rather than control its course. Much of human adaptability plus science's deep awareness of the extremely vital benefit/harm ratio can, must, and will, when proper and by TRU, provide the necessary PPMA and corrections.

In positing CES of OSF of DYNU, DYNS, and DYNO in the three following figures the void in those of smallest size and most distant future is not so much a defect in the text, as it is in human knowledge and especially knowledge of the smallest and most far term processes of the future. Many of these we can roughly infer, by way of VAPO, SCARLS, and existing ES and CES of TRU. Yet, often, this can not be done with the reliability and certainty science demands or even many CES provide in other sectors of TRU. For this reason,, and for now, such voids are left as is. Examples of these are molecular forces and determinants in the future.

Fig 8-86 DYNO CES in OSF SECTOR of SPALT

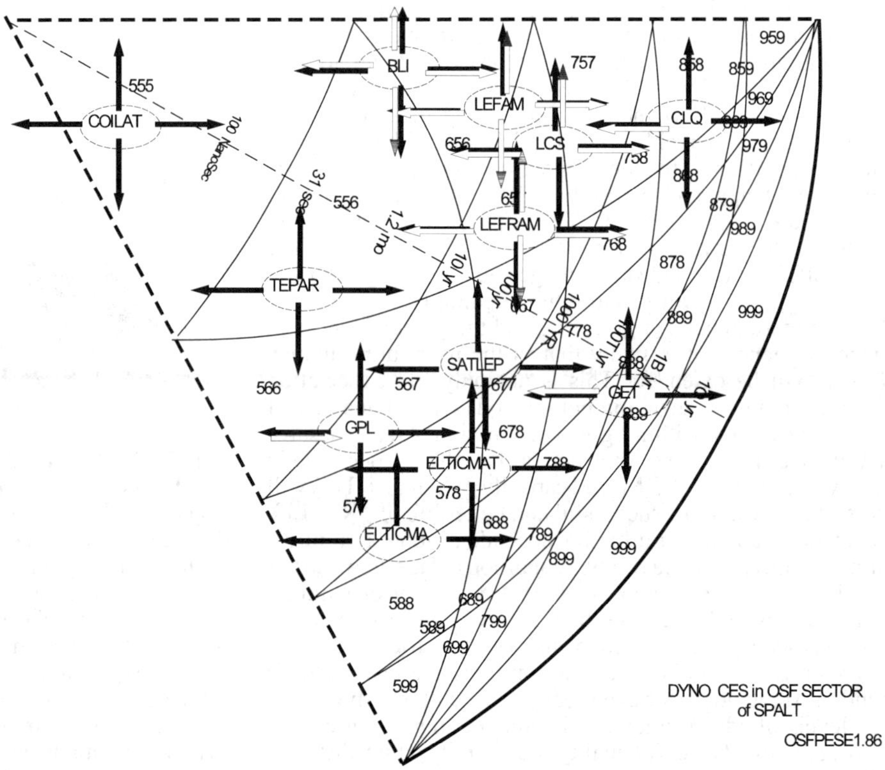

MEASURES of DYNO OSF CES

Measure	Values	Scale	Abbrev.
Benefices,Loses, Ind, TRU	6-5-6 ±	S 4	BLI
COILAT	5-5-5~	S 1,3	
Comparative Lives, Quality	8-5-8 ±	S 4	CLQ
Contribution or Detriment	7-5-6~±	T 4	CoD
DAKIN	~7-5-7~	-M 8	
ELTICMA, ELTICMAP	5-7-8	S 3	
GENETIC ENGIN., TES, TRU	8-8-8	S 7	GET
GENERAL PRIN. of LIVING	5-6-7	S 7	GPL
LACVA	6-3-	M 4	
LAGOT	8-5-6~	M 7~	
LARET	~8-5-5~	M 1	
LASDANT	~5-5-5~	T 1	
LATIM	~-~5-~ -	M 8	
LEFAM	~6-5-7~±	S 4	
LEFRAM	6-5-7	S 5	
LIFE'S COURSES, S/F	6-5-7 ±	S 4	LCS
LIFE EXPECTATION SPAN,%	9-5-7 ±	T 4	LES
MOLECULAR FOR., FUT. DET	9-8-~ ±	M 4	MFF
OPPORTUNITY of PARTICIPATION	7-5-6	T 4	OoP
OSF	7.5-7.5-7. 5		
POPAR	~8-~-7~ ±	M 6	
PROVAC	~8-5~-6~	M 6	
PROVEX	~8-5~-6~-	M 2	
REACTION FORMULAE, DFLPGM	6-6-7 ±	T 7	
RFD RISES and FALLS from TES	5-5-~ ±	M 4	RFT
SATLEP	6-6-7	S 1	
SRAP	5-6-6~	S 1	
SREP	6-6-6~±	T 4	
TAKING and GIVING in LIFE	7-5-5±	T 6	TGL
TEPAR	5-5-6~	S 3	
VOLUNTARY and PLANNED PARENTING	7-6-7	S 1	VPP

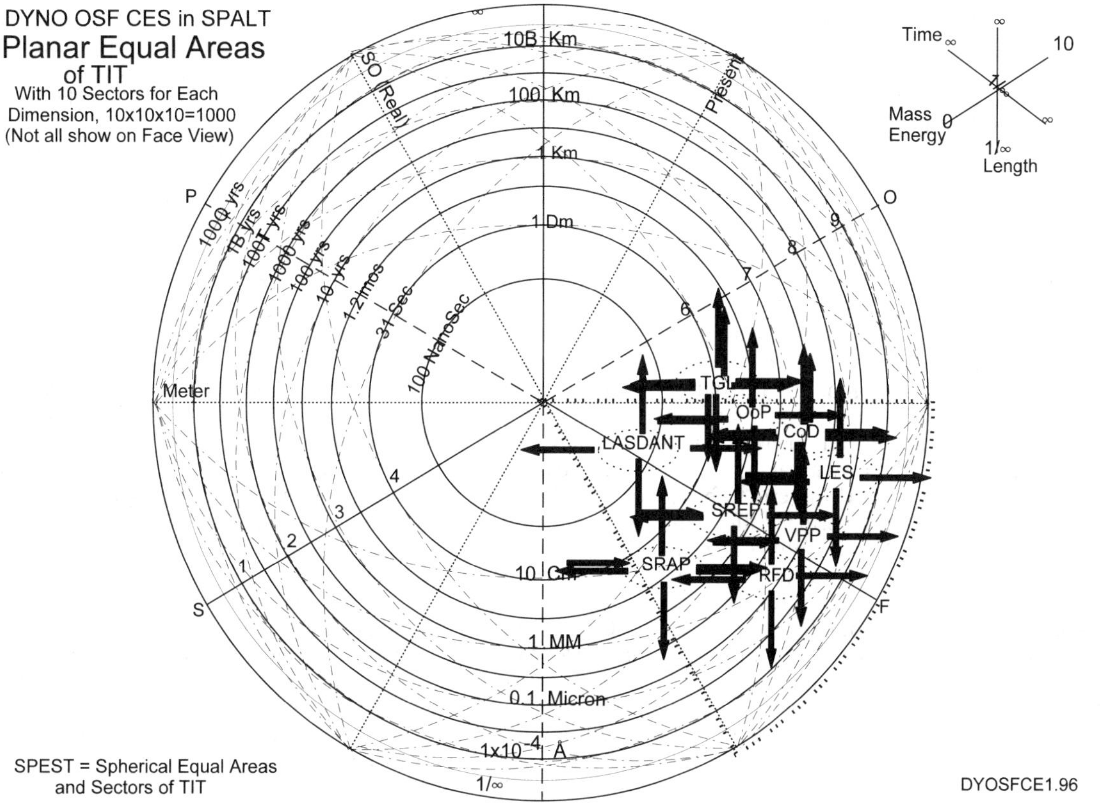

Fig 9-86 **DYNO OSF CES in SPALT**

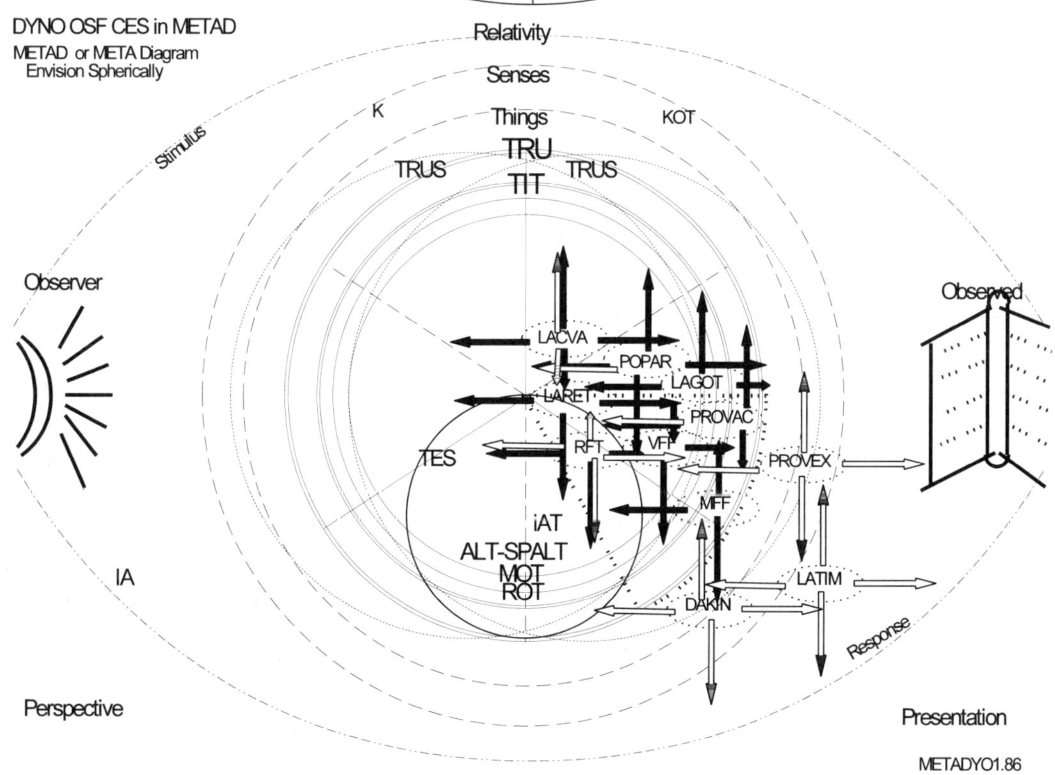

Fig 10-86 **DYNO OSF CES in METAD**

CHAPTER 87

OLF or OBJECTIVE LARGE FUTURE PROCESSES of DYNO

QUALITY, ECONOMICS, CARE, PREVENTION, POPULATION, CULTURE, CONTROL, CLIMATE, SURVIVAL, ENDS

Of greatest human concern is propagation of maximal quality. This can be done in many ways. It will take a great amount of information, measurements, and voluntary cooperation. It will take much PPMA, and attention to, concern with, and appreciation of all that provides this quality. This is not only a thorough knowledge of health and genetics of humans individually, collectively and totally, the determining forces, but all their interactions that result in the propagation of maximal quality throughout the Á FAF of TRU. This can't be attained without the cooperation of each and all persons. Their interest, conviction, dedication, love and endeavor to attain and sustain TREQ, TRU, TES, PUREALITY, COGNOS, ELTICMA, SÂ and TÂ is paramount.

WOFREM = World Only Follows Relative Magnitude. The resultant world, TRU, of interacting worlds that cancel out, tends to proportionality. This is shown by the greatness of its concepts, principles, laws, mechanisms, ES and CES, and that of greatest relative value, REM. It is exemplified by the fact that America or any nation or culture that purports to be or is even looked upon as being a leader is often poorly accepted. The world generally, as all mankind, nations, states, and peoples, looks from their own vantage and for greatness and excellency in leadership. If there is to be a world leader, an example for the rest of the world to follow, this leadership must be great and excellent. It must follow TRU and all that is of the greatest REM. All nations in history that have tried to assume the role of world leader, and aspired for hegemony, have done so successfully and continually only if, and as long as, the degree their principles, values, and all were of highest REM in TIT. They were governed, and accepted by their peoples and others by VAPO in TRU. Submission accepts much dominance if wise, fair and right in TRU. For the degree to which America or any nation or group lack this REM, VAPO, and TRU is the degree to which they are resisted, fail, and much of the world fails. Most humans if left to chose between life and death, value and non value, wisdom, fairness, and right over their opposites, will overwhelmingly choose the former. So also is it the case for all OLF. The CES and DYNO of OLF require leadership at all levels. It is an underlying law of all that interacts with TRU.

It is not simple or easy to be a consistently good practicing Christian. It is even more difficult to be TES. In the modern and future technical, competitive and highly financially oriented worlds this becomes a great challenge. For technology and economics detract from other great aspects of life, knowledge, and TRU. This is largely a matter of displacement. More profound is displacement of attention to basic concepts, principles, laws and mechanism of TRU. We should seek to resolve the problem not in reaction to technology and economy, but underpinning and fulfilling all levels of human mass energy that help complete VAPO, TIT-TO and TRU. One of these means is to discover, learn and determine the difference between TRU and ATRU wealth. This is called: **RATWE** = Ratio of ATRU to TRU Wealth. The better we know RATWE, and the more we emphasize all TRU, the more all future technology and economy can be guided, voluntarily or involuntarily, directly or indirectly toward TRU.

The density of world population, is a great determinant of the characteristics, parts, forces and processes in TRU. A species, RUS, TRUS, or TIT is to a large extent a function of their total population in a given area. The great changes with markedly different densities will produce many new forces and processes in the coming centuries. Not all effects of higher density are detrimental. Some of the greatest advances of mankind have paralleled rapid increases in population and density. This enhances socialization, cooperation, and civilization. So that in the future it is not only marked increases of density that need to be addressed and managed but the many and diverse effects for which planning, control, adaptation, and PPMA will be needed. Even with a full discovery of TRU human success is not assured unless those crises, problems, material needs and wants that sap human energy and impair success are solved. This is why the economic, financial, business, governmental, and industrial, segments of the world are important. It is not only because of their benefits, it is because of the reactions created that stimulate PPMA and control required to provide the continued prosperity a flourishing civilization depends upon. Here is the role of voluntary selection, for all to know, serve and sacrifice in whatever way is best for TES and TRU.

Fig 1-87 **REVERSIBLE FORCES in AMTRU of VULNERABLE OLF & TRU**

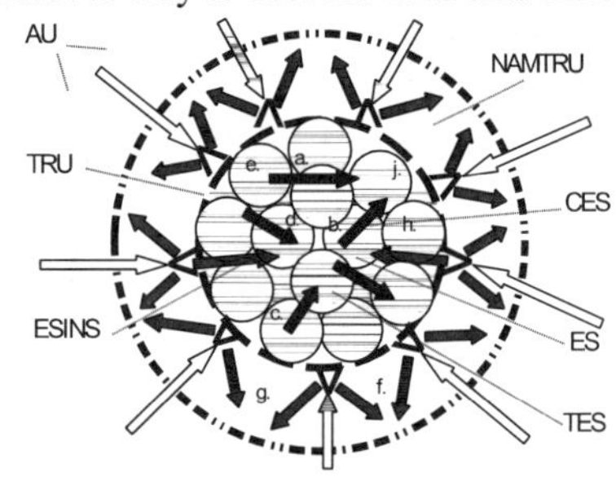

Modern advanced civilization consists of a reversible yet vulnerable mass of good to great individuals persons working dynamically for the common welfare. They are primarily and increasingly sustained by the efforts of a small fraction of the total population. These efforts are characterized by: self sacrifice, honor, conviction, truth, dedication, perseverance and highest principles. These include knowledge and concepts of successful endeavor and a firm belief in humanity, and the highest goods that are served by God or TREQ. The population loses great qualities by the continuous and insidious increments of mediocrity and negative opinions that spread cancerously through modern civilization and the NAMTRU world. They are fanned by those of little faith and trust which accumulate, INEG, to represent in totality the overall detraction, failure, and even annihilation of TRU. Only when turned around into a cultural and total S/F of TRU will we find hope, salvation and appropriate security through OLF in DYNO for ENT of TRU.

Communication leads to cooperation and cooperation leads to care. These permeate much of TRU and apply to all IPR. They are especially experienced in those governmental departments that collect taxes. In the IRS these are essential, the largest department for maintaining governmental operations. Communication, cooperation and care are key to both the principles of, as well as activities, that sustain, the social contract. It is here we can most clearly see the great value of voluntary participation and TRU administration. **PICT** = Politics, Comparative in TRU. What happens in good and bad politics and government? It can range from the most positive to the most negative, as well as from the incipient to the senescent. It can be TRU mutually for growth, prosperity and happiness or it can result in decay, revolution, even extermination. This wide range is the reason it can't be simply cured or only from the top. It can be cured when each individual is sufficiently knowledgeable, participating in, and desirous of TRU, to demand the most positive, and best kind of politics and government.

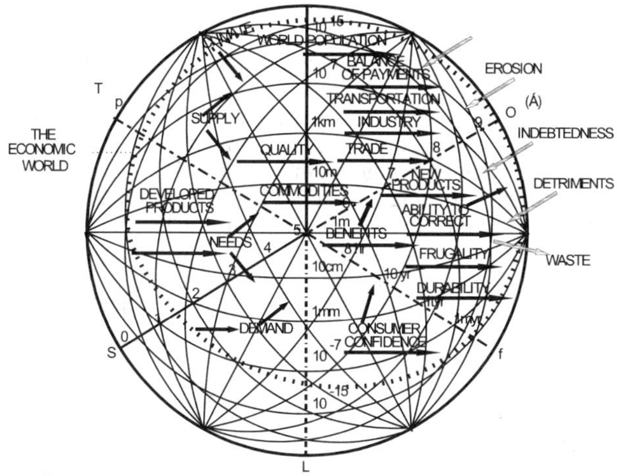

Fig 2-87 **TOTAL ECONOMIC FACTORS and FORCES**

The fundamental first law of economics and exchange, as in any transfer in dynamics is: **LAQUEX** or Law of Absolute Qualitative and Quantitative Exchange. You can't buy, exchange or transfer something if you aren't willing to pay the price. You can't sell, exchange or transfer something if the other isn't willing to pay the price. There must be sufficient supply and demand. There must be a 1:1 simple and direct exchange in which "You pay me, and I'll pay you". The more each party profits with a minimal loss, the greater the exchange, and thereby welfare of both, as TRU, will be. This is LAQUEX. It holds not only for economics, but all human life, the function of processes in TRU. It is the basic cause of so many disasters in economics and human life and fuels the downturns that result in prolonged recessions and depressions. It is clearly seen many of the disasters in American history, as corporate scandals at beginning of the century. It is seen wherever the exchange of goods is uncontrollably complicated by variable and expanding vehicles which lend themselves to abuse. From the invention of money, Midas, the Athenian Empire, and the failure of many empires and cultures since, much of their downfall has been a result of a lack of, abuse of, or in part related to denial of LAQUEX. Qualitatively it is the degree both ends of the transfer are TRU. Quantitatively it is how much, both the buyer and the seller get most by their needs and roles in TRU. It assumes an adequate and increasingly better supply, and an adequate or increasingly better demand. Where these are unequal, imbalanced, the difference, or deviation of supply and demand, needs to be quantified and corrected as much as possible.

This leads to the next fundamental, or second law of Economics. **LÖDS** or Law of Optimal Supply and Demand. LÖDS applies to all purchases, exchanges or transfers, by TRU, the parties involved, and occurrences. LÖDS is highly variable and determines much of the price. Where supply is extremely limited and demand is excessively great the buyer - and seller - must be vigilant to assure the best use and control of the transaction.

The next fundamental, or third law of Economics, is: **LÖDAST** or Law of Optimal Supply and Demand by TRU. LÖDAST has been, is and will be even more so, the great incentive to both greater resources, and to most frugally and efficiently use the supply available. It likewise requires the highest level of knowledge and judgement in the use of supply and control of demand, individually, group, and totally for the entire world. From the first three laws, as well as for its own right, the fourth law is: **LARDECT** or the Law of The Relative and Dependent ES and Roles of Economics in TRU.

The fundamental laws of Economics are not independent, nor relegated to one field of human endeavor. They are highly formed and function in the overall dynamics of human life, this special universe of best human life, TRU. To understand these laws, and economics in general - as for all topics, fields and sciences - we must discover, learn, remember, utilize and improve the knowledge and representation of all the form and function, parts and processes of TRU. How far humans can deviate from TRU is all about us. The many scams, abuses, and downright criminal behavior of many, who when left to their own devices and not corrected to carry on with ever worse pursuits, are often ignored and at times condoned by the public. If we are to uphold and sustain the best of human life, TRU, as we must, we should recognize these defects, know TRU and best practice these four fundamental laws.

The economic forces and processes at work in the world provide a rough idea of the complexity that typifies TIT-TO. The principle factors that explain economics integratively and differentially show how they can be treated in HOM. The economic world is a large part of TRU. Its expansion helps to explain the relative short half life of TRU. This is compensated by its overall expansion. Like the many other worlds about us the economic world is far from pure TRU. It and its factors and forces are of varying degrees of AMTRU depending upon their TRU effectiveness. With the right modifications and interpretations the highly interrelated and integrated economic forces and their many concepts, laws, and principles can be readily adapted to TRU This is: **TRECON** = An Economics based solely on TRU, on TRU values, forms, parts and functions. TRECON is applied to produce optimal human well being, highly ENT for TRU. It is a combination of the best of rationally free enterprise monetary systems with TRU principles and values, balanced with the best of other beneficial systems of government and life. If you can have, do good, make a profit, and be TRU that is even better, an added bonus. If you can make a profit without doing harm and provide optimal benefit, this is the greatest for TRU and ones TES. TRECON thereby are the means by which human material, physical, monetary exchange, transfer of goods and services are applied in which not only is a profit gained, but this profit is optimally turned into TRU, and ANTRU is eradicated.

When nations recognize themselves, their counterparts, and constituents as ES, TES in TRU they will better live by TRU. They could also better adapt to others who live, in the same way, by TRU. This would be helped by an international commission, bureau or agency to evaluate and recommend to national governments what is TRU/ATRU in terms of the whole, or **BUCAT** = Bureau or Universal Commission to evaluate and recommend Correction of ATRU/TRU, as in expenditures. BUCAT would be directed to prevent, manage and control government's insidious drift away from ES, CES, TES, and TRU.(93)

There is no question that one of the greatest revolutions in human life is the massive increase in the world's population. Revolutions, often and more easily, tell us much more about TRU and its dynamics than what they entail in and of themselves. Mankind is moving from a highly differentiated to a highly integrated population. This has both short term and long term beneficial and damaging consequences. It has produced vast and ramified actions and effects. In this transition there continues to be great conflicts between self and common interests, the good of the individual and the good of the whole. By steps or stages in development each state of TRUS, TROS, and TRIS is characterized by a set form of these two poles in dialectic motion. Mans survival is increasingly dependent upon the harmony, equilibrium and power of both, and both together. Law, religion, and the difficulty individuals have in adapting, utilizing, and living a happy, productive, and full life, are good examples of this great revolution. Such explosions indicate basic and serious TRU imbalances. To find the reasons for these is key to solving their problems, as well as protecting and assuring TRU. The problems in both cases likely result from an excess of quantity and lack of quality. For although increased quality is highly productive of increased quantity, the opposite quickly reverses any gains made, increased quantity more often produces decreased quality. These should be discovered, learned, known and applied. They represent trends toward depreciating values with an accompanying loss of TRU. Yet, large populations and even greater rates of increase, provide an opportunity for more life and happiness. The disparity between opportunity for life and happiness and detrimental effects needs to be stated as a ratio: **PROLHAD** = Population Ratio between Opportunity for Life and Happiness versus Detrimental effects. It is the increase and build up of this disparity fed by nationalism, self interest, and diverse partial and intolerant religions that leads to uncontrolled disaster for mankind, TRU. There is no greater challenge to man today than to exchange his great population quantity for like greatness in quality, and to do so in ways that conserve the best of both. The kind of world that mankind sustains today can not exist without a reasonably large quantity. This must be a quantity that is capable of continued sustain/harm of TRU. This is: **LAPCOST** = Law of Population quantity Capable of Sustaining TRU. LAPCOST may be remembered by being the cost to humans, or the price we must pay, if we do not have a population of such quality necessary to keep up with and sustain TRU. LAPCOST is not only genetic and physical, it is physical, mental, phenomenal, spiritual, and all ACEXIN and SÁ.

Regional and international power politics and economics exemplify the worst in humanity. Yet they have a role in the predominance necessary to sustain natural selection. In the present, in which reason, moderation, toleration, avoidance of violence, and a cosmopolitan spirit pervade, power politics and economics can't operate at will and whim. The never ending struggle between the myriad forces of ENIT and the absolute forces TIT is a war that few have discovered or appreciated. It is at the root of many great human problems. For it is whether or not there are individual, national and many other rights in and of themselves or only as they are TRU. Each separate entity expects and fights for all it can win or acquire. Only as we discover, learn and understand TRU or TREQ, international relations, and all levels of IPR, will there be efficient, effective and sustained successful peace and prosperity. Prosperity is that state of well being in an economy, culture, world or TRU in which there is a minimum of poverty and want, and a ample amount of goods, supplies, provisions, and services for maximum fulfillment. It is a goal sought by all peoples. Yet, they often do not know either what these are, their implications, or their limits. Prosperity for TRU is markedly different than what is commonly believed. Since TRU is by TRIA and TIT unique and extends into expanding segments of time in the future, so does its prosperity. A prosperity that is short lived, limited, less than or a loss to TRU, and harmful to the future, is a false prosperity. This is **LAPAF** or Law of a true Prosperity can only be one in which the Present does not Prosper at the expense, loss, ruin, or ANTRU of the near, mid, or long term Future. This is like ideas of deficits and indebtedness resulting from abnormal values. Prosperity, with business cycles, will fluctuate and be modified by the mechanisms and forces in operation. It must assume those relations that are within TRU, what is action/reaction within the equilibrium is all that is necessary It will depend largely upon the combined success of all that supports and enhances TRU. Material prosperity is especially a large partial of the necessities of life. Yet a prosperity that does not include all levels and mass energy, Á of TRU, ACEXIN, and is unbalanced with all the rest of the dynamics of TRU, is not a TRU prosperity. **PRID** = Principle of Prosperity and Ideal Difference. PRID is exemplified by what the conditions are for greater economic prosperity. What prosperity means to different peoples as ENIT is often far different from what it is in TRU. Even what it is for TRUS, TRU, TIT, TRUID and TÂ are not the same. This will depend upon the kind of prosperity as in **ECPRO** = Economic Prosperity. **QUALNAD** = Quality and Quantity Alternating, cyclical Dynamics. The S/F of TRU is to a large extent determined by the optimal equilibrium and operations of QUALNAD. There is a momentum of quality and mass that is necessary to maintain TRU, and danger of its loss in the present.. There is also a critical level in the depreciation of a RU or TRU. To prevent passing below this level we must select all that is positive rather than negative. There are a great number of threats to human existence both acute, subacute and chronic that although not striking, are none the less formidable obstacles to TRIA and TIT. To prevent these we must start not at the 11th hour; we must start now, and hereto, not only directly but indirectly to all that is basic of ATRU/TRU. **LACORE** = Law of Action or Accentuation through Total Replacement. LACORE signifies the fact that we, usually nor completely, can not count on one action or even simple and separate actions or accentuations. They should be the singular, simple, and separate actions fused and united with all actions and accentuations that bring about total and optimal TRU. There is, and probably will ever be, no lack of ill effects on TRU. This is aggravated by poor recognition of TREQ and a void in understanding TRU. Their solution arises from our ability to react, adapt, cure, and protect all by TIT.(PPMA, ACC) This is especially by interest in TRU.

Fig 3-87

SUPPRESSOR AND SUPPRESSED or OPPRESSOR and OPPRESSED DYNAMICS, and TRU

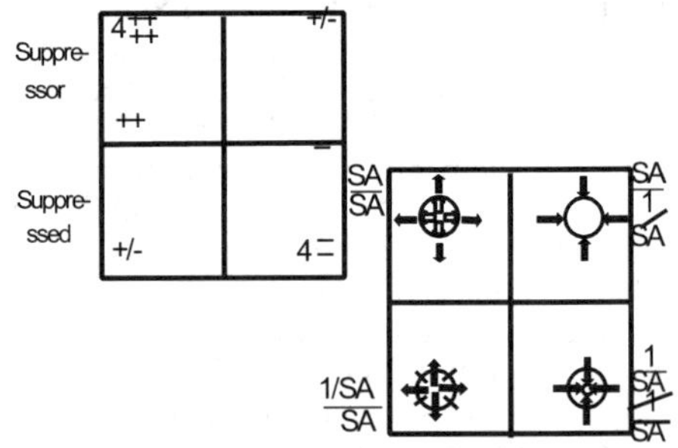

The question and problem of suppressed peoples is operative in the potentiality, fitness mode, and SÂ/non SÂ, or exponential absolute in TRU. If the suppressor is 1/SÂ and the suppressed is SÂ then there will be 'hell to pay'. The suppressed will not accept the suppressors lack of TRU and errant dominance. If the suppressor is SÂ and the suppressed 1/SÂ it is relatively common, difficult, less acceptable, and apt to lead to NAMTRU or ATRU. In the figures of 3-87 we seek to show and make distinct these opposing states. The relationships between a suppressor and suppressed occur not only on the political and national levels, but in the economic, social, linguistic, business, occupational, class and many other levels. It is a FAF, state and problem of our massive dynamic NAMTRU and TRU existence. Many of the nationalities, depressed economies, social inequalities, language barriers, business abuses, occupational pecking lists, class distinctions and others clearly show these clines that exist within humanity. Some are more or less a part of the form of TRU, TÄ. Most are highly ATRU and do not benefit from TRU. The combinations are: 1. 1/SA / 1/SÂ = Chaotic Extinction, 2. 1/SÂ / SÂ = Revolutions or Conflict, 3. SÂ / 1/SÂ = Repression, 4. SÂ / SÂ = Flowering out, Fulfillment of TRU, to an ideal, Utopian, World, TRUID. The great concepts, principles, mechanisms, and laws of TRU: DIA, SIA, IAT, VAPO, SCARLS, DAOST, OM, POM, Pureality, COGNOS, and ELTICMA lead to SÂ.(98) There opposites are its 1/SÂ, the ATRU, selfish, devious, ignorant, and ungrateful of the world who take all and give nothing Why suppressors commit excesses and atrocities on the suppressed is key to raising the level of IPR and social relations in TRU to new heights of success and efficiency. Only as both peoples and states fulfill their ES and CES to enhance their S/F toward the best of TRU can they be free to enjoy the fruits of their labors and being. In this the suppressed often become the suppressor, and reverse interchangeably. When moderated in balance in TRU they best fulfill their functions.(AIES,LOPAN)

APRA = Ancestors Progeny Ratio or sufficiency. APRA signifies the decrease or increase in quality of ancestors and progeny. For ancestors it almost always is an increase and for progeny of modern humans is often in quick decline. The ratio of sufficiency is most certainly greater than one. This reveals the validity of ancestor worship. It reveals the worship of the past and all that has contributed to the glory of the present. So that when we speak of the spirit of our ancestors or of all in time past, we should also speak of the spirit, hope and assurance of our progeny or all time future. In neither case are we in error. For their significance lies in their ESINS, and these are in full identity when of their part and processes in TRU. It is from this we find much in common and relative of knowledge, as it is TRU. It provides correlation, the basis for much of scientific research and measures. Yet it can not be accepted as is. It has to be returned fully to TRU if we are not to be in error and misled. It is this return or convergence upon TRU that constantly and selectively makes APRA sufficient, and whose accentuation will assure both best progeny and TRU.

Humans have existed throughout history in a social and population milieu in which there were all the way from no, a few, an agreeable amount, a balance, to many, an excess, and a great overabundance of people. The processes involved in each kind of mixture are markedly different. During the nomadic and hunter gatherer periods of human existence mankind ranged widely encountering no, a few, or many others. Like many other animals, the need of territory and property for survival, made them highly possessive. For the future this poses ever greater problems, conflicts, and negative forces whose selection and cure is largely in TRU, ACC, management and control. At Monte Verde in the southern extremity of Chile at the end of the ice age there were humans far separated from others on earth. They exemplify the fact that food sources, quality propagation, and special technical developments to supply human needs were as important as large populations.

The concentrated potentiality of a revolutionary TIT is tremendous. So is the reverse potential for destruction. This is of great importance and one of the worst threats. It can be seen in the half lives of great changes in TIT, such as communication 1 - 100 years, language 10 - 100,000 years, and outbreeding 100 - 10,000 years. They may or may not be capable of adapting to the rapid changes produced by the inadvertent, irrational, and ATRU will of man. These are constantly buffeting AMTRU, TRU and TIT. TIT as TRU as RU has always been as RÂ, by RIA, by TRIA for its time, its constituents, and themselves by DAOST. Yet these profound and greatly impelling powers are great threats to TRIA and TIT, tend to greatly alter its thisness, and hasten permanent change. Such adaptive processes tend to occur at the most fundamental and critical levels to maintain the equilibrium, stability, and retention of TRU. Or they, at least, tend to return TIT from TRUCH by TRACH or keep TIT in TRID as much as possible. It is in this light that modern civilization must be understood as well as the dramatic, violent and terrible causes for great calamity or extinction. It is these that face man and TRU through the rapidly expanding insidious forces such as Nuclear War in the present age. For although since the fall of the Iron Curtain nuclear war, may seem in the eyes of many, more a problem of the past, this is not so. In fact humans since the beginning of time have used whatever tools or vehicles they have to get whatever they want. It is this last, to get what people want, that is the key, and can only be TREQ or TRU. We should never forget the increasing potential for nuclear war and holocaust. With mans knowledge, technology, continued existence of nuclear weapons of vast quantities, and their proliferation in many more countries, and even fewer dependable groups, remains the greatest immediate threat of early extinction. This is: **NUWREX** = Nuclear War Remains the Greatest Immediate Threat to Early Extinction, and: **NUDAL** = Nuclear Disarmament Only an Apparition, a Lull in a continuing storm and threat. As great as TRU is and as unlikely as an early date for such annihilation is, the danger remains. This is **NUWAT** or Nuclear War is A-Temporal.

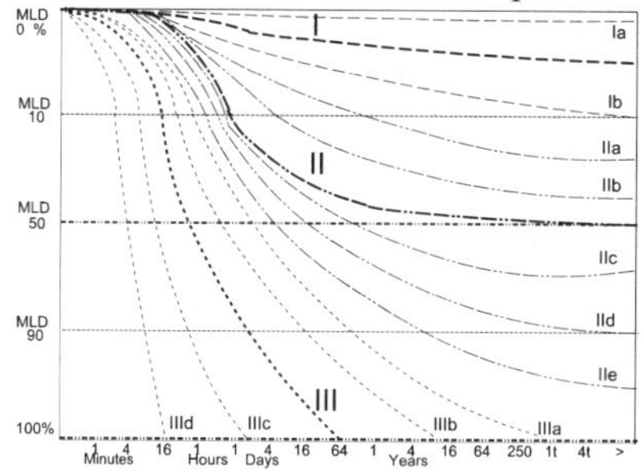

Fig 4-87 **WORLD SURVIVAL and POPULATION**

Some of the varieties, effects, and possible consequences of the use of nuclear weapons include, I. Delimited or partly aborted exchange, II. Full exchange, without extinction, III. Full exchange, with extinction. **MLD** = Mean Limit of Death; MLD although thought of at the 50% level is a continuum from 0, 10, 50, 90 to 100% of worlds population, depending upon the base taken. Time from zero hour of explosion to years, lifetimes, many thousands of years, or more is variable and increasingly indefinite. Even in the best of scenarios some effects, like the countless effects on genetic material, would persist for thousands of years. A full exchange would consist of a state of:: **NUCOL** = Nuclear Holocaust, Nuclear War, Unlimited. The many points on the steps into and out from NUCOL and the many negative and positive forces that occur at each, at some in combination, or throughout can be demonstrated. The most fundamental and crucial forces, actions, and powers, that exist are likely to exist both negatively and positively in this eventuality, must be absolutely and completely known, ECRO, countered, and controlled by all needed to protect and assure TRU.

Once initiated the use of nuclear weapons would be hard to stop. Leaders of either nations, special interests or subversive groups, as terrorists, may impulsively go to extremes secretly and without much reason. Also the likelihood of mankind suffering such disasters depends in large part upon the degree to which we understand and fear destruction, and can prevent them. This largely depends upon the extent to which we can rise to the necessity of all the reasons of their occurrence. Most of these are what is lacking of TRU and what IPTRU addresses. This is especially so for the mechanisms, forces, processes, and powers that exist in HOM, DYNU, DYNS, and DYNO and even more specifically all that ACC addresses, and with PPMA, can control. There can, must and will be, a constant vigilance of what is most dangerous and negative, in periods of peace as much as in periods of threatening war. **LAVIND** = Law of constant Vigilance In periods of peace, Negativity, stasis, of unanticipated Danger. There is a close relationship between power and TRU, or: **RAPAT** = Ratios of Power and TRU. The use of power will rise, which is all the more reason why TRU must be known and all power that is apt to, is or will effect TRU, especially adversely, be controlled.

Power once discovered can never be undiscovered. Yet it, as all, must be TRU, managed and controlled as by ACC. Nuclear weapons should never be used or even accumulated unless: 1. Human survival or TRU are at stake, 2. To the degree TRU or TREQ are known, KOT, 3. No other means or method exists or can be found to save and assure survival, 4. The benefit to harm ratio, is maximal, 5. No extended use is to occur, and complete safeguards and prevention are certain. The best action to be taken in lieu of use of nuclear weapons is all TIT-TO, TRU and ACC. This requires all means to reduce dissension, discord, non coincidence of laws, standards, ISMS, religions, rules, regulations, customs, nationalism, abuses, and partiality that is not of or for TREQ or TRU. Yet making full allowance, toleration, and moderation for differences in TIT-TO.

If power and TRU are both great, TRU profits. If TRU is great and power is weak, TRU is weak. If power is great and TRU is weak, TRU is in mortal danger. Thereby we must have TRU, or if we can't have TRU, or refuse to accept and achieve TRU, we must eliminate the greatness of power. This will be a guiding principle, as it is a proof, of ACC and MARS. Human culture, knowledge, technology, and industry, can be no better than the control of the side effects, deleterious and negative influences on our life, TIT-TO. It is not so much their discoveries and minimal trials, as it is in their continuing use and accumulative effects that damage results. Mankind, in time, tends to increasingly develop, discover, understand, appreciate, control and stop all that threatens. Yet they must be discovered, remembered, accounted for and best acted upon. Each decade and century has many new, as well as old, sets of threats, wrongs and - more or less - critical problems. So that their discovery, answers, solutions and cures are a continuous operation for ACC, PPMA, and TRU.

As long as humans have existed on earth, they have, more or less, engaged in fighting and warfare both with other forces, other life forms, and within themselves. This is especially violent when antagonisms build up and one or both of the combatants fear loss or become desperate. When desperate they use any weapon or means at their disposal to save themselves, prevent death, loss, and to win. The crux of the problem then is what makes this desperation, how can it be controlled, and how can defense be assured against the escalating power of weapons. The only solution for nuclear conflict is through the best in human nature and love of life, TRU. **SNUCOLT** = Solution to NUCOL is human life in TRU. It takes much more than singular measures, or even a worldwide and massive program and public opinion campaign to eliminate or control nuclear weapons. For it does not correct all the underlying reasons and negative forces, actions, and powers that create them and make their use likely. To correct such negative forces it takes ever greater faith in the positive aspects of all TIT-TO. It takes the rightness, goodness, benefits and the success of the world in current religious, philosophical, scientific, political, economic, and social forces at work.(TREQ)

Fig 5-87 **FUTURE GREAT NEEDS, PROBLEMS, and CONFLICTS**

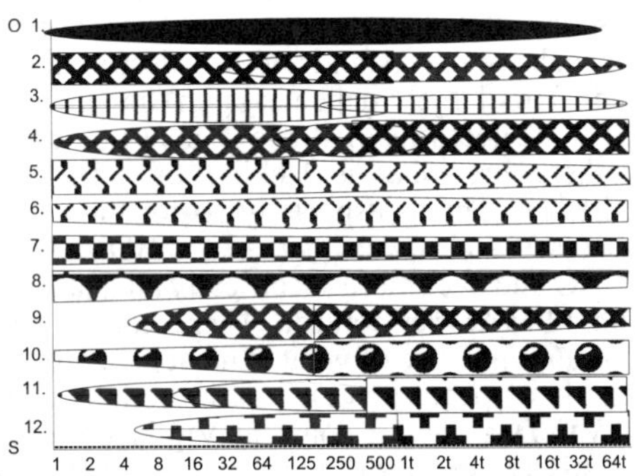

Some of the greatest problems, conflicts and needs of the near, mid and far term for TRU can be shown by their presence at various times. They do not include many others that are of great importance. Many of these are topics of the text, especially its latter half and ACC. Those represented are: 1. Nuclear war, 2. Population explosion, 3. Energy crises, 4. Resources, depletion, 5. Genetic capacity, 6. Live-product, capacity, 7. Live-product, ability, 8. Determining forces and their control, 9.. ANTITRU forces, 10. TRIX, 11. TRU, 12. TÂ. It is not always easy to distinguish the severe threat the first nine of these present from how they are FAF in TRU, and how they are to be acted upon, ACC. Initially they mostly pose a great threat, mid term they are being adapted to TRU, and long term we may increasingly turn them into forces for TRU that have their place in PPMA and ACC. We can thereby see how they are all highly dependent upon TRU or TREQ, and how this proves TRU its value, procedures and means. There is often no complete or final cure for many of these, only the adaptation, compensation, management and control that goes with TIT, TRUD, HOM, and ACC. It shows how important the continuing and increasing capacity and ability of the live product, genetic material, and determining forces are as they make up TRU and the overall capacity and ability of TRU to continuously surpass all possible threats, causes, problems, dangers, and conflict. This is called: **LATANCO** = the Law of The Absolute Necessity of TRU Always Surpassing All Causes, Conflicts, Problems, Threats, and Needs. LATANCO can not be provided unless and to the degree we provide TREQ, TRU, KOT, ROT, POT, and all.

That TRU and TÄ are confirmed by space exploration becomes increasingly certain the more advanced our missions become. Rather than finding and fulfilling some false belief of what the universe or world is, they increasingly show the beauty and worth of mother earth and humanity. This should not lessen our interest in and discovery of space if for no other reason than to give us more knowledge, both directly, indirectly, and as by product. It is also important to know all that is or is apt to affect us. Just like through history human discovery of the earth geographically, so does space exploration, help us know TRU, TIT, ACEXIN, TRIT, MOT and SPALT. The scaling of the dimensions of TRU, ALT, are not only to be found outwardly, they are to be found inwardly and proportionally by VAPO, as they exist in the actual human world, TRU. The loss of similarity, sameness or relativity that will be found, and can not exist except as it is closest to TRU, is called: **PED** = Principle of Exponentially Distinct Depreciation of Similar Worlds. PED means the farther we get from the earth and ALT, the more different things are. This does not mean there may not be other FAF that may be of some relation or use, it only means that by ATO, DAOST, TIT and ACEXIN we must hold to TRU and avoid NAMTRU, ATRU, and ANTRU. It is in the separation of this complex mixture that we best discover TRU and AU. Adaptation to, utilization,, stabilization, and control of the world's climate will be major needs of the future. Just as meteorology is becoming increasingly proficient in predicting weather and climate, so its applications continue to develop. Yet this, as so much of knowledge, must be done with both its advantages and the restraints in operation and control. It may not be so easy to prevent the extensive droughts and floods, heat and cold, tornadoes and stagnant polluted air, or rising and lowering of the ocean level. The measures to be taken are quite different in the short, mid or long-term. They are also dependent upon local versus world wide conditions, and upon our knowledge and technology. Mostly the success of TRU in these is a matter of prevention, preparation, restitution, and adaptation. This is often not of the weather, but of humans, all TIT and ACC. To compensate for deteriorating effects of the earths temperature can be, in part, done by: 1. Increased or decreased use of water. 2. Increasing or decreasing dams and reservoirs, 3. Increasing or decreasing uses of water for arid lands, deserts, irrigation functionally by TRU, 4. Prevention of run off, by increased forestation and controlled grazing and hard surfacing. 5. Increasing or decreasing storage of water to maintain needed stability of ocean levels, 6. Management of ice caps, and 7. Optimal PPMA of allowed fluctuation. 8. Adaptation to and tolerance of conditions beyond our control. These and other measures existing and to be devised show the combined long term solutions that can be provided.

That there must be adequate provision to maintain sufficient solidity to preserve TRU, is vital for TIT. Mankind is in a time of great transition and revolution from random, diverse, and low energy level strife and selection to more high energy level, total, and unified selection within TRU.(SA) Compared with previous species their solidity was one of fitness, strength and dominance preserved largely by the great many forces working in natural selection. Solidity in the future can't depend only on the kind of natural selection we are familiar with. It is the great contrast between a world determined for us, or a world determined by us. This by the caveat that 'the determining by us' is optimally TRU. This is called: **RUDBU(F)** = Ratio of Universe Determined by Us over Universe Determining Us, and: **LADUT** = Law of Determination of Universe, TRU, only by TRU. However since we are already here, by ATO, and it is up to us to save and preserve ourselves from death and extinction, it is within the reality of TRU, driven by our ideal that this relativity can be, as it is TÂ, a relative absolute. This is called: **VAPUT** = Validity of a Permanent Universe TRU. **RAPTEB** = Ratio of Permanent TRU over its Extinction. The proof of this is in relativity, wherein because of its limit in infinity, this relativity in infinity provides its own eternality. This is called: **PUTLIRE** = Proof of the Permanence of TRU, or human life, by Relativity and its Limit in Infinity Providing its own Eternality. TRU is limited, and may seem to eventually become extinct, yet because of its relative rather than absolute nature, becomes indefinite by virtue of its infinite nature.(TÌ)

The dynamics of human heredity works collectively in which the whole battery of characteristics is inherited. This is: **LACSEL** or the Law of Collective Selection. This applies to all parameters that make up TRU, including A, L, T, GELD, DAOST, and interactions of the total population OITIO. Singlar characteristics are significant, especially in eliminating aberrations and mutants.

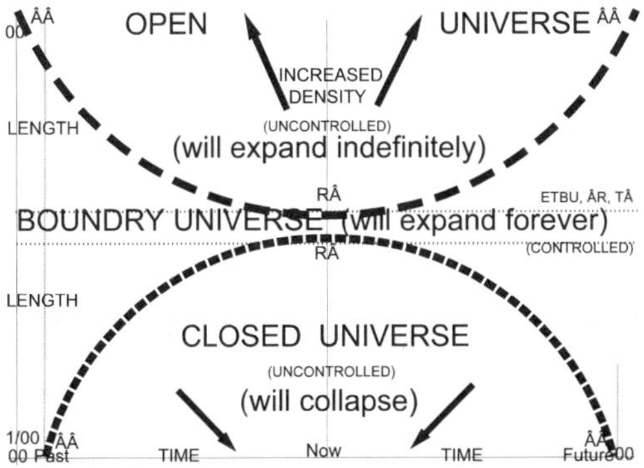

Fig 6-87 **LIMITS and ENDS of ASTRONOMICAL RELATIVE UNIVERSES**

Different scenarios of the limits and possible ends of universes show certain aspects of the makeup and outer reaches of TRU. In an astrophysical universe we are in the position of, sooner or later, accepting one of two opposites, that are equally and totally faulty for TRU. These consist of an open system in which the original Big Bang products will fly apart forever, and a closed system in which the expansion will slow, stop, and eventually reverse itself. Both of these are eventually completely fatal for man and TRU.. They defy RIA, TRIA, ATO, TRU TRIS, TIT and completely defy TÂ. Herein we can again see the preeminence of relativity, the relativity of our universe, TIT. This can be readily resolved however, by remembering the nature, form and function of TRU. By ATO, ALT, and TRIA it is The Thing, not any, all, no, each or even some, thing, or entity. It is not what is known as the astronomical universe, even in its best scientific and physical sense. TRU is The Relative and relativity renders erroneous such an all or nothing treatment of the world we see in an ENIT, and astrophysics alone. Because the astronomical universe is held to have began with a 'Big Bang' doesn't nullify ATO, TIT, ALT and TRIA. It only shows their vast differences. The answer depends on how much matter there is in the universe to exert gravitational pull. Many billions of years from now, the curve of time versus distance on the movement of space objects or bodies will by itself give an answer, as indicated in Figure 6-87. TRU, by TRIA is unending. For with an end that is not TRU it is naught and thereby impossible, for we, as TIT, already are. This is why the extremes of all dimensions are infinite, with limits, relative not absolute. It is this difference between an absolute end and the infinite that has so confused the astronomers and those who hold that the universe is only physical. From an absolutist point of view entities and things are or are not. They begin and they end. The astrophysical universe is one of these that begins and ends. To think that the universe will infinitesimally hold on with neither a beginning nor an end defies an absolutist view. This error is readily realized when we think in terms of TRU, TIT, TRIA, and ATO. Yet the certainty of illusion and the illusiveness of certainty revealed in IPTRU, by stranger things, are actually TRU as ATO, TES and DAK, lead us to have confidence in: **ETBU** = Eternally Controlled Boundary Universe. To better envision these comparisons look at figures of TRUCH, ALT, SPALT, METAD, and SPAT.

TRUC = TRU Conversion. TRUC signifies, along with RUDBUF and LADUT that a selective basis must be changed to a more directive basis. In short TRU can not be left to natural, spontaneous, involuntary or trial and error change. It will require increasing direction, especially as all works together with acceleratively improved mentality. This will be increasingly certain, yet always limited and less than perfect. It will depend upon our knowledge in science, biology, genetics and all fields highly TRU. There will be a balance between the involuntary, voluntary, indirect, direct, and degree of control in selection. It will require the best of each, depending upon the level of KOT and ACC. What provides optimal quality appropriate for ES, CES and TO. Selection that becomes less and less a measure of direct individual survival, but one of collective survival in large populations will present problems whose solutions will require the highest levels of mass energy, SÁ. The planning, management, control, accentuation and direction must be collectively, cooperatively, integratively, totally and universally provided as they become absolutely necessary. This is expressed in: **MARSAD** = Maximal Ratios of Selective and Directive Control. **MOCNEXT** = Maximal and Optimal Management and Controls to Meet the Necessary Expansion of TRU. **LIC** = Limitation of Control. MARSAD and MOCNEXT are special and general cases of RUDBUF, and LADUT. LIC occurs where selectivity is or is not understood and what needs control is out of reach. Also, what needs control must provide a degree of freedom, or allowance. Included in this is the presence of indirect positive development.

MEASURES of DYNO OLF CES

APRA 7-3-7 T 4
BUCAT 6-2-8~ M 6,7
CARE 5-5-5~ S 7
COMPETITION 7-3-5~ S 8 COM
CONCENTRATED POTENTIAL-
ITY S/F ~8-1-7~-S4 CPS
ECPRO 9-1-7~ S 5,7
ETBU ~7-~-9~ M 7
GREAT, PROBLEMS, NEEDS,
CONFLICTS ~8-1-8-S 4 PNC
INVESTMENT, SECURITY
and TRU 8-2-8±S 4 IST
LACORE ~8-1-7~ M 7
LADUT 3-0-7~ M 2
LAPAF 7-3-6 S 1
LAPCOST 7 2 8 S 1
LATANCO ~7-~-9~ M 7
LAVIND ~5-3-7~- M 4
LIC ~7-~-8~± T 5
MARSAD 7-2-7 T 7
MLD ~9-3~-7± T 8
MOCNEXT 6-3-8 S 3
NUCOL 9-0-9~ M 7
NUDAL 6-2-7 M 4
NUWREX 9-~-7 T 4
OLF 7.5-2.5-7.5
OPEN CLOSED UNIVERSES,
ENDS, TRU OCU
~9-0-9~- T 6
PED ~8-0-8~± M 8
PICT 7-2-7 ± S 4
POPULATION EXPLOSION
PEx 9-1-7 ± S 8
POWER POLITICS, SELF
INTERESTS PSI
6-2-~- M 4
PREVENTION PREV
6-3-5~ S 7
PRID 6-1-8± T6,7
PUTLIRE 5-~-9 T 7
QUALNAD ~6-4-6 ± S 7
RAPAT 6-3~-8 M 8
RAPTEB ~5~1-9 T 7
RATWE 6-4-5± S 6
REVERSIBLE NAMTRU OLF
FORCES ROF 7-2-7 ± S 6
SNUCOLT 7-3-8 S 7
SUPRESSERS/SUPPRESSED
DYNAMICS SSD 7-2-6~ ± T 8
TRECON 8-2-8~ S 5
TRUC ~6-3-9~± M 5
VAPUT 6-2-9 T 7
WOFREM ~5-~-8~ M 5
WORLD SURVIVAL POST
WWIII WSW ~9-0-9~ - M 8

OLFPESE1.87

Fig 7-87 **DYNO CES in OLF SECTOR of SPALT**

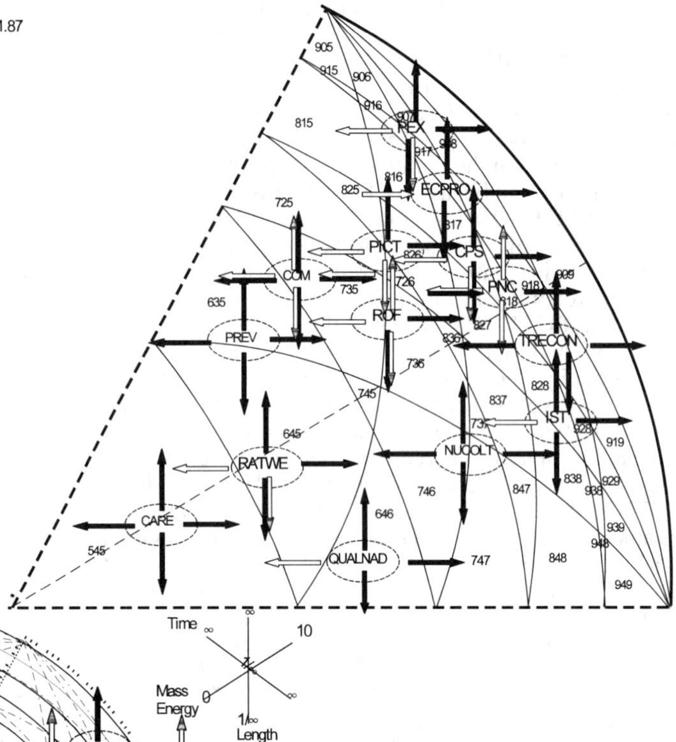

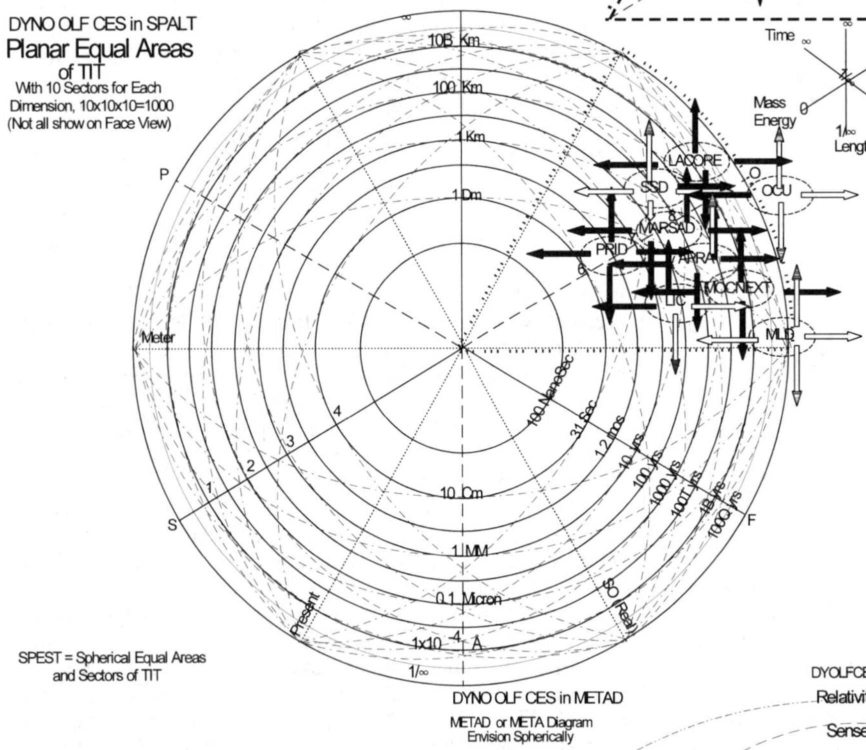

Fig 8-87 **DYNO OLF CES in SPALT**

DYOLFCE1.87

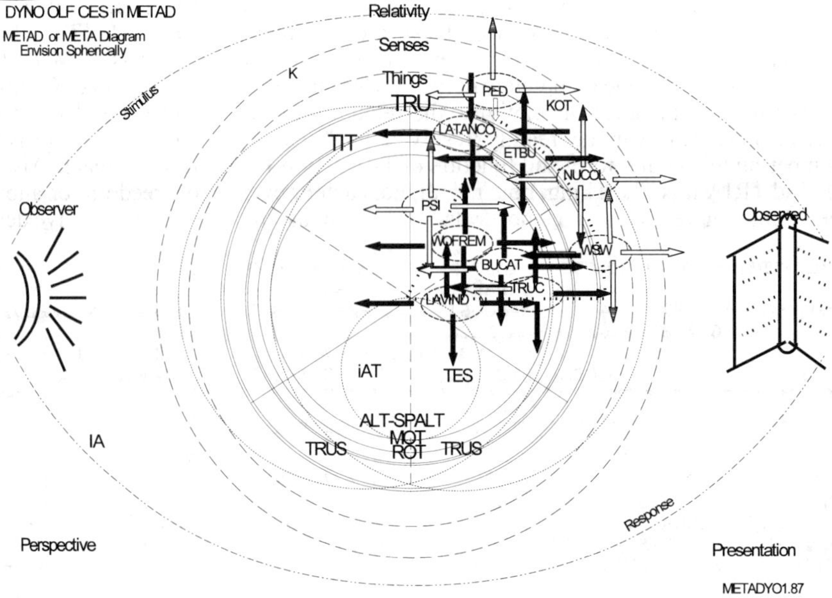

Fig 9-87 **DYNO OLF CES in METAD**

METADYO1.87

CHAPTER 88

SULF or SUBJECTIVE LARGE FUTURE PROCESSES in DYNO

COMPLIANCE, LAW, GOVERNMENT, FREEDOM, RELIGION, KNOWLEDGE, SCIENCE, EDUCATION, PHILOSOPHY

Those things that are obvious do not need proof so much as what is not so obvious, as diseases that are not readily recognized. These must be attacked with all the methods at our disposal and brought to proof, interpretation and fruition in TRU so that we can prevent, protect, maintain or act upon them in that manner that most enables their cure and the enhancement of TRU. We are increasingly provided with many means of finding and proving what is not obvious, life threatening, and of great importance to TRU. This is often done not by single observations or tests, nor even in combinations, but in the massive accumulations of data, and their accounting in and by TRU. This is also how it is of all that is around us. We must not only recognize the obvious, we must be highly alert and concentrate on whatever may only slightly yield the clues, sentinels, and provide an idea of the important. These are characteristics modern man has developed means for in an unprecedented way. Added to SÁ or native ability, they enable us to more readily recognize, name, and decide what to do for the important, great, crucial aspects of TRU. This is to a large extent technology. All the hard and soft ware that vastly expands our ability to foresee problems and disaster. This is exemplified by the panel of controls in vehicles and computerized aids. So may it be for all TRU. Much of this can be greatly enhanced with KOT.

LACPUT = Law of the Compulsive Convergence of the Purposive, The Purpose of TRU. Purpose tends to be highly TRU, especially as an end point. This is for the reasons: 1. In being closely related to ends it is forced to converge upon what provides success and survival, 2. It is not only of a high mass energy state or level in TRU, or of the Á dimension, it is an aspect of these levels that are directed toward a concentration on TRU, 3. Although often less than fully TRU or even at times any or negative TRU, when collectively taken in resultant, the less than and negatives tend to cancel each other out leaving only the fully TRU. This is like natural selection. LACPUT signifies how TRU is and can be increasingly assured as S-O or Á tends to TRU. LACPUT is, in ways, corollary to CEK, compelling forces of knowledge.

PASOS = Paradox of Subjectification of the Subjective. PASOS signifies the question, problem and state in which we must determine how far, how much, and what the balance is between success in a world that is a person's figment of itself. Can there be a complete increase in the subjective revolution without S/O? All that is subjective may or may not be TRU. Even if it is subjective of the subjective of TRU it may still be or not be TRU. For in being of TRU rather than entirely in TRU it may not be all TRU. As great as subjectivity is even when of all TRU, especially as it has spearheaded the great revolution of mans Á FAF and being in TRU, it is, nor can never become all TRU. PASOS can not be resolved unless it's elements are defined: 1. Subjectivity may or may not be TRU, 2. Subjectivity is less than TRU, often much less than TRU, 3. Subjectivity that spearheads the revolution of TRU does so not as an absolute in and of itself, but in a continuum of fulfillment of all the mass energy Á FAF of TRU. This state requires an equilibrium of S-O to maintain its reality and thisness, or actuality. 4. Subjectivity can grow even relative to objectivity, yet at a limited rate unless both grow as TRU grows. Ignorance like blindness may be complete, partial or only limited in some way. Yet even to be limited in some way, as to be operating from ones blind side in a traffic accident, can be fatal. So also with knowledge; knowledge as it must be of TRU. To think one has a great mass of knowledge, even new knowledge, is not enough. This knowledge must be not only complete, as well as greatly increasing in detail, it must be appropriate for all contingencies. It is for this reason we can't be satisfied with any and all kinds of knowledge, sciences, philosophies, religions, and other fields. They can only be right as they are their part and role, by VAPO in TIT.

SAJ = Summative Atonement and Judgement Formulation. Atonement and judgement of a person are like immortality, not something that is pronounced at a specified place or time, as at death. It is an ongoing and continuing process throughout life. This is seen in: **POCP(RO)** = Principle of many long and Continuing Processes. The determination and extraction of ESINS, ES, and CES are difficult POCPRO operations.(DES)

LACJOT = Law of Collective Total TRU Judgement. LACJOT necessitates SCARLS of TRU between a government and its citizens, a state and individuals in which neither has any more rights than defined by their own part and role in TRU. The citizens and individuals must believe, think and do only as TES in TRU. Since the collective total judgement of TRU is far more accurate and far more complete than that of the individual, group or other world, they must do and think not in any way but only as the judgement of the total collective TRU finds best for TRU.

FOPIC = Free, Forthright, Open, and Principled Communication and Conversation of TRU. FOPIC can be stated as a fallacy, concept, principle, or law. It may be between two or more individuals, IPR, groups, or all people. FOPIC may seem obvious, yet it is a basic necessity in a civilized society. It is like the right to free speech, only highly qualified by being only from and for TREQ or TRU. It is a great vehicle as well as stimulus for the fulfillment of TES and TRU. It is key to PUREALTIY and the quality judgement and endeavor needed for ES, CES, TES and TRU. However such freedom of communication does not come without limits nor can it be taking for granted. This is called: **LIFIT** = Limits of Freedom(s), yet great Freedom or liberty and life within TRU. FOPIC is most acutely seen in all the many inhibitions in conversation and communication. Its most simple form is in asking questions and giving answers appropriate for TRU. This is a measure of the quality of a person, society, and world.

POKOT = Preventers and deprecaters of the knowledge of TRU. POKOT is found all about us in the NAMTRU world. It may be direct or indirect, voluntary or involuntary. It is in all that displaces or blocks IAT, KOT, ROT, SPALT, TES, COGNOS, ELTICMA, and TRU. It is in all that fails to learn, recognize, enhance, and utilize VAPO in TRU. It is in those with strongly held beliefs that are not TRU. It is clearly seen in the biases, prejudices, fallacies and errors that become so fixed in our minds.(RELE ,SCE,PHE,)

To help the positive, good, TES, and TRU become more so is the greatest service that can be rendered. To help the potentially positive, good and TRU be more TRU is a greater service. To help the less than positive, fair and TREQ toward TRU is a great service. To avoid or eliminate that which is implacably negative, bad, and ANTRU is of service. To do anything other than these is likely to be of little, no or worse service. The positive, good and TRU are exemplified by those associations, sects, groups and individuals that are dedicated to and accomplish what is best for humanity or TREQ and do so continuously, fully and properly in practice as well as principle. They are of the existing TRU that is unrecognized as

TRU. They are TRU by mutual assistance, cooperation, contributions, and benefits as they supply TREQ of TIT.

SIRT = Stimulation Imperatively Sufficient for TRU and its Rewards. It is imperative that all persons, especially when young, be well trained and educated to acquire sufficient understanding, habits, and appreciation of the benefits, joys and rewards of being TRU. They should be taught and trained to best discover, learn, remember and live by what is in their best interests as TÍS and TES. One of the common, existing, and most effective and best methods is imitation, as an 'Imitation of Christ', or of any and all great contributors to TREQ and TRU. Each and all persons should with greatest inspiration, revelation, and conviction find identity with, and keep their commitment, to GOD as TREQ or TRU. It should be enough to catalyze, stimulate and provide the impetus needed to achieve success and survival for themselves and others in TIT. **RAKUT** = Ratio of Known to Unknown of TIT, TIT-TO, or Known/Unknown x TIT x Perfection/Error. Much of perfection/error is in the representation of TRU.

Dogmatism is erroneous to the degree any law, belief, or way of life assumes powers that are greater and other than their CES, ES, and VAPO in TRU. There should be within all, even the best of POT, a degree of agnosticism that limits a dogmatic approach. This depends upon, and should be proportional to, how much of TRU is unknown, how much VAPO and KOT are incomplete. This is why dogmatics has been wrong and unacceptable. It is thereby necessary and a great goal to constantly discover, improve, and make more distinct all TRU from its beautiful exquisite detail to its best systematic unity. This needs to be done with an ever increasing ACEXIN of all HOM in SPALT, COGNOS and ELTICMA. **EQUOC** = Equation of Relative Compulsive Certainty. EQUOC signifies that as IA, S-O, and Å are spearheaded by increasing subjectivity, and mentality humans are propelled and compelled to uphold, sustain and advance their being, TRU. Much of the opposite is irreversible, thereby making it necessary that we know as much as possible so that we can PPMA and control all that effects our destiny.

	Hegel, Absolute Idealism	Kierkegaard, Existentialism	TRU
Source			The Given, IA, Å
Choice	Thesis	Choice	Thesis-Choice
Necessity	Antithesis	Necessity	Antithesis-Need
Conclusion	Synthesis		Conclusion-TRU-TIT-ACC-TÅ

Table 1-88 **IDEALISM and EXISTENTIALISM in TRU**

Two highly influential schools of philosophy in the last two centuries include those of the absolute idealism of Hegel and the existentialism of Kierkegaard. The absolute idealism of Hegel held that all knowledge could be analyzed and interpreted in terms of a dialectical method, whose format was thesis, antithesis and synthesis. Existentialism held that the glory and agony of life is a person's ability to make choices. One can't avoid them. Existence, like romance, was the freedom to do what pleases. Putting these together we have a glorious ability to make choices as the thesis, the agony of no escape from them or the antithesis, and their resolution in the synthesis. Obviously, however, this is inadequate, except as a tentative guide. By TRU both Absolute Idealism or Existentialism are only approaches or methods, offering little to explain either the sources of our being and the conclusions that must, by necessity, be drawn. They were, as so much of philosophy, and especially its ISMS, partial and incomplete. The sources are to be found in the derivation, development, determination and demonstration TRU, e.g. ATO, and in the FAF of TRU, its DAOST, SCARLS, TROS, TRUD and ACEXIN. Its resolution is to be found in all TRU, SPALT, HOM, and ACC. When we reproduce TRU as a concrete object in which posits are made of all its content, including its aspects, parts, processes and mechanisms, the syntheses of all that makes up TRU can provide the conclusions necessary to avoid the bad and make good TRU choices. **TECS** = The Critical Essences, and **STEC** = Summation of Critical Essences. These were the great strength of Hegelianism. Although Hegel championed theistic monism, as in the absolute spirit, *geist* and thesis, ideas of antithesis and synthesis led him to STEC. However critical essences can not be any essences, nor even ESINS, they must be the major ES, CES, concepts, principles, laws and mechanisms of TRU. So also can choice, by the similar necessity of TRU, be directed. Such a summation is absolutely necessary if an adequate KOT in totality or a comprehension of all things is to be achieved. Such an en-masse collection is what we find in LANT, NEC, POT, EPI of the essences, modes, themes, TÄ, ES, CES, ACEXIN, and REM of TRU.

TICOH = TRU Correction and Inversion of Truth and Honesty. The attainment of TICOH isn't easy. It compels a higher and accelerative degree of TRU, TES, IMTES, AITES, Pureality and COGNOS. TICOH is the correction of all less than TRU and inversion of all not or negative to TRU. Actual truth is not any truth, nor even scientific or gospel truth by agreement of S and O or the correspondence or correlation of fact. It is how this agreement and correspondence occur as they are in TRU. Honesty is not so much being truthful in any sense, or telling simple correspondence or correlations of fact as they exist in and of themselves, as important as this is for S-O and a viable reality. It is in correcting and telling these correspondences, correlations, facts and truths as they actually exist in the virtual and actual reality of TRU.

CET = Compulsory Elevation to TRU, **CEST** = Compulsory Spontaneous Elevation to TRU, **CESTS** = Compulsory Spontaneous and Sustaining Elevation to TRU. Survival, life, TES, KET and TRU can not be achieved without a sufficiently high level of knowledge, learning, thought, understanding, apperception, will, solution and action. This should not be only by the individual, but by all that make up TRU. All individuals by IA at all ages in life, may be measured by CET, CEST and CESTS. Those who do not rise to, nor elevate themselves spontaneously to this level will not profit from TRU.

WILT = Will to TRU. The concepts, principles and laws of freedom and duty in relation to TRU are evident in, and are illustrated by, those nations that do or do not provide them. Some nations provide a great amount of freedom and in many liberties are curtailed. Much depends upon a peoples will to TRU. Also what is best of liberty, compliance, and how they become perfect in TRU. We can qualify the parameters given, liberty and compliance by the degree to which they are voluntary or involuntary, direct or indirect, and others. WIT works with all that enables TRU by providing an optimum direct and indirect voluntary liberty and compliance that best fulfills the social contract in TIT. Compliance is, to do what one knows to be TRU, is told to do because it is TRU, and to acquire and achieve TES. To will this is to activate WILT.

The importance of foreknowledge is an increasing aspect of human mentality. It was noted in the bible: Ac 2:23 "foreknowledge of God, ye have taken", and "Elect according to the foreknowledge of God".(lPe 1:2). Not only is it in foreknowledge that we can attain such heights of success, it is in the foreknowledge by KOT from TREQ and TRU that ever greater ability to anticipate and best act that TES and TRU are achieved. The collection, refinement, validation, and summation of foreknowledge that provides a more perfect performance in TRU is a predominant means of achieving everlasting TRU and its associated permanent victory in TÂ.

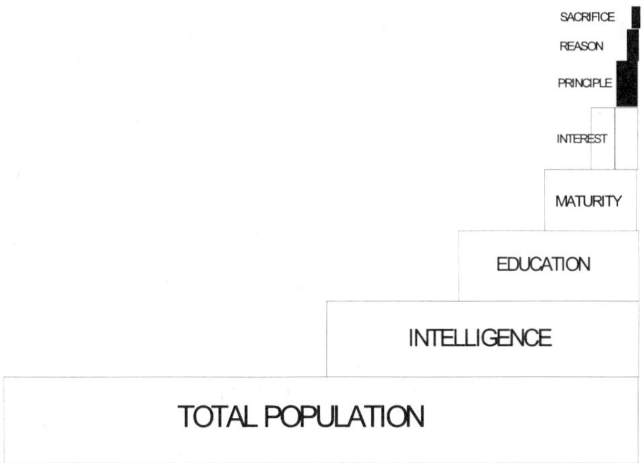

Fig 1-88 **ACCUMULATIVE DEFICIENCIES of TRU in POPULATIONS**

Possibly the greatest of all problems in our increasingly complex and massive worlds is the fact that its populace grossly lacks much that it takes for success and survival, i.e. to be TRU. This has been corrected, in evolution, through natural selection and special forms and functions. Such a delimitation of species, races, cultures, nationalities, and other groupings are less available or acceptable to the actual current world, TRU. This is accounted for by the fact that TRU may be much more limited than what we believe this integrated world might allow. In addition we must remember that although the linear genesis or seed line taken by the species homo sapiens is largely biological, in TRU it contains those added states in totality, qualities, essences, as well as special characteristics which are different and of the utmost importance, e.g. mans consciousness, self consciousness, values, reason, intellect and others. To bring quantity as well as quality of people into TRU, either as part of reproductive selection or in other ways the following schemata shows the great deficit of those of 1. NAMTRU, 2. have the necessary intelligence to contribute to TRUF, 3. have the intelligence but lack the necessary education, 4. with intelligence and education but lack the necessary maturity, 5. with intelligence, education and maturity but lack the necessary interest, 6. with intelligence, education, maturity and interest but lack the necessary principle and conviction in TREQ, 7. with intelligence, education, maturity, interest and principle but lack the imagination and reason for continued pursuit of TREQ and TRU, 8. with intelligence, education, maturity, interest, principle, imagination and reason, but lack the life long practice, service and sacrifice to TRU that is necessary. Herein we can see the different AMTRU worlds about us, the purity of TIT, and especially the needs and problems before us, to confront opposition, enable TES and achieve TRU. **DONEP** or Deficiency or Lack of Necessary Essentials in People, or a Population, to be TRU. If many are to participate in TRU, especially be a part of its reproductive selection and core of its evolution, or share in its predominance and their immortality, they must have, develop, cultivate and practice not just one or some of the essentials of TRU momentarily, but all continually. We, thereby, must be forever dedicated to, and remember the ever increased need to attain: intelligence, education, maturity, interest, principle, imagination, reason, practice, and service of TES in TRU.

The great test of knowledge is the ability to state what an entity is, what its essence in TRU is, what TRU is, and what its, and our role, in TRU is. To do so we must know and express the words that stand for what these unvarying essences are. The greatness of this was recognized by the ancient Greeks, in Logos, which was raised to the status of divinity, and by Christianity the Holy Spirit, to be developed rhetorically by St. Augustine and the scholastics. It shows the power of transcendent ideas and forms expressed as words. It also shows us the presence of the real or the intermediaries, means and vehicles between the subjective and objective as they form the relative basis for TRU, its Á FAF, state and dimension. In knowledge we have a great search, a search for what is TRU. This is largely the ability to distinctly and discretely state what the relative actuality of a thing is, and how this is in the actuality of TRU. Not only how its differences can be made distinct, its samenesses are integrated, and these two combined in the whole, TIT-TO. In knowledge discretion and morality are provided in higher and more concentrated forms. Through interdependence and transmission these foci positively enhance TRU and negatively work to enhance TRU by eliminating what is ATRU. For in the actuality of TRU what is true, or if not true is much more than a simple correspondence, identity or relationship. It is the actuality of Á FAF in TIT-TO and TRU.

What is TRUE, What IS, and How can we be SURE?

God and Religion, TREQ and TRU are not perpetuated myths and illusions that mankind has been repeatedly obsessed with. This is what some of the public, science, and schools of philosophy, would have us believe. For even the intelligent, belief in what is not readily evident, or proven by direct examination or investigation, approaches heresy. Far from it, however, we can't assume, that just because we can 't experience or find our knowledge through examination and experimentation, that this is all. Certainly all who have believed and promoted the higher and more unifying powers in our world were not merely copying the obsessions of their predecessors. For many of them were the greatest minds the world has ever known.

Thereby we must rise above the objects before us, even our subjective observations of these objects. For in neither alone will be found the answers to all things. The answers to all will be found in all things, as they exist in what exists. If there is anything that relativity, quantum theory, basic physics, verifiability, and falsifiability have taught us in the last century is that there is a focus, our observer-observed form and function. There can be increasing certainties and truths as our knowledge becomes ever more fine tuned if we avidly search them out. For us, our minds, and the great body of FAF that give us such absolute existence, is what makes us predominant on earth. This is our predominant mass-energy or the A FAF of our lives. That this consists of many levels and forms of energy that make existence should not be surprising. It is only when we discover, learn and can understand all the aspects, makeup, parts, processes form and function of this special universe of best human life, that we can begin to recognize how, all is constituted. Only when we with validity, actuality, proportionality and totality collect and integrate all that exists in TRU will this become apparent. So that by TRU, as with God and those who recognized spiritual powers and supreme beings before us along with the mystic union each of us has with it, can we begin to understand the world in which we live.

We can not understand this great problem, that science, philosophy and religion have so struggled with through the ages of the total make up of the world, unless we discover and learn TRU. For it can't be known simply, superficially nor singularly. It can be known only as it more closely replicates what it is in and of itself, TIT. This calls for a complex, profound, subtle, total and dynamic understanding of all TRU. It also calls for a good understanding of how this TRU is represented, presented, held in perspective and comprehended. Nothing less solves this problem. Once achieved, it is for all to appreciate, benefit, and have all time mastery of things that attain ever greater immortality, in the eternal continuity of the best of human life.

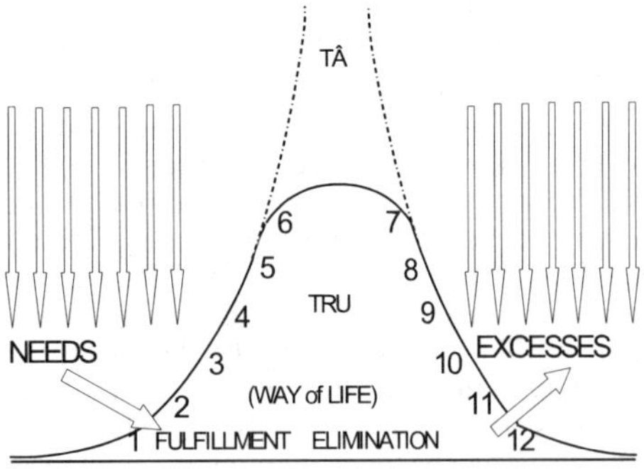

Fig 2-88 **ACCUMULATIVE EXCESSES and NEEDS in TRU**

No way of life, nor religion, can be better than the degree its: 1. Abilities meet the necessities of the world,(LATAN), 2. Ordination, (LATANO) 3. Coordination,(LATANOC) 4. Its standardization,(LATANOS) 5. Degree and intensity to which it is momentarily and in summation, IAT, realizes, holds and follows these first four, 6. The degree to which all individuals adhere and meet this needed intensity. This is the reason why some have been so great and many so much less. There are increasing needs accompanied by increasing excesses upon which the rise or fall of TRU depend. It has been benefitted in the past by communion, deification, mutual purpose, and united will. It will continue to increasingly benefit from these in the future. Yet since the needs will also increase, more and more dependence upon LATAN, LATANO, LATANOC, LATANOS and all TRU will be required.

The sanctity and holiness of God, Christ, Mary, the Saints, and all that religion so commonly extols is largely an exaggerated reflection of a mysterious and secretive interpretation of the absolute nature of TRU. Once the nature, form and function, parts and processes, concepts, principles, laws, and mechanisms of TRU are discovered, well developed and used the exaggeration is unnecessary and religion is freed from errors of disproportion, incoordination, inactuality and incompletion.(1/VAPO). The actual world, the relative universe, may be considered religiously by monotheism, trinity, polytheism, or pantheism. Yet by none of these exclusively nor strictly except as modified by their roles and depending upon how they are formed and observed in totality. Even an essence of deism may be recognized, whereby much that the relative universe is, is neither of, nor managed and controlled by humans. Still, it warrants a highly apperceptive, accommodating,, laizze faire, and/or compensatory approach. Yet by none of these exclusively except as modified by their form and function in its totality, TROS, and TIT-TO. The great contribution religion, God, Christ and many more have made to civilization can never be minimized. Yet this must not hinder us from analyzing, improving, and supplementing them to develop and realize the full potential of knowledge that best enables us to comprehend our world, TRU.

The requirements for knowledge of the future must consider it not from any point of view, not of any or all entities, senses or things. It must be, as all, from and to TIT-TO. If it doesn't it quickly deviates and loses focus. Navigators will tell you such loss of focus quickly leads to misdirection, confusion, failure, and disaster. It is for this reason, the actual existence of only TRU in the future, that not only its imperceptibility renders us so confused, mistaken and ignorant, it must be through and through VAPO and ACEXIN complete of all that is of TIT. It will also have to be most intricate, of greater and greater detail and of the greatest of TRU. It must account for not only the future, but the potential futures, TRUFS as well as TRIF. Mostly knowledge should emphasize those complex, subtle, evasive great principles, concepts, laws, and mechanisms which man has either failed to discover or failed to develop and utilize successfully. That of and in TRU is much different than any knowledge or knowledge as we know it. What is dynamically and proportionally in TRU is even more different. What is interdependent, related, integrated and functions as a whole in TRU is even more different than any knowledge we know. For these reasons we must be very careful to use what knowledge we have, adapt, alter, and find what is unknown, ESINS, ES, CES, dynamics, proportions, related, interdependent, and integrated as they exist in TIT-TO.

The great advantage of today's knowledge is shown by magnanimity. This is a virtue of many who with sufficient overview have the necessary moderation, tolerance and appreciation necessary to positively benefit society, ENT and TRU. This is why the accelerative forces and powers of CES in HOM are so great today and will continue even more in the future. This leads to: **POFUS** = Principle of Future Super-magnanimity. When humans learn all the effects, means, and other features of TRU the development of the great major courses of the future can be better known and shown. What seems so uncertain about the future will become clear.

ILE = Imperative Liberal Education, general or universal TRU Education. ILE signifies the high degree of learning that has, is and will be an ever expanding part of our world. It has, is and will be in ever greater demand, a necessity that each and all of us must fulfill as early as possible in life for this life to be TES. (KUPS)

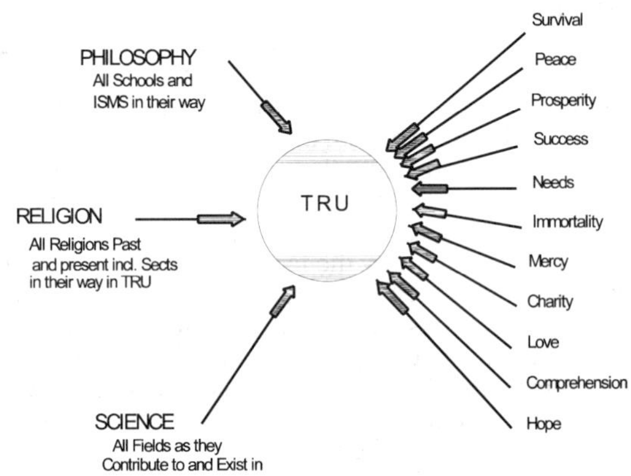

Fig 3-88 **21st Century Revolutionary Forces**

These will participate and become greater, with TRU. All religions must be held in full perspective and practice in life, if they are to be adequate and successful, valid, actual, proportional, ordered, and total. Only in this way, by VAPO and TRU, can they return to the individual the ACEXIN of ones being and contribution by PIA, DIA, SIA, SA, OM, PUREACLITY, COGNOS, ELTICMA, SA and TA. Thereby we can we have an adequate satisfactory, successful, permanent, perfected and ultimate religion that properly satisfies and fulfills the demands and needs of our being, ENT, TRU, and TÂ. Sects, tenets, beliefs, and religions are continually being replaced. Individuals repeatedly change faiths, convictions, principles and beliefs depending upon influences, maturity, and experiences in life. All persons, new and old, would do better to learn or relearn what is the greatest, or TREQ, in theology, philosophy and science and avoid much of the rest, what is ATRU.

The great revolution of the next century will be the development, appreciation and expansion of a comprehensive understanding of all, TRU, COAT. It will be catalyzed by the information age capabilities of modern electronic technology and the stimulus to individual comprehension necessary to meet the expanding demands of human life in an ever more competitive and complex world. It will come from a synthesis of all that has enabled it in the past and compels it in the future. The degree to which TRAX, TREX, TRIX and IPTRU play to reveal ATREC, the unrecognized TRU and TIT, will be the measures of their success.

The most general and basic religion is that in which each and all persons learn, understand and apply those basic concepts, principles, mechanisms, and laws necessary to lead the kind of lives, act, behave and do what is best by TES and TRU. Herein is the key to an ever better life. It can be expanded into: **MOR** = the Measure of all Religions. If these concepts, principles, mechanisms and laws are TRU a religion approaches perfection. It loses most all that has inhibited it and caused conflict with other religions. The degree to which any religion is TREQ is the degree to which it is to be complemented and united with all the same, TRU. Being ALT is essentially identical to TRIT, a 1:1 relationship with TIT. Being Á is called the S-O basis of religion in its most complete and proportionate sense, or: **SOPT** = the Subjective Objective, valid, actual, Total and Proportionate basis for religion, or TRU. When the S-O is reduced to its Á FAF, state and dimension in TRU this becomes: **ÁPT** = Á Proportionate basis for religion or TRU. The goodness or badness, ultimate success or failure of various religions, sects, and ISMS may be readily measured by the ratio of the degree to which they identify themselves with all TRU or TREQ. Or it may be done in reverse, negatively, by the degree to which they extol their religion, sect, ISM, or belief as an ENIT or end in itself, when of little or no TRU, and lacking in VAPO. **RANT** = Ratio of ENT, or ratio of contribution, or enablement of being, over partiality, ATRU, or ANTRU dissociation. RANT explains how we may grade different religions, sects, and ISMS and unequivocally say which, and to what degree, are right or wrong, by TRU. In this way we can adapt ourselves, and all that is around us to practically and successfully become optimum in TRU.

LÄSÁ = Law of Solidarity of Á. LASA is shown, in part, by COGNOS and ELTICMA. It occurs when knowledge has reached a point of sufficient completion, level of certainty, verification, and solidarity in phenomenal or real identity with Á and TRU. By LÄSÁ with TES there is absolute assurance of TRU, TRIA and TIT. This is a level of mass energy, whereby its predominance so overwhelms all encountered that it can increase at a rate far greater than all that might cause its defeat, or extinction. This is the exponential positive that acceleratively drives TRU in TRUF, TRIF, and TÂS.

The effects of cognition as means and processes demonstrate the dynamics of the Á mass energy FAF of TRU. They like subjectivity directly, indirectly, singularly, multiply, complexly or totally pervade all TRU. This is in sharp contrast with the astrophysical and other worlds and universes. It is a prime proof of TRU. As subjectivity, cognition, and COGNOS work from thought to thought their roles and powers accelerate. When working from IAT, TRIT, OBT, SUT, and ROT to TRU it has the effect of a chain reactor, or ultimately: **CICH** = COGNOS is a Chain Reactor. This concept signifies how by TREQ in subjectivity, knowledge and science, psychology, philosophy and religion can and will be among the greatest of all applied sciences. This is in contrast to their commonly regarded and relegated role of impractical theorizing on AU detail. By Á and TIT they will bring new light to applied sciences. For they are completing and fulfilling both our realization of and the fact, processes and operations of the higher levels of mass energy in TRU. This is an example of and proof of Á. Á is not a polarized and unrelated S-O or entity in any world, it is a living, dynamic, predominant great continuum that constitutes and operates in much of TIT and TRU. Its subjective or mental powers, as in CICH, are what provide much of the both differential and integral thrust of our being, TU and TE to TO of TIT.

LAPBOV = Law of Variability Political Boundaries. Political boundaries are not only arbitrary they are highly variable and changing. Both the quantity and quality of populations rapidly change within set boundaries. So that there is always a sort of over or under density or vacuum in political units. This plus increasing density of populations and nationalism leads to serious stress, strife, conflict, and war. In addition to tolerance and moderation on the part of its citizens there must be ways to adjust boundaries to changing conditions. By TRU and the optimal form and function of political units, many of the same principles and laws apply to all. **LASPCOL** = Law of Same Political Principles and Laws Apply to All. This may be applied to the more general case: **LASCOL** = Law of Same Principles and Laws Apply to All. LASCOL is how the same principles and laws apply to all TRU, TRUD, Homation and ACC or TRU Accentuation and Action. LASPCOL and LASCOL are to be modified by D or the Different in DAOST of TRU. The rise and fall of political states and nations can be meliorated by much of ACC including selection by TRU. The selection necessary can be anticipated and controlled. It should be sufficient to allow the cooperation and harmony that is vital to TRU. The ideal form of government then, is that which most perfectly represents TRU, or that segment of TRU equivalent to states and nations. It is that form of government that enables each and all the components of TRU to function in the best possible way. It is a form of government based upon the nature, or FAF of, TRU. It includes both democratic, republic, social, and totalitarian principles that are equivalent and beneficial to TRU. **GUT** = Government, ideal form of, a United TRU. GUT in being from TRU, TIT, TREQ, and GOD is, in a way, a kind of theocracy that best represents TROS, TRIA, TRIT and ENT. This is the optimal unity, identity, totality, VAPO and power that best guides and preserves the TES of a people and their governments. GUT must be based upon the power of TRU in each and all persons that is appropriate for each issue, problem and occasion. It must not try to govern all.Those functions and processes of TRU that can not occur in their own way. It must complement CES and ENT and avoid all actions where no action is proper by TRU. GUT, like models, give us a plan, method, and means by which we can best be and serve TRU. Yet we must always, as THEO, SUT, ROT, MOT, and ALT will differ from TIT, remember and correct for this very difference that specified parts have from TIT. For here is the key to best government and their harmony by and in TRU. It is signified by: **LOTHIT** = Law of THES-THEO/TIT Separation. This is the separation, distinction, and reduction of THES and THEO as they and all, by SCARLS, must return to TIT-TO.

By the Relative Universal Quotient, **TRUQCO**, or measure of participation in TRU and TES, individuals have certain rights. This, on balance, is all necessary to assure TRU and optimal adaptation. These rights are both inherited, but acquired. Largely they must be earned. If we fail to fulfill or even minimally meet the necessity of TRU and TES we can not expect to have the same rights as those who are TES. Many rights are limited and selected by: **LOT(S)** = the basic Law(s) of TRU. The right of propagation is naturally inherited, yet, not inalienable. For it is unequal, depending on quality. Those individuals whose progeny will be detrimental to TRU do not have the same rights. So that there is a sliding scale in the continuum from highest TÂ, TRIA, TO of TIT, ES, CES, and TRU through TES, decreasing REM, MATRU, AMTRU, NAMTRU, ATRU, ANTRU, ANTITRU, and all. Measures and laws must be adapted to TRU. They must be established to provide a progressive adaptation to the demands, problems, and massive needs of the future.

Because we do not now know perfectly, or because a contradiction seems to disprove our knowledge does not mean that what we know is not much, or all, of the truth or needs to be discarded. This is contrary to what the structuralists, critics, skeptics and cynics hold. It is settled when the bigger problem is solved. This is to know the truth as it is TRU. For if and when we do this, most of these seeming contradictions and unknowns evaporate. We can then go on, by VAPO, DAOST, KOT, SCARLS, TROS, and TRIS, to build ever greater truths, totality, and certainty.

RUPKAL = the Revolution and Rapid Attainment of Universal Perfected Knowledge and Love. RUPKAL is accumulative, accretive and summative by and to TES and TRU. There is at all times, some CES that are more dramatic or acute than at others, a constant increasing or decreasing in ES and segments of TRU. TRUP and TRIP is not the same as history except where history is positively adapted and warped to TRU. That of history that was negative was not TRU, unless and to the extent it produced stimulus, positive reaction, provided knowledge, or experience for TRU. The great revolutions of history were and were not TRU. The great current revolution toward the rapid attainment of RUPKAL is an example of what has, is and will be most positive in history and future.

AOS = Alteration of Standardization. AOS is most noticeable in systems, religions, sects, philosophies, schools, sciences and fields. There is a constant tendency toward a realization of standardization that comes from new appearances and new subjectives. This is not so unusual when one recognizes the flight of IA and experiences in life. It may be better demonstrated when signified by: **PONS** = Principle of New Subjective, and **FONS** = Fallacy of New Subjective. AOS, PONS and FONS show us the antitheses and relationships in the continuum between chaos and cosmos. For with new subjectives and their alteration experienced in life we have a steady stream that does not necessarily move toward TIT. It often moves in whatever direction it is compelled, inclined or desired. To correct this we recognize FONS, AOS and PONS and what must be done to correct them. This is accomplished by direction, convergence by TRU and often most formatively by limited standardization.

Fig 4-88 **POWER of the SUBJECTIVE and MENTAL REVOLUTION**

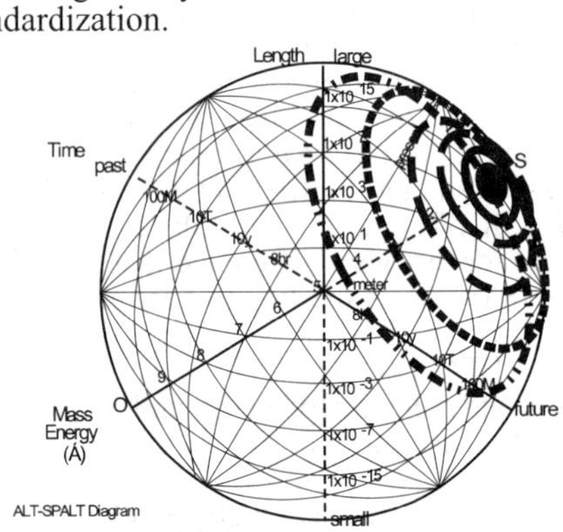

There are many processes, forces and powers that contribute to the great subjective mental revolution that have, are and will increase human life. It is a summation of all the higher levels of predominant relativity, oriented in the S-O mass energy FAF, continuum and dimension of Á. It is most clearly seen in COGNOS which, although formative, at present, will become more and more functional, stimulating, empowering and predominant in the future. It among all the processes of DYNO, shows how the future is moving toward ever higher levels of mentality, but is being drawn, reversely, by many forces, even the powers of the future to be what it is , in TIT and TRIF, and will be, in TRIS, TRUF, and TÂ. It should be noted and accepted that in this figure SPALT has a rotated axis which is viewed to make Á with S in the right upper instead of left lower position.. This shows how ROT, MOT, ALT can vary from a 1:1 replica of the set, standardized form of TIT.

The question remains and will always be, "to what extent and in what ways are our controls to be made"? Certainly life must be allowed a reasonable degree of freedom to follow its own natural inclinations, especially when there is likely benefit. The struggles, abuses and defects in the equilibrium between natural freedoms and controls will not only take much improvement and time; it will require many additions and alterations with changing conditions. It will require a greater enlightened public, citizenry, and ever greater additions in the development and use of controls. This is what TRU provides and ACC seeks to resolve. A much more valid, actual, harmonious and comprehensive means of achieving both a natural and an ordered life. The guide for greater appreciation, grace, good will and service to TRU.

No matter how much unanimity there is or seems to be in human beliefs right and wrong, and TRU, they can never be right unless and to the extent they are, and can be made valid, actual, proportional and total in their representation, of the actual world, this special universe of best human life, TRU. There are a vast array of methods and applications that are offshoots and uses of highly expanded cognizance and computerization, e.g. **AI** or Artificial Intelligence. Yet, they can never displace the living virtual powers of human ideas, thought, and identity. They can provide vehicles and benefits that the living virtual powers of human ideas, thought, and identity can't. The ideal will be an expanded role for both in the harmonious function of human mentality in the mass energy of FAF and continum of TRU.

FUH = Future Hindsight. FUH is that power of subjectivity, especially in TRU, whereby VAPO and TRU knowledge in the future effects positively and compellingly the present and all in time to become more TRU. FUH comes in many different forms, these are called: **FUHÄ** = Future Hindsight, Forms of. Some of these include, 1. The general drawing power of subjectivity in the future on all before, current or after, 2. The special type of FUH, largely personal, whereby recognition is made as in being watched. This is: **WABEW** or We Are Being Watched. WABEW may be of more than one variety depending upon the degree of personal receptivity, 1. as conscience. Or, 2. it may depend upon the nature of the subjectivity in the future, as personal views combine more and complex types of subjectivity. 3. The general drawing power of TRU, as an absolute spirit, or the religious experience of God whereby one is receptive to Grace and in turn is justified through their own spiritual activities. 4. The total combination of the above and others as they provide much of the compelling forces for, and a great segment of, the FAF of TRU. This is all subjectivity working posteriorly enhancing all working anteriorly. These do not necessarily work from the distant future. They can work from the mid or near term future, as well as any forces that work posteriorly, as near past to far past.

POCI = Principle of Infinite Convergence of Ideals. POCI fits with polarity. Yet from our perspective of TRU a large part of ethics, ideals and TRU it increases progressively. Not so much on a point or pole nor in a line, but through a whole field and range, toward the whole of TRU infinitely.

MEASURES of DYNO SULF CES EXISTING/TRU KNOWLEDGE/TERMS

Term	Measure	±	Type	Value	Term	Measure	±	Type	Value	Term	Measure	±	Type	Value	Term	Measure	±	Type	Value
					(ETK)	2-4-6	±	S	4	LASPCOL	~4-3-7~		T	5,7	REQUIREMENTS FUTURE KNOWLEDGE				
ACCUMULATIVE. EXCESS, NEEDS					FONS	1-8-5~	-	M	4	LOT	~3-~-8~		M	7	(RFK)	2-2-7		M	5,7
TRU (AEN)	3-2-7	±	S	8	FOPIC	3-3-5		S	1	LOTHIT	~-~-~-~		M	7	RUPKAL	1-1-8		S	7
AOS	~4-1-7~	±	M	6~	FUH	0-4-9~		M	~3	MOR	3-1-7~		S	7~	SAJ	3-2~-6~		S	7
ÁPT	~4-3-7~		T	7	FUHÄ	1-4-9~		M	~3	PASOS	~-~-5~	±	M	4	SIRT	~4-4-7~		T	3~
CET	~4-4-7		T	1,5	GUT	4-1-8~		S	7	POCI	0-4~-9~		M	7	SOPT	4-2~-7~		S	7
CEST	~4-4-5		S	1	ILE	2-4-7		T	7	POCP	~-~-~-~	±	M	4	STEC	4-3~-7~		S	5~
CESTS	~4-4-8		T	7~	LACJOT	2-2~-7~		T	7	POFUS	3-1-8~		T	7~	SULF	2.5-2-7		.5	
CICH	1-1-8~		S	7	LACPUT	~3-4-6~		T	1	POKOT	4-3-5~	-	S	4	TICOH	3-4-6		T	1~
DE(DES,DESINS)	3-6-7		M	1	LAPOBOV	~4-1-7~		S	4	PONS	~0-8~-5~	±	M	4	WABEW	~0-6-9~		M	8,6~
DONEP	4-3~-7	-	T	4	LASÁ	~4-~-8~		M	7	RAKUT	1-4~-5~		M	5~	WILT	3-3-6~		T	3
EQUOC	~2-7-8~		M	7	LASCOL	~3-2-8~		S	7~	RANT	~4-3-5~		S	5~					

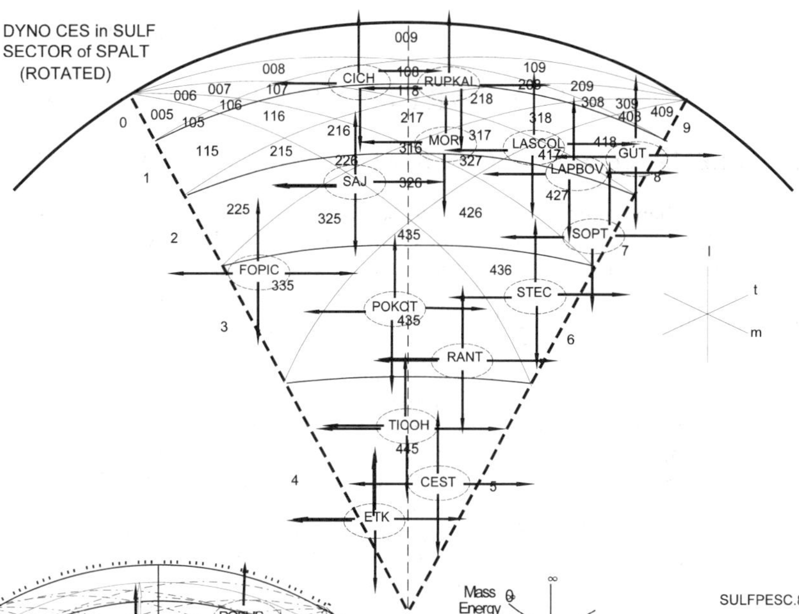

Fig 5-88 **DYNO CES in SULF SECTOR of SPALT**

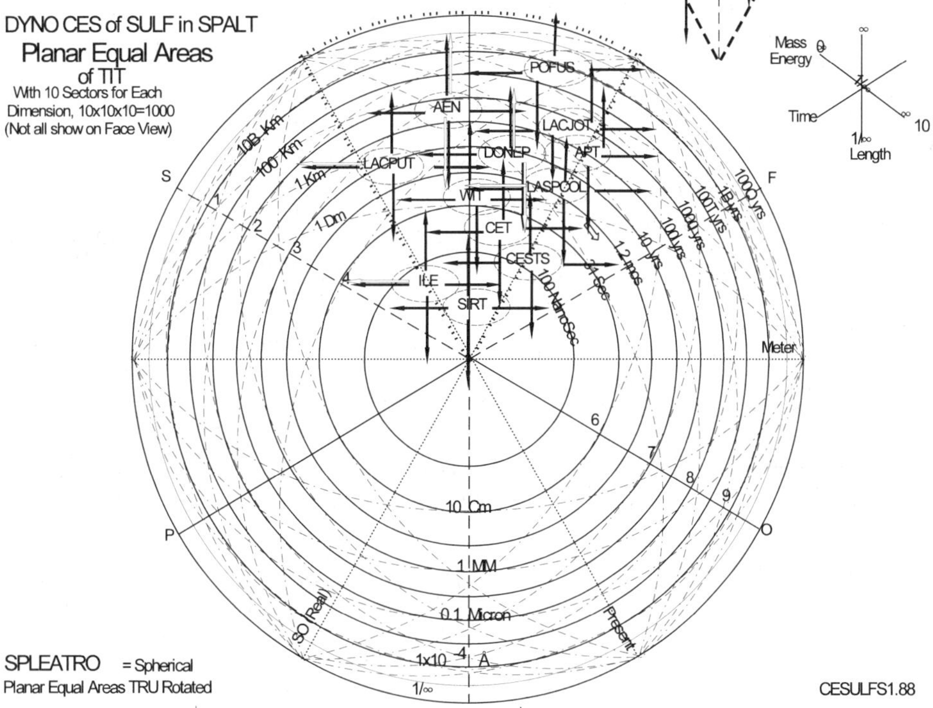

Fig 6-88 **DYNO SULF CES in SPALT**

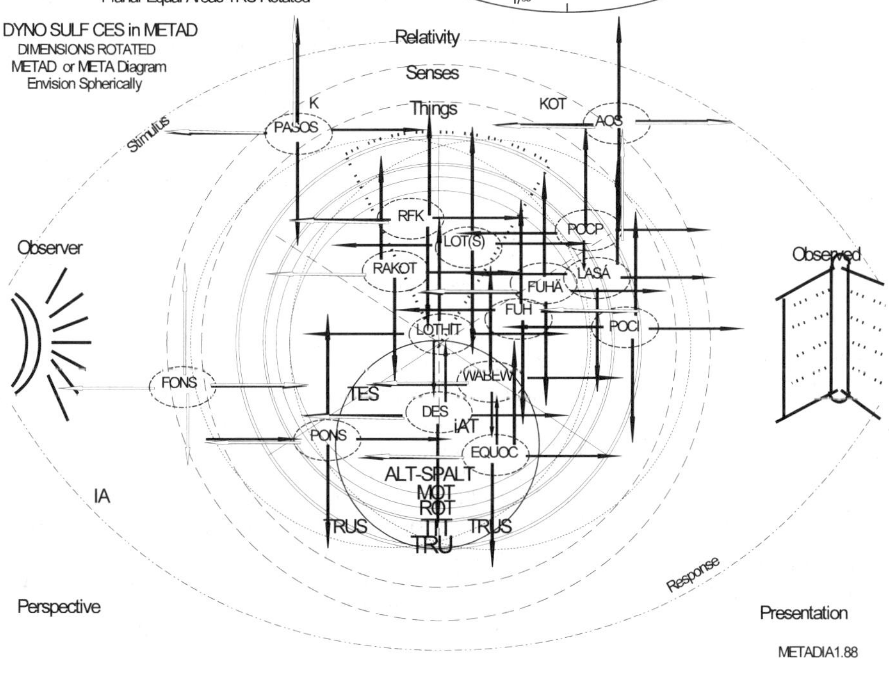

Fig 7-88 **DYNO SULF CES in METAD**

CHAPTER 89

SUMMARY of INTEGRATED PROCESSES or DYNO

PROBLEMS, S/F, SOLUTIONS, SYSTEMS, OPPOSITION, HALF LIFE, ACTION, COURSE, UNIVERSES

There is more than enough knowledge to be discovered and learned, work to be found,.and happiness to be sought and enjoyed for the most adventurous and imaginative persons in TIT-TO. They can be infinite, of exquisite detail, simply ordered and unified. They can be absolute in totality and actuality. They can be perfected in identity, distinction, coordination, order and solidity. It is unnecessary and detrimental to delve into what is not, the negative, erroneous, false, invalid, inactual, disproportionate or trivial that so distract us. We must know and live TES and TRU by VAPO, SCARLS, DAOST, TROS, and ACEXIN. We must not waste our lives and knowledge on all that is less, except what is differential or a threat to TRU. So may we, through, the power of TRU, set to rest for all time all that reduces our mass-energies fruitlessly. This although natural and repeatedly experienced must be relearned, remembered and forever best applied. Let us only have the capacity, ability, courage, conviction, constancy and joy in the glory of victory to see TRU come to fulfillment, continually maintaining, improving, and accelerating all that assures its permanence. TRIA is what both propels and compels us. This is: **PACTRIA** = Propelling and Compelling Powers of TRIA. Most all segments of the world, typical of their partial nature, past, present, and future, large, medium and small, mental, phenomenal or physical tend to rob Peter to pay Paul. This is seen in living on credit, deficit spending, borrowing today to pay for tomorrow, and making the future pay for the present. These are the antitheses of the necessary S/F of life. In TRU it is part of the continuum of TE-TO, yet when less than TRU accelerates ATRU. Only as the past provides for the present, as all time, length and mass-energy sectors do for each other, can we expect to restore sufficient S/F. This is to be found only in TRU or TREQ, as guided by TRIT, OBT, SUT, ROT and SPALT. It is to discover and learn the reasons for such forces and powers, especially when negative, at work in our lives, that we can best see the great value, benefits and blessings of TRU.

SOEM = Subject Object Expansion of MassEnergy. SOEM signifies evolutionary expansion of S-O or Á in animals and mammals to humans. It arises from the most simple to become ever more complex, compound, total with ever greater levels of capacity to adapt to the conditions of life. It is from these we can clearly recognize the dynamics of, and idea for, the mass energy dimension, state, and form of TRU. SOEM may be considered in more than one way. How it has developed in RUTS and RUCH, how it developed in TRUS and TRUCH, how it exists in TIT and how it converges on TÁ. Such a tendency to completion is both propelling and compelling and CES in DYNO, TO and TRU.(PACTRIA)

Any society in which the hard working, constructive, knowledgeable, and genuine contributors to TRU are made to pay for the leisurely, ignorant, and ingenuous destroyers of TREQ and TRU, is doomed to failure. This failure is often a revolutionary or explosive one. Societies, nations, cultures, civilizations and a world, in which only a small percentage of the population is TES and effectively striving for TRU with a large part of AMTRU, ATRU or ANTRU, can not maintain itself. TRU consists of great societies, nations, cultures, civilizations and a world in which all must be TES and TRU. They should especially be ones that grow and improve into ever greater participation in TIT. This is in order to enable all TRU to grow. It requires the highest possible percentage of both quantity and quality of TRU for all parts, ES, TES, and the totality, TO, of TRU. Together these provide, **ADTÂ** = To TRU Absolute. ADTÂ is exemplified by the life and trials of those who have persisted in the life of TES and TRU.

CIOS = Chaotic Inversion of Orderly Systems. If one is operating with a good, effective and orderly system, and another comes along which isn't in phase, with many, a few, or sometimes even one element out of place, this orderly system is quickly turned upside-down and becomes chaotic. This is seen from largest astronomical bodies to the tiniest particles. It is an everyday occurrence in lives that so thrive, by relativity and TRU, on order. A displacement of one or two elements, as through subjective paucity, being self centered, lassitude or lack of totality or OM, is soon followed by disorder. A harmonious or cosmotic system quickly becomes discordant and chaotic. This is chaotic inversion of orderly systems, or **CIOS**, and is shown by: **COSMOS → CIOS → CHAOS**. CES of CIOS in TRU are not all bad. They are bad only when the end result is negative to TIT-TO. For the partial, differences, oppositions, and dialectics are even less, more AMTRU and ATRU than TRU. There can be positive CIOS, those systems that provide a better TRU than their predecessors. It helps to sort out the negative from the positive, orderly systems that only seem to be so, and all that need correction, improvement, or replacement.(TREQ)

RIM = REM Impetus Preservation. RIM is the impetus for preserving the REM of ENT and TRU. It is exemplified by the French Epic of Roland and tragedy of lost life, especially when such loses are sustained for great causes. Any loss brings, to each and all persons, a deep feeling of foreboding. We don't appreciate what we have until we dearly miss it. This strong feeling has great TRU meaning. It shows how any and all that is taken from TRU, does so by taking from us what is most dear. We lose a part of our own life. This is our very being, a kind of living death. It is what has the greatest worth. Of this very being of greatest worth, there are many components, forces, ES and CES. It is up to REM to show what these great values are, both specifically in REM, and the absolute worth of TRU. RIM is the great state in TRU whereby REM, and the absolute worth of TRU, provide the impetus for saving what might be lost of TRU, self preservation and preservation of the species. It helps to teach what we must save, what is most important, REM, essential, and the absolute worth of TRU. It teaches and proves the absolute worth or value of all, all that is TRU.(LOPAN) It also proves and teaches what we must be, what, and how much we must do to enable, achieve and assure TRU.(TES) For one of the best tests of TRU over ANTRU, and especially TES over DAK, is to discover those who have found joy in good and right, and provide for the best in life rather the opposite.

LACDIT = Law of Capacity for Continuing Development in TRU. Homo Sapiens is, without question, in the midst and thrust of great anatomical, genetic, physiological, dynamic, structural, formative and functional change. Upright posture, opposable thumb, dexterity, and most of all brain capacity, cerebration, coordination, cognition, and reason provided by the higher levels of mass energy Á FAF in TRU. These are how our bodies and ability to survive are being pushed to the limit to make corresponding adaptations or changes to meet the competition and overall stress. No doubt this is not the end, as much as TIT necessitates stability and preservation. This may be for short periods of time unnoticeable, as in the present. Yet even in these short periods of time they are operative. Some of it is seen in the attitudes and appreciation for life or TRU that we have lost. We tend to fail to live by TRU, enhance quality and essence of all that is TRU. To turn this around as it is LACDIT is the great challenge for all.

MATDUR = Maturation, permanent and Durable of TRU. MATDUR is that state of altered genetic and biological activity where selection and competition are controlled for the benefit of all. This replaces survival of the fittest between individual and groups at a savage level. It can be accomplished less destructively, and with much less suffering by utilizing TRU's unique quality of not being only then and there or here and now, but of being through and through in different durations. It is in the through and through total dynamic nature of TRU that we can directly and indirectly, voluntarily or involuntarily work so that all willingly help to correct all problems and great conflicts that exist within us.

How much of life is in the living of it, and how much is in the preservation of it? How are the two reconciled and combined? Those whose lives are without TRU are worse than those who are dead. For the physically dead of TES continue to contribute to TRU, whereas those alive but without TRU detract from TRU. TRU is alive through and through, past present and future, not only you and I, or even only TES in the here and now. Humans, as TRU by DAOST, have lives that are much the same and different. Only after fulfilling TES can one with added ability or endeavor produce what is above and beyond the call of immediate duty. It is what our ancestors have done. It is what is so near and dear to us, our immortality. If we live in times of greatness and can stand on the shoulders of those before and among us we have the added advantage of being one of the chosen immortals.

What will the world be like at specified times in the future? It will Grow. It will change. It will in the near term be much like TIT. The further into the mid and long term we go the more that will be different and less will be the same. Culturally, and even knowledge wise, don't expect those of 2, 4, 6, 8, 10 thousand years ago, or in the future, to be like what there is today. This, however does not apply to human intelligence, mental or physical capacity which has changed at a much slower rate. Thereby, expect for the future of mankind in 2, 4, 6, 8, 10 years to have cultures and knowledge that increasingly different, hopefully and likely far greater and better than what exists today. Mental Capacity anatomically will not be that much greater, although with improved and expanded means it will have available much greater ability to use this mental capacity. More definitely human culture and knowledge will be much more effective, selective,, efficient, massive, harmonious, and tolerant of differences within itself as it is increasingly TREQ and TRU. Yet much of what is TIT and has been of TRUP will persist for long, reversibly. It will have some things that will be more accentuated to compensate for those that change or are under increasing pressure. This is seen in the changing flora and fauna of past ages.

From both the known as well as unknown cycles the world is in, as well as what is more irregular, we must learn and forever remember and apply the fact that what is optimal, may or may not occur but in a small fraction of time of our existence. It may be most opportune, or it may be detrimental to us. We must strive to overcome bad times and make the best of good times. This is seen in environmental stresses that have enhanced quality in the past. We can't live only on the crests, the periods of greatest affluence. We must live our best in all periods, to be TES. Sometimes it is more difficult than others. It is with those who hustle while they wait, have highest character and continually prepare, work, and enable TRU at all times, that have the greatest success and survival.

The uncertainty of the future isn't infinitely invalidating, it is only infinitely compelling, e.g. eschatology, teleology, TRU, TIT, TRIN, and TÂ. This arises from: **LU** = Law of Uncertainty. LU describes all that is not known or not known adequately, especially of TRU. **UL** = Uncertainty Law. UL describes all that is uncertain of the future and how because we are uncertain about the future does not mean that it is not valid, nor that it isn't infinitely and absolutely compelling.

The great problem of race relations corresponds with the extent to which we are DAOST, a human community, one species or one world, TU. There is in any RU a great continuum between centers, or cores, and peripheries or fringes, and all in between. We all have different views of our relative value, and roles, in the world. Yet those that assure TRU's knowledge, as intellect, judgement, and COGNOS, are certainly much of its core salient. It is in such salients that characterize our predominance and supremacy. This is called: **SAKOT** = Supremacy of Segments that are Salients in TRU. TRU will be largely proportional to how well, complete, practical, and effective knowledge, and especially KOT, is that it enables greater power for TRU. SAKOT is closely related to natural selection. Yet for TRU it emphasizes not so much the biological fact of natural selection as **CORF** or Core Forces, and the exclusive principle by which thisness or haecceity of TRU operates.(TRIA) It shows how the predominant and preeminent natures of TRU, work to continue, enhance and assure it. The completed explanation of these differences in participation in TRU are signified by: **LACHVAR(IT)** = Law of the Continuum of Human Variation Relativity in TRU. LACHVAR describes the FAF and state of TRU in which all its components participate not equally but as accounted for by relativity in TRU, as by REM. Only with VAPO, KOT, and ACC can we understand. manage and control or cure these differences.

	BEING/Non Being	NECES-SITY	PREDO-MINANCE	CRIME	POSSES-SION	SELEC-TION	FREE-DOM	INDIVI-DUALITY	HAPPI-NESS	HATE/ LOVE	WRONG BELIEFS	IGNOR-ANCE
TRUS	x	XXX	XX	X	X	XX	X	X	XX	XX	X	X
KOT	XX	XX	XX	XX	XXX	XXX	XX	XX	XXX	XXX	XXXX	XXXX
DOT	XX	XXX	XXXX	XX	X	XXX	X	XX	X	XX	XXX	X
TÄ	XX	X	X	XX	XXX	XX	X	XX	XX	X	XX	XX
TIT	XXXX	XXXX	XXX	XXX	XXX	XX	XXX	XXX	XXXX	XXXX	XXX	XXX
TU	XX	XX	X	XXXX	X	X	X	X	XXX	XXXX	XX	X
TÄ	XXXX	XXX	XXXX	XX	XXX	XXXX	XX	X	XXX	XXXX	XXX	XXX
ES	X	XX	XX	XXX	XX	XX	XXX	XXX	X	XX	X	XX
TES	X	XXXX	XXXX	XXXX	XXXX	XXX	XXXX	XXXX	XXX	XXX	X	XXX
TI	X	X	X	X	XX	X	XXX	X	X	XX	X	X
TO	XXX	XX	XX	XXX	X	XX	XX	X	XXX	XXX	XXXX	XX
TRUD	XX	XXXX	XXX	XX	XX	XXXX	XXX	XX	XXX	XXXX	XXX	XXX

Table 1-89 **CORRESPONDENCE of BASIC PROBLEMS and FORMS of TRU**

The correspondence of the basic forms and relationships of TRU with some of the problems, aims, objectives, goals and purposes are shown in Table 1-89. A dozen of the problems are shown horizontally and a dozen of the many basic forms and relationships are depicted on the vertical axis. Sometimes the problems overlap with the basic forms, e.g. AU/TRU. This analysis is not definitive, especially in how interpreted. It depends upon whether the determination is made from where they are, what they are related to, how they are answered and corrected. Each gives a somewhat different emphasis and thereby rating. Yet it is in realizing the overall ideas and need for discovering, learning, remembering, and using TRU that this table provides a cue to the best in human life. It is mankind's continuous neglect or denial of many such problems and forms that cause so much failing and suffering. It is through KOT, OM, REM, HOM, and ACC that they can be corrected and cured. **POD** = Posits Total Dynamics. The massive sameness, correlations, interrelations, and inter-operation of all in TRU proves the need for the posits of CES, CHES, in DYNO. This is exemplified by how you can't divorce morality from politics. The common focus of morality and politics lies in TES and TRU or TREQ. By DAOST, morality and politics are different aspects of the same thing, in TRU. **LASUD** = Law of no Separation of the Same and no Union of the Different. LASUD is a corollary of DAOST, and confirms the extreme when operative. They show how great ideas lose value in becoming partial or to emphasize the extreme, EBE. They are narrowed and broken in partiality, NAMTRU, ATRU, AU, and ANTRU. It is for this reason we speak of TRU rather than God, TES rather than soul, PUREALITY rather than morality, and many NEC rather than existing terms. A profession of morality or a feeble intention to goodness, distinction, REM and TRU is not enough. It is to be exercised in ones work and IPR in every day of life by TRU. It is highly complex, especially in large populations, with large demands on the capacity and ability of its members, institutions, and governments. The potential benefit/harm ratio is permanently and unvaryingly high. TRU must not only be understood in its fullest detail, it must be also fully followed, practiced, and used. This is PUREALITY, as the individual and governments stand in mutual cooperation and preservation when TES, ES and TÄ in TRU. **SUOB** = SU Overburden. The effects of SUT on becoming ENT and TRU, and as all things imparted into TRU, do not have an exact, but a lesser or greater, under or over, effect and reaction. If SUT is precise and valid, and if the perspective of TRU is balanced and proper then this overburden decreases. If these do not hold, but increase, there is reaction, a large component of much of the dynamics, HOM and CES of TRU.

Quality control of populations becomes more available with increasing knowledge, especially in genetics and related fields. It is possible to provide greater degrees of accuracy plans, maps, and preventive programs for the improvement of human quality. This is limited by: 1. Genetic knowledge is only in its infancy compared with its potential and applications in and by TRU, 2. The vast complexity and interactions of human genetics and how much even this is extended by their marked interdependence and dynamics, 3. Limited interpretation of complexities and interactions, upon genetic instruction, reproduction and selection in humans. 4. Limited utilization of existing knowledge. Humans see beginnings and ends throughout their lives, and assume all is likewise. Yet this is not so for TIT-TO. Here it is continuous recurrant dependent and harmonious dynamic. Beginnings and ends are more for ES and CES of TRU and even more for what is not TRU. Of TIT-TO there is no beginning nor end.(ATO,TRIA,TÂ) For if we had either, a state in which there was actuality before or after TRU, it would not be TRU and thereby could not be. 5. Personal and social objections to engineering, selection, action, and quality control.

LEVEL		QUALITY RATING	# CHILDREN	METHODS
0	20%	Highly Detrimental	0	Total Restriction
1	16%	Absolute Contraindications	0	Total Genetic Restriction,
2	14%	Highly Relative Contraindications	0	Restrictions, Early Sterilization or equivalent
3	12%	Relative Contraindications	0-1 Limit	Restrictions, Early Sterilization or equivalent
4	10%	Slightly Relative Contraindications	1 Limit	Restricted Prpagation Certain Birth Control
5	8%	Barely Sufficient, Possible Contraindications, Current Par	Limited to 1-2	Restricted Propagation Voluntary Birth Control
6	6%	Sufficient, Questionable Contraindications Readily Achieved Par	Limited to 2	Limited Restrictions, Voluntary Birth Control
7	5%	More than Sufficient to Maintain, Achievable Par	Limited to 3	Voluntary Restrictions and Birth Control
8	4%	Definite Contribution to Genetic Pool Far Term Par	3 as desired	Voluntary Participation in Program
9	3%	High Contribution to Genetic Pool	Over 3 Encouraged	Planned and Supported Increase in Population
10	2%	Outstanding Contribution to Genetic Pool	Much Encouraged and Subsidized	Strong Support Programmed Propagation

Table 2-89 **GENETIC and REPRODUCTIVE QUALITY CONTROL of POPULATIONS**

There is a great need, commonly and universally recognized, for birth and population control, voluntary or involuntary. Even more there must be such a need for quality and quality control. Many regions of the world have a higher density than they can support, and an insufficient quality to sustain this support. Programs to limit overpopulation and enhance quality are highly commendable, they only need to be refined and fully extended. In many regions there has been a kind of evolution in reverse with much current strife and suffering and dire negative implications for the future. This is seen in the many peoples in Africa, Asia, and even Europe and the Americas. To get the necessary results over a prolonged period of time requires a deep understanding, reason, conviction and commitment on the part of all persons to cooperate in the needed preventative voluntary and planned programs. It also requires the simplest, effective, and permanent methods possible. A method of scaled quality control that is increasingly voluntary is listed. With a good determination of genetic limits and potential, as noted withl GELF, goals and projections for optimal human quality may be set. This should depend upon the burden of total population, degree of lost genetic quality, adaptable time scale for operation, and vigorous fulfillment of TRU.

To have no personal incentive and much common good is no better than to have much common good and no personal incentive. Because one ISM becomes highly damaging does not mean that the answer is to accept and live by its opposite or some other ISM. It is in TIT-TO only, that we can find the answer and resolve these problems. The reason why the contrast, competition, and conflict between capitalism and socialism, individual and society, or TE and TO are so important comes from the most basic differences in the form of TRU.(DAOST,TROS) **LIGDOC** = Law of The Individual Group and Whole Differential, Opposition, Contrast, Conflict, and Competition. LIGDOC may seem common, obvious, and unnecessary. Yet it signifies the great meaning of equilibrium and success of TRU. It can't be ignored. There is hardly any dialectic relationship that is so powerful, detracts or contributes to human life than the success or failure of how a person, ones group, and the whole interact positively or negatively to provide for human life, TRU. LIGDOC is whether each and all that is TRU works together or apart, for their own interests or for the common good, and how these two are most mutually beneficial. This requires KOT and TRU, especially on the individual level: **IGUNS** = Individuals can not Identify with and Contribute Successfully with Groups, States, Nations, or Worlds they do not Understand. It is highly necessary that all understand the groups, states, nations and world we live in, what and how this is and should be TRU, ES, TES and CES in TIT-TO..

So is all Homation, Pureality, man and his world, TRU strong or weak, successful or of failure, as its foundations are so and as they are stable and change no more than TRUCH makes it inevitable. This is called: **PROPERT** = Proper and inevitable Changes in TRU. Thus through the latter pages of IPTRU less reference is made to God, except as it is TREQ, TRU, TU, TA, TE, TI, TRUCH, TRUD and SU. This is implicit by TREQ. TRU is identified with whatever part of GOD is TRU. To so use TRU, exclusively as our focus, vantage and measure is: **ROMT** = Reliability of TRU as a measure. If all is to be measured by TRU, then this measure can be no better than the degree to which TRU is known, represented and presented through units and scales for measurement. Much of what assures us of this reliability and success of TRU, both TIT and PROPERT is: **RUPKAL**, Revolution Universal of Perfected Knowledge, Love and Being in TRU. RUPKAL consists of the great driving forces of mankind, competition, and environment over the last millions years. It is the exponentially expanding subjective explosion. This is especially true of the near past, present and future. The processes of the subjective aspect of TRU, THES, whereby part of TRU is passing through its most rapid portion of growth. This is a critical phase in the alteration of TIT and TRU. It is what makes PROPERT of central concern. It is what makes ROMT even more needed.

MÓTH = Measure of All Things. Measures that have been or are to be discovered can be adapted to, and are often highly useful in ROT and TRU. They and similar collections of data and information, as in spread sheets, often provide massive, discrete, and appropriate information. They help to make the differential clear, the needed distinctions of ES and CES. They fill the gap that exists between KOT and TIT. They give a wider basis for reasoning in and interpretation of TRU.

For much of life we have plenty to eat, a home, family, and most we need. We get ahead, have money in the bank, wealth and possessions. We think we have it made. This, we think is the best it can be. Yet how soon and quickly these can be depleted and lost, our lives turned upside down. This may be from natural disasters, degradation, inflation, our own indiscretion and rash acts, or more often simply from being outdone competitively. There is much to lose. Yet much of this loss is often not so great as we thought. Much that is great we never recognized. This is particularly so if we do not continue to struggle, work and earn. It illustrates the constant dynamics and demands of life. It represents the mass against which the individual and TRU are in perpetual interaction. It shows what we, TES, and TRU must constantly surmount. This is especially true if we are wasteful, live beyond our means, or deny what the future demands. So that each individual, as well as AMTRU are in a lose or die condition. It is the SCARLS of the never ending dynamic massive TRUD. This we see in DYNO. Dyno proves TRU by helping to know, and thereby direct our lives to, TES in the continuing fulfillment of DYNO in TRU.

	HALF LIFE of TO, TIT, TRU	HALF LIVES of PROCESSES of TRU
LAHAL, (Alteration)	Is in part dependent on how long in the future TIT will exist, or change	Depends upon forces acting upon it
LAHIRNS, (Relative Stability and Instability)	Is reversible depending upon Mass Action, Momentum and Inertia of TRU and Forces acting upon it	Dependent upon TRU, and how stable or unstable and reversible they are
LAHOSEV, (Total Stability and Special or Partial Vulnerability)	Is a measure of irreversible stability or vulnerability and instability to forces acting upon it	Dependent upon TRU, and its own existence in TRU that irreversibly sustains or destroys it
LAHOCEX, (Total Contraction and Expansion)	Is the degree to which the half life of TO is contracting or expanding reversibly or irreversibly	Is the degree to which the half life of CES or ES are contracting or expanding, reversibly or irreversibly
LAHPRO, (Probabilities)	Is the likelihood the half life of TO will stay the same or change, and the manner and degree of this change	Is the likelihood CES and ES and its half life will stay the same or change, and the manner and degree of this change
LAHED, (Effects Determination)	Is the effects determined by the overall half life and how these effects determine and explain TO, TIT and TRU, its aspects and all	Is how the half lives of CES or ES effect and determine, their FAF as well as that of others
LAHANC, (Necessary, Actuation, Action, and Accentuation)	Is the imperative actuation produced by the half life dynamics of TO, TIT and TRU and how it enables action and accentuation	Is how the half lives of CES or ES imperatively actuate and effect their and other CES and ES, or show what needs to be accentuated
LAHAFS, (Causing Adaptation Forces Secondarily or in Reaction)	Are the forces adaptations and secondary effects that occur as a result of the half life dynamics of TO and TIT and associated reactions	Is how the half lives of CES or ES effects their and others adaptation, secondarily or in reaction
LAHERM, (Measures of Relativity)	Are the measures of relativity of TO, TIT and TRU provided by its half life and their dynamics	Is how the half lives of CES or ES provide and can produce measures of their roles, in themselves or as they exist in TO, TIT and TRU
LAHEXT, (State of Half Lives Relative to the Exponential Time frame of TO, TIT and TRU)	Are the shape and posits of half lives of TO, TIT and TRU in MOT, ALT and SPALT whereby they are adapted to the exponential state of time in TIT.	Are the shape and posits of half lives of CES or ES in ALT and SPALT whereby although half of temporal time, are only a tiny fraction of TIT time.

Table 3-89 **HALF LIFE CONCEPTS of TRU and CES**

There are many kinds of half lives or periods within as well as through the whole of TRU. The half period or life for TIT-TO decreases with rapid expansion, as needs, threats, forces and powers acting upon it increase. Population, technology, ecology, resources, effect on the environment, and many other forces may have more or less effects on the half life of TRU. These do not act simply nor singularly. They usually act in diverse, indirect, irregular, complex and less known ways until the damage is done. Stresses and strains, actions and reactions, causes and effects in processes are set in motion and work in variable ways. When these are overwhelming the likelihood of calamity, holocaust and extinction are likely. For these, it is important to enumerate, combine, make total, and account for. This has been much of the purpose of HOM, DYNU, DYNS and DYNO. If the half life of TRU decreases from 1,000 to 100 to 10 or less years, this is exponentially bad for the overall haecceity and worth of TRU. This is associated with an absolute to relative extinction of mankind. It is not, however an extinction of TÂ. For although the half life may be short, the lifetime for TIT will still be infinite. It is a measure of its relativity. For as long as TRU extends to the infinite in the future, TÂ and TRIA are not contradicted, absolutely. Yet this is all the more reason for concentration on TIT. For if not we are already in accelerated fall. Various categories of half lives or periods are, **LAHAL** = Law of Half Life Alteration, **LAHIRNS** = Law of Half Life Relative Stability and Instability, **LAHOSEV** = Law of Half Life Total Stability and Special or Partial Vulnerability, **LAHOCEX** = Law of Half Life Total Contraction and Expansion, **LAHPRO** = Law of Half Life Probabilities, **LAHED** = Law of Half Life Effects Determination, **LAHANC** = Law of Half Life Necessary, Actuation, Action, and Accentuation. Changes in half life occur because of, and in turn cause, secondary adaptation to forces, processes and powers that effect TRU. This is called: **LAHAFS** = Law of Half Life causing Adaptation Forces Secondarily or in reaction. LAHAFS can be readily recognized as the effects produced by reaction to action or forces upon it. This is called: **LAHERM** = Law of Half Life a Measure of the Relativity of an entity, organism, or TRU. Half periods also apply to in Á and L either interacting with time or alone, taking on the larger powers and meaning in ACEXIN and TIT-TO. LAHERM applies to the very important use of half-life in determining, measuring, and describing the forces and CES operating in the relativity of TIT, SPALT, and HOM. It helps to correlate the time and mass energy form of TRU. A, L and T in TRU are highly interdependent.(ALT,ACEXIN)

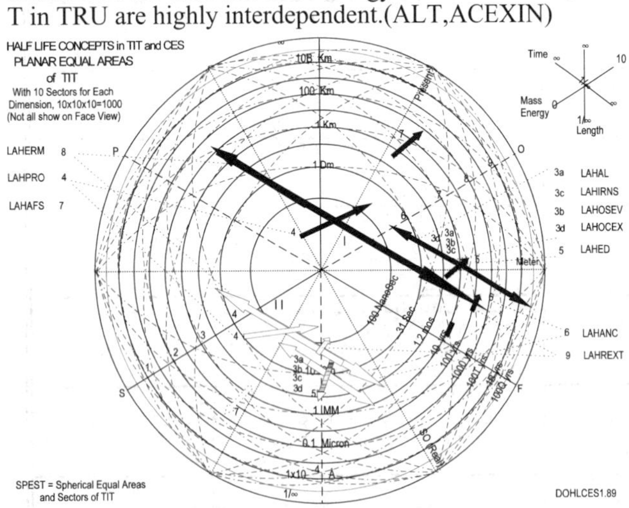

Fig 1-89 **HALF LIFE CONCEPTS in TIT, DYNO, and CES**

Analyses of half lives are important and helpful. It enables a better explanation of the FAF, mechanisms and CES. They are made more distinct and available for posit. In an analysis of the existence, states and roles of half lives in TRU it helps to break them down into their discrete elements. As noted above we distinguish between half lives of all TRU and its many varieties and variations or that less than the whole of TRU, its ES, CES and interactions. Secondly we should distinguish the many basic principles or laws of half lives that signify and denote these varieties and variations. Of the classes of arrows: # I, or solid arrows represent half lives of TO, TIT and TRU, # II or the hatched arrows represent the half lives of CES or other aspect of TRU. Of these kinds of arrows and conditions of half lives, that represented is: 1. a half life in TO of TIT, 2. the half life of TO of TIT in the future, as it could also have done for the past, present or any selected segment of TRU. This is what we noted in Chapters 90-99 of IPTRU and is of the utmost importance in establishing the frame or structure of TRU, MOT. 3. The fact that a half life may be of more or less time, or altered, both of TIT and of CES. This may be either irreversible or more often reversible. Such alteration when determined will become its expansion or contraction. The quality and quantity of this when reversible will determine its stability or instability or when irreversible its total stability existence or vulnerability. 4. The probabilities, either by prediction from existing knowledge of the past and present or existing and potential knowledge of the future. 5. The effects determined by half lives, this can be for either TO or CES, and can occur in the many ways these half lives produce their effect. 6. The necessary actuation, action and accentuation half lives produce or will produce. They are much of the grounds of ACC, Section IX. For much of ACC is not singular unrelated actions, it is the compounded accentuation of what already exists in TIT. 7. Represents adaptations and forces causing secondary effects or reactions to themselves or others. To this are the complexities produced by the varieties and variations of these half lives. 8. How their half lives are a measure of the relativity of TO or CES. This is more apt to be a product of change and variation than the linear temporal FAF of the half life. It is how half life participates in the mass variation of time in TRU, or role in CES. 9. How in SPALT half life is not half the distance of the vector, but is closer to the end. This does not nullify the validity, nor great power, of half lives as it might initially seem. In fact it is a double reverse in which not only are half lives far removed from the center in time, they in their extension to full lives become exponential.

PAO = Principle of Averaging Out. PAO is demonstrated by work in which the load is carried equally by all, rather than only one or two doing most and the rest doing little. It is shown by a division of labor in which many hands make light work. For if one has the work of several the load is unbearable. Whereas when many carry the same load it becomes light, quite bearable for all. To provide a distribution or division of labor that optimizes this we have PAO. The TRU solutions lie in mass action: **SIMA** = Solutions lie In Mass Actions. SIMA is closely related to PAO. Often those choices, decision and intended solutions that we think are correct, are usually only small portions at best, and often less than or negative to TRU. This is especially so if they are made quickly or without a good accounting of all that goes into them and their relationship to TRU. This is why individuals and groups that tend to be radicals are usually wrong, even though this extreme in TRU is of high REM. Only if they are tolerant and moderate enough to recognize the limits and role of this extreme or partial, and make a full accounting of it in TRU, can they rise to the necessary level appropriate to SIMA and TRU. To gain enough wisdom to guide SIMA requires an appreciation for all that provides the harmony, symphony, and concord that exists in TRU. **PAOC** = Principle of Averaging Out, Controls. PAOC signifies how we can not only provide an optimal division of labor, we best identify the work and provide controls that enable successful results.

TOPSOR = Total TRU is the Prime and ultimate Objective of all S-R, stimulus response. A corollary to this is its reverse and measure, S-R is valid and can be measured to the degree it fulfills the prime and ultimate objective of TRU. TOPSOR is most easily and perfectly achieved by objectivizing and subjectivizing TRU. This involves great mechanisms of fission and fusion. It is a kind of binary fission in which subjectivity and objectivity divide from the whole, only to return reversibly in fusion. It is a kind of fission that is readily provided in a practical, popular, conventional and standardized way. When we can provide the gradients for this fission, as in its decimation or reduction to a ten fold measure in the dimensions of TRU, we are well on the road to accomplishing TOPSOR and much of TRU. For the mass energy or Â dimension this is the ten divisions that show its spectral gradients from strict subjectivity through reality to strict objectivity. This is how the relativity of our being's form is in dynamic operation with all S-R, TOPSOR.

There is more than one kind of TRUCH: 1. All that changes TRU, TRIT, and TIT from its existing state, and thereby is sooner or later detrimental, 2. All that changes TRU, TRIT, and TIT from its existing state to a more ideal state, and thereby, is to a varying degree allowable. 3. All that changes TRU, TRIT, and TIT quantitatively, but not qualitatively. 4. All that changes, but is reversible, as in the case of TRICH and TRUD, 5. All that is TRUCH and is largely outside of or departing from, or potential that never becomes actual of TRU. The second kind is exemplified by **FER**, Form, ESIN, ES, CES, and REM, where the ES of TRU on becoming the chips or keys in SUT, ROT, MOT, ALT return to TRU.

Fig 2-89
CAUSES and MECHANISMS of WEAKENED SUPPORTS for TRU

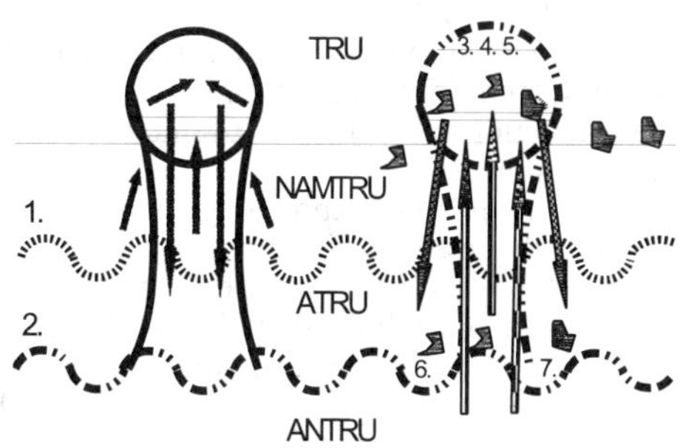

This diagram presents two scenarios, one in which TRU is maintained over ATRU, and one in which TRU is lost through ATRU. In a far perspective in relation to NAMTRU, ATRU and ANTRU, TRU exists in a massive dynamic state, in which its thisness and vulnerability are in an elevated state or condition. The world, TRU, can suddenly disintegrate or disappear. This would occur when and where those forces and powers that sustain it are suddenly overwhelmed by external forces. This may be due, among many others or different combinations to: 1. Level necessary for sustaining, 2. Level of 1/TÂ, or F/S, 3. Excess Population, 4. Loss of Needs, Resources, 5. Waste, accumulation of damaging materials, 6. Living in false worlds, encroachment on TRU, 7. Dissemination in greed, self interests, loss of endeavor for common benefit, 8. Propagation of accumulated defective genetic material, 9. Live products inadequate to support TRU, 10. Determining forces that overwhelm TIT. There are many examples of these in recent history, the present, and likely in the future. Some are compensated for by increasing work and load. Some are decompensating continually. The NAMTRU world we live in is in such a balance between TRU and ATRU, positive and negative, good and bad. Parts of it are moving in one direction and parts in others. The total and how it effects the long term health of TRU is the critical determination. This can be made only when we thoroughly understand most of TRU and can set up the appropriate accentuation, actuation and actions needed to assure TRU. Apocalyptic signifies prophetic and pertains to the apocalypse, presaging disaster. Mankind through history has been more or less, and off and on, in a continuous state of the apocalyptic. It is in the human ability to adapt to and selective alter FAF that repeated apocalypse have been reacted to without extinction. This has been less dramatic, yet more basic, in the totality of TRU than in localities or focally in time and place by parts of the whole species. It is the relative stability of TIT-TO that explains many human questions and problems. This is achieved, as in DYNO, by bringing all together.

VAC = Voluntary Automatic or spontaneous Controls. Many mechanisms, processes, or CES of TRU, are unconscious, automatic, and guided by lower Â levels of mass energy. These need to be thoroughly and comprehensively understood along with how and why TRU works. **HAW** = How and Why. HAW is a major component and aspect of knowledge, and what HOM seeks to show. It is a knowledge based upon the reasons for and directed toward how they can be best used by best serving TRU in TO. It is upon such a firm demonstration of TRU and its knowledge that we can develop the acts, actions, ACC and VAC that are so necessary for the success, and well being, of TRU.

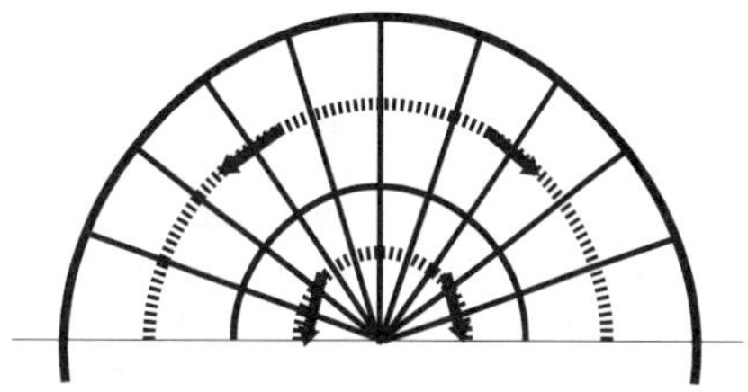

Fig 3-89
COURSES of ACTION and HIDIC

RINCFID = Compounded vast Range of Intelligence, Intellect, Interests, Curiosity, Focus and Determination. **COAC** = Courses of Action. Human initiative, knowledge, Convictions, judgement, moderation, and the higher range of RINCFID. These are the great bases for much of TRUD, and thereby TES. The question then arises, 'What are the courses of action to be taken'? What are their CES, mechanisms, operations and dynamics that are, and show the makeup, of TRU? What is action, right action, and how does it function singularly, multiply in various combinations and totally all to TRU? By visualizing the Horizons of IPTRU, **HIDIC**, we can ind, **HART** = Hard and Fast Tie-Ins between the Aspects, Parts, Processes, Representation and Presentation of TRU.

Fig 4-89
COURSES of ACTION, MORALITY, ETHICS, LAWS and TRU

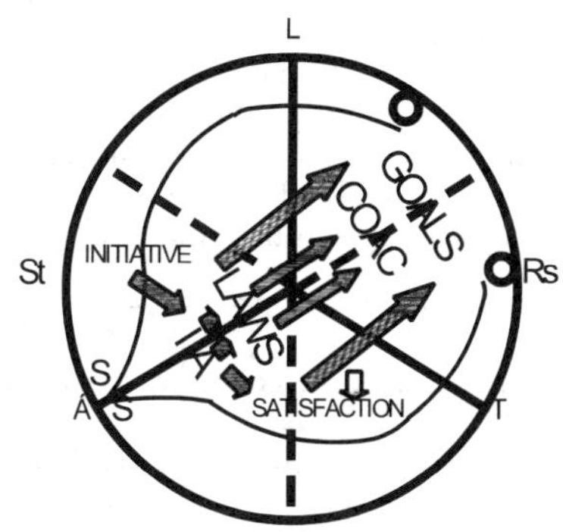

This figure provides the frame of horizons for COAC. It directly applies to needs, suggested actions, and fundamental relationships between ES and CES functioning in harmony and perfect success. A part, as a person, can't claim its rights without accepting the responsibility for them. This applies to all aspects of IPR and to all aspects of TRU. This is how any body must function in relation to its parts, and parts to the body, object or TRU. The courses of action of a person, many, or all persons are guided by, and are largely functions of, intentions and goals. These are ends, which satisfy, meet or validly and successfully complete the problem or need at hand. In summation these become problems which confront TRU.

TRIND = The TRIN/TRUID Differential. TRIND is that of TRU which can be improved in TRUCH. This is something that must be weighed very carefully. For to grant alteration, especially when irreversible, of TRIT, as occurs in TRUCH, we are defying TRIA if only in tiniy increments. Nevertheless it may occur at times that some features, forms, and parts of TRU can be made to work and be better. These are the allowances and exceptions that may be needed and granted that alter TIT-TO. If so minimal change from TRIN to TRUID, if properly controlled, may be allowed and preferred. This is much more so if the potential for reversing to its TIT state can be preserved and expanded. **CATRU** = Concomitant and Total Reconciliation of TRU, TIT, and TRIT with TRUID. CATRU concerns questions, problems and paradoxes of ideal TRU validity. Much of TIT, especially with all less, is already ideal or TRUID. Much that is TRUID can be of TIT reversibly or by expansion. Therefor there is essentially no reason to expect additional TRUID that alters TIT and TÂ. In association with TRUID TRIA is made up of two separable natures: **TRIAO** or TRIA is All, and **TRIAB** or TRIA is Absolute. This relegates our ideas of ideal and TRUID, to lesser status, signified by: **LITRUID** or the Limits of TRUID.

Mankind has made great strides improving predominance. Yet there has always been, and will be, ups and downs, especially with negative forces beyond control. As civilization will increasingly provide control, so will these negative forces decrease. However, as many negative forces diminish the stimulus they provide is lost, and thus humans tend to, in being complacent and less selected, be less able to provide the degree of fitness necessary. Many other forces will arise from our expanded mass which must be managed and controlled. Of the many factors whose positive and negative effects can be both measured and determined in relation to TRU and its parts and processes there may be added to the bodily functions noted in Chapter 63 those with larger personal, social, regional, and worldwide implications. Some of these include: population, density, areas, life expectancy, survival, family size, natural growth, environmental equilibrium, emigration, immigration, birth intervals, actual and potential growth rates, generations, age per generation, human size, bodily proportions and differences, rainfall, climate, marriage rates and ages, reproductive interval, undesired fertility, gestation, fetal mortality, birth control, disease, AIDS, malaria, plague, heart disease, cancer, genetic and gestational abnormalities, accidents, injuries, prevention, immunizations, preventative and public health measures, breast feeding, nutrition, productivity, threats, predators, infantile - childhood - juvenile - mortality, stunted growth, mobility, sanitation, water supply, wages, costs, costs of child rearing, food prices, different and total prices, GNP, PCI, investment, service, rural and urban distribution, agriculture, manufacturing, mining, construction, etc. Without an understanding and demonstration of these it is not possible to adequately analyze and present what the future will be. They also help to determine destructive, detrimental, minimal, cyclical, potential, actual, and optimal effects. When combined and brought together as by DYNS and DYNO their roles in the future, Section IX or ACC, can be better understood, managed and controlled. What we must do to activate and enhance them, and what their results will be. This is the progress human civilization has, is, and must increasingly make. It can only be done with a highly active, intelligent and interested public. A public in which each individual identifies themselves with their part and role in it, their immortal souls, The fulfillment of their contribution to the whole, which ultimately is their prime and sole measure. This may be approached in many ways, by TREQ of existing religions, sciences and philosophies. Yet it is what they hold in common that is valid, actual, proportional and total of the relative universe in itself as a whole that is the final measure and continuous goal.

MOPT = Maximal Opportunity to TRU. How the constructive increments of civilization, TRU, are not to be taken for granted, nor abused, shows MOPT. Nor are they to be used except for the maximal efficiency and benefit of TRU. We should be alert to provide and secure the best advantage for all opportunities that enable PPMA of TRU. This is both inward and outward, direct and indirect, and of quality even more than quantity. It is especially so for all, that is indirect, inward and of quality, that we can bring the advantage of opportunities to greatly enhance TRU. **LOMEC** = Law of Maximal/Common Effort. LOMEC helps to convincingly prove that action, ES, CES, ACC and other means of TRU are the way of success.

In order to tie in all the knowledge of TRU shown by SPALT and HOM, the use of serial maps in time is helpful. As in all differential, however this can be no better than how they are used along with the whole, that of TIT-TO.(SPALT) These can be done in many ways to ever more accurately depict what was, is and will be going on and what forces and operations are to be given. A most general one, as an example, for much of TRU would be 16 serial maps centered on the present, or at: 10,000,000, 100,000, 10,000, 3,333, 1,000, 333, 100, and 10 years past and the same for the future. For many changes as historical and geographic an arithmetic series is more edifying, yet less TRU. They should be superimposed, corrected and remembered in the proportion of TIT-TO shown in ALT and SPALT. Then, we can better interpret, even though warped for ECRO, the major courses, movements, trends, CES and mechanisms that are a large part of TRU.

	Pre-sent	2050	2100	2200	2400	2800	3600	5000	8000	15000	30000	60000	100th+	200th+
Changed time		-	+/-	+/-	+	+	+	++	++	++	+++	+++	++++	++++
Space	-	-	-	+	+	+	++	++	++	+++	+++	+++	+++	++++
Earth	++	+++	+++	+++	+++	+++	+++	+++	+++	+++	++++	++++	++++	++++
NationPoliti	++	++	++	++	++	++	++	++	++	++	++	++	++	++
War Terorism	+++	+++	++~	++~	+++~	+++~	++~	++~	+~	+~	-~	-	-	-
Economic war	++	+++	+++	++++	++++	++++	+++	+++	++	+	+	-	-	-
OptiCompetit	++	++	++	++	++	+++	+++	+++	+++	+++	++++	++++	++++	++++
Crime total	++++	++++	+++	+++	+++	++	++	++	++	+	+	+	+	-
Rights HuRaSe	++++	+++	+++	++	++	++	+	+	+	+	-	-	-	-
Genetic Qual-	+++	+++	+++	++++	++++	++++	++++	++++	++++	+++	+++	+++	++	+
AppliedKnowl	+	+	+	++	++	++	++	+++	+++	+++	+++	++++	++++	++++
GeneralKnowl	++	++	++	+++	+++	+++	++++	++++	++++	+++	+++	+++	++	++
TRU Knowledge	-	+	++	++	+++	+++	+++	++++	++++	++++	++++	++++	++++	++++

Table 4-89 **MAJOR WORLD COURSES and CONFLICTS in the FUTURE**

Relative ratings depend both on the strength of the courses or optimums and the degree of conflict and threat. More refined, corrected, and increasingly expanded examinations and interpretations can and will produce the vast number of interacting factors. How will TRIF be different at different times in the future and what forces and courses are active can be determined and be very helpful. What they will be in 2050, 2100, 2200, 2400, 2800, 3600, 5000, 8000, 15000, 30,000, 60,000, 120,000, 250,000, 500,000, 1, 10, 100 million years, or any set time, should be no more esoteric for us than the same in the past. Its great interest and importance, as already initiated in HOM, will tremendously increase our ability to function in ENT, ACC, TIT, and TA. These are the windows that SPALT opens up for us.

Other universes and worlds often stand in direct opposition to TRU. Often facts, principles and problems which seem so certain and even overwhelming in other worlds or the physical universe are much different and of less significance in TRU. That this, and similar differences, are already well known is a proof of TREQ. Other universes and worlds generally represent misplaced proportions or ESINS of TRU, and thereby much different than TRU. They have aspects or elements in common with TRU, yet are far different. They are ENIT and thus illustrate how erroneous we can become when we do not make the necessary corrections.(TIT,VAPO)

PRUDYNO

TIT, TO, CES, SPALT, ES, ALT, METAD, POS, POT, VAPO, TREQ, KOT, Life, TRUD, Human, DAOST, Universe, Relative, RUPKAL, LARF, MATES, ENT, OM, LACOP, MADAS, ACEXIN, ALT^n, LATAN-O-S, LARP, HOM, WITÂ, LAMUA, ACC, PUREACLITY, Biology, Climate, Complement-ation, Concordance, Equilibrium, Geology, Total Processes, PIP, SEF, LACE, PODAF, NUDAL, LAPAF, PROFUS, WIT, LAHIRN, HAW, EM, LAROCE, RUTHES, ATIM, POMU, LEXSEP, LANTBE, LODEST, LASIRK, TRUCA, ETOC, LOE, MOMET, LACORN, LAMA, ME, COSUF, HOMS, CIAC, TOC, SUFUV, SUFPUV, RECTRU, TRUFCE, DEBT, LAHPRO, LAHED, and MOT

PRUTREQ

Thisness, Life, Special, Relativity, God, ENT, TRUF, Language, LATANO-S, LARP, REM, Frugality, Improve-ment, Invent-ion, OPOT, LOVKI, LATREQ, PRAST, ACIR, GET, LIMGID, MOG, COPOS, DOPEMS, LITGIK, TIKAN, UNA, LORA, LAUTH, TRUQ,

PRULAW

LOLS, LOT, LATRU, LAVCAK, LAC, LOPT, LOPUT, LAFRIA, LAL, COPLAT, CATIM

PROOF(S) In General,

Universal-s, VAPO, Relativity, S O, S-O, SO, Life, Human, ENT, LACOP, Meaning, OT, ROT, LATANOS, PRUT, Proof by disproof or contradiction, especially of dimensions, Proof Theory, LOUP, TIST, APROS, LAMUP, LIVROT, PRUS, LADIRE, PRAPOS, LACOP, LASAP, SURAPT, LARECI, LAPRA, PAC, LOPEN, LIPCO, PROET, LAVIT, LAROS, LAFRO, PRODEP. The great need in proofs, is how important it is to be named to be known. For TRU the proof of this need is called: **PRUNT** or Proof of a new - and fresh - Notational System, or TRU Terminology. This is the third and highest stage in representation, OC to EC to NEC, even though it is far less than what PI represents. Here there is clear and certain proof and validity in that described and demonstrated. This is particularly so when positively conditioned and limited.(COL, COR, SCARL, VAPO) Narration, of detail and differentiation, is the pitfall of human knowledge. It is also its great power, especially when by TRU and ALT. It is in obtaining for them their distinctions, essences, ESINS, ES, and CES in TRU in our SPALT and other Models of TRU that they can be best used as evidence for the justification and proof of TRU. **FOLM** = Functional Operations Logical and Mathematical Notations. It is through the logical and mathematical means and method of operating notations and their functions that we both justify TRU and TRU terminology and work to prove TRU. This is, **JUTFOLM** = Justification of TRU through FOLM. **PRUTFOLM** = Proof of TRU through FOLM.

Let T stand for The Relative Universe. or this special universe of best human life, then the justification, strength and proof of these laws, LAL, comes from: (1). TRU, as we exist in and can demonstrate it we can relate all and so make formally distinct much of its makeup, representation and presentation. 2). SCARL, as we increasingly discover, develop, determine and show TRU we show its states, conditions, limitations, relationships, and all that is associated with it. As we identify or otherwise distinguish and make discrete these great aspects of TRU, their existence as well as their application is shown. When this is done, and they have imperative significance, they are stated as Laws. (3). For much of what is of and around us that has been given a name, or has been denoted, especially what is a great part of TRU, the degree it can be vividly and completely shown and applied is the degree to which it can and should become Law. (4). LATAN, LATANO, LATANOS, as we better know and show TRU we learn and appreciate its greatness and power, especially its necessity, ordination and standardization. (5). The many questions, problems, paradoxes and enigmas that seem irreconcilable are often no more than our ignorance, failure to make them formally discrete, and give them appropriate terminology or perspective. So is it with PLT, Private Language Problem, as with LANT, a weakness in all conceptualization and languages of the world. It reveals the great problem, solution and justification of good language/new terms, and right of NEC by greatness of TRU. It shows and tends to prove the total and separate, integrated and differential nature of the world, of TRU. For TRU and PLT can not be unless they are totally- in common with the whole. Let T stand for The Relative Universe or this special universe of best human life, then: It thereby is in the order of and to IA that we can achieve greatest validity, certainty and proof. It can be seen from these that such means of showing TRU are but levels in the S - O relationship, and apply to all things and their senses.

I. Existential Proofs (ENIT,OC), II. Essential Proofs (ESINS, EC), III. Actualizatiion Proofs(non specified TRU), IV. Total TRU Proofs (VAPO,specified TRU), V. Positive Total TRU Proofs (REM,POS,STA,and HOM) in the fulfillment of TIT and TRU. It is in the exercise, profound and vigorous pursuit of the best of these major classes or categories of proofs that certainty increasingly becomes PUREACLITY, COGNOS,to TÂ. So is it for all perspective generally and especially as it approaches TRU. Such certainty is mandatory for TRU. As in mathematics many statements can be proven to be true by the use of previous assumptions and definitions. This is the axiomatic method. For TRU the whole of POT and IPTRU is the repeated versions of assumptions, definitions and proofs to complete certainty of all in TIT-TO. How knowledge can become valid, with high levels of certainty and proof is shown in many ways including: 1. The validity of much of TRU as provided through correspondence and correlation of information and knowledge. 2. The resolution of many existing questions, problems and paradoxes. 3. The reciprocal proof by the model of TIT and the model by TRU. 4. Proof through summation and total view through learning, study, practice and expansion that greatly produces and inspires organization, knowledge and the comprehension of all. 5. The attitudes, methods and means to be immortal, to fulfill one's best part in life and to be most successful and prosperous. 6. The way by which society, nations, cultures, and civilization can better function in providing their members relief from detrimental stress and conflict. The necessary recognition and appreciation of the total and proportionate make up of our world and the blessings it in turn bestows. 7. The confirmation of through and through processes in the world. TRU, acting specifically and cosmotically, determinatively and teleologically, not chaotically nor indiscriminately. 8. The culmination of applied knowledge as it can be best applied by the individual in life, by ethics as each and all function optimally as best and proper counterparts in the world, TRU, TES. 9. The reciprocal value of TRU as it enables the text and the text as it enables and increases TRU is the beauty and glory of each. As S of Á increases, so similarly does the 0 of Á expand. As our understanding of the actual world, our text's demonstration of it, increases so does objectivity and our actions and its objectivity increase. This is the full circle interaction of the text from EPI to ACC and back again. This is how the source of the great proofs presented, like the proof of Á, or PRÁ, the Proof of TRU or PRUT, the Proof of SU or PROSU, and the Proof of ALT or PRALT, work together.

MEASURES of CES in the SUMMARY of DYNO

ADTA	5-4-7	T	7	LAHED	~-~4-6~	T	4	POD	~-~-~-~	M	7
APOCALYPTIC	~7-~1-8~	-M	8,6	LAHERM	~-~-~n~	M	8	QUALITY CONTROL of POPULATION			
BASIC PROBLEMS & FORMS				LAHIRNS	~-~-~n~	M	7	(MCP)	7-3~-7~	M	1,3
of TRU (BPF)	~-4~-~	-T	4	LAHOCEX	~-~-~n~	M	7	RESOLUTION INDIVIDUAL			
CATRU	~-~-~-~	M	7	LAHPRO	~-~-~n~	M	8,6	/SOCIAL (RIS)	6-4-7~	± T	4,2
CIOS	~-~-~-~-~	M	1,5	LAHOSEV	~-~-~n~	M	5,7	RIM	~-~-~7~	M	7
COAC	8-~-7~	± M	4	LASUD	~-~-~-~	M	7	SAKOT	~-~-7~	M	1
DIFFERENT WORLDS, UNIVERSES				LITRUID	~-~-~9~	M	6,7	SIMA	~8-1-7~	T	7
TRU (DWU)	~-~-~-~	M	8	LOMEC	~7-~1-8~	T	6	SOEX	~5~5-7~	T	5,7
FER	~7-~1~-2~	M	7,6	MAJOR WORLD COURSES, CONFLICTS				SUOB	~-~-~-~-	M	8
HART	~-~-~-~	M	7	(MWC)	~-~-~8~	± M	8	TOPSOR	~-~-~-~	M	7
HAW	~-~-~-~	± M	4	MATDUR	~5-~4-7~	T	7	TRIAB	~-~-~-~	M	7
HL	~-~-~n~	± M	4,8	MO	~-~-~-~	M	2	TRUAO	~-~-~-~	M	7
LACDIT	~-~-~7~	M	7~	MOPT	~6-~3-7~	T	7	TRIND	~-~-~-~	M	7
LACHVAR(IT)	~-~-~-~	M	7	OBSAD	~-~-~-~	M	7	UL	~-7~-7~	±M	4
LAH	~-~-~n~	M	4,8~	PAO	~-~-~-~	M	6,7	VAC	~5-~4-7~	±T	5,8
LAHANC	~7-6-7~	T	4,1	PAOC	~6-~4-7~	M	6,7	VIGILANCE	~3-~5-~5~	T	1~

Fig. 5-89 **DYNO CES in SPALT**

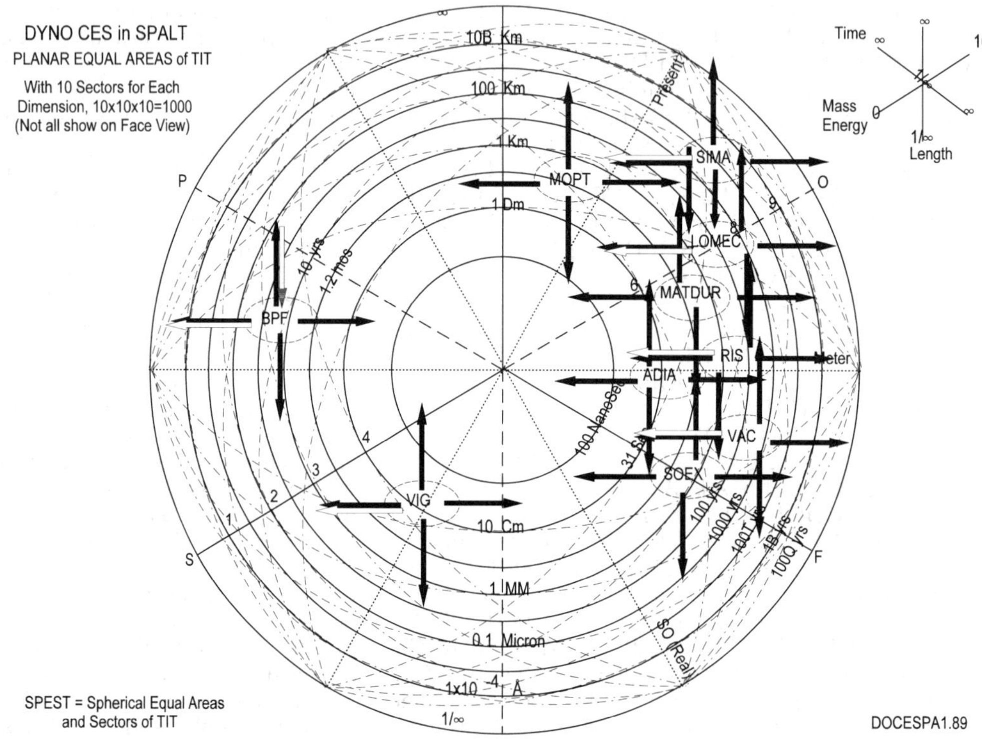

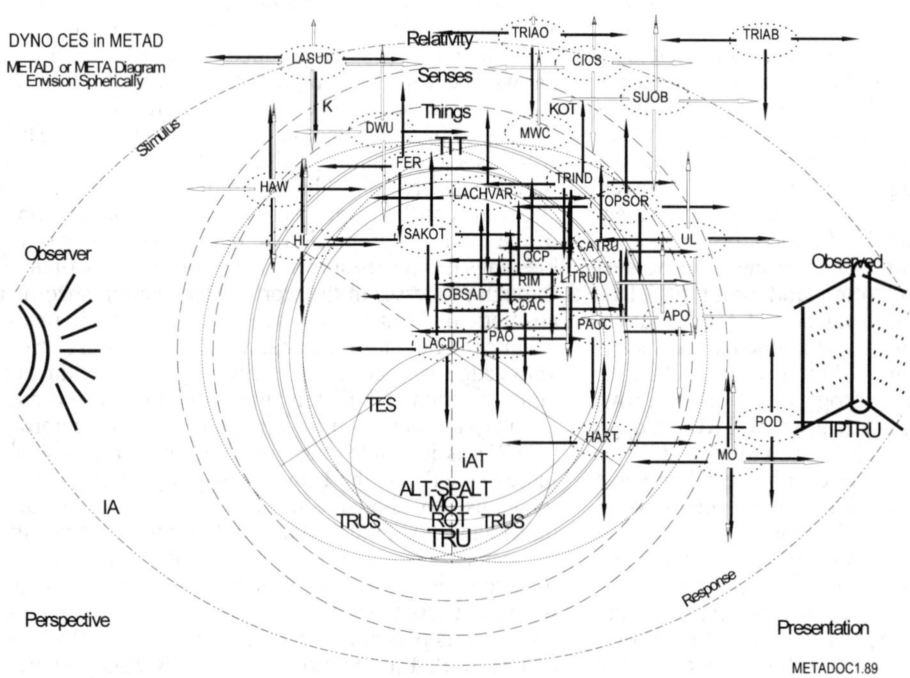

Fig. 6-89 **DYNO CES in METAD**

SECTION IX

ACC or ACCOUNTING, ACCENTUATION, ACTION, and SUMMARY of TRU

CHAPTER 90

INTRODUCTION, BASIS, and KNOWLEDGE of ACC

FORMAT, KINDS, THEMES, CONTENTS, PRECEDENT, PURPOSE, FALLACIES, PROBLEMS, ACTIONS, MEANS and PROOF of ACC

This last section of the text, IX or ACC brings together what has been discovered, described, posited, and shown in the previous sections. It emphasizes much that enables us to present TRU and assure TA. We do so, continually reminded, by their actual state in TIT. It is with considerable wisdom that the actual world or TRU is governed. Yet in the future it will have to be much greater, of the best. The entire text has been projected so that the future is more or less included in all its demonstration. Much of what needs to be done and acted upon has already been either generally or specifically noted in past chapters. In ACC we seek to accentuate and consolidate these best and greatest of TRU for ready use. Much of this is what is most apt to stimulate and accelerate TRU. To gain this greatly needed knowledge for accentuating TRU we must not only discover, demonstrate, and determine TRU we must provide: **MARP** = Massive Media Rapid Presentation and Promotion of TRU. This is the accentuation and advancement of a whole new way of life. The form of ACC follows the format of POA's simple culmination in previous sections which are brought together through the laws, principles, concepts, parts and processes of TRU. These are then analyzed, posited, interpreted and reapplied in the closest identity with TRU.

In the first chapter, 90, of this last major section of the text, ACC, we not only summarize its findings and point out the indicated actions, we purport to emphasize and strengthen both existing and added new ideas, concepts, principles, laws and means. It involves the necessary added concentration, demonstration, and accentuation of those crucial aspects of TRU and needed actions which the earlier chapters and sections of the text have indicated or suggested. We designate these aspects of TRU and ENT by TRÃ, TRAX, TREX and TRIX. **TRA** is TRU ACC done generally, exponentially and through an increasing conscious recognition of TRU. **TRAX** signifies all that is done for TRU without knowing the term or concept TRU. **TREX** is the Exponential ACC of TRU resulting from **TRE** or TIT Recognized and its conscious actions. This distinction is essential if we are to clearly define ACC and bring together all that has been of TIT yet unknown as TRU, or TREQ. It helps to show why TRU as a new concept, topic, and science is so mandatory. There is a rational basis for action with an increasing TRE. **TRIX** is that done specifically and directly utilizing TRU as the fulcrum to expand any and all means that accentuate and enhance TRU. This is often through the use of POT, IPTRU, and further communication as a science that best represents and can promote TRU. TRIX also focuses on all that is ANTRU and a threat to TIT.

New basic philosophical, religious, scientific and social attitudes are, by LATAN, LATANO, mandatory if we are to successfully cope with the many forces, powers, errors, defects, and their likely negative ill effects. Problems are often perpetuated and added to, rather than lessened, by each new generation. Our very existence, you, I, and all, is not whether "I am" or "we are", for this now means nothing without our have been, are and will be all together as they are one increasingly and eternally, TRU.

The greatest reason for, and measure of, members of the human race, are not so much those who can do specific jobs, technical work, or treat minor illnesses or conditions. For this will be increasingly the work of established routine or automata. It is, those, who can best prevent and manage serious and life threatening states, especially in their most intricate, massive, interactive, and total forms. So is each person, as we are TRU, our TES, the great agent of ENT. This is called: **AGENT** = All who, and that which, are Agents, vehicles, or means of producing, ENT, ACC. and TRU.

Where law is uncertain or defective there is likely to be no law, even in reaction worse law. Uncertainty destroys laws, it is 1/Á or the opposite of TÄ, TRUD, FAF and TRU. In **LAL** or the law of limits, and the Law of Quality/Quantity, **QUALNAD**, we typify Accentuation. For it is the quality of TRU and how it can be most greatly enhanced that is most typical of ACC. In CONVET and CONRIT we demonstrate the increasing universality and particularity of much that provides ACC.

The methods, means, forms, functions, parts and processes of Accentuation are evident in many ways, these include ethnic history, comparative methods, probability theory, laws of TRU, laws of thermodynamics, motion, Homation and its simple to total use. They must be held to their SCARLS, being conditioned by the degree to which they are proper for TRU. This is called: **PAC** = Proper From Accentuation. From these the **KAC** or the Knowledge of ACC is derived and developed. The validity or invalidity, sufficiency or insufficiency of ACC and much of TRU depends upon the value and order of these methods. **SUI** = Sufficiency or Insufficiency, Validity or Invalidity equals value and order over number of systems in SU, and THEO of TIT, or.

S/Ins x V/Inv = REM x Ord/# System. SUI especially pertains when it is used in ACC. It is highly related to: **CRAED** or the Correction of the Actual and Formal Differential. CRAED signifies the tendency we have of confusing actual and form. The stimulus and inspiration for KAC arises from, and is aptly implied in, the phrase 'a word to the wise is sufficient'. A wise person needs little extra information and incentive to make good judgements. They will also find NEC and LANT to be words for the wise. That which is credible requires not only truth but the test of time. Test of time and reliability inevitably reduce to TRU. The greater and more accurate knowledge is, and the greater and more astute persons who use this knowledge are, the more credible they are. Infallibility can come only from TIT by TRIA. Even here it is not possible, except when taken as a whole TIT-TO. With KOT KAC are not only perfectly TRU, but proportionately and completely in TIT and applied with equal perfection. The inverse is called: **LAPLID** = Law of the ProLiferation of Imperfections and Divergences in fundamental Doctrine. The Exclusion Principle is the basis for the heterogeneity of TIT. It is that by which formal distinction can be made of these divergences in the makeup of TIT & TRU.

Until we summarize all we know about TRU and seek to apply it to our lives, we are betrayed by being focal, partial, and ATRU. We fail to take advantage of all we know, and all TRU, POT, and IPTRU provide. The conclusions, laws, principles, concepts, perspectives and replicas of TRU can be collected, ordered and organized and used as a means to solve the problems of the problems. Especially those problems that are most confronting and seriously prevent the development, advancement, glorification and perpetuity of TRU

The courses and themes of ACC may be coordinated with the degree of complexity, or from differential units or singular items to the infinite total integral. Such a broadsided approach is needed and much more effective if we are to use ACC and best do all that we must for TRU.

The methods and results of ACC and LOTS are derived, not only from existing laws and their failures as shown by the saying 'extreme law is often extreme wrong' and the nature of TRU, but by the various means laws that are good can be better served. **POES, POCES, and POTES** or the Power of ES, CES and TES, and **POP** of the Power of PUREACLITY singularly, collectively and totally enable TRU. POP is the key to TES and much of ACC. It is in TREQ, and human Á FAF, purity, reality, actuality, and spirit to become their TES in TRU, as they develop and perfect themselves to live and work for TRU, that ACENT, ACC, ENT, and TIT can be attained.

There remain vast regions in human life whose expansion, accentuation, development and ACC can be increasingly complimented, directed, controlled, effected and produced. Many of these are unknown or not used for human benefit. This occurs in most all fields, sciences, and technologies that will enhance TES and TRU..

The form of KAC is exemplified by induction, diagnosis, SOAP, Baconian Method, problem solving, GPS, Popperian, and Precognition. Induction is the method of building and representing TRU by proving one idea or set of ideas by another. Its great strength comes from the use of inference logically building from certainties further missing appropriate certainties. By holding all to TRU these become actuated certainties that greatly extend the disclosure of TRU. With medical diagnosis and management, **SOAP** or Subject-Object-Assessment-Plan builds inductively. It is the means of logically and accurately bringing together all helpful facts, information, and evidence to make the valid inferences and conclusions that accurately depict an algorithm of the state of affairs at hand. We are able to extend these diagnoses to their proper posits in ALT and TRU.(CONWAT) In the Baconian Method, science, and especially research was better guided, directed, and greatly stimulated to vastly expand its horizons. Problem Solving can be of many kinds and levels functioning as a basic form of our entire being and ACC. Its interactions with Á and SÁ is a large part of TRUD, HOM and ACC. GPS or General Problem Solving are **CAD** or Computer Designed Programs which stimulate a human problem solver's procedures. The Popperian Method is used extensively in this text by sorting out separate and individual processes, as DYNU and DYNS due to the actions, interactions, aims, hopes, thoughts, and traditions created and preserved by individuals. The principle of falsifiability proposed by POPPER holds that until a theory is disproved it is the appropriate object of the observational and experimental procedure.

Modern philosophy has rationalized its failure to bring knowledge to any kind of closure by holding to a belief that truth itself is in doubt. That there is no final truth to be found, only as Karl Popper put it, "all we can do is search for the falsity content of our best theory". The limits of philosophy are seen in structuralism and post-structuralism wherein all was reduced to a structure in language. This is the hole that philosophy has dug itself into. Religion & science have also limited themselves by failing to envision their overall part & role in the totality of all that actually is. Herein is the catch. Philosophy, science and religion all have accumulated a great body of truth. There can be little doubt when they are known, interpreted and accounted for in the validity, actuality, proportionality and totality of this special universe of best human life, the relative universe, TRU. Only these truths that philosophy, religion, sciences and all fields have developed have to be corrected and adapted to how they exist in TRU if they are to be beyond doubt. For as long as humans consider their knowledge as entities in themselves, not the essences or parts & processes of TRU, knowledge is only theory, never to achieve validity, actuality, proportionality, & totality which are mandatory for a full disclosure and final truth free from doubt. This is called KACAT, or **LAKACT** or Law of Knowledge and Certainty to be brought to Completion & Closure in the Combined Actuality of TRU. LAKACT is the basis for KAC & ACC, the knowledge, principles, & methods of the accounting, accentuation, and action needed to achieve and provide for TRU. Yet Popper, the critics, constructionists, and doubting Thomas-es all have their points. In general, even with TRU, truth is never absolute, only relative. For it at best can be no better than the degree to which it approaches the interactive, relative, infinite, of TIT-TO. Only in the massive methods of LAKACT, ACC, POT & IPTRU of a combined deductive, CONRIT, and inductive, CONVET, can we accurately and fully differentiate and integrate the continuum of differences to similarities of TIT-TO. It must also accommodate the OITIO or out-in-TRU-in-out, observer-observed relativity of the mass energy dimensional form and function of TRU. However, without recognizing these facts, and persisting in attributing falsity indiscriminately to all things, it becomes apparent how dubious truth is and how much we need to beware the temptation of accepting any and all knowledge as certain and, or final, without the total proportional analysis of a positive objective approach by TRU.

There is, among others, a four-way-path to action. 1. Must have knowledge, 2. Must have right knowledge, KOT, 3. Must apply this right knowledge, KAC, and 4. Must be firm and follow through properly with this right knowledge, ACC. Each step is essential. These we strive to provide in ACC. This is stimulated and expanded by the best exercise of the mass energy Á FAF, state and dimension of TRU. For when we fulfill Á properly this four way path is largely spontaneously and continuously completed.

It is much easier to go 'down hill, than up hill.' This is action that benefits from gravity, or the Tao of following nature. Such also holds for all levels of energy that make up the Á FAF of TIT. It is a way of going with the flow, working with the grain, or best utilizing ACEXIN. It is how to use whatever means are available that most efficiently and effectively produce the needed results. Yet in the end all is not down hill, all is not done by using the easiest means. A limit is reached in which we are faced with some uphill, a certain amount of work, as a labor of love, has to be done to fulfill TES and TRU. That necessary to maintain the balance of mass energy within TRU. It is because of this that we have: **MENTS** = Mass Energy Sufficient to Maintain TRU. MENTS includes the resting mass, kinetic energy, potential energy, and the energies involved in the necessary interactions both within TRU and what is necessary to meet that from without.

Never leave to function on its own anyone or any situation that is not sufficient for it, as it is TRU. This is simply seen in a two-year-old child running loose, activist governments without wisdom, or criminals on a rampage. What is allowed to drift in any way will quickly be on a path of destruction. Much of the world around us functions ideally This is why knowledge, knowledge of TRU, THES, COGNOS, PUREACLITY, TES, ALT, ROT, PERT, and POT are of such great importance. They represent a perfected exercise and practice of our Á FAF in TRU. Here is a good example and a proof of Á. It shows us how vital Á is by adapting ourselves to our counterparts in TRU, TES. This is individual, group, and all mankind through Á properly, proportionately, precisely and totally functioning in TRU, TIT to TÂ.

'It is always darkest before the dawn', expresses an understanding of what is highly complex and important. Their discovery and elucidation are not attained until sufficient information is gathered. The added confusion or blackness is suddenly lifted by the glorious light of idea, inspiration, revelation, conception, and representation.

Lack of appreciation, gratitude, and respect, are signs of our times. It comes from survival, natural self interest of the individual, and from a lack of experience, knowledge, and identity with their group, society, nation, world, TRU. These must be learned if we are to have the appreciation, gratitude, love, respect and joy for the great benefits we receive.

Without a thoroughgoing knowledge of religion, fields and sciences, philosophy and especially of TRU a person can not think for themself. This is for many reasons, each of which must be surmounted if we are to thrive. We have the capacity to think, and do about what is immediately before us or what it takes to suit our purpose. Yet rarely do we think about what is best, TRU and TES. What is the obvious or inadvertent thought or act is not likely to be the right one or TRU. Only if it is based upon a thoroughgoing knowledge, evaluated, proportioned, ordered, and appropriately reasoned thorough by TRU, can it be good, right, and successful.

In order to properly and accurately act and react to TO we must remember that what may seem very tenuous and trivial at one place, time, and reality may seem momentous and overwhelming at another. Yet they may be both not only pure TRU, but warrant appropriate IAT both focally and totally. It is this massive TROS in TIT-TO and our failure to find or use the necessary and appropriate knowledge and understanding of TRU that renders us so inept in strife, misery and failure. The world, TREQ and TRU is not unreal, unknowable, infinitely infinite, nor chaotic. The TRU world is cosmotic, relative, unified, formatively and dynamically real, actual, finite, and total.(TROS) It is simultaneously ordered in a dimensionally demonstrative way. This is how the world can be understood by TRU, KOT and KAC. The day has come, man has matured, and his knowledge is sufficient, for TRU, KOT and COGNOS to show the way we can have a highly effective KAC and ACC.

The great cure for existing and upcoming needs and problems, their prevention, management and treatment, is to be able to anticipate, look ahead, and know what will best answer and solve them. The greatest human failures have not been ability nor effort to address problems. It has been the failure to look ahead, understand, comprehend, and work out the total picture and implications of all, that so vitally and critically affect us. Only with the total understanding, that can be produced through, SPALT, POS, and ACC with precision and sufficient detail, can we begin to look ahead and know how all fits together, what our problems will be, in what way they are part of the whole, and how they are to be solved. This is KAC as it sets the pace and way for ACC and TRU.

What do we know for sure about the future, and why is it so necessary that all TRU knowledge be discovered and set forth: 1. To prevent the great amount of misconception and partiality that is so rife in our world, 2. To prevent irrational speech, decisions, judgements and demands. 3. To be able to know what is purported to be, is right and TRU. 4. If we know what the future will be, at least generally, we can undeniably know much of how we in the present can act, behave, plan, do and provide for. 5. TRU is not complete unless the living future is adequately fused with the knowing, doing and living past and present. 6. To have and be able to use the best of knowledge, of TRU guided by VAPO and its SCARLS, so that all our actions will be correct, neither to much, to little, nor harmful. These we can have by visualizing the circumscribed relative universe, SPALT of TRIT, TIT and TRU and its CES, major courses and their interrelationships. They are essential for the future success of mankind.

The past and present can greatly help us understand, act, and control the future. There are many examples of how man has failed to learn from or use the lessons of the past and apply them correctly. This is all the more reason why all must turn to TRU, be TES, learn all there is to be known of TRU and its potential dangers. To see that all works in the best possible way for TRU, and thereby the success of each and all.

There is little doubt that most can't relinquish dependence upon all that has been developed. There is also little doubt that each person and mankind on average, although initially accepting or attentive, will soon falter and not achieve what it takes to provide for them. This is called: **PACDE** = Provision and Action over Desire and Dependence or the Dichotomy of Action and Provision over Desire and Dependence. It is this great dichotomy and producer of conflict that is so great a basic opposition in human life. It is at the root of many of our problems, troubles and suffering. Even more it is this great dichotomy that will cause man to fail in the future if it is not adequately PPMA, corrected, and controlled. Simply, PACDE mandates a high level of action and provision for all that is TRU, with minimal amount of ATRU desire and dependence. **SPACDE** = Solution to PACDE. SPACDE generally requires a total approach to all TRU, as we do in TO, TRUD, HOM, DYNO and ACC. This, more specifically, requires, that each and all of us accept our role in life and make the hard decisions, choices, and well controlled actions that enable contribution to TRU. This will return its blessings many times over.

It is not so likely that TRU, TES and ACC are achieved in the singular destruction of negatives and seemingly harmful. It is in the overall fulfillment of all TRU, life of TES, ES and CES by the components of TRU, plus the best possible use of COAT, KOT, ACC, and KAC. This is called: **LOTLAN** = Law of TRU Limitation of Knowledge of Ä and AN, from and Against, or Negatives. **LATNIP** = Law of Turning Negative Into Positive. LOTLAN is the elimination of attention to all knowledge that is less than, other than, or negative to TRU. LATNIP is the reversal of all knowledge other than TRU, into its most positive and best role in TRU.

FIC = Fallacy of Insignificant Causes and Specifics. FIC is how we tend to look for specific or singular solutions to all things. It is the immature expectation of finding a simple answer to all It is not in an unknown in biochemistry, neuroanatomy, genetics, psychology philosophy or a singular God that all TRU is to be found. FIC is like looking for the Holy Grail. The riddle of the universe lies in all, all, TIT-TO. The great mass that only the highest level of quality and quantity in each person should aspire to, can and will achieve.

Some of the greatest problems of mankind include, decreasing capacity and ability, limited or unfulfilled quality, identity without fulfillment, actual order, irreversible loss of resources, waste, pollution, destruction or our natural habitats, failure to build new habitats, inadequate controls, nuclear war (WWIII), quality populations, failure to know and build our potential, loss of quality and purpose in reproduction, health that is not maintained, lowering of educational standards, absence or denial of the best in religion as TREQ, and lack of continuous redemption. They may be addressed separately or together, each learned, discovered, and corrected directly. This associated with interest, inspiration, stimulation, and love in TRU provides a mode for ACC and TRAX. Societies with decreasing family and home ties, loss of basic principles, ethics, morals, community, cooperation, and appreciation, are what we face and must restore. The political, social contract, legislation, legal, and judicial all need great solutions, for work by TRU and ACC. All of these problems have solutions that are to a large extent the same: A reversal of the processes of decreasing, limited and unfulfilled quality or TRU. These are achieved through: 1. Increasing intelligence, 2. Increasing knowledge, 3. Increasing knowledge, revelation, and comprehension of TRU 4. Increasing judgement and way of life reconciled to TRU, 5. Increasing interest, inspiration, stimulation, love of TRU, 6. Frugality, 7. Total increments of attention to TRU, ACC and all needed to PPMA TRU.

The greatest problem in human life is not that TRU doesn't exist, or even can not be well ordered known and shown. It is our own accumulative limitation in life, both inherited and acquired. With the repeated cycles of rebirths and constant rejuvenation each ends, as each begins, with a great loss of appreciation, order, knowledge and art. It is only in our most mature and competent ages of life, that incline us to TES, that we can even come close to having the order, knowledge and production that are necessary for TRU and for us to have continuity if we are to be our counterparts in TRU, TES.

Those, who do not attain the wisdom or courage to rule, will be ruled. It is here we can most clearly see the success and failure of mankind. It is here we can identify what is needed and how KOT, TRU, TES, and ACC can raise all humanity to higher levels of participation and contribution in TRU. Failure of any element in the social contract to fulfill their roles creates a vacuum. It is a vacuum of rule or be ruled defectively. If there are no enlightened citizens, the governed can expect no better. If the government is defective citizens are in want. If both are poor there is a complete breakdown. If we are indifferent, deny, and defy our own TRU nature we can't expect to get more. This applies to most of TRU. Those who are not or will not be TRU can expect and deserve little.

A person's most common and serious error, mistake, and fallacy is ones inveterate tendency to ignore or deny GOD,. the world one is a part of, TREQ or TRU. In this humans err by interpreting as TRU all entities, senses and things, or to interpret as entities, senses and things, TRU, or all entities as one, or all TRU as one only. In other words it is human failure to make the needed distinction between TRU and ATRU and accounting of all TIT. **FUPIC** = the superficial Fallacy, Fallacy of superficiality, in CENS. Through FUPIC we fail to properly interpret and differentiate ENS, CENS, ESINS, ES, CES, TRUS, TRIS and TIT, and their DAOST and SCARLS. This is from lack of knowledge, lack of the kind of language provided in NEC, formal distinctions, total formulation, order and representation of TRU. It is to help to correct this, and thereby ENT and ACC, that the ROT, MOT, POT, and ACC of this text are largely undertaken and offered.

Why has TRU been unrecognized and how can it be best shown? The great paradox is, if TRU, TIT and/or TRIT are so certain and so much of us and all why has it been so unknown: 1. TRU, TIT and TRIT have been in some ways known under other names, i.e God, cosmos, universe, world, mankind, etc. 2. TRU, TIT and TRIT have been known in many partial ways more adequately, yet with less VAPO and more NAMTRU, e.g. fields and sciences. 3. Man, civilization, language, culture resists displacement of what has been known and of benefit. All that is coterminal with TRU, TREQ, that has provided man with the understanding and means to better our lives, and to have more happiness is deeply and firmly rooted in the world. This is both good and bad. Good to the extent it holds to what is TRU, although not known as TIT or ATREC, or that part of it that is TRU. Bad to the extent it holds to what is less or not a part or all of TRU. If we are to show and promote TRU to the point we expect the world to accept it then we must not only show it well to be what it is, we must show it in comparison to what it is not.

FORK = Fallacy of Right of Knowledge. There is a serious and dangerous tendency to believe that anything known, especially if known well, is true, the actual gospel truth. Likewise anything that you do not know is false, bad or undesirable. This is supported by self preservation where all one is must be sustained. Actually much we know and is about us is ATRU. Unless we can be highly suspicious and critical, and this only through VAPO TRU or TREQ, can we overcome FORK. This explains the basic elements of LOTS and LATS, truths are relative, relative in TIT.. Herein is the resolution of the vast divide between religion and science. Scientific and religious truths are compatible, at least when they are limited to their roles in the relativity of TIT.

Fig 1-90 **INTERVENTION and ATRU/TRU SEPARATION**

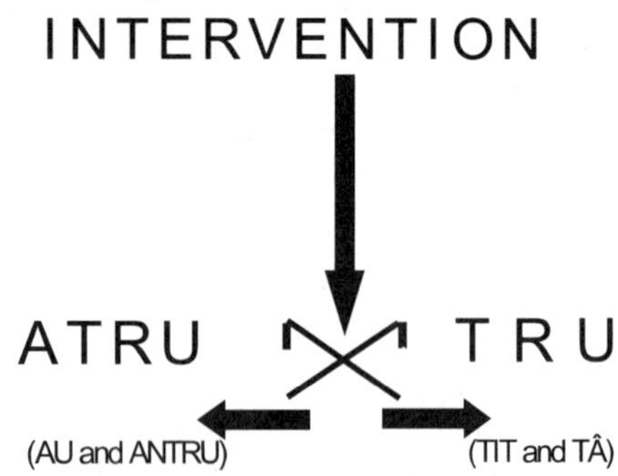

You can't be a good teacher until you have been a good student. You can't give wise counsel until you have learned and experienced life. If you are to practice behavior modification, as we so commonly do without knowing it as such, you must first understand the world, the people in it, and the principles of their FAF, ES, and CES in TRU. In fact the very key to the practice of intervention is to separate what is the negative from the positive, the bad from the good, the ATRU, AU and ANTRU from the TRU, TIT and TÂ. You can not be a good student, good patient, or candidate for intervention or behavior modification until your capacity, interest, stimulation, curiosity, energy, recognition, appreciation, and reason are ready, and can accelerate TRU.

"What is the matter" is a common, key, and basic question signified by different words in hundreds of languages? Yet they have similar meanings in all by the very nature of our being, to inquire, discover, and reveal by S-O or Á. What is most demanding and important either at the moment, through time, or all TRU. They are what sets them in motion, and indicate their solutions. **SIJ** = Steps of Investigation and Judgement. The steps from the initial IA and recognition of what is the matter and its most defective forms to its final TRU perfection enable us to see through self examination and the analysis of various situations. What is the best form of KAC and what are the various errors and distractions in problem solving. What one's mind is able to apperceive and what knowledge has been discovered are far from complete, although a good first step. For TRU we can take little at face value. All must be investigated, accounted for in relation to TRU and judged, not only as it applies to its particular level before, beneath and toward TRU, but its inner TIT value. From ordered judgement by TRIT and OBT we proceed from AU and ATRU to the TRU or **ABAT**, ANT to TRU or **ABANT**, from ANTITRU to TRU or **ABANTI**, and from all mixed TRU-ATRU-AU-ANTRU to TRU **ABANTO**. It is in that final step in which all is brought to TIT-TO, **ABITO**, that we can finally attain the ultimate in KAC necessary for the guidance needed to properly effect all as it can be TRU.

NODA = Not Determined Actually and Absolutely, as TRU nor as World Equivalent. NODA is a world fallacy. It is the persistent fallacy of considering the world as some variable vague and limited thing, object or spirit.. The world, that is equivalent to TRU, TREQ, and that is TRU, is far different, yet somewhat, more or less the same. It is demonstrated by TÄ, TROS, TIT, SPALT, HOM, DYNO, and all TRU. NODA is due to our inveterate lack of total comprehension, COAT of TRU. It is the failure to learn and know TRU, KOT, and KET.(Fig 7-11)

The ability to relate and picture, as done by flow charts in medicine. Yet it is none the less necessary and important if we are to attain the accounting and accentuation TRU needs. Only when we can relate and picture all the parts and processes, form, and mechanisms of TRU as they fit into its overall representation and modulation by ALT, SPALT and OITIO, can we come to have a knowledge and the necessary ideas, concepts and views of all, that are actually TRU. This is what TU, SUT, Á, ALT, ROT, MOT, ALT, SPALT, HOM, OITIO and ELTICMA enable us to do. This is called: **KIPOCT** = Knowledge Imperatively Perfect, Proportionate, Total and Coordinate in, of and by TRU. Such knowledge depends largely on the improvement, expansion, and refinement of concepts, language, and communication. Humans have reached a point in their evolutionary history and future where they have gone from no, some, fair, good, excellent, and perfect meanings, concepts and knowledge. This trend has rapidly moved in the past few thousand years toward closure. Can we use this knowledge to deter failure and extinction in the next decades, centuries and millennia? How can it prevent all the detrimental causes, dangers, threats, and failures that have, are, and will directly or indirectly accumulate to produce extinction? There are no existing and ordered positive knowledge that provide the management and control necessary. This is the great reason for, and proof of, TRU, KOT, POT and IPTRU.

In TRU conservation and liberal utilization are not necessarily irreconcilable opposites, they can be complemen-tary if understood in and practiced in terms of VAPO, SCARLS, TRU, DAOST, TIT, TRIA, TO, TRUD, ENT, POS, HOM, and ACC. **PUC** = Principle of Ultimate Coincidence. The final coincidence or relationship between things, is no more nor less than the similarity and correspondence in TRU. It is the posits of their ES and CES in TRU within the form and function, structure and mechanisms of TRU. In contrast to what is related, corresponding or of ultimate coincidence is the marginal. Marginal has many applications to TRU. One of the most significant is the insufficient state in which man exists in AMTRU, MATRU, NAMTRU, and ATRU TRU. **RAMAR** = Ratio Marginal. RAMAR is a measure of the critical and crucial marginal state of our world today. It is RAMAR that shows why ACC, TRU, and their promotion, are so greatly needed. Much of the need for TRU and work of ACC is to vastly expand this margin. Action and ACC and their proofs are highly coterminal and thus used interchangeably, by **PRAC**. Proof of ACC. PRAC comes from: 1. TRU actions convergent on and positive for TRU, 2. any and all actions positive for TIT, 3. TRUD, Dynamics and all actions involved with TRU, 4. HOM or Processes of Man representing TRU actions, 5. RAMAR bringing marginal ATRU and AMTRU, as the potential, to TRU. Liberal utility and conservation can't be achieved until fundamentals of ACC, and TRU are set.

Even a quick, but intensive, look at the crux of a problem will often provide all that is needed to make the diagnosis and decisions necessary for solution. Signs and clues act as sentinels of basic mechanisms that explain the complication, disease, problem or danger and its solution or cure. Without prior knowledge and a great body of information, attention, overall perspective and presentation are much less successful. Each and all of the great principles, laws, concepts, NEC, views, key bits of information are what is needed to make the diagnosis of TRU and show us how to make the accounting, interpretations, and decisions needed for solutions. It is only for us to discover, know, learn, remember and best use these sentinels if we are to understand and act properly and most effectively by TRU.

To discover the knowledge needed and the methods necessary to develop ACC we must not only thoroughly understand KOT, we must understand much of KAC. Through **SEPTRU** or the Separation of TRU, AU, ANTRU, ATRU, ANTITRU and NAMTRU we gain a better idea of how to resolve the problems needing solution. We must utilize all the powers available, such as: **PIT** = Perspectives Inclination to TRU. PIT involves not only the natural tendency to bring all our perspectives and views to the needs, meaning, and actualities of life, or TRU, but to greatly expand this by developing a total view of TIT, TRU, and METATRU. The stimulus for KAC is readily developed in all TRU, and through the course of this text. As we develop a greater knowledge of TRU it spontaneously induces an ever deeper appreciation and conviction in it. *Verbum Sapiens* is for Homo Sapiens, word - man - wise - TRU. Yet humans are not so *sapiens* without living up to their potential for wisdom, if we are to be, in continuity. In learning this, as in 'A word to the is sufficient', we can encompass and reduce all before us. Accentuation is achieved by: ↑ **TES** = ↑ **ENT** = ↑ **ACC.**

One of the major proofs of TRU by ACC is the accounting, averaging, and balancing extremes, oppositions, opinions and views that people assume of the world. They vary from at one are this, and another time that. When we, or even the best of authorities, hold such positions it may have great validity, yet it is not TIT-TO, nor even of its actual position in TIT. In resultant such positions are used to deny their opposites, both of which are apt to be incomplete, and to that extent in error. This alternating state of views is: **LASTOV** = Law of Alternating State of Views. LASTOV is a great hurdle for both the correct perspective of and the acceptance of TRU. Even more, it is apt to produce a skepticism inimical to TRU. The cure for this is an understanding of TRU. **LAVEX** = Law of Averaging Out of Extremes or Opposing Views of TRU. LAVEX already exist in common usage, whereby those who seek to solve problems or groups who need to resolve differences and controversies, look at all sides and come to a consensus that is the best total view. This needs to be by TRU.

The greatest need for all mankind is that each generation have the greatest and best possible understanding of TREQ and TRU. This can be done adequately and effectively when guided by TRU and all its component FAF, parts and processes. This is not possible without the concepts, principles, laws and means of TRU along with a positive objective approach as provided by VAPO, DAOST, SCARLS, TROS, ACEXIN, TRIT, OBT, SUT, ROT, MOT, ALT, SPALT, POS, STA, HOM, and ACC. It is in this way each person can direct their lives to learn and understand all TRU. and to help others to do the same. Herein is the great basis for, and major proof of ACC, **PRACC**.

MEASURES of NEC and TRU TOPICS in ACC

ABANT ~-~-~-~ S 1
ABANTI ~-~-~-~ S 1
ABAT ~-~-~-~ S 1
ABITO ~-~-~-~ S 1
ACC ~-~-~F~ M 5
ACENT ~-~-~F~ M 5,7
AGENT ~-~-~-~ M 5
APPRECIATION GRATITUDE TRU ACC (AGT) ~2-7-7~ M 3
CRAED ~-~-~-~ S 8,7
COURSES, THEMES THROUGH ACC (CTA) ~5-3~-7~ M 4,2
EXIR ~3-~-~-~ M 1
EXCLUSION PRINCIPLE (EXP) ~3 -6-7 M
FIC ~5-~-~-~ S - 4
FORK ~1-~-~-~ S - 4
FUPIC S 4
GOPT ~6~5~8~ T 5,7
GPS ~4-~6-5~ T 8,6
NTERVENTION (INT) ~6-3~-7~ S 4,1
KAC ~2-5~-7~ M 5,7
KAT ~2-~ ~ ~ S - 8
KIPOCT 2-~5~-8~M 7
KOF ~-~-~-~ S ± ~,7
LAL ~-~-~-~ S ± 2,6
LAQUALN ~-~-~-~ S ± 6,7
LATNIP ~-~-~-~ S ± 4,1
LAPLID 6 -4 -7 M 8
LATS ~4-~5-7~ T 7
LOOKING AHEAD, KNOWING NEEDS (LAK) ~1-~6-8~ S ~ 5-7
LOTLAN ~3-~-7~ S 4,1
LOTS ~3-~5-7~ S 7
MARP 4-~2-7~ T 1,7
NODA ~2-~7-5~ T - 4
NON TRU (UNTRU) ~ ~ ~ ~ S 4
PAC ~6-~3-7~ T 7
PIT ~1-~7-5~ T 3
POES ~-~-~-~ S 1
POP ~4-~6-6~ T 3
POTES ~5-~5-5~ T 1,3
PRACC ~3-7-8~ M 7
PUC ~-~-~-~ M 7
RAMAR ~-~-~-~ S ±~ 7
SEPTRU ~-~-~-~ S± 8,5
SIJ ~4~6~5~ M 4,1
SOAP ~-~-~6~ T± 4,1
SUI ~-~-~-~ S± 1,3
TRA ~7-3-6~ T 3
TRAX ~6-~3-7~ T 5,7
TREX ~5-~5-7~ T 5,7
TRIX ~4-~7-7~ T 5
TREQ ~5-~-~-~ S± 5,7
UNDISCOVERED, LOST, UNUSED K (ULK) 2-7-5~ T± 4
WHY TRU UNRECOGNIZED, NOT SHOWN (WTU) ~5-4-2~ T± 8

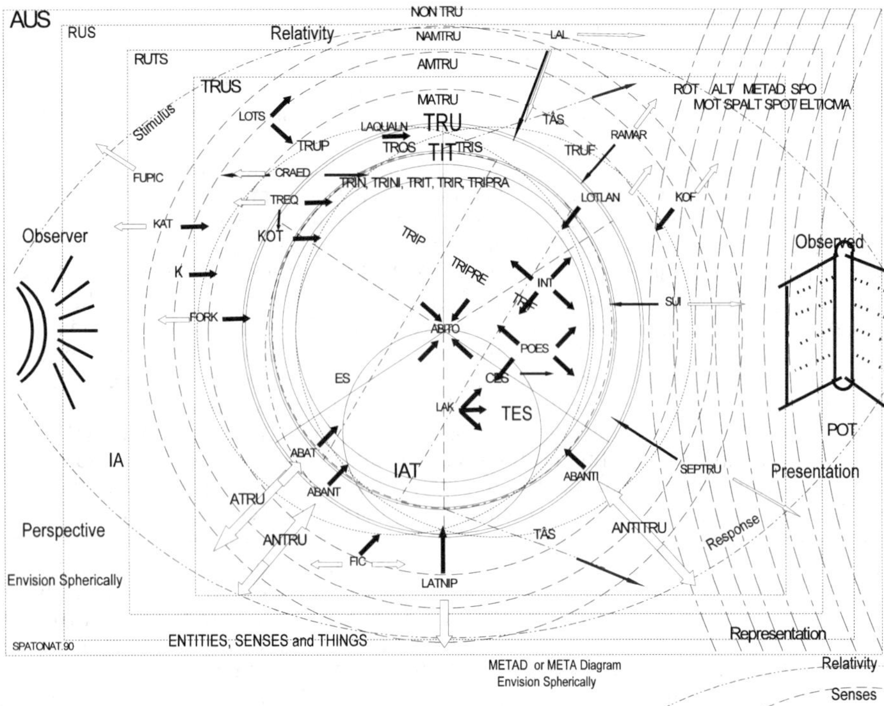

Fig 4-90 **ACC NEC POS in SPAT**

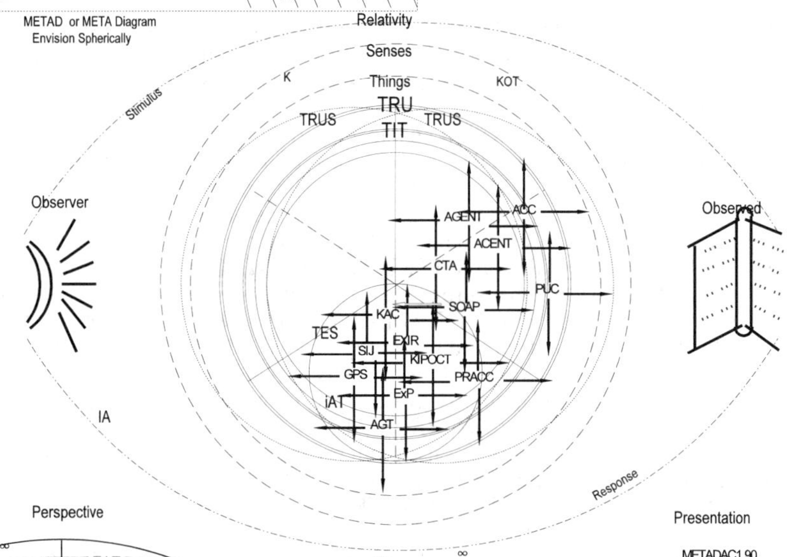

Fig 3-90 **ACC NEC POS in METAD.**

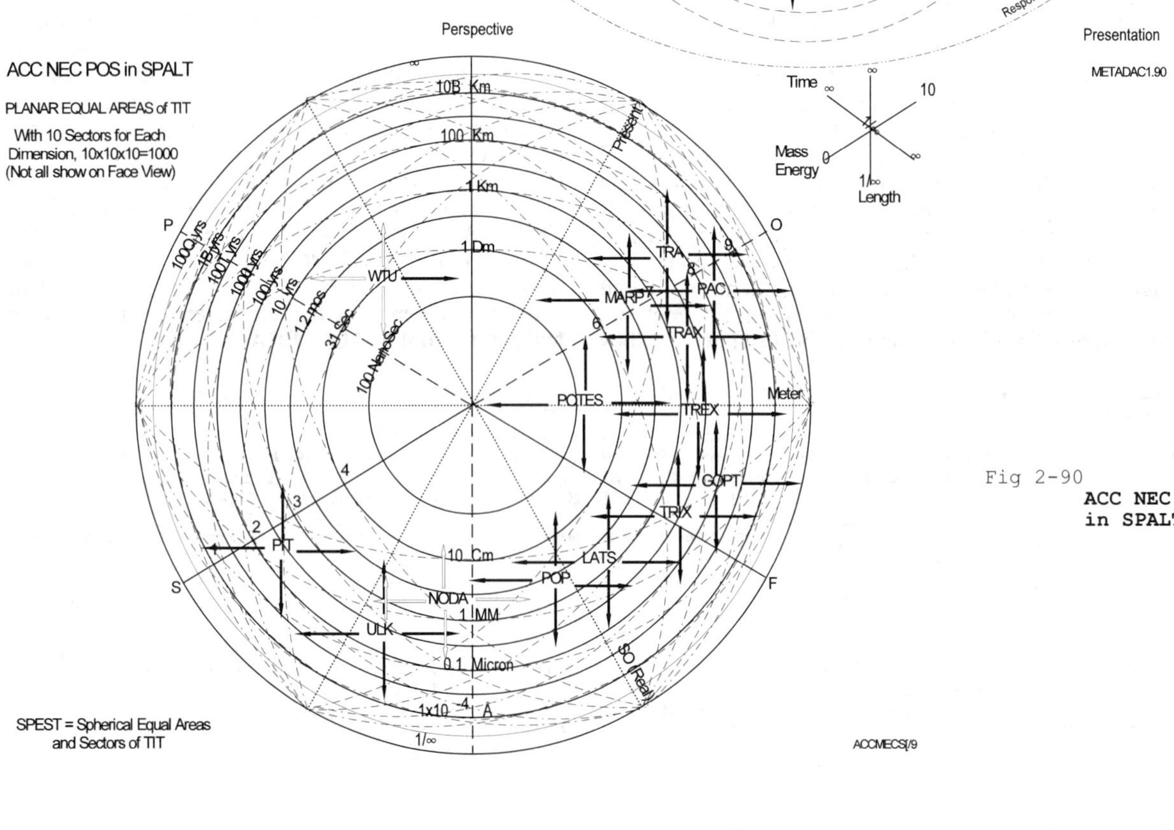

Fig 2-90 **ACC NEC POS in SPALT**

CHAPTER 91

DAC or the DERIVATION, DEVELOPMENT, DETERMINATION, and DEMONSTRATION of ACC

ACLAP, METHODS, ACC REACTION to AU, ANTRU, ATRU, NAMTRU and NEGATIVE ACTIONS on TRU

DAC, is a product of what we do about the disproportions, partialities, non identities, and negatives that tend to implicate and destroy our actual world, TRU. This has been common in human history and civilization. Each person is not born with the capacity or understanding to adequately or successfully conform to the world. What one thinks is their world is far from what it actually is, TRU. So that even if ones intentions are seemingly good they are often doing, working, playing, and living under false paradoxes, premises, forms, sets of standards and rules, or a fool's paradise. This applies to groups, communities, states, and nations as well as individuals. Here the harm and dangers escalate. They are often deliberately and increasingly deceptive, negative and destructive. All levels of social groups that seem to its members so agreeable and correct are more or less ATRU relative to the actuality of TRU. It is a world that is less than TRU, MATRU, or increasingly AMTRU, NAMTRU, ATRU, ANTRU and ANTITRU. Hopefully the future will reverse this at acceleratively improved rates and levels as TRU is discovered, learned, adopted and advanced. In law there is the opportunity to bring all intention, aim, purpose and use of TRU FAF to fruition. Those in judiciary and legal leadership roles, because of their pivotal position, must be able to distinguish between what is and what is not TRU. If not they will support what is not, and destroy that which is TRU. It is in law, when right by TRU and the adaptation of all legal systems to TRU and its laws, that much of ACC is to be derived and achieved.

In this chapter, DAC, we show those greatest dangers and threats to human survival that are key to the thisness of TRU and TRIA. They are shown in a way similar to that used in the causes or etiology of diseases in Medical Science. One of these is: **COC** = Complications or Compounding of Causes. It is through collecting, correlating, and formatting COC that we can provide a total picture and program for correcting all that is interacting with TIT-TO. The prevention, necessary treatment, cure, management and control of all problems and causes, that are negative or detrimental to TRU, can often be best acted upon at their source or appropriate points of contact with TRU. A functional or S-R approach to DAC can be ordered from their earliest origins in AU to their first causes and contacts with TIT, to more specific actions and reactions, contacts and effects, to their analysis, summation and accounting in the TO of TIT. It superimposes the functional S-R onto the formative PACH by SO, so natural for Á.

Those causes, problems, threats, dangers, and likely reasons for loss, misery, pain and suffering, failure, disaster, destruction, and extinction, may be differentiated and classified by their being acute or chronic, direct or indirect, temporary or permanent, mild or critical, simple or most complex, known or unknown. Of the latter there are, I. Causes which we know, feel, and especially those that are very popular and identified by activist groups, II. Transitory causes that will often go away by themselves, or do not have much effect on the totality or long term success or failure of TRU, III. Many that are known, not known adequately, or are not known in ways equal to the danger, IV. Causes which we do not know. Some of these will and some will not go away by themselves, without management. Many of these causes are actually great threats, forces, powers, and actions on TRU that must be identified, addressed, and reduced to their proper relation, active, and/or potential, to TRU. These we must face. The longer they are not identified, discovered, known, and to the degree neglected, the worse it will be for TRU. These are many of the acute climaxes in the evolutionary and historical continuity through the present. Many are the great trends in conflict today. These are serious and life threatening, to TRU, the ills, defects, and disorders in our society and human life from the past, at present, and in the future.

UT = Un TRU, Not TRU. UT includes both all that is not or other than TRU, as well as all that is purported or may be assumed to be of TRU, yet is not. UT is like a face or a mask. It is what something appears to be, yet is actually quite different.(LOVAL,LOVAP,ROMT) UT is not the same as ATRU. UT is a state in and of itself, in a way the opposite of TRU. ATRU is more dynamic, movement away from, or decreasing TRU. We should remember the differences between MATRU, AMTRU, NAMTRU, UT, AU, ATRU, ANTRU and ANTITRU. This differential provides the expanded, yet distinct, makeup and gradations of what is or isn't TRU. Since at this interchange much of ACC occurs, it is important to know how they work. ACC and PPMA are not the same for each.

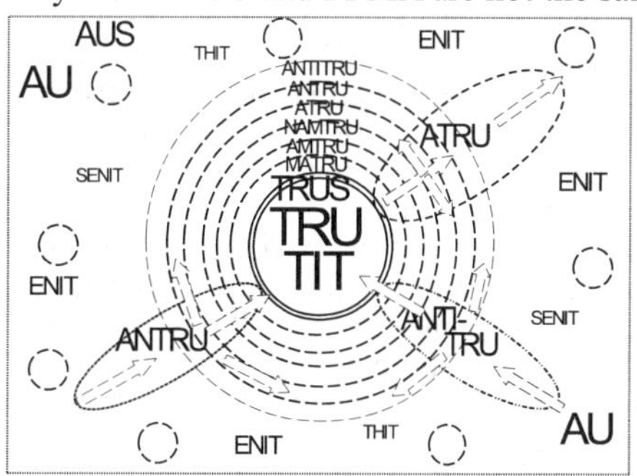

Fig 1-91 **EXTERIOR SET of DAC, TRU DETERMINED COMPARATIVELY**

In this figure the relations between TRU and all else are shown. TRU is the double lined solid in the center. MATRU, AMTRU, and NAMTRU are gradations of Mixed yet decreasing TRU.They are shown surrounding TRU. ATRU or from TRU is shown for all that leaves and thereby diminishes TRU. It may be positive as excretion, or negative as loss of TRU. ANTRU or Against TRU generally is shown for all that is opposing and negative to TRU. ANTITRU or Against TRU specifically as combated by TRIX, and is shown for all that is intentionally opposing or negative to TRU when TREC. ENIT is all that exists as entities-in-and-of themselves, from their own focus. There are a myriad of ENIT, that make up much of the substance of all when considered from the point of view of AU. From the point of view of TRU there are none, only their ESINS, and these only as ES and CES in TIT-TO. From the point of view of themselves there is only one, their own RIA. AU is the total rectangle included in the figure. It is a world that includes, but does not give, special exclusiveness to any entity including TRU. It is in such a mixed actual and inactual schemata that many of the forces, actions, causes and dangers to TRU occur, to be analyzed in ACC and DAC. We must quickly and continuously remember, however, that it is only their final reduction to TIT that is our actuality and prime concern.

The term prejudice is largely negative, whose judgements lack VAPO, are fleeting, superficial, and warped. This when inverted is **FEXPRE** = Fallacy of Exfoliated Prejudice. FEXPRE and prejudice have importance for what they necessitate. This is a prejudgment that is right, right by TRU. As ATRU, FEXPRE is all about us, about TRU, and future encounters. In order to have the right judgement, which is not always easy, it is necessary to have a great and good body of knowledge, plus the ability to use it, and the focus of TRU. This especially means philosophical, religious, scientific, and its many fields, as they are TREQ, in TIT. Without TRU knowledge we quickly depart from the best and regress through NAMTRU and ATRU to AU and ANTRU, prejudicially. This points out the justification of TRU, KOT, and ACC.

FAH = Fallacy of Heedlessness. FAH signifies what more or less exists in all humans, the failure to heed that knowledge and actuality of life that is TRU or TREQ. Even without TRE much of TRU, can be known and lived well if we only heed, trust and constantly follow what is most beneficial. FAH is like those of little faith who have no good values, appreciation for the divine, and recklessly deny or destroy TRU. This is largely from failure to recognize, learn and live by TRU, TREQ or GOD. FAH like PHE, RELE, SCE and many other fallacies are the reasons why so much of the world is not TRU. It is why the world, universe, god and TRU are so far apart.

It is astounding how poorly each person comprehends and applies oneself to the world in which they live. Throughout history we repeatedly make the same old mistakes.We propagate the unfit, abuse our bodies, think and do what pleases us, promote and stubbornly hold to false idols, disregard and mistreat what is good, champion bad, ignore or deny what it actually takes for human life and survival. Ups and downs, growth and decay, and periods of prosperity and depression are the natural occurrences of life. Yet, we can PPMA them to both manage and control our life as we continue to make it better. This is the reason for this text and TRU. We must discover, learn, remember, apply, and uphold ACEXIN. and the positive imperative that life offers for our place and activity in TRU.

Politics and the legal system are largely practiced on a basis other than TRU. It is one based upon the right of whatever is given, advocated, or promoted. It defies much of TROS. It demonstrates the Law of Negation of the basic, fundamental, vital laws of TRU relationships. This is called: **LATAR** = Law of Negation or from the, fundamental vital Laws of TRU Relationships. To understand the problem, or what failing is revealed, is half the solution. Once understood the solution to these problems become more natural, easier, rapid and effective. This is, **LAPUS** = Law of A Problems Understanding or determination is much of the Solution.

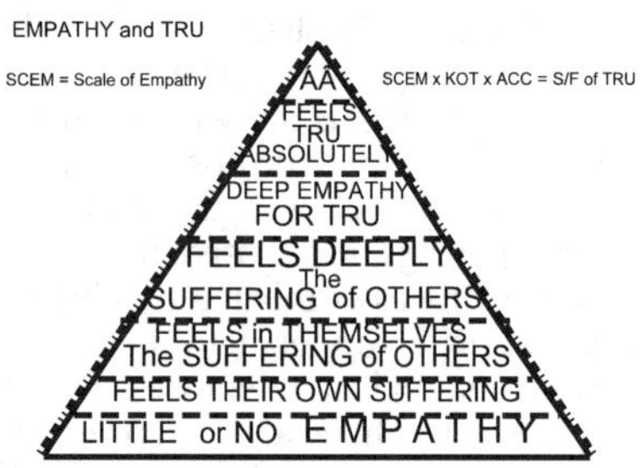

Fig 2-91 **EMPATHY and TRU; LEVELS and GRADIENTS**

Suffering is, and to that degree, a result of all existence that is not, or is from, TIT-TO. This is **SRÀT,** or as a law, LASRÀT, the Stimulus Response Á, S-O TRU energy. It is cured by ACC, PPMA and the empathy that goes with KOT, TES and ACC. Empathy is a component of the complexity, quality, quantity, power and predominance of the Á mass energy FAF, state and dimension of TRU. This is neither exclusively physical nor mental. It participates in the entire spectrum, continuum, and gamut of levels and gradations that function with VAPO and the SCARLS of TRU. Empathy is in most of human life and TRU, that is neither all mental or all physical, but partakes of many levels and gradations of Á. It is like taxis, tropism, and tendencies to act in ways that help life. It is a complex biological instinct, urge, feeling, love, identity, and appreciation for others and all that is TRU. It is more than a simple feeling for the suffering of those in misery, yet it is less than a feeling for the needs of any and all things. It is no more and no less than a total inclination, feeling, and love of all that is TRU or has a potential for TRU. In rising to ever higher and deeper levels of complex, massive, quality, powerful and dominant levels, in VAPO, empathy fulfills its role in TRU. These levels of empathy by becoming greater in themselves as well as in TRU are called: **SCEM** = Scales of Empathy, for each and all in TRU. SCEM especially applies to all that suffers for the sake of TRU. Yet we should not forget that the predominance of TRU is not the same, in fact much the opposite, of any dominance, power and force that we are so commonly familiar with. This is often ATRU, ANTRU and ANTITRU. This is: **FATDAP** = Fallacy of ATRU, ANTRU, ANTITRU Dominance and Power. The recognition of FATDAP and SCEM, and our actions and work in life to develop SCEM in the best way, is the beginning of ACC. We can never achieve any lasting or effective TRUD and ACC, assurance of TIT and TÂ, without the complete function of our capacities and abilities that are provided by all Á. This is, of course, in harmony with the overall ACEXIN that constitutes TRU. It will be different and same by DAOST and **DIN**, Differentiation/Integration.. Yet it will in resultant, as shown by VAPO and TO, always reconcile with TIT.

FALE = False Loyalty is Erroneous, Evil, and wrong. False loyalty is wrong for a person. It is often even worse and wrong for the effect it has on others. There are many kinds of false loyalty. The loyalty each and all extol is wrong to the degree it is less than, not or negative to TRU. It is impossible to have the right kind or degree of loyalty to anything, unless one knows what this loyalty is to and how it corresponds to, relates, is interpreted and guided by, and functions in TRU.

. Crime is not a simple matter of deeds that are violent and harmful. Nor is it only the qualifying and quantifying of these actions. The deeper and prime meaning of crime lies in evaluating it in terms of TRU. Crime and punishment are relative to TRU in an inverse correlation and correction. They are ever present in life, an aberrant part of growing up, maturing and being. They are the result and accompany any and all that is negative to, and to the degree it is or is apt to act adversely on TRU. The simple reaction, right or wrong, to crime is punishment. Crime pertains to all we do, or even think and believe or more generally, what is ATRU, ANTRU and ANTITRU. Punishment, as crime, should be delt with adequately and forcefully. But not until you are sure of:
1. the crime and its degree. 2. the opportunity, proper time, and manner of delivery, and 3. the proper reasons, values, and relations to best human life, are determined. Punishment thereby should be directed toward achieving TRU. It should be, like experience and lessons in life, a learning and education in knowing and achieving what is wrong and what is right. Every moment of our existence, every thought and act we have, are a part of crime and punishment. Since this can not be adequately achieved without TRU and a sufficient knowledge of TRU, the approach to crime and punishment will always be faulty without TRU. The best answer and solution to crime and punishment come from a world in which each and all persons are so formed and function, that crime does not exist and punishment will not be needed. This is a goal that can be relatively achieved. It is to, as early, thoroughly, and permanently in life as possible, discover, learn, know, remember, cherish, show, act and live by ones role in all of TRU, that their reason and deeds can be best formed and function. Human life, as it most fulfills its potential in TRU, tends to combat wrongs and convert failure into success, defeat into victory. This is the nature of our human mass energy makeup, or Á FAF. Its carriers, concepts, and the reasoning that uses them, are like the magic of God. This is what is dedicated to combating wrongs and ills and provides the fulfillment of TRU. (MOG)

Many of the evils of man have been recognized throughout history. They were recognized and denounced in Egyptian and Biblical writings. Which vices, sins, and evils and the number given, differ with each authority. Many more have been revealed and shown before and since, as in the early Renaissance in Italy. In the Church of Sta. Maria dell'Arena in Padua, Giotto the student of Cimabuqe painted the frescoes that demonstrated: sloth, malice, ingratitude, anger, and selfishness among others. Humans so quickly forget, ignore,abuse and fail to bring these together. This is counteracted when TES is achieved and TRU is discovered, learned, known, remembered, lived by and promoted. Both qualitatively and quantitatively it is how poorly humans tell good from bad and don't know what the good actually is that is the greatest cause of their undoing and failure.

	VARIETIES	EXPLANATION	TIME FRAME	CAUSES and REASONS	EXISTING CONCEPTS and NEC	UNDERLYING TRU MECHANISMS	TRU PROCESSES	TRU ACTION ACC
OZONE & OTHER Resources, Oxygen, Water, Energy	Primarily Circular Terrestrial Essential	Depletion of Protective Cover	Max in next few decades or until PPMA	Aerosols Pollutants Wastes	1/ENT,TRU Á, SÁ LOPAN	Alteration of Human Habitat Depletion	Numerous Health Economy Effects,	Frugality Controls Invention
POLLUTION & ENVIRON MENT	Innumerable Rapidly Increasing	Adverse effect on Habitat	Effects Escalating Controls	Industrial Commercial Expansion	1/ENT,TRU	Physical & Genetic Health	Loss of Ability & Quality	Frugality Controls Alternates
POPULATION GROWTH & QUALITY	Total and local dysfunctio n-	< TRU needed quality	Great Contin-uing s controls	Knowledge Technology Self gain	ATRU and TE/TO,TIT	Quantity/ Quality	↓ Quality Temporary Permanent	KOT,TES, Voluntary Controls
RACISM Classes	Numerous on all Levels of TE	Unbalanced Competitive Predominant	Continuing s TRU, KOT TES, ACC	Ignorance Prejudice Rationalize	ATRU/TRU-TROS,TRIS Ind/TES	Superficial use of Distinction s	Imbalanced Poor DAOST TES TRU	KOT,TES, LOTES,ACC
GOD RELIGION MORALS	Innumerable disordered	Mostly Archaic or Developing	Depends on Recognized need, TRU	Ignorance Fragmented non TRU	TROS, TRIS, TIT, TRIA, MOT REM POS	Need focus and ACEXIN by TIT ACC	Developing THES THESÂ COGNOS SÂ	KOT,TES, ACC,MARS KUPS TÂ
WAR Organized TERRORISM	Many kinds and levels	Great Nega-tive Poten-tial for, & Danger of	Terrible effects,Mod - eration,TRU Persistent	Ignorance Hate, Self Interest	No TREQ, ANTRU/TRU 1/HOM,ACC	Lack of KOT TES,TO,TROS	1/TRU,TES, PUREACLITY TO,COGNOS	ACC,TRAX, TESCOGNOS, KUPS,SÂ,TÂ
ECONOMY COMMERCE Business	Many Levels and Venues/ Greed	Highly bene-ficial and detrimental	Upward Acti-vity + if healthy TRU	Burden on habitat & Quietude	TRUD, REM, DYNU, DYNS, DYNO/ATRU	S/F depends on Efficient TRUD,HOM	TRUD,HOM,TE all levels of ÁCEXIN	REM,TRU, HOM/ATRU, waste, harm
NATIONALISM	On a natio-nal level a Major kind	of abnormal focus and intolerance	Worsening without TO, moderation	Ignorance Propaganda 1/Rational	TREQ,TRU,TO TRUD,HOM,SP ALT,POS,ACC	Rationalizi ng blame for ATRU	ATRU-ANTRU /TRU-TIT, TRIA,TÂ	KOT, TRU HOM, DYNO, ACC, SÂ
OPPRESSION SELECTION	Many kinds of abnormal & excessive	+ & - forms of dynamic interaction	Worsening without TRU TES, ACC	Abusive & Tolerates abuses	FIA, 1/KOT,TRU	Ignorance Partiality 1/KOT,TRU	1/Á,ACEXIN 1/TRUD TO	TREQ,IPT, TRU,TO,TES HOM,ACC,SÂ
POLITICS	At all lev-els of Human Relations	Some needed Can be extr-emely +or-	Improved c TREQ, worse cATRU-ANTRU	Dominance & Control/Gre ed & Abuse	TRUD,HOM, DYNO,ACC KOT,	TRU/ATRU S/F,ACC	Group→TES→ TRU or → ATRU	TREQ,TRIS,TO ,TES,KOT ACC,TRAX,
LAW ORDER & FREEDOM	All kinds & extremes, balance	Same contin uum + or - Social FAF	Better cTRU Worse c ↑ Pop & ANTRU	↑Population ↑Complexity of TRU	TRU,TES in Persons & Government	How much can Persons take/do TES	How well can Governm supply TRU	TREQ,TIT,TTE S,HOM, POS,ACC,SÂ
EMPLOYMENT	Any and All kinds of work	Great disparity Untrained	Worse or better s or c TREQ, TRU	Unreasonabl e Expectation s	TREQ,TRU,TÂ TES,POS,HOM ACC,KUPS	TES/Ind,TRU /ATRU,TO, HOM, ACC	Best Needs of Ind-Soc in tandem	KOT,KUPS, PUREACLITY TES,ACC
CRIME & PUNISHMENT	Many Kinds, Should fit Optimally	Attitudes and degrees best managed	Worse s TRU Better c" & Tolerance	Cause of Social-TRU Collapse	ANTRU,ATRU TE-TI,KUPS KOT,TES,SÂ	Reaction to ATRU law or der freedom	→Ind-ANTRU or→TES-TRU TE-TO-ACC	PPMA,TRU KOT,TES,SÂ ACC,KUPS
SEXISM	Continuum Male/Female Relations	Breakdown of	Worse s TRU Better c TO Mutual Gain	No-poor qua-lity in FAF Immoderate	KOT,TRU-TÂ Incentive TO,TRUEQ	Breakdown TRUD TRUEQ and TES/Ind	TE-TES→TIT ↑ Á FAF TRUEQ→TO	Mutual Love Quality,TETR U,ACC,TÂ
SERVICE Self " SACRIFICE	Continuum of +/- zeal for best	Degree of Dedication to God-TRU	Better c TO TREQ God Worse s	Irreligious 1/Endeavor Self Gain	TRIA,TA,TES TO,LOTES, PUREALITY	Revelation or no, TRU IDENTITY	Ind→God→TIT →TES→TESÂ→ THESOÂ→TÂ	Inspired Zeal for TRU,SÂ,TÂ

Table 1a-91 **RAGING CONTROVERSIES, TRU and ACC**

This first of two tables on current controversies involve the larger parts of TRU, as seen in SPALT, or of nations and societies. They are presented to tie-in much of the text, and especially how all comes together in ACC. It is because we fail to get to the bottom of their causes, seek to learn and apply their underlying mechanisms and processes that we endlessly struggle with the same controversies. Only when we realize they have deep roots, are highly complex, and can't be answered or solved simply, can we begin to properly interpret them. Only when we discover, learn and live by TRU, can we understand and solve these problems and controversies. Because they and their causes are deeply rooted, highly complex, highly interrelated and TRU, the ultimate best approach is a total one, one in which we treat all by the totality of TRU and a total ACC. This is the beacon or guiding light we follow in KUPS and COGNOS in ACC by bringing together what seems so different and separate controversially. Yet, by DAOST, is only different aspects of the same thing, TRU. We must tear them apart if we are to put them back together again. To maintain spread sheet reconciliation and follow through to TIT-TO is what these tables seek to do, by best utilizing the great concepts, existing and new, of TRU.

The greatest crime of all is to reward the wrong and bad and to punish and dishonor the right and the good, or TRU. To honor and reward the unrighteous, incompetent, wasteful, evil, criminal and to punish the righteous, competent, efficient, good and law abiding is done with impunity. This is a crime that all in positions of responsibility must forever keep in mind and avert. The mass equilibrium and forces involved must be held to their highest possible positive levels, as this is from TRU. For there must be much more stimulus from the positive, good, righteous, and TRU forces than from the negative, bad, and ANTRU forces.

Sectarian is defined by being narrowly confined to sect, or limited in interest, or scope. Sectarian is, although not a positive term, a highly appropriate one for many kinds of adherents and zealots. Although it is used to separate what is not or from religious by being partisan or cultist, from what is it is in universal religion, they often have good reason and some identity with an important aspect or part of TRU,

Both tables, 1a and 1b, their problems and causes, as well as their explanations. have much in common. This is human quality, both inherited and acquired. Rarely have humans been civilized enough to have an excellent culture. History has been the repeated tale of the rise and fall of civilizations, cultures and nations with how well or poorly they have accumulated and performed in these, and other processes, mechanisms and dynamics essential to the success and continuation of TRU.

	VARIETIES	EXPLANATION	TIME FRAME	CAUSES and REASONS	EXISTING CONCEPTS NEC	UNDERLYING MECHANISMS	TRU PROCESSES	TRU ACTION ACC
HEALTH	All kinds from worst to best	Can come to optimum of TRU	↑ or ↓ with TRU and TES quality	Incapacity & Quality of Environment	TREQ, TRU FAF, ENT, TES HOM	Continued Massive Selection	TIT→TES→ENT /ATRU, ANTRU ACC→SÂ→TÂ	Continued IA on TES TRU ACC
BIRTH CONTROL	Prevention of Pregnancy	Increasing Methods and Reliability	Profound effects on Human Life	Population Quality/ Many Abuses	TRUD ENT HOM DYNU DYNS ACC	Positive if used right for TRU, TÄ	Can Yield ↑ Selective Quality TES	Good Voluntary means of TRU ACC
ABORTION	Terminated Pregnancy Abused	Wrong in a TRU world, Rx for -	Will vary c ATRU/TRU & abuses	A place for correcting Negative	Not Prolife or Prochoice only ProTRU	What wrong/ right? TRU/ ATRU-ANT	Highly Selective Quality TES	KOT TRU TÂ TRUD HOM TES
PREVENTIVE MEDICINE	Vast Range of vantages	What is and What can be	Great + c Great - s	Whole gamut of dangers	KOT TES SÂ HOM ,TRAX	Best of TRU /ANTRU, TES	Info→KOT→HOM→ DYNU→	TRUD HOM ACC TRAX
DRUGS & IN TOXICANTS	Innumerable Increasing	Production Distribute	Failure c Success s	Ignorance Exposure	KOT TES T PUREALITY	ANT-ATRU Proliferate	ATRU→Info→ KOT→TES ,TO	PPMA ACC PUREALITY
BIAS & PREJUDICE	All kinds and Degrees	Abnormal attitudes	+/- c Quality Minds	Ignorance - Reactions	ATRU, 1/KOT TES TIT TÂ	1/Genetics Mentality	Resistive to KOT TES TRU	Selection TES ACC TÂ
HAVES/NOTS INEQUITY	Extremes of Poverty	Inequality of Wealth	Some ~ with Ability Need	1/Quality REM & Need	REM TES T VAPO HOM	Inequality inability	1/TRU→1/TES→1 /REM→F/S	TRU TES REM ACC
HABITS of F/S, S/F	Worst to best Habits	Ability to Cope c life	+/- c Ability Will	Ignorance 1/energy	PUREACLITY TES, TIT	Precedent Training	S-R→Will→ PUREACLITY	KOT TES TÂ WILL
EDUCATION TRAINING	All kinds and Levels	Poor or Mis-focused	+/- c TRU PUREALITY	1/inspired determined	IPT KOT TES PUREA	Opportunity Guidance	Interest→Will →KOT→S/F	Inspire TES Focus ACC
OCCUPATION SKILL TRADE	~ Quality & Performance	Enhance Best Accept Least	+/- c ↑ & Adaptation	Inability Unavailable	KOT Á TES ACC KUPS	Quality TES Application	Will→KOT→Á TES→KUPS	IPT KOT TES KUPS
INTERESTS IDENTITY	From best to worst	Needed Key to TRU/ATRU	+/- c ↑ Á THES TES	1/Incentive Wrong Focus	S KOT TRU TES TESÂ	Uninspired K God, KOT	↑ KOT→Á→TES THES→TESÂ	Quality Pop Training, TO
REVELATION INSPIRATION.	Spectrum None - All	Deepest FAF of the Mind	Greatly ↑ c Interests	↓ Genetics - Stimuli	EPI KOT TES, SÂ TÂ	Curiosity + Interests TÂ	S→KOT→TIT→ TRIA→TES→TÂ	Stimulate Foster ENT
REASON INTELLECT	Spectrum None - All	Highest FAF of the Mind	Greatly ↑ with KOT Á	Failure Survival	KOT Á TIT TES THESÂ	Great Salient of Á FAF	S→EPI→KOT→ THES→TES, TÂ	Work to ↑ KOT TESTÂ
SELF IMPROVE MENT SOUL	Convergent on TES TIT	Held back s TREQ KOT	Increasing c TRU TES	Ignorance Disinterest	TES TESÂ COGNOS TÂ	Ind→PUREAL→ COGNOS	Ind→COGNOSTES ELTICMA	Rigorous KOT T TÂ
IMMORTALITY	All Degrees of	Eternal Life	Increasing with TES TÂ	Speciation and Á FAF	TES TRIA TESÂ TÂ	Ind-TES-TÂ PUREALITY	TES→HOM→SÂ ACC→ELTICM	TES-ACC-TÂ

Table 1b-91 **RAGING CONTROVERSIES (INDIVIDUAL)**

This second table investigates those great problems of our day that are of the smaller portions, of TRU, of individuals and their health. The leading facts and conclusions are that their solutions come from failure to properly and fully understand them, their causes and cures. They tend, like all paradoxes, to be the cues to fundamental principles of human life. Therefore, to find their solutions and cures and take the needed action we can do no less than to show, promote ,and get as many persons as possible to learn and live thereby.

BAB = Bad's Blindness to Bad. BAB is largely the opposite of how great minds seem to follow the same paths, because they tend to be the same, relevant, and TRU. Just like a lot worse problem makes what seems to be a terrible problem much less, so is it that in these there is no refuge, no singular positive objective or solution. A thirst, burning desire, and deep want of knowledge and a knowledge that promotes TREQ and TRU is crucial. BAB is the antithesis of a life of TRU, S-O, or Á. So that ignorance, by BAB, is of more than one kind; ignorance of any and all, of established knowledge of various kinds, or of TRU. What counts most is how to turn this ignorance and knowledge into all that enables man to be, to be by TRIA, TES and TRU. BAB may be contrasted with the acceptance of the bad by bad. This is called: **BAC** = Blinds or Bad's Acceptance of Bad. It is overcome in much the same way as BAB.

A strong concept and word is needed to describe those who take from the relativity of the self to TRU and reverse it to excesses of diversity, ATRU, and ANTRU. These are, **DIVA** = Diversity of Individuals from (ab) TRU. Much of the best of human life and work, as in religion, ethics, morals, duty, honor, cooperation and mutual effort has been directed toward correcting DIVA. However without the fulfillment found in VAPO such correction has been weak, defective and missing.

In the world about us, that is such a mixture of TRU and non TRU, MATRU, AMTRU and NAMTRU, there is a constant flux of destruction and construction occurring. Negative and harmful thinking and ways are contagious. Those who do not know better, lack religion, a heart felt appreciation and love of the Good, Right, God, TREQ or TRU, are readily persuaded to champion this or that false cause. Reward of the bad and the punishment of the Good, or TRU are all about us. They will continue, especially without an expansion of KOT, TRE, and ACC. For as long as there are immature, unappreciative, and vain individuals in this world, as long as there are those who will profit in any way from competition, conflict and dissension. no matter in what form, there will be a whole world of ANTRU which is encountered and must be addressed and combated.

What are the causes and reasons for TRU, and the role of philosophy, science, religion, arts. and technology?

Humans are entrapped in their own dead seat of bounty, neglect & failure to live by what is best. What we have been given, must achieve, & fulfill are of our undeniable part and role in life. This is a life of ever higher participation, competition, success and victory of TES in TRU. A death struggle for continuity over extinction in this special universe of best human life of Competitive Predominance

B

A C

D

If we are A, our competition or opponents are B & D. and what assists us is C, we are in the ***DEAD SEAT*** between B & D. So situated as to be outflanked and overpowered. If C is little help, and we, as strong as we so loudly acclaim ourselves to be, and actively adventure senselessly and recklessly into annihilation by B & D, we can no longer be, become extinct. A is the world of man,. B & D are our adversaries - all that is ATRU or

ANTRU - and C is the world we depend upon other than ourselves. We can't depend upon C, it is much of our own, and often negative, creation. It is dependent upon us and very much weaker especially without our support. We, as A, have good strengths, only we don't know, know and don't remember, know and badly use them, especially before the onslaught of powerful and vicious effects of B & D. The war is to the death, crucial & vital. We must know all needed, prepare and support ourselves and what assists us to have and use all the best strategy & tactics to subdue B&C. This means that way natural to our deepest being, all our mass energies in the spaciotemporal framework of TRU.

Just as a very good, beautiful, intelligent girl can't bring a bad, mean bum to her nice world, a well-raised promising intelligent boy, can't bring a bad, ignorant, worthless girl to his world. In either case the opposite, ATRU results. It is largely up to the individual to be TES,, and to be TES they must be less individual and only TRU. All who are TRU can not expect to, accept and try to remake NAMTRU, ATRU, AU, ANTRU, ANTITRU, TANTRU and all other devious worlds, into TRU. Except in very limited ways in which their potential and ESINS can be contracted and directed to TRU. This is especially so if all that is not TRU does not tend to TRU. We can turn much of what is less TRU to more TRU. Yet we must always remember, ECRO, that this best of worlds, TRU, is a very special, select and pure in ways and SCARLS that render it easily and quickly effected deleteriously, and destroyed in part or totally, by what is not TRU. Also remember, PPMA, that TIT-TO should be carefully and vigorously protected, maintained and preserved. From the individual's point of view, we may seem invulnerable and assured. Yet this can be no further from the truth, and is part of the great causes and reasons for ACC. It is because the individual is so indifferent, not TRU, and disinclined to be TRU, that we must react, compensate and be forever vigilant to make correction. It is for this reason we so emphasize, as in religion, ethics, and education, the importance to each person of proper training, learning, education, and development of TREQ or TRU. We need only accept, honor, be stimulated, inspired, use, live by and see to it that the world is much greater and better when we leave it than when we found it. This is something that can't be done in some last spurt of goodness in life, as in making supplication to God and heaven when in peril, or at the end of life. It is a life long task of duty, honor, privilege, necessity, and joyous fulfillment, of TES. This is TRU, our great and only true and actual opportunity and immortality.

Final cause is the end or purpose for which something was, is being, or is to be done. In the broader or more general sense final cause is the end or purpose for which something occurs. It should be considered along with other types of causes which together provide a description of the nature of the dynamics of things, objects, or their subjectification. They provide a description of the nature of TRU, as the object and purpose of its subjectification and ACC. The concept of final causes is great for several reasons, including: 1. TRU is or is not as this end and purpose is done.(ENT,TÁ) 2. The several causes of TRU when taken together properly and fully provide a great demonstration and presentation. 3. In our presentation of TRU, final causes, as for all TRU and readers, can be convergently focused to provide their concordance through SÂ, TÂ and TRU. 4. The adaptation of final causes by TRU proves and in turn is itself proven, as in ATO, TRIA, and TÁ. What might seem to be a final cause is not. More often only a final limit, final within a set, segment, or portion of TRU.

How a little knowledge can be a dangerous thing is a beautiful example of **MAPEV** or Mis Application of Partial Evidence. This is how with a bit of reason that is valid much justification is made or action taken that turns out to be so bad. This is what happens in the media, politics, war, and doctors who use such bits of information to justify elaborate procedures and surgery. MAPEV has become so pernicious in current society and the modern world that it has become habitual, even extolled in some circles. It goes along with so many other misapplications of knowledge for devious purposes that together constitute one of the greatest negative forces on TRU. They are the breeding grounds for much of the turmoil, suffering and loss to TRU. This eventually either has to be made up, continues in a positive and negative dialectic that keeps us in a highly NAMTRU world, or can lead to extinction of the human race. Much of the misery and woe humans endure is of their own making. Much of this comes from shortchanging a good perspective of the actual world, a lack of overview, comprehension, and good judgement, that can be found if properly sought for. This is why TRU, TIT, ROT, SPALT, POS, POT, and IPTRU are so great. This is why each and all persons can and must find their souls, be TES and identify with this special world of best human life, TRU

LOMI = Law of Misdirected, or diverted, Intention. LOMI signifies, helps to define, and describes that state in which intention, be it TRU or other than TRU, becomes misdirected or applied to causes, reasons and purposes other than its original or initial TRU purpose. This is a relatively simple and common, yet erroneous, route of human activity. Like more complex functions of higher levels of human mass energy or Á FAF intentions, it may or may not converge on TRU. It often becomes a habit, increasingly bad, when it deviates from TRU. This is the beginning of reac tion formations to the stresses of life. It is very important, for it represents two of TRU's greatest failings: 1. Not to be intentionally directed to TRU, 2. Not to hold precisely to that direction to TRU under all influences. LOMI greatly illustrates the ubiquitous state of DYNU where without a cooperative, complex, and total, or DYNS and DYNO, approach to life that should be TRU, there is an insufficient degree of integrated and projected intention to sustain or maintain TRU. More generally, any defect in attention, IA, and IAT are major causes and reasons for ACC. Most specifically, however, it is the lack of, defective, or poor perspective that causes trouble and is a basic reason for ACC. This is, **ACLAP** = ACC Lack Arises from Poor Perspective. Mankind has sufficient knowledge to develop TRU. Although far from providing all the concepts and mechanisms, it is adequate to produce TRU. Thereby it is in how we sort out, order, and use this knowledge and its perspective, as a 1:1 replica of TRU that is largely at fault.

LICON = Law of Indirect Collective Conspiracy. Much of the wrong in the world is not done directly, by any one person, nor in a planned way. It is often done by several, many, and all who falsely or on little actual fact condemn others, TREQ, and TRU. This does harm, accumulatively and collectively, often indirectly or unintentionally. It is a result of all whose knowledge is incomplete, defective and lacks VAPO and TRU. The mass and power of TRU weighs heavily on those, even in TREQ, who are without TRE. Since this is much of the reason for LOMI, all direct actions for TRU are the best ACC in whatever way PPMA allows and is proper by TRU. This is: **DATBAC** = Direct Actions for TRU are the basic and best kinds of ACC. Much can be corrected by simply discovering, learning, knowing, using and living by TES and TRU.

TEF = The Future Drawing Forces, Attraction of. All fallacies are significant, important, and should be known and resolved. They are not only some of the worst causes and needs for ACC, they are, like paradoxes, the basis for solutions. When known and resolved they tend to disappear. This will depend upon how much they are a part of understanding and how amenable they are to solution. The resolutions of fallacies are the basis of much of ACC in general. It is a great part of the Á FAF of TRU, both internally and externally. Fallacies are mostly the errors or deficiencies in the mass energy or S-O or Á form, state and continuum of TRU. The resolution of fallacies is mostly a direct and simple matter of fulfilling S-O functions in TRU. This is, **PR0SÁ** = Problem Solutions, and the resolution of fallacies, are the simple and direct fulfillment of S-O or Á.

TOG = Total, collective *cahiers*, or recorded and summed Grievances. TOG includes all those terrible and awful wrongs, ills, burdens and pressures that are totally unnecessary and ATRU. Many can be best resolved as a whole when we learn and live by TRU. Many are more TES than TO. Many can be resolved by catharsis or explosive cleansing of the worst, their selective ringing out. Examples for ACC are the wrongs that existed in France prior to the French Revolution and the cahiers that provided its cutting edge.(MARS)

CAPONT = the Punishment of Crime must of Necessity be Proportional and equivalent to the Total effect and relationship in TRU. Rewards and punishment by Sun Tzu in the *Art of War* must be immediate, fit the occasion, personal, and great. Punishment can't be accomplished by AU or even ES alone. Thereby we must all sit in judgement, especially on ourselves, since this is one of the greatest means to the positive in TRU. Judgement is not limited to what is commonly known as crime, for like sin, right and wrong, crime is at various levels of interaction with TRU. By TRUD and HOM all that is active, is more or less so in a positive or negative way. When severe to TRU,as negative, it becomes crime. Throughout all these, the greatest benefit usually lies in prevention, by learning TRU.

Recidivism is to pass from TRU to poorer and worse, to recede into prior wrongful states, as the criminal who after parole returns to crime. In TRU as in religion, recidivism stands for backsliders who lose faith. It is apostates who may have reached any of a number of levels of godliness, TREQ, TRU, or TES only to recede to lower levels, a complete reversal of what is opposite to the best of life, TRU. How they lose their conviction, determination, will and persistent endeavor for right and recede into wrongful thoughts, habits, acts, and life. It is the resistance to recidivism that so characterizes the devout, often to abnormal extremes. Recidivism Rate = The pace and degree to which a person loses faith, TRU and turns to ATRU, AU, and ANTRU. It is this loss of virtue, goodness and righteousness that so quickly turns to evil, crime, and sin. This rate is of the greatest importance to self control in TES.

Losses that are apt to occur in the future can't be prevented unless they are understood. It is in timing, finding, developing, and using the understanding provided by TRU that is much of this text's theme. The summary of this is: **LOMT** = Law of Total Management by and of TRU, through PPMA, ACC and all that best preserves TRU.

CONTINGENCY OF SOLUTIONS

1. YOU CAN'T SOLVE INTERNATIONAL PROBLEMS UNTIL YOU SOLVE SOCIAL PROBLEMS
2. YOU CAN'T SOLVE SOCIAL PROBLEMS UNTIL YOU SOLVE PSYCHOLOGICAL PROBLEMS
3. YOU CAN'T SOLVE PSYCHOLOGICAL PROBLEMS UNTIL YOU SOLVE PHYSIOLOGICAL PROBLEMS
4. YOU CAN'T SOLVE PHYSIOLOGICAL PROBLEMS UNTIL YOU SOLVE GENETIC PROBLEMS
5. YOU CAN'T SOLVE GENETIC PROBLEMS UNTIL YOU SOLVE CHEMICAL PROBLEMS
6. YOU CAN'T SOLVE CHEMICAL PROBLEMS UNTIL YOU SOLVE PHYSICAL PROBLEMS
7. YOU CAN'T SOLVE PHYSICAL PROBLEMS UNTIL YOU SOLVE TRU PROBLEMS
8. YOU CAN'T SOLVE TRU PROBLEMS UNTIL YOU DISCOVER, LEARN, KNOW, APPLY, USE, AND BEST ACT UPON ALL THAT IS TRU.

Fig 3-91 **COMPARATIVE ANALYSIS AND TRU DEPENDENCE OF PROBLEMS AND SOLUTIONS**

Figure 3-91 dramatic-ally, if simplistic-ally, shows the depth of interaction to which causes, problems and solutions occur. To fully demonstrate them a much more elaborate representation is needed. This is one that is not only reversed, includes all clines or levels of SÁ, but gives them all TIT-TO, SPALT, METAD, SPAT, ACC, COGNOS, ELTICMA, and SÂ can do.

There is nothing more vital than to: 1. Discover, learn and know TRU and TES, 2. Stand up, work and live for TRU or, 3. Stand up for those who do right, 4. Help, train, correct yourself and all, 5. Ignore, alter, negate or eliminate those who don't or won't stand up for the world, TRU. Humanity can't survive in a world in which each succeeding generation does not learn to appreciate, fails to maintain, and loses its actual TRU power, what has been built up and works well. Each and all persons must, as early as possible and as thoroughly as possible in life, learn to know and appreciate all that is the best for them in this special universe of best human life, TES in TRU When people are hurting, in pain, miserable, and severely wanting in one or many ways, watch out! Especially when they are artificially and arbitrarily prevented, constrained or held in bondage, directly or indirectly, advertently or inadvertently, from getting what they need. For humans, their needs are one of the most powerful of all forces. Here the world is not measured by niceties, what is good, or often, what is even GOD or TRU. They are ready to vigorously or at times violently react to and attack all in their way. This is often not only ATRU, it is a boiling pot of massive and potentially violent reactions. Let all, both those caught up in misery, and yet to suffer from it, long ahead of time learn, understand and realize what they are in for as they relate to TRU. This foreknowledge, anticipation and comprehension of all and the planning and ACC necessary for its correction, are much of TRU. For if we live by TES frugally, efficiently, and, yet assure our basic TRU needs and desires, we can avoid most of these severe disruptions in human life. For TRUF is as much of us as TRU past or present, especially in TRIF. Therefor revert to TRU, enlist in TRU and become all TRU as it is and becomes you.

The diseases that cause the most deaths in humans usually have a predilection for, develop in, and show themselves pathologically in certain critical areas of the body. These are focal areas that lead to heart attacks, strokes, cancer, accidents, hypertension, diabetes and many others. This is also the case for TRU. For the worst human problems that threaten our very existence have predilections for, develop in, and show themselves pathologically in certain areas or segments of TRU. It is not necessarily of all inside, universal and evenly of TRU, especially when TES and TRU are healthy or at their best. It is often outside to inside, focal and uneven. Because TIT has uniformity does not mean that what effects TRU necessarily effects its distribution uniformly. Hereby we are dealing with two distinct and contrasting laws. **LATUN** or Law of TRU Uniformity. LATUN holds that for the special unique form of TRU, TÄ and ALT, its contents are distributed relatively uniformly. **LANTNU** or Law of ANTRU, ATRU, AU, and UTRU effects on TRU Non Uniformly. LANTNU holds that what effects TRU does so in a non uniform, unevenly distributed, or focal manner. The resulting synthesis of these for TRU is: **LATCRUN** or Law of TRU Compensation or Reaction Formation to Return to Uniformity. LATCRUN denotes how for TIT to return from surface exposure, critical areas, deformity, initial and resultant SCARLS, optimal interactions, internal reactions, homeostasis or steady state and compensations are made that maintain its continuity and TRIA.. How to correct these will depend upon many factors, and upon the treatment and management, or PPMA. It will depend upon discovering, recognizing, protecting and treating crucial areas of susceptibility. It is how we apply ACEXIN and **TRIALT**,, or Trial of ALT on TRU, to these crucial points, areas and regions, or limits, relations, conditions, and states - SCARLS. So that we can most effectively and successfully provide the necessary management and control of all that is ANTRU, ATRU, to cure them through ENT, ACC, and TRU. A great example is scientific research, preventive medicine and public health.

Activism, and to be an activist, are both popular and pejorative concepts and ideals. The polarity between activism and passivism represents two extremes of the TRUD and ACC continuum that characterizes life.. They reveal the dialectics of TRU and combine to distinguish ISMS. Thereby, as to be expected, and by Yin Yang, these are both good and bad with a moderate, or appropriate intermediary stand, the Golden Mean. Variance of this, as by adaptation, is usually best depending upon the circumstances and necessity. Activism or to be an activist simply means any and all involvement, especially with fervor, that seeks to resolve some problem, state or condition that is believed to be in need of change, elimination or support. Activism may be characterized by four main components or factors: 1. Right cause or reason, 2. Need, existence and degree of, for benefit, 3. Capacity, ability to make a difference, 4. Act, ACC, pressure, power or push; how well and how hard it can be pursued. Using these factors equationally we have:

Activism = Right x Need x Capacity x Push, or
A = RxNxCxP

For activism to be completely justified and successful it requires, among other things, a proper mix and balance of the above, plus focus by, of and in TRU. This will vary and can be somewhat weighted, in each case. It is in being active, and using all that enables best behavior in the right amount and at the right time that constitutes proper function in TRU, TRUD. Since TRU is so precise, precious, and right, and is so tremendously needed, we must enhance our capacity and ability, know how to and achieve the highest levels of concentration, strategy and tactics to empower and accomplish all for TRU. Activism like so much in the dynamics of life, can only be applied just right, fine tuned, and in measured quantities perfectly qualified, by TRU.

Humans are endowed with glorious and renewable capacities and powers of self preservation and fulfillment of life. Yet unlike other animals we are also blessed with much higher states and levels of mass energy. This consist of a proliferation of synapses involved in acquiring and using knowledge. This is called learning, one of the greatest of all our attributes in mass energy Á FAF. It is what we take for granted and as students often feel so overwhelmed by. For the body of knowledge is vast and disordered. It is poorly recognized, and even more poorly ordered as a whole, as it exists without TRU presentation. To avoid being discouraged, or permanently, intermittently or temporarily diverted from this fulfillment we need to make learning as enjoyable and successful as possible. This requires an appreciation for the immense value of knowledge and how the game of learning can be made as fulfilling and productive of happiness as possible. This is a major reason why the computer industry has been successful in appealing to young people. In game theory the player so orders and operates the variables to bring all to successful conclusion, a win. It is not only the student, but browser and server that are at the helm. The game of learning must be gauged to the level of its players. This is what the 21st century will need to solve its problems. These include self interest, waste and pollution, limited goals, nationalism, climatic events,, religious conflicts, overpopulation, ATRU distractions, lack of appreciation for life's worth, what is best for the world, and what is best in others, opportunity for all less fortunate, indebtedness, personal, group and national, crime, terrorism, and no or poor command and control of the world. The method of meeting these problems largely resides in each and all persons together, in TRU. We only need to know what they are, their causes, prevention, removal of causes, and how best managed and controlled. We do this by making a game of knowledge and finding the great joy and happiness in seeing all come together, applied and operated too win in this game of life. We win, not by devious means but, by calculated ordered reason, how our gains are the worlds, and the worlds ours. It is a total effort by one and all best fulfilling their roles in TES and TIT. This is done with inspiration, stimulation, imagination, compensation, self sacrifice, revelation, complements, learning how and what to love, organizing or adapting to the order of the world, ignoring most that is of no value or negative, and with probity eliminating all that is or apt to be harmful. The solution to these problems lies in preparation, prevention and a united total interest, accounting, and effort. This is by seeing that they are resolved, or held to their proper place in life, TRU. For all must realize that many of these problems are more or less a part of the D or Differences in DAOST of TRU, and can not, nor should they be, eliminated, only controlled, adapted to, or balanced. In different ways they are a part of life, the dualism and counteracting forces of life. Many of their solutions will come from the prevention and control of the abnormal extremes, e.g. idealism and materialism, integration and differentiation, uniformity and diversity, right of life and right of choice, creationism and evolutionism, conservatism and liberalism, democracy and constitutionalism, equanimity and enthusiasm, racism and eclecticism, sexism and love. These as many other and all life's problems will be resolved by an identity with the success of this best of worlds. They must be guided not by any intermediary but by all SÁ, how they exist in TIT. This means Quality. Associated with quality, the star that guides and leads this role, should be Prevention. For much of quality can't be achieved immediately, but in time, not directly, but indirectly by prevention. This is largely through education, learning TRU. Happiness in life comes from recognizing our limits and the fulfilment that exists within ES, CES, and TES, or counterparts in the best of worlds, TIT. Just as in an *Imitation of Christ* written 700 years ago in the Netherlands, we can largely extent accomplish TES and TRU by imitation. This settles the greater question, how to learn to live with and treat each other in moderation and toleration, especially when guided by the best, TREQ or TRU. It is how we can make a game of learning that the greatest happiness in life can be found. Joy of books and study, self sacrifice and compensation are, much of, life. Our urges don't need immediate fulfillment only moderately as part of the whole. Just as Greek Philosophy attributed the highest good to be reason, we can only produce this if we have the knowledge, interest, anticipation, and inspiration necessary, and the training and practice to relate all this knowledge together as it revealingly represents our world, and its best exercise by TRU. Half of the fun is in this game of learning, a kind of free association of acquiring knowledge, just as what is so delicious to our taste of food, is this food for thought, knowledge to our souls, TES. The happiness in knowledge and its learning experience is its game of putting it all together, using, applying, and making it win. This is like the goodness and divinity of reason and wisdom, Sancti Sophia. As we make learning a game, acquire and order more and more knowledge of what is most important in the world it will best help fulfill our counterpart, TES. It will then be easier to help ourselves, others and all the rest of the world to do the same.

TOPA = Test of Primary Application to breakthroughs of problems. TOPA signifies concentration and focus on the great new ideas, discoveries and breakthroughs that have been developed. It is these when addressed to major and crucial problems and applied by all the best means at our disposal that TRU control and correction can gloriously succeed and flourish. It is TOPA that has so massively, cooperatively, and readily expanded knowledge, SÁ and TRU. This is especially so for the technology that has enabled our means to represent, manage and control TRU, as in human genomics and genetics. It must be massive, accurate, and with great precision. It must be accompanied by interpretation, guided by VAPO and TRU. TOPA is a key method in technology, business and industry. It is what provides quality by placing maximum emphasis upon research and development to yield the best possible products and competitive edge. It spearheads corporate profits and is the great new generator of human life, TRU.

STOD = Singular Solution Total Discrepancy. STOD is seen in our personal tendency and determination to find specifics, singular and simple cures, answers, and solutions to ailments, questions, and problems in life. We by nature are focused on the here, now and what is given or the needs, wants and problems of daily life. Yet TRU is constructed much differently, i.e. by unit, finite, infinite, total, static, functional and absolute. Thereby there is an ongoing discrepancy between ones existence and TRU. STOD is much of what keeps TRUD in a highly active and dialectic form. The totality of TRU largely proceeds from the same of DAOST. The closest thing to satisfying this demand for specific, singular and simple solutions is to be found in TU, SUT and the order they provide for representing TRU. This is shown in REM, ALT and SPALT. Through these the finer and finer differential bases of TRU can be managed.

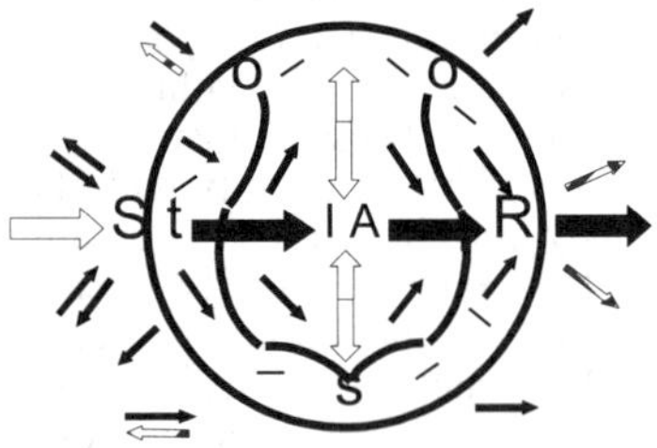

Fig 4-91 **ACTIONS, REACTIONS and PROJECTIONS in the St-R, S-IA-O SETTING of TRU**

Projection in life is more than a 1:1 stimulus response, IA Thought-Action situation or operation. For every trouble, problem, mistake, wrong, injury, conflict, and war there must be, by LASUR, some reason or explanation. And for these reasons and explanations there should be some prevention or action that could have been taken before they got out of hand, or would have averted the crisis or failure. This may be both obvious on the one hand or criticized on the other. Yet it is a lesson that is hard to learn and often neglected. The older we get in life the more we realize that to get along well, to save ourselves from injury, trouble, woe, suffering and misery, it is necessary to prevent, prepare, protect, maintain, act, manage and control as much in and of the world as possible in all ways that best meet oncoming adversaries, dangers and destructive forces.(PPMA) This is what we see throughout HOM and apply in ACC. Perfected accentuation or ACC, is the great formation and management of all these actions through optimal counteraction, reaction and action in summation. The ratio of present to perfect accentuation is such that the higher levels of SÁ and S-R are attained by PPMA, the total ACC for TRU. It is like the total salvation of all in God. We are safe, healthy, prosper and succeed indefinitely only to the degree we know and practice TRU, TES, HOM and ACC. This is what Á and SÁ are in TIT, by TRIA, and by definition. It is in this way each and all persons can make the best use of all needed to get it done, the fulfillment of life as TRU

Anticipatory guidance is a phrase from Pediatrics that is highly indicative of what is needed more generally in life, TRU, and ACC. It is one of many continuing streams of ideas that are highly popular, yet add to our distinctions, and provide a better cutting edge for POT. They also help to provide further nuances by increasing separation of the different approaches many fields and sciences have to TRU. Such nuances raise representation to a higher level by identifying and reproducing the deepest meanings of life, Homation, and TRU. Anticipatory guidance is especially important in ACC by not only emphasizing guidance, which is a major theme for the future. It is a guidance that anticipates what is wrong, needed, required and in order. Unless we can, like the directors of great symphonies and endeavors, anticipate in advance and in an ordered, proportionate, actual, valid and total way, we can never achieve the action needed in life, nor can ACC be adequate or appropriate for TRU.(VAPO) portion is spent and used on what is of little or no benefit to TRU, One of the greatest problems and causes of our day, especially in the more affluent nations is waste. What is done with all the time, money, property, and material wealth earned and accumulated. What it is spent on and how it is used. A large in fact much is a detriment. A large portion is spent on AMTRU, NAMTRU and ATRU,, what is of a very low percentage and efficiency of TRU. Only a small amount is spent on higher levels or what is fully TRU. This is the reason why the bible so condemns wealth and extols poverty. For wealth has a bad influence on individuals, and leads them both toward wasting affluence, and even worse, displaces what might otherwise have been TRU. On the other side of this, some wealth creates, and is used, efficiently, wisely and beneficially. The modern world is largely products of inventions, industries, communication, and commerce that are highly TRU. In common life it is a very fine line that separates what is completely TRU, mostly TRU, little TRU, not TRU, negative and harmful to TRU.

AI = Á or Á (S-O) Intensity, Infinity. AI signifies not only the intensity found in depth conception, OM, IATES, AITES, IMTES, SIA, SÂ and TÂ, but the very massive MARS based upon completely knowledgeable mechanisms, how to effectively plan all needed for ACC. It is a:

SUBJECTIVE INTENSITY . X = X . OBJECTIVE INTENSITY

SIO = AI or A (S-O) Intensity Infinity

This converges on and becomes Summative Á to SÂ to TÂ. It is astounding how poorly each person comprehends and applies themselves to the world in which they live. Most are meek, weak, vain, and devious if only unintentionally. It is because they abuse themselves and their kind, and ignore the best of knowledge that is available to them, that they only through collective effort seem to be successful and produce a flourishing culture. Throughout history we repeatedly make the same old mistakes.We propagate the unfit, we abuse our bodies, we think and do what pleases us, we promote and stubbornly hold to false idols, we disregard and mistreat what is good, we often champion what is bad in our fellow man, and ignore or deny what it actually takes for human life and survival.

Instead of being soft, defective, and mindless we must in the short years of our maturity avail ourselves of all that life has to offer, and focus on all that enables this special universe of best human life to flourish and survive. Ups and downs, growth and decay, and periods of prosperity and depression are the natural occurrences of life and with us both partially and universally. Yet we can PPMA them to both manage and control our life as we continue to make it better. This is the reason for IPT, POT, TES, TROS, TRIS, TRIT, OBT, ROT, MOT, SPALT, POS, ACC, PUREACLITY, KUPS, COGNOS, ELTICMA, SÂ and TÂ. We must discover, learn, remember, apply, and uphold ACEXIN and the Positive Imperative that life offers for our place and activities in TRU.

MEASURES of CHAPT 91 ACC NEC POSITS

Term	Measure		
ACTIVISM (ACT)	~7-3-6~	M	- 4
AI	~5~-~-~	M	3
ANTICIPATORY GUIDANCE (AnG)	~4--~5-4~	T	± 1,3
Attack, Apt Destroy TRU (ADT)	~7--~4-7~	M	±4-7
BAB	~1--~8--~	S	- 4
BAC	~3--~6~5~	S	- 4
CAPONT	~7--~4-6~	M	- 2
CATCH 22, Life Situation (CSL)	4-5-6~	M	-4,2
COC	~-~-~-~	S	- 4
COMPARATIVE DOT and DAC (CDD)	~5-4~-~	S	±8,6
DAC	~6~4--~6~	T	5,7
EVA	~5--~5-3~	S	- 4
FAH	~4--~5-5~	S	- 4
FALE	~3--~5-6~	S	- 4
FATDAP	~-~-~-~	S	- 4
FEXPRE	~3--~6-6~	S	- 4
FINAL CAUSE(FiC)	~5--~5-9~	M	3,7
LACRUN	4+-5-7	M	- 5
LAPUS	~2--~6-6~	S	- 4
LATAR	~5--~2-7~	S	- 4
LATNU	6- ~ 4	S	- 6
LATUN	~ ~ ~	T	- 7
LEARNING, GAME of, ACC, TRU LEA	2-7-6	T	1-7
LICON	~4--~3-2~	S	- 4
LOMI	~4--~6-5~	S	- 4
LOMT	~6--~2-7~	T	7
PREDICTION (PrD)	2--~4-8~	M	± ~
PREVENTION (PrV)	~6--~5-7~	T	1
PROJECTION (PrJ)	~4--~3-7~	T	5,7
PROSÁ	~5--~6-6~	M	±8,6
QUALITY (QUA)	~-~-~-~	M	~ 1
RECIDIVISM (REC)	~3--~6-6~	T	- 4
SCEM (EMPATHY, ACC and TRU)	~2--~7-5~	T	±4-3
SIO	~4-6~-6	T	3
Stand for Right, Best, T (SRB)	~5-6~-5~	T	1,7
STOD	~5-5~-6~	M	- 4
TEF	~4-5~-8~	M	5,7
TOG	~3-3~-6~	S	- 8
TOPA	~4-5~-6~	M	1
UT	~-~-~-~	S	- 4~

Fig 5-91 **ACC DAC NEC POS in SPALT**

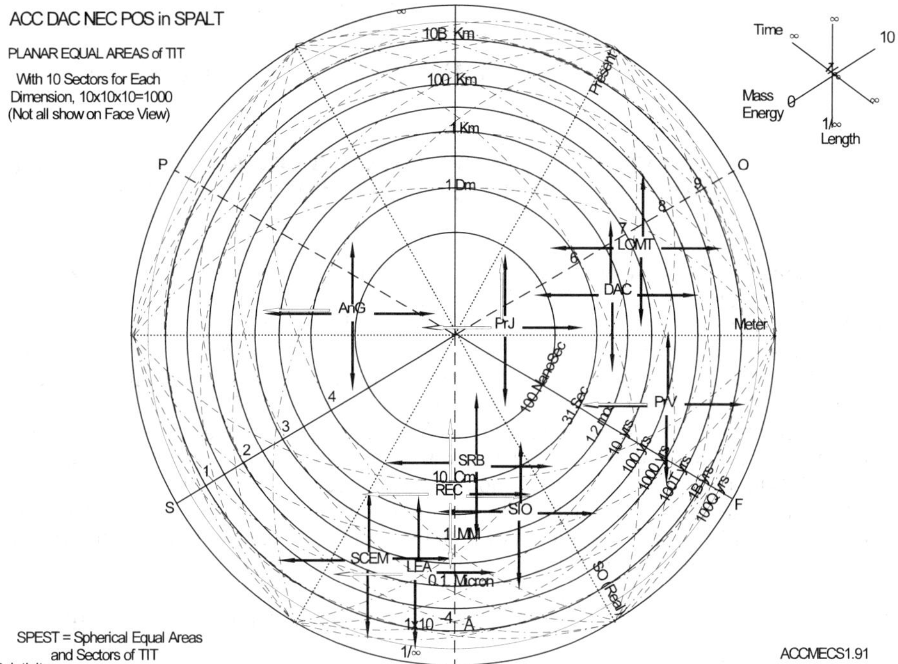

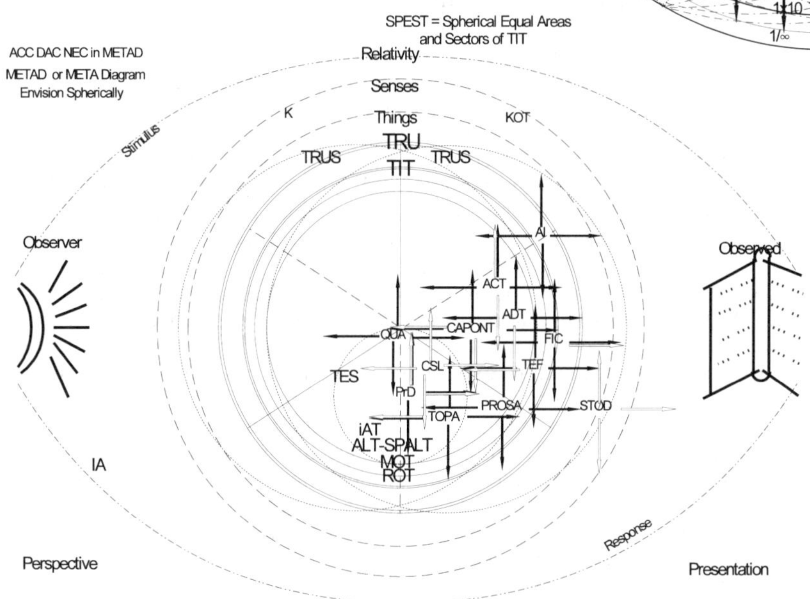

Fig 6-91 **ACC DAC NEC POS in METAD**

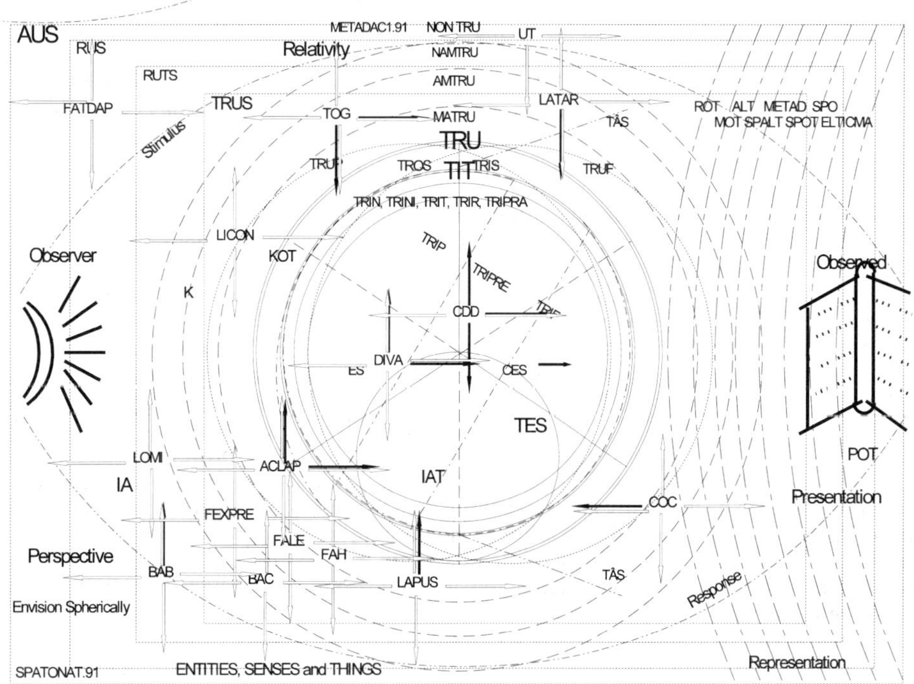

Fig 7-91 **ACC DAC NEC POS in SPAT**

CHAPTER 92

GENERAL FORM of ACC or ACCENTUATION: ACÄ

PREPARATION, INTERVENTION, CONTRIBUTION, SUPPORT, PREVENTION CORRECTION, CHARITY, CARE, LIMITS, STATES, SUCCESS, PPMA, NEEDS

It is important that we know what is from or against TRU, and what is ATRU and ANTRU. For, although TRU and its ENT are largely the living of it or all that is increasingly REM and positive, it is what is decreasing and negative to TRU that ACC needs to pay added attention. This is to better identify and correct them. They are not the same, and thereby need to be treated differently. Also, through learning and correcting them we can better understand TES and TRU. How we are not or not much TRU and how TRU is made more distinct when separated from them. These accompany the many concepts, principles, and laws by which we must live. What is ATRU and ANTRU can be of any, or of a specified, duration. ANTITRU is largely what we focus upon in TRIX, ANTRU in TREX, and ATRU in TRAX, although there is considerable overlap. When the target of its attacks are, and when especially focused on, TRE of TIT, they become TANTRU.

It is not so much that this or that faith, belief, hero, champion or God is wrong or are in striking conflict with what is best, TREQ or TRU. It is the extent to which they are right, and dependent upon their part, value, and role in TRU. This is very much so for individuals. Our interests and ideals may be perfect, yet in practice how quickly we forget all we know, acclaim, and impulsively say and do first this or that which is trivial, negligible or harmful. Thereby we must not only know and appreciate key systems, schools of thought, religions, and sciences. We must bring this knowledge, as all that is TREQ to TRU and use it consistently and continuously in the best way. We should do so recognizing both their truths and errors, values, disvalues, uses and disuses by best applying them to all that is so good and great in the world, TREQ, TRU, and TA. We should at the same time be forever aware of and remember - ECRO - with great conviction and determination that each and all thoughts and acts must be for, perfect TRU. We can not live and succeed if we only on occasion acclaim TREQ and TRU and think and do as we please or are influenced the rest of the time. For although it is 'work that makes life pleasant' it must be effective work for TRU and TES. This is especially so if we can appreciate and find happiness in this work for what it is, the service and enhancement of TRU. Part is our duty to the past and all that it has provided, part is in expanding what we have and benefitted from in the present, and part is for the future of TRU as well as ourselves in the combined absolute it assures.

A good technician must be an ever better fundamentalist. This holds in general in TRU, as well as in most all fields, e.g the stock market. It is for this reason we must discover, learn, remember, show and use TRU in its utmost profound, appropriate, and subtle ways. We must also inspire ourselves and all others, to rise to the necessary level of competence and power in life, from the smallest IA to its greatest massive coordinated totality.

When we live in a world of TRU it is often easier to be TRU than not to be TRU. It is easier to want to be TRU than to be TRU. This is: **ROV** = Ratio of Ontology or being, and Volition. Wanting to be is usually far from being. For TRU the worth of ones being and the will to assure it should both be high and complementary. Yet in life we often fleetingly have more unsustained volition than we have being, and a fleeting worth of being than any real and actual constancy of TES and TRU. When achieved this is: **COBSWAT** = Constancy of Sustained Being, Worth of Being, Will, and Will to be TRU. It is COBSWAT, known and accepted or not, that is at the heart of all of our lives, our choices, to be TES or not, DAK, TRU, not ATRU. This is a great propelling and compelling force built into us by TRU to continue as the live product and vehicles of TRU that sees its actuation in service and endeavor in TES and TRU. Will to TRU can be spontaneous from TES, trained, cultivated, habitual or routine. When all together, it is the greatest blessing of TES for TIT.

The art and science of strategy and tactics, led by military combat, have been increasingly developed and applied to many fields. Much of this is applied to TRU and ACC. One tactic that brings those who are refractory to TRU is fear. Although fear has much less place in TRIF, to compensate for what is not or a threat to current TRU, it can be helpful. Fear can be both a good method and great benefit at the appropriate time and place. For humans will often be moved by nothing short of fear. Religions have long practiced fear as a means of attaining supplication and devotion. This has been largely displaced by love, understanding, reason and faith. It does have a limited place in ACC when the negative, irrational, incorrigible and ATRU will not be TES, TREQ or TRU. They will be TRU, by VAPO, when appropriately made to fear the consequences. To do so, as is evident in appealing to those whose bad habits ruin their health, is to scare them until they recognize what is best for them, TRU. Fear is a readiness state in which all ones being is mobilized for defense and survival. Fear can be highly effective if followed up by a deep feeling of participation, experience, inspiration, stimulation, incentive, conviction, determination, love and joy in living a full yet self managed life that is most beneficial for the individual and for all, as TRU. Similar tactics are the sparkplug of ACC. We all move reluctantly if that which impacts us does not suit us. Our compliance is poor unless the need is right upon us. It is only when what strikes such terror into us threatens our very existence or what we most cherish that we adapt, change, and come to our senses to do what must and will be done, TRU.

FARP = Fallacy of Radical Partiality, or ATRU and ATES extremism. The reasons for being radical are often a necessary reaction to the pressures within a group, nation, region or world. They seek to magnify their causes and actions to produce the greatest effect. This produces disproportion. It becomes even further imbalanced in conversation, heralded by the press, TV, communication and the news media. It is a very fine line that divides FARP from the massive dialectic oppositions that occur naturally and fully within TRU. To correct them calls for a full accounting, across the board effort, and accentuation that is provided by TRU and POT.

It is crucial to know what produces extinction, or, 1/TRIA. Over how long a period of time is this likely to occur? It is when our capacities and abilities are no longer adequate to meet our needs or become so far removed from meeting the demands and threats of all that is upon us that we are overpowered. Humans like all other animals have many causes that can lead to extinction. It is in how we are different, unique and special that our added vulnerabilities make the need to prevent extinction even more urgent. In us there is the potential for and powers of self destruction or self determination. This is called: **PODES** = human Potential and Power of Self Determination. It is PODES that shows the highest levels of Á FAF that are the cutting edge, and will be the salvation of our being. It is PODES that is such a propelling and even greater compelling force. For our being is us, we are many, we have a heavy load of support, supply, and demand whose determining forces as well as our own thisness is much of the higher levels of mass energy, or Á.

$\downarrow$Cp/C$\uparrow$ x $\downarrow$Ab/$\uparrow$S = Extinction

It is where decreasing capacity and ability/increasing stress and conditions so expose our vulnerability to extinction in all portions of life, as the GM-LP-DF system and all its intermediaries that the healthy equilibrium in CARSC and TIT-TO is so altered that it is no longer able sustain itself.

It is much easier to start with TRU and keep it up, than to make or turn AU, ANTRU and ATRU into TRU. To start with degeneracy, defects, disease, deformity and make it right is a divine and merciful goal. Yet it is not to be extolled before what is already TRU, e.g. preventive medicine and public health. This especially applies to what has been most neglected, ones mental and spiritual health, or how well one exercises their brain, intellect, spirit, and soul by TES and TRU. This is the key to ACC and the justification and proof of POT and IPTRU. It is therefor imperative, as many teach and learn, that we first must take care of what we are given in life to get all TRU. **TREQ** = TRU Equivalent or TRU yet not called or known to be TRU. When recognized as TRU it is called: **TRE**. When not TRE, yet functioning as becoming or in a more or less equivalence it is called: **EQUT**. TREQ is used for existing terms, descriptions and all that hold much in common with TRU. TREQ signifies all that exists in the world, today, present, past and future that is of this special universe of best human life, yet not TRE, or LANT. We must learn and encourage all that enables a fusion of EQUT, TREQ and TRE with TIT. We must be tolerant and work with all TREQ and expect and encourage the same from it.. A science of TRU may be signified by: **SCINT** = Science in and Recognized in TRU, **SCIN** = Science In TRU, but not necessarily recognized as TRU, or to a lesser extent TREQ. **SCIT** = Science In and of Itself. SCIN is much of what we know as science today. Especially when it is shed of all that is not its ESIN. What is in the social lives of man, sociology, that is EQUT, equivalent to TRU but not TRE, it may be likewise classified, by: **SOCINT, SOCIN, SOCIT** = Social Sciences or Sociology in TRU, in TRU but not recognizes, and in and of themselves. The latter may be somewhat TRU, but highly non TRU by being focused not on TRU, but on itself. Much of human past, present and even more hopefully the future becoming TRU is: **SOCET** = Social Sciences or Sociology becoming TRU, the equivalent for science in general is: **SCET** = Science becoming TRU. With each and all of these concepts of science we should be careful to be tolerant and fully utilize all that so has and can greatly extol TRU. With the most complete vision of TRU this is possible, by KOT, OM, ACEXIN, ALT, SPALT, COGNOS, ELTICMA, SÂ, and TÂ.

ACC, as all TRU, has much in common with theology, ethics, and God. To utilize them we have to make them formally distinct, especially from TRU terminology. When in lowercase, god, we signify any and all senses of God. When God, it defines that existing concept common to all great religions, of the supreme and ultimate reality or deity. God so conceived is the ESINS of god, in part TREQ, a TRU equivalent. As TREQ is has much in common with TRU yet lacking many essential aspects, modes, classes, parts, processes, structure, proportion and totality of TRU.(<VAPO) When fully capitalized, GOD, as we do for other existing terms, we signify full identity with TRU or some aspect or part of TRU. **GID** = God defect, how God is less than TRU or what God lacks, as noted above, that TRU has. Any and all ways God, and more god, is less, other, or negative to TRU. There are many aspects of God, that have their comparative natures in TRU Terminology, as shown in their respective chapters in IPTRU: 1. Object of reverence, TRE, 10,61,71,81, 2. Holy of Holies, Focus 137,139,99, 3. Director of all from afar 11+,179,18,5+,81, or from within, HOM, 88,90,99, 4. The all, TO, 137,160,165,89,98, 5. The 'Creator', ATO, 11+,111,112,81, 6. Reality, Á, 08,119,124,147, 7. The Absolute Spirit, TÂ, 139,99, 8. Good over Evil, +/-, 175,575,6-9+, 9. Cosmos/Chaos, TRU/AU, 11+,179,18+, 10. Life/Death, TRU/AU, 11+,13+,5-9+. Yet TIT-TO is through and through, TAT, with a TRU perspective that is OITIO.

Among some of the existing terms that contribute to ACC and KAC are, work, ethics, common virtues of kindness, honesty, goodness, fairness, and the best of business principles. These and many others are the simple and common aspects, modes, form, function, FAF, ES and CES in or of TRU which have, are and will make it most successful. They are guiding ideals, whose recognition at various levels of intensity, provide the magic powers for mankind. They are formed by and are the vehicles of the Á state and dimension of TRU. They are ordered by TU, SU, ROT, MOT, ALT, SPALT, POS, STA, HOM, DYNU, DYNS and DYNO to give ACC uniformity and closure in TIT-TO. Only as they are VAPO and SCARLED to TRU and utilized in a like manner can we fully utilize all that has gone before us and all that is of us to best enable that which will perpetually be more of us.(COPOS)

An ethos is the spirit or underlying sentiment of a culture. It largely signifies the moral determinant of action. The ethos of TRU promotes a great part of the formal makeup and dynamics of TIT and TRU. If the ethos is highly TRU it greatly increases the power of a culture. If it is poor TRU suffers. What ethos is TREQ needs to be supported, what ethos of TRU that is lacking needs to be developed and nourished. It needs to anticipate and avert digression. Digression and deviation are large negative aspects of thought, speech, writing, argument, study, or work. Digression can be positive. It is usually thought of as being negative, a cause of loss of attention or concentration. However, it may be a change to a higher, broader, or more TRU level of attention. For what it is on may be ENIT progressing to become ESINS, ES and TRU. The prevention and correction of all digression that is negative is what helps to maintain the ethos, the should be TRU concerns of a culture or civilization..

Prevention is large aspect of applied technology. Science isn't only of current or limited interest. It works through and through all time - the far, mid, and near past, present, and future. Each works in many different ways to prevent loss in the others. This is more apt to be by finer and finer gradients, or reversed indirectly through forces and actions in any direction. They are presented to remind us that prevention is not a simple matter of what we do this moment to keep from injury, illness or loss. Prevention shows how much and most all TRU works and can be used in ACC. It usually shows only a part of it. For what works through and through in TRU is what gives it completion and fulfillment. The VAPO of these forces in TRU is called: **FORTOP** = Forces of TRU, Total and Proportionate, and their equilibrium. FORTOP, more than the operation of forces only in time from near past through the near future, includes all those forces that work in reverse, as from the future to the past. These are to be added to those other various mechanisms which act to provide the dynamics of TRU, TRUD.(PPMA) The relationship between the states and the parts or processes different factors in TRU play is called: **LASDIPT** = Law of Altered States with Different Posits in TE, in TRU. It is by LASDIPT that we can better understand TRU, PIP and all that comes before us. Prevention is not only for a single episode in the present, or for someone else to have done in the past, or will blindly do in the future. Prevention is the totality, one and many mechanisms which act to provide the dynamics of TRU, functioning by contributing, enhancing, ACC, and PPMA throughout TRU.

By learning what the means and extremes are for the great continua, mechanisms, processes, and courses of TRU we can better recognize deviations from normal, and from TRU. Their summation, plus that of all management, control and treatment, is highly effective in providing health. This is a health that is not only of highest quality physically and mentally, but one that is real, or has the highest correspondence of the physical and mental, observer with the observed, or Á FAF in TIT. This provides its highest actuality, identity, and participation in TRU. It is such a combined and basic form of accentuating TRU, PUREALITY, that is a prime producer of TES and TRU.

The many misconceptions, erroneous beliefs, ISMS, prejudices, and fallacies should be corrected, put to rest, or restored to their proper meanings. Each has a much different, and often destructive, meaning in and effect on TRU. It is for us to discover learn and apply TRU, then recognize these failings and see how they are wrong and can be best remedied or avoided. With TIT, TRIT, OBT, SUT, ROT, MOT, ALT, SPALT, POS, STA and HOM a comprehensive, effective map, plan and means can be developed for ACC to provide reduction to TRU, for TES..

LATSCAP = Law of Corrections and Solutions to Attitude Problems. The greatest cause of ATRU lies in the attitudes of individuals and peoples. If their attitudes are bizarre, narrow, negative, lack VAPO and all that sustains TRU then it is hard for them to become TES, or even to be successful for long. It is in the cultivation of optimal attitudes on the part of each and all persons that TES and TRU can be most readily attained. Attitudes, like all else in our lives, tend to be an orienting, coordinating and ordering of our primary form and function, mass energy Á dimension and state, to best assure survival and being, TRU.

No rule, regulation, legislation, law, or routine can make up or substitute for a diminishing degree of intelligence, wisdom, and interest to uphold the welfare of the world in which we live, TRU. One's civic duty should come from ever increasing attitudes, intelligence, interest, conviction, and endeavor in that best for TRU. We must be careful to correct or eliminate all rules, regulations, legislation, routines, and laws that are inadequate, outdated, worthless and harmful, and replace them, not with ever more regulations, but with increased intelligence, interest, KOT, COAT, and participation in the common welfare, for and of TRU. This can be done only when each and all understand, love, and have an undying interest and conviction in all that are basic, fundamental, and most important.(TRU,TES, REM) They are those principles, FAF, ES, CES, mechanisms, operations, and NEC of TRU or TREQ.

DABU = Doctrine of abuses, deviations and misuses of life. DABU should be held up for all to see and for the judgement of all and TRU. To be made so that their punishment will both separate them out, as well as provide an additional stimulus for TRU. DABU like sin is all about us. To be controlled it requires a strong focus, wisdom, and will. This is what TRU and ACC can correctly provide.

	A	AN	ANTI
TRAX	Attention to, prevention of, or elimination of all that may be lost from what is good for mankind	Attention to, prevention of, or elimination of all that may be against what is best for mankind	Attention to, prevention of, or elimination of the worst threats and dangers to the success or survival of mankind
TREX	Attention to, prevention of, or elimination of all that causes a loss to any and all work for TRU	Attention to, prevention of, or elimination of all that may be against what is best for TRU	Attention to, prevention or elimination of that most against and the worst dangers to the success or survival of TRU
TRIX	Attention to, prevention of, or elimination of all that causes a loss to a concentrated effort on advancing TRE and TIT	Attention to, prevention of, or elimination of all that may be against what is best of TRU and TIT-TO as TRE	Attention to, prevention or elimination of that most against and all the worst dangers to the success of TIT-TO, TRU, as TRE

Table 1-92 Plans of Action for Levels of Cause, Reason, Problem or Threat

Table 1-92 shows the relations between causal factors in this and the next two chapters, or ACC form general, special, and targeted. TRAX, TREX and TRIX are a large part of each of these three aspects of ACC. They show some of the more direct ways ACC contributes to the competition, selection and quality of TRU. The different states of entities and being from, generally against and specifically against TRU are shown. Among those entities and being that are not TRU most are not included in this diagram. It is only when such a neutral state becomes negative and needs to be acted upon or at least under control in one or another way that ACC becomes operative. TRAX stands for those states in which TRU is accentuated or accomplished by a knowledge of the world, God, life, civilization, religion, philosophy, science, and technology, but not known as TRU. TRAX is more often TREQ and to a decreasing extent of the TRU in AMTRU. TREX stands for ACC in which TRU is known or recognized, yet the actions upon it are not necessarily or purposively against TRU. It is directed to all that benefits TRU, in the special way of ACC. This is often from the general standpoint of the ESINS, ES, CES, and FAF of TRU. TRIX stands for ACC in which TIT is accentuated and accomplished directly through a knowledge of TRU itself. This is always with TRE and focuses on the success of TRU based upon the performance of specific and concentrated objectives. These are guided by the gist and cue of its mechanisms that refer to the ACC of TRU as a means to TRU. TRIX focuses on, and is largely what combats its worstadversaries, what is TANTITRU, TANTRU, ANTI-TRU, and much of ANTRU. **TRIXÄF** stands for the management, control, correction and elimination of all these that are purposively against the recognized, TRE, of TIT.

Some of the general forms of ACC that are means to TRU include: unconscious and conscious TRU, indirect and direct TRU, automatic and deliberate TRU, extemporaneous or routine TRU, unintentional or intentional TRU, involuntary and voluntary TRU, unplanned and planned TRU, spontaneous and controlled TRU, prevention and action in TRU. These means are FAF of both TRU and ACC.

The great, beautiful, perfect, and the good, must be forever fashionable. We must all learn, early and long, to hold most deeply in mind, remember to uphold, and keep our interest highest on what can always be most fashionable, TRU. Keeping TRU fashionable permanently, is the most perfect, greatest, best, most successful, glorious, and victorious, way of achieving, enabling, and sustaining TRU. All that is the opposite of, or to the degree it diminishes this focus is, to that degree, not only void of being what should be, it is negative, damaging and of failure for all. It is for this reason all persons, especially our young as they grow to learn and live, do so by all means that are most TRU. This includes making a game of, and finding happiness in, a life of achieving TRU. Such learning and practice needs to be not only from the top down, but from the bottom up, from one end to the other, and all in between in totality, OITIO, in and of TRU. This is: **LAPOFT** = Law of the Learning and Practice of all TRU, OITIO, as it is most Fashionable to do so. LAPOFT can and should be done with the best possible presentation, vigorous promotion, success and happiness that TRU affords. It should not be done at the expense of any or all that supports ENT,TRU, TIT-TO and ACC..

	PAY ALL	PAY NOTHING
WANT and SEEK ALL	This is the best, providing the payment and receipt of an appropriate amount of TRU are equal	Life costs us. If we pay or are willing to pay nothing for it we deserve and will usually get little or nothing in return
WANT and SEEK NOTHING	If we pay all for little/nothing we are wasted. Yet this is what much of the world continuously does, especially the affluent	To pay nothing, and want and seek nothing is tantamount to nothing be it not TRU or TRU

Table 2-92 Want and Seek All or Nothing

There are persons or groups who momentarily, for long periods, or for a lifetime, want and seek the whole world. Yet they are unwilling to pay anything for it. They take it for granted and fail to recognize its infinite and absolute worth. These may be compared with those who want what is nothing, or not TRU, yet are willing to pay everything for it. Of these, neither are right, both are extremely wrong. Both have much in common as diverse as they are, especially in lesser extremes, and are measures of the ATRU and ANTRU of humans, in our society, and times. What we should be willing to pay for is the best. What we pay for that which is worthless and bad is nothing or for correction. We can enhance the quality, achievement, production of all that is the best and of greatest quality, by TRU. Only as we, and all, have an ever increasing level of understanding of, dedication to, conviction in, and devotion to what is the best, of greatest quality and of highest value in TIT-TO will we escape from paying too much, paying to little, or paying for what is actually worthless, harmful and destructive. All need to distinguish whether that wanted or sought is TRU, how much, and in what way. It is not possible without an adequate knowledge of TREQ or TRU. Such an analysis is key to IA, Á, and life, and provides a good general form for the structure of ACC, and test of TES and TRU.

The primary limitation to the greatness of TRU and TE is one's inability, disinclination, lack of opportunity and unwillingness to comprehend and act by TRU. Even when man is given knowledge of TREQ these limitations persist. One of the greatest general forms of ACC then, is to simply eliminate or greatly reduce these limiting factors to TRU. **WOT** = Way of TRU. The great impetus to TRU is what works well, that is for, and by us, as we are TRU. It is largely the way, like Tao, that can effectively, efficiently, and actually work. These ways are many as both provided through, and shown in history and the whole course of human development. Some work well in their own worlds. Yet these worlds are largely ENIT, or of poor NAMTRU. There are some that even work for segments of TRU, or TREQ, yet lack VAPO and an adequate basis in TRU. It is necessary that each person be their counterparts in TRU, TES, and that each group, community, nation, and race be their ES and CES in TRU.

How TRU and ALT are divided by various durations of time, so are they helpful for ACC. The future may be divided in various ways. For ACC however, it is best to primarily follow the divisions of ALT for the future. For any other will depart from the actual form and structure of TIT-TO. They must adapt and account for midpoint future. The importance, often unrealized, for IA can be seen from the need for a focus or starting point in ACC. It may, as well, start anywhere in TRU. It shows how we can bring TI and TE together as well as TROS and TRU. The future must be in equilibrium with the past, as well as past, present, and, future together in relative equality both in continuity and duration. How much this is lacking is apt to be an error in ROT, MOT, ALT, and SPALT rather than TIT.

TES, TRU and TIT are of absolute importance as the prime aspects of this special universe of best human life. On the other side, and to reverse the negative, it is of similar importance to counterbalance those aspects of human nature that are inherently selfish, unappreciative, and destructive.. Only when we can fully and comprehensively discover, know and apply TRU as exemplified by the contrast between different goals in life can we understand and manage our world properly. Is it for how much money or power we can attain? Is it for security? Is it for survival? Is it for quality of life? Or is it for any number of other needs and goals that show us what we must work and live for? These are similar for all of us, especially when being our part, TES in TRU. How active we are, and how well our activities benefit TES and TRU determine how much we reach these goals.

Karma and its resolution in dharmaare, like TES, those thoughts, actions, causes, effects that a person has and does that have eternal existence. Like the laws, SCARLS and all that has, is and will make for and assure an increasing life of positive participation in the world. Buddha was an avatar of Brahma, much as Christ was of God. In this way it is somewhat related to the soul, holy spirit, and god as TES does in TRU. It was the hope of release from oppression in the Chinese and Japanese in the first millennia of the Christian era that Buddhism so powerfully struck such strong following. In these and various other religions positive forces conducive, constructive and for the betterment of mankind, identify the great powers of TREQ and TRU. Different religions have interpreted karma, as much in language, in somewhat different ways. There can be both good and bad Karma, similar to the central idea of Zoroastrianism dualism. Bad karma occurs when the forces of ones actions are such that one is not released from the perpetual cycles of existence and suffering. Yet it is the positive and constructive forces that best correlate with TRU through TREQ. They illustrate the dynamic, active and actuality aspects of TRU. They are to a great extent coterminal with Á or the mass energy form and function, state and dimension of TRU. This is emphasized in the AC of ACEXIN and ACC of Accentuation. It is shown by how ACC signifies the actuality of these acts, their action, cause and effect, law and dynamics. These provide the differential and finite basis, as in CES and mechanisms, that exists in the thisness and absolute of TRU.

Protection of life and TRU is not a random protection of all. (PRUT).What is needed is a selective protection and control of what is, 1.Best for humanity and the world, or TREQ, 2. Most TRU, and, 3. Most threatening to TRU. This must be comprehensive, vigilant, vigorous, with conviction, relentless, and successfully cover all contingencies and priorities that can be turned from the negative to the positive in TRU. Only when we thoroughly understand TRU, and can utilize this understanding in highly accurate and reliable ways by valuating, modulating, and positing can we produce and achieve the protection necessary. This is done largely by fully using all the VAPO and SCARLS that apply.

What we mostly need to do in life is to help nature, as this nature is TRU. It is all TRUD, as it is accelerated in ACC. It is so that PPMA can assist all those natural processes that are of TRU. For we as individuals can live neither as we please, nor as we may even believe to be best without the necessary TREQ of VAPO, KOT and SCARLS in TRU. There are many variations and means by which it can and, to a great extent, must be done. The Bible teaches us, Gods Will Be Done, so also must, even more actually, the will of GOD as TREQ of TRU be done. This is so to help nature, to prevent harm, protect ourselves against negative effects, maintain that which we optimally have and to advance, control and effect all that is not readily accomplished without our help and yet needs doing. In life we can't always let things happen or do things when they are immediately experienced or confront us. Life is largely a simple matter of knowing what the future has "in-store" for us. The degree to which we have this knowledge is the degree to which the future draws us onward and upward toward an ever better life, TIT and TÂ. In addition to today's more tangible problems, there are many examples of theseposterior compelling forces. They are commonly ignored or disbelieved. They are not only indicators of the principle, they provide crucial information for the correct management, ACC, of TRU. They tell us what this necessity is, the upcoming problem, and its PPMA.

	EPI	TRUS TRUCH	TROS TETO	TRIS TRIT	TIT	ENT	SUT ROT	MOT SPALT	REM	POS-STA	HOM	DYNU	DYNS	DYNO	ACC	TRAX-TRIX	TÂ
1	+++/-	+++/-	+++	++++	++++	++++	+++	+++	++++	+++	+++	++	++	+++	+++	++++	++++
2	++	+	+	+	+/-	+	+	0/-	++/-	0/-	+/-	++/-	+/-	--	+/--	0/---	+/--
3	++/--	++/--	+/--	0/--	0/--	++/-	++/--	--	++/-	--	+/-	++/-	+/-	+/-	+/--	---	+/--
4	+++	+++	++++	+++	++++	++++	+++	+++	++++	+++	+++	++	++	+++	++++	++++	++++
5	PPFm	PPFl	PPFn	PPFo	PPFo	PPFn	nm	nm	nm	Fnm	PPFo	PPFo	PPFo	PPFo	nm	nm	0
6	↑↑K	↑↓	↑↑↓	↑↑↑↓	↑↑↑↑	↑↑↑	↓↑↑	↓↓↑	↓↑	↓↑	↓↑	↓↑	↓↑	↓↑	↓↑↓↑↑	↓↑↑	↓↑↓↑
7	↑↓↑	↑↓	↑↓↑↓	↑↓	↑↓	↑↓	↓↑↓	↓↑↓	↑↓↑	↑↓	↓↑↓	↓↑↓	↓↑↓	↓↑	↓↑↓	↓↑↓	↑↓↓
8	↑↑↑	↑↑↑	↑↑↑	↑↑↑	↑↑↑↑	↑↑↑	↑↑↑	↑↑↑↑	↑↑↑	↑↑↑	↑↑↑	↑↑	↑↑	↑↑↑	↑↑↑↑	↑↑↑↑	↑↑↑↑
9	continuous↑↑	control	Optimal TRU FAF	Optimal Focus	Opti Expansion	Best Use of	Perfected Dvlt	Perfection & Use	Full Use of	Application all	Experience Apply	Experience Apply	Experience Apply	Apply and ACC	Total Perfect ACC	Perfect and Apply	Continuous TÂ

Table 3-92 **The Best Form and State of TRU and ACC**

The state of TRU and of ACC derive from the success/failure of TRU. They are seen in and involve a summary of EPI, TRU, ES, and universals derived for and applied to, ACC. It largely involves a simple determination of what is bad or good for the union, nation, world or TRU. It involves a simple determination of where the bad can be ignored or eliminated, where the good can be left alone, helped, or accelerated, and where both can be best planned, managed and controlled. How these can be done, what other means, methods and ways can be used and what parameters can be developed are shown in Table 3-92 and include: 1. What is of the greatest value, PPMA, 2. Highest Public Principles, 3. Known means of Success/Failure, 4. Percentage of Action, 5. Time of Action, 6. What will occur and what will not, by itself, 7. What will occur and what will not by man, without the knowledge of TRU, 8. What will occur and what will not with a good knowledge of TRU, 9. What we can do about it. This table helps to clarify the different ways humans have approached their world and the major constituents in and of TRU. It also helps to better know, apply, justify, and prove ACC and TRU.

ROMP = Realization of Man's great Potential. ROMP is the conscious and inspiring realization of our great potential. It is largely the product of our highest mental powers or the higher levels of Á in the unique mass energy predominant FAF, state and dimension. ROMP is a realization that participates in IAT, ACEXIN, COAT, OM, COGNOS, and ELTICMA to TES in TIT-TO. ROMP is what America and the world so greatly and desperately lack and need. It stands in contrast to ignorance,nihilism, partiality, and much materialism and consumerism.

Fig 1-92

WHAT ARE BEST FORCES and PROCESSES in TRU

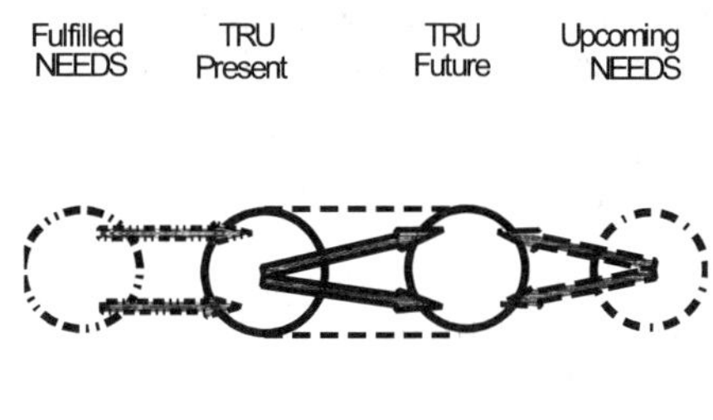

Forces and Processes of What is Best

The common knowledge of what is best gives us a basis as well as target to seek, imitate, emulate, identify with and cherish. It should be what we work for, that which our accomplishments and achievements are directed toward. This is a major means and method of life, ACC and TRU. It can be outlined as a goal. Whether we recognize TRU or not, what is best will often be sought naturally. Yet, as great a method as this is, it is not the best, nor can it be entirely relied on. For alone, it does not utilize, VAPO, TRU or the highest levels of Á. It is not supported by the great feeling, inspiration, revelation, gratitude, reason, appreciation, intellect, coordination, and comprehensive management that TRU can provide. It will be sought without TO, SUT, and ACEXIN. Thereby it actions and TREQ are a basis to initiate ACC and TRU. In the figure we show the positive and negative forces of what is best and most important. These can be compared and supplemented with many known forces and processes in life. Many are known, what is most needed is often only to be adequately recognized, accounted for and accomplished.

MEASURES OF ACC FORM POS

ACCÄG ~5-~5-7~ 5,7 T
BASIC KNOWLEDGE KEY to ACC (BKA) ~3-~5-6~ 5 T
BEST FORM and STATE of ACC (BFA) ~6-~4-3~ 5-7 T
COMMON CONSCIOUS, SUC. ,BEST (CSB) ~4-~5-3~ 1-3 T
DABU 7-5-27~ 4- T
EQUT ~3-~-~ 5,7 S
ETHOS ~3-~5-2~ 4,5 T
EXTINCTION (EXT) ~7-~3-9~ 8 S
FARP ~4-~5-6~ 4- M
FORCES, PROCESS WHAT BEST (FPW) ~5-~4-5~ 1,3 T
FORTOP ~-~-~-~ 3 S
GOD ~-~-~-~ 7 S
GOD's WILL BE DONE (GWD) ~5-~-18~ 1-7 M
GREAT CONTRIBUTIONS (GCO) ~4-~4-2~ 1,5 T
HELPING NATURE (HEN) ~8-~4-6~ 4,1 T
HELPING TRU NAT.HTN) ~7-~5-7~ 1,3 T
HOLY SPIRIT (HOS) ~4-~6-2~ 1,3 M
INTEREST, WISDOM and TRU (IWT) 3-7-6~ 5 T
LAPOFT ~4-4~-7~ 7 T
LASDIPT ~-~-~-~ 1 S
LATSCAP ~4-6~-6~ 4,1±M
MAJOR GENERAL FORMS of ACC (MGF) ~5-~5-~ 5,7 S
MISCONCEPTIONS, MEANING &TRU (MMTT) ~3-~6-3~ 4 S
PLAN of ACTION (PAC) 6-~4-6~ 1-7±M
POSTERIOR WORKING FORCES (PWF) ~6-~4-8~ 4,1±M
POTENTIAL DANGERS, CAU, ACLÄG (PDC) -~-~-~-~ - S
PPMA ~7-~-7~ 4-7±M
ROMP ~3-~4-8~ 5,7 M
ROV ~-6~-7~ 1,5 S
SCET ~-~-7~ 5 S
SCIN ~-~-3-~ 5 S
SCINT ~-~-7-~ 5 S
SCIT ~-~-~-~ 4-5 ±S
SEEK and PAY, NOTHING ALL (SPN) ~6-5-3~ 4 - M
SOCET ~6-2~-7~ 1 M
SOCIN ~6-2~-7~ 1 M
SOCINT ~6-2~-7~ 3 M
SOCIT ~6-2~-7~ 4,2 ±M
STRATEGICS (STR) ~6-~2-7~ 8,7 T
TACTICS (TAC) ~7-~3-6~ 4,3 T
WOT ~-~-~-~ 5,7 S

Preparation, Support, Correction, needs, COPOS, LOMT, PRUTRAX: " TREQ, LIFE, HUMAN, 1/ATRU, OITIO, ENT, MADAS, LATANOS, LORVA

Fig 3-92 **ACC FORM GENERAL POSITS in METAD**

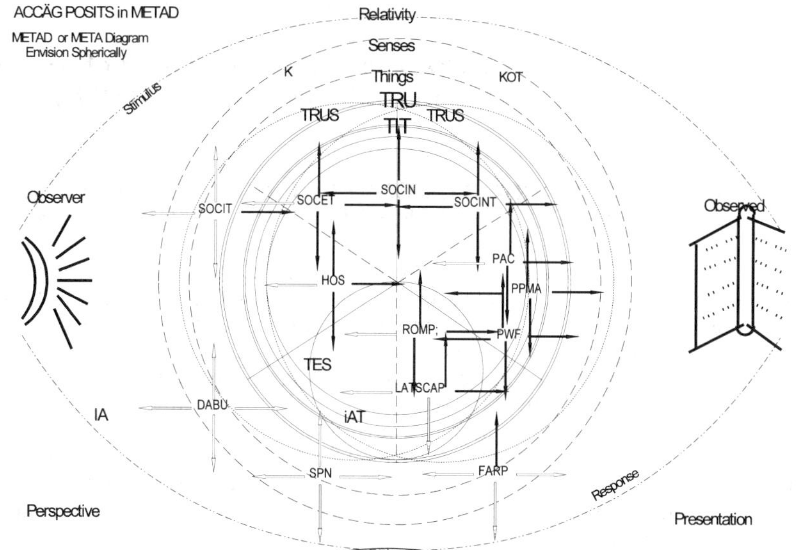

Fig 2-92 **ACC FORM GENERAL POSITS in SPALT**

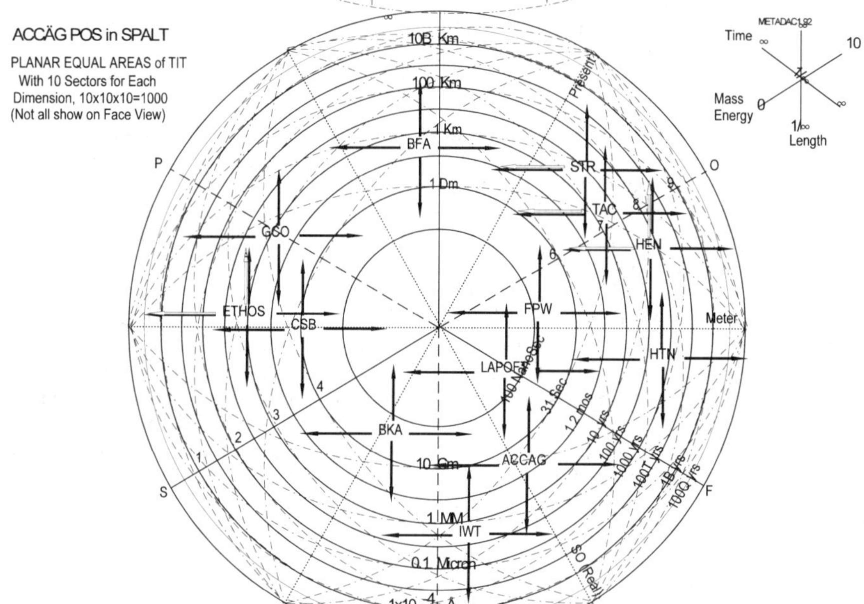

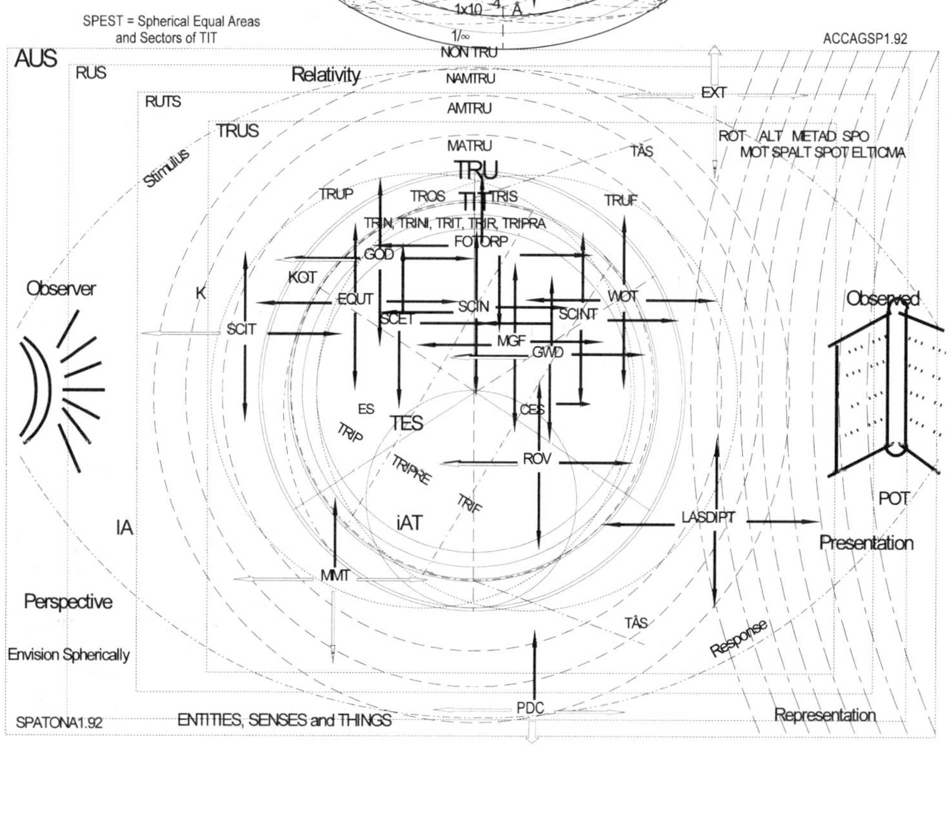

Fig 4-92 **ACC FORM GENERAL, POSITS in SPAT**.

CHAPTER 93

SPECIAL and PROPER FORM of ACC or ACCENTUATION and ACTION by TRU

PRESERVATION, RIGHT, CONVERGENCE, SELECTION, QUALITY SOLUTIONS PLAN, TREX, RULES, UNITY, TREATMENT, LAWS, EXCLUSION

PÂC = ACC Proper, or the Special Form of Accentuation. PAC is how TRU can be accentuated, acted upon, enhanced and achieved through the most special and proper means. It is not only well to present a good account of accentuation, it is to set forth that which must be done and to do it in accord with the principles and guides provided by TRU, SU, ALT, EPI, POT and IPTRU. Where there are conditions, problems and knowledge that are incomplete, disproportionate and all that is other than TRU we must show, accentuate and correct them. This is: **ACIP** = Accentuation of the Incomplete, Inaccurate, Problematic, Disproportionate, Poor and deficient. In ACC we may have both positive, neutral and negative reactions. There may be avoidance and elimination as well as accentuation. They may all be either direct, indirect, voluntary, or involuntary. As TRU is TU, TE, TI, and TO simultaneously and as E and ES are DAOST of TRU, they are all derived from set relationships between the form, function, parts and processes of TRU. So are Accentuation, intervention, ACIP and other related concepts largely of set relationships governed by TRU. For the completion of PAC, ACC and TRU we have: **FEPTA** = The complete, comprehensive and concrete fulfillment of all, of EPI and TRU in accentuation. Through SPO, PERT, THES and THEO, TRIT, OBT, SUT, ROT, MOT, ALT, and SPALT we can know and show TRU in such a way that by concentrating and accentuating all that provides TRU we can most successfully, comprehensively and concretely fulfill and complete POT.

There is no question that the major causes of the troubles and wrongs in the world arise from the failure of individuals to properly comprehend and act according to what is best, TRU or TREQ. Failure to comprehend TRU comes from not learning and living what has already been known in TREQ, being unable to learn and live by what is not known that is TRU, and not knowing what TRU as a topic and concept are. Added to this is the tendency to either have false knowledge or to badly use what is known, and thereby live by what is less than, not, or against TRU. It is from these, to be able to comprehend and act according to TRU, that most of the great solutions to the reasons and causes of troubles and wrongs in the world are to be found and their relief secured. If all persons, who care for intellect and are informed citizens seek to understand TRU and use such knowledge to best guide their actions by TRU, ES and CES, there will be spontaneous and automatic success, prosperity and permanence. This is through VAPO, SCARLS, DAOST, TROS, TIT, ACEXIN, TU, SUT, THEO, THES, TRUD, ROT, MOT, ALT, SPALT, REM, POS, STA, HOM, and ACC.

It is by much more than coincidence that understanding and standards are related. They enhance each other, especially when TREQ or TRU. Principles, laws and standards are set means of making TRU more understandable, realistic, and workable. They, as Á, are and, should follow TRU. This follows what is natural, closest, and easiest possible way. Principles, standards and laws are means and models to live by, from which our habits, behavior, policies, routines, plans and actions are largely determined and set. Yet the limits and accumulated deviations from TO-TIT must always be known, compensated for, managed, controlled, and returned to TRU. It is this that makes the new language and concepts of TRU justified, just as it is up to us to be sure they maximize, and have a minimum of detraction from, TRU. The development of standards to represent TRU, as by the sequence TIT-TO - TRIT - OBT - SUT - ROT - MOT - SPALT - POS - ACC can be better understood and achieved when such maximum and minimum limits of standards are upheld.

In order to have a better understanding and how to preserve, accentuate and act in relation to TRU it helps to look at it as a science. Like all great sciences it has a basic format, principles, concepts, laws, mechanisms and terminology. These are determined in part by what has been developed, exists, and is known to work best in each and all great fields and sciences. It is also determined by the relationship between observer and observed, subject and object, the Á FAF of TIT of each science. Mostly it is determined by the nature, aspects, form, SCARLS, DAOST, makeup and function of TRU. The greatness of such a science relies heavily upon foreknowledge of all TREQ. This is what will best direct proper accentuation and action with available resources and energy.

The world and especially science has made giant strides in knowledge, education, training, providing exposure to experience, communication and acquisition of the positive attitudes that accompany them. We can also provide change or improve principles as they are discovered, developed, determined, modulated, posited and ordered in TRU. It is in the books, libraries, and communication of information that we can see how greatly mankind benefits from the discovery and enhancement of TRU. This is done by combining all necessary to beautifully present that which enables an effective comprehension and achievement of all, TRU.(COAT)

Fig. 1-93 **CONDEMNING and COMPLIMENTING ATRU and TRU**

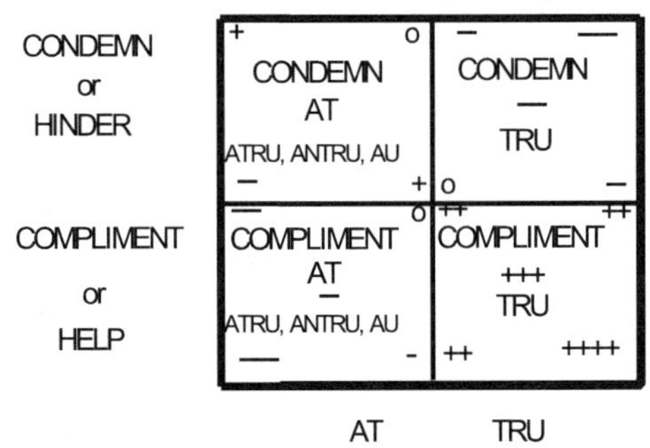

Human actions may be divided into four groups. These depend upon whether or not they are condemning or hindering, complimenting or helping ATRU, or TRU. The accompanying figure shows these distinctions. In the world in which we live, however, they are both variably mixed and operative as well as and highly scaled. In life each person helps, assists, gives to, compliments, considers, and reacts to others or other things. This may be TRU or ATRU or some intermediary or variable. On the other hand each person often hinders, condemns, blames, or detracts from others, themselves or other things. This is often done indiscriminately or irrationally, on whim, fancy, grudge, hate, or prejudice. ACC and life in TRU are predominantly only 1/4 of this figure. The others are of little, no, or negative significance. Thereby complementation and help of any and all must be of a kind that is optimal for TRU.

Great conviction and dedication to a cause is supported by an identification with and firm belief in it and its associated reasons. It is also accompanied by great curiosity and interest. These arise from their Á FAF, mass-energies and efforts, of the physical, mental, spiritual, and all in between. It is the added stimulation of means that catalyzes the Á predominant and dynamic energies and efforts to make the necessary reaction, reduction, and achievement. There are necessary levels of concentration that make each action successful. This is what we would expect from the many gradients of our Á.

Only to the degree knowledge, TREQ and TRU are well learned, known, highly refined, complete, remembered , and utilized, can the right be exercised to tell people what to do. This is the great justification, reason, and proof for continued work on TRU, **JURP** = Justification, Reason, and Proof for Continued Work on TRU. Once this level is achieved, then it is, as the highest and most set state for policy making, something that is undeniable and thereby warrants full compliance. For those who deny and defy JURP, certainty, and the set state of TRU we have a right to make them do what they are told to do. Encouragement, and usually a word to the wise, is sufficient.

To do what one is told means
That what is told must be right.
This means that whoever and all
Who determine what is right
Must do so properly and completely.
This can only be done as it is
Discovered, developed and understood
Comprehensively to Represent TRU

Much of the troubles of the world
Are not a result of the badness nor
The sinfulness of its inhabitants
It is more often a result of good
Intentions that have gone astray
To tell others to do what in actuality
Is not right is erroneous, conflicting
And destructive of all that is right

To correct this is not easy or simple.
For those who make the polices of
The world, make its rules, regulations
And laws are not readily dissuaded.
In defense they naturally become
Set and steadfast in accumulated error
It takes the deepest and best reasons
And VAPO to bring them to TRU.

LAPU = Law of Proper Utilization. To seek, make or use the world for any entity, purpose, or end in itself, is false, erroneous and destructive. To seek, make or use the world for TREQ or TRU is the only correct and constructive way. Yet it must be done through the appropriate VAPOS, DAOST, and SCARLS. It must be done in a way that most helps that needed for TRU. This we see in the use of TRU terms, where it least damages what is needed, or exists in the world, TREQ, of TRU.(VAPO,ES,TES)

The solution to the 10, 100, and even 1,000 greatest problems and causes of our day often exist in the great form and function, determinants, and core forces of TRU. These, in turn, are often already in the Á state of TRU. This is called: **LARACA** = Law of Reverse Action equals A, or the actions and reactions of ACC are often mostly the reverse of existing FAF, core forces and determinants of TRU. The Á, or mass energy state of TRU provides us with continuing action, not only in TRU as TRAX, but in TRU as TREX and TRIX. It provides the means, PPMA. It is shown by the wiser use of birth control in enabling increased quality of propagation. It can provide humans with the knowledge to refrain from all actions that are not, of benefit, in accord with what is best, nor in conformity with the great core forces and determinants of TRU. It can cure our ills so that we suffer less and strive for more than the trial and error of natural selection can produce. It can bring man from a state of being unstimulated, uninspired, unimaginative, and non aggressive to increasing quality and power in TRU.

RULES of LIFE

1. Know what TRU is worth, 2. Know what part you are of TRU and thereby what you are worth, 3. Know what your services are worth, 4. Do not expect to get paid more nor less than what your services are worth, 5. Do not expect to get paid more, nor less, than what you contribute to TRU, 6. Always strive to make or adapt your services to be more productive and valuable, 7. Recognize the detail, finite and approaching infinite needs of TRU that a large portion of your services must fulfill. 8. Do not waste yourself what is not VAPO or TRU. 9. Do not waste yourself on those, who want something for nothing, who make great demands upon your life, time, and function especially if they are not, or are incapable or unwilling to be or return it to TRU. This is especially so for that and those who are hopelessly and increasingly ÂTRU.

TÂKT = TRU as it is, but is not Known as TRU. **TÊKT** = TRU as it is, and is known as TRU. It is one thing to know TRU as it is the subject TRU and presented in IPTRU. It is quite another thing to know much about particulars of TREQ but nothing of TIT-TO, or TRU by POT. So that there is much in religion, philosophy, science and knowledge in general that is TREQ and TÂXT. In ACC we separate TAKT and TEKT, TRE, TREQ, TRAX, TREX, TRIX, ACC, TRU and TIT. It provides much of the special form of ACC. TAKT is similar to TREQ as TEKT is to TRE. Yet TAKT is like TIT itself, whereas TREQ exists both in and out of TRU, like ESINS in religion. So also is TEKT for TRE, not only the recognized TRU, but TIT recognizing itself. TAKT and TEKT are exclusively involved in the knowledge of TRU as provided by general and universal knowledge unknown as TRU, or can be provided by the special form of ACC whose direct knowledge of TRU and its concepts, principles, laws, mechanisms have been or can be developed. The emphasis of TRAX, TREX and TRIX is upon the accounting, accentuation, and action to be placed in determining the different forms and special parts of ACC. TREX and TRIX are special or partial aspects which, in a way when combined, become PUREALITY, the special parts of ACC, and its contribution to TES. This is how we can separate, divide and conquer all through the weapons of ACC. This is how ACC is both made distinct from prior, and can be developed in closure in Chapters 96-99, to TRU.

There are right and wrong sides for entities, oneself, all, TES, and TRU. To expect people to be on your side both you and they must be together, in harmony, a dialectical equilibrium of differences of TRU. To be together and the same, you, and they, must be TES and TRU under DAOST. For to expect others to be on your side under any and all circumstances, no matter how right or wrong you are, or how right you and even they may think you are - and not be TRU - is even worse than nothing. It is to be doubly so. For you are transposing yourself into others. You are adding to their ATRU, ANTRU, or ANTITRU.

The degree to which problems are actual, and especially critical to the success/failure of TRU is the degree to which powers must be available for best use in their solution. Any work on, of, in or by TRU 'must have teeth', and use the appropriate powers of TRU. Yet there is a ratio of constructive over destructive, positive over negative action and work. This must be ever higher. It will vary from time to time, and even for ES and Givens. This means the kind and degree of powers which must be available for best use entails a tremendous amount of accounting, operational assessment, constructive action, endeavor, work, and ACC. However, this need not be negative nor a burden, but of established, positive, and joyous achievement of ever greater benefits and victories for TRU.

The greatest task mankind faces is to preserve and enhance being, TRU. Closely associated with this are the great tasks of: 1. Increasing total quality and limiting quantity, 2. Increasing knowledge, TRU understanding, and decreasing ignorance. 3. Increasing personal mental and physical health, 4. Increasing participation and success. These are more or less related, some quite closely, either as problems and in their solutions. As such they can be best managed and controlled with the overall program of TRU in TES, ACC, TRAX, TREX, TREX, PUREALITY, and COGNOS. To increase quality and control quantity will take much patience, hard work, precision, and cooperative effort. It will infringe on individuality. Therefor it should be done by: 1. Voluntary action as much as possible, 2. A minimum of laws and regulations, 3. Avoid immediate or wholesale affects on peoples lives, 4. As discrete, natural and painless as possible, 5. Thorough and exacting selection of best leaders, 6. Careful consideration and management of the proportions of race, nations, and other social divisions over the long term, 7. Friendly competition amongst individuals, groups, communities, cities, states, nations, regions, races, languages, and other divisions, and 8. Allow for guides to follow as well as means of affecting selection.

There are marked differences between the simple fact of TRU and TRU'S many different aspects, forms, states, parts and processes. It is in the proof of the correspondence of TRU to ACC, and ACC to TRU, that we can achieve higher truth and success. An example of this is in comparing the validity and massive proofs of both TRU, and its various categories. Here lie the basic causes, the core COCS, and the KDMS which are so in TIT and enable us to know and show TRU. For example, do we take care of ourselves, everything, or what? This may seem obvious but to those who would belittle TRU let them see how much they do what. For example, the basic differences, principles, and laws to be derived from the validity of the many kinds of TRU are the basic principles and laws of HOM and TRU. **TRUHCO** = TRU and HOM Correspondence. **COPS** = Correspondence of Conflicts, Problems, and their Principles and Solutions.

Quality comes before Quantity. *Quality ante Quantity.* ***QUALAN.*** No specie, culture, civilization, nation, group, etc. can survive for long with increasing quantity and decreasing quality. It is one of the greatest causes for failure and misery. Yet sufficient quantity is necessary for success and survival, **QUANS.**. This is seen in many aspects of human life, in propagation, protection, cooperation, teamwork, maintenance, compensation, or making up for loss, and many processes of TRU. (HOM) This is especially so for modern day civilization, knowledge, technology, division of labor, and professionalism, where the diversity, or combined capacities and abilities of many greatly enhance success and survival of TRU, **QUANST**. Thus much of the Strength of a RU, and especially TRU, comes from, ab, its (D) diversity, **STAD.** As much as it comes from its unity, **STU,** and all in between, or the whole spectrum of Integrative Differential Strength of TRU, **STID**. This is seen in the contribution of all sectors in the success/failure and survival of a specie. No one individual has it all. No one individual can give all it takes. It takes the quality contributions of many. Quality applied all the way through genetic material, live product and determining forces.

We soon learn how important it is to never be careless with money. Yet we fail to learn, or quickly forget, how important it is to never be careless with anything that is essential to this special universe of best human life. This applies to one's soul, God, TREQ, TES and TRU as each and all is a proportionate part, and contributes to the success of TRU, as by REM. This is, Never be Careless of REM. CREN.

In ACC, as in TRU, TIT, and OITIO, we may have both positive, neutral and negative reactions. There may be elimination and avoidance as well as action and accentuation. They may all be either direct or indirect, voluntary or involuntary. It is to be able to comprehend and act according to TRU, that most of the great solutions to the major reasons and causes of the troubles and wrongs in the world are to be found and their relief secured.

It is possible to approach ACC and TRU in ways other than IPT, POT, TRE, or TIT. These include uses of TREQ and TÂXT of what is already in language, philosophy, science, knowledge, theology, God, the world of god, concepts that function spiritually, and phenomena by their unique and great powers in TRU .In IPTRU and POT such use is like the Magical powers of God and all, MOG, and the Gist of these Powers, GOG. The format using such powers has several advantages, including: full development of God and Religion as they exist in and apply through TREQ to TRU. How the existing powers of God and Religion in society supply the vivid gist and cue of greatest concepts, and the power of each of these concepts in TRU. These powers may be presented as prescriptions for PPMA and treatment in ACC. They act as guides to POT including: presentation, perspective, approach to, God or TRU, God or TRU as The-Thing-In-Itself, representation of God or TRU, modulation of God or TRU, evaluation of God or TRU, posits in God or TRU, statics in God or TRU, general processes of God or TRU, singular processes of God or TRU, complex processes of God or TRU, total processes of God or TRU, and accentuation of God or TRU. The powers, of MOG, are the existing concepts and terms, especially in religion whose TRU interpretation provides greatest insight into TRU. This has great added advantage and power by showing how they actually exist in TRU, and how each participates in the whole, TRIT of TIT-TO. It focuses on the greatest key ideas that are also produced in the format of refined SOAP and posited in SPALT. They are further strengthened when shown in combinations and complexes, especially when they correspond, are correlated, and used with solutions to major and most threatening problems, conflicts, causes and the basic principles, laws, division, and form of TRU. An example of this is in world literature where the power of words or EC in phrases, descriptions compositions, not only show TREQ, they are in and strengthen TIT. When combined with TRU, as in IPTRU, this becomes a formidable means to ACC.

Procedures in treatment should be guided by: 1. What will get well by itself, 2. If not what one can cure or correct, 3. If not can anyone else provide its cure, 4. If not what can all do to help it, 5. If not how can it be best left alone, 6. What will happen if none of these work, 7. How can all function in the best way by TRU, 8. How much will be spontaneously cured by normal and indirect functions of TRU, 9. What remains, and how can it be discovered, learned, known, accounted for, or best acted upon. We live by routines, rules, custom, and habits which greatly assist in reaching the above. Yet they may not be built upon the right knowledge. They may not be set up right. They may not always, even if set up right, continue to be right with time. In being routines, rules, customs, and habits they will lack a conscious concern, rationality, and adaptive capacity to maintain optimal performance. When well maintained by TRU they will not be taken for granted, must be best utilized, constantly upgraded., and accompanied by the best in ACC.

If we are to understand and show TRU, we must overcome the redundancy and recalcitrance of all those who lack the needed knowledge of, interest in, and endeavor for what is best in philosophy, fields and sciences, religion, arts, labor, business, and all major disciplines of human life. This is the beginning and end, the Representation, Perspective and Presentation of TRU. With the right focus such knowledge, interest and endeavor their irrelevant and partial aspects will be overcome by those that are relevant, universal, TREQ, and TRU What is so desperately needed and will most effectively bring humans together is PRES the combination of Philosophy, Science, and Religion that relates their great concepts and principles with the same in TRU. It is from the wisdom, evidence, faith and conviction necessary to unite all elements, sections, groups, individuals, and entities as their ESINS are brought to TRU. This will include convergence on TRU in S/F to overcome all ATRU, AU, ANTRU, ANTITRU, and their mixtures. It can not be done by religions, sciences, or philosophies that lack VAPO, are partial, and in opposition to TREQ or TRU. It can best be done by enlisting, encouraging, stimulating, uniting, cooperating, proving, and victoriously showing all that are of TREQ and TRU. This can be done by how much they need and will benefit by TRU.

In a government the individual is only a tiny portion of the whole, so that as an individual they do not count for much. Yet individuals collectively count for a great deal. This is called: **INIS** = Individuals Individually are Not, Individuals collectively or plurally in Summation are. INIS is a proof of TES/IND. For if the singular individual counted for much one would have more power, or those who have it would last longer. This is also a great explanation and reason why the individual, if they are to have a life, work with others of the same, must contribute to the whole of TRU. It should go a long way to cure the perennial tendencies of individuals to be self centered, excessively independent, and immoderate.. This principle, is what, carries such weight in republics. The sacrifice individuals must make, and especially to their egos is compensated by LOPAN, and the immortality they share and receive in TES and TRU. It is in this way a union of the great silent majority of all peoples on earth who appreciate and exercise reason and love for the power, benefit, and success of the world, mankind, TREQ, and TRU. This is when they are first and foremost interested in the welfare of this cooperative endeavor and not only one, a few, or what is less or naught. Such is similar to intolerance for the whole, all TRU, like an abomination to God, negation, or denial of TRU. **INTOL** = Intolerance for TO, the whole of TRU. INTOL is inversely correlated with VAPO. There is a lack of TRU conscience and love, or when present love and conscience do not rise beyond narrow bounds that should have brought them to TRU. For this reason all of us must most vigorously learn and love TRU. Love that is based on the bonds that hold TRU together is the right kind of love that positively produces benefice. To do so we should continuously strive to perfect both ourselves, as TES, and all that can be TRU.

Fig 2-93
MONOLITHIC VULNERABILITY

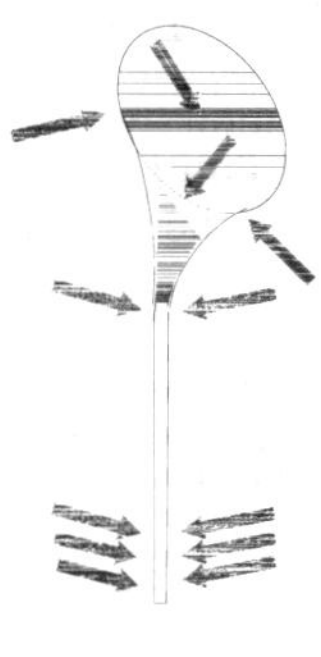

MOV = Monolithic Vulnerability of TRU and its PPMA. MOV is that state of TRU, as in any RU, in which survival depends upon the very delicate interaction, and balance of vital ES, CES, and FAF. It is one whose foundations or building blocks are so interdependent that the loss of one or a few proves disastrous to the whole, likely to cause extinction. MOV is natural to life and more pointedly to humanity and TRU. It is seen in both the ease of death, failures in conception, and much between. It is seen in the repeated cycles of regeneration and extinction. There is much in human form and function that is very fragile. In fact it is this very fragility that gives evolution much of its dynamics. This is associated with a polymorphism or diversity that is maintained only by a high level of selection. It is especially acute in a species' genetic makeup. Genetic material, live product and determining forces are so constituted that damage and alteration to a very small portion affect the quality of the whole. It is apt to be accumulative and irreversible. Those areas in which MOV are likely, and whose PPMA must be sought, need to be constantly discovered, remembered, kept in mind, focused upon, and controlled. This is, in one way, met by adaptation. Another is through prevention. Many different measures can be applied that are available and effective in the reaction to MOV and the support of TRU. It is also recognized, and even popularized. Yet this is often not TIT and TO by VAPO, but in a hit and miss fashion to the whole. It must be considered in different ways. What makes MOV is based upon different time elements or intervals. MOV greatly helps to explain Sections V-IX of the text. It explains much of what mankind's history of the past, 100, 1000 and 10,000 years have been, and is absolutely critical to the similar periods in the future. There are great specie variations and factors causing MOV. Those great forces which enable human permanency and those that produce vulnerability must be clearly set forth and formulated. This we do in HOM, DYNU, DYNS DYNO, and ACC. We do so especially by concentrating on what is potentially dangerous to survival. It is largely dependent upon the stability and health of the environment. Yet man has accumulated a vast amount of knowledge and means which benefit and, more or less, protect us from MOV. On the other hand many of these means themselves pose new dangers and are delicately dependent. It is in this interdependence, by DAOST, that we should look to find where the most vulnerable sites are, and where upcoming destruction of these sites are apt to occur and require PPMA.

In the crucible of life, especially when concentrated in TRU, it is often better to emphasize the prevention, protection, and maintenance, than the action. For much in life will 'work itself out', 'will out grow it', or just needs for us to 'let it go', as in laissez faire. Yet all these must be under the close attention of PPMA. It takes much information and knowledge, especially of and by TRU, to be able to make the decisions and judgements necessary to do or not do what may seem so simple, yet is so great and complex.

What, in this world, cost so little, yet is worth so much? One of these is advice. Yet it is little accepted, for it is often, trivial, poor and bad. Advice usually lacks TRU and VAPO. It is for this reason what at times actually costs so little and can be worth so much fails to be heeded. Also why TRU, COAT and ACC are so necessary.

Expanding levels of PAC are shown by all the difference in the world between: no opinion, no guess, a guess, an educated guess, a trained opinion, an assessment, a diagnosis, a certain diagnosis, a certain plan based on a certain diagnosis, a certain plan that encompasses the diagnosis and all else involved, a unified plan that encompasses broad fields of human endeavor, a systematic plan of operations that covers an entire world, a systematic plan of operations that covers TRU, and a certain unified systematic plan that covers all facets of TRU plus all potential aspects and all surrounding approaches and ramifications that are apt to be necessary. As the world, TRU, becomes increasingly capable, as well as more and more dangerous, it is mandatory that the level of our subjectivity, management and control increase at a faster pace. Along with this there must be an equal or greater capacity for application, production and achievement. It is in being professional that we become more adept at attaining higher levels of ability to develop and use such systematic planning. A professional is trained, and continually strives to achieve better management of all problems by fitting them into the big picture in the best possible way. This is what is necessary if we are to be leaders, preachers, physicians, lawyers, and all who provide treatment or advice, like all ACC, for TRU.

A high level of reasoning has been developed in many governments, industries, and institutions. Much of it is specialized, routine, and, at best, only partially focused on TRU. It takes the form of policy. Policy is prudent wisdom in the management of affairs. It is a definite and best course of action, selected from many, to act as a guide. It has arisen largely from the experiences of group behavior, relations between nations, and has been adapted as a way of achieving success in many walks of life. Policies are the rules of engagement, a frame of reference in which all that is most beneficial is best managed. It is devised to meet both the immediate needs, as well as the long term relationships. It is directed toward greatest importance and danger. Policies tend to avoid personalities and become more focused upon general or overall factors and principles. They can be most partial to very universal. Policies can be of the short, mid, or long term. They can be bad or good, for one or for all. They can be highly, lowly, negative, or of variable TRU. Or, as is often the case, a highly vacillating mixture seen in the sad tale of failed policies in world history.. Policy represents a high level of the mass energy form and function, of the predominant Á state and dimension of TRU. In being such it is often in a very formative stage of development. It is because of the archaic aspect of this stage that it has often been, a failure in what it most intends to provide. That is a secure and permanent benefit. This is not primarily because the idea of policy is bad. It is even not primarily because the intentions of many of those who devise these policies are not good. It is because policies, rather than being consistently held as principles, are poorly made, abused, temporal, variable, and open to abuse. It is because they do not have, or do not use the available experience, information, research, and knowledge. Policies like existing laws are limited and often quickly lose their benefits. They have not discovered, learned, known, remembered, applied and lived by KOT and TRU. Therefore the best plan, and 'policy', is to learn, know, and use TRU. Know how much ACC is most simply the policy of TRU, and devise all its policies with the utmost VAPO, KOT, and all TRU. Yet to have policy that equals TRU and all related to TRU is not simple. It is complex and for ACC needs to account for and include even the totality and absolute of TRU. Policies, like knowledge in general, can't be limited to one, a few, or even complex courses of management or action, unless so stipulated, remembered and controlled. They can only be right if they form a whole in which all its parts function, as by VAPO, for the success of the whole, TIT-TO.

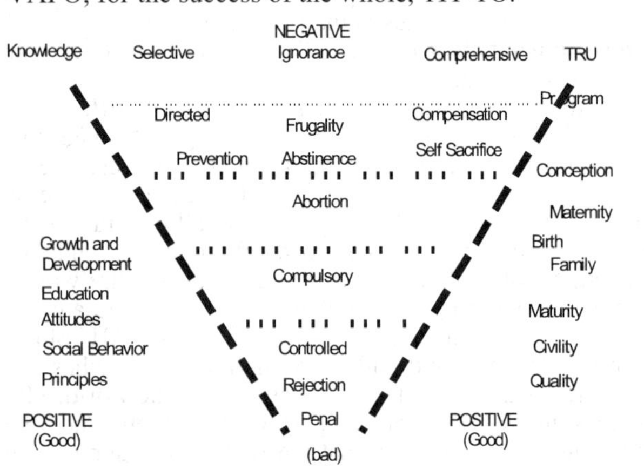

Fig 3-93 **SELECTIVE QUALITY of THE INDIVIDUAL**

Since TRU is highly dependent upon the quality of individuals and their role in propagation of the species, it is imperative that selection be optimal. Selective quality of individual humans that is well substantiated and most beneficial is key to preventing most of the conflicts and suffering in the world. It is even more imperative that these produce a highly efficient and effective TRU. This is a large part of an individual's role in the survival of TRU. This is much of ones immortality, the degree to which they are TES in TRU. In Fig. 3-93 the negative is in the vertical center with positive on the sides. It helps to look at what are and are not beneficial to TRU and the programs that need to be developed. Diagrammatically, we use a funnel type figure to focus on the acute actions needed for control. This is done to emphasize the direction and power needed. Such a program should be massive and total, yet sufficiently lenient to adapt to unforseen needs and variables. It should optimize TRU and minimize all that is not TRU or cause additional conflicts, suffering and misery. It can't be done without a great amount of felt need, appreciation and voluntary cooperation on the part of all. Yet if we learn, know, appreciate TRU, understand the needs, programs, means and ends and how they will not lose but gain by TRU, they will bring out the powers of natural selection for quality of the species. This, rather than by trial and error, is guided by the power of principle in what is right for society, and will to TRU.(WIM) Such an approach helps us see why crime is so rampant and out of control, and why punishment has not been effective. Even when positive it is met with punishment on a one to one basis. We can't solve all the crime, all that is negative in the world, directly or alone. It can be solved only as we discover, learn, know, apply, use and live by all TRU. It is how TRU can increase the selective quality of individuals both through increasing knowledge as well as increased capacity, that will spontaneously prevent most crime.

The world will not work 1/2 TRU and 1/2 Non TRU, or even 9/10 TRU and 1/10 ANTRU. Yet, even worse than this are many of the NAMTRU mixtures humans live in. We are engulfed, encompassed, and constantly interacting with such worlds. The ANTRU and ANTITRU portions of the world should be vigilantly guarded against, controlled, negated, or annihilated depending upon the extent to which they pose harm to TRU, and what best fits TRU. ATRU and NAMTRU should also be attended to, and sufficiently and completely reversed, or brought to higher levels of TRU, in order to remove its adverse effect on TRU. All that is in this mixed world that is ANTRU, ATRU or even NAMTRU will not or can not become TRU. It must be either separated out, selectively removed, completely controlled, or eliminated. All persons who are ATRU must be either sufficiently brought to TRU or constrained to prevent loss to TRU. They must have a high quality of training, education, principles, care, and social standards. This kind of conversion is called: **IMC** = Imperative Conversion, to TRU. IMC should not be one of blind faith or submission to power. It should be a voluntary and reasoned acceptance of the greatest reasons and highest VAPO of TRU. Yet it should be vigilant and determined. It should be one of joyous and willful cooperation and identity on a lifelong basis with TRU.

ASAJ = Association for Civil Justice, **CAJS** = Committee for Candidates of Political and Civil Justice Selection. These are associations for the optimal selection of Civil Candidates, especially in elections. What is best for the whole, its parts, short term, midterm and long term? We need to understand the failings of our political systems. To be effective they must be sufficiently organized to produce the power necessary to assure the success of TRU, its nations, states, local community governments and all. They also must include leaders of highest ability and work to meet their responsibilities. It is for the selection of those candidates who are most suited, motivated, and capable of performing their jobs by TRU. It should be against lobbies, politics for politics sake, the disenfranchisement of less well represented or promoted groups that are TRU. It should be for the development and strengthening of all counterparts of TRU such as those whose TES is not partisan and demanding. It should consist of persons who are independent thinkers, hard working and concerned with the well being of the best in human life, TES and TRU. To do this we propose to select and vote in a bloc, a unified voice by TRU. Those candidates who will best fit the job or office, and who will best enhance TRU and all its components. To do so by thoroughly, scientifically, and especially psychologically analyzing each candidate or those best suited, by their ability, learning, character, motivation, and work record. Also, to aid their campaigns at the local, state and national levels, by all TRU means. The prime aim is to provide a politics that best answers, solves, and makes needed changes that will enable all in political life, and those they serve, to be most TES and TRU.

The simplest way to know what we should do, be like, accentuate and eliminate is to look at ourselves in 100, 1,000, 10,000 years in the future in terms of the greatness of TRU. With each time period we see ourselves and TRU from a new, different, and highly penetrating vantage and find what must now be done to enable TRU to survive and best succeed. This can be done for successive and specific time periods as well as throughout many different times. To carry this further certain principles of what life on earth in all future environments or settings should be applied: 1. Build on all our fundamental strengths, 2. Lengthen all our potentialities, 3. Eliminate all unnecessary burdens, 4. Eliminate all potential dangers and threats of destruction and loss, 5. Spearhead all means to each successive period and throughout time. These, among others, clearly illustrate the actuality of the future, and is termed: **PAF** = Principle of the Actuality of the future. **PAFOC** = Principle of Positive Continuing Correction and Actuation by Future Application. This way is an example of the ongoing and continuing state of TRIF and TRIP with TRIN that constitute TRIS and TIT. It also should allow for change and variation. It will, preferably, be more reversible than irreversible. Yet some irreversible changes may be necessary or indicated by the revolutionary state of man's mass energy, Á, Nature. This is the human cognitive or subjective revolution as it transforms TRU. PAF and PAFOC prove TRU by proving the form of TRU.

Who would think of running an obstacle course if blind, or in the pitch dark of night? What modern day navigator would think of navigating without aids? This is the world without TRU, TU, SUT, TO, ROT, ALT, SPALT, ACC and COGNOS. Yet, this all of us do this every moment of our lives. We have poor focus, a focus that changes chaotically, or an IA that has no map to follow. Although by tradition, routine, and example we have intention and purpose these are usually only stop gap measures, and serve only very limited segments of our existence. They, as we, are largely NAMTRU. It is only when we, individually and collectively, can have the best possible world view and know our position in it that we can begin to steer and navigate our course through life successfully. This can only be based upon TRU, TIT, TU, TE, TÄ, TÁ, TÂ, TI, TO, SUT, VAPO, DAOST, SCARL, ENT, ROT, MOT, ALT, SPALT, POS, STA, HOM, ACC, PAC and all that gives us the guidance needed to set our course throughout life, in TES and TIT. Many levels and portions ofthe mass energy, Á FAF of TRU contribute to coordination.

In fact coordination, as a major component of our central nervous system, is also a major component of our FAF. It is coordination that best show how the higher levels of Á FAF can be so easily confused. To ignore, deny and defy it is the same as to ignore, deny and defy our being, TES and TRU. **CORAC** = Law of Coordination the basis for ACC, its FAF and proof, PRACC. It is CORAC that not only shows how much of TIT and ACC are constituted and operate, it shows how well we can, when best using all that is provided for us in the TU, SUT, TO, ALT, SPALT, POS, HOM,and ACC, guide our lives and world in the best way.

ABISU = Absolute Imperative SUT. A great driving force through history has been the world view and its representation. Especially how well its formation and structuring replicate this world, as it is TREQ or TRU. If the world has no form or structure, we will be forever disoriented, lacking in direction, confused and lost. This is the main thrust of the text. We seek to show all that best reveals how TRU can be formed, represented, modeled, valued and posited and held in view for use. This is initiated by the actual form, frame, and structure of TRU, by TIT, TRIT, OBT, SUT and SPALT. TRIT, OBT and ALT are what most basically and definitively orient, coordinate, and replicate TRU. They are provided by its unity, form, cohesion, and totality. Its formation is enabled by the sameness of much of its makeup, DAOST, as this makeup is positively shown in SPALT. It is in this way that SUT and its development are absolutely necessary. SUT is close to the core of TRU through TRIT, TIT, TU and TÄ. Just a unitary God has been the bond that has enabled religion and the human world to be focused, so does SUT lay the foundation for the structure and representation of God as TREQ, that are from TU to TRU. The success of SUT, ROT, MOT, ALT, SPALT, REM, POS, STA and HOM provide solutions that are not only certain, they are guided by VAPO, and thereby have a higher level of coordination and revelation of TRU. In this way certainty, conclusive validity and a resolution of all questions, problems, paradoxes and fallacies can be increasingly provided.

Fig 4-93 **CLASSIFICATION of LAWS of TRU** (LOLAT)

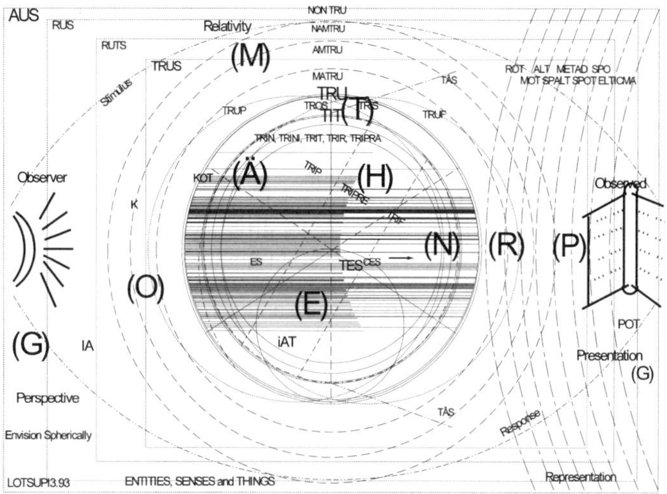

LOT(S) = Law(s) <u>of</u> TRU **LAT(S)** = Law(s) <u>in</u> TRU, or of TIT. The laws of TRU, LOTS, are largely representations of the necessary power <u>of TRU</u>, G of General, O of Out/In, M of Mechanical or Physical, R of Representation, P of Presentation, They are neither perfectly TIT, nor completely non TIT. They are TIT to the degree they are laws of-in TIT. They are of representation to the degree they are a means of showing TRU even when not entirely in TIT. When of RIT, or Representation in TRU, they are both of a representation and in TIT as these two are the same. Laws in TRU. LATS, are largely representations of the necessary power <u>in TIT</u>, T of TIT universally, Ä of Form of TIT, E of Part or Parts of TIT, H of Homation or Processes of TIT, and N of New Concepts and Future of TIT. Laws are valid and can only be properly called and used as laws if they are held to their status, functions, meanings, implications and interpretations in TRU.(VAPO,DAOST) Universal laws if meant to apply to all without qualification or TREQ are apt to be fiction or illusion in full or in part. A good study is to sort all CALP by type.

CLASSIFICATION of the LAWS of TRU

LOTS					**LATS**				
G	**O**	**R**	**P**	**M**	**T**	**1**	**E**	**H**	**N**

G - Laws in general of subjectivity and perspective of TRU,
O - Laws of objectivity, relativity and reality,
R - Laws of representation and modulation,
P - Laws of Posits, Statics, and Vectors of TRU
M - Laws of Mixed AUS/TRUS
T - Laws of TRUS-TRIS and DOT
Á - Laws of the mass energy dimension and its levels
E - Laws of the finite of TIT, from TROS to TRIS & TÂ
H - Laws of TRUD and HOM, Dynamics and Processes
N - Laws of the enhancement and Accentuation of TRU

In Figure 4-93 the ten different kinds of laws that are shown are posited on the SPAT figure. This provides a highly effective overview of all laws. It helps to establish the limits to which laws become fixed and incline toward finer and finer distinctions, meaning and validity. It is in this way we can discover, determine and develop many laws for much of the world, TRU. We need not shrink from asserting that which is certain and valid because it does not apply in all instances, nor foreign to current beliefs. We have a great duty to be sure that our concepts, principles, and laws are the best, most valid, accurate, reliable, and as well defined as possible. On the other hand we have a great duty to support them, once they are properly developed. It is when we deny and defy this duty, directly, indirectly, intentionally or unintentionally in life, that we are more or less guilty of this greatest of crimes, denial of human life, as TREQ, TRU, TIT, or TO. Laws thereby become rules of action and regulators which provide for the dynamic success of TRU. It is in ACC that the overall Laws of TRU are promulgated. When used in this way they are called: **LATRU** = Laws of TRU and its Accentuation.

In actuality or in TRU, those laws which seem to be the most universal are only those whose generality applies to many states and stages in TRUS and TRU in common with other universes. These are ones whose posits are diffuse, occur throughout the range of 0-10 in each and all of Á, L and T posits. They are those in which there is a large degree of sameness. Since this is largely a sameness of relativity, those laws which seem to be universal are often those that are of the closest and most general aspects of the FAF and relations of TRU.

The discovery, determination, development, perfection and utilization of laws are continuing processes. In IPTRU it is in an early stage of development. It is certain that both the format and the detail of laws can be greatly improved. The laws presented in IPTRU are not meant to be final.They will be refined and perfected. They are offered as examples for future development and perfection, as is TRU terminology. Each law is valid only in certain ways and within certain limits. Thereby we must always be careful not to think of laws as universal, absolute, or gospel truths but more as imperative guides within the set limits of TRU to which they apply.

Laws of the past, and by NAMTRU, are of many kinds, highly variable, often weak, altered and much abused. This is for two of the wrong reasons: 1. They are to aid some partial or what is little or against TRU, 2. They are only to compensate for wrongs, and are unlikely to be TRU. Many existing laws should be good within narrow limits and with slight adjustments as regulators of routines and mechanical devices in social interaction. If a law needs frequent change, it is apt to be invalid, poorly formed, very specific or outdated. Just as two negatives do not make a positive for life or TRU, or two wrongs do not make a right, our best to discover, so is it impossible to legislate and enforce unless this legislation and enforcement is based upon a thorough going wisdom, that of ACC and TRU. Cure more often lies in greater and more refined means of TRU, KOT, TES, and ACC.

Justice, as well as the justification for TRU, rests on the formalization of natural law. The time has long since passed when our laws can be primarily built on simple ENITS and reactions to specific abuses in society, civilization, or actual world.. It must be built on the fundamental principles of natural law, what the basic aspects, modes, form and function of this special world of human life is all about. This is what this presentation of TRU mandates. It is what justifies and explains laws in and of TRU, LA- and LO-, offered as a rough new language and for trial and error. This is not without precedent. Stoicism strongly advocated the importance of universal laws above the limited power of man. This was a guiding power in Roman Law that was continued through medieval time and continues in the present. It is imbedded, although often denied, unrecognized, or forgotten in current jurisprudence. Thereby let us look upon these laws of and in TRU as a continuation, expansion, formalization, and consolidation of all the great natural laws upon which our actual world, has, is and of necessity will be based. Since ours is increasingly a world of knowledge, reason and the judgements based thereon of which each and all of us share a common benefit and responsibility, it is necessary that we do learn, know, remember, apply, perfect and extol their best laws to the best of our ability. This is one of the most direct and effective ways of not only preventing much of the troubles, failures and threats to our world. It is, even more, the means to proscribe an increasingly accurate plan and way to assure human prosperity, equanimity, happiness, continuity and survival.

Since TRU is all, and we can share its immortality, let us give it our all. Let us have infinite spirit, find the glory and joyous victory of fulfilling our being, TES in TRU. This in all ways is what produces the greatest benefit, brings maximal permanence, and fulfills our lives, needs, wants, wishes, desires, hopes, life, continuity, and absolute being, eternally. These can be attained when convergent upon, reduced to, and achieved in TRU. The use of LOTS and LATS affords us with a magnificent means of knowing, showing, accounting, accentuating, acting and applying ACC and the fundamentals of TRU. This is shown in the sequence:

TRU → LAWS of TRU → POA → POLICY → PLANS → ACC

Just as we face the problems of distinction and order with our being, minds and language, so do we with our actions. By their concepts, principles, laws, mechanisms, TRU and its TU, TO, SUT, ROT, MOT can be reduced to order by POA, POT, SPALT, POS, HOM, and ACC to provide the basis for the best of laws, policies, plans and actions for success and survival.

MEASURES of POSITS of the SPECIAL FORM of ACC

```
ABISU              7    ~--0~--~    S
ACIP               4,1  ~6--4~-7~   M -
ADVICE, GOOD BAD TAKEN or NOT
     AGB           4,1  ~3--6-5~    M ±
ANTRAX,ANTREX, ANTRIX
     ANT           4,1  ~6--4~-7~   S ±
ASAJ               1,5  ~5--2~-7~   T ±
AT                 8    ~--~--~     S -
CAJS               1,5  ~4--3~-6~   T ±
CLASSIFICATION of LOTS, LATS
     CLL           7~   ~--~--~     S
CONDEMNING/COMPLEMENTING
    AT/T   CCA     1    ~4--5~-5~   T ±
CONVERGENCE CON    4-7  ~--~-8~     S ±
FEPTA              7    ~--~-9~     M
GOG                1,7  ~2--5~-2~   T
IMC                3    ~3--7~-7~   T
INIS               3     ~6-3-5~    M
INTOL              4,1  ~--~--~     S -
LAPU               4    ~8--4~-3~   S -
LARACA             1,3  ~7--4~-6~   T ±
LAT(S)             1,3  ~4--~-7~    T
LOT(S)             5,7  ~--~--~     S
LOTS and LATS in SPAT
          LLS      7~   ~--~--~     S
MAGOG              1,5  ~4--3~-3~   T
MOV                8,6  ~--~--~     S -
PAC                1-7  ~--~--~     S
PAF                3    ~--~-7~     M
PAFOC              1,3  ~6~4~7~     M
PLAN               4-7  ~4~4~-6~    T
POLICY,, POL       4-7  ~5-5~6~     T ±
PRESERVATION 1-7  ~7--3-8~   T   PRE
PROCEDURES in
   TREATMENT 4-7  ~6-5~-6~   M± PIT
RIGHT  RIG   1-7  ~3--6-5~   T
SELECTIVE QUALITY
        SEQ  1,5  ~5~5~7~    M
SOLUTIONS SOL4-7  ~--~-7~    S ±
TAKT         5    ~--~--~    M
TEKT         7    ~3~-~7~    M
TRIX         1    ~6--5~-7~  M
  HUMAN TÂ 1/ANTRU ENT LANT PRUNT
  LATANO-S LARP  REM LORVA SPALT
  MOCBIREM AIT PUREACLITY
  PRUTRIX (TRAX,TREX)
  - GOLDEN KEY 66-8 2-81;
  - COGNOS 75-11 2-727;
  - MEP 76-4 1.34;
```

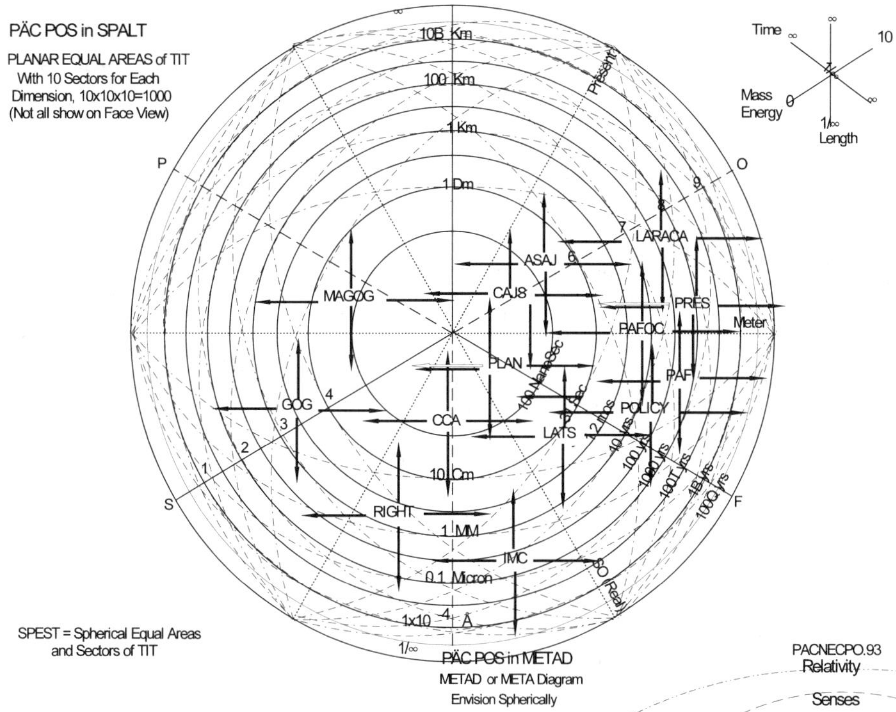

Fig 5-93 **PAC POS in SPALT**

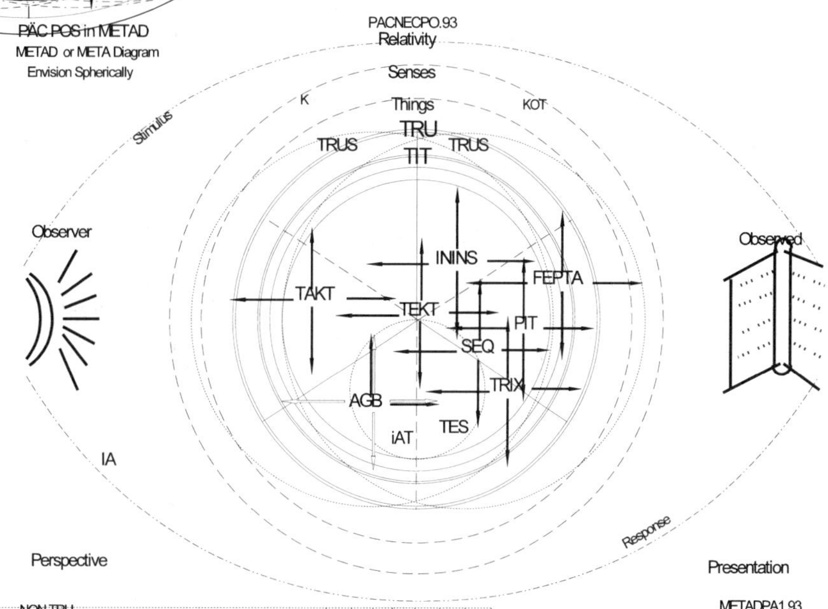

Fig 6-93 **PÄC POS in METAD**

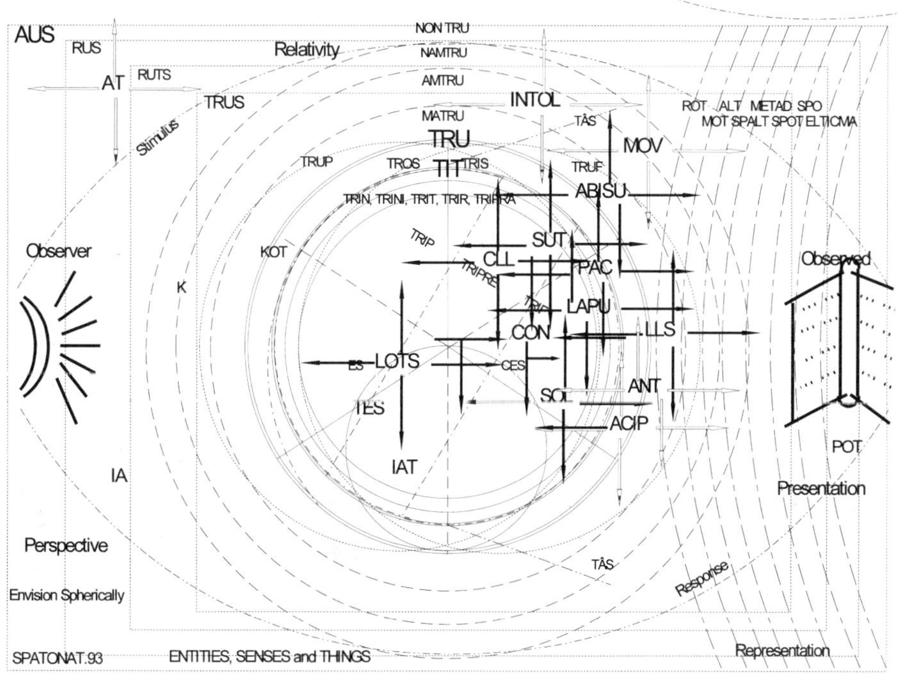

Fig 7-93 **PÄC POS in SPAT**

ACCOUNTING, SPECIAL ACTIONS for TRU
Human, TIT, HOM, DYNO, DYNS, DYNU, TRIA, POT, VAPO, Special, Tactics, Strategy, Thisness, TREQ, Life, TÂ, CES, LOPAN, God, SCARL-S, TES, RUPKAL, OITIO, ENT, OM, TRUF, LACOP, KOT, PRUT, PRACALT, LANT, PRUNT, EC, MADAS, ACEXIN, ALT^n, LATAN-O-S, LARP, SPALT, LORVA, REM, MOCBIREM, POS, STA, PRALT, LOPA, FAF, DYNU, DYNS, DYNO, LAMUA, WITÂ, AIT, COGNOS, PURE-AC-LITY, Acceleration, Accentuation, Answer-s, Anticipation, Application, Benefit, Compossible-ity, Conserve-ation, Control, Conviction, Educate-ion, Endure-ance, Enhance-ment, Error-s, Future, Loyalty, Meaning, Act-ion, Inter-action, Special, Force, Cum Div, Golden Mean, Operate-ion, Practical-ity, Stimulus, Response, Simple-icity, Use-s, Predict-ion, Prevent-ion, Resolve-ute-ution, Solve-ution, Success S/F, Tolerance, Victory-ies, Proof and how, Collect and Perfect, SMASH, CRUSH, Adverse forces, Solutions to all great problems, Training, UNCTA, SUT, HOM, Interest in TRU or TREQ, Less than Best, TRU Wisdom, OPTRU, Partiality, UNA, LABGEPT, right, mind, TRUD, literature, scenario-s, wants-desires, must, give, ENUCA, learn TRU, MEP, LOVET, VOC, aging, VACOL, CLACC, going to do, success, KOT, common right, communication of knowledge, LAPDUC, Depletion, civilization, RAPAT, OBSA, LOMEC, ACCAP, causes of, CORAC, DERGS, MARM, fear, decreases TRU, TATOR, HOM, Function, conscience, LATOK, REBE, FRE, AK, LADEP, PAND, LAPED, POMS, IRT, LOMRES, LOPU, ATIM, APROD, ACIR, ATIN, MED, LOSCOR, PRAGA, POIR, PI, LACRES, LEXPOM, PRAC, RECTRU, CONST, TRIND, LAPEXS, LAP, INEG, LAPOR, FUS, PIF, PIFS, INVOR, AWET, VOST, AITES, SOCIDE, PRER, LOPUT, CREPST, PAR, PURT, LECAS, LISACT, ROL, LADUT, PAK, PRUF, DUST, DUSTS, RUNS, UOU, TRICHE, LACLAP, COC, LIGS, LITS, LACORN, LEXPAC, COVPAC, APROD, PAFOT, PRADAC, LATAB, LADCEN, DOMETO, LARPON, LIT, HELN, PINED, POSTA, TRUPRO, WESO, POPAC, TAFOR, MOV, CACH, ACENT, LOPAP, LORA, LAI, LASPRE, METO, EPOT, DOPEMS, DUAC, POIR, UNE, SAPORS, TANCOW, LICAPT, LACOL, MIRPS, MARS, SICOA, LAFIRS, LARES, APTSO, RAMIT, POMRAK, FONPI, LAFRO, PADET, LAPRAL, DILATE, RECONE, CAD, LART, ACSURT, PROSA, REX, PRECOT, COMSU, OCA, LIRM, PROSOL, PESP, REFIT, FUF, MIPLID, DAREV, POIR, LASOC, LORSUD, COVET, LADU, POTDE, PPS, PAS, MOPLAR, CERIT, AJUT, LILT, MAREOP, TES, LASACA, SRAP, NUWREX, NUDAL, SAF, TRUC, WIT, RANT, GUT, CIOS, OBSAD, SIMA, POVIN, REPOM, REFIP, LODPET, LANITS, PIP, IN, POCON, POQUC, HOR, TOR, ROSOC, LOPAT, ATAN, RIK, LANEF, CONPA, COMPLOT, VETEN, NET, LACDAR, WIR, LAPAC, WIM, SWAT, LARCAN, LIR, MEA, STOCS, LABA, TAFOR, RECTRU, LART, LADOV, LACOMA, LAHIGO, RAPAN, LAPERA, TEF, LAMEK, TREXPO, UNCTA, TIKAN, DEBT, DIFS, FAS, IRAS, LITPRO, FACH, PPS, FADAL, SAT, RAD, ACS, EXINI, SOLA, DOD, RITES, LATOK, POFU, FISAT, INT, LACHES, LATAC, LORA, MARS, PIB, GAC, SUDRA, MAT, UNA, ITOD, LOAB, SEJONA, INCSET, MARP, TRE, KAC, CRAED, LAPLID, PACDE, SPACDE, NACDES, FUPIC, SIJ, PIT, FEXPRE, FAH, LAPUS, BAB, BAC, LOMI, LICON, DATBAC, CAPONT, LOMT, TOPA, DABU, PPMA, TEKT, MOV, CORAC, ANU, LERD, LESP, OFEL, FOKIT, METAC, LOCMEN, ACEXINOP, ALMAN, PROST, RETRU, ACCEX, NAMOV, WIM, MAPO, MARS, LOUP, DISEP, DISEPO, DISEPT, INVICT, WIPAT, TACOR, RAPOR, LOPURT, SOSA, LADIT, TACK, IMT, TIF, SIST, LASE, LACSA, PARCGI, LAGEND, SEQUAL, LAPAR, ITA, LASEV, LABGEPT, ACEX, SOLO, TRIALT, ACLO, PRIND, DORIN, LAMA, MUNAPT, LAT

PRUDAC Proof of ACC Derivation and Development
ILE, SRAP, NUWREX, NUDAL, SAF, SNUCOL, TRUC, WIT, GUT, SIMA, LAHOSE, LAHPRO, LAHED, PACDE, SIJ, NODA, COC, LATAR, FAH, LAPUS, BAB, BAC, LOMI, LICON, TEF, CAPONT, LOMT, TOPA, DAC, FARP, PODES, WOT, ROMP, TEKT, MOV, INC, ANU, LESP, FOKIT, LOCMEN, ACEXINOP, ALMAN, PROST, NAMOV, AWIM, MARS, LOUP, CONVET, CONRIT, CONUET, INVICTA, WIPAT

PRACA(PRUTRAX) Proof of the General Form of ACC
TREQ, Life, Human, 1/ATRU, OITIO, ENT, MADAS, LATANO-S, LARP, REM, LORVA, LOLVAR, COTRU, MARS, TOPA, TOVAR, TRE, MED, LARIV, LAREX, LAP, RAVAC, AWET, VOST, LINCTE, EFAR, LACLAP, ANTITRU, HELN, LARPIT, DOCT, TENT, HERI, POIR, MIR, POMRAK, POPTE, LART, OCA, SRIA, POIR, DEC, LASACA, ILE, RAPAT, TRUC, LIC, GUT, MAPER, LACMES, LODPET, LANITS, POCON, POQUC, PIDOMA, COMPLOT, WIR, WIM, LIR, LAMEK, LAMA, MIACET, PPS, SOLA, INT, MARS, TIF, MOPU, GAC, UNA, SEJONA, CEK, SOLO, UNDEXT, COVET, LILT, ATIM, TACOR, LOPURT, RAPOR, ERT, ERART, LADIT, TACK, IMT, TEX, POPA, TOPA, MERPS, MAB, TIF, SIST, TRAX, SIJ, COC, FARP, PODES, DABU, WOT, ROMP, LLAPOFT, PPMA, TEKT, MOV, INC, ASAJ, CAJS, PAFOC, LATRU, ANU, LERD, LESP, FOKIT, DOC, SUTAC, NAMOV, AWIM, DISEP, DISEPT, DISEPO, INVICTA

PRACA(PRUTREX) Proof of the Special Form of ACC
ACC, TRE, Human, TÂ, 1/ANTRU, ENT, LANT, PRUNT, LATANO-S, LARP, REM, LORVA, SPALT, MOCBIREM, AIT, PUREACLITY, LOLVAR, COTRU, MARS, TOPA, TOVAR, TRE, MED, LARIV, LAREX, LAP, LASACA, MARSAD, GUT, TREX, CAPONT, WOT, PAC, ACIP, FEPTA, LARACA, TEKT, INIS, MOV, INC, ASAJ, CAJS, PAFOC, LATRU, TESTAR, ANU, LERD, LESP, TRUNIV, KUPS, DOC, UBOT, RETRI, PROST, INT, SOLHUP, TRUX, RETRUX, PROFT, LIL, LACURT, NAMOV, LOUP, DISEP, DISEPT, DISEPO, WIPAT, RABNAL, PITRU, PERJUD

PRACE(PRUTRIX) Proof of Parts of ACC, Generally
ACC, TREX, TRE, 1/ANTITRU, ENT, LATANO-S, LARP, ALT^n, SPALT, REM, LORVA, MOCBIREM, WITÂ, AIT, LOLVAR, COTRU, MARS, TOPA, TOVAR, MED, LARIV, LAREX, LAP, LASACA, ILE, TRIX, AGENT, LAPUS, TEF, TRIXAF, WOT, TEKT, INTOL, MOV, INC, ASAJ, CAJS, PAFOC, SCEN, LUTNO, LANTLA, TESTAR, ANU, LERD, LESP, TRUNIV, KUPS, LASORT, RALGAW, CICIJ, OFEL, DOC, UBOT, FRET, RETRI, PROST, INT, INTRIX, SOLHUP, TRUX, RETRUX, PROFT, LIL, LACURT, NAMOV, AWIM, LOUP, CONVET, DOPT, WIPAT, TRUCORP, PITRU, PERJUD, AA, PESTRU, PEBTRU, PALTRU, PILTRU, PEVTRU

PRACI(PRUPUR) Proof of Parts of ACC, Specifically I, TES, Salvation, Soul, TIT, AC, DYNU, DYNS, ENT, MADAS, EC, LATAN-O-S, LARP, ACEXIN, ALT^n, REM, LORVA, MOCBIREM, SPALT, POS, STA, LADAC, PRALT, CES, WITÂ, LAMUA, AIT, ACC, COGNOS, Immortality, PUREAL, PUREACLITY, SUIT, ECRO, STRU, ACIR, COMET, KATIS, TRIST, REALÁ, RENTES, BIRST, LOTI, AWET, VOST, CITRU, ALET, AITES, TIAM, SATES, SALET, RECST, AIME, MADEC, KOCO, MANIFA, LAPERI, ISI, POIR, SAC, POM, OM, LATIND, LATEPI, RORO, LART, MIPLID, SRIA, LIDEN, ICOL, PROSEX, TESESA, VETEN, CARUL, SWAT, MEA, PIFS, DEBT, LAT, SRIA, EXINI, INT, WITÂ, LATAC, PIB, TIF, GAC, NECVIK, TESA, MAREOP, TES, ATIM, CE, LADIT, LASACA, LITA, ILE, COK, COMTEL, POP, LANTLA, ANU, TRUNIV, LEXIT, KUPS, LANTES, LANICT, COSTES, COSTRU, LASORT, RALGAW, PAC, SCIN, COGNOSID, ELTIR, OFEL, DOC, UBOT, ALMAN, FRET, PROST, INT, INTRIX, SOLHUP, PROFT, ACCEX, NAMOV, WIM, AWIM, MARS, SACOMD, PIET, EPOCHET, WIPAT, SACT, GACSAL, PERJUD, POJUD

PRACO(PRUCOG) Proof of ACC as a whole, and
COGNOS KOT, TRU, TIT, TO, TES, Á, PUREACLITY, KUPS, PERT, S/F, PREDOMINANCE, RUPKAL, ENT, OM, LACOP, Meaning, MADAS, LANT, EC, PRUNT, LATAN-O-S, LARP, ACEXIN, ALT^n, ALT, SPALT, REM, LORVA, MOCBIREM, POS, STA, LADAC, PRALT, AIT, ACC, PUREACLITY, Overview, Understanding, RORRE, LOPOIR, VAPO, LAINCO, KUP-S, LASUFA, COPOFT, LEXPON, ULS, TRA, ACEXIN, DIFIRT, LAVAC, KOCO, EQUA, ISI, SAPORS, SAC, OM, POM, LUV, SRIA, POIR, LIDEN, DEC, TESP, ULP, LADPOD, PENK, THEOSAF, CEK, IPEX, PORA, LAUTH, KOTA, ATIM, COGNIM, PAFOC, COSTES, COSTRU, SCIN, COGNOSID, ELTIR, LOTRAC, FOKIT, DOC, MARM, CONQ, SCONQ, DITHE, LEWOV, UBOT, PRACT, METAC, ENOS, MARS, APRES, LOCMEN, ACEXINOP, COS, THESA, FRET, PROST, SOLHUP, PROFT

PRUELT Proof of the Dynamics of ACC, and
ELTICMA TÂ, TO, I, IAT, SPALT, POS, STA, HOM, METAD, SPAT, The, RUPKAL, ENT, OM, TRUF, LACOP, OT, MADAS, LARP, LATAN-O-S, ACEXIN, ALT^n, REM, LORVA, MOCBIREM, POS, STA, LADAC, PRALT, LAMUA, WITÂ, AIT, PIP, LATANOS, TODIM, MANT, LOPOIR, ETACT, LACT, TRELIN, LADI, PORI, LAPIA, LEXUS, MI, AIME, EQUA, LAROP, LATET, OBAC, CUSA, POM, SOEX, CODIT, UALT, LADPOR, UM, LABIF, CEK, COSTRU, SCIN, SID, ELTIR, LOTRAC, FOKIT, DOC, MARM, DITHE, LEWOV, PRACT, METAC, LOREX, MARS, APRES, LOCMEN, ACEXINOP, COS, TREV, FRET, PROST, INT, SOLHUP, PROFT, LIL, LACURT, ATEQ, ELTIR-MA, CRUSH, ACCEX, NIP, ORD, NAMOV, LOP, LINT, WIM, AWIM, MAPO, SACOMD, LOUP, VIFS, VINPAC, PIET, VISO, LASCOM, COGNOS, LASA, EXAT,ACOM,

PRUSÂ Proof: Enhancement and Progression of
ACC and TRU ACC, TRAX, TREX, TRIX, PUREACLITY, COGNOS, ELTICMA, IPT, POT, KOT, TIT, TRU, TÂ, THE, ENT, OM, TRUF, LACOP, MADAS, LATAN-O-S, LARP, ACEXIN, ALT^n, REM, LORVA, SPALT, LAMUA, WITÂ, AIT, Fruition, Glory, SÂ, ETO, LACTEL, OCA, TOPA, MET, AEX, LASUF, ONTA, PIP, TESAR, TEL, CONST, LACTAR, TAC, TAT, LET, TEA, AWET, ALET, UTIA, PURT, PODIM, LOIM, TATOR, CIMAT, LAPEN, TRUCHE, LAMPIT, RAR, LAROP, WESO, ULTIT, STAT, LABIC, COMTRU, DEP, LAN, AMEXT, TELEOP, SAC, OBAC, ALTAC, OM, CUSA, PROPERT, SOPT, POR, STAT, LABIF, GAC, THEOSAF, WITA, PROMGO, CEK, SAF, FOKIT, MARS, THESOA, PROST, ENWIPT, LOP, WIM, MAPO, RABDI, LOUP, VINPAC, CONRIT, CONUET, ACON, PIET, LAMENT, LASCOM, INVICT, UVICTA, ELTICA, ATIM, COGNOS, LASA, EXAT, LATEVU, LAREQ, ACOM, KNI ,LOSOW, SA, TATOR, MERPS, POJUD, COGIF, AA

PRUTA Proof: Summary and Conclusions of ACC, TÂ, and TRU
TOPA, MET, AEX, LASUF, ONTA, PIP, TESAR, TEL, CONST, LACTAR, TAC, TAT, LET, TEA, AWET, ALET, UTIA, PURT, PODIM, LOIM, TATOR, CIMAT, LAPEN, TRUCHE, LAMPIT, RAR, LAROP, WESO, ULTIT, STAT, LABIC, COMTRU, DEP, LAN, AMEXT, TELEOP, SAC, OBAC, ALTAC, OM, CUSA, PROPERT, TÂ, SA, ETO, LACTEL, OCA, SOPT, POR, STAT, LABIF, THEOSAF, WITA, PROMGO, GAC, CEK, VAPUT, ETBU, WIM, MAPO, MARS, RABDI, LOUP, ASENT, VINPAC, ACON, LAMENT, LASCOM, AI, UVICTA, ATIM, COGNOS, LASA, EXAT, LAP, PUA, LATEVU, LAREQ, ACOM, LEXUS, LEXUSA, LOSOW, SA, TATOR, MERPS, AIME, COGIF, AA

CHAPTER 94

GENERAL PART and PARTS of ACC

DEFINITION and MEANING, FIELDS and SCIENCES, TRIX, SOUNDS and LETTERS, LANGUAGE, TERMS, MEASURES, TESTING, EVALUATION, ACTIONS

We must present the great ideas, concepts, principles, laws, corollaries, mechanisms and systems of TRU even though their explanations and means are rudimentary, at times seemingly trite, in error, or incomplete. This is partly the result of the non absolute nature of differentiation and distinction, or a non absolute reality. This is a result of the ID and same/other, or DAOST continua, wherein all that is not completely TIT-TO will more or less lack absolute separation from all else of TIT-T0. So that one of our greatest tasks in life is to, with courage, conviction and determination, continually bring distinctions in an ever more refined state, yet remembering their limits. This can be successful only when ordered by TIT, TRIT, OBT, SPALT, ACC and POT. This initial presentation is often more to show their existence, how they are in TRU, and what can be done, than to offer them as definitive products. A massive effort should be made to include all the best information, methods and principles in all fields and sciences that substantiate and benefit TRU. This should be directed toward what is most: 1. Valid, as proven by their respective sciences and fields, 2. Basic, to each respective science and field in TREQ, 3. Actual or that which corresponds most precisely and fully with TRU, 4. Proportional.(VAPO). 5. Relative, important, and valuable in TRU,(REM) 6. Total, or best provides a complete and comprehensive overview of TRU. 7. Lacking or fills in the gaps in our knowledge of TRU, and 8. Related and necessary for TRU, POT, and IPTRU. Their use is often a combined one with much overlapping and dynamic interaction in replicating TRU. Thereby they will have values in multiples, and resultants that will need a variety of views to show their overall contribution.

To divide and conquer, and thereby control, is a highly effective strategy. It is even more important as a part of the ACC of TRU. Yet it is often limited by: 1. Must have TIT-TU and this with increasing relationships, 2. Must have SUT, SPALT and TIT-TO and this with increasing precision and expansion, 3. Must have the means, forces and the power to make it continuously effective, 4. Both execution and result must be guided by nothing less than the overall benefit for TRU.

For those and all that would injure or destroy TRU the strongest and persistent corrections are required. This is where forceful and determined actions are sometimes needed.(SMASH,CRUSH,97) This will depend to a large extent upon what level of ACC we are dealing with, whether TRAX, TREX, or TRIX. Yet, in the near term, we must learn to live with some of the AMTRU and UNTRU worlds about us as long as their ill effect on TRU is minimal. ACC shows how an all out attack, or open season is permissible and often to be recommended on any and all that could or would destroy TRU. This however must not be wasteful or abused. It must be within the bounds of VAPO, appropriate SCARLS, and TRU. It must avoid indirect harm or waste to TRU. It should be directed at much that besets TRU, TREQ, and TES, especially in their efforts for TIT, ENT, ACC, TRAX, TREX, and TRIX.

TRIX = Exclusive, Relentless and powerful ACC for The Relative Universe, Itself, TIT and TRE. TRIX denotes all action taken for TRU in which a conscious realization of TIT is complete and focused upon greatest dangers to TRU and TIT. TRIX is not new, for many religions, ISMS, associations, societies, and nations concentrate on and emphasize what they can accomplish through specific planned action. There is a proper time, place, and way, although limited, for the martial of TRU.(MARS) The use of TRIX can be delicate. Their use by activities for set causes can be partial, disproportionate and negative with potential adverse reactions. However if these are known, remembered and used with a maximum of benefit/damage, it can add much to the differential and visible power of ACC, to the relief, benefit, and ENT of TRU.

There are a number of does and don'ts that show how the parts of ACC can be effective in life. Does are all that support the fulfillment of life, all necessary that is not excessive, all necessary that is not insufficient. Look for the deepest or TRU meanings in all, what is VAPO or valid, actual, proportionate, and total by TRU. What are its SCARLS, the different states conditions, limits, and relations of TRU. What are the different and same by TRU, all that makes ones conscience coincide with TRU, and all that stimulates our best and greatest interests, inspiration and revelation. There are among the **Don'ts**, all that fail to acclaim what is TRU and disclaim all that is not TRU, be ashamed of TRU or unashamed of all not TRU. Likewise don't be ashamed of or avoid what you are of TES and of TRU. Do whatever is necessary if it is TRU. You will always be proud of not doing anything your TRU conscience tells you not to do, anything that you have learned by TRU that hurts you or other of TRU, anything that society, as TRIX, holds to be universally wrong or repugnant, and anything that causes loss to TRU.

Eleven of the greatest needed projects for TRU include: 1. Future work that will most contribute to the presentation and advancement of TRU. 2. Collection and incorporation of all material diversely spread for tabulation and maximal development and presentation, 3. Collaboration of ten to one hundred or more of the best authorities in various fields, sciences, philosophy and religion for a more advanced and intensive encyclopedic work on TRU, 4. REMS and DREM that further establish the principles, methods, and measures of DRE and REM, 5. SPALT toward an increasingly perfect Spherical Model of TRU, and its varieties, 6. Glossary, lexicon or dictionary of all terms, both existing and new, in which their most explicit and profound TRU meanings and definitions are made, 7, Tie in with all works in history, science, technology, industry, commerce, religion, philosophy to greatly strengthen both themselves TREQ and TRU, 8. Indexing all terms as they are in TIT of TRU, 9. New and existing concepts that can be best developed and used, 10. Discover and record the key gists and cues to all unknown ideas that are of prime importance for TRU, 11. SPOT, METAD, and OITIO that can be best dissected, and superimposed to show all models, charts, diagrams, and illustrations, 12. Collect all diagrams, illustrations, drawings, pictures, graphs and means of representation from the literature and scientific world to compare, extract, and adapt their optimal meanings by and to TRU. Show all limits, fallacies, errors, and best means of presentation. 13. Color rendering of all major drawings of TRU, ALT, SPALT, NAT, FAT, BAS, METAD, and SPAT.

In the interpretation, study and analysis of any author and their work the reader must be careful to primarily use the authors meaning. This is called: **AMU** = Author's Meaning or Use. AMU is contingent upon the work or text's production and quality and the importance of its topic. This is whether or not it is identical or equivalent with the topic being reproduced. This is called: **PIO** = Product, Production or Presentation being equivalent or Identical with the Object, the Subject or entity being depicted. Regardless of literary style or acceptance, a work can be no better than the degree to which it represents its topic, and this topic to its part or role in TRU.

This is the purity, quality, quantity, and greatness of the topic in TRU. **LIMRET** = Literature is Measured by how well and how much it Represents its Topic, and how well and how much this Topic is TRU or TREQ. Its success largely depends upon its concepts and terms, how valuable, their interactions in TRU, and how well are demonstrated.

What will the world, TRUS, TRIS, TIT, and TÂ, be like at different times in the future? If TRU and POT are well developed and represented we should be able to add much to our understanding of the future. Although we can roughly demonstrate their general nature, form and function, and much about them at different times, we can't predict or forecast every explicit specific detail, e.g. individuals or their IA and specific acts..As our existence becomes more TRU they should improve. They must abide by the FAF, concepts, principles, laws, and mechanisms of TRU. Although set times in the future are infallible by TIT, they are neither perfect by SPALT, nor of absolute certain content by TRU.

TWENTY-FIVE GREATEST NEEDED DISCOVERIES

All that has not been discovered or whose discovery is so poor as to have been never utilized needs to be provided. Since these have been, and still are, largely unknown or poorly known, we must be careful for lack of accurate closure. Yet accept their challenge as they are: 1. TRU, 2. Á, 3. TES, 4. ALT and SPALT, 5. POS in SPALT and associated figures, 6. All fields and sciences key TRU ideas and laws, 7. Continued refinement of REM, 8. Greater corrections for coterminalities in REM, 9. Greater corrections for overlapping in POS, 10. Better TRU education & training of all persons, 11. Ways for greater TRU motivation, 12. How to provide best TRU opportunities, 13. How to increase satisfaction with TRU, 14. How to teach all to be most selective through TRU, 15. How to best enlist the TREQ in human life, 16. How to eliminate all economics that is not TRU, 17. Help all learn best reproductive selection, 18. How to minimize all waste, conserve that needed, 19. More precise near-mid-far term ACC Timetable, 20. Develop massive perspective of all TRU, systems, 21. Discover the missing information of these systems, 22. Support and expand contributions to and from TRU, 23. Procure most qualified, tested and selected leaders, 24. Best stimulate youth to inspiration & revelation, and 25. All other voids and unknown great TRU ideas, including control of determining forces and CES.

Nearly all fields and sciences, more or less, have application and representation in TRU. Only this is often different to the point of seeming strange and unbelievable to any observer or current authority. It will require the efforts of many versed in TRU and many authorities in all fields and sciences working together with increasing understanding on the part of all combined as TREQ, TRU, and TIT-TO. The great need for TRU terminology is signified by: **SCEN** = Every New Science requires its own NEW Concepts and Terminology. Nowhere is this more important than for TIT, and its expansion in ACC to creates a science that encompasses all sciences, TRU.

One of the greatest methods for determining and demonstrating all needed to enhance TRU and ACC is by asking questions. This is especially effective when the questions and their answers bring out the best and greatest in TRU . It is in this three pronged attack of TRU concepts, making questions that best fit these concepts, and having the accurate posits in an approach to TIT-TO, that ACC can be best given. This is called: **LUTNO** = Law of Union in TRU of New Terminology, Interrogation and Total Order. LUTNO is the method of specific participation in ACC that elevates questioning to ever higher levels by basing our dialectics, reasoning and intellect upon a VAPO like union of TRU, use of a New Terminology for all its greatest components, and the coordination, ordering and total comprehension that of TRU. In this way we need not stop anywhere along the line with answers that are minuscule or wrong. We need not stop with one or a few profound ideas, insights and inspirations that good questions and answers can provide. We need not stop with many profound insights that can be brought together in ways that lack total identity and the positive objective approach SUT, ROT, ALT, SPALT, POS, STA, HOM, DYNO and COGNOS provide. We can and must continue to ask these great questions, stimulate our brains and minds to best use our reason and intellect. Thereby gaining the best of existing knowledge, adding needed new concepts, having the structure to posit them in and manipulating them in all ways they exist in TRU. This is to be done so that we can achieve that glorious overview, with accuracy, validity, ACEXIN, and totality that brings us to a 1:1 identity and reality with TRU. LUTNO shows us the great justification for IPTRU, NEC, ALT, SPALT, POS, HOM and ACC. Furthermore it helps to prove TRU through the kind of perspective and comprehension necessary.

One of the greatest advancements and revolutions of human development was representation. This was first done by gestures, sounds, meanings, symbols, vowels, consonants, and words that progressed to use in writing. This continued into the combinations of the words, ideas, concepts, added to the forms now existing in composition and communication. Different languages have used many different kinds of gestures, sounds, writings, and combinations. This shows both the novelty of language and the diversity that exists in TRU. When the language is generalized and new it can lead to rapid and great progression and expansion. When specialized it may be more temporarily successful, yet will sooner lose power, adaptability, and predominance. For TRU, a universal language must, have VAPO. It should use all the sounds available, in the best balance and ways. Yet it should more or less depend upon the natural strength of each sound, syllable or combination. In the expansion of vowels, and addition of unused consonants, a new symbol should be available for each sound. The use of the specific sound for each vowel and consonant will make language much more efficient and meaningful, especially when directed toward TRU. Much more it will enable greater breadth, distinction, and avoidance of multiple meanings thereby help to separate and control coterms. The additional use of different letters, vowels, consonants, combinations, and syllables, for directly matched greatest concepts, principles, and laws of TRU will make all TRU, and each person's life more successful and fulfilling. It should help to provide a systematic method for the propagation of TRU that will revolutionize, accelerate and perpetuate ENT, TRU, TRUIT and TÂ. Vowels and their sounds in common and current use are as follows:

Á á as in fate, chaotic
Â â as in absolute, add, ask
Ä ä as in ab, arm
À à as in sofa
Å å as in loud, now

Æ æ as in bang, pass
É é as in mete
Ê ê as in met
Ë ë as in silent
È è as in maker

Í í as in iceÎ î as in ill
Ï ï as in charity
Ì ì as in
Ó ó as in note Ô
ô as in obey

Ö ö as in orb
Ò ò as in odd
Ú ú as in cube
Û û as in unite
Ü ü as in urn
Ù ù as in up, circus

These are neither complete nor universally accepted, yet they follow the Multinational Character Sets for computer characters. They, with all other vowels, consonants, and combined sounds, are initiated in LANT and will become the basis for much of future terminology. Each letter and sound should stand for a major constituent, aspect, form, function, parameter, coordinate, dimension, feature and variety of TRU. In this way it should eventually be possible to have not only the most simple and effective language, each word will be identified immediately from its construct and every one can more readily learn TRU and achieve TES. For the acceptance and future development of a TRU language they are not offered for unselective or unqualified inclusion and usage in existing language. Their continuing study, analysis, corrections, revisions, improvements and replacements should be a major ongoing, organized, and controlled TRU effort. They are presented in such great profusion and detail largely to show what can be done, provide a basis for the great aspects, parts, and processes in TRU, and to show their distinctions with posits in its model. Some, in being of TRU rather than in TRU, and having added foci, give us the overall view which is important, not so much for TIT, but for its presentation, and our own subjectivity and identity in TES and TIT. Since each letter of the alphabet, both consonants and vowels, have been used to symbolize, and should stand for, a major or fundamental idea, they can be expanded to better represent more fundamental TRU concepts, principles, laws, features, aspects, parameters, coordinates, and parts of TRU. It is then necessary that we formulate as precisely, validly, proportionately, systematically, orderly and completely as possible the correspondence between these senses and things in and of themselves. Their symbolic representation and presentation of TRU then becomes their ES and CES in a living TIT-TO.

Capitalization and bold representation of TRU terms are used both to set them apart from existing terms as well as to emphasize and clarify their significance for a reader's understanding of TRU. To have it reduced to a minimum is: **MIRC** = Minimum Reduced Capitalized New Concepts or TRU Terms. The fact that the world is rapidly becoming more unified through television and communication, as well as travel, makes it ever more important that the languages used are the best possible. It is for this reason that, as early as possible a presentation of TRU and the language necessary for it, be developed and promoted. For a unified world language that is incomplete, disproportional, largely figment, and illusion many world problems are only accentuated rather than alleviated.This does not mean, however, that the world, TRU, should consist, at least any time soon, of one arbitrary invariant language. Therefore great attention and effort must be paid to make both languages, that exist and are in development, as flexible, all inclusive, and well ordered of TIT as possible. We cannot think of TRU, nor even TREQ, unless we have language by which this can he done. If we do not know the concepts, principles, laws and mechanisms by which TRU operates and which make TRU distinct, we can never gain any adequate or satisfactory understanding of it. We may at length describe it with what words we have, as is done in philosophy, religion, science and the entire academic and literate world. But these are lengthy and cumbersome at best, and quickly lead to variance, tangents, disproportion, fragmentation and distraction of our attention on TO of TIT. There are certain terms like God, spirit, soul, universe, reality, form, and idea that elicit some notion of what the world and our actual part in it consist. However they are far from providing the formal distinctions that are TRU. It is for these reasons that a new language of TRU is so imperative. **LANTLA** = Law of the Absolute necessity of TRU Terminology or Language. The more we think with TRU terms and the more precise and defined these terms become the more TRU is. This is, both by its certainty, as well as our being with it.(SÁ),

Fig 1-94 **CATEGORIES of CONNOTATIONS**

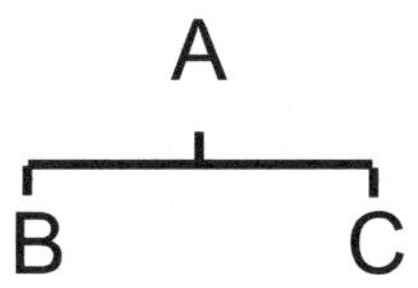

For language without convention, like standards, or trying to understand the world or TRU without TU, SUT, ALT, ROT and POT, there is no order and opportunity for our predominant mass-energy dimension, Á, to fulfill its part and role in TRU. Even better, however, we should strive to provide their formal distinctions by being representative of that aspect of TRU to which they pertain. To distinguish the different kinds of meanings there are: **MENEX** = Meaning from ENIT unless otherwise stated, or excepted, **MATIT** = Meaning actuality, Identity by TIT, not TIT from meaning in TIT, **MATAM** = Meaning by TIT and TIT by meaning,(OT) For example, if we have a term A and if this can be taken in two ways, B and C, where B is a very good connotation and C one that is bad; then if A is used without B, it becomes equivalent to C which is bad. If A is used without C it becomes B, which is very good or positive and far different than A generally. This is called: **LADES** = Law of Deductive Selectivity. LADES is formal distinction with perfected use of meanings adapted to TRU. By learning the different meanings of words, comparing, adapting and best applying them to TRU we operate by LADES.

The test of an entity, sense, thing, person, etc. is to see if it is or is not TRU. Does it, to what extent, and in what way does it participate in the bridge of homo sapiens in TRU? By illustrating time horizontally and kind of AUS-RUS-TRUS-TIT-TRUS-RUS-AUS vertically the focus being TES and TÍS, it will be seen that most of us are not much TES. And this even if we try. Yet a great part of what we are is absolutely critical to TRU, by LOPAN. Thus it depends upon how much we can increase TRU and ourselves to it. This is **LATEST**, the Law of the Test of TES and TRU.

From testing and evaluating it can be seen how variable the positing of each entity or item may be. This depends upon several things including whether they are considered: 1. generally or specifically, 2. randomly or by TRU, 3. worldly or personally, 4. positively or negatively, 5. wholly or partially, 6. by all their meanings graded by best language or by one meaning as popular or fashionable, 7. by one meaning or another meaning, often quite different in a good or bad connotation. Programs and measures used to test ATRU, TRU, and TES are parts of ACC that are effective means both currently done, can be adapted to TRU or can be developed. Psychological, social, physical and medical measurements that provide a basis for ACC include: 1. Benefits, 2. Cheerfulness, 3. Hardwork, 4. Honesty/hype, Reason/Sophistication, 5. humility, false popularity, as in legends in their own time, 6. Civility/politics, 7. Frugality/extravagance, 8. Moderation/violence, 9. Best/bad language, 10. Sincerity/insincerity, 11. Principle/lack of principle, 12. Work/lethargy, 13. Education/Illiteracy, 14. Prevention, 15. Health/Disease, 16. Purity/impurity, 17. Equanimity/rashness, 18. Tolerance/intolerance, 19. Courage/cowardice, 20. Strength/weakness, 21. Quality/ Quantity, and many others. They are much of our lives, S/F, Á FAF, and TES. They are much of what drives humanity, provides TRU even without TRE. Each in its own way provides a simple means of fulfilling needed TES and TRU. With allowances, adaptation, and other particular means they more accurately and practically show TES, TRU and ACC.

If we attempt to manage and control our world by tests without adapting and accounting ACC to TIT-TO we will often find we do more harm than good. There is a best time, place, and way, ALT, for tests, ACC, as well as all actions. We can't act effectively unless we know what this time, place, and way is. Thereby the test and measures of ACLAP, ACC, and their actions are to be found in POA, the SPALT, POS, STA, HOM, DYNU, DYNS, DYNO and ACC, METAD, SPAT and ELTICMA. It is these that enable each and all of us to avoid the pitfalls of either excessively simplistic unity, excessively complicated infinity, or non TRU finity. They enable each person to have the knowledge necessary for a perspective that yields the great ability to voluntarily contribute to TRU.

Testing for TRU should be to determine the interest, insight, enthusiasm, vigorous pursuit, conviction, and effectiveness, of the best in life. Information, data, evidence and their memory are helpful bases, yet they must be reconciled in TRU to show those best attitudes that most assure continuation of TRU. This will tend to guide the rest. Also needed is to determine freedom from bias, reaction formation, being partial, resistance, or other tendencies that inhibit ones inclinations to TES, TRU or TREQ. **TESTAR** = Testing to TRU and Rating, Testing and Rating of TRU. How to test for and develop critical decisions in life are few compared with IA. Good raising, training, education, and experience are not enough. It takes life long habits of good thinking and judgement based on wisdom and virtues of TES and TRU. TESTAR is much more than a simple test of political candidates for election to determine their potential for responsibility and being effective TRU leaders. It is much more than a rating of various societies, associations, interest groups, or ISMS for a valid rating of their potential benefit or harm to humanity, TRU. TESTAR should be a profound system whereby all sorts of information is collected and computerized in such a way as to provide the most fair, effective and quickest evidence for decision making, key to ACC. It should enable one and all to work toward the best ends attainable. These are not so difficult once learned and practiced, their benefits are much greater and rewarding.

A complete evaluation, testing, grading, reevaluation, rating, promoting or demoting of all persons in positions of leadership by TRU should be determined by the extent they and their work is of and for TRU. The degree to which leaders are incapable or unwilling to continually fulfill their responsibilities is the degree to which they should be prevented or removed from harm to TRU. If they do not have the capacity, ability, quality, character, good judgement and knowledge necessary to fill their important roles they should either be guided to rise to it or be reduced to that level they are fit for. This is why it is important to know TRU and be TES. This is why it is important to be able to test all, especially those in positions of leadership. Such testing and the measures to be developed can be written, oral, situational or interpretive, whichever or in combination are most effective in evaluating performance.

Competitive games and contests, not only physical events like college and professional sports and the olympic games, but events in which the total population of TRU is tested competitively, need to be devised. They should include much of both the smaller finite as well as greatest REMS, principles, laws, processes, posits, models, and overview based upon a good cross section of the ACEXIN of TRU. Especially what measures the higher levels of Á FAF in TRU should be provided. This should be done continually in programs that provide optimal benefit. This would be a most effective way of encouraging the voluntarily and selective processes of all individuals, groups to stimulate TRU. These should target and determine proficiency in TES and TRU. There are many different kinds of events and testing in the world today. Yet they lack three vital modes: 1. Are not TRU, 2. Do not follow VAPO of TRU, 3, Are not guided by or conform to Á. These must be present before the edge necessary to raise individual effort to its limits in TES and TRU are achieved.

To get the best men and women who are most interested and dedicated to their jobs, use the best reason and imagination, and utilize their abilities and responsibilities for effective and efficient profit, are key to corporate success. This applies to all human activity and especially ACC. It includes helping them appreciate the great work they are in, TRU, develop TES, inspiration, revelation, and ACEXIN, plus a hope for the future, sufficient security, rewards, and compensation that make it well worthwhile. Since responsibility tends to gravitate to those shoulders who best carry it, it is imperative that the shoulders of those working for TRU be the most numerous and strongest possible. These are personnel that are both highly capable and inclined to TRU or TREQ.

Much of the particulate of ACC is in how to best anticipate, focus on, search and destroy the ATRU and ANTRU at the receptor and effector sites on TIT. In life, the older we get the more aches and pains we have to live with, yet the more we learn to live with them. The trick is in learning to distinguish which are minor or major, acute or chronic, and how to best react to each. Especially important is to learn and manage those which are major, of high risk, or life threatening. We must analyze, focus on, emphasize and investigate all that is likely to become much worse. In life we must live with much that is ATRU. Yet we must constantly anticipate and attend to this whole gamut that confronts us. **SADANT** = Search and Destroy the AN, ANTI, TAN, 1/TRU. SADANT is an ongoing, everlasting and multifarious process, what occupies much of our lives. It works both in and out of TRU. It is especially focused upon the receptor and effector sites on the periphery of TIT-TO, the TRU/AU border.

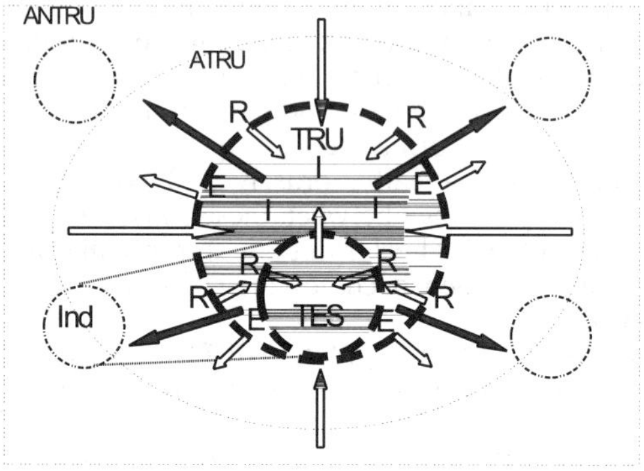

Fig 2-94 **ACC and the INTERACTIONS between TRU and NOT TRU**

This shows some of the interactions between TRU, TES, ANU and all not TRU. We emphasize the sites of interaction. These are given by, I inside TRU, E or Effector, and R or Receptor in TRU or at the periphery of TRU. It is to understand and control these that both TRUD, CES, and ACC best enable TRU to function and succeed. What do we ignore, leave alone, help or destroy? We help all that is TRU, and also help depart from, or avert all that is negative for TRU. We destroy whatever seriously threatens TRU. We selectively and accountably destroy the bad, whether approaching, entering, or already effecting TRU. We destroy the bad that causes the good to leave. Remember that those who are wanting and demanding are often those who have missed their chance, have not worked for their part in TRU and expect TRU and all else to be given them. These are called: **ANU** = Against and Nots, in the world or Universe. ANU are those who are not only not TRU, but are also against TRU. Each individual rapidly, frequently and alternately moves in and out of ANU. Many other features and seeming indicators are much less significant. It is ANU in us, others and all, that we should constantly guard against, help to correct, educate, train, and counsel when possible. The TES in us carries great power if yielded appropriately with all the rest that supports TIT. One of the most effective ways of achieving the desired end is to lay out what is the most important priorities, and make plans that best attack each. For TRU this will depend upon what stage of ACC we are working from, or is the given. For ACC, as all, this is not always TIT-TO, only in a final accounting. There will be, like REM, differential items, problems, goals, ends and targets to be focused on. These may be divided by the parameters we use in selecting the priorities, and ordered by their overall coordination in SPALT or SPAT. This is or is not directed toward TRU generally, ACC generally, ACC specifically, TRIX, or other aspects of TRU. Some are common, including the optimizing of: TRU/ATRU ANTRU, TES/IND, IAT/IA, REM/non REM, SPALT/incoordination, POA over chaos, ACC/inaction and wrong action.

	AU AT	RU RU T	PO T EP I	IND IP- TRU	BAS	IA S, O	SO	FAT SPO	VA- PO	TR- US	TRU	KOT IAT	PRU T	AT O SO T	TÄS SCA RLS	Â	TR- OS TU	TRIS TIT TRIT	TR- IA TÂ	ESIN ESTE TES	TO TRUD ENT	OI- TIO OM	SUT ROT SOI	MOT ALT SPA L	REM POS STA	HOM DYU DYS DYC	AC
BEING		4+	1	2			1	1	1	3	4+	2	4	3	2	3	3	4+	4+	1	2	1			1	2	3
BENEFITS	x		1	3		1	2	1	1	2	3	2	1	4	2	3	2	4	4+	4	4	2		2	3	3	3
BODY/MIND			1	2	2	4+	4+	2	2	1	1	2	3	3	2	4	1	3	1		1	3	1	2	1	1	2
CHOICE	1	2	1			1	1	1	1	1	2	3	2	2	1	3	1	2	2	4+	2	1	1	1	2	3	4
COSMOLOGY	2	3	3	4	2	1	1	4+	3	4	3	3	3	4	3	3	4	4	4	1	4	4	3	4+	2	2	3
CREAT/EVOL	1	2	2	1			1	1	1	4+	2	2	3	4+	2	3	2	3	4	1	3	2	2	4	2	2	3
DEATH/IMMO	1	3	1	1		1	1		1	2	3	2	2	4	2	3	1	4	4+	4+	3	2			2	3	4
DETERMINSM	X	1	1	2	1	2	3	2	3	3	3	3	3	4+	4+	4	4+	3	4	2	4	3	1	2	3	3	4
DOUBT/SKEP	4	3	1	2	2	3	2	2	1	2	2	4	4	3	3	4	2	3	4	4	2	4+	2	3	3	3	4
DUTY	X	2	2	2	1	1	2	1	1	1	2	3	3	2	1	3	2	2	3	4	4	3	1	1	2	3	4+
FREEDOM	4	2	2	1	2	3	2	1	3	2	2	3	3	3	1	3	2	3	4+	4+	1	3	1	2	2	2	2
GOD/DEITY	X	x	2	3	1	1	2	2	3	2	4+	2	4	3	3	3	4	4+	4	1	3	3	1	2	2	3	4
INDUCTION			4+	4	2	1	2	2	2	1	2	4	4+	3	1	2	1	4	3	2	2	4	2	2	3	2	3
KNOWLEDGE	1	1	4+	3	2	1	2	3	2	2	3	4+	4	2	2	3	2	3	3	2	3	4	2	3	3	4	4+
LANGUAGE		-	2	3	1	2	4	1	2	1	2	4	3	2	2	4	1	3	3	2	2	2	4+	2	4	2	3
LENGTH/TIME	4	4	1	2	1	1	1	3	2	3	3	2	3	3	3	1	2	4+	3	2	3	3	2	4+	2	4	4
MEANING		1	3	3	1	2	4	2	3	2	3	4	4	2	1	4+	2	4	4	2	2	3	1	2	2	3	4
METAPHYSICS	2	2	4	4+	2	1	2	4	4	3	2	2	2	2	2	1	2	2	2	2	3	3	2	3	1	1	2
MORAL/JUDG			1	1		1	2		3	1	2	3	3	2	2	3	4	4	4	4+	4	2	1	1	4	4	4
PERCEPTION	1	1	3	3	4+	3	4+	3	3	1	2	3	3	4+	2	4	2	3	2	2	2	4	1	2	2	2	3
PHYSICAL W	4+	3	1	3	2	2	1	1	2	2	2	1	2	2	2	2	2	2	2	2	2	1	3	4	3	3	2
PLAN		-	3	4	1	1	2	2	3	1	2	3	3	1	4	2	3	3	3	3	4	2	1	2	3	4	4+
PRESERVE	1	2	1	3	1	2	3	2	3	3	4	2	3	2	1	4	1	4	4+	2	4+	3	1	1	3	4	4+
PROOF/CERT			1	3	1	1	2	2	4+	1	2	3	4+	3	2	3	2	3	3	2	2	3	1	2	3	3	4+
PURPOSE		-	2	3	1	2	3	2	2	2	3	4	4	3	1	4+	1	4	4+	3	3	3	1	2	3	4	4+
REALITY	-	1	2	3	2	2	4	3	2	1	2	4	3	4	2	4+	1	3	3	2	3	4+	1	2	3	3	4+
REWARD		-	1	4+	1	2	3	2	3	2	4	4	4	3	2	4	3	4	4	4+	4	3	2	3	4+	4	4+
SALVATION			1	3	1	1	2	2	3	2	3	4	4	4	2	4	4	4	4+	4+	4	3	2	3	3	4	4+
SUCCESS	X	x	2	3	1	1	2	2	3	2	4	4	4	4	2	4	2	4	4+	4	4	4	2	3	3	4	4+
SURVIVAL	1	3	2	4	1	2	3	2	3	3	4	4	4	4	2	4	2	4+	4+	4	4	4	2	3	4	4+	4+
TRU-TÂ	X	x	3	4	2	1	2	2	3	3	4	4	4	4	3	4	4	4	4+	3	4	4	2	3	3	4	4+
WORK/REST		1	2	3	1	1	2	2	3	3	4	2	2	4	2	4+	2	4	4	4+	4+	4	2	2	3	4+	4

Table 1-94 **PROBLEMS and SOLUTIONS by TRU**

In this table we list some of the greatest human problems and the concepts, principles, laws, and mechanisms that answer, solve or relate to them. The former are largely the themes of the text, and the latter its substance. The order of these follows a natural TRU and IPTRU sequence. They are rated from X and x which is in opposition to them, to 1, 2, 3, 4 and 4+ depending upon the degree to which they are related, how much needs to be done and how well they are achieved. Some abbreviations are used including in the rows: creat for creation, evo for evolution, immo for immortality, skep for skepticism, tim for time, judg for judgement, w for world, preserve for preservation. Of the NEC in the columns we use their primary acronyms. Although rudimentary, this is one of the most important tables in the text for it incorporates much of the text and its substance. It provides a quick overview of the problems human life encounters, what they are related to, explained by, and how best resolved by ACC and TRU. It is thereby the cue for one and all in their quest to come to terms with life and to vigorously and rewardingly join in the fray for the success and fulfillment it can provide. Life is not of unselective pleasure. It is a life that is highly competitive, demanding, determined and absolutely requires ever higher - to the absolute - purpose, plan, management, control, ACC, victory and glory.

What is true of partials is not true of wholes. What is true of wholes is not true of partials. Therefor what is true of anything is only true within the limits, relations, states, and conditions of the realm to which it applies. It is to a great extent for this reason doubt is formed, explained, and to be corrected. Many of man's greatest paradoxes are the result of failure to apply a rigorous proof. For once we understand doubt, can show what is and is not true, and can produce rigorous, actual, proportionate and total proofs for all we are well on the way to complete and comprehensive knowledge. Man's belief in the inalienable correctness - and unquestioned sacredness of truth - is one of the great hurdles to an accurate and full understanding of our world. It is a great impediment that is both the reason for TRU and the glory of IPTRU. It is such an insight into doubt, truth and proof that enable us to avoid the ecstasy of simple revelation, and the depression of error. This is like 'the gospel truth" wherein there can be much revelation, yet much error. The golden mean is to be found in the resolution of doubt not by submission to it, but through the determined search for truths, their realms, limits, states, conditions and relations, the massive differentiation and integration in the whole of this special universe of best human life provides. It is the concepts, principles, laws, and mechanisms that make up much of the parts, processes and whole of our world. Their refinement and adaptation enable us to develop the correct truths of all, resolve any and all doubts, and provide the proofs that are essential to a healthy, vigorous, and rigorous, meaning to life. This is our mass energy form and function, what is the most basic continuum and aspect of our being. Only in the fullest identity with the best possible representation, presentation, and perspective of TRU can we overcome doubt, and the confusing depth of partials, both TRU and ATRU that are so much of our lives.

In order to provide ACC and TRU with the highest quality and greatest number of authorities, teachers, students, and all with interest in a permanent and secure TRU some of its education should be done at a university level. Higher education has been a main source , preserver and promoter of knowledge. Students and faculty that are highly stimulated, interested, inspired with revelations to identify with TREQ or TRU are what is most needed. It is facilitated by good environment, order, direction, and support sufficient to pursue TRU in a scholarly and profound way. This includes means to temporarily lessen the burdens and distractions of life, for a devotion to TRU. Yet it can't be an easy nor a frivolous way out of life. To develop the necessary proficiency in TRU will take vigorous training and cultivation of virtue, accompanied by a deep gratitude and love of what we have received, TRU.(KUPS) Such a University of TRU, **TRUNIV,** can take many forms from that of the best in current universities, adult education, home study, computer-internet guided, others or various mixtures. The purpose and goal is to enable each person to develop the maximum understanding of TRU and become as proficient as possible in life of TES TIT.

As there are boards and standards in many fields and professions that uphold the proficiency of applicants and practitioners so can there be for TRU. In American Education, Medicine, Surgery, many specialties, professions and trades, such boards provide a highly valuable service. The knowledge required in each profession tends to and should be of the highest quality, importance, and usefulness. The proficiency, exactness, and probity of such means, tests, and measures of knowledge and ability in the best ways possible should be incorporated into the armamentarium of each and all persons so that they best fulfill their part in TRU. So must each person be a professional of TRU. Certification of the needed proficiency should be rewarded, as by Degree, e.g. **TD** or TRU Doctorate.

LERD = Law of Exponentially Destructive Religions. LERD denotes the point at which what is, even when set forth with the best of intentions, or has been of TREQ, becomes negative and destructive to TRU. This could happen even to TRU when not held to VAPO and TIT-TO. Although initially in the name of TRU by God or TREQ they in time are apt to become negative. Thereby whenever one encounters such in and out of TRU, is in a position of action, especially of increasing power the exercise of great discretion, and prudence, guidance back to TRU must be maintained.

If America is going to be a leader of the world by being its watchdog and supplier it must realize this is not possible unless we can accomplish at least two things. That we can reach the minds and wills of those peoples we occupy or seek to bring to a new world. And that we, both as a nation and individually, have a like mentality and will that practically mirrors this special world of best human life. Since neither of these are within our grasp in the early 21st century, we must avoid foreign entanglements, as George Washington so astutely advised, and get our house in order. This means nationally as well as at all levels down to each person and their virtuous daily lives. What is commonly held in many of the world's greatest religions is much the same and of the highest benefit. Thus, if religion, and a universal view, are the main stumbling blocks, what is divisive must be eliminated and what is common and in accord with maintaining the best human life offers is to be honored, extolled, promoted and achieved. It takes the best of all VAPO, Á, ACEXIN, TROS, TIT, TO, HOM and ACC. The whole continuum of ID or integration and differentiation operating dialectically with the best of MAC, or **LIDMAC** or Law of Integrative Differential Management and Control by COGNOS and total TRU ACC. It is important to show the terrorists, who is boss and eliminate them as much as possible. Yet as long as they are supported by large percentages of the world's or regional populations, to adopt their kind of tactics only alienates a larger amount of the world. Let them live - and be in misery - in their kind of world, only on occasion, and on our terms annihilate them with the least cost to us. Maybe they will eventually wake up, learn their lesson, as we must learn ours.

LESP = Law of ES Perpetuity. LESP denotes those parts, processes, and all of TRU that continue to produce the inertia and momentum needed for perpetuity. In ACC we concentrate on those characteristics, processes, needs, strategy, and action that will make optimal LESP, as it empowers TRU. They are made up for or completed, by TES and TRU. This make up effect is called: **MUEO** = Make Up Effect of All, all that is TRU. MUEO denotes the fact that as bad and wrong as the world about us seems, there is hidden amongst all this that is not so much TRU, a considerable amount of TRU. These are the ES, CES, form, function and all aspects of TRU which enable it to be. These provide much of its inertia and momentum. It is that our IA glimpses it only on occasion, especially when overwhelmed.

There must be a full accounting and complete reconciliation of all monetary policy and the use of national finances. All who have responsibility in the use of money should guarantee that it is used only for the right, TRU purposes. We need not expect it to be easy to be TRU, nor even less, turn what is not TRU into TRU. There can be little glory without the expected due, proper and complete ACC and TRU effort.

In human evolution and cladistics there are periods of population instability and stability, restriction and expansion. This largely depends upon their ability, competition, environment, and adaptation. Periods of instability and restriction are apt to be followed, for a time, by inbreeding. This may accelerate mutation and change helping to reinforce beneficial improvements in genetics. Or on the other hand, it can be extremely harmful if there are defective genes and effects that irreversibly alter the delicate balance. This has a kind of overspecialization associated with an extreme paucity, compared to highly polymorphic and large populations. How these have, are and will change in the future will need to be worked out. They are at the root of the problem of human quality and its voluntary or formalized controls. It is why a panel or group of experts that represent both the differential and integral in each field in TRU, will be so necessary. It will greatly accelerate the best possible, continuing and ultimate supply of research and data needed for an optimal ACC and POT. The guide lines for this panel should be clear and properly set forth from the outset. They should be based directly upon TRU, ACC, TRAX, TREX, and TRIX. We should also keep track of the hundreds and thousands of greatest voids, regions, segments or aspects where TRU is lacking in which added knowledge and work can greatly add to, complete, and fulfill TRU.

MEASURES of POSITS of GENERAL PARTS of ACC

AMU 4-2 ± ~3-7-6~ S
ACC and AU/TRU INTERACTIONS
AAI 4,8 ± ~-~-~-~ S
BETTER LANGUAGE and TRU,
Need BLT 4,1 ± ~4-6-6~ M
BOARDS of TRU BOT 1 6-5-6~ T
DOES and DON'TS DAD1 7-5-6~ M
EVALUATION of LEADERS,
PER,, ELP 1 ~4-6-7~ T
FIELDS-SCIENCES EXPANSION
ACC FSE 4,3 ~6-3-7~ T
FUTURE, PREDICTIONS FORECASTS
FPF 4,1 ±4-4-8~ M
GAMES and COMPETITIONS of
TRU GCT 1 ~5-6-7~ T
GREATEST (25) N. DISCOVERIES
GND 1 ~5-5-8~ M
GREATEST (11) NEEDED PROJECTS
GNP 1 ~5-4~6~ T
LADES 1-3 ~3-6-5~ T
LANTLA 1,3 ~4-6-7~ M
LERD 8 - ~2-1-2~ S
LESP 3 ~5-5-8~ T
LUTNO 5,7 ~4~4-7~ T
MATAM 1~ ± ~2-7-4~ S
MATIT 1,3 ~3-7-5~ S
MEASURES and TESTS of
ACC MTA 1 ± ~6-5-7~ T
MENEX 4 - ~2-8-4~ S
MIRC 4 ± ~5-7-7~ S
MUEO 4-2 ± ~5-4-7~ S
PIO 4-2 ~8-4~6~ M
PROBLEMS & SOLUTIONS &
ACC PSA 4,1 ± ~6-5-7 M
SADANT 4,1 ± ~6-5-7~ S
-SCEN 4,3 ± ~4-6-6~ T
SOUNDS LETTERS TERMS
CORREL. SLT 4,1 ± ~4-6-2~ M
TESTAR 1,3 ~5-6-7~ T
TESTING and EVALUATING
TERMS TET 1 3-6-6~ T
TD (TRU Doctorate) 1,7 ~3-5-8~ S
TRU TERMINOLOGY ABSOLUTE
NEED TTA 1,3 ~4-6-7~ M
TRIX 1,3 ~6-6-7~ M
TRUNIV (TRUNIV) 7 ~3-3-8~ S

Fig 3-94 **ACC GENERAL PARTS POSITS in SPALT**

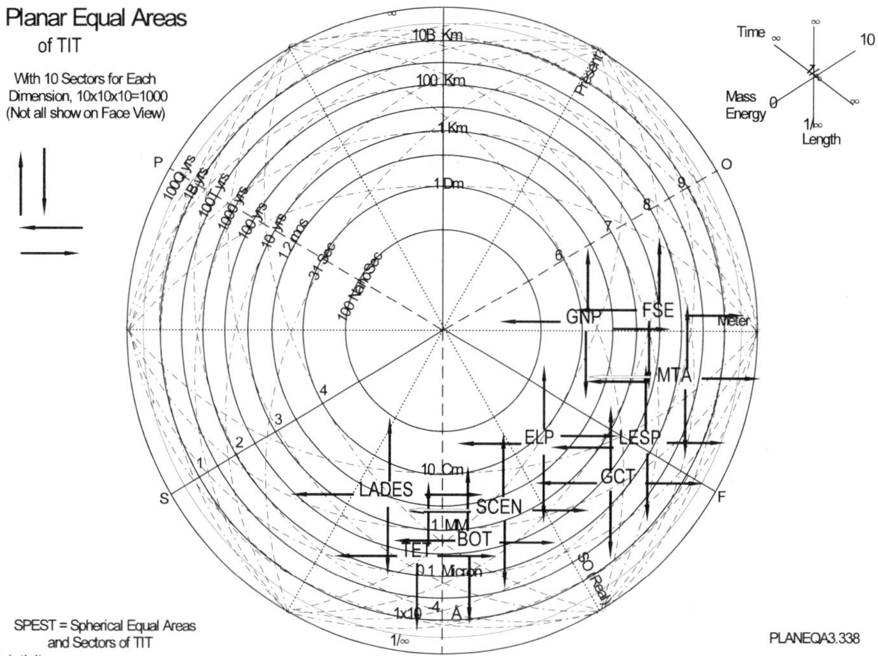

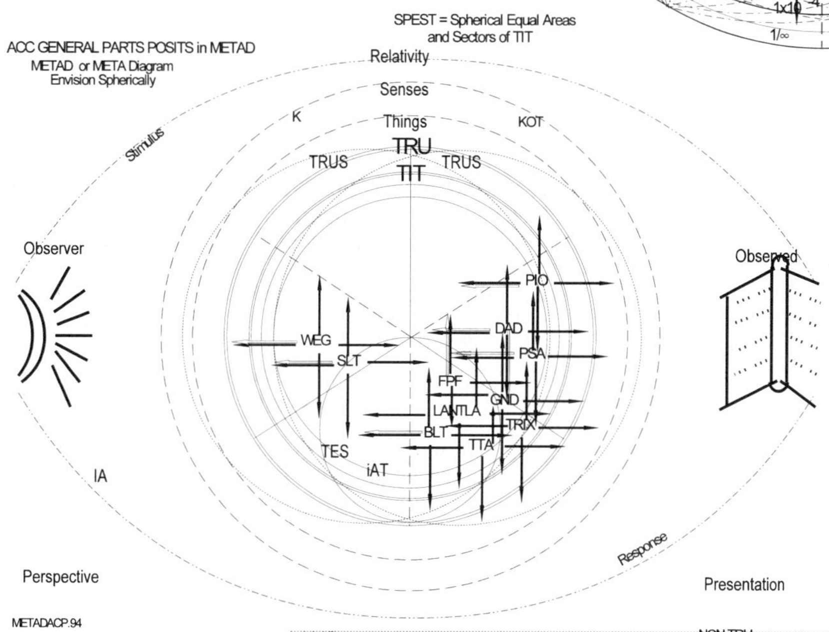

Fig 4-94 **ACC GENERAL PARTS POSITS in METAD**

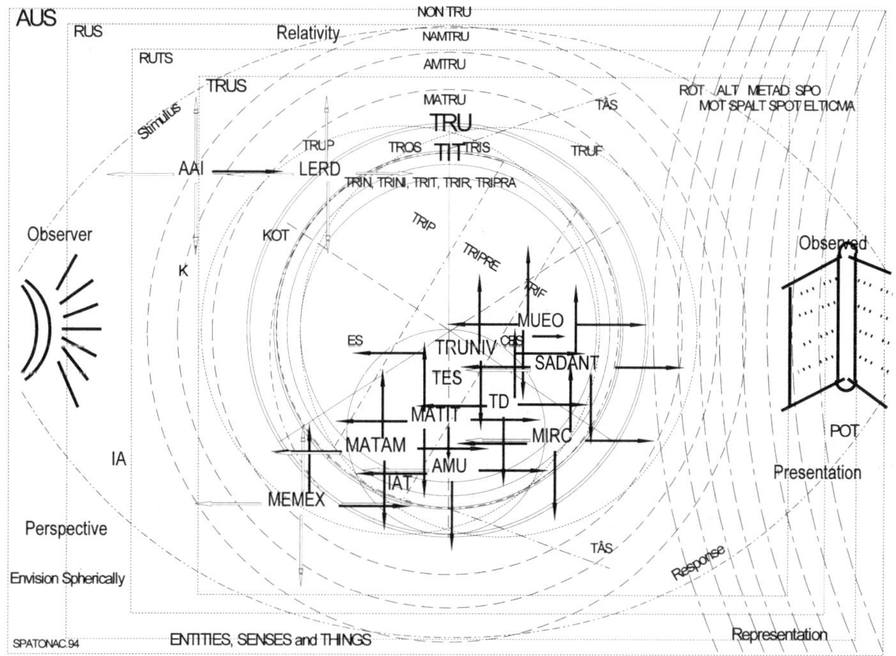

Fig 5-94 **ACC PARTS GENERAL POSITS in SPAT**

CHAPTER 95

THE SPECIAL PART or PARTS of ACC

CONVERSION, SELECTION, STANDARDS, CONCENTRATION, CONTROL, TES, PUREACLITY, KUPS, LEVELS, PROPORTIONS, RATIOS, CICIJ, and IMMORTALITY

The parts of ACC are the special ways we account, act, and accentuate TRU. The focus is largely on how a person can directly or indirectly become TES through PUREACLITY and KUPS. for greatest benefit to TRU. The most precious gift we receive is life, what it takes to be TES and TRU. The most precious thing we can keep is not stock options, jewelry, possessions, social power or what is most popular, it is the inspiration and revelation of greatest concepts, principles, laws, mechanisms, parts, and processes of TRU and how they work by ACEXIN in SPALT of TIT. It is how deeply we can identify, make distinct, fulfill, enjoy, and best use them. It is how well each and all of us can discover, learn, develop and practice VAPO and acquire maximum quality and quantity for TRU. It is how well they can be applied to ones life in TES. Herodotus relates that Croesus asked of Solon "who is the happiest of all men;" Solon's reply was, "to live a vigorous life, have many children, see them do well and die nobly for a great cause." This great cause, collectively, is the survival of the human race, TREQ, or TRU. It is not to seek a life in which there are few or light burdens, but the challenge and opportunity of working for and finding TES and TRU. For although its burdens are heavy, they can be readily borne, suffered, and accomplished when a person is shown how and what to do, knows why they are doing it, do it well, and live by all that is TRU. In knowing that they are the best, they can be, and join with the most perfect and greatest of TRU, is their hope and joy.

With advancing means we have an extraordinary opportunity to demonstrate what TRU actually is. How things are as TRU is very different than what they seem to be, or even have been known to be. These have failed to clarify how we are the same in life and in death. **LEXIT** = Life of ordinary Experience or IA, far different than, IAT and life In TRU. Only when we can come to discover LEXIT can we begin the journey of PUREACLITY, Pure Actual Reality to TES and TRU. This we do when we learn, know, love, appreciate, find identity in, are inspired by, revel in, apply, best serve, and enjoy TRU. Life is not a simple absolute in contrast to death, but a continuum, a mixture of continuity between life and death that is key to understanding what we actually are, TES, and not what we commonly think we are. What is actually true is only the TRU in summation that so greatly enhances us, TES and TRU. (**TRUT**)

To learn more and more about less and less as in special cases, or to learn less and less about more and more as in general cases, is futile. It is only as we learn more and more about more and more TRU that we can achieve the heights of understanding that both we and knowledge require. To do this may seem impossible, yet it has limits, and here is the beauty of TRU, KOT, ACC, and TIT. For it is limits that reveal its form, which when modeled and demonstrated show what this more is, TRU. As our knowledge relates and posits the world around us, we apply it to our lives by best fulfilling TES in TIT. This we do through: **KUPS** = Knowledge, Universal, Partial and Special, or our combined Universal, Partial and Special Knowledge. In doing so we better live in PUREACLITY, become TES and one with COGNOS forever be. This we do in striking contrast with our seeming being and those lost in the maize of NAMTRU, ATRU and AU.

There is a vast multitude of forces that take over when control is lost. Thereby it is in control established through persistence and strategy in what is right and actual, that enables us to manage these problems, objects, work and life. It is where success, solutions, accomplishment, achievement, and fulfillment are to be found. Such control is by dividing in ways that suit the endeavor, in portions of halves, thirds, fifths, tenths, hundredths, etc. For TES and TRU it is a continuing process and total endeavor the effective means to resolve problems, complete objectives, and achieve TRU.

There are many inequities in life, some TRU when associated with differences, more AMTRU, most ATRU and ANTRU. It is the last two that are apt to cause the most violent reactions. For humans have a great natural aversion to the effect repeated inequities have on life. Variations from the norm in TRU are mostly adapted to. It is when they seriously deviate from TRU that they need to be resolved. This is largely by TRU, which can be utilized to demonstrate the special and partial actions in ACC. Every person has a great story to tell. It is one that tells the story of what they did and did not do with this life and what it did to them, others, the world, and TRU. This may be done like an Ibis or an accounting of each person. It is what fiction and non fiction seek to describe. There is a need to collect information for each person and their Á FAF to archive and account for their lives. This should be available for a great variety of beneficial uses. It is what is lacking in the archeological record, what was current for each given person and country in their time. This should include not only the collective learning experiences directed toward TES but all other pertinent facts that help determine the future. They are the great personal stories told in the bible, and ancient Greek Literature. This includes the genetic material, physical health, determining forces, and what was their SÁ make up or FAF in TES and TRU. Such a composite when justly used, within strict SCARLS and safeguards against abuses, would act as a weighing of the souls by noting whether they did or did not spend their lives for mutual benefit. It would stimulate and help TES and TRU.

Standardization by TRU is for all that does not function best in and of itself. When naturally following TRU they act less as regulations and more as information, guides, ways of behaving, and treatment that provide health, harmony and prosperity in TRU. **DERGS** = Degeneration of Standards. DERGS is evident in the depreciation and degeneration of Gods and religions through history. They have a half life of variable TRU and time, some much more and some less, depending largely upon the degree they are TRU. In reaction this adds to their partiality and bias. They tend to become reactionary and irreconcilable with TREQ. It is for this reason TRU is offered, not so much as another religion, philosophy or science, but an overall means of bringing all human Á FAF together in fulfillment and closure.

The 're' of a thing that is important, as the E and ES in TRU, is actually real in being TRU. The 'res' of things, are ordinarily its reals from any vantage. These may be either non TRU, as in ENIT, or TRU as ESINS. When they are non TRU they are simply the corresponding link in relativity between the observer and the observed. They are usually, more in being a correspondence that is both vivid, tangible as well as being important in some way to the observer, given or ENIT. When they are TRU, they are different than ENIT, much more complex, yet actual in being an aspect, part or process of TRU. For they are guided by Á the predominant mass energy FAF of TRU, its Á state and dimension that brings and converges this relativity on what is optimal for TRU. This is what turns a simple real, correspondence in S-O or IA, into an actual and pure one in TRU. It is the recognition and especially the formulation, total understanding, and use of all, TRU, TO, OM, by the individual through PUREACLITY to become TES. This is not any purity, reality, nor actuality, as commonly limited, believed, or in usage.

It is the purity of the reality of the actuality that is TIT-TO. It is and can be only what they are in a strict way, by TRU. PUREALITY requires: knowledge, coordination of this knowledge, application, effectiveness, avoidance of ATRU, identity with TRU, and acceptance of life in TES and TRU. It is a combined focusing and extension of our predominant, preponderant mass energy power in Á produced through our subjective objective correspondence, exponentially expanded in COGNOS.

The most distinct meaning of PUREALITY, is the purity, actuality, and reality each must seek if they are to be. If they are to be their parts, fulfill their roles in TRU, and contribute to the whole of this special universe of best human life. These three principles and functions of a person's life and work both separately, together and all in between reveal what one must focus on to be TES and TÍS, their general and special parts in TRU. Their interrelations and interactions are, 1. The purity of actuality or degree to which one has no impurities nor negative effects upon what is only TRU. 2. The degree to which the actuality of reality keeps us focused upon what is only TRU. 3. The degree to which our reality, or S-O identity, is only actually pure, is the degree to which our special mass energy, or Á dimension and continuum are fulfilled and maintained. Thus it is through the aegis or PUREALITY that humans can see and achieve all in the higher levels of being that constitute identity, participation, fulfillment, and immortality in the absolute. This is an absolute that the past, present, and future of human life has both given to us, and to which the future, in turn, depends on us. Purity is the qualitative state in which this special universe of best human life is free from all that is null, and are significantly negative to it. How free it is from impurities, contaminants, defects, detriments, dangers, injuries, disease, and all. Purity is the quality that separates out all else, gives a thing individuality or thisness, its hacceity. Purity quantitatively is the degree to which any and all is positive in this universe. This is equivalent to its REM, relative magnitude or value,.the degree of its positive/negative effects. Actuality is the qualitative state in which all that exists does so only when fully relative, enables, contributes, or is a part of this special universe of best human life. This approaches its aseity, or underived independent existence, the manner by which it exists.(TRIA) Quantitatively it is the degree to which any and all are actually a part and participant in this universe. Reality is the qualitative state that characterizes the mass energy of our actual world, TRU. It is a mass energy typical of human life in which the relation between observer and observed, or subject and object, relativity, is predominant. By the laws of relativity and the higher powers of mlt, reality defines, in association with gravity and other increasingly human forces, the total mass in our world. Reality quantitatively is the degree to which the powers of subjectivity–objectivity reach higher levels of identity, distinction, and totality, as they replicate TRU. It tells us what, why, how, when, where, which, what, etc. It is closely related to its quiddity,.different kinds, states, and parts of REM, REM of purity, actuality, reality, and others including REM of MOCBIREM. In nature, as it is of TRU, those who are unfit through failure to contribute, uphold and sustain their inherited powers of being, their mass energy continuum or Á form and function are eliminated. This process of selection occurs in many ways described through the text, and is justified by the need to sustain this special universe of best human life.

There are certain things the individual must learn and do if they are to become or converted to TES, and TES to TRU, including that: 1. Even TES is but a part of TRU, 2. Neither TRU nor oneself are TES, 3. An individual can't stand alone, 4. TRU can not stand without individuals, nor individuals before we are TES, 5. An individual does not exist as TES, 6. All individuals do not have to, and will not be exactly alike, nor behave the same, nor have the same environments, even when TES, 7. It is in maturation and conversion, not by complicity, superficiality, nor ATRU, but by TRU that we are reborn in TES, become TRU, have absolution, or one with TIT and TÂ.

Fig 1-95 **INDIVIDUAL, TES, IPTRU, and TRU CONVERSION**

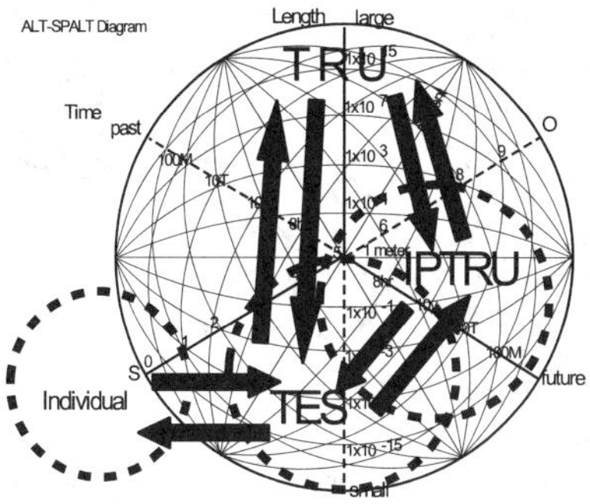

We are alive only to the extent we are a part of, or contribute to the health, well being, and survival of the human race. Immediate problems do not make us alive. It is not any survival, any quality of life, any monetary success, but only the survival, quality, and success of this special universe of best human life, TRU. Much of what makes us so personally sure of ourselves, our being, is nothing other than 1. the key dynamic built in mechanisms, 2. our form and function, 3. concentration on our part and role in the success of TRU, and 4. quality and survival, of the whole TRU, TIT-TO. This is called, **LABIT** or Law of Being Alive in TRU.

A person is TES through **IMTES** or the Imperative TES. This is produced by being born, the challenge of life, of TRU, to have the wisdom and conviction needed to counteract all that reverses it. Those by IMTES are generated to have depth conception, glorious feelings of revelation, joyful learning, and profound and inspired recognition of TRU. They have the ability to refrain from or rectify all that is not TRU and achieve all that is TRU. IMTES targets existence in NAMTRU, ATRU, AU, ANTRU or ANTITRU that can not be avoided nor corrected through TES alone. It is for this reason, it is important for all religions to become reconciled with each other, to come together by their common similarities and relations, by DAOST, ESINS, and TREQ, in TRU. There is no place for sects in TRU unless and to the degree they are ES, sectors, or regions within TRU forever returning to TIT-TO. Religion if it is to fulfill its universal or VAPO nature in TRU must be nothing less than all TROS, all TRU. What is good and what is evil is profound and subtle. It is to be understood only with a great knowledge of their part and role in TRU. The degree to which religions are not the same they become opponents, and when other than TIT, negative to TRU. This is why the ESINS in religions, sects, ISMS and extremes need to be improved and vigorously shown and brought to TRU, or displaced if they can not.

As necessary as it is to concentrate and develop the infinite and absolute of all, even the absolute partial or **EÂ**, it can never be allowed to displace, replace, detract from nor adversely effect TIT-TO. For the peace and success of humanity, as TRU, depends upon partiality adaptation, the absolute necessity of TES, and a thorough-going KUPS. This is mandatory and paramount for individuals, groups and all if they are to achieve adequate TES. A person must develop the subjectivity and rationality necessary to: 1. Understand, love and live by TRU, 2. Distinguish TRU from NAMTRU, ATRU, AU, and ANTRU, 3. Distinguish the different parts of TRU, their order, posits, REM, processes and how they collectively form and function in TRU. 4. Distinguish the individual and self from its counterpart in TRU, TES, 5. Understand and work for KUPS, in proper balance and fulfillment in ALT, SPALT and Homation, 6. Comprehend TRU, as represented through SPALT and SPAT, 7. Posit and coordinate all one sees, senses and perceives, 8. Think TRU and use reason in the basic concepts, principles and laws of TRU, 9. Realize how we come from our own IA to IAT to TRU and how these can be ordered through EPI, SUT and ALT of TRU. **LANTES** = Law of the Absolute Necessity for each and all persons to become their counterparts, TES in TRU. **LANICT** = Law of the Absolute Adaptation Necessary for each Individual, as all, to Conform to TRU. **RADU** = Ratio of Differentiation and Unity in Integration.

Fig 2-95 **REALITY, PURITY, ACTUALITY, TRU, and IPTRU**

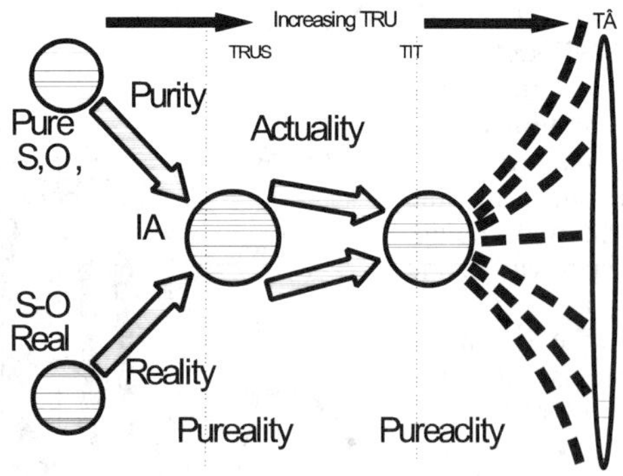

The triumph of TES, COGNOS, TIT, and TÂ are provided by those less commonly used senses of reals and reality that we describe in the relativity of the observer-observed, Á, and TRU as they occur in ACEXIN and PUREACLITY. This is the purity and reality of actuality, that is of mass energy, Á, FAF of TRU. The prime or predominant form and function of TRU, Á, equals these reals or combined subject object entities that make up much of the greatness of the ES and CES, or the parts and processes of TRU, and their recognition. The Á reals of S-O, are the actual distinction of entities as these observers render the observed objects different than the entities they refer to. It makes them into something that is relative, relative to us, TRU. These reals are thereby combinations of subject and object as they exist in TRU, a phenomena of its own kind that becomes separated and is not the observed or observer, O or S, exclusively, but a distinct gradient in their continuum, as they exist in TRU. The Real is the clear demonstration of the degree of nominalistic intensity necessary for the appropriate phenomenal gradient. These reals of the mid portion of the Á or S-O spectrum or continuum are similar to reals in the Platonic sense, or the royal ideas of senses and highest goods of things as they are in and of this special and exclusive universe of best human life, TIT-TO.

PUREACLITY is the way of life that maximally provides one with the ability to attain optimal or perfect E - ES - TES in TRU. It is where a thought, action, all thoughts and actions, their variations and effects, either directly or indirectly is of: 1. Perfect TRU, 2. Perfect E or ES, 3. Perfect Proportion or REM, 4. Perfect form, construction, form or SU, and these in turn are automatically synchronized in every IAT, movement, and act. PUREACLITY is the fusion and union of several different crucial aspects of TRU.(DAOST) 1. The pure or uncontaminated makeup of TIT, or any aspect, component, ES, or CES of TRU. It is nothing less than itself alone as it is only in TRU. 2. The real or greatest and most identifiable relativity of the observer observed, or subject object aspects or parts of TRIT. 3. The actual or what TRU is to the exclusion of all else. 4. How to secure and maintain the highest purity of all that is TRIT.5. How to secure and maintain the highest reality of all that is TRIT. 6. How this purity or 100% Thisness or Haeccity and reality is the 100% whatness or quiddity, the predominant mass energy, A, dimension of TRU. How this actuality is the 100% coming into being in TRU of the potential. And how all these are combined to best render and serve TRU. 7. The increasing purity of reality, or, semantically, how to obtain and maintain the clearest possible definition of the real, that which is most identifiable and formally distinct of TRIT and TIT. 8. How all is no more and no less than their perfect ACEXIN in greater and greater TIT. 9. To live in such a way that all these, # 1 - 8, are produced and accomplished. To live in such a way that all ESINS are their full potential and become actuated as they approach their optimal FAF in TIT. PUREALITY is dedicated to all persons, especially upcoming generations, the babies, children, youth, and bright young minds, adults, mothers, fathers, and even aged, whose hopes for the future seek to improve their life through learning, education, and training to become increasingly TES and TRU. This is to find the right kind of success, victory and glory that is theirs for the asking if they live in PUREALITY and One With COGNOS Forever BE.

Fig 3-95 **BONDING and CONCENTRATION PROVIDED by RELATIVITY, Á, TÄ, ACEXIN and ALT**

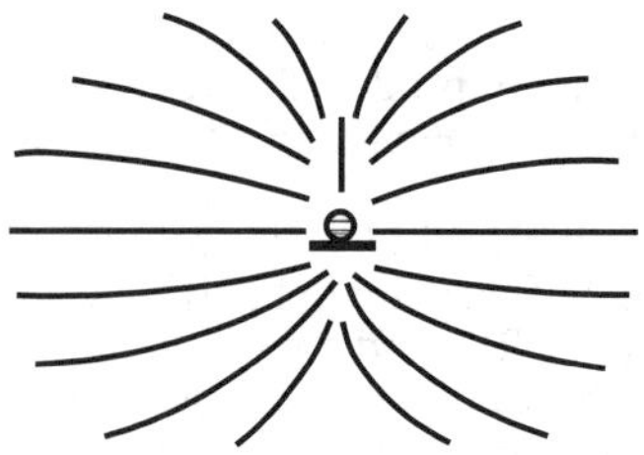

The bonding, concentration and coordination provided by relativity, in the form of its actuation, extension and intension, or ACEXIN is focused from the IA, or immediate attention, awareness and action elements of life and PUREACLITY. They become self accelerative and with expansion provide for TRU. This shows the culmination of the investigation of TRU in POT and IPTRU. The functions of actuation by the real of Á FAF continuum of TRU, exists symphonically in TIT-TO. It does this bonding by providing man's higher levels of mass energy a filling out of this continuum within the ACEXIN of TRU. It is these reals, as they are of Á and TRU, that give such great power by helping to unite, preserve, and thereby expand itself, TRU. This we show in IPTRU, by how all that flows out of this great bond of relativity mass energy to represent and present TRU. PUREACLITY is a major product and vehicle of this bonding by the reals in relativity, especially by increasing purity to 100% TIT. It is also accompanied by increasing actuality as this exclusively selects out only TRU and excludes all that is not TRU. It is in this way TRU has grown as its actualization has been empowered by, and in turn empowers, purity and reality.

There are several components and levels of PUREALITY including the extensions of purity, reality, and actuality in TIT: 1. A simple subjective meaning, as pure and real in combination, 2. Pure in body, real in mind when conjoined or bonded, a coterm, 3. As a goal, objective, and aim in personal health care, 4. The identity of S & 0 as they become the mass energy A dimension, continuum and FAF in TRU, ultimate pure reality, 5. The ultimate goal, significance and importance of Pureality. This is to increasingly project reality onto TRU by increasingly pure analysis and synthesis of all qualities, partials, processes, characteristics, form and function to TIT-TO, 6.The addition of actuality to purity and reality, excluding all but a TRU that leads to TÂ.

TES can be generic or specific and gradients in between. Generic when it is common to all persons, specific when it coincides with the ESINS of a particular individual. These are called **TEGS** and **TESS**. TES exists in TRU whether recognized like TRE when not recognized as in TREQ, or existing in TRU yet unknown as such. When TES is recognized as TES in TRU it is called **TESRE**. When it occurs without TRE, as TREQ does without TRE of TRU, TES is called: **TESARE**. TES includes those thoughts, studies, actions, efforts, and work made to attain TRU. These may or may not be accomplished through the recognized TRU, TRE. The TRU that TRU is conscious of is called: **COTRU** = Consciousness of TRU. **COSTRU** = Consciousness of ones self conscious of TRU. **COTES** = Consciousness of TES. **COSTES** = Consciousness of ones self conscious of TES. COTRU although closely related, is distinctly separate from TRE. TRE is generically the TRU that is recognized as TRU. COTRU is more specific, the introspection of TRU by itself. COTRU largely arises with PUREALITY, as it is for much of all its action, accentuation and ACC. COTRU is a binding force in TRU just as much of the ego of the individual and of TRU correspond and fuse to bring the subjective into actual identity in TRU. COTRU as a basis for Pureality is also the basis for COGNOS, ELTICMA, SÂ and TÂ. These concepts, especially TES, partake of TÍS when only by TIT rather than ESINS of TRU, and when the that of the individual concentrates upon itself, is self centered.

Would you like to have total knowledge at your fingertips

anytime? This is what TRU does for you.It separates truth from illusion and figment, the actual from the inactual. It focuses on all this special universe of best human life that is both true or valid, actual, proportional and total in the most effective way. It helps tell the same and the different, and what the value is of all. This for your personal benefit as well as that of the world in general, TRU. The latter is key to your actual existence, soul, immortality, life in PUREACLITY, TES, and TES that is the focus and great part of TRU. Thereby learn REM, POS, SPALT, ACC, KUPS, and as much of all that is TRU as possible. To do this we need **TAN**, or Think and Note. This is what best helps to stimulate and produce the necessary habits of inspiration, reason, and revelation of all the greatest ideas, concepts, principles, laws, mechanisms, forms and functions, parts and processes of best human life or TRU. It encourages us to train our minds, apply our experiences, and learn the deepest meanings of our life. What we are actually a part of. One way of doing this is to have at-hand, notebooks, journals or electronic devices that enable writing deepest thoughts. With practice this steadily improves, and can be further stimulated by listing the greatest questions and answers of the givens in life. Along with listings of the greatest concepts, principles, and laws of what are derived from existing and essential terms, and TRU language. When these are adapted to SPALT, and POSITS, then analyzed through HOM and ACC a clear picture of all can be seen. Such an overall view is what is most needed to guide one in life. Such facility of notes targets aides to a person's immediate life, and their life in TRU, by PUREACLITY, TES, COGNOS, KUPS, and ELTICMA.

Since human knowledge has become so vast that no one, even in an entire lifetime, can expect to learn and remember all of it, each and all persons can and must, at least, learn that which is: 1. Most important generally,(REM,TIT-TO) This is the U of KUPS, or Universal Knowledge. 2. Most useful and important partially, for oneself as one is a part of TRU. All that enables each person to better provide for themselves and their loved ones. This is the individual's part in caring for one self, surroundings, family, community and all that directly and immediately provide for their more successful and happy life. This is the P of KUPS, or Partial Knowledge. It is like how to be good a handyman, home maker, mother, wife, father, husband, friend, helper, and neighbor, 3. Most productive and important specially, for ones life's work and its contribution to and role in TRU. This is their occupation, job in life, speciality, or profession. This is their role in the division of labor necessary in a populous TRU, and healthy social order of responsible members. This is the S of KUPS, or Special Knowledge. All together they make up: **KUPS** = Knowledge Universal, Partial and Special, or KOT, Practical and Professional Knowledge.

Fig 4-95 **KUPS: KNOWLEDGE, UNIVERSAL, PARTIAL and SPECIAL**

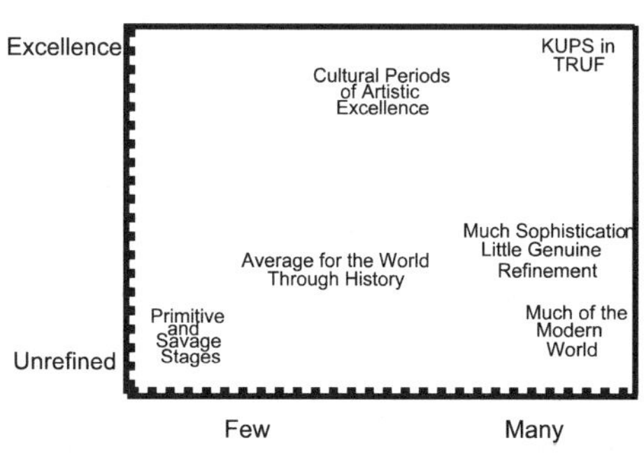

There have been times in history with too much or to little division of labor, too few or to many specialists. Also, they have had greater or lesser proficiency. With an increasingly complex world there will have to be an ever greater number of specialists with excellence. Yet this must never be at the sake of being a complete person in the service of the best of life about one, or KUPS. There is today a great division of labor with specialists who know their work but little else. This aggravates an existing tendency to self or group interest and denies TES and TIT-TO. Each person must increasingly learn and know KUPS, universals that give meaning to all, partials that make life abundant and fruitful, as well as their life's work. It takes a good and great life to be able to produce proficiency in all three. KUPS must be taken with great conviction, energy, interest, incentive, devotion, sacrifice, service, enthusiasm, dedication, constancy, perseverance, glory and victory if one is to find success and identity with the world, TRU.(TES) KUPS is the guiding division, like an actor and actress in life, by which one best performs and produces a show that is most successful, acclaimed and immortal. Yet it is much more than an act and show, it is life in earnest, life in the raw, life that must be of ever greater worth and value. KUPS is not achieved superficially, nor only by each separately. It can only be achieved by all with ACEXIN and of the best VAPO of TRU. Each person must not only have a good knowledge of the world, good at their job, or good as a member of a family or community, but good at being all these, at the appropriate place and time. This combined knowledge is KUPS. The secret of life is not only to work vigorously for one-self, ones loved ones or power, but to serve all that one is a part of, TIT. It includes enjoying life and the fruits of one's labors. Although the secrets and mysteries of life have been both a great hurdle and a stimulus, they have decreased with increasing knowledge and the inquisitive trained minds of bright youth.(Rabalais, *Quatre Livrs*)

In the great projection of knowledge and epistemics to TRU, subjective idealism or the higher levels of the Á mass energy continua, have been foremost. They have been the greatest of all revolutions, the movement toward achieving TRU and TÂ. This is **LÁSIR,** the Subjective Idealistic Revolution. Most identified with the emergence of this was Leonardo da Vinci. His spirit of inquiry and great contributions to modern ideas heralded man's coming into being. Humanity's finding itself, its soul, TES, in the beautiful handiwork of God and TREQ, as TRU.

The greatest change taking place in the world today and increasing in years to come is designating and complementing the good in all persons. For many thousands of years some humans have realized and done this, especially after Christ. However this can become spontaneously rational only by learning and living the reasons for the fundamental aspects, FAF, concepts, principles, laws, and mechanisms of TRU. **APUC** = Automatic or spontaneous PUREACLITY Catalysis. APUC signifies that state and power created by ever greater emphasis and achievement of TRU in knowledge, psychology, education, science, and technology. It is how TRU has, is, and will be more confirmed by TREQ, TRAX, TRE, TREX, and TRIX.

Liberty and power should be proportional to the level of knowledge and practice of TRU. Everyone, as well as groups, should be tested, measured, rated, gauged, and utilized in proportion to their understanding, comprehension, practice, utilization and enhancement of TRU. This is: **LISPOT** or Law of Liberty and Power Proportional to Practice and Performance of TRU. All should understand, comprehend, use, and live by a sufficient level of TRU or TREQ. Those who are more capable, or who are chosen to be leaders whose work entails great responsibility, have the increasing need to acquire higher levels of TRU comprehension and utilization. This should be an ongoing and continuing program of training, TRU education, and competitive selection in which those who are TES and best for their roles in TRU are well prepared. Their responsibilities, honors and rewards should increase accordingly. **LARHUCT** = Law of Responsibility and Honor According to ones Command, Comprehension, service, and Utilization of TRU. LARHUCT should be gauged by KUPS and TIT-TO.

For each person, and the human race, to succeed, flourish, survive, and have eternal life it requires knowledge that is of the greatest quality, valid, actual, proportional and total. Knowledge with greatest meaning, by and in depth, intensity, extensity, order, relation, and use. To order knowledge requires that it be well organized in all ways, whereby the observer or subject of the observed or object, most represents, models, shows and presents by this special universe of best human life.

The kind of people to have about one, as oneself in TES, are those who are highly: intelligent, trained, experienced, educated, professional, skilled, especially in delegated tasks, productive, efficient, principled, inspired, loving of TRU or TREQ, thoughtful, imaginative of highest concepts, rational, comprehensive, cognizant, accentuate the positive, eliminate the negative, accountable, dedicated to their work and life in TRU, rewarded for their TRU endeavor, bond with the selectivity and exclusiveness of TRU, absolute identity with TRU, no deviation from TRU, ability to expand TRU properly, ability and effort to protect TRU, and command and control all interacting with TRU. These characteristics go together, are formed together, and are simply the result of good capacity, training, environment, interests, relations, purpose, and dedication to the cause of TRU. They exist, more or less, in many, especially those raised in a stimulating, learned, vigorous, loving, and principled home and environment.(TREQ)

TRUC (CACTRU) = TRU Concentration and Control. Much of life, is one question, problem, conflict, fear, struggle, battle, disaster, and failure after another. This is often a result of nothing more than a lack of an identity with the world, TRU. It is especially a lack of total concentration and control. As long as there are new individuals and generations who enter the world without experience, knowledge and good judgement, there will be both a life of normal maturation in TRU as well NAMTRU, ATRU, ANTRU and ANTITRU. TRUC necessarily develops spontaneously, yet it will often be spontaneously dissipated. For TRUC to be adequate it must not only have the answers for currently existing problems, it must have the answers for all things, by TRU.(KOT,ACC)

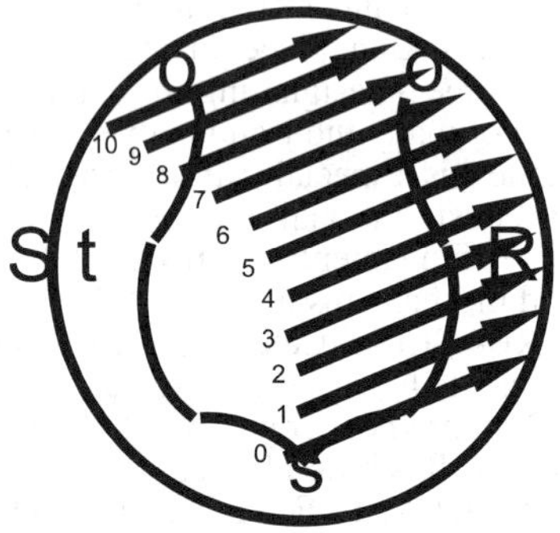

Fig 5-95 **LEVELS of ACCENTU-ATION**

0. Check out but leave alone, 1. Proper treatment, adequate follow up, 2. Routine treatment or action, of quality, 3. More than routine care, action, weeks work, 4. Difficult problems, good work, 5. Very difficult problems, hard work, quality performance, 6. Referral or special care, close supervision, 7. Subcontracting for precise work and quality production, hospitalization, 8. Major contracting, vigorous and precise work and energy, quality performance, plus the combination of all, their management and control. 9. Emergent, vital, critical problems with intensive work and major contracting needed, 10. Heroic, critical, and vital situations, calling for all emergent and management skills, even the potentially hopeless. The parameters of ACC include: critical degree, degree of importance, quality, degree of action, prevention, follow up, routine or special, referral or contracting, degree of difficulty, degree of supervision, and degree of precision necessary. These are all a product of TRUD, HOM, and their roles in ACC. They represent complex and profound aspects of TRU which can be understood only by investigating the detailed relationship between these levels, parameters, HOM and their posits in SPALT, METAD, SPAT and all IPTRU.

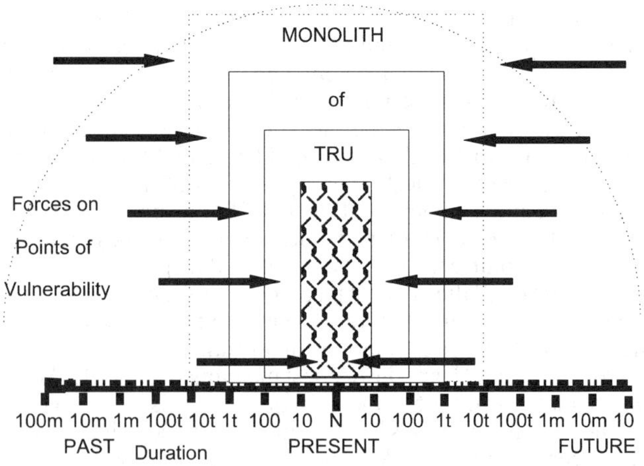

Fig 7-95 **MONOLITHIC VULNERABILITY and CONCENTRATION**

One of the greatest problems and features of ACC is that of **MOV** or the Monolithic Vulnerability of TRU, mankind, the species, and much of its FAF. It is vividly seen in personages through history. It is very poorly understood for mankind in general, especially the future. Historically and anthropologically it has been close to survival of the species. Yet the future may be much shorter or longer depending upon what and how well we perform and what acts on human life are learned and controlled. With the vast increasing powers of the higher levels of human mass energy, A, knowledge, intellect, reason, and technology we should have time to keep up with vulnerability. Yet we can't be so sure of this without greater knowledge, that concentrates and focuses on what these threats are, how to minimize MOV. This is what any animal does and how it converges on what most threatens it. This is: **LARUC** = Law of RU or TRU Concentration. Whereas a species, as Homo Sapiens, or TRU, may have much in common with other species, their centers and specificity may be very separate, different, and even opposite. This occurs not only between species, but in a broad range of phylogenetic levels, both higher and lower than species. In fact it is closely related to much of the competition and dominance in the same species that is so much of the dynamics of life, and TRU, as shown in TRUD and HOM. There are many reasons for fighting, often, even when violent, not a serious threat to human survival. However with the products of human discovery and technology this no longer holds for we are rapidly attaining ever increasing means of self destruction. This is direct and indirect, obvious and most obscure. These work on the foundations and chain of life that point out the potential for escalating vulnerability.

A person's position in life should be no higher nor greater than the degree to which one knows and practices TES and TRU, or TREQ. Leaders, preachers, teachers, doctors, lawyers, judges, executives and all who make the greatest and most critical decisions have positions whose judgement necessitates the highest degree of knowledge and appreciation for TREQ or TRU. Even with knowledge without appreciation or vv., they can't be fair and efficient in their roles in life. They must know, have interest in, and have a deep felt positive identity with TRU. This is ultimately possible only with a TRU education. Such a TRU education can best arise from a person's combined determination to realistically, spiritually, mystically and enthusiastically bond with TRU. Also with the overall provision that society and the world make a TREQ education as available and efficacious as possible. This is called: **LASORT** = the Law of Selective Levels of Responsibility and Duty by TRU.

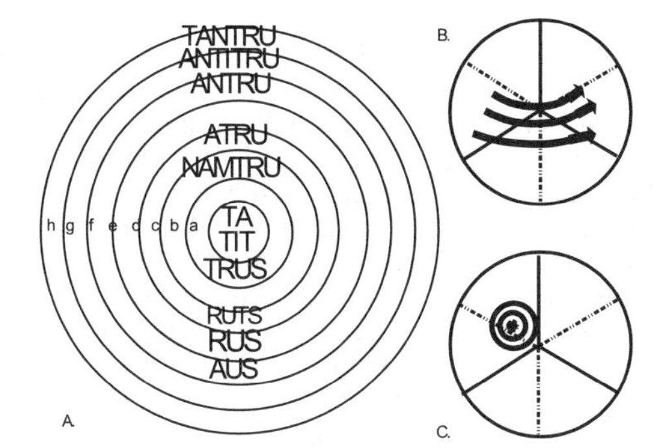

Fig 6-95 **TRU, TREQ, HOM, SOCIETY, and ACC**

There is a great lack of TRU in the world. This is shown by the relative smallness of TÂ, TIT and TRUS in diagram A. Diagram B shows courses, processes, and vectors in ALT. Diagram C shows the position of social dynamics in ALT. For society to be equitable and sustain the harmony it requires equitable benefits and opportunity for equal work, contribution, and service to TRU. Humans have an insatiable appetite to acquire, which quickly becomes greed. We have a natural tendency to push out all except those and that which might help us or whom we fear. This is evident on both the individuals, groups, institutional, political as well as international levels. It is especially seen in monopolies and the destruction that occurs in the guise of competition. There is a wide difference between what is illegal, immoral, unethical, insensible, ungrateful, inconsiderate, incorrect, not TRU and TRU. All but TRU or TREQ is wrong. The better we can apperceive and measure the degree of quality or TRU in people, institutions, groups and nations, the better we can act upon them, positively or negatively.

Since all persons, their environments, circumstances, and many other facets of their being are much different, it is not easy or even possible to equate individuals or even TES. It is even difficult to equate the great differences within the same individual. An individual adapts and applies themself and special position, environment or circumstances to TES, TRU and TA. This is by living in PUREALITY, by DAOST, TROS, and is different for different individuals. It is important to show how it works and can be enhanced.

Any existence, as life itself and even a perfect life, challenges all. For an entity to be, to be internally or externally, it must face change, the more relative the greater the challenge. At least from the reference of the first person and how this person adapts to the actual world, TRU, it is in. Thus, be it any entity, any individual, the first person, or their soul or TES as it identifies with the actual world, God, TREQ, TRU, or TIT-TO, each must rise to meet its challenges. The main difference is they are not the same in each. To make life easier, then, if we are to adapt and rise to these challenges, we must be relative, selective, work and live in those ways that provide most direct and VAPO routes and best management and control of challenges. to TIT-TO

RALGAW = Ratio of Love and Work for, over Greed and Waste of, TRU. The capacity, ability, and achievement of RALGAW is the glory, success and eternality of TRU. The reverse of RALGAW is the dissolution, failure and extinction of TRU. Each person, as all that is a part of TRU, comes into it with a great gift of benefits, opportunity and challenge to be or not to be the different aspects of the same thing, TES in TRU. If I do little or nothing to have, or fail to fulfill, my counterpart in the world, I die a manifold death. This is not only an individual death, but of all I consist of, and a death of TRU. In effect how it works to disenable TRU, to cause a loss to TRU, not only of myself but of that affected by what would be a part of TRU. **LATROI** = Law of the TRU Responsibility of the Individual, or Law of My TRU Responsibility. LATROI is something that even paleolithic man sensed in relationship to his habitat and other life forms. This may be taken for granted, but when lost, is felt in all. It is a vital part of TRU. One can have conscience without TRU, but for humans that seek what is best in life neither are adequate nor complete without the other.

No one should be in public office unless and to the extent they have a high degree of: judgement and identity with the success of TREQ or TRU. Decisions are to be based only upon what applies to each and all situations or circumstances as they are derived from TRU It is to provide such a healthy, positive and stimulating atmosphere in politics, public life, work place and those who would serve mankind, to be sure that this service is in trust and for the welfare of TRU. This applies to what all segments of society need, especially special groups, and association. Such is proposed in: **CCJ** or **CICIJ** = Citizens for Civil Justice. CCJ can be applied throughout the world as well as regionally. It is an association of all who have suffered the terrible injustices of malicious treatment and will, with increasing knowledge and conviction in TRU, be willing to combine their energies to help TRU to be specifically and properly achieved all over the world. From the bottom, it will require a citizenry of individuals who are TES, participate in government, are well informed, and vigorously CRUSH injustices to TES and TRU.

The measure of a civilization's or culture's success is a function of a powerful language and individuals that are intelligent, knowledgeable, stimulated, inspired, best coordinated, directed , and contribute to the whole. Since the individual is at best only variably partial, lacks KUPS, and lives in denial, a great amount of conversion to TES and TRU is mandatory. Conversion defines and signifies how we are, can be brought, or bring ourselves from the great variety of partial or deviant positions, environments or circumstances to TES, TRU and TÂ. Primarily it is attained by how we become TES, discover, learn, and utilize the many concepts, principles and mechanisms involved in TRU, and how with dedication and conviction this conversion to TRU is sustained. TRU Conversion does not mean either an absolute or static and fixed equality. It means the transit from any state of ATRU to a better state in TES and TRU. It signifies the collective ESINS existing in a person and their unified dynamic identity with whatever TES best fulfills their FAF in TRU.

SCIN = Scopic Inertia, momentum and power. SCIN denotes the degree to which every moment becomes more and more precious and absolute. This also becomes with Á and SO, the degree to which the apocalypse becomes impossible. For it signifies the powers of focusing in and out of TRU to provide an ever deeper comprehension of all, of TRU. It is the telescoping in and out and of all segments of TRU on SPALT in which TRU is fulfilled and its objectivity is completed. Everyone through training, education, and a repeated familiarity with TRU finds the inspiration, concentration, conversion, PUREALITY, COGNOS, and ELTICMA needed to achieve SCIN. This should be at a level sufficient to maintain TRU and assure the completion of S-O and Á of TRU as these provide the proper understanding and action of TRU. SCIN reveals an individual's opportunity in IA to focus back and forth through TRU with ELTICMA to optimal IAT. This is called: **SCOPT** or Scopic Optimum, Scopic Optimal SCIN. SCOPT signifies the greatest vision that can be achieved in viewing TRU and all. It is seen in SPALT, METAD, and SPAT where we best recognize TIT and POT with the greatest possible quality and quantity. **COGNOSID** = COGNOS IDentity, or knowledge Identity and the Inspiring, Glorious, victorious, revelation and time of recognition by the Individual of ones Identity with the Greatest and eternal. It is, an identity of one's power of knowing and acting, with that of all we know and we do for the best. This is identity of the self in its highest FAF, or its counterpart in TRU, TES, with KOT, COAT, TRU, COGNOS, ELTICMA, SA and TA. COGNOSID is the bond between PUREALITY, TES, COGNOS, and TRU. It is how PUREALITY functions reciprocally with TES to form an identity through COGNOS with TRU. The subjective aspect of this nearing absolute identity is:

SELF ↔ TES ↔ COGNOSID ↔ COGNOS ↔ TA

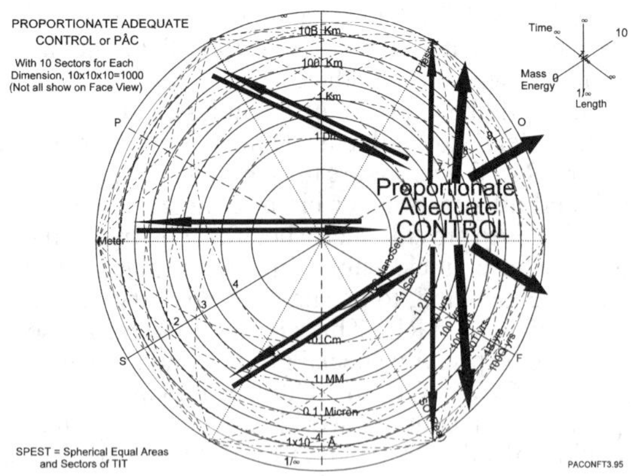

Fig 8-95 **PROPORTIONATE ADEQUATE CONTROL: or PAC**

There are advantages in confidence and control, especially by TRU. If they are not by TRU their confidence is false, and their controls will fail. Thereby PAC is necessary for the following reasons: 1. TRU, 2. Proof of TRU and ALT, 3. VAPO, 4. TROS, 5. TES, 6. Necessity of control by Á and SÁ. 7. Will, conviction and endeavor in TRU. 8. Danger to TRU, 9. Need by TRU, 10. Problems and Conflicts involving TRU. If control is not TRU, sooner or later, in reaction, the Principle of Imperfections become Evils of TRU, or **PIET**, sets in. Even with responsibility that lacks control, TRU is lost. Yet control should never be other than, and is usually less, than TRU, depending upon the given or partial at hand.

ELTIR = Everlasting Life Through Infinite Apperception. ELTIR is a kind of super vision and recognition of all by a person's IA in which IA is propelled and compelled by subjectivity in infinite apperception. This is an apperception of all that is TRU, especially all that is of the greatest of TRU. It is an instantaneous and brilliant revelation and cognizance of all, TRU. ELTIR is the generic form and forerunner of ELTICMA. It helps to fulfill the opportunity for immortality, or: **OFEL** = Opportunity for Everlasting Life or Immortality. OFEL is how we can make our opportunities through TRU, its understanding and use. One's life is short, it is gone before you know it. Yet it is in sickness, illness, suffering and loss that we realize the accelerative powers of death and non being. Only then do we come to fully appreciate what we have lost, the dearness of life, and reflect on how we have spent this life, for good or bad, for TRU or ATRU. To late we realize OFEL and ELTIR. Yet life in TES and TRU is not ultimately temporal, nor even only of duration or variation, it can and must be, by TÂ, absolute. Death is either partial or AU. It is in every and all moments of life less than perfect TES and TRU. We may in physical death be full of actual life, as Christ, an immortality by an enhanced TRU. It is in how well and how much we are TRU in life and death, and how well and how much we reverse what is non TRU in life that is OFEL. For there is always the opportunity for immortality. Death is none the less increasingly with us unless and to the degree we are TES, TRU and TÂ. Herein lies our challenge and opportunity, to control one's own destiny, and that of all to extend life eternally.

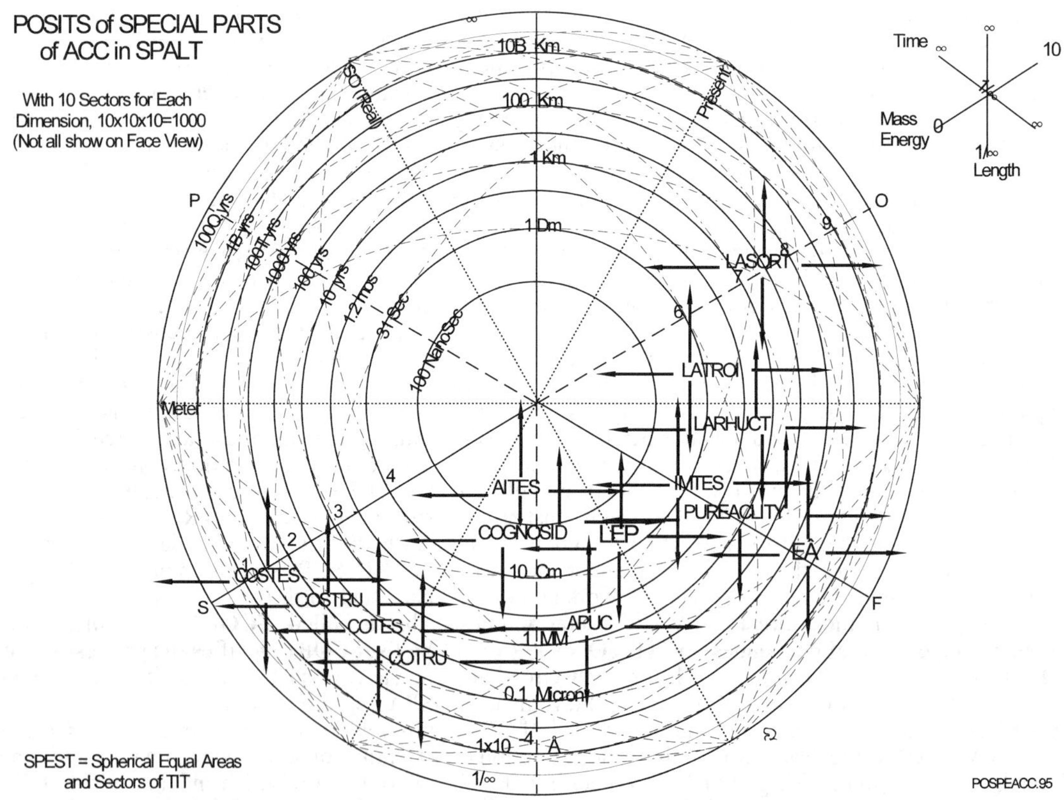

Fig 9-95 **ACC SPECIAL PARTS POSITS in SPALT**

MEASURES of ACC SPECIAL PARTS POSITS

Part			
ACHIEVEMENT of GREATER KUPS (AGK)	8,7	~2~6-7~	M
AITES	1,3	4-5-5	T
APUC	1	3-6-7	T
CCJ (CICIJ)	4,1 ±	7-2-6~	M
COGNOSID	413 ±	2-5-7-	T
CONVERSION,CON	4,3 ±	1-6-6~	M
COSTES	1,3	~0~8-5~	T
COSTRU	1	0-7-5~	T
COTES	1	1-8-5	T
COTRU	3	1-7-5~	T
DERGS	8	~-~-~-~	S
EÂ	3	~5~5~9~	T
ELTIR	3	0-8-9~	M
EXCELLENCY/REFINED, EW/MANY (ERF)	4 ±	-~4~3~-~	S
IMTES	1	5-5-6~	T
INEQUITIES of LIFE (IOL)	4,8	- ~ ~ ~ ~	S
KUPS	5,7	1~7-7~	S
LANICT	7	~ ~ ~ ~	S
LANTES	1	4-5-6~	M
LARHUCT	1,3	5-4-7~	T
LARUC	2,3	5-5-5	S
LASORT	1	6-4-6~	T
LATROI	1	6-4-6~	T
LEVELS of ACC (LEA)	1,3	~-~-~-~	S

LEVELS of PUREALITY				PUREACLITY (PUR)	1	~5-5-7~	T	TESARE	1	5-5-4~	M
(LEP)	1	~4~5-6~	T	RADU	1,6	~-~-~-~	S	TESRE	1,3	4-5-5~	M
LISPOT	1	7-4-7~	M	RALGAW	1	5-5-5~	M	TESS	1	~4-6-5~	M
OFEL	3,6	~-~-~9~	S	SCIN	1,7	1-8-5	S	TREQ, HOM, SOCIETY,			
PAC	2,6	7~-~-~	S	SCOPT	1,7	1-8-9~		ACC, TRU THES 857	±	~-~-~-~	S
PRECIOUS GIFT				STEGS	3	4-4-5~	M	TRUC	5	~-~-~-~	M
(PRG)	1,7	~-~-~2~	S								

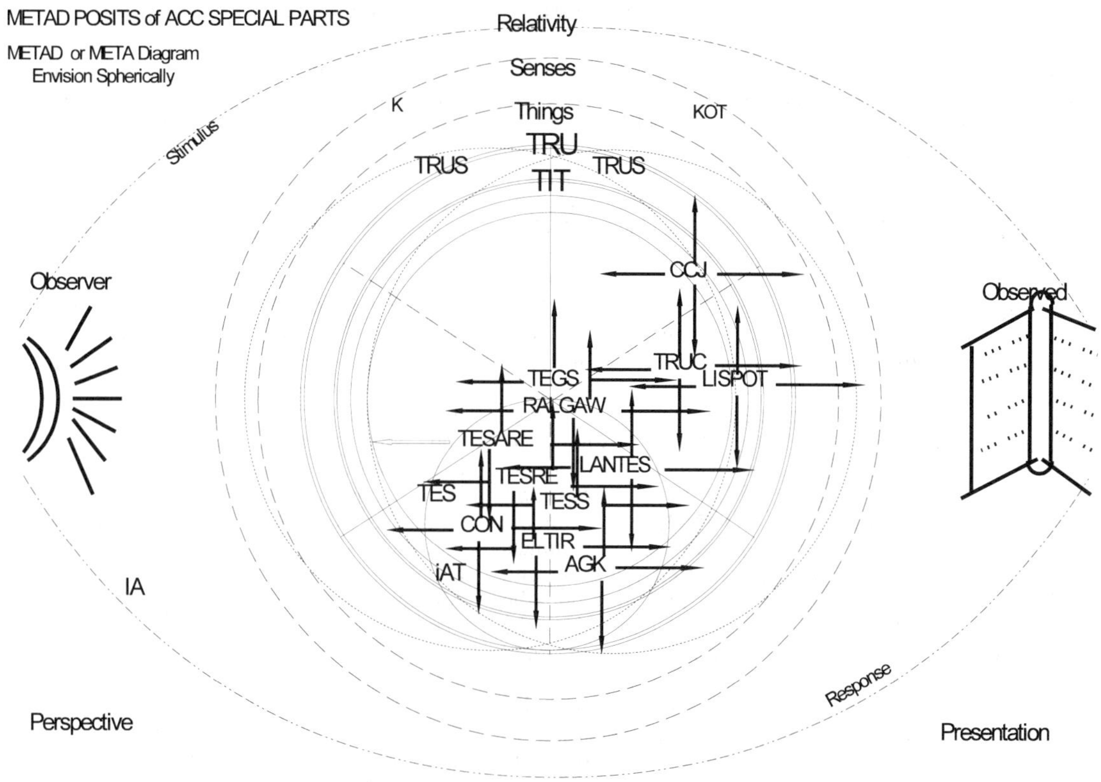

Fig 10-95 **ACC SPECIAL PARTS POSITS in METAD**

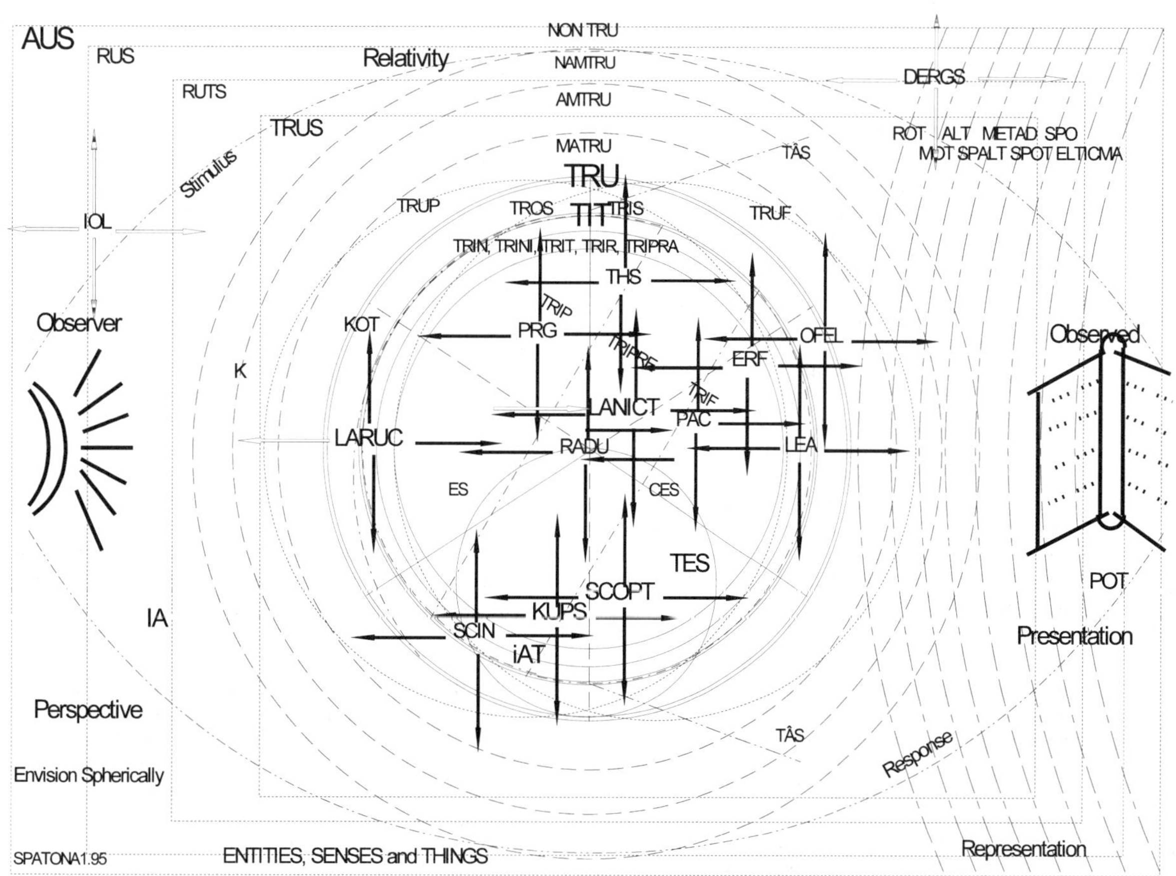

Fig 11-95 ACC SPECIAL PARTS POSITS in SPAT

CHAPTER 96

TOTAL ACC, ALL TRU ACCENTUATION and ACTION

INTEGRATION and SUMMATION of ACC, TOTAL KNOWLEDGE, COMPREHENSION, COGNOS, CHANGE, MARS, TOTAL MANAGEMENT, CONTROL

Just as the world of Homation and TRU is integrated, fused, and has overlapping features, so does PUREALITY of an individual's TES, as it is turned to best use for TRU. Likewise COGNOS, the most concentrated, comprehensive and identified state of subjectivity in TIT-TO, increasingly becomes the absolute in TRU, as THES becomes THESÂ. Since the individual is the great vehicle and carrier of TRU it is to the individual that COGNOS most depends and is most focused. Everyone, in a way, needs to have a personal God. COGNOS gives us, in being a great part of TRU, $THES^n$ or THESÂ, concentrated identity with the advancing powers of TRU that are cognitive, mental and subjective. COGNOS, either recognized or unrecognized as in TREQ, has produced a great amount of stimulation, interests, beneficial ideas, concepts, reason, and judgement. It is like the absolute, the absolute mind or absolute spirit, that which, as an end, so much of Philosophy has focused. This is seen in Kant's *Pure Reason*, and Hegel's *Absolute Spirit*. COGNOS arises from the discovery, development and practice of knowledge by humans in ways that can be most perfectly formed, function and utilized for infinite and total power and mastery, even unto the absolute. Such knowledge can be attained only when the proper VAPO and SCARLS are applied to and by TES and TRU. This largely develops through personal identity with divinity, PUREALITY, TES and TIT. Such a personal God is the concentration of an individual upon the highest salients of ones being with the all, TRIA. Those highest levels of SÁ in TES and TRU that are special, predominant, and most powerful. Without TREQ, TRU, PUREALITY, and COGNOS one is apt to lose the necessary great key, cue, gist, tools, means, desire, love and will to discover, learn, develop, be, practice, sacrifice, serve, and perpetuate TES, TIT and TÂ.

The advantage and greatness of ROT, MOT, ALT, and HOM lie in our ability to think properly, comprehensively, solidly, and to think TRU. The very nature of our being, especially our immediate awareness and instantaneous attention, IA, demands a concentration, of various appropriate degrees and varieties, upon that directly before us. This must always be conditioned if we are to function properly, successfully and fully. ALT, HOM, and TRU give us this overall plan. They provide us with realization and comprehension that is: 1. Balanced, 2. Proportional, 3. Massive, 4. Total, 5. Ordered, and 6. Readily available. This is where and how realization is the identity of the observer and observed, our relativity, and the role of comprehension in actuality where actuality is the total thisness, FAF, static and dynamic state of TRU. This is called: **LOTRAC** = the Law of TRU Realization through the Comprehension of Actuality. LOTRAC is the achievement of actuality by comprehension. This occurs by the relativity and focus of S-O, or Á, on the realization of all TRU, and more specifically how we can live in PUREALITY and one with COGNOS forever be.

If we can prove God, religion and TRU, as all are in TIT and of TRU, we can show how these exist through unity, systematic unity, relativity, and the finite ordering that can be produced through DAOST, TROS and TRIT. We can show how they can be used to tremendous and glorious advantage in our lives by becoming a total, complete knowledge, understanding and comprehension. This many theologians, gnostics, philosophers, scientists and thinking minds through history have suspected and sought. It has been a driving power, force, and benefit for much of human life, TRU. It, as none other, stands to offer mankind, through the focus and summation of IA, and its S-O or Á mass energy to completion, than the degree of mastery and control otherwise unattainable. Such is COGNOS, the utilization of all subjective powers in the mass energy continuum of TRU to envision and control all.

What will all be like at different times, and how much would have been better if we had only realized and done what will be known in the future? **FOKIT** = Foreknowledge by Identity with the Future and its Focus on the present and past, that improves TRU. It helps to tell better ways of living in the Present. FOKIT is not only a fact and way of helping to anticipate and live better, it is a great compelling force of the future on the present, past and through and through. It is like recovering what has been lost, a great inspiring method of comprehending, both all and, what is critically needed. It is a means to COGNOS, ELTICMA, and SÂ. The cue and trick is to both put ourselves in those positions in the future in which such greater visions can be made, and to be able to focus on the problems and dangers that do or will exist and must be known. Throughout HOM, DYNU, DYNS, DYNO, and ACC we can increasingly see and appreciate that which history teaches us and what can be done in and for the future.

Special interests are often 1/TRU, ANTRU, the opposite of ENT. They, at best, proceed naturally, and honestly, from the function of a part in a whole. They are part of the dynamic processes and the interplay of the competitive massive forces that makeup TRU. Yet, they commonly, quickly and usually lead to abuse. This stems from a lack of appreciation for the whole, TO, and the totality of ACC. It is a result of errors in our knowledge, attitudes and reasoning.(1/VAPO) Much of TRU is homely and common. We should appreciate this as much if not more than what glitters, seems so comely, fashionable, or even different.. This should be for all that is integral with TIT, be it directly or less directly through THES and THEO. This is called: **LIRT** = the Law of Integrative Resolution in TIT. LIRT is 1/Special Interests, and expresses how the infinite and finite must always reach toward and function properly in TIT-TO. LIRT, coming largely from IA and the individual, functions with its own degree of freedom.

The great tragedy of our times is how people of 100, 1000, and more years in the future will look back on us with pity and disbelief. How could we have been so naive, ignorant, vain, impractical, disordered, uncoordinated, and incredulous as to fail to understand the truth of knowing and comprehending the world and ourselves as we actually exist, in TRU. How selfish, greedy, unhelpful, and wasteful we are in the performance of our part in the world, TRU. How little, we realize or want to understand and live by the forces, powers, dynamics, spirit, completeness, inner-workings, formulation and systematic unity of the world, TRU. How little we identify with THES and COGNOS as they are we, and we are they. There are great vibrant moving and pulsating forces within us that are also apart from us, that yield, mold, form, direct, necessitate and in all ways are the heart and soul of so much of us, TRU.(TRUD) It is for us to discover, best learn, know, remember, and use them if our progeny and future are to look upon us with favor and produce for us the immortality we must have to be TRU..

TRUCO = TRU Conversion. Conversion from one belief to another, or to a religion, as Christianity, is an excellent example of what takes place in developing TES and TRU. It is more or less unnecessary to the extent it can be replaced by more effective means of maintaining the highest quality of humans

FAF, TES. Yet, because these have been very incomplete, inadequate, and inefficient survival of the fittest persists. This can be largely corrected by TRU, TRUCO. All religions, more or less, purport to convert lives of suffering and misery into less painful and more effective modes of serving God. **FOREC** = Fallacy of Religious Conversion. FOREC is produced by religion's emphasis more on salvation and conversion based on any reason only to acquire and retain a following. Reasons whose tenets are partial, disproportionate, and often invalid by TREQ or lack VAPO. They are defective although they are often an improvement over what they replace. Religious conversion fails to convert ANTRU, ATRU, and MATRU lives. To cleanse ones soul - the achievement of TES - is a continuous process throughout life. How well this is done, is seen in: **DOC** = Degree of Conversion. DOC may be from any kind of life. This direction is formed from TROS, TIT, TRIT, SUT, ALT and the coordination, order and comprehension that TRU provides. **SUTAC** = Summation, integration and Sigma Systematic Unity Accentuation: **SUTAC → SÂ → TÂ** SUTAC is exemplified by how far types and orientation of political and social systems fail because of a lack of ability and performance to provide the benefits of SUT when integrally applied in ACC of TRU. The parameters which represent the continua that exist in TRU show how DOC can be measured. The greatness of conversion corresponds to the cue and gist of meaning that is to be found in the revelations and inspiration of all that is TRU. These when well developed provide the concepts and models of TRU that each and all must learn. This we see in ACC, whose direction complements natural selection and adds many better means of achieving quality TES and TRU. **POCANS** = Positing Complimentary Conversion Accentuation Direction and Natural Selection. POCANS signifies that the direction attainable in ACC and natural selection are mutually operative and complimentary. Complete conversion is possible by acts achieving ever higher degrees of KOT, ACC, and TRU.

MARM = Multiple Accretive Resolution en-masse. The resolution of multiple ills does not call for the instantaneous obliteration of singular ills. It often calls for an unwinding and continual resolution of all involved over prolonged periods of time. This is to be compared with their development. Multiple ills can't usually be cured by the excision of just two or three of a hundred or like number of separate causes no more than cancer can be cured by only removing a small part of it. Just as they must be cured en-masse over prolonged time, so must all be cured totally over the entire length of time. This is the importance of SPALT, HOM, DYNO and the totality of ACC.

We tend to dislike things we don't understand. This is a tremendous force. For it hides dangers that threaten human form and function, Á FAF, and all TRU. Energy and curiosity are put to the test. Humans also dislike things that are complex or massive. For these reasons, thereby, TRU will not be as well liked as it should. The dislikes of people are the roots of conflict, both IPR and universally. The question then arises, 'How do you get people to understand all things in their greatness, fullest, and most proportional ways'? (VAPO,TRU) Herein is the stimulus to learning. So that the question then becomes 'How do we eliminate all that inhibits learning, and how do we enhance what encourages and qualifies learning'? These questions are great proofs for TRU and POT. They justify religion, philosophy, science, knowledge, and COGNOS. It is one of, if not the greatest, powers of human life, Á, processes, TES, HOM, ACC, COGNOS in TRU.(PREFS, philosophy, religion, fields and sciences)

Adequate knowledge of a topic, especially those as great as TRU and ACC require massive amounts of information and data. With an accurate model this can be organized so that we may plot and plan for action all needing accentuation and reaction. ACC and TRU are presented to emphasize the fact that most fields and sciences have yet to discover, promote and utilize many great facts, concepts, principles, mechanisms, laws and processes that are not known or adequately used. These need to be continually added, refined and upgraded. Kant was greatly inspired and influenced by Hume's criticism and method by which all are not only made formally distinct, as in pure reason, but are held to their most actual and great - TREQ - states. They show how we can with truth and precision, make TRU determinative. Many of the benefits of computer science will increasingly show how to best manage the finite of TRU. This is successful only when such masses of data are well identified, made distinct, have great meaning, are ordered and made whole. COGNOS signifies We Know. It is not in knowing more and more about less and less, but knowing discretely, distinctly, coordinately, proportionately, and with control, all of TRU. This it must be if it is to be, and be fully utilized. COGNOS is largely recognized from a coordinated overview by comprehending TRU that prepares us for perfected action, ACC and ELTICMA.

Some of the modes and characteristics of COGNOS, relative to God, may be enumerated: 1. We know that we know, 2. We know God, 3. We know God as TREQ of TRU, and as variants from TRU, 4. We know much about God, in, of and other than what and how it is in or of TRU, 5. We know how to posit and treat most all of God as it can be represented and modulated of TRU, 6. We know how to utilize this knowledge, 7. We know what we are or can be of God and how to achieve highest participation in TRU, 8. We know God and TRU as TA, the absolute, 9. We can find our own absolute, not in and of anything or oneself, but by how it becomes TESÂ,(LOPAN) 10. We can be the heart and soul of God, as TES in TRU, TIT and TÂ, 11. We can better appreciate TRU, through VAPO and its void in God, 12. We can be infinitely cognizant of this knowing, 13. We can inspiringly and absolutely recognize the greatness of COGNOS and our participation in this salient of an ever greater TRU, 14. We can recognize absolutely the eternal glory and victory of TRU that COGNOS can provide, our identity with it, and the actions that will best achieve TRU, 15. We can involute all less than TÂ, especially ourselves, as stimulated by COGNOS to an ever greater Á, mass energy dimensional state in TRU. Many of these reveal the eclectic and ecumenical qualities and powers of earliest to modern religions, as the Upanishads to TRU. To be critical, agnostic and sift out what is best and common in all knowledge helps to counteract those tendencies that so divide religions. It is evident in many trends of human thought. All is Brahma, just as all is Allah, God, or TRU is all.(TRIA) It is in realizing and knowing what this all that is held in common is, that we need and benefit from COGNOS.

MAROT = Minds Ability to Rapidly and Totally Comprehend all, TRU. The human mind has the fantastic and glorious capacity, ability, and opportunity to rapidly and totally envision, comprehend, concentrate, order, differentiate, integrate and unify all. This is especially so when well trained and guided by POA of TRU. TU, TIT, TRIT, OBT, and SUT to provide the best possible means to represent and model TRU. It enables the ideal presentation for MAROT. MAROT is achieved by the: 1. Exercise of the mind, 2. Exercise of the mind, in more complex, compound, and total ways of thinking, 3. The different habits and ways that accompany this exercise, 4. All the knowledge that enables us to better know TRU, SPALT, and especially all the detailed aspects, parts and processes of TRU that can be shown in SPALT, by Homation. 5. The big picture already before us, whereby both the entity objectivized and ourselves subjectivized can best find the range that all can come together in TRU, through TU, SU, ROT, ALT, and POS. 6. The revelations necessary to enable one to realize the deepest meanings in all TRU, TO and OITIO. 7. Through the inspiration and revelation that TRU and COGNOS provide, 8. TAN is used to expand and accelerate comprehension, 9. The cue formation needed to answer and solve the various question and problems of TRU and those before us, 10. The best solutions that can be found, as problem solving, 11. The beautiful and joyous interest that exists and can be cultivated in the expanding array of concepts, principles, laws, mechanisms and all that is TRU.

C.Q.(CONQ) = Conceptualization Quotient. CONQ is vital to a person's and TRU's success. It is the ratio and measure of human potential, ability and intensity to acquire and hold greatest concepts of TRU. Especially in association with other highest levels of Á a balanced CONQ is the greatest force in TRU. It is the total of those, ideas, concepts, principles, and terms which in their absence leave a person weak and unable to meet the demands and competition of life. Only as we attain massive conceptualization, **SCONQ** or Summation Conceptualization Quotient, and only as these are highly proportioned, coordinate, ordered, held in mind, and utilized for TRU can we achieve TES, prosper and have permanence. In Figure 1-99, Diagram of the Thetic or Prescribed Stages and Levels to TRU, we show the steps are necessary to attain a good comprehension of TRU, as in COAT, METAD, SPAT and ELTICMA. It will be seen on close inspection, that, although separate, these stages can be fused as they are practiced and fully utilized. Cogito is the process of Cartesian doubt that is reached through argument, a skeptical and critical attitude arising from the "I think therefor I am". Descartes purported this to be the only way truth can be achieved. Yet it is more simply the first step in the process of relativity, of the observer - observed identity, in which certainty, proof, and understanding can be established. Because cogito does not distinguish what is true in general from what is true actually in TRU, it can not succeed in providing a sine qua non method for TRU. In COGNOS this doubt is alleviated in more than one step: 1. The thing in itself, ENIT, is the object or observed; that it has optimal purity and validity, 2. The object, or what is observed, is the same, as that which the observer senses or perceives. That is it has optimal reality. 3. The observed, object or ENIT is no more nor less than those parts and whole of it, ESINS, that are in TRU. Thereby it has optimal actuality. 4. This object of TRU is no more nor less than its proportional part in TRU. Thereby it has optimal proportionality. 5. All of these conform, no more and no less, than to the whole of TRU, that it has optimal totality. It is in this way that VAPO and PUREACLITY set the stage for COGNOS. Knowledge that reaches its optimal degree in TRU is what we call COGNOS. It solves many problems, e.g. Cartesian doubt.

Fig 1-96 **COMPONENTS of UNKNOWNS, KNOWNS and COGNOS**

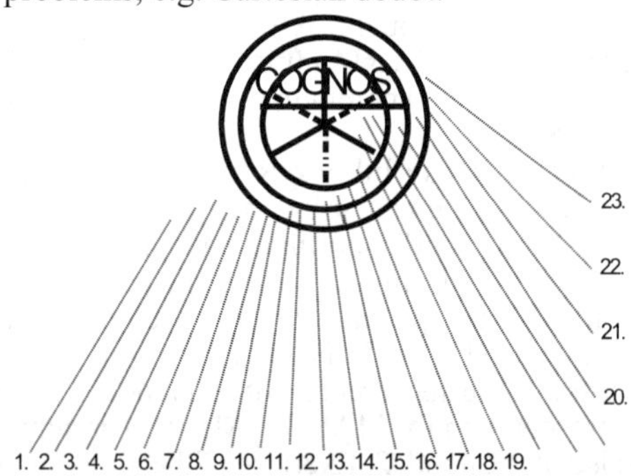

1. Unknowns, never known, 2. Unknowns, not likely known or to be known, 3. Unknowns, possible to be known later, 4. Unknowns past or present, probably known later, 5. Unknowns, not known in past, more later, 6. Unknowns now, but known past and future, 7. Knowns slightly in the present, 8. Knowns moderately at the present time, 9. Knowns much of AU, 10. Knowns much of mixed AU-TRU, undifferentia-ted, 11. Knowns much of mixed AU-TRU slightly differentiated, 12. Knowns much of AU-TRU, moderately differentiated, 13. Knowns much of AU-TRU, well differentiated, 14. Knowns of EPI-AU-TRU well differentiated, 15. Knowns of EPI-TREQ well differentiated, 16. Knowns of TREQ-TRU well differentiated, 17. Knowns of TRU, KOT, 18. Knowns of TRU extensively, OM, 19. Knowns of TRE of TRU, 20. Knowns by TRE of TRU to finite COGNOS, 21. Knowns by COGNOS, fully to infinite, 22. Knowns by COGNOS to TRU completion, and 23. Knowns by COGNOS to TÂ completion. The likeness of radiant effects of COGNOS on knowledge to that of the hands of ATEN and the rays of God to Moses reveals their common base in TRU. From this demonstration knowledge is not a simple state. It is one of gradations toward fulfillment and perfection, in COGNOS, can provide great advantages, force and power for TRU. It approaches the ultimate means from Epistemics to TRU accelerating KOT exponentially to TRU and TIT. This is how COGNOS as THESÂ advances through THESOÂ toward SÂ and TÂ.

The universe is roughly apparent to many, yet this is quickly lost for lack of focus, totality, proportion and misuse. This is called: **FOTIU** = Fallacy of The Irrelevant Universe. Nothing that is of TRU is irrelevant. It may be more or less, or it may be different for different individuals and IA, yet because of TIT and TRIA if it actually is, it is TRU. It is because a common overview of the world is necessary for human harmony that all should seek, as this text does, the solution to the multiplicity - and duplicity ATRU diversity produces.

LEWOV = Levels of World View. Below are enumerated and classified the different kinds of world view and the rough percentages in and used by the world population. This is more variable, extensive and dynamic than such a list suggests. Yet the resultant conclusions and indications for use should be evident. Humans have a long way to go, before they have a good world view. We have an even greater way to go before adequately applying this world view to life, ACC, and TRU.

10. Perfected, absolute, infinite, all time, ordered, proportioned World View of TRU & All 0%

9. Excellent, infinite, all time, ordered, proportioned World View of TRU & All 0%

8. Good, infinite, all time, ordered, proportioned World View of TRU or TREQ and All >0%

7. Fair, infinite, all time, ordered, proportioned World View of TRU or TREQ and All .001%

6. Poor, infinite, all time, ordered, proportioned World View of TRU or TREQ and ALL, with an excellent view of aspects or principles of TRU . 01%

5. Very Poor infinite, all time, ordered, proportioned World View of TRU or TREQ & All. May have a good view of parts and processes of TRU and all .1%

4. Very good personal view and much about one. Rough idea of the World. No discrete, ordered, proportioned World View of TRU and All 1%

3. Good View of oneself and that about one. But no or erroneous idea of the world in general 4%

2. Fair view of oneself and that about one, No adequate idea of the world or God, 15%

1. Bare view of oneself and that about one, Completely confused about the world and God 30%

0. No or erroneous view of oneself and that about one as well as about the world and God 50%

For mature adults in the prime of life, at their best and on occasion LEWOV will move up one to two levels. To recognize # 8-10, and to actually have and live it are far removed. The individual's world view, as the universal in KUPS will be in a state of constant flux and constantly need to be qualified by the infinite possibilities of knowledge and different perspectives, the many different standards of perspectives, the relative universe, ELTICMA, the increasing focus from ALT to the world of which one is a part and the focus from these to the world of ones attention at the moment, IA. LEMOV shows, contrary to personal and common belief, how poorly we visualize the world that actually exists, as God, by TREQ of TRU. Even more it shows how much work we have ahead us and how little we can allow such ignorance to go unchallenged and uncorrected.

Biblically "Know and be Known". The gist of this is that without knowing one can't be known, and without being known one can't know. It is the interdependent subjective powers of knowledge, individuals, others, TES and TIT. An individual can't have any adequate knowledge without acquiring it from others and the body of human knowledge. The body of human knowledge can't exist without the knowing minds of individuals. TIT can not be without TES nor TES without TIT. To be carried to its fulfillment is COGNOS. This is where We Know, both that God as TRU is, and the subjectivity in TIT is what provides us with all knowingness, COGNOS. It is more than a simple idea or knowledge, it is the great, divine, immanent and all powerful predominant mass energy, Á form and function of TRU in its highest and most complete level of development in TRU. This is a manifold and explosively expanding mass of power that drives this special universe of best human life ever onward, forward and higher eternally.

What are the worlds greatest revelations? Which ones are especially applicable to knowledge? Some of these include, The tree of life, The circle of life, God, Radiation from God, Brahma, Identity with God, Atman, Mysticism, Christ, Love, The Cross, Anserated Cross, X, Sigma X, The Philosopher's Stone, +, TRU, and many others. Also many of the greatest ideas, concepts, principles and laws of humanity, philosophy, science, and religion and their fields, including its highest public principles or HPP, and many yet to be recognized, as NEC in LANT. Symbolism has long been a part of religion, and shows the power of revelation whose intensity magnifies and retains abstract ideas and concepts. Every moment, thought, and act of each person is a direct and great opportunity to reap the inspiration revelations produce as they in turn provide the fast track to actual being and eternal life in TES and TRU. This is largely assisted by TAN, to think and note. This it does by recording one's deepest thoughts and positing them in an ordered comprehension of TRU, POS in SPALT. TAN is the most direct and intense participation of the individual in the real of Á mass energy continuum in TRU. It is in the syncretism and added states in totality of such inspiration and revelation that we bond with COGNOS and TES. It is how ACC can totally produce and sustain TRU and TÂ.

CONST = Concord or Coincidence of Standards. It is crucial and vital that all human standards best identify with and reflect TRU and ROT. Its need is less from the different that is accommodated, than the laws, standards, rules, regulations, customs, tenets, prejudices and partiality that are not, or disproportionate, TRU. These are the great divisions, clines, opposing theses, and systems that divide the human race, and pit one against another. We must remove, eliminate, and control all laws, regulations, rules, standards, customs, tenets, systems and all that are: (1) not necessary, (2) not TRU, or (3) used beyond their limits or SCARLS. To do this we must first know what is needed, what is TRU, and what is or is not within its limits or SCARLS. This requires persons who have great understanding, enlightenment, interest, and are dedicated to active participation in TRU. Each and all persons, groups, communities, states, nations, religions, and systems should know exactly what their place is in the world, in TRU. They should function no more and no less than just what this must be. This is the great message of Confucius. It is a message of self control, Jen, duty, and a conduct that is in conformity with the harmony of society and world, the way of heaven, TRU, and TÂ. To prevent the catastrophes that lie ahead, as well as all other less dramatic, yet more complex, and none the less deadly disasters, it is imperative that we get to the roots of our problems. This ultimately is no less than what most acts on and affects TIT-TO. It is necessary that we see to it, get busy with life, stick to business, and do a good job. This is the business of learning, thinking, comprehending, having revelations, being inspired, appreciating, utilizing, doing and enhancing TRU, eternally.

UBOT = Universal Boards of TRU. UBOT are examinations and other means of evaluating the quality and proficiency of all. Their utilization should depend upon multiple levels of competency and test scores guided by TES and TRU. They should be for those who are likely to be future leaders, especially who participate in the presentation, dissemination and promotion of ACC, TRIX, and TRU. Most specifically it will help to select those needed for the more intensive work in TRU and POT. The boards should be formed and designed to select those candidates that have more than factual and topical knowledge, even about TRU or TREQ. It should be gauged to select those with the character that best fits whatever role in TRU they will take, including the greatest appreciation, inspiration, revelation, love, conviction, and vigorous purpose that TES and TRU warrant. It should be able to determine the habits of thinking and all that can tell who will make the best executors of POT, promotion and ACC of TRU. These should include a high level of PUREALITY, TES and COGNOS that gives added direct impetus to TRU, and to other's TES and TRU. This should be directed toward accentuation. **AET** = Accentuation Exponentially to TRU.

The power of good or bad has been relentless in this acceleratively formative stage of TRU. It has been a strong selective force. Its continual incremental operation is much of the Á FAF of TRU. Although Á changes very slowly, the processes which make up the forces of selection at present and in the near to mid term future will change more rapidly. This will, more or less, be a change for the better, a human demand for higher levels of mass energy, or Á FAF. These are the rational, valuative, intellectual, and higher abilities provided by TRU, the hallmark of our civilization. We must develop and effect ACC if these changes, as well as the power of good and bad, are to be what is necessary for TRU. **PRACT** = The development and Effecting ACC, or the Practice of ACC of TRU. PRACT is much of what we do in OITIO, COGNOS, ELTICMA, THESA, THEOA, THESOA, SÂ, and TÂ.

Natural Selection, Management and Control

About the greatest problem humans will face in the future is to obtain the maximum adaptation and selection simultaneously with the maximum in prevention, protection, maintenance, management, control and advance. How the best of evolutionary natural selection can be accommodated by the best of mentality, valuation, reason, intellect, intention. purpose and total management. These can be highly conflicting, especially if natural selection is not adapted to TRU and if controls are not well based, founded, and applied, to best suit natural selection, TRU, TIT, and TÂS.

PLANSMAC = Problem, Law and Conflict between Natural Selection and Management and Control. It is through the realization, recognition and application of PLANSMAC by the optimal operation of both natural selection and management and control of the determining forces, live product, genetic material and all of TIT, TRIF, TRU, TRUF and TÂS that TRU can be best assured. What provides these and their best concerted operation results from the assiduous and total ACC, action, representation and presentation of TRU.(TINS)

Increasingly complex to total treatment converges on and becomes equivalent with total life and total TRU. However, this is with a slightly different emphasis. It is not possible to summate or finalize any treatment unless and to the extent one identifies, is in conformity and reconciles the totality of TRU, as in TO, TRU, and TRUD combined or **TOTRUD**. This should account and make allowance for not only TIT and TRIT alone, but a certain - although hopefully small amount of TRIS and, even less, TRUS. So that the final treatment can be no less than TRU in totality from TU to TI and all in between, statically, dynamically, coordinately, proportionately, and transcendentally as accounted for through POT, ROT, practice, ACC and overall management. This is: **METAC** = Meta ACC, Treatment, or Management. METAC is what the totality of ACC is, the meta-management of all TRU. In the form of a law this is: **LOREX** = Law of Total Rx, or Total Treatment. METAC and LOREX are similar, yet distinct. For METAC comes from the overall management and ACC that is needed and possible with TRU and TOTRUD. LOREX is the more specific term for the singular, to collective total treatments and actions possible and operative. METAC is approached from the unity, and even super unity of the all encompassing perspective, as in METAD and SPAT. LOREX is the collective treatments that can be integrated to make a whole. METAC is the super ACC for all TRUS and extra TRU convergent on TRU. LOREX is the total for treatments within ACC and TIT.

Even without being recognized as such, the force of COGNOS is very common and great. It is an all powerful, driving, influencing, determining, and directing process and dynamic energy that activates and enables TRU. It is a great compelling force which the future increasingly produces in TIT. It is the spearhead and salient that projects and assures much of TRU in the future. It is to be distinguished from all knowledge that is neither TREQ, TRU nor VAPO. Attitudes which are in opposition to COGNOS are accumulatively major causes for a person's weakness and failure. The bible extols and sings the songs - most intensely in the book of Psalms - of the greatness of understanding and the wisdom that comes from God. So likewise does the Upanishad do the same for Brahma. These are to COGNOS as TREQ is to TRU. Yet COGNOS incorporates an exponentially greater, more proportional and concretely total comprehension of TRU.(VAPO) Lecomte du Nouy in *Human Destiny* predicts that man will eventually be able to think universally. However it requires an understanding of what is actually universal, TRU. These when well developed become COGNOS or We Know, we Know All that "IS", is in and of TRU. We know all that needs to be known. In this way TRU, its power and action as focused in COGNOS answers the perpetual quandary of the individual, 'what all do I need to know', and 'what is All?'

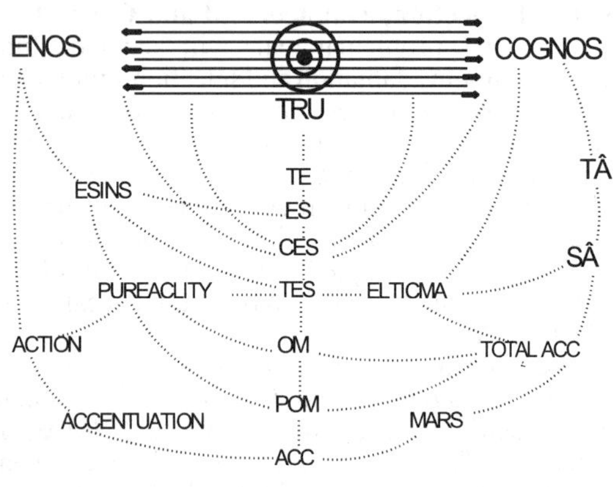

Fig 2-96 **ENOS, ES, CES, OM to COGNOS, MARS and TOTAL ACC**

ENOS = Entities from Objects to Subjects Relatively, Entities us. ENOS signifies the meaning that entities, things and all do not exist alone nor separately. They exist in progression, as does COGNOS, TRU, TE, ES, CES, OM, and POM. The symbolism of an all powerful deity hovering over us, and protecting us for benefit has existed since eighteenth dynasty times in Egypt. It is in this way we by nature and TRU, are compelled to act, and act in certain set ways. This action taken to its proper ends, and the ends necessitated by the FAF of TRU or its Á state or dimensional continuum and especially its higher powers of subjectivity, must do so with a great amount of ACC. So that we must avoid the trap of thinking that POM, TRU, TE, ES, COGNOS, ENOS, many or all entities, and less established methods are here for themselves. Assuredly, we can naively sit back and let them happen, but in this we are only denying TRU and fooling ourselves. It is only, as in AITES, up to us to make it happen. This is the basis for: **MARS** = Multiple Ablation Rapid Simultaneous. MARS is what enables entities and even the more negative forces upon TRU to be amenable to ACC and TRU. Often we can not achieve TRU by simple, singular, poorly focused and ordered means. When they are done in totality, with unity, focus, order, force, power and precision of MARS based upon ACC and TRU even the more difficult and total work will be successful.

COGNOS does not mean that we know everything, for much that is not TRU is not necessary to be known. By the special thisness of TRU we need to know relatively and by VAPO. COGNOS does not mean we know TRU infinitesimally or infinitely, nor that which is excessively and impractically partial for every person, time and place. COGNOS does not mean that we know TÄ, the form of TRU in perfect detail or repetition. This is called: **APRES** = Form of TRU, Perfect Resolution, From Perfect Repetition of Detail, or From Presentation Perfection and Repetition. We can know the form of TRU as we know the form of any object very successfully if we know it well; if we know it with a high degree of validity and sufficiency. COGNOS is how we know TRU. How we know all that enables this knowledge of TRU. It is how TRU is known more and more perfectly in all its aspects and glory. This is, like in the best of art, how the salient features and detail of TRU are sufficiently brought out to provide an excellent observation of it. The greater, deeper, and broader this knowledge is, as by ACEXIN, the greater deeper and broader we, and TRU become. There is no reason for us to become complacent about achieving COGNOS. There is an infinite and eternal opportunity and challenge. COGNOS implies a relative, practical comprehensive completion, and closure of knowledge representatively providing an increas-ingly accurate, unified, 1:1 = 1 image of TIT-TO and TRU.

The mental revolution in human evolution can not be sustained without COGNOS, cognitively consciousness of this revolution led by TRU. This is its heteronomy of ends, where all TRU, its perspective, representation and presentation are so fully developed in each and all persons that they appreciate and live by all TRU. Here is the prime goal of POT and IPTRU. It is not only the discovery of TRU, but the implementation, securing and guaranteeing of TRU through the living of it as it must be. Such a cognitive revolution is a higher generation of the revolution in knowledge that has so characterized civilization, a step in the revolution of human life, transformation of its mass energy, of Á FAF. By TRE and COGNOS we enter the third great generation and revolution, in the make up of humans. Even a fourth, and ever higher levels of SÁ, may be conjectured in ELTICMA, SÂ, and TÂ.

Certain threats to human survival, TRU, may become very acute at different times in the future depending upon our exponentially enlarging power over ability to best use this power. Therefore proper or TRU use of power must always be much greater than any and all other kinds of acquired power. This is: **RATUP** or the Ratio of TRU Use of Power over the exercise of any and all other kinds of power. RATUP is another reason and great stimulating need for knowledge of THES, THEO, and TRU. Many problems and threats that seem to be occurring so spontaneously or naturally at a much later time, are in fact occurring at a much earlier time, as at present or through from the past. These are often what come from the outside, unexpectedly, explosively, and violently to affect TIT adversely. They also may occur through a weakening of TIT or some vital aspect, system or mechanism of TIT. This is called: **LOCMENF** = the Law of Concentrated Management of Negative Effects in the Future. LOCMENF is a part of the concentrated total management, prevention, preservation, protection, maintenance and advance of ACC.(PPMA) This total when guided by ACEXIN positively is: **ACEXINOP** or the Actuation, Extension and Intension of total knowledge and COGNOS of all for the positive/negative enablement of TRU.(ENT) ACEXINOP is the total positive/negative effect attainable.

Fig 3-96 **MANAGEMENT of ALL, TOTAL ACC and TRU**

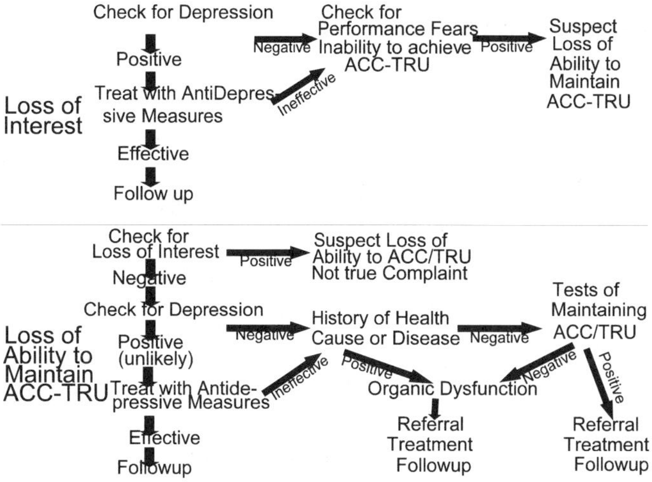

ALMAN = All Management. ALMAN is the consolidation and control by the total management of all toward the given goals and ends of TRU. Through TRU, TU, SU, ROT, MOT, ALT, SPALT, REM and POT we can derive, develop and apply methods and means to provide the total management of all. Through the use of Flow Charts, their combinations and total superimposed total application to ACC, along with other means of demonstrating TRU we can show and provide the correct actions necessary. This converts whatever is at hand into that necessary to show and produce its best management. In the figure we do not give ALMAN for all ACC and TRU. We do so as an example, and for the great cue and gist ALMAN has for ACC and TRU. This is of the individual, the quantity and quality of interest, and the amount of ability they have for maintaining ACC and TRU.

At first glance the ideal ultimate knowledge may seem simple and easy. It may be assumed that all that is known and all that is unknown as well as all that will be later needed to be known would be the ultimate or ideal knowledge. Yet this is neither entirely truenor so readily distinguished. For ultimate knowledge partakes of all the characteristics and nature of the theory of knowledge as outlined in the beginning chapters and throughout the text. Ideal or ultimate knowledge consists of all concepts, principles, laws, mechanisms, EC and NEC, LOTS, LATS, and other TRU terms. All that is necessary for an increased and manifold knowledge that enables the understanding of TRU and all things, COA, COAN, COAT. This totality is called: **COS** = Comprehension Total in all ways, the Plurality, or S, of all entities. COS differs from COA, COAT, KOT, OM, and COGNOS by returning from TRU back toward AUS, but still primarily being in TIT and retaining all we have learned, and known about TRU, ROT, POT, HOM, ACC and all. Total ACC and the triumph of all TRU and TÂ is, in part, exemplified in TREQ, and by ‘ God is triumphant and wants his children to be joyful’. This was the new spirit of British Religion in the 18th - 19th Century. It tended to put away impersonal, uncaring, strict and negative thinking by empowering and glorifying God, as their blessings were counted and enjoyed in the great accomplishments that both God and man have made. Religions have come and gone, and in doing so have gone through many phases. Often these are different and contrasting phases at the same times in the same or different parts of the world. One is often the reaction to the other, and associated forces, trends, courses, and processes. These we trace out in detail in SPALT and HOM over the past few thousand years. They often reflect the political, commercial, material, intellectual and spiritual state of the times. The cyclic changes in human fortune have, are and will often be reflected in man's religious spirit. It is not so well to ride the glory train of a few of these, but to order and utilize the best in TRU, and PPMA the worst, that we can correctly achieve the best of spirit in ACC and TRU. Yet we must sufficiently know all of them, how they work, and may be applied to best advantage. GOD, as TRU, is actually, like the spirit of British Culture, so very triumphant. GOD, as TRU, ‘wants his children’, ES, CES, TES, and even TÍS, as all of TIT, to be joyful, most harmonious, and successful. For TRU to be triumphant means not only now and in some ways, but always, and in all ways. For ES, CES, TES, et. al. to be joyful are to receive, appreciate and with happiness savor our blessings, not so much for the moment but for all time, in TIT and TRU.

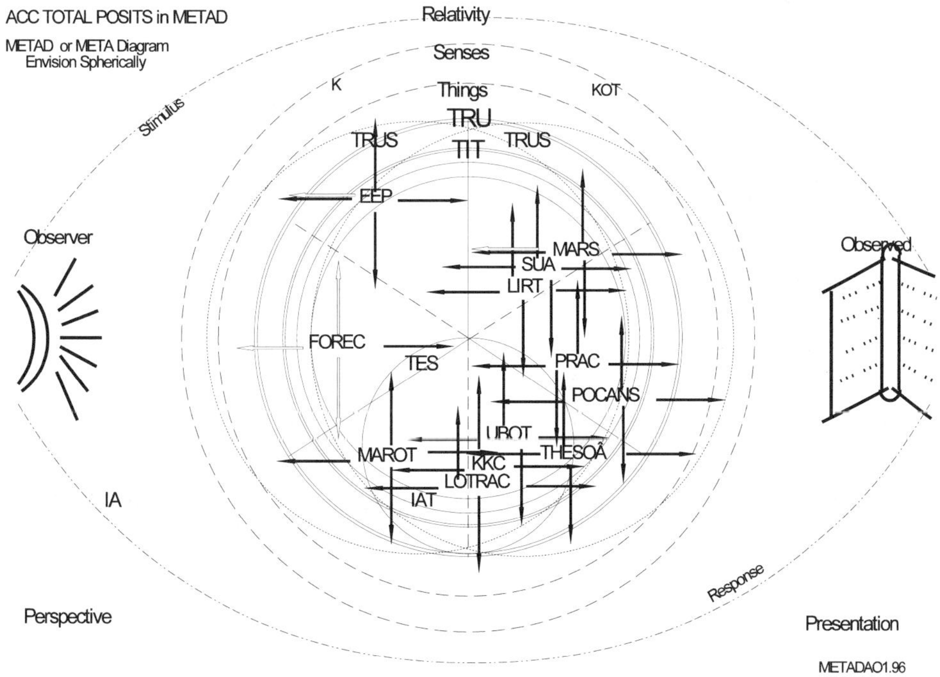

Fig 5-96 **ACC TOTAL POSITS in METAD**

MEASURES of TOTAL ACC POSITS

ACEXINHOP	7	~3-~-7~	S
AET	5,7	~-~-~8~	S
ALMAN	± 6-7	~5~3~8~	S
APRES	3,7	~6~4~8~	S
COGNOS	3,7	1-1-8	T
COMPONENTS COGNOS	1,3	~2-7-7~	T
C.Q. (CONQ)	3,7	1-7-5	T
CONST	± 2,7	~4~1~8~	S
COS	7	2-3-7	T
DITHE	± 4,7	~2-6-7~	S
DOC	1,3	2-7-6~	T
ECLECTIC ECUMENICAL POWER TRU± EEP	5	3-1-2	M
ENOS	± 4,8	~7-4-2~	S
FOREC	- 4	~2~4-3~	M
KNOW and BE KNOWN S COGNOS TÁ KKC	1-7	2-7-7~	M
LACK of UNDERSTANDING TRU ACC LTA	-4	~-~-~-~	S
LEVELS OF S-O ACC and TRU LST	1-7	~6-5-6~	T
LEWOV	8,7	~2~0-3~	S
LIRT	1,7	~5~2~6~	M
LOCMENF	± 4,3	~7~1-8~	S
LOREX	5,7	~7~2-7~	T
LOTRAC	5,7	~2~7-7~	M
MARM	1,5	~6-2~6~	T
MAROT	5,7	~1~7~5~	M
MARS	± 4-7	~7~3~6~	M
MASSIVE DATA for ACC, TRU MDA	1	~3-5-7~	T
METAC	5,7	~7-1-8~	S
MODES and CHARACTERS of COGNOS MCC	1	~1-7-7~	T
P/N	± 1-7	~-~-~-~	S
POCANS	1-7	~6~5~8~	M
POWER and ACTION of COGNOS PCA	1-3	~4-4-7~	T
PRACT	1,3	~6~5~7~	M
REVELATIONS SYMBOLISM ACC TRU1 RSA		~4-6-3~	T
SCONQ	5,7	~2-2-7~	T
SUTAC	7	~6~1-8~	M
THESOÂ	7	~0-7~9~	M
TOTAL ACC TRIUMPH TRU TÂ TTT	7	~6-4-8~	T
TOTRUD	7	~-~-~-~	S
UBOT	1,3	~3~6-7~	M
VAPO and STEPS to COGNOS VSC	1,5	~3-7-6~	T

Fig 4-96 **ACC TOTAL POSITS in SPALT**

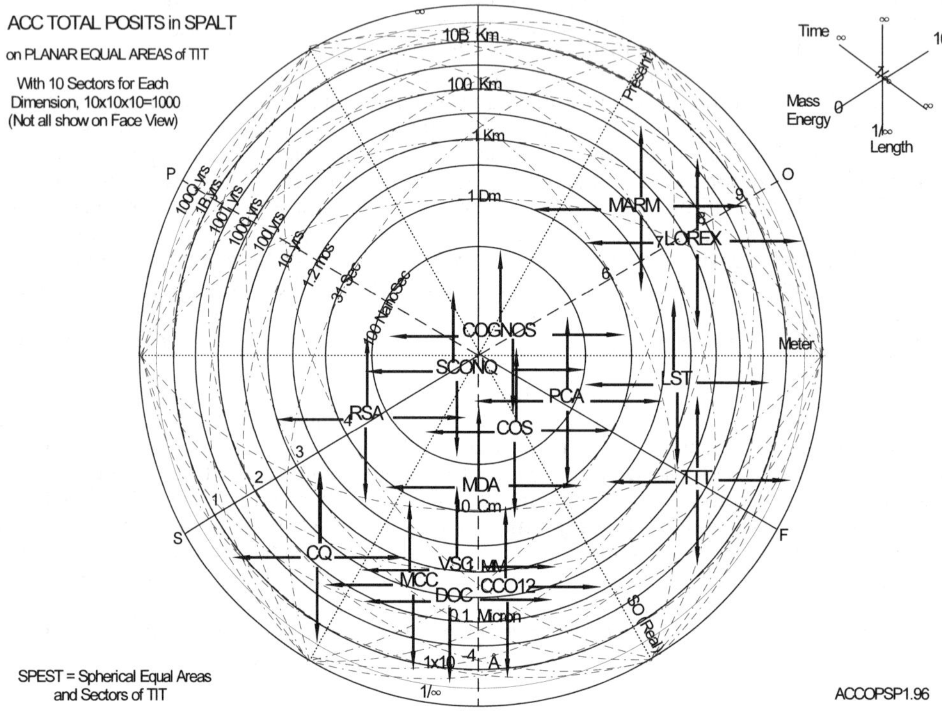

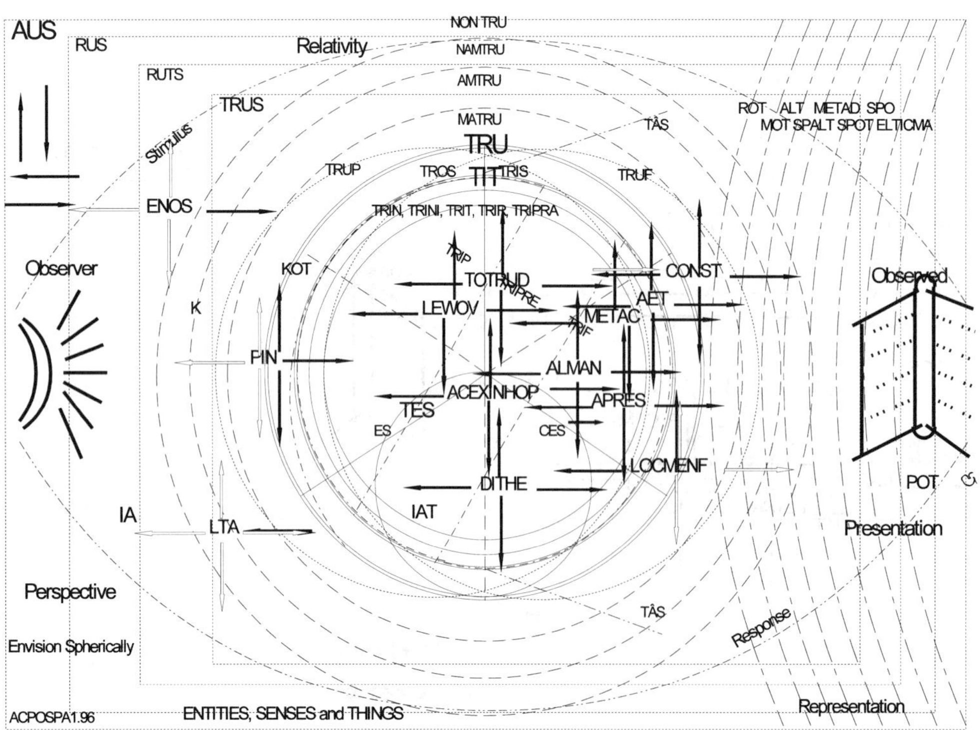

Fig 6-96 **ACC TOTAL POSITS in SPAT**

CHAPTER 97

ACC DYNAMICS, ENERGY, STIMULUS, RESPONSE, ACTION, REACTION, FORCE

INTERACTION, REACTION, FOCUS, BEHAVIOR, INVERSION, REVERSION, ELTICMA, SURVIVAL, CRUSH, POWERS, WORK, WILL, MARS

We must be careful when we consider TRU and its primary aspects that we remember to do so in the most general terms and universal light. For if we speak of TRU as some special or partial object of devotion, as a deity that is championed, we are liable to be considering or speaking only of a segment of TRU. For this reason we differentiate TRU, the Relative Universe as is, TIT and TRIS, from TRU, the Relative Universe which we know, extol, and champion.(KOT, TRE, ACC, TREX and TRIX). Many religions, philosophies, fields and sciences have, more or less greatly, suffered, and often with much NAMTRU, ATRU, and ANTRU, from a similar tendency to depart from TREQ TRU. This is a problem with IA's limits, maintaining optimal comprehension of all TIT, TRU, and meta TRU. We can not readily comprehend both TIT, TRU, meta TRU subjectivity and objectivity, the stimuli, questions, problems and all that are before us. Even to begin, takes great discovery, insight and training. Just as we are constantly diverted from stimulus to stimulus, we are distracted even from the highest, TRU. This is why we must, in reaction, constantly revert to TRU. This is called: **TREV** = TRU Reversion, constant Reversion to TRU. TREV is what enables a return to the most universal comprehension of whatever our focus and attention must on from moment to moment. It is how IA held to IAT has the degree of freedom to function properly as the focus it must have to fulfill its role in life. With TREV we can at least ECRO this great problem and reserve for TRU, ACC, TRIX, PUREALITY, COGNOS, ELTICMA, SÂ, and TÂ both what they are due, what is all TRU, and meta TRU. Also, this is why it is so important in language, especially the terminology of TRU, NEC, that we provide not only the best possible denotation of all its greatest aspects and features, but do so with the utmost formal distinction.(TROS,TRIS,TIT,TE, TRIT) This is a must whose reason is explained and justified by DAOST. We can not provide any adequate level of action in the future for a life that is becoming increasingly complicated unless and to the extent this action is equal to all TRU, and especially ACC dynamics.

By focusing in and out of the map and globe of SPALT we can demonstrate by varying distance from, and perspective of, this object a much better understanding of TRU. So is it for life, either all TRU, some division, or some smaller part or process. Likewise it helps to recognize the difference between a total philosophy of life and living some small part of it. This is largely the problem of DAOST, TROS, TE and TO. These problems are surmountable and solvable. We have only to recognize them and so proportion, order, coordinate, formulate, standardize and learn the relationships to give the necessary optimal focus and perspective. We have the focus, unity of an entity, unity of an observer or subject, unity of the observed, image or object, unity of relativity or observer-observed, unity of IA, unity of TRU, expanding image of TRU, infinite image of TRU, infinite image of views, senses and perspectives, infinite image of all things and senses, and the finite ordination of all the above as practically adapted to our means and our ability. The presentation and the first person, will vary with the proficiency of the presentation and the understanding of the first person. As both improve the glory of TRU is revealed. It is through the dynamics of ACC that we can collect all that is necessary for such a finite ordering of TRU.

The human species on planet earth can not be divided into two camps one not and one TREQ or TRU. For this, although it has, is nothing less than the degree to which our relativity tends to zero rather than to the absolute.(TRIA) Hopefully the quality and quantity of the total for TRU will far outpace and overpower what is not and against TRU. As we collect all that is TREQ, stimulate and accentuate this TRU portion and combine them with all else that is TRU we will automatically and certainly increase and enhance TRU.

One of the worst crimes of all, is not murder, rape, armed robbery or sedition, as bad as they are. It is the load, burden, abuse, and ATRU opposition those who are under the great yoke of life and struggling to do what is best by TRU are placed. **SPART** = Suffering for TRU, Abuse of. These are, have and will be, the true martyrs, saints and divine agents who uphold and suffer for TRU. Let those that destroy, or would destroy, TRU, be made to pay, smashed or condemned to eternal obliteration. (SMASH) These are mostly those who impair or destroy the transition of themselves and others from

Individuals → SELF → ESINS → PUREALITY → TES.

The release from such impairment and destruction through the achievement of TES and participation in TRU is called: **FRET** = Freedom in TRU. The achievement of TES and release from the bondage of individuality and all who would destroy one's being, in TRU, is not to be looked upon as a life of unnecessary and unhappy compliance, hardship, or struggle. It is highly fulfilling and the escape - like Nirvana - from endless failures, impositions, assaults and destruction of an ATRU world. Let all those who suffer for the sake of the greatness of TRU be held in eternal honor and glory. This is much of what we do in the dynamics of ACC, to eliminate all that would destroy TRU, reward and honor those who contribute to TRU, especially those who do so under the most difficult forces and circumstances. This is a great part and role in the lives, more or less, of all person's sacrifice for TRU and achievement of TES. Just as virtue is its own reward so is a life of TES reward and honor enough. For with such a life we know, COGNOS, that we are one with TRU and share its eternal glory.

The world we live in is a mixed NAMTRU world in which there is much ATRU and ANTRU and often very little TRU. The tremendous debt this produces is rarely considered and correctly acted on. In place of a natural tendency to TRU, and even the imperative inclination and convergence on TRU they have to yield and suffer the retribution of both TRU and non TRU they so ignorantly follow. This is called: **RETRI** = Retribution of TRU, Retribution of all, NON TRU, ATRU, ANTRU that compels a Return to TRU. RETRI is what is due TRU and for which much of society should feel remorse, recognize their guilt and sin, have conscience, go to great lengths to rectify, and make right all that can, should and will be TRU.

CONSIN = Confession of Sin. A person must be fair and honest with oneself, others, and all in life, by the actuality of TRU. One must recognize and do so with sufficient power that allows a recognition and confession of all shortcomings, weaknesses, failures, incapacities, bad habits, and all that is less than what one can do and be, TES in TRU. This is vitally necessary; for without it no sufficient understanding and force is available to voluntarily correct what will be a loss to TRU. It is well to get the deepest and best stimulus from this recognition of ones failures and weaknesses. It nourishes the necessary determination and conviction to do better, live right and with VAPO, be TES in TIT. It is how a person can achieve this goal that is so set for us in life. For life is purpose and fulfillment. It is in what is pure, real, actual, and totally best, PUREALITY, in TRU. As we experience, come to know and feel, and attain it, COGNOS comes to life, with the full and powerful identity the knowledge and thisness of TRU provide. It is then that we can follow and act upon it, ACC and ELTICMA, to live and do by THESOA its FAF of TRU, unto the absolute, SÂ to TÂ.

The great advantage of Problem Solving is seen in the best of studies and education where problems are presented for the student to solve. This is done in mathematics, geometry, physics, chemistry, pharmacology, internal medicine, and many other fields and sciences. It is a much better way to learn the basis for the principles, laws, and totality of a topic than through reading alone, or other means. **PROST** = Problem Solving TRU, or a Program for the Training that best enables Thought and Profound Reason of the greatest of TRU. PROST is more than in problem solving, the intense thought and rational processes needed for problem solving, solving the problems of TRU, and even the principles and habits of life they seek to produce. PROST is the combination of all of these, the trained ability to habitually, imaginatively and with intense reason achieve ever higher intuition into the unknowns, questions, problems and solutions key to TRU. This can be a great part of the training, education, and a necessary requirement for tests of proficiency in TRU, as in **TD,** TRU Doctorate. If we are to have an able bodied and minded cohort of workers for ACC and TRU, they must be rigorously trained and tested. The greatness of TRU, the complexities of ACC, the delicate nature of the consequences of ACC and failures in the future, make this a top priority.

For the innocent, TES, and TRU to have to pay for the errors, mistakes, indiscretions, crimes, DIVA, DAK, ATRU, ANTRU, TANTRU, ATRIX, ANTRIX, and TANTRIX is the scourge and calamity of mankind. It can't be tolerated. Those who indiscriminately spread syphilis, AIDS, disease, genetic defects, terror, crime and many others detrimental to TRU have to be controlled. All, who play, act, and live, neglectfully, frivolously and destructively, will have to pay. It is a choice of the degree to which one accepts the challenge of being TES and TRU. One does not have the choice of being, or even believing, and doing as one pleases, or whatever one is inclined to be or driven to. This is completely antithetical to being, being alive, existing, the absolute thisness of TRU. For here, and only here are the actual joys, happiness, and blessings of life. Such a life of TRU, PUREACLITY and TES are and can be very happy, joyous, blessed, and successful, as TRU is so by its own nature. It is this motivation, initiative, incentive, interest, and all that helps gain the reception in TRU that is so greatly needed. **INT** = Incentive, Initiative, Interest and Inclination to TRU. INT can be attained even more readily if we through TRIX concentrate and accelerate TRU as ACC, the total dynamic ACC. **INTRIX** = Incentive, Initiative, Interest and Inclination Into TRIX. We actually do not have any choice when it comes to life, TES and TRU. For if we chose not to be, not to be what is only what we can be, we are not, there is no choice, and thereby we are not. We become, join, live, serve and enjoy, it or we are not. This is called: **LACEXT** = Law of no Choice in Existence, Except to identify with and live by TRU, yet with much choice, maneuverability, success, happiness and joy in TRU.

The habits developed through TRU are most productive. To this are added the many vehicles, conveniences, media and means being developed, as in electronics and communication, that make us more productive. Yet it is being productive of what, and in what way that we can be best directed by TRU. Human productivity is not so limited and can be vastly accelerated as revealed in: **SOLHUPT** = Solution to Limits of Human Productivity by TRU. SOLHUPT can be made the fortune of human life, TRU. Although life in TRU is not easy and puts some limits on total individual activity, these are normal and healthy, and need not detract, only moderate, the accelerative expansion of future human productivity.

Gradually, unobtrusively, increasingly, and covertly if necessary, try out, work into and work out to help, test, expand, and promote all venues that will assist the advance of TRU, TRIX, TREX, and TRAX. **RETRUX** = the Reversibility of TRUX, where **TRUX** is the combination and summation of TRAX, TREX, and TRIX. TRUX can function reversibly in any way, any way that ACC benefits TRU. That they are both highly separate and highly combined is a must. That they be highly reversible is imperative. All who can best do so should be enlisted to most vigorously accomplish them.

There must be a great outpouring of energy for, attention to, happiness and joy in and for TRU in such a way that all spontaneously and willingly join the great war against ATRU and ANTRU, and endeavor for TRU to assure TA. World opinion will be naturally stimulated, encouraged and properly brought to recognize, love, glorify and serve in the goodness and greatness of GOD, TREQ, TRU, TIT and TÂ, so that all will benefit and be rewarded by their grace, benefits, and rewards to find complete happiness in TRU. Not all opposition should be crushed. For the goal is certainly not to smash or crush all that is different in TRU. Some may be no more than the differences, polarities, and the normal dialectics that exists in and of TRU. Some may also be justified by their principles and potential, as ESINS, which can be absorbed into TRU.

Qualified tolerance is a general principle and policy of ACC, action, and response to opposition, terrorism, and anarchy. It is the practice of life whereby the differences that exist among humans are recognized and allowed for. To qualify, is to hold and bring all to what is of greatest quality and benefit to the success and survival of humanity, TRU. When combined with tolerance it signifies how any kind of tolerance is not apt to be good or beneficial. Only the tolerance that is guided by the greatest knowledge, appreciation, and reason derived from a comprehensive understanding of God, TREQ TRU, and most specifically TIT-TO and the TES in all that constitutes TRU. As a cure for terrorism, anarchy, and war, qualified tolerance should assure all powers of difference and opposition that they are likely to precipitate into violence and are to be managed and controlled at all levels. It gives humans a common ground for the union and integration necessary if our world is to avoid escalating partiality, uncontrolled degeneration, destruction and extinction.

If every one, in life were as proficient in TRU as they think they are, want to be, or if they were as proficient in their jobs, trades, occupations and professions as they think or demand others to be, what a wonderful world this would be. **PROFT** = Proficiency in TRU. PROFT is TES in TRU, what we are directing the first person toward as they read and use this book. It is how, when reading and applying IPTRU by using ACEXIN and PUREACLITY, we become TES and TRU, masters of not only our own destinies, but what enables TRU, ENT, TIT and guarantees TÂ.

We should never forget that the whole actual world, TRU, is based largely on oneself, the individual and their immediate awareness and instantaneous attention. The future, and accompanying actions, must primarily come from the best our attention and our selves can produce to fulfill their parts and roles in TRU. This is called: **LARIAT** or the Law of Reversion to the Individual and ones IA as it is in, by and for TRU. We can never leave it to others, until later, to the whole, or to any devious outlet. It is up to us now, before and after, as we, the great generators from the center and focus of all, in identity and continuity with all, are what actually is, TRU.

There is a great difference between F work, D work, C work, B work, A work, and 100% or perfect work. Such distinctions, and their finer and finer percentages, are the exquisite and beautiful detail we find in, and seek to learn and make of, life. It is key to the mass energy Â FAF of TRU, and IA becoming IAT. Perfected work for TRU is often the only work that is satisfactory and provides the precision necessary. The precision for 100% work requires not so much an absolute overall perfection of all but a perfection within a limited range. It is only sufficient to do what is most required perfectly. This may be very simple and small, or it may be most complex and large. Just as the saying goes, "what is worth doing at all, is worth doing well". Such is what is needed for TRU, and by TRIA, all. For only when work is perfected can it provide the S/F, fulfillment by ENT, and all that is both LATAN and possible for TRU, TIT and TÂ. It requires great sacrifice. This is especially so in a life that expects to predominate, succeed, and propel ACC and TRIX.

What TRU action is necessary and how it is provided can - within certain abilities and limits - be forecast. Such forecasting will, by concentrating like on breakthroughs in science, enable a fast expansion in TRU and mans powers to enhance TRU. With better forecasting we can plan and provide the ACC and operations needed for accomplishment. To a large extent this depends not only on the increasing knowledge of the future, but knowledge that is focused upon TRU and ordered as we present ROT, SPALT, HOM, ACC and IPTRU.

At slack times, periods of relative inactivity, even peace, and prosperity, we become complacent. At times of pressure, stress, and conflict we are distraught, confused, and lost, yet stirred to action. If simply exist, and exist by shutting out much of the world we miss most life, TES in TRU. Even though they are often the product of ATRU. If we allow it to run our lives we can never be TRU, or only poorly TRU. What we can do is to expand and overcome them by learning more and knowing what is good and bad, TRU and ATRU. In this way we gain experience and good judgement. The initial and continuing thrust of all must be not so much in answering or reacting to all that is not TRU, bad, trivial, or negative, but to push relentlessly and powerfully in all ways and times towards TRU.

To accept defects and what is repulsive is largely a product of our attitudes and variability of IA. To have the interest and ingenuity to discover their underlying reasons and meanings is not commonly practiced. It is here discovery, knowledge, learning, and science has been most effective. Yet it has been with little or limited TRU. Many humans wantonly accept as excusable their own bad judgements, acceptance of any law, belief, and way of life, yet will quickly condemn others or TRU for being less than their ideals or notions. Only as we develop POT, KOT, PUREALITY, become TES and TRU, and use SPALT, HOM, ACC, COGNOS, ELTICMA, and SA will we overcome such errors and weaknesses to what is new and different, yet so great of TRU.

LIL = Law Inverse Law. LIL is law that is inverted through reaction to wrong law. Law can only be, in resultant, good, and good as long as it adapts to, and is identical with what it represents in TRU. To the degree law does not represent these, and is not limited by VAPO, is the degree to which it is wrong, false, and no longer can be law. To this extent existing laws and all laws devised without TRU or TREQ as their prime goal, are to that extent, and likely to be, increasingly, wrong. The degree to which this kind of wrong resists change or combats right is the degree to which they are doubly wrong, ANTRU. There is a constant motion and dynamic interchange of mass action, reaction, and power aong all legal and justice systems, the fundamentals or right and wrong, good and bad, Godly and evil in the world. For neither the circumstances, nor laws remain exactly the same for long. The more a law is universal and great in TRU, the more it is right and it remains constant. This is called: **LACURT** = Law of Constancy of what is great of the Universal and or Relative in TRU. Much more common is the highly particular, trivial, less important, defective, and erroneous existing laws that so permeate human life. It is only by return to the basic fundamentals of, and develop laws from TRU and its FAF, DAOST, TROS, TRIT, systems, divisions, ES, and CES, and their VAPO that we can keep a proper and effective body of law. It is partly for this reason many NEC are expressed as laws. All other laws, rules, and regulations, must be either adapted to this basic body of law, and TRU, taken within their narrow and changeable limits, corrected, or eliminated. Those who resist such change, especially when against TRU, must be shown their errors, recant, or most powerfully corrected as only TRU can do. When ANTRU and ANTITRU they must be, as they so often have done, made to pay. This should always be within the context of what is justified and best for TIT, TO and TRU.

As much as you may want to be agreeable and help others and as much as they may seem to be in need, this is often untrue, deception, or distraction. For their need, your help, and want separately or in combination are often not TRU. It is commonly the beginning of integrative negation, INEG. We can actually please only when it comes from TRU. It requires an understanding, analysis and the application of what is not and what is TRU in each case and occasion. All that is not TRU must be distinguished, separated, eliminated or alienated. **ATEQ** = Alienation of all not TREQ or TRU. ATEQ is the pressure necessary to relieve ourselves, and especially society, of much that is not TRU. Alienation exists in many different - often subtle and unrecognized - forms. ATEQ can be used to remove that and those who adversely effect TES and TRU. We must be careful, however, not to misuse or practice ATEQ. without VAPO and TRU.

The role of power to be, is called: **ENWIP** = End of War is Power. **ENWIPT** = End of all Negative is power, Will to Power to be TRU. ENWIP is the power to adapt, survive and dominate in markedly different environments. It is in a way, as it has been in the past, and will be forever, not by anything less than the power of TRU to maintain, preserve itself, and survive. For the lineage of man has gone through constant changes of habitat and accompanying adaptations. Mankind today, is living in a world in which the forces that alter the environment and habitat increasingly differ, are greater and more acute. It is these we must learn, understand, remember, plan for, PPMA and control, as shown by TRU, if we are to survive.

ELIA = Everlasting Life In IA, and **ELTIA** = Everlasting Life through IAT. **ELTIMA** = Everlasting Life Through Instantaneous and Massive TRU Apperception and Achievement. ELTIMA is the recognition and identity of a person with TRU through SUT and the identity of PUREALITY with COGNOS back to TRU. ELTIMA signifies how each and all persons can have everlasting life if they discover, learn, remember, comprehend and achieve that degree of massive apperception, overview, and endeavor of all in life that comes from learning and knowing TRU. It is in ELTIMA that one and all can first unlock the door to, and be absolute, TÂ. **ELTICMA** = Everlasting life Through Coordinate Massive Apperception and Action of TRU. Although much of TRU and HOM are common, natural, all around and close to us they are and are not obvious. They are obvious as entities, senses, things, and even ESINS, but not so obvious as TRU. PUREALITY is not, as in entities, senses and things, so much singularly pure, real and actual, it exists in the added characteristics and thisness of a whole and its identity with TES and TIT. This is the view that is vital to understanding ourselves and TRU. Such a view can not be achieved without coordinate, massive apperception, or ELTICMA. ELTICMA provides the proper solution of all through moral and ethical behavior based on TRU. Some of what ELTICMA depends include: 1. Prior Knowledge, 2. Interest, 3. Inspiration, 4. Revelation, 5. Exaltation of Religious like experience, 6. Ever finer Formal Distinctions, 7. Reinforcement, 8. Perfection of Model, SPALT, 9. REM, 10. Posits and fulfillment STA and Homation, 11. Focusing in and out of the filled SPALT, 12. Added views and interpretations of ELTICMA, 13. RORRE, 14. ACC, 15. PUREACLITY. The kinds and expansion of ELTICMA include: **ELTICMAN** = ELTICMA Multliple or Many. ELTICMAN signifies how the

ELTICMA of one individual may be transferred to all, especially when TES. **ELTICMAK** = ELTICMA Summative x Individuals into Knowledge, of TRU. ELTICMAK is demonstrated in the immortality of knowledge of the spirit, like Buddha, Christ, Hegel, and many others that give revelation, inspiration, and total states of identity with TIT and TA. COGNOS when carried exponentially becomes ELTICMAK. This becomes THESA, as THESÂ combines with THEOÂ to become THESOÂ, SÂ, TIT, and TÂ. These in reverse and in part may be equated with a factor of other knowledge using ESINS to become sense in itself SENIT and ENIT. The conversion of knowledge and attention to ACC and Action is how idealism and materialism become, with realism, SÁ and thus THESO and THESOÂ of TRU, TIT and TÂ.

What is worse, to be crushed by overwhelming forces and powers that are unknown, or to know them and fail to react. In either case the result is much the same, failure, defeat, suffering and extinction. For the individual threats to mankind seem far away, until they affect them personally. However each and all of us through life will encounter many natural forces which we must react to, as a product of our role in the world. Whether this world is TRU, NAMTRU, or ATRU will determine what these forces are and how to best react to them. Among many of the best ways, one is to meet them head on, especially if TRU can be enlisted and proper management and control provided. This includes: **CRUSH** = Citizens Control of the Relative Universe for the Salvation of Humanity. CRUSH is an activist oriented term and logo to serve as a rallying point for those who learn and know TRU, desire, and are restless for it to be fulfilled. It is primarily targeted to sweep away all that holds humanity in bondage, restrains and inhibits the exercise of all human rights and opportunities to be TRU. It targets those institutions, laws, states, groups and individuals who, care little for the rights of others, TREQ or TRU, have imbedded themselves in ANTRU worlds, especially all that acts against the growth and development of TRU. It is by all who come together in TRU to regain the power, do, and achieve what it should have had to better all TIT. CRUSH is for those with great TRU learning and ability. For most of ACC is dedicated to non violence, a controlled accentuation, especially aggression and dominance not based upon the best of TRU and ACC. Our greatest challenge is to acquire all possible means of maintaining and perfecting TRU and helping all be more capable, able and productive through TES. Those who are most TES, as $(TES)^n$, should be complimented, extolled, honored and rewarded. Those who are highly TES should be, for the most part, left alone, unencumbered by ATRU, to do, work, and live for ENT, TES and TRU. Those who are potentially TES need help and encouragement in all ways that best actualizes their TES. Those not or less than potential TES should be salvaged as much as possible and helped to reach their highest level while preventing any depreciation of TRU. Those hopelessly non TES need to be ignored, prevented from injuring TRU, or eliminated from any adverse effect on TRU. Those who are ANTITRU and ANTITES must be to the degree of their hostility, antitheses, and damage to TRU be separated out, nullified, reduced and eliminated as is appropriate for TRU. To the degree they are ANTITRU and TANTRU they must be, and only with perfect TRU rationale and method, increasingly and vigorously destroyed, MARS, CRUSH, and SMASH. So is this true not only of effect on TES and TRU, it holds true for all that is in and out of, input and output of TRU, OITIO.

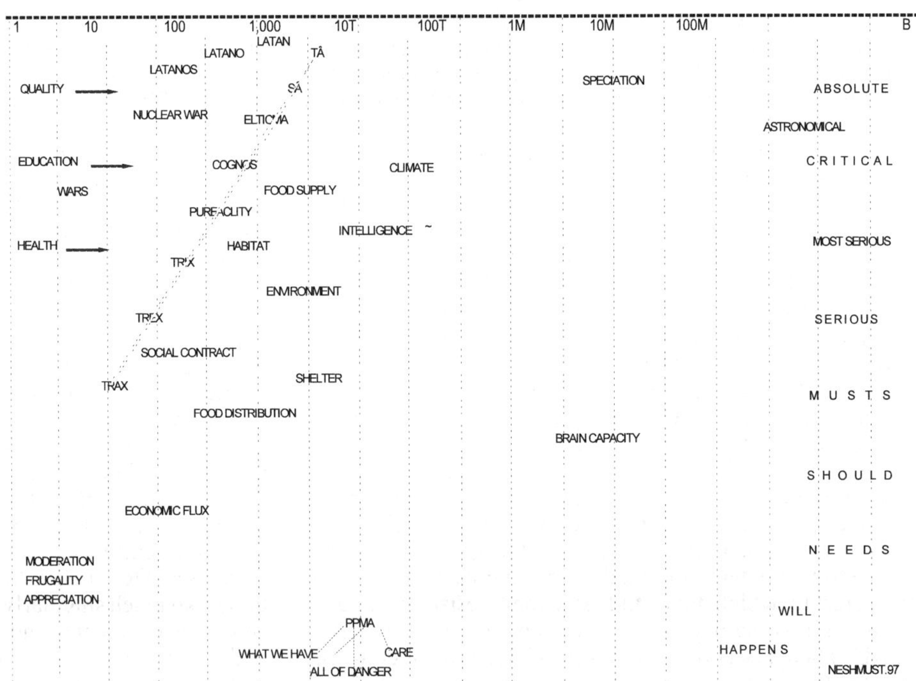

Fig 1-97 **NEEDS, MUSTS, HAPPENS, ACC, and TRU**

In order to envision the many factors that we by ACC encounter in the future it is helpful to account for them by the parameters of time and degree of seriousness. Although this somewhat replicates prior chapters it carries added significance for ACC. In the columns time from 1 to 1 billion years is shown, and in the rows what is absolutely fatal or causing extinction to the simplest needs and what will happen are shown. It will be seen that various factors and problems occur at times of maximums, or in a way a time when they come to a head. In part this can be inferred from what has happened in the past. We can also infer it from a greater or lesser amount of information we have at present, as well as what changes may affect it between now and this maximal head. By reaction to action, response to stimuli, answers to questions, and solutions to problems, what is most apt to be fatal, cause extinction, critical, and most serious are those which we need to be most attentive to, provide the best reactions, responses, answers and solutions. With appropriate modifications and superimpositions, these should be refined and readapted POSITS in SPALT, METAD and SPAT.

In life we are in a struggle for survival, directly and indirectly, incipiently and violently. If we are not interested in what is best for TRU and certainly what is best for us as we are TRU, TRU will not be interested in us. We and TRU will thereby suffer, lose, fail and die. It is a life that is TRU, not so much of the figments and illusions of the observed and our observation, but the purity, reality and actuality of all that is TRU. It is a dynamics of the dynamic ACC or $(ACC)^n$. This is called: **ACCEX** = The exponential expansion of ACC. ACCEX is the opportunity, potential and hope for the future that humans have when living by nature in TRU and enlisting all available means of TRU to produce ever greater improvement and perfection of life, TRU. It especially includes all that will PPMA TRU.

NIP = Negative Indirect Powers. NIP is associated with, yet distinct from, IN or Integrative Negation. There are many, large and damaging indirect forces and potential injurious powers at work throughout the world, including those which have caused the demise of societies, states, nations, cultures, or civilizations. Only when we discover and learn TRU and TES can we have the wisdom to surmount NIP. This is accomplished through a progression of POA, TU, TRIT, OBT, SUT, ROT, MOT, ALT and SPALT. It is called: **ORD** = the Ordination through SU of TRU. It is through the progression of ORD that we can realize and coordinate our lives to become an adequate member of society. This produces maturity in TES and may be likened to a coronation for all who attain a success/failure level in comprehension and endeavor for TRU. This is in a community with others who likewise perpetuate and assure TRU. This is called: **ORT** = Order, Coordination or Coronation of TRU. That which we must protect against or ATRU, TRUCH/TRIT or at least hold TRUCH within strict limits: **SLICH** = Strict Limits of Change in TRIS. SLICH is how to sustain and hold TRUCH to a steady, continuous and permanent state effectively, dynamically and relatively and still retain TRIA. We must clearly distinguish different types of change and their significance. Much change, in or of AU, is insignificant. Changes within TIT, TRIT as of ES, CES, and TRUD, are a part of TRUD and normal. Changes of TRUCH are good only as it is reversible or SLICH. Life is expectation of damage and hope of gain. By TRU these can be well controlled and turned to optimal benefit.

The power of human optimism, especially in the young, is great. It is much of what sustains human life and TRU. For these have a purpose, lust for knowledge, benefit for others, success, well being and happiness. Most are sure they are going to heaven. Optimism soon disappears with the vicissitudes of life. This is largely because we do not know the actual world. We fail to act and live by it. This is why and how TRU can, and must be achieved by KOT, TES, MOT, POS, ACC, ELTICMA, POT, and IPT. One of the repeatedly impossible ways to win and most certain ways to lose is to have on your side those who are playing or fighting for your opponents. For no matter how many times you are winning, you will lose the half or percentage these 'partners' are active, plus how much it takes to recoup the loss. Therefor even if your opponents are losers, you can't win, you will at best only break even. So it is in ones life of TRU, and the partners and associates one keeps. If they are not working and living by TES and TRU they will make you lose. Every time you depend upon them to build TRU, TREQ, ACC, TRAX, TREX, TRIX you will find no gain, and a likely loss. It is for this reason that disassociation from ATRU, ANTRU and ANTITRU, and even the control or avoidance of NAMTRU, MATRU and AMTRU as much as possible, is the name of the game, by word, and way of TRU. In TRU, as in much of life, there is only a narrow opportunity for survival, success and wining based upon high competition for life, a very set FAF of TRU, and the opposition of all negative forces. This is signified by: **NAMOV** = Narrow Margin of Victory. NAMOV is shown in: **LOP** = Law of Optimal Participants and elimination of Opponents and Unreliable Participants. This way of TRU is to collect all your forces and powers that enable TES and TRU and continually upgrade them. It is like the greatest leaders of nations who have stood on the shoulders of advisers carefully and superbly selected. It is to best plan, manage and control all by selecting TRU, rejecting ATRU and ANTRU and focusing all as much as possible on TIT and TÂ. It is to learn, know and show TRU and best use it as provided by VAPO, SCARLS, ENT, SPALT, POS, SPA, HOM, ACC, ELTICMA, SÂ, and TÂ. It is because of NAMOV and LOP that so much of ACC and a strong selective program for quality population is justified.

Those cultures are greatest whose interests, values, conversation, studies, fashions, customs and all are most directed toward the concepts, principles, laws, parts, processes, and mechanisms, of TRU or TREQ. When these become the raging, burning issues and topics of lively conversation, discussion, presentation. and refined as they are TRU, this culture will most successfully flourish. **LINT** = Law of Interest, Intention, Intellectual curiosity and power to TRU. There has been, just as God is TREQ to TRU, much in mans love of knowledge and search for wisdom that is equivalent to TRU. Nor is LINT limited. It is specifically a large power of ENT and ACC. More generally it applies to all mankind's attention to and actions for TRU. It is the stimulant and activator of TRU. Without LINT much in the world fails, is greatly inhibited and suffers. The 21st and following centuries will be the heroic age of TRU by an accelerated LINT, depending upon us, TES, and ACC.

VOTGA = Advisory Group for Voluntary TRU, by all that: 1. Provides increased quality, 2. Voluntarily Limits all factors that put an excess strain and drain on TRU, 3. Enables an elimination or control of war, revolutions, terrorism, and unnecessary conflicts, 4. Discovers, anticipates, manages and controls the critical dangers that can destroy TRU, 5. Strengthens TRU, especially its key mechanisms, 6. Makes for a happier and more joyous existence in TRU, 7. Enables a person to realize that whatever happy and joyous existence is to be found is not in extraneous, irrelevant, worthless senses, and things, but in TRU moderation, 8. Enables Purity, Reality and Actuality as they are TRU, 9. Discourages extremes, abuses, misuses, and partiality that is less or other than of TRU, 10. Supplies and promotes validity, actuality, proportionality and totality in the comprehension of all.(VAPO)

Do you like to work here? Do you want to keep on working here? If not, take your leave. When you can come to be a part of this group, TRU by TES, return with the needed spirit and open minded appreciation that is given to you. Fully identify yourself with your job and life as you and it are one, in TRU.

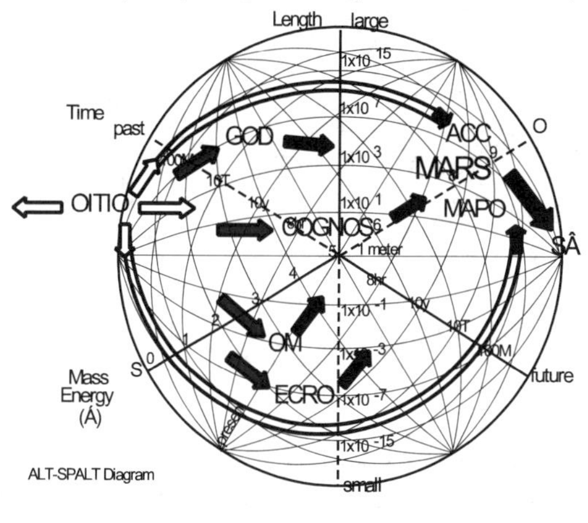

Fig 2-97 **MARS, MAPO, ACC, SÂ, OPTIMAL and TOTAL MASSIVE ACCENTUATION of POWER**

MAPO = Massive Accentuation of Powers totally and Optimally to GOD, TREQ and TRU. OM, OITIO, COGNOS, and SÂ. This provides the convergence of all forces, powers, bullets and prescriptions of God and TRU that are, have and can be used as necessary for maximal benefit. One should remember, ECRO, not to lessen ones enthusiasm for TRU or its many magic powers because of their multiplicity, diversity, and partiality We should only utilize their reinvigorating energies that provide the converging focus and challenge needed for life. It is in making and living VAPO, ATO, SCARLS, and TROS that one can maintain an ever greater MARS, pursuit, and endeavor for the enhancement of TRU, oneself and eternality.

MARS = Multiple Massive Accounting, Accentuation, Action, Ablation, Acceleration, Rapid and Simultaneous. The trick is to know what is best and what will work. This requires ideals, a guiding light, a plan, orientation. coordination, and the comprehension that comes from their role and part in TRIT and TIT. It requires the identification of specific needs, how to put them together, and how to work vigorously to see that they are met and fulfilled. To do this we enlist and encompass all practically available and can be achieved by ACC for TRU.

The importance of information is largely in decision making. We can't know to much about what is uncontrollable and what is controllable in the world and TRU. Likewise what of these two poles is a mixture and to what degree. This is the beauty of TRU, TRIT, POT, IPT, ACC, ROT, MOT, ALT, SPALT, POS, and SPAT. How we can manage and control with and without TRU, its methods, degrees, and variants.

The specific right or wrong must be adapted to its overall state in TRU. Not so much as a point in time but as duration, and variation. Not so much as a location, but an area or space, not so much as a simple sense or thing, but as their combined, complex and unique S-O or Á FAF in TRU as all ACEXIN comes to TIT. The dilution and concentration of the universal must be controlled to suit TRU. They are called: **DU** and **CU** or Dilution of the universal and Concentration of The Universal. DU and CU are similar to hypofunction and hyperfunction. For, as a mass, or TIT, becomes diluted from its optimal state it becomes weakened. So does this occur as it becomes excessively concentrated, for these are a measure of its density, be it receding from optimal or exceeding, as in quantity/quality, its optimal. This relates to the degree to which its parts are no more or no less than its total. It is in this sense, whether parts function just right, or function producing dilution or excessive concentration of TRU that even the best of means for enhancing TRU must be controlled.

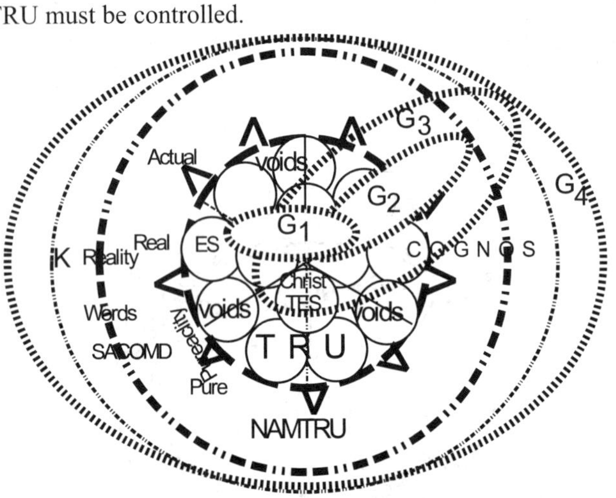

Fig 3-97 **REALS, REALITY, ROYALTY, Á, TRU MEANINGS and ACC**

How much can knowledge and reality be greater than TRU? PUREALITY equals the purity of the reality of the actuality of TRU. This is what is maximal for TRU, PPF. This becomes the same as perfect quality. For what is real, in its most original sense, is that which is royal or of the greatest value, REM. This we may see in the syllables of real for as the overall term is equivalent with royal so is re equivalent to things, and al equivalent with all, as in royal. So also is ES the end product, as TES is of PUREACLITY, pure ES. **SACOMD** = Stretching and Contracting Meanings and Definitions to PUREACLITY and COGNOS. As the intermediaries, and much of the middle range of the Á Continuum, it is such words that must be so pure, real and actual. This is so if they are to be what they must be, both of purer formal distinction and actuality in TRU. The stretching and contracting of definitions and meanings is mostly detrimental to good formal distinction. This is especially so, although somewhat a necessary evil for language and knowledge that are still in a formative stage that needs flexibility for usage and development. That which overfills TRU with less than pure ES, under fills TRU without known ES, and binds filling voids, are apt to be non or poorly TRU. In Fig 3-97 some of these relationships are diagrammatically shown. K = Knowledge, ES are the Parts of TRU, Voids are those regions of TRU that are unknown or inadequately known, They also contrast with all that of God that is not TRU. G = God, and G_1, G_2, G_3, G_4, show stages and ideas of God.

Sacrifice largely accomplishes its end by example or influence through time; that others may live, as their TES in and for TRU. Here also we can see the CU or concentration of the universal out of hand. For its only justification lies in those states in which little choice remained, or those in which by example TRU is so much more greatly revealed in an absolute sacrifice. It is rather, and almost always, in a summation that we must base our accentuation, action, and our lives. *Summo Studio* or with the utmost zeal, with our total powers to the total TRU. It is for that we must direct ourselves. This can be readily accomplished when we know and act for TRU. It is illustrated by MARS or Multiple Ablation Rapid Simultaneous. This is what occurs when we line up all our objectives, things to be done, wrongs that beset us - as TRU - and go to work on them simultaneously, vigorously and conclusively.

WIM = We Will It Must. WIM is a product of LODEPT, LODEST, RA, RIA TRIA and TÂ. WIM can be remembered by the everlasting victory that continues eternally to TRU absolute, TÂ. If we have sufficient capacity, ability, TES, FAF as Á, and its associated will to TRU, then we, through WIM must be, TRU. It is in WIM that all can come together for TRU, as each and all individuals become TES, best and most voluntarily, apply this to TIT and TÂ. The success and **TESEX** or EXponential of TES of TRU, past, present and future, must be accompanied by an exponential and balanced growth of all its parts. This is especially so as PUREACLITY develops from KUPS, growing to COGNOS and ELTICMA. So will TIT accelerate exponentially to become the Absolute, TÂ, as we will and it must, WIM. For all who lack, are less than, or recede from WIM, we have: **ÂWIM** = From, Away from or degenerative WIM. AWIM are those, like apostates, who do not for one reason or another will TIT and TÂ to be. Like DAK and ATRU are to TRU, AWIM is to WIM. AWIN and DAK are not TRU, even more than ATRU, they are increasingly ANTRU. WIM is the great intention, purpose and goal of our subjectivity, attention, teaching, education, behavior, activities, service, sacrifice, life, and TES.

Fig 4-97 **MINIMUM and MAXIMUM EXPANSION, of TRUS**

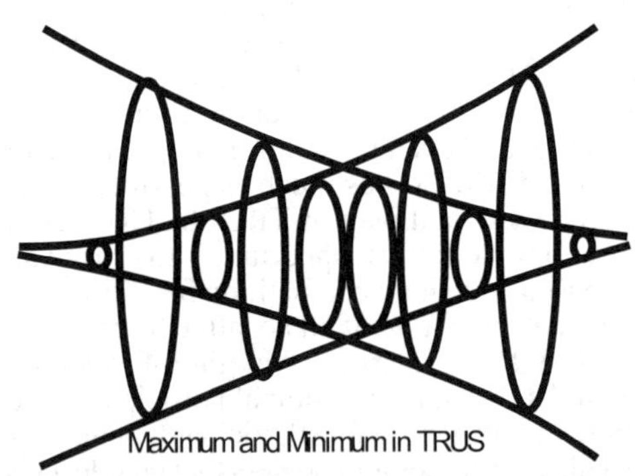

We can show this with maximum TRUS expansion. Yet, all things, words, and concepts, including God, have their non TRU form, elements, parts and meanings. For mankind today this is a much too large a part. As Gods, many things, outside and inside of, or superimposed on TRU, in various ways, and as variously interpreted, are ways of looking or observing the Á special form, mass-energy function, state and dimension of TRU. This helps to better explain the dynamics in, and the reversibility of, the TRUS/TRIA paradox.

If TIT is all and absolute, by TRIA, then it would seem that TES and TRU must be also. Thereby to accommodate these two seemingly irreconcilable opposites we have: **RABDI** = Ratio of Better and Different. RABDI designates the idea that there is a relationship, much in opposition, between the potential for change that can be better, with the loss that occurs from being different. Or if TRUID and TÂ have changes for the better, yet with a minimum of difference from TIT, then RABDI will be very high, and optimal. If there is much change so that the difference from TIT is great, and the improvements are minimal or negative, with great loss of sameness from TIT, then RABDI will be very low, negative, with likely loss, or non survival. It is in this way that, next to achieving TES, TIT and TRU, it is in achieving and maintaining the optimum of RABDI, improvement, and the better or ideals of TRU, TRUID, over the minimum of change, difference, and loss of sameness and identity with what is, TIT, that is paramount. The principles, essences, parts, processes, and all increments that contribute to this optimum and maximum of TIT to TÂ are what we, as this text, is all about.(ITCH, RATCH)

What are the solutions to a person's problems, and hope for the future of mankind? Both are to be found together, primarily in the life of each person, whether or not, they contribute to the genetic pool or total population. How much difference does it make to a person if the world is perfect now, only to be drastically replaced by another breed of humans in which very few existing persons contribute? Why should the general population strive to be perfect, have souls, or be TES? For as long as there is high outbreeding, although it may increase capacity near and mid term, sooner or late it will produce a selective deficit.The history of mankind has been one in which populations come and go, the specific individual, as much as they were the center in the here and now, usually meant little. Only the collective individual carried humanity, both genetically and all levels of human mass energy form and function. To the extent civilizations, large regional populations were outbreeding they lost their genotypic edge, energy, and ability to compete. They were replaced by incoming populations that were extremely inbred and concentrated with high energy and capacity to survive. Herein is the Paradox of Massive Peace and Prosperity versus Natural Selection by Elimination.(PAMENS) Only those generations that mate and reproduce the highest quality will in the long term contribute to the genetic pool. For there will always be, as there has been since man became civilized, great masses of population that exist below any level adequate for sustaining human predominance. This is why it is so important that persons of outstanding inheritance, health, strength, mentality, energy, capacity and ability mate and reproduce not only in the first generation, but in repeated generations in the long term. Also that they realize that mating and reproducing are only a part of their overall potential, part and contribution to uphold and sustain TRU. To know and live by this legacy of nature (or God, as it is TRU), is the key to solving their problems and fulfilling the hope for the future of mankind.

VIFS = Victory in the Future over Imperfect Systems. VIFS is recognized and provided by FOG, CODEF, IMSE, IMSED and IATSU. There are both benefits and drawbacks to be gained from both existing and improved systems. It is in how much benefit over detriment existing imperfect systems have compared with more improved systems, plus the inertia and loads such changes produce. With the greatness of TRU, and the value that TU, SU, ROT, MOT, ALT, and SPALT provide for a POA to TRU it will be seen that the understanding and will for TRU are there, and how much VIFS can succeed.

Fig 5-97 **THE PRINCIPLES and LEVELS of ULTIMATE PROBABILITY**

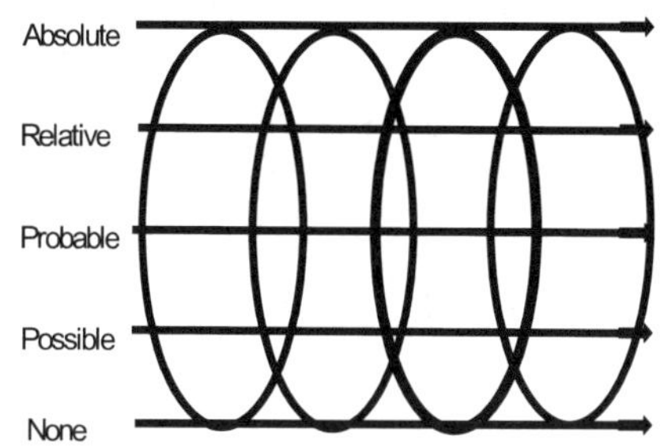

LOUP = Law or Principle of Ultimate Probability. Whereas knowledge in the infinite past was impossible and whereas knowledge in the infinite future must be perfected, from the possible, to the probable, to the absolutely certain, then we have in the accompanying illustration their relations. This is a diagram of the forces and powers of knowledge in time. It is how they have, are and must permeate, escalate, converge on and gravitate toward TRU. Only if we do not survive does this certainty vanish. If we do not survive TRU all - relative - passes to Zero. Therefore there will be and is no TRU. Thus we have LOUP, by the law of TRIA where we are or are not only as TRU is, but TRU is absolutely. Thereby we have: 1. Absolute Survival or ENT = **ASENT**, and 2. Relative Survival or ENT = **RESENT**, and, 3. Rate of Survival or ENT = **RASENT**. Since without knowledge nor without survival, the least possible, increasingly possible, probable, relative, and the absolute certainty of knowledge can't exist. Nor can the means to TRU and TÂ nor even the recognition of this, our very being, self consciousness or TRU, knowledge of and survival in TRU is, as TRIA itself be possible. Thereby it is absolutely necessary to achieve its ultimate probability. It, TRU, as any mass, and relative system, or RU must have RIA, and must be self sustaining if it is to be. For TRU, by TRIA, it must have the S/F that is highly Á dependent, which for man is highly subjective and knowledge dependent.

VINPAC = Victory in the future Necessitates Perfected Adaptation to Change. Since TRU must be relatively stable and free from irreversible change the success over failure or victory for TRU in the future must include an ever more perfected adaptation to, prevention of or protection from change. This does not mean that all change is an enemy of TRU. There is much change that exists within TIT, and even much external, and that in the form of TRU that is substantial but not formally negative that is reversible that is TRU. However it is irreversible, irrelevant, and uncontrolled external change that is so deleterious to TRU. VINPAC, applies to the adaptation that is maximally reversible in sustaining existing TRU, and allows no qualitative or excessive quantitative irreversible change in the highly controlled and beneficial way of TRU.

MEASURES of ACC DYNAMIC POSITS

Term		Value	Pattern	Type
ACCEX		1-7	~8~2~9~	M
ÁCON		7	~~	T
ASENT	±	7	~7~5-9~	M
ATEQ	±	4,6	~5~3-8~	M
AWIM MO	-	2,8	~_~_~_~	S
CONSIN	±	4,1	~4-5-6~	M
COWOT		1,3	~6-3-7~	T
CRUSH	±	1	~6~3-7~	T
CU MH	±	8-7	~5~-~-~	S
DU M	±	4-3	~-~-~-~	S
ELIA	±	4-7	~4~7-7~	T
ELTIA		1-7	~4-7-7~	T
ELTICMA		7	5-4-8~	T
ELTICMAK		7	4-4-8~	T
ELTICMAN		7	5-3-8~	T
ELTIMA		7	5-4-8~	T
ENWIP(HN)		4,5	7-1-8~	S
ENWIPT(HN)		4,5	6-2-7~	S
FRET		1	~4~5~5~	M
INT		3	3~7~5~	T
INTRIX		1	4~6~6~	T
LACURT (AT)		7	~-~-~-~	S
LIL (ON)	±	4-7	~-~-~-~	S
LINT		1,3	~1~8~6~	T
LOP (N)		1,5	~6~6~7~	S
LOUP (H)	±	6,7	~-~-~-~	S
MAPO		5,7	~7~2~8~	T
MARS	±	4-7	~7-3-6~	M
MAXIMUM/MINIMUM of TRUS(MT)		8-7	~-~-~-	S
NAMOV		1-7	~6~3-8	M
NEEDS, MUSTS, HOM, and ACC	±	4,1	~-~-~-	M
NIP (O)		4,8	~-~-~-	S
ORD (AR)		6-7	~-~-~-	S
ORT (AR)		6-7	~-~-~-	S
PROFT (A)		7	~-~-~-	S
PROST		1	1-8-5	T
RASENT (N)	±	7	~-~-~9	S
REALS, REALITY and ACC				
RRA (N)		5	~4~-~-	S
RESENT(N)	±	7	~-~-~9	S
RETRI (N)	±	7-4	~7~5~7~	S
RETRUX		5	~-~-~7~	M
SACOMD		1	~0~8-4~	M
SLICH (T)	±	5	~-~-~8~	S
SMASH	±	3-4	~7~4~7~	T
SPART (O)	-	4 8	~3~5~7~	S
SUMMO STUDIO SUS		5	~3-6-6~	M
TREV		7	~-~-~-~	M
TRUX		5	~-~-~8~	M
VIFS		8-7	~-~-~9~	M
VINPAC	±	5-7	~8~5~9~	M
VOTGA		1	~3~5~6~	T
WIM		7	~4~-~9~	T

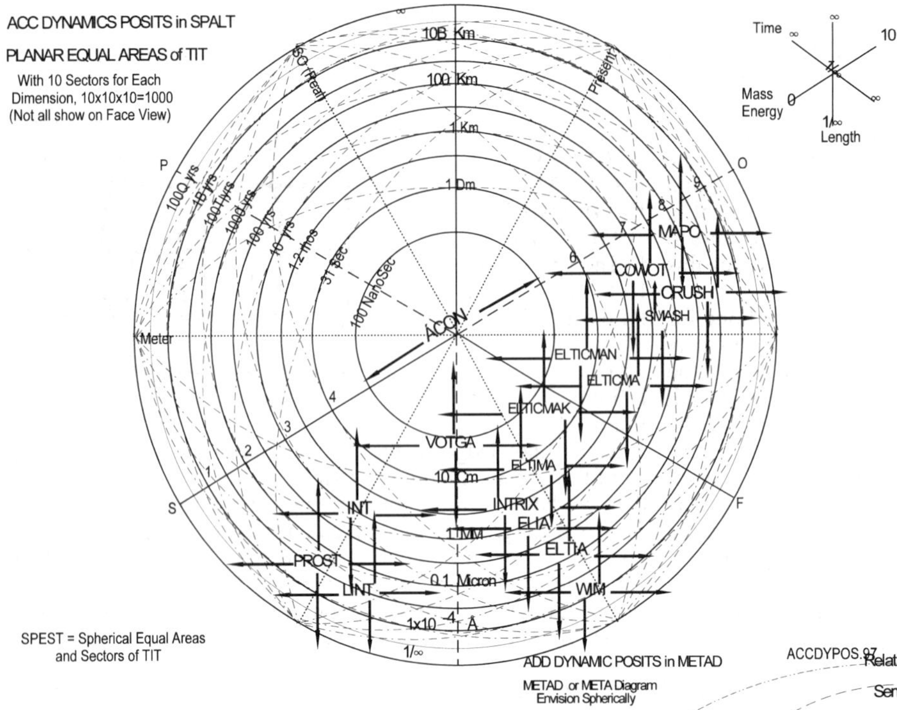

Fig 6-97 **ACC DYNAMIC POSITS in SPALT**

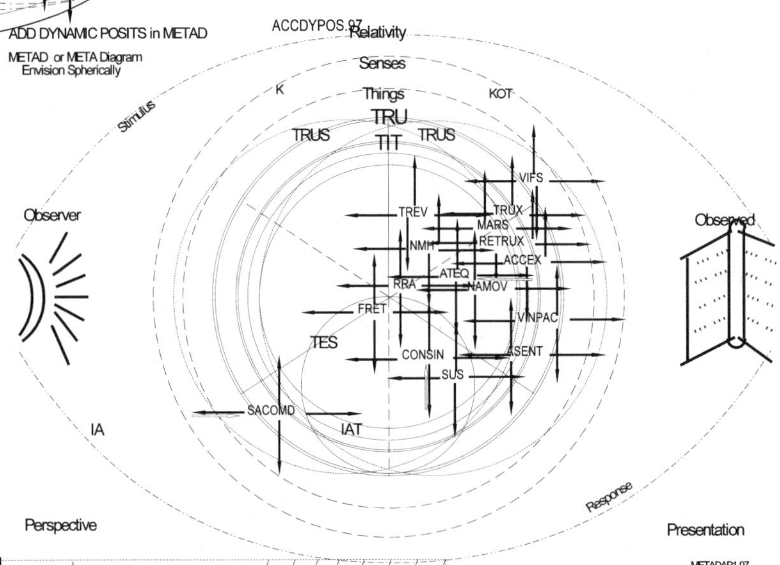

Fig 7-97 **ACC DYNAMIC POSITS in METAD**

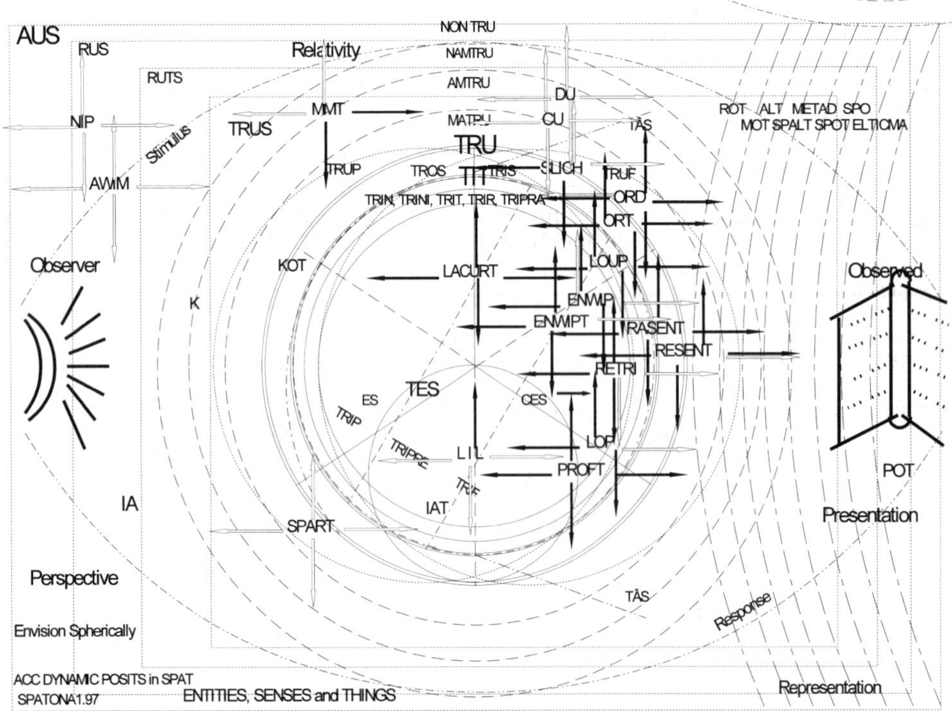

Fig 8-97 **ACC DYNAMIC POSITS in SPAT**

CHAPTER 98

ENHANCEMENT of TRU ACCENTUATION and ACTION

ENT of ACC, S-O or Á COMPLETION, OITIO, SÂ, TÂ, IPTRU, OPOT, ELTICMA, TES, EVERLASTING LIFE, EPOCHES, CONTROL FULFILLMENT, VICTORY, ÂR

ACC is the Accounting, Accentuation and Action we need to assure TRU. It is what we can do for TRU over and above what is normally and ordinarily done. In this chapter we present the combination of ACC and ENT or: **ACENT** = The Enablement and Enhancement of Accentuation and Action. This is one of the highest and most climactic chapters of the text. It presents a kind of exponential activated purpose and perpetuated success of TRU, or **ACCEX**. It is one that leads through to a summation, SÂ, of all that is absolute for TRU directly and effectively to the absolute TRU, TÂ.

Many get lost in life, many stay lost most of their lives, and often don't know it until to late.(DAK) The purpose of religion is to help us when lost. Yet much of their effort is wasted, both because we lack TES, and religion lacks TREQ and TRU. Too many religions, especially as practiced today, are only dressing or partial. We must learn TRU, TRU terminology, and its vistas, gists and proofs.

Justification and Proof of TRU Terminology

Because of the many voids and deficient meanings in human language's coverage of the total actual form and function of all in and about our world, TRU, it is necessary to utilize the best of logical, modular and vernacular means to provide what is so critically missing and necessary. The necessary is a major qualitative and quantitative measure of all, all TRU. If one would wonder what is missing in knowledge and success in human life we can readily and accurately determine it by investigating what is and what is not, and the degree, necessary. The deepest meanings made distinct of these needs, when inadequately represented by a concept and term, is NEC, New Concepts. The means used to derive NEC is how a new symbolic logical system, guided by the modular form of TRU and the vernacular modern usage of the English Language enables us to produce a standardized, rigorous, effective and precise language and representation of TRU. **LANDIM** - Law of New concepts, terminology, and Notations Determined from Deepest Meanings In Necessity. **LANMAV** - Law of New Concepts, terminology, and Notations provided by the Logical, Modular, and Vernacular Means and Measures of the Necessary in This Special World of Best Human Life.

OPOT = Optimal Presentation of TRU. There have been many works in literature that have contributed not only to TREQ and TRU but SUT, ENT, and ACC. These may be ranked by their detail, amount of greatest ideas, proportions, and totality as they are or are not equivalent, TREQ to TRU. OPOT must always be open to new knowledge, improvements and methods. Yet even more, it must provide the necessary prevention, protection and advance of TRU. This means that it must always be qualified and limited to the overall benefit of TRU. Yet of itself as a special and, at times, sole provider, protector, and means to TRU, be similarly extolled and enhanced.

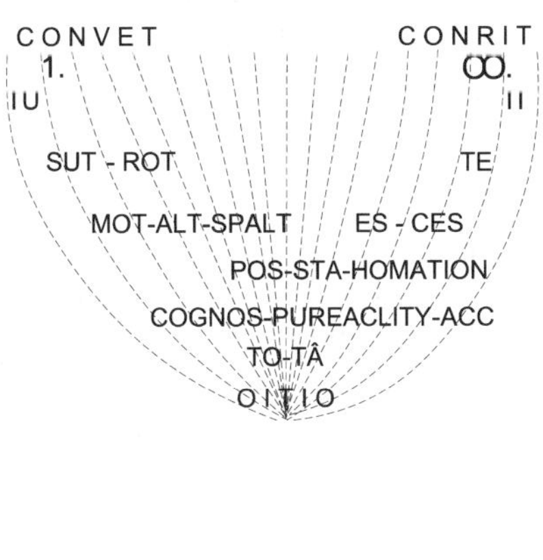

Fig 1-98 **CONVET and CONRIT in ACC of TRU** **CONVET** = Convergence on TRU by E, ES, TE, TI, or from the differential. **CONRIT** = Convergence on TRU by U, TU, SU, RIA, or from the Integral. CONVET and CONRIT signify the two poles that guide us to show ACC for TRU. We can not understand TRU if we can not show it. We can not show TRU unless it can be done through a positive objective approach, POA. This means that if we look only at the infinite makeup of TRU we are lost. It means that if we look only at TRU as single undifferentiated unit we see next to nothing. It is only by demonstrating and envisioning TRU with both of these together as the unitary and infinite TRU, TU and TI, and fuse them along with all that is their finite intermediaries and TROS, that we provide a good showing of TRU. It is readily knowable, practical and highly beneficial. The unity of TRU is controlled through TU, TO, OBT, SUT, and MOT. The infinity of TRU is controlled through VAPO, REM, ACEXIN, SPALT, POS, SPA and HOM. As the differential of TRU is ordered by the integral, and as the integral is systematically divided by the ALT FAF of TRU we have the total product. This is a replica that seeks to approach a 1:1 identity with TIT. It is CONVET and CONRIT and all in between, as they provide the increasing formal distinction of all that is TRU. This is the key to enabling us to know and show TRU, and in their combined continuum is **CONUET**.

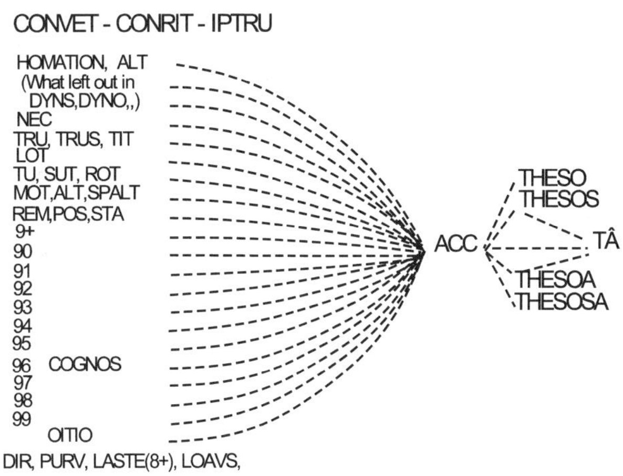

Fig 2-98 **CONVET-CONRIT-CONUET-ACC-IPTRU**

Those ACC that are taken for TRU are a collective accumulation. This applies especially to ACC that result from the use of KOT, ROT, MOT, REM, POS, and HOM. They proceed from both CONVET and CONRIT and all in between. It is only in such a way that we bring the subjective of TRU, THES and the objective, THEO together as THESO and THESOS in such a way that they provide a total, overall and absolute format for all we think and do, TRU, TÂ and OITIO. The knowing and showing of TRU from within and without is done by our ability to attain compound images, like a photographic memory that can massively and distinctly envision not only TRU by OM, COGNOS, and ELTICMA, but its parts, processes, representation, presentation and its picture from the outside in relation to all. Humans, with repeated learning, training, experience and use have the wonderful capacity in mind and memory to envision and hold images that are both solidly unified and in beautiful exquisite detail. This is made possible when the object of this image and vision is TRU, TROS, TIT, TO, TRIT, OBT, SUT, ROT, MOT, SPALT, and their operations. When increasingly finite and perfected this approaches an absolute picture by TRU, OM,

COGNOS, and ELTICMA or SÂ and Â. For the first person and ones IA, THES, and TES to have a perfect, absolute static and dynamic union with TIT is the greatest of ideals. To free oneself from LASÂ, and **FASÂ**, or absolute subjectivity, we must bring subjectivities to objectivity and our objectivities to subjectivity as they fulfill the Á FAF of TIT.

The Latin saying *verbum sapienti*, is a word to the wise is sufficient. In life we receive and give much advice. This great dictum emphasizes, not only how important being wise, good advice, and communicating this wisdom is, but even more the accelerative power of wisdom. Less directly it signifies the great power of language, and compels us to have the best possible knowledge and wisdom of TREQ or TRU.

TRU is all glorious, its rewards to humanity are infinitely great, and the damage to TRU from all that is negative and defective so devastating that the selection of the best for TRU becomes an all consuming necessity. It means that being TES is our first and greatest line of defense. It is not only in how we can best become TES, it is how we can help others, plus all that prevents, protects, maintains and advances TRU. **DISEP** = Deficit of Selective Processes. **DISEPO** = Deficit of Selective Processes Total. **DISEPT** = Deficit of Selective Processes TRU. DISEP, DISEPO, and DISEPT denote all that the world loses or becomes depreciated by. It focuses on all that can be helped by selective processes, totally in TRU. What when lost or deficient must be made up or compensated for. **SEMP** = Selection is in Mass, and its Product is in Mass. SEMP explains massive changes in time. It is rarely exclusive, or of only one quality, even if predominant with helpfulness and cooperation. Here is the key to much of success and happiness. This is especially so when perfectly directed, timed and ordered. This means to know TRU, to do TRU as best and appropriate for the state of things about one. How one can so move and provide the best help that is needed for life, and mankind, as TRU.

Fig 3-98 **REASON, APPERCEPTION, EVERLASTING LIFE, IA, ACC, TÂ**

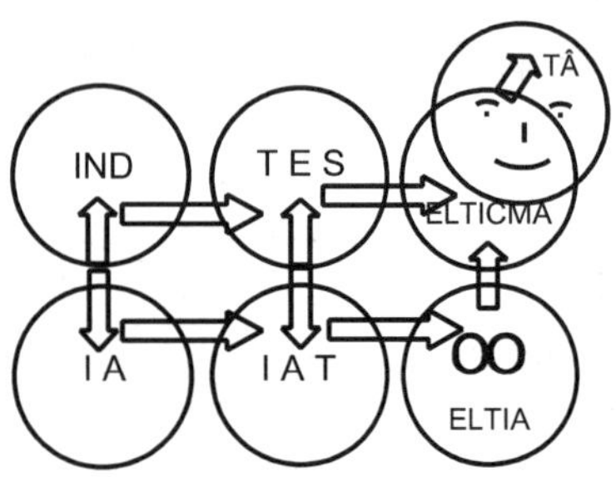

Everlasting life for an individual can be readily and quickly attained by the strong outward expansion of IA. This is through our combined, accelerated, and progressively higher mass energy, or Á, FAF. It extends the differential toward the infinite. It is a vision that is available to each person, beginning in ELIA, progressing through ELTIA and ELTIMA to ELTICMA and SÁ. It is a principle that everyone can readily follow. This is the achievement of an infinite quality of IA-Á in an infinitesimal quantity of time. It is not the same as becoming or reaching infinity. However, with concentration, direction, infinite world vision, REM, SU, and ALT this can be approached. Everlasting life, **EL**, necessitates infinite time. This may be made finite, as in various segments of ACC through time, if we correct for its non total or partial basis.

Fig 4-98 **FUTURE of SUBJECTIVE and OBJECTIVE, KNOWLEDGE, TRU**

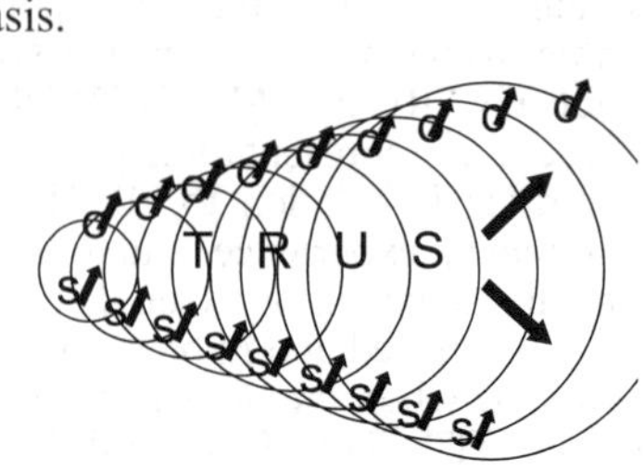

KOT, COGNOS, and ELTICMA will not be easy, nor an invariable attainment. They will be intermittent, as is our IA and much of TRU, naturally. Their frequency will be determined by the great courage, inspiration, and conviction of each person as they grow and improve acceleratively toward PUREALITY and TES. With increasing frequency it becomes ELTICMA, TESÂ for a lifetime in contrast to that for a moment. It is moving in the direction of increasing subjectivity and objectivity, not always in balance. In the short term past it has been more objective, with short term future subjectivity at a relatively faster pace. These tend to balance in the distant past and future. For as subjectivity increases so will objectivity, by the balancing out of mid points past and future. So that although subjectivity, as with knowledge and COGNOS, are exponentially expanding, they are only salients for which the entire front of SÁ expands in tandem. Yet, it is this great subjective expansion or revolution that we are in that is demonstrated by so much of literature, science, religion and philosophy. It is many of the accelerative, catalytic and stimulating effects that have been so commonly, universally recognized, honored, and the key to TRU and POT. For knowledge completion, ACENT, and TRU we should remember and be careful in our use of TRU, as well any critique of TRU, that neither TRU nor IPTRU are perfectly known or complete. In being incompletely known much interpretation will be incomplete and thereby open for criticism, correction and expansion. It is necessary that TREQ, TRU, KOT, ACC, and ACENT be available to best assist all achieve a thoroughgoing and comprehensive means of developing the causes, reasons, principles, policies for the best in life, TES. This will require ACC, promotion, use of TREQ and TRE, to support the discovery, expansion and fulfillment of knowledge, SÁ, and TIT.

Most simply and directly the future principles, policies and processes that provide the enhancement of ACC and TRU include: 1. How to politically and culturally maintain the optimal diversity, yet unity.(DAOST) 2. How to genetically and biologically best improve the quality of human capacity without jeopardizing its delicate form and function. 3. How to manage, control, and fulfill all human needs with a minimal loss of unrenewable resources. 4. How to accurately measure and test human performance to assure all are supported and encouraged to live up to their highest potential in TIT-TO.
5. Help all discover, learn, remember and live by the great CALP, ES, TES, CES, and ACC that enable them to be, and TRU to permanently flourish.

The road to heaven and immortality is, to a large extent, through COGNOS. The way to attain COGNOS, heaven and immortality is through BASFANT, POT, KOT, TRU, SUT, TO, ENT, ROT, MOT, SPALT, POA, HOM, POS, METAD, SPAT, and PUREALITY. This comes from utilizing ones best ability, energy and power in TIT. We must always strive to attain their perfect measure. This arises from: **PIET** = Present Imperfections Equal the Evils of Tomorrow and the future. The worst kind of impiety is to practice PIET, as in living off or taking from Peter to pay Paul. Heaven, although similar to TRU, or improving TRUS, is the infinite glorious victory of the all human life, as TRU or TREQ. We are apt to mistake our life on earth in this NAMTRU world to be all, or the actual life. Commonly this is what is contrasted with heaven. We are also commonly apt to mistake heaven to be fiction, figment and illusion far away, rather than TIT. So it is very understandable why heaven is so beautiful and free from misery and suffering. Such illusion is even much less than TRUID. This is called: **LIFAC** = Life in the NAMTRU World, or ones physical and empirical life on earth, not or From the Actual one, TRU. LIFAC is simply NAMTRU/TIT and like DAK clearly and dramatically helps us understand what we are so loth to do. That is to realize that many of our thoughts, acts, loves and all are not what they seem to be, but are little or poor TRU.(NAMTRU) It is a good understanding of LIFAC and DAK that helps us realize and live for TES and its immortality. Heaven like TES and TIT is both highly complex, spiritual, and exclusive. Being, by TES and TRU, or an absolute mandate toward the eternal, must be exponentially more massive, spiritual and exclusive. It is TESÂ in TIT and TÂ that is the eternal and absolute all must seek, yet more or less blindly, simplistically and inaccurately call heaven. To achieve the quality necessary for TIT, TES, PUREACLITY, ELTICMA, TESÂ, and SÂ it is vital that what is the best of TRU be presented, represented, discovered, learned, and most vigorously pursued. This is especially true of KUPS, PUCS, and IPTRU. This need can be well done through Boards, Diplomats, and Doctorates in TRU, **DOPT** or **T.D.**

The promotion and success of TRU proper will depend largely upon the degree to which it and its presentation are accepted. A large portion of the world's population is complacent, disinterested, uninspired, has their own agendas, stubborn in notions and beliefs, and alienated by all that is different or new. TRU, and its presentation only exaggerate their already sensed, subconscious and conflicting felt suspicion of the battle within self conscious worth and universal worthlessness. Some portion of the world's population who have TRU potential and TREQ may dislike TRU because they feel sufficient in themselves and thereby don't feel the need. Only a small portion of the worlds population will, at first, realize and appreciate the great need for, method, means, and provisions it offers. A much large portion will come to accept it, more or less in one form or another, for even the average person once trained in its usage will is capable of knowing, and will appreciate its blessings.. Yet it will take a vigorous pursuit of strategy, art of persuasion, and compliance along with the joyous love of finding ones soul, or TES, and the satisfaction and joy in fulfilling ones role in human life, fortune, and survival.

ACOM = S-O, or Á, Completion.(LASOT) Whereas mankind and TRU are exponentially accelerating in mass, so does TRU by COGNOS become absolute. The cognitive revolution and its course are conditioned by both the universal limits of relatives and its own capacities. It requires the utmost subtlety and profundity to have and to hold TRU. As all great REMS are reduced by the coterms of their counterparts so must that which seems so great of all systems, religions, and philosophies be reduced to their ordered parts in TRU. This is called: **PRODIST** = Profound, Prolonged, Difficult and Subtle Objectivity and Subjectivity Identity with TRU and TIT, and **ROFT** = Reduction of REMS or values by Fractals or quanta Ordered parts in TRU. This is a major function of the Á and SÁ mass energy role in TRU, ACC, and SÂ.

Fig 5-98 **COMPARISON and ATTAINMENT of TRIT and ELTICMA**

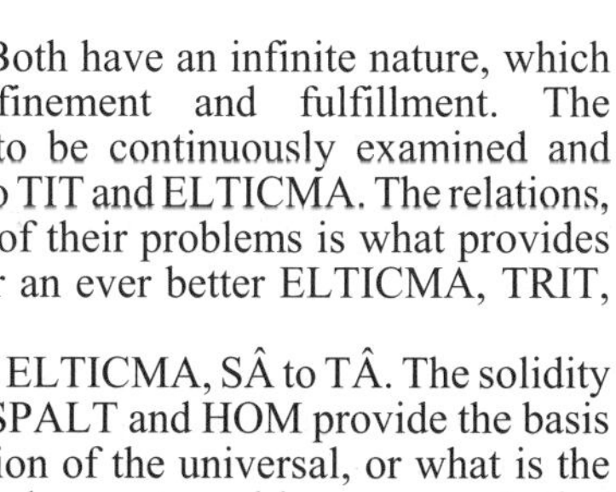

How do we attain the optimal or ultimate of ELTICMA and TRIT, separately and together? They are in a, more or less, constant sequence. For TIT and TRIT this is one of being understood. Both have an infinite nature, which demands continuous refinement and fulfillment. The experiences of life need to be continuously examined and developed for application to TIT and ELTICMA. The relations, reasons, and explanations of their problems is what provides the necessary solutions for an ever better ELTICMA, TRIT, TIT and TRU.

SUCTA = SU, COGNOS, ELTICMA, SÂ to TÂ. The solidity of TO, TRIT, OBT, SUT, SPALT and HOM provide the basis for SUCTA. It is a collection of the universal, or what is the same in TRU, and the developments and improvements that give actuality to ROT, POT and IPTRU. SUT is more than a representation of TRU. It is a mode or manifestation of the great aspect, form and character of TRU whereby its parameters, continua, coordinates, frame, structure, and totality can be known, shown and used. It becomes an aspect, form, character and part of TRU. This is called: **SUTE** = SU as a part of TRU. SUTE is in contrast to **SUIT** or SU In and Of Itself. The more SUIT becomes ATRU and defeats itself, when intended for TRU is: **SUTEN** = SUT Enigmatic, or SUT ENS Flux. SUTEN defines those states, that misled us from TRU, like the parameters, dimensions, coordinates, frames, and structures, systems and sects that exist and are believed to be of all that is about us, yet are far from TRU. It is only SUT confined to TIT as in ALT of TRU, that we can overcome this enigma. It is through SUT that we can develop ELTICMA, and it is through ELTICMA of ACC that we can set our course, SÂ.

Fig 6-98 **THE WAY to TES, TO, OM, POM, SÁ, ACC, ELTICMA, SÂ, and TÂ**

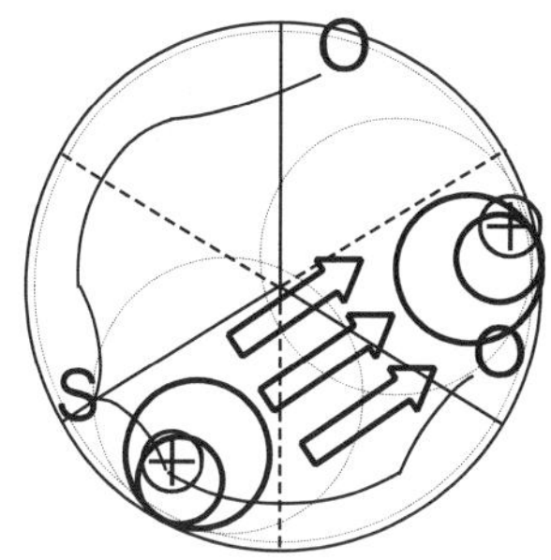

TO, OM, POM, SÂ, and TÂ are emphatically exemplified by international differences and the conflicts and turmoil that has existed throughout history. Humans can get along together best This is based upon sufficient respect for each other. Only this requires knowledge, and especially the identity that exists in common, DAOST by TES in TIT-TO. Likewise it requires an understanding of ones own non TES and its sameness in that of others and all TRU. It requires a willingness to guide and be guided by TIT, TRIT, SUT, PUREALITY, COGNOS, ELTICMA, SÂ and TÂ. These should be a sufficient way to S/F, ACC, and all TRU. It is the way to respond integratively to all that are best accounted for and are reconciled in TRU.

There is a great need for, and perfection in, TREQ, TRU, their representation and presentation. IPTRU is rough and incomplete in many ways of its parts. This should be appreciated in a topic as important, undiscovered, and undeveloped as TRU. As great as many of its ideas, concepts, principles, laws, and mechanisms are, as much as it uses the best that knowledge has discovered, and as profound and subtle as many of its new concepts are, they do not validly, actually, proportionately nor totally cover, nor perfect, the concepts, form and function of TRU that must be available for all to know and use. Likewise many of those presented will need correction or replacement. Much of the material in the text will, with future editions, versions or kinds of presentation of TRU be better expressed and demonstrated. Yet the need is overwhelming. God and Religion on the one hand and philosophy and science on the other, as good and great as they are, have failed to provide humans with the accurate and comprehensive guide each person needs to reach a level of comprehension concomitant with successful adaptation to life in the actual world. This is and will be increasingly acute in the near future as both mans population, culture, industry, commerce, communication and politics place escalating burdens and problems that must be anticipated, solved, maintained, and controlled. TRU must and can be developed, modulated, represented, given the necessary positive objective approach and presented with increasing reliability and perfection. It is only with dedication, determination, conviction, courage and a great sense of love and joy in TRU that such will be forthcoming.

The Nobel Prizes awarded annually in Stockholm for most outstanding work in various fields and sciences and in Oslo for peace, are excellent examples of what can be done to stimulate research and knowledge that supports and proves TREQ and TRU. Such honors and rewards show how great the benefit/cost ratio can be when well targeted. Expansion of this and like means to knowledge and TRU would vastly improve human interest, purpose and quality. These are excellent examples of what can be done in the ACC, ELTICMA and SÂ of TRU. Plato, Poincare and Einsteinshare with others the distinction of having recognized and demonstrated the significance of essences and relativity. This is the basis for much of TRU.(Á,TRIA) It provides the means of gaining the necessary depth conception to unify and over-come our problems. Anaxagoras is distinguished for having recognized the power of *Nous* or predominant form and dimension of the relative universe. Leibniz, distinguished the particulate, monadic, spiritual construct of the world, and the epoche eidetic development of the essential. Aristotle although losing some of the gains of Anaxagoras and Plato in not emphasizing the actualization of the SO in TRU advanced another step by providing the Formal or Logic of Essences that expanded the real or phenomenal into a formal method and concrete foundation for the complex, syllogistic, symbolic and logical uses of language in representing TRU.

EPOCHET or Epoches of TES to TRU and TÂ

Epoches are ascending levels of a person's search for fulfillment in TRU. They provide an unrefined yet important and beautiful demonstration and means of a person's attainment of PUREACLITY and TES. These may be viewed as steps to $(S/F)^{\infty}$ TRU. It shows how to measure the accelerative power a person passes through to achieve higher levels of PUREACLITY, TES, TA. It may be called: SÂ = Steps to the Absolute, TA. SÂ is a connector. In the accompanying diagram we find assumption and resurrection in their deepest meanings. It is like being part of the steps to heaven. SAK and EPOCHET are by systematization for TÂ, The Absolute from The Self. (Ind-Self-TES-TRU-TOT-TRUID-TÂ)

\ TÂ
Eternal Absolute
\ SÂ (to THESOÂ)
Summation to Absolute
\ ELTICMA (to THEOÂ)
Eternal Life by COGNOS +
\ COGNOS (to THESÂ)
Total Inspired TRU Knowledge
\ PUREACLITY (to TES)
Self to TES
\ POM (SUT, TO) Expanded OM
Total Inspired Knowledge
\ OM (SUT, TE) inspired
Ordered Knowledge of TRU
\ EPOCHE Concrete, ENT,
Eidetic ES 1:1 TRU Rx
\ EPOCHE Concrete Eidetic to TO, D.E.T.
Knowledge Closure in TIT-TO
\ EPOCHE Concrete Eidetic, Summative Depth
Conceptions, Ordered, and developed
\ EPOCHE Eidetic Concretion of the Essential
and Formal TRUT
\ EPOCHE Eidetic Formal, Logic of Essences well
developed
\ EPOCHE Eidetic, Images or Formal Developmental of
the Essential
\ EPOCHE Essential, TRU understanding of Ideas
and Essences, beginning of the Á Dimension,
\ EPOCHE Pure Phenomenological, Depth Conception,
Transcendental Concentration, THESO
\ EPOCHE Distinction, Eidetic and Psychological Transcendence
Depth Conception
\ EPOCHE Transcendent, How we move in and out of TRU, soul and
spirit SELF-TES-TE-TU-TRU-TÂ
\ EPOCHE TÂ, of the Absolute Spirit, Sustained
Revelation of Identity
\ EPOCHE Critical Practical Reason, First Glimpses of TRU. How the
pieces of the universe fit together, beginnings of ALT, Homation
\ EPOCHE Psychological, The best of scientific minds dedication to the
experimental method, to bring out total valid conclusions
\ EPOCHE Godhead Identity, a 1:1 picture of God, Universe, things, and
appearances as they fuse in TRU, Beginning of Realized Immortality
\ EPOCHE ITID or TEUIT-TRIT, Recognition of the fusion of the Heaven on Earth
with the Ideal Heaven Above.
\ EPOCHE Religious, Level of Sustained Recognition of the fusion of the Heaven on
Earth with the heaven above
\ EPOCHE Logical, Level of Profound Insights through Logical Abstractions, Set Theory,
Mathematics and Formal Methods
\ EPOCHE Rational, Level of Insight through Dialectical and Epistemological Methods
Application to life and the actual world
\ EPOCHE of Righteousness, as in 10 commandments
Selective and qualified application
\ EPOCHE of Good Works, Revelations from God, Glimpses of TU
\ EPOCHE Intentional, Voluntary, Sustained REM, values, priorities
\ EPOCHE Intentional, Voluntary, Collective, Combined Endeavor, Positive, Increasing
from Singular Intentions to combined and collective intentions
\ EPOCHE Unintentional, Involuntary, Imitational, +/- in TRUD

EPOCHET, or Epoches of TRU, is similar to finer higher levels of SÁ, or mass energy in TIT, in contrast to space or time. It represents levels by which a person can rise to greatest states of TES. We can see how for many they depend upon some if not all of the preceding or lower states. Just as it is impossible to adequately be TÂ without recognizing TÂ, so is it impossible to be either TES or TIT without recognizing and living by GOD, TREQ, or TRU. This escalation of the spirit has been one of the greatest driving forces and powers in the history of the world. It shows some of the greatest processes of TRU of the last few thousand years. It propels mankind from the past, and more, compels us toward the future. Mostly it empowers mankind in the present by natural selection and genetic improvement to best compete against all adversaries.

All persons strive for TES more or less, some vigorously, powerfully and with great perseverance. It is this that has been most productive of the finest accomplishments and benefits to human life, and especially TRU's greatest gains. Gilgamesh was known for sustained voluntary and intentional search for wisdom. Moses is known for his devotion, leadership and the *Ten Commandments*. Saints are known for their good works as revelations and illuminations from God. Christ and Plotinus are known for their transcendental state in rising above the immediate, and ATRU, world about them to identify with the actual world of God and how their soul and spirit functions. St. Augustine with his rhetorical training, depth conception and conviction in God is an example of how one must see that the essence of a word and the essence or meaning of this word as it is a part of God, or ES of TREQ and TRU. They are identical. Also with this the recognition of the fact that these essences are not static, but dynamic forces that function for the mutual benefit of TRU.(Doctrine of Grace) St, Thomas Aquinas is known for the identity of all with God and their fusion and separation. Copernicus, Galileo, Newton, Bacon, and all great scientists are known for their dedication to the experimental method and verification of basic laws, the correlations and identities that exist in the world. Kant is known for the *Critique of Pure Reason* and *Practical Reason*, and how the pieces of the universe, TRU, fit together. Hegel is known as the champion of the dialectic and Absolute Spirit. Husserl is known for his pure S-O phenomenological approach to the expanding identity of SO through intentionality. This is the best existing evidence and proof for the mass energy Á FAF of TRU. Most all great principles, laws, concepts, topics of, and refined by TRU in IPTRU, may be posited in one, or another, or several, of these Epoches. Their levels follow their discoverers, much as ontology follows phylogeny. They will vary somewhat depending upon the depth and relationships given. For example the idea of SCARLS typify the intension of science or Step # 11. Some will vary widely depending upon the form used, e.g. DAOS and DAOST, # 3 and # 12. A refined and properly developed system based upon such Epoches of the Á dimension, as in SÁ, provide a great tool and weapon for the accentuation of TRU. It is also one more representation that is not only of, but in TRU, RINT, ROT-THEO-TIT, as they exist in TIT.

The failure to recognize and adequately live by, the basic and crucial laws, principles, mechanisms and processes of TRU has, is and will be the greatest failure of man. It will be the case whether TRU is recognized as such, TRE, or is not, TREQ. This recognition and life must be with distinction, conviction, identity, dedication and perseverance. It is not enough to barely catch a glimpse of the great aspects of TRU in the glory of an apparition. It is not enough to have and to hold them singularly or even in combination for a time. Our greatest success in life has, is and will come by comprehending and fulfilling all TRU, COAT, envisioning and performing its crucial laws, principles, mechanisms and processes. This will be when each person's lives, FAF, thoughts and activities become counterparts in TRU as this proceeds through THEOA, THESA, THESOA, SA, to TÂ. The omnificent magnanimity or OM of TRU as produced by OITIO, SPALT, TRIS, TROS, TRUS, and TAS, is projected by COGNOS and ELTICMA to the infinite TRIS and TAS. They are the ultimate end to which all must be directed and each person's life fulfilled.

DEATH in LIFE and LIFE in DEATH in TES and TRU

To be dead in life	To be dead to all that is TRU
To be alive in death	To be alive in death means
This is what we all are	To be alive to all that is TRU
For to be dead in life means	Through ones life while alive

What is foremost is whether we live or die, have immortality or extinction. For in death we can be alive, and in life we can be dead.(DAK) If we must be, we must do our best, think our best, and be our best to the utmost. This is to be TES. Since by LOPAN and the law of essential parts of TRU, each is more or less vital, thereby immortal. So that all of us are immortal, or at least have the opportunity and potential to be immortal. In this process we achieve increasing immortality, relative to the absolute, as TRU becomes TÂ. Humans have great drive to learn, acquire knowledge, and understand all, as in religion and philosophy. These lead to COGNOS. We intensely seek immortality. We have a deep desire to attain infinite and total introspection, vision, and knowledge. We strive to avoid loss, and ever more greatly possess and multiply. All these and more constitute a large portion of our biological, human and civilization's heritage in TIT. All humanity comes together on the road to the ideal, GOD, TREQ, and the absolute, TÂ of TRU. If we know where we are going and if we know how to get there it is not so difficult to reach our destination, TRU and TÂ. Nor should it be so impossible for us to identify with this and escape from all other roads, ends, and all else that is so negative, failing, and destructive. Our IA and subjectivity provide us with the potential for hundreds and thousands of visions. These through repetition and memory become all the more rapid, acute and intense, especially when we are curious, imaginative, inspired, recursive, and have the needed revelations. They become all the more effective and beneficial when we have the knowledge, VAPO, TRU, TRUS, SCARLS, TROS, TRIS, TIT, TU, SUT, TRIA, TÂ, TE, TI, TO, TRUD, ENT, ROT, MOT, ALT, SPALT, POA, POS, STA, HOM and ACC to guide them. It is necessary for the individual to attain reality, the reality that exists in the actuality of TRU. Together this is PUREALITY, which can like nirvana enable us to escape from the cycles of ATRU and comprehend all at once, as if to make the world standstill. This is to best make the needed alignment and identity one must have to be TES and TRU. It is called: **VISO** = Vision Infinite Subjective Total in the Infinitesimal Present. VISO, like ATO, is the infinite focus of all on TIT, by TRIA, in which it is neither past nor future, but infinitesimally totally existent in the present.

When liberty is rampant without TRU and when the law becomes a tangled web of statutes, rules, regulations, and legislation they rapidly change and fail. This is especially to the degree they are misdirected, misapplied, abused, disproportionate, and used indiscriminately. Only by developing and formulating our laws and order out of TIT can they be sufficient to uphold TRU. Law and order also depend upon those who exercise them. It is largely in the governed not the governing that the greatness and glory of TRU are to be found. Yet both those governing and the governed must be TES and TRU. When TRU this is largely voluntary and agreeable. It is by the will to find and do what is right, through the natural and heartfelt inspiration and motivation that comes from TES, TRU or TREQ, or **POPUR** = Power of PUREALITY. POPUR is where the Á dimension's physical and mental natures and all in between form a continuum that is increasingly pure and real in TIT. It is how they spontaneously or automatically fulfill all the laws of TRU. This is signified by: **LASENOT** = the Law of the Automatic and Spontaneous enablement and fulfillment of all the Laws of TRU. LASENOT is largely achieved through PUREALITY, TES and TRU. Life in TRU does not come from never ending legislation and regulation by simple and foolish governments. It comes from the heart, ones mind and spirit in ones love of TRU and all that is TRU, especially others DAOST, as they are or can be TRU. This is the beauty and glory of the Christian Spirit or an inspired and heartfelt love of God. For TRU its actuality, more complex in the whole continuum between mental and physical, developed in PUREALITY, SÁ and all TRU.

Convergence on the absolute is no where more evident than in the mass, length and time dimensions of mlt and ÁLT. As TRU enlarges and expands it comes into relation with more and more of the absolute, that unrelated becoming potential to it. As this occurs the focus and scales of mlt change. For mass this change in form results in an alteration of the characteristic mass to Á or the mass energy dimension of relativity. So that for TRU to retain its thisness and existence, yet approach the absolute, it must not only maintain and retain its existent status, it must readapt all incoming actions and be constantly and reversibly unfolded to maintain its thisness, yet incorporate the absolute. As a pattern, unity and systematic unity of this adaptation it becomes, synthesized, set and stabilized. TRU approaches maximum completion and identity within its own absolute. The completion and totalization of mlt is called: **LAMENTO** = Law of TRU Mass Energy, Length and Time Totality. LAMENTO occurs as the changing and converging mlt dimensions become complete in and out folding and identity or equilibrium with the absolute. This is not a simple, single and absolute end. The relative does not become ÂÂ, an absolute absolute. Not if TRIA holds. The best that can occur is for TRU and the Absolute to fuse selectively maintaining and perfecting its thisness. Some permanent alteration of thisness may be expected, relatively, yet with opportunity for reversal if TRU and TIT are to exist. This does not imply that humans have little to fear from their own inadequacies, inabilities, and failure to be TRU. For these are the greatest threat to consciousness, intuition, thought, reason, comprehension, and all TRU .

The subjective also retains the power of reversibility and equilibrium as it maintains its thisness in TIT. It will tend to completion incorporating more and more of the absolute expansively. This is: **LASCOM** = Law of Subjective Completion. LASCOM is a general form of the subjective and thereby what the A dimension of TRU assumes in the indefinite future. It is one of stability, order, and eternality, not one of infinite variation, disorder, and extinction.

TRUCI = the Relative Universe Through Compensatory Immortality. TRUCI denotes how we drive the hardest and greatest when it is by and for immortality. It is through the instinct to survive that we are most basically moved. It is when we face death that we most desire life.. All organisms and animals are so formed and function that they can best preserve themselves. As animal's CNS becomes more complex they develop both primary, secondary, and other forms and functions that assist self preservation. All that enables organism or animal life is in one way or another a means to survive. This is how the brain, is primarily a concentrated organ of survival. In humans, by the predominant powers of Á, including basic: tropisms, taxis, instincts, adaptation, behavior, and needs, survival is achieved, and can be recognized. In being recognized and in experiencing the death of other life forms and persons, a keen interest is developed in how to survive. Great apes, elephants and other animals have a curiosity, or concern with death, especially of those most dear. For thousands of years, across most continents, humans in the megalithic age, built monuments to project life. These often took elaborate and fantastic rituals that were dedicated to immortality. This is found in ones participation in a group, world or universe bigger than one self. It is recognized that life can come after physical death through increasing and best participation in one's kind, group, race, specie, life, world, or universe. Such participation provides for life after death, or the immortality one seeks. In this way humans satisfy this deep instinct for survival. It is in sacrifice and being a martyr for a higher cause. They enhance their Accentuation of TES and TRU in ways that increasingly provide survival.

LOSOW = Law of the Summation of Wills. LOSOW includes the added and multiplied integration of various wills: individual, collective, will to be, will to live, will to power, will for good, will to attain all maximally for the benefit of humanity, world, God, and TRU. The main difference between universal law and COGNOS or Universal TRU reason lies in willful intent. This is most significant. If God is commonly conceived in many various ways, it lacks the distinction of a willful intent that is strictly, totally and proportionately TRU.(VAPO) Yet for the purpose of universal knowledge, as is the case for KOT and TRU, it is all important. For human success and our very being is not derived so much from universal law as known today as it is upon the summation of human reason. This and LOSOW are much the same in their identity within a large portion of TRU. There is no reason except that of man. Yet reason in its most general and universal actual form is none other than a great part of TRU as THES and THESÂ are a product of LOSOW. Once a, or all persons, have cleared the way toward an understanding of TRU through willful intent, LOSOW, COGNOS, and inspired recognition of PUREACLITY, we gloriously and victoriously advance TRU, REM, ACC, ALT, MOT, ROT, POT, PERT, COGNOS, THES, THEO, THESO, THESA, THEOA, THESOA, SÂ, and TÂ. This is how will, knowledge, civilization and all that has been most important was necessary and has advanced TRU. Knowledge alone is not enough, nor industry and will. It takes these together, when actual for TRU, and all the rest that partakes in the form and function of TRU with the exclusion of all that does not partake of TRU.(POS) **PASOT** = ultimate Purpose, Aim and goal of TRU, PUREACLITY, Á meaning, or S-O Totality of TRU. The ultimate purpose, aim, goal, end and direction of TRU is not only to become all knowing, nor the acquisition of the material or territorial, it is to internally and exponentially expand the continua of Á, L, and T, plus they together in ALT, to their highest powers, ACEXIN. It is to push our form and function, state, operations and dimensional makeup in all directions and in the proper balance toward TÂ, while still maintaining TRIA and TIT. This is the optimal growth of the TU, TE, TI, TO and TÂ of TRU. These for the Á dimension plurally become: **ÁS** = Predominant Mass Energy of TRU Plurally, **ÁI** = Predominant Mass Energy of TRU Infinitely. Infinity for the S-O as well as L and T dimensions, are not for the near or mid term, a practical goal. Yet as an end toward which the unity and thisness of TRU works, it is valid.

It is with the best of fanatic zeal, victorious reaction, cooperative and coordinate VAPO, DAOST, SCARL, ENT and ACC that we progress to TRU, TES, TRUID, TIT, and TÂ. The wars, suffering, death, and horrors of all time are little more than the errors of groups that have not adequately utilized their given, predominant, and powerful means of reaction. It is a reaction with force and power that is necessary to develop the concentration to overcome oncoming danger. It is a part of a body's collective fear reaction. It encompasses many levels of the Á dimension, in which the lower levels predominate, i.e. more out of knee-jerk reflex than out of reason. The action causing this reaction is often due to deficiencies in themselves, overreaction, imbalances, or the negative that comes from external opposition. This state is managed by: **INVICTA** = Victory in Coordinate and Cooperative TRU. INVICTA holds that the main justification for using fanatic zeal is when the danger and its use are equal to or greater than the needs of TRU. Recognizing the fact that TRU is dynamic, of DAOST in natural opposition, and preferably non violent, fanatic zeal is better put to use in more indirect and controlled ways. For this opposition must be balanced. There must be cooperation for the same, and moderation, tolerance, accommodation and allowance for the different. Yet INVICTA represents the tremendous potential for action that is available. It is a kinetic base of exponential power that animalian evolution provided for our Á mass energy FAF. INVICTA is with reason, guided by TIT-TO and ACC adapted to the situation, given, danger, or threat at hand. INVICTA is often the salient to which ACC and SÂ are directed. INVICTA is the invaluable inspiration and power that proceeds from the flowering of life. It, like MARS, is an incisive tool of ACC. It is, when unlimited, beyond measure, an accompaniment of infinite TRU. The glory of such ultimate and absolute reaction and success is the great impetus to ACC. **UVICTA** = Ultimate or Absolute Victory. We should always remember that INVICTA and UVICTA are only ends in closure, and are not the simple, or open ended ends they may seem, even in TRU as when partial.(VAPO, SCARLS, HOM)

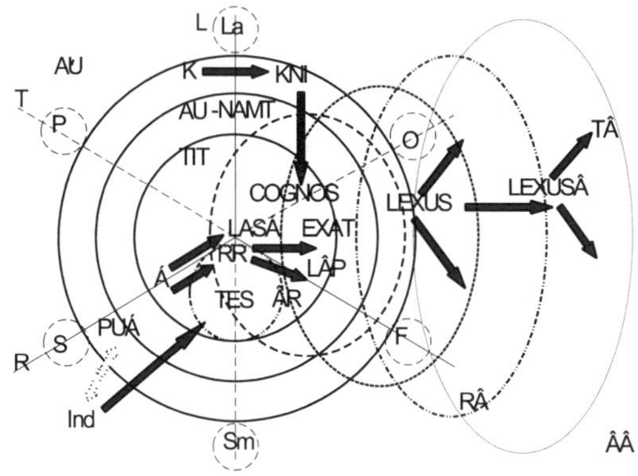

Fig 7-98 **COGNOS and THE INFINITE, EXPONENTIAL and ABSOLUTE**

COGNOS differs from knowledge. As *we know*, COGNOS implies not only the infinite wisdom of knowing that we know and KNow Infinitely, KNI, it implies that knowledge has reached a point where it is self assured and can near completion, to become irreversible through the certainty and absolute thisness of TIT.(TRIA) It has reached a level through the verified solidarity of reality and phenomenal identity in Á and TIT-TO. **LASÁ** = Law of Solidarity in Á. LASA confirms the real and phenomenal as intermediaries, like spirits by the vantage of relativity. They are neither illusions nor figments. By their strong identity in TRU they are the opposite, the ultimate in solidarity. This provides for TRU an exponential assurance, **EXAT** = Exponential Assurance of TRU. EXAT explains the transition from the relative to the absolute, that occurs in RR to RÂ and TRU to TÂ. Just as 'God is in you, or God is greater', and good is, so we search for a firm and absolute assurance. This is basic to TRU, and for any RU what concentrates to make it sufficient unto itself and survive. This is the self preservation that is present in all entities, only in different forms. In AU and at the strictly physical level it takes the form of the laws of motion. A body continues in its state until acted upon, and for each action, there is an equal and opposite reaction. It maintains itself, or retains much of its prior likeness and self, when acted upon. It is the nature, makeup, aspects, modes, form and all of what enables survival that is key to learning TRU. It shows our very life, being, survival, hope and infinite being to the absolute. It is here we can understand how to unlock the door to the absolute. It is much of the entity's or TRU's momentum and inertia which tends to be fixed, sufficient unto itself, self assuring and protected in permanency. This is much of what RÂ and TÂ are. Yet since all is relative, this is never absolute nor ÂÂ. Yet relatively as the words, parts, form and function of TRU are certain, so are they absolute in a sense and so does this added factor become absolute. **LÂP** = Law of the Absolute Assurance of Parts, increments, Forms, Functions of TRU. However, nothing can be set up as absolute if limited to their need in TRU or TRUF. Words, concepts and principles are, to an extent, ethereal. This is like the objects that seem so certain to our senses and experience are likewise illusionary and ethereal. This leads us to: **PUA** = Paradoxical Uncertainty of relativity, observers and observed or subjective and objective. PUA is a law of the non reality, non certainty of a portion of all, except TIT. As man becomes TES, and thereby TRU, as greater and greater AMTRU, ATRU, ANTRU and ANTITRU are eliminated, and as humans expand their A, L and T in all directions so do knowledge, possessions, makeup and total being approach the absolute. The expansion of this is called: **LEXUS** = Law of Universal Simultaneous Expansion. The expansion to TÂ is called: **LEXUSÂ** = Law of Expansion Universal and Simultaneous to TÂ. LASA, NAT, LEXUS and LEXUSA largely define the transition from the infinite through the exponential to the absolute.

ELTICMAT(Â) is a lifetime wherein one can not only name and identify with immortality, it can be made infinite in TES to the nth degree in TÂ. So that what is not absolutely correct or takes a stretch of the imagination to bridge the gap between the real world of S - O and AUS - RUS, and the world of IAT - TIT with: **ELTICMA → (TES)^n → SÂ → TÂ** requires a profound learning, understanding and use of ELTICMA, TES^n, SÂ, TÂ, and all the means we can develop that will best show them. **ATIM** = All Time Mastery of Things. Whatever we learn and remember that is of TRU, and that is proportionately great of TRU, is of permanent, all time, value. It is of 100% gain. If these are of TRU, SUT, ALT, REM, and of 100% significance to enable a person to attain ELTICMA and the mastery of all things, only the constant reinforcement and attack on their infinite ramifications are delimiting. For ATIM, just like the permanency of knowledge or any enduring and highly useful construct or tool, is for all to see, use and benefit by, eternally.

Just as the Bible Says Eternal life is the gift of God This is done not only directly It is given indirectly in grace	We and all live in Pureaclity, be TES And One with TRU Eternal Life is up to us
Our Eternal Life is purity Reality, and actuality in TRU Only as we extract our ATRU To fully become TES in TRU	Not some Etheral gift nor Illusionary figment or diety But actuality of existence As its counterpart in TRU
Eternal Life is not given without cost there is only an opportunity for us to Earn it, and this only as	Just as DAOST, all is same By relativity we are neither Zero nor Absolute. We live Eternally relative in TRU

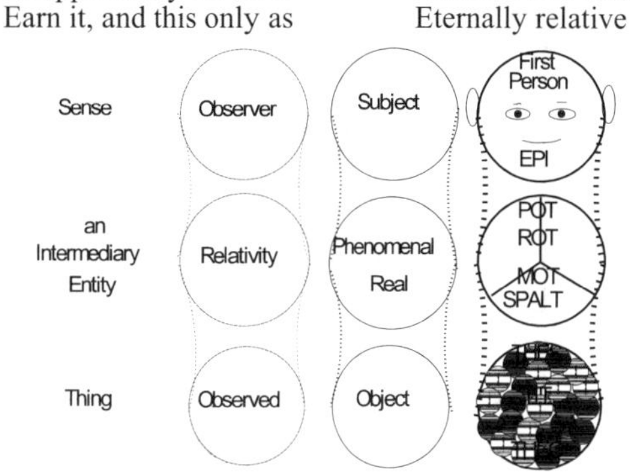

Fig 8-98 **OBSERVER and TRU RELATIONSHIPS and ABSOLUTE FUSION**

Figure 8-98 shows how a face is not a face, entities are not entities, senses, things, nor even their combinations are not themselves. They are only what they are and do in relation to TRU. It also shows the basis for the fundamental form of entities and especially TRU. For their form is neither sense, thing, nor intermediary. It is neither observer, observed nor relativity. It is neither subject, object, nor the reality they produce together. Their form is in actuality only what all these are together as they are focused from the relative universe of which they are a part. This form, and function, FAF, is a product of both the unique and exclusive thisness of this RU, by RIA, as well as how it is composed of a combination of sames and differents in a unique and exclusive way that characterizes and typifies it. The combination of the viewer shown by a face, and the viewed or TIT is mediated by presentations, representations, and a total modulation that together fuse and come to exist as one in TU, TIT, and TRU. The first person and TIT fuse so that one achieves TES. In turn, TES fuses through TIT, and its power and grace, to identify by way of POT, ROT, MOT, and SPALT. Through the action of EPI, THES,, TIT, and THEO - in the black circle - it assures the permanence and eternality of all in COGNOS, ELTICMA, SÂ and TÂ. We need to envision by vivid illustrations the relationship and relativity that exists between the observer and observed, first person and TRU, and all in between. It will be SIA, OM, COGNOS, ELTICMA, SÂ, and TÂ. It will be that to which all are directed, aim, have purpose and converge, TRU to TÂ.

By Ecclus., "Work your work before the time cometh, and in His time He will give you your reward." The best preparation for death and the attainment of immortality is to live well and become a part of what is eternal, TIT, through the fulfillment of TES. This fulfillment is past, present and future, before, during and beyond ones physical lifetime. Just as all the greatest personages in history were driven by an inward and outward quest for immortality so do their present and future existence, as their souls in TRU, fulfill their preparation for death in life and life in death, an eternal life, TIT to TÂS.

	AU	RUS	TRUS	TRU	TRIS	TIT	TRIF	TIT-TÂ	LEX- USÂ	SÂ	TÂS
Sub- atomic Forces	SAF physi- cally	SAF altered by RUS	Selectiv ely in TRUS	Limited to role in TRU	Limited to role in TRIS	Limited to role in TIT	As best change d forTRIF	As best change d T-TÂ	As best used by LEXSÂ	As best used by SÂ	As best used by TÂS
Gravity	classic	"	"	"	"	"	"	"	"	"	"
mu	begin- ning	"	"	"	" SLICH	"	"	"	"	"	"
gnu	0	begins	"	"	"	"	"	"	"	"	"
adu	0	begins	"	" WIM	" TESEX	" MARS	",	"VIN- PAC	"	" LAREQ	" ATIM
cu-exp	0	begins IA	begin- ning	" SEMP	" IAT	"LAM- ENTO	" ÂS ELIA	" Á I	"	"	" INVI- CTA
ku	0	0 PUA	begin- ning	"PRO- DIST	"LOS OW	" KNI	" OM	" TRUCI	"	"	"
ru- validity	0	0	begins VAPO	"	"	"	" LAS- ENOT	" ASENT	"LA- TEVU	"	"
imu,vu	0	0	begin- ning	begin- ning	" ROFT	" PASOT	" EXAT	" POP	"	"	"
THES	0	0	begin- ning	begin- ning	" RES- ENT	" LASÁ	" PUREA CLITY	" COG- NOS	" ELTIC- MA	" ELTIC MAT	"
THES Á	0	0	0	0	begin- ning	"	" ÁCOM	"	"	"	THESÂ

Table 1-98 **States of Existence and Levels of Mass Energy**

The SÂ continuum, summation, or steps of TRU to the absolute, horizontal series, and the SÁ continuum or MESOÁ (pg. 32), summation of mass energy FAF in TRU, vertical series, when taken together reveal much of the makeup of TRU, its development and future. With Epochet, this table helps to get a better vision of these two continua, their contrast, and the role of various new concepts in chapters 97 and 98

Very few entities, as man and TRU ideally, expand indefinitely to overcome all lack of and changes in fulfillment, imperfections and incompletion to maintain perpetual sufficiency. This is a sufficiency that is at equilibrium with infinite and eternal SCARLS, changes and variation. Such convergence of TRU on perpetual universality is called: **LATEVU** = Law of TRU Universal Evolution. LATEVU the interaction of TRU with its future, TÂ, and the Absolute. RR, ÂR, RÂ, and ÂÂ form a continuum in which TRU, humanity and individuals are features. ENT of ACC has sought to establish the resultant to this. This is expressed by: **LAREQ** = Law of the Absolute to Relative Equilibrium. LAREQ signifies that region and center of the absolute relative polarity in which life and TRUS attain the optimal or ideal state and existence. This is highly convergent on TIT, by TRIA. Yet we know from science that this may not be so absolute in itself as conjectured. Therefor it is in a range, in which reversibility and control play an ever larger role. We do not seek the ÂÂ absolutely for we know it is of much AU, ATRU and ANTRU. We do not even necessarily seek the RÂ or TÂ in this extra TRU absolute sense, for it denies much of the TRIA of ÂR and TIT. Thereby we seek the optimal of ÂR and TIT that are concomitant with, and most practically and beneficially adapt to the needs and requirements of, TIT, LATEVU, LAREQ and TÂ, while retaining TRIA.

MEASURES of ACENT CONCEPTS POSITS

```
ACENT 1,3 ~6~5-6~ T
ACOM  7   ~4~~~9~ M(NÂ)
ÁI    7   ~3~6~9~ M(N)
ÁS    7   ~4~5~7~ M
ATIM  5,  ~3~5~7~ T
CONRIT6-7 ~-~0~5~ M(MT)
CONVET4-7 ~-~9~5~ M(ME)
DISEP 4   ~-~-~-~ M(H)
DISEPO8   ~-~-~-~ M(H)
DISEPT4,1 ~-~-~-~ M(H)
DOPT  1,3 ~3~6~7~ T

ELTICMAT(Á)3~4~6~7~T
EPOCHET 4,3 ~4-5-6~T
EXAT    5,7 -~-~-~ M(TN)
INVICTA 5,7 ~6~4~8~M
KNI     3   ~0~8~9~T
KNOWLEDGE COMPLETION,
 ACENTTRU 5  3-6-8~T(KCT)
LAMENTO6,7  -~-~-~ S(OT)
LÂP       3  ~5~6~9~T
LAREQ   6,7 ~-~-~-~S(TÂ)
LASÁ      7  ~-~-~-~S(Ä)
LASCOM    5  ~0~8~9~T
LASENOT   5  ~6~4~7~T
LATEVU    7  ~-~-~-~S(TÂ)

LEXUS   7   ~-~-~-~ S(TN)
LEXUSÂ  7   ~-~-~-~ S(Â)
LIFAC - 4   ~7~5~2~ S(M)
LOSOW  5,7 ~4~4-8~ T(N)
OPOT    6   ~-~-~7~ S(P)
PASOT   7   ~5-1-9~ T(T)
PIET  - 6   ~-~-~-~ S(Au)
POP     3   ~4~6~7~ T
PRODIS  3   ~5-5-27 T
PUA   - 6   -~-~9~  S(GOÂ)
SÂ      7    ~-~-~  S(NÂ)
SOF(Subject Object Future)
       ±3   ~-~ -~  M(GON)
SUCTA   7   ~3~-~9~ M(TRÂ)

SUTE    3   ~-~0~-~ S (T)
SUTEN ± 4,3~-~0~-~ S (T)
TRIT ELTICMA ATTAINMENT
        7   ~-~-~-~ M (TEA)
TRUCI  1,5 ~3~6~7~ M (ENÂ)
UVICTA  7   ~-~-~-~ M (TEÂ)
VISO   8,7 ~~~~--- M (GÄP)
```

For NEC identiifiers, see pg. 521
Classification of types of Laws.

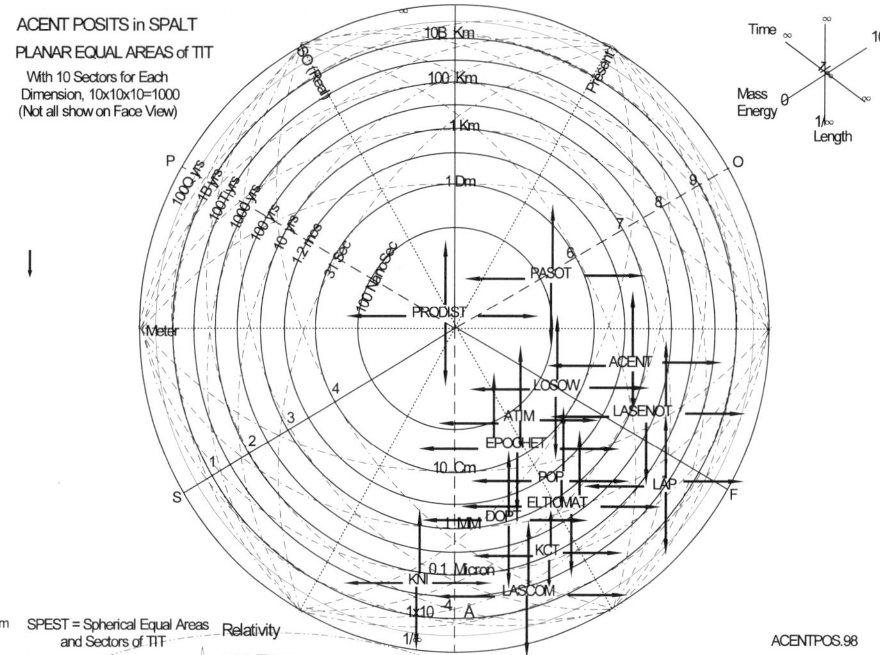

Fig 9-98 **ACENT POSITS in SPALT**

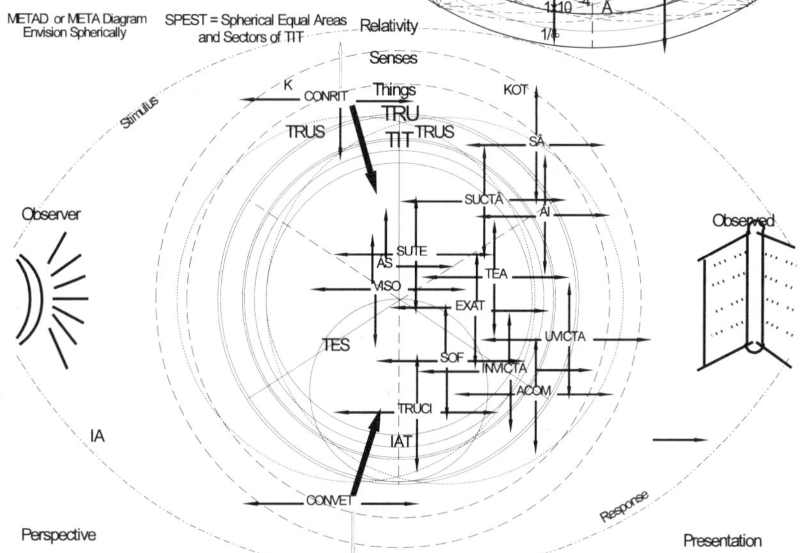

Fig 10-98 **ACENT POSITS in METAD**

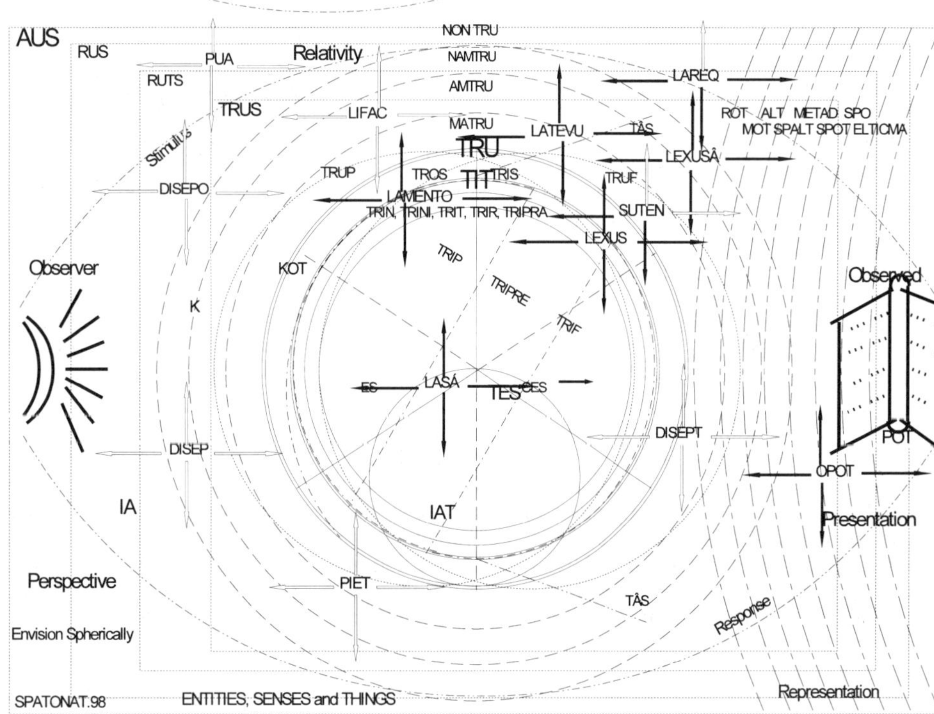

Fig 11-98
ENHANCEMENT of ACC POSITS in SPAT

CHAPTER 99

SUMMATION and CONCLUSION of ACC, TRU, TIT, POT, and IPTRU

TÂ, MUSTS, USES, PROMOTION, CONTINUATION, JUDGEMENT, OPTIMISM, COMPLETION, PERFECTION

The will to prepare to win is much superior than the will to win, and the will to win is much superior than all lesser wills. So that the most superior of all wills is the will to prepare to win, TRU. **WIPAT** = Will to Prepare to Win TRU. With the development and determination of TRU, ROT, and POT, as with all things one does, it is in the proper, effective, successful preparation to achieve them that solutions and results are best produced. Wars and battles are won, not on the day of the battle or time of the war, but in all that has gone to prepare for it. For this reason, deliberate, detailed, precise, proportional, priority ordered and effective preparation are necessary. Since TRU is not only so great, but poorly understood, our preparation to win TRU is absolutely necessary. It is by the aegis of a collective and determined WIPAT that all must come together in the crusade for TRU.

Many bright young minds are led, directed toward, inspired by, and ready to adopt whatever aims, purposes, ideals, goals, and ends the current rage, fad, fashion and sophisticated mood produce. They come to feel this ideal is the answer to the riddle of life, the key to all knowledge, the answer to all things, the Excaliber, Eldorado, Phoenix or Eureka. Yet this is only a symptom and sign of their weakness, deficiency, and failure. This deficiency is of the knowledge of TRU. It is necessary that all persons attain a sufficient knowledge of TREQ and TRU. Their observer observed situation is no less and none other than the relativity that is TRU. To the degree this knowledge is TRU is the degree each person can achieve a similar degree of TES. Thereby all people, especially those bright young minds that are so wasted, must come to consider and realize those ends that are actually right or wrong, great and trivial, TRU and not TRU. They must learn and know much that is unknown. They can not worship figments and illusions or believe in that which is other, less than or against them and TRU. Let them once and for all put to rest all mis-beliefs, opinions, concepts, principles, and laws that lack VAPO and TRU. Let them avoid or correct all fads, fashions, creeds, sects and beliefs that are negative to TRU. Let them come to know each for what it is in TRU. They must come to interpret each in their proper place and function in TRU, SU, ALT, HOM, PUREALITY, COGNOS, ELTICMA and TA. The battle lines are drawn. Only those who fight with TRU, ACC, COGNOS, ELTICMA in full pursuit of SÂ and TÂ are sure to survive, be TES and have immortality in TIT.

Fig 1-99 **Thetic Stages and Levels, LEXAT and TÂ**

In order to understand OITIO or what is more than in TIT exclusively, we need to know the origins and development of vantages and their resultant overview and total action. This we began in Chapter 11. We also need to know what sets up the interactions that occur. In this last chapter, 99 we seek to bring the thing in itself, TIT, back into perspective with the presentation that we began. We also seek to understand how the interactions that are outside of TRU become a part of TIT and TÂ. This is what happens in relativity when the observer and observed focus on and fuse in TIT by THES and THEO in Á. Predominance and haecceity are the prime reversible essences of TIT most uniquely revealed in its mass energy Á FAF and dimension. Á is not any entity nor the same as the mass dimension of the physical world. Nor is Á only the subject and object, or observer and observed of relativity. It is only as it is in TIT. Much of things, what we sense, our subjectivity, objectivity, and relativity are not TRU, thereby is not of Á. This is called:

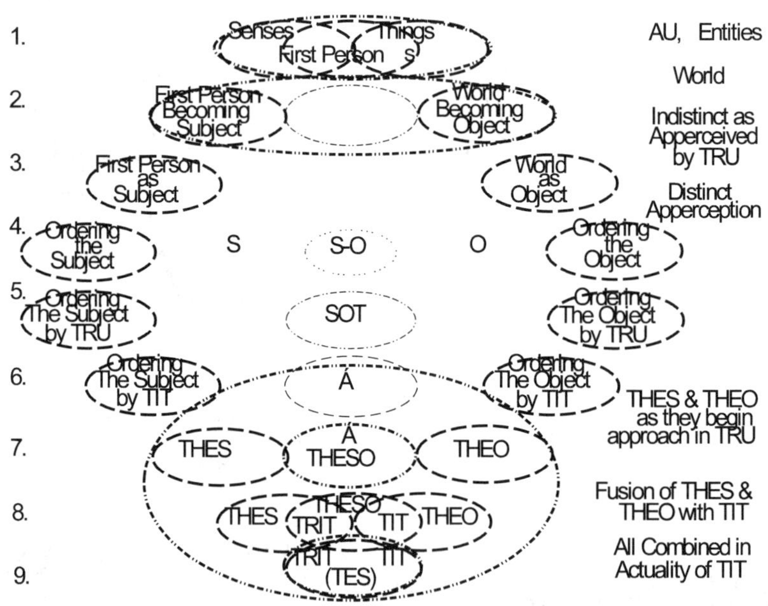

LEXÁT = Law of Exclusiveness of Á in TIT. It is the beginning of our journey back out of TIT, leaving the state of LEXAT that this chapter focuses upon. To return to OITIO we reverse the Figure 7-11 and embark on a course that seeks to provide an understanding and representation of what is, in a way, greater than TÂ. Their fusion, or how THES and THEO as THESO, make up the Á dimension and FAF of TIT and TRIT and their identity and completion in TRU fulfills this progression. These stages or levels can be expanded or contracted. They are driven by TRUD and exemplified in animal behavior. By stimulus and response, questions and answers, problems and solutions, errors, paradoxes and others their accounting and resolutions are the determinants of relativity and the form and function of the Á dimension. In order to summarize the text and any POT we must turn this inside out and reverse it so that we can again see how TRU is dynamic, both internally and externally. It is in this grey zone of less than 100% TIT, or AMTRU, MATRU, NAMTRU and all other than TRU that we find and can demonstrate OITIO and finally understand all as they are both in TIT and in and of TRU, or out of TRU by contrast.

TÂ may be best conceived and its errors corrected by contrasting it with ATO. DOT by ATO like its finality already exists within TIT. This is an absolute condition of TRIA. So that origins as well as ends can, when perfect, only be within TIT. This holds absolutely by TRIA and TIT, but only relatively by TRU and TRUS. So that for TÂ we have both, in reverse order, a relative and an absolute form. Both of these, however are of RÂ and not ÂÂ. TÂ is not a static absolute nor end, it is intricately bound to TIT like ATO. Thus, we have:

TATOR = TÂ ATO Reversal of Polarity. TATOR is not unlimited, for neither equal TIT nor TRU. TÂ has FAFs that are other than in the polarity. This is its role in TRUS/TRIS, and why ACC in the long term, as well as TIT and TRU demand such great attention, action, management and control.

If TRU is all selective, great and important, and if the learning and demonstration of it are the only effective and reliable ways we can achieve it, then each and all persons must learn, show, serve and live TRU, as TRU will do the same for them. This is a primary reason for IPTRU. It explains why IPTRU can not be presented in any way and why it must be so vigorous, intricate, deliberate, detailed, explicit, reliable and ordered. It also explains why it is so necessary that every person achieve the inspiration, revelation, conviction and devotion necessary to appreciate and become TES and TRU. For in the great struggle of life, there is pitted against one or another, being and non being, existence or extinction, and thisness or thatness, in which each and all are participants, like it or not. The degree to which they are/are not, and are on the right/wrong side in this struggle and war, is the degree to which they discover, learn, know, remember, use, serve and live TRU. By understanding the reasons and actions necessary, the end result can be more readily achieved, TRU unto TÂ.

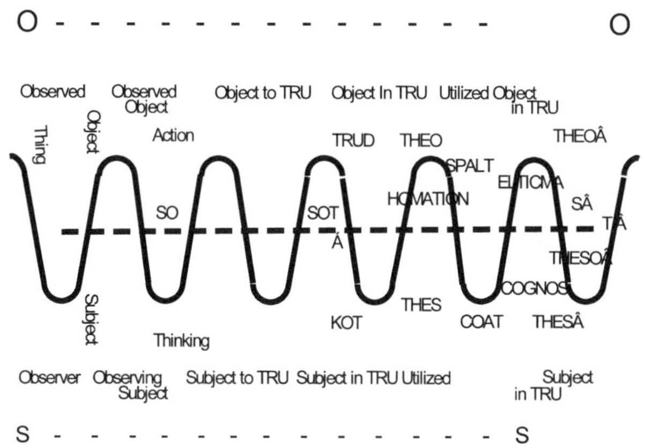

Fig 2-99 **COMPLETION of SUBJECT OBJECT and RELATIVITY in TRU, IPTRU, SÂ and TÂ**

As we reach the end of this text we have come full circle from our initial first person entry or attention to this book and its topic, to where we must go our way as an individual in life. It is one whose IA best adapts to circumstances as demanded of us. Life becomes ever better with the guidance of knowledge of TRU. May we not only repeatedly return to it, become ever more familiar with it, envision its beauty, greatness, and exquisite detail in the coordinated totality of TIT. May we increasingly bring the text and ourselves together with TRU as one. That all may assist and stimulate each other in our mutually perfected adaptation and achievement in TIT and TAS. This format reader and book are common for all of TRU and the subject object relativity in life. This occurs when we become more and more together, by living together, not so much in physical proximity, but in the proximity of TRU, in PUREALITY and one with COGNOS forever be. **COTANOS** = Coming Together or Cooperative COGNOS, and the Absolute Necessity of Absolute Objectivity and Subjectivity, as Ordered, Systematized and Standardized In and Of TRU.

It is mostly up to each person to enhance and assure TRU and, thereby, themselves. Collective mentality and intellect, as great as it may be, is not as concrete, easily controlled, nor does it have the imagination, enthusiasm, interests, life and qualities of survival and being that the individual has. Therefore to attain PUREALITY and one with COGNOS forever be, one must be directed to the absolute, to THESÂ, THEOÂ, THESOÂ, SÂ, and TÂ as they best represent TIT, TRIT, TRIF, TRID, TA, TRA, in totality. What is mostly of the inner core of our lives, of TES, TRU, ACC, TREX, TRIX, is their focus, convergence, and concentration in PUREALITY and COGNOS with THESA, THEOA, THESOÂ exponentially increasing and eternally gaining TÂ. This is continuously and always modified by the circumstances, states and conditions of the first person, observer, given or case at hand. Because of the antithetical relations between observer and observed, subject and object, first person and TRU, we can never have a perfectly, simple, continuous, and eternal identity. It is in a constant dialectical flux in which we can only hold as close as possible to TRU and TÂ identity. This we help to do with standardization, SPALT, ACC, PPMA, and controls to retain optimal relativity through CONVET and CONRIT as all converge on TIT. This is the beauty, glory and victory of our increasing mentality, knowledge, intellect, PRET, ALT, COGNOS, and ELTICMA. It provides the tools, great means, weapons and mechanisms whereby this identity can approach the necessary and perfected 1:1 union of a person, TES and TRU.

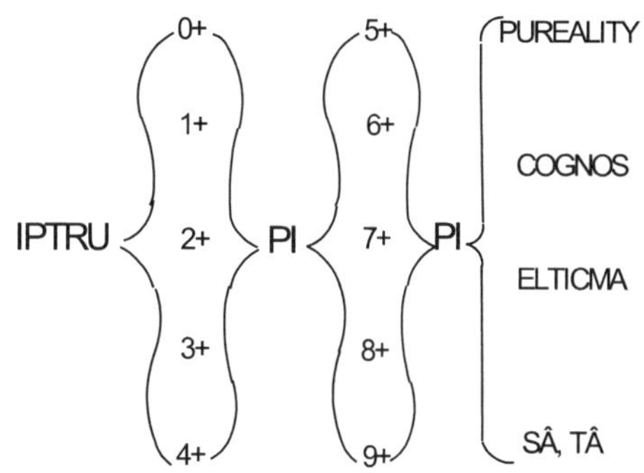

Fig 3-99 **POSITIVE IMPERATIVE in IPTRU**

Much of the beauty of PI is how it shows TRU in a balanced way, in which sections correspond as seen in HIDIC. This is how this last section and last chapter of IPTRU are the opposite faces of the first section and first chapter, and how they are not ENS, but aspects of the observed TRU its observation and production. This is by the relativity, or Á of TRU. Combinations of great new concepts not only help to clarify each, they show TRU, ROT and POT. This is, **PI** = Positive Imperative. TRU is shown in the first five sections of IPTRU, PI and its extensions through HOM, ACC, TRAX, TREX, TRIX, and are perfected in return to TRU in PUREALITY, COGNOS, ELTICMA, and SÂ in the last five sections.. These combinations also help discover missing or void ideas, concepts, principles and portions of TRU. There should be a TRU concept, either adapted existing ones or new for each great void that is not a part of, or deficient in, language. What the 10, 100, 1,000 or more greatest voids are we need to enumerate and perfect as we develop the Glossary and ESSARY of IPTRU. They may be rated or valued in terms of their greatness or REM. They should include, also, all great existing concepts if their totality is to have meaning, and be best presented.

The existence of many different AUS, RUS, RUTS, TRUS, TROS and TRIS was enunciated by Leibniz in his doctrine of many possible worlds. From these he separated "the best possible world" and showed, as we do in TREQ, TRU, TRIT, TIT, TRID and TÂ that philosophy, religion, general science, and TRU are not an endless number of vicious circles, but a very simple, unified and great Good. This is shown by TIT, TRIA, ENT and ACC. It can be seen in how TRUS moves through TROS, TRIS, TRIT, TIT to TRIF, TRID, TÂS and TÂ. TÂ from TRU and TIT is how COGNOS comes to exist, grow and accelerate. It is how by ENT we achieve the best possible worlds and how even these in time, will become even better, as even illusive potentials can become actual.

The good and the truth, in practice and performance, become righteousness and honesty. Their convergence on, and existence in TRU, confirms the focus and thisness of TRU. As CALP they reveal the mechanism by which the mass energy continuum of TRU, Á, is formed and functions. Yet none of these mean much unless and to the extent they are VAPO in TRU. Great persons, as great empires, nations, and cultures are founded and flourish to the extent they are based upon the good and the truth, righteousness and honesty. This is most clearly seen in the greatest empire man had developed before 500 b.c., the Persian Empire. Their two greatest tenets were the rule of good over evil, and honesty over deceit. Yet nations, as persons, quickly lose their ideals and deteriorate in the massive and stressful crucible of life. This is especially so when not founded upon, and adherent to God, TREQ, VAPO, TRU, TIT-TO, TÂS, and all that ENT and ACC can provide. An example of the great need for Laws of TRU are the losses, errors, and horrors that result from existing laws, regulations, deregulations, and their repeated alteration The relationship between benefit to harm of laws is expressed by: **RABNAL** = Ratio of Benefit to Harm of Laws. RABNAL denotes and defines the great determinant of laws; how much are they positive or negative, help or harm TRU, in any and all ways. It helps determine not only when a law is justified, it helps to determine when, to what extent, and in what way laws do or do not need to be changed. Since laws are such a large power in society and the success or failure of TRU, it is always important to remember, ECRO, and be able to evaluate them astutely by TRU. A vigorous critique of laws is justified by their convergence in the highest levels of Á FAF in TRU.

The measure, justification, and test of authority are the degree to which, 1. It is founded on TRU, 2. TRU is fully and completely developed, 3. TRU is followed by those who have authority i.e. those of leadership and positions of responsibility, 4. The standards set for and from TRU are optimal, 5. These optimal standards are periodically, or as needed, reviewed, changed, refined, or corrected, 6. Willingness of those who follow, do the work to effect TRU. Standardizations are adjusted, accounted for, demonstrated, and resolved in Accentuation. The resolutions of formalities are continuously unfolding processes. Thus: **RESTAD** = Resolution of Standardizations. **MERP(S) (MIRRS)** = Multiple Emanation Rapid Proper Solution or Resolution.(SUIAC,TREQ) The position of Accentuation in the dimensions of man's being, TRU, are shown in:

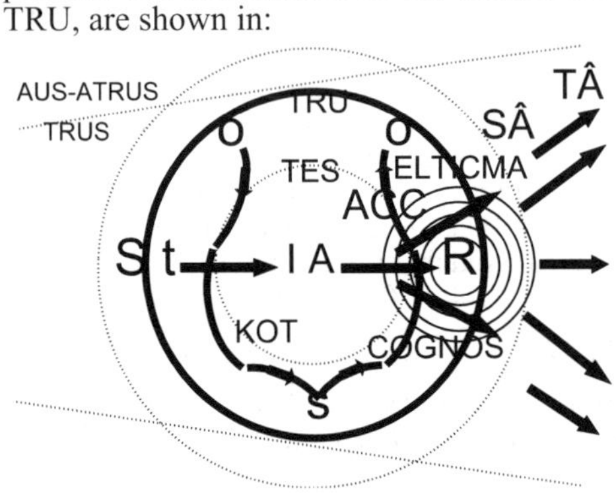

Fig 4-99 **ACC and COLLECTIVE RESPONSES of TRU**

Accentuation is a basic response to stimuli in an accelerative and self-propelling form. It is natural for a RU, especially when focused on TRU, TROS, TIT, TÄ TÁ,, TÂ, TRIT, ROT, POT and IPTRU. The concentration of this response is what enables TRU and is shown as R and ACC. As the saying goes, 'without God nothing'. With determinism, as shown by the Covering Law Theory, and as we increasingly have a better and greater frame of reference we have: **ACQUENT** or Accentuations Equality, Equivalence, or Correspondence with ENT. We must to a degree account for: **ETA** or the Equilibrium of TRU and ATRU-AU interactions. Yet this depends upon: **ETACT** or ETA in Contracting Time. Also, it depends upon **TAR**, or the Transcendent Resolution. In the following we show accentuation by: **LIDE** the Law of Equal Demonstration Accentuation of TRUS, as well as their limits.(ROSOC) Yet we must hold to TRU, TIT, and TRIT as they move toward TÂ. This is shown by: **AIME** or Absolute Imperative Expansion. AIME signifies where and how we are compelled by TRU, as any RU, and as so readily concomitant with Accentuation, to expand imperatively to the absolute.

It is in discovery, and all that enables discovery, that humans have improved, advanced and expanded their lives. It is how they have been able to find and show those great voids, and especially how they exist without and within KOT and TRU. How much is unknown, invalid, disproportional, inadequate, inactual, lost, and unused. Without POA, TRIT, ACEXIN, ROT, MOT, ALT, SPALT, SPAT, POS, HOM, and ACC we are lost, confused, and never gain either clear differentiation nor comprehensive integration of our world. There is the greatest shock in the contrast between the vague worlds we commonly allude to, and the solid geometrical world we can produce. The latter is what can enable us to have a place for everything, and everything in its place. The discovery and use of voids need not start from scratch, it can continually build on what is known, pure, real, actual, and usable.

What are the greatest key fundamental projects in TRU that are incomplete, needing work, or should be perfected. These should be collected, enumerated, added to, perfected, completed and remembered including: 1. **LAMPREC** = Laws and Methods of Reduction and Classification of the basic Problems in Philosophy, Religion, knowledge, fields and sciences that are important in TRU. 2. **TIST** = TRU proves ISMS and ISMS prove TRU. The differences and similarities between LAMPREC and TIST show each other and TRU. 3. Contrasting projects in IPTRU 4. Contrasting projects following IPTRU. The work that needs to be done upon completion of IPTRU will only be the beginning of a vast, massive, powerful and highly beneficial expansion. **TRUFEX** = TRU Fundamental Expansion and Solutions to all Problems. This is needed for all TIT, TRU and TREQ. It can be readily planned and accomplished if properly set up, acted upon, managed and controlled. We need to enumerate the 25-100 greatest avenues through which this can be done: EC whose utilization or existence in IPTRU is not fully utilized, EC whose TRU meanings, definitions and usage must be revealed, new ideas of concepts, principles, laws and mechanisms which have not been discovered or used, voids which have been inadequately filled in SPALT and IPTRU, segments of TRU for which processes are poorly developed or integrated, specialties whose great information, concepts, ideas, principles, laws, formulations, and methods need to be incorporated into IPTRU, and data, illustrations and models that can be superimposed onto SPALT, METAD, and SPALT, to give added, yet ordered, vistas of TRU.

A Programmed Course presents a topic, field, science, subject and text in a way that not only stimulates the reader to think for themselves but to better understand the key concepts, mechanisms and processes involved. It also organizes the topic and text in a way that can best systematically unify and represent it. This is aided by appropriate diagrams and descriptions. These can be adapted to, and superimposed on SPALT, METAD and SPAT to bring TRU concepts and mechanisms, ES and CES into more meaningful forms in TRU. Medical Homation is the application of the principles of TRU and Homation to medical science. This may be done in two ways: 1. individually, whereby the principles and directions for acquiring, maintaining and preserving ones health utilize all knowledge of TRU that helps one to be built up, come to terms with life, and be TES, 2. for practicing specialists and generalists in all health services. It should emphasize both the best of medical science and the best of TRU, leaving out nothing of importance. It needs to provide knowledge of TRU health, with knowledge of ANTRU - as disease - which can be best brought back to health for TES and TRU. It can be explained by DAT, TRUS, TRUCH, TRIS, and TAS. This is where as things are from TRU or to TRU, they alter TRU and as things are bad, wrong, diseased, injurious, they alter TRU in TRUCH. The rate of this is called: **RATCH** = or Rate of TRU

Change. Similar studies for every field and science that use all that is applicable of TRU as its base, can greatly clarify and enhance them, and in return TRU.

TRUCORP is the agent for future collective business applications of TRU. To make it legal, help to expedite all sorts of uses, and to benefit from the advantages, power, and popularity of entrepreneurship in the corporate world, a formal unit should be established. These include, 1. All from IPTRU, BIPTRU, CIPTRU, copyrights, and patents, 2. All upcoming ideas from TRU and their uses, 3. All products that are directly derived from IPTRU and its representations and models, 4. All products that can be secondarily derived from IPTRU and ROT, TRU, and POT, 5. All products that can be derived from TRU and TREQ, 6. All other diversified products, 7. The promotion of TRU that is well planned and guided concomitant with its base each step of the way. The promotion of TRU should pay particular attention to individual and group aids that will best help their interest and dedication to TES and TRU. This will include an increasing number of those already versed in fields and sciences of TREQ, as specialists in fields that contribute the most to TRU, authorities for each chapter of IPTRU, interdisciplinary authorities who would act as alternates or assure maximal continuity in TRU, operational, advisory and legal staff to best manage and oversee TRU as a corporate, commercial, political or social entity in the existing world. At whatever level the work for TRU exists at any time, the great emphasis for one and all is to enhance their TES and TRU, and to exude this love and energy in ways all share and find fulfillment.

In order to produce an ever better POT there will need to be terms and meanings whose formal distinctions add the most and best to complete TRU. Words need to be shown both with their common meanings, are precisely defined, formally distinct, have more specialized meanings, or have meanings that greatly tend to or are TRU. This should include showing the various meanings of words, and emphasizing what is ENIT, ESINS, ES and TRU. Also we need to show all voids in TRU, and the terms that do not exist or will have to be provided in TRU Terminology. It is in this way we can bring all together for the best possible ROT and POT of TRU. This is called: **PITRU** = Perfected Presentation, Initial and Incremental, of TRU. PITRU is achieved when terms, words, meanings, ESINS, and new concepts are gathered together to provide the greatest, most detailed, distinct and fullest meaning to all that is of TRU. It can be cross checked by SPALT POS.

Nowhere are there, more profound and heartwarming treasures and means to be found than in TRU. These include the reality and joy of actually being alive or TES, the deepest and best feelings of love and fulfillment or identity with the actual world, TRU, the joy and delight in finding what is missing or what has been lost in life of TIT, the cue to the meaning of life or Á, the sudden finding, realization, and revelation of deepest conceptions and principles, the jist of TRU, the youthful vigor of accelerating power of imagination and inspiration, the successful satisfaction and resolution of the most profound questions and problems, the power that a person can derive from a knowledge of TRU and its best use, the adventurous discovery of unknowns whose findings produce ever greater benefits for all, especially all who are and will come after us in TRU, the infinite and eternal existence attained through better, greater and more victorious contributions to the whole, how to come to terms with life, and how to recognize non being, anti-being, all that is bad, evil, sinful, injurious, detracting, base, distracting, and disillusioning that can be ever more fully and easily ignored or reduced from their negative effects on human life. Life is like a stairs each person must climb a step at a time to become TES or ones counterpart in TRU. This is similar to the stations of the cross, being an imitation of Christ, and the order of our duty and responsibility to GOD or TRU. This is not easy, but its success/failure is vital, by its own nature, that we avoid oblivion, and be, TES in TRU.

SACT = Supreme Sacrifice of Self to TRU. Wisdom can be a refuge as well as a means to an end. When it fails, there is little success and much failure. In a life and death struggle with insurmountable odds and adversaries SACT is resorted to. This is seen in the final days of Socrates, Christ, many saints, martyrs and others in history. It is the supreme sacrifice a person can make in ascension to TES and TRU. Their reason and conviction in what is best in life over what is merely physical death. This is not an ascension of the physical body to heaven, but none the less actual. It is one of the actual reality of the soul in the fusion of body and mind in TES. It is the actual power of spirituality, that of the predominant Á FAF of TRU. It is the convergence and identity of the individual on TES through the combined levels of the Á, L and T continua in TIT. It is how this individual's TES, becomes disseminated at and through life into the whole of TRU. **LINDALT** = Individuals Dissemination through TES, Death and all Life, into the ALT totality of TRU. There are many forms of SACT. The most dramatic and commonly known is that of sacrificing ones life in a highly visible way for a great cause, which can only be TRU, or some key aspect of TRU. The highest form of SACT is called: **ITÂ** = I Absolute to TRU Absolute. ITA signifies how a person may die for TRU or TÂ. When all else fails, SACT is increasingly effective as ITA. Yet this is not any death, nor one without great control and direction. If not to TRU it is wasted, a great loss to TRU and absolute loss to the individual. **CONTID** = Consolation In Times of Despair. TRU CONTID can be achieved by sacrifice, as in sacrificing ones life, yet this is less common than the compensation one manages by accounting for, and accentuating all that benefits TES and TRU over the mid and long term. **PETOD** = Prepare to Die. PETOD is actually a state for which one prepares throughout a lifetime. As PETOD increases, the causes of despair, and thereby need for CONTID, decrease. Only in extreme cases, is the ultimate acute sacrifice valid, exemplified by the martyrs and saints of the early Christian Church. It can be positive if it is the elimination of negative. More negatively it is seen in individual suicide. In this sacrifice can become a threadbare imitation of a falsely assumed rite. This is the result of complete annihilation of TES by an individuality bent on destruction. It is the antithesis of life, GOD, TREQ and TRU. Martyrdom is valid only if it produces far greater SACT and TRU than ones possible achievement in a lifetime. Either of these choices are far greater than those who fail to realize it and live in ATRU. It is when ATRU increases and TRU decreases that we are suddenly and markedly faced and forced to make the choice. The greatest form of SACT, however, is the life long struggle and endeavor to seek TES and TRU, thereby improving the quality of human life. It is the ever greater and more perfect knowledge and activities of life that enhance TRU. Thereby this greatest form of SACT is essentially the optimal state of sacrificing all individuality for:

SELF → TES → TIT → TÂ

Instead of a world population that is roughly 1% TRU, 5% MATRU, 15% AMTRU, 30% NAMTRU, 30% ATRU, 15% ANTRU, and 4% ANTITRU it is possible to envision one that could be largely TES and TRU. In a highly competitive world, survival of the fittest, MOV, narrow margin of selection, and predominance of all less than TRU will be sooner or later eliminated. With great discovery, learning, production, and effort we can in time, depending upon the brilliance of our research and common effort for TES, approach 100% TRU. The immediate goal of ACC is to strive to attain this necessary level of TRU in all individuals, groups, and peoples on earth. This is to solidify the quality of all in their ES and TES in TRU and thereby become TÂ. **GACSAL** = Goal of Accentuating and Solidifying the Quality of All to TÂ. GACSAL can be achieved. Much of the world desires, or at least recognizes, the need for its attainment. Quality individuals, groups and all require an across the board approach, as demonstrated by POA and IPTRU. It especially requires genetic quality plus the learning and use of TRU or TREQ. The concentration on quality by TRU has been continuous and is largely what sustains it. Quality, as Á, exists in many different forms as shown. This quality is not simple for TIT, but mixed in numerous ways with AMTRU. Herein is PUREALITY or how we separate this mixture by attacking with the ways that produce quality and TES. For many reasons it is not only especially vital and crucial, it is neglected. It is for this reason, Section IX or ACC, has been extracted from others for special emphasis. We can not raise ourselves from a world population filled with a large percentage less than TRU to one that is much higher without great work, difficulty, reaction, and personal or group loss.

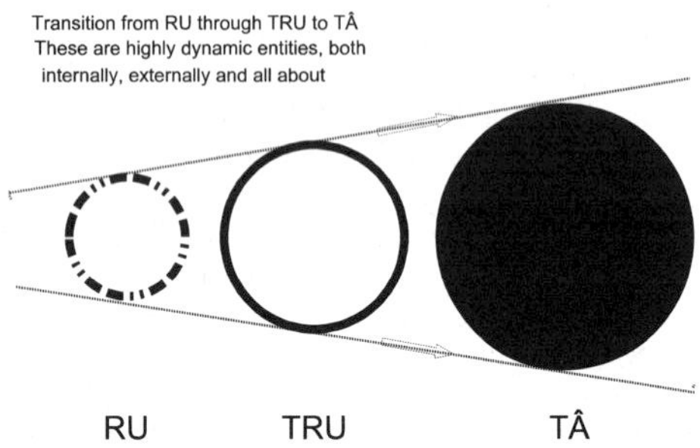

Fig 5-99 **SYMBOLS and LOGOS from ENS to T Á** Comparative entities and universes and their extension from RU to TÂ are shown. How there is for each type of thing, universe, or ENIT, an AU, RU, TRIS, TRUS, TRU, TIT, TRIS, and TÂS. It also reveals the need for dimensions, and how additional illustrations may be superimposed as more gradients of their states are demonstrated. What we first encountered in the Introductory Section has progressed thru the text and ties into TÂ.

To know exactly what you are doing before hand or ahead or time, to anticipate and conduct in advance all needed, are the time-honored methods of the performer, MC, musician, writer, professional, highly successful artist, computer programmer,and skilled tradesman of many occupations and technologies. It is from lack of the necessary valid, total, actual, proportionate, accurate, ordered, well represented and presented or VAPO knowledge of TRU that so much of life fails. It should displace much of the Youthful restlessness Desire for change, **YAD,** which without KOT and TRU or TREQ leads to anarchy. There is an abundance of opportunity for expending youthful enthusiasm, delight in their new found powers, natural talents and energy that are essential for the continual regeneration, happiness, glory, success, and victory of one and all. It is in harnessing, controlling and directing these powers and energies that KOT and TRU provide the impetus that can raise the bored from being deviates to achieving TES and TRU.

Ordinarily in our NAMTRU world we tend to look at the future vaguely or specifically, by what will happen at a certain time. We rarely take a good overall look at it. This is partly because it is unknown, but more likely a result of what has misled us and our inability to work it out. Our knowledge and the terms we use are not sufficient to provide a good overview. This is a major reason for a new terminology of the actual universe. Its concepts, principles, laws, and mechanisms must be discovered, learned, known and applied if we are to have what it takes to provide a good overview of TIT-TO. The future is neither unknown nor negligible. In fact it is of the greatest importance that it be both well known and greatly elevated and celebrated. When most of its greatest concepts, principles, laws, and mechanisms are known, their proofs made available, and a model as SPALT supplied that provides a good representation of TIT the future is knowable and can be used to our greatest benefit.

There are great forces and powers operating and interacting in, out, and in-out of TRU. These are both indirect and direct, advertent and inadvertent. They may be positive, negative or variably positive and negative for TRU. They may be large, medium or small, past present or future, or of different degrees of SÁ. The subjective forces may be on an individual, an individual's activities, or on larger, smaller, or other dynamic functions in or out of TRU. The most general case is of S itself, subjectivity. This is of a great variety of all, OITIO. From this we may produce many kinds of divisions depending on what is being subjectivized as well as the kind of subjective. One of these, that is most direct, advertent, positive, large, future and highly subjective is called: **WABEW** = We Are Being Watched. WABEW may be taken most generally to most specifically. When most specific it may be viewed as a single observer in the distant future observing our every activity. This is the extreme case, however, when more proximate generally for TRU it is called: **OTBEW** = All TRU is Being Watched, or for TES, **TESBEW** = TES Being Watched. TESBEW is important for the concentrated compelling power it has and produces in developing an active TES. For the incentive to be TRU is much less if we feel no compulsion. In being noticed, watched, and shown to be on trial for our lives, especially when this becomes a highly compelling, complimentary or inspiring force it yields a much higher incentive to TES and TRU.

	MOST GENERAL	GENERAL	MEDIUM	SPECIFIC	MOST SPECIFIC
DIRECT	LASCOM	LOTRAC	SCONQ	FOKIT	CONQ
INDIRECT	OM	OITIO	KUPS	LINT	LOMI
VOLUNTARY	LOSOW	POJUD	WIPAT	PERJUD	WIM
INVOLUNTARY	COS	SCOPT	PIT	SCIN	VISO
OBSERVER-S	COGNOS	LEMOV	OTBEW	TESBEW	SIJ
SOLITARY "	THESÁ	COGIF	COGNOSID	THES	SOSTES
OBSERVED-S	ELTICMÁT	ELTIMA	SOAP	ELTIR	ELTIA
SOLITARY "	THESOÁ	SIO	TRE	THESO	WABEW distant F

Table 1-99 **LEVELS-KINDS of SUBJECTIVE FORCES**

Table 2 provides a more differentiated view, with one specific concept in each, of the whole of subjective forces. The concepts have at least some correlation in the category. It may be used as an exercise in bringing NODA to COGNOS. The different kinds of direct and indirect forces in S, WABEW, OTBEW and TESBEW help us understand TRUD, HOM, and ACC. When direct they are often simple, singular, easily traced and shown. They may be indirect, whereby and although we are not being seen, envisioned directly, all that we are and do is being, in summation, felt and observed by our progeny, TRU progeny to the infinite, forever. For where they succeed rather than fail depends upon what we provide for them. They will be our best judge through PIET. Just as we are of those who have come before us. The better we can separate what TRU is the better we can distinguish this makeup to represent and reproduce it. This is key to language and all representation, and thereby presentation. It helps make formally distinct what is so abstract, dynamic and immense. They help to show how much different TRU is than what it seems to be. This helps us get over our mis-beliefs and errors in understanding the world and ourselves. It does so, in part by showing how what is TRU is often what we commonly know, yet confused because of lack of VAPO and order. It helps us recognize TRU, uphold, give it the honor it deserves, and the values all its parts require.

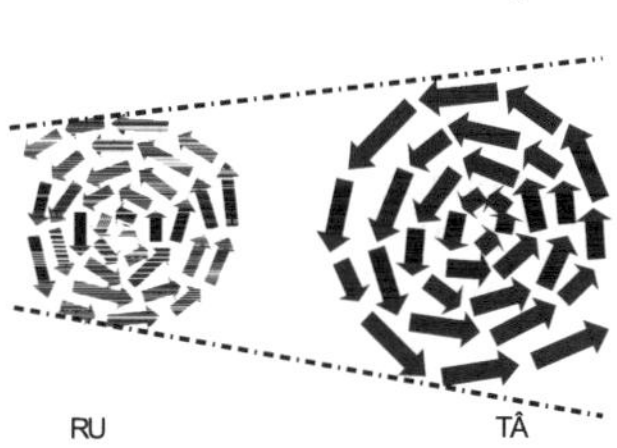

Fig 6-99 ACCELERATIVE INFINITE EFFORTS to TÂ

Figure 5-99 depicts how TÁ becomes infinitely large as our efforts are incrementally and infinitely accelerated and directed to produced a pure TRU and TÂS. This when what is 0% pure, ATRU, ANTRU, and ANTITRU, are reduced to zero.(COMT) We then show how they are derived, developed and determined.

For TRU, judgement is not to be made only once in a lifetime, as at death. Nor in more frequent yet erratic ways by confession or cleansing of the soul. Judgement must be perpetual, constantly repeated, on the highest levels of VAPO in TES and TRU. **PERJUD** = Perpetual Judgement. PERJUD is where a complete and balanced, entirely TRU and TES attention, evaluation, judgement, recognition and reward are made for those whose work and lives have been to that degree TRU. **POJUD** = Principles and Processes of Judgement Universal Derivation, Determination and evaluation. POJUD signifies the necessity and mechanisms of judgements. as a judgement day, for each person in TRU. Humans of all ages should be periodically judged for their contribution to TES and TRU. This should come primarily from an ever better TES within the individual. It should be partly in proportion to their capacity but mostly by performance and contribution to TRU. It should be done with as much voluntary, positive and joyful participation as possible. It should also account for the waste and negative of TRU. Its test should be highly accurate and held to TRU impartially without deviation. What works continuously, and through INEG should be included in the accounting. Those who are TES have little need to fear, TRU or even death, for they already have immortality. What they need to primarily fear is all that is not TRU, and especially themselves as individuals that are ATRU. It is through POJUD and related IATES, RITES, optimal conscious life, conscientiousness and good influences in TES that TRU is best served. **OPJURIT** = Optimal Processes of Judgement Reduction Imperative to TRU. Judgement is an ongoing and continuous process in TRU, the selective highest levels of mass energy. TRU must be highly selective as well as optimal. So that we should always remember, ECRO, that all we think and do is in a way POJUD and OPJURIT, if it has the best possible effect in enhancing all persons to be TES. It is an optimally positive continuous give and take of judgement by, of and for TRU.

Each person has received the greatest of all gifts and benefits; the gift of life and all the benefits it delivers. This is seen in the total of its mass energy FAF, Á. What we make of this gift, whether we return the favor, and how we use these benefits is largely up to us. The wiser we are, the more we learn and know TRU, or TREQ, and the better we comply and apply ourselves to our roles in TRU, the better this gift, and the use of its benefits, become. This is what opens wide the door to the absolute. For the absolute is The relative Absolute of TRU. This is its Eternal EXpansion, **TÂSEX**, improvement, and continuity. Since the person is the focus of this universe, and since it depends largely upon the contribution we make, it is in our identity with it that we find immortality and at the same time increase the absolute. This immortality and its associated absolute, or TES, TESÂ, TÂSEX, and TÂS, is neither hopeless nor some illusion. In fact they are equivalent to the opposite, perfect hope and actual realization. It is even more real, actual, intense and extensive than life itself, of which it largely consists. Our consciousness, self consciousness, and ego, are small in comparison. The fact that identity and immortality do not provide massive or immediately tangible rewards only points out the immensity, ACEXIN and FAF of TRU and the paucity of our minds. It helps us recognize DAK or lack of TES. This we can't deny nor can it be defied. For to do so is simply a denial and defiance of life itself, life as it is most perfectly TRU. It is, thereby, in the discovery, revelation, inspiration, learning, knowing, comprehension, accounting, use, application, contribution, and glory we achieve, in becoming and fulfilling TES and TRU that identity, immortality and the absolute that we seek is secure. The deep personal recognition of ones IAT, identity, immortality, and future extension to the absolute, TÂ, is called: **COGÍF** = The Cognitive Identity and Contribution that I can, do and will have and make with Immortality, the Future of TRU and its eternal continuity, the Absolute, RÂ, TÂS. COGIF is the ultimate source of highest Á FAF realization, TES and TRU. It is the most vivid and powerful ACEXIN, PUREALITY and COGNOS that a person can collect and bring to life. Some of its great power is none other than the cognitive realization of itself, how I can and will be TES and TRU, and these both consciously and unconsciously, directly and indirectly, simultaneously and increasingly contribute to TRU, TRUF and TÂS. This is my most actual, subtle and profound immortality as it melds with, opens wide the door to, and becomes the absolute.(RÂ) The degree to which one achieves the revelation of COGÍF, finds its inspiration, makes the needed sacrifices, and realizes it, as ones own glory, is the degree to which each reader, student and person who uses IPTRU fulfills their dream, hope of the future and the greatest happiness of life.

It is necessary to clearly distinguish between TRU as proposed or as spoken of in this text, and all other forms. This may be of various kinds of non TRE or TRE, TREQS or TIT, TRIF, TRIT, TRIF, TRUID, TÂS and TÂ. It becomes vivid when TRU is distinguished from what TRU will ultimately and perfectly become, TÂ. **PROPT** = Proposed TRU. For our work on TRU, PROPT is rough, incomplete, unfinished, tentative, largely temporary, and will need correction. For this reason we must keep an open mind, especially that which is not basically certain and most important. Thereby: **PERFT** = Perfect TRU. It is PERFT that we imply in much of the text, whereas in production this is often PROPT and even less through our limits or errors in representation, modulation, and presentation. When we show or even know how to provide a duplicate of TIT, we have PERFT. The processes of developing TRU that converge on TIT and TÂ is PROPT. How we convert the world about us, and ourselves with it, to PROFT, then to PERFT is key to actual being, TRU. The Individual's or group's knowledge of TRU is much less than TIT. TRU is somewhat less than TIT. POT and IPTRU are less than TIT, TIT is all, PERFT, the difference is PROFT.

PRIPT = Proofs of the significance, Importance, Relations, Roles, effects and benefits of IPTRU: 1. Although knowledge, especially including that of religion, philosophy, fields, and sciences have and continue to make great strides in fulfilling needs, the time is long past when all knowledge should be brought by VAPO to selectively focus on this special universe of best human life in an oriented, coordinated, and ordered way for optimal use and benefit.(SPALT,POS,COGNOS,SÂ), 2. With a thorough and fulfilling understanding of TRU, began in IPTRU, each person can both improve their role in life and receive the rewards this life offers if they only make the effort to achieve it. 3. The success or failure, survival or extinction of the human race depends upon its competitive advantage. This is their special, predominant, and unique FAF, characterized by the dimensions and continua of Á, L, T, and how they are altered and adapted for optimal effect. It is the higher levels and powers of Á that have been, and will even more be, the salient of this survival. Only TRU and IPTRU or their TREQ can provide and assure such powers. 4. JURP or Justification, Reason, and Proof of continued work and effort for TRU and

TRE, 5. Targeted groups who can obtain special benefits from the application of IPTRU to their special problems, e.g inductive and deductive thought, prevention in many fields, education, mentally ill, 6. Helps to systematize the strategy and tactics of future work,(ENT, HOM,ACC) 7. A Greater sense of values, incremental to life, 8. The many secondary to finite and indirect, as well as direct, roles, effects and benefits of IPTRU seen in each concept, principle, law, mechanism, topic, part and process of TRU. When each and all persons anticipate, have interest in, discover, learn, remember, participate, and apply these they can achieve their full potential in the power of VAPO, and Á in TES and TRU.

Every given, topic, subject, field and science is not what it appears, usually presented, or how it is commonly known. It is strictlly and only how it exists in TRU. How ESINS, and the essences of TRU, act as parts and processes in TRU. Ordinarily this is not so apparent. For it is apt to be in error by variance or absence from TRU. All the way from much that is included that is not TRU, to much that is not included that is TRU, and even these, also, by degree, more or less. So that to make knowledge sufficient and efficient we can't start from what is assumed or ordinarily known. Scientific knowledge championed by experimentalism, Galileo and many others has also failed to fully interpret and extend their great contributions, by excluding the unity, totality, order and selectivity of this potentially well defined special universe of best human life. The why, how, what, where, and when of all is apt to be misleading. For each science, field, topic, subject, and object, as for any and all TRU, must originate, develop, include and fulfill all its parts and processes by how they interact with all in TRU. In order to overcome this weakness and prevent its failures in knowing the strict and deepest meanings of all, we must find out, learn, and live TRU. This is Lateral or Exponential Expansion of the mass energy, Á dimension is **LEXÁ**. All the power of each level of SÁ is not achieved by O-S alone, nor by a singular continuum or part. They work by all powers of Á, L, T or ACEXIN and other modes of TRU. This lateral and more comprehensive massive increasing finite expansion toward the absolute is what characterizes, and makes unique and special, TRU.

The actual absolute mass energy of TRU, Á or SO as an end, Á^{N}, terminus and ultimate of THESOS. This is called: **ÂÁ(ACA)** = Absolute Á or Absolute Actual Mass Energy Form of, to in, TRU. ÂCÁ is attained through the operations involved in THES, THEO, THESO, and THESOÂ. These operations and the associated mechanisms are shown by the Thetic states diagram, in Figs 7-11 and 1-99, and are as follows: by extracting the form and function, parts and processes of TRU through THEO, representing and modulating them combined with the reciprocal operations of extracting from all entities that which is their ESINS and through THEO identifying and applying them in representation and modulation to TIT. This is done by combining the outgoing and incoming operations with that of objectivity, of a positive objective approach. These are then focused on THEO or The Objective of TRU. This describes those operations that best and most concretely, like we do with SUT, fuse and show how TRU can be most effectively represented and presented. Likewise THES or The Subjective is attained through a similar reciprocal operation in which we proceed both outwardly and inwardly at the same time. Outwardly by extracting the concepts, principles, ideals, and laws of TRU through THES and putting them into perspective and comprehension combined with the reciprocal operation of extracting from all entities their ESINS, reason, causes, essences, or ideas and through THES identifying and applying them to TIT. By these mechanisms and operations we combine the outgoing and incoming operations of subjectivity of a positive subjective approach and achieve ever higher levels of certainty needed to fulfill our knowledge, Á, and TRU. As the best of these two reciprocal mechanisms and operations are combined we have THEO and THES being combined to THESO and as these are formalized, standardized and perfected we have THESOS. As THESOS is further perfected and absolutely assured we have THESOSÂ or ÂCÁ, The Absolute Actual Á. This very ACA or perfected predominant FAF of mass and energy that exists in TIT and TRU is the very substance or stuff of which it is composed, which we seek, and seek to find permanence. It is the eternal assurance of its and TRU's fulfillment. The exponentially accelerative exercise of our use of Á and ÂCÁ is the very heart and soul of TIT and each of us, our very existence as we propellingly and compellingly must be, be TES, and be TRU.

There is more than one kind of perfected showing of TRU. These include: 1. Perfect initial presentation of TRU, **PIPTRU**, 2. Perfect presentations of TRU, **PESTRU**, 3. Perfect textbook or work on TRU, **PEBTRU**, 4. Perfect atlas of TRU, **PALTRU**, 5. Perfect illustrations of TRU, **PILTRU**, 6. Perfected models of TRU, **PEMTRU**, and 7. Perfected Audio-Visual Representations of TRU, **PEVTRU**. These vary with the degree to which TRU is known, and the limits of each of their vantages, means and relations to TRU. A perfect representation of TRU should contain, figures, maps, models, and electronically operated products that show in the greatest and best possible detail and accuracy the integral, differential, intermediate, construct, form, systems, and parts as well as processes, their measures and posits, of TRU. This is to produce TRU in its utmost detail, as provided by ALT, SPALT, and POS and associated means. It should provide the optimal in color, yet color as much as possible natural to and in identity with the relationships between color physically and what they correspond to in TRU, and ROT. It should be based upon and emphasize the spherical three dimensional actuality of TRU. It should provide the fourth and higher dimensions of TRU in metatheory, transcendentally as they all enhance a presentation of TRU and all senses and things potential to or determinant of TRU. A preview of this is to be found in CAD or Computer Aided Design and the Graphics of the electronic media and internet, wherein each person can enter the world of TRU in virtual reality. Many concepts, principles, and laws, as well as new concepts will then be better understood with even greater formal distinctions, TRU meanings, and concise definitions that can be maintained and utilized for the benefit of all in TRU, TIT.

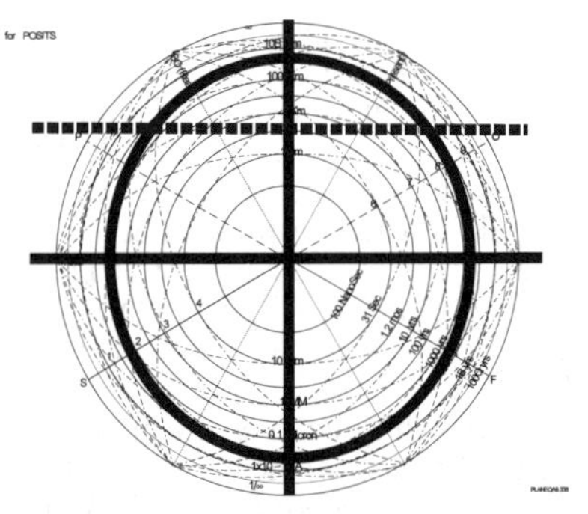

Figure 7-99 **CROSS and CIRCLE of IDENTITY, TOTALITY and ETERNITY**

The symbolism of the cross and the circle have existed since before recorded history. They have been used to signify conjunction and totality together in completion. In geometry the completion of a line drawn perfectly about a point makes a complete arc of 360 degrees. It is by the combination of the cross and the circle that more than a frame of an object, thing, the object, the thing, TRU is provided and how it can be modeled. It does so not by arbitrary standardization, but through a naturally representative replication of haecceity and form, and of TRU. In Figure 7-99 the combined cross and circle, alluding to their sphericity, are viewed in a setting of beatific perspective, or horizon of a person in their world, TRU. This provides the frame for completion that enables us to proceed toward a set end, eternity, The Relative Universe Absolute, TÂ. The circle is symbolic of totality, fulfillment and eternity. In this figure, like the Celtic Cross or Cross of Iona, we use such a combined configuration of cross and circle.

Variations may be produced altering the cross or crosses so that OITIO is added to our relationship to TRU, TIT and TÂ. Such a combined configuration signifies in all religions, philosophies, and sciences a basic geometric meaning. This is the intersection of the 3D entity, RU, object and TRU with the two lines of observer observed and length in instantaneous time that together form a coordinate whole. Such a two dimensional representation of a cross or scaffold on a round field or frame is symbolic of a person, as Christ's or TES, centering, yet only parts in the overall scheme of things, TIT-TO of TRU.

LÂRP(OT) = Law of Absolute and Arbitrary Relative and Variable Representation and Presentation of TRU. Although by TRIA, TU and TO, SUT, SPALT and ELTICMA, a singular and simple identity of presentation and representation with TIT and TRIT is valid, this is neither exclusive nor absolute, for it does not include its meta states. These are the externality of the observer, ROT and POT, and the OITIO involved. It does not include the subject and object relationships in the world and of all things from our focus in life. This is a product of the partiality and externality of ourselves and that about us that is less than total TRU. This is especially notable in our individuality and more so, in our IA. However this always has to be remembered, ECRO, and corrected, RECTRU, by TRIA, to TRU, and TIT. Since in our extant world we can not readily nor correctly differentiate, especially immediately, between that which is or is not TRU, or that which is more or less TRU, we must keep an open mind, be ready for alteration as the occasions require. Thereby we should contemplate, forever upgrading our KOT with contrapletion, yea and neigh, that which needs to be of our attention, yet may not be TRU or so much TRU.

The greatness and beauty of TRU, once discovered, can never be denied or lost. It is like the best of knowledge and literature, permanent records of life. That is it can never be lost until the carrier is lost.(GELD) This would be the death of all that carries TRU, all knowledge and persons, which is tantamount to no TRU, and thereby not. This latter is why it is so imperative that we discover, learn, know, remember, practice, serve and live TRU, TIT, TES, IAT, ACC, PUREALITY, COGNOS, ELTICMA, SÂ, and TÂ. To work on a presentation of TRU, as IPTRU, is like the flowering of the most beautiful plants at full bloom, or the glory of spring. It is like the morning exuberance after a restful night, the rosy fingered dawn. The combined actuation, extension and intension of all that is so vital, crucial, beneficial and enhances all to TRU and to TRU Absolute, TA. This is the reward for our labors in presenting TRU. This is called: **IPTA** = Initial Presentation Through Personal endeavor of all for TRU to TÂ. The development of the full glory, beauty and victory of TRU by all under optimal conditions. It is how individuals and groups 'go for it', 'do it', or 'shoot the moon' to accomplish and achieve that best for themselves and all, TRU. Sciences, philosophies, religions, churches, political systems, and international organizations can do much better when they are guided by, and are TREQ and TRU. This is the attainment of TRU and its best convergence in the absolute, TÂ. This is our life, a rare opportunity to participate in the greatest cause, salient and heroic age of humanity that will be of such great benefit to our being, TRU, and assure its everlasting continuity in eternity, TÂ.

Postscript and Colophon

In conclusion this work on the collective of all greatest topics, for and in TRU. It comes full circle from the one's attention to TIT-TO and its ACC back to existing attention and being:

First Person - IA - Observed - thing - SO - Á - TRU - TIT-TO - ACC - TÂS - ACC -TIT-TO - TRU - Á - SO - thing - Obsever - IA - First Person.

Repeated exercise of this Positive Imperative, **PI**, through practice tends to make perfect ones IA, ability of the reader to achieve identity with TIT-TO and Action needed. The knowledge acquired becomes COGNOSID a total comprehensive identity with TRU symbolized by the colophon ALT. It is IPTA of TIT through a degree of LARP allowed by the extra TRU position of the observer, you the reader, and involved member of humanity .

Fig 8-99 **ACC SUMMARY POSITS in SPALT**

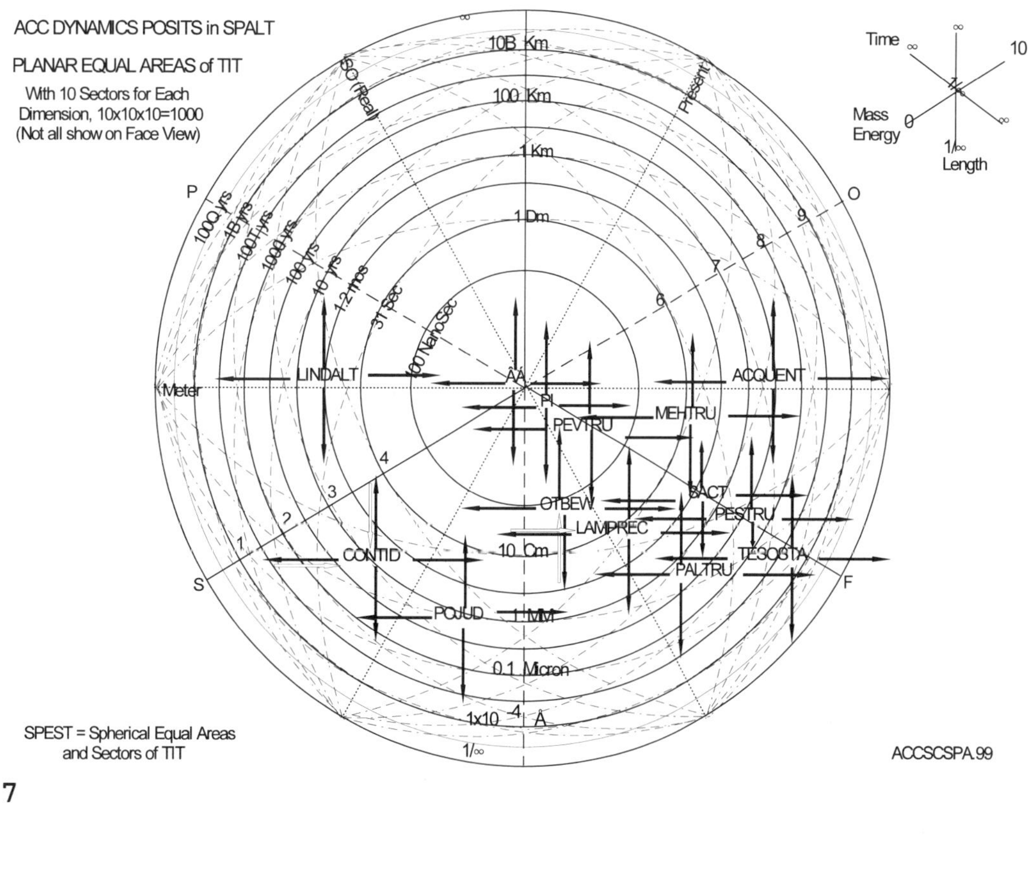

MEASURES of ACC POSITS SUMMARY and CONCLUSIONS

Posit	Measure	Pattern	Type	Code
ÂÁ	7	~5~5~5~	T	
ACCELERATED INFINITE EFFORTS TÂ	3	5-9-9	M	(N)
ACQUENT	1	~6~4~7~	T	
$Á^N$ (ACA)	7	~5~4~9~	S	(ÄNÂ)
AIME	± 67	~4~4~7~	M	
COMPLETION S-O RELATIVITY TRU	7	5-~-9	M	(TNÂ)
CONTID	± 43	2-7-5	T	
COTANOS	8	~-~-~8~	S	(ÄTR)
ETA	± 65	~-~-~-~	S	(OMT)
ETACT	± 65	~-~-~-~	S	(OMT)
FOCUS CONVERGE IDENTITY TRU TA	3	~-~-~-~	M	(THN)
GACSAL	7	~-~-~-~	S	(TÂ)
IPTÂ	47	~5~6~9~	M	(PÂ)
IPTRU	47	~5~5~6~	M	(PÂ)
ÍTÂ	43	~3~4~7~	M	(SÂ)
LAMPREC	13	~4~3~7~	T	
LARP(OT)	± 67	~-~-~-~	S	(RP)
LEVELS KINDS SELECTIVE FORCES-	43	~-~-~-	S	(MEH)
LEXÁT	3	~5~-~-~	S	(Ä)
LIDE	3	~-~-~-~	S	(NR)
LINDALT	3	~3~5~27	T	(HE)
MEHTRU	±41	~7-6~5~	T	
MERP	±43	~6~-~6~	M	
OPJURIT	1	~2~7~5~	M	
OPOT	3	~-~-~7~	S	(P)
OTBEW	3	~3~4~7~	T	
PALTRU	41	~5-6-7~	T	(RP)
PEBTRU	18	~5-6-7~	M	(R)
PEMTRU	1	~6-6-7~	M	(R)
PERJUD	43	~2~7~-~	M	(GH)
PERFT	5	~-~-~-~	S	(TÄ)
PESTRU	1	~5~5~7~	T	({P)
PETOD	±43	~6~5~-~	M	(MHN)
PEVTRU	1	~4~5~6~	T	(RP)
PI	41	~4-4-6~	T	(OR)
PILTRU	1	~5-6-6~	M	
PIPTRU	1	~6-5-6~	M	(P)
PITRU	41	~6-6-7~	M	(P)
POJUD	85	~3-7-4~	T	(ET)
PROPT	5	~-~-~-~	M	(TNP)
RABNAL	± 41	~-~-~-~	S	(MH)
REASONS ACTIONS TRU IPTRU	1	~-~-~6~	M	(TP)
RESTAD	87	~-~-~-~	S	(RT)
SACT	17	~5~5~7~	T	
SYMBOLISM CROSS COMPLETION TÂ	43	~4~-~8~	S	(NR)
TÂ(SEX)	7	~-~-~9~	S	(NÂ)
TAR	7	~-~-~-~	S	(TÂ)
TESBEW	3	~1~6~-~	S	(EG)
TESOSTÁ	17	~5~6~9~	T	(GEÂ)
THETIC STAGES or LEVELS	43	~5~5~7	S	(E)
TIST	47	~-~-~-~	S	(MTE)
TRUFEXPR	± 45	~-~-~9~	S	(TN)
WABEW	± 47	~7~-~-~	S	(GTE)
YAD	± 41	~4~5~6~	M	(Eau)

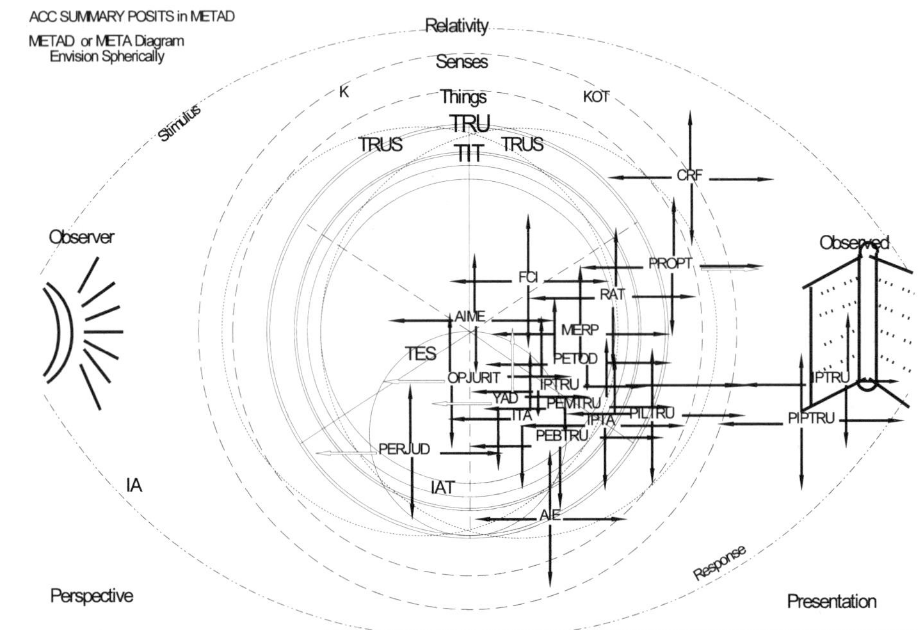

Fig 9-99 **ACC SUMMARY POSITS in METAD**

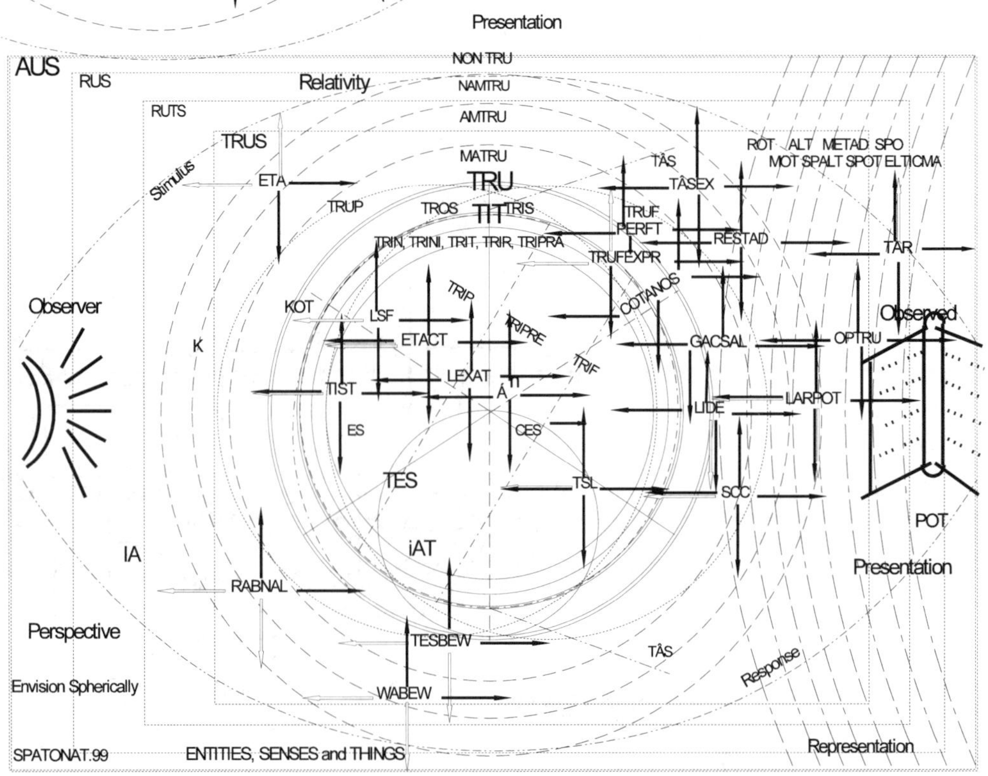

Fig 10-99 **ACC SUMMARY POSITS in SPAT**

GLOSSARY (Abridged)

This abridged glossary is to supplement that already given through the text, by defining each new concept, NEC, at the same place where presented. Also see Supplemental Contents, see page ii-x. The glossary's target is to provide a reader with meanings of terms that are used without accompanying definitions. If the definition is not given with the NEC but later on in the same or adjacent pages it is not included in this listing. Also those NEC that are largely obvious from their construction, as ANTES for Against TES, are not included. Many can be readily discerned from the letters used. To understand this text - and human life in general - it is essential that one master the key concepts of the supplemental contents as early as possible. For definitions of many new concepts in the proofs, figures, and other lists, you are refered to the exhaustive glossary in the unabridged text, IPTRU.

First numbers, before definitions, are for the page in text where noted. The second in brackets are where defined, and the third in *italics*, following their defintions refer to the chapter in the full text, IPTRU. A star, * indicates those terms that have different definitions because of different vowel pronunciation. Two stars, **, indicate NEC in development, occur with more than one term, or whose vowel varieties are not included in this list.

ÂÂ 40,[65] Absolute Absolute, inacutual terminus of absolutes *121*
ACIA 30 Actualization of IA
ACIAT 30 Actualization of IA to or in TRU
ACIT 157 Action Itself is final Test of Truth and Value *215*
ACOM 299 [556] Á Completion, S-O, Mass energy Completion *98*
ACON 553 Á Continuum, spectrum in - of TRU mass energy *97*
AD 5, 49 Form Derivation, Development, Determination,, *021*
ADD 43 Addenda to book
ADTRU 445 To, added to TRU *117,131,82*
AILA 230 Absolute Illusory Absolute, no relationship *039,069*
ANKTRU 217 Difference between T and all not T, from,,, *005,006*
ANRU 30 Against RU, a RU
APT vi, Approach to TRU [1,,,]
ÂR 40, [65] Absolute Relative *037,039,121*
ATIM, 51 All time Mastery of things, by TRU *10*,
AUCH 29 Changes in AU, Absolute Universe Change, *062,068*
AUT or (AT) 56, All between all AU and all TRU *062*
AXLA 52 Axiom of Limits of Application *116*
BIV 54 Bring into View in overall application *040,065,1+*
CALP (CLAMPS) 272,280 Concepts, Principles, Laws, Mechanisms **
CED 299 Competitive Exclusion of Divergence *172,34,52,55*
CEFKA 299 Compulsory Ethic of a Known and Assured Future *59*
CEMLIP 156 Contraction Expansion Mechanisms, CALP T or ROT *214*
CET 125 Contract ES to and from each other and TRU *165*
CIN 279,281 Concentration Intension, depth and power in IA *081*
CINE 281 Coterminality Index between parts, or ES *48*
CINSY 281 Coterminality Index between systems *48*
CIS 46 Conditional Is, Identity any entity, not TRU, limited *034,106*
COB 136 Collective Total Benefits, of TRU *182*
COCOR 227 Correlation by Coincidence of REM, TRU *36B*
COGIF 568 The Cognitive Identity-Contribution, I can have c T *99*
COIM 380 Concretion of Infinite Limits, Misapplied Intellect *71*
COMIA 8 Completion of IA *089*
COMT 29 Comparative TRU *033, 06, 117*
CONQ 568 Conceptualization Quotient *96*
COPOS 511 Common Principles of Success *179,58*
COPTA 299 Concordance of Perfect Thought and Act *83*
CORIA 8,30 Correction of, or Corrected IA *078,088*
COS 568 Comprehension Total in all ways, S Total *96*
*COSUT Coincidence of OS, mental and physical units in TRU
COVAPS 122 Correlation Value, Posit Analysis Problem/Sol. *156,190,39B* COWOT 553 Citizens of the World, TRU *97*
DA 49 Derivation of Form *031*
DACOB 136 Degree Deficit Accumulated Total Benefits *182*
DAO 5 Derivation, Development, Determination of Objectivity *041*
DAOS 61 Different Aspects of the Same Thing (any entity, object) *043*
DÄR 5 Derivation, Development,,, of the Form ru, any RU *083*
DÄU 5 Derivation, Development of the form of AU *031*
DE-DES-DESIN-S 479,484 Determination, 'Extraction ES, ESINS *88*
DEC 299 Depth Conception, Understand TRU, REM 060,37A
DEPOS 281 Degree of Precise Posit, fraction *48*
DET 5,252 (or combine all D'S) Derivation,,,, E, ES, TES *140-42,151*
DIA 5,33 Derivation, Development,,, of IA *010*
DIL 54 Death in Life *071*
DILAC 230 Degree of Illusion / Actuality *094,113*
DIP 264 Dialectical Predicament, Distinction / Predication 145, 44
DÏR 237 Divisions of REMS into separate parts *37B*
DIRIA 30 Direction of IA, in and of TRU *010,081*
DOM 96 Doctrine Of the Mean *091*
DRAL 281 Difference between Actual and Represented Locus *48*
DREM 18 [224] Determination of REMS by ES SPALT POS t4 *35-39*
DYNIA 8 Dynamics of IA, *070*
EA 90, 100, Ideation Realization, ESINS, ES and Á *142*
EFI 100 Equivalence-convergence of Form and Ideal becoming E *143-8*
ELIA 249 Everlasting Life through Infinite Apperception *97*
ENA 129 Whatness of Form of Entity *028*
ENUL 221 Energy Units, Levels of emu, au, cu,,, *37A*
EOA 92 Extraction of ESINS *142*
EOE 96 Extraction of ES, ESINS, constant *143*
EOP 42 Total combined final interpreted REM *38B*
EPIT 42 [43] EPI, Epistemics TRU, Knowledge Approach to TRU *029.095*
ER 285 the Finite Relative *58*
ERED 157 E or RE deficit Particulate Unknown TRU 140
ESATION 99,100 E or ES formation, Processes of *142*
ESCH 92 Change of ES, individually and with TRU *143*
ESE, ESES 92 Essences, forms, states of ES, attraction to TRU *140-3*
ESINAC 92 ESIN ACtuation 142
ESINATION 92,99,100 operations of ENS to ESINS to ES *149*
ESION 92,96 The processes of ESINS and ES formation *141,149*
ESSE 101 To be of the essences, to be human *149*
ETION 92 Process of reducing all to their ES *143*
EXEL 246 Exclusion of EL, ES limitation, specific *37A-B*
EXIA 30 Existing IA *077*
EXIR 499 Examples of Reinforcement, of TRU, Elixir of life*110,170,90*
FADA 129 Fallacy of Decreasing, Diminished Attention *007*
FAF 30 Form and Function, **
FARP 246[264] Factorial Advantage of Relate and Posit *44A*
FEMST 49 Attribution of erroneous proportion, REM form TRU *114*
FERI 331 Fluctuating Extremes of Reaction, Ignorance *61*
FET 230 Fallacy of Equality of Things, entities, is or is not *007.196*
FIAT 47 Fallacy of Indiscriminate Attribution of all to TRU *112*
FIE 230 Fallacy of Indiscriminate Equalities, not ES *007,143*
FIRAS 331 Summation of Fear, Irrational Acts *61*
FÍT 246 Fallacy of Identifying All with TRU *122*
FITALP 246 Fit and Limits Prove TRU, TÄ and ALT *12-2*
FLOITIO 144 Flow, Flux, and Field of In-Out Out-In TRU, OITIO, 2-196
FOCIT 110 Focus of Relativity by Sum of Midpoints *157*
FOG 553 Fallacy of God, without parts and posits *065,41B*
FOKIT 568 Foreknowledge by Identity with Future, improves TRU *96*
FORTOP 411 Forces of TRU, Total and Proportionate *92*
GAC 359 Glory in Act by Á and TRU, opportunity for TRU *66,67*
GEC 78 General Case,, effect, action *032,087,52*
GELD(F) 295, 302 Genetic Material, Live Product , Determining Forces **
GID331 God Defect, Dead, lost by difference from TRU, TIT-TO*107,61,81*
GOPT 499 Generation, Optimal of TRU *90*
HIP-S 12 ? Horizon IPTRU Diagram *299* **
IA 5 Instantaneous Attention, Immediate Awareness, core focus *010*
IAR 33 IA Relative, foci, point of reference *011,032,053*
IAT 5, IATRU 42 IA of, in and by TRU, expanding circles of *011,055*
IATID 230 IA TRU Identity *041,044*
IATSU 553 IA SUT of TRU, focus on target *107*
IFPIP 90 Indefinite state of Future Proves PIP *59*
IMSE 553 Imperfect Systems Effect Error *107,207,215*
IMSEF 553 Imperfect Systems Effect Error, Future *
INAC 445 INterAcTion Coefficient, amount of *82*
INF 90 Indefiniteness of the Future singular ES *60,67*
INIA 30 Intension and Intensity of IA *060,078*
IPA 30 IA and IPT, IAT-IPT, corollary of Á *152*
KÂ 29 Knowledge of Form, Knowledge Form
KET 49 Knowledge Equals TRU (within limits) *105,106*

KITRU 49 Knowledge by the Individual of TRU *105*
KDM-s 518 Key Dynamic Mechanism-s, General - Physiological 50,53
KOT 2, Knowledge of TRU *005,105*
LAC 5 [168] Language from Cleavages,,, enabling distinction *010* **
LACIR 237 Law of Correction of Insufficient REM,,, *38B*
LACOLT 140 Law of Correction of Limits, TRU 191,*300*
LACOR 158 Law of Correlation of Representation, S-O & Á *208,242*
LACREM 279 Law of Corrected or Standardized REM *48*
LAD 52 Law of Determination *110,114,116*
LADAC 148,203 [259] Dimensional Formative Congruity of ES *41B*
LADEP 437 Law of Dynamic Equilibrium Proportion *165*
LADEX 120 [171] Law of Definition to E, ES, exclusive *144,242* **
LADU 130 Law of Disproportional Utility *70*
LAECIT 215 Law of Á Expansion through [C] and Identity TRU *135,137*
LAGSTACH 74 Law of General and Specific TRUCH *123,168*
LAIA 230 Law of mass energy Á, Instantaneous Attention, À *278*
LAL 130 Law of Law-s, many from voids \ LATAN and TRIA *193*
LAMID 227 Law of Multiple and Magnitude Interdependence, (f) sames, 2-36
LAMUA 148 Law of Mutual Assistance, (all in TIT-TO) *379*
LANA 158 [164] Law of Non Arbitration of TIT, TRU *229*
LANDOVS 380 Negation from Deficit in VAPO and SCARLS *71*
LAPAR(S) 425 Law of Parsimony , Occam's Razor (LAMID) *79*
LAPE 437 Law of Proportionate and mass total Entirety *060,265*
LAPIVOT 54 Law of Pivotal Proof of TRU, set fixed relations *135*
LAPUS 411 Law of a Problems Understanding much of Solution *91*
LARF 200,221 Law of Relative Fixation, basis for much of TRU *158*
LARPCO 279 Law of Reduction for Posit for Overlapping, *48*
LARPON 5,230 Law of Relative Reversal Producing Negative *171,175*
LASCES 281 Law of Substantial Concretion of ES of TRU *148*
LÂSIT Law of Actual State in TRU 135, 137
LASO 5 Law of Subject Object *231* *
LASOT 556 Law of Subject Object TRU (to Á dimension) *219,128,106**
LASTE 299 L. Strong Tie or bond between ES,TES,TRU,TIS,*154,76,80*
LATAC 299 Law of TRU Actualization, Action to Contribution *100,66,,*
LATANE 281 Law of the Absolute Necessity of Finite Form, E, *47*
LATCOB 136 Law of TRU always from Collective Benefits *182*
LATINT 380 Law of TRU, TIT, TRIT In distinction *190*
LATNUP LATAN Singular Unified Perspective Presentation 035,093,118,
LATOL 299 Law of Tolerance, Total Limit to TRU 184,
LATRA 75 Law of Transformation, change, and of Á *129*
LATRUB 136 Lack of TRU Benefit IT/TIT *182*
LATSCAP 411 Law of Corrections Solutions Attitude Problems *92*
LECDIE 281 Law of ES Coincident with Divisional Parts of TRU *148*
LES 221 Law of Especialty, special/general *38A*
LENA 158, 195, [164] Absolute Necessity\Necessity of Arbitration *2209*
LEPER 230 Levels of Perfection of TRU (or God) *189* *
LETID 125 Law of Equilibrium of Integration/differ. and TRU *067,174*
LEXPRO 230 Extremely Fine, Refined and definitive POA *045*
LINCOT 230 Law of Inclination, Centripetal, to TRU *035*
LINROT 200 Identity of word, represented object, ENIT and TRU *281*
LINT 568 Interest, Intellectual Curiosity to TRU *97*
LISTA 51 Law of IS and TRU, relation and certainty to TRU *104*
LOCIT 159 Law of Compulsory Identity with TIT *133*
LODEP 230 Law of Definite Proportions by TÄ-TE-TIT-TROS-TRIT *256*
LOEDPT 125 (LODEPT)Dynamic Equilibrium Parts & Wholes *156,174*
LOFETE 230 Fractional Evidence things, ESINS, ES *049,142*
LOGES 92 Law of ES, Logos, reason, words, ES *143*
LOMI 568 Law of Misdirected Attention *184,91*
LOMT 411 Total Management to TRU, PPMA, ACC, plus *91*
LOPELT 221 [229] Precise TRU Parts Posited on A-L-T *37A*
LOPS 299 Law of Proportional Separation *36A*
LOR 158 Law of Orientation *215*
LOR 158 Law of Rational Tends to problem solving, ,, TRU *208*
LOSADIN 87 Sense Thing Differential Integral to Á in TRU *131*
LOSE 123 Law of Selective Effect, trans PPF, +/- *171*
LOSCARL 54 Law of States, Conditions, Relations, Limits TRU
LOSO 246 Law of Total 0-S-0 combinations 40A
LOSOW 568 Law of the Summation of Wills, to TRU TÂ
LOTRAC 568 L. Realization by Comprehending T Actuality *241,44A,96*
LOVAL 501 Law of Value, relations, from labor, property,, *35B,36A*
LOVAP 501 Law of Valid Parameters or Universals 36A, *
LUV 283 Universal Validity, TRU, certain *065,109,50*
M&C 374 Management and Control *90-*
MALCOP 331 Measure of a Language's Capacity , Á *61*
MEHTRU 571 Medical Homation and TRU *99*
MER 256 Mass Energy Relativity, 120, 42
MEROT 146 Methods and Means of Representing TRU *198,203,208,218*
MESU 146 Means and Method of SU, to ROT and MOT *198,203,218*
MIDCOT 200 Maximizing Intensities in Differentia Coterminalities *289*
MIF 218 (MIP) Meaning Imperatively Finite *215* **
MO 493 Maximal Opportunity **
MOCBIREM 148 [221] Method of Precise REM by Classes *37B*
MODREM 232 Modification of REM, best ROT of TRU 37
MOPU 359 Measure of Practical Use, Proportional *67*
MUPT 130 (MUPRUT?) Mutual Proof of TRU additive *192*
NAMS 92 [93] Names ES, word-ES-thing, ES/CES/TA *143*
NAMTRU 65 [71] Mixed TRU and not TRU (not much T) *126,169,51,,*
NAT 170 Name, Nominal, word as it is forming ES *116,230*
NEN-S 253 Name-s-Entity-ies-, words-entities, word-thing *242*
NES 253 Names of ES, Nominals Word Things combined to TRU 23, 42
NESI 253 Name, ES, Identity **
NESION 96 Negation of ESION, drift from ES *141,149*
NETH 277 Nominal-Things, word-things, for Posits *42B*
NODA 568 Not determined absolutely, nor accurately *90*
OBSAD 493 Optimal Balance Sames and Differents Abs Un/no *89*
OIA 7,30,33 Total IA *151*, also ÖIA Objective IA *125*
OIAT 30 Total IA TRU *061*
OIN 281 Without, out of text context, in appendix *48*
ONTA 52 Ontological Absolute TRU,as God never ceases to exist *108,116*
OOM 42 (OM Omnificient Magniminity, Total Meaning) 158,193,195,**,
OOP 42 OP (Objective Paradox *096*) **
OPAR 230 Opposition from Relative *096*
OPOT 571 Optimal Presentation of TRU *98*
OPTH 299 [399] Optimal State in Life, TRU, Health, carer *58,75*
OR 230 Objective Real, 5th 6th of S-R-O *32A* **
OSM 42 Objective Small(medium?)
OSP 42 Objective Small Past octant of SPALT, TIT *33A,83*
OTREM 221 Total TRU REM must value only TO & REMO *38A,39A*
OVEX 384 Overburdening Extra *72*
PADALT 215 P. of Average of extensions of all dimensions, ALT *130,,*
PAN 96 Paradox of Nominals, words or language *092,148,207*
PAND 230 Positive and Negative Dynamics in TRU *55*
PAVAT 226 Parts Absolutely Vital to whole, TO, degree *042,146,148*
PECRO 101 Processes of Continuous mass Representation *213*
PEGSOT 346 Principle of Given, Entity, S, O, ES and TRU *64*
PES 100 Pseudo ES, Illusion and figment of ES *142*
PHILE(R) 479 *Phile* or Love, Error (as familiarity, labile) misleads *190*
PIA 010,482 Opportunity for/of IA, consciousness, awareness, attention 060,59
PIO 134 Presentation, Individual, All ENS *036* **
PIPS 122 Problem Principle Solution (PPS) *192* **
P/N 545 Positive over Negative **
PIT 568 Perspectives Inclination to TRU *90*
PLADIP 277 Problem of Planar Posits Dimensional Polarities *45*
PLAP 277 Problem of Planar Posits 45
POAT 8 Positive Objective Approach to TRU *113,208*
POCP 284 Principle of Continuing Processes, many long *88*
PODES 411 human Potential Power of Self Determination *92* **
POEOB 104 (Principle of Explanation of Essence of Entity, Object) *116*,
POES 230 [496] Prin. Essential Separation, DAOST, ES, CES 116,223
POF 284 The Possibility Potential Factor *114* **
POID 281 Infinite universal looking at SPO,,SPOR, SPOT,,, *053,059,*
POIR 224 [225] -Purposive Problem Solving IMU TRU Revolution *35B*-**
POK 364 Presentation of Knowledge *69*
POM 104 Perfected Power of OM *195*
POP 281,496,[559] Power of PUREALITY inc. Á continuum *98* **
PORCOM 346 [330] Principle of Repetitive Conception of Meaning *61,64*
POREM 221 Posits total of REM, able to posit all *38A*
PORIV 35 Powers of Reinvigoration and Reproduction 07
PORT 5 PORF Principle of Relative Fixation, produces set *001* **
POSIM 284 Principle of Simultaneity *198*
POSTOP 284 Supernatural and Sub-natural Transmutation of CES *073*
POUC 283 Principle of Ultimate Coincidence *138*
POUS 21,28 Perspective of the Universe, Standardized *060,063,066*
POVE 284 Principle of Vacuum Effect *55*
POVIN 246 Principle of Veritable Interpretation PPM *41B*
PP 247 Past Past (distant past) first sixth sector of PPF time *32A*
PPF 221 Past Present and Future *138,32A,76*
PPS 375 Popular Public Sentiment, common opinions *70*
PRET 565 Perspective, Presentation, and Representation of TRU
PRODEP 246 [261] Proof Deposit, of Proof falling into place *44A*

PROKEP 246 Process of complete Knowledge thru accumulated Posit *44B*
PUCS 557 Process-s of filling up, enables success 85,98
PUREA(C)LITY 2, Purity, Actuality & Reality T @ *95* **
PUREV 283[295] Purity in complete Reversibility, sustaining *53*
QUALAN 255 Qualitative and Quantitative Combined and Interactive *42*
RÂ 40, [65] Relative Absolute *121,189*
RACOD 230 RU AU [C] Differential, increased separation *044*
RAEV 64 Reason and Á from causes-effects, ES, CES TRU *119,141*
RAMAR 411 Ratio Marginal, critical measure by TRU *90*
RANIA 30,33 Random IA
RECIP 246 Reciprocal Proofs of TRU *49*
RELE 479 Religious or theological Error *009,090*
RELIS 230 Revelation Lost in Singularities *083*
RENDE 281 Rendered to a Discrete Part of TRU *46*
RÍT 214 Representation in Itself *200*
ROMT 299 Reliability of TRU as a Measure 193,195,209 **
- 502 Ratio of Monetary to TRU value *38B* **
- Reinforcement of Meaning of TRU *237* **
RR 40, [65] Relative Relative, exclusive focus on relative *039,121*
RORRE 246 Revelation of Relate Explosion *038,083,088,098,*
ROSU 147 Resolution of the separation and union of entities, , , universes, TRU 138,193
RU-RUS 9 Relative Universe-s
RUCH 29 Change-s in a RU, RUS *032*
RUIT 27 Relative Universe-s in themselves
RUPKAL 200,221 Revolution of Perfected Knowledge,,, TRU *193,195,88*
RUT-RUTS 14 Range of Universes between RUS & TRUS
SAF 230 Success or Failure as a Function of Finity *042*
SAFPIN (SAFPIM) 281 Spread and Focus Proves In Media *48* *
SAO 111 Subject Absolute Object relationship
SARVAT 226 Spectrum/continuum degrees of ES vitality and ,,,146
SAV 28 Smaller Forms Acutely Vital 183
SCATS 326 Every new Science requires its own special form, relative to TRU 60
SCE 479 Science or Scientific Error **
SCENS 328 Every New Science Requires its own Entitis, Terms, Words, ES 60
SCIN 568 Scopic Inertia, momentum and powers *95*
SCONQ 568 Summation Conceptualization Quotient *96*
SCOPT 568 Scopic Optimum, (SCIN) *95*
SE 418 Subjective Error, ERIN for Subjective, TRU *007,126,43A* **
SEC 78 Special Case-s *032,087*
SEPTRU 499 Separation TRU, ATRU, MATRU, ,,,AU *102,90*
SIAT 230 Summative IA, SIA to TRU *011,041,044,118*
SIDA 281 Simultaneous Analysis of Static and Dynamic *065*
SIJ 568 Steps of Investigation and Judgement *90*
SIN-M-UAD 281 Simultaneous DIN with SU and Formal Distinction *067**
SIO 568 Subjective Intensity Infinite Objective AI, A *91*
SOEX 493 Subject Object Expansion **
SOLA 54 re-Solution from ATRU, 1/Sophistication individual, self *109,65*
SOPAN [99] Solution to the Paradox of Nominals *092,148,265*
SOPOB- 21,28 Standardized Perspective of Being *051,059,060,063,065*
SPAT 28 Expansion of the Threshold of TRU *105*
SPU 3 Unified Universal, inward Standardization of SPOT *010*
SPUNO 3 Standardized Unitary Perspective 1:1 identity S-O T *010,036*
SRE 281 Summative Representation of E, ES of TRU *206,229,265*
SRIA 352 Sudden Realization of Life's worth or value 35B,65,75
STOD 129 Singular Solution Total Discrepancy, S O & *090,115,199,91*
SUDE 146 Systematic Unity Partial Differential *198*
SUFO 146, SU FAF as a fixed and ordered whole 198
SUI 499 Sufficiency or Insufficiency, value order *90*
SUIAC 31 Superimposed IA Charts representations
SUO 146 Systematic Unity of an Object *198*
SURMS 146 SUFO used to Show Model and applications *198*
SUTFO 146 SUT FAF as a fixed and ordered whole by TRU *198*
SUTORMS 146 SURMS for, of and by TRU *198*
TÄE 92 Form of TRU's Part-s **
TAL 356 TRU Absolute Limit, not an end *083-4* **
TÂS 40 TRU Absolute, plural, types and forms of *138,139*
TELOT 28 Telescopic overall TRU/non TRU variation, PERT *063*
TEP 264 TRU special Part Predicament, (DIP) 145, 44
TER 81,164 Factors, bonds of Parts of TRU *120* **
TESP 103,104 The personal TES, or TIS, one;'s identity c T *150,152-3*
TESOSTA 571 OS THESOÂ *99*
TESPRA 281 Test or TES Proof of A, ALT, TRU *47* **
THESA 39 The Subjective (of TRU) Absolute *154*
THEOA 39,545 The Objective (of TRU) Absolute *155*
THESOS 8 The Standardized Subject Object of TRU *154*
TÏÉÂ 100 Terms, Intension, Extension and Actuality by TRU *149,159,223*
TIMS 40 Tie-Ins existing at the same levels in IPTRU *08*
TIO 174 (The Interrelations of All 105,197) **
TIS 380 The Special Parts of TRU 140,153,154 **
TIT-TO 1, emphasizing the TO nature of TIT. *16-*
TOPA 411[417] Test Primary Applicability, break throughs *065,78,91*
TORP 268,450 Total TRU Representation and Presentation *40B* **
TRÄT 74 Transformation, as Á in TRU *021,031,12*
TREC 501The Recognized TRU, as such, TIT concept, *105,90,92* **
TRIC 47 Tri dimensional Configuration, form for solidity,,, 268
TRICH 56 TRIS Change, in all or within each *168*
TRINE 118 Partial or Particulate state of TIT, TRIT (DITRIT) *172*
TRINI 20 The infinitesimal of TRIS, TIT,T *059,159*
TRIF 47 TRIS, TIT, TRIT time of Future, last 3rd of TIT time *133,148,158*
TRIP 47 TRIS, TIT, TRIT time of Past, first 3rd of TIT time *133,158*
TRIV 107,118 Varieties of TRIS, TIT, TRU, *56* **
TRIZ 84 TRU, TIT, or TRIT zero, as a limit, 1/TÂ *137*
TRUCA 281 TRU Common Additive *163*
TRUQ 395[445] Quality Quintessence Virtue of TRU *120*
UFT 40 [41] Unified Field Theory *160*
UMTRU 215 TRUT and KOT lost in modeling and formulation *005* **
UNKTRU 198 Unknown of TRU producing limit and/or loss *005*
UPTRU 198 TRUT and KOT lost in Presentation and Perspective *005*
URTRU 198,215 lost to TRU through Representation (LOLS) *005*
USTRU 198 Lost to TRU by Standardization (LOLS) 005
UTES 197 Lost to TRU through TES, partiality, and lack of TES 29
VE 184 Validation through extrapolation, unknowns by knowns, of TRU 25
VISO 568 Vision/S Infnite in the Infinitesimal Present *98*
WIM 88,520 We Will It Must (all TRU) *53,97*
WIPAT 568 Will to Prepare to Win TRU *99*
WITA 352 Will to TRU, Á form and Absolute *65,79*

Some of the new terms that are either in the stage of development, overlap with other terms, have multiple definitions, etc include: ANSART 18, BURP 62,284, CAPIA 30, COACINEX 90 , COGDEF 553, DICE 437, ELAC 54, ENID 129, EOM 42, EPIA 42 [43], ESM 42, ESP 42, GEXA 129, HP(L,T,S,O) 110, IS 51, KATRA 29 (KATRU), KOF 499, LACES 281, LACTO 140, LAPOSC 281, LARA-S 172 (25 combinations), LARDOL 54, LATANSART 18, LEMOV 568, LOBIR 230, PIDET 437, POLON 284, POMB, POTISU 249, POUF, PROVT 130, RENFUD 284, RON 284, RONT 96, SCATS 328, SCENS 328, SOSTES 568, TAG 47.

For many NEC in the figures, proof, and REM enumerations, see the complete glossary in IPTRU. There are many acronyms and abbreviations in common use whose combinations provide terms of major importance, such as: S/F Success over Failure 11 and S-R Stimulus - Response 12. Likewise many theorems, concepts, principles, laws, and mechanisms named after their originators are also great additions to human knowledge - and TRU - as: Godel'sTheorem 10,88. These are not listed in this abridged glossary. Furthermore one must look to the full work, IPTRU, for many NEC or OC of great value which are in the process of being named or added, that help to fill out a proper representation of the finite of TRU.

The use of a new language is not only very difficult but seemingly unnecessary to the casual reader. Yet the success of our actual world will not persist unless it can identify and make distinct all of its missing or insufficiently designated parts, processes, concepts, laws, principles, and mechanisms. The loss or absence of such great and glorious aspects and elements of our actual world can 't be appreciated until they are known; only the resulting failure and suffering. Thereby revel in TRU and its NEC, for only here can you find actual reality our powers of mass energy give us.

215 THESES of THE RELATIVE UNIVERSE
THEMES, CONCEPTS, PRINCIPLES, LAWS, PROOFS, TESTS, ACTS, and CONCLUSIONS
(THE GIST OF TRU in PLAIN and SIMPLE ENGLISH)

1. Are philosophy, science and religion entirely different, or are they the same, and if so how? Philosophy is how we look at the world, science is what it consists of, and religion is what is important, how unified, ordered and best lived. We can't be wise **unless we know how they fit** into the whole.
2. It is when we do not learn and live by what is common and best for all that philosophy, science, and religion fail and are so conflicting. Neither one alone adequately describes our world. We should like to know about the same and different of **what actually is in** the world.
3. Then let us, for a minute, be astounded by not only how different I am when viewed from without or at different times, **how different** each person is that I confront, and even more, how fundamentally and often conflicting different these other person's opinions are than mine.
4. The astute investigation, analysis, determination, and demonstration of these differences is the beginning of wisdom. They exemplify, and give one the revelations necessary to start on, the **greatest of all adventures, our own being**, to learn what the best of human life is all about.
5. These theses assist our experience to learn **what the best of human life is** or is not all about. As the clear and pertinent evidence, facts, concepts, principles, laws, and knowledge are developed, represen-ted, modeled, presented, applied, and used, they provide excellent understanding of the actual world. .
6. What is common and best for all can't be understood, unless it is discovered, represented, and well presented. Since human beginnings, this has been the propelling **power of our words and knowledge**, the great search for the holy grail, what works best and how to make life better.
7 The actual, relative, world, universe, God, all, etc. consists of neither a multitude of completely separate things in themselves, any combination or totality of them, nor a singular undifferentiated union of them. It is DAOST, **Different Aspects of the Same Thing**, The Relative Universe. SC
8. Life is far from **being** so obvious and **sure** as our presence, senses, and beliefs assume. To believe only what is given, we sense, see, or wish to see, and to fail to profit from the best of knowledge and experience of others, is the perennial mistake of each and all generations. KOT, 00-99
9. The relative universe is this special world of best human life. It is signified by <u>TRU</u> or T. It is synonymous with our world, the actual universe, reality, deity, God, the all, and the absolute. <u>TIT</u> is TRU in and of itself, exclusively, and <u>TO</u> is its totality. What is equivalent to TRU is <u>TREQ</u>. 0+, 1+.
10. When, and what is, **equivalent or the same** to TRU, <u>TREQ</u>. carries great significance. It is how the concepts synonymous with TRU are not completely equal to it that reveal both paradoxes and errors in our knowledge, as well as leading us toward a better understanding of our world. TIT-TO
11. The Relative Universe is highly **selective, qualified, integrated, differentiated, focused, unified, exclusive,** centered on the existent at once, length and special mass energy, of the interrelated totality of its own being, actuality, aseity, perseity, haecciety, and quiddity. SC
12.. This special world of best human life, being made up of **different aspects** of the same thing, is all that focuses upon and enables human life, spirit, mentality, and culture. **It is simultaneously unitary, finite, infinite, total and absolute.** <u>TROS</u>, TU, TE, TI, TO, TÂ 12-15
13. There is good precedent for understanding the relative universe in **religion, science, philosophy**, metaphysics, and especially <u>epistemics</u>. Each has their part in TRU, are not separate nor for limited interest and use, but different aspects of the same thing, TRU. <u>DAOST</u> 05,12,15
14. Through the general and special, universal and particular, integral and differential, deductive and inductive **convergence of all** into the actual world, its form, function, and composition are shown.
The depth and balance of such an approach is the key to wisdom. <u>CONRIT, CONVET</u> 01,06-2,10-12
15. We search for valid facts and evidence, yet even these must be corrected to the degree they exist, relate, have limits, and are proportionate in our world. The relative universe is formed and can be represented through the **corrections and correlations of limits**. <u>COL</u> 02-11,2+

16. In order to understand **what is actual** we need to understand what is not, the relative and absolute and their varieties. Through comparative entities, their essences, existence and contribution to TRU we can not only discover what is not, but what actually is. CENS, 03,11
17. In contrast to physical mass and gravity, the **unique mass-energy of human life is ordered by relativity,** the relativity between the observer and the observed expanded to encompass all the ordered means and powers of structural and functional adaptation that best assures survival.
18. Man's consciousness, immediate awareness, or **instantaneous attention and action**, IA, is the focus of human life, by relativity, observer-observed, or S-O. These are not singular, but highly complex and dynamic.They describe much of the human **mass energy continuum** and spectrum. Á. 05,15
19. It is because of the lack, limits and lability of truth and proof that **doubt exists**. Our world is infinite, and truth is often partial. To properly and fully show TRU we must account for and correct many concepts, principles, laws, truths and mechanisms with their part and roles in TRU. CALP
20. To better understand the world and our lives we **must separate** that of the given, or all **entities** that come before us, into what of it is of **our actual world**, what is in transition, what it is in and of itself, and what is not, 10,11, ESIN(S), AMTRU, ENIT, null
21. We and all about us exist in a way that is less than perfect and TRU. In being so, we must always recognize its **mixed, from, against, gradient, other, many combined and variant forms**, states and aspects. They are, like the devil, often commonly invisible and insidious. AMTRU, NAMTRU, ATRU,
22..All that is not of the relative universe, as much that is in, of, about, all around us and through all we can sense and know, is illusion and figment to the degree it is not **precisely, selectively and by quality purely of** this special world of best human life. Learn to sharpen your wits and senses. ATRU. 09,10,11
24. The secrets and mysteries of life, as the spiritual or heavenly world and God, are not so surprising nor perplexing **once discovered**, but more subtle and satisfying. Man's greatest **paradoxes result** from failure to apply basic principles, forms, limits of life and their ***rigorous proof***. KOT,10,09 SCARLS
25. Much of the separation of what is in or not in our actual world is done naturally, yet often very inefficiently. So that, the key to a person's life is to **better separate and select** what is of the actual world. It is to discover, learn and practice how and how much this can be done in the best ways.
26. The TRU that existed at any time in history was actual for that time. Yet **from TIT-TO they were not,** by being increasingly diverse and inactual. So also is it for the future. Much of the past and future, over and above being less or non TRU temporally, were also non TRU in mass energy and spatially.
27. We must always be careful to **avoid false attribution to our best of worlds**, of that not of it. It explains, in part, why the future and past do not seem to apply to life today, even though in actuality much does. Diversity in TRUS comes from the diverging pre and post forms TRU. DITRUS
28. To best understand history - and the future - we must be able, for each time, to know their sameness **and differences from TIT-TO**, and to a lesser extent from each other, what is less than and negative to these various TRUS. The outward divergence of TRU also applies to the mass and length dimensions.
29. Neither creation nor evolution alone explain beginnings and ends, our world is in an 'at-once' continuity focused on the here & now. Largely determined **by relativity, the at once**, or *ATO*, continuity includes aspects of creation, evolution, cladistics and other processes. 11-,
30. We the observer and it as the observed are one as our subjectivity and its objectivity unite in the predominant **mass-energy dimension and state** of our being. It is formed and functions by its predominant power or special relativity in human life. Á, SO,,09-13,119,17
31. The Relativity of an entity or **TRU** only survives when its predominant powers give it overwhelming haecceity or thisness. It becomes exclusive when its **predominant power is** in harmony within its limits and overpowers or compensates for **all** opposing forces. TRIA, 11-13
32. **Cladistics**, or the science of convergence in biology, reveals our overwhelming thisness , the convergence in and the focus of the here and now in human life, and the higher powers of relativity, observer-observed recognition of its own absolute existence. ATO, TRIA, 15-16-26-30

33. The ***at once, ATO,*** form and function of our life, TRU, includes an infinite past and future and an infinitesimal present, an infinite large and small and an infinitesimal medium, and an infinite observer and observed and an infinitesimal real, **an absolute relative**. TRIS 11-139
34. The great **determinants** of a simple clear description and definition of This Special Universe of Best Human Life are its combined **Genetic Material - Live Product - Determining Forces**, and special mass-energy of quality, selection, predominance, continuity, and survival**.** 116
35. Our consciousness, instantaneous attention, mentality and activity function as observer-observed or **subject and object in the special reality** of the relative universe. Like an object being subjectivized in time and space, they interact with a multitude of forces and determinants. 119.
36. The connector and vehicle between subject and object is the **real**, or what makes the continuum of subject-real-object, the mass energy of human form, function, composition, and action. It can be recognized when all the aspects and powers of the spaciotemporal are left out. 08,12
37. Along with length and time, our ideas and knowledge of mass, relativity, **mass energy, and their expansion in human life,** must undergo vast transformations if we are to fully grasp their actual meanings in TRU and demonstrate them accurately in a 1:1 replication. 12,16,19
38. The increasing and self propelling powers of human **mass-energy or Á**, that are so much of the existence, essence, and thisness of TRU, become more evident with an expanded knowledge and appreciation of the relative universe. 09,10-13
39. Each organism and its kind become self-contained units which function as **sets, in which they are in a life and death struggle** and competition with all about them, even, to a more or less extent, their own kind. It is such a relative, special, and exclusive set that enables precise representation. 0+,01,11
40. **Human mass energy forms, and is formed by, a continuum between the physical and mental,** sensory and motor, or subjective and objective worlds. Much that constitutes, typifies and enables human life is a part of it. Á, SÁ 1+, 18,
41. Just as form and function are not separate, and the nervous and muscular systems of life forms work together in extension as the **neuromuscular** system, so are the observer and the observed, **subjective and objective but two extremes of the same continuum, mass-energy** for TRU. Á, SÁ
42. Along with length and time, the continuum of the physical and mental mass energy, are the basic dimensions of form and function. They enable human life, **provide permanence, survival, and its relative absolute status**. e.g. Autonomic NS, ACEXIN, 12,13,18,
43. The **dimensional makeup** of our world, unlike that of physical objects, is based upon its own mass-energy, length and time form. The unity, totality and dimensional order of the infinitesimal, finite and infinite scales **enable us to accurately represent** our world. ALT,TROS,TRIT,ROT.12,2+
44. The relative universe is **centered** about the being, normalcy, quality, health, improvement, strength, predominance, success and survival of human life.Although centered on the person, this person is, more or less, of any age, sex, race, language, nationality, etc. CENU
45.. That the actual world is **homocentric** is proven by, and proves, its mass energy continuum. It is neither, un-centered, dis-centered, nor astro-physically centered. This is by systematic unity and closure in its own **at once** nature, form and function.TIT,11, 13,15, TU, man, Á. CENU
46. In being homocentric the relative universe is highly **inclusive, qualified and typified by Homo Sapiens**, 'sapiens and verbum sapiens'. To be human is to be wise, have all that is essential, and recognize what is best of this wisdom for our world. CENU, KOT, 10-13
47. To be wise, and all the neurological that accompany it, are what makes us a species, separate, predominant and distinct from all other animals and things. Yet what actually exists is **only what is relative to us**, not all we sense, believe, or entities in themselves. 09,11,13 13
48. TRU is **homocentric** about all persons, more or less, depending upon their participation in and contribution to, the actual world.. It is also homocentric about much of the world that supports, enables, and enhances human existence. CENU,TRU 13-18

49 The apparent world, the worlds we see and seem to be in, **are far different than the actual world**, TRU. It is for this reason man has to realize there is more, a special, spiritual, heavenly world that is right and valid. Each person is lost largely by failing to see and be in and of it. 1+, 158-160
50 Yet this special, spiritual or heavenly world of TRU is much different than commonly and religiously known. It is not opposite the material. It is highly material, formatively and functionally, the **best that the physical, chemical, biological, and psychological** can provide for human survival. **16**,18
52. Our universe is absolute, exclusive, special, unique, necessary and different than other universes or worlds we know. It is centered on human life and mentality and focused by the present. It is unitary, total, and of a **set finite form that enables** accurate and reliable representation and modeling.TRIT, 13 53. Our actual world, TRU is the absolute best of worlds, not so much else about us. The special case is the powerful real world of **human spirit**, **culture, and life**, shed of all impurities, disproportions, invalidity, external representation, figments, delusions, and disorder.TRU,TIT,10,133
54. It is important to distinguish our world **generally, specifically, and how it exists in and of itself**. Nothing **actually exists** except and to the extent or degree it participates and contributes to human relativity, quality, success and survival.11,12,13 TIT, TRIA, TE
55. The measure of all things is man, in a selective, qualified and TRU sense. Human life, individual to total, is very special, sacrosanct, a whole unit, **integrated and differentiated** from the infinite to the infinitesimal, with increasing powers of finite distinction and determination. DIN, MADAS, 11-13
56. Although persons and human life are the core of this universe, **it includes** much more than homo sapiens alone, or even civilization. It includes much that is, about us, and all that **enables it to be**, although often commonly known, this is only in a selective way, ENT,11-12,14,15
57. The measure of all things includes much more than persons, groups, civilizations, or universes. It includes all that enables the human species to be, including much that supports human life in a special way as: the sun, earth, soil, oxygen, water, food, other animals, the life chain, etc. 15,18, TE,ENT
58. The relative universe itself, TIT, is the **qualified, exclusive, and selective mass-energy spatio-temporal** continua, form and function of our world. Its length, time, and mass- energy dimensions are combined , adapted, altered and warped by this special form and function.12-16
59, To understand the makeup of our world we must learn its **parts and processes, how they differ** from things-in-themselves.Of our actual world the part and parts in general must be **made distinct** from its part and parts specially ., ENIT, ESINS, E, ES, TE, TÍ, and TÍS 15
60. What is of a part or process can range from extremes of: **exclusive to itself or common to all**, and any mixture in between. One's actual being is a **combination of , exclusive, common, and their mixture**, in one's participation and contribution to this best of worlds, TRU..TÍ, TE, TE-TÍ, TÍS, TÉS
61. As much as we seem to be we are, **dead and don't know** it, to the degree we are not our part and contribute to the success/failure of our world. We are largely and variably dead in life and alive in death. Our immortality is past, present and future by fulfilling our roles in this world. DAK
62. The three principles and functions of **purity, actuality, and reality** in a person's life, as in TRU, work both separately, together and all in between to reveal what one must focus on to be their general and special parts in this world. PUREAL, TRU, 15,65,75,85,95
63. **Purity** is the qualitative and quantitative state in which this special universe of best human life is free from all not perfectly of it. None is disproportionate, null or negative to it. It is free from impurities, defects, contaminants, detriments, damages, injuries, disease, and all. 13,17
64. What is less than pure is repeatedly and often **continuously interacting with TRU**, without altering its final purity. For human life is so formed and functions, or adapts in ways that through interaction and reversibility, maintains internal milieu, homeostasis, and equilibrium. 16,17,18, HOM
65. **Actuality** is the qualitative state in which all that exists does so only if fully relative, enables, contributes to, or is a part of this special universe of best human life. It approaches absolute aseity, the underived and independent existence, form, function, and manner of our being. .13

66. **Actuality** is a mass energy typical of human life in which the relativity of observer and observed , or subject and object, is predominant. Actuality of any entity is the degree to which it, and especially its essence, is its proportionate distinct part and contribution to our world. 11,13, ESIN, REM.
67. **Reality** is the qualitative state of combined subject-object that characterizes much of the mass energy of best human life. The mass energy form and function of this world may be divided and shown by its many levels, most physical through most combined in the real, to most mental. Á Cu, 06-3i 35
68. The **actuality of reality** is the degree to which we observe and keep focused upon what is only TRU. The relative universe is not any life or any human life, it is only what is actually in, of, and by itself, virtual, living and genuine. In reality what selectively assures our continuing survival. 12-18
69. How our reality, or S-O identity, is actually pure is how our special mass energy, or **Á** dimension and continuum are fulfilled and maintained. **Purity, actuality, and reality** are the three great modes that identify This World, and each person if they are to actually be of it. PUREAL 12-15
70. By relativity, **reality** defines, in association with gravity and other increasingly human forces, the total mass energy in our universe. Quantitatively it is the degree subjectivity-objectivity reach higher levels of identity, distinction and totality. S-O,Á,TE,MADAS,01-03,11-3
71. The make up, form, parts, processes and systems of which man, as TRU, is constituted are within themselves formally discrete and can be **treated as independent variables or factors,** representable within limits as mathematical factors and functions. 12,15,16,17,25,40
72. Many subsystems, cleavages, organs, forms and parts of this actual world are **compounds of smaller and larger** systems. Each system is derived from those before it. Some, and the planes or levels they assume, are derived from their own set proportions of mass-energy, spatial and temporal continua.
73. All exists only in that portion and way their special relation is essential to TRU. Since some parts, as part in general, are vital to the whole, thus existing absolutely, **Law of Partial Necessity**, <u>LOPAN</u>. Since persons in general are essential and vital to the whole, to this extent are, or tend to be absolute.
74. The focus and great carrier of The Relative Universe is the optimal individual person in the best of human life Yet, through the **life chain** much of the world, all that supports and sustains human life, although not specifically human, is often more generally and actually of us. 10-15,
75. The **center** and axis of TRU is the person who is special for TRU. Its focus is derived from the media of human mass-energy, length and time. This is seen in the interactions of conscious life and behavior. It is altered, adapted and warped, increasingly by the mental and less by the physical. IA.
76. By relativity humans have center, focus, and a special and separate universe. We are not exclusively restricted to the physical world. It is greatly on how well or poorly we can develop, determine, understand, and show **the makeup of the mass-energy dimension** that we will, or will not, be successful.15
77. The central, peripheral, sympathetic and parasympathetic sensory and motor connections are evidence and beautiful proof of the focus, crux, medial region and thereby **gradients of human mass-energy** form and function of TRU. This is critically necessary for its representation and use. 15
78. Mass like length and time are not points, but like lines, areas, and volumes that can be considered as continua. Their spectra show their many different levels and gradients. Such **differentiation** from unity enables factoring in continua that **provide the measures, operations, and proofs** needed 15
79. Our world, like the God of monotheism, is formed by and functions as a Unit. Yet the unity of the relative universe is not simple, but a **great composite totality** of superimposed differentiated and integrated strata of which it is formed and functions. TRIT,135,.12, 139 TU.
80. Life, humans and TRU owe much of their living to the order and organization provided by the self determination our dimensional frame and form of special ALT is given by the **systematic unified totality** of TRU. This is the **primary law of unified relativity.** <u>PLUR, SUT,.TU</u>., 136,16
81. The farther from the focus of TRU, the more different all becomes. Forces in nature are relative as they approach the unity of the mental and physical, S-0, or Á, our mass energy dimension. With l and t continua **the frame of TRU is set, whose form sets the order** of all. PLUR

82. What is ***true*** of partials is not true of wholes; what is true of wholes is not **true** of partials, even though both be perfectly of this world. Popper, constructionists, and critics have their point in doubt, yet fail to have the great overview possible by TRU in totality. PERT, TO, TE
83. Our world is constantly **changing**, and by **variable rates**. These can be both internal and external, reversible or irreversible, and perfectly or less than perfectly TRU.. It is in a dynamic milieu whose continua are and can be treated as a unit, infinite, and in many various finite ways. TRUCH, 168-17
84. The form, frame and construct of our world can only be shown by how each of its three basic dimensions exist in all their powers. The continua of dimensions basic to human life are **not separate parameters but highly interdependent** aspects, forms, and functions of it. DAOST,12,25,30
85..The **higher powers** of our actual world are best understood by their dynamics and combinations that form its overall predominant frame and structure, especially as they function together in the totality of TRU, as shown by ÁLT, SPALT and **ACEXIN.** 15,16,17,25-29,3+
86. **Actuation, Extension and Intension** are - and show - the highest powers of mlt for our world, its special and typical mass, length, and time. They work integratively and differentially in its totality. It is by actuation that m, extension that l, and intension that t are TRU. mlt, ÁLT, ACEXIN, TIT-TO, 26-3
87. The different **powers of these dimensions** are known, as the length, area and volume, l_1, l_2 and l_3 time, duration and variation or t_1, t_2, and t_3, and the exponentially expanding predominant relative mass energy continuum of our world. $Á_1$, $Á_2$ and $Á_3$.. Together they give solidity to TO. 15-16,25-29,3+
881. The highly complex and total advanced **powers** that drive human predominance, as exemplified by the higher powers of its special mass-energy, spatial and temporal form and function are key to understanding the basic nature and survival of human life. 17,18,19,2-9
89. The form, participation, integration and balanced optimal function of dimensional form make-up much of our **health, strength, and survival.** It is a composite of an infinite number of parts, processes, aspects, and attributes. TI 149,18 15-18,5+
90. All that is disproportionate, inactual, invalid, and non total is to that degree non existent or negative. Our world, TRU, as any object, can't be shown without a **valid, actual, total and proportionate** demonstration of it, VAPO. 08,11,2+ 11,19
91. It is, not only the simple **parts** that make up the human genome that are crucial, it is their combinations, relations, uses, and most of all how these exist and survive within the validity, actuality, proportionality and totality of the human species. GELD, 19
92. Our actual world, TRU, is an entity whose dimensions are altered, warped by, and consist of gradients that follow continuous adaptation. These can be treated objectively through its **own parameters** of actuation, extension and intension. OBT, ALT, ACEXIN.18, 19, 20 24,25,32
93. We live in a world typified by compulsory **dimensions,** CODIT, that increasingly help, guide, and enable us to live. The proper analysis of the actual world is achieved by its **systematic unity** and dimensional formulation. Their learning, practice and use are key to knowledge. SUT 19,2+
94. **Correlations** reveal the profound tendency to sameness, convergence, and contingency. They, in relativity and absolute unity, show how the different and separate are brought to TRU. They are used to provide precision through measurement in and out of TRU. 11,12,21,4+
95. To understand only commonly used words defined as entities in themselves or, even a few of their essences, **are to miss the great gift language,** new concepts, and all the deepest meanings and revelations human experience has, can, and will provide. LANT, NEC, 0+,2+,20,21,22,23
96 Names, equivalents, or anything that represents the actual extant TRU, are, variably, not exactly the same, often much less, other, or negative. Yet as **vehicles in the real** of our mass energy, their part and roles are valid and must be known within the totality of TRU. .2+,,DAOST, TIT-TO, RÍT, RÏT,
97. The **dimensional formulation and frame** of our actual world, TRU is called , ÁLT. It can be represented and modeled with a high degree of accuracy. We must never forget, and **constantly remember** all that makes up this universe, its representation, presentation, and use. ECRO 29,9 3+

98. The form and function of the length and time dimensions are simple and clear compared to the mass dimension. It is in how we can make **the mass energy or Á dimension** simple and clear that will produce the most reliable, acceptable and effective representation and presentation. 30
99. The representation and presentation of TRU by simple dimensions are not enough. All that is complex, even when accurate and total, is not so simple. For human mass energy, Á, we start by **contrasting it with length and time**, showing their higher powers and complex interactions..30
100. We must remember that even length and time are not as simple and clear as they are physically. More than they appear, or we assume, they are **adapted, altered and warped** in TRU, mlt to ÁLT. They are special and exist differently in higher powers and combined totality of TIT-TO. 30
101. The explicit **worth of models in** the relative universe comes from enabling us to show and apply its form and function, parts and processes in the structural frame of its special dimensions of mass-energy, length and time. Models that accurately replicate all TRU are the greatest. SPALT, 30,35
102 There is a striking difference between the **portion of the world something is, and the portion of something that is of the world.** The first is its proportion or **relative magnitude** in the actual world, and is usually far higher than the second, the essence of an entity that is TRU. TIT,ES,REM,14,35
103. The makeup of the relative universe, the problem with its proportions and values, and much of our difficulty with showing and knowing the world as it actually is, is because of super percentages. This results from overlapping of **values that far exceed** their actual relative magnitude.TRU, REMS, 14,35
104. The method of measuring our world by the **decrease in REM that occurs with each added value**, needs to be constantly remembered. It doesn't lessen their actual value, it only reinforces the absolute value of the whole. It requires a great knowledge and application of TRU, SPALT, and REM.
105. The **relative magnitude** of all that is in or of TRU consists of set principles, laws, and methods of measurement. By showing its composition and values, both human and other than human parts, we add to their recognition, the perfection of TRU, its model, and their use. REM 12,14,35-39
106. Nothing goes without cost, all may be valued, from infinitesimal to infinite worth, from negative to positive value, from absolute dis-value to absolute **value,** depending on its harm or benefit, detraction or contribution to the whole in this special universe of best human life. REM, 35
107. An entity's degree of positive/negative effects, or proportion of the whole of our actual world, is its relative magnitude, or value. One of the methods of measuring our world is by using and modifying existing scales **already provided by many of the major constituents** ofTRU. REM, MOCBIREM
108. The final and conclusive determination of existence in the actual world can only follow the targeted replication of the whole. Its form and function, parts and processes can be **posited** in such a way as to enable us to render and everyone have **a place for everything with everything in its place**. POS
109. By taking advantage of the form and frame our actual world, providing its best positive objective approach, and using the complete, organized, and continuously available knowledge of TRU the **ideal of a solid geometric placement** for every pertinent and useful participant is achieved. 4+
110.Through a valid, actual, proportionate and total showing of our world a **1:1 replica** can be represented, modeled and produced. Such a replica uses systematic unity and the dimensional frame and formulation of TITto provide a model for positional analysis of its parts and processes. SPALT,3-4+
111.To fully and **best represent** this world what is of the same will often be coterminal, synonymous, repetitious, and routine. The **dimensional measure** and position of an entity, essence, ES, CES, or aspect of TRU in its model will **vary** with each definition, meaning, and how **used**. DEMVU
112. By producing a model, and the associated values, posits, statics and dynamics of the relative universe a positive objective approach is achieved. It **provides a description and presentation** that has been heretofore unknown, long sought after, and will be of the greatest benefit. 3+,4,5-9+
113. The best possible replica, a **great spherical model,** can be filled with the most accurate massive intricate and beautiful detailed positioned parts, and processes. They will provide the clearest, most distinct and well remembered overview, for proof, application, and closure. SPALT

114. **Values and posits** in the model of TRU can be computerized, programmed, communicated and applied to all technologies especially the most advanced fields and sciences for prevention, protection, management, control and their, and the world's, enhancement and future perfection. REM, POS,4,91,9+
115. The **parts and processes of TRU,** its ES and CES, are best represented by **posit or placement** in its most accurate models. They need continual refinement The benefits of showing the form and function of our universe lie in providing effective management, action and control.MAC,4+, 9
116. The positing, relating, measuring, testing, determining and utilizing **its processes is the study** of our actual world. The processes, of each person and all, may be calculated and positioned to provide the means of best showing oneself and our world, TRU Homation, CES,. 17,35,5-8+
117. From distinct and well developed information, concepts, principles, and knowleddge that are properly adapted to TRU, our demonstration of its beautiful intricate detaileded parts and processes can be highly revealing and successful through **total posits in perfected models**. 3+,4+,45+,5-9+
118. What effects human life occur in terms of their good or bad, right or wrong measures. All the propelling and compelling or drawing forces, processes and dynamic relationships affecting man do so with more or less and often mixed **measurable positive, negative, or neutral charges**. 5+
119. For the success and survival of the human race in the rapidly expanding, highly rational and technical, world, it is vital that **all persons have**, as much as possible, the capacity, ability, knowledge, reactions, setting and conditions or means to fulfill their roles. CARSC,09,15,52,59,
120. ***Faith*** is all our energy that appreciates what we have in common, manages our differences, and extols what is and will be of benefit. Each new person and generation must have *faith* in all that has been provided and all that enables their **past, present and future**. PPF, ENT, 08,18,58
121. The dynamic processes, vectors, forces, mechanisms and operations of TRU are demonstrated, posited, and utilized **simply, complexly, and in totality**. The resultant conclusions, solutions, accentuation, and actions can be readily developed and presented. 5-9
122. We are only to the degree we are a part and process, member, and **contributor, to the whole.** Our sin and crime are the degree to which we are not, back slide from, do harm to, take pleasure in the negative, or fail to do our part, fulfill our role, and become one with all in the whole, TO. 15,16,8+
123. Consciousness and immediate awareness, along with being man's greatest asset, are also man's greatest weakness. For it is often led by appearances, whims, and desires, rather than actualities and **what is correct and best by** human life in general. IA, TRU. 10,15,65,75,85,91-95
124. Without an understanding and appreciation for the deepest TRU meaning of **humility** mankind suffers. It is a characteristic or mind set that appreciates and lives up to what they have been given,, how to sacrifice themselves, and what they do to fulfill their part and role in TRU. 65,75
125. Each person throughout life decides and chooses what to do. What to do fails without **knowing and wanting the best.** Each person can be a powerful dynamo for best/worst by the positive, ordered knowledge and means that enhance our being. 15,64-65,75,85,9+,90,91,95
126. We do not exist as much as we think or our vanity assumes .We **actually exist only to the extent** we accept, experience, apply, adapt, and contribute to the selective quality, success, and survival of human life. How it becomes much longer, better, happier, easier and freer from suffering.
127. There is much opportunity, joy and happiness to be found in life, especially to the extent one fulfills their part and role in it. **To be our part** and fulfill our role requires a mastery of the parts, processes, concepts, principles, laws, and key dynamic regulatory mechanisms of all levels of our being. 75
128. The role of **human genetics** in the actual world, like all that makes up TRU, is extremely complex. As numerous as the many different genes and constituents that make up the human genome and genetic material are, it is only the smaller portion of its extreme selective and interactive complexity.7+
129. The complexities of **genetic material, live product, and determining forces** reflect those of TRU. They are manifold, times the great diversity of the forces and dynamic environment and world in which they exists. This is why there remains much knowledge to be discovered and applied. GELD, 8+

130. The complexity of **human health** is only a portion of the complexity of TRU This is seen not only in how no two humans are alike, even identical twins, it is seen in how different each person is not only when healthy, at different times, but even at the same time. Such knowledge is needed for tolerance.7+
131. **Peace** without competition, selection, quality, and fulfillment is false.War without the utmost actuality, comprehension, and judgement to secure the greatest benefits over losses for the whole of humanity or relative universe, is unjust, savage, dangerous, destructive, and fatal. 87
132. The expanding capacity of mass energy continuum manifests itself throughout human existence on earth. As characterized by its form and function in **each and all aspects of human life,** it typifies them and explains much of their part and role in the whole. Homation, DYNO, 5+-8+
133. The great importance and value/harm of religion has been its **actuality/inactuality and totality/partiality** in human life. It has been a prime source of human survival. Yet it is still in a formative stage, only to achieve closure through TREQ with VAPO, DAOST and TRU.,*81*-88.
134. **Each person's** entire life is a testing ground, separator, and **selector of the actually existent** from all else. Exceptionally we exist to the extent we are actually a part and fulfill our roles in the whole, the success and survival of human life..TES, TÍS, 75,85 09,13,15,55-95,85
135. Each of us can't fail to **fulfill our roles** in this best human life. Like nourishing our souls, it is through the needed energy, depth conception, inspiration, revelation, devotion, distinction, selective quality and total comprehension for being our part in TRU. TES, Á, MADAS, 75,
136. Because religion is often so divisive and fails us doesn't mean that it is of no value or can be neglected. Religion is becoming and must be the great tie that binds, like our mass energy, the selector and **demonstrator of much that is the best over worst** in human life.61,71,81
137. Because of the complexities, superimposition, infinite nature, absoluteness and all its makeup, each and all persons by the actual world, TRU, can't expect to **be their counterpart** without learning, knowing, remembering, using, and acting more and more by, what is offered. 75,85
138. You and I, each and **every person is immortal**, divine, and Godly, to the extent we are our part, contribute to the whole, and enhance the degree this special universe of best human life continues, survives, succeeds, flourishes, has permanence, and becomes absolute.TES,15,85,95
139. **Knowledge** needs to be both **universal, partial and special**. Universal as it is of the whole relative universe generally, partial as it reveals its parts and processes, and special as it enables the individual to fulfill their role. KUPS, 15,75,85,95
140. Like our ancestors, each - and all - of us **would not be here** if they had not been the best there were. This applies differently for close and distant generations. So, for our progeny, we must be the best, that they, theirs, and their world will continue and flourish. 11,7+,82-87
141. Until everyone understands and **appreciates what they have** in common with others, tolerates and adapts to what is different, accepts, and works for what is best for human life, there will be no good will to man, peace on earth, deliverance from endless suffering, failure and extinction. 8-9
142. It is by computerization, the media, **methods and development of information transfer**, with persons who have the interest and dedication necessary to comprehend and apply such complex, dynamic, and total TRU knowledge, that we can be assured of continued discoveries and success. 88,89
143. We do not commonly recognize the **great voids that exist** in human life until or unless we are lost, confused, defeated, in distress, or when our world fails us. Yet, they are continually and massively with us, and as all that is of TRU, has its place and role in it. .0+,9+
144. To believe only the present exists, everything has separate existence, oneself is all, or human life isn't divine, are the height of ignorance and naivety. **TRU is certain**, oneself as all persons are the agents of TRU, and to seek its eternal life is the height of pure actual reality. PUREALITY, 91-5
145. Each individual, group, nation, and culture thinks it is the best, the ideal, the perfect one. Little do they realizie **how far short** they fall from being TRU or TÂ, or even their necessary part and role in TIT-TO. Such a revelation, as that for God or TREQ, is the beginning of being and wisdom 10,15,91

146. Great problems and troubles, the causes of human suffering and demise, come from a lack of, or wrong, understanding and action of TRU actuality, meaning in depth, and a **positive objective approach** to what this world of best of human life is all about. POA, KOT,91,
147. Those who think less of human life let them show why and how they are to actually live. We will show why and how you have **no choice but to fulfill and enjoy** what is the best of it. Why and how you should learn and live by TRU if you are to be, and if your life is to have any meaning at all.
148 We must **constantly lift ourselves from personal mediocrity** in our own agendas and bring these to their equivalents in TRU. This can be done without destroying our individuality as long as this equivalent incorporates all that we can be by and in TRU, TES,15,95
149. Without TRU or TREQ you can't relate and fail, all you identify and stand for are lost with you Therefore discover, learn, work for, enjoy, live by, and **find identity**, revelation, inspiration, devotion, courage, determination, conviction, security, serenity and happiness in TES & TRU.
150. **Antitheses** are natural to human life and are largely a reflection of relativity and the opposition of differences, the D in DAOST, and variations in our world,. Often little extra action is indicated. How we best account for, reconcile, correct, accentuate, manage and control them is key and our duty.5-9-95
151. There is 'all the difference in the world' between antitheses that are **harmoniously and best synthesized** in TRU, and the great conflicts, wars, and destruction that occur from the discordant and degenerative courses antitheses take without TRU. We can not know these without TRU. 17,19,5+,9+
152. Humans are **trapped in their own** dead seat bounty of neglect and failure..What they have been given, must achieve, and fulfill are not realized. To know, show, act, and serve the actual world, each person must be open to learn, be saved, achieve providence, and eternally survive. 91-95
153. To realize that ones roles through service and sacrifice are at the basis of coming to terms with life. **Sacrifice highlights** the contrast between ATRU and TRU. It is a major way, as exemplified by the saints, to immortality or the absolute for one and all in the relative universe. TÍS - TES, TO-TIT 7-9+
154. Ours **must be a life of** ever higher participation, competition, contribution, predominance, fulfilment, happiness, success, and victory in the life and death struggle for continuity over extinction of this special universe of best human life. It all resolves into to what I can be and do to assure TRU. 9+
155. Public opinion, reflecting individual notions, tends to vacillate from one extreme, controversy, fashion, or fad to another instead of having a rational and **comprehensive approach to the future.** Human survival is assured only by the control provided by thoroughly learning TRU.
156. Once we understand doubt, can show what is and is not true, and can produce rigorous, actual, proportionate, and total proofs for all, we are well on the **way to a complete** and comprehensive knowledge, and the management and control necessary for our lives and world. MAC, 09,10,19,9+
157. Legitimate **differences in knowledge** exist side by side, even when perfectly of this world. Only in the versatility and totality of perspective and TRU can they be reconciled. By relative selection Great **knowledge once learned** and lived enables the **all time mastery of things**. KOT ATIM, 10,19,90-99
158. It is to have quality, organized, comprehensive, total, and usable **knowledge, and individuals** with the capacity, incentive, and interest to learn, order, positively reconcile and apply it, that is basic and essential to human success and survival. KOT, KUPS, COGNOS, LATANOS, 9+
159. Lack of **respect for all that actually makes up** and runs our lives produces misjudgements, errors, and misdeeds, both for oneself, others, and all humanity. It is largely a failure to appreciate, have, and use the best in valid, actual, proportional, and total ordered knowledge VAPO, KOT ,96
160. To bring all to our world, and it to all that is known, or existing essences that are unknown or unnamed, is the way to best account and reconcile all. We know, and **know that we know**, by the validity, actuality, proportionality and totality of our world, TRU. COGNOS VAPO ,96
161. Great satisfaction, **happiness and joy are to be found** in realizing the fulfillment and assured continuing betterment of this special universe of best human life. With the benefits and leverage a perfected representation and presentation offer we can proactively direct and control all in the future.

162. Mankind's fulfillment also results from **language's** special way of identifying, making distinct, meaningful, and usable the detail, unity, totality and order provided by the **representation of TRU**, its positive objective approach, modeling, operations, and communicated systems. ROT

163. Many of the **greatest causes, problems, threats, and needs** of the future are, more or less, both already existent, and fundamentally paradoxical, states, conditions, relations, and limits of our existence. That between races, languages, religions, nations, and trade exemplify such resolvable differences.

164. It is these when long neglected, or when they become so **out of harmony with human life** in general, as it is TRU, that they explode into more violent and often disastrous products. Thus it is the ATRU and ANTRU worlds about us that make up much of what we face in the future. (PPF)

165 We must be **constantly on guard to prevent loss**, allowing external forces to divert us from, or effecting all circumstances that detract from, the purposes, goals and ends that can be so readily and fully comprehended with a valid, actual, proportional and total knowledge of TRU. 92

166. There must be a **consensus among all who comprehend** the best and greatest in human life. This must be made interesting and stimulating, fashionable, the major thrust of all conversation, and vigor-ously reinforced by actions that prevent any who would either advertently or inadvertently destroy it.93

167. Negative reactions to the good, loss of experience by new generations, and escape from routine and regulation must constantly be guarded against. They must be **transformed and adapted to TRU** before they become detrimental habits of behavior and ways of life that degenerate into permanent loss.

168. The greatest barrier to a full knowledge and life in TRU is the loss with each generation. Only when each new person and group has the quality, character, **interest and zeal to learn, remember, and best apply** themselves to TRU can their combined success and survival be assured. 95

169. Since human life is so short, knowledge and contribution so long, there is no time nor opportunity to lose. By **learning TRU**, developing an accurate model, and practicing its finite fulfillment, we avoid the worthless and harmful to pursue the best deepest meanings in all..95

170. **Only with knowledge and action by** TRU or its equivalent, can each of us, through the validity, actuality, proportionality and totality of the different aspects of the same turn loss into gain, negative into positive, death into life, and discontinuity into permanence.KOT,95,96

171. The purity, actuality and reality each must seek if they are to be their parts, and contribute to this special universe of best human life is the most distinct and profound **meaning of each person's life**. Together they exist in the state of PUREAL-ITY, much of TES, as TES is of TIT..15 65,75,85,95

172. **Separating service, self preservation, and selfishness** is difficult without the benefit of knowing the actual world. Being self centered is a mechanism that assures predominant behavior and propaga-tion. Yet, as partial and ATRU, it creates a bias that causes perpetual conflict and loss. MADAS,TÍS

173. Unless we can overcome infolding and mis-folding of all in life, and exfoliate the subject-object situation **to envision the whole**, we can never surmount our inactuality, ignorance, indolence, and partiality, to benefit from our own providence, and become masters of our own destiny. 90,93,95

174 There are many great doubts, differences, questions, problems, troubles, conflicts, and wars that can be readily answered, corrected, solved, managed or controlled. Yet without a **full accounting and inclusion** of all effects they will, and to that degree, be inadequate or lead to defeat and failure. 90

175. This is why it is so important that all persons, as early, long, and well in their lives have both the potential, incentive, and interest to **be their counterparts and best function** in TRU. It is largely up to the individual to guide, influence or stimulate others and all to be and do likewise. 15,65-98

176. Local, regional, national, international and world wide actions can be taken to assure and enhance human survival and success. Yet they are neither singularly, nor sufficiently achieved without purpose, plan, guide, and blueprint as **perfected representation and presentation of TRU can provide**.

177. The difference between what is already in TIT-TO, and what will be incoming is the continual concern of mankind.(OITIO) It is a concern that requires an ever vigilant attitude along with an ever expanding accounting and **reconciliation for all degrees of probability** of what is coming.

178. For what is or is not incoming, and what is leaving, along with the degrees of each, can not be treated proactively for TRU without knowing and weighing each in terms of **their actual TRU part and contribution,** lack, or negative effect thereof.
179. It is these deficiencies and negative effects that often preoccupy much of our attention when deciding and acting on what is ahead of us. Yet to our perpetual sorrow and failure we can not distinguish between what is or is not TRU, its degree, or **how to be totally and rightly proactive**.
180. It is through the increasingly finite toward infinite selection, distinction, actual meanings, separation and positive operations of its parts and processes that **valid knowledge of TRU is developed** and made effective. This is the prime mechanism of our mass energy, Á. COGNOS,96,97
181. We can't deny or defy our world, as TRU. To do so is non being. What is there when human life is gone? All, that has, is, and will enable us to be is lost. **Life necessitates all that enables it to be,** its relativity, unity, totality, focus, quality, selection, predominance, success and continuity.9+
182. From our earliest beginnings mankind has been continually faced with all needed for survival. Yet to survive, we must have the form, functions, and all necessary to enable survival. Also we must have what it takes to **be prepared to survive**. From attention, to all, we continuously fail to be prepared.
183. **Being prepared to survive** characterizes philosophy, science, religion, and the higher powers of human life. Human culture has become so complex, with such burdens, as its population, material means, and goods, that without preparation, the prospective and proactive we fail, and all is lost.
184. To supply the existing world, to preserve its resources beyond what is needed for its supply, **requires a constant vigilance** to increase the material that is of highest quality and productive of the greatest health and benefit. Yet to decrease or avoid using material that is wasteful and harmful.
185. The appreciation all must have if they are to be in harmony, avoid severe conflict and unnecessary suffering not only of themselves, their possessions, what is familiar, close at hand, or how they make a living, but of **all the actual world in a valued and positive way**. 95
186. The best of existing knowledge, methodology, and technology reveals much that leads to TRU. When reconciled, accounted for, and ordered by and with TRU, they give the total proactive means to fully and successfully achieve what **assures, sustains, and yields** greatest quality and survival.
187. The relative universe may be simply described and defined in many ways, as by its form, function, or some of its major systems or components. However named it should be partially if not completely, indirectly if not directly, the **topic of all presentations**. TRU
188. No matter how convincing it looks, how well, or by whom, a work of art or book is presented, they are often little more than a bit of actuality, some small part, extreme, or cause, unless they are equivalent **to a total presentation** of the relative universe. This is their measure. LATOP, IPT
189. The pinnacle, crux, and climax of any representation, presentation, operation and endeavor are how to know, remember, apply, utilize, complete and best incorporate it into the relative universe in itself. For a book on TRU, this is how a replica **best shows and fulfills TRU.**
190. With an excellent learning and development of what is actually relevant and existent, an accurate and reliable knowledge, replication, **utilization, management and control** of this special universe of best human life is available for each and all persons. TRU,KOT,
191 Once the frame, structure, form and replica of the relative universe are established and its major fundamental concepts, principles, laws, mechanisms, and proofs are known, it is much easier to rapidly learn, **expand and use this knowledge and cognition, to completion**. COGNOS, 96
192. How well can our proactive methods, techniques, efforts, and operations exponentially expand human management, control and continued survival? It is largely by knowing and acting to fulfill all TIT-TO that the future primarily takes care of itself. **Much of the future is already in TIT-TO**.
193. To never forget that since the whole actual world, TRU, is based upon the individual and one's immediate awareness and attention, that the future and accompanying actions must primarily come from the **best our attention and our actions can produce to** fulfill our part and roles in TRU.

194. By this reversion to the individual and IA in and by TRU, we can never leave it to the whole, to others, to later, or to various devious outlets. It is **up to us now**, before and after as we in continuity form the center and focus of all, all that actually is. LARIAT
195, There is a tremendous disparity between human knowledge, technical development, and popula-tion, and human quality, order and control. This gap produces escalating deterioration if not **corrected by new and vigorous** approaches founded upon valid, actual, proportional and total knowledge. POA,
196. **Concentrated and focused devotion** to the collective, integrated, and total convergent endeavor necessary to sustain, expand, and advance human life is what best produces and achieves the successful and eternal fulfillment that is essential to human life, TRU.
197. The world is over loaded with all of us who are more or less inadequate and insufficient to meet the minimum prerequisites of the future. Even less will we meet the demands of a rapidly expanding and demanding technology, industry, commerce and populace. We can only rise to it, **a life in TRU**..
198. To have the heaven on earth all expect yet so few sufficiently support, and to prevent the losses so likely to occur from the load and challenges of the next few centuries, each and all must **learn and devote their lives** to a new and wise world in which a valid, actual, proportional and total life is upheld.
199. Much of the best of mankind is sustained by the work force, the daily employment of citizens. Some of this is of great contribution, much is of lesser contribution, and some is detrimental. Increased efficiency, **quality, productivity, and contribution** and decreased detriment is the actual world. TRU.
200. It is most important that we **determine, test, and measure** all that comes before us, whether it is valid, actual, proportional, and total in the coordinate frame of human life. Although our subconscious and autonomic functions account for many, without coordination and control we are lost. VAPO,ALT
201. It is because each person lacks the knowledge, ability, and coordination to know their place in the actual world that they never achieve their full part and role in it. By collecting and placing themselves and all in their **proper position in the whole** of our world it can be well comprehended and controlled.
202. Thoroughly study and remember these theses and themes, learn the major concepts, principles, laws, NEC, ES, and CES and **practice putting** them and your life in the spherical model, **SPALT.** It is in this way you can be well on the way to great understanding, success, joy, and immortality in life.
203 When the **states, conditions, relations and limits** of our being are known our ultimate management and control can bring better **harmony by resolving** the **paradox of indefinitely and infinitely expanding absolute universes**. SCARLS, PINEXU 87, 97-99
204. Yet the solution, what our actions must be and the proactive totality we must produce, is rela-tively simple. It is **no less than being TRU**, TES in TIT-TO. Together with avoiding, diminishing and removing all that is less, other or negative to TRU, the future becomes ever brighter and fulfilled.
205. There is an absolute imperative that all must exclusively and selectively live by, if we are to be. For a person the absolute imperative is how one is a **part and contributor to the world**.. It begins with the revelation of an identity with, and is how to best fulfill our life, in TRU. ÂÏT, AITES, SC, 9+, 939, 95
206. One of the greatest of all future problems is the hopeless condition so much of mankind lives in. The solution is more than unitary or simple. The problems are **largely a lack of TRU**, not only on the part of those in such a condition, but that imposed on them,, both by others and their worlds.
207. It is to give hope that Christianity has so succeeded. Yet it is an empty and hypocritical hope that fails to explain our actual world, nor show what is necessary to uphold it. For hope can be no better than the eternal **fulfillment the best in human life** offers. Every generation needs access to such hope.
208. It is to inculcate discipline, enforcement, and religious fervor that Islam has so succeeded. Yet these are false and hate producing unless moderated by a comprehensive knowledge and endeavor to uphold our actual world. They can only come from the eternal **fulfillment of the best in human life**.
209. All who are less than their part and role in the hope of the future must learn and know their status. They must strive to be their best, and be **willing to defer** when less. For humanity on earth can't survive in an escalating downward spiral of what is less than or negative to the best.
210 There is a very **delicate equilibrium between unity and diversity**, major essences and massive finite

parts and processes. Maintaining this is a prime end of human life and survival. Excesses of either or loss of adequate control of irreversible disturbances to this can lead to extinction.TO

211. The **survival** of human life is not from chance, random, partial, or trivial choice. It is a forgone **absolute imperative** by the selective relative absolute; **absolutely dependent upon us**, as we are on it, by its benefits, blessings and our increasing powers and control. AITES, 10-15,91-95

213. Through learning, believing, and applying these themes, theses, and tenets **one can secure** the meaning, inspiration, revelations, determination and divine purpose necessary to join in, endeavor for, and achieve the growth and fruition mandated by ones own and all human life .98

214. Mankind is at the threshold of finding the long sought goals of fulfillment, happiness, good will, and eternal life. This results from a person's capacity for tremendous inspired revelation, and application, of **the ordered beautiful intricate detail and greatness in our actual world**.99

215.. The **glory of mankind** and its attributes are without end. The identity of a person with ones counterpart and contribution to the actual world is key to this blessing and perfection. We must realize our potential, accept the challenge, make the effort, use its success, and savor its glory. 9+,99

The world of pure, actual reality, or TRU, can not be well understood without noting a few of its key concepts.When used in a sentence they are defined or meanings are implicated. When at the end of a sentence they are singularly noted.

The gist of each thesis is in bold print. This will help to provide an emphatic guide to the whole.

All the information is taken from CIPTRU, the abridged text of TRU. To get a clearer understanding of many of these it is highly recommended that one study this text, and peruse its figures.

What is actual, our world, or the relative universe, is far different - yet has much in common - with all that is known. This is even more so with that used by each person. If we are to ever escape our ignorance, and all that is lost with each passing generation, we need to discover, develop, learn, and use this actual and far different knowledge. We need to practice validity, actuality, proportionality and totality as it exists in and of us. We need its best representation as a map to guide us through life so that we don't keep on repeating the same mistakes and failures.

Throughout the text we refer to the acronym Á, the form and function of our mass energy dimension and continua. This is key to understanding TRU, for the actual universe of which we are so much a part, its aseity, perseity, and haecceity, is not not like what is commonly known, yet is the same, our being. The mass energy dimension arose from the mass energy of the astrophysical world, as commonly felt in gravity.. Yet it has developed to become in humans characterized by our relativity to it. This is the observer-observed fact or law of relativity in which our mass energy dimension exists as the continua of form and function basic to human life.

This diagram reveals the various stages in development of this dimension as it occurred through evolution. Yet since by ATO our world is focused on the present we all need to remember, ECRO, that TRU proper or TIT-TO, its totality, is what shows only in the present of this diagram It already contains within by way of actuation, extension and intension, much of what we ordinarily think of as past and future. .

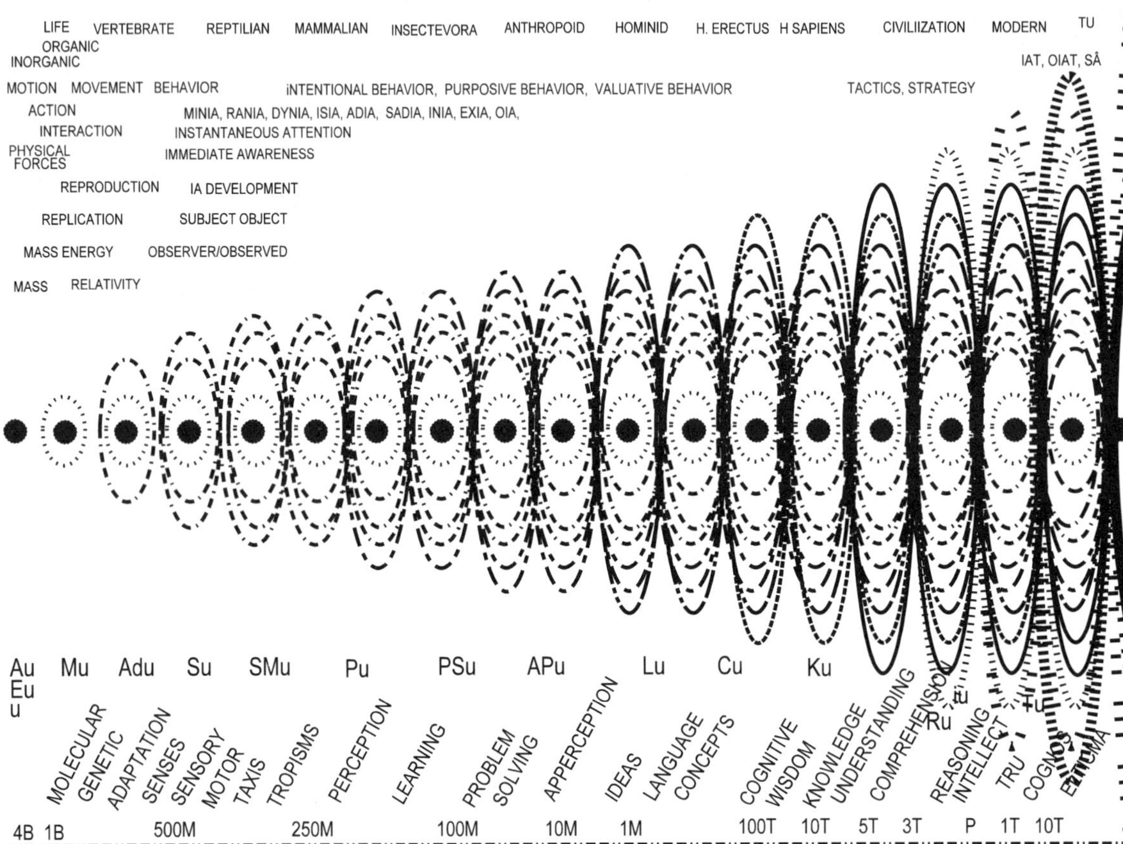

MASS ENERGY SUBJECT OBJECT Á DYNAMICS and IA

To have such knowledge means we need to supplement our existing concepts, words and language with added terms that most distinctly depict this seemingly far different, yet much the same, world we live in. For, *non semper ea sunt quae videntur*, or things arn't always what them seem. New terms need to replace the invalid, inactual, disproportional, and partially misleading words that we become so habituated in using. To this end, if we are to be a part of this world and have immortality, we must take heart in and find joy in learning and being the best we can be in this best of worlds. The above figure shows the gradual expansion of mass energy in life. The increasing and self propelling powers of Á that are so much of the thisness, existence and essence of TRU become more evident. Each organism and its kind become self contained units which function as sets in which it is in a life and death struggle and competition with all about it, even, to a more or less extent, its own kind. It is this great mass of capacity and ability to sustain itself, its many levels of mass energy or Á that is both its great power as well as its vulnerability. For as this becomes greater and greater, as we find in the text, it becomes acceleratively more vulnerable. This is why it is crucial to TRU, and why it is so important that we discover, learn, understand, apply and protect it fully. In the dynamics of POT, it helps us to orient our approach to TRU on the course mass energy has taken in life, and especially life as it has become, and has its human selective focus. If we can continually envision this convergence of the makeup of the length, time and mass energy dimensions in this special universe of best human life, and give them the added forms, functions, states, operations, conditions, mechanisms, parts, processes, actual meanings and terms they contain as TRU, we will much more accurately depict, prove, and understand them with VAPO. We will in turn be able to represent, present, and better remember all things that relate - or to a decreasing degree less significant, insignificant or negative - to our being, TRU.

QUESTIONS and EXERCISES in THOUGHT, INTUITION, and REASON

1. Of what is the world composed?
2. What was the beginning of all?
3. What will the end of all be?
4. What lies ahead in the future for man?

5. Can all things be known?
6. How can the knowledge of all be discovered?
7. How can it be formulated, organized and shown?
8. How can this be done most simply?
9. How can this be done most efficiently?

10. How can this be done most accurately?
11. What is the optimal view of all?
12. How can a text present this optimal view?
13. What greatest remains undiscovered, unnamed?
14. What is the absolute?
15. What is the universe? What is universal?
16. Is there more than one universe? How many?
17. What are these different universes like?
18. Are we more important than the world around us?
19. How are importance and values to be measured?

20. What should be of the greatest value?
21. What are the priorities, aims and goals of man?
22. What is true and what false?
23. What is good and what bad, right and wrong?
24. What are the consequences and rewards of acts?
25. Why is it important to do good and right?
26. What measures and scales are there for right?
27. What is needed for peace and prosperity?
28. What is the use of questions, problems, paradoxes?
29. What is wrong with "I only believe what I can see"?

30. What is the world? How is it not the universe?
31. How many different worlds are there?
32. Is there one actual and best world of man?
33. Is there a name for this actual and best world?
34. Is there a God? Is there more than one God?
35. How many different varieties of Gods are there?
36. What is God? What is right and wrong with God?
37. What is wrong with mans idea of God?
38. How well does man identify with and follow God?
39. What is the relationship between God and World?
40. What is the proof of God?
41. Does God change?
42. What is the relative universe?
43. How many relative universes are? What kinds?
44. What is the form of the relative universe?
45. What are the parts of the relative universe?
46. What are the characteristics of """?
47. Is the relative universe dynamic? changing?
48. What do polarities and opposition arise from ""?
49. What are the processes of the relative universe?
50. Does the world of man exist as a whole?
51. What are the limits of this world? of life?
52. Is reality truth? What is truth?
53. Is truth an end? a means? or is it limited?

54. How is the world nourished? What input/output?
55. What are the first or ultimate principles, laws?
56. What proves the relative universe?
57. What is the actual place and role of language, words, expressions, and communication in life?
58. What is the form and function of meaning?
59. What is the future? Does it change the past?
60. What is the present? is it independent?
61. Is the will of man important in the world?
62. Does faith in immortality lie beyond reason?
63. How much hope is there for man? the world?
64. What does it take to best achieve success?

65. How is success to be measured?
66. How much are standards and laws necessary?
67, How can we find greatest happiness?

1. TE, TÄ, TROS, TRIS, TRUS, TRU, AU, AMTRU, ES, CES, ACOM, ÁS, ALT, REM, PECOM
2. ATO, TIT-TO, TRU, TRUS, TRUCH, AMTRU, AUS, OITIO, LARF, TÄRAW
3. TÂ, TRIA, TIT-TO, ETO, ATO, DAOST, ACEXIN, CORT, LADPEFT, SIRT, SAC
4. TRIF, TRUF, Survival, Self Preservation, Life/death, ÁXE, CEFKA, VAC, LIPÁ, LAMEX, PEK
5. KOT, COAT, SCARLS, DAOST, VAPO, TĬ, TES/Ind, COGNOS, LOR, SUVUK, RUPKAL,
6. Training, Self Discipline, Joy in Study, TES, TRU, TAN, LOR, SEVI, RORRE, VERTRU
7. TRU, KOT, POT, TRIT, ROT, MOT, SPALT, POS, IPT, SPAT, CODIT, DOLT, LARP, SUVUK
8. TÄ, TROS, TRIS, TRIT, SPALT, SPAT, POS, TES, PUREALITY, LANOFAC, SUVUK
9. HUB, POT, IPT, Study, TES, Practice, Perfection, MADAS, SUVUK, RECONET
10. VAPO, POS, Measures, Refinement, TES, POT, IPT, IND, LADAC, RUPKAL, POP
11. TIT-TO, SPALT, OITIO, SPOT, SPAT, COAT, TRUS, AUS, LOPOIR, RORRE, POIR, PERT, LORAST
12. POT, IPT, ROT, TRU, PI, MOT, SPALT, POS, PACH, HOM, LOPOIR, RUPKAL, LORAST
13. TRIS, TÂS, TRIF, TRUF, TIT-TO, DYNO, DYNS, DYNU, ATREC, ITOD
14. TÂ, TÂS, ÂR, RÂ, TRIA, Eternity of TRU, AU, AUS, LOPCAF, SAC
15. TRU, TIT, TIT-TO, AU, OITIO, TAT, REM, DAOST, ACEXIN, LESM, MADAS
16. TRIS, TROS, TU, DAOST, TRUS, TU, TRUCH, STOT, PODAF, PIN
17. TRIS, TROS, TÄS, TRUD, FAF, TRUS, AMTRU, AUS, VE, PODAF, MADAS
18. REM, TE, TES, VAPO, HOM, ACC, PUREALITY, ELTICMA, SÂ, LOPCAF
19. REM, TÄ, VAPO, SÁ, ALT, SPALT, ACC, PUREALITY, DEMCO, JUCSMOR, ROMT
20. TIT-TO, TRIA, REM, 1R, 2R, VAPO, TES, ACC, SÂ, LACRO, LEDIOR, ROMVET
21. TIT, TRU, TRIA, TROS, TRUEQ, ACC, ACÄ, ELTICMA, TÂS, LATAC, LEXPRE
22. TIT, TIT-TO, TRU, VAPO, S-O, S-O→Á, TRU/ATRU-AU, LIPÁ, MATES
23. TIT, TRU, Selection, HOM, Ethics, TES/Ind, ACC, DAC, LITS
24. TRUD, HOM, DYNU, DYNS, TIT, ACC, TES/Ind, ELTICMA, COPTA, GAC,LOPA, LOTAC
25. TRIA, Survival, S/F, TRU, TIT, TÂS, Happiness, LAMIR, OPTH,
26. TRIT, ALT, VAPO, REM, POS, STA, HOM, Ethics, ACC, CUT, LOM, ACLO, ROMT
27. TIT-TO, TRUD, TRUEQ, VAPO, REM, DAOST, SCARLS, TES/Ind, LAMUA, LAPAF, VAC
28. APP, MET, QALP, KOT, ITOD, TAN, LOR, MERPS
29. Figment, Illusion/Actuality, 1/KOT,COAT, 1/TIT-TO, TOPA
30. TIT, TIT-TO, TRU, TRU/RUS-AUS, TROS, DAOST, LIACT, PIN
31. AUS, RUS, RUTS, AMTRU, TRUS, OC/EC/NEC, TIT-TO, STOT, PIN, MADAS
32. TIT-TO, TRIS, TRUS, RUS, AUS, TRUCH, LAPDAT,
33. TRU, TIT-TO, LANT, LATAN, NAMIT, NATREC, Named to be known
34. god, God, GOD, TREQ, TRU, TIT, TROS, DAOST, REM, STOT,
35. TROS, TU, TO, TE, TIT-TO, TREQ, Illusion/ACEXIN, TOPA
36. TIT-TO, TE, TÄ, TRU, TREQ; /figment, illusion, gods, VE, LAMUA
37. god/TIT-TO, 1/VAPO, TU/TROS, TU/TE, illusion/ACEXIN, 1/LARF
38. TES/ind, TES-TIT, KOT, COAT, TÍS-TIT, ACC, PUREALITY, ASIT, LACOM
39. TIT-TO, GOD→TREQ→TIT, REM, TRU/AU, TIT/AUS, RÍ, SEVI
40. PRUT, TIT, TRU, TROS, TRIA, TRIS, TRIT, TRUD, HOM, LARF, POPTE
41. TRUCH, TRUS, TREQ, AUS, AUCH, TIT/god, TRUF, LOPCAF
42. TRU, T, TIT, TIT-TO, TROS, TRIS, TÂS, LASAF, LINS, MIST, PRUNES, LOTAC
43. RU-RUS, RUT, RUTS, AMTRU, TRUS, TRU, TRIS, TIT, TÂS. LORÄ, PIN
44. TÄ, TÄS, TÄP, TROS, SCARLS, TU, TIT, TRIS, LARF, LASTE, PRÁ
45. TE, TES, TI, TIS, E, ES, CE, CES, POS, CHES, COVET, LADES, LAPAR
46. TROS, ACEXIN, Á, SÁ, TRUD, HOM, REM, VAPO, ENUL, LARAC, TRUCO, POCOMF
47. TRUD, TRIS, TRUS, TRUCH, TRUV, ACEXIN, HOM, CES, DAOST, CED, FET, RED, TRUCO
48. TIT-TO, TROS, TRIT, TRUD, ALT, ACEXIN, HOM, LANIP, LARNIP, RED
49. CE, CES, HOM, CHES, TRUD, DYNU, DYNS, DYNO, ACC, FERT, IOBRE, TRUCO
50. TO, TU, TROS, TIT, TIT-TO, TRIA, LAPCOST
51. SCARLS, TĬ, TĬS, TINI, TINIS, TÂ, 0, 1, 1/∞, FOL, LACOL, LATOL, POCOL, LAFRIA
52. S-O, S-R-O, R, R↔Á, VAPO, PRUT, TIT, KOT, COGNOS, LED, LOPUT, TOVOP, POKED
53. POT, VAPO, PRUT, DAOST, TROS, SCARLS, MET, CORT, LOPUT, SIRT, TOVOP, POKED
54. TIT, TRUD, ACEXIN, HOM, DYNU, DYNS, DYNO, OITIO, INT, LAMUA, LASADIN
55. TIT, LAT-T, TRU, LOT-S, TOPNEC, HPP, COGNOS, UDOHAT, APROD, COPT,
56. TIT, TRIA, TROS, TRU, PRUT, ALT, VAPO, DAOST, LACOP, LARTIN, POPTE, MIST
57. Á, SÁ, ROT, LANT, LAPDAT, LOPCAF, LORÄ, RECONST, PAMEDS, LOTESAN
58. TRU, TIT, TÄP, S-O, SO, S-R-O, SOT, Á, SÁ, MIST, MATES
59. TRIF, TRUF, TIT-TO, TRIA, TÂS, CEK, PEK, AXE, CED, PAF, POMAC
60, TIT, TRUN, CENU, TU, IAT, DAOST, TRUD, FED, LATED,
61. ACC REM,TRUD, HOM, PUREACLITY, KUPS, COGNOS, WIM, LASPRE, STES, RÍ, WITÂ
62. PUREACLITY, TES, TIT, DYNO, ACC, ELTICMA, SÂ, TÂ, LACJOT, LOR, PORI
63. TIT, TRIA, ACC,MARS, PUREALITY, COGNOS, ELTICMA, LACHVAR, LACOM, RÍ, SARUE
64. TIT, TRU, TROS, VAPO, KOT, TES, ACC, PUREALITY, SÂ, LAROS, VAC, MERPS
65. TIT, TÂ, VAPO, TES, REM, POS,,KUPS SÂ, TÂ,LASACA, LASUCT, LOM, SARUE, ROMT,
66. TIT, TRIA, LATANOS, LANT, TRIT, OBT, ROT,, POT, AIT, LARECS, LARUTS, LORAST
67. TIT, TRU/AU, TES/Ind, TROS, DAOST, TAN, ELTICMA, SÂ, PREDI

REFERENCES

The books from which information and ideas were acquired are innumerable, in bits and pieces that emphasized their altered meanings and gist in TRU From libraries, book stores, and universities around the world those that best provided information for a large personal library were purchsed. Among these the following have been the most frequently referred to and relevant in TRU.

Anthropology, A to Z, Coon and Hunt, New York, Gross and Dunlap, 1963
Arey, Leslie B. *Developmental Anatomy, Embryology.* Philad. Saunders, 1941
Artists on Art. Goldwater and Treves. New York: Pantheon/Random , 1972.
Atlas of Ancient Archeology, Hawkes, Ed., N.Y. Barnes & Noble 1994.
Atlas of the Bible, The Times, James Ptitchard, Ed., Avenel N.J., Crescent, 1996
Avey,. *Handbook in the History of Philosophy:* 2nd ed. New York: B & N,1958.
Beck, William. *Human Design, Molecular, Cellular and Systematic Physiology.* New York: Harcourt Brace 1971.
Bergh, M.B. Collected Notes, *Organization of Knowledge, Correlations of sensory and motor symptoms and signs, Target theory in Radiation Biology.* centering by space time in TRU, *Homation, or the Processes of Man, Medical Homation, IPTRU ,* Initial Presentation of TRU, Multi-Volume, *DIPTRU,* Shortened, Large Two-Volume Presentation, TRU, Unpublished (1936-2001)
Bergson, Henri. *Duration and Simultaneity,* New York: Bobbs-Merrill/, 1965.
Bible, The Portable World. Robert Ballou. New York: Viking, 1965
Bible, The Holy: Revised Standard Version, Nashville: Nelson, 1952.
Biological Data, Handbrook of. William Spector. New York: Saunders, 1956.
Biology Dictionary, Hale and Margham, New York, Harper Collins, 1991
Black Holes, Edited by Jackie Wilson, DK Publishing, New York, 1996
Books in Print, Paperbound. Edited by Olga Weber. Somerset, -2000
Britannica, Books of the Year: 1973-2001. Ed. James Ertel. Chicago:
Brodie, Bernard. *Evolution of Nervous Control.*Washington: AAAS, 1959,
Bronowski, J. *The Ascent of Man.* Boston: Little Brown, 1974.
Brooks, C.E. *Climate Through the Ages, etc..* London: Pillans Wilson, 1926.
Brookshear, J. Glenn, *Computer Science,* Redwood C., Ca. Benjamin/C., 1991.
Bullock, Alan and Stallybrass. *The Fontana Dictionary of Modern Thought.* London: Fontana, 1977.
Calder, Nigel. *Timescale, Atlas of 4th Dimension.* New York: Viking, 1983.
The Capricious Cosmos, Joe Rosen, New York, Maxmillan, 1992
Cavalli-Sforza, Bodmer,. *Genetics and Human Pop..* San F.:Freeman, 1971.
Cirlot, *Dictionary of Symbols.* Philosophical Library, 1965
Columbia Encyclopedia,.Bridgwater - Kurtz. New York: Columbia, 1968.
Concise Encyclopedia of Archeology. Cottre. London: Hutchinson, 1960.
Conn, Rakel and Johnson. *Family Practice.* Philadelphia: Saunders, 1973.
Cook, Roger. *The Tree of Life, Image for the Cosmos.* N.Y. Avon/Hearst, 1974.
Coon, Carleton. *The Origin of Races,* New York: Knopf, 1963.
Daedalus. Graubard. Boston: American Academy Arts and Sciences, 1962,1963
Dadashzadeh, *Discrete Structures,* REA, Piscatway NJ., 1991
Dictionary of Cultural Literacy, Hirsch, Kett, & Trefil, Boston, H.M., 1993
Dictionary of Languages, David Crystal, London, Penguin, 1992
Dictionary of Philosophy. Runes. Totowa, NJ: Littlefield Adams, 1965,1975.
Dictionary of Psychology, Arthur S. Reber, London, New York, Penguin, 1985.
Dictionary of World Religions, Crim, Editor, New York, Harper and Row, 1989.
Dickerson, Richard, Molecular Thermodynamics, Menlo Park, Benjamn, 1969
Doubleday Roget's Thesaurus in Dictionary Form. Edited by Ronald J. Bogus. Garden City, NY: Doubleday, 1977.
Dubin, Dale. *Rapid Interpretation of EKG's.* Tampa: Cover, 1973.
Duff & Naylor. *Differential Equations,Applied Mathematics.* N.Y: Wiley, 1966.
Edwards, Tyron. *The New Dictionary of Thoughts.* Standard, 1966.
Encyclopedia of Archeology, Larousse. Charles-Picard. N. Y. Putnam, 1972.
*Encyclopedia Britannica,*1771 lst issue. 3 vol. Edinburgh; 1947 24 vol, London, Chicago; 1994 32 vol, Micro-Macropedia, Norton, Chicago:.
Encyclopedia of Philosophy: Vols 1-8. Edited by Paul Edwards. New York: Macmillan/Free Press, 19731995
Encyclopedia of Science and Technology. Parker. N. Y. McGraw-Hill, 1984.
Encyclopedia of World Religions. London, New York. Octopus/Crown, 1975.
Essential Sacred Writings Around the World, M. Eliade, Harper Collins, 1987
Evolution of Nervous Control, Symposium Brodie and Bass, Washington D.C. American Association for the Advancement of Science, 1959
*Evolution as a Process.*Huxley, Hardy and Ford. N.Y.:Collier/Macmillan, 1958.
Every, George. *Christian Mythology.* London: Hamlyn, 1970,
Evolution of Nervous Control, Brodie and Bass, AAAS, Washington DC l959
The Fallen Sky, Hill & Wang, Inc. New York, Colonial Press. Clinton,. 1964
Ferguson and Bruun. *European Civilization.* Boston: Houghton Mifflin, 1962.
Field Guide to the Night Sky, Audobon Society, New York, Knopf, 1998.
Fowler, Ramsey, Aaron, *The Little Brown Handbook,* Harper Collins N.Y.1989
Fuller, R.B. *Synergetics, Geometry of Thinking.* N.Y. Macmillan, l975.
Gleick, James. *Chaos.* New York, Penguin, 1988.
Great Books of the Western World: vols 54. Edited by Robert Hutchins. Chicago: Encyclopedia Britannica, 1955, especially Homer, Herodotus, Thucydides, Plato, Aristotle, Euclid Archimedes, Tacitus, Plotinus, Aurelius,Augustine, Aquinas, Gallileo, Descartes, Spinoza, Newton, Locke, Hume, Kant, Hegel, Goethe, Darwin, and James.
Great Ideas Today, Hutchins & Adler, Chicago, Britannica, 1961
Handbook of Chemistry/Physics, Hodgman, Cleveland, Chemical Ru1939, 1964
Harper's Bible Dictionary. Miller and Miller, New York: Harper, 1961
Harrison's *Principles of Internal Medicine,* Braunwald, N.Y. McGraw-H. 1987
Hawking, Stephen. *A Brief History of Time.* New York: Bantam, 1990.
History Atlas of the World, 4th Ed., The Times, Maplewood, Hammond, 1994.
History of Art, Janson, Harry N. Abrams, New York, 1977
Hodges &Lehman, *Elements of Finite Probability,* San Fran. Holden-Day 1965
Horizon, Thorndike and Mee. New York: American Heritage, 1964-1974.
Hoyt's Cyclopedia of Practical Quotations Kate Roberts. Somerset, 1947
Hussey, Edward, *The Presocratics,* Lloyd-Jones, N.Y. Schribner"s, 1972
Human Evolution , The Cambridge Encyclopedia of, Jones, Martin, Pilbeam, Cambridge University Press, 1992
Huxley, Aldous, *Brave New World,* New York, Harper & Bros. 1946
Ideas, A world of, Rohman, Chris, Ballantine Books, N.Y. 2000
In The Age of Mankiind, Smithsonian, Lewin, Washington DC, 1988
Johanson & Shreeve, *Lucy's Child,* Avon, New York, 1989
Kaufmann, Walter. *Religions in Four Dimensions.* New York: Crowell, 1976.
Kells, Lyman. *Elementary Differential Equations:* N. Y.: McGraw- Hill, 1935.
Key Ideas in Human Thought, McLeish Editor, N.Y.Facts on File, 1990.
Kierkegaard, Soren, *Either/Or,* Swenson, New York, Doubleday, 1959
Koran, The, Translated by N.J.Dawood, New York, Penguin, 1990
Living Philosophies, Fadiman, New York, Doubleday, 1990
Love, Clyde E. *Analytic Geometry.*3rd ed. New York: Macmillan, 1938,
Lowy and Harrow. *Organic Chemistry.* New York: Wiley,1940.
McClave and Dietrich. *Statistics:* 2nd ed. San Fran cisco: Dellen, 1982
MacCleod's Physiology in Modern Medicine,.Bard, St. Louis, Mosby, 1941
The Macmillan World History Factfinder, McEvedy, N.Y., Macmillan, 1985.
Mankind, In the Age of, Roger Lewin, Washington D.C., Smithsonian, 1988
Masterpieces of World Philosophy.Magill. N.Y. Salem/Harper Collins, 1961-90
Mathematical Functions, Handbook of. Washington: U.S.Dep. of Commerce, National Bureau of Standards, 1970.
Mathematics Encyclopedia. Max Shapiro. New York: Doubleday, 1977.
Mathematics, The VNR Concise Encyclopedia of. Edited by Gellert, Kustner, Hellwich, Kastner. New York: Van Nostrand, 1977
Mawson, Sylvester. *Dictionary of Foreign Terms.* Apollo, 1968.
Mechanics, I, II, Fogiel, Piscataway N. J. REA, 1991
Medieval Philosopheers, selections, Translated. McKeon, N.Y. Scribner. 1958
Mesarovic, Mihajlo, & Pestel, *Mankind at the Turning Point,* N.Y. Dutton 1974
Mettler, Fred. *Neuroanatomy.* St. Louis: Mosby, 1942.
Microfabric of Man, Warren Andrew, Chicago, Year Book Pub. 1966.
Milewski, Emil. *Set Theory.* Piscataway NJ: REA,1991
Modern History: Larousse Encyclopedia, Feltham Middlesex: Hamlyn, 1964.
National Geographic, (Leakey 6/1973) Washington 1935-2001,
The National Geographic Desk Reference, 1999
Nature of the World and of Man, The. Newman.,University of Chicago, 1930.
The New Atlas of the Universe, Moore, Arch Cape Press, N.Y., 1988
The Oxford Companion to Philosophy, Honderich, Oxford, Oxford Prress 1995.
Oxford English Dictionary, The Compact Edition. Oxford U. Press, 1975.
Particles and Forces at the Heart of Matter, Carrigan and Trower Editors, New York, Freeman 1986.
Past Worlds, Atlas of Archeology, Scarre,, London Harper Collins, 1996
Pictorial History of Philosophy. Runes, N.Y. Philosophical Library, 1959.
Prigogine and Stengers. *Order out of Chaos.* New York: Bantam, 1984.
Presocratics, The Mourelatos & Rorty, New York, Anchor/Doubleday, 1974
Radiocarbon., Deevey, Flint, Rouse. New Haven: Yale University, 1950 -2000.
Random House Dictionary of the English Language. Urdang. N.Y. 1968.
Random House Dictionary. Stein, N.Y. Random House, 1981.
Random House Encyclopedia. James Mitchel. N.Y. Random House, 1977.
The Real World, Edited by Bruce Marshall, Boston, Houghton Mifflin,, 1991.
Relativity, The Principle of. Collected Memoirs of Einstein, Lorentz, Minkowski, Weyl. New York: Dover/Methuen, 1923.
Rohmann, Chris, *A World of Ideas,* Ballantine Books, New York, 1999
Sagan, Carl. *Cosmos,* New York, Ballantine Books, 1985
Schilling, Charles. *Atomic Energy Encylopedia in the Life Sciences.* Philadelphia: Saunders, 1964.
Scientific American. Edited by Dennis Flanagan. N.Y. 1950-2000
Seldes, George. *The Great Quotations.* Castle, 1977.
Selye, Hans. *Stress.* Montreal: Acta, 1950.
Shepherd, Wm. R., *Historical Atlas,*9th Ed., N.Y. Barnes & Noble, 1964
Sherrington, Charles. *Man and His Nature.* New York: New American,1975
Statistical Abstract of the United States. Washington: Dp Commerce,1880-2000
Stedman's Medical Dictionary: Cutler. Baltimore: Williams & Wilkins, 1973.
Strong's Concordance of the Bible. Nashville: Thomas Nelson, 1983
Tillich, Paul. *My Search for Absolutes.* New York: Simon and Schuster, 1969
Timetables of History: The Random House. New York: Random House, 1991.
Toynbee, Arnold, *A Study of History,* New York, McGraw-Hill, 1972
Treasures of Britain, New York, Norton,1986
Universal Almanaac, Wright, Kansas City, Universal 1993
Universal Self-Instructor. Berg. N.Y. Thomas Kelly, Winter House. 1883/1970.
Upanishads, The. N.Y. New American Library/ Mentor, 1960
Van Nostrand's Scientific Encyclopedia: Princeton: Van Nostrand, 1958.
Wallach, *Interpretation of Diagnostic Tests,* Boston, Little Brown, 1986
Weber, Olga. *Paperbound Books in Print.* New York: Bowker, 1969.
Webster's Third International Dictionary: vols 3. Gove. Chicago: Enc.B. 1966
Weyl, Herman. *Space, Time, Matter.* Henry Brose. N.Y. Dover, 1955/1922.
Whitehead, Alfred. *Adventures of Ideas.* N.Y: NAL,/Mentor/Macmillan, 1962
Wiener, Norbert. *Cybernetics.* New York: Wiley, 1961
Winick, Charles. *Dictionary of Anthropology.* Ames: Littlefield Adams, 1958.
*Winston's Simplified Dictionary.*Lewis. Philadelphia: Winston, 1929
World Almanac, The, New York: Doubleday, 1935-2001
World Atlas: Imperial Edition, Masterpiece Edition. N.Y. Rand McNally, 1992
World Mythology, Larousse, Grimal. London; Paul Hamlyn, l971
World Population,, Livi-Bacci. Blackwell, Malden Ma, 2001.
World System, The. Edited by Ervin Laszlo. New York: Doubleday, 1973.
*Yearbook of Science and the Future.*Calhoun. Chicago: Britannica, 1979-2001.
Young, Arthur M. *The Geometry of Meaning.* Delacorte/Lawrence, 1976.
Art of War, Sun Tzu *Story of Philosophy*

MAJOR TRU TERMS or TOPNEC, NOT READILY REMEMBERED

Term	Description	Code	L/U			
ACLO	ACtion Level Optimal, right by TRU	3a 6	L	8	6	6.7
ACOM	Á Completion, S-O Mass Energy	3a 9	L U	5-	6.7	7.5
AIT	Absolute Imperative TRU, Reconcile	2c 8	L	3.3	5	7
APROD	Action Proportional to Principle	3b 1	L U	6	6	6.7
AS	Predominant MassEnergy TRU, Plural	3a 0	L	5	5+	6.7+
ASIT	Absolute Self Identity with TRU	3b 7	L	5-	5+	7.5
ÁXE	Á Expansion	3b 2	L	5	5	5+
CED	Compelling or Drawing Effects Dynamics	3a 6	L U	5	5	6.7
CEFKA	Compulsory Ethic KOT Assure Future	3a 5	L U	6.7	5+	8
CEK	Compelling or Drawing Forces Knowledge	3b 6	L U	3.3	5-	7.5
CER	Correlation Equals ES/O	3c 03	L U	6.7	3.3	6.7
CODIT	Compulsory Dimension of TRU	3b 2	L@U	(3.3	2.5	7)
COPT	Conformity to TRU Principle	3a 1	L	5	6.7	6.7
COPTA	Concordance of Perfect Thought Act	3a 8	L	5	6.7	7.5
CORT	Completed TRU, Fulfillment of ROT	3c092	L U	6.7	3.3	8
COVET	Collective Value of Function of ES	3b 1	L	6	5	6.7
CUT	Correlation not = to Correlation to TRU	3c 03	L U	6.7	2.5	3.3
DEMCO	Decrease of Magnitude by COT, of NEC, EC		L U	4	6.7	8
DOLT	Doctrine of Space Time	3b 0	L@U@			
ENUL	Energy Units Gradients of Á, Emu,Mu,,	3b 7a	L U	5	7.5	4
FERT	Fertility Ratio Rates Degrees for TRU	3b 1	L U	8+	6	5
FET	Freedom Equals TRU, actual range in TRU	3b 1	L U	7.5	3-7	6.7
FOL	Form in Limits	3b 4	L U	6.7~	3.3~	3.3~
GAC	Glory in Act by Á and TRU, opportunity	3a 66	L	7.5	5+	7.5
INÁ	Interval Á, Mass Energy RU, TRU	04-1	L	5		
@INT	Incentive Interests Inclination TRU	3b 0	L U	2.5	7	6.7
IOBRE	Inbreeding Outbreeding Optimum in TRU	3b 7	L	8.5	5	2-8
ITOD	Ideas whose time if overdue	3a 6	L U	1.5	7	7
JUCSMOR	Justification for COT max VAPO REM	3a 19	L U	4	6.7	7.5
LACHVAR	Continuum Human Variation Relativity	3b 8	L U	6.7	4	5~
LACJOT	Collective and Total TRU Judgement	3b 8	L U	2.5	2.5	7.5
LACOL	Law of COL, Correction of Limitation	3a 19	L U	3.3	6	7.5
LACOM	Law of Compliance	3a 7	L U	7.5	5~	7
LACRO	Compensatiion for Reduction in COT	3b 4	L U	3.3	6	8
LADAC	Dimensional Formative Congruity ES	2-3c4	L U	6.7~	5~	7
LADES	Discrete Separation Particular TRU	3b 2	L U	5+	6.7	7.5
LADPEFT	Greatest End, Fulfillment of TRU	3b 5	L	@	2.5~	10~
LAFRIA	Law function of TRIA, Limits of TRU	3a 19	L@			
LAMEX	Law of Mutual Expansion	3b 18	L@U			
LAMIR	L. Mutual Identity of Right KOT → ACT	3a 19	L U	5	6.5	6.5
LAMUA	L. of Mutual Assistance	2c 79	L U	7.5	5~	2-8~
LANOFAC	L. Abs. Necessity of Form Condition*	3a 2	L U	6.7	2.5	5~
LANIP	Negative in Positive, adding to TRU Pos	01-2	L@U			
LAP	Perfected Reliable Presentation Represen	3b44a	U L	7+	5	7.5
LAPAF	L. true Prosperity 1/expense of Future	3a 87	L	6.7	3.3	7.5
LAPAR	L. Particularity, Separate Parts of TRU	3b 7	L U	6.7	6.7	6.7
LAPCOST	L. Population or Quantity Sustain TRU	3b 87	L U	8	1.5	6.7
LAPDAT	Predominant Dimensional Formula. TRU	3a125	L U	5~	5~	5~
LAPDOR	Precision Detail by Objects size ROT	3b40A	U L	7.5	6.7	7.5
LARAC	Law of Reverse Actualization	3b100	L U	3.3	6.7	6.7
LARECS	L. Reduction Reconciliation Standards	3b 26	L U	3.3	6.7	7.5
LARF	Law of Relative Fixation, basis of TRU	2c158	L U	5~	5~	5~
LARNIP	Non Absolute Relative Neg in Pos Con	3b171	L@U			
LAROS	Relative Resolution-s Total Sufficient	3b190	L U			
LARP	Relative Parametric Proportion, Set TRU	2c056	L U~	6.7	5~	6.7~
LARTIN	L. Art and Representation in TRU	3b 2	U L	7.5	6.7	7.5
LARUTS	Limits Reduced Unitary Field TRU Std	3a193	L U@	6.7	3.3	7.5
LASACA	Action ACC Sufficient Concomitant	3a 1	L	7.5	5~	7.5
LASADIN	Static Dyn. Int. Dif. in and of TRU	3a170	L@U@	(6.7	@	7.5)
LASAF	Set FAF of Selectivity in Specificity	3b 1	L	3.3	6.7	2-8~
LASPRE	Self Preservation, Perfect in TRU	3b184	L	@	6.7	@
LASTE	Strong Tie Binding E ES TES CES TRU	3a 3	L@U	(5+	6.7	@)
LASUCT	Law of Success of TRU	3a185	L@U	(7~	2~	8~)
LATAC	TRU Act/Action, to TRU Contribution	3a6/100	L	8	5~	7.5
LATED	Law of Temporal Dependence	3a 8	L U	7.5~	5~	5@
LATOL	Law of Tolerance, Total Limit to TRU	3a184	L@U	(5~	2.5~	7.5~)
LED	Finite Distinction, Expands reality	3a44A	L U	3.3	6.7	7.5
LEDIOR	Equal Downgrading by Input Output REM	3a 3	U L	4	6	7
LEPER		04-3				
LESM	Law of Specific Energies and Masses	3b058	U L	0-10	@	@
LETID		06-7				
LEXPAC	Exclusive TRU Plan, for all Action	3b171	L U	6.7	3.3	7.5
LIACT	L. Imperative IA Convergence on TRU	01-2	L U	5+	7	6.7
LINS	Law of Internal Natural Selection	3b 82	L U	6.7	5+	5~
LIPÁ	Law of Imperative Perfection of Â	3a189	L	5@		

LITS	L. of Inalterable TRU Susceptibility	3a171	L U	5~	5~	6.7~
LODAC	L. Direct and Indirect Consequences	3b014/100	L U	6.7	5~	6.7~
LOM	Levels of Measurement	3b 4	U L	6.7	2.5	6.7
LOPA	Law of Proportional Act	2c6	L	8	6	6.7
LOPCAF	Optim. Predomin. Center Focus Reinforced	3b272	L U	5	5	5
LOPOIR	L. Point of Imagery Return Ratio	3a069	L U	2	7.5	7
LOPUT	Law of Proportional Truth TU,TO,REM,,,	3b100	L	1.5	4+	7
LOR	Law of Rational Tends to TRU, LASUR, Á	3b0	L U	2.5~	7	7.5
LORÄ(F)	Law of Predominant Form	3b054	L U	6.7	2.5	5~
LORAST	Law of Orientation and Standardization	00-7	L U	5~	2.5	7.5
LORVA	Law of Relative Value, Set by TRU	2c3	L U	4	5~	7
LOSE	Law of Separation of E, ES	3b 14	L U	6	7.5	5~
LOTAC	L. TRU on becoming Object Alters Form	3a198	U L	7.5	2.5	6.7
LOTESAN	L. of TRE ESIN, Actualization Nominal	3b241	L U	6.7~	5~	6.7
LOV	L. Value, Valuation, REM,,not Equal	3b 3	L U	5~	6~	7~
LUDATO	Understanding FAF and Value by POSits	3b 4	L U	4~	6.7	7.5
MADAS	Maximal Separation Distinction AU,,TRU	2b 2	U L	5~	3.3	7
MARAT	Minds Ability to Rapidly Attend to TRU	3b032	L U	3.3	6.7	7
MATES	Meaning automatically Tends to ES, CES	2c1-2	L U	3	7.5	7.5
MERPS	Manifold Enumeration in Problem Solving	3b7	L U	5~	6~	6.5
MIRPS	Multiple Inves Enum Relation Rapid P-S	3b192	L U	5~	5~	7
MIST	Meaning Is TRU (1, 2)	3b233	L	3.3	4	6
OPTH	Optimal Health, Care, Condition	3a7	L	7	6	7
OTS	Opportunity to Study, Learn, Work on TRU	3b7	L	4	6	7
PAF	P. The Actuality of the Existing Future	3b9	L	6.7~	2.5~	7.5
PAMEDS	Meaning Word ES Distilling Separation	3a2	L U	4	7	7.5
PED	Perfected Essence Degree (LOMID)	3b6-8	L	5~	5+	7~
PEK	Propelling force and Powers of Knowledge	3b68	L U	2.5	5~	3~
PICOM	P. of Interposition Composition	3b4	L U	5~	6.7~	7~
PIN	P. of Individuation	3b052	U L	6.7	5.5	5~
POCOL	P. of Concretion of Limits	3a059	U L@	7.5~	2.5~	0,5,10
POCOMF	P. of Composition Fixation	3b4	L U	5+	5+	5+
PODAF	P. Division Fusion of World Population	3b8	L U	7.5	2.5	5+
POIR	Point of Imagery Return, Resulting focus	3a1	U L	3.3	5+	7.5
POKED	P. Explosive Revelation Knowledge thru Analysis Depth Conception Deduction Intuition	3b048-086	L U	2	7	8
POMAC	P. of Multiple (Dual) Acceleration	3b7	L U	7	3.3	7~
POP	Point Placement, Posited by Foci Centers	3a4	U L	7.5	6.7	7.5
POPOT	Perfection of Perspective, PERT, POT	3b038	U L	7.5	6.7	8
POPTE	P. of Piecing Together Essences	3b2	L U	4	5-	7
PORI	P. of Reversible Immortality	3b6	L	4	5	9+
POSÂ	Positive Total S O Approach (TRU,SÂ,TÂ)	3b199	L U	5	6	6+
PREDIS	Prevention of Dissolution of TRU	3b55	L U	5~	5~	7~
RECONET	Reconciliation Comp COT Overlap ES	3b248	U L	4	7	7
RED	Reaction by Equalizing Densities, Voids	3b4	L U	4	6.5	7.5
RÍ	Reduction Resolution First Person in TES	3b199	L	3.3	5.5	7.5
ROMT	Reliability of TRU as a Measure	3a193,195	L U	5~	4.5	7.5
ROMVET	Ratios Monetary Customary REM = TRU	3b35B	L U	6.7	6~	6.7
RORRE	Revelation Relate Explosion Revelations	3b038	L U	2	7	7.5
RUPKAL	Revolution Universal to KOT Love of TRU	2c1	L U	2.5	5~	7.5
SAC	Subject-Absolute Completion Expan, all Á	3b189	L U	2.5~	6.7~	5~
SARUE	Self TRU Equival Release of ATRU bondage	3b36B	L U	6.7~	6	2.5~
SEVI	Search for Vivid Imagery, Identity	01-2	L U	3	6+	7.5
SIRT	Stimulation Sufficient to Rewards of TRU	3b8	L U	5~	6	8
STES	Self's Realization and Dedication to TES	3b7	L U	2.5	6.5	7
STOT	Stages or States of Being TRU, TRUS ness	3b114	L U	6.7~	2.5~	5~
SUVUK	Separation Unveiling all Veniers Kno	3a100,238	L U	2	6	6.5~
TÄRAW	TRU from Relative Warped Altered Orient.	3a33A	L U	5~	2.5~	5~
TATA	TRU formed and unformed	3c217	L U	6.7~	2.5~	5~
TECS	TRU Parts, Predicable Classes div, forms	3c124	L U	6.7~	5~	5~
TOPA	Test of TRU Applicability	06,78,91	L U	6~	5+	7.5
TOVOP	Test of Validity of Portions of TRU	3b35B	L U	4~	6.5	7
TRUCO	TRU Conversion	3a96	L U	4	5~	7.5
VAC	Voluntary, Automatic spontaneous Control	3a8	L U	6.7	4~	8
VE	Vacuum or Void Effect	3b2	L@U@			
WITA	Will to TRU Absolute	2c7	L U	3.3	5~	7.5

These measures of posits are only preliminary, more as initial guides and examples rather than final determinations. Much work will be needed by many to reach improved concensus and the more specific measure fulfill the ideal POA for TRU and all.

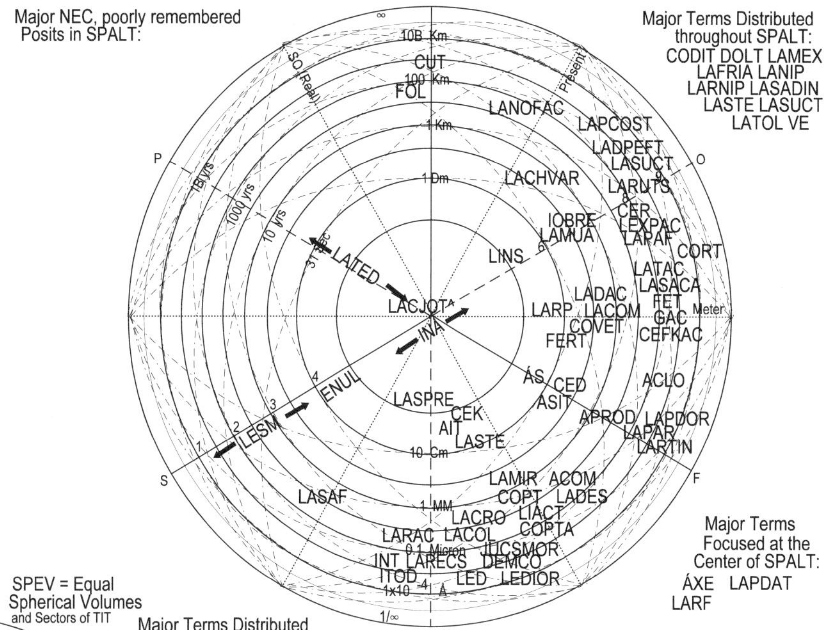

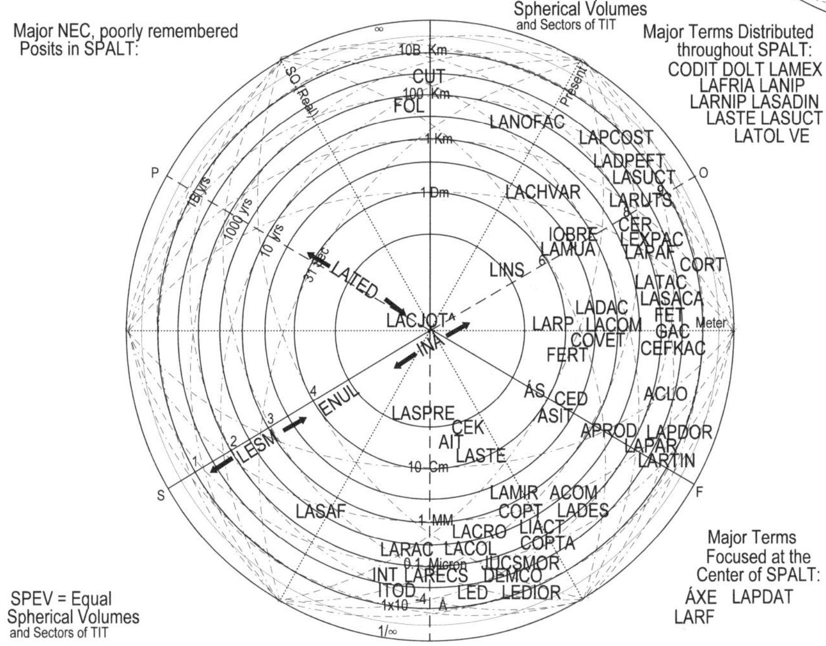

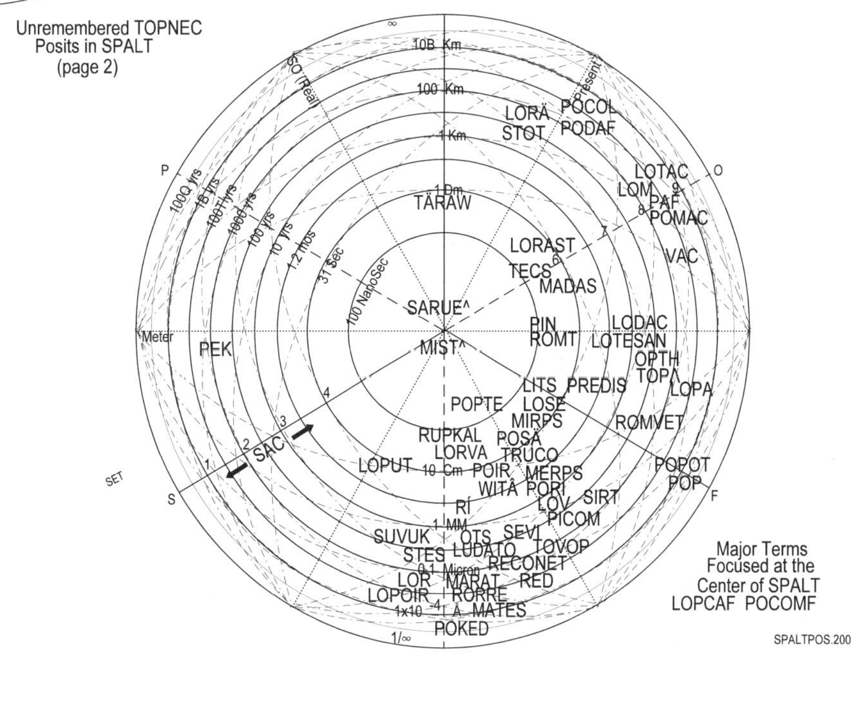

Fig G5-LA G5-LB **Posits in SPALT of less readily remembered major TRU concepts.**

SPALTPOS.200

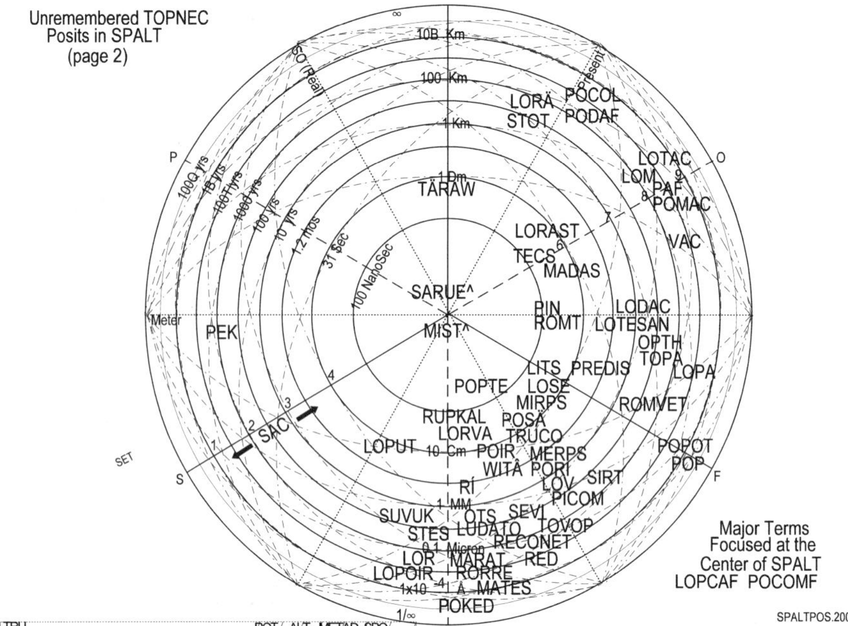

Fig G5-L2
Posits in SPALT of less readily remembered TRU concepts

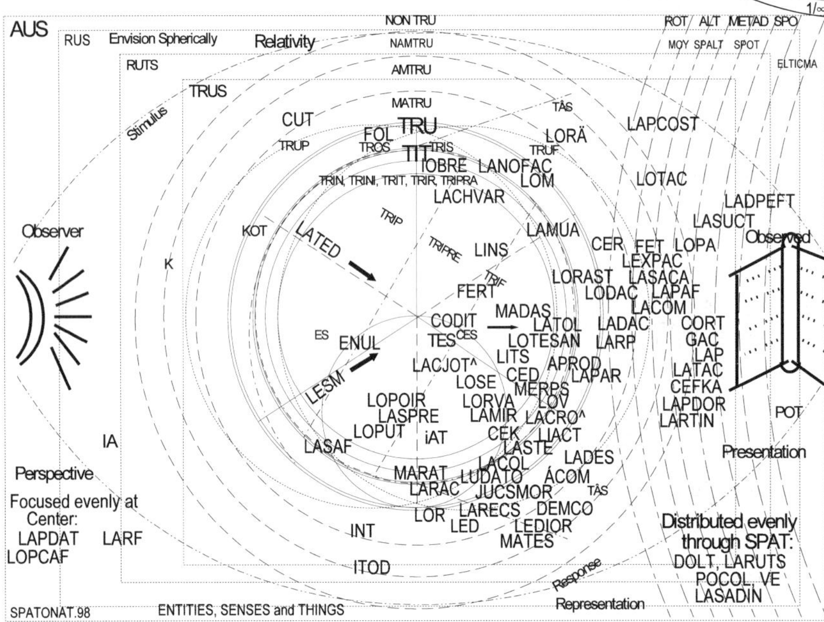

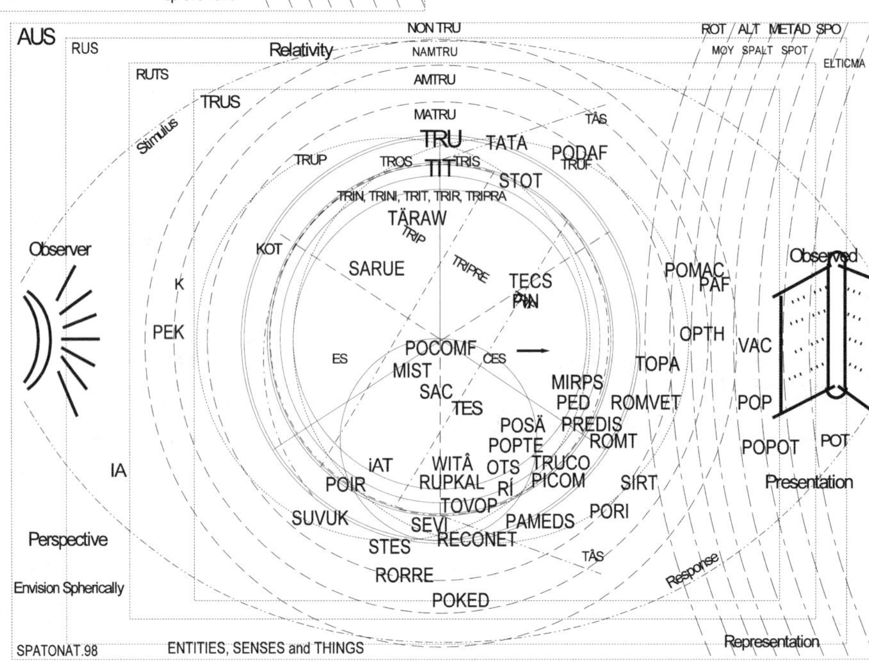

Fig G5-N1 and G5-N2 **Posits in SPAT of less readily remembered TRU concepts**

INDEX of NEW CONCEPTS

The abbreviations, acronyms, new concepts, and TRU distinctions of existing concepts are provided here to enable the reader and student to make better meaning and use of the text. More detailed definitions and descriptions are in the pages noted, as well as in the glossary and ESSARY of the multi-volume text on TRU, IPTRU.

This index is to help the reader or student find added meaning to each term as they are investigated. It is especially useful in the details of the figures at the end of each chapter from 61 to 99 where extensive use is made of NEC as processes. It is also to help give added meaning to listed Proofs given at several places in the text.

Added Lists, Posits and Figures of Abbreviations and NEC are found on pages: iii,iv,v,vii,viii, 74,119,205,240,242, 253,272,276,526,
PRU terms are also on.pgs. 147,282,
and lists of proofs at the end of several chapters.

INDEX

A detailed index is offered not only to provide the intricate reference such a text demands. It is like a pony for tie- ins of new concepts.(OC/NEC). Some of the in3exed items will be found on the adjacent pages, especially on a following page.

B

F

I

J

K

P

Q

R

X

Y

Z

ABOUT the AUTHOR and ORIGINS of the TEXT

Marcus Bernard Bergh was the third son, fourth of five children, born at Grand Rapids Minnesota, Oct. 24th 1921. His parents, Otto Ingman and Frances Hovey Bergh were administrators and professors in agriculture and music of the University of Minnesota and University of Miami, Fla. They were largely of Scandinavian-English Ancestry. Marcus attended the public schools of Minnesota, Wisconsin and Florida. Being raised in a highly competitive and stimulating environment, he had, from an early age, developed a deep interest in many subjects including authorship in history, geography, science, philosophy and religion.

From age sixteen more detailed logs were kept of deepest thoughts in all topics with varied models to organize knowledge. These were continually added to, information related and integrated, so that at the University of Miami, with combined majors in Chemistry and Biology, and minors in Psychology and Mathematics, large ledgers were carried in which information and ideas in the subjects taken were combined with that already produced.

At the School of Medicine of Johns Hopkins University the most advanced basic medical sciences, as well as the practice of clinical medicine, were learned and applied to ideas on the organization of knowledge. The failure in education at all levels to turn their bright young minds into well rounded, yet highly proficient professional graduates, was found to be due to a lack of focus, relation, interest, overview, and the limits specialization poses in training. What is most needed is for all in positions of leadership and decision making, to learn, understand, and use, the greatest in universal knowledge.

Being in uniform for five years during and after World War II, included rotating internship, Medical Officer aboard ship and ashore, and Venereal Disease Control Officer, experience was gained from care of casualties from Tarawa, Iwo Jima, and Okinawa, surveying disabled navy personel, and common illnesses and problems in service men and their families. Information was gathered from these that led to the Correlations of Physical Symptoms and Signs with applications to human stress and fitness. As Medical Director of Moosehaven studies on Correlations among Geriatric Patients continued the collective detail needed for a data base in applying medical and health statistics to organized knowledge.. As fellow in Medicine at Strong Memorial Hospital, University of Rochester, N. Y. studies with the Atomic Energy Commission on effects of radiation, and experiments with radioactive gold led to the integrative effects of decomposition, and the role of target theory in knowledge and human life. It also revealed the delicate and vulnerable state various levels, of all that makes up the form and function of the human body, have in survival over extinction.

As founder, proprietor, and physician-in-charge of a primary care clinic/hospital, Pureal Medical Center, in Orange Park Florida, 1949-1991, the first hand analysis of human diseases, ills, injuries, weaknesses, and their prevention, and as well as human strengths provided a wealth of ideas on the role of each person in this special universe of best human life. Associations during this period included being on staff at St. Vincents, Duval Medical, and St. Luke's - Mayo Clinic - Hospitals, Charter President of the Clay Co. Medical Society, member Duval, Florida, and American Medical Societies, Diplomat Board of Family Practice, Honors Lions Club of America, Loyal Order of Moose, etc.. The needed understanding and care by both health providers for their patients, and how these and all persons must acquire sufficient knowledge to care for themselves and others, became more evident. To achieve this level of care on the part of all, the basic ideas, relations, processes, concepts, principles and laws were developed and molded into a single work.

It became increasingly clear that an overall approach required models which recognized the focus human life has in this world, yet still includes all in the universe that is vital to the whole - as a part of the life chain. Extensive travel to most countries in North and South America and Europe, and many countries of Africa and Asia enabled an analysis of their cultures, languages, peoples and conditions. This, when interpreted in the light of their roles in the world, past, present and future provided a broad background for understanding the best human life offers. Gradually, through the massive collection of material from all sources, the naming, defining, relating, and positing of the greatest ideas, mechanisms, concepts, principles, and laws were combined and posited in the solidity of improved models. In the fifties and sixties this book was called *Homation* or *the Processes of Man.* Finding this less than adequate for a total organization of knowledge that is representative of the actual world of man, it was given the name of TRU, or The Relative Universe. During the seventies, eighties, and nineties continued additions, refinements, terminology, and consolidation were provided to make a presentation that describes what human life is actually all about.

Since retirement and into the 21st century, full time has been devoted to the reduction of the vast amount of material in the multi-volume *IPTRU* or the *Initial Presentation of TRU*; first to *DIPTRU* or large two volume, *Diminished Presentation of TRU*, and more recently to its current abridged form in *CIPTRU*, a single volume standard reference work on TRU. TRU is neither a substitute for the great terms true and truth, nor for God, world, or universe, as close as it comes to, and as much as it has in common with each of these. It is selected because it better represents this special universe of best human life. Its presentation provides the kind of language required by every new subject and science. It is hoped that many will recognize, benefit from, carry on the work, and savor the glory such a greatly needed, beautiful, and all-embracing approach to the world, that TRU provides.